S0-BXN-104

Dictionary of Organic Compounds

FIFTH EDITION

SEVENTH SUPPLEMENT

Dictionary of Organic Compounds

FIFTH EDITION

SEVENTH SUPPLEMENT

Ref.
QD246
D5
1982
Suppl.7
1989

LONDON   NEW YORK
CHAPMAN AND HALL
SCIENTIFIC DATA DIVISION

The Fifth Edition of the Dictionary of Organic Compounds
in seven volumes published 1982
The First Supplement published 1983
The Second Supplement published 1984
The Third Supplement published 1985
The Fourth Supplement published 1986
The Fifth Supplement (in two volumes) published 1987
The Sixth Supplement published 1988
This Seventh Supplement published 1989

Chapman and Hall
Scientific Data Division
11 New Fetter Lane, London EC4P 4EE
29 West 35th Street, New York, NY 10001

Printed in Great Britain at
the University Press, Cambridge

ISBN 0 412 17070 1

© 1989 Chapman and Hall

All rights reserved. No part of this book may be reprinted, or
reproduced or utilized in any form or by any electronic, mechanical
or other means, now known or hereafter invented, including photocopying
and recording, or in any information storage or retrieval system,
without permission in writing from Chapman and Hall.

British Library Cataloguing in Publication Data

Dictionary of organic compounds – 5th ed.
Seventh supplement
1. Organic compounds
I. Buckingham, J. (John)
547

ISBN 0-412-17070-1

Library of Congress Cataloging in Publication Data

Dictionary of organic compounds.
    Seventh supplement.—5th ed.
        p.   cm.
    "Executive editor, J. Buckingham"—P.
    Includes bibliographical references.
    ISBN 0-412-17070-1
    1. Chemistry, Organic—Dictionaries.
    I. Buckingham, J.
QD246.D5   1982   Suppl. 7
547′.003—dc20   89–23854

EXECUTIVE EDITOR
J. Buckingham

ASSISTANT EDITOR
C.M. Cooper

EDITORIAL BOARD

J.I.G. Cadogan
British Petroleum Company

R.A. Raphael
University of Cambridge

C.W. Rees
Imperial College

INTERNATIONAL ADVISORY BOARD

D.H.R. Barton
Texas A & M University

A.J. Birch
Australian National University, Canberra

J.F. Bunnett
University of California, Santa Cruz

E.J. Corey
Harvard University

P. Doyle
ICI, Plant Protection Division

Sukh Dev
Malti-Chem Research Centre, Vadodara

A. Eschenmoser
Eidgenossische Technische Hochschule, Zurich

K. Hafner
Technische Hochschule, Darmstadt

Huang Wei-yuan
Shanghai Institute of Organic Chemistry, Academia Sinica

A. Kende
University of Rochester

K. Kondo
Sagami Chemical Research Centre, Kanagawa

R.U. Lemieux
University of Alberta

J. Mathieu
Roussel Uclaf, Romainville

T. Mukaiyama
University of Tokyo

G. Ourisson
Université Louis Pasteur de Strasbourg

H.E. Simmons
E.I. du Pont de Nemours and Company, Wilmington

B.M. Trost
University of Wisconsin

SPECIAL EDITORS

J.W. Barton
University of Bristol

A.J. Boulton
University of East Anglia

L. Bretherick
BP Research Centre

B.W. Bycroft
University of Nottingham

P.M. Collins
Birkbeck College, London

J.D. Connolly
University of Glasgow

J.S. Davies
University College of Swansea

I.D. Entwistle
Shell Biosciences Laboratory

F.D. Gunstone
University of St Andrews

R.A. Hill
University of Glasgow

B.A. McAndrew
Proprietary Perfumes Limited

W.J. Oldham
BP Chemicals

G. Pattenden
University of Nottingham

ADDITIONAL COINTRIBUTORS TO THIS SUPPLEMENT
P.I.C. Berthier, P. Devi, C.M. Pattenden, A.D. Roberts

Seventh Supplement

Introduction

For detailed information about how to use DOC 5, see the Introduction in Volume 1 of the Main Work.

1. Using DOC 5 Supplements

As in the Main Work volumes, every Entry is numbered to assist ready location. The DOC Number consists of a letter of the alphabet followcd by a five-digit number. In this seventh supplement the first digit is invariably 7. Cross-references within the text to Entries having numbers beginning with zero refer to Main Work Entries and with 1, 2, 3, 4, 5 or 6 refer to the first six supplements.

Where a Supplement Entry contains additional or corrected information referring to an Entry in the Main Work or earlier supplements, the whole Entry is reprinted, with the accompanying statement "Updated Entry replacing . . .". In such cases, the new Entry contains all of the information which appeared in the former Entry, except for any which has been deliberately deleted. In such cases there is therefore no necessity for the user to consult the Main Work or previous supplements.

2. Literature Coverage

In compiling this Supplement the primary literature has been surveyed to mid-1988. A considerable number of compounds from the older literature have also been included for the first time.

3. Indexes

The indexes in the Supplement cover the Sixth and Seventh Supplements. A cumulative index volume to Supplements 1–5 inclusive was issued as part of the Fifth Supplement. In order to find a compound in DOC, look first in the Seventh Supplement, then in the Fifth Supplement cumulative indexes, then in the Main Work indexes.

Note to Readers

Always use the latest Supplements

Supplements are published in the middle of each year and contain new and updated Entries derived from the primary literature of the preceding year. Searching the entire Supplement series is facilitated by consulting first the indexes in the latest Supplement. The Supplement indexes are cumulative to facilitate the rapid location of data.

For full information on Supplements please write to:

The Marketing Manager   *or*   Chapman and Hall
Scientific Data Division   29 West 35th Street
Chapman and Hall   New York, NY 10001
11 New Fetter Lane   U.S.A.
London
EC4P 4EE

New Compounds for the DOC 5

The Editor is always pleased to receive comments on the selection policy of DOC 5, and in particular welcomes specific suggestions for compounds or groups of compounds to be considered for inclusion in the Supplements.

Write to:

The Editor
Dictionary of Organic Compounds
Scientific Data Division
Chapman and Hall
11 New Fetter Lane
London
EC4P 4EE

Specialist Dictionaries

Four important specialist publications are now available which greatly extend the coverage of the DOC databank in key specialist areas. These are as follows:

Dictionary of Alkaloids, 1989, 2 volumes, ISBN 0 412 24910 3

Dictionary of Antibiotics and Related Substances, 1987, ISBN 0 412 25450 6

Dictionary of Organophosphorus Compounds, 1987, ISBN 0 412 25790 4

Carbohydrates (a Chapman and Hall Chemistry Sourcebook), 1987,
    ISBN 0 412 26960 0

Caution

Treat all organometallic compounds as if they have dangerous properties.

The information contained in this volume has been compiled from sources believed to be reliable. However, no warranty, guarantee or representation is made by the Publisher as to the correctness or sufficiency of any information herein, and the Publisher assumes no responsibility in connection therewith.

The specific information in this publication on the hazardous and toxic properties of certain compounds is included to alert the reader to possible dangers associated with the use of these compounds. The absence of such information should not however be taken as an indication of safety in use or misuse.

Contents

Seventh Supplement Entries   *page* 1

Name Index   449

Molecular Formula Index   527

CAS Registry Number Index   615

A

Abbeymycin   **A-70001**

1,2,3,10,11,11a-Hexahydro-2-hydroxy-11-methoxy-5H-pyrrolo[2,1-c][1,4]benzodiazepin-5-one, 9CI

[108073-64-9]

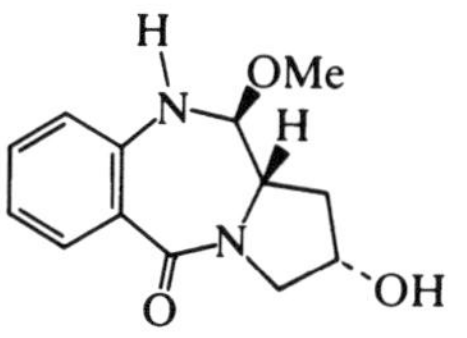

$C_{13}H_{16}N_2O_3$ M 248.281

Anthramycin-type antibiotic. Prod. by *Streptomyces* AB 999F-52. Exhibits weak antibacterial activity. Platelets ($CHCl_3$/MeOH). Mp 142-144° dec. $[\alpha]_D^{25}$ +303° (c, 0.741 in H_2O).

Hochlowski, J.E. *et al*, *J. Antibiot.*, 1987, **40**, 145 (*isol, struct, props*)

Abbottin   **A-70002**

8,9-Dihydro-5-methoxy-8-(1-methylethenyl)-2-phenyl-2H-furo[2,3-h]-1-benzopyran, 9CI

[106327-62-2]

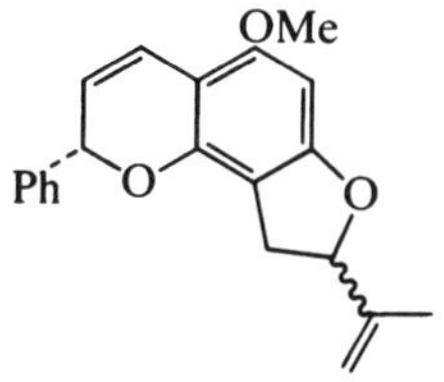

$C_{21}H_{20}O_3$ M 320.387

Isol. from *Tephrosia abbottiae*. Red viscous oil.

Gomez-Garibay, F. *et al*, *Chem. Ind. (London)*, 1986, 827 (*isol*)

Abiesolidic acid   **A-70003**

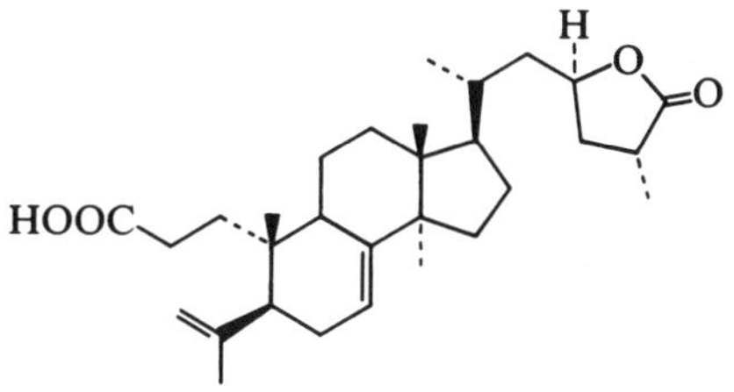

$C_{30}H_{46}O_4$ M 470.691

Constit. of *Abies sibirica*.

Me ester: [108195-55-7]. Cryst. (Et_2O/pet. ether). Mp 154-155°. $[\alpha]_D^{21}$ +10.3° (c, 3.86 in $CHCl_3$).

Raldugin, V.A. *et al*, *Khim. Prir. Soedin.*, 1986, **22**, 688; *Chem. Nat. Compd.*, p. 645 (*isol, cryst struct*)

8,11,13-Abietatriene-7,18-diol   **A-70004**

$C_{20}H_{30}O_3$ M 318.455

7α-form

Constit. of pollen grains of *Cedrus deodara*. Needles (Me_2CO). Mp 89°. $[\alpha]_D^{20}$ −3.3° (c, 0.46 in EtOH).

Ohmoto, T. *et al*, *Chem. Pharm. Bull.*, 1987, **35**, 229.

8,11,13-Abietatriene-12,18-diol   **A-70005**

18-Hydroxyferruginol

$C_{20}H_{30}O_2$ M 302.456

Constit. of *Torreya nucifera*. Cryst. (Me_2CO). Mp 180-181°. $[\alpha]_D^{25}$ +70.3° (c, 0.37 in EtOH).

18-Aldehyde: 12-Hydroxy-8,11,13-abietatrien-18-al.
18-Oxoferruginol.
$C_{20}H_{28}O_2$ M 300.440
Constit. of *Torreya nucifera*. Cryst. (hexane). Mp 139-141°. $[\alpha]_D$ +69.6° (c, 1.05 in $CHCl_3$).

Fukushima, I. *et al*, *Agric. Biol. Chem.*, 1968, **32**, 1103 (*isol, struct*)
Harrison, L.J. *et al*, *Phytochemistry*, 1987, **26**, 1211 (*deriv*)

8,11,13-Abietatrien-3-ol   **A-70006**

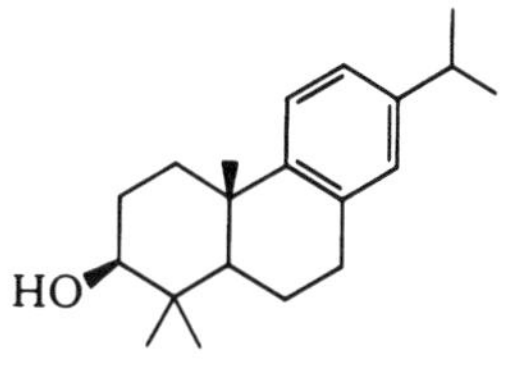

$C_{20}H_{30}O$ M 286.456

3β-form [78078-41-8]

Constit. of *Nepeta tuberosa*. Cryst. (hexane). Mp 109-111° (nat.), Mp 136.5-138° (synthetic). $[\alpha]_D$ +37.26° (c, 0.95 in $CHCl_3$), $[\alpha]_D$ +50.4° ($CHCl_3$).

Ac: Mp 112-114°. $[\alpha]_D$ +58.9° ($CHCl_3$).

Matsumoto, T. *et al*, *Bull. Chem. Soc. Jpn.*, 1981, **54**, 581 (*synth, pmr*)
Urones, J.G. *et al*, *Phytochemistry*, 1988, **27**, 523 (*isol*)

Acanthocerebroside A   **A-70007**

1-Glucopyranosyl-2-(2-hydroxytetracosanoylamino)-1,3,4-hexadecanetriol

[110744-71-3]

$R = (CH_2)_{21}CH_3$,
$R' = (CH_2)_{11}CH_3$.

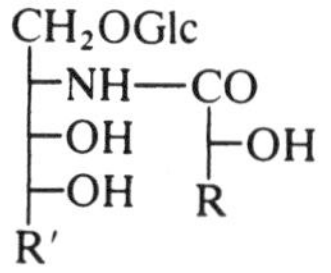

$C_{46}H_{91}NO_{10}$ M 818.226

Cerebroside from the starfish *Acanthaster planci*. Needles + $3H_2O$ (MeOH). Mp 209-210°. $[\alpha]_D$ +2.4° (c, 0.81 in propanol).

Kawano, Y. *et al*, *Justus Liebigs Ann. Chem.*, 1988, 19 (*isol, pmr, cmr, struct*)

Acanthocerebroside B   **A-70008**

1-Glucopyranosyl-2-(2-hydroxyhexadecanoylamino)-1,3,4-docosanetriol

[110744-72-4]

As Acanthocerebroside *A*, A-70007 with

$R = -(CH_2)_{13}CH_3$, $R' = (CH_2)_{17}CH_3$

$C_{44}H_{87}NO_{10}$ M 790.172

Cerebroside from the starfish *Acanthaster planci*. Needles + $1H_2O$ (MeOH). Mp 218-219°. $[\alpha]_D$ +7.8° (c, 0.24 in propanol).

Kawano, Y. *et al, Justus Liebigs Ann. Chem.*, 1988, 19 (*isol, pmr, cmr, struct*)

Acanthocerebroside *C*   A-70009

1-Glucopyranosyl-2-(2-hydroxyhexadecanoylamino)-13-docosene-1,3,4-triol

[110744-73-5]

As Acanthocerebroside *A*, A-70007 with

$$R = (CH_2)_{13}CH_3, \ R' = -(CH_2)_8CH=CH(CH_2)_7CH_3$$
(*Z*-)

$C_{44}H_{85}NO_{10}$ M 788.156

Cerebroside from the starfish *Acanthaster planci*. Needles (MeOH). Mp 203-204°. $[\alpha]_D$ +18.3° (c, 1.24 in propanol).

Kawano, Y. *et al, Justus Liebigs Ann. Chem.*, 1988, 19 (*isol, pmr, cmr, struct*)

Acanthocerebroside *D*   A-70010

[110744-74-6]

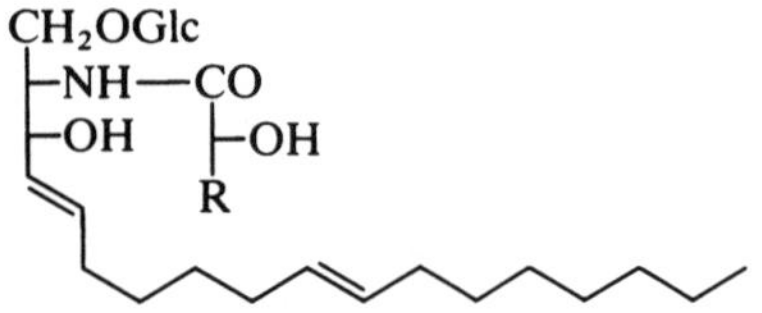

CH₂OGlc
—NH—CO
—OH ⌐OH
         R

R = −(CH₂)₁₉CH₃

$C_{46}H_{87}NO_9$ M 798.195

Cerebroside from the starfish *Acanthaster planci*. Needles + 2H₂O (MeOH). Mp 198-199°. $[\alpha]_D$ +1.1° (c, 0.3 in propanol).

Kawano, Y. *et al, Justus Liebigs Ann. Chem.*, 1988, 19 (*isol, pmr, cmr, struct*)

Acanthocerebroside *E*   A-70011

[110744-75-7]

As Acanthocerebroside *D*, A-70010 with

R = −(CH₂)₂₀CH₃

$C_{47}H_{89}NO_9$ M 812.221

Cerebroside from the starfish *Acanthaster planci*. Needles + 3H₂O (MeOH). Mp 194-195°. $[\alpha]_D$ +1.0° (c, 0.3 in propanol).

Kawano, Y. *et al, Justus Liebigs Ann. Chem.*, 1988, 19 (*isol, pmr, cmr, struct*)

Acanthocerebroside *F*   A-70012

[110744-76-8]

As Acanthocerebroside *D*, A-70010 with

R = (CH₂)₂₁CH₃

$C_{48}H_{91}NO_9$ M 826.248

Cerebroside from the starfish *Acanthaster planci*. Needles + 2H₂O (MeOH). Mp 193-194°. $[\alpha]_D$ +1.2° (c, 0.5 in propanol).

Kawano, Y. *et al, Justus Liebigs Ann. Chem.*, 1988, 19 (*isol, pmr, cmr, struct*)

1-Acenaphthenecarboxaldehyde   A-70013

Updated Entry replacing A-00061
1-Formylacenaphthene

[37977-48-3]

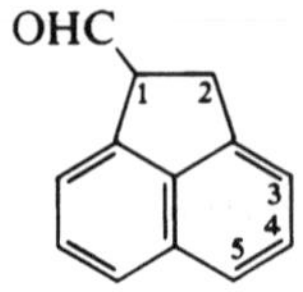

$C_{13}H_{10}O$ M 182.221

(±)-*form*

Yellowish needles (Et₂O/pet. ether). Mp 99.5-100.5°. Bp$_{0.1}$ 154°.

Oxime:
$C_{13}H_{11}NO$ M 197.236
Yellow needles (C₆H₆). Mp 152°.

Feiser, L. *et al, J. Am. Chem. Soc.*, 1940, **62**, 49 (*synth*)
Ger. Pat., 2 064 279, (*1972*); *CA*, **77**, 114126r (*synth*)
Raffaelli, A. *et al, Synthesis*, 1988, 893 (*synth, deriv, ir, pmr*)

1-Acenaphthenecarboxylic acid, 8CI   A-70014

Updated Entry replacing A-00062
1,2-Dihydro-5-acenaphthylenecarboxylic acid, 9CI. 1-Acenaphthoic acid

[6833-51-8]

$C_{13}H_{10}O_2$ M 198.221

(±)-*form*

Needles (EtOH aq. or Et₂O/pet. ether). Mp 163-164°.

Nitrile: 1-Cyanoacenaphthene.
$C_{13}H_9N$ M 179.221
Cryst. (pet. ether). Mp 68°.

Julia, M. *et al, Bull. Soc. Chim. Fr.*, 1952, 1065 (*synth*)
Canceid, J. *et al, Bull. Soc. Chim. Fr.*, 1973, 2727 (*synth*)
Fay, C.K. *et al, J. Org. Chem.*, 1973, **38**, 3122 (*pmr*)
Raffaelli, A. *et al, Synthesis*, 1988, 893 (*nitrile*)

5,6-Acenaphthenedicarboxylic acid, 8CI   A-70015

1,2-Dihydro-5,6-acenaphthylenedicarboxylic acid, 9CI

[5698-99-7]

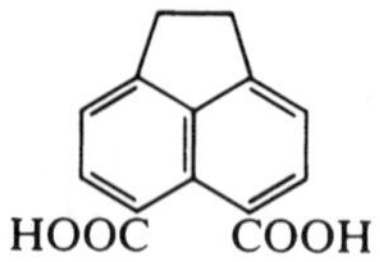

$C_{14}H_{10}O_4$ M 242.231
Mp 294°.

Dinitrile: [86528-79-2]. *5,6-Dicyanoacenaphthene.*
$C_{14}H_8N_2$ M 204.231
Cryst. (CH₂Cl₂). Mp 310-311° (charred before melting).

Bis(dimethylamide): [31458-06-7].
$C_{18}H_{20}N_2O_2$ M 296.368
Solid (EtOH aq.). Mp 112-114°.

Freund, M. *et al*, *Justus Liebigs Ann. Chem.*, 1913, **399**, 182 (*deriv, synth*)
Carpino, L.A. *et al*, *J. Org. Chem.*, 1964, **29**, 2824 (*synth*)
Trost, B.M. *et al*, *J. Am. Chem. Soc.*, 1971, **93**, 737 (*synth, ir, pmr, ms*)
Rieke, L.I. *et al*, *J. Org. Chem.*, 1983, **48**, 2949 (*deriv, synth, ir, ms*)

Acenaphtho[1,2-*b*]benzo[*d*]thiophene   A-70016

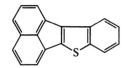

C₁₈H₁₀S   M 258.337

$C_{18}H_{10}S$ M 258.337
Solid. Mp 125°.

Kar, G.K. *et al*, *Org. Prep. Proced. Int.*, 1988, **20**, 213.

Acenaphtho[1,2-*b*]thiophene   A-70017

[1969-60-4]

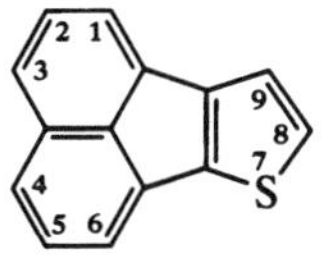

$C_{14}H_8S$ M 208.277
Yellow needles (pet. ether). Mp 73-74° (67°).

Hauptmann, S. *et al*, *J. Prakt. Chem.*, 1969, **311**, 614 (*synth, uv*)
Kar, G.K. *et al*, *Org. Prep. Proced. Int.*, 1988, **20**, 213 (*synth*)

Acephenanthrylene, 9CI   A-70018

[201-06-9]

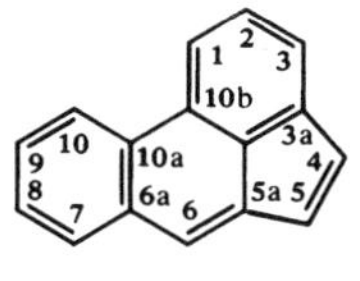

$C_{16}H_{10}$ M 202.255
Yellowish cryst. (MeOH). Mp 143-144°.

Laarhoven, W.H. *et al*, *Recl. Trav. Chim. Pays-Bas*, 1976, **95**, 165 (*synth, uv, pmr*)
Chung, Y.-S. *et al*, *J. Org. Chem.*, 1987, **52**, 1284 (*synth, pmr*)

Acetamide, 9CI   A-70019

Updated Entry replacing A-00092
[60-35-5]

$$H_3CCONH_2$$

C_2H_5NO M 59.068
Solubiliser, plasticiser, stabiliser, used industrially as solv. in molten form. Dissolves virtually all classes of organic and inorganic compds. Deliquescent, hexagonal cryst. Odourless when pure but usually has characteristic "mouse" odour. V. sol. H₂O, EtOH, sol. CHCl₃, prac. insol. Et₂O. Mp 82-83°. Bp 222°, Bp₅ 92°. Triboluminescent.

▷ Mild irritant, exp. carcinogen. AB4025000.

B,¹/₂HBr: [20731-46-8]. Reagent for brominating acid-sensitive compds. Needles. Mp 138-139°.
Picrate: Mp 117°.
N-Chloro: see N-Chloroacetamide, C-00656

N-Bromo: see N-Bromoacetamide, B-01883
N-Iodo: see N-Iodoacetamide, I-00357
N-Ac: see Diacetamide, D-00497
N-Me: [79-16-3].
  C_3H_7NO M 73.094
  Bp₈ 65°.
▷ AC5960000.
N-Di-Me: [127-19-5].
  C_4H_9NO M 87.121
  Bp 179-181°, Bp₁₀ 68-69°.
▷ AB7700000.

Org. Synth., Coll. Vol., **1**, 3 (*synth*)
Robson, J.H. *et al*, *J. Am. Chem. Soc.*, 1955, **77**, 498 (*deriv*)
Marakami, M. *et al*, *Bull. Chem. Soc. Jpn.*, 1962, **35**, 11 (*synth*)
Ueki, M. *et al*, *Bull. Chem. Soc. Jpn.*, 1971, **44**, 1108 (*deriv*)
Ottersen, T., *Acta Chem. Scand., Ser. A*, 1975, **29**, 944 (*cryst struct*)
Kerridge, D.H., *Chem. Soc. Rev.*, 1988, **17**, 181 (*rev*)
Fieser, M. *et al*, *Reagents for Organic Synthesis*, Wiley, 1967-84, **1**, 3.
Sax, N.I., *Dangerous Properties of Industrial Materials*, 5th Ed., Van Nostrand-Reinhold, 1979, 332.

1-Acetoxy-7-acetoxymethyl-3-acetoxy-methylene-11,15-dimethyl-1,6,10,14-hex-adecatetraene   A-70020

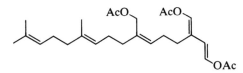

$C_{26}H_{38}O_6$ M 446.583
Compd. not named in the paper, not indexed by *CA*. Metab. of *Penicillus dumetosus*. Antibacterial and antifungal. Oil. $[\alpha]_D^{25}$ +17° (c, 1.3 in CHCl₃).

Paul, V.J. *et al*, *Tetrahedron*, 1984, **40**, 2913.

2-Acetyl-3-aminobenzofuran   A-70021

1-(3-Amino-2-benzofuranyl)ethanone, 9CI. *3-Amino-2-benzofuryl methyl ketone*
[49615-96-5]

$C_{10}H_9NO_2$ M 175.187
Cryst. (EtOH). Mp 153-154°.

Gewald, K. *et al*, *J. Prakt. Chem.*, 1973, **315**, 779 (*synth*)
Bachechi, F. *et al*, *Acta Crystallogr., Sect. C*, 1988, **44**, 1449 (*cryst struct*)

2-Acetyl-3-aminopyridine   A-70022

1-(3-Amino-2-pyridinyl)ethanone, 9CI. *3-Amino-2-pyridinyl methyl ketone*, 8CI
[13210-25-8]

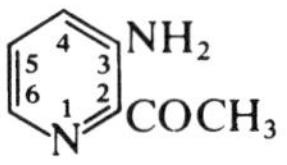

$C_7H_8N_2O$ M 136.153
Yellow plates (pet. ether). Mp 66-67° (63-64°).

2,4-Dinitrophenylhydrazone; B,HCl: [13385-49-4]. Needles (AcOH). Mp 276-277°.
Picrate: [51460-34-5]. Yellow prisms (EtOH aq.). Mp 200-201°.

Atkinson, C.M. *et al, J. Chem. Soc. (C)*, 1966, 2053 (*synth*)
Edward, J.T. *et al, J. Heterocycl. Chem.*, 1973, **10**, 1047 (*synth*)

Atkinson, C.M. *et al, J. Chem. Soc. (C)*, 1966, 2053 (*synth*)
Edward, J.T. *et al, J. Heterocycl. Chem.*, 1973, **10**, 1047 (*synth*)

2-Acetyl-5-aminopyridine   A-70023

1-(5-Amino-2-pyridinyl)ethanone, 9CI. 5-Amino-2-pyridinyl methyl ketone, 8CI

[51460-32-3]

$C_7H_8N_2O$ M 136.153

Mp 108-109°.

B,HCl: [34689-84-4]. Mp 208-210° dec.

Picrate: [51460-36-7]. Yellow needles (EtOH aq.). Mp 191-192°.

N-Ac: [31557-77-4].
  $C_9H_{10}N_2O_2$ M 178.190
  Cryst. (C_6H_6). Mp 114-116°.

Edward, J.T. *et al, J. Heterocycl. Chem.*, 1973, **10**, 1047 (*synth*)
Cooper, G.H. *et al, J. Chem. Soc. (C)*, 1971, 772, 3257 (*synth*)

3-Acetyl-2-aminopyridine   A-70024

1-(2-Amino-3-pyridinyl)ethanone, 9CI. 2-Amino-3-pyridinyl methyl ketone

[65326-33-2]

$C_7H_8N_2O$ M 136.153

Mp 136°.

Güngör, T. *et al, J. Organomet. Chem.*, 1981, **215**, 139 (*synth, pmr*)

3-Acetyl-4-aminopyridine   A-70025

1-(4-Amino-3-pyridinyl)ethanone, 9CI. 4-Amino-3-pyridinyl methyl ketone

[53277-43-3]

$C_7H_8N_2O$ M 136.153

Cryst. (C_6H_6). Mp 165-166°.

Clark, B.A.J. *et al, Tetrahedron*, 1974, **30**, 475 (*synth*)

4-Acetyl-2-aminopyridine   A-70026

1-(2-Amino-4-pyridinyl)ethanone, 9CI. 2-Amino-4-pyridinyl methyl ketone

[42182-25-2]

$C_7H_8N_2O$ M 136.153

Cryst. (toluene). Mp 133-133.5°.

Oxime: [80882-45-7]. Cryst. (EtOAc). Mp 215-217°.

N²,N²-Di-Me: [80882-53-7].
  $C_9H_{12}N_2O$ M 164.207
  Cryst. Mp 37-42°.

N²,N²-Di-Me, oxime: [80882-54-8]. Pale-yellow cryst. (toluene). Mp 145-148°.

LaMattina, J.L., *J. Heterocycl. Chem.*, 1983, **20**, 533 (*synth*)

4-Acetyl-3-aminopyridine   A-70027

1-(3-Amino-4-pyridinyl)ethanone, 9CI. 3-Amino-4-pyridinyl methyl ketone

[13210-52-1]

$C_7H_8N_2O$ M 136.153

Yellow plates (pet. ether). Mp 91-92° (87°).

2,4-Dinitrophenylhydrazone; B,HCl: [13210-54-3]. Pale-yellow needles (AcOH). Mp 287-288° dec.

Picrate: [51460-35-6]. Yellow leaflets (EtOH aq.). Mp 189-191°.

5-Acetyl-2-aminopyridine   A-70028

1-(6-Amino-3-pyridinyl)ethanone, 9CI. 6-Amino-3-pyridinyl methyl ketone, 8CI

[19828-20-7]

$C_7H_8N_2O$ M 136.153

Cryst. (Me_2CO/hexane or Et_2O). Mp 122° (89-90°).

B,HCl: [19828-21-8]. Cryst. (MeOH/Et_2O). Mp 194°.

N-Oxide: [49647-13-4].
  $C_7H_8N_2O_2$ M 152.152
  Cryst. (Me_2CO aq.) (as hydrochloride). Mp 190-191° (hydrochloride).

Swiss Pat., 452 525, (*1968*); *CA*, **69**, 96488 (*synth*)
Korytnyk, W. *et al, J. Med. Chem.*, 1973, **16**, 959 (*synth*)

2-Acetyl-3-aminothiophene   A-70029

1-(3-Amino-2-thienyl)ethanone, 9CI

[31968-33-9]

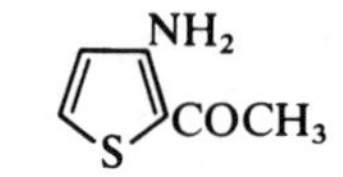

C_6H_7NOS M 141.187

Needles (pet. ether). Mp 82-83° (89-90°).

Huddleston, P.R. *et al, Synth. Commun.*, 1979, **9**, 732 (*synth*)

1-Acetylanthracene   A-70030

1-(1-Anthracenyl)ethanone, 9CI. 1-Anthryl methyl ketone, 8CI

[7396-21-6]

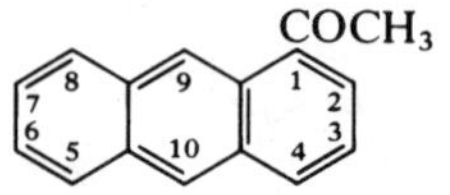

$C_{16}H_{12}O$ M 220.270

Cryst. (EtOH). Mp 110.5-111°.

2,4-Dinitrophenylhydrazone: Cryst. (Py/EtOH). Mp 260°.

Gore, P.H. *et al, J. Org. Chem.*, 1957, **22**, 135.
Bassilios, H.F. *et al, Recl. Trav. Chim. Pays-Bas*, 1963, **82**, 298 (*synth*)
Gore, P.H. *et al, J. Chem. Soc. (C)*, 1966, 1729 (*synth, deriv, uv, ir*)

2-Acetylanthracene   A-70031

1-(2-Anthracenyl)ethanone, 9CI. 2-Anthryl methyl ketone, 8CI

[10210-32-9]

$C_{16}H_{12}O$ M 220.270

Cryst. (C_6H_6). Mp 195.5-196°.

2,4-Dinitrophenylhydrazine: Cryst. (Py). Mp 297°.

Gore, P.H. *et al, J. Org. Chem.*, 1957, **22**, 135.
Gore, P.H. *et al, J. Chem. Soc. (C)*, 1966, 1729 (*synth, deriv, uv, ir*)

Acetylcedrene

A-70032

Vertofix. Lixetone

[32388-55-9]

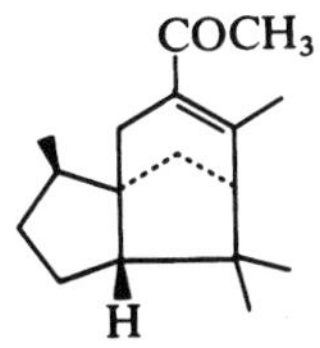

$C_{17}H_{26}O$ M 246.392

Perfumery ingredient. Bp_4 120-122°, $Bp_{0.01}$ 84-86°. $[\alpha]_D^{25}$ −38.5° (neat). n_D^{20} 1.5152.

Kitchens, G.C. *et al*, *J. Org. Chem.*, 1972, **37**, 6 (*synth, ir, ms, pmr*)

McAndrew, B.A. *et al*, *J. Chem. Soc., Perkin Trans. 1*, 1983, 1373 (*synth*)

2-Acetyl-5-chloropyridine

A-70033

1-(5-Chloro-2-pyridinyl)ethanone, 9CI

[94952-46-2]

C_7H_6ClNO M 155.584

Eur. Pat., 122 056, (*1983*); *CA*, **102**, 132042 (*synth*)

3-Acetyl-2-chloropyridine

A-70034

1-(2-Chloro-3-pyridinyl)ethanone, 9CI

[55676-21-6]

C_7H_6ClNO M 155.584

Sainsbury, M. *et al*, *J. Chem. Soc., Perkin Trans. 1*, 1975, 289.

U.S.P., 4 060 601, (*1977*); *CA*, **88**, 136458 (*synth*)

4-Acetyl-2-chloropyridine

A-70035

1-(2-Chloro-4-pyridinyl)ethanone, 9CI. 2-Chloro-4-pyridinyl methyl ketone, 8CI

[23794-15-2]

C_7H_6ClNO M 155.584

Cryst. (pet. ether). Mp 36-39°.

Ger. Pat., 1 811 833, (*1969*); *CA*, **71**, 91325 (*synth*)

LaMattina, J.L., *J. Heterocycl. Chem.*, 1983, **20**, 533 (*synth*)

4-Acetyl-3-chloropyridine

A-70036

1-(3-Chloro-4-pyridinyl)ethanone, 9CI. 3-Chloro-4-pyridinyl methyl ketone

[78790-82-6]

C_7H_6ClNO M 155.584

Liq. Bp_2 81°, $Bp_{0.1}$ 55°.

LaMattina, J.L. *et al*, *J. Org. Chem.*, 1981, **46**, 4179 (*synth*)

Marsais, F. *et al*, *J. Organomet. Chem.*, 1981, **216**, 139 (*synth*)

5-Acetyl-2-chloropyridine

A-70037

1-(6-Chloro-3-pyridinyl)ethanone, 9CI

[55676-22-7]

C_7H_6ClNO M 155.584

Mp 104°.

Phenylhydrazone: Pale-yellow cryst. Mp 164°.

Binz, A. *et al*, *Justus Liebigs Ann. Chem.*, 1931, **487**, 127 (*synth*)

Sainsbury, M. *et al*, *J. Chem. Soc., Perkin Trans. 1*, 1975, 289.

1-Acetylcyclohexene

A-70038

Updated Entry replacing A-00249

1-(1-Cyclohexen-1-yl)ethanone, 9CI. 3,4,5,6-Tetrahydroacetophenone

[932-66-1]

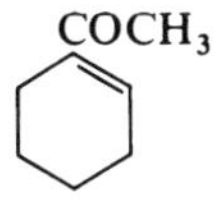

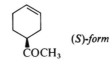

$C_8H_{12}O$ M 124.182

Bp_{22} 85-88°. n_D^{20} 1.488.

Oxime: [23042-98-0].

$C_8H_{13}NO$ M 139.197

Cryst. (C_6H_6/ligroin). Mp 99°.

2,4-Dinitrophenylhydrazone: Mp 204-205° (196-197°).

Org. Synth., Coll. Vol., 3, 22 (*synth*)

Ferres, H., *J. Chem. Soc., Perkin Trans. 2*, 1973, 936 (*synth*)

Hasbrouck, R. *et al*, *J. Org. Chem.*, 1973, **38**, 2103 (*synth*)

Zhang, P. *et al*, *Synth. Commun.*, 1986, **16**, 957 (*synth, ir, uv*)

4-Acetylcyclohexene

A-70039

3-Cyclohexenyl methyl ketone. 1′,2′,3′,6′-Tetrahydroacetophenone

$C_8H_{12}O$ M 124.182

(S)-form

$[\alpha]_D^{23}$ −116.3° (c, 1.25 in $CHCl_3$) (~99% opt. pure).

Ceder, O. *et al*, *Acta Chem. Scand., Ser. B*, 1976, **30**, 908 (*synth, ir, pmr, ms*)

Brown, H.C. *et al*, *J. Am. Chem. Soc.*, 1987, **109**, 5420 (*synth, ir, pmr*)

4-Acetyl-2,3-dihydro-5-hydroxy-2-isopropenylbenzofuran

A-70040

1-[2,3-Dihydro-5-hydroxy-2-(1-methylethenyl)-4-benzofuranyl]ethanone, 9CI. 4-Acetyl-2-isopropenyl-2,3-dihydro-5-benzofuranol

[110694-79-6]

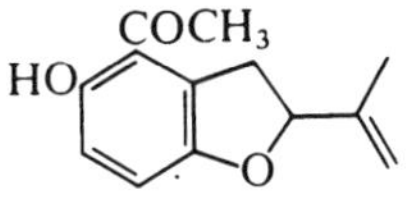

$C_{13}H_{14}O_3$ M 218.252

Isol. from *Lasiolaena morii*. Cryst. (hexane). Mp 75-76°.

Bohlmann, F. *et al*, *Phytochemistry*, 1982, **21**, 161 (*isol*)

Yamaguchi, S. *et al*, *Bull. Chem. Soc. Jpn.*, 1986, **59**, 3983 (*synth, struct*)

2-Acetyl-4,5-dihydrothiazole

A-70041

1-(4,5-Dihydro-2-thiazolyl)ethanone, 9CI. Methyl 2-thiazolin-2-yl ketone, 8CI. 2-Acetyl-2-thiazoline

[29926-41-8]

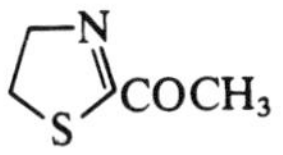

C_5H_7NOS M 129.176

Roasted meat-like flavour ingredient. Mp 25-26°. Bp_{11} 94°. n_D^{20} 1.5294.

Oxime: [37112-89-3].
$C_5H_8N_2OS$ M 144.191
Cryst. (CCl_4). Mp 178-179° dec.

Doornbos, T. *et al, Recl. Trav. Chim. Pays-Bas,* 1972, **91**, 711 (*synth, deriv, ir, pmr, uv, ms*)
U.S.P., 3 778 518, (*1973*); *CA,* **80**, 69405e (*props*)

4-Acetyl-2,2-dimethyl-1,3-dioxolane   A-70042

1-(2,2-Dimethyl-1,3-dioxolan-4-yl)ethanone, 9CI

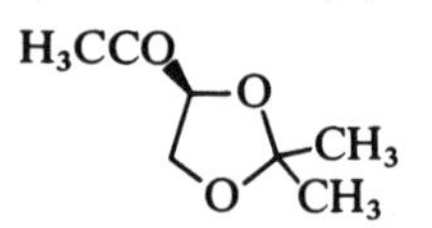

$C_7H_{12}O_3$ M 144.170

(*S*)-*form* [99566-52-6]
Chiral intermed. Oil. $[\alpha]_D$ −65° (c, 1.5 in C_6H_6), $[\alpha]_D$ −82° (c, 3.3 in $CHCl_3$). 96% enantiomeric excess. Previous syntheses claimed a lower opt. rotn.

Tanner D. *et al,* Synth. Commun.1986, **16**, 1517 (*synth, pmr, bibl*)

6-Acetylhomopterin   A-70043

6-Acetyl-2-amino-1,7,8,9-tetrahydro-4H-pyrimido[4,5-b] [1,4]diazepin-4-one, 9CI

[80003-63-0]

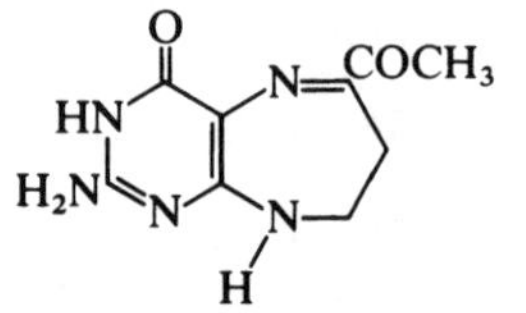

$C_9H_{11}N_5O_2$ M 221.218
Naturally occurring diazepine from *Drosophila melanogaster.* Intermed. involved in the biosynth. of Drosopterin, D-60521 . Greenish-yellow cryst.

Wiederrecht, G.J. *et al, J. Biol. Chem.,* 1981, **256**, 10399 (*isol, struct, ms, pmr*)
Jacobson, K.B. *et al, Biochemistry,* 1982, **21**, 5700 (*isol, struct, pmr, cmr, uv*)
Boyle, P.H. *et al, Tetrahedron Lett.,* 1987, **28**, 5331 (*synth*)

5-Acetyl-2-(1-hydroxy-1-methylethyl)-benzofuran   A-70044

1-[2-(1-Hydroxy-1-methylethyl)-5-benzofuranyl]-ethanone, 9CI. 5-Acetyl-2-(2-hydroxyisopropyl)-benzofuran

[64165-99-7]

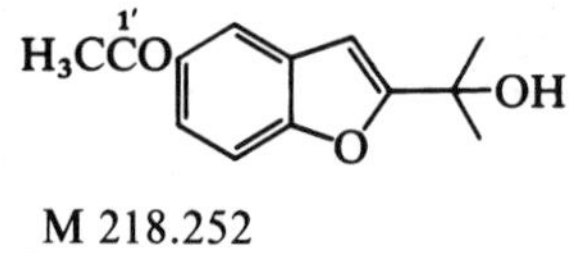

$C_{13}H_{14}O_3$ M 218.252
Isol. from *Podachaenium eminens, Fleischmanniopsis leucocephala, Smallanthus fruticosus, Stylotrichium rotundifolium* and other plants. Cryst. (Et_2O/pet. ether). Mp 73°.

1′-Alcohol: 5-(1-Hydroxyethyl)-2-(1-hydroxy-1-methylethyl)benzofuran
$C_{13}H_{16}O_3$ M 220.268
Constit. of *Smallanthus fruticosus.* Oil. No opt. rotn. reported.

Bohlmann, F. *et al, Phytochemistry,* 1977, **16**, 1304; 1978, **17**, 2101; 1980, **19**, 973; 1981, **20**, 1887 (*isol, ir, ms, struct*)

Schneiders, G.E. *et al, Synth. Commun.,* 1980, **10**, 699 (*synth*)

2-Acetyl-1*H*-indene   A-70045

1-(1H-Inden-2-yl)ethanone, 9CI. Inden-2-yl methyl ketone

[43073-11-6]

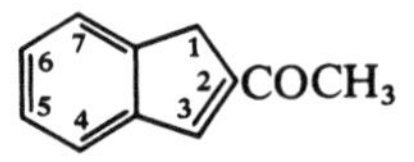

$C_{11}H_{10}O$ M 158.199
Cryst. (Et_2O/hexane). Mp 62° (56-58°). Originally reported (1921) as having Mp 122°; this was a trimer.

Doyle, P. *et al, Tetrahedron Lett.,* 1973, 2903 (*synth*)
Marx, J.N. *et al, Synth. Commun.,* 1973, **3**, 95 (*synth, pmr*)
Eliasson, B. *et al, J. Chem. Soc., Perkin Trans. 2,* 1981, 403 (*cmr*)
Satoh, T. *et al, Bull. Chem. Soc. Jpn.,* 1987, **60**, 1839 (*synth, ir, pmr, ms*)

3-Acetyl-1*H*-indene   A-70046

1-(1H-Inden-3-yl)ethanone, 9CI. 8CI. 1-Acetylindene

[28529-48-8]
$C_{11}H_{10}O$ M 158.199
Mp 65-66° (52-54°).

Uhde, W. *et al, Chem. Ber.,* 1970, **103**, 2675 (*synth, struct*)
Craig, J.C. *et al, J. Org. Chem.,* 1974, **39**, 1669 (*synth*)

6-Acetyl-1*H*-indene   A-70047

1-(1H-Inden-6-yl)ethanone, 9CI. Inden-6-yl methyl ketone
$C_{11}H_{10}O$ M 158.199
Liq. Bp_3 143-145°.
2,4-Dinitrophenylhydrazone: Cryst. (Py). Mp 265-270°.

Kumazawa, Z. *et al, Agric. Biol. Chem.,* 1961, **25**, 798 (*synth*)

2-Acetylnaphtho[1,8-*bc*]pyran   A-70048

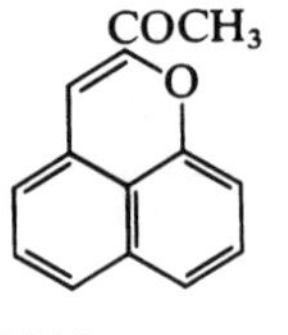

$C_{14}H_{10}O_2$ M 210.232
Yellow-orange microcryst. (pet. ether).

Buisson, J.-P. *et al, J. Heterocycl. Chem.,* 1988, **25**, 539 (*synth, pmr, uv*)

12-(Acetyloxy)-10-[(acetyloxy)methylene]-6-methyl-2-(4-methyl-3-pentenyl)-2,6,11-dodecatrienal, 9CI   A-70049

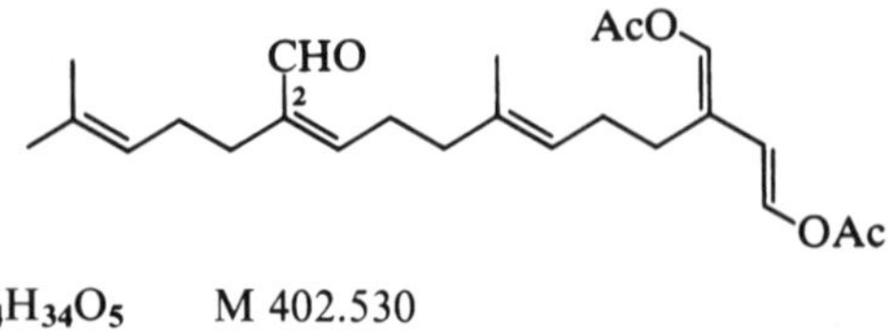

$C_{24}H_{34}O_5$ M 402.530
(*2E*)-*form*
Metab. of *Penicillus dumetosus.* Oil.

(2Z)-form

Metab. of *P. dumetosus*. Oil.

Paul, V.J. *et al*, *Tetrahedron*, 1984, **40**, 2913.

2-[2-(Acetyloxy)ethenyl]-6,10-dimethyl-2,5,9-undecatrienal, 9CI   A-70050

[93888-67-6]

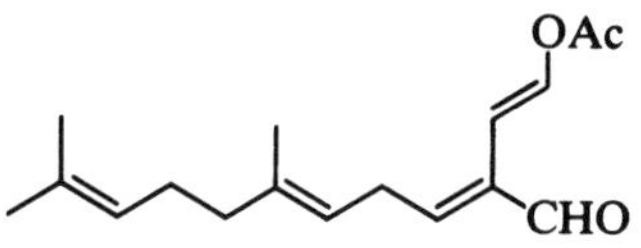

$C_{17}H_{24}O_3$ M 276.375

Metab. of algae *Penicillus capitatus* and *Udotea cyathiformis*. Antibacterial and ichthyotoxin. Oil.

Paul, V.J. *et al*, *Tetrahedron*, 1984, **40**, 2913.

2-Acetylpiperidine   A-70051

1-(2-Piperidinyl)ethanone, 9CI

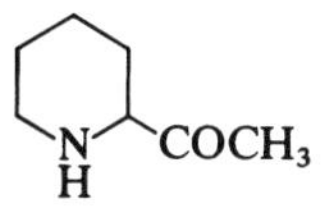

$C_7H_{13}NO$ M 127.186

(±)-form [106318-86-9]

Light-yellow oil. Unstable, lib. from HCl salt immed. prior to use.

B,HCl: [106318-66-5]. Powder. Mp 225-228° dec.

Oxime (E-): [106318-67-6].

$C_7H_{14}N_2O$ M 142.200

Solid. Mp 103-105°.

Phenylhydrazone: Bp$_{0.54}$ 143-145°.

tert-*Butylhydrazone:* Oil. Bp$_{0.5}$ 100-105°.

Ethylene ketal: [106318-87-0].

$C_9H_{17}NO_2$ M 171.239

Liq. Bp$_{0.28}$ 55-60°.

Norman, M.H. *et al*, *J. Org. Chem.*, 1987, **52**, 226 (*synth, ir, pmr, cmr*)

3-Acetylpyrrole   A-70052

Updated Entry replacing A-00414

1-(1H-Pyrrol-3-yl)ethanone, 9CI. Methyl 3-pyrryl ketone. 3-Acetopyrrole

[1072-82-8]

C_6H_7NO M 109.127

Cryst. (CHCl$_3$/pet. ether). Mp 115-116°.

Khan, M.K.A. *et al*, *Tetrahedron*, 1966, **22**, 2095 (*synth*)
Loader, C.E. *et al*, *Tetrahedron*, 1969, **25**, 3879 (*synth, pmr*)
v. Leusen, A.M. *et al*, *Tetrahedron Lett.*, 1972, 5337 (*synth*)
Anderson, H.J. *et al*, *Synth. Commun.*, 1987, **17**, 401 (*synth*)

2-Acetyl-1,2,3,4-tetrahydronaphthalene   A-70053

1-(1,2,3,4-Tetrahydro-2-naphthalenyl)ethanone, 9CI. 2-Acetyltetralin

[35060-50-5]

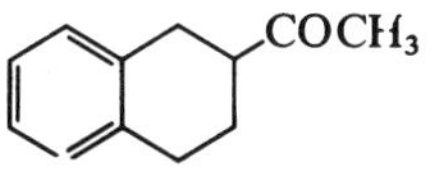

$C_{12}H_{14}O$ M 174.242

(±)-form

Oil.

2,4-Dinitrophenylhydrazone: Mp 164°.

Ravichandran, K. *et al*, *Heterocycles*, 1987, **26**, 645 (*synth, pmr*)

Achalensolide   A-70054

Updated Entry replacing A-30038

3-Oxo-4,11(13)-guaiadien-12,8β-olide

[87302-42-9]

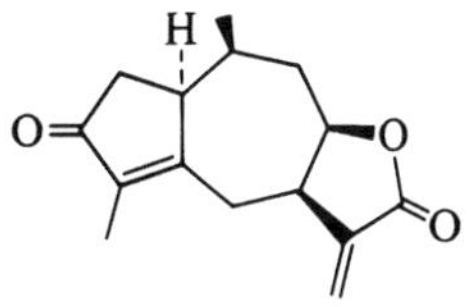

$C_{15}H_{18}O_3$ M 246.305

Constit. of *Stevia achalensis*. Cryst. (MeOH/Et$_2$O). Mp 176-177°. [α]$_D$ +226.8° (c, 0.34 in CHCl$_3$).

3-Deoxo: 4,11(13)-Guaiadien-12,8β-olide. 3-Desoxoachalensolide.

$C_{15}H_{20}O_2$ M 232.322

Constit. of *S. polyphylla*. Oil. [α]$_D^{24}$ +162° (c, 1.55 in CHCl$_3$).

3-Deoxo, 11β,13-Dihydro: 4-Guaien-12,8β-olide. 11β,13-Dihydrodesoxoachalensolide.

$C_{15}H_{22}O_2$ M 234.338

Constit. of *S. polyphylla*.

Oberti, J.C. *et al*, *J. Org. Chem.*, 1983, **48**, 4038 (*isol, struct*)
Zdero, C. *et al*, *Phytochemistry*, 1988, **23**, 2835 (*derivs*)

Aconitic acid   A-70055

Updated Entry replacing A-00441

1-Propene-1,2,3-tricarboxylic acid, 9CI. Achilleaic acid. Citridinic acid. Equisetic acid. Pyrocitric acid

[499-12-7]

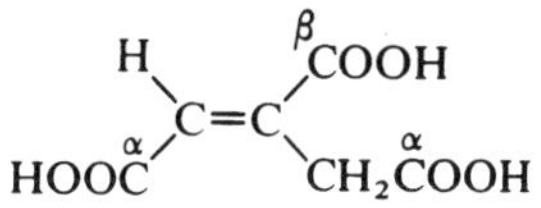

$C_6H_6O_6$ M 174.110

▷UD2380000.

(E)-form [4023-65-8]

Isol. from *Asarum europaeum*, from cane-sugar molasses and other plant sources. Used to prod. unsatd. polyesters. Leaflets (H$_2$O). Sol. H$_2$O, EtOH. Mp 194-195° dec. pK_{a1} 2.8 (25°), pK_{a2} 4.46 (25°). Decarboxylates at Mp to Itaconic acid. Mp variable with rate of htg.

α-Mono-Me ester: [65146-88-5].

$C_7H_8O_6$ M 188.137

Prisms (Me$_2$CO/C$_6$H$_6$). Mp 136-137°. Called 3-ester in CA.

β-Mono-Me ester: [65146-89-6].

$C_7H_8O_6$ M 188.137

Prisms (Me$_2$CO/C$_6$H$_6$). Mp 144-145°. Called 2-ester in CA.

γ-Mono-Me ester: [62424-07-1].

$C_7H_8O_6$ M 188.137

Prisms (Me$_2$CO/C$_6$H$_6$). Mp 154-155°. Called 1-ester in CA.

Tri-Me ester: [4271-99-2].

$C_9H_{12}O_6$ M 216.190

Sol. EtOH, Et$_2$O. Bp 270°, Bp$_{14}$ 161°.

Triamide:
$C_6H_9N_3O_3$ M 171.155
Needles. Sol. hot H_2O, insol. EtOH, Et_2O. Turns brown at 250°. Sinters without melting at 260°.
Anhydride:
$C_6H_4O_5$ M 156.095
Mp 134-135°.

(Z)-form [585-84-2]
Mp 125°. Gives (E)-form on heating.
α-Mono-Me ester: [65146-85-2]. Prisms (Me-$_2CO/C_6H_6$). Mp 101-102°.
β-Mono-Me ester: [65146-86-3]. Mp 102-104°.
γ-Mono-Me ester: [65146-87-4]. Prisms (Me-$_2CO/C_6H_6$). Mp 126-127°.
Anhydride: [31511-11-2]. Mp 74°.

Malachowski, R. *et al, Ber.*, 1928, **61**, 2521 (synth, props)
Krogh, A., *Acta Chem. Scand.*, 1969, **23**, 2932 (isol)
Org. Synth., Coll. Vol., **2**, 12 (synth)

1-Acridinecarboxaldehyde   A-70056
1-Formylacridine
[113139-13-2]

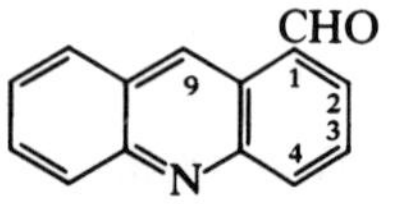

$C_{14}H_9NO$ M 207.231
Pale-green plates (hexane). Mp 118-119°.

Takahashi, K. *et al, J. Heterocycl. Chem.*, 1987, **24**, 977 (synth, pmr)

2-Acridinecarboxaldehyde   A-70057
2-Formylacridine
[113139-14-3]
$C_{14}H_9NO$ M 207.231
Straw-coloured plates (hexane). Mp 196-198°.

Takahashi, K. *et al, J. Heterocycl. Chem.*, 1987, **24**, 977 (synth, pmr)

3-Acridinecarboxaldehyde   A-70058
3-Formylacridine
[113139-15-4]
$C_{14}H_9NO$ M 207.231
Pale-yellow needles (hexanes). Mp 133-134°.

Takahashi, K. *et al, J. Heterocycl. Chem.*, 1987, **24**, 977 (synth, pmr)

4-Acridinecarboxaldehyde   A-70059
4-Formylacridine
[113139-16-5]
$C_{14}H_9NO$ M 207.231
Yellow prisms (hexane). Mp 140-142°.

Takahashi, K. *et al, J. Heterocycl. Chem.*, 1987, **24**, 977 (synth, pmr)

9-Acridinecarboxaldehyde   A-70060
[885-23-4]
$C_{14}H_9NO$ M 207.231
Yellow needles (C_6H_6/pet. ether). Mp 147°.
10-Oxide: [10228-97-4].
    $C_{14}H_9NO_2$ M 223.231

Mp 261° (200° dec.).

Chardonnens, L. *et al, Helv. Chim. Acta*, 1949, **32**, 656 (synth, bibl)
Tsuge, O. *et al, Bull. Chem. Soc. Jpn.*, 1965, **38**, 2037 (synth)
Kawashima, K. *et al, Chem. Pharm. Bull.*, 1978, **26**, 951 (oxide)
Faure, R. *et al, Chem. Scr.*, 1980, **15**, 62 (cmr)

9-Acridinecarboxylic acid, 9CI   A-70061
Updated Entry replacing A-00461
[5336-90-3]
$C_{14}H_9NO_2$ M 223.231
Yellow cryst. Spar. sol. H_2O. Mp >300°.
Me ester: [5132-81-0].
    $C_{15}H_{11}NO_2$ M 237.257
    Mp 127-128°.
    ▷AR7675000.
Ph ester:
    $C_{20}H_{13}NO_2$ M 299.328
    Fine needles (hexane/THF). Mp 189-190°.
Chloride: [5132-80-9].
    $C_{14}H_8ClNO$ M 241.676
    Mp 221-222° (as hydrochloride).
Amide: [35417-96-0].
    $C_{14}H_{10}N_2O$ M 222.246
    Yellow needles (EtOH). Mp 263-264°.
Nitrile: [42978-64-3]. 9-Cyanoacridine.
    $C_{14}H_8N_2$ M 204.231
    Mp 181°.

Lehmstedt, K. *et al, Ber.*, 1928, **61**, 2044 (deriv, synth)
Albert, A., *The Acridines*, 1951, Arnold, London, 282 (synth)
Rauhut, M.M. *et al, J. Org. Chem.*, 1965, **30**, 3587 (deriv, synth)
White, E.H. *et al, J. Am. Chem. Soc.*, 1987, **109**, 5189 (deriv, synth, ir, pmr)

β-Actinorhodin   A-70062
[109978-15-6]

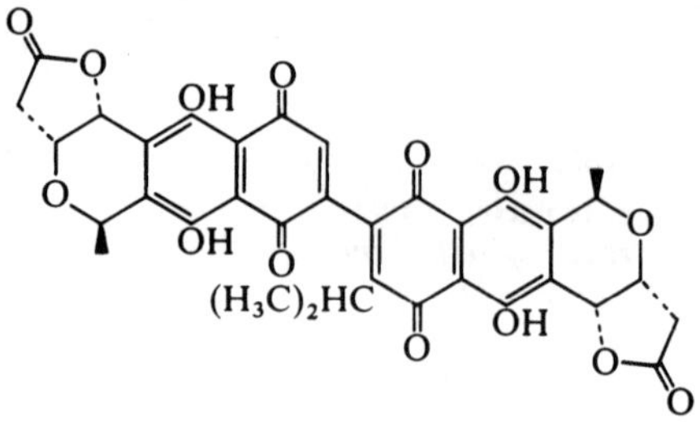

$C_{35}H_{28}O_{14}$ M 672.598
Anthraquinone. Prod. by *Streptomyces violaceoruber*.

Krone, B. *et al, Justus Liebigs Ann. Chem.*, 1987, 751 (pmr, struct)

Consult the *Dictionary of Antibiotics and Related Substances* for a fuller treatment of antibiotics and related compounds.

Acuminatolide
A-70063

Updated Entry replacing A-60063

4-(1,3-Benzodioxol-5-yl)tetrahydro-1H,3H-furo[3,4-c]furan-1-one, 9CI

[108645-28-9]

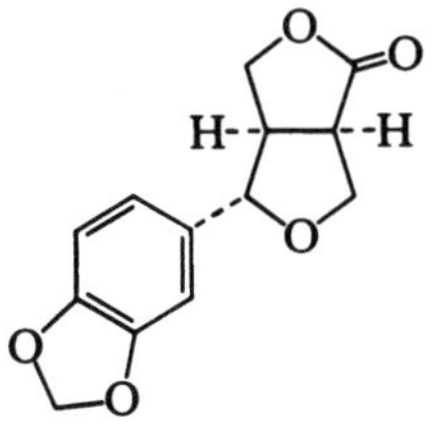

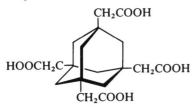

C₁₃H₁₂O₅ M 248.235

Constit. of *Helichrysum acuminatum.* Cryst. Mp 118°.

4-Epimer: [28168-96-9]. **Pluviatide.** From *Zanthoxylum alatum* and *Z. pluviatile.* Needles (MeOH). Mp 162°. [α]_D −35.5° (CHCl₃). Epimeric at the aryl group.

Corrie, J.E.T. *et al, Aust. J. Chem.,* 1970, **23**, 133 (*isol*)
Dashpande, V.H. *et al, Indian J. Chem., Sect. B,* 1977, **15**, 95 (*isol*)
Jakupovic, J. *et al, Phytochemistry,* 1987, **26**, 803 (*isol, struct*)

1,3,5,7-Adamantanetetraacetic acid
A-70064

*Tricyclo[3.3.1.1³,⁷]decane-1,3,5,7-tetraacetic acid, 9CI.
1,3,5,7-Tetrakis(carboxymethyl)adamantane*

[84782-76-3]

C₁₈H₂₄O₈ M 368.383

Cryst. (AcOH). Mp >300°.

Tetranitrile: [84782-75-2]. *Tricyclo[3.3.1.1³,⁷]decane-1,3,5,7-tetraacetonitrile. 1,3,5,7-Tetrakis(cyanomethyl)adamantane.*
C₁₈H₂₀N₄ M 292.383
Cryst. (AcOH). Mp 266-268°.

Tetrakis(dimethylamide):
C₂₆H₄₄O₄N₄ M 476.658
Oil. Bp₀.₁ 345-350° (oven).

Naemura, K. *et al, Tetrahedron,* 1986, **42**, 1763 (*synth, ir*)

Adenanthin
A-70065

[109974-30-3]

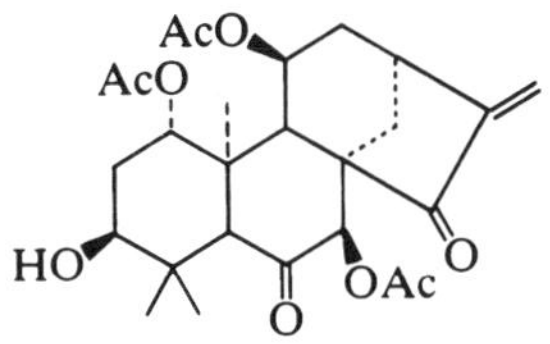

C₂₆H₃₄O₉ M 490.549

ent-Kaurene terpenoid antibiotic. Isol. from *Rabdosia adenantha.* Possesses bacteriostatic, antiinflammatory and antitumour props. Mp 255° dec. [α]_D¹³ −76° (c, 0.25 in CHCl₃).

Xu, Y.-L. *et al, Tetrahedron Lett.,* 1987, **28**, 499 (*isol, struct*)

Adonirubin
A-70066

3-Hydroxy-β,β-carotene-4,4′-dione. Phoenicoxanthin

[4418-72-8]

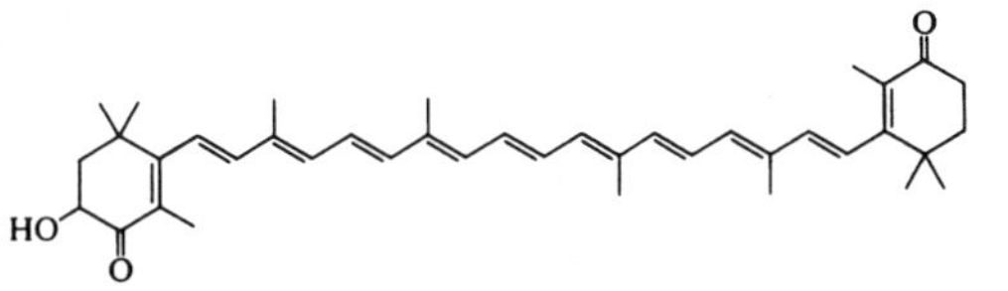

C₄₀H₅₂O₃ M 580.849

Red pigment from feathers of flamingos (e.g. *Phoenicopterus ruber*) and from flowers of *Adonis annua.* Red gum. λ_max 478 nm (C₆H₆).

Cooper, R.D.G. *et al, J. Chem. Soc., Perkin Trans. 1,* 1975, 2195.

Adoxoside
A-70067

Updated Entry replacing A-50069

[42830-26-2]

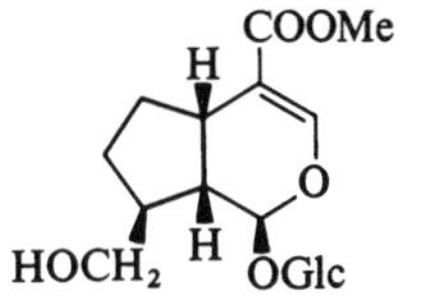

C₁₇H₂₆O₁₀ M 390.386

Constit. of *Adoxa moschatellina.* Amorph.

Penta-Ac: Cryst. Mp 140.5-141.5°. [α]_D²² −63° (c, 1 in CHCl₃).
6β-Hydroxy: [110559-86-9]. **6β-Hydroxyadoxoside.**
C₁₇H₂₆O₁₁ M 406.386
Constit. of *Castilleja integra.* Foam. [α]_D²⁷ −83.4° (c, 0.43 in MeOH).

Jensen, S.R. *et al, Biochem. Syst. Ecol.,* 1979, **7**, 103 (*isol*)
Damtoft, S. *et al, Phytochemistry,* 1981, **20**, 2717 (*cmr*)
Gardner, D.R. *et al, J. Nat. Prod.,* 1987, **50**, 485 (*isol*)

Adoxosidic acid
A-70068

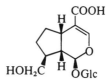

C₁₆H₂₄O₁₀ M 376.360

Constit. of *Castilleja integra* and *Fouquieria columnaris.*

Jensen, S.R. *et al, Biochem. Syst. Ecol.,* 1979, **7**, 103 (*isol*)
Jensen, S.R. *et al, Phytochemistry,* 1982, **21**, 1623 (*isol*)
Gardner, D.R. *et al, J. Nat. Prod.,* 1987, **50**, 485 (*isol*)

Adriadysiolide
A-70069

[113540-73-1]

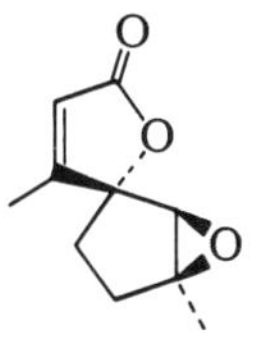

C₁₀H₁₂O₃ M 180.203

Constit. of a *Dysidea* sp. of the Adriatic Sea. Powder. Mp 76-77°. [α]₄₃₅²⁰+2.1° (c, 0.625 in MeOH).

Mancini, I. *et al*, *Helv. Chim. Acta*, 1987, **70**, 2011.

Afraglaucolide A-70070

[115346-35-5]

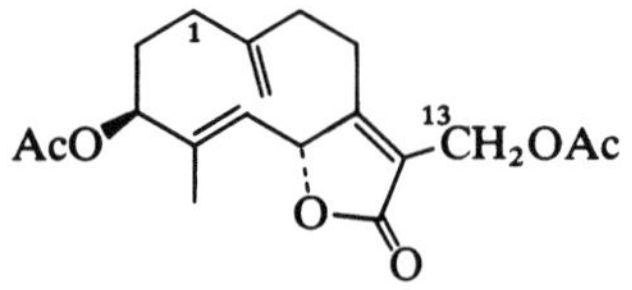

C$_{19}$H$_{24}$O$_6$ M 348.395

1α-Hydroxy: [115367-47-0]. *1α-Hydroxyafraglaucolide*.
  C$_{19}$H$_{24}$O$_7$ M 364.394
  Constit. of *Artemisia afra*.
1β-Hydroxy: [115346-39-9]. *1β-Hydroxyafraglaucolide*.
  C$_{19}$H$_{24}$O$_7$ M 364.394
  Constit. of *A. afra*.
1α-Hydroxy, 13-de-Ac: [115334-66-2]. *13-Desacetyl-1α-hydroxyafraglaucolide*.
  C$_{17}$H$_{22}$O$_6$ M 322.357
  Constit. of *A. judaica*.
1β-Hydroxy, 13-de-Ac: *13-Desacetyl-1β-hydroxyafraglaucolide*.
  C$_{17}$H$_{22}$O$_6$ M 322.357
  Constit. of *A. judaica*. Oil.

Khafagy, S.M. *et al*, *Phytochemistry*, 1988, **27**, 1125 (*isol*)
Jakupovic, J. *et al*, *Phytochemistry*, 1988, **27**, 1129 (*isol*)

Agaritine A-70071

Updated Entry replacing A-00625
β-N-(γ-Glutamyl)-4-hydroxymethylphenylhydrazine.
Glutamic acid 5-2-(α-hydroxy-p-tolyl)hydrazide, 8CI
[2757-90-6]

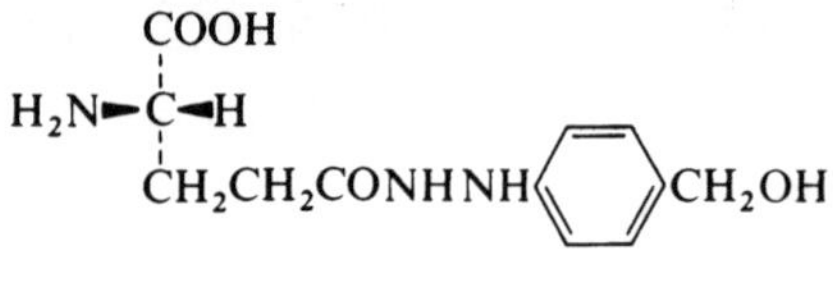

C$_{12}$H$_{17}$N$_3$O$_4$ M 267.284
▷MA1284000.
(S)-form
  L-form
  Constit. of some members of the family *Agaricaceae*,
  notably *Agaricus bisporus*. Cryst. (EtOH/butanol).
  Mp 205-209°. [α]$_D^{25}$ +7° (c, 0.8 in H$_2$O). pK_{a1} 3.4,
  pK_{a2} 8.86 (H$_2$O).

Daniels, E.G. *et al*, *J. Org. Chem.*, 1962, **23**, 3229 (*isol, synth*)
Levenberg, B., *J. Biol. Chem.*, 1964, **239**, 2267 (*isol, struct*)
Datta, S. *et al*, *Helv. Chim. Acta*, 1987, **70**, 1261 (*synth, ir, pmr, bibl*)

Agerasanin A-70072

[116397-92-3]

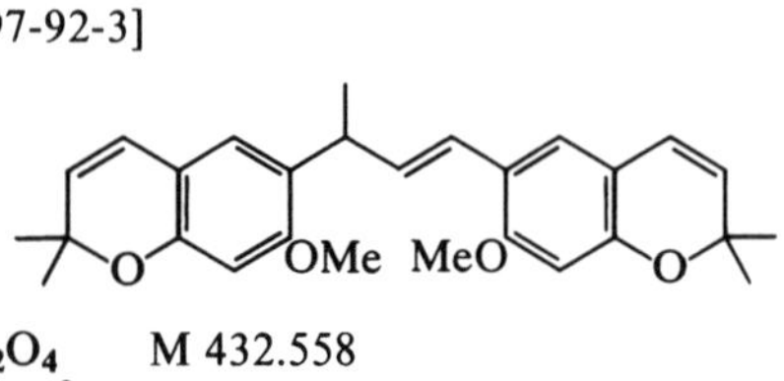

C$_{28}$H$_{32}$O$_4$ M 432.558
Constit. of *Ageratina arsenii*.

Fang, N. *et al*, *Phytochemistry*, 1988, **27**, 1902.

Aglajne 2 A-70073

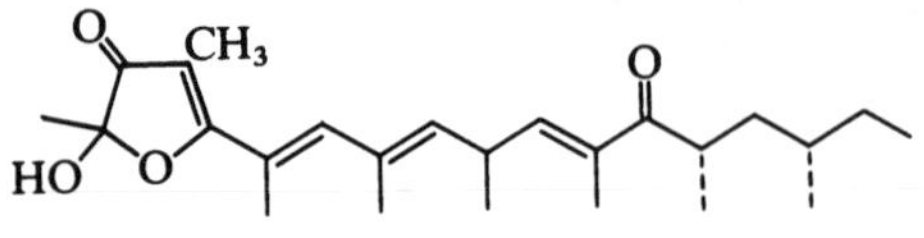

C$_{25}$H$_{38}$O$_4$ M 402.573
Constit. of the mollusc *Aglaja depicta* and its prey *Bulla striata*.
Ac: Oil. [α]$_D^{20}$ +46° (c, 1.3 in CHCl$_3$).

Cimino, G. *et al*, *J. Org. Chem.*, 1987, **52**, 5326.

Aglajne 3 A-70074

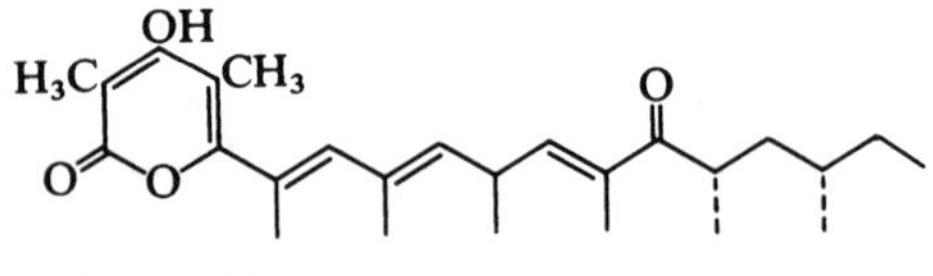

C$_{26}$H$_{38}$O$_4$ M 414.584
Constit. of the molusc *Aglaja depicta* and its prey *Bulla striata*. Oil. [α]$_D^{20}$ +105° (c, 0.8 in CHCl$_3$).

Cimino, G. *et al*, *J. Org. Chem.*, 1987, **52**, 5326.

Agropinic acid A-70075

1-(2-Carboxy-5-oxo-1-pyrrolidinyl)-1-deoxy-D-mannitol, 9CI
[74474-75-2]

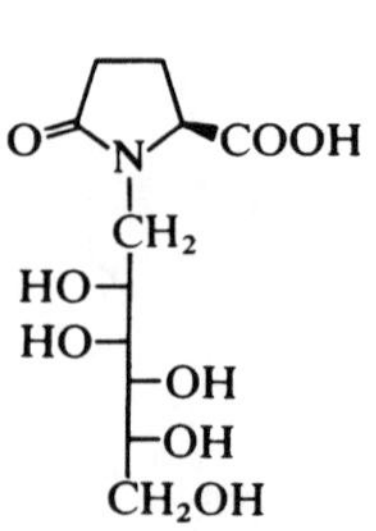

C$_{11}$H$_{19}$NO$_8$ M 293.273
Opine found in plant tumours.

Tate, M.E. *et al*, *Carbohydr. Res.*, 1982, **104**, 105 (*synth*)
Chilton, W.S. *et al*, *Biochemistry*, 1984, **23**, 3290 (*cd*)
Chilton, W.S. *et al*, *J. Bateriol.*, 1984, **158**, 650 (*deriv, struct, metab*)
Chilton, W.S. *et al*, *Phytochemistry*, 1985, **24**, 2945.

Agroskerin A-70076

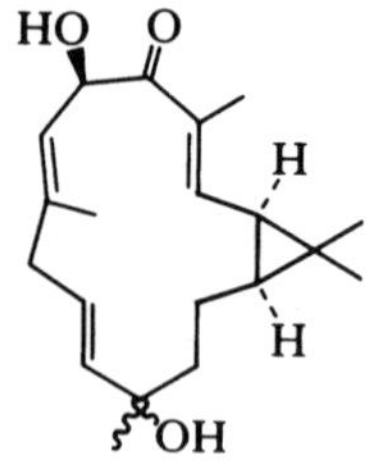

C$_{20}$H$_{30}$O$_3$ M 318.455
Constit. of *Agrostistachys hookeri*. Resin. [α]$_D$ −54° (c, 0.1 in CHCl$_3$).

Choi, Y.-H. *et al*, *J. Nat. Prod.*, 1988, **51**, 110.

Agrostophyllin   A-70077
*2,6-Dimethoxy-5H-phenanthro-[4,5-*bcd]*pyran-7-ol, 9CI*
[116174-99-3]

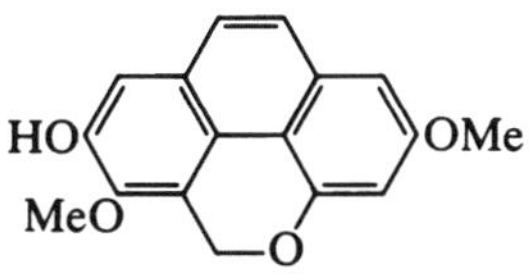

$C_{17}H_{14}O_4$ M 282.295
Constit. of *Agrostophyllum khasiyanum*. Cryst.
(EtOAc/pet. ether). Mp 86°.

Majumder, P.L. *et al, Phytochemistry*, 1988, **27**, 1899.

AH$_{10}$ A-70078
5,6,7,8-Tetrahydro-6,7,8-trihydroxy-5-[[4-oxo-2-(2-
phenylethyl)-4H-1-benzopyran-6-yl]oxy]-2-(2-pheny-
lethyl)-4H-benzopyran-4-one, 9CI
[112664-27-4]

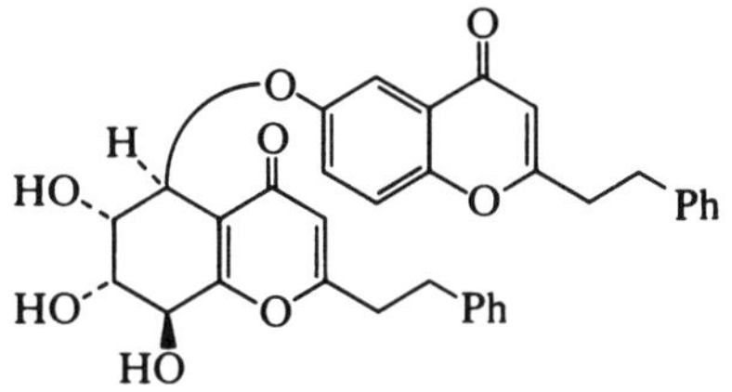

$C_{34}H_{30}O_8$ M 566.606
Isol. from Agalwood "Jinko". Powder. Mp 110°. $[\alpha]_D$
−127.7° (MeOH).
Tri-Ac: Powder. Mp 82-83°. $[\alpha]_D$ −43.6° (MeOH).

Iwagoe, K. *et al, Chem. Pharm. Bull.*, 1986, **34**, 4889 (*isol*)

Ajoene   A-70079
2-Propenyl 3-(2-propenylsulfinyl)-1-propenyl disulfide,
9CI. 4,5,9-Trithia-1,6,11-dodecatriene 9-oxide
[92285-01-3]

$$H_2C=CHCH_2S(O)CH_2CH=CHSSCH_2CH=CH_2$$

$C_9H_{14}OS_3$ M 234.389
Isol. from garlic (*Allium sativum*) extracts. Potent
inhibitor of platelet aggregation.

Block, E. *et al, J. Am. Chem. Soc.*, 1986, **108**, 7045 (*synth*)

Albomitomycin A   A-70080
[110934-24-2]

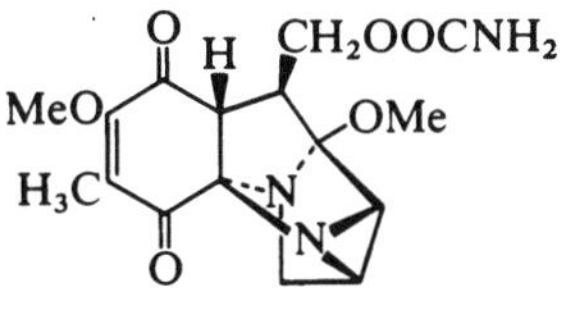

$C_{16}H_{19}N_3O_6$ M 349.343
Quinone antibiotic. Metab. of *Streptomyces caespitosus*.
Prod. by intramolecular rearrangement of Mitomycin
A. Plates (CHCl$_3$). $[\alpha]_D^{23}$ −2.7° (c, 0.5 in CHCl$_3$). Dec.
at >130°.

Kono, M. *et al, J. Am. Chem. Soc.*, 1987, **109**, 7224 (*isol,*
struct)

Alectorialic acid, 8CI   A-70081
Updated Entry replacing A-00736

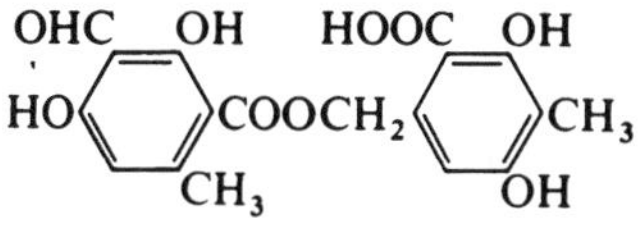

$C_{18}H_{16}O_9$ M 376.319
Constit. of *Alectoria nigricans* and *Bryoria nadvorni-*
kiana. Mp 176-179°.
Me ester: Cryst. (MeOH). Mp 202-204°.

Solberg, Y.J., *Z. Naturforsch., B*, 1967, **22**, 777.
Persson, B. *et al, Acta Chem. Scand.*, 1970, **24**, 345.
Elix, J-A. *et al. Aust. J. Chem.*, 1987, **40**, 1841 (*isol, synth*)

Alisol B   A-70082
Updated Entry replacing A-00749
24R,25-Epoxy-11β,23S-dihydroxydammar-13(17)-en-
3-one, 9CI. 24R,25-Epoxy-11β,23S-dihydroxy-13(17)-
protosten-3-one
[18649-93-9]

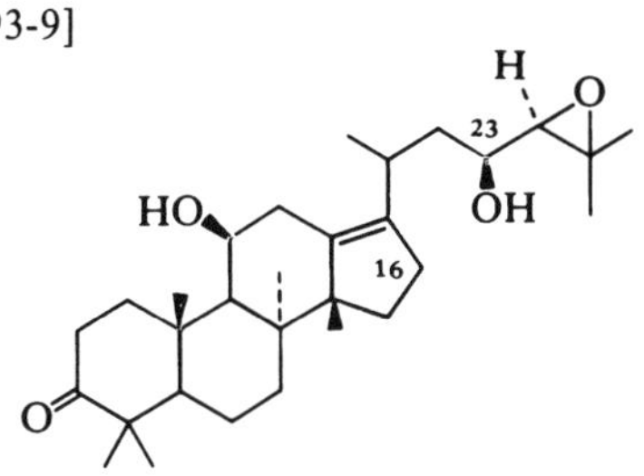

$C_{30}H_{48}O_4$ M 472.707
Constit. of rhizomes of *Alisma plantago-aquatica*. Cryst.
(EtOAc). Mp 166-168°. $[\alpha]_D$ +130° (CHCl$_3$).
23-Ac: [19865-76-0].
  $C_{32}H_{50}O_5$ M 514.744
  Constit. of *A. plantago-aquatica*. Cryst.
  (Me$_2$CO/hexane). Mp 162-163°. $[\alpha]_D$ +121°
  (CHCl$_3$).
23-Ac, 16β-hydroxy: [115346-25-3]. **16β-Hydroxyalisol**
  B monoacetate.
  $C_{32}H_{50}O_6$ M 530.743
  Constit. of *A. plantago-aquatica*. Needles
  (CH$_2$Cl$_2$/MeOH). Mp 196.5-198°. $[\alpha]_D^{25}$ +11° (c,
  0.32 in CHCl$_3$).
23-Ac, 16β-methoxy: [115333-90-9]. **16β-Methoxyalisol**
  B monoacetate.
  $C_{33}H_{52}O_6$ M 544.770
  Constit. of *A. plantago-aquatica*. Prisms
  (CH$_2$Cl$_2$/MeOH). Mp 164-166°. $[\alpha]_D^{25}$ +89.4° (c,
  0.92 in CHCl$_3$).

Murata, T. *et al, Chem. Pharm. Bull.*, 1970, **18**, 1347, 1369
  (*isol, struct*)
Pei-Wu, G. *et al, Phytochemistry*, 1988, **27**, 1161 (*derivs*)

Alliodorin   A-70083
Updated Entry replacing A-00773
[50906-59-7]

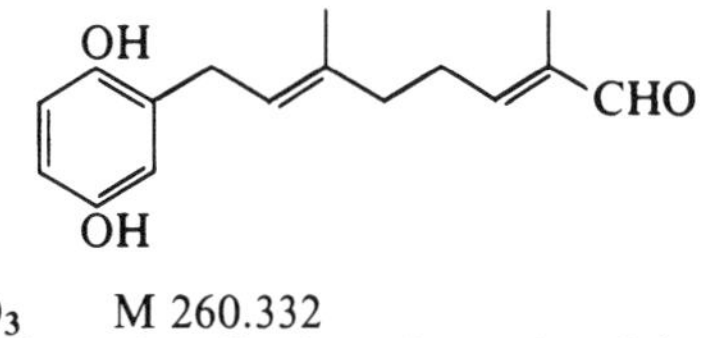

$C_{16}H_{20}O_3$ M 260.332
Constit. of *Cordia alliodora*. Cryst. (Et$_2$O/pet. ether).
Mp 87°.

Stevens, K.L. *et al, Tetrahedron*, 1976, **32**, 665 (*isol, struct*)
Tortajada, J. *et al, Bull. Soc. Chim. Belg.*, 1986, **95**, 253 (*synth*)

Altamisic acid   A-70084

Updated Entry replacing A-50095

C$_{15}$H$_{20}$O$_5$ M 280.320

Constit. of *Ambrosia tenuifolia*. Cryst. Mp 141-142°.

Me ester: **Methyl altamisate**. *1'-Noraltamisin*.
  C$_{16}$H$_{22}$O$_5$ M 294.347
  Constit. of *A. confertiflora*. Cryst.
  (Me$_2$CO/diisopropyl ether). Mp 93-95°. [α]$_D^{21}$ +34.1°
  (c, 0.2 in MeOH).

Et ester: [67604-04-0]. **Altamisin**.
  C$_{17}$H$_{24}$O$_5$ M 308.374
  Constit. of *A. cumanensis*. Cryst. (hexane). Mp 101-
  102°. [α]$_D$ +34.9°.

Borges, J. *et al, Tetrahedron Lett.*, 1978, 1513.
Oberti, J.C. *et al, Phytochemistry*, 1986, **25**, 1355.
Delgado, G. *et al, J. Nat. Prod.*, 1988, **51**, 83.

2-(Aminoacetyl)pyridine   A-70085

*2-Amino-1-(2-pyridinyl)ethanone, 9CI. Aminomethyl 2-
pyridinyl ketone. Picolinylaminomethane*
[75140-33-9]

C$_7$H$_8$N$_2$O   M 136.153

B,HCl: Prisms (EtOH). Mp 171-172° dec.
B,2HCl: Tan cryst. Mp 178-180° dec.

Clemo, G.R. *et al, J. Chem. Soc.*, 1938, 753 (*synth*)
Burrus, H.O. *et al, J. Am. Chem. Soc.*, 1945, **67**, 1468 (*synth*)

3-(Aminoacetyl)pyridine   A-70086

*2-Amino-1-(3-pyridinyl)ethanone, 9CI. Nicotinylamino-
methane. Aminomethyl 3-pyridinyl ketone*
[51941-15-2]

C$_7$H$_8$N$_2$O   M 136.153

B,HCl: [93103-00-5]. Mp 234-245° dec.
B,2HCl: Needles (Me$_2$CO/EtOH). Mp 172° dec.

Clemo, G.R. *et al, J. Chem. Soc.*, 1938, 753 (*synth*)
Burrus, H.O. *et al, J. Am. Chem. Soc.*, 1945, **67**, 1468 (*synth*)
Maeda, S. *et al, Chem. Pharm. Bull.*, 1984, **32**, 2536 (*synth*)

4-(Aminoacetyl)pyridine   A-70087

*2-Amino-1-(4-pyridinyl)ethanone, 9CI. Aminomethyl 4-
pyridinyl ketone. Isonicotinylaminomethane*
[75140-34-0]

C$_7$H$_8$N$_2$O   M 136.153

B,2HCl: [51746-83-9]. Cryst. (MeOH). Mp 240-245°
  (230-235°) dec.
Semicarbazone; B,HCl: [74783-36-1]. Cryst. Mp 216-
  218° dec.

Kofman, H. *et al, Recl. Trav. Chim. Pays-Bas*, 1953, **72**, 236
  (*synth*)
Cymerman-Craig, J. *et al, J. Chem. Soc.*, 1955, 4315 (*synth*)

10-Amino-9-anthracenecarboxylic acid   A-70088

C$_{15}$H$_{11}$NO$_2$ M 237.257

Me ester: [79693-15-5].
  C$_{16}$H$_{13}$NO$_2$ M 251.284
  Bright-orange rhombs. Mp 192° (173°). Diazotised
  with difficulty.

N-Ac, Me ester:
  C$_{18}$H$_{15}$NO$_3$ M 293.321
  Cryst. (MeOH/C$_6$H$_6$). Mp 307-308°.

N-Di-Me, Me ester: [79693-35-9].
  C$_{18}$H$_{17}$NO$_2$ M 279.338
  Bright-yellow rhombic cryst. Mp 127°.

Rigaudy, J. *et al, Bull. Soc. Chim. Fr., Part II*, 1981, 223
  (*synth, uv, pmr*)
Brown, A.T. *et al, Org. Prep. Proced. Int.*, 1985, **17**, 223 (*synth,
  ir, pmr*)

3-Amino-1,2-benzenediol, 9CI   A-70089

Updated Entry replacing D-05306
*3-Aminopyrocatechol, 8CI. 2,3-Dihydroxyaniline. 3-
Aminocatechol*
[20734-66-1]

C$_6$H$_7$NO$_2$ M 125.127

Unstable solid. Mp 142-145°.

B,HCl: [51220-97-4]. Light-green leaflets (AcOH). Mp
  198-202.5° dec.
1-Benzenesulfonyl: [36383-42-3]. Cryst.
  (trichloroethylene). Mp 130-131°.
2-Ac, 1-benzenesulfonyl: [36383-45-6]. Cryst. (C$_6$H$_6$).
  Mp 128°.
O^1-Me: [40925-71-1]. *2-Amino-6-methoxyphenol. 6-
  Aminoguaiacol.*
  C$_7$H$_9$NO$_2$ M 139.154
  Rectangular leaflets (ligroin). Mp 83.5°. Also referred
  to as 3-aminoguaiacol.
O^1-Me, 4-methylbenzenesulfonyl: Prisms (MeOH). Mp
  117.5-118.5°.
O^2-Me: *3-Amino-2-methoxyphenol. 3-Aminoguaiacol.*
  Needles (ligroin). Mp 97-100°. Rapidly turns straw-
  brown in air. Also referred to as 6-aminoguaiacol.
Di-Me ether: [6299-67-8]. *2,3-Dimethoxyaniline, 8CI.
  2,3-Dimethoxybenzenamine, 9CI. 3-Aminoveratrol. 3-
  Amino-1,2-dimethoxybenzene.*
  C$_8$H$_{11}$NO$_2$ M 153.180
  Liq. Bp$_{1.5}$ 100-101°.

Oxford, A.E., *J. Chem. Soc.*, 1926, 2004 (*synth, derivs*)
Booth, H. *et al, J. Chem. Soc.*, 1956, 940 (*synth, deriv*)
Boyer, J.H. *et al, J. Org. Chem.*, 1961, **26**, 1654 (*synth, deriv*)
Sunkel, J. *et al, Ber. Bunsenges Phys. Chem.*, 1969, **73**, 203 (*uv*)
Kampouris, J.M., *J. Chem. Soc., Perkin Trans. 1*, 1972, 1088
  (*synth, derivs*)
Stuart, J.G. *et al, J. Heterocycl. Chem.*, 1987, **24**, 1589 (*synth,
  ir, pmr, deriv*)

4-Amino-1,2-benzenediol, 9CI   A-70090

Updated Entry replacing D-05307
4-Aminopyrocatechol. 3,4-Dihydroxyaniline
[13047-04-6]

C$_6$H$_7$NO$_2$ M 125.127

Sympathomimetic in animals. Apparently unstable.
▷Nephrotoxic

B,HCl: [4956-56-3]. Mp 224-227°.
2-Benzenesulfonyl: [36383-38-7]. Flakes (C_6H_6). Mp 140-141°.
1-Ac, 2-benzenesulfonyl: [36383-41-2]. Prisms (MeOH). Mp 135-136°.
O¹-Me: [1687-53-2]. *5-Amino-2-methoxyphenol, 9CI. 5-Aminoguaiacol. 4-Aminoguaiacol.*
$C_7H_9NO_2$ M 139.154
Grey cryst. (C_6H_6). Mp 131-133° (125-127°).
O¹-Me; B,HCl: [35589-31-2]. Green prisms (HCl). Mp 160-180° dec.
O¹-Me, 2-Ac: [53606-43-2].
$C_9H_{11}NO_3$ M 181.191
Needles (EtOAc/ligroin). Mp 81°. Darkens in air.
O²-Me: [52200-90-5]. *4-Amino-2-methoxyphenol, 9CI.*
Prisms or grey cryst. (H_2O). Mp 175-177° (184°) dec.
Rapidly turns violet or brown in air.
O²-Me; B,HCl: [4956-02-9]. Green cryst. or yellow-green plates or prisms. Mp 160-180° (242°) (340°).
O²-Me, 1-Ac:
$C_9H_{11}NO_3$ M 181.191
Prisms (EtOAc/ligroin). Mp 80°. Slowly darkens in air.
Di-Me ether: [6315-89-5]. *3,4-Dimethoxyaniline, 8CI. 3,4-Dimethoxybenzenamine, 9CI. 4-Aminoveratrol. 4-Amino-1,2-dimethoxybenzene.*
$C_8H_{11}NO_2$ M 153.180
Mp 85-86°. Bp_{18} 169°.
▷BX4550000.
Di-Me ether, N-Ac: [881-70-9].
$C_{10}H_{13}NO_3$ M 195.218
Plates (EtOH). Mp 133°.

Head, F.S.H. *et al, J. Chem. Soc.,* 1931, 1241 (*synth, deriv*)
Drake, N.L. *et al, J. Am. Chem. Soc.,* 1948, **70**, 168 (*synth, deriv*)
Harman, R.E. *et al, J. Org. Chem.,* 1952, **17**, 1047 (*synth*)
Hargreaves, K.R. *et al, J. Appl. Chem.,* 1958, **8**, 273 (*synth, derivs*)
Collins, R.F. *et al, J. Chem. Soc. (C),* 1966, 366 (*synth, derivs*)
Sunkel, J., *Ber. Bunsenges Phys. Chem.,* 1969, **73**, 203 (*uv*)
Kampouris, J.M., *J. Chem. Soc., Perkin Trans. 1,* 1972, 1088 (*synth, derivs*)
Mori, M. *et al, Heterocycles,* 1984, **22**, 253 (*synth, deriv*)
Stuart, J.G. *et al, J. Heterocycl. Chem.,* 1987, **24**, 1589 (*synth, ir, pmr, deriv*)

4-Amino-1*H*-1,5-benzodiazepine-3-carbonitrile, 9CI   A-70091

4-Amino-3-cyano-1H-1,5-benzodiazepine

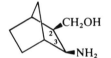

$C_{10}H_8N_4$ M 184.200
Intermed. for synth. of many benzodiazepines and related compds. esp. by Michael addn. Mp 203° dec.
B,HCl: [42510-46-3]. Mp 280° dec.

Okamoto, Y. *et al, Chem. Pharm. Bull.,* 1975, **23**, 1391 (*synth, pmr*)
Okamoto, Y. *et al, J. Heterocycl. Chem.,* 1987, **24**, 885 (*rev*)

2-Amino-4*H*-3,1-benzothiazine   A-70092

4H-3,1-Benzothiazin-2-amine, 9CI. 2-Amino-4H-benzo[d] [*1,3*]*thiazine*
[78959-46-3]

$C_8H_8N_2S$ M 164.225
Cryst. ($CHCl_3/Et_2O$). Mp 137-138°.
N-Et: [108288-44-4].
$C_{10}H_{12}N_2S$ M 192.278
Cryst. (hexane). Mp 100-101°.
N-Benzyl: [108288-46-6].
$C_{15}H_{14}N_2S$ M 254.349
Cryst. ($CHCl_3$/hexane). Mp 121-122°.
N-Ph: [16781-33-2].
$C_{14}H_{12}N_2S$ M 240.322
Cryst. ($CHCl_3$/pet. ether). Mp 199-200°.

Beilenson, B. *et al, J. Chem. Soc.,* 1942, 98 (*synth*)
Shridhar, D.R. *et al, Indian J. Chem., Sect. B,* 1981, **20**, 471 (*synth*)
Gonda, J. *et al, Collect. Czech. Chem. Commun.,* 1986, **51**, 2802, 2810 (*synth, ir, pmr, cmr, ms*)

3-Aminobicyclo[2.2.1]heptane-2-methanol, 9CI   A-70093

3-Hydroxymethylbicyclo[2.2.1]heptan-2-amine. 2-Amino-3-hydroxymethylnorbornane

$C_8H_{15}NO$ M 141.213
(*1RS,2SR,3RS*)-form [78365-98-7]
(±)-exo,exo-*form*
Mp 123-126°. Bp_4 106-108°.
Picrate: [88330-31-8]. Mp 178-180°.
N-Me: [95630-81-2].
$C_9H_{17}NO$ M 155.239
Bp_{400} 96-98°.
(*1RS,2RS,3SR*)-form [95630-79-8]
(±)-endo,endo-*form*
Bp_{400} 90-92°.
N-Me: [95630-80-1]. Oil. Bp_{400} 94-96°.

Kirmse, W. *et al, Chem. Ber.,* 1981, **114**, 1793 (*synth*)
Stajer, G. *et al, J. Heterocycl. Chem.,* 1983, **20**, 1181; 1984, **21**, 1373 (*synth, deriv*)

3-Aminobicyclo[2.2.1]hept-5-ene-2-methanol, 9CI   A-70094

3-Hydroxymethylbicyclo[2.2.1]hept-5-en-2-amine. 5-Amino-6-hydroxymethylnorborn-2-ene

$C_8H_{13}NO$ M 139.197
(*1RS,2RS,3SR*)-form [95630-78-7]
(±)-exo,exo-*form*
Liq., cryst. on standing. Bp_{400} 96-98°.
N-Me: [104770-09-4].
$C_9H_{15}NO$ M 153.224
Cryst. ($EtOH/Et_2O$) (as hydrochloride). Mp 163-165° (hydrochloride).

(*1RS,2SR,3SR*)-form [87768-94-3]
endo,endo-*form*
Bp_4 116-120°.
Picrate: [88330-30-7]. Mp 193-195°.
N-*Me:* [104770-08-3]. Cryst. (hexane). Mp 40-42°.

Stajer, G. *et al*, *J. Heterocycl. Chem.*, 1983, **20**, 1181; 1984, **21**, 1373 (*synth*)
Falop, F. *et al*, *Tetrahedron*, 1985, **41**, 5159 (*synth, deriv*)

2-Aminobicyclo[2.1.1]hexane   **A-70095**
Bicyclo[2.1.1]hexan-2-amine, 9CI

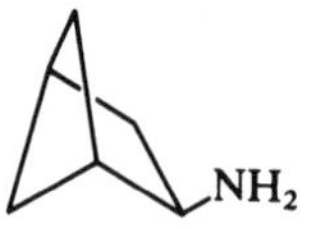

$C_6H_{11}N$ M 97.160
(±)-form
B,HCl: [102781-57-7]. Cryst. (EtOAc/MeOH). Mp 295° subl. Also known in opt. active form.

Kirmse, W. *et al*, *J. Am. Chem. Soc.*, 1986, **108**, 4912 (*synth, nmr*)

7-Amino-2,10-bisaboladiene   **A-70096**
7-Aminobisabolene

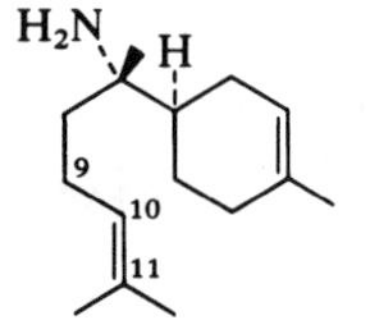

$C_{15}H_{27}N$ M 221.385
Constit. of sponge *Theonella* sp. Oil. $[\alpha]_D$ +39° (MeOH).
Δ^9-*Isomer, 11-hydroxy:* 7-*Amino-2,9-bisaboladien-11-ol.* **Aminobisabolenol.**
$C_{15}H_{27}NO$ M 237.384
Constit. of *T.* sp. Oil. $[\alpha]_D$ +29° (MeOH).
Δ^{11}-*Isomer, 10R-hydroxy:* 7-*Amino-2,11-biaboladien-10R-ol.* **Isoaminobisabolenol b.**
$C_{15}H_{27}NO$ M 237.384
Constit. of *T.* sp. Oil. $[\alpha]_D$ +40° (MeOH).
Δ^{11}-*Isomer, 10S-hydroxy:* 7-*Amino-2,11-bisaboladien-10S-ol.* **Isoaminobisabolenol a.**
$C_{15}H_{27}NO$ M 237.384
Constit. of *T.* sp. Oil. $[\alpha]_D$ +34° (MeOH).

Kitagawa, I. *et al*, *Chem. Pharm. Bull.*, 1987, **35**, 928 (*cryst struct*)

3-Amino-2-bromobenzoic acid   **A-70097**
$C_7H_6BrNO_2$ M 216.034
Me ester: [106896-48-4].
$C_8H_8BrNO_2$ M 230.061
Pale-yellow oil.

Krolski, M.E. *et al*, *J. Org. Chem.*, 1988, **53**, 1170 (*synth, deriv, ir, pmr*)

2-Amino-3-bromobutanedioic acid   **A-70098**
3-Bromoaspartic acid. 2-Amino-3-bromosuccinic acid

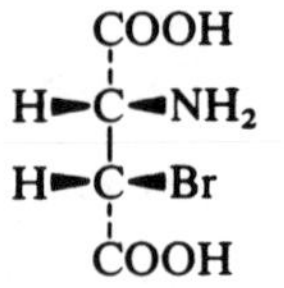

$C_4H_6BrNO_4$ M 212.000
(2S,3S)-form
L-threo-*form*
No phys. props. reported.

Akhtar, M. *et al*, *Tetrahedron*, 1987, **43**, 5899 (*synth*)

3-Amino-2-bromophenol   **A-70099**
[100367-36-0]
C_6H_6BrNO M 188.024
Cryst. Mp 88-89.5°.
Me ether: [112970-44-2]. *2-Bromo-3-methoxyaniline. 3-Amino-2-bromoanisole.*
C_7H_8BrNO M 202.050
Mp 198-200° (as hydrochloride).

Krolski, M.E. *et al*, *J. Org. Chem.*, 1988, **53**, 1170 (*synth, deriv, ir, pmr*)

2-Amino-4-bromopyrimidine, 8CI   **A-70100**
4-Bromo-2-pyrimidinamine, 9CI

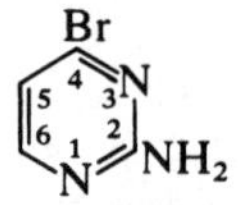

$C_4H_4BrN_3$ M 174.000
Cryst. (H_2O). Mp 237-238°. Sublimes.

Golding, D.R.V. *et al*, *J. Org. Chem.*, 1947, **12**, 293 (*synth*)
Sollazzo, R. *et al*, *Farmaco, Ed. Sci.*, 1954, **9**, 282 (*synth*)

2-Amino-5-bromopyrimidine, 8CI   **A-70101**
5-Bromo-2-pyrimidinamine, 9CI
[7752-82-1]
$C_4H_4BrN_3$ M 174.000
Lustrous plates (EtOH aq.). Mp 242-244°. Sublimes.
B,HCl: [99651-86-2]. Mp 190° dec.

Nishiwaki, T., *Tetrahedron*, 1966, **22**, 2401 (*uv*)
Khmel'nitskii, R.A. *et al*, *Chem. Heterocycles* Compd. (*Engl. Transl.*), 1973, 1528 (*ms*)
Jovanovic, M.V., *Heterocycles*, 1983, **20**, 2011 (*synth*)
Paudler, W.W. *et al*, *J. Org. Chem.*, 1983, **48**, 1064 (*synth, pmr*)
Fichter, R. *et al*, *J. Chem. Educ.*, 1985, **62**, 905 (*synth*)
Watton, H.L.L. *et al*, *Acta Crystallogr.*, Sect. C, 1988, **44**, 1857 (*cryst struct*)

4-Amino-5-bromopyrimidine, 8CI   **A-70102**
5-Bromo-4-pyrimidinamine, 9CI
[1439-10-7]
$C_4H_4BrN_3$ M 174.000
Yellowish needles (H_2O). Mp 209-211° dec.

Chesterfield, J. *et al*, *J. Chem. Soc.*, 1955, 3478 (*synth*)
Hara, H. *et al*, *J. Heterocycl. Chem.*, 1982, **19**, 1285 (*synth*)
Paudler, W.W. *et al*, *J. Org. Chem.*, 1983, **48**, 1064 (*synth, pmr*)

5-Amino-2-bromopyrimidine   A-70103

2-Bromo-5-pyrimidinamine, 9CI

[56621-91-1]

$C_4H_4BrN_3$ M 174.000

Cryst. (C_6H_6). Mp 182-183°.

Krchnak, V. *et al*, *Collect. Czech. Chem. Commun.*, 1975, **40**, 1396 (*synth, uv*)

Shkurko, O.P. *et al*, *Chem. Heterocycles* Compd. (*Engl. Transl.*), 1979, 1537 (*pmr*)

2-Amino-4-*tert*-butylcyclopentanol   A-70104

2-Amino-4-(1,1-dimethylethyl)cyclopentanol, 9CI

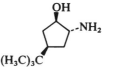

(*1RS,2RS,4RS*)-*form*

$C_9H_{19}NO$ M 157.255

(*1RS,2RS,4RS*)-*form* [37745-18-9]

(±)-trans,cis-*form*

Cryst. (pet. ether). Mp 65.5-66°.

Benzoyl:
   $C_{16}H_{23}NO_2$ M 261.363
   Cryst. (C_6H_6/pet. ether). Mp 127.5-128°.

(*1RS,2SR,4RS*)-*form* [37715-19-8]

(±)-cis,cis-*form*

Plates (pet. ether). Mp 106-106.5°.

(*1RS,2SR,4SR*)-*form* [37745-21-4]

(±)-cis,trans-*form*

Cryst. (pet. ether). Mp 88.5-89°.

(*1RS,2RS,4SR*)-*form* [37745-11-2]

(±)-trans,trans-*form*

Translucent prisms (pet. ether). Mp 68-69°.

Bernáth, G. *et al*, *Tetrahedron*, 1972, **28**, 3475 (*synth, deriv*)

2-Amino-α-carboline   A-70105

9H-Pyrido[2,3-b]indol-2-amine. 3-Amino-3-carboline
(*obsol.*)

[26148-68-5]

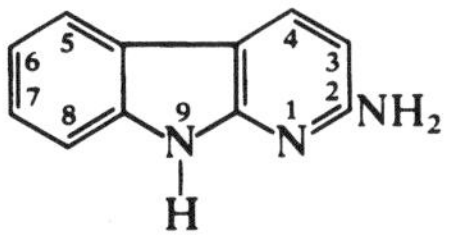

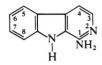

$C_{11}H_9N_3$ M 183.212

Found in cooked foods. Cryst. (EtOH). Mp 202-203°.

▷UU9351600.

N^2-*Me:* [26148-70-9]. *2-(Methylamino)-1H-pyrido[2,3-b]indole, 8CI.*
   $C_{12}H_{11}N_3$ M 197.239
   Cryst. (C_6H_6/pet. ether). Mp 156-157°.

Stephenson, L. *et al*, *J. Chem. Soc.* (*C*), 1970, 1355 (*synth*)

Matsumoto, T. *et al*, *Agric. Biol. Chem.*, 1979, **43**, 675 (*synth*)

3-Amino-α-carboline   A-70106

9H-Pyrido[2,3-b]indol-3-amine. 4-Amino-3-carboline
(*obsol.*)

[13315-72-5]

$C_{11}H_9N_3$ M 183.212

Mp 271-273°.

Nantka-Namirski, P., *Acta Pol. Pharm.*, 1966, **23**, 331 (*synth*)

Stephenson, L. *et al*, *J. Chem. Soc.* (*C*), 1970, 1355 (*synth*)

4-Amino-α-carboline   A-70107

1H-Pyrido[2,3-b]indol-4-amine, 9CI. *6-Amino-3-carboline* (*obsol.*)

[25208-34-8]

$C_{11}H_9N_3$ M 183.212

Mp 237.5-239.5°.

Nantka-Namirski, P. *et al*, *Acta Pol. Pharm.*, 1971, **28**, 219 (*synth*)

B.P., 1 268 773, (*1972*); *CA*, **76**, 140760 (*synth*)

6-Amino-α-carboline   A-70108

9H-Pyrido[2,3-b]indol-6-amine. 10-Amino-3-carboline (*obsol.*)

$C_{11}H_9N_3$ M 183.212

Yellow needles (Py). Mp 263-264° (243-245°).

U.S.P., 2 690 441, (*1954*); *CA*, **49**, 13298b (*synth*)

Saxena, J.P., *Indian J. Chem.*, 1966, **4**, 148 (*synth*)

1-Amino-β-carboline   A-70109

9H-Pyrido[3,4-b]indol-1-amine, 9CI

[30684-41-4]

$C_{11}H_9N_3$ M 183.212

Needles (EtOH aq.). Mp 200-201°.

Snyder, H.R. *et al*, *J. Am. Chem. Soc.*, 1949, **71**, 527 (*synth*)

Ho, B.T. *et al*, *J. Pharm. Sci.*, 1970, **59**, 1445 (*synth*)

3-Amino-β-carboline   A-70110

9H-Pyrido[3,4-b]indol-3-amine

[73834-77-2]

$C_{11}H_9N_3$ M 183.212

Shiny-yellow plates (EtOH). Mp 289-291°.

N^3-*Ac:* [91985-75-0].
   $C_{13}H_{11}N_3O$ M 225.249
   Mp 210°.

N^3-*Et:* [95935-52-7].
   $C_{13}H_{13}N_3$ M 211.266
   Cryst. (EtOH/hexane). Mp 143-145°.

N^3-*Formyl:* [95935-50-5].
   $C_{12}H_9N_3O$ M 211.223
   Cryst. (EtOH/hexane). Mp 266° dec.

Agarwal, S.K. *et al*, *Indian J. Chem., Sect. B*, 1980, **19**, 45 (*synth*)

Dodd, R.H. *et al*, *J. Med. Chem.*, 1985, **28**, 824 (*synth, pmr*)

4-Amino-β-carboline   A-70111

9H-Pyrido[3,4-b]indol-4-amine, 9CI

[98263-44-6]

$C_{11}H_9N_3$ M 183.212

Cryst. (EtOAc). Mp 222-224°.

Fukoda, N. *et al*, *Tetrahedron Lett.*, 1985, **26**, 2139 (*synth*)

Trudell, M.L. *et al*, *J. Org. Chem.*, 1987, **52**, 4293 (*synth, pmr*)

6-Amino-β-carboline   A-70112

9H-Pyrido[3,4-b]indol-6-amine, 9CI

[6453-27-6]

$C_{11}H_9N_3$ M 183.212

Cryst. (EtOH). Mp 302-303° (268°).

Saxena, J.P., *Indian J. Chem.*, 1966, **4**, 148 (*synth*)
Ho, B.T. *et al*, *J. Pharm. Sci.*, 1970, **59**, 1445 (*synth*)
Ger. Pat., 3 240 514, (*1984*); *CA*, **101**, 151824

8-Amino-β-carboline　　　　A-70113

9H-*Pyrido*[3,4-b]*indol-8-amine*, 9CI
[30684-51-6]
$C_{11}H_9N_3$　　M 183.212
Cryst. (EtOH/Et$_2$O). Mp 242° dec.

Ho, B.T. *et al*, *J. Pharm. Sci.*, 1970, **59**, 1445 (*synth*)

1-Amino-γ-carboline　　　　A-70114

5H-*Pyrido*[4,3-b]*indol-1-amine*, 9CI
[79642-27-6]

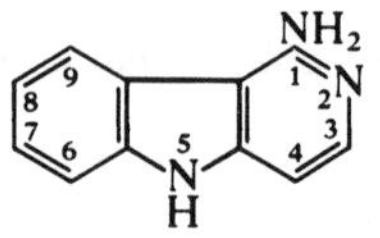

$C_{11}H_9N_3$　　M 183.212
Mp 231-232°.
N^1-*Di-Me*: *1-(Dimethylamino)-γ-carboline.*
　$C_{13}H_{11}N_3$　　M 209.250
　Mp 58-60°.
N^1-*Ph*: *1-Anilino-γ-carboline.*
　$C_{17}H_{13}N_3$　　M 259.310
　Mp 239-240°.

Lee, C.-S. *et al*, *Heterocycles*, 1981, **16**, 1081 (*synth*)

3-Amino-γ-carboline　　　　A-70115

5H-*Pyrido*[4,3-b]*indol-3-amine*, 9CI
[69901-70-8]
$C_{11}H_9N_3$　　M 183.212
B,HCl: [75240-08-3]. Cryst. (MeOH/EtOAc). Mp 266-
　267°.

Takeda, K. *et al*, *Chem. Pharm. Bull.*, 1981, **29**, 1280 (*synth*)

5-Amino-γ-carboline　　　　A-70116

5H-*Pyrido*[4,3-b]*indol-5-amine*, 9CI
[79642-25-4]
$C_{11}H_9N_3$　　M 183.212
Mp 167-168°.

Lee, C.-S. *et al*, *Heterocycles*, 1981, **16**, 1081 (*synth*)

6-Amino-γ-carboline　　　　A-70117

5H-*Pyrido*[4,3-b]*indol-6-amine*, 9CI
[79642-26-5]
$C_{11}H_9N_3$　　M 183.212
Mp 240-241°.

Lee, C.-S. *et al*, *Heterocycles*, 1981, **16**, 1081 (*synth*)

8-Amino-γ-carboline　　　　A-70118

5H-*Pyrido*[4,3-b]*indol-8-amine*, 9CI
[79642-24-3]
$C_{11}H_9N_3$　　M 183.212
Mp 230-231°.

Lee, C.-S. *et al*, *Heterocycles*, 1981, **16**, 1081 (*synth*)

8-Amino-δ-carboline　　　　A-70119

5H-*Pyrido*[3,2-b]*indol-8-amine*

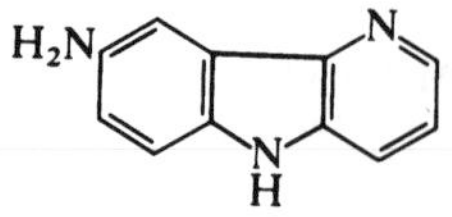

$C_{11}H_9N_3$　　M 183.212
B,HCl: Cryst. (MeOH). Mp >300°.

Abramovitch, R.A. *et al*, *Can. J. Chem.*, 1962, **40**, 864 (*synth*)

α-Amino-2-carboxy-5-oxo-1-pyrrolidinebu-　　A-70120
tanoic acid, 9CI

*1-(3-Amino-3-carboxypropyl)-5-oxo-2-pyrrolidinecar-
boxyic acid. Lp-2*
[115553-99-6]

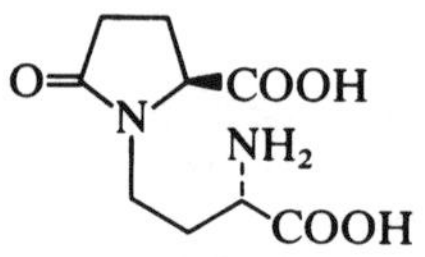

$C_9H_{14}N_2O_5$　　M 230.220
Amino acid from the basidiomyete *Lactanus piperatus*.
　Amorph. solid. [α]$_D$ −17.5° (c, 0.939 in H$_2$O).

Fushiya, S. *et al*, *Chem. Pharm. Bull.*, 1988, **36**, 1366 (*isol, ir,
　pmr, cmr, synth*)

2-Amino-3-chlorobutanedioic acid　　　　A-70121

3-Chloroaspartic acid. 2-Amino-3-chlorosuccinic acid

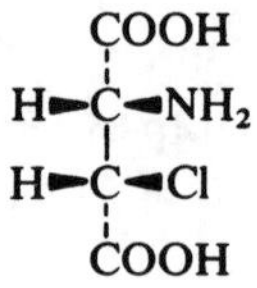

$C_4H_6ClNO_4$　　M 167.549
(*2R,3S*)-*form* [106073-70-5]
　L-threo-*form*
　Cryst. (MeOH aq.). Mp 168-171°. [α]$_D^{20}$ −37.5° (c, 0.4
　in H$_2$O).

Akhtar, M. *et al*, *Tetrahedron*, 1987, **43**, 5899 (*synth*)

2-Amino-3-chloro-1,4-naphthoquinone　　　　A-70122

2-Amino-3-chloro-1,4-naphthalenedione, 9CI
[2797-51-5]

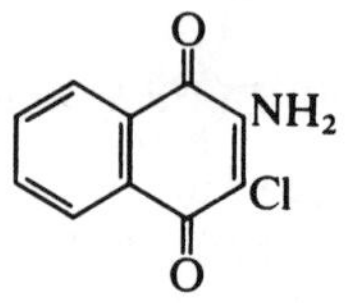

$C_{10}H_6ClNO_2$　　M 207.616
Has been used in herbicides. Yellow-brown needles
　(AcOH). Mp 196-198° (193°).
▷QL7350000.
N-*Ac*: [5397-78-4].
　$C_{12}H_8ClNO_3$　　M 249.653
　Red-brown platelets or yellow powder. Mp 219°.
　▷AB7444800.

Hoover, J.R.E. *et al*, *J. Am. Chem. Soc.*, 1954, **76**, 4148 (*synth*)
Gaultier, J. *et al*, *Acta Crystallogr.*, 1965, **19**, 585 (*cryst struct*)
Neidlein, R. *et al*, *Helv. Chim. Acta*, 1983, **66**, 2285 (*cmr*)

Hammam, A.S. *et al, Collect. Czech. Chem. Commun.,* 1985, **50**, 71 (*synth, uv*)

2-Aminocycloheptanecarboxylic acid, 9CI   A-70123

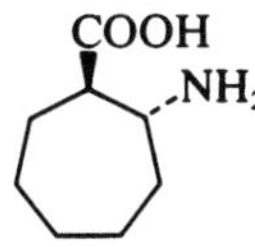

COOH
NH₂

(*1RS,2RS*)-*form*

$C_8H_{15}NO_2$ M 157.212
(**1RS,2RS**)-**form** [42418-84-8]
  (±)-*trans-form*
  Mp 282-285° dec.
(**1RS,2SR**)-**form** [42418-83-7]
  (±)-*cis-form*
  Mp 242-243° dec.
Bernáth, G. *et al, Tetrahedron,* 1973, **29**, 981.

2-Aminocycloheptanemethanol, 9CI   A-70124

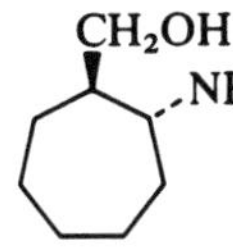

CH₂OH
NH₂

(*1RS,2RS*)-*form*

$C_8H_{17}NO$ M 143.228
(**1RS,2RS**)-**form** [72450-76-1]
  (±)-*trans-form*
  B,HCl: [42418-79-1]. Mp 103-104°.
(**1RS,2SR**)-**form** [72450-75-0]
  (±)-*cis-form*
  B,HCl: [42418-78-0]. Mp 149-150°.
Bernáth, G. *et al, Tetrahedron,* 1973, **29**, 981; 1979, **35**, 799.

3-Amino-2-cyclohexen-1-one   A-70125
[5220-49-5]

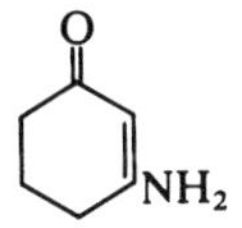

O

NH₂

C_6H_9NO M 111.143
Readily obt. useful synthetic intermed., e.g. for cyclocon-
densations leading to heterocycles. Light-yellow cryst.
(C_6H_6) or canary-yellow powder. Mp 133-135° (118-
119°).
N-Ac: [23674-56-8].
  $C_8H_{11}NO_2$ M 153.180
  Plates (C_6H_6). Mp 164-166°.
Dubas-Sluyter, M.A.T. *et al, Recl. Trav. Chim. Pays-Bas,* 1972,
  91, 157 (*synth*)
Ramalingam, K. *et al, Indian J. Chem.,* 1972, **10**, 62 (*synth, ir,
  uv*)

1-Aminocyclopropenecarboxylic acid   A-70126
[110374-54-4]

H₂N COOH

$C_4H_5NO_2$ M 99.089
B,HCl: [110374-53-3]. Cryst. solid. Mp 164° dec.
Wheeler, T.N. *et al, J. Org. Chem.,* 1987, **52**, 4875 (*synth, ir,
  pmr, cmr*)

N⁴-Aminocytidine   A-70127
Uridine 4-hydrazone, 9CI
[57294-74-3]

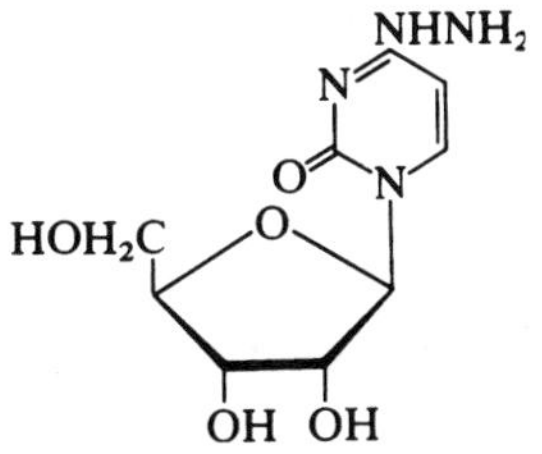

NHNH₂

HOH₂C
OH OH

$C_9H_{14}N_4O_5$ M 258.233
Mutagen causing AT⇌GC base-pair transitions in DNA.
Cryst. + ½H_2O. Freq. obt. as glass.
▷Potent mutagen. YU8055000.
Negishi, K. *et al, Chem. Pharm. Bull.,* 1987, **35**, 3884 (*synth,
  uv, bibl*)

3'-Amino-3'-deoxyadenosine, 9CI, 8CI   A-70128
Updated Entry replacing A-01352
9-(3-Amino-3-deoxy-β-D-ribofuranosyl)adenine
[2504-55-4]

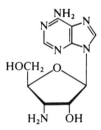

NH₂

HOCH₂ O

H₂N OH

$C_{10}H_{14}N_6O_3$ M 266.259
Nucleoside antibiotic. Obt. from the mould *Cordyceps
  militaris* and from *Helminthosporium* spp. culture
  filtrates. Antitumour agent. Mp 275-278° dec. (265-
  267°). $[\alpha]_D^{25}$ −37° (c, 2.0 in 0.1M HCl), $[\alpha]_D^{25}$ −40°
  (c, 0.4 in DMF). λ_{max} 260 nm (ε 17 300) (H_2O).
Dipicrate: Mp 222-224° dec.
3'N-Me: [25787-43-3].
  $C_{11}H_{16}N_6O_3$ M 280.286
  Mp 247-250° dec. $[\alpha]_D^{19}$ −103° (c, 1.0 in 1N NaOH).
  λ_{max} 259 (ε 15 100), 207 (20 400) (pH 7); 256 (14
  800), 206 (22 100) (pH 1); 259 nm (15 200) (pH 13).
3',6,6-Tri-N-Me: [34522-43-5].
  $C_{13}H_{20}N_6O_3$ M 308.339
  Mp 216-218°. $[\alpha]_D^{23}$ −52° (c, 0.97 in EtOH). λ_{max} 275
  (ε 19 400), 214 (17 100) (pH 7); 267 (19 200), 208 (20
  100) (pH 1); 275 nm (19 700) (pH 13).
3',3',6,6-Tetra-N-Me: [25787-42-2].
  $C_{14}H_{22}N_6O_3$ M 322.366
  Mp 184.5-186°. $[\alpha]_D^{22}$ −27° (c, 1.0 in H_2O). λ_{max} 275
  (ε 19 000), 214 (16 100) (pH 7); 267 (18 600), 209 (17
  700) (pH 1); 276 nm (18 700) (pH 13).
3'N-Ac, 3'N-Me, 2',5'-di-Ac: [25834-71-3].
  $C_{17}H_{22}N_6O_6$ M 406.397
  Mp 224-225.5°. $[\alpha]_D^{21}$ −15° (c, 1.0 in Py). λ_{max} 259 (ε
  13 900) (pH 7), 257 (13 800) (pH 1), 260 nm (14 200)
  (pH 13).
3'N-Ac, 3'N-Me, 2',5'-di-Ac, 6N-benzoyl: [36913-17-4].
  Mp 96-99.5°. $[\alpha]_D^{20.5}$ −27° (c, 1.0 in EtOH). λ_{max} 291
  (ε 26 300), 252 (11 400) (pH 1); 303 nm (13 900) (pH
  13).
Gerber, N.N. *et al, J. Org. Chem.,* 1962, **27**, 1731.
Guarino, A.J. *et al, Biochim. Biophys. Acta,* 1963, **68**, 317 (*isol*)

Lee, W.W. *et al, J. Org. Chem.*, 1970, **35**, 3808 (*synth, pmr*)
Nagasawa, H.T. *et al, J. Med. Chem.*, 1972, **15**, 177.
Azhaev, A.V. *et al, Collect. Czech. Chem. Commun.*, 1978, **43**, 1520 (*synth*)
Klimke, G. *et al, Z. Naturforsch., C*, 1979, **34**, 1075 (*conformn*)
Lichtenthaler, F.W. *et al, Chem. Ber.*, 1979, **112**, 2588 (*synth*)
Sheldrick, W.S. *et al, Acta Crystallogr., Sect. B*, 1980, **36**, 2328 (*cryst struct*)

4-Amino-2,5-dichlorophenol   A-70129

[50392-39-7]
C₆H₅Cl₂NO   M 178.018
Cryst. (C₆H₆). Mp 178-179°.

Bargellini, G. *et al, Gazz. Chim. Ital.*, 1929, **59**, 30 (*synth*)
Santilli, A.A. *et al, J. Pharm. Sci.*, 1974, **63**, 449.

4-Amino-1,7-dihydro-2H-1,3-diazepin-2-one   A-70130

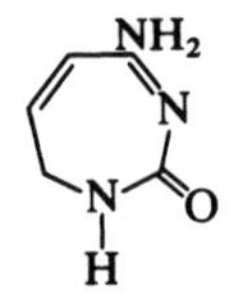

C₅H₇N₃O   M 125.130
1-β-D-Ribofuranosyl: [107711-30-8]. *Homocytidine.*
C₁₀H₁₅N₃O₅   M 257.246
Shows weak biological activity. Sl. tan foam. $[\alpha]_D^{27}$ −6.6° (c, 0.12 in H₂O).

Kim, C.-H. *et al, J. Org. Chem.*, 1987, **52**, 1979 (*deriv, synth, pmr, ms*)

4-Aminodihydro-3-methylene-2-(3H)-furanone, 9CI   A-70131

β-Amino-α-methylene-γ-butyrolactone

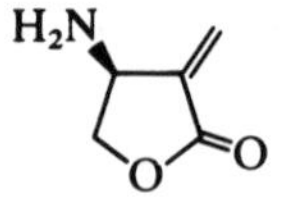

(*R*)-*form*

C₅H₇NO₂   M 113.116
(*R*)-*form* [117752-92-8]
B,HCl: Cryst. Mp >120° dec. $[\alpha]_D$ +85.7° (c, 0.5 in H₂O).
N-Ac: [117752-93-9].
C₇H₉NO₃   M 155.153
Mp 75-77°. $[\alpha]_D$ +89.8° (c, 0.4 in CHCl₃).

(*S*)-*form*
B,HCl: Mp >135° dec. $[\alpha]_D$ −85.2° (c, 0.6 in H₂O).
N-Ac: [117752-87-1]. Mp 75-77°. $[\alpha]_D$ −90.8° (c, 0.3 in CHCl₃).

Hvidt, T. *et al, Can. J. Chem.*, 1988, **66**, 779 (*synth, pmr, cmr*)

2-Amino-4,7-dihydro-4-oxo-7β-D-ribofuranosyl-1H-pyrrolo[2,3-d]pyrimidin-5-carbonitrile, 9CI   A-70132

5-Cyano-7-deazaguanosine
[61210-21-7]

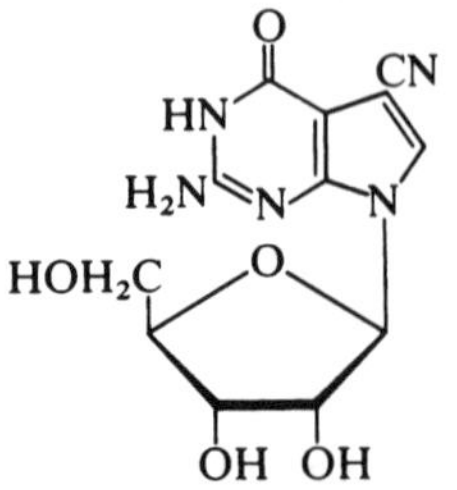

C₁₂H₁₃N₅O₅   M 307.265
Isol. from tRNA^{tyr} of a mutant of *Escherichia coli*. Pale-yellow cryst. Mp 260-270° dec. $[\alpha]_D^{15}$ −59.1° (c, 0.1 in DMSO).

Noguchi, S. *et al, Nucleic Acids Res.*, 1978, **5**, 4215.
Kondo, T. *et al, Tetrahedron*, 1986, **42**, 207 (*synth, pmr, uv, ir*)

2-Amino-1,7-dihydro-7-phenyl-6H-purin-6-one, 9CI   A-70133

7-Phenylguanine
[110718-94-0]

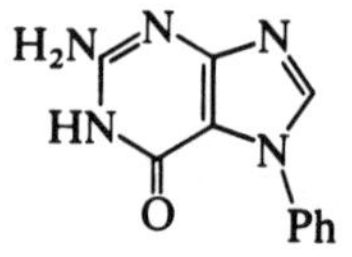

C₁₁H₉N₅O   M 227.225
Of possible significance in benzene carcinogenesis. Yellowish cryst. (EtOH aq.). Mp >300° dec.

Verkoyen, C. *et al, Justus Liebigs Ann. Chem.*, 1987, 957 (*synth, pmr, ir, ms*)

6-Amino-2,3-dihydro-1,4-phthalazinedione   A-70134

4-Aminophthalhydrazide. 4-Amino-1,2-benzenedicarbonyl hydrazide. Isoluminol
[3682-14-2]

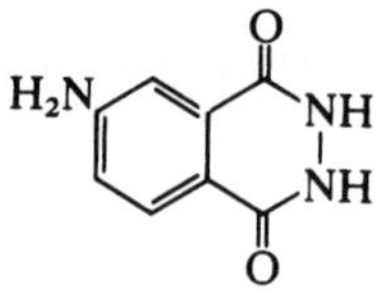

C₈H₇N₃O₂   M 177.162
Chemoluminescent compd. showing lower luminescence than Luminol, L-00472 but derivs. have advantages for luminescence immunoassay over Luminol derivs. Yellowish needles + 1H₂O. Mp >305°.

Curtius, T., *J. Prakt. Chem.*, 1907, **76**, 301 (*synth*)
Bélanger, A. *et al, Can. J. Chem.*, 1987, **65**, 1392 (*bibl*)

5-Amino-3,4-dihydro-2H-pyrrole-2-carboxylic acid, 9CI   A-70135

2-Amino-1-pyrroline-5-carboxylic acid

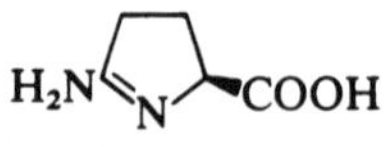

S-form

C₅H₈N₂O₂   M 128.130

(S)-form
B,HCl: [110971-81-8]. Solid. Mp 150-155°. $[\alpha]_D^{22}$ −8.2°
(c, 0.0095 in H_2O).
Me ester: [110971-80-7].
$C_6H_{10}N_2O_2$ M 142.157
Cryst. (CH_2Cl_2/hexane)(as hydrochloride). Mp 169-
171° (hydrochloride). $[\alpha]_D^{22}$ −1.6° (c, 1.8 in H_2O).
Et ester: [110971-82-9].
$C_7H_{12}N_2O_2$ M 156.184
Mp 105-109° (hydrochloride). $[\alpha]_D^{22}$ −2.9° (c, 0.0123
in H_2O).

Lee, M. et al, J. Org. Chem., 1987, **52**, 5717 (synth, pmr, ir, ms)

2-Amino-3,3-dimethylbutanoic acid   A-70136
3-Methylvaline, 9CI. tert-Butylglycine. tert-Leucine. 2-
Aminoneohexanoic acid. Bug. Pseudoleucine
[20859-02-3]

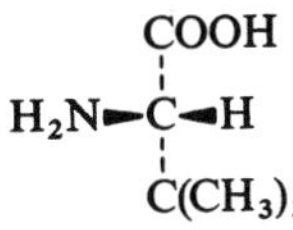

$$
\begin{array}{c}
\text{COOH} \\
| \\
H_2N\!\!-\!\!C\!\!-\!\!H \\
| \\
C(CH_3)_3
\end{array}
$$

$C_6H_{13}NO_2$ M 131.174
(S)-form [20859-02-3]
L-form
Residue from Bottromycin A. Chiral inducer in
asymmetric syntheses. Leucine analogue in modified
peptides. Mp 248° dec. $[\alpha]_D^{25}$ −9.5° (c, 3 in H_2O),
$[\alpha]_D^{25}$ +8.7° (c, 3 in 5M HCl).
N-tert-Butyloxycarbonyl, dicyclohexylamine salt: Mp
165°. $[\alpha]_D^{23}$ −7.2° (c, 1 in MeOH).
Me ester: [63038-26-6].
$C_7H_{15}NO_2$ M 145.201
Mp 170°. $[\alpha]_D^{23}$ +17.0° (c, 1 in MeOH).
Amide: [62965-57-5].
$C_6H_{14}N_2O$ M 130.189
Mp 205° (as hydrochloride). $[\alpha]_D^{23}$ +41.8° (c, 1 in
MeOH).
(±)-form [33105-81-6]
Cryst. (Me_2CO aq.). Mp 250-251° subl.

Izumiya, N. et al, J. Biol. Chem., 1953, **205**, 221 (struct)
Nakamura, S. et al, Chem. Pharm. Bull., 1965, **13**, 599 (isol)
Miyazawa, T., Bull. Chem. Soc. Jpn., 1979, **52**, 1539 (synth,
resoln)
Fauchère, J.-L., Helv. Chim. Acta, 1980, **63**, 824 (props)
Viret, J. et al, Tetrahedron Lett., 1986, **27**, 5865 (resoln, bibl)

2-Amino-4,4-dimethylpentanoic acid   A-70137
4-Methylleucine, 9CI. Neopentylglycine

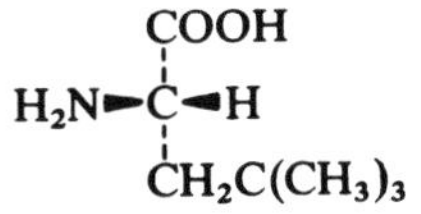

$$
\begin{array}{c}
\text{COOH} \\
| \\
H_2N\!\!-\!\!C\!\!-\!\!H \\
| \\
CH_2C(CH_3)_3
\end{array}
$$ (S)-form

$C_7H_{15}NO_2$ M 145.201
(S)-form [57224-50-7]
L-form
Mp 259-260°. $[\alpha]_D$ +14.7° (AcOH). (R)-form also
known.
N-tert-Butyloxycarbonyl: Cryst. (Et_2O/pet. ether). Mp
102-103°. $[\alpha]_D$ +14.0° (MeOH).
N-Formyl:
$C_8H_{15}NO_3$ M 173.211
Mp 175-177°. $[\alpha]_D$ −13.0° (EtOH).

N-Ac:
$C_9H_{17}NO_3$ M 187.238
Cryst. (EtOH/Et_2O). Mp 224-225°.
(±)-form [60122-72-7]
Mp 235-237° dec.
N-Formyl: Cryst. (EtOH/C_6H_6). Mp 162-163°.

Praetorius, H.J. et al, Chem. Ber., 1975, **108**, 3079 (synth)
Fauchere, J.L. et al, Int. J. Pept. Protein Res., 1981, **18**, 249
(synth)
Pospisek, J. et al, Collect. Czech. Chem. Commun., 1987, **52**,
514 (synth, resoln, abs config)

2-Amino-3-ethylbutanedioic acid   A-70138
3-Ethylaspartic acid, 9CI. 2-Amino-3-ethylsuccinic acid

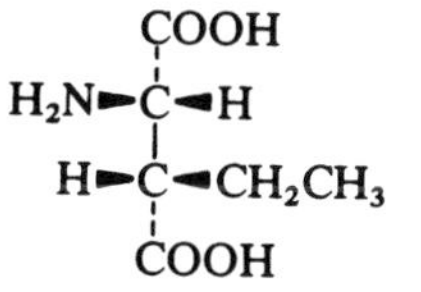

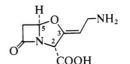 (2S,3S)-form

$C_6H_{11}NO_4$ M 161.157
(2S,3S)-form
Cryst. (EtOH). Mp 245-246°.
(2RS,3RS)-(?)-form [15383-88-7]
Mp 235-237°. A single diastereoisomer which may be the
(2RS,3SR)-form.

Winkler, M.F. et al, Biochim. Biophys. Acta, 1967, **146**, 287
(synth)
Bochenska, M. et al, Rocz. Chem., 1976, **50**, 1195 (synth)
Akhtar, M. et al, Tetrahedron, 1987, **43**, 5899 (synth)

3-(2-Aminoethylidene)-7-oxo-4-oxa-1-   A-70139
azabicyclo[3.2.0]heptane-2-carboxylic
acid, 9CI

  (2R,3Z,5R)-form

$C_8H_{10}N_2O_4$ M 198.178
(2R,3Z,5R)-form [65788-55-8]
9-Aminodeoxyclavulanic acid
Synthetic.
(2S,3Z,5S)-form [112296-12-5]
Clavaminic acid
Isol. from *Streptomyces clavuligerus*.

Ger. Pat., 2 725 203, (1977); CA, **88**, 105312 (synth)
Newall, C.E., Spec. Publ.-R. Soc. Chem., 1981, **38**, 151 (synth)
Elson, S.W. et al, J. Chem. Soc., Chem. Commun., 1987, 1736
(isol)

2-Amino-3-fluorobenzoic acid   A-70140
3-Fluoroanthranilic acid
[825-22-9]

$$
\begin{array}{c}
\text{COOH} \\
\text{6}\diagdown\text{1}\diagdown\text{NH}_2 \\
\text{5}\diagup\text{4}\diagup\text{3}\text{F}
\end{array}
$$

$C_7H_6FNO_2$ M 155.128
Mp 177°.

Holt, S.J. et al, Proc. Roy. Soc. London, Ser. B, 1958, **148**, 481
(synth)
Milner, D.J., Synth. Commun., 1985, **15**, 485 (synth)

2-Amino-4-fluorobenzoic acid   A-70141

4-Fluoroanthranilic acid

[446-32-2]

$C_7H_6FNO_2$ M 155.128

Needles (H_2O). Mp 195°.

N-*Ac:*
  $C_9H_8FNO_3$ M 197.166
  Platelets (EtOH aq.). Mp 209-209.5°.

Steck, E.A. *et al, J. Am. Chem. Soc.*, 1948, **70**, 439 (*synth, deriv*)
Holt, S.J. *et al, Proc. Roy. Soc. London, Ser. B*, 1958, **148**, 481 (*synth*)
Stiles, M. *et al, J. Am. Chem. Soc.*, 1963, **85**, 1792 (*synth*)

2-Amino-5-fluorobenzoic acid   A-70142

5-Fluoroanthranilic acid

[446-08-2]

$C_7H_6FNO_2$ M 155.128

Needles (H_2O). Mp 185° (182-183° dec.).

Me ester: [319-24-4].
  $C_8H_8FNO_2$ M 169.155
  Liq. Bp_2 105°.
N-*Ac:*
  $C_9H_8FNO_3$ M 197.166
  Cryst. (EtOH aq.). Mp 180-181°.

Fosdick, L.S. *et al, J. Am. Chem. Soc.*, 1944, **66**, 1165 (*deriv*)
Castle, R.N. *et al, J. Heterocycl. Chem.*, 1965, **2**, 459 (*synth, deriv*)
Rajšner, M. *et al, Collect. Czech. Chem. Commun.*, 1975, **40**, 719 (*synth, deriv*)

2-Amino-6-fluorobenzoic acid   A-70143

6-Fluoroanthranilic acid

[434-76-4]

$C_7H_6FNO_2$ M 155.128

Nitrile: [77326-36-4]. *2-Cyano-3-fluoroaniline.*
  $C_7H_5FN_2$ M 136.128
  Cryst. (C_6H_6/hexane). Mp 126-127.5°.

Klaubert, D.H. *et al, J. Med. Chem.*, 1981, **24**, 742 (*synth, deriv*)
Hynes, J.B. *et al, J. Heterocycl. Chem.*, 1988, **25**, 1176 (*synth, deriv, pmr*)

3-Amino-4-fluorobenzoic acid   A-70144

[2365-85-7]

$C_7H_6FNO_2$ M 155.128

Needles (H_2O). Mp 182-183°.

Me ester: [369-26-6].
  $C_8H_8FNO_2$ M 169.155
  Mp 123-126°.
N-*Ac:*
  $C_9H_8FNO_3$ M 197.166
  Needles (H_2O). Mp 245-246°.

Dunker, M.F.W. *et al, J. Am. Chem. Soc.*, 1939, **61**, 3005 (*synth, deriv*)
Fosdick, L.S. *et al, J. Am. Chem. Soc.*, 1943, **65**, 2305 (*deriv*)

4-Amino-2-fluorobenzoic acid   A-70145

[446-31-1]

$C_7H_6FNO_2$ M 155.128

Cryst. (H_2O). Mp 216-216.5°.

N-*Ac:*
  $C_9H_8FNO_3$ M 197.166
  Cryst. (H_2O). Mp 256-257°.

Schmelkes, F.C. *et al, J. Am. Chem. Soc.*, 1944, **66**, 1631 (*synth, deriv*)
Valkanas, G. *et al, J. Chem. Soc.*, 1963, 1925 (*synth*)

4-Amino-3-fluorobenzoic acid   A-70146

[455-87-8]

$C_7H_6FNO_2$ M 155.128

Cryst. (H_2O). Mp 215-216°.

▷DG2525000.

N-*Ac:*
  $C_9H_8FNO_3$ M 197.166
  Cryst. (H_2O). Mp 282-283° dec.
Nitrile: [63069-50-1]. *4-Cyano-2-fluoroaniline.*
  $C_7H_5FN_2$ M 136.128
  Cryst. (hexane). Mp 87-88°.

Tomcufcik, A.S. *et al, J. Org. Chem.*, 1961, **29**, 3351 (*synth*)
Frolov, A.N. *et al, J. Org. Chem. USSR (Engl. Transl.)*, 1977, **13**, 552 (*synth, deriv, ir, nmr*)

5-Amino-2-fluorobenzoic acid   A-70147

[56741-33-4]

$C_7H_6FNO_2$ M 155.128

Needles (H_2O). Mp 190°.

Me ester: [56741-34-5].
  $C_8H_8FNO_2$ M 169.155
  Pale-yellow needles (CCl_4). Mp 86°.
Nitrile: [53312-81-5]. *3-Cyano-4-fluoroaniline.*
  $C_7H_5FN_2$ M 136.128
  Mp 88-89°.

Drewes, S.E. *et al, J. Chem. Soc., Perkin Trans. 1*, 1975, 1283 (*synth, deriv*)
Mulvey, D.M. *et al, Tetrahedron Lett.*, 1978, 1419 (*nitrile*)
Tani, J. *et al, Chem. Pharm. Bull.*, 1982, **30**, 3530 (*synth, deriv, pmr*)

3-Amino-6-fluoropyridazine   A-70148

6-Fluoro-3-pyridazinamine, 9CI

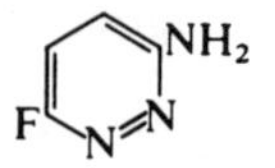

$C_4H_4FN_3$ M 113.094

Cryst. (EtOAc). Mp 177-179°.

Barlin, G.B., *Aust. J. Chem.*, 1986, **39**, 1803 (*synth, pmr, F nmr*)

2-Amino-5-guanidino-3-hydroxypentanoic   A-70149
acid

3-Hydroxyarginine, 9CI

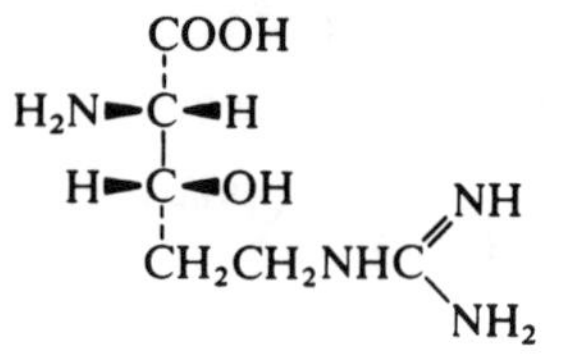

(2S,3R)-*form*

$C_6H_{14}N_4O_3$ M 190.202

(2S,3R)-form

B,HCl: [107942-05-2]. Cryst. (EtOH aq.). Mp 212° dec. $[\alpha]_D^{25}$ +10.4° (c, 2 in H_2O).

(2S,3S)-form

B,HCl: [107942-06-3]. Cryst. (EtOH aq.). Mp 204° dec. $[\alpha]_D^{25}$ −3.8° (c, 2 in H_2O).

Wityak, J. *et al, J. Org. Chem.*, 1987, **52**, 2179 (*synth*)

8-Aminoguanosine, 9CI   A-70150

*2,8-Diamino-9-β-D-ribofuranosyl-9H-purin-6(1H)-one.
8-Amino-9-β-D-ribofuranosylguanine. 2,8-
Diaminoinosine*

[3868-32-4]

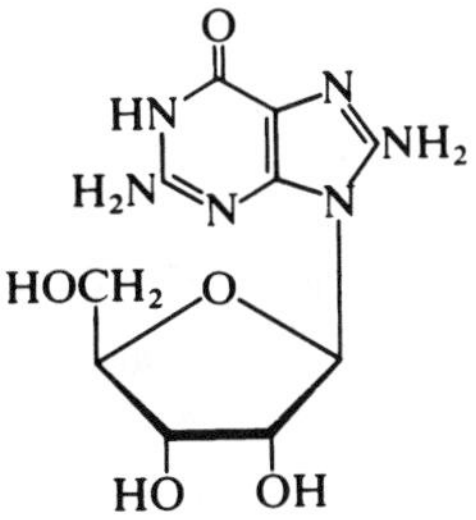

$C_{10}H_{14}N_6O_5$ M 298.258
Cryst. (H_2O). Mp >241° dec. Slowly darkens in air.

N⁸-Me: [13389-05-4]. *8-(Methylamino)guanosine, 9CI.*
   $C_{11}H_{16}N_6O_5$ M 312.285
   Cryst. (H_2O). Mp >196° dec.

N⁸,N⁸-Di-Me: [7057-52-5]. *8-(Dimethylamino)-
guanosine, 9CI.*
   $C_{12}H_{18}N_6O_5$ M 326.311
   Cryst. (H_2O). Mp >211° dec.

Holmes, R.E. *et al, J. Am. Chem. Soc.,* 1965, **87**, 1772 (*synth*)
Ikehara, M.I. *et al, Chem. Pharm. Bull.,* 1966, **14**, 46 (*derivs*)
Long, R.A. *et al, J. Org. Chem.,* 1967, **32**, 2751 (*derivs*)
Saneyoshi, M. *et al, Chem. Pharm. Bull.,* 1968, **16**, 1616 (*synth, uv*)
Jordan, F. *et al, Biochim. Biophys. Acta,* 1977, **476**, 265 (*pmr*)
Lin, T.-S. *et al, J. Med. Chem.,* 1985, **28**, 1194 (*synth*)

3-Amino-4-hydroxy-1,2-benzenedicarboxy-lic acid, 9CI   A-70151

3-Amino-4-hydroxyphthalic acid, 8CI

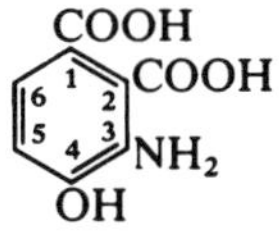

COOH
COOH
NH₂
OH

$C_8H_7NO_5$ M 197.147

*Me ether: 3-Amino-4-methoxy-1,2-benzenedicarboxylic
acid.*
   $C_9H_9NO_5$ M 211.174
   Needles (H_2O). Mp 152° dec.

Me ether, anhydride:
   $C_9H_7NO_4$ M 193.159
   Needles (AcOH). Mp 182°.

Grewe, R., *Chem. Ber.,* 1938, **71**, 907 (*synth*)

4-Amino-5-hydroxy-1,2-benzenedicarboxy-lic acid, 9CI   A-70152

4-Amino-5-hydroxyphthalic acid, 8CI

$C_8H_7NO_5$ M 197.147

Me ether, di-Me ester: [99684-26-1]. *Dimethyl 4-amino-
5-methoxyphthalate.*
   $C_{11}H_{13}NO_5$ M 239.227
   Cryst. (MeOH). Mp 149°.

King, H., *J. Chem. Soc.,* 1939, 1157 (*synth*)

4-Amino-6-hydroxy-1,3-benzenedicarboxy-lic acid, 9CI   A-70153

4-Amino-6-hydroxyisophthalic acid, 8CI

[15540-79-1]
$C_8H_7NO_5$ M 197.147
Cryst. (MeOH aq.). Mp 280° (210-212° dec.).

Di-Me ester: [71407-99-3].
   $C_{10}H_{11}NO_5$ M 225.201
   Cryst. (CHCl₃/cyclohexane). Mp 144-145°.

Beyerman, H.C. *et al, Recl. Trav. Chim. Pays-Bas,* 1950, **69**, 1021 (*synth, deriv*)
Baker, S.R. *et al, J. Chem. Soc., Perkin Trans. 1,* 1979, 677 (*synth, deriv, uv, ir, pmr*)
Schmidt, H.W. *et al, Justus Liebigs Ann. Chem.,* 1979, 2005 (*synth, ir, ms*)

5-Amino-2-hydroxy-1,3-benzenedicarboxy-lic acid   A-70154

5-Amino-2-hydroxyisophthalic acid, 8CI

$C_8H_7NO_5$ M 197.147
Mp 273° dec.

Di-Me ester: [67294-57-9].
   $C_{10}H_{11}NO_5$ M 225.201
   Yellow needles (MeOH). Mp 115-118°.

Benica, W.S. *et al, CA,* 1945, **39**, 1633 (*synth*)
Ebel, K. *et al, Chem. Ber.,* 1988, **121**, 323 (*synth, deriv, ir, pmr*)

5-Amino-4-hydroxy-1,3-benzenedicarboxy-lic acid, 9CI   A-70155

5-Amino-4-hydroxyisophthalic acid, 8CI

$C_8H_7NO_5$ M 197.147
Needles (H_2O). Mp 297° dec.

Di-Me ester:
   $C_{10}H_{11}NO_5$ M 225.201
   Mp 170°.

Me ether, di-Me ester:
   $C_{11}H_{13}NO_5$ M 239.227
   Cryst. (C_6H_6/pet. ether). Mp 63-64°.

Hunt, S.E. *et al, J. Chem. Soc.,* 1956, 3099 (*synth, uv*)

2-Amino-4-hydroxybutanoic acid   A-70156

Updated Entry replacing A-01761
Homoserine, 9CI
[498-19-1]

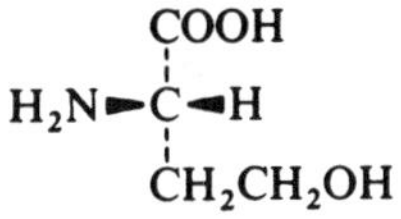

COOH
H₂N►C◄H
CH₂CH₂OH

(S)-form
Absolute
configuration

$C_4H_9NO_3$ M 119.120

(R)-form [6027-21-0]
D-form
Mp 203° dec. $[\alpha]_D^{23}$ −8° (H_2O).

(S)-form [672-15-1]
L-form
Present in peas, Jack bean seeds (*Canavalia ensiformis*)
   and the seedlings of many leguminous plants.
   Intermediate in the conversion of Aspartic acid into
   Homocysteine and α-ketobutyrate in microbia and
   fungi. Mp 203° dec. $[\alpha]_D^{23}$ +8° (H_2O).

N-Benzyloxycarbonyl, benzyl ester: Cryst. (Et₂O/pet.
   ether). Mp 52-57°. $[\alpha]_D^{25}$ 0° (c, 0.976 in CHCl₃).
Ac: see 4-Acetoxy-2-aminobutanoic acid, A-00126
Et ether: [17268-93-8]. *2-Amino-4-ethoxybutanoic acid.*
   $C_6H_{13}NO_3$ M 147.174

Prod. by *Corynebacterium ethanolaminophilum* grown in presence of ethanol. Mp 262°. $[\alpha]_D^{30}$ −14° (c, 2.5 in H_2O). Higher homologues are prod. by growing the bacterium with other alcohols present.

O-Succinyl: see O-Succinylhomoserine, S-00928

O-Oxalyl: [4096-48-4]. *O-Oxalylhomoserine.*
$C_6H_9NO_6$ M 191.140
Prod. by *Lyrus sativus*. Cryst. (H_2O).

O-(2-Amino-3-hydroxypropyl): [37662-02-5].
$C_7H_{16}N_2O_4$ M 192.214
Prod. by *Rhizobium japonicum*.

O-Phosphate: [4210-66-6]. *O-Phosphohomoserine. Phosphorylhomoserine.*
$C_4H_{10}NO_6P$ M 199.100
Major metab. of *Lactobacillus casei*. Intermed. in conversion of homoserine to threonine.

(±)-*form* [1927-25-9]
Needles (EtOH aq.). Mp 186-188° dec. Readily lactonises.

N-Benzoyl:
$C_{11}H_{13}NO_4$ M 223.228
Needles (H_2O). Mp 121° (126-127°, 140-144°).

O,N-Dibenzoyl:
$C_{18}H_{17}NO_5$ M 327.336
Plates (EtOH aq.). Mp 198°.

Lactone: see 3-Aminodihydro-2(3H)-furanone, A-60138

Ågren, G., *Acta Chem. Scand.*, 1962, **16**, 1607 (*phosphate*)
Guenther, D. et al, *Chem. Ber.*, 1964, **97**, 159 (*synth*)
Kollonitsch, A. et al, *J. Am. Chem. Soc.*, 1964, **86**, 1857 (*synth*)
Greenstein, J.P. et al, *Chemistry of the Amino Acids*, 1965, Wiley, N.Y., **3**, 2612 (*synth*)
Murooka, Y. et al, *Agric. Biol. Chem.*, 1967, **31**, 1035 (*derivs*)
Lawrence, J.M., *Phytochemistry*, 1973, **12**, 2207 (*isol*)
Miyoshi, M. et al, *Chem. Lett.*, 1973, 5 (*abs config*)
Bell, E.A., *FEBS Lett.*, 1976, **64**, 29 (*derivs*)
Baldwin, J.E. et al, *Tetrahedron Lett.*, 1987, **28**, 3167.
Baldwin, J.E. et al, *Tetrahedron*, 1988, **44**, 637 (*synth*)

1-Amino-3-(hydroxymethyl)-1-cyclobutane- carboxylic acid   A-70157

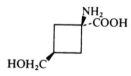

$C_6H_{11}NO_3$ M 145.158
cis-form [109794-96-9]
Isol. from seeds of *Atelia herbert-smithii*. Cryst. + $1H_2O$ (EtOH aq.). Dec. at 210°.

Austin, G.N. et al, *Tetrahedron*, 1987, **43**, 1857 (*isol, cryst struct*)

4-Amino-5-hydroxy-2-oxo-7- oxabicyclo[4.1.0]hept-3-ene-3-carboxa- mide, 9CI   A-70158

2-Amino-4,5-epoxy-3-hydroxy-6-oxo-1-cyclohexene- carboxamide. Antibiotic 2061 B
[99291-28-8]

$C_7H_8N_2O_4$ M 184.151

Quinone-related antibiotic. Isol. from *Streptomyces* sp. Weakly active against gram-positive and -negative bacteria.

5-Ketone: [63582-80-9]. **Antibiotic 2061 A**. *G 7063-2. Antibiotic G 7063-2.*
$C_7H_6N_2O_4$ M 182.135
From *S.* spp. Mod. active against gram-positive and - negative bacteria and fungi. Yellow cryst. (CH_3Cl/EtOAc). Mp >180°.
▷RN8789000.

Noble, M.F. et al, *J. Antibiot.*, 1977, **30**, 455 (*manuf*)
Reddy, G.C.S. et al, *J. Antibiot.*, 1984, **38**, 1596 (*config*)
Orezzi, G.P. et al, *Kangshengsu*, 1986, **11**, 474 (*isol*)

4-Amino-5-hydroxypentanoic acid, 9CI   A-70159

Updated Entry replacing A-01810
[40217-11-6]

$$CH_2CH_2COOH$$
$$H{-}\overset{\displaystyle |}{\underset{\displaystyle |}{C}}{-}NH_2 \qquad (S)\text{-}form$$
$$CH_2OH$$

$C_5H_{11}NO_3$ M 133.147
(S)-form
Mp 167-168°. $[\alpha]_D^{25}$ +24.2° (c, 2 in H_2O).
(±)-*form*
γ-Aminobutyric acid transaminase inactivator. Cryst. Mp 147-148°.

Prasad, B. et al, *Indian J. Chem.*, 1974, **12**, 290 (*formn*)
Silverman, R.B. et al, *J. Org. Chem.*, 1980, **45**, 815 (*synth*)
Barlos, K. et al, *J. Chem. Soc., Chem. Commun.*, 1987, 1583 (*synth, bibl*)

3-Amino-2-hydroxypropanoic acid, 9CI   A-70160

Updated Entry replacing A-01840
Isoserine. 2-Hydroxy-β-alanine. 3-Aminolactic acid
[565-71-9]

$$COOH$$
$$H{-}\overset{\displaystyle |}{\underset{\displaystyle |}{C}}{-}OH \qquad (R)\text{-}form$$
$$CH_2NH_2$$

$C_3H_7NO_3$ M 105.093
(R)-form [632-11-1]
Cryst. (EtOH aq.). Mp 199-201° dec. $[\alpha]_D^{20}$ +32.44° (H_2O).
Benzoyl:
$C_{10}H_{11}NO_4$ M 209.201
Mp 107-109°. $[\alpha]_D$ +29.46°.
(S)-form [632-13-3]
Cryst. (H_2O). Mp 199-201° dec. (188-189°). $[\alpha]_D^{20}$ −32.58° (c, 1 in H_2O).
Benzoyl: Mp 107-109°. $[\alpha]_D$ −30.03°.
(±)-*form* [632-12-2]
Prisms (EtOH aq.). Mp 248° dec. pK_{a1} 9.27, pK_{a2} 11.22.
Et ester:
$C_5H_{11}NO_3$ M 133.147
Mp 75°.
Benzoyl: Mp 151°.
Phenylurethane: Mp 183-184°.

Freudenberg, K., *Ber.*, 1914, **47**, 2027 (*abs config*)
Kenneth, C.L. et al, *J. Org. Chem.*, 1962, **27**, 438 (*synth*)

Miyazawa, T. *et al*, *Agric. Biol. Chem.*, 1976, **40**, 1651 (*synth*)
Andruszkiewicz, R. *et al*, *Synthesis*, 1983, 31 (*synth*)
Williams, T.M. *et al*, *J. Org. Chem.*, 1985, **50**, 91 (*synth*)
Dureault, A. *et al*, *Synthesis*, 1987, 491 (*synth*)

4-Amino-1*H*-imidazo[4,5-*c*]pyridine   A-70161

1H-Imidazo[4,5-c]pyridin-4-amine, *9CI*
[6811-77-4]

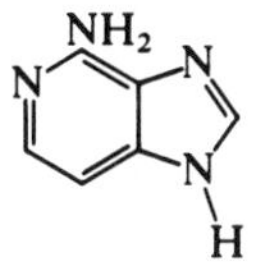

$C_6H_6N_4$ M 134.140
Powder. Mp 225°.
B,2HCl: [80639-85-6]. Cryst. Mp >250°.
4-N-Di-Me: [98858-07-2].
   $C_8H_{10}N_4$ M 162.194
   Cryst. (MeOH) (as hydrochloride). Mp >250°
   (hydrochloride).
4-N-Benzyl: [80443-18-1].
   $C_{13}H_{12}N_4$ M 224.265
   Yellow solid + ¼H_2O. Mp 60-64°.
1-β-D-Ribofuranosyl: [6736-58-9].
   $C_{11}H_{14}N_4O_4$ M 266.256
   Solid. Mp 226°. $[\alpha]_D^{20}$ −63.4° (c, 1 in DMF).
1-(2-Deoxy-β-D-ribofuranosyl): [78582-15-7].
   $C_{11}H_{14}N_4O_3$ M 250.257
   Mp 206° dec. (215°). $[\alpha]_D^{20}$ −25.5° (c, 1 in DMF).
1-(5-Deoxy-β-D-ribofuranosyl): [98858-09-4].
   $C_{11}H_{14}N_4O_3$ M 250.257
   Cryst. + 1H_2O. Mp 140°. $[\alpha]_D^{20}$ −31.9° (c, 0.5 in
   DMF).

Salemink, C.A. *et al*, *Recl. Trav. Chim. Pays-Bas*, 1949, **68**,
   1013.
Mizuno, Y. *et al*, *Chem. Pharm. Bull.*, 1964, **12**, 866; 1968, **16**,
   2011.
Montgomery, J.A. *et al*, *J. Med. Chem.*, 1965, **8**, 708.
Rousseau, R.J. *et al*, *Biochemistry*, 1966, **5**, 756.
DeRoos, K.B. *et al*, *Recl. Trav. Chim. Pays-Bas*, 1969, **88**, 1263;
   1971, **90**, 654.
May, J.A. *et al*, *J. Chem. Soc., Perkin Trans. 1*, 1975, 125.
Montgomery, J.A. *et al*, *J. Heterocycl. Chem.*, 1977, **14**, 195.
Krenitsky, T.A. *et al*, *J. Med. Chem.*, 1986, **29**, 1386 (*synth, uv,
   pmr, derivs*)

2-Amino-3-isopropylbutanedioic acid   A-70162

3-(1-Methylethyl)aspartic acid, *9CI*. *2-Amino-3-isopro-
pylsuccinic acid*. *β-Isopropylaspartic acid*

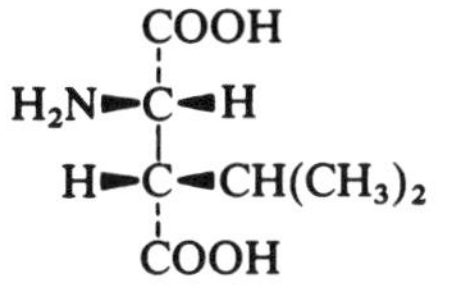

(2S,3S)-form

$C_7H_{13}NO_4$ M 175.184
(2S,3S)-form
   Cryst. (MeOH). Mp 239-242°. $[\alpha]_D^{20}$ +7.35° (c, 0.4 in
   6*M* HCl).
(2RS,3RS)-(?)-form [15383-90-1]
   Mp 235-237°. A single racemate which may be the
   (2*RS*,3*SR*)-form.

Winkler, M.F. *et al*, *Biochim. Biophys. Acta*, 1967, **146**, 287
   (*synth*)
Bochenska, M. *et al*, *Rocz. Chem.*, 1976, **50**, 1195 (*synth*)
Akhtar, M. *et al*, *Tetrahedron*, 1987, **43**, 5899 (*synth*)

2-Amino-3-mercapto-1,4-naphthoquinone,   A-70163
8CI
2-Amino-3-mercapto-1,4-naphthalenedione, *9CI*
[76148-76-0]

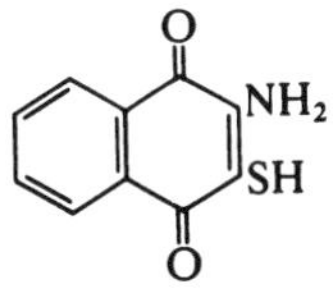

$C_{10}H_7NO_2S$ M 205.231
Orange-red needles (dioxan). Mp 293-295°.
Me ether: [72653-76-0]. *2-Amino-3-(methylthio)-1,4-
   naphthalenedione*, *9CI*.
   $C_{11}H_9NO_2S$ M 219.258
   Cryst. (EtOH). Mp 132°.

Ikeda, N., *CA*, 1956, **50**, 3358i (*deriv*)
Hammam, A.S. *et al*, *Collect. Czech. Chem. Commun.*, 1985,
   50, 71 (*synth, uv*)
Katritzky, A.R. *et al*, *J. Heterocycl. Chem.*, 1988, **25**, 901
   (*synth, uv, ir, pmr, cmr*)

2-Amino-3-methylbiphenyl   A-70164

Updated Entry replacing A-01991
3-Methyl-[1,1'-biphenyl]-2-amine, *9CI*. *3-Methyl-2-
biphenylamine*
[14294-33-8]
$C_{13}H_{13}N$ M 183.252
Cryst. (pet. ether). Mp 63.5-64° (61-62°).
N-Ac: *2-Acetamido-3-methylbiphenyl*.
   $C_{15}H_{15}NO$ M 225.290
   Mp 139-140°.

Risaliti, A. *et al*, *Ann. Chim. (Rome)*, 1967, **57**, 3; *CA*, **67**,
   11291 (*synth*)
Atwell, G.J. *et al*, *J. Med. Chem.*, 1988, **31**, 774 (*synth*)

5-Aminomethyl-3-bromoisoxazole   A-70165

3-Bromo-5-isoxazolemethanamine, *9CI*
[2763-93-1]

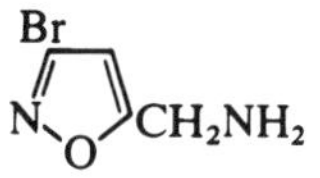

$C_4H_5BrN_2O$ M 177.000
B,HCl: Mp 170° dec.

Chiarino, D. *et al*, *Tetrahedron Lett.*, 1986, **27**, 3181 (*synth*)

2-Amino-3-methylbutanedioic acid   A-70166

Updated Entry replacing A-20122
3-Methylaspartic acid, *9CI*. *2-Amino-3-methylsuccinic
acid*
[2955-50-2]

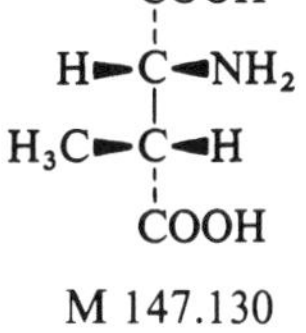

(2R,3R)-form

$C_5H_9NO_4$ M 147.130
(2R,3R)-form
   D-threo-form
   $[\alpha]_D^{25}$ +11.7° (H_2O).

(2S,3S)-form [31571-69-4]

L-threo-*form*

Prod. by *Clostridium tetanomorphum*; component of Aspartocin, Amphomycin, Glutamycin. $[\alpha]_D^{25}$ −12.4° (H_2O).

Di-Et ester:

$C_9H_{17}NO_4$ M 203.238

Prisms (as hydrochloride). Mp 118° (hydrochloride).

(2RS,3RS)-form

(±)-threo-*form*

Cryst.

(2RS,3SR)-form

(±)-erythro-*form*

Cryst.

Biochem. Prep., 1961, **8**, 93, 96 (*synth*)
Traynham, J.G. *et al*, *J. Org. Chem.*, 1962, **27**, 2959 (*synth*)
Sprecher, M. *et al*, *J. Biol. Chem*, 1966, **241**, 868.
Burrows, I.E. *et al*, *J. Chem. Soc. (C)*, 1968, 40.
Dressel, L.A. *et al*, *J. Am. Chem. Soc.*, 1976, **98**, 4150 (*resoln*)
Mori, K. *et al*, *Tetrahedron Lett.*, 1981, 1127.
Gani, D. *et al*, *J. Chem. Soc., Perkin Trans. 1*, 1985, 1363 (*synth, pmr*)
Akhtar, M. *et al*, *Tetrahedron*, 1987, **43**, 5899 (*synth, pmr*)

2-Amino-2-methylbutanoic acid   **A-70167**

Isovaline, 9CI

[465-58-7]

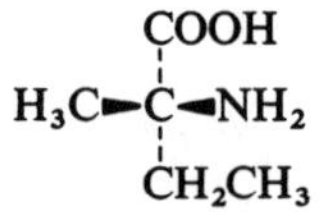

$H_3C-\overset{\overset{\displaystyle COOH}{|}}{\underset{\underset{\displaystyle CH_2CH_3}{|}}{C}}\blacktriangleleft NH_2$ (R)-*form*

$C_5H_{11}NO_2$ M 117.147

(R)-form [3059-97-0]

D-*form*

Cryst. Mp ca. 300°. $[\alpha]_D^{20}$ −9.10° (H_2O), $[\alpha]_D^{20}$ −6.11° (20% HCl). Sublimes.

(S)-form [595-40-4]

L-*form*

Needles + $1H_2O$. $[\alpha]_D^{25}$ +10.7° (H_2O). Sublimes.

(±)-form [595-39-1]

Needles. Sol. H_2O, spar. sol. EtOH. Mp 315° (305°). Subl. at 300°.

B,HCl: Mp 110-115° dec.

Et ester:

$C_7H_{15}NO_2$ M 145.201

Bp_{20} 65-66°.

Nitrile: [4475-95-0].

$C_5H_{10}N_2$ M 98.147

Bp_4 68°, Bp_{20} 72°.

Fischer, E. *et al*, *Justus Liebigs Ann. Chem.*, 1914, **406**, 5.
Bulman *et al*, *Bull. Soc. Chim. Fr.*, 1934, 1661.
Li, L. *et al*, *J. Chin. Chem. Soc.*, 1942, **9**, 1.
Greenstein, J.P. *et al*, *Chemistry of the Amino Acids*, 1961, Wiley, **3**.
Yamada, S. *et al*, *Chem. Pharm. Bull.*, 1964, **12**, 1525; 1966, **14**, 537 (*abs config*)
Bosch, R. *et al*, *Tetrahedron*, 1982, **38**, 3579 (*cryst struct, abs config*)
Schöllkopf, U. *et al*, *Synthesis*, 1983, 675 (*synth*)
Fieser, M. *et al*, *Reagents for Organic Synthesis*, Wiley, 1967-84, **4**, 274.

Handle all chemicals with care

2-(Aminomethyl)cycloheptanol, 9CI   **A-70168**

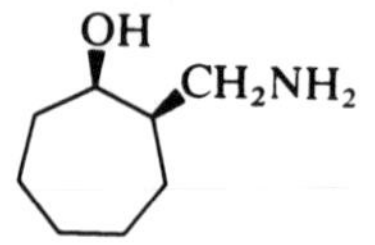

(1RS,2RS)-*form*

$C_8H_{17}NO$ M 143.228

(1RS,2RS)-form [72450-73-8]

(±)-cis-*form*

B,HCl: [42418-76-8]. Mp 170°.

(1RS,2SR)-form [72450-74-9]

(±)-trans-*form*

B,HCl: [42418-77-9]. Mp 132°.

Bernáth, G. *et al*, *Tetrahedron*, 1973, **29**, 981; 1979, **35**, 799.

2-(Aminomethyl)cyclohexanol, 9CI   **A-70169**

[60659-10-1]

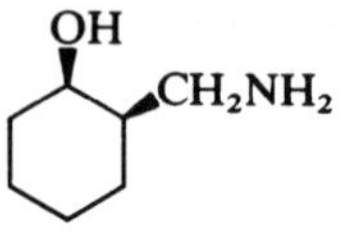

(1RS,2RS)-*form*

$C_7H_{15}NO$ M 129.202

(1RS,2RS)-form [28250-37-5]

(±)-cis-*form*

Bp_5 105-108°.

B,HCl: [24947-68-0]. Cryst. (EtOH/Et_2O). Mp 182.5-183°.

(1RS,2SR)-form [5691-09-8]

(±)-trans-*form*

$Bp_{1.1}$ 89°.

B,HCl: [24948-05-8]. Mp 151°.

N-Benzoyl: [20617-33-8].

$C_{14}H_{19}NO_2$ M 233.310

Cryst. (C_6H_6/pet. ether). Mp 120-122.5°.

Mousseron, M. *et al*, *C.R. Hebd. Seances Acad. Sci.*, 1948, **91**, 226 (*synth*)
Bernáth, G. *et al*, *Acta Chim. Acad. Sci. Hung*, 1970, **64**, 183 (*synth*)
Schwartz, L.H. *et al*, *J. Am. Chem. Soc.*, 1972, **94**, 180 (*nmr, ms, synth*)
Gorenstein, D.G. *et al*, *J. Am. Chem. Soc.*, 1979, **101**, 4925.

2-(Aminomethyl)cyclooctanol, 9CI   **A-70170**

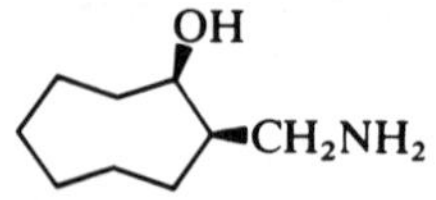

(1RS,2RS)-*form*

$C_9H_{19}NO$ M 157.255

(1RS,2RS)-form [57906-83-9]

(±)-cis-*form*

Liq. crystallising at r.t. Bp_{35} 145-155°.

B,HCl: [57906-84-0]. Mp 141-142°.

N-Benzoyl:

$C_{16}H_{23}NO$ M 245.364

Cryst. (EtOH aq.). Mp 99-101°.

(1RS,2SR)-form

(±)-trans-*form*

B,HCl: [60622-99-3]. Cryst. (EtOH/Et_2O). Mp 138-139°.

N-Benzoyl: Mp 107-108°.

Bernáth, G. *et al*, *Acta Chim. Acad. Sci. Hung.*, 1975, **86**, 187.

2-Amino-3-methyl-1,4-naphthoquinone, 8CI   A-70171

2-Amino-3-methyl-1,4-naphthalenedione, 9CI

[7427-09-0]

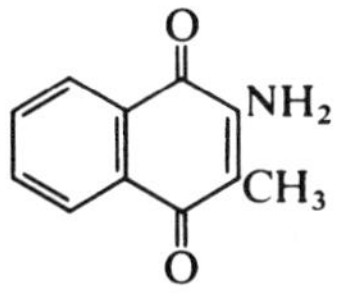

$C_{11}H_9NO_2$ M 187.198

Red cryst. (EtOH). Mp 165-167°.

Ac:

$C_{13}H_{11}NO_3$ M 229.235

Mp 157-158°.

Fieser, L.F. *et al, J. Am. Chem. Soc.*, 1948, **70**, 3212 (*synth*)
Pearce, D.S. *et al, J. Org. Chem.*, 1974, **39**, 1362 (*synth*)
Husu, B. *et al, Monatsh. Chem.*, 1988, **119**, 215 (*synth*)

1-Amino-2-naphthalenethiol   A-70172

1-Amino-2-mercaptonaphthalene. 2-Mercapto-1-naphthylamine

[63512-54-9]

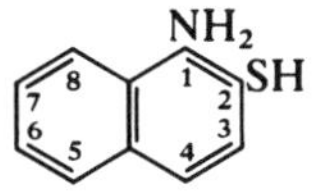

$C_{10}H_9NS$ M 175.248

Cryst. Unstable, readily oxid. to disulfide.

Ohkura, Y. *et al, Anal. Chim. Acta*, 1978, **99**, 317 (*synth, use*)

8-Amino-1-naphthalenethiol   A-70173

1-Amino-8-mercaptonaphthalene. 8-Mercapto-1-naphthylamine

[72319-64-3]

$C_{10}H_9NS$ M 175.248

Orange cryst. (Et$_2$O). Mp 75-76°. Unstable in air; turns blue-black.

Nakayama, J. *et al, J. Am. Chem. Soc.*, 1979, **101**, 7684 (*synth, ir, pmr*)

3-Amino-4-nitro-1,2-benzenedicarboxylic acid, 9CI   A-70174

3-Amino-4-nitrophthalic acid, 8CI

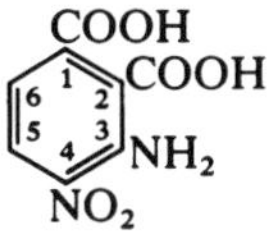

$C_8H_6N_2O_6$ M 226.145

Di-Me ester: [52412-66-5].

$C_{10}H_{10}N_2O_6$ M 254.199

Cryst. (CCl$_4$). Mp 85-86.5°.

Di-Me ester, Ac:

$C_{12}H_{12}N_2O_7$ M 296.236

Cryst. (MeOH). Mp 125.5-127°.

Williams, R.L. *et al, J. Heterocycl. Chem.*, 1973, **10**, 891 (*synth, ir, pmr*)

3-Amino-6-nitro-1,2-benzenedicarboxylic acid, 9CI   A-70175

3-Amino-6-nitrophthalic acid, 8CI

$C_8H_6N_2O_6$ M 226.145

Yellow prisms (EtOAc). Mp 242°.

Di-Me ester: [57113-93-6].

$C_{10}H_{10}N_2O_6$ M 254.199

Yellow needles (MeOH). Mp 149-151°.

Momose, T. *et al, Chem. Pharm. Bull.*, 1961, **9**, 263 (*synth*)
Nagai, U., *Chem. Pharm. Bull.*, 1975, **23**, 1841 (*synth, uv, ir, pmr*)

4-Amino-3-nitro-1,2-benzenedicarboxylic acid, 9CI   A-70176

4-Amino-3-nitrophthalic acid, 8CI

$C_8H_6N_2O_6$ M 226.145

Yellow prisms (H$_2$O). Mp 234°.

Di-Me ester: [52412-59-6].

$C_{10}H_{10}N_2O_6$ M 254.199

Cryst. (C$_6$H$_6$). Mp 122.5-124°.

Di-Me ester, Ac: [52412-56-3].

$C_{12}H_{12}N_2O_7$ M 296.236

Cryst. (CCl$_4$). Mp 125-127°.

Momose, T. *et al, Chem. Pharm. Bull.*, 1964, **12**, 14 (*synth*)
Williams, R.L. *et al, J. Heterocycl. Chem.*, 1973, **10**, 891 (*synth, ir, pmr*)

4-Amino-5-nitro-1,2-benzenedicarboxylic acid, 9CI   A-70177

4-Amino-5-nitrophthalic acid, 8CI

$C_8H_6N_2O_6$ M 226.145

Yellow needles (H$_2$O). Mp 238°.

Di Me ester, Ac: [52412-55-2].

$C_{12}H_{12}N_2O_7$ M 296.236

Cryst. (MeOH). Mp 123-124.5°.

Anhydride:

$C_8H_4N_2O_5$ M 208.130

Needles (C$_6$H$_6$). Mp 166-167°.

Goldstein, H. *et al, Helv. Chim. Acta*, 1952, **35**, 1476 (*synth*)
Momose, T. *et al, Chem. Pharm. Bull.*, 1964, **12**, 14 (*synth*)
Williams, R.L. *et al, J. Heterocycl. Chem.*, 1973, **10**, 891 (*synth, deriv, ir, pmr*)

5-Amino-3-nitro-1,2-benzenedicarboxylic acid, 9CI   A-70178

5-Amino-3-nitrophthalic acid, 8CI

$C_8H_6N_2O_6$ M 226.145

Yellow prisms (H$_2$O). Mp >300°.

Momose, T. *et al, Chem. Pharm. Bull.*, 1964, **12**, 14 (*synth*)

2-Amino-1,10-phenanthroline   A-70179

1,10-Phenanthrolin-2-amine, 9CI

[22426-18-2]

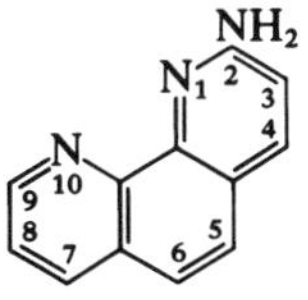

$C_{12}H_9N_3$ M 195.223

Claus, K.G. *et al, Inorg. Chem.*, 1969, **8**, 59.

8-Amino-1,7-phenanthroline   A-70180

1,7-Phenanthrolin-8-amine, 9CI

[86538-53-6]

$C_{12}H_9N_3$ M 195.223

Mp 206-208°.

Picrate: [86538-55-8]. Mp 307-310° dec.

van den Haak, H.J.W. *et al*, *J. Heterocycl. Chem.*, 1983, **20**, 447 (*synth, pmr*)

4-Amino-1,10-phenanthroline   A-70181

1,10-Phenanthrolin-4-amine

[69380-48-9]

$C_{12}H_9N_3$ M 195.223

Yellow prisms (cyclohexane). Mp 124-125°.

N-*Ph:* [69380-42-3]. *4-Anilino-1,10-phenanthroline.*

$C_{18}H_{13}N_3$ M 271.321

Yellow needles (MeOH aq.). Mp 295-300°.

Cook, M.J. *et al*, *J. Chem. Soc., Perkin Trans. 2*, 1978, 1215 (*synth, tautom, uv*)

5-Amino-1,10-phenanthroline   A-70182

1,10-Phenanthrolin-5-amine, 9CI

[54258-41-2]

$C_{12}H_9N_3$ M 195.223

Mp 259-260°.

B,HBr: [22426-20-6]. Yellow solid. Mp 302-304°.

B,2HBr: [22426-19-3]. Yellow solid. Mp 281°.

Koft, E. *et al J. Org. Chem.*, 1962, **27**, 865 (*synth*)
Gillard, R.D. *et al*, *J. Chem. Soc., Dalton Trans.*, 1974, 1217 (*deriv, ir, ms*)

1-Amino-1-phenyl-4-pentene   A-70183

α-3-Butenylbenzenemethanamine, 9CI

[109925-99-7]

$$H_2C{=}CHCH_2CH_2CH(NH_2)Ph$$

$C_{11}H_{15}N$ M 161.246

(±)-*form*

Oil. $Bp_{0.02}$ 80° (Kugelrohr).

Hippeli, C. *et al*, *Synthesis*, 1987, 77 (*synth, ir, pmr*)

1-Aminophthalazine   A-70184

1-Phthalazinamine

[19064-69-8]

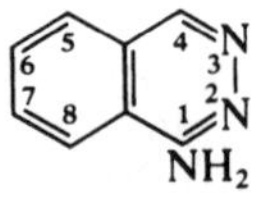

$C_8H_7N_3$ M 145.163

Needles (MeNO$_2$). Mp 212-213°.

B,MeI: Pale-yellow needles (H$_2$O).

N^1-*Ac:* *1-Acetamidophthalazine.*

$C_{10}H_9N_3O$ M 187.201

Needles (EtOH). Mp 183-184°.

N-*Me:* [39998-73-7].

$C_9H_9N_3$ M 159.190

Yellow cryst. (CHCl$_3$/pet. ether). Mp 178°.

N,N-*Di-Me:* [35552-67-1].

$C_{10}H_{11}N_3$ M 173.217

Mp 67°. $Bp_{0.5}$ 146°.

Atkinson, C.M. *et al*, *J. Chem. Soc.*, 1956, 1081 (*synth*)
Guingant, A. *et al*, *Bull. Soc. Chim. Fr.*, 1975, 2246 (*synth*)
Counotte-Potman, A. *et al*, *J. Heterocycl. Chem.*, 1983, **20**, 1259 (*synth*)

5-Aminophthalazine   A-70185

5-Phthalazinamine, 9CI

[102072-84-4]

$C_8H_7N_3$ M 145.163

Cryst. (MeOH). Mp 223-224°.

Shaikh, I.A. *et al*, *J. Med. Chem.*, 1986, **29**, 1329 (*synth, ir, pmr*)

2-Amino-3-propylbutanedioic acid   A-70186

3-Propylaspartic acid, 9CI. *2-Amino-3-propylsuccinic acid*

[69779-95-9]

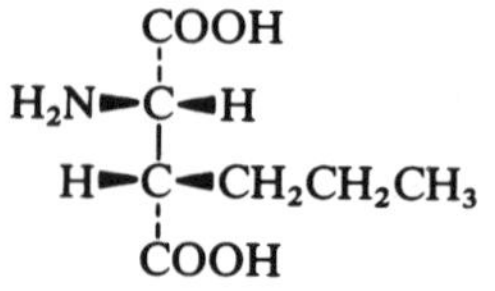

(2S,3S)-*form*

$C_7H_{13}NO_4$ M 175.184

(**2S,3S**)-*form*

Cryst. (MeOH aq.). Mp 235-237°. $[\alpha]_D^{20}$ +7.6° (c, 0.5 in 6*M* HCl).

(**2RS,3RS**)-(**?**)-*form* [15383-89-8]

Mp 224-227°. A single diastereoisomer which may be the (2*RS*,3*SR*)-form.

Winkler, M.F. *et al*, *Biochim. Biophys. Acta*, 1967, **146**, 287 (*synth*)
Bochenska, M. *et al*, *Rocz. Chem.*, 1976, **50**, 1195 (*synth*)
Akhtar, M. *et al*, *Tetrahedron*, 1987, **43**, 5899 (*synth*)

3-Amino-1*H*-pyrazole-4-carboxylic acid,   A-70187
9CI

[41680-34-6]

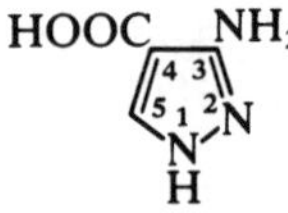

$C_4H_5N_3O_2$ M 127.102

Cryst. (H$_2$O). Mp 120° dec. 1-Unsubstituted derivs. tautomeric with the 5-amino compds.

▷UQ6390000.

Me ester: [29097-00-5].

$C_5H_7N_3O_2$ M 141.129

Mp 135-136°.

Et ester: [6994-25-8].

$C_6H_9N_3O_2$ M 155.156

Cryst. (H$_2$O or Et$_2$O). Mp 102-103°.

Amide: [52906-16-8].

$C_4H_6N_4O$ M 126.118

Cryst. (H$_2$O). Mp 222-225°.

Anilide: [90399-89-6].

$C_{10}H_{10}N_4O$ M 202.215

Mp 198-199°.

Nitrile: [16617-46-2]. *5-Amino-4-pyrazolecarbonitrile. 3-Amino-4-cyanopyrazole.*

$C_4H_4N_4$ M 108.102

Cryst. (H$_2$O). Mp 174-175°.

N(*1*)-*Me, Et ester:* [21230-43-3].

$C_7H_{11}N_3O_2$ M 169.183

Cryst. (CH$_2$Cl$_2$/pet. ether). Mp 92-93°.

N(*1*)-*Me, Nitrile:* [21230-50-2]. *3-Amino-1-methyl-4-pyrazolecarbonitrile*, 9CI. *3-Amino-4-cyano-1-methylpyrazole.*

$C_5H_6N_4$ M 122.129

Cryst. (CH$_2$Cl$_2$/pet. ether or EtOH). Mp 135-136° (126°).

Nitrile, N(*3*)-*Ac:*

$C_6H_6N_4O$ M 150.140

Cryst. (H₂O). Mp 221-222°.
Et ester, N(*3*)-*Ac:* [15250-36-9].
   C₈H₁₁N₃O₃   M 197.193
   Cryst. (EtOH aq.). Mp 200°.
N(*1*)-*Benzyl:* [99007-16-6].
   C₁₁H₁₁N₃O₂   M 217.227
   Cryst. (H₂O). Mp 176-180° dec.
N(*1*)-*Benzyl, Et ester:* [21377-11-7].
   C₁₃H₁₅N₃O₂   M 245.280
   Needles (EtOH). Mp 115-116°.

Robins, R.K., *J. Am. Chem. Soc.*, 1956, **78**, 784 (*nitrile, amide*)
Schmidt, P. *et al, Helv. Chim. Acta*, 1956, **39**, 986; 1959, **42**, 349, 763 (*synth, derivs*)
Dorn, H., *et al, Z. Chem.*, 1968, **8**, 420 (*1-benzyl derivs*)
Straley, J.M. *et al, CA*, 1970, **73**, 78529 (*Me ester*)
Ege, G. *et al, J. Heterocycl. Chem.*, 1982, **19**, 1267 (*1-Me, 1-Ph, nitriles*)
Finlander, ʹP. *et al, Chem. Scr.*, 1984, **23**, 23 (*Et ester, N-Ac, anilide*)

3-Amino-1*H*-pyrazole-5-carboxylic acid   A-70188

*5-Amino-2*H-*pyrazole-3-carboxylic acid*
C₄H₅N₃O₂   M 127.102
N(*1*)-*Me:* [117860-54-5].
   C₅H₇N₃O₂   M 141.129
   Cryst. (MeOH aq.). Mp 240.5-241.5°.
N(*1*)-*Me, Me ester:* [89088-56-2].
   C₆H₉N₃O₂   M 155.156
   Cryst. (MeOH). Mp 114-118° (112-113°).
N(*1*)-*Me, Et ester:* [89088-57-3].
   C₇H₁₁N₃O₂   M 169.183
   Mp 58-60°. Subl. at 35°/0.15 mm.

Chenard, B.L., *J. Org. Chem.*, 1984, **49**, 1224.
Lee, H.H. *et al, J. Org. Chem.*, 1989, **54**, 428.

4-Amino-1*H*-pyrazole-3-carboxylic acid,   A-70189
9CI

C₄H₅N₃O₂   M 127.102
Flakes (H₂O). Mp 212.5° dec. 1-Unsubstituted derivs. tautomeric with the 5-carboxylic acid.
Et ester: [55904-61-5].
   C₆H₉N₃O₂   M 155.156
   Mp 90-92°.
Amide: [67221-50-5].
   C₄H₆N₄O   M 126.118
   Cryst. (EtOH). Mp 181-182°.
Nitrile: [68703-67-3]. *4-Amino-3-cyanopyrazole.*
   C₄H₄N₄   M 108.102
   Prisms (H₂O or by subl.). Mp 143.5-145°.
1-Ph: [64299-26-9].
   C₁₀H₉N₃O₂   M 203.200
   Cryst. (H₂O). Mp 223-225°.
1-Ph, amide: [64299-25-8].
   C₁₀H₁₀N₄O   M 202.215
   Cryst. (propanol). Mp 189-191°.

Bertho, A. *et al, Justus Liebigs Ann. Chem.*, 1927, **457**, 278 (*synth*)
Robins, R.K. *et al, J. Am. Chem. Soc.*, 1956, **78**, 2418 (*synth, amide*)
Tam, H.-D. *et al, J. Org. Chem.*, 1975, **40**, 2825 (*Et ester*)
Gewald, K. *et al, Monatsh. Chem.*, 1977, **108**, 611 (*1-Ph derivs*)
Kagan, J. *et al, J. Heterocycl. Chem.*, 1979, **16**, 1113 (*nitrile*)

4-Amino-1*H*-pyrazole-5-carboxylic acid   A-70190

C₄H₅N₃O₂   M 127.102

1-Unsubstituted derivs. tautomeric with the 3-carboxylic acid, q.v.ʹ
N(*1*)-*Me, Amide:* [92534-73-1].
   C₄H₆N₄O   M 126.118
   Cryst. (2-propanol). Mp 174-175°.
N(*1*)-*Me; B,HCl:* [92534-70-8]. Cryst. (2-propanol). Mp 216-218° dec.

Perevalov, V.P. *et al, Zh. Org. Khim.*, 1984, **20**, 1073.

5-Amino-1*H*-pyrazole-3-carboxylic acid   A-70191

C₄H₅N₃O₂   M 127.102
N(*1*)-*Me:* [117860-53-4].
   C₅H₇N₃O₂   M 141.129
   Cryst. (AcOH/toluene). Mp 162-164°.
N(*1*)-*Me, Me ester:* [92406-53-6].
   C₆H₉N₃O₂   M 155.156
   Needles (EtOAc/Et₂O). Mp 101-102°.

Lee, H.H. *et al, J. Org. Chem.*, 1989, **54**, 428 (*derivs*)

5-Amino-1*H*-pyrazole-4-carboxylic acid   A-70192

C₄H₅N₃O₂   M 127.102
1-Unsubstituted derivs. are tautomeric with the derivs. of 3-Amino-1*H*-pyrazole-4-carboxylic acid, A-70187 .
N(*1*)-*Me, N(5)-Ac:* [92914-76-6].
   C₇H₉N₃O₃   M 183.166
   Cryst. (Me₂CO aq.). Mp 194.5-195°.
N(*1*)-*Ac, Et ester:* [14333-81-4].
   C₈H₁₁N₃O₃   M 197.193
   Cryst. (EtOH aq.). Mp 74-76°.
N(*1*)-*Me, Et ester:* [31037-02-2].
   C₇H₁₁N₃O₂   M 169.183
   Cryst. (C₆H₆/pet. ether). Mp 101°.
N(*1*)-*Me, amide:* [18213-75-7].
   C₅H₈N₄O   M 140.144
   Cryst. (H₂O). Mp 237-239°.
N(*1*)-*Me, nitrile:* [5334-41-8]. *5-Amino-4-cyano-1-methylpyrazole.*
   C₅H₆N₄   M 122.129
   Cryst. (H₂O). Mp 222-223°.
N(*1*)-*Me, N(5)-Ac, Et ester:* [88320-47-2].
   C₉H₁₃N₃O₃   M 211.220
   Mp 101-103°.
N(*1*)-*Me, N(5)-Ac, nitrile:* [5334-42-9]. *N-(4-Cyano-1-methyl-1H-pyrazol-5-yl)acetamide, 9CI.*
   C₇H₈N₄O   M 164.166
   Cryst. (toluene). Mp 138-139°.
N(*1*)-*Benzyl, amide:* [56156-22-0]. *5-Amino-1-(phenylmethyl)-1H-pyrazole-4-carboxamide, 9CI.*
   C₁₁H₁₂N₄O   M 216.242
   Cryst. (EtOH). Mp 224-225°.
N(*1*)-*Benzyl, nitrile:* [91091-13-3]. *5-Amino-1-benzyl-4-cyanopyrazole. 3-Amino-2-benzyl-4-cyanopyrazole.*
   C₁₁H₁₀N₄   M 198.227
   Cryst. (EtOH). Mp 184-185.5°.
N(*1*)-*Ph, amide:* [50427-77-5].
   C₁₀H₁₀N₄O   M 202.215
   Cryst. (H₂O). Mp 172-173°.
N(*1*)-*Ph, nitrile:* [5334-43-0].
   C₁₀H₈N₄   M 184.200
   Cryst. (H₂O). Mp 140°.

Cheng, C.C. *et al, J. Org. Chem.*, 1956, **21**, 1240 (*1-Me, 1-Ph, nitriles, amides*)
Schmidt, P. *et al, Helv. Chim. Acta*, 1959, **42**, 349 (*1-benzyl, nitrile, amide*)

Hartke, K. *et al*, *Arch. Pharm.* (*Weinheim, Ger.*), 1966, **299**, 914 (*1-Ac, Et ester*)
Finlander, P. *et al*, *Chem. Scr.*, 1983, **22**, 171 (*derivs*)
Nielsen, S.V. *et al*, *Chem. Scr.*, 1984, **23**, 240 (*derivs*)

2-Amino-3,5-pyridinedicarboxylic acid, 9CI   A-70193

$C_7H_6N_2O_4$ M 182.135
Solid. Mp 344.5-346.5° dec.
3-Mono-Et ester:
　$C_9H_{10}N_2O_4$ M 210.189
　Cryst. (dioxan aq.). Mp >300° dec.
5-Mono-Et ester:
　$C_9H_{10}N_2O_4$ M 210.189
　Cryst. (dioxan aq.). Mp 248.5-250°.
Di-Et ester:
　$C_{11}H_{14}N_2O_6$ M 270.241
　Needles (hexane). Mp 99-100°.
Diamide:
　$C_7H_8N_4O_2$ M 180.166
　Cryst. (H_2O). Mp 290.5-292°.
5-Nitrile, 3-amide: 2-*Amino-5-cyanonicotinamide.*
　$C_7H_6N_4O$ M 162.151
　Cryst. (EtOH aq.). Mp >256° dec.
Dinitrile: [78473-10-6]. 2-*Amino-3,5-dicyanopyridine.*
　$C_7H_4N_4$ M 144.135
　Cryst. (MeCN). Mp 220° dec.

Mulvey, D.M. *et al*, *J. Org. Chem.*, 1964, **29**, 2903 (*synth, deriv*)
Piper, J.R. *et al*, *J. Med. Chem.*, 1986, **29**, 1080 (*deriv, synth, pmr*)

4-Amino-2,6-pyridinedicarboxylic acid   A-70194
γ-Aminodipicolinic acid
[2683-49-0]
$C_7H_6N_2O_4$ M 182.135
Cryst. (H_2O). Mp 299°.

Di-Et ester:
　$C_{11}H_{14}N_2O_4$ M 238.243
　Mp 283°.
N-Benzoyl:
　$C_{14}H_{10}N_2O_5$ M 286.243
　Mp 309° dec.

Koenigs, E. *et al*, *Chem. Ber.*, 1924, **57**, 1172 (*synth*)
Lewis, J.C. *et al*, *J. Biol. Chem*, 1972, **247**, 1861 (*synth*)

5-Amino-3,4-pyridinedicarboxylic acid   A-70195
5-Aminocinchomeronic acid
$C_7H_6N_2O_4$ M 182.135
Pale-yellow solid + 1H_2O. Mp 243-245° dec.

Di-Me ester:
　$C_9H_{10}N_2O_4$ M 210.189
　Mp 74-76°.

van der Waal, B. *et al*, *Recl. Trav. Chim. Pays-Bas*, 1961, **80**, 203 (*synth*)
Gadekar, S.M. *et al*, *J. Med. Chem.*, 1962, **5**, 531 (*synth, uv*)

5-Amino-4(1H)-pyrimidinone, 9CI   A-70196
5-Amino-4-hydroxypyrimidine
[69785-94-0]

$C_4H_5N_3O$ M 111.103
Mp 213-216°.
B,HCl: [106913-64-8]. Cryst. (MeOH). Mp 213-217° dec.
5-N-Formyl: [106289-05-8]. N-(*1,4-Dihydro-4-oxo-5-pyrimidinyl*)*formamide*, 9CI.
　$C_5H_5N_3O_2$ M 139.113
　Mp 260-264°.

Nemeryuk, M.P. *et al*, *Collect. Czech. Chem. Commun.*, 1986, **51**, 215 (*synth*)

3-Amino-2(1H)quinolinethione, 9CI   A-70197
[71316-38-6]

$C_9H_8N_2S$ M 176.236
Fluorimetric reagent.

Yoshida, T. *et al*, *Bunseki Kagaku*, 1979, **28**, 351 (*synth*)
Yoshida, T. *et al*, *Chem. Pharm. Bull.*, 1986, **34**, 3774 (*use*)

8-Amino-9-β-D-ribofuranosyl-6H-purin-6-one   A-70198
8-Aminoinosine, 9CI
[13389-16-7]

$C_{10}H_{13}N_5O_5$ M 283.243
Cryst. (MeOH/EtOH). Mp 192° dec.
6-Hydrazone: [55627-78-6]. Cryst. (EtOH aq.). Mp 236-238°.
N^8,N^8-*Di-Me:* [13389-17-8]. 8-(*Dimethylamino*)*inosine.*
　$C_{12}H_{17}N_5O_5$ M 311.297
　Mp 209-211° dec.

Long, R.A. *et al*, *J. Org. Chem.*, 1967, **32**, 2751 (*synth*)
Long, R.A. *et al*, *J. Chem. Soc., Chem. Commun.*, 1970, 1087 (*pmr*)
Szekeres, G.L. *et al*, *J. Heterocycl. Chem.*, 1975, **12**, 15 (*deriv*)
Lin, T.-S. *et al*, *J. Med. Chem.*, 1985, **28**, 1481 (*synth, uv, pmr*)

3-Aminotetrahydro-3-thiophenecarboxylic acid, 9CI   A-70199

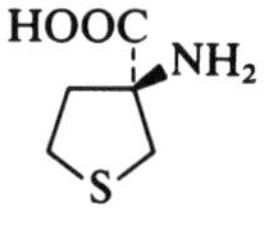

$C_5H_9NO_2S$ M 147.192

(**S**)-**form** [114715-53-6]

Solid + $1H_2O$. Mp 207-209°. $[\alpha]_D^{25}$ −6.67° (c, 1.32 in H_2O).

Morimoto, Y. *et al, Chem. Pharm. Bull.*, 1987, **35**, 3845 (*synth, pmr, cmr*)

3-Aminotetrahydro-2*H*-thiopyran   A-70200

*Tetrahydro-2*H*-thiopyran-3-amine, 9CI. 3-Aminothiacyclohexane*

[117593-38-1]

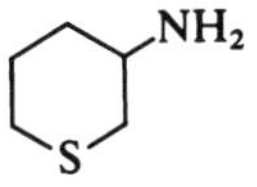

$C_5H_{11}NS$ M 117.209

(±)-**form**

Oxalate: [117593-51-8]. Mp 226°.
Picrate: Mp 175-176°.
S-Oxide:
  $C_5H_{11}NOS$ M 133.208
  Noncryst. Mixt. of diastereoisomers.
S,S-Dioxide: [117593-43-8].
  $C_5H_{11}NO_2S$ M 149.207
  Cryst. (hexane/EtOAc). Mp 83-84°.
N,N-Di-Me: [117593-44-9].
  $C_7H_{15}NS$ M 145.262
  Cryst. (EtOH) (as oxalate). Mp 111-112° (oxalate).
N-Ph: [117593-45-0].
  $C_{11}H_{15}NS$ M 193.306
  $Bp_{0.1}$ 120-140°.

Brunet, E. *et al, Tetrahedron*, 1988, **44**, 1751.

4-Amino-2,2,6,6-tetramethylpiperidine   A-70201

Updated Entry replacing T-01554
2,2,6,6-Tetramethyl-4-piperidinamine, 9CI
[36768-62-4]

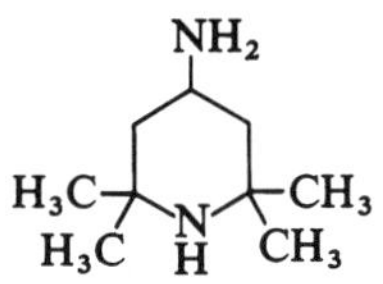

$C_9H_{20}N_2$ M 156.270

Intermed. in synth. of polymer stabilisers and pharmaceuticals. Mp 16-18°. Bp 188-189°.

N-Ac: [40908-37-0].
  $C_{11}H_{22}N_2O$ M 198.308
  Mp 205° (acetate salt).

Harries, C., *Justus Liebigs Ann. Chem.*, 1918, **417**, 107 (*synth*)
Ger. Pat., 2 412 750, (1975); *CA*, **84**, 43860g (*synth*)
Fioshin, M.Ya. *et al, Collect. Czech. Chem. Commun.*, 1987, **52**, 182 (*synth, bibl*)

5-Amino-1*H*-tetrazole   A-70202

1H-Tetrazol-5-amine, 9CI
[4418-61-5]

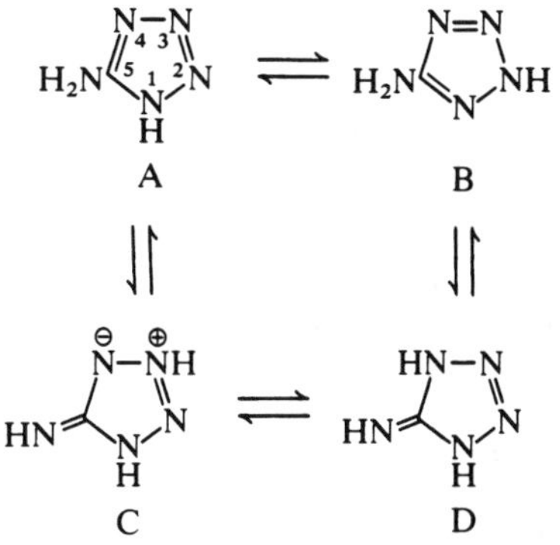

CH_3N_5 M 85.068
Cryst. Mp 206° dec. pK_a 1.675.
▷XF7465000.

Monohydrate: [15454-54-3]. Prisms (H_2O). Mp 200-204° (199°).
B,HNO_3: Cryst. (H_2O). Mp 178-179° (174°) dec.

1*H*-Amino-(A)-form

Major tautomeric form.

1-Me: [5422-44-6]. *5-Amino-1-methyl-1*H*-tetrazole.*
  $C_2H_5N_5$ M 99.095
  Needles (H_2O or $CHCl_3$). Mp 226° (220-223°).
1,5(N)-Di-Me: [17267-51-5]. *1-Methyl-5-methylamino-1*H*-tetrazole.*
  $C_3H_7N_5$ M 113.122
  Flat needles (EtOH). Mp 173-174°.
5-N-Me: [53010-03-0]. *5-Methylamino-1*H*-tetrazole.*
  $C_2H_5N_5$ M 99.095
  Cryst. (EtOH). Mp ca. 185°.
5-N-Et: *5-Ethylamino-1*H*-tetrazole.*
  $C_3H_7N_5$ M 113.122
  Cryst. (MeCN). Mp 174-175°.
1-Benzyl: [31694-90-3]. *1-(Phenylmethyl)-1*H*-tetrazol-5-amine, 9CI. 5-Amino-1-benzyl-1*H*-tetrazole.*
  $C_8H_9N_5$ M 175.193
  Cryst. (2-propanol). Mp 187-189°.
1-Benzyl, 5-N-Me: [53010-00-7]. *1-Benzyl-5-methylamino-1*H*-tetrazole.*
  $C_9H_{11}N_5$ M 189.219
  Needles (EtOH). Mp 117-118°.
5-N,N-Di-Me: [5422-45-7]. *5-Dimethylamino-1*H*-tetrazole.*
  $C_3H_7N_5$ M 113.122
  Cryst. (H_2O). Mp 235-236°.
5-N-Benzyl: [14832-58-7]. *N-(Phenylmethyl)-1*H*-tetrazol-5-amine, 9CI. 5-Benzylamino-1*H*-tetrazole.*
  $C_8H_9N_5$ M 175.193
  Cryst. (H_2O). Mp 183°.
1-Et: [65258-53-9]. *5-Amino-1-ethyl-1*H*-tetrazole.*
  $C_3H_7N_5$ M 113.122
  Cryst. (H_2O). Mp 148°.
1,5(N)-Dibenzyl: [66907-78-6].
  $C_{15}H_{15}N_5$ M 265.317
  Cryst. (EtOH). Mp 168-171°.

2*H*-Amino-(B)-form

2-Me: [6154-04-7]. *5-Amino-2-methyl-2*H*-tetrazole.*
  $C_2H_5N_5$ M 99.095

Needles (C_6H_6/pet. ether). Mp 105.5-106.5°.
2,5(N)-Di-Me: 2-Methyl-5-methylamino-2H-tetrazole.
$C_3H_7N_5$ M 113.122
Cryst. (Et_2O/pet. ether). Mp 48-49°.
2-Me, 5(N)-Ac: 5-Acetamido-2-methyl-2H-tetrazole.
$C_4H_7N_5O$ M 141.132
Prisms (MeCN). Mp 153-154°.
2-Benzyl: [31694-91-4]. *5-Amino-2-benzyl-2H-tetrazole.*
$C_8H_9N_5$ M 175.193
Needles (2-propanol). Mp 84.5-85°.
2-Me, 5-N-Benzyl:
$C_9H_{11}N_5$ M 189.219
Cryst. (pet. ether). Mp 86-87°.
2,5(N)-Dibenzyl:
$C_{15}H_{15}N_5$ M 265.317
Prisms (toluene/pet. ether). Mp 64-65°.
2-Et: 5-Amino-2-ethyl-2H-tetrazole.
$C_3H_7N_5$ M 113.122
Mp ca. 20°. Bp_1 94°.

1H,3H-5-Imino-(C)-form

1,3-Di-Me: 5-Imino-1,3-dimethyl-1H,3H-tetrazole.
$C_3H_7N_5$ M 113.122
Mp 41-43°. Bp_1 88°. pK_a ca. 11.5. Hygroscopic.
1,3-Di-Me; B,HCl: Needles or prisms (2-propanol). Mp 208-209°.
1,3-Di-Me, picrate salt: Yellow cryst. (EtOH). Mp 186-187°.
1-Me, 3-Benzyl: 3-Benzyl-5-imino-1-methyl-1H,3H-tetrazole.
$C_9H_{11}N_5$ M 189.219
Oil.
1-Me, 3-benzyl, picrate salt: Yellow needles (EtOH). Mp 152-153°. Solidifies and remelts at 163-164°.
1-Benzyl, 3-Me: 1-Benzyl-5-imino-3-methyl-1H,3H-tetrazole.
$C_9H_{11}N_5$ M 189.219
Oil.
1-Benzyl, 3-Me; B,HCl: Cryst. (2-propanol/Et_2O). Mp 180-181° dec.

1,4-Dihydro-5-imino-(D)-form

1,4-Di-Me: 1,4-Dihydro-5-imino-3,4-dimethyltetrazole.
$C_3H_7N_5$ M 113.122
Cryst. (C_6H_6). Mp 108.5-109.5°. pK_a 8.68.
1,4-Di-Me; B,HCl: Needles (2-propanol). Mp 242-244° dec.
1,4-Di-Me, picrate salt: Yellow needles (EtOH). Mp 211.5-212.5° dec.
1-Me, 4-Benzyl:
$C_9H_{11}N_5$ M 189.219
Cryst. (EtOAc/pet. ether). Mp 38-40°. Hygroscopic.
1-Me, 4-Benzyl; B,HCl: Cryst. (2-propanol aq.). Mp 216-217°.
1-Me, 4-benzyl, picrate salt: Needles (EtOH). Mp 107-108°.
1,4,5(N)-tri-Me: [35151-69-0]. *N-(1,4-Dihydro-1,4-dimethyl-5H-tetrazol-5-ylidene)methanamine,* 9CI.
$C_4H_9N_5$ M 127.149
Oil or deliquescent cryst. pK_a 9.57.
1,4,5(N)-tri-Me; B,HCl: Cryst. (2-propanol). Mp 202-203°.

Stollé, R., *Ber.*, 1929, **62**, 1118 (*synth*)
Bryden, J.H., *Acta Crystallogr.*, 1953, **6**, 669 (*cryst struct*)

Finnegan, W.G. *et al*, *J. Org. Chem.*, 1953, **18**, 779 (*derivs*)
Herbst, R.M. *et al*, *J. Org. Chem.*, 1953, **18**, 941 (*synth*)
Garbrecht, W.L. *et al*, *J. Org. Chem.*, 1953, **18**, 1003, 1014 (*derivs*)
Henry, R.A. *et al*, *J. Am. Chem. Soc.*, 1954, **76**, 923, 2894 (*1,3- and 1,4-disubst derivs*)
Murphy, D.B. *et al*, *J. Org. Chem.*, 1954, **19**, 1807 (*1,4-disubst derivs*)
Bianchi, G. *et al*, *J. Chem. Soc.* (*B*), 1971, 2355 (*tautom*)

3-Amino-1,2,4-triazine   A-70203

1,2,4-Triazin-3-amine, 9CI
[1120-99-6]

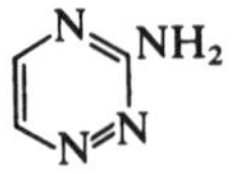

$C_3H_4N_4$ M 96.091
Mp 177-179°.
▷XY2969000.
N-*Me:* [65915-07-3].
$C_4H_6N_4$ M 110.118
Mp 90°.
N,N-*Di-Me:* [53300-17-7].
$C_5H_8N_4$ M 124.145
Yellow cryst. (MeOH). Mp 27-29°.

Daunis, J. *et al*, *Bull. Soc. Chim. Fr.*, 1969, 3675 (*synth, uv, pmr*)
Neunhoeffer, H. *et al*, *Chem. Ber.*, 1976, **109**, 1113.
Jacobsen, N.W. *et al*, *Aust. J. Chem.*, 1987, **40**, 1979 (*synth, pmr, cmr, derivs*)

Ammothamnidin   A-70204

Updated Entry replacing A-40116
[88640-94-2]

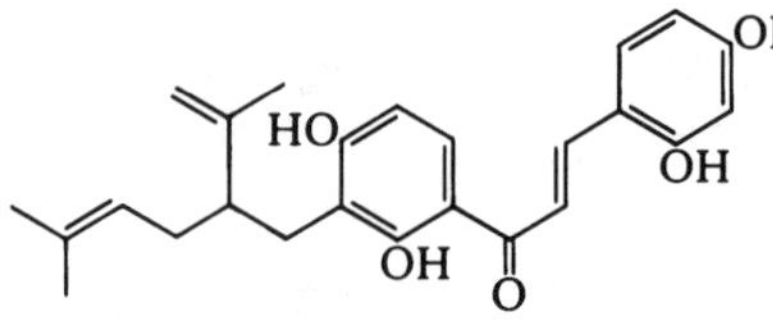

$C_{25}H_{28}O_5$ M 408.493
Constit. of *Ammothamnus lehmanni*. Cryst. (Me_2CO-/hexane). Mp 112-114°. $[\alpha]_D^{20}$ +4.5° (c, 0.22 in MeOH).

Sattikulov, A. *et al*, *Khim. Prir. Soedin.*, 1983, **19**, 445 (*isol*)
Bakirov, E.Kh. *et al*, *Khim. Prir. Soedin.*, 1987, **23**, 429 (*struct, synth*)

Amphidinolide *B*   A-70205

[110786-78-2]

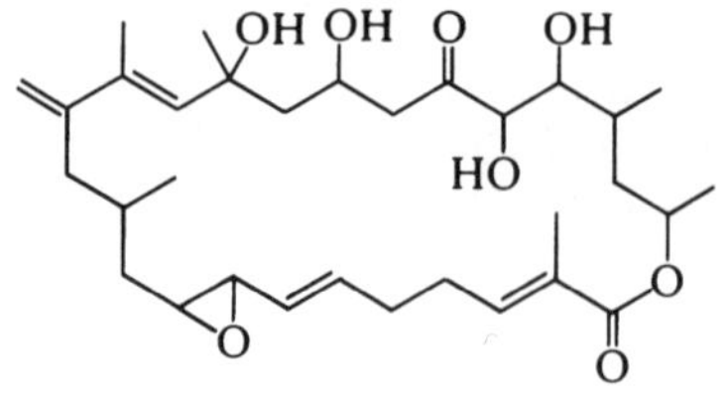

$C_{32}H_{50}O_8$ M 562.742
Macrolide antibiotic. Isol. from *Amphidinium* sp. Shows antitumour props. Amorph. $[\alpha]_D^{25}$ −45° (c, 1 in $CHCl_3$).

Ishibashi, M. *et al*, *J. Chem. Soc., Chem. Commun.*, 1987, 1127 (*isol, struct*)

Anastrephin   A-70206

Updated Entry replacing A-30163
*4-Ethenylhexahydro-4,7a-dimethyl-2(3H)-benzofuran-
one, 9CI*
[77670-94-1]

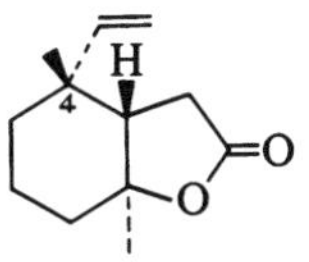

Relative
configuration

$C_{12}H_{18}O_2$ M 194.273
Abs. config. of the (−)-form now known to be as illus.

(−)-form

Component of the male pheromone of the Caribbean fruit
fly *Anastrepha suspensa* and the Mexican fruit fly *A.
ludens*. Cryst. Mp 94-95° (91-91.5°). $[\alpha]_D^{23.5}$ −50.4°
(c, 0.25 in hexane). (+)-form also known synthetically.
Props. refer to synthetic material of 100% opt. purity.
The nat. prod. is a partial racemate.

4-Epimer: [77670-93-0]. *Epianastrephin*.
$C_{12}H_{18}O_2$ M 194.273
Component of male pheromone of *A. anastrepha* and
A. ludens. Mp 49-50°. $[\alpha]_D^{23}$ −83.9° (c, 0.27 in
hexane).

(±)-form

Mp 81.5-83.5°.

4-Epimer: Mp 28-29°.

Battiste, M.A. *et al, Tetrahedron Lett.*, 1983, **24**, 2611 (*isol,
struct, synth*)
Saito, A. *et al, Chem. Lett.*, 1984, 729 (*synth*)
Mori, K. *et al, Justus Liebigs Ann. Chem.*, 1988, 167 (*synth, abs
config, ir, pmr, cmr*)

Andalucin   A-70207

[117255-08-0]

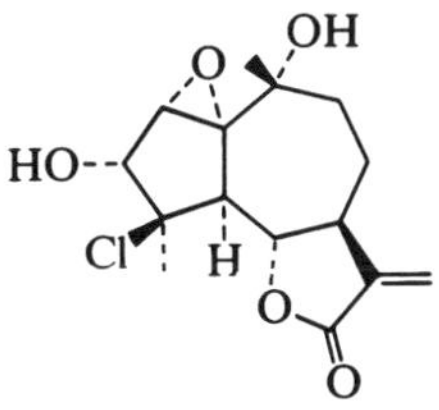

$C_{15}H_{19}ClO_5$ M 314.765
Constit. of *Artemisia lanata*. Gum.

Aguilar, J.M. *et al, Phytochemistry*, 1988, **27**, 2229.

Angoluvarin   A-70208

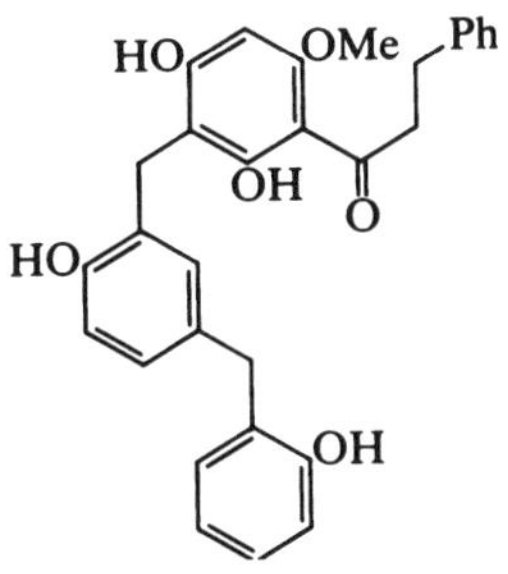

$C_{30}H_{28}O_6$ M 484.548
Constit. of *Uvaria angolensis*. Cryst. Mp 154-156°.

Hufford, C.D. *et al, J. Org. Chem.*, 1987, **52**, 5286.

Anguinomycin *A*   A-70209

[111278-01-4]

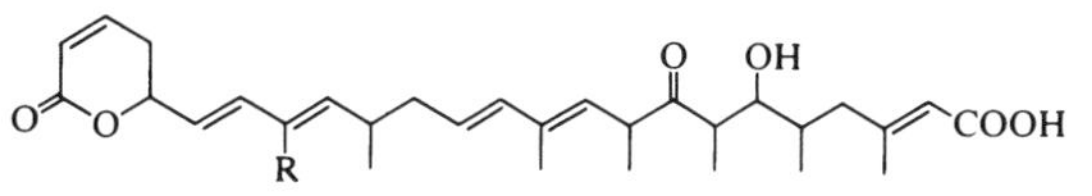

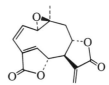

R = CH₃

$C_{31}H_{44}O_6$ M 512.685
Isol. from *Streptomyces* sp. R2827. Antitumour agent.
Visc. oil. $[\alpha]_D^{27}$ −139° (c, 0.1 in MeOH). Similar to
heptomycins.

Hayakawa, Y. *et al, J. Antibiot.*, 1987, **40**, 1349 (*isol, struct*)

Anguinomycin *B*   A-70210

[111278-00-3]

As Anguinomycin *A*, A 70209 with

$$R = -CH_2CH_3$$

$C_{32}H_{46}O_6$ M 526.712
From *Streptomyces* sp. R2827. Antitumour agent. Visc.
oil. $[\alpha]_D^{27}$ −130° (c, 0.1 in MeOH). Similar to heptomycins.

Hayakawa, Y. *et al, J. Antibiot.*, 1987, **40**, 1349 (*isol, struct*)

Anhydroscandenolide   A-70211

[114742-71-1]

$C_{15}H_{14}O_5$ M 274.273
Constit. of *Mikania urticifolia*. Cryst. Mp 73-75°.

Gutierrez, A.B. *et al Phytochemistry*, 1988, **27**, 938.

Annonacin   A-70212

[111035-65-5]

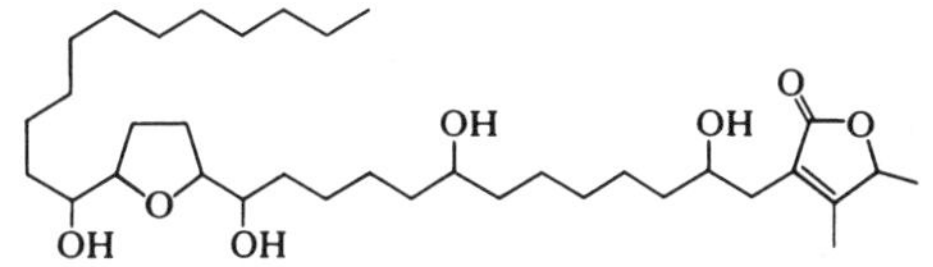

$C_{35}H_{64}O_7$ M 596.886
Closely related to Goniothalamicin, G-70027 . Constit. of
Goniothalamus giganteus and *Annona densicoma*.
Wax.

McCloud, T.G. *et al, Experientia*, 1987, **43**, 947 (*isol, struct*)
Alkofahi, A. *et al, Experientia*, 1988, **44**, 83 (*isol*)

Antheliolide *A*   A-70213
[114915-36-5]

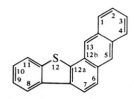

$C_{24}H_{32}O_3$ M 368.515
Novel acetoacetylated diterpenoid skeleton. Constit. of
   soft coral *Anthelia glauca*. Mp 180°. Exists as mixt. of
   two conformers.
8,23-Epoxide: [114933-19-6]. **Antheliolide B.**
   $C_{24}H_{32}O_4$ M 384.514
   Constit. of *A. glauca*. Oil.
Green, D. *et al*, *Tetrahedron Lett.*, 1988, **29**, 1605 (*struct*)

Anthra[1,2-*b*]benzo[*d*]thiophene   A-70214
[223-47-2]

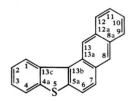

$C_{20}H_{12}S$ M 284.375
Pale-yellow leaves (C_6H_6). Mp 228°.
Tedjamulia, M.L. *et al*, *J. Heterocycl. Chem.*, 1983, **20**, 861
   (*synth, uv, pmr*)

Anthra[2,1-*b*]benzo[*d*]thiophene   A-70215
[28887-47-0]

$C_{20}H_{12}S$ M 284.375
Yellow leaves. Mp 168-169°.
Tedjamulia, M.L. *et al*, *J. Heterocycl. Chem.*, 1983, **20**, 861
   (*synth, uv, pmr*)

Anthra[2,3-*b*]benzo[*d*]thiophene   A-70216
[249-05-8]

$C_{20}H_{12}S$ M 284.375
Yellow fluorescent leaves. Mp 280° (250°).
Tedjamulia, M.L. *et al*, *J. Heterocycl. Chem.*, 1983, **20**, 861
   (*synth, uv, pmr, bibl*)

Anthra[1,2-*c*]furan   A-70217
[227-53-2]

$C_{16}H_{10}O$ M 218.254
Light-yellow prisms (CH_2Cl_2/pet. ether). Mp 169-175°
dec.
Moursounidis, J. *et al*, *Aust. J. Chem.*, 1988, **41**, 235 (*synth,
   pmr, uv, ms*)

Antibiotic 273a₂   A-70218

Antibiotic 273a₂ₐ R = H
273a₂β R = H

Glycoside-type antibiotic complex. Prod. by *Streptomyces paulus*. Active against gram-positive bacteria. Isol.
   with Paldimycin, P-70002 Related also to Paulomycin.
Antibiotic 273a₂ₐ [102326-69-2]
   $C_{39}H_{55}N_3O_{20}S_2$ M 949.992

Mp 119-150° dec. $[\alpha]_D^{25}$ −33° (c, 0.89 in MeOH).
Antibiotic 273a₂β [102301-13-3]
   $C_{38}H_{53}N_3O_{20}S_2$ M 935.965

$[\alpha]_D^{25}$ −34° (c, 0.4 in MeOH). Dec. begins at 122°.
Argoudelis, A.D. *et al*, *J. Antibiot.*, 1987, **40**, 408, 419 (*isol,
   struct, synth*)

Antibiotic A 40926   A-70219
A 40926
[102961-72-8]

A 40926A R = H
B 40926B R = CH₃

Glycopeptide antibiotic complex. Prod. by *Actinomedura*
   sp. Active against gram-positive bacteria and *Neisseria
   gonorrhoeae*. Similar to Aridicins. Other components
   isol. but not elucidated as yet.

Antibiotic A 40926A [102961-73-9]
$C_{82}H_{86}Cl_2N_8O_{29}$ M 1718.524

Same gross struct. as Parvodicin B_2.
Antibiotic A 40926B [102961-74-0]
$C_{83}H_{88}Cl_2N_8O_{29}$ M 1732.550

Same gross struct. as Parvodicin C_1.

Waltho, J.P. *et al*, *J. Chem. Soc., Perkin Trans. 1*, 1987, 2103 (*isol, struct*)

Antibiotic B 1625*FA*$_{2\beta\text{-}1}$ A-70220

*B 1625*FA$_{2\beta\text{-}1}$
[111420-30-5]

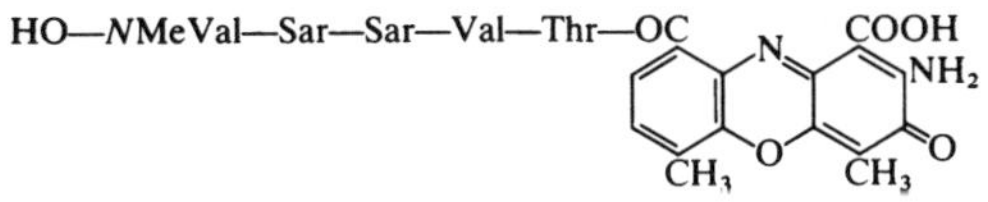

HO—*N*MeVal—Sar—Sar—Val—Thr—OC

$C_{37}H_{49}N_7O_{12}$ M 783.834
Actinomycin-related antibiotic. Isol. from *Streptomyces antibioticus*. Orange-red powder. Mp 185°. $[\alpha]_D$ +3° (CHCl$_3$).

Fujimoto, Y. *et al*, *Agric. Biol. Chem.*, 1987, **51**, 803 (*isol, struct, props*)

Antibiotic BMY 28100   A-70221

BMY 28100
[92665-29-7]

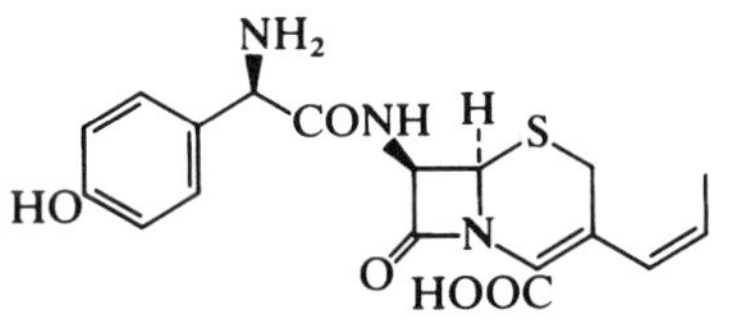

$C_{18}H_{19}N_3O_5S$ M 389.425
Cephalosporin antibiotic. Semisynthetic. Against gram-positive and -negative bacteria. Prisms. Mp 218-220° dec.

Kawaguchi, H. *et al*, *J. Antibiot.*, 1987, **40**, 991, 1175 (*synth, props*)

Antibiotic CP 54883   A-70222

CP 54883
[112396-63-1]

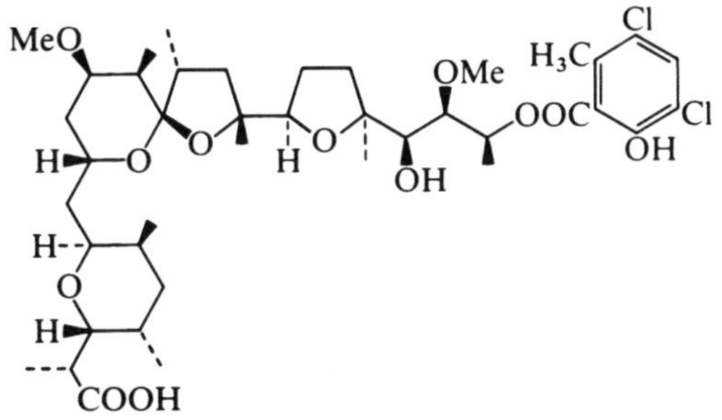

$C_{41}H_{62}Cl_2O_{12}$ M 817.840
Polyether antibiotic. Prod. by *Actinomadura routienii*. Ionophore. Active against gram-positive bacteria and *Eimeria* coccidia.

Na salt: Cryst. (Et$_2$O). Mp 330-340°. $[\alpha]_D^{25}$ +11.9° (c, 0.2 in CHCl$_3$).

Watts, P.C. *et al*, *J. Antibiot.*, 1987, **40**, 1490, 1496 (*isol, struct, props*)

Antibiotic FR 900359   A-70223

FR 900359
[107530-18-7]

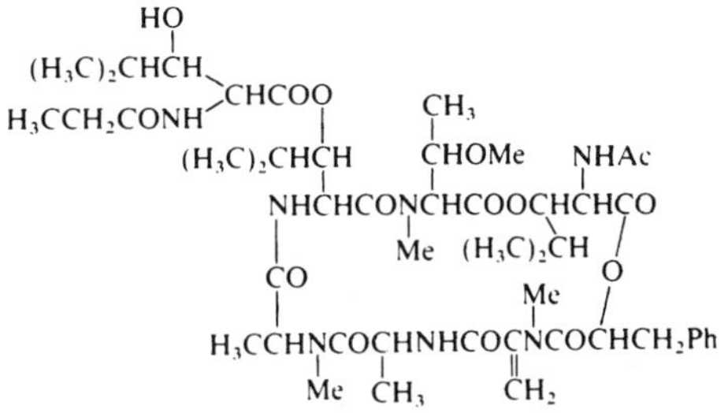

$C_{49}H_{75}N_7O_{15}$ M 1002.169
Cyclic depsipeptide antibiotic. Prod. by *Ardisia crenata*. Platelet aggregation inhibitor and hypotensive agent.

Fujioka, M. *et al*, *CA*, 1987, **106**, 133972 (*isol, struct*)

Antibiotic L 660631   A-70224

5-(1-Hydroxy-2,4,6-heptatriynyl)-2-oxo-1,3-dioxolane-4-heptanoic acid, 9CI. L 660631
[115216-83-6]

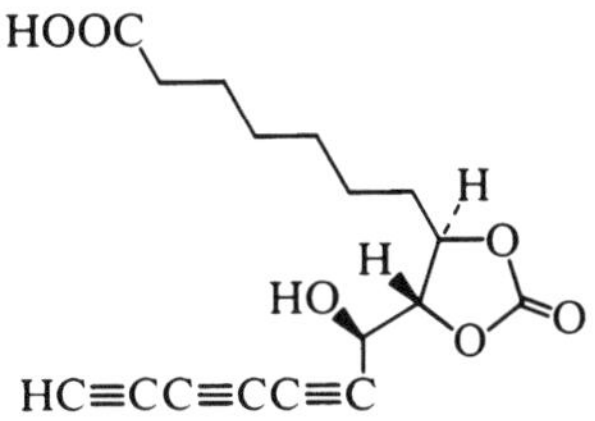

$C_{17}H_{18}O_6$ M 318.326
Isol. from *Actinomyces* sp. Active against pathogenic yeasts and filamentous fungi. No phys. props. reported.

Lewis, M.D. *et al*, *Tetrahedron Lett.*, 1987, **28**, 5129 (*abs config*)

Antibiotic PB 5266*A*   A-70225

PB 5266A

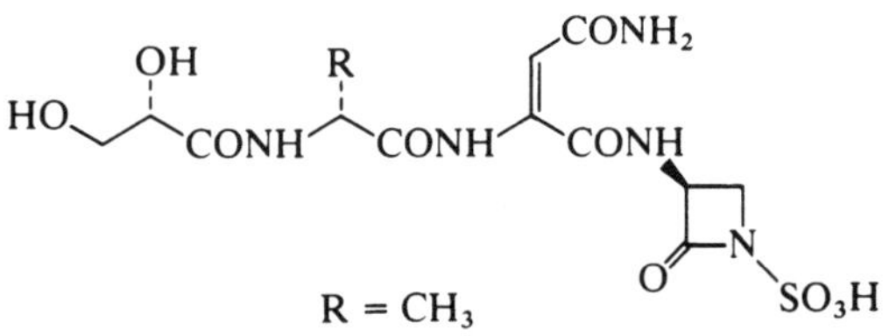

R = CH$_3$

$C_{13}H_{19}N_5O_{10}S$ M 437.381
Monobactam antibiotic. Prod. by *Cytophaga johnsonae*. Exhibits weak antibacterial activity.

K salt: Acidic powder + 2H$_2$O.

Kato, T. *et al*, *J. Antibiot.*, 1987, **40**, 135, 139 (*isol, struct, props*)

Antibiotic PB 5266*B*   A-70226

PB 5266B
[108065-95-8]
As Antibiotic PB 5266*A*, A-70225 with

$$R = CH_2OH$$

$C_{13}H_{19}N_5O_{11}S$ M 453.380
Monobactam antibiotic. Prod. by *Cytophaga johnsonae*. Exhibits weak antibacterial activity.

K salt: Acidic powder + 2H₂O.

Kato, T. *et al, J. Antibiot.*, 1987, **40**, 135, 139 (*isol, struct, props*)

Antibiotic PB 5266*C* — A-70227

PB 5266C

[108065-96-9]

As Antibiotic PB 5266*A*, A-70225 with

R = H

$C_{12}H_{17}N_5O_{10}S$ M 423.354

Monobactam antibiotic. Prod. by *Cytophaga johnsonae*. Exhibits weak antibacterial props.

K salt: Acidic powder + 2H₂O.

Kato, T. *et al, J. Antibiot.*, 1987, **40**, 135, 139 (*isol, struct, props*)

Antibiotic PC 766*B* — A-70228

PC 766B

[108375-77-5]

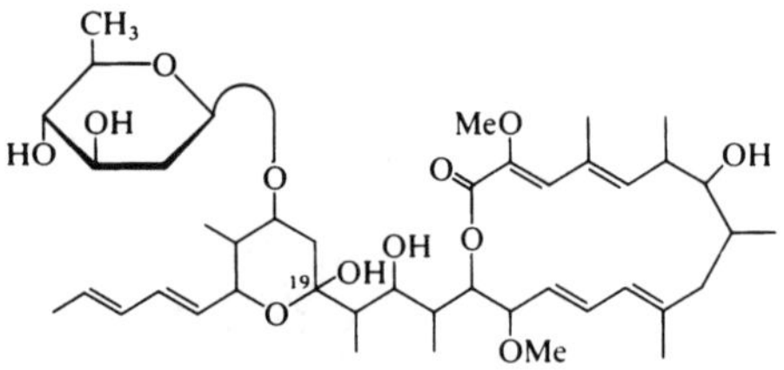

$C_{43}H_{68}O_{12}$ M 777.003

Macrolide antibiotic. Isol. from *Nocardia* sp. Related to Bafilomycins.

19-Me ether: [108351-33-3]. **Antibiotic PC 766B'**. *PC 766B'*.
$C_{44}H_{70}O_{12}$ M 791.030
From *N.* sp.

Japan. Pat., 87 05 990, (*1987*); *CA*, **106**, 212560 (*isol*)

Antibiotic PDE II — A-70229

PDE II

[62874-94-6]

As Antibiotic PDE I, A-60286 with

R = Ac

$C_{14}H_{14}N_2O_5$ M 290.275

Isol. from culture closely related to *Streptomyces griseoflavus*. Shows antitumour props. Inhibitor of cyclic adenosine-3',5'-monophosphate phosphodiesterase. Mp 253° dec.

Enomoto, Y. *et al, Agric. Biol. Chem.*, 1978, **42**, 1331, 1337; 1979, **43**, 559 (*isol, struct, synth*)
Bolton, R.E. *et al, J. Chem. Soc., Perkin Trans. 1*, 1987, 931 (*synth*)
Carter, P. *et al, J. Am. Chem. Soc.*, 1987, **109**, 2711 (*synth, bibl*)
Boger, D.L. *et al, J. Am. Chem. Soc.*, 1987, **109**, 2717 (*synth*)

Antibiotic R 20Y7 — A-70230

R 20Y7

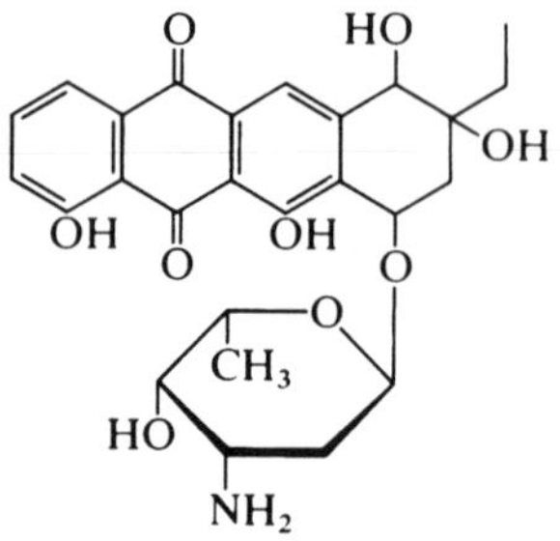

$C_{26}H_{29}NO_9$ M 499.516

Anthracycline antibiotic. Prod. by *Actinomadura roseoviolacea*. Antitumour agent. Similar to Feudomycin *A*, F-20003 .

Japan. Pat., 87 81 398, (*1987*); *CA*, **107**, 174374 (*isol, props*)

Antibiotic Ro 09-0198 — A-70231

Ro 09-0198

Large peptide antibiotic. Prod. by *Streptoverticillium griseoverticillatum*.

B,HCl: Needles (MeCN aq.).

Kessler, H. *et al, Helv. Chim. Acta*, 1987, **70**, 726 (*isol, struct*)

Antibiotic SF 2415*A*₁ — A-70232

SF 2415A₁

[110200-32-3]

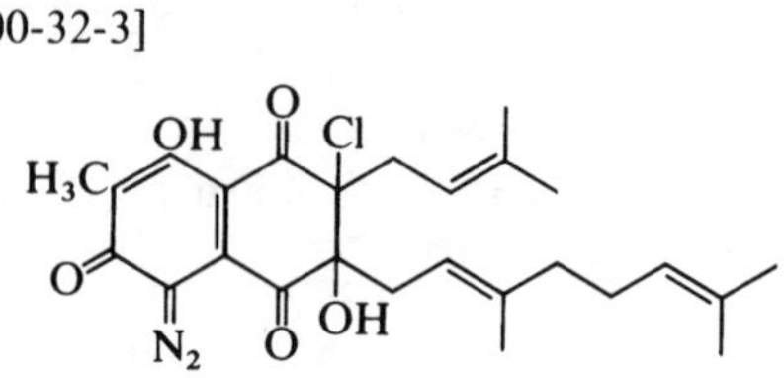

$C_{26}H_{31}ClN_2O_5$ M 486.994

Naphthoquinone-related antibiotic. Isol. from *Streptomyces aculeolatus*. Active against gram-positive bacteria. Yellow needles (hexane). Mp 89-90°. $[\alpha]_D^{22}$ +133° (c, 0.5 in MeOH).

Gomi, S. *et al, J. Antibiot.*, 1987, **40**, 732, 740 (*isol, struct, props*)

Antibiotic SF 2415*A*₂ — A-70233

SF 2415A₂

[110200-31-2]

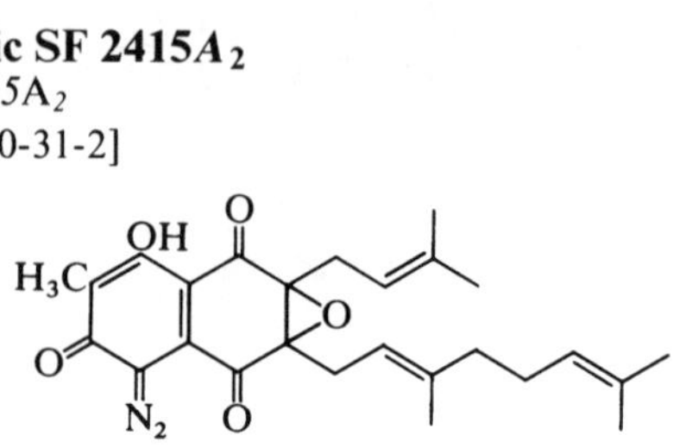

$C_{26}H_{30}N_2O_5$ M 450.533

Naphthoquinone-related antibiotic. Isol. from *Streptomyces aculeolatus*. Active against gram-positive bacteria. Red oil. $[\alpha]_D^{22}$ +49° (c, 0.5 in MeOH).

Gomi, S. *et al, J. Antibiot.*, 1987, **40**, 732, 740 (*isol, struct, props*)

Antibiotic SF 2415A_3 A-70234
SF 2415A$_3$
[110200-33-4]

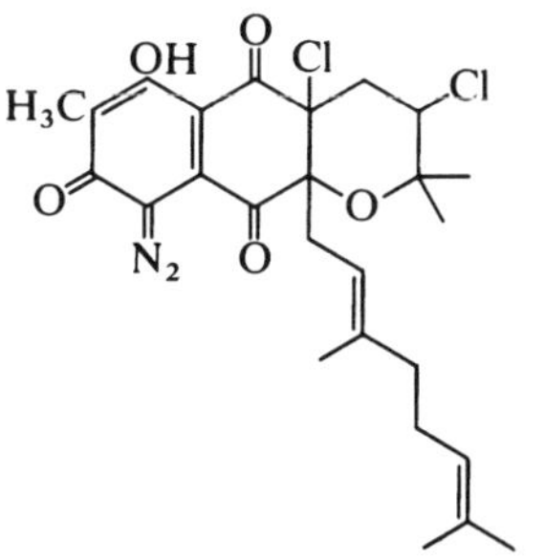

$C_{26}H_{30}Cl_2N_2O_5$ M 521.439
Naphthoquinone-related antibiotic. Isol. from
Streptomyces aculeolatus. Active against gram-
positive bacteria. Red powder. $[\alpha]_D^{22}$ +195° (c, 0.5 in
MeOH).

Gomi, S. *et al, J. Antibiot.*, 1987, **40**, 732, 740 (*isol, struct, props*)

Antibiotic SF 2415B_1 A-70235
SF 2415B$_1$
[110200-34-5]

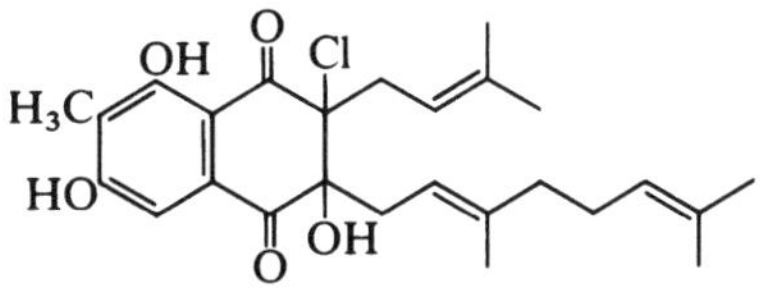

$C_{26}H_{33}ClO_5$ M 460.997
Naphthoquinone-related antibiotic. Isol. from
Streptomyces aculeolatus. Active against gram-
positive bacteria. Pale-yellow oil. $[\alpha]_D^{22}$ −82° (c, 0.5 in
MeOH).

Gomi, S. *et al, J. Antibiot.*, 1987, **40**, 732, 740 (*isol, struct, props*)

Antibiotic SF 2415B_2 A-70236
SF 2415B$_2$
[110200-30-1]

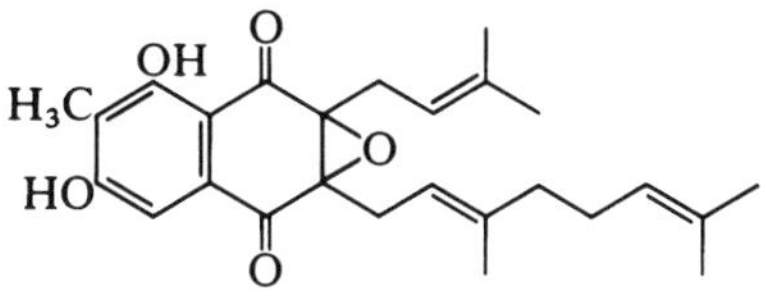

$C_{26}H_{32}O_5$ M 424.536
Naphthoquinone-related antibiotic. Isol. from
Streptomyces aculeolatus. Active against gram-
positive bacteria. Yellow oil. $[\alpha]_D^{22}$ +122° (c, 0.5 in
MeOH).

Gomi, S. *et al, J. Antibiot.*, 1987, **40**, 732, 740 (*isol, struct, props*)

Antibiotic SF 2415B_3 A-70237
7-Methylnapyradiomycin A. SF 2415B$_3$
[110200-29-8]

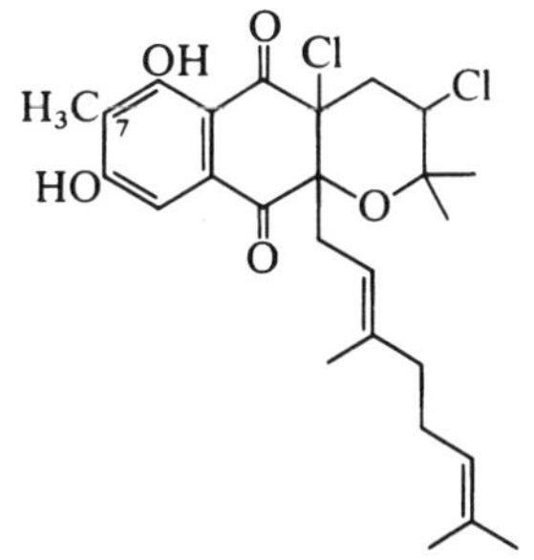

$C_{26}H_{32}Cl_2O_5$ M 495.442
Naphthoquinone-related antibiotic. Isol. from
Streptomyces aculeolatus. Active against gram-
positive bacteria. Yellow needles. $[\alpha]_D^{22}$ +33° (c, 0.5 in
MeOH). Similar to Napyradiomycin *A*, N-50076 .

Gomi, S. *et al, J. Antibiot.*, 1987, **40**, 732, 740 (*isol, struct, props*)

Antibiotic SS 21020C A-70238
*SS 21020*C
[108112-82-9]

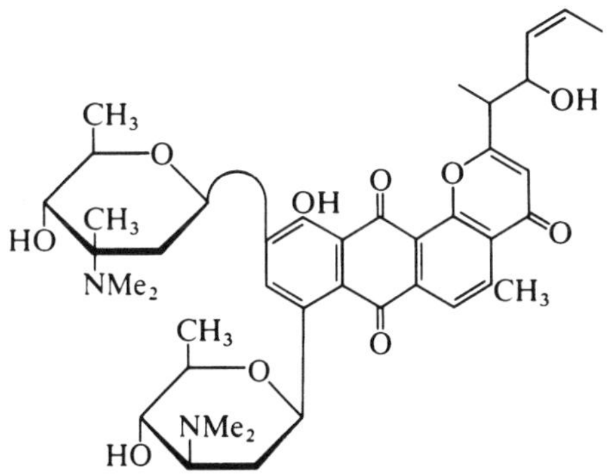

$C_{41}H_{52}N_2O_{10}$ M 732.869
Quinone antibiotic. Prod. by *Streptomyces* sp. S 21020.
Shows antitumour and antibacterial props.

Japan Pat., 86 274 693, (*1986*); *CA*, **107**, 21949

Antibiotic SS 43405D A-70239
*7,12-Dihydro-11-hydroxy-2-(1-methylpropyl)-4,7,12-
trioxo-4H-anthra[1,2-b]pyran-5-acetic acid, 9Cl. SS
43405*D
[105514-98-5]

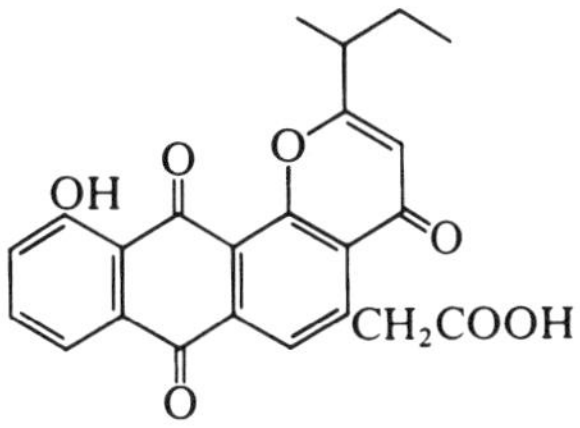

$C_{23}H_{18}O_7$ M 406.391
Quinone antibiotic. Prod. by *Streptomyces* sp. S 43405.
Yellow cryst.

Japan. Pat., 86 139 394, (*1986*); *CA*, **105**, 224599 (*isol*)

Antibiotic TMF 518*D*   A-70240
*TMF 518*D
[103528-07-0]

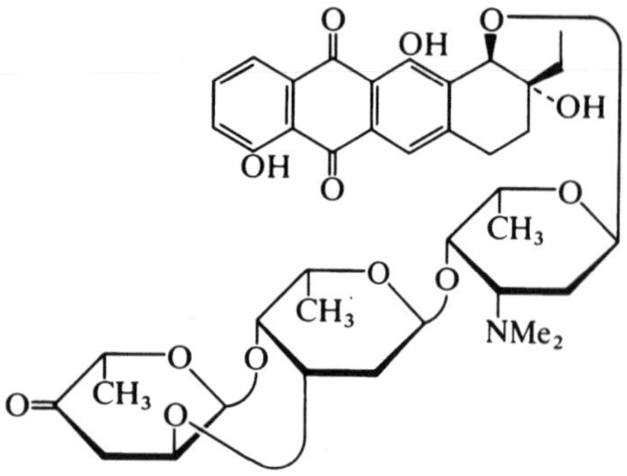

C$_{40}$H$_{49}$NO$_{13}$ M 751.826
Anthracycline antibiotic. Isol. from *Streptomyces cosmosus*. Related to Serirubicin, S-50059 .

Japan Pat., 86 05 090, (*1986*); *CA*, **106**, 65840 (*isol*)

Antibiotic W 341*C*   A-70241
*W 341*C
[110368-36-0]

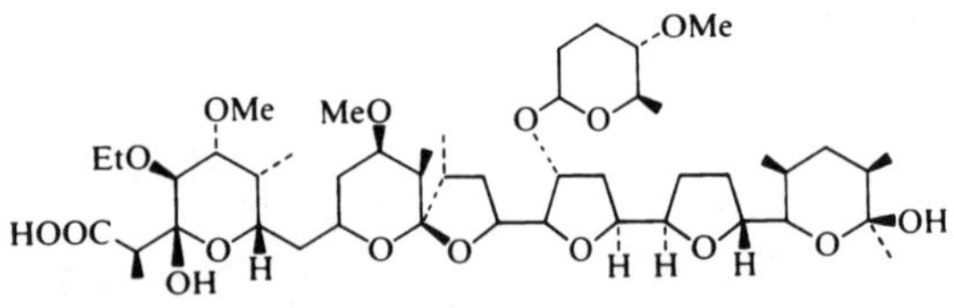

C$_{47}$H$_{80}$O$_{16}$ M 901.139
Polyether antibiotic. Prod. by *Streptomyces* W341. Anticoccidial agent. Ionophore.

Wehbie, R.S. *et al*, *J. Antibiot.*, 1987, **40**, 887 (*props, biol*)

Aphyllocladone   A-70242

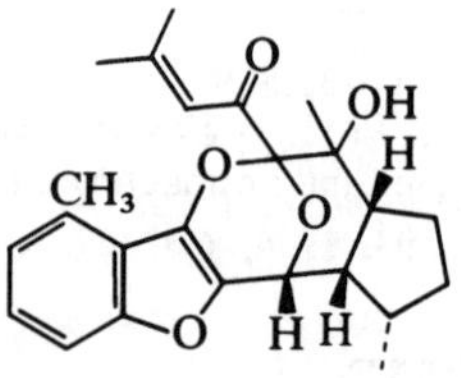

C$_{24}$H$_{28}$O$_5$ M 396.482
Constit. of *Aphyllocladus denticulatus*. Gum.

Zdero, C. *et al*, *Phytochemistry*, 1988, **27**, 1821.

Aphyllodenticulide   A-70243

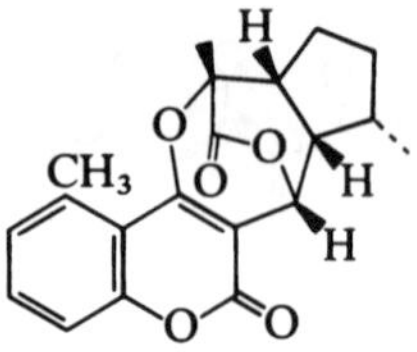

C$_{20}$H$_{20}$O$_5$ M 340.375
Constit. of *Aphyllocarpus denticulatus*. Cryst. (Et$_2$O). Mp 221-223°.

Zdero, C. *et al*, *Phytochemistry*, 1988, **27**, 1821.

Aplysiapyranoid *A*   A-70244
[109927-30-2]

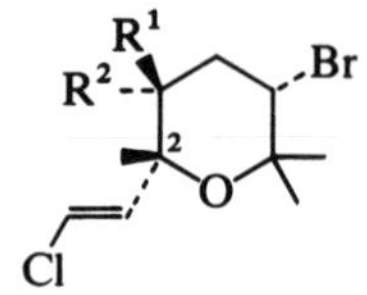

$$R^1 = Br, R^2 = H$$

C$_{10}$H$_{15}$Br$_2$ClO   M 346.489
Constit. of *Aplysia kurodai*. Oil. [α]$_D$ +4.4° (c, 1 in CHCl$_3$).
2-Epimer: [108212-12-0]. **Aplysiapyranoid B**.
  C$_{10}$H$_{15}$Br$_2$ClO   M 346.489
  Constit. of *A. kurodai*. Cryst. (EtOH). Mp 46-49°. [α]$_D$ −27° (c, 0.91 in CHCl$_3$).
Kusumi, T. *et al*, *J. Org. Chem.*, 1987, **52**, 4597.

Aplysiapyranoid *C*   A-70245
[109927-31-3]
As Aplysiapyranoid *A*, A-70244 with

$$R^1 = H, R^2 = Cl$$

C$_{10}$H$_{15}$BrCl$_2$O   M 302.038
Constit. of *Aplysia kurodai*. Oil. [α]$_D$ +52° (c, 1 in CHCl$_3$).
2-Epimer: [109927-32-4]. **Aplysiapyranoid D**.
  C$_{10}$H$_{15}$BrCl$_2$O   M 302.038
  From *A. kurodai*. Oil. [α]$_D$ +3.4° (c, 1.1 in CHCl$_3$).
Kusumi, T. *et al*, *J. Org. Chem.*, 1987, **52**, 4597.

10′-Apo-β-caroten-10′-ol   A-70246

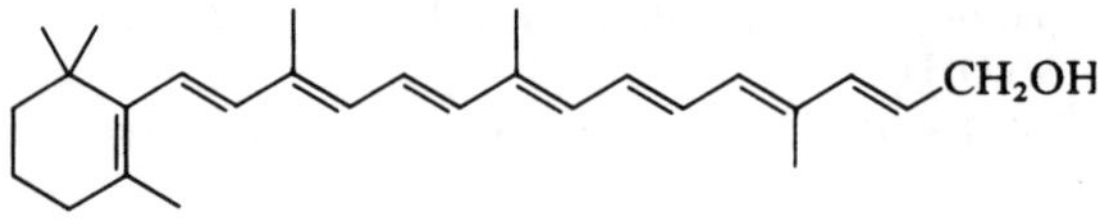

C$_{27}$H$_{38}$O   M 378.597
Constit. of rose flowers (*Rosa foetida*).

Märki-Fischer, E. *et al*, *Helv. Chim. Acta*, 1987, **70**, 1988.

Apotrisporin *E*   A-70247
Apotrisporin
[29538-78-1]

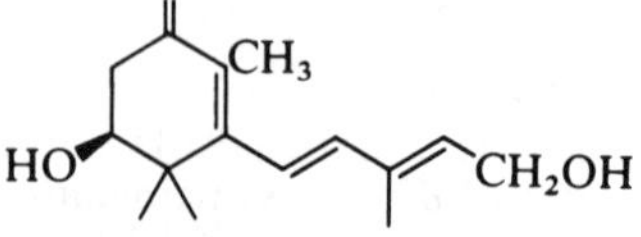

C$_{15}$H$_{22}$O$_3$ M 250.337
Isol. from *Phycomyces blakesleeanus* and *Blakeslea trispora*. λ$_{max}$ 235, 300 nm.
Deoxy: **Apotrisporin C**.
  C$_{15}$H$_{22}$O$_2$ M 234.338
  Isol. from *B. trispora*.
Sutter, R.P., *Exp. Mycol.*, 1986, **10**, 256 (*isol, struct, nmr*)

Arbutin, 9CI, 8CI   A-70248

Updated Entry replacing A-03400

4-Hydroxyphenyl β-D-glucopyranoside, 9CI, 8CI. Hydro-quinone-glucose. Arbutoside

[497-76-7]

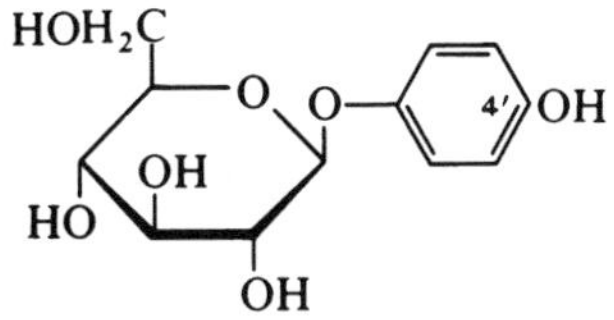

$C_{12}H_{16}O_7$ M 272.254

Glucoside in pear leaves (*Pyrus communis*). Mp 142-143°. $[\alpha]_D^{18}$ −60.34° (H_2O).

6-Ac: [10338-88-2]. **6-O-Acetylarbutin**. *Monoacetylarbutin*.
$C_{14}H_{18}O_8$ M 314.291
Constit. of immature pear (*P. communis*) and mountain cranberry (*Vaccinium vitis*). Mp 214-216°. $[\alpha]_D^{23}$ −58.8° (c, 2.0 in H_2O).

6-Benzoyl: [71202-99-8]. **Eximin**.
$C_{19}H_{20}O_8$ M 376.362
Isol. from leaves of *Protea eximia*. Mp 199-201°. $[\alpha]_D$ −48° (c, 0.94 in EtOH).

Penta-Ac: [14698-56-7]. Mp 146°. $[\alpha]_D^{20}$ −28.18° (Me_2CO).

Penta-Me ether: Mp 76-78°. $[\alpha]_D^{18}$ −48.2° (EtOH).

4,6-Bis(3,4,5-trimethoxybenzoyl): **4,6-Di-O-galloylarbutin**.
$C_{26}H_{24}O_{15}$ M 576.467
From *Bergenia purpurascens*. Amorph. powder + $1H_2O$. $[\alpha]_D$ −15.5° (c, 1 in Me_2CO).

6-(3,4,5-Trimethoxybenzoyl): **6-O-Galloylarbutin**.
$C_{19}H_{20}O_{11}$ M 424.360
From *B. purpurascens*. Needles + $1\frac{1}{2}H_2O$. Mp 250-251°. $[\alpha]_D$ −33.6° (c, 1.1 in Me_2CO aq.).

Bradfield, F.J. *et al, J. Chem. Soc.*, 1952, 4740.
Haslam, E., *J. Chem. Soc.*, 1964, 5649, 5653.
Durkee, A.B. *et al, J. Food Sci.*, 1968, **33**, 461 (*isol, deriv*)
Haslam, E. *et al, Tetrahedron*, 1968, **24**, 4015 (*synth*)
Perol, G.W. *et al, J. Chem. Soc., Perkin Trans. 1*, 1979, 239 (*synth, pmr*)
Xin-Min, C. *et al, Phytochemistry*, 1987, **26**, 515 (*derivs*)

Arctigenin   A-70249

Updated Entry replacing A-20198

4-[(3,4-Dimethoxyphenyl)methyl]dihydro-3-[(4-hydroxy-3-methoxyphenyl)methyl]-2(3H)-furanone, 9CI

[7770-78-7]

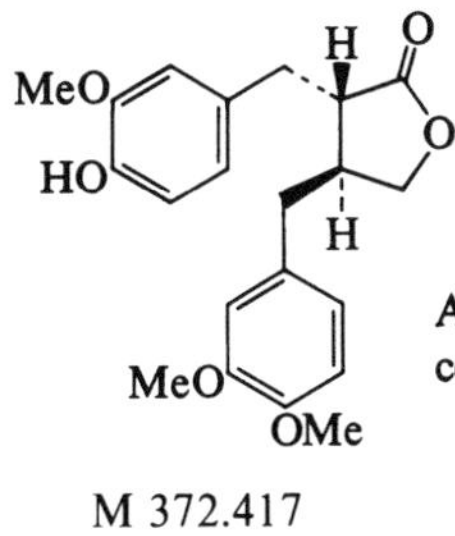

Absolute configuration

$C_{21}H_{24}O_6$ M 372.417

Constit. of *Cnicus benedictus, Forsythia viridissima, Wikstroemia viridiflora* and in the seeds of burdock (*Arctium lappa*). Plates (MeOH/Et_2O). Mp 102°. $[\alpha]_D^{23}$ +28.05° (c, 1.23 in EtOH).

Ac: Needles (EtOAc). Mp 52-60°.

Me ether: see Matairesinol, M-00185

4-β-D-Glucopyranoside: [20362-31-6]. **Arctiin**.
$C_{27}H_{34}O_{11}$ M 534.559

Constit. of the seeds of *A. lappa* and of the stems of *Trachelospermum asiaticum*, from *F.* spp. and *Centaurea imperialis*. Mp 110-112°. $[\alpha]_D^{27}$ −51.5° (c, 2 in EtOH).

4-β-Gentiobioside: [41682-24-0].
$C_{33}H_{44}O_{16}$ M 696.701
Mp 179-180°. $[\alpha]_D^{19}$ −52° (c, 0.4 in H_2O).

Ozawa, T., *J. Pharm. Soc. Jpn.*, 1952, **72**, 285, 551; *CA*, **47**, 2745, 3276 (*synth, config*)
Inagaki, I. *et al, Chem. Pharm. Bull.*, 1972, **20**, 2710 (*isol*)
Nishibe, S. *et al, Chem. Pharm. Bull.*, 1973, **21**, 2778 (*synth*)
Takaoka, D., *Bull. Chem. Soc. Jpn.*, 1977, **50**, 2821 (*isol*)
Omar, A.A., *Lloydia*, 1978, **41**, 638 (*Arctiin*)
Suzuki, H. *et al, Phytochemistry*, 1982, **21**, 1824 (*isol*)
Nishibe, S. *et al, Chem. Pharm. Bull.*, 1984, **32**, 4653 (*cmr*)

Arguticinin   A-70250

[116497-04-2]

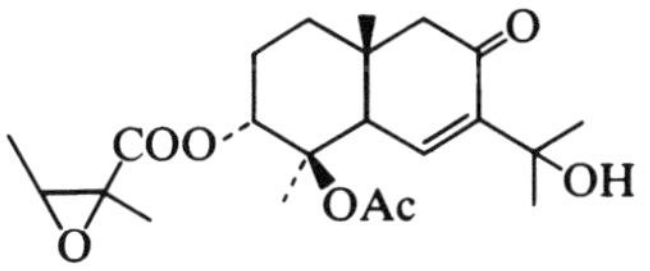

$C_{22}H_{32}O_7$ M 408.491

Constit. of *Pluchea arguta*. Gum. $[\alpha]_D^{20}$ +98° (c, 0.04 in MeOH).

4-Epimer: [68799-58-6].
$C_{22}H_{32}O_7$ M 408.491
Isol. from *P. foetida*. Oil.

Bohlmann, F. *et al, Phytochemistry*, 1978, **17**, 1189.
Ahmad, V.U. *et al, Phytochemistry*, 1988, **27**, 1861.

Aristochilone   A-70251

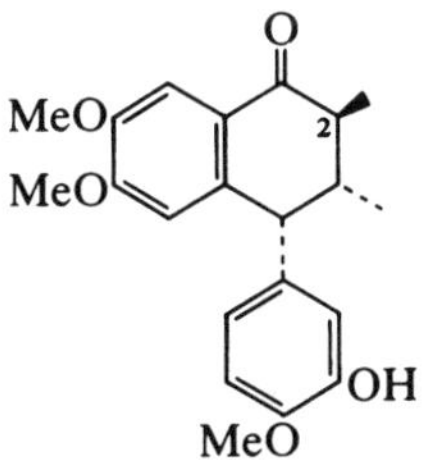

$C_{21}H_{24}O_5$ M 356.418

Constit. of *Aristolochia chilensis*. Amorph. $[\alpha]_D$ −134° (c, 1.62 in $CHCl_3$).

Me ether: **Aristoligone**.
$C_{22}H_{26}O_5$ M 370.444
Constit. of *A. chilensis*. Amorph. $[\alpha]_D$ −190° (c, 0.78 in $CHCl_3$).

Me ether, 2-epimer: **Aristosynone**.
$C_{22}H_{26}O_5$ M 370.444
Constit. of *A. chilensis*.

Urzúa, A. *et al, J. Nat. Prod.*, 1988, **51**, 117.

1(10)-Aristolen-9-one   A-70252

Gansongone

[114339-93-4]

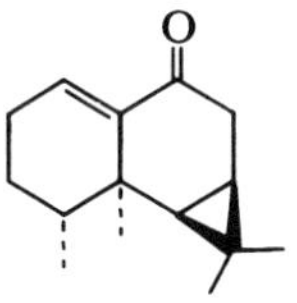

$C_{15}H_{22}O$ M 218.338

Constit. of *Nardostachys chinensis*. Oil. $[\alpha]_D^{20}$ +20° (c, 2.0 in $CHCl_3$).

Shide, L. *et al*, *Planta Med.*, 1987, **53**, 556.

Aristotetralol

A-70253

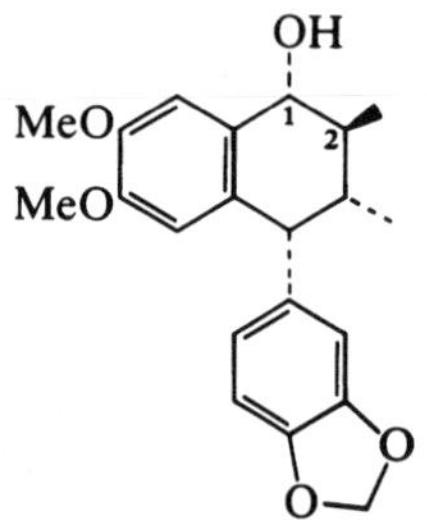

$C_{21}H_{24}O_5$ M 356.418

Constit. of *Aristolochia chilensis*. Amorph. $[\alpha]_D$ −149° (c, 1.49 in CHCl₃).

1-Ketone, 2α-hydroxy: **2-Hydroxyaristotetralone**.
$C_{21}H_{22}O_6$ M 370.401
Constit. of *A. chilensis*. Amorph. $[\alpha]_D$ −62° (c, 0.23 in CHCl₃).

1-Ketone, 2α-acetoxy: **2-Acetoxyaristotetralone**.
$C_{23}H_{24}O_7$ M 412.438
Constit. of *A. chilensis*. Amorph. $[\alpha]_D$ −62° (c, 0.58 in CHCl₃).

Urzúa, A. *et al*, *J. Nat. Prod.*, 1988, **51**, 117.

Arizonin C_1

A-70254

[108890-86-4]

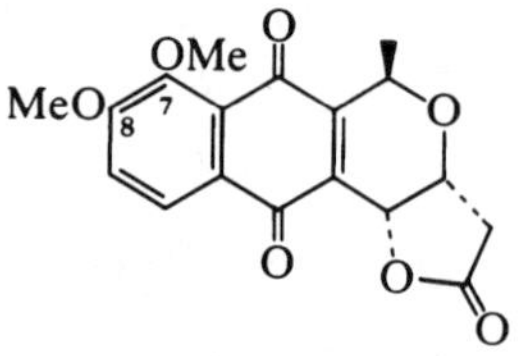

$C_{18}H_{16}O_7$ M 344.320

Naphthoquinone-type antibiotic. Prod. by *Actinoplanes arizonaensis*. Active against a number of gram-positive bacteria. Orange cryst. Mp 110-135° dec. $[\alpha]^{25}_{-84.3}$(c°, 1.017 in MeOH). Similar to Kalafungin.

7-O-De-Me: [108890-89-7]. **Arizonin B₁**.
$C_{17}H_{14}O_7$ M 330.293
From *A. arizonaensis*. Active against a number of gram-positive bacteria.

8-O-De-Me: [108890-87-5]. **Arizonin A₁**.
$C_{17}H_{14}O_7$ M 330.293
From *A. arizonaensis*. Active against a number of gram-positive bacteria. Orange cryst. Mp 233-235° dec. $[\alpha]^{25}_D$ −92° (c, 0.06 in MeOH).

Hochlowski, J.E. *et al*, *J. Antibiot.*, 1987, **40**, 401 (*isol, struct*)

Arizonin C_3

A-70255

[108925-06-0]

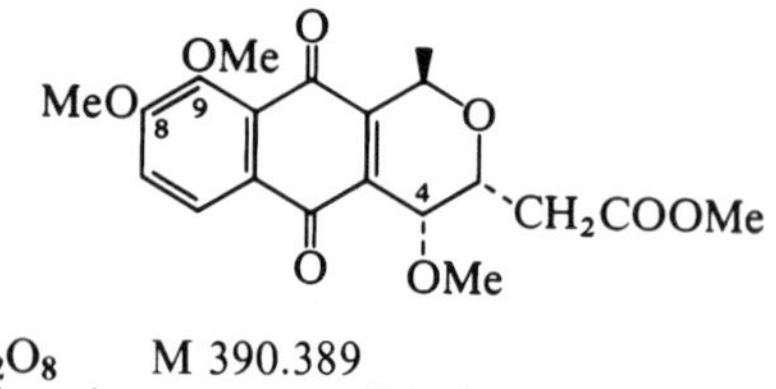

$C_{20}H_{22}O_8$ M 390.389

Naphthoquinone-type antibiotic. Prod. by *Actinoplanes arizonaensis*. Active against a number of gram-positive bacteria. Orange cryst. Mp 195-208°. $[\alpha]^{25}_D$ −104° (c, 0.969 in MeOH). Similar to Kalafungin.

4,8-Di-O-de-Me: [108890-88-6]. **Arizonin A₂**.
$C_{18}H_{18}O_8$ M 362.335
From *A. arizonaensis*. Active against a number of gram-positive bacteria. Orange cryst. $[\alpha]^{25}_D$ −149° (c, 0.203 in MeOH).

4,9-Di-O-de-Me: [108905-76-6]. **Arizonin B₂**.
$C_{18}H_{18}O_8$ M 362.335
From *A. Arizonaensis*. Active against a number of gram-positive bacteria. Orange semi-solid.

Hochlowski, J.E. *et al*, *J. Antibiot.*, 1987, **40**, 401 (*isol, struct*)

1(5),3-Aromadendradiene

A-70256

[111778-06-4]

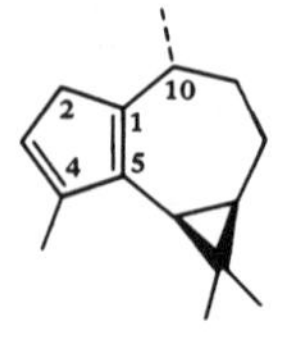

$C_{15}H_{22}$ M 202.339
Constit. of Tolu balsam (from *Balsamum tolutanum*). Oil.

Friedel, H.D. *et al*, *Helv. Chim. Acta*, 1987, **70**, 1753.

1(10),4-Aromadendradiene

A-70257

[112362-74-0]
$C_{15}H_{22}$ M 202.339
Constit. of Tolu balsam (from *Balsamum tolutanum*). Oil.

Friedel, H.D. *et al*, *Helv. Chim. Acta*, 1987, **70**, 1753.

1-Aromadendrene

A-70258

[111821-79-5]
$C_{15}H_{24}$ M 204.355
Constit. of Tolu balsam (from *Balsamum tolutanum*). Oil. $[\alpha]^{25}_D$ +42° (c, 0.2 in pentane).

Friedel, H.D. *et al*, *Helv. Chim. Acta*, 1987, **70**, 1753.

9-Aromadendrene

A-70259

[112421-20-2]

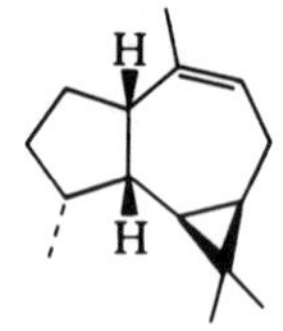

$C_{15}H_{24}$ M 204.355
Constit. of Tolu balsam (from *Balsamum tolutanum*). Oil. $[\alpha]^{25}_D$ −78° (c, 0.6 in pentane).

Friedel, H.D. *et al*, *Helv. Chim. Acta*, 1987, **70**, 1753.

1(5)-Aromadendren-7-ol

A-70260

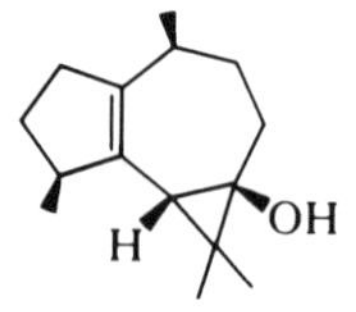

$C_{15}H_{24}O$ M 220.354
Constit. of the soft coral *Xenia novae-britanniae*. Cryst. (EtOAc). Mp 121-124°. $[\alpha]_D$ +66° (c, 0.1 in CHCl₃).

Bowden, B.F. *et al*, *Aust. J. Chem.*, 1987, **40**, 1483.

Artausin   A-70261

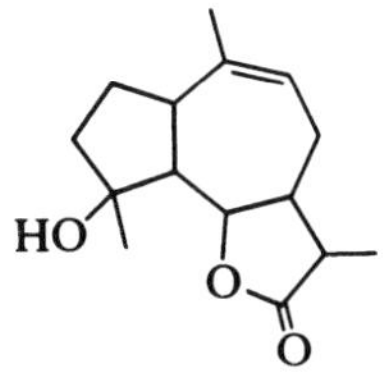

$C_{15}H_{22}O_3$ M 250.337

Constit. of *Artemisia austriaca*. Cryst. (EtOH). Mp 201-204°.

Adekenov, S.M. *et al*, *Khim. Prir. Soedin.*, 1987, **23**, 127.

Arteannuin *B*   A-70262

Updated Entry replacing A-50321

[50906-56-4]

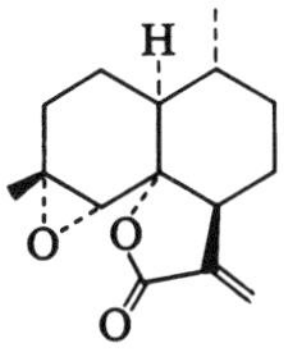

$C_{15}H_{20}O_3$ M 248.321

Constit. of *Artemisia annua*. Cryst. Mp 152°. $[\alpha]_D^{20}$ −6°.

Jeremić, D. *et al*, *Tetrahedron Lett.*, 1973, 3039 (*isol*)
Uskokovic, M.R. *et al*, *Helv. Chim. Acta*, 1974, **57**, 600 (*struct*)
Leppard, D.G. *et al*, *Helv. Chim. Acta*, 1974, **57**, 602 (*cryst struct*)
Goldberg, O. *et al*, *Helv. Chim. Acta*, 1980, **63**, 2455 (*synth*)
Yusupova, I.M. *et al*, *Khim. Prir. Soedin.*, 1986, **22**, 733 (*cryst struct*)
Lansbury, P. *et al*, *Tetrahedron Lett.*, 1986, **27**, 3967 (*synth*)
Akhida, A. *et al*, *Phytochemistry*, 1987, **26**, 1927 (*biosynth*)
Jung, M. *et al*, *J. Nat. Prod.*, 1987, **50**, 972 (*synth*)

Artedouglasiaoxide   A-70263

[86575-88-4]

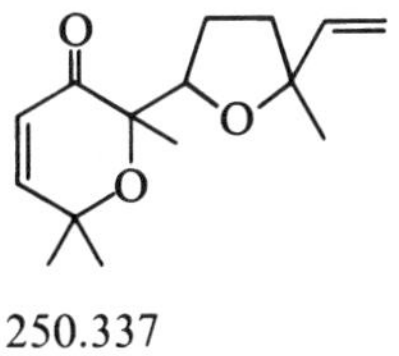

$C_{15}H_{22}O_3$ M 250.337

Isol. from *Artemisia douglasiana*. Gum.

Bohlmann, F. *et al*, *Phytochemistry*, 1982, **21**, 2691 (*isol, struct*)

Arvoside   A-70264

[112543-27-8]

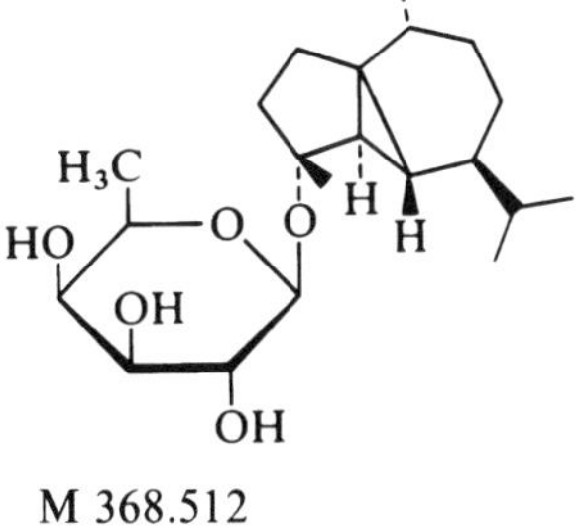

$C_{21}H_{36}O_5$ M 368.512

Constit. of *Calendula arvensis*. $[\alpha]_D$ −27.6°.

Pizza, C. *et al*, *J. Nat. Prod.*, 1987, **50**, 784.

Assafoetidin   A-70265

[115361-83-6]

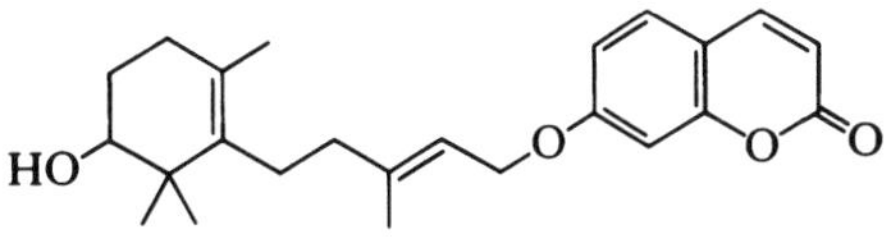

$C_{24}H_{30}O_4$ M 382.499

Constit. of gum resin of *Ferula assafoetida*. Needles. Mp 112°. $[\alpha]_D^{20}$ +11.25° (CHCl₃).

Banerji, A. *et al*, *Tetrahedron Lett.*, 1988, **29**, 1557.

ent-3β,16α,17-Atisanetriol   A-70266

[115783-44-3]

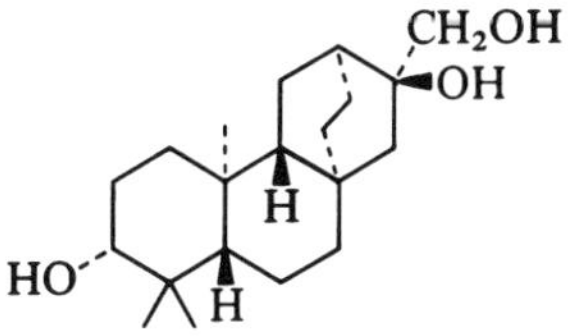

$C_{20}H_{34}O_3$ M 322.487

Constit. of *Euphorbia acaulis*. Cryst. (EtOAc). Mp 226°.

Satti, N.K. *et al*, *Phytochemistry*, 1988, **27**, 1530.

Atratogenin *A*   A-70267

15,20-Epoxy-2α,3β-dihydroxy-14,15-seco-5,15,17(20)-pregnatrien-14-one

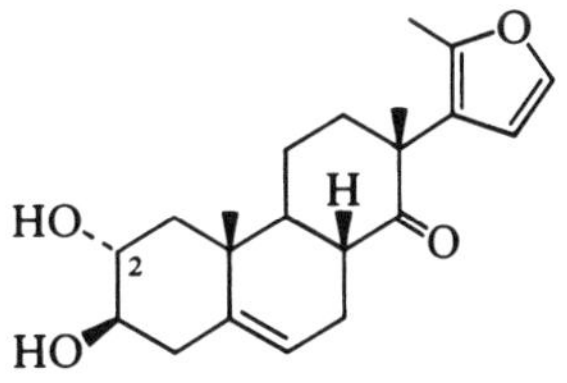

$C_{21}H_{28}O_4$ M 344.450

Constit. of *Cynanchum atratum*. Pale-yellow powder. Mp 69-73°. $[\alpha]_D$ −88.2° (c, 1 in MeOH).

*3-O-β-D-Cymaropyranosyl-(1→4)-α-L-diginopyranosyl-(1→4)-β-D-cymaropyranoside: **Atratoside A**.*
$C_{42}H_{64}O_{13}$ M 776.960
Constit. of *C. atratum*. Amorph. powder. Mp 105-110°. $[\alpha]_D$ −65.9° (c, 1.6 in MeOH).

*3-O-β-D-Glucopyranosyl-(1→4)-β-D-cymaropyranosyl-(1→4)-α-L-diginopyranosyl-(1→4)-β-D-cymaropyranoside: **Atratoside B**.*
$C_{48}H_{74}O_{18}$ M 939.102
Constit. of *C. atratum*. Amorph. powder. Mp 153-158°. $[\alpha]_D$ −48.3° (c, 2.1 in MeOH).

*2-Ketone: 15,20-Epoxy-3β-hydroxy-14,15-seco-5,15,17(20)-pregnatriene-2,14-dione. **Atratogenin B**.*
$C_{21}H_{26}O_4$ M 342.434
Constit. of *C. atratum*. Pale-yellow oil. $[\alpha]_D$ −87.4° (c, 2.1 in CHCl₃).

*3-O-β-D-Glucopyranosyl-(1→4)-β-D-cymaropyranosyl-(1→4)-α-L-diginopyranosyl-(1→4)-β-D-cymaropyranoside: **Atratoside C**.*
$C_{48}H_{72}O_{18}$ M 937.086
Constit. of *C. atratum*. Amorph. powder. Mp 148-153°. $[\alpha]_D$ −58.8° (c, 2.2 in MeOH).

Zhang, Z.-X. *et al*, *Phytochemistry*, 1988, **27**, 2935.

Auramine
A-70268

Updated Entry replacing A-10278

*4,4'-Carbonimidoylbis[N,N-dimethylbenzenamine], 9CI.
Bis(4-dimethylaminophenyl)methylenimine. 4,4'-Bisdi-
methylaminobenzophenone imide. Apyonin. Yellow
pyoctenin. CI Solvent yellow 34*

[492-80-8]

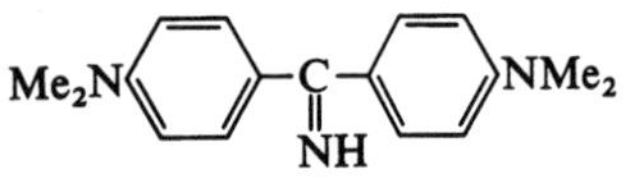

$C_{17}H_{21}N_3$ M 267.373

Imide of 4,4'-Bis(dimethylamino)benzophenone, B-10194
. Non light-fast dye for solvents, oils, waxes, laquers
and inks. Yellow plates (EtOH). Mp 136°.

▷Highly toxic by skin absorption and inhalation. Human
carcinogen, manuf. in the UK controlled by the
Carcinogenic Substances Regulations, 1967.
BY3500000.

B,HCl: [2465-27-2]. *CI Basic Yellow 2.* Dye for silk,
cotton, wool, leather, paper. Dye for inks; fluorescence
probe; has some bacteriostatic activity. Golden-yellow
plates + H_2O (H_2O). Spar. sol. cold H_2O. Mp 267°.

▷BY3675000.

Lynch, D.F.J. *et al*, *J. Am. Chem. Soc.*, 1933, **55**, 2515 (*synth*)
Hellerman, L. *et al*, *J. Am. Chem. Soc.*, 1946, **68**, 1890 (*synth*)
Muralidharan, D. *et al*, *J. Chromatogr.*, 1986, **368**, 405 (*tlc*)
Sax, N.I., *Dangerous Properties of Industrial Materials*, 5th
Ed., Van Nostrand-Reinhold, 1979, 740.
Hazards in the Chemical Laboratory, (Bretherick, L., Ed.), 3rd
Ed., Royal Society of Chemistry, London, 1981, 186.

Auraptenol
A-70269

Updated Entry replacing A-20229

*8-(2-Hydroxy-3-methyl-3-butenyl)-7-methoxy-2H-1-
benzopyran-2-one, 9CI. 8-(2-Hydroxy-3-methyl-3-bu-
tenyl)-7-methoxycoumarin*

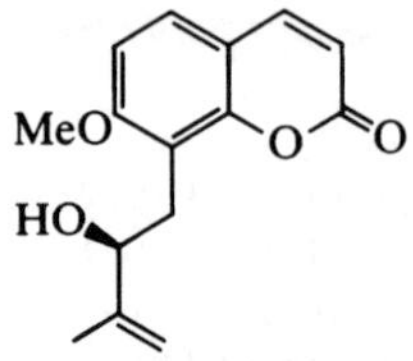

(S)-form

$C_{15}H_{16}O_4$ M 260.289

Constit. of *Cnidii monnieri* and oil of *Citrus aurantium*.
Mp 119-122° (109-110°).

(S)-form [1221-43-8]

Constit. of *Citrus aurantium*. Needles (EtOH). Mp 109-
110°. $[\alpha]_D^{26}$ +14° (c, 1 in EtOH).

Hydroperoxide: [109741-39-1]. **Peroxyauraptenol**.
$C_{15}H_{16}O_5$ M 276.288
Constit. of *Murraya exotica*. Prisms. Mp 114-116°.
$[\alpha]_D$ +3.53° (CHCl$_3$). Has −OOH replacing −OH.
The name peroxyauraptenol is somewhat misleading.

(±)-form [61235-25-4]
Cryst. (C_6H_6/hexane). Mp 109-110°.

Stanley, W.L. *et al*, *Tetrahedron*, 1965, **21**, 89 (*isol, struct*)
Bhakuni, D.S. *et al*, *Indian J. Chem., Sect. B*, 1976, **14**, 332
(*synth*)
Barik, B.R. *et al*, *Phytochemistry*, 1983, **22**, 792 (*abs config*)
Yamahara, J. *et al*, *Chem. Pharm. Bull.*, 1985, **33**, 1676 (*isol*)
Ito, C. *et al*, *Heterocycles*, 1987, **26**, 1731 (*Peroxyauraptenol*)

Austrobailignan-5
A-70270

Updated Entry replacing A-03669

*5,5'-(2,3-Dimethyl-1,4-butanediyl)bis-1,3-benzodioxole,
9CI. 2,3-Bis(3,4-methylenedioxybenzyl)butane. 2,3-Di-
methyl-1,4-dipiperonylbutane*

[55890-23-8]

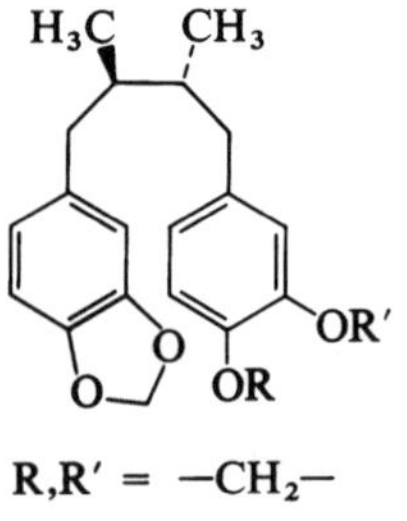

R,R' = −CH$_2$−

(2R,3R)-form
Absolute
configuration

$C_{20}H_{22}O_4$ M 326.391

(2R,3R)-form
Constit. of *Austiobaileya scandens*. Oil. Bp$_{0.02}$ 100°.
$[\alpha]_D^{25}$ −37° (c, 2.5 in CHCl$_3$).

(2RS,3RS)-form [68964-81-8]
Oil.

(2RS,3SR)-form [110269-50-6]
meso-*form*. **Machilin A**
Constit. of *Machilus thunbergii*. Needles. Mp 48-50°.

Murphy, S.T. *et al*, *Aust. J. Chem.*, 1975, **28**, 81 (*isol, struct*)
Biftu, T. *et al*, *J. Chem. Soc., Perkin Trans. 1*, 1978, 1147
(*synth*)
Mahalanabis, K.K. *et al*, *Tetrahedron Lett.*, 1982, **23**, 3975
(*synth*)
Shimomura, H. *et al*, *Phytochemistry*, 1987, **26**, 1513 (*Machilin
A*)

Austrobailignan-7
A-70271

Updated Entry replacing A-03671

*4-[5-(1,3-Benzodioxol-5-yl)tetrahydro-3,4-dimethyl-2-
furanyl]-2-methoxyphenol, 9CI*

[55890-25-0]

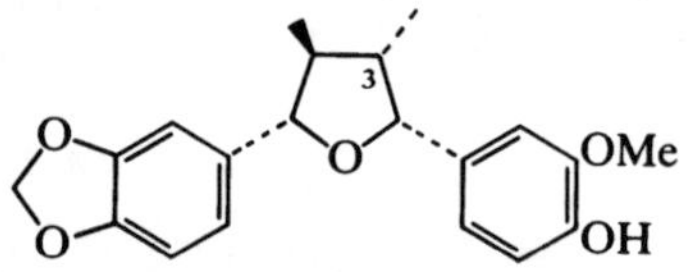

Absolute
configuration

$C_{20}H_{22}O_5$ M 342.391

Constit. of *Austrobaileya scandens* and from mace
(*Myristica fragrans*). Pale-yellow gum. Bp$_{0.02}$ 130°.
$[\alpha]_D^{25}$ +12° (c, 2.9 in CHCl$_3$).

Ac: [55730-99-9].
$C_{22}H_{24}O_6$ M 384.428
Mp 81-82°.

Me ether: [19950-67-5]. **Calopiptin**.
$C_{21}H_{24}O_5$ M 356.418
Constit. of *Piptocalyx moorei*. Cryst. (C_6H_6 or pet.
ether). Mp 95.5°. $[\alpha]_D^{14}$ +28° (c, 2 in CHCl$_3$).

3-Epimer: **Fragransin E$_1$**. **Machilin F**.
$C_{20}H_{22}O_5$ M 342.391
Constit. of *M. fragrans* and *Machilus thunbergii*. Oil.
Abs. config. unknown (quasi-meso). Descr. as
(2R,3R,4S,5R) which appears to be incorr.

3-Epimer, Me ether: [114422-19-4]. **Machilin G**.
$C_{21}H_{24}O_5$ M 356.418
Constit. of *M. thunbergii*. Oil. $[\alpha]_D^{24}$ +3.9° (c, 0.3 in
CHCl$_3$). Abs. config. unknown (quasi-meso).

McAlpine, J.B. *et al*, *Aust. J. Chem.*, 1968, **21**, 2095
(*Calopiptin*)
Murphy, S.T. *et al*, *Aust. J. Chem.*, 1975, **28**, 81 (*isol, struct*)
Hada, S. *et al*, *Phytochemistry*, 1988, **27**, 563 (*isol, deriv*)
Shimomura, H. *et al*, *Phytochemistry*, 1988, **27**, 634 (*deriv*)

Axamide-1   A-70272

Updated Entry replacing A-03701

[56012-88-5]

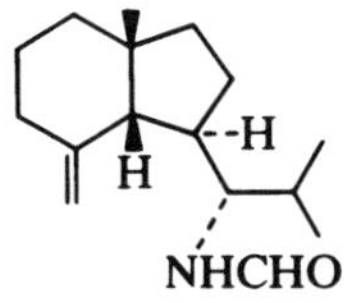

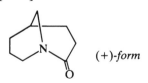

$C_{16}H_{27}NO$ M 249.395

Constit. of *Axinella cannabina*. Oil. $[\alpha]_D$ +10° (c, 1.2 in CHCl$_3$).

Fattorusso, E. *et al*, *Tetrahedron*, 1975, **31**, 269 (*isol*)
Adinolfi, M. *et al*, *Tetrahedron Lett.*, 1977, 2815 (*struct*)
Piers, E. *et al*, *Can. J. Chem.*, 1986, **64**, 2475 (*synth*)
Piers, E. *et al*, *Tetrahedron*, 1987, **43**, 5521 (*synth*)

1-Azabicyclo[3.2.0]heptane-2,7-dione, 9CI   A-70273

[115022-87-2]

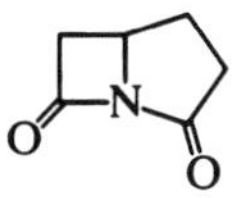

$C_6H_7NO_2$ M 125.127

(±)-*form*

Solid by subl. Mp 61-63°.

Drummond, J.T. *et al*, *Tetrahedron Lett.*, 1987, **28**, 5245 (*synth, ir*)

8-Azabicyclo[5.2.0]nonane, 9CI   A-70274

[39872-95-2]

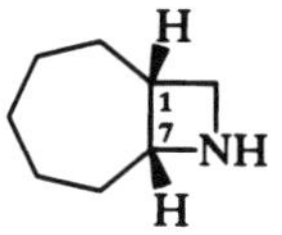

(1RS,7RS)-*form*

$C_8H_{15}N$ M 125.213

(*1RS,7RS*)-*form*

(±)-cis-*form*

Picrate: [72052-64-3]. Mp 135-136°.

(*1RS,7SR*)-*form*

(±)-trans-*form*

Picrate: [72052-66-5]. Mp 160-162°.

Gondos, G. *et al*, *J. Chem. Soc., Perkin Trans. 1*, 1979, 1770 (*synth, deriv*)

1-Azabicyclo[3.3.1]nonan-2-one   A-70275

$C_8H_{13}NO$ M 139.197

(+)-*form*

Waxy cryst. (EtOAc or by subl.). Mp 54-61°. $[\alpha]_D$ +233° (c, 1 in MeCN).

(±)-*form*

Mp 77-81°.

Steliou, K. *et al*, *J. Am. Chem. Soc.*, 1983, **105**, 7130 (*synth, pmr*)
Brehm, R. *et al*, *Helv. Chim. Acta*, 1987, **70**, 1981 (*synth, uv, cd, pmr, cmr, ms, abs config*)

2-Azabicyclo[3.3.1]nonan-7-one, 9CI   A-70276

Updated Entry replacing A-03740

[58937-88-5]

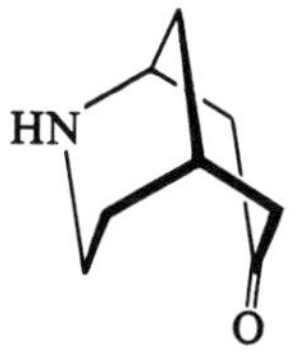

$C_8H_{13}NO$ M 139.197

Yellow oil, cryst. on standing. Bp$_{0.3}$ 80-100°. Unstable, polym. on standing.

Picrate: Mp 222-226° dec.

B,HCl: [106140-37-8]. Solid (EtOH). Mp 185-187°.

Adachi, J. *et al*, *Chem. Pharm. Bull.*, 1976, **24**, 85.
Bonjoch, J. *et al*, *J. Org. Chem.*, 1987, **52**, 267 (*synth, pmr, ir, ms*)

7-Azabicyclo[4.2.0]octane, 9CI   A-70277

[278-36-4]

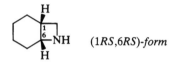   (1RS,6RS)-*form*

$C_7H_{13}N$ M 111.186

(*1RS,6RS*)-*form*

(±)-cis-*form*

Picrate: [5691-13-4]. Cryst. (EtOH/Et$_2$O). Mp 131°.

(*1RS,6SR*)-*form*

(±)-trans-*form*

Picrate: [5691-38-3]. Cryst. (EtOH/Et$_2$O). Mp 146-147°.

Wulfman, D.S. *et al*, *Tetrahedron Lett.*, 1972, 3933.
Gondos, G. *et al*, *J. Chem. Soc., Perkin Trans. 1*, 1979, 1770.

8-Azabicyclo[3.2.1]octane, 9CI   A-70278

Updated Entry replacing A-03751

Nortropane, 8CI

[280-05-7]

$C_7H_{13}N$ M 111.186

Cryst. Sol. H$_2$O. Mp ca. 60°. Bp ca. 161°. Absorbs CO$_2$ from air.

B,HCl: [6760-99-2]. Cryst. Mp 281° dec.

N-*Benzoyl:*

$C_{14}H_{17}NO$ M 215.294

Mp 94-95°. Bp$_{14}$ 204-205°.

N-*Me:* [529-17-9]. *Tropane. Dihydrotropidine.*

$C_8H_{15}N$ M 125.213

Parent substance of an important group of alkaloids.

Liq. Misc. H$_2$O. d_4^{20} 0.931. Bp 167°. n_D^{25} 1.4732.

N-*Me, picrate:* Needles. Mp 284-285° dec.

N-*Chloro:* [74327-95-0].

$C_7H_{12}ClN$ M 145.632

Oil.

Hess, K., *Ber.*, 1918, **51**, 1007, 1014 (*synth*)

v. Braun, J. et al, Ber., 1930, **63**, 496 (synth)
Kay, J.B. et al, J. Chem. Soc., 1965, 5112 (synth, derivs)
Banholzer, R. et al, Justus Liebigs Ann. Chem., 1975, 2227 (synth)
Nelsen, S.F. et al, J. Am. Chem. Soc., 1985, **107**, 3829 (deriv, synth, pmr, cmr)
Bathgate, A. et al, Tetrahedron Lett., 1987, **28**, 5937 (synth)

2-Azabicyclo[2.2.2]oct-5-ene　　　A-70279

[3693-58-1]

$C_7H_{11}N$　　M 109.171
Mp 77-79° (sealed tube). Bp$_{12}$ 29-31°.

Bussmann, R. et al, Chem. Ber., 1987, **120**, 1767 (synth, pmr, cmr, ms)

Azadirachtol　　　A-70280

Updated Entry replacing A-50360
[101508-37-6]

$C_{32}H_{46}O_6$　　M 526.712
Constit. of *Azadirachta indica*. Cryst. Mp 110-112°.
$[\alpha]_D^{25}$ +6.25° (CHCl$_3$).

Siddiqui, S. et al, Planta Med., 1985, 478 (isol)
Siddiqui, S. et al, J. Nat. Prod., 1988, **51**, 30 (rev)

Azadiradione　　　A-70281

Updated Entry replacing A-03780
[26241-51-0]

$C_{28}H_{34}O_5$　　M 450.574
Constit. of *Melia azadirachta*. Insect growth inhibitor against *Heliothis virescens*. Cryst. (MeOH). Mp 168°.
$[\alpha]_D^{20}$ +35.5° (c, 1 in CHCl$_3$).
17-Epimer: [68681-08-3]. **17-epi-Azadiradione**.
　$C_{28}H_{34}O_5$　　M 450.574
　Constit. of *A. indica*. Cryst. (MeOH). Mp 205°. $[\alpha]_D^{20}$ −97° (c, 1 in CHCl$_3$).
O^7-*De-Ac:* [101509-58-4]. **Nimbocinol**. *7-Deacetylajadiradione*.
　$C_{26}H_{32}O_4$　　M 408.536
　Constit. of *A. indica*. Active against *H. virescens*.
O^7-*De-Ac, 17β-Hydroxy:* [117659-70-8]. **7-Deacetyl-17β-hydroxyazadiradione**.
　$C_{26}H_{32}O_5$　　M 424.536

Constit. of *A. indica*. Active against *H. virescens*.

Lavie, D. et al, Tetrahedron, 1971, **27**, 3927 (isol, struct)
Kraus, W. et al, Tetrahedron Lett., 1978, 2395 (epimer)
Lee, S.M. et al, Phytochemistry, 1988, **27**, 2773 (deriv)

3-Azatetracyclo[5.3.1.1^{2,6}.0^{4,9}]dodecane　A-70282

Decahydro-3,6-methanonaphthalen-2,7-imine, 9CI. 3-Azawurtzitane

$C_{11}H_{17}N$　　M 163.262
B,HCl: [77483-35-3]. Mp >360°.
N-Me: [77483-36-4].
　$C_{12}H_{19}N$　　M 177.289
　Cryst. (CH$_2$Cl$_2$/Et$_2$O). Mp 149-150° dec.

Klaus, R.O. et al, Helv. Chim. Acta, 1980, **63**, 2559 (synth, ir, pmr, cmr, ms)

Azetidine, 9CI　　　A-70283

Updated Entry replacing A-03834
Trimethyleneimine
[503-29-7]

C_3H_7N　　M 57.095
Liq. with ammoniacal odour. Misc. H$_2$O, EtOH. d^{20} 0.844. Bp 63°. n_D^{24} 1.4287.
Picrate: Yellow needles. Mp 166-167°.
N-Nitroso: [15216-10-1].
　$C_3H_6N_2O$　　M 86.093
　▷Exp. carcinogen. CM4375000.
N-Ph: [3334-89-2].
　$C_9H_{11}N$　　M 133.193
　Oil. Bp 242-245°, Bp$_{16}$ 130-132°.
N-Benzyl: [7730-39-4].
　$C_{10}H_{13}N$　　M 147.219
　Bp$_5$ 71-75°, Bp$_4$ 65-70°.
N-Benzhydryl:
　$C_{16}H_{17}N$　　M 223.317
　Mp 107-109°.
N-(4-Methylbenzenesulfonyl): [7730-45-2]. Cryst. (EtOH). Mp 116-118°.

Vaughan, W.R. et al, J. Org. Chem., 1961, **26**, 138 (deriv, synth)
Berman, H.M. et al, J. Am. Chem. Soc., 1969, **91**, 6177 (cryst struct)
Org. Synth., 1973, **53**, 13 (synth)
White, J. et al, J. Org. Chem., 1974, **39**, 1973.
Cromwell, N.H., Chem. Rev., 1979, **79**, 331 (rev, bibl)
Szmuszkovicz, J. et al, J. Org. Chem., 1981, **46**, 3562 (synth, deriv)
Sammes, P.G. et al, J. Chem. Soc., Perkin Trans. 1, 1984, 2415 (deriv, synth, pmr)
Causey, D.H. et al, Synth. Commun., 1988, **18**, 205 (synth, deriv)
Sax, N.I., Dangerous Properties of Industrial Materials, 5th Ed., Van Nostrand-Reinhold, 1979, 866.

2-Azetidinecarboxylic acid, 9CI   A-70284

Updated Entry replacing A-03835

[2517-04-6]

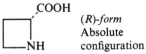

(R)-form
Absolute
configuration

$C_4H_7NO_2$ M 101.105

(R)-form [7729-30-8]

$[\alpha]_D^{20}$ +102° (c, 3.6 in H_2O).

(S)-form [2133-34-8]

Present in roots and leaves of *Convallaria majalis* (lily-of-the-valley) and other liliaceous plants and seeds of *Fritillaria meleagris*. Seedling growth inhibitor. Proline antagonist. Cryst. (MeOH). Sol. H_2O. $[\alpha]_D^{20}$ −108° (c, 3.6 in H_2O). Unstable in mineral acids. Does not melt but darkens at 270°.

▷CM4310500.

N-*Benzyloxycarbonyl:* Oil. $[\alpha]_D^{20}$ −99.0° (c, 3.9 in $CHCl_3$).

(±)-form [20063-89-2]

Darkens without melting >200°.

Berman, H.M. *et al, J. Am. Chem. Soc.*, 1969, **91**, 6177 (*cryst struct*)
Phillips, B.A. *et al, J. Heterocycles* Chem., 1973, **10**, 795 (*synth*)
Miyoshi, M. *et al, Chem. Lett.*, 1973, 5 (*abs config*)
Leete, E., *Phytochemistry*, 1975, **14**, 1983 (*biosynth*)
Cromwell, N.H., *Chem. Rev.*, 1979, **79**, 331 (*rev, bibl*)
Aldrich Atlas of NMR Spectra, 1974, **3**, 16B (*pmr*)
Aldrich Atlas of IR Spectra, 1975, 2nd Ed., 309F (*ir*)
Baldwin, J.E. *et al, Tetrahedron*, 1988, **44**, 637 (*synth*)

7-Azidobicyclo[2.2.1]hepta-2,5-diene   A-70285

7-Azidonorbornadiene

[104172-76-1]

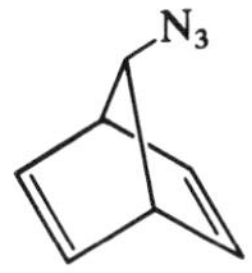

$C_7H_7N_3$ M 133.152

Bp_1 41°.

▷Explosive

Hoffmann, R.W. *et al, Chem. Ber.*, 1986, **119**, 3297 (*synth, pmr*)

4-Azidobutanal, 9CI   A-70286

[99545-47-8]

$$N_3CH_2CH_2CH_2CHO$$

$C_4H_7N_3O$ M 113.119

Gold oil.

Hudlicky, T. *et al, J. Am. Chem. Soc.*, 1986, **108**, 3755 (*synth, nmr, ir*)

2-Azidoethanol, 9CI   A-70287

Updated Entry replacing A-03852

Azidoethyl alcohol

[1517-05-1]

$$N_3CH_2CH_2OH$$

$C_2H_5N_3O$ M 87.081

Misc. H_2O. d_{24}^{24} 1.149. Bp_{20} 73°, Bp_8 60°. n_D^{25} 1.4578.

▷Explosive, particularly on distn.

Fagley, T.F. *et al, J. Am. Chem. Soc.*, 1953, **75**, 3104 (*synth, props*)
Karpenko, G.V. *et al, CA*, 1979, **90**, 103317 (*synth*)
Smith, R.H. *et al, J. Org. Chem.*, 1988, **53**, 1467 (*synth, haz*)

1-[3-Azido-4-(hydroxymethyl)cyclopentyl]-5-methyl-2,4(1H,3H)pyrimidinedione   A-70288

C-AZT

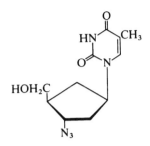

$C_{11}H_{15}N_5O_3$ M 265.271

(+)-form [114489-64-4]

Carbocyclic analogue of 3′-Azido-3′-deoxythymidine, A-50371 . $[\alpha]_D^{20}$ +14.4° (c, 2.6 in Me_2CO).

(±)-form [100020-51-7]

Cryst. Mp 154-157°.

Shealy, Y.F. *et al, J. Med. Chem.*, 1986, **29**, 483 (*synth, uv, ms, ir, pmr*)
Bodenteich, M. *et al, Tetrahedron Lett.*, 1987, **28**, 5311 (*synth*)

2-Azidopropane, 9CI   A-70289

Isopropyl azide

[691-57-6]

$$(H_3C)_2CHN_3$$

$C_3H_7N_3$ M 85.108

Bp 67°. n_D^{20} 1.4020.

Smirnov, A.N. *et al, Zh. Obshch. Khim.*, 1965, **35**, 178 (*synth*)
Roberts, B.P. *et al, J. Chem. Soc., Perkin Trans. 2*, 1982, 1549 (*use*)

3-Azido-2-propenoic acid   A-70290

$$N_3CH{=}CHCOOH$$

$C_3H_3N_3O_2$ M 113.076

(E)-form

2-Bromoethyl ester: Liq.

(Z)-form

Me ester:
    $C_4H_5N_3O_2$ M 127.102
    Mp 65°.

Et ester:
    $C_5H_7N_3O_2$ M 141.129
    Characterised spectroscopically.

2-Bromoethyl ester: Characterised spectroscopically.

Priebe, H., *Acta Chem. Scand., Ser. B*, 1987, **41**, 640.

3(5)-Azidopyrazole   A-70291
[29176-21-4]

C$_3$H$_3$N$_5$ M 109.090
Long needles (H$_2$O). Mp 55-57°. Turns brown in air.

Reimlinger, H. *et al, Chem. Ber.,* 1970, **103**, 3284 (*synth, ir*)

2*H*-Azirine, 9CI   A-70292
Updated Entry replacing A-03873
[157-16-4]

C$_2$H$_3$N   M 41.052
Polymerises rapidly at r.t.

Ford, R.G., *J. Am. Chem. Soc.,* 1977, **99**, 2389 (*synth, micro-wave*)
Nair, V., *Chem. Heterocycles* Compd., (*Hassner, A., Ed.*), 1983, **42(1)**, 215 (*rev*)
Bock, H. *et al, Chem. Ber.,* 1987, **120**, 1961 (*bibl*)
Guillemin, J-C. *et al, Tetrahedron,* 1988, **44**, 4447 (*synth, uv, pmr*)

1,4-Azulenedicarboxylic acid   A-70293
C$_{12}$H$_8$O$_4$ M 216.193
Di-Me ester: 1,4-Dicarbomethoxyazulene.
C$_{14}$H$_{12}$O$_4$ M 244.246
Constit. of *Ixiolaena leptolepis.* Violet oil.

Lehmann, L. *et al, Phytochemistry,* 1988, **27**, 2994.

2,6-Azulenedicarboxylic acid   A-70294
C$_{12}$H$_8$O$_4$ M 216.193
Di-Me ester:
  C$_{14}$H$_{12}$O$_4$ M 244.246
  Green cryst. (cyclohexane). Mp 144-144.5°.

Treibs, W., *Chem. Ber.,* 1959, **92**, 2152 (*synth, uv*)

5,7-Azulenedicarboxylic acid   A-70295
C$_{12}$H$_8$O$_4$ M 216.193
Di-Et ester: [54092-46-5].
  C$_{16}$H$_{16}$O$_4$ M 272.300
  Deep-violet needles (EtOH). Mp 82-83°.
Dinitrile: [54092-45-4]. *5,7-Dicyanoazulene.*
  C$_{12}$H$_6$N$_2$ M 178.193
  Dark-blue needles (EtOH). Mp 201-202°.

Jutz, C. *et al, Chem. Ber.,* 1974, **107**, 2956 (*synth, deriv, uv, pmr*)
Fukazawa, Y. *et al, Tetrahedron Lett.,* 1982, **23**, 2129 (*synth, deriv*)

B

Baccharinoid *B1*　　　　　　　　B-70001

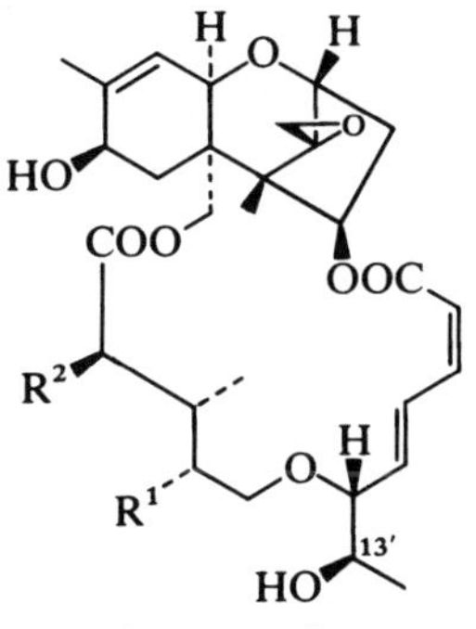

R¹ = OH, R² = H

$C_{29}H_{40}O_{10}$　　M 548.629
Constit. of *Baccharis megapotamica*. Cryst. (EtOAc/hexane). Mp 162-164°. $[\alpha]_D^{24}$ +100° (c, 1.4 in CH_2Cl_2).

13'-Epimer: **Baccharinoid B2.**
$C_{29}H_{40}O_{10}$　　M 548.629
Constit. of *B. megapotamica*. Cryst. (CH_2Cl_2/Et_2O). Mp 177-180°. $[\alpha]_D^{24}$ +116° (c, 1 in CH_2Cl_2).

Jarvis, B.B. *et al, J. Nat. Prod.*, 1987, **50**, 815.

Baccharinoid *B7*　　　　　　　　B-70002

As Baccharinoid *B1*, B-70001 with

R¹ = H, R² = OH

$C_{29}H_{40}O_{10}$　　M 548.629
Constit. of *Baccharis megapotamica*. Cryst. (EtOH). Mp 229-231°. $[\alpha]_D^{25}$ +150° (c, 0.66 in CH_2Cl_2).

13'-Epimer: **Baccharinoid B3.**
$C_{29}H_{40}O_{10}$　　M 548.629
Constit. of *B. megapotamica*. Cryst. (Me₂CO/hexane). Mp 172-180°. $[\alpha]_D^{25}$ +164° (c, 0.58 in CH_2Cl_2).

Jarvis, B.B. *et al, J. Nat. Prod.*, 1987, **50**, 815.

Baccharinol　　　　　　　　　　B-70003

Updated Entry replacing B-00011
Baccharinoid B4
[63783-94-8]

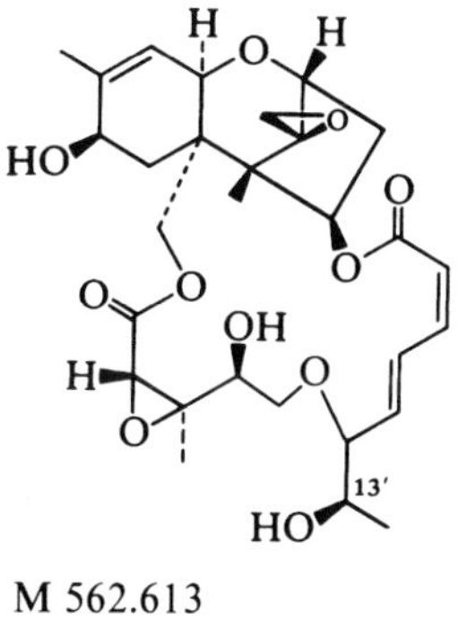

$C_{29}H_{38}O_{11}$　　M 562.613
Constit. of *Baccharis megapotamica*. Cryst. (MeOH/CHCl₃/Et₂O). Mp 259-263°. $[\alpha]_D^{24}$ +165° (c, 0.5 in MeOH). λ_{max} 260 nm (ϵ 20 400) (EtOH).

Tri-Ac: Mp 255-257°.
13'-Epimer: [63814-58-4]. **Isobaccharinol.** *Baccharinoid B6.*
$C_{29}H_{38}O_{11}$　　M 562.613
From *B. megapotamica*. Cryst. (Me₂CO/hexane). Mp 249-251°. $[\alpha]_D^{24}$ +149° (c, 0.66 in MeOH).

Kupchan, S.M. *et al, J. Org. Chem.*, 1977, **42**, 4221 (*isol, struct*)

Bacchariolide *B*　　　　　　　　B-70004

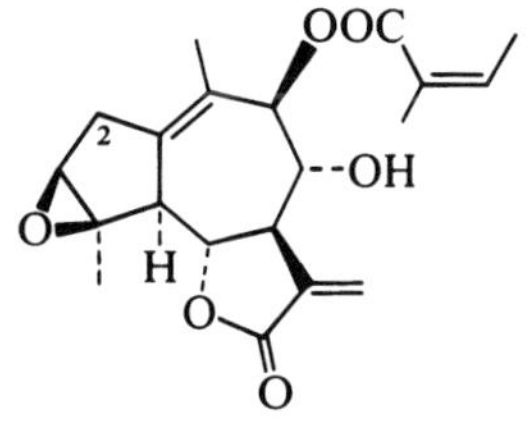

$C_{20}H_{24}O_6$　　M 360.406
Constit. of *Baccharis salicina*. Gum.

2-Hydroxy: **Bacchariolide A.**
$C_{20}H_{24}O_7$　　M 376.405
Constit. of *B. salicina*. Gum.

Parodi, F.J. *et al, Phytochemistry*, 1988, **27**, 2987.

Bacteriochlorophyll f　　　　　　B-70005

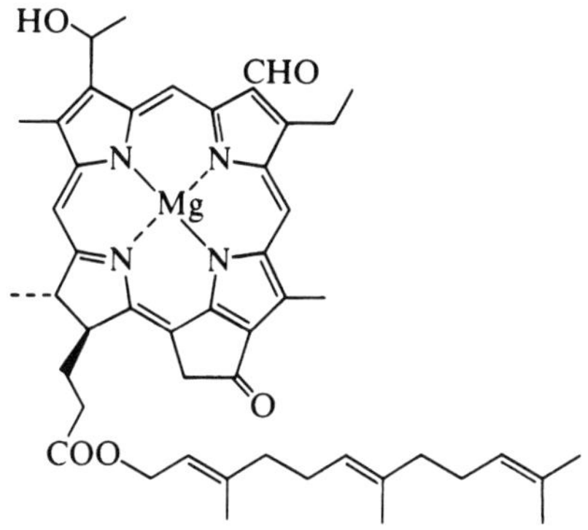

$C_{46}H_{56}MgN_4O_5$　　M 769.277
Not yet found in nature but postulated to occur naturally.
Risch, N. *et al, Justus Liebigs Ann. Chem.*, 1988, 343.

Balaenol　　　　　　　　　　　　B-70006

[115031-67-9]

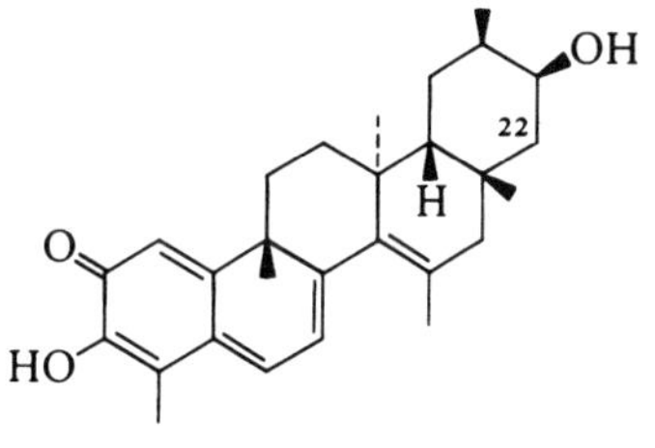

$C_{28}H_{36}O_3$　　M 420.591

45

Constit. of *Cassine balae*. Mp 139-140°.

22-Oxo: [115028-52-9]. **Balaenonol**.
$C_{28}H_{34}O_4$ M 434.574
Constit. of *C. balae*. Mp 205-208°.

Fernando, H.C. *et al, Tetrahedron Lett.*, 1988, **29**, 387 (*struct*)

Barbatic acid   B-70007

Updated Entry replacing B-00079
2-Hydroxy-4-[(2-hydroxy-4-methoxy-3,6-dimethylbenzoyl)oxy]-3,6-dimethylbenzoic acid, 9CI.
Barbatinic acid. Rhizoic acid. Coccellic acid
[17636-16-7]

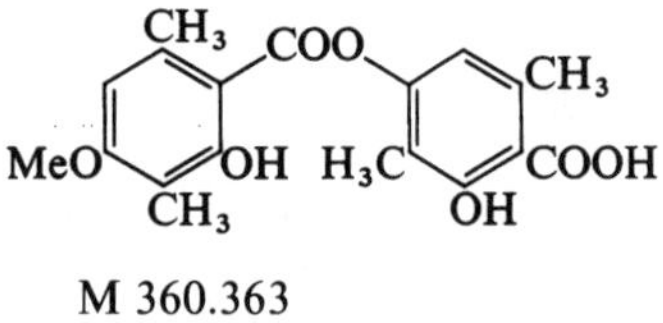

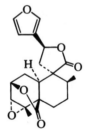

$C_{19}H_{20}O_7$ M 360.363
Constit. of *Usnea* spp. and *Alectoria ochroleuca*. Cryst.
(C_6H_6). Mp 191° (186°).
Mono-Ac: Needles (AcOH). Mp 172°.
Me ester: [5014-22-2]. Prisms (MeOH). Mp 170°.
5-Chloro, 4-O-de-Me, Me ester: **Methyl 5-chloro-4-**O-**demethylbarbatate**.
$C_{19}H_{19}ClO_7$ M 394.808
Constit. of *Erioderma* spp. Cryst. (CHCl_3). Mp 165-166°.

St. Pfau, A., *Helv. Chim. Acta*, 1928, **11**, 864 (*struct*)
Robertson, A. *et al, J. Chem. Soc.*, 1932, 1675 (*isol*)
Elix, J.A. *et al, Aust. J. Chem.*, 1975, **28**, 1113; 1987, **40**, 1581 (*synth, deriv*)

Bartemidiolide   B-70008

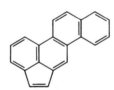

$C_{20}H_{22}O_6$ M 358.390
Constit. of *Baccharis artemisioides*. Cryst. (MeOH). Mp
224-225°. $[\alpha]_D^{25}$ +84.4° (c, 0.7 in CHCl_3).

Tonn, C.E. *et al, Phytochemistry*, 1988, **27**, 489 (*isol, cryst struct*)

Benz[*a*]aceanthrylene, 9CI   B-70009

Updated Entry replacing B-20009
Benzo[a]*fluoranthene*
[203-33-8]

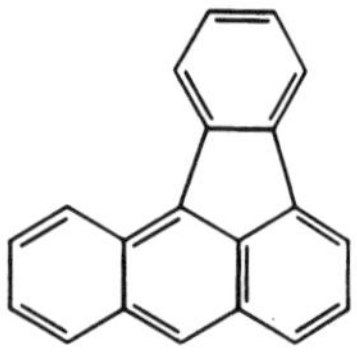

$C_{20}H_{12}$ M 252.315
Orange-yellow needles. Mp 144-144.5°.

Marks, A. *et al, J. Chem. Soc.*, 1951, 2941 (*synth*)
Stubbs, J. *et al, J. Chem. Soc.*, 1951, 2939 (*synth*)

Ray, J.K. *et al, J. Org. Chem.*, 1982, **47**, 3335 (*synth*)
Chung, Y.-S. *et al, J. Org. Chem.*, 1987, **52**, 1284 (*synth, uv, pmr, ms*)
Cho, B.P. *et al, J. Org. Chem.*, 1987, **52**, 5668 (*synth, pmr, uv*)

Benz[*e*]acephenanthrylene, 9CI   B-70010

Updated Entry replacing B-20011
Benzo[b]*fluoranthene. 3,4-Benzofluoranthene*
[205-99-2]

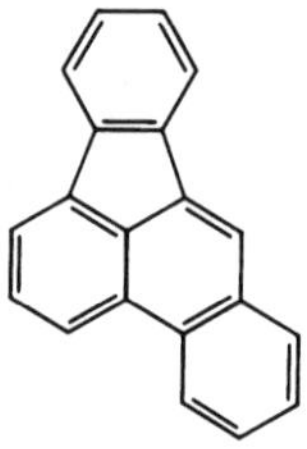

$C_{20}H_{12}$ M 252.315
Needles (C_6H_6 or EtOH). Mp 168°.
▷Exp. carcinogen. CU1400000.
Picrate: Yellow needles (EtOH). Mp 156°.
1,3,5-Trinitrobenzene complex: Orange-yellow needles
(EtOH). Mp 179°.

Tobler, R. *et al, Helv. Chim. Acta*, 1941, **24**, 100E (*synth*)
Buu-Hoï, N.P. *et al, J. Chem. Soc.*, 1959, 1845 (*synth*)
Bartle, K.D. *et al, Spectrochim. Acta, Part A*, 1964, **25**, 1603
(*nmr*)
Lavitt Lamy, D. *et al, Bull. Soc. Chim. Fr.*, 1966, 2613 (*uv*)
Jutz, C. *et al, Angew. Chem., Int. Ed. Engl.*, 1972, **11**, 315
(*synth*)
Jones, D.W. *et al, Spectrochim. Acta, Part A*, 1974, **30**, 489
(*nmr*)
Brennan, J. *et al, J. Chem. Res. (S)*, 1977, 107 (*synth*)
Praefcke, K. *et al, Justus Liebigs Ann. Chem.*, 1978, 1399
(*synth, ms*)
Amin, S. *et al, J. Org. Chem.*, 1981, **46**, 2573 (*synth, tox, bibl*)
Anthony, I.J. *et al, Aust. J. Chem.*, 1984, **37**, 1283 (*synth*)
Chung, Y.-S. *et al, J. Org. Chem.*, 1987, **52**, 1284 (*synth, uv, pmr, ms*)
Cho, B.P. *et al, J. Org. Chem.*, 1987, **52**, 5668, 5679 (*synth, pmr, cmr*)
Sax, N.I., *Dangerous Properties of Industrial Materials*, 5th
Ed., Van Nostrand-Reinhold, 1979, 406.

Benz[*j*]acephenanthrylene   B-70011

[216-48-8]

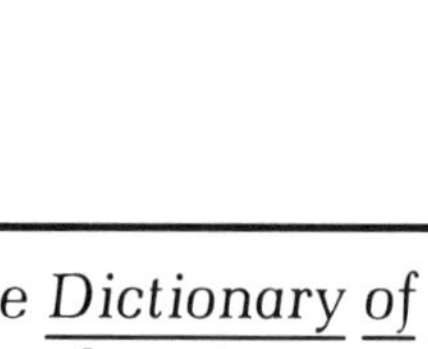

$C_{20}H_{12}$ M 252.315
Orange-yellow cryst. (hexane). Mp 170-171°.

Sangaiah, R. *et al, J. Org. Chem.*, 1987, **52**, 3205 (*synth, pmr, uv, ms, ir*)

---

*Consult the Dictionary of Alkaloids
for a comprehensive treatment of
alkaloid chemistry.*

Benz[*a*]anthracene-1,2-dione   B-70012

[82120-27-2]

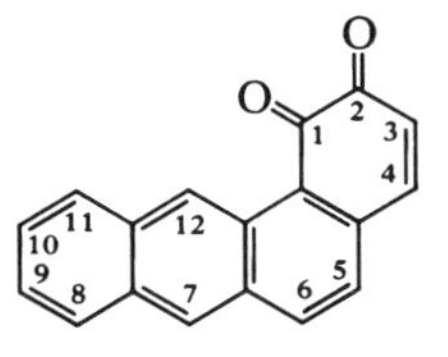

$C_{18}H_{10}O_2$ M 258.276
Mp 173°.

Platt, K.L. *et al, Tetrahedron Lett.*, 1982, **23**, 163 (*synth, uv, ir, pmr, ms*)

Benz[*a*]anthracene-3,4-dione   B-70013

[74877-25-1]

$C_{18}H_{10}O_2$ M 258.276
Deep-violet cryst. Mp 233°.

Sukumaran, K.B. *et al, J. Org. Chem.*, 1980, **45**, 4407 (*synth, pmr*)
Harvey, R.G. *et al, J. Med. Chem.*, 1988, **31**, 154 (*synth*)

Benz[*a*]anthracene-5,6-dione   B-70014

[18508-00-4]

$C_{18}H_{10}O_2$ M 258.276
Red needles ($CHCl_3$). Mp 260-262°.

Cho, H. *et al, Tetrahedron Lett.*, 1974, 1491 (*synth, pmr*)
Okamoto, T. *et al, Chem. Pharm. Bull.*, 1978, **26**, 2014 (*synth*)
Yang, D.T.C. *et al, Org. Prep. Proced. Int.*, 1982, **14**, 202 (*synth*)
Balani, S.K. *et al, J. Org. Chem.*, 1986, **51**, 1773 (*synth*)

Benz[*a*]anthracene-7,12-dione, 9CI   B-70015

Updated Entry replacing B-00155
1,2-Benzanthraquinone. Naphthanthraquinone. Tetraphene-7,12-quinone

[2498-66-0]

$C_{18}H_{10}O_2$ M 258.276
Yellow prisms (AcOH). V. sol. $CHCl_3$, C_6H_6, sol. Me_2CO, AcOH, spar. sol. EtOH, Et_2O. Mp 169°.

Heller, G. *et al, Ber.*, 1908, **41**, 3633 (*synth*)
Groggins, P.H. *et al, Ind. Eng. Chem.*, 1930, **22**, 159 (*synth*)
Arnold, R.T. *et al, J. Org. Chem.*, 1940, **5**, 250 (*synth*)
Mukherji, R.N. *et al, Tetrahedron*, 1967, **23**, 3859 (*synth, uv, ir*)
Tomaszewski, J.E. *et al, Tetrahedron Lett.*, 1977, 971 (*synth*)
Sammes, P.G. *et al, J. Chem. Soc., Chem. Commun.*, 1979, 33 (*synth*)
Kuroda, R. *et al, Acta Crystallogr., Sect. B*, 1982, **38**, 1674 (*cryst struct*)
Fujisawa, S. *et al, Bull. Chem. Soc. Jpn.*, 1985, **58**, 3356 (*synth*)

Benz[*a*]anthracene-8,9-dione   B-70016

[82120-26-1]

$C_{18}H_{10}O_2$ M 258.276
Mp 238°.

Platt, K.L. *et al, Tetrahedron Lett.*, 1982, **23**, 163 (*synth, uv, ir, pmr, ms*)

Benzeneethanethioic acid, 9CI   B-70017

Phenylthioacetic acid. Thio-α-toluic acid. Thiophenylacetic acid

[62167-00-4]

$$PhCH_2COSH \rightleftharpoons PhCH_2CSOH$$

C_8H_8OS M 152.211
Oil. Slowly oxid. in air.
O-Et ester: [10602-65-0].
    $C_{10}H_{12}OS$ M 180.264
    Bp 108-110°, $Bp_{0.2}$ 55°. n_D^{25} 1.5554, n_D^{20} 1.5518.
S-Et ester: [14476-63-2].
    $C_{10}H_{12}OS$ M 180.264
    d_{20} 1.115. Bp_8 108-109°. n_D^{20} 1.5961.
Amide: [645-54-5]. *Phenylthiolacetamide.*
    C_8H_9NS M 151.226
    Cryst. (EtOH). Mp 97.5-98°.

Sjoberg, B., *Acta Chem. Scand.*, 1959, **13**, 1036 (*synth, ir*)
Boiko, Yu.A. *et al, J. Org. Chem. USSR*, 1966, **2**, 1893 (*synth*)
Radeglia, R. *et al, Z. Naturforsch., B*, 1969, **24**, 283 (*pmr*)
Mori, M. *et al, Int. J. Sulfur Chem., Sect. A*, 1972, **2**, 79 (*synth*)
Barton, D.H.R. *et al, J. Chem. Soc., Perkin Trans. 1*, 1973, 1571 (*synth, ir, pmr*)
Kroeber, H. *et al, Int. J. Sulfur Chem.*, 1976, **8**, 611 (*synth*)
Lang, M. *et al, J. Am. Chem. Soc.*, 1979, **101**, 6296 (*synth*)
Takeda, K. *et al, J. Org. Chem.*, 1985, **50**, 273 (*synth*)

Benzenesulfonyl peroxide, 8CI   B-70018

Bis(phenylsulfonyl)peroxide, 9CI. Bis(benzenesulfonyl)-peroxide

[29342-61-8]

$$PhSO_2-O-O-SO_2Ph$$

$C_{12}H_{10}O_6S_2$ M 314.327
Easily-handled two-electron oxidant for a variety of functional groups. Waxy cryst. Mp 66° (53-54° dec.). Stability is increased by electron-withdrawing substits. in the aromatic rings and nitrobenzenesulfonyl peroxides are freq. used.
▷May explode violently on storge at r.t.

Govatt, L.W. *et al, J. Org. Chem.*, 1959, **24**, 2031 (*synth*)
Dannelly, R.L. *et al, J. Org. Chem.*, 1970, **35**, 3076 (*analogues*)
Hoffman, R.V., *Org. Prep. Proced. Int.*, 1986, **18**, 181 (*rev*)

1,3,5-Benzenetrimethanethiol, 9CI   B-70019

1,3,5-Tris(thiomethyl)benzene. 1,3,5-Tris(mercaptomethyl)benzene

[38460-57-0]

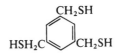

$C_9H_{12}S_3$ M 216.374
Oil. Bp_1 177° (Bp_1 90-95° Kugelrohr).

Ricci, A. *et al, J. Chem. Soc., Perkin Trans. 1*, 1976, 1691 (*synth*)
Houk, J. *et al, J. Am. Chem. Soc.*, 1987, **109**, 6825 (*synth, pmr, ir*)

7,16[1',2']-Benzeno-7,16-dihydroheptacene, **B-70020**
9CI
[110372-75-3]

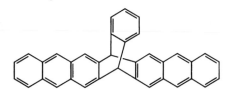

C₃₆H₂₂   M 454.570
Pale-yellow cryst. (1,3,5-trimethylbenzene).

Luo, J. *et al, J. Org. Chem.*, 1987, **52**, 4833 (*synth, pmr, cmr, ms, uv*)

Benz[*f*]imidazo[5,1,2-*cd*]indolizine, 9CI   **B-70021**
1-Azabenzo[h]*cycl*[3.2.2]*azine*
[104169-29-1]

C₁₃H₈N₂   M 192.220
Pale-yellow prisms (MeOH). Mp 90°.

Tominaga, Y. *et al, J. Heterocycl. Chem.*, 1988, **25**, 185 (*synth, ir, uv, pmr, ms*)

1*H*-Benzimidazole-2-sulfonic acid   **B-70022**

C₇H₆N₂O₃S   M 198.196
Cryst. + 1H₂O (H₂O). Mp >380°. Exists as zwitterion in crystal.

Chidambaram, S. *et al, Acta Crystallogr., Sect. C*, 1988, **44**, 898 (*synth, cryst struct*)

Benz[*def*]indeno[1,2,3-*qr*]chrysene, 9CI   **B-70023**
[111189-34-5]

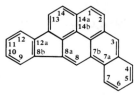

C₂₆H₁₄   M 326.397
Orange needles (C₆H₆/hexanes). Mp 241-242°.

Cho, B.P. *et al, J. Org. Chem.*, 1987, **52**, 5668 (*synth, pmr, cmr, uv*)

Benz[*def*]indeno[1,2,3-*hi*]chrysene, 9CI   **B-70024**
[111189-33-4]

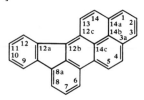

C₂₆H₁₄   M 326.397
Orange powder (xylene). Mp 213-215°.

Cho, B.P. *et al, J. Org. Chem.*, 1987, **52**, 5668 (*synth, pmr, cmr, uv*)

Benz[5,6]indeno[2,1-*a*]phenalene, 9CI   **B-70025**
[76727-41-8]

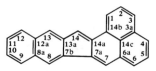

C₂₄H₁₄   M 302.375
Red flakes with coppery sheen (pet. ether/C₆H₆). Mp 284-285° dec.

Kemp, W. *et al, J. Chem. Soc., Perkin Trans. 1*, 1980, 2812 (*synth, pmr, ms*)

1*H*-Benz[*e*]indole-1,2(3*H*)-dione, 9CI   **B-70026**
Updated Entry replacing B-10033
Benz[e]*indoline-1,2-dione, 8CI. Benz*[e]*isatin. β-Naphthisatin*
[5588-87-4]

C₁₂H₇NO₂   M 197.193
Red needles (EtOH). Mp 252°.
N-*Ac:*
    C₁₄H₉NO₃   M 239.230
    Yellow plates (C₆H₆). Mp 195°.
N-*Me:*
    C₁₃H₉NO₂   M 211.220
    Red needles (AcOH). Mp 216-218°.
N-*Et:*
    C₁₄H₁₁NO₂   M 225.246
    Red needles (EtOH). Mp 173°.
N-*Ph:* [13471-31-3].
    C₁₈H₁₁NO₂   M 273.290
    Red needles (EtOH). Mp 227°.
1-Oxime:
    C₁₂H₈N₂O₂   M 212.207
    Yellow cryst. Mp 240° (186° dec.).
1-Phenylhydrazone: Orange plates (EtOH). Mp 220° dec.

Hinsberg, O. *et al, Ber.*, 1888, **21**, 110; 1898, **31**, 250 (*synth*)
Wichelhaus, H., *Ber.*, 1903, **36**, 1736 (*derivs*)
Heller, G., *Ber.*, 1926, **59**, 704 (*deriv*)
Wendelin, W. *et al, J. Heterocycl. Chem.*, 1987, **24**, 1381 (*ir, uv,*

Benz[*cd*]indol-2-(1*H*)-one, 9CI   B-70027

Updated Entry replacing N-00223
Naphthostyril. 2H-Benzo[cd]*isoindol-2-one. 8-Amino-
1-naphtholactam*
[130-00-7]

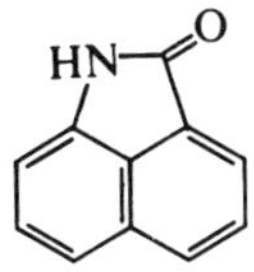

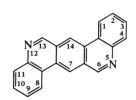

C$_{11}$H$_7$NO   M 169.182
Fine needles (EtOH). Mod. sol. boiling H$_2$O, spar. sol.
Et$_2$O. Mp 180-181°. Sol. EtOH → green fluor.
N-*Ac:* [58779-65-0].
    C$_{13}$H$_9$NO$_2$ M 211.220
    Mp 125°.
N-*Benzoyl:*
    C$_{18}$H$_{11}$NO$_2$ M 273.290
    Mp 170°.
N-*Me:* [1710-20-9].
    C$_{12}$H$_9$NO   M 183.209
    Cryst. (C$_6$H$_6$/pet. ether). Mp 78-80°.
N-*Ph:* [1830-57-5].
    C$_{17}$H$_{11}$NO   M 245.280
    Yellow needles (pet. ether). Mp 104-105°.
N-*Chloromethyl:* [114044-23-4]. *1-
Chloromethylbenz*[cd]*indol-2(1H)-one.*
    C$_{12}$H$_8$ClNO   M 217.654
    Reagent for the derivatisation and fluorometric
    determination of carboxylic acids. Cryst.
    (cyclohexane). Mp 141°.

Rule, H.G. *et al, J. Chem. Soc.*, 1934, 137; 1935, 318 (*synth*)
Dokunitchin, N.S. *et al, CA*, 1963, **58**, 6772b (*synth*)
Protiva, M. *et al, Collect. Czech. Chem. Commun.*, 1985, **50**,
    1888 (*deriv*)
Wendelin, W. *et al, J. Heterocycl. Chem.*, 1987, **24**, 1381
    (*synth, ir, uv, use, deriv*)

Benzo[1,2-*d*:4,5-*d'*]bis[1,3]dithiole-2,6-dione, 9CI   B-70028

1,3,5,7-Tetrathia-s-indacene-2,6-dione
[108970-88-3]

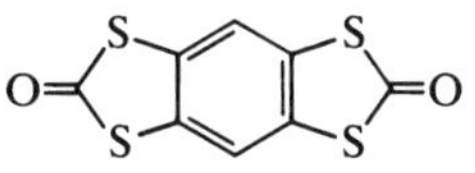

C$_8$H$_2$O$_2$S$_4$ M 258.343
Synthon for prepn. of new multisulfur tetrathiofulvalene
    donors and nickel-dithiolene complexes. Fine needles
    (toluene). Mp >320°.

Larsen, J. *et al, J. Org. Chem.*, 1987, **52**, 3285 (*synth, use*)

Benzo[5,6]cycloocta[1,2-*c*]furan, 8CI   B-70029

1,2-Benzo[5,6-c]furocyclooctatetraene
[26261-80-3]

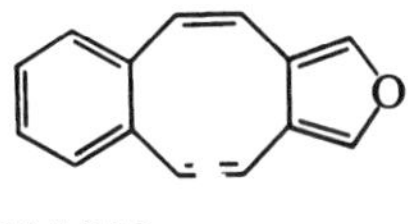

C$_{14}$H$_{10}$O   M 194.232
Prisms by subl. Mp 86-88°.

Elix, J.A. *et al, J. Am. Chem. Soc.*, 1987, **92**, 973 (*synth, pmr,
    ir, uv*)

Benzo[1,2-*f*:4,5-*f'*]diquinoline   B-70030

5,12-Diazadibenz[a,h]*anthracene*
[73513-77-6]

C$_{20}$H$_{12}$N$_2$ M 280.328
Yellow needles (C$_6$H$_6$). Mp 320-321.5°.

Klemm, L.H. *et al, J. Heterocycl. Chem.*, 1971, **8**, 763 (*synth,
    uv, pmr, ms*)

1,3-Benzodiselenole   B-70031

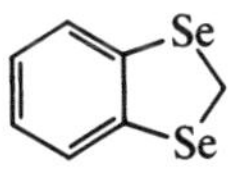

C$_7$H$_6$Se$_2$ M 248.044
Cryst. (Et$_2$O/pet. ether). Mp 30-32°.

Neidlein, R. *et al, Helv. Chim. Acta*, 1988, **71**, 1242 (*synth, ir,
    pmr, cmr, ms*)

1,3-Benzodiselenole-2-thione, 9CI   B-70032

[87143-01-9]

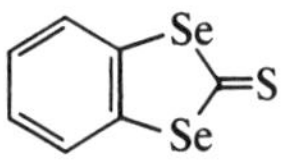

C$_7$H$_4$SSe$_2$ M 278.089
Yellow needles (CCl$_4$). Mp 147-148°.

Kistenmacher, T.J. *et al, J. Chem. Soc., Chem. Commun.*, 1983,
    294 (*synth*)
Lambert, C. *et al, Tetrahedron Lett.*, 1984, **25**, 834 (*synth*)

1,3,2-Benzodithiazol-1-ium(1+)   B-70033

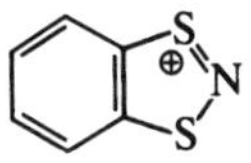

C$_6$H$_4$NS$_2$$^{\oplus}$ M 154.224 (ion)
Chloride: [91934-41-7].
    C$_6$H$_4$ClNS$_2$ M 189.677
    Yellow solid. Mp 218-220°.
Bromide: [109988-40-1].
    C$_6$H$_4$BrNS$_2$ M 234.128
    Dark-red cryst.

Morris, J.L. *et al, J. Chem. Soc., Perkin Trans. 1*, 1987, 211,
    217 (*synth, uv, ir, pmr*)

3*H*-1,2-Benzodithiole-3-selone, 9CI   B-70034

4,5-Benzo-3-selenoxo-1,2-dithiole
[51846-33-4]

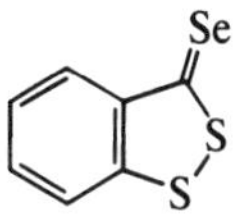

C$_7$H$_4$S$_2$Se   M 231.189
Mp 83-84°.

Tanagaki, S. *et al, Heterocycles*, 1974, 45 (*synth*)

Benzohydroxamic acid   B-70035

N-Hydroxybenzamide, 9CI

[495-18-1]

PhCONHOH

C₇H₇NO₂   M 137.138
Complexing agent. Cryst. (EtOAc). Mp 124°. pK_a 8.81
(H₂O).
▷DF9650000.

Org. Synth., Coll. Vol., 2, 67.
Trickes, G. *et al, Z. Naturforsch., A,* 1977, **32**, 956 (*cmr*)
Yamada, K. *et al, J. Am. Chem. Soc.,* 1981, **103**, 7003 (*synth*)
Ando, W. *et al, Synth. Commun.,* 1983, **13**, 1053 (*synth*)

1,2-Benzo[2.2]metaparacyclophan-9-ene   B-70036
12,15-Etheno-5,9-metheno-9H-benzocyclotridecene, 9CI
[82733-10-6]

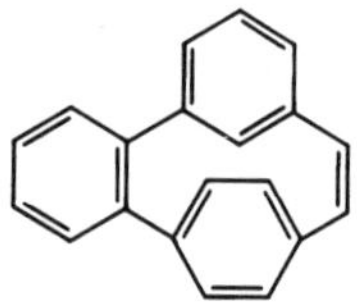

C₂₀H₁₄   M 254.331
Cryst. (hexane). Mp 90-91°.

Wong, T. *et al, Angew. Chem., Int. Ed. Engl.,* 1988, **27**, 705
(*synth, pmr*)

1*H*-Benzo[*de*][1,6]naphthyridine   B-70037
Didemethoxyaaptamine
[36917-96-1]

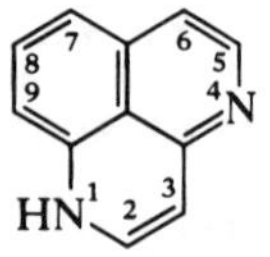

C₁₁H₈N₂   M 168.198
Yellow solid.
B,HCl: [105400-80-4]. Bright-yellow solid. Mp >250°.
Pelletier, J.C. *et al, J. Org. Chem.,* 1987, **42**, 616 (*synth, pmr, ir*)

1,2-Benzo[2.2]paracyclophane   B-70038
*9,10-Dihydro-5,8:11,14-diethenobenzocyclododecene,
9CI. [2](4,4'')-1,1':2',1''-Terphenylophane*
[64586-13-6]

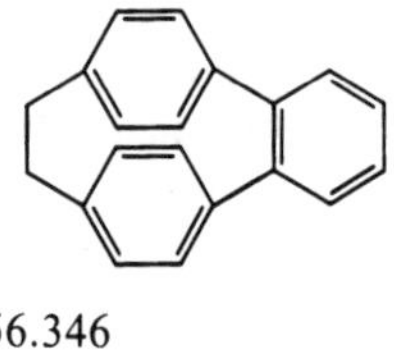

C₂₀H₁₆   M 256.346
Needles (heptane). Mp 217-219°.

Jacobsen, N. *et al, Angew. Chem., Int. Ed. Engl.,* 1978, **17**, 46
(*synth, pmr*)
Chan, C.W. *et al, J. Am. Chem. Soc.,* 1988, **110**, 462 (*synth, uv,
pmr*)

Benzo[*b*]phenanthro[1,2-*d*]thiophene   B-70039
[198-11-8]

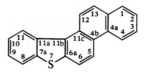

C₂₀H₁₂S   M 284.375
Cryst. (hexane). Mp 168.5° (162-164°).

Davies, W. *et al, J. Chem. Soc.,* 1957, 459 (*synth*)
Pratap, R. *et al, J. Heterocycl. Chem.,* 1982, **19**, 219 (*synth,
pmr, uv*)

Benzo[*b*]phenanthro[4,3-*d*]thiophene   B-70040
[192-88-1]

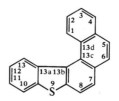

C₂₀H₁₂S   M 284.375
Cryst. Mp 133°.

Pratap, R. *et al, J. Heterocycl. Chem.,* 1982, **19**, 219 (*synth,
pmr, uv*)
Tedjamulia, M.L. *et al, J. Heterocycl. Chem.,* 1983, **20**, 861
(*synth*)

Benzophenone-2,2'-dicarboxylic acid   B-70041
Updated Entry replacing B-00449
*2,2'-Carbonylbisbenzoic acid, 9CI. 2,2'-Carbonyldiben-
zoic acid, 8CI. Diphenylmethanone-2,2'-dicarboxylic acid*
[573-32-0]

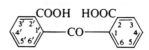

C₁₅H₁₀O₅   M 270.241
Forms anhydride at 148-150° before melting.
Di-Me ester: [65594-65-2].
    C₁₇H₁₄O₅   M 298.295
    Leaflets. Mp 85-86°.
Di-Et ester:
    C₁₉H₁₈O₅   M 326.348
    Prisms. Mp 73-74°.
Anhydride:
    C₁₅H₈O₄   M 252.226
    Leaflets. Mp 212° (203-205°). Sublimes.
Imide:
    C₁₅H₉NO₃   M 251.241
    Leaflets. Mp 251-252°.

v. Braun, J. *et al, Justus Liebigs Ann. Chem.,* 1929, **468**, 277
(*synth*)
Finkelstein, H., *Chem. Ber.,* 1959, no. 5, xxxvii (*derivs*)
Hellwinkel, D. *et al, Chem. Ber.,* 1987, **120**, 1151 (*synth, pmr*)

1*H*-2-Benzopyran-4(3*H*)-one, 9CI   B-70042
4-Isochromanone
[20924-56-5]

$C_9H_8O_2$ M 148.161
Needles (Et_2O/pet. ether). Mp 53°. Bp_1 90°. An earlier ref. to this compd. was incorrect.

2,4-Dinitrophenylhydrazone: Orange cryst. (EtOAc). Mp 250° dec.

Thibault, J., *Ann. Chim.* (*Paris*), 1971, **6**, 381 (*synth, ir, uv, pmr*)
Nilsen, B.P. *et al*, *Acta Chem. Scand., Ser. B*, 1976, **30**, 619 (*synth, pmr, uv*)

Benzo[*a*]quinolizinium(1+), 9CI   B-70043
Updated Entry replacing B-00513
[231-02-7]

$C_{13}H_{10}N^\oplus$ M 180.229 (ion)
Bromide: [31806-66-3].
   $C_{13}H_{10}BrN$ M 260.133
   Cryst. (EtOH/Me_2CO). Mp 259°.
Perchlorate: [4146-32-1].
   $C_{13}H_{10}ClNO_4$ M 279.679
   Yellow needles (H_2O or EtOH). Mp 197°.
Picrate:
   $C_{19}H_{12}N_4O_7$ M 408.326
   Yellow needles (EtOH). Mp 178°.

Glover, E.E. *et al*, *J. Chem. Soc.*, 1958, 3021.
Mörler, D. *et al*, *Justus Liebigs Ann. Chem.*, 1971, **744**, 65.
Mariano, P.S. *et al*, *J. Org. Chem.*, 1977, **42**, 1122.
Arai, S. *et al*, *J. Chem. Soc., Perkin Trans. 1*, 1987, 481 (*synth, ir, pmr*)

1,4-Benzoquinone, 8CI   B-70044
Updated Entry replacing B-00515
2,5-Cyclohexadiene-1,4-dione, 9CI. Quinone. p-Benzoquinone
[106-51-4]

$C_6H_4O_2$ M 108.096
Component of arthropod defensive secretions; isol. from the grasshopper *Romalea microptera*. Dienophile, H acceptor for Oppenauer oxidns., mild dehydrogenating agent. Yellow cryst. with penetrating chlorine-like odour (pet. ether or H_2O). V. spar. sol. H_2O. d_4^{20} 1.32. Mp 117°. Sublimes. Steam-volatile.
▷Irritant, causes dermatitis and conjunctivitis. Highly toxic. DK2625000.

Monoxime: see 4-Nitrosophenol, N-01413
Dioxime: [105-11-3].
   $C_6H_6N_2O_2$ M 138.126

Pale-yellow cryst. Mp 240° dec.
▷DK4900000.
Monosemicarbazone: Yellow cryst. Mp 178° dec.
Disemicarbazone: Red cryst. Mp 241-243° dec.
1,4-Benzenediol complex (1:1): [106-34-3].
   Quinhydrone. Red-brown needles. Mp 171°.
▷VA4550000.
Mono(ethylene ketal): [35357-34-7]. *7,10-Dioxaspiro[5.4]deca-2,5-dien-4-one*.
   $C_8H_8O_3$ M 152.149
   Needles (C_6H_6/cyclohexane). Mp 52-53°.
Bis(ethylene ketal): [35357-33-6].
   $C_{10}H_{12}O_4$ M 196.202
   Cryst. (C_6H_6). Mp 235-236°.

Evans, T.W. *et al*, *J. Am. Chem. Soc.*, 1930, **52**, 3204 (*complex*)
Org. Synth., Coll. Vol., **1**, 482 (*synth*)
Org. Synth., Coll. Vol., **2**, 553 (*synth*)
Flaig, W. *et al*, *Justus Liebigs Ann. Chem.*, 1958, **618**, 117 (*uv*)
Norris, R.K. *et al*, *Aust. J. Chem*, 1966, **19**, 617 (*pmr*)
Kikot, B.S. *et al*, *J. Gen. Chem. USSR* (*Engl. Transl.*), 1968, **38**, 883 (*ir*)
Heiss, J. *et al*, *Org. Mass. Spectrom.*, 1969, **2**, 1325 (*ms*)
Eisner, T. *et al*, *Science*, 1971, **172**, 277 (*isol*)
Berger, S. *et al*, *Tetrahedron*, 1972, **28**, 3123 (*cmr*)
Heller, J.E. *et al*, *Helv. Chim. Acta*, 1973, **56**, 272 (*deriv, synth, ir, pmr*)
v. Bolhuis, F. *et al*, *Acta Crystallogr., Sect. B*, 1978, **34**, 1015 (*cryst struct*)
Caparelli, M.P. *et al*, *J. Org. Chem.*, 1987, **52**, 5360 (*synth*)
Fieser, M. *et al*, *Reagents for Organic Synthesis*, Wiley, 1967-84, **1**, 49.
Sax, N.I., *Dangerous Properties of Industrial Materials*, 5th Ed., Van Nostrand-Reinhold, 1979, 952.
Hazards in the Chemical Laboratory, (Bretherick, L., Ed.), 3rd Ed., Royal Society of Chemistry, London, 1981, 1193.

1,2-Benzoquinone-3-carboxylic acid   B-70045
5,6-Dioxo-1,3-cyclohexadiene-1-carboxylic acid, 9CI

$C_7H_4O_4$ M 152.106
Me ester: [93081-07-3].
   $C_8H_6O_4$ M 166.133
   Formed *in situ* by the oxidn. of methyl 2,3-dihydroxybenzoate with Ag_2O. Intermed. for cycloaddition reactions. Unstable.

Forte, M. *et al*, *Gazz. Chim. Ital.*, 1985, **115**, 41 (*synth, use*)

1,2-Benzoquinone-4-carboxylic acid   B-70046
3,4-Dioxo-1,5-cyclohexadiene-1-carboxylic acid, 9CI
[75435-17-5]
$C_7H_4O_4$ M 152.106
Me ester: [93127-02-7].
   $C_8H_6O_4$ M 166.133
   Formed *in situ* by the oxidn. of methyl 3,4-dihydroxybenzoate with Ag_2O. Intermed. for cycloaddition reactions. Unstable.

Al-Hamdany, R. *et al*, *J. Iraqi Chem. Soc.*, 1982, **6-7**, 53 (*synth, use*)
Forte, M. *et al*, *Gazz. Chim. Ital.*, 1985, **115**, 41 (*synth, use*)

1,4-Benzoquinone-2,3-dicarboxylic acid   B-70047
3,6-Dioxo-1,4-cyclohexadiene-1,2-dicarboxylic acid, 9CI

$C_8H_4O_6$ M 196.116
Di-Me ester: [77220-15-6].
    $C_{10}H_8O_6$ M 224.170
    Orange needles (C_6H_6/pet. ether). Mp 155.5-157°.

Ansell, M.F. *et al, J. Chem. Soc.*, 1963, 3028 ,(synth)
Brownbridge, P. *et al, Tetrahedron Lett.*, 1980, **21**, 3423 (*synth*)
Padias, A.B. *et al, J. Org. Chem.*, 1985, **50**, 5417 (*ir*)

1,4-Benzoquinone-2,5-dicarboxylic acid   B-70048
3,6-Dioxo-1,4-cyclohexadiene-1,4-dicarboxylic acid, 9CI
[81566-86-1]
$C_8H_4O_6$ M 196.116
Di-Me ester: [81566-86-1].
    $C_{10}H_8O_6$ M 224.170
    Orange cryst. ($CHCl_2/Et_2O$). Mp 130°.

Ansell, M.F. *et al, J. Chem. Soc.*, 1961, 2908 (*synth*)
Padias, A.B. *et al, J. Org. Chem.*, 1985, **50**, 5417 (*synth, uv, ir, pmr*)

1,4-Benzoquinone-2,6-dicarboxylic acid   B-70049
2,5-Dioxo-3,6-cyclohexadiene-1,3-dicarboxylic acid, 9CI
$C_8H_4O_6$ M 196.116
Di-Et ester: [110935-41-6].
    $C_{12}H_{12}O_6$ M 252.223
    Red oil.

Ebel, K. *et al, Chem. Ber.*, 1988, **121**, 323 (*synth, ir, pmr*)

Benzoselenazole   B-70050
Updated Entry replacing B-20057
[273-91-6]

C_7H_5NSe M 182.083
Skin antimycotic. Various derivs. used as photosensi-
    tisers. Low-melting solid/liq. with quinoline-like odour.
    Mp 31-32°. Bp_{20} 125°.
B,HBr: Mp 132°.
Picrate: Small powdery cryst. Mp 180° (173°).

Ochiai, E. *et al, J. Pharm. Soc. Jpn.*, 1947, **67**, 138; *CA*, **45**,
    9539c (*synth*)
Develotte, J., *Ann. Chim.* (*Paris*), 1950, **5**, 215 (*synth*)
Croisy, A. *et al, Org. Mass Spectrom.*, 1972, **6**, 1321 (*ms*)
Bryce, M.R. *et al, Synth. Commun.*, 1988, **18**, 181 (*synth*)

1H-2-Benzoselenin-4(3H)-one, 9CI   B-70051
2-Selenochroman-4-one
[10352-22-4]

C_9H_8OSe M 211.122
Cryst. (C_6H_6/hexane). Mp 67-68°.
2,4-Dinitrophenylhydrazone: Mp 212°.

Renson, M. *et al, Bull. Soc. Chim. Belg.*, 1966, **75**, 456 (*synth*)
Hori, M. *et al, J. Org. Chem.*, 1987, **52**, 1397 (*synth*)

[1]Benzoselenopheno[2,3-b][1]-benzothiophene   B-70052
[206-76-8]

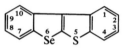

$C_{14}H_8SSe$ M 287.237
Platelets (EtOH). Mp 161°.
Picrate: [34673-69-3]. Red-brown platelets (EtOH). Mp
    142°.
Trinitrofluorenone complex: [34673-70-6]. Brown plate-
    lets (EtOH). Mp 213° dec.

Kirsch, G. *et al, C.R. Hebd. Seances Acad. Sci.*, Ser. C, 1971,
    273, 902.

1,4-Benzothiazine   B-70053

C_8H_7NS M 149.210
Parent compd. unknown. Nucleus present in compds. re-
    sponsible for formation of mammalian phaeomelanins
    by oxidative polymerisation. Readily forms trimers and
    tetramers in soln.

Chioccara, F. *et al, J. Heterocycl. Chem.*, 1987, **24**, 1741.

2H-1,4-Benzothiazine-2,3(4H)-dione, 9CI   B-70054
[73042-64-5]

$C_8H_5NO_2S$ M 179.193
Cryst. (AcOH). Mp 250° dec.

Zahn, K., *Ber.*, 1923, **56**, 578.
Dora, E.K. *et al, J. Indian Chem. Soc.*, 1979, **56**, 620 (*synth*)
Kollenz, G. *et al, Monatsh. Chem.*, 1984, **115**, 623 (*synth*)

2*H*-1,4-Benzothiazine-3(4*H*)-thione   B-70055

2H-1,4-Benzothiazine-2-thiol. 2-Mercapto-1,4-benzothiazine

[22191-30-6]

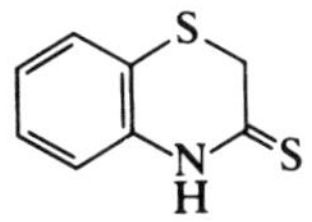

C$_8$H$_7$NS$_2$ M 181.270
Yellowish-green cryst. (EtOH). Mp 126-128°.
1,1-Dioxide: [22191-26-0].
  C$_8$H$_7$NO$_2$S$_2$ M 213.269
  Cryst. (Me$_2$CO/pet. ether). Mp 184-186°.
4-Benzyl: [22191-31-7].
  C$_{15}$H$_{13}$NS$_2$ M 271.394
  Cryst. (Et$_2$O/hexane). Mp 92-94°.

Kiprianov, A.I. *et al*, *J. Gen. Chem. USSR (Engl. Transl.)*, 1962, **32**, 3870 (*synth*)
Prasad, R.N., *J. Med. Chem.*, 1969, **12**, 290 (*synth*)
Chorvat, R.J. *et al*, *J. Med. Chem.*, 1983, **26**, 845 (*deriv, synth*)

3-(2-Benzothiazolyl)-2*H*-1-benzopyran, 9CI   B-70056

3-(2-Benzothiazolyl)coumarin

[1032-98-0]

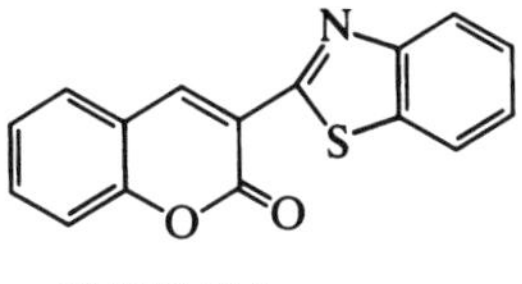

C$_{16}$H$_9$NO$_2$S   M 279.313
Substd. derivs. are patented as photographic sensitisers, fluorescent dyes, dye intermeds., etc. Cryst. (EtOAc). Mp 217-218°.

Dryanska, V., *Synth. Commun.*, 1987, **17**, 203 (*synth, bibl*)

N-2-Benzothiazolyl-*N*,*N*′,*N*″-triphenyl-guanidine, 9CI   B-70057

2N-Phenyl-N-[[(phenylamino)(phenylimino)methyl]-amino]benzothiazole. Hugershoff's base

[76402-03-4]

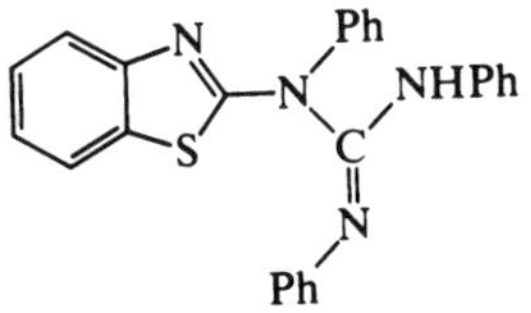

C$_{26}$H$_{20}$N$_4$S   M 420.531
(*Z*)-*form* [96816-28-3]
  Oxidn. prod. of 1,3-diphenylthiourea with H$_2$O$_2$ or Br$_2$/H$_2$O. Cryst. (EtOH). Mp 135-136°.

Hugershoff, A., *Ber.*, 1903, **36**, 3121 (*synth*)
Sureshi, K.S., *J. Indian Chem. Soc.*, 1960, **37**, 581 (*uv, ir*)
Kandror, I.I. *et al*, *Khim. Geterotsikl. Soedin.*, 1985, 206; *Chem. Heterocycles* Compd., 169 (*cryst struct*)

6*H*-[1]Benzothieno[2,3-*b*]indole   B-70058

[111550-86-8]

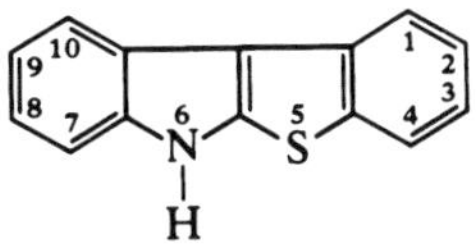

C$_{14}$H$_9$NS   M 223.292

Cryst. (MeOH). Mp 149-150°.

Levy, J. *et al*, *Bull. Soc. Chim. Fr.*, 1987, 193 (*synth, uv, ir, pmr, cmr*)

10*H*-[1]Benzothieno[3,2-*b*]indole, 9CI   B-70059

Updated Entry replacing B-00568

[248-67-9]

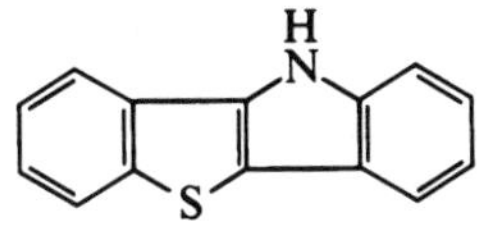

C$_{14}$H$_9$NS   M 223.292
Small prisms (MeOH). Mp 250-252°.
N-Me: [31486-46-1].
  C$_{15}$H$_{11}$NS   M 237.319
  Prisms Mp 222-224°.
N-Et: [31486-47-2].
  C$_{16}$H$_{13}$NS   M 251.345
  Cryst. (EtOH). Mp 118-120°.

Werner, L.H. *et al*, *J. Am. Chem. Soc.*, 1957, **79**, 1675.
Chippendale, K.E. *et al*, *J. Chem. Soc., Perkin Trans. 1*, 1972, 2023.
McKinnon, D.M. *et al*, *Can. J. Chem.*, 1988, **66**, 1405 (*synth*)

[1]Benzothieno[2,3-*c*][1,5]naphthyridine   B-70060

[65682-45-3]

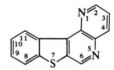

C$_{14}$H$_8$N$_2$S   M 236.291
Powder. Mp 181-183°.

Kudo, H. *et al*, *J. Heterocycl. Chem.*, 1987, **24**, 1009 (*synth, pmr*)

[1]Benzothieno[2,3-*c*][1,7]naphthyridine, 9CI   B-70061

[65682-47-5]

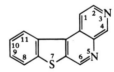

C$_{14}$H$_8$N$_2$S   M 236.291
Yellow prisms (C$_6$H$_6$). Mp 170-171°.

Kudo, H. *et al*, *J. Heterocycl. Chem.*, 1987, **24**, 1009 (*synth, pmr*)

[1]Benzothieno[2,3-*c*]quinoline   B-70062

[57289-92-6]

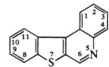

C$_{15}$H$_9$NS   M 235.303
Irregular plates (hexane). Mp 118-119°.

Zektzer, A.S. *et al*, *Magn. Reson. Chem.*, 1986, **24**, 1083 (*pmr, cmr*)

McKenney, J.D. *et al, J. Heterocycl. Chem.*, 1987, **24**, 1525 (*synth, ir, ms*)

[1]Benzothieno[3,2-*b*]quinoline   B-70063
[243-47-0]

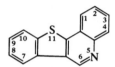

C₁₅H₉NS   M 235.303
Needles (EtOH). Mp 171.5-172.5°.

Nölting, E. *et al, Ber.*, 1910, **43**, 3512; 1911, **44**, 2585.
Klemm, L.H. *et al, J. Heterocycl. Chem.*, 1971, **8**, 383 (*synth, pmr*)
Bushkov, A.Ya. *et al, Khim. Geterotsikl Soedin.*, 1985, 273; *Chem. Heterocycles* Compd. (*Engl. transl.*), 232 (*synth, pmr*)

[1]Benzothieno[3,2-*c*]quinoline, 9CI   B-70064
[33193-41-8]

C₁₅H₉NS   M 235.303
Cryst. Mp 141°. Subl. at 135°/0.1 mm.

Neidlein, R. *et al, Synthesis*, 1977, 65.

Benzothiet-2-one   B-70065
7-Thiabicyclo[4.2.0]octa-1,3,5-trien-8-one, 9CI. 2-Thiobenzpropiolactone
[21083-36-3]

C₇H₄OS   M 136.168
Pale-yellow solid collected at −196°. Rapidly polym. >−20°.

Chapman, O.L. *et al, J. Am. Chem. Soc.*, 1970, **92**, 7001 (*synth, ir*)
Wentrup, C. *et al, J. Org. Chem.*, 1987, **52**, 3838 (*synth, ir, pmr, cmr, bibl*)

1-Benzothiopyrylium(1+), 9CI   B-70066
1-Thianaphthalenium. 1-Thionianaphthalene
[254-38-6]

C₉H₇S⊕   M 147.214 (ion)
Perchlorate: [3220-72-2].
    C₉H₇ClO₄S   M 246.665
    Greenish-yellow needles (AcOH). Mp 219-220° dec.

Bonthrone, W. *et al, Chem. Ind.* (*London*), 1960, 1192 (*synth*)
Lüttringhaus, A. *et al, Chem. Ber.*, 1960, **93**, 1525 (*synth, uv*)

2-Benzothiopyrylium(1+), 9CI   B-70067
2-Thianaphthalenium. 2-Thionianaphthalene

C₉H₇S⊕   M 147.214 (ion)
Perchlorate: [7432-88-4].
    C₉H₇ClO₄S   M 246.665
    Green needles (AcOH/Ac₂O). Mp 189-190° dec.

Lüttringhaus, A. *et al, Chem. Ber.*, 1960, **93**, 1525 (*synth, uv*)
Price, C.C. *et al, J. Am. Chem. Soc.*, 1963, **85**, 2278 (*synth, uv*)

1*H*-Benzotriazole   B-70068
Updated Entry replacing B-00589
Aziminobenzene (*obsol.*). *Azimidobenzene* (*obsol.*)

C₆H₅N₃   M 119.126
Needles (C₆H₆). Mp 100°.
▷Mod. toxic. Can detonate during vac. dist.
N-*Ac:* [18773-93-8].
    C₈H₇N₃O   M 161.163
    Mp 51°.
N-*Benzoyl:* [4231-62-3].
    C₁₃H₉N₃O   M 223.234
    Mp 112°.
N-*Hydroxy:* see 1-Hydroxybenzotriazole, H-50156
1-Me: see 1-Methyl-1*H*-benzotriazole, M-20089
2-Me: see 2-Methyl-2*H*-benzotriazole, M-20092
1-Et:
    C₈H₉N₃   M 147.179
    Bp₁₀ 150°.
1-Benzyl:
    C₁₃H₁₁N₃   M 209.250
    Plates (MeOH). Mp 114-115°.

Boyer, J.H., *Heterocycles* Compd., (*Elderfield, R.C., Ed.*), 1961, **7**, 384 (*rev*)
Carbon, J.A., *J. Org. Chem.*, 1962, **27**, 185 (*synth*)
Rees, C.W. *et al, J. Chem. Soc., Chem. Commun.*, 1971, 532.
Katritzky, A.R. *et al, J. Chem. Soc., Perkin Trans. 1*, 1987, 781 (*derivs*)
Bretherick, L., *Handbook of Reactive Chemical Hazards*, 2nd Ed., Butterworths, London and Boston, 1979, 576.
Sax, N.I., *Dangerous Properties of Industrial Materials*, 5th Ed., Van Nostrand-Reinhold, 1979, 408.
Hazards in the Chemical Laboratory, (Bretherick, L., Ed.), 3rd Ed., Royal Society of Chemistry, London, 1981, 193.

Benzo[1,2-*d*:3,4-*d'*:5,6-*d''*]tris[1,3]dithiole, 9CI   B-70069
Tris(methylenedithio)benzene
[110431-55-5]

C₉H₆S₆   M 306.506
Mp 295°.

Lapouyade, R. *et al*, *J. Chem. Soc., Chem. Commun.*, 1987, 223 (*synth, pmr*)

1,3-Benzoxathiole-2-selone   B-70070

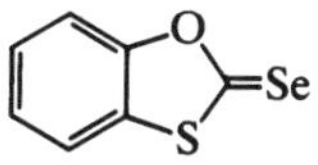

C_7H_4OSSe M 215.128
Yellow prisms (EtOH). Mp 130-131°.

Copeland, C.M. *et al*, *Aust. J. Chem.*, 1988, **41**, 549 (*synth, uv, pmr, cmr*)

Benzoylcyclohexane   B-70071

Updated Entry replacing B-00651
Cyclohexylphenylmethanone, 9CI. Cyclohexyl phenyl ketone. Hexahydrobenzophenone
[712-50-5]

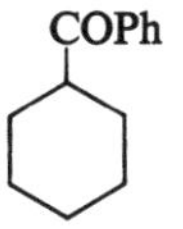

$C_{13}H_{16}O$ M 188.269
Needles (pet. ether). Mp 59-60°. Bp$_{18}$ 164-165°.
▷OB1760000.

Semicarbazone: [839-97-4]. Mp 168-169°, 175°.
(E)-*Oxime:* [23554-71-4].
   $C_{13}H_{17}NO$ M 203.283
   Prisms (EtOH). Mp 111°.
(Z)-*Oxime:* [23517-35-3]. Needles (EtOH). Mp 158°.
2,4-Dinitrophenylhydrazone: [1248-51-7]. Mp 192°, 197-198°, 201-202°.

Niichiro, T. *et al*, *Bull. Chem. Soc. Jpn.*, 1962, **35**, 1779 (*oximes*)
Rouzand, J. *et al*, *Bull. Soc. Chim. Fr.*, 1964, 2908 (*derivs*)
Aldrich Library of NMR Spectra, 1974, **6**, 9C (*pmr*)
Aldrich Library of IR Spectra, 1975, 2nd Ed., 748B (*ir*)
Huenig, S. *et al*, *Synthesis*, 1975, 180 (*synth*)
Zhang, P. *et al*, *Synth. Commun.*, 1986, **16**, 957 (*synth*)
Weiberth, F.J. *et al*, *J. Org. Chem.*, 1987, **52**, 3901 (*synth*)

1-Benzoylcyclopentene   B-70072

1-Cyclopenten-1-ylphenylmethanone, 9CI. 1-Cyclopentenyl phenyl ketone
[21573-70-6]

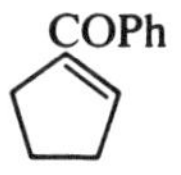

COPh

$C_{12}H_{12}O$ M 172.226
Oil. Bp$_1$ 117-118°. n_D^{20} 1.5719.

Zhang, P. *et al*, *Synth. Commun.*, 1986, **16**, 957 (*synth*)

2-Benzoyl-1,2-dihydro-1-isoquinolinecar-   B-70073
bonitrile, 9CI

1-Cyano-2-benzoyl-1,2-dihydroisoquinoline. Isoquinoline Reissert compound
[844-25-7]

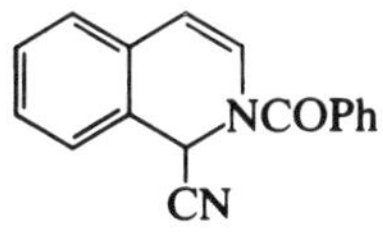

NCOPh
CN

$C_{17}H_{12}N_2O$ M 260.295
Prod. of reaction of isoquinoline with PhCOCl/KCN.
   Intermed. in synth. of 1-substd. isoquinolines. Cryst.
   (EtOH). Mp 124-126°.

Reissert, A., *Ber.*, 1905, **38**, 3427.
Padbury, J.J. *et al*, *J. Am. Chem. Soc.*, 1945, **67**, 1268.
McEwen, W.E. *et al*, *Chem. Rev.*, 1955, **55**, 511 (*rev*)
Popp, F.D., *Adv. Heterocycles* Chem., 1968, **9**, 1; 1979, **24**, 18 (*rev*)

2-Benzoylimidazole   B-70074

1H-Imidazol-2-ylphenylmethanone, 9CI. Phenyl 2-imidazolyl ketone
[38353-02-5]

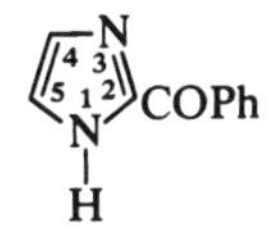

$C_{10}H_8N_2O$ M 172.186
Needles (C_6H_6). Mp 159-159.5°.

Iwasaki, S., *Helv. Chim. Acta*, 1976, **59**, 2738 (*synth, uv, ir, pmr, cmr*)

3-Benzoylisoxazole   B-70075

3-Isoxazolylphenylmethanone, 9CI
[38791-02-5]

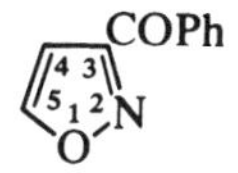

COPh

$C_{10}H_7NO_2$ M 173.171
Mp 31-32°.

p-*Nitrophenylhydrazone:* Yellow needles (EtOH). Mp 197°.
Semicarbazone: Mp 171°.

Quilico, A. *et al*, *Gazz. Chim. Ital.*, 1947, **77**, 586 (*synth*)

2-Benzoylnaphtho[1,8-bc]pyran   B-70076

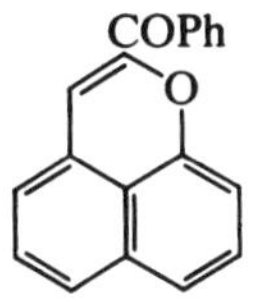

COPh

$C_{19}H_{12}O_2$ M 272.303
Fine yellow needles (EtOH). Mp 114°.

Buisson, J.-P. *et al*, *J. Heterocycl. Chem.*, 1988, **25**, 539 (*synth, pmr, uv*)

Benzoyl phenyl diselenide   B-70077

Benzenecarbo(diselenoperoxoic)acid phenyl ester, 9CI
[111931-75-0]

PhCO–Se–Se–Ph

$C_{13}H_{10}OSe_2$ M 340.141
Oil.

Ishihara, H. *et al*, *Synthesis*, 1987, 371 (*synth, ir, pmr*)

3-Benzoylpyrazole   B-70078

Phenyl-1H-pyrazol-3-ylmethanone, 9CI. Phenyl 3-pyra-zolyl ketone

[19854-92-3]

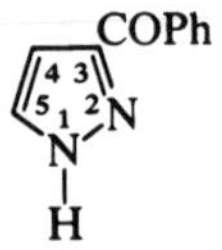

$C_{10}H_8N_2O$ M 172.186
Needles (C_6H_6). Mp 94-95°.

Doyle, M. *et al, J. Heterocycl. Chem.*, 1983, **20**, 943 (*synth, pmr*)
Schweizer, E.E. *et al, J. Org. Chem.*, 1973, **38**, 3069 (*synth*)

4-Benzoyl-1,2,3-triazole   B-70079

[14677-10-2]

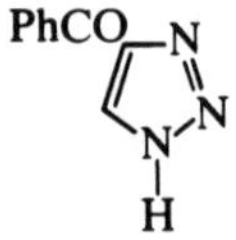

$C_9H_7N_3O$ M 173.174
Cryst. (C_6H_6 or AcOH). Mp 122-123° (118-120°).
Hydrazone:
  $C_9H_9N_5$ M 187.204
  Cryst. (EtOH aq.). Mp 158-160°.

1H-form

  1-Me:
  $C_{10}H_9N_3O$ M 187.201
  Cryst. (EtOH). Mp 150-152°.

2H-form

  2-Me:
  $C_{10}H_9N_3O$ M 187.201
  Cryst. (EtOH). Mp 70-72°.

3H-form

  3-Me:
  $C_{10}H_9N_3O$ M 187.201
  Cryst. (EtOH). Mp 101-103°.

Nesmeyanov, A.N. *et al, Dokl. Akad. Nauk SSSR, Ser. Sci. Khim.*, 1964, **158**, 408; *CA*, **61**, 14664h (*synth*)
Maiorana, S., *Ann. Chim. (Rome)*, 1966, **56**, 1531 (*synth*)
Bianchi, M. *et al, J. Heterocycl. Chem.*, 1988, **25**, 743 (*pmr, derivs*)

Benzvalenoquinoxaline   B-70080

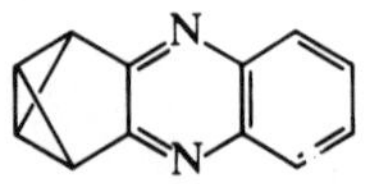

$C_{12}H_8N_2$ M 180.209
Valence isomer of Phenazine, P-00850 . Mp 122-125°.

Christl, M. *et al, Angew. Chem., Int. Ed. Engl.*, 1988, **27**, 1369 (*synth, pmr, cmr*)

Benzylidenecyclobutane   B-70081

(Cyclobutylidenemethyl)benzene, 9CI. Cyclobutylidene-phenylmethane, 8CI

[5244-75-7]

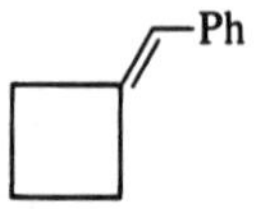

$C_{11}H_{12}$ M 144.216

Bp$_{15}$ 114°, Bp$_{0.6}$ 62-63°.

Bestmann, H.J. *et al, Angew. Chem., Int. Ed. Engl.*, 1967, **6**, 81 (*synth*)
Schweizer, E.E. *et al, J. Org. Chem.*, 1968, **33**, 3082 (*synth*)
Bestmann, H.J. *et al, Chem. Ber.*, 1969, **102**, 1802 (*synth*)
Norden, W. *et al, Chem. Ber.*, 1983, **116**, 3097 (*synth*)
Crandall, J.R. *et al, J. Org. Chem.*, 1984, **49**, 4244 (*synth*)

Benzylidenecyclohexane   B-70082

(Cyclohexylidenemethyl)benzene, 9CI. Cyclohexylidene-phenylmethane, 8CI

[1608-31-7]

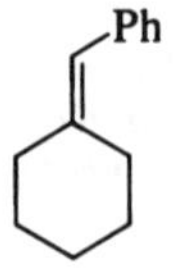

$C_{13}H_{16}$ M 172.269
Liq. Bp$_{12}$ 124.5-126.5°, Bp$_{0.04}$ 60°.

Belletire, J.L. *et al, Tetrahedron Lett.*, 1983, **24**, 1475 (*synth*)
Mitchell, T.N. *et al, J. Organomet. Chem.*, 1983, **252**, 47 (*synth*)
Norden, W. *et al, Chem. Ber.*, 1983, **116**, 3097 (*synth*)
Baldwin, J.E. *et al, J. Chem. Soc., Chem. Commun.*, 1984, 22 (*synth*)
Crandall, J.K. *et al, J. Org. Chem.*, 1984, **49**, 4244 (*synth*)

Benzylidenecyclopentane   B-70083

(Cyclopentylidenemethyl)benzene, 9CI. Cyclopentyliden-ephenylmethane, 8CI

[4410-77-9]

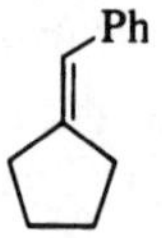

$C_{12}H_{14}$ M 158.243
Liq. Bp$_5$ 90°.

Hara, S. *et al, Tetrahedron Lett.*, 1975, 1545 (*synth*)
Norden, W. *et al, Chem. Ber.*, 1983, **116**, 3097 (*synth*)
Crandall, J.K. *et al, J. Org. Chem.*, 1984, **49**, 4244 (*synth*)
Aguero, A. *et al, J. Chem. Soc., Chem. Commun.*, 1986, 531 (*synth*)

1-Benzylidene-1H-cyclopropabenzene   B-70084

7-(Phenylmethylene)bicyclo[4.1.0]hepta-1,3,5-triene, 9CI. 1-(Phenylmethylene)-1H-cyclopropabenzene

[103322-21-0]

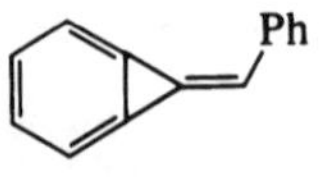

$C_{14}H_{10}$ M 178.233
Pale-yellow solid (pet. ether).

Halton, B. *et al, J. Am. Chem. Soc.*, 1986, **108**, 5949 (*synth, nmr, ir, uv*)

2-Benzyloxyacetaldehyde   B-70085

(Phenylmethoxy)acetaldehyde, 9CI

[60656-87-3]

$$PhCH_2OCH_2CHO$$

$C_9H_{10}O_2$ M 150.177

Yellow liq. Bp$_{1.2-1.5}$ 75-76°.

Garner, P. *et al*, *Synth. Commun.*, 1987, **17**, 189 (*synth, pmr, ms, bibl*)

C-Benzyloxycarbohydroxamic acid   B-70086

Benzyl hydroxycarbamate. N-
Benzyloxycarbonylhydroxylamine
[3426-71-9]

PhCH$_2$OCONHOH

C$_8$H$_9$NO$_3$ M 167.164
Plates (C$_6$H$_6$/pet. ether). Mp 71°.

Boyland, E. *et al*, *J. Chem. Soc.* (*C*), 1966, 354 (*synth*)
Frankel, M. *et al*, *J. Chem. Soc.* (*C*), 1969, 1746 (*synth*)
Briggs, M.T. *et al*, *J. Chem. Soc., Perkin Trans. 1*, 1979, 2138 (*synth*)

Benzyne   B-70087

Updated Entry replacing B-00921
1,3-Cyclohexadien-5-yne. 1,2-Dehydrobenzene
[462-80-6]

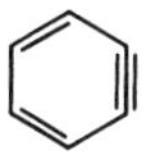

C$_6$H$_4$ M 76.098
V. reactive intermed. in the elimination and
fragmentation reactions of certain mono- and 1,2-
disubstituted benzenes. Reacts with dienes, e.g. furan,
anthracene and with nucleophiles.

Berry, R.S. *et al*, *J. Am. Chem. Soc.*, 1960, **82**, 5240; 1962, **84**,
3570; 1964, **86**, 2738; 1965, **87**, 4497 (*uv, ms*)
Hoffmann, R.W., *Dehydrobenzene and Cycloalkynes*, 1967,
Academic Press (*rev*)
Chapman, O.L. *et al*, *J. Am. Chem. Soc.*, 1973, **95**, 6134; 1975,
97, 6586 (*synth, ir*)
Laing, J.W. *et al*, *J. Am. Chem. Soc.*, 1976, **98**, 660 (*ir, struct*)
Levin, R.H., *React. Intermed.*, 1978, **1**, 1 (*rev*)
Dewar, M.J.S. *et al*, *J. Chem. Soc., Chem. Commun.*, 1985,
1243 (*pe*)
Brown, R.D. *et al*, *J. Am. Chem. Soc.*, 1986, **108**, 1296
(*microwave*)

Bharangin   B-70088

[72711-84-3]

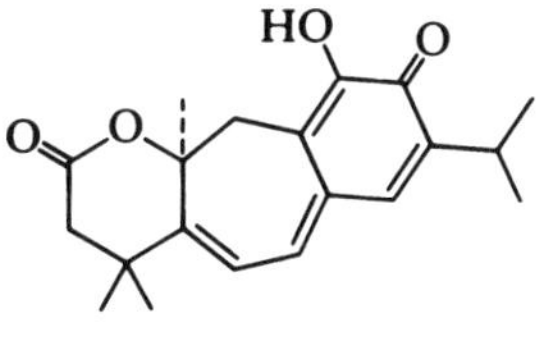

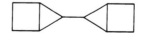

Absolute
configuration

C$_{20}$H$_{24}$O$_4$ M 328.407
Constit. of *Pygmacopremna herbaceae*. Yellow needles
(CCl$_4$). Mp 213-214°. [α]$_D^{26}$ +350.3° (c, 0.878 in
CHCl$_3$).

Sankaram, A.V.B. *et al*, *Tetrahedron Lett.*, 1988, **29**, 245
(*struct*)

Bharanginin   B-70089

7-Isopropyl-3,3,4-trimethylnaphtho[2,3-b]furan-
2,5,6(3H)-trione

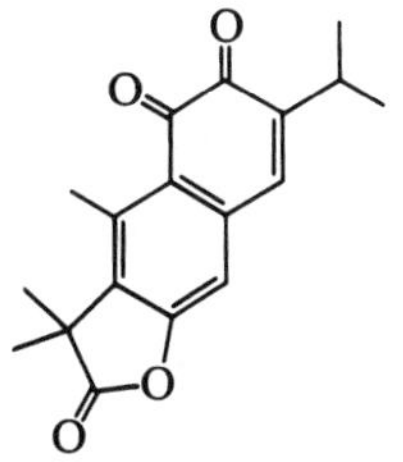

C$_{18}$H$_{18}$O$_4$ M 298.338
Constit. of *Pygmacopremna herbacea*. Orange cryst.
(pet. ether). Mp 171-173°.

Sankaram, A.V.B. *et al*, *Tetrahedron Lett.*, 1988, **29**, 4881
(*struct*)

9,9′-Biacridylidene   B-70090

9,20-Dihydro-9-(9(10H)-acridinylidene)acridine, 9CI

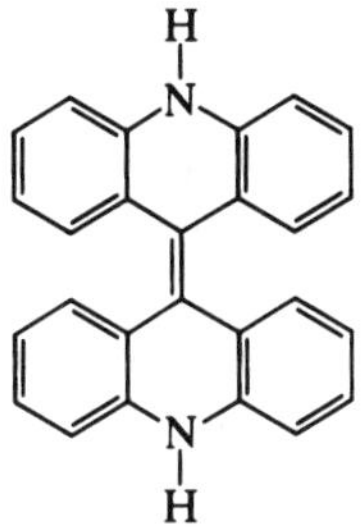

C$_{26}$H$_{18}$N$_2$ M 358.442
N,N′-*Di-Me*: [23663-77-6].
C$_{28}$H$_{22}$N$_2$ M 386.495
Mp >360°.

Surgi, M.R. *et al*, *Org. Prep. Proceed. Int.*, 1988, **20**, 295 (*synth,
ir, ms, pmr, bibl*)

Biadamantylidene epoxide   B-70091

Dispiro[tricyclo[3.3.1.1^{3,7}]decane-2,2′-oxirane-3,2″-
tricyclo[3.3.1.1^{3,7}]decane], 9CI. Adamantylideneadaman-
tane epoxide
[29186-07-0]

C$_{20}$H$_{28}$O   M 284.441
Mp 181.5-183°. V. unreactive. Unaffected by LiAlH$_4$,
butyllithium or PPh$_3$.

Wynberg, H. *et al*, *Tetrahedron Lett.*, 1970, 3613 (*synth*)
Jefford, C.W. *et al*, *Helv. Chim. Acta*, 1977, **60**, 2673.
Bartlett, P.D. *et al*, *J. Org. Chem.*, 1980, **45**, 3000 (*synth*)
Watson, W.H. *et al*, *Acta Crystallogr., Sect. C*, 1988, **44**, 1627
(*cryst struct*)

5,5′-Bibicyclo[2.1.0]pentane, 9CI   B-70092

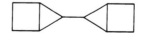

C$_{10}$H$_{14}$ M 134.221
Bp$_{14}$ 120°. Mixt. of isomers.

Adam, W. *et al*, *J. Am. Chem. Soc.*, 1987, **109**, 5250 (*synth, ir, ms, cmr, pmr*)

Bicoumol   B-70093

Updated Entry replacing B-60082
7,7'-Dihydroxy-[6,8'-bi-2H-benzopyran]-2,2'-dione.
7,7'-Dihydroxy-6,8'-bicoumarin
[15575-52-7]

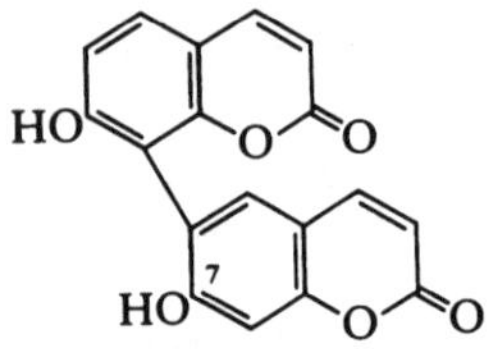

$C_{18}H_{10}O_6$ M 322.273
Struct. in 9CI and 8CI is incorrect. Isol. from ladino clover (*Trifolium repens*) and *Ruta* sp. Cryst. (EtOH/Me$_2$CO). Mp 293-294°.
Di-Ac: [18304-03-5]. Cryst. (Me$_2$CO aq.). Mp 228.5-229.5°.
7-Me ether: [89320-82-1]. **Bhubaneswin**.
  $C_{19}H_{12}O_6$ M 336.300
  Constit. of *Boenninghausenia albiflora*. Cryst. (Me$_2$CO/CHCl$_3$). Mp 320°.
Di-Me ether: [3153-73-9]. *7,7'-Dimethoxy-[6,8'-bi-2H-benzopyran]-2,2'-dione. 7,7'-Dimethoxy-6,8'-bicoumarin.* **Matsukaze lactone**.
  $C_{20}H_{14}O_6$ M 350.327
  Constit. of *B. albiflora* and *B. japonica*. Platelets (MeOH). Mp 266.5-267.5°.

Miyazaki, T. *et al*, *Chem. Pharm. Bull.*, 1964, **12**, 1236 (*isol*)
Spencer, R.R. *et al*, *J. Agric. Food Chem.*, 1967, **15**, 536 (*isol, struct*)
Kozawa, M. *et al*, *Chem. Pharm. Bull.*, 1974, **22**, 2746 (*isol*)
Gonzalez, A.G. *et al*, *An. Quim.*, 1977, **73**, 1015 (*isol*)
Basa, S.C. *et al*, *Heterocycles*, 1984, **22**, 333 (*isol, cmr, pmr*)
Basa, S.C., *Phytochemistry*, 1988, **27**, 1933 (*rev*)

Bicyclo[1.1.0]butane-1-carboxylic acid   B-70094

Updated Entry replacing B-01000
[30493-99-3]

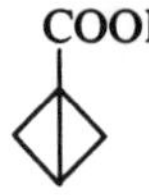

$C_5H_6O_2$ M 98.101
Cryst. (pentane). Mp ca. 51° (polymerises).
Et ester: [29820-55-1]. *1-Carbethoxybicyclo[1.1.0]-butane.*
  $C_7H_{10}O_2$ M 126.155
  Bp$_{15}$ 56-58°.
Nitrile: [16955-35-4]. *1-Cyanobicyclo[1.1.0]butane.*
  C_5H_5N M 79.101
  Bp$_4$ 38°.

Wiberg, K.B. *et al*, *J. Am. Chem. Soc.*, 1959, **81**, 5261 (*ester*)
Kirmse, W. *et al*, *Justus Liebigs Ann. Chem.*, 1970, **739**, 231 (*ester*)
Hall, H.K. *et al*, *J. Am. Chem. Soc.*, 1971, **93**, 110, 121, (*derivs, synth, ir, pmr*)

Bicyclobutylidene   B-70095

Updated Entry replacing B-01002
Cyclobutylidenecyclobutane, 9CI
[6708-14-1]

C_8H_{12} M 108.183
Liq. Bp$_{100}$ 85°.

Everett, J.W. *et al*, *J. Chem. Soc., Chem. Commun.*, 1972, 642 (*synth*)
Bee, L.K. *et al*, *J. Org. Chem.*, 1975, **40**, 2212 (*synth*)
Fitjer, L. *et al*, *Synthesis*, 1987, 299 (*synth*)

Bicyclo[8.8.2]eicosa-1(19),10(20),19-triene,   B-70096
9CI

[99605-65-9]

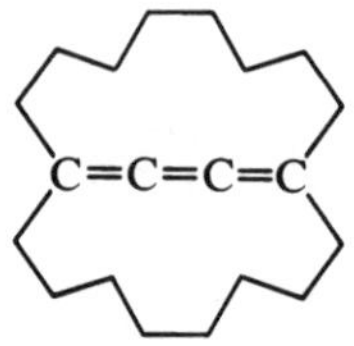

$C_{20}H_{32}$ M 272.473
First bicyclic cumulatriene. Semisolid. Mp 82-92°.

Macomber, R.S. *et al*, *J. Am. Chem. Soc.*, 1986, **108**, 343 (*synth, ir, nmr, uv*)

Bicyclo[3.2.0]hepta-2,6-diene   B-70097

[2422-86-8]

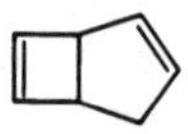

C_7H_8 M 92.140
Liq. Bp 90-91°.

Baldwin, J.E. *et al*, *J. Org. Chem.*, 1987, **52**, 4772 (*synth, pmr, ms, bibl*)

Bicyclo[2.2.1]heptan-2-one   B-70098

Updated Entry replacing B-01042
2-Norbornanone, 8CI. Norcamphor. 2-Oxonorbornane
[497-38-1]

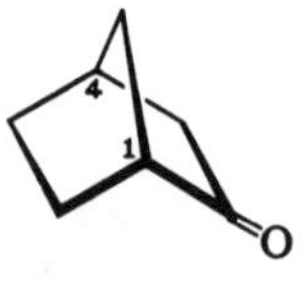

                     (1*R*,4*S*)-*form*
                     Absolute
                     configuration

$C_7H_{10}O$ M 110.155
▷RB7680000.
(**1R,4S**)-*form* [29583-35-5]
  (−)-*form*
  Cryst. Mp 93-95°. [α]$_D$ −31°.
(**1S,5R**)-*form* [2630-41-3]
  (+)-*form*
  Cryst. Mp 93-95°. [α]$_D$ +31°.
(**1RS,4SR**)-*form* [22270-13-9]
  (±)-*form*
  Waxy solid. Mp 97-98° (92.5-93.5°). Bp 172°.
  Oxime: [4576-48-1].
    $C_7H_{11}NO$ M 125.170
    Mp 46-48°. Bp$_{12}$ 114-116°.

2,4-Dinitrophenylhydrazone: [3281-03-6]. Orange needles (EtOH). Mp 132°.

Wildman, W.C. *et al, J. Org. Chem.*, 1952, **17**, 1641 (*synth*)
Zbinden, R. *et al, J. Am. Chem. Soc.*, 1960, **82**, 1215 (*ir*)
Hill, R.K. *et al, Tetrahedron*, 1965, **21**, 1501 (*abs config*)
Goto, T. *et al, Tetrahedron*, 1966, **22**, 2213 (*ms*)
McDonald, R.N. *et al, J. Am. Chem. Soc.*, 1970, **92**, 5664 (*synth*)
Lightner, D.A. *et al, J. Am. Chem. Soc.*, 1971, **93**, 2677 (*synth, cd*)
Conan, J.Y. *et al, Bull. Soc. Chim. Fr.*, 1974, 1405 (*synth, resoln*)
Marshall, J.L. *et al, J. Am. Chem. Soc.*, 1974, **96**, 6358 (*pmr*)
Org. Synth., Coll. Vol. **5**, 852 (*synth*)
Morris, D.G. *et al, J. Chem. Soc., Perkin Trans. 2*, 1976, 1579 (*cmr*)
Carlsen, P.H.J. *et al, Synth. Commun.*, 1987, **17**, 19 (*synth*)

Bicyclo[3.2.0]hept-2-ene-6,7-dione, 9CI   **B-70099**
[61149-88-0]

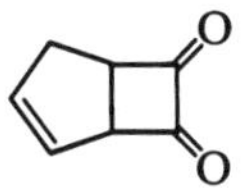

$C_7H_6O_2$ M 122.123

Rubin, M.B. *et al, J. Am. Chem. Soc.*, 1976, **98**, 5699 (*synth, pmr, ir, ms*)

Bicyclo[2.1.1]hexan-2-ol   **B-70100**
[5062-80-6]

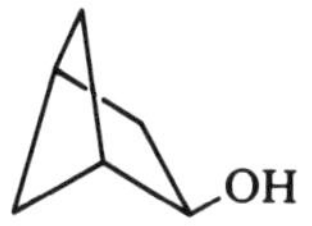

$C_6H_{10}O$ M 98.144

(−)-*form*
$[\alpha]_D^{23}$ −13.7° (c, 0.073 in pentane).
(±)-*form*
Mp 82-84° subl.

Bond, F.T. *et al, Tetrahedron Lett.*, 1965, 4685 (*synth*)
Kirmse, W. *et al, J. Am. Chem. Soc.*, 1986, **108**, 4912 (*synth, nmr*)

Bicyclo[12.4.1]nonadec-   **B-70101**
1,3,5,7,9,11,13,15,17-nonaene, 9CI
1,6-Methano[18]annulene
[110977-28-1]

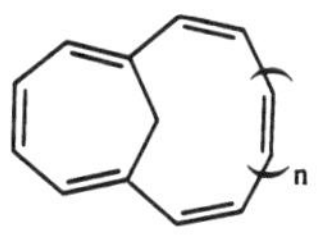

n = 4

$C_{19}H_{18}$ M 246.351
Compds. also prepd. where *n* = 5, 6 and 7. Mp 138°.
Compds. where *m* = 4 and *m* = 6 are diatropic; where *m* = 5 and 7, compds. are paratropic.

Yamamoto, K. *et al, J. Chem. Soc., Chem. Commun.*, 1987, 199 (*synth, pmr, uv*)

Bicyclo[3.3.1]nona-2,6-diene   **B-70102**
[13534-07-1]

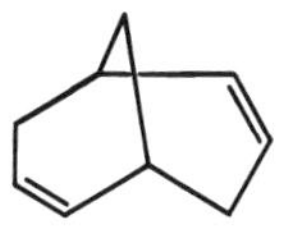

C_9H_{12} M 120.194
Liq. Bp_{110} 105-107°.

Henkel, J.G. *et al, J. Org. Chem.*, 1981, **46**, 3483 (*synth, cmr*)
Trinks, R. *et al, Chem. Ber.*, 1987, **120**, 1481 (*synth*)

Bicyclo[3.3.1]nonane-2,4-dione, 9CI   **B-70103**
[98442-42-3]

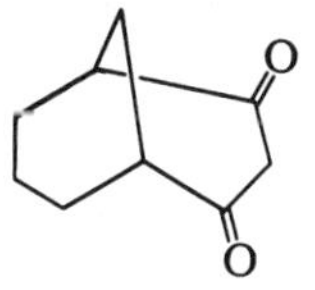

$C_9H_{12}O_2$ M 152.193
Cryst. (EtOAc). Mp 153-155° (136-138°). 64% enolised in $CDCl_3$ at r.t.

Inoye, Y. *et al, Bull. Chem. Soc. Jpn.*, 1987, **60**, 4369 (*synth, uv, ir, pmr*)
Yamazaki, T. *et al, Chem. Pharm. Bull.*, 1987, **35**, 3453 (*synth, pmr, ms*)

Bi(cyclononatetraenylidene)   **B-70104**
Nonafulvalene

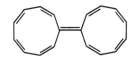

$C_{18}H_{16}$ M 232.324
Orange. Unstable, undergoes valence isomerisation at −50°.

Escher, A. *et al, Helv. Chim. Acta*, 1987, **70**, 1623 (*synth, props*)

Bicyclo[3.3.1]non-3-en-2-one, 9CI   **B-70105**
[10036-12-1]

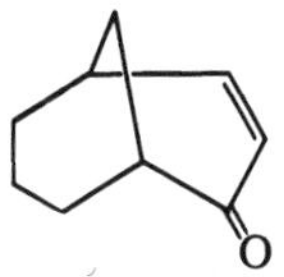

$C_9H_{12}O$ M 136.193
Cryst. (pentane). Mp 97-98°.

Schaefer, J.P. *et al, J. Org. Chem.*, 1967, **32**, 1372 (*synth*)
Inouye, Y. *et al, Bull. Chem. Soc. Jpn.*, 1987, **60**, 4369 (*synth*)

Bicyclo[4.2.1]non-2-en-9-one, 9CI   **B-70106**
[52089-56-2]

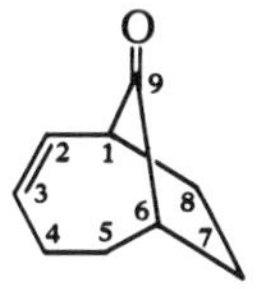

$C_9H_{12}O$ M 136.193
Mp 25-27°.

Stamm, E. *et al*, *Helv. Chim. Acta*, 1979, **62**, 2174 (*synth, uv, ir, pmr, cmr, ms*)

Bicyclo[4.2.1]non-3-en-9-one   B-70107

[40863-57-8]

$C_9H_{12}O$ M 136.193

Cryst. by subl. Mp 56-56.5° (49-50°).

Stamm, E. *et al*, *Helv. Chim. Acta*, 1979, **62**, 2174 (*synth, uv, pmr, cmr, ms*)
Föhlisch, B. *et al*, *Chem. Ber.*, 1987, **120**, 1951 (*synth, ms*)

Bicyclo[4.2.1]non-7-en-9-one, 9CI   B-70108

[42948-91-4]

$C_9H_{12}O$ M 136.193

Cryst. by subl. Mp 91-91.5°.

Stamm, E. *et al*, *Helv. Chim. Acta*, 1979, **62**, 2174 (*synth, uv, ir, pmr, cmr, ms*)

Bicyclo[2.2.2]octa-5,7-diene-2,3-dione, 9CI   B-70109

[32273-62-4]

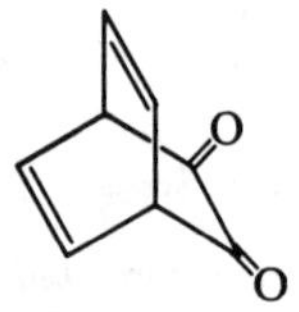

$C_8H_6O_2$ M 134.134

Golden-yellow cryst. (CCl₄ or by subl.). Mp 107-108° (100-102°).

Dioxime: [35558-00-0].
    $C_8H_8N_2O_2$ M 164.163
    Cryst. (dioxane/THF). Mp 225-227° dec.
Bisphenylhydrazone: Orange-brown cryst. (EtOH aq.). Mp 154-155°.

Scharf, H. *et al*, *Chem. Ber.*, 1972, **105**, 575 (*synth*)
McMahon, R.J. *et al*, *J. Am. Chem. Soc.*, 1987, **109**, 2456 (*synth, pmr, ir, ms*)

Bicyclo[4.2.0]octa-1,5-diene-3,4-dione, 9CI   B-70110

Benzocyclobutene-4,5-quinone. 4,5-Cyclobuta-1,2-benzoquinone

[4082-19-3]

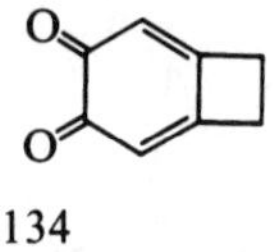

$C_8H_6O_2$ M 134.134

Bright-red leaflets (Et₂O). Mp 161° dec. (vigorous deflagration). Originally erroneously reported as the 4,5-dione. Stable at r.t.

▷Deflagrates at Mp

Lloyd, J.B.F. *et al*, *Tetrahedron*, 1965, **21**, 2281 (*synth, pmr, ir, uv*)

Bicyclo[5.1.0]octa-3,5-dien-2-one, 9CI   B-70111

Updated Entry replacing B-01114

Homotropone

[3818-97-1]

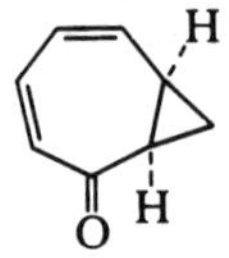

1R,7R-form

C_8H_8O M 120.151

(1R,7R)-form

$[\alpha]_D^{20}$ +48.4° (c, 1.076 in CHCl₃).

(1RS,7RS)-form [116460-77-6]

(±)-*form*

Pale-yellow oil. Bp₀.₂ 56-57°.

HSbCl₆ addn. compd.: Yellow cryst. Mp 89-90°. Dec. in moist air.
Semicarbazone: Yellow plates (EtOH aq.). Mp 146-147°.

Holmes, J.D. *et al*, *J. Am. Chem. Soc.*, 1963, **85**, 2531 (*synth*)
Oda, M. *et al*, *Synthesis*, 1974, 721 (*synth*)
Franck-Neumann, M. *et al*, *Tetrahedron Lett.*, 1975, 1759 (*synth*)
Tajiri, A. *et al*, *Tetrahedron Lett.*, 1987, **28**, 6465 (*resoln, abs config, cd*)

Bicyclo[3.2.1]octane-1-carboxylic acid, 9CI   B-70112

[2534-83-0]

$C_9H_{14}O_2$ M 154.208

Cryst. (pentane or by subl.). Mp 73.5-74.5° (70-71°).

Me ester: [81523-46-8].
    $C_{10}H_{16}O_2$ M 168.235
    Yellow oil.

Vaughan, W.R. *et al*, *J. Am. Chem. Soc.*, 1965, **87**, 2204.
Chow, A.W. *et al*, *Tetrahedron Lett.*, 1966, 5427 (*synth*)
Kropp, P.J. *et al*, *J. Am. Chem. Soc.*, 1982, **104**, 3972 (*synth, nmr, ms*)
Della, E.W. *et al*, *Aust. J. Chem.*, 1986, **39**, 1061 (*synth, pmr, cmr*)

Bicyclo[2.2.2]oct-2-ene-1-carboxylic acid, 9CI   B-70113

[2534-80-7]

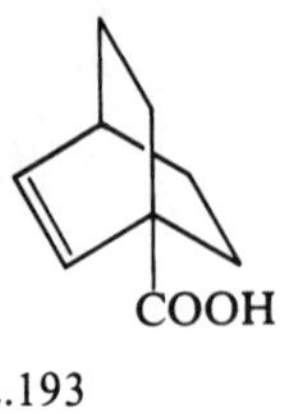

$C_9H_{12}O_2$ M 152.193

Cryst. by subl. Mp 120-121°.

Et ester: [3037-84-1].
    $C_{11}H_{16}O_2$ M 180.246
    Bp₁₀ 95-97°. nD²¹ 1.4749.
Nitrile: [90014-13-4].
    $C_9H_{11}N$ M 133.193
    Mp 42-43°. Bp₀.₁ 70°.

Grob, C.A., *Helv. Chim. Acta*, 1958, **41**, 1191 (*synth*)
Baker, F.W. *et al*, *J. Org. Chem.*, 1967, **32**, 3344 (*synth, pmr, deriv*)
Eichel, W. *et al*, *Chem. Ber.*, 1984, **117**, 1235 (*synth, ir, pmr*)

[1,1'-Bicyclopropyl]-1,1'-dicarboxylic acid, B-70114
9CI

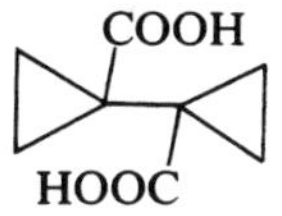

$C_8H_{10}O_4$ M 170.165
Dinitrile: [64760-88-9]. *1,1'-Dicyanobicyclopropyl.*
$C_8H_8N_2$ M 132.165
Liq.

Cobb, R.L. *et al*, *J. Org. Chem.*, 1978, **43**, 931 (*synth*)
Schrumpf, G. *et al*, *Acta Crystallogr., Sect. C*, 1987, **43**, 1594 (*cryst struct*)

Bicyclo[4.4.1]undeca-1,6-diene B-70115

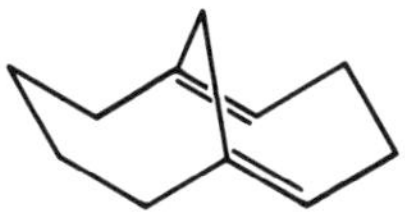

$C_{11}H_{16}$ M 148.247

Shea, K.J. *et al*, *J. Am. Chem. Soc.*, 1986, **108**, 5901 (*synth, nmr, ir*)

Bi(1,3-diselenol-2-ylidene) B-70116
Updated Entry replacing T-01744
2-(1,3-Diselenol-2-ylidene)-1,3-diselenole, 9CI.
Tetraselenafulvalene
[54489-01-9]

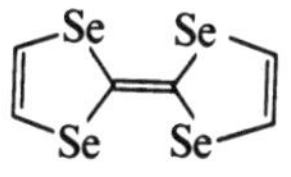

$C_6H_4Se_4$ M 391.938
Pink plates (hexane). Mp 133-134°. A labile form
(yellow needles) can be obt. at low temps.

Tetracyanoquinodimethane salt: [56003-87-3]. Charge-
transfer compd. showing metallic-type electrical
conductivity and superconducting props. Long black
needles.

Engler, E.M. *et al*, *J. Am. Chem. Soc.*, 1974, **96**, 7376 (*synth, pmr, uv*)
Lakshmikantham, M.V. *et al*, *J. Org. Chem.*, 1976, **41**, 882 (*synth*)
Narita, M. *et al*, *Synthesis*, 1976, 489 (*rev*)
Andersen, J.R. *et al*, *Org. Mass Spectrom.*, 1978, **13**, 121 (*ms*)
Johannsen, I. *et al*, *J. Chem. Soc., Chem. Commun.*, 1984, 89 (*synth*)
Jackson, Y.A. *et al*, *Tetrahedron Lett.*, 1987, **28**, 5635 (*synth*)

Bi(1,3-ditellurol-2-ylidene) B-70117
2-(1,3-Ditellurol-2-ylidene)-1,3-tellurole, 9CI.
Tetratellurafulvalene
[108418-48-0]

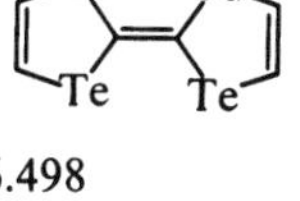

$C_6H_4Te_4$ M 586.498

McCullough, R.D. *et al*, *J. Am. Chem. Soc.*, 1987, **109**, 4115 (*synth, ms, pmr, uv, ir*)

Bi(1,3-dithiolo[4,5-*b*][1,4]dithiin-2-ylidene) B-70118
2-(1,3-Dithiolo[4,8-b] [1,4]-dithiin-2-ylidene)-1,3-dith-
iolo[4,5-b] [1,4]dithiin, 9CI. Bis(vinylenedithio)-
tetrathiafulvalene
[88682-15-9]

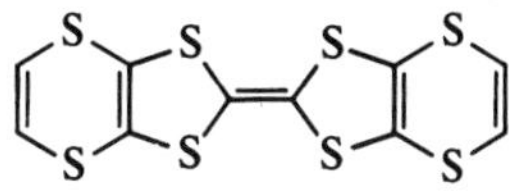

$C_{10}H_4S_8$ M 380.622
Planar molecule of interest in organic semiconducting.
Red needles. Mp 227° dec.

Nakamura, T. *et al*, *Bull. Chem. Soc. Jpn.*, 1987, **60**, 365 (*synth*)

Bi(1,3-dithiol-2-ylidene) B-70119
2-(1,3-Dithiol-2-ylidene)-1,3-dithiole, 9CI.
Tetrathiafulvalene
[31366-25-3]

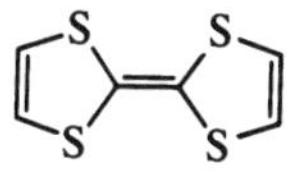

$C_6H_4S_4$ M 204.338
Long orange needles (hexane). Mp 119-119.5°.

Wudl, F. *et al*, *J. Chem. Soc., Chem. Commun.*, 1970, 1453 (*ir, pmr, ms*)
Wudl, F. *et al*, *J. Org. Chem.*, 1974, **39**, 3608 (*synth*)
Le Coustumer, G. *et al*, *J. Chem. Soc., Chem. Commun.*, 1980, 38 (*synth*)
Krief, A. *et al*, *Tetrahedron*, 1986, **42**, 1209 (*synth, rev*)

1,1'-Biindanylidene B-70120
1-(2,3-Dihydro-1H-inden-1-ylidene)-2,3-dihydro-1H-
indene, 9CI
[17666-94-3]

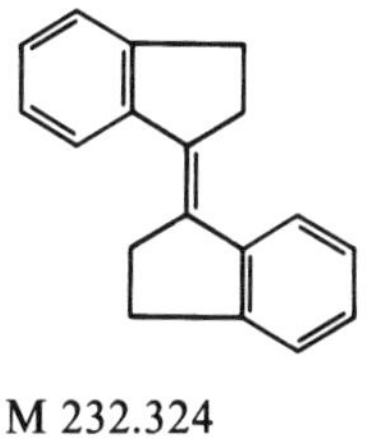

(E)-form

$C_{18}H_{16}$ M 232.324
E-form [24536-68-3]
Cryst. Mp 142-143°.

Saltiel, J. *et al*, *J. Am. Chem. Soc.*, 1972, **94**, 6445.
Yeung, L.L. *et al*, *J. Chem. Soc., Chem. Commun.*, 1987, 981 (*synth, pmr, ms*)

Biopterin B-70121
Updated Entry replacing B-60111
2-Amino-6-(1,2-dihydroxypropyl)-4(1H)-pteridinone,
9CI. 2-Amino-4-hydroxy-6-(1,2-dihydroxypropyl)pterin

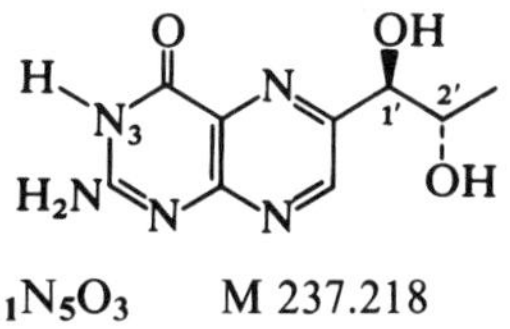

(1'R,2'S)-form

$C_9H_{11}N_5O_3$ M 237.218

(1′R,2′S)-form [22150-76-1]

L-erythro-*form*

Widely distributed in microorganisms, insects, algae, amphibia and mammals. Found in urine. Growth factor. Pale-yellow cryst. (AcOH aq.). Mp 250-280° dec. $[\alpha]_D^{20}$ −66° (c, 0.2 in 0.1M HCl). pK_{a1} 2.23, pK_{a2} 7.89.

3-Me: [111317-37-4]. **2-Amino-6-(1,2-dihydroxypropyl)-3-methylpterin-4-one**.

$C_{10}H_{13}N_5O_3$ M 251.244

Isol. from the marine anthozoan *Astroides calycularis*. Mp 229-231°. $[\alpha]_D$ −60° (c, 0.3 in 0.1M HCl).

5,6,7,8-Tetrahydro: see Pyreno[4,5-c]furan, P-70166

(1′R,2′R)-form [13019-52-8]

D-threo-*form*

Mp >300°. pK_{a1} 2.20, pK_{a2} 7.92.

(1′S,2′S)-form [13039-82-2]

L-threo-*form*. **Ciliapterin**

Isol. from the protozoan *Tetrahymena pyriformis*. Mp >300°. pK_{a1} 2.24, pK_{a2} 7.87.

(1′S,2′R)-form [13039-62-8]

D-erythro-*form*

Mp >300°. pK_{a1} 2.23, pK_{a2} 7.90.

Patterson, E.L. et al, J. Am. Chem. Soc., 1955, 77, 3167.
Kidder, G.W. et al, J. Biol. Chem, 1968, 243, 826 (Ciliapterin)
Kidder, G.W. et al, Methods Enzymol., 1971, 18B, 739 (Ciliapterin)
Rembold, H. et al, Angew. Chem., Int. Ed. Engl., 1972, 11, 1061 (biochem)
Sugimoto, T. et al, Bull. Chem. Soc. Jpn., 1975, 48, 3767 (synth)
Taylor, E.C. et al, J. Am. Chem. Soc., 1976, 98, 2301 (synth)
Schircks, B. et al, Helv. Chim. Acta, 1977, 60, 211; 1985, 68, 1639 (synth, pmr, cmr)
Armarego, W.L.F. et al, Aust. J. Chem., 1982, 35, 785 (synth)
Kappel, M. et al, Justus Liebigs Ann. Chem., 1984, 1815 (synth)
Aiello, A. et al, Experientia, 1987, 43, 950 (deriv)

2,2′,3,4-Biphenyltetrol   B-70122

2,2′,3,4-*Tetrahydroxybiphenyl*

$C_{12}H_{10}O_4$ M 218.209

Tetra-Me ether: 2,2′,3,4-*Tetramethoxybiphenyl*.

$C_{16}H_{18}O_4$ M 274.316

Cryst. (2-propanol). Mp 105°.

Itoh, Y. et al, Helv. Chim. Acta, 1988, 71, 1199 (synth, deriv, ir, pmr)

2,3,4,4′-Biphenyltetrol   B-70123

2,3,4,4′-*Tetrahydroxybiphenyl*

$C_{12}H_{10}O_4$ M 218.209

Tetra-Me ether: 2,3,4,4′-*Tetramethoxybiphenyl*.

$C_{16}H_{18}O_4$ M 274.316

Cryst. (2-propanol). Mp 75°.

Itoh, Y. et al, Helv. Chim. Acta, 1988, 71, 1199 (synth, deriv, ir, pmr)

2,3,4-Biphenyltriol, 8CI   B-70124

Updated Entry replacing B-01326

4-*Phenylpyrogallol*. 2,3,4-*Trihydroxybiphenyl*

$C_{12}H_{10}O_3$ M 202.209

Cryst. (xylene under N_2). Mp 216°.

Tri-Ac:

$C_{18}H_{16}O_6$ M 328.321

Plates (C_6H_6/pet. ether). Mp 111°.

Tri-Me ether: [95127-11-0]. 2,3,4-*Trimethoxybiphenyl*.

$C_{15}H_{16}O_3$ M 244.290

Needles (pet. ether). Mp 46.5°. $Bp_{0.1}$ 121-122°.

Bruce, J.M. et al, J. Chem. Soc., 1955, 4435 (synth)
Horner, L. et al, Chem. Ber., 1965, 98, 2016.
Banwell, M.G. et al, J. Org. Chem., 1988, 53, 4945 (synth, derivs, ir, pmr, cmr)

[1,1′-Bipyrrolidine]-2,2′-dione   B-70125

2,2′-*Dioxo*-N,N′-*bipyrrolidinyl*

[60769-64-4]

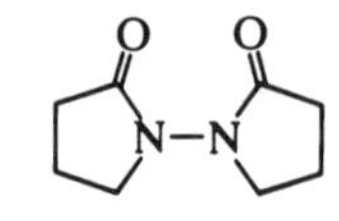

$C_8H_{12}N_2O_2$ M 168.195

Cryst. (CCl_4). Mp 125-126°.

Stetter, H. et al, Chem. Ber., 1958, 91, 1982 (synth)
Nelsen, S.F. et al, J. Am. Chem. Soc., 1987, 109, 5724 (props, cryst struct)

2-Bisabolene-1,12-diol   B-70126

[117232-52-7]

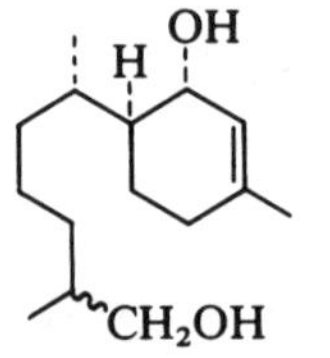

$C_{15}H_{28}O_2$ M 240.385

Constit. of *Vittadinia gracilis*. Oil. $[\alpha]_D^{24}$ +6° (c, 0.2 in $CHCl_3$).

Zdero, C. et al, Phytochemistry, 1988, 27, 2251.

1,1:2,2-Bis([10]annulene-1,6-diyl)ethylene   B-70127

11-*Bicyclo[4.4.1]undeca-1,3,5,7,9-pentaen-11-ylidenebicyclo[4.4.1]undeca-1,3,5,7,9-pentaene*, 9CI

[116204-21-8]

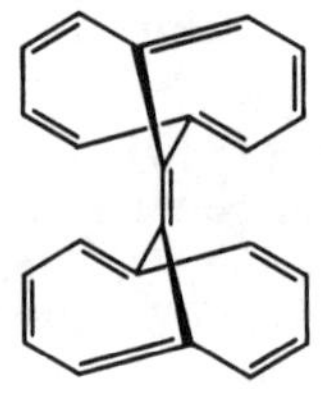

$C_{22}H_{16}$ M 280.368

Pale-yellow needles (C_6H_6/hexane). Mp 245-247°.

Yamamoto, K. et al, Bull. Chem. Soc. Jpn., 1988, 61, 1281 (synth, uv, pmr)

2,3-Bis(9-anthryl)cyclopropenone   B-70128

[78594-10-2]

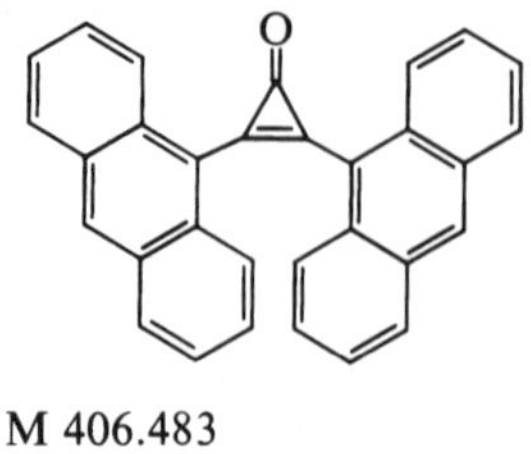

$C_{31}H_{18}O$ M 406.483

Cryst. (CH$_2$Cl$_2$). Mp ca. 284° dec. (>300°).

Wadsworth, D.H. et al, Synthesis, 1981, 285 (synth)
Becker, H.-D. et al, J. Org. Chem., 1987, **52**, 5205 (synth)

6H,24H,26H,38H,56H,58H-16,65:48,66-   B-70129
Bis([1,3]-benzenomethanothiomethano[1,3]-benzeno)-32,64-etheno-1,5:9,13:14,-18:19,23:27,31:33,37:41,45:46,50:51,-55:59,63-decemetheno-8H,40H[1,7]dithiacyclotetratriacontino[9,8-h][1,77]dithiocyclotetratriacontin, 9CI

[109978-12-3]

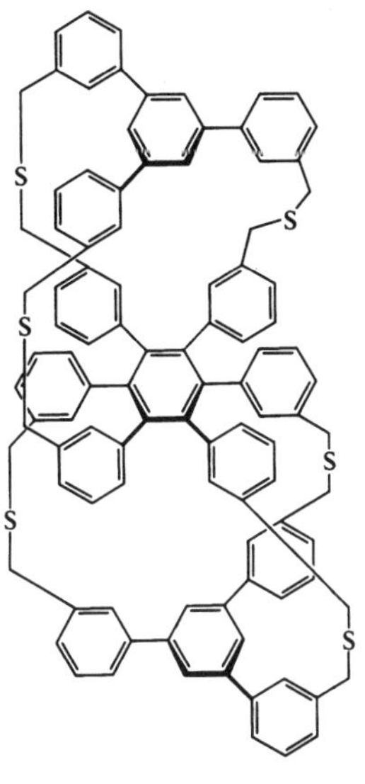

C$_{102}$H$_{78}$S$_6$ M 1496.098

Two isomers prepd., that illus. having (1α,2α-,3β,4α,5β,6β)-config. about the central ring, which is chiral and was resolved, and the symmetrical (1α,2β,3α,4β,5α,6β)-isomer.

Sendhoff, N. et al, Angew. Chem., Int. Ed. Engl., 1987, **26**, 777.

N,N'-Bis[3-(3-bromo-4-hydroxyphenyl)-2-   B-70130
oximidopropionyl]cystamine

N,N'-(Dithiodi-2,1-ethanediyl)bis[3-bromo-4-hydroxy-α-(hydroxyimino)benzenepropanamide], 9CI

[110659-91-1]

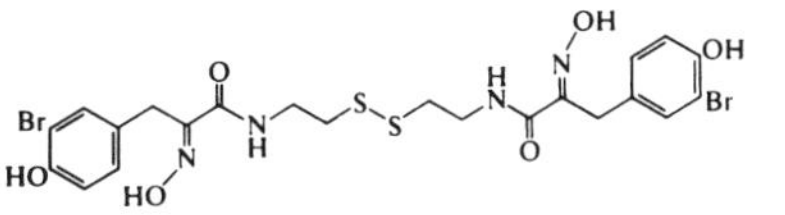

(E,E)-form

C$_{22}$H$_{24}$Br$_2$N$_4$O$_6$S$_2$ M 664.383

(E,E)-form

Major metab. from an unidentified sponge from Guam (prob. of the Verongidae family). Amorph. powder (hexane/EtOH). Mp 172-174°.

(E,Z)-form

Minor metab. from the same unidentified sponge. Powder. The (E,Z)- or (Z,Z)-isomer is poss. the nat. metab., and isomerises on storage. Isomerizes slowly to the (E,E)-form.

Arabshahi, L. et al, J. Org. Chem., 1987, **52**, 3584 (isol, uv, ir, pmr, cmr, ms, struct)

1,2-Bis(bromomethyl)-3-fluorobenzene   B-70131

[62590-16-3]

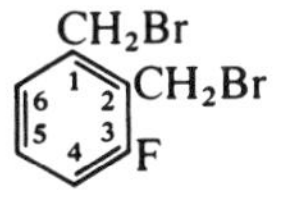

C$_8$H$_7$Br$_2$F   M 281.950
Plates (pet. ether). Mp 41-42.5°.

Rice, J.E. et al, J. Org. Chem., 1988, **53**, 1775 (synth, pmr)

1,2-Bis(bromomethyl)-4-fluorobenzene   B-70132

[55831-04-4]
C$_8$H$_7$Br$_2$F   M 281.950
Plates (Et$_2$O/pet. ether). Mp 47-48.5°.

Rice, J.E. et al, J. Org. Chem., 1988, **53**, 1775 (synth, pmr)

1,3-Bis(bromomethyl)-2-fluorobenzene   B-70133

[25006-86-4]
C$_8$H$_7$Br$_2$F   M 281.950
Prisms (hexane). Mp 90-91°.

Boekelheide, V. et al, J. Org. Chem., 1973, **33**, 3928 (synth, pmr)
Yamato, T. et al, J. Chem. Soc., Perkin Trans. 1, 1987, 1 (synth)

3,12-Bis(carboxymethyl)-6,9-dioxa-3,12-   B-70134
diazatetradecanedioic acid, 9CI

Ethylenebis(oxyethylenenitrilo)tetraacetic acid, 8CI. Ethylene glycol bis[2-[bis(carboxymethyl)amino]ethyl] ether. Ethylene glycol bis(2-aminoethyl ether)-N-tetraacetic acid. egta

[67-42-5]

(HOOCCH$_2$)$_2$NCH$_2$CH$_2$OCH$_2$CH$_2$OCH$_2$CH$_2$N-(CH$_2$COOH)$_2$

C$_{14}$H$_{24}$N$_2$O$_{10}$ M 380.351
Chelating agent, e.g. used in detn. of Ca in presence of Mg. Biochemistry much investigated.

▷LD$_{50}$ 9.43 mmoles/kg (as di-Na salt, rat). AH3760000.

B.P., 695 346, (1953); CA, **48**, 11486i (manuf)
Schmid, R.W. et al, Anal. Chem., 1957, **29**, 264 (use)
Fr. Pat., 1 381 847, (1964); CA, **62**, 9014d (manuf)
Sudmeier, J.L. et al, Anal. Chem., 1964, **36**, 1698 (pmr)
Wynn, J.E. et al, Toxicol. Appl. Pharmacol., 1970, **16**, 807 (use, tox)
Schauer, C.K. et al, Acta Crystallogr., Sect. C, 1988, **44**, 981 (cryst struct, complexes)

Bis(2,4-dichlorophenyl)ether, 8CI   B-70135

1,1'-Oxybis[2,4-dichlorobenzene], 9CI. 2,2',4,4'-Tetrachlorodiphenyl ether

[28076-73-5]

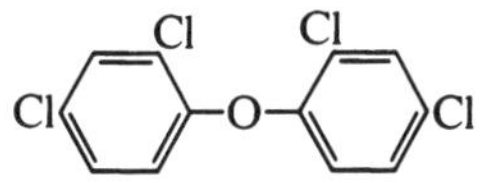

C$_{12}$H$_6$Cl$_4$O   M 307.991
Present as impurity in commercial chlorophenol preparations. Precursor to toxic tetrachlorodibenzofurans. Cryst. (Et$_2$O). Mp 70-72°. Rel. nontoxic.

Dahlgard, M. *et al*, *J. Am. Chem. Soc.*, 1958, **80**, 5861 (*ir, uv*)
Farrar, W.V., *J. Appl. Chem.*, 1960, **10**, 207 (*synth*)
Busch, M.L. *et al*, *Biomed. Mass Spectrom.*, 1979, **6**, 157 (*ms*)
Humppi, T. *et al*, *J. Chromatogr.*, 1985, **331**, 410 (*glc, ms*)
Humppi, T., *Chemosphere*, 1986, **15**, 2003 (*synth*)
Rissanen, K. *et al*, *Acta Crystallogr., Sect. C*, 1988, **44**, 1644 (*cryst struct*)

Bis(2,6-dichlorophenyl) ether   B-70136

2,2'-Oxybis[1,3-dichlorobenzene]. 2,2',6,6'-Tetrachloro-diphenyl ether

$C_{12}H_6Cl_4O$ M 307.991
Insol. pet. ether. Mp 68°.

Ger. Pat., 1 139 847, (*1962*); *CA*, **58**, 11277c (*synth*)

Bis(3,4-dichlorophenyl) ether   B-70137

1,1'-Oxybis[3,4-dichlorobenzene], 9CI. 3,3',4,4'-Tetrachlorodiphenyl ether
[56348-72-2]

$C_{12}H_6Cl_4O$ M 307.991
Cryst. (EtOH). Mp 70°. $Bp_{0.2}$ 160-162°.

Busch, M.L. *et al*, *Biomed. Mass Spectrom.*, 1979, **6**, 157 (*ms*)
Granoth, I., *J. Chem. Soc., Perkin Trans. 1*, 1979, 2166 (*synth, pmr*)
Singh, P. *et al*, *Acta Crystallogr., Sect. B*, 1980, **36**, 210 (*cryst struct*)
Ryan, J.J. *et al*, *J. Chromatogr.*, 1984, **303**, 351 (*glc*)
Humppi, T. *et al*, *J. Chromatogr.*, 1985, **331**, 410 (*glc, ms*)
Humppi, T., *Chemosphere*, 1986, **15**, 2003 (*synth*)

S,S-Bis(4,6-dimethyl-2-pyrimidinyl)-dithiocarbonate   B-70138

$C_{13}H_{14}N_4OS_2$ M 306.400
Coupling reagent for esterification of carboxylic acids. Yellow cryst. (pet. ether/CH_2Cl_2). Mp 95-96°.

Kim, S. *et al*, *Synthesis*, 1986, 1017 (*synth, ir, pmr, use*)

1,3-Bis(1,2-diphenylethenyl)benzene, 9CI   B-70139

1,3-Bis(1,2-diphenylvinyl)benzene

$C_{34}H_{26}$ M 434.579
(*Z,Z*)-**form** [109764-44-5]
Cryst. (C_6H_6/EtOH). Mp 149-150°.

Du, C.-J.F. *et al*, *J. Org. Chem.*, 1987, **52**, 4311 (*synth, pmr, ms*)

1,4-Bis(1,2-diphenylethenyl)benzene, 9CI   B-70140

1,4-Bis(1,2-diphenylvinyl)benzene

(*E,E*)-**form** [31020-17-4]
cis,cis-*form*
Mp 172-174°.
(*E,Z*)-**form** [31020-18-5]
cis,trans-*form*
Mp 109-110°.
(*Z,Z*)-**form** [31024-80-3]
trans,trans-*form*
Mp 145-145.5°.

Bertoniere, N.R. *et al*, *J. Org. Chem.*, 1971, **36**, 2956.
Du, C.-J.F. *et al*, *J. Org. Chem.*, 1987, **52**, 4311 (*synth*)

Bis(diphenylmethyl) diselenide, 9CI   B-70141

Diphenylmethyl diselenide
[24572-14-3]

$$Ph_2CHSeSeCHPh_2$$

$C_{26}H_{22}Se_2$ M 492.380
Yellow needles (pet. ether). Mp 121.5-123°.

Collard-Charon, C. *et al*, *Bull. Soc. Chim. Belg.*, 1962, **71**, 563 (*synth*)
Palmer, H.T. *et al*, *Acta Crystallogr., Sect. B*, 1969, **25**, 1090 (*cryst struct*)
Okuma, K. *et al*, *Tetrahedron Lett.*, 1987, **28**, 6649 (*synth, pmr, cmr, ir*)

1,2;3,4-Bisepoxy-11(13)-eudesmen-12,8-olide   B-70142

$C_{15}H_{18}O_4$ M 262.305
(*1α,2α,3α,4α,8β*)-**form**
Constit. of *Ferreyranthus fruticosus*. Cryst. Mp 192°.

Jakupovic, J. *et al*, *Phytochemistry*, 1988, **27**, 1113.

1,11-Bis(3-furanyl)-4,8-dimethyl-3,6,8-undecatrien-2-ol   B-70143

α-[9-(3-Furanyl)-2,6-dimethyl-1,4,6-nonatrienyl]-3-furanethanol, 9CI

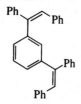

$C_{21}H_{26}O_3$ M 326.435
Ac: [97530-63-7]. *2-Acetoxy-1,11-bis(3-furanyl)-4,8-dimethyl-3,6,8-undecatriene.*
$C_{23}H_{28}O_4$ M 368.472
Constit. of nudibranch *Dendrodoris grandiflora*. $[\alpha]_D$ +12.8° (c, 0.24 in $CHCl_3$).

Cimino, G. *et al*, *Tetrahedron*, 1985, **41**, 1093.

2,6-Bis(hydroxyacetyl)pyridine   B-70144

1,1'-(2,6-Pyridinediyl)bis[2-hydroxyethanone], 9CI
[78812-65-4]

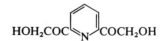

$C_9H_9NO_4$ M 195.174
Mp 120-122°.

Di-Me ether: [105729-07-5]. *2,6-Bis(methoxyacetyl)-pyridine.*
$C_{11}H_{13}NO_4$ M 223.228
Cryst. (hexane). Mp 101-102°.
Bis di-Me ketal:
$C_{13}H_{21}NO_6$ M 287.312
Mp 153-154°.

Moriarty, R.M. *et al, Tetrahedron Lett.*, 1981, **22**, 1283 (*synth, ir, pmr*)
Moriarty, R.M. *et al, J. Org. Chem.*, 1987, **52**, 150 (*deriv, synth, pmr, ir, ms*)

S,S'-**Bis(2-hydroxyethyl)-1,2-ethanedithiol**   **B-70145**

2,2'-[1,2-Ethanediylbis(thio)]bisethanol, 9CI. 2,2'-(Ethylenedithio)diethanol, 8CI. 3,6-Dithia-1,8-octane-diol. 1,2-Bis(2-hydroxyethylthio)ethane

[5244-34-8]

$$HOCH_2CH_2SCH_2CH_2SCH_2CH_2OH$$

$C_6H_{14}O_2S_2$ M 182.295
Cryst. (Me₂CO). Mp 64° (58-60°). Bp₂ 100°.
S-Tetroxide:
$C_6H_{14}O_6S_2$ M 246.293
Mp 115-116°.

Welti, D. *et al, J. Chem. Soc.*, 1962, 3955 (*synth*)
Wolf, R.E. *et al, J. Am. Chem. Soc.*, 1987, **109**, 4328 (*synth*)

1,7-**Bis(hydroxymethyl)-7-methylbicyclo[2.2.1]heptan-2-ol**   **B-70146**

2,8,10-Bornanetriol. 8,10-Dihydroxyborneol

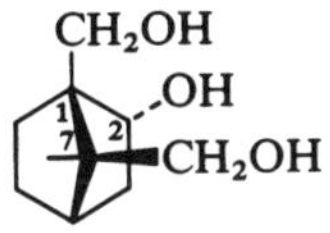

$C_{10}H_{18}O_3$ M 186.250
(1S,2R,7R)-form
Constit. of *Florestina tripteris*. Oil. $[\alpha]_D^{24}$ −10° (c, 0.36 in CHCl₃).

Domínguez, X.A. *et al, Phytochemistry*, 1988, **27**, 613.

1,2-**Bis(4-hydroxyphenyl)-1,2-propanediol,**   **B-70147**
9CI

4,4'-Dihydroxy-α-methylhydrobenzoin. **Marsupol**

[84323-33-1]

$C_{15}H_{16}O_4$ M 260.289
Constit. of *Pterocarpus marsupium*. Cryst. Mp 156°.
$[\alpha]_D$ −13.6° (c, 2.3 in MeOH).

Rao, A.V.S. *et al, Phytochemistry*, 1982, **21**, 1837.

2,2'-**Bis(2-imidazolyl)biphenyl**   **B-70148**

2,2'-[1,1'-Biphenyl]-2,2'-diylbis-1H-imidazole, 9CI

[107011-08-5]

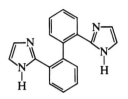

$C_{18}H_{14}N_4$ M 286.335
Bidentate donor ligand. Cryst. Mp 273-274°.

Knapp, S. *et al, J. Am. Chem. Soc.*, 1987, **109**, 1882 (*synth, use*)

1,4-**Bis(1-isocyanato-1-methylethyl)-benzene, 9CI**   **B-70149**

p-*Tetramethylxylylene diisocyanate*

[2778-41-8]

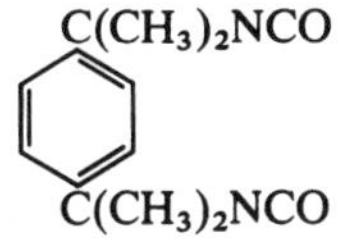

$$C(CH_3)_2NCO$$

$$C(CH_3)_2NCO$$

$C_{14}H_{16}N_2O_2$ M 244.293
Intermed. for light-stable polyurethanes. Mp 78° (74-76°). Bp₁ 110-130°.
▷Potentially hazardous synth.

Hoover, F.W. *et al, J. Org. Chem.*, 1964, **29**, 143 (*synth*)
Henderson, W.A. *et al, Org. Prep. Proced. Int.*, 1986, **18**, 149 (*synth*)

S,S'-**Bis(2-mercaptoethyl)-1,2-ethanedith-iol**   **B-70150**

2,2'-[1,2-Ethanediylbis(thio)]bisethanethiol, 9CI. 3,6-Dithia-1,8-octanedithiol

[25423-55-6]

$$HSCH_2CH_2SCH_2CH_2SCH_2CH_2SH$$

$C_6H_{14}S_4$ M 214.417
Crown ether precursor. Viscous oil that solidifies on standing. Bp₂.₆ 170-175°.

Wolf, R.E. *et al, J. Am. Chem. Soc.*, 1987, **109**, 4328 (*synth, pmr, ir*)

2,3-**Bis(2-methoxy-4-nitro-5-sulfophenyl)-5-[(phenylamino)carbonyl]-2H-tetrazoli-um hydroxide inner salt, 9CI**   **B-70151**

XTT. 3,3'-[1-[(Phenylamino)carbonyl-3,4-tetrazolium]-bis(4-methoxy-6-nitrobenzenesulfonic acid)]

[117038-70-7]

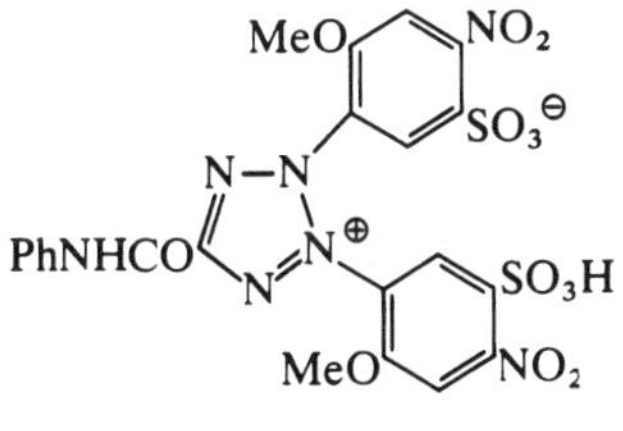

$C_{22}H_{17}N_7O_{13}S_2$ M 651.535
Other contributing forms possible. Tetrazolium salt of interest in bioreduction studies in cell culture assay systems. Unusual in that the formazan produced on

redn. is highly soluble.
Na salt: Cryst. + $1\frac{1}{2}H_2O$. Mp 285° dec.

Paull, K.D. *et al, J. Heterocycl. Chem.*, 1988, **25**, 911 (*synth, ir, pmr, uv*)

9,10-Bis(methylene)tricyclo[5.3.0.0²,⁸]deca-3,5-diene, 9CI   B-70152
[61278-86-2]

$C_{12}H_{12}$ M 156.227

Paquette, L.A. *et al, Tetrahedron Lett.*, 1987, **28**, 4965 (*synth, pmr*)

3,6-Bis(1-methylpropyl)-2,5-piperazine-dione   B-70153
Di-sec-butyldiketopiperazine. Cyclo(isoleucylisoleucyl)
[39897-59-1]

$C_{12}H_{22}N_2O_2$ M 226.318
Prod. by *Beauveria bassiana* and *Ustilago cynodontis.*
$[\alpha]_D^{22}$ −47° (c, 0.09 in EtOH). Subl. at 250° without melting.
N,N-Di-Me: Cryst. (Et₂O/pet. ether). Mp 122-124°.

Yoshimura, J. *et al, Bull. Chem. Soc. Jpn.*, 1975, **48**, 605 (*deriv*)
Suzuki, K. *et al, Pept. Chem.*, 1980, **8**, 11 (*synth*)
Grove, J.F. *et al, Phytochemistry*, 1981, **20**, 815 (*isol*)

Bis(methylsulfonyl) peroxide, 9CI   B-70154
Methanesulfonyl peroxide
[1001-62-3]

$$MeSO_2-O-O-SO_2Me$$

$C_2H_6O_6S_2$ M 190.186
Easily-handled two-electron oxidant for a variety of functional groups. Cryst. Indefinitely stable at r.t.

Myall, C.J. *et al, J. Chem. Soc., Perkin Trans. 1*, 1975, 953 (*synth, ir*)
Hoffman, R.V., *Org. Prep. Proceed. Int.*, 1986, **18**, 181 (*rev*)

S,S'-Bis(1-phenyl-1H-tetrazol-5-yl) dithio-carbonate   B-70155
Carbonodithioic acid S,S-bis(1-phenyl-1H-tetrazol-5-yl)ester, 9CI
[32276-00-9]

$C_{15}H_{10}N_8OS_2$ M 382.417
Esterification reagent. Cryst. (EtOAc/Et₂O). Mp 126° dec.

Takeda, K. *et al, Synthesis*, 1987, 560 (*synth, pmr, ir, use*)

1,2-Bis(phenylthio)cyclobutene   B-70156
[70871-28-2]

$C_{16}H_{14}S_2$ M 270.407
Mp 28°.
S-Oxide: [105858-81-9]. *1-(Phenylsulfinyl)-2-(phenylthio)cyclobutene.*
$C_{16}H_{14}OS_2$ M 286.406
Oil.
S¹,S²-Dioxide: 1,2-Bis(phenylsulfinyl)cyclobutene.
$C_{16}H_{14}O_2S_2$ M 302.405
Mp 113.5-114.5°. Mixt. of *meso*- and (±)-stereoisomers.
S-Trioxide: [105858-78-4]. *1-(Phenylsulfinyl)-2-(phenylsulfonyl)cyclobutene.*
$C_{16}H_{14}O_3S_2$ M 318.405
Mp 107-109°.
S-Tetraoxide: [105858-77-3]. *1,2-Bis(phenylsulfonyl)-cyclobutene.*
$C_{16}H_{14}O_4S_2$ M 334.404
Mp 134-135°.

Landen, H. *et al, Chem. Ber.*, 1987, **120**, 171 (*synth, uv, pmr, cmr*)

3,3-Bis(phenylthio)propanoic acid, 9CI   B-70157

$$(PhS)_2CHCH_2COOH$$

$C_{15}H_{14}O_2S_2$ M 290.394
Et ester:
$C_{17}H_{18}O_2S_2$ M 318.448
Oil.

Tilak, B.D. *et al, Tetrahedron*, 1966, **22**, 7 (*synth*)
Cohen, T. *et al, J. Am. Chem. Soc.*, 1986, **108**, 3718 (*synth*)

3,3-Bis(phenylthio)-1-propanol   B-70158
[83711-16-4]

$$(PhS)_2CHCH_2CH_2OH$$

$C_{15}H_{16}OS_2$ M 276.411
Yellow oil.

Cohen, T. *et al, J. Am. Chem. Soc.*, 1982, **104**, 7142; 1986, **108**, 3718 (*synth, nmr, ir*)

N,N'-Bis(3-sulfonatopropyl)-2,2'-bipyridin-ium   B-70159
1,1'-Bis(3-sulfopropyl)-2,2'-bipyridinium dihydroxide bis(inner salt), 9CI
[86690-04-2]

$C_{16}H_{20}N_2O_6S_2$ M 400.464
Electron acceptor. Solid.

Degani, Y. *et al, J. Am. Chem. Soc.*, 1983, **105**, 6228 (*synth, pmr, cryst struct, use*)

N,N'-**Bis(3-sulfonatopropyl)-4,4'-bipyridinium**   **B-70160**

1,1'-Bis(3-sulfopropyl)-4,4'-bipyridinium dihydroxide bis(inner salt), 9Cl. Dipropyl 4,4'-bipyridinium disulfonate. Propyl viologensulfonate

[77951-49-6]

$C_{16}H_{20}N_2O_6S_2$ M 400.464
Electron acceptor. Solid. Hygroscopic.

Willner, I. *et al, J. Phys. Chem.*, 1981, **85**, 3277 (*props, use*)
Willner, I. *et al, J. Heterocycl. Chem.*, 1983, **20**, 1113 (*synth, pmr*)

Bis[1,2,5-thiadiazolo][3,4-*b*:3',4'-*e*]-pyrazine   **B-70161**

[113055-86-0]

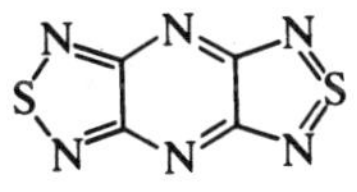

$C_4N_6S_2$ M 196.204
Hypervalent S—N heterocycle. Stable red cryst. Mp 323-329° dec.

Yamashita, Y. *et al, Angew. Chem., Int. Ed. Engl.*, 1988, **27**, 434 (*synth, uv, cryst struct*)

Bis(2,4,6-tri-*tert*-butylphenyl)ditelluride   **B-70162**

Bis[2,4,6-tris(1,1-dimethylethyl)phenyl]ditelluride, 9Cl. Bis(2,4,6-tri-tert-butylphenyl)ditellane

[99354-18-4]

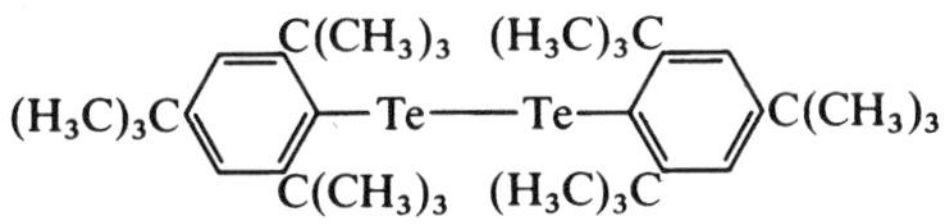

$C_{36}H_{58}Te_2$ M 746.054
Deep-red highly birefringent cryst. V. sol. THF, CCl₄. Mp 192° dec.

du Mont, W.-W. *et al, Chem. Ber.*, 1988, **121**, 11 (*synth, uv, pmr, cmr, ms, cryst struct*)

Bis(trichloromethyl) carbonate   **B-70163**

Triphosgene

[32315-10-9]

$$Cl_3C-O-CO-O-CCl_3$$

$C_3Cl_6O_3$ M 296.749
Stable crystalline source of phosgene for use in reactions. Cryst. (Et₂O). Mp 79°.

Councler, C., *Ber.*, 1880, **13**, 1697 (*synth*)
Eckert, H. *et al, Angew. Chem., Int. Ed. Engl.*, 1987, **26**, 894 (*synth, bibl, use*)

4,5-Bis(trifluoromethyl)-1,3,2-dithiazol-2-yl   **B-70164**

[71042-53-0]

$C_4F_6NS_2$ M 240.161
Poss. dimeric. Stable free radical. Black-green liq. Mp 12°. Paramagnetic at r.t. Thermally stable, air-sensitive, photochemically sensitive. Blue in the gas phase.

Fairhurst, S.A. *et al, J. Chem. Soc., Faraday Trans. 1*, 1983, **79**, 925 (*esr*)
Awere, E.G. *et al, J. Chem. Soc., Chem. Commun.*, 1987, 66 (*synth, ir, esr, uv, props*)

1,1'-Bitetralinylidene   **B-70165**

1-(3,4-Dihydro-1(2H)-naphthalenylidene)-1,2,3,4-tetrahydronaphthalene, 9Cl. 1-(1-Tetralinylidene)tetralin. 1,1'-(1,2,3,4-Tetrahydronaphthylidene)

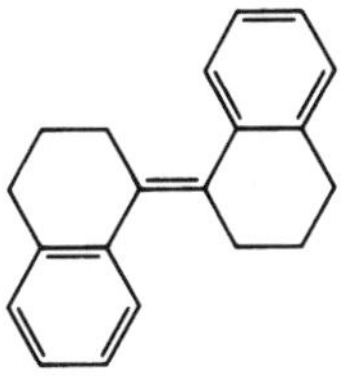

(E)-form

$C_{20}H_{20}$ M 260.378
(***E***)-*form* [91590-50-0]
   Mp 142-144° (120-123°).
(***Z***)-*form* [91573-04-5]
   Cryst. (MeOH). Mp 109-112°. Obt. only in small yield, by crystal picking.

Lemmen, P. *et al, Chem. Ber.*, 1984, **117**, 2300 (*synth, ir, pmr, ms*)
Yeung, L.L. *et al, J. Chem. Soc., Chem. Commun.*, 1987, 981 (*synth, pmr, ms*)
Ogawa, K. *et al, Bull. Chem. Soc. Jpn.*, 1988, **61**, 939 (*cryst struct*)

2,2'-Bithiazole   **B-70166**

2,2'-Dithiazolyl

[13816-21-2]

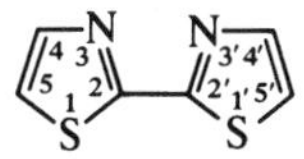

$C_6H_4N_2S_2$ M 168.231
Cryst. (pet. ether). Mp 102.5°.

Erlenmeyer, H. *et al, Helv. Chim. Acta*, 1939, **22**, 698 (*synth*)
Vernin, G. *et al, Bull. Soc. Chim. Fr.*, 1963, 2504 (*synth*)
Dondoni, A. *et al, Synthesis*, 1987, 185 (*synth, ms, pmr*)
Craig, D.C. *et al, Aust. J. Chem.*, 1988, **41**, 1625 (*synth, cryst struct*)

2,4'-Bithiazole   **B-70167**

2,4'-Dithiazolyl

[82326-33-8]
$C_6H_4N_2S_2$ M 168.231
Small spangles (H₂O). Mp 117.5-118° (114-116°).
Picrate: Needles (EtOH). Mp 173-174°.

Erlenmeyer, H. *et al, Helv. Chim. Acta*, 1948, **31**, 1142 (*synth*)
Dondoni, A. *et al, Synthesis*, 1987, 185 (*synth, pmr, ms*)

2,5'-Bithiazole   **B-70168**

2,5'-Dithiazolyl

[19960-71-5]
$C_6H_4N_2S_2$ M 168.231
Mp 31-32°. Bp₀.₁ 120° (bath).

Traupel, W. *et al, Helv. Chim. Acta*, 1950, **33**, 1960 (*synth*)
Dondoni, A. *et al, Synthesis*, 1987, 185 (*synth, ms, pmr*)

4,4'-Bithiazole   **B-70169**

4,4'-Dithiazolyl

[89324-31-2]

$C_6H_4N_2S_2$ M 168.231
Cryst. (C_6H_6). Mp 173-173.5°.

Erlenmeyer, H. *et al, Helv. Chim. Acta,* 1939, **22**, 938; 1946, **29**, 275 (*synth*)
Bayer, H. *et al, Chem. Ber.,* 1956, **89**, 2777 (*synth*)
Dondoni, A. *et al, Synthesis,* 1987, 185 (*synth, ms, pmr*)
Craig, D.C. *et al, Aust. J. Chem.,* 1988, **41**, 1625 (*cryst struct*)

4,5'-Bithiazole   B-70170

4,5'-Dithiazolyl

[111185-06-9]

$C_6H_4N_2S_2$ M 168.231
Cryst. (pet. ether). Mp 93-94° (82-84°).

Picrate: Yellow needles (EtOH). Mp 150-152°.

Erlenmeyer, H. *et al, Helv. Chim. Acta,* 1948, **31**, 26 (*synth*)
Dondoni, A. *et al, Synthesis,* 1987, 185 (*synth, ms, pmr*)

5,5'-Bithiazole   B-70171

5,5'-Dithiazolyl

[19960-72-6]

$C_6H_4N_2S_2$ M 168.231
Needles. Mp 94-95° (89-91°). Gradually becomes yellowish.

Dipicrate: Yellow needles (EtOH). Mp 169-170°.

Erne, M. *et al, Helv. Chim. Acta,* 1953, **36**, 354 (*synth*)
Dondoni, A. *et al, Synthesis,* 1987, 185 (*synth, ms, pmr*)

Blumealactone *C*   B-70172

[111545-48-3]

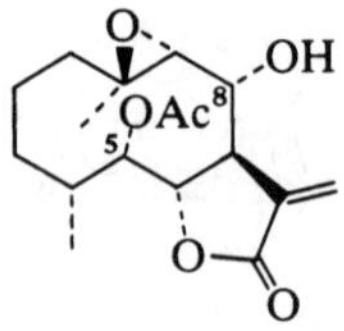

$C_{17}H_{24}O_6$ M 324.373
Constit. of *Blumea balsamifera.* Cryst. Mp 247-250°. [α]$_D$ −4.85° (c, 1.55 in $CHCl_3$).

5-De-Ac, 8-(2-methyl-2-butenoyl): [111545-46-1]. **Blumealactone A.**
$C_{20}H_{28}O_6$ M 364.438
Constit. of *B. balsamifera.* Cryst. Mp 205-208°. [α]$_D$ −45.9° (c, 0.88 in $CHCl_3$).

5-De-Ac, 8-(3-methyl-2-butenoyl): [111545-47-2]. **Blumealactone B.**
$C_{20}H_{28}O_6$ M 364.438
Constit. of *B. balsamifera.* Cryst. Mp 221-222°. [α]$_D$ −34.2° (c, 0.58 in $CHCl_3$).

Fujimoto, Y. *et al, Phytochemistry,* 1988, **27**, 1109.

Boesenboxide   B-70173

[113532-14-2]

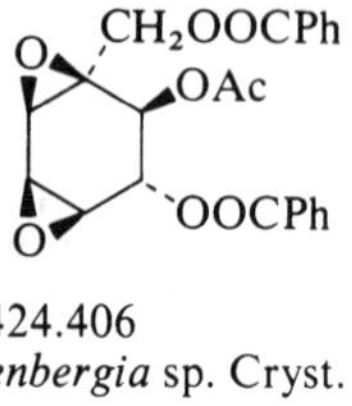

$C_{23}H_{20}O_8$ M 424.406
Constit. of a *Boesenbergia* sp. Cryst. (MeOH). Mp 171-172°.

Tuntiwachwuttikul, P. *et al, Aust. J. Chem.,* 1987, **40**, 2049.

Botrylactone   B-70174

10-Hydroxy-1,2,5,7,9-pentamethyl-3,11,12-trioxatricyclo[5.3.1.1^{2,6}]dodecan-4-one

[72579-05-6]

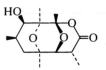

$C_{14}H_{22}O_5$ M 270.325
Prod. by *Botrytis cinerea.* Active against gram-positive bacteria. No phys. props. reported.

Ac: [72579-04-5]. Cryst. (Me_2CO/pet. ether). Mp 136-140°. [α]$_D^{20}$ +88° (c, 1 in $CHCl_3$).

Welmar, K. *et al, Chem. Ber.,* 1979, **112**, 3598 (*cryst struct, nmr, ms*)

Bowdichione   B-70175

Updated Entry replacing B-01834

2-(7-Hydroxy-4-oxo-4H-1-benzopyran-3-yl)-5-methoxy-2,5-cyclohexadiene-1,4-dione, 9CI

[53774-75-7]

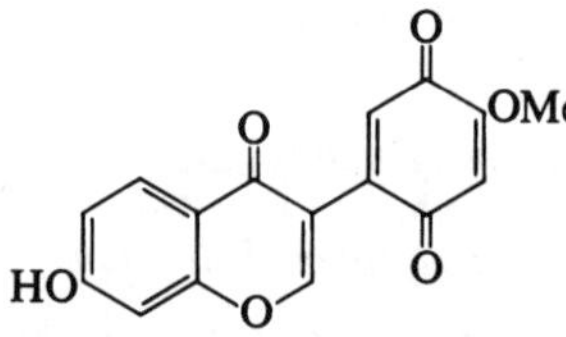

$C_{16}H_{10}O_6$ M 298.251
Constit. of *Bowdichia nitida.* Yellow needles (Me_2CO aq.). Mp 269-271°.

5-Hydroxy: [112448-38-1]. **5-Hydroxybowdichione.**
$C_{16}H_{10}O_7$ M 314.251
Constit. of *Dalbergia candenatensis.* Yellow cryst. ($CHCl_3$/MeOH). Mp 241-245° dec.

Brown, P.M. *et al, Justus Liebigs Ann. Chem.,* 1974, 1295 (*isol, struct*)
Hamburger, M. *et al, J. Nat. Prod.,* 1987, **50**, 696 (*deriv*)

Boxazomycin *A*   B-70176

5-Hydroxy-2-[5-hydroxy-2-(hydroxymethyl)-4-pyrimidinyl]-7-methyl-4-benzoxazolecarboxamide, 9CI

[107021-64-7]

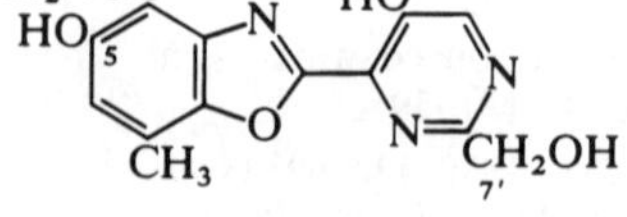

$C_{14}H_{12}N_4O_5$ M 316.273
Isol. from *Pseudonocardia* sp.

5-Deoxy: [107021-66-9]. **Boxazomycin C.**
$C_{14}H_{12}N_4O_4$ M 300.273
From *P.* sp.

7'-Deoxy: [107021-65-8]. **Boxazomycin B.**
$C_{14}H_{12}N_4O_4$ M 300.273
From *P.* sp.

Kusumi, T. *et al, CA,* 1987, **106**, 119520.

Brassicadiol   B-70177

*2,3-Dihydro-α²,α²-dimethyl-2,7-benzofurandimethanol,
9CI. 2,3-Dihydro-7-hydroxymethyl-2-(2-hydroxy-2-
propyl)-6-propylbenzofuran*
[110538-24-4]

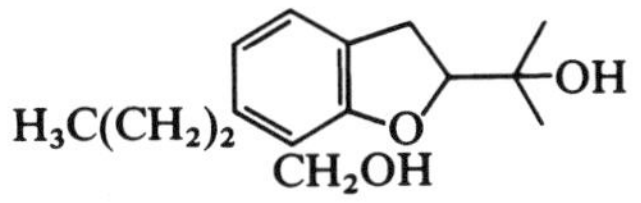

$C_{15}H_{22}O_3$ M 250.337
Metab. of *Alternaria brassicae*. Oil.

Ayer, W.A. *et al*, *J. Nat. Prod.*, 1987, **50**, 408.

Brassinolide   B-70178

Updated Entry replacing B-20147
*2α,3α,22R,23R-Tetrahydroxy-24S-methyl-B-homo-7-
oxa-5α-cholestan-6-one*
[72962-43-7]

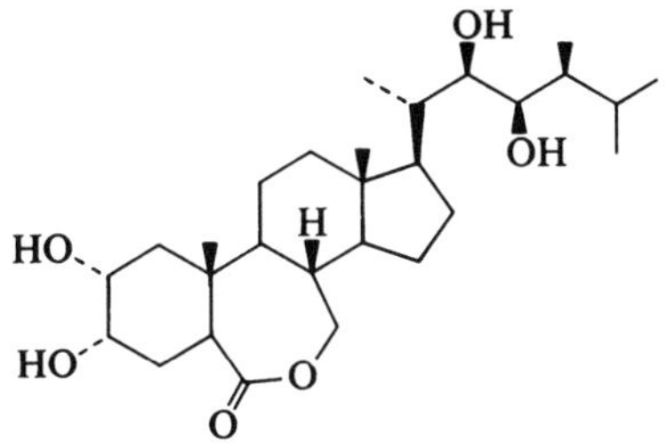

$C_{28}H_{48}O_6$ M 480.684
Constit. of bee collected rape pollen (*Brassica napus*).
Plant growth promoting substance. Cryst. Mp 274-
275°. $[\alpha]_D^{24}$ +41.9° ($CHCl_3$/MeOH, 9:1).

Grove, M.D. *et al*, *Nature* (*London*), 1979, **281**, 216 (*isol, cryst
struct*)
Mori, K. *et al*, *Tetrahedron*, 1982, **38**, 2099 (*synth*)
Sakakibara, M. *et al*, *Agric. Biol. Chem.*, 1983, **47**, 663 (*synth,
bibl*)
Donaubaurer, J.R. *et al*, *J. Org. Chem.*, 1984, **49**, 2833 (*synth*)
Adam, G. *et al*, *Phytochemistry*, 1986, **25**, 1787 (*rev*)
Singh, H. *et al*, *Indian J. Chem., Sect. B*, 1986, **25**, 989 (*rev*)
Aburatani, M. *et al*, *Agric. Biol. Chem.*, 1987, **51**, 1909 (*synth*)

Brevifloralactone   B-70179

Updated Entry replacing B-60198
[110668-26-3]

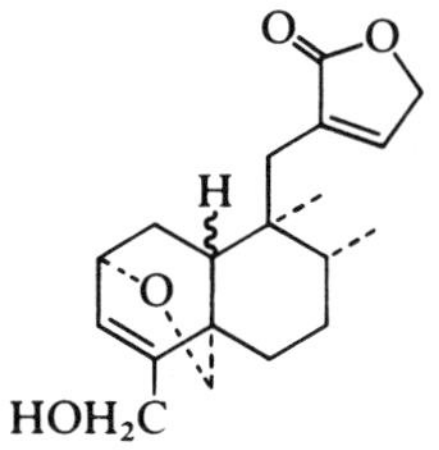

$C_{20}H_{28}O_4$ M 332.439
Constit. of *Salvia breviflora*. Cryst. Mp 126°. $[\alpha]_D^{20}$
+0.232° (c, 2.3 in MeOH).

Ac: Brevifloralactone acetate.
$C_{22}H_{30}O_5$ M 374.476
Constit. of *S. breviflora*. Cryst. Mp 112°.

Cuevas, G. *et al*, *Phytochemistry*, 1987, **26**, 2019.

3-Bromo-1,2,4-benzenetriol   B-70180

[99910-88-0]
$C_6H_5BrO_3$ M 205.008

Tri-Me ether: [25245-41-4]. *2-Bromo-1,3,4-
trimethoxybenzene.*
$C_9H_{11}BrO_3$ M 247.088
Cryst. (pentane). Mp 77-78°.

Bacon, R.G.R. *et al*, *J. Chem. Soc.* (*C*), 1969, 1978 (*synth,
deriv, pmr*)
Arias, S. *et al*, *Electrochim. Acta*, 1985, **30**, 1441.

5-Bromo-1,2,4-benzenetriol   B-70181

$C_6H_5BrO_3$ M 205.008
Tri-Ac: [23046-47-1].
$C_{12}H_{11}BrO_3$ M 283.121
Cryst. (EtOH). Mp 116.5-117.5°.
Tri-Me ether: [20129-11-7]. *1-Bromo-2,4,5-
trimethoxybenzene.*
$C_9H_{11}BrO_3$ M 247.088
Cubes (pet. ether). Mp 54-55°.

Fabinyi, R. *et al*, *Chem. Ber.*, 1910, **43**, 2676.
Blatchly, J.M. *et al*, *J. Chem. Soc.* (*C*), 1969, 1350 (*synth, pmr*)
Bacon, R.G.R. *et al*, *J. Chem. Soc.* (*C*), 1969, 1978 (*synth*)

6-Bromo-1,2,4-benzenetriol   B-70182

$C_6H_5BrO_3$ M 205.008
Amorph. powder. Mp 137-139°.

Tri-Ac: [38475-38-6].
$C_{12}H_{11}BrO_3$ M 283.121
Mp 73-74°.
Tri-Me ether: [23030-39-9]. *1-Bromo-2,3,5-
trimethoxybenzene.*
$C_9H_{11}BrO_3$ M 247.088
Oil. Bp_1 117-119°.

Hughes, G.K. *et al*, *Aust. J. Sci. Res., Ser. A*, 1950, **3**, 497
(*synth, deriv*)
Blatchly, J.M., *J. Chem. Soc.* (*C*), 1972, 2286 (*deriv*)
Sanchez, I.H. *et al*, *Tetrahedron*, 1985, **41**, 2355 (*synth, deriv,
uv, ir, pmr, cmr*)
Sinhababu, A.K. *et al*, *J. Heterocycl. Chem.*, 1988, **25**, 1155
(*synth, deriv, pmr*)

3-Bromo-4H-1-benzopyran-4-one   B-70183

3-Bromochromone
[49619-82-1]

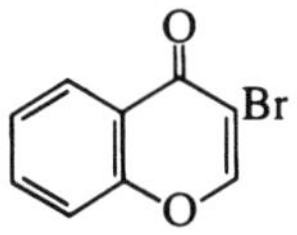

$C_9H_5BrO_2$ M 225.041
Needles (CH_2Cl_2/hexane). Mp 95-99°.

Gammill, R.B., *Synthesis*, 1979, 901 (*synth, ir, pmr*)
Davies, S.G. *et al*, *J. Chem. Soc., Perkin Trans. 1*, 1987, 2597
(*synth, ir, pmr, ms*)

1-Bromobicyclo[2.2.2]octane   B-70184

[7697-09-8]

$C_8H_{13}Br$ M 189.095
Cryst. by subl. Mp 66-68° (65°).

Honegger, E. *et al*, *Chem. Ber.*, 1987, **120**, 187 (*synth, ir, pmr,
pe*)

69

2-Bromo-3,3′-bipyridine, 9CI   B-70185

[101002-03-3]

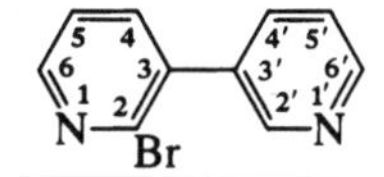

$C_{10}H_7BrN_2$ M 235.083
Cryst. Mp 97-98°.

Moran, D.B. *et al, J. Heterocycl. Chem.*, 1986, **23**, 1071.

4-Bromo-2,2′-bipyridine, 9CI   B-70186

[14162-95-9]
$C_{10}H_7BrN_2$ M 235.083
Needles (EtOH aq.). Mp 52°.
Picrate: [14097-60-0]. Needles (EtOH). Mp 174-175°.

Jones, R.A. *et al, J. Chem. Soc. (B)*, 1967, 106 (*synth*)
Cook, M.J. *et al, J. Chem. Soc., Perkin Trans. 2*, 1984, 1293 (*synth*)

5-Bromo-2,3′-bipyridine, 9CI   B-70187

[774-53-8]
$C_{10}H_7BrN_2$ M 235.083
Mp 75-77°. Bp₁ 150°.

Terashima, M. *et al, Chem. Pharm. Bull.*, 1985, **33**, 4755 (*synth, pmr, ir*)

5-Bromo-2,4′-bipyridine, 9CI   B-70188

[106047-33-0]
$C_{10}H_7BrN_2$ M 235.083
Mp 120-121°.

Terashima, M. *et al, Chem. Pharm. Bull.*, 1985, **33**, 4755 (*synth, pmr, ir*)

5-Bromo-3,3′-bipyridine, 9CI   B-70189

[15862-22-3]
$C_{10}H_7BrN_2$ M 235.083
Mp 94-95°. Bp₁ 130-131°.

Terashima, M. *et al, Chem. Pharm. Bull.*, 1985, **33**, 4755 (*synth, pmr, ir*)

5-Bromo-3,4′-bipyridine, 9CI   B-70190

[106047-38-5]
$C_{10}H_7BrN_2$ M 235.083
Cryst. (Me₂CO/Et₂O). Mp 149-150°.

Terashima, M. *et al, Chem. Pharm. Bull.*, 1985, **33**, 4755 (*synth, pmr, ir*)

6-Bromo-2,2′-bipyridine, 9CI   B-70191

[10495-73-5]
$C_{10}H_7BrN_2$ M 235.083
Cryst. (pet. ether). Mp 70-71°.

Case, F.H., *J. Org. Chem.*, 1966, **31**, 2399 (*synth*)
Keats, N.G. *et al, J. Heterocycl. Chem.*, 1979, **16**, 1431 (*ms*)

6-Bromo-2,3′-bipyridine, 9CI   B-70192

[106047-28-3]
$C_{10}H_7BrN_2$ M 235.083
Mp 73-74°. Bp₁ 150°.

Terashima, M. *et al, Chem. Pharm. Bull.*, 1985, **33**, 4755 (*synth, pmr, ir*)

6-Bromo-2,4′-bipyridine, 9CI   B-70193

[106047-29-4]
$C_{10}H_7BrN_2$ M 235.083
Cryst. (Me₂CO/hexane). Mp 110-111°.

Terashima, M. *et al, Chem. Pharm. Bull.*, 1985, **33**, 4755 (*synth, pmr, ir*)

3-Bromo-4-[(3-bromo-4,5-dihydroxyphenyl)methyl]-5-(hydroxy-methyl)-1,2-benzenediol, 9CI   B-70194

3-Bromo-2-(3-bromo-4,5-dihydroxybenzyl)-4,5-dihy-droxybenzyl alcohol. 5′-Hydroxyisoavrainvilleol
[111537-53-2]

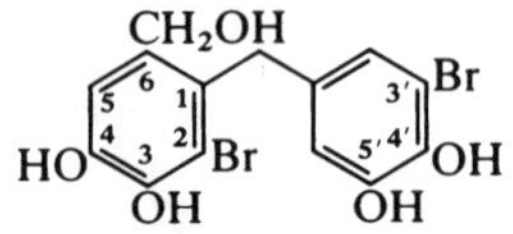

$C_{14}H_{12}Br_2O_5$ M 420.054
Constit. of *Avrainvillea nigricans*. Light-yellow oil.

Colon, M. *et al, J. Nat. Prod.*, 1987, **50**, 368.

2-[[3-Bromo-5-(1-bromopropyl)tetrahydro-2-furanyl]methyl]-5-(1-bromo-2-pro-pynyl)-tetrahydro-3-furanol, 9CI   B-70195

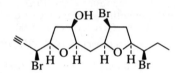

$C_{15}H_{21}Br_3O_2$ M 473.042
Ac: [94444-25-4]. *6-Acetoxy-3,10,13-tribromo-4,7:9,12-diepoxy-1-pentadecyne.*
$C_{17}H_{23}Br_3O_4$ M 531.078
Metab. of red alga *Laurencia obtusa*. Mobile oil. $[\alpha]_D$ −29.3° (c, 1.14 in CHCl₃).

Gonzalez, A.G. *et al, Tetrahedron*, 1984, **40**, 3443.

1-Bromo-1,2-butadiene   B-70196

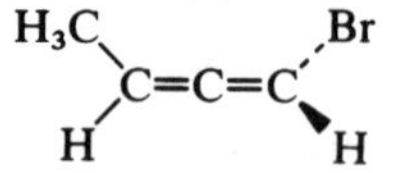

C_4H_5Br M 132.988
(R)-form [94137-75-4]
$[\alpha]_D^{25}$ −58.4° (c, 11.2 in EtOH). n_D^{20} 1.5240.
(±)-form [109279-75-6]
Liq.

Elsevier, C.J. *et al, J. Org. Chem.*, 1985, **50**, 364 (*synth, ir, pmr, cmr*)
Caporusso, A.M. *et al, J. Org. Chem.*, 1987, **52**, 3920 (*synth*)

4-Bromobutanal   B-70197

[38694-47-2]

$$BrCH_2CH_2CH_2CHO$$

C_4H_7BrO M 151.003
Bp₀.₀₇ 35-40°.

Vedejs, E. *et al, J. Org. Chem.*, 1979, **44**, 3230 (*synth, ir, pmr*)
Little, R.D. *et al, J. Org. Chem.*, 1982, **47**, 362 (*synth, ir, pmr*)

4-Bromo-1-butyne   B-70198

[38771-21-0]

$$HC\equiv CCH_2CH_2Br$$

C_4H_5Br M 132.988
Bp 109-110°.

Eglington, G. *et al, J. Chem. Soc.*, 1950, 3650 (*synth*)
Schlubach, H.H. *et al, Justus Liebigs Ann. Chem.*, 1950, **568**, 141 (*synth*)

5-Bromo-7-chlorocavernicolin   B-70199

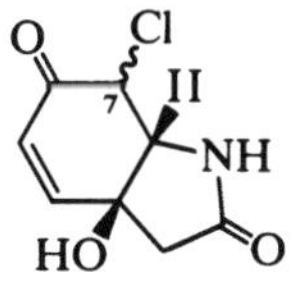

$C_8H_7BrClNO_3$ M 280.505
Metab. from the Mediterranean sponge *Aplysina cavernicola* (*Verongia cavernicola*) (Verongidae). Racemic. Isol. as a mixt. of C(7)-epimerizing monoacetates.

D'Ambrosio, M. *et al, Helv. Chim. Acta*, 1984, **67**, 1484 (*ms, struct*)

1-Bromo-6-chlorohexane, 9Cl   B-70200

[6294-17-3]

$$BrCH_2(CH_2)_4CH_2Cl$$

$C_6H_{12}BrCl$ M 199.518
Liq. Bp_{13} 95-96°.

Gautier, J.A. *et al, Bull. Soc. Chim. Fr.*, 1963, 1368 (*synth*)
Camps, F. *et al, Synthesis*, 1987, 511 (*synth, pmr, cmr*)

2-Bromo-1,3-cyclohexanedione   B-70201

2-Bromodihydroresorcinol
[60060-44-8]

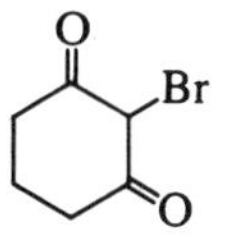

$C_6H_7BrO_2$ M 191.024
Cryst. Mp 170-171°.
Me ether: [85493-96-5]. *2-Bromo-3-methoxy-2-cyclohexen-1-one, 9Cl.*
$C_7H_9BrO_2$ M 205.051
Cryst. (EtOAc). Mp 92-94°.

Nazarov, I.N. *et al, Izv. Akad. Nauk SSSR, Ser. Khim.*, 1959, 668; *Engl. Transl.*, p. 639 (*synth*)
Shepherd, R.G. *et al, J. Chem. Soc. (A)*, 1987, 2153 (*synth, deriv, ir, pmr*)

4-Bromo-4,4-difluoro-2-butynoic acid, 9Cl   B-70202

(Bromotrifluoromethyl)propiolic acid
[82477-42-7]

$$BrF_2CC\equiv CCOOH$$

$C_4HBrF_2O_2$ M 198.952
Mp 25°. Bp_2 130°.

Rico, I. *et al, J. Chem. Soc., Perkin Trans. 1*, 1982, 1063 (*synth, pmr, F pmr*)

3-Bromo-3,3-difluoro-1-phenylpropyne   B-70203

(3-Bromo-3,3-difluoro-1-propynyl)benzene, 9Cl.
(Bromodifluoromethyl)phenylacetylene
[82477-43-8]

$$PhC\equiv CCF_2Br$$

$C_9H_5BrF_2$ M 231.039
Liq. Bp_1 50°.

Rico, I. *et al, J. Chem. Soc., Perkin Trans. 1*, 1982, 1063 (*synth, pmr, F nmr*)

3-Bromo-2,3-dihydrothiophene   B-70204

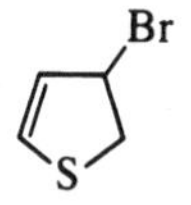

C_4H_5BrS M 165.048
(±)-*form*

1,1-Dioxide: [53336-42-8]. *4-Bromo-2-sulfolene.*
$C_4H_5BrO_2S$ M 197.046
Butadienyl cation equivalent. Large cryst. (H_2O). Mp 63.5-64°.

Bailey, W.J. *et al, J. Am. Chem. Soc.*, 1954, **76**, 1932 (*synth*)
Chou, T. *et al, J. Am. Chem. Soc.*, 1987, **52**, 3394 (*use*)

5-Bromo-3,4-dihydro-4-thioxo-2(1H)-pyrimidinone, 9Cl   B-70205

5-Bromo-4-thiouracil
[72812-45-4]

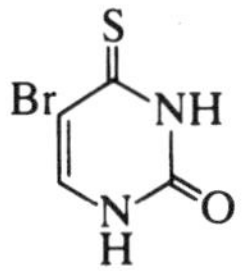

$C_4H_3BrN_2OS$ M 207.045
Cryst. Mp 244-245° dec.

White, W.E. *et al, Biochem. Pharmacol.*, 1979, **28**, 1467 (*synth*)
Ľapucha, A., *Synthesis*, 1987, 256 (*synth, ir, pmr, cmr, ms*)

1-(4-Bromo-2,5-dihydroxyphenyl)-7-hydroxy-3,7-dimethyl-2-octen-1-one   B-70206

(4-Bromo-2,5-dihydroxyphenyl)(2,6-dimethyl-6-hydroxy-2-heptenyl)ketone

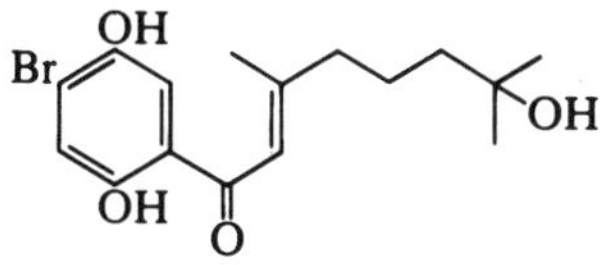

$C_{16}H_{21}BrO_4$ M 357.244
(E)-*form*
Constit. of *Cymopolia barbata*. Cryst. (hexane/CH_2Cl_2). Mp 126-128°.

Estrada, D.M. *et al, J. Nat. Prod.*, 1987, **50**, 735.

1-Bromo-3,3-dimethyl-1-butyne, 9CI B-70207
Bromo-tert-*butylacetylene*
[13601-86-0]

$$(H_3C)_3CC{\equiv}CBr$$

C_6H_9Br M 161.041
Liq. Bp_{118} 58°. Unstable.

Katritzky, A.R. *et al, J. Chem. Soc., Perkin Trans. 2*, 1974, 282 (*ir*)
Razuvaev, G.A. *et al, J. Organomet. Chem.*, 1981, **222**, 55.
Naemura, K. *et al, Tetrahedron*, 1986, **42**, 1763 (*synth*)

4-Bromo-3,5-dimethylpyrazole, 9CI B-70208
[3398-16-1]

$C_5H_7BrN_2$ M 175.028
Needles (Et_2O), cryst. (hexane or Et_2O/pet. ether). Mp 124-125°.
B,HBr: Needles. Mp 174°.
Picrate: Cryst. (EtOH). Mp 194-195°.

Grandberg, I.I. *et al, J. Gen. Chem. USSR*, 1962, **32**, 3515 (*deriv*)
Elguero, J. *et al, Bull. Soc. Chim. Fr.*, 1966, 3744 (*uv*)
Wilczynski, J.J. *et al, J. Org. Chem.*, 1974, **39**, 1909 (*pmr, synth*)
Kurihara, T. *et al, Chem. Pharm. Bull.*, 1978, **26**, 1141 (*synth*)
Faure, R. *et al, Can. J. Chem.*, 1988, **66**, 1141 (*cmr*)

3-Bromo-1,2-diphenylpropene B-70209
*1,1'-[1-(Bromomethyl)-1,2-ethenediyl]bisbenzene, 9CI.
2,3-Diphenylallyl bromide*
[70671-91-9]

(Z)-form

$C_{15}H_{13}Br$ M 273.172
$Bp_{0.01}$ 131-133°.
(Z)-form [106403-98-9]
Oil.

Lüttringhaus, R. *et al, Justus Liebigs Ann. Chem.*, 1948, **560**, 201 (*synth*)
Suárez, A.R. *et al, J. Org. Chem.*, 1987, **52**, 1145 (*synth, pmr*)

2-(2-Bromoethyl)pyridine B-70210
[39232-04-7]

C_7H_8BrN M 186.051
B,HBr: [72996-65-7]. Cryst. (2-propanol). Mp 133-135°.
Wróbel, J.T. *et al, Synthesis*, 1987, 452 (*synth, ir, pmr*)

3-(2-Bromoethyl)pyridine B-70211
C_7H_8BrN M 186.051
B,HBr: [41039-91-2]. Cryst. Mp 124-125°.
Hawes, E.M. *et al, J. Heterocycl. Chem.*, 1973, **10**, 39 (*synth, ir, pmr*)

1-Bromo-3-fluoropropane B-70212
[352-91-0]

$$BrCH_2CH_2CH_2F$$

C_3H_6BrF M 140.983
Liq. d_4^{25} 1.54. Bp 100-101.5°.

Hoffman, F.W., *J. Org. Chem.*, 1949, **14**, 105; 1950, **15**, 425 (*synth, props*)
Chi, D.Y. *et al, J. Org. Chem.*, 1987, **52**, 658 (*synth, pmr*)

2-Bromo-3-fluoropyridine B-70213
[40273-45-8]

C_5H_3BrFN M 175.988
Oil. Bp_{22} 62°.

Marsais, F. *et al, C.R. Hebd. Seances Acad. Sci., Ser. C*, 1972, **275**, 1535 (*synth*)
Marsais, F. *et al, Tetrahedron*, 1983, **39**, 2009 (*synth*)
Saari, W.S. *et al, J. Med. Chem.*, 1984, **27**, 1182 (*synth*)

3-Bromo-4-fluoropyridine B-70214
[116922-60-2]
C_5H_3BrFN M 175.988
Bp_{20} 70°.

Marsais, F. *et al, J. Heterocycl. Chem.*, 1988, **25**, 81 (*synth, pmr*)

8-Bromoguanosine, 9CI B-70215
2-Amino-8-bromo-9-β-D-ribofuranosyl-9H-purin-6(1H)-one. 8-Bromo-9-β-D-ribofuranosylguanine. 2-Amino-8-bromoinosine
[4016-63-1]

$C_{10}H_{12}BrN_5O_5$ M 362.139
Cryst. (H_2O). Mp 200° dec. (softens at 192°). $[\alpha]_D^{30}$ −25.9° (c, 2 in DMSO).
2',3',5'-Tri-Ac: Cryst. (Me_2CO aq.). Mp 216-218°.

Holmes, R.E. *et al, J. Am. Chem. Soc.*, 1964, **86**, 1242 (*synth*)
Ikehara, M. *et al, Chem. Pharm. Bull.*, 1965, **13**, 639 (*synth*)
Bugg, C.E. *et al, Biochem. Biophys. Res. Commun.*, 1969, **37**, 623 (*cryst struct*)
Srivastava, P.C. *et al, Experientia*, 1970, **26**, 220 (*synth*)
Tavale, S.S. *et al, J. Mol. Biol.*, 1970, **48**, 109 (*cryst struct*)
Uesugi, S. *et al, Chem. Pharm. Bull.*, 1978, **26**, 3040 (*cmr*)
Guschlbauer, W., *Biochem. Biophys. Acta*, 1980, **610**, 47 (*pmr, conformn*)
Lin, T.S. *et al, J. Med. Chem.*, 1985, **28**, 1194 (*synth*)

2-Bromo-1,5-hexadiene B-70216
[101933-87-3]

$$H_2C{=}CHCH_2CH_2CBr{=}CH_2$$

C_6H_9Br M 161.041
Bp$_{28}$ 43°.

Peterson, P.E. *et al*, *J. Org. Chem.*, 1986, **51**, 2381 (*synth, pmr, cmr*)

α-(1-Bromo-3-hexenyl)-5-(1-bromo-2-propynyl)tetrahydro-3-hydroxy-2-furanethanol, 9CI   B-70217

$C_{15}H_{22}Br_2O_3$ M 410.145
Ac: [94444-24-3]. *6-Acetoxy-3,10-dibromo-4,7-epoxy-12-pentadecen-1-yne.*
$C_{17}H_{24}Br_2O_4$ M 452.182
Metab. of red alga *Laurencia obtusa*. Oil. $[\alpha]_D$ −14.5° (c, 0.62 in CHCl$_3$).

Gonzalez, A.G. *et al*, *Tetrahedron*, 1984, **40**, 3443.

1-Bromo-1-hexyne   B-70218
[1119-64-8]

$$H_3C(CH_2)_3C{\equiv}CBr$$

C_6H_9Br M 161.041
Liq. Bp$_{39}$ 60°, Bp$_{15}$ 38°.

Pflaum, D.J. *et al*, *J. Am. Chem. Soc.*, 1934, **56**, 1106 (*synth*)
Ando, T. *et al*, *Agric. Biol. Chem.*, 1982, **46**, 717 (*synth, ir, pmr*)

1-Bromo-2-hexyne, 9CI   B-70219
[18495-25-5]

$$BrCH_2C{\equiv}CCH_2CH_2CH_3$$

C_6H_9Br M 161.041
Liq. Bp$_{80}$ 97-98°, Bp$_{18}$ 65-69°.

Newman, M.S. *et al*, *J. Am. Chem. Soc.*, 1949, **71**, 1292 (*synth*)
Kajiwara, T. *et al*, *Agric. Biol. Chem.*, 1977, **41**, 1481 (*synth*)
Carless, H.A.J. *et al*, *J. Chem. Soc., Perkin Trans. 1*, 1987, 1999 (*synth, ir, pmr*)

2-Bromo-1*H*-imidazole, 9CI   B-70220
[16681-56-4]

$C_3H_3BrN_2$ M 146.974
Long prismatic needles (H$_2$O). Mp 207°.
▷Violent dec. at Mp
B,HNO₃: Stout needles (H$_2$O). Mp 137° dec.
Picrate: Long glistening yellow needles (H$_2$O). Mp ca. 232° dec.

King, H. *et al*, *J. Chem. Soc.*, 1923, 621 (*synth*)

4(5)-Bromo-1*H*-imidazole   B-70221
[2302-25-2]
$C_3H_3BrN_2$ M 146.974
Plates (H$_2$O, CHCl$_3$ or C$_6$H$_6$). Mp 130°.
B,HCl: Cryst. (5*M* HCl). Mp 162-165°.
B,HNO₃: Needles (H$_2$O). Mp 135° dec.

B,(COOH)₂: Needles (H$_2$O). Mp 218° dec. (after sintering).
Picrate: Short yellow needles (H$_2$O). Mp 162°.
1-Me: *4-Bromo-1-methyl-1*H-*imidazole*.
    $C_4H_5BrN_2$ M 161.001
    Mp 155° (as nitrate).
3-Me: *5-Bromo-1-methyl-1*H-*imidazole*.
    $C_4H_5BrN_2$ M 161.001
    Mp 45-46°.
3-Me; B,HCl: Mp 155°.
1-Benzyl: [106848-38-8].
    $C_{10}H_9BrN_2$ M 237.099
    Cryst. (Et$_2$O/pet. ether). Mp 88-90°.

Balaban, E. *et al*, *J. Chem. Soc.*, 1922, 947; 1924, 1564 (*synth*)
Stensiö, K.-E. *et al*, *Acta Chem. Scand.*, 1973, **27**, 2179 (*synth, pmr*)
Iddon, B. *et al*, *J. Chem. Soc., Perkin Trans. 1*, 1987, 1453 (*deriv, synth, pmr*)

1-Bromo-2-iodo-3-nitrobenzene   B-70222
[32337-96-5]

$C_6H_3BrINO_2$ M 327.904
Cryst. Mp 120.5-121.5°.

Liedholm, B. *et al*, *Acta Chem. Scand.*, 1971, **25**, 113 (*synth*)
Liedholm. B. *et al*, *Acta Chem. Scand., Ser. A*, 1984, **38**, 555 (*synth, pmr*)

2-Bromo-1-iodo-3-nitrobenzene   B-70223
[32337-95-4]
$C_6H_3BrINO_2$ M 327.904
Cryst. (MeOH). Mp 99-100°.

Liedholm, B., *Acta Chem. Scand.*, 1971, **25**, 106 (*synth*)

4-Bromo-1(3*H*)-isobenzofuranone   B-70224
4-Bromophthalide
[102308-43-0]

$C_8H_5BrO_2$ M 213.030
Cryst. by subl. Mp 103-104°.

Soucy, C. *et al*, *J. Org. Chem.*, 1987, **52**, 129 (*synth, ir, pmr*)

7-Bromo-1(3*H*)-isobenzofuranone   B-70225
7-Bromophthalide
[105694-44-8]
$C_8H_5BrO_2$ M 213.030
Cryst. by subl. Mp 141-143°.

Soucy, C. *et al*, *J. Org. Chem.*, 1987, **52**, 129 (*synth, ir, pmr*)

2-Bromo-4-mercaptobenzoic acid   B-70226
3-Bromo-4-carboxybenzenethiol

$C_7H_5BrO_2S$ M 233.079
Cryst. (EtOH aq.). Mp 182-184°.
Me thioether: 2-*Bromo*-4-(*methylthio*)*benzoic acid.*
   $C_8H_7BrO_2S$ M 247.106
   Cryst. (EtOH aq.). Mp 181-183°.
Netherlands Pat., 6 512 881, (*1966*); *CA*, **65**, 8916d (*synth, deriv*)

2-Bromo-5-mercaptobenzoic acid   B-70227
4-Bromo-3-carboxybenzenethiol
[84889-60-1]
$C_7H_5BrO_2S$ M 233.079
Me ether: [22362-40-9]. 2-*Bromo*-5-(*methylthio*)*benzoic acid.*
   $C_8H_7BrO_2S$ M 247.106
   Prisms (Et$_2$O/pet. ether). Mp 146-147°.
Künzle, F. *et al, Helv. Chim. Acta*, 1969, **52**, 622 (*synth, deriv*)

3-Bromo-4-mercaptobenzoic acid   B-70228
2-Bromo-4-carboxybenzenethiol
[58123-70-9]
$C_7H_5BrO_2S$ M 233.079
Cryst. (EtOH). Mp 230°. pK_a 4.61 (50% EtOH aq., 25°).
Me ether: [58123-71-0]. 3-*Bromo*-4-(*methylthio*)*benzoic acid.*
   $C_8H_7BrO_2S$ M 247.106
   Cryst. (EtOH). Mp 222°. pK_a 5.33 (50% EtOH aq., 25°).
Kalfus, K. *et al, Collect. Czech. Chem. Commun.*, 1975, **40**, 3009 (*synth, deriv*)

1-(Bromomethyl)acridine   B-70229
[113139-04-1]

$C_{14}H_{10}BrN$ M 272.144
Pale-yellow needles (hexane). Mp 153-155° dec.
Takahashi, K. *et al, J. Heterocycl. Chem.*, 1987, **24**, 977 (*synth, pmr*)

2-(Bromomethyl)acridine   B-70230
[113139-05-2]
$C_{14}H_{10}BrN$ M 272.144
Prisms. Mp 159-161° dec. Not fully purified.
Takahashi, K. *et al, J. Heterocycl. Chem.*, 1987, **24**, 977 (*synth, pmr*)

3-(Bromomethyl)acridine   B-70231
[113139-06-3]
$C_{14}H_{10}BrN$ M 272.144

Pale-yellow prisms (hexane/C$_6$H$_6$). Mp 152-154° dec.
Takahashi, K. *et al, J. Heterocycl. Chem.*, 1987, **24**, 977 (*synth, pmr*)

4-(Bromomethyl)acridine   B-70232
[15787-60-7]
$C_{14}H_{10}BrN$ M 272.144
Pale-yellow needles (hexane/C$_6$H$_6$). Mp 164-165°.
Takahashi, K. *et al, J. Heterocycl. Chem.*, 1987, **24**, 977 (*synth, pmr*)

9-(Bromomethyl)acridine   B-70233
[1556-34-9]
$C_{14}H_{10}BrN$ M 272.144
Reagent for fluorimetric detn. of tertiary amines. Cryst. (CCl$_4$). Mp 169-170°.
Campbell, A. *et al, J. Chem. Soc.*, 1958, 1145 (*synth*)
Lehr, R.E. *et al, J. Pharm. Sci.*, 1975, **64**, 950 (*synth, use*)

2-Bromo-3-methylaniline   B-70234
Updated Entry replacing B-02654
2-Bromo-3-methylbenzenamine, 9CI. 2-Bromo-m-toluidine, 8CI. 3-Amino-2-bromotoluene
[54879-20-8]

C_7H_8BrN M 186.051
Cryst. or pale-yellow oil. Mp 80-81°.
N-Ac: [54879-19-5].
   $C_9H_{10}BrNO$ M 228.088
   Cryst. Mp 144.0-144.5°.
Bunnett, J.F. *et al, J. Am. Chem. Soc.*, 1961, **83**, 1691 (*synth*)
Inoue, S. *et al, J. Chem. Soc., Perkin Trans. 1*, 1974, 2097 (*synth, ir*)
Krolski, M.E. *et al, J. Org. Chem.*, 1988, **53**, 1170 (*synth, ir, pmr*)

2-(Bromomethyl)-6-chloropyridine   B-70235
[63763-79-1]

C_6H_5BrClN M 206.469
B,HBr: [32938-50-4]. Mp 95-111° dec. Unstable to recrystn.
Norton, S.J. *et al, J. Med. Chem.*, 1971, **14**, 557 (*synth*)

3-(Bromomethyl)-2-chloropyridine   B-70236
C_6H_5BrClN M 206.469
B,HBr: [32918-38-0]. Mp 98-110° dec. Unstable to recrystn.
Norton, S.J. *et al, J. Med. Chem.*, 1971, **14**, 557 (*synth*)

4-(Bromomethyl)-2-chloropyridine, 9CI   B-70237
[83004-15-3]
C_6H_5BrClN M 206.469

B,HBr: [32938-48-0]. Mp 113-119° dec. Unstable to recrystn.
1-Oxide: [70258-20-7].
C$_6$H$_5$BrClNO   M 222.469
Cryst. (THF/hexane). Mp 88-90°.
1-Oxide; B,HBr: Cryst. (MeOH). Mp 138-142°.

Norton, S.J. *et al, J. Med. Chem.*, 1971, **14**, 557 (*synth*)
Tilley, J.W. *et al, J. Heterocycl. Chem.*, 1979, **16**, 333 (*derivs*)

5-Bromomethyl)-2-chloropyridine   B-70238
C$_6$H$_5$BrClN   M 206.469
B,HBr: [32918-40-4]. Mp 92-98° dec. Unstable to re-
crystn.

Norton, S.J. *et al, J. Med. Chem.*, 1971, **14**, 557 (*synth*)

2-(Bromomethyl)cyclohexanamine, 9CI   B-70239
1-Amino-2-(bromomethyl)cyclohexane

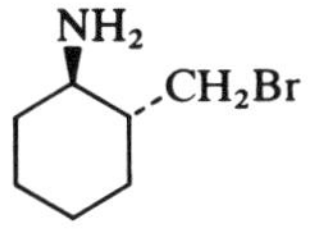

(1R*,2R*)-form

C$_7$H$_{14}$BrN   M 192.098
(1R*,2R*)-form
(+)-trans-*form*
B,HBr: [29586-62-7]. Mp 207-208°. [α]$_D^{20}$ +36.0° (c, 3
in EtOH).
(1RS,2RS)-form
(±)-trans-*form*
B,HBr: [72053-13-5]. Cryst. (EtOH/Et$_2$O). Mp 220-
221°.
(1RS,2SR)-form
(±)-cis-*form*
B,HBr: [24717-06-4]. Cryst. (EtOH/Et$_2$O). Mp 160°.

Armarego, W.L.F. *et al, J. Chem. Soc. (C)*, 1969, **12**, 1635
(*synth*)
Nohira, H. *et al, Bull. Chem. Soc. Jpn.*, 1970, **43**, 2230 (*resoln*)
Gondos, G. *et al, J. Chem. Soc., Perkin Trans. 1*, 1979, 1770
(*synth, deriv*)

2-(Bromomethyl)cyclopentanamine, 9CI   B-70240
1-Amino-2-(bromomethyl)cyclopentane

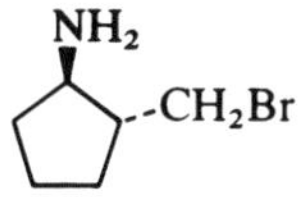

(1RS,2RS)-form

C$_6$H$_{12}$BrN   M 178.072
(1RS,2RS)-form
(±)-trans-*form*
B,HBr: [72053-12-4]. Cryst. (EtOH/Et$_2$O). Mp 177-
178°.
(1RS,2SR)-form
(±)-cis-*form*
B,HBr: [72053-11-3]. Cryst. (EtOH/Et$_2$O). Mp 155-
156°.

Armarego, W.L.F. *et al, J. Chem. Soc. (C)*, 1969, **12**, 1635
(*synth*)
Gondos, G. *et al, J. Chem. Soc., Perkin Trans. 1*, 1979, 1770
(*synth*)

(Bromomethyl)cyclopentane   B-70241
[3814-30-0]

CH$_2$Br

C$_6$H$_{11}$Br   M 163.057
Bp$_1$ 44°.

Nystrom, R.G. *et al, J. Am. Chem. Soc.*, 1947, **69**, 2548 (*synth*)
Wiley, G.A. *et al, J. Am. Chem. Soc.*, 1964, **86**, 964 (*synth*)
Samsel, E.G. *et al, J. Am. Chem. Soc.*, 1986, **108**, 4790 (*synth,
nmr*)

1-Bromo-1-methylcyclopropane   B-70242
[50915-27-0]

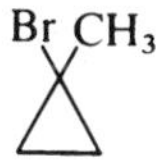

Br CH$_3$

C$_4$H$_7$Br   M 135.003
Bp 78-80°.

Anderson, B.C., *J. Org. Chem.*, 1962, **27**, 2720 (*synth, pmr*)
Deycard, S. *et al, J. Am. Chem. Soc.*, 1987, **109**, 4954 (*synth,
pmr, ms*)

2-Bromomethyl)-6-fluoropyridine   B-70243
[100202-78-6]

F—N—CH$_2$Br

C$_6$H$_5$BrFN   M 190.015
Mp 108-112° dec.
B,HBr: [31140-62-2]. Mp 108-112° dec. Unstable to re-
crystn.

Norton, S.J. *et al, J. Med. Chem.*, 1971, **14**, 211 (*synth*)

3-(Bromomethyl)-2-fluoropyridine   B-70244
C$_6$H$_5$BrFN   M 190.015
B,HBr: [31140-59-7]. Mp 105-120° dec. Unstable to re-
crystn.

Norton, S.J. *et al, J. Med. Chem.*, 1971, **14**, 211 (*synth*)

4-(Bromomethyl)-2-fluoropyridine   B-70245
[64992-03-6]
C$_6$H$_5$BrFN   M 190.015
B,HBr: [31140-60-0]. Mp 93-105° dec. Unstable to recrystn.

Norton, S.J. *et al, J. Med. Chem.*, 1971, **14**, 211 (*synth*)

5-(Bromomethyl)-2-fluoropyridine   B-70246
C$_6$H$_5$BrFN   M 190.015
B,HBr: [31140-61-1]. Mp 114-120° dec. Unstable to re-
crystn.

Norton, S.J. *et al, J. Med. Chem.*, 1971, **14**, 211 (*synth*)

1-[3-(Bromomethyl)-1-methyl-2-methylene-cyclopentyl]-4-methyl-1,4-cyclohexa-diene, 9CI   B-70247

5-Bromomethyl-2-(4-methyl-1,4-cyclohexadienyl)-2-methyl-1-methylenecyclopentane

[93395-09-6]

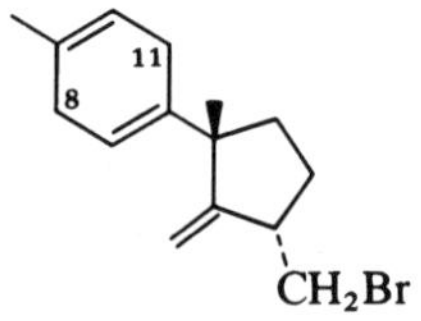

$C_{15}H_{21}Br$ M 281.235

Constit. of red alga *Laurencia pinnatifida*. Unstable oil.
$[\alpha]_D$ +2.3° (c, 2.15 in $CHCl_3$).

8,11-Didehydro: [93395-10-9]. *1-[3-(Bromomethyl)-1-methyl-2-methylenecyclopentyl]-4-methylbenzene, 9CI. 3-(Bromomethyl)-1-methyl-2-methylene-1-(4-bromophenyl)cyclopentane.*
$C_{15}H_{19}Br$ M 279.219
Constit. of *L. pinnatifida*. Oil. $[\alpha]_D$ +4.7° (c, 3.19 in $CHCl_3$).

Gonzalez, A.G. *et al, Tetrahedron*, 1984, **40**, 2751.

2-Bromo-4-methylpentanoic acid, 9CI   B-70248

2-Bromo-4-methylvaleric acid, 8CI

[49628-52-6]

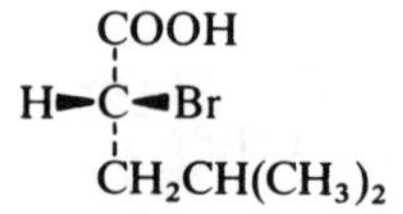

$$\underset{\underset{CH_2CH(CH_3)_2}{|}}{\overset{\overset{COOH}{|}}{H \blacktriangleright C \blacktriangleleft Br}} \qquad (R)\text{-form}$$

$C_6H_{11}BrO_2$ M 195.056

(R)-form [42990-28-3]
D-form
Oil. $Bp_{0.5}$ 88-90°. $[\alpha]_D^{22}$ +39.6° (c, 1.4 in MeOH).
(S)-form [28659-87-2]
L-form
Oil. $Bp_{0.5}$ 87-88°. $[\alpha]_D^{22}$ −38.9° (c, 1 in MeOH).

La Noce, T. *et al, Ann Chim.* (*Rome*), 1968, **58**, 393 (*synth*)
Gaffield, W. *et al, Tetrahedron*, 1971, **27**, 915 (*cd, ord, uv*)
Dutta, A.S. *et al, J. Chem. Soc., Perkin Trans. 1*, 1987, 111 (*synth*)

2-Bromomethyl-2-propenoic acid, 9CI   B-70249

Updated Entry replacing B-02838
3-Bromo-2-methylenepropanoic acid. α-Bromomethylacrylic acid

[72707-66-5]

$$H_2C=C(CH_2Br)COOH$$

$C_4H_5BrO_2$ M 164.986
Me ester: [4224-69-5].
$C_5H_7BrO_2$ M 179.013
$Bp_{8.3}$ 40-42°.
Et ester: [17435-72-2].
$C_6H_9BrO_2$ M 193.040
$Bp_{3.9}$ 59-65°.

Ferris, A.F., *J. Org. Chem.*, 1955, **20**, 780 (*synth*)
Villieras, J. *et al, Synthesis*, 1982, 924 (*synth*)
Anzeveno, P.B. *et al, Synth. Commun.*, 1986, **16**, 387 (*synth, bibl*)
Drewes, S.E. *et al, Synth. Commun.*, 1987, **17**, 291 (*synth, pmr*)

3(5)-Bromo-4-methylpyrazole, 9CI   B-70250

[5932-20-7]

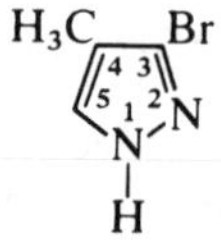

$C_4H_5BrN_2$ M 161.001
Cryst. (EtOH aq.). Mp 127-128°.
1-Me: [13745-59-0]. *3-Bromo-1,4-dimethylpyrazole.*
$C_4H_5BrN_2$ M 161.001
No phys. props. reported.

Hüttel, R. *et al, Justus Liebigs Ann. Chem.*, 1955, **593**, 179 (*synth*)
Elguero, J. *et al, Bull. Soc. Chim. Fr.*, 1966, 3727, 3744 (*synth, pmr, uv*)
Elguero, J. *et al, C.R. Hebd. Seances Acad. Sci., Ser. C*, 1966, **263**, 1456 (*synth, pmr, deriv*)
Cabildo, P. *et al, Org. Magn. Reson.*, 1984, **22**, 603 (*cmr*)

4-Bromo-3(5)-methylpyrazole, 9CI   B-70251

[13808-64-5]
$C_4H_5BrN_2$ M 161.001
Exists as 5-Me tautomer in solid state. Plant growth stimulator. Cryst. (ligroin). Mp 76-77°.
B,HBr: Needles (H_2O). Mp 135°.
Picrate: Fine pale-yellow needles. Mp 158°.
1-Me: [5775-82-6]. *4-Bromo-1,3-dimethylpyrazole.*
$C_5H_7BrN_2$ M 175.028
Mobile oil. d_4^{15} 1.498. Bp_{40} 100°. n_D^{20} 1.5285.
1-Me; B,HBr: Mp 197°.
1-Me, picrate: Light-yellow cryst. (EtOH). Mp 116-117°.
2-Me: [5775-86-0]. *4-Bromo-1,5-dimethylpyrazole.*
$C_5H_7BrN_2$ M 175.028
Mp 40-42° (38°). Bp_{21} 105°.
2-Me; B,HBr: Mp 125-126° (120-122°).
2-Me, picrate: Greenish or light-yellow cryst. (EtOH). Mp 122-122.5°.

von Auwers, K. *et al, Justus Liebigs Ann. Chem.*, 1924, **437**, 36 (*synth, deriv*)
Elguero, J. *et al, Bull. Soc. Chim. Fr.*, 1966, 293, 2727, 2832, 3727, 3744 (*synth, pmr, uv*)
Butler, R.N., *Can. J. Chem.*, 1973, **51**, 2315 (*pmr, derivs*)
Cabildo, P. *et al, Org. Magn. Reson.*, 1984, **22**, 603 (*cmr, derivs*)
Faure, R. *et al, Can. J. Chem.*, 1988, **66**, 1141 (*cmr, tautom*)

3(5)-Bromo-5(3)-methylpyrazole   B-70252

[57097-81-1]
$C_4H_5BrN_2$ M 161.001
Exists as 3-bromo-5-methyl tautomer in solid state. Cryst. solid by subl. Mp 138-139°.
1-Me: [5744-80-9]. *3-Bromo-1,5-dimethylpyrazole.*
$C_5H_7BrN_2$ M 175.028
No phys. props. reported.
2-Me: [5744-70-7]. *5-Bromo-1,3-dimethylpyrazole.*
$C_5H_7BrN_2$ M 175.028
Bp_{30} 90°.

Elguero, J. *et al, Bull. Soc. Chim. Fr.*, 1966, 293, 3727, 3744 (*synth, uv, pmr, derivs*)
Butler, R.N., *Can. J. Chem.*, 1973, **51**, 2315 (*pmr, derivs*)
Faure, R. *et al, Can. J. Chem.*, 1988, **66**, 1141 (*cmr, tautom*)
Jain, R. *et al, J. Am. Chem. Soc.*, 1988, **110**, 1356 (*synth, pmr, ms*)

2-(Bromomethyl)pyridine, 9CI B-70253

2-Pyridylmethyl bromide. 2-Picolyl bromide

[55401-97-3]

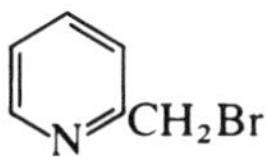

C_6H_6BrN M 172.024

B,HBr: [31106-82-8]. Needles. Mp 145-146°.
Picrate: Cryst. (Me_2CO/EtOH). Mp 152-153°.

Šorm, F. *et al, Collect. Czech. Chem. Commun.*, 1948, **13**, 289 (*synth*)
Winterfeld, K. *et al, Arch. Pharm. (Weinheim, Ger.)*, 1956, **26**, 448 (*synth*)
Hamana, M. *et al, Chem. Pharm. Bull.*, 1960, **8**, 692 (*synth*)

3-(Bromomethyl)pyridine, 9CI B-70254

3-Pyridylmethyl bromide. 3-Picolyl bromide

[69966-55-8]

C_6H_6BrN M 172.024

B,HBr: [4916-55-6]. Cryst. (MeOH). Mp 147.5-149° subl.

Coombes, R.G. *et al, J. Chem. Soc.*, 1965, 7029 (*synth*)
Fischer, A. *et al, Can. J. Chem.*, 1978, **56**, 3068 (*synth*)
Jokela, R. *et al, Heterocycles*, 1985, **23**, 1707.

4-(Bromomethyl)pyridine, 9CI B-70255

4-Pyridylmethyl bromide. 4-Picolyl bromide

[54751-01-8]

C_6H_6BrN M 172.024

B,HBr: [73870-24-3]. Needles. Mp 185-187° dec.
1-Oxide; B,HBr: Mp 161-162°.

Bixler, R.L. *et al, J. Org. Chem.*, 1958, **23**, 575 (*synth*)
Hamano, M. *et al, Chem. Pharm. Bull.*, 1962, **10**, 961 (*synth*)

2-Bromo-3-nitrobenzyl alcohol, 8CI B-70256

2-Bromo-3-nitrobenzenemethanol, 9CI

[90407-20-8]

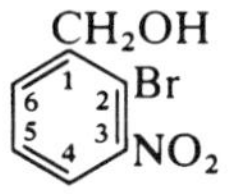

$C_7H_6BrNO_3$ M 232.033
Needles (C_6H_6). Mp 76-77°.

Rahman, L.K.A. *et al, J. Chem. Soc., Perkin Trans. 1*, 1984, 385 (*synth, ir, pmr*)

2-Bromo-3-nitrophenol B-70257

Updated Entry replacing B-03049

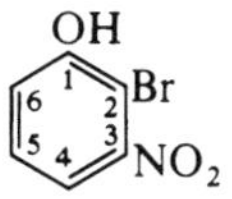

$C_6H_4BrNO_3$ M 218.007
Yellow needles (EtOH aq.). Mp 147°. Subl.

Me ether: [67853-37-6]. *2-Bromo-1-methoxy-3-nitro-benzene. 2-Bromo-3-nitroanisole.*
$C_7H_6BrNO_3$ M 232.033
Needles (EtOH). Mp 103-104°.

Et ether: 2-Bromo-1-ethoxy-3-nitrobenzene. 2-Bromo-3-nitrophenetole.
$C_8H_8BrNO_3$ M 246.060
Yellow prisms. Mp 57°.

Schlieper, F.W., *Ber.*, 1892, **25**, 552.

Ando, M. *et al, Bull. Chem. Soc. Jpn.*, 1978, **51**, 2437 (*synth, deriv*)
Krolski, M.E. *et al, J. Org. Chem.*, 1988, **53**, 1170 (*synth, Me ether, pmr, ir*)

1-Bromo-1-nitropropane B-70258

[5447-96-1]

$$H_3CCH_2CHBrNO_2$$

$C_3H_6BrNO_2$ M 167.990
(±)-*form*
Bp_{50} 82-85°.

Meyer, V. *et al, Justus Liebigs Ann. Chem.*, 1876, **180**, 112 (*synth*)
Seigle, L.W. *et al, J. Org. Chem.*, 1940, **5**, 100 (*synth*)

2-Bromo-2-nitropropane B-70259

[5447-97-2]

$$(H_3C)_2CBrNO_2$$

$C_3H_6BrNO_2$ M 167.990
Mp 33-35°. Bp_{50} 73-75°.

Seigle, L.W. *et al, J. Org. Chem.*, 1940, **5**, 100 (*synth*)
Amrollah-Madjdabadi, A. *et al, Synthesis*, 1986, 828 (*synth, ir, pmr*)

8-Bromo-1-octanol B-70260

[50816-19-8]

$$BrCH_2(CH_2)_6CH_2OH$$

$C_8H_{17}BrO$ M 209.126
Liq. $Bp_{0.01}$ 77-78°.

Maurer, B. *et al, Helv. Chim. Acta*, 1977, **60**, 1155 (*synth*)

1-Bromo-1-octene, 9CI B-70261

1-Octenyl bromide

$$H_3C(CH_2)_5CH=CHBr$$

$C_8H_{15}Br$ M 191.111
(*E*)-*form* [51751-87-2]
Bp_6 59-60°.
(*Z*)-*form* [42843-49-2]
Bp_{17} 64-70° (bath).

Brown, H.C. *et al, J. Am. Chem. Soc.*, 1973, **95**, 6456 (*synth*)
Brown, H.C. *et al, Synthesis*, 1984, 919 (*synth*)
Tamao, K. *et al, J. Org. Chem.*, 1987, **52**, 1100 (*synth, pmr*)

(3-Bromo-2-oxopropylidene)propanedioic acid B-70262

$$BrCH_2COCH=C(COOH)_2$$

$C_5H_5BrO_5$ M 224.996
Di-Me ester:
$C_8H_9BrO_5$ M 265.060
Highly reactive compd., e.g. in Michael addns., in cycloaddns. and as enophile. Microcryst. powder. Mp 30-31°.

Cameron, D.W. *et al, Aust. J. Chem.*, 1987, **40**, 1831 (*synth, ir, pmr, cmr, ms, props*)

5-Bromopentanal B-70263

δ-Bromovaleraldehyde

[1191-30-6]

$$BrCH_2(CH_2)_3CHO$$

C_5H_9BrO　　M 165.030
Bp_{14} 84°.

2,4-Dinitrophenylhydrazone: Mp 106.5-107.5°.
Ethylene acetal: [87227-41-6]. *2-(4-Bromobutyl)-1,3-dioxole.*
　　$C_7H_{12}BrO_2$　　M 208.075
　　No phys. props. given.

Ratcliffe, R. *et al, J. Org. Chem.*, 1970, **35**, 4000 (*synth*)
Andersen, N.H. *et al, Tetrahedron Lett.*, 1978, 4315 (*synth*)
Kuenhe, M.E. *et al, J. Org. Chem.*, 1981, **46**, 2002 (*synth, pmr*)
Little, R.D. *et al, J. Org. Chem.*, 1982, **47**, 362 (*synth, ir, pmr*)
Baker, J.K. *et al, J. Med. Chem.*, 1985, **28**, 46 (*synth, deriv, pmr*)
Newcomb, M. *et al, J. Am. Chem. Soc.*, 1986, **108**, 240 (*synth*)

2-Bromo-1,10-phenanthroline, 9CI　　　　B-70264

[22426-14-8]

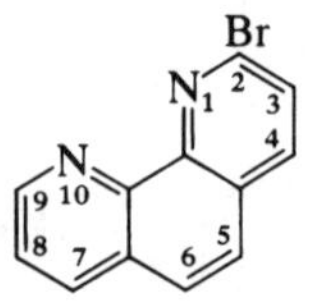

$C_{12}H_7BrN_2$　　M 259.105
Needles (C_6H_6). Mp 164-165°.

Ogawa, S. *et al, J. Chem. Soc., Perkin Trans. 1*, 1974, 976 (*synth*)

8-Bromo-1,7-phenanthroline, 9CI　　　　B-70265

[86538-52-5]
$C_{12}H_7BrN_2$　　M 259.105
Mp 146-147°.

van den Haak, H.J.W. *et al, J. Heterocycl. Chem.*, 1983, **20**, 447 (*synth, pmr*)

1-Bromophthalazine　　　　　　　　B-70266

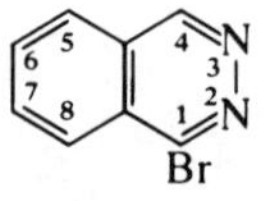

$C_8H_5BrN_2$　　M 209.045
Pale-yellow solid. Mp 175° (foams).

Hirsch, A. *et al, Can. J. Chem.*, 1965, **43**, 2708 (*synth*)

5-Bromophthalazine, 9CI　　　　　　B-70267

[103119-78-4]
$C_8H_5BrN_2$　　M 209.045

Japan. Pat., 86 37 707, (*1986*); *CA*, **105**, 37472 (*synth*)

6-Bromophthalazine, 9CI　　　　　　B-70268

[19064-74-5]
$C_8H_5BrN_2$　　M 209.045
Mp 110°.

Bowie, J.H. *et al, Aust. J. Chem.*, 1967, **20**, 2677 (*ms*)
Robev, S., *Tetrahedron Lett.*, 1981, **22**, 345 (*synth*)

2-Bromo-9-β-D-ribofuranosyl-6H-purin-6-　B-70269
one

2-Bromoinosine, 9CI

[77977-04-9]

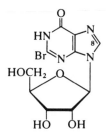

$C_{10}H_{11}BrN_4O_5$　　M 347.125
Mono-NH₄ salt: [52538-26-8]. Cryst. + ½H₂O (EtOH).
　Mp 176-178° dec.
2′,3′,5′-Tri-Ac: [41623-91-0]. Pale-yellow cryst. (EtOH).
　Mp 178-180°.

Marumoto, R. *et al, Chem. Pharm. Bull.*, 1974, **22**, 342; 1975, **23**, 759 (*synth*)
Dudycz, L.W. *et al, Nucleosides Nucleotides*, 1984, **3**, 33 (*synth*)

8-Bromo-9-β-D-ribofuranosyl-6H-purin-6-　B-70270
one

8-Bromoinosine, 9CI

[55627-73-1]
$C_{10}H_{11}BrN_4O_5$　　M 347.125
Cryst. (EtOH aq.). Mp 198-200° dec.
2′,3′,5′-Tri-Ac: Cryst. (EtOH). Mp 191-192°.
2′-(4-Methylbenzenesulfonyl): [79483-43-5]. Cryst.
　(EtOH). Mp 169-171° dec.

Holmes, R.E. *et al, J. Am. Chem. Soc.*, 1964, **86**, 1242 (*synth, uv*)
Sternglanz, H. *et al, Acta Crystallogr., Sect. B*, 1977, **33**, 2097 (*cryst struct*)
Uesugi, S. *et al, J. Am. Chem. Soc.*, 1977, **99**, 3250 (*cmr*)
Lin, T.-S. *et al, J. Med. Chem.*, 1985, **28**, 1481 (*synth, uv, pmr*)

2-Bromotetrahydro-2H-pyran　　　　B-70271

Tetrahydro-2-bromo-2H-pyran. 1,5-Epoxy-5-bromopentane

[6667-26-1]

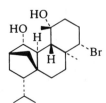

C_5H_9BrO　　M 165.030
Fuming liq. d^8_{15} 1.50. Bp_{16} 61-63°. Unstable, rapidly
　resinifies in air.

Paul, R., *Bull. Soc. Chim. Fr.*, 1934, 1397 (*synth*)
Anderson, C.B. *et al, J. Org. Chem.*, 1967, **32**, 607 (*synth*)

Bromotetrasphaerol　　　　　　　　B-70272

[104900-64-3]

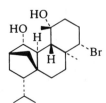

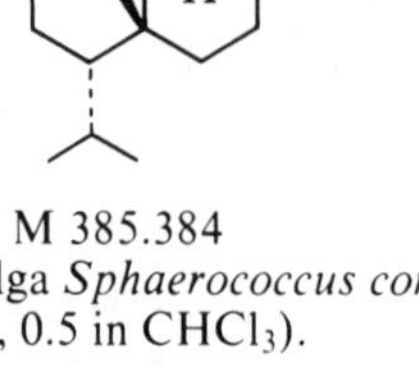

$C_{20}H_{33}BrO_2$　　M 385.384
Constit. of red alga *Sphaerococcus coronopifolius*. Oil.
　$[\alpha]_D$ −5.9° (c, 0.5 in CHCl₃).

Cafieri, F. *et al*, *Tetrahedron*, 1985, **41**, 4273.

5-Bromo-1,2,3-triazine   B-70273
[114078-88-5]

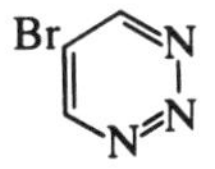

C$_3$H$_2$BrN$_3$ M 159.973
Cryst. (hexane). Mp 125-126° dec.

Kaihoh, T. *et al*, *Chem. Pharm. Bull.*, 1987, **35**, 3952 (*synth*)

1-Bromo-1,1,2,2,3,3,4,4,5,5,6,6,6-trideca-   B-70274
fluorohexane, 9CI
Perfluorohexyl bromide
[335-56-8]

F$_3$C(CF$_2$)$_4$CF$_2$Br

C$_6$BrF$_{13}$ M 398.949
X-ray contrast medium. Liq. Bp 100°.

Haszeldine, R.N., *J. Chem. Soc.*, 1952, 4259; 1953, 3761 (*synth*)

3-Bromo-3-(trifluoromethyl)-3*H*-diazirine,   B-70275
9CI
[117113-33-4]

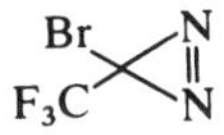

C$_2$BrF$_3$N$_2$ M 188.935
Gas.

▷Potentially explosive

Grayston, M.W. *et al*, *J. Am. Chem. Soc.*, 1976, **98**, 1278 (*synth*)
Dailey, W.P., *Tetrahedron Lett.*, 1987, **28**, 5801 (*synth*)

Bromotriphenylethylene   B-70276
1,1',1''-(1-Bromo-1,2,2-ethenediyl)trisbenzene, 9CI
[1607-57-4]

Ph$_2$C=CPhBr

C$_{20}$H$_{15}$Br   M 335.243
Cryst. (hexane). Mp 116-117°.

Kaupp, G. *et al.*, *Chem. Ber.*, 1987, **120**, 1897 (*synth, uv, pmr, ms*)

1-Bromo-2-undecene   B-70277
[67952-61-8]

H$_3$C(CH$_2$)$_7$CH=CHCH$_2$Br

C$_{11}$H$_{21}$Br   M 233.191
Liq. Bp$_{10}$ 123-124°, Bp$_{0.5}$ 87-90°.

Hiromichi, N. *et al*, *Nippon Kagaku Zasshi*, 1971, **92**, 1214; *CA*, **76**, 126325x (*synth*)
Camps, F. *et al*, *Synthesis*, 1987, 511 (*synth, pmr, cmr*)

Bromovulone I   B-70278
[105343-04-2]

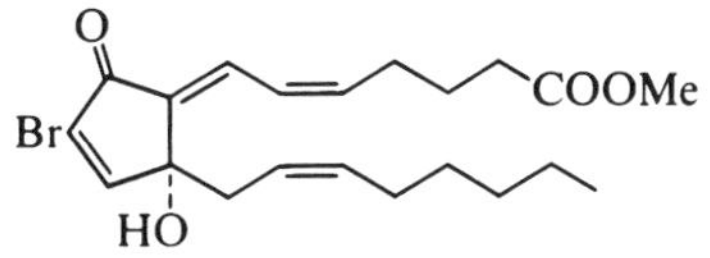

Absolute configuration

C$_{21}$H$_{29}$BrO$_4$ M 425.362
Prostanoid isol. from Japanese marine stolonifer *Clavularia viridis*. Shows antitumour activity. Oil.

Iodine analogue: [105343-03-1]. **Iodovulone I**.
   C$_{21}$H$_{29}$IO$_4$ M 472.362
   Isol. from *C. viridis*. Shows antitumour activity. Oil.

Iguchi, K. *et al*, *J. Chem. Soc., Chem. Commun.*, 1986, 981 (*isol, ms, ir, uv, pmr, abs config*)

Brosimone *A*   B-70279

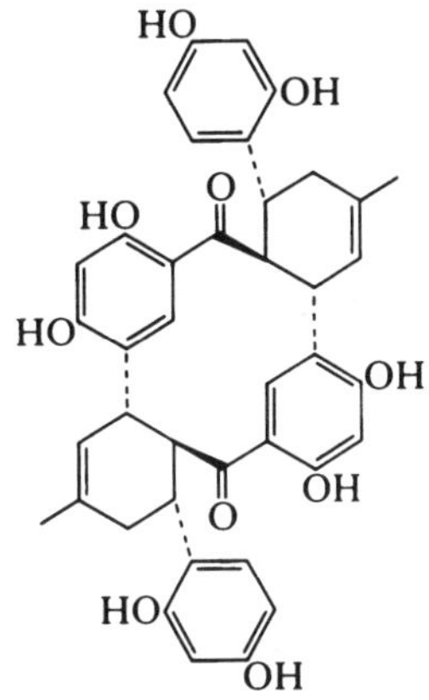

C$_{40}$H$_{36}$O$_{10}$ M 676.718
Constit. of roots of *Brosimopsis oblongifolia*. Amorph. powder. [α]$_D^{20}$ −711° (MeOH).

Messana, I. *et al*, *Tetrahedron*, 1988, **44**, 6693.

Bruceanol *C*   B-70280

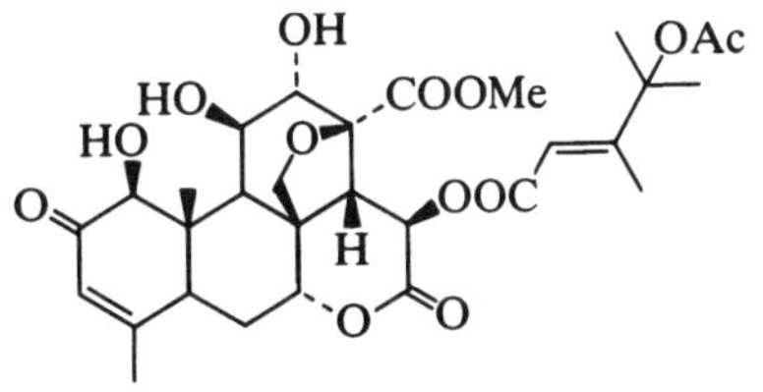

C$_{30}$H$_{38}$O$_{13}$ M 606.622
Constit. of *Brucea antidysenterica*. Shows cytotoxic activity. Amorph. powder. Mp 125-127°. [α]$_D^{25}$ +38° (c, 0.24 in EtOH).

Fukamiya, N. *et al*, *J. Nat. Prod.*, 1988, **51**, 349.

Bruceantinoside *C*   B-70281

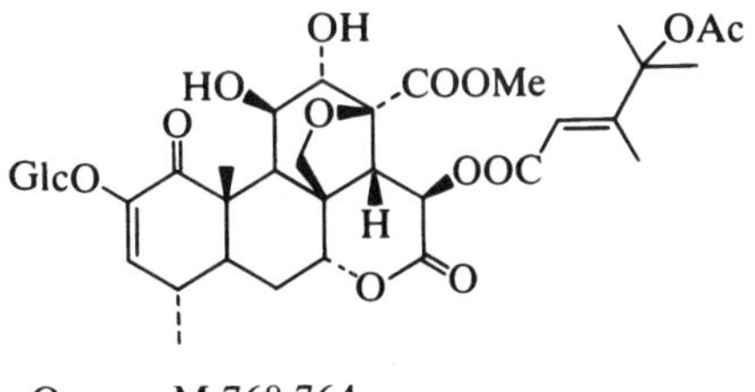

C$_{36}$H$_{48}$O$_{18}$ M 768.764

Constit. of *Brucea antidysenterica*. Cytotoxic agent.
Amorph. powder. Mp 153-155°. $[\alpha]_D^{23}$ +12.7° (c, 1.1
in EtOH).

Fukamiya, N. *et al*, *J. Nat. Prod.*, 1987, **50**, 1075.

BR-Xanthone *A*   B-70282
[112649-48-6]

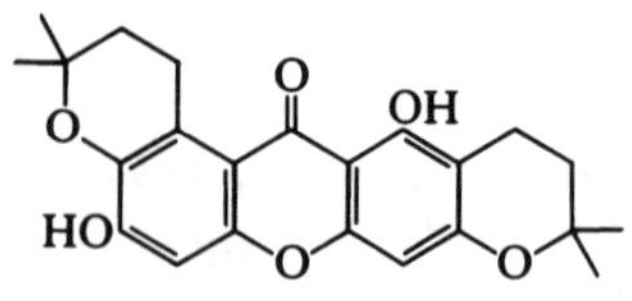

$C_{23}H_{24}O_6$ M 396.439

Constit. of the dry fruit hulls of *Garcinia mangostana*.
Pale-yellow needles (C_6H_6/pet. ether). Mp 181-182°.

Balasubramanian, K. *et al*, *Phytochemistry*, 1988, **27**, 1552.

Bryoflavone   B-70283
[111200-22-7]

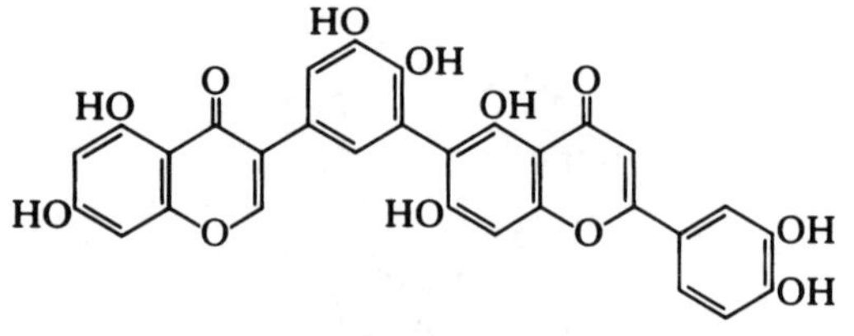

$C_{30}H_{18}O_{12}$ M 570.465

Constit. of *Bryum capillare*.

Geiger, H., *Z. Naturforsch., C*, 1987, **42**, 863.

Bryostatin 1   B-70284

Updated Entry replacing B-30333
NSC 339555
[83314-01-6]

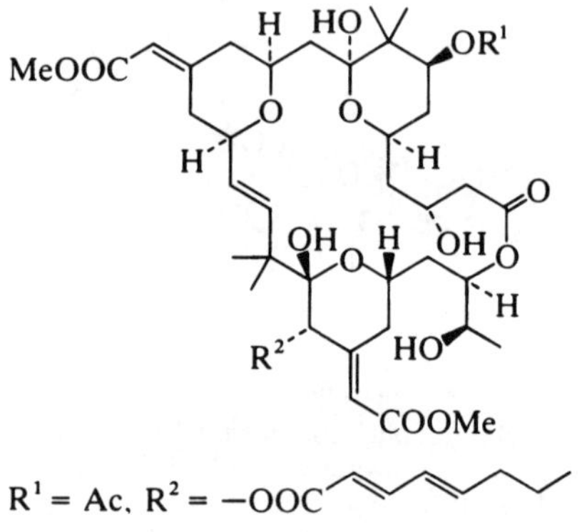

R^1 = Ac, R^2 = $-$OOC

$C_{47}H_{68}O_{17}$ M 905.044

Macrolide antibiotic. Prod. by *Bugula neritina*. Active
against M531 murine ovary sarcoma. Cryst.
(CH_2Cl_2/MeOH). Mp 230-235°. $[\alpha]_D^{25}$ +34.1° (c,
0.044 in MeOH). λ_{max} 233, 263 nm (MeOH).

O-De-Ac: [87745-28-6]. ***Bryostatin 2***.
$C_{45}H_{66}O_{16}$ M 863.007
Prod. by *B. neritina*. Shows potent antitumour props.
Fine cryst. (CH_2Cl_2/MeOH). Mp 201-203°. $[\alpha]_D^{25}$
+50° (c, 0.05 in MeOH). λ_{max} 231, 261 nm (MeOH).

Petit, G.R. *et al*, *J. Am. Chem. Soc.*, 1982, **104**, 6846 (*isol, cryst
struct, ms, uv, ir*)
Petit, G.R. *et al*, *J. Nat. Prod.*, 1983, **46**, 528 (*deriv, pmr*)

Petit, G.R. *et al*, *Tetrahedron*, 1985, **41**, 985 (*rev*)

Bryostatin 10   B-70285
[102580-65-4]
As Bryostatin 1, B-70284 with

$$R^1 = -COCH_2CH(CH_3)_2, R^2 = H$$

$C_{42}H_{64}O_{15}$ M 808.959
Isol. from *Bugula neritina*. Antineoplastic agent. Plates
(MeOH/CH_2Cl_2). Mp 161-164°. $[\alpha]_D^{27}$ +99.8° (c,
0.04 in MeOH).

Pettit, G.R. *et al*, *J. Org. Chem.*, 1987, **52**, 2848 (*isol, struct,
props*)

Bryostatin 11   B-70286
[102580-63-2]
As Bryostatin 1, B-70284 with

$$R^1 = Ac, R^2 = H$$

$C_{39}H_{58}O_{15}$ M 766.878
Isol. from *Bugula neritina*. Antineoplastic agent. Needles
(MeOH/CH_2Cl_2). Mp 171-173°. $[\alpha]_D^{27}$ +42.5° (c,
0.05 in MeOH).

Pettit, G.R. *et al*, *J. Org. Chem.*, 1987, **52**, 2848 (*isol, struct,
props*)

Bryostatin 12   B-70287
[107021-10-3]
As Bryostatin 1, B-70284 with

$$R^1 = -COCH_2CH_2CH_3, R^2 = -OOCCH=CH-$$
$$CH=CHCH_2CH_2CH_3$$

$C_{49}H_{72}O_{17}$ M 933.098
Isol. from *Bugula neritina*. Antineoplastic agent. $[\alpha]_D^{27}$
+39° (c, 0.108 in MeOH).

Pettit, G.R. *et al*, *J. Org. Chem.*, 1987, **52**, 2854 (*isol, struct,
props*)

Bryostatin 13   B-70288
[107021-11-4]
As Bryostatin 1, B-70284 with

$$R^1 = -COCH_2CH_2CH_3, R^2 = H$$

$C_{41}H_{62}O_{15}$ M 794.932
Isol. from *Bugula neritina*. Antineoplastic agent. No
phys. props. reported.

Pettit, G.R. *et al*, *J. Org. Chem.*, 1987, **52**, 2854 (*isol, struct,
props*)

*Suggestions for new DOC Entries
are welcomed.   Please write to
the Editor, DOC 5, Chapman
and Hall Ltd, 11 New Fetter Lane,
London EC4P 4EE*

Bufalin   B-70289
Updated Entry replacing B-20217
3β,14β-Dihydroxy-5β-bufa-20,22-dienolide
[465-21-4]

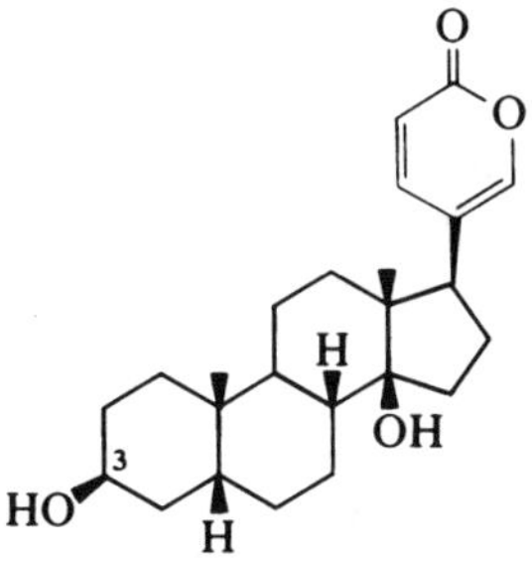

$C_{24}H_{34}O_4$ M 386.530
Constit. of toad venom. Cryst. (EtOAc). Mp 240-243°.
$[\alpha]_D$ −9° (CHCl$_3$).

▷ Highly toxic, LD$_5$ 0.137 mg/kg (cat). EI2962500.
Ac: Mp 230° and 247° (double Mp). $[\alpha]_D$ −6° (CHCl$_3$).
3-(Hydrogen suberoyl): [30219-13-7].
    $C_{32}H_{46}O_7$ M 542.711
    Constit. of Chinese toad venom drug Ch'an Su.
    Amorph.
3-(Methylsuberoyl): [20987-33-1]. Cryst. Mp 154-156°.
3-(Arginylsuberoyl): [35455-33-5]. **Bufalitoxin.**
    $C_{38}H_{58}N_4O_8$ M 698.898
    Constit. of *Bufo vulgaris formosus* venom. Cryst.
    (MeOH/Me$_2$CO). Mp 204-211° dec. $[\alpha]_D^{22}$ −5.2° (c,
    0.2 in MeOH).

Ruckstuhl, J.P. *et al, Helv. Chim. Acta*, 1957, **40**, 1270 (*isol, struct*)
Kamano, Y. *et al, Tetrahedron Lett.*, 1968, 5673 (*isol*)
Gsell, L. *et al, Helv. Chim. Acta*, 1969, **52**, 557 (*pmr*)
Kamano, Y. *et al, J. Am. Chem. Soc.*, 1972, **94**, 8592 (*synth*)
Porto, A.M. *et al, J. Steroid Biochem.*, 1972, **3**, 11 (*biosynth*)
Sondheimer, F. *et al, Tetrahedron Lett.*, 1973, 765 (*synth*)
Pettit, G.R. *et al, J. Org. Chem.*, 1974, **39**, 3003 (*isol*)
Shimada, K. *et al, Chem. Pharm. Bull.*, 1977, **25**, 714 (*isol*)
Rohrer, D.C. *et al, Acta Crystallogr., Sect. B*, 1982, **38**, 1865 (*cryst struct*)
Sen, A. *et al, J. Chem. Soc., Chem. Commun.*, 1982, 1213 (*synth*)
Tsai, T.Y.R. *et al, Can. J. Chem.*, 1982, **60**, 2161 (*synth*)
Wiesner, K. *et al, Helv. Chim. Acta*, 1983, **66**, 2632 (*synth*)
Pettit, G.R. *et al, J. Org. Chem.*, 1987, **52**, 3573 (*synth*)

6-(1,3-Butadienyl)-1,4-cycloheptadiene   B-70290
Desmarestene
[83013-90-5]

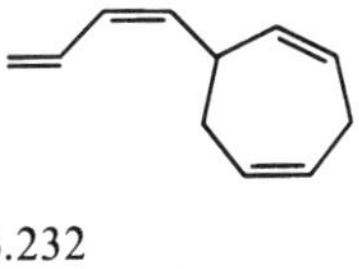

$C_{11}H_{14}$ M 146.232
Isol. from the brown algae *Desmarestia aculeata* and *D. viridis*. Algal sex attractant.

Jaenicke, L. *et al, Angew. Chem., Int. Ed. Engl.*, 1982, **21**, 643 (*rev*)
Müller, D.G. *et al, Naturwissenschaften*, 1982, **69**, 290.

3-Butene-1,2-diol, 9CI   B-70291
Updated Entry replacing B-03508
Erythrol
[497-06-3]

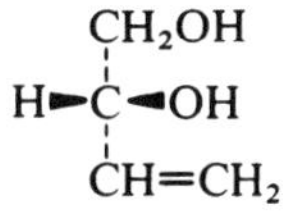

$$\begin{array}{l} CH_2OH \\ H\!\!-\!\!C\!\!-\!\!OH \qquad (R)\text{-}form \\ CH\!\!=\!\!CH_2 \end{array}$$

$C_4H_8O_2$ M 88.106

(R)-form
    Bp$_{15}$ 98-100°. $[\alpha]_D^{25}$ +40° (c, 4.6 in 2-propanol).
    1-O-Benzyl: [113426-94-1]. *1-(Phenylmethoxy)-3-bu-ten-2-ol, 9CI. 1-Benzyloxy-3-buten-2-ol.*
    $C_{11}H_{14}O_2$ M 178.230
    Chiral synthon. $[\alpha]_D$ +6.2° (c, 1.6 in CHCl$_3$).

(±)-form
    Liq. Bp 196.5°, Bp$_{16}$ 98°.
    Ac: [18085-01-3].
        $C_6H_{10}O_3$ M 130.143
        Liq. Bp 159-162°.
    Di-Ac: [18085-02-4].
        $C_8H_{12}O_4$ M 172.180
        Liq. Bp 202-203°.
    Dibenzoyl: [73318-93-1].
        $C_{18}H_{16}O_4$ M 296.322
        Liq. Bp$_6$ 199-200°.
    1-Me ether: [17687-76-2]. *1-Methoxy-3-buten-2-ol.*
        $C_5H_{10}O_2$ M 102.133
        Bp 143-144°.
    Bisphenylurethane: Cryst. Mp 125-126°.

Prevost, C., *C. R. Hebd. Seances Acad. Sci.*, 1926, **183**, 1292; 1928, **186**, 1209 (*synth*)
Hurd, C.D. *et al, J. Am. Chem. Soc.*, 1939, **61**, 1156 (*synth*)
Bessinger, W.E. *et al, J. Am. Chem. Soc.*, 1947, **69**, 2955 (*synth*)
Howes, D.A. *et al, J. Chem. Res. (S)*, 1983, 9 (*synth*)
Rao, A.V.R. *et al, Tetrahedron Lett.*, 1987, **28**, 6497 (*deriv, synth, use*)

2-Butene-1,4-dithiol   B-70292
1,4-Dimercapto-2-butene
[6725-69-5]

$$HSCH_2CH{=}CHCH_2SH$$

$C_4H_8S_2$ M 120.227

(E)-form
    Yellowish foul-smelling liq. Bp$_{11}$ 81-82°, Bp$_2$ 50-53°.
(Z)-form [55443-61-3]
    Liq. Bp$_{11}$ 80-81°, Bp$_{0.1}$ 30°.

Houk, J. *et al, J. Am. Chem. Soc.*, 1987, **109**, 6825 (*synth, ir, pmr*)

3-Buten-1-ynylcyclopropane, 9CI   B-70293
1-Cyclopropyl-2-vinylacetylene. 4-Cyclopropyl-1-buten-3-yne
[71452-17-0]

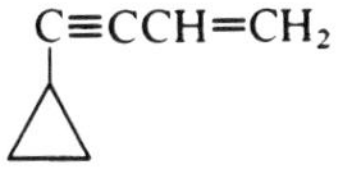

C_7H_8 M 92.140
Doering, W. von E. *et al, J. Am. Chem. Soc.*, 1987, **109**, 2697 (*synth, pmr, ir*)

3-*tert*-Butylbicyclo[3.1.0]hexane　　　B-70294

3-(1,1-Dimethylethyl)-6-oxabicyclo[3.1.0]hexane, 9CI.
4-tert-Butylcyclopentene-1,2-oxide

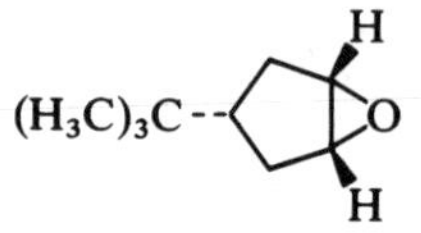

(H₃C)₃C-- 　　　(1α,3α,5α)-*form*

C₉H₁₆O　　M 140.225

(*1α,3α,5α*)-*form* [5590-96-5]
cis-*form*
Bp₁₅ 75-76°. n_D^{20} 1.4497.
(*1α,3α,5β*)-*form* [5581-98-6]
trans-*form*
Liq.

Bernáth, G. *et al*, *Tetrahedron*, 1972, **28**, 3475 (*synth*)
Sohár, P. *et al*, *Acta Chim. Acad. Sci. Hung.*, 1975, **87**, 289 (*pmr, ir*)

tert-Butyl chloromethyl ether　　　B-70295

2-(Chloromethoxy)-2-methylpropane, 9CI. Chloro-methyl tert-butyl ether
[40556-01-2]

(H₃C)₃COCH₂Cl

C₅H₁₁ClO　　M 122.594
Reagent for the introduction of the *tert*-butoxymethyl protecting group for alcohols, histidine residues, etc. Stable under N₂ at r.t. as a CCl₄ soln.

Pinnick, H.W. *et al*, *J. Org. Chem.*, 1978, **43**, 3964 (*synth*)
Jones, J.H. *et al*, *Synth. Commun.*, 1986, **16**, 1607 (*synth*)

1-*tert*-Butyl-1,3-cyclohexadiene　　　B-70296

1-(1,1-Dimethylethyl)-1,3-cyclohexadiene
[51497-33-7]

C₁₀H₁₆　　M 136.236
Liq. Bp₁₈ 88°.

Arain, M.F. *et al*, *Aust. J. Chem.*, 1988, **41**, 505 (*synth, ir, uv, pmr, cmr*)

3-*tert*-Butyl-2-cyclohexen-1-one　　　B-70297

3-(1,1-Dimethylethyl)-2-cyclohexen-1-one, 9CI.
[17299-35-3]

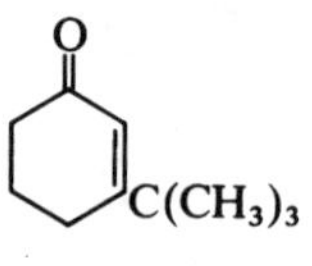

C₁₀H₁₆O　　M 152.236
Bp₀.₄₅ 85°.

Piers, E. *et al*, *J. Org. Chem.*, 1978, **40**, 2694 (*synth, pmr*)
Arain, M.F. *et al*, *Aust. J. Chem.*, 1988, **41**, 505 (*synth*)

1-*tert*-Butylcyclopentene　　　B-70298

1-(1,1-Dimethylethyl)cyclopentene, 9CI
[3419-67-8]

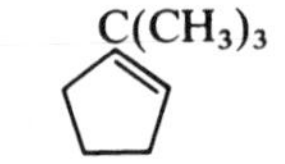

C₉H₁₆　　M 124.225
Bp 138-140°.

Brown, H.C. *et al*, *J. Org. Chem.*, 1979, **44**, 1910 (*synth, ir, pmr*)

3-*tert*-Butylcyclopentene　　　B-70299

3-(1,1-Dimethylethyl)cyclopentene, 9CI
[6189-88-4]
C₉H₁₆　　M 124.225
(±)-*form*
Liq. Bp₄₀ 52°.

Brown, H.C. *et al*, *J. Org. Chem.*, 1979, **44**, 1910 (*synth, ir, pmr*)

4-*tert*-Butylcyclopentene　　　B-70300

4-(1,1-Dimethylethyl)cyclopentene, 9CI
[5581-97-5]
C₉H₁₆　　M 124.225
Liq.

Richer, J.C. *et al*, *Can. J. Chem.*, 1965, **43**, 3419 (*synth, pmr*)
Bernáth, G. *et al*, *Tetrahedron*, 1972, **28**, 3475 (*synth*)
Brown, H.C. *et al*, *J. Org. Chem.*, 1979, **44**, 1910 (*synth*)

1,1-*tert*-Butyl-3,3-diethoxy-2-azaallen-ium(1+)　　　B-70301

[(H₃C)₃C]₂C=N⊕=C(OEt)₂

C₁₄H₂₈NO₂⊕　　M 242.381 (ion)
Representative of a new class of 2-azaallenium salts.
Tetrafluoroborate: [108593-59-5].
　　C₁₈H₂₈BF₄NO₂　　M 377.228
　　Cryst. (CH₂Cl₂/Et₂O). Mp 85-88° (sinters from 80°).
Krestel, M. *et al*, *Chem. Ber.*, 1987, **120**, 1271 (*synth, pmr, cmr, ir, cryst struct*)

5-*tert*-Butyl-1,2,3,5-dithiadiazolyl　　　B-70302

4-(1,1-Dimethylethyl)-3H-1,2,3,5-dithiadiazol-3-yl, 9CI

C₅H₉N₂S₂　　M 161.260
Stable paramagnetic free radical. Red-purple liq. Mp 20-21°.

Brooks, W.V.F. *et al*, *J. Chem. Soc., Chem. Commun.*, 1987, 69 (*synth, esr, props*)

5-*tert*-Butyl-1,3,2,4-dithiadiazolyl   B-70303

5-(1,1-Dimethylethyl)-1,3,2,4-dithiadiazol-2-yl, 9CI

[111299-76-4]

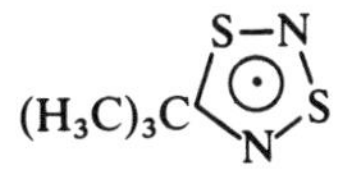

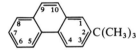

$C_5H_9N_2S_2$ M 161.260

Stable paramagnetic free radical. Brown-black liq. Mp 0-1°. Stable in dark at r.t. Photochemically isom. to 5-*tert*-Butyl-1,2,3,5-dithiadiazolyl, B-70302 .

Brooks, W.F. *et al*, *J. Chem. Soc., Chem. Commun.*, 1987, 69 (*synth, esr, props*)

3-*tert*-Butyl-4-hydroxybenzaldehyde   B-70304

3-(1,1-Dimethylethyl)-4-hydroxybenzaldehyde, 9CI

[65678-11-7]

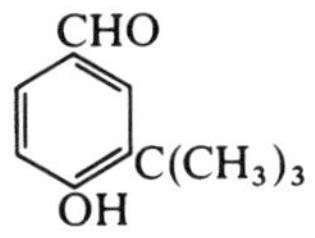

$C_{11}H_{14}O_2$ M 178.230

Cryst. (toluene). Mp 140-142°.

Hewgill, F.R. *et al*, *Aust. J. Chem.*, 1978, **31**, 907 (*synth, pmr, ir, uv*)
Katsumi, I. *et al*, *Chem. Pharm. Bull.*, 1986, **34**, 121 (*synth, ir, pmr*)

2-*tert*-Butylphenanthrene   B-70305

2-(1,1-Dimethylethyl)phenanthrene, 9CI

[66553-04-6]

$C_{18}H_{18}$ M 234.340

Pearly plates. Mp 99-100°. Brilliant-purple fluor. in uv light.

Picrate: Orange needles. Mp 130-131°.

Fieser, L.F. *et al*, *J. Am. Chem. Soc.*, 1936, **58**, 1838 (*synth*)
Imagire, Y., *Kogyo Kagaku Zasshi*, 1959, **62**, 837; *CA*, **57**, 8517b,e (*synth*)
Pataki, J. *et al*, *J. Org. Chem.*, 1982, **47**, 1133 (*synth, pmr*)

3-*tert*-Butylphenanthrene   B-70306

3-(1,1-Dimethylethyl)phenanthrene, 9CI

[33240-33-4]

$C_{18}H_{18}$ M 234.340

Long prismatic needles (EtOH). Mp 54-55° (49-50°). Bright-blue fluor. in uv light.

Picrate: Mp 142-143°.

Fieser, L.F. *et al*, *J. Am. Chem. Soc.*, 1936, **58**, 1838 (*synth*)
Cameron, D.W. *et al*, *Aust. J. Chem.*, 1977, **30**, 859 (*synth, uv, pmr*)
Bansal, R.C. *et al*, *Org. Prep. Proced. Int.*, 1988, **20**, 305 (*synth, cmr*)

9-*tert*-Butylphenanthrene   B-70307

9-(1,1-Dimethylethyl)phenanthrene, 9CI

[17024-05-4]

$C_{18}H_{18}$ M 234.340

Long needles (EtOH). Mp 64-65°.

1,3,5-Trinitrobenzene complex: Silky bright-yellow needles (MeOH). Mp 143°.

Anet, F.A.L. *et al*, *Can. J. Chem.*, 1956, **34**, 991; 1963, **41**, 2160 (*synth, uv*)
Harvey, R.G. *et al*, *J. Org. Chem.*, 1976, **41**, 3722 (*synth*)

2-Butyne-1,4-dithiol, 9CI   B-70308

1,4-Dimercapto-2-butyne

[74912-54-2]

$$HSCH_2C{\equiv}CCH_2SH$$

$C_4H_6S_2$ M 118.211

Pale-yellow foul-smelling liq. Bp_1 45-50°. Extensive polym. >100°.

Houk, J. *et al*, *J. Am. Chem. Soc.*, 1987, **109**, 6825 (*synth, pmr*)

3-Butynoic acid   B-70309

Updated Entry replacing B-03841

[2345-51-9]

$$HC{\equiv}CCH_2COOH$$

$C_4H_4O_2$ M 84.074

Mp 82-84°.

Me ester: [32804-66-3].
     $C_5H_6O_2$ M 98.101
     Liq. Bp_{10-15} 30-32°.

Wotiz, J.H. *et al*, *J. Am. Chem. Soc.*, 1951, **73**, 5503 (*synth*)
Gaudemar, M, *Ann. Chim. (Paris)*, 1956, **1**, 161
Benghiat, V. *et al*, *J. Chem. Soc., Perkin Trans. 2*, 1972, 1772 (*cryst struct*)
Crombie, L. *et al*, *J. Chem. Soc., Perkin Trans. 1*, 1975, 1081 (*synth*)
Bigley, D.B. *et al*, *Org. Mass Spectrom.*, 1976, **11**, 352 (*ms*)
Collins, P.W. *et al*, *J. Med. Chem.*, 1987, **30**, 193 (*ester, synth, pmr*)

C

Cacotheline — C-70001

2,3-Dihydro-4-nitro-2,3-dioxo-9,10-secostrychnidin-10-oic acid, 9CI

[561-20-6]

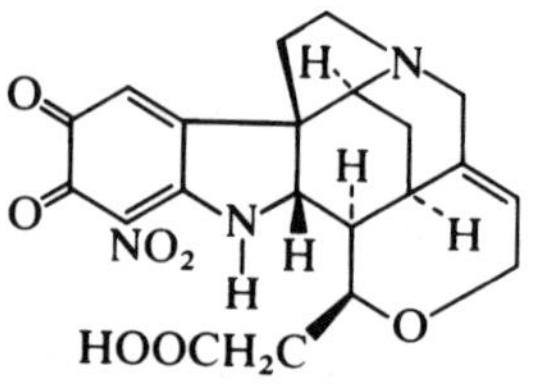

$C_{21}H_{21}N_3O_7$ M 427.413

Prod. by oxidn. and nitration of brucine. Indicator for Sn^{2+} titrations. Yellow cryst.

Leuchs, H. *et al, Ber.*, 1910, **43**, 1042 (*synth*)
Teuber, H.-J. *et al, Chem. Ber.*, 1953, **86**, 232 (*struct*)

Cadambagenic acid — C-70002

Updated Entry replacing C-00013
3β-Hydroxy-18α-olean-12-ene-27,28-dioic acid

[53318-10-8]

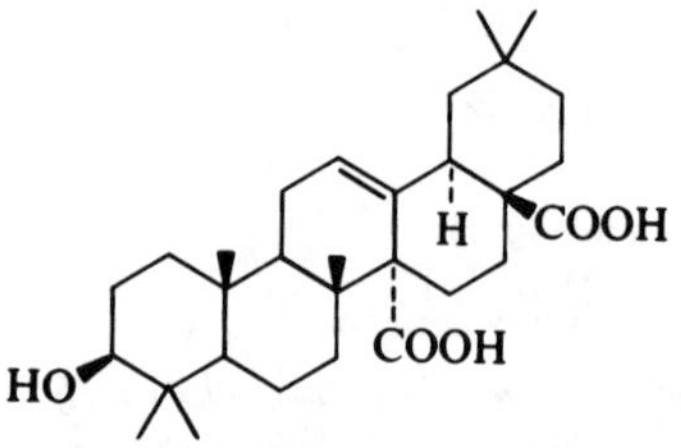

$C_{30}H_{46}O_5$ M 486.690

Constit. of *Anthocephalus cadamba*. Cryst.
(MeOH/CHCl₃). Mp 312-314°. $[\alpha]_D^{25}$ +94.4°
(EtOH).

O^3-[β-D-Glucopyranosyl(1→2)-α-L-rhamopyranosyl(1→4)-β-D-glucopyranoside]:
[61775-04-0]. **Saponin A**.
$C_{48}H_{76}O_{19}$ M 957.117

Saponin from stem-bark of *Anthrocephalus cadamba*.
Mp 240-241°. $[\alpha]_D^{28}$ +3.2° (EtOH).

Sahu, N.P. *et al, Indian J. Chem.*, 1974, **12**, 284 (*isol, struct*)
Banerji, N. *et al, Indian J. Chem., Sect. B*, 1976, **14**, 614 (*deriv*)

4(15),5-Cadinadiene — C-70003

Nephthene

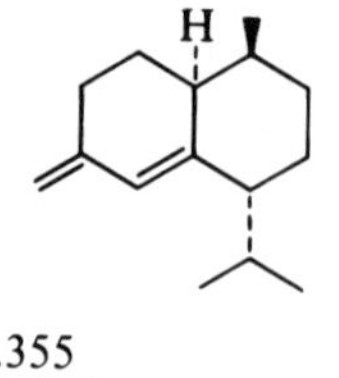

$C_{15}H_{24}$ M 204.355

Constit. of coral *Nephthea* sp. Oil. $[\alpha]_D^{20}$ −9° (c, 5.3 in CHCl₃).

Kitagawa, I. *et al, Chem. Pharm. Bull.*, 1987, **35**, 124.

4,11(13)-Cadinadien-12-oic acid — C-70004

1,2,3,4,4a,5,6,8a-Octahydro-4,7-dimethyl-α-methylene-1-naphthaleneacetic acid, 9CI. Artemisininic acid. Artemisic acid. Qing Hau acid. Arteannuic acid

[80286-58-4]

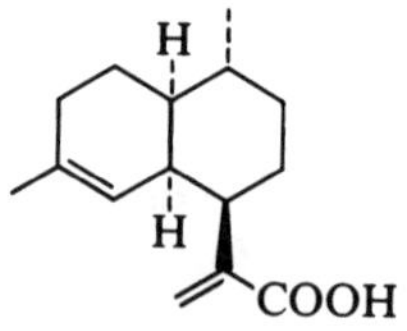

$C_{15}H_{22}O_2$ M 234.338

Isol. from *Artemisia annua*. Shows antibacterial activity.
Cubes (pet. ether). Mp 131°. $[\alpha]_D$ +36° (c, 0.01 in CHCl₃).

Tu, Y.Y. *et al, Planta Med.*, 1982, **44**, 143 (*isol*)
Zhou, W. *et al, CA*, 1985, **103**, 160709 (*config*)

3-O-Caffeoylquinic acid — C-70005

Updated Entry replacing C-00041
3-(3,4-Dihydroxycinnamoyl)quinic acid. Chlorogenic acid. Caffeylquinic acid. Caffetannic acid

[327-97-9]

$C_{16}H_{18}O_9$ M 354.313

Constit. of many plants including *Chrozophora,
Cinchona, Scabiosa, Valeriana, Senecio* and
Hypericum spp., isol. originally from Liberian coffee.
Needles + ½H₂O. Mp 208°. $[\alpha]_D^{16}$ −35.2° (H₂O).
▷GU8480000.

Tri-Ac: Cryst. (MeOH). Mp 150-152°.
Penta-Ac: Needles (EtOH aq.). Mp 181°.
3'-Me ether: **3-O-Feruloylquinic acid**.
$C_{17}H_{20}O_9$ M 368.340
Constit. of coffee beans. Also from *Lycopersicon
esculentum, Helianthus annuus* and *Nicotiana
tabacum*. Mp 196-197°. $[\alpha]_D^{28}$ −42.8° (EtOH).
O^4-(4-Hydroxy-3,5-dimethoxycinnamoyl): [110241-35-5]. **3-O-Caffeoyl-4-O-sinapoylquinic acid**.
$C_{27}H_{28}O_{13}$ M 560.510
Constit. of Gardenia fructus (*Gardenia jasminoides*).
Lipoxygenase inhibitor. Amorph. powder. $[\alpha]_D^{20}$ −252°
(c, 0.54 in MeOH).

Fischer, H.O.L. *et al, Ber.*, 1932, **65**, 1037 (*struct*)
Corse, J. *et al, Tetrahedron*, 1962, **18**, 1207 (*deriv*)
Waiss, A.C. *et al, Chem. Ind. (London)*, 1964, 1984 (*pmr*)
Schulten, H.R. *et al, Biomed. Mass Spectrom.*, 1974, **1**, 120 (*ms*)
Kelley, C.J. *et al, J. Org. Chem.*, 1976, **41**, 449 (*cmr*)
Nagels, L. *et al, Phytochemistry*, 1976, **15**, 703 (*biosynth*)
Nishizawa, M. *et al, Chem. Pharm. Bull.*, 1986, **34**, 1419; 1987, **35**, 2133 (*deriv*)

Calaminthone — C-70006

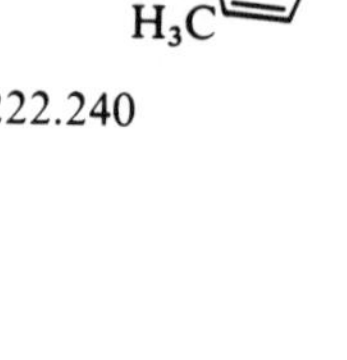

$C_{12}H_{14}O_4$ M 222.240

Constit. of *Calamintha ashei*. Oil.

Tanrisever, N. *et al*, *Phytochemistry*, 1988, **27**, 2523.

Calaxin   C-70007

Updated Entry replacing C-00061

[30412-86-3]

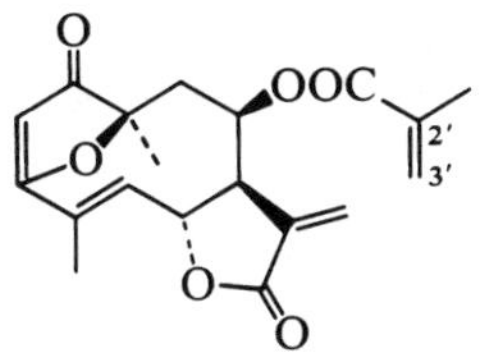

$C_{19}H_{20}O_6$ M 344.363

Constit. of *Helianthus ciliaris* and *Calea axillaris*. Cryst.
Mp 180-182°. $[\alpha]_D$ −115°.

2′,3′-Dihydro: [30412-87-4]. **Ciliarin**.
$C_{19}H_{18}O_6$ M 342.348
Constit. of *H. ciliarus*. Cryst. Mp 148°. $[\alpha]_D$ −143°.

Chawdhury, P.K. *et al*, *J. Org. Chem.*, 1980, **45**, 4993 (*struct*)

Calichemicin γ₁¹   C-70008

LL E33288γ₁. Antibiotic LL E33288γ₁

[108212-75-5]

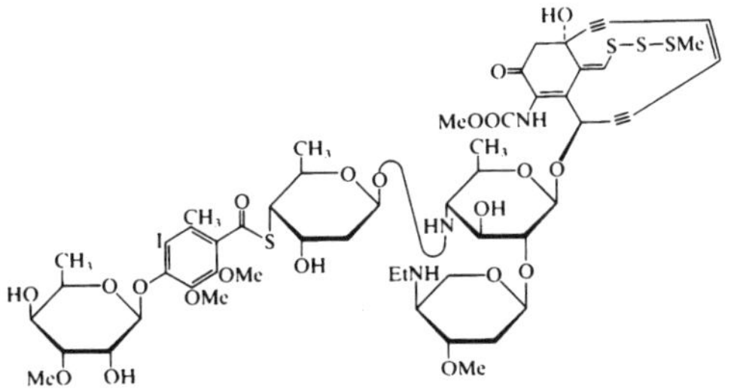

$C_{55}H_{74}IN_3O_{21}S_4$ M 1368.342

Glycosidic antibiotic. Prod. by *Micromonospora
echinospora* ssp. *calichensis*. Antitumour agent.
Amorph. $[\alpha]_D^{26}$ −124° (c, 0.98 in EtOH). Major
component of Calichemicin complex. Similar to
Esperamicins.

Lee, M.D. *et al*, *J. Am. Chem. Soc.*, 1987, **109**, 3464, 3466
(*struct*)

α-Camphorene   C-70009

Updated Entry replacing C-00106
*4-(5-Methyl-1-methylene-4-hexenyl)-1-(4-methyl-3-
pentenyl)cyclohexene, 9CI. Paracamphorene. p-
Camphorene*

[532-87-6]

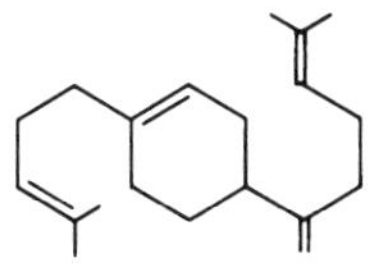

$C_{20}H_{32}$ M 272.473

Constit. of camphor oil and oil of *Humulus lupulus*. Oil.
Bp₀.₀₃ 110°.

Lammens, H. *et al*, *Bull. Soc. Chim. Belg.*, 1968, **77**, 497 (*isol*)
Eisfelder, W. *et al*, *Justus Liebigs Ann. Chem.*, 1977, 988
(*synth*)
Teresa, J. de P. *et al*, *An. Quim.*, 1978, **74**, 305 (*synth*)
Vig, O.P. *et al*, *J. Indian Chem. Soc.*, 1986, **63**, 507 (*synth*)

Camporic acid   C-70010

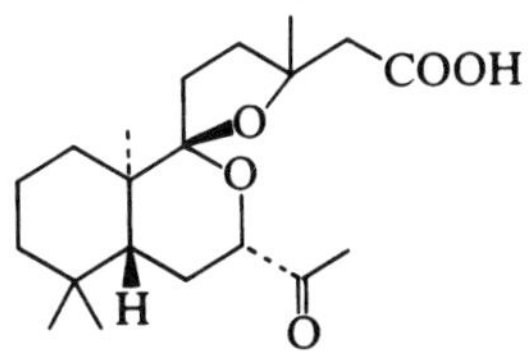

$C_{20}H_{32}O_5$ M 352.470

Constit. of *Grindelia camporum*.
Me ester: Oil. $[\alpha]_D^{25}$ −5.7° (c, 3.4 in CHCl₃).

Hoffmann, J.J. *et al*, *Phytochemistry*, 1988, **27**, 493.

Canaliculatol   C-70011

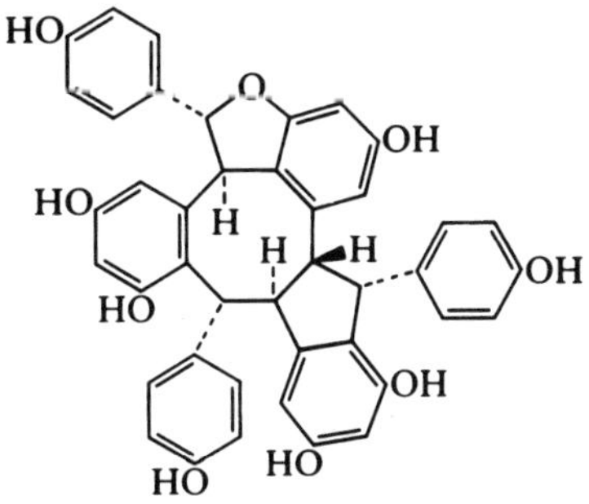

$C_{42}H_{32}O_9$ M 680.709

Constit. of the bark of *Stemonoporus canaliculatus*.
Antifungal agent. Cryst. Mp 245° dec. $[\alpha]_D$ −25.5°
(MeOH).

Bokel, M. *et al*, *Phytochemistry*, 1988, **27**, 377.

Candicanin   C-70012

[36149-85-6]

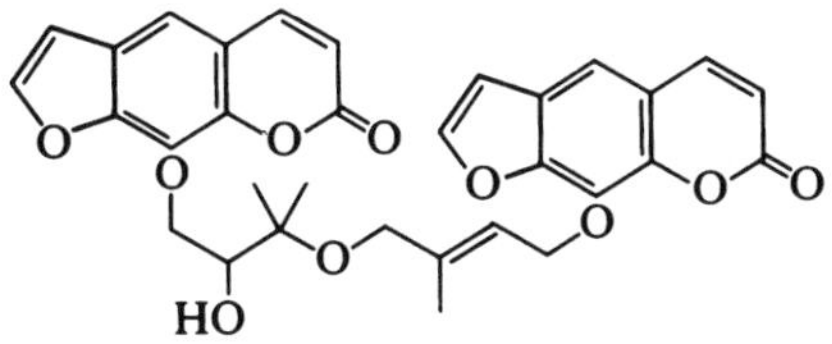

$C_{32}H_{28}O_{10}$ M 572.567

Constit. of *Heracleum candicans*. Cryst. (C₆H₆/EtOAc).
Mp 153°.

Bandopadhyay, M. *et al*, *Tetrahedron Lett.*, 1971, 4221.

Caniojane   C-70013

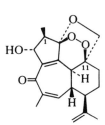

$C_{20}H_{24}O_5$ M 344.407

Constit. of *Jatropha grossidentata*. Cryst. Mp 167-168°.
$[\alpha]_D^{24}$ −348° (c, 0.13 in CHCl₃).

1,11-Diepimer: **1,11-Bisepicaniojane**.
$C_{20}H_{24}O_5$ M 344.407
Constit. of *J. grossidentata*. Gum.

Jakupovic, J. *et al*, *Phytochemistry*, 1988, **27**, 2997.

O^6-[(3-Carbamoyl-2H-azirine-2-ylidene)-amino]-1,2-O-isopropylidene-3,5-di-O-tosyl-α-D-glucofuranoside   C-70014

6-[[[3-(Aminocarbonyl)-2H-azirine-2-ylidene]amino]-oxy]-6-deoxy-1,2-O-(1-methylethylidene)-α-D-gluco-furanose 3,5-bis(4-methylbenzenesulfonate), 9CI

[115319-22-7]

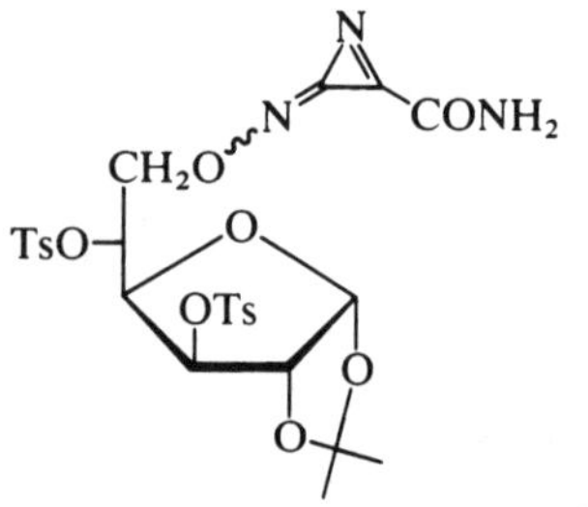

$C_{26}H_{29}N_3O_{11}S_2$ M 623.649

First known compd. contg. an azireneimine unit. Mp 130-131°. $[\alpha]_D^{20}$ +13.8° (c, 0.2 in $CHCl_3$).

Meyer zu Reckendorf, W. et al, Angew. Chem., Int. Ed. Engl., 1988, **27**, 1075 (synth, pmr, cmr, N nmr, ms)

N-Carbamoylglucosamine   C-70015

SF 1993. Antibiotic SF 1993

[71868-25-2]

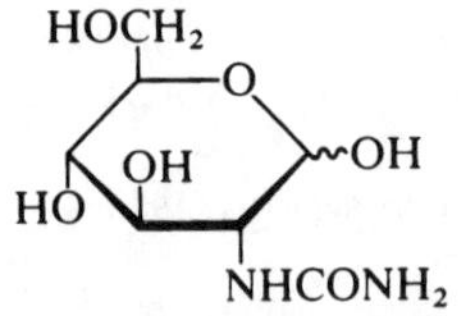

$C_7H_{14}N_2O_6$ M 222.197

Aminoglycoside antibiotic. Prod. by *Streptomyces halstedi*. Active against gram-negative bacteria and fungi. Needles. Mp 158-159°. $[\alpha]_D^{20}$ +73.6° (c, 0.98 in H_2O). Related to Antibiotic CV 1.

Michael, F. et al, Chem. Ber., 1956, **89**, 1246 (synth)
Shomura, T. et al, J. Antibiot., 1979, **32**, 427, 436 (isol, struct, props)

N^6-(Carbamoylmethyl)-2′-deoxyadenosine   C-70016

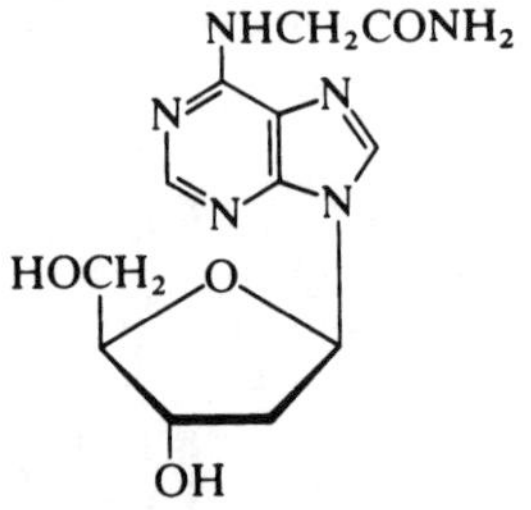

$C_{12}H_{16}N_6O_4$ M 308.296

Modified nucleoside found in *E. coli* bacteriophage Mu. Needles (MeOH). Mp 215-216° (211°).

Seela, F. et al, Helv. Chim. Acta, 1987, **70**, 1649 (synth, uv, pmr, bibl)

1-Carbethoxy-2-cyano-1,2-dihydropyridine   C-70017

[51364-89-7]

$C_9H_{10}N_2O_2$ M 178.190

Reissert compd. of the pyridine series. Red-brown oil. Stable for 2 months at 4°.

Reuss, R.H. et al, J. Org. Chem., 1974, **39**, 2027 (synth, pmr)
Cooney, J.V. et al, Org. Prep. Proced. Int., 1985, **17**, 60 (synth, pmr)

Carbodine   C-70018

4-Amino-1-[2,3-dihydroxy-4-(hydroxymethyl)-cyclopentyl]-2(1H)-pyrimidinone

[62805-43-0]

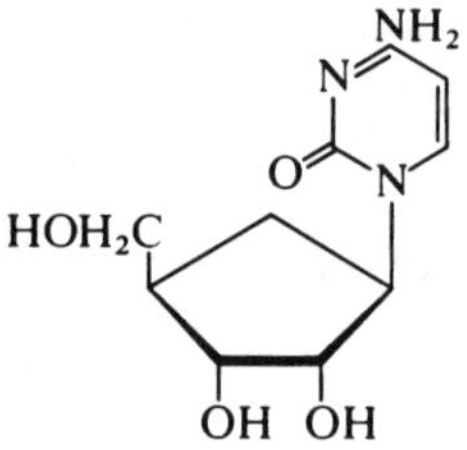

$C_{10}H_{15}N_3O_4$ M 241.246

Carbocyclic analogue of cytidine; antitumour, antiviral. Cryst. (H_2O). Mp 253-255°.

Shealy, Y.F. et al, J. Heterocycl. Chem., 1976, **13**, 1353; 1980, **17**, 353 (synth, ms, uv, pmr, cmr)
Shealy, Y.F. et al, J. Med. Chem., 1986, **29**, 1720 (biochem, bibl)

Carbonimidic acid, 9CI   C-70019

Imidocarbonic acid, 8CI

[6703-56-6]

$$HN{=}C(OH)_2$$

CH_3NO_2 M 61.040

Free acid (unknown) is a tautomer of Carbamic acid, C-00196 .

Di-Ph ester: [4513-71-7]. *Diphenyl imidocarbonate.*
$C_{13}H_{11}NO_2$ M 213.235
Cryst. (pet. ether). Mp 55°.

Houben, J. et al, Ber., 1913, **46**, 2447 (synth)
Hedayatullah, M., Bull. Soc. Chim. Fr., 1967, 416 (synth)
Kupfer, R. et al, Chem. Ber., 1986, **119**, 3236 (synth, ir, pmr, cmr)

Carbon monosulfide   C-70020

[2944-05-0]

CS

CS   M 44.071

Obs. in upper atmosphere. Gas. Rapidly forms brown-black polymer.

▷Condensed CS polymerises explosively

Klabunde, K.J. et al, Inorg. Chem., 1974, **13**, 1778 (synth, bibl)
Klabunde, K.J. et al, J. Am. Chem. Soc., 1984, **106**, 263.
Moltzen, E.K. et al, Acta Chem. Scand., Ser. B, 1986, **40**, 609 (synth)
Moltzen, E.K. et al, J. Org. Chem., 1987, **52**, 1156 (synth, bibl)

3-Carboxy-2,3-dihydro-8-hydroxy-5-meth-   **C-70021**
ylthiazolo[3,2-a]pyridinium hydroxide in-
ner salt, 9CI, 8CI
8-Hydroxy-5-methyldihydrothiazolo[3,2-a]pyridinium-
3-carboxylate
[13431-29-3]

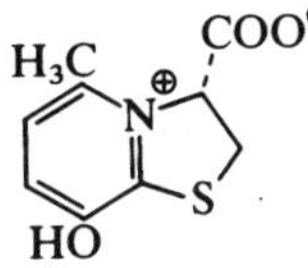

$C_9H_9NO_3S$ M 211.235
(R)-form
L-form
Fluorescent substance from acid hydrol. of bovine liver
extracts. Mp 152-153°.
(S)-form [26574-86-7]
D-form
$[\alpha]_D^{25}$ +99° (c, 0.6 in 1M NaOH).
(±)-form
Cryst. (H_2O). Mp 160-162°.
Me ester: Mp 140-145°. No counterion given in the lit.
May be a zwitterion.

Undheim, K. *et al, Acta Chem. Scand.*, 1969, **23**, 371, 1704;
1972, **26**, 2267, 2385; 1973, **27**, 1390 *(struct, uv, abs config,
ord, cd, ms, synth)*

N-(N-1-Carboxy-6-hydroxy-3-oxoheptyl)-   **C-70022**
alanyltyrosine, 9CI
WF 10129
[109075-64-1]

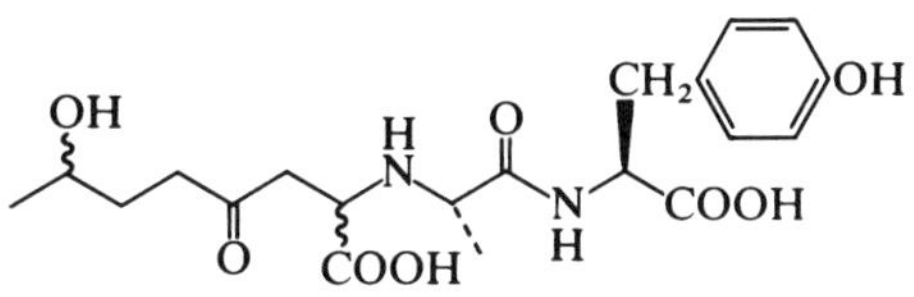

$C_{20}H_{28}N_2O_8$ M 424.450
Prod. by *Doratomyces putredinis*. Inhibitor of
angiotensin converting enzyme. Powder. Mp 90-95°.
$[\alpha]_D^{23}$ +12.9° (c, 0.375 in H_2O).

Ando, T. *et al, J. Antibiot.*, 1987, **40**, 468 *(isol, struct, props)*

Cardiophyllidin   **C-70023**
[113464-53-2]

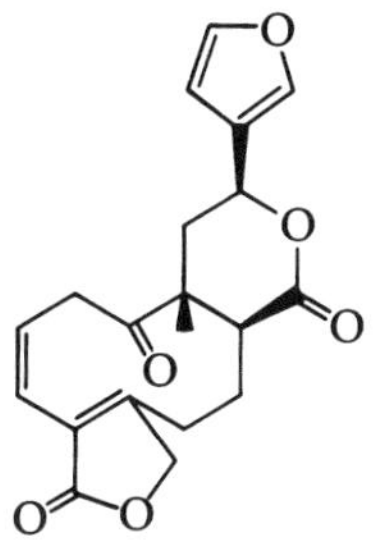

Absolute
configuration

$C_{20}H_{20}O_6$ M 356.374
Constit. of *Salvia cardiophylla*. Cryst. Mp 239-241°.

González, A.G. *et al, Tetrahedron Lett.*, 1988, **29**, 363 *(cryst
struct)*

Carnitine, INN   **C-70024**
Updated Entry replacing C-20054
3-Carboxy-2-hydroxy-N,N,N-trimethyl-1-propanamin-
ium hydroxide inner salt, 9CI. β-Hydroxy-γ-butyrotri-
methylbetaine. Novain. Vitamin B_T. Other proprietary
names
[461-06-3]

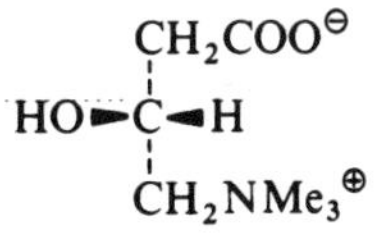

(R)-form
Absolute
configuration

$C_7H_{15}NO_3$ M 161.200
(R)-form [541-15-1]
Levocarnitine, INN, USAN
Constit. of striated muscle, liver and whey. Facilitator of
long-chain fatty acids through mitochondrial
membranes, thus allowing their metabolic oxidn.
Regulator of blood lipid levels, used in sport and infant
nutrition. Drug used to increase cardiac output and
improve myocardial function; often administered after
haemodialysis. Extremely hygroscopic solid
(EtOH/Me_2CO). Mp 196-198°. $[\alpha]_D^{25}$ −31.3° (c, 10 in
H_2O) (>99% opt. pure).
▷BP2980000.
B,HCl: Hygroscopic cryst. Mp 137-139°. $[\alpha]_D^{22}$ −20.4°
(H_2O).
Me ether:
$C_8H_{17}NO_3$ M 175.227
Prisms (Me_2CO) (as hydrochloride). Mp 178°.
Et ester: [40915-13-7]. *4-Ethoxy-2-hydroxy-N,N,N-tri-*
methyl-4-oxo-1-butanaminium, 9CI.
$C_9H_{20}NO_3^{\oplus}$ M 190.262 (ion)
Needles (Me_2CO) (as chloride). Mp 146°.
O-Ac: [3040-38-8].
$C_9H_{17}NO_4$ M 203.238
Cryst. (EtOH/Et_2O). Mp 145°. $[\alpha]_D^{20}$ −19.5°.
▷BP2990000.
O-Ac; B,HCl: Mp 188-190°.
Dimeric intermolecular ester:
$C_{14}H_{28}N_2O_5$ M 304.386
Mp 198-200°.
(S)-form [541-14-0]
Shows unfavourable physiol. props.; administration of the
L-form appears preferable to the racemate. Mp 210-
212° dec. $[\alpha]_D$ +30.9°.
B,HCl: Mp 142° dec.
(±)-form [406-76-8]
Mp 195-197° dec.
▷BP2979900.
B,HCl: Mp 196°.
Nitrile: [18933-33-0]. *3-Cyano-2-hydroxy-N,N,N-tri-*
methylpropanaminium, 9CI.
$C_7H_{15}N_2O^{\oplus}$ M 143.208 (ion)
Mp 232° dec. (as chloride).
O-Ac: [870-77-9]. Mp 187-188°.

Carter, H.E. *et al, Methods Enzymol.*, 1957, **3**, 660 *(isol)*
Strack, E. *et al, Hoppe-Seylers Z. Physiol. Chem.*, 1960, **318**,
129 *(synth)*
Wolf, G. *et al, Arch. Biochem. Biophys.*, 1961, **92**, 360
(biosynth)
Kaneko, T. *et al, Bull. Chem. Soc. Jpn.*, 1962, **35**, 1153 *(abs
config)*
Vasil'eva, E.D., *Khim. Prir. Soedin.*, 1969, 463; *CA*, **72**, 54672a.
Tomita, K. *et al, Bull. Chem. Soc. Jpn.*, 1974, **47**, 1988 *(cryst
struct)*
Bamji, M., *Biochem. Rev.*, 1980, **50**, 99 *(rev)*
Frenkel, R.A. *et al, Eds., Carnitine: Biosynth. Metab. and
Functions*, Academic Press, N.Y., 1980 *(book)*
Boch, K. *et al, Acta Chem. Scand., Ser. B*, 1983, **37**, 341 *(synth)*

Bremer, J., *Physiol. Rev.*, 1983, **63**, 1420 (*rev*)

Zhou, B.N. *et al, J. Am. Chem. Soc.*, 1983, **105**, 5925 (*synth*)

Comber, R.N. *et al, Org. Prep. Proced. Int.*, 1985, **17**, 175 (*synth*)

Reboucha, C.J. *et al, Annu. Rev. Nutr.*, 1986, **6**, 41 (*rev*)

Voeffray, R. *et al, Helv. Chim. Acta*, 1987, **70**, 2058 (*synth, ir, pmr, bibl*)

Martindale, The Extra Pharmacopoeia, 28th/29th Eds., 1982/1989, Pharmaceutical Press, London, 12528.

Negwer, M., *Organic-Chemical Drugs and their Synonyms*, 6th Ed., Akademie-Verlag, Berlin, 1987, 789 (*synonyms*)

Cassigarol *A*   C-70025

10,11-Dihydro-2,4,7,8-tetrahydroxy-10-(3,4-dihydroxy-phenyl)-5-[(3,5-dihydroxyphenyl)methyl]-5H-dibenzo[a,d]*cycloheptene*

[106387-02-4]

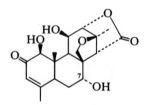

$C_{28}H_{24}O_8$ M 488.493

Isol. from heartwood of *Cassia garrettiana*. Pale-brown oil.

Octa-Ac: Cryst. Mp 189-190°.

Baba, K. *et al, Chem. Pharm. Bull.*, 1986, **34**, 4418 (*isol, struct*)

Catalpol   C-70026

Updated Entry replacing C-50061

$C_{15}H_{22}O_{10}$ M 362.333

Constit. of *Plantago lanceolata, Buddleia globosa* and *B. variabilis*. Mp 207-209° dec. $[\alpha]_D^{22}$ −122° (EtOH aq.).

10-(E)-Cinnamoyl: [1399-49-1]. **Globularin**. *Scutellarioside I*.

$C_{24}H_{28}O_{11}$ M 492.479

Constit. of *Globularia alypum* and *Scutellaria altissima*. Cryst. (EtOH) or amorph. powder. Mp 115-117°. $[\alpha]_D^{20}$ −64.9° (c, 0.81 in MeOH), $[\alpha]_D^{20}$ −76.9° (c, 0.5 in EtOH). The stereochem. of Scutellarioside I was not detd. but it is prob. identical with Globularin.

10-(Z)-Cinnamoyl: [76248-14-1]. **Globularicisin**. *Picroside I*.

$C_{24}H_{28}O_{11}$ M 492.479

Constit. of *G. alypum* and *Picrorhiza kurrooa*. Amorph. $[\alpha]_D^{20}$ −97.2° (c, 0.66 in CHCl₃).

10-(Z)-Cinnamoyl, 3,4-dihydro: [70256-08-5]. **Globularidin**.

$C_{24}H_{30}O_{11}$ M 494.494

Constit. of *G. alypum*. Amorph. $[\alpha]_D^{20}$ −57.7° (c, 0.51 in MeOH).

6-(3,4-Dihydroxybenzoyl): [50932-20-2]. **Verproside**.

$C_{22}H_{26}O_{13}$ M 498.440

Constit. of *Veronica officinalis*. Amorph. $[\alpha]_D^{20}$ −164.8° (c, 0.86 in MeOH). May be 10-substd.

6-(4-Hydroxy-3-methoxybenzoyl): [39012-20-9]. **Picroside II**.

$C_{23}H_{28}O_{13}$ M 512.466

Mp 178-179°. $[\alpha]_D^{20}$ −105° (CHCl₃). May be 10-substd.

6-(4-Hydroxy-3-methoxycinnamoyl): [64461-95-6]. **Picroside III**.

$C_{25}H_{30}O_{13}$ M 538.504

Mp 154-155°. $[\alpha]_D^{20}$ −78° (CHCl₃). May be 10-substd.

10-((E)-4-Hydroxycinnamoyl): [58286-53-6]. **Scutellarioside II**.

$C_{24}H_{28}O_{12}$ M 508.478

From *S. altissima*. Powder. $[\alpha]_D^{20}$ −80.1° (c, 0.5 in EtOH).

6-Deoxy: [99499-99-7]. **6-Deoxycatalpol**.

$C_{15}H_{22}O_9$ M 346.333

Constit. of *Utricularia australis, Catilleja miniata* and *Cistanche salsa*. Cryst. (EtOH), needles (MeOH). Mp 212.5-214° (204-206°). $[\alpha]_D^{22}$ −47° (c, 0.4 in MeOH), $[\alpha]_D^{18}$ −50.0° (c, 1.0 in MeOH). Also descr. as a foam.

O⁶-Me: **Methylcatalpol**.

$C_{16}H_{24}O_{10}$ M 376.360

Isol. from leaves of *B. globosa* and *B. variabilis*. Cryst. (H₂O). Mp 236-238°. $[\alpha]_D^{22.5}$ −122° (c, 1.64 in 90% EtOH aq.).

Duff, R.B. *et al, Biochem. J.*, 1965, **96**, 1 (*Methylcatalpol*)

Inouye, H. *et al, Chem. Pharm. Bull.*, 1971, **19**, 1438 (*abs config*)

Weinges, K. *et al, Justus Liebigs Ann. Chem.*, 1975, 2190; 1977, 1053 (*Picrosides, Scutellariosides*)

Afifi-Yazar, F.Ü. *et al, Helv. Chim. Acta*, 1980, **63**, 1905 (*Verproside*)

Chaudhuri, R.K. *et al, Helv. Chim. Acta*, 1981, **64**, 3 (*Globularin*)

Arslanian, R.L. *et al, J. Nat. Prod.*, 1985, **48**, 957 (*6-Deoxycatalpol*)

Damtoft, S. *et al, Phytochemistry*, 1985, **24**, 2281 (*6-Deoxycatalpol*)

Kobayashi, H. *et al, Chem. Pharm. Bull.*, 1985, **33**, 3645 (*6-Deoxycatalpol*)

Cedronin   C-70027

$C_{19}H_{24}O_7$ M 364.394

Constit. of *Simaba cedron*. Cryst. (MeOH). Mp 238-240°. $[\alpha]_D$ +70° (c, 0.44 in CHCl₃), $[\alpha]_D$ −14° (c, 1.2 in Py).

7-Epimer: **7-Epicedronin**.

$C_{19}H_{24}O_7$ M 364.394

Constit. of *S. cedron*. Cryst. (EtOAc/hexane). Mp 232-234°. $[\alpha]_D$ +109.3° (c, 0.57 in CHCl₃).

Jacobs, H. *et al, J. Nat. Prod.*, 1987, **50**, 700.

Celastanhydride   C-70028

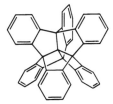

C$_{28}$H$_{36}$O$_5$ M 452.589
Constit. of *Kokoona zeylanica* and other Celastraceae spp. Unstable yellow solid. [α]$_D^{24}$ +160.8° (CHCl$_3$).

Gamlath, C.B. *et al*, *Tetrahedron Lett*., 1988, **29**, 109 (*struct*)

1,3,7,11-Cembratetraen-15-ol   C-70029

C$_{20}$H$_{32}$O   M 288.472
Constit. of *Lobophytum pauciflorum*. Cryst. (hexane). Mp 81-83°.

Bowden, B.F. *et al*, *J. Nat. Prod*., 1987, **50**, 650.

2,7,11-Cembratriene-4,6-diol   C-70030

Updated Entry replacing C-20071
4,8,13-Duvatriene-1,3-diol
[57605-80-8]

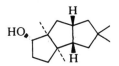

(1*S*,2*E*,4*R*,6*R*,7*E*,11*E*)-*form*

C$_{20}$H$_{34}$O$_2$ M 306.487
Cryst.
(*1S,2E,4R,6R,7E,11E*)-*form* [58190-98-0]
  β-*Cembrenediol*
  Constit. of tobacco. Plant growth inhibitor. Cryst. Mp 150-152° (123°). [α]$_D$ +40° (CHCl$_3$).
(*1S,2E,4S,6R,7E,11E*)-*form* [75282-01-8]
  α-*Cembrenediol*
  Constit. of tobacco. Mp 118-120°. [α]$_D$ +100° (CHCl$_3$).

Springer, J.P. *et al*, *Tetrahedron Lett*., 1975, 2737 (*isol*)
Chang, S.Y. *et al*, *Phytochemistry*, 1976, **15**, 961 (*isol*)
Wahlberg, I. *et al*, *Acta Chem. Scand., Ser. B*, 1982, **36**, 443 (*isol*)
Crombie, L. *et al*, *Phytochemistry*, 1988, **27**, 1685 (*biosynth*)
Begley, M.J. *et al*, *Phytochemistry*, 1988, **27**, 1695 (*cryst struct*)

Centrohexaindane   C-70031

Hexabenzohexacyclo[5.5.2.2^{4,10}.1^{1,7}.0^{4,17}.0^{10,17}]-heptadecane

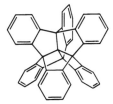

C$_{41}$H$_{24}$ M 516.641
Needles (xylene).

Kuck, D. *et al*, *Angew. Chem., Int. Ed. Engl*., 1988, **27**, 1192 (*synth, uv, ir, pmr, cmr, ms*)

Cerapicol   C-70032

1,4,4,8-Tetramethyltricyclo[6.2.1.0^{2,6}]undecan-11-ol

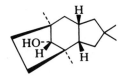

C$_{15}$H$_{26}$O   M 222.370
Metab. of fungus *Ceratocystis piceae*. Oil. [α]$_D^{20}$ +24.7° (c, 1.00 in CHCl$_3$).

Hanssen, H.-P. *et al*, *Tetrahedron*, 1988, **44**, 2175 (*struct*)

Ceratopicanol   C-70033

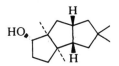

C$_{15}$H$_{26}$O   M 222.370
Metab. of fungus *Ceratocystis piceae*. Gum. [α]$_D^{20}$ +6.4° (c, 0.50 in CHCl$_3$).

Hanssen, H.-P. *et al*, *Tetrahedron*, 1988, **44**, 2175 (*struct*)

Cervicol   C-70034

8,14-Oxido-9-oxo-8,9-secodolast-1(15)-ene-7,8-diol
[103772-42-5]

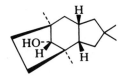

C$_{20}$H$_{32}$O$_4$ M 336.470
Constit. of brown alga *Dictyota cervicornis*.
7-Ac: Pale-yellow gum. [α]$_D^{25}$ −76.1° (c, 1.00 in CHCl$_3$).
Teixeira, V.L. *et al*, *Bull. Soc. Chim. Belg*., 1986, **95**, 263.

Cetraric acid   C-70035

9-(Ethoxymethyl)-4-formyl-3,8-dihydroxy-1,6-dimethyl-11-oxo-11H-dibenzo[b,e][1,4]dioxepin-7-carboxylic acid, 9CI
[489-49-6]

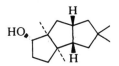

C$_{20}$H$_{18}$O$_9$ M 402.357
Isol. from the lichen *Cetraria islandica*. Dec. at ca. 250°.

Koller, G. *et al*, *Monatsh. Chem*., 1929, **53**, 931; 1934, **64**, 1 (*isol, struct*)
Asahina, Y. *et al*, *Ber*., 1933, **66**, 700 (*struct*)
Sticher, O., *Pharm. Acta Helv*., 1965, **40**, 385, 483 (*props*)

Chalaurenol   C-70036

[85359-61-1]

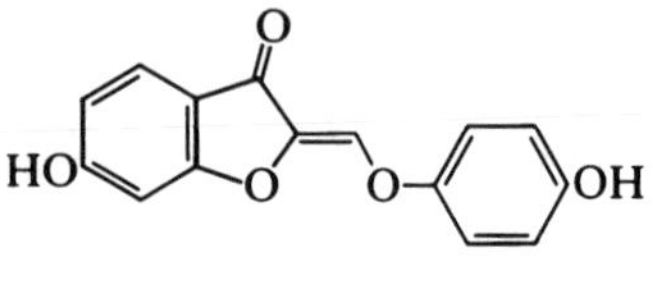

HO

O

O

OH

$C_{15}H_{10}O_5$ M 270.241

Formed from 2',4,4'-trihydroxychalcone in *Amorpha fruticosa* seedlings. Cryst. Mp 222-234°.

Begley, M.J. et al, J. Chem. Soc., Perkin Trans. 1, 1987, 2775 (isol, cryst struct)
Crombie, L. et al, J. Chem. Soc., Perkin Trans. 1, 1987, 2783 (synth)

Chapinolin   C-70037

[115547-12-1]

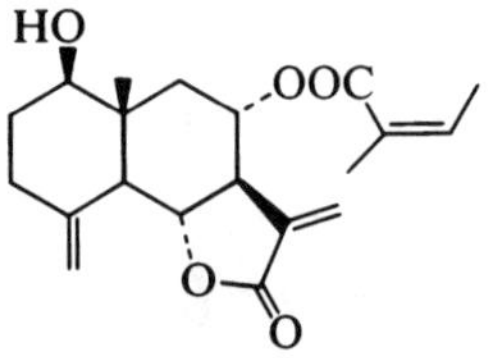

HO

OOC

O

O

$C_{20}H_{26}O_5$ M 346.422

Constit. of *Squamopappus skutchii*. Gum.

Vargas, D. et al, Phytochemistry, 1988, 27, 1413.

Chiisanogenin   C-70038

Updated Entry replacing C-30061

1R-Hydroxy-3,4-seco-4(23),20(29)-lupadiene-3,28-dioic acid 3,11α-lactone

[89353-99-1]

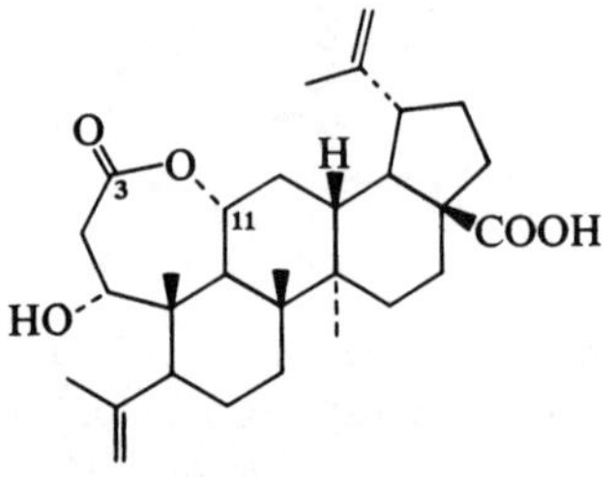

O

O

3

H

11

COOH

HO

$C_{30}H_{44}O_5$ M 484.675

Needles (Et₂O). Mp 232-234°. $[\alpha]_D^{22}$ +86.4° (c, 0.66 in MeOH).

α-L-Rhamnopyranosyl(1→4)-β-D-glucopyranosyl(1→6)β-D-glucopyranosyl ester:
[89354-01-8]. **Chiisanoside**.
$C_{48}H_{74}O_{19}$ M 955.101
Constit. of leaves and stem bark of *Acanthopanax chiisanensis*. Needles (butanol). Mp 228°. $[\alpha]_D^{14}$ +7.7° (c, 1.69 in MeOH).

β-D-Glucopyranosyl-(1→6)-β-D-glucopyranosyl ester:
Divaroside.
$C_{42}H_{64}O_{15}$ M 808.959
Constit. of leaves of *A. divaricatus*. Powder. $[\alpha]_D^{19}$ +30.0° (c, 0.59 in MeOH).

Kasai, R. et al, Chem. Pharm. Bull., 1986, 34, 3284 (isol, struct)
Matsumoto, K. et al, Chem. Pharm. Bull., 1987, 35, 413 (Divaroside)

2-Chloro-6-aminopurine   C-70039

2-Chloropurin-6-amine, 9CI. 2-Chloroadenine

[1839-18-5]

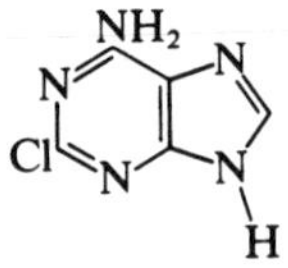

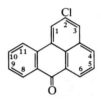

$C_5H_4ClN_5$ M 169.573

Powder.

9-β-D-Ribofuranosyl: see 2-Chloroadenosine, C-50083

Davoll, J. et al, J. Am. Chem. Soc., 1952, 74, 1563 (synth, uv)

2-Chloro-7H-benz[de]anthracen-7-one   C-70040

2-Chlorobenzanthrone. 2'-Chlorobenzanthrone (obsol.)

[82-04-2]

Cl

1 2 3
1   4
11   4
10   5
9   8   6
7
O

$C_{17}H_9ClO$ M 264.711

Yellowish-green plates (C_6H_6). Mp 191-192°.

Heilbron, I.M. et al, J. Chem. Soc., 1936, 781.
Vaidyanathan, A., Indian J. Chem., Sect. B, 1982, 21, 356 (pmr)

3-Chloro-7H-benz[de]anthracen-7-one   C-70041

3-Chlorobenzanthrone. 13-Chlorobenzanthrone (obsol.)

[6409-44-5]
$C_{17}H_9ClO$ M 264.711
Yellow needles (AcOH). Mp 182-183°.

Cahn, R.S. et al, J. Chem. Soc., 1933, 444 (synth)
Boyd, H.C. et al, Acta Crystallogr., 1954, 7, 142 (cryst struct)
Vaidyanathan, A., Indian J. Chem., Sect. B, 1982, 21, 356 (pmr)

4-Chloro-7H-benz[de]anthracen-7-one   C-70042

4-Chlorobenzanthrone. 2-Chlorobenzanthrone (obsol.)

[81-97-0]
$C_{17}H_9ClO$ M 264.711
Yellow needles (AcOH). Mp 204-205° (194.5-195°).

Lüttringhaus, A. et al, Justus Liebigs Ann. Chem., 1929, 473, 259 (synth)
Pandit, P.N. et al, Proc. Indian Acad. Sci., Sect. A, 1953, 38, 355; CA, 49, 1684f (synth)

5-Chloro-7H-benz[de]anthracen-7-one, 9CI   C-70043

5-Chlorobenzanthrone

$C_{17}H_9ClO$ M 264.711
Cryst. (C_6H_6). Mp 205-206°.

Malhotra, S.S. et al, Proc. Indian Acad. Sci., Sect. A, 1953, 38, 361; CA, 49, 1684i (synth)

6-Chloro-7*H*-benz[*de*]anthracen-7-one   C-70044
6-Chlorobenzanthrone
[51958-72-6]
C₁₇H₉ClO   M 264.711
Yellow granules (EtOH). Mp 146-147°.

Bradley, W. *et al, J. Chem. Soc.*, 1948, 1622, 1746.

8-Chloro-7*H*-benz[*de*]anthracen-7-one   C-70045
8-Chlorobenzanthrone
[30468-03-2]
C₁₇H₉ClO   M 264.711
Yellow needles. Mp 178-179°.

Bradley, W. *et al, J. Chem. Soc.*, 1948, 1746 (*synth*)
Nagai, Y. *et al, CA*, 1971, **74**, 43506q (*synth*)

9-Chloro-7*H*-benz[*de*]anthracen-7-one, 9CI   C-70046
9-Chlorobenzanthrone. 6-Chlorobenzanthrone (obsol.)
[24092-47-5]
C₁₇H₉ClO   M 264.711
Golden-yellow needles (AcOH or PhNO₂). Mp 186-187°.

Scholl, R. *et al, Ber.*, 1922, **55**, 109 (*synth*)
Pandit, P.N. *et al, Proc. Indian Acad. Sci., Sect. A*, 1953, **38**, 355; *CA*, **49**, 1684f (*synth*)

10-Chloro-7*H*-benz[*de*]anthracen-7-one   C-70047
10-Chlorobenzanthrone. 7-Chlorobenzanthrone (obsol.)
[30013-84-4]
C₁₇H₉ClO   M 264.711
Yellow cryst. Mp 188-189°.

Cahn, R.S. *et al, J. Chem. Soc.*, 1933, 447 (*synth*)
Pandit, P.N. *et al, Proc. Indian Acad. Sci., Sect. A*, 1953, **38**, 355; *CA*, **49**, 1684f (*synth*)
Tokita, S., *CA*, 1971, **75**, 151588x (*synth*)

11-Chloro-7*H*-benz[*de*]anthracen-7-one   C-70048
11-Chlorobenzanthrone
[30468-02-1]
C₁₇H₉ClO   M 264.711
Nagai, Y. *et al, CA*, 1971, **74**, 43506q (*synth*)

4-Chloro-1,3-benzenedisulfonic acid, 9CI   C-70049
[27886-55-1]

C₆H₅ClO₆S₂   M 272.675
Cryst. (pet. ether). Mp 87-88°.

Disulfonamide: [671-95-4]. **Clofenamide, INN.** *4-Chloro-1,3-benzenedisulfonamide, 9CI. Saltron. Soluran. Aquedux. Numerous proprietary names.*
C₆H₇ClN₂O₄S₂   M 270.705
Diuretic, antihypertensive agent. Needles. Mp 217-219° (206-207°).
▷CZ9100000.

Olivier, S.C.J., *Recl. Trav. Chim. Pays-Bas*, 1918, **37**, 311 (*synth, deriv*)
Davies, W. *et al, J. Chem. Soc.*, 1928, 1122 (*synth*)
Petrow, V. *et al, J. Pharm. Pharmacol.*, 1960, **12**, 648 (*synth, deriv, pharmacol*)

4-Chloro-1,2-benzisothiazole, 9CI   C-70050
[25380-61-4]

C₇H₄ClNS   M 169.628
Pale-yellow needles (pet. ether). Mp 41-42°, Mp 83°.
Bp₂ 95-100°. The higher Mp could possibly be due to confusion with 5-Chloro-1,2-benzisothiazole, C-70051 (reported Mp 82-83°).

Ricci, A. *et al, Ann. Chim. (Rome)*, 1963, **53**, 1860; *CA*, **60**, 12000g (*synth*)
von Becke, F. *et al, Justus Liebigs Ann. Chem.*, 1969, **729**, 146 (*synth*)
Markert, J. *et al, Justus Liebigs Ann. Chem.*, 1980,

5-Chloro-1,2-benzisothiazole, 9CI   C-70051
[41918-07-4]
C₇H₄ClNS   M 169.628
Yellow needles (MeOH). Mp 82-83°.

Ricci, A. *et al, Ann. Chim. (Rome)*, 1963, **53**, 577; *CA*, **59**, 8721d (*synth*)

6-Chloro-1,2-benzisothiazole, 9CI   C-70052
C₇H₄ClNS   M 169.628
Cryst. (MeOH aq.). Mp 37-38°.

Ricci, A. *et al, Ann. Chim. (Rome)*, 1963, **53**, 577; *CA*, **59**, 8721d (*synth*)

7-Chloro-1,2-benzisothiazole, 9CI   C-70053
[89583-90-4]
C₇H₄ClNS   M 169.628
Needles (MeOH aq.). Mp 49-50°.

Ricci, A. *et al, Ann. Chim. (Rome)*, 1963, **53**, 1860 (*synth*)
Rahman, L.K.A. *et al, J. Chem. Soc., Perkin Trans. 1*, 1984, 385 (*synth, pmr*)

1-Chlorobenzo[*g*]phthalazine   C-70054
[30800-68-1]

C₁₂H₇ClN₂   M 214.654
Tan cryst. (C₆H₆). Mp 178-180°.

Hill, J.H.M. *et al, J. Org. Chem.*, 1971, **36**, 3248 (*synth, props*)

4-Chlorobenzo[*g*]phthalazin-1(2*H*)-one, 9CI   C-70055
[77766-56-4]

$C_{12}H_7ClN_2O$ M 230.653
Ppt. (H_2O).

Me ether: [30800-69-2]. *1-Chloro-4-methoxybenzo*[g]-*phthalazine.* Tan cryst. (C_6H_6). Mp 134-136°.

Hill, J.H.M. *et al, J. Org. Chem.,* 1971, **36**, 3248 (*synth, deriv*)
Kormendy, K. *et al, Acta Chim. Acad. Sci. Hung.,* 1980, **105**, 175; *CA,* **94**, 208787k (*synth, use*)

2-Chlorobicyclo[2.2.1]hepta-2,5-diene   C-70056
2-Chloronorbornadiene
[2294-41-9]

C_7H_7Cl M 126.585
Bp_{14} 94.96°. n_D^{22} 1.4960. Polymerises on standing at r.t.
Adam, W. *et al, Chem. Ber.,* 1987, **120**, 531 (*synth, ir, pmr*)

1-Chlorobicyclo[2.2.2]octane   C-70057
[2064-03-1]

$C_8H_{13}Cl$ M 144.644
Mp 103-104°. Bp_{12} 70° subl.

Baker, C.W. *et al, J. Org. Chem.,* 1963, **28**, 514 (*synth*)
Becker, K.B. *et al, Synthesis,* 1973, 493 (*synth*)
Honegger, E. *et al, Chem. Ber.,* 1987, **120**, 187 (*synth, ir, pmr, pe*)

4-Chloro-1-butyne   C-70058
[51908-64-6]

$$HC{\equiv}CCH_2CH_2Cl$$

C_4H_5Cl M 88.537
Bp 86°. n_D^{22} 1.4383.

Eglington, G. *et al, J. Chem. Soc.,* 1950, 3650 (*synth*)

2[(Chlorocarbonyl)oxy]-3a,4,7,7a-tetrahydro-4,7-methano-1*H*-isoindole-1,3(2*H*)-dione, 9CI   C-70059
5-Norbornene-2,3-dicarboximido carbonochloridate
[99502-89-3]

$C_{10}H_8ClNO_4$ M 241.631

The synonym, used in the reference, is inaccurate.
Reagent for introduction of urethane protective groups via activated carbonic esters. Solid. Mp 98-100° dec. Stable in absence of air and moisture.

Henklein, P. *et al, Synthesis,* 1987, 166 (*synth, ir, use*)

1-Chloro-3-(chloromethyl)-7-methyl-2,6-octadiene, 9CI   C-70060

$C_{10}H_{16}Cl_2$ M 207.142
(*Z*)-*form* [112642-61-2]
Constit. of *Chondrococcus hornemannii.* Oil.
Coll, J.C. *et al, Aust. J. Chem.,* 1987, **40**, 1893.

3-Chloro-2-chloromethyl-1-propene, 9CI   C-70061
Updated Entry replacing C-00881
3,3′-Dichloroisobutylene
[1871-57-4]

$$H_2C{=}C(CH_2Cl)_2$$

$C_4H_6Cl_2$ M 124.997
d_4^{27} 1.18. Mp −11°. Bp_{756} 137-138°, Bp_{50} 58.8-59.8°.

Gragson, J.T. *et al, J. Am. Chem. Soc.,* 1953, **75**, 3344 (*synth*)
Latour, S. *et al, Synthesis,* 1987, 742 (*synth*)

5-Chloro-1,3-cycloheptadiene   C-70062
[107616-06-8]

C_7H_9Cl M 128.601
(±)-*form*
$Bp_{1.4}$ 66-71°. Cont. 20% 6-chloro isomer.
Mayr, H. *et al, Tetrahedron,* 1986, **42**, 6657, 6663 (*synth*)

5-Chloro-2-cyclopenten-1-one   C-70063
[19931-06-7]

C_5H_5ClO M 116.547
(±)-*form*
d_4^{20} 1.216. Bp_{18} 91°. n_D 1.506.
2,4-Dinitrophenylhydrazone: Mp 142°.

Martin, G.J. *et al, Bull. Soc. Chim. Fr.,* 1970, 3098 (*synth, ir*)
Rizzi, C.J. *et al, J. Org. Chem.,* 1987, **52**, 5280 (*synth, ir, pmr*)

5'-Chloro-5'-deoxyarabinosylcytosine   C-70064

4-Amino-1-(5-chloro-5-deoxy-β-D-arabinofuranosyl)-2(1H)-pyrimidinone

[32659-31-7]

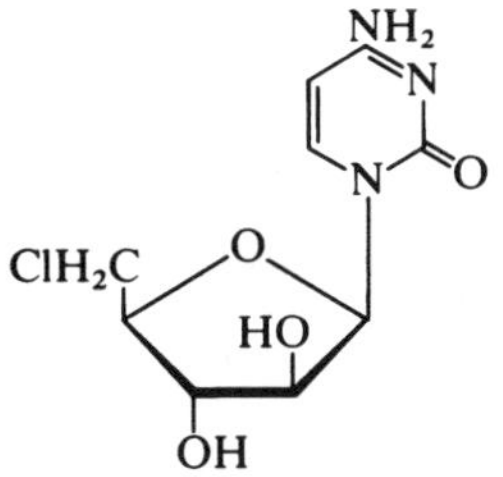

$C_9H_{12}ClN_3O_4$ M 261.665

Potent and specific inhibitor of DNA synthesis in tumour cells. Cryst. (H_2O). Mp 206-208.5° (202-204.5°). $[\alpha]_D^{25}$ +163.8° (c, 0.5 in H_2O).

Kikugawa, K. *et al, J. Org. Chem.*, 1972, **37**, 284 (*synth, uv*)
Hřebabecký, H. *et al, Collect. Czech. Chem. Commun.*, 1980, **45**, 599 (*synth*)
Birnbaum, G.I. *et al, Can. J. Chem.*, 1988, **66**, 1203 (*cryst struct, pmr, cmr, bibl*)

5-Chloro-2,3-dihydroxybenzaldehyde   C-70065

Updated Entry replacing C-50112

[73275-96-4]

$C_7H_5ClO_3$ M 172.568

3-Me ether: [7740-05-8]. *5-Chloro-2-hydroxy-3-methoxybenzaldehyde.*
   $C_8H_7ClO_3$ M 186.595
   Cryst. (CCl_4). Mp 118-119°.
Di-Me ether: [86232-28-2]. *5-Chloro-2,3-dimethoxybenzaldehyde.*
   $C_9H_9ClO_3$ M 200.621
   Cryst. (EtOH aq.). Mp 80-81°.

Daukshas, V.K. *et al, J. Org. Chem. USSR (Engl. transl.)*, 1983, **19**, 458 (*synth, deriv, uv, ir, pmr*)

12-Chlorododecanoic acid   C-70066

[22075-86-1]

$$ClCH_2(CH_2)_{10}COOH$$

$C_{12}H_{23}ClO_2$ M 234.766
Cryst. (pet. ether). Mp 43-44°.

Chloride: [78476-69-4].
   $C_{12}H_{22}Cl_2O$ M 253.211
   $Bp_{0.15}$ 139-145°.

Logemann, E. *et al, Chem. Ber.*, 1981, **114**, 2245 (*synth*)
Bidd, I. *et al, J. Chem. Soc., Perkin Trans. 1*, 1987, 2455 (*synth*)

3-Chloro-4-fluorobenzaldehyde   C-70067

Updated Entry replacing C-40077

[34328-61-5]

C_7H_4ClFO M 158.559
Liq.

French, F.A. *et al, J. Med. Chem.*, 1971, **14**, 862 (*synth*)
Yoshida, Y. *et al, Chem. Lett.*, 1988, 1355 (*synth*)

8-Chloroguanosine, 9CI   C-70068

2-Amino-8-chloro-9-β-D-ribofuranosyl-9H-purin-6(1H)-one. 8-Chloro-9-β-D-ribofuranosylguanine. 2-Amino-8-chloroinosine

[2104-68-9]

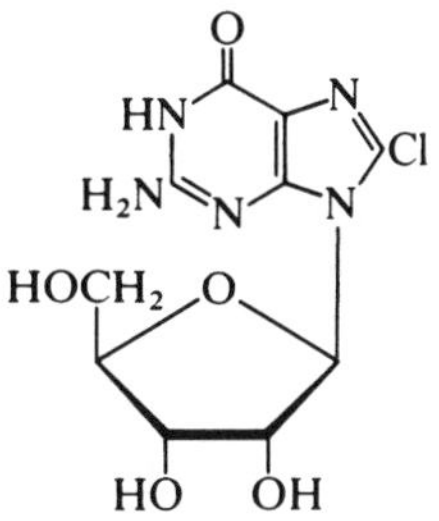

$C_{10}H_{12}ClN_5O_5$ M 317.688
Cryst. (EtOH aq.). Mp >260°.

Ryu, E.K. *et al, J. Org. Chem.*, 1981, **46**, 2819 (*synth, uv, pmr*)
Birnbaum, G.I. *et al, Biochemistry*, 1984, **23**, 5048 (*cryst struct, conformn*)
Lassota, P. *et al, Z. Naturforsch., C*, 1984, **39**, 55 (*pmr, conformn*)

7-Chloro-2-heptenoic acid   C-70069

$$ClCH_2(CH_2)_3CH{=}CHCOOH$$

$C_7H_{11}ClO_2$ M 162.616

(E)-form [75078-18-1]
   Oil. $Bp_{0.5}$ 118-121°.
Et ester: [107408-35-5].
   $C_9H_{15}ClO_2$ M 190.669
   Bp_{13} 129-131°.

Cooke, M.P. *et al, J. Org. Chem.*, 1987, **52**, 1381 (*synth, pmr*)

2-Chloro-1,5-hexadiene   C-70070

[101933-88-4]

$$H_2C{=}CHCH_2CH_2CCl{=}CH_2$$

C_6H_9Cl M 116.590
Bp_{100} 52-54°.

Peterson, P.E. *et al, J. Org. Chem.*, 1986, **51**, 2381 (*synth, pmr*)

3-Chloro-4-hydroxy-3-cyclobutene-1,2-dione   C-70071

Squaric acid monochloride

[68057-72-7]

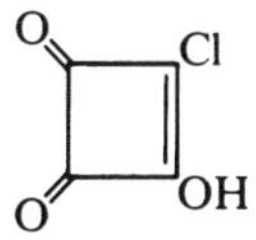

C_4HClO_3 M 132.503
Cryst. Mp 149-152°. Also descr. as yellowish oil.

Belluš, D. *et al, Helv. Chim. Acta*, 1978, **61**, 1784 (*synth, uv, ir*)
Schmidt, A.H. *et al, Synthesis*, 1987, 134 (*synth, ir, cmr*)

3-Chloro-2-hydroxypropanoic acid   C-70072

Updated Entry replacing C-20116
β-Chlorolactic acid
[1713-85-5]

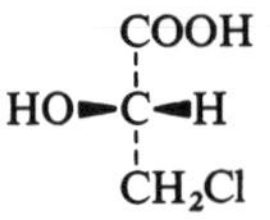

$$\begin{array}{c} COOH \\ | \\ HO\!\!-\!\!C\!\!-\!\!H \\ | \\ CH_2Cl \end{array} \qquad (R)\text{-form}$$

$C_3H_5ClO_3$ M 124.524

(R)-form [61505-41-7]
L-form
Mp 88-89°. $[\alpha]_D^{20}$ +4.14° (c, 0.91 in H_2O).

(S)-form [82079-44-5]
D-form
Mp 88-89°. $[\alpha]_D$ +3.97° (c, 0.91 in H_2O). Note sign of rotn. reported as (+) for both enantiomers.

(±)-form [50906-02-0]
Cryst. (Et_2O). Mp 78°.

Me ester: [32777-04-1].
$C_4H_7ClO_3$ M 138.551
Bp 185-187°.

Koelsch, C.F. *et al*, *J. Am. Chem. Soc.*, 1930, **52**, 1105 (synth)
Hirschbein, B.L. *et al*, *J. Am. Chem. Soc.*, 1982, **104**, 4458 (synth, abs config)
Morin, C. *et al*, *Synthesis*, 1987, 479 (synth)
Bretherick, L., *Handbook of Reactive Chemical Hazards*, 2nd Ed., Butterworths, London and Boston, 1979, 427.
Hazards in the Chemical Laboratory, (Bretherick, L., Ed.), 3rd Ed., Royal Society of Chemistry, London, 1981, 246.

4-Chloro-6-hydroxy-2(1*H*)-pyridinone, 9CI   C-70073

4-Chloro-2,6-dihydroxypyridine
[62616-12-0]

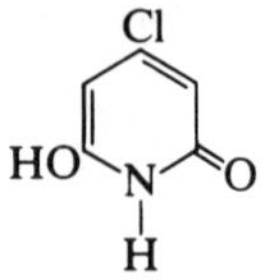

$C_5H_4ClNO_2$ M 145.545
Cryst. (AcOH). Mp 224° dec.

Di-Me ether: [62616-14-2]. *4-Chloro-2,6-dimethoxypyridine.*
$C_7H_8ClNO_2$ M 173.599
Prisms (MeOH). Mp 64-65°.

Me ether, N-Me: [62616-13-1]. *4-Chloro-6-methoxy-1-methyl-2(1H)-pyridinone, 9CI.*
$C_7H_8ClNO_2$ M 173.599
Cryst. (pet. ether). Mp 104°.

Elvidge, J.A. *et al*, *J. Chem. Soc., Perkin Trans. 1*, 1976, 2462 (synth, deriv, ir, uv, pmr)
Kaneko, C. *et al*, *Chem. Pharm. Bull.*, 1986, **34**, 3658 (synth, pmr, deriv)

6-Chloro-4-hydroxy-2(1*H*)pyridinone, 9CI   C-70074

2-Chloro-4,6-dihydroxypyridine
$C_5H_4ClNO_2$ M 145.545
Mp 233°.

4-Me ether: [108279-65-8]. *6-Chloro-4-methoxy-2(1H)-pyridinone.*
$C_6H_6ClNO_2$ M 159.572
Prisms (EtOAc). Mp 190-194°.

Di-Me ether: [108279-89-6]. *2-Chloro-4,6-dimethoxypyridine.*
$C_7H_8ClNO_2$ M 173.599

Prisms (pentane). Mp 81-83°.

Davis, S.J. *et al*, *J. Chem. Soc.*, 1962, 3638 (synth)
Elvidge, J.A. *et al*, *J. Chem. Soc., Perkin Trans. 1*, 1976, **23**, 2462 (deriv, synth)
Kaneko, C. *et al*, *Chem. Pharm. Bull.*, 1986, **34**, 3658 (synth, pmr, derivs)

5-Chloro-8-hydroxyquinoline   C-70075

Updated Entry replacing C-60103
5-Chloro-8-quinolinol, 9CI. **Cloxyquin, USAN.** *Cloxiquine, INN. Anametil. Atletol. Chlorisept. Dermofongin A*
[130-16-5]
C_9H_6ClNO M 179.606
Bactericide. Topical antifungal. Mp 129-130° (122-123°).
▷VC4590000.

B,HCl: [25395-13-5]. Yellow cryst. Mp 256-258°.
1-Oxide: [21168-34-3].
$C_9H_6ClNO_2$ M 195.605
Cryst. (ligroin). Mp 169-170°.
Ac: [10173-02-1]. *Silital. S 604.*
$C_{11}H_8ClNO_2$ M 221.643
Antibacterial, antidiarrhoeal, antifungal agent. Cryst. (C_6H_6). Mp 83°.

Das Gupta, S.J., *J. Indian Chem. Soc.*, 1952, **29**, 711 (synth)
B.P., 791 409, (1958); CA, **52**, 12931 (acetate)
Vogt, H. *et al*, *Arch. Pharm. (Weinheim, Ger.)*, 1958, **291**, 168 (acetate)
Sukhina, L.F. *et al*, *Zh. Obshch. Khim.*, 1962, **32**, 1356.
Ritter, P. *et al*, *Arzneim.-Forsch.*, 1966, **16**, 1647 (metab, acetate)
Naga, S. *et al*, *Bull. Chem. Soc. Jpn.*, 1975, **48**, 863 (ms)
Sawada, Y. *et al*, *Chem. Pharm. Bull.*, 1978, **26**, 1357 (metab)
Zuev, A.P. *et al*, *Khim.-Farm. Zh.*, 1980, **14**, 128 (synth)
Kidric, J. *et al*, *Org. Magn. Reson.*, 1981, **15**, 280 (pmr, cmr)
Banerjee, T. *et al*, *Acta Crystallogr., Sect. C*, 1986, **42**, 1408 (cryst struct)
Martindale, The Extra Pharmacopoeia, 28th/29th Eds., 1982/1989, Pharmaceutical Press, London, 15330.

2-Chloro-1*H*-imidazole   C-70076

[16265-04-6]

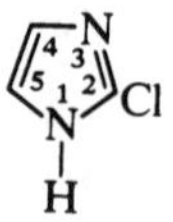

$C_3H_3ClN_2$ M 102.523
Cryst. by subl. Mp 165-166°.

Imbach, J.L. *et al*, *J. Heterocycl. Chem.*, 1967, **4**, 451 (synth)

4(5)-Chloro-1*H*-imidazole   C-70077

[15965-31-8]
$C_3H_3ClN_2$ M 102.523
Solid (H_2O). Mp 117-118°.

Lutz, A.W. *et al*, *J. Heterocycl. Chem.*, 1967, **4**, 399 (synth)

1-Chloro-2-iodoethane   C-70078

[624-70-4]

$$ClCH_2CH_2I$$

C_2H_4ClI M 190.411
Bp 137-139°.

Francis, A.W., *J. Am. Chem. Soc.*, 1925, **47**, 2340 (*synth*)
Cracknell, M.E. *et al*, *J. Chem. Soc., Perkin Trans. 1*, 1985, 115 (*synth*)

4-Chloro-1(3H)-isobenzofuranone   **C-70079**
4-Chlorophthalide
[52010-22-7]

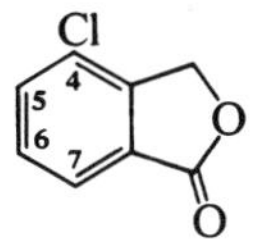

Cl

$C_8H_5ClO_2$ M 168.579
Cryst. by subl. Mp 87-88°.

Soucy, C. *et al*, *J. Org. Chem.*, 1987, **52**, 129 (*synth, pmr, ir*)

7-Chloro-1(3H)-isobenzofuranone   **C-70080**
7-Chlorophthalide
[70097-45-9]
$C_8H_5ClO_2$ M 168.579
Cryst. by subl. Mp 148-150°.

Soucy, C. *et al*, *J. Org. Chem.*, 1987, **52**, 129 (*synth, pmr, ir*)

1-Chloro-2-isothiocyanoethane, 9CI   **C-70081**
2-Chloroethyl isothiocyanate. β-Chloroethyl mustard oil
[6099-88-3]

$$ClCH_2CH_2NCS$$

C_3H_4ClNS M 121.584
Reagent for synth. of *S*-contg. heterocycles. Unpleasant-smelling liq. d_{20} 1.265. $Bp_{0.13}$ 80°.

▷Mucous membrane irritant

Brintzinger, H. *et al*, *Chem. Ber.*, 1949, **82**, 389.
Outcalt, R.J., *J. Heterocycl. Chem.*, 1987, **24**, 1425 (*use, bibl*)

2-Chloro-6-mercaptobenzoic acid   **C-70082**
2-Carboxy-3-chlorobenzenethiol
[20324-51-0]

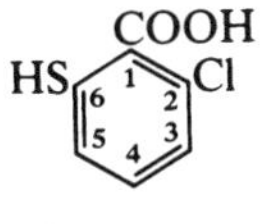

HS   COOH
        Cl

$C_7H_5ClO_2S$ M 188.628
Mp 106-108°.
Nitrile, Me thioether: [51271-34-2]. *2-Chloro-6-(methylthio)benzonitrile.*
C_8H_6ClNS M 183.655
Cryst. (EtOH). Mp 118-119°.

Amoretti, L. *et al*, *CA*, 1968, **68**, 77913m (*synth*)
Beck, J.R. *et al*, *J. Org. Chem.*, 1974, **39**, 1839 (*synth, deriv*)

3-Chloro-2-mercaptobenzoic acid   **C-70083**
3-Carboxy-2-chlorobenzenethiol
[17839-51-9]
$C_7H_5ClO_2S$ M 188.628
Yellow solid. Mp 172-173°.
Nitrile: 2-Chloro-6-cyanobenzenethiol.
C_7H_4ClNS M 169.628
Cryst. Mp 110°.

Markert, J. *et al*, *Justus Liebigs Ann. Chem.*, 1980, 768 (*nitrile*)

Rahman, L.K.A. *et al*, *J. Chem. Soc., Perkin Trans. 1*, 1984, 385 (*synth, ir*)

4-Chloro-2-mercaptobenzoic acid   **C-70084**
2-Carboxy-5-chlorobenzenethiol
[20324-49-6]
$C_7H_5ClO_2S$ M 188.628
Cryst. Mp 196°.

Amoretti, L. *et al*, *CA*, 1968, **68**, 77913m (*synth*)

5-Chloro-2-mercaptobenzoic acid   **C-70085**
2-Carboxy-4-chlorobenzenethiol
[20324-50-9]
$C_7H_5ClO_2S$ M 188.628
Cryst. (C_6H_6). Mp 193-195°.
Me thioether: [62176-39-0]. *5-Chloro-2-(methylthio)-benzoic acid.*
$C_8H_7ClO_2S$ M 202.655
Cryst. (EtOH aq.). Mp 184°.

Ruff, F. *et al*, *Tetrahedron*, 1978, **34**, 2767 (*synth, deriv*)
Šindelář, K. *et al*, *Collect. Czech. Chem. Commun.*, 1978, **43**, 471 (*synth*)
Hung, J. *et al*, *J. Heterocycl. Chem.*, 1983, **20**, 1575 (*synth*)

Chloromethanesulfonic acid   **C-70086**
[40104-07-2]

$$ClCH_2SO_3H$$

CH_3ClO_3S M 130.546
Na salt: [10352-63-3]. Cryst. (MeOH). Mp 261-262.5°.
Benzylthiuronium salt: Cryst. (H_2O). Mp 97-98°.
Chloride: [3518-65-8]. *Chloromethanesulfonyl chloride.*
$CH_2Cl_2O_2S$ M 148.992
Liq. Bp_{25} 80-81°.

▷Lacrymator. PB2800000.
Bromide: Chloromethanesulfonyl bromide.
CH_2BrClO_2S M 193.443
Oil. $Bp_{0.6}$ 50-54°.

Org. Synth. Coll. Vol. 5, 231 (*deriv, synth*)
Smith, T.L. *et al*, *J. Am. Chem. Soc.*, 1953, **75**, 3566 (*synth*)
Silhanek, J. *et al*, *Collect. Czech. Chem. Commun.*, 1977, **42**, 524 (*synth, pmr*)
Block, E. *et al*, *J. Am. Chem. Soc.*, 1986, **108**, 4568 (*deriv, synth, ir, pmr, use*)

3-Chloro-3-methoxy-3H-diazirine, 9CI   **C-70087**
[4222-27-9]

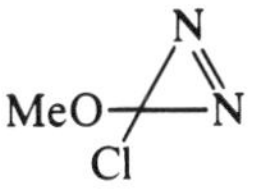

$C_2H_3ClN_2O$ M 106.512
Reagent for generation of methoxychlorocarbene.

Smith, N.P. *et al*, *J. Chem. Soc., Perkin Trans. 2*, 1979, 213, 1298 (*synth, nmr, uv, ir*)
Sheridan, R.S. *et al*, *J. Am. Chem. Soc.*, 1986, **108**, 99 (*synth*)

2-(Chloromethoxy)ethyl acetate   **C-70088**
2-(Chloromethoxy)ethanol acetate, 9CI. 1-Acetoxy-2-chloromethoxyethane

[40510-88-1]

$$AcOCH_2CH_2OCH_2Cl$$

$C_5H_9ClO_3$ M 152.577

Reagent for prepn. of acyclic nucleoside analogues. Bp_5 74-76°.

Foye, W.O. *et al, J. Heterocycl. Chem.*, 1982, **19**, 497 (*synth, pmr, use*)

Aitken, D.J. *et al, Tetrahedron Lett.*, 1986, **27**, 3417 (*use, bibl*)

3-Chloro-2-(methoxymethoxy)-1-propene,   **C-70089**
9CI

2-(Chloromethyl)-3,5-dioxa-1-hexene

[105104-40-3]

$$H_2C=C(CH_2Cl)OCH_2OMe$$

$C_5H_9ClO_2$ M 136.578

Acetonylating reagent. Liq. Bp_{35} 62-63°.

Gu, X.-P. *et al, J. Org. Chem.*, 1987, **52**, 3192 (*synth, pmr, ms, ir, use*)

2-Chloro-3-methyl-2-cyclopenten-1-one   **C-70090**

[73923-18-9]

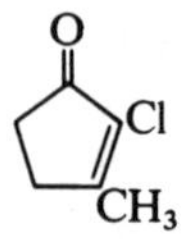

C_6H_7ClO M 130.574

Cryst. (pentane). Mp 37-38°.

Ponaras, A.A. *et al, J. Org. Chem.*, 1987, **52**, 5630 (*synth, ir, pmr*)

5-Chloro-3-methyl-2-cyclopenten-1-one   **C-70091**

[110874-85-6]

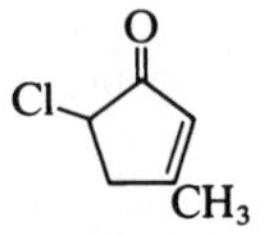

C_6H_7ClO M 130.574

(±)-*form*

Liq. $Bp_{0.35}$ 69-70°.

Ponaras, A.A. *et al, J. Org. Chem.*, 1987, **52**, 5630 (*synth, ir, pmr, cmr*)

1-(Chloromethyl)-1H-indole-2,3-dione, 9CI   **C-70092**

N-*Chloromethylisatin*

[31704-42-4]

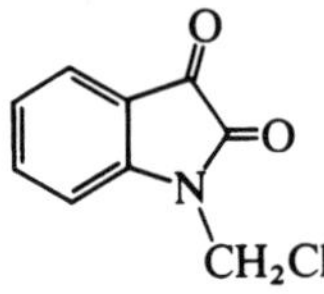

$C_9H_6ClNO_2$ M 195.605

Reagent for derivatisation and labelling of carboxylic acids. Fine light-yellow cryst. (C_6H_6/cyclohexane). Mp 126°.

Wendelin, W. *et al, Monatsh. Chem.*, 1972, **103**, 1632 (*synth, use*)

Wendelin, W. *et al, J. Heterocycl. Chem.*, 1987, **24**, 1381 (*bibl*)

2-(Chloromethyl)-3-methylpyridine   **C-70093**

[4377-43-9]

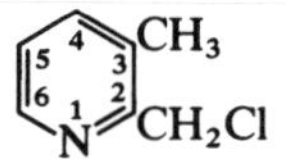

C_7H_8ClN M 141.600

$Bp_{1.4}$ 59-61°. n_D^{20} 1.5407.

B,HCl: [4370-22-3]. Cryst. (2-propanol). Mp 159° (157°).

Picrate: Mp 146-146.5°.

Mathes, W. *et al, Angew. Chem., Int. Ed. Engl.*, 1963, **2**, 144 (*synth*)

Jeromin, G.E. *et al, Chem. Ber.*, 1987, **120**, 649 (*synth, pmr*)

2-(Chloromethyl)-4-methylpyridine   **C-70094**

[38198-16-2]

C_7H_8ClN M 141.600

Oil. $Bp_{0.9}$ 51-52°. n_D^{20} 1.5326.

▷Irritant

Picrate: Mp 159-160° (138-139°).

Mathes, W. *et al, Angew. Chem., Int. Ed. Engl.*, 1963, **2**, 144 (*synth*)

Matsumura, E. *et al, Bull. Chem. Soc. Jpn.*, 1970, **43**, 3540 (*synth*)

2-(Chloromethyl)-5-methylpyridine   **C-70095**

[767-01-1]

C_7H_8ClN M 141.600

Bp_3 65-71°.

Picrate: Mp 160-161°.

Arata, Y. *et al, Yakugaku Zasshi*, 1959, **79**, 108; *CA*, **53**, 10211d (*synth*)

U.S.P., 3 501 486 (*1970*); *CA*, **72**, 121377y

2-(Chloromethyl)-6-methylpyridine   **C-70096**

[3099-29-4]

C_7H_8ClN M 141.600

Mp 18-19°. Bp_7 72-75°. n_D^{20} 1.5315. Stable on storage at 0° for 6 months.

▷Skin irritant

B,HCl: [3099-30-7]. Needles. Mp 155-156°.

Picrate: Mp 161-162°.

Baker, W. *et al, J. Chem. Soc.*, 1958, 3594 (*synth*)

Mathes, W. *et al, Angew. Chem., Int. Ed. Engl.*, 1963, **2**, 144 (*synth*)

Matsumoto, I., *Chem. Pharm. Bull.*, 1967, **15**, 1990 (*synth*)

Jeromin, G.E. *et al, Chem. Ber.*, 1987, **120**, 649 (*synth, pmr*)

3-(Chloromethyl)-2-methylpyridine   **C-70097**

C_7H_8ClN M 141.600

B,HCl: [58539-77-8]. Cryst. (C_6H_6).

Sato, Y., *Chem. Pharm. Bull.*, 1959, **7**, 241.

3-(Chloromethyl)-4-methylpyridine   **C-70098**

C_7H_8ClN M 141.600

Free base unstable.

B,HCl: Mp 175°.

Picrate: Mp 159-160°.

Japan Pat., 65 2148, (*1965*); *CA*, **62**, 14635f (*synth*)

4-(Chloromethyl)-2-methylpyridine   C-70099

[75523-42-1]

C_7H_8ClN M 141.600

Oil.

▷Irritant

Picrate: Yellow needles (MeOH). Mp 146-148°.

Furukawa, S., *Chem. Pharm. Bull.*, 1955, **3**, 413.
Pratesi, P. *et al*, *Farmaco, Ed. Sci.*, 1980, **35**, 621.

5-(Chloromethyl)-2-methylpyridine   C-70100

C_7H_8ClN M 141.600

No phys. props. reported.

Nakashima, T., *Yakugaku Zasshi*, 1958, **78**, 661; *CA*, **52**, 18399a (*synth*)

2-(Chloromethyl)phenol, 9CI   C-70101

α-Chloro-o-cresol, 8CI. 2-Hydroxybenzyl chloride

[40053-98-3]

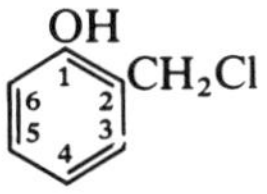

C_7H_7ClO M 142.585

Ac: [15068-08-3]. $Bp_{1.5}$ 98-100°. n_D^{25} 1.5235.

Me ether: [7035-02-1]. *1-(Chloromethyl)-2-methoxy-benzene, 9CI. 2-Methoxybenzyl chloride.*

C_8H_9ClO M 156.612

Mp 29-30°. Bp_{15} 102°, $Bp_{0.1}$ 52-54°. n_D^{22} 1.5478, 1.5492.

Grice, R. *et al*, *J. Chem. Soc.*, 1963, 1947 (*synth, deriv*)
Taylor, L.D. *et al*, *J. Org. Chem.*, 1978, **43**, 1197 (*deriv*)
Reimann, E. *et al*, *Arch. Pharm.* (*Weinheim, Ger.*), 1985, **318**, 1105 (*synth, deriv*)
Vejdělek, Z. *et al*, *Collect. Czech. Chem. Commun.*, 1987, **52**, 2545 (*deriv*)

3-(Chloromethyl)phenol   C-70102

α-Chloro-m-cresol, 8CI. 3-Hydroxybenzyl chloride

[60760-06-7]

C_7H_7ClO M 142.585

$Bp_{0.1}$ 105-110°.

Ac: [4530-44-3].

$C_9H_9ClO_2$ M 184.622

$Bp_{0.001}$ 95-98°. n_D^{19} 1.5272.

Me ether: [824-98-6]. *1-(Chloromethyl)-3-methoxyben-zene.* m-*Methoxybenzyl chloride.*

C_8H_9ClO M 156.612

Bp_{16} 102-104°, $Bp_{0.1}$ 58°. n_D^{20} 1.5427.

Grice, R. *et al*, *J. Chem. Soc.*, 1963, 1947 (*synth, deriv*)
Alvarez-Ibarra, C. *et al*, *Tetrahedron*, 1979, **35**, 1767 (*synth, deriv, ir, pmr*)
Reimann, E. *et al*, *Arch. Pharm.* (*Weinheim, Ger.*), 1985, **318**, 1105 (*synth, deriv*)
Aizpura, J.M. *et al*, *Can. J. Chem.*, 1986, **64**, 2342 (*synth, pmr*)

4-(Chloromethyl)phenol   C-70103

α-Chloro-p-cresol, 8CI. 4-Hydroxybenzyl chloride

[35421-08-0]

C_7H_7ClO M 142.585

Ac: [39720-27-9].

$C_9H_9ClO_3$ M 200.621

Liq. $Bp_{1.5}$ 104-106°, $Bp_{0.01}$ 59-60°. n_D^{25} 1.5290.

Me ether: [824-94-2]. *1-Chloromethyl-4-methoxyben-zene.* p-*Methoxybenzyl chloride.*

C_8H_9ClO M 156.612

Bp_{19} 113°, $Bp_{0.01}$ 59-60° (64-65°). n_D^{19} 1.5492.

Grice, R. *et al*, *J. Chem. Soc.*, 1963, 1947 (*synth, deriv*)
Taylor, L.D. *et al*, *J. Org. Chem.*, 1978, **43**, 1197 (*deriv*)
Alvarez-Ibarra, C. *et al*, *Tetrahedron*, 1979, **35**, 1767 (*synth, ir, pmr, deriv*)
Amin, S. *et al*, *J. Org. Chem.*, 1981, **46**, 2394 (*deriv, pmr*)
Aizpurva, J.M. *et al*, *Can. J. Chem.*, 1986, **64**, 2342 (*deriv*)

1-Chloro-4-methylphthalazine, 9CI   C-70104

[19064-68-7]

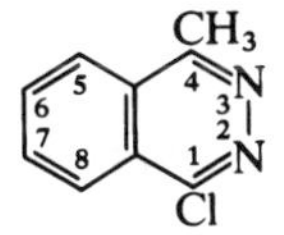

$C_9H_7ClN_2$ M 178.621

Needles (H_2O). Mp 130°.

Gabriel, S. *et al*, *Ber.*, 1893, **26**, 705; 1897, **30**, 3022 (*synth*)

2-Chloro-3-methylpyrazine, 9CI   C-70105

[95-58-9]

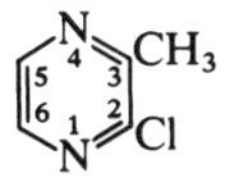

$C_5H_5ClN_2$ M 128.561

Liq. Bp_{65} 94-96°, Bp_{15} 55-65°.

1-Oxide: [61689-42-7].

$C_5H_5ClN_2O$ M 144.560

Needles (pet. ether). Mp 51-52°. Identity doubtful.

▷UQ2465000.

4-Oxide: [61689-41-6]. *3-Chloro-2-methylpyrazine 1-oxide.*

$C_5H_5ClN_2O$ M 144.560

Cryst. (C_6H_6/pet. ether). Mp 74-76°.

Karmas, G. *et al*, *J. Am. Chem. Soc.*, 1952, **74**, 1580 (*synth*)
Klein, B. *et al*, *J. Org. Chem.*, 1963, **28**, 1682 (*oxides*)

2-Chloro-5-methylpyrazine, 9CI   C-70106

[59303-10-5]

$C_5H_5ClN_2$ M 128.561

Bp_{60} 94-96°.

Karmas, G. *et al*, *J. Am. Chem. Soc.*, 1952, **74**, 1580 (*synth*)

2-Chloro-6-methylpyrazine, 9CI   C-70107

[38557-71-0]

$C_5H_5ClN_2$ M 128.561

Mp 50-51°. Bp_{40} 84-85°.

4-Oxide: [76850-13-0].

$C_5H_5ClN_2O$ M 144.560

Needles (EtOH). Mp 108-110°.

Karmas, G. *et al*, *J. Am. Chem. Soc.*, 1952, **74**, 1580 (*synth*)
Klein, B. *et al*, *J. Org. Chem.*, 1963, **28**, 1682 (*oxide*)

4-Chloro-3(5)methylpyrazole, 9CI   C-70108

[15878-08-7]

$C_4H_5ClN_2$ M 116.550

Exists as 5-Me tautomer in solid state. Plant growth stimulator. Cryst. (H_2O or pet. ether). Mp 65-66°. Bp_{12} 116-118°.

▷UQ6456000.

B,HCl: [98816-35-4]. Mp 175-177°.

Picrate: Greenish-yellow plates (Et_2O). Mp 151.5-152.5°.

von Auwers, K. *et al, J. Prakt. Chem.*, 1927, **116**, 65 (*synth*)

Hüttel, R. *et al, Justus Liebigs Ann. Chem.*, 1956, **598**, 186 (*synth*)

Elguero, J. *et al, Bull. Soc. Chim. Fr.*, 1966, 3727, 3744 (*uv, pmr*)

Franssen, M.C.R. *et al, J. Heterocycl. Chem.*, 1987, **24**, 1313 (*synth*)

Faure, R. *et al, Can. J. Chem.*, 1988, **66**, 1141 (*cmr, tautom*)

3(5)-Chloro-5(3)-methylpyrazole, 9CI   C-70109

[15953-45-4]

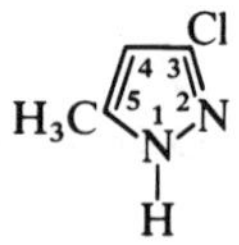

$C_4H_5ClN_2$ M 116.550

Cryst. (Et_2O/pet. ether). Mp 118-119°. Bp ~258°, Bp_{15} 138°.

B,HCl: Mp 137°.

Grandberg, I.I. *et al, J. Gen. Chem. USSR*, 1963, **33**, 511 (*uv*)

Elguero, J. *et al, Bull. Soc. Chim. Fr.*, 1966, 3727; 1968, 5019 (*synth, pmr*)

Tensmeyer, L.G. *et al, J. Org. Chem.*, 1966, **31**, 1878 (*pmr*)

Cabildo, P. *et al, Org. Magn. Reson.*, 1984, **22**, 603 (*cmr*)

3-Chloro-4-methylpyridazine, 9CI   C-70110

[68206-04-2]

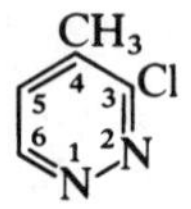

$C_5H_5ClN_2$ M 128.561

Plates (cyclohexane). Mp 46-47°.

1-Oxide:

$C_5H_5ClN_2O$ M 144.560

Needles (C_6H_6). Mp 148-149°.

Ogata, M. *et al, Chem. Pharm. Bull.*, 1963, **11**, 35 (*synth*)

3-Chloro-5-methylpyridazine, 9CI   C-70111

$C_5H_5ClN_2$ M 128.561

Prisms or needles (pet. ether). Mp 35°.

1-Oxide:

$C_5H_5ClN_2O$ M 144.560

Needles (C_6H_6). Mp 127-128°.

Takahayashi, N., *Chem. Pharm. Bull.*, 1957, **5**, 229 (*synth*)

Ogata, M. *et al, Chem. Pharm. Bull.*, 1963, **11**, 35 (*oxide*)

3-Chloro-6-methylpyridazine, 9CI   C-70112

[1121-79-5]

$C_5H_5ClN_2$ M 128.561

Needles (pet. ether). Mp 58°.

1-Oxide: [17762-03-7].

$C_5H_5ClN_2O$ M 144.560

Mp 163-164°.

2-Methochloride:

$C_6H_8Cl_2N_2$ M 179.049

Reagent for chlorodehydroxylation of alcohols. Hygroscopic cryst. (EtOAc). Mp 133-134°.

Overend, W.G. *et al, J. Chem. Soc.*, 1947, 239 (*synth*)

Kano, H. *et al, Chem. Pharm. Bull.*, 1961, **9**, 1017 (*deriv*)

Yoshihara, M. *et al, Synthesis*, 1980, 746 (*deriv, use*)

2-Chloro-4-methylpyrimidine   C-70113

[13036-57-2]

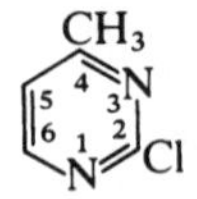

$C_5H_5ClN_2$ M 128.561

Elongated plates (pet. ether). Mp 59-60° (48°). Bp_{25} 95-100°.

Marshall, J.R. *et al, J. Chem. Soc.*, 1951, 1004 (*synth*)

Vanderhaeghe, H. *et al, Bull. Soc. Chim. Belg.*, 1957, **66**, 276 (*synth*)

Brown, D.J. *et al, J. Chem. Soc. (B)*, 1971, 2214 (*synth*)

2-Chloro-5-methylpyrimidine   C-70114

[22536-61-4]

$C_5H_5ClN_2$ M 128.561

Mp 91°.

Brown, D.J. *et al, Aust. J. Chem.*, 1968, **21**, 243; 1973, **26**, 443 (*synth*)

Brown, D.J. *et al, J. Chem. Soc. (B)*, 1971, 2214 (*pmr*)

4-Chloro-2-methylpyrimidine   C-70115

[4994-86-9]

$C_5H_5ClN_2$ M 128.561

Mp 59-60°.

Gabriel, S., *Ber.*, 1904, **37**, 3638 (*synth*)

Brown, D.J. *et al, J. Chem. Soc. (B)*, 1971, 2214 (*synth*)

Undheim, K. *et al, Acta Chem. Scand.*, 1971, **25**, 3227 (*ms*)

Busby, R.E. *et al, J. Chem. Soc., Perkin Trans. 1*, 1980, 1427 (*pmr*)

4-Chloro-5-methylpyrimidine   C-70116

[51957-32-5]

$C_5H_5ClN_2$ M 128.561

Mp 25-27°. Bp_{7-8} 62-64°.

Vanderhaeghe, H. *et al, Bull. Soc. Chim. Belg.*, 1957, **66**, 276 (*synth*)

4-Chloro-6-methylpyrimidine   C-70117

[3435-25-4]

$C_5H_5ClN_2$ M 128.561

Mp 38-39.5° (34-36°). Bp_{25} 65-70°.

1-Oxide: [52816-76-9].

$C_5H_5ClN_2O$ M 144.560

Cryst. (C_6H_6/pet. ether). Mp 110°.

Marshall, J.R. *et al, J. Chem. Soc.*, 1951, 1004 (*synth*)

Vanderhaeghe, H. *et al, Bull. Soc. Chim. Belg.*, 1957, **66**, 276 (*synth*)

Peereboom, R. *et al, Recl. Trav. Chim. Pays-Bas*, 1974, **93**, 58 (*synth, pmr, oxide*)

Busby, R.E. *et al, J. Chem. Soc., Perkin Trans. 1*, 1980, 1427 (*pmr*)

5-Chloro-2-methylpyrimidine   C-70118
[54198-89-9]
$C_5H_5ClN_2$ M 128.561
Cryst. with quinoline-like odour. Mp 56-57°.

Budĕšínský, Z., *Collect. Czech. Chem. Commun.*, 1949, **14**, 223.

5-Chloro-4-methylpyrimidine   C-70119
[54198-82-2]
$C_5H_5ClN_2$ M 128.561
Liq. or cryst. Mp 49-50°. Bp_{19} 90°.
Picrate: Yellow prisms (Et_2O/hexane). Mp 110-111°.

Brown, D.J. *et al, Aust. J. Chem.*, 1974, **27**, 2251 (*synth*)
Yamanaka, H. *et al, Chem. Pharm. Bull.*, 1987, **35**, 3119
  (*synth, pmr*)

2-(Chloromethyl)quinoxaline   C-70120
[106435-53-4]

$C_9H_7ClN_2$ M 178.621
Yellow cryst. Mp 45-46° dec.

Jeromin, G.E. *et al, Chem. Ber.*, 1987, **120**, 649 (*synth, pmr, uv*)

2-Chloro-3-methylquinoxaline, 9CI   C-70121
[32601-86-8]

$C_9H_7ClN_2$ M 178.621
Platelets (EtOH aq.). Mp 87-88°. Mps between 79° and
  92° have been quoted.
1-Oxide: [61689-44-9].
  $C_9H_7ClN_2O$ M 194.620
  Mp 105-106°.
4-Oxide: [61689-45-0].
  $C_9H_7ClN_2O$ M 194.620
  Mp 94° (89-91°).
1,4-Dioxide: [39576-78-8].
  $C_9H_7ClN_2O_2$ M 210.620
  Cryst. (MeOH). Mp 166-168°.

Munk, M. *et al, J. Am. Chem. Soc.*, 1952, **74**, 3433 (*synth*)
Hayashi, E. *et al, Yakugaku Zasshi*, 1964, **84**, 163 (*oxide*)
Abushanab, E., *J. Org. Chem.*, 1973, **38**, 3105 (*oxide*)
Mixan, C.E. *et al, J. Org. Chem.*, 1977, **42**, 1869 (*oxides*)

2-Chloro-5-methylquinoxaline, 9CI   C-70122
[61148-17-2]
$C_9H_7ClN_2$ M 178.621
Mp 95°.

Wolf, F.J. *et al, J. Am. Chem. Soc.*, 1949, **71**, 6 (*synth*)
Platt, B.C., *J. Am. Chem. Soc.*, 1949, **71**, 3247 (*struct*)

2-Chloro-6-methylquinoxaline, 9CI   C-70123
[55687-00-8]
$C_9H_7ClN_2$ M 178.621
Plates (pet. ether). Mp 105-107°.
4-Oxide: [39267-00-0]. *3-Chloro-7-methylquinoxaline
1-oxide.*
  $C_9H_7ClN_2O$ M 194.620

Needles (pet. ether). Mp 142-143°.

Platt, B.C. *et al, J. Chem. Soc.*, 1948, 1310 (*synth*)
Wolf, F.J. *et al, J. Am. Chem. Soc.*, 1949, **71**, 6 (*synth*)
Ahmad, Y. *et al, J. Org. Chem.*, 1973, **38**, 2176 (*oxide*)

2-Chloro-7-methylquinoxaline, 9CI   C-70124
$C_9H_7ClN_2$ M 178.621
Needles (EtOH aq.). Mp 77°. Steam-volatile.

Hinsberg, O., *Justus Liebigs Ann. Chem.*, 1888, **248**, 71 (*synth*)
Platt, B.C. *et al, J. Chem. Soc.*, 1948, 1310.
Platt, B.C. *et al, J. Am. Chem. Soc.*, 1949, **71**, 3247 (*synth*)

6-Chloro-7-methylquinoxaline, 9CI   C-70125
$C_9H_7ClN_2$ M 178.621
Needles (pet. ether). Mp 120-122°.
1-Oxide: [39266-96-1].
  $C_9H_7ClN_2O$ M 194.620
  Needles (EtOH). Mp 168-169°.
1,4-Dioxide: [33368-92-2].
  $C_9H_7ClN_2O_2$ M 210.620
  Mp 227°.

Landquist, J.K., *J. Chem. Soc.*, 1953, 2816 (*synth, oxides*)
Ahmad, Y. *et al, J. Org. Chem.*, 1973, **38**, 2176 (*oxides*)

2-Chloro-6-nitrobenzylamine   C-70126
2-Chloro-6-nitrobenzenemethanamine

$C_7H_7ClN_2O_2$ M 186.598
Cryst. Mp 28-30°.
B,HCl: Plates (EtOH). Mp 258-260°.

Southwick, P.L. *et al, J. Org. Chem.*, 1959, **24**, 753 (*synth*)
Stalder, H., *Helv. Chim. Acta*, 1986, **69**, 1887 (*synth, ir, pmr,
  ms*)

4-Chloro-2-nitrobenzylamine   C-70127
4-Chloro-2-nitrobenzenemethanamine, 9CI
[67567-43-5]
$C_7H_7ClN_2O_2$ M 186.598
pK_a' 8.34.
B,HCl: Mp 225-227°.

Celnik, K. *et al, Pol. J. Chem.*, 1978, **52**, 947 (*synth, ir, pmr*)
U.S.P., 4 316 900, (*1982*); *CA*, **96**, 199750 (*synth*)

4-Chloro-3-nitrobenzylamine   C-70128
4-Chloro-3-nitrobenzenemethanamine
$C_7H_7ClN_2O_2$ M 186.598
B,HCl: Mp 227-228°.
Picrate: Cryst. (EtOH). Mp 210° dec.

McKay, A.F. *et al, J. Am. Chem. Soc.*, 1959, **81**, 4328 (*synth*)

5-Chloro-2-nitrobenzylamine   C-70129
5-Chloro-2-nitrobenzenemethanamine, 9CI
[67567-44-6]
$C_7H_7ClN_2O_2$ M 186.598
pK_a' 8.19.
B,HCl: Mp 244-246°.

Celnik, K. *et al, Pol. J. Chem.*, 1978, **52**, 947 (*synth, ir, pmr*)

1-Chloro-1-nitrocyclobutane   C-70130

[109178-30-5]

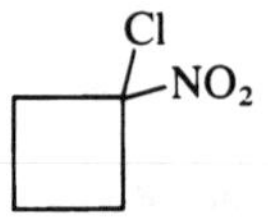

$C_4H_6ClNO_2$ M 135.550

Amrollah-Madjdabadi, A. *et al, Synthesis*, 1986, 826 (*synth, ir, pmr*)

1-Chloro-1-nitrocyclohexane   C-70131

[873-92-7]

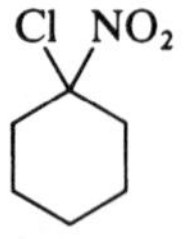

$C_6H_{10}ClNO_2$ M 163.604
Bp$_8$ 81-82°, Bp$_{0.35}$ 35-40°.

Robertson, J.A., *J. Org. Chem.*, 1948, **13**, 395 (*synth*)
Amrollah-Madjdabadi, A. *et al, Synthesis*, 1986, 828 (*synth, ir, pmr*)

2-Chloro-3-nitro-9H-fluorene   C-70132

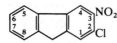

$C_{13}H_8ClNO_2$ M 245.665
Yellow needles (pet. ether). Mp 116-117°.

Ishikawa, N. *et al, CA*, 1957, **51**, 14655b (*synth*)

2-Chloro-7-nitro-9H-fluorene, 9CI   C-70133

[6939-05-5]
$C_{12}H_8ClNO_2$ M 233.654
Yellow cryst. Mp 241-241.5°.

Gutman, H.R. *et al, J. Am. Chem. Soc.*, 1951, **73**, 4033 (*synth, bibl*)
Hayashi, M. *et al, CA*, 1957, **51**, 1920f (*synth*)

9-Chloro-2-nitro-9H-fluorene, 9CI   C-70134

[63027-86-1]
$C_{12}H_8ClNO_2$ M 233.654
Yellow columns (C_6H_6). Mp 141°.

Arcus, C.L. *et al, J. Chem. Soc.*, 1954, 3977 (*synth*)

9-Chloro-3-nitro-9H-fluorene   C-70135

[73748-77-3]
$C_{12}H_8ClNO_2$ M 233.654
Needles (C_6H_6). Mp 150-151°.

Arcus, C.L. *et al, J. Chem. Soc.*, 1954, 3977 (*synth*)

Chloropentakis(dichloromethyl)benzene   C-70136

[111351-51-0]

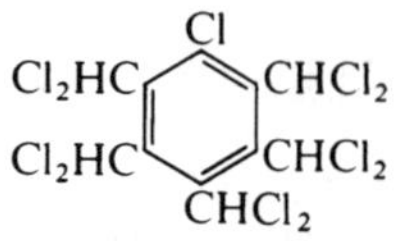

$C_{11}H_5Cl_{11}$ M 527.144

Cryst. (tetrachloroethane). Mp >300°.

Biali, S.E. *et al, J. Org. Chem.*, 1988, **53**, 135 (*synth, pmr, stereochem*)

5-Chloropentanal   C-70137

[20074-80-0]

$$ClCH_2(CH_2)_3CHO$$

C_5H_9ClO M 120.579
Oil. Bp$_{10}$ 60-66°.

Kuehne, M.E. *et al, J. Org. Chem.*, 1981, **46**, 2002 (*synth, ir, pmr*)

1-Chloro-3-pentanol, 9CI   C-70138

[32541-33-6]

$$H_3CCH_2CH(OH)CH_2CH_2Cl$$

$C_5H_{11}ClO$ M 122.594
(±)-*form*
Bp$_{15}$ 75-76°.
Ac:
    $C_7H_{13}ClO_2$ M 164.632
    Bp$_9$ 75-76°.

Bartok, M. *et al, Acta Phys. Chem.*, 1963, **9**, 25 (*synth, deriv*)

5-Chloro-1-pentyne   C-70139

[14267-92-6]

$$HC{\equiv}CCH_2CH_2CH_2Cl$$

C_5H_7Cl M 102.563
d$_4^{20}$ 0.97. Bp$_{350}$ 88°, Bp$_{145}$ 67-69°. n$_D^{20}$ 1.445.

Henne, A.L. *et al, J. Am. Chem. Soc.*, 1945, **67**, 484 (*synth*)
Eglington, G. *et al, J. Chem. Soc.*, 1950, 3650 (*synth*)

3-Chloro-3-phenyl-3H-diazirine, 9CI   C-70140

[4460-46-2]

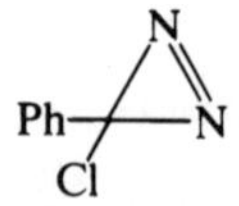

$C_7H_5ClN_2$ M 152.583
Pale-yellow, sweet-smelling liq. Bp$_{0.1}$ 30-35°.
▷Explosive, dec. on dist. >35°

Org. Synth., 1981, **60**, 53 (*synth, haz*)
McMahon, R.J. *et al, J. Am. Chem. Soc.*, 1987, **109**, 2456 (*synth, ir, uv, pmr, cmr, haz*)

1-Chloro-4-phenylphthalazine   C-70141

[10132-01-1]

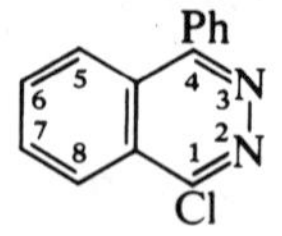

$C_{14}H_9ClN_2$ M 240.692
Plates (EtOH). Mp 160-161°.

Lieck, A., *Ber.*, 1905, **38**, 3918 (*synth*)
Hill, J.H.M. *et al, J. Org. Chem.*, 1971, **36**, 3248 (*props*)

2-Chloro-3-phenylpyrazine — C-70142

[41270-65-9]

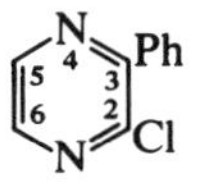

$C_{10}H_7ClN_2$ M 190.632
Mp 65-66°.

1-Oxide: [66769-58-2].
$C_{10}H_7ClN_2O$ M 206.631
Mp 188-189°.

4-Oxide: [58861-87-3].
$C_{10}H_7ClN_2O$ M 206.631
Prisms (MeOH). Mp 151-152°.

1,4-Dioxide: [58861-88-4].
$C_{10}H_7ClN_2O_2$ M 222.631
Needles (MeOH). Mp 238.5-239.5° dec.

Chillemi, F. *et al, Farmaco (Pavia),* Ed. Sci., 1963, **18**, 566; *CA,* **59**, 13976 (*synth*)
Sato, N., *J. Org. Chem.,* 1978, **43**, 3367 (*oxides*)
Ohta, A. *et al, J. Heterocycl. Chem.,* 1982, **19**, 1061 (*oxides*)

2-Chloro-5-phenylpyrazine — C-70143

[25844-73-9]
$C_{10}H_7ClN_2$ M 190.632
Cryst. (MeOH). Mp 98.5-99°.

1-Oxide: [61578-11-8].
$C_{10}H_7ClN_2O$ M 206.631
Pale-yellow needles (MeOH). Mp 148-150°.

4-Oxide: [61578-12-9].
$C_{10}H_7ClN_2O$ M 206.631
Pale-yellow needles (MeOH). Mp 139-141°.

1,4-Dioxide: [85093-93-2].
$C_{10}H_7ClN_2O_2$ M 222.631
Pale-yellow prisms (MeOH). Mp 248-250.5° dec.

Lont, P.J. *et al, Recl. Trav. Chim. Pays-Bas,* 1973, **92**, 449 (*synth*)
Ohta, A. *et al, J. Heterocycl. Chem.,* 1982, **19**, 1061 (*oxides*)

2-Chloro-6-phenylpyrazine — C-70144

[41270-62-6]
$C_{10}H_7ClN_2$ M 190.632
Needles (hexane). Mp 34-35°. Bp$_{0.3}$ 99-100°.

1-Oxide: [66769-59-3].
$C_{10}H_7ClN_2O$ M 206.631
Cryst. (EtOH). Mp 128°.

4-Oxide: [66769-60-6].
$C_{10}H_7ClN_2O$ M 206.631
Pale-yellow prisms (MeOH). Mp 115-116°.

Lont, P.J. *et al, Recl. Trav. Chim. Pays-Bas,* 1973, **92**, 449 (*synth*)
Sato, N., *J. Org. Chem.,* 1978, **43**, 3367 (*synth, oxide*)
Ohta, A. *et al, J. Heterocycl. Chem.,* 1982, **19**, 1061 (*oxides*)

2-Chloro-3-phenylquinoxaline, 9CI — C-70145

[7065-92-1]

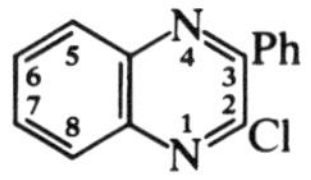

$C_{14}H_9ClN_2$ M 240.692
Needles (EtOH). Mp 130° (127-128°).

4-Oxide:
$C_{14}H_9ClN_2O$ M 256.691
Microneedles (EtOH). Mp 126-127°.

Shiho, D. *et al, J. Am. Chem. Soc.,* 1960, **82**, 4044 (*synth*)
Ahmad, Y. *et al, J. Org. Chem.,* 1966, **31**, 2613 (*synth, oxide*)

6-Chloro-2-phenylquinoxaline, 9CI — C-70146

[25187-19-3]
$C_{14}H_9ClN_2$ M 240.692
Needles (EtOH). Mp 148-149°.

4-Oxide: [57433-56-4].
$C_{14}H_9ClN_2O$ M 256.691
Mp 196-198°.

1,4-Dioxide (?): [36059-14-0].
$C_{14}H_9ClN_2O_2$ M 272.690
Cryst. (CF$_3$COOH/MeOH). Mp 216-220°. May be the 7-chloro isomer.

Mufarrij, N.A. *et al, J. Chem. Soc., Perkin Trans. 1,* 1972, 965 (*dioxide*)
Bannore, S.N. *et al, Indian J. Chem.,* 1973, **11**, 631; 1975, **13**, 609 (*synth, pmr, uv, oxide*)

7-Chloro-2-phenylquinoxaline, 9CI — C-70147

[49634-78-8]
$C_{14}H_9ClN_2$ M 240.692
Needles (EtOH). Mp 130-131°.

4-Oxide: [57433-57-5].
$C_{14}H_9ClN_2O$ M 256.691
Mp 155-156°.

1,4-Dioxide (?): see under 6-Chloro-2-phenylquinoxaline, C-70146

Bannore, S.N. *et al, Indian J. Chem.,* 1973, **11**, 631; 1975, **13**, 609 (*synth, pmr, oxide*)

1-Chlorophthalazine — C-70148

[5784-45-2]

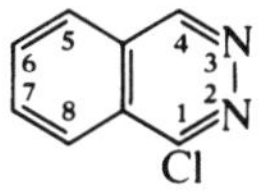

$C_8H_5ClN_2$ M 164.594
Yellow cryst. Mp 120-121°.

Gabriel, S. *et al, Ber.,* 1893, **26**, 523 (*synth*)
Brasyunas, V.B. *et al, CA,* 1959, **53**, 16144 (*synth*)
Hayashi, E. *et al, Yakugaku Zasshi,* 1962, **82**, 584 (*synth*)

5-Chlorophthalazine, 9CI — C-70149

[78032-08-3]
$C_8H_5ClN_2$ M 164.594
Mp 128°.

Robev, S., *Tetrahedron Lett.,* 1981, **22**, 345 (*synth*)

6-Chlorophthalazine, 9CI — C-70150

[78032-07-2]
$C_8H_5ClN_2$ M 164.594
Mp 139°.

Robev, S., *Tetrahedron Lett.,* 1981, **22**, 345 (*synth*)

2-Chloro-2-propen-1-amine — C-70151

3-Amino-2-chloro-1-propene. 2-Chloroallylamine
[38729-96-3]

$$H_2C{=}CClCH_2NH_2$$

C_3H_6ClN M 91.540

Bp 105-114°.

Speziale, A.J. *et al*, *J. Am. Chem. Soc.*, 1956, **78**, 2556 (*synth*)
Lattrell, R. *et al*, *Justus Liebigs Ann. Chem.*, 1974, 870 (*synth*)
Bargar, T.M. *et al*, *J. Org. Chem.*, 1987, **52**, 678 (*synth*)

1-Chloro-2-(2-propenyl)benzene, 9CI   C-70152

1-Allyl-2-chlorobenzene. 3-(2-Chlorophenyl)-1-propene
[1587-07-1]

C₉H₉Cl   M 152.623

Aresta, M. *et al*, *J. Organomet. Chem.*, 1973, **56**, 395 (*synth*)
Klabunde, K.J. *et al*, *J. Organomet. Chem.*, 1974, **71**, 309 (*synth*)
Padwa, A. *et al*, *J. Org. Chem.*, 1984, **49**, 1353 (*synth*)

1-Chloro-3-(2-propenyl)benzene, 9CI   C-70153

1-Allyl-3-chlorobenzene, 8CI. 3-(3-Chlorophenyl)-1-propene
[3840-17-3]
C₉H₉Cl   M 152.623
Oil. Bp 198-202°, Bp₃ 60-65°.

Martin, M.M. *et al*, *J. Am. Chem. Soc.*, 1964, **86**, 233 (*synth*)
Jones, L.B. *et al*, *J. Org. Chem.*, 1970, **35**, 1777 (*synth*)

1-Chloro-4-(2-propenyl)benzene, 9CI   C-70154

1-Allyl-4-chlorobenzene. 3-(4-Chlorophenyl)-1-propene
[1745-18-2]
C₉H₉Cl   M 152.623
Oil. Bp 199-201°, Bp₂₁ 91-93°.

La Combe, E.M. *et al*, *J. Am. Chem. Soc.*, 1961, **83**, 3457 (*synth*)
Martin, M.M. *et al*, *J. Am. Chem. Soc.*, 1964, **86**, 233 (*synth*)
Hartley, F.R. *et al*, *J. Polym. Sci. Polym. Chem. Ed.*, 1982, **20**, 2395 (*synth*)
Amano, T. *et al*, *Chem. Pharm. Bull.*, 1986, **36**, 4653 (*synth*)

3-Chloro-1*H*-pyrazole, 9CI   C-70155

Updated Entry replacing C-02183
[14339-33-4]

C₃H₃ClN₂   M 102.523
Cryst. (pentane). Mp 43°. Bp₀.₅ 76-77°.

1H-form
1-Me: [63425-54-7]. *3-Chloro-1-methylpyrazole.*
C₄H₅ClN₂   M 116.550
Bp₁₃ 72°. n_D²⁰ 1.4968.
1-Benzyl: [50877-43-5]. *1-Benzyl-3-chloropyrazole.*
C₁₀H₉ClN₂   M 192.648
Cryst. (cyclohexane). Mp 62°.

2H-form
2-Me: [42110-76-9]. *5-Chloro-1-methylpyrazole, 9CI.*
C₄H₅ClN₂   M 116.550

Bp 134°, Bp₁₃ 33°.
2-Benzoyl: [50877-39-9]. *1-Benzyl-5-chloropyrazole.*
C₁₀H₉ClN₂   M 192.648
Bp₀.₁ 65°.

Reimlinger, W.K. *et al*, *Chem. Ber.*, 1966, **99**, 3350.
Elguero, J. *et al*, *Bull. Soc. Chim. Fr.*, 1966, 3727 (*pmr*)
Frêche, P. *et al*, *Tetrahedron Lett.*, 1976, 1495 (*synth, pmr*)
Dorn, H., *J. Prakt. Chem.*, 1977, **319**, 281 (*synth, ir, pmr, derivs*)
Ferguson, I.J. *et al*, *J. Chem. Soc., Perkin Trans. 1*, 1977, 672 (*synth, pmr, deriv*)

4-Chloro-1*H*-pyrazole, 9CI   C-70156

Updated Entry replacing C-02184
[15878-00-9]
C₃H₃ClN₂   M 102.523
Cryst. (pet. ether). Mp 77°. Bp 220°.
▷UQ6445000.

N-Me: [35852-81-4]. *4-Chloro-1-methylpyrazole.*
C₄H₅ClN₂   M 116.550
Yellowish oil. Bp₇₅₆ 167°. Turns red in light.

Knorr, L., *Ber.*, 1895, **28**, 715 (*synth*)
Hüttel, R. *et al*, *Justus Liebigs Ann. Chem.*, 1956, **598**, 186 (*synth*)
Elguero, J. *et al*, *Bull. Soc. Chim. Fr.*, 1966, 3727 (*pmr*)
v. Thuijl, J., *Org. Mass Spectrom.*, 1970, **3**, 1549 (*ms*)
Butler, R.N., *Can. J. Chem.*, 1973, **51**, 2315 (*pmr, deriv*)
Cabildo, P. *et al*, *Org. Magn. Reson.*, 1984, **22**, 603 (*cmr, deriv*)
Franssen, M.C.R. *et al*, *J. Heterocycl. Chem.*, 1987, **24**, 1313 (*synth, deriv*)

5-Chloropyrimidine, 9CI   C-70157

Updated Entry replacing C-02202
[17180-94-8]
C₄H₃ClN₂   M 114.534
Mp 36.5°.
N-Oxide: [114969-48-1].
C₄H₃ClN₂O   M 130.534
Mp 155-156°.

Lythgoe, B. *et al*, *J. Chem. Soc.*, 1951, 2323.
Yamanaka, H. *et al*, *Chem. Pharm. Bull.*, 1987, **35**, 3119 (*synth, pmr, oxide*)

6-Chloro-2,4-(1*H*,3*H*)-pyrimidinedione, 9CI   C-70158

6-Chlorouracil, 8CI
[4270-27-3]

C₄H₃ClN₂O₂   M 146.533
Amorph. solid. Mp 298-300° dec.

Horwitz, J.P. *et al*, *J. Org. Chem.*, 1961, **26**, 3392 (*synth, nmr*)

2-Chloro-9-β-D-ribofuranosyl-9H-purine,   **C-70159**
9CI
2-Chloronebularine
[5466-11-5]

$C_{10}H_{11}ClN_4O_4$ M 286.674
Cryst. (H₂O). Mp 154-156°. $[\alpha]_D^{26}$ −30.0° (c, 0.38 in H₂O).

Schaeffer, H.J. *et al*, *J. Am. Chem. Soc.*, 1958, **80**, 4896 (*synth*)
Ger. Pat., 2 209 078, (*1972*); *CA*, **78**, 27908 (*synth*)

6-Chloro-9-β-D-ribofuranosyl-9H-purine   **C-70160**
6-Chloronebularine
[5399-87-1]
$C_{10}H_{11}ClN_4O_4$ M 286.674
Antineoplastic agent. Cryst. (MeOH aq.). Mp 165-166° dec. $[\alpha]_D^{24}$ −43.3° (H₂O).

Biochem. Prep., 1963, **10**, 145 (*synth, bibl*)
Zemlicka, J. *et al*, *Collect. Czech. Chem. Commun.*, 1965, **30**, 1880 (*synth*)
Guilford, H. *et al*, *Chem. Scr.*, 1972, **2**, 165 (*synth*)
Sternglanz, H. *et al*, *Acta Crystallogr., Sect. B*, 1975, **31**, 2888 (*cryst struct*)
Sugiyama, T. *et al*, *Agric. Biol. Chem.*, 1978, **42**, 1791 (*synth*)
Zemlicka, J. *et al*, *Nucleic Acid Chem.*, 1978, **2**, 611 (*synth, use*)
Keyser, G.E. *et al*, *Tetrahedron Lett.*, 1979, 3263 (*synth*)

2-Chloro-9-β-D-ribofuranosyl-6H-purin-6-   **C-70161**
one
2-Chloroinosine, 9CI
[13276-43-2]

$C_{10}H_{11}ClN_4O_5$ M 302.674
Prod. by *Bacillus subtilis* in medium containing 2-chlorohypoxanthine. Cryst. (H₂O). Mp 123°.

Gerster, J.F. *et al*, *J. Org. Chem.*, 1966, **31**, 3258 (*synth*)
Imai, K. *et al*, *Chem. Pharm. Bull.*, 1966, **14**, 1377 (*synth*)
Yamanoi, A. *et al*, *J. Gen. Appl. Microbiol.*, 1966, **12**, 299 (*isol*)
Yamazaki, A. *et al*, *Chem. Pharm. Bull.*, 1969, **17**, 2581 (*synth*)

5-Chloro-4-thioxo-2(1H)-pyrimidinone, 9CI   **C-70162**
5-Chloro-4-thiouracil
[63331-62-4]

$C_4H_3ClN_2OS$ M 162.594
Mp 305-306° dec.

Lapucha, A., *Synthesis*, 1987, 256 (*synth, ir, pmr, cmr, ms*)

1-Chloro-1,1,2,2,3,3,4,4,5,5,6,6,6-trideca-   **C-70163**
fluorohexane, 9CI
Perfluorohexyl chloride
[355-41-9]

$$F_3C(CF_2)_4CF_2Cl$$

C_6ClF_{13} M 354.498
d_{25}^{25} 1.705. Bp 86°. n_D^{15} 1.287.

Haszeldine, R.N., *J. Chem. Soc.*, 1952, 4259; 1953, 3761 (*synth*)
Ger. Pat., 3 338 300 (*1985*); *CA*, **103**, 14859 (*synth*)

2-Chloro-3-(trifluoromethyl)benzaldehyde   **C-70164**
[93118-03-7]

$C_8H_4ClF_3O$ M 208.567
Cryst. (hexane). Mp 43-44°.

Arrowsmith, J.E. *et al*, *J. Med. Chem.*, 1986, **29**, 1696 (*synth*)

2-Chloro-5-(trifluoromethyl)benzaldehyde,   **C-70165**
9CI
[82386-89-8]
$C_8H_4ClF_3O$ M 208.567
Oil. Bp₂.₁ 93-96°.

Sindelar, K. *et al*, *Collect. Czech. Chem. Commun.*, 1982, **47**, 967 (*synth*)

2-Chloro-6-(trifluoromethyl)benzaldehyde,   **C-70166**
9CI
[60611-22-5]
$C_8H_4ClF_3O$ M 208.567
Ger. Pat., 2 603 399, (*1976*); *CA*, **85**, 143109 (*synth*)

4-Chloro-3-(trifluoromethyl)benzaldehyde,   **C-70167**
9CI
[34328-46-6]
$C_8H_4ClF_3O$ M 208.567
Liq. $n_D^{18.5}$ 1.4964.

French, F.A. *et al*, *J. Med. Chem.*, 1971, **14**, 862 (*synth*)
Ger. Pat., 2 021 620, (*1971*); *CA*, **76**, 45911 (*synth*)

5-Chloro-2-(trifluoromethyl)benzaldehyde, 9CI   C-70168

[90381-07-0]

$C_8H_4ClF_3O$ M 208.567

Japan. Pat., 84 21 637, (*1981*); *CA*, **101**, 6825 (*synth*)

3-Chloro-3-(trifluoromethyl)-3*H*-diazirine, 9CI   C-70169

[58911-30-1]

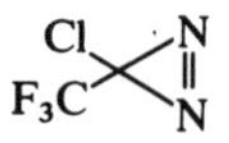

$C_2ClF_3N_2$ M 144.484

Gas. Bp −19° to −18°.

▷Potentially explosive

Grayston, M.W. *et al, J. Am. Chem. Soc.*, 1976, **98**, 1278 (*synth, ir, uv, pmr*)

2-Chloro-3,3,3-trifluoropropanoic acid   C-70170

[110230-36-9]

$$F_3CCHClCOOH$$

$C_3H_2ClF_3O_2$ M 162.496

(±)-*form*

Solid. Mp 26-27°.

Blazejewski, J.-C. *et al, J. Chem. Soc., Perkin Trans. 1*, 1987, 1861 (*synth, pmr, F nmr, ms*)

Chokorin   C-70171

[115812-55-0]

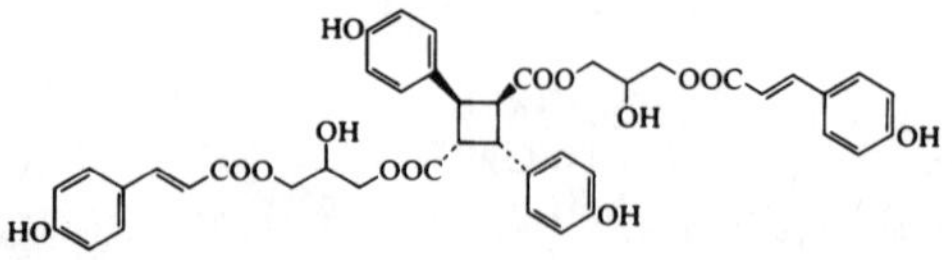

$C_{42}H_{40}O_{14}$ M 768.770

Constit. of stromata of *Epichloe typhina* on *Phleum pratense*. Cryst. Mp 226-228°.

Koshino, H. *et al, Phytochemistry*, 1988, **27**, 1333.

Cholest-5-ene-3,16,22-triol   C-70172

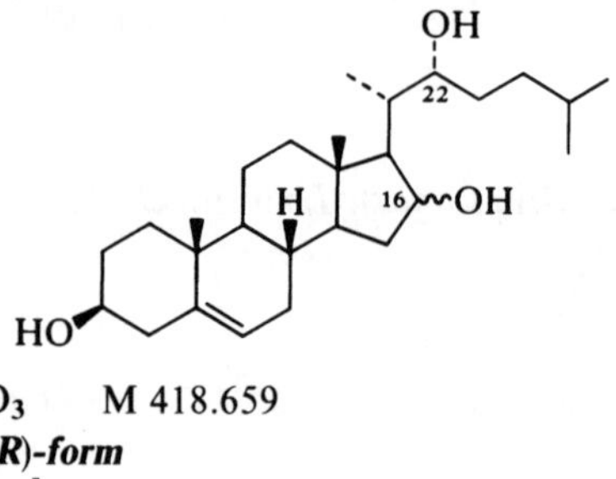

$C_{27}H_{46}O_3$ M 418.659

(3β,16ξ,22R)-*form*

Saxosterol

Constit. of *Narthecium ossifragum*. Cryst. (MeOH). Mp 182-183°. $[\alpha]_D^{20}$ −27° (c, 0.2 in CHCl₃).

(3β,16ξ,22ξ)-*form* [11003-34-2]

Calibragenin

Constit. of *Calibranus hookeri*. Cryst. (MeOH). Mp 195-196°. $[\alpha]_D^{20}$ −56° (CHCl₃).

Giral, F. *et al, Phytochemistry*, 1975, **14**, 793 (*isol, struct*)
Stabursvik, A. *et al, Phytochemistry*, 1988, **27**, 1893 (*isol*)

Chorismic acid   C-70173

Updated Entry replacing C-20182

[617-12-9]

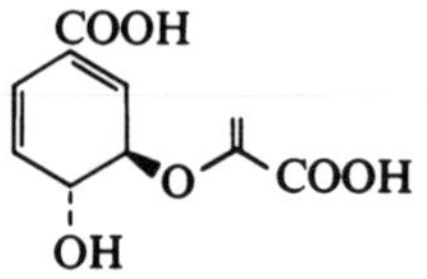

$C_{10}H_{10}O_6$ M 226.185

Intermed. in the biosynth. of aromatic compds. *via* the Shikimic acid pathway. Cryst. + 1H₂O (EtOAc/pet. ether). Mp 148-149°. $[\alpha]_D^{25}$ −295.5° (c, 0.2 in H₂O).

Edwards, J.M. *et al, Aust. J. Chem.*, 1965, **18**, 1227 (*struct, ir, uv, pmr, ms*)
Biochem. Prep., 1968, **12**, 94 (*synth*)
Hill, R.K. *et al, J. Am. Chem. Soc.*, 1969, **91**, 5893 (*biosynth*)
Pittard, J. *et al, Curr. Top. Cell. Regul.*, 1970, **2**, 29 (*rev*)
McGowan, D.A. *et al, J. Am. Chem. Soc.*, 1982, **104**, 1153, 7036 (*synth*)
Hoare, J.H. *et al, J. Am. Chem. Soc.*, 1983, **105**, 6264 (*synth*)
Pawlak, J.L. *et al, J. Org. Chem.*, 1987, **52**, 1765 (*synth*)
Posner, G.H. *et al, J. Org. Chem.*, 1987, **52**, 4836 (*synth*)

Chrysene-5,6-oxide   C-70174

Updated Entry replacing C-20187

1a,11c-Dihydrochryseno[5,6-b]*oxirene, 9CI. 5,6-Epoxy-5,6-dihydrochrysene*

[15131-84-7]

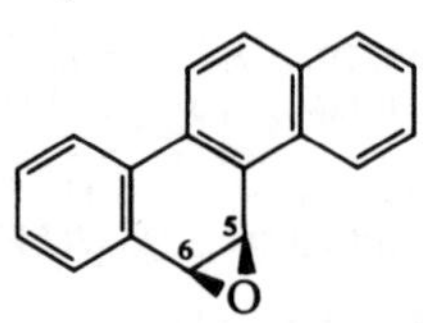

(5R,6S)-*form*

$C_{18}H_{12}O$ M 244.292

▷GC1225000.

(5R,6S)-*form* [98243-93-7]

$[\alpha]_D$ +9°.

(5S,6R)-*form* [98243-92-6]

$[\alpha]_D$ −9°.

Harvey, R.G. *et al, J. Am. Chem. Soc.*, 1975, **97**, 3468 (*synth*)
Krishnan, S. *et al, J. Am. Chem. Soc.*, 1977, **99**, 8121 (*synth*)
Balani, S.K. *et al, J. Org. Chem.*, 1987, **52**, 137 (*synth, pmr, cd, abs config*)

Chrysophyllin *B*   C-70175

3,3a,4,5-Tetrahydro-5,7-dimethoxy-2-methyl-3a-(2-propenyl)-3-(3,4,5-trimethoxyphenyl)-6(2H)-benzofuranone, 9CI

[85541-04-4]

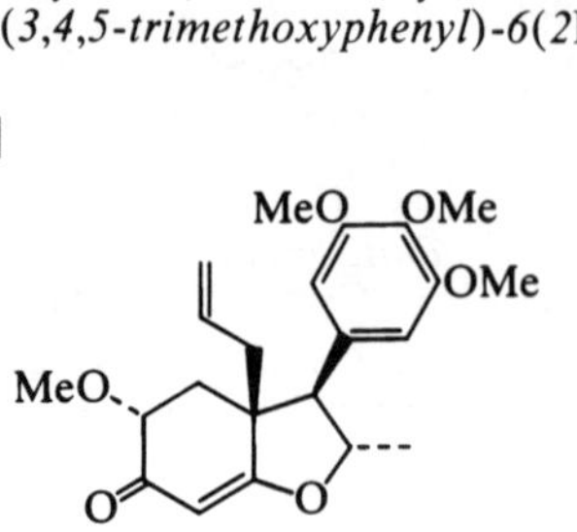

$C_{22}H_{28}O_6$ M 388.460

Constit. of *Licaria chrysophylla*.

Ferreira, Z.S. *et al, Phytochemistry*, 1982, **21**, 2756 (*isol*)

Ciclamycin O   C-70176
3′-Hydroxycinerubin X
[111233-40-0]

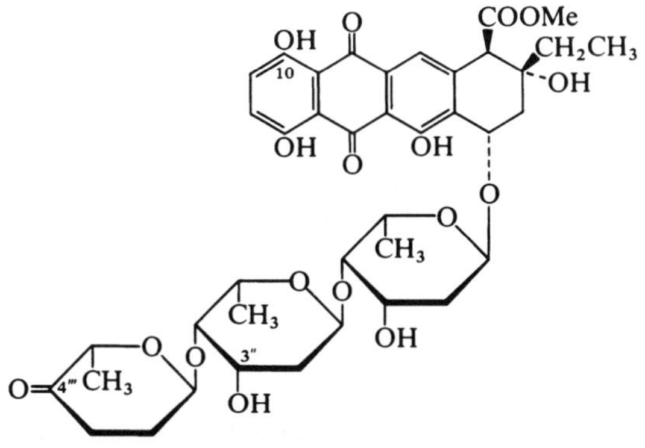

$C_{40}H_{48}O_{17}$ M 800.809

Anthracycline antibiotic. Isol. from *Streptomyces capoamus*. Antitumour agent. Red needles (MeOH). Mp 165-167°. Similar to Antibiotic MA 144U_7.

3″-Deoxy, 4‴ α-*alcohol:* [111233-39-7]. **Ciclamycin 4**.
$C_{40}H_{50}O_{16}$ M 786.825
From *S. capoamus*. Antitumour agent. Red powder (EtOH/EtOAc). Mp 143-145°.

Bieber, L.W. *et al*, *J. Antibiot.*, 1987, **40**, 1335 (*struct, bibl*)

Cinnamodial   C-70177
Updated Entry replacing C-02479
Ugandensidial. Agandencidial
[23599-45-3]

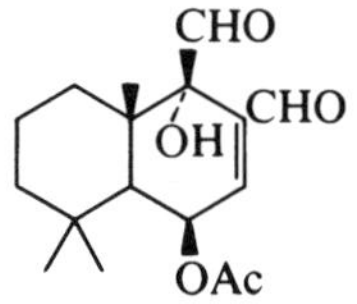

$C_{17}H_{24}O_5$ M 308.374

Constit. of *Cinnamosa fragrans*. Cryst. Mp 141-143°. $[\alpha]_D^{20}$ −421.5° (c, 1 in CHCl$_3$).

Brooks, C.J.W. *et al*, *Tetrahedron*, 1969, **25**, 2887 (*struct*)
Canonica, L. *et al*, *Tetrahedron*, 1969, **25**, 3895 (*struct*)
Burton, L.P.J. *et al*, *J. Am. Chem. Soc.*, 1981, **103**, 3226 (*synth*)
White, J.D. *et al*, *J. Org. Chem.*, 1985, **50**, 357 (*synth*)

3-Cinnolinecarboxaldehyde   C-70178
3-Formylcinnoline
[51073-57-5]

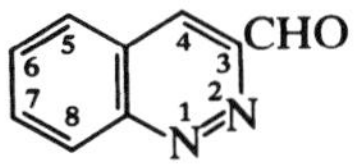

$C_9H_6N_2O$ M 158.159

Cryst. (pet. ether). Mp 119-120°.
Semicarbazone: Cryst. (AcOH). Mp 253-254°.

Haas, H.J. *et al*, *Chem. Ber.*, 1963, **96**, 2427 (*synth*)

4-Cinnolinecarboxaldehyde   C-70179
4-Formylcinnoline
$C_9H_6N_2O$ M 158.159
Yellow needles (C$_6$H$_6$/pet. ether). Mp 147-149°. Unstable on standing.

Oxime:
$C_9H_7N_3O$ M 173.174
Orange needles (EtOH). Mp 223° dec.
2,4-Dinitrophenylhydrazone: Orange needles (DMF). Mp 316-317° dec.
Semicarbazone: Yellow granules (EtOH). Mp 234-235° dec.
Hydrazone:
$C_9H_8N_4$ M 172.189
Yellow needles (DMF). Mp 297° dec.

Castle, R.N. *et al*, *J. Org. Chem.*, 1961, **26**, 4465 (*synth, ir*)

3,4-Cinnolinedicarboxaldehyde, 9CI   C-70180
3,4-Diformylcinnoline
[113290-67-8]

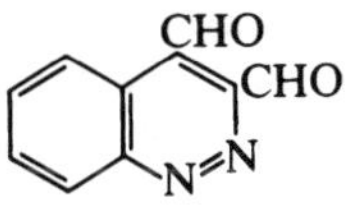

$C_{10}H_6N_2O_2$ M 186.170

Yellow needles (pet. ether). Mp 144-145°. Discoloured with time.

Barton, J.W. *et al*, *J. Chem. Soc., Perkin Trans. 1*, 1987, 1541 (*synth, ir, pmr, cmr, ms*)

Citreoviridinol A$_1$ C-70181
[94161-12-3]

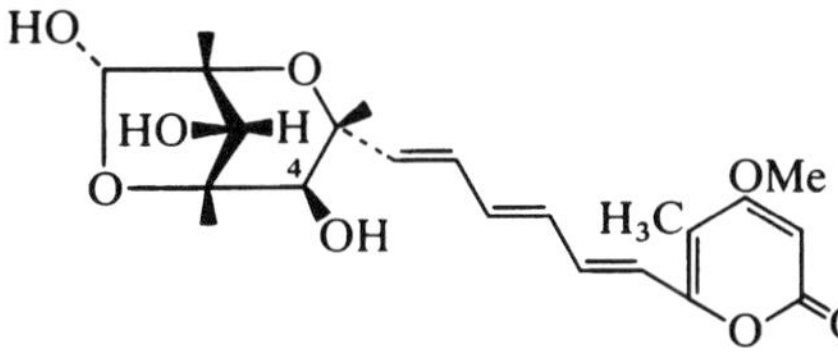

$C_{22}H_{28}O_8$ M 420.458

Prod. by *Penicillium pedemontanum*. Mycotoxin.
4-Epimer: [94122-10-8]. **Citreoviridinol A$_2$**.
$C_{22}H_{28}O_8$ M 420.458
From *P. pedemontanum*. Mycotoxin.

Rebuffat, S. *et al*, *Org. Mass. Spectrom.*, 1984, **19**, 349 (*ms*)

Clavamycin *C*   C-70182

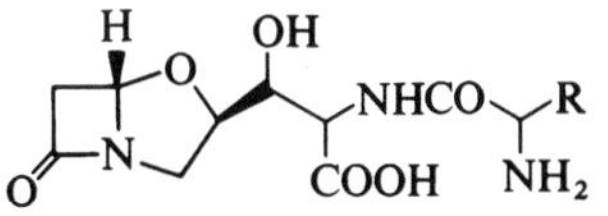

R = —CH(OH)CH(OH)CH$_2$NH$_2$

$C_{13}H_{22}N_4O_8$ M 362.339

β-Lactam antibiotic. Prod. by *Streptomyces hygroscopicus*. Shows potent antifungal props.

King, H.D. *et al*, *J. Antibiot.*, 1986, **39**, 510, 516 (*isol, struct, props*)

Clavamycin *D*   C-70183
As Clavamycin *C*, C-70182 with

$$R = -CH(CH_3)_2$$

$C_{13}H_{21}N_3O_6$ M 315.325

β-Lactam antibiotic. Prod. by *Streptomyces hygroscopicus*. Shows potent antifungal props. Cryst. (MeOH). Mp >210° dec. $[\alpha]_D^{25}$ −41.3° (c, 0.76 in H$_2$O).

Stereoisomer (?): **Antibiotic CA 146A**. *CA 146A. G 0069A. Antibiotic G 0069A.*
$C_{13}H_{21}N_3O_6$ M 315.325

Isol. from *S. lavendulae* and *S.* sp. G 0069A. Antifungal and antitumour agent. Same plane struct. as Clavamycin *D*, may be identical. Identity of CA 146A and G 0069A has not been confirmed.

King, H.D. *et al*, *J. Antibiot.*, 1986, **39**, 510, 516 (*isol, struct, props*)
Japan. Pat., 86 212 587, (*1986*); *CA*, **106**, 100846
Japan. Pat., 87 10 088, (*1987*); *CA*, **107**, 76085

Clavamycin *E*   C-70184

As Clavamycin *C*, C-70182 with

$$R = CH_3$$

$C_{11}H_{17}N_3O_6$ M 287.272
β-Lactam antibiotic. Prod. by *Streptomyces hygroscopicus*. Shows potent antifungal props.
Stereoisomer (?): [103059-96-7]. **Antibiotic CA 146B**. *CA 146B*.
$C_{11}H_{17}N_3O_6$ M 287.272
Isol. from *S. lavendulae*. Antifungal agent. No stereochem. May be identical with Clavamycin *E*.

King, H.D. *et al*, *J. Antibiot.*, 1986, **39**, 510, 516 (*isol, struct, props*)
Japan. Pat., 87 10 088, (*1987*); *CA*, **107**, 76085 (*isol*)

Cleomiscosin *D*   C-70185

[114422-15-0]

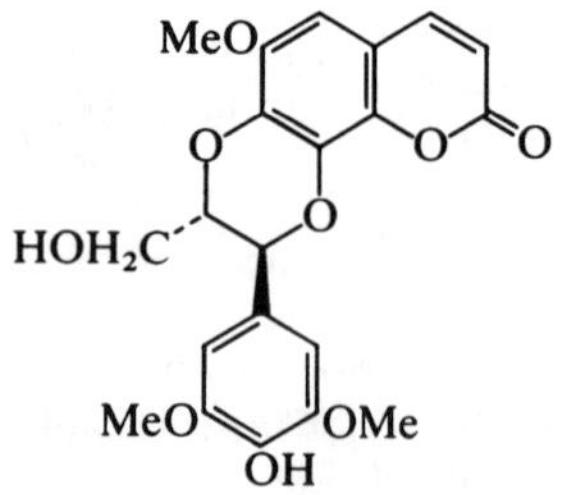

$C_{21}H_{20}O_9$ M 416.384
Constit. of *Cleome viscosa* seeds. Needles (MeOH/EtOAc). Mp 243-246°.

Kumar, S. *et al*, *Phytochemistry*, 1988, **27**, 636.

Clibadic acid   C-70186

[116310-63-5]

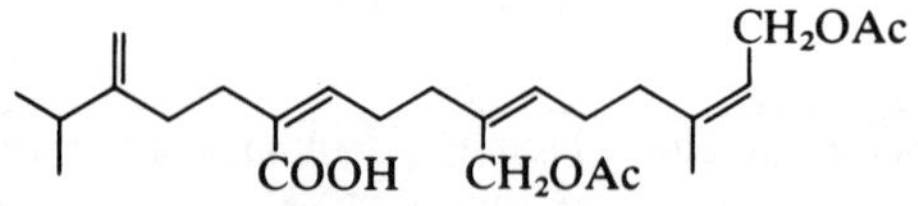

$C_{25}H_{38}O_6$ M 434.572
Constit. of *Clibadium pittierii*. Gum.

Tamayo-Castillo, G. *et al*, *Phytochemistry*, 1988, **27**, 1868.

Clibadiolide   C-70187

[116310-64-6]

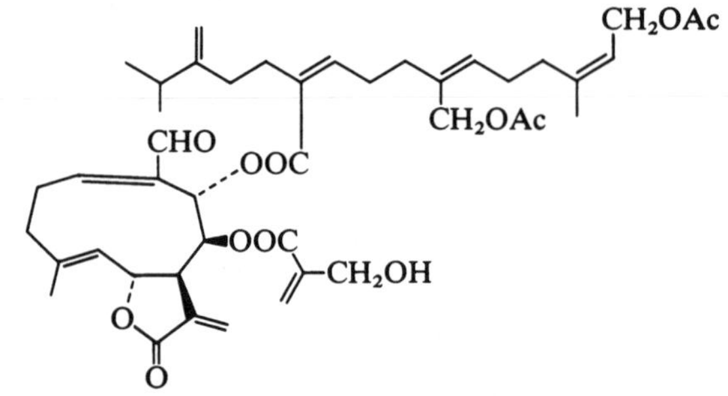

$C_{44}H_{58}O_{12}$ M 778.935
Constit. of *Clibadium pittierii*. Gum.

Tamayo-Castillo, G. *et al*, *Phytochemistry*, 1988, **27**, 1868.

Colipase   C-70188

Lipase cofactor, 9CI
[55126-92-6]

H-Val-Pro-Asp-Pro-Arg-Gly-Ile-Ile-Ile-Asn-Leu-Asp-Glu-Gly-Glu-Leu-Cys-Leu-Asn-Ser-Ala-Gln-Cys-Lys-Ser-Asn-Cys-Cys-Gln-His-Asp-Thr-Ile-Leu-Ser-Leu-[37]Ser-Arg-Cys-Ala-Leu-Lys-Ala-Arg-Glu-Asn-Ser-Glu-Cys-[50]Ser-Ala-Phe-Thr-Leu-Tyr-Gly-Val-Tyr-Tyr-Lys-Cys-Pro-Cys-Glu-Arg-Gly-Leu-Thr-Cys-Glu-Gly-Asp-Lys-Ser-Leu-Val-Gly-Ser-Ile-Thr-Asn-Thr-Asn-Phe-Gly-Ile-Cys-His-Asn-Val-Gly-Arg-Ser-Asp-Ser-OH

Struct. of porcine colipase illus. Equine colipase has sequence deletions at posns. 37 and 50. Isol. from pancreatic tissue. Assumed to act as a cofactor by anchoring lipase to insoluble triglyceride substrates.

Sakina, K. *et al*, *Chem. Pharm. Bull.*, 1988, **36**, 425 (*struct, bibl*)

Colneleic acid   C-70189

Updated Entry replacing N-01515
9-(2,4-Nonadienyloxy)-8-nonenoic acid

$$H_3C(CH_2)_4(CH=CH)_2OCH=CH(CH_2)_6COOH$$

$C_{18}H_{30}O_3$ M 294.433
(*2ʹE,4ʹZ,8E*)-form [52761-34-9]
Product of the enzymic oxidation of potato lipids. Isol. as the Me ester by glc.

Galliard, T. *et al*, *Biochem. J.*, 1972, **129**, 743 (*isol*)
Curtis, R.R., *Chem. Phys. Lipids*, 1973, **11**, 11 (*synth*)
Galliard, T. *et al*, *Chem. Phys. Lipids*, 1973, **11**, 173 (*isol*)
Corey, E.J. *et al*, *Tetrahedron Lett.*, 1987, **28**, 4917 (*synth*)
Crombie, L. *et al*, *J. Chem. Soc., Chem. Commun.*, 1987, 502, 503 (*biochem*)

Colubrinic acid   C-70190

[67594-73-4]

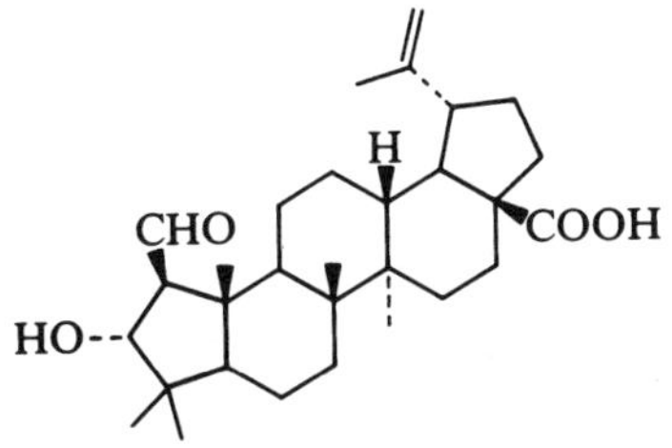

$C_{30}H_{46}O_4$ M 470.691

Constit. of *Colubrina texenis* and *C. granulosa*. Prisms (MeOH). Mp 282-284°.

Roitman, J.N. *et al, Phytochemistry*, 1978, **17**, 491 (*isol*)
Baxter, R.L. *et al, Phytochemistry*, 1988, **27**, 2350 (*isol, cryst struct*)

Comaparvin   C-70191

Updated Entry replacing C-02729
5,8-Dihydroxy-10-methoxy-2-propyl-4H-naphtho[1,2-b]pyran-4-one, 9CI

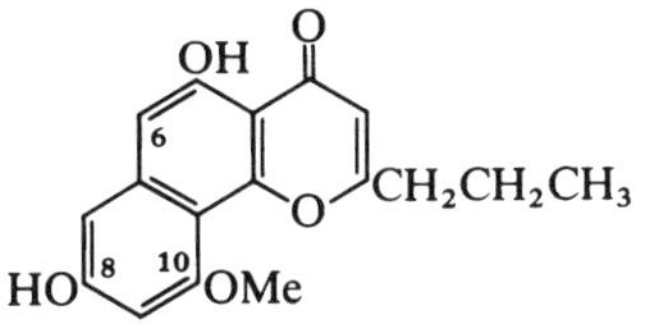

$C_{17}H_{16}O_5$ M 300.310

Pigment from the crinoid *Comanthus parvicirrus*. Yellow needles. Mp 232-233°.

Ac: Golden-yellow needles. Mp 185-186°.

8-Me ether: Pale-yellow needles. Mp 142-143°.

O-De-Me, O⁸-Me: [111397-58-1]. **5,10-Dihydroxy-8-methoxy-2-propyl-4H-naphtho[1,2-b]pyran-4-one, 9CI.**
$C_{17}H_{16}O_5$ M 300.310
Constit. of *C. parvicirrus*. Yellow needles (MeOH). Mp 184-186° dec.

6-Methoxy: [33646-78-5]. *5,8-Dihydroxy-6,10-dimethoxy-2-propyl-4H-naphtho[1,2-b]pyran-4-one, 9CI*. **6-Methoxycomaparvin**. Pigment from *C. parvicirrus*. Bright-yellow needles. Mp 200-201.5° dec. (chars from 195°).

6-Methoxy, O⁵-Me: [33646-79-6]. *8-Hydroxy-5,6,10-trimethoxy-2-propyl-4H-naphtho[1,2-b]pyran-4-one, 9CI*. **6-Methoxycomaparvin 5-methyl ether**. Constit. of *C. parvicirrus*. Yellow needles (Me₂CO). Mp 221-222° dec.

6-Methoxy, di-Me ether: [60658-78-8]. Needles (cyclohexane). Mp 93-94°.

Smith, I.R., *Aust. J. Chem.*, 1971, **24**, 1487 (*isol, struct*)
Rideout, J.A. *et al, Aust. J. Chem.*, 1976, **29**, 1087 (*synth*)
Sakuma, Y. *et al, Aust. J. Chem.*, 1987, **40**, 1613 (*deriv*)

The first digit of the DOC Number defines the Supplement in which the Entry is found.   0 indicates the Main Work.

Combretastatin   C-70192

Updated Entry replacing C-50269
3-Hydroxy-4-methoxy-α-(3,4,5-trimethoxyphenyl)-benzenemethanol, 9CI. 2-(3-Hydroxy-4-methoxyphenyl)-1-(3,4,5-trimethoxyphenyl)ethanol
[82855-09-2]

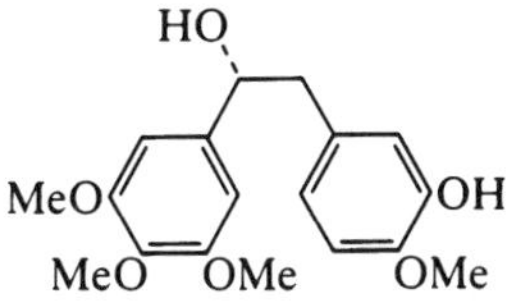

$C_{18}H_{22}O_6$ M 334.368

Isol. from the South African tree *Combretum caffrum*. Shows antitumour activity. Needles. Mp 130-131°. $[\alpha]_D^{26}$ -8.51° (c, 1.41 in CHCl₃).

Pettit, G.R. *et al, Can. J. Chem.*, 1982, **60**, 1374 (*isol, cryst struct*)
Pettit, G.R. *et al, J. Org. Chem.*, 1985, **50**, 3404 (*synth*)
Latey, P.P. *et al, Indian J. Chem., Sect. B*, 1986, **25**, 299 (*synth*)
Pettit, G.R. *et al, J. Nat. Prod.*, 1987, **50**, 386 (*abs config*)

Confertoside   C-70193

[113332-15-3]

$C_{27}H_{44}O_{15}$ M 608.636

Constit. of *Penstemon confertus*. Powder (CHCl₃/MeOH). Mp 100-101°. $[\alpha]_D^{20}$ -57° (c, 0.147 in MeOH).

Gering, B. *et al, Phytochemistry*, 1987, **26**, 3011.

Conocandin   C-70194

Updated Entry replacing C-02759
[61371-61-7]

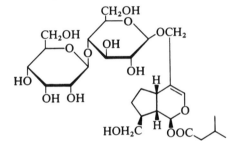

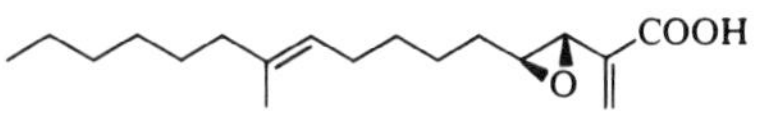

Absolute configuration

$C_{18}H_{30}O_3$ M 294.433

Metab. of *Hormococcus conorum*. Fungicidal antibiotic. Viscous oil. $[\alpha]_D^{20}$ -7° (c, 0.557 in EtOH).

▷RR0352000.

Müller, J.M. *et al, Helv. Chim. Acta*, 1976, **59**, 2506 (*isol, struct*)
Banfi, L. *et al, J. Org. Chem.*, 1987, **52**, 5452 (*synth, abs config*)

Conotoxin G1   C-70195

[76862-65-2]

Glu-Cys-Cys-Asn-Pro-Ala-Cys-Gly-Arg-His-Tyr-Ser-Cys-NH₂

Peptide neurotoxin from sea snail *Conus geographus*.

Gray, W.R. *et al, J. Biol. Chem.*, 1981, **256**, 4734 (*isol, biochem*)
Gray, W.R. *et al, Biochemistry*, 1984, **23**, 2796 (*synth, struct*)

Atherton, E. *et al*, *J. Chem. Soc., Perkin Trans. 1*, 1985, 2065 (*synth*)

Cooperin   C-70196

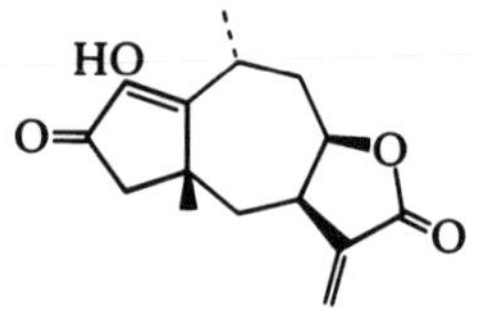

$C_{15}H_{18}O_4$ M 262.305
Constit. of *Psilostrophe cooperi*. Cryst. (CHCl$_3$/EtOAc). Mp 202°.

Stuppner, H. *et al*, *Phytochemistry*, 1988, **27**, 2681.

α-Copaene   C-70197
Updated Entry replacing C-02773
1,3-Dimethyl-8-(1-methylethyl)tricyclo[4.4.0.0^{2,7}]dec-3-ene, 9CI. 8-Isopropyl-1,3-dimethyltricyclo[4.4.0.0^{2,7}]-dec-3-ene
[3856-25-5]

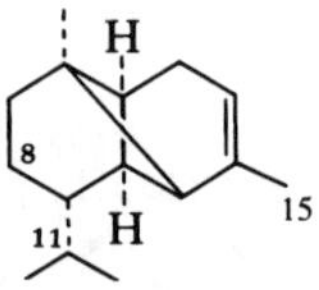

$C_{15}H_{24}$ M 204.355
$C_{15}H_{24}O$ M 220.354
Constit. of the oils of African Copaiva balsam and *Sindora wallichii*. Oil. Bp 246-251°. [α]$_D$ −13.3° (CHCl$_3$).

8α-Hydroxy: [58569-25-8]. *α-Copaen-8-ol*.
$C_{15}H_{24}O$ M 220.354
Constit. of the root oil of the *Angelica archangelica*, isol. by glc.
11-Hydroxy: [41370-56-3]. *α-Copaen-11-ol*.
$C_{15}H_{24}O$ M 220.354
Constit. of the essential oil of *Parabenzoin praecox*. Cryst. Mp 99.5-100.5°. [α]$_D$ +9.2° (CHCl$_3$).
15-Hydroxy: [41610-70-2]. *15-Copaenol*.
$C_{15}H_{24}O$ M 220.354
Constit. of *Pilgerodendron uvifera*. Oil. [α]$_D^{25}$ −28.6° (c, 0.2 in CHCl$_3$).

Kapadia, V.H. *et al*, *Tetrahedron*, *1965*, **21**, 607 (*pmr*)
de Mayo, P. *et al*, *Tetrahedron*, 1965, **21**, 619 (*ir*)
Corey, E.J. *et al*, *J. Am. Chem. Soc.*, 1973, **95**, 2303 (*synth*)
Ohara, K. *et al*, *Bull. Chem. Soc. Jpn.*, 1973, **46**, 641 (*α-Copaen-11-ol*)
Taskinen, J., *Acta Chem. Scand.*, *Ser. B*, 1975, **29**, 999 (*α-Copaen-8-ol*)
Oyarzún, M.L. *et al*, *Phytochemistry*, 1988, **27**, 1121 (*15-Copaenol*)

Coriandrin   C-70198
4-Methoxy-7-methyl-5H-furo[2,3-g][2]benzopyran-5-one, 9CI
[116408-80-1]

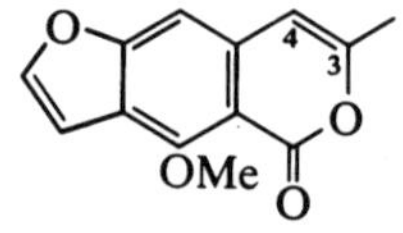

$C_{13}H_{10}O_4$ M 230.220
Constit. of *Coriandrum sativum*. Needles (MeOH). Mp 142-143°.
3,4-Dihydro: [116383-99-4]. *Dihydrocoriandrin*.

$C_{13}H_{12}O_4$ M 232.235
Constit. of *C. sativum*.

Ceska, O. *et al*, *Phytochemistry*, 1988, **27**, 2083.

Coriolin   C-70199
Updated Entry replacing C-40177

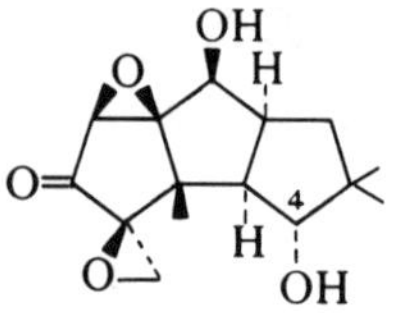

Absolute configuration

$C_{15}H_{20}O_5$ M 280.320
▷WH0203000.

(−)-*form* [33404-85-2]
Metab. of *Coriolus consors*. Antibiotic active against gram-positive bacteria, *Trichomonas* and Yoshida sarcoma cells. Cryst. Mp 175°. [α]$_D$ −20.7° (c, 0.03 in CHCl$_3$).
▷WH0203000.
4-Octanoyl: [33400-89-4]. *Coriolin B*.
$C_{23}H_{34}O_6$ M 406.518
Metab. of *C. consors*. Shows antitumour props. Cryst. Mp 215-216°.
4-(2-Hydroxyoctanoyl): [33400-90-7]. *Coriolin C*.
$C_{23}H_{34}O_7$ M 422.517
Metab. of *C. consors*.

(±)-*form* [74183-96-3]
Mp 151-154°.

Takahashi, S. *et al*, *Tetrahedron Lett.*, 1969, 4663; 1970, 1637; 1971, 1955 (*isol, struct*)
Nakamura, H. *et al*, *J. Antibiot.*, 1974, **27**, 301 (*abs config*)
Danishefsky, S. *et al*, *Tetrahedron Lett.*, 1980, 3439 (*synth*)
Shibasaki, M. *et al*, *Tetrahedron Lett.*, 1980, 3587 (*synth*)
Trost, B.M. *et al*, *J. Am. Chem. Soc.*, 1981, **103**, 7380 (*synth*)
Iseki, K. *et al*, *Tetrahedron*, 1981, **37**, 4411 (*synth*)
Tatsuta, K. *et al*, *Tetrahedron*, 1981, **37**, 4365 (*synth*)
Mehta, G. *et al*, *J. Chem. Soc., Chem. Commun.*, 1982, 540 (*synth*)
Koreeda, M. *et al*, *J. Am. Chem. Soc.*, 1983, **105**, 7203 (*synth*)
Wender, P.A. *et al*, *Tetrahedron Lett.*, 1983, **24**, 5325 (*synth*)
Demuth, M. *et al*, *Helv. Chim. Acta*, 1984, **67**, 2023 (*synth*)
Ito, T. *et al*, *Tetrahedron*, 1984, **40**, 241 (*synth*)
Schuda, P.F. *et al*, *Tetrahedron*, 1984, **40**, 2365 (*synth*)
Demuth, M. *et al*, *J. Am. Chem. Soc.*, 1986, **108**, 4149 (*bibl*)
Van Hijfte, L. *et al*, *J. Org. Chem.*, 1987, **52**, 4647 (*synth*)

Crombeone   C-70200
Updated Entry replacing C-20238
6a,12a-Dihydro-2,3,8,10-tetrahydroxy[2]-benzopyrano[4,3-b][1]benzopyran-7(5H)-one, 9CI. 3',4',5,8-Tetrahydroxypeltogynan-4-one
[30759-13-8]

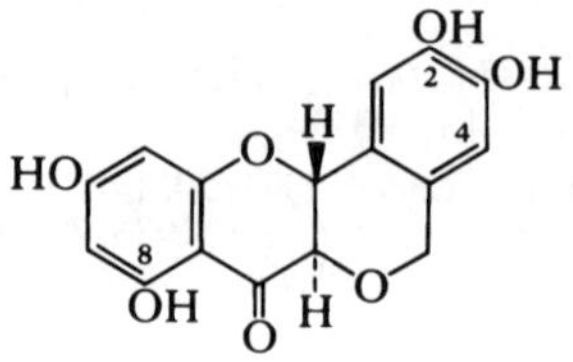

$C_{16}H_{12}O_7$ M 316.267
CA numbering shown. Flavonoid-type (peltogynan) numbering has also been used. Constit. of *Acacia crombei*, isol. with difficulty.

O^8-Me: [38279-47-9]. *6a,12a-Dihydro-2,3,10-trihydroxy-8-methoxy[2]benzopyrano[4,3-b][1]benzopyran-7(5H)-one, 9CI. 2,3,10-Trihydroxy-8-methoxy-6,12-dioxabenz[a]anthracen-7(5H,6aH,12aH)-one.*
$C_{17}H_{14}O_7$ M 330.293

Isol. from *Goniorrhachis marginata*. Cryst. Mp 260-262°.

Tri-Me ether: Amorph. powder. Mp 209° dec. $[\alpha]_D^{28}$ +255° (Py).

8-Deoxy: [38279-52-6]. *6a,12a-Dihydro-2,3,10-trihydroxy[2]benzopyrano[4,3-b][1]benzopyran-7(5H)-one, 9CI.* *2,3,10-Trihydroxy-6,12-dioxabenz[a]-anthracen-7(5H,6H,12aH)-one.*
$C_{16}H_{12}O_6$ M 300.267
Isol. from *G. marginata*. Rods (CHCl₃/MeOH). Mp 259-261°.

2-Deoxy, 4-hydroxy, O⁸-Me: [38279-50-4]. *6a,12a-Di-hydro-3,4,10-trihydroxy-8-methoxy[2]-benzopyrano[4,3-b][1]benzopyran-7(5H)-one, 9CI.* *3,4,10-Trihydroxy-8-methoxy-6,12-dioxabenz[a]-anthracen-7(5H,6H,12aH)-one.*
$C_{17}H_{14}O_7$ M 330.293
Isol. from *G. marginata*. Cryst. (MeOH). Mp 280-282°.

Brandt, E.V. *et al, J. Chem. Soc., Chem. Commun.,* 1971, 116 (*struct*)
Gottlieb, O.R. *et al, Phytochemistry,* 1972, **11**, 2841 (*derivs*)
Brandt, E.V. *et al, J. Chem. Soc., Perkin Trans. 1,* 1981, 514 (*synth*)

Cryptoresinol   C-70201

2,5-Dihydro-3,5-bis(4-hydroxyphenyl)-2-furanmeth-anol, 9CI

[115713-11-6]

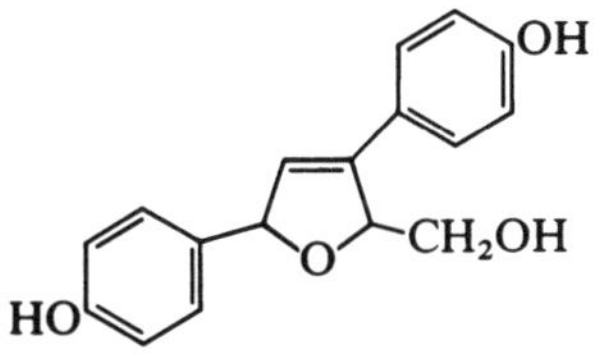

$C_{17}H_{16}O_4$ M 284.311
Constit. of *Cryptomeria japonica* heartwood. Cryst. (Et₂O/MeOH). Mp 253-255°. $[\alpha]_D^{25}$ −170.4° (c, 1.25 in MeOH).

Takahashi, K. *et al, Phytochemistry,* 1988, **27**, 1550.

Crystal violet lactone   C-70202

6-(Dimethylamino)-3,3-bis[4-(dimethylamino)phenyl]-1(3H)-isobenzofuranone, 9CI

[1552-42-7]

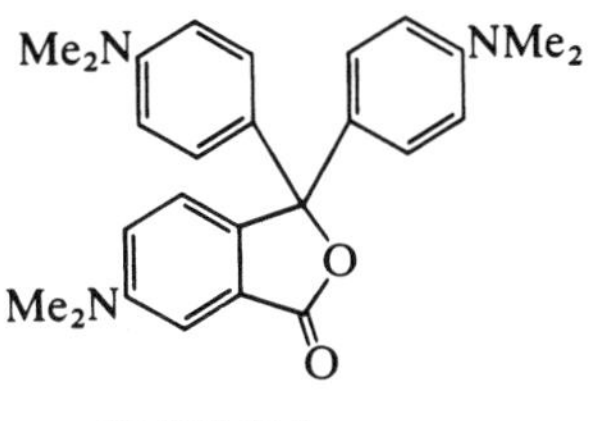

$C_{26}H_{29}N_3O_2$ M 415.534
Photographic/reprographic dye. Blue-green cryst. (CHCl₃/MeOH). Mp 162-163°.

Ger. Pat., 2 835 450, (*1980*); *CA,* **93**, 27735a (*manuf*)
Theocharis, C.R. *et al, J. Crystallogr. Spectrosc. Res.,* 1984, **14**, 121 (*cryst struct*)
U.S.P., 4 455 435, (*1984*); *CA,* **101**, 92793r (*synth*)
Chunaev, Yu.M., *Zh. Org. Khim.,* 1986, **22**, 2240 (*Engl. Transl.* p. 2012) (*synth, struct*)

Cucurbitacin *D*   C-70203

2,16,20,25-Tetrahydroxy-9-methyl-19-norlanosta-5,23-diene-3,11,22-trione, 9CI. Elatericin A

[3877-86-9]

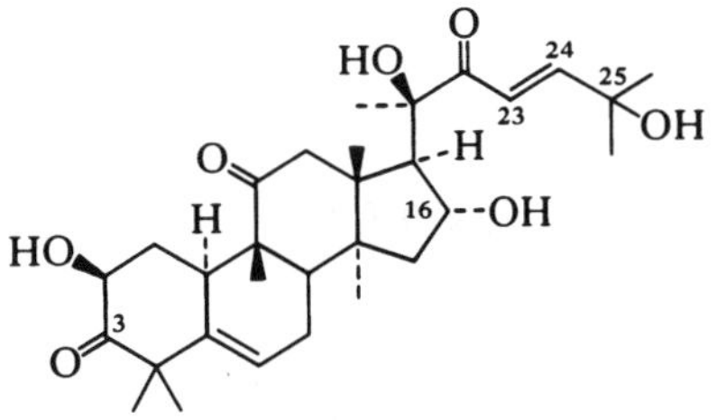

$C_{30}H_{44}O_7$ M 516.673
Isol. from plants of the cucurbitaceae. e.g. *Phormium tenax* and *Ecballium elaterium*. Cryst. + ½H₂O (EtOH). Mp 151-152°. $[\alpha]_D$ +52° (EtOH).

▷RC6170000

O¹⁶-(2,3-Di-O-acetyl-α-L-allopyranoside): [87827-49-4]. *Datiscoside* C.
$C_{40}H_{58}O_{13}$ M 746.890
Isol. from *Datisca glomerata*. Cytotoxic. Prisms (C₆H₆/Et₂O).

O¹⁶-(2-O-Acetyl-6-deoxy-α-L-ribo-hexopyranos-3-uloside): [36067-56-8]. *Datiscoside.*
$C_{38}H_{54}O_{12}$ M 702.837
Glycoside from *D. glomerata*. Antileukaemic and cytotoxic. Needles (MeOH). Mp 174-175°. $[\alpha]_D^{23}$ +26° (c, 1.04 in CHCl₃).

O¹⁶-(2,4-Di-O-acetyl-6-deoxy-α-L-allopyranoside): [87718-77-2]. *Datiscoside* E.
$C_{40}H_{58}O_{13}$ M 746.890
Isol. from *D. glomerata*. Cytotoxic. Plates (Et₂O).

O¹⁶-(2,3-Di-O-acetyl-4,6-dideoxy-α-ribo-hexopyranoside): [87710-32-5]. *Datiscoside* F.
$C_{40}H_{58}O_{12}$ M 730.891
Isol. from *D. glomerata*. Plates (Et₂O). $[\alpha]_D^{23}$ +16.6° (c, 1 in CHCl₃).

O¹⁶-β-D-Glucoside, O²-Ac: *Datiscoside* G.
$C_{38}H_{56}O_{13}$ M 720.853
Isol. from *D. glomerata*. Cytotoxic. Cryst. (Et₂O). Registered in *CA* as lacking the 2-Ac group.

23,24-Dihydro: [55903-92-9]. *Cucurbitacin* R.
$C_{30}H_{46}O_7$ M 518.689
Isol. from juice of *E. elaterium*. Mp 140-143°.

3α-Alcohol, O¹⁶-(2,3-di-O-acetyl-α-D-allopyranoside): [87827-50-7]. *Datiscoside* D.
$C_{40}H_{60}O_{13}$ M 748.906
Isol. from *D. glomerata*. Cytotoxic. Plates (Et₂O). $[\alpha]_D^{23}$ +19.3° (c, 1 in CHCl₃).

3α-Alcohol, O¹⁶-(2-O-Acetyl-6-deoxy-α-L-ribo-hexopyranos-3-uloside): [87710-34-7]. *Datiscoside* B.
$C_{38}H_{54}O_{12}$ M 702.837
Isol. from *D. glomerata*. Cytotoxic. Plates (EtOH). $[\alpha]_D^{23}$ +20.7° (c, 1 in CHCl₃).

3α-Alcohol, O¹⁶-(2-O-Acetyl-3,6-dideoxy-α-L-erythro-hexopyranos-4-uloside): [87710-33-6]. *Datiscoside* H.
$C_{38}H_{56}O_{11}$ M 688.854
Isol. from *D. glomerata*. Cytotoxic. Plates (C₆H₆/Et₂O).

O²⁵-Ac: see Cucurbitacin B, C-20247

Lavie, D. *et al, Progr. Chem. Org. Nat. Prod.,* 1971, **29**, 307 (*rev*)
Kupchan, S.M. *et al, J. Am. Chem. Soc.,* 1972, **94**, 1353 (*isol, uv, ir, pmr, Datiscoside*)
Restivo, R.J. *et al, J. Chem. Soc., Perkin Trans. 2,* 1973, 892 (*abs config, cryst struct, Datiscoside*)
Rao, M.M. *et al, J. Chem. Soc., Perkin Trans. 1,* 1974, 2552 (*Cucurbitacin R*)
Kupchan, S.M. *et al, Phytochemistry,* 1978, **17**, 767 (*isol*)

Yamada, Y. *et al*, *Phytochemistry*, 1978, **17**, 1798 (*isol*)
Sasamori, H. *et al*, *J. Chem. Soc., Perkin Trans. 1*, 1983, 1333
  (*Datiscosides*)

Bull, J.R. *et al*, *S. Afr. J. Chem.*, 1979, **32**, 27 (*cmr*)
Amonkar, A.A. *et al*, *Phytochemistry*, 1985, **24**, 1803 (*isol*)
Che, C.-T. *et al*, *J. Nat. Prod.*, 1985, **24**, 429 (*pmr*)

Cucurbitacin *E*   C-70204

Updated Entry replacing C-02978
  α-Elaterin
  [18444-66-1]

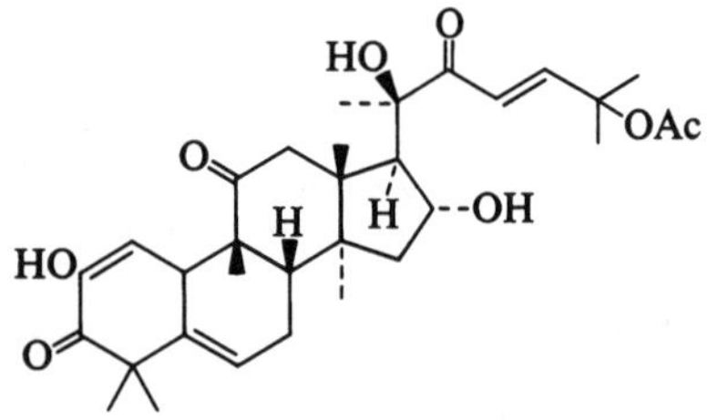

$C_{32}H_{44}O_8$ M 556.695
Constit. of *Ecballium elaterium*. Mp 234°. $[\alpha]_D^{20}$ −64.3°
  (CHCl$_3$).
▷RC6305500.

2-O-β-D-Glucosyl: [1398-78-3]. **Elaterinide**.
  Colocynthin.
  $C_{38}H_{54}O_{13}$ M 718.837
  Constit. of *Citrullus lanatus* and *C. colocynthis*. Mp
  148-150°. $[\alpha]_D^{20}$ −63.5° (CHCl$_3$).
  ▷RC6192500.

Deacetyl: [2222-07-3]. **Cucurbitacin I**. *Elatericin* B.
  $C_{30}H_{42}O_7$ M 514.658
  Constit. of seed of *Iberis amara*. Cryst. Mp 148-149°
  dec. $[\alpha]_D^{20}$ −58.5° (MeOH).
  ▷RC6200000.

Deacetyl, 2-O-β-D-glucopyranosyl: [29803-94-9].
  $C_{36}H_{52}O_{12}$ M 676.800
  Constit. of *C. lanatus*. Mp 241-243°. $[\alpha]_D^{20}$ −72.6°
  (EtOH).
  ▷RC6191000.

23,24-Dihydro, deacetyl: [1110-02-7]. **Cucurbitacin L**.
  $C_{30}H_{44}O_7$ M 516.673
  Constit. of *C. ecirrhosus* and *C. colocynthis* (also as
  glycosides), *Gratiola officinalis* and *Bryonia dioica*.
  Cryst. $+\frac{1}{2}$H$_2$O (MeOH aq.). Mp 122-127°. $[\alpha]_D^{28}$
  −41° (EtOH).

Deacetyl, 23,24-dihydro, 2-O-β-D-glucopyranoside:
  [61105-51-9]. **Bryoamaride**.
  $C_{36}H_{54}O_{12}$ M 678.815
  Constit. of *Bryonica dioica* and *C. colocynthus*. Mp
  228-235°. $[\alpha]_D^{20}$ −85.7° (c, 0.82 in EtOH).

23,24-Dihydro, 2-O-β-D-glucopyranoside: [61014-18-4].
  25-O-Acetylbryoamaride.
  $C_{38}H_{56}O_{13}$ M 720.853
  Constit. of *B. dioica*. $[\alpha]_D^{20}$ −34.3° (c, 1.04 in CHCl$_3$).

Deacetyl, 20-Ac: [38308-89-3]. **Datiscacin**.
  $C_{32}H_{44}O_8$ M 556.695
  Constit. of *Datisca glomerata*. Mp 208-212°. $[\alpha]_D^{23}$
  −18° (c, 0.87 in CHCl$_3$).

Deacetyl, 11-Deoxo: **11-Deoxocucurbitacin I**.
  $C_{30}H_{44}O_6$ M 500.674
  Constit. of *Desfontainia spinosa*. Cryst. (C$_6$H$_6$). Mp
  212-213°.

. Kupchan, S.M. *et al*, *J. Org. Chem.*, 1973, **38**, 1420 (*Datiscacin*)
Ripperger, H., *Tetrahedron*, 1976, **32**, 1567 (*isol*)
Nielsen, J.K. *et al*, *Phytochemistry*, 1977, **16**, 1519 (*isol*)
Yamada, Y. *et al*, *Phytochemistry*, 1978, **17**, 1798 (*isol*)

Cudraflavone *A*   C-70205

Updated Entry replacing C-40181
[96843-73-1]

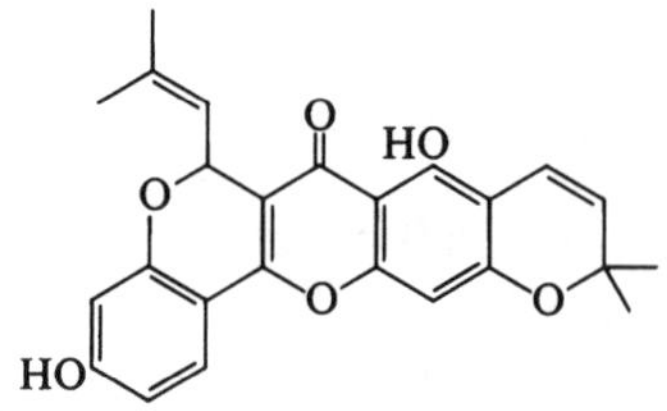

$C_{25}H_{22}O_6$ M 418.445
Constit. of *Cudrania tricuspidata*. Cryst. Mp 265-272°.
  $[\alpha]_D^{19}$ +27.3° (c, 0.147 in CHCl$_3$).

5'-Hydroxy: [114339-95-6]. **5'-Hydroxycudraflavone** A.
  $C_{25}H_{22}O_7$ M 434.445
  Constit. of *Brosimopsis oblongifolia*. Cryst.
  (CHCl$_3$/hexane). Mp 235-237°. $[\alpha]_D^{20}$ +13.8° (c, 1.1
  in Me$_2$CO).

Fujimoto, T. *et al*, *Planta Med.*, 1984, **50**, 218 (*isol, struct*)
Messana, I. *et al*, *Planta Med.*, 1987, **53**, 541 (*deriv*)

Cudraisoflavone *A*   C-70206

[114727-97-8]

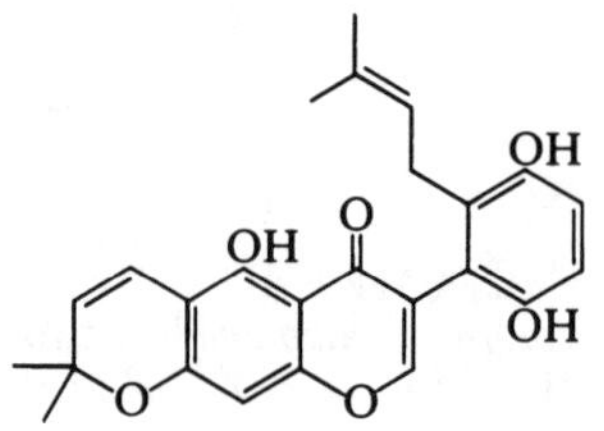

$C_{25}H_{24}O_6$ M 420.461
Constit. of *Cudrania cochinchinensis*. Yellow cryst.
  (CHCl$_3$). Mp 167-170°.

Sun, N.-J. *et al*, *Phytochemistry*, 1988, **27**, 951.

Curcudiol   C-70207

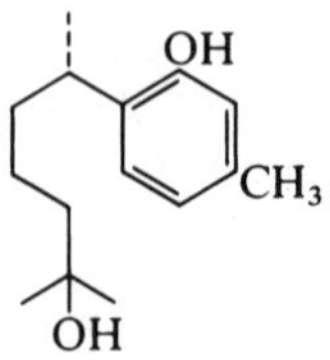

$C_{15}H_{24}O_2$ M 236.353
Constit. of the sponge *Didiscus flavus*. Oil. $[\alpha]_D^{22}$ +9.2°
  (c, 10.8 in CHCl$_3$).

Wright, A.E. *et al*, *J. Nat. Prod.*, 1987, **50**, 976.

Curcuphenol   C-70208

Updated Entry replacing C-03016

2-(1,5-Dimethyl-4-hexenyl)-5-methylphenol, 9CI

[17194-58-0]

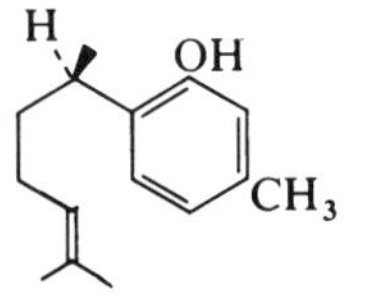

$C_{15}H_{22}O$ M 218.338

(**R**)-*form* [69301-27-5]

Constit. of *Pseudopterogorgia rigida* and *Lasianthaea podocephala*. Oil. $[\alpha]_D$ −7° (c, 3.65 in $CHCl_3$), $[\alpha]_D$ −23.6° ($CHCl_3$).

(**S**)-*form*

Constit. of the sponge *Epipolasis* sp. and of *Didiscus flavus*. Oil. $[\alpha]_D^{23}$ +29.1° (c, 3.13 in $CHCl_3$).

8,9-Didehydro (E-): **Dehydrocurcuphenol**.

$C_{15}H_{20}O$ M 216.322

Constit. of *E.* sp. Oil. $[\alpha]_D^{23}$ −1.2° (c, 0.48 in $CHCl_3$).

10,11-Dihydro, 11-hydroxy: **10,11-Dihydro-11-hydroxycurcuphenol**.

$C_{15}H_{24}O_2$ M 236.353

Constit. of *E.* sp. Oil. $[\alpha]_D^{23}$ +1.3° (c, 5.92 in $CHCl_3$).

McEnroe, F.J. *et al, Tetrahedron,* 1978, **34**, 1661 (*isol, struct*)

Ghisalberti, E. *et al, Aust. J. Chem.,* 1979, **32**, 1627 (*isol*)

Fusetani, N. *et al, Experientia,* 1987, **43**, 1234 (*isol*)

Wright, A.E. *et al, J. Nat. Prod.,* 1987, **50**, 976 (*isol*)

2-Cyanoethanethioamide, 9CI   C-70209

2-Cyanothioacetamide, 8CI

[7357-70-2]

$$NCCH_2CSNH_2$$

$C_3H_4N_2S$ M 100.138

Reactive compd. for the synth. of *N-* and *S-*heterocycles. Cryst. (EtOH or C_6H_6). Mp 121-123°. Darkens on long standing.

Riad, B.Y. *et al, Heterocycles,* 1987, **26**, 205 (*rev*)

Cyanoviridin RR   C-70210

3-L-Arginine-5-L-argininecyanoginosin LA, 9CI

[111755-37-4]

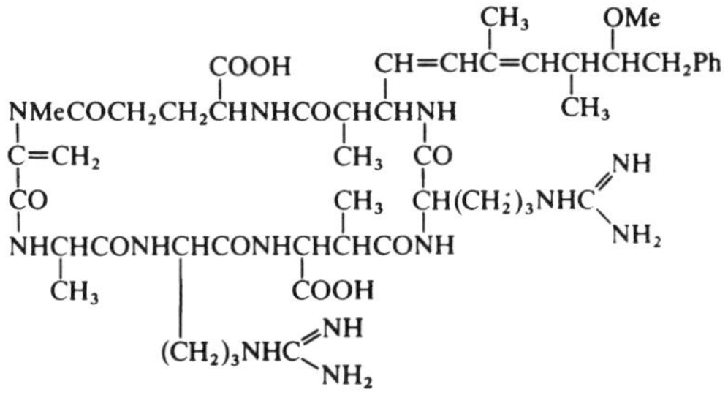

$C_{49}H_{75}N_{13}O_{12}$ M 1038.211

Cyclic peptide antibiotic. Isol. from blue-green alga *Microcystis viridis*. Toxin. Isol. with 2 other components, undergoing research.

Kusumi, T. *et al, Tetrahedron Lett.,* 1987, **28**, 4695 (*isol, struct*)

Cycloaspeptide A   C-70211

[109171-13-3]

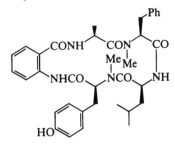

$C_{36}H_{43}N_5O_6$ M 641.766

Cyclic peptide antibiotic. Isol. from *Aspergillus* sp. NE-45. Needles + $1H_2O$ (MeOH). Mp 270-272°. $[\alpha]_D^{20}$ −228° (c, 1.01 in $CHCl_3$).

N^{Tyr}*-de-Me:* [109171-14-4]. **Cycloaspeptide B**.

$C_{35}H_{41}N_5O_6$ M 627.739

Needles + $2H_2O$ (MeOH). Mp 250-252°. $[\alpha]_D^{15}$ −138° (c, 0.1 in THF/DMF).

N^{Phe}*-de-Me:* [109171-15-5]. **Cycloaspeptide C**.

$C_{35}H_{41}N_5O_6$ M 627.739

Needles + $4H_2O$ (MeOH). Mp 281-283°. $[\alpha]_D^{15}$ −168° (c, 0.09 in THF/DMF).

Kobayashi, R. *et al, Chem. Pharm. Bull.,* 1987, **35**, 1347 (*isol, struct*)

1,3-Cyclobutadiene, 9CI   C-70212

Updated Entry replacing C-03142

[4]Annulene

[1120-53-2]

C_4H_4 M 52.076

Transient intermediate. Formed by oxidn. of cyclobutadieneiron tricarbonyl and trapped by condensation.

Watts, L. *et al, J. Am. Chem. Soc.,* 1965, **87**, 3253 (*synth*)

Dewar, M.J.S. *et al, J. Am. Chem. Soc.,* 1971, **93**, 3437 (*struct*)

Chapman, O.L. *et al, J. Am. Chem. Soc.,* 1973, **95**, 614 (*synth*)

Maier, G. *et al, Tetrahedron Lett.,* 1973, 861.

Masamune, S. *et al, Can. J. Chem.,* 1976, **54**, 2679 (*spectra*)

Maier, G. *et al, Tetrahedron Lett.,* 1984, **25**, 5645 (*synth*)

Maier, G., *Angew. Chem., Int. Ed. Engl.,* 1988, **27**, 307 (*rev*)

1,1-Cyclobutanedimethanethiol, 9CI   C-70213

1,1-Bis(mercaptomethyl)cyclobutane. 1,1-Bis(thiomethyl)cyclobutane

[110206-37-6]

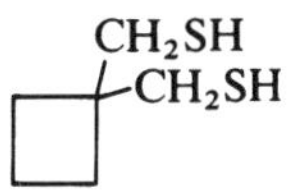

$C_6H_{12}S_2$ M 148.281

Liq. $Bp_{0.4}$ 45-50°.

Houk, J. *et al, J. Am. Chem. Soc.,* 1987, **109**, 6825 (*synth, ir, pmr*)

1,2-Cyclobutanedimethanethiol   C-70214

1,2-Bis(thiomethyl)cyclobutane. 1,2-Bis(mercaptomethyl)cyclobutane

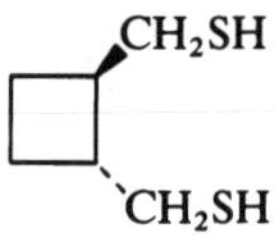

$C_6H_{12}S_2$ M 148.281

(**1RS,2RS**)-*form* [110206-40-1]

(±)-trans-*form*

Liq. Bp$_{0.5}$ 46-50°.

Houk, J. *et al, J. Am. Chem. Soc.*, 1987, **109**, 6825 (*synth, ir, pmr*)

1,1-Cyclobutanedithiol   C-70215

1,1-Dimercaptocyclobutane

[15144-23-7]

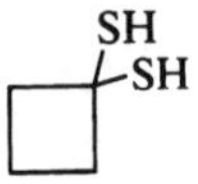

$C_4H_8S_2$ M 120.227

Liq. Bp$_5$ 51°.

Fournier, C. *et al, Org. Magn. Reson.*, 1977, **10**, 20 (*synth, ms*)
Houk, J. *et al, J. Am. Chem. Soc.*, 1987, **109**, 6825 (*synth, ir, pmr*)

Cyclocratystyolide   C-70216

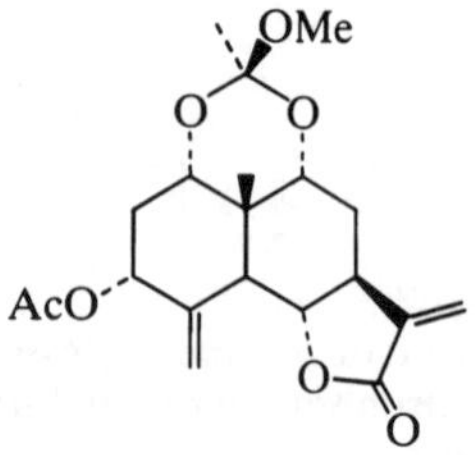

$C_{20}H_{26}O_7$ M 378.421

Constit. of *Cratystylis conocephala*. Oil.

Zdero, C. *et al, Phytochemistry*, 1988, **27**, 865.

1,4,7,10-Cyclododecatetraene   C-70217

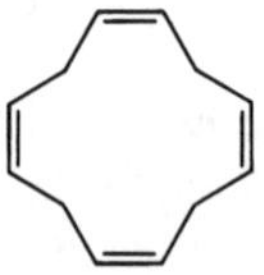

(*Z,Z,Z,Z*)-*form*

$C_{12}H_{16}$ M 160.258

(**Z,Z,Z,Z**)-*form* [112681-65-9]

Liq.

(**E,Z,Z,Z**)-*form* [112790-14-4]

Liq.

Brudermüller, M. *et al, Angew. Chem., Int. Ed. Engl.*, 1988, **27**, 298 (*synth, pmr*)

Cycloeuphordenol   C-70218

4,14-Dimethyl-9,19-cycloergost-20-en-3β-ol, 9CI.
4α,14α,24R-Trimethyl-9β,19-cyclo-20-cholesten-3β-ol

[117193-18-7]

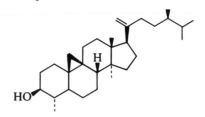

$C_{30}H_{50}O$ M 426.724

Constit. of *Euphorbia tirucalli*. Needles (MeOH). Mp 105-106°. [α]$_D$ +39° (c, 0.127 in CHCl$_3$).

Khan, A.Q. *et al, Phytochemistry*, 1988, **27**, 2279.

Cyclohepta[1,2-*b*:1,7-*b'*]bis[1,4]-benzoxazine, 9CI   C-70219

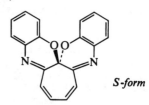

S-form

$C_{19}H_{12}N_2O_2$ M 300.316

Contains highly twisted π-electron system.

(**S**)-*form* [93380-74-6]

Yellow cryst. Mp 184-185°. [α]$_D$ −4700° (c, 0.0018 in MeOH).

(±)-*form* [93278-93-4]

Yellow needles (Et$_2$O). Mp 191°.

Someya, T. *et al, Bull. Chem. Soc. Jpn.*, 1983, **56**, 2756 (*synth, uv, pmr, cmr*)
Harada, N. *et al, J. Am. Chem. Soc.*, 1987, **109**, 1661 (*synth, uv, cd, abs config*)

Cyclohepta[*def*]carbazole   C-70220

[52879-18-2]

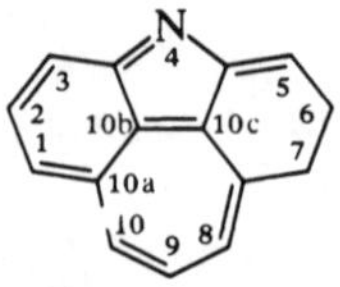

$C_{15}H_9N$ M 203.243

Currently unknown. The cation is stable in HClO$_4$ soln. but attempts at deprotonation were unsuccessful.

B,HBF$_4$: [113728-07-7]. *Cyclohepta*[def]*carbazolium tetrafluoroborate*.

$C_{15}H_{10}BF_4N$ M 291.054

Dark-blue flakes.

Freiermuth, B. *et al, Angew. Chem., Int. Ed. Engl.*, 1988, **27**, 585 (*synth, ir, pmr, uv*)

2-(2,4,6-Cycloheptatrien-1-ylidene)-1,3-dithiane, 9CI   C-70221

8,8-Dithiaheptafulvene

[60576-03-6]

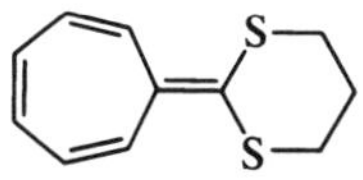

$C_{11}H_{12}S_2$ M 208.336

Dark-red cryst. (pet. ether). Mp 45°.

Rapp, K.M. *et al*, *Tetrahedron Lett.*, 1976, 2011 (*synth, ir, uv, ms, pmr*)

5-(2,4,6-Cycloheptatrien-1-ylideneethylidene)-3,6-cycloheptadiene-1,2-dione   C-70222

8-(8-Heptafulvenyl)-p-tropoquinone methide

[113789-85-8]

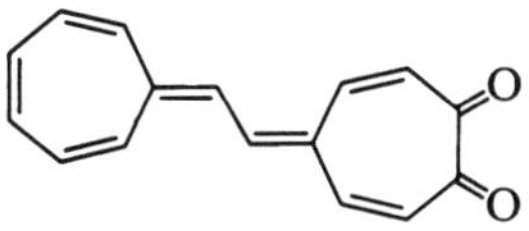

$C_{16}H_{12}O_2$ M 236.270

Dipolar delocalised forms possible. Obt. as a deep-blue soln. Solid unstable and not isol.

Takahashi, K. *et al*, *J. Chem. Soc., Chem. Commun.*, 1987, 935 (*synth, pmr, uv*)

1,1-Cyclohexanedimethanethiol, 9CI   C-70223

1,1-Bis(mercaptomethyl)cyclohexane. 1,1-Bis(thiomethyl)cyclohexane

[56472-19-6]

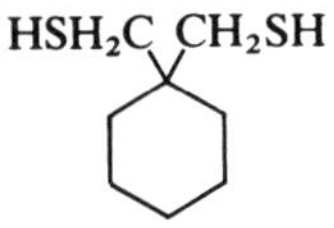

HSH₂C CH₂SH

$C_8H_{16}S_2$ M 176.334

Liq. Bp₁₃ 140°, Bp₀.₁ 70-75°.

Goor, G. *et al*, *Phosphorus Sulfur*, 1976, **1**, 81 (*synth*)
Houk, J. *et al*, *J. Am. Chem. Soc.*, 1987, **109**, 6825 (*synth, ir, pmr*)

1,2-Cyclohexanedimethanethiol, 9CI   C-70224

1,2-Bis(mercaptomethyl)cyclohexane. 1,2-Bis(thiomethyl)cyclohexane

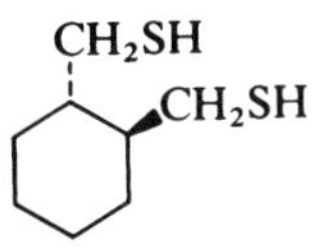

(1S,2S)-form

$C_8H_{16}S_2$ M 176.334

(1S,2S)-form [38261-87-9]

(+)-*trans-form*

Liq. Bp₂.₅ 90°, Bp₀.₂ 73-75°. [α]₂₇.₁ +67.25° (c, 2.55 in MeOH).

(1RS,2SR)-form

cis-*form*

Liq. Bp₁₂.₅ 138-142°, 4 . Bp₁ 80-85°.

Lüttringhaus, A. *et al*, *Chem. Ber.*, 1959, **92**, 2271 (*synth*)
Casey, J.P. *et al*, *J. Am. Chem. Soc.*, 1972, **94**, 6141 (*synth*)
Houk, J. *et al*, *J. Am. Chem. Soc.*, 1987, **109**, 6825 (*synth, ir, pmr*)

1,2-Cyclohexanedithiol   C-70225

1,2-Dimercaptocyclohexane

[61947-27-1]

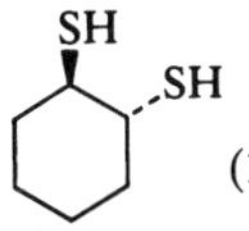

(1RS,2RS)-form

$C_6H_{12}S_2$ M 148.281

(1RS,2RS)-form [19756-06-0]

(±)-*trans-form*

Liq. Bp₁₈₋₁₉ 104-106°, Bp₀.₅ 40-45°.

(1RS,2SR)-form [2242-71-9]

cis-*form*

Liq. Bp₉ 99-100°, Bp₀.₂ 48-51°.

Iqbal, S.M. *et al*, *J. Chem. Soc.*, 1960, 1030 (*synth*)
Böhme, H. *et al*, *Chem. Ber.*, 1965, **98**, 1455 (*synth*)
Forster, R.C. *et al*, *J. Chem. Soc., Perkin Trans. 1*, 1978, 822 (*synth*)
Houk, J. *et al*, *J. Am. Chem. Soc.*, 1987, **109**, 6825 (*synth, pmr, ir, cmr*)

1,3,5-Cyclohexanetricarboxaldehyde, 9CI   C-70226

1,3,5-Triformylcyclohexane

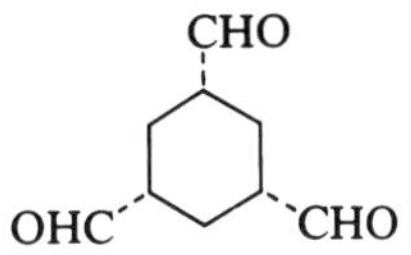

$C_9H_{12}O_3$ M 168.192

(1α,3α,5α)-form [107354-37-0]

Small chunky prisms. Mp 60-63°. Polym. on attempted recryst. from CCl₄.

Tris(phenylhydrazone): [107383-98-2]. Cryst. (2-propanol). Mp 87-89°, Mp 176-182° (dimorph.?).

Trishydrazone: [107383-99-3].
    $C_9H_{18}N_6$ M 210.281
    Mp 75-100°. Readily polym., could not be recryst.

Tris(di-Me acetal): [107384-03-2]. *1,3,5-Tris(dimethoxymethyl)cyclohexane, 9CI.*
    $C_{15}H_{30}O_6$ M 306.398
    Cryst. Mp 78-80°.

Nielsen, A.T. *et al*, *J. Org. Chem.*, 1987, **52**, 1656 (*synth, ir, pmr, ms*)

3-Cyclohexene-1-carboxaldehyde, 9CI   C-70227

Updated Entry replacing C-03370

1,2,3,6-Tetrahydrobenzaldehyde. 4-Formylcyclohexene

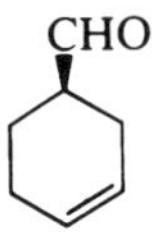

$C_7H_{10}O$ M 110.155

(R)-form [60631-76-7]

[α]D +95° (c, 3 in CHCl₃).

(±)-form [100-50-5]

Oil. Mp −96.1°. Bp 163.5-164.5°, Bp₁₃ 51-52°.

Polymerises very readily.

▷ Mod. toxic. Flammable. GW2800000.

Oxime: [4736-13-4].
    $C_7H_{11}NO$ M 125.170
    2 forms, oil and cryst. Mp 75-76° (cryst.). Bp₂₅ 128-129° (oil).

Semicarbazone: Cryst. (C_6H_6/ligroin or MeOH aq.). Mp 153.5-154.5°.

Phenylhydrazone: Bp_{22} 207-208°.

Sobecki, W., *Ber.*, 1910, **43**, 1040 (*synth*)
Diels, O. *et al*, *Justus Liebigs Ann. Chem.*, 1928, **460**, 121 (*synth*)
Org. Synth., 1971, **51**, 11 (*synth*)
Ceder, O. *et al*, *Synth. Commun.*, 1976, **6**, 381 (*synth*)
Sax, N.I., *Dangerous Properties of Industrial Materials*, 5th Ed., Van Nostrand-Reinhold, 1979, 1016.

N-Cyclohexylhydroxylamine
C-70228

Updated Entry replacing C-03407
N-*Hydroxycyclohexanamine, 9CI.*
Hydroxylaminocyclohexane

[2211-64-5]

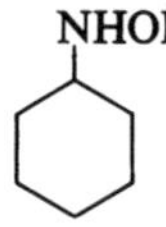

NHOH

$C_6H_{13}NO$ M 115.175

Reagent for the formn. of α,β-unsaturated acids from cyclic anhydrides, for the acylation of benzilic acid and for the prepn. of coumarin nitrones. Needles. Mp 140-141° (135-137°). Sublimes. A Mp of 102-103° is given by Fieser and Fieser but appears to be incorr.

▷NC3410400.

Gygax, P. *et al*, *Helv. Chim. Acta*, 1977, **60**, 507 (*use*)
Shridhar, D.R. *et al*, *J. Indian Chem. Soc.*, 1978, **55**, 902 (*use*)
Schmidt, J. *et al*, *Arch. Pharm.* (*Weinheim, Ger.*), 1979, **312**, 1019 (*use*)
Geffken, D., *Arch. Pharm.* (*Weinheim, Ger.*), 1980, **313**, 337 (*use*)
Varma, R.S. *et al*, *Org. Prep. Proced. Int.*, 1985, **17**, 254 (*synth*)
Fieser, M. *et al*, *Reagents for Organic Synthesis*, Wiley, 1967-84, **8**, 135 (*use*)

Cyclolycoserone
C-70229

[98094-08-7]

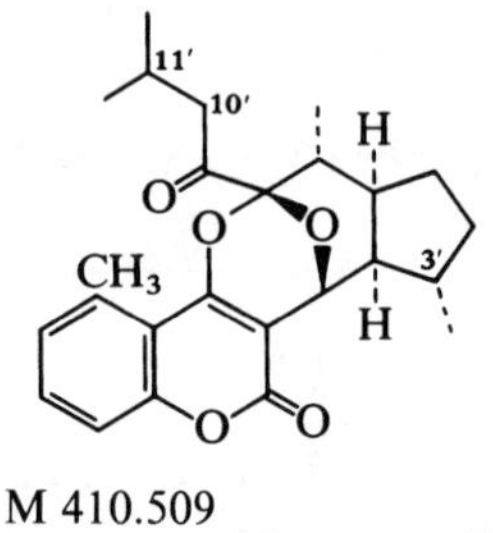

$C_{25}H_{30}O_5$ M 410.509

Constit. of *Lycoseris latifolia*. Oil. $[\alpha]_D^{24}$ +63° (c, 0.17 in $CHCl_3$).

3'-Epimer:
  $C_{25}H_{30}O_5$ M 410.509
  Constit. of *Aphyllocladus denticulatus*. Gum. $[\alpha]_D^{24}$ −78° (c, 1.62 in $CHCl_3$).

10',11'-Didehydro: **10',11'-Dehydrocyclolycoserone**.
  $C_{25}H_{28}O_5$ M 408.493
  Constit. of *A. denticulatus*. Gum.

10'-Hydroxy: **10'-Hydroxycyclolycoserone**.
  $C_{25}H_{30}O_6$ M 426.508
  Constit. of *A. denticulatus*. Gum.

Bohlmann, F. *et al*, *Justus Liebigs Ann. Chem.*, 1985, 1367 (*isol*)
Zdero, C. *et al*, *Phytochemistry*, 1988, **27**, 1821 (*isol*)

3*H*-Cyclonona[*def*]biphenylen-3-one
C-70230

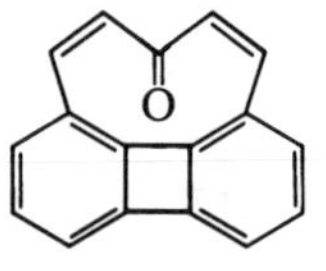

Z,Z-form

$C_{17}H_{10}O$ M 230.265

(*Z,Z*)-form [108058-71-5]
  Orange cryst. solid. Mp 200-203°. Nonplanar, showed no antiaromatic props. Slowly dec. in soln.

Wilcox, C.F. *et al*, *J. Org. Chem.*, 1987, **52**, 2635 (*synth, ir, pmr, cmr, uv, ms, cryst struct*)

Cycloocta[2,1-*b*:3,4-*b'*]di[1,8]naphthyridine
C-70231

[90134-81-9]

$C_{20}H_{12}N_4$ M 308.342
Cryst. (EtOH). Mp >280°.

Wang, X.C. *et al*, *Tetrahedron Lett.*, 1987, **28**, 5833 (*synth, pmr*)

Cycloocta[2,1-*b*:3,4-*b'*]diquinoline
C-70232

[95172-28-4]

$C_{22}H_{14}N_2$ M 306.366
Cryst. (EtOAc). Mp 261-262°.

Wang, X.C. *et al*, *Tetrahedron Lett.*, 1987, **28**, 5833 (*synth, pmr, cryst struct*)

Cycloocta[*def*]fluorene
C-70233

[60047-82-7]

$C_{17}H_{12}$ M 216.282
Planar, paratropic, antiaromatic system. Red oil.

Willner, I. *et al*, *J. Org. Chem.*, 1980, **45**, 1628 (*synth, ir, uv, ms, pmr*)

Cycloocta[a]naphthalene, 9CI   C-70234
[231-95-8]

$C_{16}H_{12}$ M 204.271
Cryst. by subl. Mp 52-54°.

Zimmerman, H.E. et al, J. Am. Chem. Soc., 1970, **92**, 4366 (synth, ir, pmr)

Cycloocta[b]naphthalene, 9CI   C-70235
[262-83-9]

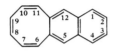

$C_{16}H_{12}$ M 204.271
Cryst. (EtOH). Mp 113-114°.

Krebs, A. et al, Justus Liebigs Ann. Chem., 1967, **707**, 66 (synth, uv, ms)

1H-Cyclooctapyrazole, 9CI   C-70236
[16767-46-7]

$C_9H_8N_2$ M 144.176
Yellow needles (CH$_2$Cl$_2$/hexane). Mp 103-105°.

Sanders, D.C. et al, J. Org. Chem., 1987, **52**, 5622 (synth, ir, pmr, uv)

1,3,5,7-Cyclooctatetraene-1-acetic acid, 9CI   C-70237
[56900-31-3]

$C_{10}H_{10}O_2$ M 162.188
Yellow oil.

Paquette, L.A. et al, J. Am. Chem. Soc., 1975, **97**, 3565 (synth, pmr)

Cyclooctatetraenecarboxaldehyde   C-70238
Formylcyclooctatetraene
[30844-12-3]

C_9H_8O M 132.162
Yellowish liq. Bp$_{0.5}$ 40-45°. Stored at −78°.

2,4-Dinitrophenylhydrazone: [110661-69-3]. Orange cryst. (CH$_2$Cl$_2$/hexane). Mp 218-219°.
4-Methylbenzenesulfonylhydrazone: [110661-70-6]. Light-yellow cryst. (CH$_2$Cl$_2$/hexane). Mp 114-116°.

Harman, C.A. et al, J. Org. Chem., 1973, **38**, 549 (synth, ir)
Paquette, L.A. et al, J. Am. Chem. Soc., 1975, **97**, 4649 (synth, pmr)
Sanders, D.C. et al, J. Org. Chem., 1987, **52**, 5622 (synth, ir, pmr)

Cyclooctatetraenecarboxylic acid   C-70239
Updated Entry replacing C-03514
[2411-95-2]

$C_9H_8O_2$ M 148.161
Yellow needles (Et$_2$O). Mp 72-73°.

Me ester: [37464-73-6].
   $C_{10}H_{10}O_2$ M 162.188
   Yellow oil. Bp$_5$ 75-76.2°.
Nitrile: [37164-17-3]. Cyanocyclooctatetraene.
   C_8H_7N M 117.150
   Yellow liq. Bp$_{1.5}$ 66°.

Cope, A.C. et al, J. Am. Chem. Soc., 1952, **74**, 173 (synth)
Antkowiak, T.A. et al, J. Am. Chem. Soc., 1972, **94**, 5366 (deriv, synth, ir)
Harman, C.A. et al, J. Org. Chem., 1973, **38**, 549 (deriv, synth)

1,3,5-Cyclooctatrien-7-yne, 9CI   C-70240
1,2-Dehydrocyclooctatetraene
[4514-69-6]

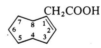

C_8H_6 M 102.135
Transient intermediate.

Krebs, A. et al, Justus Liebigs Ann. Chem., 1967, **707**, 66 (synth)
Lankey, A.S. et al, J. Org. Chem., 1971, **36**, 3339 (synth, props)

1-Cyclooctene-1-acetic acid, 9CI   C-70241
[56900-24-4]

$C_{10}H_{16}O_2$ M 168.235
Bp$_{0.1}$ 108-110°.

Paquette, L.A. et al, J. Am. Chem. Soc., 1975, **97**, 3565 (synth, pmr)

4-Cyclooctene-1-acetic acid, 9CI   C-70242
[50585-13-2]
$C_{10}H_{16}O_2$ M 168.235
(±)-**form**
Cryst. (hexane).

Moon, S. et al, J. Org. Chem., 1974, **39**, 995 (synth, ir, pmr)

1,1-Cyclopentanedimethanethiol, 9CI   C-70243

1,1-Bis(mercaptomethyl)cyclopentane
[56472-18-5]

$C_7H_{14}S_2$ M 162.308
Liq. Bp_{14} 123°, $Bp_{0.5}$ 76-79°.

Goor, G. *et al, Phosphorus Sulfur*, 1976, **1**, 81 (*synth*)
Houk, J. *et al, J. Am. Chem. Soc.*, 1987, **109**, 6825 (*synth, ir, pmr*)

4H-Cyclopenta[def]phenanthrene   C-70244

Updated Entry replacing C-20299
Phenanthrindene
[203-64-5]

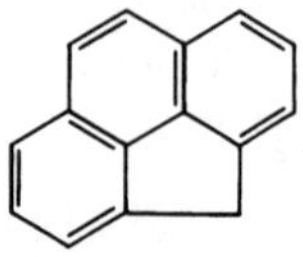

$C_{15}H_{10}$ M 190.244
Cryst. (EtOH). Mp 116°. Bp 353°. pK_a 24.5 (DMSO).
▷Exp. carcinogen. GY5280000.

Kruber, D., *Ber.*, 1934, **67**, 1000.
Medenwald, H., *Chem. Ber.*, 1953, **86**, 287.
Douris, J. *et al, Bull. Soc. Chim. Fr.*, 1971, 3365 (*nmr*)
Yoshida, M. *et al, Bull. Chem. Soc. Jpn.*, 1983, **56**, 2179 (*synth*)
Minabe, M. *et al, Bull. Chem. Soc. Jpn.*, 1988, **61**, 995 (*synth*)
Sax, N.I., *Dangerous Properties of Industrial Materials*, 5th Ed., Van Nostrand-Reinhold, 1979, 532.

4H-Cyclopenta[c]thiophen-5(6H)-one, 9CI   C-70245

[38447-47-1]

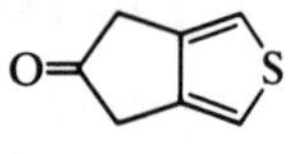

C_7H_6OS M 138.184
Yellowish cryst. Mp 106-107°.

Helmers, R., *J. Prakt. Chem.*, 1972, **314**, 334 (*synth, ir, pmr, ms*)

1H-Cyclopropabenzene   C-70246

Updated Entry replacing B-00316
Bicyclo[4.1.0]hepta-1,3,5-triene, 9CI. Benzocyclopropene
[4646-69-9]

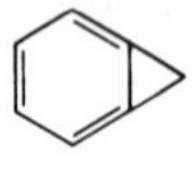

C_7H_6 M 90.124
Liq. Bp_{30} 35°.

Vogel, E. *et al, Tetrahedron Lett.*, 1965, 3625 (*uv, ir*)
Wentrup-Byrne, E. *et al, Org. Mass Spectrom.*, 1977, **12**, 636.
Billups, W.E., *Acc. Chem. Res.*, 1978, **11**, 245 (*rev*)
Neidlein, R. *et al, Angew. Chem., Int. Ed. Engl.*, 1988, **27**, 294 (*cryst struct*)
Billups, W.E. *et al, Tetrahedron*, 1988, **44**, 1305 (*rev*)
Neidlein, R. *et al, Chem. Ber.*, 1988, **121**, 1199 (*synth*)
Fieser, M. *et al, Reagents for Organic Synthesis*, Wiley, 1967-84, **6**, 33.

5H-Cyclopropa[f][2]benzothiophene, 9CI   C-70247

[113605-05-3]

C_9H_6S M 146.206
Pale-yellow solid. Unstable, solid dec. slowly.

Anthony, I.J. *et al, Tetrahedron Lett.*, 1987, **28**, 4217 (*synth, pmr, cmr, uv*)

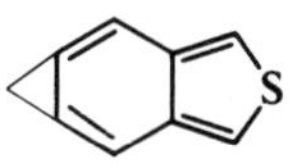

1,2-Cyclopropanedicarboxylic acid, 9CI   C-70248

Updated Entry replacing C-03708

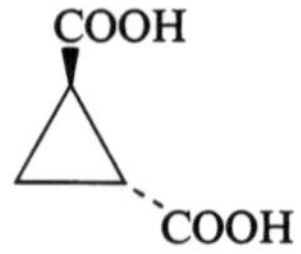

(1R,2R)-form
Absolute
configuration

$C_5H_6O_4$ M 130.100
(1R,2R)-form [34202-45-4]
    (−)-trans-*form*
    Mp 175°. $[\alpha]_D^{27}$ −84.4°.
(1S,2S)-form [14590-54-6]
    (+)-trans-*form*
    Mp 175°. $[\alpha]_D^{27}$ +84.87°.
(1RS,2RS)-form [58616-95-8]
    (±)-trans-*form*
    Needles. Mp 175°. pK_a 5.69. Heating with Ac_2O to 200° → *cis*-anhydride.
(1RS,2SR)-form [696-74-2]
    cis-*form*
    Prisms (H_2O or Et_2O). Sol. H_2O. Mp 139°. pK_a 5.40 (25°).
Di-Me ester: [826-34-6].
    $C_7H_{10}O_4$ M 158.154
    Bp 219-220°.
Anhydride: [5617-74-3]. *3-Oxabicyclo[3.1.0]hexane-2,4-dione, 9CI.*
    $C_5H_4O_3$ M 112.085
    Mp 58-60°.

Wassermann, A., *Helv. Chim. Acta*, 1930, **13**, 229 (*synth*)
Inouye, Y. *et al, Tetrahedron*, 1964, **20**, 1695 (*abs config*)
Inouye, Y. *et al, CA*, 1967, **66**, 64751r (*synth*)
Payne, G.B. *et al, J. Org. Chem.*, 1967, **32**, 3351 (*synth*)
Landor, S.R. *et al, J. Chem. Soc., Perkin Trans. 1*, 1983, 2921 (*ester, synth, ir, pmr*)
Schrumpf, G. *et al, Acta Crystallogr., Sect. C*, 1987, **43**, 1748, 1755 (*cryst struct*)

1,1-Cyclopropanedimethanethiol   C-70249

1,1-Bis(thiomethyl)cyclopropane. 1,1-Bis(mercaptomethyl)cyclopropane

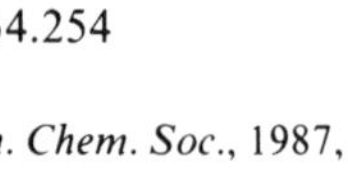

HSH₂C CH₂SH

$C_5H_{10}S_2$ M 134.254
Liq. Bp_1 45-50°.

Houk, J. *et al, J. Am. Chem. Soc.*, 1987, **109**, 6825 (*synth, ir, pmr*)

Cyclopropanehexacarboxylic acid   C-70250

C$_9$H$_6$O$_{12}$ M 306.139
Hexa-Et ester: [61936-90-1].
  C$_{21}$H$_{30}$O$_{12}$ M 474.461
  Cryst. Mp 115°. Bp$_{12}$ 197-202°. Difficult to crystallise, formerly descr. as a viscous oil.

Kötz, A. *et al, J. Prakt. Chem.*, 1903, **68**, 156 (*synth*)
Lablanche-Combier, A. *et al, Tetrahedron Lett.*, 1976, 3081 (*synth*)
Schrumpf, G. *et al, Acta Crystallogr., Sect. C*, 1987, **43**, 1758, 2015 (*cryst struct, ir, pmr, cmr, bibl*)

5*H*-Cycloprop[*f*]isobenzofuran, 9CI   C-70251
[33059-38-0]

C$_9$H$_6$O   M 130.146
Oil. Unstable, polym. rapidly when neat, dec. in soln.

Anthony, I.J. *et al, Tetrahedron Lett.*, 1987, **28**, 4217 (*synth, uv, cmr, pmr*)

1,8-Cyclotetradecadiene, 9CI   C-70252
Updated Entry replacing C-03743
[4308-14-9]

C$_{14}$H$_{24}$ M 192.344
Bp$_{0.001}$ 48°.

Hubert, A.I. *et al, Chem. Ind. (London)*, 1961, 249 (*synth*)
Bestmann, H.J. *et al, Angew. Chem., Int. Ed. Engl.*, 1972, **11**, 508 (*synth*)
Warwel, S. *et al, Synthesis*, 1987, 935 (*synth, ir, pmr, ms*)

Cyclotetradecane   C-70253
[295-17-0]

C$_{14}$H$_{28}$ M 196.375
Cryst. (EtOH aq. or by subl.). Mp 56.2°.

Ruzicka, L. *et al, Helv. Chim. Acta*, 1930, **13**, 1152 (*synth*)
Borgen, T. *et al, J. Chem. Soc., Dalton Trans.*, 1970, 1340 (*conformn*)
Drotloff, H. *et al, J. Am. Chem. Soc.*, 1987, **109**, 7797 (*synth, cmr, cryst struct*)

Cyclotriveratrylene   C-70254
Updated Entry replacing C-03764
10,15-Dihydro-2,3,7,8,12,13-hexamethoxy-5H-tribenzo[a,d,g]*cyclononene*
[1180-60-5]

C$_{27}$H$_{30}$O$_6$ M 450.530
Cryst. (C$_6$H$_6$). Mp 234°.

Umezawa, B. *et al, J. Chem. Soc. (C)*, 1970, 465 (*synth*)
Arçoleo, A. *et al, Chem. Ind. (London)*, 1976, 853 (*synth*)
Birnbaum, G.I. *et al, Can. J. Chem.*, 1985, **63**, 3258 (*cryst struct*)
Collet, A., *Tetrahedron*, 1987, **43**, 5725 (*rev*)

2-Cycloundecen-1-one   C-70255

C$_{11}$H$_{18}$O   M 166.263
Bp$_{0.4}$ 52-57°. Bp refers to mixt. of stereoisomers.
(*E*)-*form* [24593-68-8]
  Semicarbazone: Cryst. (EtOH aq.). Mp 182°.
(*Z*)-*form* [24593-67-7]
  Liq., obt. by glc.

Regitz, M. *et al, Chem. Ber.*, 1969, **102**, 3877 (*synth, ir, pmr, uv*)
Bissinger, H.-J. *et al, Justus Liebigs Ann. Chem.*, 1988, 221 (*synth, pmr, cmr*)

1-Cycloundecen-3-yne   C-70256

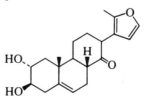   (*E*)-*form*

C$_{11}$H$_{16}$ M 148.247
(*E*)-*form* [111689-53-3]
  Oil with characteristic odour. Isom. at r.t. to *Z*-isomer with t$_{1/2}$ = 2.2h.

Bissinger, H.-J. *et al, Justus Liebigs Ann. Chem.*, 1988, 221 (*synth, pmr, cmr*)

Cynajapogenin *A*   C-70257
[107484-53-7]

C$_{20}$H$_{26}$O$_4$ M 330.423
Constit. of *Cynanchum japonicum* and *C. atratum*.
Amorph. powder. Mp 71-74°. [α]$_D$ −45.3° (c, 0.8 in MeOH).

3-O-α-D-Oleandropyranosyl-(1→4)-β-D-digitoxopyran-osyl-(1→4)-β-D-cymaropyranoside: [118002-94-1].
Atratoside D.
$C_{40}H_{60}O_{13}$ M 748.906
Constit. of *C. atratum*. Amorph. powder. Mp 92-94°.
$[\alpha]_D$ −52.3° (c, 0.8 in MeOH).

Runsink, J. *et al, Tetrahedron Lett.*, 1974, 55 (*isol, struct*)
Zhang, Z.-X. *et al, Phytochemistry*, 1988, **27**, 2935 (*isol, glycoside*)

Cyparenoic acid   C-70258

COOH

$C_{15}H_{22}O_2$ M 234.338
Constit. of *Sandwithia guyanensis*. Cryst. (Me₂CO/hex-ane). Mp 162-164°. $[\alpha]_D^{25}$ −18.8° (c, 0.08 in CHCl₃).

Jacobs, H. *et al, J. Nat. Prod.*, 1987, **50**, 835.

Cyperenol   C-70259

Updated Entry replacing C-03793
[16981-80-9]
As Cyperene, C-03792 with

$$R = CH_2OH$$

$C_{15}H_{24}O$ M 220.354
Constit. of *Cyperus scariosus*. Cryst. Mp 94°. $[\alpha]_D^{30}$ −12.1° (c, 4.3 in CHCl₃).
Ac: [115334-13-9]. **Cyperenyl acetate**.
    $C_{17}H_{26}O_2$ M 262.391
    Constit. of *Cirsium dipsacolepis*. Oil. $[\alpha]_D$ −5.1° (c, 0.8 in CHCl₃).
14-Aldehyde: [115334-16-2]. **Cyperenal**.
    $C_{15}H_{22}O$ M 218.338
    Constit. of *C. dipsacolepis*. Oil. $[\alpha]_D$ +9.6° (c, 0.4 in CHCl₃).

Nerali, S.B. *et al, Tetrahedron Lett.*, 1967, 2447 (*isol, struct*)
Takano, S. *et al, Phytochemistry*, 1988, **27**, 1197 (*deriv*)

Cystofuranoquinol   C-70260

$C_{27}H_{36}O_3$ M 408.580
5-Hydroxy: [115788-00-6]. **5-Hydroxycystofuranoquinol**.
    $C_{27}H_{36}O_4$ M 424.579
    Constit. of the brown alga *Cystoseira spinosa* var. *squarrosa*. Oil. $[\alpha]_D^{20}$ +1.2° (c, 2.8 in EtOH).
5-Oxo: [115787-99-0]. **5-Oxocystofuranoquinol**.
    $C_{27}H_{34}O_4$ M 422.563
    Constit. of *C. spinosa* var. *squarrosa*. Oil.
5-Oxo, 6Z-isomer: [115787-98-9]. **5-Oxoisocystofuranoquinol**.
    $C_{27}H_{34}O_4$ M 422.563
    Constit. of *C. spinosa* var. *squarrosa*. Oil.

Amico, V. *et al, Phytochemistry*, 1988, **27**, 1327.

D

24-Dammarene-3,7,18,20,27-pentol   D-70001

$C_{30}H_{52}O_5$ M 492.738

(3β,7β,20S,24Z)-form

Needles (CHCl$_3$). Mp 119-121°. [α]$_D^{17}$ +7.6° (c, 0.87 in MeOH).

20-O-β-D-Glucopyranoside: [108906-62-3]. **Actinostemmoside C.**
$C_{36}H_{62}O_{10}$ M 654.880
Constit. of *Actinostemma lobatum.* Needles (MeOH aq.). Mp 194-197°. [α]$_D^{17}$ +3.3° (c, 1.0 in MeOH).

Iwamoto, M. *et al, Chem. Pharm. Bull.,* 1987, **35**, 553.

24-Dammarene-3,6,20,27-tetrol   D-70002

$C_{30}H_{52}O_4$ M 476.738

(3β,6α,20S,24Z)-form

Needles (Et$_2$O). Mp 148-150°. [α]$_D^{20}$ +52.3° (c, 0.13 in MeOH).

20-O-β-D-Glucopyranoside: [108906-64-5]. **Actinostemmoside A.**
$C_{36}H_{62}O_9$ M 638.880
Constit. of *Actinostemma lobatum.* Needles (EtOH aq.). Mp 125-130°. [α]$_D^{16}$ +32.3° (c, 0.3 in MeOH).

(3β,6α,20R,24Z)-form

Needles (MeOH aq.). Mp 183-185°. [α]$_D^{17}$ +49.0° (c, 0.1 in MeOH).

20-O-α-L-Rhamnopyranosyl-(1→2)-β-D-glucopyranoside: [108906-61-2]. **Actinostemmoside D.**
$C_{42}H_{72}O_{13}$ M 785.023
Constit. of *A. lobatum.* Needles (EtOH aq.). Mp 168-171°. [α]$_D^{17}$ −2.2° (c, 1.0 in MeOH).

Iwamoto, M. *et al, Chem. Pharm. Bull.,* 1987, **35**, 553.

24-Dammarene-3,7,20,27-tetrol   D-70003

$C_{30}H_{52}O_4$ M 476.738

(3β,7β,20S,24Z)-form

Needles (Et$_2$O). Mp 180-183°. [α]$_D^{21}$ +24.5° (c, 0.1 in MeOH).

20-O-β-D-Glucopyranoside: [108906-63-4]. **Actinostemmoside B.**
$C_{36}H_{62}O_9$ M 638.880
Constit. of *Actinostemma lobatum.* Needles (EtOH aq.). Mp 142-145°. [α]$_D^{19}$ +15.4° (c, 0.5 in MeOH).

Iwamoto, M. *et al, Chem. Pharm. Bull.,* 1987, **35**, 553.

Danshenspiroketallactone   D-70004

Updated Entry replacing D-50009

[100414-80-0]

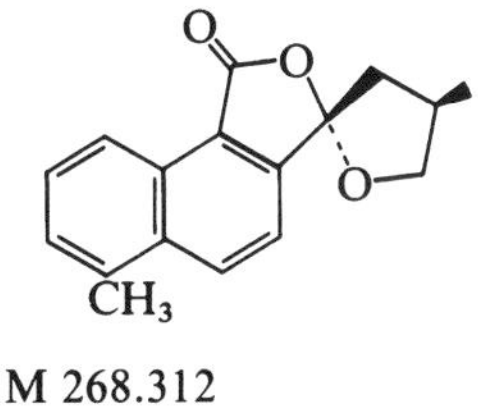

$C_{17}H_{16}O_3$ M 268.312
Constit. of *Salvia miltiorrhiza.*

13-Epimer: **Epidanshenspiroketallactone.**
$C_{17}H_{16}O_3$ M 268.312

Constit. of *S. miltiorrhiza.*

Kong, D. *et al, Acta Pharm. Sinica,* 1985, **20**, 747 (*cryst struct*)
Luo, H.W. *et al, Phytochemistry,* 1988, **27**, 290 (*isol*)

Daphnodorin *D*   D-70005

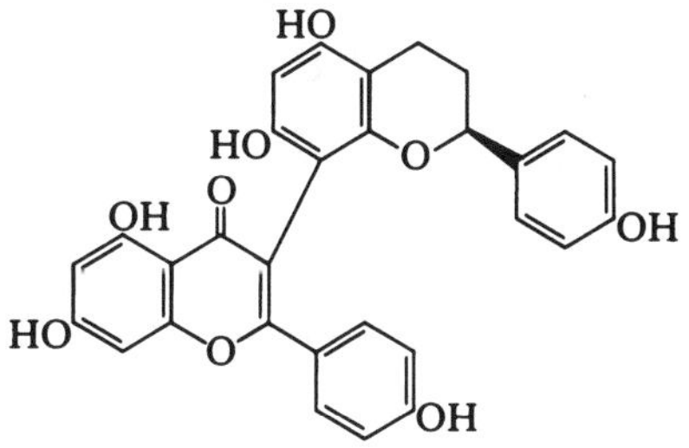

$C_{30}H_{22}O_9$ M 526.498

(S)-form [112757-06-9]

Constit. of *Daphne odora.* Cryst. Mp 212-214° dec.

Baba, K. *et al, Yakugaku Zasshi,* 1987, **107**, 863.
Baba, K. *et al, CA,* 1988, **108**, 183599t.

Daphnoretin   D-70006

Updated Entry replacing D-30004

7-Hydroxy-6-methoxy-3-[(2-oxo-2H-1-benzopyran-7-yl)oxy]-2H-1-benzopyran-2-one, 9CI

[2034-69-7]

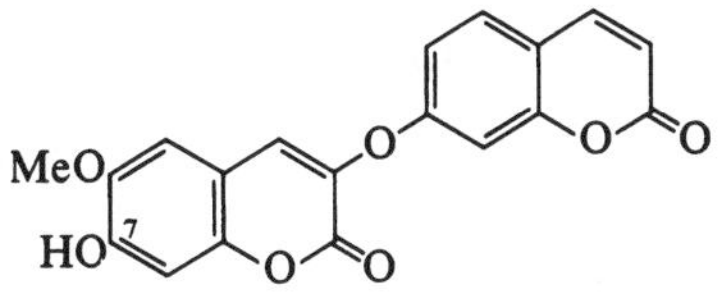

$C_{19}H_{12}O_7$ M 352.300

Constit. of *Daphne mezereum,* and *Wikstroemia viridiflora.* A common bis-coumarin found in many plants. Yellow needles (THF/MeOH) or long rods. Mp 244-247°.

Ac: **7-O-Acetyldaphnoretin.**
$C_{21}H_{14}O_8$ M 394.337
Constit. of *Edgeworthia gardneri.* Cryst. (MeOH). Mp 230-232°.

7-Glucoside: [55806-40-1]. **Daphnorin.**
$C_{25}H_{22}O_{12}$ M 514.442
Found in plants of the order Thymelaceae. Fine needles (MeOH aq.). Mp 202-204° dec. [α]$_D^{20}$ −78° (c, 0.6 in H$_2$O).

O-De-Me: [53947-90-3]. **Edgeworthin.** *6,7-Dihydroxy-3-[(2-oxo-2H-1-benzopyran-7-yl)oxy]-2H-1-benzopyran-2-one, 9CI. 6,7-Dihydroxy-3,7'-bicoumarin.* **Demethyldaphnoretin.**
$C_{18}H_{10}O_7$ M 338.273
Constit. of *E. gardneri* and *D. gnidioides.* Cryst. (THF). Mp 280-282° dec.

O-De-Me, di-Ac: Mp 189°.

Tschesche, R. *et al, Justus Liebigs Ann. Chem.,* 1963, **662**, 113; *Naturwissenschaften,* 1963, **50**, 521 (*struct*)
Kirkiacharian, B. *et al, Bull. Soc. Chim. Fr.,* 1966, 770 (*synth*)
Majumder, P.L. *et al, Phytochemistry,* 1974, **13**, 1929 (*Edgeworthin*)

Tandon, S. *et al*, *Phytochemistry*, 1977, **16**, 1991 (*isol*)
Cordell, G.A., *J. Nat. Prod.*, 1984, **47**, 84 (*pmr, cmr*)
Chakrabarti, R. *et al*, *Phytochemistry*, 1986, **25**, 557 (*isol*)
Ulubelen, A. *et al*, *J. Nat. Prod.*, 1986, **49**, 692 (*Edgeworthin*)

Daturilinol   D-70007

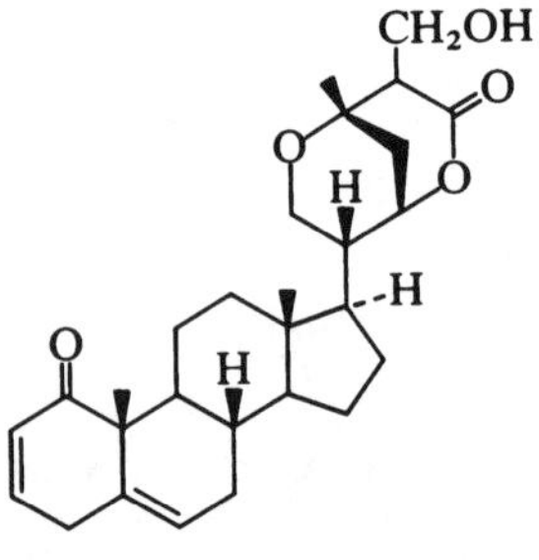

$C_{28}H_{38}O_5$ M 454.605
Constit. of *Datura metel*. Cryst. (EtOAc/MeOH). Mp
145-146°. $[\alpha]_D^{20}$ −80° (c, 0.09 in $CHCl_3$).

Mahmood, T. *et al*, *Heterocycles*, 1988, **27**, 101.

Daucene   D-70008

Updated Entry replacing D-00049
4,8-Daucadiene
[16661-00-0]

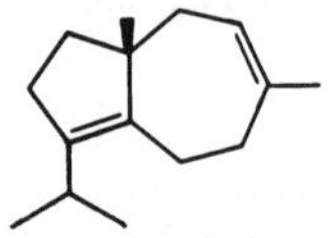

$C_{15}H_{24}$ M 204.355
Constit. of *Daucus carota*. Oil. $[\alpha]_D$ +39° (c, 0.3 in
$CHCl_3$).

Bisepoxide: Cryst. Mp 89°. $[\alpha]_D$ +25° (c, 0.4 in $CHCl_3$).

de Broissia, H. *et al*, *Bull. Soc. Chim. Fr.*, 1972, 4314.
Audenaert, F. *et al*, *Tetrahedron*, 1987, **43**, 5593 (*synth*)

8-Daucene-2,4,6,10-tetrol   D-70009

2,4,6,10-Tetrahydroxy-8-daucene

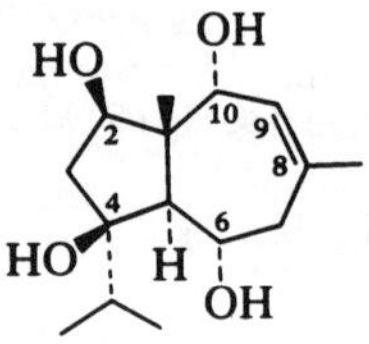

$C_{15}H_{26}O_4$ M 270.368
(2β,4β,6α,10α)-form
Cryst. by subl. Mp 243-245°.

6,10-Diangeloyl: [95342-43-1]. **Tingitanol**.
  $C_{25}H_{38}O_6$ M 434.572
  Constit. of roots of *Ferula tingitana*. Amorph.
2,6-Diangeloyl: **Desoxodehydrolaserpitin**.
  $C_{25}H_{38}O_6$ M 434.572
  Constit. of *Laserpitium latifolium*. Cryst. (pet. ether).
  Mp 59-60°.
8,9-Epoxide, 2,6-Diangeloyl: [16836-36-5].
  Isolaserpitin.
  $C_{25}H_{38}O_7$ M 450.571
  Constit. of *L. latifolium*. Cryst. Mp 157-158°. $[\alpha]_D^{20}$
  −27.5° (MeOH).

Holub, M. *et al*, *Monatsh. Chem.*, 1967, **98**, 1138 (*deriv*)
Miski, M. *et al*, *Tetrahedron*, 1984, **40**, 5197 (*cryst struct*)

Debneyol   D-70010

Updated Entry replacing D-40008
[99694-82-3]

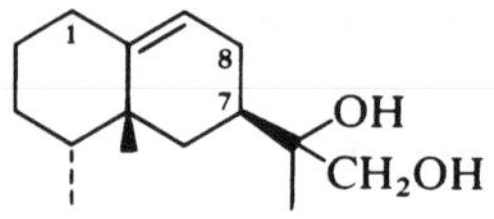

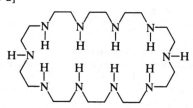

$C_{15}H_{26}O_2$ M 238.369
Phytoalexin from *Nictiana debneyi* and *N. tabacum*.

7-Epimer: **7-Epidebneyol**.
  $C_{15}H_{26}O_2$ M 238.369
  Constit. of *N. tabacum*.
1β-Hydroxy: **1-Hydroxydebneyol**.
  $C_{15}H_{26}O_3$ M 254.369
  Constit. of *N. tabacum*.
8β-Hydroxy: **8-Hydroxydebneyol**.
  $C_{15}H_{26}O_3$ M 254.369
  Constit. of *N. tabacum*.

Burden, R.S. *et al*, *Phytochemistry*, 1985, **24**, 2191 (*isol*)
Watson, D.G. *et al*, *Phytochemistry*, 1985, **24**, 2195 (*isol*)
Brooks, C.J.W. *et al*, *Phytochemistry*, 1987, **26**, 2243 (*biosynth*)
Whitehead, I.M. *et al*, *Phytochemistry*, 1988, **27**, 1365 (*isol,
    derivs*)

1,4,7,10,13,16,19,22,25,28-Decaazacyclo-   D-70011
triacontane, 9CI

[862-28-2]

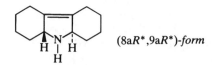

$C_{20}H_{50}N_{10}$ M 430.682
B,8HCl: Mp >260°.
Decakis(4-methylbenzenesulfonyl): Cryst. (CH_2Cl_2/hex-
    ane or CH_2Cl_2/EtOH). Mp >260°.

Hosseini, M.W. *et al*, *J. Am. Chem. Soc.*, 1987, **109**, 7047
    (*synth, pmr, cmr, ms*)

1,2,3,4,5,6,7,8,8a,9a-Decahydrocarbazole   D-70012

(8aR*,9aR*)-form

$C_{12}H_{19}N$ M 177.289
Previous syntheses were mixts. of stereoisomers and/or
    impure.

(8aR*,9aR*)-form
(+)-trans-*form*
Felted needles (Et_2O/pentane). Mp 101°. $[\alpha]_D^{22}$ +61.5°
    (c, 1.0 in CH_2Cl_2). Abs. config. not detd.
Hydrogen tartrate: Powder. Mp 217-218° dec.
(8aRS,9aRS)-form
(±)-trans-*form*
Needles (Et_2O/pentane). Mp 99.5-100°.
N-Ac:
  $C_{14}H_{21}NO$ M 219.326
  Orange granular cryst. (Et_2O/pet. ether). Mp 97-98°.

N-*Benzoyl:*
$C_{19}H_{23}NO$ M 281.397
Felted needles (EtOH aq.). Mp 166.8-167.2°.
N-*Me:*
$C_{13}H_{21}N$ M 191.316
Oil.

(8aRS,9aSR)-form
cis-*form*
Not isol. in pure form.
N-*Ac:* Almost colourless prisms (Et₂O/pentane). Mp 101-101.5°.
N-*Me:* Oil.

Anderson-McKay, J.E. *et al, Aust. J. Chem.*, 1988, **41**, 1013 (*synth, cmr, resoln, bibl*)

Decahydroquinazoline, 9CI   D-70013

Updated Entry replacing D-00126
Perhydroquinazoline
[61557-94-6]

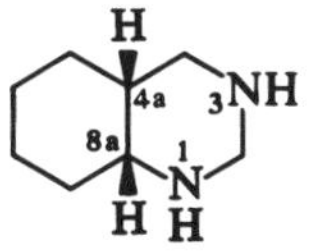

(4a*R*,8a*R*)-*form*
Absolute
configuration

$C_8H_{16}N_2$ M 140.228
(4aR,8aR)-form [26693-40-3]
(+)-cis-*form*
Mp 48-49°.
Dipicrate: Mp 183-184° dec.
(4aS,8aS)-form [26693-41-4]
(−)-cis-*form*
Mp 47-48°.
(4sS,8aR)-form [26685-89-2]
(+)-trans-*form*
Oil. $[\alpha]_D$ +3.0°.
Dipicrate: Mp 188-189.5° dec.
(4aR,8aS)-form [24716-98-1]
(−)-trans-*form*
Oil. $[\alpha]_D$ −3.0° (c, 3.3 in EtOH).

Armarego, W.L.F. *et al, J. Chem. Soc. (C)*, 1970, 1597 (*synth, abs config, ord*)
Booth, H. *et al, Tetrahedron*, 1988, **44**, 1465 (*synth, pmr, cmr*)

Decahydro-5,5,8a-trimethyl-2-naphtha-lenol   D-70014

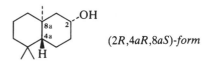

(2*R*,4a*R*,8a*S*)-*form*

$C_{13}H_{24}O$ M 196.332
(2R,4aR,8aS)-form [113667-24-6]
Cryst. (hexane). Mp 89-90°. $[\alpha]_D^{20}$ +19.6° (c, 1.01 in CHCl₃).
Ac: [113667-19-9].
$C_{15}H_{26}O_2$ M 238.369
Woody-like odorant with ionone-like undertone. $[\alpha]_D^{20}$ −16.69° (c, 1.4 in CHCl₃).
2-Ketone: [113667-25-7]. *Octahydro-5,5,8a-trimethyl-2(1H)-naphthalenone.*
$C_{13}H_{22}O$ M 194.316
Weaker woody patchouli-like odorant.
(2S,4aS,8aR)-form [58239-51-3]
Cryst. (hexane). Mp 89-90°. $[\alpha]_D^{20}$ −19.6° (c, 1.2 in CHCl₃).

Ac: [98676-96-1]. Woody odorant with less rich tone than the (+)-form.
2-Ketone: [54808-90-1]. Strong woody amber-like odourant with damp-earth overtones. Mp 87-89°. $[\alpha]_D^{20}$ −84° (c, 1.3 in CHCl₃).

Gautier, A. *et al, Helv. Chim. Acta*, 1987, **70**, 2039 (*synth, abs config, pmr, ms, props*)

1,3,5,7,9-Decapentaene   D-70015

[2423-91-8]

$$H_2C=CHCH=CHCH=CHCH=CHCH=CH_2$$

$C_{10}H_{12}$ M 132.205
Cream-coloured needles (EtOH). Mp 146-147°.

Sondheimer, F. *et al, J. Am. Chem. Soc.*, 1961, **83**, 1675 (*synth, ir, uv*)
Block, E. *et al, J. Am. Chem. Soc.*, 1986, **108**, 4568 (*synth, uv*)

3-Decenal   D-70016

Updated Entry replacing D-00163
[58474-80-9]

$$H_3C(CH_2)_5CH=CHCH_2CHO$$

$C_{10}H_{18}O$ M 154.252
(E)-form [68676-85-7]
Liq. Bp₀.₃ 70°.
2,4-Dinitrophenylhydrazone: Mp 126-127°.
(Z)-form [69891-94-7]
Liq.
Di-Me acetal: 1,1-Dimethoxy-3-decene.
$C_{12}H_{24}O_2$ M 200.320
Bp₁₁ 109-111°.

Stowell, J.C. *et al, Synthesis*, 1979, 132 (*synth*)
Achmatowicz, O. *et al, Synthesis*, 1987, 413 (*synth, ir, pmr*)

5-Decene, 9CI   D-70017

Updated Entry replacing D-00169
[19689-19-1]

$$H_3C(CH_2)_3CH=CH(CH_2)_3CH_3$$

$C_{10}H_{20}$ M 140.268
Bp₇₅₀ 170°.
(E)-form [7433-56-9]
Bp 172-173.5°. n_D^{25} 1.4213.
▷HE2080000.
(Z)-form [7433-78-5]
Bp 169-170°, Bp₁₁ 59°. n_D^{25} 1.4230.

Davis, A. *et al, J. Chem. Soc., Perkin Trans. 1*, 1972, 286 (*synth*)
Bestmann, H.J. *et al, Synthesis*, 1974, 798 (*synth*)
House, H.O. *et al, J. Org. Chem.*, 1974, **39**, 747 (*synth*)
McMurry, J.E. *et al, J. Org. Chem.*, 1975, **40**, 2555 (*synth*)
Miyaura, N. *et al, Synthesis*, 1975, 669 (*synth*)
Sonnet, P.E. *et al, J. Org. Chem.*, 1978, **43**, 1841 (*synth*)
Murahashi, S. *et al, J. Org. Chem.*, 1979, **44**, 2408 (*synth, nmr*)

8-Decene-4,6-diynoic acid, 9CI   D-70018

Dihydromatricaria acid

$$H_3CCH=CHC\equiv CC\equiv CCH_2CH_2COOH$$

$C_{10}H_{10}O_2$ M 162.188

Defensive antifeedant secretion isol. from soldier beetles.

(Z)-*form* [19949-46-3]

Me ester: [4161-50-6].
$C_{11}H_{12}O_2$ M 176.215
$Bp_{0.009}$ 65°. n_D^{20} 1.5290.

Derzhinskii, A.R. *et al, Izv. Akad. Nauk SSSR, Ser. Khim.,*
1965, **7**, 1237; *CA*, **63**, 14691g (*synth, deriv*)
Meinwald, J. *et al, Science*, 1968, **160**, 890 (*isol*)
Eisner, T. *et al, J. Chem. Ecol.*, 1981, **7**, 1149 (*isol*)

5-Decen-1-ol   D-70019

Updated Entry replacing D-00190

$$H_3C(CH_2)_3CH{=}CH(CH_2)_3CH_2OH$$

$C_{10}H_{20}O$ M 156.267

(E)-*form* [56578-18-8]
Pheromone of the moth *Anarsia lineatella*. Unpleasant
smelling liq. Bp_{14} 114-115°, $Bp_{0.15}$ 83-85°.

(Z)-*form* [51652-47-2]
Unpleasant-smelling liq. Bp_{14} 114-115°, $Bp_{0.2}$ 100-105°.
Ac: [67446-07-5].
$C_{12}H_{22}O_2$ M 198.305
Pheromone of female moth *Agrotis fucosa*.
3-Methylbutanoyl: [37616-04-9].
$C_{15}H_{28}O_2$ M 240.385
Sex pheromone from the female pine emperor moth
(*Nudaurelia cytherea cytherea*).

Henderson, H.E. *et al, J. Chem. Soc., Chem. Commun.*, 1972,
686.
Roelofs, W. *et al, Environ. Entomol.*, 1975, **4**, 580 (*isol*)
Bestmann, H.J. *et al, Angew. Chem., Int. Ed. Engl.*, 1978, **17**,
768 (*isol, synth*)
Bestmann, H.J. *et al, Justus Liebigs Ann. Chem.*, 1981, 1705
(*synth*)
Buss, A.D. *et al, J. Chem. Soc., Perkin Trans. 1*, 1987, 2569
(*synth, ir, pmr, ms*)

4-Decen-1-yne   D-70020

$$H_3C(CH_2)_4CH{=}CHCH_2C{\equiv}CH$$

$C_{10}H_{16}$ M 136.236

(Z)-*form*
$Bp_{0.15}$ 100°.

Corey, E.J. *et al, Tetrahedron Lett.*, 1987, **28**, 3547 (*synth*)

2-Dehydroarcangelisinol   D-70021

Updated Entry replacing D-50037
[96552-88-4]

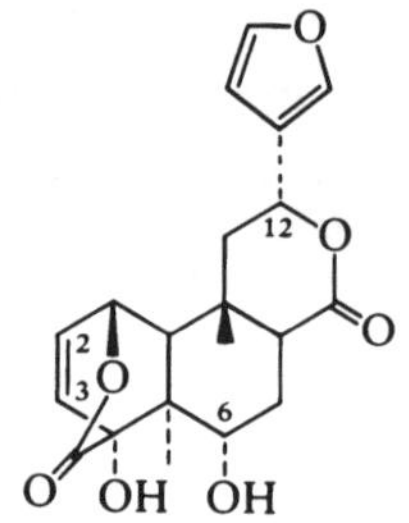

$C_{20}H_{22}O_7$ M 374.390
Constit. of *Arcangelisia flava*. Prisms (MeOH). Mp 208-
212° dec. $[\alpha]_D^{20}$ +98.8° (c, 0.5 in Py).

2β,3β-Epoxide: [96552-87-3]. **6-Hydroxyarcangelisin**.
$C_{20}H_{22}O_8$ M 390.389

Constit. of *A. flava*. Prisms (MeOH). Mp 274° dec.
$[\alpha]_D^{20}$ +55.27° (c, 0.54 in Py).
2β,3β-Epoxide, 6,12-diepimer:
$C_{20}H_{22}O_8$ M 390.389
Constit. of *Tinospora cordifolia*. Cryst. (EtOH). Mp
258-260°. $[\alpha]_D^{22}$ +10.0° (c, 0.5 in Me_2CO).

Kunii, T. *et al, Chem. Pharm. Bull.*, 1985, **33**, 479 (*isol, struct*)
Hanuman, J.B. *et al, J. Nat. Prod.*, 1988, **51**, 197 (*deriv*)

2′-Deoxy-5-azacytidine   D-70022

5-Aza-2′-deoxycytidine

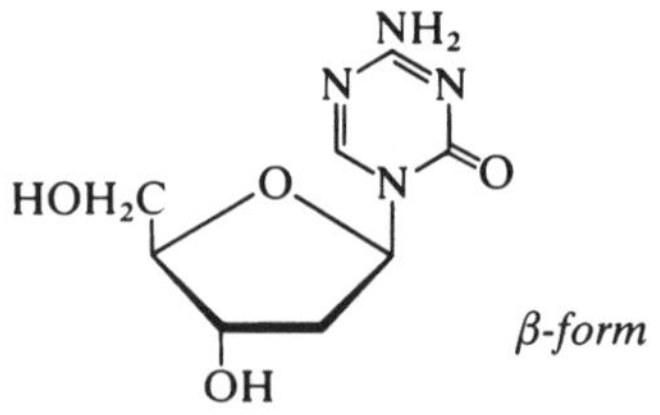

β-form

$C_8H_{12}N_4O_4$ M 228.207

α-*form* [22432-95-7]
Cryst. (MeOH). Mp 181°. $[\alpha]_D^{29}$ −40.8° (c, 1.0 in H_2O).

β-*form* [2353-33-5]
Inhibits DNA methylation. Cryst. (MeOH). Mp 191°
dec. $[\alpha]_D^{29}$ +63.8° (c, 1.00 in H_2O).
▷XZ3012000.

Winkley, M.W. *et al, J. Org. Chem.*, 1970, **35**, 491 (*synth, uv, ir,
pmr*)
Ben-Hattar, J. *et al, J. Org. Chem.*, 1986, **51**, 3211 (*synth, pmr,
bibl*)

2′-Deoxyinosine, 9CI   D-70023

*9-(2-Deoxy-β-D-erythro-pentofuranosyl)-1,9-dihydro-
6H-purin-6-one. Hypoxanthine 2′-deoxyriboside*
[890-38-0]

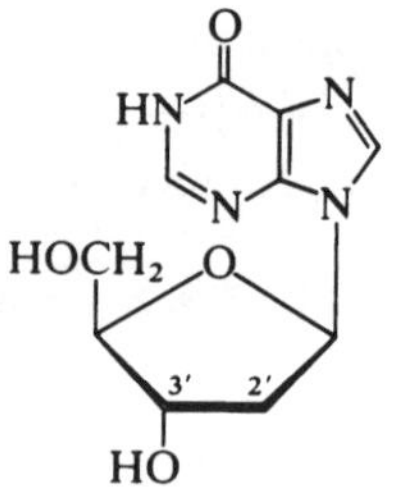

$C_{10}H_{12}N_4O_4$ M 252.229
Cryst. (H_2O). Mp 216-218°. $[\alpha]_D^{30}$ −21° (c, 1 in H_2O).
Oxime: [51385-49-0]. Mp 139-140°. $[\alpha]_D^{22}$ −21.8° (c, 1
in MeOH). Softens at 110-115° and 130-135°.

Brown, D.M. *et al, J. Chem. Soc.*, 1950, 1990 (*struct*)
Robins, M.J. *et al, Can. J. Chem.*, 1973, **51**, 3161 (*synth*)
Yamazaki, A. *et al, Chem. Pharm. Bull.*, 1973, **21**, 1143 (*synth*)
Mengel, R. *et al, Justus Liebigs Ann. Chem.*, 1977, 1585 (*synth*)

3′-Deoxyinosine, 9CI   D-70024

*9-(3-Deoxy-β-D-erythro-pentofuranosyl)-1,9-dihydro-
6H-purin-6-one. Hypoxanthine 3′-deoxyriboside*
[13146-72-0]

$C_{10}H_{12}N_4O_4$ M 252.229
Needles (EtOH aq.). Mp 202-203°. $[\alpha]_D^{30}$ −87.9° (c, 1 in
Py).

Haga, K. *et al, Bull. Chem. Soc. Jpn.*, 1970, **43**, 3922 (*synth, pmr*)
Yamazaki, A. *et al, Chem. Pharm. Bull.*, 1973, **21**, 1143 (*synth*)
Montgomery, J.A. *et al, J. Med. Chem.*, 1975, **18**, 564 (*synth*)
Mengel, R. *et al, Justus Liebigs Ann. Chem.*, 1977, 1585 (*synth*)

Desacylligulatin *C*　　　　　　　　　　D-70025

C_{15}H_{22}O_4　　M 266.336
Constit. of *Rudbeckia grandiflora*. Gum.
Di-Ac: [65179-91-1]. **Ligulatin C**.
　C_{19}H_{26}O_6　　M 350.411
　Constit. of *Parthenium argentatum* and *P.
　tomentosum*. Cryst. (Me_2CO/hexane). Mp 134-138°.
4-Ketone, 14-acetoxy, 15-Ac: [65129-90-0]. **Ligulatin A**.
　C_{19}H_{24}O_7　　M 364.394
　Constit. of *P.* spp.

Rodriguez, E., *Biochem. Syst. Ecol.*, 1977, **5**, 207 (*isol*)
Isman, M.B. *et al, Phytochemistry*, 1983, **22**, 2709 (*isol*)
Maldonado, E. *et al, Phytochemistry*, 1985, **24**, 2981 (*isol*)
Vasquez, M. *et al, Phytochemistry*, 1988, **27**, 2195 (*deriv*)

10-Desmethyl-1-methyl-1,3,5(10),11(13)-　D-70026
eudesmatetraen-12,8-olide

C_{15}H_{16}O_2　　M 228.290
8β-form
　Constit. of *Ferreyranthus fruticosus*. Cryst. Mp 122°.

Jakupovic, J. *et al, Phytochemistry*, 1988, **27**, 1113.

Desmethylzeylasterone　　　　　　　　D-70027
Updated Entry replacing D-20036
*2,3-Dihydroxy-6-oxo-D:A-friedo-24-noroleana-
1,3,5(10),7-tetraen-23,29-dioic acid, 9CI.
Demethylzeylasterone*
[87064-40-2]

C_{29}H_{36}O_7　　M 496.599
Constit. of outer stem bark of *Kokoona zeylanica*. Pale-
　yellow cryst. (MeOH aq.). Mp 190-192°. [α]_D^{27}
　−36.48° (c, 2.46 in CHCl_3).
23-Aldehyde: [107316-88-1]. **Demethylzeylasteral**.
　C_{29}H_{36}O_6　　M 480.600

Constit. of *K. zeylanica*. Yellow solid. Mp 158-160°.
　[α]_D^{27} −67.9° (c, 1.06 in CHCl_3).
23-Aldehyde, 29-Me ester: [87064-16-2]. **Zeylasteral**.
　C_{30}H_{38}O_6　　M 494.627
　Constit. of stem bark of *K. zeylanica*. Pale-yellow
　cryst. (MeOH). Mp 278-280°. [α]_D^{27} −136.0° (c, 0.86
　in CHCl_3).

Kamal, G.M. *et al, Tetrahedron Lett.*, 1983, **24**, 2025 (*isol,
　struct*)
Gamlath, C.B. *et al, J. Chem. Soc., Perkin Trans. 1*, 1987, 2849
　(*isol, struct*)

1,5-Diacetylanthracene, 8CI　　　　　D-70028
1,1′-(1,5-Anthracenediyl)bisethanone, 9CI
[10210-34-1]

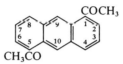

C_{18}H_{14}O_2　　M 262.307
Deep-yellow needles (AcOH). Mp 223.5-224°.

Bassilios, H.F. *et al, Recl. Trav. Chim. Pays-Bas*, 1962, **81**, 679;
　1963, **82**, 298 (*synth*)
Gore, P.H. *et al, J. Chem. Soc.* (*C*), 1966, 1729 (*synth, uv, ir*)

1,6-Diacetylanthracene, 8CI　　　　　D-70029
1,1′-(1,6-Anthracenediyl)bisethanone, 9CI
C_{18}H_{14}O_2　　M 262.307
Cryst. (EtOH). Mp 211-212°.

Gore, P.H. *et al, J. Chem. Soc.* (*C*), 1966, 1729 (*synth, ir*)

1,8-Diacetylanthracene, 8CI　　　　　D-70030
1,1′-(1,8-Anthracenediyl)bisethanone, 9CI
[10208-16-9]
C_{18}H_{14}O_2　　M 262.307
Cryst. (AcOH). Mp 183.5-184°.

Bassilios, H.F. *et al, Recl. Trav. Chim. Pays-Bas*, 1963, **82**, 298
　(*synth*)
Gore, P.H. *et al, J. Chem. Soc.* (*C*), 1966, 1729 (*synth, uv, ir*)

9,10-Diacetylanthracene, 8CI　　　　　D-70031
1,1′-(9,10-Anthracenediyl)bisethanone, 9CI
[67263-73-4]
C_{18}H_{14}O_2　　M 262.307
Yellow cryst. (EtOH). Mp 248.5-249.5°.

Duerr, B.F. *et al, J. Org. Chem.*, 1988, **53**, 2120 (*synth, pmr,
　cmr*)

1,1-Diacetylcyclopropane　　　　　　D-70032
1,1′-Cyclopropylidenebisethanone, 9CI
[695-70-5]

H_3COC　COCH_3

C_7H_{10}O_2　　M 126.155
Bp_8 74-74.5°.
2,4-Dinitrophenylhydrazone: Mp 258-260°.

Ichikawa, K. *et al, J. Org. Chem.*, 1966, **31**, 447 (*synth, ir, uv,
　nmr, deriv*)

Zefirov, N.S. *et al*, *Zh. Org. Khim.*, 1983, **19**, 541 (*synth*)
Zefirov, N.S. *et al*, *Tetrahedron*, 1986, **42**, 709 (*props*)

2,12-Di-9-acridinyl-7-phenyldibenz[*c,h*]-acridine, 9CI   D-70033

[111773-09-2]

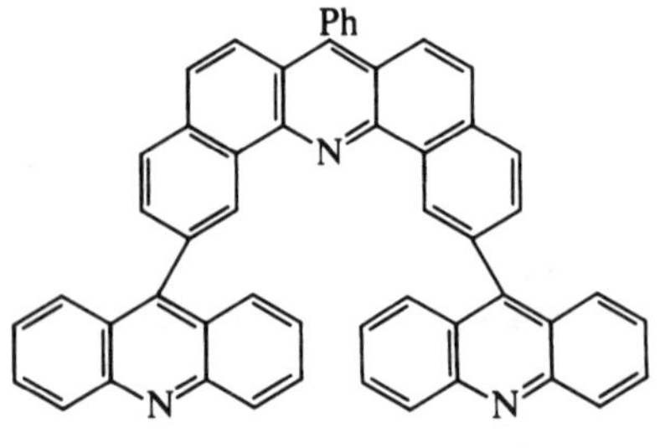

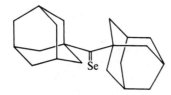

C₅₃H₃₁N₃   M 709.848

$C_{53}H_{31}N_3$ M 709.848

"Molecular tweezers"; nonmacrocyclic complexing receptor molecule.

Zimmerman, S.C. *et al*, *J. Am. Chem. Soc.*, 1987, **109**, 7894 (*synth, props*)

Di-1-adamantyl selenoketone   D-70034

$C_{21}H_{30}Se$ M 361.428
Blue cryst. (hexane). Mp 161-162°.

Ishii, A. *et al*, *Bull. Chem. Soc. Jpn.*, 1988, **61**, 861 (*synth, pmr, cmr, Se nmr, ir, uv*)

2,5-Diamino-1,4-benzenedithiol   D-70035

2,5-Dimercapto-p-*phenylenediamine*
[15657-79-1]

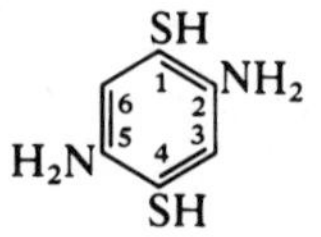

$C_6H_8N_2S_2$ M 172.263
Dyestuff intermed.

B,2HCl: [75464-52-7]. Cryst. Dec. at 200-210° without melting.

Osman, A.M. *et al*, *UAR J. Chem.*, 1971, **14**, 475; *CA*, **80**, 16426f (*synth*)
Wolfe, J.F. *et al*, *Polym. Prep.* (*ACS Div. Polym. Chem.*), 1978, **19**, 1; 1981, **22**, 60 (*synth, deriv*)
Wolfe, J.F. *et al*, *Macromolecules*, 1981, **14**, 915 (*synth, ir*)

3,6-Diamino-1,2-benzenedithiol, 9CI   D-70036

$C_6H_8N_2S_2$ M 172.263
B,2HI: [107474-49-7]. Cryst. (CH₂Cl₂/MeOH/Et₂O). Mp >150° dec.
Di-Me thioether: [107474-50-0]. *2,3-Bis(methylthio)-1,4-benzenediamine.*
$C_8H_{12}N_2S_2$ M 200.316
Cryst. (as dihydriodide). Mp >200° dec.

Lakshmikantham, M.V. *et al*, *J. Org. Chem.*, 1987, **52**, 1874 (*synth, ir*)

4,6-Diamino-1,3-benzenedithiol, 9CI   D-70037

4,6-Dimercapto-m-*phenylenediamine. 4,6-Diaminothiorecorcinol*
[17366-20-0]
Dyestuffs intermed. Prisms (H₂O). Mp 99-102°. Unstable in air, turning blue-violet.
Di-S-Me:
$C_8H_{12}N_2S_2$ M 200.316
Cryst. (EtOH aq.). Mp 105°.
Di-S-Et:
$C_{10}H_{16}N_2S_2$ M 228.370
Cryst. (pet. ether). Mp 54°.
Di-S-Ph:
$C_{18}H_{16}N_2S_2$ M 324.458
Large prisms (EtOH). Mp 94-95°.

Finzi, C. *et al*, *Gazz. Chim. Ital.*, 1959, **89**, 2543 (*synth*)
Grandolini, G., *Ann. Chim.* (*Rome*), 1961, **51**, 195 (*synth, deriv*)
Neunhoeffer, O. *et al*, *Z. Chem.*, 1961, **1**, 278 (*synth*)

1,5-Diamino-1,5-dihydrobenzo[1,2-*d*:4,5-*d'*]bistriazole   D-70038

[91477-70-2]

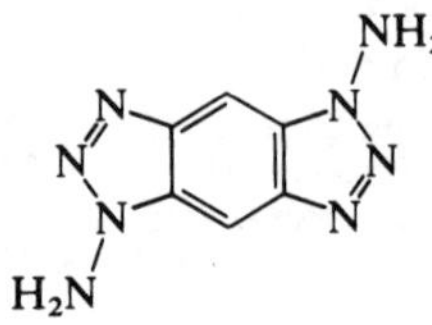

$C_6H_6N_8$ M 190.167
1,4-Benzadiyne equivalent. Mp 292° dec.

Hart, H. *et al*, *J. Org. Chem.*, 1986, **51**, 979 (*synth, pmr, cmr, ms, ir, use*)
Ok, D. *et al*, *J. Org. Chem.*, 1987, **52**, 3835 (*props*)

2,5-Diamino-3-hydroxypentanoic acid   D-70039

3-Hydroxyornithine, 9CI

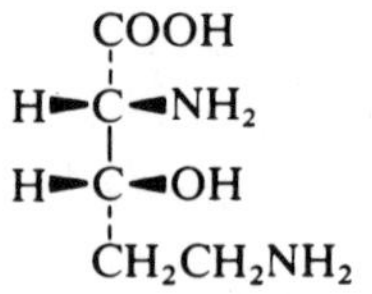

$C_5H_{12}N_2O_3$ M 148.161
(2R,3R)-*form* [86831-55-2]
D-*threo-form*
Constit. of the peptidoglycan of *Corynebacterium* sp. Co 112.
(2S,3R)-*form*
L-*erythro-form*
Degradn. prod. of Capreomycin, C-00173 .
B,HCl: Mp 123° dec. $[\alpha]_D^{25}$ +2.9° (c, 3 in H₂O), $[\alpha]_D$ +9.6° (c, 2 in 6*M* HCl).
(2S,3S)-*form*
L-*threo-form*
B,HCl: Mp 232° dec. $[\alpha]_D^{25}$ +0.4° (c, 3 in H₂O), $[\alpha]_D$ +18.0° (c, 2 in 6*M* HCl).
(2RS,3RS)-*form* [64817-94-3]
(±)-threo-*form*
B,HCl: [66954-82-3]. Cryst. (EtOH aq.). Mp 189-190°.
N⁵-*Phthaloyl:* Cryst. (H₂O). Mp 214.5° dec.
N⁵-*Benzyloxycarbonyl:* Cryst. (H₂O). Mp 212-213°.
(2RS,3SR)-*form* [64818-17-3]
(±)-erythro-*form*
B,HCl: [66954-83-4]. Mp 225° dec.

N^5-*Phthaloyl:* Cryst. (H$_2$O). Mp 245° dec.

Bey, P. *et al, J. Med. Chem.*, 1977, **21**, 50 (*synth*)
Wakamiya, T. *et al, Bull. Chem. Soc. Jpn.*, 1978, **51**, 850 (*synth*)
Schleifer, K.H. *et al, Arch. Microbiol.*, 1983, **134**, 243 (*isol*)
Wityak, J. *et al, J. Org. Chem.*, 1987, **52**, 2179 (*synth*)

4,5-Diaminoisoquinoline   D-70040

4,5-Isoquinolinediamine, 9CI
[110191-89-4]
C$_9$H$_9$N$_3$ M 159.190
Pale-yellow needles by subl. Mp 151.5-153°.

Woodgate, P.D. *et al, Heterocycles*, 1987, **26**, 1029 (*synth, pmr, ms*)

5,8-Diaminoisoquinoline   D-70041

5,8-Isoquinolinediamine, 9CI
[1127-49-7]
C$_9$H$_9$N$_3$ M 159.190
Light-brown needles. Mp 138-140° (132-133°).

Potts, K.T. *et al, J. Org. Chem.*, 1986, **51**, 2011.

1,8-Diaminonaphthalene   D-70042

Updated Entry replacing D-10056
1,8-Naphthalenediamine, 9CI
[479-27-6]
C$_{10}$H$_{10}$N$_2$ M 158.202
Antioxidant. Used for determination of nitrite, nitrate and selenium. Sol. hot H$_2$O, spar. sol. cold. Mp 66.5°. Bp$_{12}$ 205° subl. pK_a 4.61.
N,N'-Di-Me: [20734-56-9]. *1,8-Bis(methylamino)-naphthalene.*
C$_{12}$H$_{14}$N$_2$ M 186.256
Mp 103-104°. pK_a 5.61.
N,N,N'-Tri-Me: [20734-57-0].
C$_{13}$H$_{16}$N$_2$ M 200.283
Mp 29-30°. pK_a 6.43.
N,N,N',N'-Tetra-Me: [20734-58-1]. *1,8-Bis(dimethylamino)naphthalene. Proton sponge.*
C$_{14}$H$_{18}$N$_2$ M 214.310
Weakly nucleophilic, very hindered strong base. Mp 47-48°. pK_a 12.34.

Hodgson, H.H. *et al, J. Chem. Soc.*, 1945, 202 (*synth*)
Smith, J.L. *et al, Org. Mass Spectrom.*, 1971, **5**, 473 (*ms*)
Einspahr, H. *et al, Acta Crystallogr., Sect. B*, 1973, **29**, 1611 (*cryst struct*)
Seita, J. *et al, Org. Magn. Reson.*, 1978, **11**, 239 (*cmr*)
Alder, R.W. *et al, J. Chem. Soc., Perkin Trans. 1*, 1981, 2840 (*deriv*)
Benoit, R. *et al, Can. J. Chem.*, 1987, **65**, 996 (*props, deriv*)
Staab, H.A. *et al, Angew. Chem., Int. Ed. Engl.*, 1988, **27**, 865 (*rev, deriv*)
Fieser, M. *et al, Reagents for Organic Synthesis*, Wiley, 1967-84, **6**, 50 (*deriv*)

2,6-Diamino-4(1*H*)-pyrimidinone, 9CI   D-70043

Updated Entry replacing D-01065
2,4-Diamino-6-hydroxypyrimidine
[56-06-4]

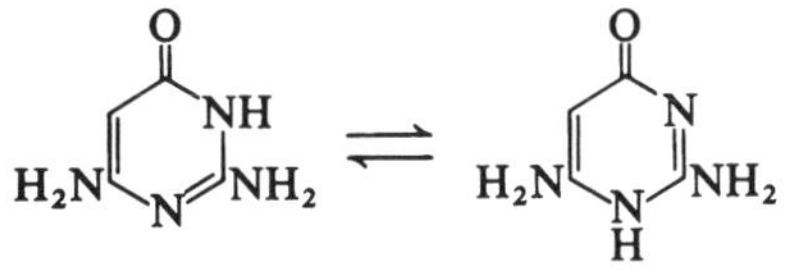

C$_4$H$_6$N$_4$O   M 126.118

Yellow needles (AcOH). Mp 260-270°.
1-Me: [51093-34-6].
C$_5$H$_8$N$_4$O   M 140.144
Cryst. Mp 284°.
2-N-Me: [89181-81-7]. *6-Amino-2-(methylamino)-4(3H)-pyrimidinone.*
C$_5$H$_8$N$_4$O   M 140.144
Cryst. + 1H$_2$O (MeOH aq.). Mp 232-234°.

Traube, W., *Ber.*, 1900, **33**, 1371 (*synth*)
Org. Synth., Coll. Vol., **4**, 245 (*synth*)
Munesada, K. *et al, J. Org. Chem.*, 1987, **52**, 5655 (*synth, deriv, cmr, ir, ms*)

4,5-Diaminoquinoline   D-70044

4,5-Quinolinediamine, 9CI
[45990-28-1]
C$_9$H$_9$N$_3$ M 159.190
Pale-yellow cryst. by subl. Mp 134-139°. Darkens on standing in air.
B,AcOH: [40107-00-4]. Mp 172-173° (sinters >100°).

Ellis, J. *et al, Aust. J. Chem.*, 1973, **26**, 907 (*synth*)
Woodgate, P.D. *et al, Heterocycles*, 1987, **26**, 1029 (*synth, pmr*)

3,6-Diamino-1,2,4,5-tetrazine   D-70045

1,2,4,5-Tetrazine-3,6-diamine, 9CI
[19617-90-4]

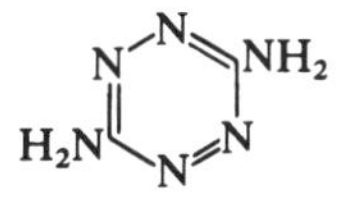

C$_2$H$_4$N$_6$ M 112.094
Orange-red microcryst. (dioxan or H$_2$O) or red amorph. solid. V. spar. sol. H$_2$O. Mp >300° (subl. above 200°).
Di-N-Ac:
C$_6$H$_8$N$_6$O$_2$ M 196.168
Orange microcryst. (EtOH). Mp 156-158°.

Lin, C.-H. *et al, J. Am. Chem. Soc.*, 1954, **76**, 427 (*synth*)
Katunina, A.B. *et al, Khim. Geterotsikl. Soedin.*, 1975, 847 (*synth*)
Krieger, C. *et al, Acta Crystallogr., Sect. C*, 1987, **43**, 1320 (*cryst struct*)

6,10-Diamino-2,3,5-trihydroxydecanoic acid   D-70046

Galantinamic acid

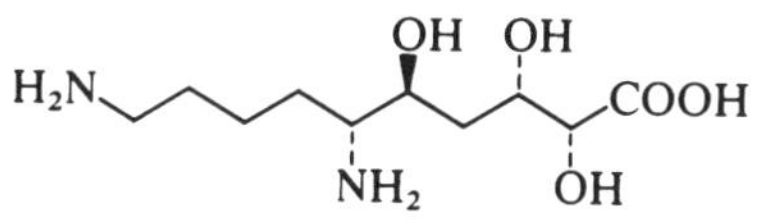

C$_{10}$H$_{22}$N$_2$O$_5$ M 250.294
Prob. abs. config. (illus.) based on unpubl. work. Component of Galantin I, G-10001 .
B,HCl: Mp 207.5-209° dec. $[\alpha]_D^{22}$ −0.4° (c, 0.5 in 1M HCl).

Wakamiya, T. *et al, Bull. Chem. Soc. Jpn.*, 1988, **61**, 1422 (*isol, struct, cmr*)

3,5-Diamino-2,4,6-trinitrobenzoic acid   D-70047
[97217-74-8]

C$_7$H$_5$N$_5$O$_8$ M 287.145
Cryst.
Nitrile: [105363-51-7].
  C$_7$H$_4$N$_6$O$_6$ M 268.145
  Orange-brown cryst. Mp 220-221° dec.

Ammon, H.L. *et al, Acta Crystallogr., Sect. C*, 1985, **41**, 921 (*cryst struct*)
Chaykovsky, M. *et al, Synth. Commun.*, 1986, **16**, 205 (*nitrile*)

3,5-Dianilino-1,2,4-thiadiazole   D-70048
N,N′-Diphenyl-1,2,4-thiadiazole-3,5-diamine, 9CI.
Dost's base
[22713-97-9]

C$_{14}$H$_{12}$N$_4$S   M 268.336
Isom. prod. of Hector's base, H-60012 . Cryst. (CHCl$_3$).
Mp 220-222°.

Dost, K., *Ber.*, 1906, **39**, 863.
Kurzer, F. *et al, J. Chem. Soc.*, 1962, 4191 (*uv*)
Akiba, K. *et al, Bull. Chem. Soc. Jpn.*, 1976, **49**, 550 (*synth*)
Butler, A. *et al, Acta Chem. Scand., Ser. B*, 1986, **40**, 779 (*cmr, struct*)

Dianthramide A   D-70049
2-[(2-Hydroxybenzoyl)amino]-4-methoxybenzoic acid,
9CI. N-Salicyl-4-methoxyanthranilic acid
[93289-90-8]

C$_{15}$H$_{13}$NO$_5$ M 287.271
Isol. from *Dianthus caryophyllus* (carnation) infected with *Phytophthora parasitica*. Phytoalexin.
O-De-Me, Me ester: [93289-91-9]. **Dianthramide B**.
Methyl 2-[(2-hydroxybenzoyl)amino]-4-hydroxyben-
zoate, 9CI. Methyl N-salicyl-4-hydroxyanthranilate.
C$_{15}$H$_{13}$NO$_5$ M 287.271
Isol. from *D. caryophyllus* and *P. parasitica*.
Phytoalexin.

Ponchet, M. *et al, Phytochemistry*, 1984, **23**, 1901 (*isol, uv, ir, ms, pmr*)

Diaporthin   D-70050
Updated Entry replacing D-60052
8-Hydroxy-3-(2-hydroxypropyl)-6-methoxy-1H-2-ben-
zopyran-1-one, 9CI. 8-Hydroxy-3-(2-hydroxypropyl)-6-
methoxyisocoumarin

C$_{13}$H$_{14}$O$_5$ M 250.251
(S)-form [10532-39-5]
  Isol. from cultures of *Endothia parasitica*. Phytotoxin.
  Needles or plates. Mp 90-92° (83-85°). [α]$_D$ +54° (c, 10.87 in CHCl$_3$).
  O-De-Me: **De-O-methyldiaporthin**.
  C$_{12}$H$_{12}$O$_5$ M 236.224
  Phytotoxin from *Drechslera siccans*. Solid. [α]$_D$ +22° (c, 0.09 in MeOH).

Hardegger, E. *et al, Helv. Chim. Acta*, 1966, **49**, 1283 (*isol, props*)
Hallock, Y.F. *et al, Phytochemistry*, 1988, **27**, 3123 (*isol*)

1,4-Diazabicyclo[2.2.2]octane   D-70051
Updated Entry replacing D-01125
Triethylenediamine. Dabco
[280-57-9]

C$_6$H$_{12}$N$_2$ M 112.174
Reagent for the cleavage of β-keto esters and geminal diesters and dehydrohalogenation reactions. Also a cyclisation catalyst and catalyst for urethane foam prepn. Nonnucleophilic base. Cryst. Mp 158°. Bp 174°. pK$_{a1}$ 3.0, pK$_{a2}$ 8.7. Extremely hygroscopic. Sublimes readily at r.t.
▷Irritant. HM0354200.

B,HCl: Long deliquescent needles. Mp 320° dec.
Br$_2$ complex: [5770-70-7]. Reagent for the oxidn. of aromatic and secondary alcohols. Yellow cryst. Mp 155-160° dec. Stable to light, air and H$_2$O.

Mann, F.G. *et al, J. Chem. Soc.*, 1957, 1881 (*synth*)
Zaugg, H.E. *et al, Tetrahedron*, 1966, **22**, 1257 (*use*)
Miles, D.H. *et al, J. Org. Chem.*, 1976, **41**, 208 (*use*)
Nimmo, J.K. *et al, Acta Crystallogr., Sect. B*, 1976, **32**, 597 (*cryst struct*)
Blair, L.K. *et al, J. Org. Chem.*, 1977, **42**, 1816 (*deriv*)
Cocivera, M. *et al, J. Org. Chem.*, 1980, **45**, 415 (*use*)
Benoit, R.L. *et al, Can. J. Chem.*, 1987, **65**, 996 (*props*)
Guzonas, D.A. *et al, Can. J. Chem.*, 1988, **66**, 1249 (*ir, raman*)
Fieser, M. *et al, Reagents for Organic Synthesis*, Wiley, 1967-84, **7**, 86; **8**, 141.
Sax, N.I., *Dangerous Properties of Industrial Materials*, 5th Ed., Van Nostrand-Reinhold, 1979, 1051.

10*b*,10*c*-Diazadicyclopenta[*ef*,*kl*]heptalene   D-70052
Diazaazupyrene
[42851-22-9]

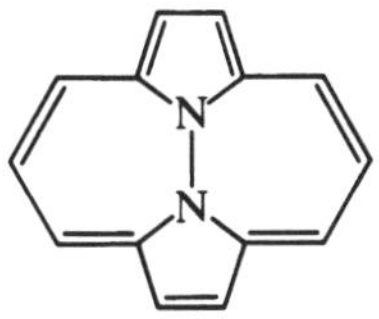

$C_{14}H_{10}N_2$ M 206.246
Cryst. Mp 295° dec. Readily oxidn. in soln. No colour mentioned but prob. red.

Flitsch, W. *et al*, *Chem. Ber.*, 1987, **120**, 1925 (*synth, ir, uv, pmr*)

3,6-Diazafluorenone   D-70053
5H-*Cyclopenta[2,1-c:3,4-c']dipyridin-5-one*, 9CI
[109528-43-0]

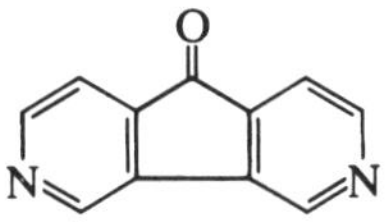

$C_{11}H_6N_2O$ M 182.181
Yellow cryst. (Me$_2$CO). Mp 167.5°.

Li, Y-Z. *et al*, *J. Org. Chem.*, 1987, **52**, 3975 (*synth, pmr*)

3,4-Diazatricyclo[3.1.0.0^{2,6}]hex-3-ene, 9CI   D-70054
3,4-Diazabenzvalene
[114068-57-4]

$C_4H_4N_2$ M 80.089
$t_{1/2}$ 22 min. at −60°.

Kaisaki, D.A. *et al*, *Tetrahedron Lett.*, 1987, **28**, 5263 (*synth, pmr, uv*)

1,2-Diazetidin-3-one, 9CI   D-70055
Updated Entry replacing D-40070
3-Oxo-1,2-diazetidine

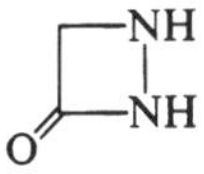

$C_2H_4N_2O$ M 72.066
4-Methylbenzenesulfonate salt: [79289-49-9]. Solid. Mp 147-149° dec.
1,2-Di-Ph: [14790-51-3]. *1,2-Diphenyl-1,2-diazetidin-3-one.*
  $C_{14}H_{12}N_2O$ M 224.262
  Mp 117-118°.
1-Ac, 2-Ph: [108511-44-0].
  $C_{10}H_{10}N_2O_2$ M 190.201
  Bright-orange oil.

Taylor, E.C. *et al*, *J. Am. Chem. Soc.*, 1981, **103**, 7743 (*synth, ir, pmr*)
Taylor, E.C. *et al*, *J. Org. Chem.*, 1983, **48**, 4567 (*use*)
Taylor, E.C. *et al*, *J. Org. Chem.*, 1987, **52**, 4107 (*deriv, synth, pmr, ir, cmr, ms*)

2,2-Diazido-1,3-indanedione, 8CI   D-70056
2,2-Diazido-1H-indene-1,3(2H)-dione, 9CI
[16291-08-0]

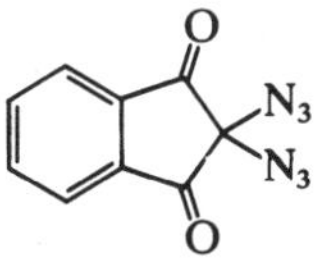

$C_9H_4N_6O_2$ M 228.170
Mp 163° dec. Dec. in light.
▷Explodes on heating

Gudrinece, E. *et al*, *Dokl. Akad. Nauk SSSR, Ser. Sci. Khim.*, 1966, **171**, 869; *CA*, **67**, 53900h.

6-Diazo-2,4-cyclohexadien-1-one, 9CI   D-70057
Updated Entry replacing D-50118
2-Hydroxybenzenediazonium hydroxide inner salt, 9CI.
1,2-Benzoquinone diazide. o-*Diazophenol*. *1,2,3-Benzoxadiazole*
[4024-72-0]

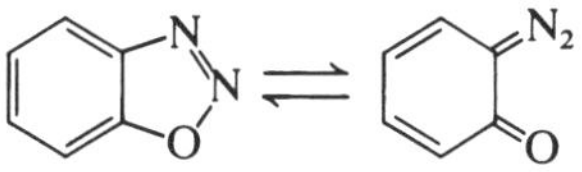

$C_6H_4N_2O$ M 120.110
Mesomeric struct., with diazide-form the main contributor. Tautomeric in soln. and in the gas phase. Yellow plates or prisms (Et$_2$O at −50°). Mp 66° dec. Dec. in a few hours in light at r.t.

Puza, M. *et al*, *Synthesis*, 1971, 481 (*synth, pmr*)
Schulz, R. *et al*, *Angew. Chem., Int. Ed. Engl.*, 1984, **23**, 509 (*uv, pe*)
Lowe-Ma, C.K. *et al*, *J. Chem. Res. (S)*, 1988, 214 (*derivs*)

2-Diazo-1,3-diphenyl-1,3-propanedione,   D-70058
9CI, 8CI
Diazodibenzoylmethane
[2085-31-6]

$$(PhCO)_2C{=}N_2$$

$C_{15}H_{10}N_2O_2$ M 250.256
Light-yellow platelets (EtOH). Mp 107° (115°).

Regitz, M. *et al*, *Chem. Ber.*, 1966, **99**, 3128 (*synth, ir, uv*)
Lauer, W. *et al*, *Chem. Ber.*, 1988, **121**, 465 (*synth, cmr*)

3-Diazo-2,4(5*H*)-furandione   D-70059
3-Diazotetronic acid
[98026-98-3]

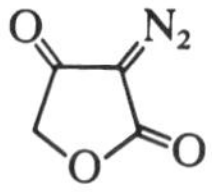

$C_4H_2N_2O_3$ M 126.071
Precursor of 1,2-Propadien-1-one, P-50249 . Solid. Mp 91-91.5°.

Chapman, O.L. *et al*, *J. Am. Chem. Soc.*, 1987, **109**, 6867 (*synth, use*)

2-Diazo-2*H*-imidazole   D-70060
[50846-98-5]

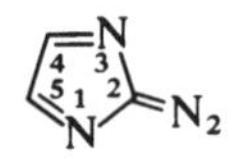

C$_3$H$_2$N$_3$ M 80.069
Used in prep. of *N*-containing heterocycles. Yellow solid.
▷Shock sensitive; used freshly prepd. in soln.

Magee, W.L. *et al, J. Org. Chem.*, 1987, **52**, 5538 (*synth, ir, use*)

4-Diazo-4*H*-imidazole   D-70061
[89108-47-4]

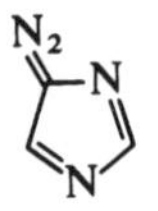

C$_3$H$_2$N$_4$ M 94.076
Used in prepn. of *N*-containing heterocycles. Cryst. solid which dec. to a black oil.
▷Potentially hazardous

Magee, W.L. *et al, J. Org. Chem.*, 1987, **52**, 5538 (*synth, ir, pmr, use*)

1-(Diazomethyl)pyrene, 9CI   D-70062
1-Pyrenyldiazomethane
[78377-23-8]

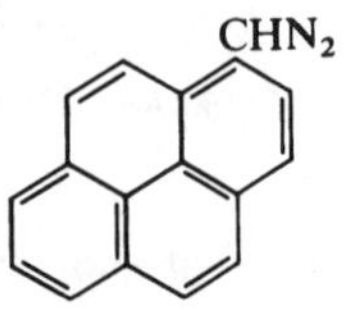

C$_{17}$H$_{10}$N$_2$ M 242.279
Reagent for prepn. of 1-pyrenylmethyl esters, photolabile protecing group. Dark-red cryst. Mp 112°. Stable in dark at 0°.

Iwamura, M. *et al, Tetrahedron Lett.*, 1987, **28**, 679 (*synth, use*)

6-Diazo-5-oxo-1,3-cyclohexadiene-1-car-   D-70063
boxylic acid, 9CI
2,3-Dihydro-2-diazo-3-oxobenzoic acid. TAN 665A. Antibiotic TAN 665A
[105918-54-5]

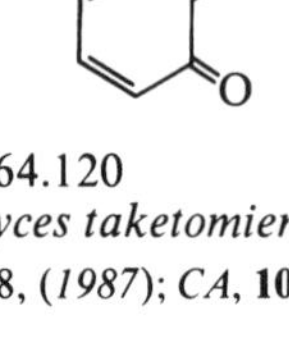

C$_7$H$_4$N$_2$O$_3$ M 164.120
Isol. from *Streptomyces taketomiensis*. Antifungal agent.
Japan. Pat., 87 132 848, (*1987*); *CA*, **107**, 196525

2-Diazo-2,2,6,6-tetramethyl-3,5-heptane-   D-70064
dione
[60681-09-6]

$$[(H_3C)_3CCO]_2C{=}N_2$$

C$_{11}$H$_{18}$N$_2$O$_2$ M 210.275

Yellow oil. Bp$_{0.1}$ 35.5°. n$_D^{20}$ 1.4755.

Korobitsyna, I.K. *et al, J. Org. Chem. USSR (Engl. Transl.)*, 1976, **12**, 1245 (*synth*)
Lauer, W. *et al, Chem. Ber.*, 1988, **121**, 465 (*synth, ir, ms, cmr*)

9-Diazo-9*H*-thioxanthene, 9CI   D-70065
[23619-77-4]

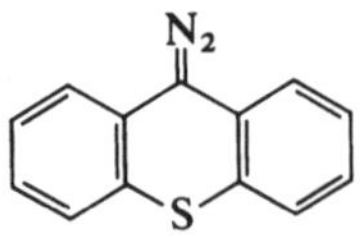

C$_{13}$H$_8$N$_2$S   M 224.280
Green cryst. (Et$_2$O). Mp 105° to give red liq.
10,10-Dioxide: [3166-17-4].
  C$_{13}$H$_8$N$_2$O$_2$S   M 256.278
  Red-brown cryst. (butanol). Mp 158-160° dec.

Schonberg, A. *et al, J. Am. Chem. Soc.*, 1959, **81**, 2259 (*synth*)
Regitz, M., *Chem. Ber.*, 1964, **97**, 2742 (*deriv, synth*)
Patrick, T.B. *et al, J. Org. Chem.*, 1978, **43**, 3303 (*props*)

Dibenz[*e,k*]acephenanthrylene, 9CI   D-70066
Updated Entry replacing D-30054
Naphtho[2,3-b]fluoranthene
[206-06-4]

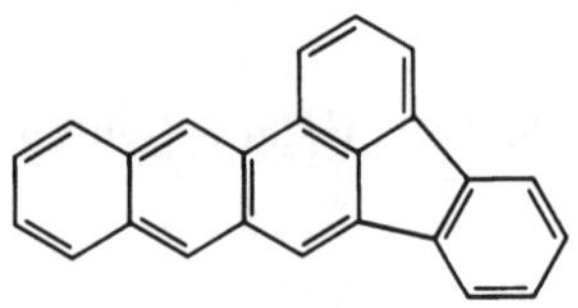

C$_{24}$H$_{14}$ M 302.375
Yellow needles with greenish-yellow fluor. (C$_6$H$_6$). Mp 229-230°. Blue fluor. in soln.

Campbell, N. *et al, J. Chem. Soc.*, 1949, 1555; 1950, 3466 (*synth*)
Clar, E. *et al, Tetrahedron*, 1969, **25**, 5639 (*synth, uv*)
Cho, B.P. *et al, J. Org. Chem.*, 1987, **52**, 5668 (*synth, pmr, uv*)

2*H*-Dibenz[*e,g*]isoindole   D-70067

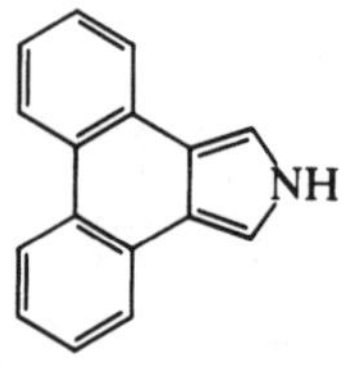

C$_{16}$H$_{11}$N   M 217.270
N-tert-Butyl: [110028-19-8].
  C$_{20}$H$_{19}$N   M 273.377
  Cryst. Mp 144°.
N-Benzyl: [110028-20-1].
  C$_{23}$H$_{17}$N   M 307.394
  Mp 177-179°.

Kreher, R.P. *et al, Chem. Ber.*, 1988, **121**, 81 (*synth, ir, uv, pmr, cmr, ms*)

Dibenzo-54-crown-18   D-70068

[108332-47-4]

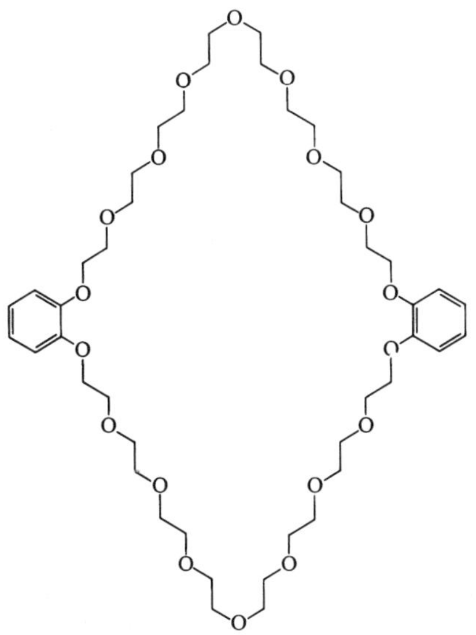

$C_{44}H_{72}O_{18}$ M 889.042
Cryst. Mp 78-80°.

Talma, A.G. *et al*, *Synthesis*, 1986, 680 (*synth*)
van Eerden, J. *et al*, *Acta Crystallogr., Sect. C*, 1987, **43**, 799 (*cryst struct, pmr*)

5*H*,7*H*-Dibenzo[*b,g*][1,5]dithiocin, 9CI   D-70069

Updated Entry replacing D-50134

[60418-10-2]

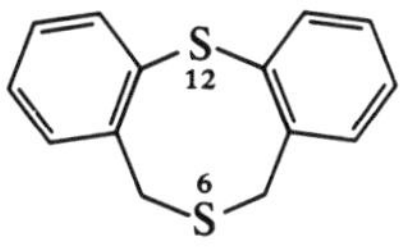

$C_{14}H_{12}S_2$ M 244.369
Cryst. (EtOH). Mp 128-130°.
6-Oxide: [103896-92-0].
    $C_{14}H_{12}OS_2$ M 260.368
    Mp 178-180°.
12-Oxide: [103896-93-1].
    $C_{14}H_{12}OS_2$ M 260.368
    Mp 195-196°.
12,12-Dioxide: [60418-11-3].
    $C_{14}H_{12}O_2S_2$ M 276.368
    Cryst. (EtOH). Mp 160-166°.
6,6,12,12-Tetroxide: [60418-12-4].
    $C_{14}H_{12}O_4S_2$ M 308.366
    Cryst. (EtOH). Mp 293°.

Gellatly, R.P. *et al*, *J. Chem. Soc., Perkin Trans. 1*, 1976, 913.
Ohkata, K. *et al*, *Tetrahedron Lett.*, 1985, **26**, 4491 (*conformn*)
Akasaka, T. *et al*, *Tetrahedron Lett.*, 1985, **26**, 5049 (*props*)
Fujihara, H. *et al*, *J. Org. Chem.*, 1987, **52**, 4254 (*monoxides*)

6*H*,12*H*-Dibenzo[*b,f*][1,5]dithiocin, 9CI   D-70070

[263-06-9]

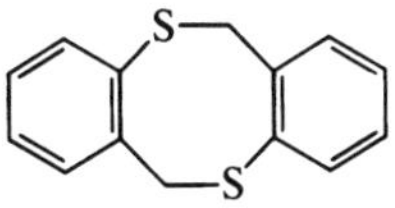

$C_{14}H_{12}S_2$ M 244.369

Cryst. by subl. Mp 174-176°.
S-Oxide: [105440-04-8].
    $C_{14}H_{12}OS_2$ M 260.368
    Mp 119°.

Stacey, G.W. *et al*, *J. Org. Chem.*, 1965, **30**, 4074 (*synth*)
Ollis, W.D. *et al*, *J. Chem. Soc., Perkin Trans. 1*, 1978, 1421 (*synth*)
Fujihara, H. *et al*, *J. Org. Chem.*, 1987, **52**, 4254 (*deriv, synth, ir*)

1,2:9,10-Dibenzo[2.2]metaparacyclophane   D-70071

5,8-Etheno-17,13-metheno-13H-dibenzo[a,g]-*cyclotridecene, 9CI*

[114032-09-6]

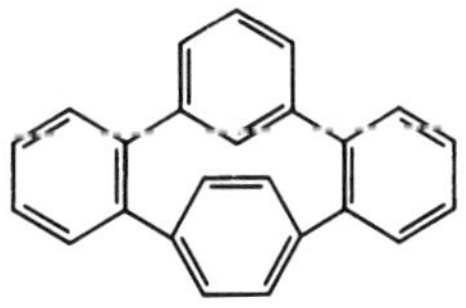

$C_{24}H_{16}$ M 304.390
Cryst. (hexane). Mp ~250° dec.

Wong, T. *et al*, *Angew. Chem., Int. Ed. Engl.*, 1988, **27**, 705 (*synth, pmr*)

1,2:7,8-Dibenzo[2.2]paracyclophane   D-70072

Updated Entry replacing D-40079
5,8:13,16-Diethenodibenzo[a,g]*cyclododecene, 9CI.*
1,2:9,10-Dibenzo[2.2]*paracyclophanediene*

[97315-26-9]

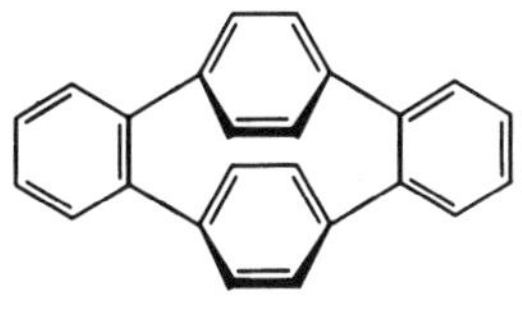

$C_{24}H_{16}$ M 304.390
Cryst. Subl. >275°.

Wong, H.N.C. *et al*, *J. Am. Chem. Soc.*, 1985, **107**, 4790 (*synth, pmr*)
Wong, H.N.C. *et al*, *Acta Crystallogr., Sect. C*, 1986, **42**, 703 (*synth, pmr, cryst struct*)
Reiser, O. *et al*, *Angew. Chem., Int. Ed. Engl.*, 1987, **26**, 1277 (*synth*)
Yang, Z. *et al*, *Helv. Chim. Acta*, 1987, **70**, 299 (*pe*)
Chan, W.C. *et al*, *J. Am. Chem. Soc.*, 1988, **110**, 462 (*synth, uv, pmr*)

Dibenzo[*a,j*]perylene-8,16-dione, 9CI   D-70073

Updated Entry replacing D-01307
1,2:7,8-Dibenzoperylene-3,9-quinone.
Heterocoerdianthrone

[5737-94-0]

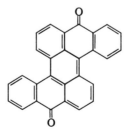

$C_{28}H_{14}O_2$ M 382.417

Reddish-violet microcryst. (PhCl or Py). Mp 363-365°.

Clar, E. *et al*, *Chem. Ber.*, 1949, **82**, 46.
Scholl, R. *et al*, *Justus Liebigs Ann. Chem.*, 1952, **494**, 201.
Hirakawa, K. *et al*, *Bull. Chem. Soc. Jpn.*, 1987, **60**, 2292 (*synth, pmr, bibl*)

Dibenzo[2,3:10,11]perylo[1,12-*bcd*]-thiophene   D-70074

Flavophene

[196-23-6]

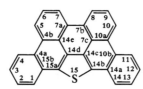

$C_{28}H_{14}S$ M 382.479

Fine yellow needles (PhNO$_2$ or PhCOOEt). Mp 378-380° (evac. tube) (391-392°).

Badger, G.M. *et al*, *J. Chem. Soc.*, 1957, 4417 (*struct, uv, bibl*)
Zander, M. *et al*, *Chem. Ber.*, 1973, **106**, 2752 (*synth*)
Riepe, W. *et al*, *Org. Mass Spectrom.*, 1979, **14**, 455 (*ms*)

Dibenzo[*f,h*]quinoline   D-70075

1-Azatriphenylene. Triphenylidene (*obsol.*)

[217-65-2]

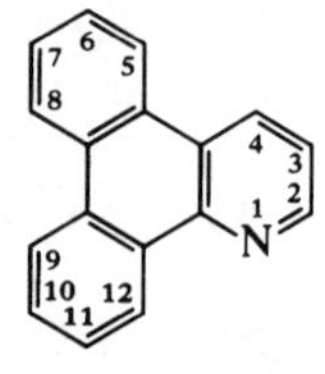

$C_{17}H_{11}N$ M 229.281

Yellow or colourless needles (C$_6$H$_6$). Mp 174° (171°).

Herschmann, F., *Ber.*, 1908, **41**, 1998 (*synth*)
Krueger, J.W. *et al*, *J. Org. Chem.*, 1940, **5**, 313 (*synth*)
Geerts-Evrard, F. *et al*, *Tetrahedron, Suppl. 7*, 1966, 287 (*synth*)

Dibenzo[*a,h*]quinolizinium(1+), 9CI   D-70076

Benzo[a]*phenathridizinium*

[340-45-4]

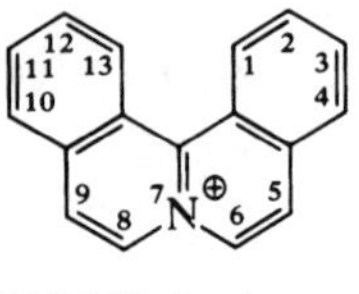

$C_{17}H_{12}N^{\oplus}$ M 230.289 (ion)

Bromide: [1556-78-1].
    $C_{17}H_{12}BrN$ M 310.193
    Cryst. (MeOH/Me$_2$CO). Mp 338.8-339.9°.
Perchlorate: [1556-79-2].
    $C_{17}H_{12}ClNO_4$ M 329.739
    Cryst. (MeOH/Me$_2$CO). Mp 268.0-269.5°.
Picrate: Yellow needles (MeOH). Mp 224.5-226.5°.

Bradsher, C.K. *et al*, *J. Org. Chem.*, 1965, **30**, 1846 (*synth, uv*)
Arai, S. *et al*, *J. Chem. Soc., Perkin Trans. 1*, 1987, 481 (*synth, ir, pmr*)

Dibenzotetraselenofulvalene   D-70077

2-(1,3-Benzodiselenol-2-ylidene)-1,3-benzodiselenole, 9CI

[82452-81-1]

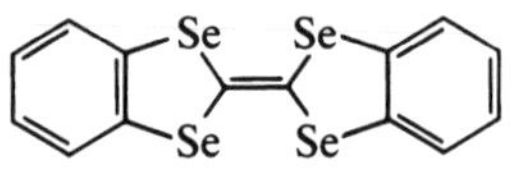

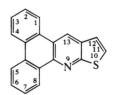

$C_{14}H_8Se_4$ M 492.057

Red platelets (CCl$_4$). Mp 288-291°.

Kistenmacher, T.J. *et al*, *J. Chem. Soc., Chem. Commun.*, 1983, 294 (*synth*)
Johannsen, I. *et al*, *J. Chem. Soc., Chem. Commun.*, 1983, 295 (*synth, ir, uv, pmr*)
Lambert, C. *et al*, *Tetrahedron Lett.*, 1984, **25**, 833 (*synth*)

Dibenzo[*b,i*]thianthrene-5,7,12,14-tetrone, 9CI   D-70078

[21634-42-4]

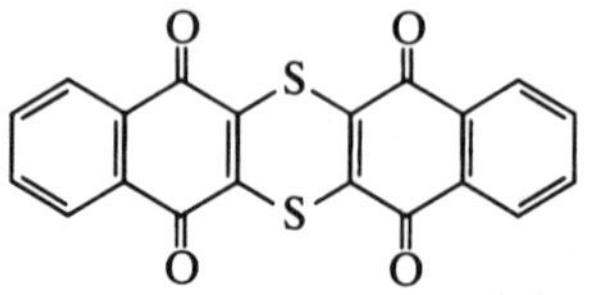

$C_{20}H_8O_4S_2$ M 376.401

Purple powder. Mp 306-308° (302°).

Brass, K. *et al*, *Ber.*, 1922, **55**, 2543 (*synth*)
Katritzky, A.R. *et al*, *J. Heterocycl. Chem.*, 1988, **25**, 901 (*synth, ir, uv*)

Dibenzo[*f,h*]thieno[2,3-*b*]quinoline, 9CI   D-70079

Phenanthro[9,10-e]*thieno*[2,3-b]*pyridine*

[109216-79-7]

$C_{19}H_{11}NS$ M 285.363

Fluffy, cryst. solid. Mp 147-150°.

1-Oxide: [109216-80-0].
    $C_{19}H_{11}NOS$ M 301.362
    Solid. Low sol. in org. solvents, could not be purified by cryst.

Taylor, E.C. *et al*, *J. Org. Chem.*, 1987, **52**, 4280 (*synth, ir, pmr, cmr*)

6H-Dibenzo[*b,d*]thiopyran-6-one, 9CI   D-70080

3,4-Benzothiocoumarin

[4445-36-7]

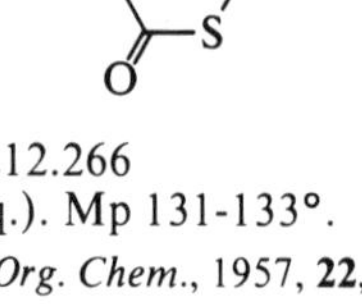

$C_{13}H_8OS$ M 212.266

Needles (EtOH aq.). Mp 131-133°.

Gilman, H. *et al*, *J. Org. Chem.*, 1957, **22**, 851 (*synth*)

Dibenzo[*b*,*d*]thiopyrylium(1+), 9CI   D-70081

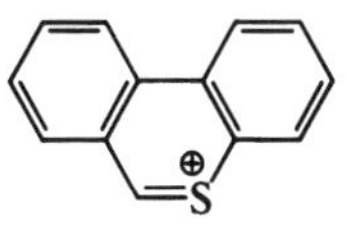

$C_{13}H_9S^{\oplus}$ M 197.274 (ion)

Perchlorate: [7432-94-2].

$C_{13}H_9ClO_4S$ M 296.725

Lemon-yellow cryst. Mp 201°.

Lüttringhaus, A. *et al*, *Z. Naturforsch., B*, 1961, **16**, 762 (*synth, uv*)

Hori, M. *et al*, *J. Org. Chem.*, 1987, **52**, 3668 (*synth*)

Dibenzo[*b*,*m*]triphenodithiazine-5,7,9,14,16,18(8*H*,17*H*)-tetrone, 9CI   D-70082

[116489-73-7]

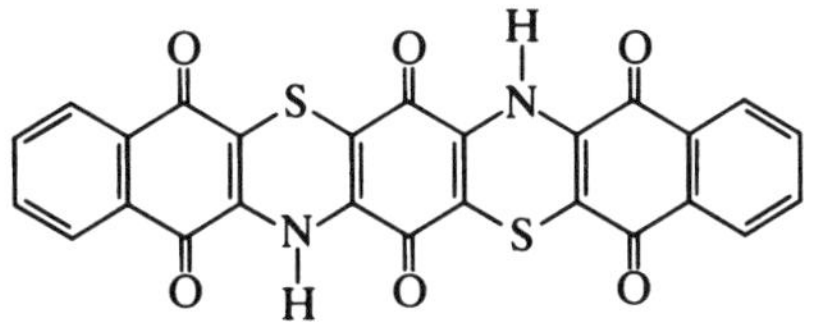

$C_{26}H_{10}N_2O_6S_2$ M 510.495

Red powder. Mp 275-278°.

Katritzky, A.R. *et al*, *J. Heterocycl. Chem.*, 1988, **25**, 901 (*synth, ir, uv*)

2,3-Dibenzoylbutanedioic acid, 9CI   D-70083

2,3-Dibenzoylsuccinic acid, 8CI

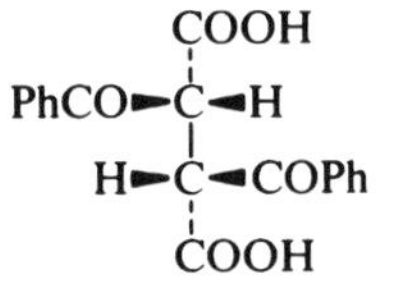

(2RS,3RS)-form

$C_{18}H_{14}O_6$ M 326.305

No evidence for significant conc. of enolic forms of esters in soln., though they interconvert, presumably by enolisation. Early refs. refer to enol. esters.

(2RS,3RS)-form

(±)-threo-*form*

Et ester: [81581-30-8]. Prisms. Mp 75°. Originally referred to as the γ-ester.

Dibutyl ester: [76695-74-4]. Could not be cryst. (ir, uv, ms, pmr, data).

(2RS,3SR)-form

erythro-*form*

Di-Et ester: [67560-61-6].

$C_{22}H_{22}O_6$ M 382.412

Prisms (EtOH). Mp 128-130°. Originally referred to as the β-ester.

Dibutyl ester: [76695-73-3].

$C_{26}H_{30}O_6$ M 438.519

Cryst. (hexane). Mp 89-90°.

Suehiro, T. *et al*, *Nippon Kagaku Zasshi*, 1958, **79**, 457; *CA*, **54**, 4486i (*synth*)

Mordecai, B.R. *et al*, *J. Chem. Soc., Perkin Trans. 1*, 1980, 2670 (*synth, deriv, ir, uv, ms, pmr*)

Stetter, H. *et al*, *Synthesis*, 1981, 626 (*synth, ir, pmr*)

Pelter, A. *et al*, *J. Chem. Soc., Perkin Trans. 1*, 1982, 183 (*synth, ir, pmr*)

1,3-Dibromoadamantane, 9CI   D-70084

Updated Entry replacing D-01420

1,3-Dibromotricyclo[3.3.1.1³,⁷]decane, 9CI

[876-53-9]

$C_{10}H_{14}Br_2$ M 294.029

Mp 112-113°.

Stetter, H. *et al*, *Chem. Ber.*, 1960, **93**, 1366 (*synth*)

Baughman, G.L., *J. Org. Chem.*, 1964, **29**, 238 (*synth, pmr*)

Fort, R.C. *et al*, *J. Org. Chem.*, 1965, **30**, 789 (*nmr*)

Lightner, D.A. *et al*, *Tetrahedron*, 1987, **43**, 4905 (*synth, ir, pmr*)

2,3-Dibromoanthracene   D-70085

$C_{14}H_8Br_2$ M 336.025

Light-yellow cryst. (C_6H_6). Mp 270°.

Lin, C.T. *et al*, *Synthesis*, 1988, 628 (*synth, uv, ir, pmr, ms*)

1,4-Dibromobicyclo[2.2.2]octane   D-70086

[10364-04-2]

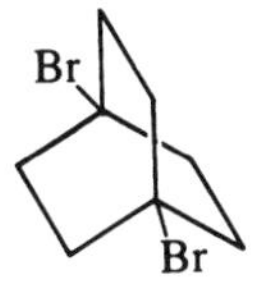

$C_8H_{12}Br_2$ M 267.991

Cryst. (Me₂CO). Mp 253-256° (248-249°).

Honegger, E. *et al*, *Chem. Ber.*, 1987, **120**, 187 (*synth, ir, pmr, cmr, pe, bibl*)

1,4-Dibromo-2-butyne   D-70087

[2219-66-1]

$$BrCH_2C{\equiv}CCH_2Br$$

$C_4H_4Br_2$ M 211.884

Liq. Bp₁₅ 92°, Bp₀.₁ 42°. n_D^{20} 1.5877.

▷Strongly lachrymatory

Valette, A., *Ann. Chim. (Paris)*, 1948, **3**, 644 (*synth*)

Eglington, G. *et al*, *J. Chem. Soc.*, 1950, 3650 (*synth*)

7,8-Dibromo-6-(chloromethylene)-2-methyl-2-octene, 9CI   D-70088

[112642-60-1]

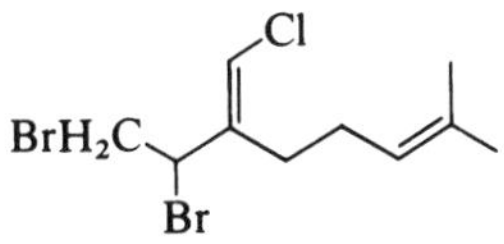

$C_{10}H_{15}Br_2Cl$ M 330.490

Constit. of *Chondrococcus hornemannii*. Oil. [α]_D +12.5° (c, 0.015 in CHCl₃).

Coll, J.C. *et al*, *Aust. J. Chem.*, 1987, **40**, 1893.

1,3-Dibromo-2,3-dimethylbutane, 9CI   D-70089

Updated Entry replacing D-01661

[49623-54-3]

$$(H_3C)_2CBrCH(CH_3)CH_2Br$$

$C_6H_{12}Br_2$ M 243.969

(±)-*form*

Bp₁₆.₅ 88-89°.

Bergmann, A.G., *CA*, 1923, **17**, 1420 (*synth*)

1,5-Dibromo-3,3-dimethylpentane   D-70090

[37746-17-1]

BrH$_2$CCH$_2$C(CH$_3$)$_2$CH$_2$CH$_2$Br

C$_7$H$_{14}$Br$_2$ M 257.996
Bp$_8$ 116-117°, Bp$_3$ 98-99°.

Schmerling, L. *et al, J. Am. Chem. Soc.*, 1952, **74**, 2885 (*synth*)
Cartledge, F.K. *et al, J. Org. Chem.*, 1986, **51**, 2206 (*synth, pmr*)

1,2-Dibromo-1,2-diphenylethane   D-70091

Updated Entry replacing D-01716
1,1'-(1,2-Dibromo-1,2-ethanediyl)bisbenzene, 9CI. α,α'-Dibromobibenzyl, 8CI. Stilbene dibromide. α,β-Dibromodibenzyl
[5789-30-0]

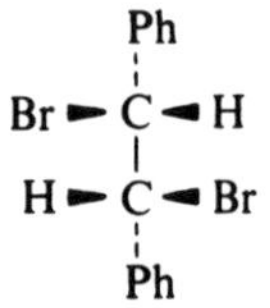

(1RS,2RS)-form
Relative configuration

C$_{14}$H$_{12}$Br$_2$ M 340.057
(*1RS,2RS*)-**form** [13027-48-0]
(±)-*form*
Cryst. (EtOH). Mp 113-114°.
(*1RS,2SR*)-**form** [13440-24-9]
meso-*form*
Cryst. Mp 237°.

Buckles, R.E. *et al, J. Am. Chem. Soc.*, 1950, **72**, 2496 (*synth*)
Hartshorn, M.P. *et al, Aust. J. Chem.*, 1973, **26**, 917 (*synth, pmr*)
Kaupp, G. *et al, Chem. Ber.*, 1987, **120**, 1897 (*synth, pmr, cmr, ms*)

1,20-Dibromoicosane   D-70092

[14296-16-3]

BrCH$_2$(CH$_2$)$_{18}$CH$_2$Br

C$_{20}$H$_{40}$Br$_2$ M 440.344
Mp 62-64°. Bp$_{0.05}$ 170-180°.

Woolford, R.G., *Can. J. Chem.*, 1962, **40**, 1846 (*synth*)
Natrajan, A. *et al, J. Am. Chem. Soc.*, 1987, **109**, 7477 (*synth, pmr, cmr*)

2,4(5)-Dibromo-1*H*-imidazole, 9CI   D-70093

[64591-03-3]

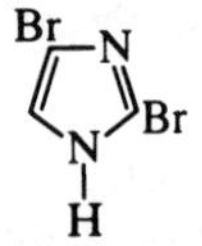

C$_3$H$_2$Br$_2$N$_2$ M 225.870
Needles (H$_2$O). Mp 193°.
B,HCl: Cryst. + 1H$_2$O (5M HCl). Mp 120-130° (sinters >100°).

Balaban, I.E. *et al, J. Chem. Soc.*, 1922, 947 (*synth*)
USSR Pat., 558 913, (*1979*); *CA*, **87**, 184497t (*synth*)

4,4-Dibromo-3-methyl-3-buten-2-one   D-70094

[103367-81-3]

Br$_2$C═C(CH$_3$)COCH$_3$

C$_5$H$_6$Br$_2$O   M 241.910
Light-green liq. Bp$_{5.5}$ ca. 63°.
Di-Me acetal: [103367-82-4]. *1,1-Dibromo-3,3-dimethoxy-2-methyl-1-butene.*
C$_7$H$_{12}$Br$_2$O$_2$ M 287.979

Liq. Bp$_1$ 50°.

Boeckman, R.K. *et al, J. Am. Chem. Soc.*, 1986, **108**, 5549 (*synth, nmr*)

6,7-Dibromo-1-naphthol   D-70095

6,7-Dibromo-1-naphthalenol, 9CI
[117157-37-6]
C$_{10}$H$_6$Br$_2$O   M 301.965
Needles. Mp 143-144°.

Ashnagar, A. *et al, J. Chem. Soc., Perkin Trans. 1*, 1988, 559 (*synth, ir, pmr*)

6,7-Dibromo-1,4-naphthoquinone   D-70096

6,7-Dibromo-1,4-naphthalenedione, 9CI
C$_{10}$H$_4$Br$_2$O$_2$ M 315.948
Orange-yellow needles. Mp 171-172°.

Ashnagar, A. *et al, J. Chem. Soc., Perkin Trans. 1*, 1988, 559 (*synth, ir, pmr*)

6,7-Dibromo-1-naphthylamine   D-70097

6,7-Dibromo-1-naphthalenamine, 9CI. 1-Amino-6,7-dibromonaphthalene
[117157-36-5]
C$_{10}$H$_7$Br$_2$N   M 300.980
Needles (EtOH aq.). Mp 123-124°.

Ashnagar, A. *et al, J. Chem. Soc., Perkin Trans. 1*, 1988, 559 (*synth, ir, pmr*)

6,7-Dibromo-1-nitronaphthalene, 9CI   D-70098

2,3-Dibromo-5-nitronaphthalene (incorr.)
[117157-35-4]
C$_{10}$H$_5$Br$_2$NO$_2$ M 330.963
Yellow needles (AcOH). Mp 186-187°.

Ashnagar, A. *et al, J. Chem. Soc., Perkin Trans. 1*, 1988, 559 (*synth, ir, pmr*)

1,8-Dibromopyrene, 9CI   D-70099

[38303-35-4]
C$_{16}$H$_8$Br$_2$ M 360.047
Needles (C$_6$H$_6$/hexane). Mp 210-211°.

Grimshaw, J. *et al, J. Chem. Soc., Perkin Trans. 1*, 1972, 1622 (*synth, pmr*)

4,5-Dibromo-1,2,3-triazine   D-70100

[114078-89-6]

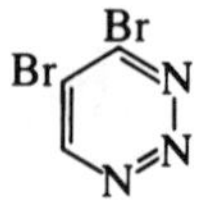

C$_3$HBr$_2$N$_3$ M 238.869
Cryst. (Et$_2$O/hexane). Mp 138-139° dec.

Kaihoh, T. *et al, Chem. Pharm. Bull.*, 1987, **35**, 3952 (*synth*)

2,2-Di-*tert*-butyl-3-(di-*tert*-butylmethylene)cyclopropanone   D-70101

[108263-91-8]

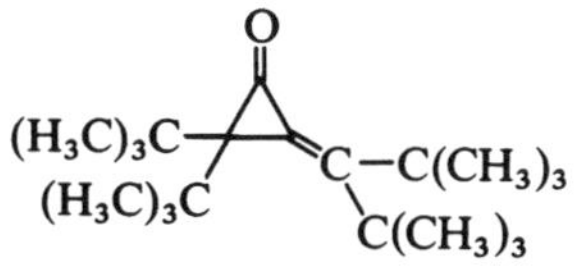

C$_{20}$H$_{36}$O   M 292.504

Stable methylenecyclopropanone. Yellow solid. Mp 56-57°.

Crandall, J.K. *et al*, *J. Am. Chem. Soc.*, 1987, **109**, 4338 (*synth, ir, pmr, cmr, uv, ms*)

2,7-Di-*tert*-butyldicyclopenta[*a,e*]-cyclooctene   D-70102

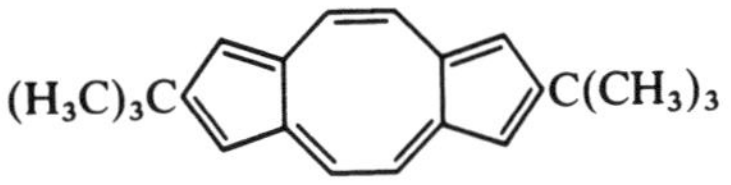

C$_{22}$H$_{26}$ M 290.447

Reddish-brown cryst. Mp 199°. Air-stable.

Hafner, K. *et al*, *Angew. Chem., Int. Ed. Engl.*, 1988, **27**, 1191 (*synth, pmr, cmr, uv, cryst struct*)

2,5-Di-*tert*-butyl-2,5-dihydrobenzo[*e*]-pyrrolo[3,4-*g*]isoindole   D-70103

[111558-02-2]

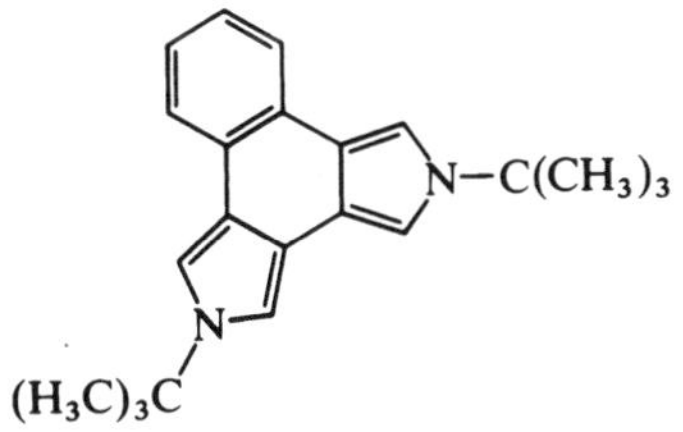

C$_{22}$H$_{26}$N$_2$ M 318.461

Mp 203-204°.

Kreher, R.P. *et al*, *Angew. Chem., Int. Ed. Engl.*, 1987, **26**, 1262 (*synth, pmr*)

2,7-Di-*tert*-butyl-2,7-dihydroisoindolo[5,4-*e*]isoindole   D-70104

[111558-01-1]

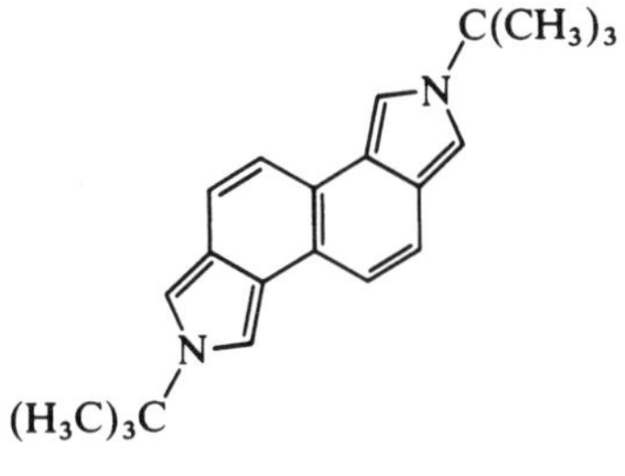

C$_{22}$H$_{26}$N$_2$ M 318.461

Mp 276-277°.

Kreher, R.P. *et al*, *Angew. Chem., Int. Ed. Engl.*, 1987, **26**, 1262 (*synth, pmr*)

5-(3,5-Di-*tert*-butyl-4-oxo-2,5-cyclohexadienylidene)-3,6-cycloheptadiene-1,2-dione   D-70105

5-[3,5-Bis(1,1-dimethylethyl)-4-oxo-2,5-cyclohexadien-1-ylidene]-3,6-cycloheptadiene-1,2-dione, 9CI

[112292-35-0]

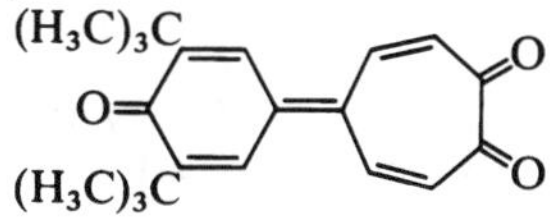

C$_{21}$H$_{24}$O$_3$ M 324.419

Reddish-purple needles. Mp 149.5-150.5°.

Iyoda, M. *et al*, *Tetrahedron Lett.*, 1987, **28**, 625 (*synth, ms, pmr, cmr, ir, uv*)

3,4-Di-*tert*-butylpyrazole, 8CI   D-70106

3,4-Bis(1,1-dimethylethyl)pyrazole

[16867-95-1]

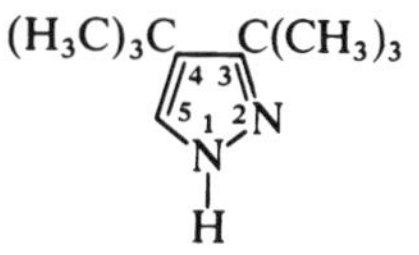

C$_{11}$H$_{20}$N$_2$ M 180.292

Cryst. (MeOH aq.). Mp 129-130°.

De Groot, A. *et al*, *J. Org. Chem.*, 1968, **33**, 3337 (*synth, ir, pmr*)

3,5-Di-*tert*-butylpyrazole, 8CI   D-70107

3,5-Bis(1,1-dimethylethyl)pyrazole, 9CI

[1132-14-5]

C$_{11}$H$_{20}$N$_2$ M 180.292

Cryst. Mp 193°.

Bertini, V. *et al*, *Gazz. Chim. Ital.*, 1964, **94**, 915.
Elguero, J. *et al*, *Bull. Soc. Chim. Fr.*, 1966, 3727; 1968, 707 (*synth, uv, pmr*)
Cabildo, P. *et al*, *Org. Magn. Reson.*, 1984, **22**, 603 (*cmr*)
Faure, R. *et al*, *Can. J. Chem.*, 1988, **66**, 1141 (*cmr*)

3,6-Di-*tert*-butylpyrrolo[3,2-*b*]pyrrole   D-70108

3,6-Bis(1,1-dimethylethyl)pyrrolo[3,2-b]pyrrole, 9CI.
*3,6-Di-*tert-*butyl-1,4-diazapentalene*

[115271-20-0]

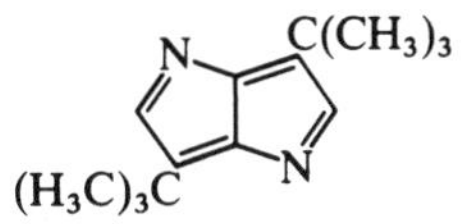

C$_{14}$H$_{20}$N$_2$ M 216.325

First azapentalene not stabilised by conjugative substituents. Fine red-brown needles. Mp 93°. Mod. unstable in air at r.t.

Tanaka, S. *et al*, *Angew. Chem., Int. Ed. Engl.*, 1988, **27**, 1061 (*synth, ir, pmr, cmr, uv*)

3,4-Di-*tert*-butyl-2,2,5,5-tetramethylhexane   D-70109

*3,4-Bis(1,1-dimethylethyl)-2,2,5,5-tetramethylhexane, 9CI. 1,1,2,2-Tetra-*tert-*butylethane*

[62850-21-9]

$$[(H_3C)_3C]_2CHCH[C(CH_3)_3]_2$$

$C_{18}H_{38}$ M 254.498
Conformational analysis extensively studied. Mp 168-172°.

Mendenhall, G.D. *et al*, *J. Am. Chem. Soc.*, 1974, **96**, 2441 (*synth, conformn*)
Brownstein, S. *et al*, *J. Am. Chem. Soc.*, 1977, **99**, 2073 (*synth, conformn*)
Beckhaus, H.D. *et al*, *Chem. Ber.*, 1978, **111**, 72 (*synth, conformn*)
Osawa, E. *et al*, *J. Am. Chem. Soc.*, 1979, **101**, 4824 (*conformn*)
Hellmann, G. *et al*, *Chem. Ber.*, 1982, **115**, 3364 (*conformn*)

3,4-Di-*O*-caffeoylquinic acid   D-70110

Updated Entry replacing D-02268
[2271-12-7]

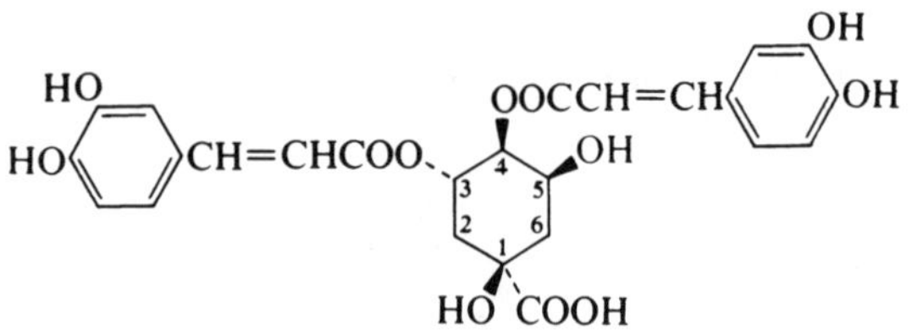

$C_{25}H_{24}O_{12}$ M 516.457
Isol. from coffee and maté. $[\alpha]_D^{28}$ −225° (MeOH).
O^5-(*3-Hydroxy-3-methylglutaroyl*): [103744-76-9].
$C_{31}H_{32}O_{16}$ M 660.584
Isol. from fruit of *Gardenia jasminoides*. Lipoxygenase inhibitor. $[\alpha]_D^{18}$ −169.8° (c, 0.95 in MeOH).

Scarpati, M.L., *Tetrahedron Lett.*, 1964, 2851 (*isol*)
Taylor, A.O., *Phytochemistry*, 1968, **7**, 63 (*isol*)
Nishizawa, M. *et al*, *Chem. Pharm. Bull.*, 1986, **34**, 1419 (*deriv*)

Dicarbonic acid, 9CI   D-70111

Updated Entry replacing D-60084
Oxydiformic acid, 8CI. *Pyrocarbonic acid*
[503-81-1]

HOOC−O−COOH

$C_2H_2O_5$ M 106.035
Di-Me ester: [4525-33-1]. *Dimethyl dicarbonate. Dimethyl pyrocarbonate. Velcorin.*
$C_4H_6O_5$ M 134.088
Yeast inhibitor and preservative for alcoholic beverages. Bp_5 44-47°. n_D^{20} 1.3948.
Di-Et ester: [1609-47-8]. *Diethyl dicarbonate. Diethyl pyrocarbonate. Baycovin.*
$C_6H_{10}O_5$ M 162.142
Preservative for beverages, food additive. Bp_3 58.5-62°.
▷Poss. protocarcinogen. LQ9350000.
Di-tert-butyl ester: Di-tert-butyl dicarbonate. Di-tert-butyl pyrocarbonate.
$C_{10}H_{20}O_5$ M 220.265
Coupling reagent in organic synthesis. Liq. Mp 21-22°. Bp_3 73-75°. n_D^{25} 1.4071, 1.4085.
Dibenzyl ester: [31139-36-3]. *Dibenzyl dicarbonate. Dibenzyl pyrocarbonate.*
$C_{16}H_{14}O_5$ M 286.284
Reagent for prepn. of *N*-benzyloxycarbonyl protected amino acids. Cryst. (hexane). Mp 28°.
Di(2-propenyl) ester: [115491-93-5]. *Diallyl dicarbonate. Diallyl pyrocarbonate.*
$C_8H_{10}O_5$ M 186.164
Reagent for protection of amino-sugars, aminoacids and nucleosides. $Bp_{0.05}$ 65°. Stable at r.t.

Fr. Pat., 1 542 382, (1968); CA, **71**, 123555h (*manuf, dimethyl ester*)
Brysova, V.P. *et al*, *J. Org. Chem. USSR*, 1974, **10**, 2551 (*synth, dimethyl ester*)
Turczan, J.W. *et al*, *J. Agric. Food Chem.*, 1977, **25**, 594 (*pmr, diethyl ester*)
Org. Synth., 1977, **57**, 45 (*di-tert-*butyl ester)
Pauli, G.H., *J. Chem. Educ.*, 1984, **61**, 332 (*rev, diethyl ester*)
Sennyey, G. *et al*, *Tetrahedron Lett.*, 1986, **27**, 5375; 1987, **28**, 5809 (*Diallyl, dibenzyl esters*)

α-Diceroptene   D-70112

Updated Entry replacing I-40045
Isoceroptene
[98094-16-7]

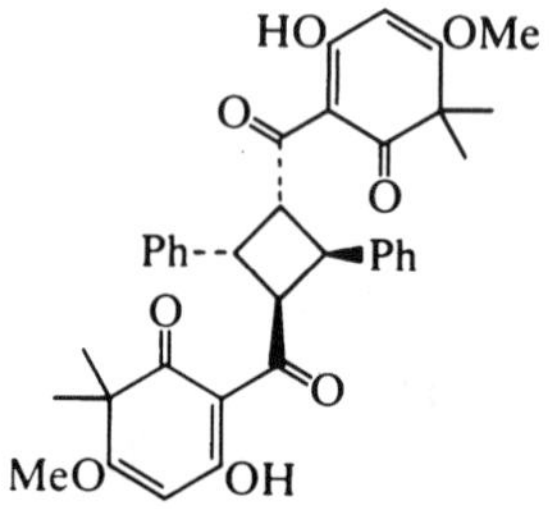

$C_{36}H_{36}O_8$ M 596.676
Struct. revised in 1987-formerly considered to be a C_{18} compd. Constit. of *Pityrogramma triangularis*. Cryst. Mp 218-220°.

Markham, K.R. *et al*, *Z. Naturforsch., C*, 1985, **40**, 317 (*isol*)
Vilain, C. *et al*, *Z. Naturforsch., C*, 1987, **42**, 849 (*struct*)

2,2-Dichloro-1,3-benzodioxole, 9CI   D-70113

Updated Entry replacing D-30142
1,2-[(Dichloromethylene)dioxy]benzene, 8CI. *Catechol dichloromethylene ether*
[2032-75-9]

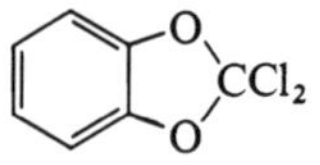

$C_7H_4Cl_2O_2$ M 191.013
Bp_{12} 82-89°.

Gross, H. *et al*, *Chem. Ber.*, 1961, **49**, 544; 1963, **96**, 1382 (*synth, use*)
Mayr, H. *et al*, *Chem. Ber.*, 1988, **121**, 339 (*synth, ir, pmr, cmr, ms, bibl*)
Fieser, M. *et al*, *Reagents for Organic Synthesis*, Wiley, 1967-84, **4**, 70.

1,4-Dichlorobenzo[*g*]phthalazine, 9CI   D-70114

[30800-67-0]

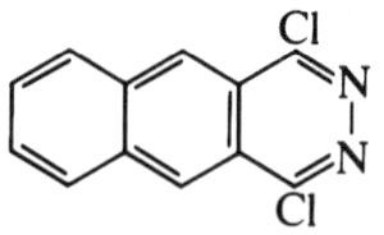

$C_{12}H_6Cl_2N_2$ M 249.099
Needles ($CHCl_3$/hexane). Mp 258-261° dec.

Hill, J.H.M. *et al*, *J. Org. Chem.*, 1971, **36**, 3248 (*synth, props*)

134

1,4-Dichlorobicyclo[2.2.2]octane　　D-70115
[1123-39-3]

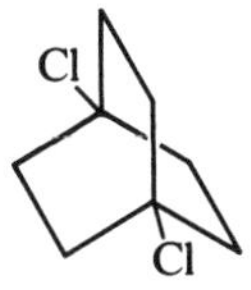

$C_8H_{12}Cl_2$　　M 179.089
Cryst. (hexane). Mp 234-235° (228-231°).

Becker, K.B. *et al*, *Synthesis*, 1973, 493 (*synth*)
Honegger, E. *et al*, *Chem. Ber.*, 1987, **120**, 187 (*synth, ir, pmr, cmr, pe*)

1,8-Dichloro-6-chloromethyl-2-methyl-2,6-octadiene, 9CI　　D-70116
[112642-59-8]

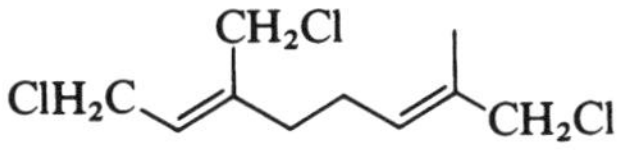

$C_{10}H_{15}Cl_3$　　M 241.588
Constit. of *Chondrococcus hornemannii*. Oil.

Coll, J.C. *et al*, *Aust. J. Chem.*, 1987, **40**, 1893.

3,8-Dichloro-6-chloromethyl-2-methyl-1,6-octadiene, 9CI　　D-70117
[112642-62-3]

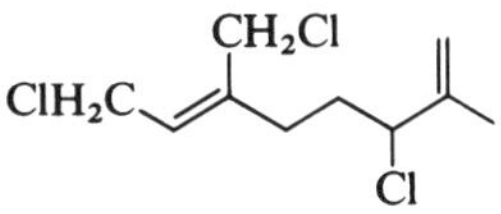

$C_{10}H_{15}Cl_3$　　M 241.588
Constit. of *Chondrococcus hornemannii*. Oil. $[\alpha]_D$ −33.7° (c, 0.002 in $CHCl_3$).

Coll, J.C. *et al*, *Aust. J. Chem.*, 1987, **40**, 1893.

1,3-Dichloro-2-(chloromethyl)propane, 9CI　　D-70118
Tris(chloromethyl)methane
[66703-69-3]

$$HC(CH_2Cl)_3$$

$C_4H_7Cl_3$　　M 161.458
Liq. Bp$_{50}$ 90°.

Latour, S. *et al*, *Synthesis*, 1987, 742 (*synth, ms, ir, pmr*)

4,6-Dichloro-5-[(dimethylamino)-methylene]-3,6-cyclohexadiene-1,3-dicarboxaldehyde, 9CI　　D-70119
[101392-95-4]

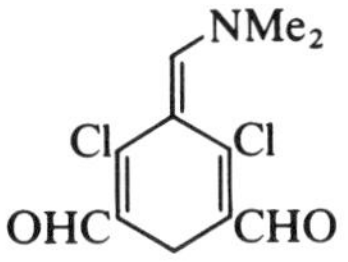

$C_{11}H_{11}Cl_2NO_2$　　M 260.119
Remarkably stable nonbenzenoid tautomer of an aromatic compd. Orange prisms ($CHCl_3$/pet. ether). Mp 149-150°. Stable for several weeks at r.t. Substituting Ph for one Me favours the aromatic tautomer.

Katritzky, A.R. *et al*, *Tetrahedron Lett.*, 1985, **26**, 4715 (*synth, cmr*)

Katritzky, A.R. *et al*, *Tetrahedron*, 1988, **44**, 3209 (*cryst struct, pmr, cmr*)

2,5-Dichloro-4-hydroxybenzaldehyde　　D-70120
[27164-10-9]
$C_7H_4Cl_2O_2$　　M 191.013
Cryst. (CCl_4). Mp 159-160°.

Knuutinen, J.S. *et al*, *J. Chem. Eng. Data*, 1983, **28**, 139 (*synth, pmr, cmr, ms*)

2,6-Dichloro-4-hydroxybenzaldehyde　　D-70121
[60964-09-2]
$C_7H_4Cl_2O_2$　　M 191.013
Cryst. (toluene). Mp 223.5-224.5°.

Me ether: [82772-93-8]. *2,6-Dichloro-4-methoxybenzaldehyde.*
$C_8H_6Cl_2O_2$　　M 205.040
Needles (MeOH). Mp 109-111°.

Baldwin, J.J. *et al*, *J. Med. Chem.*, 1979, **22**, 687 (*synth*)
Knuutinen, J.S. *et al*, *J. Chem. Eng. Data*, 1983, **28**, 139 (*pmr, cmr, ms*)
Karl, J. *et al*, *J. Med. Chem.*, 1988, **31**, 72 (*synth, deriv*)

4,5-Dichloro-1H-imidazole　　D-70122
[15965-30-7]

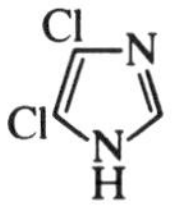

$C_3H_2Cl_2N_2$　　M 136.968
Cryst. (H_2O). Mp 179-180°.

Lutz, A.W. *et al*, *J. Heterocycl. Chem.*, 1967, **4**, 399 (*synth, ir*)

5,6-Dichloro-1H-indole-3-acetic acid　　D-70123
[98640-00-7]

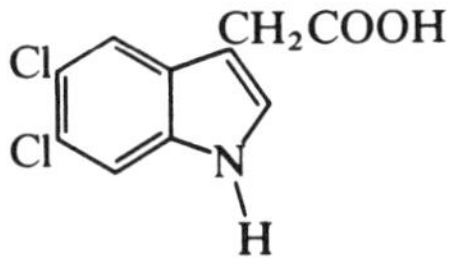

$C_{10}H_7Cl_2NO_2$　　M 244.077
Most potent auxin of all investigated natural and synthetic compds. Mp 189-191°.

Hatano, T. *et al*, *Experientia*, 1987, **43**, 1237 (*synth, pmr, ir, ms*)

6,7-Dichloro-1,4-naphthoquinone　　D-70124
Updated Entry replacing D-02932
6,7-Dichloro-1,4-naphthalenedione
[577-67-3]
$C_{10}H_4Cl_2O_2$　　M 227.046
Yellow plates (EtOH). Mp 186-187° (199°).

Babu Rao, K. *et al*, *J. Sci. Ind. Res., Sect. B*, 1958, **17**, 225 (*synth*)
Bluestone, H. *et al*, *J. Org. Chem.*, 1961, **26**, 346 (*synth*)
Ashnagar, A. *et al*, *J. Chem. Soc., Perkin Trans. 1*, 1988, 559 (*synth, ir, pmr*)

2,3-Dichloro-4-nitrophenol, 9CI   D-70125
[59384-57-5]

$$C_6H_3Cl_2NO_3 \qquad M\ 208.001$$

Me ether: [105630-54-4]. *2,3-Dichloro-1-methoxy-4-nitrobenzene,* *9CI. 2,3-Dichloro-4-nitroanisole.*
$C_7H_5Cl_2NO_3 \qquad M\ 222.027$
Pale-yellow prisms (hexane). Mp 81-82°.

Belg. Pat., 872 530, (*1979*); *CA*, **91**, 91347m
Hine, J. *et al*, *J. Org. Chem.*, 1987, **52**, 2089 (*synth, pmr*)

1,1-Dichlorooctane, 9CI   D-70126
[20395-24-8]

$$H_3C(CH_2)_6CHCl_2$$

$C_8H_{16}Cl_2 \qquad M\ 183.120$
Bp 95-99°.

Ransley, D.L., *J. Org. Chem.*, 1969, **34**, 2618 (*synth*)

1,2-Dichlorooctane, 9CI   D-70127
[21948-46-9]

$$H_3C(CH_2)_5CHClCH_2Cl$$

$C_8H_{16}Cl_2 \qquad M\ 183.120$
(±)-form
Bp_9 87°, Bp_4 67-71°. n_D^{27} 1.4503.

Kharasch, M.S. *et al*, *J. Am. Chem. Soc.*, 1951, **73**, 964 (*synth*)
Beringer, F.M. *et al*, *J. Am. Chem. Soc.*, 1955, **77**, 5533 (*synth*)
Okamoto, Y. *et al*, *Bull. Chem. Soc. Jpn.*, 1970, **43**, 2613 (*synth*)
Kovacic, P. *et al*, *J. Org. Chem.*, 1971, **36**, 3566 (*synth*)

1,3-Dichlorooctane, 9CI   D-70128
[5799-71-3]

$$H_3C(CH_2)_4CHClCH_2CH_2Cl$$

$C_8H_{16}Cl_2 \qquad M\ 183.120$
(±)-form
Bp_{12} 97-99°, $Bp_{0.4}$ 48-50°. n_D^{20} 1.4554.

Goldwhite, H. *et al*, *Tetrahedron*, 1964, **20**, 1613; 1965, **21**, 2743 (*synth, ir, nmr*)

2,2-Dichlorooctane, 9CI   D-70129
[73642-95-2]

$$H_3C(CH_2)_5CCl_2CH_3$$

$C_8H_{16}Cl_2 \qquad M\ 183.120$
$Bp_{0.1}$ 75°.

Dehmlow, E.V. *et al*, *Justus Liebigs Ann. Chem.*, 1980, 1 (*synth, pmr*)

2,3-Dichlorooctane, 9CI   D-70130

$$\begin{array}{c} CH_3 \\ Cl\text{—}\overset{2}{C}\text{—}H \\ H\text{—}\overset{3}{C}\text{—}Cl \\ (CH_2)_4CH_3 \end{array} \qquad (2RS,3RS)\text{-}form$$

$C_8H_{16}Cl_2 \qquad M\ 183.120$
(2RS,3RS)-form [37464-71-4]
(±)-threo-*form*
Bp_{25} 104°.
(2RS,3SR)-form
(±)-erythro-*form*
Bp 206.1-206.2°, Bp_{10} 78°. n_D^{20} 1.4523.

Hoff, M.C. *et al*, *J. Am. Chem. Soc.*, 1951, **73**, 3333 (*synth*)
Beringer, F.M. *et al*, *J. Am. Chem. Soc.*, 1955, **77**, 5533 (*synth*)
Uemura, S. *et al*, *Bull. Chem. Soc. Jpn.*, 1972, **45**, 1482 (*synth, pmr*)

4,5-Dichlorooctane, 9CI   D-70131

$$\begin{array}{c} CH_2CH_2CH_3 \\ Cl\text{—}C\text{—}H \\ H\text{—}C\text{—}Cl \\ CH_2CH_2CH_3 \end{array} \qquad (4RS,5RS)\text{-}form$$

$C_8H_{16}Cl_2 \qquad M\ 183.120$
(4RS,5RS)-form
(±)-erythro-*form*
Liq.
(4RS,5SR)-form [51149-23-6]
meso-*form*
d_4^{20} 1.0175. Bp 202.5°. n_D^{20} 1.4538.

Kharasch, M.S. *et al*, *J. Am. Chem. Soc.*, 1951, **73**, 632 (*synth*)
Hoff, M.C. *et al*, *J. Am. Chem. Soc.*, 1951, **73**, 3329 (*synth*)
Sicher, J. *et al*, *Helv. Chim. Acta*, 1973, **56**, 1630 (*synth*)

1,4-Dichlorophenanthrene   D-70132

$C_{14}H_8Cl_2 \qquad M\ 247.123$
Cryst. (MeOH aq.). Mp 78°.

Greenhalgh, N., *J. Chem. Soc.*, 1954, 4699 (*synth*)

2,4-Dichlorophenanthrene   D-70133
[7299-38-9]
$C_{14}H_8Cl_2 \qquad M\ 247.123$
Cryst. (MeOH aq.). Mp 87-88°.

Greenhalgh, N., *J. Chem. Soc.*, 1954, 4699 (*synth*)
Letcher, R.M., *Org. Magn. Reson.*, 1981, **16**, 220 (*pmr, cmr*)

3,6-Dichlorophenanthrene   D-70134
[20851-90-5]
$C_{14}H_8Cl_2 \qquad M\ 247.123$
Needles (toluene). Mp 167.5-169°.

Weis, C.D., *Helv. Chim. Acta*, 1968, **51**, 1572 (*synth, uv, pmr*)

3,9-Dichlorophenanthrene   D-70135
$C_{14}H_8Cl_2 \qquad M\ 247.123$

Needles (dioxan/MeOH). Mp 125-125.5°.

Schultz, J. *et al, J. Org. Chem.*, 1946, **11**, 320 (*synth*)

4,5-Dichlorophenanthrene   D-70136

[108665-39-0]

$C_{14}H_8Cl_2$ M 247.123

Prisms (MeOH). Mp 126-127°.

Cosmo, R. *et al, Aust. J. Chem.*, 1987, **40**, 2137 (*synth, uv, ir, pmr, ms*)

9,10-Dichlorophenanthrene   D-70137

[17219-94-2]

$C_{14}H_8Cl_2$ M 247.123

Needles (EtOH). Mp 160-160.5°. Bp$_{0.1}$ 175-185°.

Carey, J.G. *et al, J. Chem. Soc.*, 1959, 3144 (*synth*)
Cameron, D.W. *et al, Aust. J. Chem.*, 1977, **30**, 859 (*uv, pmr*)
Ding, M. *et al, Macromolecules*, 1983, **16**, 839 (*synth*)

4-(2,5-Dichlorophenylhydrazono)-5-methyl-   D-70138
2-phenyl-3*H*-pyrazol-3-one

*4-[(2,5-Dichlorophenyl)azo]-2,4-dihydro-5-methyl-2-phenyl-3*H-*pyrazol-3-one, 9*CI*. 3-Methyl-1-phenylpyra-zole-4,5-dione 4-(2,5-dichlorophenyl)hydrazone. CI Pigment yellow 10*

[6407-75-6]

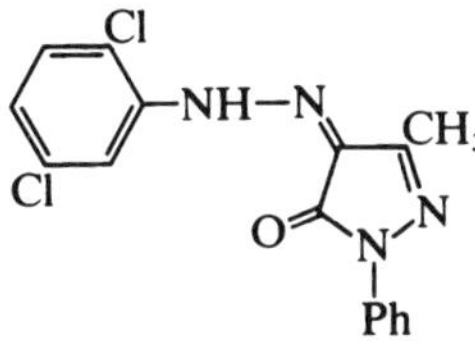

$C_{16}H_{12}Cl_2N_4O$ M 347.203

Hydrazone tautomer shown predominates over poss. azo tautomer in cryst. state. Yellow pigment used in printing inks, emulsion paints and resins. Yellow cryst. (toluene), orange cryst. (PhNO$_2$). Mp 224°.

Cassino, D., *CA*, 1951, **45**, 7353g (*synth*)
Whitaker, A., *Acta Crystallogr., Sect. C*, 1988, **44**, 1767 (*cryst struct*)

1,5-Dichloroxanthone   D-70139

*1,5-Dichloro-9*H-*xanthen-9-one, 9*CI

[76093-26-0]

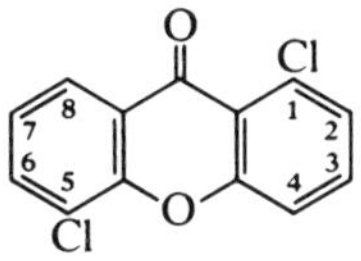

$C_{13}H_6Cl_2O_2$ M 265.095

Needles (EtOH). Mp 192°.

Okabayashi, I. *et al, Chem. Pharm. Bull.*, 1980, **28**, 2831 (*synth, ir*)

1,7-Dichloroxanthone   D-70140

*1,7-Dichloro-9*H-*xanthen-9-one, 9*CI

[76093-30-6]

$C_{13}H_6Cl_2O_2$ M 265.095

Needles (EtOH). Mp 172-173.5°.

Okabayashi, I. *et al, Chem. Pharm. Bull.*, 1980, **28**, 2831 (*synth, ir*)

2,6-Dichloroxanthone   D-70141

*2,6-Dichloro-9*H-*xanthen-9-one, 9*CI

[1556-62-3]

$C_{13}H_6Cl_2O_2$ M 265.095

Needles (EtOH). Mp 219-220°.

McNelis, E., *J. Org. Chem.*, 1963, **28**, 3188 (*synth*)
Okabayashi, I. *et al, Chem. Pharm. Bull.*, 1980, **28**, 2831 (*synth, ir*)

2,7-Dichloroxanthone   D-70142

*2,7-Dichloro-9*H-*xanthen-9-one, 9*CI

[55103-01-0]

$C_{13}H_6Cl_2O_2$ M 265.095

Mp 228-230° (219°).

Granoth, I. *et al, J. Org. Chem.*, 1975, **40**, 2088.
Kimura, M. *et al, Chem. Pharm. Bull.*, 1987, **35**, 136 (*synth, ir, pmr, cmr*)

3,6-Dichloroxanthone   D-70143

*3,6-Dichloro-9*H-*xanthen-9-one, 9*CI

$C_{13}H_6Cl_2O_2$ M 265.095

Mp 190-191° (184-186°).

Goldberg, A.A. *et al, J. Chem. Soc.*, 1958, 4234.
Kimura, M. *et al, Chem. Pharm. Bull.*, 1987, **35**, 136 (*synth, ir, pmr, cmr*)

Dicranolomin   D-70144

2′,6″-Biluteolin

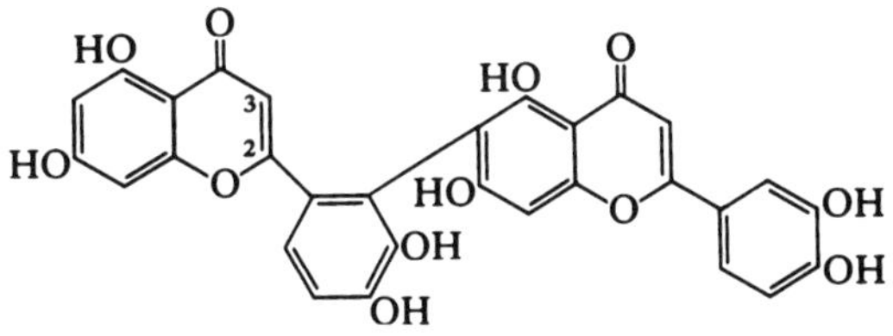

$C_{30}H_{18}O_{12}$ M 570.465

Constit. of *Dicranoloma robustum*.

2,3-Dihydro: **2,3-Dihydrodicranolomin**. *2,3-Dihydro-2′,6″-biluteolin.*
$C_{30}H_{20}O_{12}$ M 572.481
Constit. of *D. robustum*.

Markham, K.R. *et al, Phytochemistry*, 1988, **27**, 1745.

Dictyodial *A*   D-70145

Updated Entry replacing D-30183

[70552-61-3]

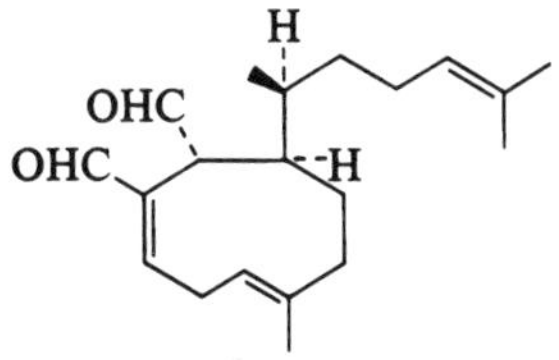

$C_{20}H_{30}O_2$ M 302.456

Constit. of *Dictyota crenulata* and *D. flabellata*. Oil. $[\alpha]_D^{25}$ −95° (c, 1.2 in CHCl$_3$).

4β-Hydroxy: [89482-11-1].
$C_{20}H_{30}O_3$ M 318.455
From *D. crenulata*. Oil. $[\alpha]_D^{25}$ −121° (c, 0.33 in EtOH).

4β-Hydroxy, 18,O-Dihydro, 18-Ac: [84164-85-2].
$C_{22}H_{34}O_4$ M 362.508

From *D. crenulata*. Oil. $[\alpha]_D^{25}$ −187° (c, 0.8 in EtOH).

Finer, J. *et al*, *J. Org. Chem.*, 1979, **44**, 2044 (*isol*)
Kirkup, M.P. *et al*, *Phytochemistry*, 1983, **22**, 2539 (*isol*)
Nagaoka, H. *et al*, *Tetrahedron Lett.*, 1988, **29**, 5945 (*abs config*)

Dictyone   D-70146

[84164-88-5]

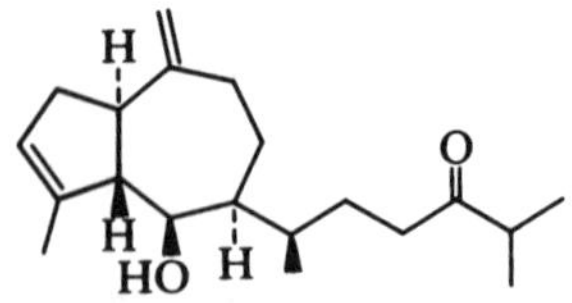

$C_{20}H_{32}O_2$ M 304.472

Constit. of brown alga *Dictyota dichotoma*. Oil. $[\alpha]_D^{28}$ +48.5° (c, 1.15 in CHCl$_3$).

Enoki, N. *et al*, *Chem. Lett.*, 1982, 1837.

Dictyotriol A   D-70147

[93710-24-8]

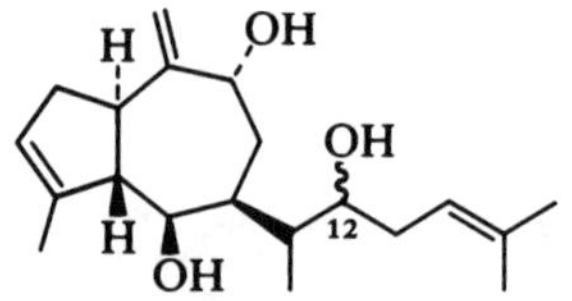

$C_{20}H_{32}O_3$ M 320.471

Constit. of the brown alga *Glossophora kuntii*. Amorph. solid. $[\alpha]_D^{25}$ +24.9° (c, 0.47 in CHCl$_3$).

12-Ac:
$C_{22}H_{34}O_4$ M 362.508
From *G. kuntii*. Oil. $[\alpha]_D^{25}$ +7.8° (c, 0.017 in CHCl$_3$).

Rivera, A.P. *et al*, *J. Nat. Prod.*, 1987, **50**, 965.

Dictytriene A   D-70148

[84164-86-3]

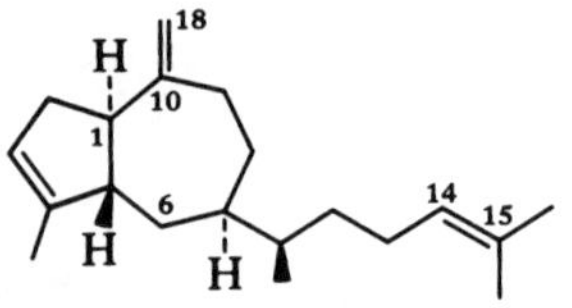

$C_{20}H_{32}$ M 272.473

Constit. of brown alga *Dictyota dichotoma*. Oil. $[\alpha]_D^{22}$ +37° (c, 0.28 in cyclohexane).

$\Delta^{1(10)}$*-Isomer:* [84164-87-4]. **Dictytriene B**.
$C_{20}H_{32}$ M 272.473
Constit. of *D. dichotoma*. Oil. $[\alpha]_D^{22}$ +54.4° (c, 0.29 in cyclohexane).

14,15-Dihydro, 6β,14R,15-trihydroxy: **Dictytriol**.
$C_{20}H_{34}O_3$ M 322.487
Constit. of *D. dichotoma*. Cryst. Mp 92-93°. $[\alpha]_D^{24}$ +80.7° (c, 0.63 in CHCl$_3$).

14,15-Dihydro, 6β,14S,15-trihydroxy: **Isodictytriol**.
$C_{20}H_{34}O_3$ M 322.487
Constit. of *D. dichotoma*. Cryst. Mp 125-126°.

Enoki, N. *et al*, *Chem. Lett.*, 1982, 1837 (*isol*)
Kusumi, T. *et al*, *Chem. Lett.*, 1986, 1241 (*cryst struct, deriv*)

Dicyanoacetic acid   D-70149

(NC)$_2$CHCOOH

$C_4H_2N_2O_2$ M 110.072

Me ester: [2040-70-2]. *Methyl dicyanoacetate.*
$C_5H_4N_2O_2$ M 124.099
Cryst. (CHCl$_3$). Mp 59-60° (65°).

Et ester: [74908-84-2]. *Ethyl dicyanoacetate.*
$C_6H_6N_2O_2$ M 138.126
Cryst. Mp 36-37°. Bp$_{0.2}$ 30-50°.

Arndt, F. *et al*, *Justus Liebigs Ann. Chem.*, 1935, **521**, 95 (*synth*)
Gano, J.E. *et al*, *J. Org. Chem.*, 1987, **52**, 2102 (*synth, pmr, cmr, ir*)

Dicyclohexylmethanol, 8CI   D-70150

α-*Cyclohexylcyclohexanemethanol*, 9CI.
Dicyclohexylcarbinol

[4453-82-1]

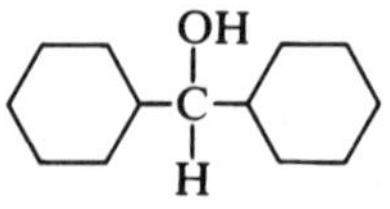

$C_{13}H_{24}O$ M 196.332
Mp 66° (62-63°). Bp$_{12}$ 154°.

Phenylurethane: Cryst. (pet. ether). Mp 157°.
Acid phthalate: Mp 136°.

Neunhoeffer, O., *Justus Liebigs Ann. Chem.*, 1934, **509**, 115 (*synth*)
Brown, H.C. *et al*, *J. Org. Chem.*, 1985, **50**, 4032 (*synth*)
Sgarabotto, P. *et al*, *Acta Crystallogr., Sect. C*, 1988, **44**, 671 (*cryst struct*)

5,6:12,13-Diepoxy-5,6,12,13-tetrahydrodibenz[*a,h*]anthracene   D-70151

1a,6b,7a,12b-Tetrahydrodibenz[3,4:7,8]anthra[1,2-b:5,6-b']bisoxirene, 9CI. *Dibenz*[a,h]*anthracene-5,6:11,12-dioxide*

[55400-87-8]

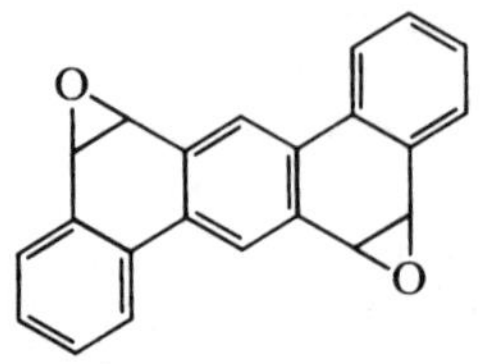

$C_{22}H_{14}O_2$ M 310.351
Needles (C$_6$H$_6$/hexane). Mp 192-193° dec. Light sensitive.

Agarwal, S.C. *et al*, *J. Org. Chem.*, 1975, **40**, 2307 (*synth, ir, uv*)

4,5:9,10-Diepoxy-4,5,9,10-tetrahydropyrene   D-70152

3b,4a,7b,8a-Tetrahydropyreno[4,5-b:9,10-b']bisoxirene, 9CI. *Pyrene-4,5:9,10-dioxide*

[55400-88-9]

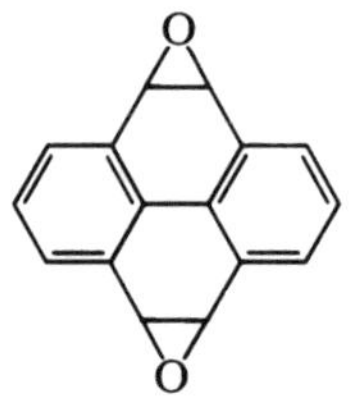

$C_{16}H_{10}O_2$ M 234.254

Needles (dioxan). Mp 206-208° dec.

Agarwal, S.C. *et al*, *J. Org. Chem.*, 1975, **40**, 2307 (*synth*, *ir*, *uv*)

Moriarty, R.M. *et al*, *Tetrahedron Lett.*, 1975, 2557 (*synth*)

Diethoxyacetaldehyde D-70153

Updated Entry replacing D-10251
Glyoxal diethylacetal
[5344-23-0]

$$(EtO)_2CHCHO$$

$C_6H_{12}O_3$ M 132.159
Bp_{10} 50-55°. Readily forms hydrate.

Stetter, H. *et al*, *Synthesis*, 1981, 129 (*synth*, *pmr*)
Bahler, J.H., *et al*, *Synth. Commun.*, 1987, **17**, 77 (*synth*, *pmr*)

1,2-Diethylidenecyclohexane, 9CI D-70154

1,2-Bis(ethylidene)cyclohexane

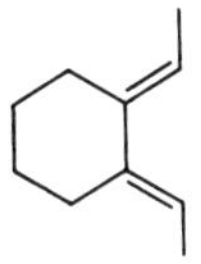

$C_{10}H_{16}$ M 136.236
(*E,E*)-*form* [92013-62-2]
Liq. Bp_{12} 67-69°.

Nugent, W.A. *et al*, *J. Am. Chem. Soc.*, 1987, **109**, 2788 (*synth*, *pmr*)

N,N'-Diethylthiourea, 9CI D-70155

Updated Entry replacing D-03450
N,N'-*Diethylthiocarbamide*
[105-55-5]

$$EtNHCSNHEt$$

$C_5H_{12}N_2S$ M 132.223
Cryst. Sol. H_2O, EtOH. Mp 77°.
▷YS9800000.

Hofman, A.W., *Ber.*, 1868, **1**, 27 (*synth*)
Martin, D. *et al*, *CA*, 1965, **62**, 3960 (*synth*)
B.P., 1 173 521, (*1969*); *CA*, **72**, 54821 (*synth*)

1,3-Diethynylbenzene, 9CI D-70156

Updated Entry replacing D-03457
[1785-61-1]
$C_{10}H_6$ M 126.157
Starting compound for many polymers. Mp 2.5°. Bp_{15} 78°.

Deluchat, R., *C. R. Hebd. Seances Acad. Sci.*, 1931, **192**, 1387 (*synth*)
Brogli, F. *et al*, *Helv. Chim. Acta*, 1975, **58**, 2620.
King, G.W. *et al*, *J. Mol. Spectrosc.*, 1978, **70**, 53 (*ir*, *raman*)
Neenan, T.X. *et al*, *J. Org. Chem.*, 1988, **53**, 2489 (*synth*, *ir*, *pmr*)

1,4-Diethynylbicyclo[2.2.2.]octane D-70157

[105858-44-4]

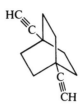

$C_{12}H_{14}$ M 158.243
Cryst. by subl. Mp 162-164°.

Honegger, E. *et al*, *Chem. Ber.*, 1987, **120**, 187 (*synth*, *ir*, *pmr*, *cmr*, *pe*)

2,7-Diethynylnaphthalene D-70158

[113705-27-4]
$C_{14}H_8$ M 176.217
Cryst. (MeOH). Mp 135-137°.

Neenan, T.X. *et al*, *J. Org. Chem.*, 1988, **53**, 2489 (*synth*, *ir*, *pmr*)

1,3-Diethynyl-2,4,5,6-tetrafluorobenzene D-70159

[113705-24-1]

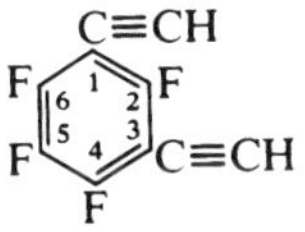

$C_{10}H_2F_4$ M 198.119
Liq. Bp_1 55°.

Neenan, T.X. *et al*, *J. Org. Chem.*, 1988, **53**, 2489 (*synth*, *ir*, *pmr*)

1,4-Diethynyl-2,3,5,6-tetrafluorobenzene D-70160

[38002-32-3]
$C_{10}H_2F_4$ M 198.119
Cryst. ($CHCl_3$/MeOH). Mp 132-134°.

Waugh, F. *et al*, *J. Organomet. Chem.*, 1972, **39**, 275 (*synth*, *uv*, *ir*)
Neenan, T.X. *et al*, *J. Org. Chem.*, 1988, **53**, 2489 (*synth*, *ir*, *pmr*, *ms*)

9,10-Difluoroanthracene, 9CI D-70161

Updated Entry replacing D-03485
[1545-69-3]
$C_{14}H_8F_2$ M 214.214
Yellow needles. Mp 164-165°.

Wittig, G. *et al*, *Justus Liebigs Ann. Chem.*, 1959, **623**, 17 (*synth*)
Logothelis, A.L. *et al*, *J. Org. Chem.*, 1966, **31**, 3686 (*synth*)
Anderson, G. *et al*, *J. Am. Chem. Soc.*, 1971, **93**, 6984 (*pmr*)
Stock, L.M. *et al*, *J. Org. Chem.*, 1976, **41**, 1660 (*synth*)
Duerr, B.F. *et al*, *J. Org. Chem.*, 1988, **53**, 2120 (*synth*, *pmr*)

1,1-Difluoro-1,3-butadiene D-70162

[590-91-0]

$$H_2C=CHCH=CF_2$$

$C_4H_4F_2$ M 90.072
Bp 3.5-5°.

Tarrant, P. *et al*, *J. Am. Chem. Soc.*, 1954, **76**, 944, 3466 (*synth*)

Dolbier, W.R. *et al*, *J. Org. Chem.*, 1987, **52**, 1872 (*pmr, F nmr*)

2,2-Difluorobutanedioic acid, 9CI   D-70163

Updated Entry replacing D-03502
2,2-Difluorosuccinic acid
[665-31-6]

$$HOOCCF_2CH_2COOH$$

$C_4H_4F_2O_4$ M 154.070
Cryst. (MeNO$_2$). Mp 144-146°.

Org. Synth., Coll. Vol., **5**, 393 (*synth*)
Ovenall, D.W. *et al*, *J. Magn. Reson.*, 1977, **25**, 361 (*cmr*)
Akhtar, M. *et al*, *Tetrahedron*, 1987, **43**, 5899 (*synth, ir, pmr*)

3,3-Difluorocyclobutanecarboxylic acid   D-70164

[107496-54-8]

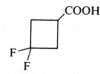

$C_5H_6F_2O_2$ M 136.098
Needlelike cryst. Mp 48-50°. Bp$_{4-5}$ 85-90°.

Dolbier, W.R. *et al*, *J. Org. Chem.*, 1987, **52**, 1872 (*synth, ir, pmr, Fnmr, ms*)

3,3-Difluorocyclobutene   D-70165

[29507-09-3]

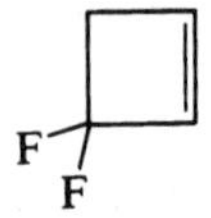

$C_4H_4F_2$ M 90.072

Dolbier, W.R. *et al*, *J. Org. Chem.*, 1987, **52**, 1872 (*synth, pmr, F nmr, ms*)

4,5-Difluorophenanthrene   D-70166

[108665-38-9]

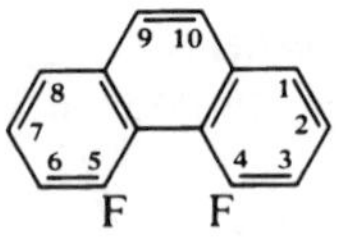

$C_{14}H_8F_2$ M 214.214
Prisms (pet. ether). Mp 105°.

Cosmo, R. *et al*, *Aust. J. Chem.*, 1987, **40**, 2137 (*synth, uv, ir, pmr, ms*)

9,10-Difluorophenanthrene   D-70167

[56830-34-3]
$C_{14}H_8F_2$ M 214.214

Zupan, M. *et al*, *J. Org. Chem.*, 1975, **40**, 3794 (*F nmr*)

1,3-Di(2-furyl)benzene   D-70168

2,2′-(1,3-Phenylene)bisfuran, 9CI

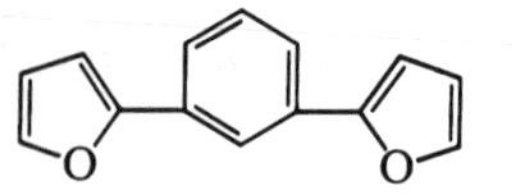

$C_{14}H_{10}O_2$ M 210.232
Cryst. Mp 37-38°.

Pelter, A. *et al*, *Tetrahedron Lett.*, 1987, **28**, 5213 (*synth, uv*)

1,4-Di(2-furyl)benzene   D-70169

2,2′-(1,4-Phenylene)bisfuran, 9CI
[34121-64-7]
$C_{14}H_{10}O_2$ M 210.232
Cryst. by subl. Mp 148-150°. Rapidly oxidised.

Pelter, A. *et al*, *Synthesis*, 1987, 51 (*synth, pmr, cmr, ms*)
Pelter, A. *et al*, *Tetrahedron Lett.*, 1987, **28**, 5213 (*synth, uv*)

6b,7a-Dihydroacenaphth[1,2-b]oxirene,   D-70170
9CI

*1,2-Epoxyacenaphthylene, 8CI. 1,2-Epoxyacenaphthene.
Acenaphthylene oxide*
[22058-69-1]

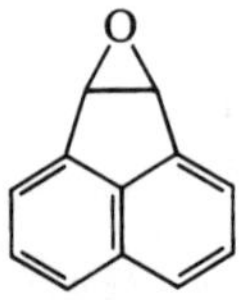

$C_{12}H_8O$ M 168.195
Cryst. (pet. ether), plates (CCl$_4$). Mp 83-84°. Stable on storage.

Kinstle, T.H. *et al*, *J. Org. Chem.*, 1970, **35**, 257 (*synth, pmr*)
Hunter, D.H. *et al*, *J. Am. Chem. Soc.*, 1973, **95**, 8333 (*synth*)

1,2-Dihydro-3H-azepin-3-one, 9CI   D-70171

1H-Azepin-3(2H)-one

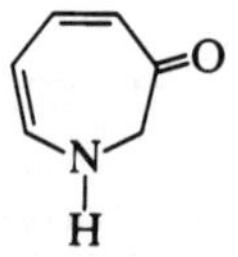

C_6H_7NO M 109.127
1-Me: [110561-67-6].
     C_7H_9NO M 123.154
     Characterised spectroscopically.
1-Ph: [110561-68-7].
     $C_{12}H_{11}NO$ M 185.225
     Characterised spectroscopically.

McNab, H. *et al*, *J. Chem. Soc., Chem. Commun.*, 1987, 140, 141 (*synth, pmr, cmr, props*)

2,3-Dihydroazete, 9CI   D-70172

1-Azetine
[6788-85-8]

C_3H_5N M 55.079
Polymerises rapidly at r.t.

Bock, H. *et al*, *Chem. Ber.*, 1987, **120**, 1961 (*synth, pe, bibl*)
Guillemin, J-C. *et al*, *Tetrahedron*, 1988, **44**, 4447 (*synth, uv, pmr*)

1,4-Dihydrobenzocyclooctatetraene, 9CI   **D-70173**

Bicyclo[6.4.0]dodeca-1(8),2,4,6,10-pentaene

[23709-78-6]

$C_{12}H_{12}$ M 156.227

Yellow oil.

Elix, J.A. *et al, J. Am. Chem. Soc.*, 1969, **91**, 4734 (*synth, uv, ir, pmr, ms*)
Streitweiser, A. *et al, Organometallics*, 1983, **2**, 1873 (*synth, pmr*)

1,8-Dihydrobenzo[2,1-*b*:3,4-*b'*]dipyrrole   **D-70174**

[112149-08-3]

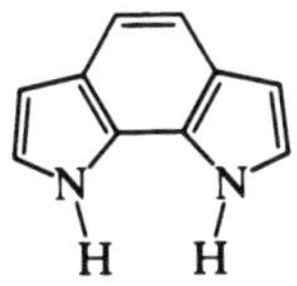

$C_{10}H_8N_2$ M 156.187

Cryst. Mp 217° dec. $Bp_{0.01}$ 180° subl.

Berlin, A. *et al, J. Chem. Soc., Chem. Commun.*, 1987, 1176 (*synth, pmr*)

3,4-Dihydro-2*H*-1-benzopyran-6-ol   **D-70175**

6-Chromanol. 6-Hydroxychroman

[5614-78-8]

$C_9H_{10}O_2$ M 150.177

Cryst. (hexane). Mp 99-100°.

Ac:
   $C_{11}H_{12}O_3$ M 192.214
   Cryst. (pentane). Mp 51-52°.

Me ether: 3,4-Dihydro-6-methoxy-2H-1-benzopyran. 6-Methoxychroman.
   $C_{10}H_{12}O_2$ M 164.204
   d_4^{23} 1.11. $Bp_{0.05}$ 92°. n_D^{23} 1.5471.

Willhalm, B. *et al, Tetrahedron*, 1964, **20**, 1185 (*synth, ms*)
Shiratsuchi, M. *et al, Chem. Pharm. Bull.*, 1987, **35**, 632 (*synth, pmr*)

1,4-Dihydro-2(3*H*)-benzopyran-3-one, 9CI   **D-70176**

Updated Entry replacing D-10300

3-Isochromanone

[4385-35-7]

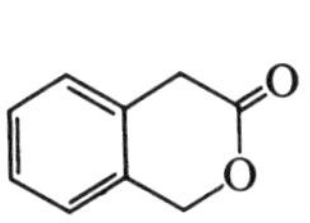

$C_9H_8O_2$ M 148.161

Mp 82-83°. Bp_1 130°.

Mann, F.G. *et al, J. Chem. Soc.*, 1954, 2819.
Spangler, R.J. *et al, Synthesis*, 1973, 107 (*synth*)
Garwood, R.F. *et al, J. Chem. Soc., Perkin Trans. 1*, 1975, 2471 (*synth*)
Chatterjee, A. *et al, Synthesis*, 1981, 818 (*synth*)
Cottet, F. *et al, Synthesis*, 1987, 497 (*synth*)

1,2-Dihydrocyclobuta[*a*]naphthalene-2-carboxylic acid, 9CI   **D-70177**

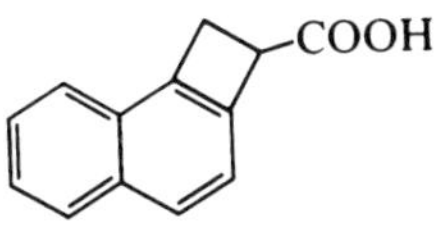

$C_{13}H_{10}O_2$ M 198.221

(±)-*form* [113340-31-1]

Needles (hexane/Et_2O). Mp 181-182°.

Me ester: [113340-32-2].
   $C_{14}H_{12}O_2$ M 212.248
   Synthon for steroid synth. Cryst. (pentane). Mp 48-49°.

Amide: [85180-89-8].
   $C_{13}H_{11}NO$ M 197.236
   Needles (MeOH). Mp 215-219°.

Nitrile: [113340-30-0]. *2-Cyano-1,2-dihydrocyclobuta[a]naphthalene.*
   $C_{13}H_9N$ M 179.221
   Synthon for steroid synth. Needles (hexane/Et_2O). Mp 105-106°.

Sato, M. *et al, Chem. Pharm. Bull.*, 1987, **35**, 3647 (*synth, pmr*)

3,8-Dihydrocyclobuta[*b*]quinoxaline-1,2-dione, 9CI, 8CI   **D-70178**

1,2,3,8-Tetrahydrocyclobuta[b]quinoxaline-1,2-dione

[20420-52-4]

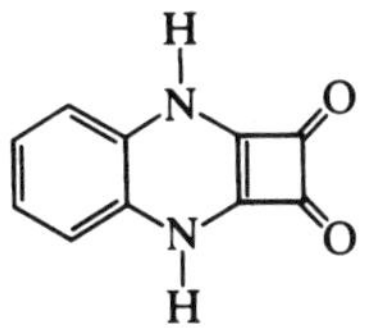

$C_{10}H_6N_2O_2$ M 186.170

Red-brown to orange-red needles and spars. or dark-red solid. Mp 330° (303°) dec.

Skujins, S. *et al, J. Chem. Soc., Chem. Commun.*, 1968, 598 (*synth, pmr*)
Ried, W. *et al, Chem. Ber.*, 1969, **102**, 1439 (*synth, ir*)
Griffiths, G.T. *et al, J. Mol. Struct.*, 1971, **9**, 333 (*uv*)

5,6-Dihydro-4*H*-cyclopenta-1,2,3-thiadiazole, 9CI   **D-70179**

Cyclopenteno-1,2,3-thiadiazole

[56382-73-1]

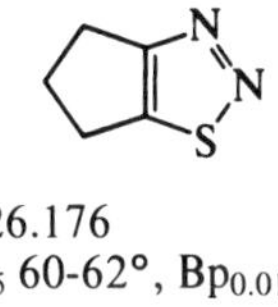

$C_5H_6N_2S$ M 126.176

Sl. yellow liq. $Bp_{0.05}$ 60-62°, $Bp_{0.01}$ 42-46°.

2-Oxide:
   $C_5H_6N_2OS$ M 142.175
   Mp 123°.

Braun, H.P. *et al, Tetrahedron*, 1975, **31**, 637 (*synth, pmr, cmr, uv*)
Spencer, H.K. *et al, J. Org. Chem.*, 1976, **41**, 730 (*synth*)

5,6-Dihydro-4H-cyclopenta[b]thiophene,   **D-70180**
9CI

Cyclopenta[b]*thiophene*
[5650-50-0]

C₇H₈S   M 124.200
Liq. Mp ca. −17°. Bp₁₅ 71-72°.

MacDowell, D.W.H. *et al, J. Org. Chem.*, 1967, **32**, 1226 (*synth, ir, pmr*)

5,6-Dihydro-4H-cyclopenta[c]thiophene,   **D-70181**
9CI

Cyclopenta[c]*thiophene*
[7690-98-4]

C₇H₈S   M 124.200
Liq. Bp₂₀ 81-82°.

MacDowell, D.W.H. *et al, J. Org. Chem.*, 1967, **32**, 1226 (*synth, ir, uv, pmr*)

4,5-Dihydro-6H-cyclopenta[b]thiophen-6-   **D-70182**
one

6-Oxocyclopenta[b]*thiophene*
[5650-52-2]

C₇H₆OS   M 138.184
Solid by subl. Mp 92-93°. Bp₀.₉ 97-99°.

2,4-Dinitrophenylhydrazone: Red prisms (EtOAc). Mp 255-255.5°.

MacDowell, D.W.H. *et al, J. Org. Chem.*, 1967, **32**, 1226 (*synth, uv, ir, pmr*)

4,6-Dihydro-5H-cyclopenta[b]thiophen-5-   **D-70183**
one

5-Oxocyclopenta[b]*thiophene*
[33449-51-3]

C₇H₆OS   M 138.184
Faint-yellow cryst. (hexane). Mp 100-101°.

Skramstad, J., *Acta Chem. Scand.*, 1971, **25**, 1287 (*synth, pmr, ir, ms*)

5,6-Dihydro-4H-cyclopenta[c]thiophen-4-   **D-70184**
one, 9CI

4-Oxocyclopenta[c]*thiophene*
[7687-82-3]

C₇H₆OS   M 138.184
Needles (hexane or by subl.). Mp 81-82°.

MacDowell, D.W.H. *et al, J. Org. Chem.*, 1967, **32**, 1226 (*synth, uv, ir, pmr*)

10,11-Dihydro-5H-dibenz[b,f]azepine, 9CI   **D-70185**

Updated Entry replacing D-50320
o-*Imidodibenzyl. Iminodibenzyl*
[494-19-9]

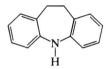

C₁₄H₁₃N   M 195.263
Imipramine metab. Pale-yellow cryst. by subl.; prisms (pet. ether). Mp 110°. pKₐ −0.7 (15.75% EtOH, 20°).
▷HN8950000.

5-Ac: [13080-75-6].
   C₁₆H₁₅NO   M 237.301
   Cryst. (EtOH). Mp 96-97°.
5-Me: [4513-01-3]. *10,11-Dihydro-5-methyl-5H-dibenz*[b,f]*azepine, 9CI.*
   C₁₅H₁₅N   M 209.290
   Needles (MeOH). Mp 106-107°.
5-Ph: [78943-60-9]. *10,11-Dihydro-5-phenyl-5H-dibenz*[b,f]*azepine, 9CI.*
   C₂₀H₁₅N   M 269.345
   Prisms (hexane). Mp 86-88°.

Thiele, J. *et al, Justus Liebigs Ann. Chem.*, 1899, **305**, 96 (*synth*)
Reynaud, R. *et al, Bull. Soc. Chim. Fr.*, 1963, 1805 (*props*)
Fr. Pat., 1 351 837, (*1964*); *CA*, **60**, 15847d (*synth*)
Itier, J. *et al, Bull. Soc. Chim. Fr.*, 1969, 2355 (*uv*)
Profitt, J.A. *et al, J. Org. Chem.*, 1979, **44**, 3972 (*synth*)
Ohta, T. *et al, J. Am. Chem. Soc.*, 1980, **102**, 6385 (*synth*)
Ohta, T. *et al, Chem. Pharm. Bull.*, 1981, **29**, 1221 (*derivs*)
Caranoni, C. *et al, J. Appl. Crystallogr.*, 1983, **16**, 649 (*cryst struct*)
Hallberg, A. *et al, J. Heterocycl. Chem.*, 1984, **21**, 197 (*pmr, cmr*)
Renfroe, B. *et al, Chem. Heterocycles* Compd., 1984, **43**, 1 (*rev*)
Al-Showaier, I. *et al, J. Heterocycl. Chem.*, 1986, **23**, 731 (*ms*)

6,7-Dihydro-5H-dibenzo[a,c]cycloheptene,   **D-70186**
9CI

Dibenzocyclohepta-1,3-diene
[1015-80-1]

C₁₅H₁₄   M 194.276
Mp 40°, Mp 55.5-55° (dimorph.). Bp₀.₃ 99-100°.

Rapoport, H. *et al, J. Am. Chem. Soc.*, 1964, **71**, 1774 (*synth*)
Griffin, G.W. *et al, Org. Prep. Proced. Int.*, 1985, **17**, 187 (*synth*)

5,7-Dihydro-6H-dibenzo[a,c]cyclohepten-   **D-70187**
6-one, 9CI

1,2,3,4-Dibenzcyclohepta-1,3-diene-6-one
[1139-82-8]
C₁₅H₁₂O   M 208.259
Cryst. (MeOH aq.). Mp 78-79.5°.

Kenner, J. *et al, J. Chem. Soc.*, 1911, 2101 (*synth*)
Cope, A.C. *et al, J. Am. Chem. Soc.*, 1956, **78**, 1012 (*synth*)
Mislow, K. *et al, J. Am. Chem. Soc.*, 1962, **84**, 1455 (*synth, uv, pmr*)

Satyanarayana, N. *et al*, *Tetrahedron Lett.*, 1987, **28**, 2633 (*synth, pmr*)

6,7-Dihydro-5*H*-dibenzo[*a,c*]cyclohepten-5-one, 9CI   D-70188

Dibenzo[a,c] *[1,3]cycloheptadien-5-one. 1,2,3,4-Dibenz-cyclohepta-1,3-dien-5-one*

[53137-51-2]

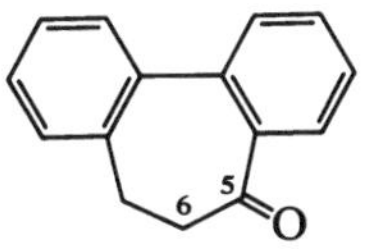

$C_{15}H_{12}O$ M 208.259

Cryst. (MeOH aq.). Mp 85-86°.

Cook, J.W. *et al*, *J. Am. Chem. Soc.*, 1947, 746 (*synth*)
Rapoport, H. *et al*, *J. Am. Chem. Soc.*, 1949, **71**, 1774 (*synth*)
Coburn, T.T. *et al*, *J. Am. Chem. Soc.*, 1974, **96**, 5218 (*synth*)

5,7-Dihydrodibenzo[*c,e*]selenepin, 9CI   D-70189

5H,7H-Dibenzo[c,e]*selenepin*

[33948-83-3]

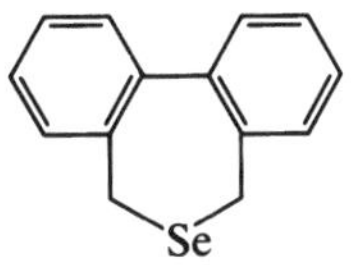

$C_{14}H_{12}Se$ M 259.209

Yellow solid (hexane); cryst. (pet. ether). Mp 65-67°.

Truce, W.E. *et al*, *J. Am. Chem. Soc.*, 1956, **78**, 6130 (*synth*)
Williams, D.J. *et al*, *Org. Prep. Proced. Int.*, 1985, **17**, 270 (*synth, ir, pmr, bibl*)

5,7-Dihydrodibenzo[*c,e*]thiepin, 9CI   D-70190

[6672-64-6]

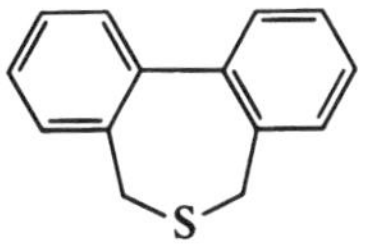

$C_{14}H_{12}S$ M 212.309

Cryst. (MeOH). Mp 90-91°.

S-*Oxide:*
  $C_{14}H_{12}OS$ M 228.308
  Cryst. (EtOH). Mp 129-130°.

S-*Dioxide:*
  $C_{14}H_{12}O_2S$ M 244.308
  Cryst. (EtOH). Mp 210-211°.

Truce, W.E. *et al*, *J. Am. Chem. Soc.*, 1956, **78**, 6130 (*synth, uv*)
Šindelář, K. *et al*, *Collect. Czech. Chem. Commun.*, 1986, **51**, 2848 (*synth*)

2,3-Dihydro-3,5-dihydroxy-2-methyl-naphthoquinone   D-70191

Dihydrodroserone. PZ 5

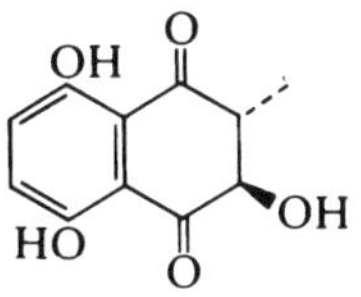

$C_{11}H_{10}O_4$ M 206.198

Isol. from *Plumbago zeylanica*. Mp 67°.

Dinda, B. *et al*, *Chem. Ind. (London)*, 1986, 823 (*isol*)

1,4-Dihydro-3,6-dimethoxy-1,2,4,5-tetra-zine   D-70192

[81930-32-7]

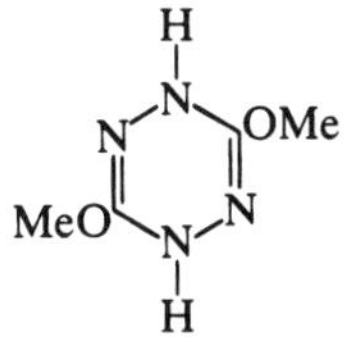

$C_4H_8N_4O_2$ M 144.133

Cryst. (CH_2Cl_2/C_6H_6). Mp 160° dec. Forms a radical anion.

Neugebauer, F.A. *et al*, *Justus Liebigs Ann. Chem.*, 1982, 387 (*synth, pmr*)

2,3-Dihydro-2,2-dimethyl-4*H*-1-benzothio-pyran, 9CI   D-70193

2,2-Dimethyl-4-oxothiochroman

[28035-02-1]

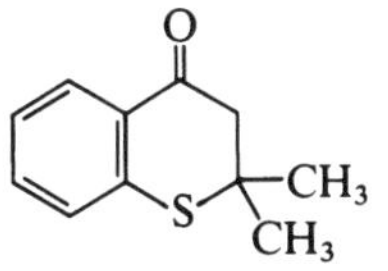

$C_{11}H_{12}OS$ M 192.275

Mp 66-68° (63-65°). $Bp_{0.3}$ 115°.

Kurth, H.J. *et al*, *Justus Liebigs Ann. Chem.*, 1977, 1141 (*synth, pmr, ir*)
Tercio, J. *et al*, *Synthesis*, 1987, 149 (*synth, pmr, ir*)

9,10-Dihydro-9,10-dimethyleneanthracene   D-70194

9,10-Dihydro-9,10-bis(methylene)anthracene. 9,10-Anthraquinodimethane

[105020-57-3]

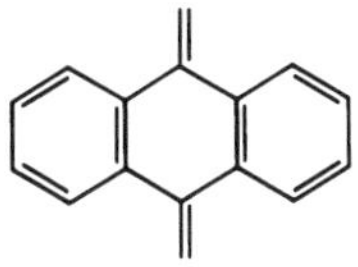

$C_{16}H_{12}$ M 204.271

Unstable compd. stabilised as Fe complex.

Williams, D.J. *et al*, *J. Am. Chem. Soc.*, 1970, **92**, 1436; 1971, **93**, 5034 (*pmr, uv*)
Koray, A.R. *et al*, *Angew. Chem., Int. Ed. Engl.*, 1985, **24**, 521.

2,3-Dihydro-2,3-dimethylenethiophene   D-70195

2,3-Dihydro-2,3-bis(methylene)thiophene, 9CI

[99646-78-3]

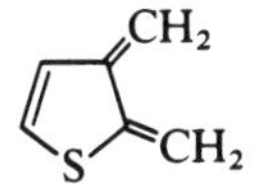

C_6H_6S M 110.173

Intermed. prepd. in soln. and trapped as Diels-Alder adduct.

Chadwick, D.J. *et al*, *Tetrahedron Lett.*, 1987, **28**, 6085 (*synth*)

Dihydro-4,4-dimethyl-2,3-furandione, 9CI   **D-70196**

α-Keto-β,β-dimethyl-γ-butyrolactone. Ketopantolactone

[13031-04-4]

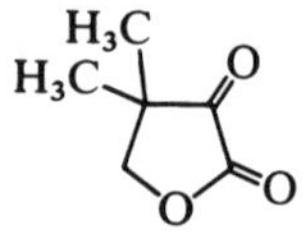

$C_6H_8O_3$ M 128.127

Cryst. (Et₂O). Mp 66-68° (60°). Bp₁₂ 120-125°.

Hydrazone:
  $C_6H_{10}N_2O_2$ M 142.157
  Needles (EtOH). Mp 200-202°.
Phenylhydrazone: Needles (CHCl₃). Mp 169-171°.
2,4-Dinitrophenylhydrazone: Yellow cubes (AcOH). Mp 242-243.5°.

Kuhn, R. *et al*, *Ber.*, 1942, **75**, 121 (*synth*)
Lipton, S.H. *et al*, *J. Am. Chem. Soc.*, 1949, **71**, 2364 (*synth*)
Nagase, O. *et al*, *Chem. Pharm. Bull.*, 1969, **17**, 398 (*synth, ir*)

Dihydro-4,4-dimethyl-5-methylene-2(3H)-   **D-70197**
furanone, 9CI

4,4-Dimethyl-5-methylene-γ-butyrolactone

[65371-43-9]

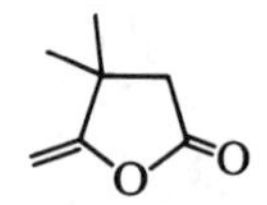

$C_7H_{10}O_2$ M 126.155

Synthon for pyrethroids and other compds. Liq. Bp₆ 51-55°.

Joshi, G.S. *et al*, *Synth. Commun.*, 1988, **18**, 559 (*synth, ir, pmr, bibl*)

4,5-Dihydro-4,4-dimethyloxazole, 9CI   **D-70198**

4,4-Dimethyl-2-oxazoline

[30093-99-3]

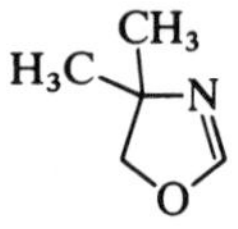

C_5H_9NO M 99.132

Mobile liq. Bp 99-100°.

B,MeI: Solid. Mp 215° dec. Slightly hygroscopic.

Meyers, A.I. *et al*, *J. Am. Chem. Soc.*, 1970, **92**, 6676 (*synth, ir, pmr*)
Org. Synth., 1974, **54**, 42 (*synth, use*)

3a,6a-Dihydro-3a,6a-dimethyl-1,6-penta-   **D-70199**
lenedione

1,5-Dimethylbicyclo[3.3.0]octa-3,6-diene-2,8-dione

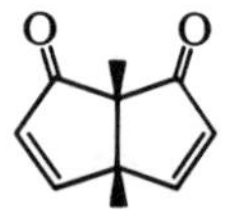

$C_{10}H_{10}O_2$ M 162.188

(3aRS,6aSR)-form [109308-71-6]
cis-*form*
Cryst. (CH₂Cl₂). Mp 203-205° (198-199°).

Quast, H. *et al*, *Justus Liebigs Ann. Chem.*, 1987, 965 (*synth, pmr, cmr*)

3,4-Dihydro-2,2-dimethyl-2H-pyrrole, 9CI   **D-70200**

5,5-Dimethyl-Δ¹-pyrroline

[2045-76-3]

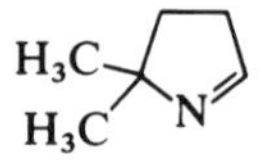

$C_6H_{11}N$ M 97.160

Bp 104-108°.

1-Oxide: [3317-61-1].
  $C_6H_{11}NO$ M 113.159
  Bp₀.₆ 66-67°. Hygroscopic.
Picrate: Yellow prisms (EtOH/Et₂O). Mp 120-124°.

Bonnett, R. *et al*, *J. Chem. Soc.*, 1959, 2087, 2094 (*synth, ir*)

4,5-Dihydro-5,5-dimethyl-3H-1,2,4-tria-   **D-70201**
zol-3-one, 9CI

5,5-Dimethyl-Δ¹-1,2,4-triazolin-3-one

[112700-85-3]

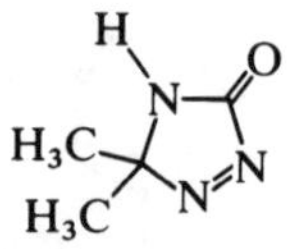

$C_4H_7N_3O$ M 113.119

Cryst. (Et₂O). Sol. H₂O. Mp 84-87°.

Schantl, J.G. *et al*, *Heterocycles*, 1987, **26**, 1439 (*synth*)
Schantl, J.G. *et al*, *J. Heterocycl. Chem.*, 1987, **24**, 1401 (*ir, uv, pmr, cmr, ms, cryst struct*)

5,6-Dihydro-11,12-diphenyldibenzo[a,e]-   **D-70202**
cyclooctene

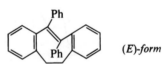

(E)-form

$C_{28}H_{22}$ M 358.482

(E)-form [114491-65-5]
First example of a *Z,E,Z*-cyclooctatriene deriv. Yellow cryst. (EtOH). Mp 173-174°.
(Z)-form [114504-92-6]
No phys. props. reported.

Hannemann, K. *et al*, *Angew. Chem., Int. Ed. Engl.*, 1988, **27**, 853 (*synth, uv, pmr, cmr*)

2,3-Dihydro-5,6-diphenyl-1,4-oxathiin, 9CI   **D-70203**

[58041-19-3]

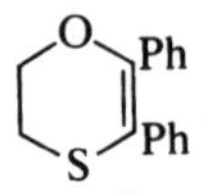

$C_{16}H_{14}OS$ M 254.346

Mp 59-60°.

▷RP5320000.

Mattay, J. *et al*, *J. Org. Chem.*, 1986, **51**, 1894 (*synth, ir, pmr, cmr*)

2,5-Dihydro-3,6-diphenylpyrrolo[3,4-*c*]- pyrrole-1,4-dione, 9CI   D-70204

3,6-Diphenyl-2,5-diaza-1,4(2H,5H)-pentalenedione

[54660-00-3]

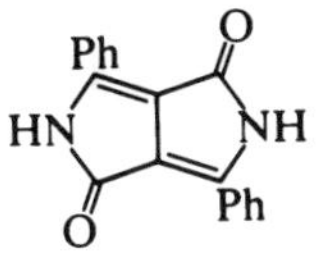

$C_{18}H_{12}N_2O_2$ M 288.305

Pigment. Red powder.

Farnum, D.G. *et al, Tetrahedron Lett.*, 1974, 2549.
Closs, F. *et al, Angew. Chem., Int. Ed. Engl.*, 1987, **26**, 552 (*synth, uv*)

2,5-Dihydro-3,4-diphenylselenophene, 9CI   D-70205

[113495-60-6]

$C_{16}H_{14}Se$ M 285.247

Cryst. Thermally stable.

Nakayama, J. *et al, J. Chem. Soc., Chem. Commun.*, 1987, 1072 (*synth*)

9,10-Dihydro-9,10-epidithioanthracene, 9CI   D-70206

9,10-Epidithio-9,10-dihydroanthracene

[116195-01-8]

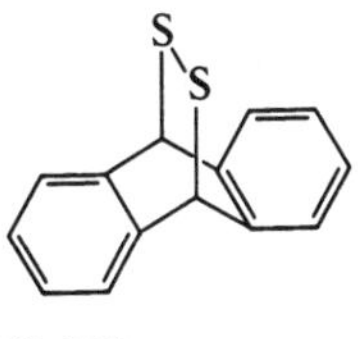

$C_{14}H_{10}S_2$ M 242.353

Unstable intermed.; source of singlet diatomic *S* which can be trapped with conj. dienes.

Ando, W. *et al, Tetrahedron Lett.*, 1987, **28**, 6653 (*synth*)

9,10-Dihydro-9,10-ethanoanthracene, 9CI   D-70207

Dibenzobicyclo[2.2.2]octadiene

[5675-64-9]

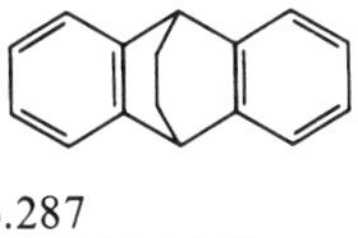

$C_{16}H_{14}$ M 206.287

Cryst. (pentane). Mp 145-146°.

Tanida, H. *et al, Tetrahedron Lett.*, 1964, 807.
De Wit, J. *et al, Tetrahedron*, 1973, **29**, 1399 (*synth*)
Adcock, W. *et al, J. Organomet. Chem.*, 1975, **102**, 297 (*synth, nmr*)
Baldwin, J.E. *et al, Tetrahedron*, 1986, **42**, 3943 (*synth, uv, nmr*)

9,10-Dihydro-9,10-ethenoanthracene- 11,12-dicarboxylic acid, 9CI   D-70208

[1625-81-6]

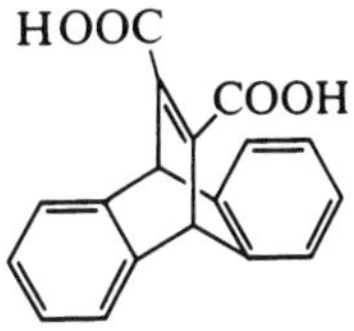

$C_{18}H_{12}O_4$ M 292.290

Mp 215.5-216°.

Di-Me ester: [1625-82-7].
   $C_{20}H_{16}N_2$ M 284.360
   Mp 160-161°.
Dinitrile: [19067-49-3].
   $C_{18}H_{10}N_2$ M 254.290
   Cryst. (MeCN). Mp 267-268°.

Weis, C.D., *J. Org. Chem.*, 1963, **28**, 74 (*nitrile*)
Figeys, H.P. *et al, Tetrahedron*, 1972, **28**, 3031 (*synth*)
Oliver, S.W. *et al, Acta Crystallogr., Sect. C*, 1986, **42**, 927 (*synth, cryst struct, nitrile*)

3,4-Dihydro-3-fluoro-2*H*-1,5-benzodioxe- pin, 9CI   D-70209

[107408-92-4]

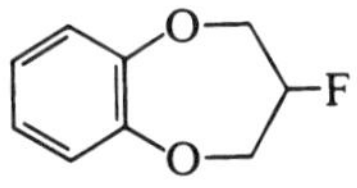

$C_9H_9FO_2$ M 168.167

Cl, Br and I analogues also prepd. Solid. Mp 40°.

Dionne, P. *et al, J. Am. Chem. Soc.*, 1987, **109**, 2616 (*synth, pmr, ms, cmr, conformn*)

Dihydro-2(3*H*)-furanthione, 9CI   D-70210

Thionobutyrolactone

[39700-44-2]

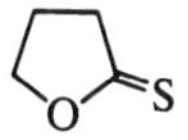

C_4H_6OS M 102.151

Kaloustian, M.K. *et al, J. Am. Chem. Soc.*, 1986, **108**, 6683 (*synth, ir, pmr*)

2,3-Dihydrofuro[2,3-*b*]pyridine, 9CI   D-70211

7-Azacoumaran

[27038-50-2]

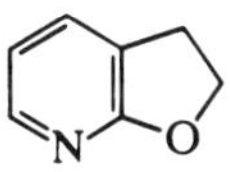

C_7H_7NO M 121.138

Bp_{15} 90°.

Sliwa, H., *Bull. Soc. Chim. Fr.*, 1970, 646 (*synth, ir, pmr*)
Frissen, A.E. *et al, Tetrahedron Lett.*, 1987, **28**, 1589 (*synth, pmr*)

4,5-Dihydro-3-hydroxy-5-(hydroxymethyl)-   D-70212
2(3H)-furanone

2-Hydroxy-4-hydroxymethyl-4-butanolide

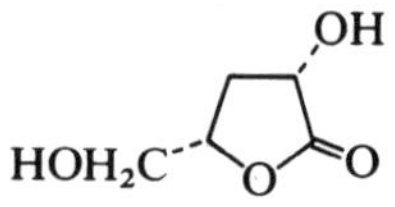

$C_5H_8O_4$ M 132.116

(3S,5S)-form

Isol. from human and rat blood serum. Shows appetite-stimulant props. Oil. $[\alpha]_D^{20}$ +23.2° (c, 3.61 in MeOH). Other stereoisomers also prepd.

Uchikawa, O. *et al*, *Bull. Chem. Soc. Jpn.*, 1988, **61**, 2025 (*synth, ir, pmr, cmr, bibl*)

2,3-Dihydro-3-hydroxy-2-(1-methyleth-   D-70213
enyl)-5-benzofurancarboxaldehyde, 9CI

5-Formyl-3-hydroxy-2-isopropenyl-2,3-dihydrobenzofuran

[115699-93-9]

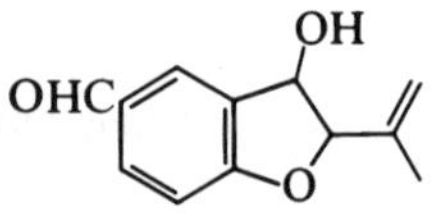

$C_{12}H_{12}O_3$ M 204.225

Metab. of *Heterobasidion annosum*. Oil. $[\alpha]_D^{20}$ +77.7° (c, 5.5 in CHCl₃).

Donnelly, D.M.X. *et al*, *Phytochemistry*, 1988, **27**, 2709.

2,3-Dihydro-2-(1-hydroxy-1-methylethyl)-   D-70214
5-benzofurancarboxaldehyde, 9CI

5-Formyl-2-(1-hydroxyisopropyl)-2,3-dihydrobenzofuran

[115699-94-0]

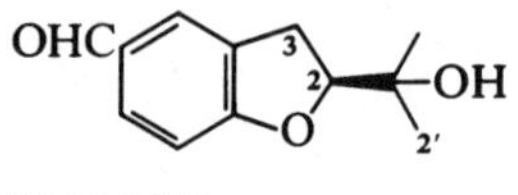

$C_{12}H_{14}O_3$ M 206.241

Metab. of *Heterobasidion annosum*. Cryst. (CHCl₃/hexane). Mp 83-86°. $[\alpha]_D^{21}$ +109.1° (c, 1.4 in CHCl₃).

2'-Hydroxy: [115699-95-1]. *2-(1,2-Dihydroxy-1-methylethyl)-2,3-dihydro-5-benzofurancarboxaldehyde, 9CI. 5-Formyl-2-(1,2-dihydroxyisopropyl)-2,3-dihydrobenzofuran.*
$C_{12}H_{14}O_4$ M 222.240
Metab. of *H. annosum*. Oil. $[\alpha]_D^{16}$ +35.14° (c, 1.1 in CHCl₃).

2,3-Didehydro: [116919-57-4]. *2-(1-Hydroxy-1-methylethyl)-5-benzofurancarboxaldehyde, 9CI. 5-Formyl-2-(1-hydroxyisopropyl)benzofuran.*
$C_{12}H_{12}O_3$ M 204.225
Metab. of *H. annosum*. Oil.

Donnelly, D.M.X. *et al*, *Phytochemistry*, 1988, **27**, 2709.

1,5-Dihydro-6H-imidazo[4,5-c]pyridazine-   D-70215
6-thione, 9CI

6-Mercaptoimidazo[4,5-c]pyridazine. Imidazo[4,5-c]-pyridazine-6-thiol

[17645-02-2]

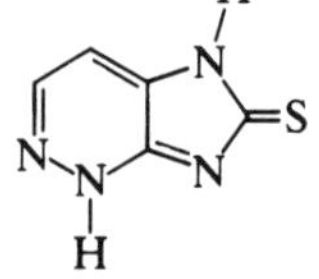

$C_5H_4N_4S$ M 152.173
Complex tautomerism possible. Cryst. (EtOH). Mp 360°.

S-Me: [17645-03-3]. *6-(Methylthio)-5H-imidazo[4,5-c]pyridazine.*
$C_6H_6N_4S$ M 166.200
Mp ca. 245° (212-213°).

N^1,*S-Di-Me:* [79690-93-0].
$C_7H_8N_4S$ M 180.227
Cryst. (cyclohexane). Mp 114-116°.

N^2,*S-Di-Me:* [79690-90-7].
$C_7H_8N_4S$ M 180.227
Cryst. (C₆H₆). Mp 145-146°.

N^5,*S-Di-Me:* [79690-91-8].
$C_7H_8N_4S$ M 180.227
Cryst. (C₆H₆). Mp 207-208°.

N^7,*S-Di-Me:* [79690-92-9].
$C_7H_8N_4S$ M 180.227
Cryst. (cyclohexane). Mp 150-153°.

Murakani, H. *et al*, *J. Heterocycl. Chem.*, 1967, **4**, 555 (*synth, uv, ir*)
Barlin, G.B. *Aust. J. Chem.*, 1981, **34**, 1361 (*deriv, synth, pmr*)

2,3-Dihydroimidazo[1,2-a]pyridine   D-70216

[27578-93-4]

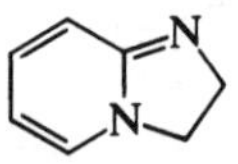

$C_7H_8N_2$ M 120.154
Yellow prisms (C₆H₆). Mp 64-65°.
Picrate: Cryst. (EtOH). Mp 213°.

Bremer, O., *Justus Liebigs Ann. Chem.*, 1935, **521**, 286 (*synth*)
Munavalli, S. *et al*, *Chem. Ind.* (*London*), 1987, 243 (*synth*)

5,6-Dihydro-4H-indene   D-70217

[108835-42-3]

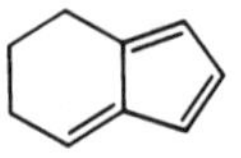

C_9H_{10} M 118.178
Orange fluorescent oil. Bp₁.₀ 40°.

Paquette, L.A. *et al*, *J. Org. Chem.*, 1987, **52**, 3250 (*synth, ir, pmr, cmr*)

5,10-Dihydroindeno[2,1-a]indene, 9CI   D-70218

[6543-29-9]

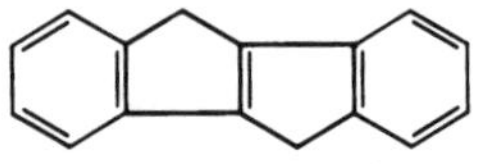

$C_{16}H_{12}$ M 204.271
Cryst. (EtOH). Mp 210°.

Brand, K. *et al*, *Ber.*, 1922, **55**, 601 (*synth*)

4*b*,9*a*-Dihydroindeno[1,2-*a*]indene-9,10-dione   D-70219

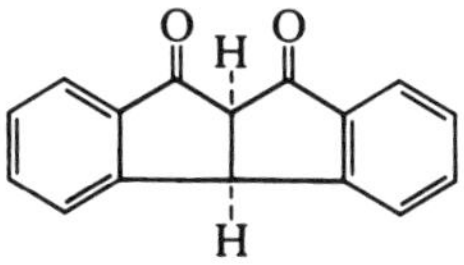

C$_{16}$H$_{10}$O$_2$ M 234.254
cis-form [29746-51-8]
Cryst. (Me$_2$CO aq.). Mp 257-259°.

Cristol, S.J. *et al*, *J. Am. Chem. Soc.*, 1970, **92**, 4013 (*synth*, *pmr*)

4*b*,9*b*-Dihydroindeno[2,1-*a*]indene-5,10-dione, 9CI   D-70220

[50703-54-3]

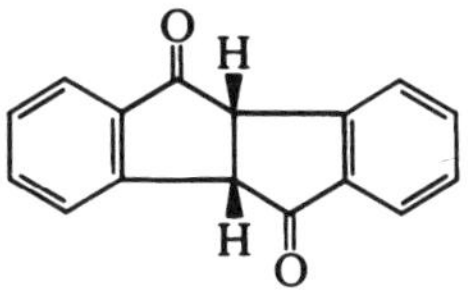

4bS,9bS-form

C$_{16}$H$_{10}$O$_2$ M 234.254
(4*bS*,9*bS*)-form [74431-35-9]
(+)-*cis-form*
Cryst. (EtOH). Mp 253.5-254.5° (subl. at ∼240°). [α]$_D$ +51.1° (c, 0.057 in EtOH).
(4*bRS*,9*bRS*)-form [16293-80-4]
(±)-*cis-form*
Solid (Et$_2$O/EtOAc). Mp 206-207°.

Dioxime:
   C$_{16}$H$_{12}$N$_2$O$_2$ M 264.283
   Small needles. Mp 254° dec.

Roser, W., *Justus Liebigs Ann. Chem.*, 1888, **247**, 152 (*synth*)
Mittal, R.S.D. *et al*, *Tetrahedron*, 1973, **29**, 1321 (*synth*, *ir*, *pmr*, *ms*)
Ogura, F. *et al*, *Bull. Chem. Soc. Jpn.*, 1980, **53**, 291 (*synth*, *ir*, *uv*, *cd*, *abs config*)
Papageorgiou, C. *et al*, *J. Org. Chem.*, 1987, **52**, 4403 (*synth*, *pmr*)

2,3-Dihydro-1*H*-indole-2-sulfonic acid   D-70221
Indoline-2-sulfonic acid

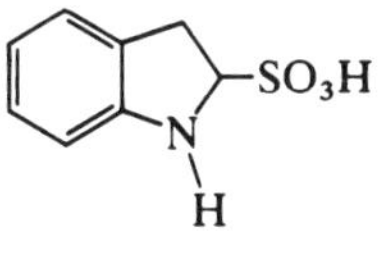

C$_8$H$_9$NO$_3$S   M 199.224
Na salt: [26807-68-1]. Readily obt. by reacn. of indole with NaHSO$_3$. Solid + 1H$_2$O. Mp 340-360° dec. Readily hydrol. by its water of crystallisation back to indole.
N-*Ac:* [26807-69-2].
   C$_{10}$H$_{11}$NO$_4$S   M 241.261
   Source of 5-subst. indoles by subn. reacns. Solid + ½H$_2$O (as Na salt). Mp >300° dec.

Russell, H.F. *et al*, *Org. Prep. Proced. Int.*, 1985, **17**, 391 (*synth*, *ir*, *pmr*, *use*, *bibl*)

4,5-Dihydro-3-(methoxymethylene)-2(3*H*)-furanone   D-70222
α-Methoxylidene-γ-butyrolactone

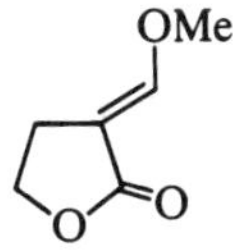

C$_6$H$_8$O$_3$ M 128.127
(*E*)-form
Needles. Mp 117-118°. Homolgoues with *O*-Et, *O*-benzyl and *O*-allyl groups also prepd.

Murray, A.W. *et al*, *Synth. Commun.*, 1986, **16**, 853 (*synth*, *pmr*)

8,8*a*-Dihydro-3-methoxy-5-methyl-1*H*,6*H*-furo[3,4-*e*][1,3,2]-dioxaphosphepin 3-oxide, 9CI   D-70223
CGA 134736
[113266-71-0]

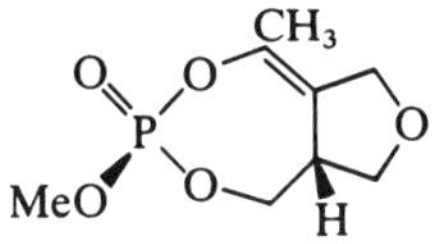

C$_8$H$_{13}$O$_5$P   M 220.161
Prod. by *Streptomyces antibioticus* strain DSM 1951. Potent insecticide, first known natural organophosphorus insecticide. Accompanied by the homologue CGA 134735 having propyl in place of the *C*-Me group.

Neumann, R. *et al*, *Experientia*, 1987, **43**, 1235.

4,5-Dihydro-2-(1-methylpropyl)thiazole   D-70224
2-sec-Butyl-2-thiazoline
[56367-27-2]

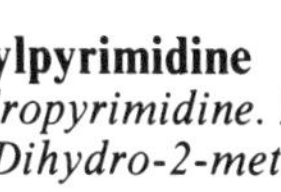

C$_7$H$_{13}$NS   M 143.246
Pheromone of the common mouse *Mus musculus*. Androgen-dependent substance eliciting male aggressive behaviour.

Fomum, Z.T. *et al*, *Tetrahedron Lett.*, 1975, 1101 (*synth*)
Novotny, M. *et al*, *Proc. Natl. Acad. Sci. USA*, 1985, **82**, 2059.

1,4-Dihydro-2-methylpyrimidine   D-70225
2-Methyl-1,6-dihydropyrimidine. 2-Methyl-1,4-dihydropyrimidine. 1,6-Dihydro-2-methylpyrimidine

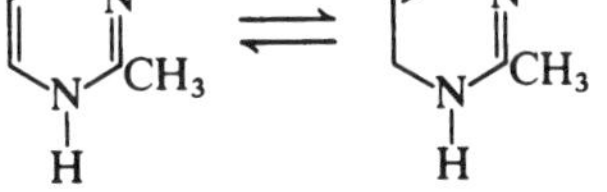

C$_5$H$_8$N$_2$ M 96.132
Needles by subl. Mp 82°. Mixt. of tautomers.

Weis, A.L. *et al*, *J. Org. Chem.*, 1987, **52**, 3421 (*synth*, *ir*, *pmr*, *ms*)

3,4-Dihydro-5-methyl-2*H*-pyrrole, 9CI   D-70226

2-Methyl-1-pyrroline, 8CI

[872-32-2]

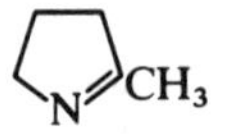

C_5H_9N M 83.133

Liq. Bp 103-105°.

B,MeClO₄: [2730-96-3]. Cryst. (EtOH). Mp 230° dec.

Picrate: Cryst. (C_6H_6). Mp 122°.

Cervinka, O. *et al, Collect. Czech. Chem. Commun.,* 1965, **30**, 1736 (*pmr*)
Etienne, A. *et al, Bull. Soc. Chim. Fr.,* 1969, 3704 (*synth*)
Bielawski, J. *et al, J. Heterocycl. Chem.,* 1978, **15**, 97 (*synth, pmr*)
Borg, R.M. *et al, J. Am. Chem. Soc.,* 1987, **109**, 2728 (*deriv, synth, pmr*)

2,3-Dihydro-2-methylthiophene   D-70227

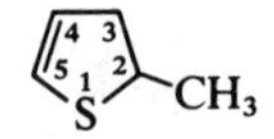

C_5H_8S M 100.178

(±)-*form*

 1,1-Dioxide:
  $C_5H_8O_2S$ M 132.177
  d_4^{20} 1.25. Mp 25-26°. Bp_1 75-77.5°. n_D^{20} 1.4949.

 Krug, R.C. *et al, J. Org. Chem.,* 1958, **23**, 1697 (*deriv, synth, ir*)

2,3-Dihydro-4-methylthiophene, 8CI   D-70228

C_5H_8S M 100.178

Bp 139.4°.

 1,1-Dioxide: [872-94-6]. *α-Isoprene sulfone.*
  $C_5H_8O_2S$ M 132.177
  Long needles (C_6H_6/pet. ether). Mp 77-78°.

 Birch, F.F. *et al, J. Chem. Soc.,* 1951, 3411 (*synth, deriv*)

2,3-Dihydro-5-methylthiophene   D-70229

C_5H_8S M 100.178

Bp 127-128°, Bp_{18} 35-36°.

 1,1-Dioxide:
  $C_5H_8O_2S$ M 132.177
  Needles (EtOH). Mp 128.0-128.5°.

 Birch, F.F. *et al, J. Chem. Soc.,* 1951, 3411 (*synth*)
 Bacchetti, T. *et al, Gazz. Chim. Ital.,* 1953, **83**, 1037 (*synth*)
 Krug, R.C. *et al, J. Org. Chem.,* 1958, **23**, 1697 (*deriv, synth*)

2,5-Dihydro-2-methylthiophene   D-70230

[55273-87-5]

C_5H_8S M 100.178

(±)-*form*

 1,1-Dioxide: [6007-71-2]. *2-Methyl-3-sulfolene.*
  $C_5H_8O_2S$ M 132.177
  Oil. d_4^{20} 1.25. Bp_7 85°. n_D^{20} 1.4929.

 Krug, R.C. *et al, J. Org. Chem.,* 1958, **23**, 1697; 1962, **27**, 1305 (*deriv, synth, ir*)
 McIntosh, J.M. *et al, Can. J. Chem.,* 1983, **61**, 1872 (*synth, pmr, cmr*)

Chou, T.S. *et al, J. Chem. Soc., Perkin Trans. 1,* 1985, 515 (*deriv, synth, ir, pmr*)
Chou, T.S. *et al, J. Org. Chem.,* 1987, **52**, 3394 (*deriv, synth*)

2,5-Dihydro-3-methylthiophene   D-70231

[42855-50-5]

C_5H_8S M 100.178

Liq. Bp 147.5°. n_D^{20} 1.5196.

 1,1-Dioxide: [1193-10-8]. *Isoprene cyclic sulfone.*
  $C_5H_8O_2S$ M 132.177
  Thick plates (MeOH). Mp 63.5-64°.

Org. Synth., Coll. Vol., 3, 499 (*deriv, synth*)
Birch, S.F. *et al, J. Chem. Soc.,* 1951, 3411 (*synth, deriv*)
McIntosh, J.M. *et al, J. Org. Chem.,* 1974, **39**, 202 (*synth*)

2,3-Dihydro-6-methyl-2-thioxo-4(1*H*)-pyrimidinone, 9CI   D-70232

6-Methyl-2-thiouracil, 8CI. *Methylthiouracil,* INN. *Muracin. Prostrumyl. Strumacil. Thimecil. Thiothymin. Thyreostat*

[56-04-2]

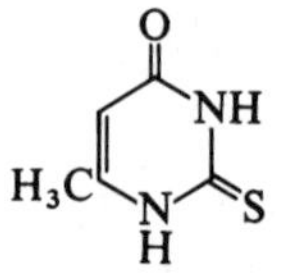

$C_5H_6N_2OS$ M 142.175

Thyroid inhibitor. Mp 326-331°. Subl. readily. Numerous tradenames.

▷YR0875000.

List, G., *Justus Liebigs Ann. Chem.,* 1886, **236**, 1 (*synth*)
Galimberti, P. *et al, Ann. Chim.* (*Paris*), 1958, **48**, 457 (*uv*)
Negwer, M., *Organic-Chemical Drugs and their Synonyms,* 6th Ed., Akademie-Verlag, Berlin, 1987, 281 (*synonyms*)
Martindale, The Extra Pharmacopoeia, 28th/29th Eds., 1982/1989, Pharmaceutical Press, London, 836.

3,4-Dihydro-5-methyl-4-thioxo-2(1*H*)-pyrimidinone, 9CI   D-70233

5-Methyl-4-thiouracil. 4-Thiothymine

[35455-79-9]

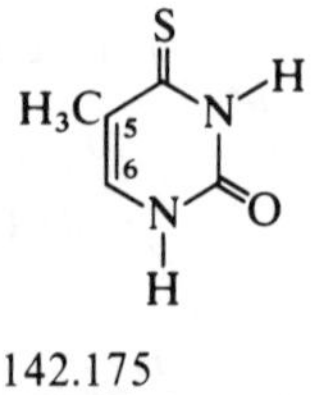

$C_5H_6N_2OS$ M 142.175

Fine-yellow needles (H_2O). Mp 306-309°.

Veda, T. *et al, J. Med. Chem.,* 1963, **6**, 697 (*synth*)
Lamon, R.W., *J. Heterocycl. Chem.,* 1968, **5**, 837 (*synth, uv*)
Lapucha, A., *Synthesis,* 1987, 256 (*synth, ir, pmr, cmr, ms*)

3,4-Dihydro-6-methyl-4-thioxo-2(1*H*)-pyrimidinone, 9CI   D-70234

6-Methyl-4-thiouracil

[638-13-1]

$C_6H_5N_2OS$ M 153.178

Yellow cryst. (H_2O). Mp 338-341° dec. (>250° dec.).

Wheeler, H.L. *et al, Am. Chem. J.,* 1909, **42**, 431 (*synth*)

Brown, D.J. *et al, Aust. J. Chem.*, 1980, **33**, 1147 (*synth*)
Łapucha, A., *Synthesis*, 1987, 256 (*synth, ir, pmr, cmr, ms*)

2,3-Dihydronaphtho[2,3-*b*]furan, 9CI   D-70235

[7193-15-9]

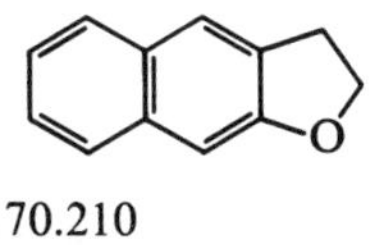

$C_{12}H_{10}O$ M 170.210
Cryst. (EtOH). Mp 104°.

Narasimhan, N.S. *et al, Indian J. Chem.*, 1969, **7**, 536 (*synth*)
Ghera, E. *et al, Tetrahedron Lett.*, 1987, **28**, 709 (*synth, pmr*)

2,5-Dihydro-3-nitrofuran, 9CI   D-70236

[111286-34-1]

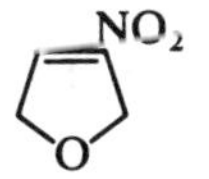

$C_4H_5NO_3$ M 115.088
Oil solidifying at 0°. Unstable.

Bitha, P. *et al, J. Heterocycl. Chem.*, 1988, **25**, 1035 (*synth, pmr*)

9,11-Dihydro-22,25-oxido-11-oxoholothur-   D-70237
inogenin

22,25-Epoxy-3β,17α-dihydroxy-11-oxo-7-lanosten-18,20S-olide

[112570-87-3]

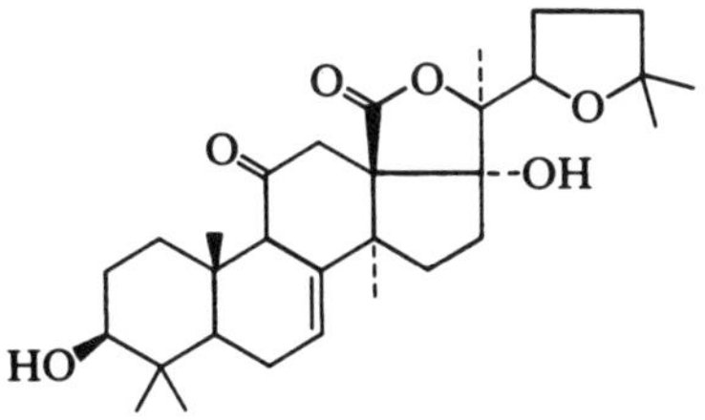

$C_{30}H_{44}O_6$ M 500.674
Aglycone from *Holothuria atra* and *H. scabra*. Cryst. (MeOH). Mp 280-282°. [α]$_D$ +16.6°.

Sarma, N.S. *et al, Indian J. Chem., Sect. B*, 1987, **26**, 715.

7,8-Dihydro-8-oxoguanosine, 9CI   D-70238

8-Hydroxyguanosine. 2-Amino-9-β-D-ribofuranosyl-6,8(1H,9H)-purinedione, 8CI. 2-Amino-8-hydroxy-9-β-D-ribofuranosyl-9H-purin-6(1H)-one. 8-Hydroxy-9-β-D-ribofuranosylguanine

[3868-31-3]

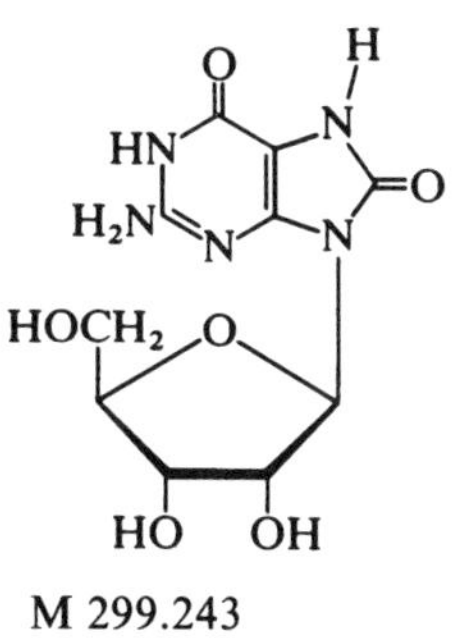

$C_{10}H_{13}N_5O_6$ M 299.243
Cryst. (H$_2$O). Mp 229-231° dec.
8-Me ether: [7057-53-6]. *8-Methoxyguanosine, 9CI.*
   $C_{11}H_{15}N_5O_6$ M 313.269

Cryst. (H$_2$O). Mp >216° dec.
8-Benzyl ether: [3868-36-8]. Cryst. (EtOH aq.). Mp 170-171°.

Holmes, R.E. *et al, J. Am. Chem. Soc.*, 1965, **87**, 1772 (*deriv*)
Ikehara, M. *et al, Chem. Pharm. Bull.*, 1966, **14**, 46 (*deriv*)
Ogilvie, K.K. *et al, J. Org. Chem.*, 1971, **36**, 2556 (*synth*)
Uesugi, S. *et al, J. Am. Chem. Soc.*, 1977, **99**, 3250 (*cmr*)
Lin, T.-S. *et al, J. Med. Chem.*, 1985, **28**, 1194 (*synth, pmr*)

Dihydroparthenolide   D-70239

Updated Entry replacing D-03986

[66397-64-6]

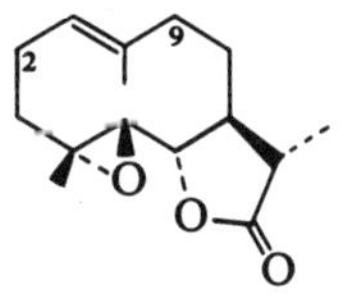

$C_{15}H_{22}O_3$ M 250.337
Constit. of *Michelia lanuginosa* and *M. compressa*. Cryst. (Et$_2$O/hexane). Mp 127°. [α]$_D$ −68.5°.
2α-Hydroxy: [114076-72-1]. *2α-Hydroxydihydroparthenolide.*
   $C_{15}H_{22}O_4$ M 266.336
Constit. of *Paramichelia baillonii*. Oil.
9α-Hydroxy: [114076-73-2]. *9α-Hydroxydihydroparthenolide.*
   $C_{15}H_{22}O_4$ M 266.336
Constit. of *P. baillonii*. Oil. [α]$_D^{22}$ −60° (CHCl$_3$).

Ogura, M. *et al, Phytochemistry*, 1978, **17**, 957 (*isol*)
Ruangrungsi, N. *et al, J. Nat. Prod.*, 1988, **51**, 163 (*derivs*)

1,2-Dihydropentalene   D-70240

Bicyclo[3.3.0]octa-1,3,5-triene

[30294-54-3]

C_8H_8 M 104.151
Yellow oil. Air-sensitive, thermolabile.

Kaiser, R. *et al, Angew. Chem., Int. Ed. Engl.*, 1970, **9**, 892 (*synth, uv, pmr*)
Meier, H. *et al, Chem. Ber.*, 1987, **120**, 1607 (*synth, pmr, cmr, struct*)

1,4-Dihydropentalene   D-70241

Bicyclo[3.3.0]octa-1(5),2,6-triene

[61771-84-4]

C_8H_8 M 104.151
Obt. only in admixture with 1,6-Dihydropentalene, D-70243 . V. unstable at r.t.

Baird, M.S. *et al, Tetrahedron Lett.*, 1976, 2895.
Meier, H. *et al, Chem. Ber.*, 1987, **120**, 1607 (*synth, pmr, cmr, props*)

1,5-Dihydropentalene   D-70242

Bicyclo[3.3.0]octa-1,4,6-triene

[33284-11-6]

C_8H_8 M 104.151
Oil. Bp$_{16}$ 50-54°.

Katz, T.J. *et al, J. Am. Chem. Soc.*, 1964, **86**, 249 (*synth, uv, pmr*)
Brinker, U.H. *et al, Chem. Lett.*, 1984, 45 (*synth*)
Meier, H. *et al, Chem. Ber.*, 1987, **120**, 1607 (*synth, pmr, cmr, props*)

1,6-Dihydropentalene   **D-70243**
Bicyclo[3.3.0]octa-1(5),2,7-triene
[89654-25-1]
C_8H_8 M 104.151
Obt. only in admixture with 1,4-Dihydropentalene, D-70241 . V. unstable at r.t., even in soln.

Meier, H. *et al, Chem. Ber.,* 1987, **120**, 1607 (*synth, pmr, cmr, props*)

1,6a-Dihydropentalene   **D-70244**
Bicyclo[3.3.0]octa-1,3,7-triene
C_8H_8 M 104.151

Baird, M.S. *et al, Tetrahedron Lett.,* 1976, 2895.

Dihydro-5-pentyl-2(3H)-furanone, 9CI, 8CI   **D-70245**
Updated Entry replacing P-00569
γ-*Nonalactone. 4-Hydroxynonanoic acid γ-lactone. γ-Amylnonalactone. 1,4-Nonanolide. 5-Pentyldihydro-2(3H)-furanone*
[104-61-0]

$$H_3C(CH_2)_4 \underset{O}{\overset{}{\diagup}} O \qquad (R)\text{-}form$$

$C_9H_{16}O_2$ M 156.224
Isol. from ethereal oil of *Rosa rugosa.*
▷LU3675000.
(R)-form [63357-96-0]
$[\alpha]_D^{24}$ +47.5° (c, 1.82 in MeOH).
(S)-form [63357-97-1]
$[\alpha]_D^{24}$ −47.9° (c, 1.70 in MeOH).
(±)-form
Perfumery and flavouring ingredient. Bp_{12} 134-134.5°, $Bp_{1.5}$ 93-94°.

Terai, Y. *et al, Bull. Chem. Soc. Jpn.,* 1956, **29**, 822 (*synth*)
Sakai, T. *et al, CA,* 1962, **58**, 8845 (*isol*)
Goldwhite, H. *et al, Tetrahedron,* 1964, **20**, 1657 (*synth, ir*)
Honkanen, E. *et al, Acta Chem. Scand.,* 1965, **19**, 370 (*ms*)
Oser, B.L. *et al, Food Cosmet. Toxicol.,* 1965, **3**, 563 (*rev*)
Sataro, M. *et al, Bull. Chem. Soc. Jpn.,* 1980, **53**, 770 (*synth*)
Miyashita, Y. *et al, Agric. Biol. Chem.,* 1981, **45**, 2521 (*synth*)
Ballini, R. *et al, Synthesis,* 1987, 711 (*synth, ir, pmr*)

9,10-Dihydro-2,3,4,6,7-phenanthrenepentol   **D-70246**
2,3,4,6,7-Pentahydroxy-9,10-dihydrophenanthrene.
9,10-Dihydro-2,3,4,6,7-pentahydroxyphenanthrene

$C_{14}H_{12}O_5$ M 260.246
2,3,4-Tri-Me ether: [39500-00-0]. *9,10-Dihydro-5,6,7-trimethoxy-2,3-phenanthrenediol, 9CI. 9,10-Dihydro-6,7-dihydroxy-2,3,4-trimethoxyphenanthrene.*
$C_{17}H_{18}O_5$ M 302.326
Constit. of *Combretum caffrum.* Cryst. (MeOH). Mp 212-214°.
3,4,6-Tri-Me ether: [39499-93-9]. *9,10-Dihydro-2,7-dihydroxy-3,4,6-trimethoxyphenanthrene. 9,10-Dihydro-3,4,6-trimethoxy-2,7-phenanthrenediol, 9CI.*
$C_{17}H_{18}O_5$ M 302.326

Constit. of *C. caffrum.* Yellow needles ($CHCl_3$/hexane). Mp 112-114°.
2,3,4,6-Tetra-Me ether: [116963-91-8]. *9,10-Dihydro-7-hydroxy-2,3,4,6-tetramethoxyphenanthrene. 9,10-Dihydro-3,4,6,7-tetramethoxy-2-phenanthrenol, 9CI.*
$C_{18}H_{20}O_5$ M 316.353
Constit. of *C. caffrum.* Cryst. ($CHCl_3$/hexane). Mp 176-178°.

Pettit, G.R. *et al, Can. J. Chem.,* 1988, **66**, 406.

9,10-Dihydro-2,4,5,6-phenanthrenetetrol   **D-70247**
2,4,5,6-Tetrahydroxy-9,10-dihydrophenanthrene

$C_{14}H_{12}O_4$ M 244.246
2,4-Di-Me ether: [58115-33-6]. *9,10-Dihydro-5,6-dihydroxy-2,4-dimethoxyphenanthrene. 9,10-Dihydro-5,7-dimethoxy-3,4-phenanthrenediol, 9CI.*
$C_{16}H_{16}O_4$ M 272.300
Constit. of the rhizomes of *Dioscorea prazeri.* Needles. Mp 144-145°.
2,4-Di-Me ether, di-Ac: Mp 147-149°.

Rajaraman, K. *et al, Indian J. Chem.,* 1975, **13**, 1137.

2,3-Dihydro-2-phenyl-4H-1-benzothio-   **D-70248**
pyran-4-one, 9CI
Updated Entry replacing D-04008
2,3-Dihydro-2-phenyl-4H-benzo[b]thiin-4-one.
Thioflavanone
[5962-00-5]

$C_{15}H_{12}OS$ M 240.319
(+)-form [110318-34-8]
Mp 61°. $[\alpha]_D$ +61° (c, 0.7 in $CHCl_3$).
(−)-form [110318-35-9]
Mp 61°. $[\alpha]_D^{20}$ −56° (c, 0.8 in $CHCl_3$).
(±)-form [110318-31-5]
Needles (EtOH or pet. ether/CS_2). Mp 55-56°.
1-Oxide: [58109-92-5].
$C_{15}H_{12}O_2S$ M 256.319
Cryst. (C_6H_6/pet. ether). Mp 148-151°.
1,1-Dioxide: [58109-94-7].
$C_{15}H_{12}O_3S$ M 272.318
Cryst. (2-propanol). Mp 133-134°, Mp 156-157°.

Arndt, F., *Ber.,* 1923, **56**, 1274 (*synth*)
Chauhan, M.S. *et al, Can. J. Chem.,* 1975, **53**, 2880 (*cmr*)
Bognar, R. *et al, Justus Liebigs Ann. Chem.,* 1977, 1529 (*synth*)
Tökés, A.L., *Justus Liebigs Ann. Chem.,* 1987, 1007 (*resoln, pmr*)

1,4-Dihydro-2-phenylpyrimidine   D-70249

2-Phenyl-1,6-dihydropyrimidine. 2-Phenyl-1,4-dihydro-pyrimidine. 1,6-Dihydro-2-phenylpyrimidine

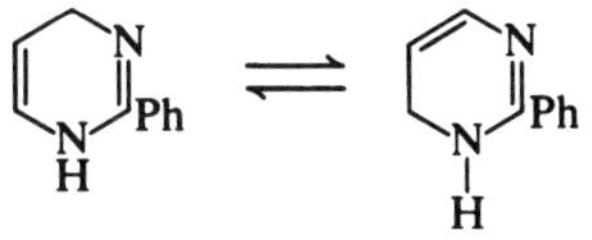

$C_{10}H_{10}N_2$ M 158.202

Yellow needles (hexane). Mp 131-132°. Mixt. of tautomers.

Weis, A.L. *et al, J. Org. Chem.*, 1987, **52**, 3421 (*synth, pmr, ir, uv, ms*)

5,6-Dihydro-2-phenyl-4(1H)-pyrimidinone,   D-70250
9CI

6-Oxo-2-phenyl-1,4,5,6-tetrahydropyrimidine. 4,5-Di-hydro-2-phenyl-6(1H)-pyrimidinone

[20456-88-6]

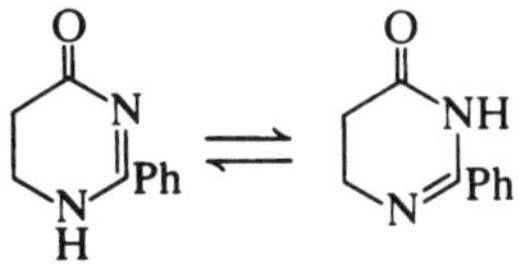

$C_{10}H_{10}N_2O$ M 174.202

Named by the authors as a 4(1H)-pyrimidinone, but their evidence supports the 6(1H) (or 4(3H))-form. Cryst. Mp 145°.

Weis, A.L. *et al, J. Org. Chem.*, 1987, **52**, 3421 (*synth, pmr*)

3,4-Dihydro-5-phenyl-2H-pyrrole, 9CI   D-70251

2-Phenyl-1-pyrroline, 8CI

[700-91-4]

$C_{10}H_{11}N$ M 145.204

Liq. Bp 245-247°, Bp_{18} 120-130°, Bp_5 104-106°.

Picrate: Cryst. (C_6H_6). Mp 200°.

Craig, L.C. *et al, J. Am. Chem. Soc.*, 1931, **53**, 1831 (*synth*)
Etienne, A. *et al, Bull. Soc. Chim. Fr.*, 1969, 3704 (*synth*)
Bielawski, J. *et al, J. Heterocycl. Chem.*, 1978, **15**, 97 (*synth, pmr*)

3,4-Dihydro-2H-pyran-5-carboxaldehyde,   D-70252
9CI

[25090-33-9]

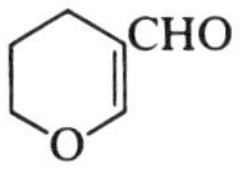

$C_6H_8O_2$ M 112.128

Liq. Bp_{16} 82-84°, Bp_3 68°.

Skattebøl, L., *J. Org. Chem.*, 1970, **35**, 3200 (*synth, ir, uv, pmr*)
Piancatelli, G. *et al, Ann. Chim.* (Rome), 1972, **62**, 394 (*synth*)
Breitmaier, E. *et al, Synthesis*, 1987, 1 (*synth*)

4,5-Dihydro-1H-pyrazole, 9CI   D-70253

Updated Entry replacing P-02762

2-Pyrazoline, 8CI

[109-98-8]

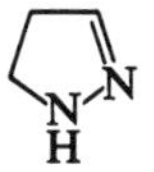

$C_3H_6N_2$ M 70.094

Fluorescence brightener. Liq. with faint amine odour. Misc. H_2O, EtOH. d_4^{20} 1.02. Bp 144°, Bp_{40} 63-64°. n_D^{20} 1.4782. Volatile in steam and Et_2O.

Picrate: Yellow needles. Mp 130°.

1-Ph: [936-53-8].
$C_9H_{10}N_2$ M 146.191
Plates (ligroin). Mp 52°. Bp_{754} 273°, Bp_{14} 151°.

Wirsing, F., *J. Prakt. Chem.*, 1894, **50**, 531 (*synth*)
v. Auwers, K. *et al, Justus Liebigs Ann. Chem.*, 1927, **458**, 175 (*props*)
Nardelli, M. *et al, Ric. Sci.*, 1960, **30**, 904 (*cryst struct*)
Gol'ding, G.S. *et al, Khim. Geterotsikl. Soedin.*, 1970, 429; *CA*, **73**, 25347p (*synth*)
Faure, R. *et al, Spectrosc. Int. J.*, 1983, **2**, 381 (*cmr*)
Ger. Pat., 3 245 109, (*1984*); *CA*, **101**, 211136r (*manuf*)
El-Rayyes, N.R. *et al, Synthesis*, 1985, 1028 (*rev, synth, derivs*)

7,12-Dihydropyrido[3,2-b:5,4-b']diindole,   D-70254
9CI

[98263-45-7]

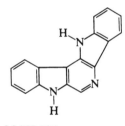

$C_{17}H_{11}N_3$ M 257.294

Cryst. (MeOH/MeCN). Mp >350°.

Trudell, M.L. *et al, J. Org. Chem.*, 1987, **52**, 4293 (*synth, ir, pmr*)

1,4-Dihydropyrimidine   D-70255

Dihydropyrimidine, 9CI. 1,6-Dihydropyrimidine

[27790-74-5]

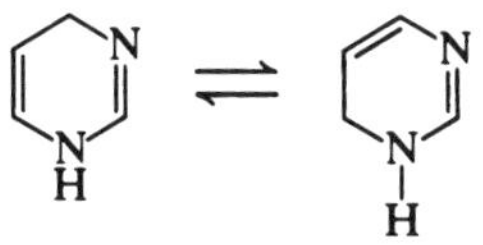

$C_4H_6N_2$ M 82.105

Obt. only 85% pure, unstable, quickly dec. or polym.

Weis, A.L. *et al, J. Org. Chem.*, 1987, **52**, 3421 (*synth, pmr*)

Dihydro-4,6-(1*H*,5*H*)-pyrimidinedione, 9CI   D-70256

2,5-Dihydro-4,6-pyrimidinediol. 4,6-Dihydroxy-2,5-dihydropyrimidine

[21215-00-9]

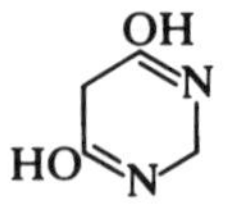

$C_4H_6N_2O_2$ M 114.104

O-Alkyl derivs. descr. are first examples of stable 2,5-dihydropyrimidines. No evidence for tautomerism of derivs.

Di-Me ether: [97984-28-6]. *2,5-Dihydro-4,6-dimethoxypyrimidine.*
$C_6H_{10}N_2O_2$ M 142.157
Cryst. Mp 48-50°.

Di-Et ether: [97984-29-7]. *4,6-Diethoxy-2,5-dihydropyrimidine.*
$C_8H_{14}N_2O_2$ M 170.211
Cryst. Mp 72-73°.

Di-tert-butyl ether: [104995-43-9].
$C_{12}H_{22}N_2O_2$ M 226.318
Cryst. Mp 63-64°.

Kheifels, G.M. *et al, Tetrahedron*, 1967, **23**, 1197 (*pmr, struct, derivs*)
Weis, A.L. *et al, J. Org. Chem.*, 1986, **51**, 4623 (*synth, pmr, ir, cryst struct*)

2,5-Dihydro-4,6-pyrimidinedithiol   D-70257

2,5-Dihydro-4,6-dimercaptopyrimidine

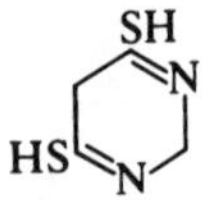

$C_4H_6N_2S_2$ M 146.225

Parent compd. unknown. Derivs. are first examples of 2,5-dihydropyrimidines.

Di-Et thioether: [97984-32-2]. *4,6-Bis(ethylthio)-2,5-dihydropyrimidine.*
$C_8H_{14}N_2S_2$ M 202.332
Cryst. Mp 39-40°.

Di-Ph thioether: [104995-46-2].
$C_{16}H_{14}N_2S_2$ M 298.420
Cryst. Mp 152-153°. Exists as mixt. of 1,2- and 2,5-dihydro forms.

Weis, A.L. *et al, J. Org. Chem.*, 1986, **51**, 4623 (*synth, pmr, ir*)

6,7-Dihydro-5*H*-1-pyrindine, 9CI   D-70258

Pyrindane. 1-Pyrindane. 2,3-Cyclopentenopyridine. 2,3-Trimethylenepyridine. 2,3-Pyrindane

[533-37-9]

C_8H_9N M 119.166

Found in coal tar and tobacco smoke condensates. Bp$_{11}$ 87-88°.

Picrate: Mp 179-181.5°.

Thummel, R.P. *et al, J. Org. Chem.*, 1977, **42**, 2742; 1978, **43**, 4882 (*synth, pmr, cmr, uv*)
Irie, H. *et al, Heterocycles*, 1979, **12**, 771 (*synth*)
Stefaniak, L. *et al, Org. Magn. Reson.*, 1984, **22**, 201 (*N nmr*)

6,7-Dihydro-5*H*-2-pyrindine, 9CI   D-70259

2-Pyrindane. 3,4-Cyclopentenopyridine. 3,4-Trimethylenepyridine. 3,4-Pyrindane

[533-35-7]

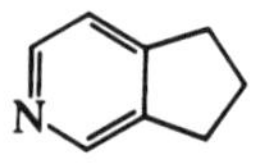

C_8H_9N M 119.166

Found in coal tar fractions.

Picrate: Mp 139-140.5°.

Thummel, R.P. *et al, J. Org. Chem.*, 1977, **42**, 2742; 1978, **43**, 4882 (*synth, pmr, cmr, uv*)

Dihydro-1*H*-pyrrolizine-3,5(2*H*,6*H*)-dione, 9CI   D-70260

Updated Entry replacing D-20211

2,5-Pyrrolizidinedione. 3,4-Dioxopyrrolizidine. **Rolziracetanal,** *BAN, INN. PD 105587. CI 911. Rolziracetam. Lukes-Šorm dilactam*

[18356-28-0]

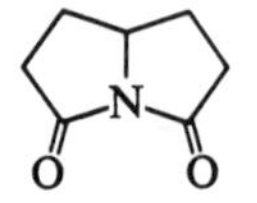

$C_7H_9NO_2$ M 139.154

Note various numbering systems used for pyrrolizines.
Shows amnesia-reversing props. in memory-impaired and demented patients. Cryst. by subl., needles (EtOH). Mp 181°.

Lukes, R. *et al, Collect. Czech. Chem. Commun.*, 1947, **12**, 278 (*synth*)
Micheel, F. *et al, Chem. Ber.*, 1955, **88**, 509; 1956, **89**, 129 (*synth*)
Butler, D.E. *et al, J. Med. Chem.*, 1987, **30**, 498 (*synth, pharmacol, bibl*)
Alonso, R. *et al, Synth. Commun.*, 1988, **18**, 37 (*synth, pmr*)

2,3-Dihydro-1*H*-pyrrolizin-1-ol   D-70261

[18377-79-2]

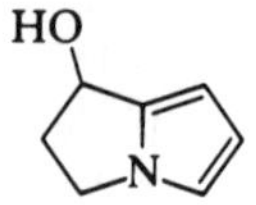

C_7H_9NO M 123.154

(±)-**form**
Cryst. by subl. Mp 66-67°.

Karchesy, J.J. *et al, J. Org. Chem.*, 1987, **52**, 3867 (*synth, pmr, ms*)

1,2-Dihydro-3*H*-pyrrol-3-one, 9CI   D-70262

Updated Entry replacing D-50398

1H-Pyrrol-3-ol, 9CI. 3(2H)-Pyrrolone. 3-Hydroxypyrrole

[5860-48-0]

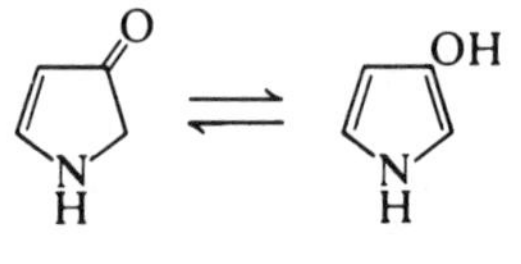

C_4H_5NO M 83.090

Parent compd. not known. *N*-substd. derivs. adopt oxo-form in nonpolar, *OH*-form in polar solvs.

N-*Formyl:* [114049-72-8]. *2,3-Dihydro-3-oxo-1*H-*pyr-role-1-carboxaldehyde*, *9CI.*
$C_5H_5NO_2$ M 111.100
Cryst.

N-*Ac:* [114049-73-9]. *1-Acetyl-1,2-dihydro-3*H-*pyrrol-3-one*, *9CI.*
$C_6H_7NO_2$ M 125.127
Cryst.

N-*Me:* [54630-44-3]. *1-Methyl-1,2-dihydro-3*H-*pyrrol-3-one.*
C_5H_7NO M 97.116
Air-sensitive oil. Neat compd. shows keto:enol = 14:3.

N-*Ph:* [65172-10-3]. *1,2-Dihydro-1-phenyl-3*H-*pyrrol-3-one. 1-Phenyl-4-oxo-2-pyrroline.*
$C_{10}H_9NO$ M 159.187
Air-sensitive oil.

Momose, T. *et al*, *Heterocycles*, 1977, **6**, 1827 (*synth, props*)
Momose, T. *et al*, *Chem. Pharm. Bull.*, 1979, **27**, 1448 (*synth, ir, uv, ms, pmr*)
McNab, H. *et al*, *J. Chem. Soc., Chem. Commun.*, 1985, 213 (*synth, props, bibl*)
Flitsch, W. *et al*, *Tetrahedron Lett.*, 1987, **28**, 4397 (*derivs, synth, pmr*)

2,3-Dihydro-1*H*-pyrrolo[2,3-*b*]pyridine,   D-70263
9CI

2,3-Dihydro-7-azaindole. 7-Azaindoline
[10592-27-5]

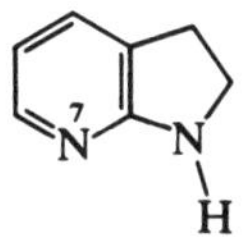

$C_7H_8N_2$ M 120.154
Cryst. Mp 83-85° (78°).
B,HCl: Needles. Mp 212°.
Picrate: Yellow cryst. (C_6H_6). Mp 224°.
7-Oxide:
$C_7H_8N_2O$ M 136.153
Thick needles (EtOAc). Mp 155-156°.

Kruber, O., *Ber.*, 1943, **76**, 128 (*synth*)
Robison, M.M. *et al*, *J. Am. Chem. Soc.*, 1957, **79**, 2573; 1959, **81**, 743 (*synth, struct, props*)
Taylor, E.C. *et al*, *Tetrahedron Lett.*, 1987, **28**, 379 (*synth*)

5,14-Dihydroquinoxalino[2,3-*b*]phenazine   D-70264
Updated Entry replacing D-40198
Fluorindine
[531-47-5]

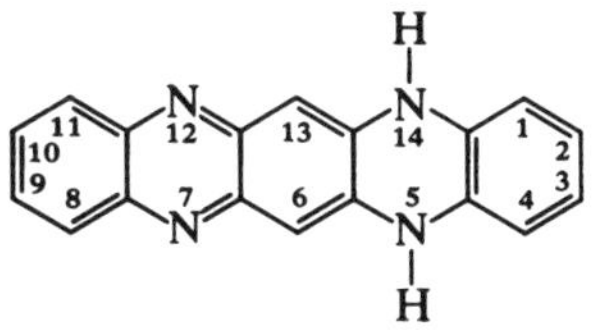

$C_{18}H_{12}N_4$ M 284.320
Formerly assigned the 5,12-dihydro struct. Dark-blue, purple or green needles or powder (AcOH). Red fluorescence in soln.

Badger, G.M. *et al*, *J. Chem. Soc.*, 1951, 3211 (*synth, props*)
Wanzlick, H.W. *et al*, *Chem. Ber.*, 1968, **101**, 3744 (*synth*)
Armand, J. *et al*, *Can. J. Chem.*, 1987, **65**, 1619 (*pmr, struct*)

Dihydroserruloside   D-70265
[113332-14-2]

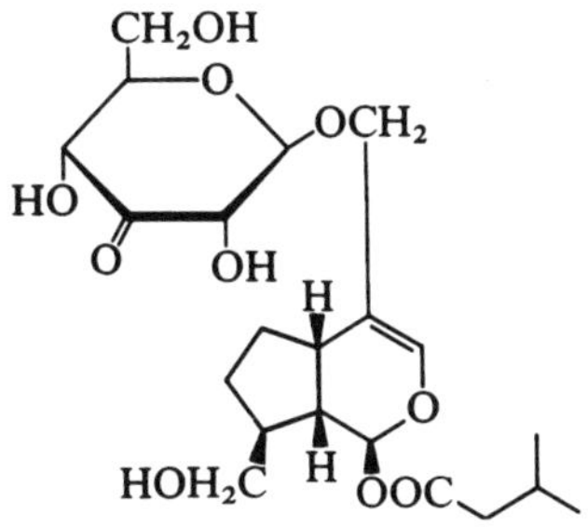

$C_{21}H_{32}O_{10}$ M 444.478
Constit. of *Penstemon confertus*. Yellow amorph. powder (CHCl$_3$/MeOH).

Gering, B. *et al*, *Phytochemistry*, 1987, **26**, 3011.

4,7-Dihydro-1,1,3,3-tetramethyl-1*H*,3*H*-   D-70266
3*a*,7*a*-episeleno-4,7-epoxyisobenzofuran,
9CI
[112520-11-3]

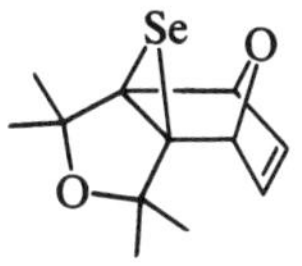

$C_{12}H_{16}O_2Se$ M 271.217
Stable selenirane.

Ando, W. *et al*, *Tetrahedron Lett.*, 1987, **28**, 2867 (*synth, pmr, cmr, ms*)

3,6-Dihydro-3,3,6,6-tetramethyl-2(1*H*)-   D-70267
pyridinone
[109070-40-8]

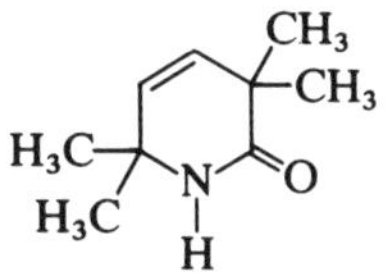

$C_9H_{15}NO$ M 153.224
Cryst. Mp 132°.

Grimaldi, J. *et al*, *Tetrahedron Lett.*, 1986, **27**, 5089 (*synth*)

1,2-Dihydro-1,2,4,5-tetrazine-3,6-dione,   D-70268
9CI

1,2,4,5-Tetrazine-3,6-diol. 6-Hydroxy-1,2,4,5-tetrazin-3(2H)-one. 3,6-Dihydroxy-1,2,4,5-tetrazine
[81930-28-1]

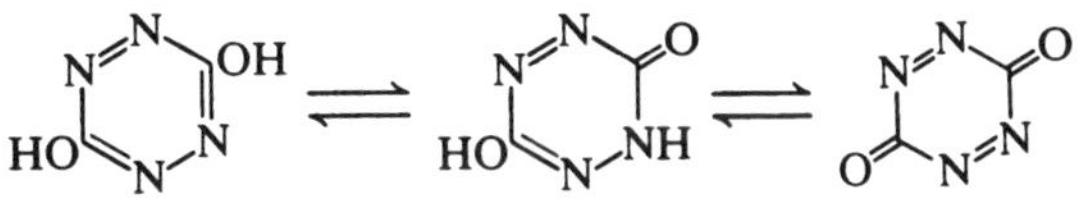

$C_2H_2N_4O_2$ M 114.063
Orange-brown cryst. (Me$_2$CO/pet. ether). Mp 168° (explodes).
Di-Me ether: [81930-31-6]. *3,6-Dimethoxy-1,2,4,5-tetrazine.*
$C_4H_6N_4O_2$ M 142.117
Red-violet needles (C_6H_6). Mp 62-63°.

Neugebauer, F.A. *et al*, *Justus Liebigs Ann. Chem.*, 1982, 387 (*synth, ir, uv, pmr*)

Krieger, C. *et al, Acta Crystallogr., Sect. C*, 1987, **43**, 1412 (*cryst struct, deriv*)

2,3-Dihydrothieno[2,3-*b*]pyridine   D-70269
[103020-20-8]

C_7H_7NS M 137.199
$Bp_{0.4}$ 60-70°.
1-Oxide: [109216-74-2].
    C_7H_7NOS M 153.198
    Cryst. solid. Mp 96.0-97.5°.

Seitz, G. *et al, Tetrahedron Lett.*, 1985, **26**, 4355 (*synth, pmr*)
Frissen, A.E. *et al, Tetrahedron Lett.*, 1987, **28**, 1589 (*synth, pmr*)
Taylor, E.C. *et al, J. Org. Chem.*, 1987, **52**, 4280 (*deriv, synth, ir, pmr, cmr*)

2,5-Dihydrothiophene   D-70270
[1708-32-3]

C_4H_6S M 86.151
Liq. d^{20} 1.06. Bp 122.3°.
1,1-Dioxide: see 3-Sulfolene, S-00999
Birch, S.F. *et al, J. Chem. Soc.*, 1951, 2556 (*synth*)

2,3-Dihydro-4-(trichloroacetyl)furan   D-70271
2,2,2-Trichloro-1-(4,5-dihydro-3-furanyl)ethanone
[83124-80-5]

$C_6H_5Cl_3O_2$ M 215.463
$Bp_{11.8}$ 123.3°.
Hojo, M. *et al, Synthesis*, 1986, 1016 (*synth, ir, pmr*)

3,4-Dihydro-5-(trichloroacetyl)-2*H*-pyran   D-70272
*2,2,2-Trichloro-1-(3,4-dihydro-2*H-*pyan-5-yl)ethanone*
[83124-87-2]

$C_7H_7Cl_3O_2$ M 229.490
Bp_{10} 132°.
Hojo, M. *et al, Synthesis*, 1986, 1016 (*synth, ir, pmr*)

Dihydrotrichostin   D-70273
[115713-37-6]

$C_{21}H_{24}O_7$ M 388.416
Constit. of *Piper trichostachyon*. Semi-solid. $[\alpha]_D^{20}$ −13.3° (c, 0.6 in MeOH).
Koul, S.K. *et al, Phytochemistry*, 1988, **27**, 1479.

2,3-Dihydro-4-(trifluoroacetyl)furan   D-70274
1-(4,5-Dihydro-3-furanyl)-2,2,2-trifluoroethanone
[109317-75-1]

$C_6H_5F_3O_2$ M 166.100
Bp_{10} 48.5°.
Hojo, M. *et al, Synthesis*, 1986, 1016 (*synth, ir, pmr*)

3,4-Dihydro-5-(trifluoroacetyl)-2*H*-pyran   D-70275
*1-(3,4-Dihydro-2*H*-pyran-5-yl)-2,2,2-trifluoroethan-one, 9CI*
[109317-74-0]

$C_7H_7F_3O_2$ M 180.126
$Bp_{10.5}$ 64.6°.
Hojo, M. *et al, Synthesis*, 1986, 1016 (*synth, ir, pmr*)

5,6-Dihydro-6-(2,4,6-trihydroxyheneico-syl)-2*H*-pyran-2-one, 9CI   D-70276
7,9,11-Trihydroxy-2-hexacosen-1,5-olide
[116561-97-8]

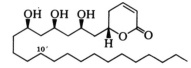

$C_{26}H_{48}O_5$ M 440.662
Constit. of *Eupatorium pilosum*. Cryst. (MeOH). Mp 101°.
10′-Hydroxy: [69616-80-4]. *5,6-Dihydro-6-(2,4,6,10-te-trahydroxyheneicosyl)-2*H*-pyran-2-one. 7,9,11,15-Tetrahydroxy-2-hexacosen-1,5-olide.*
$C_{26}H_{48}O_6$ M 456.662
Constit. of *E. pilosum*. Needles (MeOH). Mp 109°. $[\alpha]_D^{22}$ −33.6° (c, 0.0250 in MeOH).
Herz, W. *et al, Phytochemistry*, 1978, **17**, 1327 (*isol*)
Nakata, T. *et al, Tetrahedron Lett.*, 1987, **28**, 5661 (*synth, abs config*)

7,7a-Dihydro-3,6,7-trihydroxy-1a-(3-methyl-2-butenyl)naphth[2,3-b]oxiren-2(1aH)-one, 9CI   D-70277

2,3-Epoxy-4,5,8-trihydroxy-2-prenyl-1-tetralone

[111261-13-3]

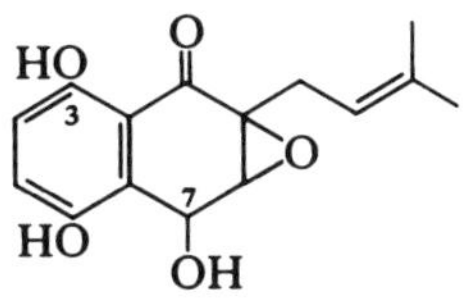

$C_{15}H_{16}O_5$ M 276.288

Constit. of *Sesamum angolense*.

3-O-β-D-Glucopyranoside: [111261-15-5].
$C_{21}H_{26}O_{10}$ M 438.430
Constit. of *S. angolense*.

7-Ketone: [111261-12-2]. *3,6-Dihydroxy-1a-(3-methyl-2-butenyl)naphth[2,3-b]-2,7(1aH,7aH)-dione*.
$C_{15}H_{14}O_5$ M 274.273
Constit. of *S. angolense*.

7-Ketone, 3-O-β-D-glucopyranoside: [111261-14-4].
$C_{21}H_{24}O_{10}$ M 436.415
Constit. of *S. angolense*.

Potterat, O. *et al, Helv. Chim. Acta*, 1987, **70**, 1551.

1,4-Dihydro-4,6,7-trimethyl-9H-imidazo[1,2-a]purin-9-one, 9CI   D-70278

7-Methylwye

[96881-39-9]

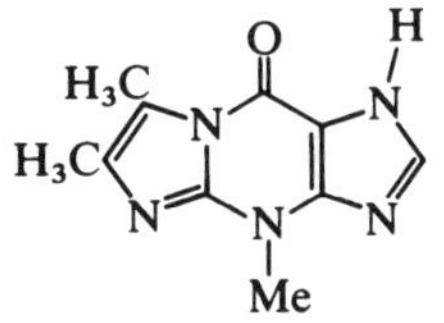

$C_{10}H_{11}N_5O$ M 217.230

Found in tRNA of highly thermophilic archaebacteria. Monohydrate. Mp >300°.

Itaya, T. *et al, Tetrahedron Lett.*, 1985, **26**, 347 (*synth, pmr*)
Milloskey, J.A. *et al, Nucleic Acids Res.*, 1987, **15**, 683.

Dihydro-3-vinyl-2(3H)-furanone   D-70279

3-Ethenyldihydro-2(3H)-furanone. 2-Vinyl-γ-butyrolactone

[43142-60-5]

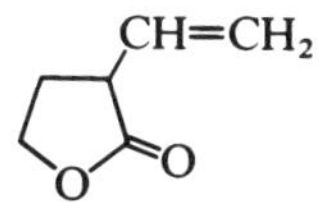

$C_6H_8O_2$ M 112.128

(±)-*form*
Bp$_{0.5}$ 41-42°.

Hénin, F. *et al, Synthesis*, 1983, 1019 (*synth, ir, pmr*)
Tamaru, Y. *et al, Tetrahedron Lett.*, 1987, **28**, 3497 (*synth*)

7,15-Dihydroxy-8,11,13-abietatrien-18-oic acid   D-70280

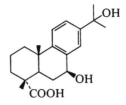

$C_{20}H_{28}O_4$ M 332.439

7β-form
Constit. of pollen of *Cedrus deodara*. Needles (CHCl$_3$/C$_6$H$_6$). Mp 166-168°. [α]$_D^{20}$ +24.1° (c, 0.28 in EtOH).

Ohmoto, T. *et al, Chem. Pharm. Bull.*, 1987, **35**, 229.

2,12-Dihydroxy-8,11,13-abietatrien-7-one   D-70281

$C_{20}H_{28}O_3$ M 316.439

2α-form
2α-Hydroxysugiol
Constit. of *Salvia cardiophylla*. Cryst. (EtOAc/pet. ether). Mp 269-271°.

González, A.G. *et al, Phytochemistry*, 1988, **27**, 1540.

2,11-Dihydroxy-7,9(11),13-abietatrien-12-one   D-70282

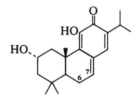

$C_{20}H_{28}O_3$ M 316.439

2α-form [116425-75-3]
6-Deoxo-2α-hydroxytaxodione
Constit. of *Salvia texana*. Brownish-yellow oil.

6-Oxo: [116425-76-4]. *2α,11-Dihydroxy-7,9(11),13-abietatriene-6,12-dione. 2α-Hydroxytaxodione*.
$C_{20}H_{26}O_4$ M 330.423
Constit. of *S. texana*. Reddish-brown oil.

6α-Hydroxy: [116425-77-5]. *2α,6α,11-Trihydroxy-7,9(11),13-abietatrien-12-one. 2α-Hydroxytaxodone*.
$C_{28}H_{28}O_4$ M 428.527
Constit. of *S. texana*. Yellow oil.

6α,7-Dihydroxy: [116425-78-6]. *2α,6α,7,11-Tetrahydroxy-7,9(11),13-abietatrien-12-one. 2α,7-Dihydroxytaxodone*.
$C_{20}H_{28}O_5$ M 348.438
Constit. of *S. texana*. Yellow oil.

González, A.G. *et al, Phytochemistry*, 1988, **27**, 1777.

12,13-Dihydroxy-1(10)-aromadendren-2-one   D-70283

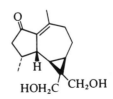

$C_{15}H_{22}O_3$ M 250.337

Di-Ac: 12,13-Diacetoxy-1(10)-aromadendren-2-one.
$C_{19}H_{26}O_5$ M 334.411
Constit. of *Gnephosis brevifolia*. Oil.

Jakupovic, J. *et al, Phytochemistry*, 1988, **27**, 3181.

ent-16α,17-Dihydroxy-3-atisanone   D-70284

ent-3-Oxo-16α,17-atisanediol

$C_{20}H_{32}O_3$ M 320.471
Constit. of *Euphorbia acaulis*. Cryst. (EtOAc). Mp
138°. $[\alpha]_D^{27}$ −30.8° (c, 1 in CHCl₃).

Satti, N.K. *et al, J. Nat. Prod.*, 1987, **50**, 790.

2,3-Dihydroxybenzaldehyde, 9CI, 8CI   D-70285

Updated Entry replacing D-20219
[24677-78-9]

$C_7H_6O_3$ M 138.123
Yellow needles. Mp 108°. Bp₁₆ 119-120°.

Phenylhydrazone: Yellow needles (EtOH). Mp 167°.
2-Me ether: 3-Hydroxy-2-methoxybenzaldehyde.
  $C_8H_8O_3$ M 152.149
  Needles (C₆H₆/hexane). Mp 113-115° (108-110°).
3-Me ether: [148-53-8]. *2-Hydroxy-3-methoxybenzalde-
hyde. o-Vanillin. 3-Formylguaiacol.*
  $C_8H_8O_3$ M 152.149
  Yellow needles (H₂O). Mp 44-45°. Bp₁₀ 128°.
▷CU6530000.
*Di-Me ether: 2,3-Dimethoxybenzaldehyde. o-
Veratraldehyde.* Needles (EtOH, Et₂O or pet. ether).
Mp 54°. Bp₇₄₅ 256°, Bp₁₂ 137°.
Di-Me ether, oxime: [5470-95-1].
  $C_9H_{11}NO_3$ M 181.191
  Needles (EtOH aq.). Mp 98-99°.
*Methylene ether: See 2,3-Methylenedioxybenzaldehyde,
M-01733*

Pauly, H. *et al, Justus Liebigs Ann. Chem.*, 1911, **383**, 312
  (*synth*)
U.S.P., 1 345 649, (*1920*); *CA*, **14**, 2644 (*synth*)
Kratzl, K. *et al, Monatsh. Chem.*, 1960, **91**, 219.
Neville, G.A., *Org. Magn. Reson.*, 1972, **4**, 633 (*pmr, derivs*)
Santavy, F. *et al, Collect. Czech. Chem. Commun.*, 1972, **37**,
  1825 (*uv, derivs*)
Huneck, S., *Tetrahedron*, 1976, **32**, 109 (*synth, deriv*)
Smidrkal, J., *Collect. Czech. Chem. Commun.*, 1982, **47**, 2140
  (*synth*)
Akashi, T. *et al, Chem. Pharm. Bull.*, 1986, **34**, 2024 (*deriv*)

1-[3-(2,4-Dihydroxybenzoyl)-4,6-dihy-   D-70286
droxy-2-(4-hydroxyphenyl)-7-benzofur-
anyl]-3-(4-hydroxyphenyl)-2-propen-1-
one, 9CI

*4,6-Dihydroxy-3-(2,4-dihydroxybenzoyl)-7-(4-hydrox-
ycinnamoyl)-2-(4-hydroxyphenyl)benzofuran*
[88901-97-7]

$C_{30}H_{20}O_9$ M 524.483
Orange pigment from *Brackenridges zanguebarica*.
  Orange-red rosettes (Me₂CO/pet. ether). Mp 252-
  253°.

2,3-Dihydro (trans-): [88901-98-8].
  $C_{30}H_{22}O_9$ M 526.498
  Constit. of *Brackenridges zanguebarica*.
2,3-Dihydro, hexa-Me ether: Cryst. (MeOH). Mp 66°.

Drewes, S.E. *et al, J. Chem. Soc., Perkin Trans. 1*, 1987, 2809.

2,5-Dihydroxybenzyl alcohol, 8CI   D-70287

Updated Entry replacing D-04307
*2,5-Dihydroxybenzenemethanol, 9CI. α,2,5-Trihydroxy-
toluene. 2-Hydroxymethyl-1,4-benzenediol. Gentisyl alcohol. Salirepol. Gentisin alcohol*
[495-08-9]

$C_7H_8O_3$ M 140.138
Phytotoxic metab. of *Phoma* spp. and *Penicillium
  roqueforti*. Red-brown cryst. (Et₂O/pet. ether). Mp
  104-105°. Readily oxidised.

Tribenzoyl: Mp 124.5-125.5°.
O²-β-D-Glucoside: [26652-12-0]. *Salirepin.*
  $C_{13}H_{18}O_8$ M 302.280
  Obt. by debenzoylation of Salireposide. Mp 165-167°.
2-Glucoside, hexa-Ac: Mp 117°. $[\alpha]_D^{27}$ −25.5° (CHCl₃).
O¹′-Benzoyl, O²-β-D-glucoside: [16955-55-8].
  Salireposide.
  $C_{20}H_{22}O_9$ M 406.388
  Constit. of bark of *Salix repens* and *S. purpurea* var.
  angustifolia. Fine needles. Mp 206-207°. $[\alpha]_D$ −37°
  (Py/Me₂CO).

Pearl, I.A. *et al, Tetrahedron Lett.*, 1967, 1869.
Séquin-Frey, M. *et al, Helv. Chim. Acta*, 1971, **54**, 851.
Harwig, J. *et al, CA*, 1978, **89**, 159844 (*isol*)
Burton, G. *et al, Can. J. Chem.*, 1980, **58**, 1839 (*cmr*)
Casiraghi, G. *et al, Synthesis*, 1980, 124 (*synth*)

3-(3,4-Dihydroxybenzyl)-7-hydroxy-4-   D-70288
chromanone
3-[(3,4-Dihydroxyphenyl)methyl]-2,3-dihydro-7-
hydroxy-4H-1-benzopyran-4-one, 9CI
[102067-88-9]

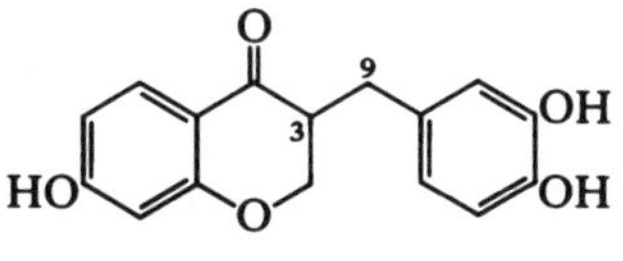

$C_{16}H_{14}O_5$ M 286.284
Constit. of dried heartwood of *Caesalpinia sappan*
(Sappan Lignum). Needles. Mp 192-194°. $[\alpha]_D^{24}$
−10.3° (c, 1.00 in MeOH).
3,9-Didehydro(E-): [104778-14-5]. *3-(3,4-Dihydroxy-*
benzylidene)-7-hydroxy-4-chromanone.
$C_{16}H_{12}O_5$ M 284.268
Constit. of Sappan Lignum (*C. sappan*). Yellow
needles. Mp 220-221°. $[\alpha]_D$ ±0°.
3-Hydroxy: [104778-15-6]. *3-(3,4-Dihydroxybenzyl)-*
3,7-dihydroxy-4-chromanone.
$C_{16}H_{14}O_6$ M 302.283
Constit. of Sappan Lignum (*C. sappan*). Powder. $[\alpha]_D^{24}$
+51.6° (c, 1.00 in MeOH).
3-Hydroxy, 4-alcohol: [102067-87-8]. *3-(3,4-Dihydrox-*
ybenzyl-3,4,7-trihydroxychroman.
$C_{16}H_{16}O_6$ M 304.299
Constit. of Sappan Lignum (*C. sappan*). Needles. Mp
157-160°. $[\alpha]_D^{29}$ +3.7° (c, 0.50 in MeOH).
3-Hydroxy, 4-alcohol, O^4-Me: 3-(3,4-Dihydroxyben-
zyl)-3,7-dihydroxy-4-methoxychroman.
$C_{17}H_{18}O_6$ M 318.326
Constit. of Sappan Lignum (*C. sappan*). Powder. $[\alpha]_D^{29}$
+53.6° (c, 1.50 in MeOH).

Saitoh, T. *et al, Chem. Pharm. Bull.*, 1986, **34**, 2506.

6,8'-Dihydroxy-2,2'-bis(2-phenylethyl)[5,5'-   D-70289
bi-4H-1-benzopyran]-4,4'-dione, 9CI
6',8-Dihydroxy-2,2'-bis(2-phenylethyl)-5,5'-bichro-
mone. AH_{11}
[112649-07-7]

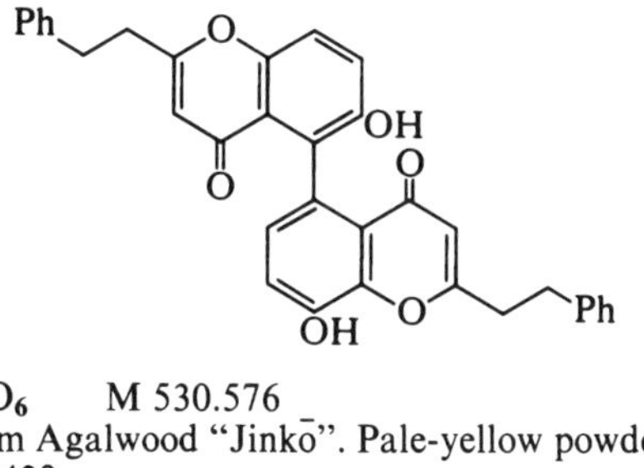

$C_{34}H_{26}O_6$ M 530.576
Isol. from Agalwood "Jinkō". Pale-yellow powder. Mp
239-242°.
Di-Ac: Needles. Mp 240°.

Iwagoe, K. *et al, Chem. Pharm. Bull.*, 1986, **34**, 4889 (*isol*)

2,3-Dihydroxy-2-cyclopropen-1-one, 9CI   D-70290
Updated Entry replacing D-04362
Deltic acid
[54826-91-4]

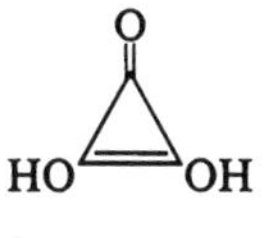

$C_3H_2O_3$ M 86.047

Solid.
Di-Me ether (ester): Dimethoxycyclopropenone.
$C_5H_6O_3$ M 114.101
Liq.
Diisopropyl ether (ester): [111870-18-9]. *2,3-*
Diisopropoxycyclopropenone.
$C_9H_{14}O_3$ M 170.208
Bp$_{0.05}$ 60-65°.
Di-tert-butyl ether (ester): [66478-67-9]. *2,3-Di-tert-*
butoxycyclopropenone.
$C_{11}H_{18}O_3$ M 198.261
Mp 80-82° (77-79°).

Pericas, M.A. *et al, Tetrahedron Lett.*, 1977, 4437 (*synth*)
Städeli, W. *et al, Helv. Chim. Acta*, 1977, **60**, 948 (*struct, cmr*)
Lautié, A. *et al, Can. J. Chem.*, 1985, **63**, 1394 (*ir*)
Semmingsen, D. *et al, J. Am. Chem. Soc.*, 1987, **109**, 7238
(*cryst struct*)
Dehmlow, E.V. *et al, Chem. Ber.*, 1988, **121**, 569 (*esters*)

4',5-Dihydroxy-3',7-dimethoxyflavone, 8CI   D-70291
Updated Entry replacing D-04383
5-Hydroxy-2-(4-hydroxy-3-methoxyphenyl)-7-
methoxy-4H-1-benzopyran-4-one, 9CI. **Velutin.** *Flavoya-*
dorigenin B
[25739-41-7]

$C_{17}H_{14}O_6$ M 314.294
Constit. of the leaves of *Ceanothus velutinus* and
Vernonia flexuosa. Mp 225-227°.
Di-Ac: Mp 207°.
4'-Glucoside: [30271-21-7]. **Flavoyadorinin B**.
$C_{23}H_{24}O_{11}$ M 476.436
Constit. of leaves of *Viscum album*. Pale-yellow
needles (EtOH). Mp 203-206°.

Ohta, N. *et al, Agric. Biol. Chem.*, 1970, **34**, 900
(*Flavoyadorinin B*)
Chirikdjian, J.J. *et al, Sci. Pharm.*, 1971, **39**, 65 (*uv*)
Jain, A.C. *et al, Phytochemistry*, 1973, **12**, 1455 (*synth*)
Kisiel, W., *Pol. J. Pharmacol. Pharm.*, 1975, **27**, 339 (*isol, pmr*)

5,8-Dihydroxy-6,10-dimethoxy-2-propyl-   D-70292
4H-naphtho[2,3-b]pyran-4-one
[111397-47-8]

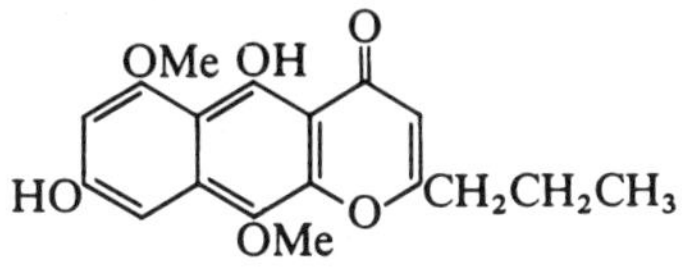

$C_{18}H_{18}O_6$ M 330.337
Constit. of the crinoid *Comanthus parvicirrus*. Orange-
red needles (Me$_2$CO/EtOH). Mp 214-214.5°.

Sakuma, Y. *et al, Aust. J. Chem.*, 1987, **40**, 1613.

2,6-Dihydroxy-3,5-dimethylbenzoic acid   D-70293
$C_9H_{10}O_4$ M 182.176
Di-Me ether: [91971-25-4]. *2,6-Dimethoxy-3,5-dimeth-*
ylbenzoic acid.
$C_{11}H_{14}O_4$ M 210.229
Mp 77-79° (71-72°).

Zbiral, E. *et al, Monatsh. Chem.*, 1962, **93**, 15.
de Paulis, T. *et al, J. Med. Chem.*, 1986, **29**, 61.

2,7-Dihydroxy-1,6-dimethyl-5-vinylphen-anthrene   D-70294

5-Ethenyl-1,6-dimethyl-2,7-phenanthrenediol.
Dehydrojuncusol

$C_{18}H_{16}O_2$ M 264.323
Constit. of *Juncus roemerianus*. Yellow cryst.
(EtOAc/hexane). Mp 241-243°.

Sarkar, H. *et al, Phytochemistry*, 1988, **27**, 3006.

3,11-Dihydroxy-7-drimen-6-one   D-70295

$C_{15}H_{24}O_3$ M 252.353
3β-form
Deoxyuvidin B
Metab. of *Alternaria brassicae*. Powder.

Ayer, W.A. *et al, J. Nat. Prod.*, 1987, **50**, 408.

5,20-Dihydroxy-6,8,11,14-eicosatetraenoic acid   D-70296

$C_{20}H_{32}O_4$ M 336.470
(5S,6E,8Z,11Z,14Z)-form [107373-23-9]
Hydroxylated eicosanoid metab. $[\alpha]_D^{23}$ +10.2° (c, 1.53 in
Me$_2$CO).

Lumin, S. *et al, J. Chem. Soc., Chem. Commun.*, 1987, 389
(*synth, pmr*)

15,20-Dihydroxy-5,8,11,13-eicosatetraen-oic acid, 9CI   D-70297

$C_{20}H_{32}O_4$ M 336.470
(5Z,8Z,11Z,13E,15S)-form [112168-45-3]
Hydroxylated eicosanoid metabolite. $[\alpha]_D^{23}$ +12.0° (c,
1.76 in Me$_2$CO).

Lumin, S. *et al, J. Chem. Soc., Chem. Commun.*, 1987, 389
(*synth, pmr*)

1,5-Dihydroxy-2,4(15),11(13)-eudesma-trien-12,8-olide   D-70298

$C_{15}H_{18}O_4$ M 262.305

(1α,5α,8β)-form
Constit. of *Ferreyranthus fruticosus*. Oil.

Jakupovic, J. *et al, Phytochemistry*, 1988, **27**, 1113.

1,3-Dihydroxy-4-eudesmen-12,6-olide   D-70299

$C_{15}H_{22}O_4$ M 266.336
(1α,3α,6α)-form [71241-93-5]
Alkhanol
Constit. of *Artemisia fragrans*. Cryst. Mp 179-181°.
(1β,3α,6α)-form [85847-69-4]
1-Epialkhanol
Constit. of *Picris aculeata*. Gum.

Serkerov, S.V., *Khim. Prir. Soedin.*, 1979, **15**, 282, 424 (*isol,
struct*)
Bruno, M. *et al, Phytochemistry*, 1988, **27**, 1201 (*isol*)

2′,8-Dihydroxyflavone   D-70300

$C_{15}H_{10}O_4$ M 254.242
Constit. of *Primula pulverulenta*. Yellow powder
(EtOH). Mp >250°.

Wollenweber, E. *et al, Phytochemistry*, 1986, **27**, 1483.

21,27-Dihydroxy-3-friedelanone   D-70301

Updated Entry replacing D-40233
21,27-Dihydroxy-D:A-friedooleanan-3-one

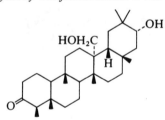

$C_{30}H_{50}O_3$ M 458.723
21α-form [72183-91-6]
Kokoondiol
Constit. of *Salacia reticulata* var. *diandra* and *Kokoona
zeylanica*. Cryst. (pet. ether). Mp 271-273°. $[\alpha]_D^{27}$
−29° (CHCl$_3$), $[\alpha]_D$ +18.0° (CHCl$_3$). Gunatikala *et
al* give both positive and negative opt. rotns. for
Kokoondiol.
21-Deoxy: [72183-92-7]. *27-Hydroxy-3-friedelanone.*
Kokoonol.
$C_{30}H_{50}O_2$ M 442.724
Isol. from *K. zeylanica*. Mp 272°. $[\alpha]_D$ −28.5°
(CHCl$_3$).
21-Ketone: [72183-90-5]. *27-Hydroxy-3,21-friedelane-
dione. Kokoononol.*
$C_{30}H_{48}O_3$ M 456.707
Isol. from *K. zeylanica*. Mp >325°. $[\alpha]_D$ +100.0°
(CHCl$_3$).

Gunatikala, A.A.L. *et al, J. Chem. Soc., Perkin Trans. 1*, 1983,
2459 (*isol*)
Kumar, V. *et al, Phytochemistry*, 1985, **24**, 2067 (*isol*)
Tanaka, R. *et al, Tetrahedron Lett.*, 1988, **29**, 4751 (*struct*)

2,8-Dihydroxy-1(10),4,11(13)-germacra-trien-12,6-olide   D-70302

Updated Entry replacing D-04491

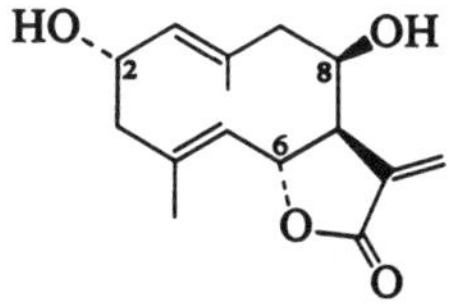

$C_{15}H_{20}O_4$ M 264.321

(*1(10)E,2α,4E,6α,8β*)-*form* [72229-33-5]
Constit. of *Eupatorium mikanioides*. Cryst. (CHCl$_3$/MeOH). Mp 184-186°. [α]$_D$ +196.5° (c, 0.1 in Py).

8-(2R,3R-Epoxy-2-methylbutanoyl): [72229-34-6].
Mollisorin B.
$C_{20}H_{26}O_6$ M 362.422
Constit. of *E. mikanioides* and *Helianthus mollis*. Gum or cryst. Mp 165-166°. [α]$_D$ +67.7° (c, 0.2 in CHCl$_3$). Identity of samples not established.

8-(2-Hydroxymethyl-2Z-butenoyl): [38456-39-2].
Deacetyleupaserrin.
$C_{20}H_{26}O_6$ M 362.422
Constit. of *E. mikanioides* and *E. semiserratum*. Cryst. (EtOAc/hexane) (also descr. as unstable amorph. foam). Mp 132-135°. [α]$_D^{25}$ +75° (c, 0.9 in MeOH).

8-(2R,3R-Epoxy-2-hydroxymethylbutanoyl): [72229-35-7].
$C_{20}H_{26}O_7$ M 378.421
Constit. of *E. mikanioides*. Cryst. (CHCl$_3$/MeOH). Mp 170-172°. [α]$_D$ +97.8° (c, 0.027 in Py).

8-(2S-Hydroxy-2-hydroxymethyl-3S-mercaptobutanoyl): [72229-36-8].
$C_{20}H_{28}O_7S$ M 412.497
Constit. of *E. mikanioides*. Cryst. (CHCl$_3$/MeOH). Mp 180-181°. [α]$_D$ +102° (c, 0.02 in Py).

8-(3S-Hydroxy-2-methylenebutanoyl): [72229-37-9].
$C_{20}H_{26}O_6$ M 362.422
Constit. of *E. mikanioides*. Gum. [α]$_D$ +80.5° (c, 0.23 in CHCl$_3$).

8-(2-Methyl-2Z-butenoyl): [72704-04-2]. **Mollisorin A**.
$C_{20}H_{26}O_5$ M 346.422
Constit. of *H. mollis*. Oil.

8-(2-Acetoxymethyl-2Z-butenoyl): [38456-36-9].
Eupaserrin.
$C_{22}H_{28}O_7$ M 404.459
Constit. of *E. semiserratum*. Shows tumour-inhibitory props. Mp 153-154°. [α]$_D^{25}$ +71.2° (c, 0.94 in MeOH).

Herz, W. *et al*, *J. Org. Chem.*, 1973, **38**, 1260; 1980, **45**, 489.

3,13-Dihydroxy-1(10),4,7(11)-germacra-trien-12,6-olide   D-70303

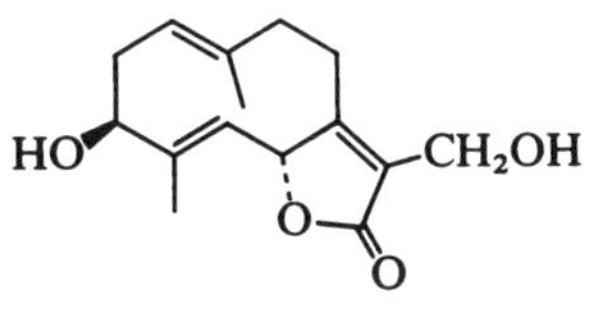

$C_{15}H_{20}O_4$ M 264.321
Di-Ac: [115346-31-1]. **Artemisiaglaucolide**.
$C_{19}H_{24}O_6$ M 348.395
Constit. of *Artemisia afra*. Oil.

Jakupovic, J. *et al*, *Phytochemistry*, 1988, **27**, 1129.

2,4-Dihydroxy-5,11(13)-guaiadien-12,8-olide   D-70304

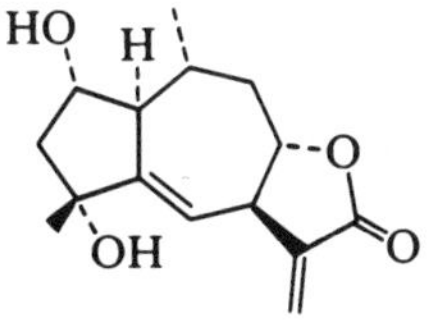

$C_{15}H_{20}O_4$ M 264.321
(*2α,4α,8α*)-*form* [117634-61-4]
Constit. of *Gaillardia pulchella*.

Yu, S. *et al*, *Phytochemistry*, 1988, **27**, 2887.

3,9-Dihydroxy-10(14)-guaien-12,6-olide   D-70305

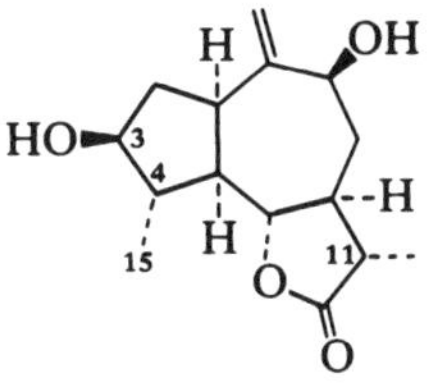

$C_{15}H_{22}O_4$ M 266.336
(*3β,4α,6α,9β,11α*)-*form* [88010-68-8]
9β-Hydroxy-4β,11β,13,15-tetrahydrozaluzanin C
Constit. of *Liabum floribundum*. Oil.

4,15-Didehydro: [116360-07-7]. *3β,9β-Dihydroxy-4(15),10(14)-guaiadien-12,6α-olide. 9β-Hydroxy-11β,13-dihydrozaluzanin* C.
$C_{15}H_{20}O_4$ M 264.321
Constit. of *L. floribundum*. Oil.

3-Ketone: [78012-02-9]. *9β-Hydroxy-3-oxo-10(14)-guaien-12,6α-olide. 9β-Hydroxy-4β,11β,13,15-tetrahydrodehydrozaluzanin* C.
$C_{15}H_{20}O_4$ M 264.321
Constit. of *L. floribundum* and *Aretolis grandis*. Oil.

(*3β,4α,6α,9β,11β*)-*form* [116360-06-6]
9β-Hydroxy-4β,11α,13,15-tetrahydrozaluzanin C
Constit. of *L. floribundum*. Oil. [α]$_D^{24}$ +27° (c 0.57 in CHCl$_3$).

4,15-Didehydro: [116360-08-8]. Constit. of *L. floribundum*. Oil.

3-Ketone: [116404-82-1]. Constit. of *L. floribundum*. Gum. [α]$_D^{24}$ +83° (c, 0.24 in CHCl$_3$).

Halim, A.F. *et al*, *Phytochemistry*, 1980, **19**, 2767 (*isol*)
Jakupovic, J. *et al*, *Phytochemistry*, 1988, **27**, 1771 (*isol*)

1,3-Dihydroxy-2-hydroxymethylanthra-quinone   D-70306

Updated Entry replacing D-50466
1,3-Dihydroxy-2-(hydroxymethyl)-9,10-anthracene-dione, 9CI. **Lucidin**
[478-08-0]

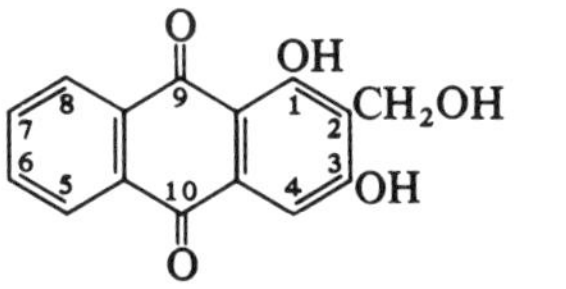

$C_{15}H_{10}O_5$ M 270.241
Obt. from *Coprosma*, *Morinda* and *Rubia* spp. Yellow cryst. (dioxan). Mp >330°. pK_a 8.11 (H$_2$O, 20°). Prob. artefact.

▷CB6712000.

Tri-Ac:
$C_{21}H_{16}O_8$　　M 396.353
Cryst. (EtOH). Mp 175-178°.
1-Me ether: [477-83-8]. *3-Hydroxy-2-hydroxymethyl-1-methoxyanthraquinone.* **Damnacanthol.**
$C_{16}H_{12}O_5$　　M 284.268
Obt. from roots of *M. cirrifolia, Damnacanthus* spp. and bark of *Coprosma rotundifolia.* Mp 288-290°.
1,1'-Di-Me ether: *3-Hydroxy-1-methoxy-2-(methoxymethyl)anthraquinone.*
$C_{17}H_{14}O_5$　　M 298.295
Isol. from *M. lucida.* Prob. artefact.
1-Me, $O^{1'}$-Et: *2-Ethoxymethyl-3-hydroxy-1-methoxyanthraquinone.*
$C_{18}H_{16}O_5$　　M 312.321
Isol. from *Plocama pendula.*

▷Prob. artefact
Tri-Me ether:
$C_{18}H_{16}O_5$　　M 312.321
Cryst. (EtOH). Mp 173°.
$O^{1'}$*-Et:* [17526-17-9]. *1,3-Dihydroxy-2-(ethoxymethyl)-anthraquinone.* **Ibericin.**
$C_{17}H_{14}O_5$　　M 298.295
Isol. from roots of *Rubia iberica.* Citron-yellow cryst. (C_6H_6). Mp 182-183° dec.
▷CB6705000.
$O^{1'}$*-Et, Di-Ac:* Mp 163-164°.

Agyangar, N.R. *et al, J. Sci. Ind. Res., Sect. B,* 1956, **15**, 359 (*synth*)
Bloom, H. *et al, J. Chem. Soc.,* 1959, 178 (*ir*)
Stikhin, V.A. *et al, Khim. Prir. Soedin.,* 1966, **2**, 12; 1967, **3**, 276; *Chem. Nat. Compd.,* pp. 9, 230 (*Ibericin*)
Hirose, Y. *et al, Chem. Pharm. Bull.,* 1973, **21**, 2790 (*synth*)
Leistner, E., *Planta Med.,* 1975 (*Suppl.*), 214 (*biosynth*)
Briggs, L.H. *et al, J. Chem. Soc., Perkin Trans. 1,* 1976, 1789 (*isol, pmr*)
Castonguay, A. *et al, Can. J. Chem.,* 1977, **55**, 1324 (*deriv*)
González, A.G. *et al, An. Quim.,* 1977, **73**, 869 (*isol, deriv*)
Roberts, J.L. *et al, J. Chem. Soc., Perkin Trans. 1,* 1977, **30**, 1553 (*synth*)
Demagos, G.P. *et al, Z. Naturforsch., B,* 1981, **36**, 1180 (*isol, deriv*)

5,6-Dihydroxyindole　　　　D-70307

Updated Entry replacing D-50478
1H-Indole-5,6-diol, *9CI*
[3131-52-0]
$C_8H_7NO_2$　　M 149.149
Intermed. in the tyrosinase-cat. biosynth. pathway from tyrosine to melanin. Needles (C_6H_6/pet. ether). Sol. hot H_2O, insol. pet. ether. Mp 140° dec. Unstable on storage, undergoes spont. oxidn. and condensation to melanin.

O,O-Di-Ac: [15069-79-1].
$C_{12}H_{11}NO_4$　　M 233.223
Prisms (C_6H_6/pet. ether). Mp 135-136°.
N-Me: [4821-00-5].
$C_9H_9NO_2$　　M 163.176
Cryst. (pet. ether). Mp 134°.
N-Me, di-O-Ac: [13988-19-7].
$C_{13}H_{13}NO_4$　　M 247.250
Mp 109-110° (101°).
O^5*-Me:* [2380-82-7]. *5-Methoxy-1H-indol-6-ol. 6-Hydroxy-5-methoxyindole.*
$C_9H_9NO_2$　　M 163.176
Mp 110-112°.
O^6*-Me:* [2380-83-8]. *6-Methoxy-1H-indol-5-ol. 5-Hydroxy-6-methoxyindole.*
$C_9H_9NO_2$　　M 163.176

Mp 113°.
Di-Me ether: [14430-23-0]. *5,6-Dimethoxyindole.*
$C_{10}H_{11}NO_2$　　M 177.202
Leaflets (EtOH/pet. ether), rods (C_6H_6), needles (EtOH aq.). Mp 154-155°. Bp$_8$ 198°. Darkens rapidly.
Di-Me ether, N-Ac:
$C_{12}H_{13}NO_3$　　M 219.240
Yellow-brown leaflets (pet. ether). Mp 150-152°.
Di-Me ether, N-Me: [80639-40-3].
$C_{11}H_{13}NO_2$　　M 191.229
Needles (heptane). Mp 134-136°.

Oxford, A.E. *et al, J. Chem. Soc.,* 1927, 417.
Raper, H.S., *Biochem. J.,* 1927, **21**, 89.
Beer, R.J.S. *et al, J. Chem. Soc.,* 1948, 2223; 1951, 2029.
Harley-Mason, J., *J. Chem. Soc.,* 1950, 1276; 1953, 200.
Heacock, P.A. *et al, J. Am. Chem. Soc.,* 1963, **85**, 1825.
Murphy, B.P. *et al, J. Org. Chem.,* 1985, **50**, 2790 (*synth, uv, pmr, cmr, ir*)
Corradini, M.G. *et al, Tetrahedron,* 1986, **42**, 2083 (*deriv, synth, props*)
Rogers, C.B. *et al, J. Heterocycl. Chem.,* 1987, **24**, 941 (*derivs*)

5,6-Dihydroxy-1H-indole-2-carboxylic acid, 9CI　　　　D-70308

Updated Entry replacing D-04529
[4790-08-3]

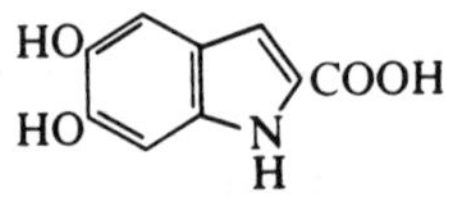

$C_9H_7NO_4$　　M 193.159
Degradn. prod. of melanin. Likely precursor in biosynth. of eumelanins. Cryst. Mp 240°.

Et ester:
$C_{11}H_{11}NO_4$　　M 221.212
Yellow prisms (EtOH). Mp 172°.
Di-Me ether:
$C_{11}H_{11}NO_4$　　M 221.212
Plates (C_6H_6). Mp 202-203° dec. H_2SO_4 → orange soln.

Perkin, W.H. *et al, J. Chem. Soc.,* 1926, 360.
Oxford, A.E. *et al, J. Chem. Soc.,* 1927, 420.
Raper, H.S., *Biochem. J.,* 1927, **21**, 89.
Benigni, J.D. *et al, J. Heterocycl. Chem.,* 1965, **2**, 387 (*synth, uv*)
d'Ischia, M. *et al, Tetrahedron Lett.,* 1985, **26**, 2801 (*isol*)
Palumbo, P. *et al, Tetrahedron,* 1987, **43**, 4203 (*biochem*)

2,5-Dihydroxy-4-isopropylbenzyl alcohol　　D-70309
2-(Hydroxymethyl)-5-(1-methylethyl)-1,4-benzenediol.
p-Cymene-2,5,7-triol, *8CI*

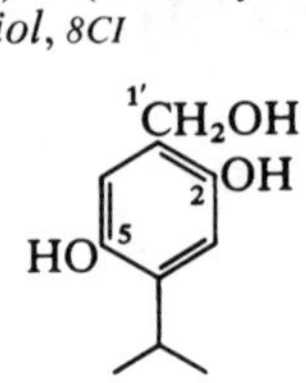

$C_{10}H_{14}O_3$　　M 182.219
2,5-Di-Me ether, 1'-(2-methylpropanoyl): [34272-56-5]. *1-Isobutyryloxymethyl-4-isopropyl-2,5-dimethoxy-benzene. Isobutyric acid 4-isopropyl-2,5-dimethoxy-benzyl ester,* *8CI.*
$C_{16}H_{24}O_4$　　M 280.363
Constit. of the root of *Inula salicifolia.* Oil. Bp$_{0.1}$ 95-105°.

Anthonsen, T. *et al*, *Acta Chem. Scand.*, 1971, **25**, 390.

3,15-Dihydroxy-8(17),13-labdadien-19-al   D-70310

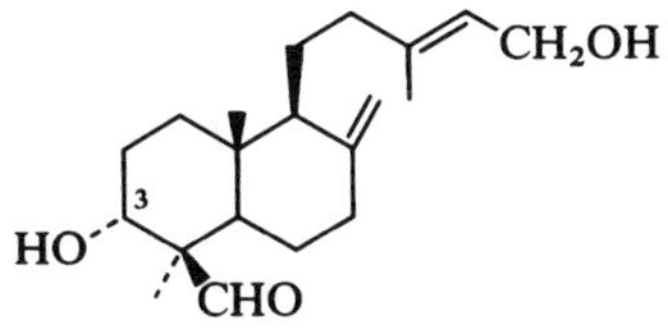

$C_{20}H_{32}O_3$ M 320.471

(3α,13E)-form [117254-99-6]
3α-Hydroxyisoagatholal
Constit. of *Juniperus thurifera*. Cryst. (CH_2Cl_2/hexane).
Mp 129°. $[\alpha]_D^{23}$ −0.7° ($CHCl_3$).

(3β,13E)-form [117255-05-7]
3β-Hydroxyisoagatholal
Cryst. (CH_2Cl_2/hexane). Mp 134°. $[\alpha]_D^{23}$ +22.3°
($CHCl_3$).

3-Ac: [117232-44-7].
$C_{22}H_{34}O_4$ M 362.508
Constit. of *J. thurifera*. Oil. $[\alpha]_D^{23}$ −14.5° ($CHCl_3$).

San Feliciano, A. *et al*, *Phytochemistry*, 1988, **27**, 2241.

8,15-Dihydroxy-13-labden-19-al   D-70311

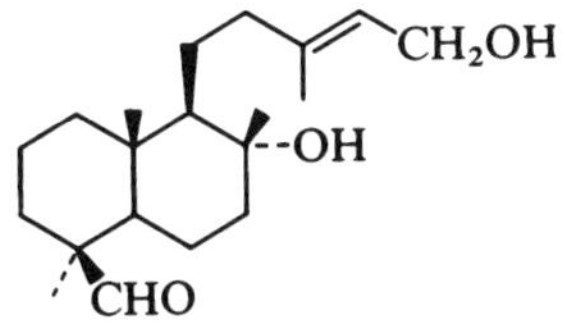

$C_{20}H_{34}O_3$ M 322.487

(8α,E)-form [117232-45-8]
Constit. of *Juniperus thurifera*. Cryst. (CH_2Cl_2/hexane).
Mp 134°. $[\alpha]_D^{23}$ +17.8° ($CHCl_3$).

San Feliciano, A. *et al*, *Phytochemistry*, 1988, **27**, 2241.

3,8-Dihydroxylactariusfuran   D-70312

[59684-34-3]

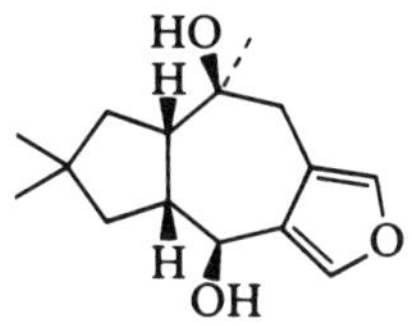

$C_{15}H_{22}O_3$ M 250.337
Constit. of *Lactarius* spp. Oil.

Nozoe, S. *et al*, *Tetrahedron Lett.*, 1971, 3125 (*isol, struct*)
Sterner, O. *et al*, *Acta Chem. Scand.*, *Ser. B*, 1988, **42**, 43 (*isol*)

4,8-Dihydroxylactariusfuran   D-70313

[114728-44-8]

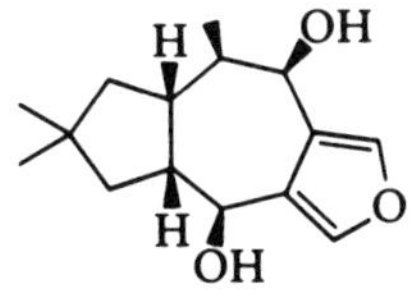

$C_{15}H_{22}O_3$ M 250.337

Constit. of *Lactarius* spp. Oil. $[\alpha]_D^{24}$ +4° (c, 0.7 in
$CHCl_3$).
Di-Ac: Cryst. Mp 70-75°. $[\alpha]_D^{24}$ +56° (c, 0.6 in $CHCl_3$).
Sterner, O. *et al*, *Acta Chem. Scand.*, *Ser. B*, 1988, **42**, 43.

3,15-Dihydroxy-7,9(11),24-lanostatrien-26-   D-70314
oic acid

Updated Entry replacing D-60344
$C_{30}H_{46}O_4$ M 470.691
The ganoderic acid synonyms are confused owing to mul-
tiple use by different workers. See also refs. under
3,15,22-Trihydroxy-7,9(11),24-lanostatrien-26-oic
acid, T-70266 .

(3α,5α,15α,24E)-form [112430-67-8]
Ganodermic acid Ja
Metab. of *Ganoderma lucidum*.

3-Ac: [108026-94-4]. *3α-Acetoxy-15α-hydroxy-
7,9(11),24E-lanostatrien-26-oic acid. Ganoderic acid
Mf.*
$C_{32}H_{48}O_5$ M 512.728
Metab. of *Ganoderma lucidum*. Syrup. $[\alpha]_D^{24}$ +42° (c,
0.2 in $CHCl_3$).

15-Ac: [86377-53-9]. *15α-Acetoxy-3α-hydroxy-
7,9(11),24E-lanostatrien-26-oic acid. Ganoderic acid
X.*
$C_{32}H_{48}O_5$ M 512.728
Metab. of *G. lucidum*.

15-Ac, Me ester: Mp 161-163°. $[\alpha]_D$ +76°.

Di-Ac: [108026-93-3]. *3α,15α-Diacetoxy-7,9(11),24E-
lanostatrien-26-oic acid. Ganderic acid Me. Ganoderic
acid R.*
$C_{34}H_{50}O_6$ M 554.765
Metab. of *G. lucidum*. Syrup or amorph. powder
($CHCl_3$/hexane). Mp 126-129°. $[\alpha]_D^{24}$ +53° (c, 0.26 in
MeOH).

(3β,15α,24E)-form [112430-68-9]
Ganodermic acid Jb
Metab. of *G. lucidum*. Cryst. Mp 201-202°.

Di-Ac: [112430-63-4]. *3β,15α-Diacetoxy-7,9(11),24E-
lanostatrien-26-oic acid. Ganoderic acid S.*
$C_{34}H_{50}O_6$ M 554.765
Metab. of *G. lucidum*. Needles ($CHCl_3$/hexane). Mp
123-124°.

Nishitoba, T. *et al*, *Agric. Biol. Chem.*, 1987, **51**, 619 (*isol*)
Shiao, M.-S. *et al*, *J. Nat. Prod.*, 1987, **50**, 886 (*isol*)

7,13-Dihydroxy-2-longipinen-1-one   D-70315

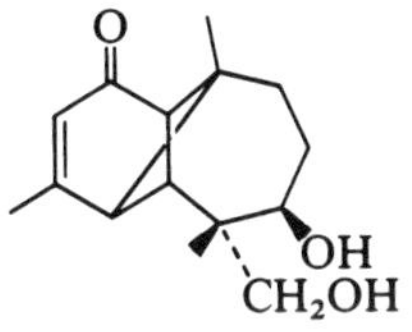

$C_{15}H_{22}O_3$ M 250.337

7β-form [100045-39-4]
Constit. of *Stevia achalensis*. Oil. $[\alpha]_D^{24}$ +64° (c, 0.65 in
$CHCl_3$).

Bohlmann, F. *et al*, *Justus Liebigs Ann. Chem.*, 1985, 1764;
1986, 799 (*isol*)

3,23-Dihydroxy-20(30)-lupen-29-al   D-70316
3,23-Dihydroxy-20(29)-lupen-30-al

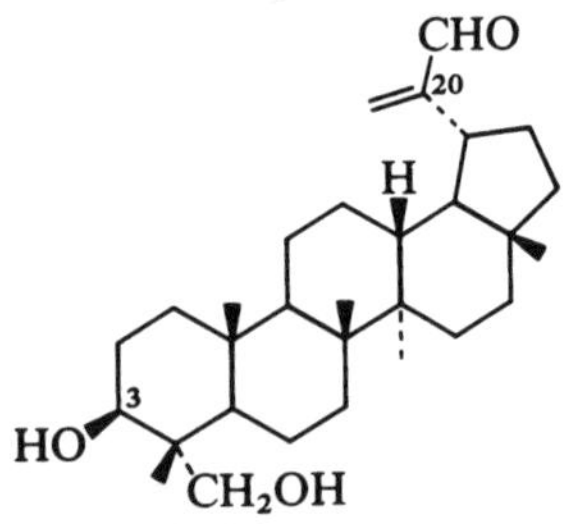

$C_{30}H_{48}O_3$ M 456.707

3β-form

Di-Ac: [116199-61-2]. *3β,23-Diacetoxy-20(30)-lupen-29-al.* **Skimmial.**
$C_{34}H_{52}O_5$ M 540.782
Constit. of *Skimmia laureola.* Needles (C_6H_6/pet. ether). Mp 218-220°. $[\alpha]_D^{25}$ +23.8° (c, 2 in EtOH).

3-Ketone, Ac: [116175-10-1]. *23-Acetoxy-3-oxo-20(30)-lupen-29-al.* **Skimmianone.**
$C_{32}H_{48}O_4$ M 496.729
Constit. of *S. laureola.* Needles (MeOH/pet. ether). Mp 187-188°. $[\alpha]_D^{25}$ +33.2° (c, 1.05 in EtOH).

Razdan, T.K. *et al, Phytochemistry,* 1988, **27**, 1890.

3,23-Dihydroxy-20(29)-lupen-28-oic acid   D-70317
$C_{30}H_{48}O_4$ M 472.707

3β-form

Constit. of *Paeonia japonica* callus tissue. Cryst. (MeOH/CHCl₃). Mp 263-267°. $[\alpha]_D^{21}$ +88° (c, 0.075 in Py).

Ikuta, A. *et al, Phytochemistry,* 1988, **27**, 2813.

23,30-Dihydroxy-20(29)-lupen-3-one   D-70318
$C_{30}H_{48}O_3$ M 456.707

23-Ac: [116175-04-3]. *23-Acetoxy-30-hydroxy-20(29)-lupen-3-one.* **Skimmiol.**
$C_{32}H_{50}O_4$ M 498.745
Constit. of *Skimmia laureola.* Amorph. Mp 192-193°. $[\alpha]_D^{25}$ +30.2° (c, 1.2 in EtOH).

Razdan, T.K. *et al, Phytochemistry,* 1988, **27**, 1890.

5,7-Dihydroxy-4'-methoxyflavone, 8CI   D-70319
Updated Entry replacing D-04595
5,7-Dihydroxy-2-(4-methoxyphenyl)-4H-1-benzopyran-4-one, 9CI. Apigenin 4'-methyl ether. **Acacetin.** *Linarigenin.* **Buddleoflavonol**

[480-44-4]

$C_{16}H_{12}O_5$ M 284.268
Constit. of *Robina pseudoacacia* (common acacia) and *Ammi visnaga.* Antiinflammatory, capillary protective and spasmolytic. Pale-yellow needles (EtOH). Mp 263°.

▷DJ3002000.

Di-Ac: [5892-39-7]. Needles (EtOH). Mp 203°.

7-β-D-Galactoside: [35013-09-3].
$C_{22}H_{22}O_{10}$ M 446.410
Mp 259° dec. $[\alpha]_D^{25}$ −36.6° (c, 0.55 in DMF).

7-β-D-Glucurono-β(1→2)-D-glucuronide: [36730-68-4].
Constit. of leaves of *Clerodendron trichotomum.* Mp 191-205° dec. $[\alpha]_D^{22}$ −48° (c, 1.3 in Py).

7-β-D-Glucoside: **Tilianin.**
$C_{22}H_{22}O_{10}$ M 446.410
Obt. from leaves of *Tilia japonica.* Cryst. + 2.5H₂O (MeOH). Mp 259-260° (anhyd.). $[\alpha]_D^{20}$ −63.3° (7:3 Py/EtOH).

7-β-Rutinoside: **Linarin.** *Acaciin.* **Buddleoflavonoloside.**
$C_{28}H_{32}O_{14}$ M 592.552
Occurs in *Linaria vulgaris, Robinia* and *Micromeria* spp. and other plants. Mp 272-276°. $[\alpha]_D^{24}$ −90.3° (Py).

7-(2,4-Di-O-acetyl-α-L-rhamnopyranosyl(1→6)-β-D-glucopyranosyl(1→2)-β-D-glucopyranosyl(1→2)-3-O-acetyl-β-D-glycopyranoside): [59985-51-2]. **Coptiside** I.
$C_{46}H_{58}O_{27}$ M 1042.948
Constit. of leaves of *Coptis japonica.* Pale-yellow amorph. powder. Mp 190-191°. $[\alpha]_D^{25}$ −93° (Py).

7-Neohesperidoside: [20633-93-6]. **Fortunellin.**
$C_{28}H_{32}O_{14}$ M 592.552
Constit. of *Fortunella* spp. Cryst. (MeOH aq.). Mp 214-216°.

Robinson, R. *et al, J. Chem. Soc.,* 1926, **128**, 2344 (*synth*)
Shibata, S. *et al, Yakugaku Zasshi,* 1960, **80**, 620; *CA,* **54**, 21488 (*pharmacol*)
Nakaoki, T. *et al, CA,* 1961, **55**, 10803 (*deriv*)
Nogradi, M. *et al, Chem. Ber.,* 1967, **100**, 2783 (*deriv*)
Bowie, J.H. *et al, J. Chem. Soc.* (*B*), 1969, **2**, 89 (*ms*)
Khadzhai, Y. *et al, Farmakol. Toksikol.* (*Moscow*), 1969, **32**, 451 (*pharmacol*)
Okigawa, M. *et al, Tetrahedron Lett.,* 1970, 2935 (*deriv*)
Wagner, H. *et al, Chem. Ber.,* 1969, **102**, 1445, 2083; 1970, **103**, 851 (*derivs*)
Fujiwara, H. *et al, Chem. Pharm. Bull.,* 1976, **24**, 407 (*Coptiside*)
Wagner, H. *et al, Tetrahedron Lett.,* 1976, 1799 (*nmr*)

2,3-Dihydroxy-5-methyl-1,4-benzoquinone   D-70320
Updated Entry replacing D-04702
2,3-Dihydroxy-5-methyl-2,5-cyclohexadiene-1,4-dione, 9CI. 5,6-Dihydroxy-p-toluquinone. 3,4-Dihydroxy-2,5-toluquinone

[825-33-2]

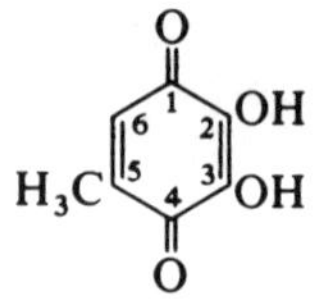

$C_7H_6O_4$ M 154.122
Prod. by *Aspergillus fumigatus* and *Penicillium spinulosum.* Amorph. violet substance. Mp 157-159°. λ_{max} 279, 385 nm (CHCl₃).

2-Me ether: [484-89-9]. *3-Hydroxy-2-methoxy-5-methyl-1,4-benzoquinone.* **Fumigatin.**
$C_8H_8O_4$ M 168.149
Prod. by *Aspergillus fumigatus.* Inhibits gram-positive and -negative bacteria. Maroon cryst. (pet. ether). Mp 114°. Subl. undec.

3-Me ether: [2446-67-5]. *2-Hydroxy-3-methoxy-5-methyl-1,4-benzoquinone.*
$C_8H_8O_4$ M 168.149
Prod. by *A. fumigatus.* Yellow oil.

2-Me ether, 3-Ac:
$C_{10}H_{10}O_5$ M 210.186
Canary-yellow cryst. (pet. ether). Mp 96-97°.

3-Me ether, 2-Ac: [2698-71-7].
$C_{10}H_{10}O_5$ M 210.186

Yellow needles (pet. ether). Mp 48°.

Di-Me ether: [605-94-7]. *2,3-Dimethoxy-5-methyl-1,4-benzoquinone. Fumigatin methyl ether.*
$C_9H_{10}O_4$ M 182.176
Occurs in defensive secretions of the millipedes *Rhapidostreptus innominatus, Pachybolus brachysternus, Aphistreptus levis* and *Metiche tanganyicense.* Intermed. in Coenzyme *Q* synth. Deep-crimson needles (pet. ether). Mp 58-59°.

2-Me ether, 5,6-epoxide: [1716-19-4]. *3-Hydroxy-4-methoxy-1-methyl-7-oxabicyclo[4.1.0]hept-3-ene-2,5-dione, 8CI.* **Fumigatin epoxide.** *Fumigatin oxide.*
$C_8H_8O_5$ M 184.148
From *A. fumigatus.* Unstable cryst. Mp 74°. $[\alpha]_D^{14}$ +28.5° (EtOH).

Anslow, W.K. *et al, Biochem. J.,* 1938, **32**, 687 (*isol*)
Baker, W. *et al, J. Chem. Soc.,* 1941, 670 (*synth*)
Seshadri, T.R. *et al, J. Chem. Soc.,* 1959, 1660 (*synth*)
Pettersson, G., *Acta Chem. Scand.,* 1963, **17**, 1771; 1964, **18**, 1428; 1965, **19**, 543, 1016 (*isol, biosynth*)
Imada, I. *et al, Chem. Pharm. Bull.,* 1965, **13**, 130 (*synth, ir*)
Packter, N.M., *Biochem. J.,* 1965, **97**, 321 (*ir*)
Yamamoto, Y. *et al, Chem. Pharm. Bull.,* 1965, **13**, 935; 1974, **22**, 83 (*isol, deriv*)
Simonart, P. *et al, Bull. Soc. Chim. Biol.,* 1966, **48**, 943 (*biosynth*)
Weinstock, L.M. *et al, J. Chem. Eng. Data,* 1967, **12**, 154 (*synth*)
Yamamoto, Y. *et al, Chem. Pharm. Bull.,* 1967, **15**, 427; 1972, **20**, 931; 1974, **22**, 83 (*isol, synth, biosynth*)
Koshi, K. *et al, Chem. Pharm. Bull.,* 1968, **16**, 2343.
Wilczynski, J.J. *et al, J. Am. Chem. Soc.,* 1968, **90**, 5593 (*nmr*)
Bowie, J.H. *et al, J. Chem. Soc.* (*B*), 1969, 89 (*ms*)
Silverman, J. *et al, Acta Crystallogr., Sect. B,* 1971, **27**, 1846 (*cryst struct*)
Sheehan, J.C. *et al, J. Med. Chem.,* 1974, **17**, 371 (*biosynth*)
Keinan, E. *et al, J. Org. Chem.,* 1987, **52**, 3872 (*synth, deriv*)
Cole, R.J. *et al, Handbook of Toxic Fungal Metabolites,* Academic Press, N.Y. 1981, 773.

1,3-Dihydroxy-4-methyl-6,8-decadien-5-one   D-70321

SS 7313A. E2Z4

[108605-55-6]

$$H_3C(CH{=}CH)_2COCH(CH_3)CH(OH)CH_2CH_2OH$$

$C_{11}H_{18}O_3$ M 198.261
Isol. from *Streptomyces cavourensis.* Immunoregulator.

Japan. Pat., 87 00 434, (*1987*); *CA,* **107**, 5734 (*isol*)

2′,5′-Dihydroxy-4′-nitroacetophenone   D-70322

1-(2,5-Dihydroxy-4-nitrophenyl)ethanone. 2-Acetyl-5-nitroquinol

[64481-14-7]

$C_8H_7NO_5$ M 197.147

Di-Me ether: [70313-21-2]. *2′,5′-Dimethoxy-4′-nitroacetophenone.*
$C_{10}H_{11}NO_5$ M 225.201
Yellow needles (MeOH). Mp 122-123°.

Howe, C.A. *et al, J. Org. Chem.,* 1960, **25**, 1245; 1962, **27**, 1923.

4,6-Dihydroxy-20-nor-2,7-cembradien-12-one   D-70323

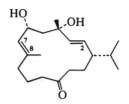

$C_{19}H_{32}O_3$ M 308.460

(1S,2E,4S,6R,7E)-form
Constit. of tobacco. Oil. $[\alpha]_D$ +1.5° (c, 0.46 in CHCl₃).

7R,8R-Epoxide: 4,6-Dihydroxy-7,8-epoxy-20-nor-2-cembren-12-one.
$C_{19}H_{32}O_4$ M 324.459
Constit. of tobacco. Cryst. Mp 133-134°. $[\alpha]_D$ −42° (c, 0.27 in CHCl₃).

Wahlberg, I. *et al, Acta Chem. Scand., Ser. B,* 1987, **41**, 749.

3,23-Dihydroxy-30-nor-12,20(29)-oleana-dien-28-oic acid   D-70324

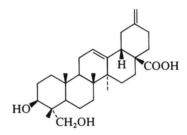

$C_{29}H_{44}O_4$ M 456.664

3β-form [117654-06-5]
Constit. of *Paeonia japonica* callus tissue. Cryst. (MeOH/EtOAc). Mp 249°. $[\alpha]_D^{21}$ +109.6° (c, 0.31 in Py).

Ikuta, A. *et al, Phytochemistry,* 1988, **27**, 2813.

1,3-Dihydroxy-12-oleanen-29-oic acid   D-70325

$C_{30}H_{48}O_4$ M 472.707

(1α,3β)-form [114076-53-8]
Imberbic acid
Constit. of *Cambretum imberbe.* Cryst. (EtOH). Mp 286-288°. $[\alpha]_D$ +70.0° (c, 1 in Py).

Rogers, C.B. *et al, Phytochemistry,* 1988, **27**, 531.

3,6-Dihydroxy-12-oleanen-29-oic acid   D-70326

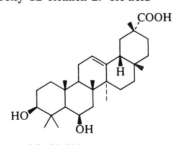

$C_{30}H_{48}O_4$ M 472.707

(3β,6β)-form [113762-83-7]
Myrianthinic acid
Constit. of *Myrianthus arboreus.* Cryst. (CH₂Cl₂). Mp 290-292°.

Ngounou, F.N. *et al, Phytochemistry*, 1988, **27**, 301.

3,19-Dihydroxy-12-oleanen-28-oic acid   D-70327

Updated Entry replacing S-00399

$C_{30}H_{48}O_4$ M 472.707

(*3α,19α*)-form [56586-60-8]

3-Episiaresinolic acid

Constit. of *Gardenia latifolia*.

3-Ac: Mp 266-268°. $[\alpha]_D$ +13.6° (CHCl$_3$).

(*3β,19α*)-form [511-77-3]

Siaresinolic acid

Isol. from Siam benzoin (siamese gum) and of *Osteospermum corymbosum*. Cryst. (MeOH/Et$_2$O). Mp 274-275°. $[\alpha]_D^{24}$ +98.5° (EtOH).

3-O-β-D-Xylopyranoside: [108529-23-3]. ***Ilexoside* A**. $C_{35}H_{56}O_8$ M 604.823

Constit. of leaves of *Ilex chinensis*. Cryst. (MeOH). Mp 264-267°. $[\alpha]_D$ +7.8° (c, 0.20 in MeOH).

Me ester: Cryst. Mp 198-200°.

3-Ketone: **19α-Hydroxy-3-oxo-12-oleanen-28-oic acid**. $C_{30}H_{46}O_4$ M 470.691

Constit. of *O. corymbosum*.

3-Ketone, Me ester: Cryst. Mp 187-190°.

(*3β,19β*)-form [32205-23-5]

Spinosic acid

Constit. of *Randia spinosa*. Cryst. (CHCl$_3$/MeOH). Mp 270-272°. $[\alpha]_D$ +12.5° (c, 0.42 in CHCl$_3$).

Barton, D.H.R. *et al, J. Chem. Soc.*, 1952, 78 (*struct*)
Aplin, R.T. *et al, J. Chem. Soc. (C)*, 1971, 1067 (*isol*)
Reddy, G.C.S. *et al, Indian J. Chem.*, 1975, **13**, 749 (*isol*)
Inada, A. *et al, Chem. Pharm. Bull.*, 1987, **35**, 841 (*Ilexoside A*)
Jakupovic, J. *et al, Phytochemistry*, 1988, **27**, 2881 (*isol*)

4,5-Dihydroxy-3-oxo-1-cyclohexenecarboxylic acid   D-70328

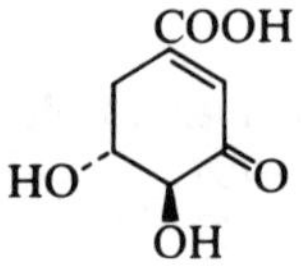

$C_7H_8O_5$ M 172.137

(*4S,5R*)-form [2922-42-1]

3-Dehydroshikimic acid. 5-Dehydroshikimic acid (*obsol.*)

Isol. from culture filtrates of an *E. coli* mutant; widespread compd. Immediate precursor of Shikimic acid in the shikimate pathway to aromatic aminoacids in plants and microorganisms. Needles (EtOAc). Mp 146-147°.

Salamon, J.J. *et al, J. Am. Chem. Soc.*, 1953, **75**, 5567 (*isol*)
Weiss, U. *et al, J. Am. Chem. Soc.*, 1953, **75**, 5572 (*isol, struct, uv, ir*)
Haslam, G. *et al, Methods Enzymol.*, 1963, **6**, 498.
McKittrick, B. *et al, J. Org. Chem.*, 1985, **50**, 5897 (*synth*)
Abell, C. *et al, Acta Crystallogr., Sect. C*, 1988, **44**, 1290 (*cryst struct*)

1,5-Dihydroxy-2-oxo-3,11(13)-eudesmadien-12,8-olide   D-70329

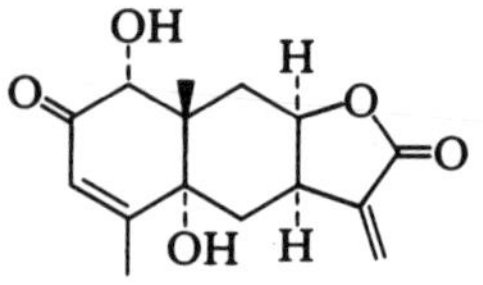

$C_{15}H_{18}O_5$ M 278.304

(*1α,5α,8β*)-form

1α,5α-Dihydroxypinnatifidin

Constit. of *Ferreyranthus fruticosus*. Oil.

Jakupovic, J. *et al, Phytochemistry*, 1988, **27**, 1113.

3,15-Dihydroxy-23-oxo-7,9(11),24-lanostatrien-26-oic acid   D-70330

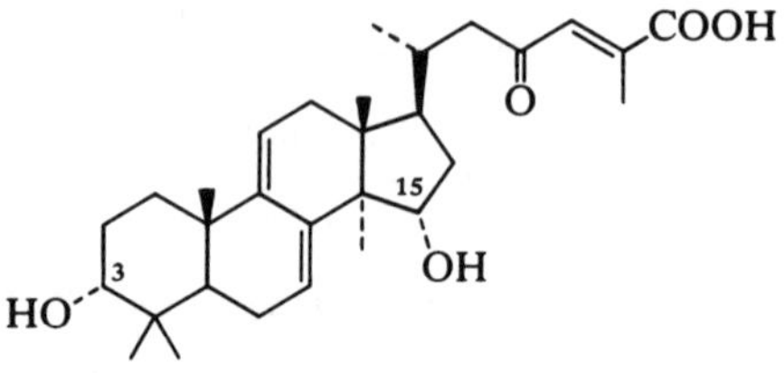

$C_{30}H_{44}O_5$ M 484.675

(*3α,15α,24E*)-form

3-Ac: [117383-37-6]. *3α-Acetoxy-15α-hydroxy-23-oxo-7,9(11),24E-lanostatrien-26-oic acid.* $C_{32}H_{46}O_6$ M 526.712

Constit. of *Ganoderma lucidum*.

15-Ac: [117383-35-4]. *15α-Acetoxy-3α-hydroxy-23-oxo-7,9(11),24E-lanostatrien-26-oic acid.* $C_{32}H_{46}O_6$ M 526.712

Constit. of *G. lucidum*.

Di-Ac: [117383-36-5]. *3α,15α-Diacetoxy-23-oxo-7,9(11),24E-lanostatrien-26-oic acid.* $C_{34}H_{48}O_7$ M 568.749

Constit. of *G. lucidum*.

Shiao, M.-S. *et al, Phytochemistry*, 1988, **27**, 2911.

19,29-Dihydroxy-3-oxo-12-oleanen-28-oic acid   D-70331

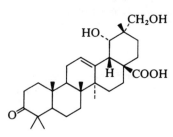

$C_{30}H_{46}O_5$ M 486.690

19α-form

Constit. of *Osteospermum corymbosum*.

Me ester: [117615-11-9]. Cryst. Mp 190-193°.

Jakupovic, J. *et al, Phytochemistry*, 1988, **27**, 2881.

3,6-Dihydroxy-2-(1-oxo-10-tetradecenyl)-   **D-70332**
2-cyclohexen-1-one, 9CI

[112727-26-1]

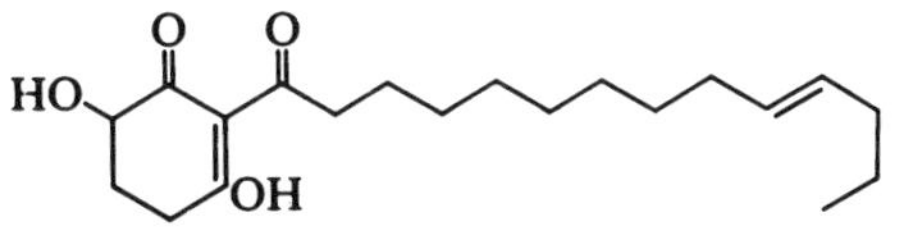

$C_{20}H_{32}O_4$ M 336.470
Constit. of *Corythucha ciliata*. Oil.

Lusby, W.R. *et al, J. Nat. Prod.*, 1987, **50**, 1126.

22,23-Dihydroxy-3-oxo-12-ursen-30-oic   **D-70333**
acid

$C_{30}H_{46}O_5$ M 486.690

22α-form

Me ester: [109974-22-3]. **Regelinol.**
$C_{31}H_{48}O_5$ M 500.717
Constit. of *T. regelii*. Cryst. (MeOH). Mp 210-212°.
$[\alpha]_D^{26}$ +14° (c, 0.5 in CHCl₃).

Hori, H. *et al, Chem. Pharm. Bull.*, 1987, **35**, 2125.

4-(3,4-Dihydroxyphenyl)-5,7-dihydroxy-   **D-70334**
2H-1-benzopyran-2-one

4-(3,4-Dihydroxyphenyl)-5,7-dihydroxycoumarin

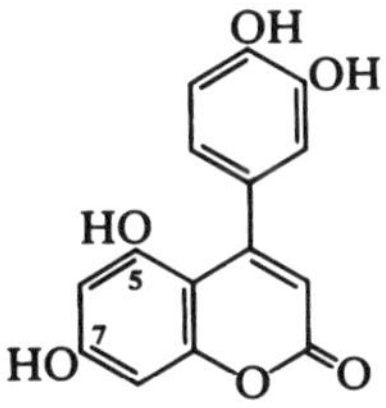

$C_{15}H_{10}O_6$ M 286.240

7-Me ether: Cryst. (Me₂CO). Mp 138-140°.
7-Me ether, 5-O-β-D-galactopyranosyl: [112078-67-8].
$C_{22}H_{22}O_{11}$ M 462.409
Constit. of *Exostema caribaeum*. Cryst.
(Me₂CO/EtOH). Mp 228-231°.

Mata, R. *et al, J. Nat. Prod.*, 1987, **50**, 866.

1-(2,4-Dihydroxyphenyl)-2-(3,5-   **D-70335**
dihydroxyphenyl)ethylene

4-[2-(3,5-Dihydroxyphenyl)ethenyl]-1,3-benzenediol,
9CI. 2,3′,4,5′-Stilbenetetrol, 8CI. 2,3′,4,5′-Tetrahydroxys-
tilbene. **Oxyresveratrol**

[4721-07-7]

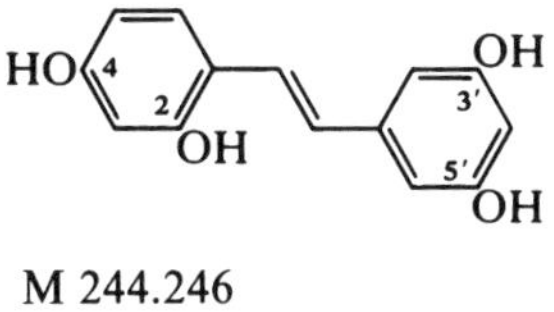

$C_{14}H_{12}O_4$ M 244.246

(E)-form [29700-22-9]
Constit. of *Morus* spp., *Artocarpus* spp. and *Chlorophora
regia*. Shows antifungal activity. Mp 205°.
Tetra-Ac: Mp 142°.
Tetra-Me ether: Mp 84°.
3′,4-Di-O-β-D-Glucopyranoside: [102841-42-9]. **Mul-
berroside A.**
$C_{26}H_{32}O_{14}$ M 568.530
From *M. lhou*. Amorph. powder. $[\alpha]_D^{25}$ −78° (c, 0.88
in MeOH).

Barnes, R.A. *et al, J. Am. Chem. Soc.*, 1955, **77**, 3259 (*isol*)
Mongolsuk, S. *et al, J. Chem. Soc.*, 1957, 2231 (*isol*)
Reimann, E., *Justus Liebigs Ann. Chem.*, 1971, **750**, 109 (*synth*)
Takasugi, M. *et al, Chem. Lett.*, 1978, 1241 (*isol, deriv*)
Hirakura, K. *et al, Chem. Pharm. Bull.*, 1985, **33**, 1088 (*cmr,
    deriv*)
Hirakura, K. *et al, J. Nat. Prod.*, 1986, **49**, 218 (*deriv*)

1-(2,4-Dihydroxyphenyl)-3-(3,4-   **D-70336**
dihydroxyphenyl)propane

Updated Entry replacing D-30381
4-[3-[(2,4-Dihydroxyphenyl)propyl]]-1,2-benzenediol

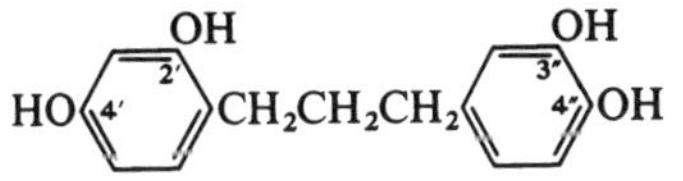

$C_{15}H_{16}O_4$ M 260.289

2′,3″-Di-Me ether: [57430-10-1]. *1-(4-Hydroxy-2-meth-
oxyphenyl)-3-(4-hydroxy-3-methoxyphenyl)propane.*
Broussonin F.
$C_{17}H_{20}O_4$ M 288.343
Constit. of *Broussonetia papyrifera* and wood of
Iryanthera coriacea. Antifungal phytoalexin. Oil.
4′,4″-Di-Me ether: [90902-21-9]. *1-(2-Hydroxy-4-meth-
oxyphenyl)-3-(3-hydroxy-4-methoxyphenyl)propane.*
Broussonin E.
$C_{17}H_{20}O_4$ M 288.343
Constit. of *B. papyrifera*. Phytoalexin showing
antifungal activity. Cryst. Mp 72-74°.
3″,4′-Di-Me ether: *1-(2-Hydroxy-4-methoxyphenyl)-3-
(4-hydroxy-3-methoxyphenyl)propene.*
$C_{17}H_{20}O_4$ M 288.343
Constit. of *Virola calophylloidea*. Cryst.
(MeOH/Me₂CO). Mp 87-88°.

Alves de Lima, R. *et al, Phytochemistry*, 1975, **14**, 1831 (*isol*)
Takasugi, M. *et al, Chem. Lett.*, 1984, 689 (*isol*)
Martinez, V.J.C. *et al, J. Nat. Prod.*, 1987, **50**, 1045 (*isol*)

3,5-Dihydroxy-4-phenylisoxazole   **D-70337**

*5-Hydroxy-4-phenyl-3(2H)-isoxazolone, 9CI. 3-
Hydroxy-4-phenyl-5(2H)-isoxazolone, 9CI. 2-Phenyl-
3,5-isoxazolidinedione, 8CI. 4-Phenyl-3,5-isoxazolediol.
4-Phenyldisic acid*

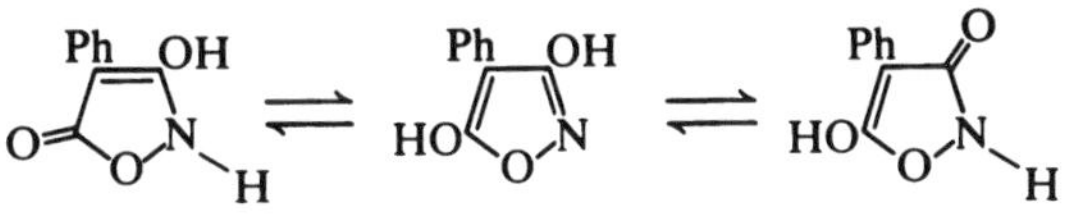

$C_9H_7NO_3$ M 177.159
Complex tautomerism including dipolar forms (not
illus.). Cryst. + ½H₂O (EtOAc/CCl₃), light-yellow
solid (anhyd.). Mp 150-152°. pK_a −1.86. Turns yellow
above 135°.

Zvilichovsky, G., *Tetrahedron*, 1975, **32**, 1861 (*synth, pmr, ir,
    uv*)
Zvilichovsky, G., *J. Heterocycl. Chem.*, 1987, **24**, 465 (*cryst
    struct, uv, tautom*)

1-(2,6-Dihydroxyphenyl)-1-tetradecanone,   **D-70338**
9CI

2-Tetradecanoyl-1,3-benzenediol

[113201-68-6]

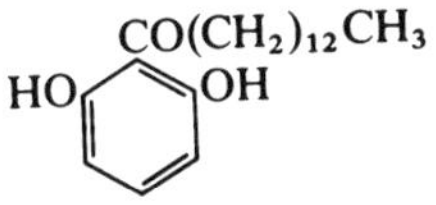

$C_{20}H_{32}O_3$ M 320.471

Constit. of the stem bark of *Myristica dactyloides*. Pale-yellow needles (CH₂Cl₂/pet. ether). Mp 91-92°.

Mono-Me ether: [114226-24-3]. *1-(2-Hydroxy-6-methoxyphenyl)-1-tetradecanone.*
$C_{21}H_{34}O_3$ M 334.498
Constit. of the stem bark of *M. dactyloides*. Needles (MeOH). Mp 51-52°.

Kumar, N.S. *et al, Phytochemistry*, 1988, **27**, 465.

3-(3,4-Dihydroxyphenyl)-1-(2,4,5-trihy-   **D-70339** droxyphenyl)-2-propen-1-one

2′,3,4,4′,5′-Pentahydroxychalcone. Stillopsidin. Neoplathymenin

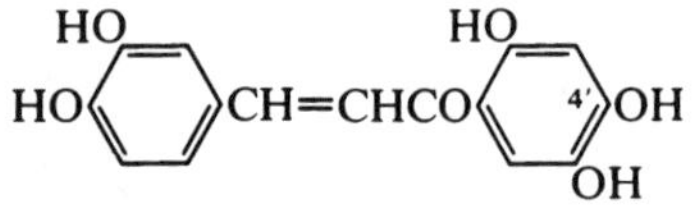

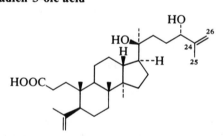

$C_{15}H_{12}O_6$ M 288.256

(E)-form

Orange plates (MeOH). Mp 226-228°.

4′-β-D-Glucopyranoside: [72241-26-0]. **Stillopsin**.
$C_{21}H_{22}O_{11}$ M 450.398
Pigment from *Coreopsis stillmanii* and *C. bigelovii*. Rare oxygenation pattern.

4′-β-D-Glucopyranoside, Octa-Ac: Yellow-tinted needles (MeOH). Mp 196.5-197.5°.

Seikel, M.K. *et al, J. Am. Chem. Soc.*, 1950, **72**, 5720; 1955, **77**, 1196 (*isol, uv, struct, deriv*)
Laumas, K.R. *et al, Proc. Indian Acad. Sci.*, 1957, **46**, 343; *CA*, **52**, 14569i (*synth*)
Nicholls, K.W. *et al, Phytochemistry*, 1979, **18**, 1076 (*isol*)

5-(1,3-Dihydroxypropyl)-2-isopropenyl-2,3-   **D-70340** dihydrobenzofuran

1-[2,3-Dihydro-2-(1-methylethenyl)-5-benzofuranyl]-1,2-propanediol, 9CI

[115699-96-2]

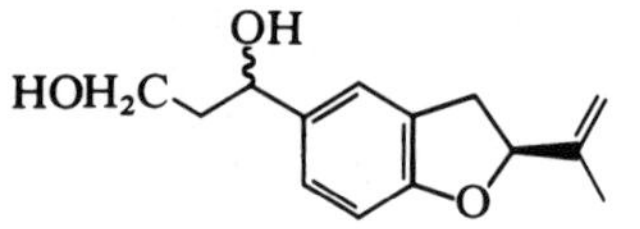

$C_{14}H_{18}O_3$ M 234.294
Constit. of *Heterobasidion annosum*. Oil.

Donnelly, D.M.X. *et al, Phytochemistry*, 1988, **27**, 2709.

4,6-Dihydroxy-2(1H)-pyridinone, 9CI   **D-70341**

2,4,6-Pyridinetriol, 8CI. 2,4,6-Trihydroxypyridine

[626-47-1]

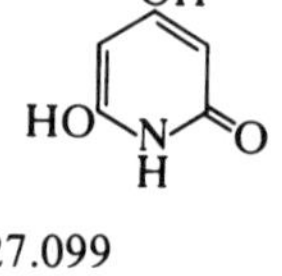

$C_5H_5NO_3$ M 127.099

4,6-Di-Me ether: [108279-66-9]. *4,6-Dimethoxy-2(1H)-pyridinone.*
$C_7H_9NO_3$ M 155.153
Cryst. (EtOAc). Mp 158-160°.

Tri-Me ether: [91591-88-7]. *2,4,6-Trimethoxypyridine.*
$C_8H_{11}NO_3$ M 169.180

Prisms (pentane). Mp 47-48°.

Sterk, H. *et al, Monatsh. Chem.*, 1967, **98**, 1763 (*ir, uv*)
Kaneko, C. *et al, Chem. Pharm. Bull.*, 1986, **34**, 3658 (*derivs*)

3,4-Dihydroxy-2-pyrrolidinecarboxylic   **D-70342** acid

Updated Entry replacing D-50547
3,4-Dihydroxyproline, 9CI

[63121-50-6]

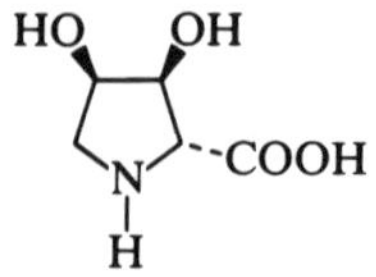

(2R,3S,4R)-form

$C_5H_9NO_4$ M 147.130

(2R,3S,4R)-form [105118-17-0]
Isol. from diatom cell walls. Cryst. (EtOH aq. or Me₂CO aq.). $[\alpha]_D^{20}$ −6.8° (c, 0.43 in H₂O). Dec. at 247°.

(2S,3R,4R)-form [74644-88-5]
Found in toxic peptides of *Amanita virosa*. Potent and specific β-D-glucuronidase inhibitor. $[\alpha]_D^{19}$ −12.6° (c, 0.53 in H₂O).

(2S,3S,4S)-form [23161-63-9]
Isol. from diatom cell walls. $[\alpha]_D^{22}$ −63° (c, 0.8 in H₂O).

Karle, I.L. *et al, Science*, 1969, **164**, 1401 (*ms, cryst struct*)
Adams, E. *et al, Annu. Rev. Biochem.*, 1980, **49**, 1005 (*rev*)
Kahl, J.-U. *et al, Justus Liebigs Ann. Chem.*, 1981, 1445 (*synth*)
Lindblad, W.J. *et al, J. Chromtogr.*, 1984, **315**, 447 (*synth*)
Ohfune, Y. *et al, Tetrahedron Lett.*, 1985, **26**, 5307 (*synth*)
Rule, C.J. *et al, Tetrahedron Lett.*, 1985, **26**, 5379 (*props*)
Baird, P.D. *et al, J. Chem. Soc., Perkin Trans. 1*, 1987, 1785 (*synth, bibl*)

20,24-Dihydroxy-3,4-seco-4(28),23-dam-   **D-70343** maradien-3-oic acid

$C_{30}H_{50}O_4$ M 474.723

(20S,24S)-form
Constit. of *Alnus japonica* flowers.

Aoki, T. *et al, Phytochemistry*, 1988, **27**, 2915.

20,25-Dihydroxy-3,4-seco-4(28),23-dam-   **D-70344** maradien-3-oic acid

$C_{30}H_{50}O_4$ M 474.723

(20S,23E)-form
Constit. of *Alnus japonica* flowers.

Aoki, T. *et al, Phytochemistry*, 1988, **27**, 2915.

3,19-Dihydroxy-13(16),14-spongiadien-2-one   D-70345

Updated Entry replacing D-05065

$C_{20}H_{28}O_4$ M 332.439

3α-form [71302-28-8]
Spongiadiol
Constit. of *Spongia* spp. Cryst. (EtOAc). Mp 181-183°.
[α]$_D^{21}$ +73.6° (c, 1 in CHCl$_3$).

Di-Ac: [71302-24-4].
$C_{24}H_{32}O_6$ M 416.513
Constit. of *S.* spp. Cryst. (MeOH). Mp 131.5-133°.
[α]$_D^{25}$ +14.5° (c, 1 in CHCl$_3$).

3β-form [71302-26-6]
Episongiadiol
Constit. of *S.* spp. and *Hyatella intestinalis.* Cryst.
(EtOAc). Mp 157-158.5°. [α]$_D^{21}$ +18.7° (c, 1 in CHCl$_3$).

3-Ac: 3β-Acetoxy-19-hydroxy-13(16),14-spongiadien-2-one.
$C_{22}H_{30}O_5$ M 374.476
Constit. of *H. intestinalis.* Cryst. (Me$_2$CO/hexane).
Mp 216-218°. [α]$_D$ +34.0° (c, 0.15 in CHCl$_3$).

Di-Ac: [71302-30-2]. Constit. of *S.* spp. and *H. intestinalis.* Cryst. (MeOH). Mp 195-198°. [α]$_D^{21}$ +45.2° (c, 1 in CHCl$_3$).

Kazlauskas, R. *et al*, *Aust. J. Chem.*, 1979, **32**, 867 (*isol, struct*)
Cambie, R.C. *et al*, *J. Nat. Prod.*, 1988, **51**, 293 (*isol, deriv*)

7,12-Dihydroxysterpurene   D-70346

Updated Entry replacing D-60374

$C_{15}H_{24}O_2$ M 236.353
Metabolite of fungus *Stereum purpureum.*

Abell, C. *et al*, *Tetrahedron Lett.*, 1987, **28**, 4887; 1988, **29**, 1985 (*isol, abs config*)

9,12-Dihydroxysterpurene   D-70347

1-Sterpurene-9,12-diol
$C_{15}H_{24}O_2$ M 236.353
Metab. of *Stereum purpureum.*

Abell, C. *et al*, *Tetrahedron Lett.*, 1988, **29**, 4337 (*struct, biosynth*)

Consult the *Dictionary of Organophosphorus Compounds* for organic compounds containing phosphorus.

1-(5,7-Dihydroxy-2,2,6-trimethyl-2*H*-1-benzopyran-8-yl)-2-methyl-1-propanone, 9CI   D-70348

5,7-Dihydroxy-8-isobutyryl-2,2,6-trimethylchromene
[111983-96-1]

$C_{16}H_{20}O_4$ M 276.332
Constit. of *Hypericum revolutum.* Antifungal agent.
Yellow prisms. Mp 79-81°.

3'-Methyl: [111983-97-2]. **1-(5,7-Dihydroxy-2,2,6-trimethyl-2H-1-benzopyran-8-yl)-2-methyl-1-butanone, 9CI.**
$C_{17}H_{22}O_4$ M 290.358
Constit. of *H. revolutum.* Yellow oil.

Décosterd, L.A. *et al*, *Helv. Chim. Acta*, 1987, **70**, 1694.

2,3-Dihydroxy-12-ursen-28-oic acid   D-70349

Updated Entry replacing D-05105

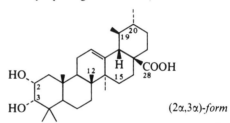

(2α,3α)-form

$C_{30}H_{48}O_4$ M 472.707
(2α,3α)-form [52213-27-1]
Constit. of *Prunus serotina, P. lusitanica* and *Pseudopanax arboreum.*

Me ester:
$C_{31}H_{50}O_4$ M 486.734
Cryst. (MeOH aq.). Mp 196-198°. [α]$_D^{20}$ +52° (c, 1.2 in CHCl$_3$).

(2α,3β)-form [4547-24-4]
2α-Hydroxyursolic acid
Constit. of *Chamaenerion angustifolium.* Cryst. (EtOH).
Mp 243-245° dec. [α]$_D$ +42.1° (c, 1 in Py).

Me ester: Cryst. (pet. ether). Mp 212-214°. [α]$_D$ +54.5° (c, 0.88 in CHCl$_3$).

3-(4-Hydroxycinnamoyl): [63303-42-4]. **Jacoumaric acid.** Constit. of *Jacaranda caucana.*

3-(4-Hydroxycinnamoyl), Me ester: Cryst. (MeOH).
Mp 297°.

2-(4-Hydroxy-E-cinnamoyl): [112693-17-1]. **Neriucoumaric acid.**
$C_{39}H_{54}O_6$ M 618.852
Constit. of *Nerium oleander* leaves. Needles (MeOH).
Mp 120-121°. [α]$_D^{24}$ +16.66° (CHCl$_3$).

2-(4-Hydroxy-Z-cinnamoyl): **Isoneriucoumaric acid.**
$C_{39}H_{54}O_6$ M 618.852
Constit. of *N. oleander* leaves. Needles (MeOH). Mp 208-209°. [α]$_D^{24}$ +50.0° (CHCl$_3$).

Glen, A.T. *et al*, *J. Chem. Soc. (C)*, 1967, 510 (*isol*)
Biessels, H.W.A. *et al*, *Phytochemistry*, 1974, **13**, 203 (*isol*)
Bowden, B.F. *et al*, *Aust. J. Chem.*, 1975, **28**, 91 (*isol*)
Seo, S. *et al*, *J. Chem. Soc., Chem. Commun.*, 1975, 270, 954 (*biosynth, pmr*)

Ogura, M. *et al*, *Phytochemistry*, 1977, **16**, 286 (*deriv*)
Siddiqui, S. *et al*, *Planta Med.*, 1987, **53**, 424 (*deriv*)

3,19-Dihydroxy-12-ursen-28-oic acid　　　D-70350

Updated Entry replacing D-10432

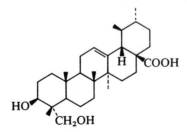

C$_{30}$H$_{48}$O$_4$　　M 472.707
(*3β,19α*)-*form* [13849-91-7]
Pomolic acid. *Benthamic acid*
Constit. of apple peel, *Micromeria benthami* and
Euscaphis japonica. Cryst. Mp 301-303°. [α]$_D^{20}$ +37°
(c, 2.0 in THF).
Me ester: Cryst. (pet. ether). Mp 127-129°. [α]$_D^{20}$ +39°
(CHCl$_3$).
3-O-β-D-Xylopyranoside: [108544-40-7]. **Ilexoside B**.
C$_{35}$H$_{56}$O$_8$　　M 604.823
Constit. of leaves of *Ilex chinensis*. Cryst. (as Me
ester). Mp 155-158° (Me ester). [α]$_D$ +15.9° (c, 0.15
in MeOH).
3-O-[β-D-Glucopyranosyl-(1→2)-α-L-
arabinopyranosyl]: [83725-19-3]. **Ilexside I**.
C$_{41}$H$_{66}$O$_{13}$　　M 766.965
Constit. of *I. cornuta*. Cryst. (MeOH) (as Me ester).
Mp 240-243° (Me ester). [α]$_D^{20}$ +15.6° (c, 0.80 in
MeOH).
3-O-[β-D-Glucopyranosyl-(1→2)-α-L-
arabinopyranosyl],28-O-β-D-glucopyranosyl: [88255-
95-2]. **Ilexside II**.
C$_{47}$H$_{76}$O$_{18}$　　M 929.107
Constit. of *I. cornuta*. Powder. Mp 268-270°. [α]$_D^{20}$
−15.0° (c, 0.71 in MeOH).
3-Ketone: 19α-Hydroxy-3-oxo-12-ursen-28-oic acid.
Pomonic acid.
C$_{30}$H$_{46}$O$_4$　　M 470.691
Isol. as the Me ester from *Pyrus malus* (apple).
3-Ketone, Me ester: Mp 204°. [α]$_D^{20}$ +50° (c, 2.0 in
CHCl$_3$).

Brieskorn, C.H. *et al*, *Chem. Ber.*, 1967, **100**, 1252 (*isol*)
Takahashi, K. *et al*, *Chem. Pharm. Bull.*, 1974, **72**, 650 (*isol*)
Nakanishi, T. *et al*, *Phytochemistry*, 1982, **21**, 1373 (*isol*)
Inada, A. *et al*, *Chem. Pharm. Bull.*, 1987, **35**, 841 (*Ilexoside B*)

3,23-Dihydroxy-12-ursen-28-oic acid　　　D-70351

C$_{30}$H$_{48}$O$_4$　　M 472.707
3β-form [94414-19-4]
Constit. of *Cigarrilla mexicana*. Needles. Mp 266-268°.
Mata, R. *et al*, *Phytochemistry*, 1988, **27**, 1887.

4,6-Dihydroxy-5-vinylbenzofuran　　　D-70352

5-Ethenyl-4,6-benzofurandiol

C$_{10}$H$_8$O$_3$　　M 176.171
Di-Me ether: 4,6-Dimethoxy-5-vinylbenzofuran. *5-Eth-*
enyl-4,6-dimethoxybenzofuran.
C$_{12}$H$_{12}$O$_3$　　M 204.225
Constit. of *Dorstenia barnimiana*. Oil.

Woldu, Y. *et al*, *Phytochemistry*, 1988, **27**, 1227.

1,2-Dihydroxyxanthone　　　D-70353

Updated Entry replacing D-05112
1,2-Dihydroxy-9H-xanthen-9-one, 9CI
[17623-69-7]

C$_{13}$H$_8$O$_4$　　M 228.204
Yellow needles (EtOH). Mp 170.5-171.5°.
1-Me ether: 2-Hydroxy-1-methoxyxanthone.
C$_{14}$H$_{10}$O$_4$　　M 242.231
Isol. from trunk wood of *Kielmeyera speciosa*. Cryst.
Mp 169-171°.

Davies, J.S.H. *et al*, *J. Chem. Soc.*, 1958, 1790 (*synth*)
de Oliveira, G.G. *et al*, *An. Acad. Bras. Cienc.*, 1966, **38**, 421;
　CA, **67**, 108528 (*isol, deriv*)
Arends, P. *et al*, *Dan. Tidsskr. Farm.*, 1972, **46**, 133 (*synth,*
　nmr)

Di-1H-imidazol-2-ylmethanone, 9CI　　　D-70354

Bis(2-imidazolyl)ketone
[64269-79-0]

C$_7$H$_6$N$_4$O　　M 162.151
Cryst. (H$_2$O). Mp 285° dec. An earlier ref. to this
compd. is erroneous.
Hydrazone: Mp 86-88°.
1,1′-Di-Me: [62366-40-9].
　C$_9$H$_{10}$N$_4$O　　M 190.204
　Powder (Me$_2$CO). Mp 151°.

Joseph, M. *et al*, *Synthesis*, 1977, 459 (*synth*)
Regel, E. *et al*, *Justus Liebigs Ann. Chem.*, 1977, 145 (*deriv,*
　synth)
Gorun, S.M. *et al*, *J. Am. Chem. Soc.*, 1987, **109**, 4244 (*deriv,*
　synth)

9,10-Diiodoanthracene　　　D-70355

[113705-11-6]

C$_{14}$H$_8$I$_2$　　M 430.026
Yellow needles (CCl$_4$). Mp 254-255°.

Duerr, B.F. *et al*, *J. Org. Chem.*, 1988, **53**, 2120 (*synth, pmr*)

2,6-Diiodobenzoic acid   D-70356

$C_7H_5I_2O_2$ M 374.924

Diethylamide: [97567-50-5].
  $C_{11}H_{13}I_2NO$ M 429.039
  Needles (hexane). Mp 116.5-117.5°.

Mills, R.J. *et al*, *Tetrahedron Lett.*, 1985, **26**, 1145 (*synth, deriv*)
Eaton, P.E. *et al*, *J. Org. Chem.*, 1988, **53**, 2728 (*synth, deriv, pmr*)

1,3-Diiodobicyclo[1.1.1]pentane, 9CI   D-70357

[105542-98-1]

$C_5H_6I_2$ M 319.911

Wiberg, K.B. *et al*, *Tetrahedron Lett.*, 1986, 1553 (*synth*)

1,3-Diiodo-2-(iodomethyl)propane, 9CI   D-70358

Tris(iodomethyl)methane
[66587-78-8]

$$HC(CH_2I)_3$$

$C_4H_7I_3$ M 435.813
Pale-rose liq. Bp$_{0.35}$ 106°.

Latour, S. *et al*, *Synthesis*, 1987, 742 (*synth, ms, ir, pmr*)

1,4-Diiodopentacyclo[4.2.0.0²,⁵.0³,⁸.0⁴,⁷]-octane, 9CI   D-70359

1,4-Diiodocubane
[97229-08-8]

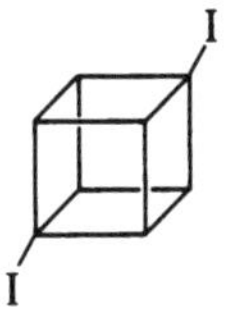

$C_8H_6I_2$ M 355.944
Cryst. (hexane). Mp 226-227°.

Honegger, E. *et al*, *Helv. Chim. Acta*, 1985, **68**, 23 (*synth, pmr, ir, ms*)
Moriarty, R.M. *et al*, *J. Chem. Soc., Chem. Commun.*, 1987, 675 (*synth*)

2,7-Diiodophenanthrene   D-70360

[62325-31-9]

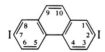

$C_{14}H_8I_2$ M 430.026
Needles (CCl₄). Mp 231°.

Wirth, H.O. *et al*, *Makromol. Chem.*, 1963, **63**, 53 (*synth*)
Bochenkov, V.N., *J. Org. Chem. USSR (Engl. Transl.)*, 1976, **12**, 2355 (*synth, uv*)

3,6-Diiodophenanthrene   D-70361

[835-05-2]

$C_{14}H_8I_2$ M 430.026
Needles (EtOH aq.). Mp 224°.

Staab, H.A. *et al*, *Chem. Ber.*, 1968, **101**, 879 (*synth*)

Diisocyanogen   D-70362

[114861-40-4]

$$CN—NC$$

C_2N_2 M 52.035
Dec. > −30° with formation of brown polymer.

van der Does, T. *et al*, *Angew. Chem., Int. Ed. Engl.*, 1988, **27**, 936 (*synth, ms*)

6,6′;7,3a′-Diligustilide   D-70363

[88182-33-6]

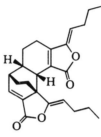

$C_{24}H_{28}O_4$ M 380.483
Constit. of *Ligusticum porteri*. Cryst. Mp 123°.

Delgado, G. *et al*, *Heterocycles*, 1988, **27**, 1305.

3,4-Dimercapto-3-cyclobutene-1,2-dithione, 9CI   D-70364

Tetrathiosquaric acid. Tetrathioquadratic acid

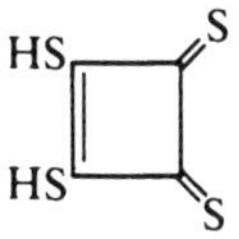

$C_4H_2S_4$ M 178.300
Di-K salt: [56284-07-2]. Yellow cryst. + 1H₂O.

Allmann, R. *et al*, *Chem. Ber.*, 1976, **109**, 2209 (*synth, ir, uv, cmr, cryst struct*)
Grenz, R. *et al*, *Chem. Ber.*, 1986, **119**, 1217 (*complexes*)

6,7-Dimethoxy-2,2-dimethyl-2*H*-1-benzopyran, 9CI   D-70365

Updated Entry replacing D-05343
6,7-Dimethoxy-2,2-dimethyl-3-chromene. **Precocene II**
[644-06-4]

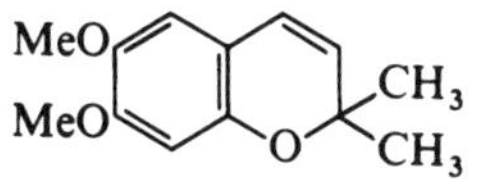

MeO, MeO — 2,2-dimethyl benzopyran with CH₃, CH₃, O

$C_{13}H_{16}O_3$ M 220.268
From *Ageratum houstonianum, A. conyzoides* and *Verbesina* spp. Antiallotropin (induces premature ecdysis in insects). Pale-yellow needles (MeOH) or oil. Mp 49-50°. Bp₄ 145-150°.

O^6-*De-Me*, O^6-*Et:* [65383-74-6]. *6-Ethoxy-7-methoxy-2,2-dimethyl-2H-1-benzopyran. Precocene III.*
  $C_{14}H_{18}O_3$ M 234.294
  Synthetic. Proallatocidin; the most potent precocene analogue having ~10 times the activity of Precocene II. Oil. Bp₀.₁ 109°.

Hlubucek, J.R. *et al, Aust. J. Chem.*, 1971, **24**, 2347 (*synth*)
Kasturi, T.R. *et al, Indian J. Chem.*, 1973, **11**, 91 (*isol*)
Suga, T. *et al, Phytochemistry*, 1975, **14**, 306 (*isol, ir, uv, pmr*)
Bowers, W.S. *et al, Science*, 1976, **193**, 542 (*isol, synth, use*)
Ohta, T. *et al, Chem. Pharm. Bull.*, 1977, **28**, 2788 (*synth*)
Ohta, T. *et al, J. Agric. Food Chem.*, 1977, **25**, 478 (*metab*)
Schwartz, M., *J. Heterocycles* Chem., 1977, **14**, 333 (*synth*)
Bohlmann, F. *et al, Chem. Ber.*, 1978, **111**, 263 (*isol*)
Kiehlmann, E. *et al, Org. Prep. Proced. Int.*, 1982, **14**, 337 (*Precocene III*)
Timár, T. *et al, J. Heterocycl. Chem.*, 1988, **25**, 871 (*synth, pmr, cmr, ms, bibl*)

1,3-Dimethyladamantane   **D-70366**

Updated Entry replacing D-05453
1,3-Dimethyltricyclo[3.3.1.1^{3,7}]decane, 9CI
[702-79-4]

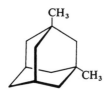

$C_{12}H_{20}$ M 164.290
Bp_{10} 86°.

Schleyer, P.v.R. *et al, Tetrahedron Lett.*, 1961, 305 (*synth*)
Schneider, A. *et al, J. Am. Chem. Soc.*, 1964, **86**, 5365 (*synth*)
Lightner, D.A. *et al, Tetrahedron*, 1987, **43**, 4905 (*synth, ir, pmr*)
Dooley, T.J. *et al, Org. Prep. Proced. Int.*, 1988, **20**, 293 (*synth*)

1-Dimethylamino-4-nitro-1,3-butadiene   **D-70367**

N,N-Dimethyl-4-nitro-1,3-butadien-1-amine, 9CI
[66085-83-4]

$$Me_2NCH\!=\!CHCH\!=\!CHNO_2$$

$C_6H_{10}N_2O_2$ M 142.157
Nitrodienamine useful in synthesis. Yellow-brown flaky cryst. Mp 92-93° (87°).

Severin, T. *et al, Chem. Ber.*, 1978, **111**, 692.
Takeuchi, N. *et al, Chem. Pharm. Bull.*, 1988, **36**, 481 (*synth, ir, pmr, use*)

N-(7,7-Dimethyl-3-azabicyclo[4.1.0]hepta-1,3,5-trien-4-yl)benzamide, 9CI   **D-70368**

[108869-44-9]

$C_{15}H_{14}N_2O$ M 238.288
First known cyclopropapyridine. Cryst. (Et$_2$O/pentane). Mp 114-116°. Acid-sensitive, stable under Ar at r.t.

Bambal, R. *et al, Angew. Chem., Int. Ed. Engl.*, 1987, **26**, 668 (*synth, uv, pmr, cmr*)

N,N-Dimethylazidochloromethyleniminium(1+)   **D-70369**

N-(*Azidochloromethylene*)-N-*methylmethanaminium*, *9CI. Azidophosgeniminium*

$$N_3CCl\!=\!NMe_2^{\oplus}$$

$C_3H_6ClN_4^{\oplus}$ M 133.560 (ion)
Diazo transfer reagent.
Chloride: [57020-32-3].
  $C_3H_6Cl_2N_4$ M 169.013
  Solid.

Viehe, H.G. *et al, Chimia*, 1975, **29**, 209 (*synth, cmr*)
Kokel, B. *et al, J. Heterocycl. Chem.*, 1987, **24**, 1493 (*synth, use, bibl*)

2,3-Dimethyl-1,4-benzodioxin   **D-70370**

[79792-92-0]

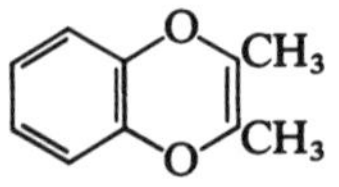

$C_{10}H_{10}O_2$ M 162.188
Cryst. (hexane). Mp 37.5-38.5°.

Adam, W. *et al, Synthesis*, 1982, 322 (*synth, ms, ir, pmr*)
Kashima, C. *et al, J. Org. Chem.*, 1987, **52**, 5616 (*synth, ir, pmr, cmr*)

2,5-Dimethyl-4H-1-benzopyran-4-one   **D-70371**

2,5-Dimethylchromone
[96574-27-5]

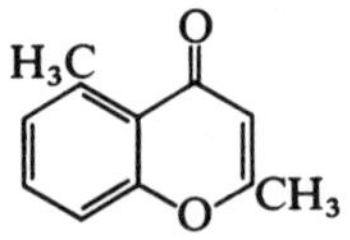

$C_{11}H_{10}O_2$ M 174.199
Constit. of the ant *Rhytidoponera metallica*. Cryst. Mp 67-69°.

Brophy, J.J. *et al, J. Nat. Prod.*, 1988, **51**, 99.

3-(2,2-Dimethyl-2H-1-benzopyran-6-yl)-2-propenoic acid, 9CI   **D-70372**

[104387-05-5]

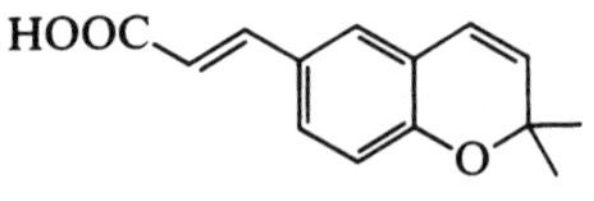

$C_{14}H_{14}O_3$ M 230.263
Isol. from *Baccharis* spp. Mp 191-192°.
Me ester: [92632-00-3].
  $C_{15}H_{16}O_3$ M 244.290
  From *Werneria stuebellii*. Cryst. (pet. ether). Mp 64°.

Bohlmann, F. *et al, Phytochemistry*, 1984, **23**, 1135 (*deriv*)
Labbe, C. *et al, J. Nat. Prod.*, 1986, **49**, 516 (*isol, struct*)

2,2-Dimethyl-2H-1-benzothiopyran, 9CI   **D-70373**

2,2-Dimethyl-2H-thiochromene
[19114-85-3]

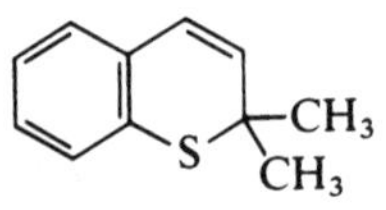

$C_{11}H_{12}S$ M 176.276
Liq. $Bp_{0.3}$ 50°.

Tercio, J. *et al, Synthesis*, 1987, 149 (*synth, ir, pmr, cmr*)

1,2-Dimethylcarbazole   D-70374

[18992-67-1]

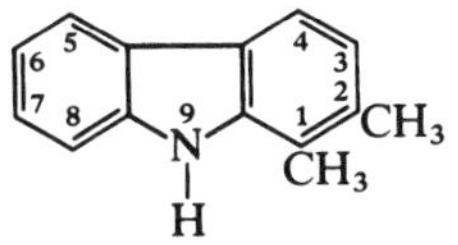

$C_{14}H_{13}N$ M 195.263
Mp 151° (140-143°).

Kuroki, M. *et al*, *J. Heterocycl. Chem.*, 1981, **18**, 709, 715
(*synth, uv, bibl*)
Bergman, J. *et al*, *Tetrahedron*, 1988, **44**, 5215 (*synth, ir, pmr*)

1,3-Dimethylcarbazole   D-70375

Updated Entry replacing D-05841

[18992-68-2]
$C_{14}H_{13}N$ M 195.263
Cryst. powder (ligroin). Mp 93-95°.

N-*Me*: *1,3,9-Trimethylcarbazole.*
$C_{15}H_{15}N$ M 209.290
Cryst. (pet. ether). Mp 100-101°.

Ullmann, F. *et al*, *Justus Liebigs Ann. Chem.*, 1904, **332**, 82.
Sundberg, R.J. *et al*, *J. Org. Chem.*, 1974, **39**, 2546 (*synth*)
Kuroki, M. *et al*, *J. Heterocycl. Chem.*, 1981, **18**, 709, 715
(*synth, uv, bibl*)
Clancy, M.G. *et al*, *J. Chem. Soc., Perkin Trans. 1*, 1984, 429
(*synth*)
Kulagowski, J.J. *et al*, *J. Chem. Soc., Perkin Trans. 1*, 1985,
2725 (*synth, deriv, pmr, ms*)

1,4-Dimethylcarbazole   D-70376

Updated Entry replacing D-05842

[18028-55-2]
$C_{14}H_{13}N$ M 195.263
Cryst. (EtOH aq.). Mp 97-98°.

Robinson, R. *et al*, *J. Chem. Soc.*, 1952, 976 (*synth*)
Cranwell, P.A. *et al*, *J. Chem. Soc.*, 1962, 3482 (*synth*)
Ahond, A. *et al*, *Tetrahedron*, 1978, **34**, 2385 (*cmr*)
Fusco, R. *et al*, *Tetrahedron Lett.*, 1978, 4827 (*synth, pmr*)
Kuroki, M. *et al*, *J. Heterocycl. Chem.*, 1981, **18**, 709, 715
(*synth, uv, bibl*)
Clancy, M.G. *et al*, *J. Chem. Soc., Perkin Trans. 1*, 1984, 429
(*synth*)

1,5-Dimethylcarbazole   D-70377

[51640-60-9]
$C_{14}H_{13}N$ M 195.263
Mp 137-138°.

Kuroki, M. *et al*, *J. Heterocycl. Chem.*, 1981, **18**, 709, 715
(*synth, uv*)

1,6-Dimethylcarbazole   D-70378

[78787-77-6]
$C_{14}H_{13}N$ M 195.263
Mp 174.5-175.5°.

Kuroki, M. *et al*, *J. Heterocycl. Chem.*, 1981, **18**, 709, 715
(*synth, uv*)

1,7-Dimethylcarbazole   D-70379

[78787-78-7]
$C_{14}H_{13}N$ M 195.263

Mp 166-168°.
Kuroki, M. *et al*, *J. Heterocycl. Chem.*, 1981, **18**, 709, 715
(*synth, uv*)

1,8-Dimethylcarbazole   D-70380

Updated Entry replacing D-05843

[6558-83-4]
$C_{14}H_{13}N$ M 195.263
Mp 178-180° (175-176°).

Pausacker, K.H. *et al*, *J. Chem. Soc.*, 1949, 1384 (*synth*)
Bajwa, G.S., *Can. J. Chem.*, 1970, **48**, 2293 (*synth*)
Kuroki, M. *et al*, *J. Heterocycl. Chem.*, 1981, **18**, 709, 715
(*synth, uv*)
Kulagowski, J.J. *et al*, *J. Chem. Soc., Perkin Trans. 1*, 1985,
2725 (*synth*)

2,3-Dimethylcarbazole   D-70381

Updated Entry replacing D-05844

[18992-70-6]
$C_{14}H_{13}N$ M 195.263
Mp 251-253°.

Campbell, N. *et al*, *J. Chem. Soc.*, 1950, 2870 (*synth*)
Noland, W.E. *et al*, *Tetrahedron Lett.*, 1962, 589 (*synth*)
Kuroki, M. *et al*, *J. Heterocycl. Chem.*, 1981, **18**, 709, 715
(*synth, uv*)

2,4-Dimethylcarbazole   D-70382

Updated Entry replacing D-05845

[18992-71-7]
$C_{14}H_{13}N$ M 195.263
Needles (pet. ether). Mp 137-138.5° (124-126°).

N-*Nitroso:*
$C_{14}H_{12}N_2O$ M 224.262
Mp 135°.
N-*Me*: *2,4,9-Trimethylcarbazole.*
$C_{15}H_{15}N$ M 209.290
Cryst. (EtOH). Mp 134°.

Horning, E.C. *et al*, *J. Am. Chem. Soc.*, 1948, **70**, 3935 (*synth*)
Smolinsky, G., *J. Am. Chem. Soc.*, 1960, **82**, 4717 (*deriv*)
Fusco, R. *et al*, *Tetrahedron Lett.*, 1978, 4827 (*synth, pmr*)
Kuroki, M. *et al*, *J. Heterocycl. Chem.*, 1981, **18**, 709, 715
(*synth, uv*)

2,5-Dimethylcarbazole   D-70383

[78787-79-8]
$C_{14}H_{13}N$ M 195.263
Mp 140.5-142°.

Kuroki, M. *et al*, *J. Heterocycl. Chem.*, 1981, **18**, 709, 715
(*synth, uv*)

2,7-Dimethylcarbazole   D-70384

Updated Entry replacing D-05847

[18992-65-9]
$C_{14}H_{13}N$ M 195.263
Mp 288-289° (283°).

v. Niementowski, S., *Ber.*, 1901, **34**, 3325 (*synth*)
Kuroki, M. *et al*, *J. Heterocycl. Chem.*, 1981, **18**, 709, 715
(*synth, uv*)

3,4-Dimethylcarbazole   D-70385

[18992-72-8]

$C_{14}H_{13}N$ M 195.263
Mp 168.5-169° (159-160°).

Anderson, G. et al, J. Chem. Soc., 1950, 2855 (synth)
Kuroki, M. et al, J. Heterocycl. Chem., 1981, **18**, 709, 715 (synth, uv)

3,5-Dimethylcarbazole   D-70386

[78787-81-2]
$C_{14}H_{13}N$ M 195.263
Mp 154-156°.

Kuroki, M. et al, J. Heterocycl. Chem., 1981, **18**, 709, 715 (synth, uv)

3,6-Dimethylcarbazole   D-70387

Updated Entry replacing D-05848
[5599-50-8]
$C_{14}H_{13}N$ M 195.263
Needles (EtOH). Mp 219°.

N-*Ac*:
    $C_{16}H_{15}NO$ M 237.301
    Mp 129°.
N-*Nitroso*:
    $C_{14}H_{12}N_2O$ M 224.262
    Mp 106°.

Vaniček, V. et al, Collect. Czech. Chem. Commun., 1955, **20**, 996.
Buu-Hoi, N.P. et al, J. Chem. Soc. (C), 1966, 924.
Wentrup, C. et al, Helv. Chim. Acta, 1971, **54**, 2108 (synth)
Kuroki, M. et al, J. Heterocycl. Chem., 1981, **18**, 709, 715 (synth, uv)
Filimonov, V.D. et al, Khim. Geterotsikl Soedin., 1984, 204 (cmr)

4,5-Dimethylcarbazole   D-70388

[18992-66-0]
$C_{14}H_{13}N$ M 195.263
Mp 177-177.5°.

Kuroki, M. et al, J. Heterocycl. Chem., 1981, **18**, 709, 715 (synth, uv)

24,24-Dimethyl-3-cholestanol   D-70389

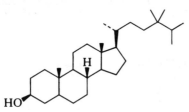

$C_{29}H_{52}O$ M 416.729
(*3β,5α*)-*form*
    Constit. of *Gynostemma pentaphyllum*.
    Akihisa, T. et al, Phytochemistry, 1988, **27**, 2931.

Dimethyl *N*-cyanoimidocarbonate   D-70390

Dimethyl cyanocarbonimidate, 9CI
[24771-25-3]

$$(MeO)_2C{=}NCN$$

$C_4H_6N_2O_2$ M 114.104
Reagent for synth. of guanidines and heterocyclic compds. Cryst. (pet. ether). Mp 53-55°.

Zmitek, J. et al, Org. Prep. Proced. Int., 1985, **17**, 256 (synth, bibl)

2,7-Dimethyl-2,4,6-cycloheptatrien-1-one,   D-70391
9CI

2,7-Dimethyltropone
[49747-09-3]

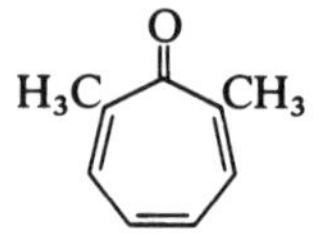

$C_9H_{10}O$ M 134.177
Bp$_{0.001}$ 50-70°.

Closs, G.L. et al, J. Am. Chem. Soc., 1961, **83**, 599 (synth, uv, ir)
Mayor, C. et al, J. Org. Chem., 1978, **43**, 4498 (synth, pmr)
Takaya, H. et al, J. Am. Chem. Soc., 1978, **100**, 1778 (synth, pmr)
Föhlisch, B. et al, Justus Liebigs Ann. Chem., 1987, 1 (synth, cmr)

5,5-Dimethyl-2-cyclohexene-1,4-dione, 9CI   D-70392

[45731-99-5]

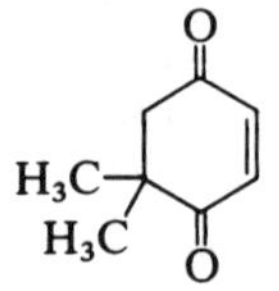

$C_8H_{10}O_2$ M 138.166
Yellow plates (hexane). Mp 39-39.5°.

Yates, P. et al, Can. J. Chem., 1987, **65**, 69 (synth, pmr, cmr)

2,2-Dimethyl-1,3-cyclopentanedione   D-70393

[3883-58-7]

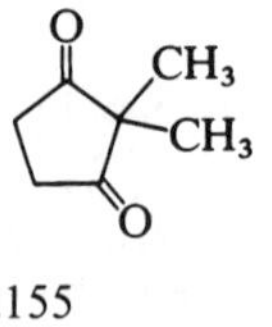

$C_7H_{10}O_2$ M 126.155
Mp 45-47°. Bp$_{0.15}$ 110°.

Agosta, W.C. et al, J. Org. Chem., 1970, **35**, 3856 (synth, ir, pmr)
Crispin, D.J. et al, J. Chem. Soc. (C), 1970, 10 (synth, ir)
Doyle, J.D. et al, Synthesis, 1986, 845 (synth)

4,4-Dimethyl-1,3-cyclopentanedione   D-70394

5,5-Dimethyl-1-hydroxy-1-cyclopenten-3-one
[4683-51-6]

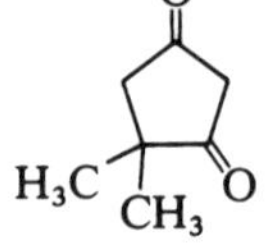

$C_7H_{10}O_2$ M 126.155
Largely enolised. Large prisms (H_2O). Mp 97°.

Farmer, E.H. et al, J. Chem. Soc., 1922, 128 (synth)
Guntrum, E. et al, Synthesis, 1986, 921 (synth, ir, pmr, cmr)

2,2-Dimethyl-4-cyclopentene-1,3-dione   **D-70395**
[26154-22-3]

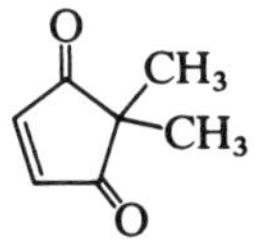

$C_7H_8O_2$ M 124.139
Yellow liq. Bp_{44} 90°.

Agosta, W.C. *et al, J. Org. Chem.*, 1970, **35**, 3856 (*synth, ir, uv, pmr*)

2,2-Dimethylcyclopropanemethanol, 9CI   **D-70396**
2-Hydroxymethyl-1,1-dimethylcyclopropane
[930-50-7]

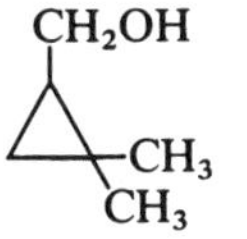

$C_6H_{12}O$ M 100.160
(±)-*form*
Bp_{118} 95-96°, Bp_{12} 66°.
Trimethylsilyl ether: Bp_{80} 110°.

Bly, R.S. *et al, J. Org. Chem.*, 1965, **30**, 10.
Wilcox, C.F. *et al, J. Am. Chem. Soc.*, 1972, **94**, 8232 (*deriv*)
Zimmerman, H.E. *et al, Tetrahedron*, 1978, **34**, 1775 (*synth*)
Laurie, D. *et al, Tetrahedron*, 1986, **42**, 1035 (*synth, pmr, deriv*)

2,3-Dimethylcyclopropanol   **D-70397**
Updated Entry replacing D-05982
[13830-35-8]

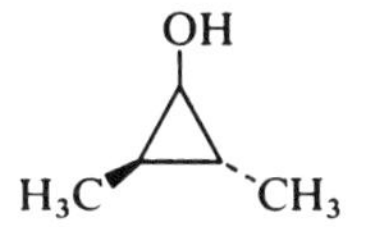

$(2R^*,3R^*)$-*form*

$C_5H_{10}O$ M 86.133
3 Stereoisometric forms possible, one of which (illus.) is known (as derivs.) in both opt. active and racemic form. The $(1\alpha,2\alpha,3\alpha)$-form appears to be unknown.
$(2R^*,3R^*)$-*form*
  Ac: [66769-49-1].
    $C_7H_{12}O_2$ M 128.171
    Bp_{30} 46-48°. $[\alpha]_D^{22}$ −44.9° (c, 0.009 in EtOH).
  Me ether: [66791-94-4]. *1-Methoxy-2,3-dimethylcyclopropane.*
    $C_6H_{12}O$ M 100.160
    $[\alpha]_{237}^{22}$+13.4° (c. 0.095 in EtOH).
$(2RS,3RS)$-*form* [13830-35-8]
  (±)-*trans-form.* $(2\alpha,3\beta)$-*form.*
  $Bp_{0.35}$ 60-61°.
$(1\alpha,2\beta,3\beta)$-*form*
  trans,cis-*form*
  $Bp_{0.14}$ 40-42°.

Schöllkopf, U. *et al, Chem. Ber.*, 1966, **99**, 3391 (*synth, pmr*)
Schöllkopf, U., *Angew. Chem., Int. Ed. Engl.*, 1968, **7**, 588 (*synth*)
Longone, D.T. *et al, Tetrahedron Lett.*, 1969, 2859 (*synth*)
Andrist, A.H. *et al, J. Org. Chem.*, 1978, **43**, 3422 (*synth, resoln*)

2,3-Dimethyl-3,4-dihydro-2H-pyran, 9CI   **D-70398**

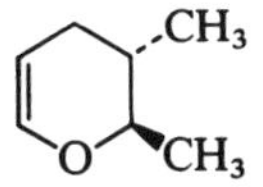

$C_7H_{12}O$ M 112.171
(*2R,3S*)-*form* [101978-00-1]
trans-*form*
Oil. Bp 135-136°.

Kocieński, P.J. *et al, J. Chem. Soc., Perkin Trans. 1*, 1987, 2183 (*synth, pmr, ms*)

Dimethyldioxirane, 9CI   **D-70399**
[74087-85-7]

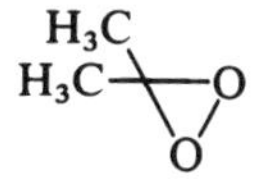

$C_3H_6O_2$ M 74.079
Oxygen-atom transfer reagent. Obt. in ≤0.1 M soln. Volatile.

Murray, R.W. *et al, J. Org. Chem.*, 1985, **50**, 2847 (*synth, uv, ir, pmr, cmr*)
Baumstark, A.L. *et al, Tetrahedron Lett.*, 1987, **28**, 3311 (*use, bibl*)

2,2-Dimethyl-1,3-dioxolane-4-carboxalde-   **D-70400**
hyde, 9CI
2,3-O-Isopropylideneglyceraldehyde. Glyceraldehyde acetonide. Acetoneglyceraldehyde
[5736-03-8]

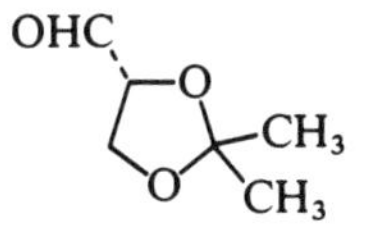

$C_6H_{10}O_3$ M 130.143
Unstable in pure form. Important starting compd. for prep. of many chiral C_3 compds.
(*R*)-*form* [15186-48-8]
  D-*form*
  Bp_{17} 53°. $[\alpha]_D$ +64.9°.
(*S*)-*form* [22323-80-4]
  L-*form*
  Bp_{35} 64-66°. $[\alpha]_D^{20}$ −63.5° (c, 8 in C_6H_6).

Baer, E. *et al, J. Am. Chem. Soc.*, 1939, **61**, 761 (*synth*)
Ichimura, K., *Bull. Chem. Soc. Jpn.*, 1970, **43**, 2501 (*synth*)
Pfander, H. *et al, Helv. Chim. Acta*, 1983, **66**, 814 (*synth*)
Hubschwerlen, C., *Synthesis*, 1986, 962 (*synth*)
Jurczak, J. *et al, Tetrahedron*, 1986, **42**, 448 (*rev*)
Jackson, D.Y., *Synth. Commun.*, 1988, **18**, 337 (*synth*)
Marco, J.L. *et al, Tetrahedron Lett.*, 1988, **29**, 1997 (*synth*)

2,2-Dimethyl-4,4-diphenyl-3-butenal, 9CI   **D-70401**
[5820-02-0]

$$Ph_2C{=}CHC(CH_3)_2CHO$$

$C_{18}H_{18}O$ M 250.340
$Bp_{0.2}$ 135°, $Bp_{0.02}$ 110-112°.
Oxime: [111220-88-3].
  $C_{18}H_{19}NO$ M 265.354
  Cryst. (MeOH). Mp 69-70°.
Semicarbazone: Cryst. (MeOH). Mp 145°.
2,4-Dinitrophenylhydrazone: Cryst. Mp 186°.

Julia, M. *et al, Bull. Soc. Chim. Fr.*, 1966, 734 (*synth*)
Zimmerman, H.E. *et al, J. Am. Chem. Soc.*, 1970, **92**, 6259
  (*synth, pmr*)
Armesto, D. *et al, J. Chem. Soc., Perkin Trans. 1*, 1987, 743
  (*deriv, synth, ir, pmr*)

2,2-Dimethyl-4,4-diphenyl-3-butenoic acid,   D-70402
9CI

[110835-94-4]

$$Ph_2C{=}CHC(CH_3)_2COOH$$

$C_{18}H_{18}O_2$ M 266.339
Cryst. (EtOH). Mp 95-97°.
Nitrile: [73050-40-5].
  $C_{18}H_{17}N$ M 247.339
  Cryst. (EtOH). Mp 108-109°.

Pratt, A.C. *et al, J. Chem. Soc., Perkin Trans. 1*, 1987, 359
  (*synth, ir, pmr*)
Armesto, D. *et al, J. Chem. Soc., Perkin Trans. 1*, 1987, 743
  (*deriv, synth, ir, uv, pmr*)

2,5-Dimethyl-3,4-diphenyl-2,4-cyclopenta-   D-70403
dien-1-one, 9CI
Hemicyclone
[26307-17-5]

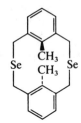

$C_{19}H_{16}O$ M 260.335
Red. Forms dimer on htg.
2,4-Dinitrophenylhydrazone: Red cryst. (butanol). Mp
  242°.
Dimer: [38883-84-0].
  $C_{38}H_{32}O_2$ M 520.670
  Cryst. (2-methoxyethanol). Mp 181-182°.

Allen, C.F.H. *et al, J. Am. Chem. Soc.*, 1942, **64**, 1260; 1950,
  72, 5165.
Fuchs, B. *et al, J. Chem. Soc., Chem. Commun.*, 1977, 537
  (*synth, ir, uv*)
Warrener, R.N, *et al, Aust. J. Chem.*, 1977, **30**, 1481 (*dimer,
  synth, pmr*)

9,18-Dimethyl-2,11-diselena[3.3]-   D-70404
metacyclophane

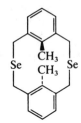

$C_{18}H_{20}Se_2$ M 394.276
anti-form

Mp 215-216°. Previously reported compd. of this struct.
  was incorrect.

Mitchell, R.H. *et al, Tetrahedron Lett.*, 1987, **28**, 5119 (*synth,
  cryst struct, pmr, cmr, Se nmr*)

2,3-Dimethylenebicyclo[2.2.1]heptane   D-70405
2,3-Bis(methylene)bicyclo[2.2.1]heptane, 9CI. 2,3-
Dimethylenenorbornane
[36439-78-8]

C_9H_{12} M 120.194
Bp$_{50}$ 67°.

Butler, D.N. *et al, Can. J. Chem.*, 1972, **50**, 795 (*synth, ir, uv,
  pmr*)
Clennan, E.L. *et al, J. Org. Chem.*, 1987, **52**, 3483 (*synth, pmr*)
Sustmann, R. *et al, Chem. Ber.*, 1987, **120**, 1323 (*struct, props*)

2,3-Dimethylenebicyclo[2.2.3]nonane   D-70406
6,7-Bis(methylene)bicyclo[3.2.2]nonane, 9CI
[36439-81-3]

$C_{11}H_{16}$ M 148.247
Bp$_{30}$ 103°.

Butler, D.N. *et al, Can. J. Chem.*, 1972, **50**, 795 (*synth, ir, uv,
  pmr*)
Clennan, E.L. *et al, J. Org. Chem.*, 1987, **52**, 3483 (*synth, pmr*)

7,8-Dimethylenebicyclo[2.2.2]octa-2,5-di-   D-70407
ene
7,8-Bis(methylene)bicyclo[2.2.2]octa-2,5-diene, 9CI
[51698-73-8]

$C_{10}H_{10}$ M 130.189
Needles by subl. Mp 40°.

Butler, D.N. *et al, Can. J. Chem.*, 1974, **52**, 447 (*synth, uv, ir,
  pmr*)
Pfeffer, H.U. *et al, Chem. Ber.*, 1975, **108**, 2923 (*pe*)
Azzena, U. *et al, Synth. Commun.*, 1988, **18**, 351 (*synth*)

2,3-Dimethylenebicyclo[2.2.2]octane   D-70408
2,3-Bis(methylene)bicyclo[2.2.2]octane
[36439-79-9]

$C_{10}H_{14}$ M 134.221
Bp$_{50}$ 73°.

Butler, D.N. *et al, Can. J. Chem.*, 1972, **50**, 795 (*synth, ir, uv,
  pmr*)
Clennan, E.L. *et al, J. Org. Chem.*, 1987, **52**, 3483 (*synth, pmr*)

Dimethylenebutanedioic acid   D-70409

Updated Entry replacing D-10512
Bis(methylene)butanedioic acid, 9CI. 1,3-Butadiene-2,3-dicarboxylic acid. Dimethylenesuccinic acid. Fulgenic acid
[488-20-0]

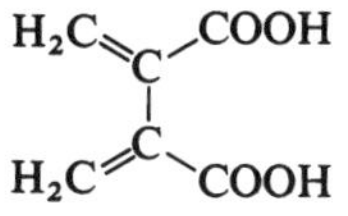

$$H_2C=C-COOH$$
$$H_2C=C-COOH$$

$C_6H_6O_4$ M 142.111
Di-Me ester: [38818-30-3]. *2,3-Dicarbomethoxy-1,3-butadiene.*
      $C_8H_{10}O_4$ M 170.165
      Yellow oil.
Dinitrile: [19652-57-4]. *Bis(methylene)butanedinitrile, 9CI. 2,3-Dicyano-1,3-butadiene.*
      $C_6H_4N_2$ M 104.111
      Needles. Mp 118-120°.

Belluš, B. *et al, Helv. Chim. Acta,* 1973, **56**, 3004.
Hamon, P.G. *et al, Synthesis,* 1981, 873.
Grundke, C. *et al, Chem. Ber.,* 1987, **120**, 1461 (*synth, ir, pmr, ester*)
Tarnchompoo, B. *et al, Tetrahedron Lett.,* 1987, **28**, 6671 (*synth, use*)

5,6-Dimethylene-1,3-cyclohexadiene   D-70410

Updated Entry replacing D-06105
1,2-Phenylenebismethyl, 9CI. 1,2-Benzoquinonedimethide. o-*Xylylene.* o-*Quinodimethane*
[14690-58-5]

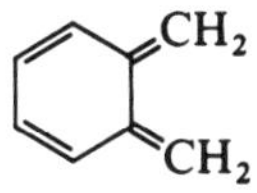

C_8H_8 M 104.151
Reactive Diels-Alder diene. Reactive intermed. Dimerises readily.

Bauld, N.L. *et al, Tetrahedron Lett.,* 1972, 2443 (*synth*)
Michl, J. *et al, J. Am. Chem. Soc.,* 1973, **95**, 5802 (*synth, props, bibl*)
Flynn, C.R. *et al, J. Am. Chem. Soc.,* 1974, **96**, 3280 (*synth*)
Ito, Y. *et al, J. Am. Chem. Soc.,* 1982, **104**, 7609 (*synth, use, bibl*)
Trahanovsky, W.S. *et al, J. Am. Chem. Soc.,* 1986, **108**, 6820 (*synth, props, bibl*)

6,6'-(1,2-Dimethyl-1,2-ethanediyl)bis[1-phenazinecarboxylic acid]   D-70411

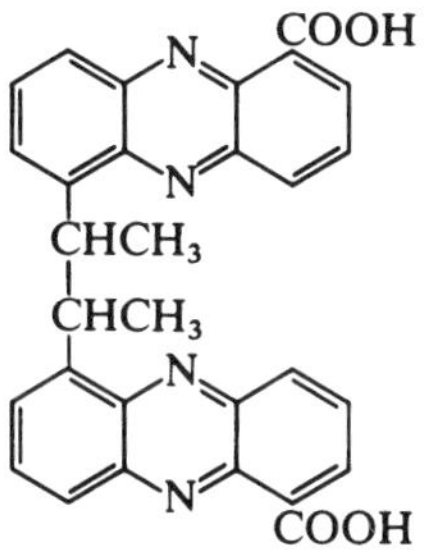

$C_{30}H_{22}N_4O_4$ M 502.528
Di-Me ester: [73649-05-5].
      $C_{32}H_{26}N_4O_4$ M 530.582
      Prod. by *Streptomyces* sp. Phosphodiesterase inhibitor.

Japan. Pat., 80 3 733, (*1980*); *CA,* **92**, 213541 (*isol*)

2,5-Dimethyl-3-furancarboxaldehyde   D-70412

3-Formyl-2,5-dimethylfuran
[54583-69-6]

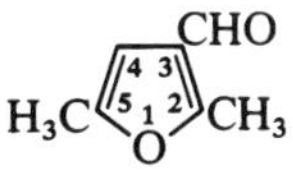

$C_7H_8O_2$ M 124.139
Oil. Bp$_{21}$ 84°. Unstable, readily oxidized.
2,4-Dinitrophenylhydrazone: Cryst. Mp 228-230°.

Baxter, G.J. *et al, Aust. J. Chem.,* 1974, **27**, 2605 (*synth, ir, pmr*)
Comins, D.L. *et al, J. Org. Chem.,* 1987, **52**, 104 (*synth, pmr*)

4,5-Dimethyl-2-furancarboxylic acid   D-70413

[89639-83-8]
$C_7H_8O_3$ M 140.138
Cryst. (toluene). Mp 157-158°.

Reichstein, T. *et al, Helv. Chim. Acta,* 1933, **16**, 28 (*synth*)
Klein, L.L., *Synth. Commun.,* 1986, **16**, 431 (*synth, pmr*)

2,2-Dimethyl-4-hexen-3-one, 9CI   D-70414

tert-*Butyl-1-propenyl ketone*
[81925-83-9]

$$(H_3C)_3CCOCH=CHCH_3$$

$C_8H_{14}O$ M 126.198
Liq.

Bienvenüe, A., *J. Am. Chem. Soc.,* 1973, **95**, 7345 (*uv, ir, config*)
Saegusa, T. *et al, J. Am. Chem. Soc.,* 1977, **99**, 1487 (*synth*)
Snowden, R.L. *et al, Tetrahedron Lett.,* 1982, **23**, 335 (*synth*)
Bretsch, W. *et al, Justus Liebigs Ann. Chem.,* 1987, 175 (*synth, pmr, ir*)

5,5-Dimethyl-2-hexynedioic acid, 9CI   D-70415

$$HOOCC\equiv CCH_2C(CH_3)_2COOH$$

$C_8H_{10}O_4$ M 170.165
Di-Me ester: [90171-34-9].
      $C_9H_{12}O_4$ M 184.191
      Liq. Bp$_{1.0}$ 84-86°.

Crimmins, M.T. *et al, J. Am. Chem. Soc.,* 1986, **108**, 800 (*synth, nmr*)

1,1-Dimethylhydrazine, 9CI   D-70416

Updated Entry replacing D-06269
Dimazine. UDMH
[57-14-7]

$$Me_2NNH_2$$

$C_2H_8N_2$ M 60.099
Rocket propellant. Protecting group for carbonyl compds., reagent for conversion of aldehydes to nitriles. Liq. with ammoniacal odour. Sol. H_2O, EtOH. d^{22} 0.791. Mp −58°. Bp 63°. n_D^{22} 1.4075. V. hygroscopic.
▷Highly corrosive to eyes, skin and mucous membranes. Convulsant poison. TLV 1. Suspected carcinogen. Ignites violently with oxidants. MV2450000.
B,HCl: [593-82-8]. Hygroscopic cryst. (EtOH). Mp 83°.
      ▷MV2900000.

Oxalate: Mp 142-143°.
Picrate: Mp 145-146° dec.

Org. Synth., 1936, **16**, 22 (*synth*)
Rowe, R.A. *et al*, *J. Am. Chem. Soc.*, 1956, **78**, 563 (*synth*)
Ohme, R. *et al*, *Justus Liebigs Ann. Chem.*, 1968, **713**, 74 (*synth*)
Schantl, J. *et al*, *Monatsh. Chem.*, 1978, **109**, 1481 (*pmr*)
Zenkovich, I.G. *et al*, *J. Org. Chem. USSR*, 1978, **14**, 1047 (*ms*)
Sisler, H.H. *et al*, *Inorg. Chem.*, 1980, **19**, 2846 (*synth*)
Mathis, R. *et al*, *Spectrochim. Acta, Part A*, 1981, **37**, 677; 1982, **38**, 133; 1986, **42**, 519 (*ir*)
Kirk-Othmer. Encycl. Chem. Technol., 3rd Ed., 1984, **2**, 279; **12**, 739, 766 (*rev*)
Riggs, N.V. *et al*, *Aust. J. Chem.*, 1987, **40**, 1783 (*struct, conformn*)
Fieser, M. *et al*, *Reagents for Organic Synthesis*, Wiley, 1967-84, **8**, 192.
Bretherick, L., *Handbook of Reactive Chemical Hazards*, 2nd Ed., Butterworths, London and Boston, 1979, 402.
Sax, N.I., *Dangerous Properties of Industrial Materials*, 5th Ed., Van Nostrand-Reinhold, 1979, 607.
Hazards in the Chemical Laboratory, (Bretherick, L., Ed.), 3rd Ed., Royal Society of Chemistry, London, 1981, 304.

1,4-Dimethyl-4-[4-hydroxy-2-(hydroxymethyl)-1-methyl-2-cyclopentenyl]-2-cyclohexen-1-ol   D-70417

FS 2

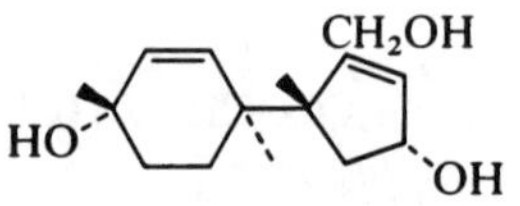

$C_{15}H_{24}O_3$ M 252.353

Isol. from *Fusarium sporotrichioides*. Mycotoxin. No phys. props. reported. Related to Trichodiol.

Corley, D.G. *et al*, *J. Org. Chem.*, 1987, **52**, 4405 (*isol, struct*)

O,N-Dimethylhydroxylamine   D-70418

Updated Entry replacing D-06279
N-*Methoxymethanamine, 9CI*
[1117-97-1]

MeNHOMe

C_2H_7NO M 61.083

Reagent for synth. of aldehydes, esp. sensitive ones *via N*-methoxy-*N*-methylamides. Liq. with sweet odour. Bp 42.2-42.6°. Does not reduce Fehling's soln.

B,HCl: [6638-79-5]. Plates. Mp 115-116°.

Bissot, T.C. *et al*, *J. Am. Chem. Soc.*, 1957, **79**, 796.
Goel, O.P. *et al*, *Org. Prep. Proced. Int.*, 1987, **19**, 75 (*synth, ir, pmr, bibl*)

2,2-Dimethyl-4-methylene-1,3-dioxolane, 9CI   D-70419

[19358-05-5]

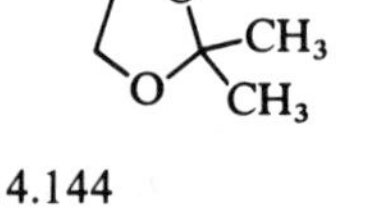

$C_6H_{10}O_2$ M 114.144

Starting material for a 2-lithiopropenal equivalent. Liq. Bp$_{95-98}$ 49-51°. Stable to storage at −10°.

Salomaa, P. *et al*, *Acta Chem. Scand.*, 1967, **21**, 2479 (*synth*)
Tius, M.A. *et al*, *J. Org. Chem.*, 1987, **52**, 2625 (*use*)

1,3-Dimethyl-2-methyleneimidazolidine   D-70420

[68738-47-6]

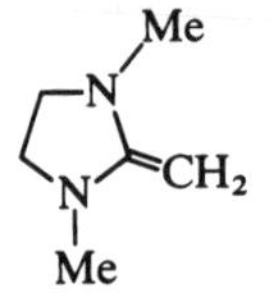

$C_6H_{12}N_2$ M 112.174

pK_a 26.6 (in MeCN). Various homologues also prepd.

Gruseck, U. *et al*, *Chem. Ber.*, 1987, **120**, 2053, 2065 (*synth, uv, pmr, cmr, use*)

8,8-Dimethyl-1,4,5(8*H*)-naphthalenetrione, 9CI   D-70421

[82208-38-6]

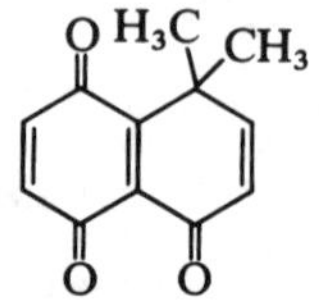

$C_{12}H_{10}O_3$ M 202.209

Red-orange cryst. (cyclohexane). Mp 146-147°.

Cassis, R. *et al*, *J. Chem. Soc., Perkin Trans. 1*, 1987, 2855 (*synth, ir, pmr*)

2,6-Dimethyl-3,7-octadiene-2,6-diol, 9CI   D-70422

[13741-21-4]

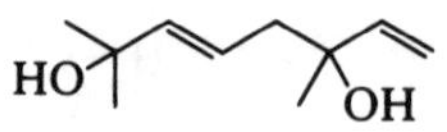

$C_{10}H_{18}O_2$ M 170.251

Constit. of *Achillea ligustica* and *Cinnamomum camphora*. Detected in many fruits and aromatic oils and in wines. Photooxidn. prod. of linalool. Oil. $[\alpha]_D^{24}$ +3.6° (c, 0.38 in CHCl₃).

Takaoka, D. *et al*, *Phytochemistry*, 1976, **15**, 330 (*isol, pmr*)
Bruno, M. *et al*, *Phytochemistry*, 1988, **27**, 1871 (*isol*)

3-(3,7-Dimethyl-2,6-octadienyl)-4-hydroxybenzoic acid   D-70423

3-Geranyl-4-hydroxybenzoic acid
[68631-48-1]

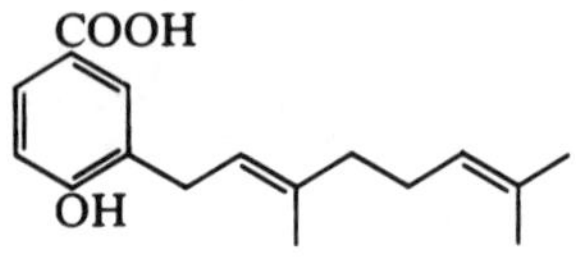

$C_{17}H_{22}O_3$ M 274.359

Isol. from cell cultures of *Lithospermum erythrorhiza*. Intermed. in biosynth. of Shikonin, S-70038 . Oil.

Inouye, H. *et al*, *Phytochemistry*, 1979, **18**, 1301.
Yazaki, K. *et al*, *Chem. Pharm. Bull.*, 1986, **34**, 2290.

4,6-Dimethyl-4-octen-3-one   D-70424

Updated Entry replacing D-50653

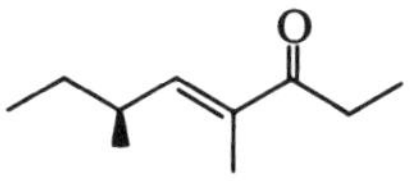

$C_{10}H_{18}O$ M 154.252

(S,E)-form [60132-36-7]
Manicone
Pheromone from the mandibular glands of the ants *Manica mutica, M. bradleyi* and *M. rubida.* Oil.

Blanco, L. *et al, Tetrahedron Lett.*, 1981, 645 (*isol, struct*)
Alexakis, A. *et al, Tetrahedron*, 1984, **40**, 715 (*synth*)
Bestmann, H.J. *et al, Angew. Chem., Int. Ed. Engl.*, 1987, **26**, 784 (*abs config*)

6,6-Dimethyl-3-oxabicyclo[3.1.0]hexan-2-one, 9CI   D-70425

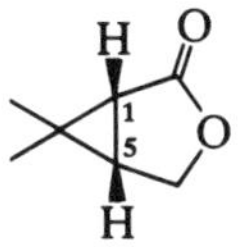

(1R,5S)-form

$C_7H_{10}O_2$ M 126.155

(1R,5S)-form [82442-72-6]
(−)-cis-*form*
Bp$_{10}$ 100°. $[\alpha]_D^{25}$ −36.6° (c, 1.4 in CHCl$_3$), $[\alpha]_D^{25}$ −72.8° (c, 1.4 in CHCl$_3$).
(1RS,5SR)-form [62222-77-9]
(±)-cis-*form*
No phys. props. reported.

Jakovac, I.J. *et al, J. Am. Chem. Soc.*, 1982, **104**, 4659 (*synth, pmr, ir, abs config*)
Mukaiyama, T. *et al, Chem. Lett.*, 1983, 385 (*synth*)
Org. Synth., 1985, **63**, 10 (*synth*)
Mandal, A.K. *et al, Tetrahedron*, 1986, **42**, 5715 (*synth, ir, pmr*)
Sabbioni, G. *et al, J. Org. Chem.*, 1987, **52**, 4565 (*synth, pmr, ir, cmr*)

3,3-Dimethyloxiranemethanol, 9CI   D-70426

2,3-Epoxy-3-methyl-1-butanol. 2-Hydroxymethyl-1,1-dimethyloxirane

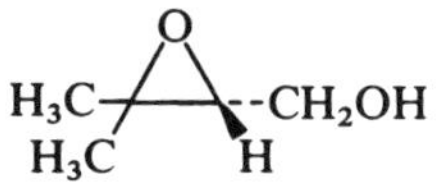

(R)-form

$C_5H_{10}O_2$ M 102.133

(R)-form [62748-09-8]
Liq. $[\alpha]_D^{25}$ +21.0° (c, 1.71 in CHCl$_3$). >93% opt. pure.
(S)-form
Liq. Bp$_7$ 58-60°. $[\alpha]_D^{25}$ −20.1° (c, 0.42 in CHCl$_3$).

Dumont, R. *et al, Helv. Chim. Acta*, 1983, **66**, 814 (*synth, ir, pmr, ms*)
Suga, T. *et al, J. Chem. Soc., Perkin Trans. 1*, 1987, 2845 (*synth, ir, pmr, ms*)

3,3-Dimethyl-1,5-pentanedithiol   D-70427

1,5-Dimercapto-3,3-dimethylpentane
[80980-56-9]

$$HSCH_2CH_2C(CH_3)_2CH_2CH_2SH$$

$C_7H_{16}S_2$ M 164.323

Oil. Bp$_{0.1}$ 50-55°.

Houk, J. *et al, J. Am. Chem. Soc.*, 1987, **109**, 6825 (*synth, ir, pmr*)

4,4-Dimethylpentanoic acid, 9CI   D-70428

4,4-Dimethylvaleric acid, 8CI
[1118-47-4]

$$(H_3C)_3CCH_2CH_2COOH$$

$C_7H_{14}O_2$ M 130.186
Liq. d^{20} 0.91. Mp −3° to −1°. Bp 211-4° (205-211°), Bp$_{15}$ 170°, Bp$_{12}$ 75-78°.
Me ester: [15673-17-3].
　$C_8H_{16}O_2$ M 144.213
　Bp$_{40}$ 77-84°.
Et ester: [10228-99-6].
　$C_9H_{18}O_2$ M 158.240
　Bp$_8$ 60-62°.
Chloride: [5699-78-5].
　$C_7H_{13}ClO$ M 148.632
　Bp$_{37}$ 40-51°.
Amide: [15672-96-5].
　$C_7H_{15}NO$ M 129.202
　Cryst. (H$_2$O). Mp 139.5-140°.
Nitrile: [15673-05-9].
　$C_7H_{13}N$ M 111.186
　Liq. Bp 165-170°, Bp$_8$ 57-59°.

Moureau, C. *et al, Bull. Soc. Chim. Fr.*, 1903, **29**, 664 (*synth*)
Spindt, R.S. *et al, J. Am. Chem. Soc.*, 1951, **73**, 3693 (*synth*)
Brändström, A., *Acta Chem. Scand.*, 1959, **13**, 613 (*synth, deriv*)
Whitesides, G.M. *et al, J. Am. Chem. Soc.*, 1967, **89**, 1135 (*synth, deriv, pmr, ir*)
Romanuk, M. *et al, Collect. Czech. Chem. Commun.*, 1972, **37**, 1755 (*synth*)
Brown, H.C. *et al, J. Org. Chem.*, 1984, **49**, 892 (*synth*)
Cooke, M.P., *J. Org. Chem.*, 1987, **52**, 5729 (*synth*)

3,3-Dimethyl-4-penten-2-one, 9CI   D-70429

[4181-07-1]

$$H_2C{=}CHC(CH_3)_2COCH_3$$

$C_7H_{12}O$ M 112.171
V. mobile liq. Bp 129.5°, Bp$_{55}$ 52-55°.
Oxime: [57606-91-4].
　$C_7H_{13}NO$ M 127.186
　Liq. Bp$_1$ 57-58°.
2,4-Dinitrophenylhydrazone: Mp 117°.

Cywinski, N.F. *et al, J. Org. Chem.*, 1965, **30**, 3814 (*synth*)
House, H.O. *et al, J. Org. Chem.*, 1976, **41**, 863 (*synth, deriv, ir, pmr*)
Cazes, B. *et al, Bull. Soc. Chim. Fr.*, 1977, 925 (*synth, ir, ms, pmr, deriv*)
Bretsch, W. *et al, Justus Liebigs Ann. Chem.*, 1987, 175 (*synth, pmr, ir*)

3,4-Dimethyl-5-pentylidene-2(5H)-furanone, 9CI   D-70430

Bovolide
[774-64-1]

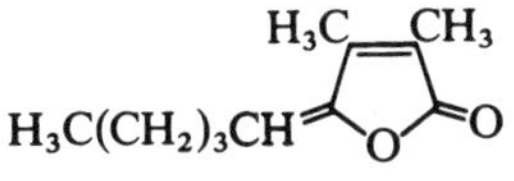

$C_{11}H_{16}O_2$ M 180.246
Constit. of tobacco and peppermint oil.
▷LU4025000.

Demole, E. *et al, Helv. Chim. Acta*, 1972, **55**, 1866 (*isol*)

Katsuhiro, T. *et al. Agric. Biol. Chem.*, 1980, **44**, 1535 (*isol*)
Wulff, W.D. *et al, J. Am. Chem. Soc.*, 1986, **108**, 520 (*synth*)

4,5-Dimethyl-2-phenyl-1,3-dioxol-1-ium, 9CI   D-70431

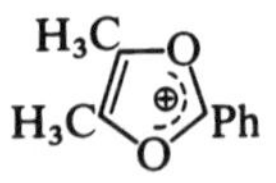

$C_{11}H_{11}O_2^{\oplus}$ M 175.207 (ion)
Trifluoromethanesulfonate: [106016-52-8].
  $C_{12}H_{11}F_3O_5S$ M 324.271
  Moisture-sensitive solid. Mp 138-140°.

Lorenz, W. *et al, J. Org. Chem.*, 1987, **52**, 375 (*synth, ir, pmr, cmr*)

4,4-Dimethyl-2-phenyl-4H-imidazole, 9CI   D-70432

[89002-63-1]

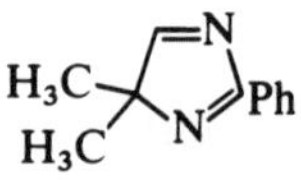

$C_{11}H_{12}N_2$ M 172.229
Cryst. (pet. ether at −78°). Mp 40-44°.

Casey, M. *et al, J. Chem. Soc., Perkin Trans. 1*, 1987, 1389 (*synth, ir, uv, pmr, cmr, ms*)

2,2-Dimethyl-1,3-propanedithiol, 9CI   D-70433
1,3-Dimercapto-2,2-dimethylpropane
[53555-42-3]

$(H_3C)_2C(CH_2SH)_2$

$C_5H_{12}S_2$ M 136.270
Bp_{12} 72°, Bp_2 42-44°.

Backer, H.J., *Recl. Trav. Chim. Pays-Bas*, 1938, **57**, 1183 (*synth*)
Houk, J. *et al, J. Am. Chem. Soc.*, 1987, **109**, 6825 (*synth, ir, pmr*)

3-(1,1-Dimethyl-2-propenyl)-4-hydroxy-6-   D-70434
phenyl-2H-pyran-2-one

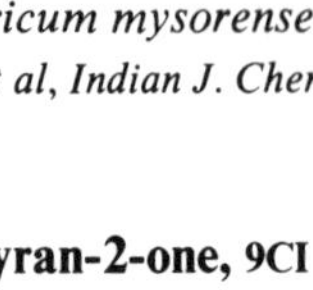

$C_{16}H_{16}O_3$ M 256.301
O-(*3-Hydroxy-3-methylbutanoyl*): *4-(3-Hydroxy-3-methylbutanoyloxy)-3-(1,1-dimethyl-2-propenyl)-6-phenyl-2H-pyran-2-one.*
  $C_{21}H_{24}O_5$ M 356.418
  Constit. of *Hypericum mysorense*. Oil.

Vishwakarma, R.A. *et al, Indian J. Chem., Sect. B*, 1987, **26**, 486.

4,6-Dimethyl-2H-pyran-2-one, 9CI   D-70435
Updated Entry replacing D-06954
4,6-Dimethyl-α-pyrone. 4,6-Dimethylcoumalin. Mesitene lactone
[675-09-2]
$C_7H_8O_2$ M 124.139
Mp 50-51°. Bp_{35} 140-142°.
▷UQ0700000.

Org. Synth., 1952, **32**, 57 (*synth*)
Izumi, T. *et al, Bull. Chem. Soc. Jpn.*, 1975, **48**, 1673 (*synth*)
Dieter, R.K. *et al, J. Org. Chem.*, 1988, **53**, 2031 (*synth, uv, ir. pmr, cmr*)

4,5-Dimethylpyrimidine, 9CI   D-70436
Updated Entry replacing D-07003
[694-81-5]
$C_6H_8N_2$ M 108.143
Needles. Misc. H_2O. Mp 3°. Bp 177°.
Picrate: Mp 162°.
1-Oxide: [114969-53-8].
  $C_6H_8N_2O$ M 124.142
  Cryst. Mp 139-140°.
3-Oxide:
  $C_6H_8N_2O$ M 124.142
  Hygroscopic semisolid.

Schlenker, J., *Ber.*, 1901, **34**, 2812.
Bredereck, H. *et al, Chem. Ber.*, 1960, **93**, 230, 1402.
Riand, J. *et al, Tetrahedron Lett.*, 1974, 3123 (*cmr*)
Yamanaka, H. *et al, Chem. Pharm. Bull.*, 1987, **35**, 3119 (*synth, pmr, oxides*)

2,5-Dimethylpyrrolidine   D-70437
Updated Entry replacing D-07040
[3378-71-0]

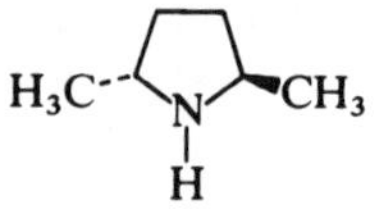

(2R,5R)-form

$C_6H_{13}N$ M 99.175
(2R,5R)-form
  (+)-trans-*form*
  B,HCl: [70144-18-2]. Solid. Mp 197-200°. $[\alpha]_D$ +5.47° (c, 3.0 in CH_2Cl_2).
(2RS,5RS)-form [39713-72-9]
  (±)-trans-*form*
  Bp 108-109°. n_D^{23} 1.4291.
  Picrate: Mp 130-131°.
  N-*Me:* [18887-22-4]. *1,2,5-Trimethylpyrrolidine.*
    $C_7H_{15}N$ M 113.202
    Mp 229-230° (as picrate).
(2RS,5SR)-form [39713-71-8]
  cis-*form*
  d_4^{20} 0.821. Bp 106-106.7°. n_D^{25} 1.4276.
  Picrate: Mp 120-121°.
  N-*2,4-Dinitrophenyl:* Mp 227°.
  N-*Me:* [18887-24-6]. Bp 120-130° (bath).
  N-*Me, picrate:* [18887-25-7]. Mp 213-214°.

Merling, G., *Justus Liebigs Ann. Chim.*, 1891, **264**, 328 (*synth*)
de Jong, H. *et al, Recl. Trav. Chim. Pays-Bas*, 1930, **49**, 245 (*synth*)
Evans, G.G., *J. Am. Chem. Soc.*, 1951, **73**, 5230 (*synth*)
Overberger, C.G. *et al, J. Am. Chem. Soc.*, 1955, **77**, 4100 (*synth*)
Ohki, S. *et al, Chem. Pharm. Bull.*, 1968, **16**, 269 (*deriv*)
Schlessinger, R.H. *et al, Tetrahedron Lett.*, 1987, **28**, 2083 (*synth, ir, pmr, ms*)

2,5-Dimethyl-3-thiophenecarboxaldehyde   D-70438
3-Formyl-2,5-dimethylthiophene

[26421-44-3]

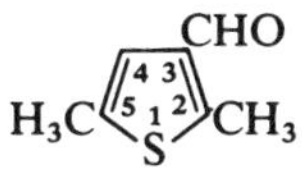

H$_3$C—[5 4 3 2]—CH$_3$, CHO, S

C_7H_8OS M 140.200
Light-yellow oil. Bp$_{25}$ 116-117°.

Phenylhydrazone: Cryst. (EtOH aq.). Mp 95-96°.

Weston, W. *et al*, *J. Am. Chem. Soc.*, 1950, **72**, 1422 (*synth*)
Comins, D.L. *et al*, *J. Org. Chem.*, 1987, **52**, 104 (*synth, pmr*)

3,5-Dimethyl-2-thiophenecarboxaldehyde   D-70439
2-Formyl-3,5-dimethylthiophene

[85895-83-6]

C_7H_8OS M 140.200
Light-yellow oil.

Hydrazone:
    $C_7H_9N_2S$ M 153.222
    Mp 125° dec.
4-Nitrophenylhydrazone: Red needles (EtOH). Mp 221-223°.

Sice, J., *J. Org. Chem.*, 1954, **19**, 70.
Comins, D.L. *et al*, *J. Org. Chem.*, 1987, **52**, 104 (*synth, pmr*)

1,2-Dimethylthioxanthone   D-70440
1,2-Dimethyl-9H-thioxanthen-9-one, *9CI*

[88488-54-4]

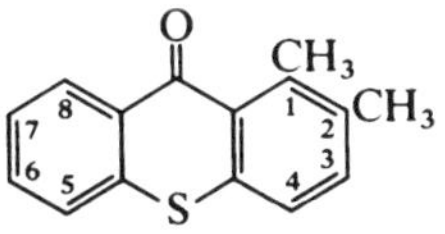

$C_{15}H_{12}OS$ M 240.319
Plates. Mp 103-104°.

Honek, J.F. *et al*, *Synth. Commun.*, 1983, **13**, 977 (*synth*)
Okabayashi, I. *et al*, *Chem. Pharm. Bull.*, 1987, **35**, 2545 (*synth, pmr*)

1,3-Dimethylthioxanthone   D-70441
1,3-Dimethyl-9H-thioxanthen-9-one, *9CI*

[76529-35-6]

$C_{15}H_{12}OS$ M 240.319
Needles (EtOH aq.). Mp 110-111°.

Ternay, A.L. *et al*, *J. Org. Chem.*, 1981, **46**, 1793.
Okabayashi, I. *et al*, *Chem. Pharm. Bull.*, 1987, **35**, 2545 (*synth, pmr*)

1,4-Dimethylthioxanthone   D-70442
1,4-Dimethyl-9H-thioxanthen-9-one, *9CI*

[25942-61-4]

$C_{15}H_{12}OS$ M 240.319
Cryst. (EtOH). Mp 112-113°.

Dioxide: [14936-99-3].
    $C_{15}H_{12}O_3S$ M 272.318
    Cryst. (EtOH). Mp 149-150°.

Kaiser, E.T. *et al*, *J. Am. Chem. Soc.*, 1967, **89**, 5179 (*synth, deriv*)
Ternay, A.L. *et al*, *J. Org. Chem.*, 1974, **39**, 2941 (*synth*)
Honek, J.F. *et al*, *Synth. Commun.*, 1983, **13**, 977 (*synth*)

2,3-Dimethylthioxanthone   D-70443
2,3-Dimethyl-9H-thioxanthen-9-one, *9CI*

[81877-48-7]

$C_{15}H_{12}OS$ M 240.319
Pale-yellow needles. Mp 176-177°.

Honek, J.F. *et al*, *Synth. Commun.*, 1983, **13**, 977 (*synth*)
Okabayashi, I. *et al*, *Chem. Pharm. Bull.*, 1987, **35**, 2545 (*synth, pmr*)

2,4-Dimethylthioxanthone   D-70444
2,4-Dimethyl-9H-thioxanthen-9-one, *9CI*

[76293-13-5]

$C_{15}H_{12}OS$ M 240.319
Yellow needles (EtOH). Mp 144.5-145°

Dioxide: [7741-54-0].
    $C_{15}H_{12}O_3S$ M 272.318
    Cryst. (EtOH). Mp 181-182°.

Kaiser, E.T. *et al*, *J. Am. Chem. Soc.*, 1967, **89**, 5179 (*synth, deriv*)
Ternay, A.L. *et al*, *J. Chem. Soc., Chem. Commun.*, 1980, **17**, 846.
Okabayashi, I. *et al*, *Chem. Pharm. Bull.*, 1987, **35**, 2545 (*synth, pmr*)

3,4-Dimethylthioxanthone   D-70445
3,4-Dimethyl-9H-thioxanthen-9-one, *9CI*

[81877-47-6]

$C_{15}H_{12}OS$ M 240.319
Pale-brown needles. Mp 196-198°.

Okabayashi, I. *et al*, *Chem. Pharm. Bull.*, 1987, **35**, 2545 (*synth, pmr*)

2,3-Dimethyl-5-(2,6,10-trimethylundecyl)-   D-70446
thiophene, *9CI*

[99835-10-6]

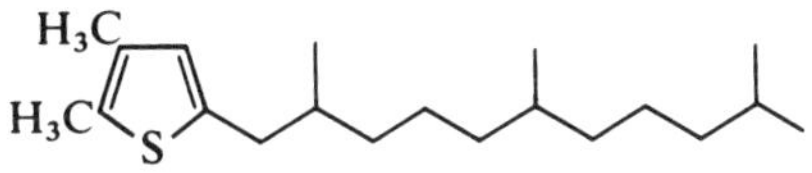

$C_{20}H_{36}S$ M 308.564
Major constituent of certain petroleums and sediment extracts. Liq.

Sinninghe Damsté, J.S. *et al*, *Tetrahedron Lett.*, 1987, **28**, 957 (*synth, pmr, cmr, ms*)

1,1-Di-4-morpholinylethene   D-70447
4,4'-Ethenylidenebismorpholine, *9CI*

[14212-87-4]

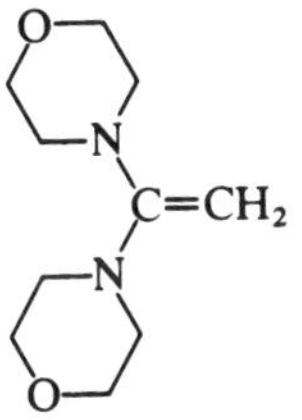

$C_{10}H_{18}N_2O_2$ M 198.264
Basic, electron-rich alkene with synthetic uses. Cryst. mass. Mp 59-60°. Bp$_{0.2}$ 90-92°. Darkens in air.

Böhme, H. *et al*, *Chem. Ber.*, 1961, **94**, 3109 (*synth*)
Baganz, H. *et al*, *Chem. Ber.*, 1962, **95**, 2095 (*synth*)
Gandhi, S.S. *et al*, *Can. J. Chem.*, 1987, **65**, 2717 (*bibl*)

7*H*,9*H*,16*H*,18*H*-Dinaphtho[1,8-*cd*:1′,8′-*ij*][1,7]dithiacyclododecin, 9CI   D-70448
[76727-31-6]

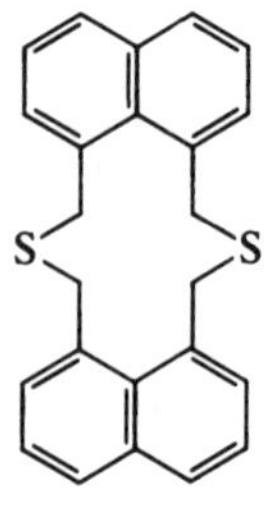

$C_{24}H_{20}S_2$ M 372.542
Feathery needles (CHCl$_3$). Mp 290-291° dec.

Kemp, W. *et al, J. Chem. Soc., Perkin Trans. 1*, 1980, 2812 (*synth, ir, pmr, ms*)

4,5-Dinitro-1,2-benzenediol, 9CI   D-70449
Updated Entry replacing D-07357
4,5-Dinitrocatechol
[77400-30-7]
$C_6H_4N_2O_6$ M 200.107
Yellow cryst. (C$_6$H$_6$/EtNO$_2$). Mp 166.5-167.5°.

Mono-Me ether: 2-Methoxy-4,5-dinitrophenol. 4,5-Dinitroguaiacol.
$C_7H_6N_2O_6$ M 214.134
Yellow cryst. (EtOH or xylene). Mp 177°.

Mono-Me ether, Ac:
$C_9H_8N_2O_7$ M 256.171
Mp 123-124°.

Di-Me ether: [3395-03-7]. *1,2-Dimethoxy-4,5-dinitrobenzene. 4,5-Dinitroveratrole.*
$C_8H_8N_2O_6$ M 228.161
Yellow needles (EtOH). Mp 130-132°.

Ehrlich, J. *et al, J. Org. Chem.*, 1947, **12**, 522.
Wulfman, D.S. *et al, Synthesis*, 1978, 924.
Eswaran, S.V. *et al, J. Heterocycl. Chem.*, 1988, **25**, 803 (*synth, ir, uv, pmr, deriv*)

4,6-Dinitro-1,3-benzenediol, 9CI   D-70450
Updated Entry replacing D-07359
4,6-Dinitroresorcinol, 8CI
[616-74-0]
$C_6H_4N_2O_6$ M 200.107
Pale-yellow cryst. Mp 215°. pK_a 3.98.

Di-Ac:
$C_{10}H_8N_2O_8$ M 284.182
Mp 139°.

Me ether: [51652-35-8]. *5-Methoxy-2,4-dinitrophenol.*
$C_7H_6N_2O_6$ M 214.134
Mp 113°.

Et ether: 5-Ethoxy-2,4-dinitrophenol.
$C_8H_8N_2O_6$ M 228.161
Mp 77°.

Di-Me ether: [1210-96-4]. *1,5-Dimethoxy-2,4-dinitrobenzene, 9CI, 8CI.*
$C_8H_8N_2O_6$ M 228.161
Mp 157° (154°).

Di-Et ether: 1,5-Diethoxy-2,4-dinitrobenzene.
$C_{10}H_{12}N_2O_6$ M 256.215
Mp 133°.

Pantlischenko, M. *et al, Monatsh. Chem.*, 1950, **81**, 293 (*synth*)
Rauner, W. *et al, Z. Chem.*, 1968, **8**, 338 (*synth*)
Granzhan, V.A. *et al, Zh. Strukt. Khim.*, 1971, **12**, 809; *CA*, **76**, 24579 (*ir*)

Crampton, M.R. *et al, J. Chem. Soc., Perkin Trans. 2*, 1972, 1178 (*deriv*)
Schmitt, R.J. *et al, J. Org. Chem.*, 1988, **53**, 5568 (*synth*)

1,2-Dinitrocyclohexene   D-70451
[107400-83-9]

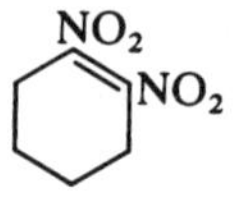

$C_6H_8N_2O_4$ M 172.140
Unstable yellow oil.

Boyer, J.H. *et al, Org. Prep. Proced. Int.*, 1986, **18**, 363 (*synth, pmr, cmr*)

3,5-Dinitroisoxazole   D-70452
[42216-62-6]

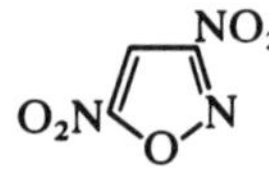

$C_3HN_3O_5$ M 159.058
Cryst. (CCl$_4$). Mp 53.4-54.5°.

Golod, E.L. *et al, J. Org. Chem. USSR (Engl. Transl.)*, 1973, **9**, 1139 (*synth*)
Cromer, D.T. *et al, Acta Crystallogr., Sect. C*, 1987, **43**, 2011 (*cryst struct*)

2,4-Dinitropentane   D-70453

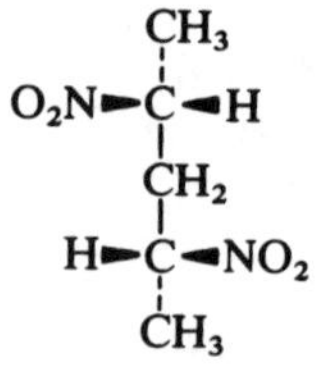

(2*RS*,4*RS*)-*form*

$C_5H_{10}N_2O_4$ M 162.145
▷Vigorous dec. can occur during synth. or dist.
(2*RS*,4*RS*)-*form*
(±)-*form*
Cryst. Mp 46-47.5°.
(2*RS*,4*SR*)-*form*
meso-*form*
Oil.

Bachman, G.B. *et al, J. Am. Chem. Soc.*, 1956, **78**, 484 (*synth*)
Wade, P.A. *et al, J. Am. Chem. Soc.*, 1987, **109**, 5452 (*synth, pmr, haz*)

1,1-Dinitro-2-propanone   D-70454
1,1-Dinitroacetone. Acetyldinitromethane

H$_3$CCOCH(NO$_2$)$_2$

$C_3H_4N_2O_5$ M 148.075
O-*Acetyloxime:* [108560-93-6].
$C_5H_7N_3O_6$ M 205.127
Cryst. (Et$_2$O/pet. ether). Mp 88-89° dec.

Soderquist, J.A. *et al, J. Org. Chem.*, 1987, **52**, 3441 (*deriv, synth, ir, nmr*)

1,3-Dinitro-5-(trifluoromethyl)benzene   D-70455

α,α,α-Trifluoro-3,5-dinitrotoluene. 3,5-Dinitrobenzotrifluoride

[401-99-0]

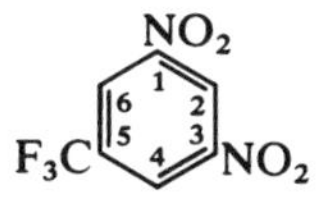

$C_7H_3F_3N_2O_4$ M 236.107
Cryst. (MeOH). Mp 49-50°.

Finger, G.C. *et al, J. Am. Chem. Soc.*, 1944, **66**, 1972 (*synth*)
Benkeser, R.A. *et al, J. Am. Chem. Soc.*, 1952, **74**, 3011 (*synth*)

2,4-Dinitro-1-(trifluoromethyl)benzene   D-70456

α,α,α-Trifluoro-2,4-dinitrotoluene. 2,4-Dinitrobenzotrifluoride

[30287-26-4]

$C_7H_3F_3N_2O_4$ M 236.107
Cryst. (MeOH). Mp 49-50°.

Malichenko, B.F. *et al, J. Org. Chem. USSR (Engl. Transl.)*, 1970, **6**, 2302 (*synth*)
Kondratenko, N.V. *et al, Synthesis*, 1980, 932 (*synth*)
Clark, J.H. *et al, J. Chem. Soc., Chem. Commun.*, 1988, 638 (*synth*)

14,15-Dinor-13-oxo-7-labden-17-oic acid   D-70457
Havardic acid E

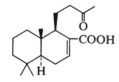

$C_{18}H_{28}O_3$ M 292.417
Constit. of *Grindelia havordii*.

Jolad, S.D. *et al, Phytochemistry*, 1987, **26**, 483.

1,6-Dioxacyclodeca-3,8-diene   D-70458

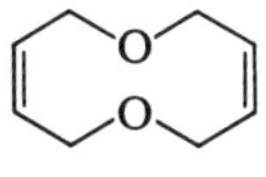

$C_8H_{12}O_2$ M 140.182

Barnes, J.C. *et al, Acta Crystallogr., Sect. C*, 1987, **43**, 2245 (*cryst struct*)

1,13-Dioxa-4,7,10,16,20,24-hexaazacyclo-hexacosane, 9CI   D-70459

[105763-01-7]

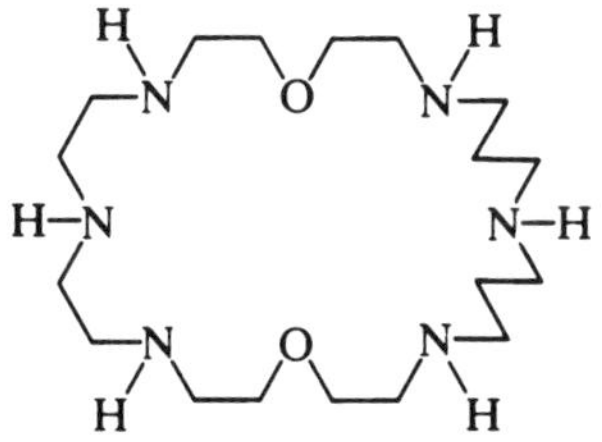

$C_{18}H_{42}N_6O_2$ M 374.569
B,6HCl: [110661-83-1]. Cryst. (EtOH aq.). Mp 252-254°.

Hexakis(4-methylbenzenesulfonyl): Glass.

Hosseini, M.W. *et al, J. Am. Chem. Soc.*, 1987, **109**, 7047 (*synth, pmr, cmr*)

1,13-Dioxa-4,7,10,16,19,22-hexaazacyclo-tetracosane, 9CI   D-70460
1,3-Dioxahexaaza-24-crown-8

[43090-52-4]

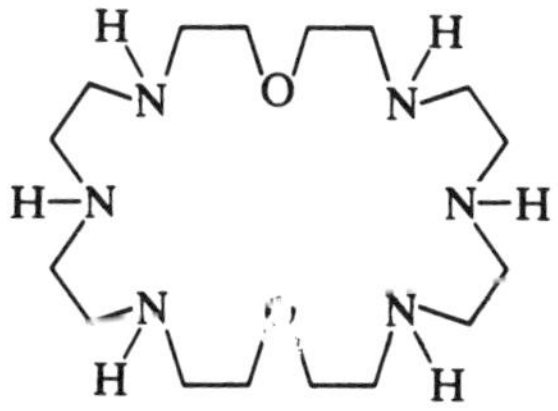

$C_{16}H_{38}N_6O_2$ M 346.515
Receptor mol. forming complexes with transition metal ions, polyprotonated form binds anions. Catalyst in transformation reactions of ATP and acetyl phosphate; shows protokinase and protophosphatase activity. Oil, cryst. on standing at −35°.

4,10,16,22-Tetrakis(4-methylbenzenesulfonyl): [64844-66-2]. Solid + 1H₂O.
7,19-Bis(aminoethyl); B,HCl: [107098-23-7]. Cryst. (MeOH). Hygroscopic.

Comarmond, J. *et al, J. Am. Chem. Soc.*, 1982, **104**, 6330 (*synth, pmr, cmr*)
Blackburn, G.M. *et al, Tetrahedron Lett.*, 1987, **28**, 2779 (*use*)
Hosseini, M.W. *et al, J. Am. Chem. Soc.*, 1987, **109**, 537 (*biochem*)
Hosseini, M.W. *et al, J. Org. Chem.*, 1987, **52**, 1662 (*deriv, synth, use, bibl*)

11,12-Dioxahexacyclo[6.2.1.1³,⁶.0²,⁷.0⁴,¹⁰.0⁵,⁹]-dodecane   D-70461

Hexahydro-2,6,3,5-ethanediylidene-2H,3H-1,4-dioxacyclopenta[cd]pentalene, 9CI. Dioxa-1,3-bishomopentaprismane

[106139-18-8]

$C_{10}H_{10}O_2$ M 162.188
Cryst. (CH₂Cl₂/hexane). Mp 271-273°.

Mehta, G. *et al, J. Org. Chem.*, 1987, **52**, 460 (*synth, pmr, cmr*)

3,9-Dioxaspiro[5.5]undecane   D-70462

[180-47-2]

$C_9H_{16}O_2$ M 156.224
Oil. Bp₄.₅ 88-89°, Bp₁.₀₋₁.₅ 65-66°.

Rice, L.M. *et al, J. Org. Chem.*, 1967, **32**, 1966 (*synth*)
Newkome, G.R. *et al, J. Org. Chem.*, 1987, **52**, 5480 (*synth, pmr*)

5,12-Dioxatetracyclo[7.2.1.0^{4,11}.0^{6,10}]-dodeca-2,7-diene   D-70463

2a,3a,5a,6a,6b,6c-Hexahydro-3,6-dioxacyclopenta[cd,gh]*pentalene, 9CI*

[106139-20-2]

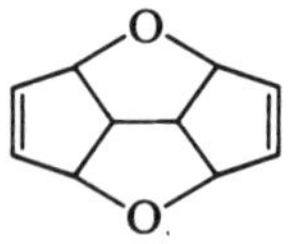

C$_{10}$H$_{10}$O$_2$ M 162.188

Cryst. (CH$_2$Cl$_2$/hexane). Mp 136-137°.

Mehta, G. *et al, J. Org. Chem.*, 1987, **52**, 460 (*synth, pmr, cmr*)

2*H*-1,5,2-Dioxazine-3,6(4*H*)-dione, 9CI   D-70464

[110434-71-4]

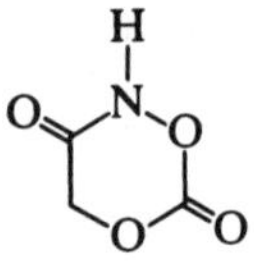

C$_3$H$_3$NO$_4$ M 117.061

Mp 75° dec. Liquifies after 3 months at r.t. due to isomerisation.

Schwarz, G. *et al, Justus Liebigs Ann. Chem.*, 1988, 35 (*synth, ir, pmr*)

1,5-Dioxiranyl-1,2,3,4,5-pentanepentol   D-70465

1,2:8,9-Dianhydrononitol, 9CI. WF 3405. Antibiotic FR 68504. Antibiotic WF 3405. FR 68504

[108354-43-4]

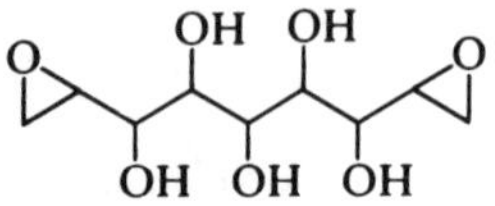

C$_9$H$_{16}$O$_7$ M 236.221

Prod. by *Amauroascus aureus*. Antitumour agent. Prisms (MeOH aq.). Mp 157-159°. [α]$_D^{23}$ −5.8° (c, 1 in H$_2$O).

Eur. Pat., 187 528, (*1986*); *CA*, **105**, 132146 (*isol*)

Kiyoto, S. *et al, J. Antibiot.*, 1987, **40**, 290 (*isol, struct, props*)

1,1′-(1,2-Dioxo-1,2-ethanediyl)bis-1*H*-indole-2,3-dione, 9CI   D-70466

1,1′-Oxalylbisisatin

[69314-18-7]

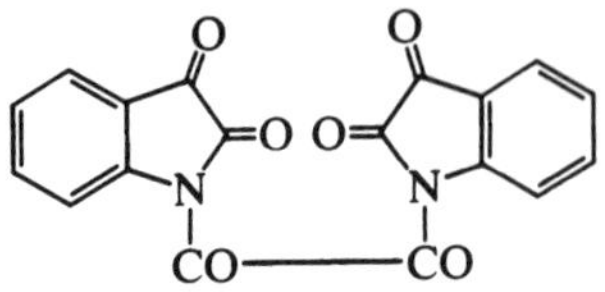

C$_{18}$H$_8$N$_2$O$_6$ M 348.271

Highly reactive towards nucleophiles, reactivity resembles that of an acid chloride.

Black, D.St.C. *et al, Tetrahedron Lett.*, 1978, 2837.

Black, D.St.C. *et al, Aust. J. Chem.*, 1987, **40**, 1755 (*cryst struct*)

1,2-Dioxolane   D-70467

[4362-13-4]

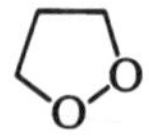

C$_3$H$_6$O$_2$ M 74.079

Salomon, M.F. *et al, J. Am. Chem. Soc.*, 1979, **101**, 4290 (*synth, nmr*)

Bloodworth, A.J. *et al, J. Org. Chem.*, 1986, **51**, 2110 (*synth, cmr, pmr*)

1,3-Dioxolane-2-methanol   D-70468

2-Hydroxymethyl-1,3-dioxolane

[5694-68-8]

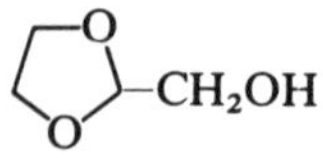

C$_4$H$_8$O$_3$ M 104.105

Ethylene acetal of Hydroxyacetaldehyde, H-01166 . Liq. Bp$_{15}$ 83-84°.

4-Nitrobenzoyl: Yellowish needles (pet. ether). Mp 58-59.5°.

3,5-Dinitrobenzoyl: Cryst. (Et$_2$O or MeOH aq.). Mp 106-107°.

Späth, E. *et al, Monatsh. Chem.*, 1947, **76**, 65 (*synth, deriv*)

Sanderson, J.R. *et al, J. Org. Chem.*, 1987, **52**, 3243 (*synth, pmr, cmr, ir, ms*)

3,7-Dioxo-12-oleanen-28-oic acid   D-70469

Rubonic acid

[113738-79-7]

C$_{30}$H$_{44}$O$_4$ M 468.675

Constit. of *Rubus moluccanus*. Cryst. (CHCl$_3$/MeOH). Mp 282-283°.

Me ester: Cryst. (C$_6$H$_6$). Mp 256-257°.

Shaw, A.K. *et al, Indian J. Chem., Sect. B*, 1987, **26**, 896.

Diphenanthro[2,1-*b*:1′,2′-*d*]furan   D-70470

[194-54-7]

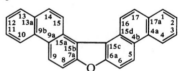

C$_{28}$H$_{16}$O   M 368.434

Yellow or colourless needles. Mp 228-229°.

Crawford, M. *et al, J. Chem. Soc.*, 1962, 674 (*synth, uv*)

Diphenanthro[9,10-*b*:9′,10′-*d*]furan, 9CI   D-70471

Tetraphenylenefurfuran

[202-71-1]

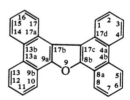

C$_{28}$H$_{16}$O   M 368.434

Silky needles. Mp 306°.

Japp, F.R. *et al*, *J. Chem. Soc.*, 1897, **71**, 1115 (*synth*)
Weitzberg, M. *et al*, *J. Heterocycl. Chem.*, 1983, **20**, 1019 (*pmr*)

Di-9-phenanthrylamine, 8CI   D-70472

N-9-Phenanthrenyl-9-phenanthrenamine
[16269-40-2]

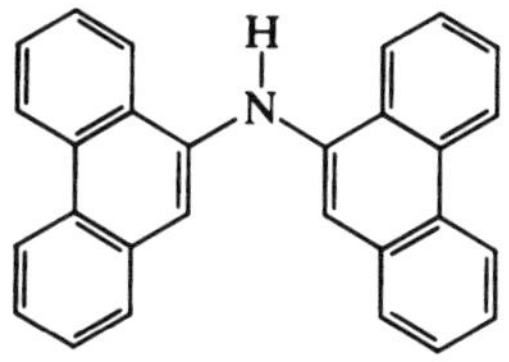

$C_{28}H_{19}N$ M 369.465
Cryst. (Et$_2$O). Mp 228-229°. A previously reported synthesis (1910) giving a prod. of Mp 360° was evidently erroneous.

Altiparmakian, R.H. *et al*, *J. Chem. Soc. (C)*, 1967, 1818.
Weitzburg, M. *et al*, *J. Heterocycl. Chem.*, 1983, **10**, 1019 (*synth, pmr, uv*)

7,7-Diphenylbenzo[*c*]fluorene   D-70473

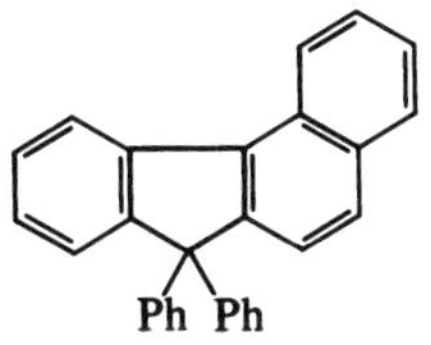

$C_{29}H_{20}$ M 368.477
Cryst. Mp 203-205.5°.

Walsh, T.D., *J. Am. Chem. Soc.*, 1987, **109**, 1511 (*synth, ms, uv*)

1,3-Diphenylbenzo[*c*]thiophene, 9CI   D-70474

[16587-39-6]

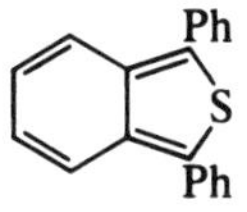

$C_{20}H_{14}S$ M 286.391
Yellow cryst. with green fluorescence. Mp 120° (118-119°). Forms a dianion which can be characterized by pmr and cmr at −20°.

Lepage, L. *et al*, *J. Heterocycl. Chem.*, 1978, **15**, 1185 (*synth, uv*)
Cohen, Y. *et al*, *J. Chem. Soc., Chem. Commun.*, 1987, 1538 (*synth, pmr, cmr*)

1,1′-Diphenylbicyclohexyl, 8CI   D-70475

(1,1′-Bicyclohexyl)-1,1′-diylbisbenzene, 9CI
[59358-71-3]

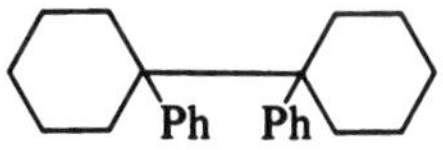

$C_{24}H_{30}$ M 318.501
Cryst. (EtOH). Mp 182-184°.

Beckhaus, H.D. *et al*, *Chem. Ber.*, 1976, **109**, 1369 (*synth*)

1,1-Diphenyl-1,3-butadiene   D-70476

1,1′-(1,3-Butadienylidene)bisbenzene, 9CI

[4165-81-5]

$$Ph_2C{=}CHCH{=}CH_2$$

$C_{16}H_{14}$ M 206.287
Cryst. d$_4^{17.5}$ 1.02. Mp 35-37°. Bp$_{0.2}$ 130°. Partially polym. on dist. Unstable in air.

Normant, H. *et al*, *Bull. Soc. Chim. Fr.*, 1956, 951 (*synth*)
Holm, T., *Acta Chem. Scand.*, 1963, **17**, 2437 (*synth, uv*)
Mitsudo, T. *et al*, *J. Org. Chem.*, 1987, **52**, 1695 (*synth, pmr, cmr, ir*)

3,6-Diphenyl-2,5-dihydropyrrolo[3,4-*c*]-pyrrole-1,4-dione   D-70477

[54660-00-3]

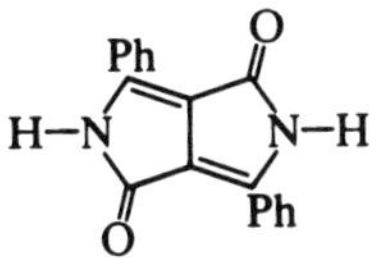

$C_{18}H_{12}N_2O_2$ M 288.305
Readily obt. from PhCN and diethyl succinate. Parent substance of a family of fluorescent dyes. Cryst. (toluene). Mp >360°. No colour mentioned, but has λ$_{max}$ 505 nm.

Potrawa, T. *et al*, *Chem. Ber.*, 1987, **120**, 1075 (*synth, uv, pmr*)

2,5-Diphenyl-1,3-dioxol-1-ium, 9CI   D-70478

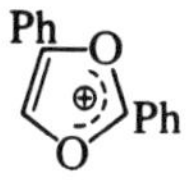

$C_{15}H_{11}O_2^{\oplus}$ M 223.251 (ion)
Trifluoromethanesulfonate: [106016-54-0].
$C_{16}H_{11}F_3O_5S$ M 372.315
V. moisture-sensitive cryst. powder. Mp 149-150° dec.

Lorenz, W. *et al*, *J. Org. Chem.*, 1987, **52**, 375 (*synth, ir, pmr, cmr*)

2,11-Diphenyldipyrrolo[1,2-*a*:2′,1′-*c*]-quinoxaline   D-70479

[111557-80-3]

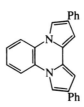

$C_{26}H_{18}N_2$ M 358.442
Yellow cryst. Mp 222°.

Kaupp, G. *et al*, *Angew. Chem., Int. Ed. Engl.*, 1987, **26**, 1280 (*synth, uv, pmr, ms*)

1,2-Diphenyl-1,2-ethanediamine, 9CI   **D-70480**

Updated Entry replacing D-07887

1,2-Diphenylethylenediamine, 8CI. 1,2-Diamino-1,2-di-phenylethane. Stilbenediamine

[5700-60-7]

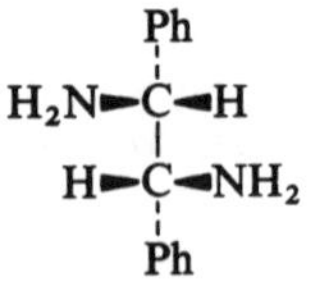

$$\begin{array}{c} \text{Ph} \\ \text{H}_2\text{N}\!-\!\!\text{C}\!-\!\!\text{H} \\ \text{H}\!-\!\!\text{C}\!-\!\!\text{NH}_2 \\ \text{Ph} \end{array}$$

(1*R*,2*R*)-form
Absolute configuration

$C_{14}H_{16}N_2$ M 212.294

(1*R*,2*R*)-form [29841-69-8]

Cryst. Mp 85-86.5° (80°). $[\alpha]_D^{22}$ +106.9° (c, 1.07 in MeOH), $[\alpha]_D^{18}$ +82.6° (c, 0.97 in Et$_2$O) (opt. pure). Has been assigned the opposite (incorrect) abs. config.

(1*S*,2*S*)-form [35132-20-8]

Mp 85.5-86°. $[\alpha]_D^{22}$ −83° (c, 0.1 in Et$_2$O).

(1*RS*,2*RS*)-form [16635-95-3]

(±)-*form*

Cryst. (ligroin). Mp 90-92° (76-77°, 83°).

B,2HCl: Cryst. Mp 251° dec.

Picrate: Cryst. Mp 220°.

N,N′-*Di-Me:* [60508-97-6].
  $C_{16}H_{20}N_2$ M 240.347
  Oil.

N-*Tetra-Me:* [94533-52-5].
  $C_{18}H_{24}N_2$ M 268.401
  Cryst. (hexane). Mp 104.5-106.5°.

N-*Tetra-Et:* [94533-54-7].
  $C_{22}H_{32}N_2$ M 324.508
  Cryst. (MeOH aq.). Mp 74.5-77°.

(1*RS*,2*SR*)-form [951-87-1]

meso-*form*

Leaflets (Et$_2$O). Mp 121°.

B,2HCl: Cryst. Mp 256° dec.

Picrate: Cryst. Mp 225°.

N,N′-*Di-Me:* [60509-62-8]. Cryst. (hexane). Mp 132-133.5°.

N-*Tetra-Me:* [94533-51-4]. Cryst. (hexane). Mp 195-197°.

N-*Tetra-Et:* [94533-53-6]. Cryst. (MeOH aq.). Mp 85-86°.

Feist, F. *et al, Ber.*, 1895, **28**, 3169 (*synth*)
Biltz, H. *et al, Justus Liebigs Ann. Chem.*, 1912, **391**, 208 (*synth*)
Okaku, N., *Bull. Chem. Soc. Jpn.*, 1967, **40**, 2326 (*synth*)
Meric, R. *et al, Tetrahedron Lett.*, 1974, 2059 (*abs config*)
Chang, C.A. *et al, Inorg. Chem.*, 1979, **18**, 1266 (*synth, resoln*)
Saigo, K. *et al, Bull. Chem. Soc. Jpn.*, 1986, **59**, 931 (*resoln, props*)
Betschart, C. *et al, Helv. Chim. Acta*, 1987, **70**, 2215 (*synth, ir, pmr, derivs*)

1,7-Diphenyl-3,5-heptanediol   **D-70481**

3,5-Dihydroxy-1,7-diphenylheptane

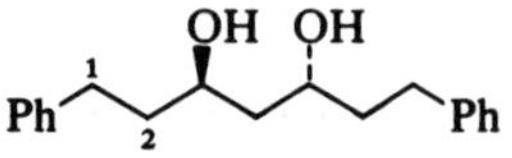

$$\text{Ph}\!-\!\!\overset{1}{\underset{2}{\frown}}\!-\!\!\overset{\text{OH}}{\frown}\!-\!\!\overset{\text{OH}}{\frown}\!-\!\!\text{Ph}$$

(3*R*,5*R*)-form

$C_{19}H_{24}O_2$ M 284.397

(3*R*,5*R*)-form [103729-38-0]

Yashabushidiol B

Constit. of male flowers of *Alnus sieboldiana*. Cryst. Mp 92-93°. $[\alpha]_D$ +7.2° (CHCl$_3$).

(3*S*,5*S*)-form [103729-39-1]

Cryst. Mp 91-92°. $[\alpha]_D$ −7.3° (CHCl$_3$).

3-Ketone: [24192-01-6]. *5-Hydroxy-1,7-diphenyl-3-*

heptanone, 9CI. **Dihydroyashabushiketol.**
  $C_{19}H_{22}O_2$ M 282.382
  Constit. of male flowers of *A. sieboldiana*. Cryst. Mp 72.5-73.0°. $[\alpha]_D^{25}$ −2.7° (c, 0.67 in MeOH), $[\alpha]_D$ +14.0° (c, 0.67 in CHCl$_3$).

1,2-Didehydro, 3-ketone: [24192-00-5]. *5-Hydroxy-1,7-diphenyl-1-hepten-3-one.* **Yashabushiketol.** Constit. of male flowers of *A. sieboldiana*. Needles (hexane). Mp 59.5-60.5°. $[\alpha]_D^{20}$ +29.0° (c, 1.05 in CHCl$_3$).

(3*RS*,5*SR*)-form [52393-53-0]

meso-*form.* **Yashabushidiol A**

Constit. of male flowers of *A. sieboldiana*. Cryst. Mp 80-81°.

Suga, T. *et al, Bull. Chem. Soc. Jpn.*, 1983, **56**, 3353 (*abs config*)
Hashimoto, T. *et al, Chem. Pharm. Bull.*, 1986, **34**, 1846.

2,3-Diphenyl-1*H*-inden-1-one, 9CI   **D-70482**

2,3-Diphenylindone, 8CI

[1801-42-9]

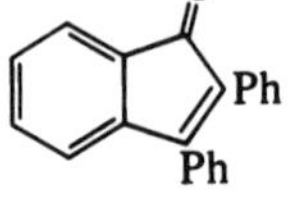

$C_{21}H_{14}O$ M 282.341

Red cryst. Mp 152-153°.

Rockett, B.W. *et al, J. Am. Chem. Soc.*, 1963, **85**, 3491 (*synth*)
Clark, T.J., *J. Chem. Educ.*, 1971, **48**, 554 (*synth*)
Ivanov, C. *et al, Rev. Roum. Chim.*, 1975, **20**, 547 (*ir, uv, pmr*)
Liebeskind, L.S. *et al, J. Org. Chem.*, 1980, **45**, 5426 (*synth*)
Jens, K.J. *et al, Chem. Ber.*, 1984, **117**, 2469 (*cryst struct*)
Butler, I.R. *et al, J. Chem. Soc., Chem. Commun.*, 1987, 439 (*synth*)
Kobayashi, T. *et al, Bull. Chem. Soc. Jpn.*, 1987, **60**, 3062 (*synth, ir, pmr*)
Watson, W.H. *et al, Acta Crystallogr., Sect. C*, 1987, **43**, 2444 (*cryst struct*)

2,3-Diphenylindole   **D-70483**

[3469-20-3]

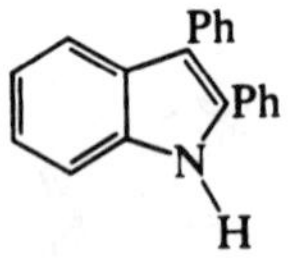

$C_{20}H_{15}N$ M 269.345

Cryst. (EtOH aq.). Mp 122-123°.

Picrate: Mp 155°.

Allen, C.F.H. *et al, J. Am. Chem. Soc.*, 1951, **73**, 5850 (*synth*)
Kamlet, M.J. *et al, J. Org. Chem.*, 1961, **26**, 220 (*uv*)
Das, H. *et al, Recl. Trav. Chim. Pays-Bas*, 1965, **84**, 965 (*synth*)
Powers, J.C. *et al, J. Org. Chem.*, 1968, **33**, 2044 (*ms*)
Schmelter, B. *et al, Acta Crystallogr., Sect. B*, 1973, **29**, 971 (*cryst struct*)
Mudry, C.A. *et al, Tetrahedron*, 1974, **30**, 2983 (*pmr, ms*)
Erra-Balsells, R., *J. Heterocycl. Chem.*, 1988, **25**, 221, 1059 (*ms, cmr*)

1-(Diphenylmethylene)-1*H*-cyclopropabenzene   **D-70484**

7-(Diphenylmethylene)bicyclo[4.1.0]hepta-1,3,5-triene, 9CI

[92012-54-9]

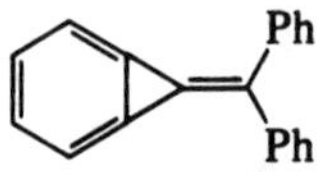

$C_{20}H_{14}$ M 254.331

Yellow needles (pet. ether). Mp 89-91°.

Halton, B. *et al*, *J. Am. Chem. Soc.*, 1986, **108**, 5949 (*synth, nmr, ir, uv*)

**1-(Diphenylmethylene)-1*H*-cyclopropa[*b*]-   D-70485
naphthalene, 9CI**
[92012-57-2]

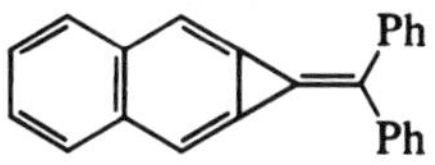

C$_{24}$H$_{16}$ M 304.390
Yellow needles (pet. ether). Mp 110-111°.

Halton, B. *et al*, *J. Am. Chem. Soc.*, 1986, **108**, 5949 (*synth, nmr, ir, uv*)

2,2-Diphenyl-4-methylene-1,3-dioxolane   D-70486

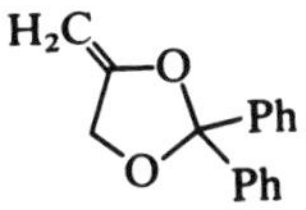

C$_{16}$H$_{14}$O$_2$ M 238.285
Mp 39.0-40°. Bp$_{0.06}$ 82°.

Hiraguri, Y. *et al*, *J. Am. Chem. Soc.*, 1987, **109**, 3779 (*synth, pmr, ir*)

**4-(Diphenylmethylene)-1(4*H*)-naphthalen-   D-70487
one, 9CI**
p-*Naphthofuchsone*
[5690-41-5]

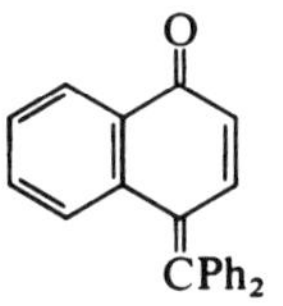

C$_{23}$H$_{16}$O   M 308.379
Cryst. (Me$_2$CO). Mp 182°.

Gomberg, M. *et al*, *J. Am. Chem. Soc.*, 1920, **42**, 1867 (*synth*)
Pisova, M. *et al*, *Collect. Czech. Chem. Commun.*, 1976, **41**, 2919 (*synth*)
Koutek, B. *et al*, *Collect. Czech. Chem. Commun.*, 1982, **47**, 1645 (*spectra*)

**7-(Diphenylmethylene)-2,3,5,6-   D-70488
tetramethylenebicyclo[2.2.1]heptane**
7,7-Diphenyl[2.2.1]hericene

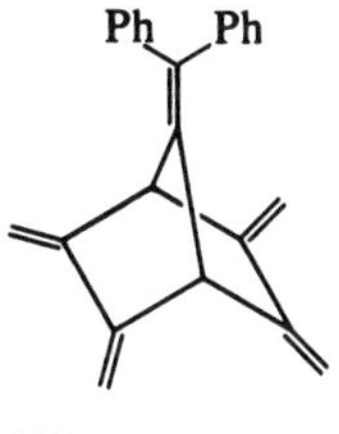

C$_{24}$H$_{20}$ M 308.422
Powder (hexane). Mp 131-133° dec.

Rubello, A. *et al*, *Helv. Chim. Acta*, 1988, **71**, 1268 (*synth, uv, ir, pmr, cmr, ms*)

1,1-Diphenyl-1,3-pentadiene   D-70489
1,1'-(1,3-Pentadienylidene)bisbenzene, 9CI

[39129-27-6]

Ph$_2$C=CHCH=CHCH$_3$

C$_{17}$H$_{16}$ M 220.313
(***E***)-*form* [15295-32-6]
Oil. Bp$_{0.1}$ 100°.

Cheminat, B., *Bull. Soc. Chim. Fr.*, 1972, 3415 (*synth, uv, ir, pmr*)
Mitsudo, T. *et al*, *J. Org. Chem.*, 1987, **52**, 1695 (*synth, pmr, cmr, ir*)

1,5-Diphenyl-1,3-pentanediol, 9CI, 8CI   D-70490
Updated Entry replacing D-60487

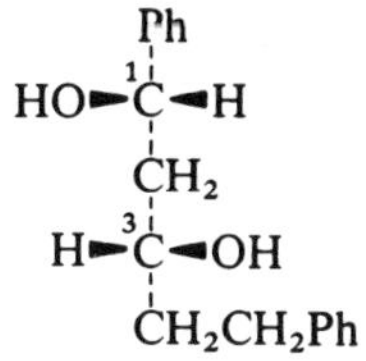

(1*S*,3*S*)-*form*

C$_{17}$H$_{20}$O$_2$ M 256.344
(***1S,3S***)-*form* [109007-67-2]
(−)-erythro-*form*
Constit. of the wood of *Flindersia laevicarpa*. Mp 89-90°. [α]$_D$ −19° (c, 3 in EtOH).
(***1R,3S***)-*form*
(+)-threo-*form*
Isol. from *F. laevicarpa* leaves. [α]$_D$ +29.8° (c, 1 in MeOH).
(***1S,3R***)-*form* [109007-68-3]
(−)-threo-*form*
Cryst. (hexane/Et$_2$O). Mp 91-92°. [α]$_D$ −23.3° (c, 1.0 in MeOH).
(***1RS,3RS***)-*form*
(±)-erythro-*form*
Needles (MeOH). Mp 86-87°.
(***1RS,3SR***)-*form*
(±)-threo-*form*
Needles (MeOH). Mp 79°.

Breen, G.J.W., *Aust. J. Chem.*, 1962, **15**, 819 (*isol, uv, ir, synth*)
Picker, K. *et al*, *Aust. J. Chem.*, 1976, **29**, 2023 (*isol, ir, uv*)
Niwa, M. *et al*, *Chem. Pharm. Bull.*, 1987, **35**, 108 (*synth, abs config, pmr, cmr*)

1,2-Diphenyl-3-(phenylimino)cyclopropene   D-70491
N-(*2,3-Diphenyl-2-cyclopropen-1-ylidene*)benzenamine,
9CI. 2,3,N-Triphenylcyclopropenoneimine. Diphenylcyclopropenone anil
[48194-29-2]

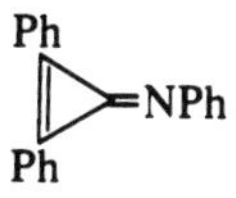

C$_{21}$H$_{15}$N   M 281.356
Cryst. Mp 103-105° dec.
B,HBF$_4$: [113880-22-1]. Cryst. Mp 241-243° dec.

Eicher, T. *et al*, *Synthesis*, 1987, 887 (*synth, uv, ir, pmr, ms*)

**(Diphenylpropadienylidene)propanedioic   D-70492
acid**

Ph$_2$C=C=C=C(COOH)$_2$

$C_{18}H_{12}O_4$ M 292.290
Di-Et ester: Ethyl 2-ethoxycarbonyl-5,5-diphenyl-2,3,4-
pentatrienoate.
$C_{22}H_{20}O_4$ M 348.398
Yellow needles. Mp 114-115°.

Browne, N.R. *et al, Aust. J. Chem.*, 1987, **40**, 1675 (*synth, uv, pmr, cmr, ms*)

2,2-Diphenylpropanoic acid, 8CI   D-70493

Updated Entry replacing D-08069
α-Methyl-α-phenylbenzeneacetic acid, 9CI. α-Phenylhy-
dratropic acid
[5558-66-7]

$$H_3CCPh_2COOH$$

$C_{15}H_{14}O_2$ M 226.274
Mp 177.5-178°. pK_a 5.77 (45% EtOH aq.), 7.2 (76% EtOH aq.).
Me ester: [50354-48-8].
$C_{16}H_{16}O_2$ M 240.301
Bp_3 149-152°.
Chloride: [40997-78-2].
$C_{15}H_{13}ClO$ M 244.720
Mp 95-96°.
Amide: [54561-74-9].
$C_{15}H_{15}NO$ M 225.290
Mp 149°.
Nitrile: 1-Cyano-1,1-diphenylethane.
$C_{15}H_{13}N$ M 207.274
Bp_2 145°.

Bateman, D.E. *et al, J. Am. Chem. Soc.*, 1927, **49**, 2914 (*synth*)
Steyard, O.W. *et al, J. Chem. Soc. (A)*, 1968, 3119 (*synth*)
Lu, A.C. *et al, J. Med. Chem.*, 1987, **30**, 273 (*synth, nitrile, pmr, ir*)

2,4-Diphenylpyrimidine   D-70494

[25095-48-1]

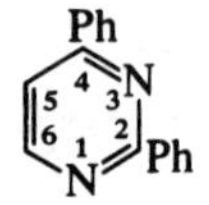

$C_{16}H_{12}N_2$ M 232.284
Needles (hexane). Mp 71°.

Wagner, R.M. *et al, Chem. Ber.*, 1971, **104**, 2975 (*synth, pmr*)

4,5-Diphenylpyrimidine   D-70495

[25922-93-4]
$C_{16}H_{12}N_2$ M 232.284
Mp 130-131°.
1-Oxide: [114769-49-2].
$C_{16}H_{12}N_2O$ M 248.284
Mp 128°.

Padwa, A. *et al, J. Am. Chem. Soc.*, 1973, **95**, 1954 (*synth, ir, pmr*)
Yamanaka, H. *et al, Chem. Pharm. Bull.*, 1987, **35**, 3119 (*synth, pmr, oxide*)

4,6-Diphenylpyrimidine   D-70496

[3977-48-8]
$C_{16}H_{12}N_2$ M 232.284
Cryst. (pet. ether). Mp 102-103°.
B,HCl: Mp 196°.
Picrate: Mp 168-169°.

N-Oxide: [60545-98-4].
$C_{16}H_{12}N_2O$ M 248.284
Cryst. (pet. ether). Mp 107-108°.

Bredereck, H. *et al, Chem. Ber.*, 1957, **90**, 942 (*synth*)
Roeterdink, F. *et al, J. Chem. Soc., Perkin Trans. 1*, 1976, 1202 (*synth, pmr, oxide*)

3,4-Diphenyl-1,2,5-selenadiazole   D-70497

[19768-00-4]

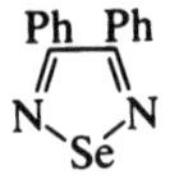

$C_{14}H_{10}N_2Se$ M 285.206
Cryst. (EtOH). Mp 145°.

Bertini, V., *Gazz. Chim. Ital.*, 1967, **97**, 1870 (*synth*)
Neidlein, R. *et al, Helv. Chim. Acta*, 1987, **70**, 1076 (*synth, uv, ir, pmr, cmr, ms*)

2,5-Diphenyl-1,2,4,5-tetraazabicyclo[2.2.1]heptane, 9CI   D-70498

[19437-32-2]

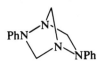

$C_{15}H_{16}N_4$ M 252.318
Principal prod. of condensation of phenylhydrazine and formaldehyde. Cryst. (EtOH). Mp 186°.

Sim, G.A., *Acta Crystallogr., Sect. C*, 1987, **43**, 2178 (*cryst struct, bibl*)

N,N′-Diphenylurea, 9CI   D-70499

Updated Entry replacing D-08221
Carbanilide
[102-07-8]

$$PhNHCONHPh$$

$C_{13}H_{12}N_2O$ M 212.251
Prisms. Mp 239-240°. Bp 260°.
▷FD9800000.
Mono-N-Ac:
$C_{15}H_{14}N_2O_2$ M 254.288
Mp 106°.
N-Me: [612-01-1].
$C_{14}H_{14}N_2O$ M 226.277
Cryst. (EtOH or xylene). Mp 106°. Bp 203-205° dec.
N,N′-Di-Me: [611-92-7]. N,N′-*Dimethyl*-N,N′-*diphenylurea. Centralite II.*
$C_{15}H_{16}N_2O$ M 240.304
Explosion regulator. Mp 122°.
▷FE0600000.
N-Et, N′-Me: [4474-03-7]. N-*Ethyl*-N′-*methyl*-N,N′-*diphenylurea. Centralite III.*
$C_{16}H_{18}N_2O$ M 254.331
Explosion regulator. Mp 58-59°.
N,N′-Di-Et: [85-98-3]. N,N′-*Diethyl*-N,N′-*diphenylurea. Centralite I.*
$C_{17}H_{20}N_2O$ M 268.358
Explosion regulator. Mp 72-73°.
▷FE0350000.

Org. Synth., Coll. Vol., **1**, 453 (*synth*)
Baker, J.W. *et al, J. Chem. Soc.*, 1957, 4652 (*synth*)

Novacek, A., *Collect. Czech. Chem. Commun.*, 1967, **32**, 1712 (*synth*)
Knobler, Y. *et al*, *Isr. J. Chem.*, 1971, **9**, 165 (*synth*)
Ayyangar, N.R. *et al*, *Chem. Ind.* (*London*), 1988, 599 (*synth, bibl, derivs*)

1,1-Di-1-piperidinylethene　　　　　D-70500
1,1'-Ethenylidenebispiperidine, 9CI
[42259-31-4]

$C_{12}H_{22}N_2$　　M 194.319
Basic electron-rich alkene with synthetic uses. Liq. Bp_{12} 123-124°. n_D^{25} 1.5056. Darkens in air. Immisc. H_2O but readily dissolves with hydrol.

Böhme, H. *et al*, *Chem. Ber.*, 1961, **94**, 3109 (*synth, ir*)
Baganz, H. *et al*, *Chem. Ber.*, 1962, **95**, 2095 (*synth*)
Gandhi, S.S. *et al*, *Can. J. Chem.*, 1987, **65**, 2717 (*bibl*)

Dipyrido[1,2-a:1',2'-c]imidazol-10-ium(1+), 9CI　　　　　　D-70501
Updated Entry replacing D-08274
Imidazo[a,c]*dipyridinium*
[245-75-0]

$C_{11}H_9N_2^{\oplus}$　　M 169.206 (ion)
Bromide:
　$C_{11}H_9BrN_2$　　M 249.110
　Mp 150-152°.
Iodide:
　$C_{11}H_9IN_2$　　M 296.110
　Mp 252-253°.
Picrate: Mp 217-218°.
Perchlorate:
　$C_{11}H_9ClN_2O_4$　　M 268.656
　Green-yellow microplates (MeOH). Mp 200-201.5°.

Yoshida, H. *et al*, *Chem. Pharm. Bull.*, 1963, **11**, 694.
Jones, T.R. *et al*, *J. Chem. Soc., Perkin Trans. 1*, 1987, 2585 (*synth, pmr*)

Di-2-pyridyl ketone　　　　　D-70502
Di-2-pyridinylmethanone, 9CI. *2-Pyridyl ketone*, 8CI
[19437-26-4]

$C_{11}H_8N_2O$　　M 184.197
Prisms (pet. ether). Mp 54-55°. $Bp_{0.4}$ 132-135°.
Picrate: Yellow cryst. (EtOAc). Mp 181-182°.
Semicarbazone: Plates or needles (EtOH). Mp 220-221° (217-218°).
Oxime: [1562-95-4].
　$C_{11}H_9N_3O$　　M 199.212

Reagent for est. of metal ions (Pd(II), Fe(II), Co(III) etc.). Needles (MeOH aq. or EtOAc). Mp 141-142.5°.
Ethylene ketal: [42772-86-1]. *2,2'-(1,3-Dioxan-2-ylidene)bispyridine*, 9CI.
$C_{13}H_{12}N_2O_2$　　M 228.250
Needles (pet. ether). Mp 164-166°.

Wibaut, J.P. *et al*, *Recl. Trav. Chim. Pays-Bas*, 1951, **70**, 989, 1054 (*synth, derivs*)
Leete, E. *et al*, *Can. J. Chem.*, 1952, **30**, 563 (*synth, derivs*)
Holland, W.J. *et al*, *J. Anal. Chem.*, 1968, **40**, 433 (*oxime, use*)
Newkome, G.R. *et al*, *Tetrahedron Lett.*, 1973, 1599 (*acetal*)

Di-3-pyridyl ketone　　　　　D-70503
Di-3-pyridinylmethanone. 3-Pyridyl ketone
[35779-35-2]
$C_{11}H_8N_2O$　　M 184.197
Cryst. (pet. ether). Mp 117.6-118.8°. Dec. in light.
Dipicrate: Yellow cryst. (H_2O). Mp 208° dec.
Phenylhydrazone: Pale-yellow cryst. (EtOH/Py). Mp 170-171°.
Oxime:
　$C_{11}H_9N_3O$　　M 199.212
　Cryst. (EtOAc). Mp 160-165°.

Wibaut, J.P. *et al*, *Recl. Trav. Chim. Pays-Bas*, 1951, **70**, 1054 (*synth, deriv*)
Niemers, E. *et al*, *Synthesis*, 1976, 593 (*deriv*)

Di-4-pyridyl ketone　　　　　D-70504
Di-4-pyridinylmethanone, 9CI. *4-Pyridyl ketone*, 8CI
[6918-15-6]
$C_{11}H_8N_2O$　　M 184.197
Needles (pet. ether/Py). Mp 141-141.5° (136.2-137.5°).
B,MeI: Brown-red cryst. (butanone). Mp 175-177° dec.
B,2MeI: Orange-red cryst. (butanone). Mp 246-248°.
Phenylhydrazone: Yellow needles (EtOH/Py). Mp 250.3-250.8°.
Oxime: [58088-22-5].
　$C_{11}H_9N_3O$　　M 199.212
　Mp 189-191°.

Wibaut, J.P. *et al*, *Recl. Trav. Chim. Pays-Bas*, 1955, **74**, 1003 (*synth, deriv*)
Minn, F.L. *et al*, *J. Am. Chem. Soc.*, 1970, **92**, 3600 (*synth, deriv, uv, props*)

Disalicylaldehyde　　　　　D-70505
Updated Entry replacing D-08303
6,12-Epoxy-6H,12H-dibenzo[b,f][1,5]dioxocin, 9CI.
3,4:7,8-Dibenz-2,6,9-bisdioxan
[252-72-2]

$C_{14}H_{10}O_3$　　M 226.231
Anhydro dimer of 2-Hydroxybenzaldehyde, H-01251.
Needles (EtOH). Mp 130°.

Adams, R. *et al*, *J. Am. Chem. Soc.*, 1922, **44**, 1126 (*synth, bibl*)

Jones, P.R. *et al, J. Org. Chem.*, 1981, **46**, 194 (*uv, pmr, cmr, ms, struct, bibl*)
Kulkarni, V.S. *et al, Synth. Commun.*, 1986, **16**, 191 (*synth*)

Disnogamycin   **D-70506**
Nogamycin
[64267-46-5]

$C_{37}H_{47}NO_{14}$ M 729.777
Anthracycline antibiotic. Semisynthetic. Antitumour agent. Cryst. (MeOH/EtOAc). Mp 210-215° dec. [α]$_D$ +273° (c, 0.923 in CHCl$_3$).
▷KD1146500.

Wiley, P.F. *et al, J. Org. Chem.*, 1979, **44**, 4030 (*synth*)
Wiley, P.F. *et al, J. Med. Chem.*, 1982, **25**, 560 (*props*)

1,3-Dithiane-2-selone   **D-70507**

$C_4H_6S_2Se$ M 197.171
Mod. stable red needles (CHCl$_3$/2-propanol/pet. ether at −78°). Mp 91-92°.

Copeland, C.M. *et al, Aust. J. Chem.*, 1988, **41**, 549 (*synth, uv, pmr, cmr*)

Dithiatopazine   **D-70508**
Dodecahydro-6H,13H-5a,12a-epidithiopyrano[3,2-b]-pyrano[2′,3′:6,7]oxepino[2,3-f]oxepin

$C_{16}H_{24}O_4S_2$ M 344.483
First stable 1,2-dithietane. Yellow-orange topazlike cryst. (hexane). Mp 134-135°.

Nicolaou, K.C. *et al, J. Am. Chem. Soc.*, 1987, **109**, 3801 (*synth, uv, ir, pmr, cmr, ms, cryst struct*)

Dithieno[2,3-*b*:2′,3′-*d*]pyridine   **D-70509**

$C_9H_5NS_2$ M 191.265
Mp 92-93°.

Gronowitz, S. *et al, Chem. Scr.*, 1988, **28**, 281 (*synth, pmr*)

Dithieno[2,3-*b*:3′,2′-*d*]pyridine   **D-70510**

$C_9H_5NS_2$ M 191.265
Mp 115-116° (110-111°).

Yang, Y. *et al, Chem. Scr.*, 1988, **28**, 275 (*synth, ir, pmr*)

Dithieno[2,3-*b*:3′,4′-*d*]pyridine   **D-70511**

$C_9H_5NS_2$ M 191.265
Mp 90-92°.

Yang, Y. *et al, Chem. Scr.*, 1988, **28**, 275 (*synth, ir, pmr*)

Dithieno[3,2-*b*:2′,3′-*d*]pyridine   **D-70512**

$C_9H_5NS_2$ M 191.265
Mp 78-80° (75.5-77°).

Gronowitz, S. *et al, Chem. Scr.*, 1988, **28**, 281 (*synth, pmr*)

Dithieno[3,2-*b*:3′,4′-*d*]pyridine   **D-70513**

$C_9H_5NS_2$ M 191.265
Mp 70-72°.

Yang, Y. *et al, Chem. Scr.*, 1988, **28**, 275 (*synth, ir, pmr*)

Dithieno[3,4-*b*:2′,3′-*d*]pyridine   **D-70514**

$C_9H_5NS_2$ M 191.265
Mp 80-81.5°.

Gronowitz, S. *et al, Chem. Scr.*, 1988, **28**, 281 (*synth, pmr*)

Dithieno[3,4-*b*:3′2′-*d*]pyridine   **D-70515**

$C_9H_5NS_2$ M 191.265
Mp 105-107°.

Yang, Y. *et al, Chem. Scr.*, 1988, **28**, 275 (*synth, ir, pmr*)

Dithieno[3,4-*b*:3',4'-*d*]pyridine   D-70516

C₉H₅NS₂   M 191.265
Mp 60-62°.

Yang, Y. *et al, Chem. Scr.*, 1988, **28**, 275 (*synth, ir, pmr*)

1,2-Dithiepane, 9CI, 8CI   D-70517

Updated Entry replacing D-08400
1,2-Dithiacycloheptane
[6008-51-1]

C₅H₁₀S₂   M 134.254
Oil. Bp₅ 57-60°.
1,1-Dioxide: [18321-17-0].
    C₅H₁₀O₂S₂   M 166.253
    Solid (Et₂O). Mp ca. 25°, Mp 50-51.5°. Slowly dec. at
    −20°.
1,1,2,2-Tetraoxide: [18321-20-5].
    C₅H₁₀O₄S₂   M 198.252
    Cryst. (CH₂Cl₂/Et₂O). Mp 159-160°.

Field, L. *et al, J. Org. Chem.*, 1969, **34**, 36 (*synth, pmr, ir, uv*)
Macke, J.D. *et al, J. Org. Chem.*, 1988, **53**, 396 (*synth, dioxide, pmr, cmr*)

1,4-Dithiino[2,3-*b*:5,6-*c'*]dipyridine, 9CI   D-70518
9,10-Dithia-1,6-diazaanthracene
[113372-47-7]

C₁₀H₆N₂S₂   M 218.291
Mp 120-121°.

Lindsay, C.M. *et al, J. Heterocycl. Chem.*, 1987, **24**, 1357
    (*synth, pmr, cmr, ms*)

[1,4]Dithiino[2,3-*b*:6,5-*b'*]dipyridine, 9CI   D-70519
9,10-Dithia-1,8-diazaanthracene
[113372-44-4]

C₁₀H₆N₂S₂   M 218.291
Off-white cryst. (CHCl₃). Mp 184-186°.

Lindsay, C.M. *et al, J. Heterocycl. Chem.*, 1987, **24**, 1357
    (*synth, pmr, cmr, ms*)

1,4-Dithiino[2,3-*b*:6,5-*c'*]dipyridine, 9CI   D-70520
9,10-Dithia-1,7-diazaanthracene
[113372-48-8]

C₁₀H₆N₂S₂   M 218.291
Mp 119-120°.

Lindsay, C.M. *et al, J. Heterocycl. Chem.*, 1987, **24**, 1357
    (*synth, pmr, cmr, ms*)

[1,4]Dithiino[2,3-*b*]-1,4-dithiin, 9CI   D-70521

Updated Entry replacing D-08408
1,4,5,8-Tetrathianaphthalene
[255-55-0]

C₆H₄S₄   M 204.338
Pale yellow needles. Mp 126-127°.

Mizuno, M. *et al, J. Org. Chem.*, 1976, **41**, 1484.
Varma, K.S. *et al, J. Heterocycl. Chem.*, 1988, **25**, 783 (*synth, ms, pmr, cryst struct*)

2,2'-Dithiobis-1*H*-isoindole-1,3(2*H*)-dione,   D-70522
9CI
Dithiobisphthalimide
[7764-30-9]

C₁₆H₈N₂O₄S₂   M 356.370
Sulfur transfer reagent. Heavy prisms (CHCl₃/MeOH).
    Mp 225°.

Kalnins, M.V., *Can. J. Chem.*, 1966, **44**, 2111 (*synth, ir*)
Harpp, D.N. *et al, J. Am. Chem. Soc.*, 1978, **100**, 1222 (*synth*)
Huang, N.-Z. *et al, J. Org. Chem.*, 1987, **52**, 169 (*synth, use*)

1,1'-Dithiobispiperidine, 9CI   D-70523
Di-1-piperidinyl disulfide
[10220-20-9]

C₁₀H₂₀N₂S₂   M 232.401
Antioxidant for mineral oils, polymer additive.
▷TM7900000.

Hatch, C.E., *J. Org. Chem.*, 1978, **43**, 3953 (*synth*)
Furukawa, M. *et al, Chem. Lett.*, 1982, 2007 (*props*)
Kivekäs, R. *et al, Acta Chem. Scand., Ser. B*, 1986, **41**, 213
    (*cryst struct*)

1,1'-Dithiobis-2,5-pyrrolidinedione, 9CI   D-70524

N,N'-Dithiobissuccinimide

[34251-41-7]

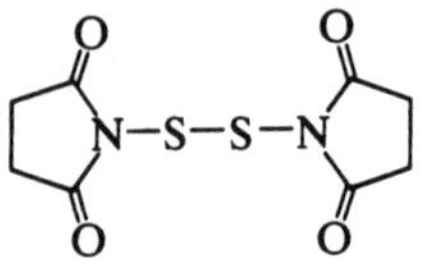

C$_8$H$_8$N$_2$O$_4$S$_2$ M 260.282

Sulfur transfer reagent. Cryst. (CH$_2$Cl$_2$/hexane). Mp 187-190° (163.5-165.0°).

Haque, M.-U. et al, J. Chem. Soc., Perkin Trans. 2, 1974, 1459 (synth, cryst struct)

Harpp, D.N. et al, J. Am. Chem. Soc., 1978, 100, 1222 (synth, ir, pmr, ms)

Huang, N.-Z. et al, J. Org. Chem., 1987, 52, 169 (synth, use)

1,3-Dithiolane-2-selone   D-70525

[57560-02-8]

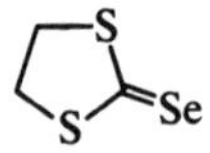

C$_3$H$_4$S$_2$Se   M 183.145

Red needles (Et$_2$O/hexane). Mp 44.5-45°.

Lakshmikantham, M.V. et al, J. Org. Chem., 1976, 41 879.

Copeland, C.M. et al, Aust. J. Chem., 1988, 41, 549 (synth, uv, cmr)

1,5-Dithionane, 9CI   D-70526

1,5-Dithiacyclononane

[6573-48-4]

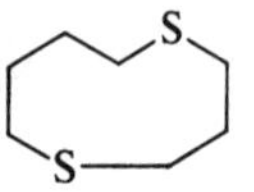

C$_7$H$_{14}$S$_2$ M 162.308

Mp 58.5-60°. Bp$_{0.6}$ 78-80°.

1-Oxide: [67463-84-7].
    C$_7$H$_{14}$OS$_2$ M 178.307
    Mp 75.0-76.5°.

Musker, W.K. et al, J. Am. Chem. Soc., 1978, 100, 6416 (synth, pmr, cmr)

1,4-Dithioniabicyclo[2.2.0]hexane, 9CI   D-70527

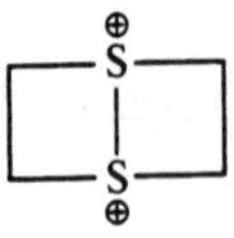

C$_4$H$_8$S$_2$$^{\oplus\oplus}$ M 120.227 (ion)

Bis(trifluoromethanesulfonate): [113504-10-2].
    C$_6$H$_8$F$_6$O$_6$S$_4$ M 418.356
    Mp 135° dec.

Fujihara, H. et al, J. Chem. Soc., Chem. Commun., 1987, 930 (synth, pmr)

Divaronic acid   D-70528

[103538-05-2]

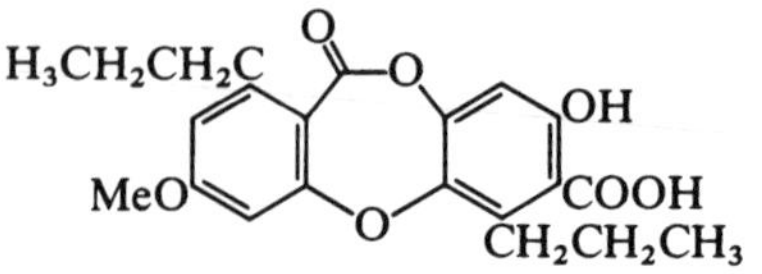

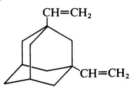

C$_{21}$H$_{22}$O$_7$ M 386.401

Constit. of Paraparmelia mongaensis. Prisms. Mp 110°, Mp 160-162° (double Mp).

Culberson, C.F. et al, Bryologist, 1985, 88, 380 (isol)

Elix, J.A. et al, Aust. J. Chem., 1987, 40, 1451 (synth)

1,3-Divinyladamantane   D-70529

1,3-Diethenyltricyclo[3.3.1.1^{3,7}]decane, 9CI

[103724-51-2]

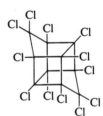

C$_{14}$H$_{20}$ M 188.312

Oil.

Majerski, Z. et al, Synth. Commun., 1986, 16, 51 (synth, pmr, cmr, ms)

Dodecachloropentacyclo[5.3.0.0^{2,6}.0^{3,9}- .0^{5,8}]decane   D-70530

1,1a,2,2,3,3a,4,5,5,5a,5b,6-Dodecachlorocyclooctahydro-1,3,4-metheno-1H-cyclobuta[cd]pentalene, 9CI. Perchloropentacyclo[5.3.0.0^{2,6}.0^{3,9}.0^{5,8}]decane. Dodecachlorodihomocubane. Mirex. Perchlordecone. Dechlorane

[2385-85-5]

C$_{10}$Cl$_{12}$ M 545.546

Superseded organochlorine insecticide used against ants in particular. Common lipophilic environmental pollutant found in animal fats. Used in flame-retardant coatings. Cryst. solid. Mp 485°.

▷PC8225000.

U.S.P., 2 996 553, (1961); CA, 56, 1363f (synth)

Verschueren, K., Handbook of Environmental Data on Organic Compounds, 2nd Ed., Van Nostrand Reinhold, 1983, 878.

Haley, T.J., Dangerous Prop. Ind. Mater. Rep., 1986, 6, 2 (rev)

1,3,5,7,9,11-Dodecahexaene   D-70531

[2423-92-9]

H$_2$C=CHCH=CHCH=CHCH=CHCH=CH-CH=CH$_2$

C$_{12}$H$_{14}$ M 158.243

Sondheimer, F. et al, J. Am. Chem. Soc., 1961, 83, 1675 (synth, uv)

Christensen, R.L. et al, J. Am. Chem. Soc., 1980, 102, 1777 (isom)

Block, E. *et al*, *J. Am. Chem. Soc.*, 1986, **108**, 4568 (*synth, ir, uv, cmr, pmr*)

Dodecahydrocarbazole D-70532

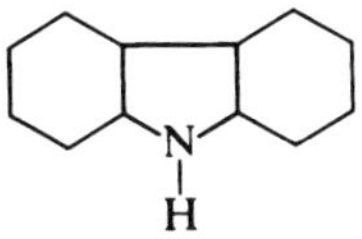

$C_{12}H_{21}N$ M 179.305

Several preparations reported but no information on configs. of products. 4 racemates and 2 *meso*-forms possible.

Anderson-McKay, J.E. *et al*, *Aust. J. Chem.*, 1988, **41**, 1013 (*bibl*)

Dodecahydro-3*a*,6,6,9*a*-tetramethyl-naphtho[2,1-*b*]furan, 9CI D-70533

Updated Entry replacing D-60515
1,1,4a,6-Tetramethyl-5-ethyl-6,5-oxidodecahydron-aphthalene. Bicyclofarnesyl epoxide
[6790-58-5]

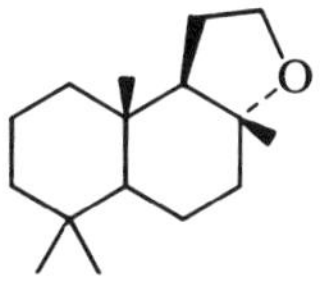

$C_{16}H_{28}O$ M 236.397

Volatile constit. of Ambergris tincture. Important amber perfumery ingredient.

(*3aα,5aβ,6α,9aα,9bβ*)-*form* [6790-58-5]
Ambroxide. Ambrox. n-*Epoxide*
Constit. of ambergris. Cryst. (pet. ether). Mp 75-76°.

(*3aα,5aα,9aβ,9bα*)-*form* [68365-88-8]
Isoambrox. Isoepoxide
Mp 60°.

Stoll, M. *et al*, *Helv. Chim. Acta*, 1950, **33**, 1251; 1951, **34**, 1664 (*synth, config*)
Hinder, M. *et al*, *Helv. Chim. Acta*, 1953, **36**, 1995 (*synth, ir*)
Lucius, G. *et al*, *Arch. Pharm. (Weinheim, Ger.)*, 1958, **291**, 57; *CA*, **52**, 14635e (*config*)
Torii, S. *et al*, *J. Org. Chem.*, 1978, **43**, 4600 (*synth*)
Ohloff, G. *et al*, *Helv. Chim. Acta*, 1985, **68**, 2022.
Decorzaut, R. *et al*, *Tetrahedron*, 1987, **43**, 1871 (*synth*)
Gonzalez-Sierra, M. *et al*, *Heterocycles*, 1987, **26**, 2801 (*synth*)
Koyama, H. *et al*, *Tetrahedron Lett.*, 1987, **28**, 2863 (*synth*)

2,6-Dodecanedione D-70534

[112497-53-7]

$$H_3CCO(CH_2)_3CO(CH_2)_5CH_3$$

$C_{12}H_{22}O_2$ M 198.305
Oil.

Ono, N. *et al*, *Synthesis*, 1987, 532 (*synth, ir, pmr*)

4,7,10,15,18,21,24,27,30,33,36,39-Dodecaoxatricyclo[11.9.9.9^{2,12}]-tetraconta-1,12-diene, 9CI D-70535

[110512-27-1]

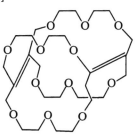

$C_{28}H_{48}O_{12}$ M 576.680

Three-dimensional analogue of 18-Crown-6, C-10319 Behaves as a hydrophilic cylinder. Cryst. solid. Mp 118-123°.

Walba, D.M. *et al*, *J. Am. Chem. Soc.*, 1987, **109**, 7081 (*synth, pmr, cmr, ir, cryst struct, props*)

3-(1,3,5,7,9-Dodecapentaenyloxy)-1,2-pro-panediol, 9CI D-70536

Updated Entry replacing D-40428
Fecapentaene-12
[84000-59-9]

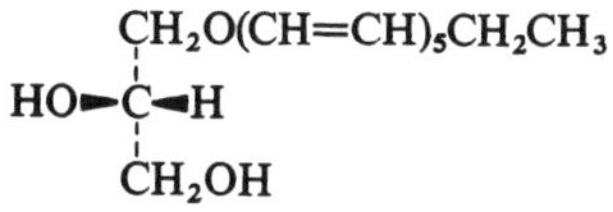

$$\begin{array}{c} CH_2O(CH=CH)_5CH_2CH_3 \\ HO-C-H \\ CH_2OH \end{array}$$

$C_{15}H_{22}O_3$ M 250.337

Natural fecapentaene-12 is a mixt. of geom. isomers in which the all *E*-form predominates.

(*R,all-E*)-*form*
Mp 190.5-191° (sealed tube under Ar). $[\alpha]_D^{20}$ +9° (c, 0.5 in DMSO).

(*S,all-E*)-*form* [83248-46-8]
1-O-(1,3,5,7,9-dodecapentaenyl)-sn-glycerol
Mutagen with suspected implication in human colonic cancer. Pale-yellow plates. Mp 191-192° (sealed tube under Ar). $[\alpha]_D^{20}$ −7° (c, 0.5 in DMSO). This is the main biological isomer. It is labile and unstable and many biol. studies may have been carried out on isomeric mixts. Mutagenicity is however independent of chirality.

▷Highly mutagenic

((±)-*all-trans*)-*form*
Mp 194.5-197.5° (sealed tube under Ar).

Hirai, N. *et al*, *J. Am. Chem. Soc.*, 1982, **104**, 6149 (*struct, ms, uv, pmr*)
Bruce, W.R. *et al*, *Naturwissenschaften*, 1982, **69**, 557.
Gunatilaka, A.A.L., *et al*, *Biochemistry*, 1983, **22**, 241.
Guntilaka, A.A.L. *et al*, *Tetrahedron Lett.*, 1983, **24**, 5457 (*synth*)
de Wit, P.P. *et al*, *Recl. Trav. Chim. Pays-Bas*, 1985, **104**, 307 (*synth*)
Pfaendler, H.R. *et al*, *J. Am. Chem. Soc.*, 1986, **108**, 1338 (*synth, pmr*)
Baptista, J. *et al*, *Angew. Chem., Int. Ed. Engl.*, 1987, **26**, 1186 (*config*)
Pfaendler, R. *et al*, *Justus Liebigs Ann. Chem.*, 1988, 499 (*synth, ir, pmr, cmr, ms*)

9-Dodecen-1-ol, 9CI D-70537

Updated Entry replacing D-08585

$$H_3CCH_2CH=CH(CH_2)_7CH_2OH$$

$C_{12}H_{24}O$ M 184.321

(E)-form [35237-62-8]
Pheromone for various insect spp.
Ac: [35148-19-7].
  $C_{14}H_{26}O_2$ M 226.358
  Pheromone for various insect spp.

(Z)-form [35148-18-6]
Pheromone for varous insect spp. $Bp_{0.01}$ 83°.
Ac: [16974-11-1]. Pheromone for various insect spp. $Bp_{0.1}$ 101-102°.
Formyl: [56218-77-0].
  $C_{13}H_{24}O_2$ M 212.331
  Pheromone for various insect spp. $Bp_{0.05}$ 78-80°.
Propanoyl: [56218-84-9].
  $C_{15}H_{28}O_2$ M 240.385
  Pheromone for various insect spp. $Bp_{0.05}$ 96-97°.

Kovaleva, A.S. *et al, Zh. Org. Khim.,* 1974, **10**, 696.
Bestmann, H.J. *et al, Chem. Ber.,* 1975, **108**, 3582; 1976, **109**, 1694 (synth)
Barabas, A. *et al, Tetrahedron,* 1978, **34**, 2191 (nmr)
Michelot, D., *Synthesis,* 1983, 130 (synth)
Körblová, E. *et al, Collect. Czech. Chem. Commun.,* 1985, **50**, 2284 (synth, ir)
Szurdoki, F. *et al, Org. Prep. Proced. Int.,* 1988, **20**, 475 (synth, pmr, cmr, bibl)

5-Dodecen-7-yne   D-70538

$$H_3C(CH_2)_3CH{=}CHC{\equiv}C(CH_2)_3CH_3$$

$C_{12}H_{20}$ M 164.290

(E)-form [16336-82-6]
Liq. $Bp_{0.3}$ 63-64°.

Zweifel, G. *et al, Synthesis,* 1977, 52 (synth)
Hoshi, M. *et al, Bull. Chem. Soc. Jpn.,* 1983, **56**, 2855 (synth, pmr, ir)
Stille, J.K. *et al, J. Am. Chem. Soc.,* 1987, **109**, 2138 (synth, pmr, cmr, ir)

5-Dodecyn-1-ol   D-70539

[88109-70-0]

$$H_3C(CH_2)_5C{\equiv}C(CH_2)_3CH_2OH$$

$C_{12}H_{22}O$ M 182.305
Liq. $Bp_{0.4}$ 99-102°.

Stille, J.K. *et al, J. Am. Chem. Soc.,* 1987, **109**, 2138 (synth, pmr, cmr, ir)

11-Dodecyn-1-ol   D-70540

[18202-10-3]

$$HC{\equiv}C(CH_2)_9CH_2OH$$

$C_{12}H_{22}O$ M 182.305
Waxy solid. Mp 28-30°. $Bp_{0.2}$ 50-60° subl., $Bp_{0.05}$ 83-86°.

Ac:
  $C_{14}H_{24}O_2$ M 224.342
  Oil.

Christie, W.W. *et al, Chem. Phys. Lipids,* 1967, **1**, 407 (synth)
Fyles, T.M. *et al, Can. J. Chem.,* 1977, **55**, 4135 (synth, ir, pmr)
Stille, J.K. *et al, J. Am. Chem. Soc.,* 1987, **109**, 2138 (synth, pmr, cmr, ir)

3,7-Dolabelladiene-6,12-diol   D-70541
6,12-Dihydroxy-3,7-dolabelladiene

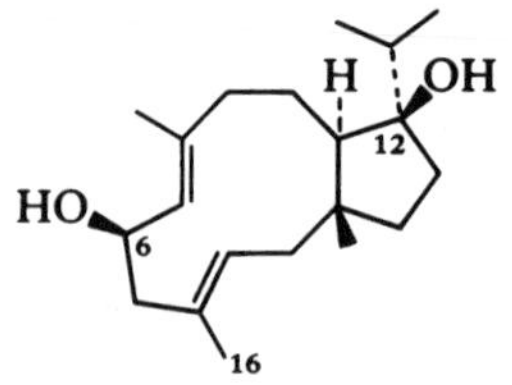

$C_{20}H_{34}O_2$ M 306.487

(1R,3E,6R,7E,11R,12R)-form [115404-62-1]
Constit. of *Odontoschisma denudatum.* Cryst. Mp 139-140°. $[\alpha]_D$ +56.2° (c, 1.3 in $CHCl_3$).
6-Ac: [90375-51-2]. **Acetoxyodontoschismenol.**
  $C_{22}H_{36}O_3$ M 348.525
  Constit. of *O. denudatum.* Cryst. Mp 75-76°. $[\alpha]_D$ +68.2° (c, 0.9 in $CHCl_3$).
16-Hydroxy, 6-Ac: [110505-66-3]. *6-Acetoxy-3,7-dolabelladiene-12,16-diol.*
  $C_{22}H_{36}O_4$ M 364.524
  Constit. of *O. denudatum.* Cryst. Mp 103.5-104.5°. $[\alpha]_D$ +36° (c, 1.4 in $CHCl_3$).
16-Hydroxy, 6,16-Di-Ac: [115334-52-6]. *6,16-Diacetoxy-3,7-dolabelladien-12-ol.*
  $C_{24}H_{38}O_5$ M 406.561
  Constit. of *O. denudatum.* Cryst. Mp 76-77°. $[\alpha]_D$ +2.9° (c, 1.1 in $CHCl_3$).
16-Oxo, 3S,4S-epoxide, 6-Ac: *6-Acetoxy-3,4-epoxy-12-hydroxy-7-dolabellen-16-al.*
  $C_{22}H_{34}O_5$ M 378.508
  Constit. of *O. denudatum.* Oil. $[\alpha]_D$ −23.3° (c, 2.2 in $CHCl_3$).

Matsuo, A. *et al, Phytochemistry,* 1988, **27**, 1153.

Dolastatin 10   D-70542
[110417-88-4]

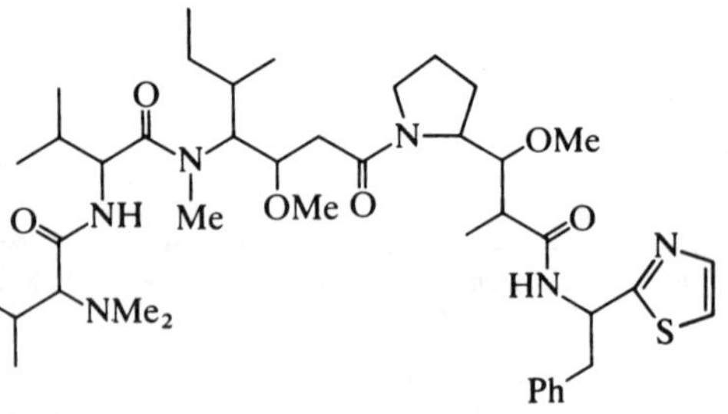

$C_{42}H_{68}N_6O_6S$ M 785.096
Peptide antibiotic. Isol. from *Dolabella auricularia.* Antineoplastic agent. Amorph. powder (MeOH/CH_2Cl_2). Mp 107-112°. $[\alpha]_D^{29}$ −68° (c, 0.01 in MeOH).

Pettit, G.R. *et al, J. Am. Chem. Soc.,* 1987, **109**, 6883 (isol, struct)

1(15)-Dolastene-4,8,9,14-tetrol   D-70543

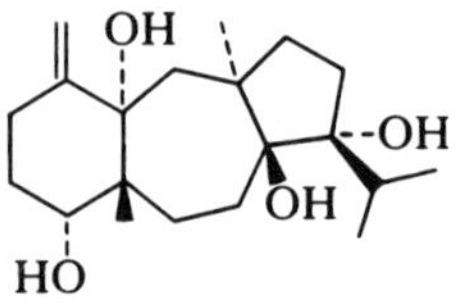

$C_{20}H_{34}O_4$ M 338.486
Constit. of *Dictyota cervicornis.* Cryst. Mp 196-199°. $[\alpha]_D$ −80.7° (c, 1 in $CHCl_3$).

Kelecom, A. *et al, Phytochemistry,* 1988, **27**, 2907.

Dolichosterone   D-70544
Updated Entry replacing D-20482
2α,3α,22R,23R-Tetrahydroxy-5α-ergost-24(28)-en-6-one

$C_{28}H_{46}O_5$ M 462.668
Constit. of *Dolichos lablab* seed. Cryst. Mp 233-237°.
6-Deoxo: ***6-Deoxodolichosterone***.
$C_{28}H_{48}O_4$ M 448.685
Constit. of *Phaseolus vulgaris*.

Baba, J. *et al*, *Agric. Biol. Chem.*, 1983, **47**, 659.
Yukota, T. *et al*, *Agric. Biol. Chem.*, 1983, **47**, 2149.
Adam, G. *et al*, *Phytochemistry*, 1986, **25**, 1787 (*rev*)
Singh, H. *et al*, *Indian J. Chem., Sect. B*, 1986, **25**, 989 (*rev*)

α-Doradexanthin   D-70545
[29125-77-7]

HO — OH
(Absolute configuration)

$C_{40}H_{54}O_3$ M 582.865
Carotenoid from the goldfish *Carassius auratus*.
3'-Epimer: [68474-14-6]. ***Fritschiellaxanthin***.
$C_{40}H_{54}O_3$ M 582.865
Isol. from the green alga *Fritschiella tuberosa*.

Buchecker, R. *et al*, *Helv. Chim. Acta*, 1978, **61**, 1962.

Dracaenone   D-70546

$C_{16}H_{14}O_2$ M 238.285
Parent compd. unknown.
10-Hydroxy, 11-methoxy: [113459-56-6]. ***10-Hydroxy-11-methoxydracaenone***.
$C_{17}H_{16}O_4$ M 284.311
Constit. of *Dracaena loureiri*. Cryst. (CHCl₃/pet. ether). Mp 263-265°. [α]$_D^{23}$ −411.3° (c, 0.025 in MeOH).
7,10-Dihydroxy, 11-methoxy: [113477-35-3]. ***7,10-Dihydroxy-11-methoxydracaenone***.
$C_{17}H_{16}O_5$ M 300.310
From *D. loureiri*. Cryst. Mp 262°. [α]$_D^{23}$ −465.9° (c, 0.0088 in MeOH).

Meksuriyen, D. *et al*, *J. Nat. Prod.*, 1987, **50**, 1118.

6-Drimene-8,9,11-triol   D-70547
8,9,11-Trihydroxy-6-drimene

$C_{15}H_{26}O_3$ M 254.369
(8β,9α)-form
Isoalbrassitriol
Metab. of *Alternaria brassicae*. Oil.

Ayer, W.A. *et al*, *J. Nat. Prod.*, 1987, **50**, 408.

7-Drimene-6,9,11-triol   D-70548
6,9,11-Trihydroxy-7-drimene
$C_{15}H_{26}O_3$ M 254.369
(7α,9α)-form
Albrassitriol
Metab. of *Alternaria brassicae*. Amorph. powder. Mp 100-104°.

Ayer, W.A. *et al*, *J. Nat. Prod.*, 1987, **50**, 408.

Duramycin   D-70549
[1391-36-2]

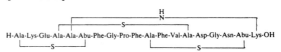

H-Ala-Lys-Glu-Ala-Ala-Abu-Phe-Gly-Pro-Phe-Ala-Phe-Val-Ala-Asp-Gly-Asn-Abu-Lys-OH

Large peptide antibiotic. Prod. by *Streptomyces cinnamomeous* f. *azacoluta*. Active against gram-positive bacteria and some yeasts and fungi. Cryst.

Shotwell, O.L. *et al*, *J. Am. Chem. Soc.*, 1958, **80**, 3912 (*isol, props*)
Kessler, H. *et al*, *Helv. Chim. Acta*, 1987, **70**, 726 (*struct*)

Duryne   D-70550
4,15,26-Triacontatriene-1,29-diyne-3,28-diol
[108641-87-8]

$C_{30}H_{48}O_2$ M 440.708
Isol. from marine sponge *Cribrochalina dura*. Potent cytotoxic metabolite. Solid. Mp 44-45°. Nomenclature error in CA.

Wright, A.E. *et al*, *Tetrahedron Lett.*, 1987, **28**, 1377 (*isol, ir, ms, pmr, cmr*)

E

Egonol E-70001

Updated Entry replacing E-10003

2-(1,3-Benzodioxol-5-yl)-7-methoxy-5-benzofuranpropanol, 9CI. 5-(2-Hydroxyphenyl)-7-methoxy-2-(3,4-methylenedioxyphenyl)benzofuran

[530-22-3]

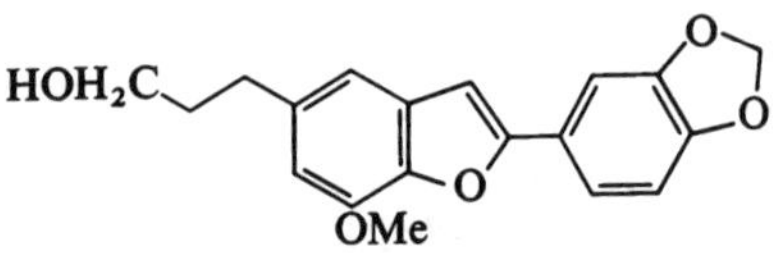

$C_{19}H_{18}O_5$ M 326.348

Constit. of *Styrax japonicum*. Plates (butanol). Mp 117.5-118°.

Ac: [15434-00-1].
  $C_{21}H_{20}O_6$ M 368.385
  Constit. of *S. obassia*. Cryst. (EtOH). Mp 107°.

β-D-Glucopyranoside: [77690-83-6]. *Egonol glucoside.*
  $C_{25}H_{28}O_{10}$ M 488.490
  Constit. of *S. obassia*. Brown needles (EtOH). Mp 169-170°.

O-(2-Methylbutanoyl): [115334-06-0]. *Egonol 2-methylbutanoate.*
  $C_{24}H_{26}O_6$ M 410.466
  Constit. of *S. obassia*. Pale-yellow oil.

O-(2-Methylbutanoyl), 7-demethoxy: [115334-07-1]. *Demethoxyegonol 2-methylbutanoate.*
  $C_{23}H_{24}O_5$ M 380.440
  Constit. of *S. obassia*. Needles (MeOH). Mp 55.5-56°.

Schreiber, F.G. *et al, J. Chem. Soc., Perkin Trans. 1*, 1976, 1514 (*bibl*)
Takanashi, M. *et al, Phytochemistry*, 1988, **27**, 1224 (*derivs*)

Elatol E-70002

10R-Bromo-4-chloro-9S-hydroxy-3,7(14)-chamigradiene

[55303-97-4]

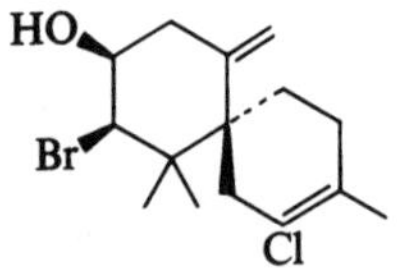

$C_{15}H_{22}BrClO$ M 333.695

Constit. of *Laurencia elata*. Oil. $[\alpha]_D$ +83.5° (c, 0.365 in MeOH).

Ac: Cryst. (hexane). Mp 157-158°.

Debromo: [61661-37-8]. **4-Chloro-9R-hydroxy-3,7(14)-chamigradiene**.
  $C_{15}H_{23}ClO$ M 254.799
  Constit. of *L. obtusa*. Oil. $[\alpha]_D$ +98°.

Sims, J.J. *et al, Tetrahedron Lett.*, 1974, 3487 (*cryst struct, abs config*)
Gonzalez, A.G. *et al, Tetrahedron Lett.*, 1976, 3051 (*isol*)

Eldanolide E-70004

Updated Entry replacing E-50013

Dihydro-4-methyl-5-(3-methyl-2-butenyl)-2(3H)-furanone, 9CI. 3,7-Dimethyl-6-octen-4-olide

[84107-93-7]

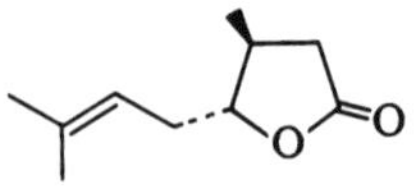

Absolute configuration

$C_{10}H_{16}O_2$ M 168.235

Wing gland phereomone of the male African sugar-cane borer *Eldona saccharina*. Oil. Bp$_{2.5}$ 75°. $[\alpha]_D^{23}$ +51.5° (c, 1.15 in MeOH).

Kunesch, G. *et al, Tetrahedron Lett.*, 1981, **22**, 5271 (*isol, struct*)
Uematsu, T. *et al, Agric. Biol. Chem.*, 1983, **47**, 597 (*synth*)
Yokoyama, Y. *et al, Chem. Lett.*, 1983, 1245 (*synth*)
Chakraborty, T.K. *et al, Tetrahedron Lett.*, 1984, **25**, 2891 (*synth*)
Vigneron, J.P. *et al, Tetrahedron*, 1984, **40**, 3521 (*struct, synth, abs config*)
Davies, H.G. *et al, J. Chem. Soc., Chem. Commun.*, 1985, 1166 (*synth*)
Suzuki, K. *et al, Tetrahedron Lett.*, 1985, **26**, 861 (*synth*)
Frauenrath, H. *et al, Tetrahedron*, 1986, **42**, 1135 (*synth*)
Butt, S. *et al, J. Chem. Soc., Perkin Trans. 1*, 1987, 903 (*synth*)
Ortuño, R.M. *et al, Tetrahedron*, 1987, **43**, 4497 (*synth*)

Elemasteiractinolide E-70005

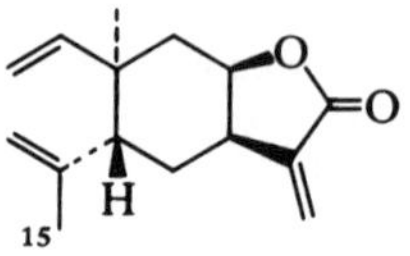

$C_{15}H_{20}O_2$ M 232.322

15-Hydroxy: [100045-45-2].
  $C_{15}H_{20}O_3$ M 248.321
  Constit. of *Stevia achalensis*. Oil. $[\alpha]_D^{24}$ +76° (c, 0.38 in CHCl$_3$).

11β,13-Dihydro, 15-hydroxy: [100045-46-3].
  $C_{15}H_{22}O_3$ M 250.337
  Constit. of *S. achalensis*. Oil.

Bohlmann, F. *et al, Justus Liebigs Ann. Chem.*, 1986, 799.

1,3,7(11)-Elematriene   E-70006

γ-*Elemene*

[29873-99-2]

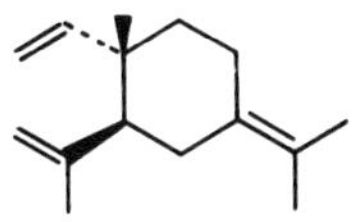

$C_{15}H_{24}$ M 204.355

Constit. of Gurjun balsam oil. Oil. $Bp_{0.01}$ 44°. $[\alpha]_D$ +3.93° (c, 1 in $CHCl_3$).

Gough, J.H. *et al*, *Aust. J. Chem.*, 1964, **17**, 1270 (*isol*)
Ganter, C. *et al*, *Helv. Chim. Acta*, 1971, **54**, 183 (*synth*)
Thomas, A.F., *Helv. Chim. Acta*, 1972, **55**, 2429 (*synth*)
Kim, D. *et al*, *J. Org. Chem.*, 1987, **52**, 4633 (*synth*)

Ellagic acid, INN   E-70007

Updated Entry replacing E-60003
2,3,7,8-Tetrahydroxy[1]benzopyrano[5,4,3-cde][1]-benzopyran-5,10-dione, 9CI. Benzoaric acid. Elagostasine. Lagiotase
[476-66-4]

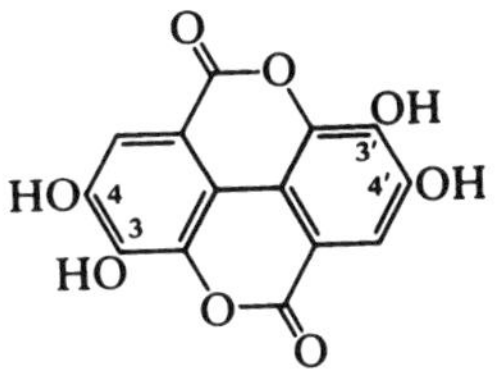

$C_{14}H_6O_8$ M 302.197

Occurs free and combined in galls. Isol. from the leaves of *Castanopsis* spp. and several other spp. Astringent (intestinal). Haemostatic. Needles + 2Py (Py). Spar. sol. H_2O, EtOH, insol. Et_2O. Mp >360°.

▷DJ2620000.

Tetra-Ac: Mp 317-319°, 343-346°.
3,3'-Di-Me ether: Nasutin C.
  $C_{16}H_{10}O_8$ M 330.250
  Constit. of the haemolymph of the termite *Nasutitermes exitiosus*. Pale-cream needles (Me_2CO). Mp 336-338° dec.
3,3'-Di-Me ether, 4-glucoside:
  $C_{22}H_{20}O_{13}$ M 492.392
  Occurs in *Terminalia paniculata*. Prisms (MeOH aq.). Mp 214-215°. $[\alpha]_D^{30}$ +79° (c, 0.504 in MeOH).
3,3',4-Tri-Me ether: Nasutin B.
  $C_{17}H_{12}O_8$ M 344.277
  Constit. of the haemolymph of *N. exitiosus*. Pale-cream platelets (EtOH). Mp 298° dec.
Tetra-Me ether: Mp 355° dec.
O^4-β-*Gentiobioside: Amritoside*.
  $C_{26}H_{26}O_{18}$ M 626.481
  Isol. from *Psidium guajava*. Platelets (EtOH). Mp 248-250°. $[\alpha]_D^{25}$ −414° (Py).

Perkin, A.G., *J. Chem. Soc.*, 1905, **87**, 1415 (*synth*)
Nierenstein, M., *Helv. Chim. Acta*, 1931, **14**, 912 (*isol*)
Zetzsche, F., *Helv. Chim. Acta*, 1931, **14**, 240 (*isol*)
Row, L.R., *Tetrahedron*, 1962, **18**, 357 (*isol*)
Moore, B.P., *Aust. J. Chem.*, 1964, **17**, 901 (*derivs*)
Seshadri, T.R. *et al*, *Phytochemistry*, 1965, **4**, 317, 989 (*Amritoside*)
Bhargava, U.C. *et al*, *J. Pharm. Sci.*, 1968, **57**, 1728 (*pharmacol*)
Arthur, H.R., *Aust. J. Chem.*, 1969, **22**, 597 (*isol*)
Doyle, B. *et al*, *Xenobiotica*, 1980, **10**, 247 (*metab*)
Sato, T., *Phytochemistry*, 1987, **26**, 2124 (*synth, cmr*)
Martindale, The Extra Pharmacopoeia, 28th/29th Eds., 1982/1989, Pharmaceutical Press, London, 309.

Encecanescin   E-70008

Updated Entry replacing E-20012
6,6'-(Oxydiethylidene)bis[7-methoxy-2,2-dimethyl-2H-1-benzopyran], 9CI
[87592-85-6]

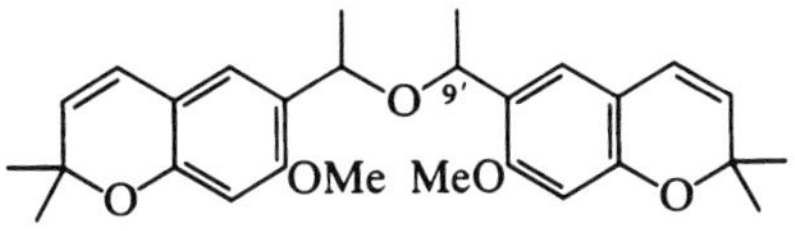

$C_{28}H_{34}O_5$ M 450.574

Constit. of *Encelia canescens*. Cryst. Mp 162-163°. $[\alpha]_D^{24}$ −113° (c, 0.35 in $CHCl_3$).

9'-Epimer: [87592-86-7]. *9'-Epiencecanescin*.
  $C_{28}H_{34}O_5$ M 450.574
  Constit. of *E. canescens*. Gum.

Bohlmann, F. *et al*, *Phytochemistry*, 1983, **22**, 557 (*isol*)
Fang, N. *et al*, *Phytochemistry*, 1988, **27**, 1902 (*struct*)

Entandrophragmin   E-70009

Updated Entry replacing E-00150
[11013-05-1]

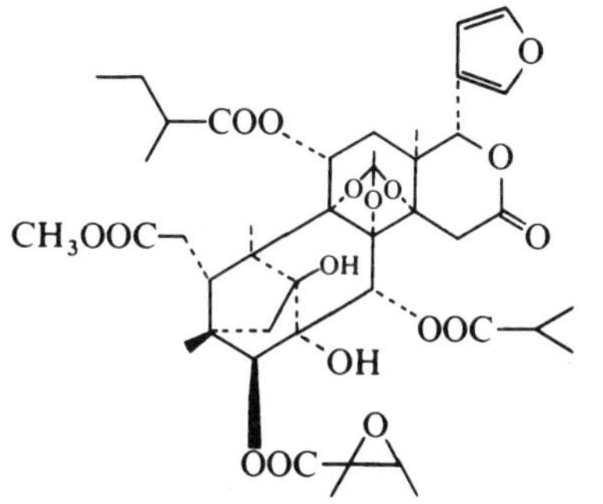

$C_{43}H_{56}O_{17}$ M 844.905

Constit. of *Entandrophragma* spp. Cryst. (MeOH aq.). Mp 256°. $[\alpha]_D^{20}$ −4°.

Deepoxy: [11048-67-2]. *Candollein*.
  $C_{43}H_{58}O_{16}$ M 830.922
  Minor constit. of *E. candollei* and *E. cylindricum*. Mp 254-256°. Conts. a 2-methylbutanoyl residue in place of 2,3-Epoxy-2-methylbutanoyl. Same ir spectrum and Mp as Entandrophragminin, no depression of Mp.
Deepoxy, hydroxy: [64504-48-9]. β-*Dihydroentandrophragmin*.
  $C_{43}H_{58}O_{17}$ M 846.921
  Constit. of *E. cylindricum*. Conts. a 2-hydroxy-2-methylbutanoyl residue in place of 2,3-epoxy-2-methylbutanoyl.

Halsall, T.G. *et al*, *J. Chem. Res. (S)*, 1977, 154

Epidermin, 9CI   E-70010

[99165-17-0]

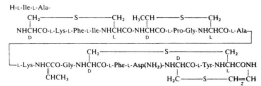

$C_{98}H_{141}N_{25}O_{23}S_4$ M 2165.586

Cyclic peptide antibiotic. Isol. from *Staphlococcus epidermidis*. Active against staphylocci, streptococci and *Propionibacterium acnes*. No phys. props. reported.

Allgaier, H. *et al, Angew. Chem., Int. Ed. Engl.,* 1985, **24**, 1051 (*isol, struct*)
Allgaier, H. *et al, Eur. J. Biochem.,* 1986, **160**, 9 (*struct*)

2,3-Epoxy-2,3-dihydro-1,4-naphthoquin-one, 8CI   **E-70011**

1a,7a-Dihydronaphth[2,3-b]oxirene-2,7-dione, 9CI
[15448-58-5]

$C_{10}H_6O_3$ M 174.156
Needles (AcOH aq.), prisms (C_6H_6/pet. ether). Mp 136°.

Zincke, T., *Ber.,* 1892, **25**, 3599 (*synth*)
Williams, D.H. *et al, J. Org. Chem.,* 1968, **33**, 998 (*pmr*)
Kerr, K.A., *Acta Crystallogr., Sect. C,* 1987, **43**, 956 (*cryst struct*)

11,12-Epoxy-2,14-dihydroxy-5,8(17)-briar-adien-18,7-olide   **E-70012**

$C_{20}H_{28}O_5$ M 348.438
O^2-*Butanoyl*, O^{14}-*Ac:* [112781-24-5]. *14-Acetoxy-2-bu-tanoyloxy-5,8(17)-briaradien-18,7-olide.*
$C_{26}H_{36}O_7$ M 460.566
Constit. of *Briareum steckei.* Plates (CH_2Cl_2/hexane). Mp 158-160°. $[\alpha]_D$ +55.2° (c, 0.27 in $CHCl_3$).
3α-Acetoxy, O^2-*butanoyl*, O^{14}-*Ac:* [112781-23-4]. *3,14-Diacetoxy-2-butanoyloxy-5,8(17)-briaradien-18,7-olide.*
$C_{28}H_{38}O_9$ M 518.603
Constit. of *B. steckei.* Glass. $[\alpha]_D$ +155° (c, 0.62 in $CHCl_3$).
4β-Acetoxy, O^2-*butanoyl*, O^{14}-*Ac:* [112781-25-6]. *4,14-Diacetoxy-2-butanoyloxy-5,8(17)-briaradien-18,7-olide.*
$C_{28}H_{38}O_9$ M 518.603
Constit. of *B. steckei.* Cryst. Mp 183-186°. $[\alpha]_D$ +106° (c, 0.24 in $CHCl_3$).
4β-Acetoxy, O^2-*propanoyl*, O^{14}-*Ac:* [112781-26-7]. *4,14-Diacetoxy-2-propanoyloxy-5,8(17)-briaradien-18,7-olide.*
$C_{27}H_{36}O_9$ M 504.576
Constit. of *B. steckei.*

Bowden, B.F. *et al, Aust. J. Chem.,* 1987, **40**, 2085.

24,25-Epoxy-7,26-dihydroxy-8-lanosten-3-one   **E-70013**

$C_{30}H_{48}O_4$ M 472.707
(*7α,24S,25S*)-*form* [114020-56-3]
Epoxyganoderiol A
Metab. of *Ganoderma lucidum.* Gummy solid. $[\alpha]_D^{23}$ +65° (c, 0.2 in $CHCl_3$).

Nishitoba, T. *et al, Agric. Biol. Chem.,* 1988, **52**, 211.

8,16-Epoxy-8H,16H-dinaphtho[2,1-b:2′,1′-f][1,5]dioxocin, 9CI   **E-70014**

[52243-98-8]

$C_{22}H_{14}O_3$ M 326.351
Formed by dimerisation of 2-Hydroxy-1-naphthaldehyde, H-02660 . Cryst. (AcOH). Mp 242-244°.

Andrieux, J. *et al, Bull. Soc. Chim. Fr.,* 1973, 3421.
Jones, P.R. *et al, J. Org. Chem.,* 1981, **46**, 194 (*synth, ms, uv, pmr, cmr*)
Bachet, B. *et al, Acta Crystallogr., Sect. C,* 1986, **42**, 1630 (*cryst struct*)
Hosangadi, B.D. *et al, Synth. Commun.,* 1986, **16**, 191 (*synth*)

22,25-Epoxy-2,3,6,26-furostanetetrol   **E-70015**

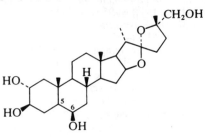

$C_{27}H_{44}O_6$ M 464.641
(*2α,3β,5α,6β,20S,22S,25S*)-*form*
Constit. of *Nothoscordum gramineum* var. *philippianum.*
Tetra-Ac: Cryst. (MeOH). Mp 214-218°.
6-Deoxy, 5,6-didehydro: 22,25-Epoxy-5-furostene-2α,3β,26-triol.
$C_{27}H_{42}O_5$ M 446.626
Constit. of *N. gramineum* var. *philippianum.*
6-Deoxy, 5,6-didehydro, Tri-Ac: Cryst. (MeOH). Mp 187-190°.

Brunengo, M.C. *et al, Phytochemistry,* 1988, **27**, 2943.

14,15-Epoxy-5-hydroxy-6,8,10,12-eicosa-tetraenoic acid   E-70016

5-Hydroxy-13(3-pentyloxiranyl)-6,8,10,12-tridecate-traenoic acid, 9CI. 5-Hydroxy-14,15-LTA₄

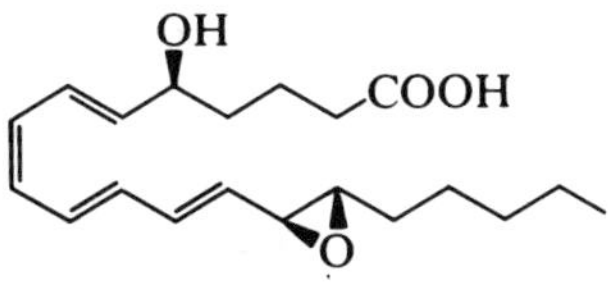

$C_{20}H_{30}O_4$ M 334.455

(5S,6E,8Z,10E,12E)-form [114144-28-4]
Intermed. in lipoxin biosynth. $t_{1/2}$ 2 min. in pH 7.4 buffer.

Leblanc, Y. *et al, Tetrahedron Lett.*, 1987, **28**, 3449 (*synth, biochem*)

5,6-Epoxy-20-hydroxy-8,11,14-eicosa-trienoic acid   E-70017

3-(14-Hydroxy-2,5,8-tetradecatrienyl)oxiranebutanoic acid, 9CI

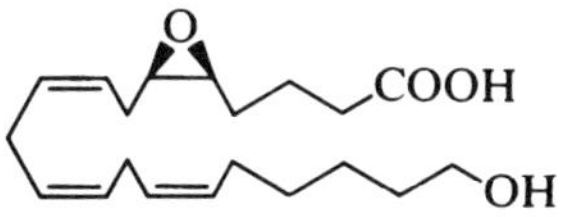

$C_{20}H_{32}O_4$ M 336.470

(5S,6R,8Z,11Z,14Z)-form [112168-43-1]
Hydroxylated eicosanoid metab. $[\alpha]_D^{23}$ −3.77° (c, 1.28 in Me₂CO).

Lumin, S. *et al, J. Chem. Soc., Chem. Commun.*, 1987, 389 (*synth, pmr*)

14,15-Epoxy-20-hydroxy-5,8,11-eicosa-trienoic acid   E-70018

13-[3-(5-Hydroxypentyl)oxiranyl]-5,8,11-tridecatrien-oic acid, 9CI

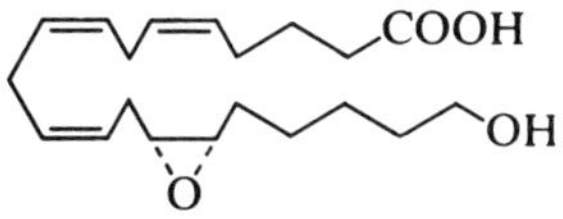

$C_{20}H_{32}O_4$ M 336.470

(5Z,8Z,11Z,14R,15S)-form [112168-44-2]
Hydroxylated eicosanoid metabolite. $[\alpha]_D^{23}$ +4.66° (c, 1.01 in Me₂CO).

Lumin, S. *et al, J. Chem. Soc., Chem. Commun.*, 1987, 389 (*synth, pmr*)

1,2-Epoxy-3-hydroxy-4,11(13)-eudesma-dien-12,8-olide   E-70019

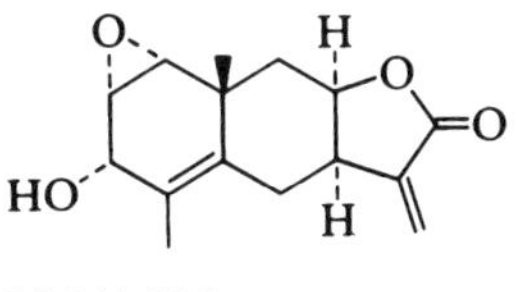

$C_{15}H_{18}O_4$ M 262.305

(1α,2α,3α,8β)-form
Constit. of *Ferreyranthus fruticosus*. Cryst. Mp 166°.

Jakupovic, J. *et al, Phytochemistry*, 1988, **27**, 1113.

ent-8,13-Epoxy-14,15,19-labdanetriol   E-70020

ent-*14,15,19-Trihydroxymanoyl oxide*

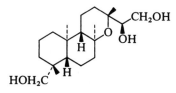

$C_{20}H_{36}O_4$ M 340.502
Constit. of *Florestina tripteris*. Gum.

Domínguez, X.A. *et al, Phytochemistry*, 1988, **27**, 613.

8,13-Epoxy-3-labdanol   E-70021

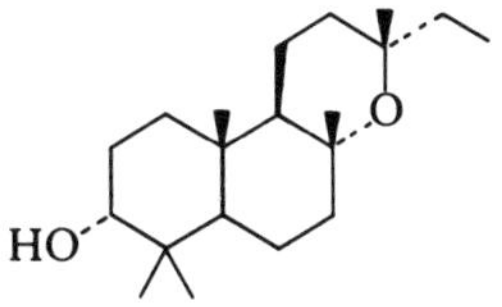

$C_{20}H_{36}O_2$ M 308.503

(3α,8α,13S)-form
3α-Hydroxy-14,15-dihydromanoyl oxide
Constit. of *Juniperus pseudosabina*. Cryst. Mp 86-87°.

Pandita, K. *et al, Indian J. Chem., Sect. B*, 1987, **26**, 453.

ent-15,16-Epoxy-7,13(16),14-labdatriene   E-70022

[117249-10-2]

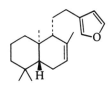

$C_{20}H_{30}O$ M 286.456
Constit. of *Acrisione denticulata*. Oil.

Aal, M.A. *et al, Phytochemistry*, 1988, **27**, 2599.

8,13-Epoxy-14-labdene   E-70023

Updated Entry replacing E-00269

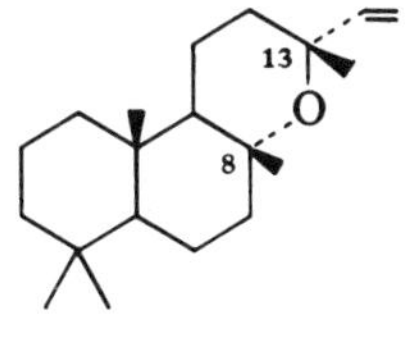

$C_{20}H_{34}O$ M 290.488

(8α,13R)-form [596-84-9]
Manoyl oxide
Constit. of *Dacrydium colensoi, Cupressus sempervirens* and *Juniperus oxycedrus*. Cryst. (MeOH aq.). Mp 29°. Bp₀.₃ 135-137°. $[\alpha]_D^{13}$ +19.6° (EtOH).
19-Hydroxy: **19-Hydroxymanoyl oxide.**
$C_{20}H_{34}O_2$ M 306.487
Constit. of *Polemonium viscosum*. Cryst. Mp 47-48°. $[\alpha]_D$ +13.1° (c, 0.013 in CHCl₃).
19-Acetoxy: **19-Acetoxymanoyl oxide.**
$C_{22}H_{36}O_3$ M 348.525
Constit. of *P. viscosum*. Oil. $[\alpha]_D$ +31.5° (c, 0.003 in CHCl₃).

13-Epimer, 12α-hydroxy: **12α-Hydroxy-13-epi-(+)-manoyl oxide**.
$C_{20}H_{34}O_2$ M 306.487
Isol. from Turkish tobacco (*Nicotiana tabacum*). Mp 141-142°. $[\alpha]_D^{25}$ +44° (CHCl$_3$).

(8β,13R)-form
  8-epi-Manoyl oxide
  Constit. of *Chamaecyparis nootkatensis*. Cryst. Mp 44-45°. $[\alpha]_D^{23}$ −9.5° (c, 2.1 in CHCl$_3$).
(8β,13S)-form [28235-38-3]
  8,13-diepi-Manoyl oxide
  Constit. of *C. nootkatensis*. Cryst. Mp 79-84°. $[\alpha]_D^{23}$ +23.4° (c, 0.6 in CHCl$_3$).
(8α,13S)-form [1227-93-6]
  13-epi-Manoyl oxide
  Constit. of Jack pine bark. Cryst. (MeOH). Mp 94-96°. $[\alpha]_D^{20}$ +37.5° (c, 10 in CHCl$_3$).
(ent-8α,13S)-form [27642-41-7]
  Olearyl oxide
  Metab. of *Gibberella fujikuroi* and constit. of *Olearia paniculata* and *Solidago missouriensis*. Cryst. (MeOH). Mp 98-99.5°. $[\alpha]_D^{24}$ −37° (c, 0.25 in EtOH).

Giles, J.A. *et al, Tetrahedron*, 1962, **18**, 169 (*12α-Hydroxy-13-epimanoyl oxide*)
Bower, G.L. *et al, Phytochemistry*, 1967, **6**, 151 (*isol*)
Cheng, Y.S. *et al, Phytochemistry*, 1970, **9**, 2517 (*isol*)
Serebryakov, E.P. *et al, Tetrahedron*, 1970, **26**, 5215 (*synth*)
Hanson, J.R. *et al, Phytochemistry*, 1972, **11**, 703 (*biosynth*)
Anthonsen, T. *et al, Acta Chem. Scand.*, 1973, **27**, 1073 (*isol*)
Almqvist, S., *Acta Chem. Scand., Ser. B*, 1975, **29**, 695 (*cmr*)
Tabacchi, R. *et al, Helv. Chim. Acta*, 1975, **58**, 1184 (*isol*)
Stierle, D.B. *et al, Phytochemistry*, 1988, **27**, 517 (*19-Hydroxymanoyl oxide, 19-Acetoxymanoyl oxide*)

ent-9,13-Epoxy-7-labdene-15,17-dioic acid   E-70024
17-Carboxygrindelic acid

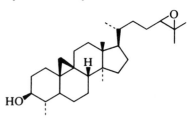

$C_{20}H_{30}O_5$ M 350.454
Constit. of *Grindelia camporum*.
Di-Me ester: [114420-29-0]. Oil. $[\alpha]_D^{25}$ −70.3° (c, 2.5 in CHCl$_3$).
Hoffmann, J.J. *et al, Phytochemistry*, 1988, **27**, 493.

24,25-Epoxy-7,9(11)-lanostadiene-3,26-diol   E-70025

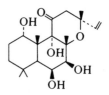

$C_{30}H_{48}O_3$ M 456.707
(3β,24S,25S)-form [114020-58-5]
  Epoxyganoderiol C
  Metab. of *Ganoderma lucidum*. Syrup. $[\alpha]_D^{23}$ +43° (c, 0.2 in EtOH).
  3-Ketone: [114020-57-4]. *24S,25S-Epoxy-26-hydroxy-7,9(11)-lanostadien-3-one.* **Epoxyganoderiol B**.
  $C_{30}H_{46}O_3$ M 454.692

Metab. of *G. lucidum*. Syrup. $[\alpha]_D^{23}$ +35° (c, 0.4 in EtOH).
Nishitoba, T. *et al, Agric. Biol. Chem.*, 1988, **52**, 211.

24,25-Epoxy-29-nor-3-cycloartanol   E-70026

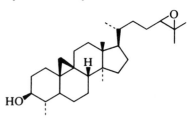

$C_{29}H_{48}O_2$ M 428.697
3β-form
  Constit. of *Aglaia roxburghiana*. Cryst. (EtOAc/C$_6$H$_6$). Mp 110°.
Ac: Amorph. powder. $[\alpha]_D^{20}$ +35° (c, 1.5 in CHCl$_3$).
Vishnoi, S.P. *et al, Planta Med.*, 1988, **54**, 40.

8,13-Epoxy-1,6,7,9-tetrahydroxy-14-labden-11-one   E-70027
Updated Entry replacing E-00354
[66428-84-0]

$C_{20}H_{32}O_6$ M 368.469
(1α,6β,7β,8α,9α,13R)-form [64657-20-1]
  Constit. of *Coleus forskohlii*. Cryst. Mp 177-180°.
  6-Ac: [64657-21-2].
    Coleonal B
    $C_{22}H_{34}O_7$ M 410.506
    Constit. of *C. forskohlii*. Cryst. Mp 208-210°.
  7-Ac: [6675-29-9]
    Coleonal. Forskolin. Colforsin, USAN
    $C_{22}H_{34}O_7$ M 410.506
    Constit. of *C. forskohlii*. Antihypertensive agent. Used to treat glaucoma. Cryst. Mp 230-232°. $[\alpha]_D^{25}$ −26.2° (c, 1.68 in CHCL$_3$).

(1α,6β,7α,9α,13S)-form
  6-Ac: [67921-06-6]. **Coleonol C**. From *C. forskohlii*. Cryst. (Me$_2$CO/hexane). Mp 205-206°. $[\alpha]_D$ −6.7° (c, 1 in CHCl$_3$).

Bhat, S.V. *et al, Tetrahedron Lett.*, 1977, 1669 (*isol*)
Tandon, J.S. *et al, Indian J. Chem., Sect. B*, 1977, **15**, 880; 1978, **16**, 341.
Lindner, E. *et al, Arzneim.-Forsch.*, 1978, **28**, 284; 1983, **33**, 1436 (*pharmacol*)
De Souza, N.J. *et al, Med. Res. Rev.*, 1983, **3**, 201 (*rev, pharmacol*)
Ramakumar, S. *et al, Z. Kristallogr., Kristallgeom., Kristallphys., Kristallchem.*, 1985, **173**, 81 (*cryst struct*)
Saksena, A.K. *et al, J. Chem. Soc., Chem. Commun.*, 1985, 1748 (*synth, bibl*)
Saksena, A.K. *et al, Tetrahedron Lett.*, 1985, **26**, 551 (*struct, coleonol B, bibl*)
Prakash, O. *et al, J. Chem. Soc., Perkin Trans. 2*, 1986, 1779 (*pmr, cmr, config, Colforsin*)

Hrib, N.J. *et al, Tetrahedron Lett.*, 1987, **28**, 19 (*synth, Colforsin*)
Ziegler, F. *et al, J. Am. Chem. Soc.*, 1987, **109**, 8115 (*synth, Colfasin*)
Hashimoto, S. *et al, J. Am. Chem. Soc.*, 1988, **110**, 3670 (*synth*)
Martindale, The Extra Pharmacopoeia, 28th/29th Eds., 1982/1989, Pharmaceutical Press, London, 16542.

12,13-Epoxytrichothec-9-en-3-ol, 9CI   E-70028

3-Hydroxytrichothecene

[104155-10-4]

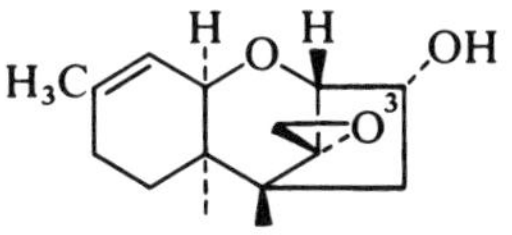

$C_{15}H_{22}O_3$ M 250.337

Isol. from *Fusarium* sp. Dec. prod. of trichotriol.
Mycotoxin. No phys. props. reported.

Ac: [91423-90-4]. **Isotrichodermin**.
  $C_{17}H_{24}O_4$ M 292.374
  From *F. culmorum* and *F. roseum*. Mycotoxin.

Greenhalgh, R. *et al, J. Agric. Food Chem.*, 1984, **32**, 1261; 1986, **34**, 98 (*deriv*)
Corley, D.G. *et al, J. Org. Chem.*, 1987, **52**, 4405 (*isol, struct*)
Greenhalgh, R. *et al, Bioact. Mol.*, 1987, **1**, 125, 137 (*isol*)

9,11(13)-Eremophiladien-12-oic acid   E-70029

[69905-01-7]

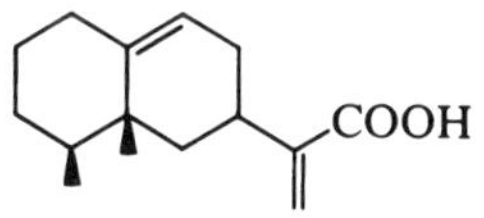

$C_{15}H_{22}O_2$ M 234.338

Constit. of *Athanasia thodei*. Oil.

Me ester: Oil. $[\alpha]_D$ −0.9° (c, 1.74 in CHCl₃).
2β-Hydroxy: [100045-53-2].
  $C_{15}H_{22}O_3$ M 250.337
  Constit. of *Stevia achalensis*. Oil. $[\alpha]_D^{24}$ −16° (c, 0.25 in CHCl₃).
3β-Hydroxy: [100045-54-3].
  $C_{15}H_{22}O_3$ M 250.337
  Constit. of *S. achalensis*. Oil. $[\alpha]_D^{24}$ +13° (c, 1.41 in CHCl₃).
3-Oxo: [100045-55-4].
  $C_{15}H_{20}O_3$ M 248.321
  Constit. of *S. achalensis*. Oil.

Bohlmann, F. *et al, Phytochemistry*, 1978, **17**, 1595.
Bohlmann, F. *et al, Justus Liebigs Ann. Chem.*, 1986, 799.

11-Eremophilene-2,8,9-triol   E-70030

Decahydro-4a,5-dimethyl-3-(1-methylethenyl)-1,2,7-naphthalenetriol, 9CI. 2,8,9-Trihydroxy-11-eremophilene

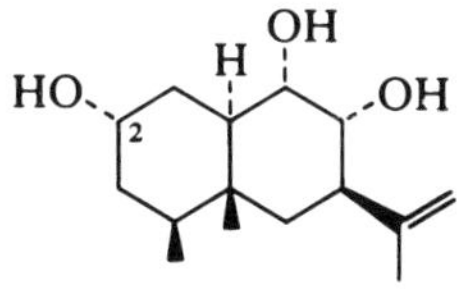

$C_{15}H_{26}O_3$ M 254.369

(2α,8α,9α,10αH)-form [53820-53-4]
Lateriflorol

Occurs as mixt. of esters in *Euryopa lateriiflorus* and *E. imbricatus*. Cryst. (Et₂O/pet. ether). Mp 158°.

2-Ketone: 8α,9α-Dihydroxy-10βH-eremophil-11-en-2-one.
  $C_{15}H_{24}O_3$ M 252.353
  Occurs as mixt. of esters in *E.* spp. Oil.

Bohlmann, F. *et al, Chem. Ber.*, 1974, **107**, 2730 (*isol, struct*)
Bohlmann, F. *et al, Phytochemistry*, 1978, **17**, 1135 (*isol, struct*)
Bohlmann, F. *et al, Justus Liebigs Ann. Chem.*, 1984, 1785 (*synth*)

1(10)-Eremophilen-7-ol   E-70031

1(10)-Valencen-7-ol

[117005-44-4]

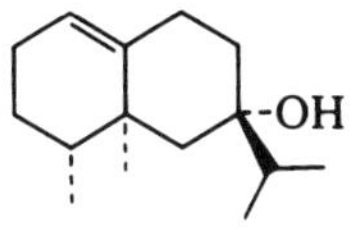

$C_{15}H_{26}O$ M 222.370

Constit. of *Bazzania fauriana*. Oil. $[\alpha]_D$ +75° (c, 0.36 in CHCl₃).

Toyota, M. *et al, Phytochemistry*, 1988, **27**, 2155.

Eriodermic acid   E-70032

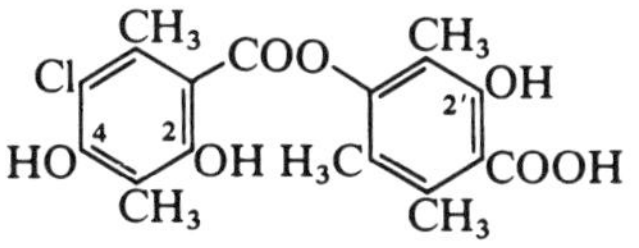

$C_{19}H_{19}ClO_7$ M 394.808

Me ester: [111258-41-4]. *Methyl eriodermate.*
  $C_{20}H_{21}ClO_7$ M 408.835
  Constit. of *Erioderma* spp. Prisms (CH₂Cl₂/pet. ether). Mp 165°.
2-Me ether, Me ester: [111258-43-6]. *Methyl 2-O-methyleriodermate.*
  $C_{21}H_{23}ClO_7$ M 422.862
  Constit. of *E.* spp. Prisms (CH₂Cl₂/pet. ether). Mp 167°.
2′-Me ether, Me ester: [111258-42-5]. *Methyl 2′-O-methyleriodermate.*
  $C_{21}H_{23}ClO_7$ M 422.862
  Constit. of *E.* spp. Prisms (CH₂Cl₂/pet. ether). Mp 172°.
4-Me ether, Me ester: [111258-44-7]. *Methyl 4-O-methyleriodermate.*
  $C_{21}H_{23}ClO_7$ M 422.862
  Constit. of *E.* spp. Amorph. solid (CH₂Cl₂/pet. ether). Mp 124°.
2,2′-Di-Me ether, Me ester: [111258-45-8]. *Methyl 2,2′-di-O-methyleriodermate.*
  $C_{22}H_{25}ClO_7$ M 436.888
  Constit. of *E.* spp. Needles (CH₂Cl₂/pet. ether). Mp 168°.

Elix, J.A. *et al, Aust. J. Chem.*, 1987, **40**, 1581.

Erivanin   E-70033

Updated Entry replacing E-00419

[25645-08-3]

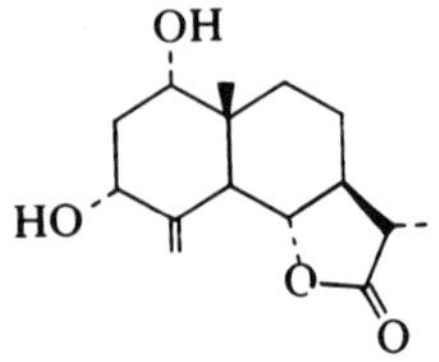

$C_{15}H_{22}O_4$ M 266.336

Constit. of *Artemisia fragrans.* Cryst. Mp 203-205°.
$[\alpha]_D^{20}$ +112° (c, 3.9 in EtOH).

Di-Ac: Cryst. Mp 192-193.5°.

1-Epimer: [32203-46-6]. **1-Epierivanin**.
$C_{15}H_{22}O_4$ M 266.336
Constit. of *Picris aculeata.* Gum.

Evstratova, R.I. *et al, Khim. Prir. Soedin.*, 1969, **5**, 239; *CA*, **72**, 32051v (*isol*)

Samek, Z. *et al, Collect. Czech. Chem. Commun.*, 1975, **40**, 2676 (*struct*)

Arias, J.M. *et al, J. Chem. Soc., Perkin Trans. 1*, 1987, 471 (*synth*)

Bruno, M. *et al, Phytochemistry*, 1988, **27**, 1201 (*deriv*)

Erycristin   E-70034

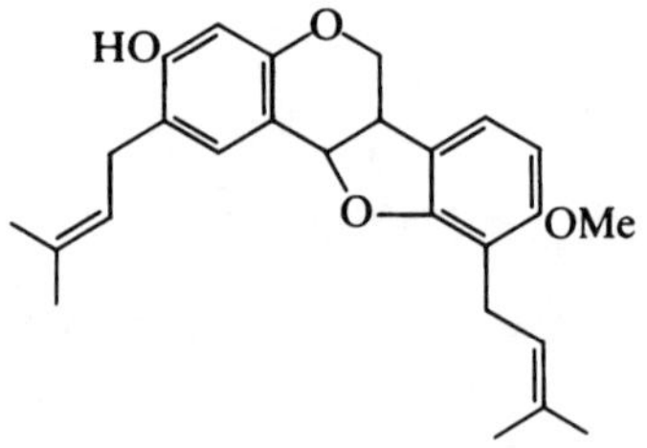

$C_{26}H_{30}O_4$ M 406.521

Constit. of the stem bark of *Erythrina crista-galli.* Anti-microbial agent. Cryst. (cyclohexane). Mp 120-121°.
$[\alpha]_D$ −140° (c, 1.964 in MeOH).

Mitscher, L.A. *et al, Phytochemistry*, 1988, **27**, 381.

Esperamicin *X*   E-70035

[107175-47-3]

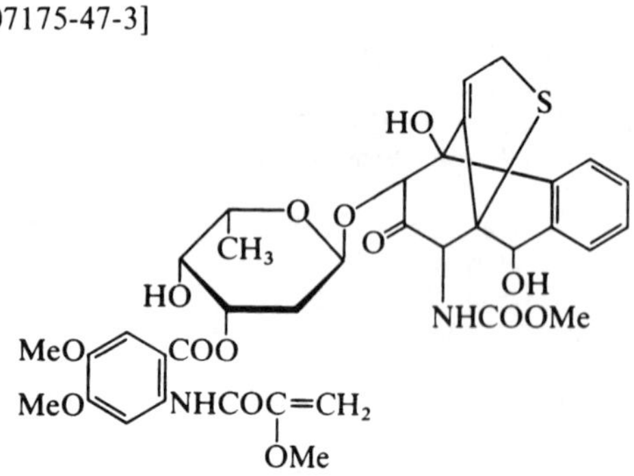

$C_{36}H_{40}N_2O_{14}S$ M 756.777

Prod. by *Actinomadura verrucosospora.* Biol. inactive component of esperamicin complex. Cryst. Mp 182-184°. $[\alpha]_D$ −36° (c, 0.5 in CHCl₃).

Golik, J. *et al, J. Am. Chem. Soc.*, 1987, **109**, 3461 (*struct*)

5-Ethenylbicyclo[2.2.1]hept-2-ene, 9CI   E-70036

Updated Entry replacing E-00516

5-Vinyl-2-norbornene

[3048-64-4]

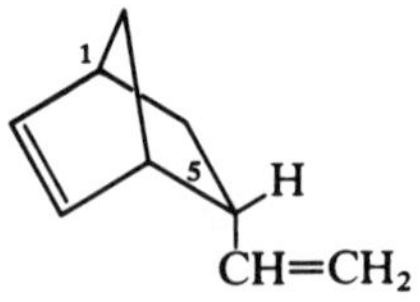

(1RS,5RS)-form

C_9H_{12} M 120.194

Major prod. of Diels-Alder reacn. between cyclopentadiene and butadiene. Used in polymerisations. Bp 137°.

▷RC0350000.

(**1RS,5RS**)-**form** [25093-48-5]

(±)-endo-*form*

Liq. Bp₂₀ 51-55°. n_D^{20} 1.4806.

(**1RS,5SR**)-**form** [23890-32-6]

(±)-exo-*form*

Liq. Bp₂₅ 55-62°. n_D^{20} 1.4791.

Inoue, Y., *Bull. Chem. Soc. Jpn.*, 1987, **60**, 1954 (*synth, pmr*)

2-Ethoxy-2-propenoic acid, 9CI   E-70037

α-Ethoxyacrylic acid. Pyruvic acid ethyl enol ether

[32821-76-4]

$$H_2C{=}C(OEt)COOH$$

$C_5H_8O_3$ M 116.116

Cryst. solid. Mp 57-58°.

Me ester:
$C_6H_{10}O_3$ M 130.143
Bp₂₅ 77-78°.

Amide: [34068-59-2].
$C_5H_9NO_2$ M 115.132
Cryst. solid (diisopropyl ether). Mp 68-69°.

4-Nitrobenzyl ester: [108818-44-6]. Cryst. solid (cyclohexane). Mp 91-92°.

Divers, G.A. *et al, Synth. Commun.*, 1977, **7**, 43 (*synth, ir, pmr*)

LaMattina, J.L. *et al, J. Org. Chem.*, 1987, **52**, 3479 (*synth*)

3-(2-Ethyl-2-butenyl)-9,10-dihydro-1,6,8-trihydroxy-9,10-dioxo-2-anthracenecarboxylic acid   E-70038

K 259-2

[102819-46-5]

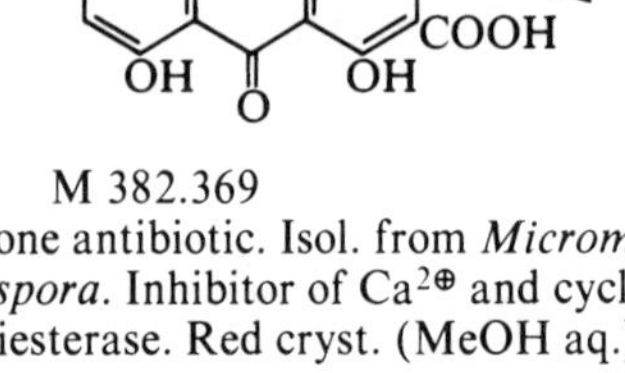

$C_{21}H_{18}O_7$ M 382.369

Anthraquinone antibiotic. Isol. from *Micromonospora divasterospora.* Inhibitor of $Ca^{2\oplus}$ and cyclic nucleotide phosphodiesterase. Red cryst. (MeOH aq.). Mp 140-145° dec.

Matsuda, Y. *et al, J. Antibiot.*, 1987, **40**, 1092, 1101 (*isol, struct, props*)

3-Ethylcyclohexene, 9CI   E-70039
Updated Entry replacing E-60047
[2808-71-1]

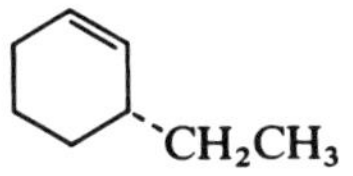

(R)-form

C_8H_{14} M 110.199
(R)-form [76152-65-3]
$[\alpha]_D^{25}$ +77° (c, 1 in toluene).
(±)-form
Liq. Bp 160°, Bp$_{30}$ 42-44°.

Bailey, B. et al, J. Chem. Soc., 1954, 967 (synth)
Bellassoued, M. et al, Synthesis, 1977, 205 (synth, pmr)
Barillier, D. et al, Tetrahedron, 1983, 39, 767 (cmr)
Rang, S. et al, Org. Mass Spectrom., 1984, 19, 193 (ms)
Buono, G. et al, J. Org. Chem., 1985, 50, 1781 (synth)
Pearson, A.J. et al, J. Org. Chem., 1986, 51, 2505 (synth, ir, pmr)
Goering, H.L. et al, J. Org. Chem., 1986, 51, 2884 (synth)

3-Ethyl-5,7-dihydroxy-4H-1-benzopyran-   E-70040
4-one, 9CI
3-Ethyl-5,7-dihydroxychromone. **Lathodoratin**
[76693-50-0]

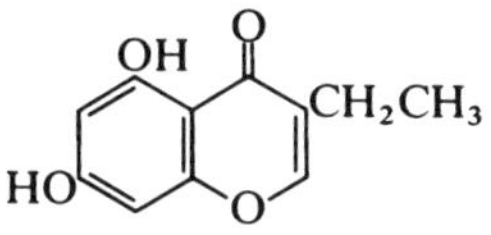

$C_{11}H_{10}O_4$ M 206.198
Phytoalexin from Lathyrus odoratus. Cryst. (MeOH).
Mp 205-206°.

Robeson, D.J. et al, Phytochemistry, 1980, 19, 2171 (isol, struct)
Al-Douri, N.A. et al, Phytochemistry, 1988, 27, 775 (biosynth)

6-Ethyl-2,4-dihydroxy-3-methylbenzoic   E-70041
acid

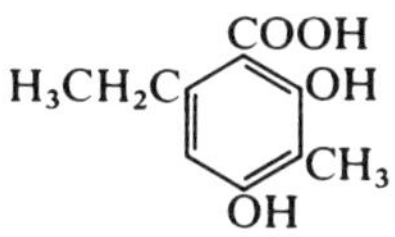

$C_{10}H_{12}O_4$ M 196.202
Me ester: [110925-97-8].
   $C_{11}H_{14}O_4$ M 210.229
   Metab. of Aspergillus ustus when grown in the presence of methanol. Cryst. (C_6H_6/hexane). Mp 104-106°.
Et ester: [110925-96-7].
   $C_{12}H_{16}O_4$ M 224.256
   Metab. of A. ustus when grown in the presence of ethanol. Cryst. (C_6H_6/hexane). Mp 95-97°.

De Jesus, A.E. et al, J. Chem. Soc., Perkin Trans. 1, 1987, 2253 (isol, struct, biosynth)

3-Ethyl-2,19-dioxabicyclo[16.3.1]docosa-   E-70042
3,6,9,18(22),21-pentaen-12-yn-20-one,
9CI
[106001-28-9]

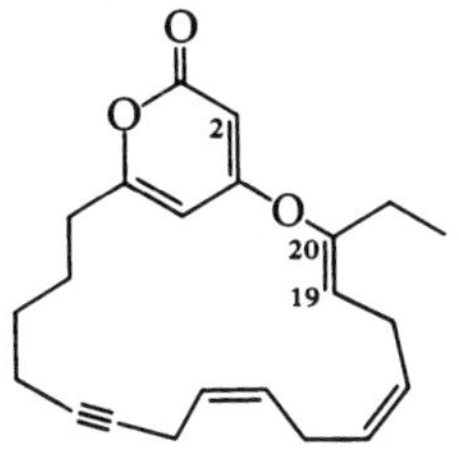

$C_{22}H_{26}O_3$ M 338.446
Authors' numbering shown. Metab. of red alga Phacelo-carpus labillardieri.

19,20-Dihydro, 2,19-dibromo: [106009-90-9]. 4,21-Di-bromo-3-ethyl-2,19-dioxabicyclo[16.3.1]docosa-6,9,18(22),21-tetraen-12-yn-20-one, 9CI.
   $C_{22}H_{26}Br_2O_3$ M 498.254
   Metab. of P. labillardieri.

Shin, J. et al, Tetrahedron Lett., 1986, 27, 5189.

2-Ethyl-1,6-dioxaspiro[4.4]nonane   E-70043
Updated Entry replacing E-30058
Chalcogran
[38401-84-2]

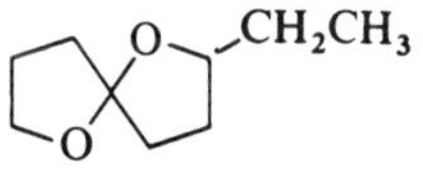

$C_9H_{16}O_2$ M 156.224
Principal aggregation pheromone of the bark beetle Pityogenes chalcographus. Ident. by glc/ms. Has been synthesised in racemic and opt. active forms; the nat. pheromone is a mixt. of diastereoisomers.

Francke, W. et al, Naturwissenschaften, 1977, 64, 590 (isol)
Jacobson, R. et al, J. Org. Chem., 1982, 47, 3140 (synth)
Redlich, H. et al, Justus Liebigs Ann. Chem., 1982, 708 (synth)
Rosini, G. et al, Angew. Chem., Int. Ed. Engl., 1986, 25, 941.
Högberg, H.E. et al, Acta Chem. Scand., Ser. B, 1987, 41, 694 (synth)

5-Ethyl-2-hydroxy-2,4,6-cycloheptatrien-   E-70044
1-one, 9CI
5-Ethyltropolone
[7159-85-5]

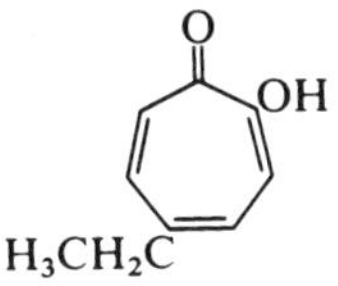

$C_9H_{10}O_2$ M 150.177
Constit. of the heartwood of Libocedrus formosana.
Prisms by subl. Mp 81-82° (78°).

Lin, Y.-T. et al, Experientia, 1966, 22, 141 (isol, uv)
Krajniak, E.R. et al, Aust. J. Chem., 1973, 26, 1337 (synth, uv, pmr)

Ethylnylcyclopropane, 9CI   E-70045
Cyclopropylacetylene
[6746-94-7]

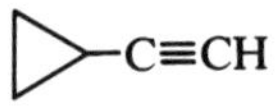

C₅H₆   M 66.102
Liq.

Slobodin, Ya.M. *et al, Zh. Org. Khim. USSR*, 1969, **5**, 1315
   (*synth*)
Schrumpf, G. *et al, Spectrochim. Acta, Part A*, 1985, **41**, 1251
   (*synth, ir*)
Hopf, H. *et al, Chem. Ber.*, 1987, **120**, 1259 (*props*)

1-Ethylnylnaphthalene, 8CI   E-70046
Updated Entry replacing N-00271
   (*1-Naphthyl*)*acetylene*
[15727-65-8]

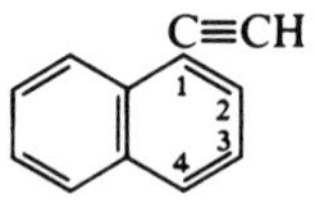

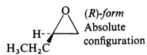

C₁₂H₈   M 152.195
Oil. d⁰ 1.066. Bp₄ 92°. n_D^{20} 1.6360. Polymerises to
   maroon polymer.

Bertin, D., *C. R. Hebd. Seances Acad. Sci.*, 1949, **229**, 660.
Atkinson, R.E. *et al, J. Chem. Soc. (C)*, 1969, 2173 (*synth*)
Okamoto, Y. *et al, J. Org. Chem.*, 1972, **37**, 3185 (*nmr*)
Neenan, T.X. *et al, J. Org. Chem.*, 1988, **53**, 2489 (*synth, ir,
   pmr*)

2-Ethyloxirane   E-70047
Updated Entry replacing E-10115
   1,2-Epoxybutane, 9CI. 1-Butene oxide
[3760-95-0]

H–⟨△⟩O   (*R*)-*form*
H₃CH₂C   Absolute
            configuration

C₄H₈O   M 72.107
▷Irritant, highly flammable fl. p. −15°
(*R*)-*form*
   Bp 59-62°. [α]$_D^{16}$ +12.4° (c, 5.98 in dioxan).
(*S*)-*form*
   Bp 72°. [α]$_D^{16}$ −12.25° (c, 6 in dioxan).
(±)-*form*
   Liq. Sol. H₂O. d$_{20}^{20}$ 0.831. Bp 63°.

Coke, J.L. *et al, J. Org. Chem.*, 1973, **38**, 2210 (*synth*)
Schmidt, U. *et al, Synthesis*, 1986, 986 (*synth, pmr*)
Chenault, H.K. *et al, J. Org. Chem.*, 1987, **52**, 2608 (*synth,
   resoln, pmr*)
Sax, N.I., *Dangerous Properties of Industrial Materials*, 5th
   Ed., Van Nostrand-Reinhold, 1979, 641.

4-Ethynylbenzaldehyde   E-70048
Updated Entry replacing E-10120
   [63697-96-1]
   C₉H₆O   M 130.146
   Yellow solid. Mp 88-90°.

Austin, W.B. *et al, J. Org. Chem.*, 1981, **46**, 2281 (*synth, ir,
   pmr, ms*)
Furber, M. *et al, J. Chem. Soc., Perkin Trans. 1*, 1987, 1573
   (*synth, ir, pmr*)

1-Ethynylbicyclo[2.2.2]octane   E-70049
[96454-73-8]

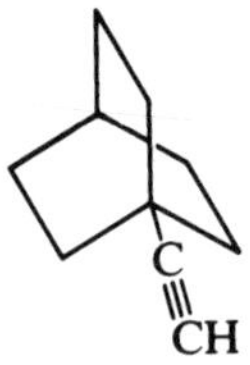

C₁₀H₁₄   M 134.221
Volatile liq.

Adcock, W. *et al, J. Org. Chem.*, 1985, **50**, 2551.
Honegger, E. *et al, Chem. Ber.*, 1987, **120**, 187 (*synth, ir, pmr,
   cmr, pe*)

1-Ethynylcycloheptene   E-70050
[2809-83-8]

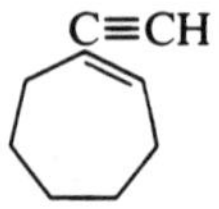

C₉H₁₂   M 120.194
Liq. Bp₁₀ 65°, Bp₁.₀ 50-52°.

Quang, L.V. *et al, Bull. Soc. Chim. Fr.*, 1965, 1525 (*synth, ir,
   uv*)
Carlson, R.G. *et al, J. Org. Chem.*, 1977, **42**, 2382 (*synth, ir,
   pmr*)

1-Ethynylcyclohexene, 9CI   E-70051
[931-49-7]

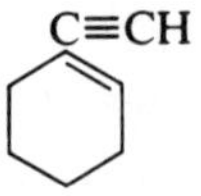

C₈H₁₀   M 106.167
Liq. Bp₆₀ 71-72°, Bp₄₀ 53-56°.

Hamlet, J.C. *et al, J. Chem. Soc.*, 1951, 2652 (*synth*)
Carlson, R.G. *et al, J. Org. Chem.*, 1977, **42**, 2382 (*synth, ir,
   pmr*)
Stille, J.K. *et al, J. Am. Chem. Soc.*, 1987, **109**, 2138 (*synth,
   pmr*)

1-Ethynylcyclopentene, 9CI   E-70052
[1610-13-5]

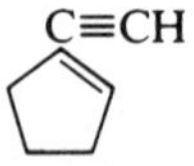

C₇H₈   M 92.140
Liq. Bp₁₀₀ 60-62°, Bp₁₅ 44-46°.

Roumestant, M.L. *et al, Bull. Soc. Chim. Fr.*, 1972, 591 (*synth,
   pmr, ir*)
Carlson, R.G. *et al, J. Org. Chem.*, 1977, **42**, 2382 (*synth, ir,
   pmr*)

Ethynylpentafluorobenzene   E-70053
(*Pentafluorophenyl*)*acetylene*
[5122-07-6]

C₆F₅C≡CH

C₇HF₅   M 180.077
Liq. Bp 130-131°, Bp₃₇ 50-52°.

Coe, P.L. *et al, J. Chem. Soc. (C)*, 1966, 597 (*synth, uv, ir, pmr*)

Waugh, F. *et al*, *J. Organomet. Chem.*, 1972, **39**, 275 (*synth, uv, ir, pmr*)

2-Ethynyl-1,3-propanediol, 9CI   E-70054

2-(Hydroxymethyl)-3-butyn-1-ol

[102573-92-2]

$$HC \equiv CCH(CH_2OH)_2$$

$C_5H_8O_2$ M 100.117

Bates, H.A. *et al*, *J. Org. Chem.*, 1986, **51**, 2637 (*synth, ir, pmr*)

2-Ethynylthiophene   E-70055

Updated Entry replacing E-01257

[4298-52-6]

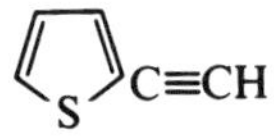

C_6H_4S M 108.158

Liq. Bp_{20} 54-60°. n_D^{24} 1.5882.

Patrick, T.B. *et al*, *J. Org. Chem.*, 1972, **37**, 4467 (*synth, ir, pmr*)
Okuhara, K., *J. Org. Chem.*, 1976, **41**, 1487 (*synth*)
Bery, J.-P. *et al*, *J. Org. Chem.*, 1982, **47**, 2201 (*synth, pmr*)
Kagan, J. *et al*, *J. Org. Chem.*, 1983, **48**, 703 (*synth, pmr*)
Carpita, A. *et al*, *Tetrahedron*, 1985, **41**, 1919 (*synth, pmr*)
Stille, J.K. *et al*, *J. Am. Chem. Soc.*, 1987, **109**, 2138 (*synth, pmr, cmr, ir*)

Euchrenone A₁   E-70056

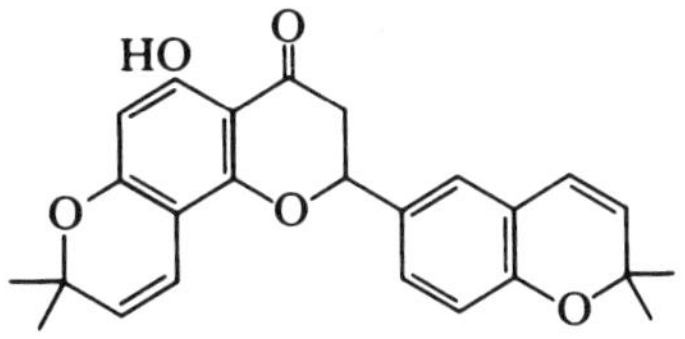

$C_{25}H_{24}O_5$ M 404.462

Constit. of *Euchresta japonica*. Pale-orange rectangles (EtOAc/hexane). Mp 120-122°.

Mizuno, M. *et al*, *Phytochemistry*, 1988, **27**, 1831.

Euchrenone A₂   E-70057

[116310-62-4]

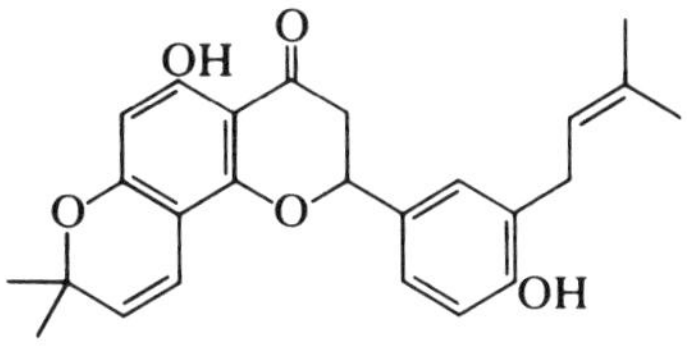

$C_{25}H_{26}O_5$ M 406.477

Constit. of *Euchresta japonica*. Pale-yellow plates (hexane). Mp 145-146°.

Mizuno, M. *et al*, *Phytochemistry*, 1988, **27**, 1831.

Euchrenone b₁   E-70058

4',5,7-Trihydroxy-3',6,8-triprenylisoflavone

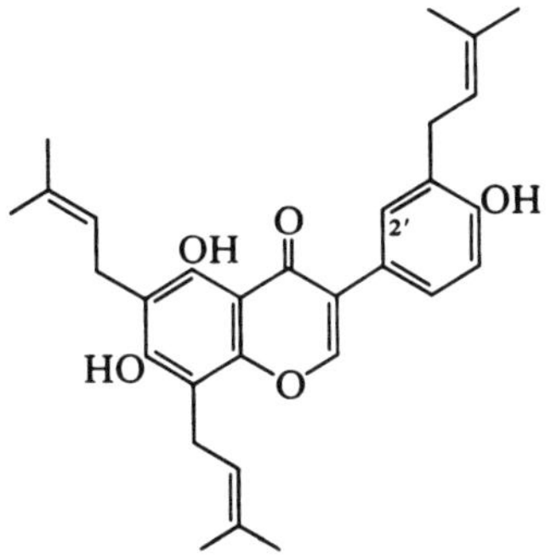

$C_{30}H_{34}O_5$ M 474.596

Constit. of *Eucresta japonica*.

2'-Hydroxy: **Eucrenone b₂**. *2',4',5,7-Tetrahydroxy-3',6,8-triprenylisoflavone.*
$C_{30}H_{34}O_6$ M 490.595
Constit. of *E. japonica*.

Mizuo, M. *et al*, *Phytochemistry*, 1988, **27**, 2975.

Euchrenone b₃   E-70059

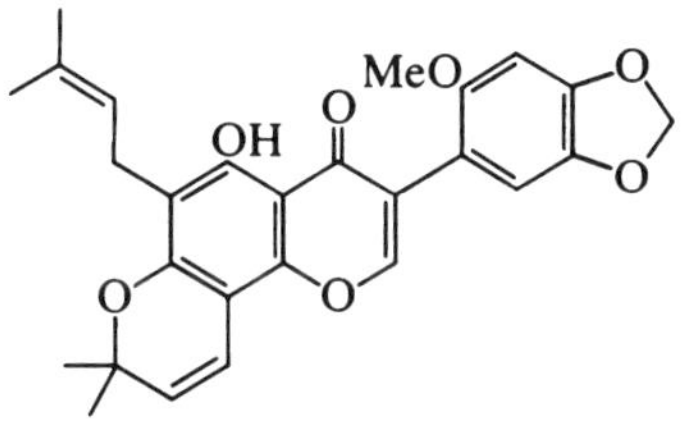

$C_{27}H_{26}O_7$ M 462.498

Constit. of *Eucresta japonica*.

Mizuo, M. *et al*, *Phytochemistry*, 1988, **27**, 2975.

3,11(13)-Eudesmadien-12-oic acid   E-70060

[28399-17-9]

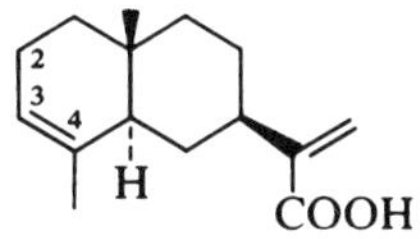

$C_{15}H_{22}O_2$ M 234.338

Constit. of *Stevia achalensis* and *Schistostephium crataegifolium*.

▷ LE3100000.

2α-Hydroxy: [88153-64-4].
$C_{15}H_{22}O_3$ M 250.337
Constit. of *S. crataegifolium*. Gum. $[\alpha]_D^{24}$ +11° (c, 0.1 in $CHCl_3$).
2-Oxo: [88153-65-5].
$C_{15}H_{20}O_3$ M 248.321
Constit. of *S. crataegifolium*. Gum. $[\alpha]_D^{24}$ +31° (c, 0.08 in $CHCl_3$).

Bohlmann, F. *et al*, *Phytochemistry*, 1983, **22**, 1632.
Bohlmann, F. *et al*, *Justus Liebigs Ann. Chem.*, 1986, 799.

4,11(13)-Eudesmadien-12-oic acid   E-70061

Isocostic acid

[69978-82-1]

$C_{15}H_{22}O_2$ M 234.338

Constit. of *Ageratina glabrata*. Oil.

3β-Hydroxy: [100045-52-1].
$C_{15}H_{22}O_3$ M 250.337

Constit. of *Stevia achalensis*. Oil.
3-Oxo: [62458-42-8]. *3-Oxo-4,11(13)-eudesmadien-12-oic acid.* **3-Oxoisocostic acid**.
$C_{15}H_{20}O_3$ M 248.321
Constit. of *S. achalensis* and *A. glabrata*. Cryst.
(Et$_2$O/pet. ether). Mp 153°.

Bohlmann, F. *et al, Chem. Ber.*, 1977, **110**, 301.
Bohlmann, F. *et al, Justus Liebigs Ann. Chem.*, 1986, 799.

1,11-Eudesmanediol   E-70062

1,11-Selinanediol
$C_{15}H_{28}O_2$ M 240.385
1β-form [71748-38-4]
   Balanitol
   Constit. of bark of *Balanites roxburghii*. Needles (C$_6$H$_6$).
   Mp 158-160°. $[\alpha]_D^{25}$ +10.5° (c, 0.84 in CHCl$_3$).

Cordano, E. *et al, J. Indian Chem. Soc.*, 1978, **55**, 1148.
Anglea, T.A. *et al, Tetrahedron*, 1987, **43**, 5537.

5,11-Eudesmanediol   E-70063

$C_{15}H_{28}O_2$ M 240.385
(4R,5S,7R,10S)-form
   Constit. of *Alpina japonica*. Needles (hexane). Mp 130-132°. $[\alpha]_D$ −21.8° (c, 0.12 in CHCl$_3$).

Itokawa, H. *et al, Chem. Pharm. Bull.*, 1987, **35**, 1460.

4(15)-Eudesmene-1,11-diol   E-70064

Updated Entry replacing E-10131
4(15)-Selinene-1,11-diol

$C_{15}H_{26}O_2$ M 238.369
1β-form [83217-89-4]
   Constit. of *Pterocarpus marsupium*. Cryst. Mp 156-157°. $[\alpha]_D^{31}$ +56.4° (c, 1.5 in CHCl$_3$).

Cordano, G. *et al, J. Indian Chem. Soc.*, 1978, **55**, 1148 (*isol*)
Adinarayana, D. *et al, Phytochemistry*, 1982, **21**, 1083.
Anglea, T.A. *et al, Tetrahedron*, 1987, **43**, 5537 (*synth*)

3-Eudesmen-6-ol   E-70065

$C_{15}H_{26}O$ M 222.370
6β-form [74144-55-1]
   Constit. of *Bazzania fauriana*. Oil. $[\alpha]_D$ −35° (c, 1.1 in CHCl$_3$).

Toyota, M. *et al, Phytochemistry*, 1988, **27**, 2155.

4(15)-Eudesmen-11-ol   E-70066

Updated Entry replacing E-01311
β-Eudesmol. 4(15)-Selinen-11-ol
[473-15-4]
$C_{15}H_{26}O$ M 222.370
Constit. of eucalyptus oils. Cryst. by subl. Mp 80-82°.
α-L-Arabopyranoside: [21615-76-9].
   $C_{20}H_{34}O_5$ M 354.486
   Constit. of *Machaeranthera tanacetifolia*. Cryst.
   (Et$_2$O/Me$_2$CO). Mp 129-130°. $[\alpha]_D^{24}$ +37.1° (c, 0.25 in EtOH).

Yoshioka, H. *et al, J. Org. Chem.*, 1969, **34**, 3697 (*deriv*)
Miller, R.B. *et al, J. Org. Chem.*, 1973, **38**, 4424 (*synth*)
Bhedi, D.N. *et al, Indian J. Chem., Sect. B*, 1976, **14**, 22 (*synth*)
Wijnberg, J.B.P.A. *et al, J. Org. Chem.*, 1983, **48**, 4380 (*synth*)
Schwartz, M.A. *et al, J. Org. Chem.*, 1985, **50**, 1359 (*synth*)
Vite, G.D. *et al, J. Org. Chem.*, 1988, **53**, 2555 (*synth*)

F

Fasciculatin
F-70001

Updated Entry replacing F-00034

[37905-12-7]

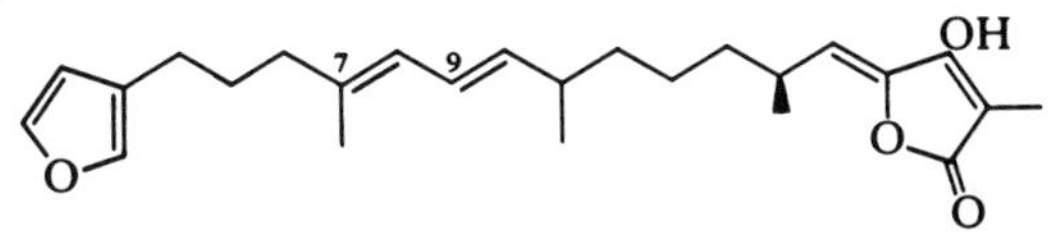

$C_{25}H_{34}O_4$ M 398.541

Constit. of the sponge, *Irania fasciculata*. Oil. $[\alpha]_D$ −15.6° (c, 0.5 in CHCl₃). The name has also been given to a flavonoid glycoside from *Veronia fasciculata*.

$\Delta^{8,10}$-*Isomer:* [94936-00-2]. **Isofasciculatin**.
  $C_{25}H_{34}O_4$ M 398.541
  Constit. of the sponge *Cacospongia scalaris*. Inhibits cell division of fertilised starfish eggs.
$\Delta^{8,10}$-*Isomer, Ac:* $[\alpha]_D$ −34.7°.

Cafieri, F. *et al*, *Tetrahedron*, 1972, **28**, 1579 (*struct, pmr, uv*)
Alfano, G. *et al*, *Experientia*, 1979, **35**, 1136 (*stereochem*)
Fusetani, N. *et al*, *Tetrahedron Lett.*, 1984, **25**, 4941 (*Isofasciculatin*)

Fauronol
F-70002

[25146-17-2]

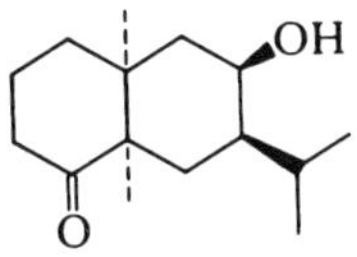

$C_{15}H_{26}O_2$ M 238.369

Ac: [2658-82-4]. *Fauronyl acetate.*
  $C_{17}H_{28}O_3$ M 280.406
  Constit. of root of *Valeriana officinalis* var. *latifolia*. Mp 85-86°. $[\alpha]_D$ −77.6°.
Ac, semicarbazone: Mp 218.5-220°.

Hikino, H. *et al*, *Chem. Pharm. Bull.*, 1965, **13**, 631; 1966, **14**, 735 (*isol, struct, abs config*)
Sammes, P.G. *et al*, *J. Chem. Soc., Perkin Trans. 1*, 1986, 281 (*synth*)

Ferprenin
F-70003

[114727-96-7]

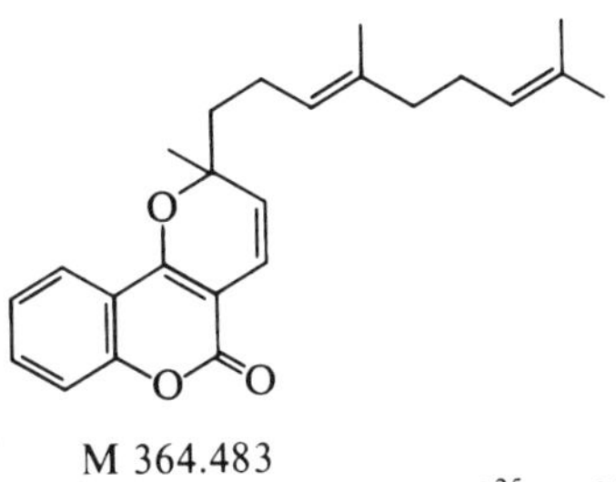

$C_{24}H_{28}O_3$ M 364.483

Constit. of *Ferula communis*. Oil. $[\alpha]_D^{25}$ +10° (c, 0.9 in CHCl₃).

Appendino, G. *et al*, *Phytochemistry*, 1988, **27**, 944.

Ferreyrantholide
F-70004

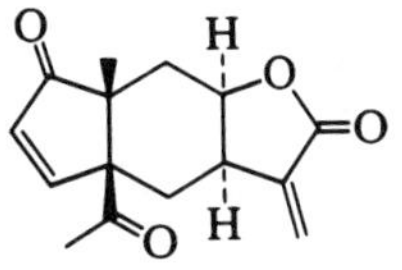

$C_{15}H_{16}O_4$ M 260.289

Constit. of *Ferreyranthus fruticosus*. Oil.

Jakupovic, J. *et al*, *Phytochemistry*, 1988, **27**, 1113.

Ferulide
F-70005

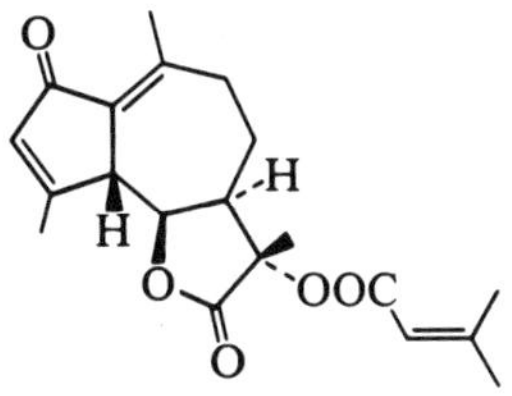

$C_{20}H_{24}O_5$ M 344.407

Constit. of *Ferula penninervis*. Cryst. Mp 135-137°.

Abdullaev, N.D. *et al*, *Khim. Prir. Soedin.*, 1987, **23**, 200.

Ferulidin
F-70006

[30860-27-6]

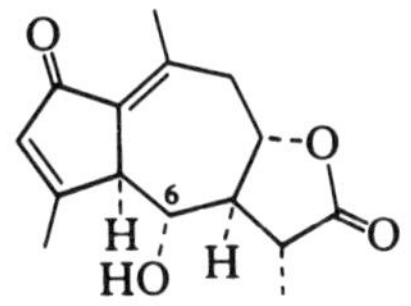

$C_{15}H_{18}O_4$ M 262.305

Constit. of *Ferula oopoda*. Mp 170-172°.

Ac: [117232-60-7]. **6-Acetylferulidin**.
  $C_{17}H_{20}O_5$ M 304.342
  Constit. of *Artemisia lanata*. Cryst. (EtOAc). Mp 207-210°.
6-Epimer: [117255-07-9]. **Carmenin**.
  $C_{15}H_{18}O_4$ M 262.305
  Constit. of *A. lanata*. Cryst. (CHCl₃/MeOH). Mp 273-275°.

Serkerov, S.Y. *et al*, *Khim. Prir. Soedin.*, 1970, **6**, 428 (*struct*)
Aguilar, J.M. *et al*, *Phytochemistry*, 1988, **27**, 2229 (*isol, struct*)

Feruone
F-70007

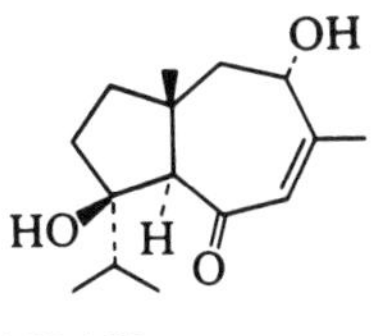

$C_{15}H_{24}O_3$ M 252.353

Constit. of *Ferula jaeschkeana*. Viscous mass. $[\alpha]_D^{20}$ +1.5° (c, 0.15 in MeOH).

Garg, S.N. *et al*, *Planta Med.*, 1987, **53**, 341.

Fibrostatins   F-70008

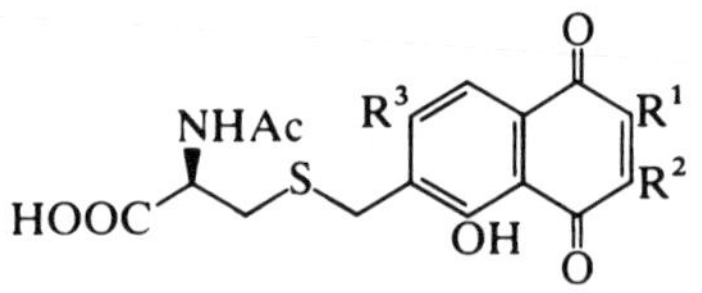

Fibrostatin *A* $R^1 = H, R^2 = CH_3, R^3 = OMe$
       B $R^1 = R^3 = OMe, R^2 = CH_3$
       C $R^1 = R^3 = OMe, R^2 = H$
       D $R^1 = OMe, R^2 = CH_3, R^3 = OH$
       E $R^1 = H, R^2 = CH_2OH, R^3 = OMe$
       F $R^1 = R^3 = OMe, R^2 = CH_2OH$

Naphthoquinone antibiotics. Isol. from *Streptomyces catenulae* ssp. *griseospora*. Prolyl hydroxylase inhibitors.

Fibrostatin A [91776-42-0]
N-*Acetyl*-S-[(*5,8-dihydro-1-hydroxy-3-methoxy-7-methyl-5,8-dioxo-2-naphthalenyl)methyl*]-L-*cysteine*, *9CI*. *P 23924*A
$C_{18}H_{19}NO_7S$ M 393.411
Orange-yellow cryst. Mp 186-188°. $[\alpha]_D^{23}$ −91° (c, 0.5 in MeOH).

Fibrostatin B [91776-48-6]
*P 23924*B
$C_{19}H_{21}NO_8S$ M 423.437
Yellowish-orange cryst. Mp 200-202°. $[\alpha]_D^{23}$ −90° (c, 0.51 in MeOH).

Fibrostatin C [91776-47-5]
*P 23924*C
$C_{18}H_{19}NO_8S$ M 409.410
Yellowish-orange cryst. Mp 187-190°. $[\alpha]_D^{23}$ −93° (c, 0.51 in MeOH).

Fibrostatin D [91776-46-4]
*P 23924*D
$C_{18}H_{19}NO_8S$ M 409.410
Yellowish-orange cryst. Mp 205-207°. $[\alpha]_D^{23}$ −62° (c, 0.51 in MeOH).

Fibrostatin E [91776-44-2]
*P 23924*E
$C_{18}H_{19}NO_8S$ M 409.410
Orange-yellow cryst. Mp 174-176°. $[\alpha]_D^{23}$ −86° (c, 0.5 in MeOH).

Fibrostatin F [91776-45-3]
*P 23924*F
$C_{19}H_{21}NO_9S$ M 439.436
Yellowish-orange cryst. Mp 191-195°. $[\alpha]_D^{23}$ −91° (c, 0.51 in MeOH).

Ishimaru, T. *et al*, *J. Antibiot.*, 1987, **40**, 1231, 1239 (*isol*, *struct*, *props*)

Flabellata secoclerodane   F-70009
[116369-02-9]

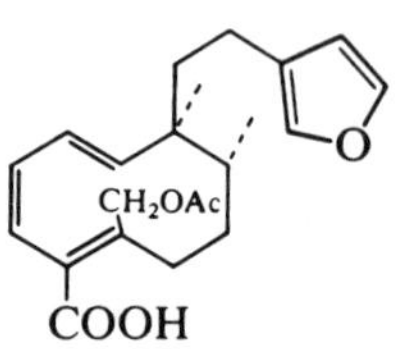

$C_{22}H_{28}O_5$ M 372.460
Constit. of *Baccharis flabellata*. Oil. $[\alpha]_D$ −33.8° (c, 4.5 in CHCl₃).

Saad, J.R. *et al*, *Phytochemistry*, 1988, **27**, 1884.

Flaccidin   F-70010
9,10-Dihydro-7-methoxy-5H-phenanthro[4,5-bcd]-pyran-2,8-diol

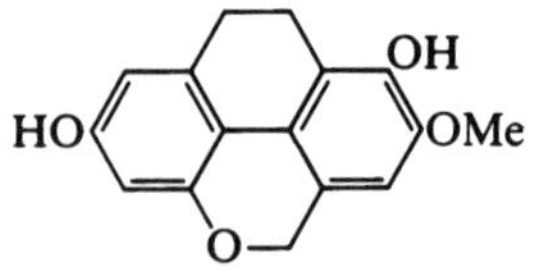

$C_{16}H_{14}O_4$ M 270.284
Constit. of *Coelogyne flaccida*. Cryst. (CHCl₃). Mp 200°.

Majumder, P.L. *et al*, *Phytochemistry*, 1988, **27**, 899.

Flavanthrin   F-70011

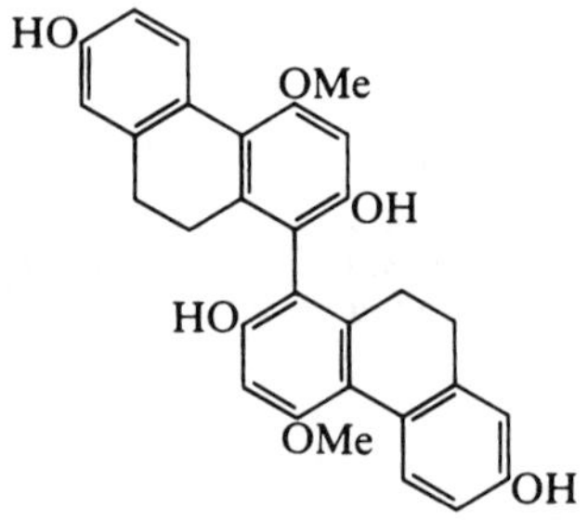

$C_{30}H_{26}O_6$ M 482.532
Constit. of *Eria flava*. Cryst. (EtOAc/pet. ether). Mp 285°.

Majumder, P.L. *et al*, *Tetrahedron*, 1988, **44**, 7303 (*struct*)

Flexilin   F-70012
Updated Entry replacing F-00196
[69625-33-8]

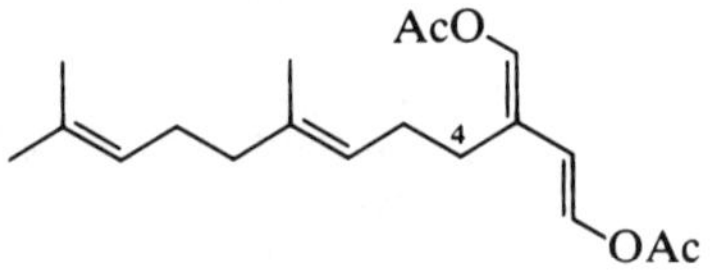

$C_{19}H_{28}O_4$ M 320.428
Constit. of *Caulerpa flexilis*. Bp₀.₁ 100°.

4-Acetoxy: [93888-65-4]. ***4-Acetoxyflexilin***.
$C_{21}H_{30}O_6$ M 378.464
Metab. of algae *Penicillus capitatus* and *Udotea cyathiformis*. Antibacterial and ichthyotoxin. Oil. $[\alpha]_D^{25}$ +18.0° (c, 1.2 in CHCl₃).

Blackman, A.J. *et al*, *Tetrahedron Lett.*, 1978, 3063 (*isol*)
Paul, V.J. *et al*, *Tetrahedron*, 1984, **40**, 2913 (*isol*, *deriv*)

Consult the *Dictionary of Alkaloids* for a comprehensive treatment of alkaloid chemistry.

Flindissol   F-70013

Updated Entry replacing F-00203
21,23-Epoxylanosta-7,24-diene-3,21-diol, *9CI*
[6805-37-4]

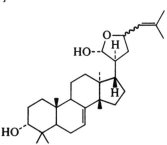

$C_{30}H_{48}O_3$ M 456.707
Constit. of *Flindersia* spp. Cryst. (Me$_2$CO). Mp 198°.
$[\alpha]_D^{24}$ −46° (c, 1.1 in CHCl$_3$).

3-Ketone: **Flindissone**.
  $C_{30}H_{46}O_3$ M 454.692
  Constit. of *Aucoumea klaineana*. Plates (EtOAc/pet.
  ether). Mp 127-130°.
21-Oxo: **Flindissol lactone**.
  $C_{30}H_{46}O_3$ M 454.692
  Constit. of *A. klaineana*. Plates (EtOAc/pet. ether).
  Mp 229-234°. $[\alpha]_D$ −50° (c, 1.18 in CHCl$_3$).
3-Ketone, 21-oxo: **Flindissone lactone**. Constit. of *A.
  klaineana*. Cryst. (EtOAc/pet. ether). Mp 193-195°.
  $[\alpha]_D$ −68° (c, 0.21 in CHCl$_3$).

Birch, A.J. *et al*, *J. Chem. Soc.*, 1963, 2762 (*isol*)
Bevan, C.W.L. *et al*, *J. Chem. Soc.* (*C*), 1967, 820 (*synth*)
Guang-Li, L. *et al*, *Phytochemistry*, 1988, **27**, 2283 (*isol, deriv*)

9*H*-Fluorene-1-carboxaldehyde   F-70014

1-Formylfluorene
[95264-32-7]

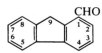

$C_{14}H_{10}O$ M 194.232
Needles (EtOH). Mp 90°.

Oxime:
  $C_{14}H_{11}NO$ M 209.247
  Needles (pet. ether). Mp 138.5-139°.
2,4-Dinitrophenylhydrazone: Orange cryst. (PhNO$_2$).
  Mp 284-286° (262°).

Mulholland, T.P.C. *et al*, *J. Chem. Soc.*, 1954, 4676 (*synth,
  deriv*)
Bergmann, E.D. *et al*, *J. Am. Chem. Soc.*, 1956, **78**, 2821
  (*synth, deriv*)
Grunewald, G.L. *et al*, *J. Med. Chem.*, 1988, **31**, 60 (*synth, pmr,
  cmr*)

9*H*-Fluorene-9-carboxaldehyde, *9CI*   F-70015

9H-Fluoren-9-ylidenemethanol, *9CI*. *9-Formylfluorene*
[20615-64-9]

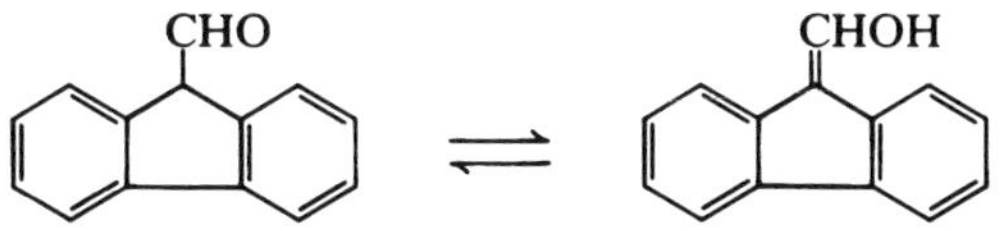

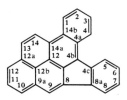

$C_{14}H_{10}O$ M 194.232
Mostly enolised (4% aldehyde in aq. soln. at r.t.). Bp$_2$
169-172°. Polymerises on storage.

▷Skin irritant
Phenylhydrazone: Mp 126°.

Von, I. *et al*, *J. Org. Chem.*, 1944, **9**, 155 (*synth, bibl*)
Harcourt, M.P. *et al*, *Bull. Soc. Chim. Fr.*, 1988, 407 (*tautom*)

Fluoreno[3,2,1,9-*defg*]chrysene, *9CI*   F-70016

Benz[a]*indeno*[*1,2,3*-cd]*pyrene*
[192-35-8]

$C_{26}H_{14}$ M 326.397
Fine burgundy or orange-red needles (xylene). Mp 261-
262°.

Aitken, I.A. *et al*, *J. Chem. Soc.*, 1960, 663 (*synth*)
Cho, B.P. *et al*, *J. Org. Chem.*, 1987, **52**, 5668 (*synth, pmr, uv*)

Fluoreno[9,1,2,3-*cdef*]chrysene, *9CI*   F-70017

[193-37-3]

$C_{26}H_{14}$ M 326.397
Cryst. (C$_6$H$_6$/hexane). Mp 239-240°.

Cho, B.P. *et al*, *J. Org. Chem.*, 1987, **52**, 5668 (*synth, pmr, uv*)

4-Fluoro-1,3-benzenediol, *9CI*   F-70018

Updated Entry replacing F-20029
4-Fluororesorcinol, *8CI*

$C_6H_5FO_2$ M 128.103

Di-Me ether: [17715-70-7]. *1-Fluoro-2,4-
  dimethoxybenzene*.
  $C_8H_9FO_2$ M 156.156
  Liq. Bp 210°.

Durrani, A.A. *et al*, *J. Chem. Soc.*, *Perkin Trans. 1*, 1980, 1658
  (*synth, pmr*)
Bélanger, P.C. *et al*, *Can. J. Chem.*, 1988, **16**, 1479 (*synth, pmr*)

5-Fluoro-3,4-dihydro-4-thioxo-2(1*H*)-pyri-   F-70019
midinone, *9CI*

5-Fluoro-4-thiouracil
[671-41-0]

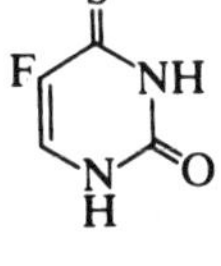

$C_4H_3FN_2OS$ M 146.139
Mp 277-278° dec.

Ueda, T. *et al*, *J. Med. Chem.*, 1963, **6**, 697 (*synth*)
Łapucha, A., *Synthesis*, 1987, 256 (*synth, ir, pmr, cmr, ms*)

2-Fluoro-6-iodobenzoic acid   F-70020

$C_7H_4FIO_2$ M 266.010

Nitrile: [79544-29-9]. *2-Cyano-1-fluoro-3-iodobenzene.*
C_7H_3FIN M 247.010
Mp 48.5-50.5°.

Hynes, J.B. *et al, J. Heterocycl. Chem.*, 1988, **25**, 1173 (*synth, deriv, nmr*)

3-Fluoro-2-iodobenzoic acid   F-70021

$C_7H_4FIO_2$ M 266.010
Cryst. (H_2O). Mp 152-153°.

Me ester:
$C_8H_6FIO_2$ M 280.037
Liq. Bp_4 127-128°.

Stanley, W.M. *et al, J. Am. Chem. Soc.*, 1933, **55**, 706 (*synth*)

3-Fluoro-4-iodobenzoic acid   F-70022

$C_7H_4FIO_2$ M 266.010
Cryst. (EtOH aq.). Mp 229-230° dec.

Tomcufcik, A.S. *et al, J. Org. Chem.*, 1961, **26**, 3351 (*synth*)

4-Fluoro-2-iodobenzoic acid   F-70023

[56096-89-0]
$C_7H_4FIO_2$ M 266.010
Cryst. (C_6H_6). Mp 147-149°.

Rajšner, M. *et al, Collect. Czech. Chem. Commun.*, 1975, **40**, 719 (*synth, uv, ir*)

4-Fluoro-3-iodobenzoic acid   F-70024

$C_7H_4FIO_2$ M 266.010
Cryst. Mp 175-176°.

Mittlestaedt, S.G. *et al, J. Am. Pharm. Assoc.*, 1950, **39**, 4; *CA*, **44**, 3465e (*synth*)

5-Fluoro-2-iodobenzoic acid   F-70025

[52548-63-7]
$C_7H_4FIO_2$ M 266.010
Cryst. (C_6H_6). Mp 149-150°.

Rajšner, M. *et al, Collect. Czech. Chem. Commun.*, 1975, **40**, 719 (*synth, uv, ir*)

1-Fluoro-3-iodopropane   F-70026

[462-40-8]

$$FCH_2CH_2CH_2I$$

C_3H_6FI M 187.983
Liq. Bp_{742} 127.5-127.8°, Bp_{95} 68-69°.

Pattison, F.L. *et al, J. Org. Chem.*, 1956, **21**, 748 (*synth*)
Chi, D.Y. *et al, J. Org. Chem.*, 1987, **52**, 658 (*synth, pmr*)

4-Fluoro-1(3H)-isobenzofuranone   F-70027
4-Fluorophthalide
[2211-81-6]

$C_8H_5FO_2$ M 152.125
Cryst. by subl. Mp 99-100°.

Bunnett, J.F. *et al, J. Am. Chem. Soc.*, 1965, **87**, 2214.
Soucy, C. *et al, J. Org. Chem.*, 1987, **52**, 129 (*synth, pmr, ir*)

7-Fluoro-1(3H)-isobenzofuranone   F-70028
7-Fluorophthalide
[2211-82-7]
$C_8H_5FO_2$ M 152.125
Cryst. by subl. Mp 166-168°.

Bunnett, J.F. *et al, J. Am. Chem. Soc.*, 1965, **87**, 2214.
Soucy, C. *et al, J. Org. Chem.*, 1987, **52**, 129 (*synth, pmr, ir*)

2-Fluoro-4-mercaptobenzoic acid   F-70029
4-Carboxy-3-fluorobenzenethiol

$C_7H_5FO_2S$ M 172.174

Nitrile: [101187-91-1]. *4-Cyano-3-fluorobenzenethiol.*
C_7H_4FNS M 153.174
Intermed. for liquid-crystal material synthesis.

Japan. Pat., 85 60 163 857, (*1985*); *CA*, **104**, 148557y (*synth*)

2-Fluoro-6-mercaptobenzoic acid   F-70030
2-Carboxy-3-fluorobenzenethiol
$C_7H_5FO_2S$ M 172.174

Nitrile, Me ether: 2-Fluoro-6-(*methylthio*)*benzonitrile.*
C_8H_6FNS M 167.201
Mp 63-64.5°.

Hynes, J.B. *et al, J. Heterocycl. Chem.*, 1988, **25**, 1173 (*synth, deriv, nmr*)

4-Fluoro-2-mercaptobenzoic acid   F-70031
2-Carboxy-5-fluorobenzenethiol
[81223-43-0]
$C_7H_5FO_2S$ M 172.174
Cryst. (C_6H_6). Mp 169-172° subl.

Me ester: [81223-44-1].
$C_8H_7FO_2S$ M 186.201
Mp 45-47°.

Polívka, Z. *et al, Collect. Czech. Chem. Commun.*, 1981, **46**, 2222 (*synth, deriv, uv, ir*)

(Fluoromethyl)oxirane, 9CI F-70032

Updated Entry replacing F-00484
1,2-Epoxy-3-fluoropropane, 8CI. Epifluorohydrin
[503-09-3]

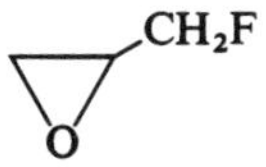

C$_3$H$_5$FO   M 76.070

▷TZ3325000.

(±)-form [103129-23-3]
Liq. d$_4^{20}$ 1.067. Bp 84-85°. n_D^{25} 1.3679. (*R*)-form also
known, no phys. props. reported.

▷Lachrymator, corrosive. Flash pt. 4°

McBee, E.T. *et al, J. Am. Chem. Soc.*, 1952, **74**, 3022 (*synth*)
Pattison, F.L.M. *et al, J. Am. Chem. Soc.*, 1957, **79**, 2311
    (*synth*)
Thomas, W.A., *J. Chem. Soc. (B)*, 1968, 1187 (*pmr, conformn*)
MacDonald, C.J. *et al, Can. J. Chem.*, 1970, **48**, 1046 (*pmr, bibl*)
Charles, S.W. *et al, J. Mol. Struct.*, 1974, **20**, 83 (*ir, conformn*)
Fujiwara, F.G. *et al, J. Mol. Struct.*, 1977, **41**, 169 (*bibl, microwave*)
Shapiro, M., *J. Org. Chem.*, 1977, **42**, 1434 (*cmr, conformn*)
Kalasinsky, V.F. *et al, J. Raman Spectrosc.*, 1980, **9**, 45 (*ir, raman*)
Martin, J.C. *et al, J. Org. Chem.*, 1982, **47**, 3531 (*R-form*)
Aldrich Library of FT-IR Spectra, 1st Ed., **1**, 231C (*ir*)
Aldrich Library of NMR Spectra, 2nd Ed., **1**, 194B (*pmr*)
Sigma-Aldrich Library of Chemical Safety Data, 1st Ed., 854C
    (*haz*)

3-Fluoro-4-nitrobenzoic acid F-70033

[403-21-4]
C$_7$H$_4$FNO$_4$ M 185.111
Cryst. (H$_2$O). Mp 174-175°.

Schmelkes, F.C. *et al, J. Am. Chem. Soc.*, 1944, **66**, 1631
    (*synth*)
Henkin, J. *et al, J. Med. Chem.*, 1983, **26**, 1193 (*synth, pmr*)

1-Fluoro-1-octene, 9CI F-70034

[72011-63-3]

H$_3$C(CH$_2$)$_5$CH=CHF

C$_8$H$_{15}$F   M 130.205

Burton, D.J. *et al, J. Org. Chem.*, 1975, **40**, 2796 (*synth*)
Barnette, W.E., *J. Am. Chem. Soc.*, 1984, **106**, 452 (*synth*)
Schwartz, J. *et al, J. Am. Chem. Soc.*, 1986, **108**, 2445 (*synth, nmr*)

4-Fluoro-1-phenyl-1-butene F-70035

(*4-Fluoro-1-butenyl*)*benzene, 9CI*

PhCH=CHCH$_2$CH$_2$F

C$_{10}$H$_{11}$F   M 150.195

(*E*)-form [110653-03-7]
Oil.

Kanemoto, S. *et al, Tetrahedron Lett.*, 1987, **28**, 663 (*synth*)

3-Fluoro-3-phenyl-3*H*-diazirine F-70036

Updated Entry replacing F-40038
[87282-19-7]

C$_7$H$_5$FN$_2$ M 136.128
Fluorocarbene precursor. Bp$_{14}$ 45-50°.

▷Can dec. explosively on dist. >35°

Cox, D.P. *et al, J. Am. Chem. Soc.*, 1983, **105**, 6513 (*synth, ir, uv, pmr, fmr, cmr*)
Moss, R.A. *et al, J. Org. Chem.*, 1984, **49**, 3828 (*use*)
McMahon, R.J. *et al, J. Am. Chem. Soc.*, 1987, **109**, 2456
    (*synth, ir, uv, pmr, cmr, haz*)

3-Fluoro-3-(trifluoromethyl)-3*H*-diazirine, F-70037
9CI

[117113-32-3]

F$_3$C $\overset{F}{\underset{}{\diagup}}$ ⟨N‖N⟩

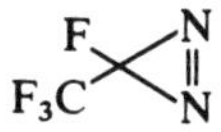

C$_2$F$_4$N$_2$ M 128.029
Source of trifluoroethylidene. Gas.

▷Potentially explosive

Dailey, W.P., *Tetrahedron Lett.*, 1987, **28**, 5801 (*synth, ir, uv, ms, F nmr*)

2-(2-Formyl-3-hydroxymethyl-2-cyclopen- F-70038
tenyl)-6,10-dimethyl-5,9-undecadienal

α-(4,8-Dimethyl-3,7-nonadienyl)-2-formyl-3-(hydroxy-methyl)-2-cyclopentene-1-acetaldehyde, 9CI
[93888-62-1]

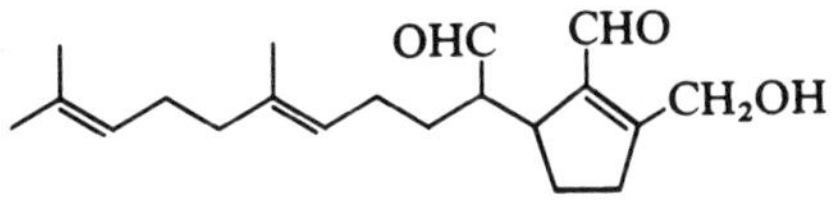

C$_{20}$H$_{30}$O$_3$ M 318.455
Metab. of *Udotea flabellum*. Antifungal, antibacterial,
    ichthyotoxin. Oil. [α]$_D^{25}$ −26.4° (c, 0.9 in CHCl$_3$).

Paul, V.J. *et al, Tetrahedron*, 1984, **40**, 2913.

2-Formylpropanoic acid, 9CI F-70039

H$_3$CCH(CHO)COOH

C$_4$H$_6$O$_3$ M 102.090
tert-*Butyl ester:*
    C$_8$H$_{14}$O$_3$ M 158.197
    Oil. ca. 60% enolised.
tert-*Butyl ester, semicarbazone:* Cryst. (Et$_2$O/hexane).
    Mp 126-128°.

Sato, M. *et al, Chem. Pharm. Bull.*, 1986, **34**, 4577.

Fragransin D_1 F-70040

[114394-21-7]

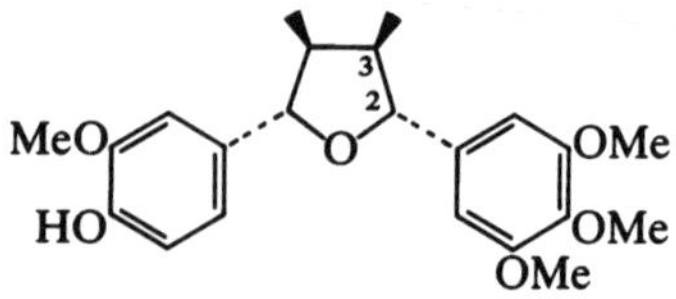

$C_{22}H_{28}O_6$ M 388.460
Constit. of *Myristica fragrans*. Oil. $[\alpha]_D$ +18.38° (c,
0.136 in $CHCl_3$).

3-Epimer: [114422-24-1]. **Fragransin D_3**.
  $C_{22}H_{28}O_6$ M 388.460
  Constit. of *M. fragrans*. Oil. $[\alpha]_D$ +11.45° (c, 0.262 in
  $CHCl_3$).
2,3-Diepimer: [114422-23-0]. **Fragransin D_2**.
  $C_{22}H_{28}O_6$ M 388.460
  Constit. of *M. fragrans*. Oil. $[\alpha]_D$ +30.49° (c, 0.106 in
  $CHCl_3$).

Hada, S. *et al*, *Phytochemistry*, 1988, **27**, 563.

Fragransol A   F-70041

[114394-19-3]

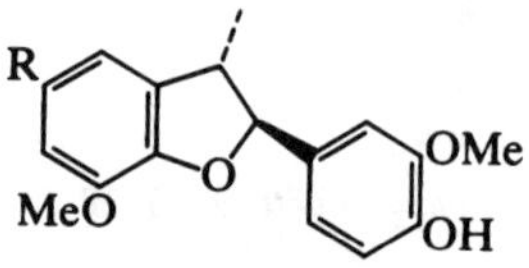

$$R = -CH(OMe)CH(OH)CH_3$$

$C_{21}H_{26}O_6$ M 374.433
Constit. of *Myristica fragrans*. Oil.
Hada, S. *et al*, *Phytochemistry*, 1988, **27**, 563.

Fragransol B   F-70042

[114394-20-6]
As Fragransol A, F-70041 with

$$R = -CH_2CH_2OH$$

$C_{19}H_{22}O_5$ M 330.380
Constit. of *Myristica fragrans*. Oil.
Hada, S. *et al*, *Phytochemistry*, 1988, **27**, 563.

Fridamycin A   F-70043

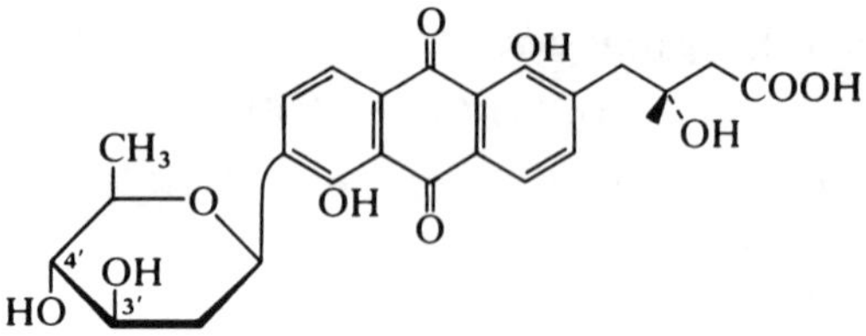

$C_{25}H_{26}O_{10}$ M 486.474
Anthracycline antibiotic. Prod. by *Streptomyces parvulus*. Related to Vineomycin B_2, V-20018 .

Aglycone: [116120-54-8]. **Fridamycin E**.
  $C_{19}H_{16}O_7$ M 356.331
  From *S. parvulus*.
3′,4′-Diepimer: **Fridamycin B**.
  $C_{25}H_{26}O_{10}$ M 486.474
  From *S. parvulus*.

Krohn, K. *et al*, *Tetrahedron*, 1988, **44**, 49.

Fridamycin D   F-70044

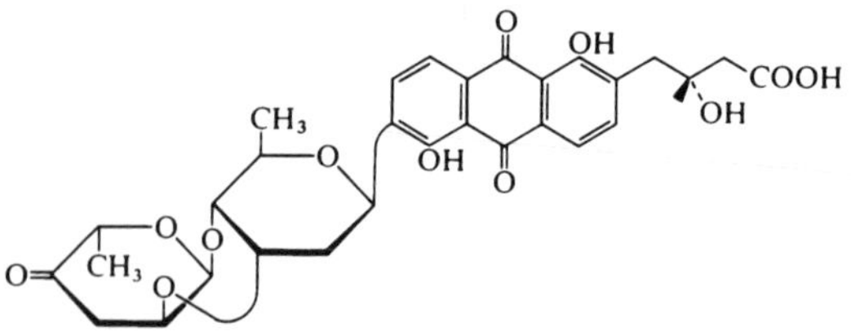

$C_{31}H_{32}O_{12}$ M 596.587
Anthracycline antibiotic. Prod. by *Streptomyces parvulus*.

Krohn, K. *et al*, *Tetrahedron*, 1988, **44**, 49.

Fruticolide   F-70045

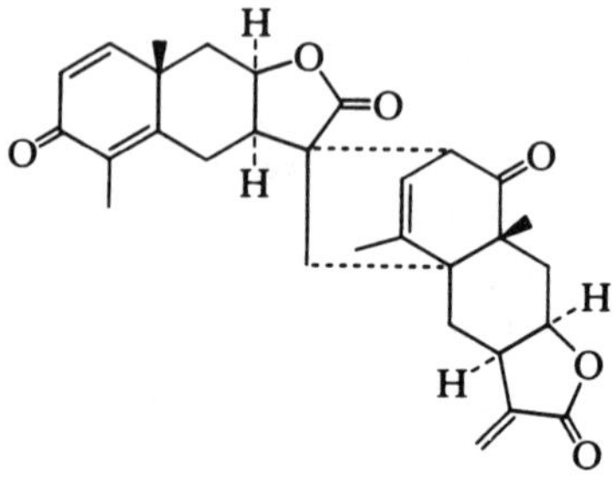

$C_{30}H_{32}O_6$ M 488.579
Constit. of *Ferreyranthus fruticosus*. Oil.
Jakupovic, J. *et al*, *Phytochemistry*, 1988, **27**, 1113.

Funadonin   F-70046

*6-(2-Hydroxy-3-oxobutyl)-7-methoxy-2H-1-benzo-
pyran-2-one, 9CI*
[117597-78-1]

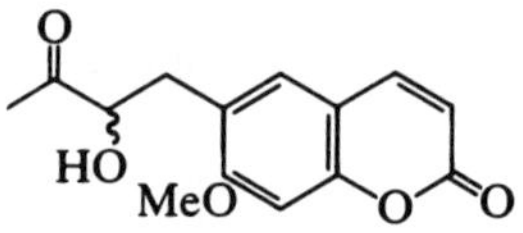

$C_{14}H_{14}O_5$ M 262.262
Constit. of *Citrus funadoko*. Oil. $[\alpha]_D$ +9.43° (c, 0.003
in $CHCl_3$).

Ju-ichi, M. *et al*, *Heterocycles*, 1988, **27**, 1451.

Furanoeremophilane-2,9-diol   F-70047

Updated Entry replacing F-00835
2,9-Dihydroxyfuranoeremophilane

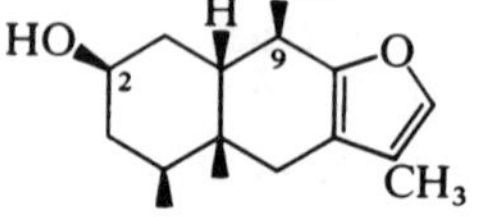

$C_{15}H_{22}O_3$ M 250.337
(2β,9β)-form
  Furanopetasol
  O^2-*Angeloyl:* [6902-62-1]. **Furanopetasin**.
    $C_{20}H_{28}O_4$ M 332.439
    Constit. of *Petasites officianalis*. Cryst. Mp 105-106°.

Novotný, L. *et al*, *Collect. Czech. Chem. Commun.*, 1964, **29**,
  1922; 1987, **52**, 1786 (*isol, struct, cd*)

5-(2-Furanyl)oxazole, 9CI
F-70048

2-(5-Oxazolyl)furan

[70380-67-5]

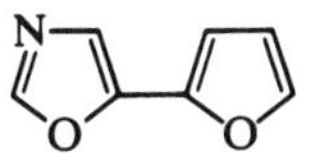

$C_7H_5NO_2$ M 135.122

Pale-yellow oil. Bp$_{0.3}$ 61-63°.

Saikachi, H. *et al*, *Chem. Pharm. Bull.*, 1979, **27**, 793 (*synth*, *uv*, *pmr*)

14-(3-Furanyl)-3,7,11-trimethyl-7,11-tetra-decadienoic acid, 9CI
F-70049

[82124-06-9]

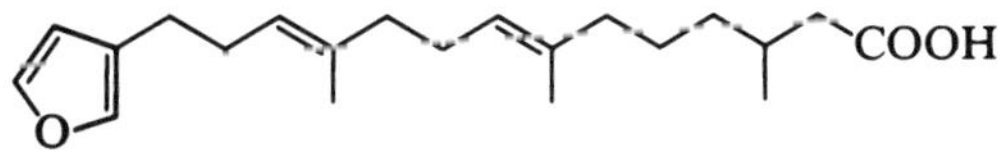

$C_{21}H_{32}O_3$ M 332.482

Constit. of a *Sarcotragus* sp. Oil.

Barrow, C.J. *et al*, *J. Nat. Prod.*, 1988, **51**, 275.

5-[13-(3-Furanyl)-2,6,10-trimethyl-6,8-tri-decadienyl]-4-hydroxy-3-methyl-2(5H)-furanone
F-70050

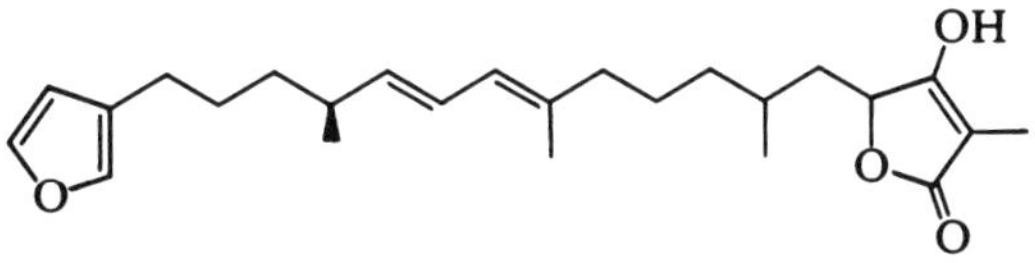

$C_{25}H_{36}O_4$ M 400.557

Constit. of a sponge *Ircinia* sp. Oil. [α]$_D$ +41° (c, 3.7 in CHCl$_3$).

Capon, R.J. *et al*, *Aust. J. Chem.*, 1987, **40**, 1327.

Furazano[3,4-b]quinoxaline
F-70051

[1,2,5]Oxadiazolo[3,4-b]quinoxaline, 9CI

[67506-48-3]

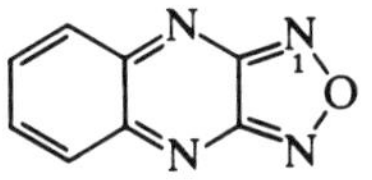

$C_8H_4N_4O$ M 172.146

Mp 181-182°. Coloured but colour not specified.

1-Oxide: [79421-50-4]. Furoxano[3,4-b]quinoxaline.
$C_8H_4N_4O_2$ M 188.145
Mp 161-162°. Coloured but no colour mentioned.

Nicolaides, D.N. *et al*, *Synthesis*, 1981, 638 (*synth*, *uv*, *ir*, *pmr*, *ms*)

Furcellataepoxylactone
F-70052

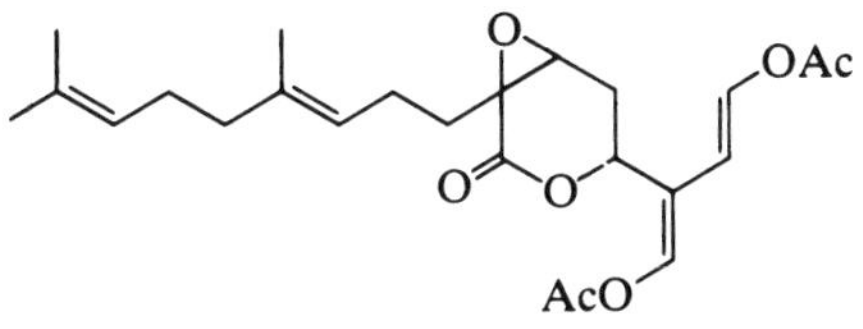

$C_{24}H_{32}O_7$ M 432.513

Constit. of *Pseudochlorodesmis furcellata*. Oil. [α]$_D^{25}$ +1.6° (c, 0.7 in CHCl$_3$).

Paul, V.J. *et al*, *Phytochemistry*, 1988, **27**, 1011.

Furlone yellow
F-70053

[6007-50-7]

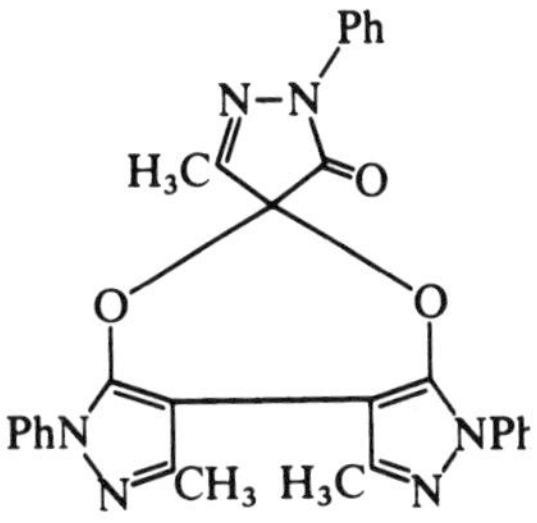

$C_{30}H_{24}N_6O_3$ M 516.558

Trimer of 3-methyl-1-phenyl-2-pyrazolin-5-one. Yellow cryst. (EtOH). Mp 158° dec.

Westöö, G., *Acta Chem. Scand.*, 1953, **7**, 360 (*synth*, *uv*)
Mann, G. *et al*, *Tetrahedron Lett.*, 1979, 4645 (*pmr*, *cmr*)
Rissanen, K. *et al*, *Acta Crystallogr., Sect. C*, 1988, **44**, 845 (*cryst struct*)

Furodysinin hydroperoxide
F-70054

[103202-15-9]

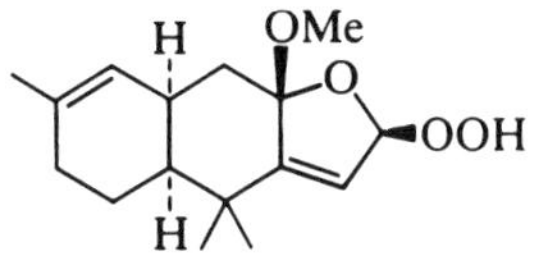

$C_{16}H_{24}O_4$ M 280.363

Metab. of the nudibranch *Chromodoris funerea*. Cryst. (Et$_2$O/hexane). Mp 142-143°. [α]$_D$ −63.4° (c, 0.5 in CHCl$_3$).

Faulkner, D.J. *et al*, *J. Org. Chem.*, 1986, **51**, 3528 (*isol*, *ir*, *pmr*, *cmr*, *cryst struct*)

3-(Furo[3,4-b]furan-4-yl)-2-propenenitrile, 9CI
F-70055

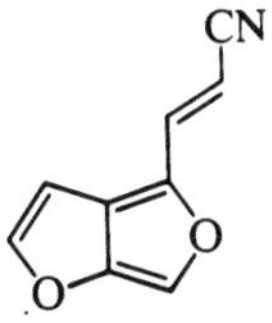

(*E*)-*form*

$C_9H_5NO_2$ M 159.144

First known furo[3,4-b]furan.

(*E*)-*form* [113893-50-8]
Stable yellow cryst. (Et$_2$O/pet. ether). Mp 144°.

(*Z*)-*form* [113893-52-0]
Characterised by pmr.

Eberbach, W. *et al*, *Angew. Chem., Int. Ed. Engl.*, 1988, **27**, 568 (*synth*, *ir*, *uv*, *pmr*)

Furoixiolal
F-70056

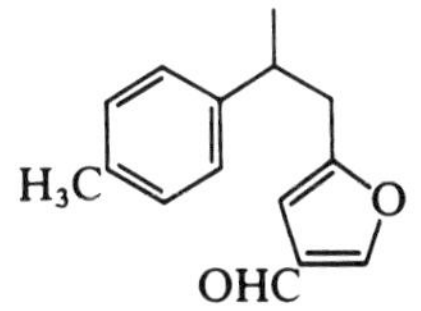

$C_{15}H_{16}O_2$ M 228.290

Constit. of *Ixiolaena leptolepis*. Oil.

Lehmann, L. *et al*, *Phytochemistry*, 1988, **27**, 2994.

Furoscalarol   F-70057

17a,21-Epoxy-4,4,8-trimethyl-D(17a)-homopregna-17,20-diene-12α,16β-diol 12-acetate, 9CI

[64285-85-4]

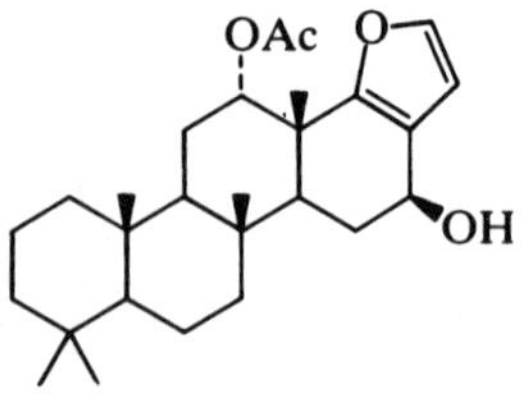

$C_{27}H_{40}O_4$ M 428.611

Isol. from the sponge *Cacospongia mollior*. Cryst. (pet. ether). Mp 181-183°. $[\alpha]_D$ +14.7° (c, 1 in $CHCl_3$). Stereochem. shown is obt. from CA; not detd. in the ref. quoted.

Cafieri, F. *et al, Gazz. Chim. Ital.*, 1977, **107**, 71.

Fusarubinoic acid   F-70058

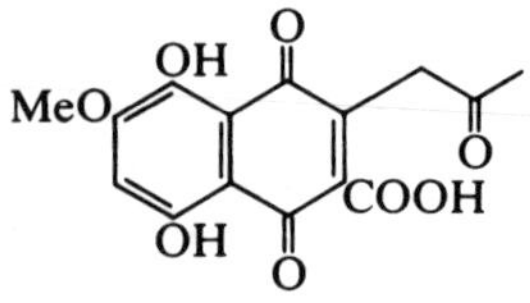

$C_{15}H_{12}O_8$ M 320.255

Metab. of *Nectria haematococca*. Red cryst. (CH_2Cl_2). Mp 200-210°.

Me ester: Red needles (MeOH). Mp 187-190°.

Parisot, D. *et al, Phytochemistry*, 1988, **27**, 3002.

G

Gallicadiol
G-70001

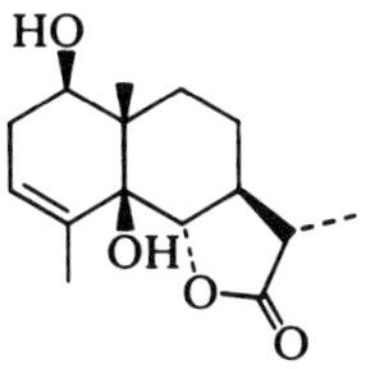

C$_{15}$H$_{22}$O$_4$ M 266.336

Constit. of *Artemisia maritima gallica*. Cryst. (CH$_2$Cl$_2$/hexane). Mp 219-221°. [α]$_D$ −11.7° (c, 0.2 in CHCl$_3$). Rare *cis*-eudesmanolide.

Gonzalez, A.G. *et al*, *Tetrahedron*, 1988, **44**, 6750.

Galloxanthin
G-70002

10'-Apo-β-carotene-3,10'-diol

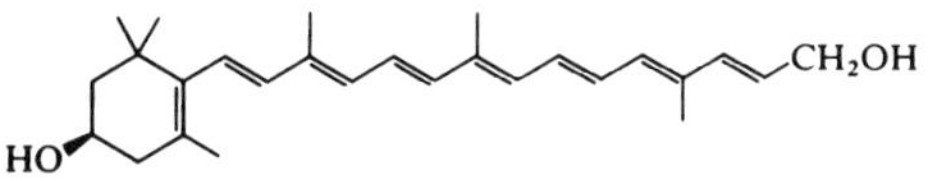

C$_{27}$H$_{38}$O$_2$ M 394.596

(**R**)-**form** [113724-67-7]

Constit. of rose flowers and retina of chicken.

Märki-Fischer, E. *et al*, *Helv. Chim. Acta*, 1987, **70**, 1988.

Galtamycin
G-70003

[103735-89-3]

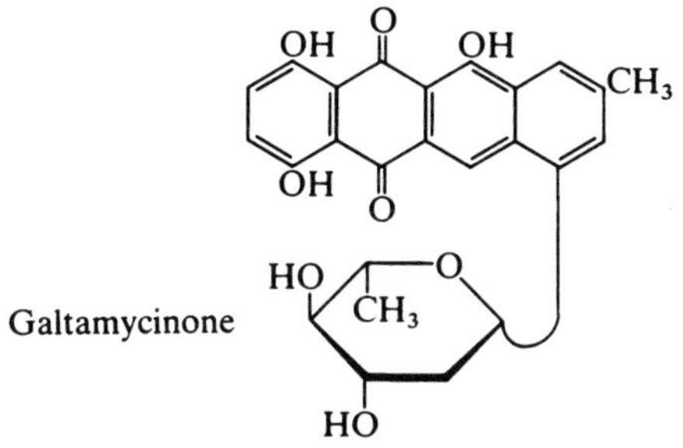

Galtamycinone

Anthracycline antibiotic of partially unknown struct. Isol. from *Streptomyces* sp. Shows low antitumour activity.

Hydrolysis prod.: [105997-04-4]. *Galtamycinone*.
C$_{25}$H$_{22}$O$_8$ M 450.444
Contains 2 less sugar residues than parent.

Murenets, N.V. *et al*, *Antibiot. Med. Bioteknol.*, 1986, **31**, 428, 431 (*isol, struct, props*)

Garcinone *D*
G-70004

[107390-08-9]

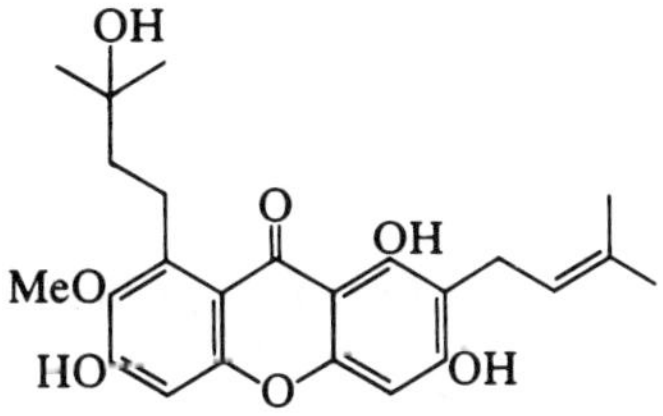

C$_{24}$H$_{28}$O$_7$ M 428.481

Constit. of *Garcinia mangostana*. Light-yellow needles (MeOH). Mp 202-204°.

Sen, A.K. *et al*, *Indian J. Chem., Sect. B*, 1986, **25**, 1157.

GB 1
G-70005

Updated Entry replacing H-00267

2,2',3,3'-Tetrahydro-3',5,5',7,7'-pentahydroxy-2,2'-bis(4-hydroxyphenyl)[3,8'-bi-4H-1-benzopyran]-4,4'-dione, 9Cl. 3'',4',4''',5,5'',7,7''-Heptahydroxy-3,8''-biflavanone, 8Cl

[14736-58-4]

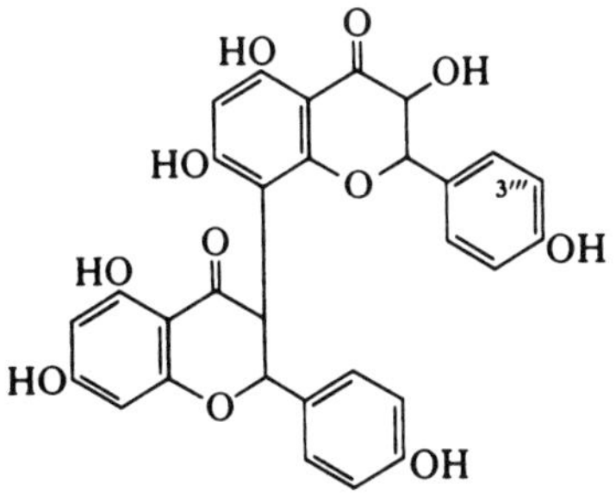

C$_{30}$H$_{22}$O$_{11}$ M 558.497

Constit. of *Garcinia buchananii* (heartwood), *G. terpnophylla* (timber and bark) and *G. kola* (nuts), also of *G. buchananii*. Component of Kolaviron. Kolaviron shows antihepatotoxic props. Cryst. (C$_6$H$_6$/Me$_2$CO) or amorph. powder. Mp 210°.

3'''-Hydroxy: [18913-18-3]. *3'',3''',4',4''',5,5'',7,7''-Octa-hydroxy-3,8''-biflavanone, 8Cl.* **GB-2**.
C$_{30}$H$_{22}$O$_{12}$ M 574.497
From *G. buchananii*, *G. kola* and *G. terpnophylla*. Amorph. solid. Mp 230°.

3''-Deoxy: [19360-72-6]. **GB1a**.
C$_{30}$H$_{22}$O$_{10}$ M 542.498
Constit. of the nuts of *G. kola*, *G. buchananii* and *G. spicata*. Amorph.

3'-Deoxy, 3'''-hydroxy: [18412-96-9]. **GB2a**.
C$_{30}$H$_{22}$O$_{11}$ M 558.497
Constit. of *G. kola* and *G. spicata*.

O$^{4''}$-Me, 3'''-hydroxy: [68705-66-8]. **Kolaflavanone**.
C$_{31}$H$_{24}$O$_{12}$ M 588.523
Constit. of the nuts of *G. kola*. Amorph. powder.

Pelter, A., *Tetrahedron Lett.*, 1967, 1767 (*struct*)
Kanoshima, M. *et al*, *Tetrahedron Lett.*, 1970, 4203 (*isol*)
Jackson, B. *et al*, *J. Chem. Soc. (C)*, 1971, 3791 (*isol, ir, uv, pmr, ms*)
Bandaranyake, W.M. *et al*, *Phytochemistry*, 1975, **14**, 1878 (*isol, ir, uv, pmr, ms*)

Cotterill, P.J. *et al, J. Chem. Soc., Perkin Trans. 1*, 1978, 532 (*isol, ir, uv, pmr, cmr*)
Duddeck, H. *et al, Phytochemistry*, 1978, **17**, 1369 (*cd*)
Iwu, M. *et al, J. Nat. Prod.*, 1982, **45**, 650 (*isol*)

Gedunin   G-70006

Updated Entry replacing G-00069

[2753-30-2]

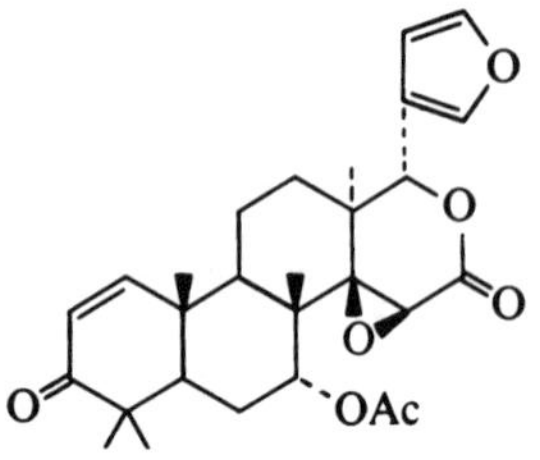

$C_{28}H_{34}O_7$ M 482.572

Constit. of *Entandrophragma angolense*. Cryst. (MeOH). Mp 157° and 218° (double Mp). $[\alpha]_D^{20}$ −44° (CHCl₃).

▷ Exp. neoplastic agent

1,2-Dihydro: [2629-11-0]. ***1,2-Dihydrogedunin***.
$C_{28}H_{36}O_7$ M 484.588
Constit. of *Guaren thompsonii*. Cryst. (CHCl₃/MeOH). Mp 237-238°. $[\alpha]_D$ +3.7° (CHCl₃).
▷ WH1318000.

6α-Hydroxy: [39838-58-9]. ***6α-Hydroxygedunin***.
$C_{28}H_{34}O_8$ M 498.572
Constit. of *Carapa guianensis*. Cryst. (CHCl₃/hexane). Mp 175-177°. $[\alpha]_D$ +100° (c, 0.1 in CHCl₃).

6α-Acetoxy: [39724-61-3]. ***6α-Acetoxygedunin***.
$C_{30}H_{36}O_9$ M 540.609
From *C. guianensis*. Cryst. (Me₂CO/MeOH). Mp 270-273°. $[\alpha]_D$ +141° (c, 1 in CHCl₃).

6β-Hydroxy: Cryst. (C₆H₆). Mp 212-220°. $[\alpha]_D^{20}$ +41°

11β-Acetoxy: [6042-73-5]. ***11β-Acetoxygedunin***.
$C_{30}H_{36}O_9$ M 540.609
Constit. of *Carapa guianensis*. Needles (CHCl₃/Et₂O/pet. ether). Mp 176-178°. $[\alpha]_D$ +33° (c, 1.2 in CHCl₃).

6α,11β-Diacetoxy: ***6α,11β-Diacetoxygedunin***.
$C_{32}H_{38}O_{11}$ M 598.646
Constit. of *C. guianensis*. Cryst. (Me₂CO/Et₂O/pet. ether). Mp 184-188°. $[\alpha]_D$ +120° (c, 1.46 in CHCl₃).

7-Deacetoxy, 7-ketone: [13072-74-7]. ***7-Deacetoxy-7-oxogedunin***.
$C_{26}H_{30}O_6$ M 438.519
Constit. of *Cedrela odorata* and *Carapa guayanensis*. Prisms (C₆H₆/pet. ether). Mp 262-265°. $[\alpha]_D^{20}$ −50° (CHCl₃).

Housley, J.R. *et al, J. Chem. Soc.*, 1962, 5095 (*isol*)
Bevan, C.W.L. *et al, J. Chem. Soc.*, 1963, 980 (*7-Deacetoxy-7-oxogedunin*)
Connolly, J.D. *et al, Tetrahedron*, 1966, **22**, 891 (*11-Acetoxygedunin, 6,11-Diacetoxygedunin*)
Baldwin, M.A. *et al, J. Chem. Soc.* (*C*), 1967, 1026 (*ms*)
Ohochucku, N.S. *et al, J. Chem. Soc.* (*C*), 1969, 864 (*pmr*)
Ohochucku, N.S. *et al, J. Chem. Soc.* (*C*), 1970, 421 (*deriv*)
Lavie, D. *et al, Bioorg. Chem.*, 1972, **2**, 59 (*deriv*)
Taylor, D.A.H. *et al, J. Chem. Soc., Perkin Trans. 1*, 1974, 437 (*cmr*)
Sax, N.I., *Dangerous Properties of Industrial Materials*, 5th Ed., Van Nostrand-Reinhold, 1979, 703.

Geodoxin   G-70007

[428-21-7]

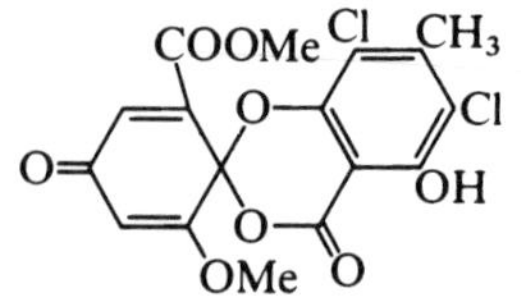

$C_{17}H_{12}Cl_2O_8$ M 415.183

Metab. of *Aspergillus terreus*. Yellow needles (CHCl₃/Et₂O or EtOAc/pet. ether). Mp 216-217° dec. (211-212°). $[\alpha]_D$ 0° (c, 1 in CHCl₃). Closely related to Geodin, G-00111 .

Hassall, C.H. *et al, J. Chem. Soc.*, 1959, 2831; 1961, 2312 (*isol, struct, uv, synth, biosynth*)
Ballantine, J.A. *et al, Org. Mass Spectrom.*, 1969, **2**, 1145 (*ms*)

4-Geranyl-3,4′,5-trihydroxystilbene   G-70008

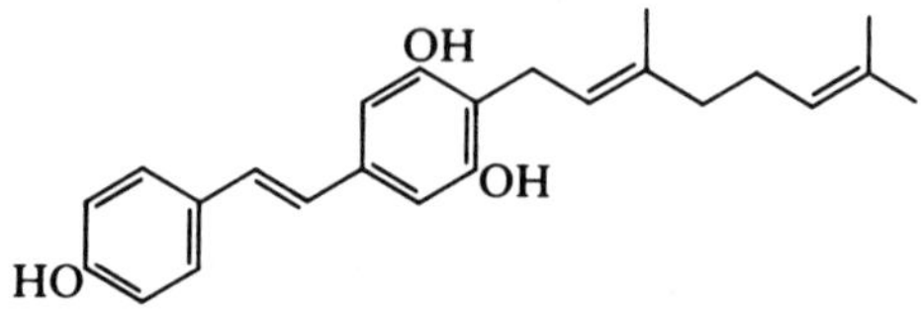

$C_{24}H_{28}O_3$ M 364.483

Constit. of *Chlorophora excelsa*. Cryst. Mp 145-150°. Related to Chlorophorin, C-02143 .

Christensen, L.P. *et al, Phytochemistry*, 1988, **27**, 3014.

5,10(14)-Germacradiene-1,4-diol   G-70009

Updated Entry replacing G-20018

4-Methyl-10-methylene-7-(1-methylethyl)-5-cyclodecene-1,4-diol, 9Cl. 7-Isopropyl-4-methyl-10-methylene-5-cyclodecene-1,4-diol. 1,4-Dihydroxy-5,10(14)-germacradiene

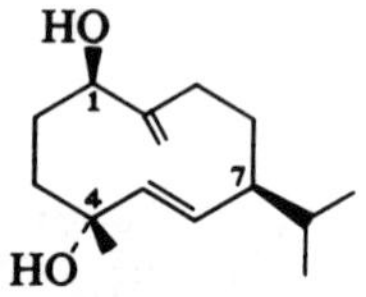

$C_{15}H_{26}O_2$ M 238.369

(1S,3S)-form [63181-39-5]

Isol. from *Laurencia subopposita, Parthenium lozanianum* and *Wikstroemea sikokiana*. Cryst. (Et₂O). Mp 118-120°. $[\alpha]_D^{20}$ +55° (c, 2.5 in CHCl₃), $[\alpha]_D^{24}$ −13° (c, 1.3 in CHCl₃). Discrepancy in opt. rotn. of the *P. lozanianum* isolate is unexplained.

1-Ac: [63181-40-8].
$C_{17}H_{28}O_3$ M 280.406
Constit. of soft coral *Lemnalia africana*. Oil. $[\alpha]_D$ +23° (c, 1.39 in CHCl₃).

Wratten, S.J. *et al, J. Org. Chem.*, 1977, **42**, 3343 (*isol*)
Izac, R.R. *et al, Tetrahedron*, 1982, **38**, 301 (*isol*)
Jakupovic, J. *et al, Phytochemistry*, 1987, **26**, 761 (*isol*)
Kitagawa, I. *et al, Chem. Pharm. Bull.*, 1987, **35**, 124 (*isol*)

1(10),4,11(13)-Germacratrien-8,12-olide G-70010

Updated Entry replacing G-30004
Desacetoxylaurenobiolide

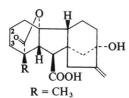

$C_{15}H_{20}O_2$ M 232.322

8α-form
Constit. of a *Ferreyanthus* sp. Oil.
8β-form
Constit. of *Stevia polyphylla*. Gum.

Bohlmann, F. *et al*, *Phytochemistry*, 1984, **23**, 1669.
Zdero, C. *et al*, *Phytochemistry*, 1988, **27**, 2835.

Gibberellin A₂₀ G-70011

Updated Entry replacing G-20027
ent-*10β,13-Dihydroxy-20-nor-16-gibberellene-7,19-dioic acid 19,10-lactone. Pharbitis gibberellin*
[19143-87-4]

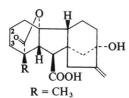

R = CH₃

$C_{19}H_{24}O_5$ M 332.396

See also Gibberellins, G-20021 . Constit. of *Pharbitis nil*.
    Cryst. Mp 232-233°.

3β-Hydroxy: [29774-53-6]. **Gibberellin A₂₉**.
    $C_{19}H_{24}O_6$ M 348.395
3β-Methoxy: Cryst. Mp 197-200°.
3-O-β-D-Glucopyranosyloxy: [30046-29-8]. **3-O-Glucosylgibberellin A₂₉**.
    $C_{25}H_{34}O_{11}$ M 510.537
    Constit. of the immature seeds of *P. nil.*.
3-O-β-D-Glucopyranonyloxy, penta-Ac: Mp 269-272°.
15β-Hydroxy: [105593-16-6]. **Gibberellin A₆₇**.
    $C_{19}H_{24}O_6$ M 348.395
    Constit. of seeds of *Helianthus annuus*. Cryst.
    (Me₂CO/pet. ether). Mp 135-137°.
3β,15β-Dihydroxy: **Gibberellin A₇₂**.
    $C_{19}H_{24}O_7$ M 364.394
    Constit. of seeds of *H. annuus*.

MacMillan, J. *et al*, *Tetrahedron Lett.*, 1968, 1357 (*struct*)
Murofushi, N. *et al*, *Agric. Biol. Chem.*, 1968, **32**, 1239 (*struct*)
Yokota, T. *et al*, *Agric. Biol. Chem.*, 1971, **35**, 583 (*deriv*)
Railton, I.D. *et al*, *Phytochemistry*, 1974, **13**, 793 (*biosynth*)
Bearder, J.R. *et al*, *Phytochemistry*, 1975, **14**, 1741 (*biosynth*)
Durley, R.C. *et al*, *Plant Physiol.*, 1979, **64**, 214 (*biosynth*)
Duri, Z.J. *et al*, *J. Chem. Soc., Perkin Trans. 1*, 1981, 161
    (*synth*)
Kirkwood, P.S. *et al*, *J. Chem. Soc., Perkin Trans. 1*, 1982, 699
    (*synth*)
Kamiya, Y. *et al*, *Phytochemistry*, 1983, **22**, 681 (*biosynth*)
Dolan, S.C. *et al*, *J. Chem. Soc., Perkin Trans. 1*, 1986, 2741
    (*Gibberellin A₆₇*)
Hutchison, M. *et al*, *Phytochemistry*, 1988, **27**, 2695 (*Giberellin A₇₂*)

Gibberellin A₆₄ G-70012

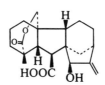

$C_{20}H_{26}O_5$ M 346.422
Constit. of seeds of *Helianthus annuus*.

Hutchison, M. *et al*, *Phytochemistry*, 1988, **27**, 2695.

Gibberellin A₆₅ G-70013

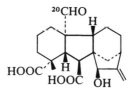

$C_{20}H_{26}O_6$ M 362.422
Constit. of seeds of *Helianthus annuus*.

20-Carboxylic acid: **Gibberellin A₆₆**.
    $C_{20}H_{26}O_7$ M 378.421
    Constit. of seeds of *H. annuus*.

Hutchison, M. *et al*, *Phytochemistry*, 1988, **27**, 2695.

Glabrachromene II G-70014

2-(1,3-Benzodioxol-5-yl)-1-(5-hydroxy-2,2-dimethyl-2H-1-benzopyran-6-yl)-2-propen-1-one, 9CI. 5-Hydroxy-2,2-dimethyl-6-(3,4-methylenedioxystyryl)-chromene
[51848-09-0]

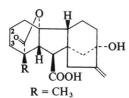

$C_{21}H_{18}O_5$ M 350.370
Minor constit. of *Pongamia glabra*. Deep-yellow needles
    (EtOH) (synthetic). Mp 145-146°.

Sharma, P. *et al*, *Indian J. Chem.*, 1973, **11**, 985 (*occur, synth, pmr*)

Glabranin G-70015

2,3-Dihydro-5,7-dihydroxy-8-(3-methyl-2-butenyl)-2-phenyl-4H-1-benzopyran-4-one, 9CI. 5,7-Dihydroxy-8-prenylflavanone
[41983-91-9]

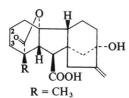

$C_{20}H_{20}O_4$ M 324.376
Isol. from *Glycyrrhiza lepidota*. Shows sl. activity against
    gram-positive bacteria. Cryst. (C₆H₆/pet. ether). Mp
    169-170°.

7-Me ether: [75350-44-6]. **Tephrinone**. Glabranin 7-methyl ether.
    $C_{21}H_{22}O_4$ M 338.402
    Isol. from *Tephrosia villosa* and other *T.* spp. Cryst.
    (cyclohexane). Mp 123-124°.

Rao, P.P. *et al, Curr. Sci.*, 1981, **50**, 319 (*isol, deriv*)
Mitscher, L.A. *et al, Phytochemistry*, 1983, **22**, 573 (*isol*)
Gomez, F. *et al, Phytochemistry*, 1983, **22**, 1305; 1985, **24**, 1057 (*isol*)

Glabridin   G-70016

4-[3,4-Dihydro-8,8-dimethyl-2H,8H-benzo[1,2-b:3,4-b']dipyran-3-yl]-1,3-benzenediol

[59870-68-7]

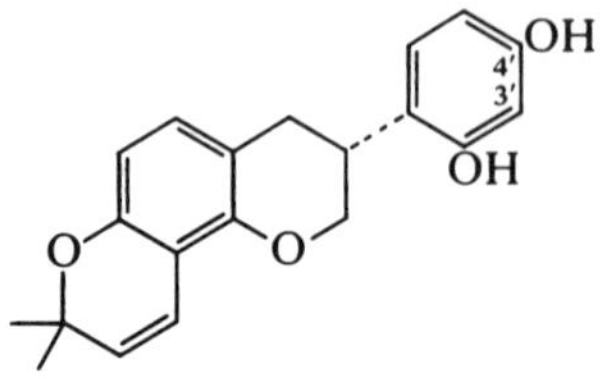

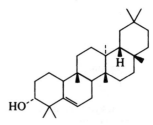

$C_{20}H_{20}O_4$ M 324.376

Isol. from *Glycyrrhiza glabra*. Possesses antimicrobial props. Plates (C_6H_6). Mp 154-155°. $[\alpha]_D^{20}$ +8.2° (c, 2.07 in $CHCl_3$).

Di-Ac: Prisms (EtOH). Mp 164-166°.
4'-Me ether: [68978-09-6]. **4'-O-Methylglabridin**.
  $C_{21}H_{22}O_4$ M 338.402
  From *G. glabra*. Possesses antimicrobial props. Needles (cyclohexane). Mp 120-121°. $[\alpha]_D^{28}$ +10.2° (c, 1.04 in $CHCl_3$).
Di-Me ether: [59870-70-1]. Needles (EtOH). Mp 110-111°. $[\alpha]_D^{28}$ +11.8° (c, 1.02 in $CHCl_3$).
3'-Methoxy: [74046-05-2]. **3'-Methoxyglabridin**.
  $C_{21}H_{22}O_5$ M 354.402
  From *G. glabra*. Cryst. (cyclohexane). Mp 104-105°. $[\alpha]_D^{25}$ +10.28° (c, 1.41 in $CHCl_3$).

Saitoh, T. *et al, Chem. Pharm. Bull.*, 1976, **24**, 752 (*isol, struct*)
Mitscher, L.A. *et al, J. Nat. Prod.*, 1980, **43**, 259 (*isol, derivs*)
Castro, O. *et al, J. Nat. Prod.*, 1986, **49**, 680 (*deriv*)

Glaucin *B*   G-70017

[115458-72-5]

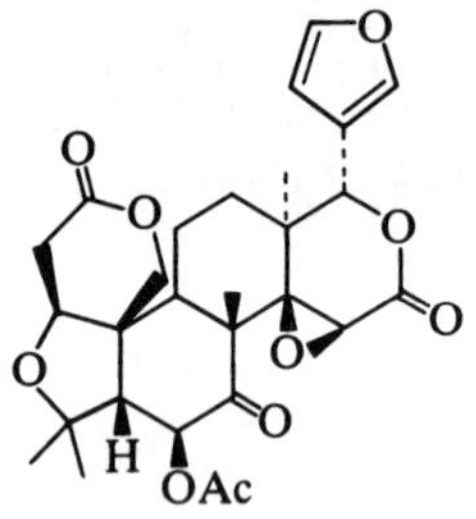

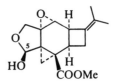

$C_{28}H_{32}O_{10}$ M 528.555

Constit. of *Evodia glauca*. Powder (MeOH). Mp 228-231°. $[\alpha]_D^{15}$ +29° (c, 0.001 in MeOH).

Nakatani, M. *et al, Phytochemistry*, 1988, **27**, 1429.

Glomelliferonic acid   G-70018

[113706-23-3]

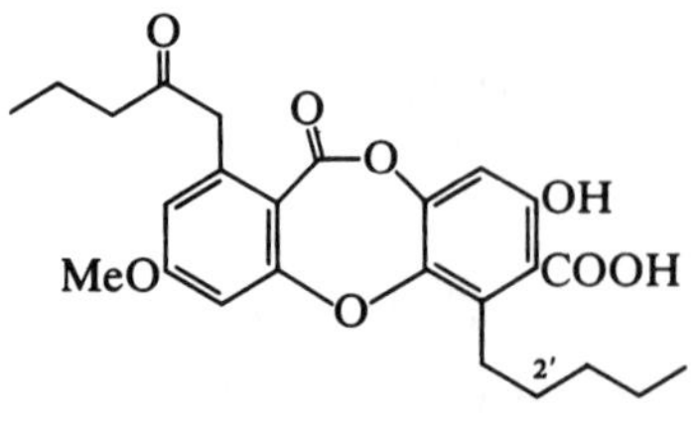

$C_{25}H_{28}O_8$ M 456.491

Metab. of *Neofuscelia subincerta*. Cryst. (EtOAc/pet. ether). Mp 184-185°.

2'-Oxo: [113706-22-2]. **Glomellonic acid**.
  $C_{25}H_{26}O_9$ M 470.475
  Metab. of *N. subincerta*. Cryst. (EtOAc/pet. ether). Mp 149-151°.

Elix, J.A. *et al, Aust. J. Chem.*, 1987, **40**, 2031.

N-[15-(β-D-Glucopyranosyloxy)-8-hydroxypalmitoyl]taurine   G-70019

2-[[15-(β-D-Glucopyranosyloxy)-8-hydroxy-1-oxohexadecyl]amino]ethanesulfonic acid, 9CI

[107959-84-2]

$$H_3CCH(OGlc)(CH_2)_6CH(OH)(CH_2)_6CONHCH_2CH_2\text{-}SO_3H$$

$C_{24}H_{47}NO_{11}S$ M 557.695

Oviposition-deterring pheromone of the European cherry fruit fly *Rhagoletis cerasi*.

Hurter, J. *et al, Experientia*, 1987, **43**, 157.

5-Gluten-3-ol   G-70020

Updated Entry replacing G-00304
D:B-Friedoolean-5-en-3-ol, 9CI

$C_{30}H_{50}O$ M 426.724

Mp 210.5-211.5° (203-205°). $[\alpha]_D$ +64° (+61°) ($CHCl_3$).

3α-form [14554-13-3]
  Alnusenol. Glutinol.
  Constit. of *Euphorbia cyparissias* and *E. royleana*. Cryst. (MeOH). Mp 203-205°. $[\alpha]_D$ +61° (c, 1 in $CHCl_3$).
3β-form [545-24-4]
  Constit. of *E. spp*. Cryst. (MeOH/$CHCl_3$). Mp 210-213°. $[\alpha]_D^{23}$ +63.3° (c, 0.71 in $CHCl_3$).
  Ac:
    $C_{32}H_{52}O_2$ M 468.762
    Constit. of *E. maculata*. Needles (MeOH/$CHCl_3$). Mp 190-191.5°. $[\alpha]_D^{23}$ +76.8° (c, 0.5 in $CHCl_3$).

Sengupta, P. *et al, J. Indian Chem. Soc.*, 1965, **42**, 543 (*isol*)
Starratt, A.N., *Phytochemistry*, 1966, **5**, 1341 (*isol*)
Matsunaga, S. *et al, Phytochemistry*, 1988, **27**, 535 (*isol*)

Glutinopallal   G-70021

$C_{16}H_{20}O_5$ M 292.331

O⁵-Octadecanoyl: Stearylglutinopallal.
  $C_{34}H_{54}O_6$ M 558.797
  Metab. of *Lactarius glutinopallens*.
O⁵-Hexadecanoyl: Palmitylglutinopallal.
  $C_{32}H_{50}O_6$ M 530.743

Metab. of *L. glutinopallens*.

Favre-Bonvin, J. *et al, Phytochemistry*, 1988, **27**, 286.

Glycinoeclepin *A*   G-70022

Updated Entry replacing G-50029
[83216-10-8]

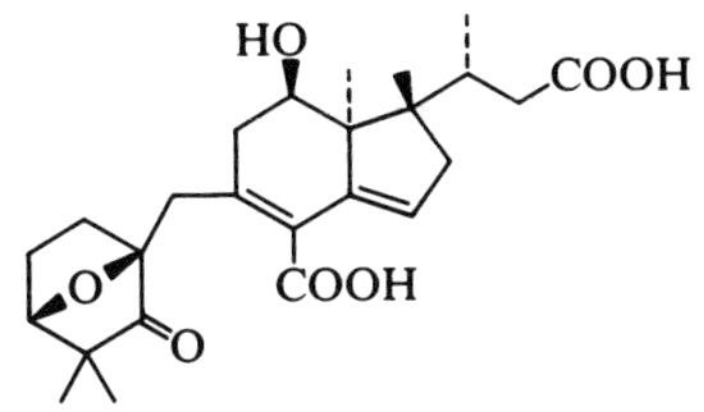

$C_{25}H_{34}O_7$ M 446.539

A natural hatching stimulus for soybean cyst nematode. Struct. is incorrectly drawn in the reference.

Fukuzawa, A. *et al, J. Chem. Soc., Chem. Commun.*, 1985, 222.

Gnetin A   G-70023

[82084-87-5]

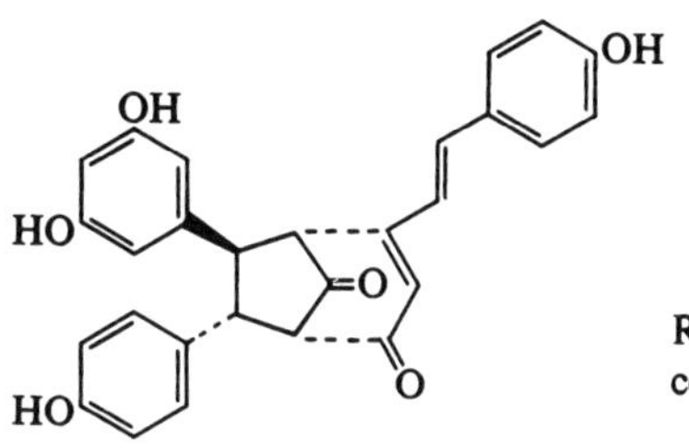

Relative configuration

$C_{28}H_{22}O_6$ M 454.478

Constit. of *Gnetum* spp. Yellow cryst. Mp 179-180°.

Lins, A.P. *et al, J. Nat. Prod.*, 1982, **45**, 754.

Gochnatolide   G-70024

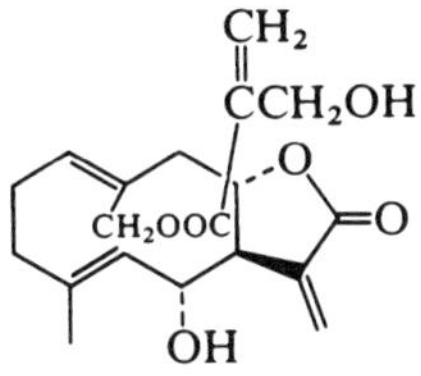

$C_{19}H_{24}O_6$ M 348.395

Constit. of *Gochnatia hypoleuca*. Cryst. (Me$_2$CO/hexane). Mp 107-109°. $[\alpha]_D^{25}$ +8.89° (c, 0.169 in CHCl$_3$).

Maldonado, E. *et al, Phytochemistry*, 1988, **27**, 861.

Gomisin *D*   G-70025

[60546-10-3]

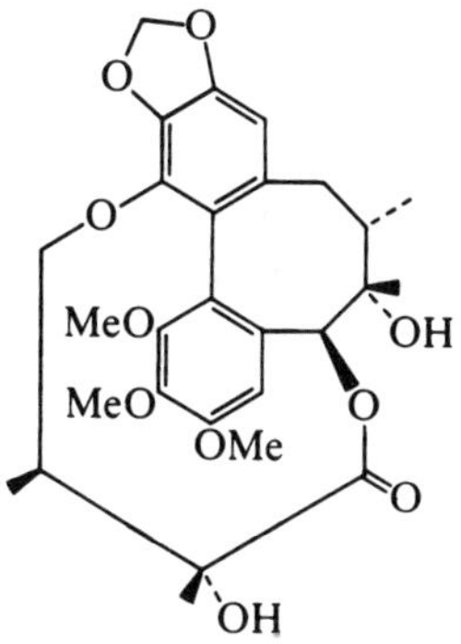

Absolute configuration

$C_{28}H_{34}O_{10}$ M 530.571

Lignan from fruits of *Schizandra chinensis*. Prisms (hexane/Et$_2$O). Mp 194°. $[\alpha]_D^{25}$ −58.8° (c, 0.265 in CHCl$_3$).

Ikeya, Y. *et al, Tetrahedron Lett.*, 1976, 1359 (*isol, uv, ir, pmr*)

Gomphidic acid   G-70026

4-Hydroxy-α-[3-hydroxy-5-oxo-4-(3,4,5-trihydroxyphenyl)-2(5H)-furanylidene]benzeneacetic acid, 9CI
[25328-77-2]

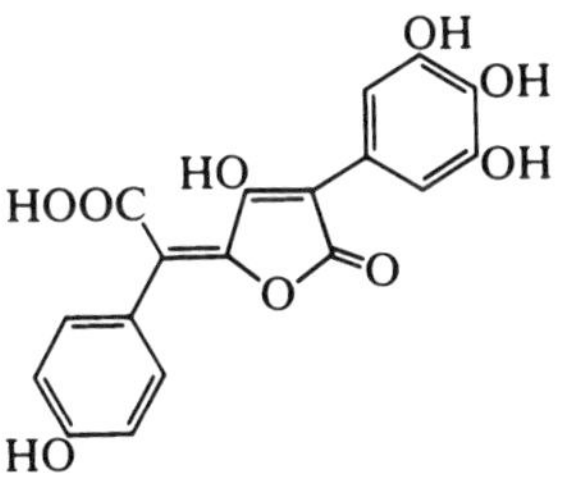

$C_{18}H_{12}O_9$ M 372.287

Pigment from the lichen *Gomphidus glutinosus*. Orange-red. Mp 300-302° dec. (unsharp). Closely related to Variegatic acid, V-00044 and Xerocomic acid, X-30005.

O-*Tetra-Ac:* Mp 198°.

Steglich, W. *et al, Z. Naturforsch., B*, 1969, **24**, 941; 1974, **29**, 96 (*isol, uv, ir, struct, pmr*)

Goniothalamicin   G-70027

[113817-64-4]

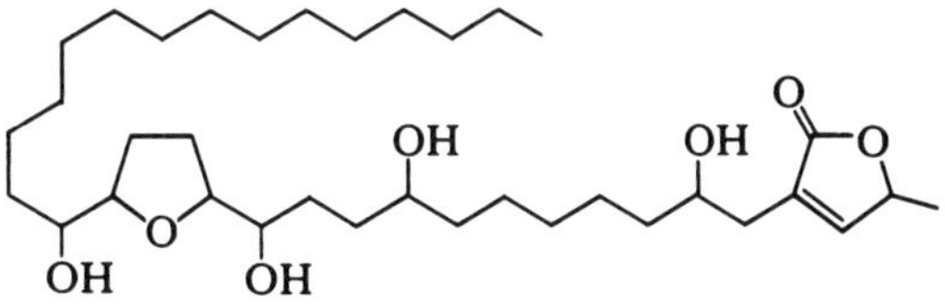

$C_{35}H_{64}O_7$ M 596.886

Closely related to Annonacin, A-70212. Constit. of *Goniothalamus giganteus*. Wax. Mp 86-88°. $[\alpha]_D$ +1.6°.

Alkofahi, A. *et al, Experientia*, 1988, **44**, 83.

Grasshopper ketone   G-70028

Updated Entry replacing D-05098

4-(2,4-Dihydroxy-2,6,6-trimethylcyclohexylidene)-3-buten-2-one, 9CI. 2,6,6-Trimethyl-1-(3-oxo-1-butenylidene)-2,4-cyclohexanediol

[41703-38-2]

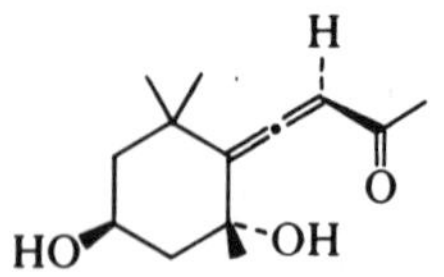

$C_{13}H_{20}O_3$ M 224.299

Isol. from the grasshopper *Romalae microptera*. Needles (Me₂CO/C₆H₆). Mp 134-136°. $[\alpha]_D^{25}$ −63.0° (c, 1.15 in MeOH).

3-O-β-D-Glucopyranoside: [109062-00-2]. **Icariside B₁.**
$C_{19}H_{30}O_8$ M 386.441
Constit. of *Epimedium grandiflorum* var *thunbergianum*. Amorph. powder. $[\alpha]_D^{25}$ −73.5° (c, 1.00 in MeOH).

Meinwald, J. *et al, Tetrahedron Lett.*, 1968, 2959 (*isol*)
de Ville, T.E. *et al, J. Chem. Soc., Chem. Commun.*, 1970, 1231 (*cryst struct*)
Hlubucek, J.R. *et al, J. Chem. Soc., Perkin Trans. 1*, 1974, 848 (*synth*)
Mori, K., *Tetrahedron*, 1974, 1065 (*synth, abs config*)
Miyase, T. *et al, Chem. Pharm. Bull.*, 1987, **35**, 1109 (*deriv*)

Grenoblone   G-70029

Updated Entry replacing G-50039

[104021-41-2]

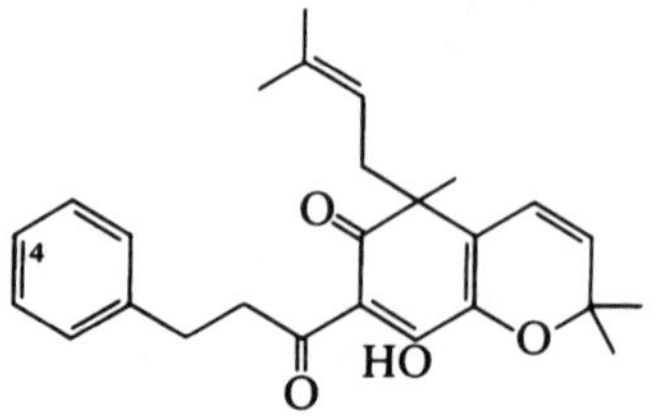

$C_{26}H_{30}O_4$ M 406.521

Constit. of *Platanus acerifolia*. Yellow oil.

4-Hydroxyphenyl analogue: **4-Hydroxygrenoblone**.
$C_{26}H_{30}O_5$ M 422.520
From *P. acerifolia*. Yellow oil.

Kaouadji, M., *J. Nat. Prod.*, 1986, **49**, 500, 508.

Grimaldone   G-70030

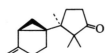

Absolute configuration

$C_{15}H_{22}O$ M 218.338

Constit. of *Mannia fragrans*. Cryst. (pentane). Mp 91-92°. $[\alpha]_D^{24}$ +49.6° (c, 0.635 in CHCl₃).

Huneck, S. *et al, Phytochemistry*, 1988, **27**, 1405.

1(5),6-Guaiadiene   G-70031

1,2,3,4,5,6-Hexahydro-1,4-dimethyl-7-(1-methylethyl)-azulene, 9CI. 1,2,3,4,5,6-Hexahydro-7-isopropyl-1,4-dimethylazulene

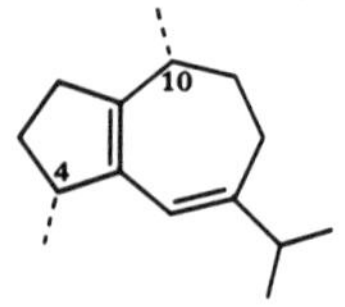

(4α,10α)-*form*

$C_{15}H_{24}$ M 204.355

(4α,10α)-*form* [111900-51-7]
Constit. of Tolu balsam (from *Balsamum tolutanum*). Oil. $[\alpha]_D^{25}$ +80° (c, 0.25 in CHCl₃).

(4β,10α)-*form* [111900-50-6]
Constit. of Tolu balsam (*B. tolutanum*). Oil. $[\alpha]_D^{25}$ +36.2° (c, 0.16 in CHCl₃).

Friedel, H.D. *et al, Helv. Chim. Acta*, 1987, **70**, 1616.

1(10),3-Guaiadien-12,8-olide   G-70032

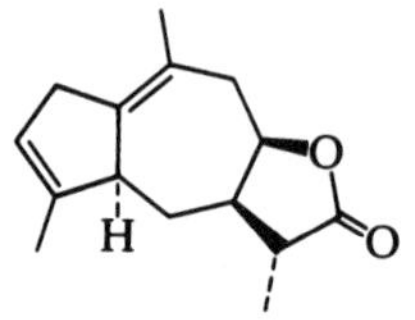

$C_{15}H_{20}O_2$ M 232.322

8β-form
Constit. of *Stevia samaipatensis*. Oil.

Zdero, C. *et al, Phytochemistry*, 1988, **27**, 2835.

Guaiaretic acid   G-70033

[500-40-3]

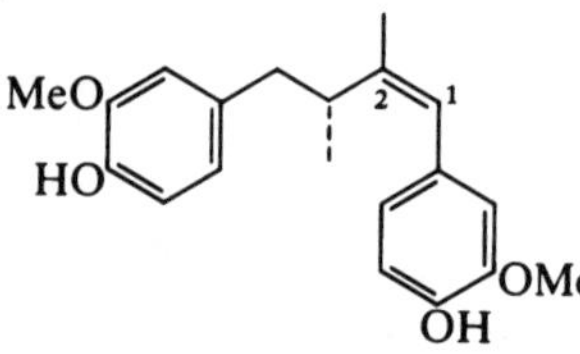

$C_{20}H_{24}O_4$ M 328.407

Constit. of resin of *Guaiacum officinale*. Key compd. in elucidation of lignan abs. configs. Cryst. (EtOH). Mp 99-100.5°. $[\alpha]_D$ −94° (EtOH).

1,2S-Dihydro: **Dihydroguaiaretic acid**.
$C_{20}H_{26}O_4$ M 330.423
From resin of *G. officinale*. Cryst. (MeOH aq.). Mp 88-88.5°.

Hearon, W.M. *et al, Chem. Rev.*, 1955, **55**, 957 (*rev*)
Carnmalm, B., *Chem. Ind.* (*London*), 1956, 1093 (*abs config*)
Schrecker, A.W. *et al, J. Org. Chem.*, 1956, **21**, 381 (*abs config*)
King, F.E. *et al, J. Chem. Soc.*, 1964, 4011 (*isol*)

α-Gurjunene   G-70034

Updated Entry replacing G-00751

[489-40-7]

$C_{15}H_{24}$ M 204.355

Constit. of, *inter alia*, gurjun balsam from *Dipterocarpus dyeri*. Oil. Bp$_{10}$ 114-116°, Bp$_3$ 75-77°. [α]$_D^{20}$ −227° (c, 1.2 in cyclohexane).

1-Epimer: [112421-19-9]. **1-Epi-α-gurjunene**.
$C_{15}H_{24}$ M 204.355
Constit. of Tolu balsam (from *Balsamum tolutanum*). Oil. [α]$_D^{25}$ −68° (c, 0.2 in pentane).

Streith, J. *et al*, *Bull. Soc. Chim. Fr.*, 1963, 1960 (*isol, struct*)
Friedel, H.D. *et al*, *Helv. Chim. Acta*, 1987, **70**, 1753 (*isol, cmr*)

Gymnomitrol   G-70035

Updated Entry replacing G-20066
[41410-53-1]

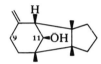

$C_{15}H_{24}O$ M 220.354
Constit. of *Gymnomitrion obtusum*. Cryst. Mp 114-116°. [α]$_D$ +7° (c, 2.3 in CHCl$_3$).

*9α-Hydroxy, 11-Ac: **9α-Hydroxygymnomitryl acetate**.*
$C_{17}H_{26}O_3$ M 278.391
Constit. of *Plagiochila trabeculata*. Oil.
*9α-Hydroxy, 11-cinnamoyl: **9α-Hydroxygymnomitryl cinnamate**.*
$C_{24}H_{30}O_3$ M 366.499
Constit. of *P. trabeculata*. Oil.
*9-Oxo, 11-Ac: **9-Oxogymnomitryl acetate**.*
$C_{17}H_{24}O_3$ M 276.375
Constit. of *P. trabeculata*. Oil.

Connolly, J.D. *et al*, *J. Chem. Soc., Perkin Trans. 1*, 1974, 2487 (*isol*)
Coates, R.M. *et al*, *J. Am. Chem. Soc.*, 1982, **104**, 2198 (*synth, bibl*)
Toyota, M. *et al*, *Phytochemistry*, 1988, **27**, 2161 (*derivs*)

Gypopinifolone   G-70036

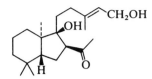

$C_{20}H_{34}O_3$ M 322.487
Constit. of *Gypothamnium pinifolium*. Oil.

Zdero, C. *et al*, *Phytochemistry*, 1988, **27**, 2953.

H

Hamachilobene A — H-70001

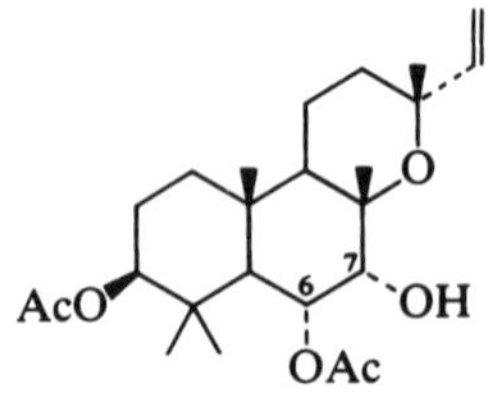

$C_{24}H_{38}O_6$ M 422.561

Constit. of *Frullania hamachiloba*. Cryst. Mp 136-138°. $[\alpha]_D$ +38.6° (c, 0.36 in $CHCl_3$).

7-Epimer: **Hamachilobene** C.
$C_{24}H_{38}O_6$ M 422.561
Constit. of *F. hamachiloba*. Cryst. Mp 94-95°. $[\alpha]_D$ +19.1° (c, 1.46 in $CHCl_3$).

7-Ketone: **Hamachilobene** B.
$C_{24}H_{36}O_6$ M 420.545
Constit. of *F. hamachiloba*. Cryst. Mp 154-156°. $[\alpha]_D$ +37.8° (c, 0.26 in $CHCl_3$).

7-Epimer, 7-Ac, O^6-De-Ac: **Hamachilobene** D.
$C_{24}H_{38}O_6$ M 422.561
Constit. of *F. hamachiloba*. Cryst. Mp 144-146°. $[\alpha]_D$ +46.4° (c, 0.58 in $CHCl_3$).

O^3-De-Ac, 6-deoxy, 7-Ac: **Hamachilobene** E.
$C_{22}H_{36}O_4$ M 364.524
Constit. of *F. hamachiloba*. Cryst. Mp 90-92°. $[\alpha]_D$ 0° (c, 0.76 in $CHCl_3$).

Toyota, M. *et al, Phytochemistry*, 1988, **27**, 1789.

Hardwickiic acid — H-70002

Updated Entry replacing H-60007

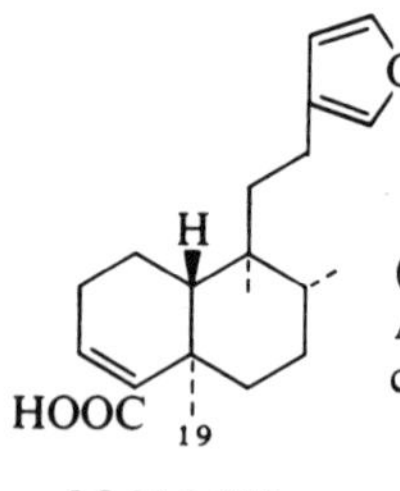

(−)-*form*
Absolute
configuration

$C_{20}H_{28}O_3$ M 316.439

(+)-*form* [24470-47-1]
Constit. of *Copaifera officinalis*. Mp 104-106°. $[\alpha]_D^{20}$ +125°.

(−)-*form* [1782-65-6]
Constit. of, *inter alia, Hardwickia pinnata*. Cryst. Mp 106-107°. $[\alpha]_D$ −114.7° ($CHCl_3$).

19-Hydroxy: [18411-75-1]. **Hautriwaic acid**.
$C_{20}H_{28}O_4$ M 332.439
Constit. of *Dodonnea viscosa* and *D. attenuata*. Also isol. as various esters from *Conyza scabrida*. Cryst. Mp 183-184°. $[\alpha]_D$ −105°.

19-Acetoxy: Hautriwaic acid acetate.
$C_{22}H_{30}O_5$ M 374.476
Constit. of *Baccharis macraei*. Gum. $[\alpha]_D^{25}$ −85° (c, 0.9 in $CHCl_3$).

19-Acetoxy, 1,2-didehydro: [88378-97-6]. **19-O-Acetyl-1,2-dehydrohautriwaic acid**.
$C_{22}H_{28}O_5$ M 372.460

Isol. from *C. scabrida* (as Me ester, after methylation of extract). Also from *Baccharis flabellata*. Oil. $[\alpha]_D$ −122.3° (c, 0.62 in $CHCl_3$).

19-Oxo:
$C_{20}H_{26}O_4$ M 330.423
Isol. from *C. scabrida* as Me ester, after methylation.

19-Oxo, Me ester: $[\alpha]_D^{24}$ −67° (c, 0.15 in $CHCl_3$).

2β,19-Dihydroxy: **2-Hydroxyhautriwaic acid**.
$C_{20}H_{28}O_5$ M 348.438
Constit. of *B. sarothroides*. Cryst. (C_6H_6). Mp 188-189°.

2α-Acetoxy: **2α-Acetoxyhardwickiic acid**.
$C_{22}H_{30}O_5$ M 374.476
Constit. of *Grangea maderaspatana*.

2α-Acetoxy, Me ester: Gum. $[\alpha]_D$ −1.1° (c, 2.07 in $CHCl_3$).

15,16-Dihydro, 16-oxo: **15,16-Dihydro-16-oxohardwickiic acid**.
$C_{20}H_{28}O_4$ M 332.439
Constit. of *G. maderaspatana*.

15,16-Dihydro, 16-oxo, Me ester: Gum. $[\alpha]_D$ −62° (c, 1.38 in $CHCl_3$).

15,16-Dihydro, 15-methoxy, 16-oxo: **15,16-Dihydro-15-methoxy-16-oxohardwickiic acid**.
$C_{21}H_{30}O_5$ M 362.465
Constit. of *G. maderaspatana*.

15,16-Dihydro, 15-methoxy, 16-oxo, Me ester: Gum. $[\alpha]_D$ −30.8° (c, 0.37 in $CHCl_3$).

2β-Hydroxy: **2β-Hydroxyhardwickiic acid**.
$C_{20}H_{28}O_4$ M 332.439
Constit. of *C. coulteri*.

Cocker, W. *et al, Tetrahedron Lett.*, 1965, 1983 (*isol*)
Bohlmann, F. *et al, Chem. Ber.*, 1972, **105**, 3123 (*deriv*)
Payne, T.G. *et al, Tetrahedron*, 1973, **29**, 2575 (*struct*)
Ferguson, G. *et al, J. Chem. Soc., Chem. Commun.*, 1975, 299 (*stereochem*)
Misra, R. *et al, Tetrahedron*, 1979, **35**, 2301 (*abs config*)
Bohlmann, F. *et al, Justus Liebigs Ann. Chem.*, 1983, 2008 (*derivs*)
Arriaga-Giner, F.J. *et al, Phytochemistry*, 1986, **25**, 719 (*2-Hydroxyhautriwaic acid*)
Gambaro, V. *et al, Phytochemistry*, 1987, **26**, 475 (*isol, derivs*)
Jolad, S.D. *et al, Phytochemistry*, 1988, **27**, 1211 (*2-Hydroxyhardwickiic acid*)
Singh, P. *et al, Phytochemistry*, 1988, **27**, 1537 (*derivs*)
Saad, J.R. *et al, Phytochemistry*, 1988, **27**, 1884 (*19-O-Acetyl-1,2-dehydrohautriwaic acid*)

7(18)-Havannachlorohydrin — H-70003

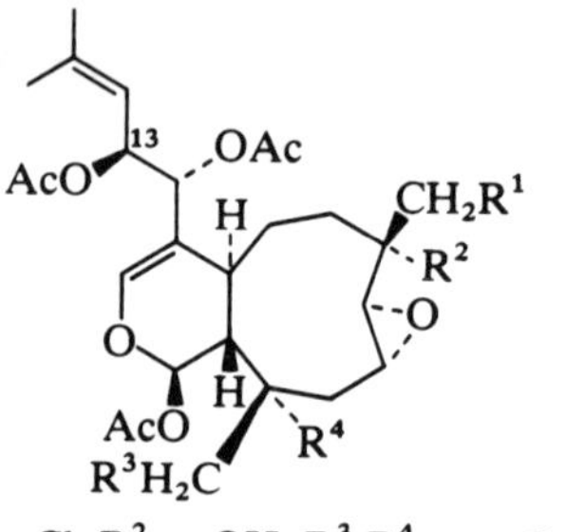

$R^1 = Cl, R^2 = OH, R^3, R^4 = —O—$

$C_{26}H_{35}ClO_{10}$ M 543.010
Constit. of *Xenia membranacea*. Amorph. $[\alpha]_D$ +22° (c, 1.4 in $CHCl_3$).

Almourabit, A. *et al*, *J. Nat. Prod.*, 1988, **51**, 283.

11(19)-Havannnachlorohydrin   H-70004

As 7(18)-Havannnachlorohydrin, H-70003 with

$$R^1R^2 = -O-, R^3 = Cl, R^4 = OH$$

$C_{26}H_{35}ClO_{10}$ M 543.010
Constit. of *Xenia membranacea*. Cryst. (EtOH). Mp
145°. $[\alpha]_D$ −1° (c, 2.4 in $CHCl_3$).
O¹³-De-Ac: 13-Deacetyl-11(9)-havannnachlorohydrin.
$C_{24}H_{33}ClO_9$ M 500.972
Constit. of *X. membranaceae*. Amorph.

Almourabit, A. *et al*, *J. Nat. Prod.*, 1988, **51**, 283.

7(18),11(19)-Havannnadichlorohydrin   H-70005

As 7(18)-Havannnachlorohydrin, H-70003 with

$$R^1 = R^3 = Cl, R^2 = R^4 = OH$$

$C_{20}H_{36}Cl_2O_{10}$ M 507.404
Constit. of *Xenia membranacea*. Amorph. $[\alpha]_D$ −5° (c,
1.6 in $CHCl_3$).

Almourabit, A. *et al*, *J. Nat. Prod.*, 1988, **51**, 283.

Heavenly blue anthocyanin   H-70006

[79620-69-2]

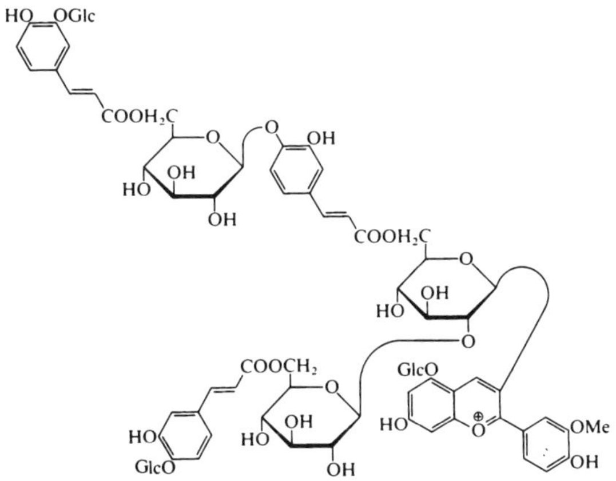

$C_{79}H_{91}O_{45}^{\oplus}$ M 1760.561 (ion)
Constit. of *Ipomoea tricolor*. Isol. as chloride.

Kondo, T. *et al*, *Tetrahedron Lett.*, 1987, **28**, 2273.

Heliopsolide   H-70007

[117274-13-2]

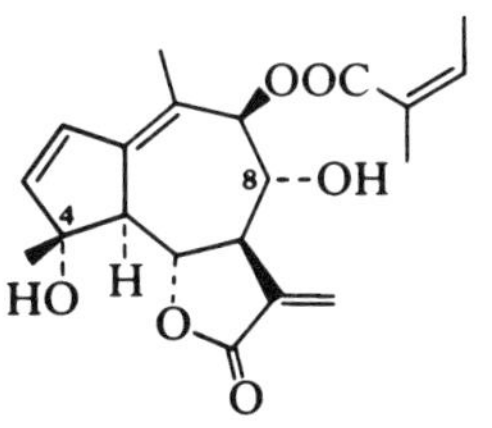

$C_{20}H_{24}O_6$ M 360.406
Constit. of *Heliopsis helianthoides*. Oil. Several related
esters also isol. from this sp.
8-Ac: [117274-04-1].
$C_{22}H_{26}O_7$ M 402.443
Constit. of *H. helianthoides*. Oil.
4-Epimer, 8-Ac: [117274-11-0].
$C_{22}H_{26}O_7$ M 402.443

Constit. of *H. helianthoides*. Oil. $[\alpha]_D^{24}$ +10° (c, 0.39
in $CHCl_3$).
2α,3α-Epoxide, 8-Ac:
$C_{22}H_{26}O_8$ M 418.443
Constit. of *H. helianthoides*. Oil.

Jakupovic, J. *et al*, *Phytochemistry*, 1988, **27**, 2235.

Helogynic acid   H-70008

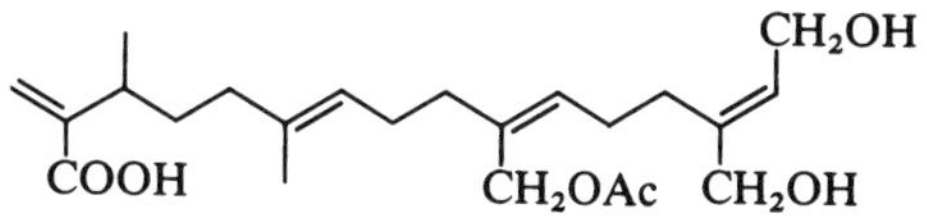

$C_{23}H_{36}O_6$ M 408.534
Constit. of *Helogyne apaloidea*. Oil

Zdero, C. *et al*, *Phytochemistry*, 1988, **27**, 616.

26-Henpentacontanone   H-70009
Cerotone
[542-48-3]

$$H_3C(CH_2)_{24}CO(CH_2)_{24}CH_3$$

$C_{51}H_{102}O$ M 731.366
Mp 93°.

Oura, H. *et al*, *Yakugaku Zasshi*, 1958, **78**, 141; *CA*, **52**, 10885f
(*synth*)

1,14,34,35,36,37,38-Heptaazaheptacy-clo[12.12.7.1³,⁷.1⁸,¹².1¹⁶,²⁰.1²¹,²⁵.1²⁸,³²]-octatriaconta-3,5,7(38),8,10,12-(37),16,18,20(36),21,23,25(35),28,30,32-(34)pentadecane, 9CI   H-70010

$C_{31}H_{27}N_7$ M 497.601
Cryptand.
NaBr complex: [117500-97-7]. Mp >260°.

Alpha, B. *et al*, *Helv. Chim. Acta*, 1988, **71**, 1042 (*synth, uv,
pmr, cmr*)

Heptacyclo[7.7.0.0²,⁶.0³,¹⁵.0⁴,¹².0⁵,¹⁰.0¹¹,¹⁶]-hexadeca-7,13-diene   H-70011

$C_{16}H_{16}$ M 208.302
Cryst. by subl. Mp 95-96°.
Tetrahydro: Heptacyclo[7.7.0.0²,⁶.0³,¹⁵.0⁴,¹².0⁵,¹⁰.0¹¹,¹⁶]-hexadecane.
$C_{16}H_{20}$ M 212.334

Mp 67-68°.

Osawa, E. *et al*, *J. Org. Chem.*, 1980, **45**, 2985 (*deriv*)
Barden, T.J. *et al*, *Aust. J. Chem.*, 1988, **41**, 817 (*synth, ir, pmr, cmr, ms*)

Heptacyclo[6.6.0.0^{2,6}.0^{3,13}.0^{4,11}.0^{5,9}.0^{10,14}]-tetradecane   H-70012

Updated Entry replacing H-40011

Decahydro-1,3,4,6-ethanediylidenecyclopenta[cd,gh]-pentalene, 9CI. Heptacyclo[5.5.1.1^{4,10}.0^{2,6}.0^{3,11}.0^{5,9}.0^{8,12}]-tetradecane

[17872-39-8]

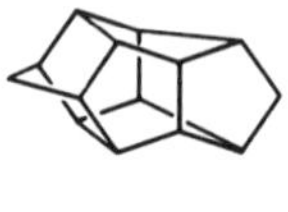

C$_{14}$H$_{16}$ M 184.280

Dimerization product of norbornadiene. Cryst. by subl. Mp 165-165.5°. Thermally stable >500°.

Lemal, D. *et al*, *Tetrahedron Lett.*, 1961, 368 (*synth, ms, pmr, uv, ir*)
Bird, C.W. *et al*, *Tetrahedron Lett.*, 1961, 373 (*synth, ir, pmr, uv*)
Scharf, H.D. *et al*, *Tetrahedron Lett.*, 1967, 4227 (*synth, pmr*)
Acton, N. *et al*, *J. Am. Chem. Soc.*, 1972, **94**, 5446.
Chow, T.J. *et al*, *J. Chem. Soc., Chem. Commun.*, 1985, 700 (*cryst struct*)
Chow, T.J. *et al*, *J. Am. Chem. Soc.*, 1987, **109**, 797 (*synth, pmr*)

Heptacyclo[9.3.0.0^{2,5}.0^{3,13}.0^{4,8}.0^{6,10}.0^{9,12}]-tetradecane   H-70013

Dodecahydro-2,6,3,5-ethanediylidenecyclobut[jkl]-as-indacene, 9CI. 1,4-Bishomohexaprismane. Garudane

[93569-21-2]

C$_{14}$H$_{16}$ M 184.280

Face to face dimer of Bicyclo[2.2.1]hepta-2,5-diene, B-20088 . Volatile waxy solid by subl. Mp 180°.

Mehta, G. *et al*, *J. Am. Chem. Soc.*, 1987, **109**, 7230 (*synth, pmr, cmr*)

2,4-Heptadien-1-ol   H-70014

[62488-55-5]

H$_3$CCH$_2$CH=CHCH=CHCH$_2$OH

C$_7$H$_{12}$O   M 112.171

(**2E,4E**)-*form* [33467-79-7]
Bp$_{15}$ 93-95°.

(**2E,4Z**)-*form* [70979-88-3]
Bp$_{14}$ 80°. n_D^{25} 1.4898.

Jaenicke, L. *et al*, *Chem. Ber.*, 1975, **108**, 225 (*synth*)
Bestmann, H.J. *et al*, *Tetrahedron Lett.*, 1979, **26**, 2467 (*synth, pmr*)
Kuroda, S. *et al*, *Tetrahedron Lett.*, 1987, **28**, 803 (*synth, pmr*)

1,5-Heptadien-3-one   H-70015

H$_2$C=CHCOCH$_2$CH=CHCH$_3$

C$_7$H$_{10}$O   M 110.155

(**Z**)-*form* [33698-68-9]
Liq. Bp$_{20}$ 58-60°.

Gibson, T.W. *et al*, *J. Org. Chem.*, 1972, **37**, 1148 (*synth, uv, pmr*)
Schröder, C. *et al*, *J. Am. Chem. Soc.*, 1987, **109**, 5491 (*synth*)

1,5-Heptadien-4-one   H-70016

H$_2$C=CHCH$_2$COCH=CHCH$_3$

C$_7$H$_{10}$O   M 110.155

(**E**)-*form* [33698-63-4]
Liq. Bp$_{16}$ 42-44°.

Gibson, T.W. *et al*, *J. Org. Chem.*, 1972, **37**, 1148 (*synth, uv, ir*)

1,6-Heptadien-3-one   H-70017

[33698-60-1]

H$_2$C=CHCOCH$_2$CH$_2$CH=CH$_2$

C$_7$H$_{10}$O   M 110.155
Liq. Bp$_{20}$ 35-40°.

Gibson, T.W. *et al*, *J. Org. Chem.*, 1972, **37**, 1148 (*synth, uv, pmr*)

4,5-Heptadien-3-one   H-70018

[108425-53-2]

H$_3$CCH=C=CHCOCH$_2$CH$_3$

C$_7$H$_{10}$O   M 110.155
Bp$_{34}$ 86-96°.

Sugita, T. *et al*, *J. Org. Chem.*, 1987, **52**, 3789 (*synth, pmr, cmr, ir*)

1,2,3,4,5,6,7-Heptafluoronaphthalene   H-70019

1H-Heptafluoronaphthalene

[7539-68-6]

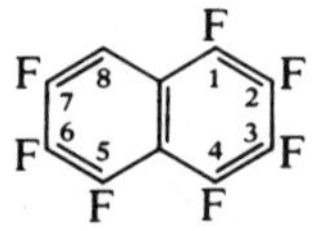

C$_{10}$HF$_7$ M 254.107
Cryst. (MeOH). Mp 38-38.5°.

Yakobson, G.G. *et al*, *CA*, 1966, **64**, 14142c (*synth*)

1,2,3,4,5,6,8-Heptafluoronaphthalene   H-70020

2H-Heptafluoronaphthalene

[784-00-9]

C$_{10}$HF$_7$ M 254.107
Cryst. (MeOH). Mp 63-64.5°.

Gething, B. *et al*, *J. Chem. Soc.*, 1962, 186, 190 (*synth*)
Yakobson, G.G. *et al*, *CA*, 1966, **64**, 14142e (*synth*)
Bolton, R. *et al*, *J. Chem. Soc., Perkin Trans. 2*, 1978, 746 (*F nmr*)

3,3',4',5,5',7,8-Heptahydroxyflavone, 8CI   H-70021

Updated Entry replacing H-40020

3,5,7,8-Tetrahydroxy-2-(3,4,5-trihydroxyphenyl)-4H-1-benzopyran-4-one, 9CI. **Hibiscetin**

[489-35-0]

$C_{15}H_{10}O_9$ M 334.239
Cryst. (EtOH). Mp 350° dec.

Hepta-Ac: Mp 242-244°.
Glycoside: **Hibiscitin**.
$C_{21}H_{20}O_{14}$ M 496.381
Glycoside from *Hibiscus sabdariffa*. Cryst. (MeOH). Struct. unknown.

3,4',8-Tri-Me ether: [99816-54-3]. *3',5,5',7-Tetrahydroxy-3,4',8-trimethoxyflavone.*
$C_{18}H_{16}O_9$ M 376.319
From *Gutierrezia microcephala*.

3,4',5',8-Tetra-Me ether: [21634-47-9]. *3',5,7-Trihydroxy-3,4',5',8-tetramethoxyflavone.*
$C_{19}H_{18}O_9$ M 390.346
Isol. from leaves and stems of *Beyeria brevifolia*. Yellow prisms (MeOH). Mp 214-216°.

3',4',7,8-Tetra-Me ether: [71149-60-5]. *3',5,5'-Trihydroxy-3,4',7,8-tetramethoxyflavone.*
$C_{19}H_{18}O_9$ M 390.346
Isol. from leaves of *Solanum* spp.

3',4',5',7,8-Penta-Me ether: [72620-08-7]. *3,5-Dihydroxy-3',4',5',7,8-pentamethoxyflavone.*
$C_{20}H_{20}O_9$ M 404.373
Cryst. (EtOAc/MeOH). Mp 218-219°.

3,3',4',5',8-Penta-Me ether: [62953-00-8]. *5,7-Dihydroxy-3,3',4',5',8-pentamethoxyflavone.*
$C_{20}H_{20}O_9$ M 404.373
Constit. of *Heteromma simplicifolium*. Cryst. Mp 203-204°.

Hepta-Me ether: [21634-52-6]. *3,3',4',5,5',7,8-Heptamethoxyflavone.*
$C_{22}H_{24}O_9$ M 432.426
Isol. from the fruit of *Murraya exotica*.

Rao, P.S. *et al, Proc. Indian Acad. Sci., Sect. A*, 1942, **15**, 148 (*isol*)
Rao, P.S. *et al, Proc. Indian Acad. Sci., Sect. A*, 1948, **27**, 209.
Chow, P.W. *et al, Aust. J. Chem.*, 1968, **21**, 2529 (*isol, uv*)
Nakayama, M. *et al, Bull. Chem. Soc. Jpn.*, 1970, **43**, 3276 (*synth, uv*)
Chakraborty, J. *et al, Indian Chem. Soc.*, 1971, **48**, 80.
Bohlmann, F. *et al, Phytochemistry*, 1979, **18**, 1081 (*isol, uv, ir, pmr*)
Whalen, M.P. *et al, Phytochemistry*, 1983, **22**, 2107 (*isol, pmr, uv, ms*)
Fang, N. *et al, Phytochemistry*, 1985, **24**, 2693; 1986, **25**, 927 (*isol*)
Horie, T. *et al, Phytochemistry*, 1988, **27**, 1491 (*synth, struct*)

4,6,8,10,12,14,16-Heptamethyl-6,8,11-octadecatriene-3,5,13-trione, 9CI   H-70022

Updated Entry replacing H-40023
Aglajne I
[98571-24-5]

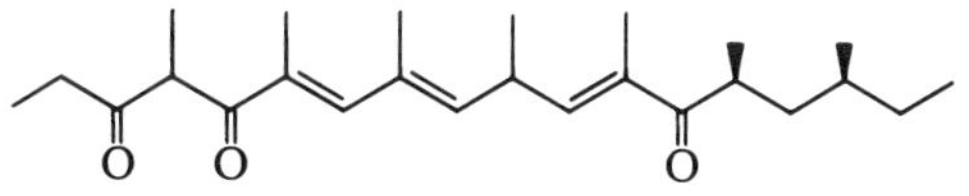

$C_{25}H_{40}O_3$ M 388.589
Metab. of mollusc *Aglaja depicta* and its prey *Bulla striata*. Oil. $[\alpha]_D$ +72° (c, 2.6 in $CHCl_3$).

Cimino, G. *et al, Tetrahedron Lett.*, 1985, **26**, 3389 (*isol*)
Cimino, G. *et al, J. Org. Chem.*, 1987, **52**, 5326 (*struct*)

1,7-Heptanedithiol   H-70023

1,7-Dimercaptoheptane
[62224-02-6]

$$HSCH_2(CH_2)_5CH_2SH$$

$C_7H_{16}S_2$ M 164.323
Liq. Bp_{14} 127°, $Bp_{0.1}$ 80-85°.

Houk, J. *et al, J. Am. Chem. Soc.*, 1987, **109**, 6825 (*synth, ir, pmr*)

5-Heptenoic acid, 9CI   H-70024

Updated Entry replacing H-00340
[3593-00-8]

$$H_3CCH=CH(CH_2)_3COOH$$

$C_7H_{12}O_2$ M 128.171
(E)-form [18776-90-4]
Bp_{16} 120-121°.

Nitrile: [555999-04-7]. *5-Heptenenitrile, 9CI. 6-Cyano-2-pentene.*
$C_7H_{11}N$ M 109.171
Bp_{25} 90°.
4-Bromophenacyl ester: Plates (EtOH aq.). Mp 62-63°.

(Z)-form
Bp_{13} 116-117°.
4-Bromophenacyl ester: Cryst. (MeOH). Mp 49-50°.

Ansell, M.F. *et al, J. Chem. Soc.*, 1958, 1788 (*synth*)
Ansell, M.F. *et al, J. Chem. Soc. (C)*, 1968, 217 (*synth*)
Chiusoli, G.P. *et al, Gazz. Chim. Ital.*, 1973, **103**, 569.
Denmark, S.E. *et al, J. Org. Chem.*, 1987, **52**, 877 (*deriv, synth, pmr*)

6-Heptyn-2-one   H-70025

[928-39-2]

$$HC{\equiv}C(CH_2)_3COCH_3$$

$C_7H_{10}O$ M 110.155
Oil. Bp_{11} 53°.

2,4-Dinitrophenylhydrazone: Mp 118-119° (115-116°).

Schreiber, J. *et al, Helv. Chim. Acta*, 1967, **50**, 2101 (*synth*)
Fülöp, F. *et al, Tetrahedron*, 1986, **42**, 2345.
Bierer, D.E. *et al, Org. Prep. Proced. Int.*, 1988, **20**, 63 (*synth, pmr, cmr, ir*)

Herpetetrone   H-70026

[112899-36-2]

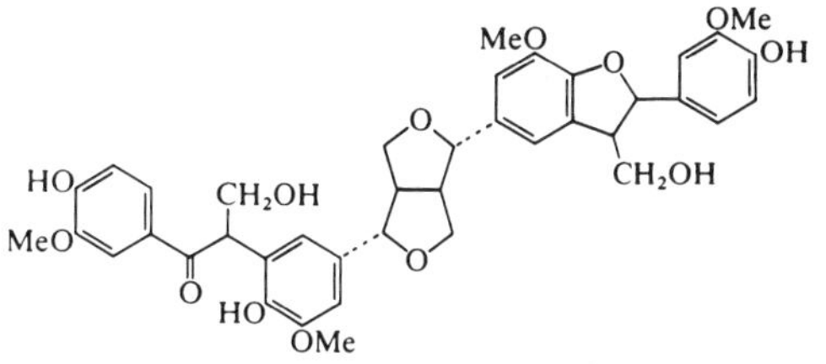

$C_{40}H_{42}O_{13}$ M 730.764
Constit. of *Herpetospermum caudigerum* seeds. Amorph. powder.

Kaouadji, M. *et al, J. Nat. Prod.*, 1987, **50**, 1089.

Heterobryoflavone   H-70027

[111200-23-8]

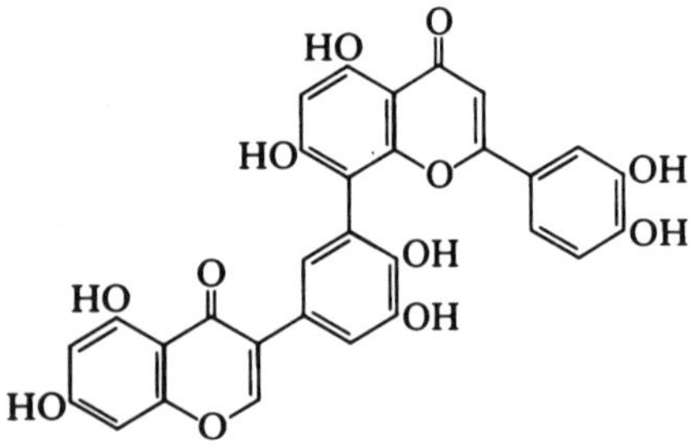

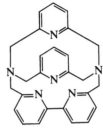

$C_{30}H_{18}O_{12}$ M 570.465
Constit. of *Bryum capillare*.

Geiger, H., *Z. Naturforsch., C*, 1987, **42**, 863.

1,4,7,12,15,18-Hexaazacyclodocosane   H-70028

[58512-71-3]

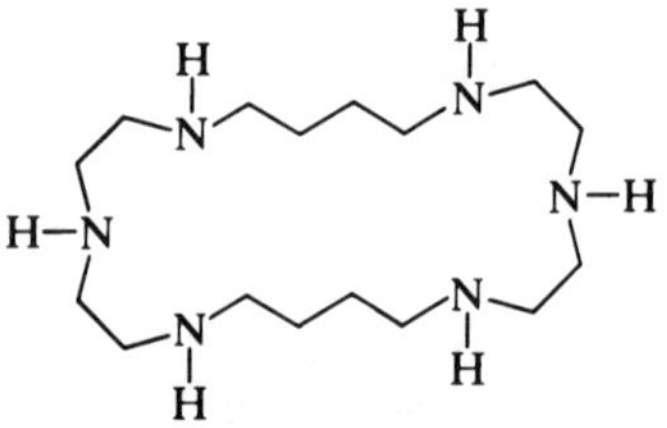

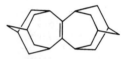

$C_{16}H_{38}N_6$ M 314.516
Hygroscopic solid. Mp 110.0-111.0°.

B,6HCl: Mp >260°.
Hexakis(4-methylbenzenesulfonyl): Solid. Mp 249°.

Martin, A.E. *et al*, *J. Org. Chem.*, 1982, **47**, 415 (*synth, pmr, cmr, ir*)
Hosseini, M.W. *et al*, *J. Am. Chem. Soc.*, 1987, **109**, 7047 (*synth, pmr, cmr*)

1,4,7,13,16,19-Hexaazacyclotetracosane   H-70029

[56187-15-6]

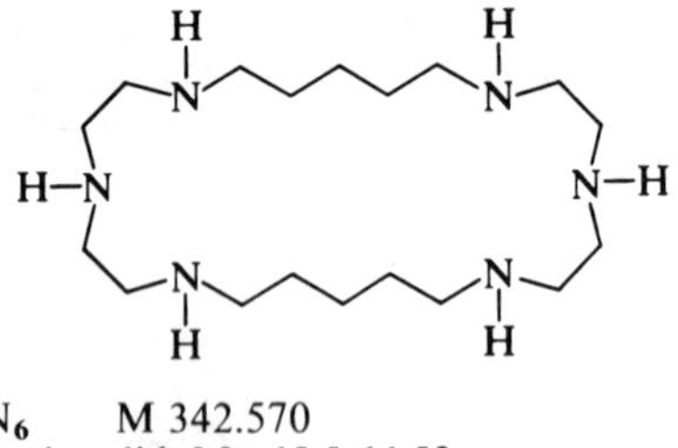

$C_{18}H_{42}N_6$ M 342.570
Hygroscopic solid. Mp 65.5-66.5°.

B,6HCl: Mp >260°.
Hexakis(4-methylbenzenesulfonyl): Cryst. (CHCl₃/Me-₂CO/hexane/DMF). Mp 205°.

Martin, A.E. *et al*, *J. Org. Chem.*, 1982, **47**, 415 (*synth, pmr, cmr, ir*)
Hosseini, M.W. *et al*, *J. Am. Chem. Soc.*, 1987, **109**, 7047 (*synth, pmr, cmr*)

1,14,29,30,31,32-Hexaazahexacyclo-[12.7.7.1³,⁷.1⁸,¹².1¹⁶,²⁰.1²³,²⁷]dotriaconta-3,5,7(32),8,10,12(31),16,18,20-(30),23,25,27(29)-dodecaene, 9CI   H-70030

$C_{26}H_{24}N_6$ M 420.516
Cryptand.

LiBr complex: [117472-88-5]. Monohydrate. Mp >260°.

Alpha, B. *et al*, *Helv. Chim. Acta*, 1988, **71**, 1042 (*synth, uv, pmr, cmr*)

Hexacyclo[11.3.1.1³,⁷.1⁵,⁹.1¹¹,¹⁵.0²,¹⁰]eicos-2-ene   H-70031

1,2,3,4,5,6,7,8,9,10,11,12,13,14-Tetradecahydro-1,5:3,7:8,12:10,14-tetramethanononalene, 9CI.
Bis(homoadamantane)
[30614-34-7]

$C_{20}H_{28}$ M 268.441
Mp 202-204° (199-201°). Chemically inert.

Gill, G.B. *et al*, *Tetrahedron Lett.*, 1971, 181 (*synth, raman*)
Boelema, E. *et al*, *Tetrahedron Lett.*, 1971, 4029 (*synth*)
Watson, W.H. *et al*, *Acta Crystallogr., Sect. C*, 1987, **43**, 2465 (*cryst struct*)

Hexacyclo[4.4.0.0²,⁵.0³,⁹.O⁴,⁸.0⁷,¹⁰]decane, 9CI   H-70032

Updated Entry replacing P-00506
Pentaprismane
[4572-17-2]

$C_{10}H_{10}$ M 130.189
Cryst. Mp 127.5-128.5°.

Eaton, P.E. *et al*, *J. Am. Chem. Soc.*, 1981, **103**, 2134 (*synth, ir, nmr*)
Dauben, W.G. *et al*, *J. Org. Chem.*, 1983, **48**, 2842 (*synth*)
Eaton, P.E. *et al*, *Tetrahedron*, 1986, **42**, 1621 (*synth, ir, pmr, cmr*)

11,13-Hexadecadienal   H-70033

Updated Entry replacing H-20036

$$H_3CCH_2CH{=}CHCH{=}CH(CH_2)_9CHO$$

$C_{16}H_{28}O$ M 236.397
Oil.

(E,E)-form [73264-91-2]
Sex pheromone of the cabbage webworm *Hellulla undalis fabricius*.
(Z,Z)-form [71317-73-2]
Pheromone of the Navel Orangeworm *Pamyelois transitella*. Liq. $Bp_{0.1}$ 140° (bath).

Bishop, C.E. *et al, J. Org. Chem.*, 1983, **48**, 657 (*synth*)
Michelot, D., *Synthesis*, 1983, 130 (*synth*)
Lo, V.M. *et al, Synth. Commun.*, 1986, **16**, 1647 (*synth, pmr, bibl*)
Takayama, H. *et al, J. Org. Chem.*, 1986, **51**, 4934 (*synth, ms, pmr, cmr*)

8,9-Hexadecanedione   H-70034

[18229-29-3]

$$H_3C(CH_2)_6COCO(CH_2)_6CH_3$$

$C_{16}H_{30}O_2$ M 254.412
Yellow plates (MeOH). Mp 48.5-49.5°.

Srinivasan, N.S. *et al, J. Org. Chem.*, 1979, **44**, 1574 (*synth, ir, pmr*)
Fatiadi, A.J. *et al, Synthesis*, 1987, 85 (*synth*)

7-Hexadecenal   H-70035

$$H_3C(CH_2)_7CH=CH(CH_2)_5CHO$$

$C_{16}H_{30}O$ M 238.412
(Z)-form [56797-40-1]
Active component of trail-following pheromone of the Argentine ant *Iridomyrmex humilis*. Liq. $Bp_{0.05}$ 110-112°. n_D^{20} 1.4525.

Van Vorhiskey, S.E. *et al, J. Chem. Ecol.*, 1982, **8**, 3.
Brown, H.C. *et al, J. Org. Chem.*, 1986, **51**, 4518 (*synth, pmr, cmr*)

1,5-Hexadiene-3,4-diol, 9CI   H-70036

Updated Entry replacing H-00594
Divinylethylene glycol. 1,2-Divinylglycol
[1069-23-4]

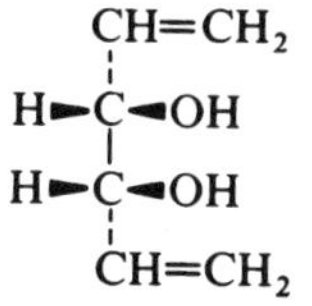

$$\begin{array}{c} CH{=}CH_2 \\ | \\ H{-}C{-}OH \\ | \\ H{-}C{-}OH \\ | \\ CH{=}CH_2 \end{array} \quad \textit{Meso-form}$$

$C_6H_{10}O_2$ M 114.144
▷MM2100000.
(3RS,4RS)-form [19700-97-1]
(±)-form
Mp 21.7°. Bp_8 90.5°.
(3RS,4SR)-form [19700-96-0]
meso-*form*
Cryst. (ligroin). Mp 88°.
Di-Et ether: 3,4-Diethoxy-1,5-hexadiene.
$C_{10}H_{18}O_2$ M 170.251
Bp 224-226°, Bp_{10} 111-113°.

Young, W.G. *et al, J. Am. Chem. Soc.*, 1943, **65**, 1245 (*synth*)
Braun, R.A., *J. Org. Chem.*, 1963, **28**, 1383 (*synth*)
Wiemann, J. *et al, Bull. Soc. Chim. Fr.*, 1967, 3293 (*synth*)
Galaj, S. *et al, Bull. Soc. Chim. Fr.*, 1972, 3979 (*synth*)
Rao, A.V.R. *et al, Tetrahedron Lett.*, 1987, **28**, 2183 (*synth*)

3,4-Hexadienoic acid   H-70037

$$H_3CCH=C=CHCH_2COOH$$

$C_6H_8O_2$ M 112.128

Me ester: [81981-05-7].
$C_7H_{10}O_2$ M 126.155
Liq.

Lewis, F.D. *et al, J. Am. Chem. Soc.*, 1986, **108**, 3016 (*synth, pmr, ir*)

3,5-Hexadienoic acid, 9CI   H-70038

[29949-29-9]

$$H_2C=CHCH=CHCH_2COOH$$

$C_6H_8O_2$ M 112.128
(E)-form [73670-87-8]
Bp_{20} 79-80°.
Me ester: [32775-94-3].
$C_7H_{10}O_2$ M 126.155
Bp_{17} 66-68°.
Et ester: [74054-58-3].
$C_8H_{12}O_2$ M 140.182
Bp_{30} 87-88°.
Chloride: [73670-84-5].
C_6H_7ClO M 130.574
Liq. Bp_{50} 79-81°.
p-*Nitrobenzyl ester:* [87785-65-7]. Yellow cryst. (Et_2O-/pentane). Mp 40-41°.
(Z)-form [2196-20-5]
Oil. Bp_{70} 100-103°.

Torssell, K., *Tetrahedron Lett.*, 1974, 623 (*synth, deriv*)
Stevens, R.V. *et al, J. Am. Chem. Soc.*, 1976, **98**, 6317 (*synth, deriv, ir, pmr*)
Martin, S.F. *et al, J. Am. Chem. Soc.*, 1980, **102**, 5274 (*deriv, synth, ir, pmr, ms*)
Ueda, Y. *et al, Can. J. Chem.*, 1983, **61**, 1996 (*synth, deriv, ir, pmr*)

3,5-Hexadien-1-ol, 9CI   H-70039

[5747-07-9]

$$H_2C=CHCH=CHCH_2CH_2OH$$

$C_6H_{10}O$ M 98.144

Howden, M.E.H. *et al, J. Am. Chem. Soc.*, 1966, **88**, 1732 (*synth*)
Martin, S.F. *et al, J. Org. Chem.*, 1983, **48**, 5170 (*synth, pmr, cmr, ms*)
Brandsma, L. *et al J. Chem. Soc., Chem. Commun.*, 1984, 735 (*synth*)

1,2-Hexadien-5-yne   H-70040

Propargylallene
[33142-15-3]

$$H_2C=C=CHCH_2C≡CH$$

C_6H_6 M 78.113
d_4 0.80. Bp 89-90°.

Peiffer, G. *et al, Bull. Soc. Chim. Fr.*, 1962, 776 (*synth*)
Hopf, H., *Chem. Ber.*, 1971, **104**, 1499 (*props*)
Bischof, P. *et al, J. Am. Chem. Soc.*, 1975, **97**, 5467 (*pe*)

1,3-Hexadien-5-yne   H-70041

Ethynylbutadiene
[10420-90-3]

$$H_2C=CHCH=CHC≡CH$$

C_6H_6 M 78.113
d_4^{20} 0.77. Bp_{756} 82-83°. Readily polym. in air. Mixt. of (E)- and (Z)-forms.

Sondheimer, F. *et al, J. Am. Chem. Soc.*, 1961, **83**, 1682 (*synth, uv*)
Ben-Efraim, D.A. *et al, Tetrahedron*, 1969, **25**, 2837 (*synth, ir, uv*)
Baldwin, J.E. *et al, J. Am. Chem. Soc.*, 1987, **109**, 805 (*synth, pmr*)

2-(2,4-Hexadiynylidene)-1,6-dioxaspira[4.5]dec-3-en-8-ol   H-70042

Updated Entry replacing H-00608
Spiroketal enol ether polyyne

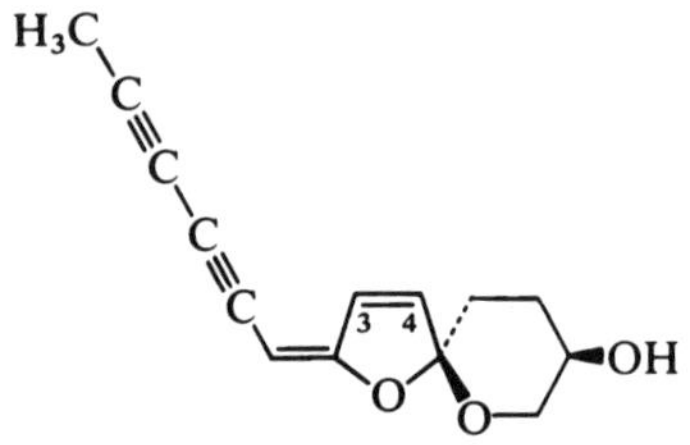

(*E*)-form

C$_{14}$H$_{14}$O$_3$ M 230.263

(*E*)-*form* [3306-40-9]
Constit. of *Chrysanthemum* spp. and of *Artemesia princeps*. Mp 83°.

Ac: 8-*Acetoxy*-2-(*2,4-hexadiynylidene*)-*1,6-dioxaspiro[4.5]dec-3-ene.*
C$_{16}$H$_{16}$O$_4$ M 272.300
Isol. from roots of *C. arcticum.* λ$_{max}$ 240, 254, 310, 318 nm. Not obt. pure.

O-(*3-Methylbutanoyl*): 8-*Isovaleryloxy*-2-(*2,4-hexadiynylidene*)-*1,6-dioxaspiro[4.5]dec-3-ene.*
C$_{19}$H$_{22}$O$_4$ M 314.380
Isol. from roots of *A. pedemontana.* Cryst. (pet. ether).
Mp 93-94°. [α]$_D^{23}$ +7.7° (c, 1.71 in Et$_2$O).

Deoxy: [3306-40-9]. 2-(*2,4-Hexadiynylidene*)-*1,6-dioxaspiro[4.5]dec-3-ene. Spiroketal enol ether polyyne.*
C$_{14}$H$_{14}$O$_2$ M 214.263
Constit. of *C.* spp. and *A. princeps,* also of roots of *Tanacetum vulgare.* Mp 83°. [α]$_{546}^{20}$+29° (c, 2.0 in Et$_2$O). Also descr. as resinous liq.

(*Z*)-*form* [5535-87-5]
Constit. of the roots of *C. arcticum* and *A. princeps.* Mp 78°. Both isomers also descr. as resinous liquids.

Ac: Isol. from roots of *C. arcticum, C. serotinum* and *A. pedemontana.* Cryst. (Et$_2$O/pet. ether). Mp 127°.
[α]$_D^{23}$ +15.2° (c, 3.08 in Et$_2$O).

O-(*3-Methylbutanoyl*): Isol. from roots of *A. pedemontana.* Cryst. (pet. ether). Mp 87-88°. [α]$_D^{23}$ −35.1° (c, 2.45 in Et$_2$O).

Deoxy: [5535-87-5]. Constit. of the roots of *C. arcticum, C. serotinum, C. pyrethrum, A. princeps* and *A. pedemontana.* Mp 78°. [α]$_D$ ±0°. Also descr. as resinous liq.

Deoxy, 3,4-epoxy: 2-(*2,4-Hexadiynylidene*)-*3,4-epoxy-1,6-dioxaspiro[4.5]decane.*
C$_{14}$H$_{14}$O$_3$ M 230.263
Isol. from roots of *C. pyrethrum.* λ$_{max}$ 224, 264, 276.5, 291.5 nm. No stereochem.

Bohlmann, F. *et al, Chem. Ber.*, 1960, **93**, 1937; 1961, **94**, 3193; 1963, **96**, 226; 1964, **97**, 1179; 1966, **99**, 990, 1830, 2416.
Yano, K. *et al, Phytochemistry*, 1972, **11**, 2577.

Hexaethylbenzene, 9CI, 8CI   H-70043

Updated Entry replacing H-60042
[604-88-6]

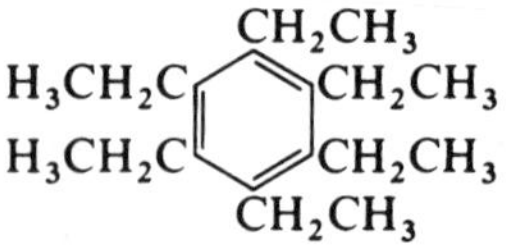

$$H_3CH_2C \quad \overset{CH_2CH_3}{\underset{CH_2CH_3}{\bigcirc}} \quad CH_2CH_3$$

C$_{18}$H$_{30}$ M 246.435
Cryst. (EtOH), needles (heptane). Mp 130°. Bp 298°.
▷DA3000000.

Wertyporoch, E. *et al, Justus Liebigs Ann. Chem.*, 1933, **500**, 287 (*synth*)
Smith, L. *et al, J. Am. Chem. Soc.*, 1940, **62**, 2631 (*synth*)
Richards, R. *et al, Proc. R. Soc. London, Ser. A*, 1948, **195**, 1 (*ir*)
Jhingan, A.K. *et al, J. Org. Chem.*, 1987, **52**, 1161 (*synth, pmr, cmr, ms*)

1,2,3,4,5,6-Hexafluoronaphthalene   H-70044

[68395-57-3]

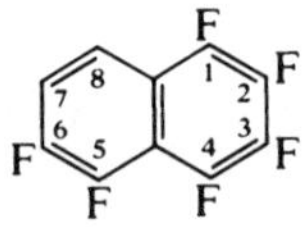

C$_{10}$H$_2$F$_6$ M 236.116
Oil.

Bolton, R. *et al, J. Chem. Soc., Perkin Trans. 2*, 1978, 746 (*synth, F nmr*)

1,2,4,5,6,8-Hexafluoronaphthalene   H-70045

C$_{10}$H$_2$F$_6$ M 236.116
Cryst. Mp 72-73°.

Chambers, R.D. *et al, J. Chem. Soc., Perkin Trans. 1*, 1988, 257 (*synth, pmr, F nmr*)

Hexahydroazirino[2,3,1-*hi*]indol-2(1*H*)-one   H-70046

[113466-89-0]

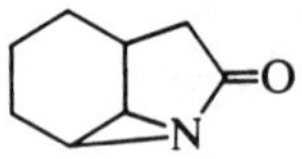

C$_8$H$_{11}$NO   M 137.181
Mp 112-114°.

Knapp, S. *et al, Tetrahedron Lett.*, 1987, **28**, 3213 (*synth, ir*)

Hexahydro-4(1*H*)-azocinone   H-70047

1-Aza-4-cyclooctanone

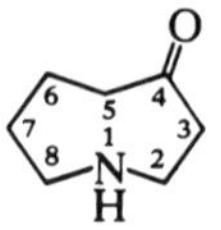

C$_7$H$_{13}$NO   M 127.186

B,HClO$_4$: [16803-16-0]. Flakes (EtOH/Et$_2$O). Mp 87-89°.

Ethylene acetal: [16853-11-5]. *1,4-Dioxa-8-azaspiro[4.7]dodecane.*
C$_9$H$_{17}$NO$_2$ M 171.239
Oil. Bp$_{0.1}$ 68-69°.

Johnson, R.A. *et al, J. Org. Chem.*, 1968, **33**, 3187 (*synth, ir*)

Hexahydro-5(2*H*)-azocinone, 9CI   H-70048
1-Aza-5-cyclooctanone
$C_7H_{13}NO$ M 127.186
1-Me: [71512-38-4].
   $C_8H_{15}NO$ M 141.213
   Mp 45°.
1-Me; B,HClO₄: Prisms (EtOH/Et₂O). Mp 260-261°.
1-Me, picrate: Needles (EtOH aq.). Mp ca. 300° dec.
1-Et: [37727-90-5].
   $C_9H_{17}NO$ M 155.239
   Bp 225°.
1-tert-Butyl: [108696-10-2].
   $C_{11}H_{21}NO$ M 183.293
   Mp 41°.
1-tert-Butyl, picrate: Cryst. (EtOH). Mp 164° dec.
Ethylene ketal: [16803-13-7]. *1,4-Dioxa-9-azaspiro[4.7]dodecane.*
   $C_9H_{17}NO_2$ M 171.239
   Oil. Bp₀.₂ 70-75°.

Leonard, N.J. *et al, J. Am. Chem. Soc.*, 1955, **77**, 6234.
Johnson, R.A. *et al, J. Org. Chem.*, 1968, **33**, 3187.
Spanka, G. *et al, J. Org. Chem.*, 1987, **52**, 3362 (*ir, pmr, cmr*)

Hexahydro-2(3*H*)-benzofuranone, 9CI   H-70049
2-Hydroxycyclohexaneacetic acid lactone
[6051-03-2]

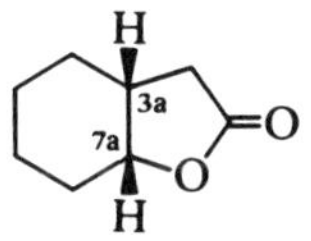

(*3aR,7aR*)-*form*

$C_8H_{12}O_2$ M 140.182
(*3aR,7aR*)-*form* [43119-27-3]
   (+)-cis-*form*
   Oil. $[\alpha]_D^{27}$ +45.5° (c, 0.43 in MeOH), $[\alpha]_D^{20}$ +41.9° (c, 10.3 in CHCl₃).
(*3aR,7aS*)-*form* [74741-45-0]
   (−)-trans-*form*
   $[\alpha]_D^{24}$ −42° (c, 1.6 in MeOH), $[\alpha]_D^{22.8}$ −77.6° (c, 4.6 in CHCl₃).
(*3aS,7aS*)-*form* [74708-16-0]
   (−)-cis-*form*
   $[\alpha]_D^{24}$ −86° (c, 0.3 in MeOH), $[\alpha]_D^{24.3}$ −40.3° (c, 8 in CHCl₃).
(*3aS,7aR*)-*form* [74708-14-8]
   (+)-cis-*form*
   $[\alpha]_D^{23}$ +78.5° (c, 2.9 in CHCl₃).
(*3aRS,7aRS*)-*form* [24871-12-3]
   (±)-cis-*form*
   Liq. d_4^{20} 1.09. Bp₂₀ 140-145°.
(*3aRS,7aSR*)-*form* [27345-71-7]
   (±)-trans-*form*
   Liq. d_4^{20} 1.09. Bp₆ 118-119°, Bp₀.₀₅ 90-95°.

Newman, M.S. *et al, J. Am. Chem. Soc.*, 1945, **67**, 233 (*synth*)
Klein, J., *J. Am. Chem. Soc.*, 1959, **81**, 3611 (*synth, ir*)
Corey, E.J. *et al, J. Org. Chem.*, 1974, **39**, 256 (*synth, ord, abs config, ir, pmr*)
Pirkle, W.H. *et al, J. Org. Chem.*, 1980, **45**, 4111 (*synth, abs config, ir, pmr*)
Bestmann, H.J. *et al, Tetrahedron Lett.*, 1987, **28**, 2111 (*synth*)
Wimmer, Z. *et al, Collect. Czech. Chem. Commun.*, 1987, **52**, 2326 (*synth, pmr, abs config, cd, ir*)

2,3,6,7,10,11-Hexahydrobenzo[1,2-*b*:3,4-*b'*:5,6-*b"*]tris[1,4]dithiin, 9CI   H-70050
Tris(ethylenedithio)benzene
[110431-54-4]

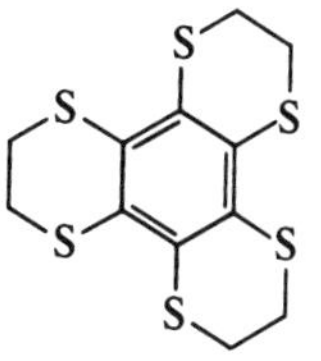

$C_{12}H_{12}S_6$ M 348.587
Mp 310°.

Tetrafluoroborate: [110431-67-9]. Black platelets. Radical cation.

Lapouyade, R. *et al, J. Chem. Soc., Chem. Commun.*, 1987, 223 (*synth, pmr, ms*)

2,3,6,7,10,11-Hexahydrobenzo[1,2-*b*:3,4-*b'*:5,6-*b"*]tris[1,4]oxathiin, 9CI   H-70051
[110448-11-8]

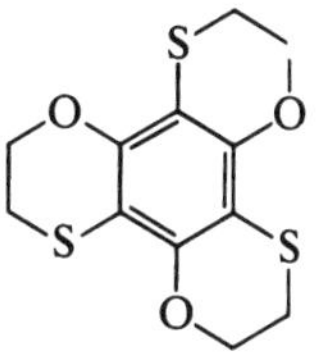

$C_{12}H_{12}O_3S_3$ M 300.405
Mp 305°.

Lapouyade, R. *et al, J. Chem. Soc., Chem. Commun.*, 1987, 223 (*synth, pmr, ms*)

Hexahydro-2(3*H*)-benzoxazolone, 9CI   H-70052
Cyclohexano[b]-2-oxazolidone
[17539-96-7]

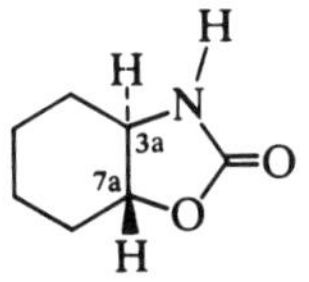

(3a*RS*,7a*RS*)-*form*

$C_7H_{11}NO_2$ M 141.169
(*3aRS,7aRS*)-*form* [7480-31-1]
   (±)-trans-*form*
   Cryst. (Et₂O or by subl.). Mp 100-102°.
   N-*Me:*
     $C_8H_{13}NO_2$ M 155.196
     Cryst. (EtOAc/pet. ether). Mp 51-52°.
(*3aRS,7aSR*)-*form* [7480-30-0]
   (±)-cis-*form*
   Cryst. (Et₂O/pet. ether). Mp 55-56°. Bp₀.₀₂ 130-135°.
   N-*Me:* Liq. Bp₀.₀₂ 110-115°.

Mousseron, M. *et al, Bull. Soc. Chim. Fr.*, 1953, 737 (*synth*)
Foglia, T.A. *et al, J. Org. Chem.*, 1967, **32**, 75 (*synth, ir*)
Herweh, J.E. *et al, J. Org. Chem.*, 1968, **33**, 4029 (*synth, pmr*)
Wright, J.J.K. *et al, J. Chem. Soc., Chem. Commun.*, 1976, 668 (*synth*)
Knapp, S. *et al, Tetrahedron Lett.*, 1987, **28**, 5399 (*synth*)
Marais, P.C. *et al, J. Chem. Soc., Perkin Trans. 1*, 1987, 1553 (*synth, ms, ir, pmr*)

Hexahydro-2,5-bis(trichloromethyl)-　　H-70053
furo[3,2-b]furan-3a,6a-diol
Chloretyl

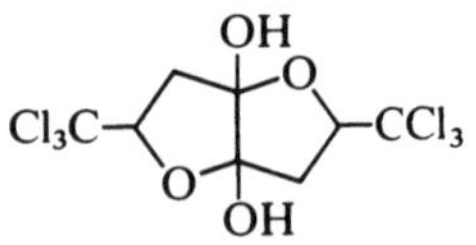

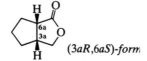

$C_8H_8Cl_6O_4$　　M 380.867

Two stereoisomers obt., configs. not certain. Prod. of reacn. of chloral and biacetyl.

α-form
Shows strong reversible knock-down insecticidal props. Mp 175° dec., Mp 206° dec. (dimorph.). Major isomer.

β-form
Mp 175-177°.

Schenk, H., *Chem. Ber.*, 1952, **85**, 901 (*synth*)
Leonard, N.J. et al, *J. Am. Chem. Soc.*, 1957, **79**, 2642.

Hexahydro-1H-cyclopenta[c]furan-1-one,　　H-70054
9CI

Updated Entry replacing O-50052
3-Oxabicyclo[3.3.0]octan-2-one. 2-Hydroxymethylcyclopentanecarboxylic acid lactone

(3aR,6aS)-form

$C_7H_{10}O_2$　　M 126.155

(3aR,6aS)-form [75658-84-3]
(+)-cis-*form*
$Bp_{0.25}$ 60°. $[\alpha]_D^{25}$ +96.9° (c, 1 in $CHCl_3$).
(3aRS,6aSR)-form [82442-63-5]
(±)-cis-*form*
Bp_{19} 127-128°.

Cope, A.C. et al, *J. Am. Chem. Soc.*, 1960, **82**, 4656 (*synth, ir*)
Jakovac, I.J. et al, *J. Am. Chem. Soc.*, 1982, **104**, 4659 (*synth, ir, pmr, abs config*)
Org. Synth., 1985, **63**, 10 (*synth*)
Kraus, G.A. et al, *Tetrahedron*, 1985, **41**, 4039 (*synth, pmr, ir, ms*)
Sabbioni, G. et al, *J. Org. Chem.*, 1987, **52**, 4565 (*synth, ir, pmr, cmr*)

Hexahydro-3,7a-dihydroxy-3a,7-dimethyl-　　H-70055
1,4-isobenzofurandione, 9CI

(3R,3aR,7R,7aR)-form

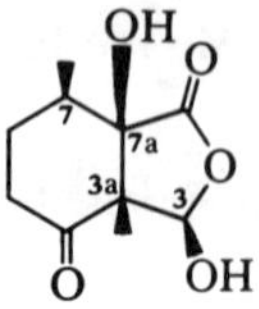

$C_{10}H_{14}O_5$　　M 214.218

(3R,3aR,7R,7aR)-form
Di-Ac: [115783-47-6]. *3,7a-Diacetoxyhexahydro-2-oxa-3a,7-dimethyl-1,4-indanedione.*
$C_{14}H_{18}O_7$　　M 298.292
Constit. of *Nepeta tuberosa*. Oil. $[\alpha]_D^{25}$ +113.8° (c, 4.9 in $CHCl_3$).
(3R,3aR,7S,7aR)-form
Di-Ac: [115783-49-8]. Constit. of *N. tuberosa*. Oil. $[\alpha]_D^{25}$ +34.8° (c, 2.4 in $CHCl_3$).

(3S,3aR,7R,7aR)-form
Di-Ac: [115783-48-7]. Constit. of *N. tuberosa*. Oil. $[\alpha]_D^{25}$ +21.35° (c, 2.13 in $CHCl_3$).
(3S,3aR,7S,7aR)-form
Di-Ac: [115783-50-1]. Constit. of *N. tuberosa*. Oil. $[\alpha]_D^{25}$ +17.47° (c, 1.75 in $CHCl_3$).

Urones, J.G. et al, *Phytochemistry*, 1988, **27**, 1525.

1,2,3,6,7,8-Hexahydro-3a,5a-ethenopyr-　　H-70056
ene, 9CI
[62665-02-5]

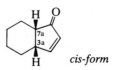

$C_{18}H_{18}$　　M 234.340
Platelets ($CHCl_3$/EtOH). Mp 130.5-132°.

Longone, D.T. et al, *Tetrahedron Lett.*, 1976, 4559 (*synth, uv, ir, pmr, ms*)
Mori, N. et al, *J. Chem. Soc., Chem. Commun.*, 1988, 575 (*synth*)

2,3,4,7,8,8a-Hexahydro-8-(5-hydroxy-4-　　H-70057
methylpentyl)-6,8-dimethyl-1H-3a,7-
methanoazulene-3-carboxylic acid, 9CI
[101467-55-4]

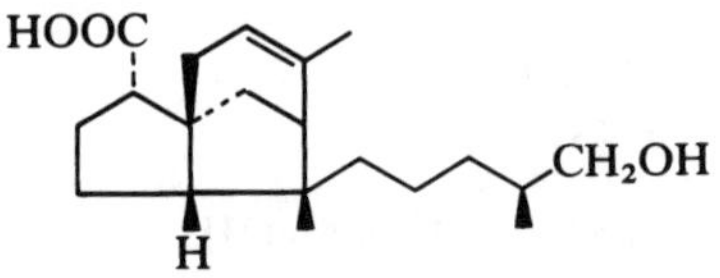

$C_{20}H_{32}O_3$　　M 320.471
New diterpene class representing an isoprenologue of the cedranes. Constit. of *Eremophila georgii*. Cryst. (Et_2O/pentane). Mp 90-91°. $[\alpha]_D$ −83.8° (c, 0.3 in $CHCl_3$).

Forster, P.G. et al, *Tetrahedron*, 1986, **42**, 215 (*isol, cryst struct*)

3a,4,5,6,7,7a-Hexahydro-1H-inden-1-one,　　H-70058
9CI
Bicyclo[4.3.0]non-8-en-7-one

$C_9H_{12}O$　　M 136.193

(3aRS,7aSR)-form [81255-91-6]
(±)-cis-*form*
Oil. $Bp_{0.08}$ 100°.

Karpf, M. et al, *Helv. Chim. Acta*, 1979, **62**, 852 (*synth, uv, cmr, pmr, ir, ms*)
Jones, T.K. et al, *Helv. Chim. Acta*, 1983, **66**, 2377.
Rizzo, C.L. et al *J. Org. Chem.*, 1987, **52**, 5280 (*synth, ir, pmr*)

2,3,5,6,7,8-Hexahydro-1H-indolizinium(1+)   H-70059

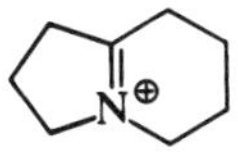

$C_8H_{14}N^{\oplus}$ M 124.205 (ion)

Perchlorate: [14594-57-1].
   $C_8H_{14}ClNO_4$ M 223.656
   Flakes (EtOH/Et$_2$O). Mp 227-228° (218-219° dec.).

Leonard, N.J. *et al, J. Org. Chem.*, 1956, **21**, 344 (*synth*)
Johnson, R.A. *et al, J. Org. Chem.*, 1968, **33**, 3187 (*synth, ir*)

1,3,3a,4,5,7a-Hexahydro-2H-indol-2-one, 9CI   H-70060

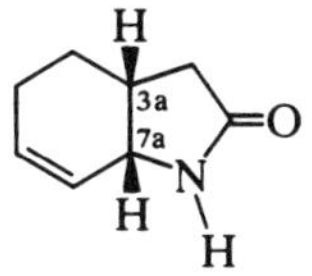

$C_8H_{11}NO$ M 137.181

(*3aRS,7aSR*)-*form* [113466-87-8]
   (±)-cis-*form*
   Oil.

Knapp, S. *et al, Tetrahedron Lett.*, 1987, **28**, 3213 (*synth, ir*)

Hexahydro-1(3H)-isobenzofuranone   H-70061

Updated Entry replacing H-50071
   Hexahydrophthalide. 8-Oxabicyclo[4.3.0]nonan-7-one
   [2611-01-0]

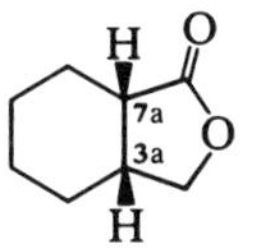

(*3aR,7aS*)-*form*

$C_8H_{12}O_2$ M 140.182

(*3aR,7aS*)-*form* [82442-61-3]
   (+)-cis-*form*
   Bp$_2$ 86°. [α]$_D^{25}$ +48.8° (c, 0.5 in CHCl$_3$).
(*3aRS,7aRS*)-*form*
   (±)-trans-*form*
   Bp$_{20}$ 140°.
   Hydrazide: Cryst. (EtOH/C$_6$H$_6$). Mp 176°.
(*3aRS,7aSR*)-*form* [6939-71-5]
   (±)-cis-*form*
   Bp$_{18}$ 131-132°, Bp$_{0.2}$ 70°.
   Hydrazide: Cryst. (C$_6$H$_6$/pet. ether). Mp 105-106°.

Christol, H. *et al, Bull. Soc. Chim. Fr.*, 1966, 2535 (*synth, ir*)
Bloomfield, J.J. *et al, J. Org. Chem.*, 1967, **32**, 3919 (*synth, ir, pmr*)
Jakovac, I.J. *et al, J. Org. Chem.*, 1982, **104**, 4659 (*synth, ir, pmr, abs config*)
Anzalone, L. *et al, J. Org. Chem.*, 1985, **50**, 2128 (*synth, pmr*)
Org. Synth., 1985, **63**, 10 (*synth*)
Murahashi, S.-H. *et al, J. Org. Chem.*, 1987, **52**, 4319 (*synth, ir, pmr, ms*)
Sabbioni, G. *et al, J. Org. Chem.*, 1987, **52**, 4565 (*synth, ir, pmr, cmr*)

3a,4,5,6,7,7a-Hexahydro-3-methylene-2-(3H)-benzofuranone   H-70062

Updated Entry replacing H-40055
   2-(2-Hydroxycyclohexyl)propenoic acid lactone. 3-Methyleneperhydrobenzofuran-2-one. 4-Methylene-2-oxabicyclo[3.3.0]octan-3-one
   [53387-38-5]

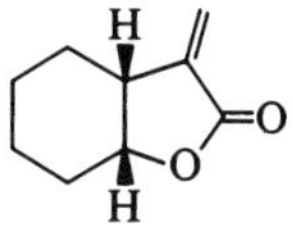

$C_9H_{12}O_2$ M 152.193

(*3aRS,7aRS*)-*form* [16822-06-3]
   (±)-cis-*form*
   Bp$_{0.06}$ 60° (bath).
(*3aRS,7aSR*)-*form* [3727-53-5]
   (±)-trans-*form*
   Mp 39-41° (35-37°). Bp$_1$ 60°.

Marshall, J.A. *et al, J. Org. Chem.*, 1965, **30**, 3475 (*synth, ir, pmr*)
Minato, H. *et al, J. Chem. Soc. (C)*, 1967, 1575 (*synth, ir*)
Behare, E.S. *et al, J. Chem. Soc., Chem. Commun.*, 1970, 402 (*synth*)
Greene, A.E. *et al, Tetrahedron Lett.*, 1972, **24**, 2489, 3375 (*synth, bibl*)
Grieco, P.A. *et al, J. Chem. Soc., Chem. Commun.*, 1972, 1317 (*synth, ir, pmr*)
Grieco, P.A. *et al, J. Org. Chem.*, 1974, **39**, 120 (*synth, ir, pmr*)
Murray, T.F. *et al, J. Am. Chem. Soc.*, 1981, **103**, 7520 (*synth, ir, pmr*)
Kozikowski, A.P. *et al, Tetrahedron Lett.*, 1983, **26**, 2623 (*synth*)
Nishitani, K. *et al, Tetrahedron Lett.*, 1987, **28**, 655 (*synth, ir, pmr*)
Srikishna, A., *J. Chem. Soc., Chem. Commun.*, 1987, 587 (*synth*)

1,3,4,5,7,8-Hexahydro-2,6-naphthalenedione, 9CI   H-70063

Bicyclo[4.4.0]dec-1(6)-ene-3,8-dione. Δ$^{4a(8a)}$-Octalin-2,6-dione. 3,4,7,8-Tetrahydro-2,6(1H,5H)-naphthalenedione

[84213-17-2]

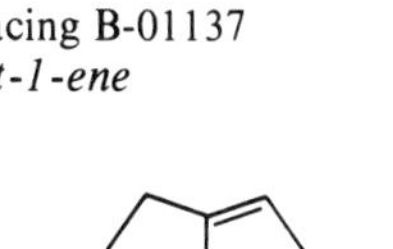

$C_{10}H_{12}O_2$ M 164.204
Mp 96°.

Jones, J.B. *et al, Can. J. Chem.*, 1987, **65**, 2397 (*synth*)

1,2,3,3a,4,5-Hexahydropentalene, 9CI   H-70064

Updated Entry replacing B-01137
   Bicyclo[3.3.0]oct-1-ene
   [694-73-5]

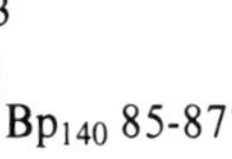

C_8H_{12} M 108.183
(±)-*form* [105601-68-1]
   Bp 124-129° (139°), Bp$_{140}$ 85-87°.

Kettey, A.D. *et al, J. Org. Chem.*, 1965, **30**, 940 (*synth*)
Becker, K.B., *Helv. Chim. Acta*, 1977, **60**, 68 (*synth, ir, pmr, cmr*)
Whitesell, J.K. *et al, J. Org. Chem.*, 1977, **42**, 3878 (*cmr*)

Crandall, J.K. *et al, J. Org. Chem.*, 1982, **47**, 5372 (*synth, pmr*)

1,2,3,3*a*,4,6*a*-Hexahydropentalene, 9CI   H-70065

Bicyclo[3.3.0]oct-2-ene

[5549-09-7]

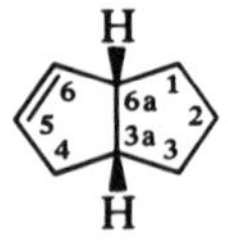

C$_8$H$_{12}$ M 108.183

(3*a*RS,6*a*RS)-form

(±)-cis-*form*

Cryst. (MeOH). Mp 159-160° dec.

Murahashi, S.-I. *et al, Bull. Chem. Soc. Jpn.*, 1974, **47**, 2420 (*synth*)
Whitesell, J.K. *et al, J. Org. Chem.*, 1977, **42**, 3878 (*cmr*)
Evans, W.J., *J. Org. Chem.*, 1981, **46**, 3925 (*synth*)
Brown, H.C. *et al, J. Org. Chem.*, 1985, **50**, 5574 (*synth*)
Behr, A. *et al, J. Mol. Catal.*, 1986, **35**, 9 (*synth*)
Benn, R. *et al, Magn. Reson. Chem.*, 1987, **25**, 653 (*cmr*)

1,2,3,4,5,6-Hexahydropentalene, 9CI   H-70066

Bicyclo[3.3.0]oct-1(5)-ene

[6491-93-6]

C$_8$H$_{12}$ M 108.183

Liq. with heavy sweet odour. Bp$_{707}$ 141-142°. n_D^{25} 1.4803; n_D^{20} 1.4836.

Corey, E.J. *et al, J. Org. Chem.*, 1969, **34**, 1233 (*synth, pmr*)
Becker, K.B., *Helv. Chim. Acta*, 1977, **60**, 68 (*cmr*)
Whitesell, J.K. *et al, J. Org. Chem.*, 1977, **42**, 3878 (*cmr*)
Kopecky, K.R. *et al, Can. J. Chem.*, 1981, **59**, 851 (*synth*)
Okarma, P.J. *et al, Org. Prep. Proced. Int.*, 1985, **17**, 212 (*synth, pmr, ms*)
Orendt, A.M. *et al, J. Am. Chem. Soc.*, 1988, **110**, 3386 (*cmr*)

Hexahydro-1*H*-pyrazolo[1,2-*a*]pyridazine, 9CI   H-70067

9-Azaindolizidine. 1,5-Diazabicyclo[4.3.0]nonane

[5721-43-7]

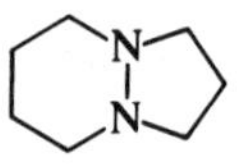

C$_7$H$_{14}$N$_2$ M 126.201

Bp$_{25}$ 71-73°.

Monopicrate: Cryst. (EtOH). Mp 200-202° dec.

Stetter, H. *et al, Chem. Ber.*, 1958, **91**, 1982 (*synth*)

1,2,3,3*a*,4,5-Hexahydropyrene   H-70068

as-*Hexahydropyrene*

[5385-37-5]

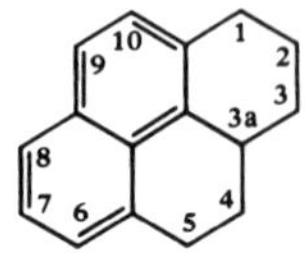

C$_{16}$H$_{16}$ M 208.302

Leaflets (EtOH). Mp 105-105.5°.

Picrate: Reddish-orange needles (EtOH). Mp 147.5-148°.

Cook, J.W. *et al, J. Chem. Soc.*, 1933, 398 (*synth*)

1,2,3,6,7,8-Hexahydropyrene, 9CI   H-70069

s-*Hexahydropyrene*

[1732-13-4]

C$_{16}$H$_{16}$ M 208.302

Long needles (EtOH). Mp 132-133°.

Picrate: Needles. Mp 119°.

Kagehira, I., *Bull. Chem. Soc. Jpn.*, 1931, **6**, 241 (*synth*)
Cook, J.W. *et al, J. Chem. Soc.*, 1933, 398 (*synth*)

Hexahydropyridazino[1,2-*a*]pyridazine-1,4-dione, 9CI   H-70070

1,6-Diazabicyclo[4.4.0]decane-2,5-dione. 1,2-Succinoylpiperidazine

[3661-10-7]

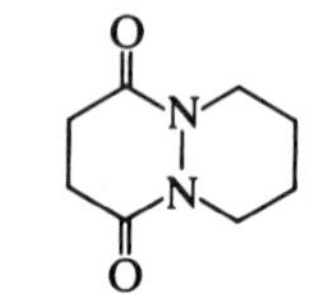

C$_8$H$_{12}$N$_2$O$_2$ M 168.195

Cryst. (CCl$_4$). Mp 179-180°.

Stetter, H. *et al, Chem. Ber.*, 1958, **91**, 1982 (*synth*)
Agmon, I. *et al, J. Am. Chem. Soc.*, 1986, **108**, 4477 (*cryst struct*)
Nelsen, S.F. *et al, J. Am. Chem. Soc.*, 1987, **109**, 5724 (*props*)

1,2,3,5,6,7-Hexahydropyrrolizinium(1+)   H-70071

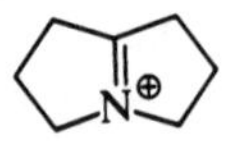

C$_7$H$_{12}$N$^{\oplus}$ M 110.179 (ion)

Perchlorate: [16853-10-4].
    C$_7$H$_{12}$ClNO$_4$ M 209.629
    Cryst. (EtOH). Mp 239-241° dec.

Johnson, R.A. *et al, J. Org. Chem.*, 1968, **33**, 3187 (*synth, ir*)

1,2,3,5,10,10*a*-Hexahydropyrrolo[1,2-*b*]-isoquinoline, 9CI   H-70072

1,2,3,5,10,10a-Hexahydrobenz[f]indolizine. Benzo[c]-1-azabicyclo[4.3.0]nonane

[54436-60-1]

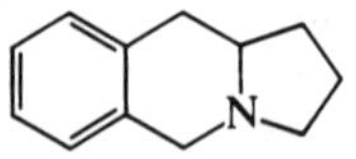

C$_{12}$H$_{15}$N   M 173.257

(±)-*form*

Liq., cryst. on long standing at 0°.

Picrate: Yellow needles. Mp 186.5-188°.

Leonard, N.J. *et al, J. Am. Chem. Soc.*, 1954, **76**, 3193 (*synth*)
Rigo, B. *et al, J. Heterocycl. Chem.*, 1983, **20**, 893 (*synth*)

The first digit of the DOC Number defines the Supplement in which the Entry is found. 0 indicates the Main Work.

Hexahydropyrrolo[1,2-*a*]pyrazine-1,4-dione, 9CI   H-70073

Cyclo(prolylglycine). 3,6-Dioxohexahydropyrrolo[1,2-a]pyrazine (incorr.)
[19179-12-5]

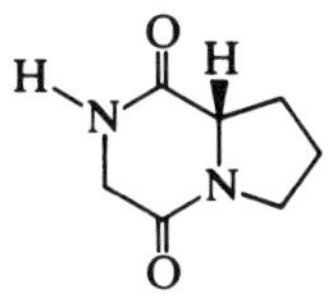

C₇H₁₀N₂O₂   M 154.168

(*S*)-form [3705-27-9]
L-form
Prod. by cultures of *Fusarium* spp. Isol. from the echinoderm *Luidia clathrata*. Alkaloid from *Pinellia pedatisecta* (Araceae). Cryst. (MeOH/Me₂CO). Mp 216-218°. [α]$_D^{20}$ −217.4° (c, 7.5 in H₂O).

Vičar, J. *et al, Collect. Czech. Chem. Commun.*, 1972, **37**, 4060; 1973, **38**, 1940, 1957 (*synth, ir, pmr, bibl*)
White, E.P., *N. Z. J. Sci.*, 1972, **15**, 178; *CA*, **77**, 98731b (*isol*)
Pettit, G.R. *et al, Experientia*, 1973, **29**, 521 (*isol, ms, struct*)
Von Dreele, R.B., *Acta Crystallogr., Sect. B*, 1975, **31**, 966 (*cryst struct*)
Qin, W. *et al, Zhongcaoyao*, 1984, **15**, 490; *CA*, **102**, 109774f (*isol*)

1,2,3,6,7,9*a*-Hexahydro-4(1*H*)-quinolizin-one   H-70074

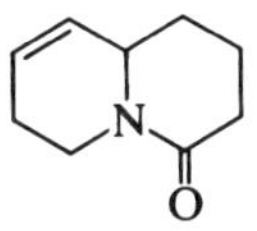

C₉H₁₃NO   M 151.208

(±)-*form* [87682-75-5]
Liq. Bp₁ 120°.

Flann, C. *et al, J. Am. Chem. Soc.*, 1987, **109**, 6097 (*synth, pmr, cmr, ir, ms*)

Hexahydro-1,5-thiazonin-6(7*H*)-one, 9CI   H-70075

5-Aza-6-oxo-1-thiacyclononane
[53579-96-7]

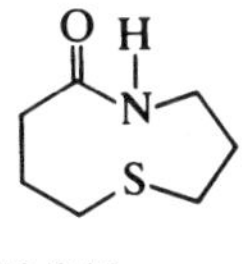

C₇H₁₃NOS   M 159.246
Cryst. (2-propanol). Mp 92-93°.

Wise, L.D. *et al, J. Med. Chem.*, 1974, **17**, 1232 (*synth*)
Doi, J.T. *et al, J. Org. Chem.*, 1987, **52**, 2581 (*synth, pmr, ir, cryst struct*)

2,3,4,5,8,9-Hexahydrothionin, 9CI   H-70076

Thiacyclonon-4-ene
C₈H₁₄S   M 142.259

(*E*)-form [68013-79-6]
Bp₂ 72°.
1-Oxide: [104808-79-9].
C₈H₁₄OS   M 158.258
Bp₀.₅ 126°.

Fava, A. *et al, J. Org. Chem.*, 1978, **43**, 4826 (*synth, ir, pmr, cmr*)
Sandri, E. *et al, J. Org. Chem.*, 1986, **51**, 4880 (*synth, deriv, pmr, cmr*)

6*bH*-2*a*,4*a*,6*a*-Hexahydrotriazacyclopenta[*cd*]-pentalene, 9CI   H-70077

[67705-38-8]

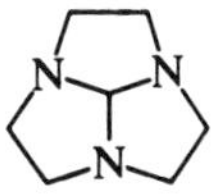

C₇H₁₃N₃   M 139.200

Atkins, T.J., *J. Am. Chem. Soc.*, 1980, **102**, 6364 (*synth, pmr*)
Erhardt, J.M. *et al, J. Am. Chem. Soc.*, 1980, **102**, 6365 (*synth, pmr, props*)
Weisman, G.R. *et al, Tetrahedron Lett.*, 1981, **21**, 3635 (*synth, pmr, cmr*)
Weisman, G.R. *et al, J. Chem. Soc., Chem. Commun*, 1987, 886 (*synth, pmr, cmr*)

1*H*,4*H*,7*H*,9*bH*-2,3,5,6,8,9-Hexahydro-3*a*,6*a*,9*a*-triazaphenalene, 9CI   H-70078

[67705-41-3]

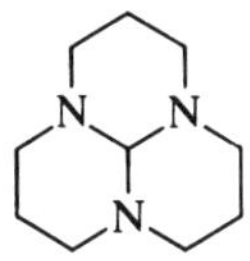

C₁₀H₁₉N₃   M 181.280

Atkins, T.J., *J. Am. Chem. Soc.*, 1980, **102**, 6364 (*synth, pmr*)
Erhardt, J.M. *et al, J. Am. Chem. Soc.*, 1980, **102**, 6365 (*synth, pmr, props*)
Weisman, G.R. *et al, Tetrahedron Lett.*, 1981, **21**, 3635 (*synth, pmr, cmr*)
Weisman, G.R. *et al, J. Chem. Soc., Chem. Commun.*, 1987, 886 (*synth, pmr*)

2′,3,3′,5,7,8-Hexahydroxyflavone   H-70079

C₁₅H₁₀O₈   M 318.239
3,7,8-Tri-Me ether: 2′,3′,5-Trihydroxy-3,7,8-trimethoxyflavone.
C₁₈H₁₆O₈   M 360.320
Constit. of *Notholaena aliena*. Cryst. (EtOH). Mp 203°.

Wollenweber, E. *et al, Phytochemistry*, 1988, **27**, 2673.

2′,3,5,5′,7,8-Hexahydroxyflavone   H-70080

C₁₅H₁₀O₈   M 318.239
3,7,8-Tri-Me ether: 2′,5,5′-Trihydroxy-3,7,8-trimethoxyflavone.
C₁₈H₁₆O₈   M 360.320
Constit. of *Notholaena californica*. Cryst. (AcOH aq.). Mp 230-231°.
3,7,8-Tri-Me ether, 2′-Ac: 2′-Acetoxy-5,5′-dihydroxy-3,7,8-trimethoxyflavone.
C₂₀H₁₈O₉   M 402.357
Constit. of *N. aliena*. Cryst. (C₆H₆/pet. ether). Mp 189-191°.
2′,3,7,8-Tetra-Me ether: 5′,5′-Dihydroxy-2′,3,7,8-tetramethoxyflavone.
C₁₉H₁₈O₈   M 374.346
Constit. of *N. californica*. Cryst. (EtOH). Mp 165-166°.

Wollenweber, E. *et al, Phytochemistry*, 1988, **27**, 2673.

3′,4′,5,5′,6,7-Hexahydroxyflavone   H-70081

Updated Entry replacing H-50088
C₁₅H₁₀O₈   M 318.239

6,7-Di-Me ether: 3′,4′,5,5′-*Tetrahydroxy-6,7-dimethoxyflavone.*
$C_{17}H_{14}O_8$　　M 346.293
Constit. of *Gutierrezia sphaerocephala.*

5′,6-Di-Me ether: 3′,4′,5,7-*Tetrahydroxy-5′,6-dimethoxyflavone.*
$C_{17}H_{14}O_8$　　M 346.293
Constit. of *Artemisia hispanica.* Yellow plates (MeOH). Mp 267-269°.

5′,6,7-Tri-Me ether: 3′,4′,5-*Trihydroxy-5′,6,7-trimethoxyflavone.*
$C_{18}H_{16}O_8$　　M 360.320
Constit. of *A. hispanica.* Yellow solid.

4′,5′,6-Tri-Me ester: 3′,5,7-*Trihydroxy-4′,5′,6-trimethoxyflavone.*
$C_{18}H_{16}O_8$　　M 360.320
Constit. of *Artemisia frigida* and *A. hispanica.* Yellow needles (MeOH). Mp 243-245°.

3′,4′,5,6-Tetra-Me ester: [68710-17-8]. 5,7-*Dihydroxy-3′,4′,5′,6-tetramethoxyflavone.* **Arteanoflavone**.
$C_{19}H_{18}O_8$　　M 374.346
Constit. of *Artemisia hispanica* and *A. anomala.* Yellow needles (MeOH). Mp 173-175°.

3′,5′,6,7-Tetra-Me ether: [83133-17-9]. 4′,5-*Dihydroxy-3′,5′,6,7-tetramethoxyflavone.*
$C_{19}H_{18}O_8$　　M 374.346
Constit. of *Artemisia mesatlantica.* Yellow plates (MeOH). Mp 240-242°.

3′,5,6,7-Tetra-Me ether, 4′,5′-methylene ether: [89029-10-7]. 5,5′,6,7-*Tetramethoxy-3′,4′-methylenedioxyflavone.*
$C_{20}H_{18}O_8$　　M 386.357
Constit. of the root of *Bauhinia championii.* Prisms (MeOH). Mp 200-201°.

3′,5,5′,6,7-Penta-Me ether: [93124-90-4]. 4′-*Hydroxy-3′,5,5′,6,7-pentamethoxyflavone.* **Ageconyflavone** C.
$C_{20}H_{22}O_8$　　M 390.389
From *Ageratum conyzoides.* Yellow solid.

Hexa-Me ether: [29043-07-0]. 3′,4′,5,5′,6,7-*Hexamethoxyflavone.*
$C_{21}H_{22}O_8$　　M 402.400
Constit. of *B. championii.* Prisms. Mp 116-117°.

Bouzid, N. *et al, Phytochemistry,* 1982, **21**, 803 (*isol*)
Chen, C.-C. *et al, Chem. Pharm. Bull.,* 1984, **32**, 166 (*isol*)
Vyas, A.V. *et al, Phytochemistry,* 1986, **25**, 2625 (*isol*)
Li, R. *et al, Phytochemistry,* 1988, **27**, 1556 (*isol*)

1,1,2,2,4,4-Hexamethyl-3,5-bis(methylene)-cyclopentane　　H-70082

[107010-25-3]

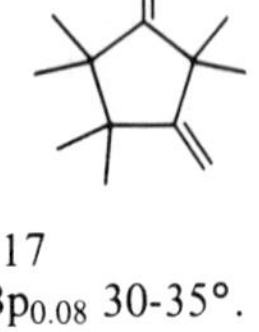

$C_{13}H_{22}$　　M 178.317
Liq. Mp ca. −20°. $Bp_{0.08}$ 30-35°.

Mayr, H. *et al, J. Org. Chem.,* 1987, **52**, 1342 (*synth, ir, pmr, cmr, ms*)

2,2,3,4,5,5-Hexamethyl-3-hexene, 9CI　　H-70083
Updated Entry replacing H-00763

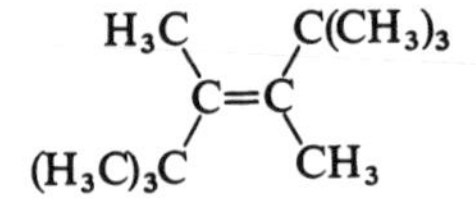

$C_{12}H_{24}$　　M 168.322
(**E**)-*form* [54290-40-3]
　Bp 142°.
(**Z**)-*form* [54429-93-5]
　Slowly isomerises to the *E*-form on pyrolysis in tetradecane. Isomers sepd. by glc.

Lenoir, D., *Chem. Ber.,* 1978, **111**, 411.
Gano, J.E. *et al, J. Org. Chem.,* 1987, **52**, 5636 (*synth, cmr, pmr, ir, ms*)

2-[3,7,11,15,19,23-Hexamethyl-25-(2,6,6-trimethyl-2-cyclohexenyl)pentacosa-2,14,18,22-tetraenyl]-3-methyl-1,4-naphthoquinone　　H-70084

[106611-74-9]

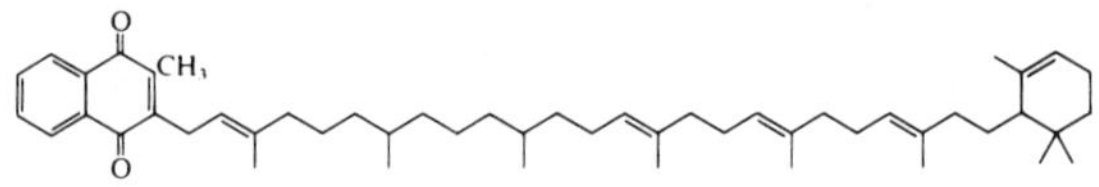

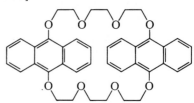

$C_{51}H_{76}O_2$　　M 721.160
Vitamin K-like compd. Isol. from *Nocardia* sp. No phys. props. reported.

Howarth, O.W. *et al, Biochem. Biophys. Res. Commun.,* 1986, **140**, 916 (*isol, struct*)

1,4,7,10,25,28,31,34-Hexaoxa[10.10](9,10)anthracenophane　　H-70085
7,8,10,11,13,14,23,24,26,27,29,30-Dodecahydro-5,3,2[1′,2′]:16,21[1″,2″]-dibenzenodibenzo[l,z] [1,4,7,10,15,18,21,24]-octaoxacyclooctacosin, 9CI

[89317-94-2]

$C_{40}H_{40}O_8$　　M 648.751
Mp 175°.

Castellan, A. *et al, Tetrahedron Lett.,* 1983, **24**, 5215 (*synth*)
Guinand, G. *et al, Acta Crystallogr., Sect. C,* 1986, **42**, 835 (*cryst struct*)

1,4,7,22,25,28-Hexaoxa[7.7](9,10)-anthracenophane   H-70086

7,8,10,11,20,21,23,24-Octahydro-5,26[1':2']:13,18[1'',2'']-dibenzenodibenzo[i,t] [1,4,7,12,15,18]-hexaoxacyclodocosin, 9CI

[89317-93-1]

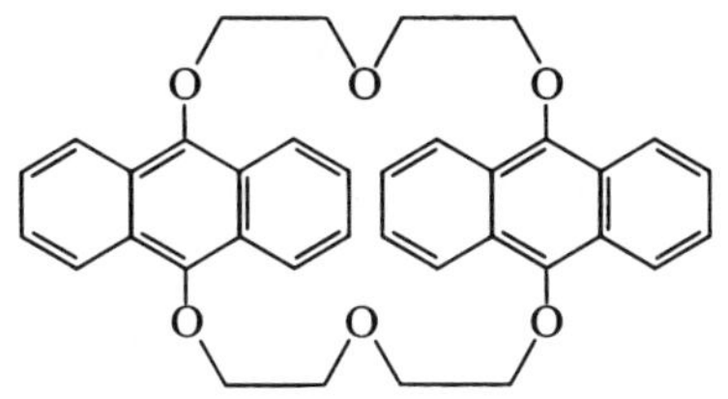

$C_{36}H_{32}O_6$ M 560.645
Mp 240°.

Castellan, A. *et al, Tetrahedron Lett.*, 1983, **24**, 5215 (*synth*)
Guinand, G. *et al, Acta Crystallogr., Sect. C*, 1986, **42**, 835 (*cryst struct*)

1,4,7,10,13,16-Hexathiacyclooctadecane   H-70087

Hexathia-18-crown-6. S_6-Ethano-18

[296-41-3]

$C_{12}H_{24}S_6$ M 360.682
Needles (hexane/Me$_2$CO). Mp 90-91°.

Black, D.St.C. *et al, Tetrahedron Lett.*, 1969, 3961 (*synth*)
Ochrymowycz, L.A. *et al, J. Org. Chem.*, 1974, **39**, 2079 (*synth*)
Wolf, R.E. *et al, J. Am. Chem. Soc.*, 1987, **109**, 4328 (*synth, pmr, ir, cryst struct, bibl*)

4-Hexen-1-amine, 9CI   H-70088

6-Amino-2-hexene

[60168-05-0]

$$H_3CCH=CHCH_2CH_2CH_2NH_2$$

$C_6H_{13}N$ M 99.175
(*E*)-*form* [55108-01-5]
Bp$_{11}$ 29°.

Venanzi, L.M. *et al, J. Organomet. Chem.*, 1979, **181**, 255 (*synth, ir, uv, pmr*)
Venanzi, L.M. *et al, J. Am. Chem. Soc.*, 1983, **105**, 6877 (*pmr*)

1-Hexen-5-yn-3-ol   H-70089

Propargylvinylcarbinol

[1573-66-6]

$$H_2C=CHCH(OH)CH_2C\equiv CH$$

C_6H_8O M 96.129
(±)-*form*
Bp$_{200}$ 105°, Bp$_{20}$ 53-54°.

Viola, A. *et al, J. Am. Chem. Soc.*, 1968, **90**, 6141 (*synth, ir*)
Schröder, C. *et al, J. Am. Chem. Soc.*, 1987, **109**, 5491 (*synth*)
Baldwin, J.E. *et al, J. Am. Chem. Soc.*, 1987, **109**, 8051 (*synth, pmr*)

3-Hexyne-2,5-dione, 9CI   H-70090

Diacetylacetylene

[54415-31-5]

$$H_3CCOC\equiv CCOCH_3$$

$C_6H_6O_2$ M 110.112
Oil. Bp$_{0.2}$ 25-50°.

Bis(2,4-dinitrophenylhydrazone): Orange cryst. Mp 230-232°.

Goldschmidt, S. *et al, Chem. Ber.*, 1961, **94**, 169 (*synth*)
Dunn, P.J. *et al, J. Chem. Soc., Perkin Trans. 1*, 1987, 1579 (*synth, pmr*)

Hipposterol   H-70091

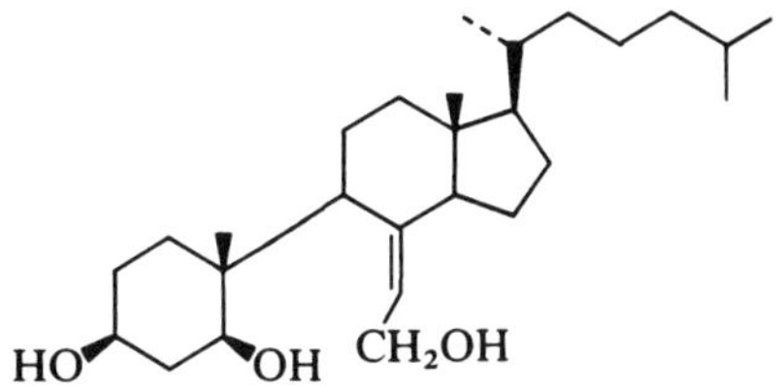

$C_{27}H_{48}O_3$ M 420.674
Constit. of *Hippospongia communis*. Cryst. Mp 85-87°.
[α]$_D$ +71.9° (c, 0.3 in CHCl$_3$).

Madaio, A. *et al, Tetrahedron Lett.*, 1988, **29**, 5999 (*struct, synth*)

Honyucitrin   H-70092

5,7-Dihydroxy-2-[4-hydroxy-3,5-bis(3-methyl-2-butenyl)phenyl]-4H-1-benzopyran-4-one, 9CI. 4',5,7-Trihydroxy-3',5'-diprenylflavone

[114542-44-8]

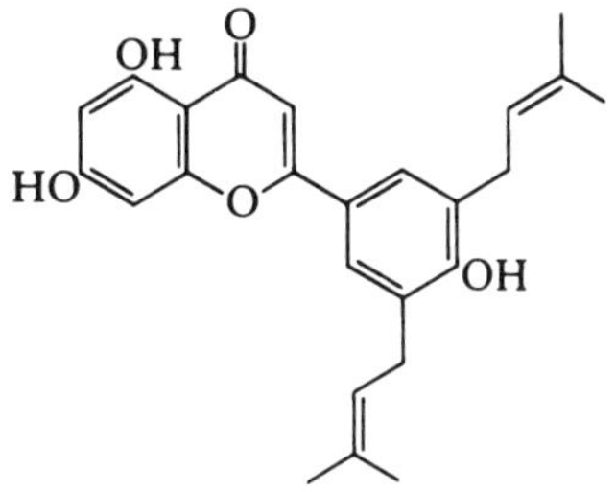

$C_{25}H_{26}O_5$ M 406.477
Constit. of *Citrus grandis*. Pale-yellow powder (diisopropyl ether). Mp 199.5-201°.

Wu, T.-S. *et al, Phytochemistry*, 1988, **27**, 585.

Honyudisin   H-70093

5-Hydroxy-8,8-dimethyl-6-(3-methyl-2-butenyl)-2H,8H-benzo[1,2-b:3,4-b']dipyran-2-one, 9CI

[114542-45-9]

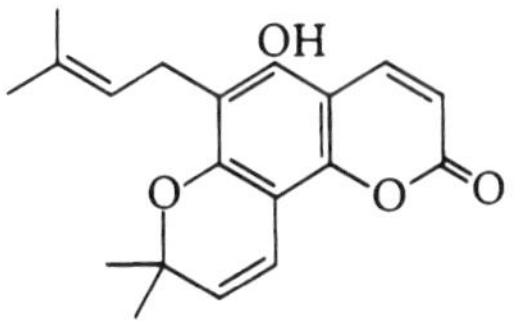

$C_{19}H_{20}O_4$ M 312.365
Constit. of *Citrus grandis*. Yellowish-green granules (EtOAc/hexane).

Wu, T.-S. *et al, Phytochemistry*, 1988, **27**, 585.

Hydnowightin   H-70094
[71392-06-8]

C₃₅H₃₀O₁₂   M 642.615
Constit. of *Hydnocarpus wightiana*. Cryst. Mp 239-241°.
[α]_D +40° (MeOH).

Sharma, D.K. *et al, Planta Med.*, 1979, **37**, 79.

2-Hydrazinobenzothiazole   H-70095

C₇H₇N₃S   M 165.212
Needles (EtOH). Mp 197-199°.

Boggust, W.A. *et al, J. Chem. Soc.*, 1949, 355 (*synth*)
Peet, N.P. *et al, J. Heterocycl. Chem.*, 1988, **25**, 543 (*synth, ir, pmr*)

2-Hydrazinoethylamine   H-70096
2-Hydrazinoethanamine, 9CI. (2-Aminoethyl)hydrazine, 8CI. 1-Amino-2-hydrazinoethane
[14478-61-6]

$$H_2NNHCH_2CH_2NH_2$$

C₂H₉N₃   M 75.113
Bp₁₅ 81-84°. Ca. 90% pure. Unstable at r.t., stored under N₂ <−20°.

Ger. Pat., 1 108 233, (*1961*); *CA*, **56**, 14080b (*synth, deriv*)
Grudzinski, S. *et al Acta Pol. Pharm.*, 1969, **26**, 217; *CA*, **72**, 43001e (*synth, deriv*)
Ropenga, J. *et al, Acta Pol. Pharm.*, 1984, **41**, 579; *CA*, **103**, 53776j (*synth*)
Showalter, H.D.H. *et al, J. Heterocycl. Chem.*, 1986, **23**, 1491 (*synth*)

5-Hydroperoxy-4(15)-eudesmen-11-ol   H-70097

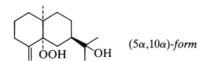

(5α,10α)-form

C₁₅H₂₆O₃   M 254.369
(5α,10α)-form
    Constit. of *Albinia japonica*. Needles (hexane). Mp 146-147°. [α]_D −48.6° (c, 0.21 in CHCl₃).
(5β,10α)-form
    Constit. of *A. japonica*. Needles (hexane). Mp 131-133°. [α]_D −52.0° (c, 0.2 in CHCl₃).

Itokawa, H. *et al, Chem. Pharm. Bull.*, 1987, **35**, 1460.

7'-Hydroxyabscisic acid   H-70098
Updated Entry replacing H-50141
Nigellic acid
[91897-25-5]

C₁₅H₂₀O₅   M 280.320
Metab. of *Xanthium strumarium* and constit. of leaves of *Vicia faba*.

Boyer, G.L. *et al, Phytochemistry*, 1986, **25**, 1103.
Lehmann, H. *et al, Phytochemistry*, 1988, **27**, 677 (*isol*)

8'-Hydroxyabscisic acid   H-70099
[25841-53-6]

C₁₅H₂₀O₅   M 280.320
Constit. of *Vigna unguiculata* fruits.

Adesomoju, A.A. *et al, Phytochemistry*, 1980, **19**, 223 (*isol*)

4-Hydroxyacenaphthylene   H-70100
4-Acenaphthylenol, 9CI
[111013-09-3]

C₁₂H₈O   M 168.195
Yellow cryst. Mp 148-150°. Dec. rapidly in polar solvs.
Ac: [111013-18-4].
    C₁₄H₁₀O₂   M 210.232
    Yellow cryst. Mp 54.5-55.5°.
Me ether: [111013-14-0]. *4-Methoxyacenaphthylene.*
    C₁₃H₁₀O   M 182.221
    Yellow cryst. Mp 46-48°.

Brown, R.F.C. *et al, Aust. J. Chem.*, 1987, **40**, 107 (*synth, ir, pmr, ms, cmr*)

2-(Hydroxyacetyl)pyridine   H-70101
2-Hydroxy-1-(2-pyridinyl)ethanone, 9CI
[95836-52-5]

C₇H₇NO₂   M 137.138
Mp 70-71°.
Hydrazone: [95836-53-6].
    C₇H₉N₃O   M 151.168
    Light-straw needles (C₆H₆). Mp 113-115°.
Me ether: [105729-06-4]. *2-(Methoxyacetyl)pyridine. 2-Methoxy-1-(2-pyridinyl)ethanone.*
    C₈H₉NO₂   M 151.165
    Bp₀.₀₅ 82-83° part. dec.

Klayman, D.L. *et al, J. Pharm. Sci.*, 1984, **73**, 1763 (*synth, deriv, use*)
Moriarty, R.M. *et al, Synthesis*, 1985, 943 (*synth, ir, pmr, ms*)

Moriarty, R.M. *et al, J. Org. Chem.*, 1987, **52**, 150 (*deriv, synth, ir, pmr, ms*)

3-(Hydroxyacetyl)pyridine   H-70102

2-Hydroxy-1-(3-pyridinyl)ethanone, 9CI
[104501-59-9]
$C_7H_7NO_2$ M 137.138
Cryst. (Me$_2$CO). Mp 112-113°.

Moriarty, R.M. *et al, Synthesis*, 1985, 943 (*synth, ir, pmr, ms*)

2-(Hydroxyacetyl)thiophene   H-70103

2-[(Hydroxymethyl)carbonyl]thiophene

$C_6H_6O_2S$ M 142.172
Me ether: [105729-09-7]. *2-(Methoxyacetyl)thiophene.*
2-Methoxy-1-(2-thienyl)ethanone.
$C_7H_8O_2S$ M 156.199
Bp$_{0.01}$ 59-62°.

Moriarty, R.M. *et al, J. Org. Chem.*, 1987, **52**, 150 (*deriv, synth, pmr, ir*)

3-Hydroxybenz[a]anthracene   H-70104

Benz[a]anthracen-3-ol, 9CI
[4834-35-9]
$C_{18}H_{12}O$ M 244.292
Cryst. (C$_6$H$_6$/hexane). Mp 209-210°.
Ac: [70092-10-3].
$C_{20}H_{14}O_2$ M 286.329
Cryst. (C$_6$H$_6$/hexane). Mp 165-166°.
Me ether: [69847-25-2]. *3-Methoxybenz[a]anthracene.*
$C_{19}H_{14}O$ M 258.319
Cryst. (C$_6$H$_6$/hexane). Mp 160-161°.

Fu, P.P. *et al, J. Org. Chem.*, 1979, **44**, 4265 (*synth, deriv, pmr*)
Harvey, R.G. *et al, J. Med. Chem.*, 1988, **31**, 154 (*synth, deriv, uv, pmr*)

2-Hydroxy-1,4-benzenedicarboxylic acid,   H-70105
9CI

Updated Entry replacing H-10089
Hydroxyterephthalic acid
[636-94-2]
$C_8H_6O_5$ M 182.132
Mp 332° dec.
Di-Me ester: [6342-72-9].
$C_{10}H_{10}O_5$ M 210.186
Mp 94°.
Ac, di-Me ester:
$C_{12}H_{12}O_6$ M 252.223
Mp 73-74°.
Me ether: [5156-00-3]. *2-Methoxy-1,4-benzenedicar-boxylic acid.*
$C_9H_8O_5$ M 196.159
Mp 296-297° subl.

Ruoff, P.M. *et al, J. Org. Chem.*, 1961, **26**, 939 (*synth*)
Grossa, M. *et al, Monatsh. Chem.*, 1966, **97**, 570 (*synth*)
Field, L. *et al, J. Org. Chem.*, 1970, **35**, 3647 (*synth*)
Weinges, K. *et al, Justus Liebigs Ann. Chem.*, 1987, 833 (*synth, pmr, cmr, derivs*)
Miura, Y. *et al, J. Org. Chem.*, 1988, **53**, 439 (*synth*)

4-Hydroxybenzofuran   H-70106

Updated Entry replacing H-01301
4-Benzofuranol, 9CI. Karanjol
[480-97-7]

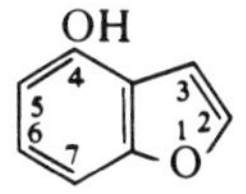

$C_9H_8O_2$ M 148.161
Mp 55-56°, 58°.
Me ether: [18014-96-5]. *4-Methoxybenzofuran.*
$C_9H_8O_2$ M 148.161
Bp 220-222°.

Adel-Wahab, S.M. *et al, J. Chem. Soc. (C)*, 1968, 67 (*deriv*)
Rene, C. *et al, Bull. Soc. Chim. Fr.*, 1973, 2355 (*synth*)
Kneen, G. *et al, Synth. Commun.*, 1986, **16**, 1635 (*synth*)

1-(4-Hydroxy-5-benzofuranyl)-3-phenyl-2-   H-70107
propen-1-one

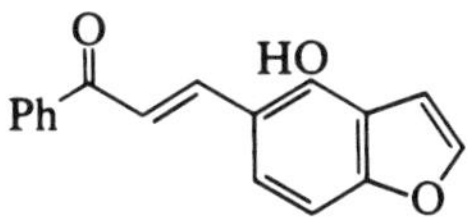

$C_{17}H_{12}O_3$ M 264.280
Constit. of *Milletia ovalifolia*. Yellow needles (EtOH).
Mp 124°.

Saxena, D.B. *et al, Indian J. Chem., Sect. B*, 1987, **26**, 704.

2-Hydroxybenzoic acid, 9CI   H-70108

Salicylic acid. Numerous proprietary names
[69-72-7]

$C_7H_6O_3$ M 138.123
Occurs in form of esters in essential oils and plant prods. eg. oil of wintergreen. Used as an antiseptic and antifungal agent, and for various skin conditions. Kerolytic. Simple esters are perfumery and flavouring ingredients. Needles (H$_2$O). Mod. sol. H$_2$O. Mp 159°. Bp$_{20}$ 211°. pK_a 3.0. Steam volatile. Subl. *in vacuo.*
▷VO 052000. VO0525000.

Diethylamine salt (1:1): [4419-92-5]. *Aciphen. Alesal. Gallisal. Mysal. Ponostop. Rheumatidermol.*
Antirheumatic. Antipyretic. Mp 87-90°.
Imidazole salt (1:1): [36364-49-5]. *Imidazole salicylate,* INN. *Imidazate. Salizolo. Selezen. ITF 182.*
Antiinflammatory, analgesic, antipyretic and antithrombotic.
Lysine salt (1:1): [57282-48-1]. *Verpyran.* Antipyretic.
N-Methylglucamine salt (1:1): [23277-50-1]. *Meglumine salicylate. PFA 186.* Analgesic.
2,2',2''-Nitrilotriethanol salt (1:1): [2174-16-5]. *Aspercreme. Aspergel. Mobisyl. Myoflex. Royflex. SalRub.* Antirheumatic and analgesic.
2-Pyridinecarboxamide salt (1:1): [66634-12-6]. *Nicotinamide salicylate. Niksalin.* Analgesic, antiinflammatory and antipyretic. Mp 132-134°.
Na salt: [54-21-7]. *Numerous proprietary names.* Antipyretic and antineuralgic. Cryst. Other metal salts of salicylic acid are used; mainly as analgesics and antiinflammatory agents.
▷VO5075000.

Me ester: see Methyl 2-hydroxybenzoate, M-02063
Et ester: [118-61-6]. **Ethyl salicylate**. *Mesotol. Sal ethyl.*
　Salstan.
　$C_9H_{10}O_3$　　M 166.176
　Has been used as a vet. counterirritant. Antirheumatic.
　Liq. Mp 1.3°. Bp 231-235°.
▷VO3000000.
4-(Acetylamino)phenyl ester: [118-57-0]. **Acetaminosa-**
　lol, INN. *Acetamidosalol. Acetaminosal. Acetylpara-*
　minosalol. Asalphen. Cetosal. Cetosalol. Phenetsal.
　Phenosol. Salophen.
　$C_{15}H_{13}NO_4$　　M 271.272
　Analgesic, antipyretic and antirheumatic. Cryst.
　(EtOH). Mp 187°.
2-Carboxyphenyl ester: see Salicyloylsalicylic acid, S-
　00024
Choline ester: [2016-36-6]. **Choline salicylate, INN**. *Oth-*
　er proprietary names.
　$C_{12}H_{19}NO_4$　　M 241.286
　Analgesic, antipyretic. Mp 49.5-50°.
▷GA6475000.
2,4-Dinitrobenzyl ester: Mp 168°.
2-Hydroxyethyl ester: [87-28-5]. **Glycol salicylate**. *Dolo-*
　gel. Dolunguent. Espirosal. Glysal. Kytta-Gel. Rheu-
　macyl. Saliment. Salocolum. Sarocol. Spirosal. GL 7.
　$C_9H_{10}O_4$　　M 182.176
　Counterirritant and antirheumatic. Bp$_{12}$ 169-172°.
Methoxymethyl ester: [575-82-6]. **Methoxymethyl salic-**
　ylate. *Ericin. Mesotan. Metoxal. Salimethyl.*
　Salmester.
　$C_9H_{10}O_4$　　M 182.176
　Antirheumatic and counterirritant. Bp$_{32}$ 153°, Bp$_{0.01}$
　65°.
Tetrahydrofurfuryl ester: [2217-35-8]. **Thurfyl**
　salicylate.
　$C_{12}H_{14}O_4$　　M 222.240
　Topical analgesic. Bp$_{0.05}$ 106°.
Chloride: [1441-87-8].
　$C_7H_5ClO_2$　　M 156.568
　Needles. Mp 19-19.5°. Bp$_{15}$ 92°.
Amide: [65-45-2]. *2-Hydroxybenzamide, 9CI*. **Salicyla-**
　mide, INN. *Numerous proprietary names.*
　$C_7H_7NO_2$　　M 137.138
　Analgesic. White or sl. pink cryst. Mp 140°.
▷VN6475000.
Amide, Et ether: [938-73-8]. *2-Ethoxybenzamide, 9CI*.
　Ethenzamide, INN. *Etenzamide. Other proprietary*
　names.
　$C_9H_{11}NO_2$　　M 165.191
　Analgesic. Cryst. (EtOAc/hexane). Mp 132-134°.
▷CV4900000.
Amide, Ethoxyethyl ether: [15302-15-5]. *2-(2-*
　Ethoxyethoxy)benzamide, 9CI. **Etosalamide, INN**.
　Ethosalamide. Ap 2.
　$C_{11}H_{15}NO_3$　　M 209.244
　Analgesic. Cryst. (2-propanol). Mp 70-71.5°.
Anilide: [87-17-2]. *2-Hydroxy-N-phenylbenzamide, 9CI*.
　Salicylanilide, INN. *Ansadol. Hyanilid. Salifebrin.*
　Salinidol. ASK.
　$C_{13}H_{11}NO_2$　　M 213.235
　Antipyretic and topical antifungal agent. Prisms
　(H$_2$O). Mp 135.8-136.2°.
Nitrile: [611-20-1]. *2-Cyanophenol.*
　C_7H_5NO　　M 119.123
　Prisms (C$_6$H$_6$/pet. ether). Mp 98°.
Ac: see 2-Acetoxybenzoic acid, A-00129
Benzoyl: [4578-66-9].
　$C_{14}H_{10}O_4$　　M 242.231

Needles. Mp 132°.
4-Nitrobenzyl: Yellow cryst. (MeOH). Mp 205°.
Me ether: see 2-Methoxybenzoic acid, M-30038
Et ether: [134-11-2]. *2-Ethoxybenzoic acid.*
　$C_9H_{10}O_3$　　M 166.176
　Mp 19.3-19.5°.
Ph ether: see 3-Phenoxybenzoic acid, P-00889

Org. Synth., Coll. Vol. **4**, 178 (*derivs*)
Brewster, R.Q., *J. Am. Chem. Soc.*, 1918, **40**, 1136
　(*Acetaminosalol*)
Weizmann, C. *et al*, *J. Org. Chem.*, 1948, **13**, 796 (*Glycol*
　salicylate)
Reichert, B., *Arzneim.-Forsch.*, 1953, **3**, 255 (*Salicylamide*)
B.P., 774 635, (*1957*); *CA*, **51**, 16564 (*Etosalamide*)
Liberman, S. *et al*, *Bull. Soc. Chim. Fr.*, 1958, 185
　(*Salicylanilide*)
Ritschel, W., *Pharm. Acta Helv.*, 1959, **34**, 195 (*Diethylamine*
　salt)
Shapiro, S.L. *et al*, *J. Am. Chem. Soc.*, 1959, **81**, 3728
　(*Ethenzamide*)
Kurbjuweit, H.G. *et al*, *Arzneim.-Forsch.*, 1960, **10**, 820
　(*Ethenzamide*)
Winstead, M.B. *et al*, *J. Chem. Eng. Data*, 1962, **7**, 265
　(*Etosalamide*)
Bouvet, P. *et al*, *Ann. Pharm. Fr.*, 1963, **21**, 87 (*Choline*
　salicylate)
Cummings, A.J. *et al*, *J. Pharm. Pharmacol.*, 1963, **15**, 212
　(*Etosalamide*)
Netherlands Pat., 6 504 517, (*1965*); *CA*, **64**, 8097
　(*Acetaminosalol*)
Wragg, R.T., *J. Chem. Soc.*, 1965, 7162 (*Thurfyl salicylate*)
Kametani, T. *et al*, *Chem. Pharm. Bull.*, 1967, **15**, 613 (*Choline*
　salicylate)
Smith, M.J.H. *et al*, *The Salicylates: a Critical Biographical*
　Review, 1967, Wiley, London.
B.P., 1 220 447, (*1971*); *CA*, **75**, 35462 (*Methoxymethyl*
　salicylate)
Ger. Pat., 1 493 954, (*1972*); *CA*, **77**, 114803 (*Meglumine salt*)
Gopalakrishna, E.M. *et al*, *Acta Crystallogr.*, Sect. B, 1972, **28**,
　2917 (*Ethenzamide*)
Dardoize, F. *et al*, *Synthesis*, 1977, 567 (*Methoxymethyl*
　salicylate)
Legheand, J. *et al*, *Ann. Pharm. Fr.*, 1977, **35**, 387
　(*Ethenzamide*)
Goldsmith, L.A., *Int. J. Dermatol.*, 1979, **18**, 32 (*rev,*
　pharmacol)
Levy, G., *Drug Metab. Rev.*, 1979, **9**, 1 (*rev, metab*)
Erickson, S.H., *Kirk-Othmer Encycl. Chem. Technol. 3rd Ed.*,
　1982, **20**, 500 (*rev*)
Pagella, P.G. *et al*, *Arzneim.-Forsch.*, 1983, **33**, 716 (*Imidazole*
　salt)
Brune, K. *et al*, *Arzneim.-Forsch.*, 1984, **34**, 1060 (*rev,*
　pharmacol)
Shibasaki, J. *et al*, *J. Pharmacobio.-Dyn.*, 1984, **7**, 804
　(*Salicylamide, Ethenzamide*)
Kashino, S. *et al*, *Acta Crystallogr.*, Sect. C, 1986, **42**, 457
　(*Salicylanilide*)
Piccolo, O. *et al*, *Tetrahedron*, 1986, **42**, 885 (*Salicylanilide*)
Szirmai, E. *et al*, *CA*, 1986, **104**, 141908 (*Niksalin*)
Hazards in the Chemical Laboratory, (*Bretherick, L., Ed.*), 3rd
　Ed., Royal Society of Chemistry, London, 1981, 188
Negwer, M., *Organic-Chemical Drugs and their Synonyms*, 6th
　Ed., Akademie-Verlag, Berlin, 1987, 368, 645, 663, 1211
　(*synonyms*)
Martindale, The Extra Pharmacopoeia, 28th/29th Eds.,
　1982/1989, Pharmaceutical Press, London, 2602, 2625, 2641,
　2642, 2656, 2663, 2664, 2700, 2701, 2705, 2708, 12707.
Sax, N.I., *Dangerous Properties of Industrial Materials*, 6th
　Ed., Van Nostrand-Reinhold, 1984, 400, 962, 989.

6-Hydroxy-2*H*-1-benzopyran-2-one, 9CI　　**H-70109**

Updated Entry replacing H-01314
6-Hydroxycoumarin

[6093-68-1]

$C_9H_6O_3$　　M 162.145
Mp 250°.

Me ether: [17372-53-1]. *6-Methoxy-2*H-*1-benzopyran-2-one. 6-Methoxycoumarin.*
$C_{10}H_8O_3$ M 176.171
Mp 102-103°.

Das Gupta, A.K. *et al, J. Chem. Soc. (C),* 1969, 29 (*synth*)
De Graw, J.I. *et al, J. Chem. Eng. Data,* 1969, **14**, 509 (*synth*)
Günther, H. *et al, Org. Magn. Reson.,* 1975, **7**, 339 (*cmr, deriv*)
Joseph-Nathan, P. *et al, J. Heterocycl. Chem.,* 1984, **21**, 1141 (*pmr, deriv*)
Sato, K. *et al, J. Chem. Soc., Perkin Trans. 1,* 1987, 1753 (*synth*)
Ziegler, T. *et al, Chem. Ber.,* 1987, **120**, 373 (*synth, deriv*)

4-Hydroxy-2*H*-1-benzothiopyran-2-one,   H-70110
9CI

*2-Hydroxy-4*H-*1-benzothiopyran-4-one. 4-Hydroxythiocoumarin. 2-Hydroxythiochromone*
[107514-60-3]

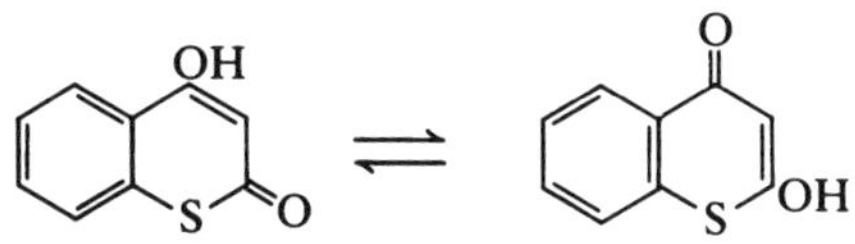

$C_9H_6O_2S$ M 178.205
The authors refer to the 4-one form, but the 2-one form is more probable. Mp 206-208°.

Lau, C.K. *et al, J. Org. Chem.,* 1987, **52**, 1670 (*synth, pmr*)

4′-Hydroxy-2-biphenylcarboxylic acid, 8CI   H-70111

Updated Entry replacing H-01382
2-(4-Hydroxyphenyl)benzoic acid
$C_{13}H_{10}O_3$ M 214.220
Pale-yellow cryst. (H_2O). Mp 206.5°.
Me ether: [18110-71-9]. *4′-Methoxy-2-biphenylcarboxylic acid.*
$C_{14}H_{12}O_3$ M 228.247
Cryst. (EtOH aq.). Mp 146-148°.
Me ether, Me ester:
$C_{15}H_{14}O_3$ M 242.274
Isol. from *Trifolium repens.* Cryst. (EtOAc). Mp 132-134°.

Drapala, T. *et al, Rocz. Chem.,* 1960, **34**, 1371; 1972, **46**, 9 (*synth, uv*)
Hey, D.H. *et al, J. Chem. Soc.,* 1961, 232 (*synth*)
Brown, P.M. *et al, J. Chem. Soc. (C),* 1968, 842.
Swenton, J.S. *et al, J. Org. Chem.,* 1973, **38**, 1157.
Ghosal, S. *et al, J. Chem. Res. (S),* 1988, 196 (*deriv, isol, uv, ir, pmr*)

2-Hydroxybutanal, 9CI   H-70112
α-Hydroxybutyraldehyde
[37428-67-4]

$$H_3CCH_2CH(OH)CHO$$

$C_4H_8O_2$ M 88.106
(±)-*form*
Cryst.
Me ether: [107847-11-0]. *2-Methoxybutanal.*
$C_5H_{10}O_2$ M 102.133
Liq. Bp 106-108°.
Di-Me acetal: 1,1-Dimethoxy-2-butanol.
$C_6H_{14}O_3$ M 134.175
Bp_1 39-42°.
4-Nitrophenylhydrazone: Cryst. Mp 135°.

Dworzak, R. *et al, Monatsh. Chem.,* 1929, **52**, 143 (*synth*)

Stevens, C.L. *et al, J. Am. Chem. Soc.,* 1954, **76**, 2695 (*deriv, synth*)
Masamune, T. *et al, Bull. Chem. Soc. Jpn.,* 1975, **48**, 2294 (*ir, pmr*)
Lodge, E.P. *et al, J. Am. Chem. Soc.,* 1987, **109**, 3353 (*deriv, synth, ir, pmr, cmr*)

2-Hydroxy-4-*tert*-butylcyclopentanecar-   H-70113
boxylic acid
4-(1,1-Dimethylethyl)-2-hydroxycyclopentanecarboxylic acid, 9CI

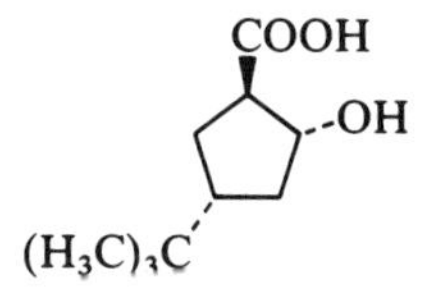

$C_{10}H_{18}O_3$ M 186.250
(**1RS,2RS,4RS**)-*form* [37715-20-1]
(±)-trans,trans-*form*
Fine needles (C_6H_6/pet. ether). Mp 90-91°.
Hydrazide: Cryst. (EtOAc). Mp 134-134.5°.
(**1RS,2RS,4SR**)-*form*
(±)-trans,cis-*form*
Hydrazide: Cryst. (EtOAc). Mp 155.5-156.5°.
(**1RS,2SR,4SR**)-*form* [37715-16-5]
(±)-cis,cis-*form*
Mp 119-120°.
Hydrazide: Plates (EtOH). Mp 196-197°.

Bernáth, G. *et al, Tetrahedron,* 1972, **28**, 3475 (*synth*)

8-Hydroxy-1(6),2,4,7(11)-cadinatetraen-   H-70114
12,8-olide
[100324-68-3]

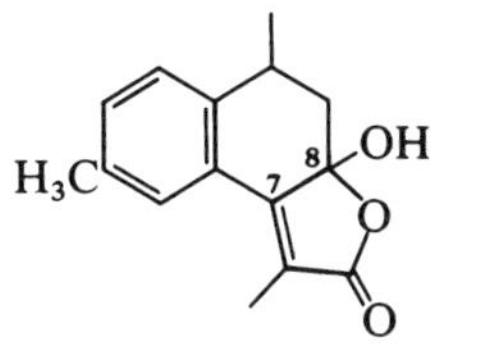

$C_{15}H_{16}O_3$ M 244.290
Constit. of *Chromolaena laevigata.* Oil.
O^8-*Me:* [100324-69-4]. *8-Methoxy-1(6),2,4,7(11)-cadinatetraen-12,8-olide.*
$C_{16}H_{18}O_3$ M 258.316
Constit. of *C. laevigata.* Oil.

Misra, L.N. *et al, Tetrahedron,* 1985, **41**, 5353.

7-Hydroxy-13(17),15-cleistanthadien-18-   H-70115
oic acid

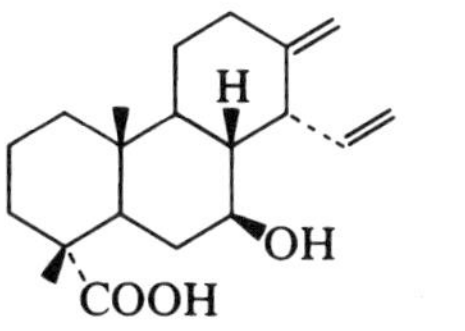

$C_{20}H_{30}O_3$ M 318.455
7β-form [114742-92-6]
Constit. of *Pogostemon auricularis.* Amorph.
Ac: [114742-93-7]. *7β-Acetoxy-13(17),15-cleistanthadien-18-oic acid.*
$C_{22}H_{32}O_4$ M 360.492
Constit. of *P. auricularis.* Cryst. Mp 178°.

Hussaini, F.A. *et al*, *J. Nat. Prod.*, 1988, **51**, 212.

7-Hydroxy-3,13-clerodadiene-16,15;18,19-diolide   H-70116

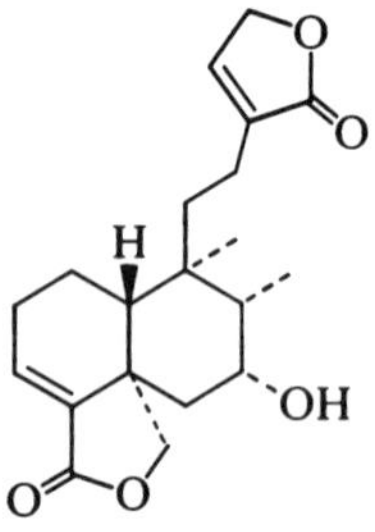

$C_{20}H_{26}O_5$ M 346.422

7α-form

Constit. of *Salvia melissodora*. Cryst. Mp 178-180°. $[\alpha]_D^{20}$ −154° (c, 0.165 in $CHCl_3$).

Esquivel, B. *et al*, *Phytochemistry*, 1988, **27**, 2903.

ent-7β-Hydroxy-3,13-clerodadiene-15,16;18,19-diolide   H-70117

[100343-68-8]

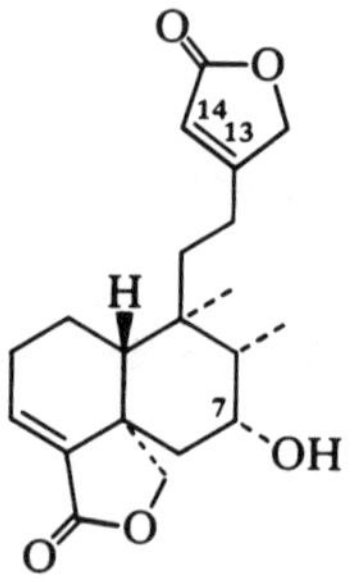

$C_{20}H_{26}O_5$ M 346.422

Constit. of *Baccharis genistelloides*. Prisms (EtOAc). Mp 198-202°. $[\alpha]_D^{20}$ −150.9° (c, 2.0 in MeOH).

13,14-Dihydro: [100343-70-2]. *ent*-7β-*Hydroxy-3-clero-dene-15,16:18,19-diolide*.
$C_{20}H_{28}O_5$ M 348.438
Constit. of *B. genistelloides*. Prisms (EtOAc/CHCl3).
Mp 199-206°. $[\alpha]_D$ −152.0° (c, 1.5 in $CHCl_3$).

13,14-Dihydro, 7-ketone: [100343-69-9]. *ent*-7-*Oxo-3-clerodene-15,16:18,19-diolide*.
$C_{20}H_{26}O_5$ M 346.422
Constit. of *B. genistelloides*. Viscous oil. $[\alpha]_D^{20}$ −128.2° (c, 1.2 in $CHCl_3$).

Herz, W. *et al*, *J. Org. Chem.*, 1977, **42**, 3913 (*isol*)
Bohlmann, F. *et al*, *Phytochemistry*, 1979, **18**, 1011 (*isol*)
Kuroyanagi, M. *et al*, *Chem. Pharm. Bull.*, 1985, **33**, 5075 (*isol*)

16-Hydroxy-3,13-clerodadien-15,16-olide   H-70118

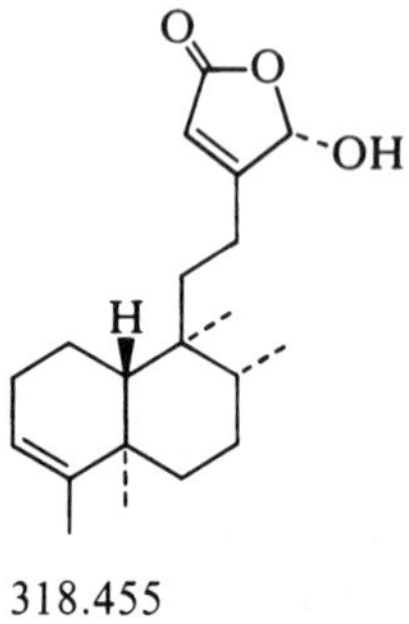

$C_{20}H_{30}O_3$ M 318.455

Constit. of *Polyalthia longifolia*. Gum. $[\alpha]_D^{26}$ −70.58° (c, 0.01 in MeOH).

Ac: Cryst. (EtOAc/pet. ether). Mp 175°. $[\alpha]_D^{26}$ −24.24° (c, 0.066 in MeOH).

Phadnis, A.P. *et al*, *Phytochemistry*, 1988, **27**, 2899.

2-Hydroxycyclohexanecarboxylic acid, 9CI   H-70119

Updated Entry replacing H-01475
Hexahydrosalicylic acid. Cyclohexanol-2-carboxylic acid
[609-69-8]

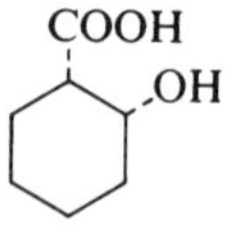

(1S,2R)-form
Absolute
configuration

$C_7H_{12}O_3$ M 144.170

(1R,2R)-form [1654-67-7]
(−)-*trans-form*
Mp 110-112°. $[\alpha]_D^{21.5}$ −51.8° (c, 3.517 in $CHCl_3$).

(1R,2S)-form [1655-00-1]
(+)-*cis-form*
Bp0.1 96.8°. $[\alpha]_D^{24}$ +33.0° (c, 1.12 in Et_2O).
Me ester: [13375-12-7]. Bp15 110-111°. $[\alpha]_D^{24}$ +31.7° (c, 4.242 in Et_2O).

(1S,2R)-form [1655-01-2]
(−)-*cis-form*
Bp0.01 98-100°. $[\alpha]_D^{25}$ −34.7° (c, 1.74 in Et_2O).
Me ester: [13375-11-6].
$C_8H_{14}O_3$ M 158.197
Bp25 119-120°. $[\alpha]_D^{23}$ −33.9° (c, 2.443 in Et_2O).

(1RS,2SR)-form [3749-17-5]
(±)-*cis-form*
Mp 82-83° (76-78°). Bp17 180°.
Ac:
$C_9H_{14}O_4$ M 186.207
Mp 221°.
Amide: [73045-98-4].
$C_7H_{13}NO_2$ M 143.185
Mp 113.7-114.7°.

(1RS,2RS)-form [17502-32-8]
(±)-*trans-form*
Needles (EtOAc). Sol. H_2O, spar. sol. C_6H_6. Mp 111°.
Ac: Mp 101-102°.
Me ester: [936-04-9]. Bp10 108-109°.
Hydrazide:
$C_7H_{14}N_2O_2$ M 158.200
Mp 208°.

Einhorn, A. *et al*, *Ber.*, 1894, **27**, 2466 (*synth*)
Marshall, E.R. *et al*, *J. Org. Chem.*, 1942, **7**, 444.
Mousseron, M. *et al*, *C. R. Hebd. Seances Acad. Sci.*, 1951, **232**, 637.
Torne, P.G., *CA*, 1967, **66**, 55082w (*synth, abs config*)
Chilina, K. *et al*, *Biochemistry*, 1969, **8**, 2846 (*abs config*)
Kay, J.B. *et al*, *J. Chem. Soc. (C)*, 1969, 248 (*resoln*)
Robinson, J.B., *J. Pharm. Pharmacol.*, 1970, **22**, 222 (*abs config*)
Schwartz, L.H. *et al*, *J. Am. Chem. Soc.*, 1972, **94**, 180 (*synth*)

2-Hydroxycyclohexanemethanol, 9CI   H-70120

2-Hydroxymethylcyclohexanol

[27583-43-3]

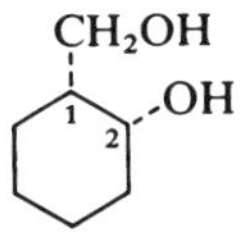

(1R,2R)-form

$C_7H_{14}O_2$ M 130.186

(1R,2R)-form [51606-50-9]

(−)-cis-*form*

$[\alpha]_D^{20}$ −36.4° (c, 0.24 in H_2O).

(1RS,2RS)-form [4187-60-4]

(±)-cis-*form*

Mp 49-50°. Bp_2 124°.

(1RS,2SR)-form [4187-59-1]

(⊥)-cis-*form*

$Bp_{0.9}$ 103-104°.

Bernáth, G. *et al, Acta Chim. Acad. Sci. Hung.,* 1970, **64**, 183.
Ferrand, G. *et al, Bull. Soc. Chim. Fr.,* Part 2, 1975, 356 (synth)
Trigalo, F. *et al, J. Chem. Soc., Perkin Trans.* 1, 1982, **8**, 1733 (synth)
Masamune, S. *et al, J. Org. Chem.,* 1983, **48**, 4441 (synth, nmr, ir)

2-Hydroxycyclooctanecarboxylic acid, 9CI   H-70121

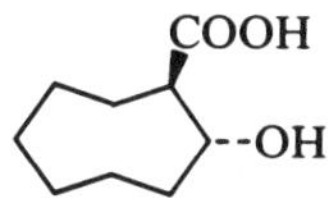

(1RS,2RS)-form

$C_9H_{16}O_3$ M 172.224

(1RS,2RS)-form [53242-29-8]

(±)-trans-*form*

Me ether, Me ester: [35193-37-4].

$C_{11}H_{20}O_3$ M 200.277

Bp_2 84°.

(1RS,2SR)-form [19297-14-4]

Mp 70°. $Bp_{0.8}$ 160-161°.

Et ester: [19297-13-3].

$C_{11}H_{20}O_3$ M 200.277

Bp_5 122-123°.

Saharia, G.S. *et al, Indian J. Chem.,* 1968, **6**, 69 (synth, deriv)
Bernáth, G. *et al, Acta Chim. Acad. Sci. Hung.,* 1972, **74**, 471; 1977, **92**, 175 (deriv, nmr, ir)

4'-Hydroxydehydrokawain   H-70122

Updated Entry replacing H-01512

6-[2-(4-Hydroxyphenyl)ethenyl]-4-methoxy-2H-pyran-2-one, 9CI. 6-(4-Hydroxystyryl)-4-methoxy-2-pyrone. Noryangonin

[7639-27-2]

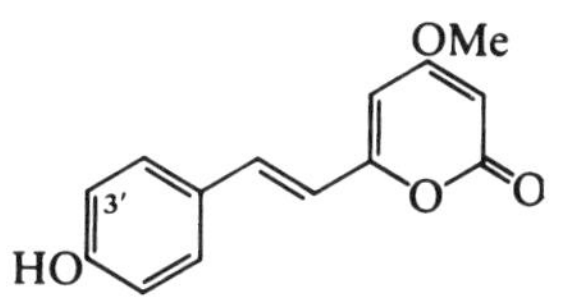

$C_{14}H_{12}O_4$ M 244.246

(E)-form [39986-86-2]

Constit. of *Anaphalis adnata.* Greenish-yellow needles (Me_2CO/C_6H_6). Mp 268° dec.

Ac: Mp 178°.

Me ether: **Yangonin.**

$C_{15}H_{14}O_4$ M 258.273

Constit. of Kawa root. Greenish-yellow cryst. with blue fluor. (MeOH or Me_2CO). Mp 153-154°.

3'-Methoxy: 6-[2-(*4-Hydroxy-3-methoxyphenyl)-ethenyl*]-4-methoxy-2H-*pyran-2-one, 9CI.* 6-(*4-Hydroxy-3-methoxystyryl*)-4-methoxy-2-*pyrone.* **11-Methoxynoryangonin.**

$C_{15}H_{14}O_5$ M 274.273

Isol. from a *Piper* sp. from New Guinea. Yellow cryst. (EtOH aq.). Mp 218-219°.

3'-Methoxy, Me ether: 6-[2-(*3,4-Dimethoxyphenyl)-ethenyl*]-4-methoxy-2H-*pyran-2-one.* 6-(*3,4-Dimethoxystyryl*)-4-methoxy-2-*pyrone.* **11-Methoxyyangonin.**

$C_{16}H_{16}O_5$ M 288.299

Isol. from *P. methysticum* and another *P.* spp. Yellow needles (EtOH). Mp 162-164°.

Chmielewska, I. *et al, Tetrahedron,* 1958, **4**, 36.
Hänsel, R. *et al, Arch. Pharm. (Weinheim, Ger.),* 1966, **299**, 503, 507 (11-*Methoxynoryangonin,* 11-*Methoxyyangonin*)
Harris, T.M. *et al, J. Org. Chem.,* 1968, **33**, 2399 (synth)
Engel, P. *et al, Z. Kristallogr. Kristallgeom., Kristallphys., Kristallchem.,* 1971, **134**, 180 (cryst struct)
Pradhan, D.K. *et al, Indian J. Chem., Sect. B,* 1976, **14**, 300 (isol)
Hipolito, E. *et al, Rev. Latinoam. Quim.,* 1977, **8**, 79; *CA,* **87**, 53025q (isol, synth)

2-Hydroxy-1,2-diphenyl-1-propanone, 9CI   H-70123

Updated Entry replacing H-01744

2-Hydroxy-2-phenylpropiophenone, 8CI. Benzoylmethylphenylmethanol. α-Methylbenzoin

[5623-26-7]

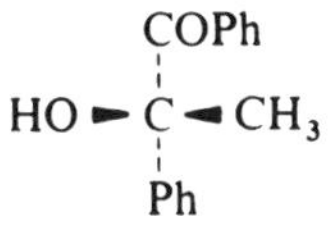

$C_{15}H_{14}O_2$ M 226.274

(S)-form

Oil. $[\alpha]_D^{24}$ −171° (EtOH).

(±)-form

Needles (pet. ether). Mp 65-66°.

Roger, R., *J. Chem. Soc.,* 1925, **127**, 518.
Stühmer, W. *et al, Arch. Pharm. (Weinheim, Ger.),* 1953, **286**, 26
Cram, D.J., *J. Am. Chem. Soc.,* 1959, **81**, 2752, 5754 (abs config)
Wu, T.C. *et al, J. Org. Chem.,* 1988, **53**, 2381 (synth, ir, pmr, cmr)

2-Hydroxy-1,3-diphenyl-1-propanone, 9CI   H-70124

Updated Entry replacing H-01745

2-Hydroxy-3-phenylpropiophenone, 8CI. Benzoylbenzylcarbinol. β-Hydroxy-β-benzylacetophenone

[5381-83-9]

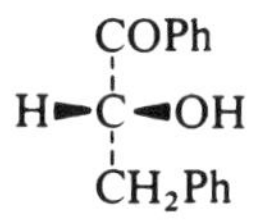

(R)-form

$C_{15}H_{14}O_2$ M 226.274

(R)-form [69897-44-5]

Prisms (pet. ether). Mp 75.5-76.5°. $[\alpha]_D^{17.5}$ +12.6° (Me_2CO).

(S)-form [69897-45-6]

$[\alpha]_D^{12.1}$ −19.3° (EtOH).

(±)-**form**

Needles (H₂O). Mp 65-66°.

Benzoyl:
   C₂₂H₁₈O₃   M 330.382
   Cryst. (EtOH). Mp 109-110°.

Oxime:
   C₁₅H₁₅NO₂   M 241.289
   Mp 117°.

McKenzie, A. *et al*, *J. Chem. Soc.*, 1914, **105**, 1583 (*synth, resoln*)

Tishchenko, I.G. *et al*, *CA*, 1973, **78**, 4165 (*synth*)

Marsman, B. *et al*, *J. Org. Chem.*, 1979, **44**, 2312 (*abs config*)

10-Hydroxy-7,11,13,16,19-docosapentaen-oic acid   H-70125

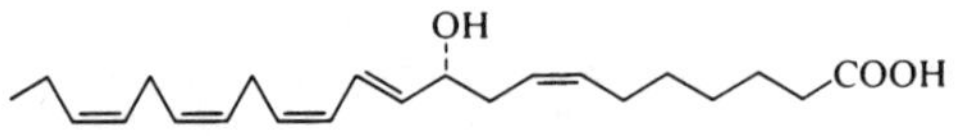

C₂₂H₃₄O₃   M 346.509

(7Z,10R,11E,13Z,16Z,19Z)-form [117332-99-7]
Leiopathic acid

Isol. from the antipatharian hexacoral, *Leiopathes* sp. [α]²⁰_D +3.1° (c, 1.85 in CHCl₃).

Et ester: [117333-00-3].
   C₂₄H₃₈O₃   M 374.562
   Isol. from *L*. sp. [α]²⁰_D +5.3° (c, 0.53 in CHCl₃).

Guerriero, A. *et al*, *Helv. Chim. Acta*, 1988, **71**, 1094 (*isol, struct, uv, pmr, cmr, ms, abs config*)

8-Hydroxy-5,9,11,14,17-eicosapentaenoic acid, 9CI   H-70126

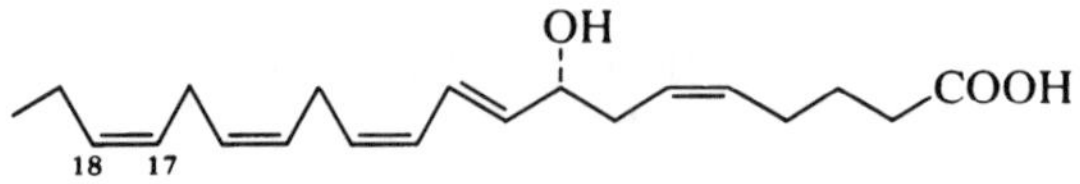

C₂₀H₃₀O₃   M 318.455

(5Z,8R,9E,11Z,14Z,17Z)-form [117407-06-4]

Constit. of the antipatharian hexacoral *Leiopathes* sp. [α]²⁰_D +3.0° (c, 0.43 in CHCl₃).

Et ester: [117333-01-4].
   C₂₂H₃₄O₃   M 346.509
   Constit. of *L*. sp. [α]²⁰_D +3.7° (c, 0.33 in CHCl₃).

17,18-Dihydro: [105500-09-2]. *8-Hydroxy-5,9,11,14-icosatetraenoic acid.*
   C₂₀H₃₂O₃   M 320.471
   Isol. from *L*. sp. [α]²⁰_D +4.0° (c, 0.48 in CHCl₃).

17,18-Dihydro, Et ester: [117333-02-5].
   C₂₂H₃₆O₃   M 348.525
   Isol. from *L*. sp. [α]²⁰_D +4.5° (c, 0.32 in CHCl₃).

Guerriero, A. *et al*, *Helv. Chim. Acta*, 1988, **71**, 1094 (*isol, struct, uv, pmr, cmr, abs config*)

11-Hydroxy-12,13-epoxy-9,15-octadeca-dienoic acid   H-70127

11-Hydroxy-1-[3-(2-pentenyl)oxiranyl]-9-undecenoic acid, 9CI

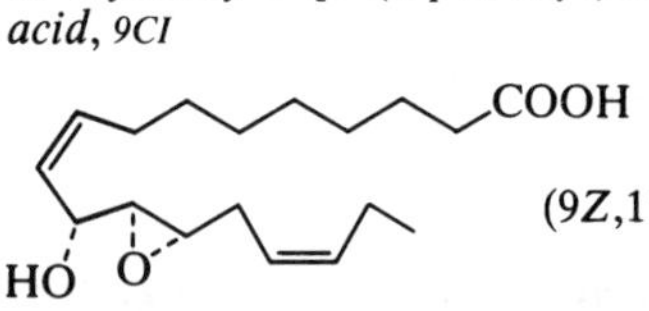

(9Z,11R,12S,13S,15Z)-form

C₁₈H₃₀O₄   M 310.433

(9Z,11R,12S,13S,15Z)-form [106034-49-5]

Isol. from rice plants with rice blast disease (*Pyricularia oryzae*).

O-(p-*Bromophenacyl*): [α]_D +26.0° (c, 0.493 in CHCl₃).

(9Z,11S,12S,13S,12Z)-form [105977-35-3]

Isol. from rice plants with rice blast disease (*P. oryzae*).

O-(p-*Bromophenacyl*): [α]_D −41.2° (c, 0.131 in CHCl₃).

Kato, T. *et al*, *J. Chem. Soc., Chem. Commun.*, 1986, 743 (*isol, ms, pmr, abs config*)

7-Hydroxy-4,11(13)-eudesmadien-12,6-olide   H-70128

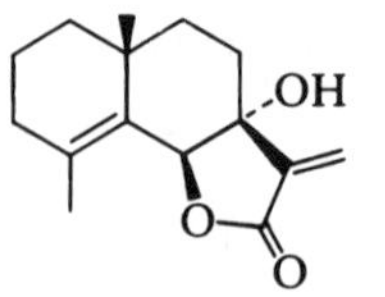

C₁₅H₂₀O₃   M 248.321

(6β,7α)-**form**

Constit. of *Sphaeranthus indicus*. Oil. [α]_D −56.8° (c, 0.43 in CHCl₃).

Sohoni, J.S. *et al*, *J. Chem. Soc., Perkin Trans. 1*, 1988, 157.

1-Hydroxy-2,4(15),11(13)-eudesmatrien-12,8-olide   H-70129

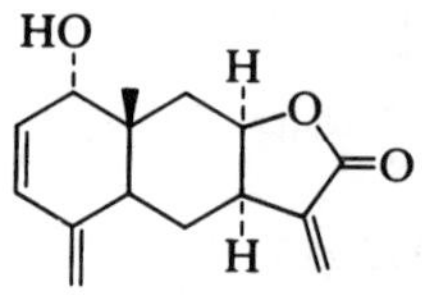

C₁₅H₁₈O₃   M 246.305

(1α,8β)-**form**

Constit. of *Ferreyranthus fruticosus*. Oil.

Jakupovic, J. *et al*, *Phytochemistry*, 1988, **27**, 1113.

6-Hydroxyflavone, 8CI   H-70130

Updated Entry replacing H-60143
6-Hydroxy-2-phenyl-4H-1-benzopyran-4-one, 9CI. 6-Hydroxy-2-phenylchromone

[6665-83-4]

C₁₅H₁₀O₃   M 238.242

Yellow needles (EtOH aq.). Mp 235.5-236.5° (231-232°).

Me ether: [26964-24-9]. *6-Methoxyflavone.*
   C₁₆H₁₂O₃   M 252.269
   Mp 163-164°.

Kostanecki, S. *et al*, *Ber.*, 1899, **32**, 326 (*synth*)

Simonis, H. *et al*, *Ber.*, 1926, **59**, 2914 (*deriv*)

Looker, J.H. *et al*, *J. Org. Chem.*, 1962, **27**, 381 (*synth, ir*)

Fozdar, B.I. *et al*, *Chem. Ind. (London)*, 1986, 586 (*synth*)

7-Hydroxyflavone   H-70131

Updated Entry replacing H-60144
7-Hydroxy-2-phenyl-4H-1-benzopyran-4-one, 9CI. 7-Hydroxy-2-phenylchromone

[6665-86-7]

C₁₅H₁₀O₃   M 238.242

Needles (EtOH aq.). Mp 244° (240°).

Me ether: [22395-22-8]. *7-Methoxyflavone.*
   C₁₆H₁₂O₃   M 252.269

Mp 110-111°.

Emilewicz, T. et al, Ber., 1899, **32**, 309 (deriv)
Looker, J.H. et al, J. Org. Chem., 1962, **27**, 381 (synth, ir)
Audier, H., Bull. Soc. Chim. Fr., 1966, 2892 (ms)
Naik, G.N. et al, Indian J. Chem., 1966, **4**, 273 (pmr)
Fozdar, B.I. et al, Chem. Ind. (London), 1986, 586 (synth)

8-Hydroxyflavone H-70132

Updated Entry replacing H-60145
[77298-64-7]
$C_{15}H_{10}O_3$ M 238.242
Mp 250-252° (249-250°).
▷LK8650100.

Me ether: [26964-26-1]. *8-Methoxyflavone.*
$C_{16}H_{12}O_3$ M 252.269
Mp 200-201°.

Gupta, D.S. et al, CA, 1955, **49**, 1713 (synth)
Looker, J.H. et al, J. Org. Chem., 1962, **27**, 381 (synth, ir)
Fozdar, B.I. et al, Chem. Ind. (London), 1986, 586 (synth)

10-Hydroxy-2H,13H-furo[3,2-c:5,4-h']-bis[1]benzopyran-2,13-dione, 9CI H-70133

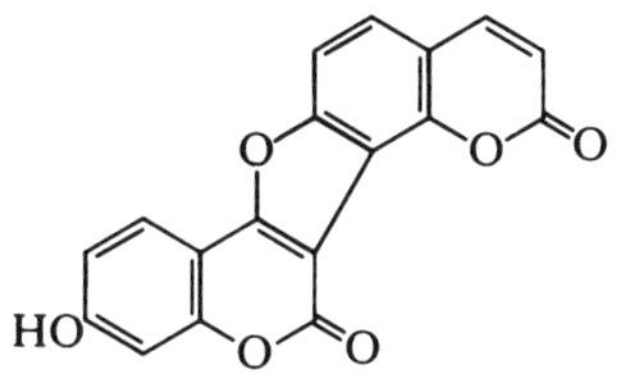

$C_{18}H_8O_6$ M 320.258
O-(α-L-Rhamnoside): [80680-23-5]. **Eriocephaloside.**
$C_{24}H_{18}O_{10}$ M 466.400
Constit. of *Lasiosiphon eriocephalus*. Cryst. (MeOH).
Mp 350° dec.

Bhandari, P. et al, Phytochemistry, 1981, **20**, 2044.

6-Hydroxy-1(10),4,11(13)-germacratrien-12,8-olide H-70134

Updated Entry replacing L-20028
[35001-25-3]

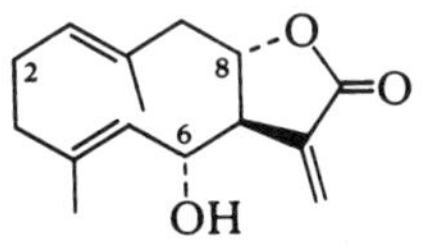

(6α,8α)-form

$C_{15}H_{20}O_3$ M 248.321
(6α,8α)-form [35001-24-2]
Desacetyllaurenobiolide
$C_{15}H_{20}O_3$ M 248.321
Constit. of *Artemisia* spp. Gum. $[\alpha]_D$ +34.5° (c, 1.6 in CHCl₃).
Ac: [35001-25-3]. **Laurenobiolide.**
$C_{17}H_{22}O_4$ M 290.358
Constit. of *Laurus nobilis*. Cryst. (Et₂O/hexane). Mp 101-103°. $[\alpha]_D$ +17.1° (EtOH).
Ac,2α-Hydroxy: 6α-Acetoxy-2α-hydroxy-1(10),4,11(13)-germacratrien-12,8-olide. 2α-Hydroxylaurenobiolide.
$C_{15}H_{18}O_3$ M 246.305
Constit. of *Gnephosis arachnoidea*. Gum. $[\alpha]_D^{24}$ +49° (c, 0.21 in CHCl₃).
Ac,2α-Acetoxy: 2α-Acetoxylaurenobiolide.
$C_{19}H_{24}O_6$ M 348.395

Constit. of *Mikania grazielae*. Gum. $[\alpha]_D^{24}$ +56° (c, 0.37 in CHCl₃).
(6β,8α)-form
6-Epidesacetyllaurenobiolide
Constit. of *Montanoa grandiflora*. Cryst. (CHCl₃/pet. ether). Mp 117-118°.

Tada, H. et al, J. Chem. Soc., Chem. Commun., 1971, 1391 (struct)
Shafizadeh, F. et al, Phytochemistry, 1973, **12**, 857 (isol)
Tada, H. et al, Chem. Pharm. Bull., 1976, **24**, 667 (isol)
Tori, K. et al, Tetrahedron Lett., 1976, 387 (cmr)
Bohlmann, F. et al, Phytochemistry, 1982, **21**, 1169 (isol)
Quijano, L. et al, Phytochemistry, 1984, **23**, 1971 (isol)
Jakupovic, J. et al, Phytochemistry, 1988, **27**, 3181 (2α-Hydroxylaurenobiolide)

3-Hydroxy-4,11(13)-guaiadien-12,8-olide H-70135

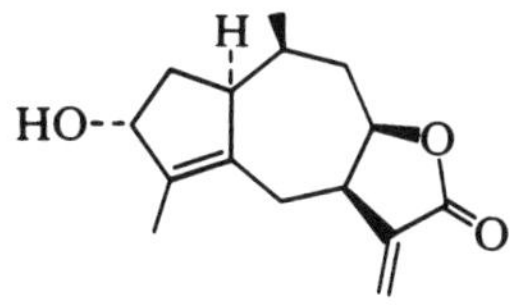

$C_{15}H_{20}O_3$ M 248.321
(3α,8β)-form
3α-Hydroxydesoxoachalensolide
Constit. of *Stevia polyphylla*. Oil.
3-Me ether:
$C_{16}H_{22}O_3$ M 262.348
Constit. of *S. polyphylla*. Oil.
(3β,8β)-form
3β-Hydroxydesoxoachalensolide
Constit. of *S. polyphylla*. Cryst. (Et₂O). Mp 128-129°. $[\alpha]_D^{24}$ +116° (c, 0.5 in CHCl₃).
Ac:
$C_{17}H_{22}O_4$ M 290.358
Constit. of *S. polyphylla*. Oil. $[\alpha]_D^{24}$ +139° (c, 2.2 in CHCl₃).
Me ether:
$C_{16}H_{22}O_3$ M 262.348
Constit. of *S. polyphylla*. Oil.
11β,13-Dihydro: 3β-Hydroxy-11β,13-dihydrodesoxoachalensolide. 3β-Hydroxy-4-guaien-12,8β-olide.
$C_{15}H_{22}O_3$ M 250.337
Constit. of *S. polyphylla*. Cryst. (Et₂O). Mp 116°.

Zdero, C. et al, Phytochemistry, 1988, **27**, 2835.

15-Hydroxy-3,10(14)-guaiadien-12,8-olide H-70136

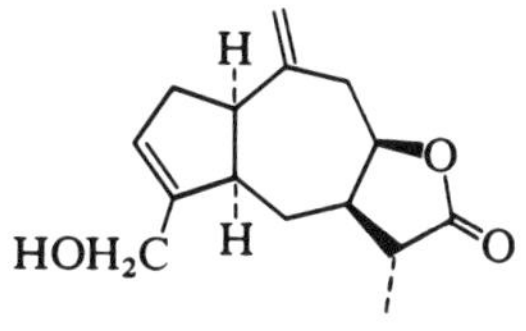

$C_{15}H_{20}O_3$ M 248.321
8β-form
Constit. of *Stevia samaipatensis*. Oil.
Ac:
$C_{17}H_{22}O_4$ M 290.358
Constit. of *S. samaipatensis*. Oil. $[\alpha]_D$ +125° (c, 3.6 in CHCl₃).

Zdero, C. et al, Phytochemistry, 1988, **27**, 2835.

5-Hydroxy-3,11(13)-guaiatrien-12,8-olide H-70137

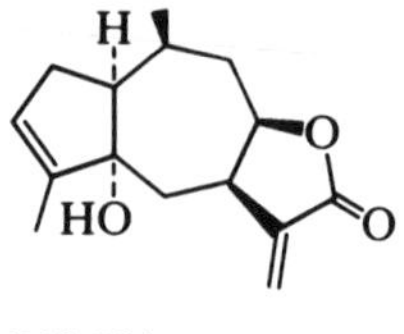

$C_{15}H_{20}O_3$ M 248.321

(5α,8β)-form

Constit. of *Stevia polyphylla*. Oil.

Zdero, C. *et al*, *Phytochemistry*, 1988, **27**, 2839.

8-Hydroxy-4(15),11(13),10(14)-guaiatrien-12,6-olide H-70138

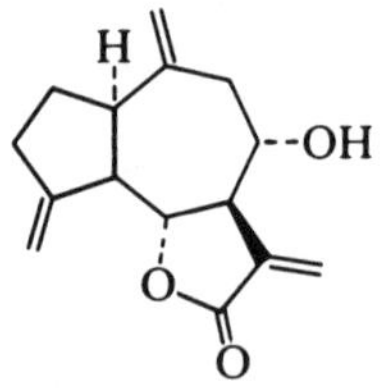

$C_{15}H_{18}O_3$ M 246.305

Cryst. (EtOAc/pet. ether). Mp 106-108°. [α]$_D$ +61° (CHCl$_3$).

8-(2-Hydroxymethylpropanol): [81421-79-6]. **Subexpinnatin**.

$C_{19}H_{22}O_5$ M 330.380

Constit. of *Centaurea canariensis*. Oil. [α]$_D$ +62° (c, 3 in CHCl$_3$).

Gonzalez, A.G. *et al*, *Tetrahedron Lett.*, 1982, **23**, 895.

7-Hydroxy-6-(hydroxymethyl)-2H-1-benzopyran-2-one H-70139

7-Hydroxy-6-hydroxymethylcoumarin

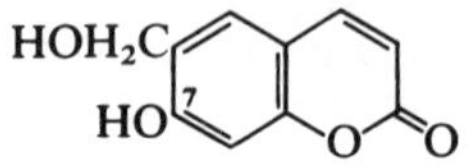

$C_{10}H_8O_4$ M 192.171

O^7-*Me:* [117597-79-2]. *6-(Hydroxymethyl)-7-methoxy-2H-1-benzopyran-2-one, 9CI*. **6-Hydroxymethylherniarin**.

$C_{11}H_{10}O_4$ M 206.198

Constit. of *Citrus funadoko*. Oil.

Ju-ichi, M. *et al*, *Heterocycles*, 1988, **27**, 1451.

1-[3-Hydroxy-4-(hydroxymethyl)-cyclopentyl]-2,4(1H,3H)-pyrimidine-dione, 9CI H-70140

Carbocyclic 2'-deoxyuridine

[62102-28-7]

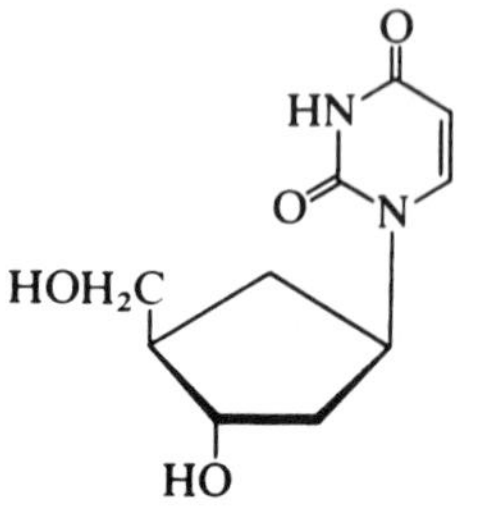

$C_{10}H_{14}N_2O_4$ M 226.232

Cryst. (EtOH). Mp 160-163° dec. Stereoisomers also prepd.

Shealy, Y.F. *et al*, *J. Heterocycl. Chem.*, 1976, **13**, 1015 (*synth, uv, ms*)
Cookson, R.C. *et al*, *J. Chem. Soc., Perkin Trans. 1*, 1986, 399, 405 (*synth, conformn, uv, ir, pmr, ms*)
Ravenscroft, P. *et al*, *Tetrahedron Lett.*, 1986, **27**, 747 (*synth, uv*)

2-Hydroxy-2-(hydroxymethyl)-2H-pyran-3(6H)-one H-70141

Microthecin

[73033-01-9]

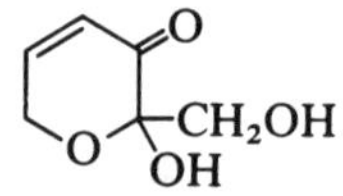

$C_6H_8O_4$ M 144.127

Prod. by *Melanospora ornata*. Cryst. (EtOAc).

Japan. Pat., 79 122 796, (*1979*); *CA*, **92**, 126906 (*isol, struct*)

3-Hydroxy-5-(4-hydroxyphenyl)pentanoic acid H-70142

β,4-Dihydroxybenzenepentanoic acid. **Niduloic acid**

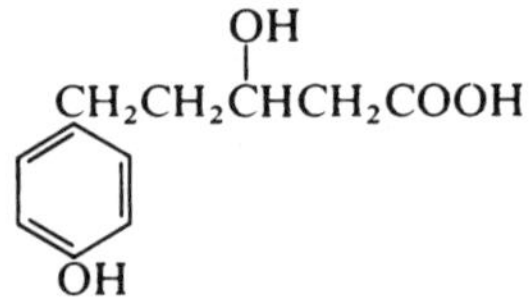

$C_{11}H_{14}O_4$ M 210.229

(+)-*form* [78472-11-4]

Prod. by *Nidula niveo-tomentosa*. Cryst. + ½H$_2$O. Mp 151-152°. [α]$_D^{25}$ +10.8° (MeOH).

Ayer, W.A. *et al*, *Phytochemistry*, 1980, **19**, 2717 (*isol, struct, nmr*)

2-Hydroxy-5-iodobenzaldehyde, 9CI H-70143

Updated Entry replacing H-01990

5-Iodosalicylaldehyde, 8CI

[1761-62-2]

$C_7H_5IO_2$ M 248.020

Pale-yellow needles (EtOH). Mp 102°. pK_a 8.98 (30°, 75% v/v dioxan aq.).

Oxime:

$C_7H_6INO_2$ M 263.034

Light-yellow needles (EtOH). Mp 135°.

Me ether: [42298-41-9]. *5-Iodo-2-methoxybenzaldehyde.*

$C_8H_7IO_2$ M 262.047

Needles (EtOH). Mp 144.5-145.5°.

Mathai, K.P. *et al*, *J. Indian Chem. Soc.*, 1964, **41**, 347 (*synth*)
Tozuka, Z. *et al*, *CA*, 1973, **79**, 66092.

7-Hydroxy-4-isopropyl-3-methoxy-6-methylcoumarin H-70144

7-Hydroxy-3-methoxy-6-methyl-4-(1-methylethyl)-2H-1-benzopyran-2-one, 9CI

[111394-32-2]

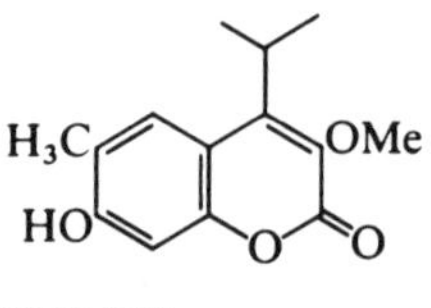

$C_{14}H_{16}O_4$ M 248.278

Constit. of *Macrothelypteris torresiana*. Cryst. Mp 197-199°.

Hori, K. *et al*, *Yakugaka Zasshi*, 1987, **107**, 491.

15-Hydroxy-8(17),12-labdadien-16-al   H-70145

$C_{20}H_{32}O_2$ M 304.472
Constit. of rhizomes of *Alpinia formosana*. Oil. $[\alpha]_D^{20}$ +20° (c, 0.1 in CHCl₃).

Itokawa, H. *et al*, *Phytochemistry*, 1988, **27**, 435.

19-Hydroxy-8(17),13-labdadien-15-al   H-70146

$C_{20}H_{32}O_2$ M 304.472
(*E*)-*form* [10266-89-4]
Constit. of *Juniperus thurifera*. Cryst. Mp 112°. $[\alpha]_D^{23}$ +26.8° (CHCl₃).
(*Z*)-*form* [117404-55-4]
Constit. of *J. thurifera*. Cryst. Mp 115°. $[\alpha]_D^{23}$ +14.1° (CHCl₃).

San Feliaciano, A. *et al*, *Phytochemistry*, 1988, **27**, 2241.

2-Hydroxy-8(17),13-labdadien-15-oic acid   H-70147
$C_{20}H_{32}O_3$ M 320.471
(*2β,13E*)-*form*
2-Ac: 2β-*Acetoxy-8(17),13E-labdadien-15-oic acid.*
$C_{22}H_{34}O_4$ M 362.508
Constit. of *Nolana filifolia*.
2-Ac, Me ester: Amorph. powder (EtOAc/pet. ether). $[\alpha]_D^{25}$ −20.1° (c, 1.5 in CHCl₃).
(*2β,13Z*)-*form*
2-Ac: 2β-*Acetoxy-8(17),13Z-labdadien-15-oic acid.*
$C_{22}H_{34}O_4$ M 362.508
Constit. of *N. filifolia*.
2-Ac, Me ester: Cryst. (EtOAc/pet. ether). Mp 101-103°. $[\alpha]_D^{25}$ −13.5° (c, 1.11 in CHCl₃).

Garbarino, J.A. *et al*, *Phytochemistry*, 1988, **27**, 1795.

ent-6β-Hydroxy-8(17),13-labdadien-15-oic   H-70148
acid

$C_{20}H_{32}O_3$ M 320.471
Constit. of *Aristolochia cymbifera*.

Me ester: Oil. $[\alpha]_D^{25}$ −11° (c, 0.35 in CHCl₃).

Lopes, L.M.X. *et al*, *Phytochemistry*, 1986, **27**, 2265.

8-Hydroxy-13-labden-15-al   H-70149

$C_{20}H_{34}O_2$ M 306.487
(*8α,E*)-*form* [89576-50-1]
Constit. of *Juniperus thurifera*. Cryst. Mp 105°. $[\alpha]_D^{23}$ +12.1° (CHCl₃).
(*8α,Z*)-*form* [89576-51-2]
Constit. of *J. thurifera*. Cryst. (CH₂Cl₂/hexane). Mp 109°. $[\alpha]_D^{23}$ +5.2° (CHCl₃).

San Feliciano, A. *et al*, *Phytochemistry*, 1988, **27**, 2241.

15-Hydroxy-8(17)-labden-19-al   H-70150
Updated Entry replacing H-02129
$C_{20}H_{34}O_3$ M 322.487
(*13R*)-*form*
Imbricatolal. *15-Hydroxyimbricatolal (incorr.)*
Constit. of *Araucaria imbricata*.
Ac: **Acetylimbricatolol.**
$C_{22}H_{26}O_3$ M 338.446
Constit. of *A. imbricata*. $[\alpha]_D^{20}$ +20°.
O-*Formyl:* **15-Formylimbricatolal.** *15-Formyloxyimbricatolal (incorr.).*
$C_{21}H_{34}O_3$ M 334.498
Constit. of *A. araucana*. Oil. $[\alpha]_D^{25}$ +17.6° (c, 1.2 in CHCl₃).

Bruns, K., *Tetrahedron*, 1968, **24**, 3417 (*isol, struct*)
Garbarino, J.A. *et al*, *J. Nat. Prod.*, 1987, **50**, 935 (*isol*)

15-Hydroxy-7-labden-3-one   H-70151

$C_{20}H_{34}O_2$ M 306.487
Constit. of *Stevia samaipatensis*. Oil.

Zdero, C. *et al*, *Phytochemistry*, 1988, **27**, 2835.

3-Hydroxy-8,24-lanostadien-21-oic acid   H-70152
Updated Entry replacing H-02139

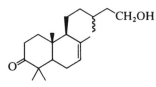

$C_{30}H_{48}O_3$ M 456.707
3β-form [24160-36-9]
Trametenolic acid
Metab. of *Trameles odorata*. Cryst. Mp 258-261°. $[\alpha]_D$ +40.2°.

Halsall, T.G. *et al*, *J. Chem. Soc.*, 1959, 2036 (*isol*)
Batta, A.K. *et al*, *J. Chem. Soc., Perkin Trans. 1*, 1975, 451 (*isol*)

7-Hydroxy-6-methoxy-2*H*-1-benzopyran-2-one, 9CI   H-70153

Updated Entry replacing H-02170
7-Hydroxy-6-methoxycoumarin. **Scopoletin**. *Aesculetin 6-methyl ether. Chrysatropic acid. Gelseminic acid. β-Methylaesculetin*
[92-61-5]

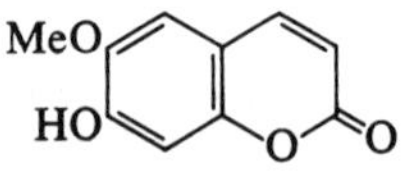

$C_{10}H_8O_4$ M 192.171
Occurs widely in the plant world, for example, the root of *Gelsemium sempervirens*, *Atropa belladonna*, *Convolvulus scammonia*, *Ipomaea orizabensis*, *Prunus serotina*, *Fabiana imbricata* and also *Diospyros* spp., *Peucedanum* spp., *Heracleum* spp., *Skimmea* spp. Also occurs in the Chinese crude drug Toki (from *Angelica acutiloba*). Needles or prisms (EtOH). Mp 204°. Reduces Fehling's and Tollen's reagents.
▷GN6930000.
Ac: Mp 177°.
O-*β-D-Glucoside:* **Fabiatrin**.
  $C_{16}H_{18}O_9$ M 354.313
  Isol. from leaves of *Fabiana imbricata*. Needles + 2H_2O (H_2O). Mp 226-228°.
O-*(2,3-Dihydroxy-3-methylbutenoyl):* **Obtsusinin**.
  $C_{15}H_{18}O_6$ M 294.304
  Isol. from *Haplophyllum obtusifolium*. Mp 136-138°.

Head, F.G.H. *et al*, *J. Chem. Soc.*, 1931, 1241 (*synth*)
Seka, R. *et al*, *Ber.*, 1931, **64**, 909.
Gonzalez, A.G. *et al*, *An. Quim.*, 1973, **69**, 1013 (*pmr*)
Tanaka, S. *et al*, *Arzneim.-Forsch.*, 1977, **27**, 2039 (*isol*)
Herath, W.H.M. *et al*, *Phytochemistry*, 1978, **17**, 1007 (*isol*)
Ishibura, N. *et al*, *Z. Naturforsch., C*, 1979, **34**, 628.
Ahluwalia, V.K. *et al*, *Monatsh. Chem.*, 1982, **113**, 197 (*Obtusinin*)
Abu-Eittah, R.H. *et al*, *Can. J. Chem.*, 1985, **63**, 1173 (*uv*)
Koul, S.K. *et al*, *Indian J. Chem., Sect. B*, 1987, **26**, 574 (*synth*)

4-Hydroxy-3-methoxy-2-methyl-9*H*-carbazole-1-carboxaldehyde, 9CI   H-70154

Carbazomycin E. *Carbazomycinal*
[103744-20-3]

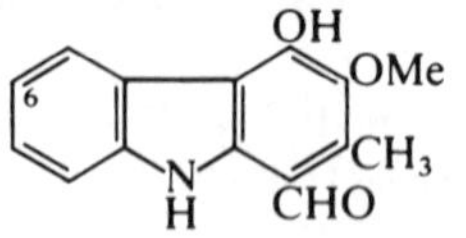

$C_{15}H_{13}NO_3$ M 255.273
Isol. from *Streptoverticillium ehimense*. Pale-yellow needles (EtOAc/hexane).
6-Methoxy: [103744-21-4]. *4-Hydroxy-3,6-dimethoxy-2-methyl-9H-carbazole-1-carboxaldehyde, 9CI.* **Carbazomycin** F. *6-Methoxycarbazomycinal.*
  $C_{16}H_{15}NO_4$ M 285.299
  From *S. ehimense*. Pale-yellow needles (EtOAc/hexane).

Kondo, S. *et al*, *J. Antibiot.*, 1986, **39**, 727 (*isol, struct*)
Naid, T. *et al*, *J. Antibiot.*, 1987, **40**, 157 (*isol*)

5-Hydroxy-3-methoxy-7-oxabicyclo[4.1.0]-hept-3-en-2-one, 9CI   H-70155

Updated Entry replacing E-10046
5,6-Epoxy-4-hydroxy-2-methoxy-2-cyclohexen-1-one. **Chaloxone**
[78472-10-3]

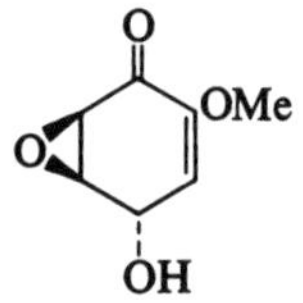

$C_7H_8O_4$ M 156.138
Metab. of the fungus *Chalara microspora*. Mp 142-143°. $[\alpha]_D^{20}$ +271° (c, 0.56 in EtOH).

Fex, T. *et al*, *Acta Chem. Scand., Ser. B*, 1981, **35**, 97 (*isol, struct, spectra*)
Fex, T., *Tetrahedron Lett.*, 1981, **22**, 2707 (*synth*)

2-(2-Hydroxy-4-methoxyphenyl)-6-methoxy-3-methylbenzofuran   H-70156

5-Methoxy-2-(6-methoxy-3-methyl-2-benzofuranyl)-phenol, 9CI
[3207-47-4]

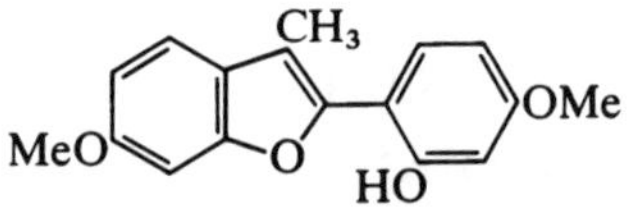

$C_{17}H_{16}O_4$ M 284.311
Constit. of *Indigofera microcarpa*. Prisms (MeOH). Mp 111° (109°). Wrongly named in one ref. as the 5-methoxy isomer.

Bevan, C.W.L. *et al*, *J. Chem. Soc.*, 1964, 5991 (*synth*)
De Moraes E Souza, M.A. *et al*, *Phytochemistry*, 1988, **27**, 1817 (*isol*)

2-(2-Hydroxy-4-methoxyphenyl)-3-methyl-5,6-methylenedioxybenzofuran   H-70157

5-Methoxy-2-(7-methylfuro[2,3-f]-1,3-benzodioxol-6-yl)phenol, 9CI
[3207-48-5]

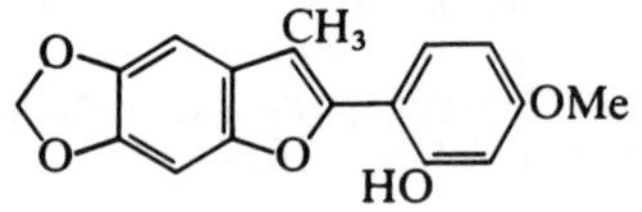

$C_{17}H_{14}O_5$ M 298.295
Constit. of *Indigofera microcarpa*. Needles (MeOH). Mp 123°.

De Moraes E Souza, M.A. *et al*, *Phytochemistry*, 1988, **27**, 1817.

2′-(Hydroxymethyl)acetophenone, 8CI   H-70158

1-[2-(Hydroxymethyl)phenyl]ethanone, 9CI
[32521-21-4]

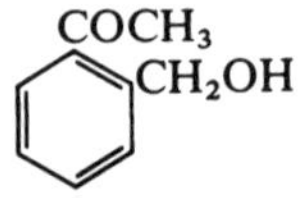

$C_9H_{10}O_2$ M 150.177
Oil with strong aromatic odour. Bp_{0.05} 48°.

Wulff, G. *et al*, *Chem. Ber.*, 1986, **119**, 1876 (*synth, pmr, cmr*)

4′-(Hydroxymethyl)acetophenone   H-70159

1-[4-(Hydroxymethyl)phenyl]ethanone

$C_9H_{10}O_2$ M 150.177
Fine needles (pet. ether). Mp 54°.

Schmid, L. *et al, Monatsh. Chem.*, 1952, **83**, 185 (*synth*)

2-Hydroxy-4-methylbenzaldehyde   H-70160

Updated Entry replacing H-02272
2-Hydroxy-p-tolualdehyde. m-Cresol-6-aldehyde. 4-
Methylsalicylaldehyde. 2,4-Cresotaldehyde. 2-Formyl-
5-methylphenol
[698-27-1]
$C_8H_8O_2$ M 136.150
Needles (H_2O or EtOH). Mp 60-61°. Bp_{726} 219-221°.
Steam-volatile.

Oxime:
$C_8H_9NO_2$ M 151.165
Plates (EtOH aq.). Mp 108.5-109°.
Semicarbazone: Mp 268°.
2,4-Dinitrophenylhydrazone: [973-82-0]. Mp 266-267°.
Me ether: [57415-35-7]. *2-Methoxy-4-*
methylbenzaldehyde.
$C_9H_{10}O_2$ M 150.177
Needles. Mp 42-43°. Bp_{720} 263-264°. Steam-volatile.

Anselmino, O., *Ber.*, 1917, **50**, 395.
Duff, J.C., *J. Chem. Soc.*, 1941, 547.
Gross, H. *et al, Chem. Ber.*, 1963, **96**, 308.
Casnati, G. *et al, Tetrahedron Lett.*, 1965, 243 (*synth*)
Bruce, J.M. *et al, J. Chem. Soc., Perkin Trans. 1*, 1974, **2**, 288 (*synth*)
Hauser, F.M. *et al, Synthesis*, 1987, 723 (*deriv, synth, ir, pmr, ms*)

4-Hydroxy-2-methylbenzaldehyde   H-70161

Updated Entry replacing H-02278
4-Hydroxy-o-tolualdehyde. m-Cresol-4-aldehyde. 4,2-
Cresotaldehyde. 4-Formyl-3-methylphenol
[41438-18-0]
$C_8H_8O_2$ M 136.150
Plates (H_2O). Mp 110°.

Oxime:
$C_8H_9NO_2$ M 151.165
Mp 151-152°.
Me ether: [52289-54-0]. *4-Methoxy-2-methylbenzalde-*
hyde. 2-Methyl-p-anisaldehyde.
$C_9H_{10}O_2$ M 150.177
Cryst. (MeOH). Bp 257°, Bp_{24} 150°.

Gatterman, L., *Justus Liebigs Ann. Chem.*, 1907, **357**, 313.
Gross, H. *et al, Chem. Ber.*, 1963, **96**, 308.
Hauser, F.M. *et al, Synthesis*, 1987, 723 (*deriv, synth, ir, pmr, ms*)

5-Hydroxy-3-methyl-1,2-benzenedicarbox-   H-70162
ylic acid, 9CI

Updated Entry replacing H-02291
5-Hydroxy-3-methylphthalic acid, 8CI. β-Coccinic acid
$C_9H_8O_5$ M 196.159
Mp 155-157°.

Anhydride:
$C_9H_6O_4$ M 178.144
Mp 166-168°.
Me ether:
$C_{10}H_{10}O_5$ M 210.186
Mp 184° dec.

Me ether, 2-Me ester: [116913-58-7].
$C_{11}H_{12}O_5$ M 224.213
Cryst. (EtOAc/pet. ether). Mp 146-147°.
Me ether, di-Me ester: [108298-35-7].
$C_{12}H_{14}O_5$ M 238.240
Oil.

Meldrum, A.N., *J. Chem. Soc.*, 1911, **99**, 1712.
Meldrum, A.N. *et al, CA*, 1935, **29**, 5430.
Tam, T.F. *et al, Synthesis*, 1988, 383 (*synth, deriv, ir, pmr*)

2-Hydroxy-4-methylbenzoic acid   H-70163

Updated Entry replacing H-02317
2-Hydroxy-p-toluic acid. 2,4-Cresotic acid. 4-Methylsa-
licylic acid. m-Cresol-6-carboxylic acid
[50-85-1]
$C_8H_8O_3$ M 152.149
Needles (H_2O), plates ($CHCl_3$). Mod. sol. H_2O. Mp 172-
174°, 177° subl. pK_a 3.16 (25°).

Ac: [14504-07-5].
$C_{10}H_{10}O_4$ M 194.187
Needles (H_2O or C_6H_6). Mp 125-126°, 139°.
Me ester: [4670-56-8].
$C_9H_{10}O_3$ M 166.176
Cryst. Mp 27-28°. Bp 242-244°.
Et ester: [60770-00-5].
$C_{10}H_{12}O_3$ M 180.203
d_4^{23} 1.095. Bp 254°.
Me ether: [704-45-0]. *2-Methoxy-4-methylbenzoic acid.*
$C_9H_{10}O_3$ M 166.176
Plates (H_2O). Mp 74°.

King, C. *et al, J. Am. Chem. Soc.*, 1945, **67**, 2080.
Prelog, V. *et al, Helv. Chim. Acta*, 1947, **30**, 675.
Yamazaki, M. *et al, CA*, 1973, **80**, 47030b (*cmr*)
Hauser, F.M. *et al, Synthesis*, 1987, 723 (*deriv, synth, ir, pmr, ms*)

4-Hydroxy-2-methylbenzoic acid   H-70164

Updated Entry replacing H-02323
4-Hydroxy-o-toluic acid. m-Cresol-4-carboxylic acid.
4,2-Cresotic acid
[578-39-2]
$C_8H_8O_3$ M 152.149
Needles + $^{1}/_2H_2O$ (H_2O). Mp 177-178° (anhyd.).

Et ester: [57081-00-2].
$C_{10}H_{12}O_3$ M 180.203
Needles (ligroin). Mp 98°. Bp 300°.
Me ether: [6245-57-4]. *4-Methoxy-2-methylbenzoic*
acid.
$C_9H_{10}O_3$ M 166.176
Needles (H_2O). Mp 176°. Bp_{16} 143°.

Gomberg, M. *et al, J. Am. Chem. Soc.*, 1917, **39**, 1679.
Rosenmund, K.W., *Justus Liebigs Ann. Chem.*, 1928, **460**, 56.
King, C. *et al, J. Am. Chem. Soc.*, 1945, **67**, 2089.
Fujio, M. *et al, Bull. Chem. Soc. Jpn.*, 1975, **48**, 2127 (*pmr, ester*)
Comins, D.L. *et al, J. Org. Chem.*, 1986, **51**, 3566 (*synth, deriv*)
Hauser, F.M. *et al, Synthesis*, 1987, 723 (*deriv, synth, ir, pmr, ms*)

8-Hydroxy-4-methyl-2*H*-1-benzopyran-2-   H-70165
one, 9CI

Updated Entry replacing H-02350
8-Hydroxy-4-methylcoumarin
[53081-42-8]

$C_{10}H_8O_3$ M 176.171
Cryst. (MeOH). Mp 188-190°.

Me ether: 8-Methoxy-4-methyl-2H-1-benzopyran-2-one. 8-Methoxy-4-methylcoumarin.
$C_{11}H_{10}O_3$ M 190.198
Isol. from wood of *Ekebergia senegalensis*. Needles (MeOH). Mp 137-138°, Mp 165°.

Bevan, C.W.L. *et al*, *Chem. Ind.* (*London*), 1965, 383 (*isol, deriv*)
Venturella, P. *et al*, *Heterocycles*, 1974, **2**, 345 (*synth, pmr*)

Hydroxymethylbilane   H-70166

Updated Entry replacing H-02361
Preuroporphyrinogen
[73023-76-4]

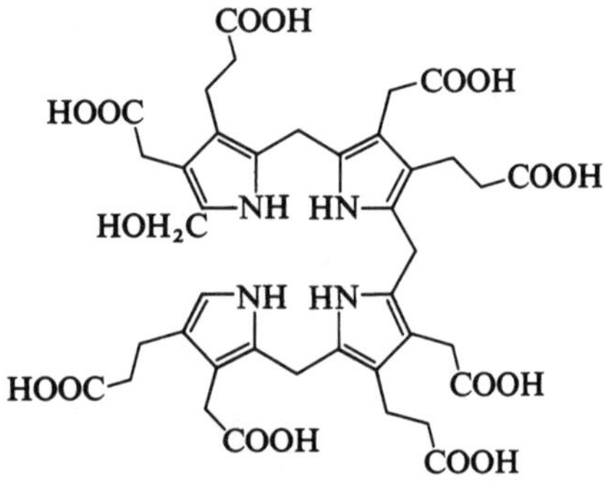

$C_{40}H_{42}N_4O_{17}$ M 850.788
Isol. by incubating Porphobilinogen with porphobilinogen deaminase enzyme system. Intermed. in the biosynthetic pathway to natural Uroporphyrinogens. Unstable.

Battersby, A.R. *et al*, *J. Chem. Soc., Chem. Commun.*, 1979, 1155 (*synth*)
Scott, A.I. *et al*, *J. Am. Chem. Soc.*, 1979, **101**, 3114.
Battersby, A.R. *et al*, *Nature* (*London*), 1980, **285**, 17
Battersby, A.R. *et al*, *Angew. Chem., Int. Ed. Engl.*, 1981, **20**, 293.
Schauder, J.-R. *et al*, *J. Chem. Soc., Chem. Commun.*, 1987, 436 (*synth, biosynth*)

1-[4-Hydroxy-3-(3-methyl-1,3-butadienyl)-   H-70167
phenyl]-2-(3,5-dihydroxyphenyl)ethylene

5-[2-[4-Hydroxy-3-(3-methyl-1,3-butadienyl)phenyl]-ethenyl]-1,3-benzenediol. 3',4',5'-Trihydroxy-3-isopentadienylstilbene

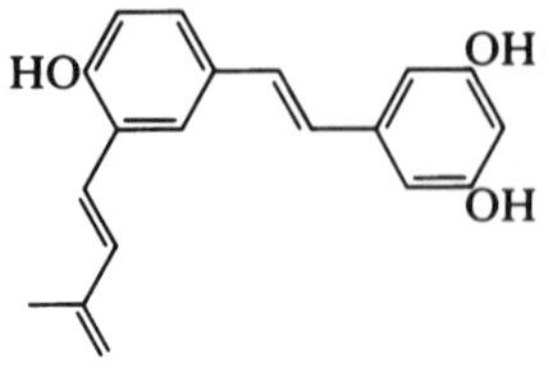

$C_{19}H_{18}O_3$ M 294.349
(E)-form
Constit. of *Arachis hypogaea*. Antifungal phytoalexin.
Cooksey, C.J. *et al*, *Phytochemistry*, 1988, **27**, 1015.

2-Hydroxy-2-methylcyclobutanone   H-70168

[25733-27-1]

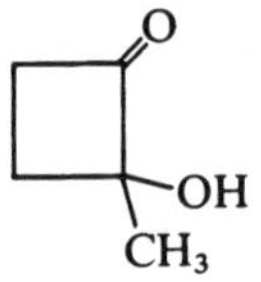

$C_5H_8O_2$ M 100.117
Isol. from the latex of the rubber tree *Hevea brasiliensis* (no opt. rotn. reported).

(±)-form
Liq. Bp$_6$ 75°. n_D^{20} 1.4524. Dimerises on standing.
p-*Nitrophenylhydrazone:* Mp 173-174°.
2,4-Dinitrophenylhydrazone: Mp 205-207°.

Urry, W.H. *et al*, *Tetrahedron Lett.*, 1962, 609 (*synth*)
Nishimura, H. *et al*, *Phytochemistry*, 1977, **16**, 1048 (*isol*)

4-Hydroxy-4-methyl-2,5-cyclohexadien-1-   H-70169
one, 9CI

[23438-23-5]

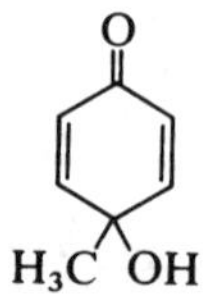

$C_7H_8O_2$ M 124.139
Needles (CH$_2$Cl$_2$/pet. ether). Mp 77.5-79° (74-75°).

Bamberger, E., *Ber.*, 1903, **36**, 2028 (*synth*)
Adam, W. *et al*, *Chem. Ber.*, 1988, **121**, 21 (*synth, ir, pmr, cmr*)

5-(Hydroxymethyl)-1,2,3,4-cyclohexanete-   H-70170
trol

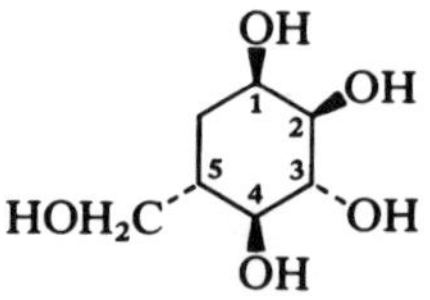

(1R,2R,3R,4S,5S)-form

$C_7H_{14}O_5$ M 178.185
(1R,2R,3R,4S,5S)-form
Pseudo-α-L-glucopyranose
$[\alpha]_D^{21}$ −67° (c, 1.5 in MeOH).
2,3,4,5-Tetra-O-Ac: $[\alpha]_D^{20}$ −56° (c, 1.1 in CHCl$_3$).
(1S,2S,3S,4R,5R)-form [94943-08-5]
Pseudo-α-D-glucopyranose. 1,2-Dideoxy-2-(hydroxymethyl)-D-chiro-inositol, 9CI. 1D-(1,2,4/3,5)-5-C-Hydroxymethyl-1,2,3,4-cyclohexanetetrol
Syrup. $[\alpha]_D$ +67° (+30°) (c, 0.5 in MeOH).
2,3,4,5-Tetra-O-Ac: Syrup. $[\alpha]_D$ +57° (+37°) (c, 0.90 in CHCl$_3$).
Penta-Ac: [94943-06-3].
$C_{17}H_{24}O_{10}$ M 388.371
$[\alpha]_D^{20}$+6.30° (c, 1.00 in CHCl$_3$), $[\alpha]_D^{20}$ +37° (c, 0.79 in CHCl$_3$).
(1R,2S,3S,4R,5R)-form [86117-82-0]
Pseudo-β-D-glucopyranose. 2,3-Dideoxy-3-(hydroxymethyl)-D-myo-inositol, 9CI. 1L-(1,3,5/2,4)-5-C-Hydroxymethyl-1,2,3,4-cyclohexanetetrol
Oil. $[\alpha]_D^{20}$ +13.0° (c, 2.1 in H$_2$O).
Monohydrate: $[\alpha]_D^{20}$ +10.9° (c, 0.83 in H$_2$O).

Penta-Ac: [90695-22-0]. Mp 115-116°. $[\alpha]_D^{20}$ +8.9° (c,
1.01 in CHCl₃), $[\alpha]_D^{20}$ +13.8° (c, 1 in CHCl₃).

(1R,2S,3S,4R,5S)-form [105017-74-1]
*Pseudo-α-ʟ-idopyranose. 1,6-Dideoxy-1-(hydroxy-
methyl)-ʟ-chiro-inositol, 9CI. 1L-(1,2,4/3,5)-1-C-Hy-
droxymethyl-2,3,4,5-cyclohexanetetrol*
Cryst. + 1H₂O (EtOH). Mp 127-128°. $[\alpha]_D^{20}$ −45.7° (c,
1.10 in H₂O).
Penta-Ac: [105017-73-0]. Cryst. (EtOAc/pet. ether).
Mp 86-87°. $[\alpha]_D^{20}$ −36.1° (c, 1.08 in CHCl₃).

(1S,2S,3R,4R,5S)-form
*Pseudo-β-ʟ-idopyranose. 2,3-Dideoxy-2-(hydroxy-
methyl)-ᴅ-epi-inositol, 9CI. 1D-(1,2,4,5/3)-5-C-Hy-
droxymethyl-1,2,3,4-cyclohexanetetrol*
Monohydrate. $[\alpha]_D^{20}$ +8.5° (c, 1.02 in H₂O).
Penta-Ac: [105017-80-9]. Cryst. (EtOAc/pet. ether).
Mp 115-117°. $[\alpha]_D^{20}$ +14.4° (c, 1.08 in CHCl₃).

(1RS,2RS,3RS,4SR,5SR)-form [95043-48-4]
Pseudo-α-ᴅʟ-glucopyranose
Penta-Ac: [77209-45-1]. Cryst. (EtOH/Et₂O). Mp 110-
111°.

(1RS,2SR,3SR,4RS,5SR)-form [74560-71-7]
Pseudo-α-ᴅʟ-idopyranose
Cryst. (EtOH). Mp 155-156°.
Penta-Ac: [74560-70-6]. Mp 106°.

Ogawa, S. *et al, Bull. Chem. Soc. Jpn.,* 1980, **53,** 1121; 1986,
59, 2956 (*synth, pmr*)
Ogawa, S. *et al, Chem. Lett.,* 1984, 355, 1919 (*synth*)
Paulsen, H. *et al, Justus Liebigs Ann. Chem.,* 1984, 433; 1987,
125, 133 (*synth, pmr*)
Ogawa, S. *et al, Carbohydr. Res.,* 1985, **136,** 77 (*synth*)

5-Hydroxy-2-methylcyclohexanone, 9CI   **H-70171**

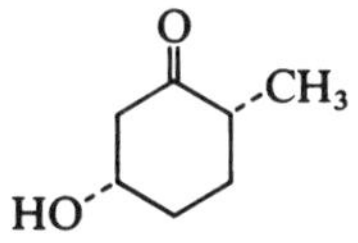

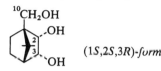

$(2R,5S)$-*form*

C₇H₁₂O₂   M 128.171
(2R,5S)-form [66901-51-7]
(−)-cis-*form*
$[\alpha]_D^{21}$ −3.2° (c, 1.1 in CHCl₃). The two racemates were
prepd. and characterised spectroscopically by Semet, *et
al.*

Solladié, G. *et al, J. Org. Chem.,* 1987, **52,** 3560 (*synth, ir, pmr*)
Corey, E.J. *et al, Chem. Ber.,* 1978, **111,** 1337 (*synth*)
Semet, R. *et al, Bull. Soc. Chim. Fr., Part II,* 1978, 185 (*synth*)

5-(Hydroxymethyl)-5-cyclohexene-1,2,3,4-   **H-70172**
tetrol, 9CI
3,4,5,6-Tetrahydroxy-1-cyclohexene-1-methanol

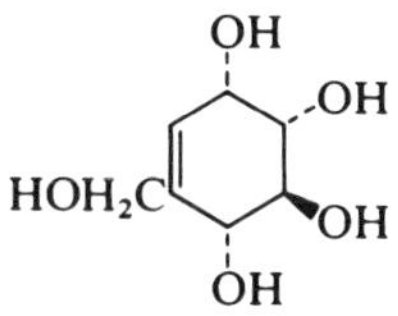

$(1S,2S,3S,4R)$-*form*

C₇H₁₂O₅   M 176.169
(1S,2S,3S,4R)-form [111136-25-5]
Streptol
Isol. from *Streptomyces* sp. 1409. Plant growth regulator.

(1RS,2RS,3RS,4SR)-form [86195-44-0]
Cryst. (EtOH). Mp 139.5-141°.
(1RS,2SR,3SR,4SR)-form [86195-45-1]
Cryst. (EtOH aq.). Mp 150.5-151.5°.

Toyokuni, T. *et al, Bull. Chem. Soc. Jpn.,* 1983, **56,** 505 (*synth*)
Isogai, A. *et al, Agric. Biol. Chem.,* 1987, **51,** 2277 (*isol, struct*)

1-(Hydroxymethyl)-7,7-   **H-70173**
dimethylbicyclo[2.2.1]heptane-2,3-diol,
9CI
2,3,10-Bornanetriol. 3,10-Dihydroxyborneol

$(1S,2S,3R)$-*form*

C₁₀H₁₈O₃   M 186.250
(1S,2S,3R)-form [114437-17-1]
Constit. of *Florestina tripteris.* Oil. $[\alpha]_{546}^{24}$ −40° (c, 0.5 in
CHCl₃).
(1S,2S,3S)-form [114489-85-9]
Constit. of *F. tripteris.* Oil. $[\alpha]_D^{24}$ +21° (c, 0.1 in CHCl₃).
3-Ketone: [114437-19-3]. *3-Hydroxy-4-(hydroxy-
methyl)-7,7-dimethylbicyclo[2.2.1]heptan-2-one, 9CI.
2,10-Dihydroxy-3-bornanone. 10-Hydroxy-3-
oxoborneol.*
C₁₀H₁₆O₃   M 184.235
Constit. of *F. tripteris.* Oil.

Domínguez, X.A. *et al, Phytochemistry,* 1988, **27,** 613.

3-Hydroxymethyl-7,11-dimethyl-2,6,11-do-   **H-70174**
decatriene-1,5,10-triol

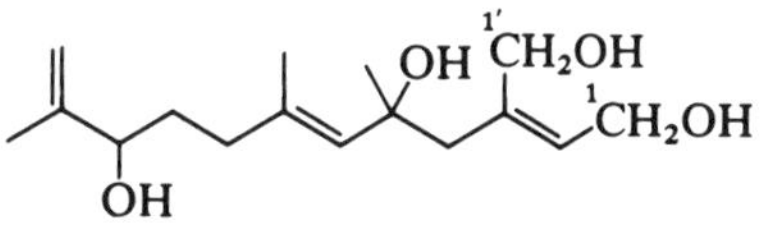

C₁₅H₂₆O₄   M 270.368
1,1′-Di-Ac:
C₁₉H₃₀O₆   M 354.442
Constit. of *Calea hispida.* Gum. $[\alpha]_D$ −7° (c, 0.1 in
CHCl₃).

Bohlmann, F. *et al, Phytochemistry,* 1982, **21,** 2899.

2-Hydroxy-4,5-methylenedioxybenzalde-   **H-70175**
hyde
*6-Hydroxy-1,3-benzodioxole-5-carboxaldehyde, 9CI. 2-
Hydroxypiperonal. 4,5-Methylenedioxysalicylaldehyde*
[4720-68-7]

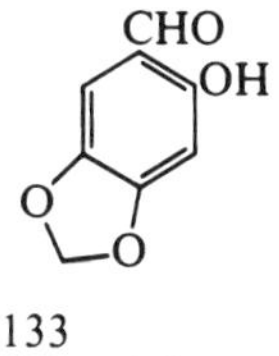

C₈H₆O₄   M 166.133
Possesses antimicrobial activity against bacteria and
fungi. Needles (EtOH). Mp 125-126°.
Ac:
C₁₀H₈O₅   M 208.170
Needles (EtOH). Mp 126-127°.
Me ether: [5780-00-7]. *2-Methoxy-4,5-methylenedioxy-
benzaldehyde. 6-Methoxy-1,3-benzodioxole-5-car-
boxaldehyde, 9CI.*
C₉H₈O₄   M 180.160

Isol. from *Piper lenticellosum*. Cryst. (EtOH). Mp 115-117° (111.5-112°).

Campbell, K.N. *et al, J. Org. Chem.*, 1951, **16**, 1736 (*synth*)
Kenjii, F. *et al, Bull. Chem. Soc. Jpn.*, 1962, **35**, 1321 (*synth*)
Scharf, H.-D. *et al, Justus Liebigs Ann. Chem.*, 1978, 573 (*deriv*)
Ishii, H. *et al, Chem. Pharm. Bull.*, 1983, **31**, 3024 (*deriv*)
Devakumar, C. *et al, Agric. Biol. Chem.*, 1985, **49**, 725 (*deriv, props*)
Diaz, P.P. *et al, J. Nat. Prod.*, 1986, **49**, 690 (*isol, deriv*)

1-(Hydroxymethyl)fluorene   H-70176
9H-Fluorene-1-methanol
[73728-55-9]

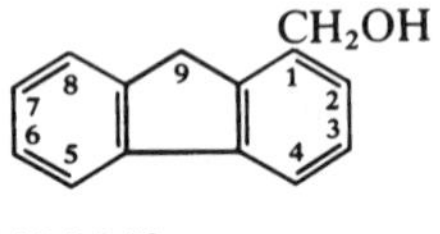

$C_{14}H_{12}O$ M 196.248
Tan needles (CH$_2$Cl$_2$). Mp 147-148.5°.
▷LL7900000.

Grunewald, G.L. *et al, J. Med. Chem.*, 1988, **31**, 60 (*synth, pmr, cmr, ms*)

5-Hydroxymethyl-2-furancarboxaldehyde, 9CI   H-70177

Updated Entry replacing H-02434
5-Hydroxymethylfurfural. 2-Formyl-5-hydroxymethylfuran
[67-47-0]

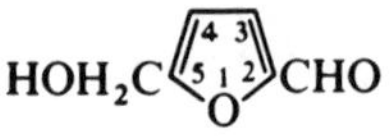

$C_6H_6O_3$ M 126.112
Obtainable from various carbohydrates. Needles. Mp 35-35.5° (31.5°). Bp$_{.02}$ 110°.
▷LT7031100.

Oxime: [37110-03-5].
  $C_6H_7NO_3$ M 141.126
  Cryst. Mp 108° (*E*-), Mp 77-78° (*Z*-).
Phenylhydrazone: Cryst. Mp 140°.
Semicarbazone: Cryst. Mp 192° (166-167°).

Teunissen, H.P., *Recl. Trav. Chim. Pays-Bas*, 1930, **49**, 784 (*bibl*)
Haworth, W.N. *et al, J. Chem. Soc.*, 1944, 667 (*synth*)
Org. Synth., Coll. Vol., **4**, 919.
Harris, D.W. *et al, J. Org. Chem.*, 1974, **39**, 724 (*synth*)
Hearn, M.T.W., *Aust. J. Chem.*, 1976, **29**, 107 (*cmr*)
Gaset, A. *et al, Inf. Chim.*, 1981, **212**, 179; **214**, 203 (*revs*)
El Hajj, T. *et al, Bull. Soc. Chim. Fr.*, 1987, 855 (*synth, pmr, ms*)

1-[2-(Hydroxymethyl)-3-methoxyphenyl]-1,5-heptadiene-3,4-diol, 9CI   H-70178
2-Methoxy-6-(3,4-dihydroxy-1,5-heptadienyl)benzyl alcohol
[72330-50-8]

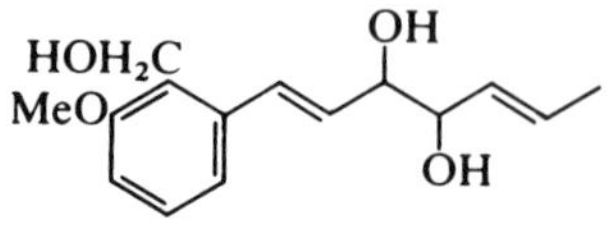

$C_{15}H_{20}O_4$ M 264.321
Prod. by *Aspergillus veisicolor*. Cryst. (CHCl$_3$). Mp 132-133°.

Dunn, A.W. *et al, J. Chem. Soc., Perkin Trans. 1*, 1979, 2122 (*isol, struct*)

6-(Hydroxymethyl)-10-methyl-2-(4-methyl-3-pentenyl)-2,6,10-dodecatriene-1,12-diol, 9CI   H-70179
18,19-Dihydroxynerylgeraniol
[117232-54-9]

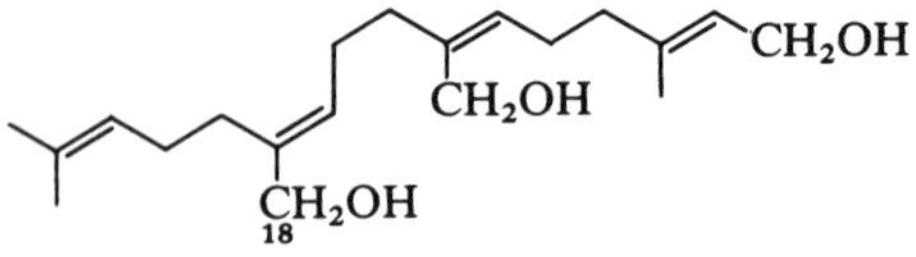

$C_{20}H_{34}O_3$ M 322.487
Constit. of *Vittandinia gracilis*. Oil.

18-Aldehyde: [117255-04-6]. *12-Hydroxy-6-(hydroxymethyl)-10-methyl-2-(4-methyl-3-pentenyl)-2,6,10-dodecatrienal, 9CI. 18-Oxo-19-hydroxynerylgeraniol.*
  $C_{20}H_{32}O_3$ M 320.471
  Constit. of *V. gracilis*. Oil.

Zdero, C. *et al, Phytochemistry*, 1988, **27**, 2251.

2-Hydroxy-5-methyl-3-nitrobenzaldehyde   H-70180

Updated Entry replacing H-02510
6-Hydroxy-5-nitro-m-tolualdehyde. 3-Nitro-2,5-cresotaldehyde
[66620-31-3]
$C_8H_7NO_4$ M 181.148
Yellow needles (EtOH aq.). Mp 141°.

Oxime:
  $C_8H_8N_2O_4$ M 196.162
  Yellow leaflets (EtOH). Mp 214-216°.
Phenylhydrazone: Mp 164-166°.

Borsche, W., *Ber.*, 1917, **50**, 1339.
Cuevas, J.C. *et al, J. Org. Chem.*, 1988, **53**, 2055 (*synth, ir, pmr, ms*)

3-(Hydroxymethyl)-7-oxabicyclo[4.1.0]-hept-3-ene-2,5-dione, 9CI   H-70181

Updated Entry replacing H-02536
5,6-Epoxy-2-hydroxymethyl-2-cyclohexene-1,4-dione.
Phyllostine

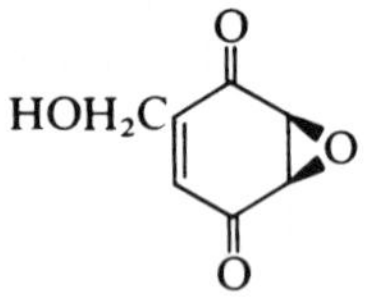

$C_7H_6O_4$ M 154.122
(−)-*form* [27270-89-9]
  Metab. of *Phyllosticta* spp. Wilting toxin. Mp 56°. $[\alpha]_D^{20}$ −105.6° (EtOH).
(±)-*form* [32180-43-1]
  Mp 48.0°.

Sakamura, S. *et al, Agric. Biol. Chem.*, 1970, **34**, 153; 1971, **35**, 105; 1974, **38**, 163 (*isol, synth*)
Ichihara, A. *et al, Tetrahedron Lett.*, 1972, 5105; 1976, 4741 (*synth*)
Sekiguchi, J. *et al, Biochemistry*, 1978, **17**, 1785.
Ichihara, A. *et al, Agric. Biol. Chem.*, 1982, **46**, 1879 (*synth*)

5-Hydroxy-3-methylpentanoic acid, 9CI   H-70182

Updated Entry replacing H-02547
5-Hydroxy-3-methylvaleric acid
[151-11-1]

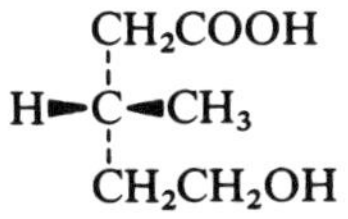

$$CH_2COOH$$
$$H \blacktriangleright C \blacktriangleleft CH_3 \quad (R)\text{-}form$$
$$CH_2CH_2OH$$

$C_6H_{12}O_3$ M 132.159

(R)-form

Lactone: [61898-55-3]. *Tetrahydro-4-methyl-2H-pyran-2-one. 3-Methyl-5-pentanolide. 3-Methyl-δ-valerolactone.*
$C_6H_{10}O_2$ M 114.144
$[\alpha]_D$ +27.1° (c, 1 in CHCl₃).

(S)-form

Lactone: [61898-56-4]. $[\alpha]_D$ −27.0° (c, 1 in CHCl₃).

(±)-form

Lactone: [62989-38-2]. Bp_{14} 107°.

Longley, R.I. et al, J. Am. Chem. Soc., 1952, **74**, 2012 (synth)
Irwin, A.J. et al, J. Am. Chem. Soc., 1977, **99**, 556 (synth, ir, pmr, abs config)
Kocienski, P.J. et al, J. Chem. Soc., Perkin Trans. 1, 1978, 834 (synth)
Enders, D. et al, Chem. Ber., 1987, **120**, 1223 (synth)
Terunuma, D. et al, J. Org. Chem., 1987, **52**, 1630 (synth)

4-Hydroxy-5-methyl-3-(3,8,11-trimethyl-8-   H-70183
oxo-2,6,10-dodecatrienyl)-2H-1-benzo-
pyran-2-one

4-Hydroxy-5-methyl-3-(8-oxofarnesyl)coumarin

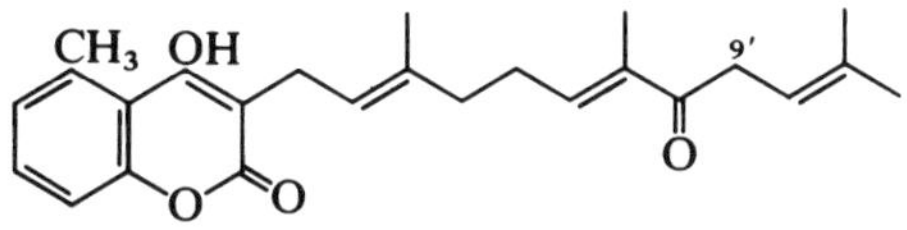

$C_{25}H_{30}O_4$ M 394.510
Constit. of *Gypothamnium pinifolium*. Oil.
9′-Hydroxy: 4-Hydroxy-3-(9-hydroxy-8-oxofarnesyl)-5-methylcoumarin.
$C_{25}H_{30}O_5$ M 410.509
Constit. of *G. pinifolium*. Oil.

Zdero, C. et al, Phytochemistry, 1988, **27**, 2953.

6-Hydroxy-1-naphthaldehyde, 8CI   H-70184

6-Hydroxy-1-naphthalenecarboxaldehyde, 9CI
$C_{11}H_8O_2$ M 172.183
Me ether: [3597-42-0]. *6-Methoxy-1-naphthaldehyde.*
$C_{12}H_{10}O_2$ M 186.210
Needles (CCl₄). Mp 82-83°. Bp_4 160-162°.
Me ether, oxime: [36112-60-4].
$C_{12}H_{11}NO_2$ M 201.224
Cryst. (EtOH aq.). Mp 114-115°.

Kessar, S.V. et al, J. Chem. Soc. (C), 1971, 262 (synth)
Gaur, S.P. et al, Indian J. Chem., Sect. B, 1982, **21**, 46 (synth, pmr)
MJustus Liebigs Ann. Chem., S. et al, J. Org. Chem., 1982, **47**, 5021 (synth)

6-Hydroxy-2-naphthaldehyde   H-70185

Updated Entry replacing H-02666
6-Hydroxy-2-naphthalenecarboxaldehyde, 9CI

$C_{11}H_8O_2$ M 172.183
2,4-Dinitrophenylhydrazone: Mp 270-271°.
Me ether: [3453-33-6]. *6-Methoxy-2-naphthaldehyde.*
$C_{12}H_{10}O_2$ M 186.210
Cryst. (pet. ether). Mp 65-66°.
Me ether, oxime:
$C_{12}H_{11}NO_2$ M 201.224
Mp 154-155°.

Horeau, A., C. R. Hebd. Seances Acad. Sci., 1953, **236**, 826.
Gandhi, R., J. Chem. Soc., 1955, 2530.
Harvey, R.G. et al, J. Med. Chem., 1988, **31**, 1308 (synth, deriv, pmr)
Lee, H. et al, J. Org. Chem., 1988, **53**, 4587 (synth, deriv, pmr)

7-Hydroxy-2-naphthaldehyde, 8CI   H-70186

7-Hydroxy-2-naphthalenecarboxaldehyde, 9CI
$C_{11}H_8O_2$ M 172.183
Me ether: 7-Methoxy-2-naphthaldehyde.
$C_{12}H_{10}O_2$ M 186.210
Prisms (Et₂O/pet. ether). Mp 56-58°.

Moffat, J.S., J. Chem. Soc. (C), 1966, 734 (synth, deriv)

Hydroxyniranthin   H-70187

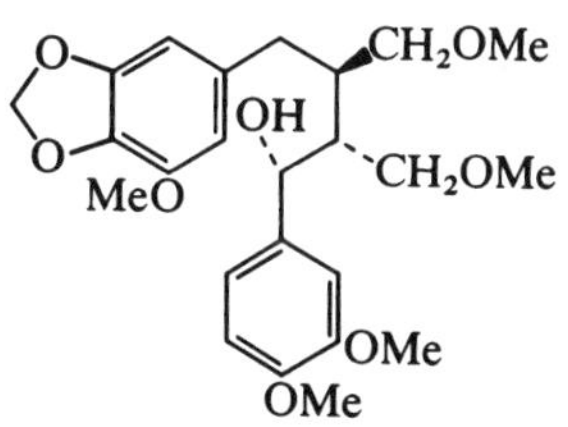

$C_{24}H_{32}O_8$ M 448.512
Constit. of *Phyllanthus niruri*. Oil. $[\alpha]_D^{28}$ −50° (CHCl₃).

Satyanarayana, P. et al, J. Nat. Prod., 1988, **51**, 44.

2-Hydroxy-5-nitro-1,3-benzenedicarboxy-   H-70188
lic acid, 9CI

Updated Entry replacing H-02778
2-Hydroxy-5-nitroisophthalic acid, 8CI
[67294-53-5]

$C_8H_5NO_7$ M 227.130
Needles + H₂O (H₂O). Mp 213-214° (anhyd.).
Di-Et ester: [72078-90-1].
$C_{12}H_{13}NO_7$ M 283.237
Needles (EtOH). Mp 61°.
Nitrile: 3-Cyano-2-hydroxy-5-nitrobenzoic acid.
$C_8H_4N_2O_5$ M 208.130
Yellow cryst. (H₂O). Mp 205°.

Meldola, R. et al, J. Chem. Soc., 1917, **111**, 533.
Jones, E.C.S. et al, J. Chem. Soc., 1931, 1842 (synth, deriv)
Fox, J.J. et al, J. Org. Chem., 1982, **47**, 1081 (uv, pmr)
Ebel, K. et al, Chem. Ber., 1988, **121**, 323 (deriv)

3-Hydroxy-6-nitro-1,2-benzenedicarboxy-   H-70189
lic acid, 9CI

Updated Entry replacing H-02779
3-Hydroxy-6-nitrophthalic acid, 8CI

$C_8H_5NO_7$　　M 227.130
Prisms (EtOAc/C_6H_6 or Et_2O/pet. ether). Mp 204°.
Me ether: 3-Methoxy-6-nitro-1,2-benzenedicarboxylic
　acid.
　$C_9H_7NO_7$　　M 241.157
　Prisms (Et_2O/pet. ether). Mp 179-180° (163°) dec.
Me ether, dinitrile: 2,3-Dicyano-1-methoxy-4-
　nitrobenzene.
　$C_9H_5N_3O_3$　　M 203.157
　Mp 198°.
Anhydride:
　$C_8H_3NO_6$　　M 209.115
　Cryst. (C_6H_6). Mp 163°.

Thomson, R.H. *et al*, *J. Chem. Soc.*, 1947, 350.
Gundermann, K.D. *et al*, *Chem. Ber.*, 1962, **95**, 2018 (*deriv*)
Momose, T., *Yakugaku Zasshi*, 1963, **83**, 724; *CA*, **59**, 12693f
　(*synth*)

4-Hydroxy-3-nitro-1,2-benzenedicarboxy-　　**H-70190**
　lic acid

Updated Entry replacing H-02780
　4-Hydroxy-3-nitrophthalic acid
$C_8H_5NO_7$　　M 227.130
Di-Et ester:
　$C_{12}H_{13}NO_7$　　M 283.237
　Needles (EtOH). Mp 93-94°.
Me ether: 4-Methoxy-3-nitro-1,2-benzenedicarboxylic
　acid.
　$C_9H_7NO_7$　　M 241.157
　Light-yellow plates (H_2O). Mp 224-225° dec.
Me ether, 2-Me ester:
　$C_{10}H_9NO_7$　　M 255.184
　Prisms (H_2O). Mp 186-187°.

Cain, J.C. *et al*, *J. Chem. Soc.*, 1914, **105**, 156.
King, H., *J. Chem. Soc.*, 1939, 1157.
Smith, P.A.S. *et al*, *J. Org. Chem.*, 1961, **26**, 27 (*deriv*)

4-Hydroxy-5-nitro-1,2-benzenedicarboxy-　　**H-70191**
　lic acid

Updated Entry replacing H-02781
　4-Hydroxy-5-nitrophthalic acid
$C_8H_5NO_7$　　M 227.130
Pale-yellow cryst. (H_2O). Mp 180-182° dec.
Di-Me ester:
　$C_{10}H_9NO_2$　　M 175.187
　Needles (EtOH). Mp 115°.
Anhydride:
　$C_8H_3NO_6$　　M 209.115
　Cryst. (C_6H_6). Mp 146-147°.
Me ether: 4-Methoxy-5-nitro-1,2-benzenedicarboxylic
　acid.
　$C_9H_7NO_7$　　M 241.157
　Needles (C_6H_6/Me_2CO). Mp 197-198° dec.
Me ether, di-Me ester:
　$C_{11}H_{11}NO_7$　　M 269.210
　Needles. Mp 118°.
Me ether, anhydride:
　$C_9H_5NO_6$　　M 223.142
　Cryst. (C_6H_6). Mp 181-182°.

Cain, J.C. *et al*, *J. Chem. Soc.*, 1914, **105**, 156.
Goldstein, H. *et al*, *Helv. Chim. Acta*, 1952, **35**, 1476.

4-Hydroxy-5-nitro-1,3-benzenedicarboxy-　　**H-70192**
　lic acid

Updated Entry replacing H-02782
　4-Hydroxy-5-nitroisophthalic acid

[22235-27-4]
$C_8H_5NO_7$　　M 227.130
Cryst. (EtOH). Mp 237-238°.
Di-Me ester: [22235-28-5].
　$C_{10}H_9NO_7$　　M 255.184
　Mp 105.5°.
Me ether: 4-Methoxy-5-nitro-1,3-benzenedicarboxylic
　acid.
　$C_9H_7NO_7$　　M 241.157
　Needles (H_2O). Mp 225-226°.
Me ether, di-Me ester:
　$C_{11}H_{11}NO_7$　　M 269.210
　Needles (pet. ether). Mp 85°.

Ger. Pat., 555 410, (*1932*); *CA*, **26**, 5105
Hunt, S.E. *et al*, *Chem. Ind.* (London), 1955, 417
Welch, D.E. *et al*, *J. Med. Chem.*, 1969, **12**, 299.

4-Hydroxy-3-nitrosobenzoic acid　　**H-70193**

[32749-97-6]

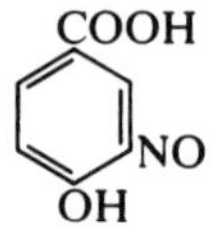

$C_7H_5NO_4$　　M 167.121
Fe complex (*2:1*): **Actinoviridin** A.
　$C_{14}H_8FeN_2O_8$　　M 388.073
　Isol. from *Streptomyces griseus*. Dark-green solid. Mp
　>300°.

Cronheim, G., *J. Org. Chem.*, 1947, **12**, 7 (*synth*)
Lazar, J. *et al*, *Acta Chim. Acad. Sci. Hung.*, 1978, **98**, 327
　(*synth*)
Kurobane, I. *et al*, *J. Antibiot.*, 1987, **40**, 1131 (*isol, struct, bibl*)

9-Hydroxy-5-nonenoic acid　　**H-70194**

$HOCH_2CH_2CH_2CH{=}CH(CH_2)_3COOH$

$C_9H_{16}O_3$　　M 172.224
(**Z**)-**form** [109997-30-0]
　Yellow oil.
　Me ester: [75964-05-5].
　　$C_{10}H_{18}O_3$　　M 186.250
　　Oil.

Bergman, N.-A. *et al*, *J. Org. Chem.*, 1987, **52**, 4449 (*synth*,
　pmr)

6-Hydroxy-7,9-octadecadiene-11,13,15,17-　　**H-70195**
　tetraynoic acid, 9CI

$H(C{\equiv}C)_4CH{=}CHCH{=}CHCH(OH)(CH_2)_4COOH$

$C_{18}H_{16}O_3$　　M 280.323
Exhibits potent antimicrobial activity.
(**7E,9E**)-**form** [111051-87-7]
　Caryoynencin A
　Isol. from *Pseudomonas caryophilli*.
(**7E,9Z**)-**form** [111138-75-1]
　Caryoynencin C
　From *P. caryophilli*.
(**7Z,9E**)-**form** [111138-74-0]
　Caryoynencin B
　From *P. caryophilli*.

Kusumi, T. *et al*, *Tetrahedron Lett.*, 1987, **28**, 3981 (*isol, struct*)

13-Hydroxy-9,11-octadecadienoic acid, 9CI   H-70196

Updated Entry replacing H-02922

[5204-88-6]

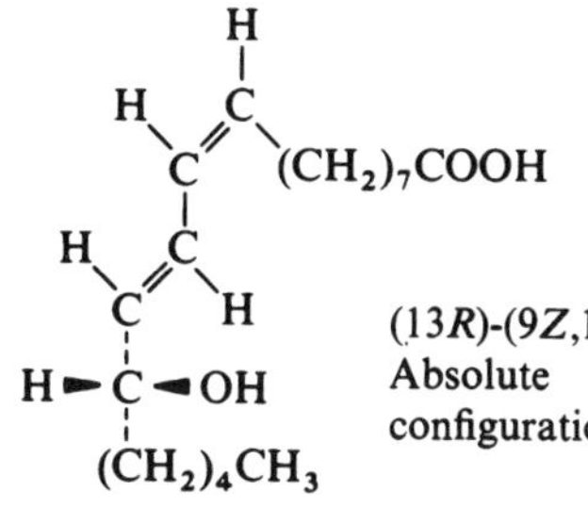

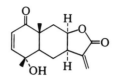

(13R)-(9Z,11E)-form
Absolute
configuration

$C_{18}H_{32}O_3$ M 296.449

(9Z,11E,13R)-form

(−)-Coriolic acid

Occurs in *Coriaria nepalensis* and *Xeranthemum anuum* seed oils. $[\alpha]_{400}-0.5°$ (c, 1.6 in hexane).

Me ester:

$C_{19}H_{34}O_3$ M 310.476

Oil. $[\alpha]_D^{23}-7.5°$ (c, 2.3 in hexane).

(9Z,11E,13S)-form [29623-28-7]

(+)-Coriolic acid

Occurs in *Mannina emarginata*.

Me ester: $[\alpha]_{546}^{20}+15.2°$ (0.86 in EtOH).

(±)-(9E,11E)-form [32819-31-1]

Occurs in *Absinthium* oils.

Tallent, W.H. *et al*, *Tetrahedron Lett.*, 1966, 4329 (*struct, uv*)
Powell, R.C. *et al*, *J. Org. Chem.*, 1967, **32**, 1442.
Conacher, H.B.S. *et al*, *Chem. Phys. Lipids*, 1969, **3**, 191 (*synth*)
Kleiman, R. *et al*, *J. Am. Oil Chem. Soc.*, 1973, **50**, 31 (*ms*)
Chan, H.W.S. *et al*, *Lipids*, 1977, **12**, 99 (*synth*)
Moustakis, C.A. *et al*, *Tetrahedron Lett.*, 1986, **27**, 303 (*synth*)
Rao, A.V.R. *et al*, *J. Org. Chem.*, 1986, **51**, 4158 (*synth, ir, pmr, ms*)
Kobayashi, Y. *et al*, *Tetrahedron Lett.*, 1987, **28**, 3959 (*synth, biochem*)
Tsuboi, S. *et al*, *Bull. Chem. Soc. Jpn.*, 1987, **60**, 1103 (*synth*)

3-Hydroxy-12-oleanen-29,22-olide   H-70197

Updated Entry replacing H-20230

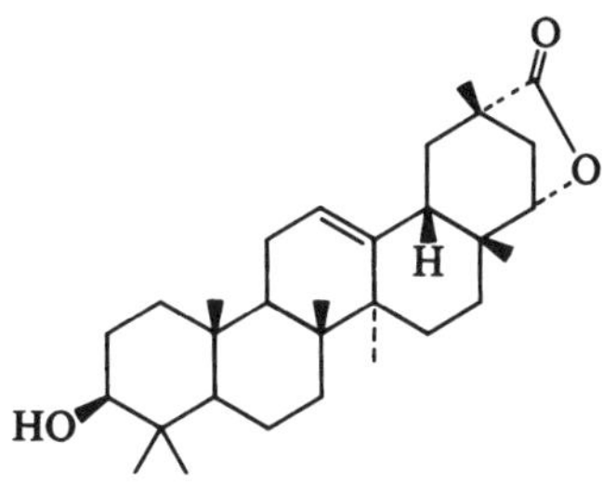

$C_{30}H_{46}O_3$ M 454.692

(3β,22α)-form [84104-71-2]

Abruslactone A. *Regelide. Wilforlide* A

Constit. of *Abrus precatorius* and *Tripterygium wilfordii* and *T. regelii*. Cryst. (CHCl₃/MeOH). Mp 329-330°.

Chang, H.-M. *et al*, *J. Chem. Soc., Chem. Commun.*, 1983, 1197 (*isol*)
Hori, H. *et al*, *Chem. Pharm. Bull.*, 1987, **35**, 2125 (*cryst struct*)

4-Hydroxy-1-oxo-2,11(13)-eudesmadien-12,8-olide   H-70198

$C_{15}H_{18}O_4$ M 262.305

(4α,8β)-form

Constit. of *Ferreyranthus fruticosus*. Cryst. Mp 117°.

Jakupovic, J. *et al*, *Phytochemistry*, 1988, **27**, 1113.

8-Hydroxy-3-oxo-1,4,11(13)-eudesmatrien-12-oic acid   H-70199

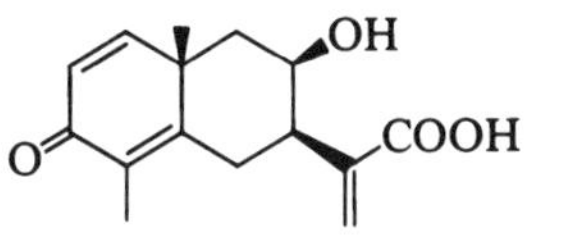

$C_{15}H_{18}O_4$ M 262.305

8β-form

Yomoginic acid

Me ester: **Methyl yomoginate**.

$C_{16}H_{20}O_4$ M 276.332

Constit. of *Ferreyranthus fruticosus*. Oil.

Jakupovic, J. *et al*, *Phytochemistry*, 1988, **27**, 1113.

7-Hydroxy-3-oxo-8,24-lanostadien-26-al   H-70200

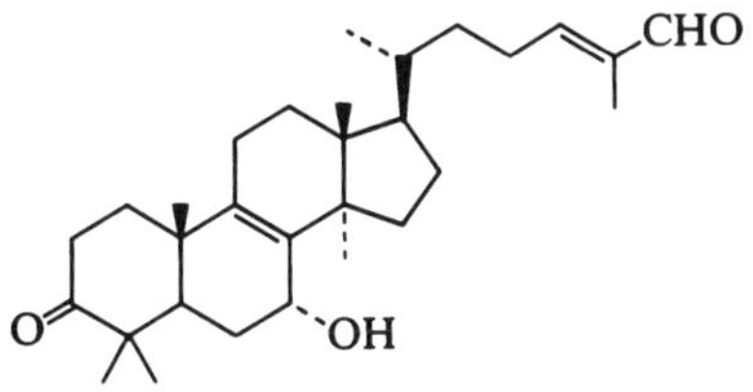

$C_{30}H_{46}O_3$ M 454.692

(7α,24E)-form

Ganoderal B

Metab. of *Ganoderma lucidum*. Syrup. $[\alpha]_D^{23}+94°$ (c, 0.1 in EtOH).

Nishitoba, T. *et al*, *Agric. Biol. Chem.*, 1988, **52**, 211.

12-Hydroxy-3-oxo-28,13-oleananolide   H-70201

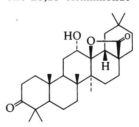

$C_{30}H_{46}O_4$ M 470.691

(12α,13β)-form

Hydroxyoleanonic lactone

Constit. of dammar resin. Cryst. (CH₂Cl₂/MeOH). Mp 304-306°. $[\alpha]_D+60.4°$ (c, 1.08 in CHCl₃).

Poehland, B.L. *et al*, *J. Nat. Prod.*, 1987, **50**, 706.

22-Hydroxy-3-oxo-12-ursen-30-oic acid   H-70202

C$_{30}$H$_{46}$O$_4$ M 470.691

22α-form

Me ester: [109974-21-2]. **Regelin**.
C$_{31}$H$_{48}$O$_4$ M 484.718
Constit. of roots of *Tripterygium regelii*. Cytotoxic.
Cryst. (CH$_2$Cl$_2$/MeOH). Mp 221-223°. [α]$_D^{27.6}$ +20°
(c, 0.9 in CHCl$_3$).

Hori, H. *et al, Chem. Pharm. Bull.*, 1987, **35**, 2125 (*cryst struct*)

(2-Hydroxyphenoxy)acetic acid   H-70203

Catecholacetic acid

[6324-11-4]

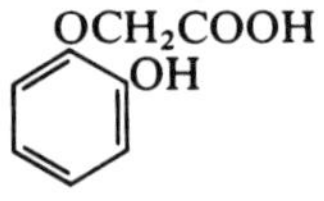

C$_8$H$_8$O$_4$ M 168.149
Plates (H$_2$O), prisms (Et$_2$O). Mp 133° (130-131°).

Me ester:
C$_9$H$_{10}$O$_4$ M 182.176
Long needles + 1H$_2$O. Mp 59°.
Et ester: Mp 53°. Bp$_{30}$ 155-156°.
O-Ac:
C$_{10}$H$_{10}$O$_5$ M 210.186
Needles (C$_6$H$_6$). Mp 110°.
Amide:
C$_8$H$_9$NO$_3$ M 167.164
Prisms (EtOH). Mp 108° (monohydrate), Mp 130°
(anhyd.).
Me ether: [1878-85-9]. (*2-Methoxyphenoxy)acetic acid*.
C$_9$H$_{10}$O$_4$ M 182.176
Needles (H$_2$O). Mp 120°.
Me ether, Et ester:
C$_{11}$H$_{14}$O$_4$ M 210.229
Bp$_{27}$ 175-179°.
Me ether, amide:
C$_9$H$_{11}$NO$_3$ M 181.191
Needles. Mp 138°.

Cutolo, A., *Gazz. Chim. Ital.*, 1894, **24**, 63 (*deriv, synth*)
Mouren, C., *Bull. Soc. Chim. Fr.*, 1899, **21**, 107 (*synth*)
Ludewig, H., *J. Prakt. Chem.*, 1900, **61**, 345 (*synth, deriv*)
Bischoff, C.A. *et al, Ber.*, 1907, **40**, 2779 (*synth*)

(3-Hydroxyphenoxy)acetic acid, 9CI   H-70204

Resorcinolacetic acid

[1878-83-7]
C$_8$H$_8$O$_4$ M 168.149
Needles (toluene). Mp 158-159°.

Et ester:
C$_{10}$H$_{12}$O$_4$ M 196.202
Cryst. (C$_6$H$_6$). Mp 55°.
Me ether: [2088-24-6]. (*3-Methoxyphenoxy)acetic acid*.
C$_9$H$_{10}$O$_4$ M 182.176
Needlelike cryst. (H$_2$O). Mp 118° (softens at 115°).
Me ether, Et ester:
C$_{11}$H$_{14}$O   M 162.231
Oil. Bp$_{24}$ 170°.

Carter, W. *et al, J. Chem. Soc.*, 1900, 1222 (*synth*)
Gilbody, W. *et al, J. Chem. Soc.*, 1901, 1409 (*deriv, synth*)

Bischoff, C.A. *et al, Ber.*, 1907, **40**, 2790 (*synth*)

(4-Hydroxyphenoxy)acetic acid, 9CI   H-70205

[1878-84-8]
C$_8$H$_8$O$_4$ M 168.149
Needles (toluene). Mp 152°.

Me ether: [1877-75-4]. (*4-Methoxyphenoxy)acetic acid*, *9CI*.
C$_9$H$_{10}$O$_4$ M 182.176
Cryst. (MeOH aq.). Mp 108-110°.
Me ether, Et ester:
C$_{11}$H$_{14}$O$_4$ M 210.229
Oil. Bp$_{0.01}$ 102-105°.
Anilide:
C$_{14}$H$_{13}$NO$_3$ M 243.262
Prisms (C$_6$H$_6$). Mp 101°.

Carter, W. *et al, J. Chem. Soc.*, 1900, 1222 (*synth, deriv*)
Koelsch, C.F., *J. Am. Chem. Soc.*, 1931, **53**, 304 (*deriv, synth*)
Sankararaman, S. *et al, J. Am. Chem. Soc.*, 1987, **109**, 5235
    (*deriv, synth, pmr, ir, cmr, ms*)

2-Hydroxy-2-phenylacetaldehyde   H-70206

α-Hydroxybenzeneacetaldehyde, 9CI. Phenylglycolaldehyde. Mandelaldehyde

[34025-29-1]

PhCH(OH)CHO

C$_8$H$_8$O$_2$ M 136.150

(±)-form

Oil. Insol. H$_2$O. Unstable, readily rearr. to 2-
Hydroxyacetophenone, H-01168 .
Semicarbazone: Cryst. Mp 222° dec.
2,4-Dinitrophenylhydrazone: Cryst. (Py). Mp 280-281°.

Evans, W.L. *et al, J. Am. Chem. Soc.*, 1913, **35**, 1770 (*synth*)
Rinkes, I.J. *et al, Recl. Trav. Chim. Pays-Bas*, 1920, **39**, 708
    (*synth*)
Staab, H.A. *et al, Justus Liebigs Ann. Chem.*, 1962, **654**, 119
    (*synth*)
Lodge, E.P. *et al, J. Am. Chem. Soc.*, 1987, **109**, 3353 (*deriv, synth, ir, pmr, cmr*)

3-(3-Hydroxyphenyl)-1*H*-2-benzopyran-1-one   H-70207

Updated Entry replacing H-01014
3-(3-Hydroxyphenyl)isocoumarin

[56383-84-7]

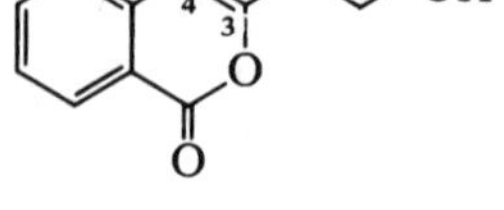

C$_{15}$H$_{10}$O$_3$ M 238.242
Needles. Mp 186-187°.

O-β-D-Glucopyranoside: [56383-84-7]. **Homalicine**.
C$_{21}$H$_{20}$O$_8$ M 400.384
Isol. from *Homalium zeylanicum*. Mp 241-243° dec.
O-β-D-Glycopyranoside, tetra-Ac: Mp 153-155°. [α]$_D$
−31.88° (c, 0.69 in CHCl$_3$).
Me ether: Mp 107-109°.
*3S,4-Dihydro: 3,4-Dihydro-3-(3-hydroxyphenyl)-1H-2-
benzopyran-1-one.*
C$_{15}$H$_{12}$O$_3$ M 240.258
Mp 179-181°. [α]$_D$ −144° (c, 1.7 in EtOH).
Racemate also known (Mp 150-151°).

3S,4-Dihydro, O-β-D-Glucopyranosyl:
Dihydrohomalicine.
$C_{21}H_{22}O_8$ M 402.400
Constit. of *H. zeylanicum*. Prisms (DMF/EtOH). Mp
260-265° dec. $[\alpha]_D$ −99.6° (c, 1 in Py).
3S,4-Dihydro, O-β-D-glucopyranosyl, tetra-Ac: Needles
(MeOH). Mp 200-202°. $[\alpha]_D$ −98.59° (c, 0.71 in
$CHCl_3$).

Govindachari, T.R. *et al, Indian J. Chem.,* 1975, **13**, 537 (*isol,
uv, ir, pmr, cd, abs config, synth*)
Narasimhan, N.S. *et al, Indian J. Chem., Sect. B,* 1983, **22**, 850
(*synth*)

1-Hydroxy-2-phenyldiazene 2-oxide   H-70208

N-*Hydroxy*-N-*nitrosobenzenamine, 9CI.* N-*Nitroso-N-
phenylhydroxylamine, 8CI.* N-*Hydroxy-N-nitrosoaniline*
[148-97-0]

$$PhN^{\oplus}(O^{\ominus}){=}NOH(Z\text{-}) \rightleftharpoons PhN(OH)NO$$

$C_6H_6N_2O_2$ M 138.126
Frequently represented as nitrosophenylhydroxylamine,
but "nitrobenzene *N*-oxime" tautomer predominates.
Cryst. (Me_2CO). V. unstable in air and organic
solvents.
▷NC4550000.
NH₄ salt: [135-20-6]. *Cupferron.*
$C_6H_9N_3O_2$ M 155.156
Reagent for separating Cu and Fe from other metals,
and quantitative analysis. Needles (H_2O). Mp 163-
164°. Dec. on heating to NH_3 and NO_x.
▷Highly toxic, exp. carcinogen. NC4725000.
4-Methylbenzenesulfonyl: [16046-28-9]. *Benzenediazo-
p-toluenesulfonate* N*-oxide, 8CI. Phenylnitrosohy-
droxylamine tosylate.*
$C_{13}H_{12}N_2O_4S$ M 292.309
Plates (EtOAc), cryst. ($CHCl_3$/hexane). Mp 136-137°
dec.
Me ether: [25370-94-9]. *1-Methoxy-2-phenyldiazene 2-
oxide, 9CI. 1-Methoxy-2-phenyldiimine 2-oxide, 8CI.*
$C_7H_8N_2O_2$ M 152.152
Prisms (pet. ether). Mp 37-38°.
Benzyl ether: [25370-92-7]. *1-Benzyloxy-2-phenyldia-
zene 2-oxide. 1-Benzyloxy-2-phenyldiimide 2-oxide,
8CI.*
$C_{13}H_{12}N_2O_2$ M 228.250
Cryst. (Et_2O/hexane). Mp 80-81°.

Bamberger, E. *et al, Ber.,* 1896, **29**, 2412 (*deriv*)
Org. Synth., Coll. Vol., **1**, 177 (*synth, deriv*)
George, M.V. *et al, Can. J. Chem.,* 1959, **37**, 679 (*deriv*)
Stevens, T.E., *J. Org. Chem.,* 1964, **29**, 311 (*deriv*)
White, E.H. *et al, Tetrahedron Lett.,* 1970, 4467 (*deriv, struct*)
Yoshimura, T. *et al, Bull. Chem. Soc. Jpn.,* 1972, **45**, 1424 (*ir,
uv, synth*)
Hickmann, E. *et al, Tetrahedron Lett.,* 1979, 2457 (*struct*)

(4-Hydroxyphenyl)ethylene   H-70209

Updated Entry replacing H-60211
4-Ethenylphenol, 9CI. p-*Vinylphenol, 8CI. 4-
Hydroxystyrene*
[2628-17-3]
C_8H_8O M 120.151
Mp 68-69°.
▷SN3800000.

Ac: [2628-16-2].
$C_{10}H_{10}O_2$ M 162.188
Fp 7.4° and 8.2° (dimorph.). Bp₃ 99-100°, Bp₀.₆ 73-
75°.
Benzoyl: [32568-59-5].
$C_{15}H_{12}O_2$ M 224.259
Needles (MeOH). Mp 75.5-76.5°.
4-O-(β-D-Xylopyranosyl(1→6)-β-D-glucopyranoside):
[90899-20-0]. **Ptelatoside A.**
$C_{19}H_{26}O_{10}$ M 414.408
Constit. of bracken fern *Pteridium aquilinum.* Cryst.
(Me_2CO aq.). Mp 183-185°. $[\alpha]_D^{22}$ −104° (c, 0.68 in
H_2O).
*4-O-(α-L-Rhamnopyranosyl(1→2)-β-D-
glucopyranoside):* [90852-99-6]. **Ptelatoside B.**
$C_{20}H_{28}O_{10}$ M 428.435
Constit. of *P. aquilinum.* Amorph. $[\alpha]_D^{23}$ −94.8° (c, 1
in H_2O).
Me ether: [637-69-4]. *1-Ethenyl-4-methoxybenzene, 9CI.
(4-Methoxyphenyl)ethylene.* p-*Methoxystyrene.*
$C_9H_{10}O$ M 134.177
Oil. Bp₁₄ 85°, Bp₃.₅ 74-75°.

Saunders, W.J. *et al, J. Am. Chem. Soc.,* 1957, **79**, 3712 (*deriv,
synth*)
Corson, B.B. *et al, J. Org. Chem.,* 1958, **23**, 544 (*synth*)
Arshady, R. *et al, J. Polym. Sci., Polym. Chem. Ed.,* 1974, **12**,
2017 (*synth*)
Hatakeyama, H. *et al, Polymer,* 1978, **19**, 593 (*synth*)
Ojika, M. *et al, Chem. Lett.,* 1984, 397.
Echavarren, A.M. *et al, J. Am. Chem. Soc.,* 1987, **109**, 5478
(*deriv, synth, ir, pmr*)
Ojika, M. *et al, Tetrahedron,* 1987, **43**, 5275 (*deriv, synth*)

2-(2-Hydroxyphenyl)-2-oxoacetic acid   H-70210

Updated Entry replacing H-40202
*2-Hydroxy-α-oxobenzeneacetic acid, 9CI. 2-Hydroxy-
benzoylformic acid.* o-*Hydroxyphenylglyoxylic acid*
[17392-16-4]

$C_8H_6O_4$ M 166.133
Yellow plates (C_6H_6/pet. ether). Mp 56-57°.
Et ester: [40785-55-5].
$C_{10}H_{10}O_4$ M 194.187
Mp 15°. Dec. on dist. *in vacuo.*
Ac:
$C_{10}H_8O_5$ M 208.170
Needles + $1H_2O$ (H_2O). Mp 101-106°.
Amide:
$C_8H_7NO_3$ M 165.148
Prisms (EtOH). Mp 170° dec.
Nitrile:
$C_8H_5NO_2$ M 147.133
Plates (AcOH). Mp 110-111°. Bp₁₄ 145-151°.

Seidel, P., *Chem. Ber.,* 1950, **83**, 20 (*synth*)
Logemann, W., *Chem. Ber.,* 1963, **96**, 2248 (*synth*)

2-(3-Hydroxyphenyl)-2-oxoacetic acid   H-70211

Updated Entry replacing H-40203
3-Hydroxy-α-oxobenzeneacetic acid, 9CI. (m-
Hydroxyphenyl)glyoxylic acid, 8CI. 3-Hydroxybenzoyl-
formic acid*

[61561-22-6]

$C_8H_6O_4$ M 166.133

Cryst. (C_6H_6). Mp 104-105°.

Me ether, nitrile: [23194-66-3].

   $C_9H_7NO_2$ M 161.160

   Mp 111-112°.

Sprenger, R.D. *et al, J. Am. Chem. Soc.*, 1950, **72**, 2874 (*synth*)

2-(4-Hydroxyphenyl)-2-oxoacetic acid   H-70212

Updated Entry replacing H-40204

4-Hydroxy-α-oxobenzeneacetic acid, 9CI. (p-*Hydroxyphenyl*)*glyoxylic acid, 8CI. 4-Hydroxybenzoylformic acid*

[15573-67-8]

$C_8H_6O_4$ M 166.133

Needles (Et_2O/C_6H_6/pet. ether). Sol. H_2O, EtOH, Et_2O, spar. sol. $CHCl_3$, C_6H_6. Mp 177-178°.

Me ether: [7099-91-4]. *4-Methoxy-α-oxobenzeneacetic acid, 9CI. Anisoylformic acid.*

   $C_9H_8O_4$ M 180.160

   Needles (C_6H_6). Mp 89°.

Me ether, oxime:

   $C_9H_9NO_4$ M 195.174

   Mp 145-146°.

Me ether, amide: [1823-76-3].

   $C_9H_9NO_3$ M 179.175

   Needles (C_6H_6). Mp 151-152°.

Me ether, nitrile: [14271-83-1]. *4-Methoxy-α-oxobenzeneacetonitrile, 9CI. 4-Methoxybenzoyl cyanide.*

   $C_9H_7NO_2$ M 161.160

   Needles (C_6H_6/pet. ether). Mp 63-64°.

Vorlander, D., *Ber.*, 1911, **44**, 2464 (*synth*)
Barnish, I.T. *et al, J. Med. Chem.*, 1981, **24**, 399 (*synth*)

5-Hydroxy-2-(phenylthio)-1,4-naphthoquin-one   H-70213

5-Hydroxy-2-(phenylthio)-1,4-naphthalenedione. 2-(Phenylthio)juglone

[71700-93-1]

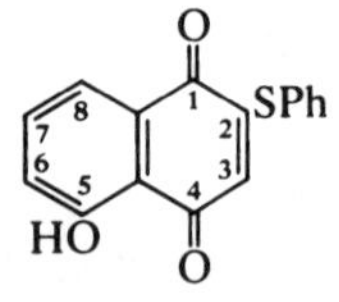

$C_{16}H_{10}O_3S$ M 282.313

Intermed. in anthracyclinone synthesis. Orange-red solid. Mp 143°.

Ac: [71700-92-0].

   $C_{18}H_{12}O_4S$ M 324.350

   Yellow-orange needles (EtOH). Mp 188°.

Me ether: [105259-49-2]. *5-Methoxy-2-(phenylthio)-1,4-naphthoquinone.*

   $C_{17}H_{12}O_3S$ M 296.340

   Mp 100-103°.

Laugraud, S. *et al, J. Org. Chem.*, 1988, **53**, 1557 (*synth, deriv, ir, pmr, cmr*)

5-Hydroxy-3-(phenylthio)-1,4-naphthoquin-one   H-70214

5-Hydroxy-3-(phenylthio)-1,4-naphthalenedione. 3-(Phenylthio)juglone

[112740-62-2]

$C_{16}H_{10}O_3S$ M 282.313

Intermed. in anthracyclinone synthesis. Orange-red solid (EtOH). Mp 155°.

Ac: [112740-63-3].

   $C_{18}H_{12}O_4S$ M 324.350

   Yellow-orange needles (EtOH). Mp 156°.

Me ether: [105245-47-4]. *5-Methoxy-3-(phenylthio)-1,4-naphthoquinone.*

   $C_{17}H_{12}O_3S$ M 296.340

   Golden-yellow needles (cyclohexane/EtOH). Mp 186-187°.

Laugraud, S. *et al, J. Org. Chem.*, 1988, **53**, 1557 (*synth, deriv, ir, pmr, cmr*)

1-(3-Hydroxyphenyl)-2-(2,4,5-trihydroxyphenyl)ethylene   H-70215

5-[2-(3-Hydroxyphenyl)ethenyl]-1,2,4-benzenetriol, 9CI. 2,3′,4,5-Tetrahydroxystilbene. 2,3′,4,5-Stilbenetetrol.

Roxburghin

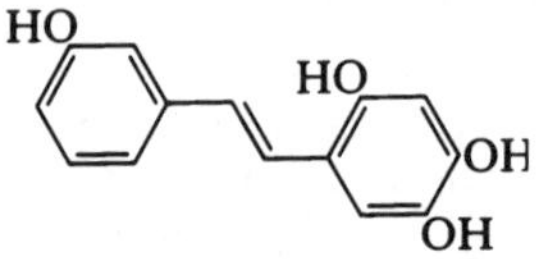

$C_{14}H_{12}O_4$ M 244.246

(*E*)-*form* [113866-80-1]

Constit. of *Cassia roxburghii*. Cryst. ($CHCl_3/Me_2CO$). Mp 224°.

Ashok, D. *et al, J. Indian Chem. Soc.*, 1987, **64**, 559.

9-Hydroxypinoresinol   H-70216

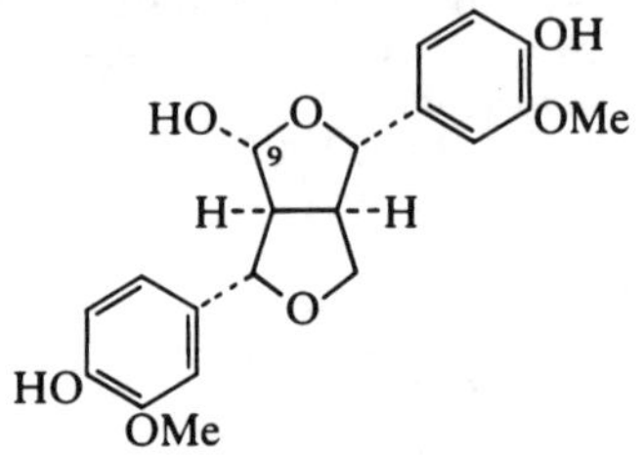

$C_{20}H_{22}O_7$ M 374.390

9α-form

Constit. of *Allamanda neriifolia*. Amorph. $[\alpha]_D^{26}$ +55.0° (c, 0.95 in MeOH).

9β-form [96738-89-5]

Constit. of *Lonicera hypoleuca*.

Khan, K.A. *et al, Phytochemistry*, 1985, **24**, 628 (*isol*)
Abe, F. *et al, Phytochemistry*, 1988, **27**, 575 (*isol, cmr*)

1-Hydroxy-2(1*H*)-pyridinethione, 9CI   H-70217

Pyrithione

[1121-30-8]

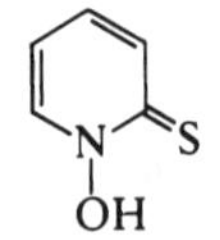

C_5H_5NOS M 127.161

Many metal complexes are used in cosmetics and pharmaceuticals. Cryst. (C_6H_6/pet. ether). Mp 68°.

▷UT9625000.

Abramovitch, R.A. *et al, J. Heterocycl. Chem.*, 1969, **6**, 989; 1975, **12**, 683 (*synth*)

Black, J.G. *et al*, *Clin. Toxicol.*, 1978, **13**, 1 (*rev, tox*)
Seymour, M.D. *et al*, *J. Chromatogr.*, 1981, **206**, 301 (*chromatog*)
Barton, D.H.R. *et al*, *Tetrahedron*, 1985, **41**, 3901; 1987, **43**, 2733 (*synth*)

ent-15-Hydroxy-8,9-seco-13-labdene-8,9-dione   H-70218

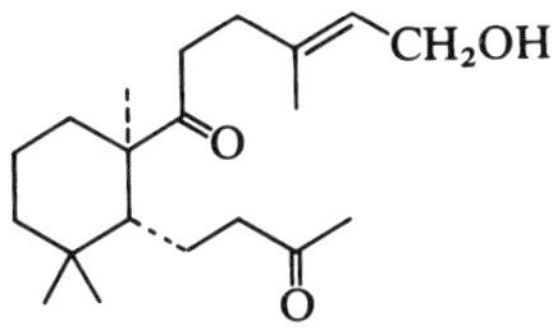

$C_{20}H_{34}O_3$ M 322.487
Constit. of *Gypothamnium pinifolium*.

Zdero, C. *et al*, *Phytochemistry*, 1988, **27**, 2953.

6-Hydroxy-3-spirostanone   H-70219

$C_{27}H_{42}O_4$ M 430.626
(5α,6α,22R,25R)-form
Constit. of *Solanum scorpioideum*. Cryst. (MeOH). Mp 208-211°. $[\alpha]_D^{21.5}$ −29.6° (c, 1.98 in CHCl$_3$).
(5α,6α,22R,25S)-form
Constit. of *S. scorpioideum*. Cryst. (MeOH). Mp 215-219°. $[\alpha]_D^{21.5}$ −52.7° (c, 4.21 in CHCl$_3$).

Usubillaga, A.N., *J. Nat. Prod.*, 1987, **50**, 636.

19-Hydroxy-13(16),14-spongiadien-3-one   H-70220

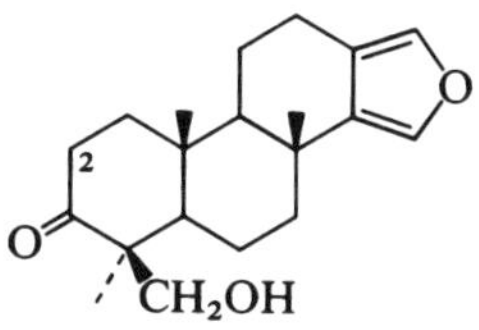

$C_{20}H_{28}O_3$ M 316.439
Constit. of *Hyatella intestinalis*. Cryst. (Me$_2$CO/hexane). Mp 143-144°. $[\alpha]_D^{21}$ +18.8° (c, 0.6 in CHCl$_3$).
2α-Hydroxy: 2α,19-Dihydroxy-13(16),14-spongiadien-3-one.
$C_{20}H_{28}O_4$ M 332.439
Constit. of *H. intestinalis*. Cryst. (MeOH aq.). Mp 180-182°. $[\alpha]_D$ −48° (c, 0.1 in CHCl$_3$).

Cambie, R.C. *et al*, *J. Nat. Prod.*, 1988, **51**, 293.

3-Hydroxytaraxastan-28,20-olide   H-70221

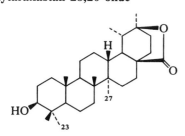

$C_{30}H_{48}O_3$ M 456.707
(3β,20β)-form
Constit. of *Pittosporum phillyraeoides*. Cryst. (MeOH). Mp 294-297°.
23,27-Dihydroxy: 3β,23,27-Trihydroxy-28,20-taraxastanolide.
$C_{30}H_{48}O_5$ M 488.706

Constit. of *P. phillyraeoides*. Cryst. (MeOH). Mp 195-196°. $[\alpha]_D$ +39° (c, 0.28 in CHCl$_3$).

Errington, S.R. *et al*, *Phytochemistry*, 1988, **27**, 543.

6-Hydroxy-2,4,8-tetradecatrienoic acid, 9ci   H-70222

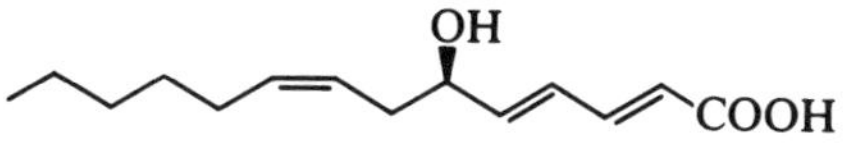

$C_{14}H_{22}O_3$ M 238.326
(2E,4E,6R,8Z)-form
Et ester: [82493-54-7].
$C_{16}H_{26}O_3$ M 266.380
Important intermed. in synth. of leukotrienes. $[\alpha]_D$ +11.4° (c, 1.31 in CDCl$_3$).
Et ester, tert-butyldiphenylsilyl ether: [82498-75-7]. Oil. $[\alpha]_D^{25}$ +40.6° (c, 2.0 in CHCl$_3$).

Guindon, Y. *et al*, *Tetrahedron Lett.*, 1982, **23**, 739 (*synth*)
Sih, C.J. *et al*, *J. Org. Chem.*, 1986, **51**, 1253 (*synth, ir, pmr*)

12-Hydroxy-3,7,11,15-tetramethyl-2,6,10,14-hexadecatetraenoic acid, 9ci   H-70223

[117007-28-0]

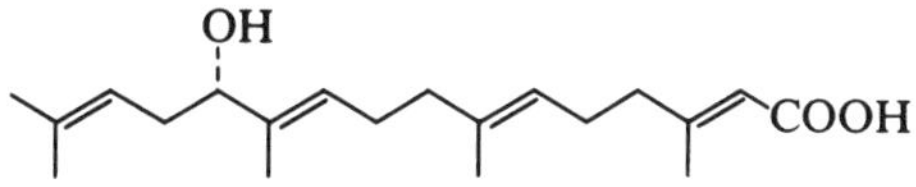

$C_{20}H_{32}O_3$ M 320.471
Constit. of *Bifurcaria bifurcata*. Oil. $[\alpha]_D$ +4.19° (c, 8.1 in CH$_2$Cl$_2$).

Semmak, L. *et al*, *Phytochemistry*, 1988, **27**, 2347.

16-Hydroxy-2,6,10,14-tetramethyl-2,6,10,14-hexadecatetraenoic acid   H-70224

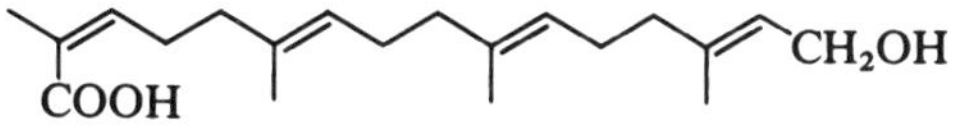

$C_{20}H_{32}O_3$ M 320.471
Ac: [80368-54-3]. **Koanoadmantic acid**.
$C_{22}H_{34}O_4$ M 362.508
Constit. of *Koanophyllon admantium*.
Ac, Me ester: Gum.

Bohlmann, F. *et al*, *Phytochemistry*, 1981, **20**, 1903 (*struct*)
Kramp, W. *et al*, *Justus Liebigs Ann. Chem.*, 1986, 226 (*synth*)

2-(5-Hydroxy-3,7,11,15-tetramethyl-2,6,10,14-hexadecatetraenyl)-6-methyl-1,2-benzenediol, 9ci   H-70225

[115788-01-7]

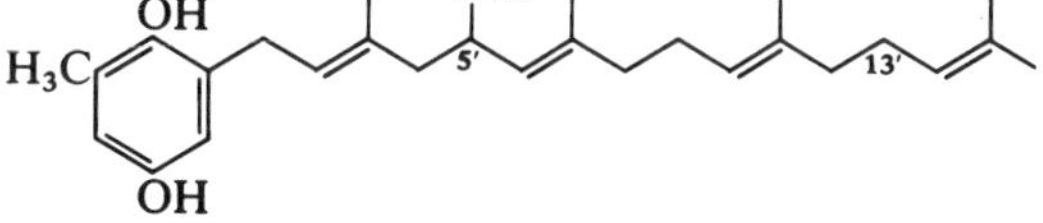

$C_{27}H_{40}O_3$ M 412.611
Constit. of the brown alga *Cystoseira spinosa* var. *squarrosa*. Yellow oil. $[\alpha]_D^{20}$ +1.0° (c, 2.3 in EtOH).
17′-Hydroxy: [115788-02-8]. 2-(5,16-Dihydroxy-3,7,11,15-tetramethyl-2,6,10,14-hexadecatetraenyl)-6-methyl-1,4-benzenediol.
$C_{27}H_{40}O_4$ M 428.611

Constit. of *C. spinosa* var. *squarrosa*.
5′-Ketone, 13′-oxo: [115788-03-9]. *16-(2,5-Dihydroxy-3-methylphenyl)-2,6,10,19-tetramethyl-2,6,10,14-hexadecatetraene-4,12-dione, 9CI. 2-(3,7,11,15-Tetra-methyl-5,13-dioxo-2,6,10,14-hexadecatetraenyl)-6-methylhydroquinone.*
$C_{27}H_{36}O_4$ M 424.579
Constit. of *C. spinosa* var. *squarrosa*. Oil.

Amico, V. *et al*, *Phytochemistry*, 1988, **27**, 1327.

ent-7β-Hydroxy-13,14,15,16-tetranor-3-   H-70226
cleroden-12-oic acid 18,19-lactone
3,5,6,6a,7,8,9,10-Octahydro-9-hydroxy-7,8-dimethyl-3-oxo-1H-naphtho[1,8a-c]furan-7-acetic acid, 9CI
[100343-67-7]

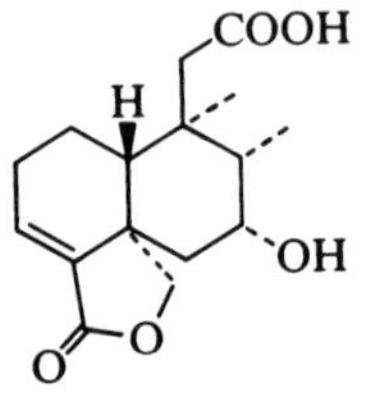

$C_{16}H_{22}O_5$ M 294.347
Compd. not named in the paper. Constit. of *Baccharis genistelloides*. Needles. Mp 253-258°. $[\alpha]_D^{20} -122.4°$ (c, 1.2 in MeOH).

Kuroyanagi, M. *et al*, *Chem. Pharm. Bull.*, 1985, **33**, 5075.

22-Hydroxy-7,24-tirucalladiene-3,23-dione   H-70227
[117073-97-9]

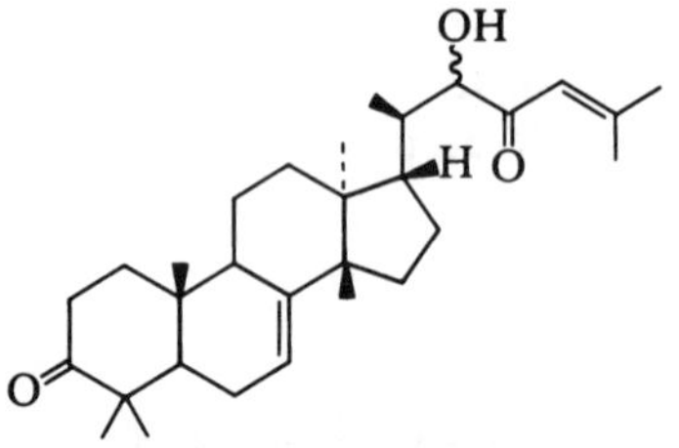

$C_{30}H_{46}O_3$ M 454.692
Constit. of *Aucoumea klaineana*. Cryst. (EtOAc/pet. ether). Mp 186-192°. $[\alpha]_D$ +37° (c, 0.3 in $CHCl_3$).

Guang-Li, L. *et al*, *Phytochemistry*, 1988, **27**, 2283.

4-Hydroxy-2-(trifluoromethyl)-   H-70228
benzaldehyde
[58914-34-4]

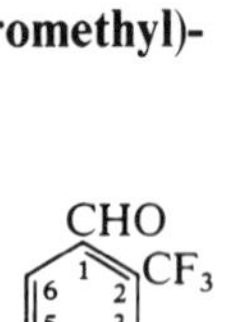

$C_8H_5F_3O_2$ M 190.122
Me ether: [106312-36-1]. *4-Methoxy-2-(trifluoromethyl)benzaldehyde. 2-(Trifluoromethyl)-anisaldehyde.*
$C_9H_7F_3O_2$ M 204.148
Needles (Et_2O/pet. ether). Mp 38-39°.

Karl, J. *et al*, *J. Med. Chem.*, 1988, **31**, 72 (*synth*)

4-Hydroxy-3-(trifluoromethyl)-   H-70229
benzaldehyde
$C_8H_5F_3O_2$ M 190.122
Me ether: [50823-87-5]. *4-Methoxy-3-(trifluoromethyl)benzaldehyde. 3-(Trifluoromethyl)-anisaldehyde.*
$C_9H_7F_3O_2$ M 204.148
Liq. Bp_6 120°.

Stogryn, E.L., *J. Med. Chem.*, 1973, **16**, 1399 (*synth*)

Hydroxyversicolorone   H-70230

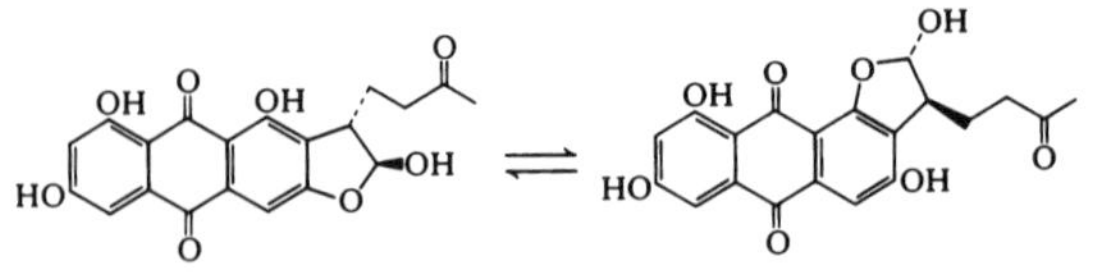

$C_{20}H_{16}O_8$ M 384.342
Metab. of *Aspergillus parasiticus*. Orange powder (Me_2CO). Mp 247-249°.

Townsend, C.A. *et al*, *J. Org. Chem.*, 1988, **53**, 2472.

6-Hydroxy-5-vinylbenzofuran   H-70231
5-Ethenyl-6-benzofuranol

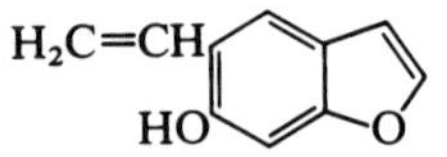

$C_{10}H_8O_2$ M 160.172
Me ether: [115333-98-7]. *6-Methoxy-5-vinylbenzofuran.*
$C_{11}H_{10}O_2$ M 174.199
Constit. of *Dorstenia barnimiana*. Oil.

Woldu, Y. *et al*, *Phytochemistry*, 1988, **27**, 1227.

3-(4-Hydroxy-2-vinylphenyl)propanoic acid   H-70232
2-Ethenyl-4-hydroxybenzenepropanoic acid, 9CI

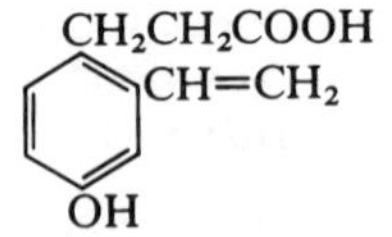

$C_{11}H_{12}O_3$ M 192.214
Me ether, Me ester: [99624-47-2]. *Methyl 3-(4-methoxy-2-vinylphenyl)propanoate.*
$C_{13}H_{16}O_3$ M 220.268
Constit. of *Gutierrezia solbrigii*. Oil.

Jakupovic, J. *et al*, *Tetrahedron*, 1985, **41**, 4537.

Hyperevoline   H-70233

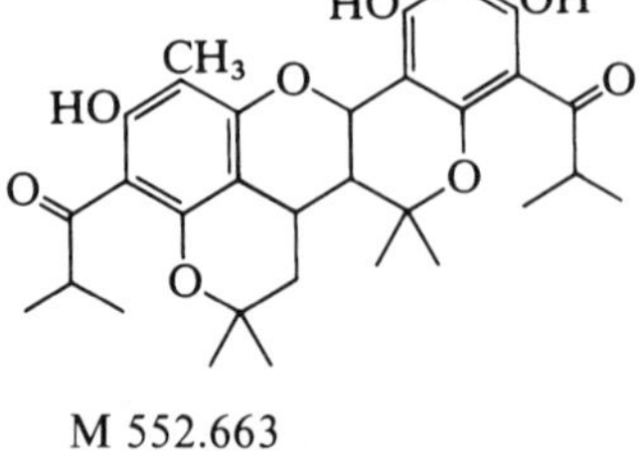

$C_{32}H_{40}O_8$ M 552.663
Constit. of *Hypericum revolutum*. Pale-yellow cryst. (Et_2O/hexane). Mp 206-210°.

Décosterd, L.A. *et al*, *Helv. Chim. Acta*, 1987, **70**, 1694.

Hypericorin   H-70234

[110901-51-4]

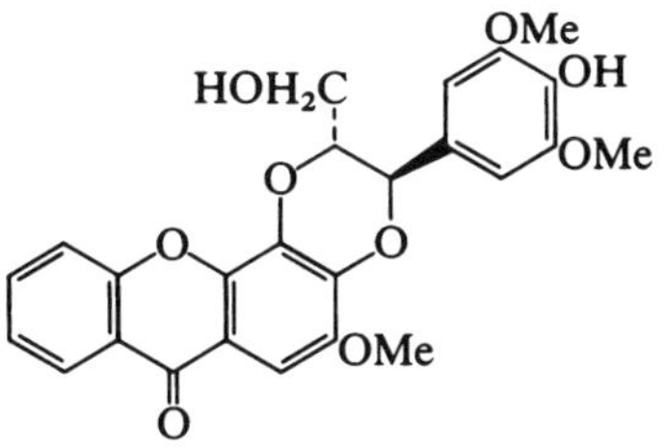

C$_{25}$H$_{22}$O$_9$ M 466.443

Constit. of *Hypericum mysorense*. Cryst. (EtOAc). Mp 202°.

Vishwakarma, R.A. *et al, Indian J. Chem., Sect. B*, 1986, **25**, 1155.

Hypocretenoic acid   H-70235

Updated Entry replacing H-20265

5β-Hydroxy-2-oxo-1(10),3,11(13)-guaiatrien-12-oic acid

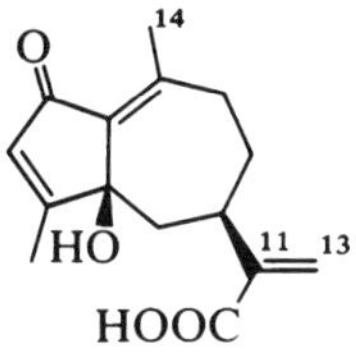

C$_{15}$H$_{18}$O$_4$ M 262.305

Me ester: [84813-80-9].
  C$_{16}$H$_{20}$O$_4$ M 276.332
  Constit. of *Hypochoeris cretensis*. Gum. [α]$_D^{24}$ +35° (c, 0.1 in CHCl$_3$).
Lactone: [84813-70-7]. **Hypocretenolide**.
  C$_{15}$H$_{16}$O$_3$ M 244.290
  Constit. of *H. cretensis*. Gum.
Lactone, 14-hydroxy: **14-Hydroxyhypocretenolide**.
  C$_{15}$H$_{16}$O$_4$ M 260.289
  Constit. of *Hedypnois cretica*. Oil.
Lactone, 14-β-D-Glucopyranosyloxy:
  C$_{21}$H$_{26}$O$_9$ M 422.431
  Constit. of *H. cretica*.
Lactone, 11α,13-Dihydro, 14-hydroxy: **14-Hydroxy-11,13-dihydrohypocretenolide**.
  C$_{15}$H$_{18}$O$_4$ M 262.305
  Constit. of *H. cretica*. Oil.
Lactone, 11α,13-Dihydro, 14-β-D-glucopyranosyloxy:
  C$_{21}$H$_{28}$O$_9$ M 424.447
  Constit. of *H. cretica*.

Bohlmann, F. *et al, Phytochemistry*, 1982, **21**, 2119 (*isol, struct*)
Harraz, F.M. *et al, Phytochemistry*, 1988, **27**, 1866 (*derivs*)

Hyptolide   H-70236

Updated Entry replacing H-03600

5,6-Dihydro-6-[3,5,6-tris(acetyloxy)-1-heptenyl]-2H-pyran-2-one, 9CI. 5,8,10,11-Tetrahydroxy-2,6-dodeca-dienoic acid δ-lactone triacetate

[908-21-4]

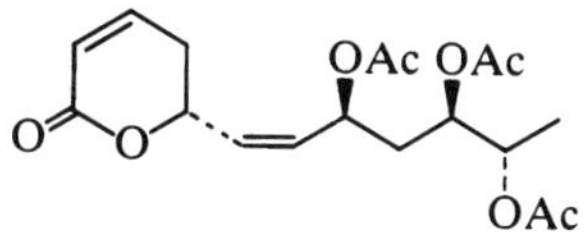

C$_{18}$H$_{24}$O$_8$ M 368.383

Bitter principle from leaves of *Hyptis pectinata*. Cryst. (Et$_2$O). Mp 88.5°. [α]$_D^{23}$ +7.43° (EtOH).

Birch, A.J. *et al, J. Chem. Soc.*, 1964, 4167 (*struct*)
Achmad, S. *et al, Acta Chem. Scand., Ser. B*, 1987, **41**, 599 (*cryst struct*)

I

Ichangensin

I-70001

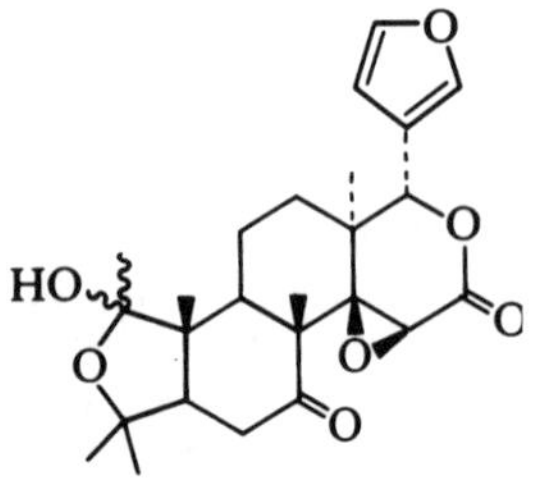

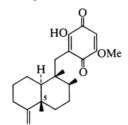

C$_{25}$H$_{32}$O$_7$ M 444.524
Constit. of seeds of *Citrus ichangensis*. Cryst.
(CH$_2$Cl$_2$/hexane). Mp 175-178°.

Bennett, R.D. *et al*, *Phytochemistry*, 1988, **27**, 1543.

Ilimaquinone

I-70002

Updated Entry replacing I-40003
[71678-03-0]

HO

O

OMe

O

H

5

Absolute
configuration

C$_{22}$H$_{30}$O$_4$ M 358.477
Constit. of *Hippiospongia metachromia*. Cryst. (hexane).
Mp 113-114°. [α]$_D^{23}$ −23.2° (c, 1.12 in CHCl$_3$).

5-Epimer: [96806-31-4]. **5-Epiilimaquinone**.
C$_{22}$H$_{30}$O$_4$ M 358.477
Constit. of a *Fenestraspongia* sp.

Luibrand, R.T. *et al*, *Tetrahedron*, 1979, **35**, 609 (*isol*)
Carté, B. *et al*, *J. Org. Chem.*, 1985, **50**, 2785 (*Epiilimaquinone*)
Capon, R.J. *et al*, *J. Org. Chem.*, 1987, **52**, 5059 (*abs config*)

Imbricatonol

I-70003

7,7′-Diacetyl-1′,8,8′-trihydroxy-6,6′-dimethyl[2,2′-bin-aphthalene]-1,4-dione, 9CI
[113540-84-4]

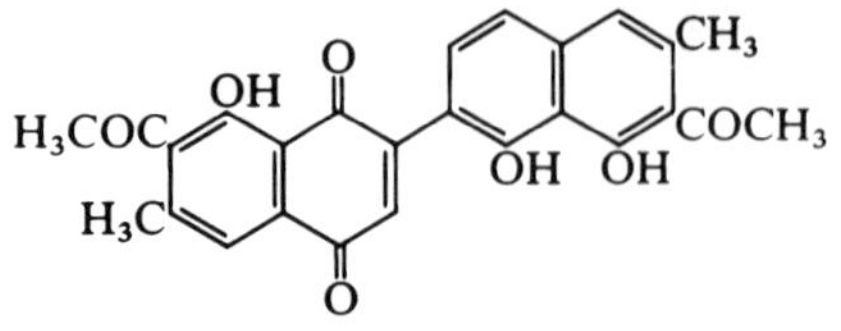

H$_3$COC

OH

O

CH$_3$

COCH$_3$

H$_3$C

OH OH

O

C$_{26}$H$_{20}$O$_7$ M 444.440
Constit. of *Stypandra imbricata* and *Dianella revoluta*.
Green solid (CHCl$_3$/Et$_2$O). Mp 240-241.5° dec.

Byrne, L.T. *et al*, *Aust. J. Chem.*, 1987, **40**, 1315.

Imidazo[5,1,2-cd]indolizine, 9CI

I-70004

Updated Entry replacing I-00108
1-Azacycl[3.2.2]azine
[209-83-6]

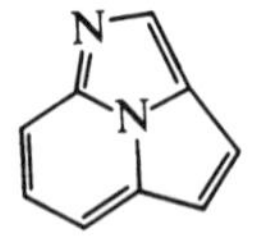

N

C$_9$H$_6$N$_2$ M 142.160
Low-melting solid.

Picrate: Mp 218-220°.
B,MeI: [58374-94-0]. Red solid. Mp 142-144°.

de Pompei, M. *et al*, *J. Org. Chem.*, 1976, **41**, 1661 (*synth*)
Tominaga, Y. *et al*, *J. Heterocycl. Chem.*, 1988, **25**, 185 (*synth, ir, uv, pmr*)

2-Imidazolidinone, 9CI, 8CI

I-70005

Updated Entry replacing I-40004
Ethyleneurea. Tetrahydro-2-oxoglyoxaline
[120-93-4]

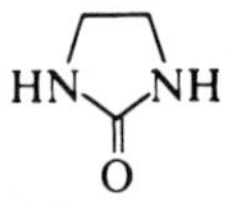

HN   NH

O

C$_3$H$_6$N$_2$O   M 86.093
Used in the manuf. of plastics, plasticizers, lacquers and
finishing agents. Many derivs. used industrially. Cryst.
+ 1/2 H$_2$O (H$_2$O), Needles (CHCl$_3$). Sol. H$_2$O, hot
EtOH, spar. sol. Et$_2$O. Mp 133.7°. Bp$_{16}$ 192°, Bp$_3$
163°. Hemihydrate is efflorescent. There is a
metastable cryst. form (transition at 80°).

▷NJ0570000.

1-Nitroso: [3844-63-1].
C$_3$H$_5$N$_3$O$_2$ M 115.091
Mp 101.5-101.8°.

▷Exp. neoplastic agent. NJ0950000.

1,3-Di-Ac: [5391-40-2].
C$_7$H$_{10}$N$_2$O$_3$ M 170.168
Mp 126.7-127.5°.

1-Me: [694-32-6].
C$_4$H$_8$N$_2$O   M 100.120
Cryst. (Ethyl acetoacetate or C$_6$H$_6$). Mp 113-114°.

1,3-Di-Me: [80-73-9]. *1,3-Dimethylimidazolidinone. Di-methylethyleneurea. DMI.*
C$_5$H$_{10}$N$_2$O   M 114.147
Solvent, replacement for the Carcinogenic
Hexamethylphosphoric triamide, H-00765 . d^{25} 1.052.
Mp 8.2°. Bp$_{11}$ 94-95°. n$_D^{20}$ 1.4720.

▷NJ0660000.

1-tert-Butyl: [92075-16-6].
C$_7$H$_{14}$N$_2$O   M 142.200
Cryst. (Et$_2$O/hexane). Mp 135-136°.

Schweitzer, C.E., *J. Org. Chem.*, 1950, **15**, 471 (*synth, bibl*)
Hofmann, K., *Imidazole and its Derivatives*, 1953, Part 1, 226,
Interscience, New York (*rev, bibl*)
Sonoda, N. *et al*, *J. Am. Chem. Soc.*, 1971, **93**, 6344 (*synth*)
Cumper, C.W.N. *et al*, *J. Chem. Soc., Perkin Trans. 2*, 1972,
2045 (*derivs*)

Butler, A.R. *et al, J. Chem. Soc., Perkin Trans. 2,* 1981, 317 (*synth*)
Quast, H. *et al, Chem. Ber.,* 1984, **117**, 2761 (*deriv*)
Dehmlow, E.V. *et al, Synth. Commun.,* 1988, **18**, 48 (*synth, props, Dimethylethylideneurea*)
Sax, N.I., *Dangerous Properties of Industrial Materials,* 5th Ed., Van Nostrand-Reinhold, 1979, 867.

1*H*-Imidazo[2,1-*b*]purin-4(5*H*)-one, 9CI   I-70006

N²,3-*Ethenoguanine*
[62962-42-9]

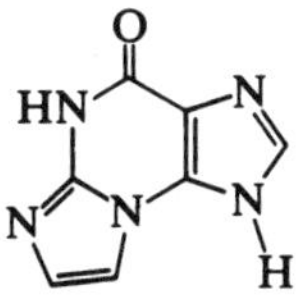

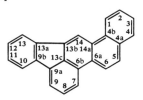

C₇H₆N₅O   M 175.149
Small needles (MeOH aq.). Mp >360°. Cryst. change at 290°.

1-β-D-Ribofuranosyl: [108060-85-1]. N²,3-*Ethenoguanosine.*
C₁₂H₁₃N₅O₅   M 307.265
Mp 320°. Hydrate softens at 170°, forms needles at 250°.

Kuśmierek, J.T. *et al, J. Org. Chem.,* 1987, **52**, 2374 (*synth, deriv, pmr, uv, ms*)

1*H*-Imidazo[4,5-*f*]quinoline, 9CI   I-70007

[233-55-6]

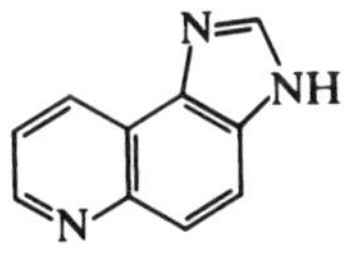

C₁₀H₇N₃   M 169.185
Mp 212-214°.

Milata, V. *et al, Collect. Czech. Chem. Commun.,* 1988, **53**, 1068 (*synth, ir, uv, pmr, cmr, bibl*)

1*H*-Imidazo[4,5-*h*]quinoline, 9CI   I-70008

[14993-03-4]

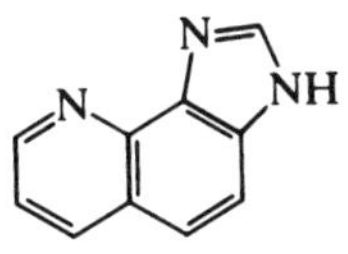

C₁₀H₇N₃   M 169.185
Mp 209-212°.

Milata, V. *et al, Collect. Czech. Chem. Commun.,* 1988, **53**, 1068 (*synth, ir, uv, pmr, cmr, bibl*)

5-(1*H*-Imidaz-2-yl)-1*H*-tetrazole, 9CI   I-70009

[63927-96-8]

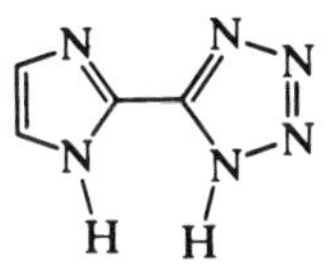

C₄H₄N₆   M 136.116
Cryst. (MeOH/Et₂O). Mp 318-320°.

N1-Benzyl: [63927-95-7]. **Bentemazole, INN**. 5-(*1-Phenylmethyl-1H-imidaz-2-yl)-1H-tetrazole, 9CI.*
C₁₁H₁₀N₆   M 226.240

Analgesic, antiuric agent. Cryst. (DMF). Mp 276-278°.

Ger. Pat., 2 550 959, (*1977*); *CA*, **87**, 102341a (*synth*)
Ger. Pat., 2 651 580, (*1978*); *CA*, **89**, 1800005z (*synth, deriv, pharmacol*)

Indeno[1,2,3-*hi*]chrysene, 9CI   I-70010

[111189-32-3]

C₂₄H₁₄   M 302.375
Long greenish-yellow needles (EtOAc/hexane). Mp 188-189°.

Cho, B.P. *et al, J. Org. Chem.,* 1987, **52**, 5668, 5679 (*synth, pmr, cmr*)

Indeno[1,2-*b*]fluorene-6,12-dione   I-70011

trans-*Fluorenacenedione*
[5695-13-6]

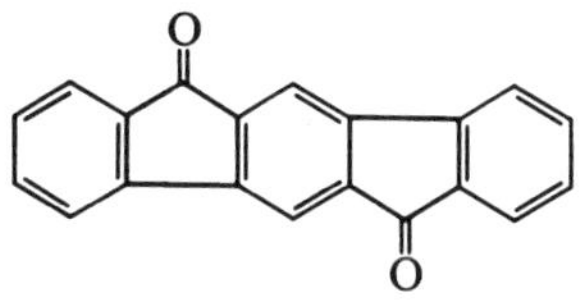

C₂₀H₁₀O₂   M 282.298
Red-violet cryst. (bromobenzene or by subl.). Mp 345-346° (341-342°).

Bis(2,4-dinitrophenylhydrazone): Orange-yellow cryst. (DMF/MeOH). Mp 277-278° dec.

Deuschel, W., *Helv. Chim. Acta,* 1951, **34**, 2403 (*synth*)
Ebel, F. *et al, Chem. Ber.,* 1956, **89**, 2794 (*synth*)

Indeno[1,2,3-*cd*]pyrene, 9CI   I-70012

Updated Entry replacing I-00240
[193-39-5]

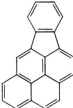

C₂₂H₁₂   M 276.337
Component of industrial smog, cigarette smoke, car exhausts, etc. Yellow cryst. (cyclohexane), bright-yellow plates (pet. ether/C₆H₆). Mp 161-163.5°. Green-yellow fluor. in soln.

▷NK9300000.

Aitkin, J.M. *et al, J. Chem. Soc.,* 1956, 3487 (*synth*)
Studt, P., *Justus Liebigs Ann. Chem.,* 1978, 528 (*synth*)
Cho, B.P. *et al, J. Org. Chem.,* 1987, **52**, 5668, 5679 (*synth, pmr, cmr*)

1*H*-Indole-3-acetic acid, 9CI   I-70013

Updated Entry replacing I-00255
3-Indolylacetic acid. Heteroauxin

[87-51-4]
$C_{10}H_9NO_2$ M 175.187
Isol. from the marine alga *Undaria pinnatifida*. Widely
distributed in higher plants. Plant growth hormone.
Cryst. (CHCl₃). Mp 164-165°.
▷NL3150000.

Me ester, picrate: [1912-33-0]. Mp 125°.
Chloride: [50720-05-3].
C_9H_6ClNO M 179.606
Flakes (Et₂O). Mp 63-65°.
Amide: [879-37-8].
$C_{10}H_{10}N_2O$ M 174.202
Mp 150-151°.
Hydrazide: [5448-47-5].
$C_{10}H_{11}N_3O$ M 189.216
Mp 138-139°.
▷NL3600000.
Nitrile: [771-51-7]. **1H-*Indole*-3-*acetonitrile***. *3-Cyano-
methyl-1H-indole.*
$C_{10}H_8N_2$ M 156.187
Naturally occurring plant growth hormone, found in
cabbage and other crucifers. Mp 36.5-37°. Bp₀.₂ 157°.
Functions by enzymic hydrol. to the acid.

▷AM0700000.
N-Me: [1912-48-7].
$C_{11}H_{11}NO_2$ M 189.213
Mp 128°. Bp₀.₁ 130-150°.

Henbest, H.B. *et al, J. Chem. Soc.*, 1953, 3796 (*isol, nitrile*)
Stowe, B.B., *Fortschr. Chem. Org. Naturst.*, 1959, **17**, 248 (*bibl*)
Johnson, H.E. *et al, J. Org. Chem.*, 1963, **28**, 1246.
Stefanescu, P., *Rev. Chim.*, 1969, **20**, 65; *CA*, **69**, 30311 (*synth*)
Abe, H. *et al, Agric. Biol. Chem.*, 1972, **36**, 2259 (*isol, uv, ms*)
Savorov, N.N. *et al, CA*, 1973, **79**, 146320 (*synth*)
Chandrasekhar, K. *et al, Acta Crystallogr., Sect. B*, 1982, **38**,
2534 (*cryst struct*)
Raucher, S. *et al, J. Am. Chem. Soc.*, 1987, **109**, 442 (*chloride*)

8*H*-Indolo[3,2,1-*de*]acridine   I-70014

[203-35-0]

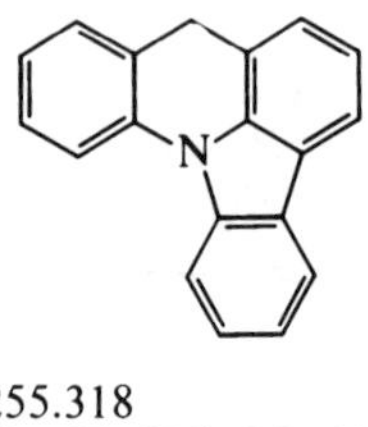

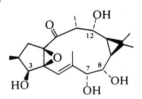

$C_{19}H_{13}N$ M 255.318
Light-yellow needles (EtOH). Mp 92.5-93.5°. Grad.
oxid. in air.

Hellwinkel, D. *et al, Chem. Ber.*, 1974, **107**, 616 (*synth, pmr*)

Indolo[2,3-*b*][1,4]benzodiazepin-12(6*H*)-one   I-70015

[116070-47-4]

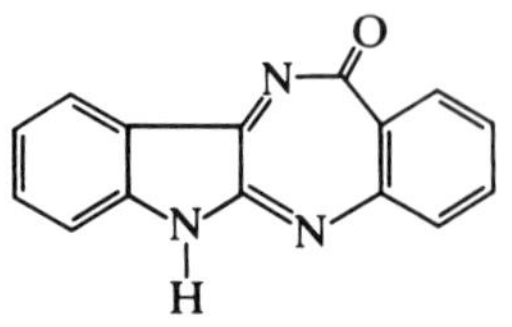

$C_{15}H_9N_3O$ M 247.256
9CI Name refers to the 5*H*-form. Authors refer to the
1*H*-form. Violet powder (toluene). Mp 238-240°.

Black, D.St.C. *et al, Aust. J. Chem.*, 1987, **40**, 1965 (*synth, uv,
ir, ms*)

10*H*-Indolo[3,2-*b*]quinoline   I-70016

Updated Entry replacing I-00286
Quindoline
[243-58-3]

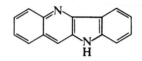

$C_{15}H_{10}N_2$ M 218.257
Alkaloid from the roots of *Cryptolepis sanguinolenta*
(Periplocaceae). Needles (EtOH) (isol. from the plant
as amorph. solid). Mp 247-248°. Blue fluor. in soln.
Subl. with part. dec.
B,HCl: Yellow cryst. H₂O, EtOH.
5-Oxide: [13220-52-5].
$C_{15}H_{10}N_2O$ M 234.257
Yellow prisms (EtOH). Mp 294-296°.
5-Me: [480-26-2]. *5-Methyl-5H-quindoline.*
Cryptolepine. Alkaloid from *C. sanguinolenta*
(Periplocaceae). Deep-purple needles. Mp 175-178°.
5-Me, monohydrate: Mp 166-169°.
5-Me; B,HCl: Mp 263-265°.
5-Me; B,HCl: Mp 263-265°.

F*Inorg. Chem.*, F. *et al, Ber.*, 1906, **39**, 3940; 1910, **43**, 3490
(*synth*)
Armit, J.N. *et al, J. Chem. Soc.*, 1922, **121**, 836 (*synth*)
Holt, S.J. *et al, J. Chem. Soc.*, 1947, 607 (*synth*)
Gellert, E. *et al, Helv. Chim. Acta*, 1951, **34**, 642 (*isol, uv,
struct, Cryptolepine*)
Schulte, K.E. *et al, Arch. Pharm.* (*Weinheim, Ger.*), 1972, **305**,
523 (*synth*)
Dwuma-Badu, D. *et al, J. Pharm. Sci.*, 1978, **67**, 433 (*isol, uv,
ir, pmr, ms*)

Ingol   I-70018

Updated Entry replacing I-10037
[51847-86-0]

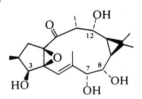

$C_{20}H_{30}O_6$ M 366.453
Hydrol. prod. from *Euphorbia ingens*. Antineoplastic
agent.

---

*Consult the Dictionary of
Antibiotics and Related Substances
for a fuller treatment of antibiotics
and related compounds.*

Tetra-Ac: [51906-02-6].
  $C_{28}H_{38}O_{10}$ M 534.602
  Constit. of *E. kamerunica*. Cryst. (isopropyl ether).
  Mp 172-175°.
3,7,8-Tri-Ac, 12-tigloyl: [81427-04-5].
  $C_{31}H_{42}O_{10}$ M 574.667
  Constit. of *E. kamerunica*.
3,7-Di-Ac, 12-tigloyl: [81980-37-2].
  $C_{29}H_{40}O_9$ M 532.630
  Constit. of *E. kamerunica*.
7-Tigloyl, O^8-Me, 3,12-di-Ac: [89984-06-5].
  $C_{30}H_{42}O_9$ M 546.656
  Constit. of latex of *E. kamerunica*.
O^7-Benzoyl, O^8-Me, 3,12-di-Ac: [89984-05-4].
  $C_{32}H_{40}O_9$ M 568.663
  Constit. of *E. kamerunica*.
7-Tigloyl, 3,8,12-tri-Ac: [92910-93-5].
  $C_{31}H_{42}O_{10}$ M 574.667
  Constit. of *E. kamerunica*.
7-Angeloyl, 3,8,12-tri-Ac: [92998-77-1].
  $C_{31}H_{42}O_{10}$ M 574.667
  Constit. of *E. kamerunica*.
7-Angeloyl, O^8-Me, 3,12-Di-Ac: [90027-10-4].
  $C_{30}H_{42}O_9$ M 546.656
  Constit. of *E. kamerunica*.

Lotter, H. *et al*, *Tetrahedron Lett.*, 1979, 77 (*cryst struct*)
Abo, K. *et al*, *Phytochemistry*, 1981, **20**, 2535 (*derivs*)
Connolly, J.D. *et al*, *Tetrahedron Lett.*, 1984, **25**, 3773 (*struct, deriv*)

myo-Inositol-1,3,4,5-tetraphosphate   I-70019

myo-Inositol-1,3,4,5-tetrakis(*dihydrogen phosphate*), *9CI*

[111929-99-8]

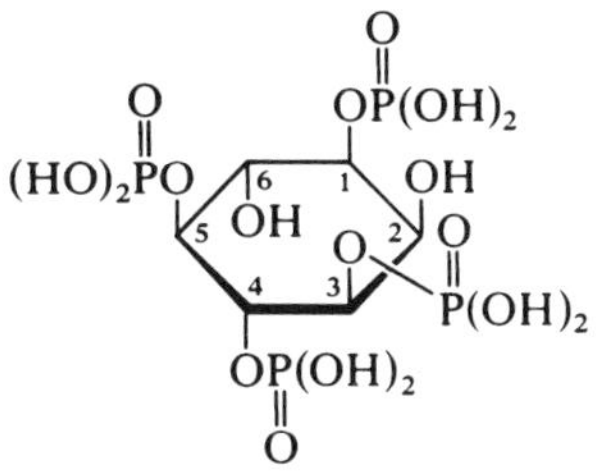

(+)-*form*

$C_6H_{16}O_{18}P_4$ M 500.077
Appears to be a cellular second messenger formed *in vivo*
from myoinositol 1,4,5-triphosphate.

(+)-*form*
1L-form
Pentakis(cyclohexylamine) salt: Mp 169.8-171.8°. $[\alpha]_D^{25}$
  +2.6° (c, 1 in H_2O).

(−)-*form*
Cyclohexylamine salt: Mp 173.7-175.7°. $[\alpha]_D^{25}$ −2.5° (c,
  1 in H_2O). Contained 4.5 mol. cyclohexylamine (cf.
  (+)-form)).

(±)-*form*
Cyclohexylammonium salt: Cryst. (Me_2CO aq.). Mp
  175-177°.

Billington, D.C. *et al*, *J. Chem. Soc., Chem. Commun.*, 1987, 1011 (*synth, pmr*)
de Solms, S.J. *et al*, *Tetrahedron Lett.*, 1987, **28**, 4503 (*synth*)
Ozaki, S. *et al*, *Tetrahedron Lett.*, 1987, **28**, 4691 (*synth*)
Baudin, G. *et al*, *Helv. Chim. Acta*, 1988, **71**, 1367 (*synth, ir, pmr, cmr, P nmr*)

myo-Inositol-1,3,4-triphosphate   I-70020
  $C_6H_{15}O_{15}P_3$ M 420.097

Secondary metab. prob. arising from the 1,4,5-triphos-
phate *in vivo*.

De Solms, S.J. *et al*, *Tetrahedron Lett.*, 1987, **28**, 4503 (*synth, bibl*)

myo-Inositol-2,4,5-triphosphate   I-70021
myo-Inositol-2,4,5-tris(*dihydrogen phosphate*), *9CI*

[91840-07-2]

$C_6H_{15}O_{15}P_3$ M 420.097
Nat. occurence not yet documented. Causes Ca^{2+} release
from tissues.

de Solms, S.J. *et al*, *Tetrahedron Lett.*, 1987, **28**, 4503 (*synth, bibl*)

Inunal   I-70022
Updated Entry replacing I-40028
[99103-26-1]

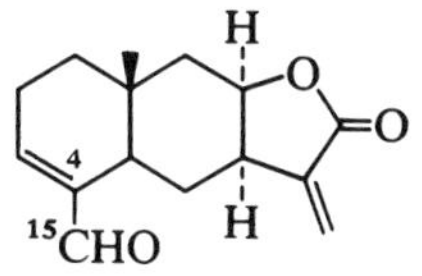

$C_{15}H_{18}O_3$ M 246.305
Constit. of *Inula racemosa*. Cryst. Mp 160°.
Δ^4-Isomer: [116965-35-6]. **Isoinunal**.
  $C_{15}H_{18}O_3$ M 246.305
  Constit. of *I. racemosa*. Plant growth regulator. Cryst.
  Mp 150°.
Δ^4-Isomer, 15-alcohol: [108560-05-0].
  $C_{15}H_{20}O_3$ M 248.321
  Constit. of *I. racemosa*. Cryst. Mp 217°.

Kaur, B. *et al*, *Phytochemistry*, 1985, **24**, 2007.
Kalsi, P.S. *et al*, *Phytochemistry*, 1988, **27**, 2079.

Inuviscolide   I-70023
Updated Entry replacing I-60037
[63109-30-8]

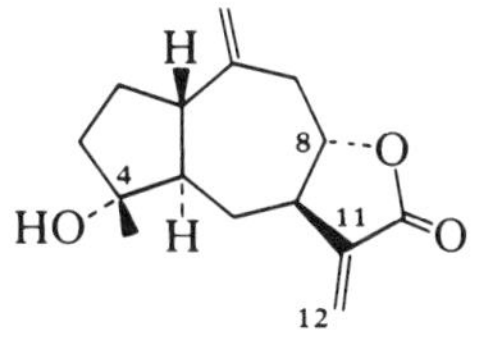

$C_{15}H_{20}O_3$ M 248.321
Constit. of *Inula viscosa*. Oil. $[\alpha]_D^{24}$ −18.6° (c, 0.35 in
$CHCl_3$).
11β,13-Dihydro: [84093-50-5]. **11β,13-
  Dihydroinuviscolide**.
  $C_{15}H_{22}O_3$ M 250.337
  Constit. of *Geigeria aspera*. Gum. $[\alpha]_D^{24}$ +28° (c, 0.13
  in $CHCl_3$).
1-Epimer: **1-epi-Inuviscolide**.
  $C_{15}H_{20}O_3$ M 248.321
  Constit. of *Dittrichia graveolens*. Oil.
8-Epimer: [108885-57-0]. **8-epi-Inuviscolide**.
  $C_{15}H_{20}O_3$ M 248.321
  Constit. of *Bedfordia arborescens*. Oil.
4,8-Diepimer: [108885-58-1]. **4,8-Bis-epi-inuviscolide**.
  $C_{15}H_{20}O_3$ M 248.321
  Constit. of *B. arborescens*. Oil.
1-Epimer, 11β,13-Dihydro: **11β,13-Dihydro-1-epi-
  inuviscolide**.
  $C_{15}H_{22}O_3$ M 250.337

Constit. of *D. graveolens*. Oil.
11α,13-Dihydro: **11α,13-Dihydroinuviscolide**.
$C_{15}H_{22}O_3$ M 250.337
Constit. of *Gnephosis brevifolia*. Gum.

Bohlmann, F. *et al*, *Chem. Ber.*, 1977, **110**, 1330 (*isol, struct*)
Bohlmann, F. *et al*, *Phytochemistry*, 1982, **21**, 1679 (*isol*)
Zdero, C. *et al*, *Phytochemistry*, 1987, **26**, 1207 (*isol, struct*)
Rustaiyan, A. *et al*, *Phytochemistry*, 1987, **26**, 2603 (*deriv*)
Jakupovic, J. *et al*, *Phytochemistry*, 1988, **27**, 3183 (*isol*)

4-Iodo-1-butyne   I-70024

[43001-25-8]

$$HC{\equiv}CCH_2CH_2I$$

C_4H_5I M 179.988
Bp_{80} 61°. n_D^{19} 1.5504.

Eglington, G. *et al*, *J. Chem. Soc.*, 1950, 3650 (*synth*)

2-Iodocycloheptanone   I-70025

[77256-26-9]

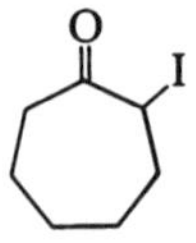

$C_7H_{11}IO$ M 238.068
(±)-*form*
Bp_2 51°.

Rubottom, G.M. *et al*, *J. Org. Chem.*, 1981, **46**, 2717 (*synth, ir, pmr, ms*)

1-Iodocyclohexene   I-70026

[17497-53-9]

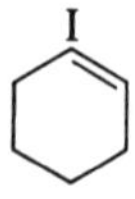

C_6H_9I M 208.042
Liq. $Bp_{0.4}$ 40°.

Pross, A. *et al*, *Aust. J. Chem.*, 1970, **23**, 989 (*synth, ir, pmr*)
Stille, J.K. *et al*, *J. Am. Chem. Soc.*, 1987, **109**, 2138 (*pmr, cmr, ir*)

2-Iodocyclooctanone   I-70027

[63641-49-6]

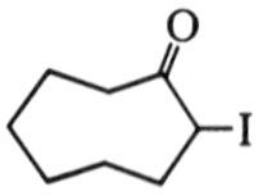

$C_8H_{13}IO$ M 252.095
(±)-*form*
$Bp_{1.5}$ 58°.

Cardillo, G. *et al*, *J. Org. Chem.*, 1977, **42**, 4268 (*synth, ir, pmr, ms*)
Rubottom, G.M. *et al*, *J. Org. Chem.*, 1979, **44**, 1731 (*synth*)
D'Auria, M. *et al*, *Synth Commun.*, 1982, **12**, 1127 (*synth, pmr*)

2-Iodo-4,5-dimethylaniline   I-70028

2-Iodo-4,5-dimethylbenzenamine, 9CI. 6-Iodo-3,4-xylidine

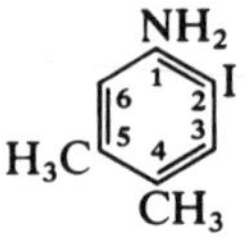

$C_8H_{10}IN$ M 247.078
Cryst. (MeOH aq.). Mp 51-52.5°.

Kajigaeshi, S. *et al*, *Bull. Chem. Soc. Jpn.*, 1988, **61**, 600 (*synth, pmr*)

4-Iodo-2,3-dimethylaniline   I-70029

4-Iodo-2,3-dimethylbenzenamine, 9CI. 4-Iodo-2,3-xylidine
$C_8H_{10}IN$ M 247.078
Cryst. (MeOH aq.). Mp 56°.

Kajigaeshi, S. *et al*, *Bull. Chem. Soc. Jpn.*, 1988, **61**, 600 (*synth, pmr*)

4-Iodo-2,5-dimethylaniline   I-70030

4-Iodo-2,5-dimethylbenzenamine, 9CI. 4-Iodo-2,5-xylidine
$C_8H_{10}IN$ M 247.078
Cryst. (MeOH aq.). Mp 69°.

Kajigaeshi, S. *et al*, *Bull. Chem. Soc. Jpn.*, 1988, **61**, 600 (*synth, pmr*)

4-Iodo-3,5-dimethylaniline   I-70031

Updated Entry replacing I-00442
4-Iodo-3,5-dimethylbenzenamine. 4-Iodo-3,5-xylidine
$C_8H_{10}IN$ M 247.078
Cryst. (MeOH aq.). Mp 112-113°.
N,N-*Di-Me:*
     $C_{10}H_{14}IN$ M 275.132
     Cryst. (EtOH). Mp 54.5-55.5°.

Tomaschewski, G., *J. Prakt. Chem.*, 1966, **33**, 168.
Kajigaeshi, S. *et al*, *Bull. Chem. Soc. Jpn.*, 1988, **61**, 600 (*synth, pmr*)

3-Iodo-4,5-dimethylpyrazole, 9CI   I-70032

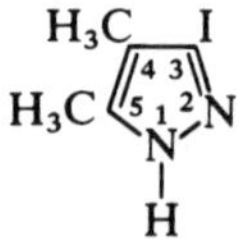

$C_5H_7IN_2$ M 222.028
Needles (MeOH aq. or by subl.). Mp 140.5-141°.

Hüttel, R. *et al*, *Justus Liebigs Ann. Chem.*, 1955, **593**, 200 (*synth*)

4-Iodo-3,5-dimethylpyrazole, 9CI   I-70033

[2033-45-6]
$C_5H_7IN_2$ M 222.028
Needles with characteristic odour (EtOH). Mp 137-138°.
Picrate: Cryst. (EtOH). Mp 191-192°.

Scott, F.L. *et al*, *J. Am. Chem. Soc.*, 1952, **74**, 4562 (*synth, deriv*)
Grandberg, I.I. *et al*, *J. Gen. Chem. USSR*, 1963, **33**, 519 (*uv*)
Orazi, O.O. *et al*, *J. Org. Chem.*, 1965, **30**, 1101 (*synth*)

Elguero, J. *et al*, *Bull. Soc. Chim. Fr.*, 1966, 3727 (*pmr*)
Faure, R. *et al*, *Can. J. Chem.*, 1988, **66**, 1141 (*cmr*)

8-Iodoguanosine, 9CI   I-70034

2-Amino-8-iodo-9-β-D-ribofuranosyl-9H-purin-6(1H)-one. 8-Iodo-9-β-D-ribofuranosylguanine
[18438-99-8]

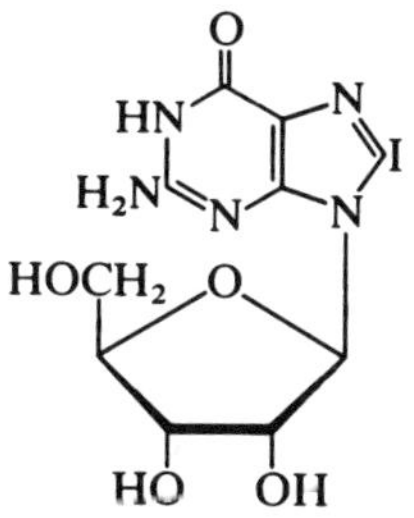

$C_{10}H_{12}IN_5O_5$ M 409.140
Cryst. (H$_2$O). Mp 203-204° dec. $[\alpha]_D$ −14.2° (c, 2 in DMSO). Darkens at 190°.

Lipkin, D. *et al*, *J. Biol. Chem*, 1963, **238**, 2249 (*synth*)
Holmes, R.E. *et al*, *J. Am. Chem. Soc.*, 1964, **86**, 1242 (*synth*)
Al-Mukhtar, J.H. *et al*, *Acta Crystallogr.*, Sect. B, 1978, **34**, 337 (*cryst struct*)

6-Iodo-1-heptene   I-70035

[13389-36-1]
$C_7H_{13}I$ M 224.084
(±)-*form*
Bp$_{25}$ 77-79°. n_D^{25} 1.5042.

Ashby, E.C. *et al*, *J. Org. Chem.*, 1984, **49**, 3545; 1987, **52**, 4554 (*synth, pmr,ms*)

7-Iodo-1-heptene   I-70036

[107175-49-5]

$$H_2C{=}CH(CH_2)_4CH_2I$$

$C_7H_{13}I$ M 224.084
Liq. Bp$_{8.0}$ 79-80°.

Bailey, W.F. *et al*, *J. Am. Chem. Soc.*, 1987, **109**, 2442 (*synth, pmr*)

1-Iodo-1-hexene   I-70037

$$H_3C(CH_2)_3CH{=}CHI$$

$C_6H_{11}I$ M 210.057
(*E*)-*form* [16644-98-7]
Liq. Bp$_5$ 85°, Bp$_3$ 50-52°.
(*Z*)-*form* [16538-47-9]
No phys. props. reported.

Zweifel, G. *et al*, *J. Am. Chem. Soc.*, 1967, **89**, 2753 (*synth*)
Dieck, H.A. *et al*, *J. Org. Chem.*, 1975, **40**, 1083 (*synth, pmr*)
Ando, T. *et al*, *Agric. Biol. Chem.*, 1982, **46**, 717 (*synth*)
Stille, J.K. *et al*, *J. Am. Chem. Soc.*, 1987, **109**, 2138 (*synth, pmr, cmr, ir*)

6-Iodo-1-hexene   I-70038

[18922-04-8]

$$H_2C{=}CH(CH_2)_3CH_2I$$

$C_6H_{11}I$ M 210.057
Bp$_{0.05}$ 45-52°. n_D^{25} 1.5106.

Garst, J.F. *et al*, *J. Am. Chem. Soc.*, 1976, **98**, 1520 (*synth*)
Albright, M.J. *et al*, *J. Organomet. Chem.*, 1977, **125**, 1 (*synth*)
Ashby, E.C. *et al*, *J. Org. Chem.*, 1984, **49**, 3545 (*synth, pmr, ms*)
Samsel, E.G. *et al*, *J. Am. Chem. Soc.*, 1986, **108**, 4790 (*synth, nmr*)

6-Iodo-5-hexen-1-ol, 9CI   I-70039

$$ICH{=}CH(CH_2)_3CH_2OH$$

$C_6H_{11}IO$ M 226.057
(*Z*)-*form* [106335-91-5]
Light-yellow oil. Bp$_{0.1}$ 90°. Darkened on standing.

Cooke, M.P. *et al*, *J. Org. Chem.*, 1987, **52**, 1381 (*synth*)
Stille, J.K. *et al*, *J. Am. Chem. Soc.*, 1987, **109**, 813 (*synth, pmr, cmr, ir*)

1-Iodo-1-hexyne   I-70040

Butyliodoacetylene
[1119-67-1]

$$H_3C(CH_2)_3C{\equiv}CI$$

C_6H_9I M 208.042
Liq. d$_{25}$ 1.56. Bp$_{30}$ 82.5°, Bp$_6$ 52.5°.

Vaughn, T.H. *et al*, *J. Am. Chem. Soc.*, 1933, **55**, 2150 (*synth*)

6-Iodo-5-hexyn-1-ol, 9CI   I-70041

[106335-92-6]

$$IC{\equiv}C(CH_2)_3CH_2OH$$

C_6H_9IO M 224.041
Bp$_{0.2}$ 90°.

Stille, J.K. *et al*, *J. Am. Chem. Soc.*, 1987, **109**, 813 (*synth, pmr, cmr, ir*)

4(5)-Iodo-1*H*-imidazole   I-70042

$C_3H_3IN_2$ M 193.975
Needles (CHCl$_3$ or H$_2$O). Mp 135-136°. An earlier ref. wrongly assigned this compd. as 2-iodoimidazole.
B,HNO$_3$: Cryst. (H$_2$O). Mp 136°.
B,(COOH)$_2$: Needles (H$_2$O). Mp 220°.
Picrate: Large yellow cryst. (EtOH), fine needles (H$_2$O). Mp 185°.

Bensusan, H.B. *et al*, *Biochemistry*, 1967, **6**, 12 (*synth*)
Naidu, N.S.R. *et al*, *J. Org. Chem.*, 1968, **33**, 1307 (*pmr, struct*)

7-Iodo-1*H*-indole   I-70043

[89976-15-8]
C_8H_6IN M 243.047
Plates (hexane). Mp 55-56°.

Somei, M. *et al*, *Chem. Pharm. Bull.*, 1987, **35**, 3146 (*synth, pmr*)

4-Iodo-1(3*H*)-isobenzofuranone   I-70044

4-Iodophthalide

[105694-45-9]

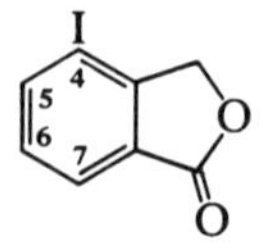

$C_8H_5IO_2$ M 260.031

Cryst. by subl. Mp 134°.

Soucy, C. *et al, J. Org. Chem.*, 1987, **52**, 129 (*synth, pmr, ir*)

7-Iodo-1(3*H*)-isobenzofuranone   I-70045

7-Iodophthalide

[105694-46-0]

$C_8H_5IO_2$ M 260.031

Cryst. by subl. Mp 137°.

Soucy, C. *et al, J. Org. Chem.*, 1987, **52**, 129 (*synth, pmr, ir*)

3-Iodo-4-mercaptobenzoic acid   I-70046

4-Carboxy-2-iodobenzenethiol

[58123-73-2]

$C_7H_5IO_2S$ M 280.080

Cryst. (EtOH). Mp 208-210°. pK_a 5.01 (50% EtOH aq., 25°).

Me ether: 3-Iodo-4-(methylthio)benzoic acid.
$C_8H_7IO_2S$ M 294.107
Cryst. (EtOH). Mp 279-281°. pK_a 5.35 (50% EtOH aq., 25°).

Benjamins, H. *et al, Can. J. Chem.*, 1974, **52**, 597 (*synth*)
Kalfus, K. *et al, Collect. Czech. Chem. Commun.*, 1975, **40**, 3009 (*synth, deriv*)

(Iodomethyl)cyclopentane   I-70047

[27935-87-1]

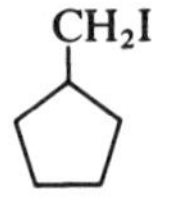

$C_6H_{11}I$ M 210.057

Samsel, E.G. *et al, J. Am. Chem. Soc.*, 1986, **108**, 4790 (*synth, nmr*)

3(5)-Iodo-4-methylpyrazole, 8CI   I-70048

[24086-18-8]

$C_4H_5IN_2$ M 208.001

Fine needles (MeOH). Mp 159°.

Hüttel, R. *et al, Justus Liebigs Ann. Chem.*, 1955, **593**, 200 (*synth*)
Bouchet, J. *et al, C.R. Hebd. Seances Acad. Sci., Ser. C*, 1969, **269**, 570 (*synth, pmr*)

4-Iodo-3(5)-methylpyrazole   I-70049

[15802-75 2]

$C_4H_5IN_2$ M 208.001

Exists as 5-Me tautomer in solid state. Polyhedra (ligroin); needles (EtOH). Mp 110.5° (185-187°). n_D^{20} 1.5670.

1-Me: [6647-97-8]. *4-Iodo-1,3-dimethylpyrazole.*
$C_5H_7IN_2$ M 222.028
Yellow oil. Bp$_{12}$ 107-108°. n_D^{20} 1.5670. Giles *et al* incorrectly identify this as 4-iodo-1,5-dimethylpyrazole (see Sinyakov *et al*).
2-Me: [6647-96-7]. *4-Iodo-1,5-dimethylpyrazole.*
$C_5H_7IN_2$ M 222.028
Needles (EtOH), cryst. (CCl$_4$). Mp 113-113.5°. Confused with 4-iodo-1,3-dimethylpyrazole. See note above.

Morgan, G.T. *et al, J. Chem. Soc.*, 1923, **123**, 1308 (*synth*)
Hüttel, R. *et al, Justus Liebigs Ann. Chem.*, 1955, **593**, 200 (*synth*)
Elguero, J. *et al, Bull. Soc. Chim. Fr.*, 1966, 3727 (*pmr*)
Giles, D. *et al, J. Chem. Soc. (C)*, 1966, 1179 (*derivs*)
Sinyakov, A.N. *et al, Bull. Acad. Sci. USSR, Ser. Chem.*, 1977, 2142 (*derivs*)
Faure, R. *et al, Can. J. Chem.*, 1988, **66**, 1141 (*cmr, tautom*)

3(5)-Iodo-5(3)-methylpyrazole   I-70050

[93233-21-7]

$C_4H_5IN_2$ M 208.001

2-Me: [85779-98-2]. *5-Iodo-1,3-dimethylpyrazole, 9CI.*
$C_5H_7IN_2$ M 222.028
Cryst. (hexane). Mp 58-59°.

Valievskii, S.F. *et al, Bull. Acad. Sci. USSR, Ser. Chem.*, 1983, 626 (*synth, pmr, deriv*)
Cabildo, P. *et al, Org. Magn. Reson.*, 1984, **22**, 603 (*cmr*)

1-Iodo-1-octene   I-70051

1-Octenyl iodide

$$H_3C(CH_2)_5CH=CHI$$

$C_8H_{15}I$ M 238.111

(*E*)-form [42599-17-7]
Bp$_5$ 120-140° (bath).

(*Z*)-form [52356-93-1]
Bp$_5$ 120-140° (bath).

Hudrlik, P.F. *et al, Tetrahedron*, 1983, **39**, 877 (*synth*)
Tamao, K. *et al, J. Org. Chem.*, 1987, **52**, 1100 (*synth, pmr, ms*)

5-Iodo-4-pentenal   I-70052

$$ICH=CHCH_2CH_2CHO$$

C_5H_7IO M 210.014

(*Z*)-form [107408-21-9]
2,4-Dinitrophenylhydrazone: Mp 116.5-117.5°.
Di-Me acetal: [107408-20-8]. *1-Iodo-5,5-dimethoxypentene, 9CI.*
$C_7H_{13}IO_2$ M 256.083
Bp$_{0.05}$ 68-70°.

Cooke, M.P. *et al, J. Org. Chem.*, 1987, **52**, 1381 (*synth, ir, pmr, cmr*)

5-Iodo-1-penten-4-yne, 9CI   I-70053

[113333-46-3]

$$H_2C=CHCH_2C\equiv CI$$

C_5H_5I M 191.999
Liq. Bp_{71} 80°.

Barluenga, J. et al, Synthesis, 1987, 661 (synth, ir, pmr, cmr)

5-Iodo-1-pentyne   I-70054

[2468-55-5]

$$HC\equiv CCH_2CH_2CH_2I$$

C_5H_7I M 194.015
Bp_{43} 84-89°. n_D^{17} 1.5351.

Eglington, G. et al, J. Chem. Soc., 1950, 3650 (synth)

1-Iodo-2-phenylacetylene   I-70055

Updated Entry replacing I-00728
Iodoethynylbenzene, 9CI. 1-Iodo-2-phenylethyne
[932-88-7]

$$PhC\equiv CI$$

C_8H_5I M 228.032
Sweet-smelling oil. d^{20} 1.66. Bp_{16} 115-117°.
▷DA3400000.

Aniline addn. cmpd.: Mp 44°.

Vaughn, T.H. et al, J. Am. Chem. Soc., 1933, 55, 2150 (synth)
Cleveland, F.F. et al, J. Am. Chem. Soc., 1940, 62, 3185 (raman)
Gavrilenko, V.V. et al, Zh. Obshch. Khim., 1967, 37, 550 (synth)
Delavarenne, S.Y. et al, Chemistry of Acetylenes, 1969, Marcel Dekker, New York.
Griebel, A. et al, J. Electron Spectrosc., 1974, 4, 185 (uv)
Cohen, M.J. et al, J. Org. Chem., 1984, 49, 515 (synth, ir, pmr)
Barluenga, J. et al, Synthesis, 1987, 661 (synth, ir, pmr, cmr)

5-Iodo-2,4(1H,3H)-pyrimidinedione, 9CI   I-70056

5-Iodouracil, 8CI
[696-07-1]

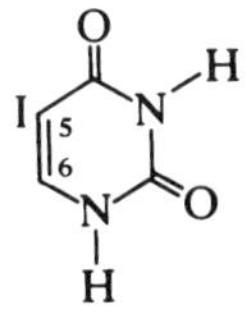

$C_4H_3IN_2O_2$ M 237.984
Needles (H_2O). Mp 270-273° dec.
▷YR0525000.

Nishiwaki, T., Tetrahedron, 1966, 22, 2401 (synth)
Sternglanz, H. et al, Acta Crystallogr., Sect. B, 1975, 31, 1393 (cryst struct)

6-Iodo-2,4-(1H,3H)-pyrimidinedione, 9CI   I-70057

6-Iodouracil
[4269-94-7]
$C_4H_3IN_2O_2$ M 237.984
Prisms (EtOH). Mp 279-280° dec.

1,3-Di-Me: [21428-19-3]. 6-Iodo-1,3-dimethyluracil.
$C_6H_7IN_2O_2$ M 266.038
Mp 174°.

Horwitz, J.P. et al, J. Org. Chem., 1961, 26, 3392 (synth, uv)
Brown, D.J. et al, Aust. J. Chem., 1973, 26, 443 (synth)
Saito, I. et al, J. Org. Chem., 1986, 51, 5148 (deriv, synth, pmr, uv, ir)

γ-Ionone   I-70058

Updated Entry replacing I-60080
4-(2,2-Dimethyl-6-methylenecyclohexyl)-3-buten-2-one, 9CI
[79-76-5]

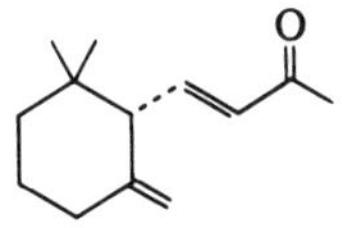

$C_{13}H_{20}O$ M 192.300
(R)-form [24190-32-7]
Oil. $[\alpha]_D^{20}$ −15.6° (EtOH).
(S)-form [64129-96-0]
Constit. of Tamarindus indica. Oil. $[\alpha]_D^{20}$ +19.5° (EtOH).

Semicarbazone: Cryst. Mp 144-144.5°.

Buchecker, R. et al, Helv. Chim. Acta, 1973, 56, 2548 (abs config)
Mukaiyama, T. et al, Chem. Lett., 1976, 1033 (synth)
Leyendecker, F. et al, Tetrahedron, 1987, 43, 85 (synth)
Oritani, T. et al, Agric. Biol. Chem., 1987, 51, 1271 (synth, abs config)

ε-Isoactinorhodin   I-70059

[109978-16-7]

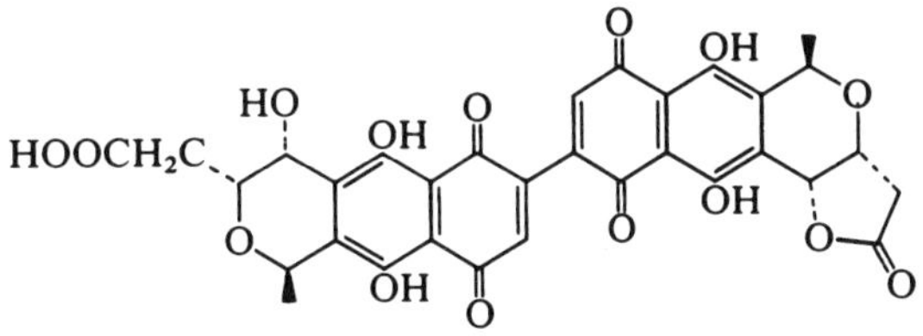

$C_{32}H_{24}O_{15}$ M 648.533
Anthraquinone antibiotic. Prod. by Streptomyces violaceoruber.

Krone, B. et al, Justus Liebigs Ann. Chem., 1987, 751 (cd, pmr, struct)

Isoafraglaucolide   I-70060

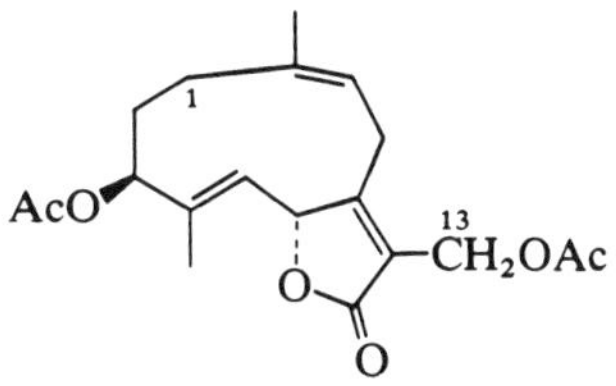

$C_{19}H_{24}O_6$ M 348.395
1α-Hydroxy: [115346-38-8]. 1α-Hydroxyisoafraglaucolide.
$C_{19}H_{24}O_7$ M 364.394
Constit. of Artemisia afra.
1β-Hydroxy, 13-de-Ac: [115338-11-9]. 13-Desacetyl-1β-hydroxyisoafraglaucolide.
$C_{17}H_{22}O_6$ M 322.357
Constit. of A. judaica. Cryst. Mp 207°. $[\alpha]_D^{24}$ +52° (c, 0.25 in $CHCl_3$).

Khafagy, S.M. et al, Phytochemistry, 1988, 27, 1125 (isol)
Jakupovic, J. et al, Phytochemistry, 1988, 27, 1129 (isol)

Isoapressin   I-70061

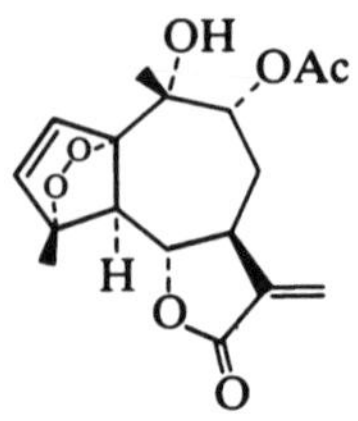

$C_{17}H_{20}O_7$ M 336.341
Constit. of *Achillea ligustica*.

Bruno, M. *et al, Phytochemistry*, 1988, **27**, 1871.

1(3*H*)-Isobenzofuranimine, 9CI   I-70062

2-(*Hydroxymethyl*)*benzonitrile*. *ψ-Phthalimidine*.
Pseudophthalimidine
[21251-69-4]

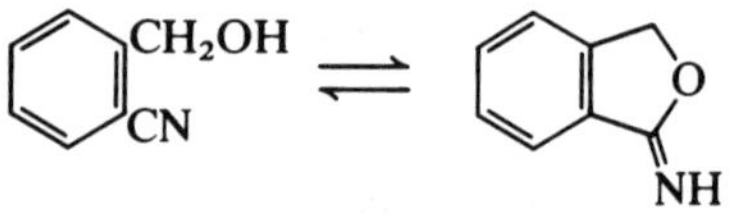

C_8H_7NO M 133.149
Tautomeric system. Oil. $Bp_{0.8}$ 118°. Unstable in moist air
which converts it into Phthalide, P-01795 .

B,HCl: Cryst., dec. on attempted recryst.

Gabriel, S. *et al, Ber.*, 1898, **31**, 2732 (*synth*)
Renson, M. *et al, Bull. Soc. Chim. Belg.*, 1964, **73**, 491 (*synth*)
Sargent, M.V., *J. Chem. Soc., Perkin Trans. 1*, 1987, 231.

1(3*H*)-Isobenzofuranthione, 9CI   I-70063

1-Thiophthalide
[4741-68-8]

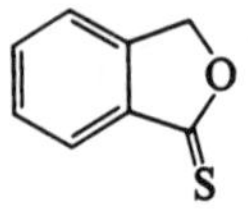

C_8H_6OS M 150.195
Green cryst. (EtOH). Mp 108-109°.

Prey, V. *et al, Monatsh. Chem.*, 1958, **89**, 505 (*synth*)
McMurray, W.J. *et al, J. Heterocycles* Chem., 1972, **9**, 1093
(*ms*)

Isochamaejasmin   I-70064

3,3′-Bi(4′,5,7-trihydroxyflavanone)
[93859-63-3]

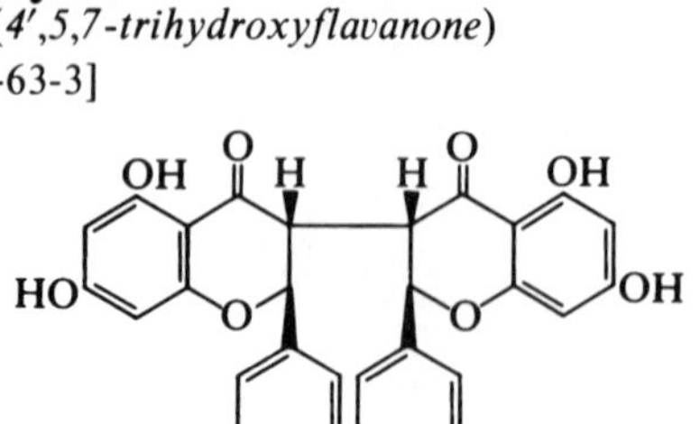

$C_{30}H_{22}O_{10}$ M 542.498
Constit. of *Thymeliacea* spp. and *Brackenridges
zanguebarica*.

Hexa-Me ether: Cryst. Mp 86°.

Niwa, M. *et al, Chem. Lett.*, 1984, 1587 (*isol, struct*)
Drewes, S. *et al, J. Chem. Soc., Perkin Trans. 1*, 1987, 2809
(*isol, pmr*)

3-Isocyanatopyridine, 9CI   I-70065

3-Pyridyl isocyanate
[15268-31-2]

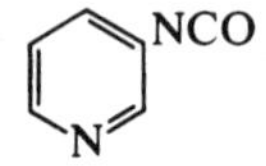

$C_6H_4N_2O$ M 120.110
Yellow solid. Mp 91-93°. Bp 183-185°.

Hyden, S. *et al, Chem. Ind.* (*London*), 1967, 1406 (*synth*)
L'abbé, G., *Synthesis*, 1987, 525 (*synth, ir*)

Isoerivanin   I-70066

Updated Entry replacing I-01029
1α,3α-Dihydroxy-4-eudesmen-12,6α-olide
[71241-93-5]

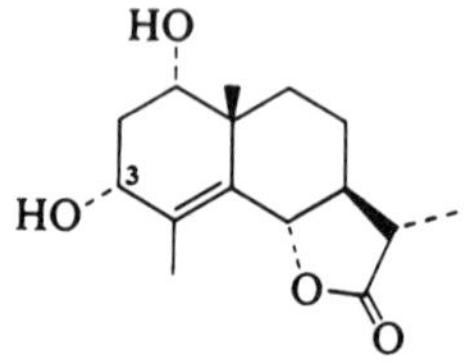

$C_{15}H_{22}O_4$ M 266.336
Constit. of *Balsamita major*. Cryst. Mp 172-174°.

3-Ketone: [71327-31-6]. **Dehydroisoerivanin**.
$C_{15}H_{20}O_5$ M 280.320
Constit. of *B. major*. Cryst. Mp 189-191°.

1-Epimer: [85847-69-4]. *1β,3α-Dihydroxy-4-eudesmen-
12,6α-olide*. **1-Epiisoerivanin**.
$C_{15}H_{22}O_4$ M 266.336
Constit. of *Artemisia judaica*. Oil. $[\alpha]_D^{24}$ +39° (c. 0.93
in $CHCl_3$).

Samek, Z. *et al, Collect. Czech. Chem. Commun.*, 1979, **44**,
1468 (*isol, struct*)
Khafagy, S.M. *et al, Phytochemistry*, 1988, **27**, 1125 (*epimer*)

Isolaureatin   I-70067

Updated Entry replacing I-01080
*7-Bromo-4-(1-bromopropyl)-2-(2-penten-4-ynyl)-3,9-
dioxabicyclo[4.2.1]nonane, 9CI. 9,13-Dibromo-6,12;7,10-
diepoxypentadec-3-en-1-yne*
[19897-64-4]

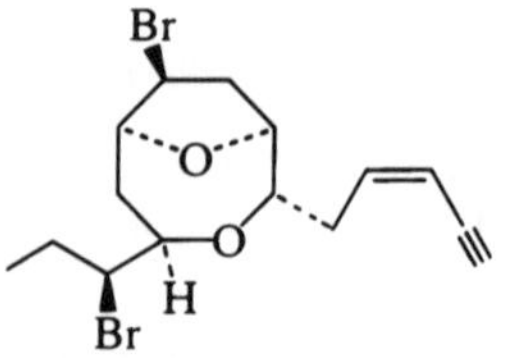

Absolute
configuration

$C_{15}H_{20}Br_2O_2$ M 392.130
Constit. of *Laurencia nipponica*. Mp 83-84°. $[\alpha]_D$ +40°
(CCl_4).

E-Isomer: [112710-66-4]. (*3E*)-*Isolaureatin*.
$C_{15}H_{20}Br_2O_2$ M 392.130
Constit. of *L. nipponica*. Oil. $[\alpha]_D^{15}$ −8.73° (c, 1.11 in
$CHCl_3$).

Irie, T. *et al, Tetrahedron Lett.*, 1968, 2735 (*isol, pmr, ms*)
Kurosawa, E. *et al, Tetrahedron Lett.*, 1973, 3857 (*cryst struct*)
Suzuki, M. *et al, Bull. Chem. Soc. Jpn.*, 1987, **60**, 3791 (*isol*)

Isolycoserone   I-70068

[116384-09-9]

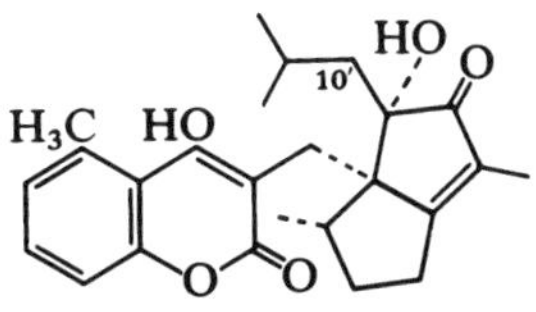

$C_{25}H_{30}O_5$ M 410.509

Constit. of *Aphyllocladus denticulatus*. Gum.

10'S-Hydroxy: [116384-12-4]. ***10'-Hydroxyisolycoserone***.

$C_{25}H_{30}O_6$ M 426.508

Constit. of *A. denticulatus*. Gum.

Zdero, C. *et al*, *Phytochemistry*, 1988, **27**, 1821.

Isomammeigin   I-70069

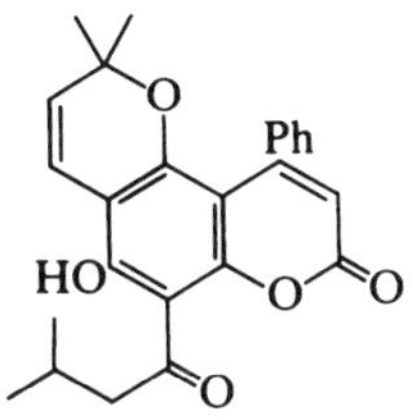

$C_{25}H_{24}O_5$ M 404.462

Constit. of *Kielmeyera pumila*. Pale-yellow prisms (Me$_2$CO). Mp 173-175°.

Nagem, T.J. *et al*, *Phytochemistry*, 1988, **27**, 2961.

Isomitomycin A   I-70070

[91717-64-5]

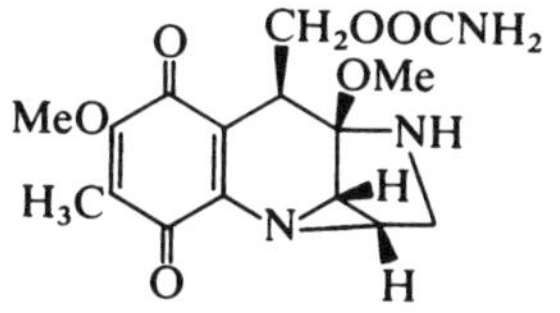

$C_{16}H_{19}N_3O_6$ M 349.343

Quinone antibiotic. Metab. of *Streptomyces caespitosus*. Prod. by intramolecular rearrangement of Mitomycin A. Yellow-orange needles (CHCl$_3$). Mp 78-80°. $[\alpha]_D^{23}$ −273° (c, 0.17 in CHCl$_3$).

Kono, M. *et al*, *J. Am. Chem. Soc.*, 1987, **109**, 7224 (*isol, struct*)
Fukuyama, T. *et al*, *J. Am. Chem. Soc.*, 1987, **109**, 7881 (*synth*)

Isomurralonginol   I-70071

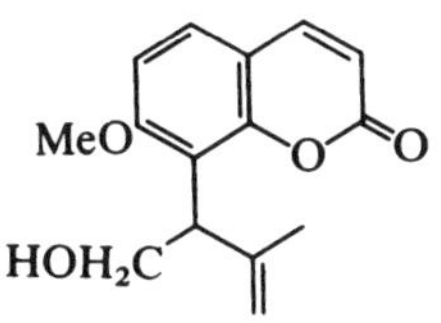

$C_{15}H_{16}O_4$ M 260.289

Ac:

    $C_{17}H_{18}O_5$ M 302.326

    Constit. of *Murraya exotica*.

3-Pyridinecarboxylate:

    $C_{21}H_{19}NO_5$ M 365.385

    Constit. of *M. paniculata*. Yellow oil. $[\alpha]_D$ +31.8° (c, 0.145 in CHCl$_3$).

Ito, C. *et al*, *Heterocycles*, 1987, **26**, 2959.

Isonimbinolide   I-70072

[108044-16-2]

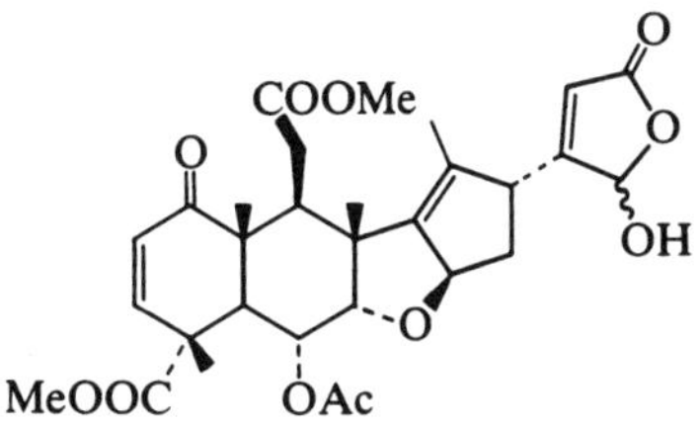

$C_{30}H_{36}O_{11}$ M 572.608

Constit. of *Azadirachta indica*. Cryst. (MeOH). Mp 172-173°.

Ara, I. *et al*, *Phytochemistry*, 1988, **27**, 1801.

Isonuezhenide   I-70073

[112693-22-8]

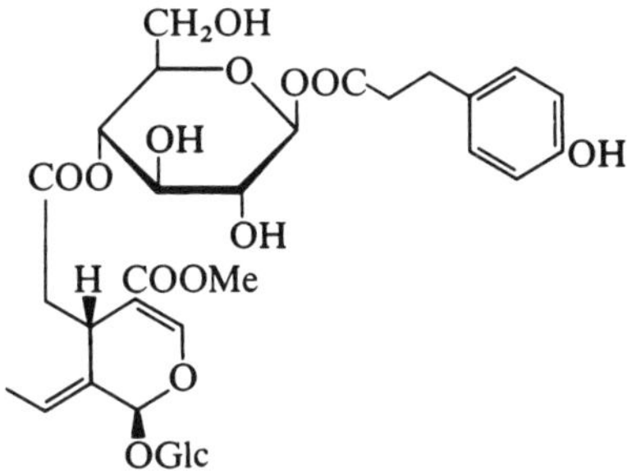

$C_{32}H_{44}O_{18}$ M 716.689

Constit. of *Ligustrum japonicum*. Powder. $[\alpha]_D$ −85° (c, 1.15 in EtOH).

Fukuyama, Y. *et al*, *Planta Med.*, 1987, **53**, 427.

Isoobtusadiene   I-70074

2-Bromo-1,1,9-trimethyl-5-methylenespiro[5.5]undeca-7,9-dien-3-ol, 9CI

[113262-62-7]

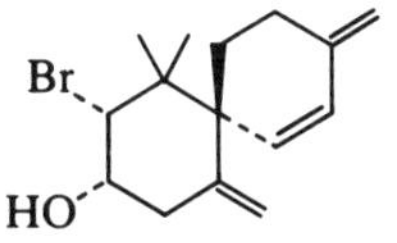

$C_{15}H_{21}BrO$ M 297.234

Constit. of the red alga *Laurencia obtusa*. Oil. $[\alpha]_D^{25}$ −11.7° (c, 0.7 in CHCl$_3$).

Gerwick, W.H. *et al*, *J. Nat. Prod.*, 1987, **50**, 1131.

Isoperezone   I-70075

5-(1,5-Dimethyl-4-hexenyl)-3-hydroxy-2-methyl-2,5-cyclohexadiene-1,4-dione, 9CI

[13120-66-6]

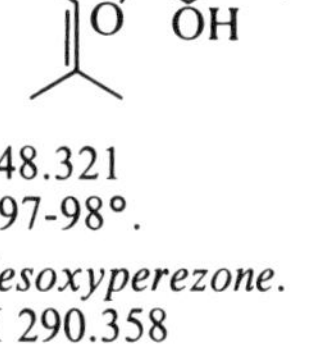

$C_{15}H_{20}O_3$ M 248.321

Yellow plates. Mp 97-98°.

Ac: 2-Acetoxy-2-desoxyperezone.

    $C_{17}H_{22}O_4$ M 290.358

    Constit. of *Coreopsis longipes*. Yellow oil. Descr. as the (*R*)-form in *CA*, but no evidence for this in the lit.

Bohlmann, F. *et al*, *Phytochemistry*, 1983, **22**, 2858 (*isol*)
Garcia, G.E. *et al*, *J. Nat. Prod.*, 1987, **50**, 1055 (*synth*)

7,15-Isopimaradiene-3,18-diol   I-70076

$C_{20}H_{32}O_2$ M 304.472

3β-form [117232-46-9]

Constit. of *Juniperus thurifera*. Cryst. (C_6H_6). Mp 121°.
$[\alpha]_D^{23}$ −31.5° (CHCl₃).

San Feliciano, A. *et al*, *Phytochemistry*, 1988, **27**, 2241.

7,15-Isopimaradiene-3,19-diol   I-70077

Updated Entry replacing V-00254

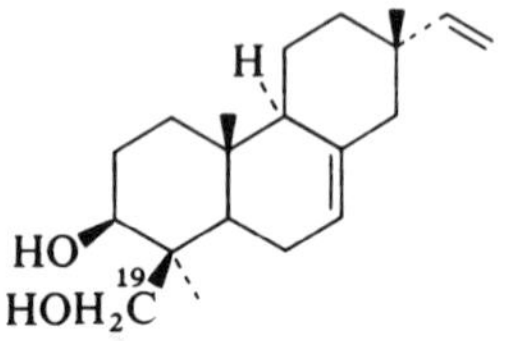

$C_{20}H_{32}O_2$ M 304.472

3β-form [22343-47-1]
Virescenol B

Cryst. (Et_2O/pet. ether). Mp 146-147°. $[\alpha]_D$ −25° (c,
0.74 in CHCl₃).

19-O-β-D-Altropyranoside: [28251-74-3]. *Virescenoside*
B. Metab. of *Oospora virescens*. Amorph. Mp 110°.
$[\alpha]_D$ −32.3° (c, 1.05 in MeOH).

19-O-β-D-Altropyranosiduronic acid: [34212-90-3].
Virescenoside G.
$C_{26}H_{40}O_8$ M 480.597
Metab. of *O. virescens*. Cryst. (EtOAc/MeOH). Mp
192-194°. $[\alpha]_D$ −85° (c, 1.11 in MeOH).

Cagnoli-Bellavita, N. *et al*, *Gazz. Chim. Ital.*, 1969, **99**, 1354;
   1977, **107**, 51 (*isol, struct*)
Polonsky, J. *et al*, *Bull. Soc. Chim. Fr.*, 1970, 1912 (*isol, struct*)
Ceccherelli, P. *et al*, *Tetrahedron*, 1973, **29**, 449 (*deriv*)
Polonsky, J. *et al*, *Tetrahedron Lett.*, 1975, 481 (*cmr, biosynth*)
Ceccherelli, P. *et al*, *J. Org. Chem.*, 1977, **42**, 3438 (*synth*)
Bellavita, N. *et al*, *J. Am. Chem. Soc.*, 1980, **102**, 17 (*nmr*)

7,15-Isopimaradiene-18,19-diol   I-70078

[117232-47-0]

$C_{20}H_{32}O_2$ M 304.472

Constit. of *Juniperus thurifera*. Cryst. (C_6H_6). Mp 108°.
$[\alpha]_D^{23}$ −10.5° (CHCl₃).

San Feliciano, A. *et al*, *Phytochemistry*, 1988, **27**, 2241.

8,15-Isopimaradiene-7,18-diol   I-70079

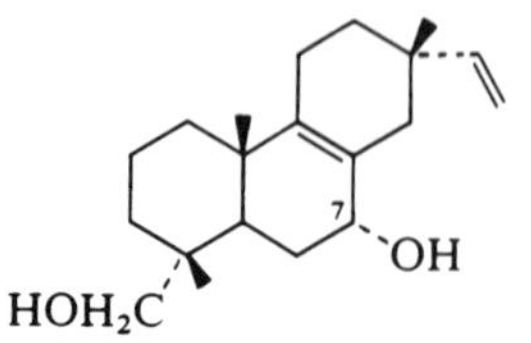

$C_{20}H_{32}O_2$ M 304.472

7α-form [114191-58-1]
Constit. of *Nepeta tuberosa*. Oil. $[\alpha]_D$ +54.5° (c, 2.3 in
CHCl₃).
7β-form [114191-60-5]
Constit. of *N. tuberosa*. Cryst. (Et_2O). Mp 95-97°. $[\alpha]_D$
+29.5° (c, 0.9 in CHCl₃).

Urones, J.G. *et al*, *Phytochemistry*, 1988, **27**, 523.

8(14),15-Isopimaradiene-3,18-diol   I-70080

8(14),15 Sandaracopimaradiene-3,18-diol

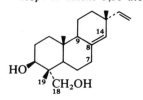

$C_{20}H_{32}O_2$ M 304.472

3β-form [59219-64-6]

Constit. of *Juniperus thurifera*. Cryst. (C_6H_6). Mp 151°.
$[\alpha]_D$ −17° (CHCl₃).

San Feliciano, A. *et al*, *Phytochemistry*, 1988, **27**, 2241.

7,15-Isopimaradiene-2,3,19-triol   I-70081

Updated Entry replacing V-00253

$C_{20}H_{32}O_3$ M 320.471

(2α,3β)-form [22343-46-0]
Virescenol A

Cryst. Mp 149-150°. $[\alpha]_D$ −44° (CHCl₃).

19-β-D-Altropyranoside: [28251-73-2]. *Virescenoside* A.
$C_{26}H_{42}O_8$ M 482.613
Metab. of *Acremonium luzulae* (*Oospora virescens*).
Amorph. Mp 130°. $[\alpha]_D$ −42.7° (c 1.03 in MeOH).

19-β-D-Altropyranosiduronic acid: [34212-88-9]. *Vires-
cenoside* F. Metab. of *A. luzulae*. Cryst.
(EtOAc/MeOH). Mp 188-190°. $[\alpha]_D$ −82° (c, 1.13 in
MeOH).

19-O-β-D-theo-Hex-4-enodialdo-1,5-pyranoside:
[34212-94-7]. *Virescenoside* E.
$C_{26}H_{38}O_7$ M 462.582
Isol. from *A. luzulae*. $[\alpha]_D$ −113° (c, 0.97 in MeOH).

19-O-β-D-threo-4-Deoxyhex-4-enopyranoside: [63758-
60-1]. *Virescenoside* L.
$C_{26}H_{40}O_7$ M 464.598
Metab. of *A. luzulae*. Microcryst. (EtOAc). $[\alpha]_D$
−154° (c, 0.57 in CHCl₃).

Cagnoli-Bellavita, N. *et al*, *Gazz. Chim. Ital.*, 1969, **99**, 1354
   (*isol*)
Polonsky, J. *et al*, *Bull. Soc. Chim. Fr.*, 1970, 1912 (*isol, struct*)
Ceccherelli, P. *et al*, *Tetrahedron*, 1973, **29**, 449 (*deriv*)
Polonsky, J. *et al*, *Tetrahedron Lett.*, 1975, 481 (*cmr, biosynth*)
Bellavita, N. *et al*, *J. Am. Chem. Soc.*, 1980, **102**, 17 (*nmr*)

8,15-Isopimaradiene-3,7,19-triol   I-70082

$C_{20}H_{32}O_3$ M 320.471

(3β,7α)-form [117255-00-2]
Constit. of *Juniperus thurifera*.

San Feliciano, A. *et al*, *Phytochemistry*, 1988, **27**, 2241.

> *Consult the Dictionary of
> Organophosphorus Compounds for
> organic compounds containing
> phosphorus.*

Isopristimerin III   **I-70083**
Updated Entry replacing I-30099
[87161-46-4]

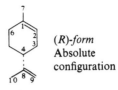

$C_{30}H_{40}O_4$ M 464.644
Rearrangement prod. of pristimerin. Needles (Et₂O/pet.
ether). Mp 230-235°.
23-Oxo: [87174-13-8]. **23-Oxoisopristimerin III**.
$C_{30}H_{38}O_5$ M 478.627
Constit. of *Kokoona zeylanica*. Cryst. Mp 157-160°.

Johnson, A.W. *et al, J. Chem. Soc.*, 1963, 2884 (*isol*)
Kamal, G.M. *et al, Tetrahedron Lett.*, 1983, **24**, 2799 (*isol*)
Gamlath, C.B. *et al, J. Chem. Soc., Perkin Trans. 1*, 1987, 2849
   (*isol, struct*)

24-Isopropenyl-7-cholestene-3,6-diol   **I-70084**
$C_{30}H_{50}O_2$ M 442.724
(3β,6α)-form
Constit. of *Dysidea herbacea*. Cryst. Mp 198-199°.
Rambabu, M. *et al, Indian J. Chem., Sect. B*, 1987, **26**, 1156.

24-Isopropenyl-7-cholesten-3-ol   **I-70085**
$C_{30}H_{50}O$ M 426.724
3β-form
Constit. of *Dysidea herbacea*. Cryst. Mp 149°.
Rambabu, M. *et al, Indian J. Chem., Sect. B*, 1987, **26**, 1156.

4-Isopropenyl-1-methyl-1,2-cyclohexane-diol   **I-70086**
1-Methyl-4-(1-methylethenyl)-1,2-cyclohexanediol, 9CI.
p-Menth-8-ene-1,2-diol

$C_{10}H_{18}O_2$ M 170.251
(1R*,2S*)-form
   Needles (MeOH). Mp 74°. [α]$_D^{25}$ +29° (c, 2 in MeOH).
Garg, S.N. *et al, Phytochemistry*, 1988, **27**, 936.

4-Isopropenyl-1-methylcyclohexene   **I-70087**
Updated Entry replacing I-01176
1,8-p-Menthadiene. **Limonene**. *Carvene. Citrene. He-*
speridene. Diisoprene. Terpilene. Kautschine. Cinene.
Cynene. Cajeputene. Isoterebentine

$C_{10}H_{16}$ M 136.236
The most important and widespread terpene known.
   Extensively used in the perfumery and flavour
   industries and in manuf. of polymers and adhesives.
(R)-form [5989-27-5]
   Major constit. of oil of orange rind, dill oil, oil of cumin,
      neroli, bergamot, caraway and lemon. Obt. comly.
      from orange oil on a large scale (ca. 50000 T/a).
      Important chiral starting material for org. synth.
      Shows insecticidal props. Oil. Bp₂₀ 71°. [α]$_D^{20}$ +126.8°.
▷GW6360000.
(S)-form [5989-54-8]
   Constit. of pine needle oil. Oil. Bp 177.6-177.8°. [α]$_D$
      −122.6°.
▷OS8350000.
(±)-form [7705-14-8]
   Dipentene
   Constit. of *Oleum cinae*, pine needle oil and many
      essential oils. Obt. comly. as by prod. in hydration of
      turpentine. Oil. Bp 178°.
▷Irritant, flammable

Pawson, B.A. *et al, J. Chem. Soc., Chem. Commun.*, 1968, 1057
   (*abs config*)
Verghese, J., *Perfum. Essent. Oil. Rec.*, 1968, **59**, 439 (*rev*)
Nijaura, N. *et al, Chem. Lett.*, 1974, 1411 (*synth*)
Org. Synth., 1977, **56**, 101 (*synth*)
Thomas, A.F. *et al, Nat. Prod. Rep.*, 1989, **6**, 291 (*rev*)
Sax, N.I., *Dangerous Properties of Industrial Materials*, 5th
   Ed., Van Nostrand-Reinhold, 1979, 621, 772.
Hazards in the Chemical Laboratory, (Bretherick, L., Ed.), 3rd
   Ed., Royal Society of Chemistry, London, 1981, 313.

6-Isopropenyl-3-methyl-2-cyclohexen-1-one   **I-70088**
3-Methyl-6-(1-methylethyl)-2-cyclohexen-1-one, 9CI.
p-Mentha-1,8-dien-3-one, 8CI. **Isopiperitenone**
[529-01-1]

$C_{10}H_{14}O$ M 150.220
(S)-form [16750-82-6]
   Alarm pheromone of the acarid mite *Tyrophagus similis*.
      Oil. d₄²⁰ 0.961. [α]$_D^{20}$ +39.1° (c, 0.169 in hexane), [α]$_D$
      +83° (synthetic). n_D^{20} 1.5037.
2,4-Dinitrophenylhydrazone: Mp 162-163°.

Schenk, G.O. *et al, Justus Liebigs Ann. Chem.*, 1965, **687**, 26
   (*synth*)
Kuwahara, Y. *et al, Agric. Biol. Chem.*, 1987, **51**, 3441 (*isol*)

6-Isopropenyl-3-methyl-9-decen-1-ol   I-70089
3-Methyl-6-(1-methylethenyl)-9-decen-1-ol

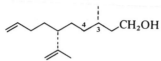

C$_{14}$H$_{26}$O   M 210.359

(3S,6R)-form

Ac: [67601-06-3].
C$_{16}$H$_{28}$O$_2$ M 252.396
Component of the sex pheromone of the California red scale insect.
3,4-Didehydro, Ac: 6-Isopropenyl-3-methyl-3,9-decadien-1-ol acetate.
C$_{16}$H$_{26}$O$_2$ M 250.380
Component of the sex pheromone of the California red scale insect.

Anderson, R.J. *et al, J. Org. Chem.,* 1980, **45**, 2229 (*synth, pmr, cmr*)
Becker, D. *et al, Tetrahedron,* 1988, **44**, 4541 (*synth, pmr, bibl*)

2-Isopropylcycloheptanone   I-70090
2-(1-Methylethyl)cycloheptanone, 9CI
[20036-72-0]

C$_{10}$H$_{18}$O   M 154.252

(±)-form
Bp$_{18}$ 93-95°.
Semicarbazone: Plates (EtOH aq.). Mp 175-176°.
Cook, J.W. *et al, J. Chem. Soc.,* 1951, 695 (*synth*)

4-Isopropylcycloheptanone   I-70091
4-(1-Methylethyl)cycloheptanone, 9CI.
Hexahydronezukone
[13656-84-3]
C$_{10}$H$_{18}$O   M 154.252

(±)-form
Bp$_{18}$ 115-117°, Bp$_1$ 62°.
Semicarbazone: Plates (EtOH aq.) (2 isomers). Mp 135-137°, Mp 180-182°.
2,4-Dinitrophenylhydrazone: Yellow plates + 4H$_2$O (EtOH). Mp 93-95°.

Cook, J.W. *et al, J. Chem. Soc.,* 1951, 695 (*synth, ir*)
Hirose, Y. *et al, Tetrahedron Lett.,* 1966, 5875 (*synth, ir, pmr*)

Isopropylcyclopentane   I-70092
(1-Methylethyl)cyclopentane, 9CI. 2-Cyclopentylpropane
[3875-51-2]

C$_8$H$_{16}$ M 112.214
Liq. d$_4^{20}$ 0.78. Bp 126.4° (118-120°).
Crane, G. *et al, J. Am. Chem. Soc.,* 1945, **67**, 1237 (*synth*)
Bailey, W.F. *et al, J. Am. Chem. Soc.,* 1987, **109**, 2442 (*synth, pmr*)

4-Isopropyl-1-methyl-3-cyclohexenene-1,2-diol   I-70093
1-Methyl-4-(1-methylethyl)-3-cyclohexane-1,2-diol,
9CI. p-Menth-3-ene-1,2-diol

C$_{10}$H$_{18}$O$_2$ M 170.251

(1R*,2S*)-form [98857-38-6]
Constit. of oil from *Ferula jaeschkeana*. Viscous mass.
[α]$_D^{25}$ −2.5° (c, 2 in CHCl$_3$).

Garg, S.N. *et al, Phytochemistry,* 1988, **27**, 936.

4-Isopropyl-1-methyl-4-cyclohexene-1,2-diol   I-70094
1-Methyl-4-(1-methylethyl)-4-cyclohexene-1,2-diol,
9CI. p-Menth-4-ene-1,2-diol

C$_{10}$H$_{18}$O$_2$ M 170.251

(1R*,2R*)-form [66965-46-6]
Constit. of oil from *Ferula jaeschkeana*. Viscous mass.
[α]$_D^{25}$ +4.8° (c, 4 in MeOH).

Garg, S.N. *et al, Phytochemistry,* 1988, **27**, 936.

3-Isopropyl-2-naphthol   I-70095
3-(1-Methylethyl)-2-naphthalenol, 9CI
[60683-46-7]
C$_{13}$H$_{14}$O   M 186.253
Cryst. (hexane). Mp 81-82°.

Matsumoto, T. *et al, Bull. Chem. Soc. Jpn.,* 1988, **61**, 911 (*synth, pmr*)

4-Isopropyl-1-naphthol   I-70096
Updated Entry replacing I-01474
4-(Methylethyl)-1-naphthalenol, 9CI
C$_{13}$H$_{14}$O   M 186.253
Needles. Mp 72° (65-66°). Bp 304-309°, Bp$_5$ 190°.

Meyer, H. *et al, Monatsh. Chem.,* 1929, **53-54**, 743.
Matsumoto, T. *et al, Bull. Chem. Soc. Jpn.,* 1988, **61**, 911 (*synth, pmr*)

3-Isopropyloxindole   I-70097
1,3-Dihydro-3-(1-methylethyl)-2H-indol-2-one, 9CI

C$_{11}$H$_{13}$NO   M 175.230

(±)-form [108665-93-6]
Cryst. (Et₂O/hexane). Mp 107-108°.

Schwarz, H., *Monatsh. Chem.*, 1903, **24**, 568 (*synth*)
Wenkert, E. *et al*, *J. Org. Chem.*, 1987, **52**, 3404.

Isopseudocyphellarin *A*   I-70098
[113706-24-4]

$C_{21}H_{22}O_8$ M 402.400
Metab. of *Pseudocyphellaria pickeringii*. Prisms. Mp 201° dec.
2'-Me ether: **2'-*O*-*Methylisopseudocyphellarin* A**.
$C_{22}H_{24}O_8$ M 416.427
Metab. of *P. pickeringii*. Plates. Mp 171°.

Elix, J.A. *et al*, *Aust. J. Chem.*, 1987, **40**, 2023.

1,3(2*H*,4*H*)-Isoquinolinedione, 9CI   I-70099
Homophthalimide
[4456-77-3]

$C_9H_7NO_2$ M 161.160
Cryst. (AcOH). Mp 230-233°.
Dioxime: [41536-78-1].
$C_9H_9N_3O_2$ M 191.189
Cryst. (MeOH aq.). Mp 223-225° dec.

Harriman, B.R. *et al*, *J. Am. Chem. Soc.*, 1945, **67**, 1481 (*synth*)
Bailey, A.S. *et al*, *J. Chem. Soc.*, 1956, 2477 (*synth*)
Barot, N.R. *et al*, *J. Chem. Soc.*, *Perkin Trans. 1*, 1973, 607 (*synth, deriv, ir, pmr, uv*)

Isosarcophytoxide   I-70100
Updated Entry replacing I-01554
2,16:11,12-Diepoxy-1(15),8E,12E-cembratriene
[70645-61-3]

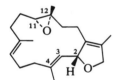

(2*R*,11*R*,12*R*)-form

$C_{20}H_{30}O_2$ M 302.456
(2*R*,11*R*,12*R*)-form
Constit. of *Sarcophyton* spp. Cryst. (pet. ether). Mp 67-69°. [α]_D −166.1° (c, 0.15 in CHCl₃).
3,4-Epoxide: [70645-59-9].
$C_{20}H_{30}O_3$ M 318.455
Constit. of *S.* spp. Oil. [α]_D −46.7° (c, 0.09 in CHCl₃).
(2*S*,11*R*,12*R*)-form
Constit. of *S. birklandi*. Cryst. Mp 70-71°. [α]_D −64° (c, 0.6 in CHCl₃).

Bowden, B.F. *et al*, *Aust. J. Chem.*, 1979, **32**, 653 (*isol, struct*)
Bowden, B.F. *et al*, *J. Nat. Prod.*, 1987, **50**, 650 (*isol, cryst struct*)

Isoschizandrin   I-70101
[114422-18-3]

$C_{24}H_{32}O_7$ M 432.513
Constit. of the fruits of *Schizandra chinensis*. Amorph. powder. [α]_D^{25} +92° (c, 1.22 in CHCl₃). Diastereoisomeric with Schizandrin, S-00159 .

Ikeya, Y. *et al*, *Phytochemistry*, 1988, **27**, 569.

Isosilybin   I-70102
[72581-71-6]

$C_{25}H_{22}O_{10}$ M 482.443
Constit. of *Silybum marianum*. Cryst. Mp 239-241°. [α]_D +16.9° (Me₂CO).

Arnone, A. *et al*, *J. Chem. Soc.*, *Chem. Commun.*, 1979, 696 (*isol, struct*)
Merlini, L. *et al*, *J. Chem. Soc.*, *Perkin Trans. 1*, 1980, 775 (*synth*)

Isosilychristin   I-70103
[77182-66-2]

$C_{25}H_{22}O_{10}$ M 482.443
Constit. of *Silybum marianum*. Cryst. Mp 155-157°. [α]_D +245° (Py).

Kalonga, M., *Z. Naturforsch.*, *B*, 1981, **36**, 262.

Isothamnosin A   I-70104
[67979-24-2]

$C_{30}H_{28}O_6$ M 484.548
Constit. of a *Ruta* sp. Cryst. (MeOH). Mp 223-226°.
Stereoisomer: [67965-45-1]. **Isothamnosin B**.
$C_{30}H_{28}O_6$ M 484.548

Constit. of a *R.* sp. Cryst. (MeOH). Mp 213-214°.

Gonzalez, A.G. *et al, An. Quim.,* 1977, **73**, 1510.

2-Isothiocyanato-2-methylpropane, 9CI   I-70105

Updated Entry replacing B-03694

Isothiocyanic acid, tert-*butyl ester,* 8CI. tert-*Butyl isothiocyanate.* tert-*Butyl mustard oil*

[590-42-1]

$$(H_3C)_3CNCS$$

C_5H_9NS M 115.193

d 0.91. Mp 10.5-11.5°. Bp 142°, Bp_{10} 30.5-32°.

▷Highly toxic, lachrymator

Kharasch, M.S. *et al, J. Am. Chem. Soc.,* 1937, **59**, 1580 (*synth*)
Schmidt, E. *et al, Justus Liebigs Ann. Chem.,* 1950, **568**, 192 (*synth*)
Nair, G.V., *Indian. J. Chem.,* 1966, **4**, 516 (*synth*)
Bäuerlein, E. *et al, Justus Liebigs Ann. Chem.,* 1978, 675 (*synth*)

2-Isothiocyanatopyridine, 9CI   I-70106

2-Pyridyl isothiocyanate

[52648-45-0]

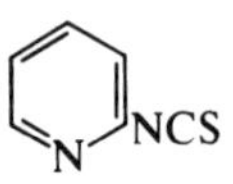

$C_6H_4N_2S$ M 136.171

Pale-yellow liq. $Bp_{0.1}$ 55-60°. Solidifies at r.t. to give brick-red cryst. of a dimer which dissociates reversibly in soln.

Fairfull, A.E.S. *et al, J. Chem. Soc.,* 1955, 796 (*synth*)
Schultz, O.E. *et al, Arch. Pharm. (Weinheim, Ger.),* 1962, **295**, 146 (*synth*)
L'abbé, G., *Synthesis,* 1987, 525 (*rev*)

3-Isothiocyanatopyridine, 9CI   I-70107

3-Pyridyl isothiocyanate

[17452-27-6]

$C_6H_4N_2S$ M 136.171

Pungent oil. Bp 231-233°, Bp_{11} 106-108°.

Fairfull, A.E.S. *et al, J. Chem. Soc.,* 1955, 796 (*synth*)
Knott, E.B., *J. Chem. Soc.,* 1956, 1644 (*synth*)
Jochims, J.C., *Chem. Ber.,* 1968, **101**, 1746 (*synth*)
L'abbé, G., *Synthesis,* 1987, 525 (*rev*)

Ivangulic acid   I-70108

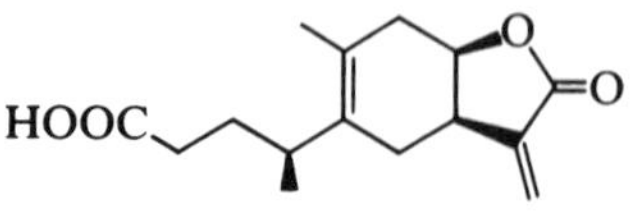

$C_{15}H_{20}O_4$ M 264.321

Constit. of *Gnephosis arachnoidea.* Gum.

Jakupovic, J. *et al, Phytochemistry,* 1988, **27**, 3181.

Ivangustin   I-70109

Updated Entry replacing I-20092

1β-Hydroxy-4,11(13)-eudesmadien-12,8β-olide. 1β,8β-Dihydroxy-4,11(13)eudesmadien-12-oic acid γ-lactone, 8CI

[14164-59-1]

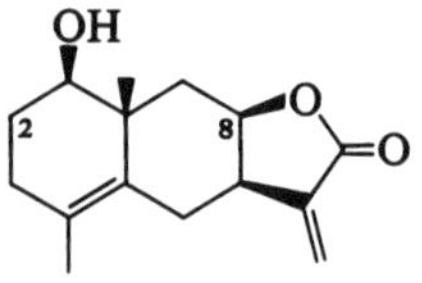

$C_{15}H_{20}O_3$ M 248.321

Isol. from *Iva angustifolia.* Cryst. ($CHCl_3$/pet. ether).
Mp 120-122°. $[\alpha]_D^{27}$ +85° (c, 1.05 in $CHCl_3$).

O-(*2-Methylbutanoyl*):
  $C_{20}H_{28}O_4$ M 332.439
  Constit. of *Wunderlichia mirabilis.* Oil. $[\alpha]_D^{24}$ +46° (c, 0.5 in $CHCl_3$).

2α-Hydroxy: **2α-*Hydroxyivangustin*.**
  $C_{15}H_{20}O_4$ M 264.321
  Constit. of *Gnaphosis arachnoidea.* Gum. $[\alpha]_D^{24}$ +39° (c, 1.09 in $CHCl_3$).

3α,6β-Dihydroxy: 1β,3α,6β-Trihydroxy-4,11(13)-eudesmadien-12,8β-olide. **3α,6β-*Dihydroxyivangustin*.**
  $C_{15}H_{20}O_5$ M 280.320
  Occurs as various esters in *Wedelia hookeriana.*

4,5α-Dihydro,4α-hydroxy, O^1-(*2-methylbutanoyl*):
  $C_{20}H_{30}O_5$ M 350.454
  Constit. of *W. mirabilis.* Oil. $[\alpha]_D^{24}$ +26° (c, 0.18 in $CHCl_3$).

8-Epimer, 2α-acetoxy: [80931-30-2]. **2α-*Acetoxy-8-epiivangustin*.**
  $C_{17}H_{22}O_5$ M 306.358
  Constit. of *Eriophyllum lanatum.* Gum. $[\alpha]_D^{24}$ +50° (c, 0.03 in $CHCl_3$).

8-Epimer, 6α-hydroxy: **6α-*Hydroxy-8-epiivangustin*.**
  $C_{15}H_{20}O_4$ M 264.321
  Constit. of *G. arachnoidea.* Gum.

8-Epimer, 6α-acetoxy: **6α-*Acetoxy-8-epiivangustin*.**
  $C_{17}H_{22}O_5$ M 306.358
  Constit. of *G. arachnoidea.* $[\alpha]_D^{24}$ −25° (c, 0.58 in $CHCl_3$).

8-Epimer, 6α-acetoxy, Ac:
  $C_{19}H_{24}O_6$ M 348.395
  Constit. of *G. arachnoidea.* Gum. $[\alpha]_D^{24}$ −20° (c, 0.58 in $CHCl_3$).

Herz, W. *et al, J. Org. Chem.,* 1967, **32**, 3658 (*isol*)
Bohlmann, F. *et al, Phytochemistry,* 1981, **20**, 2239 (*isol*)
Bohlmann, F. *et al, Phytochemistry,* 1982, **21**, 2329 (*derivs*)
Bohlmann, F. *et al, Justus Liebigs Ann. Chem.,* 1984, 228 (*derivs*)
Marco, J.A. *et al, Can. J. Chem.,* 1987, **65**, 630 (*synth*)
Jakupovic, J. *et al, Phytochemistry,* 1988, **27**, 3181 (*derivs*)

J

Jaeschkeanadiol   J-70001

Updated Entry replacing J-00012
[41690-67-9]

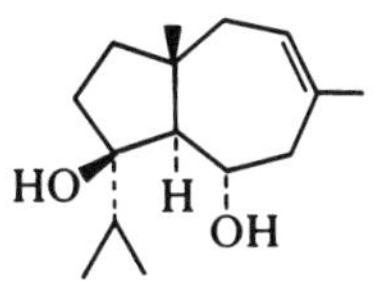

C$_{15}$H$_{26}$O$_2$ M 238.369
Constit. of *Ferula jaeschkeana*. Cryst. (pet. ether). Mp
91-92°. [α]$_D$ +38.8° (CHCl$_3$).

8α,9α-Epoxide, 6-(4-hydroxybenzoyl): [96853-63-3].
Epoxyjaeschkeanadiol p-hydroxybenzoate.
C$_{22}$H$_{30}$O$_5$ M 374.476
Constit. of *F. orientalis*.
8α,9α-Epoxide, 6-(4-hydroxy-3-methoxybenzoyl):
[112501-40-3]. **Epoxyjaeschkeanadiol vanillate**.
C$_{23}$H$_{32}$O$_6$ M 404.502
Constit. of *F. orientalis*. [α]$_D^{23}$ +52.9° (c, 1.4 in
CHCl$_3$).

Sriraman, M.C. *et al, Tetrahedron*, 1973, **29**, 985 (*isol, struct*)
Miski, M. *et al, J. Nat. Prod.*, 1987, **50**, 829 (*deriv*)

Jatrogrossidione   J-70002

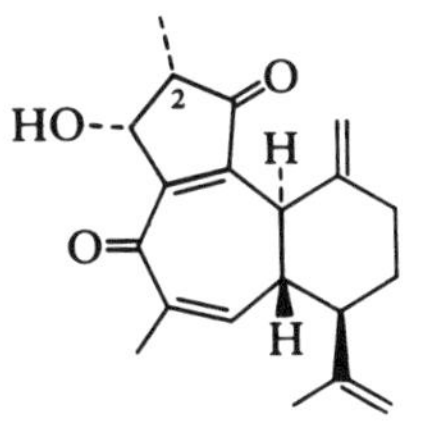

C$_{20}$H$_{24}$O$_3$ M 312.408
Constit. of *Jatropha grossidentata*. Gum.

2-Epimer: **2-Epijatrogrossidione**.
C$_{20}$H$_{24}$O$_3$ M 312.408
Constit. of *J. grossidentata*. Gum.

Jakupovic, J. *et al, Phytochemistry*, 1988, **27**, 2997.

Jewenol A   J-70003
[117590-96-2]

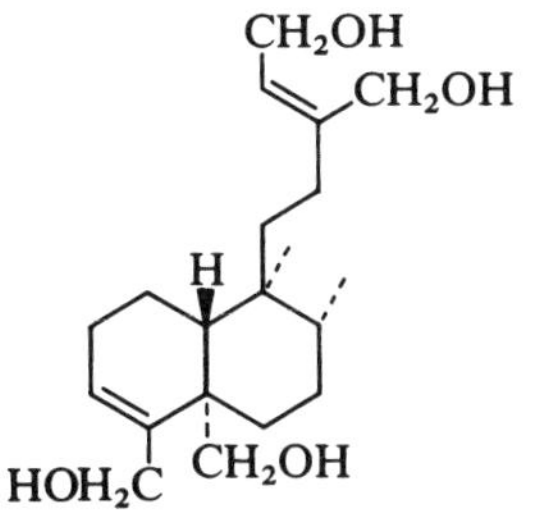

C$_{20}$H$_{34}$O$_4$ M 338.486
Constit. of *Portulaca* cv. jewel. Oil. [α]$_D^{31}$ −87.6° (c, 0.37
in EtOH).

Ohsaki, A. *et al, Phytochemistry*, 1988, **27**, 2171.

Jewenol B   J-70004
[117590-97-3]

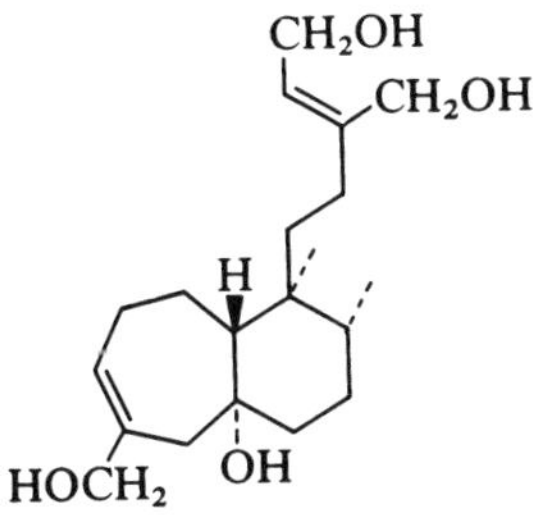

C$_{20}$H$_{34}$O$_4$ M 338.486
Constit. of *Portulaca* cv. jewel. Cryst. (MeOH aq.). Mp
123.5-124.5°. [α]$_D^{31}$ −59.2° (c, 0.52 in EtOH).

Ohsaki, A. *et al, Phytochemistry*, 1988, **27**, 2171.

Juglorin   J-70005
[109521-81-5]

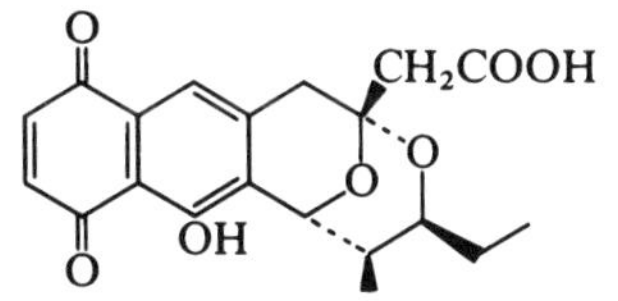

Relative
configuration

C$_{20}$H$_{20}$O$_7$ M 372.374
Quinone antibiotic. Isol. from *Streptomyces malachiti-
cus*. Spermidine synthase inhibitor. No phys. props. reported.

Hamaguchi, K. *et al, J. Antibiot.*, 1987, **40**, 717 (*isol, struct,
props*)

273

K

K 13
K-70001

[108890-90-0]

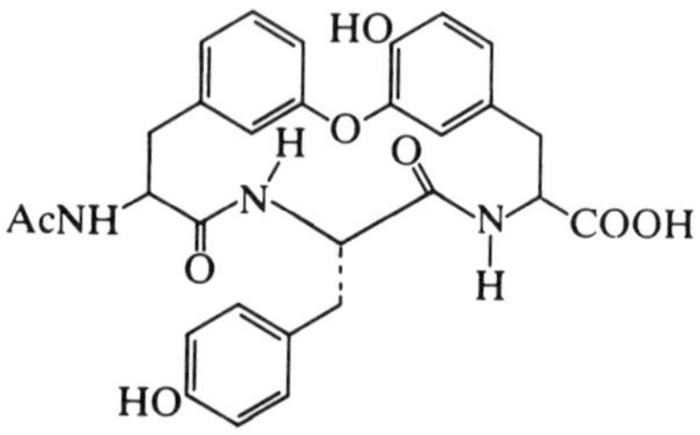

C$_{29}$H$_{29}$N$_3$O$_8$ M 547.563

Isol. from *Micromonospora halophytica* ssp. *exilisia*. Inhibitor of angiotensin I converting enzyme. Powder. Mp 265-270° dec. $[\alpha]_D^{19}$ −3.4° (c, 0.6 in MeOH). Stereochem. not fully elucidated. Related to Biphenomycin.

Kase, H. *et al*, *J. Antibiot.*, 1987, **40**, 450 (*isol, props*)
Yasuzawa, T. *et al*, *J. Antibiot.*, 1987, **40**, 455 (*pmr, cmr, struct*)

Kamebacetal A
K-70002

[73981-36-9]

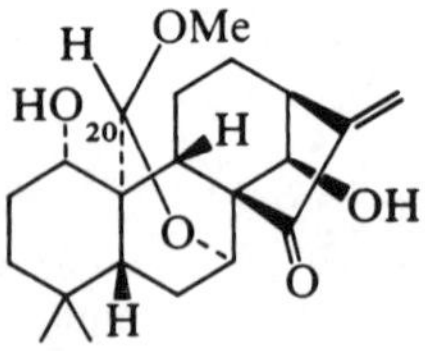

C$_{21}$H$_{30}$O$_5$ M 362.465

Constit. of *Rabdosia umbrosa*. Needles. Mp 253-256°. $[\alpha]_D^{19}$ −40° (c, 1.1 in MeOH).

20-Epimer: [73981-35-8]. **Kamebacetal B**.
C$_{21}$H$_{30}$O$_5$ M 362.465
Constit. of *R. umbrosa*. Needles. Mp 230-232°. $[\alpha]_D^{26}$ −58° (c, 0.43 in MeOH).

Takeda, Y. *et al*, *J. Chem. Soc., Perkin Trans. 1*, 1987, 2403.

Kanjone
K-70003

6-Methoxy-2-phenyl-4H-furo[2,3-h]-1-benzopyran-4-one, 9CI. 6-Methoxyfurano[4″,5″:8,7]flavone

[1094-12-8]

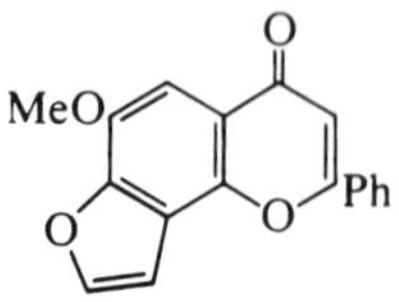

C$_{18}$H$_{12}$O$_4$ M 292.290

Constit. of *Pongamia glabra* leaves. Needles (CHCl$_3$/pet. ether). Mp 189°.

Malik, S.B. *et al*, *Indian J. Chem., Sect. B*, 1977, **15**, 536 (*isol, uv, pmr, struct*)
Talapatra, S.K. *et al*, *J. Indian Chem. Soc.*, 1982, **59**, 534 (*cmr*)

Kanshone A
K-70004

[115356-18-8]

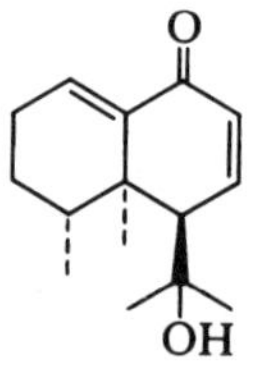

C$_{15}$H$_{22}$O$_2$ M 234.338

Constit. of *Nardostachys chinensis*. Oil. $[\alpha]_D$ −147.8° (c, 0.35 in CHCl$_3$).

Bagchi, A. *et al*, *Phytochemistry*, 1988, **27**, 1199.

Kanshone B
K-70005

[115370-61-1]

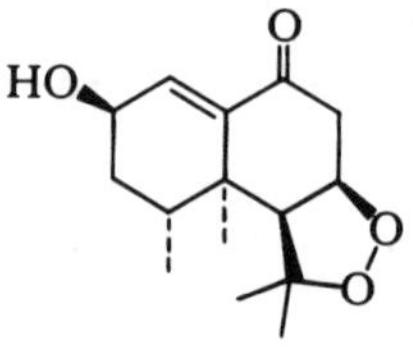

C$_{15}$H$_{22}$O$_4$ M 266.336

Constit. of *Nardostachys chinensis*. Needles. Mp 137-138°. $[\alpha]_D$ +133.8° (c, 1.1 in CHCl$_3$).

Bagchi, A. *et al*, *Phytochemistry*, 1988, **27**, 1199.

Kanshone C
K-70006

[117634-64-7]

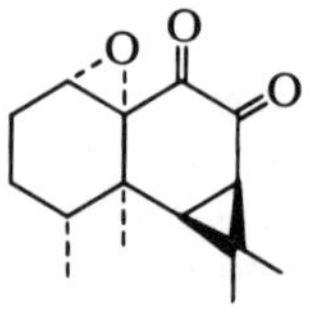

C$_{15}$H$_{20}$O$_3$ M 248.321

Constit. of *Nardostachys chinensis*. Yellow needles. Mp 136-137°. $[\alpha]_D$ −13.4° (c, 0.4 in CHCl$_3$).

Bagchi, A. *et al*, *Phytochemistry*, 1988, **27**, 2877.

Kasuzamycin B
K-70007

Hydroxyleptomycin A

[107140-30-7]

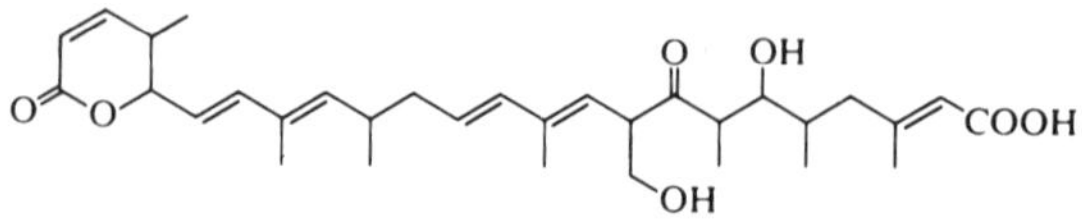

C$_{32}$H$_{46}$O$_7$ M 542.711

Isol. from *Streptomyces* sp. 81-484. Antitumour agent. Pale-yellow powder. Mp 53-55°. $[\alpha]_D^{20}$ −152° (c, 0.77 in CHCl$_3$).

Funaishi, K. *et al*, *J. Antibiot.*, 1987, **40**, 778 (*isol, struct, props*)

ent-16*S*-Kauran-17-al   K-70008

Updated Entry replacing K-10003

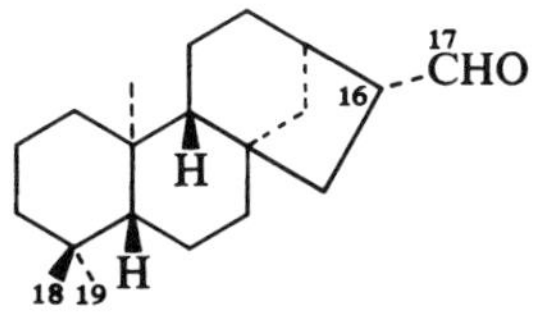

$C_{20}H_{32}O$ M 288.472

Constit. of *Baccharis minutiflora*. Gum. $[\alpha]_D^{24}$ −71° (c, 1.06 in $CHCl_3$).

19-Hydroxy: ent-*19-Hydroxy-16S-kauran-17-al.*
$C_{20}H_{32}O_2$ M 304.472
Constit. of *B. minutiflora*. Gum.

19-Aldehyde: ent-*16S-Kaurane-17,19-dial.*
$C_{20}H_{30}O_2$ M 302.456
Constit. of *B. minutiflora*. Gum. $[\alpha]_D^{24}$ −63° (c, 0.74 in $CHCl_3$).

17-Carboxylic acid: ent-*16S-Kauran-17-oic acid.*
$C_{20}H_{32}O_2$ M 304.472
Constit. of *B. minutiflora*.

17-Carboxylic acid, Me ester: Gum. $[\alpha]_D^{24}$ −56° (c, 1 in $CHCl_3$).

17-Carboxylic acid, 16-epimer: ent-*16R-Kauran-17-oic acid.*
$C_{20}H_{32}O_2$ M 304.472
Constit. of *Croton lacciferus*. Mp 215-217°. $[\alpha]_D$ −65.7° (c, 0.7 in $CHCl_3$).

Henrick, C.A. *et al, Aust. J. Chem.*, 1964, **17**, 915 (*synth*)
Bohlmann, F. *et al, Phytochemistry*, 1982, **21**, 399 (*isol*)
Bandara, B.M.R. *et al, Phytochemistry*, 1988, **27**, 869 (*isol*)

ent-16-Kaurene-7α,15β-diol   K-70009

$C_{20}H_{32}O_2$ M 304.472

Constit. of *Plagiochila pulcherrima*. Needles (MeOH). Mp 206-208°. $[\alpha]_D^{24}$ −35.8° (c, 0.56 in $CHCl_3$).

Fukuyama, Y. *et al, Phytochemistry*, 1988, **27**, 1425.

ent-16-Kauren-3β-ol   K-70010

[19891-40-8]

$C_{20}H_{32}O$ M 288.472

Constit. of *Phyllanthus flexuosus*. Needles (MeOH). Mp 177-178°. $[\alpha]_D^{23}$ −69.2° (c, 0.32 in $CHCl_3$).

Tanaka, R. *et al, Phytochemistry*, 1988, **27**, 2273.

Kerlinic acid   K-70011

[112606-14-1]

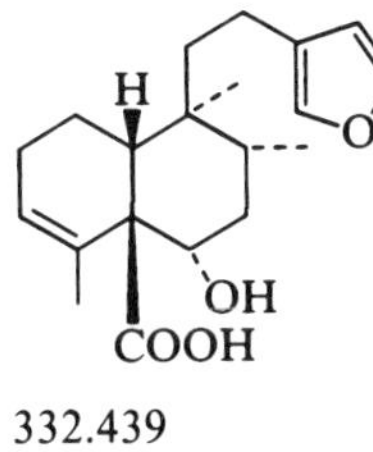

$C_{20}H_{28}O_4$ M 332.439

Constit. of *Salvia keerlii*. Cryst. (Me_2CO/pet. ether). Mp 183-185°. $[\alpha]_D^{25}$ −236.8° (c, 0.19 in $CHCl_3$).

Rodríguez-Hahn, L. *et al, Can. J. Chem.*, 1987, **65**, 2687.

Ketosantalic acid   K-70012

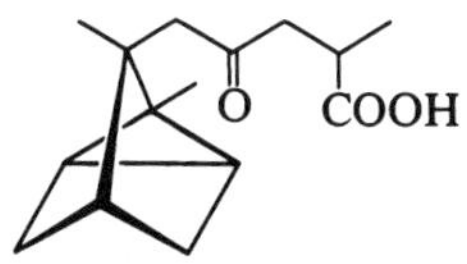

$C_{15}H_{22}O_3$ M 250.337

Constit. of Indian sandalwood oil. Oil. $Bp_{0.3}$ 180°. $[\alpha]_D^{29}$ −30° (c, 1.4 in $CHCl_3$).

Me ester, semicarbazone: Cryst. (EtOH). Mp 183-185°.

Ranibai, P. *et al, Indian J. Chem., Sect. B*, 1986, **25**, 1006.

Khellactone   K-70013

Updated Entry replacing K-00096

9,10-Dihydro-9,10-dihydroxy-8,8-dimethyl-2II,8H-benzo[1,2-b:3,4-b']dipyran-2-one, 9CI. 3',4'-Dihydro-3',4'-dihydroxyseselin

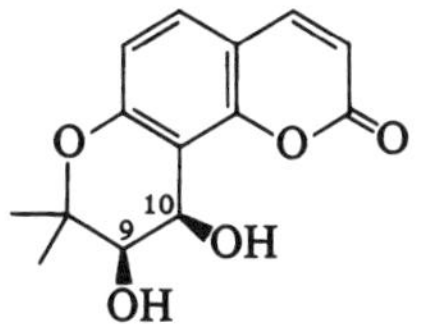

$C_{14}H_{14}O_5$ M 262.262

Abs. configs. given here are based on the review by Murray.

(9R,10R)-form [24144-61-4]

(+)-cis-*form*
Isol. from *Seseli gummiferum*. Cryst. (C_6H_6/pet. ether). Mp 174-175°. $[\alpha]_D^{22}$ +104° (c, 0.864 in EtOH), $[\alpha]_D^{20}$ +80.9° (c, 1.041 in $CHCl_3$).

Di-Ac: Cryst. (Me_2CO aq.). Mp 132-134°. $[\alpha]_D^{24}$ −17.4° (c, 0.743 in $CHCl_3$).

O,O-Diangeloyl: [4970-26-7]. **Anomalin.**
$C_{24}H_{26}O_7$ M 426.465
Isol. from *Angelica anomala, Bupleurum falcatum, Musineon divaricatum* and other plants. Mp 173-174°. $[\alpha]_D^{27}$ −78.4° (−15.5°) (EtOH).

O^{10}-Angeloyl:
$C_{19}H_{20}O_6$ M 344.363
Isol. from *A. ursina* and *Libanotis lehmanniana*. Powder. $[\alpha]_D$ −54.5° ($CHCl_3$).

O^9-Ac, O^{10}-(3-methylbutanoyl): [53023-17-9]. **Suksdorfin.**
$C_{21}H_{24}O_7$ M 388.416
Constit. of *Lomatium suksdorfi*. Mp 140.5-141°. $[\alpha]_D^{24}$ +4° (c, 0.5 in EtOH).

O^9-Ac, O^{10}-angeloyl: [13161-75-6]. **Pteryxin.**
$C_{21}H_{22}O_7$ M 386.401
Isol. from *Pteryxia terebinthina*. Mp 86-86.5°. $[\alpha]_D$ +12.5° (EtOH), $[\alpha]_D$ −4° ($CHCl_3$).

O^9-Ac, O^{10}-(3-methyl-2-butenoyl): [53023-18-0]. **Isosamidin.** *3'-Acetoxy-4'-senecioyloxy-3',4'-dihydroseselin.*
$C_{21}H_{22}O_7$ M 386.401
From roots of *S. libanotis*. Mp 120.5-121°. $[\alpha]_D$ −11.9° (c, 0.4 in EtOH).

O^9-Ac, O^{10}-(2,3-epoxy-2-methylbutanoyl): **Epoxypteryxin.**
$C_{21}H_{22}O_8$ M 402.400
Isol. from *Laserpitium archangelica*. Oil.

O^9-(3-Methylthio-2-propenoyl), O^{10}-angeloyl: [54963-31-4]. **Isofloroseselin.**
$C_{23}H_{24}O_7S$ M 444.498
Isol. from *S. coronatum*. Mp 122-123°. $[\alpha]_D$ −87.3° (EtOH). *cis-*. Abs. config. not certain, may belong to the (−)-form.

O-*Angeloyl*, O-*2-methylpropanoyl*:
$C_{24}H_{28}O_7$ M 428.481
Isol. from *S. libanotis*. Posn. of acyl residues uncertain.
O^9-(S-*2-Methylbutanoyl*), O^{10}-*Ac*: [477-32-7].
Visnadin.
$C_{21}H_{24}O_7$ M 388.416
Isol. from *Ammi visnaga*. Mp 84-86°. $[\alpha]_D$ +38°
(dioxan).

▷ET5600000.

O^9-(*3-Methylbutanoyl*), O^{10}-*Ac*: **Dihydrosamidin**.
$C_{21}H_{24}O_7$ M 388.416
Isol. from *A. visnaga*. Mp 117-119°. $[\alpha]_D$ +63°
(dioxan).
O^9-(*3-Methylbutanoyl*), O^{10}-*angeloyl*:
$C_{24}H_{28}O_7$ M 428.481
Isol. from *S. gummiferum*. Glass. $[\alpha]_D$ +21.4°
(CHCl$_3$).
O-*Angeloyl*, O-*3-methylbutanoyl*:
$C_{24}H_{28}O_7$ M 428.481
Isol. from *S. libanotis*. Placement of acyl residues
uncertain.
O^9-*Angeloyl*:
$C_{19}H_{20}O_6$ M 344.363
Isol. from *Angelica ursina* and *Libanotis lehmanniana*.
Mp 148-149°. $[\alpha]_D$ −91.8° (CHCl$_3$).
O^9-*Angeloyl*, O^{10}-*Ac*: [14017-71-1]. **Isopteryxin**.
$C_{21}H_{22}O_7$ M 386.401
Isol. from *Pteryxia terebinthina*. Mp 135-135.5°. $[\alpha]_D$
−45° (CHCl$_3$), $[\alpha]_D$ −39° (EtOH).
O^9-(*3-Methyl-2-butenoyl*), O^{10}-*Ac*: [477-33-8].
Samidin.
$C_{21}H_{22}O_7$ M 386.401
Isol. from *Ammi visnaga*. Mp 135-137°. $[\alpha]_D$ +100°
(dioxan).

▷GQ5725000.

Di-O-*Angeloyl*: [4970-26-7]. (−)-**Anomalin**.
$C_{24}H_{26}O_7$ M 426.465
Isol. from roots of *Angelica anomala*. Needles
(EtOH). Mp 173-174°. $[\alpha]_D$ −37° (CHCl$_3$).
Di-O-(*3-Methyl-2-butenoyl*): [14017-72-2].
Calipteryxin.
$C_{24}H_{26}O_7$ M 426.465
Isol. from *Pteryxia terebinthina*. Mp 147.5-148°. $[\alpha]_D$
−55° (CHCl$_3$).
O^9-(*3-Methyl-2-butenoyl*), O^{10}-*angeloyl*:
$C_{24}H_{26}O_7$ M 426.465
Isol. from *S. gummiferum*. Mp 78-79.5°. $[\alpha]_D$ +17.9°
(CHCl$_3$).
Di-O-(*3-Methyl-2-butenoyl*): 3′,4′-*Disenecioyloxy*-3′,4′-
dihydroseselin. Isol. from roots of *S. libanotis*. Cryst.
(cyclohexane/pet. ether). Mp 108-108.5°. $[\alpha]_D$ +15.8°
(c, 0.3 in CHCl$_3$).
Di-O-(*2,3-Epoxy-2-methylbutanoyl*):
$C_{24}H_{26}O_9$ M 458.464
Isol. from *Laserpitium archangelica*. Mp 179°, Mp
211°. 2 Stereoisomers isol., of unknown configs.

(9S,10S)-form
(−)-*cis-form*
Bis(*3-methylbutanoyl*): [54676-87-8].
$C_{24}H_{30}O_7$ M 430.497
Constit. of *P. japonicum*. Mp 88-89°. $[\alpha]_D$ −38.8° (c,
0.1 in CHCl$_3$).
Bis(*3-methyl-3-butenoyl*): [54676-88-9].
$C_{24}H_{26}O_7$ M 426.465
Constit. of *P. japonicum*. Mp 112-113°. $[\alpha]_D$ −47.7°
(c, 0.1 in CHCl$_3$).
O^9-*Angeloyl*, O^{10}-(*3-Methylthio-2-butenoyl*): [32686-
09-2]. **Floroselin**.
$C_{23}H_{24}O_7S$ M 444.498

Isol. from *S. sessiliflorum*. Mp 161.5-163°. $[\alpha]_D$
−121.2° (CHCl$_3$).
Di-O-*Angeloyl*: (+)-**Anomalin**.
$C_{24}H_{26}O_7$ M 426.465
Isol. from *Peucedanum formosanum*. Mp 171-172°.
$[\alpha]_D$ +34.6° (CHCl$_3$).
O^9-*Angeloyl*, O^{10}-(*3-methyl-2-butenoyl*): [6468-80-0].
Peuformosin.
$C_{24}H_{26}O_7$ M 426.465
Isol. from *P. formosanum*. Mp 155-156°. $[\alpha]_D$ +67.3°
(CHCl$_3$).
O^9-*Angeloyl*, O^{10}-(*2,3-Epoxy-2-methylbutanoyl*):
$C_{24}H_{26}O_8$ M 442.465
Isol. from *Laserpitium archangelica*. Abs. config.
uncertain, may belong to the (+)-form.
Di-O-(*3-Methyl-2-butenoyl*):
$C_{24}H_{26}O_7$ M 426.465
Isol. from *P. japonicum*. Mp 112-113°. $[\alpha]_D$ −47.7°
(CHCl$_3$).

(9RS,10RS)-form
(±)-cis-*form*
Cryst. (C$_6$H$_6$). Mp 160-161°.
Di-Ac: [35930-25-7]. Needles (C$_6$H$_6$/pet. ether). Mp
162-162.5°.

(9R,10S)-form [23458-04-0]
(−)-trans-*form*
Isol. from *Angelica purpureafolia*. Cryst. (Me$_2$CO/pet.
ether). Mp 185-186°. $[\alpha]_D^{23}$ −6.4° (c, 0.838 in EtOH),
$[\alpha]_D^{21}$ −18.0° (c, 0.901 in CHCl$_3$).
Di-Ac: [24144-59-0]. Cryst. (Me$_2$CO aq.). Mp 162-163°.
$[\alpha]_D^{21}$ −8.4° (c, 1.083 in CHCl$_3$).

(9RS,10SR)-form
(±)-trans-*form*
Cryst. (Me$_2$CO/pet. ether). Mp 184.5-186°.
Di-Ac: [15575-69-6]. Cryst. (Me$_2$CO aq. or C$_6$H$_6$/pet.
ether). Mp 157-161°, Mp 170-178°.

Schroeder, H.D. et al, Chem. Ber., 1959, **92**, 2338 (*synth, isol*)
Soine, T.O. et al, J. Pharm. Sci., 1962, **51**, 149 (*Pteryxin,
    Suksdorfin*)
Shanbhag, S.N. et al, Tetrahedron, 1965, **21**, 3591; 1967, **23**,
    1235 (*isol*)
Hata, K. et al, Chem. Pharm. Bull., 1966, **14**, 94 (*Anomalin*)
Lemmich, J. et al, Acta Chem. Scand., 1966, **20**, 2497 (*isol,
    synth, derivs*)
Battacharyya, S.C. et al, Tetrahedron, 1967, **23**, 1235
    (*Suksdorfin*)
Bernotat-Wulf, H. et al, Helv. Chim. Acta, 1969, **52**, 1165
    (*cryst struct, abs config*)
Yamada, Y. et al, Tetrahedron Lett., 1974, 2513 (*isol*)
Nielsen, B.E. et al, Phytochemistry, 1976, **15**, 1049 (*derivs*)
Shaath, N.A. et al, J. Pharm. Sci., 1976, **65**, 1028 (*ms*)
Murray, R.D.H. et al, Fortschr. Chem. Org. Naturstoffe, 1978,
    35, 200 (*rev, bibl*)
Swager, T.M. et al, Phytochemistry, 1985, **24**, 805 (*Anomalin*)

Kickxioside　　　　　　　　　　　　　**K-70014**

[110906-83-7]

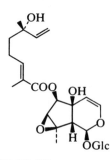

$C_{25}H_{36}O_{12}$ M 528.552

Constit. of *Kickxia spuria*. [α]$_D$ −63.2° (c, 0.7 in MeOH).

Nicoletti, M. *et al*, *Planta Med.*, 1987, 295.

Kurospongin   K-70015

[115722-54-8]

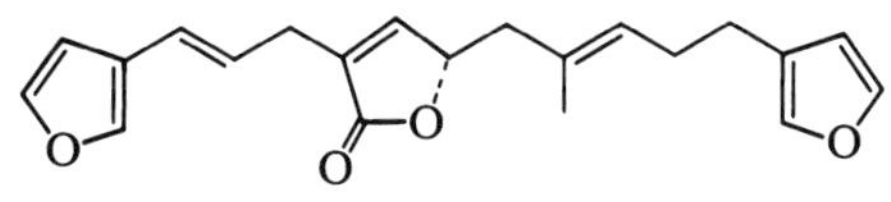

C$_{21}$H$_{22}$O$_4$ M 338.402

Constit. of *Spongia* sp. Ichthyotoxic and fish antifeedant. Yellow unstable oil. [α]$_D$ −16.8° (c, 1.24 in CHCl$_3$).

Tanaka, J. *et al*, *Tetrahedron*, 1988, **44**, 2805 (*struct, abs config*)

Kuwanone *L*   K-70016

Updated Entry replacing K-30023

[88524-65-6]

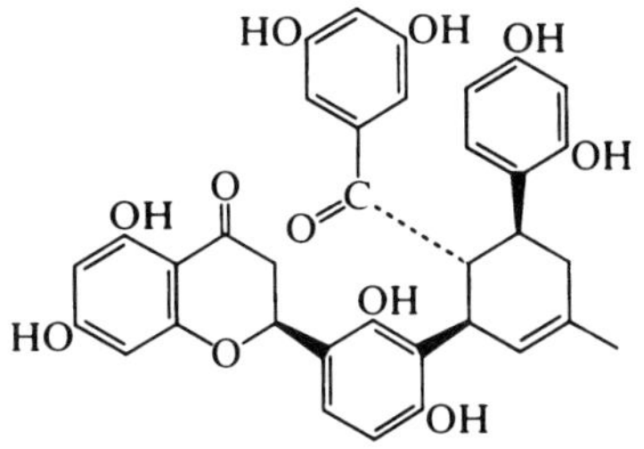

Absolute configuration

C$_{35}$H$_{30}$O$_{11}$ M 626.615

Constit. of *Morus alba*. Amorph. [α]$_D^{22}$ −227° (c, 0.04 in MeOH).

2,3-Didehydro, 3-(3-methyl-2-butenyl): [88524-66-7].
   ***Kuwanone* K**.
   C$_{40}$H$_{36}$O$_{11}$ M 692.718
   From *M. alba*. Amorph. [α]$_D$ −218° (c, 0.016 in MeOH).

Nomura, T. *et al*, *Planta Med.*, 1983, **47**, 151 (*isol*)
Hano, Y. *et al*, *Heterocycles*, 1988, **27**, 75 (*abs config*)

L

8(17),13-Labdadiene-3,15-diol   **L-70001**

$C_{20}H_{34}O_2$ M 306.487

(3α,13Z)-form

Constit. of *Juniperus pseudosabina*. Gum.

Pandita, K. *et al*, *Indian J. Chem., Sect. B*, 1987, **26**, 453.

8(17),14-Labdadiene-3,13-diol   **L-70002**

$C_{20}H_{34}O_2$ M 306.487

(3α,13R)-form

3α-Hydroxymanool

Constit. of *Juniperus pseudosabina*. Cryst.
(hexane/C_6H_6). Mp 75-76°. $[\alpha]_D^{30}$ +14.6° (c, 0.8 in
$CHCl_3$).

3-Ketone: 13R-Hydroxy-8(17),14-labdadien-3-one.

$C_{20}H_{32}O_2$ M 304.472

From *J. pseudosabina*. Cryst. (MeOH). Mp 72-73°.
$[\alpha]_D^{30}$ +25.7° (c, 1.628 in $CHCl_3$).

Pandita, K. *et al*, *Indian J. Chem., Sect. B*, 1987, **26**, 453.

ent-7,14-Labdadiene-2α,13-diol   **L-70003**

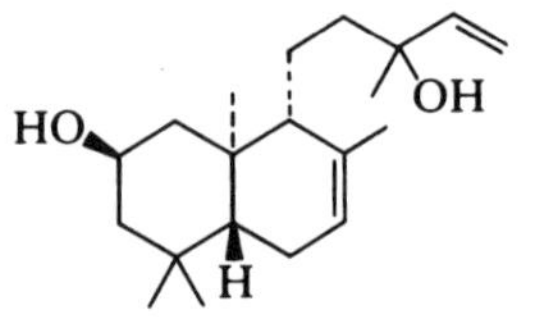

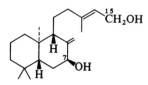

$C_{20}H_{34}O_2$ M 306.487

Constit. of *Ixiolaena leptolepis*. Oil.

Lehmann, L. *et al*, *Phytochemistry*, 1988, **27**, 2994.

ent-8,13-Labdadiene-7α,15-diol   **L-70004**

$C_{20}H_{34}O_2$ M 306.487

Constit. of *Gypothamnium pinifolium*.

7-Epimer: ent-8,13-Labdadiene-7β,15-diol.

$C_{20}H_{34}O_2$ M 306.487

Constit. of *G. pinifolium*.

7-Ketone: ent-15-Hydroxy-8,13-labdadien-7-one.

$C_{20}H_{32}O_2$ M 304.472

Constit. of *E. pinifolium*.

Zdero, C. *et al*, *Phytochemistry*, 1988, **27**, 2953.

ent-8,13(16)-Labdadiene-14,15-diol   **L-70005**

$C_{20}H_{34}O_2$ M 306.487

Constit. of *Gypothamnium pinifolium*. Oil. $[\alpha]_D^{24}$ −61°
(c, 0.82 in $CHCl_3$).

Zdero, C. *et al*, *Phytochemistry*, 1988, **27**, 2953.

ent-8(17),13-Labdadiene-7α,15-diol   **L-70006**

$C_{20}H_{34}O_2$ M 306.487

(13E)-form

7-Ac: [74635-75-9].

$C_{22}H_{36}O_3$ M 348.525

Constit. of *Austroeupatorium chaparense*. Oil.

Bohlmann, F. *et al*, *Phytochemistry*, 1980, **19**, 111.

ent-8(17),13-Labdadiene-9α,15-diol   **L-70007**

$C_{20}H_{34}O_2$ M 306.487

Constit. of *Gypothamnium pinifolium*.

Zdero, C. *et al*, *Phytochemistry*, 1988, **27**, 2953.

11,13(16)-Labdadiene-6,7,8,14,15-pentol   **L-70008**

6,7,8,14,15-Pentahydroxy-11,13(16)-labdadiene

$C_{20}H_{34}O_5$ M 354.486

(6α,7β,8α,11E,14ξ)-form

Sterebin G

Constit. of *Stevia rebaudiana*. Powder. $[\alpha]_D$ −5.9° (c,
0.5 in MeOH).

14-Epimer: Sterebin H.

$C_{20}H_{34}O_5$ M 354.486

From *S. rebaudiana*. Powder. $[\alpha]_D$ −6.7° (c, 0.5 in
MeOH).

Oshima, Y. *et al*, *Phytochemistry*, 1988, **27**, 624.

11,13-Labdadiene-6,7,8,15-tetrol   **L-70009**

6,7,18,15-Tetrahydroxy-11,13-labdadiene

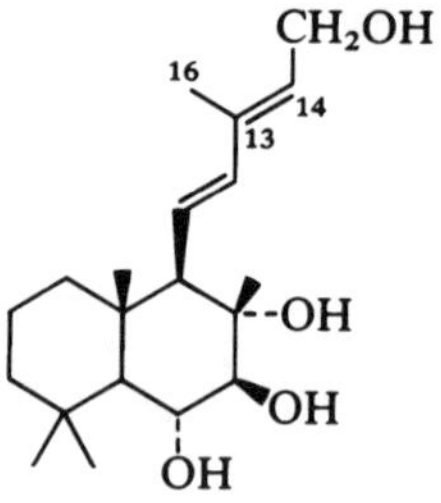

$C_{20}H_{34}O_4$ M 338.486

(6α,7β,8α,11E,13E)-form

Sterebin E

Constit. of *Stevia rebaudiana*. Powder. $[\alpha]_D$ +29.9° (c,
0.17 in MeOH).

(6α,7β,8α,11E,13Z)-form

Sterebin F

Constit. of *S. rebaudiana*. Powder. $[\alpha]_D$ +40.3° (c, 0.16
in MeOH).

Oshima, Y. *et al*, *Phytochemistry*, 1988, **27**, 624.

ent-7,14-Labdadiene-2α,3α,13-triol   **L-70010**

$C_{20}H_{34}O_3$ M 322.487

Constit. of *Ixiolaena leptolepis*. Oil.

Lehmann, L. *et al*, *Phytochemistry*, 1988, **27**, 2994.

ent-7,14-Labdadiene-2α,13,20-triol   L-70011

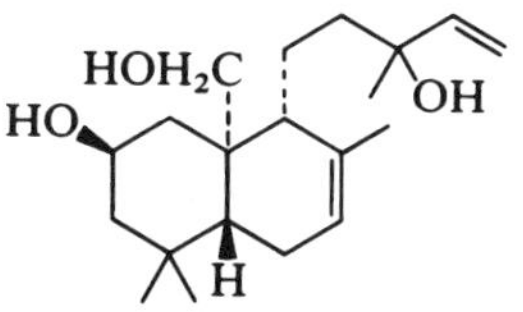

$C_{20}H_{34}O_3$ M 322.487

Constit. of *Ixiolaena leptolepis*. Oil.

Lehmann, L. *et al*, *Phytochemistry*, 1988, **27**, 2994.

8(17),12-Labdadien-3-ol   L-70012

$C_{20}H_{34}O$ M 290.488

(3α,12E)-form

Constit. of *Juniperus pseudosabina*. Oil.

Pandita, K. *et al*, *Indian J. Chem., Sect. B*, 1987, **26**, 453.

8-Labdene-3,7,15-triol   L-70013

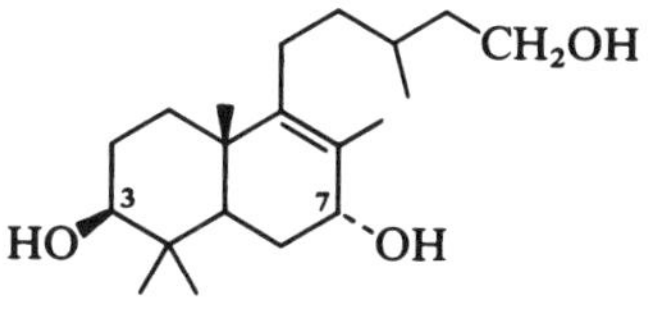

$C_{20}H_{36}O_3$ M 324.503

(3β,7α)-form

7-Me ether: [114359-99-8]. *7α-Methoxy-8-labdene-3β,15-diol.*
$C_{21}H_{38}O_3$ M 338.529
Constit. of *Halimium viscosum*.
7-Me ether, 3,15-Di-Ac: Oil. $[\alpha]_D^{22}$ +24.3° (c, 0.8 in $CHCl_3$).
7-Ketone: [114360-00-8]. *3β,15-Dihydroxy-8-labden-7-one.*
$C_{20}H_{34}O_3$ M 322.487
Constit. of *H. viscosum*.
7-Ketone, 3,15-Di-Ac: Oil. $[\alpha]_D^{22}$ +26.2° (c, 0.72 in $CHCl_3$).

Urones, J.G. *et al*, *Phytochemistry*, 1988, **27**, 501.

ent-14-Labden-8β-ol   L-70014

[117101-04-9]

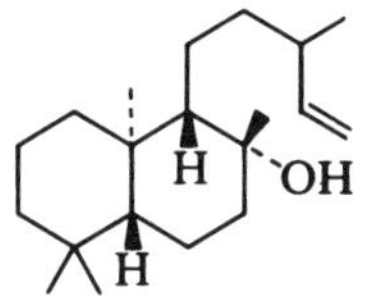

$C_{20}H_{36}O$ M 292.504

Constit. of *Aristolochia cymbifera*. Oil. $[\alpha]_D^{25}$ −5° (c, 0.60 in $CHCl_3$).

Lopes, L.M.X. *et al*, *Phytochemistry*, 1988, **27**, 2265.

Lachnellulone   L-70015

[98036-41-0]

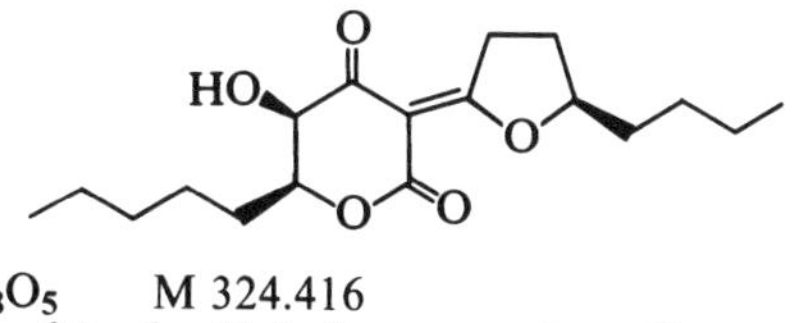

$C_{18}H_{28}O_5$ M 324.416

Metab. of *Lachnellula fuscosanguinea*. Shows antifungal props. Cryst. (Me_2CO/pet. ether). Mp 126-127°. $[\alpha]_D$ +35.0° (c, 1 in MeOH).

Ayer, W.A. *et al*, *Can. J. Chem.*, 1988, **66**, 506.

Lactarorufin *D*   L-70016

[94346-70-0]

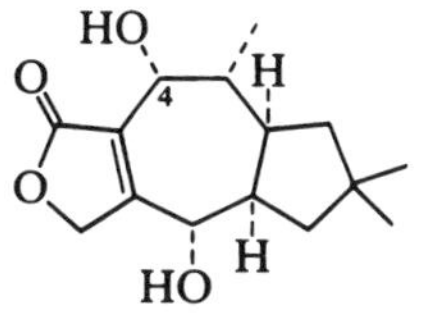

$C_{15}H_{22}O_4$ M 266.336

Constit. of *Lactarius necator*. Needles ($CHCl_3$). Mp 160-162°. $[\alpha]_D^{20}$ +93° (c, 1.0 in $CHCl_3$).

4-Epimer: [94286-87-0]. **Lactarorufin F**.
$C_{15}H_{22}O_4$ M 266.336
Constit. of *L. necator*. Needles ($CHCl_3$). Mp 125-130°. $[\alpha]_D^{20}$ +58° (c, 1.0 in $CHCl_3$).

Daniewski, W.M. *et al*, *Tetrahedron*, 1984, **40**, 2757 (*cryst struct*)

8,24-Lanostadiene-3,21-diol   L-70017

$C_{30}H_{50}O_2$ M 442.724

3β-form

21-Hydroxylanosterol
Constit. of *Uvariastrum zenkeri*. Cryst. (EtOAc). Mp 145-147°.

Muhammad, J. *et al*, *Indian J. Chem., Sect. B*, 1987, **26**, 722.

Lappaphen a   L-70018

[110325-70-7]

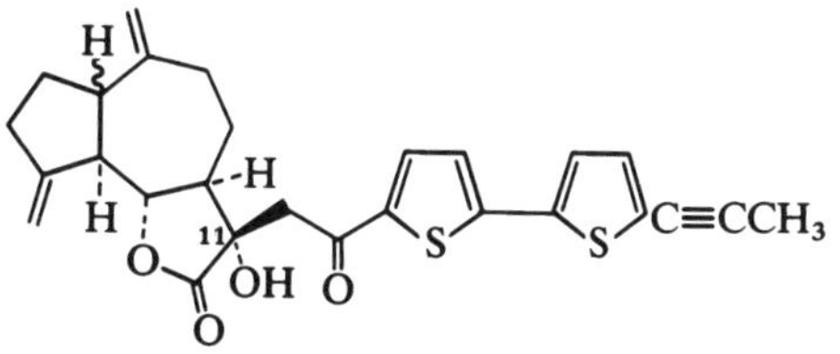

$C_{27}H_{26}O_4S_2$ M 478.620

Constit. of *Arctium lappa*. Yellow needles (EtOAc/hexane). Mp 191.5°. $[\alpha]_D^{20}$ +96° (c, 0.67 in $CHCl_3$).

11-Epimer: [110415-32-2]. **Lappaphen b**.
$C_{27}H_{26}O_4S_2$ M 478.620
Constit. of *A. lappa*. Yellow needles (EtOAc/hexane). Mp 146°. $[\alpha]_D^{20}$ −11° (c, 0.23 in $CHCl_3$).

Washino, T. *et al*, *Agric. Biol. Chem.*, 1987, **51**, 1475.

Lariciresinol   L-70019

Updated Entry replacing L-00144
Tetrahydro-2-(4-hydroxy-3-methoxyphenyl)-4-[(4-hydroxy-3-methoxyphenyl)methyl]-3-furanmethanol, 9CI

[27003-73-2]

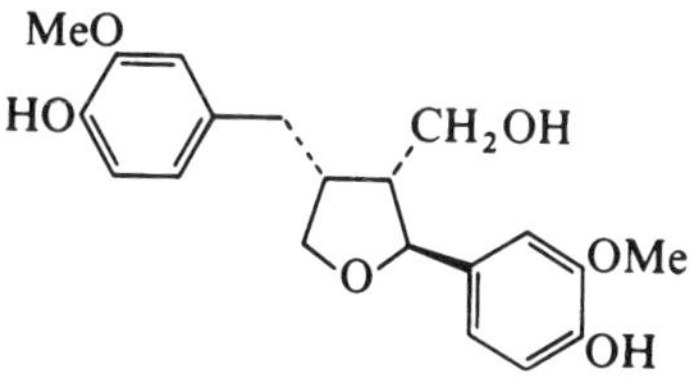

Absolute configuration

$C_{20}H_{24}O_6$ M 360.406

Constit. of the wound resin of the European larch (*Larix decidua*). Also found in *Araucaria angustifolia*. Needles (MeOH). Mp 167-168° (162-164°). $[\alpha]_D^{25}$ +18° (c, 1.0 in Me_2CO).

Di-Me ether: Prisms (Et₂O). Mp 79-80°. $[\alpha]_D^{14}$ +22°
(+12°) (Me₂CO).
Tri-Ac: Viscous oil. $[\alpha]_D^{25}$ +8° (c, 1.0 in CHCl₃).
5,5'-Dimethoxy: [116498-58-9]. **5,5'-
Dimethoxylariciresinol**.
C₂₂H₂₈O₈   M 420.458
Constit. of *Bauhinia manca.* Cryst. Mp 124-126°.
$[\alpha]_D^{21}$ +5° (c, 0.27 in MeOH).

Haworth, R.D., *J. Chem. Soc.*, 1937, 384, 1645; 1939, 1054
   (*isol*)
Ekman, R., *Holzforschung*, 1976, **30**, 79 (*isol*)
Fonseca, F.S. *et al*, *Phytochemistry*, 1978, **17**, 499 (*isol, struct*)
Achenbach, H. *et al*, *Phytochemistry*, 1988, **27**, 1835 (*isol,
   deriv*)

Larreantin   L-70020

[114094-46-1]

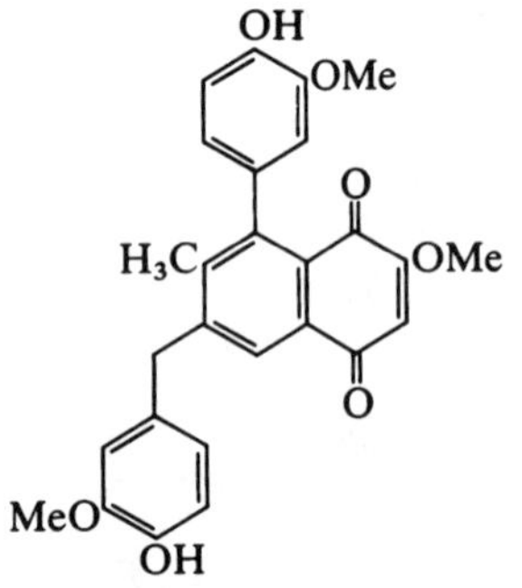

C₂₇H₂₄O₇   M 460.482
Constit. of *Larrea tridentata.* Yellow needles. Mp 204-
206°.

Luo, Z. *et al*, *J. Org. Chem.*, 1988, **53**, 2183.

Latrunculin *A*   L-70021

[76343-93-6]

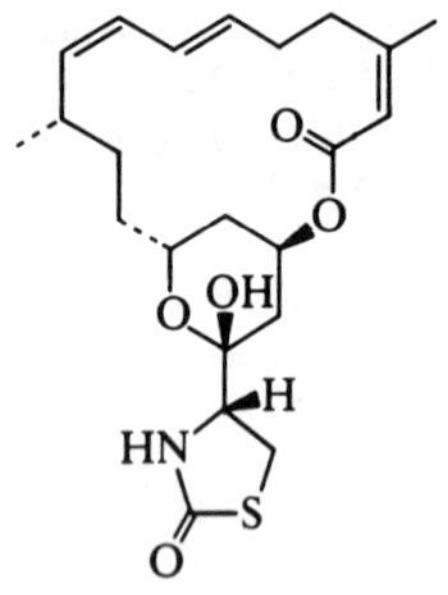

C₂₂H₃₁NO₅S   M 421.551
Constit. of sponge *Latrunculia magnifica.* Fish toxin.
   Foam. $[\alpha]_D^{24}$ +152° (c, 1.2 in CHCl₃).

Kashman, Y. *et al*, *Tetrahedron Lett.*, 1980, **21**, 3629.
Groweiss, A. *et al*, *J. Org. Chem.*, 1983, **48**, 3512.
Spector, I. *et al*, *Science*, 1983, **219**, 493 (*pharmacol*)
Kashman, Y. *et al*, *Tetrahedron*, 1985, **41**, 1905 (*abs config*)

Latrunculin *B*   L-70022

[76343-94-7]

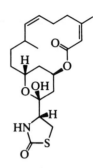

C₂₀H₂₉NO₅S   M 395.513
Constit. of sponge *Latrunculia magnifica.* Cryst. $[\alpha]_D^{24}$
+112° (c, 0.48 in CHCl₃).

Kashman, Y. *et al*, *Tetrahedron*, 1985, **41**, 1905.

Latrunculin *D*   L-70023

Updated Entry replacing L-50032
[76343-94-7]

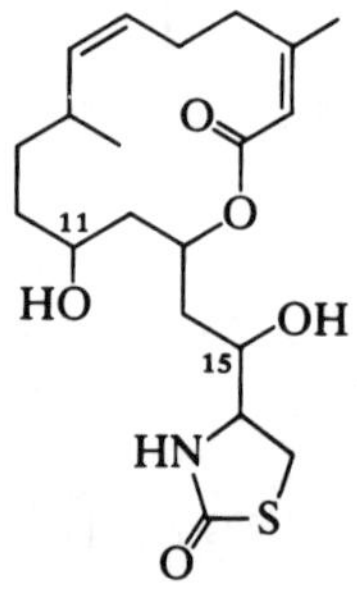

C₂₀H₃₁NO₅S   M 397.529
Macrolide. Isol. from the Red Sea Sponge *Latrunculia
   magnifica.* Ichthyotoxin. Oil.
15-Ketone, O¹¹-Me: [76376-32-4]. **Latrunculin** C.
   C₂₁H₃₁NO₅S   M 409.540
   From *L. magnifica.* Ichthyotoxin. Oil.

Kashman, Y. *et al*, *Tetrahedron*, 1985, **41**, 1905 (*isol, uv, ir,
   pmr, cmr, ms, struct*)

Laurencenone *A*   L-70024

[116384-28-2]

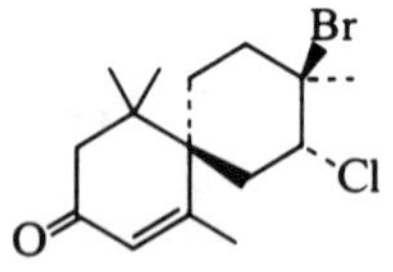

C₁₅H₂₂BrClO   M 333.695
Metab. of *Laurencia obtusa.* Oil.

Kennedy, D.J. *et al*, *Phytochemistry*, 1988, **27**, 1761.

Laurencenone *B*   L-70025

[108925-15-1]

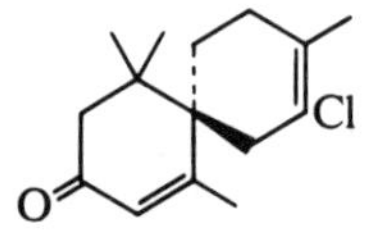

C₁₅H₂₁ClO   M 252.783
Metab. of *Laurencia obtusa.* Oil.
Dechloro: **Laurencenone** C. *1,5,5,9-
   Tetramethylspiro[5.5]undeca-1,8-dien-3-one, 9CI.*
   C₁₅H₂₂O   M 218.338

Metab. of *L. obtusa*. Oil.

Kennedy, D.J. *et al*, *Phytochemistry*, 1988, **27**, 1761.

Laurencenone *D*　　　　　　　　　L-70026

[116384-28-2]

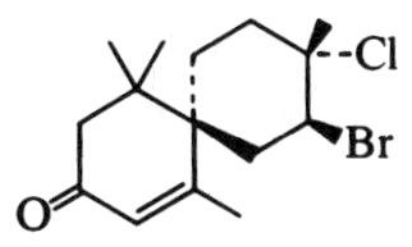

C$_{15}$H$_{22}$BrClO　　M 333.695

Metab. of *Laurencia obtusa*. Solid.

Kennedy, D.J. *et al*, *Phytochemistry*, 1988, **27**, 1761.

Lehmannin　　　　　　　　　　　　L-70027

Lehmanin

[112613-52-2]

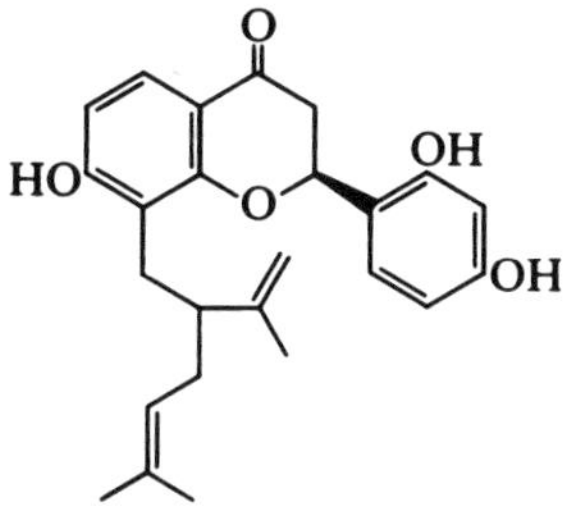

C$_{25}$H$_{28}$O$_5$　　M 408.493

Constit. of *Ammothamnus lehmannii*. Cryst. Mp 102-103°. [α]$_D$ −63.7° (c, 0.43 in MeOH).

Bakirov, E.Kh. *et al*, *Khim. Prir. Soedin.*, 1987, **23**, 429.

Lepidopterans　　　　　　　　　　L-70028

[91196-41-7]

H-Arg-Trp-Lys-Ile-Phe-Lys-Lys-Ile-Glu-Lys-Met-Gly-
Arg-Asn-Ile-Arg-Asp-Gly-Ile-Val-X-Ala-Gly-Pro-
Ala-Ile-Glu-Val-Leu-Gly-Ser-Ala-Lys-Ala-Ile-NH$_2$;
Lepidopteran A, X = Lys; Lepidopteran B, X = δ-Hy-
droxylysine

Self-defence antibacterial peptides prod. by silkworm.

Kikuchi, M., *Tetrahedron*, 1986, **42**, 829 (*struct*)

Leucettidine　　　　　　　　　　　L-70029

Updated Entry replacing L-60023

6-(1-Hydroxypropyl)-1-methyllumazine

[79121-29-2]

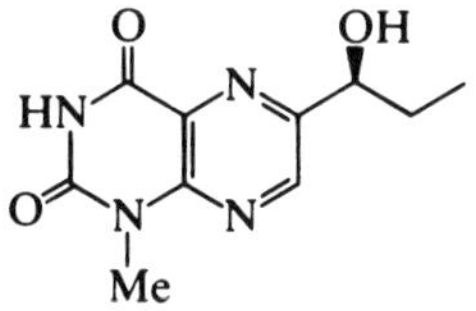

(S)-form

C$_{10}$H$_{12}$N$_4$O$_3$　　M 236.230

Struct. revised in 1988.

(S)-form

Isol. from the calcareous sponge *Leucetta microraphis*.
[α]$_D^{21}$ −35.9° (c, 1.26 in MeOH).

(±)-form

Cryst. Mp 195°.

Cardellina, J.H. *et al*, *J. Org. Chem.*, 1981, **46**, 4782 (*isol, uv, ir, pmr, ms*)

Pfleiderer, W., *Tetrahedron*, 1988, **44**, 3373 (*struct, synth, uv, pmr*)

Linderatone　　　　　　　　　　　L-70030

Updated Entry replacing L-50049

[98155-84-1]

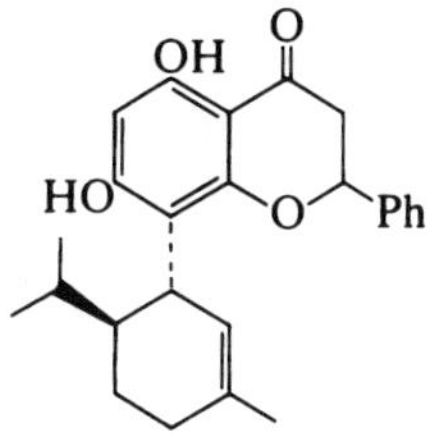

C$_{25}$H$_{28}$O$_4$　　M 392.494

Constit. of leaves of *Lindera umbellata*. Viscous oil. [α]$_D^{25}$ −25.6° (c, 0.5 in CHCl$_3$).

O^7-*Me:* [111786-26-6]. **Methyllinderatone**.
C$_{26}$H$_{30}$O$_4$　　M 406.521
Constit. of *L. umbellata* var. *membranacea*. Viscous oil. [α]$_D$ +68.6° (c, 0.35 in CHCl$_3$).

4″-*Epimer:* [111822-11-8]. **Isolinderatone**.
C$_{25}$H$_{28}$O$_4$　　M 392.494
Constit. of *L. umbellata* var. *membranacea*. Viscous oil. [α]$_D$ −67.1° (c, 1.25 in CHCl$_3$).

Tanaka, H. *et al*, *Chem. Pharm. Bull.*, 1985, **33**, 2602 (*isol, struct*)

Ichino, K. *et al*, *Chem. Pharm. Bull.*, 1987, **35**, 920 (*deriv*)

Linderazulene　　　　　　　　　　L-70031

Updated Entry replacing L-20044

3,5,8-Trimethylazuleno[6,5-b]furan, 9CI

[489-79-2]

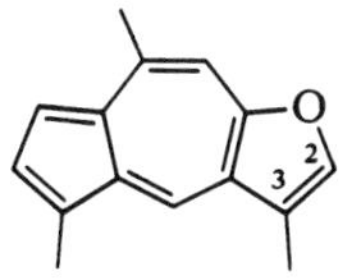

C$_{15}$H$_{14}$O　　M 210.275

Pigment of the sea gorgonian *Paramuricea chamaeleon*. Dehydrogenation prod. of sesquiterpenoids. Lustrous violet-black plates (2-propanol). Mp 106-107°.

2,3-Dihydro: [110207-64-2]. *2,3-Dihydrolinderazulene.*
11,12-Dihydrolinderazulene.
C$_{15}$H$_{16}$O　　M 212.291
Constit. of an *Acalycigorgia* sp. Purple oil. [α]$_D$ +800° (c, 0.05 in CHCl$_3$).

Takeda, K. *et al*, *J. Chem. Soc.*, 1964, 2591 (*synth, bibl*)

Imre, S. *et al*, *Experientia*, 1981, **37**, 442 (*isol*)

Sakemi, S. *et al*, *Experientia*, 1987, **43**, 624 (*deriv*)

Lipid *A*　　　　　　　　　　　　L-70032

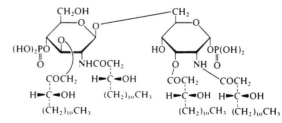

C$_{68}$H$_{130}$N$_2$O$_{23}$P$_2$　　M 1405.722

Struct. of lipid *A* from a *Salmonella* mutant is shown. Structs. of lipid *A*'s from *E. coli*, *S. minnesota* and *Proteus mirabilis* have been detd. and are derived from the struct. shown by acylation at two or more of the

3-hydroxytetradecanoyl residues. Component of gram-
negative bacteria lipopolysaccharide. Plays an important role in toxicity, pyrogenicity and adjuvant activity.

Ikeda, K. *et al*, *Chem. Pharm. Bull.*, 1987, **35**, 4436, 4517 (*bibl*)

Lipoxamycin   L-70033

Updated Entry replacing L-00362
*2-Amino-N,3-dihydroxy-N-(14-methyl-3,10-
dioxopentadecyl)propanamide, 9CI. 2-Amino-1,4-dihy-
droxy-18-methyl-4-aza-3,7,14-nonadecanetrione.
Neoenactin M$_1$*

[32886-15-0]

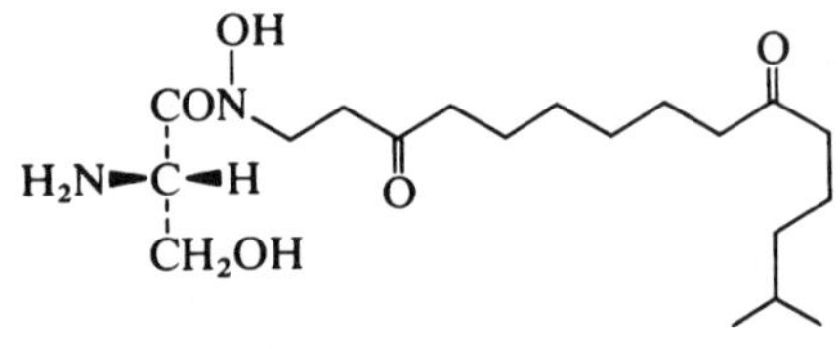

$C_{19}H_{36}N_2O_5$ M 372.504
Isol. from *Streptomyces virginiae* and *S. olivoreticuli* ssp.
neoenactus. Shows antifungal activity. Mp 68-70°.
Related to Neoenactins.

▷MU5230000.

B,H$_2$SO$_4$: Mp 155°.

Whaley, M.A., *J. Am. Chem. Soc.*, 1971, **93**, 3767 (*isol, struct*)
Roy, S.K. *et al*, *J. Antibiot.*, 1987, **40**, 266 (*isol*)

Lithospermoside   L-70034

Updated Entry replacing L-30029
[63492-69-3]

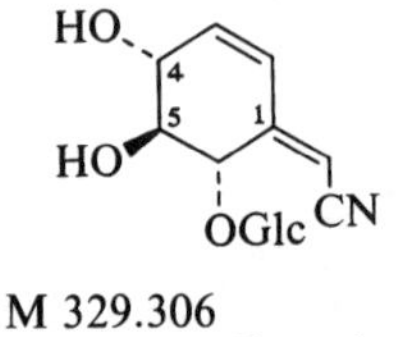

$C_{14}H_{19}NO_8$ M 329.306
Constit. of *Lithospermum officinale* and *L. caerulum*.
Cryst. (MeOH aq.). Mp 278-279°. [α]$_D^{20}$ −156° (c,
0.99 in H$_2$O).

5-Epimer: [84799-31-5].
$C_{14}H_{19}NO_8$ M 329.306
Constit. of fruits of *Ilex warburgii*. Cryst. (MeOH
aq.). Mp 221-223°. [α]$_D^{25}$ −247° (c, 0.611 in MeOH).

5-Deoxy: [67765-58-6]. **Menisdaurin**.
$C_{14}H_{19}NO_7$ M 313.307
Isol. from *I. warburgii* and *Menispermum dauricum*.
Plates. Mp 175-176°. [α]$_D^{15}$ −185.4° (c, 1.00 in
MeOH).

4,5,6-Triepimer: **Griffonin**.
$C_{14}H_{19}NO_8$ M 329.306
Constit. of roots of *Griffonia simplicifolia*. Feathery
cryst. (MeOH). Mp 263-265° dec. [α]$_D^{21}$ +6° (c, 0.5 in
H$_2$O). Griffonin is given the same reg. no. as
Lithospermoside in CA but appears to be a distinct
diastereoisomer. The opt. rotns. are v. different.

Dwuma-Badu, D. *et al*, *Lloydia*, 1976, **39**, 213 (*Griffonin*)
Sosa, A. *et al*, *Phytochemistry*, 1977, **16**, 707 (*isol, struct*)
Takahashi, K. *et al*, *Chem. Pharm. Bull.*, 1978, **26**, 1677
    (*Menisdaurin*)
Ueda, K. *et al*, *Chem. Lett.*, 1983, 149 (*isol, struct*)

Longikaurin A   L-70035

Updated Entry replacing L-00424
[75207-67-9]

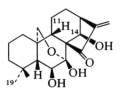

$C_{20}H_{28}O_5$ M 348.438
Constit. of *Rabdosia longituba*. Needles (MeOH). Mp
223-225°. [α]$_D$ −91.1° (c, 0.21 in Py).

19-Acetoxy: [75207-66-8]. **Longikaurin B**.
$C_{22}H_{30}O_7$ M 406.475
Constit. of *R. longituba*. Needles (CHCl$_3$/MeOH).
[α]$_D^{25}$ −115.9° (c, 0.12 in Py).

19-Acetoxy, 14-deoxy: [77284-05-0]. **Longikaurin C**.
$C_{22}H_{30}O_6$ M 390.475
Constit. of *R. longituba*. Cryst. Mp 248-250°. [α]$_D^{25}$
−137.5° (c, 0.12 in Py).

19-Acetoxy, 14-deoxy, 11α-hydroxy: [77967-61-4]. **Longikaurin D**.
$C_{22}H_{30}O_7$ M 406.475
Constit. of *R. longituba*. Mp 262-264°. [α]$_D^{25}$ −109.0°
(c, 0.13 in Py).

11α-Acetoxy, 14-deoxy: [77949-42-9]. **Longikaurin E**.
$C_{22}H_{30}O_6$ M 390.475
Constit. of *R. longituba*. Mp 252-254°. [α]$_D^{25}$ −78.6°
(c, 0.21 in Py).

11α,19-Diacetoxy, 14-deoxy: [77967-62-5]. **Longikaurin
F**.
$C_{24}H_{32}O_8$ M 448.512
Constit. of *R. longituba*. Mp 249-251°. [α]$_D^{25}$ −120.4°
(c, 0.11 in Py).

Fujita, T. *et al*, *Heterocycles*, 1981, **16**, 227 (*isol*)
Takeda, Y. *et al*, *J. Chem. Soc., Perkin Trans. 1*, 1988, 379
    (*isol*)

Longirabdosin   L-70036

[116425-02-6]

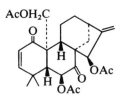

$C_{26}H_{32}O_8$ M 472.534
Constit. of *Rabdosia longituba*. Needles. Mp 196-198°.
[α]$_D^{27}$ −150.0° (c, 0.16 in MeOH).

Ichihara, T. *et al*, *Phytochemistry*, 1988, **27**, 2261.

Loroglossin   L-70037

[58139-22-3]

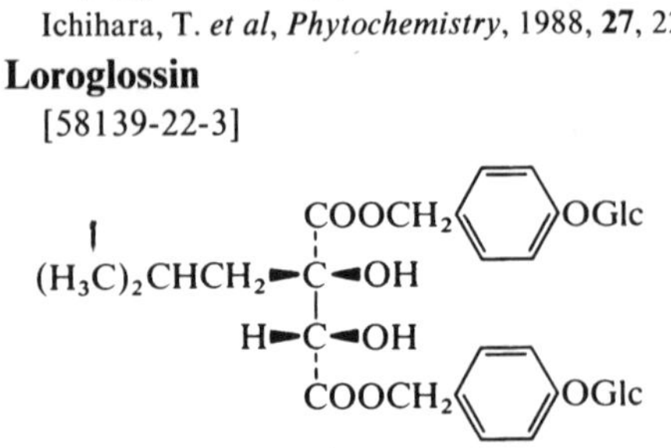

$C_{34}H_{46}O_{18}$ M 742.727
Isol. from many plants in the Orchidaceae, incl. *Orchis,
Cephalanthera, Epipactis, Goodyera, Limodorum,
Listera, Loroglossum, Platanthera* and *Spiranthes*.
Mp 151.5-153° dec. (149-151°). [α]$_D$ −45.65° (H$_2$O).

Aasen, A. *et al*, *Acta Chem. Scand., Ser. B*, 1975, **29**, 1002.
Gray, R.W. *et al*, *Helv. Chim. Acta*, 1977, **60**, 1304 (*struct, abs config, uv, ir, pmr, bibl*)

Lotisoflavan   L-70038

4-(3,4-Dihydro-5,7-dimethoxy-2H-1-benzopyran-3-yl)-1,3-benzenediol, 9CI. 5,7-Dimethoxy-2',4'-dihydroxyisoflavan

[77370-02-6]

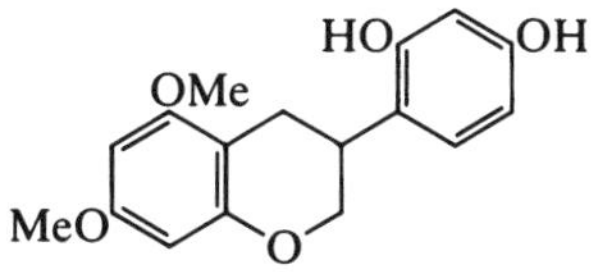

$C_{17}H_{18}O_5$ M 302.326

Isol. from *Lotus angustissimus* and *L. edulis* inoculated with *Helminthosporium carbonum*. Phytoalexin. Cryst. (EtOH aq.). Mp 135-137°.

Ingham, J.L. *et al*, *Phytochemistry*, 1980, **19**, 2799 (*isol, struct*)

Loxodellonic acid   L-70039

[113689-49-9]

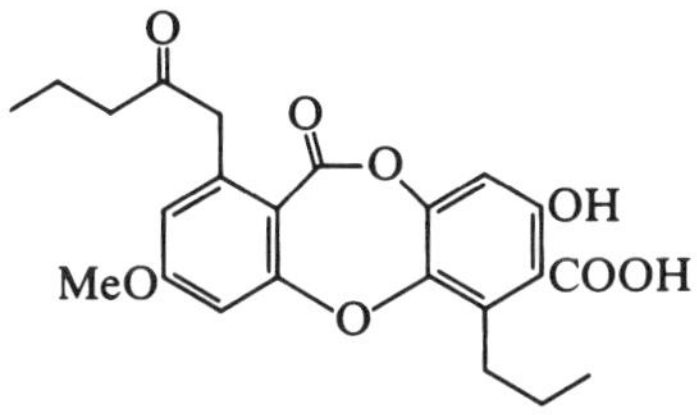

$C_{23}H_{24}O_8$ M 428.438

Constit. of *Neofuscelia subinorta*. Cryst. (EtOAc/pet. ether). Mp 159-161°.

Elix, J.A. *et al*, *Aust. J. Chem.*, 1987, **40**, 2031.

Luffariellolide   L-70040

[111149-87-2]

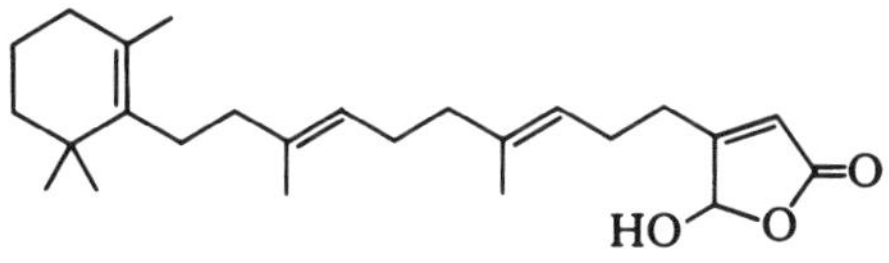

$C_{25}H_{38}O_3$ M 386.573

Constit. of a Palavan sponge *Luffariella* sp. Oil.

Albizati, K.F. *et al*, *Experientia*, 1987, **43**, 949.

Lumiyomogin   L-70041

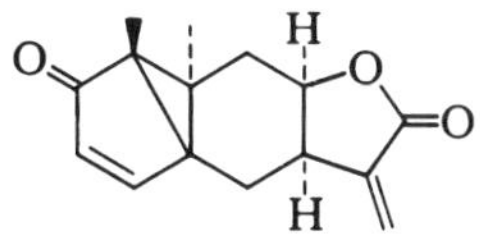

$C_{15}H_{16}O_3$ M 244.290

Constit. of *Ferreyranthus fruticosus*. Cryst. Mp 94°.

Jakupovic, J. *et al*, *Phytochemistry*, 1988, **27**, 1113.

12,20(29)-Lupadiene-3,27,28-triol   L-70042

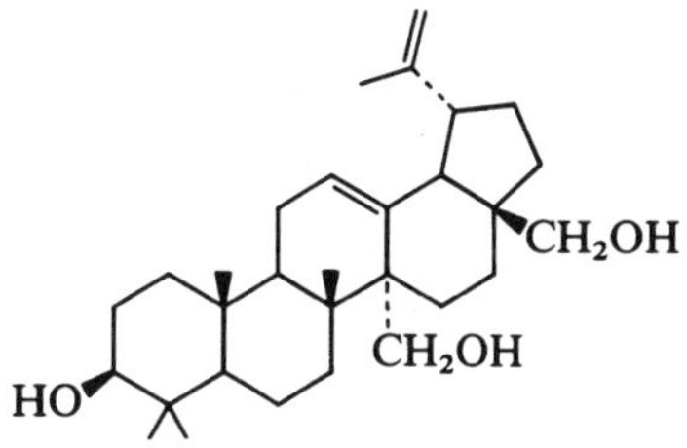

$C_{30}H_{48}O_3$ M 456.707

3β-form [114906-01-3]

Oleanderol

Constit. of *Nerium oleander*. Plates (CHCl$_3$). Mp 206-208°. $[\alpha]_D^{24}$ +6.15° (c, 0.3 in CHCl$_3$).

Siddiqui, S. *et al*, *J. Nat. Prod.*, 1988, **51**, 229.

13(18)-Lupen-3-ol   L-70043

$C_{30}H_{50}O$ M 426.724

3β-form [107159-83-1]

Constit. of *Swertia petiolata*. Cryst. Mp 268°.

Bhan, S. *et al*, *Phytochemistry*, 1988, **27**, 539.

Lycopersene   L-70044

Updated Entry replacing L-00529

2,6,10,14,19,23,27,31-Octamethyl-2,6,10,14,18,22,26,30-dotriacontaoctaene

[502-62-5]

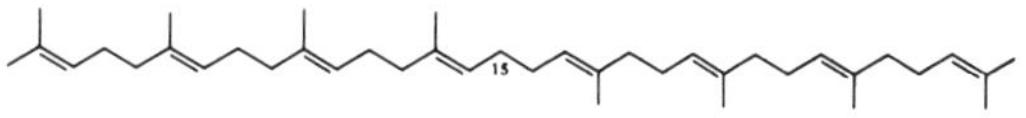

$C_{40}H_{66}$ M 546.961

Oil. Bp$_{0.02}$ 225-228°. Probably not a nat. prod.

15-Hydroxy: **15-Hydroxylycopersene**.

$C_{40}H_{66}O$ M 562.961

Constit. of *Myriophyllum verticillatum*. $[\alpha]_D$ +6.2° (c, 0.7 in CHCl$_3$). Carotenoid numbering.

Chishti, N.H. *et al*, *Tetrahedron Lett.*, 1975, 1025 (*isol*)
Lanzetta, R. *et al*, *Phytochemistry*, 1988, **27**, 887 (*deriv*)

Lycoserone   L-70045

[98094-09-8]

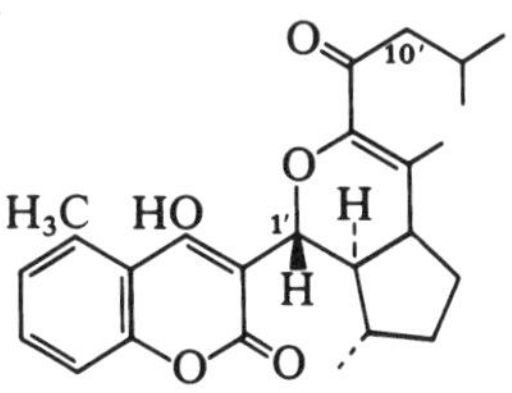

$C_{25}H_{30}O_5$ M 410.509

Constit. of *Lycoseris latifolia*.

1'-Epimer: **1'-Epilycoserone**.

$C_{25}H_{30}O_5$ M 410.509

Constit. of *L. latifolia*. Oil.

1'-Epimer, 10'-Hydroxy: **10'-Hydroxy-1'-epilycoserone**.

$C_{25}H_{30}O_6$ M 426.508

Constit. of *Aphyllocladus denticulatus*.

Bohlmann, F. *et al*, *Justus Liebigs Ann. Chem.*, 1985, 1367 (*isol*)
Zdero, C. *et al*, *Phytochemistry*, 1988, **27**, 1821 (*isol*)

Lycoxanthol L-70046

Updated Entry replacing L-10060

1,3,4,8,9,11b-Hexahydro-5,7,11-trihydroxy-4,4,8,11b-tetramethylphenanthro[3,2-b]furan-6(2H)-one, 9Cl

[33853-96-2]

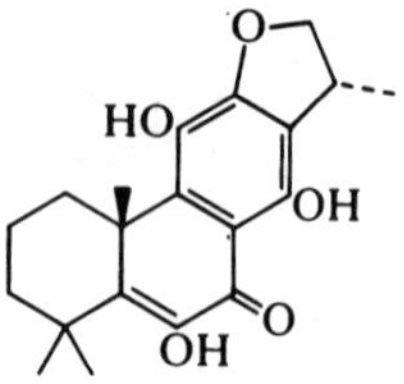

C$_{20}$H$_{24}$O$_5$ M 344.407

Pigment from *Lycopodium lucidulum.* Mp 249°.

Burnell, R.H. et al, *J. Chem. Soc., Chem. Commun.*, 1971, 897
(*struct, uv, ir*)

Burnell, R.H., *Phytochemistry*, 1972, **11**, 2815 (*struct*)

Matsumoto, T. et al, *Bull. Chem. Soc. Jpn.*, 1987, **60**, 3639
(*synth, abs config*)

M

Maackiain
M-70001

Updated Entry replacing M-20001

6a,12a-*Dihydro-6H-[1,3]dioxolo[5,6]benzofuro[3,2-c][1]benzopyran-3-ol, 9CI. 3-Hydroxy-8,9-methylenedioxypterocarpan. 7-Hydroxy-4′,5′-methylenedioxypterocarpan (obsol.). Demethylpterocarpin. Inermin*

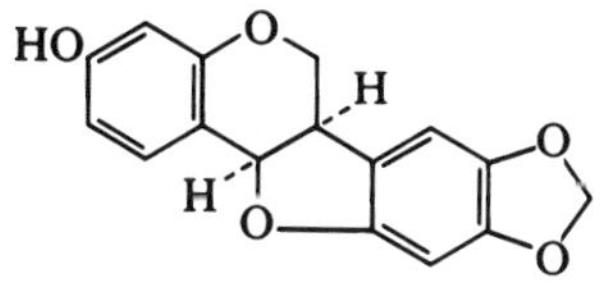

(−)-*form*
Absolute configuration

$C_{16}H_{12}O_5$ M 284.268

(−)-form [2035-15-6]

Constit. of *Maackia amurensis* heartwood and *Sophora tormentosa* aerial parts. Also from *Andira inermis*, *Swartzia madagascariensis* and *Trifolia pratense*. Shows antifungal props. Leaflets (MeOH aq.). Mp 179-181°. $[\alpha]_D^{22}$ −260° (c, 1.0 in Me$_2$CO).

O-β-D-*Glucoside:* [6807-83-6]. **Trifolirhizin**.
$C_{22}H_{22}O_{10}$ M 446.410
Constit. of red clover *T. pratense*. Shows antifungal props. Rods + 1MeOH (MeOH). Mp 142-144° dec. $[\alpha]_D^{20}$ −183° (c, 1.5 in EtOH).

O-(6-O-*Acetyl-β-D-glucoside):* [60679-70-1]. *6′-Acetyltrifolirhizin*. Constit. of *Sophora subprostrata*. Mp 223-225°. $[\alpha]_D^{22}$ −175° (AcOH).

Me ether: [524-97-0]. **Pterocarpin**.
$C_{17}H_{14}O_5$ M 298.295
Isol. from red sandalwood, *Swartzia madagascariensis* and *Flemingia chappar*. Plates (pet. ether or EtOH). Mp 168-169° (165°). $[\alpha]_D$ −214.5° (c, 0.53 in CHCl$_3$).

(+)-form

O-β-D-*Glucoside:* **Sophojaponicin**.
$C_{22}H_{22}O_{10}$ M 446.410
Isol. from roots of *Sophora japonica*. Prisms or needles (MeOH). Mp 202-204° dec. $[\alpha]_D^{17}$ −104° (c, 0.70 in AcOH).

(±)-form [19908-48-6]

Constit. of *Sophora japonica* and *Dalbergia spruceana*. Needles or plates (MeOH). Mp 196-196°.

Me ether: Leaflets (Me$_2$CO or MeOH). Mp 185-186°.

Suginome, H., *Experientia*, 1962, **18**, 161 (isol, ir, uv, pmr)
Shibata, S. et al, *Chem. Pharm. Bull.*, 1963, **11**, 167 (isol, uv, ir)
Ito, S. et al, *J. Chem. Soc., Chem. Commun.*, 1965, 595 (abs config)
Suginome, H., *Bull. Chem. Soc. Jpn.*, 1966, **39**, 1544 (Sophojaponicin)
Pachler, K.G.R. et al, *Tetrahedron*, 1967, **23**, 1817 (nmr)
Kukui, K. et al, *Experientia*, 1968, **24**, 536 (synth, ir, uv)
Harper, S.H. et al, *J. Chem. Soc. (C)*, 1969, 1109 (isol)
Dewick, P.M., *Phytochemistry*, 1975, **14**, 979 (isol)
Horino, H. et al, *J. Chem. Soc., Chem. Commun.*, 1976, 500 (synth)
Komatsu, M. et al, *Phytochemistry*, 1976, **15**, 1089 (isol, deriv)
Pelter, A. et al, *J. Chem. Soc., Perkin Trans. 1*, 1976, 2475 (cmr)
Dewick, P.M., *Phytochemistry*, 1977, **16**, 93 (biosynth)
Dewick, P.M. et al, *Phytochemistry*, 1978, **17**, 1751 (isol)
Komatsu, M. et al, *Chem. Pharm. Bull.*, 1978, **26**, 1274 (isol)

Al-Ani, H.A.M. et al, *J. Chem. Soc., Perkin Trans. 1*, 1984, 2831 (biosynth)

Machilin *H*
M-70002

[112503-91-0]

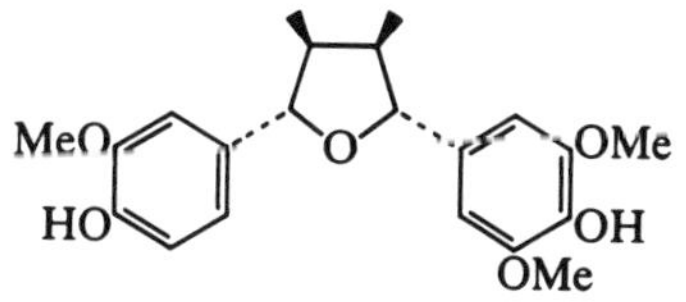

$C_{21}H_{26}O_6$ M 374.433
Constit. of *Machilus thunbergii*. Oil. $[\alpha]_D$ +8.8° (c, 0.37 in CHCl$_3$).

Shimomura, H. et al, *Phytochemistry*, 1988, **27**, 634.

Machilin *I*
M-70003

4,4′-(*Tetrahydro-3,4-dimethyl-2,5-furandiyl)bis[2-methoxyphenol], 9CI. Tetrahydro-2,5-bis(4-hydroxy-3-methoxyphenyl)-3,4-dimethylfuran*

[114422-21-8]

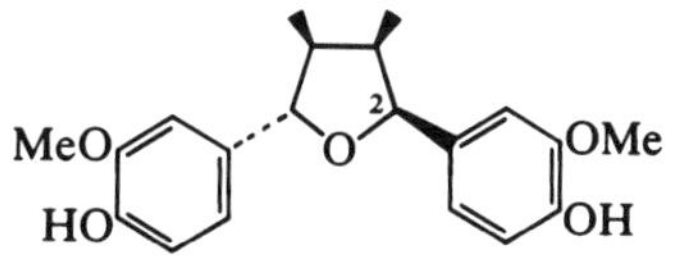

$C_{20}H_{24}O_5$ M 344.407
Constit. of *Machilus thunbergii*. Oil. $[\alpha]_D^{25}$ −93.0° (c, 0.13 in CHCl$_3$).

2-*Epimer:* [74683-16-2]. **Nectandrin B**.
$C_{20}H_{24}O_5$ M 344.407
Lignan from *Nectandra ridiga, M. thunbergii* and *Myristica fragrans*. Oil. Opt. inactive (meso-).

Le Quesne, P.W. et al, *J. Nat. Prod.*, 1980, **43**, 353.
Hattori, M. et al, *Chem. Pharm. Bull.*, 1987, **35**, 3315.
Shimomura, H. et al, *Phytochemistry*, 1988, **27**, 634.

Macrophylloside *C*
M-70004

[113270-96-5]

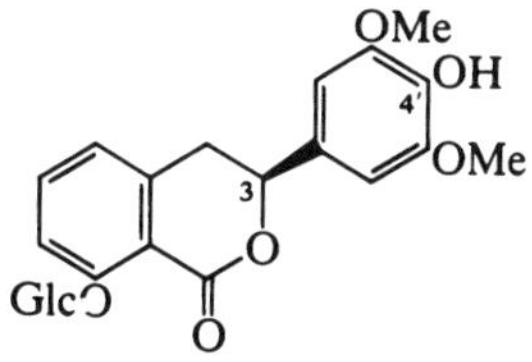

$C_{23}H_{26}O_{11}$ M 478.452
Constit. of *Hydrangea macrophylla* subsp. *serrata*.

Penta-Ac: Cryst. (Et$_2$O/EtOAc). Mp 237-238°. $[\alpha]_D$ +46.8° (c, 0.62 in CHCl$_3$).

3-*Epimer:* [113296-38-1]. **Macrophylloside B**.
$C_{23}H_{26}O_{11}$ M 478.452
Constit. of *H. macrophylla*.

3-*Epimer, Penta-Ac:* Cryst. (Et$_2$O/EtOAc). Mp 237-238°. $[\alpha]_D$ −131.0° (c, 0.45 in CHCl$_3$).

4′-*Me ether:* [113270-95-4]. **Macrophylloside A**.
$C_{24}H_{28}O_{11}$ M 492.479

Constit. of *H. macrophylla*. Cryst. (MeOH/Et$_2$O).
Mp 110-112°. [α]$_D$ −134.4° (c, 1.22 in MeOH).

Hashimoto, T. *et al*, *Phytochemistry*, 1987, **26**, 3323.

Majucin   M-70005

[114687-97-7]

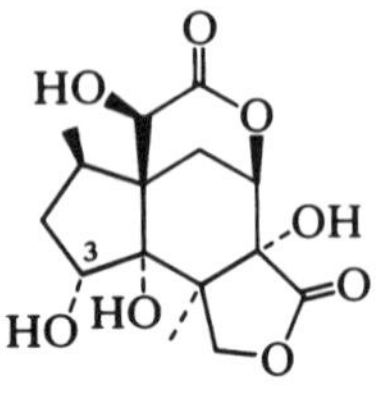

$C_{15}H_{20}O_8$ M 328.318
Constit. of *Illicium majus*. Needles (MeOH). Mp 251-
252°. [α]$_D^{24}$ −74° (c, 0.15 in dioxan).

3-Deoxy: [114687-98-8]. **Neomajucin**.
$C_{15}H_{20}O_7$ M 312.319
Constit. of *I. majus*. Octahedra (EtOAc). Mp 220-
222°. [α]$_D^{24}$ −75° (c, 0.25 in dioxan).

Yang, C.-S. *et al*, *Tetrahedron Lett.*, 1988, **29**, 1165 (*cryst struct*)

Majusculone   M-70006

1,5,5-Trimethylspiro[5.5]undeca-1,7-diene-3,9-dione,
9CI

[112642-49-6]

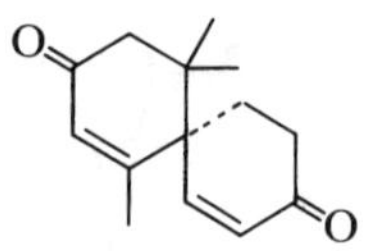

$C_{14}H_{18}O_2$ M 218.295
Constit. of the red alga *Laurencia majuscula*. Cryst.
(diisopropyl ether). Mp 91.1-92.0°. [α]$_D^{19}$ +145° (c,
0.965 in CHCl$_3$).

Suzuki, M. *et al*, *Bull. Chem. Soc. Jpn.*, 1987, **60**, 3795.

Malabaricone A   M-70007

Updated Entry replacing M-00059
1-(2,6-Dihydroxyphenyl)-9-phenyl-1-nonanone, *9CI*
[63335-23-9]

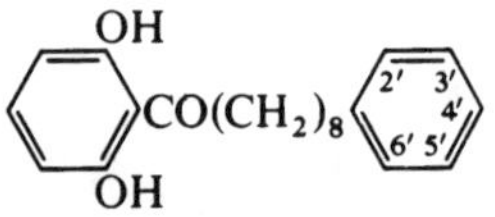

$C_{21}H_{26}O_3$ M 326.435
Constit. of *Myristica malabarica* fruit rind. Yellow cryst.
(Et$_2$O/hexane). Mp 81-82°.

4′-Hydroxy: [63335-24-0]. *1-(2,6-Dihydroxyphenyl)-9-
(4-hydroxyphenyl)-1-nonanone, 9CI*. **Malabaricone B**.
$C_{21}H_{26}O_4$ M 342.434
Constit. of *M. malabarica*. Cryst. (C$_6$H$_6$). Mp 102°.

3′,4′-Dihydroxy: [63335-25-1]. *1-(2,6-Dihydroxy-
phenyl)-9-(3,4-dihydroxyphenyl)-1-nonanone, 9CI*.
Malabaricone C.
$C_{21}H_{26}O_5$ M 358.433
Constit. of *M. malabarica*. Pale-yellow cryst. (CHCl$_3$).
Mp 123-124°.

3′,4′-(Methylenedioxy): [63335-26-2]. *9-(1,3-Benzo-
dioxol-5-yl)-1-(2,6-dihydroxyphenyl)-1-nonanone,
9CI. 9-(3,4-Dihydroxyphenyl)-1-(3,4-methylenedioxy-
phenyl)-1-nonanone.* **Malabaricone D**.
$C_{22}H_{26}O_5$ M 370.444

Constit. of *M. malabarica*. Cryst. (C$_6$H$_6$/hexane). Mp
90-91°.

*3′,4′-(Methylenedioxy), 2-Me ether: 1-(6-Hydroxy-2-
methoxyphenyl)-9-(3,4-methylenedioxyphenyl)-1-
nonanone.*
$C_{22}H_{28}O_5$ M 372.460
Constit. of the stem bark of *M. dactyloides*. Yellow
cryst. (CHCl$_3$/MeOH). Mp 51-52°.

*4′-Hydroxy, 2-Me ether: 1-(6-Hydroxy-2-methoxy-
phenyl)-9-(4-hydroxyphenyl)-1-nonanone.*
$C_{22}H_{28}O_4$ M 356.461
Constit. of the stem bark of *N. dactyloides*. Pale-
yellow cryst. (Me$_2$CO/hexane). Mp 65-66°.

Purushothaman, K.K. *et al*, *J. Chem. Soc., Perkin Trans. 1*,
1977, 587 (*isol, ir, uv, pmr, cmr, ms*)
Parthasarathy, M.R. *et al*, *Indian J. Chem., Sect. B*, 1985, **24**,
965 (*synth*)
Kumar, N.S. *et al*, *Phytochemistry*, 1988, **27**, 465 (*isol*)

Malabarolide   M-70008

$C_{18}H_{24}O_7$ M 352.383
Constit. of *Tinospora malabarica*. Cryst. (EtOH). Mp
199-200°. [α]$_D$ −4.48° (c, 1.97 in MeOH).

Atta-ur-Rahman *et al*, *Tetrahedron Lett.*, 1988, **29**, 4241 (*cryst
struct*)

Mannopinic acid   M-70009

1-Deoxy-1-[(1,3-dicarboxypropyl)amino]mannitol, 9CI
[74524-18-8]

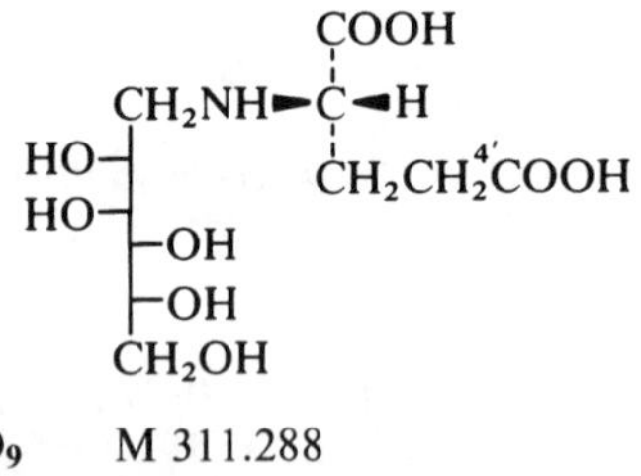

$C_{11}H_{21}NO_9$ M 311.288
4′-Amide: [87084-52-4]. **Mannopine**.
$C_{11}H_{22}N_2O_8$ M 310.303
Opine found in plant tumours.

Tempe, J. *et al*, *C.R. Hebd. Seances Acad. Sci., Ser. D*, 1980,
290, 1173 (*synth*)
Chilton, W.S. *et al*, *J. Bacteriol.*, 1984, **158**, 650.

Manwuweizic acid   M-70010

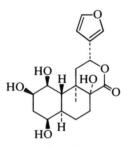

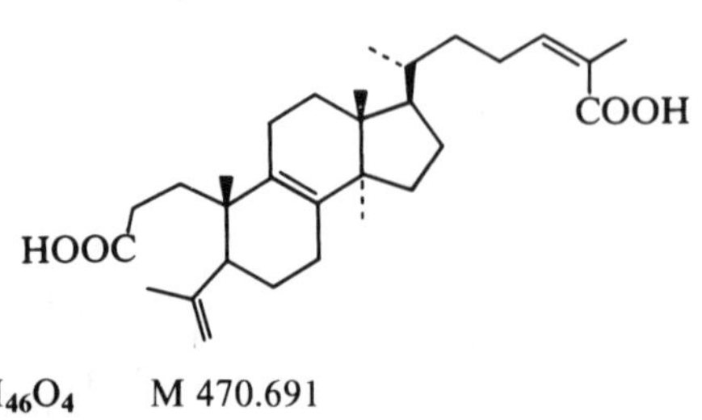

$C_{30}H_{46}O_4$ M 470.691

Constit. of *Schisandra propinqua*. Shows cytotoxic activity. Cryst (CHCl$_3$/pet. ether). Mp 191-193°. [α]$_D^{15}$ +54.3° (c, 0.291 in CHCl$_3$).

Liu, J.-S. *et al, Can. J. Chem.*, 1988, **66**, 414.

Maragenin I   M-70011

Updated Entry replacing M-00147
3β-Hydroxy-28-norolean-12-en-16-one
[71545-19-2]

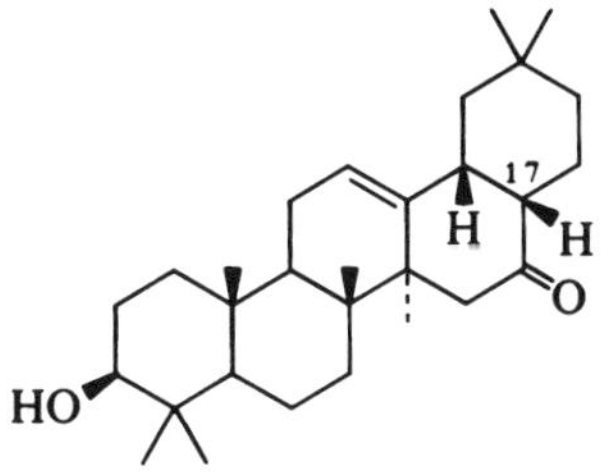

C$_{29}$H$_{46}$O$_2$ M 426.681
Constit. of *Marah macrocarpus*. Cryst. (EtOAc). Mp 218-220°.
Ac: Cryst. Mp 215-217°.
12,13-Dihydro: [71545-21-6]. *28-Nor-16-oxo-17-oleanen-3β-ol.* **Maragenin III**.
C$_{29}$H$_{48}$O$_2$ M 428.697
Constit. of *M. macrocarpus*.
12,13-Dihydro, Ac: Cryst. Mp 216-218°.
17,18-Didehydro: [71545-20-5]. *3β-Hydroxy-28-noroleana-12,17-dien-16-one.* **Maragenin II**.
C$_{29}$H$_{44}$O$_2$ M 424.665
Constit. of *M. macrocarpus*.
17,18-Didehydro, Ac: Mp 215-217°.

Hylands, P.J. *et al, Tetrahedron*, 1979, **35**, 417.

Mariesiic acid *A*   M-70012

3,23-Dihydroxy-4,4,17-trimethylcholesta-7,14,24-trien-26-oic acid, 9CI. 3α,23R-Dihydroxy-17,14-friedo-9β-lanosta-7,14,24E-trien-26-oic acid
[99624-26-7]

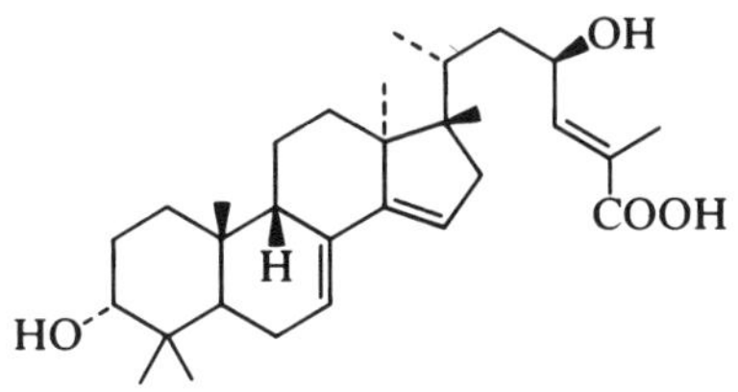

C$_{30}$H$_{46}$O$_4$ M 470.691
Constit. of seeds of *Abies mariesii*. Antimicrobial. Needles (EtOAc). Mp 197-200°. [α]$_D$ +116° (c, 1.73 in Me$_2$CO).
23-Ketone: [113105-30-9]. *23-Oxomariesiic acid* A.
C$_{30}$H$_{44}$O$_4$ M 468.675
Constit. of *A. mariesii* and *A. firma*.
23-Ketone, Me ester: Gum. [α]$_D$ +53.2° (c, 1.28 in CHCl$_3$).

Hasegawa, S. *et al, Chem. Lett.*, 1985, 1589 (*cryst struct*)
Hasegawa, S. *et al, Tetrahedron*, 1987, **43**, 1775 (*isol, deriv*)

Marrubiin   M-70013

Updated Entry replacing M-00172
[465-92-9]

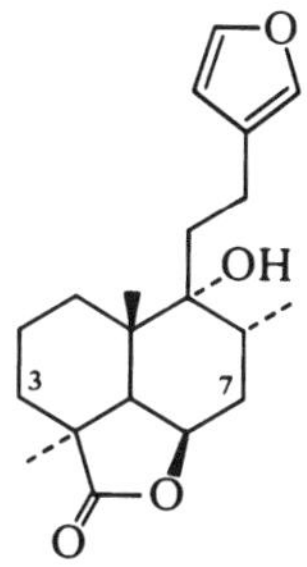

C$_{20}$H$_{28}$O$_4$ M 332.439
Constit. of *Marrubium vulgare* and *Leonotis leonurus*. Cryst. (EtOH). Mp 160°. [α]$_D$ +33.3° (c, 1 in CHCl$_3$).
3-Oxo: [19898-90-9]. **Peregrinone**. *Peregrinin*.
C$_{20}$H$_{26}$O$_5$ M 346.422
Isol. from *M. peregrinum* and *M. incanum*. Cryst. Mp 172-173°. [α]$_D$ +48° (c, 1 in MeOH).
7-Oxo: [61289-05-2]. **Ballotinone**.
C$_{20}$H$_{26}$O$_5$ M 346.422
Constit. of *Ballota nigra*. Cryst. (EtOAc/pet. ether). Mp 194°. [α]$_D^{20}$ +57° (c, 0.3 in CHCl$_3$).

Abbondanza, A. *et al, Tetrahedron Lett.*, 1965, 4337 (*biosynth*)
Canonica, L. *et al, Tetrahedron Lett.*, 1968, 3149 (*Perigrinone*)
Stephens, L.J. *et al, Tetrahedron*, 1970, **26**, 1561 (*struct*)
Mangoni, L. *et al, Tetrahedron*, 1972, **28**, 611 (*synth*)
Almqvist, S. *et al, Acta Chem. Scand., Ser. B*, 1975, **29**, 695 (*cmr*)
Savona, G. *et al, J. Chem. Soc., Perkin Trans. 1*, 1976, 1607 (*Ballotinone*)

Mbamichalcone   M-70014

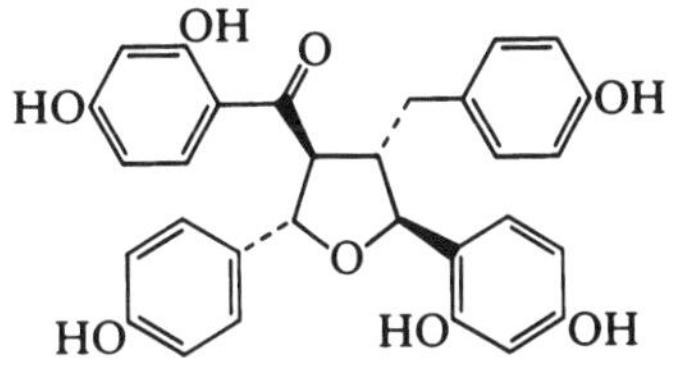

C$_{30}$H$_{26}$O$_8$ M 514.531
Constit. of *Lophira alata*. Amorph. solid. [α]$_D^{25}$ −47.8° (c, 0.65 in MeOH).

Tih, A.E. *et al, Tetrahedron Lett.*, 1988, **29**, 5797 (*struct*)

Medigenin   M-70015

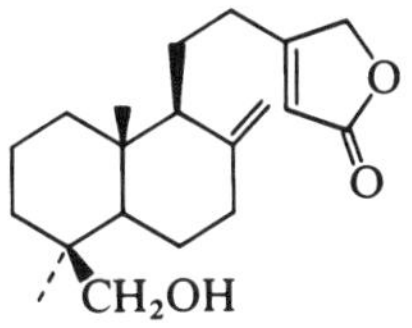

C$_{20}$H$_{30}$O$_3$ M 318.455
Constit. of *Melodinus monogynus*. Cryst. (Me$_2$CO/pentane). Mp 165-167°. [α]$_D^{25}$ −45° (c, 1 in MeOH).
β-Cellobioside: **Medinin**.
C$_{32}$H$_{50}$O$_{13}$ M 642.739
Constit. of *M. monogynus*. Cryst. (CHCl$_3$/Me$_2$CO aq.). Mp 120-122°. [α]$_D$ −26° (c, 1 in MeOH).

Sethi, A. *et al, Phytochemistry*, 1988, **27**, 2255.

Medioresinol

M-70016

[40957-99-1]

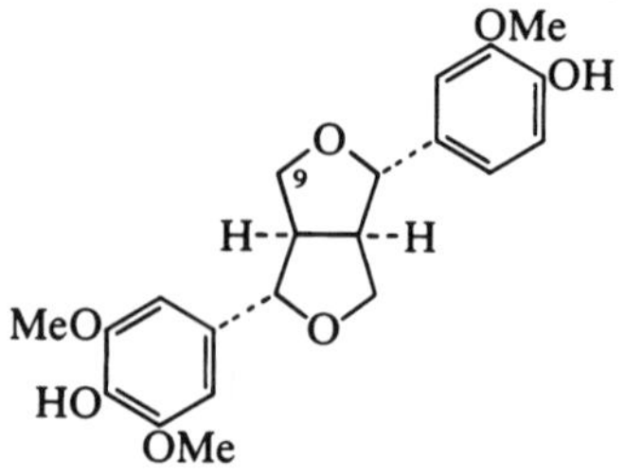

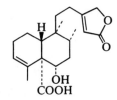

$C_{21}H_{24}O_7$ M 388.416

Constit. of *Allamanda neriifolia*. Prisms (MeOH). Mp 170-172°. $[\alpha]_D^{28}$ +77.7° (c, 0.69 in MeOH).

9α-Hydroxy: **9α-Hydroxymedioresinol**.

$C_{21}H_{24}O_8$ M 404.416

Constit. of *A. neriifolia*. Prisms (MeOH). Mp 210-213°. $[\alpha]_D^{18}$ +44.0° (c, 0.49 in MeOH).

Abe, F. *et al, Phytochemistry*, 1988, **27**, 575.

Meijicoccene

M-70017

[114066-81-8]

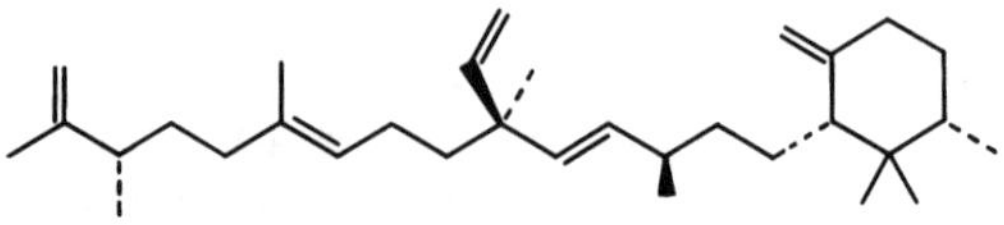

$C_{32}H_{54}$ M 438.779

Constit. of *Botryococcus braunii*. Oil.

Murakami, M. *et al, Phytochemistry*, 1988, **27**, 455.

Melisodoric acid

M-70018

Updated Entry replacing M-00280

Melissodoric acid

[54153-70-7]

$C_{20}H_{28}O_5$ M 348.438

Constit. of *Salvia melissodora*.

Rodriguez-Hahn, L. *et al, Rev. Latinoam. Quim.*, 1973, **4**, 93 (*isol*)

Rodríguez-Hahn, L. *et al, Can. J. Chem.*, 1987, **65**, 2687 (*struct*)

Melittoside

M-70019

Updated Entry replacing M-50044

[19467-03-9]

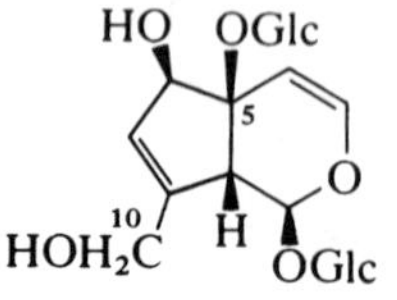

$C_{21}H_{32}O_{15}$ M 524.475

Constit. of *Melittis melissophyllum*. Prisms. Sol. H_2O. Mp 167-168°. $[\alpha]_D^{17}$ −29° (c, 1.6 in H_2O). Also exists in an amorph. form, v. sol. H_2O.

Deca-Ac: Mp 149-149.5°.

O-Deglucosyl: [20633-72-1]. **Monomelittoside**.

$C_{15}H_{22}O_{10}$ M 362.333

Isol. from *M. melissophyllum*. Amorph. powder V. sol. H_2O. $[\alpha]_D^{18}$ −180° (c, 0.7 in H_2O).

O-Deglucosyl, hepta-Ac: Cryst. (EtOH). Mp 158-160°.

10-Deoxy: **10-Deoxymelittoside**.

$C_{21}H_{32}O_{14}$ M 508.475

Constit. of *Lamiastrum galeobdolon*. Amorph. powder. $[\alpha]_D$ −60.5° (c, 1 in MeOH).

O-Deglucosyl, 6-epimer: **6-Epimonomelittoside**.

$C_{15}H_{22}O_{10}$ M 362.333

Constit. of *Tecoma heptaphylla*. Amorph. $[\alpha]_D^{20}$ −47° (c, 1.0 in MeOH).

Scarpati, M.L. *et al, Gazz. Chim. Ital.*, 1967, **97**, 1209 (*isol, struct*)

Scarpati, M.L. *et al, Ric. Sci.*, 1967, **37**, 840 (*Monomelittoside*)

Bianco, A. *et al, Phytochemistry*, 1983, **22**, 1189; 1986, **25**, 1981 (*derivs*)

2-Mercaptobenzophenone, 8CI

M-70020

(*2-Mercaptophenyl*)*phenylmethanone*, *9CI*. *2-Benzoylthiophenol*

$C_{13}H_{10}OS$ M 214.281

Cryst. (EtOH). Mp 43-45°.

Me thioether: [1620-95-7]. *2-(Methylthio)-benzophenone. [2-(Methylthio)phenyl]-phenylmethanone.*

$C_{14}H_{12}OS$ M 228.308

Yellow oil. $Bp_{0.35}$ 162-167°.

Schellenberg, K.A. *et al, J. Org. Chem.*, 1965, **30**, 1859 (*synth, uv, ir, pmr*)

Sauter, F. *et al, Monatsh. Chem.*, 1969, **100**, 905 (*synth, deriv*)

McKinnon, D.M. *et al, Can. J. Chem.*, 1988, **66**, 1405 (*synth, deriv, ir, pmr*)

3-Mercaptobenzophenone, 8CI

M-70021

(*3-Mercaptophenyl*)*phenylmethanone*, *9CI*. *3-Benzoylthiophenol*

$C_{13}H_{10}OS$ M 214.281

Me ether: 3-(*Methylthio*)*benzophenone*. [*3-(Methylthio)phenyl*]*phenylmethanone*.

$C_{14}H_{12}OS$ M 228.308

Liq. $Bp_{0.4}$ 160-170°.

Seidlova, V. *et al, CA*, 1965, **62**, 11710a (*synth, deriv*)

4-Mercaptobenzophenone, 8CI

M-70022

(*4-Mercaptophenyl*)*phenylmethanone*, *9CI*. *4-Benzoylthiophenol*

[1620-94-6]

$C_{13}H_{10}OS$ M 214.281

Cryst. Mp 71-72°.

Me ether: [23405-48-3]. 4-(*Methylthio*)*benzophenone*. [*4-(Methylthio)phenyl*]*phenylmethanone*.

$C_{14}H_{12}OS$ M 228.308

Cryst. Mp 77-78.5°.

Walker, D. *et al, J. Org. Chem.*, 1963, **28**, 3077 (*synth*)

Schellenberg, K.A. *et al, J. Org. Chem.*, 1965, **30**, 1859 (*synth, uv, ir, pmr*)

Ashby, E.C. *et al, J. Org. Chem.*, 1970, **35**, 1034 (*synth, deriv*)

3-Mercaptobutanoic acid, 9CI   M-70023

Updated Entry replacing M-00333
[26473-49-4]

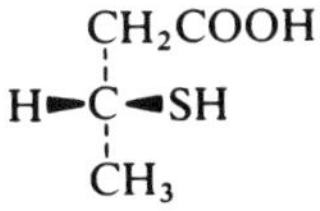

$$CH_2COOH$$
$$H \blacktriangleright \overset{|}{\underset{|}{C}} \blacktriangleleft SH$$
$$CH_3 \qquad (R)\text{-}form$$

$C_4H_8O_2S$ M 120.166

(R)-form
Liq. $Bp_{0.15}$ 120-125°. $[\alpha]_D$ +101.5° (c, 0.62 in CH_2Cl_2).
S-*Ph*: 3-*(Phenylthio)butanoic acid.*
  $C_{10}H_{12}O_2S$ M 196.264
  Oil. $Bp_{0.3}$ 160-165°. $[\alpha]_D$ -24.3° (c, 0.28 in CH_2Cl_2).
S-*Ph*, S-*oxide*: 3-*(Phenylsulfinyl)butanoic acid.*
  $C_{10}H_{12}O_3S$ M 212.263
  Needles. Mp 130-131°. $[\alpha]_D$ +119.5° (c, 2.46 in
  McOH). A single diastereoisomer obt. by fractional
  crystallisation and having $(R)_C(S)_S$-config.

(±)-form
Liq. Spar. sol. H_2O. d_4^{20} 1.14. Bp_{10} 110-111°, $Bp_{2.5}$ 87-
88°. n_D^{20} 1.4782.

Et ester:
  $C_6H_{12}O_2S$ M 148.220
  Liq. Bp_{50} 95-100°.

Lovén, J.M. *et al, Ber.*, 1915, **48**, 1257.
Johansson, H., *CA*, 1917, **11**, 2577.
Griesbeck, A. *et al, Helv. Chim. Acta*, 1987, **70**, 1326 (*synth, ir, pmr, cmr*)

2-Mercaptocyclohexanol, 9CI   M-70024

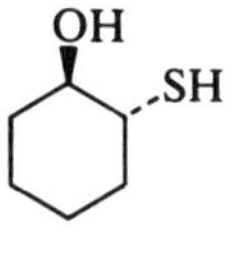

$$(1RS,2RS)\text{-}form$$

$C_6H_{12}OS$ M 132.220

(1RS,2RS)-form [60861-06-5]
(±)-trans-*form*
Bp_{15} 97-99°, $Bp_{0.015}$ 48-49°.
(1RS,2SR)-form [68972-89-4]
(±)-cis-*form*
$Bp_{0.13}$ 42-43.5°.

Culvenor, C.C.J. *et al, J. Chem. Soc.*, 1949, 278 (*synth*)
Bordwell, F.G. *et al, J. Am. Chem. Soc.*, 1953, **75**, 4959 (*synth*)
Evans, S.A. *et al, J. Org. Chem.*, 1977, **42**, 438 (*synth, ir*)
Forster, R.C. *et al, J. Chem. Soc., Perkin Trans. 1*, 1978, 822
  (*synth*)

2-Mercaptocyclohexanone   M-70025

[42904-05-2]

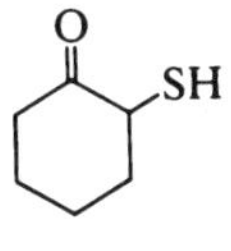

$C_6H_{10}OS$ M 130.204

(±)-form
S-*Me*: [52190-35-9]. 2-*(Methylthio)cyclohexanone.*
  $C_7H_{12}OS$ M 144.231
  Bp_6 71-72°, Bp_1 54°.
S-*Me*, S,S-*Dioxide*: [16096-71-2]. 2-*(Methylsulfonyl)-
cyclohexanone.*
  $C_7H_{12}O_3S$ M 176.230

Cryst. (hexane). Mp 55-56°.

Truce, W.E. *et al, J. Am. Chem. Soc.*, 1955, **77**, 5063 (*synth*)
Trost, B.M. *et al, J. Am. Chem. Soc.*, 1976, **98**, 4887 (*synth, ir, pmr*)
Wah, H.K. *et al, J. Chem. Soc., Perkin Trans. 1*, 1976, 651
  (*synth, ir*)
Caubère, P. *et al, J. Org. Chem.*, 1986, **57**, 1419 (*synth*)
Carreno, M.C. *et al, J. Org. Chem.*, 1987, **52**, 3619 (*synth, cmr, ms*)

3-Mercaptocyclohexanone, 9CI   M-70026

[33449-52-4]
$C_6H_{10}OS$ M 130.204

(±)-form
Not purified.
Ethylene acetal: [20051-55-2]. 1,4-*Dioxaspiro[4.5]-
decane-7-thiol.*
  $C_8H_{14}O_2S$ M 174.257
  Bp_{15} 124°.
S-*Me*: [22842-45-1]. 3-*(Methylthio)cyclohexanone.*
  $C_7H_{12}OS$ M 144.231
  $Bp_{0.1}$ 55°.

Batemen, L. *et al, J. Chem. Soc.*, 1955, 1996 (*derivs, synth*)
Napier, R.P. *et al, Int. J. Sulfur Chem., A*, 1971, **1**, 62 (*synth, ir, pmr*)

4-Mercaptocyclohexanone   M-70027

$C_6H_{10}OS$ M 130.204

S-*Me*: [23510-98-7]. 4-*Methylthiocyclohexanone.*
  $C_7H_{12}OS$ M 144.231
  Pale-yellow oil.

Gray, R.T. *et al, J. Org. Chem.*, 1970, **35**, 1525 (*deriv, synth, ir, ms*)

2-Mercapto-1,1-diphenylethanol   M-70028

α-*(Mercaptomethyl)*-α-*phenylbenzenemethanol*, 9CI
[58898-04-7]

$$Ph_2C(OH)CH_2SH$$

$C_{14}H_{14}OS$ M 230.324
Cryst. Mp 50-52°.

Ogura, K. *et al, Synthesis*, 1976, 202 (*synth, ir, pmr*)
Katritzky, A.R. *et al, J. Chem. Soc., Perkin Trans. 1*, 1987, 769
  (*synth, ir, pmr*)

8-Mercaptoguanosine, 8CI   M-70029

7,8-*Dihydro-8-thioxoguanosine*, 9CI. 2-*Amino-8-mer-
capto-9-β-D-ribofuranosyl-9H-purin-6(1H)-one. 8-
Mercapto-9-β-D-ribofuranosylguanine*
[26001-38-7]

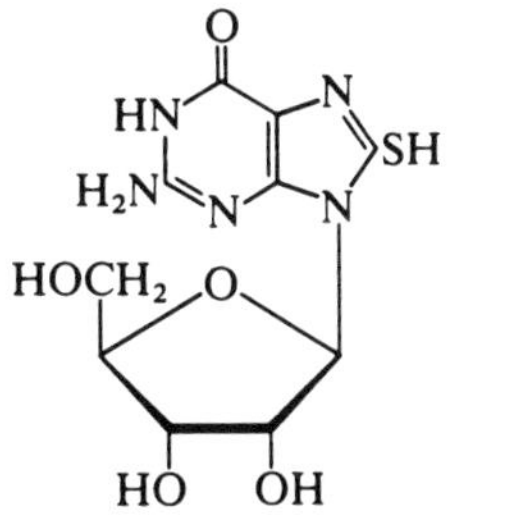

$C_{10}H_{13}N_5O_5S$ M 315.303
Cryst. (H_2O). Mp >220° dec.
S-*Me*: [2104-66-7]. 8-*(Methylthio)guanosine.*
  $C_{11}H_{15}N_5O_5S$ M 329.330

Cryst. (H$_2$O). Mp >201° dec.
S-Me, S,S-dioxide: [7057-50-3]. *8-(Methylsulfonyl)-guanosine.*
C$_{11}$H$_{15}$N$_5$O$_7$S   M 361.329
Cryst. (H$_2$O). Mp >206° dec.

Holmes, R.E. *et al, J. Am. Chem. Soc.*, 1964, **86**, 1242 (*synth*)
Ikehara, M. *et al, Chem. Pharm. Bull.*, 1966, **14**, 46 (*deriv*)
Uesugi, S. *et al, J. Am. Chem. Soc.*, 1977, **99**, 3250 (*cmr*)
Lin, T.-S. *et al, J. Med. Chem.*, 1985, **28**, 1194 (*synth*)

2-Mercaptomethyl-2-methyl-1,3-propane-dithiol, 9CI   M-70030

1,1,1-Tris(mercaptomethyl)ethane
[39597-87-0]

$$H_3CC(CH_2SH)_3$$

C$_5$H$_{12}$S$_3$ M 168.330
Pale-yellow liq. Bp$_{18}$ 114°, Bp$_4$ 70°.

Franzen, G.R. *et al, J. Am. Chem. Soc.*, 1973, **95**, 175 (*synth*)
Houk, J. *et al, J. Am. Chem. Soc.*, 1987, **109**, 6825 (*synth, ir, pmr*)

5-Mercapto-1,3,4-thiadiazoline-2-thione   M-70031

1,3,4-Thiadiazolidine-2,5-dithione, 9CI. 2,5-Mercapto-1,3,4-thiadiazole
[1072-71-5]

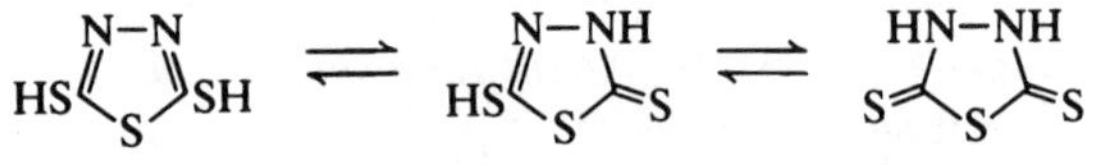

C$_2$H$_2$N$_2$S$_3$ M 150.231
5-Mercapto-2-thione-form predominates in solid phase.
Yellow cryst. (MeOH). Mp 168°.

▷Irritant. XI3850000.
Di-K salt: [4628-94-8]. Reagent for det. of Bi, Cu and Pb. Cryst. Mp 274-276° dec.
S-Me: [6264-40-0]. *5-Methylthio-1,3,4-thiadiazole-2(3H)-thione, 9CI. 2-Mercapto-5-(methylthio)-1,3,4-thiadiazole.*
C$_3$H$_4$N$_2$S$_3$ M 164.258
Cryst. (H$_2$O). Mp 136-137°.
S-Di-Me: [7653-69-2]. *2,5-Bis(methylthio)-1,3,4-thiadiazole, 9CI.*
C$_4$H$_6$N$_2$S$_3$ M 178.285
Bp$_{0.3}$ 95°.
3-Me: [29546-26-7]. *3-Methyl-1,3,4-thiadiazolidine-2,5-dithione, 9CI.*
C$_3$H$_4$N$_2$S$_3$ M 164.258
Cryst. (Et$_2$O). Mp 65-66°.
N^3,S-Di-Me: [33682-80-3]. *3-Methyl-5-(methylthio)-1,3,4-thiadiazole-2(3H)-thione, 9CI.*
C$_4$H$_6$N$_2$S$_3$ M 178.285
Cryst. (EtOH). Mp 81-82°.

Busch, M., *Ber.*, 1894, **27**, 2507 (*synth*)
Busch, M. *et al, J. Prakt. Chem.*, 1899, **60**, 25 (*synth, deriv*)
Losanitsch, S.M., *J. Chem. Soc.*, 1922, 2542 (*synth, deriv*)
Thorn, G.D., *Can. J. Chem.*, 1960, **38**, 1439 (*tautom, deriv*)
Stanovnik, B. *et al, Croat. Chim. Acta*, 1965, **37**, 17 (*tautom, deriv, uv, ir*)
Bats, J.W., *Acta Crystallogr., Sect. B*, 1976, **32**, 2866 (*cryst struct*)
Bottini, F. *et al, Org. Magn. Reson.*, 1981, **16**, 1 (*deriv, synth, struct*)
Pappalardo, S. *et al, J. Org. Chem.*, 1987, **52**, 405 (*deriv, pmr, cmr, props*)

Mesuaferrol   M-70032

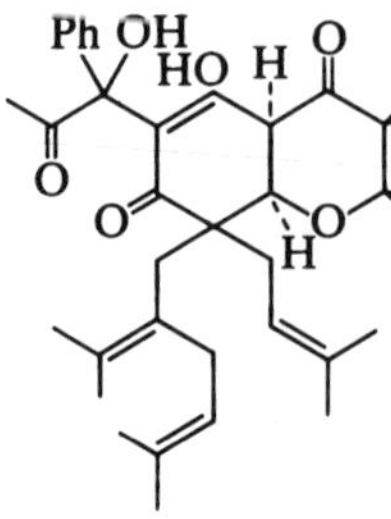

C$_{35}$H$_{46}$O$_6$ M 562.745
Constit. of *Mesua ferrea.* Amorph. powder. Mp 75°.
[α]$_D^{25}$ +27.5° (MeOH).

Dennis, T.J. *et al, Phytochemistry*, 1988, **27**, 2325.

Metachromin A   M-70033

[114466-74-9]

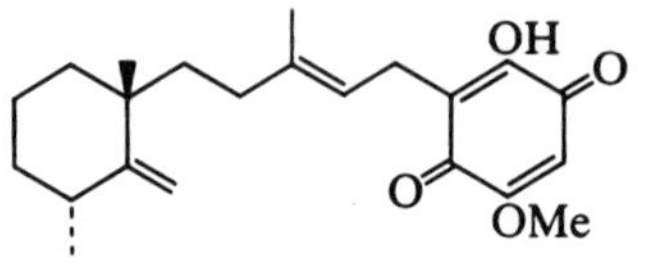

C$_{22}$H$_{30}$O$_4$ M 358.477
Isol. from the sponge *Hippospongia* cf. *metachromia.* Orange cryst. (hexane). Mp 80-82°. [α]$_D^{27}$ −11° (c, 1 in CHCl$_3$).

Ishibashi, M. *et al, J. Org. Chem.*, 1988, **53**, 2855.

Metachromin B   M-70034

[114466-75-0]

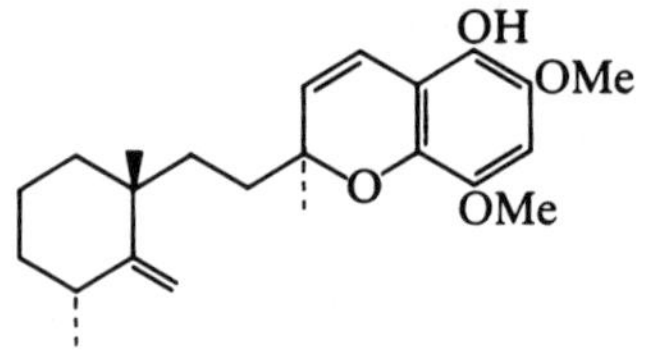

C$_{23}$H$_{32}$O$_4$ M 372.503
Isol. from the sponge *Hippospongia* cf. *metachromia.* Oil. [α]$_D^{24}$ +8° (c, 1 in CHCl$_3$).

Ishibashi, M. *et al, J. Org. Chem.*, 1988, **53**, 2855.

Methaneselenoic acid, 9CI   M-70035

Selenoformic acid

$$HC(Se)OH$$

CH$_2$OSe   M 108.986
tert-Butyl ester: [102737-62-2].
    C$_5$H$_{10}$OSe   M 165.093
    Cryst. (hexane at −30°). Mp 50.5-51.5°.
Amide: [92609-97-7]. *Methaneselenoamide, 9CI.*
    CH$_3$NSe   M 108.001
    Yellow needles (Et$_2$O). Mp 35-37°.
Dimethylamide: [31646-15-8]. N,N-*Dimethylmethaneselenoamide.*
    C$_3$H$_7$NSe   M 136.055
    Mp 2-4°. Bp$_{0.4}$ 79°.

Collard-Charon, C. *et al, Bull. Soc. Chim. Belg.*, 1963, **72**, 304 (*synth, ir*)
Henriksen, L., *Synthesis*, 1974, 501 (*ir, pmr, uv*)
Geisler, K. *et al, Z. Chem.*, 1984, **24**, 99 (*amide*)
Ishii, A. *et al, Bull. Chem. Soc. Jpn.*, 1986, **59**, 2529 (*synth, pmr, ms*)

Methanetetrapropanoic acid   M-70036

4,4-Bis(2-carboxyethyl)heptanedioic acid, 9CI. *Tetrakis(2-carboxyethyl)methane*

[10428-69-0]

$$C(CH_2CH_2COOH)_4$$

$C_{13}H_{20}O_8$ M 304.296
Cryst. (H_2O). Mp 262-263°.
Tetra-Et ester: [10428-70-3].
    $C_{21}H_{36}O_8$ M 416.511
    $Bp_{0.1}$ 187-192°.
Tetranitrile: [10428-68-9]. *Tetrakis(2-cyanoethyl)-methane.*
    $C_{13}H_{16}N_4$ M 228.296
    Cryst. (MeCN). Mp 179-180°.

Rice, L.M. *et al, J. Org. Chem.,* 1966, **32**, 1966 (*synth*)

Methanethial, 9CI   M-70037

Updated Entry replacing M-00471
Thioformaldehyde, 8CI
[865-36-1]

$$H_2C{=}S$$

CH_2S M 46.087
An interstellar molecule. Unstable. Polymerises readily.
S-Oxide: [40100-16-1].
    CH_2OS M 62.086
    Unstable.

Evans, N.J. *et al, Science,* 1970, **169**, 679 (*synth*)
Johns, J.W.L. *et al, J. Mol. Spectrosc.,* 1971, **39**, 479 (*synth, ir*)
Solouki, B. *et al, J. Am. Chem. Soc.,* 1976, **98**, 6054 (*synth*)
Vallee, Y. *et al, Can. J. Chem.,* 1987, **65**, 290 (*synth, pe, oxide*)

Methanimine   M-70038

[74-89-5]

$$H_2C{=}NH$$

CH_3N M 29.041
Interstellar molecule.
▷PF6300000.

Bock, H. *et al, Chem. Ber.,* 1987, **120**, 1961 (*synth, pe, bibl*)

10,11-Methano-1*H*-benzo[5,6]-cycloocta[1,2,3,4-*def*]fluorene-1,14-dione, 9CI   M-70039

[106880-56-2]

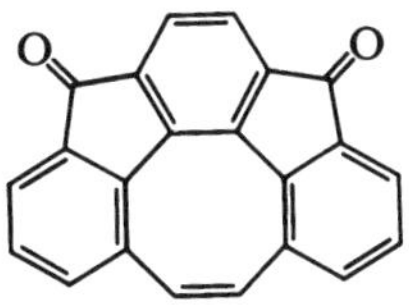

$C_{22}H_{10}O_2$ M 306.320
Conts. extremely stable planar conjugated 8-membered ring. Red needles ($CHCl_3$). Mp 305-310°.

Hou, X.L. *et al, J. Am. Chem. Soc.,* 1987, **109**, 1868 (*synth, pmr, uv, ir*)

4*b*,10*b*-Methanochrysene, 9CI   M-70040

2,3:7,8-Dibenzotricyclo[4.4.1.0^{1,6}]undeca-4,9-diene

[71949-03-6]

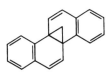

$C_{19}H_{14}$ M 242.320
Shown to have this struct. rather than the valence-isomeric dibenzomethano[10]annulene. Cryst. (EtOH). Mp 146.5-148°.

Hill, R.K. *et al, J. Am. Chem. Soc.,* 1988, **110**, 497 (*synth, uv, ir, pmr, cmr, ms*)

3*a*,7*a*-Methano-1*H*-indole, 9CI   M-70041

9-Azatricyclo[4.3.1.0^{1,6}]deca-2,4,7-triene

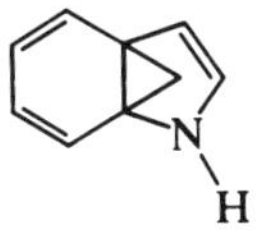

C_9H_9N M 131.177
N-Ac: [106881-87-2].
    $C_{11}H_{11}NO$ M 173.214
    Yellow oil. Contains minor equilibrium quantity of the bicyclic methanoazulene tautomer.

Tokitoh, N. *et al, J. Am. Chem. Soc.,* 1987, **109**, 1856 (*synth, pmr, cmr, ms*)

6*a*,9*b*-Methano-1*H*-phenalen-1-one   M-70042

8,1-[1]Propen[1]yl[3]ylidene-1H-benzocyclohepten-4(9H)-one, 9CI, 8CI. *2,14-Dihydrotricyclo[7.3.2.0^{5,13}]-tetradeca-1,3,5,7,9,11-hexaen-2-one. 1H-Cyclohexa[4,5,6-de]methano[10]annulen-1-one. 4,10b-Methano-8H-benzo[ab]cyclodecen-8-one. Tricyclo[7.3.2.0^{5,13}]dodeca-3,5,7,9,11(13)-hexaen-2-one. Pseudophenalenone. ψ-Phenalenone*

[32936-71-3]

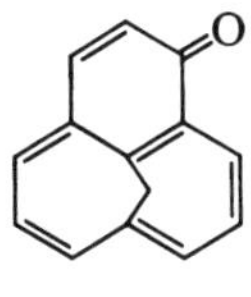

$C_{14}H_{10}O$ M 194.232
(±)-*form* [39623-20-6]
    Bridged phenalenone of theoretical and spectroscopic interest. Reddish-orange prisms. Mp 86-87.5°. Also known in opt. active form.

Murata, I. *et al, Tetrahedron Lett.,* 1971, 1789 (*synth, uv, pmr*)
Kuffner, U. *et al, Monatsh. Chem.,* 1972, **103**, 1320 (*synth*)
Yahashi, R. *et al, Bull. Chem. Soc. Jpn.,* 1973, **46**, 1441 (*cryst struct*)
Neidlein, R. *et al, Monatsh. Chem.,* 1982, **113**, 1151 (*synth, pmr, uv*)

Methionine sulfoximine   M-70043

Updated Entry replacing M-20053

S-(*3-Amino-3-carboxypropyl*)-S-*methylsulfoximine*, 9CI. *2-Amino-4-(S-methylsulfonimidoyl)butanoic acid*

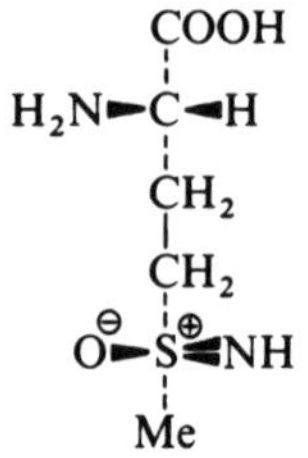

(*S*)$_c$,(*S*)$_s$-*form*
Absolute
configuration

$C_5H_{12}N_2O_3S$ M 180.221

Found in flours treated with NCl_3 as a prod. of NCl_3 action on wheat proteins.

(*S*)$_C$(*S*)$_S$-**form** [21752-32-9]
Cnestine. Glabrin
Isol. from *Cnestis glabra*, *C. polyphylla* and *Rourea orientalis*. Mp 239°. [α]$_D^{22}$ +34° (c, 2 in HCl).

▷Neurotoxin, convulsive agent, potent glutamine synthetase inhibitor

(*S*)$_C$(*R*)$_S$-**form** [21752-31-8]
Mp 235°. [α]$_D^{22}$ +39° (c, 2 in N HCl).

Christensen, B.W. *et al*, *J. Chem. Soc., Chem. Commun.*, 1969, 169 (*cryst struct, abs config*)
Sugiyama, Y. *et al*, *Tetrahedron Lett.*, 1983, **24**, 1471.
Chevrier, B. *et al*, *Acta Crystallogr., Sect. C*, 1986, **42**, 1632 (*cryst struct, bibl*)

C-Methoxycarbohydroxamic acid   M-70044

Methyl hydroxycarbamate
[584-07-6]

MeOCONHOH

$C_2H_5NO_3$ M 91.066
Hygroscopic cryst. Mp 50°.

Major, R.T. *et al*, *J. Org. Chem.*, 1959, **24**, 431.
Boyland, E. *et al*, *Analyst*, 1964, **89**, 520.
Defoin, A. *et al*, *Helv. Chim. Acta*, 1987, **70**, 554 (*synth*)

8-Methoxychlorotetracycline   M-70045

Sch 36969. Antibiotic Sch 36969
[110298-63-0]

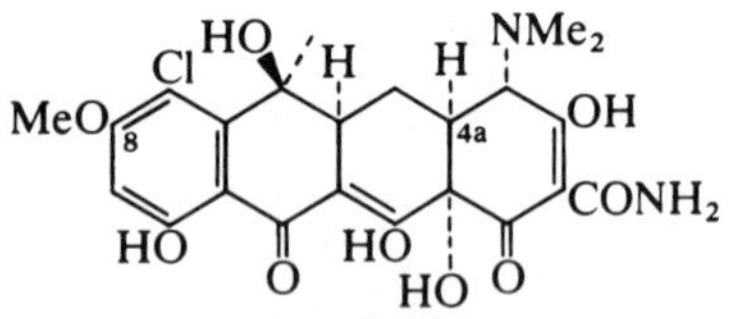

$C_{23}H_{25}ClN_2O_9$ M 508.912
Tetracycline antibiotic. Prod. by *Actinomadura brunnea*. Active against gram-positive and -negative bacteria. No phys. props. reported.

N-*Me:* [110298-65-2]. *8-Methoxy-*N-*methylchlorotetracycline. Sch 33256. Antibiotic Sch 33256.*
$C_{24}H_{27}ClN_2O_9$ M 522.938
From *A. brunnea*. Primarily active against gram-positive bacteria. Yellow cryst. Mp 180-185° dec. [α]$_D^{26}$ −105.5° (c, 0.5 in MeOH).

4aα-*Hydroxy:* [110298-64-1]. *4a-Hydroxy-8-methoxychlorotetracycline. Sch 34164. Antibiotic Sch 34164.*
$C_{23}H_{25}ClN_2O_{10}$ M 524.911
Prod. by *Dactylosporangium* sp. Active against gram-positive and -negative bacteria.

Miller, G.H. *et al*, *J. Antibiot.*, 1987, **40**, 1408, 1414, 1419, 1426 (*isol, struct, props*)

1-Methoxy-1-hexene   M-70046

[37617-02-0]

$H_3C(CH_2)_3CH{=}CHOMe$

$C_7H_{14}O$ M 114.187
Bp$_{27}$ 48-50°.

Callot, H.J. *et al*, *Bull. Soc. Chim. Fr.*, 1983, 317 (*synth*)
Newcomb, M. *et al*, *J. Am. Chem. Soc.*, 1986, **108**, 240 (*synth*)

4-Methoxy-3-(3-methyl-2-butenyl)-5-phenyl-2(5*H*)-furanone   M-70047

3-Methoxy-4-phenyl-2-prenyl-2-butenolide

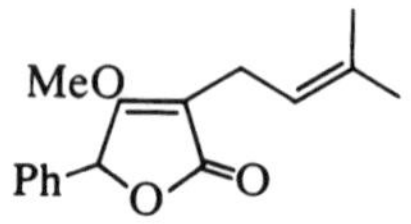

$C_{16}H_{18}O_3$ M 258.316
Constit. of *Hypericum mysorense*. Cryst. (Me$_2$CO/hexane). Mp 65°.

Vishwakarma, R.A. *et al*, *Indian J. Chem., Sect. B*, 1987, **26**, 486.

3-(2-Methoxy-4,5-methylenedioxyphenyl)-propenal   M-70048

3-(6-Methoxy-1,3-benzodioxol-5-yl)-2-propenal, 9CI. *3-Methoxy-4,5-methylenedioxycinnamaldehyde*
[67323-02-8]

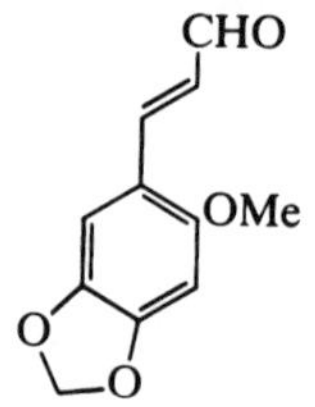

$C_{11}H_{10}O_4$ M 206.198
Isol. from *Piper lenticellosum*. Cryst. (CCl$_4$ or EtOH). Mp 171° (151-152°).

Scharf, H.-D. *et al*, *Justus Liebigs Ann. Chem.*, 1978, 573 (*synth*)
Diaz, P.P. *et al*, *J. Nat. Prod.*, 1986, **49**, 690 (*isol, struct*)

5-Methoxy-3-(8,11,14)pentadecatrienyl)-1,2,4-benzenetriol   M-70049

1,2,4-Trihydroxy-5-methoxy-3-(8,11,14-pentadecatrienyl)benzene

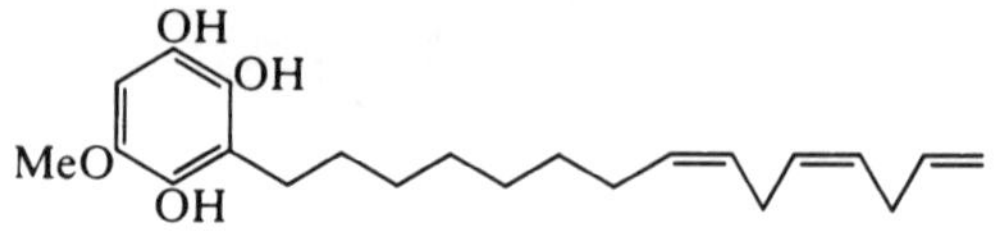

$C_{22}H_{32}O_4$ M 360.492

(*8Z,11Z*)-*form* [105018-77-7]
Germination stimulant for *Striga asiatica*, an obligate parasitic plant. V. unstable, rapidly converts to the 1,4-benzoquinone.

Lynn, D.G. *et al*, *J. Am. Chem. Soc.*, 1986, **108**, 7858 (*isol, nmr*)

2-Methyl-1,3-benzenedimethanethiol, 9CI   M-70050

2,6-Bis(thiomethyl)toluene. 2,6-Bis(mercaptomethyl)-toluene

[41563-67-1]

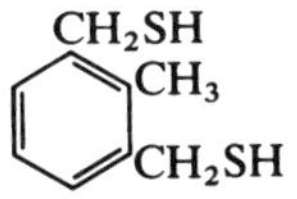

$C_9H_{12}S_2$ M 184.314

Needles (C_6H_6/hexane). Mp 40-41°.

Mitchell, R.M. *et al*, *J. Am. Chem. Soc.*, 1974, **96**, 1547 (*synth, pmr, ms, ir*)

3-Methyl-4,5-benzofurandione, 9CI   M-70051

[113297-21-5]

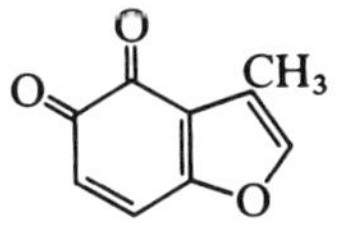

$C_9H_6O_3$ M 162.145

Dienophile. Red solid. Mp 92-93°.

Lee, J. *et al*, *Tetrahedron Lett.*, 1987, **28**, 3427 (*synth, ir, uv, pmr, cmr, use*)

3-Methyl-1*H*-2-benzopyran-1-one   M-70052

3-Methylisocoumarin

[29539-21-7]

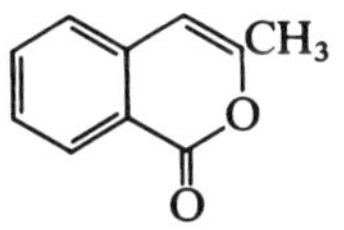

$C_{10}H_8O_2$ M 160.172

Mp 73° (67-68°).

Bhide, B.H. *et al*, *Chem. Ind.* (*London*), 1980, 84 (*synth*)
Sakamoto, T. *et al*, *Chem. Pharm. Bull.*, 1986, **34**, 2754 (*synth, pmr*)

3-Methylbicyclo[1.1.0]butane-1-carboxylic acid   M-70053

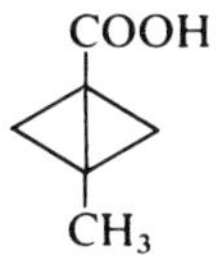

$C_6H_8O_2$ M 112.128

Amide:
  C_6H_9NO M 111.143
  Plates (EtOAc). Mp 144-144.5° dec.
Nitrile: [694-25-7]. *1-Cyano-3-methylbicyclo[1.1.0]-butane.*
  C_6H_7N M 93.128
  Bp_{100} 95-96.5°.

Blanchard, E.P. *et al*, *J. Am. Chem. Soc.*, 1966, **88**, 487 (*deriv, synth, pmr*)
Hoz, S. *et al*, *J. Org. Chem.*, 1986, **51**, 4537 (*deriv, synth*)

4-[5′-Methyl[2,2′-bithiophen]-5-yl]-3-butyne-1,2-diol, 9CI   M-70054

5-(3,4-Dihydroxy-1-butynyl)-5′-methyl-2,2′-bithiophene

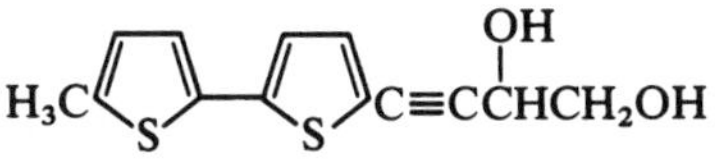

$C_{13}H_{12}O_2S_2$ M 264.357

Di-Ac: [58930-57-7]. *5-(3,4-Diacetoxy-1-butynyl)-5′-methyl-2,2′-bithiophene.*
$C_{17}H_{16}O_4S_2$ M 348.431
Constit. of *Dyssodia setifolia*. Oil.

Bohlmann, F. *et al*, *Chem. Ber.*, 1976, **109**, 901 (*isol, uv, ir, pmr, ms*)

3-Methyl-2-butenal, 9CI   M-70055

Updated Entry replacing M-01206
3-Methylcrotonaldehyde, 8CI. 3,3-Dimethylacrolein. Senecioaldehyde. Senecialdehyde

[107-86-8]

$$(H_3C)_2C{=}CHCHO$$

C_5H_8O M 84.118

d_4^{20} 0.872. Bp 132-133°. n_D^{20} 1.4526. Readily oxid.

Di-Et acetal: [1740-74-5]. *4,4-Diethoxy-2-methyl-2-butene.*
$C_9H_{18}O_2$ M 158.240
Bp 163-165°. Easily dec.
Semicarbazone: Cryst. (MeOH). Mp 221-222°.
Phenylhydrazone: Reddish-violet cryst. Mp 161-162°.

Wittig, G. *et al*, *Tetrahedron, Suppl.*, 1966, No. 8, 347 (*synth*)
Meyers, A.I. *et al*, *J. Am. Chem. Soc.*, 1969, **91**, 764 (*synth*)
Cainelli, G. *et al*, *J. Am. Chem. Soc.*, 1976, **98**, 6737 (*synth*)
Cardillo, G. *et al*, *Tetrahedron*, 1976, **32**, 107 (*synth*)
Erman, M.B. *et al*, *Tetrahedron Lett.*, 1976, 2981 (*synth*)
Pauling, H. *et al*, *Helv. Chim. Acta*, 1976, **59**, 1233 (*synth*)
Schuda, P.F. *et al*, *J. Org. Chem.*, 1987, **52**, 1972 (*synth, pmr, ir*)
Klusener, P.A.A. *et al*, *J. Org. Chem.*, 1987, **52**, 5261 (*synth, pmr*)
Sax, N.I., *Dangerous Properties of Industrial Materials*, 5th Ed., Van Nostrand-Reinhold, 1979, 967.

6-Methylcholanthrene   M-70056

1,2-Dihydro-6-methylbenz[j]aceanthrylene, 9CI

[29873-25-4]

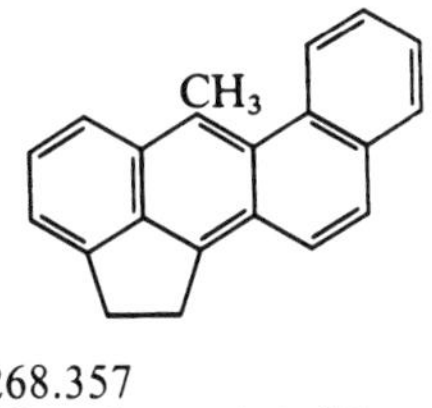

$C_{21}H_{16}$ M 268.357

Mp 143-144°. Mp refers to 1,1-dideutero analgoue.

▷Carcinogenic

Harvey, R.G. *et al*, *J. Org. Chem.*, 1986, **52**, 283 (*synth, pmr*)

24-Methyl-7,22-cholestadiene-3,5,6-triol   M-70057

Updated Entry replacing M-01271
7,22-Ergostadiene-3,5,6-triol

$C_{28}H_{46}O_3$ M 430.670

(3β,5α,6β,22E)-form [516-37-0]
Cerevisterol
Constit. of *Amanita phalloides* and *Fusarium moniliforme*. Cryst. (MeOH). Mp 254-256°. $[\alpha]_D^{21}$ −79.9° (c, 1.35 in Py).

O^6-*Me*: [117585-50-9]. *6β-Methoxy-7,22-ergostadiene-3β,5α-diol. 3β,5α-Dihydroxy-6β-methoxy-7,22-ergostadiene.*
$C_{29}H_{48}O_3$ M 444.696
Constit. of *Agaricus blazei*. Syrup. $[\alpha]_D^{20}$ −61° (c, 1.19 in $CHCl_3$).

Serebryakov, E.P. *et al*, *Tetrahedron*, 1970, **26**, 5215 (*isol*)
Kawagishi, H. *et al*, *Phytochemistry*, 1988, **27**, 2777 (*isol*)

6-Methyl-2-cyclohexen-1-ol, 9CI   M-70058

Updated Entry replacing M-01419
[3718-56-7]

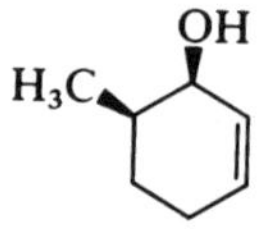

(1R,6R)-form

$C_7H_{12}O$ M 112.171

(1R,6R)-form [109428-38-8]
(+)-*cis-form*
d^{20} 0.96. $[\alpha]_D^{21}$ +41° (neat).
(1RS,6SR)-form [40523-67-9]
(±)-*trans-form*
Bp_{45} 93.3-93.8°.

Stork, G. *et al*, *J. Am. Chem. Soc.*, 1953, **75**, 4119; 1956, **78**, 4604 (*synth*)
Beckwith, A.L.J. *et al*, *Aust. J. Chem.*, 1976, **29**, 1277 (*synth*)
Solladié, G. *et al*, *J. Org. Chem.*, 1987, **52**, 3560 (*synth, ir, pmr*)

6-Methyl-2-cyclohexen-1-one, 9CI   M-70059

Updated Entry replacing M-30076
[6610-21-5]

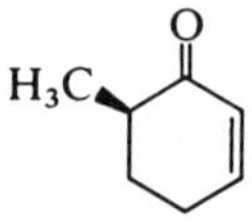

(R)-form

$C_7H_{10}O$ M 110.155

(R)-form [62392-84-1]
$Bp_{2.5}$ 33-35°. $[\alpha]_D^{21}$ +70° (c, 3 in $CHCl_3$), $[\alpha]_D^{21}$ +91° (neat).
(±)-form
Liq. with pleasant odour. Bp 172-173°, Bp_{15} 67-70°. Steam-volatile.

Semicarbazone: Cryst. (MeOH). Mp 177-178°.
2,4-Dinitrophenylhydrazone: [52456-88-9]. Mp 122-126°.

Kötz, A. *et al*, *Justus Liebigs Ann. Chem.*, 1911, **379**, 17.
Birch, A.J., *J. Chem. Soc.*, 1946, 593 (*synth*)
Mousseron, M. *et al*, *Bull. Soc. Chim. Fr.*, 1954, 1246.
McMurry, J.E. *et al*, *J. Org. Chem.*, 1973, **38**, 4367 (*synth*)
Goyal, S.C. *et al*, *Ann. Chim. (Paris)*, 1977, **2**, 57 (*derivs*)
Ryu, I. *et al*, *Tetrahedron Lett.*, 1978, 3455 (*synth*)
Marino, J.P. *et al*, *J. Am. Chem. Soc.*, 1982, **104**, 3165 (*synth*)
Solladié, G. *et al*, *J. Org. Chem.*, 1987, **52**, 3560 (*synth, pmr*)

N-Methyldiphenylamine, 8CI   M-70060

Updated Entry replacing M-01618
N-*Methyl*-N-*phenylbenzenamine*, 9CI.
Diphenylmethylamine

[552-82-9]

Ph_2NMe

$C_{13}H_{13}N$ M 183.252
Oil. d_{20}^{20} 1.06. Bp 291°, Bp_{12} 148-149°.

Ullmann, F., *Justus Liebigs Ann. Chem.*, 1903, **327**, 104.
Wieland, H., *Ber.*, 1919, **52**, 886.
Lugovkin, B.P. *et al*, *Zh. Obshch. Khim.*, 1952, **22**, 2041; *CA*, **47**, 9284c.
Dannley, R.L. *et al*, *J. Org. Chem.*, 1955, **20**, 92.
Gribble, G.W. *et al*, *Synthesis*, 1987, 709 (*synth, pmr*)

5-Methyl-1,3,2,4-dithiazolium(1+)   M-70061

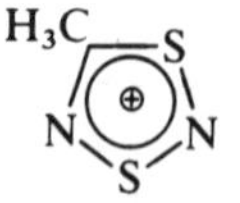

$C_2H_3N_2S_2^{\oplus}$ M 119.179 (ion)
Chloride: [115125-38-7].
   $C_2H_3ClN_2S_2$ M 154.632
   Orange solid. Mp 199°.
Hexafluoroarsenate: [88047-46-5].
   $C_2H_3AsF_6N_2S_2$ M 308.091
   Colourless needles (sic). Mp 192-194° dec.

MacLean, G.K. *et al*, *J. Chem. Soc., Dalton Trans.*, 1985, 1405 (*synth, ms, ir, cryst struct*)
Applett, A. *et al*, *J. Chem. Soc., Chem. Commun.*, 1987, 1889 (*synth, ms, pmr, ir*)

7-Methylenebicyclo[3.3.1]nonan-3-one   M-70062

7-Methylidenebicyclo[3.3.1]nonan-3-one
[17933-29-8]

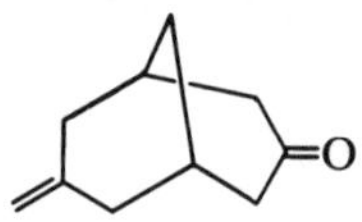

$C_{10}H_{14}O$ M 150.220
Plates (hexane). Mp 160-162.5°.

Stetter, H. *et al*, *Chem. Ber.*, 1963, **96**, 694 (*synth*)
Gagneux, A.R. *et al*, *Tetrahedron Lett.*, 1969, 1365 (*synth*)
Fischer, W. *et al*, *Helv. Chim. Acta*, 1976, **59**, 1953.
Adcock, W. *et al*, *J. Org. Chem.*, 1987, **52**, 356 (*synth, cmr*)

2,2'-Methylenebis[8-hydroxy-3-methyl-1,4-naphthalenedione], 9CI   M-70063

3,3'-Methylenediplumbagin
[58275-00-6]

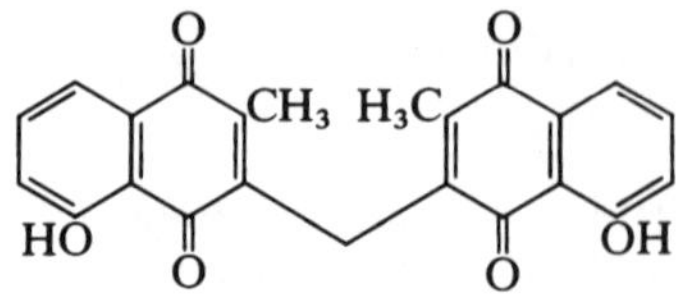

$C_{23}H_{16}O_6$ M 388.376
Constit. of *Plumbago zeylanica*. Orange-yellow needles ($CHCl_3$/pet. ether). Mp 208-210°.

Gunaherath, G.M.K.B. *et al*, *J. Chem. Soc., Perkin Trans. 1*, 1988, 407.

2,2'-Methylenebis-1*H*-imidazole, 9CI   M-70064

Di-2-imidazolylmethane

[64269-81-4]

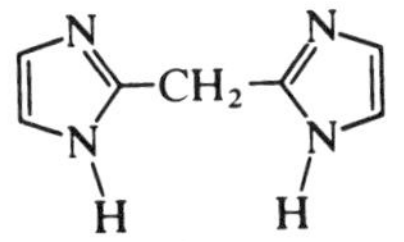

$C_7H_8N_4$ M 148.167

Needles (H_2O). Mp 285° dec.

B,2HCl: Cryst. (H_2O). Mp >190° dec.

Joseph, M. *et al, Synthesis*, 1977, 459 (*synth*)

1,1'-Methylenebispyridinium(1+), 9CI   M-70065

1,1-Methylenedipyridinium, 8CI. Dipyridiniomethane

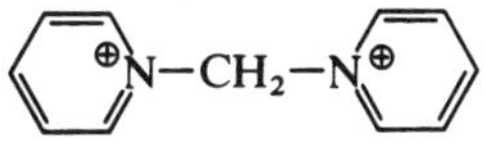

$C_{11}H_{12}N_2^{\oplus}$ M 172.229 (ion)

Insect chemosterilant; also shows fungicidal props.

Dibromide: [40032-49-3].

$C_{11}H_{12}Br_2N_2$ M 332.037

Cryst. Mp 255-258°.

Diiodide: [32405-50-8].

$C_{11}H_{12}I_2N_2$ M 426.038

Light-yellow needles. Mp 220°.

Kröhnke, F., *Chem. Ber.*, 1955, **88**, 851.
Scheibe, G. *et al, Angew. Chem.*, 1961, **73**, 736 (*synth*)
Brüdgam, I. *et al, Acta Crystallogr., Sect. C*, 1986, **42**, 866 (*cryst struct, diiodide*)

Methylenecyclobutane, 9CI   M-70066

Updated Entry replacing M-01711

[1120-56-5]

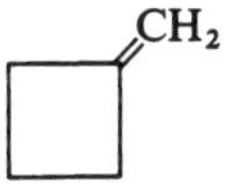

C_5H_8 M 68.118

Antiknock agent for fuels. Fp −134.68°. Bp 42°. n_D^{20} 1.4210.

Shand, W. *et al, J. Am. Chem. Soc.*, 1944, **66**, 636 (*synth*)
Roberts, J.D. *et al, J. Am. Chem. Soc.*, 1949, **71**, 3925 (*synth*)
Conia, J.M. *et al, Bull. Soc. Chim. Fr.*, 1961, 1803 (*synth*)
D'yachencko, A.I. *et al, Izv. Akad. Nauk SSSR, Ser. Khim.*, 1966, 2237 (*synth*)
Cole, K.C. *et al, Can. J. Chem.*, 1976, **54**, 657 (*nmr*)
Wickham, G. *et al, J. Org. Chem.*, 1985, **50**, 3485 (*synth*)
Fitjer, L. *et al, Synthesis*, 1987, 299 (*synth*)

2-Methylenecyclopropaneacetic acid   M-70067

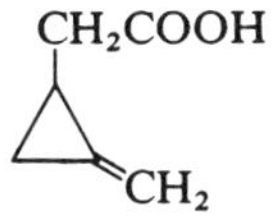

$C_6H_8O_2$ M 112.128

(±)-*form* [107617-00-5]

Oil.

Baldwin, J.E. *et al, J. Org. Chem.*, 1987, **52**, 1475 (*synth, pmr*)

6,7-Methylenedioxy-2*H*-1-benzopyran-2-one   M-70068

Updated Entry replacing M-01740

6H-1,3-Dioxolo[4,5-g][1]benzopyran-6-one, 9CI. 6,7-Methylenedioxycoumarin. **Ayapin**. *Aiapin*

[494-56-4]

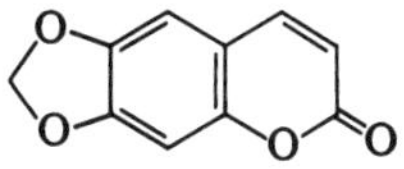

$C_{10}H_6O_4$ M 190.155

Constit. of *Eupatorium ayapena*. Cryst. (MeOH). Mp 231-232° (224°).

Bose, P.K. *et al, Ber.*, 1937, **70**, 702 (*isol, synth*)
Dieterman, L.J. *et al, Arch. Biochem. Biophys.*, 1964, **106**, 275 (*isol*)
Kelkar, S.L. *et al, Indian J. Chem., Sect. B*, 1984, **23**, 458 (*synth*)
Ziegler, T. *et al, Chem. Ber.*, 1987, **120**, 373 (*synth, pmr*)

2-Methylene-1,3-diselenole, 9CI   M-70069

[75619-06-6]

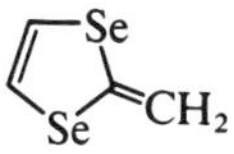

$C_4H_4Se_2$ M 209.996

Shiny plates (hexane). Mp 58-60°. Stable at −20°, dec. on standing at r.t.

Jackson, Y.A. *et al, Tetrahedron Lett.*, 1987, **28**, 5635 (*synth, pmr, uv*)

4-Methylene-1,2,5,6-heptatetraene   M-70070

$$H_2C=C(CH=C=CH_2)_2$$

C_8H_8 M 104.151

Lehrich, F. *et al, Tetrahedron Lett.*, 1987, **28**, 2697 (*synth, ir, uv, ms*)

4-Methylene-1,2,5-hexatriene   M-70071

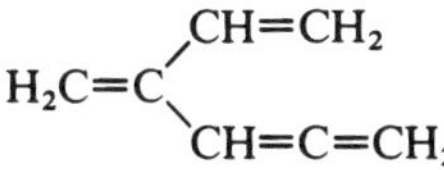

$$H_2C=C\begin{matrix} CH=CH_2 \\ CH=C=CH_2 \end{matrix}$$

C_7H_8 M 92.140

Lehrich, F. *et al, Tetrahedron Lett.*, 1987, **28**, 2697 (*synth, ir, uv, ms*)

24-Methylene-7,9(11)-lanostadien-3-ol   M-70072

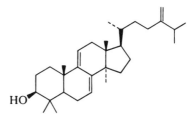

$C_{31}H_{50}O$ M 438.735

3β-form

Constit. of *Artabotrys odorotissimus*. Needles (CHCl₃/EtOAc). Mp 168-173°. $[\alpha]_D^{25}$ +51.3° (c, 0.078 in CHCl₃).

Hasan, C.M. *et al, J. Nat. Prod.*, 1987, **50**, 762.

2-Methylene-7-oxabicyclo[2.2.1]heptane, 9CI   M-70073

[111515-49-2]

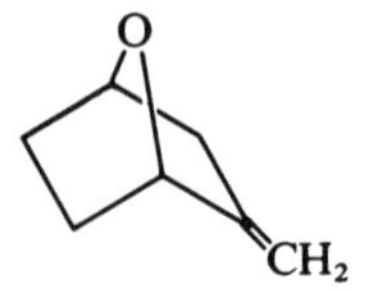

$C_7H_{10}O$ M 110.155
Liq. Bp_{40} 37°.

Senda, Y. et al, Bull. Chem. Soc. Jpn., 1987, **60**, 613 (synth, pmr, cmr)

2-Methylene-4-pentenal, 9CI   M-70074

2-Formyl-1,4-pentadiene

[17854-46-5]

$$H_2C{=}C(CHO)CH_2CH{=}CH_2$$

C_6H_8O M 96.129
Oil. Bp 115-120°.

Block, E. et al, J. Am. Chem. Soc., 1986, **108**, 7045 (synth, ir, pmr, cmr)

3-Methylene-4-penten-2-one, 9CI   M-70075

2-Acetylbutadiene

[93198-79-9]

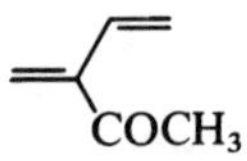

C_6H_8O M 96.129
Synthon. Readily dimerises, obt. as $Fe(CO)_3$ complex.

Franck-Neumann, M. et al, Angew. Chem., Int. Ed. Engl., 1981, **20**, 864.
Honek, J.F. et al, Synth. Commun., 1984, 483.
Hoffmann, H.M.R. et al, Angew. Chem., Int. Ed. Engl., 1987, **26**, 1015.

2-Methylene-1,3-propanedithiol, 9CI   M-70076

2-(Mercaptomethyl)-1-propene-3-thiol. 3-Mercapto-2-(mercaptomethyl)-1-propene

[52342-51-5]

$$H_2C{=}C(CH_2SH)_2$$

$C_4H_8S_2$ M 120.227
Foul-smelling liq. Bp_{13} 65-66°, $Bp_{0.3}$ 44-46°.

Schulze, K. et al, Z. Chem., 1975, **15**, 302 (synth, ir, pmr)
Houk, J. et al, J. Am. Chem. Soc., 1987, **109**, 6825 (synth, ir, pmr)

1-Methylenepyrrolidinium(1+)   M-70077

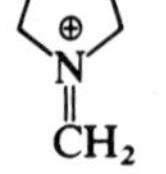

$C_5H_{10}N^{\oplus}$ M 84.141 (ion)
Chloride: [52853-03-9].
   $C_5H_{10}ClN$ M 119.594
   Solid. Extremely hygroscopic, gradually dec. to formaldehyde and pyrrolidine hydrochloride in moist air.

Mills, J.E. et al, J. Org. Chem., 1987, **52**, 1857 (synth, pmr, cmr)

24-Methylene-3,4-seco-4(28)-cycloarten-3-oic acid   M-70078

[109576-20-7]

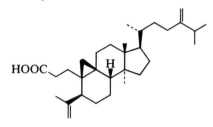

$C_{31}H_{50}O_2$ M 454.735
Constit. of needles of *Abies sibirica.*
Me ester: [109576-21-8]. Cryst. (MeCN). Mp 31-32°. $[\alpha]_D^{25}$ +61° (c, 2.83 in $CHCl_3$).

Raldugin, V.A. et al, Khim. Prir. Soedin., 1987, **23**, 259.

Methylenolactocin   M-70079

Tetrahydro-4-methylene-5-oxo-2-pentyl-3-furancarboxylic acid, 9CI

[116943-38-5]

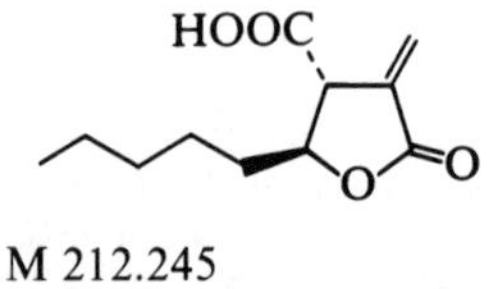

$C_{11}H_{16}O_4$ M 212.245
Metab. of *Penicillium* sp. 24-4. Antitumour antibiotic. Leaflets (EtOAc/hexane). Mp 82.5-83.5°. $[\alpha]_D^{26}$ −2.37° (c, 3.0 in MeOH).

Park, B.K. et al, Agric. Biol. Chem., 1987, **51**, 3443.

(1-Methylethenyl)oxirane, 9CI   M-70080

3,4-Epoxy-2-methyl-1-butene. Isopropenyloxirane

[7437-61-8]

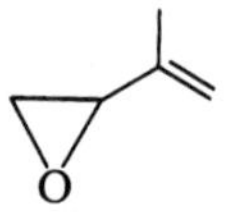

C_5H_8O M 84.118
C_5 synthon, esp. for terpene alcohols.

(±)-form
   Bp 91-93°.

   Suzuki, S. et al, Synth. Commun., 1986, **16**, 491 (synth, pmr, ir, ms, bibl)

5-Methyl-2(3H)-furanone, 9CI   M-70081

Updated Entry replacing M-01847
4-Hydroxy-3-pentenoic acid γ-lactone. α-Angelica lactone. Δ²-Angelica lactone. Angelic lactone

[591-12-8]

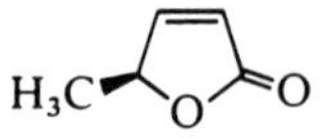

$C_5H_6O_2$ M 98.101
▷LU5075000.

(S)-form
   $[\alpha]_D^{20}$ +123° (c, 1.7 in $CHCl_3$).

(±)-form

Volatile needles. $d_4^{20.4}$ 1.091. Mp 18°. Bp 167°, Bp$_{12}$ 53°. Readily isomerizes to the 2(5H)-isomer, 5-Methyl-2(5H)-furanone, M-60083 .

▷Carcinogenic in animals

Haynes, L.J. et al, J. Chem. Soc., 1946, 954 (synth)
Kuehl, F.A. et al, J. Chem. Soc., 1950, 2213 (synth)
Wineburg, J.P. et al, J. Heterocycles Chem., 1975, 12, 749 (synth)
Kiyoshi, I., Bull. Chem. Soc. Jpn., 1977, 50, 242 (synth)
Brandänge, S. et al, Acta Chem. Scand., Ser. B, 1987, 41, 736 (synth, bibl)

2-Methyl-3-furanthiol, 9CI   **M-70082**

3-Mercapto-2-methylfuran

[28588-74-1]

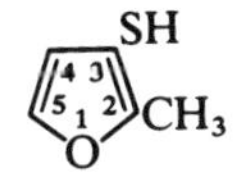

C_5H_6OS M 114.162

Organoleptic compd. with beef broth aroma. Well patented.

▷LU6235000.

S-Me: [63012-97-5].
C_6H_8OS M 128.189
Compd. with intense roasted meat aroma.

U.S.P., 4 020 175, (1977); CA, 87, 135020y (synth, deriv)
Moran, E.J. et al, Drug. Chem. Toxicol., 1980, 3, 249 (tox)
Tressel, R. et al, J. Agric. Food Chem., 1981, 29, 1078.
MacLeod, G. et al, Chem. Ind. (London), 1986, 175 (isol, deriv)

2-Methylfuro[3,2-c]quinoline   **M-70083**

[100633-68-9]

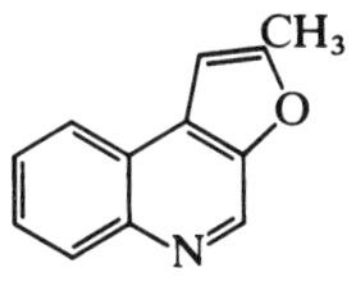

$C_{12}H_9NO$ M 183.209
Mp 180°.

Godard, A. et al, J. Heterocycl. Chem., 1988, 25, 1053 (synth, pmr)

1-Methylguanosine, 9CI   **M-70084**

2-Amino-1-methyl-9-β-D-ribofuranosyl-9H-purin-6(1H)-one. 1-Methyl-9-β-D-ribofuranosylguanine

[2140-65-0]

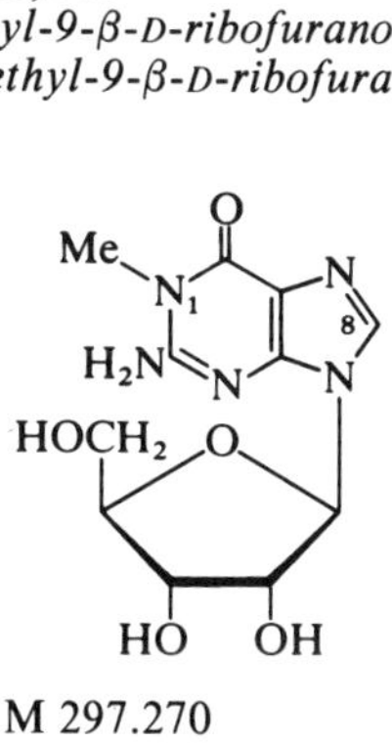

$C_{11}H_{15}N_5O_5$ M 297.270
Cryst. (MeOH). Mp 225-227°.

Bredereck, H. et al, Chem. Ber., 1947, 80, 401 (synth)
Broom, A.D. et al, Biochemistry, 1964, 3, 495 (synth)
Hecht, S.M. et al, Anal. Biochem., 1969, 30, 249 (ms)
Yamauchi, K. et al, J. Chem. Soc., Perkin Trans. 1, 1980, 2787 (synth)
Yamauchi, K. et al, Synthesis, 1980, 852 (synth)

Chang, C.-J. et al, Biochemistry, 1981, 20, 2657 (cmr)
Chang, C.-J. et al, Org. Magn. Reson., 1984, 22, 671 (cmr)

8-Methylguanosine, 9CI   **M-70085**

2-Amino-8-methyl-9-β-D-ribofuranosyl-9H-purin-6(1H)-one. 8-Methyl-9-β-D-ribofuranosylguanine

[36799-17-4]

$C_{11}H_{15}N_5O_5$ M 297.270
Needles + 1H$_2$O (H$_2$O). Mp 185°. Dec. at 230°.

Pleiderer, W. et al, Chem. Ber., 1972, 105, 1497 (synth, uv, cd)
Maeda, M. et al, Tetrahedron, 1974, 30, 2677 (synth, uv)
Uesugi, S. et al, J. Am. Chem. Soc., 1977, 99, 3250 (cmr)
Nguyen, C.-D. et al, Tetrahedron Lett., 1979, 3159 (synth)
Zady, M.F. et al, J. Org. Chem., 1980, 45, 2373.

5-Methyl-2,4-hexadienoic acid, 9CI   **M-70086**

Updated Entry replacing M-01989

5-Methylsorbic acid, 8CI. 4-Methyl-1,3-pentadiene-1-carboxylic acid. 4-Isopropylidenecrotonic acid

[19932-33-3]

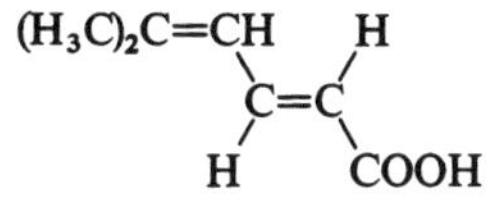

$$(H_3C)_2C=CH \quad H$$
$$C=C$$
$$H \quad COOH$$ **(E)-form**

$C_7H_{10}O_2$ M 126.155

(E)-form [79695-53-7]

Needles (EtOH aq.). Mp 113° (105°).

Me ester: [52148-91-1]. Pale-yellow oil. Bp 89°.
4-Bromophenacyl ester: Cryst. (EtOH aq.). Mp 121°.

(Z)-(?)-form

Mp 17°. Bp$_9$ 123-124°. This form reported only once and identity dubious.

Et ester:
$C_9H_{14}O_2$ M 154.208
Bp$_9$ 82-83°.

Fischer, F.G. et al, Justus Liebigs Ann. Chem., 1932, 494, 263 (synth)
Reid, E.B. et al, J. Chem. Soc., 1954, 516 (synth)
Crombie, L. et al, J. Chem. Soc., 1958, 4417 (synth, config, pmr)
Julia, M. et al, Bull. Soc. Chim. Fr., 1960, 28; 1976, 525 (synth)
Volkov, Y.P. et al, Zh. Org. Khim., 1968, 4, 763; CA, 69, 18551 (synth)
Ploquin, J. et al, Ann. Chim. (Paris), 1970, 5, 143 (synth)
Dieck, H.A. et al, J. Am. Chem. Soc., 1974, 96, 1133 (synth)
Heck, R.F. et al, J. Am. Chem. Soc., 1974, 96, 1133; Acc. Chem. Res., 1979 , 12, 146 (synth)
Boyd, J. et al, J. Chem. Soc., Chem. Commun., 1976, 380 (synth)
Lewis, F.D. et al, J. Am. Chem. Soc., 1986, 108, 3016 (synth, nmr)

5-Methyl-3,4-hexadienoic acid   **M-70087**

$$(H_3C)_2C=C=CHCH_2COOH$$

$C_7H_{10}O_2$ M 126.155
Me ester: [101934-34-3].
$C_8H_{12}O_2$ M 140.182
Bright-yellow oil.

Lewis, F.D. et al, J. Am. Chem. Soc., 1986, 108, 3016 (synth, nmr, ir)

Methyl hydrogen 2-(*tert*-butoxymethyl)-2-methylmalonate　　M-70088

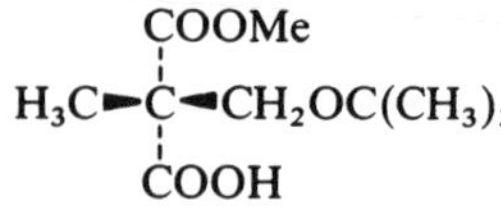

$$H_3C-\!\!-C\!\!-\!\!-CH_2OC(CH_3)_3$$

with COOMe above and COOH below the central C.

C$_{10}$H$_{18}$O$_5$　　M 218.249

(*R*)-form

Readily accessible chiral synthon. Mp 45-47°. [α]$_D$ +6.78° (c, 3.11 in MeOH).

Luyten, M. *et al*, *Helv. Chim. Acta*, 1987, **70**, 1250 (*synth, ir, pmr, ms*)

Methyl hydrogen 3-hydroxyglutarate　　M-70089

Monomethyl 3-hydroxypentanedioate

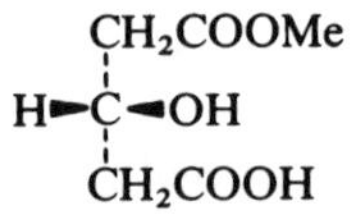

$$H-\!\!-C\!\!-\!\!-OH$$

with CH$_2$COOMe above and CH$_2$COOH below the central C.　　(*R*)-form

C$_6$H$_{10}$O$_5$　　M 162.142

Reagent for synth. of chiral hydroxyacids and related compds.

(*R*)-form [87118-53-4]

Oil. [α]$_D^{22}$ −1.7° (CHCl$_3$).

Morrison, M.A. *et al*, *J. Org. Chem.*, 1983, **48**, 4421 (*synth*)
Rosen, T. *et al*, *J. Org. Chem.*, 1984, **49**, 3657 (*purifn*)
Brooks, D.W. *et al*, *J. Org. Chem.*, 1987, **52**, 192 (*synth*)
Mohr, P. *et al*, *Helv. Chim. Acta*, 1987, **70**, 142 (*use*)

3-Methyl-1*H*-indole-2-carboxaldehyde　　M-70090

2-Formyl-3-methylindole

[5257-24-9]

C$_{10}$H$_9$NO　　M 159.187

Mp 142-143°.

2,4-Dinitrophenylhydrazone: Red cryst. (Py/MeOH). Mp 306-309°.

1-Me: [1971-44-4]. *1,3-Dimethyl-1*H*-indole-2-carboxaldehyde.*

C$_{11}$H$_{11}$NO　　M 173.214

Cryst. (EtOH aq.). Mp 36-37°.

Taylor, W.I., *Helv. Chim. Acta*, 1950, **33**, 164 (*synth, uv*)
Chatterjee, A. *et al*, *J. Org. Chem.*, 1973, **38**, 4002 (*synth*)
Itahara, T. *et al*, *Bull. Chem. Soc. Jpn.*, 1982, **55**, 3861 (*synth, ir, pmr*)
Benzies, D.W.M. *et al*, *Synth. Commun.*, 1986, **16**, 1799 (*synth, pmr, cmr*)
Comins, D.L. *et al*, *J. Org. Chem.*, 1987, **52**, 104 (*deriv, synth, pmr*)

9-Methyllongipesin　　M-70091

*4-Hydroxy-5-(1-hydroxyethyl)-3-(3-methyl-2-butenyl)-2*H-1-benzopyran-2-one, 9Cl*

[110024-38-9]

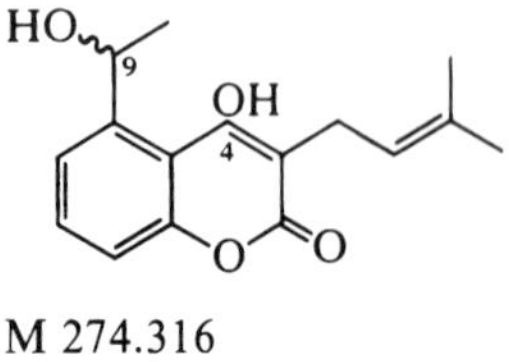

C$_{16}$H$_{18}$O$_4$　　M 274.316

Constit. of *Bothriocline longipes*.

O^9-*Ac:* [110024-40-3].

C$_{18}$H$_{20}$O$_5$　　M 316.353

Constit. of *B. longipes*.

O^9-*Ac, 4-Me ether:* Cryst. Mp 116°.

9-Propanoyl: [110024-43-6].

C$_{19}$H$_{22}$O$_5$　　M 330.380

Constit. of *B. longipes*.

9-Propanoyl, 4-Me ether: Oil. [α]$_D^{24}$ −11° (c, 0.92 in CHCl$_3$).

Jakupovic, J. *et al*, *Phytochemistry*, 1987, **26**, 1069.

4-Methyl-6,7-methylenedioxy-2*H*-1-benzo-pyran-2-one　　M-70092

*8-Methyl-6*H-1,3-benzodioxolo[4,5-g] [1]benzopyran-6-one, 9Cl, 8Cl. 4-Methyl-6,7-methylenedioxycoumarin. 4-Methylayapin*

[15071-04-2]

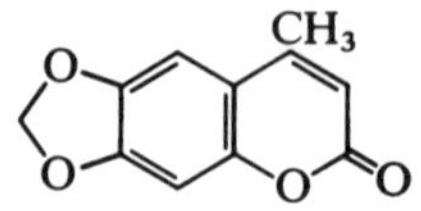

C$_{11}$H$_8$O$_4$　　M 204.182

Constit. of *Achillea schischkinii*. Cryst. (synthetic), amorph. powder (nat.). Mp 183°.

Woods, L.L. *et al*, *J. Chem. Eng. Data*, 1971, **16**, 101 (*synth, uv*)
Ulubelen, A. *et al*, *Planta Med.*, 1987, **53**, 507 (*isol*)

2-Methyl-6-methylene-2,7-octadien-4-ol　　M-70093

Updated Entry replacing M-20139

Ipsdienol

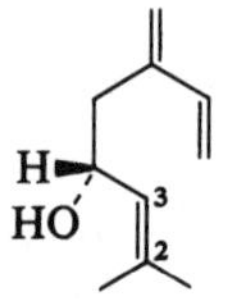

C$_{10}$H$_{16}$O　　M 152.236

A major component of the floral fragrance of several species of orchids. Racemic ipsdienol attracts various species of euglossine bees.

(*R*)-form [60894-97-5]

Pheromone of the bark beetle, *Ips confusus*, that bores into *Pinus ponderosa*. Oil. [α]$_D$ −13.6° (c, 1 in MeOH).

2,3-Dihydro: [35628-05-8]. *2-Methyl-6-methylene-7-octen-4-ol. Ipsenol.*

C$_{10}$H$_{18}$O　　M 154.252

Pheromone of *I. confusus* and *I. paraconfusus*. Oil. Bp$_{15}$ 86-88°. [α]$_D$ −17.5° (c, 1 in MeOH).

(*S*)-form [35628-00-3]

Pheromone of *I. paraconfusus*. Oil. [α]$_D^{21}$ +11.9° (c, 0.26 in MeOH).

Baekström, P. *et al*, *Acta Chem. Scand., Ser. B*, 1983, **37**, 1 (*synth, bibl*)
Sakurai, H. *et al*, *Tetrahedron*, 1983, **39**, 883 (*synth*)
Yamamoto, H. *et al*, *J. Am. Chem. Soc.*, 1986, **108**, 483 (*synth*)
Backström, P. *et al*, *Acta Chem. Scand., Ser. B*, 1987, **41**, 442 (*synth*)
Klusener, P.A.A. *et al*, *J. Org. Chem.*, 1987, **52**, 5261 (*synth*)
Whitten, W.M. *et al*, *Phytochemistry*, 1988, **27**, 2759 (*isol*)

Methyl 4-methylphenyl sulfoxide M-70094

Updated Entry replacing M-02282

1-Methyl-4-(methylsulfinyl)benzene, 9CI. Methyl p-tolyl sulfoxide

[934-72-5]

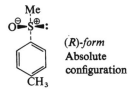

(*R*)-form
Absolute
configuration

$C_8H_{10}OS$ M 154.226

(*R*)-**form** [1519-39-7]

Orthorhombic cryst. Mp 74-76°. $[\alpha]_D^{20}$ +146° (Me_2CO).

(*S*)-**form** [5056-07-5]

Mp 74.5°. $[\alpha]_D^{20}$ −146° (Me_2CO).

(±)-**form** [39066-80-3]

Mp 42-43°, 50-54°. Bp_{38} 168°. pK_a −2.22 (H_2SO_4 or $HClO_4$).

Cerniani, A. *et al, Gazz. Chim. Ital.*, 1959, **89**, 843 (*synth*)
Mislow, K. *et al, J. Am. Chem. Soc.*, 1965, **87**, 1958 (*synth, uv, ord, cd, abs config*)
Landini, D. *et al, J. Am. Chem. Soc.*, 1969, **91**, 6703.
de la Camp, U., *Acta Crystallogr., Sect. B*, 1970, **26**, 846 (*abs config*)
Mangini, A. *et al, Int. J. Sulphur Chem., Part A*, 1972, **2**, 69 (*uv*)
Cutress, N.C. *et al, J. Chem. Soc., Perkin Trans. 2*, 1974, 268 (*ir*)
Kuneida, N. *et al, Bull. Chem. Soc. Jpn.*, 1976, **49**, 256.
Solladié, G. *et al, Synthesis*, 1987, 173 (*synth, use*)

2-Methyl-2*H*-naphtho[1,8-*de*]triazine, 9CI M-70095

Updated Entry replacing M-02362

2-Methyl-1H-naphtho[1,8-de]-1,2,3-triazinium hydroxide inner salt, 9CI

[40237-01-2]

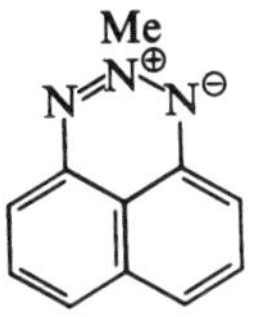

$C_{11}H_9N_3$ M 183.212

Photooxidation inhibitor, free-radical trap. Blue-black needles (pet. ether). Mp 132°.

1,3,5-Trinitrobenzene adduct: Mp 154°.

Perkins, M.J., *J. Chem. Soc.*, 1964, 3005 (*synth*)
Tavs, P. *et al, Justus Liebigs Ann. Chem.*, 1967, **704**, 150, 172 (*synth*)
Gait, S.F. *et al, J. Chem. Soc., Perkin Trans. 1*, 1975, 556 (*synth*)
Gieren, A. *et al, Z. Naturforsch., B*, 1984, **39**, 975 (*cryst struct*)
Kaupp, G. *et al, Angew. Chem., Int. Ed. Engl.*, 1986, **25**, 828 (*uv*)

2-Methyl-4-nitrobenzaldehyde M-70096

4-Nitro-o-tolualdehyde

[72005-84-6]

$C_8H_7NO_3$ M 165.148

Cryst. (EtOH). Mp 66-68°.

Goh, S.H. *et al, J. Chem. Soc., Perkin Trans. 1*, 1979, 1625 (*synth, ir, pmr*)

2-Methyl-4-nitro-1,3-benzenediol M-70097

2-Methyl-4-nitroresorcinol

$C_7H_7NO_4$ M 169.137

Yellow needles (C_6H_6). Mp 128-129°.

1-Me ether: 3-Methoxy-2-methyl-6-nitrophenol.

$C_8H_9NO_4$ M 183.163

Yellow prisms (pet. ether). Mp 68-69°.

Raphael, R.A. *et al, J. Chem. Soc., Perkin Trans. 1*, 1988, 1823 (*synth, ir, pmr, ms*)

3-Methyl-2-nitrobiphenyl M-70098

[82617-45-6]

$C_{13}H_{11}NO_2$ M 213.235

Needles (pet. ether). Mp 85-85.5°.

Atwell, G.J. *et al, J. Med. Chem.*, 1988, **31**, 774 (*synth*)

3-Methyl-4-nitropyridine, 9CI M-70099

Updated Entry replacing M-02695

4-Nitro-3-picoline, 8CI. 4-Nitro-β-picoline

[1678-53-1]

$C_6H_6N_2O_2$ M 138.126

Mp 28-29°. $Bp_{1.5}$ 67-69°.

N-*Oxide:* [1074-98-2].

$C_6H_6N_2O_3$ M 154.125

Cryst. (Me_2CO). Mp 136-137°.

▷Synth. of oxide has led to explosion. UT5775000.

Brown, E.V., *J. Am. Chem. Soc.*, 1954, **76**, 3167.
Herz, L.T., *J. Am. Chem. Soc.*, 1954, **76**, 4184.
Taylor, E.C. *et al, J. Org. Chem.*, 1954, **19**, 1633.
Org. Synth., Coll. Vol., **4**, 654 (*deriv*)
Essery, J.M. *et al, J. Chem. Soc.*, 1963, 2225 (*uv*)
Shiro, M. *et al, Acta Crystallogr., Sect. B*, 1977, **33**, 1549 (*cryst struct*)
Kuilen, A. *et al, Chem. Eng. News*, Jan. 30, 1989, 2 (*haz*)

Methyloxirane, 9CI M-70100

Updated Entry replacing M-02834

Propylene oxide. 1,2-Epoxypropane

[75-56-9]

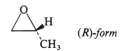

(*R*)-form

C_3H_6O M 58.080

▷Mod. toxic by inhalation and skin absorption, TLV 50. Extremely flammable, flash p. −37°. TZ2975000.

(*R*)-**form** [15448-47-2]

Liq. Bp 36.5-38°. $[\alpha]_D^{25}$ +14.0° (c, 0.5 in Et_2O), $[\alpha]_D^{22}$ +18.0° (c, 5.73 in CCl_4).

(*S*)-**form** [16088-62-3]

Liq. Bp 35°. $[\alpha]_D^{22}$ −8.21° (c, 5.04 in $CHCl_3$), −16.6° (c, 5.69 in Et_2O).

(±)-**form** [16033-71-9]

Intermed. in manuf. of polyols used for example in polyurethane foams, lubricants, hydraulic fluids and nonionic detergents. Liq. Misc. H_2O, EtOH, Et_2O. d^0 0.859. Bp 35°.

Abderhalden, E. *et al, Ber.*, 1918, **51**, 1318 (*synth*)
Fickett, W. *et al, J. Am. Chem. Soc.*, 1951, **73**, 5063 (*synth, abs config*)
Ghirardelli, R.G., *J. Am. Chem. Soc.*, 1973, **95**, 4987 (*synth*)
Golding, B.T. *et al, J. Chem. Soc., Perkin Trans. 1*, 1973, 1214 (*synth, glc*)
Pirkle, W.H. *et al, J. Org. Chem.*, 1978, **43**, 3803 (*synth*)
Schurig, V. *et al, Angew. Chem., Int. Ed. Engl.*, 1978, 937 (*synth, abs config*)

Mori, K. *et al*, *Tetrahedron*, 1985, **41**, 541 (*synth*)
Simon, E.S. *et al*, *J. Org. Chem.*, 1987, **52**, 4042 (*synth*)

5-(2-Methyloxiranyl)benzofuran   M-70101

5-(1,2-Epoxypropyl)benzofuran
[31823-03-7]

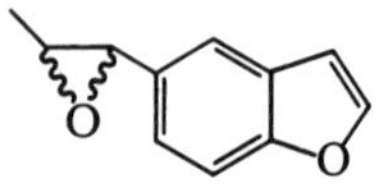

$C_{11}H_{10}O_2$ M 174.199
Metab. of *Stereum subpileatum*. Oil.

Bu'Lock, J.D. *et al*, *Phytochemistry*, 1971, **10**, 1037 (*isol, uv, ms, struct*)

2-Methyl-3-oxobutanal, 9CI   M-70102

3-Formyl-2-butanone
[22428-91-7]

$$H_3CCOCH(CH_3)CHO$$

$C_5H_8O_2$ M 100.117
Oil. Unstable. Isol. and handled as stable Na salt.

Tracy, A.H. *et al*, *J. Org. Chem.*, 1941, **6**, 63 (*synth*)
Falk, H. *et al*, *Monatsh. Chem.*, 1973, **104**, 925 (*synth*)
Paine, J.B., *J. Heterocycl. Chem.*, 1987, **24**, 351 (*synth*)

3-Methyl-4-oxo-2-butenoic acid, 9CI   M-70103

Updated Entry replacing M-10257
3-Formylcrotonic acid, 8CI. 3-Formyl-2-butenoic acid. 3-Formyl-3-methylacrylic acid

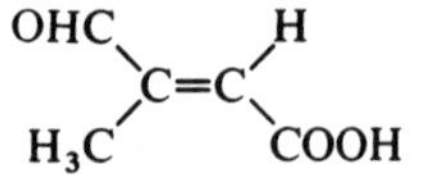

$C_5H_6O_3$ M 114.101
(*E*)-*form* [54168-84-2]
Cryst. (Et₂O). Mp 67-68°.
Me ester: [40835-18-5]. *Methyl 3-formylcrotonate. Methyl 3-formyl-3-methylacrylate.*
$C_7H_8O_3$ M 140.138
Yellow oil. Bp₇ 56-57°. Easily autooxidised.
Me ester, 2,4-dinitrophenylhydrazone: Cryst. Mp 224-225°.
Me ester, semicarbazone: Cryst. Mp 223° dec.
Me ester, di-Et acetal: [26586-00-5]. Oil. Bp₁₂ 116-122°.
Et ester: [62054-49-3].
$C_7H_{10}O_3$ M 142.154
Synthon for retinoids. Light-yellow oil.
Et ester, 2,4-dinitrophenylhydrazone: Cryst. Mp 199-200°.
Nitrile: [78843-78-4]. *3-Formyl-2-butenenitrile. 3-Cyano-2-methyl-2-propenoic acid.*
C_5H_5NO M 95.101
Bp₁₁ 71°.
Nitrile, 2,4-dinitrophenylhydrazone: Mp 276°.
(*Z*)-*form* [70143-04-3]
2,4-Dinitrophenylhydrazone: Cryst. (EtOAc). Mp 248°.
Me ester, di-Et acetal:
$C_{10}H_{18}O_4$ M 202.250
Bp₁₂ 116-122°.

Lactol-form
5-Hydroxy-4-methyl-2(5H)-furanone. 4-Hydroxy-3-methylbut-2-enolide
Prisms (Et₂O/ligroin). Sol. Et₂O, CCl₄. Bp₀.₀₁ 106-110°.

Pommer, H., *Angew. Chem.*, 1960, **72**, 811 (*esters*)
Sisido, K. *et al*, *J. Am. Chem. Soc.*, 1960, **82**, 2286 (*esters*)
Conradie, W.J. *et al*, *J. Chem. Soc.*, 1964, 594 (*synth*)
Pattenden, G. *et al*, *J. Chem. Soc.* (*C*), 1968, 1984, 1997.
Pattenden, G. *et al*, *J. Chem. Soc.* (*C*), 1970, 235 (*synth*)
Akhtar, M. *et al*, *J. Chem. Soc., Perkin Trans. 1*, 1978, 1511.
Clough, J.M. *et al*, *Tetrahedron Lett.*, 1978, 4159 (*synth*)
Chen, S., *Experientia*, 1981, **37**, 543 (*nitrile*)
Curley, R.W. *et al*, *J. Org. Chem.*, 1986, **51**, 256 (*synth, deriv, ir, uv, nmr*)
Curley, R.W. *et al*, *Synth. Commun.*, 1986, **16**, 627 (*synth*)

3-Methyl-4-oxo-2-pentenal   M-70104

Syoyualdehyde

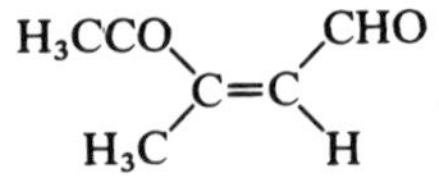

$C_6H_8O_2$ M 112.128
(*Z*)-*form* [58921-44-1]
Odoriferous constit. of soya. Bp₆ 71-100°.

Nakajima, M. *et al*, *J. Chem. Soc. Jpn.*, 1949, **70**, 47; *CA*, **45**, 7015.
Rice, K.C. *et al*, *J. Heterocycl. Chem.*, 1975, **12**, 1325 (*synth, pmr*)

3-Methyl-2-pentanone, 9CI   M-70105

Updated Entry replacing M-02910
Methyl sec-butyl ketone. 2-Acetylbutane
[565-61-7]

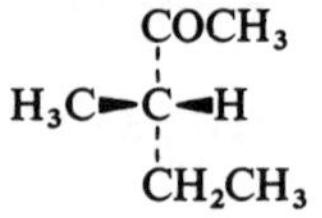

$C_6H_{12}O$ M 100.160
(*S*)-*form* [2695-53-6]
$[\alpha]_D^{25}$ +8.15° (neat).
(±)-*form* [55156-16-6]
Bp 118°.
Semicarbazone: Mp 95°.
2,4-Dinitrophenylhydrazone: Mp 72°.

Wislicenus, J., *Justus Liebigs Ann. Chem.*, 1883, **219**, 308 (*synth*)
Courtot, A., *Bull. Soc. Chim. Fr.*, 1906, **35**, 981 (*synth*)
Bartlett, P.D. *et al*, *J. Am. Chem. Soc.*, 1935, **57**, 2580 (*synth, abs config*)
Baker, R. *et al*, *J. Chem. Soc., Perkin Trans. 1*, 1987, 1613 (*synth*)
Brown, H.C. *et al*, *J. Am. Chem. Soc.*, 1987, **109**, 5420 (*synth, pmr, ir*)

2-Methyl-2-pentenal, 9CI   M-70106

Updated Entry replacing M-02912
3-Ethyl-2-methylacraldehyde. 2-Propylidenepropionaldehyde
[623-36-9]

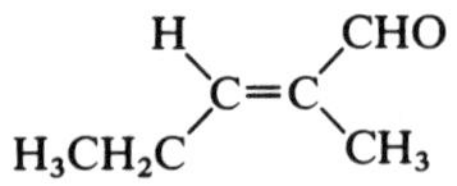

$C_6H_{10}O$ M 98.144
▷SB2100000.

(***E***)-*form* [14250-96-5]
Bp$_{25}$ 38-39°.
2,4-Dinitrophenylhydrazone: [1572-64-1]. Carmine
cryst. Mp 160°.
(***Z***)-*form* [16958-22-8]
Characterized spectroscopically.

Grignard, V. *et al, Bull. Soc. Chim. Fr.*, 1910, **7**, 642 (*synth*)
v. Auwers, K. *et al, Ber.*, 1925, **58**, 1979 (*synth*)
Häusermann, M., *Helv. Chim. Acta*, 1951, **34**, 1482 (*synth*)
Green, M.B. *et al, J. Chem. Soc.*, 1957, 3262 (*synth*)
Jewell, A. *et al, J. Org. Chem.*, 1968, **33**, 3382 (*synth, nmr, uv,
ir*)
Büchi, G. *et al, J. Am. Chem. Soc.*, 1972, **94**, 9128 (*synth*)
Woo, E.P. *et al, J. Org. Chem.*, 1986, **51**, 3704 (*synth, pmr,
cmr*)

4-Methyl-1-penten-3-ol   M-70107

Isopropyl vinyl carbinol
[4798-45-2]

$$(H_3C)_2CHCH(OH)CH{=}CH_2$$

C$_6$H$_{12}$O   M 100.160
(±)-*form*
Liq. Bp 122-127°.

Denmark, S.E. *et al, J. Org. Chem.*, 1987, **52**, 4031 (*synth*)

3-Methylphenanthrene   M-70108

Updated Entry replacing M-02959
[832-71-3]
C$_{15}$H$_{12}$ M 192.260
Cryst. (EtOH). Mp 62-63°. Bp$_6$ 140-150°.
Picrate: Cryst. (EtOH). Mp 137-138°.

Haworth, R.D. *et al, J. Chem. Soc.*, 1932, 1125.
Cook, J.W. *et al, J. Chem. Soc.*, 1944, 553.
Dalling, D.K. *et al, J. Am. Chem. Soc.*, 1977, **99**, 7142 (*cmr*)
Bansal, R.C. *et al, Org. Prep. Proced. Int.*, 1988, **20**, 305 (*synth,
pmr, cmr*)

3-Methyl-2-phenylbutanal   M-70109

α-(*1-Methylethyl)benzeneacetaldehyde*, 9CI
[2439-44-3]

$$(H_3C)_2CHCHPhCHO$$

C$_{11}$H$_{14}$O   M 162.231
(±)-*form* [107847-15-4]
Liq. Bp$_{15}$ 107-112°.

Lodge, E.P. *et al, J. Am. Chem. Soc.*, 1987, **109**, 3353 (*synth, ir,
pmr, cmr*)

2-Methyl-4-phenyl-3-butyn-2-ol, 9CI   M-70110

[1719-19-3]

$$(H_3C)_2C(OH)C{\equiv}CPh$$

C$_{11}$H$_{12}$O   M 160.215
Cryst. with rose odour. Mp 53°. Bp$_{1.5}$ 96°.
Ac:
C$_{13}$H$_{14}$O$_2$ M 202.252
Bp$_{10}$ 130-135°.

Wilson, F.J. *et al, J. Chem. Soc.*, 1923, 2612 (*synth*)
Kriz, J. *et al, Collect. Czech. Chem. Commun.*, 1967, **32**, 398
(*synth*)
Singelenberg, F.A.J. *et al, Acta Crystallogr., Sect. C*, 1987, **43**,
693 (*cryst struct*)

2-Methyl-4-phenylpyrimidine, 9CI   M-70111

[21203-79-2]

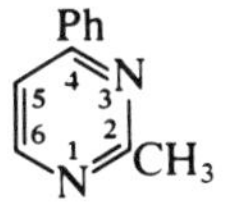

C$_{11}$H$_{10}$N$_2$ M 170.213
Leaflets (hexane or by subl.). Mp 54-55°.
Monopicrate: [21203-80-5]. Mp 205-206°.
1-Oxide: [69098-69-7].
C$_{11}$H$_{10}$N$_2$O   M 186.213
Needles (C$_6$H$_6$). Mp 153-154°.

Streef, J.W. *et al, Recl. Trav. Chim. Pays-Bas*, 1969, **88**, 1391
(*synth*)
Kato, T. *et al, Org. Mass Spectrom.*, 1974, **9**, 981 (*ms*)
Zhdanova, M.P. *et al, Chem. Heterocycles* Compd., 1978, **14**,
371 (*synth, ir*)
Yamanaka, H. *et al, Chem. Pharm. Bull.*, 1980, **28**, 1526 (*ox-
ide*)
Burdeska, K. *et al, Helv. Chim. Acta*, 1981, **64**, 113 (*pmr*)

2-Methyl-5-phenylpyrimidine, 9CI   M-70112

[34771-47-6]
C$_{11}$H$_{10}$N$_2$ M 170.213
Prisms (hexane). Mp 69.5-70°.

Wagner, R.M. *et al, Chem. Ber.*, 1971, **104**, 2975 (*synth, pmr*)
Staedeli, W. *et al, Org. Magn. Reson.*, 1981, **15**, 106; 1981, **16**,
170 (*nmr*)
Mikhaleva, M.A. *et al, Khim. Geterotsikl. Soedin.*, 1986, 380
(*synth*)

4-Methyl-2-phenylpyrimidine, 9CI   M-70113

[34771-48-7]
C$_{11}$H$_{10}$N$_2$ M 170.213
Light-yellow to colourless oil or cryst. Mp 25° (21-22°).
Bp$_{0.2}$ 82°, Bp$_{0.05}$ 92-94°.

Wagner, R.M. *et al, Chem. Ber.*, 1971, **104**, 2975 (*synth, pmr*)
Kato, T. *et al, Org. Mass Spectrom.*, 1974, **9**, 981 (*ms*)
Burdeska, K. *et al, Helv. Chim. Acta*, 1981, **64**, 113 (*synth,
pmr*)
Goerdeler, J. *et al, Chem. Ber.*, 1986, **119**, 3737 (*synth*)

4-Methyl-5-phenylpyrimidine, 9CI   M-70114

[57562-58-0]
C$_{11}$H$_{10}$N$_2$ M 170.213
Present as an impurity in illicit amphetamine samples.
Colourless or pale-yellow needles (ligroin). Mp 76-
76.5°.
1-Oxide:
C$_{11}$H$_{10}$N$_2$O   M 186.213
Viscous oil.
1-Oxide, picrate: Cryst. (Me$_2$CO/hexane). Mp 132-133°
dec.
3-Oxide: [114969-95-8].
C$_{11}$H$_{10}$N$_2$O   M 186.213
Mp 86.5-88°.

Koyama, T. *et al, Chem. Pharm. Bull.*, 1975, **23**, 2029, 2158
(*synth, pmr*)
Lambrechts, M. *et al, J. Chromatogr.*, 1984, **284**, 499; 1985,
331, 339; 1986, **369**, 365 (*hplc, chromatog, ms*)
Yamanaka, H. *et al, Chem. Pharm. Bull.*, 1987, **35**, 3119
(*synth, pmr, oxides*)

4-Methyl-6-phenylpyrimidine, 9CI   M-70115

[17759-27-2]
C$_{11}$H$_{10}$N$_2$ M 170.213

Cryst. (ligroin). Mp 45-48°. Bp$_{12}$ 145-150°.
B,HCl: Mp 236°.
Picrate: Mp 203-204°.
1-Oxide: [17759-10-3].
  C$_{11}$H$_{10}$N$_2$O   M 186.213
  Mp 135-137°.
3-Oxide: [14161-43-4].
  C$_{11}$H$_{10}$N$_2$O   M 186.213
  Pale-yellow needles; cryst. (C$_6$H$_6$/pet. ether). Mp 139-141°. Note near-identity of Mp's of 1- and 3-oxides.
3-Oxide, picrate: Mp 141°.

Bredereck, H. *et al, Chem. Ber.,* 1957, **90**, 942; 1958, **91**, 2832 (*synth, oxide*)
Kato, T. *et al, Yakugaku Zasshi,* 1967, **87**, 1096; 1977, **97**, 676; *CA,* **68**, 87268a; **87**, 95448k.
Kasuga, K. *et al, Chem. Pharm. Bull.,* 1974, **22**, 1814 (*synth, ir, ms, pmr, oxide*)
Kato, T. *et al, Org. Mass Spectrom.,* 1974, **9**, 981 (*ms*)
Konno, S. *et al, Heterocycles,* 1984, **22**, 1331 (*synth*)

5-Methyl-2-phenylpyrimidine, 9CI   M-70116

[77232-48-5]
C$_{11}$H$_{10}$N$_2$ M 170.213
Cryst. (MeOH aq. or by subl.). Mp 69-70°. Bp$_{0.1}$ 107-109°.
Picrate: Cryst. (EtOH). Mp 154°.

Bredereck, H. *et al, Chem. Ber.,* 1960, **93**, 1208 (*synth*)
Robba, M. *et al, Bull. Soc. Chim. Fr.,* 1960, 1648 (*synth*)
Burdeska, K. *et al, Helv. Chim. Acta,* 1981, **64**, 113 (*pmr*)

5-Methyl-4-phenylpyrimidine   M-70117

[57270-07-2]
C$_{11}$H$_{10}$N$_2$ M 170.213
Found in illicit amphetamine samples. Has some anti-inflammatory activity. Mp 29-31°. Bp$_{15}$ 153-156°, Bp$_{0.1}$ 100-105°.
Picrate: Cryst. (MeOH/EtOAc). Mp 140°.
1-Oxide: [57270-08-3].
  C$_{11}$H$_{10}$N$_2$O   M 186.213
  Cryst. by subl. Mp 151-153°.
3-Oxide: [57270-09-4].
  C$_{11}$H$_{10}$N$_2$O   M 186.213
  Cryst. by subl. Mp 146-148°.
1,3-Dioxide: [56642-52-5].
  C$_{11}$H$_{10}$N$_2$O$_2$ M 202.212
  Cryst. by subl. Mp 225-227°.

Bredereck, H. *et al, Chem. Ber.,* 1960, **93**, 1402 (*synth*)
Tikhonov, A.Ya. *et al, Tetrahedron Lett.,* 1975, 2721 (*synth, ir, pmr, uv, oxides*)
Volodarskii, L.B. *et al, Bull. Acad. Sci. USSR, Ser. Chem.,* 1975, 1122 (*synth, uv, pmr, dioxide*)
Bennett, G.B. *et al, J. Med. Chem.,* 1978, **21**, 623 (*synth*)
Yamanaka, H. *et al, Chem. Pharm. Bull.,* 1987, **35**, 3119 (*synth, pmr, oxide*)

3-(2-Methylpropyl)-6-methyl-2,5-piperazinedione   M-70118

Cyclo(leucylalanyl)
[1803-60-7]

C$_9$H$_{16}$N$_2$O$_2$ M 184.238

Diketopiperazine.
(3S,6S)-form [24676-83-3]
L-L-form
Prod. by *Aspergillus phoenicis* and isol. from the tubers of *Pinellia pedatisecta.* Shows virustatic activity. Mp 259°.

Caesar, F. *et al, Pharm. Acta Helv.,* 1969, **44**, 676 (*isol*)
Oya, M. *et al, Pept. Chem.,* 1977, **15**, 55 (*synth*)
Kricheldorf, H., *Org. Magn. Reson.,* 1980, **13**, 52 (*synth, cmr, N nmr*)
Czech. Pat., 210 383, (*1983*); *CA,* **101**, 17326 (*synth, props*)
Qin, W. *et al, Zhongcaoyao,* 1984, **15**, 490; *CA,* **102**, 109774 (*isol*)

(1-Methylpropyl)propanedioic acid   M-70119

2-Methylbutane-1,1-dicarboxylic acid. sec-*Butylmalonic acid*

H$_3$CCH$_2$CH(CH$_3$)CH(COOH)$_2$

C$_7$H$_{12}$O$_4$ M 160.169
(±)-form
Plates. Mp 76°.
Di-Me ester: [39520-23-5].
  C$_9$H$_{16}$O$_4$ M 188.223
  Liq. Bp$_{748}$ 217-218°.
Diamide: Cryst. (EtOH aq.). Mp 242°.

Ehrlich, F., *Ber.,* 1908, **41**, 1455 (*synth*)
Dox, A.W. *et al, J. Am. Chem. Soc.,* 1922, **44**, 1564 (*derivs*)

2-Methylpteridine, 8CI   M-70120

[2432-20-4]

C$_7$H$_6$N$_4$ M 146.151
Yellow leaflets (pet. ether). Mp 141-142°.

Albert, A. *et al, J. Chem. Soc.,* 1954, 3832 (*synth*)
Goto, T. *et al, J. Chem. Soc.,* 1963, 1773 (*pmr, ms*)
Goto, T. *et al, J. Org. Chem.,* 1965, **30**, 1844 (*pmr, ms*)
Albert, A. *et al, J. Chem. Soc. (C),* 1971, 2357 (*synth*)

4-Methylpteridine, 9CI   M-70121

[2432-21-5]
C$_7$H$_6$N$_4$ M 146.151
Yellow needles (pet. ether). Mp 152-153°.

Goto, T. *et al, J. Chem. Soc.,* 1963, 1773 (*pmr, ms*)
Goto, T. *et al, J. Org. Chem.,* 1965, **30**, 1844 (*pmr, ms*)
Albert, A. *et al, J. Chem. Soc. (C),* 1968, 1181 (*synth*)
Spanget-Larsen, J. *et al, Chem. Ber.,* 1986, **119**, 1275 (*pe*)

6-Methylpteridine, 9CI   M-70122

[24192-73-2]
C$_7$H$_6$N$_4$ M 146.151
Mp 130°.

Albert, A. *et al, J. Chem. Soc. (C),* 1970, 1540 (*synth*)

7-Methylpteridine, 9CI   M-70123

[936-40-3]
C$_7$H$_6$N$_4$ M 146.151
Pale-yellow needles (pet. ether). Mp 133-134°.

Albert, A. *et al, J. Chem. Soc.,* 1954, 3832 (*synth*)
Albert, A. *et al, J. Chem. Soc. (C),* 1970, 1540 (*pmr*)
Geerts, J.P. *et al, Org. Magn. Reson.,* 1976, **8**, 607 (*cmr*)

Zav'yalov, S.I. *et al, Izv. Akad. Nauk SSSR, Ser. Khim.*, 1980, 2575 (*synth*)

6-Methyl-3-pyridazinone   M-70124

1H-form   2H-form   OH-form

$C_5H_6N_2O$ M 110.115
2H-Form predominates. Needles (Me₂CO). Mp 130°.
Monohydrate: Cryst. (H₂O). Mp 119-123°.

1H-form
Minor tautomer.
1-Me: [57766-27-5]. *2,3-Dihydro-1,6-dimethyl-3-oxo-pyridazinium hydroxide, inner salt, 9CI.*
$C_6H_8N_2O$ M 124.142
Buff needles (CHCl₃/pet. ether). Mp 171-172°.

2H-form [13327-27-0]
2-Me: [50500-59-9].
$C_6H_8N_2O$ M 124.142
Prisms (pet. ether or EtOH/Et₂O). Mp 50-51°.
2-Ph: [38154-50-6].
$C_{11}H_{10}N_2O$ M 186.213
Cryst. (cyclohexane). Mp 79-80°.

OH-form
3-Hydroxy-6-methylpyridazine. 6-Methyl-3-pyridazinol
Minor tautomer.
Me ether: [17644-83-6]. *3-Methoxy-6-methylpyridazine.*
$C_6H_8N_2O$ M 124.142
Oil. Bp 210°.
Me ether; B,HCl: Plates (EtOH). Mp 131-132°.
Me ether, 1-Oxide: [1074-48-2].
$C_6H_8N_2O_2$ M 140.141
Mp 98-99°.

Overend, W.G. *et al, J. Chem. Soc.*, 1947, 239, 549; 1948, 2191; 1950, 3500 (*synth, deriv*)
Nakagone, T., *J. Pharm. Soc. Jpn.*, 1961, **81**, 1048 (*deriv*)
Dennis, N. *et al, J. Chem. Soc., Perkin Trans. 1*, 1975, 1506 (*deriv*)
McNab, H., *J. Chem. Soc., Perkin Trans. 1*, 1983, 1203 (*cmr*)

4-Methyl-3(2H)-pyridazinone, 9CI   M-70125
[33471-40-8]

$C_5H_6N_2O$ M 110.115
Prisms (EtOAc), cryst. (C₆H₆). Mp 165-166° (158-159°).

McMillan, F.H. *et al, J. Am. Chem. Soc.*, 1956, **78**, 407 (*synth*)
Ogata, M. *et al, Chem. Pharm. Bull.*, 1963, **11**, 35 (*synth*)
McNab, H. *et al, J. Chem. Soc., Perkin Trans. 1*, 1982, 1845; 1983, 1203 (*synth, cmr*)

5-Methyl-3(2H)-pyridazinone, 9CI   M-70126
[54709-94-3]
$C_5H_6N_2O$ M 110.115
Prisms (EtOAc). Mp 160-161° (151-153°).
2-Ph: [68143-02-2].
$C_{11}H_{10}N_2O$ M 186.213

Cryst. (cyclohexane or Et₂O/pet. ether). Mp 84°.

Schmidt, P. *et al, Helv. Chim. Acta*, 1954, **37**, 1467 (*synth*)
Wiley, R.H. *et al, J. Am. Chem. Soc.*, 1955, **77**, 403 (*deriv*)
Ogata, M. *et al, Chem. Pharm. Bull.*, 1963, **11**, 35 (*synth*)
McNab, H. *et al, J. Chem. Soc., Perkin Trans. 1*, 1982, 1845; 1983, 1203 (*synth, deriv, cmr*)

5-Methyl-2,4-(1H,3H)-pyrimidinedithione, 9CI   M-70127
5-Methyl-2,4-dithiouracil. Dithiothymine
[6217-61-4]

$C_5H_6N_2S_2$ M 158.236
Mp 283-285° (281°).
▷UV9135000.

Wheeler, H.L. *et al, Am. Chem. J.*, 1910, **43**, 19 (*synth*)
Elion, G.B. *et al, J. Am. Chem. Soc.*, 1947, **69**, 2138 (*synth*)
Łapucha, A., *Synthesis*, 1987, 256 (*synth, ir, pmr, cmr, ms*)

6-Methyl-2,4(1H,3H)-pyrimidinedithione, 9CI   M-70128
6-Methyl-2,4-dithiouracil. 2,4-Dimercapto-6-methylpyrimidine
[6308-38-9]
$C_5H_6N_2S_2$ M 158.236
Light-yellow cryst. Mp 354-356° dec. (darkens at 290°).

Brown, D.J., *J. Appl. Chem.*, 1959, **9**, 203 (*synth*)
Łapucha, A., *Synthesis*, 1987, 256 (*synth, ir, pmr, cmr, ms*)

5-Methyl-1H-pyrrole-2-carboxaldehyde, 9CI   M-70129
Updated Entry replacing M-03525
2-Formyl-5-methylpyrrole
[1192-79-6]
C_6H_7NO M 109.127
Mp 69°.
Oxime:
$C_6H_8N_2O$ M 124.142
Needles (EtOH). Mp 153°.
1-Me: [1193-59-5]. *1,5-Dimethyl-1H-pyrrole-2-carboxaldehyde.*
C_7H_9NO M 123.154
Light-yellow oil.
1-Me, 2,4-dinitrophenylhydrazone: Mp 227-228°.

Gronowitz, S. *et al, CA*, 1962, **56**, 10075 (*nmr*)
Streith, J. *et al, Tetrahedron Lett.*, 1966, 1347 (*synth*)
Comins, D.L. *et al, J. Org. Chem.*, 1987, **52**, 104 (*deriv, synth, pmr, ir*)

2-Methyl-9-β-D-ribofuranosyl-9*H*-purine   M-70130
2-Methylnebularine
[42890-34-6]

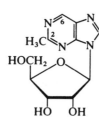

C$_{11}$H$_{14}$N$_4$O$_4$ M 266.256
Cryst. (EtOH). Mp 177-180°. [α]$_D^{25}$ −75.0° (c, 1 in 0.1*M* NaOH).

Yamazaki, A. *et al*, *Chem. Pharm. Bull.*, 1973, **21**, 692 (*synth*)

6-Methyl-9-β-D-ribofuranosyl-9*H*-purine   M-70131
6-Methylnebularine
[14675-48-0]
C$_{11}$H$_{14}$N$_4$O$_4$ M 266.256
Cryst. (EtOH). Mp 209-210°. [α]$_D^{23}$ −52.0° (c, 1 in MeOH).

2′,3′,5′-Tribenzoyl: [14985-78-5]. Mp 55°.

Gordon, M.P. *et al*, *J. Am. Chem. Soc.*, 1957, **79**, 3245 (*synth*)
Montgomery, J.A. *et al*, *J. Med. Chem.*, 1967, **11**, 48 (*synth*)
Takeda, T. *et al*, *Acta Crystallogr., Sect. B*, 1975, **31**, 1202 (*cryst struct*)
Shimazaki, M. *et al*, *Chem. Pharm. Bull.*, 1983, **31**, 3104 (*synth*)

[2-(Methylseleno)ethenyl]benzene, 9CI   M-70132
Methyl styryl selenide

PhCH=CHSeMe

C$_9$H$_{10}$Se   M 197.138
(*E*)-form [95391-78-9]
  S-Oxide: [108415-80-1]. [2-(*Methylseleninyl*)ethenyl]-benzene, 9CI. Methyl styryl selenoxide.
  C$_9$H$_{10}$OSe   M 213.137
  Hygroscopic solid.
  S-Dioxide: [108415-84-5]. [2-(*Methylselenonyl*)-ethenyl]benzene, 9CI.
  C$_9$H$_{10}$O$_2$Se   M 229.137
  Cryst. Mp 137-139°.

(*Z*)-form [95391-81-4]
  S-Oxide: [108415-82-3]. Hygroscopic solid.
  S-Dioxide: [108415-86-7]. Cryst. Mp 103-105°.

Tiecco, M. *et al*, *Tetrahedron Lett.*, 1984, **25**, 4975 (*synth, nmr*)
Tiecco, M. *et al*, *Tetrahedron*, 1986, **42**, 4889 (*deriv, synth, nmr*)

(Methylsulfonyl)[(trifluoromethyl)sulfonyl]-methane, 9CI   M-70133
Mesyltriflone
[93916-15-5]

F$_3$CSO$_2$CH$_2$SO$_2$Me

C$_3$H$_5$F$_3$O$_4$S$_2$ M 226.185
Plates. Mp 115-116°.

Hendrickson, J.B. *et al*, *J. Am. Chem. Soc.*, 1986, **108**, 2358 (*synth, nmr*)

3-Methyl-*N*-(5*a*,6*a*,7,8-tetrahydro-4-methyl-1,5-dioxo-1*H*,5*H*-pyrrolo[1,2-c][1,3]oxazepin-3-yl)butanamide, 9CI   M-70134
7,8,9,9a-Tetrahydro-3-[(3-methyl-1-oxobutyl)amino]-4-methyl-1H,5H-pyrrolo[1,2-c][1,3]oxazepine-1,5-dione. Herbicidal substance 1328-2
[102719-89-1]

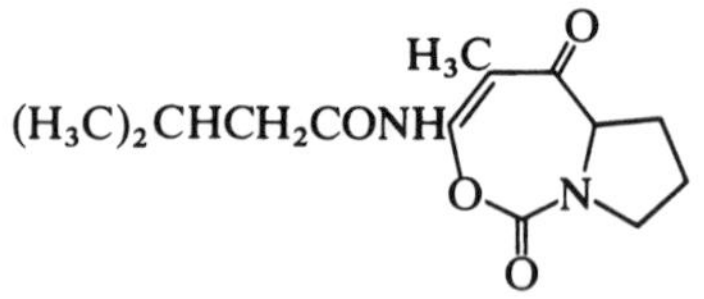

C$_{14}$H$_{20}$N$_2$O$_4$ M 280.323
Prod. by *Streptoverticillium* sp. Herbicide.
N-Hexanoyl analogue: [102719-90-4]. N-(5a,6,7,8-Te-trahydro-4-methyl-1,5-dioxo-1H,5H-pyrrolo[1,2-c][1,3]oxazepin-3-yl)hexanamide, 9CI. 7,8,9,9a-Te-trahydro-3-[(1-oxohexyl)amino]-4-methyl-1H,5H-pyrrolo[1,2-c][1,3]oxazepine-1,5-dione. Herbicidal substance 1328-3.
  C$_{15}$H$_{22}$N$_2$O$_4$ M 294.350
  From *S.* sp. Herbicide.

Japan. Pat., 87 72 691, (*1987*); *CA*, **107**, 38125k (*isol*)

2-Methyl-2-thiocyanatopropane   M-70135
Thiocyanic acid 1,1-dimethylethyl ester, 9CI. tert-*Butyl thiocyanate*
[37985-18-5]

(H$_3$C)$_3$CSCN

C$_5$H$_9$NS   M 115.193
Bp$_{10}$ 40°. Readily isom. to 1-Isothiocyanatobutane, B-03693 on heating. Obt. as a mixt. in prev. synth.

Ando, T. *et al*, *J. Org. Chem.*, 1987, **52**, 681 (*synth, cmr*)

2-(Methylthio)cyclopentanone   M-70136
[52190-34-8]

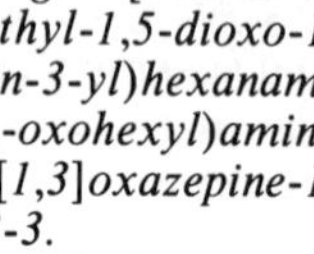

C$_6$H$_{10}$OS   M 130.204
Bp$_5$ 65-66°.
(±)-form
Bp$_{10.5}$ 84°.

Seebach, D. *et al*, *Chem. Ber.*, 1976, **109**, 1601 (*synth, nmr, ir*)
Caubère, P. *et al*, *J. Org. Chem.*, 1986, **51**, 1419 (*synth*)

[2-(Methylthio)ethenyl]benzene, 9CI   M-70137
Methyl styryl sulfide. β-(Methylthio)styrene
[7715-02-8]

PhCH=CHSMe

C$_9$H$_{10}$S   M 150.238
(*Z*)-form [35822-50-5]
Oil. Bp$_{0.5}$ 101.5°.

Caserio, M. *et al*, *J. Am. Chem. Soc.*, 1966, **88**, 5745 (*synth, nmr*)
Tiecco, M. *et al*, *J. Org. Chem.*, 1983, **48**, 4795 (*synth, nmr*)
Tiecco, M. *et al*, *Tetrahedron*, 1985, **41**, 1401.

(Methylthio)methanol   M-70138

$$MeSCH_2OH$$

C_2H_6OS M 78.129
Stable liq. Bp$_{13}$ 41°.
p-*Nitrobenzoyl:* Mp 50-52°.

Jones, J.H. *et al, Synth. Commun.,* 1986, **16**, 1607 (*synth, pmr*)

2-Methyl-3-thiophenecarboxaldehyde   M-70139
3-Formyl-2-methylthiophene
[84815-20-3]

C_6H_6OS M 126.173
Light-yellow oil. Bp$_{10}$ 80.5-82.0°.
Semicarbazone: Mp 190-191°.

Gronowitz, S. *et al, Ark. Kemi,* 1961, **17**, 165 (*synth*)
Comins, D.L. *et al, J. Org. Chem.,* 1987, **52**, 104 (*synth, pmr*)

3-Methyl-2-thiophenecarboxaldehyde   M-70140
2-Formyl-3-methylthiophene
[5834-16-2]
C_6H_6OS M 126.173
Bp$_{15}$ 100-101°, Bp$_{3.5}$ 84-85°.
Semicarbazone: Mp 194.5-195.0°.

Campaigne, E. *et al, J. Am. Chem. Soc.,* 1953, **75**, 989 (*synth*)
Gronowitz, S. *et al, Ark. Kemi,* 1961, **17**, 165 (*synth*)

4-Methyl-2-thiophenecarboxaldehyde   M-70141
2-Formyl-4-methylthiophene
[6030-36-0]
C_6H_6OS M 126.173
Liq. d$_4^{20}$ 1.160. Bp$_8$ 84-86°.
4-Nitrophenylhydrazone: Red needles (EtOH). Mp 179-181°.
Ethylene acetal:
    $C_8H_{10}O_2S$ M 170.226
    d$_4^{22}$ 1.20. Bp$_{12}$ 121-122°.

Sicé, J., *J. Org. Chem.,* 1954, **19**, 70 (*synth*)

4-Methyl-3-thiophenecarboxaldehyde   M-70142
3-Formyl-4-methylthiophene
C_6H_6OS M 126.173
Bp$_{13}$ 82-84°.
Semicarbazone: Mp 194-195°.

Gronowitz, S. *et al, Ark. Kemi,* 1961, **17**, 165 (*synth*)

5-Methyl-2-thiophenecarboxaldehyde   M-70143
2-Formyl-5-methylthiophene
[13679-70-4]
C_6H_6OS M 126.173
Light-yellow oil. Bp$_{10}$ 93-93.5°, Bp$_{3.5}$ 84-85°.

Campaigne, E. *et al, J. Am. Chem. Soc.,* 1953, **75**, 989 (*synth*)
Hoffman, R.A. *et al, Ark. Kemi,* 1960, **16**, 563 (*pmr*)
Comins, D.L. *et al, J. Org. Chem.,* 1987, **52**, 104 (*synth, pmr*)

5-Methyl-3-thiophenecarboxaldehyde   M-70144
4-Formyl-2-methythiophene

[29421-72-5]
C_6H_6OS M 126.173
Light-yellow oil. Bp$_{13}$ 91-100°.
Semicarbazone: Cryst. (EtOAc). Mp 206-207.5°.

Goldfarb, Y.L. *et al, J. Gen. Chem USSR (Engl. Transl.),* 1964, **34**, 961 (*synth*)
Comins, D.L. *et al, J. Org. Chem.,* 1987, **52**, 104 (*synth, pmr*)

Methyl trifluoromethyl sulfide, 8CI   M-70145
Trifluoro(methylthio)methane, 9CI
[421-16-9]

$$MeSCF_3$$

$C_2H_3F_3S$ M 116.101
Bp$_{750}$ 11.5-11.7°.
S-Oxide: [2697-49-6]. *Trifluoro(methylsulfinyl)-methane, 9CI. Methyl trifluoromethyl sulfoxide.*
    $C_2H_3F_3OS$ M 132.100
    Liq. d$_{20}$ 1.452. Mp −28 to −27°. Bp 115-116°. n$_D^{20}$ 1.3678-1.3684.
S,S-Dioxide: [421-82-9]. *Trifluoro(methylsulfonyl)-methane, 9CI. Methyl trifluoromethyl sulfone.*
    $C_2H_3F_3O_2S$ M 148.100
    Odourless liq. d$_4^{20}$ 1.514. Fp 14°. Bp 128-129.5°. n$_D^{20}$ 1.3486.

Truce, W.E. *et al, J. Am. Chem. Soc.,* 1952, **74**, 3594 (*synth, sulfone*)
Yu, S.-L. *et al, Inorg. Chem.,* 1974, **13**, 484 (*synth, ir, nmr, pmr, sulfide, sulfone, sulfoxide*)
Zack, N.R. *et al, J. Fluorine Chem.,* 1975, **5**, 153 (*ms*)
Haley, B. *et al, J. Chem. Soc., Perkin Trans. 1,* 1976, 525 (*synth, sulfoxide*)
Hendrickson, J.B. *et al, Tetrahedron,* 1976, **32**, 1627 (*synth, ir, pmr, sulfone*)
Yagupol'skii, L.M. *et al, J. Org. Chem. USSR,* 1976, **12**, 2197 (*synth, ir, pmr, nmr, ms, sulfoxide*)
Nwaukwai, S.O. *et al, Synth. Commun.,* 1986, **16**, 309 (*synth, ir, pmr, sulfone*)

5-Methyl-5-vinyl-1,3-cyclopentadiene   M-70146
5-Ethenyl-5-methyl-1,3-cyclopentadiene

$$H_3C \quad CH=CH_2$$

C_8H_{10} M 106.167
Liq.

Burger, U. *et al, Helv. Chim. Acta,* 1988, **71**, 389 (*synth, uv, pmr, cmr, ms*)

Microphyllinic acid   M-70147
Updated Entry replacing M-20222
2-Hydroxy-4-[[2-hydroxy-4-methoxy-6-(2-oxoheptyl)-benzoyl]oxy]-6-(2-oxoheptyl)benzoic acid, 9CI. Microphyllic acid
[491-46-3]

$C_{29}H_{36}O_9$ M 528.598
Lichen acid from *Centaria collata.* Needles (C$_6$H$_6$/pet. ether). Mp 116°. Shows ring-chain tautom. in soln.

Me ester: Needles (EtOH). Mp 118°.
2-Me ether: [79579-62-7]. **2'-O-Methylmicrophyllinic acid**.
$C_{30}H_{38}O_9$ M 542.625
Metab. of *Lecidea ferax*. Needles (EtOAc/pet. ether). Mp 147-148°.
Di-Me ether: Cryst. (EtOH). Mp 89-90°.
O^4-*De-Me:* [63744-96-7]. **4-O-Demethylmicrophyllinic acid**.
$C_{28}H_{34}O_9$ M 514.571
Metab. of *Parmotrema demethylmicrophyllinicum*. Prisms (Me₂CO/cyclohexane). Mp 162° dec.

Asahina, Y. *et al, Chem. Ber.,* 1935, **68**, 81, 2022.
Elix, J.A. *et al, Aust. J. Chem.,* 1974, **27**, 2403; 1987, **40**, 1851 (*tautom, pmr, deriv, synth*)
Chester, D.O. *et al, Aust. J. Chem.,* 1981, **34**, 1507 (*deriv, synth*)

Mikanifuran   M-70148

3-Methyl-2-(3,7,11-trimethyl-2,6,10-dodecatrienyl)-furan, 9Cl
[80442-35-9]

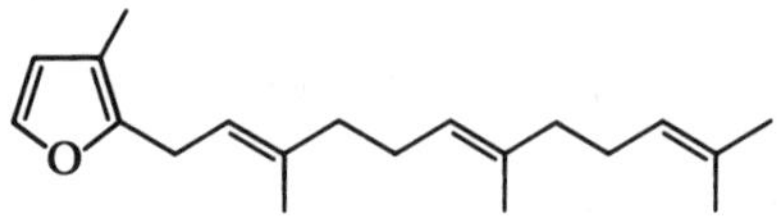

$C_{20}H_{30}O$ M 286.456
Constit. of *Mikania sessilifolia*. Oil.

Bohlmann, F. *et al, Phytochemistry,* 1981, **20**, 1899 (*struct*)
Kramp, W. *et al, Justus Liebigs Ann. Chem.,* 1986, 226 (*synth*)

Miltiodiol   M-70149

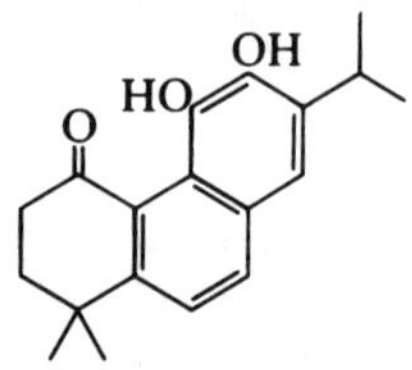

$C_{19}H_{22}O_3$ M 298.381
Constit. of Chinese drug Dan-Shen (*Salvia miltiorrhiza*).
Ginda, H. *et al, Tetrahedron Lett.,* 1988, **29**, 4603 (*struct*)

Mitsugashiwalactone   M-70150

Updated Entry replacing M-50244
Hexahydro-7-methylcyclopenta[c]pyran-1(3H)-one, 9Cl
[60363-05-5]

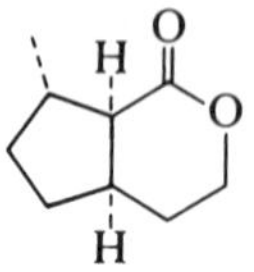

$C_9H_{14}O_2$ M 154.208
Constit. of *Boshniakia rossica*. Cat attractant. Oil.
7-Epimer: [60363-04-4]. **Onikulactone**. Isol. from *Menyanthes trifoliata*. Cat attractant. Oil.

Sakan, T. *et al, J. Chem. Soc. Jpn.,* 1969, **90**, 507 (*isol*)
Ohta, H. *et al, J. Org. Chem.,* 1977, **42**, 1231 (*synth, ir, pmr, ms*)
Nugent, W.A. *et al, J. Org. Chem.,* 1986, **51**, 3376.
Amri, M. *et al, Tetrahedron Lett.,* 1987, **28**, 5521 (*synth*)

Moluccanin   M-70151

[116521-73-4]

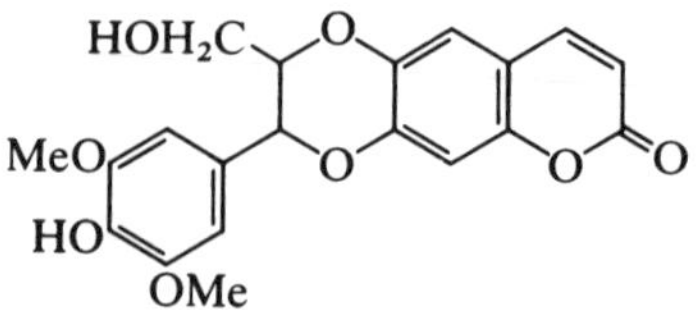

$C_{20}H_{18}O_8$ M 386.357
Constit. of *Aleurites moluccana*. Cryst. (Me₂CO-/MeOH). Mp 220°.

Shamsuddin, T. *et al, Phytochemistry,* 1988, **27**, 1908.

Monoalide   M-70152

[75088-80-1]

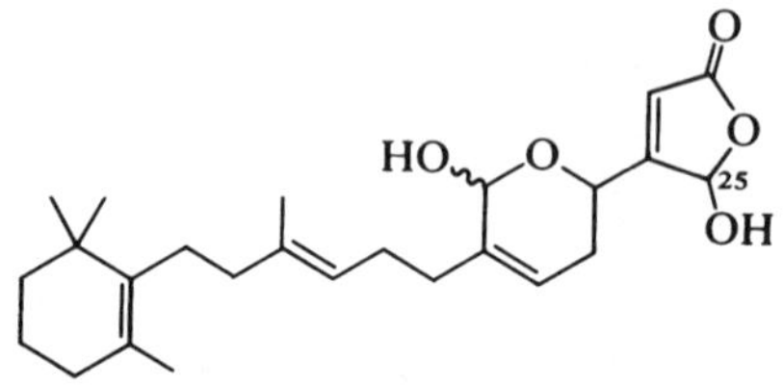

$C_{25}H_{36}O_5$ M 416.556
Constit. of *Luffariella variabilis*. Amorph.
25-Ac: Monoalide 25-acetate.
$C_{27}H_{38}O_6$ M 458.594
Constit. of *Thorectandra excavatus*. Flakes (CH₂Cl₂/hexane). Mp 117-119°. $[\alpha]_D^{17} +34°$ (c, 1 in CHCl₃).

De Silva, E.D. *et al, Tetrahedron Lett.,* 1980, 1611 (*isol, struct*)
Cambie, R.C. *et al, J. Nat. Prod.,* 1988, **51**, 331 (*deriv*)

Montanin *G*   M-70153

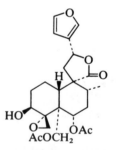

$C_{24}H_{30}O_9$ M 462.496
Constit. of *Teucrium montanum* var. *scorpilii*. Cryst. (Me₂CO/pet. ether). Mp 239-241°. $[\alpha]_D^{20} +11.2°$ (c, 0.24 in Me₂CO).

Malakov, P.Y. *et al, Z. Naturforsch., B,* 1987, **42**, 1000.

Montanolide   M-70154

[30557-67-6]

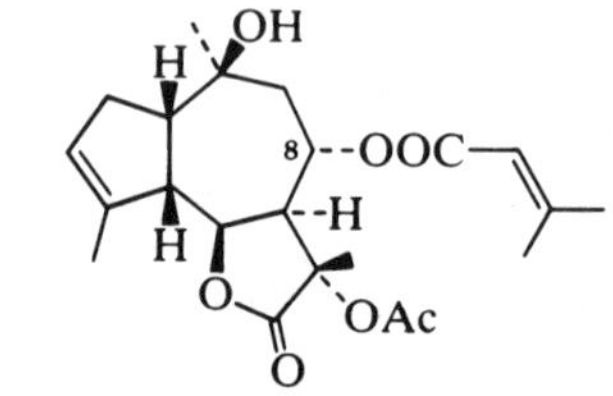

$C_{22}H_{30}O_7$ M 406.475
Isol. from *Laserpitium siler* and *L. archangelica*. Mp 132-133°. $[\alpha]_D^{20} -72.2°$ (c, 3.5 in CHCl₃).

O^8-*Deacyl, O-2-(Methyl-2-butenoyl):* [38114-47-5].
Isomontanolide.
$C_{22}H_{30}O_7$ M 406.475
Isol. from fruits of *L. siler*. Mp 176°. $[\alpha]_D^{20}$ −25.2°.
O^8-*Deacyl, O-(2-methyl-2-butenoyl), Ac:*
Acetylisomontanolide.
$C_{24}H_{32}O_8$ M 448.512
Isol. from roots and rhizomes of *L. siler*. Mp 134°.
$[\alpha]_D^{20}$ −78.2°.

Holub, M. *et al, Collect. Czech. Chem. Commun.,* 1970, **35**, 3296; 1972, **37**, 1186; 1978, **43**, 2444.

3-Morpholinecarboxylic acid, 9CI   M-70155

4-Oxapipecolic acid
[77873-76-8]

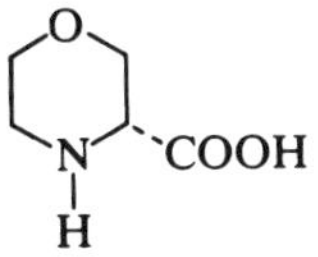

(*R*)-*form*

$C_5H_9NO_3$ M 131.131
(*R*)-*form*
 D-form
 Mp 216-218° dec. $[\alpha]_D^{25}$ +13.8° (c, 1 in 1M HCl).
(*S*)-*form*
 L-form
 Cryst. (MeOH aq.). Mp 218° dec. $[\alpha]_D^{25}$ −14.4° (c, 2 in 1M HCl).

Asher, V. *et al, Tetrahedron Lett.,* 1981, **22**, 141 (*synth*)
Kogami, Y. *et al, Bull. Chem. Soc. Jpn.,* 1987, **60**, 2963 (*synth*)

Muamvatin   M-70156

[104013-99-2]

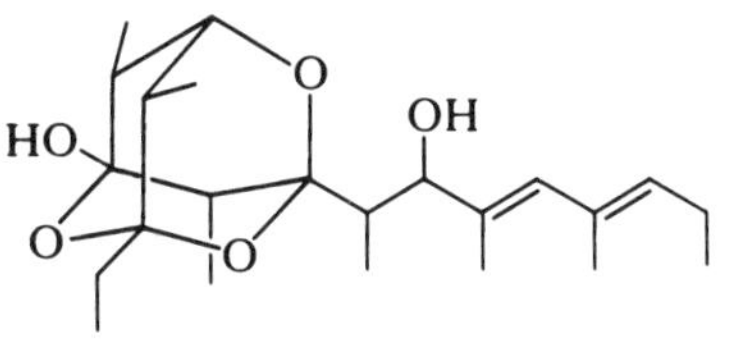

$C_{23}H_{38}O_5$ M 394.550
Isol. from the Fijian mollusc *Siphonaria normalis*. Oil.
$[\alpha]_D^{25}$ +61.1° (c, 0.175 in CH_2Cl_2).

Roll, D.M. *et al, J. Am. Chem. Soc.,* 1986, **108**, 6680 (*occur, pmr, cmr, ir, uv*)

Suggestions for new DOC Entries are welcomed. Please write to the Editor, DOC 5, Chapman and Hall Ltd, 11 New Fetter Lane, London EC4P 4EE

Mulundocandin   M-70157

1-[4,5-Dihydroxy-N-(12-methyl-1-oxotetradecyl)-ornithine]-5-serineechinocandin B, 9CI
[108351-20-8]

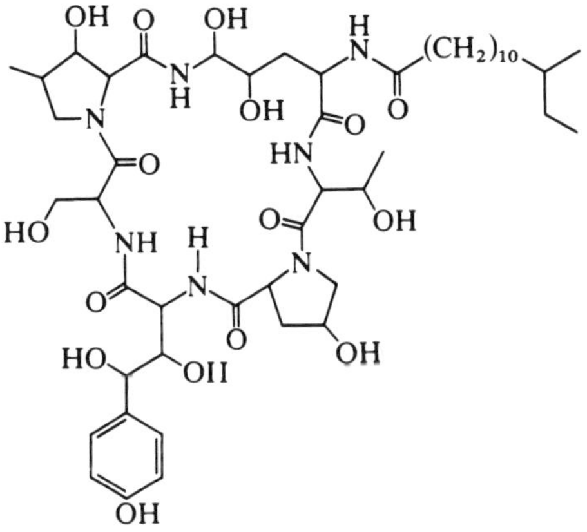

$C_{48}H_{77}N_7O_{16}$ M 1008.174
Cyclic peptide antibiotic. Prod. by *Aspergillus sydowi* var. *mulundensis*. Active against filamentous fungi and yeasts. Amorph. powder. Mp 225°. $[\alpha]_D^{25}$ −42.77° (c, 1.6 in MeOH).

Mukhopadhyay, T. *et al, J. Antibiot.,* 1987, **40**, 275, 281 (*isol, struct, props*)

Murraculatin   M-70158

5,7-Dimethoxy-α,α-dimethyl-2-oxo-2H-1-benzopyran-8-propanoic acid, 9CI
[117007-26-8]

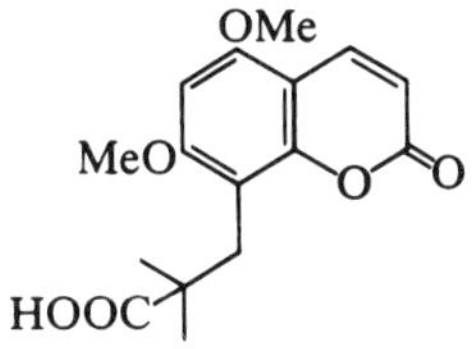

$C_{16}H_{18}O_6$ M 306.315
Constit. of *Murraya paniculata*. Needles (Me_2CO). Mp 217-218°.

Wu, T.-S., *Phytochemistry,* 1988, **27**, 2357.

Murralongin   M-70159

Updated Entry replacing M-10408
[53011-72-6]

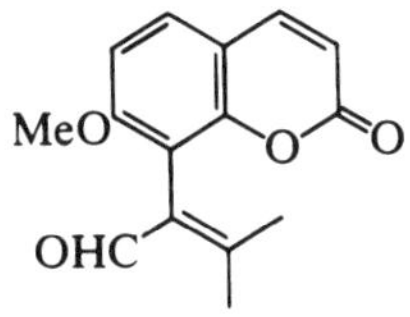

$C_{15}H_{14}O_4$ M 258.273
Constit. of the leaves of *Murraya elongata* and of *M. paniculata*. Needles (Et_2O). Mp 135°.

Talapatra, S.K. *et al, Tetrahedron Lett.,* 1973, 5005 (*isol*)
Raj, K. *et al, Phytochemistry,* 1976, **15**, 1787 (*isol*)
Imai, F. *et al, Chem. Pharm. Bull.,* 1986, **34**, 3978 (*struct*)

Murrayanone   M-70160

5,6,7-Trimethoxy-8-(3-methyl-2-oxobutyl)-2H-1-benzopyran-2-one

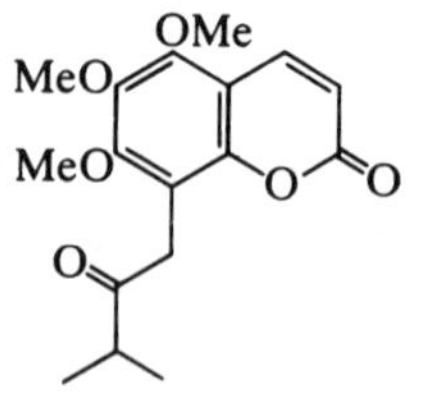

$C_{17}H_{20}O_6$ M 320.341
Constit. of *Murraya paniculata*. Syrup.
Wu, T.-S., *Phytochemistry*, 1988, **27**, 2357.

Mutisifurocoumarin   M-70161

8,9-Dihydroxy-1-methyl-6H-benzofuro[3,2-c][1]-benzopyran-6-one, 9CI

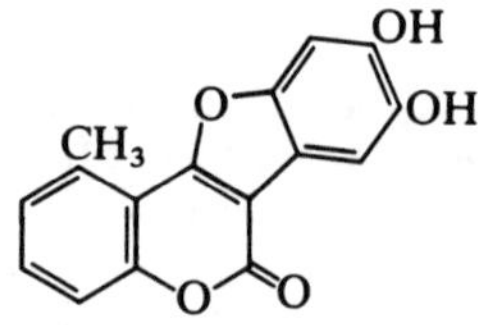

$C_{16}H_{10}O_5$ M 282.252
Constit. of *Mutisia orbignyana*.
Di-Ac: [115532-08-6]. Cryst. Mp 243-245°.
Zdero, C. *et al*, *Phytochemistry*, 1988, **27**, 891.

Myricanol   M-70162

Updated Entry replacing M-04165
16,17-Dimethoxytricyclo[12.3.1.1²,⁶]nonadeca-1(18),2,4,6(19),14,16-hexaene-3,9,15-triol, 9CI
[33606-81-4]

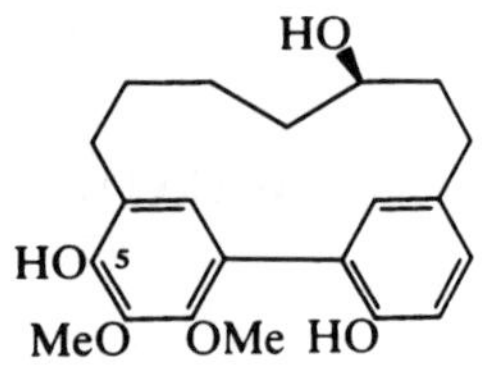

Absolute configuration

$C_{21}H_{26}O_5$ M 358.433
Constit. of the stem bark of *Myrica nagi*. Needles
(Et₂O). Mp 105-125°. $[\alpha]_D^{27.5}$ −65.5° (CHCl₃).
Ketone: [32492-74-3]. **Myricanone**.
$C_{21}H_{24}O_5$ M 356.418
Constit. of the stem bark of *M. nagi*. Cryst. (EtOH).
Mp 194-196°.
Ketone, 5-deoxy: [110007-10-8]. **5-Deoxymyricanone**.
$C_{21}H_{24}O_4$ M 340.418
Isol. from bacterial galls of *M. rubra*. Amorph.

Campbell, R.V.M. *et al*, *J. Chem. Soc., Chem. Commun.*, 1970, 1206 (*isol*)
Begley, M.J. *et al*, *J. Chem. Soc. (C)*, 1971, 3634 (*cryst struct*)
Whiting, D.A., *Tetrahedron Lett.*, 1978, 2335 (*synth*)
Takeda, Y. *et al*, *Chem. Pharm. Bull.*, 1987, **35**, 2569 (*5-Deoxymyricanone*)
Sun, D. *et al*, *Phytochemistry*, 1988, **27**, 579 (*cmr*)

Myristargenol A   M-70163

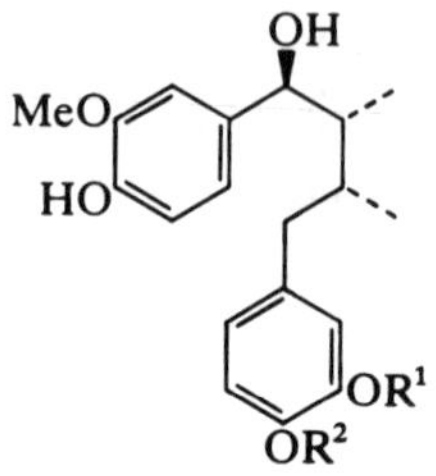

$R^1, R^2 = -CH_2-$

$C_{20}H_{24}O_5$ M 344.407
Constit. of *Myristica argentea*. Needles. Mp 133°. $[\alpha]_D^{24}$ +9.9° (c, 0.81 in EtOH).
Nakatani, N. *et al*, *Phytochemistry*, 1988, **27**, 3127.

Myristargenol B   M-70164

As Myristargenol A, M-70163 with

$R^1 = Me, R^2 = H$

$C_{20}H_{26}O_5$ M 346.422
Constit. of *Myristica argentea*. Cryst. Mp 93°. $[\alpha]_D^{24}$ +14.2° (c, 0.79 in CHCl₃).
Nakatani, N. *et al*, *Phytochemistry*, 1988, **27**, 3127.

Myrocin C   M-70165
[113122-50-2]

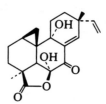

$C_{20}H_{24}O_5$ M 344.407
Metab. of *Myrothecium* sp. strain 55. Cryst. (EtOAc/hexane). Mp 178-180°. $[\alpha]_D^{20}$ +43.5° (c, 0.35 in MeOH).
Hsu, Y.-H. *et al*, *Agric. Biol. Chem.*, 1987, **51**, 3455.

Mzikonone   M-70166
[117278-48-5]

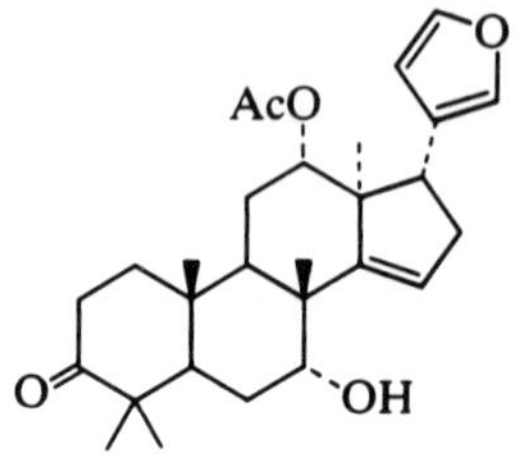

$C_{28}H_{38}O_5$ M 454.605
Constit. of *Turraea robusta*. Cryst. Mp 99-101°.
Rajab, M.S. *et al*, *Phytochemistry*, 1988, **27**, 2353.

N

Naematolin N-70001

Updated Entry replacing N-50001

[11054-16-3]

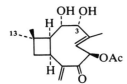

Absolute configuration

$C_{17}H_{24}O_5$ M 308.374

Metab. of *Hypholoma elongatipes, H. fasciculare, H. sublateritium* and of the fungus *Naematoloma fasciculare*. Cytotoxic. Cryst. Mp 144-145°. $[\alpha]_D^{22}$ −370° (c, 1.0 in $CHCl_3$), $[\alpha]_D^{22}$ −330°. Rare example of *cis*-fused caryophyllane.

▷QI1275000.

3-Ketone: [92121-62-5]. **Naematolone**.
$C_{17}H_{22}O_5$ M 306.358
Metab. of *H. elongatipes, H. capnoides* and *H. sublateritium*. Cytotoxic. Oil. $[\alpha]_D^{22}$ +116° (c, 0.46 in $CHCl_3$).

13-Hydroxy: [102167-68-0]. **Naematolin B**.
$C_{17}H_{24}O_6$ M 324.373
Metab. of *N. fasciculare*. Cryst. Mp 122-123°. $[\alpha]_D^{23}$ −352°.

Ito, Y. *et al, Chem. Pharm. Bull.*, 1967, **15**, 2009 (*isol*)
Backens, S. *et al, Justus Liebigs Ann. Chem.*, 1984, 1332 (*struct*)
Tsuboyama, S. *et al, Bull. Chem. Soc. Jpn.*, 1986, **59**, 1921 (*cryst struct*)

1,2-Naphthalenedithiol N-70002

1,2-Dimercaptonaphthalene

[53944-71-1]

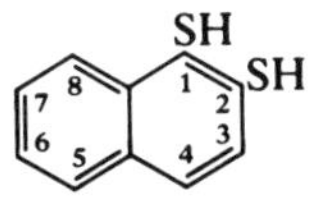

$C_{10}H_8S_2$ M 192.293

Stepanov, B.I. *et al, CA*, 1974, **81**, 135789g (*synth, deriv*)

1,8-Naphthalenedithiol, 9CI, 8CI N-70003

Updated Entry replacing N-00096

1,8-Dimercaptonaphthalene

[25079-77-0]

$C_{10}H_8S_2$ M 192.293
Reagent for determination of Pd(II). Yellow plates (EtOH/HCl). Mp 113-114°.

Price, W.B. *et al, J. Chem. Soc.*, 1928, 2373 (*synth*)
Pacault, A. *et al, Bull. Soc. Chim. Fr.*, 1952, 141 (*struct*)
Desai, H.S. *et al, J. Sci. Ind. Res., Sect. B*, 1960, **19**, 390 (*synth*)
Yui, K. *et al, Bull. Chem. Soc. Jpn.*, 1988, **61**, 953 (*synth, ir, pmr*)

2,3-Naphthalenedithiol N-70004

2,3-Dimercaptonaphthalene

[99643-52-4]

$C_{10}H_8S_2$ M 192.293
Powder (toluene under N_2). Mp 162-163°.

Di-S-benzyl: [99643-51-3].
$C_{24}H_{20}S_2$ M 372.542
Powder (EtOAc). Mp 151-152°.

Gleiter, R. *et al, J. Org. Chem.*, 1986, **51**, 370 (*synth, pmr*)

[2](1,5)Naphthaleno[2](2,7)(1,6-methano[10]annuleno)phane N-70005

15,20-Methano-1,5-(ethano[1,6]cyclodecethano)-naphthalene, 9CI

[66478-35-1]

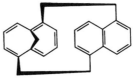

$C_{25}H_{22}$ M 322.449
Sl. yellow prisms (C_6H_6). Mp 320-321.5°.

Matsumoto, M. *et al, Tetrahedron Lett.*, 1977, 4425 (*synth, uv*)

[2](2,6)-Naphthaleno[2]paracyclophane-1,11-diene N-70006

2,6-(Etheno[1,4]benzenoetheno)naphthalene, 9CI

[117054-70-3]

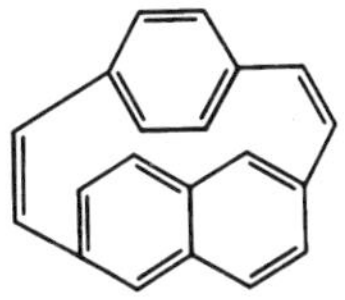

$C_{20}H_{14}$ M 254.331
(±)-*form* [116504-33-7]
Mp 124°.

Blank, N.E. *et al, Angew. Chem., Int. Ed. Engl.*, 1988, **27**, 1064 (*synth, uv, pmr, cryst struct*)

Naphtho[8,1,2-cde]cinnoline, 9CI N-70007

4,5-Diazapyrene

[63285-40-5]

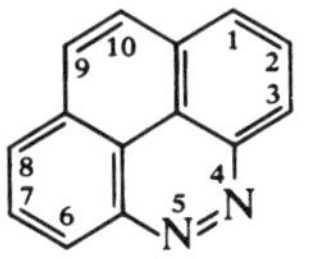

$C_{14}H_8N_2$ M 204.231
Mp 150°.

N-Oxide: [63285-41-6].
$C_{14}H_8N_2O$ M 220.230
Mp 237°.

Mugnier, Y. *et al, J. Heterocycl. Chem.*, 1977, **14**, 351 (*synth, ms, pmr, deriv*)
Mugnier, Y. *et al, Bull. Soc. Chim. Fr., Part II*, 1978, 39 (*synth*)

Naphtho[1,2-*c*]furan

N-70008

[232-74-6]

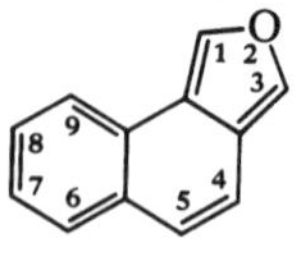

C$_{12}$H$_8$O   M 168.195
Cryst. Mp 61-63°.

Moursounidis, J. *et al*, *Aust. J. Chem.*, 1988, **41**, 235 (*synth, pmr, uv, ms*)

Naphtho[1,8-*bc*]pyran, 9CI

N-70009

Updated Entry replacing N-00204
1-Oxaphenalene. Benzo[de]*chromene*
[203-91-8]

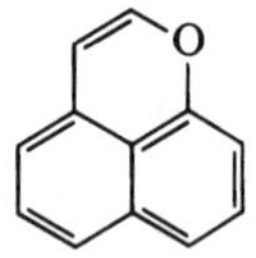

C$_{12}$H$_8$O   M 168.195
Cryst. (pet. ether). Mp 55°. Cryst. with great difficulty.
Light-sensitive.

O'Brien, S. *et al*, *J. Chem. Soc.*, 1963, 2907.
Narasimhan, N.S. *et al*, *Synthesis*, 1975, 796.
Buisson, J.-P. *et al*, *J. Heterocycl. Chem.*, 1988, **25**, 539 (*synth, uv, pmr*)

Naphtho[1,8-*bc*]pyran-2-carboxylic acid

N-70010

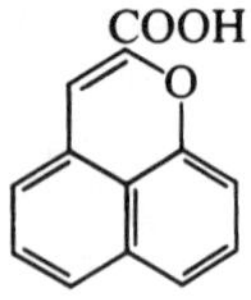

C$_{13}$H$_8$O$_3$ M 212.204
Brown microcryst. (EtOH). Mp 240°. Microanalysis
gave consistently low C.

Me ester:
C$_{14}$H$_{10}$O$_3$ M 226.231
Orange microcryst. Mp 105°.
Et ester:
C$_{15}$H$_{12}$O$_3$ M 240.258
Yellow microcryst. Mp 110°.

Buisson, J.-P. *et al*, *J. Heterocycl. Chem.*, 1988, **25**, 538 (*synth, pmr*)

2*H*-Naphtho[1,2-*b*]pyran-2-one, 9CI

N-70011

Updated Entry replacing N-00207
Benzo[h]*coumarin. 1-Hydroxy-2-naphthaleneacrylic
acid δ-lactone*
[2147-34-4]

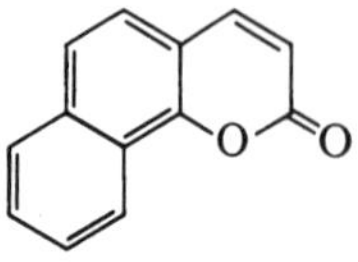

C$_{13}$H$_8$O$_2$ M 196.205
Pale-yellow needles (EtOH). Spar. sol. Et$_2$O, insol. H$_2$O.
Mp 141-142°. Greenish-yellow soln. with blue fluor. in
conc. H$_2$SO$_4$.

Koelsch, C.F. *et al*, *J. Am. Chem. Soc.*, 1953, **75**, 3596.

Das Gupta, A.K. *et al*, *J. Chem. Soc.* (*C*), 1969, 2618.
Ziegler, T. *et al*, *Chem. Ber.*, 1987, **120**, 373 (*synth*)

Naphtho[1,2-*a*]quinolizinium(1+)

N-70012

[195-40-4]

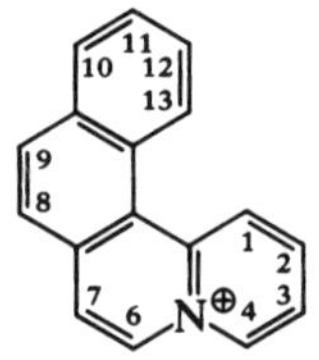

C$_{17}$H$_{12}$N$^⊕$ M 230.289 (ion)
Bromide: [112107-62-7].
C$_{17}$H$_{12}$BrN   M 310.193
Yellow needles (MeOH/EtOH).
Perchlorate: [112107-61-6].
C$_{17}$H$_{12}$ClNO$_4$ M 329.739
Pale-yellow needles (MeOH). Mp 248.4-249.3°.

Arai, S. *et al*, *J. Chem. Soc., Perkin Trans. 1*, 1987, 481 (*synth, ir*)

Naphtho[2,1-*b*]thiet-1-one

N-70013

Updated Entry replacing N-20022
Naphtho[2,1-b]*thiet-2-one*
[85601-43-0]

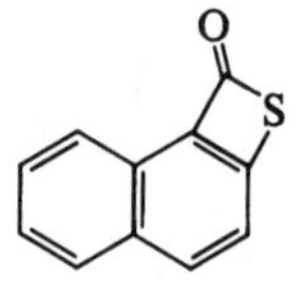

C$_{11}$H$_6$OS   M 186.228
Bright-orange cryst. Stable at −18°, polymerises in soln.
at r.t. or on heating >40°.

Wentrup, C. *et al*, *J. Org. Chem.*, 1987, **52**, 3838 (*synth, ir, pmr, cmr, ms*)

Naphtho[2,3-*b*]thiophene-2,3-dione

N-70014

[109151-44-2]

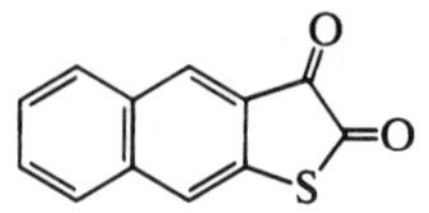

C$_{12}$H$_6$O$_2$S   M 214.238
Red plates by subl. Mp 173°.

Wentrup, C. *et al*, *J. Org. Chem.*, 1987, **52**, 3838 (*synth, ir, pmr, cmr*)

1-(1-Naphthyl)ethanol

N-70015

Updated Entry replacing N-00326
*α-Methyl-1-naphthalenemethanol, 9CI, 8CI. Methyl-1-
naphthylcarbinol. α-Hydroxy-1-ethylnaphthalene*
[1517-72-2]

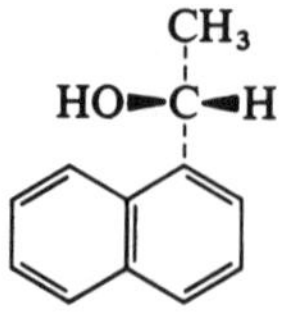

(*R*)-*form*

C$_{12}$H$_{12}$O   M 172.226

(R)-form [42177-25-3]

$[\alpha]_D^{25}$ +82.1° (c, 1.0 in Et$_2$O). Opt. pure.

(S)-form [15914-84-8]

Cryst. Mp 47° (43.5-45°). Bp$_{11}$ 166°.

(±)-form [57605-95-5]

Cryst. (pet. ether). Mp 66°. Bp$_{15}$ 178°.

Balfe, M.P. et al, J. Chem. Soc., 1946, 797 (resoln)

Prelog, V. et al, Helv. Chim. Acta, 1956, **39**, 1086 (abs config)

Mitsui, S. et al, Tetrahedron, 1967, **23**, 4271.

Cervinka, O., Collect. Czech. Chem. Commun., 1973, **38**, 441 (abs config)

Seebach, D. et al, Chem. Ber., 1974, **107**, 1748 (synth)

Sugimoto, T. et al, J. Chem. Soc., Chem. Commun., 1978, 926 (synth)

Theisen, P.D. et al, J. Org. Chem., 1988, **53**, 2374 (synth, resoln, ir, pmr, cmr)

Napyradiomycin A_2 N-70016

[111216-62-7]

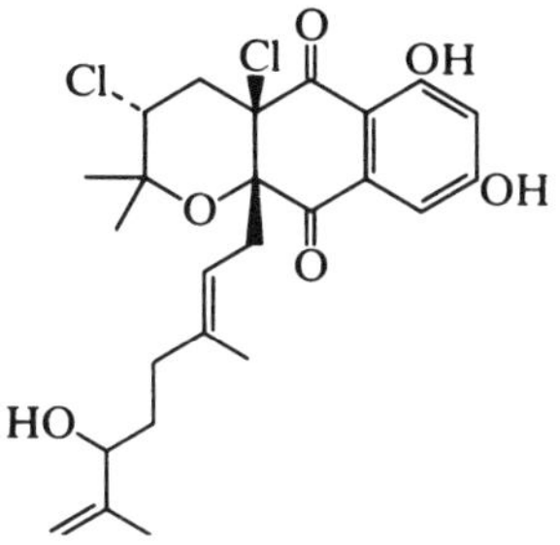

$C_{25}H_{30}Cl_2O_6$ M 497.414

Quinone antibiotic. Isol. from *Chainia rubra*. Active against gram-positive bacteria. Brownish-yellow powder. Mp 62-67°. $[\alpha]_D^{28}$ +20° (c, 0.3 in EtOH).

Shiomi, K. et al, J. Antibiot., 1987, **40**, 1213 (isol, struct)

Napyradiomycin B_4 N-70017

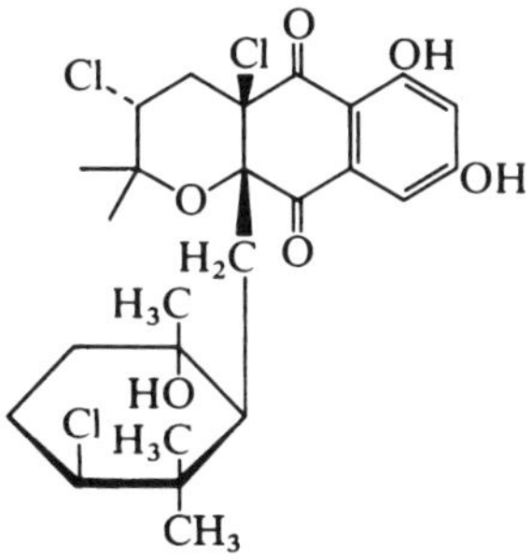

$C_{25}H_{31}Cl_3O_6$ M 533.875

Isol. from *Chainia rubra*. Active against gram-positive bacteria. Pale-yellow prisms. Mp 168-171°. $[\alpha]_D^{28}$ −81° (c, 0.2 in EtOH/C$_6$H$_6$).

Shiomi, K. et al, J. Antibiot., 1987, **40**, 1213 (isol, struct)

Nardonoxide   N-70018

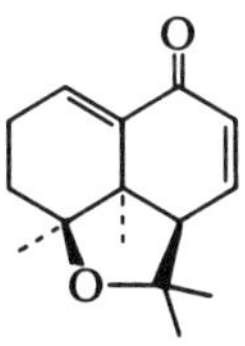

$C_{15}H_{20}O_2$ M 232.322

Constit. of *Nordostachys chinensis*. Cryst. (MeOH). Mp 62-64°. $[\alpha]_D^{20}$ −85° (c, 0.65 in CHCl$_3$).

Shide, L. et al, Planta Med., 1987, **53**, 332.

Nardostachin   N-70019

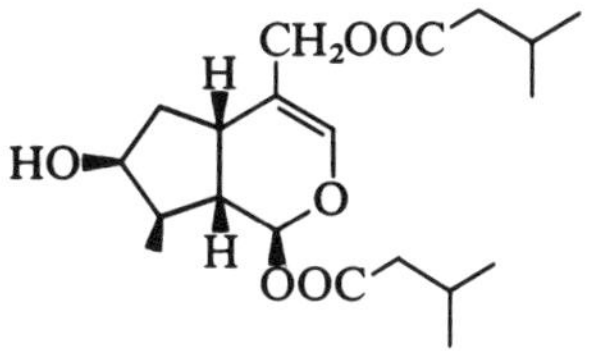

$C_{20}H_{32}O_6$ M 368.469

Constit. of *Nardostachys chinensis*. Oil. $[\alpha]_D$ −80.9° (c, 0.4 in CHCl$_3$).

Bagchi, A. et al, Planta Med., 1988, **54**, 87.

Nemorosonol   N-70020

[117193-22-3]

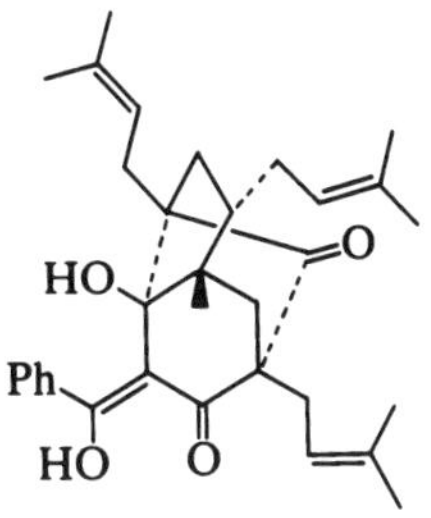

$C_{33}H_{42}O_4$ M 502.692

Constit. of *Clusia nemorosa*. Cubes (hexane). Mp 83-85°. $[\alpha]_D$ +203° (c, 0.7 in CHCl$_3$).

Delle Monache, F. et al, Phytochemistry, 1988, **27**, 2305.

Neocorymboside   N-70021

6-C-β-L-Arabinopyranosyl-8-C-β-D-galactopyranosylapigenin

[117065-26-6]

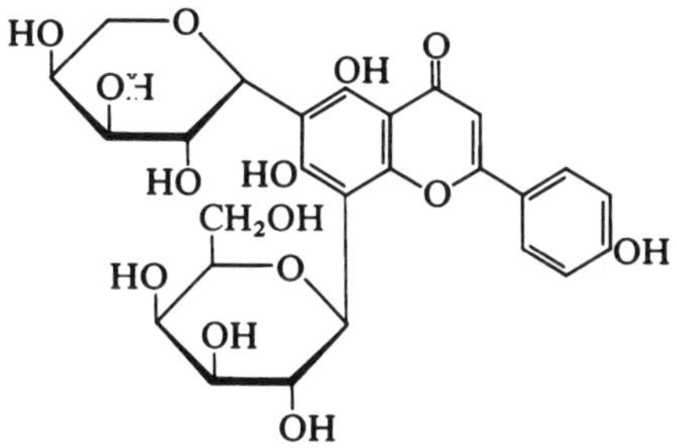

$C_{26}H_{28}O_{14}$ M 564.499

Constit. of *Atractylis gummifera*.

Chaboud, A. et al, Phytochemistry, 1988, **27**, 2360.

Neodolabellenol   N-70022

[69010-14-6]

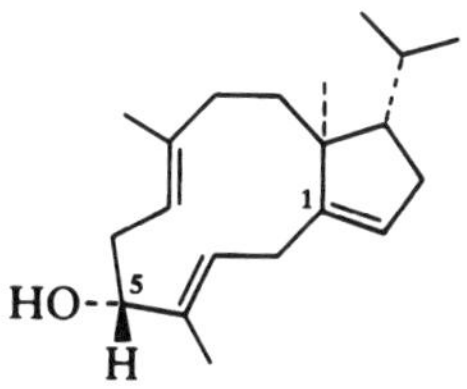

$C_{20}H_{32}O$ M 288.472

Constit. of soft corals *Clavularia koellikeri* and *C. inflata*. Needles. Mp 158-160°.

5-Epimer: [68978-05-2].
C$_{20}$H$_{32}$O   M 288.472
Constit. of soft corals. Cryst. (MeCN). Mp 108-109°.
[α]$_D$ +65.7° (c, 0.4 in CHCl$_3$).

Bowden, B.F. *et al, Aust. J. Chem.,* 1978, **31**, 2039; 1980, **33**,
927 (*isol, cryst struct*)
Kobayashi, M. *et al, Chem. Pharm. Bull.,* 1986, **34**, 2306 (*isol*)

Neoechinulin *E*   N-70023

[[2-(1,2-Dimethyl-2-propenyl)-1H-indol-3-yl]-
methylene]*piperazinetrione, 9CI*

[57944-03-3]

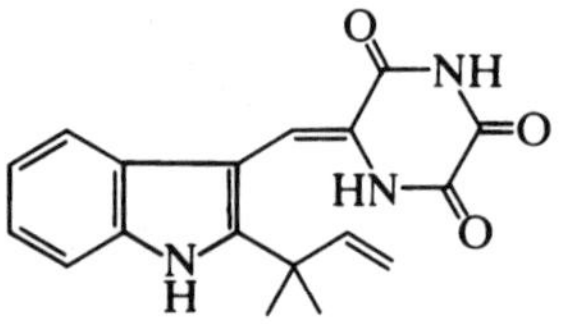

C$_{18}$H$_{17}$N$_3$O$_3$ M 323.351
Metab. of *Aspergillus amstelodami.* Orange-red cryst.
(MeOH). Mp 275°.

Marchelli, R. *et al, J. Chem. Soc., Perkin Trans. 1,* 1977, 713
(*isol, uv, ir, pmr, ms, struct*)

Neokadsuranin   N-70024

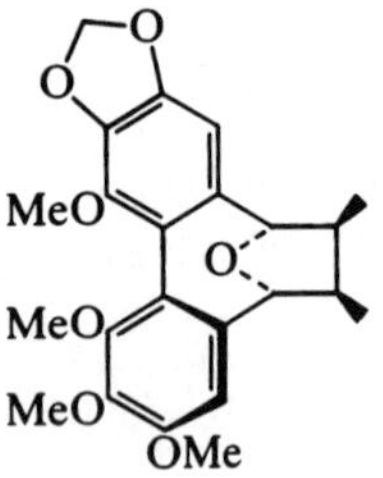

C$_{23}$H$_{26}$O$_7$ M 414.454
Constit. of *Kadsura coccinea.* Cryst. (Et$_2$O). Mp 157-
159°.

Lian-niang, L. *et al, Planta Med.,* 1988, **54**, 45.

Neokyotorphin   N-70025

[83759-54-0]

L-Thr-L-Ser-L-Lys-L-Thr-L-Arg

C$_{28}$H$_{47}$N$_9$O$_9$ M 653.734
Isol. from bovine brain and human lung carcinoma. Exhibits analgesic props. A variety of D-isomers have also
been synthesised and characterized.

B,2AcOH: Fluffy powder + 3H$_2$O. [α]$_D^{25}$ −21.7° (c, 1.2
in AcOH aq.).

Kitagawa, K. *et al, Chem. Pharm. Bull.,* 1983, **31**, 2349; 1985,
33, 377 (*synth, analogs, bibl*)
Zhu, Y.X. *et al, FEBS Lett.,* 1986, **208**, 253 (*isol*)

Neoliacinic acid   N-70026

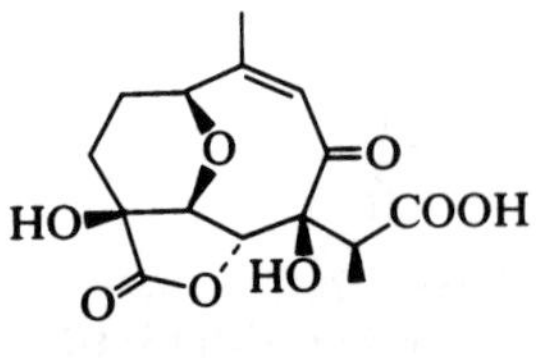

C$_{15}$H$_{18}$O$_8$ M 326.302

Constit. of *Neolitsea acciculata.*

Me ester: [116428-57-0]. Plates (EtOAc). Mp 214.5-
215.5° dec. [α]$_D$ −108° (c, 0.3 in EtOAc).

Takaoka, D. *et al, J. Chem. Soc., Chem. Commun.,* 1987, 1861.

Neonepetalactone   N-70027

Updated Entry replacing N-00499
[24190-25-8]

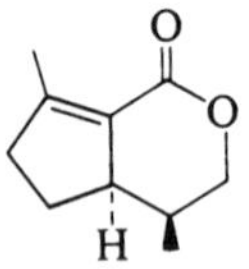

Absolute
configuration

C$_{10}$H$_{14}$O$_2$ M 166.219
Constit. of *Actinidia polygama.* Attractive to cats. Oil.
[α]$_D^{23}$ −166.8° (c, 0.31 in CHCl$_3$).

4-Epimer: [76549-18-3]. **Isoneonepetalactone.**
C$_{10}$H$_{14}$O$_2$ M 166.219
Constit. of *A. polygama.* Oil. [α]$_D$ −66° (CHCl$_3$).
4a-Epimer: [114298-64-5]. Oil. [α]$_D$ +55° (c, 0.55 in
CHCl$_3$).

Wolinsky, J. *et al, Tetrahedron,* 1969, **25**, 3767 (*isol*)
Sakai, T. *et al, Bull. Chem. Soc. Jpn.,* 1980, **53**, 3683 (*isol*)
Sakai, T. *et al, Tetrahedron,* 1980, **36**, 3115 (*abs config*)
Taber, D.F. *et al, J. Org. Chem.,* 1988, **53**, 2984 (*synth*)

Neopulchellin   N-70028

Updated Entry replacing N-00505
[28230-81-1]

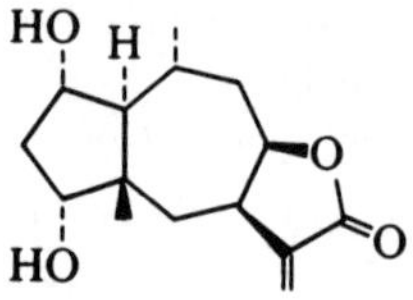

C$_{15}$H$_{22}$O$_4$ M 266.336
Constit. of *Gaillardia pulchella.* Cryst. Mp 166.5-167.5°.
[α]$_D^{25}$ +43° (c, 1 in CHCl$_3$).

4-Epimer: **4-Epineopulchellin.**
C$_{15}$H$_{22}$O$_4$ M 266.336
Constit. of *G. pulchella.* Oil.

Yanagita, M. *et al, Tetrahedron Lett.,* 1970, 3007 (*isol, struct*)
Harimaya, K. *et al, Heterocycles,* 1988, **27**, 83 (*epimer*)

Neovasinin   N-70029

7-(1,3-Dimethyl-1-pentenyl)-7,8-dihydro-4,8-dihy-
droxy-3,8-dimethyl-2H,5H-pyrano[4,3-b]pyran-2-one,
9CI

[111897-37-1]

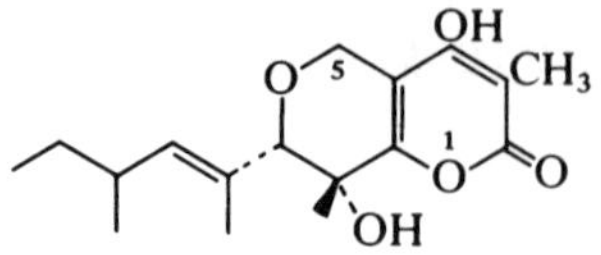

C$_{17}$H$_{24}$O$_5$ M 308.374
Prod. by *Neocosmospora vasinfecta.* Plant-growth
regulator. Plates (EtOAc/hexane). Mp 204-206°. [α]$_D$
−105° (c, 0.2 in MeOH).

5-Oxo: [108885-72-9]. **Neovasinone.**
C$_{17}$H$_{22}$O$_6$ M 322.357
From *N. vasinfecta.* Plant growth regulator. Plates
(EtOAc/hexane). Mp 193-195°. [α]$_D^{25}$ −90° (c, 0.2 in
MeOH).

Nakajima, H. *et al*, *Agric. Biol. Chem.*, 1987, **51**, 1221, 2831 (*isol, struct*)

Neoverrucosanol N-70030

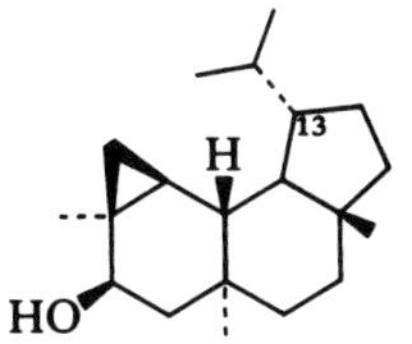

$C_{20}H_{34}O$ M 290.488

5β-form [76235-09-1]
Constit. of *Mylia verrucosa*. Cryst. Mp 174-175°. $[\alpha]_D$ −10°.

13-Epimer: **13-Epi-5β-ncoverrucosanol**.
$C_{20}H_{34}O$ M 290.488
Constit. of *Plagiochila stephensoniana*. Prisms (hexane). Mp 151-153.5°. $[\alpha]_D$ +45.5° (c, 0.8 in $CHCl_3$).

Matsuo, A. *et al*, *J. Chem. Soc., Chem. Commun.*, 1980, 822 (*isol*)
Fukuyama, Y. *et al*, *Phytochemistry*, 1988, **27**, 1795 (*isol*)

Nepetaside N-70031

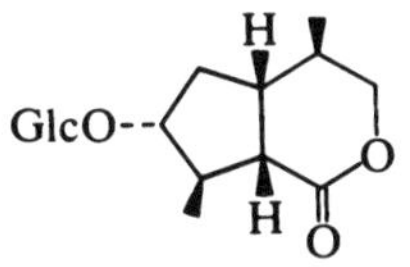

$C_{16}H_{26}O_8$ M 346.377
Constit. of *Nepeta cataria*. Needles (EtOH). Mp 204-205°. $[\alpha]_D^{22}$ −52.5° (c, 0.6 in MeOH).

Xie, S. *et al*, *Phytochemistry*, 1988, **27**, 469.

Nephroarctin N-70032

Updated Entry replacing N-00532
3,5-Diformyl-2,4-dihydroxy-6-methylbenzoic acid 3-methoxy-2,5,6-trimethylphenyl ester, 8CI. 3-Methoxy-2,5,6-trimethylphenyl 3,5-diformyl-2,4-dihydroxy-6-methylbenzoate
[23004-60-6]

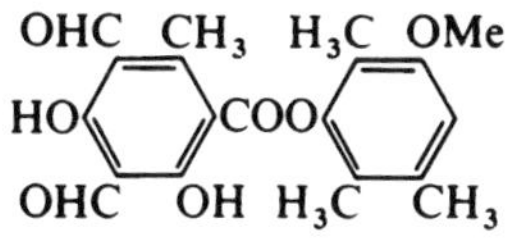

$C_{20}H_{20}O_7$ M 372.374
Isol. from *Nephroma arcticum* and *Pseudocyphellaria pickeringii*. Prisms (Me_2CO).

4'-Chloro: [113689-53-5]. **4'-Chloronephroarctin**. *1'-Chloronephroarctin*.
$C_{20}H_{19}ClO_7$ M 406.819
Constit. of *P. pickeringii*. Plates. Mp 181°.

Nuno, M. *et al*, *J. Chem. Soc., Chem. Commun.*, 1969, 78 (*isol, ir, pmr, ms, struct*)
Bruun, T., *Acta Chem. Scand.*, 1971, **25**, 2831 (*isol, uv, pmr, ms*)
Hamilton, R.J. *et al*, *J. Chem. Soc., Perkin Trans. 1*, 1976, 943 (*synth, pmr*)
Elix, J.A. *et al*, *Aust. J. Chem.*, 1987, **40**, 2023 (*isol, deriv, synth*)

Neurotoxin NSTX 3 N-70033
NSTX 3
[107288-22-2]

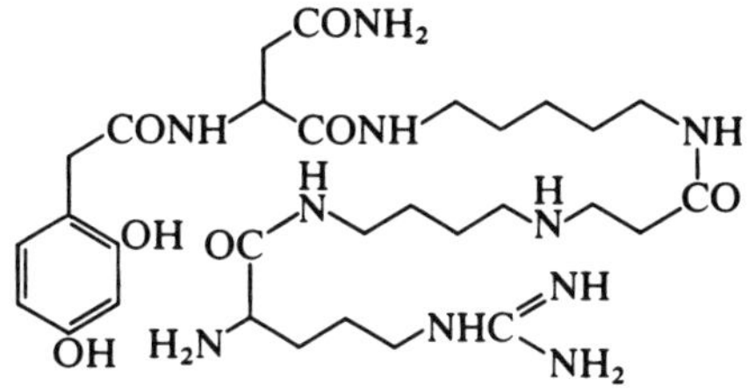

$C_{30}H_{52}N_{10}O_7$ M 664.804
Toxic principle in the venom of *Nephila maculata*.

Teshima, T. *et al*, *Tetrahedron Lett.*, 1987, **28**, 3509 (*synth*)

Nicotinuric acid N-70034
Updated Entry replacing N-00577
N-(*3-Pyridinylcarbonyl)glycine, 9CI*. N-*Nicotinoylglycine, 8CI. 3-(Pyridylformamido)acetic acid*
[583-08-4]

$CONHCH_2COOH$

$C_8H_8N_2O_3$ M 180.163
Metab. of niacin, found in urine. Also found in plants. Shows various pharmacol. props. notably depression of serum cholesterol. Cryst. (H_2O). Mp 242-244° dec. (240°).

Fox, S.W. *et al*, *J. Biol. Chem.*, 1943, **147**, 651 (*synth*)
Rohrlich, M., *Arch. Pharm. (Weinheim, Ger.)*, 1951, **284**, 6; *CA*, **45**, 10245c (*synth*)
Krishnaswamy, S. *et al*, *Acta Crystallogr., Sect. C*, 1987, **43**, 728 (*cryst struct*)

Nidoresedic acid N-70035
Nidoresedaic acid
[70377-99-0]

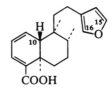

$C_{20}H_{26}O_3$ M 314.424
Constit. of *Grangea maderaspatana*. Cryst. (EtOAc). Mp 82-84°. $[\alpha]_D$ −139° (c, 0.49 in $CHCl_3$).

Me ester: **Methyl nidoresedate**.
$C_{21}H_{28}O_3$ M 328.450
Constit. of *Nidorella auriculata*. Oil.

10-Epimer: **10-Epinidoresedic acid**.
$C_{20}H_{26}O_3$ M 314.424
Constit. of *G. maderaspatana*.

10-Epimer, Me ester: Gum. $[\alpha]_D$ −66.1° (c, 1 in $CHCl_3$).

15,16-Dihydro, 15-methoxy, 16-oxo: **15,16-Dihydro-15-methoxy-16-oxonidoresedic acid**.
$C_{21}H_{28}O_5$ M 360.449
Constit. of *G. maderaspatana*.

15,16-Dihydro, 15-methoxy, 16-oxo, Me ester: Gum. $[\alpha]_D$ −120.8° (c, 0.23 in $CHCl_3$).

Bohlmann, F. *et al*, *Phytochemistry*, 1978, **17**, 1769 (*isol*)
Pandey, U.C. *et al*, *Phytochemistry*, 1984, **23**, 391 (*isol*)
Singh, P. *et al*, *Phytochemistry*, 1988, **27**, 1537 (*isol*)

Niloticin N-70036

24,25-Epoxy-23-hydroxy-7-tirucallen-3-one
[115404-57-4]

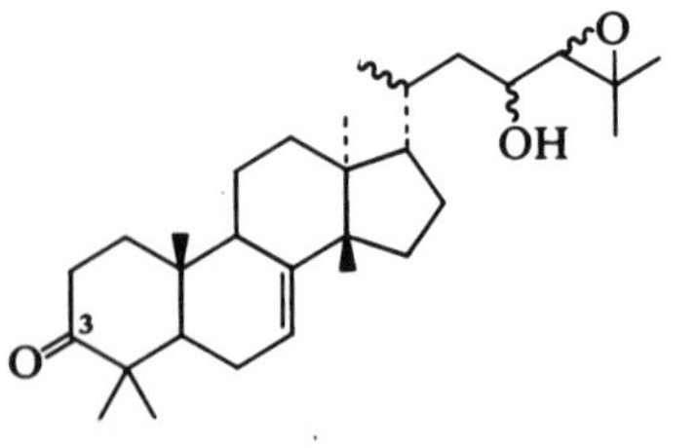

$C_{30}H_{48}O_3$ M 456.707
Constit. of *Phellodendron chinense* and *Turraea nilotica*.
Needles. Mp 147°. $[\alpha]_D$ −62° (c, 0.08 in CHCl$_3$).
Ac: [116425-97-9]. **Niloticin acetate.**
  $C_{32}H_{50}O_4$ M 498.745
  Constit. of *P. chinense*. Needles. Mp 157°. $[\alpha]_D$ −75°
  (c, 0.035 in CHCl$_3$).
3β-Alcohol: [115334-05-9]. **Dihydroniloticin.**
  $C_{30}H_{50}O_3$ M 458.723
  Constit. of *P. chinense* and *T. nilotica*. Needles
  (MeOH). Mp 174°. $[\alpha]_D$ −47° (c, 0.075 in CHCl$_3$).

Gray, A.I. *et al, Phytochemistry*, 1988, **27**, 1805.

Nimbinone N-70037

[116425-92-4]

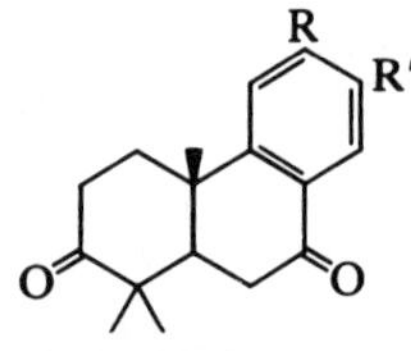

R = OH, R′ = CH$_3$

$C_{18}H_{22}O_3$ M 286.370
Constit. of *Azadirachta indica*. Cryst. (MeOH). Mp 124-
125°.

Ara, I. *et al, Phytochemistry*, 1988, **27**, 1801.

Nimbione N-70038

[116425-93-5]
As Nimbinone, N-70037 with

R = CH$_3$, R′ = OH

$C_{18}H_{22}O_3$ M 286.370
Constit. of *Azadirachta indica*. Cryst. (MeOH). Mp 102-
103°.

Ara, I. *et al, Phytochemistry*, 1988, **27**, 1801.

2,2′,2″-Nitrilotrisphenol N-70039

2,2′,2″-Trihydroxytriphenylamine
[7288-07-5]

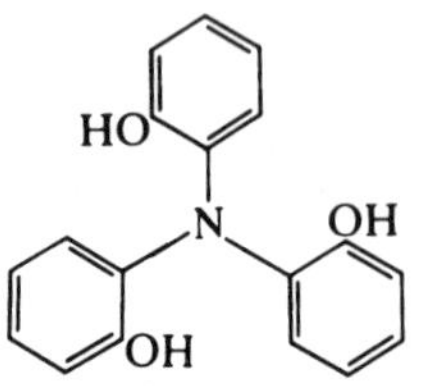

$C_{18}H_{15}NO_3$ M 293.321
Complexing agent (tetradentate). Cryst. (CH$_2$Cl$_2$ or
MeOH aq.). Mp 171-174°.

Frye, C. *et al, J. Am. Chem. Soc.*, 1966, **88**, 2727 (*synth, ir*)
Müller, E. *et al, Helv. Chim. Acta*, 1987, **70**, 499, 511, 520
(*complexes*)

4,4′,4″-Nitrilotrisphenol, 9CI N-70040

4,4′,4″-Trihydroxytriphenylamine
[25926-14-1]

$C_{18}H_{15}NO_3$ M 293.321
Needles (MeCN aq.). Mp 238-241°.

U.S.P., 3 508 888, (*1970*); *CA*, **73**, 4750 (*synth*)
Linkletter, S.J.G. *et al, J. Am. Chem. Soc.*, 1977, **99**, 5269
(*synth*)

5-Nitro-1,2-benzisothiazole, 9CI N-70041

[60768-66-3]

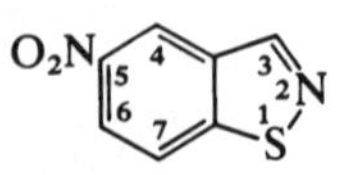

$C_7H_4N_2O_2S$ M 180.181
Cryst. (EtOH aq.). Mp 154°.

Ricci, A. *et al, Ric. Sci., Rend. Ser. B2*, 1962, 177; *CA*, **58**,
11340c (*synth*)
Hagen, H. *et al, Justus Liebigs Ann. Chem.*, 1980, **5**, 768
(*synth*)

7-Nitro-1,2-benzisothiazole, 9CI N-70042

[89641-97-4]
$C_7H_4N_2O_2S$ M 180.181
Yellow needles (Et$_2$O). Mp 162-163°.

Ricci, A. *et al, Ann. Chim. (Rome)*, 1963, **53**, 1860 (*synth*)
Rahman, L.K.A. *et al, J. Chem. Soc., Perkin Trans. 1*, 1984,
385 (*synth*)

1-Nitro-2-butanone N-70043

[22920-77-0]

$$H_3CCH_2COCH_2NO_2$$

$C_4H_7NO_3$ M 117.104
Liq. d$_{20}^{20}$ 1.16. Mp 12-15°. Bp$_{28}$ 118°, Bp$_{0.19}$ 52-54°. n_D^{20}
1.4410.
2,4-Dinitrophenylhydrazone: Cryst. (EtOH or EtOA-
c/EtOH). Mp 107-109°.

Hurd, C.D. *et al, J. Org. Chem.*, 1955, **20**, 927 (*synth*)
Jung, M.E. *et al, J. Org. Chem.*, 1987, **52**, 4570 (*synth, pmr*)

3-Nitro-2-butanone N-70044

[13058-70-3]

$$H_3CCH(NO_2)COCH_3$$

$C_4H_7NO_3$ M 117.104
Liq. d$_{20}^{20}$ 1.16. Bp$_{25}$ 92°. n_D^{20} 1.4349.
2,4-Dinitrophenylhydrazone: Cryst. (EtOH or EtOH/E-
tOAc). Mp 124-125°.

Hurd, C.D. *et al, J. Org. Chem.*, 1955, **20**, 927 (*synth*)

4-Nitro-2-butanone N-70045

[58935-95-8]

$$O_2NCH_2CH_2COCH_3$$

$C_4H_7NO_3$ M 117.104

Liq. Bp$_2$ 90-93°.

Just, G. *et al, Can. J. Chem.*, 1980, **58**, 2349 (*synth, ir, pmr*)

2-Nitrocyclodecanone N-70046

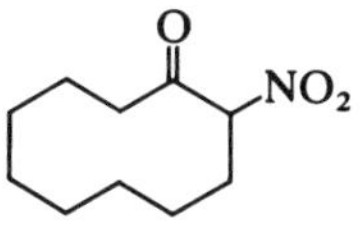

$C_{10}H_{17}NO_3$ M 199.249
Yellow oil. Bp$_{0.2}$ 105° (bulb). n_D^{22} 1.5200.

Stanchev, S. *et al, Helv. Chim. Acta*, 1987, **70**, 1389 (*synth, ir, pmr*)

1-Nitro-1,3-cyclohexadiene N-70047
[76356-92-8]

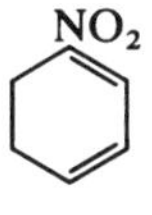

$C_6H_7NO_2$ M 125.127
Liq. Bp$_{0.5}$ 70° (oven).

Bloom, A.J. *et al, J. Chem. Soc., Perkin Trans. 1*, 1987, 2737 (*synth, ir, pmr, cmr*)

2-Nitro-1,3-diphenylpropane N-70048
1,1'-(2-Nitro-1,3-propanediyl)bisbenzene, 9CI
[69957-35-3]

$$PhCH_2CH(NO_2)CH_2Ph$$

$C_{15}H_{15}NO_2$ M 241.289
Cryst. (CH$_2$Cl$_2$/hexane). Mp 103-104°.

Battersby, A.R. *et al, J. Chem. Soc., Perkin Trans. 1*, 1987, 2027 (*synth, ir, pmr, ms*)

4-(2-Nitroethenyl)-1,2-benzenediol, 9CI N-70049
1-(3,4-Dihydroxyphenyl)-2-nitroethylene

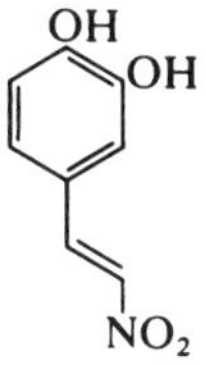

$C_8H_7NO_4$ M 181.148
(**E**)-*form* [108074-44-8]
SL 1 pigment
Prod. by *Streptomyces lavendulae*. Shows mod. activity against certain gram-positive bacteria. Active against some dermatophytes and exhibits mod. cytocidal props. Prisms (EtOAc). Mp 157-159° dec. Yellow or orange in acid/neutral solns.; wine-red to red-brown in alkali.
Di-Ac: Mp 113.5-115°.

Mikami, Y. *et al, J. Antibiot.*, 1987, **40**, 385 (*isol, props*)

4-(2-Nitroethyl)phenol, 9CI N-70050
Updated Entry replacing N-50126
[37567-58-1]

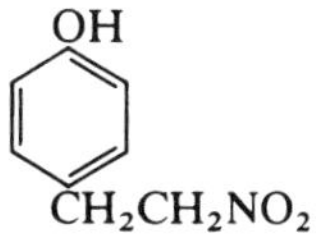

$C_8H_9NO_3$ M 167.164
O-β-D-Glucoside: [74213-96-0]. **Thalictoside**.
    $C_{14}H_{19}NO_8$ M 329.306
    Constit. of *Thalictrum aquilegifolium* and *Epimedium grandiflorum* var. *thunbergiunum*. Cryst. (MeOH/CHCl$_3$). Mp 138-139° (102-103°). [α]$_D^{25}$ −48.8° (c, 1.1 in McOH).

Chirkunov, E.V. *et al, CA*, 1972, **77**, 126172r (*synth*)
Ina, M. *et al, Chem. Pharm. Bull.*, 1986, **34**, 726 (*isol, struct, synth, deriv*)
Miyase, T. *et al, Chem. Pharm. Bull.*, 1987, **35**, 1109 (*deriv*)

1-Nitro-3-hexene N-70051

$$H_3CCH_2CH=CHCH_2CH_2NO_2$$

$C_6H_{11}NO_2$ M 129.158
(**Z**)-*form* [109178-83-8]
Corresponds to 3-hexen-1-yl synthon. Liq. Bp$_{1.6}$ 53°.

Ballini, R. *et al, Synthesis*, 1986, 849 (*synth, ir, pmr*)

5-Nitro-1H-indole N-70052
$C_8H_6N_2O_2$ M 162.148
Yellow solid or needles (C$_6$H$_6$/cyclohexane). Mp 140-141° (134-137°).

Thesing, J. *et al, Chem. Ber.*, 1962, **95**, 2205 (*synth, uv*)
Russell, H.F. *et al, Org. Prep. Proced. Int.*, 1985, **17**, 391 (*synth*)

(Nitromethyl)benzene, 9CI N-70053
Updated Entry replacing N-01027
Nitrophenylmethane. α-Nitrotoluene, 8CI. Phenylnitromethane. ω-Nitrotoluene
[622-42-4]

$$PhCH_2NO_2$$

$C_7H_7NO_2$ M 137.138
Tautomeric. Yellow liq. d 1.176. Mp 225-227°. Bp$_{35}$ 141-142°.
aci-form [622-43-5]
Phenylisonitromethane
Cryst. (Et$_2$O/pet. ether). Mp 84°.

Watarai, S. *et al, Bull. Chem. Soc. Jpn.*, 1967, **40**, 1448 (*ir*)
Neilson, A.T. *et al, J. Org. Chem.*, 1969, **34**, 984 (*synth*)
Danilova, V.I. *et al, CA*, 1973, **78**, 71211 (*tautom, uv*)
Shapiro, M.J., *J. Org. Chem.*, 1976, **41**, 3197 (*nmr*)
Chizhov, O.S. *et al, Org. Mass Spectrom.*, 1978, **13**, 611 (*ms*)
Bellamy, A.J., *Acta Chem. Scand., Ser. B*, 1979, **33**, 208 (*synth*)
Hauser, F.M. *et al, J. Org. Chem.*, 1988, **53**, 2872 (*synth, ir, pmr, cmr*)

3-Nitro-2-naphthaldehyde, 8CI N-70054
3-Nitro-2-naphthalenecarboxaldehyde, 9CI
[73428-05-4]
$C_{11}H_7NO_3$ M 201.181
Cryst. (EtOAc/hexane). Mp 124°.

Kienzle, F., *Helv. Chim. Acta*, 1980, **63**, 2364 (*synth*)
Wani, M.C. *et al*, *J. Med. Chem.*, 1980, **23**, 554 (*synth, ir, pmr*)

Nitropeptin N-70055

N-*Leucyl-3-nitroglutamic acid*, *9CI*
[109792-56-5]

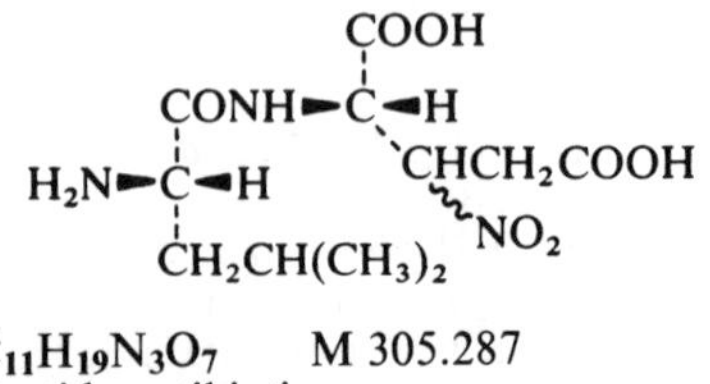

(*S,S*)-*form*

$C_{11}H_{19}N_3O_7$ M 305.287
Peptide antibiotic.

(*S,S*)-form

L,L-form

Prod. by *Streptomyces xanthochromogenus*. Active
against rice blast disease. Shows little antibacterial
props.

Mono-Na salt: Amorph. powder + $\frac{1}{2}H_2O$. Mp 166-168°
dec. $[\alpha]_D^{22}$ −27.2° (c, 1 in H_2O).

Ohba, K. *et al*, *J. Antibiot.*, 1987, **40**, 709 (*isol, struct, props*)

(2-Nitrophenyl)oxirane, 9CI N-70056

Updated Entry replacing N-01252

o-*Nitrostyrene oxide*. o-*Nitrophenylethylene oxide*.
Nitraldin
[39830-70-1]

$C_8H_7NO_3$ M 165.148

(±)-*form*

Yellow plates (MeOH). Mp 65°. Bp_{15} 150°.

▷ Dec. vigorously above 200°. Deflagrates violently on
contact with conc. H_2SO_4

Guss, C.O., *J. Org. Chem.*, 1952, **17**, 678.
Rafizadeh, K. *et al*, *Org. Prep. Proced. Int.*, 1985, **17**, 140
(*synth, bibl*)

(3-Nitrophenyl)oxirane, 9CI N-70057

Updated Entry replacing N-01253

m-*Nitrostyrene oxide*. m-*Nitrophenylethylene oxide*
[20697-05-6]

$C_8H_7NO_3$ M 165.148

(±)-*form*

Yellow oil. Bp_4 140-141°.

Guss, C.O., *J. Org. Chem.*, 1952, **17**, 678.
Rafizadeh, K. *et al*, *Org. Prep. Proced. Int.*, 1985, **17**, 140
(*synth, bibl*)

(2-Nitrophenyl)phenylethanedione, 9CI N-70058

Updated Entry replacing N-01258

2-Nitrobenzil, *8CI*. α-*2-Nitrophenyl-β-phenylglyoxal*
[35010-10-7]

$C_{14}H_9NO_4$ M 255.229
Yellow needles with green reflex (EtOH). Mp 102°.

1-Oxime: α-*Oxime*.
  $C_{14}H_{10}N_2O_4$ M 270.244
  Needles or plates (EtOH). Mp 185° dec.
2-Oxime: β-*Oxime*.
  $C_{14}H_{10}N_2O_4$ M 270.244
  Mp 265° dec.
Dioxime:
  $C_{14}H_{11}N_3O_4$ M 285.259
  Yellowish prisms (EtOH). Mp 244° dec.

Ruggli, P. *et al*, *Helv. Chim. Acta*, 1939, **22**, 147.
McKillop, A. *et al*, *J. Am. Chem. Soc.*, 1973, **95**, 1296 (*synth*)
Bowie, J.H. *et al*, *Aust. J. Chem.*, 1975, **25**, 2169 (*synth, ms*)
Tatsugi, J. *et al*, *J. Chem. Res. (S)*, 1988, 356 (*synth*)

(4-Nitrophenyl)phenylethanedione, 9CI N-70059

Updated Entry replacing N-01260

4-Nitrobenzil, *8CI*. α-*4-Nitrophenyl-β-phenylglyoxal*
[22711-24-6]

$C_{14}H_9NO_4$ M 255.229
Yellow plates or needles (EtOH). Mp 142°.

Monoxime: [31390-84-8].
  $C_{14}H_{10}N_2O_4$ M 270.244
  Mp 140°.
Dioxime: [43084-62-4].
  $C_{14}H_{11}N_3O_4$ M 285.259
  Two forms known. Mp 225° dec. (228-230°) and 185°
  (dimorph.).
Phenylhydrazone: Two forms, yellow plates or orange
  plates. Mp 200° (yellow form), 162° (orange form).

Womack, E. *et al*, *J. Chem. Soc.*, 1938, 1402.
McKillop, A. *et al*, *J. Am. Chem. Soc.*, 1973, **95**, 1296 (*synth*)
Kuse, S. *et al*, *Anal. Chim. Acta*, 1974, **70**, 65 (*deriv*)
Dillard, J.G. *et al*, *Org. Mass Spectrom.*, 1975, **10**, 728 (*ms*)
Tatsugi, J. *et al*, *J. Chem. Res. (S)*, 1988, 356 (*synth*)

1-Nitro-2-(phenylthio)ethane N-70060

[(*2-Nitroethyl*)*thio*]*benzene*, *9CI*. *2-Nitroethyl phenyl
sulfide*
[52809-70-8]

$$PhSCH_2CH_2NO_2$$

$C_8H_9NO_2S$ M 183.225
Liq. $Bp_{0.3}$ 103° (130°).

S-Oxide: [86530-73-6]. *1-Nitro-2-(phenylsulfinyl)-
ethane*. [(*2-Nitroethyl*)*sulfinyl*]*benzene*, *9CI*. *2-Ni-
troethyl phenyl sulfoxide*.
  $C_8H_9NO_3S$ M 199.224
  Reagent for prepn. of nitroethylene. Cryst. Mp 64°.

Cann, P.F. *et al*, *J. Chem. Soc., Perkin Trans. 2*, 1974, 820
(*synth, pmr, props*)

Ranganathan, S. *et al, Tetrahedron Lett.*, 1987, **28**, 2893 (*synth, pmr, use*)

1-Nitro-2-(phenylthio)ethylene　　　　　N-70061

[(*2-Nitroethenyl*)*thio*]*benzene*, *9CI*

$$PhSCH=CHNO_2$$

$C_8H_7NO_2S$　　M 181.209

(*E*)-*form* [101933-27-1]

Oil.

S-*Oxide:* [101933-28-2]. *1-Nitro-2-(phenylsulfinyl)-ethylene*. [(*2-Nitroethenyl*)*sulfinyl*]*benzene*, *9CI*.
$C_8H_7NO_3S$　　M 197.208
Cryst. (EtOH). Mp 100-115°.

S-*Dioxide:* [101933-29-3]. *1-Nitro-2-(phenylsulfonyl)-ethylene*. [(*2-Nitroethenyl*)*sulfonyl*]*benzene*, *9CI*.
$C_8H_7NO_4S$　　M 213.208
Cryst. (EtOH). Mp 148-150°.

Ono, N. *et al, J. Org. Chem.*, 1986, **51**, 2139 (*synth, ir, nmr*)

5-Nitrophthalazine, 9CI　　　　　N-70062

[89898-86-2]

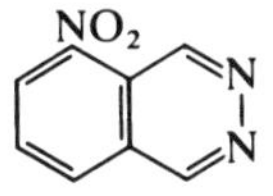

$C_8H_5N_3O_2$　　M 175.146

Yellow prismatic cryst. (EtOH). Mp 188-189°.

Kanahara, S., *Yakugaku Zasshi*, 1964, **84**, 483 (*synth*)
Shaikh, I.A. *et al, J. Med. Chem.*, 1986, **29**, 1329 (*synth, ir, pmr*)

3-Nitro-4(1*H*)-pyridone　　　　　N-70063

Updated Entry replacing N-01363
4-Hydroxy-3-nitropyridine. 3-Nitro-4-pyridinol

[15590-90-6]

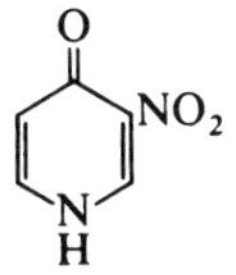

$C_5H_4N_2O_3$　　M 140.098

Pyridone form predominates. Yellow cryst. Mp 278-279° dec. (269-270°).

N-*Me:* [64761-30-4].
$C_6H_6N_2O_3$　　M 154.125
Cryst. (MeOH or Py/Et$_2$O). Mp 233° (220°).

N-*Benzyl:*
$C_{12}H_{10}N_2O_3$　　M 230.223
Mp 113-114°.

OH-form

Me *ether:* [31872-62-5]. *4-Methoxy-3-nitropyridine*.
$C_6H_6N_2O_3$　　M 154.125
Pale-yellow prisms (Et$_2$O/pet. ether). Mp 72-75°.

Et *ether:* [1796-84-5]. *4-Ethoxy-2-nitropyridine*.
$C_7H_8N_2O_3$　　M 168.152
Mp 49-50°.

Koenigs, W. *et al, Ber.*, 1924, **57**, 1187 (*synth, ethoxy*)
Takahashi, T. *et al, J. Pharm. Soc. Jpn.*, 1959, **79**, 1123; *CA*, **54**, 3418 (*N-Me*)
Raczka, A. *et al, Acta Pol. Pharm.*, 1963, **20**, 155; *CA*, **62**, 1630 (*Benzyl*)

Brignell, P.J. *et al, J. Chem. Soc.* (*B*), 1968, 1477 (*methoxy*)
Houston, D.M. *et al, J. Med. Chem.*, 1985, **28**, 467 (*ethoxy*)
Rasala, D. *et al, Org. Prep. Proced. Int.*, 1985, **17**, 409 (*synth*)

1-Nitro-2-(trifluoromethyl)benzene, 9CI　　N-70064

$α,α,α$-*Trifluoro-o-nitrotoluene*, *8CI*. o-*Nitrobenzotrifluoride*

[384-22-5]

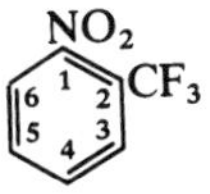

$C_7H_4F_3NO_2$　　M 191.109
Yellow cryst. Mp 32°. Bp$_{15}$ 100-102°.

▷XT3480000.

Maginnity, P.M. *et al, J. Am. Chem. Soc.*, 1951, **73**, 3579 (*synth*)
Forbes, E.J. *et al, Tetrahedron*, 1960, **8**, 67 (*synth*)
Clark, J.H. *et al, J. Chem. Soc., Chem. Commun.*, 1988, 638 (*synth*)

1-Nitro-3-(trifluoromethyl)benzene　　　N-70065

$α,α,α$-*Trifluoro-m-nitrotoluene*. m-*Nitrobenzotrifluoride*

[98-46-4]

$C_7H_4F_3NO_2$　　M 191.109
Liq. Bp 201.5°, Bp$_{10}$ 74-75°.

▷XT3500000.

Drake, N.L. *et al, J. Am. Chem. Soc.*, 1946, **68**, 1602 (*synth*)
Porai-Koshits, A.E. *et al, CA*, 1956, **50**, 4881b (*synth*)
Wiemers, D.M. *et al, J. Am. Chem. Soc.*, 1986, **108**, 832 (*synth*)

1-Nitro-4-(trifluoromethyl)benzene　　　N-70066

$α,α,α$-*Trifluoro-p-nitrotoluene*. p-*Nitrobenzotrifluoride*

[402-54-0]

$C_7H_4F_3NO_2$　　M 191.109
Cryst. Mp 41-43°. Bp$_{10}$ 81-83°.

▷XT3510000.

Drake, N.L. *et al, J. Am. Chem. Soc.*, 1946, **68**, 1602 (*synth*)
Hasek, W.R. *et al, J. Am. Chem. Soc.*, 1960, **82**, 543 (*synth*)
Wiemers, D.M. *et al, J. Am. Chem. Soc.*, 1986, **108**, 832 (*synth*)

2-Nitrotryptophan, 9CI　　　　　N-70067

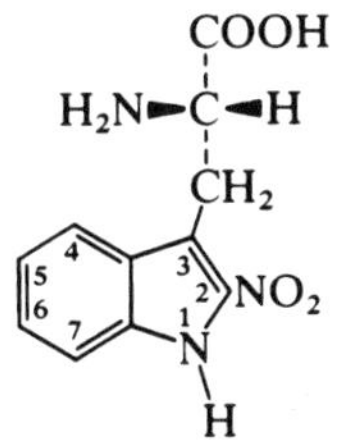

$C_{11}H_{11}N_3O_4$　　M 249.226

(*S*)-*form* [116857-23-9]
L-form
Cryst. + 1H$_2$O (MeOH aq.). Mp 240° dec. [$α$]$_D$ −38.8° (c, 0.32 in 1M HCl).

Phillips, R.S. *et al, J. Heterocycl. Chem.*, 1988, **25**, 191 (*synth, uv, pmr, ms*)

2-Nonacosyl-3-tridecene-1,10-diol, 9CI   N-70068

12-(Hydroxymethyl)-10-hentetraconten-4-ol. **Drechslerol A**

[108793-61-9]

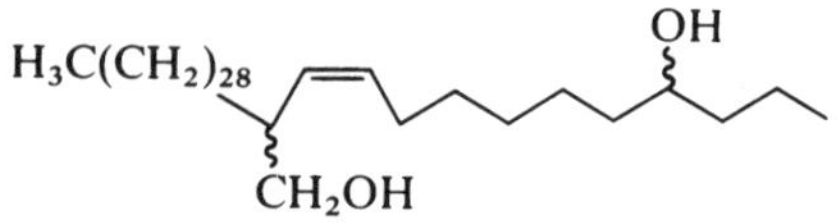

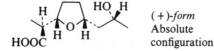

$C_{42}H_{84}O_2$ M 621.124

Prod. by *Drechslera maydis*. Phytotoxin.

Shukla, R.S. *et al*, *Plant Sci. (Limerick, Irel.)*, 1987, **48**, 159 (*isol*)

Nonactinic acid   N-70069

Updated Entry replacing N-01489

Tetrahydro-5-(2-hydroxypropyl)-α-methyl-2-furanacetic acid, 9CI. Nonactic acid

[60761-12-8]

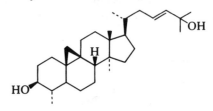

(+)-*form*
Absolute
configuration

$C_{10}H_{18}O_4$ M 202.250

Hydrol. prod. of Nonactin and other *Streptomyces* metabolites.

(+)-*form*

Me ester: $[\alpha]_D^{20}$ +16.2° (c, 2.87 in $CHCl_3$).

(−)-*form*

Me ester: $Bp_{0.03}$ 80°. $[\alpha]_D^{20}$ −17.2° (c, 1.8 in $CHCl_3$).

(±)-*form* [55220-86-5]

$Bp_{0.01}$ 120°.

Me ester: [74892-41-4]. $Bp_{0.01}$ 85°.

Bech, J. *et al*, *Helv. Chim. Acta*, 1962, **45**, 620 (*synth, nmr, ir, isol*)
Gerlach, H. *et al*, *Justus Liebigs Ann. Chem.*, 1963, **669**, 121 (*config*)
Beck, G. *et al*, *Chem. Ber.*, 1971, **104**, 21 (*synth*)
Gerlach, H. *et al*, *Helv. Chim. Acta*, 1974, **57**, 2306 (*synth, ir, uv, nmr*)
Arco, M.J. *et al*, *J. Org. Chem.*, 1976, **41**, 2075 (*synth, ir, nmr, ms*)
Schmidt, U. *et al*, *Chem. Ber.*, 1976, **109**, 2628 (*synth, nmr, ir, uv, ms*)
Baldwin, S.W. *et al*, *J. Org. Chem.*, 1987, **52**, 320 (*synth, bibl*)

Nonacyclo[10.8.0.0²,¹¹.0⁴,⁹.0⁴,¹⁹.0⁶,¹⁷.-0⁷,¹⁶.0⁹,¹⁴.0¹⁴,¹⁹]icosane   N-70070

Dodecahydro-1,7a:2,3b:3a,5:6,7b-tetramethanodicyclobuta[b,h]biphenylene, 9CI

[110318-58-6]

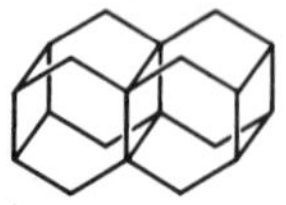

$C_{20}H_{24}$ M 264.410

Needles (MeOH). Mp 230-235°.

Hoffmann, V.T. *et al*, *Angew. Chem., Int. Ed. Engl.*, 1987, **26**, 1006 (*synth, pmr, cmr*)

1,3,5,7-Nonatetraene   N-70071

$$H_2C=CHCH=CHCH=CHCH=CHCH_3$$

C_9H_{12} M 120.194

(E,E,E)-*form* [81129-96-6]

Cryst. solid. Mp 140-147° dec. (softens from 130°).

Spangler, C.W. *et al*, *J. Chem. Soc., Perkin Trans. 1*, 1982, 2379 (*synth, nmr, uv*)
Block, E. *et al*, *J. Am. Chem. Soc.*, 1986, **108**, 4568 (*synth, nmr, ir, uv*)

4-Nonen-2-ynoic acid   N-70072

$$H_3C(CH_2)_3CH=CHC\equiv CCOOH$$

$C_9H_{12}O_2$ M 152.193

(E)-*form*

Cryst. (pentane). Mp 45-47°.

Vlahov, R. *et al*, *Synth. Commun.*, 1986, **16**, 509 (*synth, ir, pmr*)

29-Nor-23-cycloartene-3,25-diol   N-70073

$C_{29}H_{48}O_2$ M 428.697

(3β,23E)-*form*

Constit. of *Aglaia roxburghiana*. Cryst. (EtOAc/C_6H_6). Mp 155°. $[\alpha]_D^{20}$ +23° (c, 0.85 in $CHCl_3$).

Vichnoi, S.P. *et al*, *Planta Med.*, 1988, **54**, 40.

Nordracorhodin   N-70074

5-Methoxy-2-phenyl-7H-1-benzopyran-7-one, 9CI

[35290-21-2]

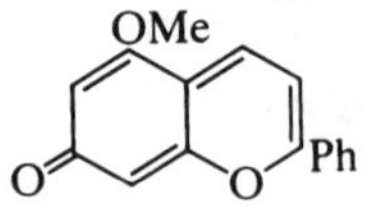

$C_{16}H_{12}O_3$ M 252.269

Pigment from Dragon's blood resin (*Sanguis draconis*). Red. Mp 120-125°.

Cardillo, G. *et al*, *J. Chem. Soc. (C)*, 1971, 3967 (*isol, uv, pmr, ms, synth*)

Nordracorubin   N-70075

[35290-22-3]

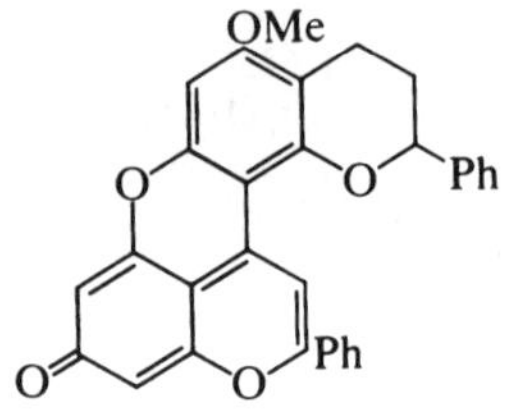

$C_{31}H_{22}O_5$ M 474.512

(−)-*form* [35290-22-3]

Pigment from Dragon's blood resin (*Sanguis draconis*). Red solid. Mp 255-260°. $[\alpha]_D^{20}$ −77.5° (c, 0.024 in MeOH).

(±)-*form*

Red plates. Mp 285° dec.

Cardillo, G. *et al*, *J. Chem. Soc. (C)*, 1971, 3967 (*isol, uv, ms, pmr, struct*)

Agbakwuru, E.O.P. *et al*, *J. Chem. Soc., Perkin Trans. 1*, 1976, 1392 (*synth*)

2-Norerythromycin   N-70076

2-Demethylerythromycin, 9CI
[111010-24-3]

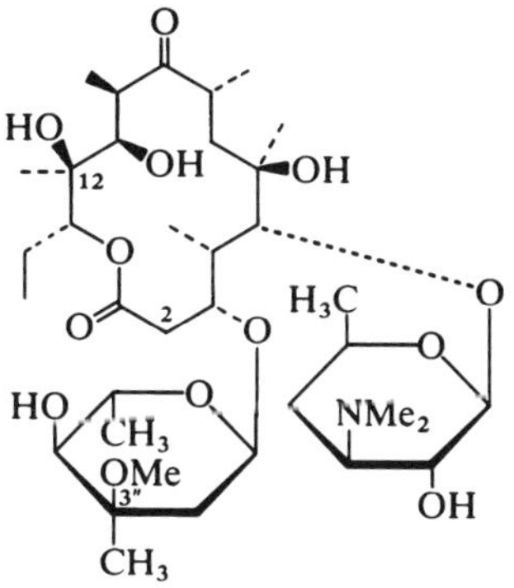

$C_{36}H_{65}NO_{13}$ M 719.908

Macrolide antibiotic. Isol. from genetically engineered actinomycetes. Mainly active against gram-positive bacteria. Foam. $[\alpha]_D^{26}$ −64° (c, 0.87 in MeOH).

12-Deoxy: [110978-90-0]. **2-Norerythromycin B**. *2-Demethyl-12-deoxyerythromycin, 9CI. 2-Demethylerythromycin* B.
$C_{36}H_{65}NO_{12}$ M 703.909
From genetically engineered actinomycetes. Similar biol. props. as parent. Foam. $[\alpha]_D^{26}$ −100° (c, 0.29 in MeOH).

O³″-De-Me: [33442-56-7]. **2-Norerythromycin C**. *3″-O-Demethyl-12-deoxyerythromycin, 9CI. 2-Demethylerythromycin* C.
$C_{35}H_{63}NO_{13}$ M 705.882
From genetically engineered actinomycetes. Similar biol. props. as parent. Glass. $[\alpha]_D^{26}$ −58° (c, 0.32 in MeOH).

12-Deoxy, O³″-de-Me: [110978-89-7]. **2-Norerythromycin D**. *2-Demethyl-3″-O-demethyl-12-deoxyerythromycin, 9CI. 2-Demethylerythromycin* D.
$C_{35}H_{63}NO_{12}$ M 689.882
From genetically engineered actinomycetes. Similar biol. props. as parent. Foam.

McAlpine, J.B. *et al*, *J. Antibiot.*, 1987, **40**, 1115 (*isol, struct, props*)

Norhardwickiic acid   N-70077

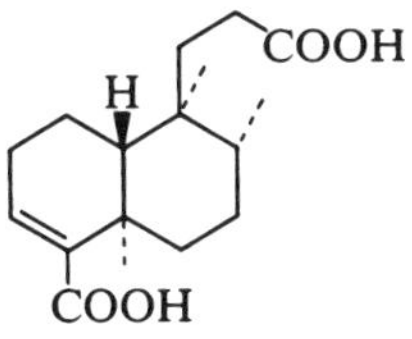

$C_{17}H_{26}O_4$ M 294.390
Constit. of *Grangea maderaspatana*.
Di-Me ester: [74284-66-5]. Gum. $[\alpha]_D$ −36° (c, 0.1 in CHCl₃).

Singh, P. *et al*, *Phytochemistry*, 1988, **27**, 1537.

Norpatchoulenol   N-70078

[41429-52-1]

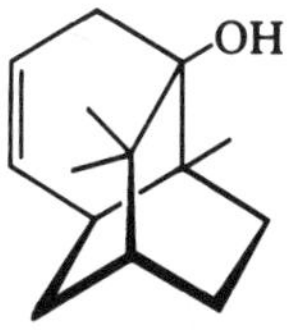

$C_{14}H_{22}O$ M 206.327
Odiferous constit. of patchouli oil. Cryst. by subl. Mp 155-160°.

Niwa, H. *et al*, *Tetrahedron Lett.*, 1984, **25**, 2797 (*synth*)

Norsalvioxide   N-70079

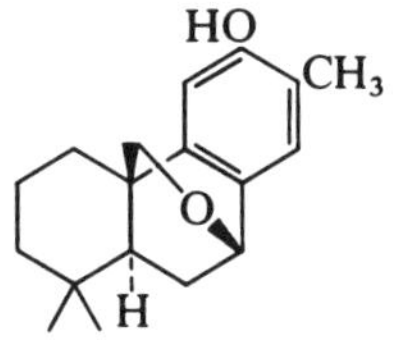

$C_{18}H_{24}O_2$ M 272.386
Constit. of Chinese drug Dan-Shen (*Salvia miltiorrhiza*).

Ginda, H. *et al*, *Tetrahedron Lett.*, 1988, **29**, 4603 (*struct*)

Norstrictic acid   N-70080

[115783-40-9]

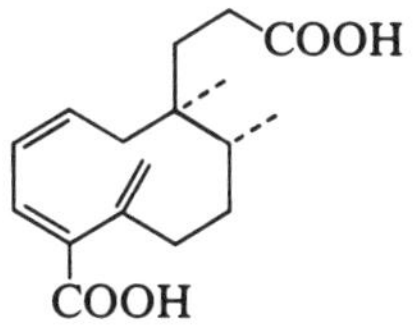

$C_{17}H_{24}O_4$ M 292.374
Constit. of *Grangea maderaspatana*.
Me ester: Gum. $[\alpha]_D$ −135° (c, 0.24 in CHCl₃).

Singh, P. *et al*, *Phytochemistry*, 1988, **27**, 1537.

Notatic acid   N-70081

Updated Entry replacing N-01689
8-Hydroxy-3-methoxy-1,4,6-trimethyl-11-oxo-11H-dibenzo[b,e] [1,4]dioxepin-7-carboxylic acid, 9CI
[38636-86-1]

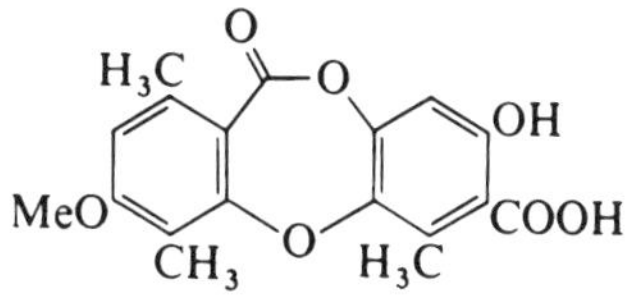

$C_{18}H_{16}O_7$ M 344.320
Isol. from *Parmelia notata*. Needles (Me₂CO). Mp 225-226°.

Me ester: [38629-30-0]. Cryst. (C₆H₆). Mp 203-204°.
O-De-Me: [50489-46-8]. **Nornotatic acid**.
$C_{17}H_{14}O_7$ M 330.293
Isol. from *P. weberi* and prob. also in *Ocellularia* and *Thelometra* spp. Needles (Me₂CO aq.). Mp 207-209°.

Cresp, T.M. *et al*, *Aust. J. Chem.*, 1972, **25**, 2167; 1975, **28**, 2417 (*isol, synth*)
Djuva, P. *et al*, *Aust. J. Chem.*, 1977, **30**, 1293 (*synth, struct, pmr, ms, deriv*)

Nuciferol   **N-70082**

Updated Entry replacing N-10128
   [39599-18-3]
   As Nuciferal, N-01699 with

$$R = CH_2OH$$

$C_{15}H_{22}O$ M 218.338

Constit. of *Torreya nucifera*. Bp$_{0.05}$ 131-132°. $[\alpha]_D^{20}$
   +41.1°.

Sakai, T. *et al*, *Tetrahedron Lett.*, 1963, 1171 (*struct*)
Evans, D.A. *et al*, *Tetrahedron Lett.*, 1973, 1389 (*synth*)
Depezay, J.-C. *et al*, *Tetrahedron Lett.*, 1974, 2755 (*synth*)
Depezay, J.-C. *et al*, *Bull. Soc. Chim. Fr.*, 1981, II, 306 (*synth*)
Takano, S. *et al*, *Tetrahedron Lett.*, 1982, **23**, 5567 (*synth*)
Yoneda, R. *et al*, *Chem. Pharm. Bull.*, 1987, **35**, 913 (*synth*)

O

Obtusadiene
O-70001

2-Bromo-1,1,9-trimethylspiro[5.5]undeca-7,9-dien-3-ol, 9CI

[113262-62-7]

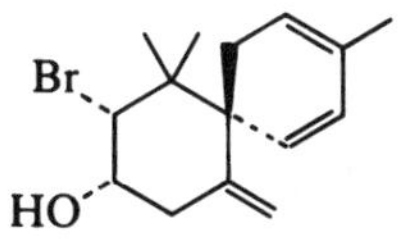

$C_{15}H_{21}BrO$ M 297.234

Constit. of the red alga *Laurencia obtusa*. Oil. $[\alpha]_D^{25}$ −52.0° (c, 2.8 in $CHCl_3$).

Gerwick, W.H. *et al, J. Nat. Prod.*, 1987, **50**, 1131.

11,11,12,12,13,13,14,14-Octacyano-1,4:5,8-anthradiquinotetramethane
O-70002

2,2′,2″,2‴-(1,4,5,8-Anthracenetetraylidene)-tetrakispropanedinitrile, 9CI

[113446-75-6]

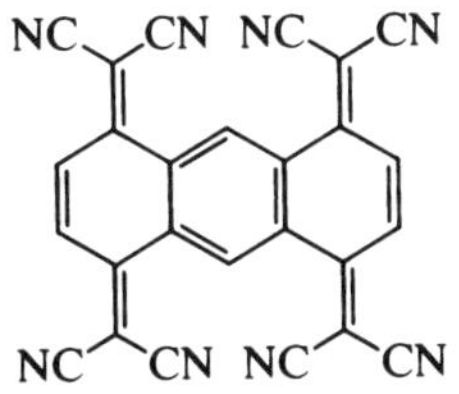

$C_{26}H_6N_8$ M 430.387

Fine golden-yellow plates (MeCN). Mp >400°.

Tetraethylammonium salt: [113446-81-4].
Semiconductor. Black needles. Radical anion.

Mitsuhashi, T. *et al, J. Chem. Soc., Chem. Commun.*, 1987, 810 (*synth, ir, cryst struct, esr, props*)

2,4-Octadecadienoic acid, 9CI
O-70003

[76282-17-2]

$$H_3C(CH_2)_{12}CH=CHCH=CHCOOH$$

$C_{18}H_{32}O_2$ M 280.450

(*E,E*)-form [59404-49-8]

Me ester: [86120-41-4].
$C_{19}H_{34}O_2$ M 294.476
Oil.

Vig, O.P. *et al, Indian J. Chem.*, 1975, **13**, 1358 (*synth, deriv*)
Akimoto, A. *et al, Angew. Chem., Int. Ed. Engl.*, 1981, **20**, 90 (*synth*)
Cardillo, G. *et al, Tetrahedron*, 1986, **42**, 917 (*deriv, synth, ir, pmr, cmr, ms*)

Octadecahydrotriphenylene
O-70004

Updated Entry replacing H-00525
Perhydrotriphenylene

[15074-91-6]

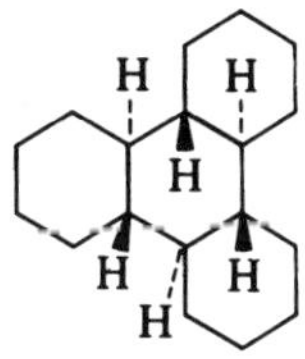

$C_{18}H_{30}$ M 246.435

Forms alkene inclusion complexes.

(*R*, *all-trans*)-form [16069-13-9]
Cryst. Mp 144°. $[\alpha]_D^{25}$ −93°.

(*S*, *all-trans*)-form
Cryst. (butanone). $[\alpha]_D^{25}$ +90° (butanone).

(±)-form
Cryst. Mp 124°.

Farina, M., *J. Am. Chem. Soc.*, 1964, **86**, 516; *J. Chem. Soc.* (*B*), 1967, 1020 (*cryst struct*)
Farina, M., *Tetrahedron*, 1970, **26**, 1839 (*synth, abs config*)

2-Octadecylicosanoic acid
O-70005

Dioctadecylacetic acid

[108293-08-9]

$$[H_3C(CH_2)_{17}]_2CHCOOH$$

$C_{38}H_{76}O_2$ M 565.017

Cryst. (EtOAc). Mp 81-83°.

Adam, N.K. *et al, J. Chem. Soc.*, 1925, 70 (*synth*)
Belletire, J.L. *et al, J. Org. Chem.*, 1987, **52**, 2549 (*synth, pmr, cmr, ir, ms*)

5,6,10,11,16,17,21,22-Octadehydro-7,9,18,20-tetrahydrodibenzo[*e,n*][1,10]-dioxacyclooctadecin, 9CI
O-70006

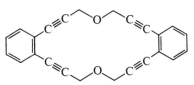

$C_{24}H_{16}O_2$ M 336.389

Mp 181-183°. *S*-analogue also prepd.

Just, G. *et al, Tetrahedron Lett.*, 1987, **28**, 5981 (*synth, pmr, ms*)

2,4-Octadien-1-ol
O-70007

$$H_3CCH_2CH_2CH=CHCH=CHCH_2OH$$

$C_8H_{14}O$ M 126.198

(*2E,4E*)-form [18409-20-6]
Mobile liq. with pleasant fruity odour. Bp$_2$ 70-73°. n_D^{25} 1.4865.

(2E,4Z)-form [56904-85-9]
Bp$_{0.1}$ 48°. n_D^{20} 1.4912.

Jacobson, M., *J. Am. Chem. Soc.*, 1956, **78**, 5084 (*synth*)
Commerçon, A. *et al, Tetrahedron*, 1980, **36**, 1215 (*synth, ir, pmr*)
Kuroda, S. *et al, Tetrahedron Lett.*, 1987, **28**, 803 (*synth, pmr*)

Octafluoronaphthalene   O-70008

[313-72-4]

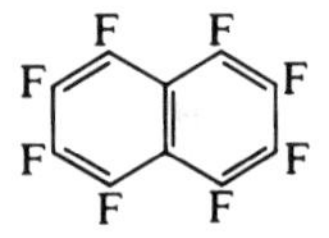

C$_{10}$F$_8$ M 272.097
Cryst. (C$_6$H$_6$). Mp 87-88°. Bp$_{10}$ 92-93°.

Gething, B. *et al, J. Chem. Soc.*, 1962, 186 (*synth, uv*)
Fuller, G., *J. Chem. Soc.*, 1965, 6264 (*synth*)
Cassidei, L. *et al, Spectrochim. Acta, Part A*, 1982, **38**, 755 (*F nmr*)

Octafluoro[2.2]paracyclophane   O-70009

5,6,11,12,13,14,15,16-Octafluorotricyclo[8.2.2.2^{4,7}]-hexadeca-4,6,10,12,13,15-hexaene, 9CI

[1785-64-4]

C$_{16}$H$_8$F$_8$ M 352.226
Solid (methylcyclohexane). Mp >205° subl.

Filler, R. *et al, Chem. Ind. (London)*, 1965, 767 (*synth, uv, ir*)
Filler, R. *et al, J. Fluorine Chem.*, 1986, **30**, 399 (*synth, ir, pmr, F nmr, cmr*)

Octahydroazocine, 9CI   O-70010

Updated Entry replacing O-00290
Perhydroazocine. Azacyclooctane. Heptamethyleneimine
[1121-92-2]

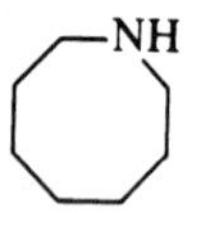

C$_7$H$_{15}$N   M 113.202
d$_4^{21}$ 0.896. Bp$_{30}$ 75-77°. pK_a 9.77. n_D^{21} 1.4740.
▷CN4825000.
B,HCl: Cryst. (Me$_2$CO). Mp 177°.
4-Methylbenzenesulfonyl: Mp 82.5°.
1-Me: [19719-81-4].
   C$_8$H$_{17}$N   M 127.229
   Mp 215° (as picrate).
1-Et: [73676-23-0].
   C$_9$H$_{19}$N   M 141.256
   Mp 163° (as picrate).
1-Isopropyl: [108696-11-3].
   C$_{10}$H$_{21}$N   M 155.283
   Mp 141° (as picrate).
1-tert-Butyl: [108696-12-4].
   C$_{11}$H$_{23}$N   M 169.309

Mp 155° (as picrate).

Blicke, F.F. *et al, J. Am. Chem. Soc.*, 1954, **76**, 2317 (*synth*)
Edwards, O.E. *et al, Can. J. Chem.*, 1972, **50**, 1167 (*synth, use*)
Anet, F.A.L. *et al, J. Org. Chem.*, 1978, **43**, 3021 (*pmr*)
Spanka, G. *et al, J. Org. Chem.*, 1987, **52**, 3362 (*deriv, synth, ir, pmr, cmr, cryst struct*)

Octahydro-2*H*-1,3-benzoxazine, 9CI   O-70011

5,6-Tetramethylenetetrahydro-1,3-oxazine

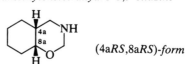

(4a*RS*,8a*RS*)-form

C$_8$H$_{15}$NO   M 141.213
(4aRS,8aRS)-form
   (±)-cis-*form*
   N-*Me:* [84808-91-3].
     C$_9$H$_{17}$NO   M 155.239
     Cryst. (EtOH/Et$_2$O) (as hydrochloride). Mp 216-219° (hydrochloride).
   N-*Benzyl:* [84909-67-1].
     C$_{15}$H$_{21}$NO   M 231.337
     Cryst. (EtOH/Et$_2$O) (as picrate). Mp 126-128° (picrate).
(4aRS,8aSR)-form
   (±)-trans-*form*
   N-*Me:* [84909-66-0]. Cryst. (EtOH/Et$_2$O) (as hydrochloride). Mp 213-216° (hydrochloride).
   N-*Benzyl:* [92597-03-0]. Cryst. (EtOH/Et$_2$O) (as picrate). Mp 150-153° (picrate).

Danilova, O.I. *et al, Zh. Org. Khim.*, 1984, **20**, 2323 (*nmr, conformn*)
Fulöp, F. *et al, Tetrahedron*, 1984, **40**, 2053 (*synth, deriv*)
Sohar, P. *et al, Org. Magn. Reson.*, 1984, **28**, 527 (*nmr, conformn*)

Octahydro-2*H*-3,1-benzoxazine, 9CI   O-70012

4,5-Tetramethylenetetrahydro-1,3-oxazine

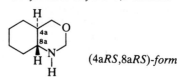

(4a*RS*,8a*RS*)-form

C$_8$H$_{15}$NO   M 141.213
(4aRS,8aRS)-form
   (±)-trans-*form*
   N-*Me:* [92597-07-4]. *Octahydro-1-methyl-2H-3,1-benzoxazine, 9CI.*
     C$_9$H$_{17}$NO   M 155.239
     Cryst. (EtOH/Et$_2$O) (as picrate). Mp 166-170° (picrate).
   N-*Benzyl:* [92597-11-0].
     C$_{15}$H$_{21}$NO   M 231.337
     Cryst. (EtOH/Et$_2$O) (as picrate). Mp 200-201° (picrate).
(4aRS,8aSR)-form
   (±)-cis-*form*
   N-*Me:* [92597-05-2]. Cryst. (EtOH/Et$_2$O) (as picrate). Mp 166-167° (picrate).
   N-*Benzyl:* [92597-09-6]. Cryst. (EtOH/Et$_2$O) (as picrate). Mp 141-143° (picrate).

Danilova, O.I. *et al, Zh. Org. Khim.*, 1984, **20**, 2323 (*nmr, conformn*)
Fulöp, F. *et al, Tetrahedron*, 1984, **40**, 2053 (*synth*)
Sohar, P. *et al, Org. Magn. Reson.*, 1984, **28**, 527 (*nmr, conformn*)

6H,13H-Octahydrodipyridazino[1,2-a:1′,2′-d][1,2,4,5]tetrazine, 9CI   O-70013

6H,13H-Octahydrobis[pyridazino[1,2-a:1′,2′-d]-s-tetrazine]

[5767-20-4]

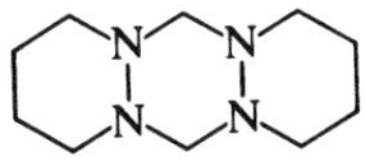

$C_{10}H_{20}N_4$ M 196.295

Mp 170-171°.

▷JM5550000.

Snyder, H.R. et al, J. Org. Chem., 1963, **28**, 1144 (synth)
Baker, V.J. et al, J. Am. Chem. Soc., 1976, **98**, 5748 (pmr, cmr, conformn)
Katritzky, A.R. et al, J. Chem. Soc., Perkin Trans. 2, 1980, 1733 (cryst struct)

Octahydro-5H,10H-dipyrrolo[1,2-a:1′,2′-d]pyrazine-5,10-dithione, 9CI   O-70014

1,7-Diazatricyclo[7.3.0.0^{3,7}]decane-2,8-dithione. Cyclo (ProtProt)

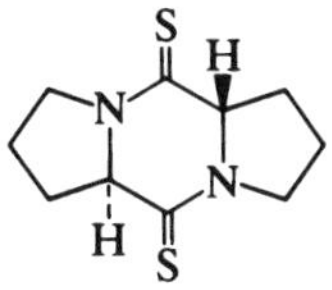

$C_{10}H_{14}N_2S_2$ M 226.354

(S,S)-form [112613-42-0]

L-L-form

Violet cryst.

Párkányi, L. et al, Acta Crystallogr., Sect. C, 1987, **43**, 2356 (synth, cryst struct)

Octahydroimidazo[4,5-d]imidazole, 9CI   O-70015

2,4,6,8-Tetraazabicyclo[3.3.0]octane

$C_4H_{10}N_4$ M 114.150

(3aRS,6aRS)-form

cis-form

1,3,4,6-Tetra-Ac: [106780-32-9].
  $C_{12}H_{18}N_4O_4$ M 282.299
  Cryst. (MeOH). Mp 243-245°.
1,3,4,6-Tetra-Me: [106780-28-3]. 2,4,6,8-Tetramethyl-2,4,6,8-tetraazabicyclo[3.3.0]octane.
  $C_8H_{18}N_4$ M 170.257
  Bp$_{42}$ 107-111°. No stereochem. specified.
1,3,4,6-Tetraisopropyl: [106780-30-7].
  $C_{16}H_{34}N_4$ M 282.471
  Unstable low-melting solid.

Nelsen, S.F. et al, J. Am. Chem. Soc., 1972, **94**, 7114 (deriv, synth, pmr, ir)
Koppes, W.M. et al, J. Org. Chem., 1987, **52**, 1113 (deriv, synth, pmr, ms, cryst struct)

Octahydropyridazino[1,2-a]pyridazine, 9CI   O-70016

1,6-Diazabicyclo[4.4.0]decane. 10-Azaquinolizidine

[3661-15-2]

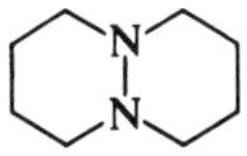

$C_8H_{16}N_2$ M 140.228

Liq. Bp$_{18}$ 79-80°.

Dipicrate: Cryst. (EtOH). Mp 155-158° dec.

Stetter, H. et al, Chem. Ber., 1958, **91**, 1982 (synth)

Octahydropyrrolo[3,4-c]pyrrole, 9CI   O-70017

3,7-Diazabicyclo[3.3.0]octane

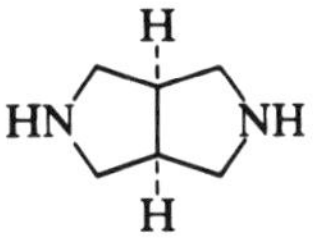

$C_6H_{12}N_2$ M 112.174

(3aRS,6aSR)-form [19885-60-0]

cis-form

Bp$_{0.7}$ 56-60°.

Dipicrate: [19885-61-1]. Mp 270-272° dec.
N-Me: [86732-28-7].
  $C_7H_{14}N_2$ M 126.201
  Bp$_{20}$ 80-81°.
N-(3,5-Dinitrobenzoyl): [19885-32-6]. Mp 267-268°.
N-Benzyl:
  $C_{13}H_{18}N_2$ M 202.299
  Bp$_{0.5}$ 126-128°.
N-Ph:
  $C_{12}H_{16}N_2$ M 188.272
  Bp$_{0.6}$ 146-148°.
N-Ph, succinate: Cryst. (MeOH). Mp 160-163° dec.

Weinges, K. et al, Chem. Ber., 1968, **101**, 3010.
Ohnmacht, C.J. et al, J. Heterocycl. Chem., 1983, **20**, 321 (deriv)

3,3′,4′,5,5′,6,7,8-Octahydroxyflavone   O-70018

Updated Entry replacing O-60023

$C_{15}H_{10}O_{10}$ M 350.238

3,6,8-Tri-Me ether: [99816-52-1]. 3′,4′,5,5′,7-Pentahydroxy-3,6,8-trimethoxyflavone.
  $C_{18}H_{16}O_{10}$ M 392.318
  Constit. of Gutierrezia grandis.
3,4′,7,8-Tetra-Me ether: [78516-78-6]. 3′,5,5′,6-Tetrahydroxy-3,4′,7,8-tetramethoxyflavone.
  $C_{19}H_{18}O_{10}$ M 406.345
  Constit. of G. microcephala.
3,3′,4′,6,8-Penta-Me ether: [96887-20-6]. 3′,5,7-Trihydroxy-3,4′,5′,6,8-pentamethoxyflavone.
  $C_{20}H_{20}O_{10}$ M 420.372
  From G. microcephala. Yellow needles (MeOH). Mp 207-210°.
3,3′,5′,6,8-Penta-Me ether: [96887-19-3]. 4′,5,7-Trihydroxy-3,3′,5′,6,8-pentamethoxyflavone.
  $C_{20}H_{20}O_{10}$ M 420.372
  From G. grandis and G. microcephala. Yellow needles (MeOH aq.). Mp 176-177°.
3,3′,4′,5′,6,8-Hexa-Me ether: [96887-18-2]. 5,7-Dihydroxy-3,3′,4′,5′,6,8-hexamethoxyflavone.
  $C_{21}H_{22}O_{10}$ M 434.399

From *G. grandis* and *G. microcephala*. Yellow needles (MeOH). Mp 178-180°.

3,3′,4′,5′,6,7,8-Hepta-Me ether: [95165-17-6]. *5-Hydroxy-3,3′,4′,5′,6,7,8-heptamethoxyflavone.*
$C_{22}H_{24}O_{10}$ M 448.426
From *G. microcephala*.

Roitman, J.N. *et al*, *Phytochemistry*, 1985, **24**, 835 (*isol*)
Fang, N. *et al*, *Phytochemistry*, 1985, **24**, 2693; 1986, **25**, 927 (*isol*)

Octaleno[3,4-c]furan, 8CI   O-70019
Furo[3,4-c]octalene
[16401-28-8]

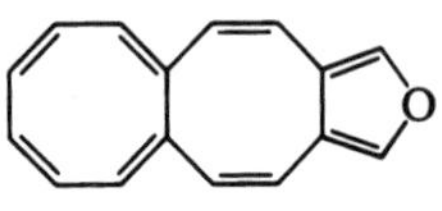

$C_{16}H_{12}O$ M 220.270
Pale-yellow liq. Unstable in air.

Elix, J.A. *et al*, *J. Am. Chem. Soc.*, 1970, **92**, 973 (*synth, ir, uv, pmr*)

2,6,10,14,19,23,27,31-Octamethyl-   O-70020
2,6,10,14,17,22,26,30-dotriacontaoc-
taen-19-ol
Isohydroxylycopersene

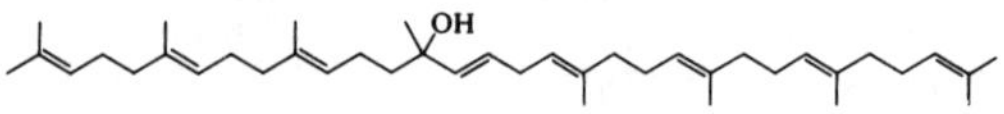

$C_{40}H_{66}O$ M 562.961
Constit. of *Myriophyllum verticillatum*. $[\alpha]_D$ +3.5° (c, 0.5 in $CHCl_3$).

Lanzetta, R. *et al*, *Phytochemistry*, 1988, **27**, 887.

1,2,2,3,3,4,4,5-Octamethyl-6-   O-70021
oxabicyclo[3.1.0]hexane, 9CI
[107010-23-1]

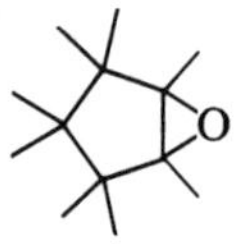

$C_{13}H_{24}O$ M 196.332
Prisms (pentane). Mp 43-46°.

Mayr, H. *et al*, *J. Org. Chem.*, 1987, **52**, 1342 (*synth, ir, pmr, cmr, ms*)

1,3,5-Octatriene   O-70022
Updated Entry replacing O-20023

$$H_3CCH_2CH{=}CHCH{=}CHCH{=}CH_2$$

C_8H_{12} M 108.183
(3E,5Z)-form [40087-61-4]
Fucoserratene
Female sex attractant from the ova of *Fucus serratus* and *F. vesiculosus*. Bp_{40} 56°.
(3Z,5Z)-form
No phys. props. reported.

Jaenicke, L. *et al*, *Chem. Ber.*, 1975, **108**, 225 (*synth*)
Mueller, D.G. *et al*, *Naturwissenschaften*, 1978, **65**, 389 (*isol*)
Janicke, L. *et al*, *Angew. Chem., Int. Ed. Engl.*, 1982, **21**, 643 (*rev*)
Baldwin, J.E. *et al*, *J. Am. Chem. Soc.*, 1987, **109**, 8051 (*synth, pmr, props*)

2,4,6-Octatriene, 9CI   O-70023
Updated Entry replacing O-00388
[764-75-0]

$$H_3CCH{=}CHCH{=}CHCH{=}CHCH_3$$

C_8H_{12} M 108.183
Bp_{100} 84-88°, Bp_{44} 66°. Geom. isomers have been characterised.

Cardenas, C.G., *J. Org. Chem.*, 1970, **35**, 264 (*synth*)
Hasiak, B. *et al*, *Bull. Soc. Chim. Fr.*, 1974, 2917 (*synth*)
Musco, A. *et al*, *J. Organomet. Chem.*, 1975, **88**, C41 (*synth*)
Rennekamp, M.E. *et al*, *Org. Mass Spectrom.*, 1975, **10**, 1075 (*synth, isom*)
Baldwin, J.E. *et al*, *J. Org. Chem.*, 1988, **53**, 1129 (*synth, pmr, isom*)

3-Octenal   O-70024
[60671-71-8]

$$H_3C(CH_2)_3CH{=}CHCH_2CHO$$

$C_8H_{14}O$ M 126.198
Used in pheromone synthesis.
(E)-form [76595-71-6]
Liq. $Bp_{0.02}$ 60°.
2,4-Dinitrophenylhydrazone: Cryst. (EtOH). Mp 80-81°.
(Z)-form [78693-34-2]
Liq. Bp_{14} 65-75° (bath).

Bestmann, H.J. *et al*, *Justus Liebigs Ann. Chem.*, 1981, 1705 (*synth, ir, pmr*)
Achmatowicz, O. *et al*, *Synthesis*, 1987, 413 (*synth, ir, pmr*)

4-Octenedial, 9CI   O-70025
[51097-60-0]

$$OHCCH_2CH_2CH{=}CHCH_2CH_2CHO$$

$C_8H_{12}O_2$ M 140.182
(Z)-form [52153-48-7]
$Bp_{3.5}$ 90-92°, $Bp_{0.3}$ 61°.
Bis(dimethyl acetal): [63284-78-6]. *1,1,8,8-Tetramethoxy-4-octene.*
$C_{12}H_{24}O_4$ M 232.319
$Bp_{1.5}$ 91-92°.
Bis-2,4-dinitrophenylhydrazone: [52096-54-5]. Mp 173° (169-170°).

Nagarkatti, J.P. *et al*, *Tetrahedron Lett.*, 1973, 4599 (*synth*)
Odinokov, V.N. *et al*, *J. Org. Chem. USSR (Engl. transl.)*, 1977, **13**, 485 (*synth, ir, pmr*)

4-Octene-1,8-diol   O-70026
[56506-08-2]

$$HOCH_2CH_2CH_2CH{=}CHCH_2CH_2CH_2OH$$

$C_8H_{16}O_2$ M 144.213
(Z)-form [62422-45-1]
Intermed. in nat. prod. synth. $Bp_{0.001}$ 78°.
Di-Ac: [67161-27-7].
$C_{12}H_{20}O_4$ M 228.288
Bp_1 102-103°.

Nagarkatti, J.P. *et al*, *J. Indian Chem. Soc.*, 1976, **53**, 666 (*synth*)
Tolstikov, G.A. *et al*, *Tetrahedron Lett.*, 1978, 1857 (*synth*)

Odinokov, V.N. *et al, J. Org. Chem. USSR (Engl. transl.)*,
  1979, **15**, 1252 (*synth, ir, pmr*)
Raederstorff, D. *et al, J. Org. Chem.*, 1987, **52**, 2337 (*synth,
  pmr, ms*)

3-Octen-2-one   O-70027

Updated Entry replacing O-50028
[1669-44-9]

$$H_3C(CH_2)_3CH=CHCOCH_3$$

$C_8H_{14}O$ M 126.198
Vanilla volatile, mushroom flavour component.
(E)-form [18402-82-9]
Oil. Bp$_{50}$ 100°.
2,4-Dinitrophenylhydrazone: Mp 93°.
(Z)-form [51193-77-2]
Oil. Bp$_{50}$ 68-69°.
2,4-Dinitrophenylhydrazone: Mp 93°.

Heilman, R. *et al, Bull. Soc. Chim. Fr.*, 1957, 119 (*synth, uv, ir*)
Clinet, J.C. *et al, Tetrahedron Lett.*, 1978, 1137 (*synth*)
Beak, P. *et al, J. Am. Chem. Soc.*, 1980, **102**, 3848 (*synth*)
Grayson, D.H. *et al, J. Chem. Soc., Perkin Trans. 1*, 1986, 2137
  (*synth, ir, pmr*)

5-Octen-2-one   O-70028

[36359-70-3]

$$H_3CCH_2CH=CHCH_2CH_2COCH_3$$

$C_8H_{14}O$ M 126.198
(E)-form [19093-20-0]
  Bp 169-173°, Bp$_{11}$ 59-61.5°.
Semicarbazone: Cryst. (pet. ether). Mp 62.5-63.5°.
2,4-Dinitrophenylhydrazone: Mp 65.5-66°.
(Z)-form [22610-86-2]
  Bp$_{20}$ 68-70°, Bp$_{10}$ 54-57°.
Semicarbazone: Leaflets (EtOH aq.). Mp 114.0-114.5°.
2,4-Dinitrophenylhydrazone: Plates (EtOH aq.). Mp
  67.5°.

Crombie, L. *et al, J. Chem. Soc.*, 1952, 869 (*synth*)
Harper, S.H. *et al, J. Chem. Soc.*, 1955, 1512 (*synth*)
Crombie, L. *et al, J. Chem. Soc. (C)*, 1969, 1016 (*synth, ir,
  pmr*)
Ballini, R. *et al, Synthesis*, 1986, 849 (*synth, ir, pmr*)

3-Octen-1-yne   O-70029

$$HC\equiv CCH=CH(CH_2)_3CH_3$$

C_8H_{12} M 108.183
(E)-form [42104-42-7]
Liq. Bp$_{640}$ 80-85°, Bp$_{50}$ 60-62°.

Mesnard, D. *et al, J. Organomet. Chem.*, 1976, **117**, 99 (*synth*)
King, A.O. *et al, J. Chem. Soc., Chem. Commun.*, 1977, 683
  (*synth*)
Stille, J.K. *et al, J. Am. Chem. Soc.*, 1987, **109**, 2138 (*synth,
  pmr, cmr*)

Okilactomycin   O-70030
YP 02908L-A. *Antibiotic YP 02908L*-A
[111367-04-5]

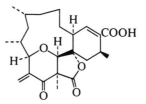

$C_{24}H_{32}O_6$ M 416.513
Cyclic lactone antibiotic. Prod. by *Streptomyces
  griseoflavus* ssp. *zamamiensis*. Exhibit weak activity
  against gram-positive bacteria and limited antitumour
  props. Prisms (CH$_2$Cl$_2$). Mp 161°. [α]$_D^{20}$ +34° (c, 1 in
  MeOH).

Imai, H. *et al, Antibiot.*, 1987, **40**, 1475, 1483 (*isol, struct,
  props*)

9(11),12-Oleanadien-3-ol   O-70031
$C_{30}H_{48}O$ M 424.709
3β-form
Constit. of *Phyllanthus flexuosus*. Cryst.
  (MeOH/CHCl$_3$). Mp 219-221°. [α]$_D^{23}$ +318° (c, 0.56
  in CHCl$_3$).

Tanaka, R. *et al, Phytochemistry*, 1988, **27**, 2273.

11,13(18)-Oleanadien-3-ol   O-70032
$C_{30}H_{48}O$ M 424.709
3β-form [117073-98-0]
Constit. of *Phyllanthus flexuosus*. Needles
  (MeOH/CHCl$_3$). Mp 232-234.5°. [α]$_D^{23}$ −61.3° (c,
  0.58 in CHCl$_3$).

Tanaka, R. *et al, Phytochemistry*, 1988, **27**, 2273.

12-Oleanene-3,15-diol   O-70033
$C_{30}H_{50}O_2$ M 442.724
(3β,15α)-form
15α-Hydroxy-β-amyrin
Constit. of *Baccharis magellanica*. Cryst. (MeOH aq.).
  Mp 230-232°. [α]$_D^{20}$ +68.4° (c, 1.12 in CHCl$_3$).

Rivera, A.P. *et al, J. Nat. Prod.*, 1988, **51**, 155.

11(12)-Oleanene-3,13,23,28-tetrol   O-70034
$C_{30}H_{50}O_4$ M 474.723
(3β,13ξ)-form
Verbascogenin
Not isol. Hydrol. of Verbascosaponin led to dehydration.
*3-O-[α-L-Rhamnopyranosyl-(1→4)-β-D-
  glucopyranosyl-(1→4)-[β-D-glucopyranosyl-(1→2)]-
  β-D-fucopyranoside]:* [74163-66-9]. **Verbascosaponin**.
  $C_{54}H_{90}O_{22}$ M 1091.292
  Saponin from *Verbascum phlomoides*. Mp 263-268°.

Tschesche, R. *et al, Chem. Ber.*, 1980, **113**, 1754 (*isol, cmr,
  struct*)

13(18)-Oleanene-3,16,28-triol   O-70035
3,16,28-Trihydroxy-13(18)-oleanene
$C_{30}H_{50}O_3$ M 458.723

(*3β,16β*)-*form* [26540-64-7]
Heliantriol A₁. *Coflotriol*
Isol. from flowers of *Helianthus annuus* and *Calendula officinalis*.
Tri-Ac: Cryst. (MeOH). Mp 229°.

St. Pyrek, J., *Pol. J. Chem.*, 1979, **53**, 2465.

13(18)-Oleanen-3-ol   O-70036
Updated Entry replacing A-02785
$C_{30}H_{50}O$ M 426.724
3α-form [52647-56-0]
Gymnorhizol
Isol. from leaves of *Bruguera gymnorhiza*. Mp 208-210°. $[\alpha]_D$ −48°.

3-Ketone: [20248-08-2]. *13(18)-Oleanen-3-one*. δ-
Amyrone. Constit. of *Alnus* spp. Cryst. (EtOH). Mp 198-201°. $[\alpha]_D^{20}$ −12°.

3β-form [508-04-3]
δ-**Amyrin**
Constit. of *Spartium junceum*. Cryst. (MeOH). Mp 212-213.5°. $[\alpha]_D^{20}$ −50.5°.

Musgrave, O.C. *et al*, *J. Chem. Soc.*, 1952, 4393 (*isol*)
v. Tamelen, E.E. *et al*, *J. Am. Chem. Soc.*, 1972, **94**, 8229 (*synth*)
Wollenweber, E., *Z. Naturforsch., C*, 1974, **29**, 362 (*isol*)
Sarkar, A. *et al*, *Indian J. Chem., Sect. B*, 1978, **16**, 742 (*Gymnorhizol*)

Oleonuezhenide   O-70037
[112693-21-7]

$C_{49}H_{66}O_{28}$ M 1103.044
Constit. of *Ligustrum japonicum*. Amorph. powder. $[\alpha]_D$ +62.3° (c, 0.83 in EtOH).

Fukuyama, Y. *et al*, *Planta Med.*, 1987, **53**, 427.

Oleuroside   O-70038
[116383-31-4]

$C_{25}H_{32}O_{13}$ M 540.520
Constit. of *Olea europaea*. Powder. $[\alpha]_D^{17}$ −83.7° (c, 1.03 in MeOH).
Hexa-Ac: Needles (EtOH). Mp 122-123°. $[\alpha]_D^{17}$ −82.8° (c, 1 in CHCl₃).

Kuwajima, H. *et al*, *Phytochemistry*, 1988, **27**, 1757.

Oligomycin *E*   O-70039
26-Hydroxy-28-oxooligomycin A, *9CI*
[110231-34-0]

$C_{45}H_{72}O_{13}$ M 821.056
Macrolide antibiotic. Isol. from *Streptomyces* sp. MCI-2225. Active against gram-positive bacteria. Shows cytotoxic props. Plates. Mp 120.5-121.5°. $[\alpha]_D^{25}$ −49.1° (c, 1.05 in dioxan).

Kobayashi, K. *et al*, *J. Antibiot.*, 1987, **40**, 1053 (*isol, struct, props*)

Ophiobolin *A*   O-70040
Updated Entry replacing O-00572
Cochliobolin A. *Ophiobolin. Cochliobolin. Ophiobalin*
[4611-05-6]

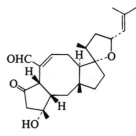

$C_{25}H_{36}O_4$ M 400.557
Sesterterpene. Metab. of *Ophiobolus miyabeanus* and *Helminthosporium* sp. and of *O. heterostrophus*, *Helminthosporium oryzae*, *H. turcicum*, *H. panici-miliacei* and *H. zizaniae*. Phytotoxin and photosynth. inhibitor. Cryst. Mp 182°. $[\alpha]_D^{29}$ +270° (CHCl₃).
▷RL1576000.

6-Epimer: [24034-72-8]. **6-Epiophiobolin** A.
$C_{25}H_{36}O_4$ M 400.557
Phytotoxin from *Drechslera maydis*. Phytotoxin and photosynth. inhibitor.
6-Epimer, 3-deoxy-, 3,4-didehydro: [90411-20-4]. **3-An-hydro-6-epiophiobolin** A.
$C_{25}H_{34}O_3$ M 382.542
Phytotoxin from *D. maydis*.

Nozoe, S. *et al*, *J. Am. Chem. Soc.*, 1965, **87**, 4968 (*struct*)
Canonica, L. *et al*, *Tetrahedron Lett.*, 1967, 3371, 4657 (*biosynth*)
Bose, A.K. *et al*, *Experientia*, 1971, **27**, 1403 (*biosynth*)
Radics, L. *et al*, *Tetrahedron Lett.*, 1975, 4415 (*cmr*)
Senter, P.D., *Diss. Abstr. Int. B*, 1982, **42**, 3696 (*synth*)
Canales, M.W., *Diss. Abstr. Int. B*, 1983, **44**, 772 (*isol, deriv*)
Kim, J.M. *et al*, *Agric. Biol. Chem.*, 1984, **48**, 803 (*isol, props*)
Muskopf, J.W., *Diss. Abstr. Int. B*, 1984, **45**, 194 (*synth*)
Canales, M.W. *et al*, *Phytochemistry*, 1988, **27**, 1653 (*isol, deriv*)

Ophiobolin J   O-70041

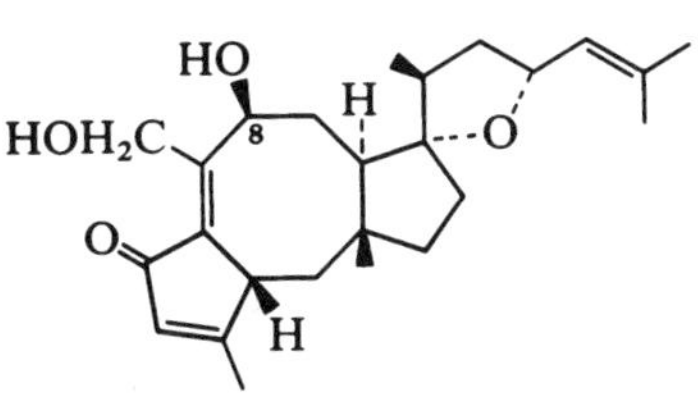

$C_{25}H_{36}O_4$ M 400.557
Metab. of *Drechslera oryzae*. Phytotoxic to rice. Oil.
$[\alpha]_D$ +48° (c, 1.7 in CHCl$_3$).
8-Deoxy: 8-Deoxyophiobolin J.
   $C_{25}H_{36}O_3$ M 384.558
   Metab. of *D. oryzae*. Phytotoxic. Cryst. Mp 138-140°.
   $[\alpha]_D$ +8° (c, 0.15 in CHCl$_3$).

Sugawara, F. *et al*, *J. Org. Chem.*, 1988, **53**, 2170.

4,10(14)-Oplopadiene-3,8,9-triol   O-70042
*1-Ethylideneoctahydro-4-methylene-7-(1-methylethyl)-
1H-indene-2,5,6-triol, 9CI. 3(14),8(10)-Oplapadiene-
2,6,7-triol (incorr.)*

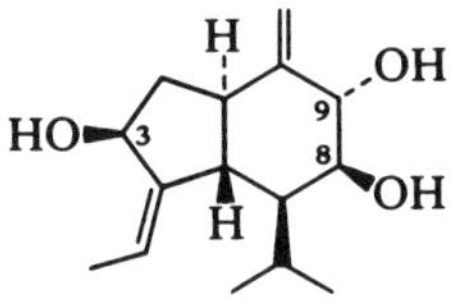

$C_{15}H_{24}O_3$ M 252.353
(3β,4Z,8β,9α)-form
   3-Ac: [117176-26-8]. *3β-Acetoxy-4,10(14)-oplopadiene-
   8β,9α-diol. 2β-Acetoxy-3(14),8(10)-oplopadiene-
   6β,7α-diol (incorr.).*
   $C_{17}H_{26}O_4$ M 294.390
   Constit. of *Acrisione denticulata*. Oil.
   3-Ac, 8-angeloyl: [117176-24-6].
   $C_{22}H_{32}O_5$ M 376.492
   Constit. of *A. denticulata*. Oil.
   3-Ac, 9-angeloyl: [117183-11-6].
   $C_{22}H_{32}O_5$ M 376.492
   Constit. of *A. denticulata*. Oil. $[\alpha]_D^{24}$ +27° (c, 0.69 in
   CHCl$_3$).
   3-Ac, 9-(3-Methyl-2-pentenoyl): [117176-25-7].
   $C_{23}H_{34}O_5$ M 390.519
   Constit. of *A. denticulata*. Oil. $[\alpha]_D^{24}$ +20° (c, 0.49 in
   CHCl$_3$).
   3-Ketone: 8β,9α-Dihydroxy-4,10(14)-oplopadien-3-one.
   $C_{15}H_{22}O_3$ M 250.337
   Various esters found in *A. denticulata*.

Aal, M.A. *et al*, *Phytochemistry*, 1988, **27**, 2599.

Oplopanone   O-70043
Updated Entry replacing O-10052

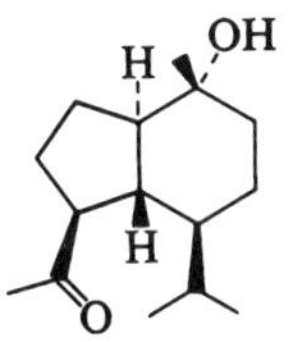

$C_{15}H_{26}O_2$ M 238.369
(−)-form [1911-78-0]
   Constit. of *Oplopanax japonicus*. Cryst. (Et$_2$O/pet.
   ether). Mp 96-97°. $[\alpha]_D^{25}$ −20° (c, 0.6 in dioxan).

(+)-form
ent-*Oplopanone*
   Constit. of soft coral *Nephthea* sp. Needles (pet. ether).
   Mp 83-84°. $[\alpha]_D^{22}$ +19° (c, 0.8 in dioxan).

Takeda, K. *et al*, *Tetrahedron, Suppl.*, 1966, **7**, 219 (*isol, ir,
   pmr, ord, struct*)
Caine, D. *et al*, *J. Org. Chem.*, 1973, **38**, 3663 (*isol*)
Taber, D.F. *et al*, *J. Org. Chem.*, 1978, **43**, 4925 (*synth*)
Koster, F.-H. *et al*, *Tetrahedron Lett.*, 1981, **22**, 3937 (*synth*)

Oreojasmin   O-70044
*6,7-Dimethoxy-8-[(2-oxo-2H-1-benzopyran-7-yl)oxy]-
2H-1-benzopyran-2-one, 9CI*
[59096-04-7]

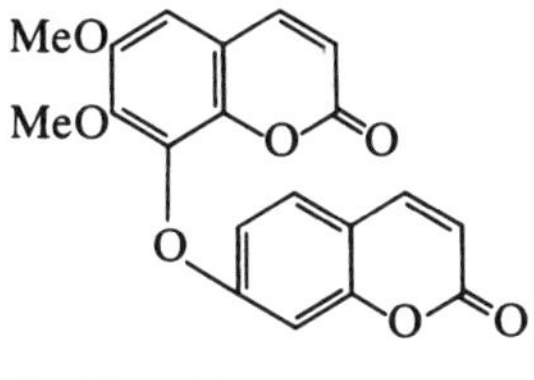

$C_{20}H_{14}O_7$ M 366.326
Constit. of *Ruta oreojasma*. Cryst. Mp 238-239°.

Gonzalez, A.G. *et al*, *An. Quim.*, 1975, **71**, 842.

Oriciopsin   O-70045
Updated Entry replacing O-20041

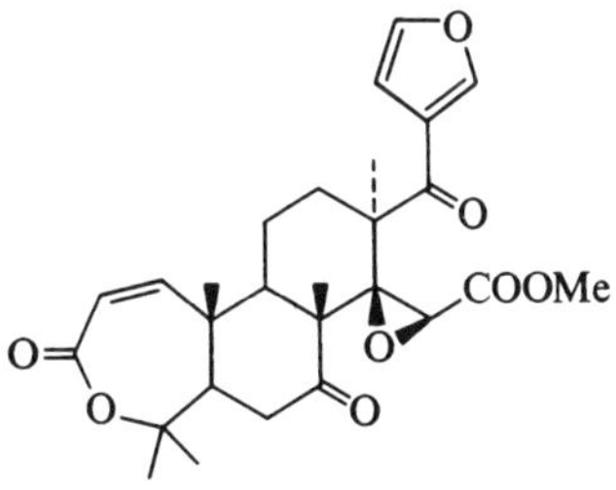

$C_{27}H_{32}O_8$ M 484.545
Constit. of *Oriciopsis glaberrima*. Rods (CHCl$_3$/Et$_2$O).
Mp 242-244°. $[\alpha]_D^{23}$ −60.5° (c, 0.8 in Me$_2$CO).
5,6-Didehydro: 5-Dehydrooriciopsin.
   $C_{27}H_{30}O_8$ M 482.529
   Constit. of *Harrisonia abyssinica*. Oil.

Ayafor, J.F. *et al*, *Phytochemistry*, 1982, **21**, 2602.
Balde, A.M. *et al*, *Phytochemistry*, 1988, **27**, 942 (*deriv*)

Orlandin   O-70046
*7,7'-Dihydroxy-4,4'-dimethoxy-5,5'-dimethyl-[8,8'-bi-
2H-1-benzopyran]-2,2'-dione, 9CI. 8,8'-Bi[7-hydroxy-4-
methoxy-5-methylcoumarin]*
[69975-77-5]

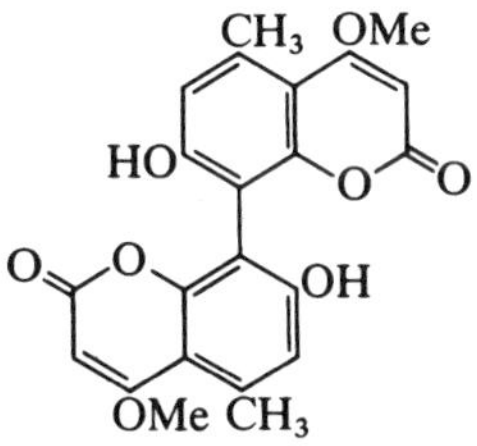

$C_{22}H_{18}O_8$ M 410.379
Metab. of *Aspergillus niger*. Cryst. Mp 285°.

Cutler, H.G. *et al*, *J. Agric. Food Chem.*, 1979, **27**, 592.

N-Ornithylglycine　　　　　　　　O-70047

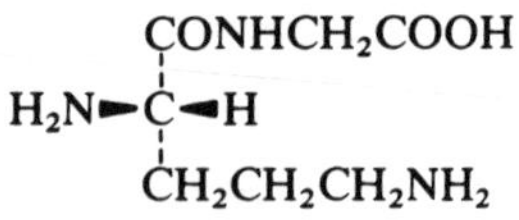

$$CONHCH_2COOH$$
$$H_2N\!-\!C\!-\!H$$
$$CH_2CH_2CH_2NH_2$$

$C_7H_{15}N_3O_3$　　M 189.214

(*S*)-form
L-form
B,HCl: [90970-33-5]. Hygroscopic cryst. + H₂O. [α]_D
+46.2° (c, 1 in H₂O).

Tada, M. *et al*, *J. Agric. Food Chem.*, 1984, **32**, 992 (*synth*)
Huynh-ba, T. *et al*, *J. Agric. Food Chem.*, 1987, **35**, 165 (*synth*, *cmr*)

Osbeckic acid　　　　　　　　　　O-70048
5-Carboxy-α-hydroxy-2-furanacetic acid, 9CI
[112923-64-5]

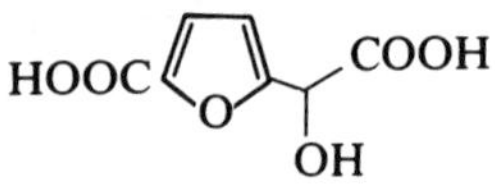

$C_7H_6O_6$　　M 186.121
Constit. of *Osbeckia chinensis*. Antioxidative synergist.
Oil. [α]_D^{25} +83.5° (c, 0.2 in MeOH).

Su, J.D. *et al*, *Agric. Biol. Chem.*, 1987, **51**, 3449.

Osthenone　　　　　　　　　　　O-70049

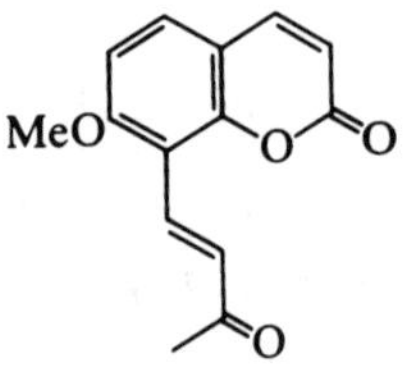

$C_{14}H_{12}O_4$　　M 244.246

(*E*)-form
Constit. of *Murraya exotica*. Oil.
(*Z*)-form
Constit. of *M. paniculata*. Oil.

Ito, C. *et al*, *Heterocycles*, 1987, **26**, 2959.

Ovothiol *A*　　　　　　　　　　O-70050
1-Methyl-4-mercaptohistidine
[108418-13-9]

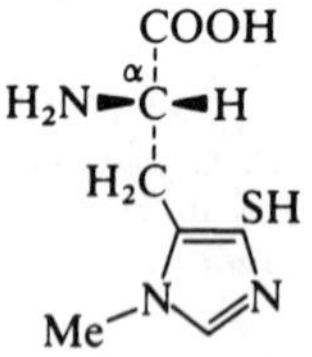

$C_7H_{11}N_3O_2S$　　M 201.243
Isol. from eggs of the starfish *Evasterias troschelii*.
Redox-active compd.
N^α-*Me*: [108418-14-0]. **Ovothiol B**.
　$C_8H_{13}N_3O_2S$　　M 215.270
　Isol. from the scallop *Chlamys hastata*.
N^α,N^α-*Di-Me*: [105496-34-2]. **Ovothiol C**.
　$C_9H_{15}N_3O_2S$　　M 229.296
　Isol. from the sea urchin *Stronglylocentrotus purpuratus*.

Turner, E. *et al*, *Biochemistry*, 1987, **26**, 4028 (*isol*, *pmr*, *struct*)

2-Oxa-3-azabicyclo[2.2.2]octane, 9CI　　O-70051
[280-50-2]

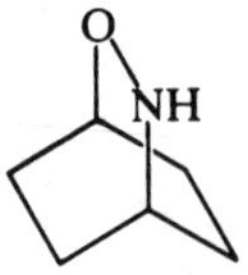

$C_6H_{11}NO$　　M 113.159
B,HCl: [110590-22-2]. Cryst. (EtOH/Et₂O). Mp 190°
dec.
N-*Formyl*: [87013-24-9].
　$C_7H_{11}NO_2$　　M 141.169
　Mp 105-106.5°.
N-*Cyano*: [110590-01-7].
　$C_7H_{10}N_2O$　　M 138.169
　Mp 35-36°.

Nelson, S.F. *et al*, *J. Am. Chem. Soc.*, 1987, **109**, 7128 (*synth*, *deriv*, *pmr*, *cmr*, *cryst struct*, *esr*, *pe*, *props*)

1-Oxa-2-azaspiro[2,5]octane, 9CI　　O-70052
Cyclohexanone isoxime. 3,3-Pentamethyleneoxaziridine
[185-80-8]

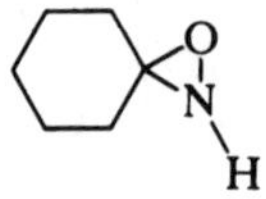

$C_6H_{11}NO$　　M 113.159
Obt. in soln., readily polym.
N-*Methanesulfonyl*: Cryst. Mp 44-46°.
N-tert-*Butyl*:
　$C_{10}H_{19}NO$　　M 169.266
　Bp_{0.7} 47-48°.
N-*Ac*: [16192-49-7].
　$C_8H_{13}NO_2$　　M 155.196
　Liq. Bp_{0.3} 51-53°.
N-*Carbamoyl*: [2289-95-4].
　$C_7H_{12}N_2O_2$　　M 156.184
　Cryst. (C₆H₆/pet. ether). Mp 128°.
N-*Benzoyl*: [2289-83-0].
　$C_{13}H_{15}NO_2$　　M 217.267
　Cryst. (Et₂O). Mp 70°.

Schmitz, E. *et al*, *Chem. Ber.*, 1964, **97**, 2521; 1967, **100**, 2593 (*synth*, *uv*, *ir*)
Jennings, W. *et al*, *J. Am. Chem. Soc.*, 1987, **109**, 8099 (*synth*, *deriv*, *pmr*, *cmr*)

3-Oxabicyclo[3.2.0]heptan-2-one, 9CI　　O-70053
2-Hydroxymethylcyclobutanecarboxylic acid lactone
[2983-95-1]

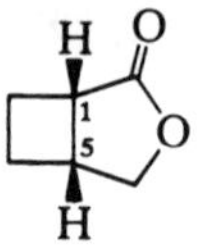

(1*S*,5*R*)-form

$C_6H_8O_2$　　M 112.128
(1*S*,5*R*)-form [75658-85-4]
(+)-cis-*form*
Bp_{20} 120°, Bp_{10} 100°. [α]_D^{25} +106.7° (c, 3.1 in CHCl₃),
[α]_D^{25} +118.7° (c, 10 in CHCl₃).

(1R,5S)-form [88335-95-9]
(−)-cis-*form*
Bp$_{20}$ 120°. [α]$_D^{25}$ −107.1° (c, 1.1 in CHCl$_3$).
(1RS,5SR)-form [14764-52-4]
(±)-cis-*form*
Bp$_{15}$ 109-113°.

Bloomfield, J.J. et al, J. Org. Chem., 1967, **32**, 3919 (synth, ir, pmr)
Jakovac, I.J. et al, J. Am. Chem. Soc., 1982, **104**, 4659 (synth, ir, pmr, abs config)
Org. Synth., 1985, **63**, 13 (synth)
Sabbioni, G. et al, J. Org. Chem., 1987, **52**, 4565 (synth, ir, pmr, cmr)

7-Oxabicyclo[2.2.1]hept-2-ene   O-70054
3,6-Epoxy-1-cyclohexene
[6705-50-6]

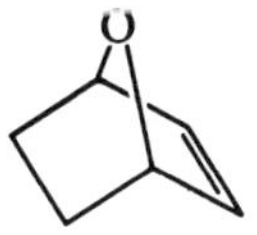

C$_6$H$_8$O   M 96.129
Liq. Bp 118-122°. n$_D^{20}$ 1.4629.

Nudenberg, W. et al, J. Am. Chem. Soc., 1944, **66**, 307 (synth)
Bain, A. et al, J. Am. Chem. Soc., 1973, **95**, 291 (pe)
Tam, N.T.T. et al, Tetrahedron, 1980, **36**, 2793 (O nmr)
Senda, Y. et al, Bull. Chem. Soc. Jpn., 1987, **60**, 613 (cmr)

3-Oxabicyclo[3.1.0]hexan-2-one, 9CI   O-70055
2-Hydroxymethylcyclopropanecarboxylic acid lactone

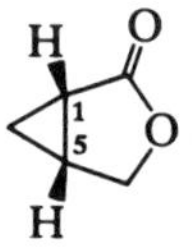

(1S,5R)-*form*

C$_5$H$_6$O$_2$ M 98.101
(1S,5R)-form [75658-86-5]
(−)-cis-*form*
Bp$_{10}$ 100°, Bp$_1$ 60°. [α]$_D^{25}$ −60° (c, 0.4 in CHCl$_3$).
(1R,5S)-form [89395-28-8]
(+)-cis-*form*
Bp$_1$ 60°. [α]$_D^{25}$ +61.1° (c, 0.72 in CHCl$_3$).
(1RS,5SR)-form [82442-65-7]
(±)-cis-*form*
Bp$_{10}$ 100°.

Jakovac, I.J. et al, J. Am. Chem. Soc., 1982, **104**, 4659 (synth, ir, pmr, abs config)
Org. Synth., 1985, **63**, 10 (synth)
Sabbioni, G. et al, J. Org. Chem., 1987, **52**, 4565 (synth, ir, pmr, cmr)

2-Oxabicyclo[3.1.0]hex-3-ene   O-70056
Homofuran
[6664-26-2]

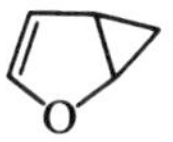

C$_5$H$_6$O   M 82.102
Bp$_{732}$ 76-78°. Polym. on exp. to light and oxygen.

Müller, E. et al, Tetrahedron Lett., 1963, 1047 (synth, uv, ir, pmr)
Herges, R., Synthesis, 1986, 1059 (synth)

9-Oxabicyclo[4.2.1]nona-2,4,7-triene, 9CI   O-70057
Updated Entry replacing O-00669
[7140-63-8]

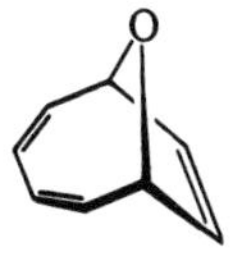

C$_8$H$_8$O   M 120.151
Bp$_{0.8}$ 35-40°. λ$_{max}$ 253 (ε 4 200), 263 (3 800) and 273 nm (2 400) (hexane).

Aumann, R. et al, J. Organomet. Chem., 1975, **85**, C4-6.
Anastassiou, A.G. et al, J. Am. Chem. Soc., 1976, **98**, 8267 (cmr)
Schmidt, H. et al, Tetrahedron, 1976, **32**, 2239.

9-Oxabicyclo[6.1.0]non-4-ene, 9CI   O-70058
1,2-Epoxy-5-cyclooctene. 1,5-Cyclooctadiene monoxide
[637-90-1]

C$_8$H$_{12}$O   M 124.182
(1RS,2SR,4Z)-form [19740-90-0]
(Z-cis)-*form*
Liq. Bp$_{40}$ 96.5°, Bp$_1$ 55°.

Traynham, J.G. et al, J. Am. Chem. Soc., 1964, **86**, 2657 (synth)
Davies, S.G. et al, J. Chem. Soc., Perkin Trans. 1, 1986, 1277 (synth, pmr, ir)
Raederstorff, D. et al, J. Org. Chem., 1987, **52**, 2337 (synth, pmr)

5-Oxabicyclo[2.1.0]pent-2-ene   O-70059
Dewar furan
[74496-19-8]

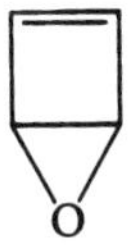

C$_4$H$_4$O   M 68.075
Transient intermed., trapped by isobenzofuran.

Pitt, I.G. et al, J. Am. Chem. Soc., 1985, **107**, 7176 (synth)

1,2,4-Oxadiazolidine-3,5-dione, 9CI   O-70060
3,5-Dioxo-1,2-oxadiazolidine. 3-Hydroxy-1,2-oxa-diazolidin-5-one. 3,5-Dihydroxy-1,2-oxadiazole
[24603-68-7]

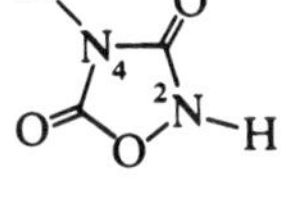

C$_2$H$_2$N$_2$O$_3$ M 102.049
Cryst. (CHCl$_3$/pet. ether). Mp 103-105°.
Dioxo-form
4-Me: [27268-68-4].
     C$_3$H$_4$N$_2$O$_3$ M 116.076
     Cryst. (CHCl$_3$/pet. ether). Mp 90-92°.
4-Et: [27280-21-3].
     C$_4$H$_6$N$_2$O$_3$ M 130.103

Bp$_{0.4}$ 102-104°.
2,4-Di-Me: [5302-11-4].
    C$_4$H$_6$N$_2$O$_3$ M 130.103
    Cryst. (EtOH). Mp 54-56°.
2-Ph: [33101-88-1].
    C$_8$H$_6$N$_2$O$_3$ M 178.147
    Cryst. (Et$_2$O/pet. ether). Mp 138-139°.
4-Ph: [5302-20-5].
    C$_8$H$_6$N$_2$O$_3$ M 178.147
    Cryst. (CHCl$_3$/pet. ether). Mp 134°.

OH-form

3-Me ether, 4-Me: [27268-89-9]. *3-Methoxy-4-methyl-*
1,2,4-oxadiazolidin-5(4H)-one.
    C$_4$H$_6$N$_2$O$_3$ M 130.103
    Cryst. (Et$_2$O). Mp 103-104°.

Zinner, G. *et al, Arch. Pharm.* (*Weinheim, Ger.*), 1969, **302**,
    691; 1970, **303**, 139; 1981, **314**, 294 (*synth, ir, deriv*)
Vsevolozhskaya, N.B. *et al, J. Org. Chem. USSR* (*Engl.*
    Transl.), 1971, **7**, 939 (*deriv*)

1,3,4-Oxadiazolidine-2,5-dione, 9CI   O-70061

[5649-86-5]

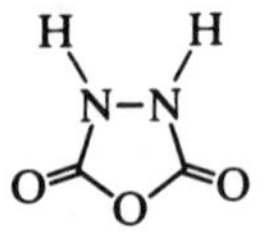

C$_2$H$_2$N$_2$O$_3$ M 102.049
Obt. as a photodecomp. prod. of Methidathion, M-00495
    Not synthesised. Sol. H$_2$O. Mp 97-98°.

Dejonckheere, W.P. *et al, J. Agric. Food Chem.*, 1974, **22**, 959
    (*ir, ms, pmr*)

2H-[1,2,4]Oxadiazolo[2,3-a]pyridine-2-   O-70062
thione, 9CI

[65478-59-3]

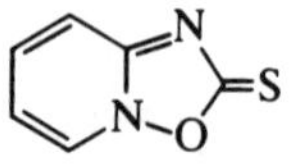

C$_6$H$_4$N$_2$OS   M 152.170
Cryst. (DMSO). Mp 134° (explodes).
▷Explodes on heating

Rousseau, D. *et al, Can. J. Chem.*, 1977, **55**, 3736 (*synth, pmr,*
    uv, ir, haz)

1,3,4-Oxadiazolo[3,2-a]pyridin-2(3H)-one   O-70063

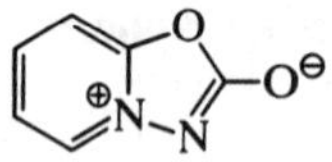

C$_6$H$_4$N$_2$O$_2$ M 136.110
Cryst. (MeOH). Mp 225°.

Hoegerle, K., *Helv. Chim. Acta*, 1958, **41**, 548 (*synth, uv, ir*)

2H-[1,2,4]Oxadiazolo[2,3-a]pyridin-2-one,   O-70064
9CI

Pyridooxadiazolone
[5678-33-1]

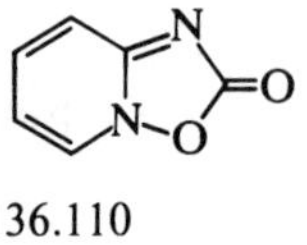

C$_6$H$_4$N$_2$O$_2$ M 136.110

Needles (EtOH). Mp 203-205° (185-187°).

Katritzky, A.R., *J. Chem. Soc.*, 1956, 2063 (*synth*)
Boyer, J.H. *et al, J. Am. Chem. Soc.*, 1957, **79**, 678 (*synth*)
Hoegerle, K., *Helv. Chim. Acta*, 1958, **41**, 548 (*synth*)

[1,4]Oxaselenino[2,3-b:5,6-b']dipyridine   O-70065
1,5-Diaza-9-oxa-10-selenaanthracene. 1,6-
Diazaphenoxaselenine

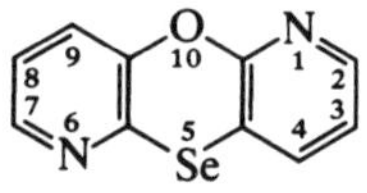

C$_{10}$H$_6$N$_2$OSe   M 249.130
Cryst. (EtOAc/cyclohexane). Mp 125-127°.

Smith, K. *et al, J. Chem. Soc., Perkin Trans. 1*, 1987, 2839
    (*synth, ir, uv, pmr, ms*)

[1,4]Oxaselenino[3,2-b:5,6-b']dipyridine   O-70066
1,8-Diaza-10-oxa-9-selenaanthracene. 1,9-
Diazaphenoxaselenine
[115855-37-3]

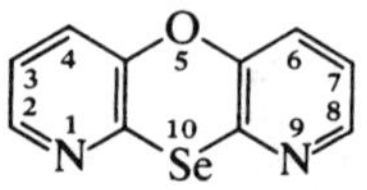

C$_{10}$H$_6$N$_2$OSe   M 249.130
Cryst. (EtOAc/cyclohexane). Mp 104.5-105.5°.

Smith, K. *et al, J. Chem. Soc., Perkin Trans. 1*, 1987, 2839
    (*synth, ir, uv, pmr, ms*)

[1,4]Oxaselenino[3,2-b:5,6-c']dipyridine   O-70067
17-Diaza-10-oxa-9-selenaanthracene. 1,8-
Diazaphenoxaselenine
[115857-55-1]

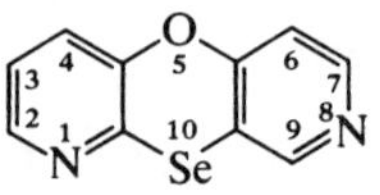

C$_{10}$H$_6$N$_2$OSe   M 249.130
Cryst. solid (Et$_2$O). Mp 96-97°.
8-Oxide: [115855-39-5].
    C$_{10}$H$_6$N$_2$O$_2$Se   M 265.130
    Cryst. (CHCl$_3$/hexane). Mp 223-224°.

Smith, K. *et al, J. Chem. Soc., Perkin Trans. 1*, 1987, 2839
    (*synth, ir, uv, pmr, ms*)

[1,4]Oxaselenino[3,2-b:6,5-c']dipyridine   O-70068
1,6-Diaza-10-oxa-9-selenaanthracene. 1,7-
Diazaphenoxaselenine
[115055-41-9]

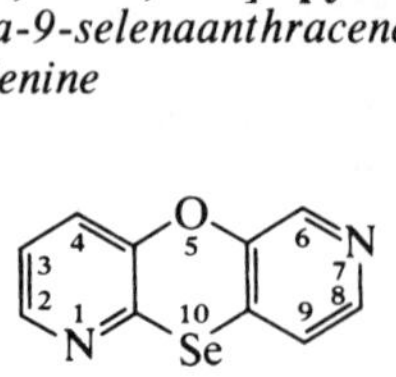

C$_{10}$H$_6$N$_2$OSe   M 249.130
Cryst. (Et$_2$O). Mp 125-126°.
7-Oxide: [115855-40-8].
    C$_{10}$H$_6$N$_2$O$_2$Se   M 265.130
    Cryst. (CHCl$_3$/hexane). Mp 187-189°.

Smith, K. *et al, J. Chem. Soc., Perkin Trans. 1*, 1987, 2839
    (*synth, ir, uv, pmr, ms*)

3-Oxatetracyclo[5.3.1.1²,⁶.0⁴,⁹]dodecane   O-70069

*Octahydro-2,7:3,6-dimethano-2H-1-benzopyran, 9CI.
Oxaiceane. 3-Oxawurtzitane*
[55092-18-7]

$C_{11}H_{16}O$ M 164.247
Mp 312-313°.

Klaus, R.O. *et al, Helv. Chim. Acta*, 1974, **57**, 2517 (*synth, ir,
    pmr, cmr*)
Hamon, D.P.G. *et al, Aust. J. Chem.*, 1977, **30**, 589 (*synth, ir,
    pmr, cmr, ms*)

1,2,4-Oxazolidine-3,5-dione   O-70070

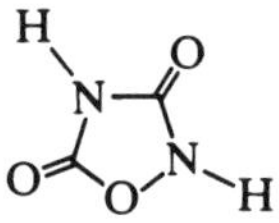

$C_2H_2N_2O_3$ M 102.049
Cryst. + 1H_2O.

Begum, A.J. *et al, Acta Crystallogr., Sect. C*, 1988, **44**, 195
    (*cryst struct*)

2-(2-Oxazolyl)pyridine, 9CI   O-70071

2-(2-Pyridyl)oxazole

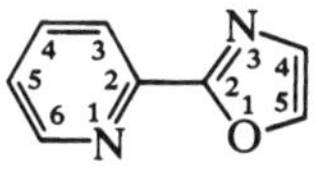

$C_8H_6N_2O$ M 146.148
Bp$_{0.1}$ 70-75°.

Dadkhah, M. *et al, Helv. Chim. Acta*, 1962, **45**, 375 (*synth*)

2-(5-Oxazolyl)pyridine, 9CI   O-70072

5-(2-Pyridyl)oxazole
[70380-73-3]
$C_8H_6N_2O$ M 146.148
Pale-yellow oil. Bp$_{16}$ 120-121°, Bp$_{0.15}$ 95-98°.
Picrate: Yellow cryst. (EtOH). Mp 165-167°.

Saikachi, H. *et al, Chem. Pharm. Bull.*, 1979, **27**, 793 (*synth,
    uv, pmr*)

3-(2-Oxazolyl)pyridine   O-70073

2-(3-Pyridyl)oxazole
$C_8H_6N_2O$ M 146.148
Cryst. (diisopropyl ether). Mp 72-75° (81-82°).
B,HCl: Cryst. (2-propanol). Mp 187-189°.
Picrate: Cryst. (EtOH). Mp 174-177°.

Merz, K.W. *et al, Arch. Pharm. (Weinheim, Ger.)*, 1960, **293**,
    92 (*synth*)
Dadkhah, M. *et al, Helv. Chim. Acta*, 1962, **45**, 375 (*synth*)

3-(5-Oxazolyl)pyridine, 9CI   O-70074

5-(3-Pyridyl)oxazole
[70380-74-4]
$C_8H_6N_2O$ M 146.148

Pale-yellow needles (cyclohexane/pet. ether). Mp 64-
    65°.

Saikachi, H. *et al, Chem. Pharm. Bull.*, 1979, **27**, 793 (*synth,
    uv, pmr*)

4-(2-Oxazolyl)pyridine, 9CI   O-70075

2-(4-Pyridyl)oxazole
[5998-92-5]
$C_8H_6N_2O$ M 146.148
Plates (Et$_2$O). Mp 104-105°.
B,HCl: Cryst. (2-propanol). Mp 221° dec.
Picrate: Cryst. (EtOH). Mp 174-178°.

Merz, K.W. *et al, Arch. Pharm. (Weinheim, Ger.)*, 1960, **293**,
    92 (*synth*)
Brufani, M. *et al, Gazz. Chim Ital.*, 1961, **91**, 767 (*synth, cryst
    struct*)
Dadkhah, M. *et al, Helv. Chim. Acta*, 1962, **45**, 375 (*synth*)

4-(5-Oxazolyl)pyridine, 9CI   O-70076

5-(4-Pyridyl)oxazole
[70380-75-5]
$C_8H_6N_2O$ M 146.148
Pale-yellow needles (cyclohexane/pet. ether). Mp 128-
    129°.

Saikachi, H. *et al, Chem. Pharm. Bull.*, 1979, **27**, 793 (*synth,
    uv, pmr*)

2-Oxepanethione, 9CI   O-70077

ε-Thionocaprolactone
[72037-38-8]

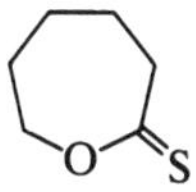

$C_6H_{10}OS$ M 130.204
Khouri, F.F. *et al, J. Am. Chem. Soc.*, 1986, **108**, 6683 (*synth,
    nmr, ir*)

2-Oxiranylfuran, 9CI   O-70078

*Furfurylethylene oxide. 2-Furyloxirane. 2-Vinylfuran
oxide*
[2745-17-7]

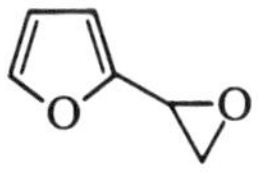

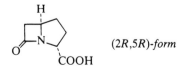

$C_6H_6O_2$ M 110.112
(±)-form
Liq. with peculiar sharp odor. d^{20} 1.411. Bp$_{14}$ 56°. n_D^{20}
    1.4845.

Novitskii, K.Yu. *et al, J. Gen. Chem. USSR (Engl. Transl.)*,
    1963, **33**, 1031 (*synth*)
Borredon, E. *et al, Bull. Soc. Chim. Fr.*, 1987, 1073 (*synth*)

7-Oxo-1-azabicyclo[3.2.0]heptane-2-car-   O-70079
boxylic acid, 9CI

Carbapenam-3-carboxylic acid

(2R,5R)-form

$C_7H_9NO_3$ M 155.153

(2R,5R)-form [112419-10-0]
Isol. from *Serratia* sp. and *Erwinia* sp.
(2S,5R)-form [112283-40-6]
From *S.* sp. and *E.* sp.

Bateson, J.H. *et al, J. Chem. Soc., Perkin Trans. 1*, 1981, 3242 (*esters*)
Berryhill, S.R. *et al, J. Org. Chem.*, 1983, **48**, 158 (*esters*)
Bycroft, B.W. *et al, J. Chem. Soc., Chem. Commun.*, 1987, 1623 (*isol*)

3-Oxobutanethioic acid, 9CI   O-70080

Thioacetoacetic acid. Acetothioacetic acid

$$H_3CCOCH_2COSH \rightleftharpoons H_3CCOCH_2CSOH$$

$C_4H_6O_2S$ M 118.150
Enol forms also possible. Esters exist as mixt. of keto and enol-forms.

(OH)-form
Et ester: [41500-02-1]. O-*Ethyl 3-oxobutanethioate.*
$C_6H_{10}O_2S$ M 146.204
Light-yellow liq. Bp_{21} 93-94°.

(SH)-form
Et ester: [3075-23-8]. S-*Ethyl 3-oxobutanethioate.*
$C_6H_{10}O_2S$ M 146.204
Bp_{1-2} 59-61°.
Isopropyl ester: [15780-62-8].
$C_7H_{12}O_2S$ M 160.231
Bp_{2-3} 68-70°.
Butyl ester: [15780-63-9]. S-*Butyl 3-oxobutanethioate.*
$C_8H_{14}O_2S$ M 174.257
Liq.
tert-*Butyl ester:* [15925-47-0]. S-tert-*Butyl 3-oxobutanethioate.*
$C_8H_{14}O_2S$ M 174.257
Pale-red oil.
Ph ester: [40053-29-0]. S-*Phenyl 3-oxobutanethioate. S-Acetoacetylthiophenol.*
$C_{10}H_{10}O_2S$ M 194.248
Liq. Dec. on dist.

Sheehan, J.C. *et al, J. Am. Chem. Soc.*, 1955, **77**, 4875 (*synth*)
Kurgane, B.V. *et al, J. Org. Chem. USSR*, 1974, **10**, 2306; *Eng. Trans.*, p. 2321 (*synth, ir, tautom*)
Bridges, A.J. *et al, J. Chem. Soc., Perkin Trans. 1*, 1975, 1603 (*synth, ir, pmr, tautom*)
Yaggi, N.F. *et al, J. Chem. Soc., Chem. Commun.*, 1977, 609 (*synth, ir, pmr, bibl*)
Kobuke, Y. *et al, Tetrahedron Lett.*, 1978, 367 (*synth, ir, pmr*)
Booth, P.M. *et al, J. Chem. Soc., Perkin Trans. 1*, 1987, 121 (*synth, pmr, ir*)

6-Oxo-3,13-clerodadien-15-oic acid   O-70081

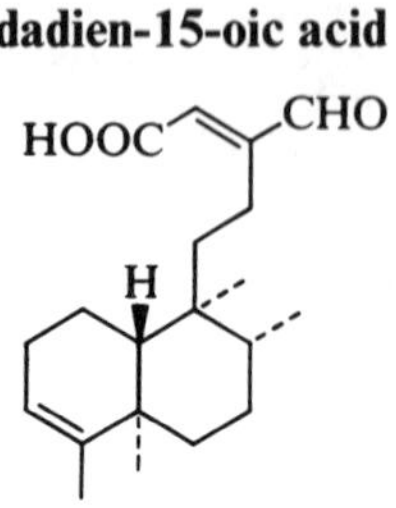

$C_{20}H_{30}O_3$ M 318.455
Constit. of *Polyalthia longifolia*. Gum. $[\alpha]_D^{26}$ −70.58° (c, 0.01 in MeOH).

Phadnis, A.P. *et al, Phytochemistry*, 1988, **27**, 2899.

4-Oxo-1-cycloheptene-1-carboxylic acid, 9CI   O-70082

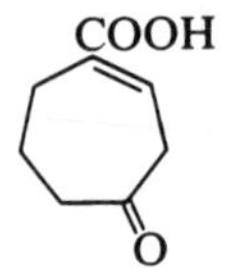

$C_8H_{10}O_3$ M 154.165
Me ester: [110129-67-4].
$C_9H_{12}O_3$ M 168.192
Yellow oil.

Van Beek, G. *et al, Tetrahedron*, 1986, **42**, 5111 (*deriv, synth, pmr, ir*)

3-Oxocyclopentaneacetic acid, 9CI, 8CI   O-70083

Updated Entry replacing O-30087
[3128-05-0]

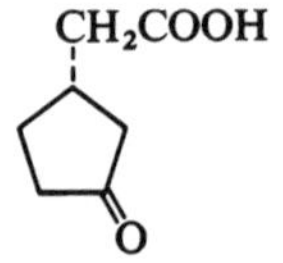

$C_7H_{10}O_3$ M 142.154
(R)-form
$Bp_{0.01}$ 90°. $[\alpha]_D$ +47.6° (c, 1.75 in CHCl$_3$).
(S)-form [2630-37-7]
Cryst. (Me$_2$CO). Mp 169-171°. $[\alpha]_D^{23}$ −115.5° (c, 1.42 in CHCl$_3$).
Me ester: [2630-38-8].
$C_8H_{12}O_3$ M 156.181
Oil. Bp_3 100°. $[\alpha]_D$ −119° (c, 0.89 in CHCl$_3$). ca. 100% opt. pure.
(±)-form
Bp_3 152-157°, $Bp_{0.001}$ 76-78°.
Phenylhydrazone: Mp 139-141° dec.
2,4-Dinitrophenylhydrazone: Mp 168-169°.
Et ester:
$C_9H_{14}O_3$ M 170.208
Mp 135-137°.
Et ester, semicarbazone: Mp 147.5-148.5°.

Demole, E. *et al, Helv. Chim. Acta*, 1962, **45**, 692 (*synth*)
Hill, R.K. *et al, Tetrahedron*, 1965, **21**, 1501 (*resoln, abs config*)
Harmon, R.E. *et al, J. Chem. Soc. (C)*, 1971, 3645 (*synth*)
Kuritani, H. *et al, J. Org. Chem.*, 1979, **44**, 452 (*synth, ir, pmr*)
Brehm, R. *et al, Helv. Chim. Acta*, 1987, **70**, 1981 (*synth*)
Hua, D.H. *et al, J. Org. Chem.*, 1987, **52**, 719 (*synth, pmr, cmr*)

2-Oxocyclopentanecarboxylic acid, 9CI   O-70084

Updated Entry replacing O-00800
Cyclopentanone-2-carboxylic acid
[50882-16-1]

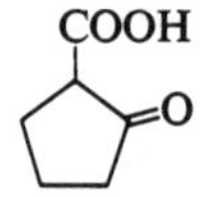

$C_6H_8O_3$ M 128.127
(±)-form
Me ester: [10472-24-9].
$C_7H_{10}O_3$ M 142.154
Bp_{19} 105°.

Me ester, semicarbazone: Cryst. Mp 167°.
Et ester: [53229-92-8]. *2-Ethoxycarbonylcyclopentan-one. Dieckmann ester.*
$C_8H_{12}O_3$ M 156.181
Bp_6 86-87°.
Nitrile: [2941-29-9]. *2-Cyanocyclopentanone.*
C_6H_7NO M 109.127
$Bp_{1.0}$ 93°.

Org. Synth., Coll. Vol., **2**, 116.
Rhoads, S.J. *et al, Tetrahedron*, 1963, **19**, 1625.
Bloomfield, J.J. *et al, Tetrahedron Lett.*, 1964, 2273.
Zupancic, B.G. *et al, Monatsh. Chem.*, 1967, **98**, 369.
Korobitsyna, I.K. *et al, Zh. Org. Khim.*, 1976, **12**, 1251; *CA*, **85**, 142560.
Coates, R.M. *et al, J. Am. Chem. Soc.*, 1987, **109**, 1160 (*nitrile*)

2-Oxo-1,3-cyclopentanediglyoxylic acid,   O-70085
8CI

2-Oxocyclopentanediylbisglyoxylic acid

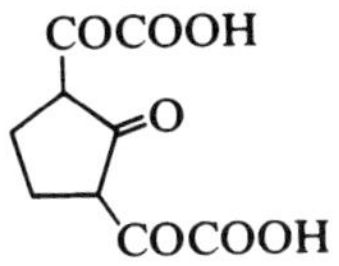

COCOOH

O

COCOOH

$C_9H_8O_7$ M 228.158
Di-Et ester: [22358-21-0].
$C_{13}H_{16}O_7$ M 284.265
Complexing agent for metals. Yellow cryst. (EtOH). Mp 108-109°.

Graddon, D.P. *et al, Aust. J. Chem.*, 1969, **22**, 505.

α-Oxo-3-cyclopenteneacetaldehyde, 9CI   O-70086

3-Cyclopentene-1-glyoxal
[80344-66-7]

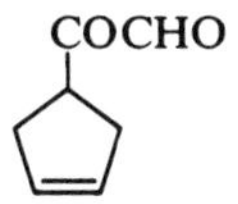

COCHO

$C_7H_8O_2$ M 124.139
2,4-Dinitrophenylhydrazone: Mp 248-250° dec.

Cook, J.M. *et al, J. Org. Chem.*, 1986, **51**, 2436 (*synth, ir, pmr, cmr*)

2-Oxodecanal   O-70087

2-Hydroxy-2-dodecenal
[110045-38-0]

$H_3C(CH_2)_9COCHO$

$C_{12}H_{22}O_2$ M 198.305
Wax.

Amon, C.M. *et al, J. Org. Chem.*, 1987, **52**, 4851 (*synth, pmr, cmr, ms, uv*)

4-Oxodecanal   O-70088

[43160-78-7]

$H_3C(CH_2)_5COCH_2CH_2CHO$

$C_{10}H_{18}O_2$ M 170.251
Solid (hexane). Mp 66-68°.

Larson, G.L. *et al, J. Org. Chem.*, 1986, **51**, 2039 (*synth, ir, pmr, cmr*)

9-Oxo-2-decenoic acid, 9CI   O-70089

Updated Entry replacing O-00812
[2575-01-1]

$H_3CCO(CH_2)_5CH{=}CHCOOH$

$C_{10}H_{16}O_3$ M 184.235
(*E*)-*form* [334-20-3]
Queen substance
Queen substance of honey bee *Apis mellifera*. Cryst. (MeOH aq. or Me_2CO). Mp 54-55.5°. $Bp_{0.001}$ 37-39°.
Me ester: [1189-64-6].
$C_{11}H_{18}O_3$ M 198.261
$Bp_{0.6}$ 107.5-108°.
2,4-Dinitrophenylhydrazone: Mp 126-130°.
Semicarbazone: Cryst. (MeOH aq.). Mp 165-168°.
(*Z*)-*form* [29282-32-4]
Oil. $Bp_{0.2}$ 110-150°.

Van der Plas, H.C. *et al, Recl. Trav. Chim. Pays-Bas*, 1964, **83**, 701 (*synth, bibl*)
Cromer, D.T. *et al, Acta Crystallogr., Sect. B*, 1972, **28**, 2128 (*cryst struct*)
Trost, B.M. *et al, J. Org. Chem.*, 1975, **40**, 148 (*synth*)
Tsuji, J. *et al, Tetrahedron Lett.*, 1977, 2267 (*synth*)
Lombardi, L. *et al, Synthesis*, 1978, 131.
Subramaniam, C.S. *et al, Indian J. Chem., Sect. B*, 1978, **16**, 318 (*synth, ir, pmr*)
Hase, T.A. *et al, Synth. Commun.*, 1979, **9**, 63 (*synth*)
Tamaru, Y. *et al, Tetrahedron*, 1979, **35**, 329 (*synth, ir, pmr*)
Villemin, D., *Chem. Ind.* (*London*), 1986, 69 (*synth*)
Dhokte, V.P. *et al, Synth. Commun.*, 1987, **17**, 355 (*synth*)

2-Oxo-3,10(14),11(13)-guaiatrien-12,6-olide   O-70090

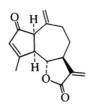

$C_{15}H_{16}O_3$ M 244.290
6α-form
2-Oxodesoxyligustrin
Constit. of *Stevia yacomensis*. Cryst. Mp 185°. $[\alpha]_D^{24}$ +250° (c, 0.62 in $CHCl_3$).

Zdero, C. *et al, Phytochemistry*, 1988, **27**, 2835.

5-Oxohexanal   O-70091

γ-Acetylbutyraldehyde. Soyanal
[505-03-3]

$H_3CCOCH_2CH_2CH_2CHO$

$C_6H_{10}O_2$ M 114.144
Poss. obt. from soya. d^{18} 1.024. Bp_{18} 94-95°. n_D^{28} 1.4615 (n_D^{23} 1.4320).
Mono-2,4-dinitrophenylhydrazone: Yellow leaflets. Mp 110-111° (107-108° dec.).
Bis-2,4-dinitrophenylhydrazone: Orange needles. Mp 142-143° dec., Mp 184°.

Nakajima, M. *et al, J. Chem. Soc. Jpn.*, 1949, **70**, 49; *CA*, **45**, 7015c.
Colonge, J. *et al, Bull. Soc. Chim. Fr.*, 1959, 370 (*synth*)
Garnick, R.L. *et al, J. Am. Chem. Soc.*, 1978, **100**, 4213 (*synth, pmr*)

5-Oxo-3-hexenoic acid, 9CI   O-70092
Updated Entry replacing O-00857

$$H_3CCOCH{=}CHCH_2COOH$$

$C_6H_8O_3$ M 128.127

(E)-form
2,4-Dinitrophenylhydrazone: [28845-69-4]. Orange needles. Mp 177.5-178°.
(Z)-form [113219-94-6]
No phys. props. reported.

Harris, C.M. *et al, J. Org. Chem.*, 1971, **36**, 2181 (*synth*)
Dieter, R.K. *et al, J. Org. Chem.*, 1988, **53**, 2031 (*synth, pmr*)

10-Oxo-6-isodaucen-14-al   O-70093
*1,2,3,3a,6,7,8,8a-Octahydro-8a-methyl-3-(1-methy-
lethyl)-8-oxo-5-azulenecarboxaldehyde, 9CI*
[100324-71-8]

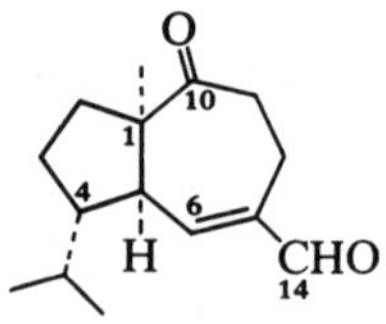

$C_{15}H_{22}O_2$ M 234.338
Constit. of *Chromolaena laevigata*. Oil.

10α-Alcohol: [100428-90-8]. *10α-Hydroxy-6-isodau-
cen-14-al.*
$C_{15}H_{24}O_2$ M 236.353
Constit. of *C. laevigata*. Apparently the enantiomer of
Aphanamol II, A-40163 although no opt. rotn. is given.

Misra, L.N. *et al, Tetrahedron*, 1985, **41**, 5353.

7-Oxo-8(14),15-isopimaradien-18-oic acid   O-70094
7-Oxosandaracopimaric acid

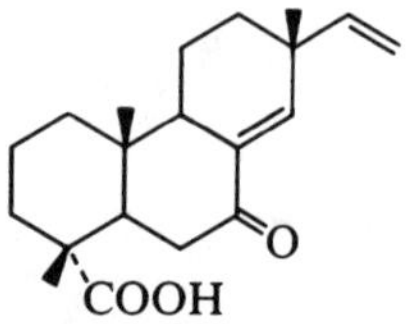

$C_{20}H_{28}O_3$ M 316.439
Constit. of *Salvia microphylla*.

*Me ester: Methyl 7-oxo-8(14),15-isopimaradien-18-
oate.*
$C_{21}H_{30}O_3$ M 330.466
Constit. of *S. microphylla*. Cryst. (EtOAc/pet. ether).
Mp 87-90°. $[\alpha]_D^{20}$ −40° (c, 0.15 in CHCl₃).

Esquivel, B. *et al, J. Nat. Prod.*, 1987, **50**, 738.

7-Oxo-8,13-labdadien-15-oic acid   O-70095
Rhinocerotinoic acid
[115374-33-9]

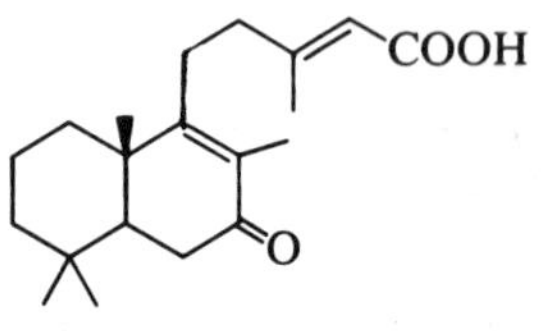

$C_{20}H_{30}O_3$ M 318.455
Constit. of *Elytropappus rhinocerotis*. Cryst. Mp 186-
189°. $[\alpha]_D^{22.5}$ +42° (c, 1 in CHCl₃).

Dekker, T.G. *et al, S. Afr. J. Chem.*, 1988, **41**, 33.

3-Oxo-8,24-lanostadien-26-oic acid   O-70096
Anwuweizonic acid

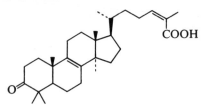

$C_{30}H_{46}O_3$ M 454.692
Constit. of *Schisandra propinqua*. Cryst. (Et₂O/pet.
ether). Mp 126-128°. $[\alpha]_D^{15}$ +63.8° (c, 0.273 in
CHCl₃).

Liu, J.-S. *et al, Can. J. Chem.*, 1988, **66**, 414.

4-Oxo-2-nonenal, 9CI   O-70097
[89186-13-0]

$$H_3C(CH_2)_4COCH{=}CHCHO$$

$C_9H_{14}O_2$ M 154.208

(E)-form [103560-62-9]
Yellow, low melting solid. Stable at −10°.

Gunn, B.P., *Heterocycles*, 1985, **23**, 3061 (*synth, pmr, cmr, ir,
ms*)
Comasseto, J.V. *et al, Synthesis*, 1987, 146 (*synth*)

4-Oxo-2-pentenoic acid, 9CI, 8CI   O-70098
Updated Entry replacing O-00927
*3-Acetylacrylic acid. 5-Hydroxy-5-methyl-2(5H)-
furanone*
[4743-82-2]

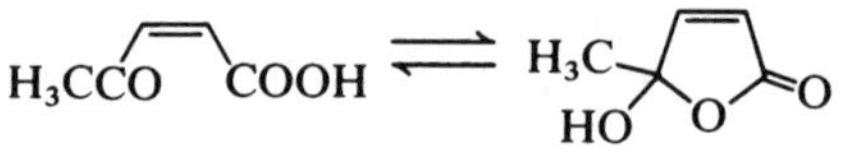

$C_5H_6O_3$ M 114.101
Prod. of metab. of phenol by *Trichosporon cutaneum*.
▷SB3360000.

(E)-form [2833-28-5]
Needles (CHCl₃), plates (C₆H₆). Mod. sol. H₂O, sol.
EtOH, Et₂O, insol. pet. ether. Mp 125°. pK_a* 5.35
(MCS). Sublimes.

Me ester: [2833-24-1].
$C_6H_8O_3$ M 128.127
Plates. Mp 60.5°. Bp₁₄ 85°.
Phenylhydrazone: Yellow needles (H₂O). Mp 169°.
Me ester, semicarbazone: Mp 196°.
Et ester:
$C_7H_{10}O_3$ M 142.154
Bp₁₂ 86-87°.

(Z)-form [4188-88-9]
Mp 33.8-36.5°. Has pseudo-struct. The anion is acyclic.
Polymerises in light.

Et ester: [2833-26-3]. Bp₀.₅ 55°. n_D^{25} 1.4432. Has
pseudo-struct.

Wolff, L., *Justus Liebigs Ann. Chem.*, 1891, **264**, 229 (*synth*)
Scheffold, R. *et al, Helv. Chim. Acta*, 1967, **50**, 798 (*synth, uv,
ir, pmr*)
Kuchar, M., *Collect. Czech. Chem. Commun.*, 1968, **33**, 880
(*synth*)
Seltzer, S. *et al, J. Org. Chem.*, 1968, **33**, 2708 (*synth, pmr, uv,
struct, bibl*)

Badovskaya, L.A. *et al*, *Zh. Org. Khim.*, 1979, **15**, 50 (*synth, ir, uv*)
Gaal, A. *et al*, *J. Bacteriol.*, 1979, **137**, 13 (*isol*)

(2-Oxo-2-phenylethyl)carbamic acid, 9CI   O-70099
Phenacylcarbamic acid, 8CI

$$PhCOCH_2NHCOOH$$

$C_9H_9NO_3$ M 179.175
Me ester: [22741-45-3]. *Methyl N-phenacylcarbamate.*
$C_{10}H_{11}NO_3$ M 193.202
Plates (AcOEt). Mp 97.4-97.8°.
Et ester: [59840-70-9]. *Ethyl N-phenacylcarbamate.*
$C_{11}H_{13}NO_3$ M 207.229
Mp 120-122°.

Fujita, S. *et al*, *Tetrahedron*, 1970, **26**, 4347 (*synth, ir, nmr*)
Lociuro, S. *et al*, *Tetrahedron Lett.*, 1983, **24**, 593 (*synth, pmr, cmr*)

(2-Oxopropylidene)propanedioic acid, 9CI   O-70100
(Acetylmethylene)malonic acid

$$H_3CCOCH{=}C(COOH)_2$$

$C_6H_6O_5$ M 158.110
Di-Me ester: [64677-33-4].
$C_8H_{10}O_5$ M 186.164
Highly reactive in Michael additions, ene-reacns. and Diels-alder reacns. Bp₃ 110-111°.

Ouali, M.S. *et al*, *Synthesis*, 1977, 626 (*synth*)
Cameron, D.W. *et al*, *Aust. J. Chem.*, 1987, **40**, 1831 (*props*)

α-Oxo-1H-pyrrole-3-acetic acid   O-70101
Pyrrole-3-glyoxylic acid

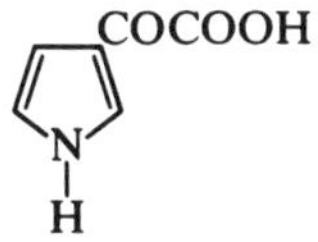

$C_6H_5NO_3$ M 139.110
Cryst. (2-propanol/Et₂O/pet. ether). Mp 154-156° dec.
Dipropylamide: Mp 83-85°.

Demopoulos, V.J., *J. Heterocycl. Chem.*, 1988, **25**, 635 (*synth, ir, pmr*)

5-Oxo-2-pyrrolidinecarboxylic acid   O-70102
Updated Entry replacing O-10080
5-Oxoproline, 9CI. Glutiminic acid. Pyroglutamic acid. Glutimic acid. 2-Pyrrolidone-5-carboxylic acid. α-Aminoglutaric acid lactam. Glutamic acid lactam. **Pidolic acid, INN**

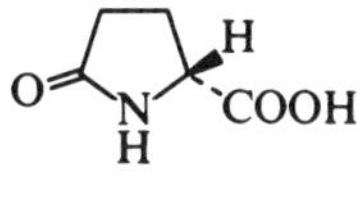

(R)-form
Absolute
configuration

$C_5H_7NO_3$ M 129.115
(R)-form [4042-36-8]
D-form
Cryst. (H₂O). Mp 182°. [α]_D +10.7° (H₂O).
Et ester: [68766-96-1].
$C_7H_{11}NO_3$ M 157.169

Mp 50°.
Amide:
$C_5H_8N_2O_2$ M 128.130
Mp 165°.
(S)-form [98-79-3]
L-form
Found in vegetables, fruits, grasses and molasses. Used in resoln. of amines. Cryst. (H₂O). Mp 156-157°. [α]$_D^{20}$ −11.45° (c, 4.44 in H₂O).
Ca salt (2:1): [31377-05-6]. *Calciopor. Calcium pidolate. Calpyrodil. Efical. Lircal.* Calcium replenisher.
Mg salt (2:1): [62003-27-4]. *Bomag. Magnesone. Solumag. MAG 2.* Magnesium source.
Et ester: [7149-65-7]. Bp₃ 161°. [α]_D −8.6° (H₂O).
Amide: [16395-57-6]. Mp 165°.
Anilide:
$C_{11}H_{12}N_2O_2$ M 204.228
Plates (EtOH). [α]$_D^{15}$ +17.9° (80% EtOH).
(±)-form [149-87-1]
Mp 178-179°.
Me ester: [54571-66-3].
$C_6H_9NO_3$ M 143.142
Mp 21-23°. Bp₂₅ 180°.
Et ester: Mp 52-54°. Bp₀.₂ 141-143°.
tert-Butyl ester:
$C_9H_{15}NO_3$ M 185.222
Mp 109-110°.
Benzyl ester:
$C_{12}H_{13}NO_3$ M 219.240
Cryst. (Et₂O). Mp 50-52°. Bp₀.₁ 185°.
Me ester, N-formyl:
$C_7H_9NO_4$ M 171.152
Cryst. (Et₂O/THF). Mp 48°. Bp₀.₁ 112°.
Me ester, N-Ac:
$C_8H_{11}NO_4$ M 185.179
Cryst. (Et₂O/pet. ether). Mp 41°.
Me ester, N-benzoyl:
$C_8H_{11}NO_4$ M 185.179
Cryst. (MeOH). Mp 141-143°.
Amide: [5626-52-8]. Mp 220-221°.

Hardegger, E. *et al*, *Helv. Chim. Acta*, 1955, **38**, 312 (*abs config*)
Gibian, H. *et al*, *Justus Liebigs Ann. Chem.*, 1961, **640**, 145 (*synth, bibl*)
Japan. Pat., 70 11 144, (*1970*); *CA*, **73**, 35763 (*synth*)
Orlowski, M. *et al*, *The Enzymes*, 1971, **4**, Academic Press, N.Y., 3rd Ed..
Cervinka, O. *et al*, *Collect. Czech. Chem. Commun.*, 1973, **38**, 897 (*cd*)
Pattabhi, V. *et al*, *J. Chem. Soc., Perkin Trans. 2*, 1974, 1085 (*cryst struct*)
Voelter, W. *et al*, *Monatsh. Chem.*, 1974, **105**, 1110 (*cmr*)
Taira, Z. *et al*, *Acta Crystallogr., Sect. B*, 1977, **33**, 3823 (*cryst struct*)
Hardy, P.M., *Synthesis*, 1978, 290 (*synth*)
Schmidt, U. *et al*, *Synthesis*, 1978, 752 (*synth*)
Silverman, R.B. *et al*, *J. Org. Chem.*, 1980, **45**, 815 (*deriv*)
Abraham, G.N. *et al*, *Mol. Cell Biochem.*, 1981, **38**, 181 (*rev*)
Amstutz, R. *et al*, *J. Med. Chem.*, 1985, **28**, 1760 (*deriv*)
Rigo, B. *et al*, *J. Heterocycl. Chem.*, 1988, **25**, 49 (*esters*)
Martindale, The Extra Pharmacopoeia, 28th/29th Eds., 1982/1989, Pharmaceutical Press, London, 13256.

Handle all chemicals with care

1-Oxo-1,2,3,4-tetrahydrocarbazole   O-70103
2,3,4,9-Tetrahydro-1H-carbazol-1-one, 9CI. 3,4-Dihy-drocarbazol-1(2H)-one, 8CI

[3456-99-3]

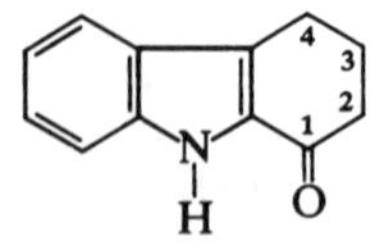

$C_{12}H_{11}NO$ M 185.225
Large flat needles (EtOH). Mp 169-170°.
Azine: Citron yellow cryst. solid. Mp 258-260° dec.
(E)-*Oxime:* [23240-52-0].
   $C_{12}H_{12}N_2O$ M 200.240
   Mp 129-136°.
(Z)-*oxime:* [23240-49-5].
   $C_{12}H_{12}N_2O$ M 200.240
   Mp 175.5-176.5°.
N-*Me:* [1485-19-4].
   $C_{13}H_{13}NO$ M 199.252
   Needles (C_6H_6/pet. ether). Mp 102.5-103°.
N-*Me, oxime:* [3449-55-6].
   $C_{13}H_{14}N_2O$ M 214.266
   Mp 184-185°.

Coffey, S., *Recl. Trav. Chim. Pays-Bas*, 1923, **42**, 528 (*synth*)
Teuber, H.-J. *et al, Justus Liebigs Ann. Chem.*, 1964, **671**, 127 (*deriv, synth*)
Dolby, L.J. *et al, J. Org. Chem.*, 1965, **30**, 1550 (*deriv*)
Hester, J.B., *J. Org. Chem.*, 1970, **35**, 875 (*oximes*)
Bailey, A.S. *et al, J. Chem. Soc. (C)*, 1971, 2479 (*derivs*)

2-Oxo-1,2,3,4-tetrahydrocarbazole   O-70104
1,3,4,9-Tetrahydro-2H-carbazol-2-one, 9CI. 3,4-Dihy-drocarbazol-2(1H)-one, 8CI

[40429-00-3]
$C_{12}H_{11}NO$ M 185.225
Cryst. (cyclohexane). Mp 131-133°.

Oxime:
   $C_{12}H_{12}N_2O$ M 200.240
   Cryst. (EtOH). Mp 192° dec.

Teuber, H.-J. *et al, Justus Liebigs Ann. Chem.*, 1964, **671**, 127 (*synth*)

3-Oxo-1,2,3,4-tetrahydrocarbazole   O-70105
1,2,4,9-Tetrahydro-3H-carbazol-3-one, 9CI. 2,4-Dihy-drocarbazol-3(1H)-one, 8CI

[51145-61-0]
$C_{12}H_{11}NO$ M 185.225
Cryst. (EtOH). Mp 156°.

Oxime:
   $C_{12}H_{12}N_2O$ M 200.240
   Cryst. (EtOH). Mp 191°.

Teuber, H.-J. *et al, Justus Liebigs Ann. Chem.*, 1964, **671**, 127 (*synth*)

4-Oxo-1,2,3,4-tetrahydrocarbazole   O-70106
1,2,3,9-Tetrahydro-4H-carbazol-4-one, 9CI. 2,3-Dihy-drocarbazol-4(1H)-one, 8CI

[15128-52-6]
$C_{12}H_{11}NO$ M 185.225
Prisms (EtOH). Mp 223°.

2,4-Dinitrophenylhydrazone: Dark-maroon needles (AcOH). Mp 288° dec.
Picrate: Bright-red rhombs (EtOH). Mp 167°.
N-*Ac:*
   $C_{14}H_{13}NO_2$ M 227.262
   Cryst. (EtOH aq.). Mp 136°.
Oxime: [14362-50-6].
   $C_{12}H_{12}N_2O$ M 200.240
   Cryst. Mp 208° dec.

Clemo, G.R. *et al, J. Chem. Soc.*, 1951, 700 (*synth*)
Teuber, H.-J. *et al, Justus Liebigs Ann. Chem.*, 1964, **671**, 127 (*deriv, synth*)

Oxysporone   O-70107
3a,7a-Dihydro-4-hydroxy-4H-furo[2,3-b]pyran-2(3H)-one, 9CI

[73343-08-5]

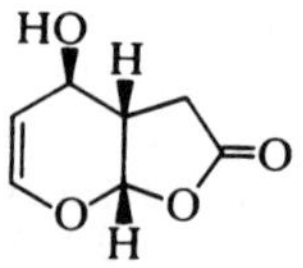

$C_7H_8O_4$ M 156.138
Prod. by *Fusarium oxysporum*. Antibiotic. Oil.

Adesogan, E.K. *et al, Phytochemistry*, 1979, **18**, 1886 (*isol, struct*)

P

Cambie, R.C. *et al*, *J. Nat. Prod.*, 1987, **50**, 948.

Pachytriol   P-70001

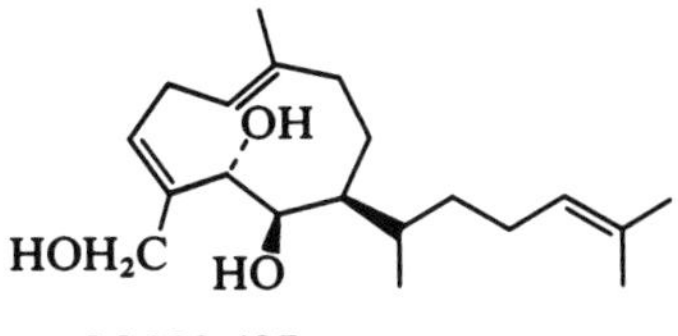

$C_{20}H_{34}O_3$ M 322.487
Constit. of *Dictyota dichotoma*. Oil. $[\alpha]_D$ −46.7° (c, 1.49 in CHCl$_3$).

González, A.G *et al*, *J. Nat. Prod.*, 1987, **50**, 500.

Paldimycin   P-70002
Antibiotic 273a$_1$

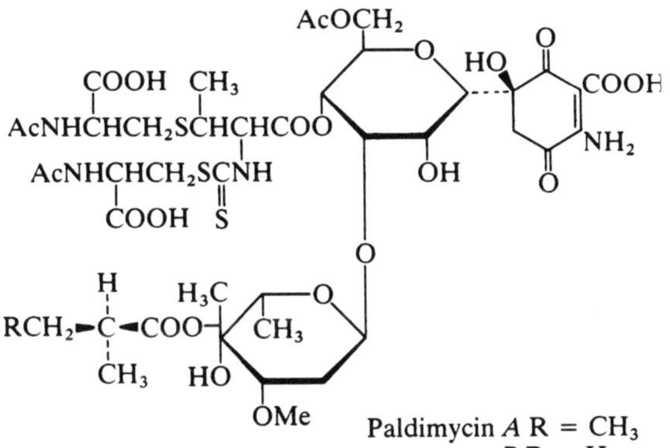

Paldimycin *A* R = CH$_3$
     B R = H

Glycoside-type antibiotic complex. Prod. by *Streptomyces paulus*. Active against gram-positive bacteria. Isol. with Antibiotic 273a$_2$, A-70218 Related also to Paulomycin.

Paldimycin A [101411-70-5]
Antibiotic 273a$_{1\alpha}$
     $C_{44}H_{64}N_4O_{23}S_3$ M 1113.183
     $[\alpha]_D$ −31° (c, 0.9 in MeOH). Dec. begins at ca. 120°.
Paldimycin B [101411-71-6]
Antibiotic 273a$_{1\beta}$
     $C_{43}H_{62}N_4O_{23}S_3$ M 1099.156
     $[\alpha]_D^{25}$ −35° (c, 0.9 in MeOH). Dec. begins at ca. 120°.

Argoudelis, A.D. *et al*, *J. Antibiot.*, 1987, **40**, 408, 419 (*isol, struct, synth*)
Brumfitt, W. *et al*, *J. Antimicrob. Chemother.*, 1987, **19**, 405 (*props*)
Rolston, K.V. *et al*, *Antimicrob. Agents Chemother.*, 1987, **31**, 653 (*props*)

Pallescensone   P-70003

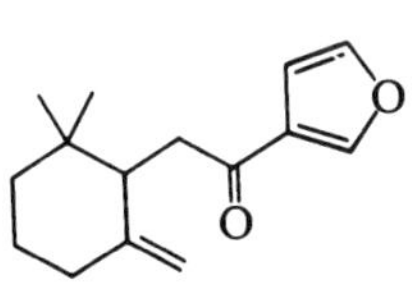

$C_{15}H_{20}O_2$ M 232.322
Constit. of the sponge *Dictyodendrilla cavernosa*. Cryst. Mp 42.5-43°. $[\alpha]_D$ +36° (c, 1 in CHCl$_3$).

Pamamycin-607   P-70004
[100905-89-3]

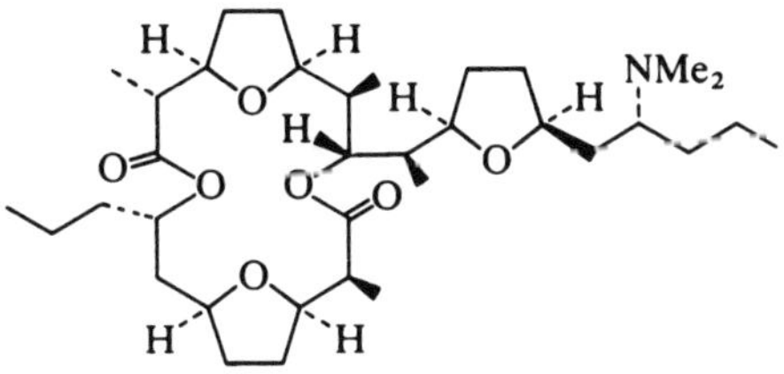

$C_{35}H_{61}NO_7$ M 607.869
Macrodiolide antibiotic. Prod. by *Streptomyces alboniger*. Ionophore. Active against gram-positive bacteria and pathogenic fungi. Oil. $[\alpha]_D^{33}$ +22.8° (c, 0.26 in MeOH). Related to Pamamycin of unknown struct.

Kondo, S. *et al*, *Tetrahedron Lett.*, 1987, **28**, 5861 (*isol, struct*)

Panasinsanol A   P-70005
[80374-27-2]

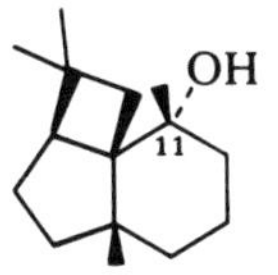

$C_{15}H_{26}O$ M 222.370
Constit. of *Panax ginseng*. Oil. $[\alpha]_D^{25}$ −51.9° (c, 0.54 in CHCl$_3$).
11-Epimer: [109785-99-1]. **Panasinsanol B**.
     $C_{15}H_{26}O$ M 222.370
     Constit. of *P. ginseng*. Oil. $[\alpha]_D^{25}$ −44.3° (c, 0.70 in CHCl$_3$).

Iwabuchi, H. *et al*, *Chem. Pharm. Bull.*, 1987, **35**, 1975.

Panial   P-70006

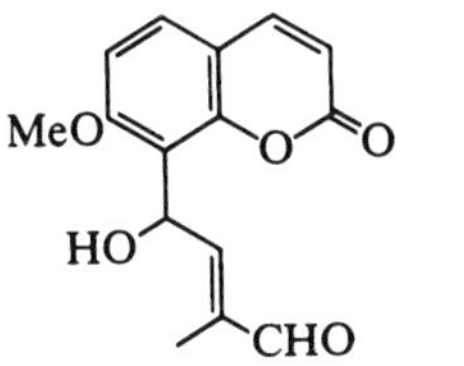

$C_{15}H_{14}O_5$ M 274.273
Constit. of *Murraya paniculata*. Oil. $[\alpha]_D$ −6.8° (c, 0.074 in CHCl$_3$).

Ito, C. *et al*, *Heterocycles*, 1987, **26**, 2959.

[2.2]Paracyclophadiene   P-70007

Updated Entry replacing P-00104

Tricyclo[8.2.2.2^{4,7}]hexadeca-2,4,6,8,10,12,13,15-octaene, 9CI. [2.2]Paracyclophan-1,9-diene

[6572-60-7]

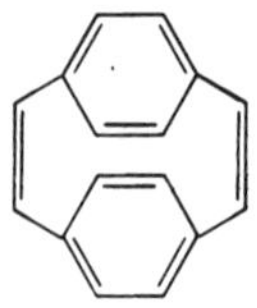

$C_{16}H_{12}$ M 204.271

Cryst. (C_6H_6/hexane). Mp 230-231°.

Dewhurst, K.C. *et al, J. Am. Chem. Soc.*, 1958, **80**, 3115 (*synth*)
Coulter, C.L. *et al, Acta Crystallogr.*, 1963, **16**, 667 (*struct*)
Otsubo, T. *et al, Tetrahedron Lett.*, 1975, 3881 (*synth*)
Stöbbe, M. *et al, Chem. Ber.*, 1987, **120**, 1667 (*synth, pmr, cmr*)
Yang, Z. *et al, Helv. Chim. Acta*, 1987, **70**, 299 (*pe*)

[14.0]Paracyclophane   P-70008

Updated Entry replacing P-00112

Tricyclo[18.2.2.2^{2,5}]hexacosa-2,4,20,22,23,25-hexaene, 9CI. 4,4′-Tetradecamethylenebiphenyl

[2013-43-6]

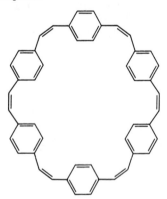

$C_{26}H_{36}$ M 348.570

Cryst. Mp 113-115°.

Nakazaki, M. *et al, Chem. Ind.* (*London*), 1965, 468 (*synth*)
Bates, R.B. *et al, Acta Crystallogr., Sect. C*, 1987, **43**, 517 (*cryst struct*)

[2₆]Paracyclophene   P-70009

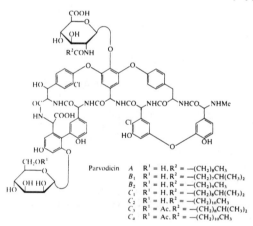

$C_{48}H_{36}$ M 612.812

4 isomers prepd.; (all-*Z*)(illus.), (*Z,Z,Z,Z,Z,E*), (*Z,Z,E,Z,Z,E*) and (*Z,E,Z,E,Z,E*).

Sundahl, M. *et al, Acta Chem. Scand., Ser. B*, 1988, **42**, 367 (*synth, pmr, uv*)

[6]-Paracycloph-3-ene   P-70010

Bicyclo[6.2.2]dodeca-4,8,10,12-tetraene, 9CI

[101055-48-5]

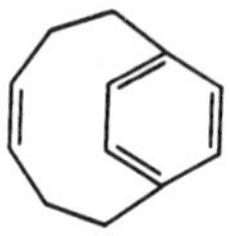

$C_{12}H_{14}$ M 158.243

Solid. Mp 29-30°.

Tobe, Y. *et al, Angew. Chem., Int. Ed. Engl.*, 1986, **25**, 369 (*synth, uv, pmr, ir, cmr*)
Tobe, Y. *et al, J. Am. Chem. Soc.*, 1987, **109**, 1136.

Paramicholide   P-70011

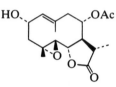

$C_{17}H_{24}O_6$ M 324.373

Constit. of *Paramichelia baillonii*. Oil.

Ruangrungsi, N. *et al, J. Nat. Prod.*, 1988, **51**, 163.

Parvodicin   P-70012

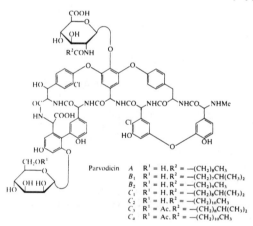

Glycopeptide antibiotic complex. Prod. by *Actinomadura parvosata*. Active against gram-positive bacteria. Similar to Aridicins.

Parvodicin A [110882-81-0]
   $C_{81}H_{84}Cl_2N_8O_{29}$ M 1704.497
Parvodicin B$_1$ [110882-82-1]
   $C_{82}H_{86}Cl_2N_8O_{29}$ M 1718.524
Parvodicin B$_2$ [110882-83-2]
   $C_{82}H_{86}Cl_2N_8O_{29}$ M 1718.524

Same gross struct. as Antibiotic A 40926*A*.
Parvodicin C$_1$ [110882-84-3]
   $C_{83}H_{88}Cl_2N_8O_{29}$ M 1732.550

Some gross struct. as Antibiotic A 40926*B*.
Parvodicin C$_2$ [110882-85-4]
   $C_{83}H_{88}Cl_2N_8O_{29}$ M 1732.550
Parvodicin C$_3$ [110882-86-5]
   $C_{85}H_{90}Cl_2N_8O_{30}$ M 1774.588

Parvodicin C$_4$

$C_{85}H_{90}Cl_2N_8O_{30}$ M 1774.588

Christensen, S.B. *et al, J. Antibiot.*, 1987, **40**, 970 (*isol, struct, props*)

Patagonic acid   P-70013

[114020-67-6]

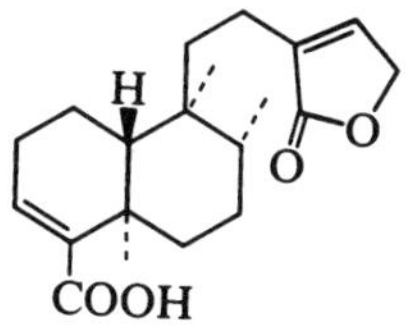

$C_{20}H_{28}O_4$ M 332.439

Constit. of *Baccharis patagonica*. Cryst. Mp 100-103°.
[α]$_D$ −65.9° (c, 2.65 in CHCl$_3$).

Rivera, A.P. *et al, J. Nat. Prod.*, 1988, **51**, 155.

Patrinalloside   P-70014

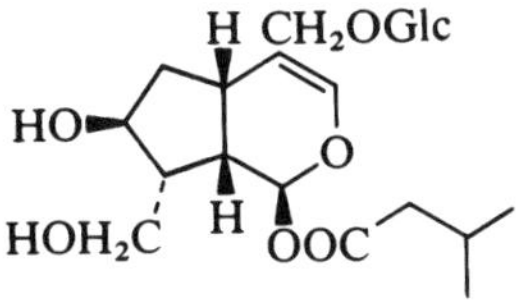

$C_{21}H_{34}O_{11}$ M 462.493

Constit. of *Penstemon hirsutus*. Powder
(CHCl$_3$/MeOH). Mp 63°. [α]$_D^{20}$ −69° (c, 0.64 in
MeOH).

Gering, B. *et al, J. Nat. Prod.*, 1987, **50**, 1048.

Pedicellosine   P-70015

[110941-52-1]

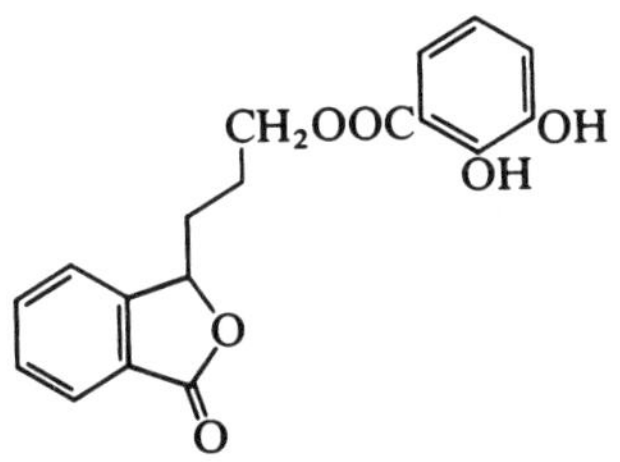

$C_{18}H_{16}O_6$ M 328.321

Constit. of *Gentiana pedicellata*. Oil.

Mpondo, E.M. *et al, Planta Med.*, 1987, 297.

Peltigerin   P-70016

Tenuiorin

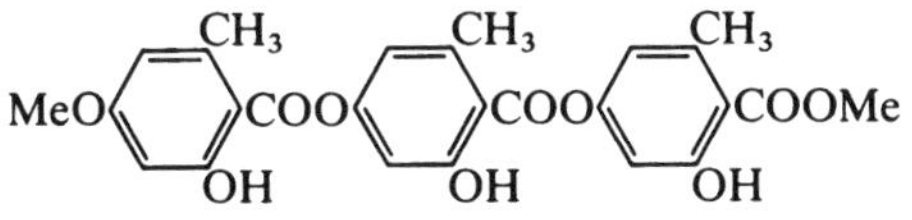

$C_{26}H_{24}O_{10}$ M 496.470

Isol. from *Peltigera polydactyla* and *Lobaria
pulmonaria*. Platelets (Me$_2$CO/C$_6$H$_6$). Mp 178-180°
(cloudy melt).

Huneck, S. *et al, Justus Liebigs Ann. Chem.*, 1965, **685**, 128
(*isol, ms, struct, bibl*)

Peltochalcone   P-70017

Updated Entry replacing P-00195

(*6,7-Dihydroxy-1*H*-2-benzopyran-3-yl*)(*2,4-
dihydroxyphenyl*)*methanone, 9CI. 3-(2,4-Dihydroxyben-
zoyl)-6,7-dihydroxy-1*H*-2-benzopyran*

[53766-30-6]

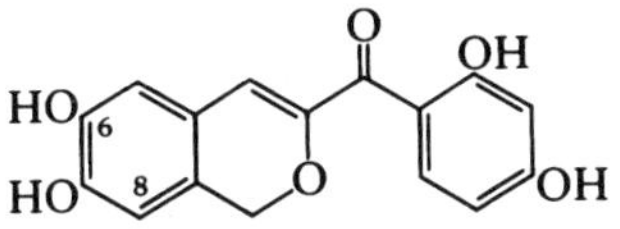

$C_{16}H_{12}O_6$ M 300.267

From heartwood of *Goniorrhachis marginata*. Yellow
cryst. (CHCl$_3$/MeOH). Mp 125° and 232-236°
(double Mp). λ$_{max}$ 255 (ε 900), 350 infl (12 000) and
402 nm (15 300) (EtOH).

4′,6,7-Tri-Me ether: [53802-73-6]. Yellow needles
(MeOH). Mp 126°.

O^6-Deoxy, O^8-hydroxy: [38279-57-1]. **3-(2,4-Dihydroxy-
benzoyl)-7,8-dihydroxy-1**H**-2-benzopyran**. (*7,8-Dihy-
droxy-1*H*-2-benzopyran-3-yl*)(*2,4-dihydroxyphenyl*)-
*methanone, 9CI. 2,4-Dihydroxyphenyl 1-hydro-7,8-di-
hydroxy-2-oxa-3-naphthyl ketone.*
$C_{16}H_{12}O_6$ M 300.267
Isol. from *Goniorrhachis marginata*. Intensely-yellow
needles. Mp 246-248°.

Gottleib, O.R. *et al, Phytochemistry*, 1972, **11**, 2841 (*isol, uv, ir,
ms, pmr, deriv*)
Gottlieb, O.R. *et al, Phytochemistry*, 1974, **13**, 1229 (*ir, ms,
pmr, uv*)

Pentachlorocyclopropane, 9CI   P-70018

Updated Entry replacing P-00251

[6262-51-7]

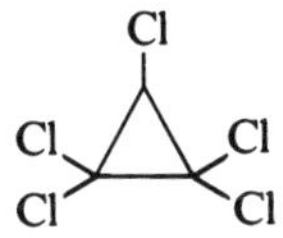

C_3HCl_5 M 214.306

Liq. Bp$_7$ 55.5-56°. $n_D^{27.5}$ 1.5170.

Tobey, S.W. *et al, J. Am. Chem. Soc.*, 1966, **88**, 2478 (*synth*)
Glück, C. *et al, Synthesis*, 1987, 260 (*synth*)

Pentacyclo[6.3.1.0^{2,4}.0^{5,10}.0^{7,8}]dodecane   P-70019

Decahydro-2,5:3,4-dimethanocyclopropa[de]-
naphthalene, 9CI. 3,12-Cycloiceane

[108643-89-6]

$C_{12}H_{16}$ M 160.258

Yamaguchi, R. *et al, J. Chem. Soc., Chem. Commun.*, 1987, 83
(*synth, cmr*)

Pentacyclo[11.3.0.0^{1,5}.0^{5,9}.0^{9,13}]hexadecane   P-70020

Octahydro-1H,4H,7H,10H-cyclobuta[1,2:1,4:2,3:3,4]-tetracyclopentene, 9CI. [4.5]*Coronane*
[108451-23-6]

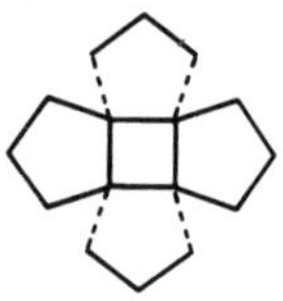

C$_{16}$H$_{24}$ M 216.366
Mp 186-188°.

Fitjer, L. *et al, Angew. Chem., Int. Ed. Engl.,* 1987, **26**, 1023.

Pentacyclo[8.4.0.0^{2,7}.0^{3,12}.0^{6,11}]tetradeca-   P-70021
4,8,13-triene

3a,4,6a,7,9a,9b-Hexahydro-1,4,7-methano-1H-phenalene, 9CI. D$_3$-*Tritwistatriene*

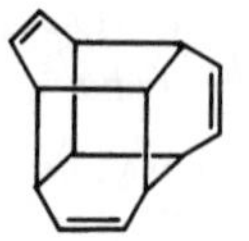

(–)-*form*

C$_{14}$H$_{14}$ M 182.265

(–)-*form*
 [α]$_D$ –414° (cyclohexane), [α]$_D$ –754° (CHCl$_3$). (+)-form also prepd.
 Hexahydro: see Pentacyclo[8.4.0.0^{2,7}.0^{3,12}.0^{6,11}]-tetradecane, P-10023

(±)-*form* [116280-37-6]
 Mp 106-107°.
 Hexahydro: see Pentacyclo[8.4.0.0^{2,7}.0^{3,12}.0^{6,11}]-tetradecane, P-10023

Müller, H. *et al, Angew. Chem., Int. Ed. Engl.,* 1988, **27**, 1103 (*synth, pmr, cmr, abs config*)

Pentacyclo[6.2.1.0^{2,7}.0^{4,10}.0^{5,9}]undecane-   P-70022
3,6-dione

Hexahydro-1,2,4-ethanylylidene-1H-cyclobuta[cd]-pentalene-5,7(1aH)dione, 9CI. Cookson's diketone
[2958-72-7]

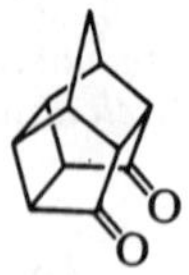

C$_{11}$H$_{10}$O$_2$ M 174.199
Cryst. by subl. Mp 245°.

Cookson, R.C. *et al, J. Chem. Soc.,* 1964, 3062 (*synth, pmr, uv*)
Bishop, R., *Aust. J. Chem.,* 1984, **37**, 319 (*cmr*)
Galkin, E.G. *et al, J. Org. Chem. USSR,* 1984, **20**, 286 (*ms*)
Marchand, A.P. *et al, Tetrahedron Lett.,* 1984, **25**, 795 (*pe*)

Pentacyclo[5.4.0.0^{2,6}.0^{3,10}.0^{5,9}]undecane-   P-70023
4,8,11-trione

Hexahydro-1,2,4-ethanylylidene-1H-cyclobuta[cd]-pentalene-3,5,7-trione, 9CI
[110243-21-5]

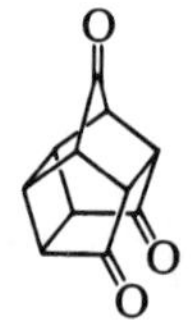

C$_{11}$H$_8$O$_3$ M 188.182
Microcryst. solid. Mp 265-270°.

Marchand, A.P. *et al, J. Org. Chem.,* 1987, **52**, 4784 (*synth, ir, pmr, cmr, ms*)

2,4-Pentadecadienal   P-70024

$$H_3C(CH_2)_9CH{=}CHCH{=}CHCHO$$

C$_{15}$H$_{26}$O   M 222.370
(*2E,4Z*)-*form* [85769-29-5]
 β-Triticene
 Antifungal constit. of *Triticium aestivum*. Oil.

Spendley, P.J. *et al, Phytochemistry,* 1982, **21**, 2403.

Pentafluorophenyl formate   P-70025

Pentafluorophenol formate, 9CI
[111333-97-2]

$$C_6F_5OCHO$$

C$_7$HF$_5$O$_2$ M 212.076
Formylating reagent. Oil. d^{20} 1.67. Slowly dec. at r.t.

Kisfaludy, L. *et al, Synthesis,* 1987, 510 (*synth, use*)

3,3',4',5,7-Pentahydroxyflavan   P-70026

Updated Entry replacing C-00436

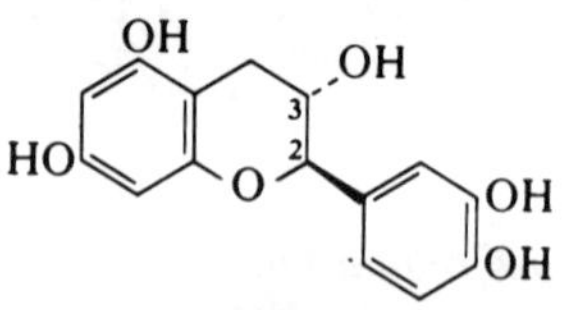

(2R,3S)-form

C$_{15}$H$_{14}$O$_6$ M 290.272
(*2R,3S*)-*form* [154-23-4]
 (+)-trans-*form*. **Catechin**. *Catechol*
 Widespread in plants. Cryst. + 4H$_2$O (AcOH aq.). Mp 93-96°, 175-177° (anhyd.). [α]$_D$ +17°. Not to be confused with 1,2-Benzenediol, B-00180 .
 ▷DJ3450000.

Penta-Ac: Cryst. (EtOH). Mp 131-132°. [α]$_D$ +40.6°.
7-L-Arabinopyranosyl: [29289-02-9]. **Polydine**.
 C$_{20}$H$_{22}$O$_{10}$ M 422.388
 Obt. from rhizomes of *Polypodium vulgare*. Mp 191-193°. [α]$_D^{20}$ –121.6°.
3-O-(3,4,5-Trihydroxybenzoyl): 3-*Galloylcatechin*.
 C$_{22}$H$_{18}$O$_{10}$ M 442.378
 Constit. of *Bergenia crassifolia* and *B. cordifolia*.
 Amorph. [α]$_D^{26}$ +56° (c, 0.9 in EtOH).
3-O-(3,4,5-Trihydroxybenzoyl), hepta-Me ether: Cryst. Mp 131-132°. [α]$_D^{20}$ –29° (c, 0.9 in CHCl$_3$).

O^3-Me: [65350-86-9]. *Meciadanol*, INN.
$C_{16}H_{16}O_6$ M 304.299
Immunomodulating agent.
$O^{4'}$-Me: *3,3',5,7-Tetrahydroxy-4'-methoxyflavan.*
$C_{16}H_{16}O_6$ M 304.299
Constit. of *Cinnamomum cassia*. Needles (H_2O). Mp
152°. $[\alpha]_D^{26}$ +6.7° (c, 0.85 in Me_2CO).
$O^{4'},O^7$-Di-Me: *3,3',5-Trihydroxy-4',7-dihydroxyflavan.*
$C_{17}H_{18}O_6$ M 318.326
Constit. of *C. cassia*. Prisms (EtOH). Mp 142-144°.
$[\alpha]_D^{26}$ −3.8° (c, 0.87 in MeOH).
3',5,7-Tri-Me ether: 3,4'-Dihydroxy-3',5,7-
trimethoxyflavan.
$C_{18}H_{20}O_6$ M 332.352
Constit. of *Viguiera quinqueradiata*. Cryst. (MeOH).
Mp 158-159°. $[\alpha]_D^{25}$ −12.8° (c, 0.18 in MeOH).
$O^{4'},O^5,O^7$-Tri-Me: *3,3'-Dihydroxy-4',5,7-*
trimethoxyflavan.
$C_{18}H_{20}O_6$ M 332.352
Constit. of *C. cassia*. Needles (hexane/EtOAc). Mp
125°. $[\alpha]_D$ −8.2° (c, 1.03 in Me_2CO).
Penta-Me ether: 3,3',4',5,7-Pentamethoxyflavan.
$C_{20}H_{24}O_6$ M 360.406
Cryst. (EtOH). Mp 95°. $[\alpha]_D$ +8.3°.

(2R,3R)-form

(−)-cis-*form.* **Epicatechin**. *Epicatechol*
Widespread. Cryst. Mp 242°. $[\alpha]_D$ −68° (EtOH).
Penta-Ac: Cryst. Mp 151-152°. $[\alpha]_D$ −15°.
3,5-Bis(3,4,5-trihydroxybenzoyl): 3,5-
Digalloylepicatechin.
$C_{29}H_{22}O_{14}$ M 594.484
Constit. of green tea leaves. Amorph. powder.
$O^{3'}$-Me:
$C_{16}H_{16}O_6$ M 304.299
Constit. of *Cinnamomum cassia, C. obtusifolium* and
Lindera umbellata var. *membranacea*. Prisms (H_2O).
Mp 237-238°. $[\alpha]_D^{28}$ −56.1° (c, 1.05 in Me_2CO).
$O^{3'},O^{4'}$-Di-Me:
$C_{17}H_{18}O_6$ M 318.326
Constit. of *L. umbellata* var. *membranacea*. Needles
(H_2O). Mp 199-201°. $[\alpha]_D^{23}$ −59.3° (c, 1.13 in
Me_2CO).
$O^{3'},O^5$-Di-Me:
$C_{17}H_{18}O_6$ M 318.326
Constit. of *C. cassia, C. obtusifolium* and *L. umbellata*
var. *membranacea*. Needles (hexane/EtOAc). Mp
206-208°. $[\alpha]_D^{28}$ −66.4° (c, 1.15 in Me_2CO).
$O^{3'},O^5,O^7$-Tri-Me:
$C_{18}H_{20}O_6$ M 332.352
Constit. of *C. cassia, C. obtusifolium* and *L. umbellata*
var. *membranacea*. Needles (C_6H_6). Mp 139°. $[\alpha]_D^{28}$
−61.2° (c, 1.22 in Me_2CO).
Penta-Me ether: Cryst. Mp 93-94°. $[\alpha]_D$ −94°.

(2RS,3SR)-form

(±)-*Catechin*
Cryst. + $3H_2O$. Mp 212-214°.

(2RS,3RS)-form

(±)-*Epicatechin*
Prisms + $4H_2O$ or needles + $1H_2O$. Mp 224-226°.

Bokaida, M.M. *et al, Proc. Chem. Soc., London*, 1960, 280
(*stereochem*)
Batterham, T.J. *et al, Aust. J. Chem.*, 1964, **17**, 428 (*pmr*)
Clark-Lewis, J.W. *et al, Aust. J. Chem.*, 1968, **21**, 3025 (*ms*)
Haslam, E., *J. Chem. Soc.* (*C*), 1969, 1824 (*isol*)
Weinges, K. *et al, Justus Liebigs Ann. Chem.*, 1970, **234**, 46
(*deriv*)
Coxon, D.T. *et al, Tetrahedron*, 1972, **28**, 2819 (*isol*)
Markham, K.R. *et al, Tetrahedron*, 1976, **32**, 2607 (*cmr*)
Jacques, D. *et al, J. Chem. Soc., Perkin Trans. 1*, 1977, 1637
(*biosynth*)
Eur. Pat., 37 800, (*1981*); *CA*, **96**, 40929y (*synth, pharmacol,*
Mediadanol)

Delgado, G. *et al, Phytochemistry*, 1984, **23**, 675 (*deriv*)
Morimoto, S. *et al, Chem. Pharm. Bull.*, 1985, **33**, 2281 (*deriv*)

3,3',4',5,7-Pentahydroxyflavanone   P-70027

Updated Entry replacing P-20041
Dihydroquercetin. Taxifolin. Distylin
[480-18-2]

$C_{15}H_{12}O_7$ M 304.256
▷LK6920000.

(2R,3R)-form [17654-26-1]

Isol. from Douglas fir sapwood and heartwood, *Pinus*
radiata, Cudrania javanensis, Urginea maritima and
Equisetum spp. Antifungal agent. Cryst. (EtOH or
H_2O). Mp 240-242°. $[\alpha]_D^{24}$ +44° (c, 1.03 in 50%
Me_2CO aq.).
Penta-Ac: [6685-67-2]. Cryst. Mp 88-89°. $[\alpha]_D^{24}$ +11.6°
(c, 1.2 in Me_2CO).
3-O-Rhamnoside: [29838-67-3]. **Astilbin**.
$C_{21}H_{22}O_{12}$ M 466.398
Constit. of *Taxillus kaempferi*.
3-O-β-D-Glucopyranoside: [83680-48-2]. **Glucodistylin**.
$C_{21}H_{22}O_{12}$ M 466.398
Constit. of *Chamaecyparis obtusa*. Mp 325° dec.
3-O-(6-Gallyl-β-D-glucopyranoside): [66656-93-7].
Taxillusin.
$C_{28}H_{26}O_{16}$ M 618.504
Constit. of *T. kaempferi*.
3-O-β-D-Xylopyranoside:
$C_{20}H_{20}O_{11}$ M 436.371
Constit. of leaves of *Thujopsis dolobrata*. Amorph.
powder. $[\alpha]_D$ −8.4° (c, 1.00 in Me_2CO).
4'-O-β-D-Glucopyranoside: **Taxifolin 4'-glucoside**.
$C_{21}H_{22}O_{12}$ M 466.398
Isol. from *Petunia hybrida*. Fine needles (MeOH aq. or
H_2O). Mp 218°. $[\alpha]_D^{21}$ −40.1° (MeOH).

(2S,3R)-form

Epitaxifolin
Amorph. $[\alpha]_D^{20}$ +58.8° (c, 0.26 in Me_2CO).
3-O-β-D-Xylopyranoside: Constit. of *T. dolobrata*.
Amorph. $[\alpha]_D^{24}$ +2.3° (c, 0.25 in Me_2CO).

(2R,3S)-form

Amorph. $[\alpha]_D^{20}$ −59.5° (c, 0.13 in Me_2CO).
3-O-β-D-Glucopyranoside: [83648-98-0].
Isoglucodistylin.
$C_{21}H_{22}O_{12}$ M 466.398
Constit. of *T. kaempferi*. Amorph. powder. Mp 169-
171°. $[\alpha]_D^{27}$ +25° (c, 0.3 in EtOH).
3-O-β-D-Xylopyranoside: Constit. of *T. dolobrata*.
Amorph. $[\alpha]_D^{21}$ −93.3° (c, 0.89 in Me_2CO aq.).

(2S,3S)-form

Needles. Mp 219-220°. $[\alpha]_D^2$ −20.6° (c, 0.32 in
Me_2CO).
3-O-β-D-Xylopyranoside: Constit. of *T. dolobrata*.
Amorph. powder. $[\alpha]_D^{21}$ −122.2° (c, 1.00 in Me_2CO).

(2RS,3RS)-form

Mp 234-236°.

Clark-Lewis, J.W. *et al, J. Chem. Soc.*, 1958, 2367 (*struct*)

Aft, H. *et al*, *J. Org. Chem.*, 1961, **26**, 1958 (*isol, deriv*)
Birkofer, L. *et al*, *Z. Naturforsch., B*, 1962, **17**, 359 (*Taxifolin 4'-glucoside*)
Fukui, Y. *et al*, *Yakugaku Zasshi*, 1966, **86**, 184 (*isol*)
Fernandez, M. *et al*, *Phytochemistry*, 1972, **11**, 1534 (*isol*)
Murti, V.V.S. *et al*, *Phytochemistry*, 1972, **11**, 2089 (*isol*)
Gupta, S.R. *et al*, *Indian J. Chem.*, 1975, **13**, 868 (*isol*)
Syrchina, A.I. *et al*, *Khim. Prir. Soedin.*, 1975, 424 (*isol*)
Nonaka, G. *et al*, *Chem. Pharm. Bull.*, 1987, **35**, 1105 (*derivs*)
Sukurai, A. *et al*, *Bull. Chem. Soc. Jpn.*, 1982, **55**, 3051 (*isol, struct*)

3',4',5,6,7-Pentahydroxyflavanone   P-70028

Updated Entry replacing P-60046
$C_{15}H_{12}O_7$ M 304.256
6-Me ether: 3',4',5,7-Tetrahydroxy-6-methoxyflavanone.
   $C_{16}H_{14}O_7$ M 318.282
   Constit. of *Eupatorium subhastatum*. Cryst. (MeOH). Mp 218-220°.
7-Me ether: 3',4',5,6-Tetrahydroxy-7-methoxyflavanone.
   $C_{18}H_{18}O_7$ M 346.336
   Constit. of *Holocarpha obconica*.
3',6-Di-Me ether: 4',5,7-Trihydroxy-3',6-dimethoxyflavanone.
   $C_{17}H_{16}O_7$ M 332.309
   Constit. of *Gutierrezia sphaerocephala*.
4',6,7-Tri-Me ether: 3',5-Dihydroxy-4',6,7-trimethoxyflavanone.
   $C_{16}H_{14}O_7$ M 318.282
   Constit. of *Vitex negundo*. Yellow cryst. (CHCl$_3$/pet. ether). Mp 136-138°.

Achari, B. *et al*, *Phytochemistry*, 1984, **23**, 703 (*isol*)
Chandra, S. *et al*, *Indian J. Chem., Sect. B*, 1987, **26**, 82 (*synth*)
Crins, W.J. *et al*, *Phytochemistry*, 1987, **26**, 2128 (*isol*)
Ferraro, G. *et al*, *Phytochemistry*, 1987, **26**, 3092 (*deriv*)
Mizuno, M. *et al*, *J. Nat. Prod.*, 1987, **50**, 751 (*isol, synth*)
Li, R. *et al*, *Phytochemistry*, 1988, **27**, 1556 (*isol*)

2',3,5,7,8-Pentahydroxyflavone   P-70029

Updated Entry replacing P-20043
8-Hydroxydatriscetin
$C_{15}H_{10}O_7$ M 302.240
Cryst. Mp 283-284°.
7-Me ether: 2',3,5,8-Tetrahydroxy-7-methoxyflavone.
   $C_{16}H_{12}O_7$ M 316.267
   Powder. Mp 249-250°.
7-Me ether, 8-Ac: 8-Acetoxy-2',3,5-trihydroxy-7-methoxyflavone.
   $C_{18}H_{14}O_8$ M 358.304
   Isol. from the fern *Notholaena sulpurea*. Cryst. Mp >300°.
2',3,7,8-Tetra-Me ether: 5-Hydroxy-2',3,7,8-tetramethoxyflavone.
   $C_{19}H_{18}O_7$ M 358.347
   Constit. of *Andrographis paniculata*. Yellow plates (MeOH). Mp 209-211°.
Penta-Me ether: 2',3,5,7,8-Pentamethoxyflavone.
   $C_{20}H_{20}O_7$ M 372.374
   Cryst. (Me$_2$CO). Mp 152-154°.

Jain, A.C. *et al*, *Proc. Indian Acad. Sci., Sect. A*, 1952, **36**, 217 (*synth*)
Gupta, K.K. *et al*, *Phytochemistry*, 1983, **22**, 314 (*isol*)
Arriaga-Giner, F.J. *et al*, *Z. Naturforsch., C*, 1987, **42**, 1063 (*deriv*)

2',3',4',5,6-Pentahydroxyflavone   P-70030

$C_{15}H_{10}O_7$ M 302.240
Penta-Me ether: 2',3',4',5,6-Pentamethoxyflavone.
   $C_{20}H_{20}O_7$ M 372.374
   Constit. of *Ardisia floribunda*. Cryst. (CHCl$_3$/MeOH). Mp 124-125°.

Parveen, M. *et al*, *Indian J. Chem., Sect. B*, 1987, **26**, 894.

3,5,6,7,8-Pentahydroxyflavone   P-70031

Updated Entry replacing P-30043
$C_{15}H_{10}O_7$ M 302.240
3,6,8-Tri-Me ether: [71479-92-0]. 5,7-Dihydroxy-3,6,8-trimethoxyflavone. **Araneol**.
   $C_{18}H_{16}O_7$ M 344.320
   Constit. of *Anaphalis araneosa*. Orange needles (CHCl$_3$/pet. ether). Mp 180-182°.
Penta-Me ether: 3,5,6,7,8-Pentamethoxyflavone.
   $C_{20}H_{20}O_7$ M 372.374
   Constit. of *Helichrysum nitens*.

Ali, E. *et al*, *Phytochemistry*, 1979, **18**, 356 (*isol, struct*)
Dutta, P.K. *et al*, *Indian J. Chem., Sect. B*, 1982, **21**, 1037 (*synth*)
Tomás-Barberán, F.A. *et al*, *Phytochemistry*, 1988, **27**, 753 (*isol*)

3',4',5,6,7-Pentahydroxyisoflavone   P-70032

Updated Entry replacing P-60049
$C_{15}H_{10}O_7$ M 302.240
6,7-Di-Me ether: 3',4',5-Trihydroxy-6,7-dimethoxyisoflavone. 6-Methoxyorobol 7-methyl ether.
   $C_{17}H_{14}O_7$ M 330.293
   Constit. of *Wyethia reticulata*.
5,6-Di-Me ether, 3',4'-methylene ether: 7-Hydroxy-5,6-dimethoxy-3',4'-methylenedioxyisoflavone. **Dipteryxin**.
   $C_{18}H_{14}O_7$ M 342.304
   Constit. of *Dipteryx odorata*. Cryst. (CHCl$_3$/Et$_2$O). Mp 235-237°. Not the same as 7,8-Dihydroxy-4',6-dimethoxyisoflavone, D-20232 .
5,6,7-Tri-Me ether, 3',4'-methylene ether: 5,6,7-Trimethoxy-3',4'-methylenedioxyisoflavone. **Odoratin**.
   $C_{19}H_{16}O_7$ M 356.331
   Constit. of *D. odorata*. Cryst. (CHCl$_3$/Et$_2$O). Mp 172-174°. Not the same as Odoratin, O-00475 or 1-(2,3,4,6-Tetrahydroxyphenyl)-3-(4-hydroxyphenyl)-2-propen-1-one, T-01302 .

Nakano, T. *et al*, *J. Chem. Soc., Perkin Trans. 1*, 1979, 2107 (*isol, struct*)
Bhardwaj, D.K. *et al*, *Proc. Indian Natl. Sci. Acad., Part A*, 1983, **49**, 168 (*struct*)
Chatterjea, J.N. *et al*, *J. Indian Chem. Soc.*, 1985, **62**, 990 (*synth*)
McCormick, S.P. *et al*, *Phytochemistry*, 1987, **26**, 2421 (*isol*)

2,3,5,6,8-Pentahydroxy-1,4-naphthoquinone   P-70033

Updated Entry replacing P-00414
2,3,5,6,8-Pentahydroxy-1,4-naphthalenedione, 9CI. **Spinochrome D**
[1143-11-9]

$C_{10}H_6O_7$ M 238.153
Tautomeric with the 2,5,6,7,8-pentahydroxy isomer.
Pigment from the sea urchins *Pseudocentrotus depressus* and *Strongylocentrotus droebachiensis*. Red cryst. Sublimes at 285-90° without melting.

Penta-Ac:
$C_{20}H_{16}O_{12}$ M 448.339
Yellow cryst. (MeOH). Mp 179-180°.
2,3,6-Tri-Me ether: [3560-71-2]. *5,8-Dihydroxy-2,3,6-trimethoxy-1,4-benzoquinone.* **Tricrozarin B**.
$C_{13}H_{12}O_7$ M 280.234
Constit. of *Tritonia crocosmaeflora*. Potent antitumour agent. Potentially tautomeric with the 5,8-dihydroxy-2,6,7-trimethoxy struct.

Kuroda, C. *et al, Proc. Imp. Acad. (Tokyo)*, 1944, **20**, 23 (isol)
Anderson, H.A. *et al, J. Chem. Soc.*, 1965, 2141 (synth)
Singh, I. *et al, J. Am. Chem. Soc.*, 1965, **87**, 4023 (synth)
Kol'tsova, E.A. *et al, Khim. Prir. Soedin.*, 1978, 438 (isol)
Masuda, K. *et al, J. Nat. Prod.*, 1987, **50**, 958 (isol)

1,3,4,5,6-Pentahydroxyphenanthrene   P-70034
1,3,4,5,6-Phenanthrenepentol

$C_{14}H_{10}O_5$ M 258.230
1,3,4-Tri-Me ether, 9,10-dihydro: [58115-32-5]. *9,10-Dihydro-5,6-dihydroxy-1,3,4-trimethoxyphenanthrene. 9,10-Dihydro-5,6,8-trimethoxy-3,4-phenanthrenediol.*
$C_{17}H_{18}O_5$ M 302.326
Constit. of the rhizomes of *Dioscorea prazeri*. Needles. Mp 154-156°.
1,3,4-Tri-Me ether, 9,10-dihydro, di-Ac: Mp 115-117°.

Rajaraman, K. *et al, Indian J. Chem.*, 1975, **13**, 1137.

2,3,4,7,9-Pentahydroxyphenanthrene   P-70035
2,3,4,7,9-Phenanthrenepentol
$C_{14}H_{10}O_5$ M 258.230
2,3,4-Tri-Me ether: [113476-61-2]. *7,9-Dihydroxy-2,3,4-trimethoxyphenanthrene. 5,6,7-Trimethoxy-2,10-phenanthrenediol, 9CI.* **Gymnopusin**.
$C_{17}H_{16}O_5$ M 300.310
Constit. of *Bulbophyllum gymnopus*. Cryst. (EtOAc/pet. ether). Mp 192°.
Penta-Me ether: [113476-63-4]. *2,3,4,7,9-Pentamethoxyphenanthrene.*
$C_{19}H_{20}O_5$ M 328.364
Cryst. (EtOAc/pet. ether). Mp 113°.

Majumder, P.L. *et al, Phytochemistry*, 1988, **27**, 245.

1,2,3,6,8-Pentahydroxyxanthone   P-70036
Updated Entry replacing P-00422
$C_{13}H_8O_7$ M 276.202
2,6-Di-Me ether: [116368-97-9]. *1,3,8-Trihydroxy-2,6-dimethoxyxanthone.*
$C_{15}H_{12}O_7$ M 304.256
Constit. of *Ixanthus viscosus*. Yellow needles (EtOAc). Mp 231-233°.
1,2,8-Tri-Me ether: [58785-54-9]. *3,6-Dihydroxy-1,2,3-trimethoxyxanthone.*
$C_{16}H_{14}O_7$ M 318.282
Constit. of the timber of *Mesua ferrea*. Cryst. + 2H$_2$O (MeOH aq.). Mp 285-287°.
2,3,6-Tri-Me ether: [34318-16-6]. *1,8-Dihydroxy-2,3,6-trimethoxyxanthone.*
$C_{16}H_{14}O_7$ M 318.282
Constit. of *I. viscosus*. Yellow needles (Me$_2$CO). Mp 177-178°.

Gunasekera, S.P. *et al, J. Chem. Soc., Perkin Trans. 1*, 1975, 2447 (isol, uv, pmr)
Ortega, E.P. *et al, Phytochemistry*, 1988, **27**, 1912 (isol, pmr, cmr)

Pentalenene   P-70037
Updated Entry replacing P-40042
1,2,3,3a,5a,6,7,8-Octahydro-1,4,7,7-tetramethylcyclopenta[c]pentalene, 9CI
[73306-73-7]

$C_{15}H_{24}$ M 204.355
Metab. of *Streptomyces griseochromogenes*. Oil.

Seto, H. *et al, J. Antibiot.*, 1980, **33**, 92 (isol)
Paquette, L.A. *et al, J. Am. Chem. Soc.*, 1983, **105**, 7358 (synth)
Cane, D.E. *et al, Bioorg. Chem.*, 1984, **12**, 312 (biosynth)
Piers, E. *et al, J. Chem. Soc., Chem. Commun.*, 1984, 959 (synth)
Mehta, G. *et al, J. Am. Chem. Soc.*, 1986, **108**, 8015 (synth)
Pattenden, G. *et al, Tetrahedron*, 1987, **43**, 5637 (synth)
Hudlicky, T. *et al, Tetrahedron*, 1987, **43**, 5685 (synth)

(Pentamethylphenyl)acetic acid   P-70038
Pentamethylbenzeneacetic acid

$C_{13}H_{18}O_2$ M 206.284
Cryst. (EtOH aq.). Mp 199-200°.

Mindl, J. *et al, Collect. Czech. Chem. Commun.*, 1987, **52**, 2936 (synth, pmr)

2,4-Pentanediamine
P-70039

Updated Entry replacing P-00472
2,4-Diaminopentane
[591-05-9]

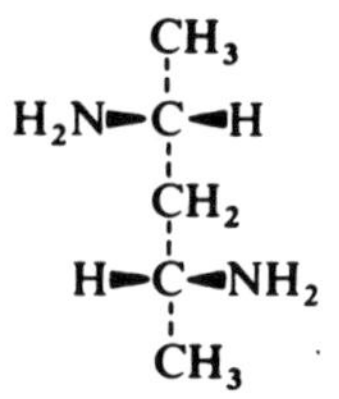

CH₃
|
H₂N—C—H
|
CH₂
|
H—C—NH₂   (2*RS*,4*RS*)-form
|
CH₃

$C_5H_{14}N_2$ M 102.179
(2R,4R)-form [34998-98-6]
$[\alpha]_D$ −8.33°.
(2S,4S)-form [34998-99-7]
$[\alpha]_D$ +8.46°.
(2RS,4RS)-form [29745-97-9]
(±)-*form.* α-*form*
Bp₁₁₋₁₂ 43-44°. Assigned the incorrect config. by Dippel, shown by Appleton *et al* to be resolveable.
B,2HNO₃: Mp 196°.
N²,N⁴-*Di-Ac*:
  $C_9H_{18}N_2O_2$ M 186.253
  Mp 168°.
N²,N⁴-*Dibenzoyl*:
  $C_{19}H_{22}N_2O_2$ M 310.395
  Mp 193-194°.
(2RS,4SR)-form [29745-96-8]
meso-*form.* β-*form*
Bp 120-140°, Bp₂₀ 46-47°.
B,2HNO₃: Mp 165°.
N²,N⁴-*Di-Ac*: Mp 168°.
N²,N⁴-*Dibenzoyl*: Mp 193-194°.

Harries, C. *et al*, *Ber.*, 1898, **31**, 550; 1899, **32**, 1193 (*synth*)
Dippel, C.J. *et al*, *Recl. Trav. Chim. Pays-Bas*, 1931, **50**, 525 (*resoln*)
Appleton, T.G. *et al*, *Inorg. Chem.*, 1970, **9**, 1800 (*synth, config*)
Murano, M. *et al*, *Macromolecules*, 1970, **3**, 605 (*resoln*)
Mizukami, F. *et al*, *Bull. Chem. Soc. Jpn.*, 1971, **44**, 3051 (*abs config*)

1,2,3,4-Pentatetraene-1,5-dione
P-70040

Pentacarbon dioxide
[51799-36-1]

O=C=C=C=C=C=O

C_5O_2 M 92.054
Yellow solid. Polymerises above −90°.

Maier, G. *et al*, *Angew. Chem., Int. Ed. Engl.*, 1988, **77**, 566 (*synth, cmr, ir, uv*)

3-Pentyn-2-ol, 9CI
P-70041

Updated Entry replacing P-00596
4-Hydroxy-2-pentyne
[27301-54-8]

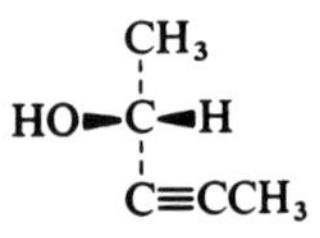

CH₃
|
HO—C—H   (*R*)-form
|
C≡CCH₃

C_5H_8O M 84.118
(R)-form [57984-70-0]
$[\alpha]_D^{20}$ +20° (c, 7.2 in CHCl₃).

(±)-form
Bp₅₉ 75-77°.
Ac:
  $C_7H_{10}O_2$ M 126.155
  Bp₂₀ 80°.

Frankenfeld, J.W. *et al*, *J. Org. Chem.*, 1971, **36**, 2110 (*synth*)
Bach, R.D. *et al*, *J. Org. Chem.*, 1975, **40**, 2559 (*nmr, abs config*)
Kosoha, K. *et al*, *Bull. Soc. Chim. Fr.*, 1975, 1291 (*synth*)
Fleming, I. *et al*, *J. Chem. Soc., Perkin Trans. 1*, 1987, 2269 (*synth, ir, pmr*)

Perfluoromethanesulfonimide
P-70042

N,*1,1,1-Tetrafluoro*-N-[(*trifluoromethyl*)*sulfonyl*]-*methanesulfonamide, 9CI*
[108388-06-3]

$(F_3CSO_2)_2NF$

$C_2F_7NO_4S_2$ M 299.135
Reagent for aromatic fluorinations. Liq. Mp −69.8°. Stable at r.t.

Singh, S. *et al*, *J. Am. Chem. Soc.*, 1987, **109**, 7194 (*synth, use*)

1*H*-Perimidin-2(3*H*)-one, 9CI
P-70043

Updated Entry replacing P-00630
3H-Benzo[de]quinazolin-2-one. Perimidone
[5157-11-9]

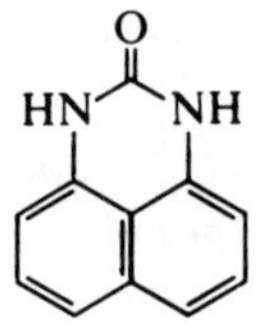

$C_{11}H_8N_2O$ M 184.197
Cryst. (EtOH or C_6H_6). Spar. sol. most org. solvents. Mp 304-305°.

Sachs, F., *Justus Liebigs Ann. Chem.*, 1909, **365**, 135 (*synth*)
Herbert, J.M. *et al*, *Heterocycles*, 1987, **26**, 1043 (*synth, ir, pmr*)

Petrosynol
P-70044

4,15,26-Triacontatriene-1,12,18,29-tetrayne-3,14,17,28-tetrol
[111554-19-9]

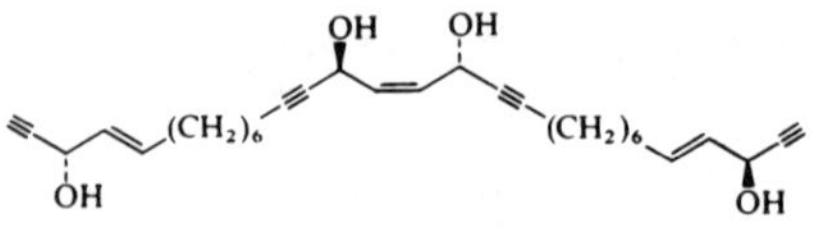

$C_{30}H_{40}O_4$ M 464.644
Isol. from marine sponge *Petrosia* sp. Exhibits antifungal props. Oil. $[\alpha]_D^{23}$ +107° (c, 0.37 in CHCl₃).
Tetraketone: [111397-55-8]. *4,15,26-Tricontatrien-1,12,18,29-tetrayne-3,14,17,28-tetrone, 9CI.*
Petrosynone.
$C_{30}H_{32}O_4$ M 456.580
From *P.* sp. Active against *Bacillus subtilis*. Yellow oil.

Fusetani, N. *et al*, *Tetrahedron Lett.*, 1987, **28**, 4313 (*isol, config*)

2-Phenacylcyclohexanone   P-70045

2-(2-Oxo-2-phenylethyl)cyclohexanone, 9CI. α-(2-*Oxocyclohexyl)acetophenone*
[33553-23-0]

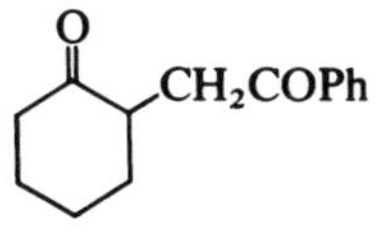

$C_{14}H_{16}O_2$ M 216.279

(±)-*form*

Prisms (Et$_2$O/pet. ether). Mp 47-48° (44-46°). Bp$_{0.05}$ 149-151°.

Baumgarten, H.E. *et al, J. Am. Chem. Soc.*, 1958, **80**, 6609 (*synth*)
Chiu, P.-K. *et al, Tetrahedron*, 1988, **44**, 3531 (*synth, pmr*)

1-Phenanthreneacetic acid   P-70046

[42050-06-6]

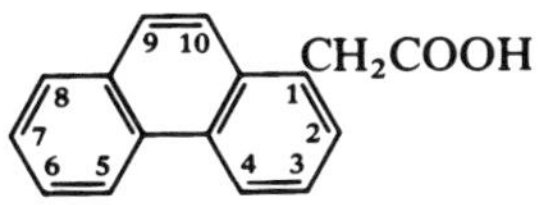

$C_{16}H_{12}O_2$ M 236.270
Cryst. (pet. ether). Mp 190-191°.

Chloride:
   $C_{16}H_{11}ClO$ M 254.715
   Light-yellow solid (Et$_2$O/pet. ether). Mp 81-82°.
Nitrile: [42155-49-7]. *1-(Cyanomethyl)phenanthrene.*
   $C_{16}H_{11}N$ M 217.270
   Needles (C$_6$H$_6$/pet. ether). Mp 130-131°.

Harmon, R.E. *et al, J. Chem. Soc., Perkin Trans. 1*, 1973, 1160 (*synth, deriv, ir, pmr*)

9-Phenanthreneacetic acid   P-70047

Updated Entry replacing P-00743
[25177-46-2]
$C_{16}H_{12}O_2$ M 236.270
Leaflets. Mp 224-225°.

Me ester: [21802-18-6].
   $C_{17}H_{14}O_2$ M 250.296
   Mp 75-75.5°.
Amide:
   $C_{16}H_{13}NO$ M 235.285
   Leaflets. Mp 250-252°.
Nitrile: [50781-52-7]. *9-Cyanomethylphenanthrene.*
   $C_{16}H_{11}N$ M 217.270
   Mp 96.5-97°.

Willgerodt, O. *et al, J. Prakt. Chem.*, 1911, **84**, 387.
Mosettig, E. *et al, J. Am. Chem. Soc.*, 1933, **55**, 2998.
Ford, W.T. *et al, J. Am. Chem. Soc.*, 1973, **95**, 6277 (*synth, deriv, uv, ir, pmr*)
Amin, S. *et al, J. Org. Chem.*, 1985, **50**, 4642 (*synth, deriv, pmr*)

4,5-Phenanthrenediol   P-70048

4,5-Dihydroxyphenanthrene
[10127-55-6]
$C_{14}H_{10}O_2$ M 210.232
Prisms (C$_6$H$_6$). Mp 182.5-183.5°.

Di-Me ether: [10127-54-5]. *4,5-Dimethoxyphenanthrene.*
   $C_{16}H_{14}O_2$ M 238.285

Plates (pet. ether). Mp 127-128°.

Newman, M.S. *et al, J. Org. Chem.*, 1967, **32**, 62 (*synth, deriv, uv*)
Cosmo, R. *et al, Aust. J. Chem.*, 1987, **40**, 2137 (*synth, pmr*)

9-Phenanthrenethiol, 9CI   P-70049

9-Mercaptophenanthrene
[61708-79-0]
$C_{14}H_{10}S$ M 210.293
Cryst. (EtOH). Mp 93.5-96.5°.

S-Ph: [16336-55-3]. *9-(Phenylthio)phenanthrene.*
   $C_{20}H_{14}S$ M 286.391
   Pale-yellow cryst. (toluene/EtOH). Mp 143-144°.

Groen, S.H. *et al, J. Org. Chem.*, 1968, **33**, 2218 (*deriv, synth*)
Castle, R.N. *et al, J. Heterocycl. Chem.*, 1982, **19**, 439 (*synth*)
Dent, B.R. *et al, Aust. J. Chem.*, 1986, **39**, 1789 (*deriv*)
Šindelář, V. *et al, Collect. Czech. Chem. Commun.*, 1986, **51**, 2848 (*synth, uv, ir, ms*)

Phenanthro[1,2-c]furan   P-70050

[219-20-5]

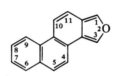

$C_{16}H_{10}O$ M 218.254
Prisms (CH$_2$Cl$_2$/pet. ether). Mp 138-143° dec.

Moursounidis, J. *et al, Aust. J. Chem.*, 1988, **41**, 235 (*synth, pmr, uv, ms*)

Phenanthro[3,4-c]furan   P-70051

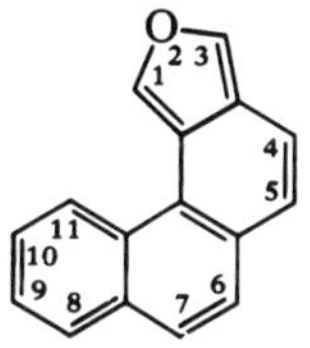

$C_{16}H_{10}O$ M 218.254
Plates (CH$_2$Cl$_2$/pet. ether). Mp 112-118° dec.

Moursounidis, J. *et al, Aust. J. Chem.*, 1988, **41**, 235 (*synth, pmr, uv, ms*)

1,7-Phenanthroline-5,6-dione, 9CI   P-70052

[82701-91-5]

$C_{12}H_6N_2O_2$ M 210.192
Orange-brown cryst. (MeOH). Mp 258-262° dec.

Eckert, T.S. *et al, Proc. Nat. Acad. Sci. USA*, 1982, **79**, 2533 (*synth*)

Phenanthro[9,10-c][1,2,5]selenadiazole, 9CI   P-70053

[236-08-8]

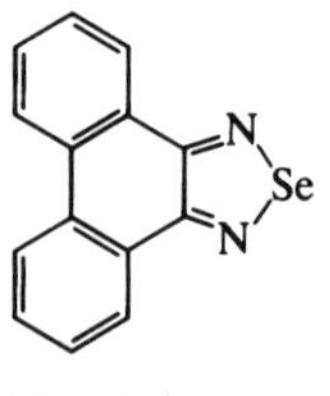

$C_{14}H_8N_2Se$ M 283.191

Light-yellow needles (EtOH/toluene). Mp 209°.

Neidlein, R. et al, Helv. Chim. Acta, 1987, **70**, 1076 (synth, uv, pmr, cmr, ms)

Phenarctin   P-70054

Updated Entry replacing P-00845

3,5-Diformyl-2,4-dihydroxy-6-methylbenzoic acid 3-hydroxy-4-(methoxycarbonyl)-2,5,6-trimethylphenyl ester, 9CI

[27839-39-0]

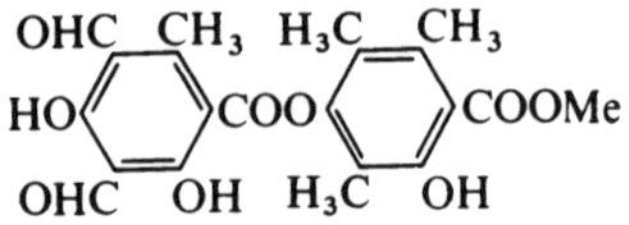

$C_{21}H_{20}O_9$ M 416.384

From *Nephroma arcticum* and *Pseudocyphellaria pickeringii*. Mp 167-168°.

2'-Me ether: [113689-52-4]. *2'-O-Methylphenarctin.*
$C_{22}H_{22}O_9$ M 430.410
Metab. of *P. pickeringii*. Plates. Mp 166°.

Bruun, T., Acta Chem. Scand., 1971, **25**, 2831 (isol, ms, pmr, uv)

Hamilton, R.J. et al, J. Chem. Soc., Perkin Trans. 1, 1976, 943 (synth)

Elix, J.A. et al, Aust. J. Chem., 1987, **40**, 2023 (isol, deriv, synth)

10H-Phenothiazine-1-carboxylic acid, 9CI   P-70055

[4182-55-2]

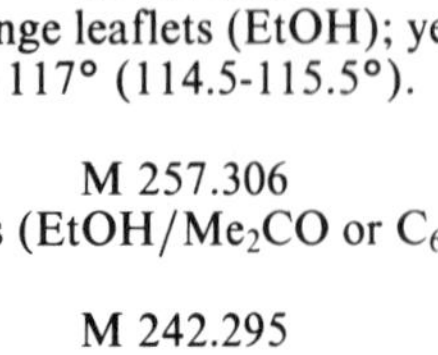

$C_{13}H_9NO_2S$ M 243.280

Orange prisms (EtOH/Me$_2$CO). Mp 265-267° (280°).

Me ester: [4063-33-6].
$C_{14}H_{11}NO_2S$ M 257.306
Yellow or orange leaflets (EtOH); yellow needles (CHCl$_3$). Mp 117° (114.5-115.5°).

N-Me:
$C_{14}H_{11}NO_2S$ M 257.306
Yellow prisms (EtOH/Me$_2$CO or C$_6$H$_6$). Mp 160°.

Amide:
$C_{13}H_{10}N_2OS$ M 242.295
Bright-yellow needles (MeOH).

N-Me, Me ester:
$C_{15}H_{13}NO_2S$ M 271.333
Yellow plates (EtOAc/Me$_2$CO). Mp 92°.

N-Et:
$C_{15}H_{13}NO_2S$ M 271.333
Orange needles (EtOH or C$_6$H$_6$). Mp 163°.

N-Et, Me ester:
$C_{16}H_{15}NO_2S$ M 285.360

Pale-yellow prisms. Mp 89°.

Me ester, 5-oxide: [96120-85-3].
$C_{14}H_{11}NO_3S$ M 273.306
Yellow needles (CHCl$_3$). Mp 167-169.2°. Racemate.

N-*Me*, 5,5-*dioxide:*
$C_{14}H_{11}NO_4S$ M 289.305
Cryst. (MeOH/EtOH). Mp 245°.

Cauquil, G. et al, Bull. Soc. Chim. Fr., 1960, 1049 (synth, derivs)

Hallberg, A. et al, J. Chem. Soc., Perkin Trans. 1, 1985, 969 (synth)

Shindo, K. et al, Bull. Chem. Soc. Jpn., 1985, **58**, 165 (oxide)

10H-Phenothiazine-2-carboxylic acid   P-70056

2-Carboxyphenothiazine

[25234-50-8]

$C_{13}H_9NO_2S$ M 243.280

Yellow cryst. (EtOH). Mp 276-277° (270-274° dec.).

Me ester: [15918-19-1].
$C_{14}H_{11}NO_2S$ M 257.306
Mp 166-167° (153°).

Et ester:
$C_{15}H_{13}NO_2S$ M 271.333
Mp 151-152°.

Amide:
$C_{13}H_{10}N_2OS$ M 242.295
Yellow needles (EtOH/2-propanol). Mp 232.5-233.5°, Mp 264-265°.

Dimethylamide:
$C_{15}H_{14}N_2OS$ M 270.348
Yellow cryst. (Me$_2$CO). Mp 176° (161-163.5°).

N-*Ac:*
$C_{15}H_{11}NO_3S$ M 285.317
Cryst. (EtOH). Mp 218°.

5,5-*Dioxide:* [54304-97-8].
$C_{13}H_9NO_4S$ M 275.278
Cryst. (DMF aq.). Mp 370°.

Me ester, 5,5-*dioxide:* [54304-03-9].
$C_{14}H_{11}NO_4S$ M 289.305
Cream cryst. (AcOH aq.). Mp 254-256°.

N-*Me:* [25244-60-4].
$C_{14}H_{11}NO_2S$ M 257.306
Yellow-green cryst. Mp 244-245°.

N-*Me, Me ester:* [54304-04-0].
$C_{15}H_{13}NO_2S$ M 271.333
Yellow leaflets (Me$_2$CO). Mp 94° (90-91°).

Nitrile: [38642-74-9]. *2-Cyano-10H-phenothiazine.*
$C_{13}H_8N_2S$ M 224.280
Mp 207-208° (202-203°).

N-*Me*, 5,5-*dioxide:* [54303-98-9].
$C_{14}H_{11}NO_4S$ M 289.305
Cryst. (EtOH). Mp 262°.

Burger, A. et al, J. Org. Chem., 1954, **19**, 1113 (synth, derivs)

Massie, S.P. et al, J. Org. Chem., 1956, **21**, 1006; 1959, **24**, 251 (synth, derivs)

Craig, P.N. et al, J. Org. Chem., 1961, **26**, 1138 (derivs)

de Antoni, J., Bull. Soc. Chim. Fr., 1964, 2214 (derivs)

Oprean, I. et al, Stud. Univ. Babes-Bolyai, Ser. Chem., 1974, **19**, 25 (derivs)

10H-Phenothiazine-3-carboxylic acid   P-70057

[68230-59-1]

$C_{13}H_9NO_2S$ M 243.280

Nitrile: [28140-93-4]. *3-Cyano-10H-phenothiazine.*
$C_{13}H_8N_2S$ M 224.280

Mp 182-184°.

▷SN6900000.

N-*Me:*
$C_{14}H_{11}NO_2S$ M 257.306
Yellow needles (EtOH/Me$_2$CO). Mp 260° (244-245°).

N-*Me, Me ester:*
$C_{15}H_{13}NO_2S$ M 271.333
Yellow cryst. (EtOH/Me$_2$CO). Mp 94°.

N-*Me, 5,5-dioxide:* [4997-39-1].
$C_{14}H_{11}NO_4S$ M 289.305
Cryst. (EtOH/Me$_2$CO). Mp 290°.

N-*Me, Me ester, 5,5-dioxide:*
$C_{15}H_{13}NO_4S$ M 303.332
Needles (EtOH/Me$_2$CO). Mp 190°.

N-*Me, nitrile:* [52805-44-4]. *3-Cyano-10-methylphenothiazine.*
$C_{14}H_{10}N_2S$ M 238.306
Yellow granules (EtOH aq.). Mp 121-122°.

N-*Et:*
$C_{15}H_{13}NO_2S$ M 271.333
Yellow needles (EtOH). Mp 200°.

N-*Et, 5,5-dioxide:*
$C_{15}H_{13}NO_4S$ M 303.332
Prisms (EtOH/Me$_2$CO). Mp 184°.

N-*Et, Me ester:*
$C_{16}H_{15}NO_2S$ M 285.360
Yellow needles (EtOH/Me$_2$CO). Mp 52°.

N-*Ph:*
$C_{19}H_{13}NO_2S$ M 319.377
Yellow needles (EtOH). Mp 220°.

N-*Ph, Me ester:*
$C_{20}H_{15}NO_3S$ M 349.403
Yellow needles (EtOH/Me$_2$CO). Mp 130°.

N-*Ph, 5,5-dioxide:*
$C_{19}H_{13}NO_4S$ M 351.376
Needles (EtOH/Me$_2$CO). Mp 295°.

N-*Ph, 5,5-dioxide, Me ester:*
$C_{20}H_{15}NO_4S$ M 365.403
Needles (EtOH/Me$_2$CO). Mp 197°.

Burger, A. *et al, J. Org. Chem.,* 1954, **19**, 1841 (*derivs*)
Cauquil, G. *et al, C.R. Hebd. Seances Acad. Sci.,* 1955, **240**, 1997 (*derivs*)
Clarke, D. *et al, J. Chem. Soc., Perkin Trans. 2,* 1977, 517 (*derivs, esr*)

10H-Phenothiazine-4-carboxylic acid   **P-70058**

$C_{13}H_9NO_2S$ M 243.280

N-*Me:*
$C_{14}H_{11}NO_2S$ M 257.306
Yellow microcryst. Mp 250°.

N-*Me, Me ester:*
$C_{15}H_{13}NO_2S$ M 271.333
Yellow prisms (EtOH/Me$_2$CO). Mp 80°. Bp$_{1.5}$ 230°.

N-*Et:*
$C_{15}H_{13}NO_2S$ M 271.333
Yellow prisms or needles (EtOH). Mp 182° (178-179°).

N-*Et, Me ester:*
$C_{16}H_{15}NO_2S$ M 285.360
Pale-yellow plates. Mp 113°.

N-*Me, 5,5-dioxide:*
$C_{14}H_{11}NO_4S$ M 289.305
Needles (MeOH). Mp 300°.

N-*Me, 5,5-dioxide, Me ester:*
$C_{15}H_{13}NO_4S$ M 303.332
Prisms (MeOH/Me$_2$CO). Mp 240°.

Cauquil, G. *et al, Bull. Soc. Chim. Fr.,* 1960, 1049.

Phenylacetohydroxamic acid   **P-70059**

N-*Hydroxybenzeneacetamide.* N-*Hydroxyphenylacetamide*
[5330-97-2]

$$PhCH_2CONHOH$$

$C_8H_9NO_2$ M 151.165
Cryst. Mp 150-152°.

N-*Me:* [72229-75-5]. N-*Methylphenylacetohydroxamic acid.*
$C_9H_{11}NO_2$ M 165.191
Prod. by *Pseudomonas mildenbergii.*

N-*Me; B,HCl:* Mp 205°.
O-*Benzyl:* Mp 78-80°.

Wise, W.M. *et al, J. Am. Chem. Soc.,* 1955, **77**, 1058 (*synth, props*)
Winternitz, F. *et al, Bull. Soc. Chim. Fr.,* 1960, 509 (*derivs*)
Kjaer, A. *et al, Acta Chem. Scand., Ser. B,* 1977, **31**, 415 (*deriv*)
Hulcher, F.H., *Biochemistry,* 1982, **21**, 4491 (*isol, deriv*)
Defoin, A. *et al, Helv. Chim. Acta,* 1987, **70**, 554 (*synth*)

2-Phenyl-2-adamantanol, 8CI   **P-70060**

2-Phenyltricyclo[3.3.1.1^{3,7}]decan-2-ol, 9CI
[29480-18-0]

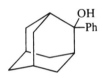

$C_{16}H_{20}O$ M 228.333
Cryst. (pet. ether). Mp 80-81°.

Tanida, H. *et al, J. Am. Chem. Soc.,* 1970, **92**, 3397 (*synth*)
Gill, G.B. *et al, J. Chem. Soc., Perkin Trans. 1,* 1980, **2**, 410 (*synth, nmr*)
Org. Synth., 1981, **60**, 104 (*synth, ir, nmr*)
Singelenberg, F.A.J. *et al, Acta Crystallogr., Sect. C,* 1987, **43**, 309 (*cryst struct*)

1-Phenyl-3-buten-1-ol   **P-70061**

Updated Entry replacing P-01060
α-*2-Propenylbenzenemethanol,* 9CI. *Allylphenylcarbinol*
[936-58-3]

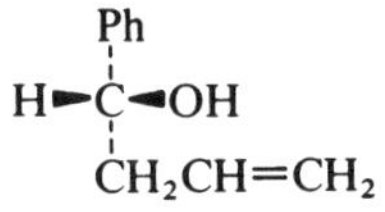

$$H \blacktriangleright \underset{\underset{CH_2CH=CH_2}{|}}{\overset{\overset{Ph}{|}}{C}} \blacktriangleleft OH \qquad (R)\text{-}form$$

$C_{10}H_{12}O$ M 148.204

(**R**)-*form* [85551-57-1]
$[\alpha]_D^{18}$ +48.3° (C$_6$H$_6$).
Hydrogen 3-nitrophthalate: Plates (C$_6$H$_6$/pet. ether).
Mp 126-127°. $[\alpha]_D^{18}$ +17.9° (CHCl$_3$).
Me ether: [86766-46-3]. *4-Methoxy-4-phenyl-1-butene.*
$C_{11}H_{14}O$ M 162.231
Bp$_{0.1}$ 35-40°. $[\alpha]_D^{22.5}$ +35.02° (c, 2.17 in CDCl$_3$).

(**S**)-*form* [77118-87-7]
Bp$_{15}$ 105-110°. $[\alpha]_D^{23}$ -44.92° (c, 7.38 in C$_6$H$_6$) (96% e.e.).

(±)-*form* [80735-94-0]
d^{18} 1.004. Bp 228-229°, Bp$_3$ 88-90°. n$_D^{21.5}$ 1.5289.

Ac: [2833-34-3].
$C_{12}H_{14}O_2$ M 190.241

Bp 239-240°.

Holding, A.F.L. *et al, J. Chem. Soc.*, 1954, 115 (*synth*)
Ayikama, S. *et al, Tetrahedron Lett.*, 1973, 4115 (*synth*)
Hathaway, S.J. *et al, J. Org. Chem.*, 1983, **48**, 3351 (*synth*)
Mashraqui, S.H. *et al, J. Org. Chem.*, 1984, **49**, 2513 (*synth*)
Jadhav, P.K. *et al, J. Org. Chem.*, 1986, **51**, 432 (*synth*)

1-Phenyl-2-butyn-1-ol, 8CI   P-70062

α-1-Propynylbenzenemethanol, 9CI

[32398-66-6]

$$H_3CC{\equiv}CCH(OH)Ph$$

C$_{10}$H$_{10}$O   M 146.188

(±)-form

Plates. Mp 37°. Bp$_{0.8}$ 110°.

4-Nitrobenzoyl: Long prisms (MeOH aq.). Mp 107°.

Braude, E.A. *et al, J. Chem. Soc.*, 1951, 2078 (*synth, uv*)
Fleming, I. *et al, J. Chem. Soc., Perkin Trans. 1*, 1987, 2269 (*synth, ir, pmr*)

4-Phenyl-3-butyn-2-ol, 9CI   P-70063

[5876-76-6]

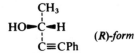

(*R*)-form

C$_{10}$H$_{10}$O   M 146.188

(*R*)-form [73922-81-3]

[α]$_D$ +51.8°. 72% e.e.

(±)-form

Liq. Bp$_{3.5}$ 95-97°.

Pitman, C.U. *et al, J. Am. Chem. Soc.*, 1965, **87**, 5632 (*synth, ir, pmr*)
Midland, M.M. *et al, J. Am. Chem. Soc.*, 1980, **102**, 867 (*synth*)
Fleming, I. *et al, J. Chem. Soc., Perkin Trans. 1*, 1987, 2269 (*synth, ir, pmr*)

3-Phenyl-2-cyclobuten-1-one, 9CI   P-70064

[38425-47-7]

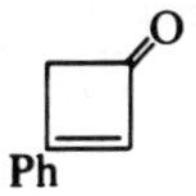

C$_{10}$H$_8$O   M 144.173

Cryst. (Et$_2$O or by subl.). Mp 51-52°. Dec. slowly in moist air.

▷Skin irritant

2,4-Dinitrophenylhydrazone: Fine red cryst. (EtOAc). Insol. most solvs. Mp 240.9-241.5°.

Manatt, S.L. *et al, J. Am. Chem. Soc.*, 1964, **86**, 2645 (*synth, ir, uv, pmr, props, haz*)
Wasserman, H.H. *et al, J. Org. Chem.*, 1973, **38**, 1451 (*synth*)
Danheiser, R.L. *et al, Tetrahedron Lett.*, 1987, **28**, 3299 (*synth*)

2-Phenylcyclohexanecarboxylic acid, 9CI   P-70065

Updated Entry replacing P-10107

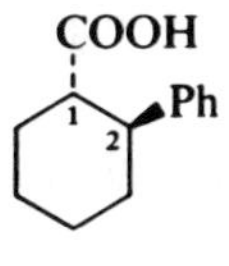

(1*S*,2*S*)-form

C$_{13}$H$_{16}$O$_2$ M 204.268

(1*S*,2*S*)-form [37982-24-4]

(+)-trans-*form*

Mp 87-88°. [α]$_D^{27}$ +68° (c, 0.12 in MeOH).

Chloride: [34713-97-8].

C$_{13}$H$_{15}$ClO   M 222.714

Mp 51-52°. [α]$_D^{27}$ +31° (c, 0.014 in heptane).

(1*S*,2*R*)-form [113215-85-3]

(+)-cis-*form*

Cryst. (hexane). Mp 56-58°. [α]$_D^{21}$ +90.8° (c, 0.13 in MeOH).

(1*R*,2*S*)-form [113215-84-2]

(−)-cis-*form*

Oil that slowly solidifies. [α]$_D^{22}$ −73.8° (c, 0.4 in MeOH).

(1*RS*,2*RS*)-form [64396-70-9]

(±)-trans-*form*

Mp 110°.

(1*RS*,2*SR*)-form

(±)-cis-*form*

Cryst. (pet. ether). Mp 76-77°.

Gutsch, C.D., *J. Am. Chem. Soc.*, 1948, **70**, 4150 (*config*)
Ziegler, F.E. *et al, J. Org. Chem.*, 1971, **36**, 3707 (*synth*)
Verbit, L. *et al, J. Am. Chem. Soc.*, 1972, **94**, 5143 (*abs config*)
Youssef, A.A. *et al, Egypt J. Chem.*, 1975, **18**, 231 (*synth, cd, abs config*)
Hawkes, G. E. *et al, J. Chem. Soc., Perkin Trans. 2*, 1976, 1709 (*synth*)
Schultz, A.G. *et al, J. Org. Chem.*, 1988, **53**, 2456 (*synth, pmr, ir, ms, abs config*)

1-Phenylcyclopropanol, 9CI, 8CI   P-70066

Updated Entry replacing P-01137

[29526-96-3]

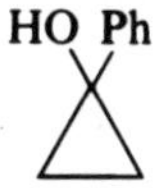

C$_9$H$_{10}$O   M 134.177

Bp$_{0.02}$ 46-48°.

Wasserman, H.H. *et al, Tetrahedron Lett.*, 1964, 341 (*synth*)
Martinez, A.M. *et al, J. Am. Chem. Soc.*, 1975, **97**, 6502 (*synth*)
Barluenga, J. *et al, Synthesis*, 1987, 584 (*synth, ir, pmr, cmr, ms*)

3-Phenyl-3*H*-diazirine   P-70067

[42270-91-7]

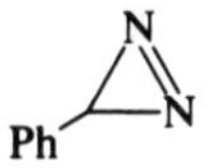

C$_7$H$_6$N$_2$ M 118.138

Pale-yellow oil.

Smith, R. *et al, J. Chem. Soc., Perkin Trans. 2*, 1975, 686 (*synth, pmr, uv, ir*)
McMahon, R.J. *et al, J. Am. Chem. Soc.*, 1987, **109**, 2456 (*synth, ir, pmr, uv*)

1-Phenylethanesulfonic acid   P-70068

α-Methylbenzeneethanesulfonic acid, 9CI

$$PhCH(CH_3)SO_3H$$

C$_8$H$_{10}$O$_3$S   M 186.225

(+)-*form* [86963-40-8]
Resolving agent. Pasty cryst. $[\alpha]_D^{25}$ +21.9° (c, 3 in DMSO).

(±)-*form*
Hemi-Ba salt: [111853-12-4]. Cryst. + $\frac{1}{2}H_2O$.
Phenylhydrazine salt: Mp 115° (rapid htg.).

Kharasch, M.S. *et al*, *J. Org. Chem.*, 1938, **3**, 178 (*synth*)
Yoshioka, R. *et al*, *Bull. Chem. Soc. Jpn.*, 1987, **60**, 649 (*synth, resoln, pmr, use*)

2-Phenylethanethiol   P-70069

Benzeneethanethiol, 9CI. β-Phenylethylmercaptan. 1-Mercapto-2-phenylethane
[4410-99-5]

$$PhCH_2CH_2SH$$

$C_8H_{10}S$ M 138.227
Bp_{23} 105°.

S-2,4-Dinitrophenyl: Mp 89-90°.

v. Braun, J., *Ber.*, 1912, **45**, 1564 (*synth*)
Bost, R.W. *et al*, *J. Am. Chem. Soc.*, 1932, **54**, 1985 (*deriv, synth*)
Katritzky, A.R. *et al*, *J. Chem. Soc.*, *Perkin Trans. 1*, 1987, 769 (*synth*)

[(2-Phenylethenyl)seleno]benzene, 9CI   P-70070

1-Phenylseleno-2-phenylethene. Phenyl styryl selenide

$$PhCH{=}CHSePh$$

$C_{14}H_{12}Se$ M 259.209

(*E*)-*form* [60466-40-2]
$Bp_{0.05}$ 100°.
S-Oxide: [108415-79-8]. *Phenyl styryl selenoxide.*
   $C_{14}H_{12}OSe$ M 275.208
   Cryst. Mp 100-101°.
S-Dioxide: [108415-83-4]. *Phenyl styryl selenone.* [(2-Phenylethenyl)selenonyl)benzene, 9CI.*
   $C_{14}H_{12}O_2Se$ M 291.208
   Mp 92-94°.

(*Z*)-*form*
$Bp_{0.05}$ 100°.
S-Oxide: [108415-81-2]. Hygroscopic solid. Mp not determined.
S-Dioxide: [108415-85-6]. Cryst. Mp 96-98°.

Petragnani, H. *et al*, *J. Organomet. Chem.*, 1976, **114**, 281 (*synth*)
Raucher, S. *et al*, *J. Org. Chem.*, 1978, **43**, 4885 (*synth*)
Tiecco, M. *et al*, *Tetrahedron Lett.*, 1984, **25**, 4975 (*synth, nmr*)
Tiecco, M. *et al*, *Tetrahedron*, 1986, **42**, 4889 (*deriv, synth, pmr, cmr*)

2-(2-Phenylethyl)phenol, 9CI   P-70071

o-Phenethylphenol, 8CI. 1-(2-Hydroxyphenyl)-2-phenylethane. 2-Hydroxybibenzyl
[7294-84-0]

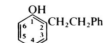

$C_{14}H_{14}O$ M 198.264
Plates (EtOH aq.). Mp 82-83°.

4-Methylbenzenesulfonyl: Prisms (pet. ether). Mp 56-57°.

Huisgen, R. *et al*, *Chem. Ber.*, 1964, **97**, 2884 (*synth*)
Bowden, B.F. *et al*, *Aust. J. Chem.*, 1975, **28**, 65 (*synth, deriv, ir, uv, pmr*)

3-(2-Phenylethyl)phenol, 9CI   P-70072

m-Phenethylphenol, 8CI. 1-(3-Hydroxyphenyl)-2-phenylethane. 3-Hydroxybibenzyl
[33675-75-1]

$C_{14}H_{14}O$ M 198.264
Cryst. (C_6H_6/pet. ether). Mp 75-76°.

Bruce, J.M. *et al*, *J. Chem. Soc. (C)*, 1971, 3749 (*synth, ir, uv, pmr*)
Bates, R.B. *et al*, *J. Org. Chem.*, 1986, **51**, 1432 (*synth*)

4-(2-Phenylethyl)phenol, 9CI   P-70073

p-Phenethylphenol, 8CI. 1-(4-Hydroxyphenyl)-2-phenylethane. 4-Hydroxybibenzyl
[6335-83-7]

$C_{14}H_{14}O$ M 198.264
Mp 103°.

Me ether:
   $C_{15}H_{16}O$ M 212.291
   Mp 63°.

Camaioni, D.M. *et al*, *J. Org. Chem.*, 1984, **49**, 1607 (*synth*)

2-(Phenylethynyl)benzoic acid, 9CI   P-70074

o-Carboxydiphenylacetylene
[1084-95-3]

$C_{15}H_{10}O_2$ M 222.243
Prisms. Mp 126-127°.

Hemihydrate: Cryst. (DMF aq.). Mp 116-118°.
Et ester:
   $C_{17}H_{14}O_2$ M 250.296
   Bp_3 175°.
Amide: [80221-08-5].
   $C_{15}H_{11}NO$ M 221.258
   Needles (EtOAc). Mp 155-157°.
Nitrile:
   $C_{15}H_9N$ M 203.243
   Bp_4 175°.

Castro, C.E. *et al*, *J. Am. Chem. Soc.*, 1964, **86**, 4358 (*deriv*)
Sakamoto, T. *et al*, *Chem. Pharm. Bull.*, 1986, **34**, 2754 (*synth, pmr*)

4-(Phenylethynyl)benzoic acid, 9CI   P-70075

p-Carboxydiphenylacetylene
[25739-23-5]

$C_{15}H_{10}O_2$ M 222.243
Needles ($CHCl_3$/pet. ether). Mp 221-222°.

Stephens, R.D. *et al*, *J. Org. Chem.*, 1963, **28**, 3313 (*synth*)
Bumagin, N.A. *et al*, *Dokl. Akad. Nauk SSSR, Ser. Sci. Khim.*, 1982, **265**, 1138 (*deriv*)
Suzuki, N. *et al*, *J. Chem. Soc., Chem. Commun.*, 1984, **22**, 1523 (*deriv*)

2-Phenyl-3-furancarboxylic acid, 9CI   P-70076
2-Phenyl-3-furoic acid
[57697-76-4]

$C_{11}H_8O_3$ M 188.182
Cryst. (diisopropyl ether). Mp 186°.

Novak, J.J.K., *Collect. Czech. Chem. Commun.*, 1975, **40**, 2855
   (*synth, pmr*)
Tius, M.A. *et al*, *Tetrahedron Lett.*, 1985, **26**, 3635 (*synth*)

4-Phenyl-2-furancarboxylic acid, 9CI   P-70077
4-Phenyl-2-furoic acid
[86471-28-5]
$C_{11}H_8O_3$ M 188.182
Cryst. (hexane/Et$_2$O). Mp 172.5-173.0°.

Alonso, M.E. *et al*, *J. Org. Chem.*, 1983, **48**, 3047 (*synth, ir, ms*)

5-Phenyl-2-furancarboxylic acid, 9CI   P-70078
5-Phenyl-2-furoic acid
[52938-97-3]
$C_{11}H_8O_3$ M 188.182
Cryst. (H$_2$O). Mp 152-155°. Bp$_{30}$ 120-122°.
Me ester: [52939-03-4].
   $C_{12}H_{10}O_3$ M 202.209
   Cryst. (hexane). Mp 65-66°.

Krutosikova, A. *et al*, *Collect. Czech. Chem. Commun.*, 1974,
   39, 767 (*synth, uv, ir, deriv*)
Masamune, T. *et al*, *Bull. Chem. Soc. Jpn.*, 1975, **48**, 491
   (*synth, ir, pmr*)
Furakawa, H. *et al*, *Chem. Pharm. Bull.*, 1981, **29**, 2420.

5-Phenyl-2(3H)-furanone   P-70079
5-Phenyl-2(3H)-butenolide
[1955-39-1]

$C_{10}H_8O_2$ M 160.172
Cryst. by subl., plates (pet. ether). Mp 91-92°. Bp$_{0.005}$
75° subl.

Kugel, M., *Justus Liebigs Ann. Chem.*, 1898, **299**, 50.
Nineham, A.W. *et al*, *J. Chem. Soc.*, 1949, 118 (*synth*)
Pelter, A. *et al*, *Tetrahedron Lett.*, 1987, **28**, 1203 (*synth*)

1-Phenyl-6-hepten-1-one   P-70080
6-Benzoyl-1-hexene
[15177-05-6]

$$H_2C{=}CH(CH_2)_4COPh$$

$C_{13}H_{16}O$ M 188.269
Oil. Bp$_6$ 148-150°, Bp$_{0.8}$ 93-95°.

Lewis, F.D. *et al*, *J. Am. Chem. Soc.*, 1974, **96**, 6090 (*synth,*
   pmr, ms)
Eddaif, A. *et al*, *J. Org. Chem.*, 1987, **52**, 5548 (*synth, ir, pmr*)

1-Phenyl-5-hexen-1-one   P-70081
5-Benzoyl-1-pentene
[22524-25-0]

$$H_2C{=}CHCH_2CH_2CH_2COPh$$

$C_{12}H_{14}O$ M 174.242
Oil. Bp$_{0.45}$ 83-84°.

Eddaif, A. *et al*, *J. Org. Chem.*, 1987, **52**, 5548 (*synth, ir, pmr*)

1-Phenyl-1H-indole, 9CI   P-70082
Updated Entry replacing P-01291
N-Phenylindole
[16096-33-6]

$C_{14}H_{11}N$ M 193.248
Yellowish oil. Insol. H$_2$O. d$_{25}$ 1.126. Bp 326-327°, Bp$_{11}$
   179-180°. Steam-volatile. n_D^{25} 1.6555.

Pfülf, A., *Justus Liebigs Ann. Chem.*, 1887, **239**, 221 (*synth*)
Pozharskii, A.F. *et al*, *Zh. Obshch. Khim.*, 1963, **33**, 1005
   (*synth*)
Powers, J.C., *J. Org. Chem.*, 1968, **33**, 2044 (*ms*)
Nishio, T., *J. Org. Chem.*, 1988, **53**, 1323 (*synth, ir, pmr, cmr*)

3-Phenyl-1,4-isoquinolinedione, 9CI, 8CI   P-70083
3-Phenyl-2-aza-1,4-naphthoquinone
[23994-23-2]

$C_{15}H_9NO_2$ M 235.242
Blood-red cryst. Mp 109-112°.

Felner, I. *et al*, *Helv. Chim. Acta*, 1969, **52**, 1810 (*synth, ir,*
   pmr)
Moore, H.W. *et al*, *Tetrahedron Lett.*, 1971, 1621 (*synth*)

9-Phenylnonanoic acid   P-70084
Benzenenonanoic acid, 9CI
[16269-06-0]

$$Ph(CH_2)_8COOH$$

$C_{15}H_{22}O_2$ M 234.338
Plates (pet. ether). Mp 35-37°.

Jefferson, A. *et al*, *Aust. J. Chem.*, 1988, **41**, 19 (*synth, pmr*)

1-Phenyl-1,2-pentadiene   P-70085
1,2-Pentadienylbenzene, 9CI
[2327-97-1]

$$H_3CCH_2CH{=}C{=}CHPh$$

$C_{11}H_{12}$ M 144.216
Bp$_1$ 55-56°.

Tolstikov, G.A. *et al*, *Bull. Acad. Sci. USSR (Engl. Trans.)*,
   1983, 569 (*synth, ir, pmr*)
Fujisawa, T. *et al*, *Tetrahedron Lett.*, 1984, **25**, 4007 (*synth*)

Caporusso, A.M. *et al*, *J. Org. Chem.*, 1987, **52**, 3920 (*synth*, *ir*, *pmr*, *ms*)

5-Phenyl-2,4-pentadienoic acid   P-70086

[1552-94-9]

PhCH=CHCH=CHCOOH

C$_{11}$H$_{10}$O$_2$ M 174.199
(*E,E*)-*form* [28010-12-0]
Mp 166-167°. (*E,Z*)- and (*Z,Z*)-isomers also known.
Me ester: [24196-39-2].
  C$_{12}$H$_{12}$O$_2$ M 188.226
  Solid. Mp 103°.

Bansal, B.S. *et al*, *J. Indian Chem. Soc.*, 1947, **24**, 443 (*synth*)
Lewis, F.D. *et al*, *J. Am. Chem. Soc.*, 1986, **108**, 3016 (*synth*, *nmr*)

1-Phenyl-1,4-pentadiyn-3-one   P-70087

[54668-28-9]

PhC≡CCOC≡CH

C$_{11}$H$_6$O   M 154.168
Cryst. (Et$_2$O/hexane). Mp 47-48.5°.

Wadworth, D.H. *et al*, *J. Org. Chem.*, 1987, **52**, 3662 (*synth*, *ir*, *pmr*)

5-Phenyl-4-penten-2-one, 9CI   P-70088

PhCH=CHCH$_2$COCH$_3$

C$_{11}$H$_{12}$O   M 160.215
(*E*)-*form* [42762-56-1]
Pale-yellow oil.

Bowers, K.G. *et al*, *J. Chem. Soc., Perkin Trans. 1*, 1987, 1657 (*synth*, *ir*, *pmr*, *ms*)

3-Phenyl-1-pentyne   P-70089

(*1-Ethyl-2-propynyl*)*benzene*, 9CI

H$_3$CCH$_2$CHPhC≡CH

C$_{11}$H$_{12}$ M 144.216
(±)-*form* [109182-83-4]
Bp$_{17}$ 84°.

Caporusso, A.M. *et al*, *J. Org. Chem.*, 1987, **52**, 3920 (*synth*, *ir*, *pmr*)

3-Phenyl-2-propenoyl isothiocyanate, 9CI   P-70090

Cinnamoyl thiocarbimide. Cinnamoyl isothiocyanate
[19495-08-0]

PhCH=CHCONCS

C$_{10}$H$_7$NOS   M 189.231
Intermed. for heterocyclic compds. Mp 41-43°.

Dixon, A.E., *J. Chem. Soc.*, 1895, **67**, 1040 (*synth*)
Kristian, P. *et al*, *Chem. Zvesti*, 1969, **23**, 173; *CA*, **71**, 38558s (*synth*)
Dzurilla, M. *et al*, *Collect. Czech. Chem. Commun.*, 1976, **41**, 742 (*conformn*)
Hafez, E.A.A. *et al*, *Justus Liebigs Ann. Chem.*, 1987, 65 (*use*)

1-Phenyl-3-pyrazolidinone, 9CI   P-70091

Phenidone
[92-43-3]

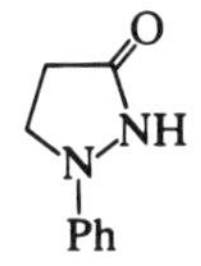

C$_9$H$_{10}$N$_2$O   M 162.191
Used in photography as developer. Inhibitor of the arachidonic acid cascade. Antiinflammatory. Mp 120-121°.
▷UQ8750000.
  B,HCl: [14776-33-1]. Mp 162-163°.

Duffin, G.F. *et al*, *J. Chem. Soc.*, 1954, 408 (*synth*)
Mikhailova, V.N. *et al*, *Zh. Obshch. Khim.*, 1958, **28**, 3112 (*synth*)
Simonova, N.I., *CA*, 1964, **60**, 4127b (*synth*)
Sbit, M. *et al*, *Acta Crystallogr., Sect. C*, 1987, **43**, 718 (*cryst struct*)

1-Phenyl-1*H*-pyrazolo[3,4-*e*]indolizine, 9CI   P-70092

[112633-64-4]

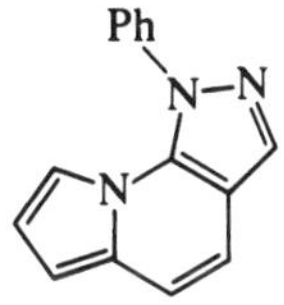

C$_{15}$H$_{11}$N$_3$ M 233.272
Cryst. (pet. ether). Mp 94-95°.

Stefancich, G. *et al*, *J. Heterocycl. Chem.*, 1987, **24**, 1199 (*synth*, *pmr*)

5-Phenylpyrimidine, 9CI   P-70093

Updated Entry replacing P-20107
[34771-45-4]
C$_{10}$H$_8$N$_2$ M 156.187
Needles (hexane). Mp 40-41°.
Picrate: Cryst. (EtOH). Mp 120°.
N-*Oxide:*
  C$_{10}$H$_8$N$_2$O   M 172.186
  Mp 105-107°.

Russel, P.B. *et al*, *J. Chem. Soc.*, 1954, 2951 (*uv*)
Wagner, R.M. *et al*, *Chem. Ber.*, 1971, **104**, 2975 (*synth*)
Oostveen, E.A. *et al*, *Recl. Trav. Chim. Pays-Bas*, 1974, **93**, 233 (*synth*)
Allen, D.W. *et al*, *J. Chem. Soc., Perkin Trans. 1*, 1977, 621 (*synth*, *pmr*)
Yamanaka, H. *et al*, *Chem. Pharm. Bull.*, 1987, **35**, 3119 (*synth*, *pmr*, *oxide*)

2-Phenylsulfonyl-1,3-butadiene   P-70094

[109802-71-3]

H$_2$C=C(SO$_2$Ph)CH=CH$_2$

C$_{10}$H$_{10}$O$_2$S   M 194.248
Versatile organic synthon. Readily dimerizes at r.t., fairly stable at −20°.

Bäckvall, J.E. *et al*, *J. Am. Chem. Soc.*, 1987, **109**, 6396 (*synth*, *pmr*, *use*)

2-(Phenylsulfonyl)-1,3-cyclohexadiene   P-70095

[102860-22-0]

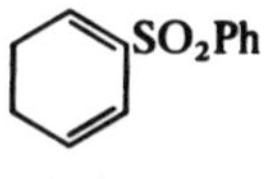

$C_{12}H_{12}O_2S$ M 220.286
Versatile organic synthon. Cryst. Mp 61-63°.

Bäckvall, J.-E. et al, J. Am. Chem. Soc., 1987, **109**, 6396 (synth, ir, pmr, cmr, use)

11-Phenylundecanoic acid   P-70096

Benzeneundecanoic acid, 9CI

[3343-24-6]

$$Ph(CH_2)_{10}COOH$$

$C_{17}H_{26}O_2$ M 262.391
Spars (pet. ether). Mp 44-45°.

Jefferson, A. et al, Aust. J. Chem., 1988, **41**, 19 (synth, pmr)

Phlebiakauranol aldehyde   P-70097

9,12,13,16-Tetrahydroxy-11-oxokauran-17-al, 9CI

[57743-92-7]

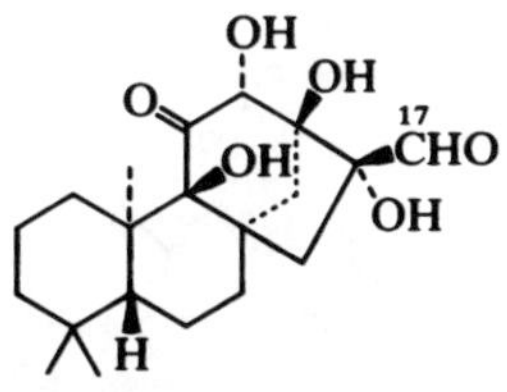

$C_{20}H_{30}O_6$ M 366.453
Terpenoid antibiotic. Prod. by Punctularia atropurpurascens. Shows antifungal, antibacterial and cytotoxic props. Cryst. (CHCl₃). Mp 198-200°. $[\alpha]_D^{22}$ +120° (c, 1.1 in DMSO).

17-Alcohol: [108708-02-7]. **Phlebiakauranol alcohol**.
$C_{20}H_{32}O_6$ M 368.469
From P. atropurpurascens. Shows weak antifungal, antibacterial and mod. cytotoxic props. Powder. Mp 206° dec. $[\alpha]_D^{22}$ +96° (c, 1.2 in DMSO).

Lisy, J.M. et al, J. Chem. Soc., Chem. Commun., 1975, 406.
Anke, H. et al, J. Antibiot., 1987, **40**, 443 (isol, struct, props)

Phleichrome   P-70098

Updated Entry replacing P-40131
4,9-Dihydroxy-1,12-bis(2-hydroxypropyl)-2,6,7,11-tetramethoxy-3,10-perylenedione, 9CI. Pleichrome

[56974-44-8]

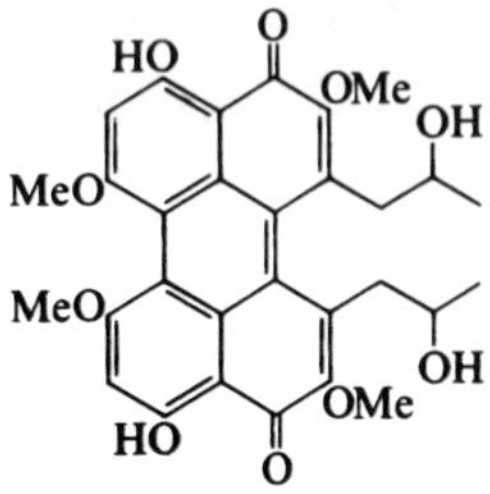

$C_{30}H_{30}O_{10}$ M 550.561
Isol. from Cladosporium phlei. Phytotoxin. Red powder or red cryst. (CHCl₂/hexane). Mp 205-210°. Tautomeric, forms two dimethyl ethers, one red and one yellow. On heating forms Isophleichrome, an atropisomer with opposite chirality of the biaryl system (cf. Cercosporin, C-20075 .

$O^2,O^{2''}$-Bis(3-hydroxybutanoyl): [11023-64-6]. **Cladochrome**. Cladochrome A.
$C_{38}H_{42}O_{14}$ M 722.741
Isol. from Cladosporium cucumerinum parasitic on cucumber seedlings. Mp 197-199° (softens at 194°). λ_{max} 478 nm. Originally assigned the MF $C_{38}H_{38}O_{12}$.

Overeem, J.C. et al, Phytochemistry, 1967, **6**, 99 (Cladochrome)
Yoshihara, T. et al, Agric. Biol. Chem., 1975, **39**, 1683 (isol, pmr, ord, ms, ir, struct)
Macri, F. et al, Agric. Biol. Chem., 1980, **44**, 2967 (isol)
Arnone, A. et al, J. Chem. Soc., Perkin Trans. 1, 1985, 1387 (cd, pmr, cmr, struct)
Thomson, R.H., Naturally Occurring Quinones, Recent Advances, Chapman and Hall, London, 1987, 593 (Cladochrome)

Phlogantholide A   P-70099

Updated Entry replacing P-40132

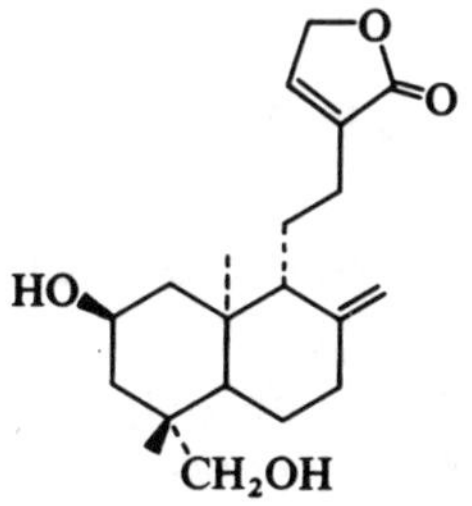

$C_{20}H_{30}O_4$ M 334.455
Constit. of Phlogacanthus thyrsiflorus. Cryst. (CHCl₃/pet. ether). Mp 144-146°. $[\alpha]_D^{26}$ −39.1° (CHCl₃).

19-O-β-D-Glucopyranoside: **Phloganthoside**.
$C_{26}H_{40}O_9$ M 496.597
Constit. of P. thyrsiflorus. Amorph.

19-O-β-D-Glucopyranoside, Penta-Ac: Cryst. Mp 193-194°. $[\alpha]_D^{26}$ −22.9° (CHCl₃).

Barua, A.K. et al, Phytochemistry, 1985, **24**, 2037; 1987, **26**, 491.

Phthalimidine   P-70100

Updated Entry replacing P-01797
2,3-Dihydro-1H-isoindol-1-one, 9CI. 1-Isoindolinone. Benzylamine-o-carboxylic lactam

[480-91-1]

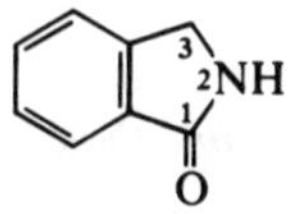

C_8H_7NO M 133.149
Needles (H₂O). Mp 155-156° (151°).

B,HCl: Mp 150° dec.

N-Ac: [3006-64-2].
$C_{10}H_9NO_2$ M 175.187
Mp 151°.

N-Me: [5342-91-6].
C_9H_9NO M 147.176
Mp 120°. Bp 300°.

N-Et: [23967-95-5].
$C_{10}H_{11}NO$ M 161.203
Mp 45°.

N-Ph: [5388-42-1]. N-Phenylphthalamidine. Phthalidanil.
$C_{14}H_{11}NO$ M 209.247
Leaflets (EtOH). Mp 162-163°.

▷TI5696000.

N-*Benzyl:* [13380-32-0].
$C_{15}H_{13}NO$ M 223.274
Mp 90-91°.

enol-form

Me ether: [49619-48-9]. *3-Methoxy-1H-isoindole.*
C_9H_9NO M 147.176
Oil. Bp$_{0.5}$ 64-65°.
Et ether: [49619-49-0]. *3-Ethoxy-1H-isoindole.*
$C_{10}H_{11}NO$ M 161.203
Oil, cryst. on standing. Mp 21°. Bp$_{0.01}$ 72°. n_D^{20}
1.5147.

Danishefsky, S., *et al, J. Org. Chem.*, 1975, **40**, 796 (*synth*)
Thompson, J.M. *et al, J. Org. Chem.*, 1975, **40**, 2667 (*spectra*)
Hennige, H. *et al, Chem. Ber.*, 1988, **121**, 243 (*synth, pmr, cmr, uv, ms, derivs*)

Piperazinone, 9CI

P-70101

2-Ketopiperazine
[5625-67-2]

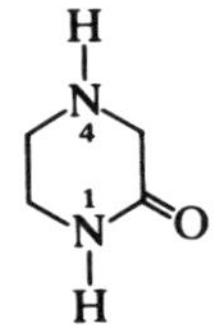

$C_4H_8N_2O$ M 100.120
Cryst. (Me$_2$CO/pet. ether). Mp 136°.
N^1-*Ph:* *1-Phenyl-2-piperazinone.*
$C_{10}H_{12}N_2O$ M 176.218
Cryst. (Et$_2$O/hexane). Mp 93-94°.
N^4-*Benzyl:* [13754-41-1].
$C_{11}H_{14}N_2O$ M 190.244
Prisms (C$_6$H$_6$). Mp 155-157° (149-150°).

Aspinall, S.R., *J. Am. Chem. Soc.*, 1940, **62**, 1202 (*synth*)
Uchida, H. *et al, Bull. Chem. Soc. Jpn.*, 1973, **46**, 3612 (*synth, benzyl*)
Nate, H. *et al, Chem. Pharm. Bull.*, 1987, **35**, 2825 (*synth, ir, pmr, deriv*)

3-Piperidinecarboxylic acid, 9CI

P-70102

Updated Entry replacing P-01954
Hexahydronicotinic acid. Nipecotic acid
[498-95-3]

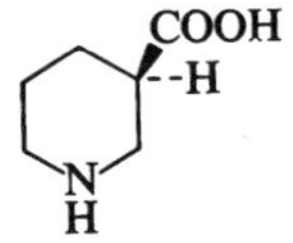

(*R*)-*form*
Absolute
configuration

$C_6H_{11}NO_2$ M 129.158
(*R*)-*form* [25137-00-2]
Mp 259-260°. $[\alpha]_D^{23}$ −3.4° (c, 5 in H$_2$O).
(*S*)-*form* [59045-82-8]
Mp 259-260°. $[\alpha]_D^{21}$ +3.6° (c, 5 in H$_2$O).
Et ester: [37675-18-6].
$C_8H_{15}NO_2$ M 157.212
$[\alpha]_D^{20}$ +1.75° (neat).
(±)-*form* [60252-41-7]
Cryst. Mp 261° dec.
B,HCl: Cryst. Mp 240-241° dec.
Me ester: [50585-89-2].
$C_7H_{13}NO_2$ M 143.185
Needles. Mp 215-217°.
Me ester; B,HCl: Mp 131-132°.
N-*Ac:* [2637-76-5].
$C_8H_{13}NO_3$ M 171.196

Mp 289-290° dec.
N-*Ac, 4-Methylbenzenesulfonyl:* Mp 167°.
N-*Nitroso:* [75822-15-0].
$C_6H_{10}N_2O_3$ M 158.157
Prisms (H$_2$O). Mp 111-112°.
Amide: [4138-26-5]. *3-Piperidinecarboxamide, 9CI. Nipecotamide, 8CI.*
$C_6H_{12}N_2O$ M 128.174
GABA and cholinesterase inhibitor. Mp 110-111°.
Amide, picrate: Cryst. (EtOH/C$_6$H$_6$). Mp 192-195°.
Diethylamide: [3367-95-1]. *N,N-Diethyl-3-piperidine-carboxamide, 9CI. N,N-Diethylnipecotamide.*
$C_{10}H_{20}N_2O$ M 184.281
Bp$_1$ 118-121°.

Freudenberg, K., *Ber.*, 1918, **51**, 1668 (*synth*)
McElvain, S.M. *et al, J. Am. Chem. Soc.*, 1923, **45**, 2738 (*synth*)
Freifelder, M. *et al, J. Org. Chem.*, 1961, **26**, 3805; 1962, **27**, 284 (*derivs*)
Barbieri, W. *et al, Tetrahedron*, 1965, **21**, 2453 (*ir, conformn, amide*)
Quan, P.M. *et al, J. Org. Chem.*, 1966, **31**, 2487 (*synth, derivs*)
Ripperger, H. *et al, Chem. Ber.*, 1969, 2864 (*abs config*)
Bettoni, G. *et al, Gazz. Chim. Ital.*, 1972, **102**, 189 (*abs config*)
Brehm, L. *et al, Acta Chem. Scand., Ser. B*, 1976, **30**, 542 (*cryst struct*)

3-Piperidineethanol

P-70103

3-(2-Hydroxyethyl)piperidine

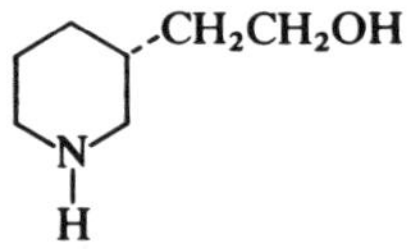

(*R*)-*form*

$C_7H_{15}NO$ M 129.202

(R)-form

Oil. $[\alpha]_D$ +2.1° (c, 8.1 in CHCl$_3$).
N-*Benzoyloxycarbonyl:* Bp$_{0.005}$ 145-150°. $[\alpha]_D$ +10.8° (c, 8.1 in C$_6$H$_6$).

(±)-form

Oil. Bp$_{0.001}$ 85-87°.

Brehm, R. *et al, Helv. Chim. Acta*, 1987, **70**, 1981 (*synth, bibl*)

Pitumbin

P-70104

[116528-02-0]

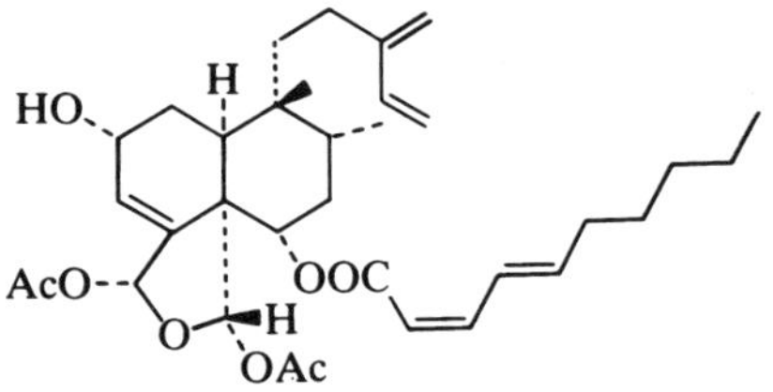

$C_{34}H_{48}O_8$ M 584.748
Constit. of seeds of *Casearia pitumba*. Oil. $[\alpha]_D$ −71.6° (c, 1 in CHCl$_3$).

Guittet, E. *et al, Tetrahedron*, 1988, **44**, 2893 (*struct*)

Platanetin P-70105

3,5,7,8-Tetrahydroxy-6-(3-methyl-2-butenyl)-2-phenyl-4H-1-benzopyran-4-one, 9CI. 3,5,7,8-Tetrahydroxy-6-prenylflavone

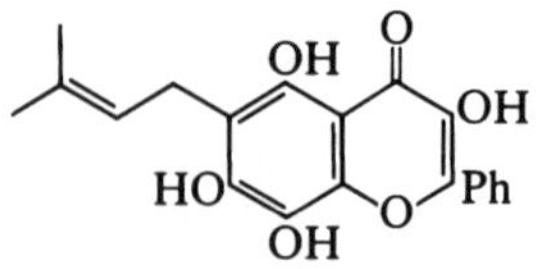

$C_{20}H_{18}O_6$ M 354.359

Constit. of *Platanus acerifolia*. Amorph. powder.

Kaouadji, M. *et al, J. Nat. Prod.*, 1988, **51**, 353.

Plumbazeylanone P-70106

Updated Entry replacing P-40172

[94410-15-8]

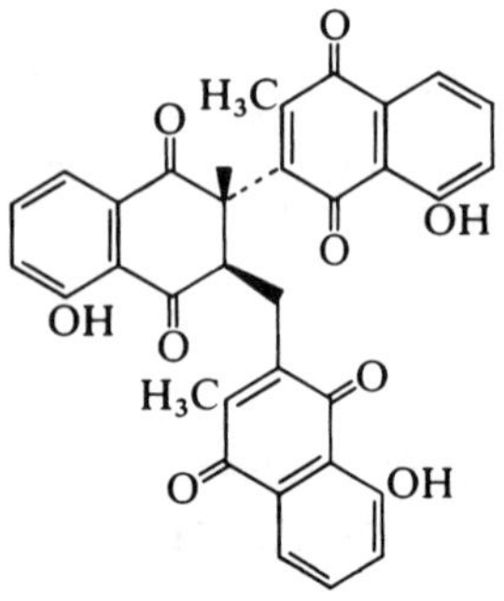

$C_{34}H_{24}O_9$ M 576.558

Constit. of roots of *Plumbago zeylanica*. Bright-orange cryst. Mp 246-248°. $[\alpha]_D$ 0° (c, 0.9 in CHCl₃).

Gunaherath, G.M.K.B. *et al, Tetrahedron Lett.*, 1984, **25**, 4801 (*isol*)

Kamal, G.M. *et al, Tetrahedron Lett.*, 1988, **29**, 719 (*cryst struct*)

13,17,21-Polypodatriene-3,8-diol P-70107

Decahydro-2,5,5,8a-tetramethyl-1-(4,8,12-trimethyl-3,7,11-tridecatrienyl)-2,6-naphthalenediol, 9CI

[89362-84-5]

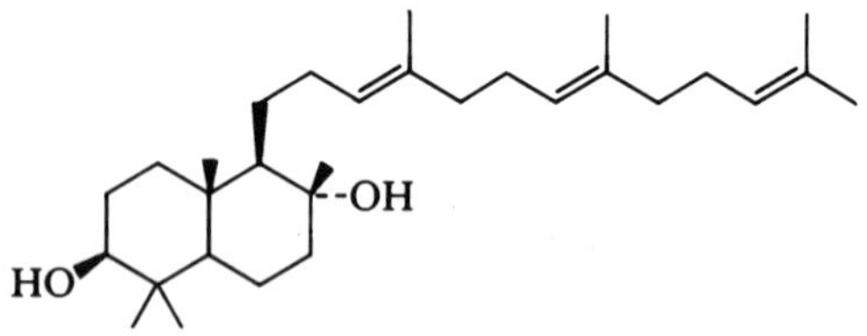

$C_{30}H_{52}O_2$ M 444.740

(*3β,8α*)-*form*

Constit. of resin of *Pistacia lentiscus*. Gum.

Boar, R.B. *et al, J. Am. Chem. Soc.*, 1984, **106**, 2476.

Pongone P-70108

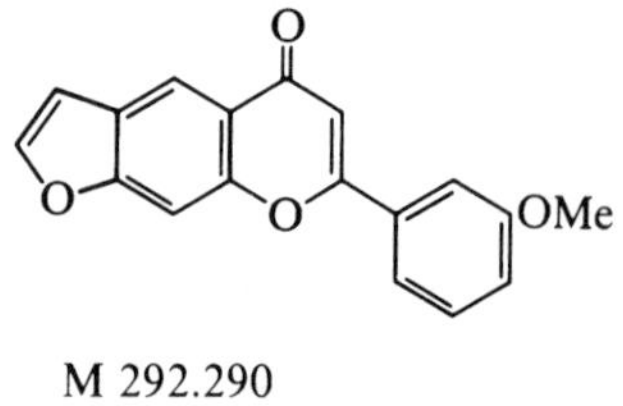

$C_{18}H_{12}O_4$ M 292.290

Constit. of *Pongamia glabra*. Cryst. (C₆H₆/pet. ether). Mp 186-188°.

Ganguly, A. *et al, Planta Med.*, 1988, **54**, 90.

21*H*,23*H*-Porphyrazine, 9CI P-70109

Tetraazaporphin

[500-77-6]

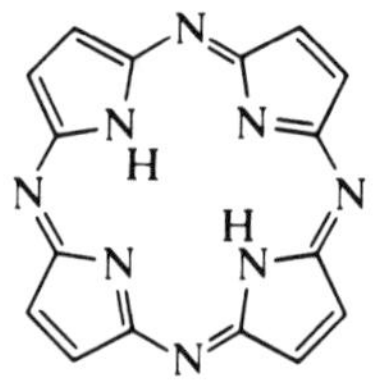

$C_{16}H_{10}N_8$ M 314.309

Forms metal complexes. Dark-purple needles. Struct. and bonding much studied.

Linstead, R.P. *et al, J. Chem. Soc.*, 1952, 4839 (*synth, uv*)
Borkovitch-Yellin, Z. *et al, J. Am. Chem. Soc.*, 1981, **103**, 6066 (*struct*)

Porphyrexide P-70110

4-Amino-2,5-dihydro-2-imino-5,5-dimethyl-1H-imidazol-1-yloxy, 9CI. 2,4-Diimino-5,5-dimethyl-1-imidazolinyloxy, 8CI

[15622-62-5]

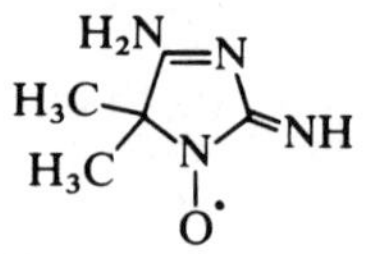

$C_5H_9N_4O$ M 141.152

Other tautomers possible. First stable organic free radical to be obt. Red cryst. + 3H₂O. Mp 248-250° (sinters at ~110°). Weak base.

B,HCl: Sol. H₂O. Mp 272° dec.

Piloty, O. *et al, Ber.*, 1901, **34**, 1870 (*synth*)
Kuhn, R. *et al, Ber.*, 1935, **68**, 1528 (*synth, struct*)
Aurich, H.G. *et al, Tetrahedron*, 1974, **30**, 2515 (*esr, tautom*)

Portmicin P-70111

Antibiotic 6270

[103521-25-1]

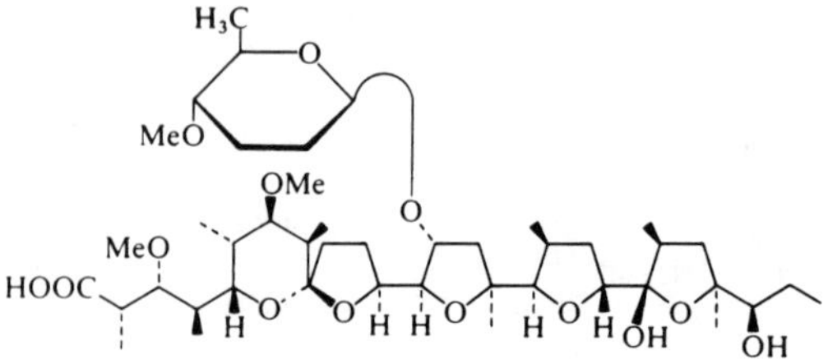

$C_{44}H_{76}O_{14}$ M 829.076

Polyether antibiotic. Prod. by *Nocardiopsis* sp. 6270. Active against gram-positive bacteria, incl. mycobacteria. Possesses coccidiostatic activity. Needles + ½H₂O (EtOAc/hexane). Mp 115-118°. $[\alpha]_D^{20}$ −11.5° (c, 1 in MeOH). Related to Cationomycin.

Kusakabe, Y. *et al, J. Antibiot.*, 1987, **40**, 237 (*isol, props*)
Seto, H. *et al, Tetrahedron Lett.*, 1987, **28**, 3357 (*pmr, cmr, struct*)

Praderin   P-70112

[116481-69-7]

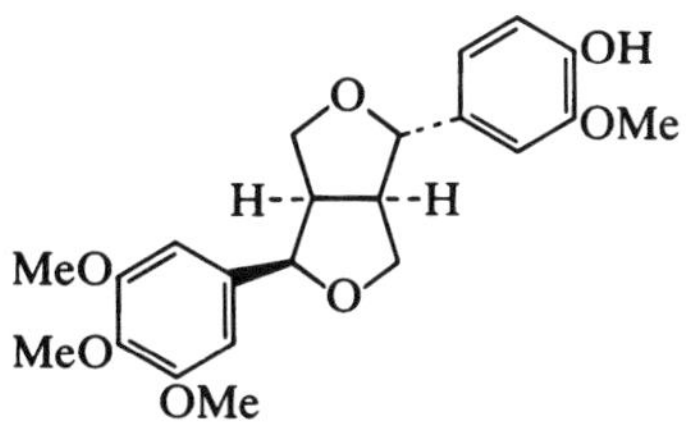

C$_{22}$H$_{26}$O$_7$ M 402.443
Constit. of *Lindera praecox*. Prisms (MeOH). Mp 155-156°. [α]$_D$ +113.8° (c, 0.05 in CHCl$_3$).

Ichino, K. *et al*, *Phytochemistry*, 1988, **27**, 1906.

Praecansone *B*   P-70113

[74517-75-2]

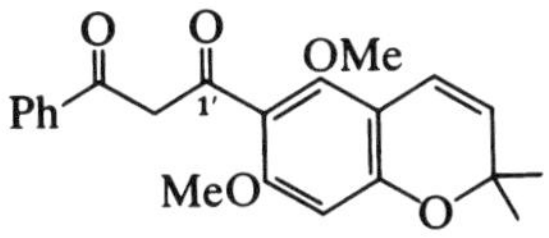

C$_{22}$H$_{22}$O$_5$ M 366.413
Enolised β-diketone. Constit. of *Tephrosia procumbens*. Yellow oil.

O$^{1'}$-Me: [74517-76-3]. **Praecansone A**.
C$_{23}$H$_{24}$O$_5$ M 380.440
Constit. of *T. procumbens*. Yellow oil. Enol ether of a tautomeric form of Praecansone B.

Camele, G. *et al*, *Phytochemistry*, 1980, **19**, 707 (*isol*)
Venkataratnam, G. *et al*, *J. Chem. Soc.*, *Perkin Trans. 1*, 1987, 2723 (*struct*)

Premnaspirodiene   P-70114

Updated Entry replacing P-20157
6,10-Dimethyl-2-(1-methylethenyl)spiro[4.5]dec-6-ene, 9CI. *2-Isopropenyl-6,10-dimethylspiro[4.5]dec-6-ene*
[82189-85-3]

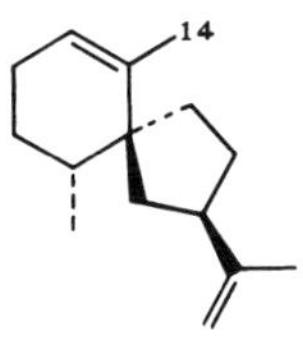

C$_{15}$H$_{24}$ M 204.355
Constit. of *Premna latifolia* and *P. integrifolia*.

14-Aldehyde: [82178-33-4]. **Premnaspiral**.
C$_{15}$H$_{22}$O   M 218.338
Constit. of *P. latifolia*. Oil.

14-Aldehyde, 2,4-dinitrophenylhydrazone: Cryst. Mp 160°.

5-Epimer: [90760-94-4]. **Agarospirene**. *Spirovetivene*.
C$_{15}$H$_{24}$ M 204.355
From the liverwort *Scapania robusta*.

Bheemasankara, C. *et al*, *Indian J. Chem.*, *Sect. B*, 1982, **21**, 267 (*isol*)
Wu, C.L. *et al*, *CA*, 1984, **101**, 51670; 1985, **102**, 3225 (*Agoraspirene*)
Rao, C.B. *et al*, *Indian J. Chem.*, *Sect. B*, 1985, **24**, 403 (*isol*)

Prephytoene alcohol   P-70115

[38005-60-6]

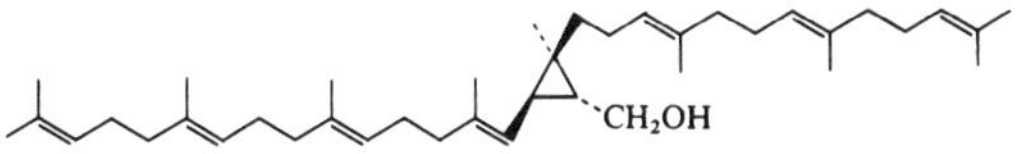

C$_{40}$H$_{66}$O   M 562.961
Constit. of *Myriophyllum verticullatum*. Biosynthetic precursor of the carotenoids. Oil. [α]$_D$ +37.5° (c, 0.7 in CHCl$_3$).

Monaco, P. *et al*, *Phytochemistry*, 1988, **27**, 2355.

Proclavaminic acid   P-70116

α-(*3-Amino-1-hydroxypropyl)-2-oxo-1-azetidineacetic acid*, 9CI. *5-Amino-3-hydroxy-2-(2-oxo-1-azetidinyl)-pentanoic acid*
[112240-59-2]

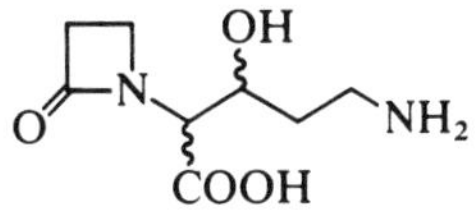

C$_8$H$_{14}$N$_2$O$_4$ M 202.210
Monocyclic β-lactam. Isol. from *Streptomyces clavuligerus*.

Elson, S.W. *et al*, *J. Chem. Soc.*, *Chem. Commun.*, 1987, 1736, 1738 (*isol, synth*)

1,2-Propadiene-1-thione, 9CI   P-70117

[83797-23-3]

$$H_2C{=}C{=}C{=}S$$

C$_3$H$_2$S   M 70.109
Detected by microwave spectroscopy.

Brown, R.D. *et al*, *J. Chem. Soc.*, *Chem. Commun.*, 1987, 573 (*synth*)
Brown, R.D. *et al*, *J. Am. Chem. Soc.*, 1988, **110**, 789 (*synth*, *props*)

1,2-Propanediol, 9CI   P-70118

Updated Entry replacing P-02351
α-*Propylene glycol*
[57-55-6]

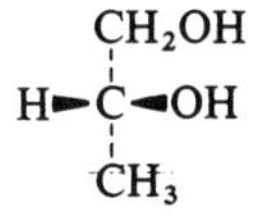

C$_3$H$_8$O$_2$ M 76.095
▷TY2000000.

(*R*)-*form* [4254-14-2]
Bp$_{12}$ 85-91°. [α]$_D$ −15.5° (neat).
Bisphenylurethane: Mp 146-147°.
2-O-Benzyl: [87037-69-2]. *2-Benzyloxy-1-propanol*.
C$_{10}$H$_{14}$O$_2$ M 166.219
[α]$_D^{20}$ −47° (c, 1 in CHCl$_3$).

(*S*)-*form* [4254-15-3]
Bp$_{12}$ 82-83°. [α]$_D^{20}$ +16.6° (neat).
1-(4-Methylbenzenesulfonyl): Mp 48-50°. [α]$_D$ +9.8° (c, 4.72 in CHCl$_3$).
2-O-Benzyl: [α]$_D^{20}$ +43° (c, 1 in CHCl$_3$).

(±)-*form* [4254-16-4]
Used in polyester resin manuf. and in the food industry as a humectant, preservative and solvent. Viscous oil with sweet taste. Misc. H$_2$O. d$^{19.4}$ 1.040. Bp 188-189°, Bp$_{21}$ 96-98°.

▷Simple ethers are mod. toxic
Di-Ac: [623-84-7].
$C_7H_{12}O_4$ M 160.169
Bp 190-191°.
▷TY4900000.
1-Me ether: [107-98-2]. *1-Methoxy-2-propanol.*
$C_4H_{10}O_2$ M 90.122
Bp 126-127°.
▷Flammable. UB7700000.
1-Me ether, Ac: [108-65-6]. *2-Acetoxy-1-methoxypropane.*
$C_6H_{12}O_3$ M 132.159
Bp 147°.
▷AI8925000.

Org. Synth., Coll. Vol., **2**, 545 (*synth*)
Baer, E. *et al, J. Am. Chem. Soc.*, 1948, **70**, 609 (*abs config*)
Guette, J.P. *et al, Bull. Soc. Chim. Fr.*, 1972, 4217 (*synth, props*)
Norula, J.L., *Chem. Era*, 1975, **11**, 21 (*resoln*)
Gombos, J. *et al, Chem. Ber.*, 1976, **109**, 2645 (*synth*)
Melchiorre, C., *Chem. Ind.* (*London*), 1976, 218 (*synth*)
Fuganti, C. *et al, J. Org. Chem.*, 1984, **49**, 543 (*deriv, synth, pmr*)
Pautard, A.M. *et al, J. Org. Chem.*, 1988, **53**, 2300 (*synth, pmr, cmr*)
Sax, N.I., *Dangerous Properties of Industrial Materials*, 5th Ed., Van Nostrand-Reinhold, 1979, 840, 943.

Propenal, 9CI   **P-70119**

Updated Entry replacing P-02385
Acraldehyde. Acrolein
[107-02-8]

$$H_2C{=}CHCHO$$

C_3H_4O M 56.064
Present in fruit aromas, black tea and prod. by microorganisms, e.g. *Clostridium perfringens*. Synthetic reagent. used in manuf. of methionine, glycerol and glutaraldehyde. Aquatic herbicide and algicide used in water treatment. Strongly lachrymatory liq. V. sol. H_2O. d_4^{20} 0.841. Mp −87°. Bp 52°. Unstable, polymerises to a white translucent solid Disacryl. n_D^{20} 1.3998.
▷Highly toxic, irritant, TLV 0.25. Highly flammable, fl. p. −26°. AS1050000.
Trimer: Metacrolein.
$C_9H_{12}O_3$ M 168.192
Plates (EtOH). Sol. EtOH, Et_2O. Mp 50°.
Di-Me acetal: [6044-68-4]. *3,3-Dimethoxypropene.*
$C_5H_{10}O_2$ M 102.133
Bp_{120} 40°.
▷UC8500000.
Cyanhydrin: [5809-59-6]. *2-Hydroxy-3-butenenitrile. Vinylglycollic nitrile. 1-Cyano-2-propen-1-ol.*
C_4H_5NO M 83.090
Bp_{17} 94°.
▷Liable to polymerise explosively. EM8225000.
Oxime: [5314-33-0].
C_3H_5NO M 71.079
Bp_9 35°.
Semicarbazone: Cryst. (EtOH/Et_2O). Mp 171°.
2,4-Dinitrophenylhydrazone: Reddish-orange cryst. (EtOH). Mp 165°.
Ethylene acetal: [3984-22-3]. *2-Vinyl-1,3-dioxolane.*
$C_5H_8O_2$ M 100.117
Bp 114-116°.

Ethylenedithioacetal: 2-Vinyl-1,3-dithiolane.
$C_5H_8S_2$ M 132.238
Bp_{14} 88-91°, $Bp_{0.5}$ 46-48°.

Doebner, O., *Ber.*, 1902, **35**, 1137 (*synth*)
B.P., 569 625, (*1945*); *CA*, **41**, 6275 (*synth*)
Bremner, J.G.M. *et al, J. Chem. Soc.*, 1946, 1018 (*synth*)
Fischer, R.F. *et al, J. Org. Chem.*, 1960, **25**, 319 (*synth*)
Lipnick, R.L., *Tetrahedron Lett.*, 1973, 931 (*struct*)
Miyajima, G. *et al, Org. Magn. Reson.*, 1974, **6**, 413 (*cmr*)
Coates, R.M. *et al, J. Org. Chem.*, 1984, **49**, 140 (*acetal, synth, ir, pmr*)
Oida, T. *et al, J. Chem. Soc., Perkin Trans. 1*, 1986, 1715 (*deriv, synth, pmr*)
Lou, J-D. *et al, Chem. Ind.* (*London*), 1987, 531 (*synth*)
Fieser, M. *et al, Reagents for Organic Synthesis*, Wiley, 1967-84, **3**, 5.
Bretherick, L., *Handbook of Reactive Chemical Hazards*, 2nd Ed., Butterworths, London and Boston, 1979, 164, 469.
Sax, N.I., *Dangerous Properties of Industrial Materials*, 5th Ed., Van Nostrand-Reinhold, 1979, 342.
Hazards in the Chemical Laboratory, (Bretherick, L., Ed.), 3rd Ed., Royal Society of Chemistry, London, 1981, 357, 429.

2-Propenethioamide, 9CI   **P-70120**

Thioacrylamide
[40620-23-3]

$$H_2C{=}CHCSNH_2$$

C_3H_5NS M 87.139
Yellow. Stable at −20° for several weeks.
N-Me:
C_4H_7NS M 101.166
Stable at −20° for several weeks.
N,N-Di-Me:
C_5H_9NS M 115.193
Stable at −20° for several weeks.

Khalid, M. *et al, Chem. Ind.* (*London*), 1988, 123 (*synth, pmr, cmr, uv, ms*)

2-Propenoic acid, 9CI   **P-70121**

Updated Entry replacing P-02390
Acrylic acid
[79-10-7]

$$H_2C{=}CHCOOH$$

$C_3H_4O_2$ M 72.063
Used widely for polymerisations, incl. prodn. of polyacrylates. Misc. H_2O. d_4^{16} 1.062. Mp 13°. Bp 141° (polymerises). pK_a 4.25 (25°). n_D^{20} 1.4424.
▷Toxic, irritant, causes burns, TLV 3.0. Exp. teratogen. AS4375000.
Me ester: [96-33-3]. *Methyl acrylate.*
$C_4H_6O_2$ M 86.090
Dienophile. Bp 85°. Polymerises on long standing.
▷Mod. toxic, irritant. Highly flammable. Forms peroxide, can polymerise violently. AT2800000.
Et ester: [140-88-5]. *Ethyl acrylate.* Polymerisation feedstock, odorant for natural gas, synthetic reagent. Liq. with acid odour. d_4^{20} 0.923. Mp −71.2°. Bp 99.8°. n_D^{20} 1.4068.
▷Skin and eye irritant, TLV 5ppm. LD_{50} 420 mg/kg. Threshold carcinogen. Flammable, Fl.p. 15°. AT0700000.
Chloride: [814-68-6].
C_3H_3ClO M 90.509
Bp 75°.
▷AT7350000.

Amide: [79-06-1]. *Acrylamide.*
C_3H_5NO M 71.079
Leaflets (C_6H_6). Sol. H_2O, EtOH, Et_2O, $CHCl_3$. Mp
85°. Polymerises on heating.
▷Highly toxic, irritant. AS3325000.
Nitrile: [107-13-1]. *Vinyl cyanide. Cyanoethylene.
Acrylonitrile.*
C_3H_3N M 53.063
Monomer, cyanoethylating reagent. Sol. H_2O. Bp 78°.
▷Highly toxic, TLV (skin) 4.5. Exp. carcinogen. Highly
flammable, fl. p. 0°. AT5250000.
Anhydride:
$C_6H_6O_3$ M 126.112
Bp_{35} 97°.

Wohlk, A., *J. Prakt. Chem.*, 1900, **61**, 212.
van der Burg, J.H.N., *Recl. Trav. Chim. Pays-Bas*, 1922, **41**, 21
    (*synth*)
Ratchford, W P. *et al, J. Am. Chem. Soc.*, 1944, **66**, 1864
    (*synth*)
Bowles, A.J. *et al, Org. Mass Spectrom.*, 1969, **2**, 809 (*ms*)
Sasaki, Y. *et al, Chem. Pharm. Bull.*, 1970, **18**, 1478 (*pmr,
    ester, nitrile*)
Katritzky, A.R. *et al, J. Am. Chem. Soc.*, 1970, **92**, 6861 (*ir*)
George, W.O. *et al, J. Chem. Soc., Perkin Trans. 2*, 1972, 400
    (*ir, conformn, esters*)
Austin, G.T., *Chem. Eng. (N.Y.)*, 1974, **81**, 86 (*rev, manuf*)
Miyajima, G. *et al, Org. Magn. Reson.*, 1974, **6**, 413 (*cmr*)
Villieras, J. *et al, Synthesis*, 1984, 406 (*synth, ester*)
Fieser, M. *et al, Reagents for Organic Synthesis*, Wiley, 1967-
    84, **5**, 439.
Bretherick, L., *Handbook of Reactive Chemical Hazards*, 2nd
    Ed., Butterworths, London and Boston, 1979, 423, 388.
Sax, N.I., *Dangerous Properties of Industrial Materials*, 5th
    Ed., Van Nostrand-Reinhold, 1979, 342, 343, 805.
Hazards in the Chemical Laboratory, (Bretherick, L., Ed.), 3rd
    Ed., Royal Society of Chemistry, London, 1981, 165, 166,
    478.

2-(2-Propenyl)benzaldehyde   P-70122

Updated Entry replacing P-60176
o-*Allylbenzaldehyde*
[62708-42-3]

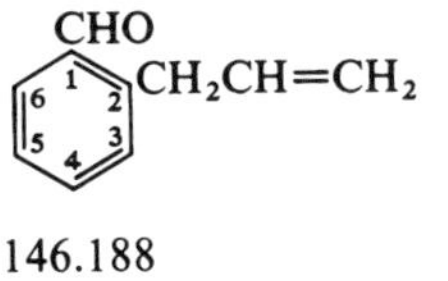

CHO
CH_2CH=CH_2

$C_{10}H_{10}O$ M 146.188
Oil. $Bp_{0.35}$ 51-52°.

Semmelhack, M.F. *et al, J. Am. Chem. Soc.*, 1983, **105**, 2034
    (*synth, spectra*)
Kampmeier, J.A. *et al, J. Org. Chem.*, 1984, **49**, 621 (*synth, ir,
    pmr, ms*)
Hickey, D.M.B. *et al, J. Chem. Soc., Perkin Trans. 1*, 1986,
    1113 (*synth, pmr*)
Ashby, E.C. *et al, J. Org. Chem.*, 1987, **52**, 4079 (*synth*)

Propenylsulfenic acid   P-70123

Updated Entry replacing P-02418
1-Propene-1-sulfenic acid
[3736-99-0]

$$H_3CCH=CHSOH$$

C_3H_6OS M 90.140

(*E*)-form

Lachrymatory factor present in onion (*Allium cepa*).

Virtanen, A.I. *et al, Acta Chem. Scand.*, 1963, **17**, 461 (*isol,
    struct, synth*)

Carson, J.F. *et al, J. Org. Chem.*, 1966, **31**, 1634 (*config*)

Protetrone   P-70124

[19556-33-3]

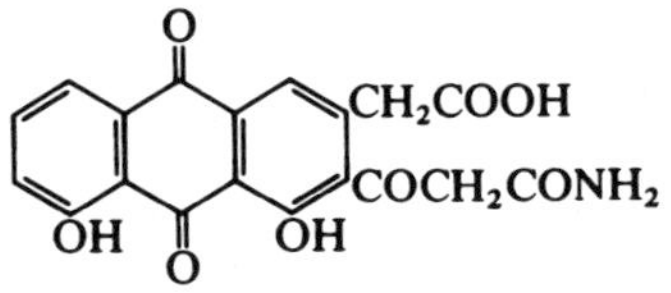

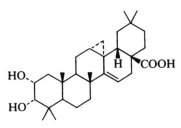

$C_{19}H_{13}NO_8$ M 383.314
Antraquinone antibiotic. Isol. from a *Streptomyces aur-
eofaciens* mutant. Biogenetically related to the tetracycline antibiotics. Orange cryst. Mp 186-190° dec. Similar to Ekatetrone.

McCormick, J.R.D. *et al, J. Am. Chem. Soc.*, 1968, **90**, 7126
    (*isol, struct*)
Prikrylova, V. *et al, J. Antibiot.*, 1978, **31**, 855.

Przewanoic acid *A*   P-70125

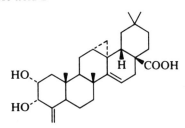

$C_{30}H_{46}O_4$ M 470.691
Constit. of *Salvia przewalskii*. Needles (Me_2CO). Mp
269-270°. $[\alpha]_D$ +125° (c, 0.08 in MeOH).

Wang, N. *et al, Phytochemistry*, 1988, **27**, 299.

Przewanoic acid *B*   P-70126

HO
HO
H
COOH

$C_{29}H_{42}O_4$ M 454.648
Constit. of *Salvia przewalskii*. Cryst. (MeOH). Mp 258-
259°. $[\alpha]_D$ +103° (c, 0.465 in MeOH).

Wang, N. *et al, Phytochemistry*, 1988, **27**, 299.

Pseudocyphellarin *B*   P-70127

Updated Entry replacing P-30197
[90685-96-4]

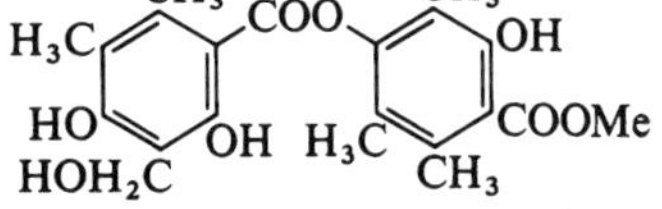

CH_3 COO CH_3
H_3C OH
HO   OH H_3C COOMe
HOH_2C CH_3

$C_{21}H_{24}O_8$ M 404.416
Constit. of *Pseudocyphellaria endochrysea* and *P. pick-
eringii*. Needles. Mp 168-169° dec.
Aldehyde: [90685-95-3]. ***Pseudocyphellarin* A.**
$C_{21}H_{22}O_8$ M 402.400
Isol. from *P. endochrysea*. Prisms (Me_2CO). Mp 173-
175°.

Aldehyde, 2'-Me ether: [113689-51-3]. **2'-O-Methylp-seudocyphellarin A.**
$C_{22}H_{24}O_8$ M 416.427
Constit. of *P. pickeringii*. Plates. Mp 190°.

Elix, J.A. *et al, Aust. J. Chem.*, 1984, **37**, 2153 (*synth*)
Huneck, S., *Phytochemistry*, 1984, **23**, 431 (*isol*)
Pulgarin, C. *et al, Helv. Chim. Acta*, 1985, **68**, 1948 (*synth*)
Elix, J.A. *et al, Aust. J. Chem.*, 1987, **40**, 2023 (*isol, deriv, synth*)

Pseudopterogorgia diterpenoid B   P-70128

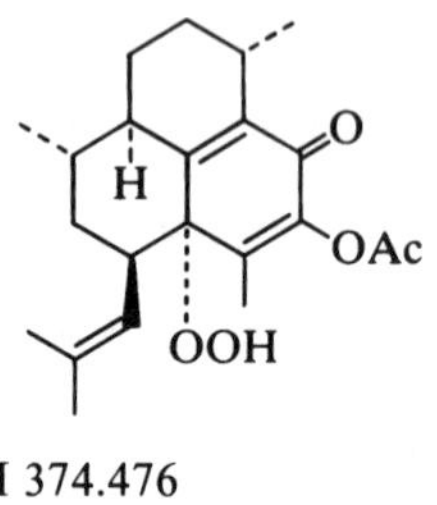

$C_{22}H_{30}O_5$ M 374.476
Constit. of *Pseudopterogorgia* sp. Yellow oil. $[\alpha]_D$ −11.3° (c, 0.006 in MeOH).

Harvis, C.A. *et al, Tetrahedron Lett.*, 1988, **29**, 4361 (*struct*)

Pseudopterogorgia diterpenoid C   P-70129

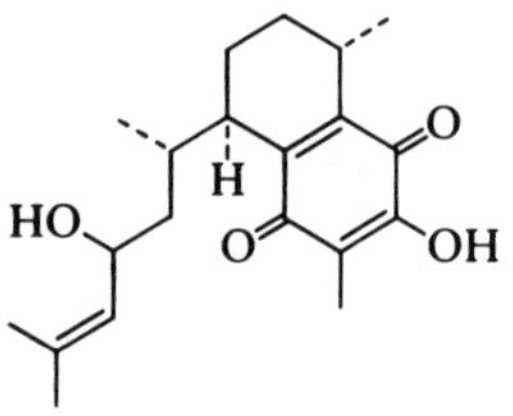

$C_{20}H_{28}O_4$ M 332.439
Constit. of *Pseudopterogorgia* sp. Orange oil. Has skeleton of aglycone of seco-pseudopterosins.

Harvis, C.A. *et al, Tetrahedron Lett.*, 1988, **29**, 4361 (*struct*)

Pseudrelone B   P-70130

Updated Entry replacing P-02608
[73702-69-9]

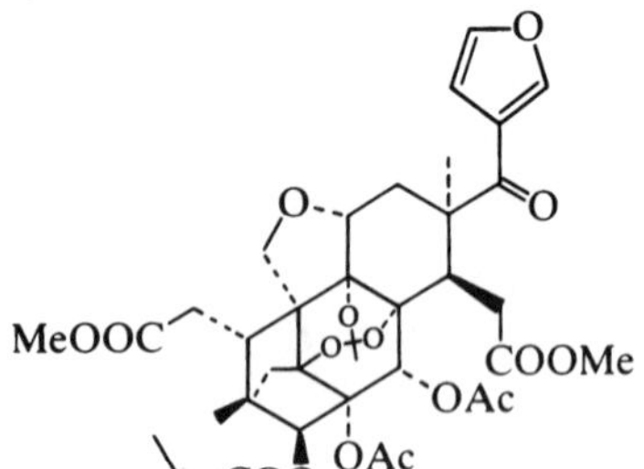

$C_{38}H_{46}O_{16}$ M 758.772
Constit. of *Pseudocedrela kotschyii*. Cryst. Mp 255-257°. $[\alpha]_D^{20}$ −70°.

Taylor, D.A.H., *Phytochemistry*, 1979, **18**, 1574 (*isol*)
Niven, M.L. *et al, Phytochemistry*, 1988, **27**, 1542 (*cryst struct*)

Psilostachyin   P-70131

Updated Entry replacing P-02613
[3533-47-9]

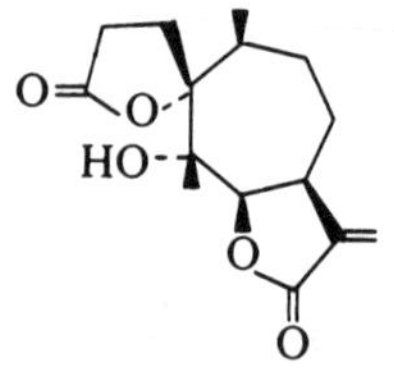

$C_{15}H_{20}O_5$ M 280.320
Constit. of *Ambrosia psilostachyia*. Cryst. (CH$_2$Cl$_2$/isopropyl ether). Mp 215°. $[\alpha]_D^{24}$ −125.2° (c, 4.76 in CHCl$_3$).

11R,13-Dihydro: [114246-84-3]. **11,13-Dihydropsilostachyin.**
$C_{15}H_{22}O_5$ M 282.336
Constit. of *A. confertiflora*. Cryst. Mp 225-226°. $[\alpha]_D^{25}$ −70° (c, 1 in MeOH).

Mabry, T.J. *et al, Tetrahedron*, 1966, **22**, 1139 (*isol, struct*)
Delgado, G. *et al, J. Nat. Prod.*, 1988, **51**, 83 (*deriv*)

Psorospermin   P-70132

Updated Entry replacing P-60189
[74045-97-9]

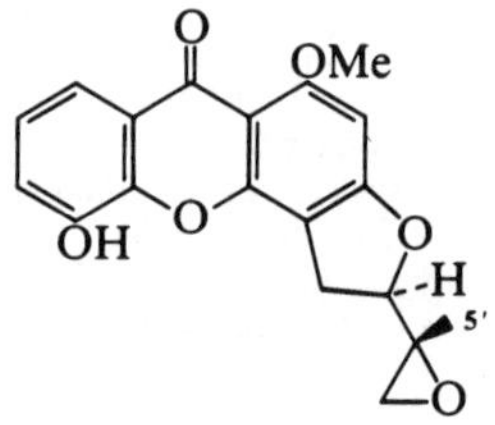

$C_{19}H_{16}O_6$ M 340.332
Constit. of *Psorospermum febrifugum*. Needles (Me$_2$CO/hexane). Mp 229-230°.

5'-Hydroxy: **5'-Hydroxypsorospermin.**
$C_{19}H_{16}O_7$ M 356.331
Constit. of *P. febrifugum*. Cryst. (MeOH). Mp 255-256°.

Kupchan, S.M. *et al, J. Nat. Prod.*, 1980, **43**, 296 (*isol*)
Ho, D.K. *et al, J. Org. Chem.*, 1987, **52**, 342 (*synth*)
Habib, A.M. *et al, J. Org. Chem.*, 1987, **52**, 412 (*isol, struct*)
Dixon, R.A. *et al, Phytochemistry*, 1988, **27**, 2801 (*deriv*)

Psorospermindiol   P-70133

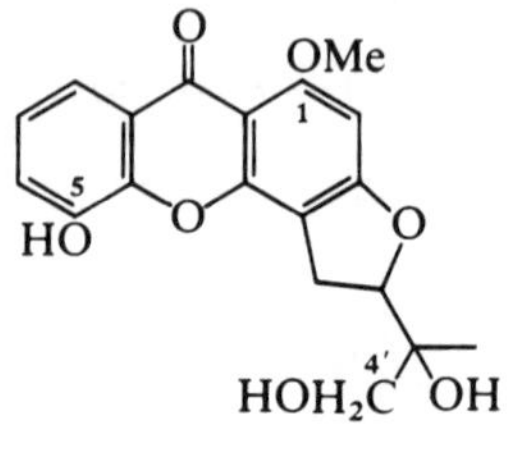

$C_{19}H_{18}O_7$ M 358.347
Constit. of *Psorospermum febrifugum*. Related to Psorospermin, P-70132 .

O-De-Me: **1-O-Demethylpsorospermindiol.**
$C_{18}H_{16}O_7$ M 344.320
Constit. of *P. febrifugum*. Cryst. (MeOH). Mp 220-221°. $[\alpha]_D^{20}$ −50° (c, 0.01 in MeOH).

O-De-Me, epimer:
$C_{18}H_{16}O_7$ M 344.320
Constit. of *P. febrifugum*. Cryst. (MeOH). Mp 222-223°. $[\alpha]_D^{20}$ +65.4° (c, 0.01 in MeOH).

O^5-Me: **5-O-Methylpsorospermindiol**.
$C_{20}H_{20}O_7$ M 372.374
Constit. of *P. febrifugum*. Cryst. (MeOH). Mp 235-236°. $[\alpha]_D^{20}$ −111° (c, 0.1 in MeOH).
$O^{4'}$-Me: **4'-O-Methylpsorospermindiol**.
$C_{20}H_{20}O_7$ M 372.374
Constit. of *P. febrifugum*. Cryst. (MeOH). Mp 243-245°.

Abou-Shoer, M. *et al*, *Phytochemistry*, 1988, **27**, 2795.

2,4,6(1H,3H,5H)-Pteridinetrione   P-70134

1,5-Dihydro-2,4,6(3H)-pteridinetrione, 9CI. *6-Oxolumazine*
[2577-35-7]

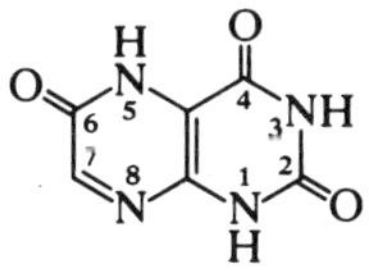

$C_6H_4N_4O_3$ M 180.123
Isol. from honey bee (*Apis mellifera*).
1-Me: [50996-37-7].
   $C_7H_6N_4O_3$ M 194.149
   Yellow-brown cryst. (H₂O). Mp 335-338° dec.
3-Me: [58947-87-8].
   $C_7H_6N_4O_3$ M 194.149
   Yellow-brown cryst. Mp >350°.
7-Me:
   $C_7H_6N_4O_3$ M 194.149
   Yellow-brown cryst. (H₂O). Mp >340°.
1,3-Di-Me: [61846-18-2].
   $C_8H_8N_4O_3$ M 208.176
   Yellowish cryst. (H₂O). Mp 300°.
1,7-Di-Me:
   $C_8H_8N_4O_3$ M 208.176
   Cryst. (H₂O). Mp 315° dec.
3,7-Di-Me:
   $C_8H_8N_4O_3$ M 208.176
   Mp >340°.
1,3,5-Tri-Me:
   $C_9H_{10}N_4O_3$ M 222.203
   Yellow needles (H₂O). Mp 180-181°.

Pfleiderer, W., *Chem. Ber.*, 1957, **90**, 2604, 2617 (*synth*)
McNutt, W.S. *et al*, *Biochemistry*, 1962, **1**, 1161 (*metab*)
Dustmann, J.H. *et al*, *Hoppe-Seylers Z. Physiol. Chem.*, 1971, **352**, 1599 (*isol*)

2,4,7(1H,3H,8H)-Pteridinetrione, 9CI, 8CI   P-70135

Isoxantholumazine. Violapterin. 7-Oxolumazine
[2577-38-0]

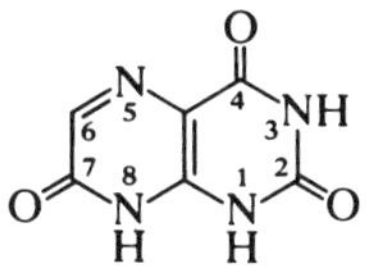

$C_6H_4N_4O_3$ M 180.123
Prod. by *Physarum polycephalum* and the ant *Formica polyctena*. Cryst. (H₂O). Mp >350°.
1-Me: [2614-44-0].
   $C_7H_6N_4O_3$ M 194.149
   Mp >340°.
3-Me: [2622-65-3].
   $C_7H_6N_4O_3$ M 194.149
   Cryst. (H₂O). Mp >340°.
6-Me: [31053-46-0].
   $C_7H_6N_4O_3$ M 194.149

Cryst. + 1H₂O (H₂O). Mp >340°.
8-Me: see *4-Oxopentanal, O-60083*
1,3-Di-Me: [2614-43-9].
   $C_8H_8N_4O_3$ M 208.176
   Mp 264°.
1,6-Di-Me:
   $C_8H_8N_4O_3$ M 208.176
   Cryst. (H₂O). Mp 330° dec.
1,8-Di-Me: [70916-42-6].
   $C_8H_8N_4O_3$ M 208.176
   Cryst. (EtOH). Mp 211°.
3,6-Di-Me:
   $C_8H_8N_4O_3$ M 208.176
   Cryst. (H₂O). Mp >340°.
3,8-Di-Me: [70916-39-1].
   $C_8H_8N_4O_3$ M 208.176
   Cryst. Mp >350°. Darkens at 300°.
6,8-Di-Me: [6743-25-5].
   $C_8H_8N_4O_3$ M 208.176
   Cryst. Mp >350°.
1,3,6-Tri-Me: [2625-21-0].
   $C_9H_{10}N_4O_3$ M 222.203
   Needles (H₂O). Mp 308°.
1,3,8-Tri-Me: [70674-02-1].
   $C_9H_{10}N_4O_3$ M 222.203
   Pale-yellow cryst. (EtOH). Mp 220°.
1,6,8-Tri-Me: [70916-41-5].
   $C_9H_{10}N_4O_3$ M 222.203
   Cryst. (EtOH). Dec. at 317°.
3,6,8-Tri-Me: [6743-26-6].
   $C_9H_{10}N_4O_3$ M 222.203
   Cryst. (EtOH). Mp 331-334°.
1,3,6,8-Tetra-Me:
   $C_{10}H_{12}N_4O_3$ M 236.230
   Pale-yellow needles (EtOH aq.). Mp 253°.

Johnson, T.B. *et al*, *Recl. Trav. Chim. Pays-Bas*, 1930, **49**, 197 (*synth*)
Pfleiderer, W., *Chem. Ber.*, 1957, **90**, 2588; 1958, **91**, 1671 (*synth, derivs*)
McNutt, W.S., *J. Biol. Chem*, 1963, **238**, 1116.
Schmidt, G.H., *Z. Naturforsch., B*, 1969, **24**, 1153 (*isol*)
Bergmann, F. *et al*, *J. Chem. Soc., Perkin Trans. 2*, 1979, 35 (*pmr, deriv*)
Loidl, P. *et al*, *CA*, 1984, **100**, 20127 (*isol*)

Ptilosarcenone   P-70136

[64597-87-1]

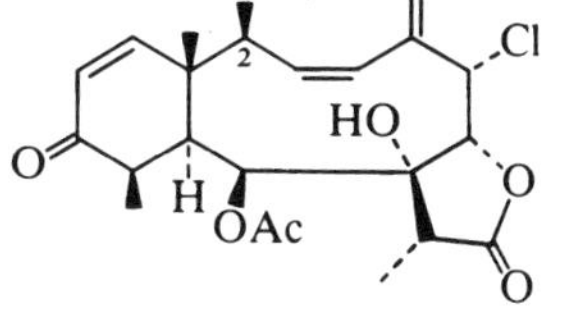

$C_{24}H_{29}ClO_8$ M 480.941
Constit. of *Ptilosarcus gurneyi* and *Tochuina tetraquetra*. Oil.

2-De-Ac, 2-butanoyl:
   $C_{26}H_{33}ClO_8$ M 508.995
   Constit. of *T. tetraquetra*. Oil. $[\alpha]_D^{25}$ −49.2° (c, 0.41 in CH₂Cl₂).

Wratten, S.J. *et al*, *Tetrahedron Lett.*, 1977, 1559 (*isol, struct*)
Williams, D.E. *et al*, *Can. J. Chem.*, 1987, **65**, 2244 (*isol*)

Puerol *A*   P-70137

Updated Entry replacing P-50268

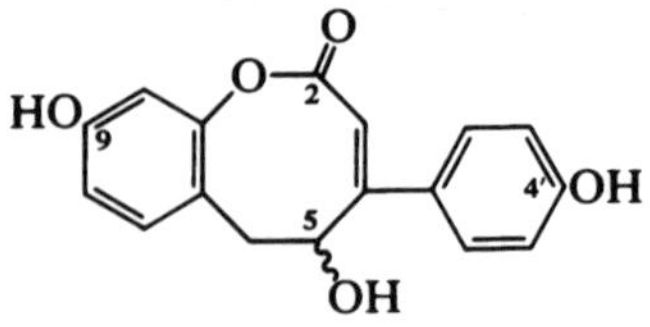

C$_{17}$H$_{14}$O$_5$ M 298.295

Constit. of roots of *Sophora japonica*. Needles. Mp 226-228°. [α]$_D$ −75.8° (MeOH). Aglucone isol. as racemate.

5-O-(α-L-Rhamnopyranosyl(1→6)-β-D-glucopyranoside): [100692-52-2]. **Pueroside A.**
C$_{29}$H$_{34}$O$_{14}$ M 606.579
Constit. of Puerariae Radix (*Pueraria lobata*). Needles. Mp 183-185°. [α]$_D$ −107.5° (MeOH).

O^9-Me: [112343-17-6]. **Puerol B.**
C$_{18}$H$_{16}$O$_5$ M 312.321
Plates. Mp 238-240°. [α]$_D$ +68.2° (MeOH).

O^9-Me, 5-O-β-D-glucopyranoside: [112343-16-5]. **Sophoraside A.**
C$_{24}$H$_{26}$O$_{10}$ M 474.463
Constit. of *S. japonica*. Needles. Mp 234-236°. [α]$_D^{25}$ +32° (c, 1.0 in DMSO).

O^9-Me, 4′,5-di-O-β-D-glucopyranoside: [100692-54-4]. **Pueroside B.**
C$_{30}$H$_{36}$O$_{15}$ M 636.605
Constit. of *P. lobata*. Needles. Mp 227-229°. [α]$_D$ −37.6° (MeOH).

Kinjo, J. *et al*, *Tetrahedron Lett.*, 1985, **26**, 6101 (*isol, struct*)
Shirataki, Y. *et al*, *Chem. Pharm. Bull.*, 1987, **35**, 1637 (*isol*)

Pulchellin *A*   P-70138

Updated Entry replacing P-40208
Pulchellin
[6754-35-4]

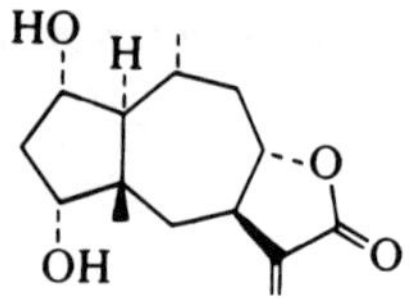

C$_{15}$H$_{22}$O$_4$ M 266.336
Constit. of *Gaillardia pulchella*. Cryst. (EtOAc). Mp 166-167°. [α]$_D^{26}$ −36.2° (c, 2.43 in CHCl$_3$).

2,4-Di-Ac: [23754-36-1]. **Pulchellin diacetate.**
C$_{19}$H$_{26}$O$_6$ M 350.411
From *Loxochysanus sinuatus*. Oil.

2-O-(3-Methylbutanoyl): **Pulchellin 2-O-isovalerate.**
C$_{20}$H$_{32}$O$_5$ M 352.470
From *L. sinuatus*. Oil. [α]$_D^{26}$ −13° (c, 0.43 in CHCl$_3$).

2-Tigloyl: [97605-30-6]. **Pulchellin 2-O-tiglate.**
C$_{20}$H$_{28}$O$_5$ M 348.438
From *L. sinuatus*. Oil.

6β-Hydroxy, 2-(3-Methylbutanoyl): [97605-31-7]. **6β-Hydroxypulchellin 2-O-isovalerate.**
C$_{20}$H$_{30}$O$_6$ M 366.453
From *L. sinuatus*. Cryst. Mp 127°.

6β-Hydroxy, 2-Ac: [97643-92-0]. **6β-Hydroxypulchellin 2-O-acetate.**
C$_{17}$H$_{24}$O$_6$ M 324.373
From *L. sinuatus*. Cryst. Mp 87°. [α]$_D^{24}$ +14° (c, 0.97 in MeOH).

4-Epimer: **4-Epipulchellin.**
C$_{15}$H$_{24}$O$_4$ M 268.352
Constit. of *G. pulchella*. Oil.

4-Epimer, di-Ac: Cryst. Mp 133-134°.

6α-Hydroxy: **6α-Hydroxypulchellin.**
C$_{15}$H$_{22}$O$_5$ M 282.336
Constit. of *G. pulchella*.

6α-Hydroxy, O^4-angeloyl: **6α-Hydroxypulchellin 4-O-angelate.**
C$_{20}$H$_{28}$O$_6$ M 364.438
Constit. of *G. pulchella*.

6β-Acetoxy, O^4-angeloyl: **6β-Acetoxypulchellin 4-O-angelate.**
C$_{22}$H$_{30}$O$_7$ M 406.475
Constit. of *G. pulchella*.

Herz, W. *et al*, *Tetrahedron*, 1963, **19**, 483 (*isol, uv, ir, pmr*)
Aota, K. *et al*, *J. Org. Chem.*, 1970, **35**, 1448 (*isol*)
Bohlmann, F. *et al*, *Phytochemistry*, 1985, **24**, 1021 (*isol*)
Inayama, S. *et al*, *Heterocycles*, 1985, **23**, 377 (*synth*)
Harimaya, K. *et al*, *Heterocycles*, 1988, **27**, 83 (*epimer*)
Yu, S. *et al*, *Phytochemistry*, 1988, **27**, 2887 (*derivs*)

Pulchelloside I   P-70139

Updated Entry replacing P-10266
[67244-49-9]

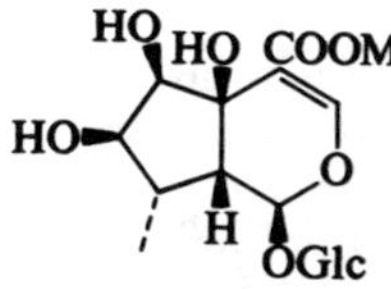

C$_{17}$H$_{26}$O$_{12}$ M 422.385
Constit. of *Verbena pulchella*. Amorph. solid. [α]$_D^{21}$ −148° (c, 1 in EtOH).

Hexa-Ac: Cryst. Mp 166-169°.

5-Deoxy: **5-Deoxypulchelloside I.**
C$_{17}$H$_{26}$O$_{11}$ M 406.386
Constit. of *Citharexylum fruticosum*.

Milz, S. *et al*, *Tetrahedron Lett.*, 1978, 895 (*isol, struct*)
Bianco, A. *et al*, *Gazz. Chim. Ital.*, 1981, **111**, 201 (*cmr*)
Ganapaty, S. *et al*, *Planta Med.*, 1988, **54**, 42 (*deriv*)

Pumilaisoflavone *A*   P-70140

[115712-89-5]

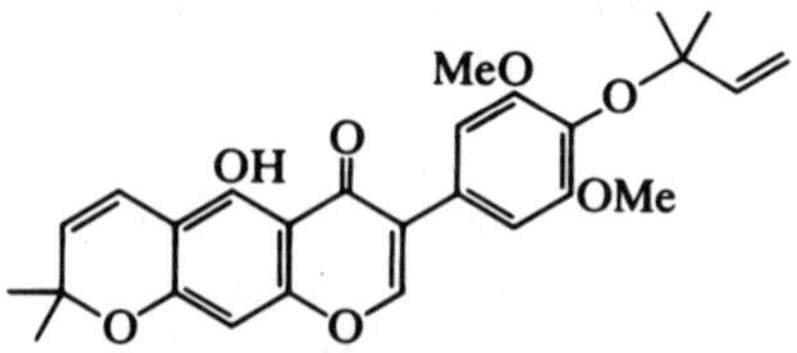

C$_{27}$H$_{28}$O$_7$ M 464.514
Constit. of *Tephrosia pumila*. Needles. Mp 197-200°.

Dagne, E. *et al*, *Phytochemistry*, 1988, **27**, 1503.

Pumilaisoflavone *B*   P-70141

[115712-90-8]

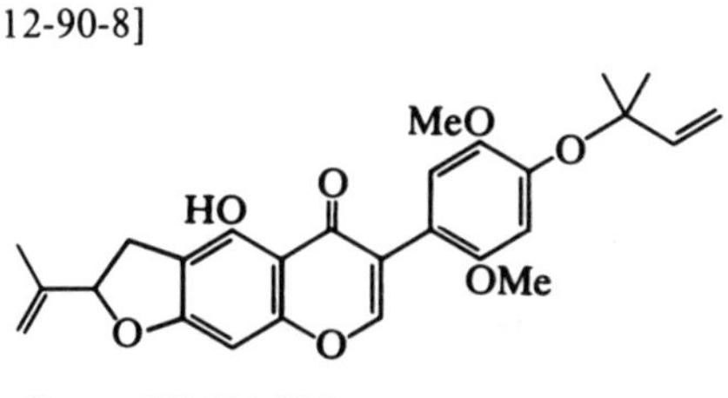

C$_{27}$H$_{28}$O$_7$ M 464.514

Constit. of *Tephrosia pumila*. Cubes (CHCl₃). Mp 126-129°.

Dagne, E. *et al*, *Phytochemistry*, 1988, **27**, 1503.

Punaglandin 1   P-70142
5,6,7-Tris(acetyloxy)-10-chloro-12-hydroxy-9-oxo-prosta-10,14,17-trien-1-oic acid methyl ester, 9Cl
[96055-63-9]

$C_{27}H_{37}ClO_{10}$ M 557.036
Isol. from Japanese octocoral *Telesto riisei*. Possesses potent antitumour props.

17,18-Dihydro: [96055-64-0]. ***Punaglandin 2***.
   $C_{27}H_{39}ClO_{10}$ M 559.052
   From *T. riisei*. Possesses potent antitumour props.

Baker, B.J. *et al*, *J. Am. Chem. Soc.*, 1985, **107**, 2976 (*isol, struct*)
Sasai, H. *et al*, *Tetrahedron Lett.*, 1987, **28**, 333.

Punaglandin 3   P-70143
5,6-Bis(acetyloxy)-10-chloro-12-hydroxy-9-oxoprosta-7,10,14,17-tetraen-1-oic acid methyl ester, 9Cl
[96055-65-1]

$C_{25}H_{33}ClO_8$ M 496.984
Isol. from Japanese octocoral *Telesto riisei*. Possesses potent antitumour props. Oil. The 12-config. in Punaglandins 3 and 4 is the opposite of that originally assigned and the opposite of that in Punaglandin 1, P-70142.

17,18-Dihydro: [96055-66-2]. ***Punaglandin 4***.
   $C_{25}H_{35}ClO_8$ M 499.000
   From *T. riisei*. Possesses potent antitumour props. Oil. $[\alpha]_D^{24}$ +72.3° (c, 0.52 in CHCl₃) (synthetic). n_D^{24} 1.4967.

17,18-Dihydro, 14E-isomer: [96055-66-2]. *(E)-Punaglandin 4.* Isol. from *T. riisei*. Oil.

Baker, B.J. *et al*, *J. Am. Chem. Soc.*, 1985, **107**, 2976 (*isol*)
Nagaoka, H. *et al*, *J. Am. Chem. Soc.*, 1986, **108**, 5019, 5021 (*synth, struct*)
Sasai, H. *et al*, *Tetrahedron Lett.*, 1987, **28**, 333 (*synth*)
Mori, K. *et al*, *Tetrahedron*, 1988, **44**, 333 (*synth, ir, pmr, ms, cd*)
Suzuki, M. *et al*, *J. Org. Chem.*, 1988, **53**, 286 (*synth, uv, pmr, ms, abs config, bibl*)

Pygmaeocine *E*   P-70144
7-Hydroxy-5,5,9-trimethyl-(1-methylethyl)-1,2,6(5H)-anthracenetrione, 9Cl. 7-Hydroxy-3-isopropyl-5,5,9-trimethyl-1,2,6(5H)-anthracenetrione
[115333-92-1]

$C_{20}H_{20}O_4$ M 324.376
Constit. of *Pygmaeopremna herbacea*. Brownish-red prisms (CHCl₃/MeOH). Mp 192-193°.

Meng, Q. *et al*, *Phytochemistry*, 1988, **27**, 1151 (*isol, cryst struct*)

Pygmaeoherin   P-70145

$C_{17}H_{18}O_4$ M 286.327
Constit. of *Pygmaeopremna herbacea*. Needles (CHCl₃/MeOH). Mp 198-200°.

Meng, Q. *et al*, *Planta Med.*, 1988, **54**, 48.

Pyoverdin *C*   P-70146
[104022-78-8]

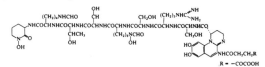

$C_{56}H_{85}N_{17}O_{23}$ M 1364.388
Oligopeptide antibiotic. Isol. from *Pseudomonas aeruginosa*. Siderophore. No phys. props. reported.

Wendenbaum, S. *et al*, *Tetrahedron Lett.*, 1983, **24**, 4877 (*isol*)
Hider, R.C., *Structure and Bonding*, 1984, **58**, 26 (*rev*)
Briskot, G. *et al*, *Z. Naturforsch., C*, 1986, **41**, 497 (*isol, struct*)

Pyoverdin *D*   P-70147
[104022-79-9]
As Pyoverdin *C*, P-70146 with

$$R = COOH$$

$C_{55}H_{85}N_{17}O_{22}$ M 1336.377
Isol. from *Pseudomonas aeruginosa*. Siderophore.

Amide: [88966-86-3]. ***Pyoverdin E***. *Pyoverdin* Pa.
   $C_{55}H_{86}N_{18}O_{21}$ M 1335.392
   From *P. aeruginosa*. Siderophore.

Wendenbaum, S. *et al*, *Tetrahedron Lett.*, 1983, **24**, 4877 (*isol*)
Hider, R.C., *Structure and Bonding*, 1984, **58**, 26 (*rev*)
Briskot, G. *et al*, *Z. Naturforsch., C*, 1986, **41**, 497 (*isol, struct*)

Pyoverdin I, 9CI   P-70148

[114616-35-2]

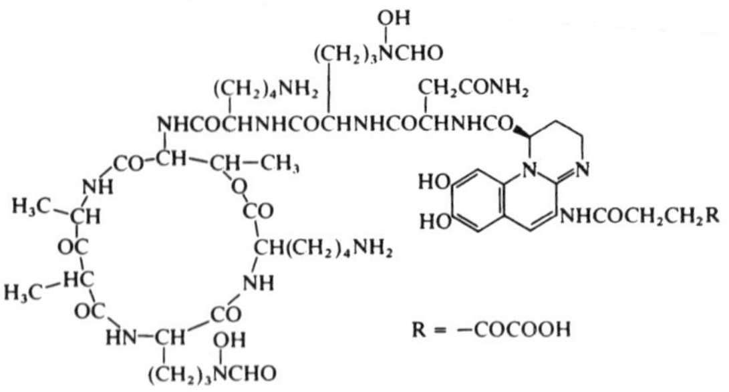

$C_{56}H_{84}N_{16}O_{21}$ M 1317.374

Cyclic depsipeptide antibiotic complex. Isol. from *Pseudomonas fluorescens*. Siderophore.

Poppe, K. *et al*, *Tetrahedron*, 1987, **43**, 2261 (*struct, bibl*)

Pyoverdin II   P-70149

[114587-58-5]

As Pyoverdin I, P-70148 with

$$R = -COOH$$

$C_{55}H_{84}N_{16}O_{20}$ M 1289.364

Isol. from *Pseudomonas fluorescens*. Siderophore.

Amide: [114587-59-6]. **Pyoverdin III**.
  $C_{55}H_{85}N_{17}O_{19}$ M 1288.379
  From *P. fluorescens*. Siderophore.

Poppe, K. *et al*, *Tetrahedron*, 1987, **43**, 2261 (*struct, bibl*)

4,4-Pyrandiacetic acid   P-70150

[110796-49-1]

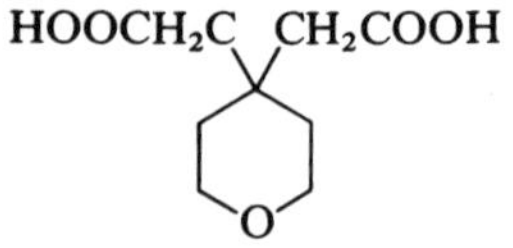

$$HOOCH_2C \quad CH_2COOH$$

$C_9H_{14}O_5$ M 202.207

Microcryst. Mp 113-115°.

Di-Me ester: [110796-54-8].
  $C_{11}H_{18}O_5$ M 230.260
  Microcryst. Mp 197-197.5°.

Di-Et ester: [110796-53-7].
  $C_{13}H_{22}O_5$ M 258.314
  Bp$_{0.5-0.8}$ 103-105°.

Newkome, G.R. *et al*, *J. Org. Chem.*, 1987, **52**, 5480 (*synth, ir, pmr, cmr*)

Pyranthrene, 9CI   P-70151

[128-70-1]

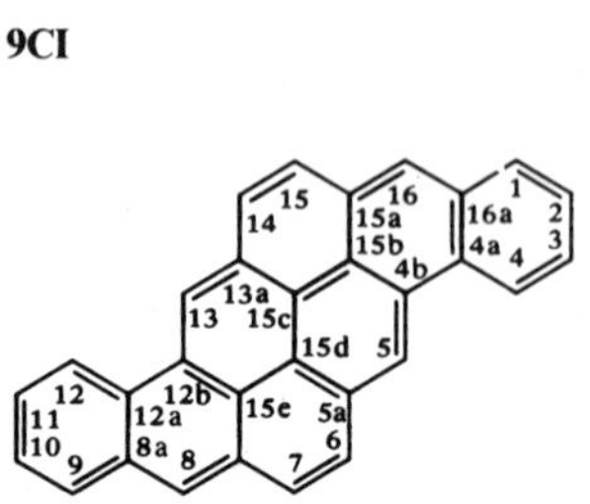

$C_{30}H_{16}$ M 376.456

Red-brown needles (xylene or by subl.). Mp 372-373°.

Clar, E., *Ber.*, 1943, **76**, 328 (*synth, uv*)
Kelly, W. *et al*, *Aust. J. Chem.*, 1960, **13**, 103 (*synth*)
Lacassagne, A. *et al*, *C.R. Hebd. Seances Acad. Sci.*, 1961, **252**, 826 (*tox*)

1H-Pyrazole, 9CI   P-70152

Updated Entry replacing P-02754
1,2-Diazole
[288-13-1]

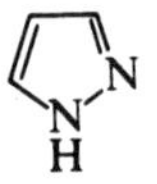

$C_3H_4N_2$ M 68.078

Needles or prisms (pet. ether). Sol. H_2O, EtOH, Et_2O, C_6H_6. Mp 69.5-70°. Bp$_{757.9}$ 186-188°. pK_a 2.52. No annular tautomerism present in cryst. state at r.t.

▷ UQ4900000.

B,HCl: Hygroscopic needles. Mp 104°.
B$_2$,H$_2$SO$_4$: Cryst. (EtOH). Mp 134°.
B,HNO$_3$: Needles. Mp 148°.
N-Ac:
  $C_5H_6N_2O$ M 110.115
  Oil. Bp$_{744}$ 155-156°.
N-Benzoyl:
  $C_{10}H_8N_2O$ M 172.186
  Mp 46°. Bp$_{747}$ 281°, Bp$_{11}$ 140°.
N-2,4-Dinitrophenyl: Cryst. (Me$_2$CO aq.). Mp 104-106°.
N-Me: see 1-Methylpyrazole, M-03446
N-COOR: see 1H-Pyrazole-1-carboxylic acid, P-02755

v. Pechmann, H., *Ber.*, 1898, **31**, 2950 (*synth*)
Wilshire, J.F.K., *Aust. J. Chem.*, 1966, **19**, 1935 (*pmr*)
Wiley, R.H., Ed., *Pyrazoles, Pyrazolines, Pyrazolidines, Indazoles, and Condensed Ring Systems*, Interscience, 1967 (*rev*)
Larsen, F.K. *et al*, *Acta Chem. Scand.*, 1970, **24**, 3248 (*cryst struct*)
Kasai, P.H. *et al*, *J. Am. Chem. Soc.*, 1973, **95**, 27 (*esr*)
Dumanovic, D. *et al*, *Talanta*, 1975, **22**, 819 (*uv*)
Litchman, W.M., *J. Am. Chem. Soc.*, 1979, **101**, 545 (*cmr*)
Schuster, I.I. *et al*, *J. Org. Chem.*, 1979, **44**, 1765 (*nmr*)
Faure, R. *et al*, *Can. J. Chem.*, 1988, **66**, 1141 (*cmr, tautom*)
Sax, N.I., *Dangerous Properties of Industrial Materials*, 5th Ed., Van Nostrand-Reinhold, 1979, 868.

Pyrazole blue   P-70153

4-(1,5-Dihydro-3-methyl-5-oxo-1-phenyl-4H-pyrazol-4-ylidene)-2,4-dihydro-5-methyl-2-phenyl-3H-pyrazol-3-one, 9CI. 3-Methyl-4-[(5-oxo-3-phenyl-2-pyrazolin-4-ylidene)methyl]-1-phenyl-2-pyrazolin-5-one, 8CI. 3,3'-Di-methyl-1,1'-diphenyl[Δ$^{4,4'}$-bi-2-pyrazoline]-5,5'-dione
[6334-24-3]

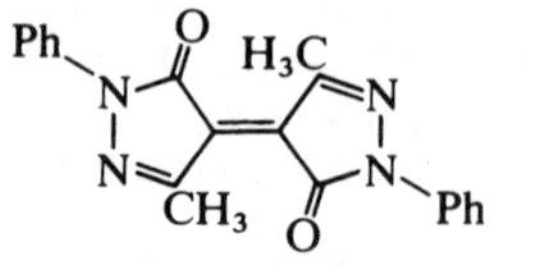

$C_{20}H_{16}N_4O_2$ M 344.372

There are also naphthalene dyes called pyrazol blues.

(E)-form [74451-57-3]
  Violet needles (EtOH/Et$_2$O). Mp 241-242° (230°) dec. Conts. reactive central double bond.

Knorr, L., *Justus Liebigs Ann. Chem.*, 1887, **238**, 137 (*synth*)
Westöö, G., *Acta Chem. Scand.*, 1953, **7**, 355 (*synth*)
Russell, G.A. *et al*, *Tetrahedron*, 1970, **26**, 3449 (*synth*)
Das, N.B. *et al*, *J. Indian Chem. Soc.*, 1978, **55**, 829 (*synth, uv, pmr*)
Mann, G. *et al*, *Tetrahedron Lett.*, 1979, 4645.

10*H*-Pyrazolo[5,1-*c*][1,4]benzodiazepine, 9CI   P-70154

[87592-08-3]

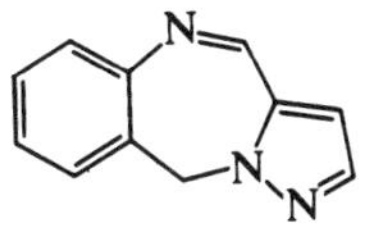

$C_{11}H_9N_3$ M 183.212
Cryst. (H_2O). Mp 105-109°.

Cecchi, L. *et al*, *J. Heterocycl. Chem.*, 1983, **20**, 871 (*synth, ir, pmr, ms*)

1*H*-Pyrazolo[3,4-*b*]pyridine, 9CI   P-70155

Updated Entry replacing P-02768
1,2,7-Triazaindene
[271-73-8]

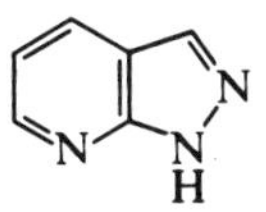

$C_6H_5N_3$ M 119.126
Mp 98-99°.

Hatt, T.L.P. *et al*, *J. Chem. Soc., Chem. Commun.*, 1966, 293 (*synth, uv*)
Khan, M.A. *et al*, *J. Heterocycl. Chem.*, 1970, **7**, 247 (*synth*)
Stanovnik, B. *et al*, *Synthesis*, 1979, 194 (*synth*)
Lynch, B.M. *et al*, *Can. J. Chem.*, 1988, **66**, 420 (*synth, pmr,*

1*H*-Pyrazolo[3,4-*d*]pyrimidine, 9CI   P-70156

[271-80-7]

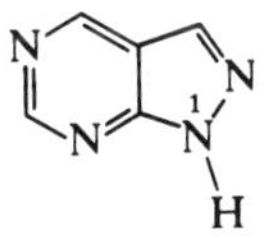

$C_5H_4N_4$ M 120.113
Deriv. show anti-tumour activity. Needles. Mp 213-214°.
1-Me: [6288-86-4].
     $C_6H_6N_4$ M 134.140
     Needles (C_6H_6). Mp 125-126°.
1-Ph: [53645-79-7].
     $C_{11}H_8N_4$ M 196.211
     Needles (pet. ether). Mp 79-80°.
1-Me, 5-oxide: [62564-63-0].
     $C_6H_6N_4O$ M 150.140
     Yellow needles (MeOH). Mp 187-189°.
1-Ph, 5-oxide: [62564-64-1].
     $C_{11}H_8N_4O$ M 212.210
     Yellow needles (EtOH). Mp 195-196°.

Robins, R.K., *J. Am. Chem. Soc.*, 1956, **78**, 784 (*synth*)
Robins, R.K. *et al*, *J. Org. Chem.*, 1956, **21**, 1240 (*deriv, synth, uv*)
Higashino, T. *et al*, *Yakugaku Zasshi*, 1974, **96**, 666 (*deriv, synth*)
Higashino, T. *et al*, *Chem. Pharm. Bull.*, 1976, **24**, 3120 (*deriv, synth*)

1*H*-Pyrazolo[5,1-*c*]-1,2,4-triazole   P-70157

[29176-18-9]

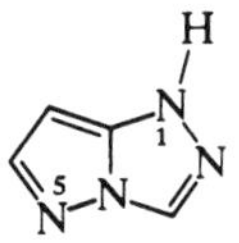

$C_4H_4N_4$ M 108.102

Tautomeric in soln. with 5*H*-form. Derivs. are intermeds. for colour photographic dyes. Many patented. Cryst. (C_6H_6). Mp 173-178°. Obt. in only 1% yield.

Reimlinger, H. *et al*, *Chem. Ber.*, 1970, **103**, 3284 (*synth, ir, pmr*)

2-(1*H*-Pyrazol-1-yl)pyridine, 9CI   P-70158

1-(2-Pyridyl)pyrazole
[25700-11-2]

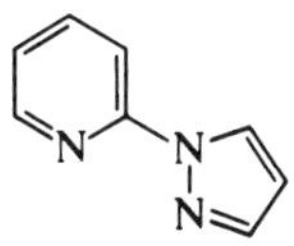

$C_8H_7N_3$ M 145.163
Complexing agent. Mp 38-40°.
B,HCl: Cryst. (EtOH). Mp 121-125°.

Polya, J.B. *et al*, *J. Chem. Soc. (C)*, 1970, 85 (*synth*)
Kauffmann, T. *et al*, *Angew. Chem., Int. Ed. Engl.*, 1972, **11**, 846 (*synth*)
Khan, M.A. *et al*, *Monatsh. Chem.*, 1980, **111**, 883 (*synth*)
Canty, A.J. *et al*, *Organometallics*, 1982, **1**, 1063.
Steel, P.J. *et al*, *Inorg. Chem.*, 1983, **22**, 1488.
Visalakshi, R. *et al*, *Inorg. Chim. Acta*, 1986, **118**, 119 (*ir, pmr, complexes*)
Sugiyarto, K.H. *et al*, *Aust. J. Chem.*, 1988, **41**, 1645 (*synth, complexes*)

2-(1*H*-Pyrazol-3-yl)pyridine, 9CI   P-70159

3-(2-Pyridyl)pyrazole
[75415-03-1]
$C_8H_7N_3$ M 145.163
Tautomeric with 5-pyrazolyl isomer. Cryst. ($CHCl_3$/hexane). Mp 124-128° (119-120°).

Sugiyarto, K.H. *et al*, *Aust. J. Chem.*, 1988, **41**, 1645 (*synth*)

3-(1*H*-Pyrazol-1-yl)pyridine, 9CI   P-70160

1-(3-Pyridyl)pyrazole
[25700-12-3]
$C_8H_7N_3$ M 145.163
Mp 30-31°.

Khan, M.A. *et al*, *J. Chem. Soc. (C)*, 1970, 85 (*synth, ir, uv*)

3-(1*H*-Pyrazol-3-yl)pyridine, 9CI   P-70161

3-(3-Pyridyl)pyrazole
[45887-08-9]
$C_8H_7N_3$ M 145.163
Tautomeric with 5-pyrazolyl isomer. Yellow oil.
B,HCl: [51747-02-5]. Mp 213-214°.
Monopicrate: [27509-30-4]. Cryst. (EtOH), thin plates or needles (H_2O). Mp 194-196°.
B,MeI: Tablets (MeOH). Mp 217.5°.

Gough, G.A.C. *et al*, *J. Chem. Soc.*, 1933, 350 (*synth*)
Terent'ev, P.B. *et al*, *Chem. Heterocycles* Compd. (*Engl. Transl.*), 1970, **6**, 460 (*synth, deriv*)
Schunack, W., *Arch. Pharm.* (*Weinheim, Ger.*), 1973, **306**, 934 (*synth, pharmacol*)

4-(1*H*-Pyrazol-1-yl)pyridine, 8CI   P-70162

1-(4-Pyridyl)pyrazole
[25700-13-4]
$C_8H_7N_3$ M 145.163

Mp 84-86°.

Khan, M.A. *et al*, *J. Chem. Soc. (C)*, 1970, 85 (*synth, ir, uv*)

4-(1*H*-Pyrazol-3-yl)pyridine, 9CI   P-70163

3-(*4-Pyridyl*)*pyrazole*

[17784-60-0]

$C_8H_7N_3$ M 145.163

Tautomeric with 5-pyrazolyl isomer. Cryst. (Me$_2$CO).
Mp 157-158° (151-153°).

Picrate: Mp 220°.

Fabrini, L., *Farmaco, Ed. Sci.*, 1954, **9**, 603 (*synth, deriv*)
U.S.P., 3 341 413, (*1967*); *CA*, **68**, 95812g (*synth, pharmacol*)
Bauer, V.J. *et al*, *J. Med. Chem.*, 1968, **11**, 981 (*synth, pharmacol*)

4-(1*H*-Pyrazol-4-yl)pyridine, 9CI   P-70164

4-(*4-Pyridyl*)*pyrazole*

[19959-71-8]

$C_8H_7N_3$ M 145.163

Cryst. (H$_2$O). Mp 198-199°. Can be sublimed.

Arnold, Z., *Collect. Czech. Chem. Commun.*, 1963, **28**, 863 (*synth*)
Bauer, V.J. *et al*, *J. Med. Chem.*, 1968, **11**, 981 (*synth, pharmacol*)

Pyreno[1,2-*c*]furan   P-70165

[56555-55-6]

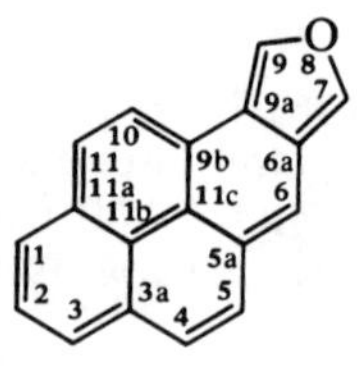

$C_{18}H_{10}O$ M 242.276

Light-yellow solid. Mp 94-98° (dec. from 63°). Highly
reactive.

Moursounidis, J. *et al*, *Aust. J. Chem.*, 1988, **41**, 235 (*synth, pmr, uv, ms*)

Pyreno[4,5-*c*]furan   P-70166

[15123-40-7]

$C_{18}H_{10}O$ M 242.276

Cryst. (CH$_2$Cl$_2$). Mp 218° dec. (darkens from 142°).

Moursounidis, J. *et al*, *Aust. J. Chem.*, 1988, **41**, 235 (*synth, pmr, uv, ms*)

Pyrenophorin   P-70167

Updated Entry replacing P-50286

8,16-Dimethyl-1,9-dioxacyclohexadeca-3,11-diene-2,5,10,13-tetrone, 9CI

[5739-85-5]

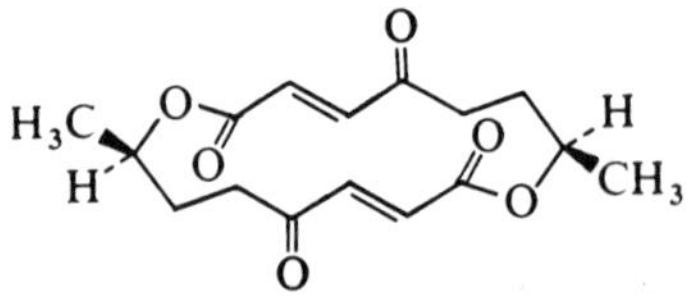

$C_{16}H_{20}O_6$ M 308.330

Macrolide antibiotic. Metab. of the plant pathogenic
fungus *Pyrenophora avenae* and of *Stemphylium
radicinium*. Antifungal agent. Needles (EtOH or pet.
ether). Mp 176°. $[\alpha]_D^{20}$ −47° (c, 0.36 in Me$_2$CO).

▷JG7524000.

Tetrahydro: Mp 156°.

Colvin, E.W. *et al*, *J. Chem. Soc., Perkin Trans. 1*, 1976, 1718 (*synth*)
Seuring, B. *et al*, *Justus Liebigs Ann. Chem.*, 1978, 2044 (*synth, abs config*)
Mali, R.S. *et al*, *Justus Liebigs Ann. Chem.*, 1981, 2272 (*synth*)
Breuilles, P. *et al*, *Tetrahedron Lett.*, 1984, **25**, 5759 (*synth*)
Dergiuni, F. *et al*, *Tetrahedron Lett.*, 1984, **25**, 5763 (*synth*)
Wakamatsu, T. *et al*, *Tetrahedron Lett.*, 1985, **26**, 1989 (*synth, bibl*)
Hatakeyama, S. *et al*, *Tetrahedron Lett.*, 1987, **28**, 2717 (*synth*)

3,6-Pyridazinedicarboxylic acid   P-70168

[57266-70-3]

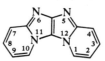

$C_6H_4N_2O_4$ M 168.109

Pale-yellow cryst. + 2H$_2$O (H$_2$O). Mp 230°. Hydrate ef-
floresces at r.t.

Di-Me ester: [2166-24-7].
  $C_8H_8N_2O_4$ M 196.162
  Yellow cryst. (MeOH). Mp 202°.
Di-Et ester: [113631-50-8].
  $C_{10}H_{12}N_2O_4$ M 224.216
  Yellow cryst. (EtOH). Mp 114°.
Dihydrazide: [113631-52-0].
  $C_6H_8N_6O_2$ M 196.168
  Yellow cryst. Mp >290° dec.

Sueur, S. *et al*, *J. Heterocycl. Chem.*, 1987, **24**, 1285 (*synth, spectra, cryst struct*)

Pyridazino[1″,6″:1′,2′]imidazo[4′,5′:4,5]-   P-70169
imidazo[1,2-*b*]pyridazine, 9CI

1,5,6,10,10a,10c-Hexaazadibenzo[a,f]*pentalene*

[117052-59-2]

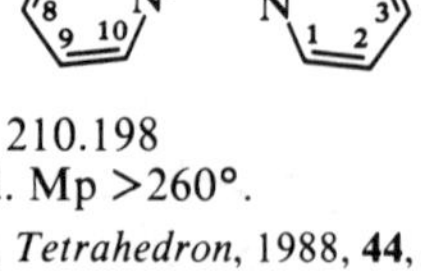

$C_{10}H_6N_6$ M 210.198

Pale-yellow solid. Mp >260°.

Pereira, D.E. *et al*, *Tetrahedron*, 1988, **44**, 3149 (*synth, ir, uv, pmr*)

2(1*H*)-Pyridineselone, 9CI   P-70170

2-Selenopyridine
[2044-26-0]

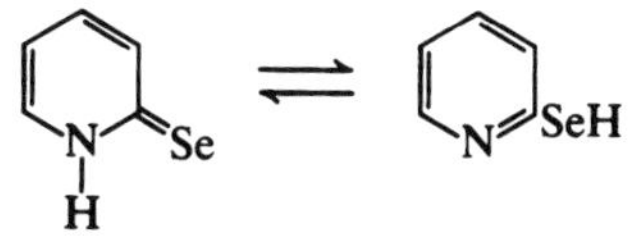

C₅H₅NSe   M 158.061

NH-form predominates. Yellow needles (C₆H₆). Mp 132-137°. pK_a 9.36. Sensitive to oxidn.

N-Oxide:
   C₅H₅NOSe   M 174.061
   Cryst. (EtOH). Mp 72.5-73°.
N-Me:
   C₆H₇NSe   M 172.088
   Cryst. (C₆H₆/pet. ether). Mp 68-78°. Air-sensitive.

SeH-form

Se-*Me:* 2-(*Methylseleno*)*pyridine.*
   C₆H₇NSe   M 172.088
   Light-yellow liq. with unpleasant odour. Bp₀.₂₅ 43-44°. n_D^{22} 1.6190.

Mautner, H.G. *et al, J. Org. Chem.*, 1962, **27**, 3671 (*synth, uv*)
Srikrishnan, T., *Acta Crystallogr., Sect. C*, 1988, **44**, 290 (*cryst struct, deriv*)

2-Pyridinesulfonic acid, 9CI   P-70171

Updated Entry replacing P-02835
[15103-48-7]

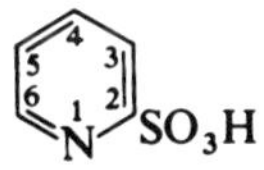

C₅H₅NO₃S   M 159.159
Needles (EtOH). V. sol. H₂O. Mp 251-252° (247-248°).
NH₄ salt: Mp 274-275°.
Ag salt: Yellow cryst. (H₂O). V. sol. hot H₂O. Mp 290°.
Chloride:
   C₅H₄ClNO₂S   M 177.605
   Cryst. below 0°. Loses SO₂ >0°.
Hydrazide:
   C₅H₇N₃O₂S   M 173.189
   Cryst. Mp 86-87° dec.

den Hertog, H.J. *et al, Recl. Trav. Chim. Pays-Bas*, 1958, **77**, 963 (*synth, uv*)
Evans, R.F. *et al, J. Org. Chem.*, 1962, **27**, 1329 (*synth*)
Hanessian, S. *et al, Synthesis*, 1987, 409 (*deriv, synth*)

N-Pyridinium-2-benzimidazole, 9CI   P-70172

Updated Entry replacing P-02854
Pyridinium 2-benzimidazolylide, 8CI
[105958-17-6]

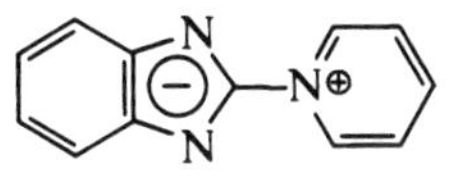

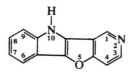

C₁₂H₉N₃   M 195.223
Bright-yellow needles (H₂O or EtOH aq.). Mp 269-270° (ca. 260° dec.). Stable to light and air.
B,HCl: [105958-33-6]. Cryst. (EtOH/H₂O/HCl). Mp 229-230°.

Boyd, G.V., *Tetrahedron Lett.*, 1966, 3369 (*synth, uv*)
Alcalde, E. *et al, J. Org. Chem.*, 1987, **52**, 5009 (*synth, props, cryst struct*)

2-Pyridinyl trifluoromethanesulfonate, 9CI   P-70173

2-(Trifluoromethylsulfonyloxy)pyridine
[65007-00-3]

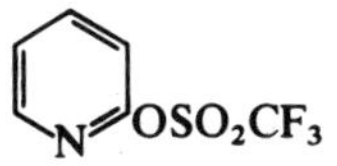

C₆H₄F₃NO₃S   M 227.158
Reagent for ketone synth. by acylation. Liq. Bp₂₅ 108°.

Keumi, T. *et al, Bull. Chem. Soc. Jpn.*, 1988, **61**, 455 (*synth, ir, pmr, ms, use*)

Pyrido[3',4':4,5]furo[3,2-*b*]indole, 9CI   P-70174

[110167-42-5]

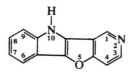

C₁₃H₈N₂O   M 208.219
Cryst. (THF). Mp 240-242°.

Koreňová, A. *et al, Collect. Czech. Chem. Commun.*, 1987, **52**, 192.

Pyrido[1'',2'':1',2']imidazo[4',5':4,5]-imidazo[1,2-*a*]pyridine, 9CI   P-70175

5,6,10a,10c-Tetraazadibenzo[a,f]*pentalene*
[100460-12-6]

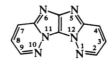

C₁₂H₈N₄   M 208.222
Pale-yellow cryst. Mod. sol. H₂O. Mp >300°.

Cruickshank, K.A. *et al, Tetrahedron Lett.*, 1985, **26**, 2723 (*synth, pmr, cmr, uv, ms*)

Pyrido[1',2':3,4]imidazo[1,2-*a*]pyrimidin-5-ium(1+), 9CI   P-70176

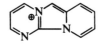

C₁₀H₈N₃⊕   M 170.193 (ion)
Perchlorate: [115879-75-9].
   C₁₀H₈ClN₃O₄   M 269.644
   Yellow cryst. (EtOH aq.). Mp 238-239° dec.

Jones, T.R. *et al, J. Chem. Soc., Perkin Trans. 1*, 1987, 2585 (*synth, ir, uv, pmr*)

6*H*-Pyrido[3',4':4,5]thieno[2,3-*b*]indole, 9CI   P-70177

[111550-91-5]

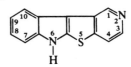

C₁₃H₈N₂S   M 224.280
Cryst. (MeOH). Mp 278-280°.

Levy, J. *et al*, *Bull. Soc. Chim. Fr.*, 1987, 193 (*synth, uv, pmr, cmr, cryst struct*)

(2-Pyridyl)(1-pyrazolyl)methane   P-70178

2-[(1H-Pyrazol-1-yl)methyl]pyridine, 9CI
[105575-75-5]

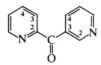

$C_9H_9N_3$ M 159.190
Complexing agent. Oil.

House, D.A. *et al*, *Aust. J. Chem.*, 1986, **39**, 1525 (*synth, pmr, cmr*)

2-Pyridyl 3-pyridyl ketone, 8CI   P-70179

2-Pyridinyl-3-pyridinylmethanone, 9CI
[56970-91-3]

$C_{11}H_8N_2O$ M 184.197
Cryst. (pet. ether). Mp 71°.

Oxime: [58088-16-7].
    $C_{11}H_9N_3O$ M 199.212
    Cryst. (EtOH). Mp 190-192°. Mixt. of stereoisomers.

Trovato, S. *et al*, *J. Mol. Struct.*, 1975, **28**, 216 (*props*)
Niemers, E. *et al*, *Synthesis*, 1976, 593 (*deriv*)
Trovato, S. *et al*, *Spectrochim. Acta, Part A*, 1976, **32**, 351 (*synth, uv*)

2-Pyridyl 4-pyridyl ketone   P-70180

2-Pyridinyl-4-pyridinylmethanone
[56970-92-4]
$C_{11}H_8N_2O$ M 184.197
Needles (pet. ether). Mp 121-122°.

Picrate: Yellow cryst. (EtOH). Mp 173.4-174.5°.
Semicarbazone: Cryst. (EtOH). Mp 241.5-242.5°.
Oxime: [58088-17-8].
    $C_{11}H_9N_3O$ M 199.212
    Cryst. (EtOH). Mp 121-122.5°. Mixt. of stereoisomers.

Wibaut, J.P. *et al*, *Recl. Trav. Chim. Pays-Bas*, 1951, **70**, 1054 (*synth, derivs*)
Niemers, E. *et al*, *Synthesis*, 1976, 593 (*deriv*)

3-Pyridyl 4-pyridyl ketone   P-70181

3-Pyridinyl-4-pyridinylmethanone
[56970-93-5]
$C_{11}H_8N_2O$ M 184.197
Cryst. (pet. ether). Mp 124°.

Oxime: [58088-20-3].
    $C_{11}H_9N_3O$ M 199.212
    Cryst. (EtOAc). Mp 150-153°. Mixt. of stereoisomers.

Niemers, E. *et al*, *Synthesis*, 1976, 593 (*deriv*)
Trovato, S. *et al*, *Spectrochim. Acta, Part A*, 1976, **32**, 351 (*synth, uv*)

1H-Pyrrole-3-carboxaldehyde, 9CI   P-70182

Updated Entry replacing P-02968
Pyrrole-3-aldehyde. 3-Formylpyrrole

[7126-39-8]
C_5H_5NO M 95.101
Cryst. (CCl$_4$/pet. ether). Mp 64°. Bp$_{0.05}$ 98-100°.
2,4-Dinitrophenylhydrazone: Yellow needles (EtOH). Mp 204°.
N-Ethoxycarbonyl:
    $C_8H_9NO_3$ M 167.164
    Oil. Bp$_{0.08}$ 68-70°.

Kahn, M.K.A. *et al*, *Tetrahedron*, 1966, **22**, 2095 (*synth, ir, pmr*)
Loader, C.E. *et al*, *Tetrahedron*, 1969, **25**, 3879 (*synth, ir, uv, pmr*)
Plieninger, H. *et al*, *Synthesis*, 1973, 422 (*synth, pmr, derivs*)
Demopolos, V.J., *Org. Prep. Proced. Int.*, 1986, **18**, 278 (*synth*)
Anderson, H.J. *et al*, *Synth. Commun.*, 1987, **17**, 401 (*synth*)

1H-Pyrrole-3,4-dicarboxylic acid, 9CI   P-70183

Updated Entry replacing P-02978
[935-72-8]
$C_6H_5NO_4$ M 155.110
Mp >300° dec., 290-297° dec.

Me ester: [41969-74-8].
    $C_7H_7NO_4$ M 169.137
    Cryst. (EtOH). Mp 247-248°.
Di-Me ester: [2818-06-6].
    $C_8H_9NO_4$ M 183.163
    Needles (xylene). Mp 243-245°.
Di-Et ester: [41969-71-5].
    $C_{10}H_{13}NO_4$ M 211.217
    Plates (EtOH aq.). Mp 153-154°.
Dimethylamide: [86492-08-2].
    $C_{10}H_{15}N_3O_2$ M 209.247
    Needles (THF/CH$_2$Cl$_2$). Mp 206-207°.
Diethylamide: [86492-07-1].
    $C_{14}H_{23}N_3O_2$ M 265.355
    Cryst. (toluene). Mp 123-124°.
1-Hydroxy, Di-Me ether:
    $C_8H_9NO_5$ M 199.163
    Oil.
1-Methoxy, Di-Me ester:
    $C_9H_{11}NO_5$ M 213.190
    Microcryst. (Et$_2$O/hexane) or yellow oil.
1-Methoxy, di-Et ester:
    $C_{11}H_{15}NO_5$ M 241.243
    Microcryst. (Et$_2$O/hexane). Mp 57-58°.

Gabel, N.W., *J. Org. Chem.*, 1962, **27**, 301 (*synth*)
Groves, J.K. *et al*, *Can. J. Chem.*, 1973, **51**, 1089 (*synth, ir, uv, nmr*)
Farnier, M. *et al*, *Bull. Soc. Chim. Fr.*, 1975, 2335 (*synth*)
Kaesler, R.W. *et al*, *J. Org. Chem.*, 1983, **48**, 4399 (*deriv, synth, pmr, cmr, ms*)
Keana, J.F.W. *et al*, *J. Org. Chem.*, 1988, **53**, 2268 (*derivs*)

2-Pyrrolidinecarboxaldehyde   P-70184

2-Formylpyrrolidine. Prolinal
[61480-98-6]

C_5H_9NO M 99.132
(S)-form [88218-12-6]
    Diethyl dithioacetal: [105089-88-1].
    $C_9H_{19}NS_2$ M 205.376

Pale-yellow oil. $[\alpha]_D^{23}$ −31.8° (c, 0.434 in CHCl₃).
N-*Benzyloxycarbonyl:* [71461-30-8].
C₁₃H₁₅NO₃   M 233.266
$[\alpha]_D^{24}$ −74.42° (c, 0.138 in CHCl₃).

Frick, L. *et al, Biochim. Biophys. Acta,* 1985, **529**, 311 (*synth, biochem*)
Langley, D.R. *et al, J. Org. Chem.,* 1987, **52**, 91 (*deriv, synth, pmr, ir, ms*)

2-Pyrrolidinemethanethiol, 9CI   P-70185
2-(Mercaptomethyl)pyrroldine

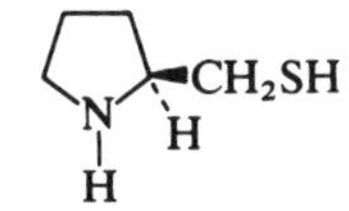

C₅H₁₁NS   M 117.209
(S)-form [85657-09-6]
L-form

Me thioether: [106865-55-8]. *2-[(Methylthio)methyl]-pyrrolidine, 9CI.*
C₆H₁₃NS   M 131.235
Liq. Bp₃₀ 60-63°.
Ph thioether: [106865-52-5]. *2-[(Phenylthio)methyl]-pyrrolidine.* Liq. Bp₀.₁ 89-91°. $[\alpha]_D^{20}$ +16.8° (c, 0.56 in CHCl₃).

(±)-form
Solid by subl. Mp 65-66°.
1-Et:
C₇H₁₅NS   M 145.262
Bp₉.₅ 65°.

Searles, S. *et al, J. Org. Chem.,* 1965, **30**, 3443 (*synth, pmr, ir*)
Otto, H.-H. *et al, Tetrahedron Lett.,* 1982, **23**, 5389 (*synth*)
Dieter, R.K. *et al, J. Am. Chem. Soc.,* 1987, **109**, 2040 (*deriv, synth, ir, pmr, cmr*)

2-Pyrrolidinethione, 9CI   P-70186
Thiopyrrolidone
[2295-35-4]

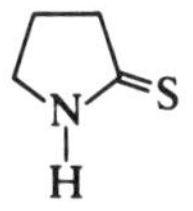

C₄H₇NS   M 101.166
Cryst. (xylene). Sol. warm H₂O with slow hydrol. Mp 106°.

N-*Me:* [10441-57-3].
C₅H₉NS   M 115.193
Oil. Bp₀.₀₈ 100°.

Tafel, J. *et al, Ber.,* 1907, **40**, 2842 (*synth*)
Mecke, R. *et al, Chem. Ber.,* 1957, **90**, 975 (*ir, raman*)
Rae, I.D., *Aust. J. Chem.,* 1979, **32**, 567 (*cmr*)
Andreocci, M.V. *et al, J. Mol. Struct.,* 1980, **69**, 151 (*pe*)
Wipf, P. *et al, Helv. Chim. Acta,* 1987, **70**, 1001 (*synth, pmr, ms, deriv*)

3-Pyrrolidinol   P-70187
3-Hydroxypyrrolidine
[40499-83-0]

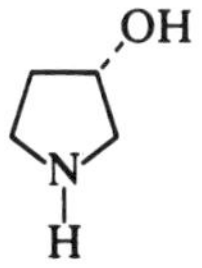

(S)-form

C₄H₉NO   M 87.121
(S)-form [100243-39-8]
Prepd. but not fully characterised.
N-*Benzyloxycarbonyl:* $[\alpha]_D^{28}$ +23.85° (c, 0.65 in MeOH). Opt. pure.
N-tert-*Butyloxycarbonyl:* Cryst. Mp 60-62°. $[\alpha]_D^{27}$ +22.75° (c, 1.02 in CHCl₃). Opt. pure.
(±)-form [83220-72-8]
N-(*Benzyloxycarbonyl):* Liq. n_D^{20} 1.5406.

Brown, H.C. *et al, J. Org. Chem.,* 1985, **50**, 1582 (*synth, ir, pmr*)
Harris, B.D. *et al, Synth. Commun.,* 1986, **16**, 1815 (*synth, ir, pmr*)

2-Pyrrolidinone, 9CI   P-70188
Updated Entry replacing P-02989
2-Oxopyrrolidine. Butyrolactam. 2-Pyrrolidone. Piperidinic lactam. 4-Aminobutanoic acid lactam
[616-45-5]

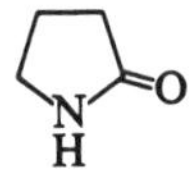

C₄H₇NO   M 85.105
Cryst. (pet. ether). V. sol. H₂O, most org. solvs., spar. sol. pet. ether. Mp 24.6°. Bp 245°, Bp₁₂ 133°. Spar. steam-volatile. Forms a monohydrate in moist air, Mp 35° (30°).
▷UY5715000.
N-*Ac:* [932-17-2].
C₆H₉NO₂   M 127.143
Bp 231°.
▷UY5717000.
N-*Me:* see *1-Methyl-2-pyrrolidinone, M-03538*
N-*Et:* see *1-Ethyl-2-pyrrolidinone, E-01176*
N-*Ph:* see *1-Phenyl-2-pyrrolidone, P-01608*

enol-form

Me ether: [5264-35-7]. *3,4-Dihydro-5-methoxy-2H-pyrrole. 2-Methoxy-1-pyrroline.*
C₅H₉NO   M 99.132
Liq. Bp 118-120°.
Et ether: [931-46-4]. *5-Ethoxy-3,4-dihydro-2H-pyrrole, 9CI. 2-Ethoxy-1-pyrroline.*
C₆H₁₁NO   M 113.159
Liq. Bp 133-137°.

Gabriel, S., *Ber.,* 1889, **22**, 3338 (*synth*)
Metzger, H. *et al, Angew. Chem., Int. Ed. Engl.,* 1963, **2**, 624 (*synth*)
Zeifman, V.I. *et al, CA,* 1963, **59**, 5110 (*synth*)
Etienne, A. *et al, Bull. Soc. Chim. Fr.,* 1969, 3704 (*derivs*)
Fronza, G. *et al, J. Chem. Soc., Perkin Trans. 2,* 1977, 1746 (*nmr*)
Pellegata, R. *et al, Synthesis,* 1978, 614 (*synth*)
Sax, N.I., *Dangerous Properties of Industrial Materials,* 5th Ed., Van Nostrand-Reinhold, 1979, 949.

5*H*-Pyrrolo[1,2-*a*]azepine, 9CI   P-70189
3a-*Azaazulene*
[276-16-4]

C_9H_9N M 131.177
Oil.

Jones, G. *et al*, *J. Chem. Soc., Perkin Trans. 1*, 1982, 1123
  (*synth, uv, pmr, ms*)
Flitsch, W., *Adv. Heterocycles* Chem., 1988, **43**, 35 (*rev*)

3*H*-Pyrrolo[1,2-*a*]azepin-3-one, 9CI   P-70190
3a-*Azaazulen-3-one*
[114709-99-8]

C_9H_7NO M 145.160
Highly coloured. Unstable to O_2.

Flitsch, W. *et al*, *Tetrahedron Lett.*, 1987, **28**, 4397 (*synth, pmr*)

5*H*-Pyrrolo[1,2-*a*]azepin-5-one, 9CI   P-70191
Updated Entry replacing P-03004
3a-*Azaazulen-4-one*
[42793-20-4]
C_9H_7NO M 145.160
Cryst. (C_6H_6). Mp 49-51°.

Flitsch, W. *et al Chem. Ber.*, 1973, **106**, 1993 (*synth*)
Flitsch, W. *et al*, *Chem. Ber.*, 1978, **111**, 2607 (*pmr*)

7*H*-Pyrrolo[1,2-*a*]azepin-7-one, 9CI   P-70192
3a-*Azaazulen-6-one*
[82218-01-7]

C_9H_7NO M 145.160
Yellow needles (hexane). Mp 122° (119-120°).

Jones, G. *et al*, *J. Chem. Soc., Perkin Trans. 1*, 1982, 1123
  (*synth, ir, uv, pmr, ms*)
Flitsch, W. *et al*, *Justus Liebigs Ann. Chem.*, 1988, 275 (*synth*)

9*H*-Pyrrolo[1,2-*a*]azepin-9-one, 9CI   P-70193
3a-*Azaazulen-8-one*
[67542-95-4]

C_9H_7NO M 145.160
Cryst. (C_6H_6/EtOAc). Mp 106-107°.

Flitsch, W. *et al*, *Chem. Ber.*, 1976, **111**, 2407 (*synth, ir, uv, pmr, ms*)

5*H*-Pyrrolo[1,2-*a*]imidazol-5-one   P-70194
[51789-99-2]

$C_6H_4N_2O$ M 120.110
Orange-red cryst. by subl. Mp 93-95°.

McNab, H., *J. Chem. Soc., Perkin Trans. 1*, 1987, 653, 657
  (*synth, pmr, uv, ir, ms, nmr*)

5*H*-Pyrrolo[1,2-*c*]imidazol-5-one, 9CI   P-70195
[111573-52-5]

$C_6H_4N_2O$ M 120.110
Yellow cryst. solid. Mp >300° (part. dec. >100°).

McNab, H., *J. Chem. Soc., Perkin Trans. 1*, 1987, 653, 657
  (*synth, pmr, uv, ir, ms, nmr*)

Pyrrolo[3,2,1-*kl*]phenothiazine   P-70196
Updated Entry replacing P-03019
[19609-07-5]

$C_{14}H_9NS$ M 223.292
Plates (hexane). Mp 117-118°.

Hollins, R.A. *et al*, *J. Heterocycles* Chem., 1978, **15**, 711.
Hallberg, A. *et al*, *Heterocycles*, 1982, **19**, 75; *J. Heterocycl. Chem.*, 1983, **20**, 37 (*synth, pmr*)
Vishwakarma, L.C. *et al*, *J. Heterocycl. Chem.*, 1983, **20**, 995
  (*synth, ms, pmr*)

Pyrrolo[1,2-*a*]pyrazine   P-70197
3a,6-*Diazaindene*
[274-45-3]

$C_7H_6N_2$ M 118.138
Component of roast beef aroma and tobacco smoke condensate. Oil. Bp_2 71°. n_D^{20} 1.6176. Rapidly darkens in air.

Picrate: Mp 212°.

Herz, W. *et al*, *J. Am. Chem. Soc.*, 1955, **77**, 6355 (*synth, uv*)
Paudler, W.W. *et al*, *J. Heterocycl. Chem.*, 1965, **2**, 410 (*synth, pmr*)
Kuhla, D.E. *et al*, *Adv. Heterocycles* Chem., 1977, **21**, 1 (*rev*)
Corbet, J.P. *et al*, *Tetrahedron Lett.*, 1982, **23**, 3565 (*synth*)
Cobb, J. *et al*, *Magn. Reson. Chem.*, 1986, **24**, 231 (*pmr, cmr*)

Pyrrolo[1,2-*a*]pyrazin-1(2*H*)-one, 9CI   P-70198

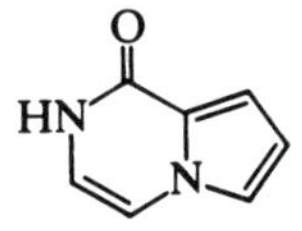

$C_7H_6N_2O$ M 134.137

N-*Me:* [116212-47-6].
   $C_8H_8N_2O$ M 148.164
   Needles (hexane/Et_2O). Mp 118-119°.

Brimble, M.A. *et al*, *Aust. J. Chem.*, 1988, **41**, 1583 (*synth, uv, pmr, cmr, ms*)

6*H*-Pyrrolo[1,2-*b*]pyrazol-6-one, 9CI   P-70199

[111573-53-6]

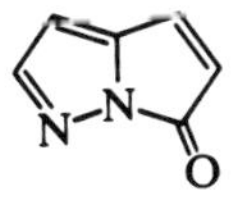

$C_6H_4N_2O$ M 120.110
Yellow solid. Mp 36-38°.

McNab, H., *J. Chem. Soc., Perkin Trans. 1*, 1987, 653, 657 (*synth, pmr, uv, ir, ms, nmr*)

2*H*-Pyrrolo[3,4-*c*]pyridine   P-70200

2,5-Diazaindene
[270-70-2]

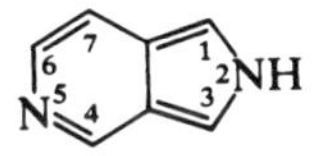

$C_7H_6N_2$ M 118.138
No phys. props. reported.

2-*Me:* [38070-67-6].
   $C_8H_8N_2$ M 132.165
   Cryst. by subl. Mp 97-98°.

2-*Me, picrate:* [38070-68-7]. Cryst. (EtOH). Mp 194.5-195.5°.

2-*Me; B,MeI:* Cryst. (EtOH). Mp 140° dec.

Armarego, W.L.F. *et al*, *J. Chem. Soc., Perkin Trans. 1*, 1972, 2485 (*deriv, synth, ir, pmr, ms, uv*)
Ahmed, I. *et al*, *Tetrahedron*, 1979, **35**, 1145 (*deriv, synth, ir, uv,pmr*)
Tsai, C.-Y. *et al*, *Tetrahedron Lett.*, 1987, **28**, 1419 (*synth, pmr, uv, cmr*)

6*H*-Pyrrolo[3,4-*b*]pyridine, 9CI   P-70201

2,4-Diazaindene
[271-01-2]

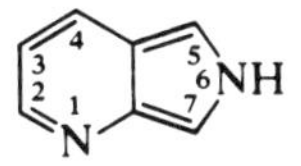

$C_7H_6N_2$ M 118.138
No phys. props. reported.

6-*Me:* [38070-65-4].
   $C_8H_8N_2$ M 132.165
   Pale-yellow needles (cyclohexane). Mp 63°.

6-*Me, picrate:* [38070-66-5]. Cryst. (MeOH). Mp 216° dec.

Armarego, W.L.F. *et al*, *J. Chem. Soc., Perkin Trans. 1*, 1972, 2485 (*deriv, synth, ir, pmr, ms, uv*)
Tsai, C.-Y. *et al*, *Tetrahedron Lett.*, 1987, **28**, 1419 (*synth, pmr, uv, cmr*)

3-(1*H*-Pyrrol-2-yl)-2-propenal, 9CI   P-70202

[6249-29-2]

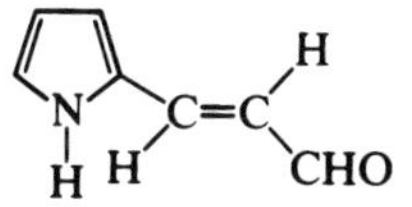

C_7H_7NO M 121.138

(*E*)-*form* [49616-64-0]
Yellow cryst. Mp 114-115° (109-111°).

Weedon, B.C.L. *et al*, *J. Chem. Soc., Perkin Trans. 1*, 1973, 1416 (*synth, pmr, uv, ms*)
Hinz, W. *et al*, *Tetrahedron*, 1986, **42**, 3753 (*synth, pmr, cmr*)

Pyrroxamycin   P-70203

Dioxapyrrolomycin. Al-R 2081. LL F42248α. SS 46506A. Antibiotic Al-R 2081. Antibiotic LL F42248α. Antibiotic SS 46506A
[105888-54-8]

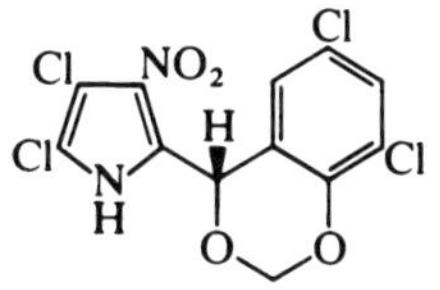

$C_{12}H_6Cl_4N_2O_4$ M 384.002
Pyrrole antibiotic. Prod. by *Streptomyces* spp. Active against gram-positive bacteria and dermophytes. Pale-yellow cryst. (EtOH) or (C_6H_6). Mp 216-220° dec. (200-207°). $[\alpha]_D^{25}$ −110° (c, 0.5 in EtOH), $[\alpha]_D^{25}$ −37.2° (c, 1 in Me_2CO).

Carter, G.T. *et al*, *J. Antibiot.*, 1987, **40**, 233 (*isol, struct, props*)
Nakamura, H. *et al*, *J. Antibiot.*, 1987, **40**, 899 (*isol, struct, props*)
Rengaraju, S. *et al*, *CA*, 1987, **106**, 152653 (*isol*)
Yano, K. *et al*, *J. Antibiot.*, 1987, **40**, 961 (*isol, struct, props*)

Q

Quercetol A   Q-70001

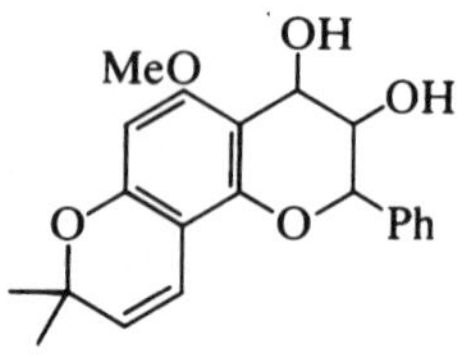

$C_{21}H_{22}O_5$ M 354.402

Constit. of *Tephrosia quercetorum*. Needles. Mp 99-100°. $[\alpha]_D$ −2.7° (c, 0.22 in CHCl$_3$).

Gómez-Garibay, F. *et al, Phytochemistry*, 1988, **27**, 2971.

Quercetol B   Q-70002

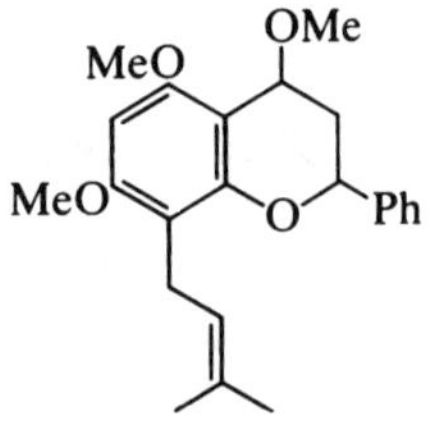

$C_{23}H_{28}O_4$ M 368.472

Constit. of *Tephrosia quercetorum*. Oil. $[\alpha]_D$ −42.2° (c, 0.19 in CHCl$_3$).

Gómez-Garibay, F. *et al, Phytochemistry*, 1988, **27**, 2971.

Quercetol C   Q-70003

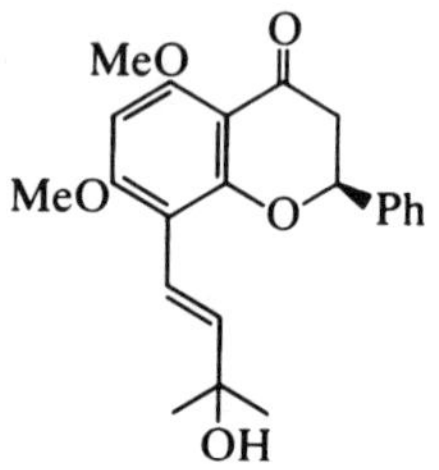

$C_{22}H_{24}O_5$ M 368.429

Constit. of *Tephrosia quercetorum*. Needles. Mp 198-200°. $[\alpha]_D$ −66° (c, 0.1 in CHCl$_3$).

Gómez-Garibay, F. *et al, Phytochemistry*, 1988, **27**, 2971.

Queuine   Q-70004

2-Amino-5-[[(4,5-dihydroxy-2-cyclopenten-1-yl)-amino]methyl]-1,7-dihydro-4H-pyrrolo[2,3-d]-pyrimidin-4-one, 9CI. 2-Amino-5-(4,5-dihydroxycyclopent-1-en-3-ylaminomethyl)pyrrolo[2,3-d]pyrimidine. 7-(4,5-Dihydroxy-2-cyclopent-1-en-3-ylaminomethyl)-7-deazaguanine. Q base

[72496-59-4]

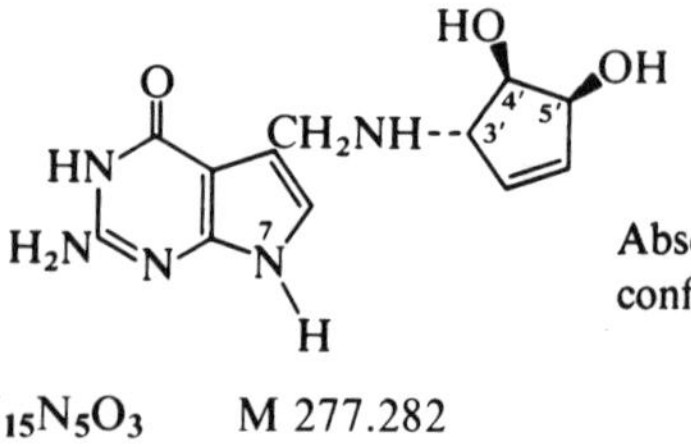

Absolute
configuration

$C_{12}H_{15}N_5O_3$ M 277.282

B,2HCl: [86496-18-6]. Prisms. Mp 230-235° dec. $[\alpha]_D^{26}$ +113° (c, 0.3 in H$_2$O).

7-β-D-Ribofuranosyl: [57072-36-3]. **Queuosine**. *Nucleoside Q.*

$C_{17}H_{23}N_5O_7$ M 409.398

Hypermodified nucleoside occurring in the first anticodon posn. in tRNA's of plants and animals. Pale-yellow cryst. (H$_2$O). Mp 225-230° dec. $[\alpha]_D$ +43.4° (c, 0.1 in H$_2$O).

7-β-D-Ribofuranosyl; B,HCl: [71050-05-0]. Glassy solid.

Ohgi, T. *et al, J. Am. Chem. Soc.*, 1979, **101**, 3629 (*deriv, synth, abs config, bibl*)
Kondo, T. *et al, Chem. Lett.*, 1983, 419 (*synth*)
Akimoto, H. *et al, J. Med. Chem.*, 1986, **29**, 1749 (*biochem, ir, pmr, bibl*)
Kondo, T. *et al, Tetrahedron*, 1986, **42**, 207 (*synth, uv, pmr, ir*)

Quinic acid   Q-70005

Updated Entry replacing Q-50003
(1α,3α,4α,5β)-1,3,4,5-Tetrahydroxycyclohexanecarboxylic acid. Hexahydro-1,3,4,5-tetrahydroxybenzoic acid
[36413-60-2]

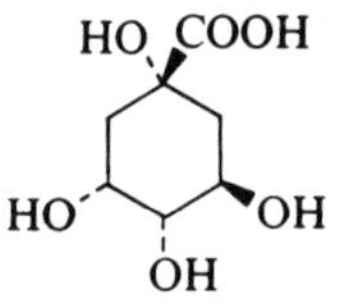

(−)-*form*
Absolute
configuration

$C_7H_{12}O_6$ M 192.168

(−)-*form* [77-95-2]

Occurs in cinchona bark, coffee beans, tobacco leaves and many other plant sources. Mp 162-163°. $[\alpha]_D^{26}$ −42.1° (H$_2$O).

▷GU8650000.

Tetra-Ac:

$C_{15}H_{20}O_{10}$ M 360.317

Mp 132-136°. $[\alpha]_D^{20}$ −22.5° (EtOH).

3-(3,4-Dihydroxycinnamyl): see 3-*O*-Caffeoylquinic acid, C-70005

Amide, tetra-Ac: Mp 186-187° dec. $[\alpha]_D$ −28.6° (C$_2$H$_2$Cl$_4$).

Lactone: [27783-00-2]. *1,3,4-Trihydroxy-6-oxabicyclo[3.2.1]octan-7-one, 9CI. Quinide.*

$C_7H_{10}O_5$ M 174.153

Mp 187°. $[\alpha]_D^{23}$ −16.75° (c, 8.3 in H$_2$O). Lactonised onto the C-3 OH group.

(+)-*form*

Mp 164°. $[\alpha]_D^{20}$ +44° (H$_2$O).

(±)-*form*

Cryst. (EtOH). Mp 155-156° (149°).

Lactone: Mp 200° (180-196°).

v. Lippmann, E., *Ber.*, 1901, **34**, 1159.
Gorter, K., *Justus Liebigs Ann. Chem.*, 1908, **359**, 217.
Fischer, H.O.L. *et al, Ber.*, 1932, **65**, 1009 (*struct*)
Wolinsky, J. *et al, J. Org. Chem.*, 1964, **29**, 3596 (*synth, bibl*)
Haslam, E. *et al, J. Chem. Soc.* (C), 1971, 1496 (*pmr*)
De Pooter, H. *et al, Bull. Soc. Chim. Belg.*, 1975, **84**, 835 (*deriv*)
Kelley, C.J. *et al, J. Org. Chem.*, 1976, **41**, 449 (*cmr*)

Abell, C. *et al, Acta Crystallogr., Sect. C*, 1988, **44**, 1287 (*cryst struct, bibl, abs config*)

5*H*,9*H*-Quino[3,2,1-*de*]acridine, 9CI   Q-70006
[33080-41-0]

C$_{20}$H$_{15}$N   M 269.345
Dark-yellow cryst. powder (EtOH). Mp 161-164° dec.
Hellwinkel, D. *et al, Chem. Ber.*, 1974, **107**, 616 (*synth, pmr*)

Quino[7,8-*h*]quinoline   Q-70007
Updated Entry replacing Q-50004
1,12-Diazabenzo[a]*phenanthrene*
[195-41-5]

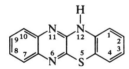

C$_{16}$H$_{10}$N$_2$ M 230.268
Claimed syntheses prior to 1987 were erroneous. Shows
proton-sponge props. Mp 196-197°. pK_a ~12.8.
Zirnstein, M.A. *et al, Angew. Chem., Int. Ed. Engl.*, 1987, **26**,
460 (*synth, pmr*)
Staab, H.A. *et al, Angew. Chem., Int. Ed. Engl.*, 1988, **27**, 865
(*rev*)

12*H*-Quinoxalino[2,3-*b*][1,4]benzothiazine   Q-70008
[258-17-3]

C$_{14}$H$_9$N$_3$S   M 251.305
Mp 280°.
N-*Ac:* [64329-58-4].
    C$_{16}$H$_{11}$N$_3$OS   M 293.342
    Mp 196°.
N-*Benzoyl:* [110744-44-0].
    C$_{21}$H$_{13}$N$_3$OS   M 355.413
    Pale-yellow cryst. Mp 194°.
Agarwal, N.L. *et al, Justus Liebigs Ann. Chem.*, 1987, 921
(*synth, ir, pmr, ms*)

12*H*-Quinoxalino[2,3-*b*][1,4]benzoxazine   Q-70009
[258-16-2]

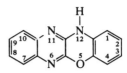

C$_{14}$H$_9$N$_3$O   M 235.245
Mp >300°.
N-*Ac:* [110744-41-7].
    C$_{16}$H$_{11}$N$_3$O$_2$ M 277.282
    Mp 170°.
Agarwal, N.L. *et al, Justus Liebigs Ann. Chem.*, 1987, 921
(*synth, ir, pmr, ms*)

2,2':5',2'':5'',2''':5''',2''''-Quinquethiophene,   Q-70010
9CI
α-Quinquethiophene
[5660-45-7]

C$_{20}$H$_{12}$S$_5$ M 412.615
Mp 253°.
Nakagama, J. *et al, Heterocycles*, 1987, **26**, 939 (*synth*)

Quisquagenin   Q-70011
20,24-Epoxy-9,19-cyclolanostane-3,16,25-triol, 9CI
[112709-68-9]

C$_{30}$H$_{50}$O$_4$ M 474.723
Constit. of *Astragalus quisqualis*. Cryst. (MeOH). Mp
232.5-234.5°. [α]$_D^{25}$ +36.6° (c, 0.745 in CHCl$_3$).
Kholzineva, L.A. *et al, Khim. Prir. Soedin.*, 1987, **23**, 439.

R

Resorcylide
R-70001

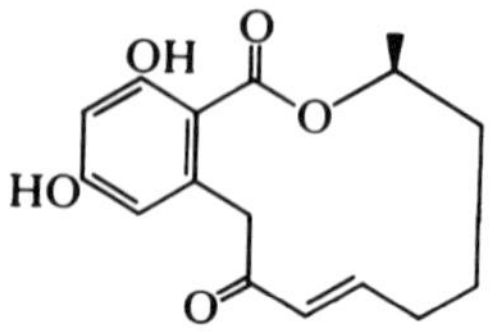

(S,E)-form

$C_{16}H_{18}O_5$ M 290.315

(S,E)-form [69483-32-5]

Prod. by *Penicillium* sp. Plant growth inhibitor. Shows antimicrobial and cytotoxic activities. Solid. $[\alpha]_D$ +78° (MeOH).

(S,Z)-form [69433-66-5]

From *P.* sp. Similar biol. props. to (*E*)-form. Cryst. Mp 215-216°. $[\alpha]_D$ +5.0° (MeOH).

Oyama, H. *et al, Agric. Biol. Chem.*, 1978, **42**, 2407 (*struct*)
Nukina, M. *et al, CA*, 1980, **92**, 177095 (*isol, struct, props*)

ε-Rhodomycinone
R-70002

Updated Entry replacing R-00109

Methyl 2-ethyl-1,2,3,4,6,11-hexahydro-2,4,5,7,12-pentahydroxy-6,11-dioxo-1-naphthacenecarboxylate, 9CI

[21288-60-8]

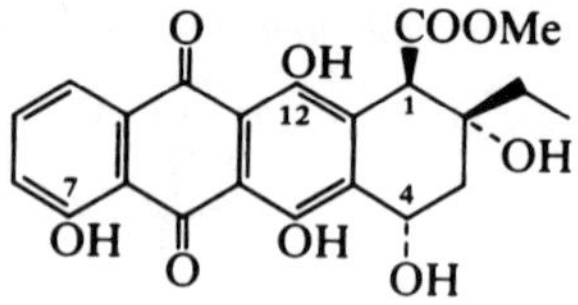

$C_{22}H_{20}O_9$ M 428.395

Pigment from *Streptomyces purpurascens*. Red needles (cyclohexane). Mp 220° dec.

Tetra-Ac: Pale-yellow needles. Mp 218-219° dec.

7-Glucoside:

$C_{28}H_{28}O_{15}$ M 604.520

Prod. by *S. peucetius* var. *vinaceus*. Cryst. Mp 168-169° dec.

4-Deoxy: [21179-17-9]. *ζ-Rhodomycinone*.

$C_{22}H_{20}O_8$ M 412.395

From *S. purpurascens*. Red cryst. Mp 275°.

4-Epimer: θ-Rhodomycinone.

$C_{22}H_{20}O_9$ M 428.395

Aglycone of antibiotics from *S.* spp. Red-brown needles (CHCl₃ or THF). Mp 220° dec. $[\alpha]_D^{25}$ +191.5° (c, 0.223 in THF).

Brockmann, H. *et al, Chem. Ber.*, 1961, **94**, 2681.
Brockmann, H. *et al, Naturwissenschaften*, 1961, **48**, 161 (*isol, struct*)
Bowie, J.H. *et al, J. Chem. Soc.*, 1964, 3927 (*θ-Rhodomycinone*)
Brockmann, H. *et al, Tetrahedron Lett.*, 1968, 4719 (*abs config*)
Biedermann, E. *et al, CA*, 1973, **78**, 82888t (*props*)
Doyle, T.W. *et al, J. Am. Chem. Soc.*, 1979, **101**, 7041 (*nmr*)
Krone, K. *et al, Tetrahedron Lett.*, 1981, **22**, 3219 (*synth*)
Kraus, G.A. *et al, J. Org. Chem.*, 1983, **48**, 3265 (*synth*)
Cassinelli, G. *et al, J. Antibiot.*, 1987, **40**, 1071 (*glucosides*)

Rhodonocardin A
R-70003

[110429-58-8]

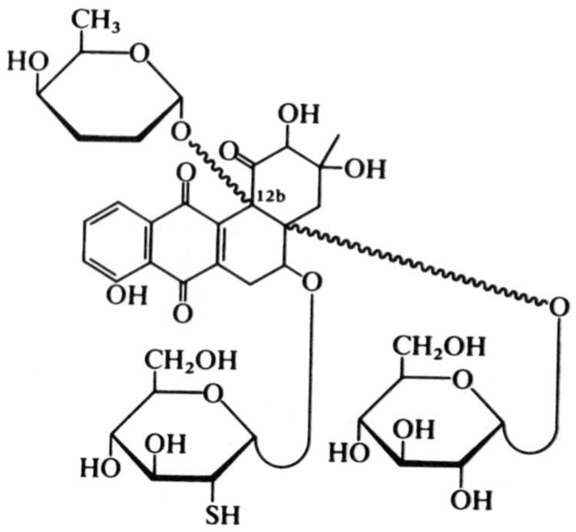

$C_{37}H_{46}O_{20}S$ M 842.818

Anthraquinone-type compd. Isol. from *Nocardia* sp. 53. Shows no antimicrobial props. Mp 214-215°. $[\alpha]_D^{20}$ +29° (c, 0.1 in MeOH). Related to Sakyomicin *A*.

De-O¹²ᵇ-rhodinosyl: [110429-59-9]. **Rhodonocardin B**.

$C_{31}H_{36}O_{18}S$ M 728.675

From *N.* sp. 53. No antimicrobial props. Mp 223°. $[\alpha]_D^{20}$ +29° (c, 0.1 in MeOH).

Etoh, H. *et al, Agric. Biol. Chem.*, 1987, **51**, 1819 (*isol, struct, props*)

Rhodotorucin A
R-70004

[66106-15-8]

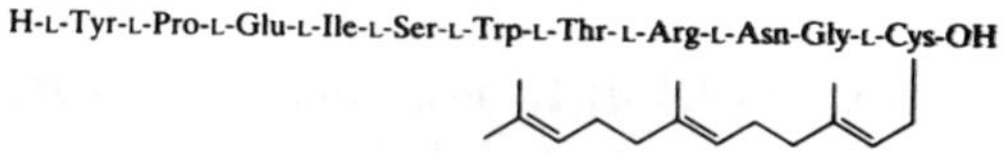

H-L-Tyr-L-Pro-L-Glu-L-Ile-L-Ser-L-Trp-L-Thr-L-Arg-L-Asn-Gly-L-Cys-OH

$C_{73}H_{108}N_{16}O_{18}S$ M 1529.813

Lipopeptide antibiotic. Prod. by *Rhodosporidium toruloides*. Fungal mating hormone. $[\alpha]_D^{25}$ −49.7° (c, 0.1 in MeOH/butanol/H₂O).

Kitada, C. *et al, Experientia*, 1979, **35**, 1275 (*synth*)
Japan. Pat., 80 39 713, (*1980*); *CA*, **93**, 43929 (*isol*)
Kamiya, Y. *et al, Naturwissenschaften*, 1981, **68**, 128 (*rev*)

Riccardin D
R-70005

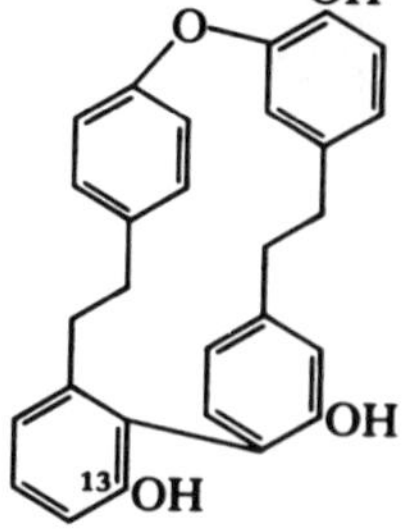

$C_{28}H_{24}O_4$ M 424.495

Constit. of *Monoclea forsteri*. Cryst. Mp 187.5-189°.

O^{13}-Me: **Riccardin E.**
C$_{29}$H$_{26}$O$_4$ M 438.522
Constit. of *M. forsteri*. Viscous oil.

Toyta, M. *et al*, *Phytochemistry*, 1988, **27**, 2605.

Richardianidin 1   R-70006
[117232-63-0]

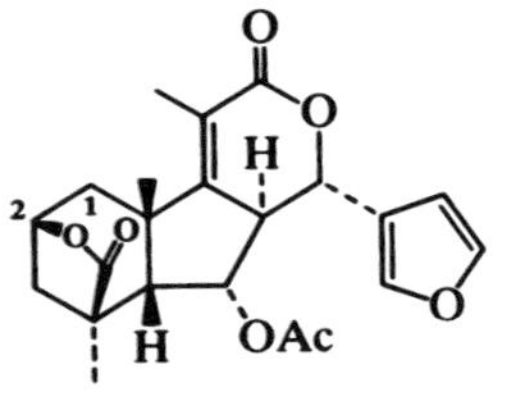

C$_{22}$H$_{24}$O$_7$ M 400.427
Novel 6,7-seco-6,11-cyclolabdane skeleton. Constit. of
Cluytia richardiana. Cryst. (MeOH). Mp 220-222°.
[α]$_D$ +8° (c, 0.4 in CHCl$_3$).
Isomer: [117210-47-6]. **Richardianidin 2.**
C$_{22}$H$_{24}$O$_7$ M 400.427
Constit. of *C. richardiana*. Mp 266-268°. [α]$_D$ −0.02°
(c, 0.14 in CHCl$_3$). Lactonised onto C(1) instead of
C(2).

Mossa, J.S. *et al*, *Tetrahedron Lett.*, 1988, **29**, 3627 (*cryst struct*)

Ridentin   R-70007
Updated Entry replacing R-00155
[28148-84-7]

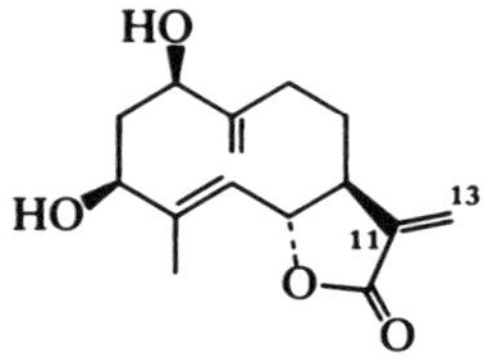

C$_{15}$H$_{20}$O$_4$ M 264.321
Constit. of *Artemisia* spp. Cryst. Mp 215-218° dec. [α]$_D^{25}$
−113° (c, 0.46 in CHCl$_3$).
11β,13-Dihydro: [117255-06-8]. **Dihydroridentin.**
C$_{15}$H$_{22}$O$_4$ M 266.336
Constit. of *A. rupicola* and *A. tripartita*. Cryst. Mp
193-194°.
11α,13-Dihydro: [117232-58-3]. **11-Epidihydroridentin.**
C$_{15}$H$_{22}$O$_4$ M 266.336
Constit. of *A. lanata*. Gum.

Irwin, M.A. *et al*, *Phytochemistry*, 1969, **8**, 2009 (*isol*)
Aguilar, J.M. *et al*, *Phytochemistry*, 1988, **27**, 2229 (*isol, struct*)

Consult the *Dictionary of Alkaloids*
for a comprehensive treatment of
alkaloid chemistry.

Roridin D   R-70008
*7'-Deoxo-2'-deoxy-2',3'-epoxy-7'-(1-hydroxyethyl)-
verrucarin A, 9CI*
[14682-29-2]

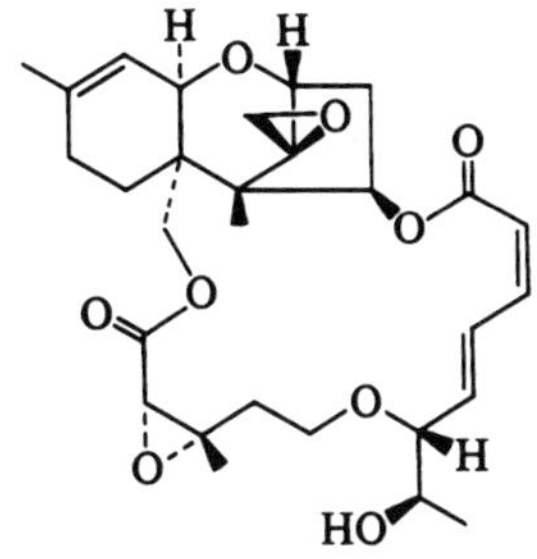

C$_{29}$H$_{38}$O$_9$ M 530.614
Trichothecene. Isol. from *Myrothecium verrucaria* and
M. roridum. Needles (Me$_2$CO/Et$_2$O). Mp 232-235°.
[α]$_D^{23}$ +29° (c, 2.71 in CHCl$_3$).
9β,10β-Epoxide: **Baccharinoid B17.**
C$_{29}$H$_{38}$O$_{10}$ M 546.613
Constit. of *Baccharis megapotamica*. Cryst.
(CH$_2$Cl$_2$/Et$_2$O). Mp >300°. [α]$_D^{25}$ +11° (c, 0.68 in
CH$_2$Cl$_2$).
8β-Hydroxy: **Baccharinoid B21.**
C$_{29}$H$_{38}$O$_{10}$ M 546.613
From *B. megapotamica*. Cryst. (CH$_2$Cl$_2$/Et$_2$O). Mp
259-260°. [α]$_D^{25}$ +73.5° (c, 0.68 in CH$_2$Cl$_2$).
3-Hydroxy: **Baccharinoid B12.**
C$_{29}$H$_{38}$O$_{10}$ M 546.613
From *B. megapotamica*. Cryst. (Me$_2$CO/hexane). Mp
248-250°. [α]$_D^{25}$ +12.6° (c, 0.54 in MeOH).

Böhner, B. *et al, Helv. Chim. Acta*, 1965, **48**, 1079 (*isol, uv, ir*)
Snatzke, G. *et al, Helv. Chim. Acta*, 1967, **50**, 1618 (*cd*)
Breitenstein, W. *et al, Helv. Chim. Acta*, 1975, **58**, 1172 (*cmr*)
Jarvis, B.B. *et al, J. Org. Chem.*, 1987, **52**, 45 (*isol, struct, derivs*)
Cole, R.J. *et al, Handbook of Toxic Fungal Metabolites*,
Academic Press, N.Y. 1981, 236.

Roseofungin   R-70009
*16,34-Didemethyl-21,27,29,31-tetradeoxy-12,13,32,33-
tetrahydro-13-hydroxy-2-methyl-21,27-dioxodermosta-
tin A, 9CI*
[12687-98-8]

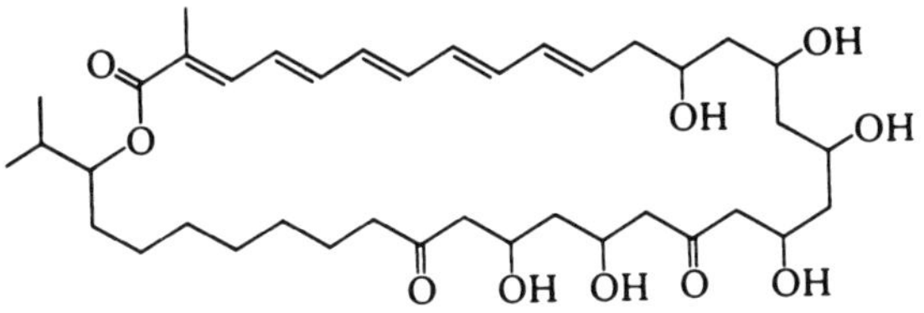

C$_{39}$H$_{62}$O$_{10}$ M 690.913
Pentaene antibiotic. Isol. from *Streptomyces roseoflavus
roseofungini*. Antifungal agent. Dec. at 97-106°.
▷VL0456000.

Vetlugina, L.A. *et al, Antibiotiki (Moscow)*, 1968, **13**, 992;
1973, **18**, 774 (*isol, props*)
Daurenbekova, *Antibiot. Med. Biotechnol.*, 1985, **30**, 525 (*isol*)
Shenin, Y.D. *et al, Antibiot. Med. Biotekhnol.*, 1986, **31**, 341
(*struct*)

Rosmanol R-70010

Updated Entry replacing R-40029

7β,11,12-Trihydroxy-8,11,13-abietatrien-20,6β-olide

[80225-53-2]

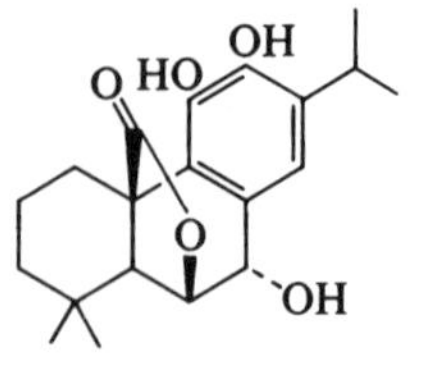

$C_{20}H_{26}O_5$ M 346.422

Isol. from *Rosmarinus officinalis*. Antioxidant. Cryst. (Me₂CO). Mp 241°. $[\alpha]_D^{18}$ −34.3° (c, 0.7 in EtOH).

7-Epimer: [93380-12-2]. **Epirosmanol**.

$C_{20}H_{26}O_5$ M 346.422

From *R. officinalis*. Cryst. (Me₂CO). Mp 221-225°.

7-Me ether: [113085-62-4]. **7-Methoxyrosmanol**.

$C_{21}H_{28}O_5$ M 360.449

From *R. officinalis*. Pale-brown powder. $[\alpha]_D^{23}$ −99.2° (c, 0.5 in EtOH).

Nakatani, N. *et al*, *Agric. Biol. Chem.*, 1981, **45**, 2385 (*isol*)
Nakatani, N. *et al*, *Agric. Biol. Chem.*, 1984, **48**, 2081 (*Epirosmanol*)
Fraga, B.N. *et al*, *Phytochemistry*, 1985, **24**, 1853 (*cryst struct*)
Arisawa, M. *et al*, *J. Nat. Prod.*, 1987, **50**, 1164 (*Methoxyrosmanol*)

Rubiflavin C₁ R-70011

11-Hydroxy-5-methyl-2-(1-methyl-1,3-pentadienyl)-8-[2,3,6-trideoxy-3-(dimethylamino)-β-D-arabino-hexopyranosyl]-10-[2,3,6-trideoxy-3-(dimethylamino)-3-C-methyl-α-L-lyxo-hexopyranosyl]-4H-anthra[1,2-b]-pyran-4,7,12-trione

[111058-14-1]

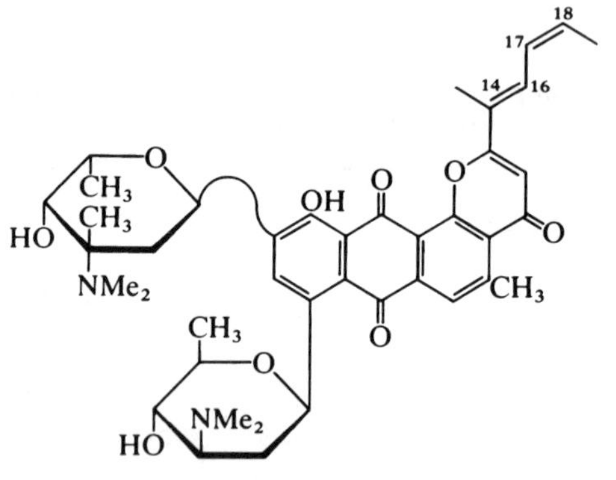

$C_{41}H_{50}N_2O_9$ M 714.854

Anthraquinone-type antibiotic. Prod. by *Streptomyces griseus*. Orange solid. Similar to Pluramycin.

17E-Isomer: [110954-32-0]. **Rubiflavin C₂**.

$C_{41}H_{50}N_2O_9$ M 714.854

From *S. griseus*. Orange solid.

17,18-Dihydro: [110954-33-1]. **Rubiflavin D**.

$C_{41}H_{52}N_2O_9$ M 716.870

Prod. by *S. griseus*. Orange solid.

14,16-Dihydro, 14-hydroxy: [104855-82-5]. **Rubiflavin E**.

$C_{41}H_{52}N_2O_{10}$ M 732.869

Prod. by *S. griseus*. Orange solid. Same gross struct. as Antibiotic SS 21020B.

Nadig, H. *et al*, *Helv. Chim. Acta*, 1987, **70**, 1217 (*struct, bibl*)

Rubiflavin *F* R-70012

Isokidamycin

[58809-18-0]

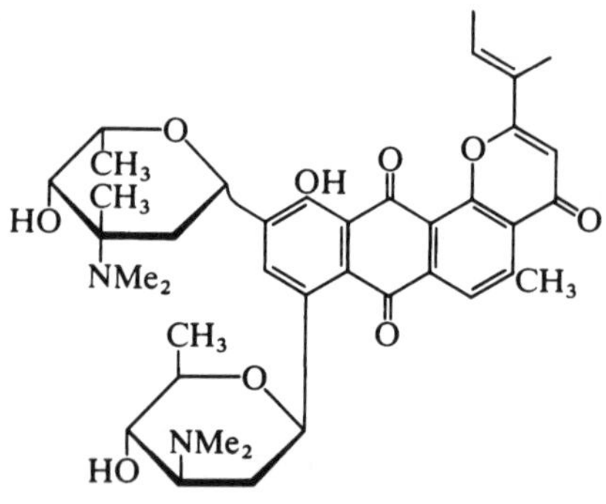

$C_{39}H_{48}N_2O_9$ M 688.816

Anthraquinone-type antibiotic. Prod. by *Streptomyces griseus*. Orange cryst. + ½H₂O (EtOAc). Mp 200-201°. $[\alpha]_D^{25}$ +62.4° (c, 1.1 in CHCl₃). Related to Pluramycin.

Furukawa, M. *et al*, *Tetrahedron*, 1975, **31**, 2989 (*synth, struct*)
Sequin, U. *et al*, *Tetrahedron*, 1978, **34**, 3623 (*cmr, conformn*)
Nadig, H. *et al*, *Helv. Chim. Acta*, 1987, **70**, 1217 (*isol, struct*)

Rudbeckin *A* R-70013

[117047-27-5]

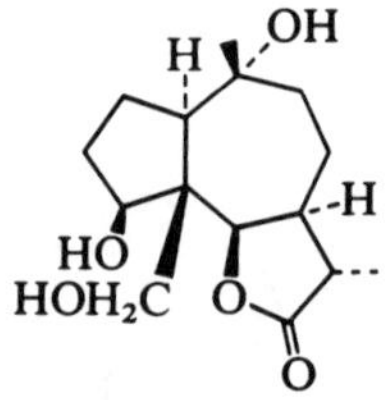

$C_{15}H_{24}O_5$ M 284.352

Constit. of *Rudbeckia grandiflora*. Gum.

Vasquez, M. *et al*, *Phytochemistry*, 1988, **27**, 2195.

Rutaevinexic acid R-70014

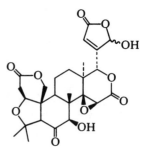

$C_{26}H_{30}O_{11}$ M 518.516

Constit. of *Tetradium glabrifolium*. Needles (EtOAc). Mp 253°. $[\alpha]_D$ −148° (c, 0.062 in MeOH).

Ng, K.M. *et al*, *J. Nat. Prod.*, 1987, **50**, 1160.

Rutalpinin   R-70015

6-(1,1-Dimethyl-2-propenyl)-8H-1,3-dioxolo[4,5-h] [1]benzopyran-8-one, 9CI. 4-(1,1-Dimethylpropenyl)-7,8-methylenedioxycoumarin

[114216-82-9]

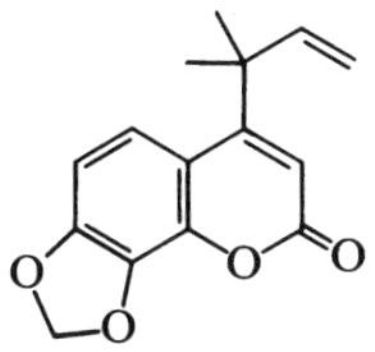

$C_{15}H_{14}O_4$ M 258.273

Constit. of *Ruta chalepensis.*

Ulubelen, A. *et al, Phytochemistry,* 1988, **27**, 650.

Rzedowskin *A*   R-70016

[117610-40-9]

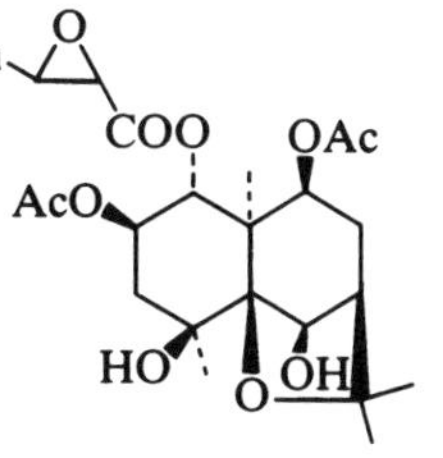

$C_{28}H_{36}O_{10}$ M 532.586

Constit. of *Rzedowskia tolantonguensis.* Cryst. (Me₂CO-/hexane). Mp 210°. Rzedowskin *B* and Rzedowskin *C* are mixtures of related esters isol. from the same source.

Jiménez, M. *et al, Phytochemistry,* 1988, **27**, 2213.

S

Safflor Yellow B
S-70001

[91574-92-4]

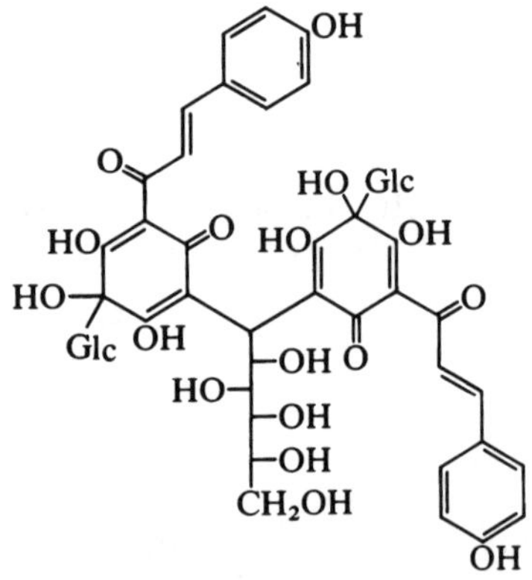

C$_{48}$H$_{54}$O$_{27}$ M 1062.938

Pigment from the flowers of *Carthamus tinctorius*.

Takahashi, Y. *et al*, *Tetrahedron Lett.*, 1984, **24**, 2471.

Salicortin
S-70002

Updated Entry replacing S-00023

2-[[[(1-Hydroxy-6-oxo-2-cyclohexen-1-yl)carbonyl]-oxy]methyl]phenyl β-D-glucopyranoside, 9CI

[29836-41-7]

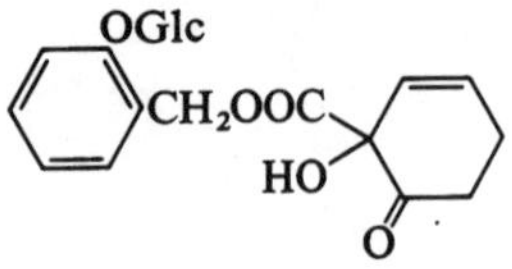

C$_{20}$H$_{24}$O$_{10}$ M 424.404

Constit. of the bark of *Salix* and *Populus* spp. Mp 135-137°.

O$^{2'}$-*Benzoyl:* [29836-40-6]. **Tremulacin.**

C$_{27}$H$_{28}$O$_{11}$ M 528.512

Constit. of the bark of *P. tremula* and *P. tremuloides*. Cryst. (EtOAc/pet. ether). Mp 122-123°. Subst. in the glucosyl residue.

Thieme, H., *Pharmazie*, 1964, **19**, 725 (*isol*)
Thieme, H., *Planta Med.*, 1967, **15**, 35 (*isol*)
Pearl, I.A., *Tetrahedron Lett.*, 1970, 3827 (*struct*)
Pearl, I.A., *Phytochemistry*, 1971, **10**, 3161 (*struct*)

Salvianolic acid *B*
S-70003

[115939-25-8]

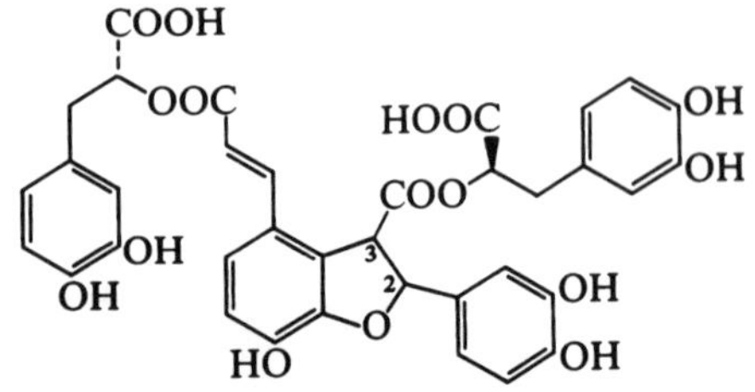

C$_{36}$H$_{30}$O$_{16}$ M 718.623

Constit. of *Salvia miltiorrhiza*. Amorph. yellow powder. [α]$_D^{18}$ +92° (c, 0.07 in EtOH).

3-Deacyl, 2,3-didehydro: [115841-09-3]. **Salvianolic acid C.**

C$_{26}$H$_{20}$O$_{10}$ M 492.438

Constit. of *S. miltiorrhiza*. Amorph. yellow powder. [α]$_D^{14}$ +70° (c, 0.102 in EtOH).

Chen, Z.-X. *et al*, *Chin. Pharm. Bull.*, 1981, **16**, 534 (*isol*)
Ai, C.-B. *et al*, *J. Nat. Prod.*, 1988, **51**, 145 (*struct, isol*)

Salvicanaric acid
S-70004

[112470-96-9]

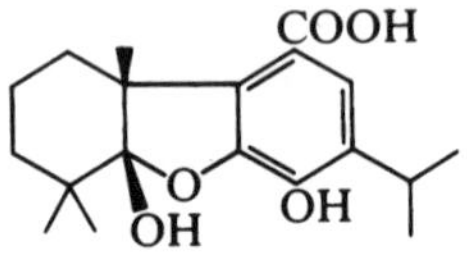

C$_{19}$H$_{26}$O$_{5}$ M 334.411

Constit. of *Salvia canariensis*. Foam.

Gonzalez, A.G. *et al*, *J. Nat. Prod.*, 1987, **50**, 341.

Salvileucolidone
S-70005

[116408-25-4]

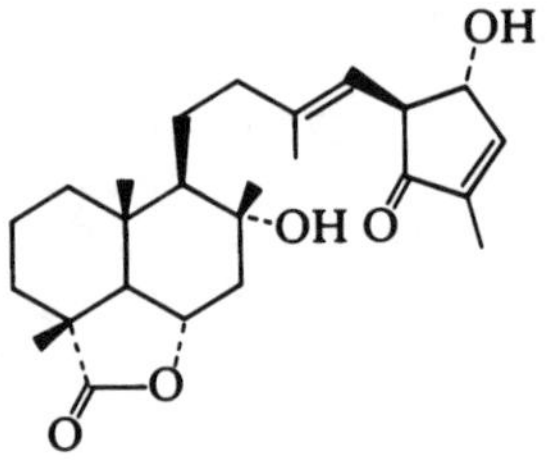

C$_{25}$H$_{36}$O$_{5}$ M 416.556

Constit. of *Salvia hypoleuca*. Gum. Various epoxide and hydroperoxide derivs. also isolated.

Rustaiyan, A. *et al*, *Phytochemistry*, 1988, **27**, 1767.

Salviolone
S-70006

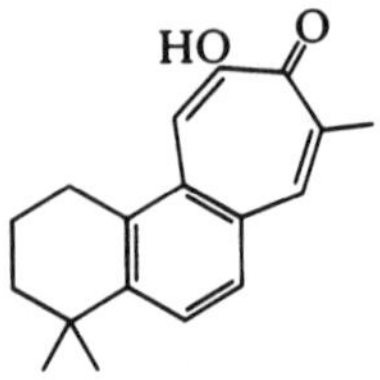

C$_{18}$H$_{20}$O$_{2}$ M 268.355

Constit. of Chinese drug Dan-Shen (*Salvia miltiorrhiza*). Cytotoxic. Pale-yellow oil.

Ginda, H. *et al*, *Tetrahedron Lett.*, 1988, **29**, 4603 (*struct*)

Sambucoin   S-70007

Updated Entry replacing S-30009
[90044-34-1]

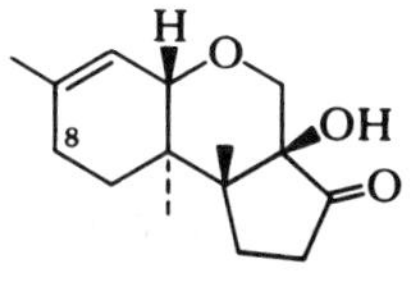

C₁₅H₂₂O₃   M 250.337

$C_{15}H_{22}O_3$ M 250.337
Trichothecene antibiotic. Metab. of *Fusarium sambucinum*. Shows possible cystatic activity. Cryst. (Me₂CO/pet. ether). Mp 205-210°. Related to Trichothecenes.

8α-Hydroxy: [112468-60-7]. *8α-Hydroxysambucoin*.
$C_{15}H_{22}O_4$ M 266.336
Metab. of *F. sporotrichioides*. Glass.

8β-Hydroxy: [112531-13-2]. *8β-Hydroxysambucoin*.
$C_{15}H_{22}O_4$ M 266.336
Metab. of *F. sporotrichioides*. Glass.

Mohr, P. *et al, Helv. Chim. Acta*, 1984, **67**, 406 (*isol, struct*)
Corley, D.G. *et al, J. Nat. Prod.*, 1987, **50**, 897 (*derivs*)

Sanadaol   S-70008

Updated Entry replacing S-60005
[83643-92-9]

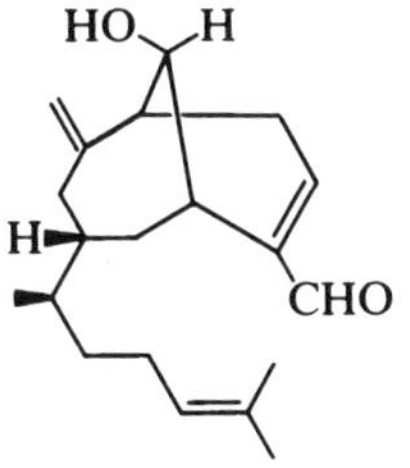

Absolute configuration

$C_{20}H_{30}O_2$ M 302.456
Constit. of brown alga *Pachydictyon coriaceum*. Oil.
[α]ᴅ +74.8° (c, 1.33 in CHCl₃). May be artifact produced from Dictyodial *A*, D-70145 .

Ac: [83643-93-0]. *Acetylsanadaol*.
$C_{22}H_{32}O_3$ M 344.493
Constit. of *P. coriaceum*. Oil. [α]ᴅ +42.5° (c, 0.89 in CHCl₃).

Ishitsuka, M. *et al, Tetrahedron Lett.*, 1982, **23**, 3179 (*isol*)
Nagaoka, H. *et al, Tetrahedron Lett.*, 1987, **28**, 2021; 1988, **29**, 5945 (*synth, abs config*)

Saptarangiquinone *A*   S-70009

[32988-67-3]

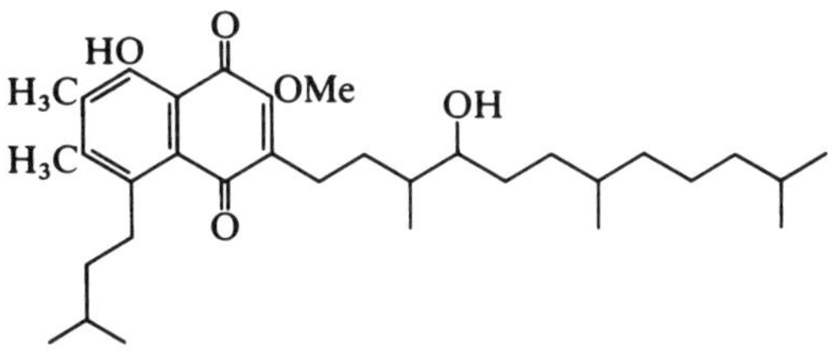

$C_{33}H_{52}O_5$ M 528.771
Isol. from the root bark of *Salacia macrosperma*. Orange plates (MeOH/CHCl₃). Mp 176-180°. 9CI name and MF incorrect.

Krishnan, V. *et al, Indian J. Chem.*, 1971, **9**, 117.

Sarcodictyin *A*   S-70010

[113540-81-1]

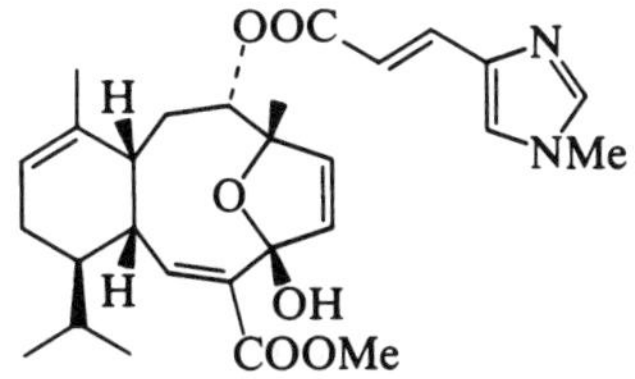

$C_{28}H_{36}N_2O_6$ M 496.602
Constit. of *Sarcodictyon roseum*. Powder (MeOH). Mp 219-222°. [α]ᴅ²⁰ −15.2° (c, 1.12 in EtOH).

Et ester analogue: [113555-26-3]. *Sarcodictyin* **B**.
$C_{29}H_{38}N_2O_6$ M 510.629
Constit. of *S. roseum*. Powder (MeOH). Mp 219-222°. [α]ᴅ −15.2° (c, 1.12 in EtOH).

D'Ambrosio, M. *et al, Helv. Chim. Acta*, 1987, **70**, 2019.

Sarcophytoxide   S-70011

Updated Entry replacing S-50016
2,16;7,8-Diepoxy-1(15),3,11-cembratriene
[70748-49-1]

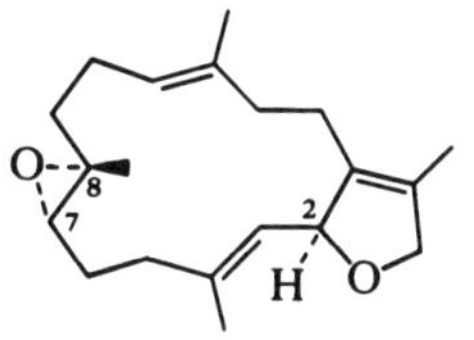

(2S,7S,8S)-form

$C_{20}H_{30}O_2$ M 302.456
(2S,7S,8S)-form
Constit. of *Sarcophyton trocheliophorum*. [α]ᴅ +157°.
(2R,7R,8R)-form
Constit. of *S. birklandi*. [α]ᴅ −191°.
(2R,7S,8S)-form
Constit. of *Lobophytum pauciflorum*. Cryst. Mp 52-56°. [α]ᴅ −64° (c, 0.6 in CHCl₃).

Tursch, B., *Pure Appl. Chem.*, 1976, **48**, 1 (*isol*)
Kobayashi, J. *et al, Experientia*, 1983, **39**, 67 (*cryst struct*)
Bowden, B.F. *et al, J. Nat. Prod.*, 1987, **50**, 650 (*isol, bibl*)

Saroaspidin *A*   S-70012

[112663-69-1]

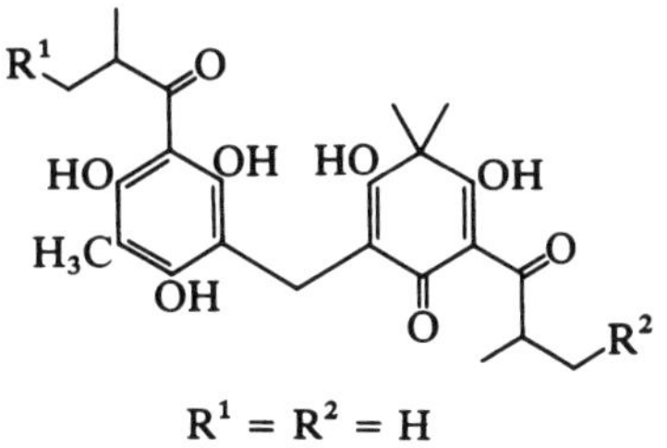

R¹ = R² = H

$C_{24}H_{30}O_8$ M 446.496
Constit. of *Hypericum japonicum*. Antibiotic. Yellow powder. Mp 205-216°.

Ishiguro, K. *et al, Planta Med.*, 1987, **53**, 415.

Saroaspidin *B*   S-70013

[112663-68-0]
As Saroaspidin *A*, S-70012 with

R¹ = H, R² = CH₃

C$_{25}$H$_{32}$O$_8$　　M 460.523
Constit. of *Hypericum japonicum*. Antibiotic. Yellow
powder. Mp 201-203°.

Ishiguro, K. *et al*, *Planta Med.*, 1987, **53**, 415.

Saroaspidin *C*　　　　　　　　　　S-70014

[112663-70-4]

As Saroaspidin *A*, S-70012 with

$$R^1 = R^2 = CH_3$$

C$_{26}$H$_{34}$O$_8$　　M 474.550
Constit. of *Hypericum japonicum*. Yellow powder. Mp
165-172°.

Ishiguro, K. *et al*, *Planta Med.*, 1987, **53**, 415.

Sauvagine　　　　　　　　　　　　S-70015

[74434-59-6]

Pyr-Gly-Pro-Pro-Ile-Ser-Ile-Asp-Leu-Ser-Leu-Glu-Leu-
Leu-Arg-Lys-Met-Ile-Glu-Ile-Glu-Lys-Gln-Glu-Lys-
Glu-Lys-Gln-Gln-Ala-Ala-Asn-Asn-Arg-Leu-Leu-
Leu-Asp-Thr-Ile-NH$_2$

Peptide from the skin of the South American frog *Phyl-
lomedusa sauvagei*. Shows hypotensive action. Inhibits
prolactin thyrotropin and growth hormone release,
causes corticotropin and endorphin release. Fluffy powder. [α]$_D^{20}$ −66.2° (c, 0.3 in 1*M* AcOH).

Montecuchi, P.C. *et al*, *Int. J. Pept. Protein Res.*, 1980, **16**, 191;
1981, **18**, 113 (*isol, struct*)
Nomizu, M. *et al*, *Chem. Pharm. Bull.*, 1988, **36**, 123 (*synth,
bibl*)

Sawaranin　　　　　　　　　　　　S-70016

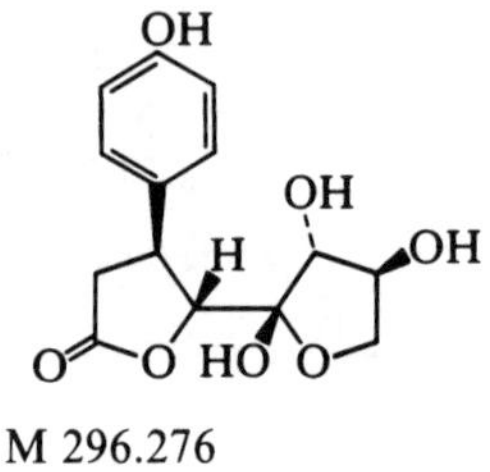

C$_{14}$H$_{16}$O$_7$　　M 296.276
Constit. of *Chamaecyparis pisifera*. Cryst. (H$_2$O). Mp
214-222° dec.

Hasegawa, S. *et al*, *Phytochemistry*, 1988, **27**, 2703.

Scapaniapyrone *A*　　　　　　　　S-70017

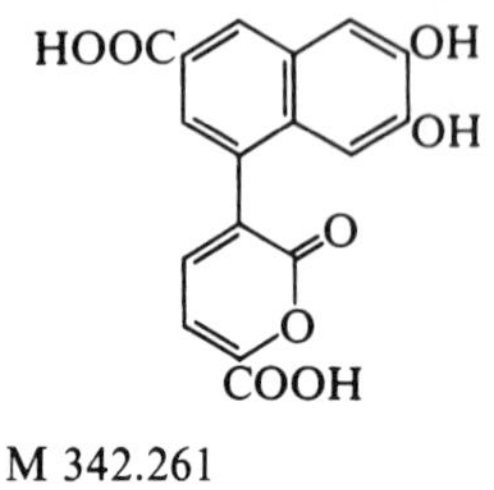

C$_{17}$H$_{10}$O$_8$　　M 342.261
Constit. of liverwort *Scapania undulata*. Yellow solid.

Mues, R. *et al*, *Tetrahedron Lett.*, 1988, **29**, 6793 (*struct*)

Sclareol　　　　　　　　　　　　　S-70018

Updated Entry replacing S-10021
14-Labdene-8α,13S-diol
[515-03-7]

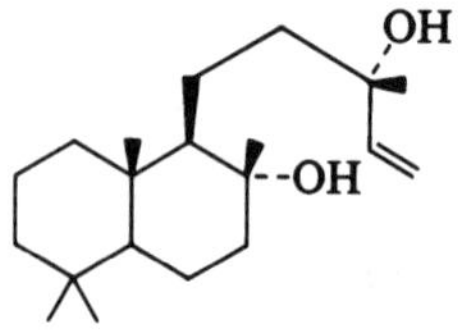

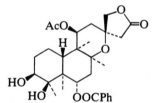

C$_{20}$H$_{36}$O$_2$　　M 308.503
Constit. of *Salvia sclarea*. Intermed. used in perfumery
industry for manuf. of synthetic ambergris odourants.
Cryst. (pet. ether). Mp 105.5-106.0°. [α]$_D^{18}$ −6.25°
(EtOH).
6′-Deoxy-α-L-idopyranoside: [77451-20-8].
　C$_{26}$H$_{46}$O$_6$　　M 454.646
　Constit. of *Aster spathulifolius*. Cryst. [α]$_D$ −33.2°
　(MeOH).
6α-Angeloyloxy:
　C$_{25}$H$_{42}$O$_3$　　M 390.605
　Constit. of *Stevia monardaefolia*. Cryst. Mp 109-110°.
　[α]$_D$ −26.4° (CHCl$_3$).
13-Epimer: 13-Episclareol. 14-Labdene-8α,13R-diol.
　C$_{20}$H$_{36}$O$_2$　　M 308.503
　Constit. of *Polemonium viscosum*. Needles (MeOH
　aq.). Mp 125-127°. [α]$_D$ +6.6° (c, 0.016 in CHCl$_3$).

Almqvist, S. *et al*, *Acta Chem. Scand., Ser. B*, 1975, **29**, 695
(*isol, struct*)
Uchio, Y. *et al*, *Tetrahedron Lett.*, 1980, 3775 (*isol, deriv*)
Quijano, L. *et al*, *Phytochemistry*, 1982, **21**, 1369 (*isol, deriv*)
Stierle, D.B. *et al*, *Phytochemistry*, 1988, **27**, 517 (*13-
Episclareol*)

Scutellone *C*　　　　　　　　　　S-70019

[114489-73-5]

C$_{29}$H$_{38}$O$_9$　　M 530.614
Constit. of *Scutellaria rivularis*. Needles (Me$_2$CO). Mp
228-230°. [α]$_D^{20}$ −20.0° (c, 1 in CHCl$_3$).

Lin, Y.-L. *et al*, *Heterocycles*, 1988, **27**, 779.

Scutellone *F*　　　　　　　　　　S-70020

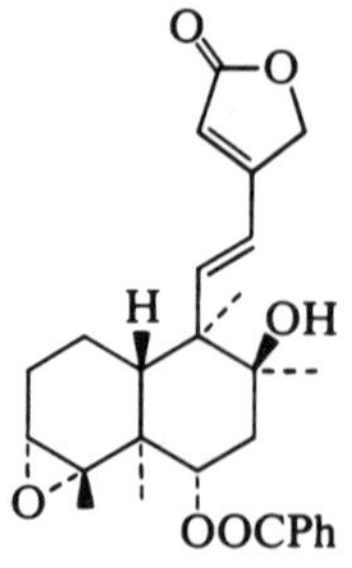

C$_{27}$H$_{32}$O$_6$　　M 452.546
Constit. of *Scutellaria rivularis*. Needles (Me$_2$CO). Mp
213-215°. [α]$_D^{20}$ +54.9° (c, 1 in CHCl$_3$).

Lin, Y.-L. *et al*, *Heterocycles*, 1988, **27**, 779.

3,4-Seco-4(28),20,24-dammaratrien-3,26-dioic acid   S-70021

[117732-40-8]

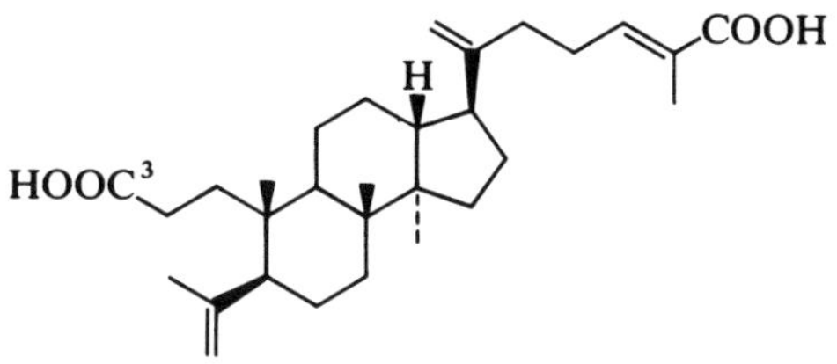

C$_{30}$H$_{46}$O$_4$ M 470.691

Constit. of *Alnus japonica* flowers. Needles (EtOAc/hexane). Mp 168.5-169.5°. [α]$_D^{25}$ +44.0° (c, 0.45 in CHCl$_3$).

3-Me ester:
C$_{32}$H$_{48}$O$_4$ M 496.729
Constit. of *A. japonica* flowers. Oil. [α]$_D^{25}$ +61.2° (c, 0.72 in CHCl$_3$).

Aoki, T. *et al*, *Phytochemistry*, 1988, **27**, 2915.

Secofloribundione   S-70022

[116374-12-0]

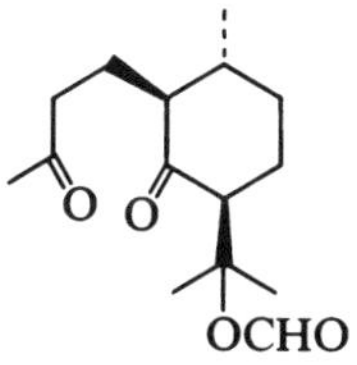

C$_{15}$H$_{24}$O$_4$ M 268.352
Constit. of *Liabum floribundum*. Oil.

Jakupovic, J. *et al*, *Phytochemistry*, 1988, **27**, 1771.

Seco-4-hydroxylintetralin   S-70023

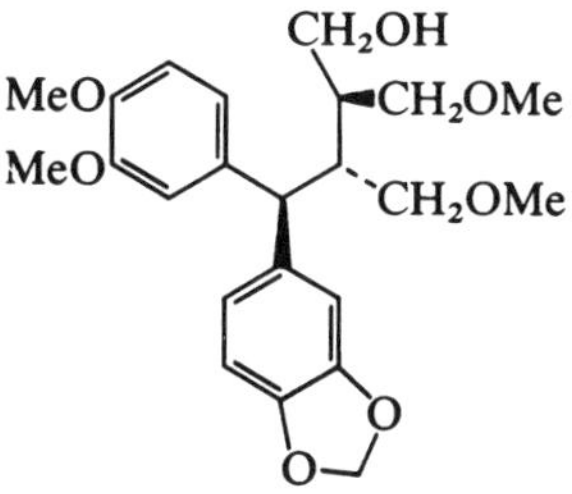

C$_{23}$H$_{30}$O$_7$ M 418.486
Constit. of *Phyllanthus niruri*. Oil. [α]$_D^{28}$ −151° (CHCl$_3$).

Satyanarayana, P. *et al*, *J. Nat. Prod.*, 1988, **51**, 44.

Secoisoerivanin pseudoacid   S-70024

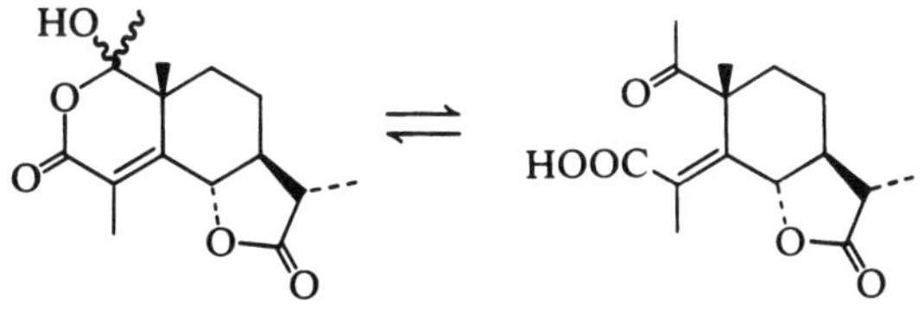

C$_{15}$H$_{20}$O$_5$ M 280.320
Constit. of *Artemisia judaica*. Oil.

Khafagy, S.M. *et al*, *Phytochemistry*, 1988, **27**, 1125.

4,5-Seconeopulchell-5-ene   S-70025

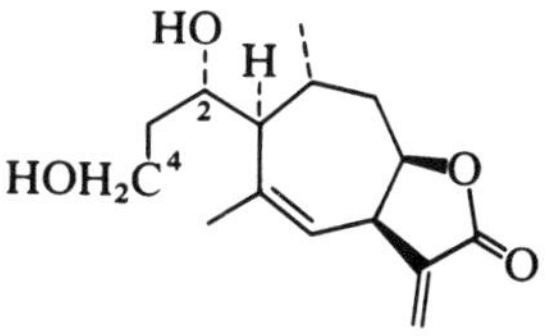

C$_{15}$H$_{22}$O$_4$ M 266.336
2-Ac: [93373-41-2].
   C$_{17}$H$_{24}$O$_5$ M 308.374
   Constit. of *Gaillardia pulchella*.
4-Ac: [117634-70-5].
   C$_{17}$H$_{24}$O$_5$ M 308.374
   Constit. of *G. pulchella*.

Bohlmann, F. *et al*, *Phytochemistry*, 1984, **23**, 1979 (*isol*)
Yu, S. *et al*, *Phytochemistry*, 1988, **23**, 2887 (*isol*)

Secopseudopterosin A   S-70026

[111466-65-0]

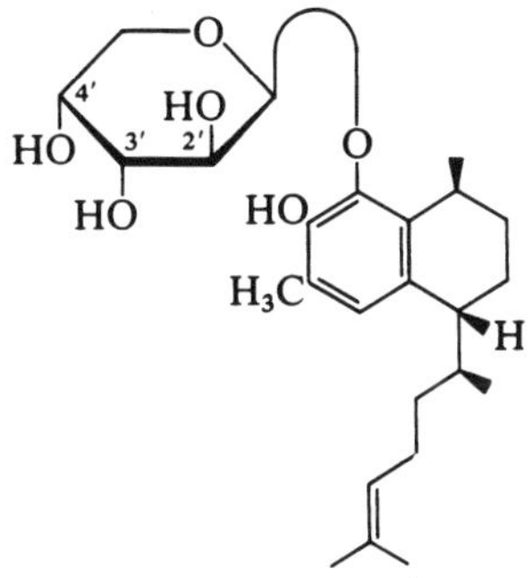

C$_{25}$H$_{38}$O$_6$ M 434.572
Constit. of sea whip *Pseudopterogorgia* sp. Analgesic and antiinflammatory. Amorph. solid. [α]$_D$ −118° (c, 1.7 in CHCl$_3$).

2′-Ac: [111397-51-4]. **Secopseudopterosin B**.
C$_{27}$H$_{40}$O$_7$ M 476.609
Constit. of *P.* sp. Unstable.
3′-Ac: [111466-66-1]. **Secopseudopterosin C**.
C$_{27}$H$_{40}$O$_7$ M 476.609
Constit. of *P.* sp. Analgesic and antiinflammatory.
Amorph. solid. [α]$_D$ −89° (c, 0.58 in CHCl$_3$).
4′-Ac: [111466-67-2]. **Secopseudopterosin D**.
C$_{27}$H$_{40}$O$_7$ M 476.609
Constit. of *P.* sp. Analgesic and antiinflammatory.
Amorph. solid. [α]$_D$ −139° (c, 0.6 in CHCl$_3$).

Look, S.A. *et al*, *Tetrahedron*, 1987, **43**, 3363.

Secothujene   S-70027

2-Acetyl-1-(1-methylethyl)cyclopropaneacetic acid, 9CI.
2-Acetyl-1-isopropylcyclopropaneacetic acid

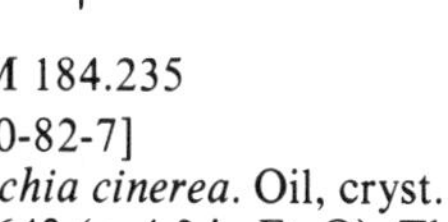

C$_{10}$H$_{16}$O$_3$ M 184.235
(1S,2S)-form [4700-82-7]
   Constit. of *Brocchia cinerea*. Oil, cryst. on standing. Mp 62°. [α]$_D$ −164° (c, 4.2 in Et$_2$O). The nat. isolate was descr. as an oil, no opt. rotn. given. Prob. identical with the semisynthetic prod. obt. by oxidn. of thujene.

Shrivastava, R.D. *et al*, *CA*, 1965, **64**, 6696g (*synth*)

Jakupovic, J. et al, Phytochemistry, 1988, 27, 2219 (isol, ms, pmr, cmr)

Seiricuprolide   S-70028

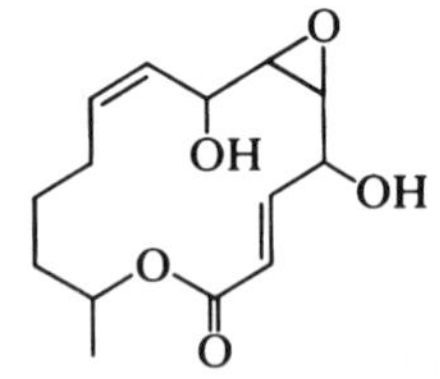

$C_{14}H_{20}O_5$ M 268.309

Metab. of *Seiridium cupressi*. Cryst. (EtOAc/pet. ether). Mp 128-130°. $[\alpha]_D^{25}$ +67.2° (c, 1.45 in CHCl₃).

Ballio, A. et al, Phytochemistry, 1988, 27, 3117.

Selenoformaldehyde   S-70029

Methaneselenal, 9CI

[6596-50-5]

$$H_2C{=\!=}Se$$

CH₂Se   M 92.987

V. unstable, detected by various spectroscopic techniques. Stabilised as complexes.

▷LP9800000.

Trimer: see 1,3,5-Triselenane, T-04418

Hofmann, L. et al, J. Organomet. Chem., 1983, 255, C41.
Bock, H. et al, Chem. Ber., 1984, 117, 187.
Judge, R.H. et al, J. Am. Chem. Soc., 1984, 106, 5406.
Glinski, R.J. et al, J. Am. Chem. Soc., 1986, 108, 531.

Selenolo[3,4-b]thiophene   S-70030

[74070-00-1]

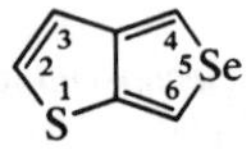

C_6H_4SSe M 187.118

Light-yellow viscous oil. Could not be recryst.

Konar, A. et al, Chem. Scr., 1982, 19, 176 (synth, pmr)
Gleiter, R. et al, Chem. Ber., 1987, 120, 1917 (pe)

Selenourea, 9CI   S-70031

Updated Entry replacing S-00256

[630-10-4]

$$(H_2N)_2C{=\!=}Se$$

CH₄N₂Se   M 123.016

Prisms or needles (H₂O). Mp 200° dec. (slow heat), 235° dec. Air and light sensitive.

▷YU1820000.

B,MeI: Methylisoselenourea iodide. Yellow cryst. Mp 187-188° dec.

B,MeHSO₄: Methylisoselenourea hydrogen sulfate. Yellow cryst. Mp 67-70° dec. Structure: [H₂N—C(SeMe)=NH₂]HSO₄.

N-*Me:* [5533-49-3].
  $C_2H_6N_2Se$ M 137.043
  Mp 156-157°.

N¹,N¹-*Di-Me:* [5117-16-8].
  $C_3H_8N_2Se$ M 151.070
  Mp 98-100°.

N¹,N³-*Di-Me:* [5533-46-0].
  $C_3H_8N_2Se$ M 151.070
  Mp 110-111°.

N-*Tetra-Me:* [5943-53-3].
  $C_5H_{12}N_2Se$ M 179.123
  Mp 79-81°.

N-*Et:* [33251-42-2].
  $C_3H_8N_2Se$ M 151.070
  Mp 125°.

N-*Ph:* [6124-02-3].
  $C_7H_8N_2Se$ M 199.114
  Mp 191-192°.

N¹,N¹-*Di-Ph:* [21347-28-4].
  $C_{13}H_{12}N_2Se$ M 275.211
  Mp 205-207° dec.

N¹,N³-*Di-Ph:* [16519-43-0].
  $C_{13}H_{12}N_2Se$ M 275.211
  Mp 190-192° (178-182° dec.).

Backer, H.J. et al, Recl. Trav. Chim. Pays-Bas, 1943, 62, 580 (synth)
Dunbar, P.E. et al, J. Am. Chem. Soc., 1947, 69, 1833 (synth)
Hope, H., Acta Chem. Scand., 1964, 18, 1800 (synth)
Jensen, K.A. et al, Acta Chem. Scand., 1966, 20, 597 (derivs)
Guedicelli, J.F. et al, Bull. Soc. Chim. Fr., 1968, 1099 (derivs)
Klayman, D.L. et al, J. Org. Chem., 1969, 34, 3549 (derivs)
Duncan, J.L. et al, J. Chem. Soc. (A), 1971, 2695 (ir, raman)
Walter, W. et al, Tetrahedron, 1972, 28, 3233 (pmr)

Semecarpetin   S-70032

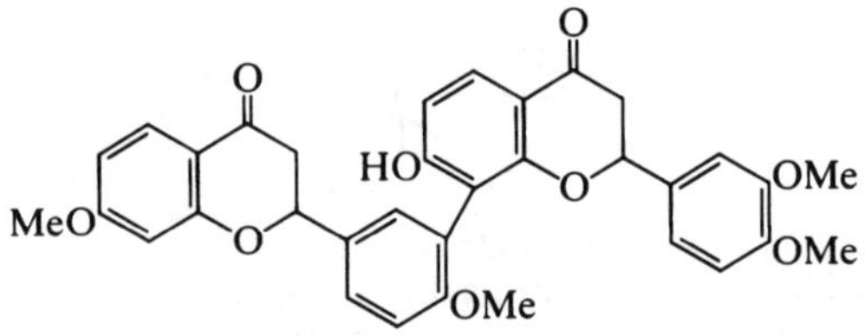

$C_{34}H_{30}O_9$ M 582.606

Constit. of *Semecarpus anacardium*.

Me ether: Cryst. Mp 138°.

Murthy, S.S.N., Phytochemistry, 1988, 27, 3020.

Senburiside II   S-70033

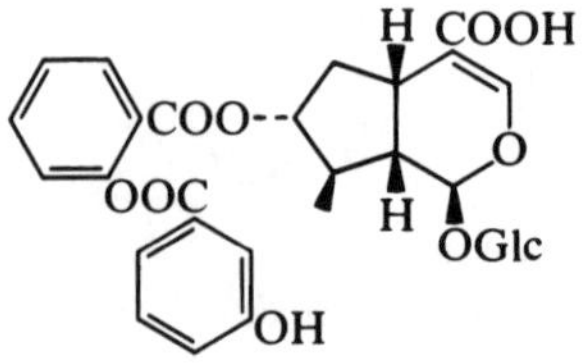

$C_{30}H_{32}O_{14}$ M 616.574

Constit. of *Swertia japonica*. $[\alpha]_D^{22}$ −88.6° (c, 0.79 in MeOH).

Ikeshiro, Y. et al, Planta Med., 1987, 158.

Senegalensein   S-70034

2,3-Dihydro-5,7-dihydroxy-2-(4-hydroxyphenyl)-6,8-bis(3-methyl-2-butenyl)-2H-1-benzopyran-4-one, 9CI.
4′,5,7-Trihydroxy-6,8-diprenylflavanone

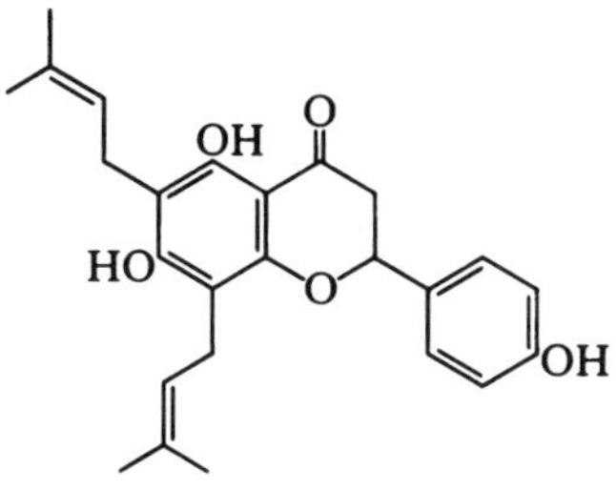

$C_{25}H_{28}O_5$ M 408.493

Constit. of the stem bark of *Erythrina senegalensis*. Pale-yellow oil. $[\alpha]_D$ −6.7° (c, 0.3 in CHCl₃).

Fomum, Z.T. *et al, J. Nat. Prod.*, 1987, **50**, 921.

Sepiapterin, 8CI   S-70035

2-Amino-7,8-dihydro-6-(2-hydroxy-1-oxopropyl)-4(1H)-pteridinone, 9CI. 6-Lactoyl-7,8-dihydropterin

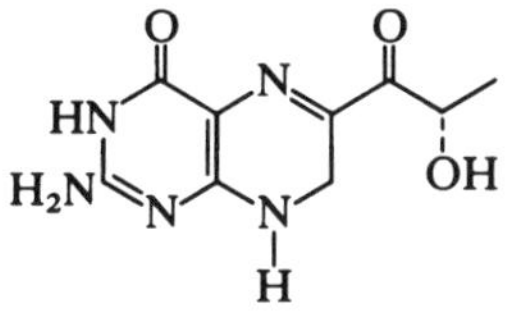

$C_9H_{11}N_5O_3$ M 237.218

(**S**)-*form* [17094-01-8]

L-form

Eye pigment from *Drosophila melanogaster* mutant *sepia*. Yellow powder or cryst.

Deoxy: [1797-87-1]. **Deoxysepiapterin.** *Isosepiapterin.*
$C_9H_{11}N_5O_2$ M 221.218
Eye pigment from *D. melanogaster*, more readily obt. from the blue-green alga *Anacystis nidulans*. Yellow cryst. (EtOH aq.). pK_{a1} 1.35, pK_{a2} 10.05.

Forrest, H.S. *et al, J. Am. Chem. Soc.*, 1954, **76**, 5656, 5658 (*isol, uv*)
Forrest, H.S. *et al, Arch. Biochem. Biophys.*, 1959, **83**, 508 (*isol*)
Nawa, S., *Bull. Chem. Soc. Jpn.*, 1960, **33**, 1555 (*struct, uv*)
Schircks, B. *et al, Helv. Chim. Acta*, 1978, **61**, 2731 (*synth, uv, cd, bibl*)
Pfleiderer, W., *Chem. Ber.*, 1979, **112**, 2750 (*synth, abs config*)
Baur, R. *et al, Helv. Chim. Acta*, 1988, **71**, 531 (*synth, uv, pmr, deriv*)

Sesamoside   S-70036

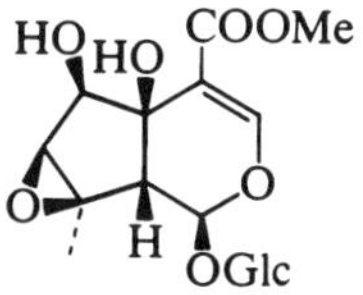

$C_{17}H_{24}O_{12}$ M 420.369

Constit. of *Sesamum angolense*. Amorph. powder.
Penta-Ac: Prisms (CH₂Cl₂/Et₂O/hexane). Mp 151-153°.

Potterat, O. *et al, Phytochemistry*, 1988, **27**, 2677.

Sesquisabinene   S-70037

Updated Entry replacing S-30035
1-(1,5-Dimethyl-4-hexenyl)-4-methylenebicyclo[3.1.0]-hexane, 9CI
[58319-04-3]

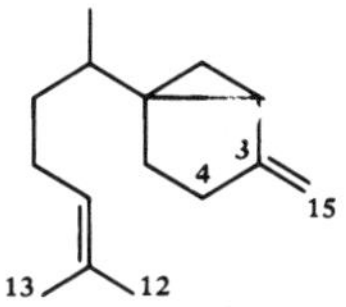

$C_{15}H_{24}$ M 204.355

Constit. of *Piper nigrum*. Isol. by glc.

$\Delta^{3,4}$-*Isomer:* [58319-06-5]. **Sesquithujene.**
$C_{15}H_{24}$ M 204.355
Isol. from *Zingiber officinale*.

3,15-Dihydro, 3-hydroxy (cis-): **Sesquisabinene hydrate.**
$C_{15}H_{24}O$ M 220.354
Isol. from essential oil of *Z. officinale*.

12-Acetoxy: **12-Sesquisabinenol acetate.** *12-Acetoxysesquisabinene.*
$C_{17}H_{26}O_2$ M 262.391
Constit. of *Arctotis grandis*, *Haplocarpha scaposa* and *H. lyrata*. Oil.

12-Oxo: [83161-51-7]. **12-Sesquisabinenal.** *2-Methyl-6-(4-methylenebicyclo[3.1.0]hex-1-yl)-2-heptenal, 9CI.*
$C_{15}H_{22}O$ M 218.338
Constit. of *H. scaposa* and *H. lyrata*. Oil.

13-Acetoxy: **13-Sesquisabinenol acetate.** *13-Acetoxysesquisabinene.*
$C_{17}H_{26}O_2$ M 262.391
Constit. of *H. scaposa* and *H. lyrata*. Oil. $[\alpha]_D^{24}$ −39.5° (c, 0.2 in CHCl₃).

Terhune, S.J. *et al, Can. J. Chem.*, 1975, **53**, 3285 (*isol, struct*)
Bohlmann, F. *et al, Phytochemistry*, 1982, **21**, 1157 (*isol, struct*)
Sharma, P.K. *et al, Phytochemistry*, 1988, **27**, 3471 (*biosynth*)

Shikonin   S-70038

Updated Entry replacing S-20053
5,8-Dihydroxy-2-(1-hydroxy-4-methyl-3-pentenyl)-1,4-naphthalenedione, 9CI

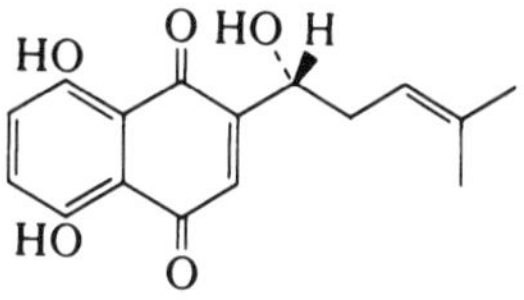

(*R*)-*form*

$C_{16}H_{16}O_5$ M 288.299

Shows antitumour activity.

(**R**)-*form* [517-89-5]
Isol. from *Lithiospermum erythrorhizon*, *L. officinale*, *L. euchromum*, *Arnebia nobilis* and *A. tibetana*. Red-brown cryst. (C₆H₆). Mp 148° (143°).
▷QL8000200.

O¹-Ac: **Shikonin acetate.**
$C_{18}H_{18}O_6$ M 330.337
Isol. from *L. erythrorhizon*, *L. euchromum* and *Jatropha glandulifera*. Red prisms or reddish-violet needles. Mp 85-86°, Mp 106-107°. $[\alpha]_D^{20}$ +26° (CHCl₃).

O¹-(3-Hydroxy-3-methylbutanoyl): β-**Hydroxyisovalerylshikonin.**
$C_{21}H_{24}O_7$ M 388.416
Isol. from *L. erythrorhizon* and *L. euchromum*. Red-violet cryst. Mp 90-92°. $[\alpha]_{600}^{15}$ +128° (EtOH).

O$^{1'}$-(*3,4-Dimethyl-3-pentenoyl*): **Teracrylshikonin**.
$C_{23}H_{26}O_6$ M 398.455
Isol. from *L. euchromum*. Amorph. $[\alpha]_{600}^{17}$ −92°
(EtOH).

O$^{1'}$-(*3-Methyl-2-butenoyl*): [5162-01-6]. *β,β*-
Dimethylacrylshikonin.
$C_{21}H_{22}O_6$ M 370.401
Isol. from *L. erythrospermum*, *L. euchromum* and
Jatropha glandulifera. Red prisms or red-violet
needles. Mp 113-114°. $[\alpha]_{600}^{22}$ +222° (EtOH).

O$^{1'}$-(*3-Methylbutanoyl*): **Shikonin isobutyrate**.
$C_{21}H_{24}O_6$ M 372.417
Isol. from *L. erythrorhizon* and *L. euchromum*. Red-
violet needles. Mp 89-90°. $[\alpha]_{600}^{23}$ +125° (EtOH).

O-(*2-Methylpropanoyl*): **Isobutyrylshikonin**.
$C_{20}H_{22}O_6$ M 358.390
Isol. from roots of *L. erythrorrhizon*. Mp 89-90°.
$[\alpha]_{600}^{23}$ +125° (EtOH).

(*S*)-form [517-88-4]
Alkannin. *Arnebin IV*
Constit. of *Alkanna tinctoria* and *Onosma echoides*.
Used in the treatment of ulcus oruris. Reddish-bronze
needles (Et$_2$O/EtOH). Mp 116-117° (149°). $[\alpha]_{Cd}^{20}$-
−157° (C_6H_6).

Mono-Ac: [38222-13-8]. **Alkannin acetate**. *Arnebin III*.
$C_{18}H_{18}O_6$ M 330.337
Constit. of the roots of *A. nobilis*. Mp 104-105°.

Tri-Ac: Yellow cryst. Mp 132°. $[\alpha]_{Cd}^{20}$−110° (C_6H_6).

O$^{1'}$-(*3-Methyl-2-butenoyl*): [5162-01-6]. *β,β*-**Dimethyla-
crylalkannin**. *Arnebin I*.
$C_{21}H_{22}O_6$ M 370.401
Constit. of the roots of *A. nobilis* and *A. tinctoria*. Red
needles (hexane). Mp 116-117°.

Me ether: Brownish-red. Mp 105° (109°).

Angeloyl: **O-Angeloylalkannin**.
$C_{21}H_{22}O_6$ M 370.401
Isol. from *A. nobilis*. Deep-red oil.

(±)-form [11031-58-6]
Shikalkin
Isol. from roots of *A. hispidissima*. Also isol. with no opt.
rotn. reported, from *Macrotomia ugamensis*, *Onosma
caucasicum* and *Echium rubrum*. Mp 148°.

Raudnitz, H. *et al*, *Ber.*, 1934, **67**, 1955; 1935, **68**, 1479 (*isol*,
struct)
Brockmann, H., *Justus Liebigs Ann. Chem.*, 1935, **521**, 1 (*isol*,
struct)
Morimoto, I. *et al*, *Tetrahedron Lett.*, 1965, 3677, 4737; 1966,
3677 (*isol*, *derivs*)
Schmid, H.V. *et al*, *Tetrahedron Lett.*, 1971, 4151 (*biosynth*)
Shcherbanovskii, L.R. *et al*, *Khim. Prir. Soedin.*, 1971, 517;
1972, 666 (*isol*)
Shukla, Y.N. *et al*, *Phytochemistry*, 1971, **10**, 1909 (*isol*,
Arnebins)
Afzal, M. *et al*, *J. Chem. Soc.*, *Perkin Trans. 1*, 1975, 1334
(*isol*)
Mizukami, H. *et al*, *Phytochemistry*, 1978, **17**, 95 (*isol*)
Papageorgiou, V.P. *et al*, *Experientia*, 1978, **34**, 1499; *Planta
Med.*, 1979, **35**, 56 (*isol*, *struct*, *use*)
Inouye, H. *et al*, *Phytochemistry*, 1979; **18**, 1301 (*biosynth*)
Sankawa, U. *et al*, *Chem. Pharm. Bull.*, 1981, **29**, 116
(*pharmacol*)
Terada, A. *et al*, *J. Chem. Soc.*, *Chem. Commun.*, 1983, 987
(*synth*)
Tanoue, Y. *et al*, *J. Org. Chem.*, 1987, **52**, 1437 (*synth*)
Terada, A. *et al*, *Bull. Chem. Soc. Jpn.*, 1987, **60**, 205 (*synth*)

Shiromodiol   S-70039

Updated Entry replacing S-00389
4α,5α-Epoxy-1(10)E-germacrene-6β,8β-diol
[20071-60-7]

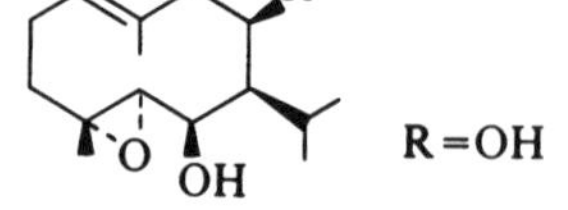

R = OH

$C_{15}H_{26}O_3$ M 254.369
Cryst. Mp 89°.

8-Ac: [20071-59-4].
$C_{17}H_{28}O_4$ M 296.406
Constit. of the leaves of *Parabenzoin trilobum*.
Antifeeding activity for the insect *Prodenia litura*.
Cryst. Mp 80°. $[\alpha]_D^{25}$ −44.8° (c, 0.34 in CHCl$_3$).

Di-Ac: [20071-58-3].
$C_{19}H_{30}O_5$ M 338.443
Constit. of the leaves of *P. trilobium*. Shows
antifeeding activity towards *P. litura* and *Trimeresia
miranda*. Cryst. Mp 112°. $[\alpha]_D^{25}$ −61.9° (c, 1.06 in
CHCl$_3$).

8-(4-Hydroxybenzoyl): [96853-63-3]. **8-p-
Hydroxybenzoylshiromodiol**.
$C_{22}H_{30}O_5$ M 374.476
Constit. of *Ferula orientalis*.

8-(4-Hydroxy-3-methoxybenzoyl): [112501-40-3]. **8-
Vanilloylshiromodiol**.
$C_{23}H_{32}O_6$ M 404.502
From *F. orientalis*.

Wada, K. *et al*, *Tetrahedron Lett.*, 1968, 4673, 4677 (*isol*,
struct, *abs config*)
Miski, M. *et al*, *J. Nat. Prod.*, 1987, **50**, 829 (*derivs*)

Shonachalin *D*   S-70040

[109972-13-6]

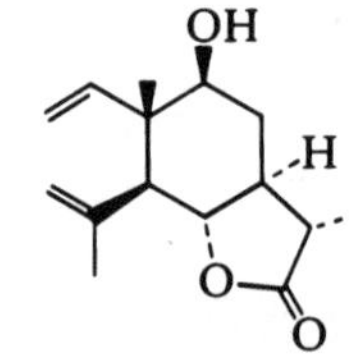

$C_{15}H_{22}O_3$ M 250.337
Constit. of *Artemisia fragrans*. Cryst. (EtOH aq.). Mp
110-112°.

Serkerov, S.V. *et al*, *Khim. Prir. Soedin.*, 1987, **23**, 84.

Sigmoidin *B*   S-70041

Updated Entry replacing S-30050
*2-[3,4-Dihydroxy-5-(3-methyl-2-butenyl)phenyl]-2,3-
dihydro-5,7-dihydroxy-4H-1-benzopyran-4-one, 9CI.
3',4',5,7-Tetrahydroxy-5'-C-prenylflavanone*
[87746-47-2]

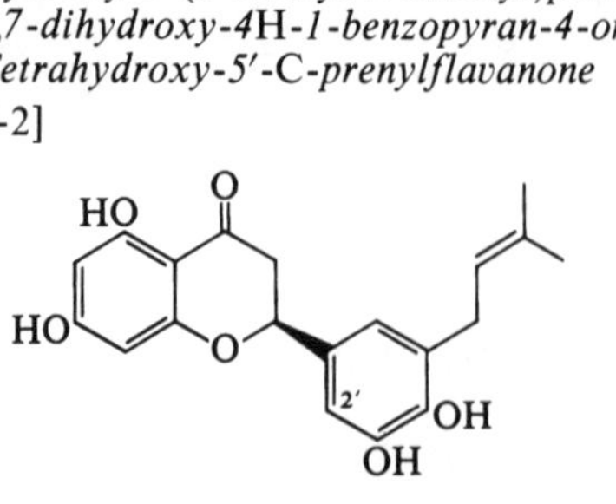

$C_{20}H_{20}O_6$ M 356.374
Constit. of *Erythrina sigmoidea*. Shows antibacterial
activity. Cryst. Mp 217-218°. $[\alpha]_D^{28}$ −54° (c, 3 in
MeOH).

2'-(3-Methyl-2-butenyl): [87746-48-3]. **Sigmoidin A.**
$C_{20}H_{28}O_6$ M 364.438
Constit. of *E. sigmoidea*. Shows antibacterial activity.
Cryst. (CHCl$_3$/MeOH). Mp 181-182°. [α]$_D^{28}$ −82° (c,
2 in MeOH).
3'-Me ether: [114340-00-0]. **4',5,7-Trihydroxy-3'-
methoxy-5'-prenylflavanone**.
$C_{21}H_{22}O_6$ M 370.401
Constit. of *E. berteroana*. Amorph. powder. Mp 123-
125°.

Fomum, Z.T. *et al, J. Chem. Soc., Perkin Trans. 1*, 1986, 33
(*isol, struct*)
Maillard, M. *et al, Planta Med.*, 1987, **53**, 563 (*deriv*)

Silandrin   S-70042
[70815-32-6]

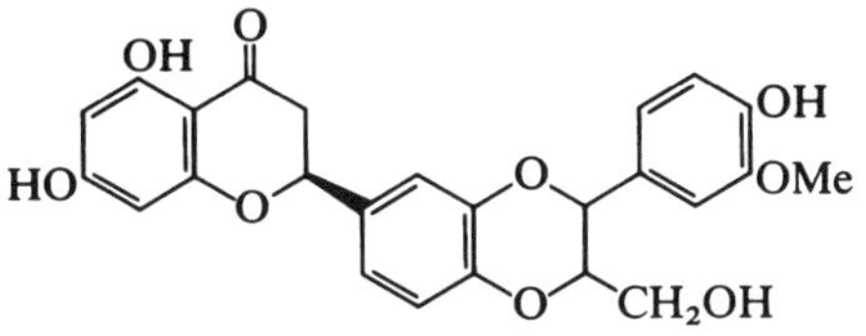

$C_{25}H_{22}O_9$ M 466.443
Constit. of *Silybum marianum*. Cryst. Mp 234-236°.
[α]$_D$ −42.7°.

Szilágyi, I. *et al, Herba Hung.*, 1978, **17**, 65.

6-Siliphiperfolene   S-70043
Updated Entry replacing S-00417
[74284-56-3]
$C_{15}H_{24}$ M 204.355
Constit. of *Siliphium perfoliatum*. Oil. [α]$_D^{24}$ −92.8° (c,
0.8 in CHCl$_3$).

Bohlmann, F. *et al, Phytochemistry*, 1980, **19**, 259.
Curran, D.P. *et al, Tetrahedron*, 1987, **43**, 5653 (*synth*)
Meyers, A.I. *et al, Tetrahedron*, 1987, **43**, 5663 (*synth*)

Silydianin   S-70044
Updated Entry replacing S-00420
Silidianin, INN
[29782-68-1]

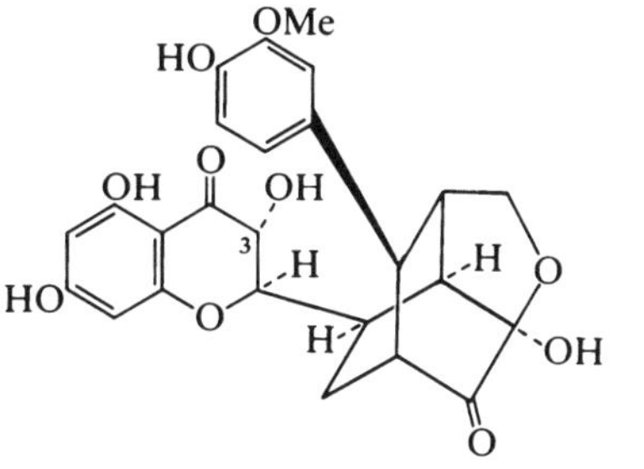

$C_{25}H_{22}O_{10}$ M 482.443
Constit. of *Silybum marianum*. Antihepatotoxic agent,
plant growth regulator. Mp 191°. [α]$_D^{24}$ +175°
(Me$_2$CO), [α]$_D$ +218°.
3-Deoxy: [70815-31-5]. **Silymonin.**
$C_{25}H_{22}O_9$ M 466.443
Constit. of *S. marianum*. Cryst. Mp 258-260°. [α]$_D$
+127°.

Janiak, B. *et al, Planta Med.*, 1960, **8**, 71 (*isol*)
Abraham, D.J. *et al, Tetrahedron Lett.*, 1970, 2675 (*struct*)

Hoelzl, J., *Z. Naturforsch., C*, 1974, **29**, 82 (*biosynth*)
Wagner, H. *et al, Z. Naturforsch., B*, 1976, **31**, 876 (*isol, ms,
pmr, cryst struct, abs config*)
Szilágyi, I. *et al, Herba Hung.*, 1978, **17**, 65 (*isol*)

Silyhermin   S-70045
[96238-88-9]

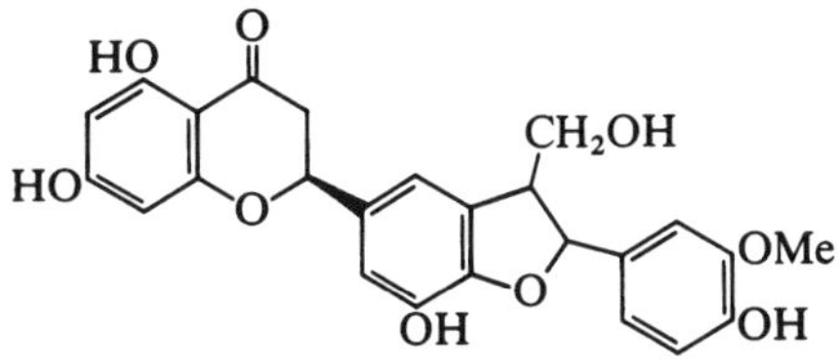

$C_{25}H_{22}O_9$ M 466.443
Constit. of *Silybum marianum*.
Penta-Ac: Cryst. Mp 93-95°. [α]$_D$ +29.9° (CHCl$_3$).

Fiebig, M. *et al, Planta Med.*, 1984, **51**, 310.

Sinapyl alcohol   S-70046
*4-(3-Hydroxy-1-propenyl)-2,6-dimethoxyphenol, 9CI. 3-
(4-Hydroxy-3,5-dimethoxyphenyl)-2-propen-1-ol, 8CI.
Syringenin*
[537-33-7]

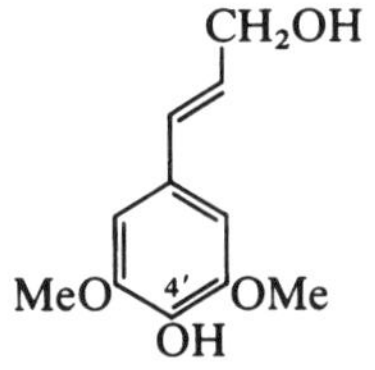

$C_{11}H_{14}O_4$ M 210.229
(E)-form
Principal building block of angiosperm wood lignin.
Needles (pet. ether). Mp 63-65°.
O$^{4'}$-β-D-Glucoside: [118-34-3]. **Syringin.** Lilacin.
Methoxyconiferin.
$C_{17}H_{24}O_9$ M 372.371
Isol. from *Syringa vulgaris, Ligustrum* spp.,
Jasminum spp., *Phyllyrea latifolia, P. decora,
Paulownia tomentosa, Forsythia suspensa, Fraxinus*
spp. and others. Needles or prisms (H$_2$O). Mp 191-
192°. [α]$_D$ −18° (H$_2$O).

Plouvier, V., *C.R. Hebd. Seances Acad. Sci.*, 1947, **224**, 670;
1948, **227**, 604; 1951, **232**, 1013; 1952, **234**, 1577; 1953, **237**,
1761; 1954, **238**, 1835 (*occur, deriv*)
Freudenberg, K. *et al, Chem. Ber.*, 1951, **84**, 67.
Aulin-Erdtman, G. *et al, Acta Chem. Scand.*, 1968, **22**, 1187
(*uv*)
Zanarotti, A., *Tetrahedron Lett.*, 1982, **23**, 3815 (*synth*)

Sintenin   S-70047
[116988-15-9]

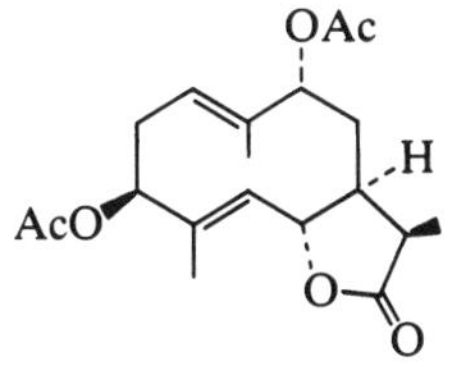

$C_{19}H_{26}O_6$ M 350.411
Constit. of *Achillea sintenisii*. Amorph.

Gören, N. *et al, Phytochemistry*, 1988, **27**, 2346.

Sirutekkone   S-70048
[117590-99-5]

C$_{20}$H$_{24}$O$_4$ M 328.407
Constit. of *Premna herbacea*. Cryst. (EtOAc/hexane).
Mp 214°. [α]$_D^{25}$ +168° (CHCl$_3$).

Sandhya, G. *et al, Phytochemistry*, 1988, **27**, 2249.

Skutchiolide A   S-70049
[117274-10-9]

C$_{22}$H$_{26}$O$_9$ M 434.442
Constit. of *Squamopappus skutchii*. Gum.

Vargas, D. *et al, Phytochemistry*, 1988, **27**, 1413.

Skutchiolide B   S-70050
[115547-14-3]

C$_{22}$H$_{24}$O$_8$ M 416.427
Constit. of *Squamopappus skutchii*. Gum.

Vargas, D. *et al, Phytochemistry*, 1988, **27**, 1413.

Solavetivone   S-70051

Updated Entry replacing S-50093
1(10),11-Spirovetivadien-2-one
[54878-25-0]

C$_{15}$H$_{22}$O   M 218.338
Stress metab. from tubers of *Solanum tuberosum*. Oil.
[α]$_D$ −119° (EtOH).
15-Hydroxy: [103573-06-4]. **Oxysolavetivone**. *15-Hydroxysolavetivone.*
C$_{15}$H$_{22}$O$_2$ M 234.338
Intermediate metabolite in the pathway to the phytoalexins of potato. Oil. [α]$_D^{21}$ −91.6° (c, 1.00 in EtOH).
13-Hydroxy: **Aglycone A$_3$**.
C$_{15}$H$_{22}$O$_2$ M 234.338
Constit. of flue-cured Virginia tobacco.

Coxon, D.T. *et al, Tetrahedron Lett.*, 1974, 2921 (*isol, struct*)

Anderson, R.C. *et al, J. Chem. Soc., Chem. Commun.*, 1977, 27 (*abs config*)
Yamada, K. *et al, J. Chem. Soc., Chem. Commun.*, 1977, 554 (*synth*)
Murai, A. *et al, J. Chem. Soc., Chem. Commun.*, 1982, 32 (*biosynth*)
Murai, A. *et al, Bull. Chem. Soc. Jpn.*, 1984, **57**, 2276, 2282 (*synth*)
Murai, A. *et al, Chem. Lett.*, 1986, 771 (*derivs*)
Iwata, C. *et al, Chem. Pharm. Bull.*, 1987, **35**, 544 (*synth*)

Solenolide A   S-70052
[114094-31-4]

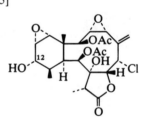

H$_3$C(CH$_2$)$_4$COO

C$_{28}$H$_{41}$ClO$_9$ M 557.080
Constit. of a *Solenopodium* sp. Shows antiviral and antiinflammatory props. Cryst. (Et$_2$O). Mp 132-133°.
[α]$_D^{20}$ −56° (c, 0.63 in CHCl$_3$).
12-Deacyl, 12-Ac: [114058-42-3]. **Solenolide B**.
C$_{24}$H$_{33}$ClO$_9$ M 500.972
Constit. of a *S.* sp. Oil. [α]$_D^{20}$ −5° (c, 1.02 in CHCl$_3$).

Groweiss, A. *et al, J. Org. Chem.*, 1988, **53**, 2401.

Solenolide C   S-70053
[114094-32-5]

C$_{24}$H$_{31}$ClO$_{10}$ M 514.956
Constit. of a *Solenopodium* sp. Amorph. solid. Mp 196-198° dec. [α]$_D^{20}$ −25° (c, 0.76 in MeOH).
12-Ac: [114058-43-4]. **Solenolide D**.
C$_{26}$H$_{33}$ClO$_{11}$ M 556.993
Constit. of a *S.* sp. Amorph. solid. Mp 183-185°. [α]$_D^{20}$ −16° (c, 0.89 in CHCl$_3$).

Groweiss, A. *et al, J. Org. Chem.*, 1988, **53**, 2401.

Solenolide E   S-70054
[114058-44-5]

C$_{22}$H$_{29}$ClO$_7$ M 440.920
Constit. of a *Solenopodium* sp. Oil. [α]$_D^{20}$ +11° (c, 0.5 in CHCl$_3$).

Groweiss, A. *et al, J. Org. Chem.*, 1988, **53**, 2401.

Solenolide F   S-70055
[114094-33-6]

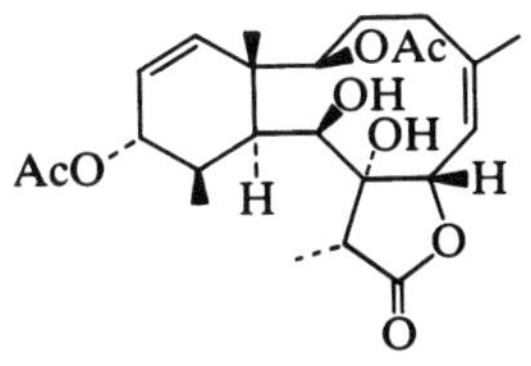

$C_{24}H_{34}O_8$ M 450.528
Constit. of a *Solenopodium* sp. Oil. $[\alpha]_D^{20}$ −45° (c, 1.52 in $CHCl_3$).

Groweiss, A. *et al*, *J. Org. Chem.*, 1988, **53**, 2401.

Sordariol   S-70056
5-[3 Hydroxy-2-(hydroxymethyl)phenyl]-4-pentene-2,3-diol, 9CI. 2-Hydroxy-6-(3,4-dihydroxy-1-pentenyl)-benzyl alcohol
[115873-03-5]

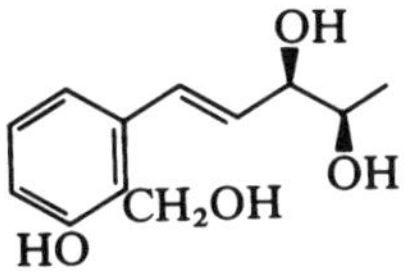

$C_{12}H_{16}O_4$ M 224.256
Metab. of *Sordaria macrospora*. Amorph. powder.

Bouillant, M.L. *et al*, *Phytochemistry*, 1988, **27**, 1517.

Soulattrone *A*   S-70057

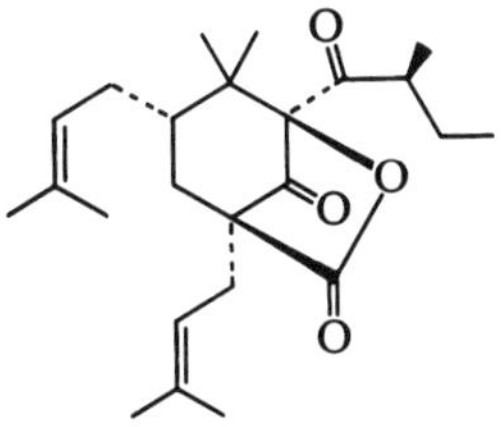

$C_{24}H_{36}O_4$ M 388.546
Constit. of the bark of *Calophyllum soulattri*. Cryst. (MeOH/CH_2Cl_2). Mp 113-114°. $[\alpha]_D^{22}$ +157.3° (c, 0.19 in EtOH).

Nigam, S.K. *et al*, *Phytochemistry*, 1988, **27**, 527 (*isol, cryst struct*)

13,17-Spatadien-10-ol   S-70058
Updated Entry replacing S-10068
$C_{20}H_{32}O$ M 288.472
Constit. of *Dilophus marginatus* and *D. okamurai*. Foam. $[\alpha]_D^{20}$ +96° (c, 0.8 in CCl_4). The spectroscopic properties of the compds. from the two spp. are not in complete agreement.

Ravi, B.N. *et al*, *Aust. J. Chem.*, 1982, **35**, 129 (*isol*)
Kurata, K. *et al*, *Phytochemistry*, 1988, **27**, 1321 (*isol*)

Specionin   S-70059

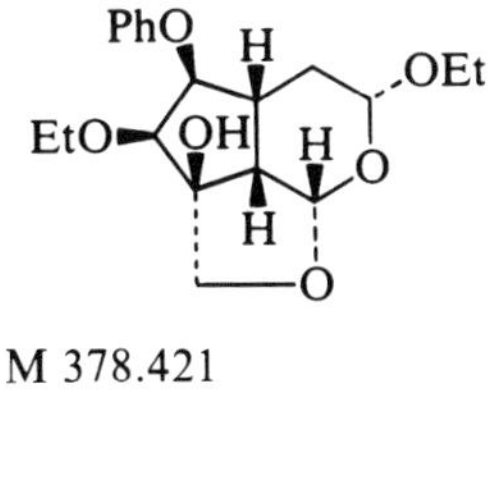

$C_{20}H_{26}O_7$ M 378.421

Isol. from *Catalpa speciosa*. Insect antifeedant.

Chang, C.C. *et al*, *J. Chem. Soc., Chem. Commun.*, 1983, 605.

Sphaeropyrane   S-70060
[115610-51-0]

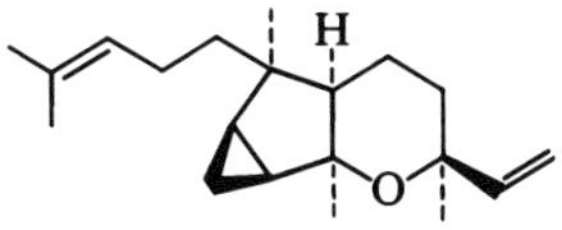

$C_{20}H_{32}O$ M 288.472
Constit. of *Sphaerococcus coronopifolius*. Cryst. Mp 136-138°. $[\alpha]_D$ −43.9° (c, 0.8 in $CHCl_3$).

Cafieri, F. *et al*, *Phytochemistry*, 1988, **27**, 621.

Sphaeroxetane   S-70061
[116310-65-7]

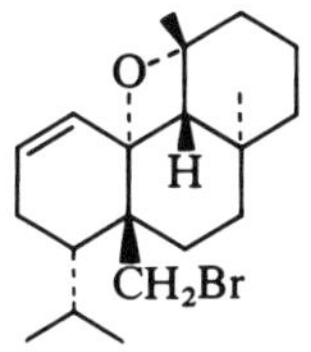

$C_{20}H_{31}BrO$ M 367.368
Constit. of *Sphaerococcus coronopifolius*. Oil. $[\alpha]_D$ −9° (c, 0.57 in $CHCl_3$).

De Rosa, S. *et al*, *Phytochemistry*, 1988, **27**, 1875.

2,2′-Spirobi[2H-1-benzopyran], 9CI   S-70062
2,2′-Spirobichromene. Diphenospiropyran
[178-30-3]

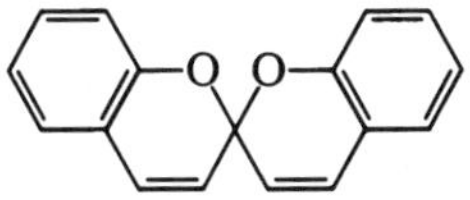

$C_{17}H_{12}O_2$ M 248.281
(±)-*form*
Cryst. (EtOH). Mp 107°. Shows photochromism.
Tetrahydro: see 3,3′,4,4′-Tetrahydro-2,2′-spirobi[2H-1-benzopyran], T-70094

Mora, P.T. *et al*, *J. Am. Chem. Soc.*, 1950, **72**, 3009 (*synth, bibl*)
Davin-Pretelli, E. *et al*, *Helv. Chim. Acta*, 1977, **60**, 215 (*ir, raman*)

Spiro[3,4-cyclohexano-4-hydroxybicyclo[3.3.1]nonan-9-one-2,1′-cyclohexane]   S-70063
Updated Entry replacing S-10075
Cornubert's ketone
[78549-00-5]

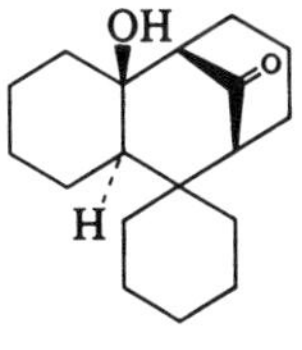

$C_{18}H_{28}O_2$ M 276.418
Product obt. in 40% yield from cyclohexanone by treatment with NaOMe/DMF at r.t. Needles (C_6H_6). Mp 186.5-187°.

Pettit, G.R. *et al, J. Org. Chem.*, 1981, **46**, 4167 (*synth, spectra*)
Pettit, G.R. *et al, Can. J. Chem.*, 1982, **60**, 629 (*struct*)
Cocker, W. *et al, Chem. Ind.* (*London*), 1988, 339 (*stereochem*)

25(27)-Spirostene-2,3,6-triol   S-70064

$C_{27}H_{42}O_5$ M 446.626

(2α,3β,5α,6β,20S,22R)-form

Sapogenin from *Nothoscordum gramineum* var. *philippianum*.

Tri-Ac: Cryst. (MeOH). Mp 125-130°.

Brunengo, M.C. *et al, Phytochemistry*, 1988, **27**, 2943.

25(27)-Spirosten-3-ol   S-70065

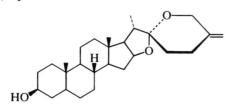

$C_{27}H_{42}O_3$ M 414.627

(3β,5α,20S,22R)-form

Constit. of *Tristagma uniflorum*.

Ac: Cryst. (Me₂CO/MeOH). Mp 182-185°.

Brunengo, M.C. *et al, Phytochemistry*, 1988, **27**, 2943.

Spiro[5.5]undecane-1,9-dione   S-70066

[32257-52-6]

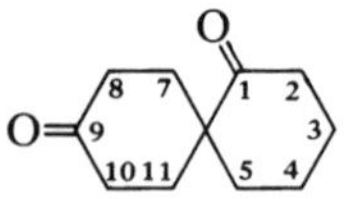

$C_{11}H_{16}O_2$ M 180.246

Mobile oil. Bp₀.₁ 112-116°, Bp₀.₀₁ 96°. n_D^{21} 1.5030.

Bis(2,4-dinitrophenylhydrazone): Yellow plates (CHCl₃/MeOH). Mp 180-183°.

Askam, V. *et al, J. Chem. Soc.* (*C*), 1971, 1524 (*synth, ir*)
Dave, V. *et al, J. Chem. Soc., Perkin Trans. 1*, 1973, 393 (*synth, ir*)

Spiro[5.5]undecane-3,8-dione   S-70067

$C_{11}H_{16}O_2$ M 180.246

Cryst. (EtOAc/pet. ether or by subl.).

Farges, G. *et al, Helv. Chim. Acta*, 1966, **49**, 552 (*synth, ir, pmr*)

Spiro[5.5]undecane-3,9-dione   S-70068

[5607-35-2]

$C_{11}H_{16}O_2$ M 180.246

Cryst. (EtOAc). Mp 112-113°.

Dioxime: [10428-73-6].

$C_{11}H_{18}N_2O_2$ M 210.275

Cryst. (THF). Mp 262-263°.

Farges, G. *et al, Helv. Chim. Acta*, 1966, **49**, 552 (*synth, ir, pmr*)
Rice, L.M. *et al, J. Org. Chem.*, 1966, **32**, 1966 (*synth*)

Sporol   S-70069

Updated Entry replacing S-50109

[101401-88-1]

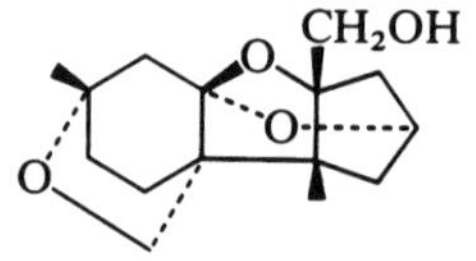

$C_{15}H_{22}O_4$ M 266.336

Mycotoxin from *Fusarium sporotrichioides* MC-72083.

Corley, D.G. *et al, Tetrahedron Lett.*, 1986, **27**, 427 (*isol*)
Ziegler, F.E. *et al, Tetrahedron Lett.*, 1988, **29**, 1665 (*struct*)

Squarrofuric acid   S-70070

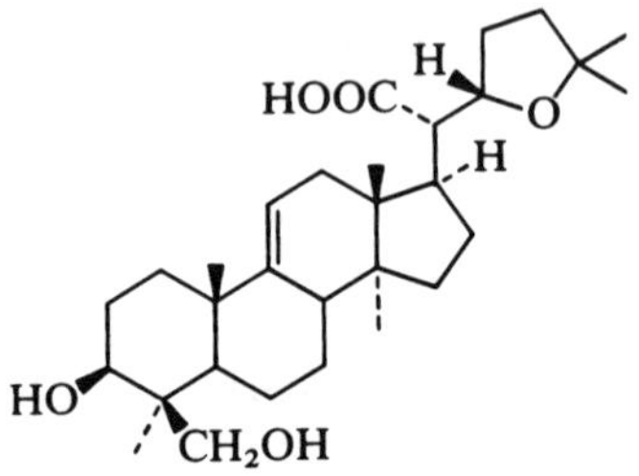

$C_{30}H_{48}O_5$ M 488.706

Sapogenin from *Thalictrum squarrosum*. Cryst. (CHCl₃/MeOH). Mp 298-300°. [α]$_D^{20}$ +20.1° (c, 0.5 in Py).

Gromova, A.S. *et al, Khim. Prir. Soedin.*, 1987, **23**, 310, 533; *Chem. Nat. Compd.*, 444 (*isol, cryst struct*)

Stemodin   S-70071

Updated Entry replacing S-20075

[41943-79-7]

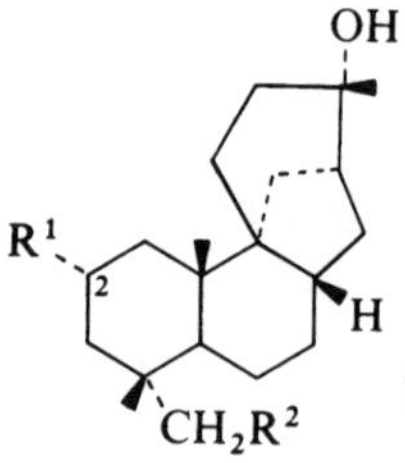

$R^1=OH, R^2=H$

$C_{20}H_{34}O_2$ M 306.487

Constit. of the leaves of *Stemodia maritima*. Cryst. Mp 196-197°. [α]$_D$ −2.6° (c, 1.07 in Py).

2-Ketone: [41943-80-0]. **Stemodinone**.

$C_{20}H_{32}O_2$ M 304.472

Constit. of *S. maritima*. Cryst. Mp 215-216°. [α]$_D$ +14.3° (c, 1 in CHCl₃).

Manchand, P.S. *et al, J. Am. Chem. Soc.*, 1973, **95**, 2705.
Corey, E.J. *et al, J. Am. Chem. Soc.*, 1980, **102**, 7612 (*synth*)
Bettolo, R.M. *et al, Helv. Chim. Acta*, 1983, **66**, 760 (*synth*)
Lupi, A. *et al, Helv. Chim. Acta*, 1984, **67**, 2261 (*synth*)
Piers, E. *et al, Can. J. Chem.*, 1985, **63**, 3418 (*synth*)
Hufford, C.D., *J. Nat. Prod.*, 1988, **51**, 367 (*pmr, cmr*)

Stenosporonic acid
S-70072

[103538-04-1]

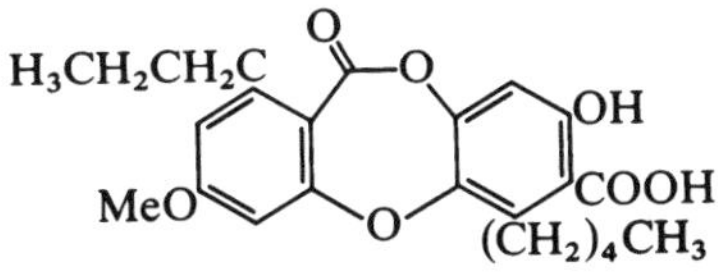

$C_{23}H_{26}O_7$ M 414.454

Constit. of *Paraparmelia mongaensis*. Needles. Mp 139-140°.

Culberson, C.F. *et al, Bryologist*, 1985, **88**, 380 (*isol*)
Elix, J.A. *et al, Aust. J. Chem.*, 1987, **40**, 1451 (*synth*)

Stevisamolide
S-70073

1(10),2,4-Guaiatrien-12,8β-olide

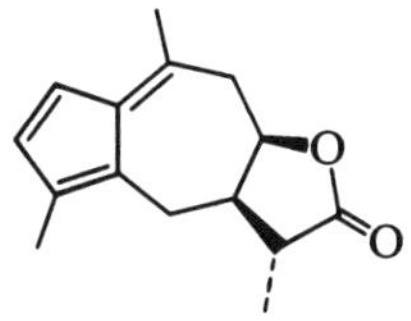

$C_{15}H_{18}O_2$ M 230.306

Constit. of *Stevia samaipatensis*. Yellow cryst. (Et$_2$O/pet. ether). Mp 96-97°. $[\alpha]_D^{24}$ −279° (c, 0.71 in CHCl$_3$).

Zdero, C. *et al, Phytochemistry*, 1988, **27**, 2835.

3,6-Stigmastanediol
S-70074

$C_{29}H_{52}O_2$ M 432.729

(*3α,6α*)-*form*

Constit. of *Spatholobus suberetus*. Fibrinolytic agent. Prisms. Mp 207-209°. $[\alpha]_D^{20}$ +41.8° (c, 0.74 in EtOH).

Fukuyama, Y. *et al, Planta Med.*, 1988, **54**, 34.

24(28)-Stigmastene-2,3,22,23-tetrol, 9CI
S-70075

24-Ethylidene-2,3,22,23-cholestanetetrol

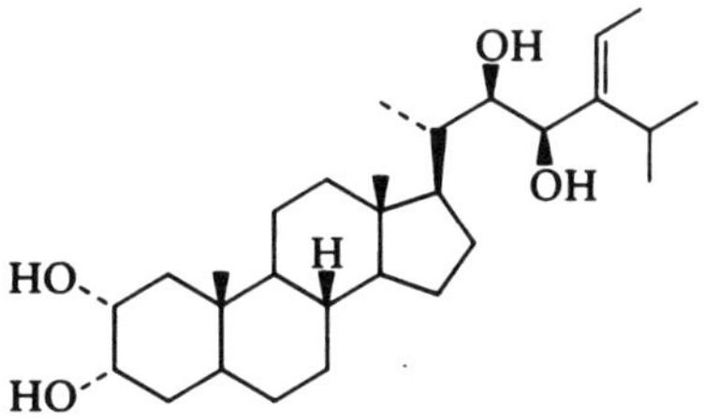

$C_{29}H_{50}O_4$ M 462.712

(*2α,3α,5α,22R,23R,24(28)E*)-*form* [110345-06-7]

6-Deoxodihydrohomodolichosterone

Constit. of *Phaseolus vulgaris* seed.

Yokota, T. *et al, Agric. Biol. Chem.*, 1987, **51**, 1625.

Strictaketal
S-70076

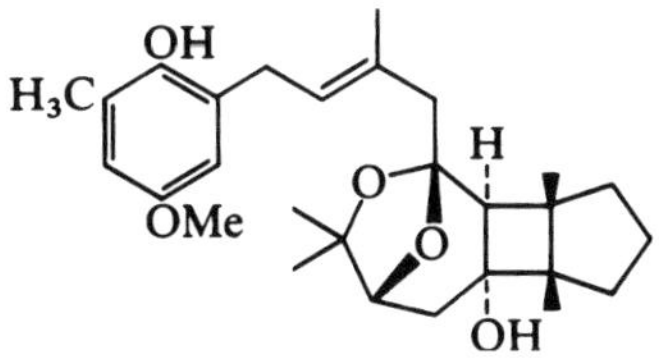

$C_{28}H_{40}O_5$ M 456.621

Constit. of *Cystoseira stricta*. Oil. $[\alpha]_D$ +21.7° (c, 0.9 in EtOH).

Amico, V. *et al, J. Nat. Prod.*, 1987, **50**, 449.

Strictic acid
S-70077

Updated Entry replacing S-30090

Conyzic acid. Seconidoresedic acid

[56317-16-9]

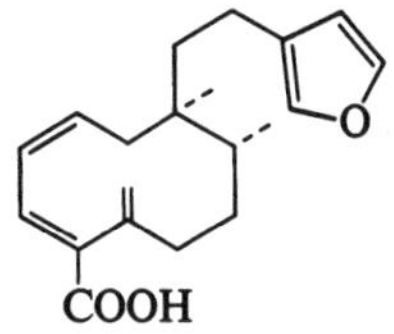

$C_{20}H_{26}O_3$ M 314.424

Constit. of *Conyza stricta*. Cryst. (MeOH). Mp 160°. $[\alpha]_D$ −190° (c, 1.1 in CHCl$_3$).

15,16-Dihydro, 15-methoxy, 16-oxo: **15,16-Dihydro-15-methoxy-16-oxostrictic acid.**
$C_{21}H_{28}O_5$ M 360.449
Constit. of *Grangea maderaspatana*.

15,16-Dihydro, 15-methoxy, 16-oxo, Me ester: Gum. $[\alpha]_D$ −135° (c, 0.24 in CHCl$_3$).

Pandey, U.C. *et al, Phytochemistry*, 1984, **23**, 391 (*isol, struct*)
Singh, P. *et al, Phytochemistry*, 1988, **27**, 1537 (*deriv*)

Subcordatolide D
S-70078

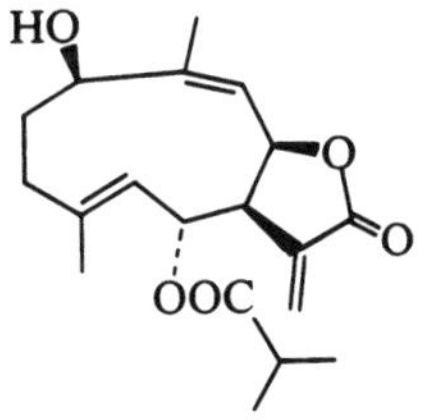

$C_{19}H_{26}O_5$ M 334.411

Constit. of *Calea subcordata*. Cryst. (CHCl$_3$/hexane). Mp 246° dec.

Ober, A.G. *et al, J. Nat. Prod.*, 1987, **50**, 604.

Subcordatolide E
S-70079

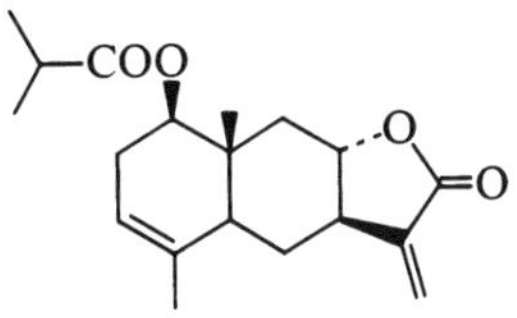

$C_{19}H_{26}O_4$ M 318.412

Constit. of *Calea subcordata*. Gum.

Ober, A.G. *et al, J. Nat. Prod.*, 1987, **50**, 604.

Suberenol
S-70080

Updated Entry replacing S-20088

6-(3-Hydroxy-3-methyl-1-butenyl)-7-methoxy-2H-1-benzopyran-2-one, 9CI

[38409-30-2]

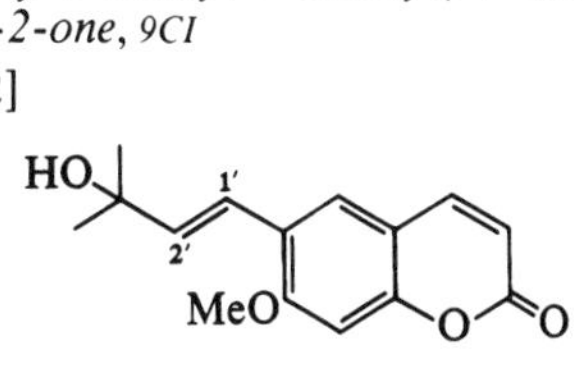

$C_{15}H_{16}O_4$ M 260.289

(E)-form [18529-47-0]
Isol. from *Zanthoxylum suberosum*, *Limonia acidissima*, *Citrus nobilis*, etc. V. pale-yellow prisms (MeOH). Mp 173°.

1′,2′-Dihydro: [81892-79-7]. **Dihydrosuberenol**.
$C_{15}H_{18}O_4$ M 262.305
Isol. from *L. acidissima*. Mp 105°.
Me ether: [117597-80-5]. (**E**)-**Methylsuberenol**.
$C_{16}H_{18}O_4$ M 274.316
Constit. of *Citrus funadoko*. Oil.

(Z)-form
Constit. of *C. funadoko*. Oil.

Me ether: [117597-81-6]. (**Z**)-**Methylsuberenol**.
$C_{16}H_{18}O_4$ M 274.316
Constit. of *C. funadoko*. Oil.

Guise, G.B. *et al*, *Aust. J. Chem.*, 1967, **20**, 2429 (*isol, struct*)
Reisch, J. *et al*, *Planta Med.*, 1980 (suppl.), 56 (*isol*)
Burke, B.A. *et al*, *Heterocycles*, 1981, **16**, 897 (*isol*)
Ghosh, P. *et al*, *Phytochemistry*, 1982, **21**, 240 (*isol*)
Ju-Ichi, M. *et al*, *Heterocycles*, 1988, **27**, 1451 (*isol, derivs*)

Subergorgic acid   S-70081

Updated Entry replacing S-40061
1,3a,4,5,5a,6,7,8-Octahydro-1,3a,6-trimethyl-8-oxocyclopenta[c]pentalene-2-carboxylic acid, 9CI
[97718-45-1]

$C_{15}H_{20}O_3$ M 248.321
Constit. of coral *Supergorgia suberosa*. Shows cardiotoxic props. Cryst. Mp 179-180°, Mp 200-202°. $[\alpha]_D$ −23° (c, 0.7 in CHCl$_3$).

Wu, Z. *et al*, *CA*, 1982, **98**, 68827 (*isol*)
Groweiss, A. *et al*, *Tetrahedron Lett.*, 1985, **26**, 2379 (*cryst struct*)
Iwata, C. *et al*, *J. Org. Chem.*, 1988, **53**, 1623 (*synth*)

Sulfuretin   S-70082

2-[(3,4-Dihydroxyphenyl)methylene]-6-hydroxy-3(2H)-benzofuranone, 9CI. 2-(3,4-Dihydroxybenzylidene)-6-hydroxy-3(2H)-benzofuranone, 8CI. 3′,4′,6-Trihydroxyaurone. 3′,4′,6-Trihydroxybenzalcoumaranone. Sulphuretin
[120-05-8]

$C_{15}H_{10}O_5$ M 270.241
Isol. from *Bidens tripartita*. Deep orange-yellow prisms (EtOH). Mp 310-312° dec. (280-285°).

6-Glucoside: [531-63-5]. **Sulfurein**. *Sulphurein*.
$C_{21}H_{20}O_{11}$ M 448.382
Glycoside from flowers of *Cosmos sulphureus*. Orange-red cryst. (H$_2$O).
3′,6-Diglucoside: [494-49-5]. **Palasitrin**.
$C_{27}H_{30}O_{15}$ M 594.525
Glycoside from flowers of *Butea frondosa*. Long brownish-red prisms + 1H$_2$O (butanol). Mp 199-200° dec. (sinters at 125°).

Shimokoriyama, M. *et al*, *J. Am. Chem. Soc.*, 1953, **75**, 1900 (*Sulfurein*)
Puri, B. *et al*, *J. Chem. Soc.*, 1955, 1589 (*Palasitrin*)
Geissman, T.A. *et al*, *J. Am. Chem. Soc.*, 1956, **28**, 832 (*uv*)
Huke, M. *et al*, *Arch. Pharm.* (*Weinheim, Ger.*), 1969, **302**, 401, 423 (*uv, pmr*)
Serbin, A.G. *et al*, *Khim. Prir. Soedin.*, 1972, **8**, 440; *Chem. Nat. Compd.*, 439 (*isol*)

Sulochrin   S-70083

Updated Entry replacing S-00966
Methyl 2-(2,6-dihydroxy-4-methylbenzoyl)-5-hydroxy-3-methoxybenzoate. 2′,4,6′-Trihydroxy-6-methoxy-4′-methyl-2-benzophenonecarboxylic acid methyl ester
[519-57-3]

$C_{17}H_{16}O_7$ M 332.309
Constit. of mycelium of *Oospora sulphurea-ochracea*, metab. of *Penicillium frequentans* and *Aspergillus terreus*. Yellow needles (Me$_2$CO aq.). Insol. H$_2$O, C$_6$H$_6$. Mp 262°.

Tri-Ac: Cryst. (EtOH). Mp 164°.
2′-Me ether: **Monomethylsulochrin**. *Methylsulochrin*.
$C_{18}H_{18}O_7$ M 346.336
Metab. of *A. fumigatus* when grown on Raulin-Thom medium. Elongated plates (EtOAc/pet. ether). Mp 198-199°.
5-Me ether: [77282-69-0]. **5-O-Methylsulochrin**.
$C_{18}H_{18}O_7$ M 346.336
Prod. by *A. wentii*.
Di-Me ether: Plates (MeOH). Mp 158°.
3′-Chloro: Methyl 2-(3-chloro-2,6-dihydroxy-4-methyl-benzoyl)-5-hydroxy-3-methoxybenzoate, 9CI. **Monochlorosulochrin**.
$C_{17}H_{15}ClO_7$ M 366.754
Metab. of *A. terreus* var. *aureus*. Active against *Trichophyton mentagrophytes*. Pale-yellow needles (EtOAc). Mp 243-250°.
O³-De-Me: [57459-06-0]. **3-O-Demethylsulochrin**.
$C_{16}H_{14}O_7$ M 318.282
Prod. by *A. wentii*.
O³-De-Me, O⁵-Me: [77282-68-9]. **Isosulochrin**.
$C_{17}H_{16}O_7$ M 332.309
Prod. by *A. wentii*.

Turner, W.B., *J. Chem. Soc.*, 1965, 6658.
Curtis, R.F. *et al*, *J. Chem. Soc.* (*C*), 1966, 168 (*biosynth, bibl*)
Afzal, M. *et al*, *J. Chem. Soc.* (*C*), 1969, 1721 (*synth*)
Gatenbeck, S. *et al*, *Acta Chem. Scand.*, 1969, **23**, 3493 (*biosynth*)
Curtis, R.F. *et al*, *J. Chem. Soc., Perkin Trans. 1*, 1972, 240 (*biosynth*)
Kiriyama, N. *et al*, *Chem. Pharm. Bull.*, 1977, **25**, 2593 (*isol, pmr, ir, uv*)
Assante, G. *et al*, *Gazz. Chim. Ital.*, 1980, **110**, 629 (*derivs*)
Inamori, Y. *et al*, *Chem. Pharm. Bull.*, 1983, **31**, 4543 (*deriv*)

Suspensolide   S-70084

4,8-Dimethyl-3E,8E-decadien-10-olide

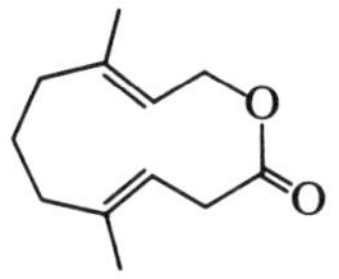

$C_{12}H_{18}O_2$ M 194.273

Not the same as Suspensolide *A*, S-10121 . Component of volatiles of male *Anastrepha suspensa* flies.

Chuman, T. *et al*, *Tetrahedron Lett.*, 1988, **29**, 6561 (*struct*)
Battiste, M.A. *et al*, *Tetrahedron Lett.*, 1988, **29**, 6565 (*synth*)
Mori, K. *et al*, *Justus Liebigs Ann. Chem.*, 1988, 162 (*synth, ir, pmr, cmr*)

Suvanine   S-70085

[94203-53-9]

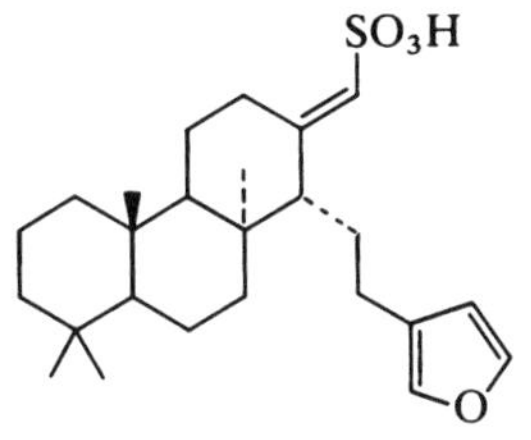

$C_{25}H_{38}O_4S$ M 434.633

Constit. of sponges of the genus *Coscinoderma*. Powder (as *N,N*-dimethylguanidine salt). Mp 224° (dimethylguanidine salt). $[\alpha]_D$ +9.5° (MeOH).

Manes, L.V. *et al*, *J. Org. Chem.*, 1985, **50**, 284; 1988, **53**, 570 (*struct*)

Swalpamycin   S-70086

[112008-27-2]

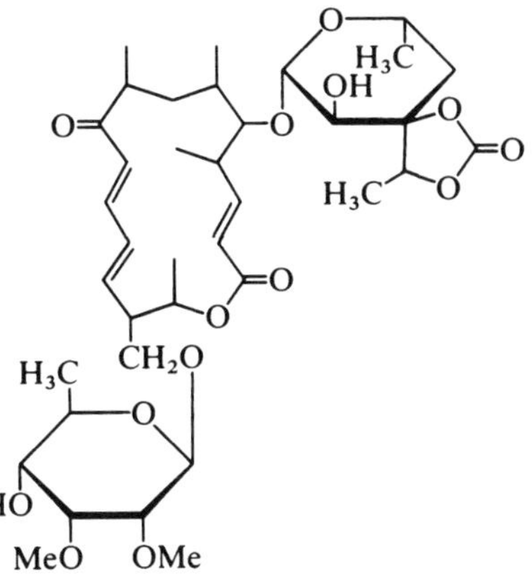

$C_{37}H_{56}O_{14}$ M 724.841

Macrolide antibiotic. Prod. by *Streptomyces* sp. Y-84030967. Active against gram-positive bacteria. Amorph. Mp 126-129°. $[\alpha]_D^{20}$ −3.8° (c, 2.1 in CHCl$_3$).

Franco, C.M.M. *et al*, *J. Antibiot.*, 1987, **40**, 1361, 1368 (*isol, struct*)

Sylvestroside I   S-70087

Updated Entry replacing S-01037
[71431-22-6]

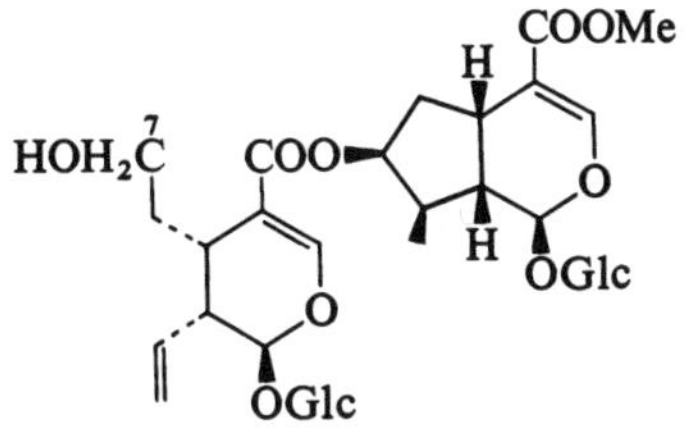

$C_{33}H_{48}O_{19}$ M 748.731

Constit. of *Dipsacus sylvestris*. Syrup. $[\alpha]_D^{21}$ −106° (c, 0.4 in EtOH).

7-Ac: [71431-23-7]. **Sylvestroside II**.
   $C_{35}H_{50}O_{20}$ M 790.768
   Constit. of *D. sylvestris*. Foam. $[\alpha]_D^{20}$ −99° (c, 1.5 in MeOH).

Nona-Ac: Cryst. (EtOH). Mp 154-155°. $[\alpha]_D^{20}$ −85° (c, 0.4 in CHCl$_3$).

Jensen, S.R. *et al*, *Phytochemistry*, 1979, **18**, 273.

Sylvestroside III   S-70088

Updated Entry replacing S-01038
[71431-24-8]

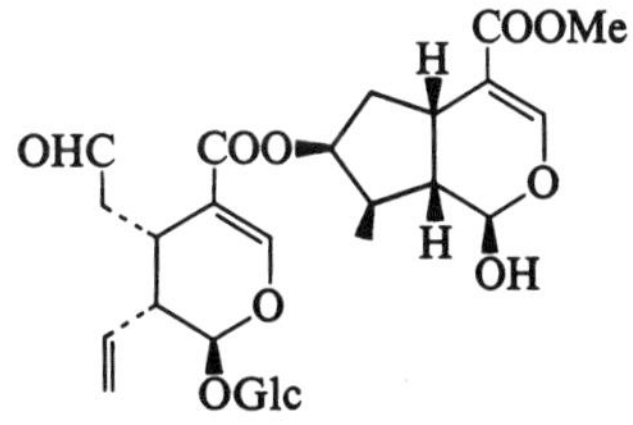

$C_{27}H_{36}O_{14}$ M 584.573

Constit. of *Dipsacus sylvestris*. Foam. $[\alpha]_D^{20}$ −85° (c, 0.4 in MeOH).

Jensen, S.R. *et al*, *Phytochemistry*, 1979, **18**, 273.

Sylvestroside IV   S-70089

Updated Entry replacing S-01039
[71431-25-9]

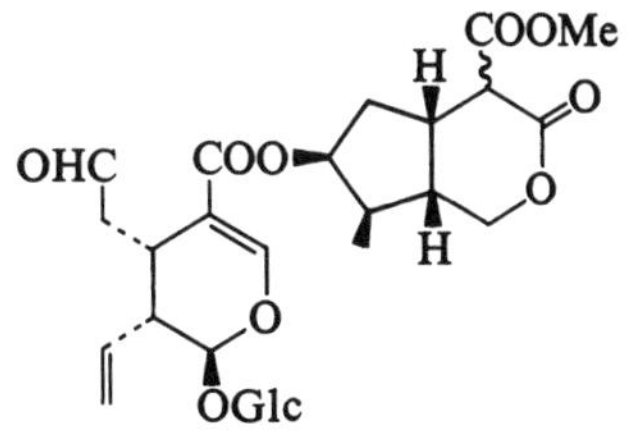

$C_{27}H_{36}O_{14}$ M 584.573

Constit. of *Dipsacus sylvestris*. Foam. $[\alpha]_D^{20}$ −57° (c, 0.4 in MeOH).

Tetra-Ac: Cryst. (EtOH). Mp 137-139°. $[\alpha]_D^{22}$ −60° (c, 0.4 in CHCl$_3$).

Jensen, S.R. *et al*, *Phytochemistry*, 1979, **18**, 273.

T

Taccalonolide *B*　　　　　　　　　　　　　　T-70001

Updated Entry replacing T-60001
　[108885-69-4]

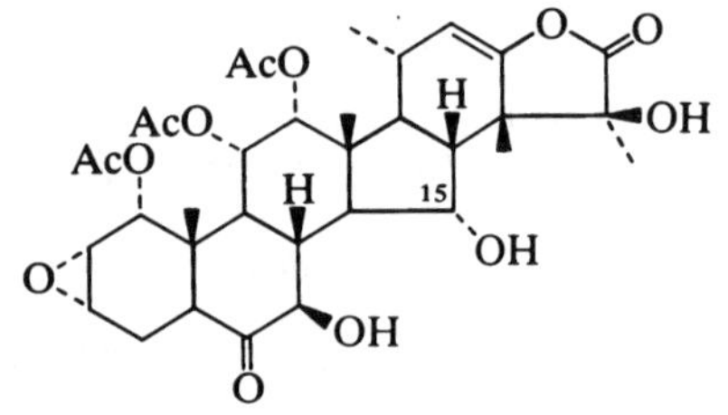

C$_{34}$H$_{44}$O$_{13}$　　M 660.714
Constit. of *Tacca plantaginea*. Cryst. Mp 266°. [α]$_D^{14}$
　+15.9° (c, 0.019 in CHCl$_3$).

7-Ac: **Taccanolide D**.
　C$_{36}$H$_{46}$O$_{14}$　　M 702.751
　Constit. of *T. plantaginea*. Cryst. Mp 284°. [α]$_D^{24}$
　+31° (c, 0.032 in CHCl$_3$).

15-Ac: [108885-68-3]. **Taccalonolide A**.
　C$_{36}$H$_{46}$O$_{14}$　　M 702.751
　Constit. of *T. plantaginea*. Mp 215°. [α]$_D^{14}$ +39.5° (c,
　0.025 in CHCl$_3$).

Chen, Z. *et al*, *Tetrahedron Lett.*, 1987, **28**, 1673 (*cryst struct*)
Chen, Z. *et al*, *Phytochemistry*, 1988, **27**, 2999.

Taccalonolide *C*　　　　　　　　　　　　　　T-70002

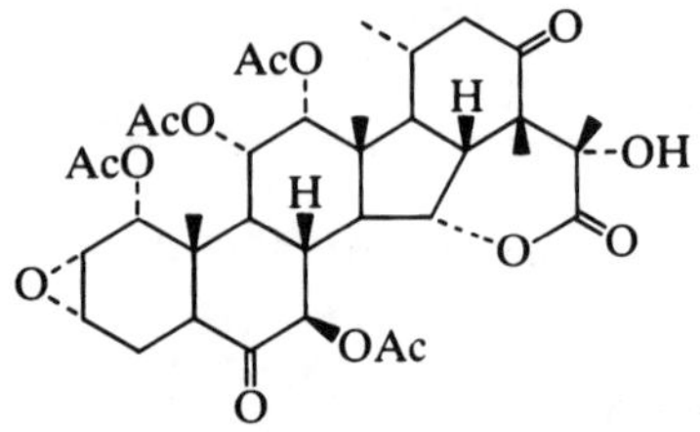

C$_{36}$H$_{46}$O$_{14}$　　M 702.751
Constit. of *Tacca plantaginea*. Cryst. Mp 305°.

Chen, Z.-L. *et al*, *Phytochemistry*, 1988, **27**, 2999.

Tagitinin *E*　　　　　　　　　　　　　　　T-70003

Updated Entry replacing T-00018
　[59979-58-7]

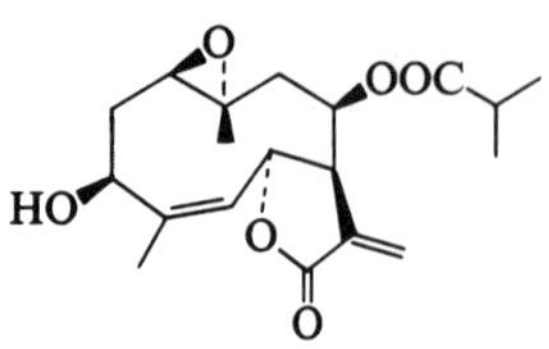

C$_{19}$H$_{26}$O$_6$　　M 350.411
Constit. of *Tithonia rotundifolia*. Cryst. (CHCl$_3$/hex-
　ane). Mp 210-211°. [α]$_D$ −101.4°.

8-Epimer: **Deacetylviguiestin**.
　C$_{19}$H$_{26}$O$_6$　　M 350.411
　Isol. from *Viguiera stenoloba*. Mp 212-214°.

8-Epimer, Ac: [69440-10-4]. **Viguiestin**.
　C$_{21}$H$_{28}$O$_7$　　M 392.448

From *V. stenoloba*.

Pal, R. *et al*, *Indian J. Chem., Sect. B*, 1977, **15**, 533 (*isol*)
Barauh, N.C. *et al*, *J. Org. Chem.*, 1979, **44**, 1831 (*struct*)
Chowdury, P.K. *et al*, *J. Org. Chem.*, 1980, **45**, 535 (*struct*)

Talaromycin *A*　　　　　　　　　　　　　　T-70004

Updated Entry replacing T-60003
*9-Ethyl-4-hydroxy-1,7-dioxaspiro[5.5]undecane-3-
methanol*, 9CI
　[83720-10-9]

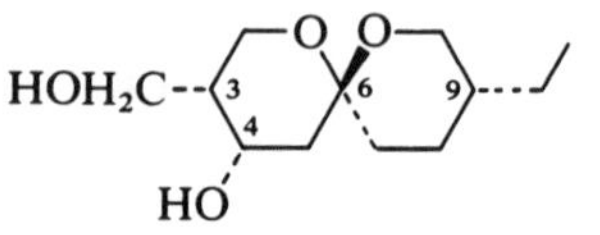

C$_{12}$H$_{22}$O$_4$　　M 230.303
Spiroketal antibiotic. Metab. of *Talaromyces stipitatus*.
　Mycotoxin. Oil.

3-Epimer: [83780-27-2]. **Talaromycin B**.
　C$_{12}$H$_{22}$O$_4$　　M 230.303
　From *T. stipitatus*. Oil.

4-Epimer: [89885-86-9]. **Talaromycin C**.
　C$_{12}$H$_{22}$O$_4$　　M 230.303
　From *T. stipitatus*. Mycotoxin. No phys. props.
　reported.

3,4-Diepimer: [111465-42-0]. **Talaromycin E**.
　C$_{12}$H$_{22}$O$_4$　　M 230.303
　From *T. stipitatus*. Mycotoxin. No phys. props.
　reported.

6-Epimer: [111465-44-2]. **Talaromycin F**.
　C$_{12}$H$_{22}$O$_4$　　M 230.303
　From *T. stipitatus*. Mycotoxin. No phys. props.
　reported.

3,6-Diepimer: [111465-43-1]. **Talaromycin D**.
　C$_{12}$H$_{22}$O$_4$　　M 230.303
　From *T. stipitatus*. Mycotoxin. No phys. props.
　reported.

4,6-Diepimer: **Talaromycin G**.
　C$_{12}$H$_{22}$O$_4$　　M 230.303
　From *T. stipitatus*. Mycotoxin. No phys. props.
　reported.

Lynn, D.G. *et al*, *J. Am. Chem. Soc.*, 1982, **104**, 7319 (*isol,
　struct*)
Kay, I.T. *et al*, *Tetrahedron Lett.*, 1984, **25**, 2035 (*synth*)
Kozikowski, A.P. *et al*, *J. Am. Chem. Soc.*, 1984, **106**, 353
　(*synth*)
Smith, A.B. *et al*, *J. Org. Chem.*, 1984, **49**, 1469 (*synth*)
Midland, M.M. *et al*, *J. Org. Chem.*, 1985, **50**, 1143 (*synth*)
Schreiber, S.L. *et al*, *Tetrahedron Lett.*, 1985, **26**, 17 (*synth*)
Iwata, C. *et al*, *Tetrahedron Lett.*, 1987, **28**, 3135 (*synth*)
Mori, K. *et al*, *Tetrahedron*, 1987, **43**, 45 (*synth*)
Phillips, N.J. *et al*, *Tetrahedron Lett.*, 1987, **28**, 1619 (*epimers*)
Whitby, R. *et al*, *J. Chem. Soc., Chem. Commun.*, 1987, 906
　(*synth*)

Tartaric acid
T-70005

Updated Entry replacing T-00079

2,3-Dihydroxybutanedioic acid, 9CI

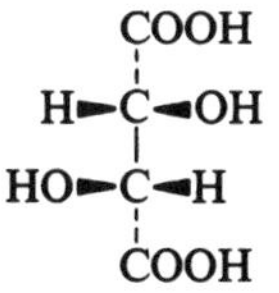

$$\begin{array}{c} \text{COOH} \\ | \\ \text{H} \blacktriangleright \text{C} \blacktriangleleft \text{OH} \\ | \\ \text{HO} \blacktriangleright \text{C} \blacktriangleleft \text{H} \\ | \\ \text{COOH} \end{array} \quad \textit{(2R,3R)-form}$$

$C_4H_6O_6$ M 150.088

DL-Nomenclature, although frequently used, is ambiguous when applied to Tartaric acid.

(2R,3R)-form [87-69-4]

L-form. L-Threaric acid

Occurs in many plants and fruit. Coml. available from the K,H salt deposited in fermenting grape juice. Acidulant for soft drinks and fruit jellies. Used to clean metals for plating, as a mordant for dyeing and in calico printing. Resolving agent for bases. Pharmaceutical aid (buffering agent). Mp 169-170°. $[\alpha]_D^{20}$ +12.0° (c, 20.0 in H_2O). pK_{a1} 2.98, pK_{a2} 4.34 (25°). Cream of tartar (the K,H salt) is used in baking powder; Rochelle salt (the K,Na salt) is used in electroplating and medicinally as a mild saline cathartic preparation; tartar emetic (the KSb salt) is used in low dose as an expectorant in cough syrups and in large dose as an emetic.

▷WW7875000.

Mono-Na salt: [526-94-3]. $[\alpha]_D^{19}$ +21.8° (H_2O).
Di-NH₄ salt: [3164-29-2]. $[\alpha]_D^{15}$ +34.6° (H_2O).

▷WW8050000.

Di-Me ester: [608-68-4]. *Dimethyl tartrate.*
    $C_6H_{10}O_6$ M 178.141
    Mp 48°, Mp 62° (dimorph.). $Bp_{11.5}$ 165-166°. $[\alpha]_D$ +6.2° ($CHCl_3$).
Di-Et ester: [87-91-2]. *Diethyl tartrate.*
    $C_8H_{14}O_6$ M 206.195
    Mp 17°. Bp_{15} 155-156°. $[\alpha]_D^{20}$ +7.5° ($CHCl_3$).
Di-Ph ester: Diphenyl tartrate.
    $C_{16}H_{14}O_6$ M 302.283
    Needles. Mp 101-102°.
Diamide: [634-63-9].
    $C_4H_8N_2O_4$ M 148.118
    Mp 195-196°. $[\alpha]_D$ +106.5° (H_2O).
Di-Ac: [51591-38-9]. *2,3-Di-O-acetyltartaric acid. Diacetoxysuccinic acid.*
    $C_8H_{10}O_8$ M 234.162
    Cryst. + $3H_2O$ (Et_2O). Sol. H_2O. Mp 58°. $[\alpha]_D^{22}$ −23.04° (H_2O).
Dibenzoyl: 2,3-Di-O-benzoyltartaric acid.
    $C_{18}H_{14}O_8$ M 358.304
    Mp 45.5°. Bp_7 234°. $[\alpha]_{546}^{17.5}$ −78.16° (Py).
Anhydride, di-Ac: 3,4-Diacetoxy-3,4-dihydro-2,5(2H,5H)-furandione, 9CI.
    $C_8H_8O_7$ M 216.147
    Mp 133-134°. $[\alpha]_D^{20}$ +97.2° (c, 0.47 in $CHCl_3$).
Bis-4-methylbenzoyl: Di-p-toluoyl tartrate.
    $C_{20}H_{18}O_8$ M 386.357
    Resolving agent. Mp 169-171°. $[\alpha]_D$ −141° (EtOH).
Mono-tert-butyl ester:
    $C_8H_{14}O_4$ M 174.196
    Cryst. ($CHCl_3/C_6H_6$ <50°). Mp 85°. $[\alpha]_{546}^{22}$+9.88° (c, 1.0 in Me_2CO).
Di-tert-butyl ester:
    $C_{12}H_{22}O_4$ M 230.303
    Mp 91°. $[\alpha]_{546}^{22}$+12.2° (c, 1.0 in Me_2CO).
Mono-tert-butyl ether:
    $C_8H_{14}O_4$ M 174.196

Mp 67°. $[\alpha]_D^{22}$ +46° (c, 1.0 in Me_2CO).
Di-tert-butyl ether:
    $C_{12}H_{22}O_4$ M 230.303
    Mp 134°. $[\alpha]_{546}^{22}$+54.7° (c, 1 in Me_2CO).

(2S,3S)-form [147-71-7]

D-form. D-Threaric acid

Found only in fruits and leaves of the West African tree *Bankinia reticulata.* Mp 169-170°. $[\alpha]_D^{20}$ −20.0° (c, 20.0 in H_2O). Not used commercially.

(2RS,3RS)-form [133-37-9]

(±)-form

Formed by heating the (+)-acid. Mp 205-206°. The first racemate to be separated into its antipodes.

Di-Me ester: Mp 84°, Mp 90° (dimorph.). Bp 282°, Bp_{12} 158°.
Dinitrile, Di-Ac:
    $C_8H_8N_2O_6$ M 228.161
    Mp 97-98°.

(2RS,3SR)-form [147-73-9]

meso-form. Racemic acid

Mp 159-160°. Not found in nature.

Di-Me ester: [5057-96-5]. Mp 111°.
Di-Et ester: Mp 55°.
Dinitrile:
    $C_4H_4N_2O_2$ M 112.088
    Plates or prisms (Et_2O). Sol. H_2O. Mp ca. 131° dec. Unstable.
Anhydride, dibenzoyl:
    $C_{18}H_{12}O_7$ M 340.289
    Mp 207-208°.

Pasteur, L., *Ann. Chim. Phys.,* 1850, **28**, 79.
Bijvoet, J.M. *et al, Nature* (London), 1951, **171**, 168 (*cryst struct*)
Bijvoet, J.M. *et al, Acta Crystallogr.,* 1958, **11**, 61 (*cryst struct*)
Org. Synth., Coll. Vol., **4**, 242 (*anhydride*)
Handbook of Biochemistry and Molecular Biology; Lipids, Carbohydrates, Steroids, (Fasman, G.D., Ed.), 1975, 169
Stothes, J.B. *et al, Can. J. Chem.,* 1977, **55**, 841 (*cmr*)
Albertsson, J. *et al, J. Appl. Crystallogr.,* 1979, **12**, 537 (*cryst struct*)
Kirk-Othmer Encycl. Chem. Technol., 3rd Ed., 1981, **13**, 103.
Hawthorne, F.C. *et al, Acta Crystallogr., Sect. B,* 1982, **38**, 2461 (*cryst struct*)
Buding, H. *et al, Angew. Chem., Int. Ed. Engl.,* 1985, **24**, 513 (*abs config, bibl*)
Uray, G. *et al, Tetrahedron,* 1988, **44**, 4357 (*tert-Butyl esters and ethers*)
Fieser, M. *et al, Reagents for Organic Synthesis,* Wiley, 1967-84, **1**, 351 (*deriv*)
Martindale, The Extra Pharmacopoeia, 28th/29th Eds., 1982/1989, Pharmaceutical Press, London, 755, 756, 1327.
Bretherick, L., *Handbook of Reactive Chemical Hazards,* 2nd Ed., Butterworths, London and Boston, 1979, 486.
Sax, N.I., *Dangerous Properties of Industrial Materials,* 6th Ed., Van Nostrand-Reinhold, 1984, 2498.

Tecomaquinone III
T-70006

[114216-80-7]

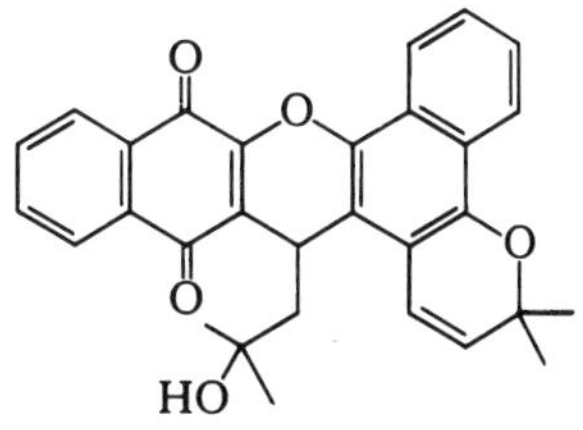

$C_{30}H_{26}O_5$ M 466.532

Constit. of *Tabebuia pentaphylla.* Dark-violet cryst. (Me_2CO/pet. ether). Mp 219-222°.

Sharma, P.K. *et al, Phytochemistry*, 1988, **27**, 632.

Tephrobbottin   T-70007

3,4-Dihydro-5-methoxy-8,8-dimethyl-2-phenyl-2H,8H-benzo[1,2-b:3,4-b']dipyran-4-ol, 9CI

[104777-98-2]

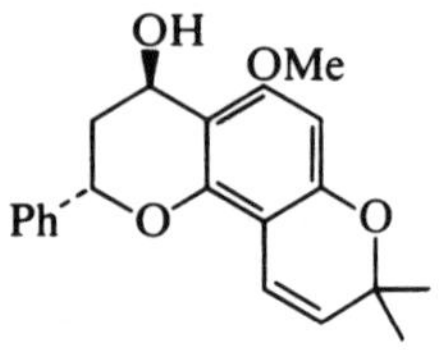

$C_{21}H_{22}O_4$ M 338.402

Isol. from *Tephrosia abbottiae*. Needles. Mp 154-155°.

Gomez-Garibay, F. *et al, Chem. Ind. (London)*, 1986, 827 (*isol*)

Ternatin   T-70008

3-[2-[2-(5-Hydroxy-4-methyl-3-pentenyl)-2-methyl-2H-1-benzopyran-6-yl]ethyl]-1,2-benzenediol, 9CI. 6-(2,3-Dihydroxyphenylethyl)-2-[(4-hydroxymethyl)-3-pentenyl]-2-methylchromene

[106982-90-5]

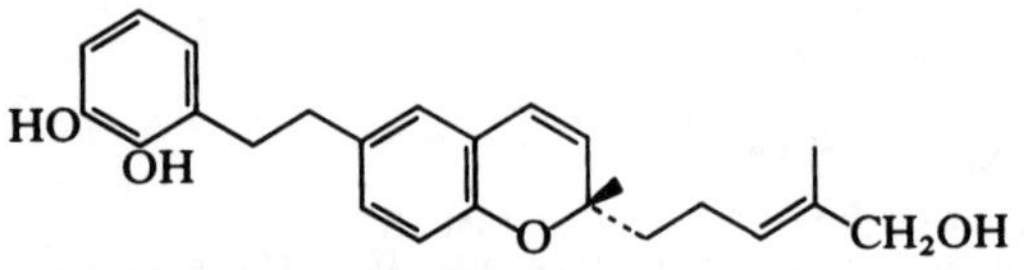

$C_{24}H_{28}O_4$ M 380.483

See also 4′,5-Dihydroxy-3,3′,7,8-tetramethoxyflavone, D-50552 . Constit. of fronds of *Sceptridium ternatum*. Oil. $[\alpha]_D^{20}$ +62° (c, 1 in MeOH).

Tanaka, N. *et al, Chem. Pharm. Bull.*, 1986, **34**, 3727 (*isol*)

Terremutin   T-70009

Updated Entry replacing T-00169

4,5-Dihydroxy-3-methyl-7-oxabicyclo[4.1.0]hept-3-en-2-one, 9CI. 5,6-Epoxy-3,4-dihydroxy-2-methyl-2-cyclohexen-1-one

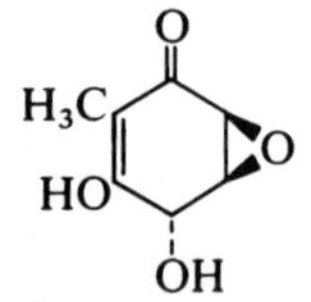

$C_7H_8O_4$ M 156.138

(−)-*form* [18746-82-2]

Metabolite of *Aspergillus terreus*. Cryst. (Et$_2$O or EtOAc). Mp 144-146° dec. $[\alpha]_D^{27.5}$ −269° (c, 1 in MeOH).

▷RN8931000.

Py complex: Mp 182-183.5° dec. $[\alpha]_D^{24.5}$ −174° (c, 1 in MeOH).

(±)-*form* [28966-95-2]

Mp 163-165°.

Py complex: Mp 185-187°.

Miller, M.W., *Tetrahedron*, 1968, **24**, 4839 (*isol, struct*)
Read, G. *et al, J. Chem. Soc., C*, 1970, 1945.
Kiriyama, N., *Chem. Pharm. Bull.*, 1977, **25**, 1265 (*biosynth*)

25,26,27,28-Tetraazapentacyclo-[19.3.1.1³,⁷.1⁹,¹³.1¹⁵,¹⁹]octacosa-1(25),3,5,7(28),9,11,13(27),15,17,19(26),-21,23-dodecaene-2,8,14,20-tetrone, 9CI   T-70010

[112634-75-0]

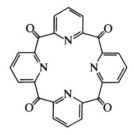

$C_{24}H_{12}N_4O_4$ M 420.383

Cryst. (CHCl$_3$).

Newkome, G.R. *et al, J. Chem. Soc., Chem. Commun.*, 1987, 854 (*synth, cryst struct, cmr*)

9H-Tetrabenzo[*a,c,g,i*]carbazole   T-70011

[88090-59-9]

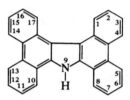

$C_{28}H_{17}N$ M 367.449

Pale-yellow cryst. Mp 242-244°.

Weitzberg, M. *et al, J. Heterocycl. Chem.*, 1983, **20**, 1019 (*synth, uv, pmr*)

Tetrabenzo[*b,h,n,t*]tetraphenylene   T-70012

9,10,23,24-Tetradehydro-5,8:11,14:19,22:25,28-tetraethenodibenzo[a,m]cyclotetracosene, 9CI

[111615-79-3]

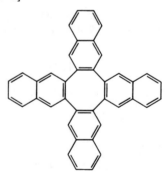

$C_{40}H_{24}$ M 504.630

Forms clathrates. Cryst. Mp >300°.

Wong, H.N.C. *et al, Tetrahedron Lett.*, 1987, **28**, 6359 (*synth, pmr, cryst struct*)

Tetra-*tert*-butoxycyclopentadienone   T-70013

2,3,4,5-Tetrakis(1,1-dimethylethoxy)-2,4-cyclopenta-dien-1-one, 9CI

[84890-11-9]

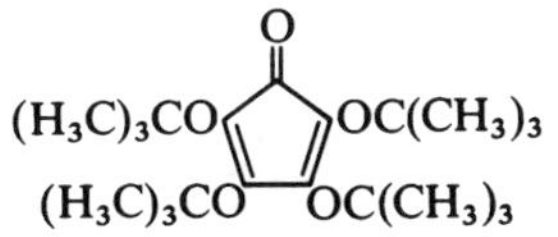

$C_{21}H_{36}O_5$ M 368.512

Ligand and synthetic intermediate. Red cryst. solid. Mp 44°.

Fornals, D. *et al, J. Chem. Soc., Perkin Trans. 1*, 1987, 2749 (*synth, use, ir, pmr*)

Tetra-*tert*-butylcyclobutadiene   T-70014

1,2,3,4-Tetrakis(1,1-dimethylethyl)-1,3-cyclobutadiene

[66809-05-0]

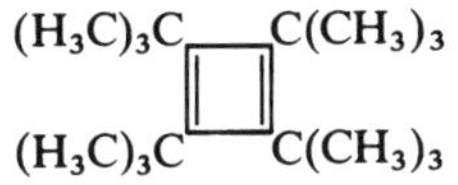

$C_{20}H_{36}$ M 276.504

V. oxygen-sensitive.

Maier, G. *et al, Angew. Chem., Int. Ed. Engl.*, 1978, **17**, 520 (*synth*)
Dunitz, J.D. *et al, Angew. Chem., Int. Ed. Engl.*, 1988, **27**, 387 (*cryst struct*)

Tetra-*tert*-butyltetrahedrane   T-70015

Updated Entry replacing T-50040
Tetrakis(1,1-dimethylethyl)tricyclo[1.1.0²,⁴]butane

[66809-06-1]

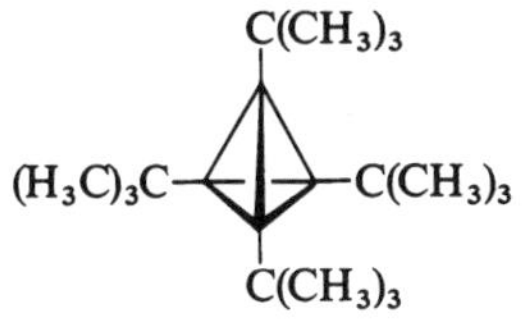

$C_{20}H_{36}$ M 276.504

Needles (MeOH). Mp 135° dec.

Bos, H.J.T., *Chem. Weekbl.*, 1978, 571 (*rev*)
Maier, G. *et al, Angew. Chem., Int. Ed. Engl.*, 1978, **17**, 520 (*synth*)
Dehnicke, K. *et al, Chem. Ber.*, 1981, **114**, 3965, 3988 (*synth, spectra*)
Loerzer, T. *et al, Angew. Chem., Int. Ed. Engl.*, 1983, **22**, 878 (*cmr*)
Irngartinger, H. *et al, Angew. Chem., Int. Ed. Engl.*, 1984, **27**, 993 (*cryst struct*)
Maier, G., *Angew. Chem., Int. Ed. Engl.*, 1988, **27**, 309 (*rev*)

1,2,3,4-Tetrachlorodibenzo-*p*-dioxin   T-70016

[30746-58-8]

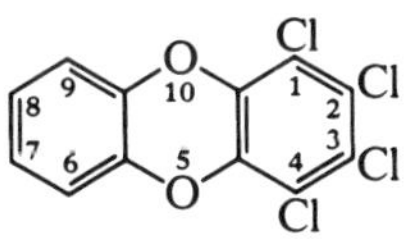

$C_{12}H_4Cl_4O_2$ M 321.974

▷HP3493000.

Rissanen, K. *et al, Acta Crystallogr., Sect. C*, 1987, **43**, 488 (*cryst struct*)

1-Tetracosyne   T-70017

[61847-84-5]

$$H_3C(CH_2)_{21}C{\equiv}CH$$

$C_{24}H_{46}$ M 334.627
Cryst. Mp 47.5-49°.

Igner, E. *et al, J. Chem. Soc., Perkin Trans. 1*, 1987, 2447 (*synth*)

12-Tetracosyne   T-70018

[113308-92-2]

$$H_3C(CH_2)_{10}C{\equiv}C(CH_2)_{10}CH_3$$

$C_{24}H_{46}$ M 334.627
Mp 33-34°. Bp$_{0.05}$ 148-151°.

Igner, E. *et al, J. Chem. Soc., Perkin Trans. 1*, 1987, 2447 (*synth*)

Tetracyclo[5.3.0.0²,⁴.0³,⁵]deca-6,8,10-triene   T-70019

5,6-Dihydro-4,5,6-metheno-4H-indene, 9CI. Azulvalene

[92622-71-4]

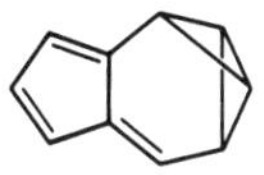

$C_{10}H_8$ M 128.173
Orange oil.

Murata, I. *et al, Tetrahedron*, 1986, **42**, 1745 (*synth, pmr, cmr, uv*)

Tetracyclo[6.4.0.0⁴,¹².0⁵,⁹]dodec-10-ene   T-70020

$C_{12}H_{16}$ M 160.258
Cryst. (MeOH). Mp 84°. Volatile.

Akhtar, J.A. *et al, J. Chem. Soc. (C)*, 1968, 812 (*synth*)
Raasch, M.S., *J. Org. Chem.*, 1980, **45**, 856 (*di- and tetrahalo derivs*)
Grimme, W. *et al, Tetrahedron Lett.*, 1987, **28**, 6035 (*synth, pmr, cmr*)

Tetracyclo[6.2.1.1³,⁶.0²,⁷]dodec-2(7)-ene   T-70021

Updated Entry replacing T-60033
1,2,3,4,5,6,7,8-Octahydro-1,4:5,8-dimethanonaphtha-lene, 9CI. Sesquinorbornene

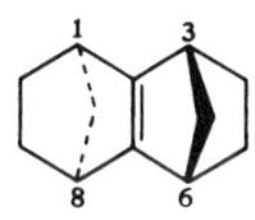

$C_{12}H_{16}$ M 160.258

(*1α,3β,6β,8α*)-*form* [73679-39-7]
anti-*form*
Mp 64-65°. Bp$_{0.6}$ 40°. *Syn*-form also known but not fully descr.

Paquette, L.A. *et al, J. Am. Chem. Soc.*, 1980, **102**, 1186 (*synth*)
Bartlett, P.D. *et al, J. Am. Chem. Soc.*, 1980, **102**, 1383 (*synth, pmr, cmr*)

Bartlett, P.D. *et al, J. Org. Chem.*, 1984, **49**, 1875 (*synth*)
Kopecky, K.R. *et al, Can. J. Chem.*, 1984, **62**, 1840 (*synth*)
DeLucchi, O. *et al, Tetrahedron Lett.*, 1986, **27**, 4347 (*synth*)

Tetracyclo[3.3.0.0^{2,8}.0^{4,6}]octane   T-70022

Octahydrodicyclopropa[cd,gh]*pentalene, 9CI, 8CI.*
Dihydrocuneane

[765-72-0]

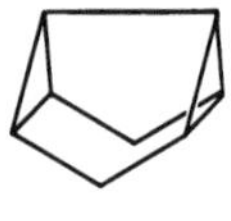

C$_8$H$_{10}$ M 106.167
The name dihydrocuneane is strictly incorrect. Liq. Bp
138-139°.

Freeman, P.K. *et al, Tetrahedron Lett.*, 1965, 3301 (*synth*)
LeBel, N.A. *et al, J. Am. Chem. Soc.*, 1965, **87**, 4301 (*synth, ir*)
Volger, H.C. *et al, J. Am. Chem. Soc.*, 1969, **91**, 218 (*synth*)
Hassenrück, K. *et al, Chem. Ber.*, 1988, **121**, 373 (*pe*)

Tetracyclo[4.2.0.0^{2,5}.0^{3,8}]octane, 9CI, 8CI   T-70023

Secocubane

[3104-90-3]

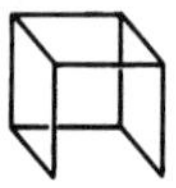

C$_8$H$_{10}$ M 106.167
Cryst. Mp 101-103°.

Scherer, K.V. *et al, Tetrahedron Lett.*, 1965, 1199 (*deriv, pmr, uv*)
Gasteiger, J. *et al, Tetrahedron*, 1978, **34**, 2939.
Stober, R. *et al, Tetrahedron*, 1986, **42**, 1757 (*synth, pmr, cmr, ir*)

3-(1,3,5,7,9-Tetradecapentaenyloxy)-1,2-propanediol, 9CI   T-70024

Updated Entry replacing T-40021
1-O-(1,3,5,7,9-Tetradecapentaenyl)glycerol. Fecapentaene-14

[84000-58-8]

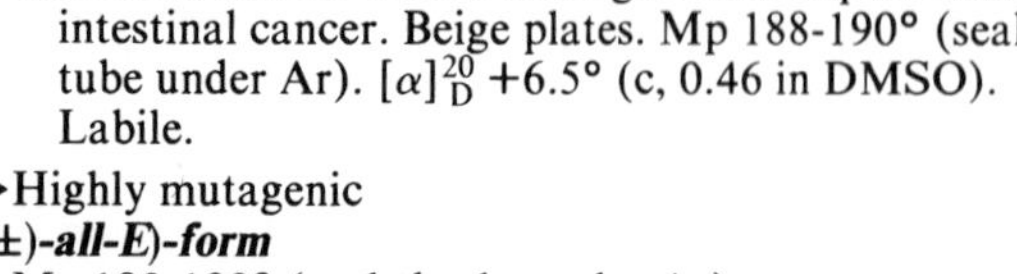

$$\text{CH}_2\text{O} \diagup\diagdown\diagup\diagdown\diagup\diagdown\diagup\diagdown\diagup\diagup$$
$$\text{HO} \text{—C} \blacktriangleleft\text{H}$$
$$\text{CH}_2\text{OH} \qquad (S, \text{ all-}E)\text{-form}$$

C$_{17}$H$_{26}$O$_3$ M 278.391
See also refs. under 3-(1,3,5,7,9-Dodecapentaenyloxy)-
1,2-propanediol, D-70536 .

(S,all-E)-form

Isol. from human faeces. Mutagen with implications for
intestinal cancer. Beige plates. Mp 188-190° (sealed
tube under Ar). [α]$_D^{20}$ +6.5° (c, 0.46 in DMSO).
Labile.

▷Highly mutagenic

((±)-all-E)-form

Mp 183-190° (sealed tube under Ar).

▷Highly mutagenic

Gupta, I. *et al, Science*, 1984, **225**, 521.
Baptista, J. *et al, Angew. Chem., Int. Ed. Engl.*, 1987, **26**, 1186
  (*config*)
Pfaendler, R. *et al, Justus Liebigs Ann. Chem.*, 1988, 449
  (*synth, ir, pmr, cmr, ms*)

1,3,5,7-Tetraethynyladamantane   T-70025

1,3,5,7-Tetraethynyltricyclo[3.3.1.1^{3,7}]decane, 9CI

[84782-82-1]

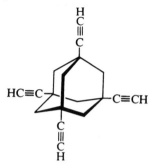

C$_{18}$H$_{16}$ M 232.324
Solid by subl. Mp 156-157°.

Naemura, K. *et al, Tetrahedron*, 1986, **42**, 1763 (*synth, pmr*)

2,3,4,5-Tetrafluorobenzaldehyde   T-70026

[16583-06-5]

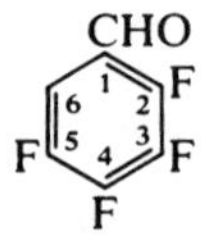

C$_7$H$_2$F$_4$O   M 178.086
Liq. Bp$_{62}$ 80-82°.

2,4-Dinitrophenylhydrazone: [19842-75-2]. Cryst.
  (EtOH). Mp 229-231°.

Belf, L.J. *et al, Tetrahedron*, 1967, **23**, 4719 (*synth, ir*)
Castle, M.D. *et al, J. Chem. Soc. (C)*, 1968, 1225 (*synth*)

2,3,4,6-Tetrafluorobenzaldehyde   T-70027

[19842-78-5]
C$_7$H$_2$F$_4$O   M 178.086
Liq. Bp 178°.

2,4-Dinitrophenylhydrazone: [19842-78-5]. Cryst.
  (AcOH). Mp 202-203°.

Castle, M.D. *et al, J. Chem. Soc. (C)*, 1968, 1225 (*synth*)

2,3,5,6-Tetrafluorobenzaldehyde   T-70028

[19842-76-3]
C$_7$H$_2$F$_4$O   M 178.086
Liq. Bp 178°, Bp$_7$ 59.5-60.5°.

2,4-Dinitrophenylhydrazone: [19842-79-6]. Cryst.
  (AcOH). Mp 204° dec.
Di-Et acetal:
  C$_{11}$H$_{11}$F$_4$O$_2$ M 251.200
  Liq. Bp$_5$ 87-89°.

Castle, M.D. *et al, J. Chem. Soc. (C)*, 1968, 1225 (*synth*)
Vysochin, V.I. *et al, J. Gen. Chem. USSR (Engl. Transl.)*, 1969,
  39, 1576 (*synth, ir*)

2,3,5,6-Tetrafluoro-1,4-benzenedicarboxylic acid, 9CI   T-70029

Tetrafluoroterephthalic acid, 8CI

[652-36-8]

COOH
F — F
F — F
COOH

$C_8H_2F_4O_4$ M 238.095
Cryst. (H_2O). Mp 283-284°.
Di-Me ester: [727-55-9].
  $C_{10}H_6F_4O_4$ M 266.149
  Cryst. (MeOH). Mp 79-80°.

Gething, B. *et al, J. Chem. Soc.*, 1961, 1574 (*synth, deriv, ir*)

2,4,5,6-Tetrafluoro-1,3-benzenedicarboxylic acid, 9CI   T-70030

Tetrafluoroisophthalic acid, 8CI

[1551-39-9]

$C_8H_2F_4O_4$ M 238.095
Cryst. Mp 203° dec.

Robson, P. *et al, J. Chem. Soc.*, 1964, 5748 (*synth*)

3,4,5,6-Tetrafluoro-1,2-benzenedicarboxylic acid, 9CI   T-70031

Tetrafluorophthalic acid, 8CI

[652-03-9]

$C_8H_2F_4O_4$ M 238.095
Cryst. (xylene). Mp 153-154°.
Anhydride: [652-12-0]. *4,5,6,7-Tetrafluoro-1,3-isobenzofurandione, 9CI.*
  $C_8F_4O_3$ M 220.080
  Cryst. by subl. Mp 94-95.5°. Hydrolyses rapidly.

Gething, B. *et al, J. Chem. Soc.*, 1961, 1574 (*synth, deriv, ir*)

2,3,4,5-Tetrafluorobenzoic acid   T-70032

[1201-31-6]

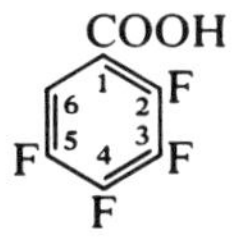

COOH
F — F
F — F
F

$C_7H_2F_4O_2$ M 194.085
Cryst. (pet. ether). Mp 85.5-87°.
Me ester: [5292-42-2].
  $C_8H_4F_4O_2$ M 208.112
  Liq. Bp 187-188°.
Nitrile: 2,3,4,5-Tetrafluorobenzonitrile. 1-Cyano-2,3,4,5-tetrafluorobenzene.
  C_7HF_4N M 175.085
  Liq. Bp 164-165°.

Yakobson, G.G. *et al, J. Gen. Chem. USSR (Engl. Transl.)*, 1966, **36**, 144 (*synth, deriv*)
Belf, L.J. *et al, Tetrahedron*, 1967, **23**, 4719 (*synth, deriv*)
Tamborski, C. *et al, J. Organomet. Chem.*, 1969, **17**, 185 (*synth*)

2,3,4,6-Tetrafluorobenzoic acid   T-70033

[32890-92-9]
$C_7H_2F_4O_2$ M 194.085

Cryst. (pet. ether). Mp 98-99.5°.

Deacon, G.B. *et al, Aust. J. Chem.*, 1978, **31**, 1709 (*synth, F nmr*)

2,3,5,6-Tetrafluorobenzoic acid   T-70034

[652-18-6]
$C_7H_2F_4O_2$ M 194.085
Cryst. (toluene). Mp 154°.
Nitrile: [5216-17-1]. *2,3,5,6-Tetrafluorobenzonitrile. 3-Cyano-1,2,4,5-tetrafluorobenzene.*
  C_7HF_4N M 175.085
  Liq. Bp_{45} 88-90°.

Alsop, D.J. *et al, J. Chem. Soc.*, 1962, 1801 (*synth*)
Vysochin, N.N. *et al, J. Gen. Chem. USSR (Engl. Transl.)*, 1969, **39**, 1607 (*synth*)

4,5,6,7-Tetrafluoro-1H-benzotriazole   T-70035

[26888-72-2]

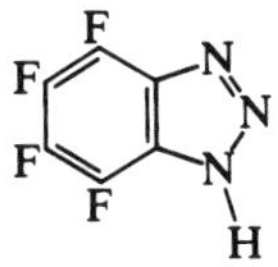

$C_6HF_4N_3$ M 191.088
Plates (C_6H_6). Mp 160-162°.
1-Ac: [26888-73-3].
  $C_8H_3F_4N_3O$ M 233.125
  Needles (pet. ether). Mp 112-113°.

Birchall, J.M. *et al, J. Chem. Soc. (C)*, 1970, 1519 (*synth, F nmr, ir*)

2,3,4,5-Tetrahydro-1H-1-benzazepine, 9CI   T-70036

Homotetrahydroquinoline

[1701-57-1]

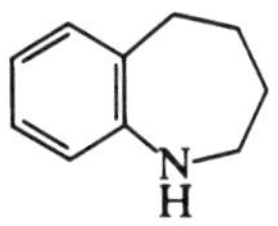

$C_{10}H_{13}N$ M 147.219
Mp 32°. Bp 253-255°, Bp_{16} 131-133°.
B,HCl: Cryst. (EtOH). Mp 186°.
Picrate: Mp 179-181°.
1-Me:
  $C_{11}H_{15}N$ M 161.246
  Oil. Bp_{15} 110-115° (bath).
N-Benzyl: [60740-67-2].
  $C_{17}H_{19}N$ M 237.344
  Oil. $Bp_{0.3}$ 127°.

v. Braun, J. *et al, Ber.*, 1912, **45**, 3378 (*synth*)
Astill, B.D. *et al, J. Am. Chem. Soc.*, 1955, **77**, 4079 (*synth, deriv*)
Thon, D. *et al, Chem. Ber.*, 1976, **109**, 2743 (*deriv, synth, pmr*)

4,5,6,7-Tetrahydro-1,2-benzisoxazol-3(2H)-one   T-70037

4,5,6,7-Tetrahydro-1,2-benzisoxazol-3-ol. Tetrahydroindoxazen-3-one

[29598-09-2]

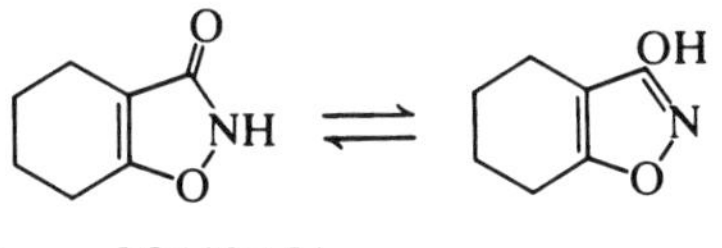

$C_7H_9NO_2$ M 139.154

Cryst. Mp 130°.

Jacquier, R. *et al*, *Bull. Soc. Chim. Fr.*, 1970, 1978 (*synth*, *ir*, *uv*, *pmr*)

4,5,6,7-Tetrahydro-2,1-benzisoxazol-   **T-70038**
3(1*H*)-one, 9CI

*4,5,6,7-Tetrahydro-2,1-benzisoxazol-3(3*aH*)-one, 9CI.*
3,4-Tetramethyleneisoxazolin-5-one. Tetrahydroanth-
ranil-3-one

[29879-50-3]

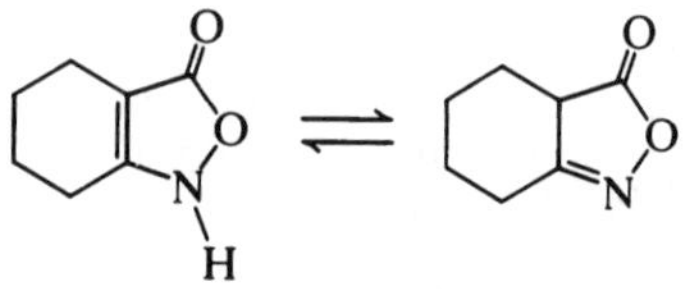

$C_7H_9NO_2$ M 139.154

NH-form predominates. Cryst. (Et$_2$O). Mp 65-69°.
Unstable even at −15°.

Katritzky, A.R. *et al*, *Tetrahedron*, 1962, **18**, 777 (*synth*, *ir*, *tautom*)
Jacquier, R. *et al*, *Bull. Soc. Chim. Fr.*, 1970, 2690 (*synth*, *ir*, *uv*, *pmr*, *tautom*)
Krogsgaard-Larsen, P. *et al*, *Acta Chem. Scand.*, 1972, **27**, 2802 (*synth*, *uv*)

2,3,8,9-Tetrahydrobenzo[1,2-*b*:3,4-*b'*]-   **T-70039**
bis[1,4]dithiin, 9CI

Bis(ethylenedithio)benzene

[110431-56-6]

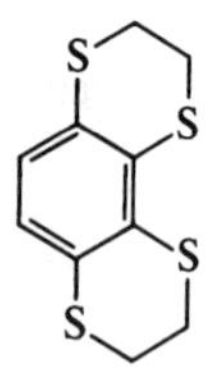

$C_{10}H_{10}S_4$ M 258.429
Oil.

Lapouyade, R. *et al*, *J. Chem. Soc., Chem. Commun.*, 1987, 223 (*synth*, *pmr*, *ms*)

2,3,8,9-Tetrahydrobenzo[2,1-*b*:3,4-*b'*]-   **T-70040**
bis[1,4]oxathiin, 9CI

Bis(ethyleneoxythio)benzene

[110431-57-7]

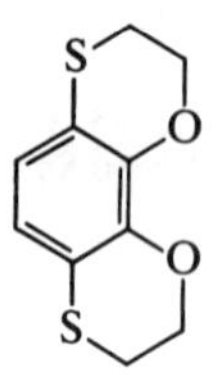

$C_{10}H_{10}O_2S_2$ M 226.308
Mp 138°.

Lapouyade, R. *et al*, *J. Chem. Soc., Chem. Commun.*, 1987, 223 (*synth*, *pmr*, *ms*)

7,8,8*a*,9*a*-Tetrahydrobenzo[10,11]-   **T-70041**
chryseno[3,4-*b*]oxirene-7,8-diol, 9CI

9,10-Epoxy-7,8,9,10-tetrahydro-7,8-
dihydroxybenzo[a]*pyrene*

[55097-80-8]

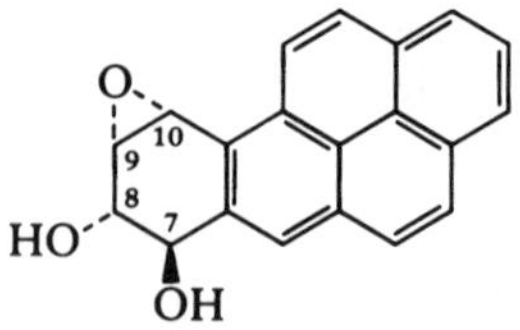

 (7R,8S,9S,10R)-form

$C_{20}H_{14}O_3$ M 302.329

(*7R,8S,9S,10R*)-*form* [63357-09-5]
  (+)-*DE-2*
  Major metabolite responsible for carcinogenic and
    mutagenic props. of benzo[*a*]pyrene. Highly
    tumorigenic to mice. Cryst. (EtOAc). Mp 214°. [α]$_D^{23}$
    +72° (THF).

(*7R,8S,9R,10S*)-*form* [63323-31-9]
  (−)-*DE-1*
  Minor metabolite responsible for carcinogenic and
    mutagenic props. of benzo[*a*]pyrene. Mp 226-228°
    dec.

Thakken, D.R. *et al*, *Proc. Natl. Acad. Sci. USA*, 1976, **73**, 3381 (*biochem*)
Yagi, H. *et al*, *J. Am. Chem. Soc.*, 1977, **99**, 1604, 2358 (*synth*, *abs config*)
Gupta, S.C. *et al*, *J. Org. Chem.*, 1987, **52**, 3812 (*props*, *bibl*)

4,5,6,7-Tetrahydrobenzothiadiazole, 9CI   **T-70042**

Cyclohexeno-1,2,3-thiadiazole

[56382-72-0]

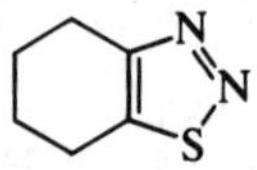

$C_6H_8N_2S$ M 140.203
Liq. Bp$_{0.05}$ 70-71°, Bp$_{0.01}$ 51-55°.
2-Oxide: [56382-76-4].
  $C_6H_8N_2OS$ M 156.202
  Mp 85°.
1,1,2-Trioxide: [56382-80-0].
  $C_6H_8N_2O_3S$ M 188.201
  Mp 140°.

Braun, H.P. *et al*, *Tetrahedron*, 1975, **31**, 637 (*synth*, *pmr*, *cmr*, *uv*)
Spencer, H.K. *et al*, *J. Org. Chem.*, 1976, **41**, 730 (*synth*)

1,2,4,5-Tetrahydro-3-benzoxepin   **T-70043**

5-Oxabenzocycloheptene

$C_{10}H_{12}O$ M 148.204
Liq. Bp$_{12}$ 103°. n_D 1.5450.

Dimroth, K. *et al*, *Chem. Ber.*, 1966, **99**, 634 (*synth*)
Canuel, L. *et al*, *Can. J. Chem.*, 1974, **52**, 3581 (*synth*, *pmr*, *conformn*)

1,3,4,5-Tetrahydro-2-benzoxepin, 9CI   T-70044

Homoisochroman

[5698-85-1]

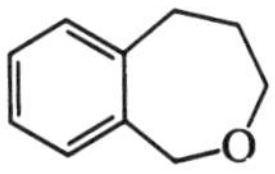

$C_{10}H_{12}O$ M 148.204

Mp 23.5-24°. Bp$_{0.13}$ 55°.

Rieche, A. *et al, Chem. Ber.*, 1962, **95**, 91 (*synth*)

Désilets, S. *et al, J. Am. Chem. Soc.*, 1987, **109**, 1641 (*synth, pmr, cmr*)

2,3,4,5-Tetrahydro-1-benzoxepin   T-70045

3-Oxabenzocycloheptene

[6169-78-4]

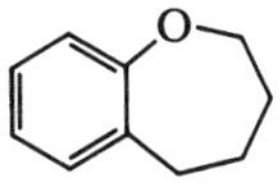

$C_{10}H_{12}O$ M 148.204

Oil.

Lachapelle, A. *et al, Can. J. Chem.*, 1987, **65**, 2575 (*synth, pmr, cmr, conformn*)

Tetrahydro-3,4-bis(methylene)furan, 9CI   T-70046

3,4-Dimethyleneoxolane. 3,4-Dimethylenetetrahydrofuran

[50521-40-9]

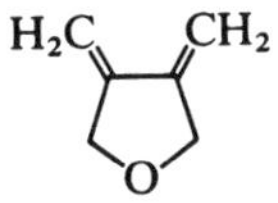

C_6H_8O M 96.129

Liq. Bp$_{20}$ 50-55° (bath).

Bailey, W.J. *et al, J. Org. Chem.*, 1963, **28**, 802 (*synth, uv, pmr*)

Gaoni, Y. *et al, J. Org. Chem.*, 1980, **45**, 870 (*synth, uv, pmr, ir*)

7,8,9,10-Tetrahydro-6H-cyclohepta[b]-naphthalene, 9CI   T-70047

2,3-Cycloheptenonaphthalene

[7092-91-3]

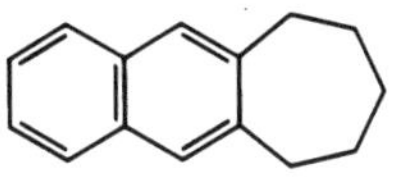

$C_{15}H_{16}$ M 196.291

Mp 104-105°.

Grice, P. *et al, Tetrahedron Lett.*, 1979, 2563 (*synth, cmr*)

Tius, M.A. *et al, Tetrahedron Lett.*, 1986, **27**, 2571 (*synth, cmr, pmr, ir, ms*)

3,3a,4,6a-Tetrahydrocyclopenta[b]pyrrol-2(1H)-one, 9CI   T-70048

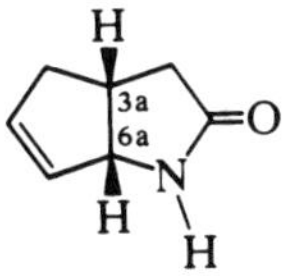

C_7H_9NO M 123.154

(3aRS,6aSR)-form [113466-88-9]

(±)-cis-*form*

Oil.

Knapp, S. *et al, Tetrahedron Lett.*, 1987, **28**, 3213 (*synth, ir*)

1,2,3,4-Tetrahydrocyclopent[b]indole, 9CI   T-70049

[2047-91-8]

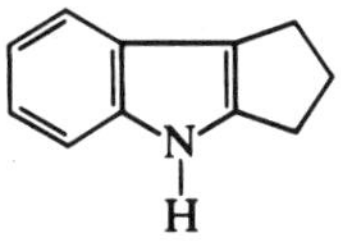

$C_{11}H_{11}N$ M 157.215

Mp 108°.

N-*Me:* [52751-31-2].

   $C_{12}H_{13}N$ M 171.241

   Cryst. (EtOH). Mp 43°.

Wender, P.A. *et al, Tetrahedron*, 1983, **39**, 3767.

Acheson, R.M. *et al, J. Chem. Soc., Perkin Trans. 1*, 1981, 3141 (*synth, pmr, ms*)

Wender, P.A. *et al, Tetrahedron*, 1986, **42**, 2985 (*synth, pmr, ir, ms*)

3,4,5,6-Tetrahydro-1H-cyclopropa[b]-naphthalene   T-70050

[112504-82-2]

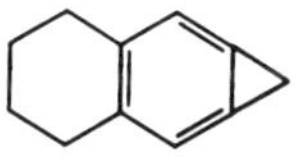

$C_{11}H_{12}$ M 144.216

Yellow oil.

Billups, W.E. *et al, J. Org. Chem.*, 1988, **53**, 1312 (*synth, ir, pmr*)

4,5,6,7-Tetrahydro-1H-cyclopropa[a]-naphthalene   T-70051

[112504-80-0]

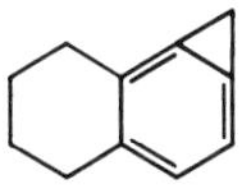

$C_{11}H_{12}$ M 144.216

Oil.

Billups, W.E. *et al, J. Org. Chem.*, 1988, **53**, 1312 (*synth, ir, pmr*)

1,3,4,5-Tetrahydrocycloprop[f]indene   T-70052

[112504-81-1]

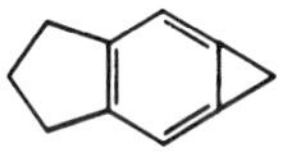

$C_{10}H_{10}$ M 130.189

Pale-yellow oil.

Billups, W.E. *et al, J. Org. Chem.*, 1988, **53**, 1312 (*synth, ir, pmr*)

1,4,5,6-Tetrahydrocycloprop[*e*]indene   T-70053
[5266-64-8]

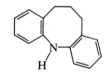

C$_{10}$H$_{10}$ M 130.189
Oil.

Billups, W.E. *et al*, *J. Org. Chem.*, 1988, **53**, 1312 (*synth, ir, pmr*)

5,6,7,8-Tetrahydrodibenz[*c,e*]azocine   T-70054
[6196-54-9]

C$_{15}$H$_{15}$N   M 209.290
Mp 119-120°.

B,HCl: [6196-36-7]. Cryst. Mp 321-322° dec.

Jeffs, P.W *et al*, *J. Am. Chem. Soc.*, 1967, **89**, 2798 (*synth, uv, pmr*)

5,6,7,12-Tetrahydrodibenz[*b,e*]azocine   T-70055
[1527-07-7]

C$_{15}$H$_{15}$N   M 209.290
Cryst. (pet. ether). Mp 59-60°. Bp$_{1.5}$ 152-155°.

B,HCl: [1527-08-8]. Cryst. (EtOH/Et$_2$O). Mp 263-266°.

Jílek, J.O. *et al*, *Monatsh. Chem.*, 1965, **96**, 182 (*synth, ir, uv*)

5,6,7,12-Tetrahydrodibenz[*b,g*]azocine   T-70056
[1639-73-2]

C$_{15}$H$_{15}$N   M 209.290
Mp 58-60°.

Fouche, J.C., *CA*, 1969, **70**, 68105p (*synth*)

5,6,11,12-Tetrahydrodibenz[*b,f*]azocine   T-70057
[5697-88-1]

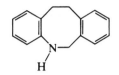

C$_{15}$H$_{15}$N   M 209.290
Cryst. (EtOH). Mp 130-132°.

N-*Ac:*
    C$_{17}$H$_{17}$NO   M 251.327
    Needles. Mp 91-94°.

Monro, A.M. *et al*, *J. Med. Chem.*, 1963, **6**, 255 (*synth, uv*)
Schindler, O. *et al*, *Helv. Chim. Acta*, 1966, **49**, 985 (*synth, ir, deriv*)

4*b*,8*b*,13,14-Tetrahydrodiindeno[1,2-*a*:2′,1′-*b*]indene, 9CI   T-70058
2,2′-o-Phenylene-1,1′-spirobiindan.
Tribenzotricyclo[6.3.0.0^{1,5}]undecane
[91158-94-0]

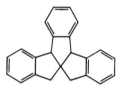

C$_{23}$H$_{18}$ M 294.395
Cryst. (hexane or EtOH/EtOAc). Mp 107.5-108°, Mp 148°.

Ten Hoeve, W. *et al*, *J. Org. Chem.*, 1980, **45**, 2930 (*synth, pmr, cmr, uv*)
Kuck, D., *Angew. Chem., Int. Ed. Engl.*, 1984, **23**, 508 (*synth*)

Tetrahydro-5,6-dimethyl-2*H*-pyran-2-one, 9CI   T-70059
Updated Entry replacing T-20061
γ,δ-Dimethyl-δ-valerolactone

H$_3$C — (ring structure) (5*S*,6*R*)-*form*
H$_3$C

C$_7$H$_{12}$O$_2$ M 128.171

(5*S*,6*R*)-form [82467-25-2]
    (+)-trans-*form*
    Liq. Bp$_1$ 80-82°, Bp$_{0.15}$ 70° (bath). [α]$_D^{21}$ +22.3° (c, 4.66 in CHCl$_3$).
    Di-Et ketal: [101977-88-2]. *2,2-Diethoxytetrahydro-5,6-dimethyl-2H-pyran, 9CI.*
    C$_{11}$H$_{22}$O$_3$ M 202.293
    Liq. Bp$_{0.08}$ 70° (bath). [α]$_D^{20}$ +22.2° (c, 1.03 in CHCl$_3$).

(5*RS*,6*RS*)-form [24405-15-0]
    (±)-cis-*form*
    Unstable liq. Bp$_5$ 99-102°.

(5*RS*,6*SR*)-form [82045-40-7]
    (±)-trans-*form*
    Mp 26-28° subl. Bp$_5$ 100-102°.

Cooke, E. *et al*, *Can. J. Chem.*, 1982, **60**, 29 (*synth, spectra*)
Williams, D.R. *et al*, *J. Am. Chem. Soc.*, 1982, **104**, 4708 (*synth*)
Schow, S.R. *et al*, *J. Am. Chem. Soc.*, 1986, **108**, 2662 (*synth, pmr*)
Kocieński, P.J. *et al*, *J. Chem. Soc., Perkin Trans. 1*, 1987, 2171 (*synth, ir, pmr*)

1,2,3,6-Tetrahydro-2,6-dioxo-4-pyrimidineacetic acid, 9CI   T-70060

Uracil-4-acetic acid

[4628-39-1]

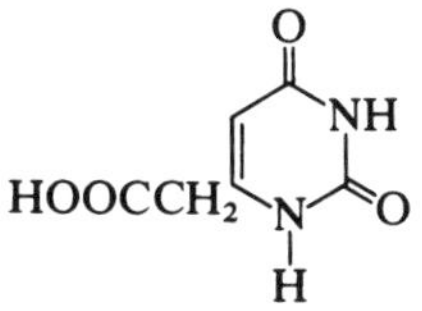

$C_6H_6N_2O_4$ M 170.124

Fine needles (H_2O). Mp >300°.

Monohydrate: Large prisms. Mp >300°.

Me ester: [6535-93-9].

$C_7H_8N_2O_4$ M 184.151

Large, elongated prisms (H_2O). Mp 216-218°.

Et ester: [6426-84-2].

$C_8H_{10}N_2O_4$ M 198.178

Cryst. by subl. Mp 189-191°.

N-*Ethylamide:* [106780-46-5].

$C_8H_{11}N_3O_3$ M 197.193

Solid. Mp 235-240° dec.

Wheeler, H.L. *et al, J. Am. Chem. Soc.,* 1908, **30**, 1156 (*synth*)
Hilbert, G.E., *J. Am. Chem. Soc.,* 1932, **54**, 2077 (*synth*)
Dietz, T.M. *et al, J. Am. Chem. Soc.,* 1987, **109**, 1793 (*deriv, synth, ir, pmr, ms*)

7,8,9,10-Tetrahydrofluoranthene, 9CI   T-70061

1,2-Cyclohexenoacenaphthylene

[42429-92-5]

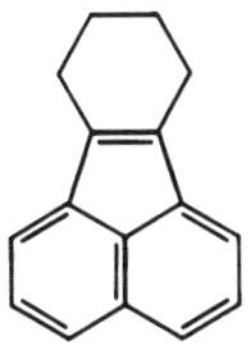

$C_{16}H_{14}$ M 206.287

Orange liq. Bp_2 165°, $Bp_{0.1}$ 132-133°.

Cristol, H. *et al, Bull. Soc. Chim. Fr.,* 1960, 1573 (*synth*)
Jackson, D.A. *et al, J. Chem. Soc., Perkin Trans. 1,* 1987, 2437 (*synth, uv, pmr*)

Tetrahydro-3,4-furandiamine, 9CI   T-70062

3,4-Diaminotetrahydrofuran

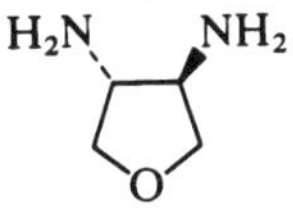

$C_4H_{10}N_2O$ M 102.136

(3RS,4RS)-form [117180-87-7]

(±)-*trans-form*

B,2HBr: [117180-88-8]. Cryst.

Feit, P.W. *et al, J. Med. Chem.,* 1967, **10**, 927; 1970, **13**, 447.
Bitha, P. *et al, J. Heterocycl. Chem.,* 1988, **25**, 1035 (*synth, pmr*)

1,2,3,4-Tetrahydro-1-hydroxy-2-naphthalenecarboxylic acid, 9CI   T-70063

1,2,3,4-Tetrahydro-1-hydroxy-2-naphthoic acid, 8CI

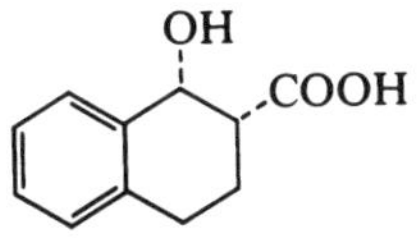

(1R,2R)-form

$C_{11}H_{12}O_3$ M 192.214

(1R,2R)-form [38157-19-6]

(+)-cis-*form*

Mp 98-102°. $[\alpha]_D^{22}$ +49.7° (c, 2 in EtOH).

Me ester: [38157-20-9].

$C_{12}H_{14}O_3$ M 206.241

$[\alpha]_D^{22}$ +50.4° (c, 2 in EtOH).

(1S,2S)-form

(−)-cis-*form*

Cryst. (EtOH). Mp 105°. $[\alpha]_D^{22}$ −102° (c, 2 in EtOH).

(1R,2S)-form [38157-17-4]

(−)-trans-*form*

Mp 126°. $[\alpha]_D^{21}$ −37.1° (c, 1.5 in CHCl₃). Not opt. pure.

Me ester: [61474-12-2]. Mp 40-41°. $[\alpha]_D^{22}$ −27.7° (c, 1.5 in CHCl₃).

(1S,2R)-form

(+)-trans-*form*

Mp 128.5-130°. $[\alpha]_D^{22}$ +52.0° (c, 1.5 in CHCl₃).

(1RS,2RS)-form

(±)-cis-*form*

Cryst. (Et₂O). Mp 130-131°.

Me ester: [32093-23-5]. Cryst. (Et₂O). Mp 75-75.5°.

(1RS,2SR)-form

(±)-trans-*form*

Mp 157-158°.

Me ester: [32093-24-6]. Mp 70.5-71°.

Bernáth, G. *et al, Acta Chim. Acad. Sci. Hung.,* 1970, **64**, 81 (*synth, ir, deriv*)
Sohar, P. *et al, Magy. Kem. Foly.,* 1970, **76**, 577; *CA,* **74**, 63815t (*nmr*)
Schoofs, A. *et al, Bull. Soc. Chim. Fr.,* 1976, 1215 (*synth, nmr, ir*)

1,2,3,4-Tetrahydro-1-hydroxy-2-naphthalenemethanol, 9CI   T-70064

2-Hydroxymethyl-1-tetralol

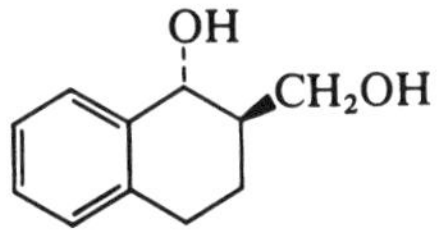

(1R,2R)-form

$C_{11}H_{14}O_2$ M 178.230

(1R,2R)-form [61474-13-3]

(−)-trans-*form*

Mp 74-81°. $[\alpha]_D^{22}$ −47.3° (c, 1.5 in EtOH).

(1R,2S)-form [38157-12-9]

(+)-cis-*form*

Mp 85-90°. $[\alpha]_D^{22}$ +40.1° (c, 2 in EtOH).

1′-(4-Methylbenzenesulfonyl): [38746-33-7]. $[\alpha]_D^{22}$ +23.8° (c, 2 in EtOH).

(1RS,2RS)-form [28053-21-6]

(±)-trans-*form*

Mp 83-83.5°.

(1RS,2SR)-form [28060-21-1]

(±)-cis-*form*

Mp 102.5-103°.

Bernáth, G. *et al, Acta Chim. Acad. Sci. Hung.*, 1970, **64**, 81 (*synth*)

Schoofs, A. *et al, Bull. Soc. Chim. Fr., Part II*, 1976, 1215 (*abs config*)

1,2,5,6-Tetrahydro-5-hydroxy-3-pyridine-carboxylic acid, 9CI   T-70065

$C_6H_9NO_3$ M 143.142

(±)-*form*

B,HBr: [86447-28-1]. Cryst. (AcOH). Mp 236-238° dec.

Allan, R.D. *et al, Aust. J. Chem.*, 1983, **36**, 601 (*synth*)

Tetrahydroimidazo[4,5-d]imidazole-2,5(1H,3H)-dione, 9CI, 8CI   T-70066

Updated Entry replacing T-00878

2,4,6,8-Tetraazabicyclooctane-3,7-dione. Hexahydro-2,6-dioxoimidazolo[4,5-d]imidazole. Glycoluril. Acetylenediurene. Glyoxaldiurene

[496-46-8]

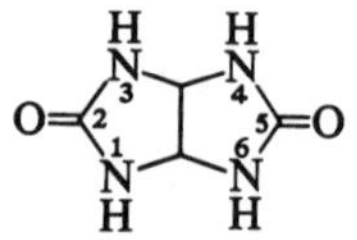

$C_4H_6N_4O_2$ M 142.117

Needles or prisms (H_2O). Mp 300° dec.

1,4-Di-Me: [17754-76-6].
  $C_6H_{10}N_4O_2$ M 170.171
  Needles (H_2O). Mp 285-287°.

1,6-Di-Me: [3720-98-7].
  $C_6H_{10}N_4O_2$ M 170.171
  Needles (H_2O). Mp 230-232°.

1,3,4,6-Tetra-Me: [10095-06-4].
  $C_8H_{14}N_4O_2$ M 198.224
  Needles (EtOH). Mp 217°.

1,3,4,6-Tetra-Ac: [10543-60-9].
  $C_{12}H_{14}N_4O_6$ M 310.266
  Needles (EtOH). Mp 236-238° dec.

1,3,4,6-Tetraisopropyl:
  $C_{16}H_{30}N_4O_2$ M 310.439
  Solid (diisopropyl ether). Mp 150°.

Petersen, H., *Synthesis*, 1973, 243 (*rev, synth*)

Hase, C. *et al, Justus Liebigs Ann. Chem.*, 1975, 95 (*synth*)

Koppes, W.M. *et al, J. Org. Chem.*, 1987, **52**, 1113 (*deriv, synth, pmr*)

2,3,5,6-Tetrahydro-s-indacene-1,7-dione, 9CI   T-70067

s-*Hydrindacene-1,7-dione*

[100939-69-3]

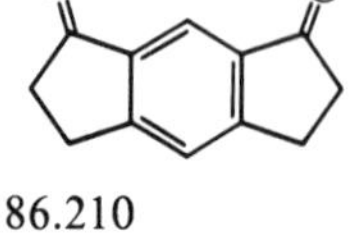

$C_{12}H_{10}O_2$ M 186.210

Cryst. solid (EtOH).

Seeger, D.E. *et al, J. Am. Chem. Soc.*, 1986, **108**, 1251 (*synth, ir, pmr, cmr*)

2,3,6,7-Tetrahydro-s-indacene-1,5-dione, 9CI   T-70068

s-*Hydrindacene-1,5-dione*

[81423-50-9]

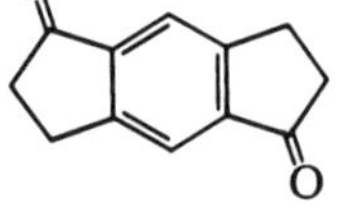

$C_{12}H_{10}O_2$ M 186.210

Mp 132.5-134°.

Seeger, D.E. *et al, J. Am. Chem. Soc.*, 1986, **108**, 1251 (*synth, ir, pmr, cmr*)

2,3,6,7-Tetrahydro-as-indacene-1,8-dione, 9CI   T-70069

as-*Hydrindacene-1,8-dione*

[17833-66-8]

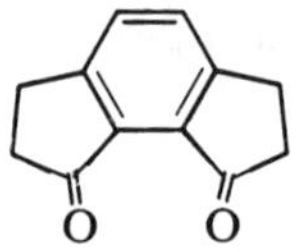

$C_{12}H_{10}O_2$ M 186.210

Mp 206-208.5°.

Wasserman, H.H. *et al, J. Am. Chem. Soc.*, 1962, **84**, 4611 (*synth, pmr*)

Seeger, D.E. *et al, J. Am. Chem. Soc.*, 1986, **108**, 1251 (*synth*)

3,5,6,7-Tetrahydro-4H-inden-4-one, 9CI   T-70070

[108835-41-2]

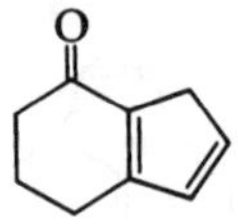

$C_9H_{10}O$ M 134.177

Thermolabile air-sensitive yellow oil.

Paquette, L.A. *et al, J. Org. Chem.*, 1987, **52**, 3250 (*synth, ir, pmr*)

1,5,6,8a-Tetrahydro-3(2H)-indolizinone   T-70071

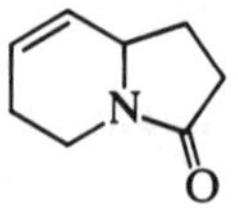

$C_8H_{11}NO$ M 137.181

(±)-*form* [71779-54-9]

Liq. Bp_1 110°.

Flann, C. *et al, J. Am. Chem. Soc.*, 1987, **109**, 6097 (*synth, ir, pmr, cmr*)

3a,4,7,7a-Tetrahydro-1(3H)-isobenzofuranone, 9CI   T-70072

Δ^4-*Tetrahydrophthalide. 3-Oxabicyclo[4.3.0]non-7-en-2-one*

[77210-71-0]

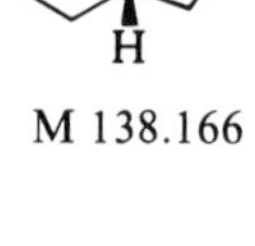

(3aR,7aS)-*form*

$C_8H_{10}O_2$ M 138.166

(3aR,7aS)-form [82442-62-4]
(−)-cis-*form*
Bp$_{0.15}$ 90°. [α]$_D^{25}$ −67.1° (c, 1 in CHCl$_3$), [α]$_D^{25}$ −71.0° (c, 0.9 in MeOH).
(3aRS,7aSR)-form [2744-05-0]
(±)-cis-*form*
Bp$_{0.1}$ 85°.
(3aRS,7aRS)-form [14679-41-5]
(±)-trans-*form*
Cryst. (C$_6$H$_6$/hexane). Mp 102.0-102.8°.

Bloomfield, J.J. et al, J. Org. Chem., 1967, **32**, 3919 (synth, ir, pmr)
Jakovac, I.J. et al, J. Org. Chem., 1982, **104**, 4659 (synth, pmr, ir, abs config)

4,5,6,7-Tetrahydro-1-methylene-1H-indene, 9CI   T-70073

[107575-50-8]

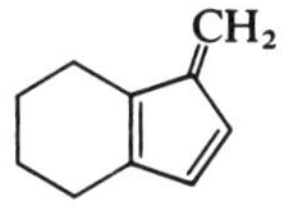

C$_{10}$H$_{12}$ M 132.205
Yellow liq. Unstable.

Kostermans, G.B.M. et al, J. Am. Chem. Soc., 1987, **109**, 2855.

Tetrahydro-6-methylene-2H-pyran-2-one   T-70074
6-Methylidenetetrahydro-2-pyrone

[5636-66-8]

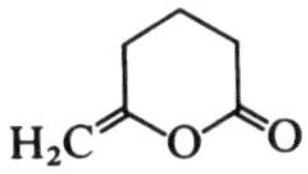

C$_6$H$_8$O$_2$ M 112.128
Oil.

Krafft, G.A. et al, J. Am. Chem. Soc., 1981, **103**, 5459 (synth, pmr)
Chan, D.M.T. et al, J. Am. Chem. Soc., 1987, **109**, 6385 (synth)

1,2,3,3a-Tetrahydropentalene   T-70075
Bicyclo[3.3.0]octa-1,3-diene

[50874-54-9]

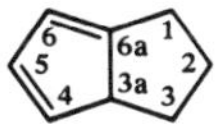

C$_8$H$_{10}$ M 106.167
Not well characterised, known as intermed.

Pauli, A. et al, Chem. Ber., 1987, **120**, 1617 (bibl)

1,2,3,4-Tetrahydropentalene   T-70076
Bicyclo[3.3.0]octa-1(5),2-diene

[50874-54-9]
C$_8$H$_{10}$ M 106.167
Liq., purified by glc.

de Meijere, A. et al, Chem. Ber., 1977, **110**, 2561 (synth, pmr)
Pauli, A. et al, Chem. Ber., 1987, **120**, 1617 (pmr, cmr, bibl)

1,2,3,5-Tetrahydropentalene   T-70077
Bicyclo[3.3.0]octa-1,4-diene

C$_8$H$_{10}$ M 106.167
Liq. Bp ~150-200°. Turns yellow and resinifies on standing.

Süs, O. et al, Justus Liebigs Ann. Chem., 1956, **593**, 91.
de Meijere, A. et al, Chem. Ber., 1977, **110**, 2561 (synth, pmr)
Pauli, A. et al, Chem. Ber., 1987, **120**, 1617 (pmr, cmr, bibl)

1,2,4,5-Tetrahydropentalene   T-70078
Bicyclo[3.3.0]octa-1,5-diene

C$_8$H$_{10}$ M 106.167
Little-known.

Pauli, A. et al, Chem. Ber., 1987, **120**, 1617 (bibl)

1,2,4,6a-Tetrahydropentalene   T-70079
Bicyclo[3.3.0]octa-1,6-diene

C$_8$H$_{10}$ M 106.167
Liq.

Murahashi, S. et al, Bull. Chem. Soc. Jpn., 1974, **47**, 2420 (synth, uv, pmr)
Eisenhuth, L. et al, Tetrahedron Lett., 1976, 1265 (synth, uv)
Freeman, P.K. et al, J. Org. Chem., 1977, **42**, 3882 (synth)
Pauli, A. et al, Chem. Ber., 1987, **120**, 1617 (bibl)

1,2,6,6a-Tetrahydropentalene   T-70080
Bicyclo[3.3.0]octa-1,7-diene

C$_8$H$_{10}$ M 106.167
Unstable liq.

Fickes, G.N. et al, J. Org. Chem., 1972, **37**, 2898 (synth, pmr, uv)
Murahashi, S. et al, Tetrahedron Lett., 1973, 4197; 1977, 3281 (synth, pmr)
Eisenhuth, L. et al, Tetrahedron Lett., 1976, 1265 (synth, pmr, uv)
Pauli, A. et al, Chem. Ber., 1987, **120**, 1617 (bibl)

1,3a,4,6a-Tetrahydropentalene   T-70081
Bicyclo[3.3.0]octa-2,6-diene

[41527-66-6]
C$_8$H$_{10}$ M 106.167
Liq.

Doering, W. von E. et al, Tetrahedron, 1963, **19**, 715 (synth, ir)
Baldwin, J.E. et al, J. Am. Chem. Soc., 1971, **93**, 3969 (synth)
Pauli, A. et al, Chem. Ber., 1987, **120**, 1617 (bibl)

1,3a,6,6a-Tetrahydropentalene   T-70082
Bicyclo[3.3.0]octa-2,7-diene

[41164-14-1]
C$_8$H$_{10}$ M 106.167
Liq.

Doering, W. von E. et al, Tetrahedron, 1963, **19**, 715 (synth, ir)
Freeman, P.K. et al, J. Org. Chem., 1973, **38**, 3635 (synth, pmr)
Pauli, A. et al, Chem. Ber., 1987, **120**, 1617 (bibl)

3,3a,6,6a-Tetrahydro-1(2H)-pentalenone, 9CI   T-70083

Updated Entry replacing B-60099
Bicyclo[3.3.0]oct-6-en-2-one
[35200-11-4]

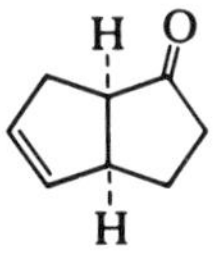

C$_8$H$_{10}$O   M 122.166

(3aRS,6aSR)-form [97563-40-1]
(±)-cis-*form*
Intermed. for polycyclopentenoid nat. prods. No phys.
   props. given.

Hudlicky, T. *et al, J. Org. Chem.*, 1980, **45**, 5020 (*synth, ir, pmr,*
   cmr, ms)
Hashimoto, S. *et al, Tetrahedron Lett.*, 1986, **27**, 2885 (*synth*)

4,5,6,6a-Tetrahydro-2(1H)-pentalenone,   T-70084
9CI

Updated Entry replacing B-60098
Bicyclo[3.3.0]oct-1-en-3-one
[72200-41-0]

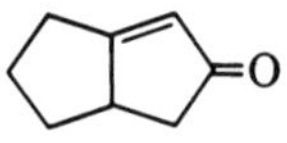

$C_8H_{10}O$ M 122.166
Sweet-smelling liq.

Begley, M.J. *et al, Tetrahedron Lett.*, 1981, 257 (*synth, spectra*)
Klipa, D.K. *et al, J. Org. Chem.*, 1981, **46**, 2815 (*synth*)
Davidsen, S.K. *et al, Synthesis*, 1986, 842 (*synth, pmr, cmr, ir*)

Tetrahydro-2H-pyran-2-thione   T-70085
δ-*Thionovalerolactone*
[72037-37-7]

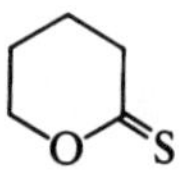

C_5H_8OS M 116.178

Kaloustian, M.K. *et al, J. Am. Chem. Soc.*, 1986, **108**, 6683
   (*synth, nmr, ir*)

Tetrahydro-1H,5H-pyrazolo[1,2-a]-   T-70086
pyrazole

1,5-Diazabicyclo[3.3.0]octane. 8-Azapyrrolizidine. 1,2-
Trimethylenepyrazolidine. Perhydro-3a,6a-
diazapentalene
[5397-67-1]

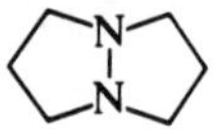

$C_6H_{12}N_2$ M 112.174
Liq. Mp 1.5-2.5°. Bp 173°, Bp_{26} 74-75°, Bp_{15} 59-62°.
Monopicrate: Cryst. (EtOH). Mp 161°.

Buhle, E.L. *et al, J. Am. Chem. Soc.*, 1943, **65**, 29 (*synth*)
Stetter, H. *et al, Chem. Ber.*, 1965, **98**, 3228 (*synth*)

Tetrahydro-1H,5H-pyrazolo[1,2-a]-   T-70087
pyrazole-1,5-dione, 9CI

1,5-Diazabicyclo[3.3.0]octane-4,8-dione
[19720-72-0]

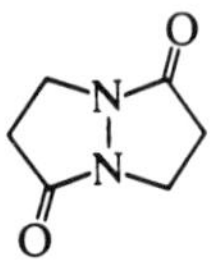

$C_6H_8N_2O_2$ M 140.141
Mp 161°. Bp_3 140-150°.

Stetter, H. *et al, Chem. Ber.*, 1965, **98**, 3228 (*synth*)
Nelsen, S.F. *et al, J. Am. Chem. Soc.*, 1987, **109**, 5724 (*props,*
   cryst struct)

Tetrahydro-1H-pyrazolo[1,2-a]pyridazine-   T-70088
1,3(2H)-dione, 9CI

[69386-75-0]

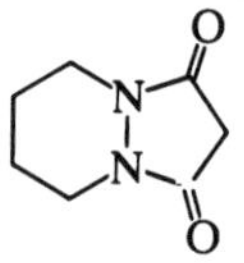

$C_7H_{10}N_2O_2$ M 154.168
Plates (EtOAc/cyclohexane). Mp 119.5°.

Lawton, G. *et al, J. Chem. Soc., Perkin Trans. 1*, 1987, 877
   (*synth, ir, pmr*)

1,2,3,6-Tetrahydropyridine, 9CI   T-70089

Updated Entry replacing T-01082
Δ^3-*Piperideine*
[694-05-3]

C_5H_9N M 83.133
Bp 113.5°. Free base polymerises readily.
▷UT8226000.

Picrate: Mp 160°.
B,HCl: [18513-79-6]. Cryst. (EtOH/Et_2O). Mp 192-
   193°.
1-Ph: [87682-63-1].
   $C_{11}H_{13}N$ M 159.230
   Pale-yellow solid. Bp_4 75°.
1-Ph; B, (COOH)$_2$: [109720-43-6]. Cryst. (2-propanol).
   Mp 143°.
1-Propyl: [53385-78-7].
   $C_8H_{15}N$ M 125.213
   Liq.
1-Propyl; B, (COOH)$_2$: [109720-41-4]. Cryst. (2-
   propanol). Mp 177-179°.

Renshaw, R.R. *et al, J. Am. Chem. Soc.*, 1938, **60**, 745 (*synth*)
Morlacchi, F. *et al, Ann. Chim. (Rome)*, 1967, **57**, 1456 (*synth*)
Ferles, M. *et al, Adv. Heterocycles* Chem., 1970, **12**, 43 (*rev*)
Flann, C. *et al, J. Am. Chem. Soc.*, 1987, **109**, 6097 (*derivs,*
   synth, ir, pmr, cmr, ms)

2,3,4,5-Tetrahydropyridine, 9CI   T-70090

Updated Entry replacing T-01083
Δ^1-*Piperideine*
[505-18-0]

C_5H_9N M 83.133
Unstable, readily trimerises.
Trimer: see Tripiperidein, T-04365

Quick, J. *et al, Synthesis*, 1976, 745 (*synth*)
Org. Synth., 1977, **56**, 118 (*synth*)
Quick, J. *et al, J. Org. Chem.*, 1979, **44**, 573 (*synth*)
Bock, H. *et al, Chem. Ber.*, 1987, **120**, 1961 (*synth, pe, bibl*)
Guittemin, J-C. *et al, Tetrahedron*, 1988, **44**, 4447 (*synth, pmr*)

6,7,8,9-Tetrahydro-5*H*-pyrido[2,3-*b*]-azepine   T-70091

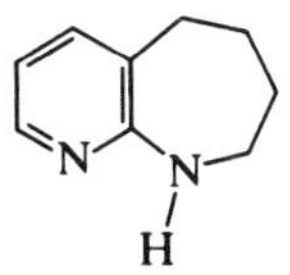

$C_9H_{12}N_2$ M 148.207
Cryst. (Et$_2$O). Mp 43-44°. Bp$_{14}$ 145-151°, Bp$_2$ 125-127°.

Hawes, E.M. *et al, J. Heterocycl. Chem.*, 1973, **10**, 39 (*synth, ir, pmr*)
Jössang-Yanagida, A. *et al, J. Heterocycl. Chem.*, 1978, **15**, 249 (*synth*)

6,7,8,9-Tetrahydro-5*H*-pyrido[3,2-*b*]-azepine, 9CI   T-70092

[67203-48-9]

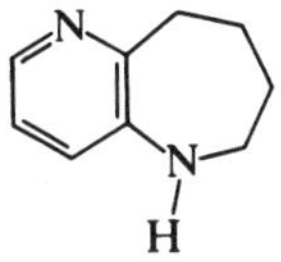

$C_9H_{12}N_2$ M 148.207
Cryst. (hexane). Mp 80.5-81°. Bp$_{14}$ 158°.

Jössang-Yanagida, A. *et al, J. Heterocycl. Chem.*, 1978, **15**, 249 (*synth, ir, pmr*)

1,2,3,4-Tetrahydroquinoline   T-70093

Updated Entry replacing T-01095
[635-46-1]

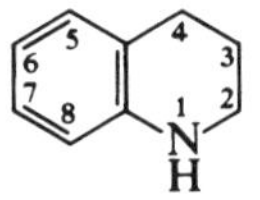

$C_9H_{11}N$ M 133.193
Bp$_{755}$ 249-250°.
B,HCl: [2739-17-5]. Prisms (EtOH). Mp 180-181°.
Picrate: Mp 142-143°.
N-Ac: [4169-19-1].
   $C_{11}H_{13}NO$ M 175.230
   Bp 295°.
N-Me: see *1,2,3,4-Tetrahydro-1-methylquinoline, T-00947*
N-Benzyl: [21863-32-1].
   $C_{16}H_{17}N$ M 223.317
   Needles (EtOH). Mp 36-37°. Bp$_{38}$ 218-222°.

Skita, A. *et al, Ber.*, 1912, **45**, 3594.
Kikugawa, Y., *Chem. Pharm. Bull.*, 1973, **21**, 1914.
Ginos, J.Z. *et al, J. Org. Chem.*, 1975, **40**, 1191 (*pmr*)
Nose, A. *et al, Chem. Pharm. Bull.*, 1984, **32**, 2421 (*synth*)
Mwahashi, S.-I. *et al, Tetrahedron Lett.*, 1987, **28**, 77 (*synth*)

3,3′,4,4′-Tetrahydro-2,2′-spirobi[2*H*-1-benzopyran], 9CI   T-70094

2,2′-Spirobichroman. Tetrahydrodiphenospiropyran
[5732-37-6]

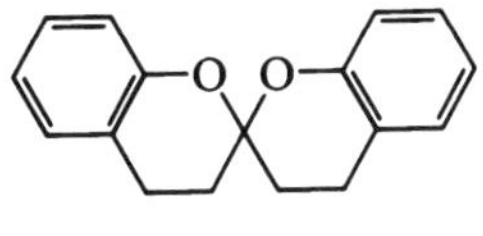

$C_{17}H_{16}O_2$ M 252.312

(±)-*form*

Cryst. (EtOH). Mp 108°.

Mora, P.T. *et al, J. Am. Chem. Soc.*, 1950, **72**, 3009 (*synth*)
Tanaka, T. *et al, Heterocycles*, 1987, **25**, 463 (*synth*)

5,6,8,13-Tetrahydro-1,7,9,11-tetrahy-droxy-8,13-dioxo-3-(2-oxopropyl)-benzo[*a*]naphthacene-2-carboxylic acid, 9CI   T-70095

KS 619-1
[103370-21-4]

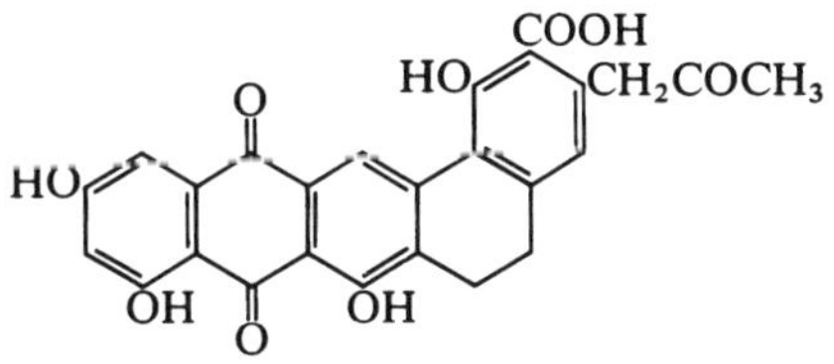

$C_{26}H_{18}O_9$ M 474.423
Anthraquinone-type antibiotic. Isol. from *Streptomyces californicus*. Inhibitor of Ca$^{2⊕}$ and cyclic nucleotide phosphodiesterase. Dark-red powder. Mp 198-200° dec.

Matsuda, Y. *et al, J. Antibiot.*, 1987, **40**, 1104, 1111 (*isol, struct, props*)

Tetrahydro-1,3,4,6-tetranitroimidazo[4,5-*d*]imidazole-2,5-(1*H*,3*H*)-dione, 9CI   T-70096

Tetranitroglycoluril. Sorguyl
[55510-03-7]

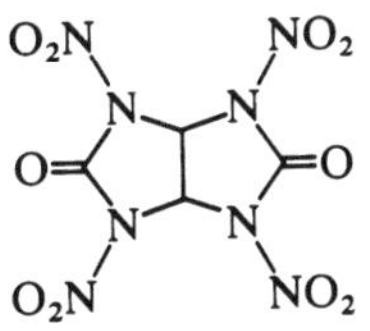

$C_4H_2N_8O_{10}$ M 322.107
High explosive.

Ger. Pat., 2 435 651, (*1975*); *CA*, **83**, 30483r (*synth*)
Boileau, J. *et al, Propellants, Explos. Pyrotech.*, 1985, **10**, 118.

Tetrahydro-1,2,4,5-tetrazine-3,6-dione, 9CI   T-70097

Updated Entry replacing T-01122
Hexahydro-s-tetrazine-3,6-dione. p-Urazine. Dicarba-mide. Diurea
[624-40-8]

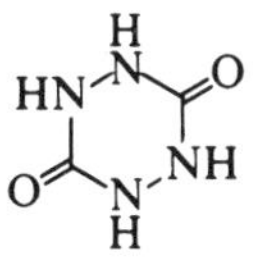

$C_2H_4N_4O_2$ M 116.079
Reports of the prepn. of this compd. prior to 1982 were incorrect. Cryst. Spar. sol. hot H$_2$O. Mp 266°. Does not melt, rearranges <200°.
▷Potentially explosive
1,5-Di-Ac: [81930-26-9].
   $C_6H_8N_4O_4$ M 200.154
   Cryst. Mp 187-189° dec.
1,2,4,5-Tetra-Ac:
   $C_{10}H_{12}N_4O_6$ M 284.228

Cryst. (DMF/EtOH). Mp 186-190° dec.
1,2,4,5-Tetrabenzyl: [81930-23-6].
$C_{30}H_{28}N_4O_2$ M 476.577
Prisms (MeOH). Mp 116-117°.

Neugebauer, F.A. *et al, Justus Liebigs Ann. Chem.*, 1982, 387.

Tetrahydro-1,3,5,7-tetrazocine-2,6(1H,3H)-dione, 9CI   T-70098

Dimethylenediurea
[51137-13-4]

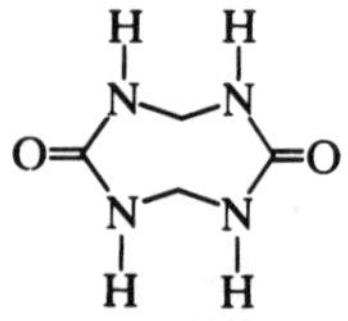

O=C...

$C_4H_8N_4O_2$ M 144.133

N-*Tetra-Me:* *1,3,5,7-Tetramethyltetrahydro-1,3,5,7-tetrazocine-2,6(1H,3H)dione.*
Tetramethyldimethylenediurea.
$C_8H_{16}N_4O_2$ M 200.240
Condensation prod. of 1,3-dimethylurea with
formaldehyde. Needles (EtOH). Mp 258°.

Kadowaki, H., *Bull. Chem. Soc. Jpn.*, 1936, **11**, 248 (*synth*)
Ebisuno, T. *et al, Bull. Chem. Soc. Jpn.*, 1988, **61**, 2191 (*cryst struct*)

Tetrahydro-2-thiophenecarboxylic acid, 9CI   T-70099

Updated Entry replacing T-50131
2-Thiolanecarboxylic acid
[19418-11-2]

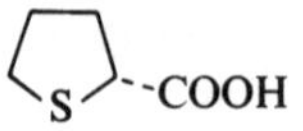

S —COOH   (*R*)-form

$C_5H_8O_2S$ M 132.177

(*R*)-form
$[\alpha]_D^{20}$ −14.2° (c, 13 in EtOH).
(±)-form
Mp 51°. Bp$_{0.4}$ 110-112°.
Chloride:
C_5H_7ClOS M 150.623
Liq. Bp$_{25}$ 113-114°.
Nitrile: 2-Cyanotetrahydrothiophene.
C_5H_7NS M 113.177
Liq. Bp$_{15}$ 110-112°.
Nitrile, 1,1-dioxide:
$C_5H_7NO_2S$ M 145.176
Cryst. (C$_6$H$_6$). Mp 47-48°.
Anilide:
$C_{11}H_{15}NOS$ M 209.306
Cryst. (EtOH). Mp 125-126°.

Putochin, N.I. *et al, Zh. Obshch. Khim.*, 1948, **18**, 1866 (*synth*)
Červinka, O. *et al, Collect. Czech. Chem. Commun.*, 1986, **51**, 405 (*resoln, abs config*)
Wróbel, J.T. *et al, Synthesis*, 1987, 452 (*synth, deriv, pmr, cmr, ir, ms*)

1,2,3,4-Tetrahydro-2-thioxo-5H-1,4-benzodiazepin-5-one, 9CI   T-70100

[111424-67-0]

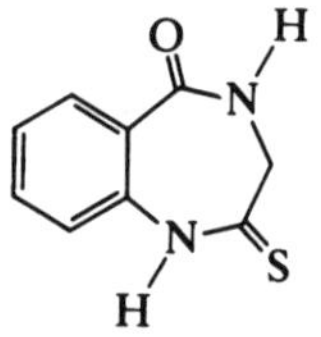

$C_9H_8N_2OS$ M 192.235
Pale-yellow cryst. (DMF/MeOH). Mp 277-283°.

Goel, O.P. *et al, Synthesis*, 1987, 162 (*synth, ir, pmr*)

Tetrahydro-1,6,7-tris(methylene)-1H,4H-3a,6a-propanopentalene, 9CI   T-70101

2,8,9-Trimethylene[3.3.3]propellane
[58461-87-3]

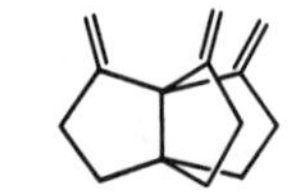

$C_{14}H_{18}$ M 186.296

Drouin, J. *et al, Tetrahedron Lett.*, 1975, 4053 (*synth, pmr*)
Paquette, L.A. *et al, Tetrahedron Lett.*, 1981, 291 (*synth, pmr*)

4,4′,5,5′-Tetrahydroxy-1,1′-binaphthyl   T-70102

1,1′-Binaphthalene-4,4′,5,5′-tetrol

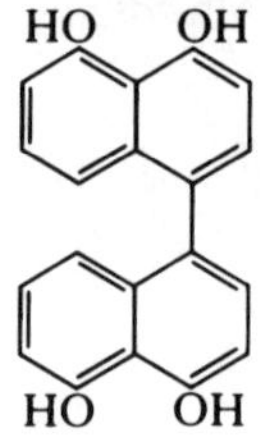

$C_{20}H_{14}O_4$ M 318.328
Metab. of *Daldinia concentrica*. Pale-yellow cryst. Mp
225-230° dec. Easily oxidized.
Tetra-Ac:
$C_{28}H_{22}O_8$ M 486.477
Cryst. Mp 245°.

Allport, D.C. *et al, J. Chem. Soc.*, 1958, 4090; 1960, 654 (*isol, synth, biosynth*)

3,4,6,8-Tetrahydroxy-11,13-clerodadien-15,16-olide   T-70103

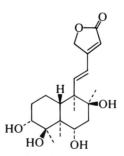

$C_{20}H_{30}O_6$ M 366.453

(3α,4β,6α,8β,11E)-form

O^6-Benzoyl: **Scuterivulactone D.**
$C_{27}H_{34}O_7$ M 470.561
Constit of crude Chinese drug "Ban Zhi Lian"
(*Scutellaria rivularis*). Needles (Me$_2$CO). Mp 260-
262°. [α]$_D^{26}$ +57.5° (MeOH).

Kizu, H. *et al, Chem. Pharm. Bull.*, 1987, **35**, 1656.

2',5,7,8-Tetrahydroxyflavanone   T-70104

Updated Entry replacing T-50148
$C_{15}H_{12}O_6$ M 288.256

(S)-form

7-Me ether: *2',5,8-Trihydroxy-7-methoxyflavanone.*
$C_{16}H_{14}O_6$ M 302.283
Constit. of rhizomes of *Iris spuria*. Cream needles
(EtOAc/pet. ether). Mp 204-205°.
2',8-Di-Me ether: [100079-33-2]. *5,7-Dihydroxy-2',8-
dimethoxyflavanone.*
$C_{17}H_{16}O_6$ M 316.310
Constit. of *Scutellaria discolor*. Needles (MeOH). Mp
208° dec.
2',5,8-Tri-Me ether: [100079-34-3]. *7-Hydroxy-2',5,8-
trimethoxyflavanone.*
$C_{18}H_{18}O_6$ M 330.337
Constit. of *S. discolor*. Needles (MeOH). Mp 214°
dec.

Tomimori, T. *et al, Chem. Pharm. Bull.*, 1985, **33**, 4457 (*isol*)
Shawl, A.S. *et al, Phytochemistry*, 1988, **27**, 1559 (*isol*)

3',4',5,7-Tetrahydroxyflavanone   T-70105

Updated Entry replacing T-01223
*2-(3,4-Dihydroxyphenyl)-2,3-dihydro-5,7-dihydroxy-
4H-1-benzopyran, 9CI.* **Eriodictyol**
[4049-38-1]
$C_{15}H_{12}O_6$ M 288.256

(S)-form [552-58-9]

Occurs in leaves of *Eriodictyon californicum, E.
glutinosum* and in several other spp. Plates (EtOH).
Mp 267°.
Ac: Mp 137°.
7-O-β-D-Glucoside: **Pyracanthoside.**
$C_{21}H_{22}O_{11}$ M 450.398
Constit. of the bark of *Malus communis*. Also from
Crataegus pyracantha and *Pyrus malus*. Mp 175-
177°, Mp 215-216°. [α]$^{.°}$ −40(60%° EtOH aq.).
3',5-Di-O-β-D-Glucoside:
$C_{27}H_{32}O_{16}$ M 612.540
Constit. of the leaves of *Crataegus phenophyrum*.
Cryst. (EtOH aq.). Mp 194-196°.
3',7-Di-Me ether, O^4-(3,6-dimethyl-2-heptenyl): **4'-
(3,6-Dimethyl-2-heptenyloxy-5-hydroxy-3',7-
dimethoxyflavanone.**
$C_{26}H_{32}O_6$ M 440.535
Minor constit. of peel of *Melicope sarcococca*. Cryst.
(MeOH). Mp 163-164°.

Power, F.B., *J. Chem. Soc.*, 1907, **91**, 895 (*struct*)
Tutin, F., *J. Chem. Soc.*, 1910, **97**, 2054 (*struct*)
Shinoda, J., *CA*, 1929, **23**, 4210 (*synth, struct*)
Brune, W. *et al, Aust. J. Chem.*, 1965, **18**, 1649 (*deriv*)
Hörhammer, L. *et al, Tetrahedron Lett.*, 1966, 5133 (*deriv*)
Kowalewski, Z., *Planta Med.*, 1971, **19**, 311 (*deriv*)
Wagner, H. *et al, Tetrahedron Lett.*, 1976, 1799 (*nmr*)
Rani, I., *Indian J. Chem., Sect. B*, 1986, **25**, 1251 (*synth*)
Babber, S. *et al, Indian J. Chem., Sect. B*, 1987, **26**, 797 (*synth*)

3',4',6,7-Tetrahydroxyflavanone   T-70106

$C_{15}H_{12}O_6$ M 288.256
6,7-Di-Me ether, 3',4'-methylene ether: *6,7-Dimethoxy-
3',4'-methylenedioxyflavanone.*
$C_{18}H_{16}O_6$ M 328.321
Constit. of *Macaranga indica*. Cryst. (C$_6$H$_6$/pet.
ether). Mp 178-179°. [α]$_D$ −54°.

Sultana, S. *et al, Indian J. Chem., Sect. B*, 1987, **26**, 801.

2',5,6,6'-Tetrahydroxyflavone   T-70107

*2-(2,6-Dihydroxyphenyl)-5,6-dihydroxy-4H-1-benzo-
pyran-4-one, 9CI*
[14813-19-5]
$C_{15}H_{10}O_6$ M 286.240
Formerly assigned the 2',5,6,7-tetrahydroxy struct.
2',6,6'-Tri-Me ether: [14813-20-8]. *5-Hydroxy-2',6,6'-
trimethoxyflavone.* **Zapotinin.**
$C_{18}H_{16}O_6$ M 328.321
Constit. of *Casimiroa edulis*. Yellow needles (MeOH).
Mp 224-225°.
Tetra-Me ether: [14813-19-5]. *2',5,6,6'-Tetramethoxy-
flavone.* **Zapotin.**
$C_{19}H_{18}O_6$ M 342.348
Constit. of the bark of *C. edulis*. Prisms (Me$_2$CO or
MeOH). Mp 150-151°.

Kincl, F.A. *et al, J. Chem. Soc.*, 1956, 4163 (*isol*)
Iriarte, J. *et al, J. Chem. Soc.*, 1956, 4170 (*isol*)
Sondheimer, F., *Tetrahedron*, 1960, **9**, 139 (*isol*)
Garrett, P.J. *et al, Tetrahedron*, 1967, **23**, 2413 (*struct*)
Dreyer, D.L., *Tetrahedron*, 1967, **23**, 4607 (*struct*)
Farkas, L. *et al, Tetrahedron Lett.*, 1968, 3993 (*synth*)

3',4',5,6-Tetrahydroxyflavone   T-70108

$C_{15}H_{10}O_6$ M 286.240
Tetra-Me ether: [94303-31-8]. *3',4',5,6-
Tetramethoxyflavone.*
$C_{19}H_{18}O_6$ M 342.348
Constit. of *Ardisia floribunda*. Cryst.
(CHCl$_3$/MeOH). Mp 171-172°.

Parveen, M. *et al, Indian J. Chem., Sect. B*, 1987, **26**, 894 (*isol*)

2,3,4,5-Tetrahydroxy-9H-fluoren-9-one   T-70109

Updated Entry replacing D-50439

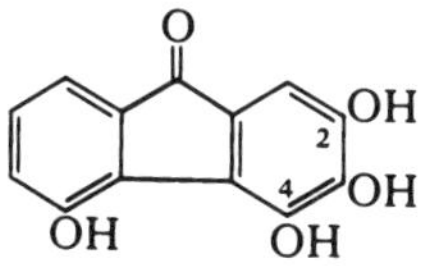

$C_{13}H_8O_5$ M 244.203
2,4-Di-Me ether: [98665-64-6]. *3,5-Dihydroxy-2,4-di-
methoxy-9H-fluoren-9-one.* **Dengibsinin.**
$C_{15}H_{12}O_5$ M 272.257
Constit. of *Dendrobium gibsonii*. Orange-red needles
(CHCl$_3$/pet. ether). Mp 227°. Struct. revised in 1987.

Talapatra, S.K. *et al, Tetrahedron*, 1985, **41**, 2765 (*isol*)
Sargent, M.V., *J. Chem. Soc., Perkin Trans. 1*, 1987, 2553
(*synth, struct*)

2',5,7,8-Tetrahydroxyisoflavone   T-70110

$C_{15}H_{10}O_6$ M 286.240

7,8-Di-Me ether: [115713-23-0]. *2',5-Dihydroxy-7,8-dimethoxyisoflavone.*
$C_{17}H_{14}O_6$ M 314.294
Constit. of rhizomes of *Iris spuria.* Yellow needles (MeOH). Mp 164-166°.

Shawl, A.S. *et al, Phytochemistry,* 1988, **27**, 1559.

3',4',6,7-Tetrahydroxyisoflavone   T-70111

Updated Entry replacing T-01381
3-(3,4-Dihydroxyphenyl)-6,7-dihydroxy-4H-1-benzopyran-4-one, 9CI
3',4'-Methylenedioxy, 6,7-di-Me ether: [55303-89-4]. *6,7-Dimethoxy-3',4'-methylenedioxyisoflavone.* **Milletenin C.**
$C_{18}H_{14}O_6$ M 326.305
Constit. of *Milletia ovalifolia.* Needles (EtOH/dioxan). Mp 252-253° (245-246°).
3',4'-Methylenedioxy, 6-Me ether: [38965-66-1]. *7-Hydroxy-6-methoxy-3',4'-methylenedioxyisoflavone.* **Fujikinetin.**
$C_{17}H_{12}O_6$ M 312.278
Constit. of bark of *Cladrastis platycarpa.* Mp 279-281° dec.
3',4'-Methylenedioxy, 6-Me ether, 7-glucoside: [38965-67-2]. **Fujikinin.**
$C_{23}H_{22}O_{11}$ M 474.420
Constit. of *C. platycarpa.* Mp 155-160°.
Tetra-Me ether: [24126-93-0]. *3',4',6,7-Tetramethoxyisoflavone.*
$C_{19}H_{18}O_6$ M 342.348
Constit. of *Pterodon apparicioi, P. pubescens* and *Cordyla africana.* Cryst. (MeOH). Mp 187-188°.

Campbell, R.V.M. *et al, J. Chem. Soc. (C),* 1969, 1787 (*isol, struct, synth*)
Braz Filho, R. *et al, Phytochemistry,* 1971, **10**, 2835 (*isol*)
Imamura, H. *et al, Mokuzai Gakkaishi,* 1972, **18**, 325 (*isol, derivs*)
Khan, H. *et al, Tetrahedron,* 1974, **30**, 2811 (*Milletenin C*)
de Almeida, M.E.L., *Phytochemistry,* 1975, **14**, 2716 (*isol*)
Antus, S. *et al, Chem. Ber.,* 1975, **108**, 3883 (*synth*)

3',4',7,8-Tetrahydroxyisoflavone   T-70112

Updated Entry replacing T-30094
$C_{15}H_{10}O_6$ M 286.240
4'-Me ether: 3',7,8-Trihydroxy-4'-methoxyisoflavone.
$C_{16}H_{12}O_6$ M 300.267
Constit. of *Xanthocercis zambesiaca.*
8-Me ether, 3',4'-methylene ether: 7-Hydroxy-8-methoxy-3',4'-methylenedioxyisoflavone. **Maximaisoflavone E.**
$C_{17}H_{12}O_6$ M 312.278
From *T. maxima.* Cryst. (EtOH). Mp 268-270°.
3',4'-Di-Me ether, 7,8-methylene ether: 3',4'-Dimethoxy-7,8-methylenedioxyisoflavone. **Maximaisoflavone D.**
$C_{18}H_{14}O_6$ M 326.305
Constit. of *Tephrosia maxima.* Cryst. (CHCl₃/MeOH). Mp 223-224°.

Rao, E.V. *et al, Phytochemistry,* 1984, **23**, 1493.
Jain, A.C. *et al, Indian J. Chem., Sect. B,* 1987, **26**, 1143 (*synth*)
Parmar, V.S. *et al, Phytochemistry,* 1987, **26**, 484 (*synth*)
Rani, I., *Indian J. Chem., Sect. B,* 1987, **26**, 1080 (*synth*)
Bezuidenhout, S.C. *et al, Phytochemistry,* 1988, **27**, 2329.

4',6,7,8-Tetrahydroxyisoflavone   T-70113

$C_{15}H_{10}O_6$ M 286.240

6-Me ether: 4',7,8-Trihydroxy-6-methoxyisoflavone.
$C_{16}H_{12}O_6$ M 300.267
Constit. of *Wisteria brachybotrys.* Powder (MeOH). Mp >300°.
6-Me ether, tri-Ac: Needles (MeOH). Mp 233-235°.

Kaneko, M. *et al, Phytochemistry,* 1988, **27**, 267.

ent-1β,7β,14S,20-Tetrahydroxy-16-kauren-15-one   T-70114
Kamebakaurin

[73981-34-7]

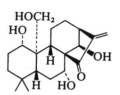

$C_{20}H_{30}O_5$ M 350.454
Constit. of *Rabdosia umbrosa.* Needles (MeOH). Mp 232-234°. $[\alpha]_D^{21.5} -107°$ (c, 1 in MeOH).

Takeda, Y. *et al, J. Chem. Soc., Perkin Trans. 1,* 1987, 2403.

ent-7β,11α,14S,20-Tetrahydroxy-16-kauren-15-one   T-70115
Kamebakaurinin

[74144-54-0]
$C_{20}H_{30}O_5$ M 350.454
Constit. of *Rabdosia umbrosa.* Needles (CH₂Cl₂/MeOH). Mp 267-269°. $[\alpha]_D^{25} -101°$ (c, 0.92 in Py).

Takeda, Y. *et al, J. Chem. Soc., Perkin Trans. 1,* 1987, 2403.

2,3,22,23-Tetrahydroxy-24-methyl-6-cholestanone   T-70116

Updated Entry replacing T-30096
2,3,22,23-Tetrahydroxy-6-ergostanone, 9CI

$C_{28}H_{48}O_5$ M 464.684
(2α,3α,5α,22R,23R,24S)-form [80736-41-0]
Castasterone
Constit. of chestnut insect galls. Shows plant hormone activity. Cryst. (MeCN aq.). Mp 259-261°.
6-Deoxo: [87833-54-3]. *6-Deoxocastasterone.*
$C_{28}H_{50}O_4$ M 450.701
Constit. of *Phaseolus vulgaris.*
2-Deoxy: [87734-68-7]. *3α,22R,23R-Trihydroxy-24S-methyl-5α-cholestan-6-one. 2-Deoxycastasterone.* **Typhasterol.**
$C_{28}H_{48}O_4$ M 448.685
Constit. of *Typha latifolia.* Plant growth promoting substance. Cryst. (CHCl₃/MeOH). Mp 227-230°.

Yokota, T. *et al, Tetrahedron Lett.,* 1982, **23**, 1275 (*isol*)

Anastasia, M. *et al*, *J. Chem. Soc., Perkin Trans. 1*, 1983, 383 (*synth*)
Schneider, J.A. *et al*, *Tetrahedron Lett.*, 1983, **24**, 3859 (*deriv*)
Yokota, T. *et al*, *Agric. Biol. Chem.*, 1983, **47**, 2149 (*isol*)
Adam, G. *et al*, *Phytochemistry*, 1986, **25**, 1787 (*rev*)
Singh, H. *et al*, *Indian J. Chem., Sect. B*, 1986, **25**, 989 (*rev*)
Takatsuto, S., *J. Chem. Soc., Perkin Trans. 1*, 1986, 1833 (*synth*)
Aburatani, M. *et al*, *Agric. Biol. Chem.*, 1987, **51**, 1909 (*synth*)

3,5,6,7-Tetrahydroxy-8-methylflavone   T-70117

3,5,6,7-Tetrahydroxy-8-methyl-2-phenyl-4H-1-benzo-pyran-4-one. **Isoplatanin**

[114567-38-3]

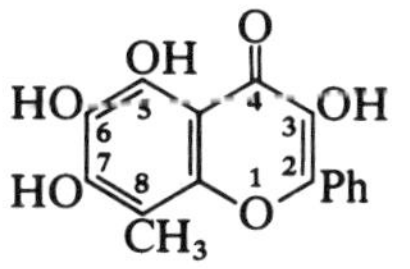

$C_{16}H_{12}O_6$ M 300.267

Constit. of *Platanus acerifolia*. Dark red-orange amorph. powder.

Kaouadji, M. *et al*, *J. Nat. Prod.*, 1988, **51**, 353.

3,5,7,8-Tetrahydroxy-6-methylflavone   T-70118

3,5,7,8-Tetrahydroxy-6-methyl-2-phenyl-4H-1-benzo-pyran-4-one. **Platanin**

$C_{16}H_{12}O_6$ M 300.267

Constit. of *Platanus acerifolia*. Amorph. powder.

Kaouadji, M. *et al*, *J. Nat. Prod.*, 1988, **51**, 353.

2,3,19,23-Tetrahydroxy-12-oleanen-28-oic acid   T-70119

Updated Entry replacing T-10100

$C_{30}H_{48}O_6$ M 504.706

(2α,3β,19α)-form [58880-25-4]
Arjungenin
Constit. of *Terminalia arjuna*. Cryst. (EtOH aq.). Mp 293-294° dec. [α]$_D$ +29° (c, 2.6 in EtOH).

O^{28}-β-D-*Glucopyranosyl ester:* [62319-70-4]. **Arjunglu-coside I.** Isol. from *T. arjuna*. Needles (MeOH aq.). Mp 231°. [α]$_D$ +17° (EtOH).

3,28-O-Bis-β-D-glucopyranosyl: [82843-99-0].
$C_{42}H_{68}O_{16}$ M 828.990
Constit. of stem bark of *Symplocos spicata*. Needles (butanol aq.). Mp 246-249°. [α]$_D$ +7.4° (c, 1.4 in MeOH).

Honda, T. *et al*, *Bull. Chem. Soc. Jpn.*, 1976, **49**, 3213 (*isol*)
Higuchi, R. *et al*, *Phytochemistry*, 1982, **21**, 907 (*isol*)

1-(2,3,4,5-Tetrahydroxyphenyl)-3-(2,4,5-trihydroxyphenyl)-1,3-propanedione   T-70120

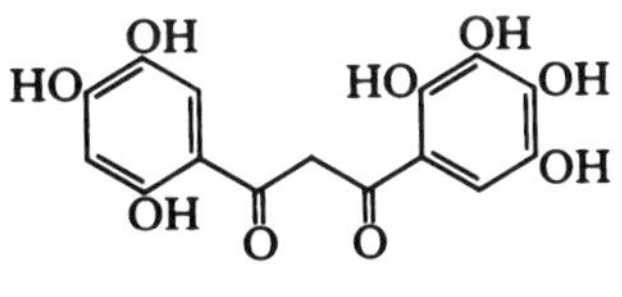

$C_{15}H_{12}O_9$ M 336.254

Hepta-Me ether: 1-(2,3,4,5-Tetramethoxyphenyl)-3-(2,4,5-trimethoxyphenyl)-1,3-propanedione.
$C_{22}H_{26}O_9$ M 434.442

Constit. of *Polygonum nepalense*. Viscous oil.

Rathone, A. *et al*, *J. Nat. Prod.*, 1987, **50**, 357.

1,2,3,19-Tetrahydroxy-12-ursen-28-oic acid   T-70121

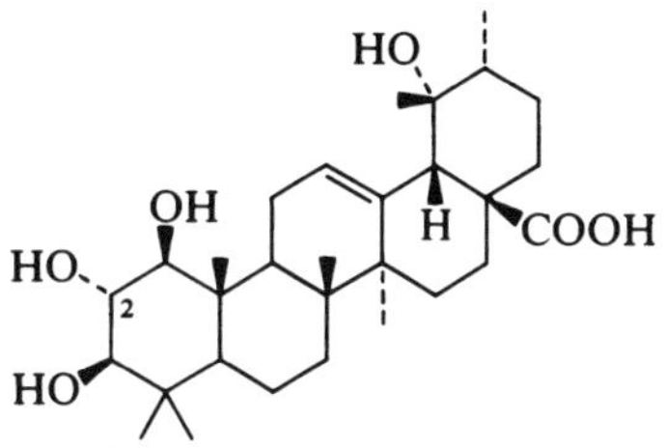

$C_{30}H_{48}O_6$ M 504.706

(1β,2α,3β,19α)-form
Constit. of *Agrimonia pilosa*.
(1β,2β,3β,19α)-form
Constit. of *A. pilosa*.

Kouno, I. *et al*, *Phytochemistry*, 1988, **27**, 297.

1,2,4,5-Tetrahydroxyxanthone   T-70122

1,2,4,5-Tetrahydroxy-9H-xanthen-9-one
$C_{13}H_8O_6$ M 260.203
1-Me ether: [115713-07-0]. *2,4,5-Trihydroxy-1-methoxyxanthone.* **BR-Xanthone B.**
$C_{14}H_{10}O_6$ M 274.229
Constit. of the dry fruit hulls of *Garcinia mangostana*. Greenish-yellow cryst. Mp 308-310°.

Balasubramanian, K. *et al*, *Phytochemistry*, 1988, **27**, 1552.

1,3,5,8-Tetrahydroxyxanthone   T-70123

Updated Entry replacing T-20103
1,3,5,8-Tetrahydroxy-9H-xanthen-9-one, 9CI.
Demethylbellidifolin

[2980-32-7]

$C_{13}H_8O_6$ M 260.203
Constit. of *Gentiana bellidifolia* and of *Swertia* spp. Cryst. (EtOH). Mp 293-295° (315-320°).

8-Glucoside: [54954-12-0].
$C_{19}H_{18}O_{11}$ M 422.345
Constit. of the leaves of *G. campestris*. Cryst. (MeOH). Mp 241°.

3-Me ether: [2798-25-6]. *1,5,8-Trihydroxy-3-methoxyxanthone.* **Bellidifolin.**
$C_{14}H_{16}O_6$ M 280.277
Constit. of *G. bellidifolia*. Yellow needles (Me$_2$CO or EtOH). Mp 270-271°.

3-Me ether, 8-glucoside: [23445-00-3].
$C_{20}H_{20}O_{11}$ M 436.371
Constit. of *G. campestris*. Cryst. (MeOH). Mp 199°.

5-Me ether: [552-00-1]. *1,3,8-Trihydroxy-5-methoxyxanthone.* **Isobellidifolin.** Isol. from *G. bellidifolia*. Cryst. (EtOAc). Mp 263-264°.

3,5-Di-Me ether: [521-65-3]. *1,8-Dihydroxy-3,5-dimethoxyxanthone.* **Swerchirin.**
$C_{15}H_{12}O_6$ M 288.256
Pigment from *S.* and *G.* spp. and *Frasera* spp. Yellow needles. Mp 185-186°.

3,5-Di-Me ether, 8-O-β-D-glucopyranoside:
$C_{21}H_{22}O_{12}$ M 466.398
Constit. of *S. speciosa*. Yellow cryst. Mp 249-252°.

5,8-Di-Me ether: [114567-42-9]. *1,3-Dihydroxy-5,8-dimethoxyxanthone.*
$C_{15}H_{12}O_6$ M 288.256

Constit. of *S. petiolata*. Cryst. (EtOAc/hexane). Mp 192-194.5°.

1,3,5-Tri-Me ether: [5557-27-7]. *8-Hydroxy-1,3,5-trimethoxyxanthone.*
$C_{16}H_{14}O_6$ M 302.283
Constit. of *S. bimaculata*. Yellow needles. Mp 215-216°.

Tetra-Me ether: [54954-13-1]. *1,3,5,8-Tetramethoxyxanthone.*
$C_{17}H_{16}O_6$ M 316.310
Mp 209-210°.

Markham, K.R., *Tetrahedron*, 1964, **20**, 991; 1965, **21**, 1449 (*isol, struct*)
Stout, G.H. *et al*, *Tetrahedron*, 1969, **25**, 1947, 1961 (*deriv*)
Kaldas, M. *et al*, *Helv. Chim. Acta*, 1974, **57**, 2557 (*isol, struct*)
Ghosal, S. *et al*, *Phytochemistry*, 1975, **14**, 1393, 2671 (*rev*)
Hostettman-Kaldas, M. *et al*, *Phytochemistry*, 1978, **17**, 2083 (*rev*)
Khetwal, K.S. *et al*, *Phytochemistry*, 1988, **27**, 1910 (*deriv*)
Kulanthaivel, P. *et al*, *J. Nat. Prod.*, 1988, **51**, 379 (*deriv*)

Tetrakis(2-bromoethyl)methane   T-70124

1,5-Dibromo-3,3-bis(2-bromoethyl)pentane
[5794-98-9]

$$C(CH_2CH_2Br)_4$$

$C_9H_{16}Br_4$ M 443.841
Long needles (EtOH). Mp 181-182°.

Rice, L.M. *et al*, *J. Org. Chem.*, 1967, **32**, 1966 (*synth*)
Newkome, G.R. *et al*, *J. Org. Chem.*, 1987, **52**, 5480 (*synth, pmr, cmr*)

1,3,5,7-Tetrakis(diethylamino)pyrrolo[3,4-f]isoindole   T-70125

N,N,N′,N′,N″,N″,N‴,N‴-Octaethylbenzo[1,2-c:4,5-c′]dipyrrole-1,3,5,7-tetramine, 9CI. 1,3,5,7-Tetrakis-(diethylamino)-2,6-diaza-s-indacene
[110568-96-2]

$C_{26}H_{42}N_6$ M 438.658
Violet rods with green-gold lustre. Mp >120° dec.

Closs, F. *et al*, *Angew. Chem., Int. Ed. Engl.*, 1987, **26**, 1037 (*synth, uv, pmr, cryst struct*)

5,8,14,17-Tetrakis(dimethylamino)[3.3]-paracyclophene   T-70126

[105562-32-1]

$C_{26}H_{40}N_4$ M 408.629
Electron donor compd. Cryst. (Et₂O).

Staab, H.A. *et al*, *Chem. Ber.*, 1987, **120**, 269 (*synth,pmr, cryst struct*)

1,3,4,6-Tetrakis(dimethylamino)-pyrrolo[3,4-c]pyrrole   T-70127

N,N,N′,N′,N″,N″,N‴,N‴-Octamethylpyrrolo[3,4-c]-pyrroleteramine, 9CI. Tetrakis(dimethylamino)-2,5-diazapentalene
[114491-81-5]

$C_{14}H_{24}N_6$ M 276.384
Orange-red needles. Mp 160-162°. Other derivs. of this ring-system prepd.

Closs, F. *et al*, *Angew. Chem., Int. Ed. Engl.*, 1988, **27**, 842 (*synth, uv, pmr, cmr, cryst struct*)

Tetrakis(diphenylmethylene)cyclobutane   T-70128

Octaphenyl[4]radialene
[51445-93-3]

$C_{56}H_{40}$ M 712.932
Iyoda, M. *et al*, *J. Am. Chem. Soc.*, 1986, **108**, 5371 (*synth, nmr, cryst struct*)

1,3,5,7-Tetrakis[2-D_3-trishomocubanyl-1,3-butadiynyl]adamantane   T-70129

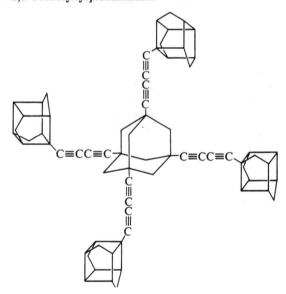

$C_{70}H_{68}$ M 909.307
(+)-*form*
Claimed to be the first resolvable molecule having T symmetry. Silky cryst. (pentane). Mp >350°. $[\alpha]_D^{27}$ +65.3° (c, 0.101 in CHCl₃).

Naemura, K. *et al*, *Tetrahedron*, 1986, **42**, 1763 (*synth*)

Tetramercaptotetrathiafulvalene   T-70130
2-(4,5-Dimercapto-1,3-dithiol-2-ylidene)-4,5-dimer-
capto-1,3-dithiole, 9CI

$C_6H_4S_8$ M 332.578
Tetrakis(S-Me): [51501-77-0].
  $C_{10}H_{12}S_8$ M 388.685
  Pale-yellow needles (MeCN). Mp 94.5-96.0°.
Tetrakis (S-Ph): [108666-51-9].
  $C_{30}H_{20}S_8$ M 636.968
  Red-orange solid (EtOAc/hexane). Mp 167-169°.

Moses, P.R. *et al, J. Am. Chem. Soc.*, 1974, **96**, 945 (*synth, ms, pmr, props*)
Hsu, S.-Y. *et al, J. Org. Chem.*, 1987, **52**, 3444 (*synth, pmr, ir*)

2,2,5,5-Tetramethylbicyclo[4.1.0]hept-1(6)-   T-70131
en-7-one
[54283-14-6]

$C_{11}H_{16}O$ M 164.247
V. stable cyclopropenone. Mp 89-89.5°.

Suda, M. *et al, J. Chem. Soc., Chem. Commun.*, 1974, 504 (*synth, ir, uv, cmr*)
Sander, W. *et al, Angew. Chem., Int. Ed. Engl.*, 1988, **27**, 398.

3,3,6,6-Tetramethylcyclohexyne   T-70132
[37494-11-4]

$C_{10}H_{16}$ M 136.236
Obt. in Ar matrix. Dimerises at 45 K.

Sander, W. *et al, Angew. Chem., Int. Ed. Engl.*, 1988, **27**, 398 (*synth, ir*)

2,2,3,3-Tetramethylcyclopropanecarboxal-   T-70133
dehyde, 9CI
[16340-68-4]

$C_8H_{14}O$ M 126.198
Bp_{60} 82°, Bp_{13} 54°. Unstable.

Gunther, H. *et al, Chem. Ber.*, 1971, **104**, 3898.
Baeckstrom, P. *et al, Tetrahedron*, 1978, **34**, 3331 (*synth*)
Laurie, D. *et al, Tetrahedron*, 1986, **42**, 1035 (*synth, pmr*)

2,2,3,3-Tetramethylcyclopropanemethanol,   T-70134
9CI
[2415-96-5]

$C_8H_{16}O$ M 128.214
Bp_{12} 72°.
Trimethylsilyl ether: Bp_{20} 90°.

Gunther, H. *et al, Chem. Ber.* 1971, **104**, 3898.
Laurie, D. *et al, Tetrahedron*, 1986, **42**, 1035 (*synth, pmr, deriv*)

3,4,7,11-Tetramethyl-1,3,6,10-dodecte-   T-70135
traene
Updated Entry replacing T-01491
Homofarnesene

$C_{16}H_{26}$ M 218.381
(3Z,6E)-form [79383-34-9]
  Trail pheromone of *Solenopsis invicta.*
(3Z,6Z)-form [79383-33-8]
  Trail pheromone of *S. invicta.*

van der Meer, R.K. *et al, Tetrahedron Lett.*, 1981, 1651 (*isol, struct*)
Alvarez, F.M. *et al, Tetrahedron*, 1987, **43**, 2897 (*synth*)

2,3,5,6-Tetramethylenebicyclo[2.2.1]-   T-70136
heptan-7-one
7-Oxa[2.2.1]hericene

$C_{11}H_{10}O$ M 158.199
Cryst. (pentane). Mp 56-57°.

Rubello, A. *et al, Helv. Chim. Acta*, 1988, **71**, 1268 (*synth, uv, pmr*)

3,7,11,15-Tetramethyl-2,6,10,13,15-hexa-   T-70137
decapentaen-1-ol, 9CI
[116988-16-0]

$C_{20}H_{32}O$ M 288.472
Constit. of *Bifurcaria bifurcata*. Oil.

Semmak, L. *et al, Phytochemistry*, 1988, **27**, 2347.

7,11,15,19-Tetramethyl-3-methylene-2-(3,7,11-trimethyl-2-dodecenyl)-1-eicosanol, 9CI   T-70138

13-Hydroxymethyl-2,6,10,18,22,26,30-heptamethyl-14-methylene-10-hentriacontene

[91914-33-9]

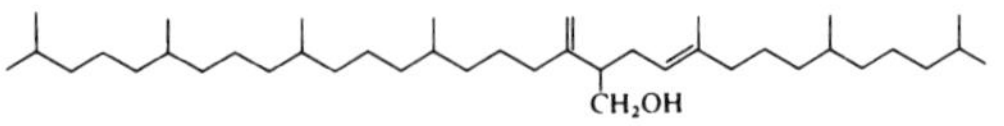

$C_{40}H_{78}O$ M 575.056

Constit. of *Elodea canadensis*. Oil. $[\alpha]_D$ −31° (CHCl₃).

Mangoni, L. *et al, Tetrahedron Lett.*, 1984, **25**, 2597.

2,5,7,7-Tetramethyloctanal   T-70139

$(H_3C)_3CCH_2CH(CH_3)CH_2CH_2CH(CH_3)CHO$

$C_{12}H_{24}O$ M 184.321

New type of odorant with interesting props. Liq. $Bp_{0.3}$ 68°. Mixt. of diastereoisomers.

Gramlick, W.R. *et al, Justus Liebigs Ann. Chem.*, 1988, 487 (*synth, ir, pmr, cmr, ms*)

2,5,8,11-Tetramethylperylene   T-70140

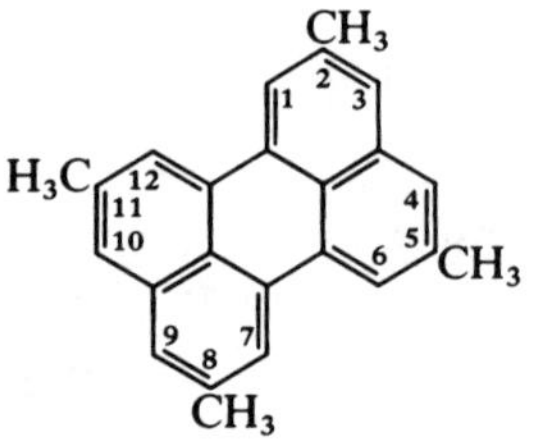

$C_{24}H_{20}$ M 308.422

Cryst. (C₆H₆). Mp 319-320°.

Michel, P. *et al, Synthesis*, 1988, 894 (*synth, pmr, cmr*)

3,4,9,10-Tetramethylperylene   T-70141

$C_{24}H_{20}$ M 308.422

Cryst. (C₆H₆). Mp 314-315°.

Michel, P. *et al, Synthesis*, 1988, 894 (*synth, pmr, cmr*)

2,4,6,7-Tetramethylpteridine, 9CI   T-70142

[19899-62-8]

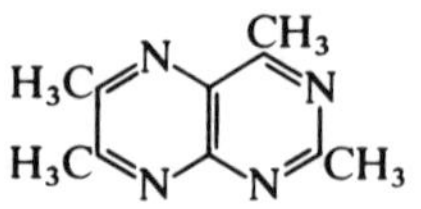

$C_{10}H_{12}N_4$ M 188.232

Pale-yellow needles (pet. ether). Mp 124-125°.

Albert, A. *et al, J. Chem. Soc.* (*C*), 1968, 2292 (*synth*)
Spanget-Larsen, J. *et al, Chem. Ber.*, 1986, **119**, 1275 (*pe*)

5,5,9,9-Tetramethyl-5*H*,9*H*-quino[3,2,1-*de*]acridine, 9CI   T-70143

[52066-62-3]

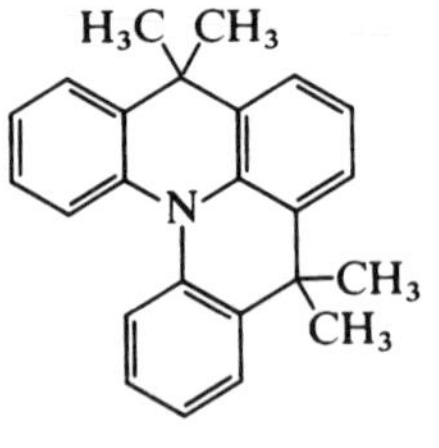

$C_{24}H_{23}N$ M 325.452

Contains a flattened planar sp³ N atom. Cryst. (2-propanol or by subl.). Mp 196-197°.

Hellwinkel, D. *et al, Chem. Ber.*, 1974, **107**, 616 (*synth, pmr*)
Fox, J.L. *et al, Org. Prep. Proced. Int.*, 1985, **17**, 169 (*synth, pmr*)

3,4,8,9-Tetraoxa-1,6-diazabicyclo[4.4.2]-dodecane, 9CI   T-70144

Hexamethylene diperoxide diamine

[112204-35-0]

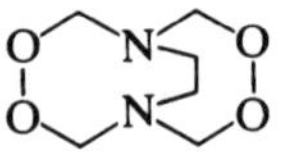

$C_6H_{12}N_2O_4$ M 176.172

Cryst. (pentyl acetate).

Fourkas, J.T. *et al, Acta Crystallogr., Sect. C*, 1987, **43**, 2160 (*synth, cryst struct*)

1,4,10,13-Tetraoxa-7,16-diazacyclooctadecane, 9CI   T-70145

Cryptand 22

[23978-55-4]

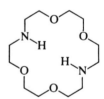

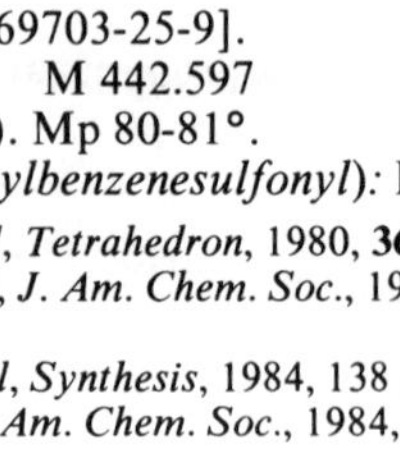

$C_{12}H_{26}N_2O_4$ M 262.348

Complexing agent.

N,N′-*Dibenzyl*: [69703-25-9].
    $C_{26}H_{38}N_2O_4$ M 442.597
    Cryst. (hexane). Mp 80-81°.

N,N′-*Bis(4-methylbenzenesulfonyl)*: Mp 164-165°.

Bradshaw, J.S. *et al, Tetrahedron*, 1980, **36**, 461 (*rev*)
Foerster, H.G. *et al, J. Am. Chem. Soc.*, 1980, **102**, 6984 (*N nmr*)
Bogatskii, A.V. *et al, Synthesis*, 1984, 138 (*synth*)
Gatto, V.J. *et al, J. Am. Chem. Soc.*, 1984, **106**, 8240 (*synth*)

Tetraoxaporphycene   T-70146
21,22,23,24-Tetraoxapentacyclo[16.2.1.1²,⁵.1⁸,¹¹.1¹²,¹⁵]-tetracosa-1,3,5,7,9,1,13,15,17,19-decaene, 9CI
[113088-07-6]

$C_{20}H_{12}O_4$ M 316.312
Violet leaflets (Et₂O/hexane). Mp 270-271°. Stable to isoln. in air. Forms a stable aromatic dication.

Vogel, E. *et al, Angew. Chem., Int. Ed. Engl.*, 1988, **27**, 411 (*synth, ir, uv, pmr, cmr, cryst struct*)

Tetraoxaporphyrinogen   T-70147
21,22,23,24-Tetraoxapentacyclo[16.2.1.1³,⁶.1⁸,¹¹.1¹³,¹⁶]-tetracoṣa-3,5,8,10,13,15,18,20-octaene, 9CI
[281-97-0]

$C_{20}H_{16}O_4$ M 320.344
Needles (EtOH or by subl.). Mp 158-159°. Stable towards O₂, can be oxid. to dication with HNO₃.

Haas, W. *et al, Angew. Chem., Int. Ed. Engl.*, 1988, **27**, 409 (*synth, ir, pmr, cmr*)

2,2,4,4-Tetraphenyl-3-butenal   T-70148
α-(2,2-Diphenylethenyl)-α-phenylbenzeneacetaldehyde, 9CI
[41326-74-3]

$$Ph_2C{=}CHCPh_2CHO$$

$C_{28}H_{22}O$ M 374.481
Cryst. (hexane). Mp 141-142.5°.

Zimmerman, H.E. *et al, J. Am. Chem. Soc.*, 1973, **95**, 2155 (*synth, pmr, ir, uv*)

1,2,3,4-Tetraphenylcyclopentadiene   T-70149
Updated Entry replacing T-01672
1,1′,1″,1‴-(1,3-Cyclopentadiene-1,2,3,4-tetrayl)-tetrakisbenzene, 9CI
[15570-45-3]

$C_{29}H_{22}$ M 370.493
Needles (EtOH/C₆H₆). Mp 177-178°.

Ziegler, K. *et al, Justus Liebigs Ann. Chem.*, 1925, **445**, 266 (*synth*)
Sonntag, N.O.V. *et al, J. Am. Chem. Soc.*, 1953, **75**, 2283 (*synth, uv*)
Cava, M.P. *et al, J. Org. Chem.*, 1964, **34**, 3641 (*synth*)
Evrard, G. *et al, Acta Crystallogr., Sect. B*, 1971, **27**, 661 (*struct*)

Bandara, H.M.N. *et al, Tetrahedron*, 1974, **30**, 2587 (*synth*)
Schumann, H. *et al, J. Organomet. Chem.*, 1987, **330**, 347 (*ir*)
Schumann, H. *et al, Chem. Ber.*, 1988, **121**, 207 (*raman, pmr, cmr, ms*)

1,1,1′,1′-Tetraphenyldimethylamine, 8CI   T-70150
N-(Diphenylmethyl)-α-phenylbenzenemethanamine, 9CI.
Dibenzhydrylamine. Bis(diphenylmethyl)amine
[5350-71-0]

$$Ph_2CHNHCHPh_2$$

$C_{26}H_{23}N$ M 349.474
Cryst. (EtOH/CHCl₃). Mp 138.5-140.5°. Resistant to conventional nitrosation.

N-*Nitroso:* [100461-65-2].
   $C_{26}H_{22}N_2O$ M 378.473
   Cryst. Mp 96-97°.

Vejdělek, Z.J. *et al, Collect. Czech. Chem. Commun.*, 1950, **15**, 671 (*synth*)
Hauser, C.R. *et al, J. Am. Chem. Soc.*, 1956, **78**, 1653 (*synth*)
Baldwin, J.R. *et al, Org. Prep. Proced. Int.*, 1985, **17**, 261 (*synth, pmr*)

9,11,20,22-Tetraphenyltetrabenzo[a,c,l,n]-pentacene-10,21-dione   T-70151
[110321-62-5]

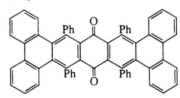

$C_{62}H_{36}O_2$ M 812.965
Bright-yellow prisms (toluene). Mp >360° (460-461°).

Pascal, R.A. *et al, Tetrahedron Lett.*, 1987, **28**, 293 (*synth, struct*)

1,3,4,6-Tetraphenylthieno[3,4-c]thiophene-5-S^{IV}, 8CI   T-70152
[23386-93-8]

$C_{30}H_{20}S_2$ M 444.608
The parent heterocycle currently (1988) unknown. Purple needles. Mp 245-248°.

Cava, M.P. *et al, J. Am. Chem. Soc.*, 1973, **95**, 2561 (*synth, uv, pmr*)
Potts, K.T. *et al, J. Am. Chem. Soc.*, 1974, **96**, 4268 (*synth*)
Horner, C.J. *et al, Tetrahedron Lett.*, 1976, 2581 (*synth*)
Gleiter, R. *et al, J. Org. Chem.*, 1978, **43**, 3893 (*pe, esr, uv*)
Miller, K.J. *et al, J. Am. Chem. Soc.*, 1983, **105**, 1705.

3,3,5,5-Tetraphenyl-1,2,4-trithiolane T-70153
[4198-12-3]

$C_{26}H_{20}S_3$ M 428.624
Cryst. Mp 128°.

Staudinger, H. *et al, Ber.*, 1928, **61**, 1836 (*synth*)
Schönberg, A. *et al, Chem. Ber.*, 1968, **101**, 725 (*synth*)
Linn, W.J. *et al, J. Org. Chem.*, 1969, **34**, 2146 (*synth*)
Campbell, M.M. *et al, J. Chem. Soc., Perkin Trans. 1*, 1973, 2862 (*synth*)
Huisgen, R. *et al, J. Am. Chem. Soc.*, 1987, **109**, 902 (*synth*)

1,3,5,7-Tetravinyladamantane T-70154
1,3,5,7-Tetraethenyltricyclo[3.3.1.1^{3,7}]decane, 9CI.
1,3,5,7-Tetraethenyladamantane
[84782-80-9]

$C_{18}H_{24}$ M 240.388
Bp$_{0.2}$ 98-100°.

Nakazaki, M. *et al, J. Chem. Soc., Chem. Commun.*, 1982, 1245 (*synth*)
Naemura, K. *et al, Tetrahedron*, 1986, **42**, 1763 (*synth, ir*)

Tetrazolo[5,1-a]phthalazine, 9CI T-70155
[234-82-2]

$C_8H_5N_5$ M 171.161
Cryst.

Haegel, K.D. *et al, J. Chromatogr.*, 1976, **126**, 517 (*synth, ms*)
Noda, A. *et al, Chem. Pharm. Bull.*, 1979, **27**, 1938, 2820.

Tetrazolo[1,5-b]quinoline-5,10-dione, 8CI T-70156
[33023-90-4]

$C_9H_4N_4O_2$ M 200.156
Mp 216° (violent dec.).

Moore, H.W. *et al, Tetrahedron Lett.*, 1971, 1621 (*synth, ir, uv, pmr*)

Teuchamaedryn A T-70157
[75443-75-3]

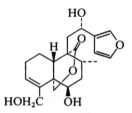

$C_{19}H_{20}O_5$ M 328.364
Isol. from aerial parts of *Teucrium chamaedrys*. Prisms (Et$_2$O/diisopropyl ether). Mp 159-160°. $[\alpha]_D^{20}$ +25°.

Papanov, G.Y. *et al, Z. Naturforsch., B*, 1980, **35**, 764.

Teulamifin B T-70158

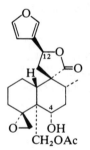

$C_{20}H_{26}O_6$ M 362.422
Constit. of *Teucrium lamiifolium* and *T. polium*. Cryst. (Me$_2$CO/pet. ether). Mp 207-209°. $[\alpha]_D^{20}$ −49.3° (c, 0.23 in Me$_2$CO).

Malakov, P.Y. *et al, Phytochemistry*, 1988, **27**, 1141.

Teupolin I T-70159
Updated Entry replacing T-60183
[72948-20-0]

$C_{22}H_{28}O_7$ M 404.459
Constit. of *Teucrium polium*. Cryst. (Et$_2$O/CHCl$_3$). Mp 211-213°. $[\alpha]_D^{20}$ +60° (c, 0.2 in CH$_2$Cl$_2$).
Ac: [67987-83-1]. **Montanin C**.
 $C_{24}H_{30}O_8$ M 446.496
 Constit. of *T. montanum*. Cryst. Mp 181-183°. $[\alpha]_D^{20}$ +8.4° (c, 0.262 in CHCl$_3$).
De-O-Ac, O^4-Ac: [72948-21-1]. **Teupolin II**.
 $C_{22}H_{28}O_7$ M 404.459
 Isol. from *T. polium*. Cryst. (Et$_2$O/Me$_2$CO). Mp 187-189°. $[\alpha]_D^{20}$ +52° (c, 0.2 in CH$_2$Cl$_2$).
De-O-Ac, O^4-Ac, 12-epimer: 12-Epiteupolin II.
 $C_{22}H_{28}O_7$ M 404.459
 Constit. of *T. lamifolium*. Cryst. (Me$_2$CO/pet. ether). Mp 173-175°. $[\alpha]_D^{20}$ +14° (c, 0.33 in Me$_2$CO).

Malakov, P.Y. *et al, Z. Naturforsch., B*, 1978, **33**, 789; 1979, **34**, 1570 (*isol*)

Gács-Baitz, E. *et al*, *Heterocycles*, 1982, **19**, 539 (*cmr*)
Fayos, J. *et al*, *J. Org. Chem.*, 1984, **49**, 1789 (*struct*)
Gács-Baitz, E. *et al*, *Phytochemistry*, 1987, **26**, 2110 (*struct, bibl*)
Boneva, I.M. *et al*, *Phytochemistry*, 1988, **27**, 295 (*deriv*)

2-Thiabicyclo[2.2.1]hept-5-ene   T-70160

Updated Entry replacing T-01847

[6841-59-4]

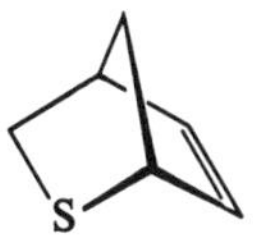

C₆H₈S   M 112.189
Cryst. by subl. or waxy solid. Mp 60-63°.
exo-*2-Oxide:*
    C₆H₈OS   M 128.189
    Pale-yellow oil.
endo-*2-Oxide:*
    C₆H₈OS   M 128.189
    Dark-brown oil. Thermally unstable, cont. 10% of *exo*-form.
2,2-Dioxide:
    C₆H₈O₂S   M 144.188
    Solid. Mp 55°.

Johnson, C.R. *et al*, *J. Org. Chem.*, 1969, **34**, 860.
Dice, D.R. *et al*, *Can. J. Chem.*, 1974, **52**, 3518.
Ohishi, J. *et al*, *J. Org. Chem.*, 1978, **43**, 4013 (*synth, nmr, ms*)
Block, E. *et al*, *J. Org. Chem.*, 1987, **52**, 809 (*oxides, synth, ir, pmr, cmr*)

2-Thiabicyclo[3.1.0]hex-3-ene, 9CI   T-70161

Homothiophene

[6573-95-1]

C₅H₆S   M 98.162
Bp₇₃₀ 130-132°, Bp₂₀ 30°. Polym. at r.t., stable at −80°.

Müller, E. *et al*, *Tetrahedron Lett.*, 1963, 1047 (*synth, uv, ir, pmr*)
Herges, R., *Synthesis*, 1986, 1059 (*synth*)

1,2,5-Thiadiazole-3,4-dicarboxaldehyde, 9CI   T-70162

3,4-Diformylthiadiazole

[85122-38-9]

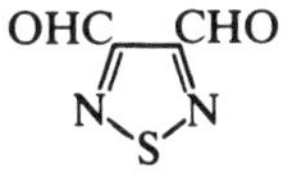

C₄H₂N₂O₂S   M 142.132
Mp 40-42°. Bp₅ 100°.

Dunn, P.J. *et al*, *J. Chem. Soc., Perkin Trans. 1*, 1987, 1579 (*synth, uv, ir, pmr, ms*)

[1,2,5]Thiadiazolo[3,4-d]pyridazine, 9CI   T-70163

[273-14-3]

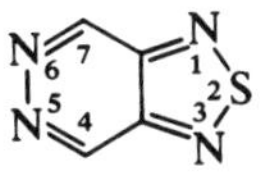

C₄H₂N₄S   M 138.147

Low-melting solid. Bp₅ 80°.

Dunn, P.J. *et al*, *J. Chem. Soc., Perkin Trans. 1*, 1987, 1579 (*synth, uv, pmr, ir, ms*)

[1,2,5]Thiadiazolo[3,4-d]pyridazin-4(5H)-one   T-70164

[33097-28-8]

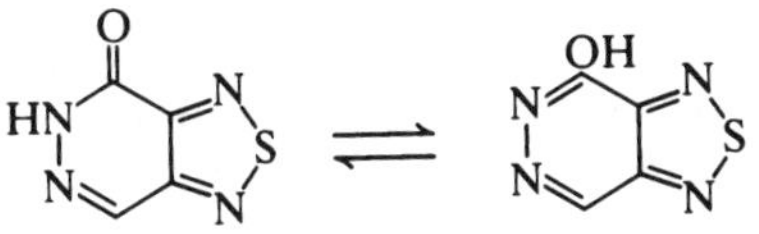

C₄H₂N₄OS   M 154.146
Exists mainly in oxo-form in solid state and in hydroxy-form in DMSO. Powder. Mp 223-225°.

Dunn, P.J. *et al*, *J. Chem. Soc., Perkin Trans. 1*, 1987, 1579 (*synth, uv, ir, pmr, ms*)

3-Thiatricyclo[2.2.1.0²,⁶]heptane, 9CI   T-70165

[53703-55-2]

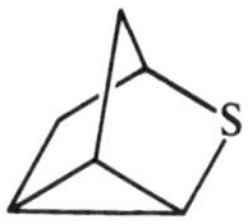

C₆H₈S   M 112.189
S,S-Dioxide: [112897-61-7].
    C₆H₈O₂S   M 144.188
    Mp 148°.

Kirmse, W. *et al*, *J. Chem. Soc., Chem. Commun.*, 1987, 709 (*synth, pmr*)

Thiazolo[4,5-b]pyridine   T-70166

[273-98-3]

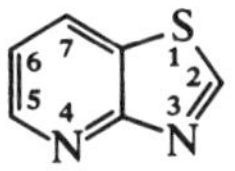

C₆H₄N₂S   M 136.171
Parent heterocycle apparently unknown, many substd. derivs. known.

Okafor, C.O., *Int. J. Sulfur Chem., Part B*, 1972, **7**, 121 (*rev*)

Thiazolo[4,5-c]pyridine   T-70167

[273-75-6]

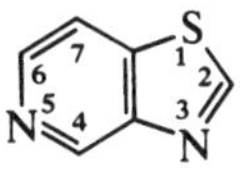

C₆H₄N₂S   M 136.171
Cryst. (Et₂O). Mp 105-106°.

Takahashi, T. *et al*, *Chem. Pharm. Bull.*, 1954, **2**, 196 (*synth*)
Ueda, K., *Chem. Pharm. Bull.*, 1956, **4**, 220 (*uv*)
Okafor, C.O., *Int. J. Sulfur Chem., Part B*, 1972, **7**, 121 (*rev*)
Attanasi, O. *et al*, *Tetrahedron*, 1976, **32**, 399 (*synth, ir*)

Thiazolo[5,4-*b*]pyridine   T-70168
Pyrido[3,2-d]thiazole
[273-84-7]

C₆H₄N₂S   M 136.171
Cryst. by subl. Mp 52°. Bp₁.₄ 71°.

Fridman, S.G. *et al, Zh. Obshch. Khim.*, 1956, **26**, 864 (*synth*)
Okafor, C.O., *Int. J. Sulfur Chem., Part B*, 1972, **7**, 121 (*rev*)
Stanovnik, B. *et al, Synthesis*, 1974, 120 (*synth*)

Thiazolo[5,4-*c*]pyridine   T-70169
[273-70-1]

C₆H₄N₂S   M 136.171
Prisms (pet. ether). Mp 105-106°.

Ueda, K. *et al, Chem. Pharm. Bull.*, 1956, **4**, 216, 220 (*synth, uv*)
Okafor, C.O., *Int. J. Sulfur Chem., Part B*, 1972, **7**, 121 (*rev*)

Thieno[3,4-*c*]thiophene-1,3,4,6-tetrathiol   T-70170
*1,3,4,6-Tetramercaptothieno[3,4-*c]thiophene

C₆H₄S₆   M 268.458
Parent compd. (and also the parent heterocycle)
currently unknown. All uncharged resonance
contributors contain hypervalent sulfur.

S-*Tetra-Et*: 1,3,4,6-Tetrakis(*ethylthio*)thieno[3,4-b]-
thiophene.
C₁₄H₂₀S₆   M 380.672
Red needles (MeOH). Mp 50.5-51°.
S-*Tetraisopropyl*: 1,3,4,6-Tetrakis(*isopropylthio*)-
*thieno[3,4-*c]thiophene.
C₁₈H₂₈S₆   M 436.779
Red needles (MeOH). Mp 116.5-117°.
S-*Tetra-tert-butyl*: 1,3,4,6-Tetrakis(tert-*butylthio*)-
*thieno[3,4-*c]thiophene.
C₂₂H₃₆S₆   M 492.886
Red needles (C₆H₆/hexane). Mp 199.5-200°.

Yoneda, S. *et al, J. Heterocycl. Chem.*, 1988, **25**, 559 (*synth, pmr, cmr, uv*)

5-(2-Thienyl)oxazole, 9CI   T-70171
2-(5-Oxazolyl)thiophene
[70380-70-0]

C₇H₅NOS   M 151.183
Pale-yellow oil. Bp₁₆ 108-110°, Bp₀.₃ 81-84°.

Saikachi, H. *et al, Chem. Pharm. Bull.*, 1979, **27**, 793 (*synth, uv, pmr*)

3-(3-Thienyl)-2-propenoic acid   T-70172
3-Thiopheneacrylic acid

C₇H₆O₂S   M 154.183
(*E*)-*form* [102696-71-9]
Mp 153-153.5°.
Et ester:
C₉H₁₀O₂S   M 182.237
d₄²⁰ 1.16. Bp₁₇ 154-155°.

Raich, W.J. *et al, J. Am. Chem. Soc.*, 1957, **79**, 3800 (*synth*)
Freeman, F. *et al, J. Am. Chem. Soc.*, 1986, **108**, 4506 (*props*)

3-(2-Thienyl)-2-propynoic acid   T-70173
2-Thienylpropiolic acid
[4843-44-1]

C₇H₄O₂S   M 152.167
Mp 135-136°.

Osbahr, A.J. *et al, J. Am. Chem. Soc.*, 1955, **77**, 1911 (*synth*)
Gosselck, J. *et al, Angew. Chem., Int. Ed. Engl.*, 1965, **4**, 1080 (*synth*)
Brown, E.J.D. *et al, J. Chem. Soc. (C)*, 1966, 1390 (*synth, ir*)
Wadsworth, D.H. *et al, J. Org. Chem.*, 1987, **52**, 3662 (*synth, ir, pmr, ms*)

2,2′-Thiobisbenzoic acid, 9CI   T-70174
*2,2′-Thiodibenzoic acid, 8CI. Diphenyl sulfide-2,2′-
dicarboxylic acid*
[22219-02-9]

C₁₄H₁₀O₄S   M 274.291
Dark-yellow needles. Mp 227-230°.

Di-Me ester: [49590-24-1].
C₁₆H₁₄O₄S   M 302.344
Colourless cryst. Mp 78-78.5° (84°).
Dichloride:
C₁₄H₈Cl₂O₂S   M 311.182
Light-yellow needles. Mp 100-102°.

Bihari, M. *et al, Adv. Mass Spectrom.*, 1978, **7**, 1362 (*ms*)
Hellwinkel, D. *et al, Chem. Ber.*, 1987, **120**, 1151 (*synth, pmr, bibl*)

4,4′-Thiobisbenzoic acid, 9CI   T-70175
*4,4′-Thiodibenzoic acid, 8CI. Diphenyl sulfide 4,4′-
dicarboxylic acid*
[4919-48-6]
C₁₄H₁₀O₄S   M 274.291
Mp 317-319°.

Di-Me ester: [14387-31-6].
C₁₆H₁₄O₄S   M 302.344
Cryst. (EtOH). Mp 122-124°.
Dichloride:
C₁₄H₈Cl₂O₂S   M 311.182
Mp 85-87°.
S-*Dioxide:* [2449-35-6]. *4,4′-Sulfonylbisbenzoic acid.
4,4′-Sulfonyldibenzoic acid.*
C₁₄H₁₀O₆S   M 306.289
Cryst. (DMF). Mp 370°.
Di-Me ester, S-dioxide: [3965-53-5].
C₁₆H₁₄O₆S   M 334.343
Cryst. (C₆H₆). Mp 193°.

Ivanova, V.M. *et al, Zh. Org. Khim.*, 1967, **3**, 46.

Bihari, M. *et al*, *Adv. Mass Spectrom.*, 1978, **7**, 1362 (*ms*)

2,2'-Thiobis-1*H*-isoindole-1,3(2*H*)-dione, 9CI

T-70176

N,N'-*Thiobisphthalimide*

[7764-29-6]

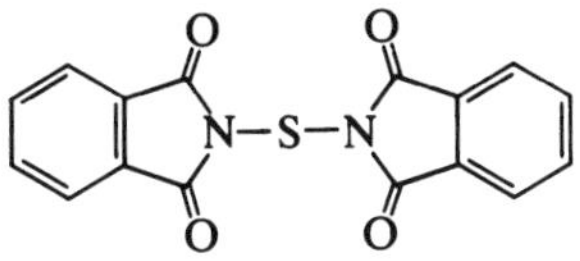

$C_{16}H_8N_2O_4S$ M 324.310

Sulfur transfer reagent. Cryst. (CHCl₃). Mp 315-317°.

Kalnins, M.V., *Can. J. Chem.*, 1966, **44**, 2111 (*synth, ir*)
Harpp, D.N. *et al*, *Tetrahedron Lett.*, 1972, 1481 (*use*)
Harpp, D.N. *et al*, *J. Am. Chem. Soc.*, 1978, **100**, 1222 (*synth*)

1-Thiocyanatoadamantane

T-70177

Thiocyanic acid tricyclo [3.3.1.1³,⁷]decyl ester, 9CI. Adamantyl thiocyanate

[39825-84-8]

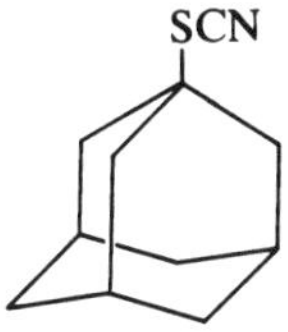

$C_{11}H_{15}NS$ M 193.306

Cryst. solid. Mp 65-66°.

S-*Oxide*: [55305-24-3]. *Tricyclo[3.3.1.1³,⁷]decyl-1-sul-
finylcyanide*, 9CI. 1-Adamantylsulfinyl cyanide.
$C_{11}H_{15}NOS$ M 209.306
Cryst. (hexane). Mp 69-70°.

Boerma-Markerink, A. *et al*, *Synth. Commun.*, 1975, **5**, 147
(*synth, deriv, ir, pmr*)
Ando, T. *et al*, *J. Org. Chem.*, 1987, **52**, 681 (*synth, pmr, cmr*)

Thiofurodysin

T-70178

[70913-81-4]

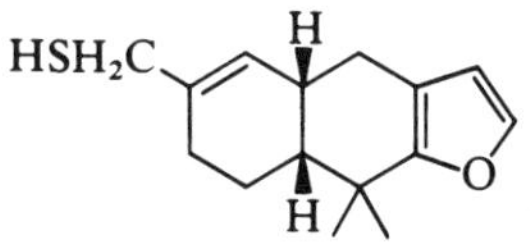

$C_{15}H_{20}OS$ M 248.382

Ac: Thiofurodysin acetate.
$C_{17}H_{22}O_2S$ M 290.420
Constit. of a *Dysidea* sp. Oil.

Kazlauskas, R. *et al*, *Tetrahedron Lett.*, 1978, 4951.

Thiofurodysinin

T-70179

[70913-80-3]

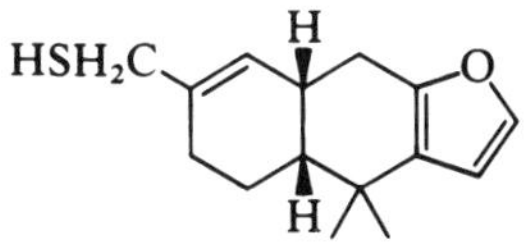

$C_{15}H_{20}OS$ M 248.382

Constit. of the sponge *Dysidea avara*. Oil.

Ac: Thiofurodysinin acetate.
$C_{17}H_{22}O_2S$ M 290.420

Constit. of a *D*. sp. Oil. [α]_D +49.3° (c, 12 in CHCl₃).

Kazlauskas, R. *et al*, *Tetrahedron Lett.*, 1978, 4951 (*isol*)
Capon, R.J. *et al*, *J. Nat. Prod.*, 1987, **50**, 1136 (*isol*)

6-Thioinosine, 9CI

T-70180

9-β-D-*Ribofuranosyl-9H-purine-6-thiol*, 8CI. 6-Mer-
capto-9-β-D-ribofuranosyl-9H-purine. Thioinosine.
Tioinosine

[574-25-4]

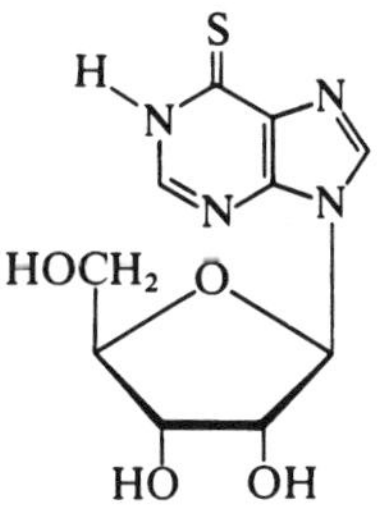

$C_{10}H_{12}N_4O_4S$ M 284.289

Antileukaemic agent. Pale-yellow cryst. Mp 210-211°
dec. [α]²⁵_D −73.0° (c, 1 in 0.1M NaOH).

▷UP0710000.

2′,3′,5′-*Tribenzoyl*: [6741-90-8]. Mp 228-230° (sealed
tube).
S-*Me*: [342-69-8]. 6-(Methylthio)-9-β-D-ribofuranosyl-
9H-purine, 9CI. 6-(Methylthio)inosine.
$C_{11}H_{14}N_4O_4S$ M 298.316
Antileukaemic agent. Cryst (EtOH). Mp 165-167°.

▷UO8985000.

Biochem. Prep., 1963, **10**, 148 (*synth, bibl*)
Shimano, S. *et al*, *Curr. Ther. Res.*, 1974, **16**, 15 (*props*)
Chenon, M.T. *et al*, *J. Am. Chem. Soc.*, 1975, **97**, 4627 (*cmr*)
Paterson, A.R.P. *et al*, *Handb. Exp. Pharmacol.*, 1975, **38**, 384
(*pharmacol*)
Roemming, C. *et al*, *Acta Chem. Scand., Ser. B*, 1976, **30**, 716
(*cryst struct, deriv*)
Keyser, G.E. *et al*, *Tetrahedron Lett.*, 1979, 3263 (*synth, deriv*)
Saneyoshi, M. *et al*, *Chem. Pharm. Bull.*, 1979, **27**, 2518 (*deriv*)
Lin, T.-S. *et al*, *J. Med. Chem.*, 1985, **28**, 1481 (*synth, props*)
Martindale, The Extra Pharmacopeia, 28th/29th Eds.,
1982/1989, Pharmaceutical Press, London, 1870.

Thiomelin

T-70181

2,4-*Dichloro-1,8-dihydroxy-5-methoxy-6-
methylxanthone*

[113734-83-1]

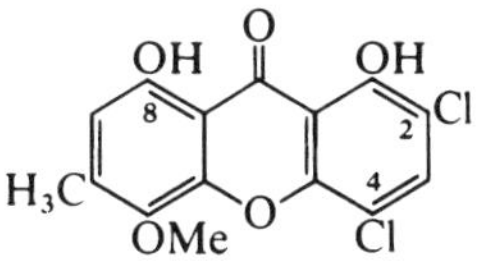

$C_{15}H_{10}Cl_2O_5$ M 341.147

Constit. of *Rinodina thiomela* and *R. lepida*. Fine yellow
threads (CH₂Cl₂/pet. ether). Mp 185°.

8-*Me ether*: [113734-85-3]. 2,4-Dichloro-1-hydroxy-
5,8-dimethoxyxanthone. 8-O-Methylthiomelin.
$C_{16}H_{12}Cl_2O_5$ M 355.174
Constit. of *R. thiomela*. Fine yellow threads
(CH₂Cl₂/pet. ether). Mp 219°.
4-*Dechloro*: [113734-84-2]. 2-Chloro-1,8-dihydroxy-5-
methoxy-6-methylxanthone.
$C_{15}H_{11}ClO_5$ M 306.702
Constit. of *R. thiomela*. Fine yellow needles
(CH₂Cl₂/pet. ether). Mp 224-227°.

4-Dechloro, 8-Me ether: [113734-86-4]. *2-Chloro-1-hydroxy-5,8-dimethoxy-6-methylxanthone. 4-Dechloro-8-O-methylthiomelin.*
$C_{16}H_{13}ClO_5$ $\qquad$ M 320.729
Constit. of *R. thiomela.* Yellow solid.

2-Dechloro, 8-Me ether: [113996-94-4]. *5-Chloro-8-hydroxy-1,4-dimethoxy-3-methylxanthone. 2-Dechloro-8-O-methylthiomelin.*
$C_{16}H_{13}ClO_5$ $\qquad$ M 320.729
Constit. of *R. thiomela.* Yellow solid.

Elix, J.A. *et al, Aust. J. Chem.,* 1987, **40**, 1169.

3-Thiomorpholinecarboxylic acid, 9CI $\qquad$ T-70182

Updated Entry replacing T-01905
Tetrahydro-2H-1,4-thiazine-3-carboxylic acid. 1,4-Thiazane-3-carboxylic acid. Deoxychondrine. Chordarine
[20960-92-3]

(*R*)-*form*
Absolute configuration

$C_5H_9NO_2S$ $\qquad$ M 147.192

(*R*)-*form* [65527-54-0]
L-form
Isol. from *Heterochordaria abietina.* Tiny needles (Me$_2$CO aq.). Mp 270-271° dec. (sealed tube). $[\alpha]_D^{25.5}$ −54.03° (c, 1.6 in H$_2$O).

S-Oxide: see Chondrine, C-02396

(±)-*form*
Cryst. (EtOH aq.), prisms (Me$_2$CO aq.). Mp 263-265° (sealed tube).

Tominaga, F. *et al, J. Biochem. Jpn.,* 1963, **54**, 222.
Carson, J.F. *et al, J. Org. Chem.,* 1964, **29**, 2203 (*synth, bibl*)
Kogami, Y. *et al, Bull. Chem. Soc. Jpn.,* 1987, **60**, 2963 (*synth*)

2-Thiophenecarboxaldehyde, 9CI $\qquad$ T-70183

Updated Entry replacing T-02066
Thiophene-2-aldehyde. 2-Formylthiophene. 2-Thienal. α-Thenaldehyde. 2-Thienylaldehyde
[98-03-3]

C_5H_4OS $\qquad$ M 112.146
Flavour ingredient. Liq. Bp$_{20}$ 93-94°, Bp$_1$ 44-45°. n_D^{20} 1.5917. Rapidly oxid. in air.
▷XM8225000.

Oxime:
C_5H_5NOS $\qquad$ M 127.161
Mp 111°.
Semicarbazone: Mp 224° dec.
Phenylhydrazone: [39677-96-8]. Cryst. (EtOH). Mp 137-139°.
2,4-Dinitrophenylhydrazone: [24383-66-2]. Red cryst. (Py). Mp 242°.

Campaigne, E. *et al, J. Am. Chem. Soc.,* 1953, **75**, 989 (*synth*)
Org. Synth., Coll. Vol., **4**, 915 (*synth, bibl*)
Archer, W.J. *et al, J. Chem. Soc., Perkin Trans. 2,* 1983, 813 (*synth*)
Satonaka, H., *Bull. Chem. Soc. Jpn.,* 1983, **56**, 2463 (*pmr*)
Casarini, D. *et al, J. Chem. Soc., Perkin Trans. 2,* 1985, 1839 (*cmr*)

Tenhosaari, A., *Org. Mass Spectrom.,* 1988, **23**, 236 (*ms*)

2,5-Thiophenedimethanethiol, 9CI $\qquad$ T-70184

2,5-Bis(thiomethyl)thiophene. 2,5-Bis(mercaptomethyl)thiophene
[14282-62-3]

$$HSH_2C\underset{S}{\diagdown\diagup}CH_2SH$$

$C_6H_8S_3$ $\qquad$ M 176.309
Liq. Bp$_1$ 90-95°.

Weber, E. *et al, Justus Liebigs Ann. Chem.,* 1976, 891 (*synth, pmr*)
Houk, J. *et al, J. Am. Chem. Soc.,* 1987, **109**, 6825 (*synth, pmr*)

2,4(3H,5H)-Thiophenedione, 9CI $\qquad$ T-70185

Thiotetronic acid
[51338-33-1]

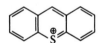

$C_4H_4O_2S$ $\qquad$ M 116.134
Exists in enol-form in solid state and in trifluoroacetic acid soln. Some dioxo-form present in etheral solvs. Other possible enol-forms not detected. Needles (Et$_2$O/pet. ether). Mp 116-119°, Mp 242-248° dec. (double Mp). pK_a 3.92.

Benary, E., *Ber.,* 1913, **46**, 2103 (*synth*)
Macierewicz, B., *Rocz. Chem.,* 1973, **47**, 1735 (*synth, uv, ir, tautom*)

Thioxanthylium(1+), 9CI $\qquad$ T-70186

[261-32-5]

$C_{13}H_9S^{\oplus}$ $\qquad$ M 197.274 (ion)
Perchlorate: [2567-20-6].
$C_{13}H_9ClO_4S$ $\qquad$ M 296.725
Red plates (AcOH). Mp 229° (217-219°) dec.

Bonthrone, W. *et al, Chem. Ind. (London),* 1960, 1192 (*synth*)
Price, C.C. *et al, J. Am. Chem. Soc.,* 1963, **85**, 2278 (*synth, ir, uv*)

2-Thioxo-4,5-imidazolidinedione, 9CI $\qquad$ T-70187

2-Thioimidazolidinetrione, 8CI. Thioparabanic acid
[496-89-9]

$C_3H_2N_2O_2S$ $\qquad$ M 130.121
Yellow cryst. (H$_2$O). Mp 174-175° dec.

Laursen, P.H. *et al, J. Org. Chem.,* 1957, **22**, 275 (*synth*)
Yonezawa, T. *et al, Bull. Chem. Soc. Jpn.,* 1969, **42**, 2323 (*uv*)

2-Thioxo-4-imidazolidinone, 9CI   T-70188

Updated Entry replacing T-30180
4-Oxo-2-imidazolidinethione. 2-Thiohydantoin
[503-87-7]

$C_3H_4N_2OS$ M 116.137
Needles (EtOH aq.). V. sol. hot Et_2O, EtOH, alkalis, spar. sol. H_2O. Mp 228° dec.
▷MU4200000.
1-N-Ac: [584-26-9].
  $C_5H_6N_2O_2S$ M 158.175
  Plates (EtOH). Mp 175-176°.
  ▷MT8575000.
1-N-Benzoyl: [577-47-9].
  $C_{10}H_8N_2O_2S$ M 220.245
  Prisms (EtOH). Mp 165°.
  ▷MT9100000.
1-N-Me:
  $C_4H_6N_2OS$ M 130.164
  Mp 227-230°.
3-N-Me: [694-68-8].
  $C_4H_6N_2OS$ M 130.164
  Needles ($CHCl_3$/ligroin). V. sol. hot H_2O, EtOH, Et_2O, spar. sol. $CHCl_3$, insol. ligroin. Mp 161°.
3-N-Ph: [2010-15-3].
  $C_9H_8N_2OS$ M 192.235
  Yellow plates (EtOH). Mp 240-242° dec.
  ▷MU4025000.
1,3-N-Di-Me: [1801-62-3].
  $C_5H_8N_2OS$ M 144.191
  Cryst. (C_6H_6/pet. ether). V. sol. hot H_2O, sol. EtOH, Et_2O, $CHCl_3$, C_6H_6. Mp 94.5°.

Johnson, T.B. *et al, J. Am. Chem. Soc.*, 1915, **37**, 2406 (*synth*)
Dalgliesh, C.E. *et al, J. Chem. Soc.*, 1947, 559 (*synth*)
Edward, J.T., *Org. Sulphur Compd.*, 1966, **2**, 287 (*rev*)
Walker, L.A., *Acta Crystallogr.*, Sect. B, 1969, **25**, 88 (*cryst struct*)
Rowley, G.L. *et al, J. Am. Chem. Soc.*, 1971, **93**, 5542 (*deriv, synth*)
Thielman, H., *Z. Chem.*, 1978, **18**, 174 (*synth*)
Blotny, G., *Synthesis*, 1983, 391 (*deriv*)
Reddick, R.E. *et al, J. Am. Chem. Soc.*, 1987, **109**, 4380 (*deriv, synth*)

Thorectolide   T-70189

$C_{25}H_{36}O_5$ M 416.556
25-Ac: Thorectolide 25-acetate.
  $C_{27}H_{38}O_6$ M 458.594
  Constit. of *Thorectandra excavatus*. Wax. $[\alpha]_D^{23}$ +33° (c, 1 in $CHCl_3$).

Cambie, R.C. *et al, J. Nat. Prod.*, 1988, **51**, 331.

Thujin   T-70190

3,8-Dihydro-3,3,4,8,8,9-hexamethylbenzo[1,2-c:4,5-c']-dipyran-1,6-dione, 9CI
[114191-55-8]

$C_{18}H_{20}O_4$ M 300.354
Constit. of *Thuja plicata*. Cryst. (EtOAc/heptane). Mp 88-91°.

Jin, L. *et al, Can. J. Chem.*, 1988, **66**, 51.

Thuriferic acid   T-70191

$C_{22}H_{20}O_8$ M 412.395
Constit. of leaves of *Juniperus thurifera*. Oil. $[\alpha]_D^{23}$ −179.6° ($CHCl_3$).

San Feliciano, A. *et al, Tetrahedron*, 1988, **44**, 7255 (*struct*)

Thymopoietin, 9CI   T-70192

Updated Entry replacing T-60219
Thymin (hormone)
[60529-76-2]

H-W-X-Phe-Leu-Glu-Asp-Pro-Ser-Val-Leu-Thr-Lys-Glu-Lys-Leu-Lys-Ser-Glu-Leu-Val-Ala-Asn-Asn-Val-Thr-Leu-Pro-Ala-Gly-Glu-Gln-Arg-Lys-Y-Val-Tyr-Val-Glu-Leu-Tyr-Leu-Gln-Z-Leu-Thr-Ala-Leu-Lys-Arg-OH. Thymopoietin I: W = Gly, X = Gln, Y = Asp, Z = His, II: W = Pro, X = Glu, Y = Asp, Z = Ser, III: W = Pro, X = Glu, Y = Glu, Z = His

Polypeptide hormone discovered by its effect on neuromuscular transmission. Induces T-lymphocyte differentiation and affects immunoregulatory balance.
Thymopoietin I [66943-28-0]
  1-Glycine-2-glutamine-43-histidinethymopoietin II, 9CI
  $C_{250}H_{410}N_{68}O_{75}$ M 5568.400
  Isol. from bovine thymus. Fluffy powder. $[\alpha]_D^{21}$ −78.4° (c, 0.3 in AcOH aq.).
Thymopoietin II, 9CI [56996-26-0]
  $C_{251}H_{412}N_{64}O_{78}$ M 5574.398

From bovine thymus.
Thymopoietin III [79103-34-7]
  34-Glutamic acid-43-histidinethymopoietin II, 9CI
  $C_{255}H_{416}N_{66}O_{77}$ M 5638.487
  From bovine spleen. Fluffy powder. $[\alpha]_D^{21}$ −81.2° (c, 0.3 in AcOH aq.).

Audhya, T. *et al, Biochemistry*, 1981, **20**, 6195 (*struct, bibl*)
Abiko, T. *et al, Chem. Pharm. Bull.*, 1985, **33**, 1583; 1986, **34**, 2133 (*synth*)

Audhya, T. *et al, Proc. Natl. Acad. Sci. USA*, 1987, **84**, 3545 (*isol*)

Thyrsiferol   T-70193

Updated Entry replacing T-60220
[66873-39-0]

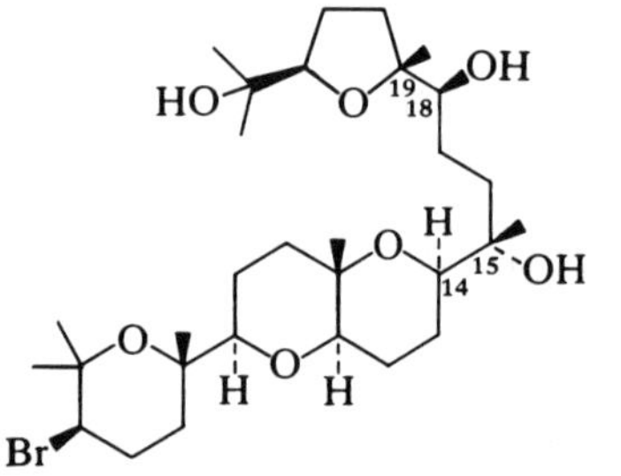

$C_{30}H_{53}BrO_7$ M 605.649

Config. at C-14 and C-15 wrongly drawn in early references. Metab. of *Laurencia thyrsifera*.

23-Ac: [96304-95-9].
    $C_{32}H_{55}BrO_8$ M 647.686
    Constit. of *L. obtusa*. Cytotoxic agent. Cryst. (MeOH aq.). Mp 118-119°. $[\alpha]_D^{29}$ +1.99° (c, 4.4 in CHCl$_4$).
18,19-Diepimer: [105880-10-2]. **Venustatriol.**
    $C_{30}H_{53}BrO_7$ M 605.649
    From *L. venusta*. Antiviral agent. Cryst. Mp 161.5°. $[\alpha]_D^{20}$ +9.4° (c, 3.2 in CHCl$_3$).

Blunt, J.W. *et al, Tetrahedron Lett.*, 1978, 69 (*isol, struct*)
Suzuki, T. *et al, Tetrahedron Lett.*, 1985, **26**, 1329 (*isol*)
Sakami, S. *et al, Tetrahedron Lett.*, 1986, **27**, 4287 (*isol, cryst struct*)

Tiacumicin *A*   T-70194

[109713-94-2]

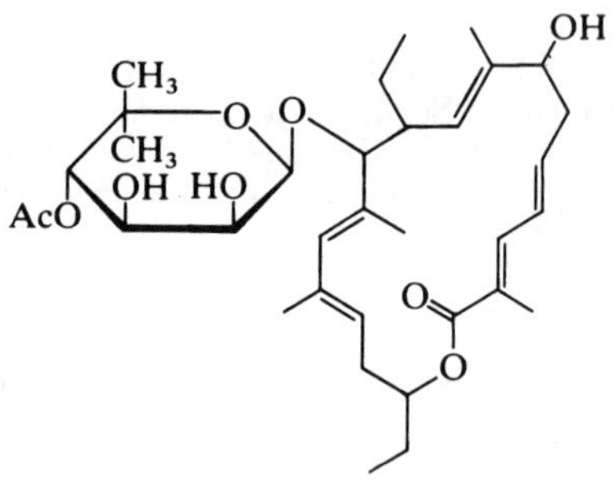

$C_{34}H_{52}O_9$ M 604.779

Macrolide antibiotic. Prod. by *Dactylosporangium aurantiacum* ssp. *hamdenensis*. V. weak activity against some gram-positive bacteria. Oil. $[\alpha]_D^{25}$ +41° (c, 0.1 in MeOH). Similar to Clostamicins and Lipiarmycin.

Theriault, R.J. *et al, J. Antibiot.*, 1987, **40**, 567 (*isol, props*)
Hochlowski, J.E. *et al, J. Antibiot.*, 1987, **40**, 575 (*pmr, cmr, ms, struct*)

Tiacumicin *C*   T-70195

Clostomicin B$_2$
[106008-70-2]

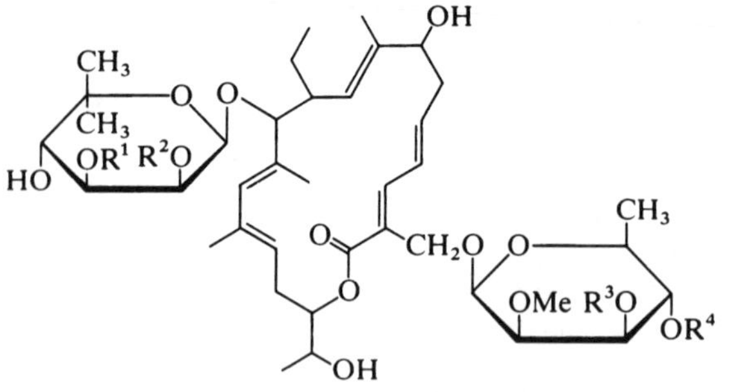

$R^1 = -COCH(CH_3)_2, R^2 = R^3 = H$

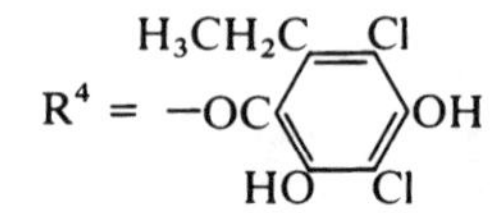

$$R^4 = -OC...$$

$C_{52}H_{74}Cl_2O_{18}$ M 1058.052

Macrolide antibiotic. Isol. from *Dactylosporangium aurantiacum* ssp. *hamdenensis* and *Micromonospora echinospora* ssp. *armeniaca*. Active against gram-positive bacteria. Mp 142-143°. $[\alpha]_D^{25}$ −8.6° (c, 15.8 in MeOH). Similar to Lipiarmycin.

Omura, S. *et al, J. Antibiot.*, 1986, **39**, 1407, 1413 (*Clostomicins*)
Theriault, R.J. *et al, J. Antibiot.*, 1987, **40**, 567 (*isol, props*)
Hochlowski, J.E. *et al, J. Antibiot.*, 1987, **40**, 575 (*pmr, cmr, struct*)

Tiacumicin *D*   T-70196

[109679-45-0]

As Tiacumicin *C*, T-70195 with

$R^1 = R^4 = H, R^2 = -COCH(CH_3)_2, R^3 = $ 3,5-Dichloro-2-ethyl-4,6-dihydroxybenzoyl

$C_{52}H_{74}Cl_2O_{18}$ M 1058.052

Macrolide antibiotic. Isol. from *Dactylosporangium aurantiacum* ssp. *hamdenensis*. Active against gram-positive bacteria. Mp 141-145°. $[\alpha]_D^{25}$ −5.6° (c, 0.47 in MeOH). Related to Clostomicins and Lipiarmycin.

Theriault, R.J. *et al, J. Antibiot.*, 1987, **40**, 567 (*isol, props*)
Hochlowski, J.E. *et al, J. Antibiot.*, 1987, **40**, 575 (*pmr, cmr, struct*)

Tiacumicin *E*   T-70197

[109679-44-9]

As Tiacumicin *C*, T-70195 with

$R^1 = R^3 = H, R^2 = -COCH_2CH_3, R^4 = $ 3,5-Dichloro-2-ethyl-4,6-dihydroxybenzoyl

$C_{51}H_{72}Cl_2O_{18}$ M 1044.025

Macrolide antibiotic. Isol. from *Dactylosporangium aurantiacum* ssp. *hamdenensis*. Active against gram-positive bacteria. Mp 138-141°. $[\alpha]_D^{25}$ −3.2° (c, 1.3 in MeOH). Related to Clostomicins and Lipiarmycin.

Theriault, R.J. *et al, J. Antibiot.*, 1987, **40**, 567 (*isol, props*)
Hochlowski, J.E. *et al, J. Antibiot.*, 1987, **40**, 575 (*pmr, cmr, struct*)

Tiacumicin *F*   T-70198

Clostomicin A

[106008-69-9]
As Tiacumicin *C*, T-70195 with

$R^1 = -COCH(CH_3)_2$, $R^2 = R^3 = H$, $R^4 = 3,5$-Dichloro-
2-ethyl-4,6-dihydroxybenzoyl

$C_{52}H_{74}Cl_2O_{18}$ M 1058.052
Macrolide antibiotic. Isol. from *Dactylosporangium aurantiacum* ssp. *hamdenensis* and *Micromonospora
echinospora* ssp. *armeniaca*. Active against gram-positive bacteria. Mp 141-143°. $[\alpha]_D^{20} +9.8°$ (c, 1 in
MeOH). Similar to Lipiarmicin.

Omura, S. *et al*, *J. Antibiot.*, 1986, **39**, 1407, 1413 (*Clostomicins*)
Therisult, R.J. *et al*, *J. Antibiot.*, 1987, **40**, 567 (*isol, props*)
Hochlowski, J.E. *et al*, *J. Antibiot.*, 1987, **40**, 575 (*pmr, cmr,
struct*)

Tifruticin   T-70199

Updated Entry replacing T-02175
[56377-69-6]

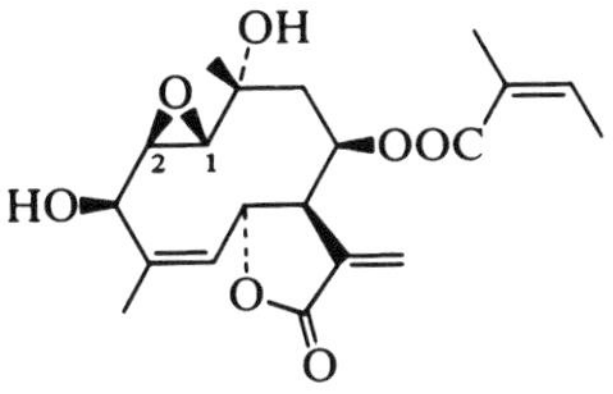

$C_{20}H_{26}O_7$ M 378.421
Constit. of *Tithonia fruticosa*. Cryst. (EtOAc). Mp 141°.
$[\alpha]_D^{22} -22°$ (c, 1.1 in CHCl_3).
Deepoxy, 1,2-didehydro: [56377-63-0]. **Deoxytifruticin**.
$C_{20}H_{26}O_6$ M 362.422
Minor constit. of *T. fruticosa*.

Herz, W. *et al*, *J. Org. Chem.*, 1975, **40**, 3118.
Baruah, N.C. *et al*, *J. Org. Chem.*, 1979, **44**, 1831 (*config*)

Tinospora clerodane   T-70200

[115334-17-3]

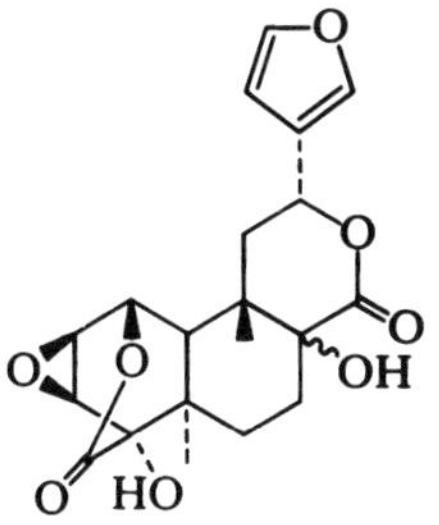

$C_{20}H_{22}O_8$ M 390.389
Constit. of *Tinospora cordifolia*. Cryst. (MeOH). Mp
231-233° dec. $[\alpha]_D^{20} +28.2°$ (c, 0.62 in DMSO).

Bhatt, R.K. *et al*, *Phytochemistry*, 1988, **27**, 1212.

7-Tirucallene-3,23,24,25-tetrol   T-70201

Updated Entry replacing T-60221

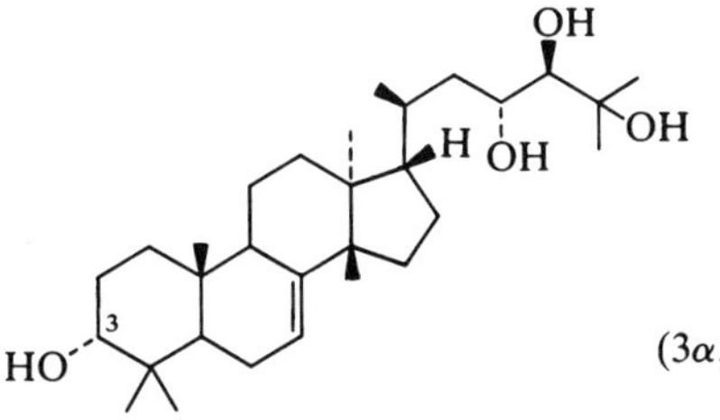

$(3\alpha,23R,24R)$-*form*

$C_{30}H_{52}O_4$ M 476.738
(3α,23R,24R)-form [78739-37-4]
Hispidol A
Constit. of *Trichilia hispida*. Cryst. (MeOH/MeCN).
Mp 118°. $[\alpha]_D^{25} -80°$ (Py).
(3β,23R,24R)-form [78739-39-6]
Hispidol B
From *T. hispida*. Cryst. (MeOH/CH_2Cl_2). Mp 252-
253°. $[\alpha]_D^{25} -57°$ (Py).

Jolad, S.D. *et al*, *J. Org. Chem.*, 1981, **46**, 4085 (*isol*)
Arisawa, M. *et al*, *Phytochemistry*, 1987, **26**, 3301 (*cryst struct*)

Tochuinol   T-70202

*1,2-Dimethyl-2-(4-methylphenyl)cyclopentanemethanol,
9CI*

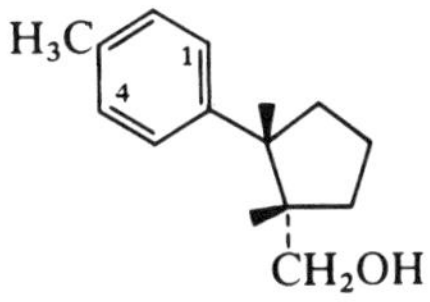

$C_{15}H_{22}O$ M 218.338
Ac: [111621-35-3]. **Tochuinyl acetate**.
$C_{17}H_{24}O_2$ M 260.375
Constit. of *Tochuina tetraqueta*. Oil. $[\alpha]_D^{25} -42.5°$ (c,
1.09 in CH_2Cl_2).
1,4-Dihydro, Ac: [111621-36-4]. **Dihydrotochuinyl
acetate**.
$C_{17}H_{26}O_2$ M 262.391
Constit. of *T. tetraqueta*. Oil. $[\alpha]_D^{25} -29.3°$ (c, 1.11 in
CH_2Cl_2).

Williams, D.E. *et al*, *Can. J. Chem.*, 1987, **65**, 2244.

Toddasin   T-70203

Mexolide
[75775-35-8]

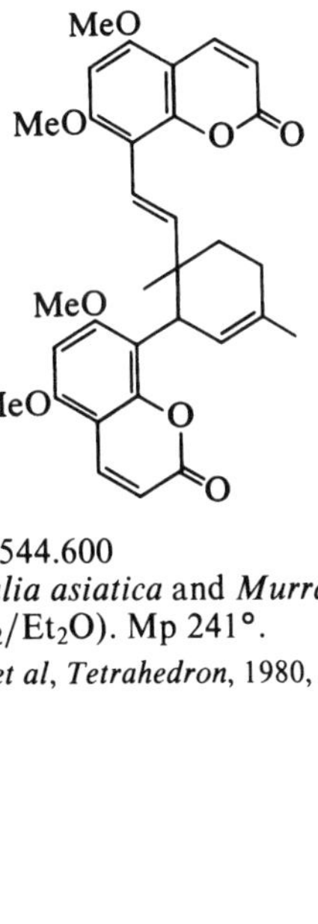

$C_{32}H_{32}O_8$ M 544.600
Constit. of *Toddalia asiatica* and *Murraya exotica*.
Cryst. (CH_2Cl_2/Et_2O). Mp 241°.

Chakraborty, D.P. *et al*, *Tetrahedron*, 1980, **36**, 3563 (*isol*)

Sharma, P.N. *et al*, *Phytochemistry*, 1980, **19**, 1258 (*isol*)

Totarol   T-70204

Updated Entry replacing T-02275
14-Isopropyl-8,11,13-podocarpatrien-13-ol
[511-15-9]

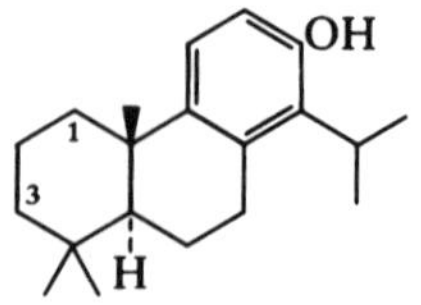

Absolute
configuration

$C_{20}H_{30}O$ M 286.456
Constit. of *Podocarpus* spp. Cryst. (pet. ether). Mp 132°.
$[\alpha]_D^{20}$ +42.5° (EtOH).
Ac: Cryst. (EtOH). Mp 125°. $[\alpha]_D^{18}$ +44.6° (Et₂O).
Me ether: Cryst. (CHCl₃/EtOH). Mp 92-92.5°. $[\alpha]_D$
+41.95° (Et₂O).
3-Oxo: [6755-93-7]. **Totarolone**.
$C_{20}H_{28}O_2$ M 300.440
Constit. of *Tetraclinis articulata* and *Juniperus
conferta*. Cryst. Mp 188-189°. $[\alpha]_D$ +102° (EtOH).
3-Oxo, Ac: Mp 113-115°. $[\alpha]_D$ +96.6° (EtOH).
1,3-Dioxo, Me ether: **1,3-Dioxototaryl methyl ether**.
$C_{21}H_{28}O_3$ M 328.450
Isol. from *Cupressus sempervirens*. Cryst. (hexane).
Mp 191-192°. $[\alpha]_D$ +125° (c, 1.0 in CHCl₃).

Barltrop, J.A. *et al*, *J. Chem. Soc.*, 1958, 2566 (*synth*)
Chow, Y.-L. *et al*, *Acta Chem. Scand.*, 1962, **16**, 1305 (*struct*)
Mangoni, L. *et al*, *Tetrahedron Lett.*, 1964, 2643 (*1,3-
Dioxototaryl methyl ether*)
Enzell, C.R., *Tetrahedron Lett.*, 1966, 2135 (*ms*)
Nishida, T. *et al*, *Org. Magn. Reson.*, 1977, **9**, 203 (*cmr*)
Matsumoto, T. *et al*, *Bull. Chem. Soc. Jpn.*, 1979, **52**, 1450
(*synth*)

1,2,12-   T-70205
Triazapentacyclo[6.4.0.0^{2,17}.0^{3,7}.0^{4,11}]-dodecane

*Hexahydro-6,1,3[1]-propan[3]ylidene-1H-
cyclopenta[c]triazirino[a]pyrazole, 9CI*
[115492-54-1]

$C_9H_{13}N_3$ M 163.222
Cryst. Mp 214° dec.

Marterer, W. *et al*, *Tetrahedron Lett.*, 1987, **28**, 5497 (*synth,
uv, N nmr, pmr, cmr, ms*)

3,5,12-Triazatetracyclo[5.3.1.1^{2,6}.0^{4,9}]-   T-70206
dodecane

*Decahydro-2,7-imino-3,6-methano-1,8-naphthyridine,
9CI*

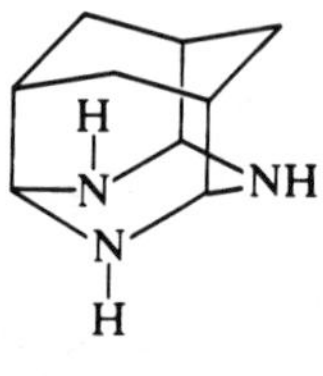

$C_9H_{15}N_3$ M 165.238
Obs. only in soln.

3,5,12-Tri-Me: [107383-92-6].
$C_{12}H_{21}N_3$ M 207.318
Long needles. Mp 64-75°. Dec. on attempted recryst.
Slowly becomes oily on standing.
3,5,12-Tri-Et: [107383-93-7].
$C_{15}H_{27}N_3$ M 249.398
Mp 50-60°. Slowly dec. on standing.

Nielsen, A.T. *et al*, *J. Org. Chem.*, 1987, **52**, 1656 (*synth, pmr,
cryst struct*)

1*H*-1,2,3-Triazolo[4,5-*f*]quinoline, 9CI   T-70207
[233-62-5]

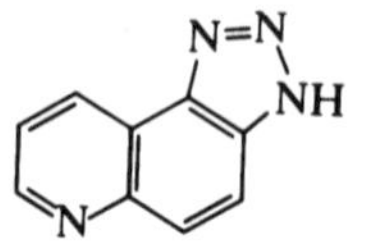

$C_9H_6N_4$ M 170.173
Mp 242-244°.

Milata, V. *et al*, *Collect. Czech. Chem. Commun.*, 1988, **53**,
1068 (*synth, ir, uv, pmr, cmr, bibl*)

1*H*-1,2,3-Triazolo[4,5-*h*]quinoline, 9CI   T-70208
[233-92-1]

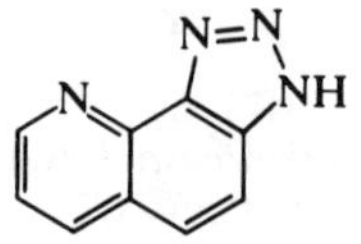

$C_9H_6N_4$ M 170.173
Mp 240-244°.

Milata, V. *et al*, *Collect. Czech. Chem. Commun.*, 1988, **53**,
1068 (*synth, ir, uv, pmr, cmr, bibl*)

[1,2,4]Triazolo[1,2-*a*][1,2,4]triazole-   T-70209
1,3,5,7(2*H*,6*H*)-tetrone

Urazourazole
[113893-26-8]

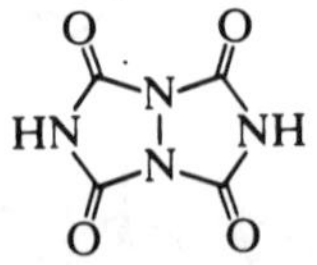

$C_4H_2N_4O_4$ M 170.084
Mp 340° dec.

Nachbaur, E. *et al*, *Angew. Chem., Int. Ed. Engl.*, 1988, **27**, 701
(*synth, ir, cmr, cryst struct*)

The first digit of the DOC Number
defines the Supplement in which
the Entry is found.   0 indicates the
Main Work.

Tribenzotritwistatriene   T-70210

4b,9b,10,14b,14c,15-Hexahydro-5,10,15-metheno-5H-benzo[a]*naphth*[*1,2,3-de*]*anthracene, 9CI*

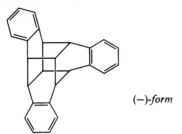

(−)-*form*

$C_{26}H_{20}$ M 332.444

(−)-*form*

$[\alpha]_D$ −493° (cyclohexane), $[\alpha]_D$ −1638° (CHCl₃). (+)-form also prepd.

(±)-*form*

Mp 215-216°. Also prepd. were mono- and dibenzannellated derivs. intermed. between this compd. and Pentacyclo[8.4.0.0^{2,7}.0^{3,12}.0^{6,11}]tetradeca-4,8,13-triene, P-70021 .

Müller, H. *et al, Angew. Chem., Int. Ed. Engl.*, 1988, **27**, 1103 (*synth, pmr, cmr, abs config*)

1,2,3-Tribromocyclopropane   T-70211

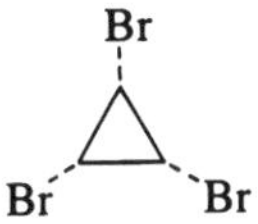

$C_3H_3Br_3$ M 278.769

(*1α,2α,3α*)-*form*

cis-*form*

Cryst. (EtOH).

Schrumpf, G. *et al, Acta Crystallogr., Sect. C*, 1987, **43**, 1188 (*synth, cryst struct*)

[(Tribromomethyl)thio]benzene   T-70212

Phenyl tribromomethyl sulfide, 8CI

[16936-63-3]

PhSCBr₃

$C_7H_5Br_3S$ M 360.889

Mp 89-90°.

S-*Oxide:* [52703-34-1]. [(*Tribromomethyl)sulfinyl*]-*benzene, 9CI. Phenyl tribromomethyl sulfoxide.*
$C_7H_5Br_3OS$ M 376.888
Props. not accessible.

S-*Dioxide:* [17025-47-7]. [(*Tribromomethyl)sulfonyl*]-*benzene, 9CI. Phenyl tribromomethyl sulfone.*
$C_7H_5Br_3O_2S$ M 392.887
Needles (EtOH aq.). Mp 144-145°.

Farrar, W.V., *J. Chem. Soc.*, 1956, 508 (*synth, deriv*)
Yagupol'skii, L.M. *et al, J. Gen. Chem. USSR (Engl. Transl.)*, 1967, **37**, 1686 (*synth*)
Del Buttero, P. *et al, Gazz. Chim. Ital.*, 1973, **103**, 809 (*synth, deriv*)
Fields, D.L. *et al, J. Org. Chem.*, 1986, **51**, 3369 (*synth, ir, pmr*)

Tribromophenylselenium, 9CI   T-70213

Phenylselenium tribromide

[38927-01-4]

PhSeBr₃

$C_6H_5Br_3Se$ M 395.778

Rose-red needles. Mp 105-106°.

Foster, D.G., *J. Am. Chem. Soc.*, 1933, **55**, 822 (*synth*)

1,3,5-Tribromo-2,4,6-trifluorobenzene   T-70214

[2368-49-2]

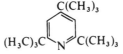

$C_6Br_3F_3$ M 368.773

Cryst. (EtOH). Mp 94.5 96°.

Finger, G.C. *et al, J. Am. Chem. Soc.*, 1951, **73**, 153 (*synth*)
Fuller, G., *J. Chem. Soc.*, 1965, 6264 (*synth*)
Bolton, R. *et al, J. Chem. Soc., Perkin Trans. 2*, 1978, 137 (*synth*)

1,3,5-Tributylbenzene   T-70215

[841-07-6]

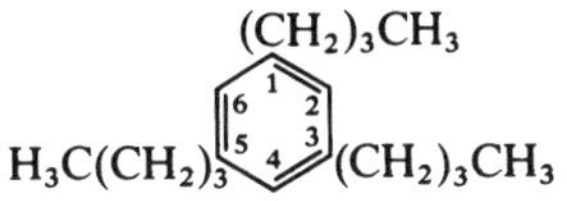

$C_{18}H_{30}$ M 246.435

Liq. Bp₀.₀₃ 100-102°.

Eapen, K.C. *et al, J. Org. Chem.*, 1988, **53**, 5564 (*synth, ms*)

1,3,5-Tri-*tert*-butyl-2-nitrobenzene   T-70216

1,3,5-Tris(1,1-dimethylethyl)-2-nitrobenzene, 9CI

[4074-25-3]

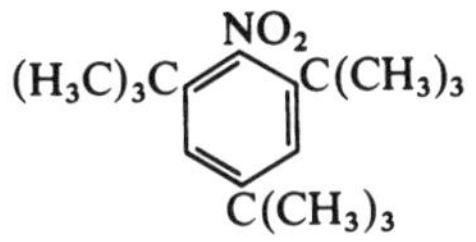

$C_{18}H_{29}NO_2$ M 291.433

Solid (pet. ether). Mp 204-205°.

Bartlett, P.D. *et al, J. Am. Chem. Soc.*, 1954, **76**, 2349 (*synth*)
Betts, E.E. *et al, Can. J. Chem.*, 1955, **33**, 1768 (*synth*)

2,4,6-Tri-*tert*-butylpyridine   T-70217

Updated Entry replacing T-02606

2,4,6-Tris(1,1-dimethylethyl)pyridine, 9CI

[20336-15-6]

$C_{17}H_{29}N$ M 247.423

Cryst. (EtOH aq.). Mp 71°. Bp₂₀ 115-120°.

Dimroth, K. *et al, Angew. Chem., Int. Ed. Engl.*, 1968, **7**, 460.
Scalzi, F.V. *et al, J. Org. Chem.*, 1971, **36**, 2541.
Böhm, S. *et al, Collect. Czech. Chem. Commun.*, 1987, **52**, 1305 (*synth, ms*)

Trichadenal
T-70218

Updated Entry replacing T-02619
3β-Hydroxy-27-friedelanal
[50465-22-0]

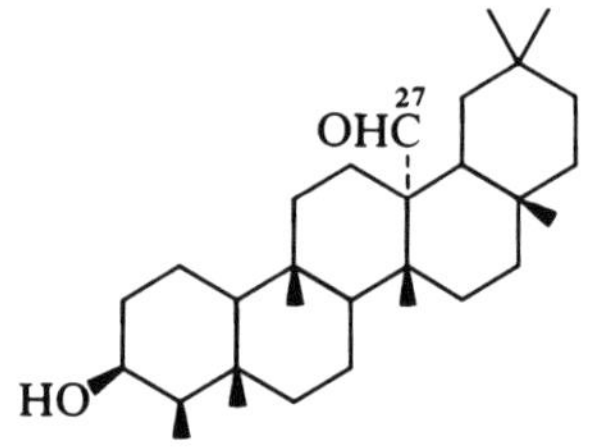

$C_{30}H_{50}O_2$ M 442.724

Note that *CA* calls the 27-posn. 26-. Constit. of bark of *Trichadenia zeylanica*. Cryst. (pet. ether). Mp 300-301°. $[\alpha]_D^{26}$ +8.7° (CHCl₃).

Ac: Cryst. (pet. ether). Mp 246-247°. $[\alpha]_D^{26}$ +13° (CHCl₃).

27-Carboxylic acid: [50656-70-7]. *3β-Hydroxy-27-frie-delanoic acid.* **Trichadenic acid B**.
$C_{30}H_{50}O_3$ M 458.723
Constit. of the bark of *Hydnocarpus octandra*. Cryst. (pet. ether). Mp 333-334°. $[\alpha]_D^{26}$ +40.1° (CHCl₃).

27-Carboxylic acid, Ac: [50464-85-2].
$C_{32}H_{52}O_4$ M 500.760
Constit. of bark of *T. zeylanica*. Cryst. (pet. ether). Mp 267-268°. $[\alpha]_D^{26}$ +56° (CHCl₃).

3-Epimer, 27-carboxylic acid: [50656-69-4]. *3α-Hydroxy-27-friedelanoic acid.* **Trichadenic acid A**.
$C_{30}H_{50}O_3$ M 458.723
Constit. of bark of *T. zeylanica*. Cryst. (pet. ether). Mp 292-293°. $[\alpha]_D^{26}$ +25° (CHCl₃).

3-Ketone, 27-carboxylic acid: [51024-97-6]. *3-Oxo-27-friedelanoic acid.* **Trichadonic acid**.
$C_{30}H_{48}O_3$ M 456.707
Constit. of the bark of *T. zeylanica*. Cryst. (pet. ether). Mp 245-246°. $[\alpha]_D^{26}$ +3° (CHCl₃).

Gunasekera, S.P. *et al, J. Chem. Soc., Perkin Trans. 1*, 1977, 418, 483 (*isol*)

Tanaka, R. *et al, Tetrahedron Lett.*, 1988, **29**, 4751 (*struct*)

2,2,2-Trichloroacetophenone
T-70219

2,2,2-Trichloro-1-phenylethanone, 9CI. Phenyl trichloromethyl ketone. Benzoyltrichloromethane
[2902-69-4]

$$PhCOCCl_3$$

$C_8H_5Cl_3O$ M 223.486

Bp₂₅ 145°, Bp₃.₅ 99-102.5°, Bp₀.₆ 75-76°.

Cohen, S.G. *et al, J. Am. Chem. Soc.*, 1950, **72**, 3952 (*synth*)

Villieras, J. *et al, Bull. Soc. Chim. Fr.*, 1970, 1189 (*synth*)

1,2,3-Trichlorocyclopropane
T-70220

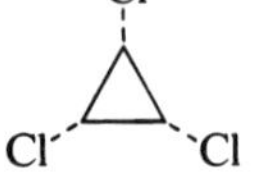

$C_3H_3Cl_3$ M 145.416

(*1α,2α,3α*)-*form* [88489-55-8]
cis-*form*
Prisms (AcOH aq.).

Schrumpf, G. *et al, J. Mol. Struct.*, 1983, **102**, 209 (*synth, ir*)

Schrumpf, G. *et al, Acta Crystallogr., Sect. C*, 1987, **43**, 1182 (*cryst struct*)

Trichlorocyclopropenylium
T-70221

$C_3Cl_3^\oplus$ M 142.392 (ion)

Tetrachloroaluminate: [10438-65-0].
C_3AlCl_7 M 311.186
Tan amorph. powder.

Hexachloroantimonate: [10421-73-5].
C_3Cl_9Sb M 476.860
Amorph. powder.

West, R. *et al, J. Am. Chem. Soc.*, 1966, **88**, 2488 (*synth, Raman, ir*)

Weiss, R. *et al, J. Org. Chem.*, 1976, **41**, 2258 (*use*)

Musigmann, K. *et al, Tetrahedron Lett.*, 1987, **28**, 4517 (*use*)

1,1,2-Trichloro-2-(diethylamino)ethylene
T-70222

Updated Entry replacing T-02726
1,2,2-Trichloro-N,N-diethylethanamine, 9CI.
Diethyl(trichlorovinyl)amine
[686-10-2]

$$Cl_2C{=}CClNEt_2$$

$C_6H_{10}Cl_3N$ M 202.511

Converts carboxylic acids to acid chlorides, alcohols to alkyl chlorides and amines to amidines. Liq. Bp₂₈ 29°.

Speziale, A.J. *et al, J. Am. Chem. Soc.*, 1960, **82**, 909 (*synth*)

Fieser, M. *et al, Reagents for Organic Synthesis*, Wiley, 1967-84, **2**, 134.

Trichloromethyl chloroformate, 8CI
T-70223

Updated Entry replacing T-02786
Trichloromethyl carbonochloridate, 9CI. Diphosgene.
Perchloromethyl chloroformate. Superpalite. Green
Cross
[503-38-8]

$$ClCOOCCl_3$$

$C_2Cl_4O_2$ M 197.833

Safer substitute for phosgene in synthesis (but see also Bis(trichloromethyl) carbonate, B-70163 . Used in WW1 as poison gas. Oily liq. giving toxic, asphyxiating vapour. d¹⁴ 1.65. Bp 128°, Bp₅₀ 49°.

▷Extremely harmful vapour. Emits toxic and corrosive fumes on heating or contact with H₂O

Ramperger, H. *et al, J. Am. Chem. Soc.*, 1933, **55**, 214.

Yura, S., *J. Chem. Soc. Jpn.*, 1948, **51**, 157; *CA*, **45**, 547 (*synth*)

Fieser, M. *et al, Reagents for Organic Synthesis*, Wiley, 1967-84, **8**, 214.

Skorna, G. *et al, Angew. Chem., Int. Ed. Engl.*, 1977, **16**, 259 (*use*)

Org. Synth., 1979, **59**, 195 (*synth, use*)

Sax, N.I., *Dangerous Properties of Industrial Materials*, 5th Ed., Van Nostrand-Reinhold, 1979, 626.

Trichlorophenylselenium, 9CI
T-70224

Phenylselenium trichloride
[42572-42-9]

$$PhSeCl_3$$

$C_6H_5Cl_3Se$ M 262.425

Synthetic intermediate. Pale-yellow solid. Unstable; can be stored at −20°.

Behagel, O. *et al, Ber.*, 1933, **66**, 708 (*synth*)
Foster, D.G., *J. Am. Chem. Soc.*, 1933, **55**, 822 (*synth*)
Engman, L., *J. Org. Chem.*, 1987, **52**, 4086 (*synth, use*)

Trichostin   T-70225

[115713-34-3]

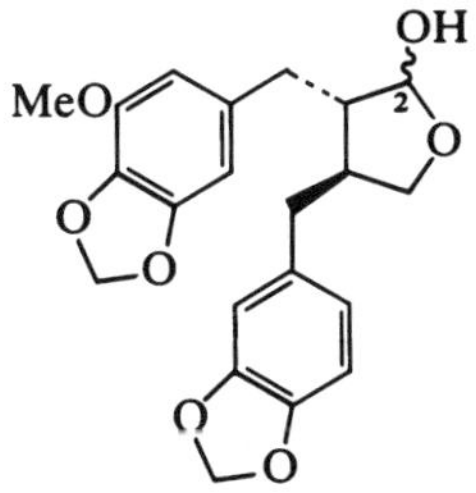

$C_{21}H_{22}O_7$ M 386.401

Constit. of *Piper trichostachyon*. Gum. $[\alpha]_D^{20}$ −62.25° (c, 0.8 in MeOH).

2-Oxo: [115794-38-2]. **Dehydrotrichostin**.
   $C_{21}H_{20}O_7$ M 384.385
   Constit. of *P. trichostachyon*. Semi-solid. $[\alpha]_D^{20}$ +10.2° (c, 0.2 in MeOH).

Koul, S.K. *et al, Phytochemistry*, 1988, **27**, 1479.

Tricyclo[4.2.2.0¹,⁶]deca-7,9-diene, 9CI   T-70226

[4.2.2]Propella-7,9-diene
[53922-55-7]

$C_{10}H_{12}$ M 132.205

Precursor for [4]Paracyclophane. Oil. Mp ca. −6°.

Landheer, I.J. *et al, Tetrahedron Lett.*, 1974, 2813 (*synth, pmr*)
Weinges, K. *et al, Chem. Ber.*, 1974, **107**, 1915 (*synth, ms, pmr, cmr*)
Kostermans, G.B.M. *et al, J. Am. Chem. Soc.*, 1987, **109**, 2471 (*use*)

Tricyclo[5.3.0.0²,⁸]deca-3,5,9-triene, 9CI   T-70227

[66039-62-1]

$C_{10}H_{10}$ M 130.189

Isom. rapidly in soln. at r.t. to Tricyclo[5.3.0.0²,¹⁰]deca-3,5,8-triene, T-02949 and Tricyclo[5.3.0.0⁴,⁸]deca-2,5,9-triene, T-50394 .

Dressel, J. *et al, J. Am. Chem. Soc.*, 1987, **109**, 2857 (*synth, pmr, cmr*)

Tricyclo[2.2.1.0²,⁶]heptan-3-one   T-70228

Updated Entry replacing T-02972
   3-Nortricyclanone
[695-05-6]

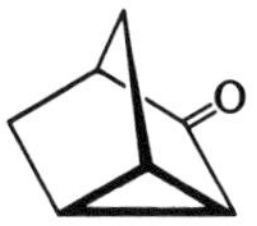

C_7H_8O M 108.140

Liq. Bp_{77} 103-105°, Bp_{24} 78-79°. n_D^{25} 1.4878.

Oxime:
   C_7H_9NO M 123.154
   Liq.
2,4-Dinitrophenylhydrazone: Cryst. (EtOH). Mp 188.2-189.6°.

Roberts, J.D. *et al, J. Am. Chem. Soc.*, 1950, **72**, 3116 (*synth, ir*)
Org. Synth., 1965, **45**, 77 (*synth*)
Chizhov, A.O. *et al, J. Org. Chem.*, 1987, **52**, 5647 (*synth, deriv, ms, ir*)

Tricyclo[3.1.0.0²,⁶]hexanedione   T-70229

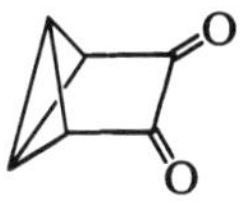

$C_6H_4O_2$ M 108.096

Valence isomer of *o*-Benzoquinone. Yellow cryst. Mp 145-155° dec. (sealed tube).

Christl, M. *et al, Angew. Chem., Int. Ed. Engl.*, 1988, **27**, 1369 (*synth, uv, pmr, cmr*)

Tricyclo[3.3.1.0²,⁸]nona-3,6-diene, 9CI, 8CI   T-70230

Updated Entry replacing T-30245
   Barbaralane
[14693-11-9]

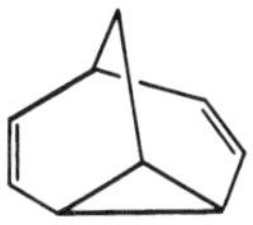

C_9H_{10} M 118.178

Cryst. by subl. Mp 46°.

Biethan, U. *et al, Angew. Chem., Int. Ed. Engl.*, 1967, **6**, 176 (*synth, nmr*)
Doering, W. v. E. *et al, Tetrahedron*, 1967, **23**, 3943 (*synth, nmr, ir*)
Henkel, J.G. *et al, J. Org. Chem.*, 1983, **48**, 3858 (*synth*)
Trinks, R. *et al, Chem. Ber.*, 1987, **120**, 1481 (*synth*)

Tricyclo[2.1.0.0¹,³]pentane   T-70231

C_5H_6 M 66.102

Strong evidence for existence as an intermed. at ca. −50°.

Wiberg, K.B. *et al, Tetrahedron Lett.*, 1987, **28**, 5411.

Tricyclo[4.4.4.0^{1,6}]tetradeca-2,4,7,9,10,12-hexaene   T-70232

4a,8a-[1,3]Butadienonaphthalene, 9*CI*. [4.4.4]-*Propellahexaene*

[108189-46-4]

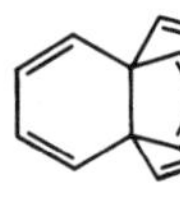

$C_{14}H_{12}$ M 180.249
Needles. Mp 48°. Stable in soln. at 105°.

Waykole, L. *et al, J. Am. Chem. Soc.*, 1987, **109**, 3174 (*synth, pmr, uv*)

Tricyclo[4.3.1.1^{3,8}]undecane-2,7-dione, 9CI   T-70233

Homoadamantane-2,7-dione

[19027-02-2]

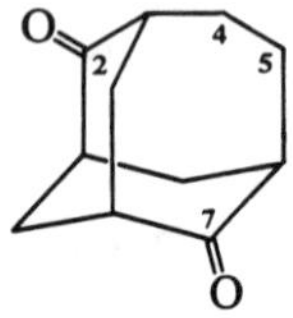

$C_{11}H_{14}O_2$ M 178.230
Cryst. (Et$_2$O/pet. ether). Mp 302-303°.

Vogt, B.R., *Tetrahedron Lett.*, 1968, 1579 (*synth*)
Dance, I.G. *et al, J. Chem. Soc., Perkin Trans. 2*, 1986, 1299 (*synth, ir, pmr, cmr*)

Tricyclo[4.3.1.1^{3,8}]undecane-4,5-dione, 9CI   T-70234

Homoadamantane-4,5-dione

[26775-76-8]

$C_{11}H_{14}O_2$ M 178.230
Cryst. (pet. ether). Mp 287° dec.

Schlatman, J.L. *et al, Tetrahedron*, 1970, **26**, 949 (*synth, ir, uv, pmr*)

2,4,6-Triethyl-1,3,5-triazine, 9CI   T-70235

[1009-74-1]

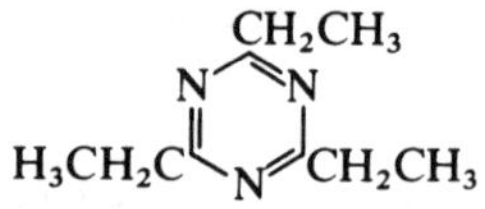

$$H_3CH_2C \overset{N\quad N}{\underset{N}{\bigtriangleup}} CH_2CH_3$$

CH$_2$CH$_3$

$C_9H_{15}N_3$ M 165.238
Mp 24-25°. Bp$_{10}$ 120°.

Kurabayashi, K., *Bull. Chem. Soc. Jpn.*, 1971, **44**, 3413.
Forsberg, J.H. *et al, J. Heterocycl. Chem.*, 1988, **25**, 767 (*synth, pmr*)

Trifarin   T-70236

Updated Entry replacing T-03155

[69625-34-9]

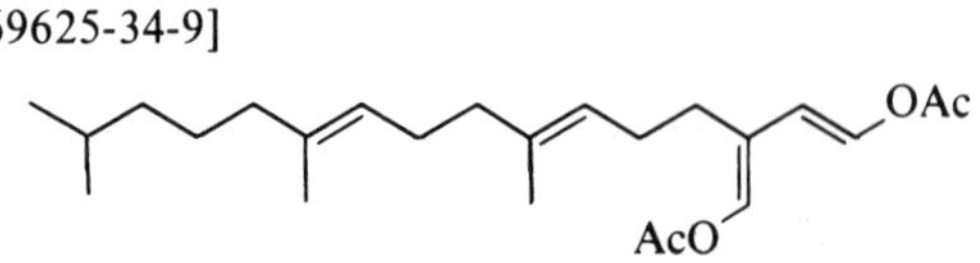

$C_{24}H_{38}O_4$ M 390.562
Constit. of *Caulerpa flexilis*. Oil.

6,7,10,11-Tetrahydro: **Bisdihydrotrifarin**.
$C_{24}H_{42}O_4$ M 394.593

Constit. of *Pseudochlorodesmis furcellata*. Oil. [α]$_D^{25}$ +1.6° (c, 0.68 in CHCl$_3$).

Blackman, A.J. *et al, Tetrahedron Lett.*, 1978, 3063 (*isol*)
Paul, V.J. *et al, Phytochemistry*, 1988, **27**, 1011 (*deriv*)

Triflorestevione   T-70237

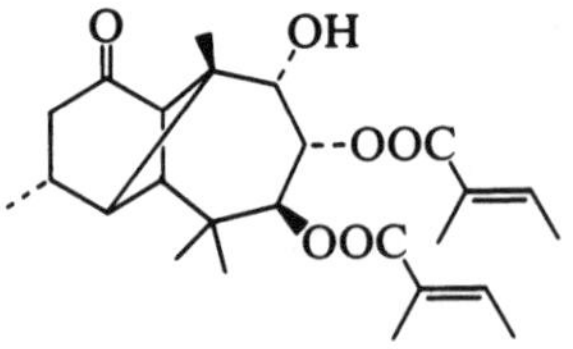

$C_{25}H_{36}O_6$ M 432.556
Constit. of *Stevia triflora*. Oil.

Amaro, J.M. *et al, Phytochemistry*, 1988, **27**, 1409.

2,3,5-Trifluorobenzoic acid   T-70238

$C_7H_3F_3O_2$ M 176.095
Mp 106°.

Finger, G.C. *et al, J. Am. Chem. Soc.*, 1959, **81**, 94 (*synth*)

2,3,6-Trifluorobenzoic acid   T-70239

[2358-29-4]
$C_7H_3F_3O_2$ M 176.095
Needles (C$_6$H$_6$). Mp 130-131° (125-126°).

Holland, D.G. *et al, J. Org. Chem.*, 1964, **29**, 3042 (*synth*)

2,4,5-Trifluorobenzoic acid   T-70240

[446-17-3]
$C_7H_3F_3O_2$ M 176.095
Cryst. Mp 97-98°.

Me ester: [20372-66-1].
   $C_8H_5F_3O_2$ M 190.122
   Liq. Bp$_{100}$ 130-135°.
Nitrile: [98349-22-5]. *2,4,5-Trifluorobenzonitrile. 1-Cyano-2,4,5-trifluorobenzene.*
   $C_7H_2F_3N$ M 157.095
   Liq. d 1.373.

Finger, G.C. *et al, CA*, 1955, **50**, 9312i (*synth*)
De Graw, J.I. *et al, J. Chem. Eng. Data*, 1968, **13**, 587 (*synth, deriv, pmr*)
Sanchez, J.P. *et al, J. Med. Chem.*, 1988, **31**, 983 (*synth, deriv, pmr*)

Trifluoromethanesulfonic acid, 9CI   T-70241

Updated Entry replacing T-03184
   Triflic acid
[1493-13-6]

$$F_3CSO_3H$$

CHF_3O_3S M 150.072
Catalyst for isomerisations, alkylations etc. Liq. Misc. H$_2$O. d$_4^{20}$ 1.69. Bp 162°. Fumes in moist air; very hygroscopic, forming monohydrate, Mp 34°. Very strong acid.

Me ester: [333-27-7].
   $CH_2F_3NO_2S$ M 149.088

Important reagent for methylations. d_4^{20} 1.45. Bp 99°.

▷Corrosive, severe poison

Et ester: [425-75-2].
 $C_3H_5F_3O_3S$ M 178.126
 Bp 115°.

Trifluoromethyl ester: [3582-05-6]. *Trifluoromethyl trifluoromethanesulfonate.*
 $C_2F_6O_3S$ M 218.071
 Trifluoromethylating agent. Liq.

Trifluoroacetyl ester: [68602-57-3]. *Trifluoroacetyl trifluoromethanesulfonate.*
 $C_3F_6O_4S$ M 246.081
 Trifluoroacetylating reagent. Bp 62.5°.

Trimethylsilyl ester: [27607-77-8].
 $C_4H_9F_3O_3SSi$ M 222.254
 Silylating agent for carbonyl compds.

Fluoride: [335-05-7].
 CF_4O_2S M 152.063
 Gas. Bp −21.7°. Hydrol. by H_2O only v. slowly at r.t., slowly at 100°.

Chloride: [421-83-0].
 $CClF_3O_2S$ M 168.518
 Bp 31.6°.

Anhydride: [358-23-6].
 $C_2F_6O_5S_2$ M 282.129
 Reagent for synth. of triflates and dehydration of aldoximes. Bp 84°.

Amide: [421-85-2].
 $CH_2F_3NO_2S$ M 149.088
 Mp 117-118°.

Diethylamide: [28048-17-1].
 $C_3H_6F_3NO_2S$ M 177.141
 Bp_7 55°.

Haszeldine, R.N. *et al, J. Chem. Soc.*, 1954, 4228 (*synth*)
Gramstad, T. *et al, J. Chem. Soc.*, 1956, 173 (*derivs*)
Burdon, J. *et al, J. Chem. Soc.*, 1957, 2574 (*synth*)
Robinson, E.A., *Can. J. Chem.*, 1961, **39**, 247 (*ir*)
Beard, C.D. *et al, J. Org. Chem.*, 1973, **38**, 3673 (*synth*)
Howells, R.D. *et al, Chem. Rev.*, 1977, **77**, 69 (*rev*)
Forbus, T.R. *et al, J. Org. Chem.*, 1979, **44**, 313; 1987, **52**, 4156 (*derivs*)
Taylor, S.L. *et al, J. Org. Chem.*, 1987, **52**, 4147 (*deriv, synth, use, bibl*)
Fieser, M. *et al, Reagents for Organic Synthesis*, Wiley, 1967-84, **6**, 406; **7**, 390.

2-(Trifluoromethyl)benzenesulfonic acid   T-70242
α,α,α-Trifluoro-o-toluenesulfonic acid

$C_7H_5F_3O_3S$ M 226.170
Chloride:
 $C_7H_4ClF_3O_2S$ M 244.616
 Bp_5 123-126°.
Amide:
 $C_7H_6F_3NO_2S$ M 225.185
 Cryst. (EtOH aq.). Mp 186-188°.

Yale, H.L. *et al, J. Org. Chem.*, 1960, **25**, 1824 (*synth*)

4-(Trifluoromethyl)benzenesulfonic acid   T-70243
α,α,α-Trifluoro-p-toluenesulfonic acid
$C_7H_5F_3O_3S$ M 226.170
Na salt: Solid.

Chloride: [2991-42-6].
 $C_7H_4ClF_3O_2S$ M 244.616
 Cryst.
Amide:
 $C_7H_6F_3NO_2S$ M 225.185
 Cryst. (EtOH aq.). Mp 176-177°.

Yale, H.L. *et al, J. Org. Chem.*, 1960, **25**, 1824 (*synth*)

1,1,1-Trifluoro-4-phenyl-3-butyn-2-one, 9CI   T-70244
[58518-08-4]

$$PhC{\equiv}CCOCF_3$$

$C_{10}H_5F_3O$ M 198.144
Liq. Bp_{24} 93-94°.

Kitazume, T. *et al, J. Fluorine Chem.*, 1985, **30**, 189 (*synth, pmr, F nmr, ir*)
Linderman, R.J. *et al, Tetrahedron Lett.*, 1987, **28**, 5271 (*use*)

Trifluoro-1,2,3-triazine   T-70245
[112291-51-7]

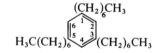

$C_3F_3N_3$ M 135.048
Liq.

Chambers, R.D. *et al, Tetrahedron*, 1988, **44**, 2583 (*synth, cmr, F nmr, ms, uv*)

1,3,5-Triheptylbenzene   T-70246
[29536-29-6]

$$\begin{array}{c}(CH_2)_6CH_3\\ H_3C(CH_2)_6 \quad (CH_2)_6CH_3\end{array}$$

$C_{27}H_{48}$ M 372.676
Liq. $Bp_{0.02}$ 152-154°.

Eapen, K.C. *et al, J. Org. Chem.*, 1984, **49**, 478; 1988, **53**, 5564 (*synth, ir, pmr, ms*)

1,3,5-Trihexylbenzene   T-70247
[29536-28-5]

$$\begin{array}{c}(CH_2)_6CH_3\\ H_3C(CH_2)_5 \quad (CH_2)_5CH_3\end{array}$$

$C_{24}H_{42}$ M 330.596
Liq. $Bp_{0.03}$ 138-140°.

Eapen, K.C. *et al, J. Org. Chem.*, 1984, **49**, 478; 1988, **53**, 5564 (*synth, ir, pmr, ms*)

2′,4′,6′-Trihydroxyacetophenone, 8CI   T-70248
Updated Entry replacing T-03241
1-(2,4,6-Trihydroxyphenyl)ethanone, 9CI. Phloracetophenone. Acetophloroglucinol
[480-66-0]
$C_8H_8O_4$ M 168.149
Isol. from bark of *Prunus domestica*. Needles (H_2O). Mp 219° (anhyd).

2',4'-Di-Me ether: [90-24-4]. 2'-Hydroxy-4',6'-dimeth-
oxyacetophenone. **Xanthoxylin**. Brevifolin.
$C_{10}H_{12}O_4$ M 196.202
Obt. from Xanthoxylum piperitum, X. alatum
(Rutaceae), Artemisia brevifolia (Compositae),
Hippomane mancinella and Sapium sebiferun
(Euphorbiaceae). Cryst. (EtOH). Mp 85-88°.
2',4'-Di-Me ether, oxime:
$C_{10}H_{13}NO_4$ M 211.217
Mp 108-110°.
2',4'-Di-Me ether, Ac: [59263-72-8].
$C_{12}H_{14}O_5$ M 238.240
Prisms (EtOH). Mp 107°.
Tri-Me ether: [832-58-6]. 2',4',6'-
Trimethoxyacetophenone.
$C_{11}H_{14}O_4$ M 210.229
Isol. from tubers of Lycoris radiata. Prisms. Mp 103°.
2'-Me ether, 4'-O-(3-methyl-2-butenyl): 1-Acetyl-4-iso-
pentenyl-6-methylphloroglucinol.
$C_{14}H_{18}O_4$ M 250.294
Isol. from Leucanthemopsis pulverentula. Mp 60-61°.
4'-Me ether: [7507-89-3]. 2',6'-Dihydroxy-4'-methoxya-
cetophenone. 4-O-Methylphloracetophenone.
$C_9H_{10}O_4$ M 182.176
Isol. from the bark of P. domestica. Mp 141-142°.
4-Me ether, O-β-D-glucoside: [24587-97-1].
Domesticoside.
$C_{15}H_{20}O_7$ M 312.319
Glycoside from the bark of P. domestica. Needles
(MeOH). Mp 189-191° (sinters at 183°).

v. Kostanecki, S. et al, Ber., 1899, 32, 2262.
Shinoda, J. et al, CA, 1928, 22, 2947.
Howells, H.P. et al, J. Am. Chem. Soc., 1932, 54, 2452.
Org. Synth., Coll. Vol., 2, 522.
Dean, F.M. et al, J. Chem. Soc., 1953, 1241.
Crow, W.D., Aust. J. Chem., 1976, 29, 2525.
Nagarajan, G.R. et al, Indian J. Chem., Sect. B, 1977, 15, 955
(isol, struct, derivs)
Nagarajan, G.R. et al, Phytochemistry, 1977, 16, 614.
de Pascual-Teresa, J. et al, Phytochemistry, 1982, 21, 791
(deriv)

2,3,9-Trihydroxy-6H-benzofuro[3,2-c][1]-   T-70249
benzopyran-6-one, 9CI

2,3,9-Trihydroxycoumestone

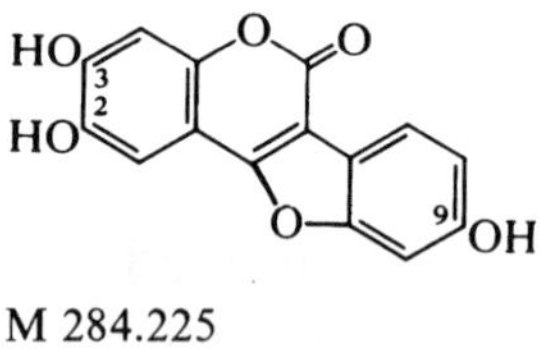

$C_{15}H_8O_6$ M 284.225
2-Me ether: [114216-99-8]. 3,9-Dihydroxy-2-
methoxycoumestone.
$C_{16}H_{10}O_6$ M 298.251
Constit. of Tephrosia hamiltonii. Cryst.
(C_6H_6/EtOAc). Mp 318-320°.
Tri-Me ether: Cryst. (CHCl3). Mp 255°.

Rajani, P. et al, Phytochemistry, 1988, 27, 648.

2,3,5-Trihydroxybenzoic acid, 9CI, 8CI   T-70250

Updated Entry replacing T-03276
[33580-60-8]
$C_7H_6O_5$ M 170.121
Cryst. (AcOH). Mp 234.5-235.5°.
Me ester: [110361-76-7].
$C_8H_8O_5$ M 184.148

Yellow cryst. solid (EtOAc/hexane). Mp 63-64°.
2,3-Di-Me ether: 5-Hydroxy-2,3-dimethoxybenzoic
acid.
$C_9H_{10}O_5$ M 198.175
Cryst. (H_2O). Mp 186°.
Tri-Me ether: [36873-96-8]. 2,3,5-Trimethoxybenzoic
acid.
$C_{10}H_{12}O_5$ M 212.202
Plates (H_2O). Mp 105°, Mp 141-143°.
Tri-Me ether, nitrile: 2,3,5-Trimethoxybenzonitrile. 1-
Cyano-2,3,5-trimethoxybenzene.
$C_{10}H_{11}NO_3$ M 193.202
Needles (CH_2Cl_2/pet. ether). Mp 61-62.5°.

Schock, R.U. et al, J. Org. Chem., 1951, 16, 1772.
Kreuchunas, A., J. Org. Chem., 1956, 21, 910 (synth)
Hill, R.A. et al, J. Chem. Soc., Perkin Trans. 1, 1987, 2209
(deriv, synth, ir, pmr)
Rizzacasa, M.A. et al, J. Chem. Soc., Perkin Trans. 1, 1988,
2425 (synth, deriv, pmr)

2,3,24-Trihydroxy-13,27-cyclo-11-oleanen-   T-70251
28-oic acid

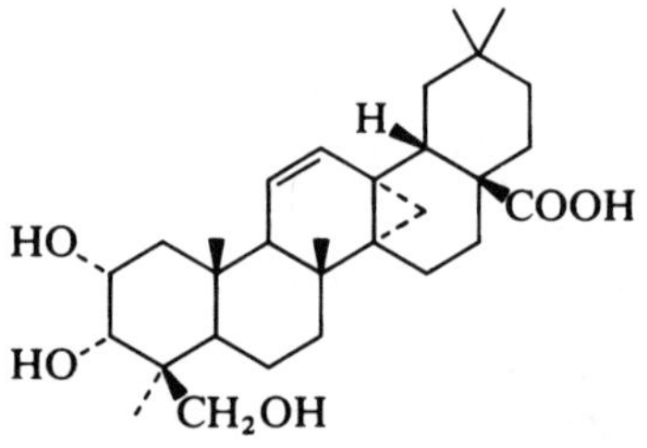

$C_{30}H_{46}O_5$ M 486.690
(2α,3α,13α)-form
Constit. of Prunella vulgaris.
Me ester: Cryst. (MeOH). Mp >310°. $[\alpha]_D^{24}$ +21.2° (c,
0.17 in CHCl3).

Kojima, H. et al, Phytochemistry, 1988, 27, 2921.

2,3,24-Trihydroxy-12,27-cyclo-14-tarax-   T-70252
eren-28-oic acid

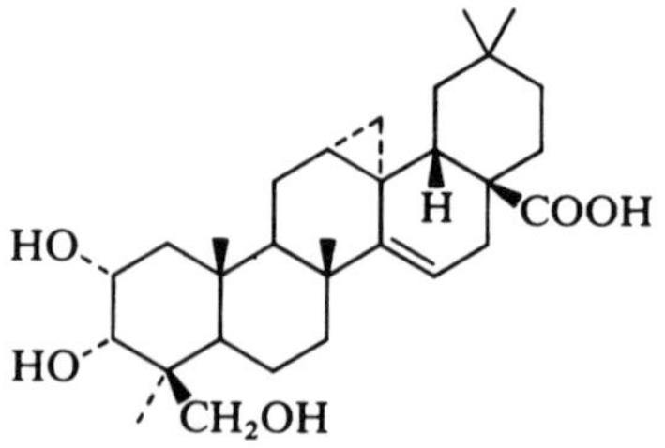

$C_{30}H_{46}O_5$ M 486.690
(2α,3α,12α)-form
Constit. of Prunella vulgaris.
Me ester: Cryst. (MeOH). Mp 276-278°. $[\alpha]_D^{24}$ +24° (c,
0.5 in CHCl3).

Kojima, H. et al, Phytochemistry, 1988, 27, 2921.

3,4,6-Trihydroxy-1,2-dimethylcarbazole   T-70253

1,2-Dimethyl-9H-carbazole-3,4,6-triol, 9CI

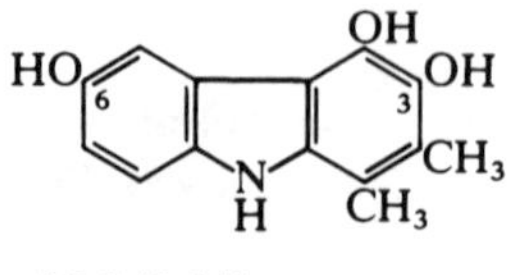

$C_{14}H_{13}NO_3$ M 243.262

3,6-Di-Me ether: [108073-62-7]. *4-Hydroxy-3,6-dimethoxy-1,2-dimethylcarbazole.* **Carbazomycin C.**
$C_{16}H_{17}NO_3$ M 271.315
Isol. from *Streptoverticillium ehimense.* Pale-yellow prisms (EtOAc/hexane). Mp 198-198.5°.

Tri-Me ether: [108073-63-8]. *3,4,6-Trimethoxy-1,2-dimethylcarbazole.* **Carbazomycin D.**
$C_{17}H_{19}NO_3$ M 285.342
From *S. ehimense.* Needles (CH₂Cl₂/hexane). Mp 129.5-130°.

Naid, T. *et al, J. Antibiot.,* 1987, **40,** 157 (*isol, struct, props*)

3,7,15-Trihydroxy-11,23-dioxo-8,20(22)-lanostadien-26-oic acid   T-70254

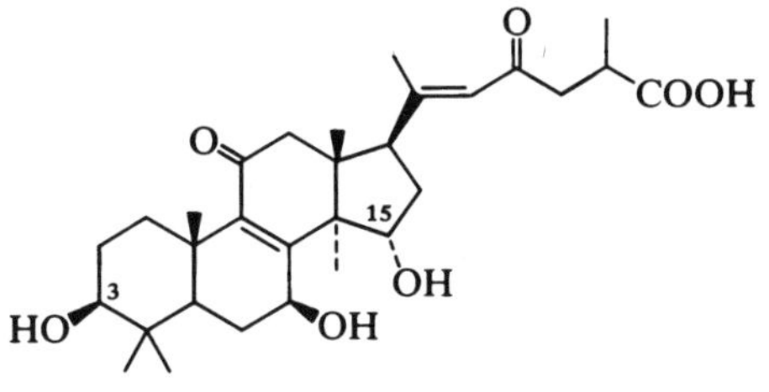

$C_{30}H_{44}O_7$ M 516.673
(3β,7β,15α,20(22)E)-form [100665-42-7]
Ganoderenic acid C
Constit. of fungus *Ganoderma lucidum.* Amorph. powder. $[\alpha]_D^{22}$ +66.2° (c, 0.15 in CHCl₃/MeOH).

3-Ketone: [100665-40-5]. *7β,15α-Dihydroxy-3,11,23-trioxo-8,20(22)E-lanostadien-26-oic acid.* **Ganoderenic acid A.**
$C_{30}H_{42}O_7$ M 514.658
Constit. of *G. lucidum.* Amorph. powder. $[\alpha]_D^{22}$ +127.8° (c, 0.43 in CHCl₃).

15-Ketone: [100665-41-6]. *3β,7β-Dihydroxy-11,15,23-trioxo-8,20(22)E-lanostadien-26-oic acid.* **Ganoderenic acid B.**
$C_{30}H_{42}O_7$ M 514.658
Constit. of *G. lucidum.* Plates (EtOAc/MeOH). Mp 211-214°. $[\alpha]_D^{22}$ +102.9° (c, 0.30 in CHCl₃).

3,15-Diketone: [100665-43-8]. *7β-Hydroxy-3,11,15,23-tetraoxo-8,20(22)-lanostadien-26-oic acid.* **Ganoderenic acid D.**
$C_{30}H_{40}O_7$ M 512.642
Constit. of *G. lucidum.* Needles (EtOH aq.). Mp 214-216°. $[\alpha]_D^{22}$ +163.4° (c, 0.34 in CHCl₃).

Komoda, Y. *et al, Chem. Pharm. Bull.,* 1985, **33,** 4829.

5,6,15-Trihydroxy-7,9,11,13,17-eicosapentaenoic acid, 9CI   T-70255

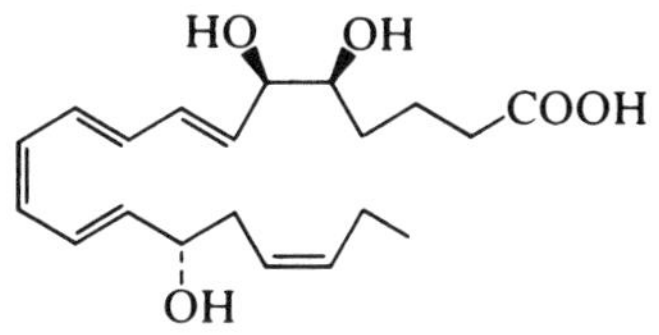

$C_{20}H_{30}O_5$ M 350.454
(5S,6R,7E,9E,11Z,13E,15S,17Z)-form [110657-98-2]
Lipoxin A₅. LXA₅
Isol. from incubation mixt. of porcine leucocytes and 15-hydroperoxyicosapentaenoic acid. Prob. stereochem. by analogy.
Me ester: $[\alpha]_D^{23}$ +11.2° (c, 1.76 in CHCl₃).

Wong, P.Y.-K. *et al, Biochem. Biophys. Res. Commun.,* 1985, **126,** 763 (*isol*)

Nicolaou, K.C. *et al, Angew. Chem., Int. Ed. Engl.,* 1987, **26,** 1019 (*synth, ir, pmr, uv*)

5,14,15-Trihydroxy-6,8,10,12,17-eicosapentaenoic acid, 9CI   T-70256

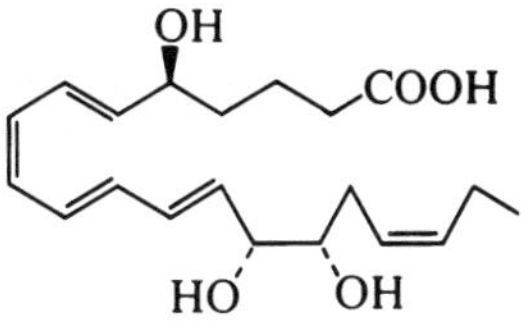

$C_{20}H_{30}O_5$ M 350.454
(5S,6E,8Z,10E,12E,14R,15S,17Z)-form [110657-99-3]
Lipoxin B₅. LXB₅
Isol. from incubation of porcine leucocytes with 15-hydroperoxyicosapentaenoic acid. Prob. stereochem. by analogy.
Me ester: $[\alpha]_D^{23}$ +14° (c, 0.570 in MeOH).

Wong, P.Y.-K. *et al, Biochem. Biophys. Res. Commun.,* 1985, **126,** 763 (*isol*)
Nicolaou, K.C. *et al, Angew. Chem., Int. Ed. Engl.,* 1987, **26,** 1019 (*synth, ir, pmr, uv*)

1,3,9-Trihydroxy-4(15),11(13)-eudesmadien-12,6-olide   T-70257

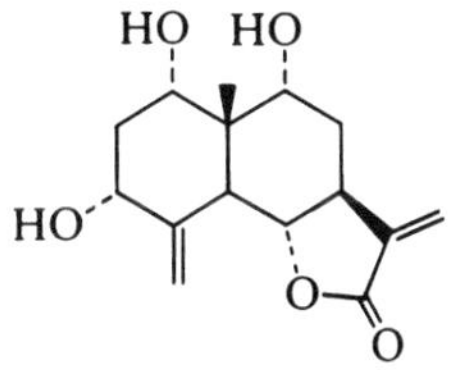

$C_{15}H_{20}O_5$ M 280.320
(1α,3α,6α,9α)-form
Cratystyolide
3-Ac:
$C_{17}H_{22}O_6$ M 322.357
Constit. of *Cratystylis conocephala.* Cryst. Mp 222-223°. $[\alpha]_D^{24}$ +133° (c, 0.25 in CHCl₃).
1,9-Di-Ac:
$C_{19}H_{24}O_7$ M 364.394
Constit. of *C. conocephala.* Cryst. Mp 236°.
Tri-Ac:
$C_{21}H_{26}O_8$ M 406.432
Constit. of *C. conocephala.* Cryst. Mp 196-198°. $[\alpha]_D^{24}$ +112° (c, 0.73 in CHCl₃).

Zdero, C. *et al, Phytochemistry,* 1988, **27,** 865.

1,3,13-Trihydroxy-4(15),7(11)-eudesmadien-12,6-olide   T-70258

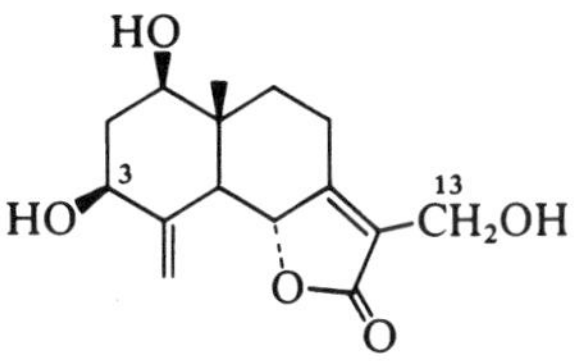

$C_{15}H_{20}O_5$ M 280.320
(1β,3β,6α)-form
3-Ac: [115338-10-8]. *13-Desacetyleudesmaafraglaucolide.*
$C_{17}H_{22}O_6$ M 322.357
Constit. of *Artemisia judaica.* Oil. $[\alpha]_D^{24}$ +82° (c, 0.88 in CHCl₃).

3,13-Di-Ac: [115346-36-6]. **Eudesmaafraglaucolide**.
$C_{19}H_{24}O_7$ M 364.394
Constit. of *A. afra*. Oil.

Jakupovic, J. *et al, Phytochemistry*, 1988, **27**, 1129 (*isol*)
Khafagy, S.M. *et al, Phytochemistry*, 1988, **27**, 1125 (*isol*)

1,4,6-Trihydroxy-11(13)-eudesmen-12,8-olide   T-70259

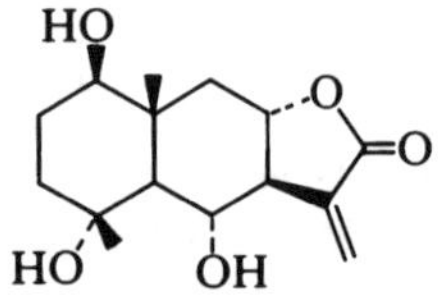

$C_{15}H_{22}O_5$ M 282.336

(1β,4α,5α,8α)-form

1,6-Di-Ac: 1β,6α-Diacetoxy-4α-hydroxy-11-eudesmen-12,8α-olide.
$C_{19}H_{26}O_7$ M 366.410
Constit. of *Gnephosis arachnoidea*. Gum. $[\alpha]_D^{24}$ −7°
(c, 0.25 in CHCl$_3$).

Jakupovic, J. *et al, Phytochemistry*, 1988, **27**, 3181.

3′,4′,7-Trihydroxyflavan   T-70260

3,4-Dihydro-2-(3,4-dihydroxyphenyl)-7-hydroxy-2H-1-benzopyran

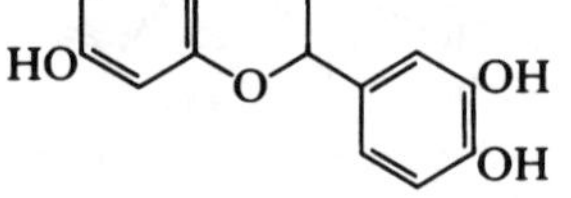

$C_{15}H_{14}O_4$ M 258.273

3′,7-Di-Me ether: [113236-00-3]. 4′-Hydroxy-3′,7-dimethoxyflavan.
$C_{17}H_{18}O_4$ M 286.327
Constit. of *Virola calophylloidea*. Cryst. (Me$_2$CO).
Mp 104-105°.

Martinez V., J.C. *et al, J. Nat. Prod.*, 1987, **50**, 1045.

2,4,5-Trihydroxy-9H-fluoren-9-one   T-70261

Updated Entry replacing D-50499

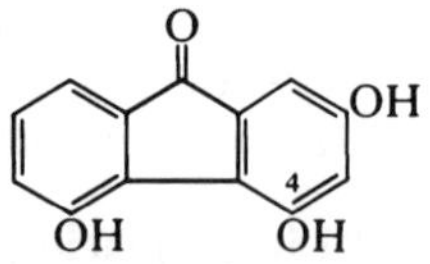

$C_{13}H_8O_4$ M 228.204

4-Me ether: [97915-33-8]. 2,5-Dihydroxy-4-methoxy-9H-fluoren-9-one. **Dengibsin**.
$C_{14}H_{10}O_4$ M 242.231
Constit. of *Dendrobium gibsonii*. Red needles
(Me$_2$CO/pet. ether). Mp 238-240°. Struct. revised in 1987.

Talapatra, S.K. *et al, Tetrahedron*, 1985, **41**, 2765 (*isol*)
Sargent, M.V., *J. Chem. Soc., Perkin Trans. 1*, 1987, 2553 (*synth, struct*)

2,5,7-Trihydroxy-3-(5-hydroxyhexyl)-1,4-naphthoquinone   T-70262

PO 1. Antibiotic PO1
[111897-38-2]

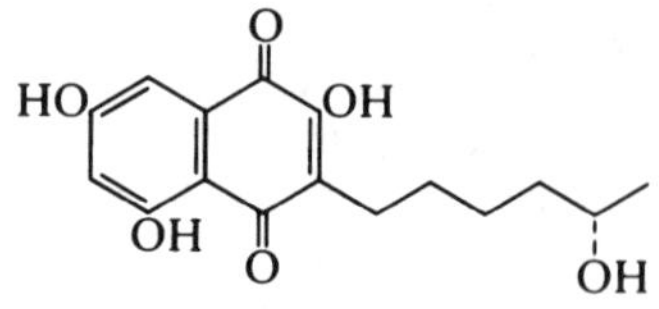

$C_{16}H_{18}O_6$ M 306.315
Naphthoquinone antibiotic. Isol. from *Penicillium* sp.
511. Shows antifungal props. Red needles. Mp 166-168°. $[\alpha]_D^{20}$ −19° (c, 1 in EtOH). Biosynth. related to Curvularin.

Kobayashi, A. *et al, Agric. Biol. Chem.*, 1987, **51**, 2857 (*isol, struct, props*)

3′,4′,7-Trihydroxyisoflavanone   T-70263

$C_{15}H_{12}O_5$ M 272.257
4′-Me ether: [67492-31-3]. 3′,7-Dihydroxy-4′-methoxyisoflavanone.
$C_{16}H_{14}O_5$ M 286.284
Constit. of *Myroxylon balsamum*. Cryst. (MeOH).
Mp 185-188°.

Alaide, B. *et al, Phytochemistry*, 1978, **17**, 593 (*isol, struct*)
Jain, A.C. *et al, Indian J. Chem., Sect. B*, 1987, **26**, 628 (*synth*)

2′,4′,7-Trihydroxyisoflavone   T-70264

3-(2,4-Dihydroxyphenyl)-7-hydroxy-4H-1-benzopyran-4-one, 9Cl

[7678-85-5]

$C_{15}H_{10}O_5$ M 270.241
Constit. of *Phaseolus vulgaris*. Cryst. (CHCl$_3$/MeOH).
Mp 275° dec.

Tri-Me ether: 2′,4′,7-Trimethoxyisoflavone.
$C_{18}H_{16}O_5$ M 312.321
Cryst. Mp 148-149°.

Woodward, M.D., *Phytochemistry*, 1980, **19**, 921 (*isol*)
Brady, W.T. *et al, J. Org. Chem.*, 1988, **53**, 1353 (*synth*)

4′,7,8-Trihydroxyisoflavone   T-70265

7,8-Dihydroxy-3-(4-hydroxyphenyl)-4H-1-benzopyran-4-one

$C_{15}H_{10}O_5$ M 270.241
4′-Me ether: [37816-19-6]. 7,8-Dihydroxy-4′-methoxyisoflavone.
$C_{16}H_{12}O_5$ M 284.268
Plates (CHCl$_3$). Mp 168-169°.

Tri-Me ether: 4′,7,8-Trimethoxyisoflavone.
$C_{18}H_{16}O_5$ M 312.321
Needles. Mp 168-169°.

Jain, A.C. *et al, Indian J. Chem., Sect. B*, 1987, **26**, 1143.

3,15,22-Trihydroxy-7,9(11),24-lanosta-trien-26-oic acid   T-70266

Updated Entry replacing T-60329

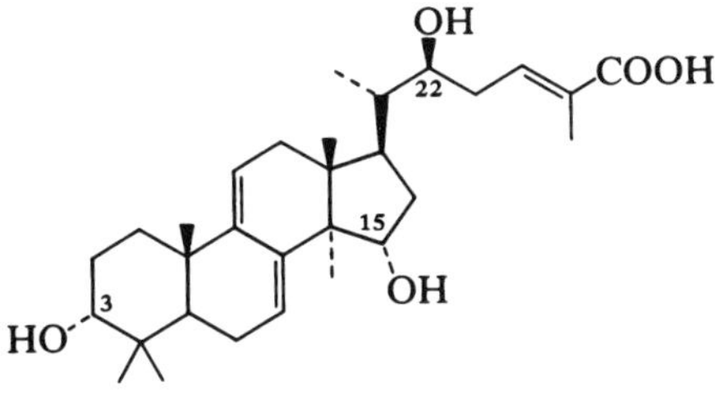

$C_{30}H_{46}O_5$ M 486.690

The Ganoderic/Ganodermic synonyms are best avoided as many of them have been duplicated. Use systematic names.

(3α,15α,22S,24E)-form

3-Ac: 3α-*Acetoxy-15α,22S-dihydroxy-7,9(11),24E-lan-ostatrien-26-oic acid.*
$C_{32}H_{48}O_6$ M 528.728
Constit. of *Ganoderma lucidum.*

3,22-Di-Ac: [110024-14-1]. 3α,22S-*Diacetoxy-15α-hydroxy-7,9(11),24E-lanostatrien-26-oic acid. Ganoderic acid* Q. *Ganodermic acid* P1.
$C_{34}H_{50}O_7$ M 570.765
Constit. of the cultured mycelium of *G. lucidum.*
Cryst. Mp 131-132°. Also descr. as a resin.

15,22-Di-Ac: [112667-14-8]. 15α,22S-*Diacetoxy-3α-hydroxy-7,9(11),24E-lanostatrien-26-oic acid. Ganoderic acid* P.
$C_{34}H_{50}O_7$ M 570.765
Constit. of the cultured mycelium of *G. lucidum.*
Cryst. Mp 211-212.5°.

Tri-Ac: [103992-91-2]. 3α,15α,22S-*Triacetoxy-7,9(11),24-lanostatrien-26-oic acid. Ganoderic acid* T.
$C_{36}H_{52}O_8$ M 612.802
Constit. of cultured mycelium of *G. lucidum.* Cryst.
Mp 200-202°. [α]$_D$ +23° (c, 0.13 in CHCl$_3$).

(3α,15α,22ξ,24E)-form

3,22-Di-Ac: 3α,22ξ-*Diacetoxy-15α-hydroxy-7,9(11),24-lanostatrien-26-oic acid. Ganoderic acid* Mk.
$C_{34}H_{50}O_7$ M 570.765
Constit. of *G. lucidum.* Syrup. [α]$_D^{23}$ +23° (c, 0.2 in MeOH). Not clear whether this is identical with Ganoderic acid Q above.

(3β,15α,22S,24E)-form

15,22-Di-Ac: [112430-69-0]. 15α,22S-*Diacetoxy-3β-hydroxy-7,9(11),24E-lanostatrien-26-oic acid. Gan-dodermic acid* P2.
$C_{34}H_{50}O_7$ M 570.765
Metab. of *G. lucidum.* Resin.

Tri-Ac: [112430-70-3]. 3β,15α,22S-*Triacetoxy-7,9(11),24-lanostatrien-26-oic acid.*
$C_{36}H_{52}O_8$ M 612.802
Metab. of *G. lucidum.*

Toth, J.O. *et al, Tetrahedron Lett.,* 1983, **24**, 1081 (*isol, deriv*)
Hirotani, M. *et al, Chem. Pharm. Bull.,* 1986, **34**, 2282 (*isol*)
Hirotani, M. *et al, Phytochemistry,* 1987, **26**, 2797 (*isol*)
Nishitoba, T. *et al, Agric. Biol. Chem.,* 1987, **51**, 1149 (*isol*)
Shiao, M.-S. *et al, Phytochemistry,* 1988, **27**, 873, 2911 (*isol*)

3,12,17-Trihydroxy-9(11)-lanosten-18,20-olide   T-70267

Updated Entry replacing T-50477

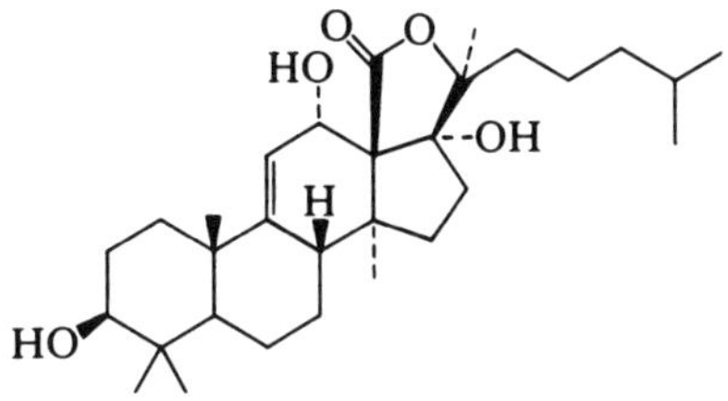

$C_{30}H_{48}O_5$ M 488.706
(3β,12α,17α,20S)-form illus.

(3β,12α,17α,20S)-form

Genuine aglycone of Echinosides *A* and *B,* antifungal gly-cosides from the sea cucumber *Actinopyga echinites.*

(3β,12β,17α,20S)-form

7,8-Dihydro-12β-hydroxyholothurinogenin
12-Me ether:
$C_{31}H_{50}O_5$ M 502.733
Aglycone from *Holothuria atra* and *H. scabra.*
12-Me ether, 3-Ac: Cryst. (MeOH). Mp 264-268°. [α]$_D$ –26°.

Kitagawa, I. *et al, Chem. Pharm. Bull.,* 1985, **33**, 5214 (*isol*)
Sarma, N.S. *et al, Indian J. Chem., Sect. B,* 1987, **26**, 715 (*isol*)

1,2,7-Trihydroxy-6-methylanthraquinone   T-70268

Updated Entry replacing T-50482

1,2,7-Trihydroxy-6-methyl-9,10-anthracenedione. 3,5,6-Trihydroxy-2-methylanthraquinone
$C_{15}H_{10}O_5$ M 270.241
Isol. from tissue cultures of *Morinda citrifolia.* Orange needles. Mp >300°. The same struct. was assigned (Keimatsu, 1929) to the substance Chrysarone. Mp 165-166°, isol. from *Rheum rhaponticum* by Hesse in 1908. In view of the Mp discrepancy this is unlikely.

O^2-β-*Primeveroside:*
$C_{26}H_{28}O_{14}$ M 564.499
Isol. from *M. citrifolia.* Yellow needles + 1H$_2$O. Mp 269-271°. [α]$_D$ –36.4° (DMF).

Inoue, K. *et al, Phytochemistry,* 1981, **20**, 1693 (*isol, uv, ir, pmr, synth*)

1,6,8-Trihydroxy-2-methylanthraquinone   T-70269

1,3,8-Trihydroxy-7-methylanthraquinone (*incorr.*)

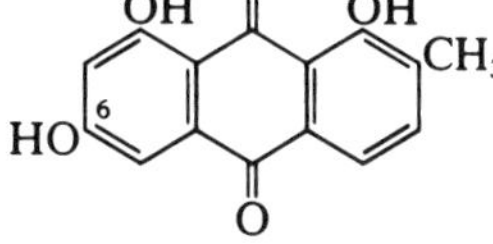

$C_{15}H_{10}O_5$ M 270.241

6-Me ether: [111228-19-4]. *1,8-Dihydroxy-6-methoxy-2-methylanthraquinone. 1,8-Dihydroxy-3-methoxy-7-methylanthraquinone* (*incorr.*).
$C_{16}H_{12}O_5$ M 284.268
Constit. of *Cassia mimosoides.* Yellow leaflets (AcOH). Mp 160-162°.

Mukherjee, K.S. *et al, J. Indian Chem. Soc.,* 1987, **64**, 130.

1,6,8-Trihydroxy-3-methylbenz[*a*]-anthracene-7,12-dione, 9CI   T-70270

Dehydrorabelomycin
[30954-70-2]

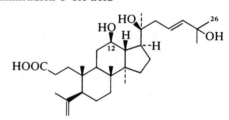

C$_{19}$H$_{12}$O$_5$ M 320.301
Anthraquinone antibiotic. Prod. by *Streptomyces murayamaensis*. Intermed. in biosynth. of Kinamycin *D*. Mp 201-203° (204-205°).

Tri-Ac: Mp 237-239°.

Liu, W.-C. *et al*, *J. Antibiot.*, 1970, **23**, 437.
Seaton, P.J. *et al*, *J. Am. Chem. Soc.*, 1987, **109**, 5282 (*isol*)

2,3,23-Trihydroxy-12-oleanen-28-oic acid   T-70271

Updated Entry replacing T-03447
C$_{30}$H$_{48}$O$_5$ M 488.706

(2α,3α)-form
Constit. of fruit of *Pseudopanax arboreum*.
Me ester: [55226-63-6]. Cryst. (EtOH). Mp 263-268°. [α]$_D^{20}$ +43° (c, 0.4 in CHCl$_3$).

(2α,3β)-form [465-00-9]
Arjunolic acid
Constit. of *Terminalia arjuna* and *Tristania conferta*.
Cryst. (Me$_2$CO). Mp 337-340°. [α]$_D^{19}$ +63.5° (c, 0.5 in EtOH).
β-D-Glucopyranosyl ester: [62369-72-6]. **Arjunglucoside II**.
C$_{36}$H$_{58}$O$_{10}$ M 650.848
Isol. from *T. arjuna*. Cryst. (MeOH). Mp 252°. [α]$_D$ +37° (c, 2.3 in EtOH).

(2β,3β)-form [6989-24-8]
Bayogenin
Sapogenin from bark of *Castanospermum australe* and other woods. Cryst. (dioxan). Mp 328-330°. [α]$_D^{27}$ +98° (c, 0.37 in dioxan).
Tri-Ac: Cryst. (EtOH). Mp 257-259°. [α]$_D^{23}$ +86° (c, 1.01 in CHCl$_3$).
Me ester: Cryst. (EtOH aq.). Mp 242-244°. [α]$_D^{24}$ +84° (c, 1.01 in CHCl$_3$).

Eade, R.A. *et al*, *Aust. J. Chem.*, 1963, **16**, 900; 1969, **22**, 2703 (*isol, struct*)
Cheung, H.T. *et al*, *J. Chem. Soc. (C)*, 1968, 1047 (*isol*)
Bowden, B.F. *et al*, *Aust. J. Chem.*, 1975, **28**, 91 (*isol*)
Honda, T. *et al*, *Bull. Chem. Soc. Jpn.*, 1976, **49**, 3213 (*Arjunglucoside II*)

5,8,9-Trihydroxy-10(14)-oplopen-3-one   T-70272

6,7,14-Trihydroxy-8(10)-oplopen-2-one (*incorr.*)

C$_{15}$H$_{24}$O$_4$ M 268.352

(8β,9α)-form
5-Ac, 9-angeloyl: 5-Acetoxy-9α-angeloyloxy-8β-hydroxy-10(14)-oplopen-3-one. 14-Acetoxy-7α-angeloyloxy-6β-hydroxy-8(10)-oplopen-2-one (*incorr.*).
C$_{22}$H$_{32}$O$_6$ M 392.491

Constit. of *Acrisione denticulata*. Oil. [α]$_D^{24}$ +8° (c, 0.76 in CHCl$_3$).

Aal, M.A. *et al*, *Phytochemistry*, 1988, **27**, 2599.

5,9,10-Trihydroxy-2*H*-pyrano[2,3,4-*kl*]-xanthen-2-one, 9CI   T-70273

7-Hydroxy-4-(3,4-dihydroxyphenyl)-2',5-oxidocoumarin
[87865-19-8]

C$_{15}$H$_8$O$_6$ M 284.225
Coumarinoid numbering shown. Constit. of *Exostema caribaeum*. Cryst. Mp 350° dec.
7-Me ether: [89783-74-4]. *9,10-Dihydroxy-5-methoxy-2H-pyrano[2,3,4-kl]xanthen-9-one, 9CI.*
C$_{16}$H$_{10}$O$_6$ M 298.251
From *E. caribaeum*. Cryst. (EtOH). Mp 273-274°.

Mata, R. *et al*, *J. Nat. Prod.*, 1987, **50**, 866.

12,20,25-Trihydroxy-3,4-seco-4(28),23-dammaradien-3-oic acid   T-70274

C$_{30}$H$_{50}$O$_5$ M 490.722
(12β,20S,23E)-form [117732-44-2]
Constit. of *Alnus japonica* flowers.
Aoki, T. *et al*, *Phytochemistry*, 1988, **27**, 2915.

20,25,26-Trihydroxy-3,4-seco-4(28),23-dammaradien-3-oic acid   T-70275

C$_{30}$H$_{50}$O$_5$ M 490.722
(20S,23E)-form [117732-43-1]
Constit. of *Alnus japonica* flowers.
Aoki, T. *et al*, *Phytochemistry*, 1988, **27**, 2915.

2,4,5-Triiodo-1*H*-imidazole, 9CI   T-70276

[1746-25-4]

C$_3$HI$_3$N$_2$ M 445.768
Cryst. (EtOH). Mp 190-191° (180-183°).
▷NI8680000.
B,HCl: Cryst. Mp 220°.
1-Me: [37067-97-3].
C$_4$H$_3$I$_3$N$_2$ M 459.795
Cryst. (EtOH aq.). Mp 93-94°.

1-Et:
   $C_5H_5I_3N_2$ M 473.821
   Cryst. (EtOH). Mp 141-142°.

Pauly, H. *et al, Ber.*, 1908, **41**, 3999; 1910, **43**, 2243 (*synth*)
Brunings, K.J., *J. Am. Chem. Soc.*, 1947, **69**, 205 (*synth*)
Iddon, B. *et al, J. Chem. Soc., Perkin Trans. 1*, 1983, 735 (*synth, props*)

Trimethidinium(2+)   **T-70277**
1,3,8,8-Tetramethyl-3-[3-(trimethylammonio)propyl]-3-azoniabicyclo[3.2.1]octane, 9CI. N-(γ-Trimethylammoniopropyl)-N-methylcamphidinium
[2624-50-2]

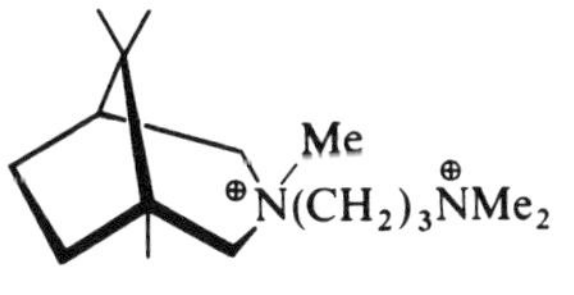

$C_{17}H_{36}N_2^{\oplus\oplus}$ M 268.485 (ion)
Ganglion blocking agent, antihypertensive.
Diiodide: [58594-48-2]. *Camphonium.*
   $C_{17}H_{36}I_2N_2$ M 522.294
   Mp 269-271°.
B,2Me₂SO₄: [14149-43-0]. **Trimethidinium methosulphate, BAN.** *Baratol. Camphidorium. Euprex. Ostensin. Ostensol. HA 106. Wy 1395.*
   $C_{19}H_{42}N_2O_8S_2$ M 490.669
   Cryst. Mp 192-193°.
▷CO1747500.

Bilecki, G., *Med. Klin.*, 1956, **51**, 1516 (*pharmacol*)
Rice, L.M. *et al, J. Org. Chem.*, 1956, **22**, 185 (*synth*)
Klupp, H., *Arzneim.-Forsch.*, 1957, **7**, 123 (*pharmacol*)
Ger. Pat., 1 086 703, (*1960*); *CA*, **55**, 27379 (*synth, pharmacol*)
Martindale, The Extra Pharmacopoeia, 28th/29th Eds., 1982/1989, Pharmaceutical Press, London, 912.

3-(2,4,5-Trimethoxyphenyl)-2-propenal, 9CI   **T-70278**
2,4,5-Trimethoxycinnamaldehyde
[106128-88-5]

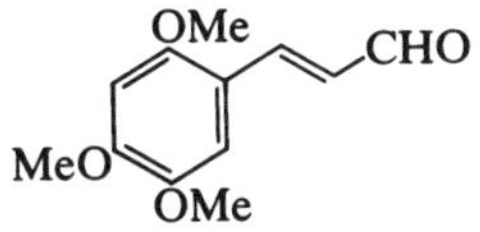

$C_{12}H_{14}O_4$ M 222.240
Constit. of *Caesulia axillaris*. Yellow cryst. Mp 140-142°.

Kulkarni, M.M. *et al, Indian J. Chem., Sect. B*, 1986, **25**, 981.

1,2,4-Trimethoxy-5-(2-propenyl)benzene   **T-70279**
1-Allyl-2,4,5-trimethoxybenzene. γ-Asarone
[5353-15-1]

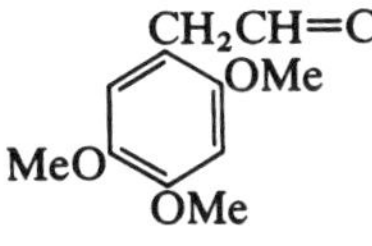

$C_{12}H_{16}O_3$ M 208.257
Constit. of *Caesulia axillaris* and *Acorus calamus*.
   Shows insect antifeedant props. Oil. Bp₂ 145-147°.

Devgan, O.N. *et al, Aust. J. Chem.*, 1968, **21**, 3001 (*isol, struct*)
Matsui, K. *et al, Agric. Biol. Chem.*, 1976, **40**, 1045 (*isol*)

Kulkarni, M.M. *et al, Indian J. Chem., Sect. B*, 1986, **25**, 981 (*isol, bibl*)

1,7,7-Trimethylbicyclo[2.2.1]heptane-2,5-   **T-70280**
diol
Updated Entry replacing T-20287
2,5-Bornanediol. 5-Hydroxyborneol

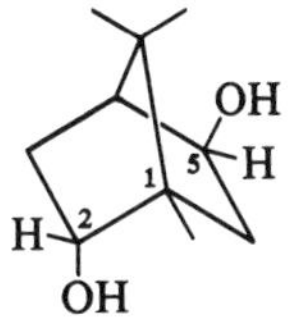

$C_{10}H_{18}O_2$ M 170.251
(1S,2R,5S)-form [64911-68-8]
   Angelicoidenol
   Constit. of *Pleurospermum angelicoides* and *Hedycarya angustifolia*. Cryst. (CHCl₃). Mp 255-257°. $[\alpha]_D^{25}$ −16.12° (c, 0.46 in MeOH).

Marquet, A. *et al, Bull. Soc. Chim. Fr.*, 1967, 128 (*synth*)
Allen, M.S. *et al, Can. J. Chem.*, 1979, **57**, 733 (*synth*)
Mahmood, U. *et al, Phytochemistry*, 1983, **22**, 774 (*isol*)
Gunawardana, Y.A.G.P. *et al, J. Nat. Prod.*, 1988, **51**, 142 (*isol, cmr, pmr*)

2,5,5-Trimethyl-2-cyclopenten-1-one   **T-70281**
[71221-73-3]

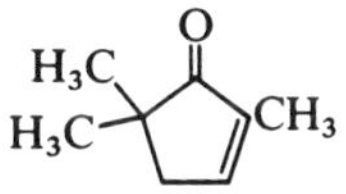

$C_8H_{12}O$ M 124.182
Liq. Bp₅₀ 110-130°.

Edwards, O.E. *et al, Can. J. Chem.*, 1963, **41**, 1592 (*synth, ir, uv*)
Adam, W. *et al, J. Am. Chem. Soc.*, 1986, **108**, 4556 (*synth*)

3,7,11-Trimethyl-1-dodecanol   **T-70282**
Hexahydrofarnesol
[6750-34-1]

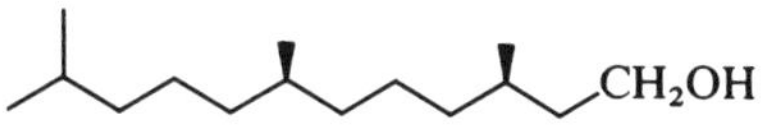

$C_{15}H_{32}O$ M 228.417
Bp 98-100°.
(3R,7R)-form [54154-26-6]
   Vitamin E side-chain. Oil. Bp₀.₁ 104-110°. $[\alpha]_D^{25}$ +3.58° (c, 3.075 in CHCl₃).
(3R,7S)-form [86238-50-8]
   $[\alpha]_D^{23}$ +3.56° (c, 1.805 in CHCl₃).

McCarthy, E.D. *et al, Tetrahedron*, 1967, **23**, 2609 (*synth, deriv*)
Koreeda, M. *et al, J. Org. Chem.*, 1983, **48**, 2122 (*synth*)
Rowland, S.J. *et al, J. Chromatogr. Sci.*, 1983, **21**, 298 (*resoln*)
Gramatica, P. *et al, Tetrahedron*, 1986, **42**, 6687 (*synth, ir, cmr, ms, pmr, bibl*)

2,6,10-Trimethyl-2,6,11-dodecatriene-1,10-   **T-70283**
diol

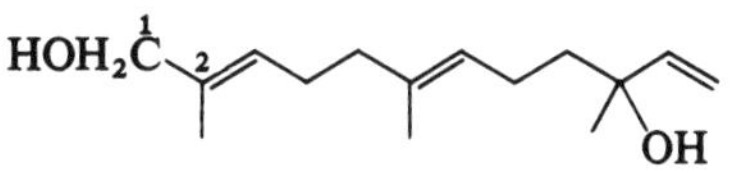

$C_{15}H_{26}O_2$ M 238.369

(2E,6E)-form [75628-02-3]
12-Hydroxynerolidol
Constit. of *Ursinia alpina*. Oil.
(2Z,6E)-form [75628-01-2]
13-Hydroxynerolidol
Constit. of *U. anethoides*. Oil.
1-Angelyl: [75628-00-1]. Constit. of *U. anethoides*. Oil.
Bohlmann, F. *et al*, *Phytochemistry*, 1980, **19**, 587.

3,7,11-Trimethyl-1,6,10-dodecatriene-3,9-diol   T-70284

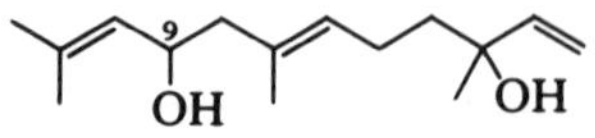

$C_{15}H_{26}O_2$ M 238.369

(E)-form
O^9-(6-O-*Acetyl-β-D-glucopyranoside*): [117479-91-1].
Gaillardoside.
$C_{23}H_{38}O_8$ M 442.548
Constit. of *Gaillardia coahuilensis*. Gum.
Gao, F. *et al*, *Phytochemistry*, 1988, **27**, 2685.

2,4,6-Trimethyl-3-nitroaniline   T-70285

Updated Entry replacing T-03976
2,4,6-Trimethyl-3-nitrobenzenamine, 9CI. 3-Nitromesidine
[1521-60-4]
$C_9H_{12}N_2O_2$ M 180.206
Golden-yellow needles (EtOH). Mp 75°.
N-*Ac:*
  $C_{11}H_{14}N_2O_3$ M 222.243
  Mp 75°.
N-*Benzoyl:*
  $C_{16}H_{16}N_2O_3$ M 284.314
  Mp 168.5°.
Benzenesulfonyl: Mp 162-163°.

Finger, G.C. *et al*, *J. Am. Chem. Soc.*, 1951, **73**, 149 (*synth*)
Adams, R. *et al*, *J. Am. Chem. Soc.*, 1957, **79**, 5525 (*synth*)
Ramaswami, R. *et al*, *Org. Prep. Proceed. Int.*, 1986, **18**, 361 (*synth, pmr, cmr, ms*)

2,6,7-Trimethylpteridine, 9CI   T-70286

[23767-00-2]

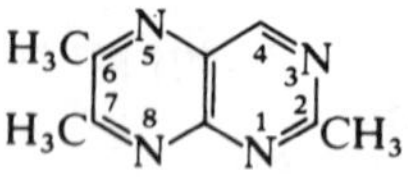

$C_9H_{10}N_4$ M 174.205
Yellow needles. Mp 134-135°.

Albert, A. *et al*, *J. Chem. Soc.*, 1954, 3832 (*synth*)
Albert, A. *et al*, *J. Chem. Soc.* (*C*), 1968, 2292 (*pmr*)

4,6,7-Trimethylpteridine, 9CI   T-70287

[19899-61-7]
$C_9H_{10}N_4$ M 174.205
Pale-yellow needles (pet. ether). Mp 139-140°.
Albert, A. *et al*, *J. Chem. Soc.* (*C*), 1968, 2292 (*synth, pmr*)

2,2′,2″-Trimethyltriphenylamine   T-70288

*2-Methyl-N,N-bis(2-methylphenyl)benzenamine, 9CI.
Tris(o-tolyl)amine. Tri-o-tolylamine*
[7287-73-2]

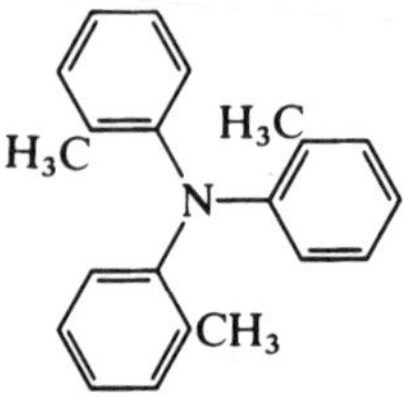

$C_{21}H_{21}N$ M 287.404
Cryst. (EtOH). Mp 103-104.5°.

Frye, C.L. *et al*, *J. Am. Chem. Soc.*, 1966, **88**, 2727 (*synth, pmr, cmr*)
Gauthier, S. *et al*, *Synthesis*, 1987, 383 (*synth*)

3,3′,3″-Trimethyltriphenylamine   T-70289

*3-Methyl-N,N-bis(3-methylphenyl)benzenamine, 9CI.
Tri-m-tolylamine, 8CI. Tris(m-tolyl)amine*
[20676-79-3]
$C_{21}H_{21}N$ M 287.404
Cryst. Mp 72°.

Cumper, C.W.N. *et al*, *J. Chem. Soc.* (*B*), 1971, 422 (*synth, props*)

4,4′,4″-Trimethyltriphenylamine   T-70290

*4-Methyl-N,N-bis(4-methylphenyl)benzenamine, 9CI.
Tri-p-tolylamine, 8CI. Tris(p-tolyl)amine*
[1159-53-1]
$C_{21}H_{21}N$ M 287.404
Forms v. stable radical cation. Cryst. powder. Mp 117°.
B,HClO₄: [34729-49-2]. Red-bronze cryst. Forms blue
  solns. in $CHCl_3$, Me_2CO and EtOH. Mp 129-130°.

Wieland, H., *Ber.*, 1907, **40**, 4260 (*synth*)
Weitz, E. *et al*, *Ber.*, 1926, **59**, 2307.
Granick, S., *J. Am. Chem. Soc.*, 1940, **62**, 2241 (*radical*)
Walter, R.I., *J. Am. Chem. Soc.*, 1955, **77**, 5999 (*synth, deriv, bibl*)
Iwaizumi, M. *et al*, *Bull. Chem. Soc. Jpn.*, 1965, **38**, 501 (*esr*)

1,2,3-Trimethyl-4,5,6-triphenylbenzene   T-70291

4′,5′,6′-Trimethyl-2′-phenyl-1,1′:3′,1″-terphenyl, 9CI
[20025-03-0]

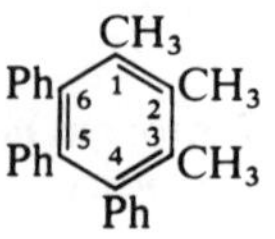

$C_{27}H_{24}$ M 348.487
Cryst. Mp 155°.

Dietl, H. *et al*, *J. Am. Chem. Soc.*, 1970, **92**, 2276 (*synth, pmr*)

1,2,4-Trimethyl-3,5,6-triphenylbenzene   T-70292

3′,4′,6′-Trimethyl-5′-phenyl-1,1′:3′,1″-terphenyl, 9CI
[10467-11-5]
$C_{27}H_{24}$ M 348.487

Brown cryst. (CHCl₃/MeOH). Mp 223°.

Hübel, W. *et al*, *Chem. Ber.*, 1960, **93**, 103 (*synth*)
Dietl, H. *et al*, *J. Am. Chem. Soc.*, 1970, **92**, 2276.
Jhingan, A.K. *et al*, *J. Org. Chem.*, 1987, **52**, 1161 (*synth, pmr, cmr*)

1,3,5-Trimethyl-2,4,6-triphenylbenzene   T-70293

2′,4′,6′-Trimethyl-5′-phenyl-1,1′:3′,1″-terphenyl, 9CI

[6231-26-1]

C₂₇H₂₄   M 348.487

Cryst. (CHCl₃/C₆H₆/MeOH). Mp 320°.

Hübel, W. *et al*, *Chem. Ber.*, 1960, **93**, 103 (*synth*)
Dietl, H. *et al*, *J. Am. Chem. Soc.*, 1970, **92**, 2276 (*synth, pmr*)
Jhingan, A.K. *et al*, *J. Org. Chem.*, 1987, **52**, 1161 (*synth, pmr, cmr*)

1,3,5-Trimorpholinobenzene   T-70294

4,4′,4″-(1,3,5-Benzenetriyl)trismorpholine, 9CI

[16857-97-9]

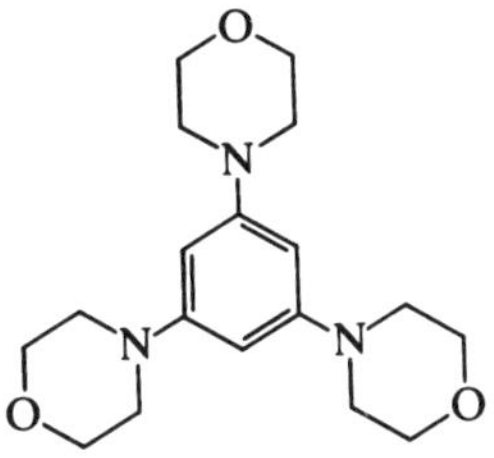

C₁₈H₂₇N₃O₃   M 333.430

Cryst. (DMF). Mp 308-312°. First protonation occurs partly on C to form a σ-complex, 2nd and 3rd protonations occur at N.

Effenberger, F. *et al*, *Chem. Ber.*, 1968, **101**, 3787 (*synth, uv, pmr*)
Knoche, W. *et al*, *Bull. Soc. Chim. Fr.*, 1988, 377 (*props, uv*)

1(15),8(19)-Trinervitadiene-2,3-diol   T-70295

Updated Entry replacing T-04164

2,3-Trinervidiol

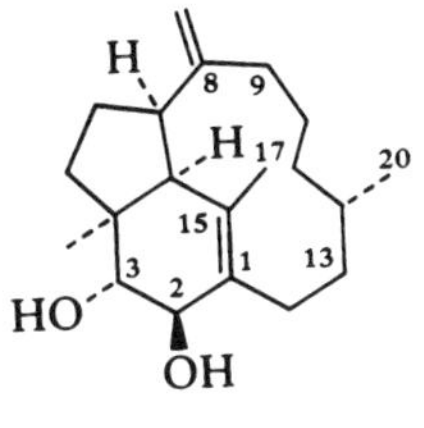

C₂₀H₃₂O₂   M 304.472

(*2β,3α*)-*form* [60753-26-6]

Constit. of the defence secretion of *Trinervitermes gratiosus* and *Nasutitermes rippertii*. Cryst. Mp 174-176°.

Prestwich, G.D. *et al*, *J. Am. Chem. Soc.*, 1976, **98**, 6062 (*isol, struct*)
Vrkoč, J. *et al*, *Collect. Czech. Chem. Commun.*, 1978, **43**, 1125 (*isol*)

1(15),8(19)-Trinervitadiene-2,3,9,20-tetrol   T-70296

C₂₀H₃₂O₄   M 336.470

(*2β,3α,9β*)-*form*

9,20-Di-Ac: [104793-17-1]. *1(15),8(19)-Trinervita-diene-2β,3α,9β,20-tetrol 9,20-diacetate. 9β,20-Diace-toxy-2β,3α-dihydroxy-1(15),8(19)-trinervitadiene.*
C₂₄H₃₆O₆   M 420.545
Constit. of secretion of *Nasutitermes* sp. Oil. Drawing error in the paper.

Braekman, J.C. *et al*, *Bull. Soc. Chim. Belg.*, 1986, **95**, 915.

11,15(17)-Trinervitadiene-3,9,13-triol   T-70297

Updated Entry replacing T-03534

3,9,13-Trihydroxy-11,15(17)-trinervitadiene

C₂₀H₃₂O₃   M 320.471

(*1β,3α,8β,9β,13α*)-*form*

Cryst. (EtOH aq.). Mp 277-278°.

3,9,13-Tripropanoyl:
C₂₉H₄₄O₆   M 488.663
Constit. of *Nasutitermes* spp.
9-Ac, 3,13-dipropanoyl:
C₂₈H₄₂O₆   M 474.636
Constit. of *N.* spp.

Prestwich, G.D. *et al*, *Tetrahedron Lett.*, 1981, 1563.

1(15),8(19)-Trinervitadiene-2,3,9-triol   T-70298

Updated Entry replacing T-04166

2,3,9-Trinervitriol

C₂₀H₃₂O₃   M 320.471

(*2β,3α,9α*)-*form*

9-Ac: [60791-30-2].
C₂₂H₃₄O₄   M 362.508
Constit. of defence secretion of *Trinervitemes gratiosus*.
2,3-Di-Ac: [60791-29-9].
C₂₄H₃₆O₅   M 404.545
Constit. of *T. gratiosus*.

Prestwich, G.D. *et al*, *J. Am. Chem. Soc.*, 1976, **98**, 6061, 6062 (*isol, cryst struct*)

1(15),8(19)-Trinervitadiene-2,3,20-triol   T-70299

2,3,20-Trihydroxy-1(15),8(19)-trinervitadiene

C₂₀H₃₂O₃   M 320.471

(*2β,3α*)-*form*

20-Ac: [104993-16-0]. *20-Acetoxy-2β,3α-dihydroxy-1(15),8(19)-trinervitadiene.*
C₂₂H₃₄O₄   M 362.508
Constit. of secretion of *Nasutitermes* sp. Drawing error in the paper.

Braekman, J.C. *et al*, *Bull. Soc. Chim. Belg.*, 1986, **95**, 915.

1(15),8(19)-Trinervitadien-9-ol   T-70300

Updated Entry replacing T-04165

Trinerviol

C₂₀H₃₂O   M 288.472

9β-form [60753-28-8]

Constit. of defence secretion of *Trinervitermes gratiosus* and *Nasutitermes rippertii*. Cryst. Mp 128-131°.

Prestwich, G.D. *et al*, *J. Am. Chem. Soc.*, 1976, **98**, 6062.

Vrkoč, J. *et al, Collect. Czech. Chem. Commun.*, 1978, **43**, 1125.

2,4,6-Trinitroaniline　　　　　T-70301

Updated Entry replacing T-04171
*Picramide. 2-Amino-1,3,5-trinitrobenzene. 2,4,6-
Trinitrobenzenamine*
[489-98-5]
$C_6H_4N_4O_6$　　M 228.121
Prisms (AcOH). Mp 192-195°.
▷Explosive
　N-*Ac:* [16400-86-5]. *2,4,6-Trinitroacetanilide.*
　　$C_8H_6N_4O_7$　　M 270.158
　　Needles (AcOH aq.). Mp ~230°.
　N-*Me:* [1022-07-7].
　　$C_7H_6N_4O_6$　　M 242.148
　　Yellow needles (EtOH). Mp 115°.
　N-*Di-Me:* [2493-31-4].
　　$C_8H_8N_4O_6$　　M 256.174
　　Yellow cryst. (C_6H_6). Mp 138°.

Hollemann, A.F., *Recl. Trav. Chim. Pays-Bas*, 1930, **49**, 112 (*synth*)
Satake, K. *et al, Bull. Chem. Soc. Jpn.*, 1961, **34**, 1346 (*synth*)
Holden, J.R. *et al, J. Phys. Chem.*, 1972, **76**, 3597 (*cryst struct*)
Lyčka, A. *et al, Collect. Czech. Chem. Commun.*, 1987, **52**, 2946 (*pmr, cmr, N nmr, bibl*)
Sax, N.I., *Dangerous Properties of Industrial Materials*, 5th Ed., Van Nostrand-Reinhold, 1979, 1064.

2,4,6-Trinitrobenzenesulfonic acid, 9CI, 8CI　　T-70302

Updated Entry replacing T-04179
[2508-19-2]

$C_6H_3N_3O_9S$　　M 293.164
Reagent for characteristation of amino acids. Cryst. (1M HCl). Mp 180°.
▷DB8400000.

Golumbic, C. *et al, J. Org. Chem.*, 1946, **11**, 518 (*synth*)
Okuyama, T. *et al, J. Biochem. (Tokyo)*, 1960, **47**, 454 (*use*)
Lyčka, A. *et al, Collect. Czech. Chem. Commun.*, 1987, **52**, 2946 (*pmr, cmr, N nmr*)

2,4,6-Trinitrobenzoic acid, 9CI, 8CI　　T-70303

Updated Entry replacing T-04185
[129-66-8]
$C_7H_3N_3O_8$　　M 257.116
Rhombohedra (H_2O). Mp 228°. Decarboxylates at Mp.
▷DI0920000.
　Me ester: [15012-38-1].
　　$C_8H_5N_3O_8$　　M 271.143
　　Orange-yellow plates (EtOH aq.). Mp 160-161°.
　Chloride: [7500-86-9].
　　$C_7H_2ClN_3O_7$　　M 275.562
　　Plates (C_6H_6). Mp 163°.
　▷DM6825000.
　Amide: [51226-42-7].
　　$C_7H_4N_4O_7$　　M 256.131
　　Cryst. (Me_2CO/pet. ether). Mp 264°.
　Anhydride:
　　$C_{14}H_4N_6O_{15}$　　M 496.217

Needles. Mp 270°.

Org. Synth., Coll. Vol., **1**, 528 (*synth, bibl*)
Lyčka, A. *et al, Collect. Czech. Chem. Commun.*, 1987, **52**, 2946 (*pmr, cmr, N nmr, bibl*)
Bretherick, L., *Handbook of Reactive Chemical Hazards*, 2nd Ed., Butterworths, London and Boston, 1979, 616.
Sax, N.I., *Dangerous Properties of Industrial Materials*, 5th Ed., Van Nostrand-Reinhold, 1979, 1064.

1,5,9-Trioxacyclododecane　　　　　T-70304

12-Crown-3

$C_9H_{18}O_3$　　M 174.239
Complexes alkali and alkaline earth metal cations. Liq. Bp_{12} 101-102°.

Dale, J. *et al, J. Chem. Soc., Chem. Commun.*, 1987, 1391 (*synth*)

1,1,4-Triphenyl-1,3-butadiene　　　　　T-70305

1,1′,1″-(1,3-Butadien-1-yl-4-ylidene)trisbenzene, 9CI
[18720-11-1]

$$Ph_2C{=}CHCH{=}CHPh$$

$C_{22}H_{18}$　　M 282.384
(*E*)-*form* [20235-61-4]
　Needles. Mp 96.5-97.5°. A Mp of 102-104° reported for a prepn. of undefined config., which was prob. also the (*E*)-form.

Bergmann, F. *et al, J. Chem. Soc.*, 1952, 2522 (*synth*)
Israelashvili, S. *et al, J. Chem. Soc.*, 1953, 1070 (*synth*)
Mitsudo, T. *et al, J. Org. Chem.*, 1987, **52**, 1695 (*synth, ir, pmr, cmr*)

1,2,3-Triphenylcyclopent[a]indene　　　　　T-70306

1,2,3-Triphenylbenzopentalene

$C_{30}H_{20}$　　M 380.488
Green cryst. (pet. ether). Mp 186-187°.

Le Goff, E., *J. Am. Chem. Soc.*, 1962, **84**, 1505 (*synth, uv, pmr*)
Brown, R.F.C. *et al, Tetrahedron Lett.*, 1988, **29**, 6791 (*synth*)

2,4,5-Triphenyl-1,3-dioxol-1-ium, 9CI　　　　　T-70307

$C_{21}H_{15}O_2^{\oplus}$　　M 299.348 (ion)
Trifluoromethanesulfonate: [106016-50-6].
　$C_{22}H_{15}F_3O_5S$　　M 448.413
　Moisture-sensitive yellow solid. Mp 171°.

Lorenz, W. *et al, J. Org. Chem.*, 1987, **52**, 375 (*synth, ir, pmr, cmr*)

1,2-Triphenylenedione   T-70308

1,2-Triphenylenequinone

[82120-28-3]

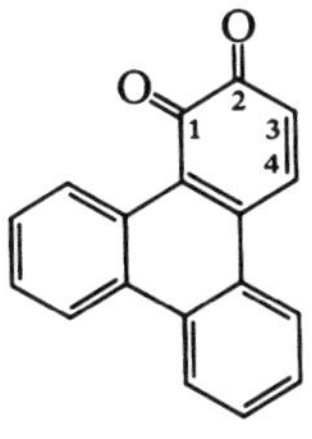

$C_{18}H_{10}O_2$ M 258.276

Red needles. Mp 188-190°. Rapidly darkens in air.

Teuber, H.J. *et al*, *Chem. Ber.*, 1959, **92**, 932.
Platt, K.L. *et al*, *Tetrahedron Lett.*, 1982, **23**, 163 (*synth*)

1,4-Triphenylenedione   T-70309

1,4-Triphenylenequinone

[76902-44-8]

$C_{18}H_{10}O_2$ M 258.276

Orange needles (Me$_2$CO). Mp 198.5-199.5°.

Klemm, L.H. *et al*, *J. Heterocycl. Chem.*, 1987, **24**, 1749 (*synth, uv, pmr, ms*)

Triphenyleno[1,12-*bcd*]thiophene   T-70310

Triphenyleno[4,5-bcd]thiophene

[68558-73-6]

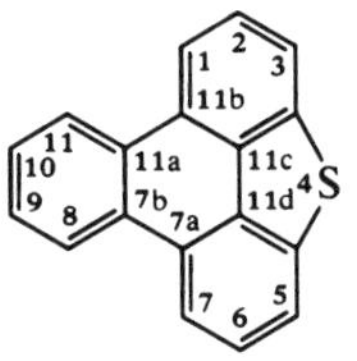

$C_{18}H_{10}S$ M 258.337

Mp 185-189°.

S-*Oxide:* [70583-30-1].

   $C_{18}H_{10}OS$ M 274.336

   Needles (MeOH). Mp 223-225°.

S-*Dioxide:* [70583-31-2].

   $C_{18}H_{10}O_2S$ M 290.336

   Platelets (MeOH/CHCl$_3$). Mp 325-327°.

Klemm, L.H. *et al*, *J. Heterocycl. Chem.*, 1979, **16**, 599; 1987, **24**, 1749 (*synth, uv, pmr*)

1,1,2-Triphenyl-1,2-ethanediol, *9CI*   T-70311

Triphenylethyleneglycol

[6296-95-3]

Ph$_2$COH
|
H—C—OH   (*R*)-*form*
|
Ph

$C_{20}H_{18}O_2$ M 290.361

(*R*)-*form* [95061-46-4]

Cryst. (CH$_2$Cl$_2$). Mp 126°. [α]$_D^{25}$ +214° (c, 1 in EtOH).

O^2-*Ac:* [95061-47-5].

   $C_{22}H_{20}O_3$ M 332.398

   Cryst. by subl. Mp 239°. [α]$_D^{25}$ +213° (c, 1 in Py).

(±)-*form*

Cryst. (Et$_2$O/hexane). Mp 165-167°.

McKenzie, A. *et al*, *J. Chem. Soc.*, 1910, **97**, 473 (*synth*)
Roger, R. *et al*, *J. Chem. Soc.*, 1931, 2229 (*props, bibl*)
Katritzky, A.R. *et al*, *Synthesis*, 1987, 415 (*synth, pmr*)
Devant, R. *et al*, *Chem. Ber.*, 1988, **121**, 397 (*synth, ir, pmr, ms*)

Triphenylphenylazomethane   T-70312

Phenyl(triphenylmethyl)diazene, 9CI

[981-18-0]

$$PhN=NCPh_3$$

$C_{25}H_{20}N_2$ M 348.446

Radical initiator. Yellow cryst. (EtOH). Mp 110-112° dec.

Gomberg, M., *J. Am. Chem. Soc.*, 1898, **20**, 773 (*synth*)
Cohen, S.G. *et al*, *J. Am. Chem. Soc.*, 1953, **75**, 5504 (*synth*)
Sigman, M.E. *et al*, *J. Org. Chem.*, 1987, **52**, 1165 (*use, bibl*)

1,1,3-Triphenyl-2-propen-1-ol   T-70313

α-Phenyl-α-(2-phenylethenyl)benzenemethanol, 9CI.
Diphenylstyrylcarbinol

[4663-36-9]

$$PhCH=CHCPh_2OH$$

$C_{21}H_{18}O$ M 286.373

Mp 110-112°.

▷GE2335000.

Lüttringhaus, A., *Ber.*, 1934, **67**, 1602 (*synth*)
Michael, A. *et al*, *J. Org. Chem.*, 1943, **8**, 60 (*synth*)
Hou, Z. *et al*, *J. Org. Chem.*, 1987, **52**, 3524 (*synth, ir, pmr*)

1,3,5-Tripiperidinobenzene   T-70314

1,1′,1″-(1,3,5-Benzenetriyl)trispiperidine, 9CI

[16857-95-7]

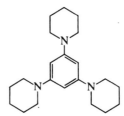

$C_{21}H_{33}N_3$ M 327.512

Cryst. (Me$_2$CO). Mp 183-184°. pK_{a1} 6.30, pK_{a2} 3.50, pK_{a3} 0.2. Protonates exclusively at N (distinction from 1,3,5-Tripyrrolidinobenzene, T-70319.

Effenberger, F. *et al*, *Chem. Ber.*, 1968, **101**, 3787 (*synth, uv, pmr*)
Knoche, W. *et al*, *Bull. Soc. Chim. Fr.*, 1988, 377 (*uv, props*)

Tri(2-propenyl)amine   T-70315

N,N-Di-2-propenyl-2-propen-1-amine, 9CI. Triallyla-mine, 8CI

[102-70-5]

$$(H_2C=CHCH_2)_3N$$

$C_9H_{15}N$ M 137.224

Oily liq. with unpleasant odour. d$_4^{14}$ 0.81. Bp 155-156°. pK_a 8.31 (20°).

▷Highly toxic vapour, flammable. XX5950000.

Butler, G.B. *et al*, *J. Chem. Educ.*, 1951, **28**, 191 (*synth*)
Sax, N.I., *Dangerous Properties of Industrial Materials*, 5th Ed., Van Nostrand-Reinhold, 1979, 1039.

2,4,6-Tri-2-pyridinyl-1,3,5-triazine   T-70316

Updated Entry replacing T-30365
[3682-35-7]

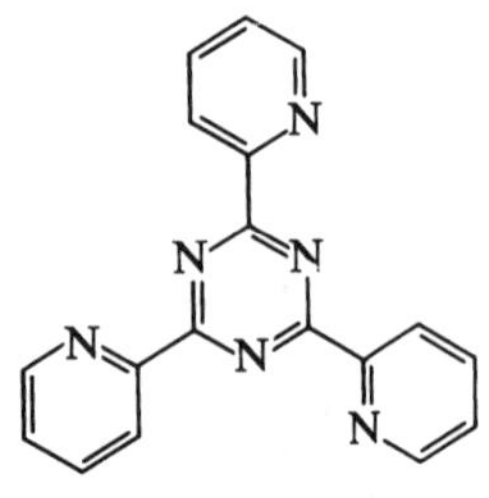

$C_{18}H_{12}N_6$ M 312.333
Forms complexes with transition metals. Anal. reagent for Fe. Cryst. + 3 or $4H_2O$. Mp 243-245°.

Case, F.H. *et al*, *J. Am. Chem. Soc.*, 1959, **81**, 905 (*synth*)
Vagg, R.S. *et al*, *Aust. J. Chem.*, 1967, **20**, 1841 (*synth, use*)
Barclay, G.A. *et al*, *Acta Crystallogr.*, Sect. B, 1977, **33**, 3487 (*cryst struct, bibl*)
Llobera, A. *et al*, *Synthesis*, 1985, 95 (*synth*)
Forsberg, J.H. *et al*, *J. Heterocycl. Chem.*, 1988, **25**, 767 (*synth*)

2,4,6-Tri-3-pyridinyl-1,3,5-triazine, 9CI   T-70317

[42333-76-6]
$C_{18}H_{12}N_6$ M 312.333
Cryst. (Py). Mp 311°.

Kurabayashi, M. *et al*, *Bull. Chem. Soc. Jpn.*, 1971, **44**, 3413 (*synth*)

2,4,6-Tri-4-pyridinyl-1,3,5-triazine, 9CI   T-70318

[42333-78-8]
$C_{18}H_{12}N_6$ M 312.333
Cryst. (Py). Mp 374°.

Kurabayashi, K., *Bull. Chem. Soc. Jpn.*, 1971, **44**, 3413 (*synth*)
Forsberg, J.H. *et al*, *J. Heterocycl. Chem.*, 1988, **25**, 767 (*synth*)

1,3,5-Tripyrrolidinobenzene   T-70319

1,1′,1″-(1,3,5-Benzenetriyl)trispyrrolidine
[16857-93-5]

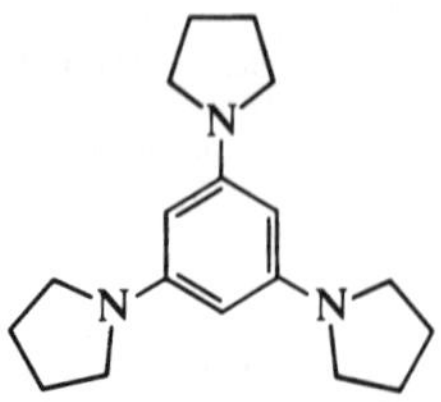

$C_{18}H_{27}N_3$ M 285.431
Almost colourless cryst. (Me_2CO aq.). Mp 179-181°.
pK_{a1} 9.62, pK_{a2} 0.80. Protonation occurs at C to give a σ-complex.

Effenberger, F. *et al*, *Chem. Ber.*, 1968, **101**, 3787 (*synth, uv, pmr*)
Effenberger, F., *Angew. Chem., Int. Ed. Engl.*, 1972, **11**, 61, 911.

Knoche, W. *et al*, *Bull. Soc. Chim. Fr.*, 1988, 377 (*uv, props*)

Tris[(2,2′-bipyridyl-6-yl)methyl]amine   T-70320

N,N-*Bis([2,2′-bipyridin]-6-ylmethyl)[2,2′-bipyridine]-6-methanamine, 9CI*
[113139-56-3]

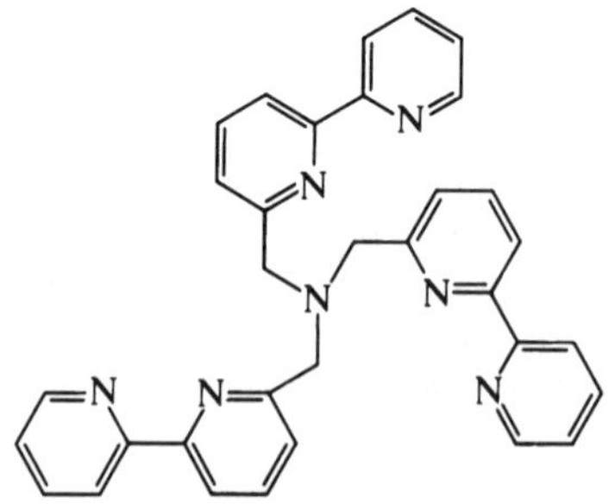

$C_{33}H_{27}N_7$ M 521.623
Ligand. Cryst. (CH_2Cl_2/hexane). Mp 142°.

Lehn, J.-M. *et al*, *J. Chem. Soc., Chem. Commun.*, 1987, 1292 (*synth, use*)

2,4,5-Tris(dicyanomethylene)-1,3-cyclopentanedione   T-70321

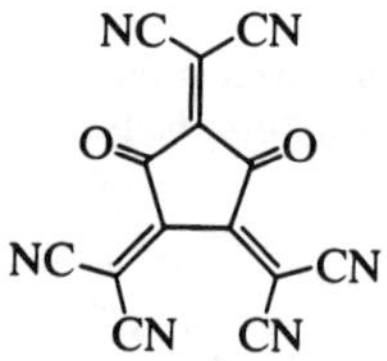

$C_{14}N_6O_2$ M 284.193
Di-K salt: [72893-87-9]. Croconate blue.
   $C_{14}K_2N_6O_2$ M 362.390
   Green metallic cryst. + 2½H_2O (H_2O). Mp >350° (darkens >180°).
Bis(tetraphenylphosphonium salt): [108816-39-3].
   $C_{62}H_{40}N_6O_2P_2$ M 962.985
   Violet tetragonal cryst. with metallic sheen. Mp 218°.

Auch, J. *et al*, *Chem. Ber.*, 1987, **120**, 1691 (*synth, ir, uv, cmr, cryst struct*)

Tris(9-fluorenylidene)cyclopropane   T-70322

9,9′,9″-(1,2,3-Cyclopropanetriylidene)tris-9H-fluorene, 9CI
[115533-27-2]

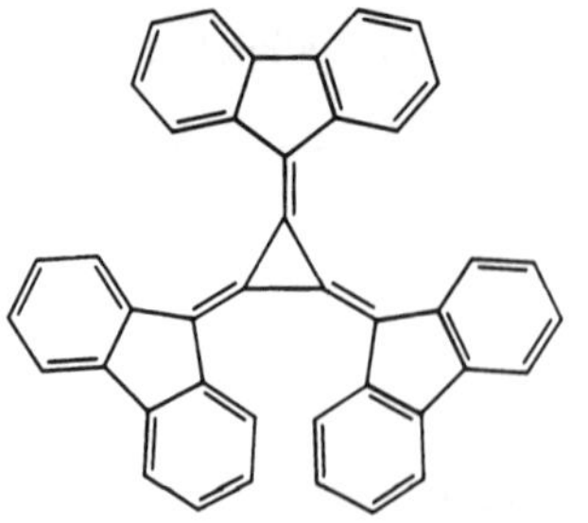

$C_{42}H_{24}$ M 528.652
Reddish-black cryst. with metallic lustre (hexane/CH_2Cl_2). Mp 300°.

Iyoda, M. *et al*, *Angew. Chem., Int. Ed. Engl.*, 1988, **27**, 1080 (*synth, pmr, cmr, uv, ir*)

Tris[(2-hydroxyethoxy)ethyl]amine   T-70323

Updated Entry replacing T-50555

2,2′,2″-[Nitrilotris(2,1-ethanediyloxy)]trisethanol, 9CI.
Tris(3,6-dioxahexyl)amine

[54384-48-4]

$$(HOCH_2CH_2OCH_2CH_2)_3N$$

$C_{12}H_{27}NO_6$ M 281.348

Phase-transfer catalyst. Alternative to toxic crown ethers.
$Bp_{0.5}$ 160°. Comly. available.

Tri-Me ether: [70384-51-9]. *2-(2-Methoxyethoxy)-*
N,N-bis[2-(2-methoxyethoxy)ethyl]ethanamine, 9CI.
Tris(2-methoxyethoxyethyl)amine. Tris(3,6-
dioxaheptyl)amine.
$C_{15}H_{33}NO_6$ M 323.429
Phase-transfer catalyst. $Bp_{0.5}$ 160°.

Soula, G., J. Org. Chem., 1985, **50**, 3717 (synth, use)
Bartsch, R.A. et al, Synth. Commun., 1986, **16**, 1777 (use)
Bose, A.K. et al, Tetrahedron Lett., 1987, **28**, 2503 (use)

Tris(methylphenylamino)methane   T-70324

N,N′,N″-Trimethyl-N,N′,N″-triphenylmethanetria-
mine, 9CI

[4960-31-0]

$$HC(NMePh)_3$$

$C_{22}H_{25}N_3$ M 331.460

Highly reactive orthoamide undergoing a variety of
reacns. Mp 268-269° (263-265° dec.).

Clemens, D.H. et al, J. Am. Chem. Soc., 1961, **83**, 2588 (synth)
Scheeren, J.W. et al, Recl. Trav. Chim. Pays-Bas, 1969, **88**, 289
(synth, struct)
Schönberg, A. et al, Chem. Ber., 1987, **120**, 1581, 1589 (props)

Trispiro[2.0.2.0.2.0]nonane, 9CI   T-70325

Updated Entry replacing T-04468

Tricyclopropylidene. [3]Rotane

[31561-59-8]

C_9H_{12} M 120.194
Mp 29-31°.

Fitjer, L., Chem. Ber., 1982, **115**, 1047 (synth, spectra)
Erden, I., Synth. Commun., 1986, **16**, 117 (synth)

Tris(thioxanthen-9-ylidene)cyclopropane   T-70326

9,9′,9″-(1,2,3-Cyclopropanetriylidene)tris-9H-thiox-
anthene, 9CI

[115591-79-2]

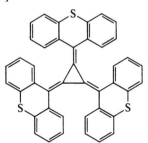

$C_{42}H_{24}S_3$ M 624.832

Blue cryst. $(CH_2Cl_2/Me_2CO/MeOH)$. Mp 269-270°.
Forms a radical cation.

Sugimoto, T. et al, Angew. Chem., Int. Ed. Engl., 1988, **27**, 1078
(synth, uv, cmr)

1,3λ⁴,δ²,5,2,4-Trithiadiazine   T-70327

[114078-94-3]

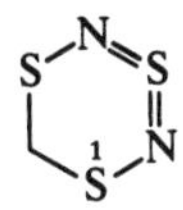

$CH_2N_2S_3$ M 138.220
Stable red cryst. solid. Mp 43-45°.

1-Oxide: [114102-66-8].
$CH_2N_2OS_3$ M 154.220
Cryst. Mp 96-97°.

Bannister, R.M. et al, J. Chem. Soc., Chem.
Commun., 1987, 1546 (synth, uv, ir, pmr, cryst struct)

2,8,17-Trithia[4⁵,¹²][9]metacyclophane   T-70328

5,11,16-Trithia[6.6.4.1³,¹³]nonadeca-1,3(19),13-triene,
9CI

[61582-73-8]

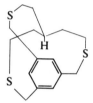

$C_{16}H_{22}S_3$ M 310.530

The "H-inside"-form (illus.) alone has been made. Cryst.
(EtOAc). Mp 300-350° dec. (softens at 170-180°)
(176°).

Trisulfone:
$C_{16}H_{22}O_6S_3$ M 406.526
Cryst. Mp >290° dec. (discolours >250°).

Ricci, A. et al, J. Chem. Soc., Perkin Trans. 1, 1976,
1691 (synth, pmr)
Pascal, R.A. et al, J. Am. Chem. Soc., 1987, **109**, 6878 (sulfone)
Pascal, R.A. et al, J. Org. Chem., 1987, **52**, 4616 (synth, struct,
pmr, ms, ir)

[3](2,7)Troponophane   T-70329

Bicyclo[4.3.1]deca-1,3,5-trien-10-one, 9CI

[112038-99-0]

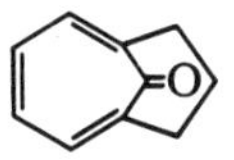

$C_{10}H_{10}O$ M 146.188
Prisms. Mp 56-58°.

Mazaki, Y. et al, Tetrahedron Lett., 1987, **28**, 977 (synth,
struct, pmr, uv, ir)

[5](2,7)Troponophane   T-70330

Bicyclo[5.4.1]dodeca-7,9,11-trien-12-one, 9CI

[112039-01-7]
$C_{12}H_{14}O$ M 174.242
Prisms. Mp 52.5-53.5°.

Mazaki, Y. et al, Tetrahedron Lett., 1987, **28**, 977 (synth, pmr,
cmr, ir, uv, ms)

Trypanothione   T-70331

L-γ-Glutamyl-L-cysteinyl-N-*[3-[4-[[[N-(N-L-γ-gluta-myl-L-cysteinyl)glycyl]amino]butyl]amino]propyl]-glycinamide cyclic(2→3)disulfide, 9CI.* N*¹*,N*⁸*-*Bis(glutathionyl)spermidine*
[96304-42-6]

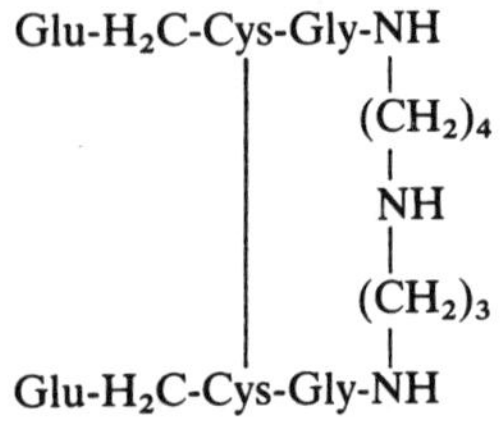

Glu-H₂C-Cys-Gly-NH
|
(CH₂)₄
|
NH
|
(CH₂)₃
|
Glu-H₂C-Cys-Gly-NH

$C_{27}H_{47}N_9O_{10}S_2$ M 721.843
Trypanosomatid metabolite.

Fairbanks, A.H. *et al, Science*, 1985, **227**, 1485 (*isol, synth*)
Henderson, G.B. *et al, J. Chem. Soc., Chem. Commun.*, 1986, 593 (*synth*)
Shames, S.L. *et al, Biochemistry*, 1986, **25**, 3519 (*biochem*)

Tubipofuran   T-70332

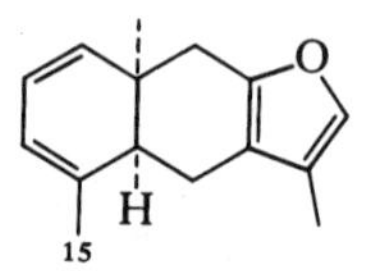

$C_{15}H_{18}O$ M 214.307
Isol. from *Tubipora musica*. Shows cytotoxic and ichthyotoxic props. Oil. [α]$_D$ +5.7° (c, 0.6 in CHCl₃).

15-Acetoxy: **15-Acetoxytubipofuran**.
$C_{17}H_{20}O_3$ M 272.343
From *T. musica*. Shows cytotoxic and ichthyotoxic props. Oil. [α]$_D$ +10.7° (c, 0.5 in CHCl₃).

Iguchi, R. *et al, Chem. Lett.*, 1986, 1789 (*isol*)

Tubocapsigenin *A*   T-70333

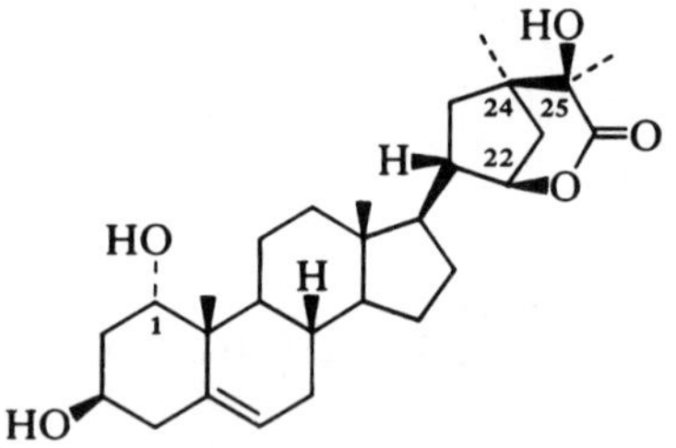

$C_{28}H_{42}O_5$ M 458.637
1,3-Di-Ac: Needles. Mp 273-275°. [α]$_D$ +17.0° (CHCl₃).
1-Ac, 3-O-β-D-glucopyranosyl(1→6)-β-D-glucopyrano-side: [107484-59-3]. **Tubocapside A**.
$C_{42}H_{64}O_{16}$ M 824.958
Constit. of berries of *Tubocapsicum anomalum*. Needles. Mp 195-200°. [α]$_D$ −44.2° (Py).
1-Ac, 3-O-[[β-D-glucopyranosyl(1→6)-β-D-glucopyranosyl(1→2)]-β-D-glucopyranoside]: [107484-60-6]. **Tubocapside B**.
$C_{48}H_{74}O_{21}$ M 987.100
Constit. of *T. anomalum*. Powder. [α]$_D$ −25.0° (Py).

Yoshida, K. *et al, Tetrahedron Lett.*, 1988, **29**, 673 (*cryst struct*)

Tumidulin   T-70334

Updated Entry replacing M-01542
3,5-Dichloro-2,4-dihydroxy-6-methylbenzoic acid 4-carbomethoxy-3-hydroxy-5-methylphenyl ester. Methyl 3,5-dichlorolecanorate

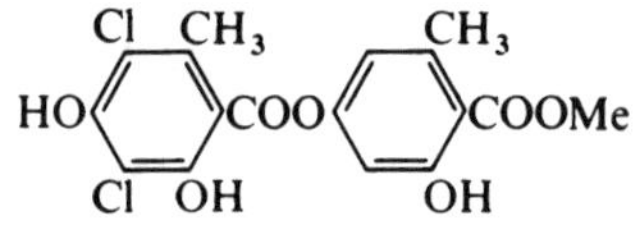

$C_{17}H_{14}Cl_2O_7$ M 401.199
Depside present in lichen of *Ramalina* spp. Cryst. (dry C₆H₆ with exclusion of light). Mp 177-177.5°.

Bendz, G. *et al, Acta Chem. Scand.*, 1965, **19**, 1188 (*isol, synth, ir, uv, pmr*)
Huneck, S. *et al, Z. Naturforsch., B*, 1965, **20**, 611; 1966, **21**, 90, 713 (*isol, struct*)
Huneck, S. *et al, Chem. Ber.*, 1966, **99**, 1106 (*uv, ir, pmr, ms*)

Tumour necrosis factor (human)   T-70335

Differentiation inducing factor. TNF. Cachectin
[94948-59-1]
Polypeptide prod. by the immune system in response to bacterial infection. Antitumour and antibacterial agent.

Ruff, M.R. *et al, Lymphokines (N.Y.)*, 1981, **2**, 235 (*rev*)
Nedwin, G.E. *et al, Nucleic Acids Res.*, 1985, **13**, 6361 (*struct*)
Shalaby, M.R. *et al, Springer Semin. Immunopathol.*, 1986, **9**, 33 (*rev*)
Takeda, K. *et al, Nature (London)*, 1986, **323**, 338

Tunichrome B-1   T-70336

3,5-Dihydroxytyrosyl-α,β-didehydro-3,5-dihydroxy-N-[2-(3,4,5-trihydroxyphenyl)ethenyl]tyrosinamide, 9CI
[97689-87-7]

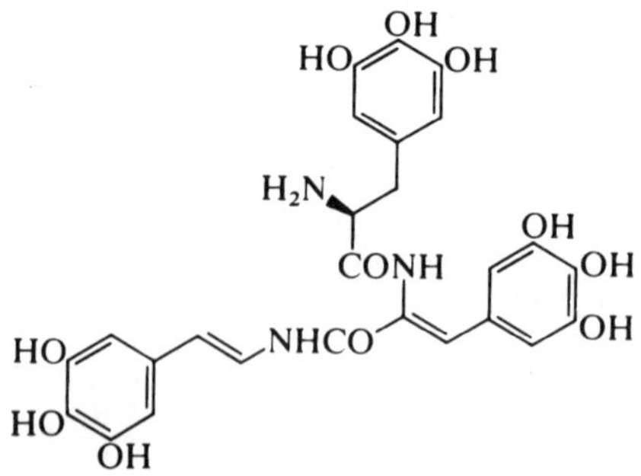

$C_{26}H_{25}N_3O_{11}$ M 555.497
Isol. from *Ascidia nigra*. Blood pigment of the sea squirt which selectively accumulates vanadium. Dec. in air or on warming.

Bruening, R.C. *et al, J. Nat. Prod.*, 1986, **49**, 193 (*isol, struct*)

U

Umbraculumin *A* — U-70001

Glycerol 2-(3-methyl-2-butenoate) 1-(2,4,11-tridecatrienoate)

[116944-99-1]

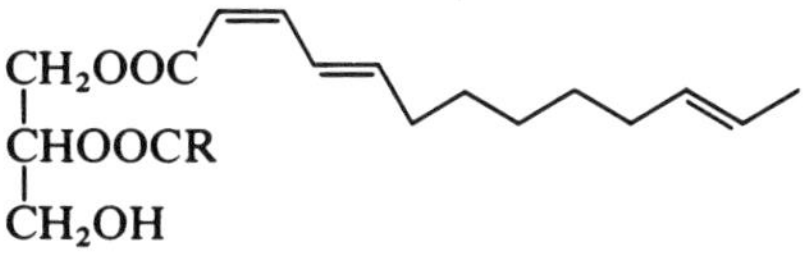

CH$_2$OOC
|
CHOOCR
|
CH$_2$OH

R = —CH=C(CH$_3$)$_2$

C$_{21}$H$_{32}$O$_5$ M 364.481

Constit. of mollusc *Umbraculum mediterraneum*.
Ichthyotoxin. [α]$_D$ −24.3° (c, 0.8 in CHCl$_3$).

Cimino, G. *et al*, *Tetrahedron Lett.*, 1988, **29**, 3613 (*struct*)

Umbraculumin *B* — U-70002

Glycerol 2-(3-methylthio-2-propenoate) 1-(2,4,11-tridecatrienoate)

[116963-50-9]

As Umbraculumin *A*, U-70001 with

R = —CH=CHSMe (*E*−)

C$_{20}$H$_{30}$O$_5$S   M 382.514

Constit. of mollusc *Umbraculum mediterraneum*.
Ichthyotoxin. [α]$_D$ +7° (c, 0.5 in CHCl$_3$).

Cimino, G. *et al*, *Tetrahedron Lett.*, 1988, **29**, 3613 (*struct*)

2,4,6,8-Undecatetraene — U-70003

[31699-37-3]

H$_3$CCH$_2$CH=CHCH=CHCH=CHCH=CHCH$_3$

C$_{11}$H$_{16}$ M 148.247

(2Z,4Z,6E,8Z)-form [109138-61-6]
Giffordene
Odoriferous substance from the brown alga *Giffordia mitchellae*. The (2E,4E,6E,8Z), (2E,4Z,6E,8Z), (2E,4Z,6E,8E), (2Z,4E,6E,8Z) and (2Z,4Z,6E,8E) isomers also occur in minor amts.

Boland, W. *et al*, *Experientia*, 1987, **43**, 466.

1,3,5-Undecatriene — U-70004

Updated Entry replacing U-00062

H$_3$C(CH$_2$)$_4$CH=CHCH=CHCH=CH$_2$

C$_{11}$H$_{18}$ M 150.263

Occurs in Galbanum extract (*Ferula galbaniflua*) and in algae.

(3E,5E)-form [19883-29-5]
Bp$_{10}$ 100-105°.
(3E,5Z)-form
Bp$_{10}$ 100-105°.
(3Z,5Z)-form [19883-26-2]
Oil. Bp$_6$ 100°. Thermally unstable.

Teisseire, P. *et al*, *Recherches*, 1967, **16**, 5 (*synth*)
Yvonne, C.B. *et al*, *Bull. Soc. Chim. Fr.*, 1967, 97 (*isol*)
Naef, F. *et al*, *Helv. Chim. Acta*, 1975, **58**, 1016 (*synth*)
Giraudi, E. *et al*, *Tetrahedron Lett.*, 1983, **24**, 489 (*synth, pmr*)

Hayashi, T. *et al*, *Tetrahedron Lett.*, 1983, **24**, 2665 (*synth*)
Andreini, B.P. *et al*, *Tetrahedron*, 1987, **43**, 4591 (*synth, pmr, ir, ms*)
Ratovelomanana, V. *et al*, *Bull. Soc. Chim. Fr.*, 1987, 174 (*synth, ir, cmr, ms, bibl*)

Urocanic acid — U-70005

Updated Entry replacing U-00136

3-(1H-Imidazol-4-yl)-2-propenoic acid, 9CI. *4-Imidazoleacrylic acid*, 8CI. *Urocaninic acid*

[104-98-3]

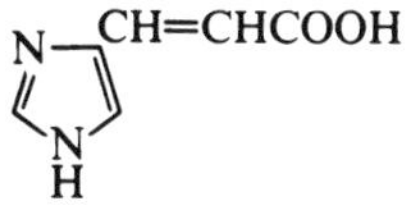

C$_6$H$_6$N$_2$O$_2$ M 138.126
Mp 243-245°.

▷N13425200. NI3425200.

(E)-form [3465-72-3]
Nat. occurring metab.; degradn. prod. of histidine, obt. by fermentation. Isol. from *Bacillus* spp., *Acromobacter liquidium* and *Micrococcus lysodeikticus*. Shows antitumour activity. Used as a sunscreen. Cryst. (dihydrate). Mp 225° (218-224°).

B,HCl: Mp 189-191°.
Picrate: Mp 218-220°.
Me ester: [70346-51-9].
    C$_7$H$_8$N$_2$O$_2$ M 152.152
    Mp 79-81°. Also descr. as viscous liq.
N^1-(1,1-Dimethyl-3-oxobutyl): [117082-14-1]. *3-[1-(1,1-Dimethyl-3-oxobutyl)imidazol-4-yl]-2-propenoic acid*, 9CI.
    C$_{12}$H$_{16}$N$_2$O$_3$ M 236.270
    Isol. from rabbit skin tissue. Cryst. (EtOH). Mp 148-149° dec. Poss. artifact.

(Z)-form [7699-35-6]
Mp 175-176°.

Me ester: [88181-49-1]. Cryst. (toluene). Mp 100-101°.

Edlbacher, S. *et al*, *Hoppe-Seylers Z. Physiol. Chem.*, 1942, **276**, 126; 1943, **279**, 63 (*isol, biosynth*)
Biochem. Prep., 1955, **4**, 50 (*synth*)
Gregoire, J. *et al*, *Bull. Soc. Chim. Belg.*, 1958, **40**, 767 (*isol*)
Japan. Pat., 61 23 094, (*1961*); *CA*, **60**, 2302 (*isol*)
Baden, H. *et al*, *Nature (London)*, 1966, **210**, 732 (*pmr*)
Shibatari, T. *et al*, *Appl. Microbiol.*, 1974, **27**, 688 (*biosynth*)
Svinning, T., *Acta Crystallogr., Sect. B*, 1979, **35**, 2813 (*cryst struct*)
Quinn, R. *et al*, *J. Am. Chem. Soc.*, 1984, **106**, 4136 (*synth, esters*)
Ienaga, K. *et al*, *J. Heterocycl. Chem.*, 1988, **25**, 1037 (*deriv*)

Urodiolenone — U-70006

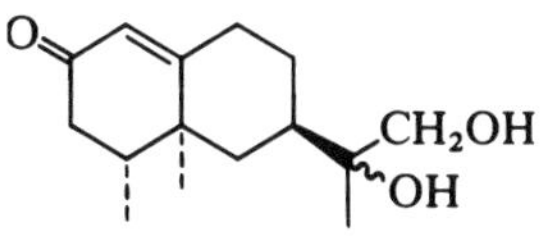

C$_{15}$H$_{24}$O$_3$ M 252.353
Found in the urine of hypertensive humans and in grapefruit. Oil.

Chayen, R. *et al*, *Phytochemistry*, 1988, **27**, 369.

9(11),12-Ursadien-3-ol   U-70007

$C_{30}H_{48}O$ M 424.709

3β-form [41753-44-0]

Constit. of *Euphorbia maculata*. Needles (MeOH/CHCl$_3$). Mp 150-151.5°. $[\alpha]_D^{20.5}$ +360° (CHCl$_3$).

Matsunaga, S. *et al*, *Phytochemistry*, 1988, **27**, 535.

V

Valerenol — V-70001

[84249-42-3]

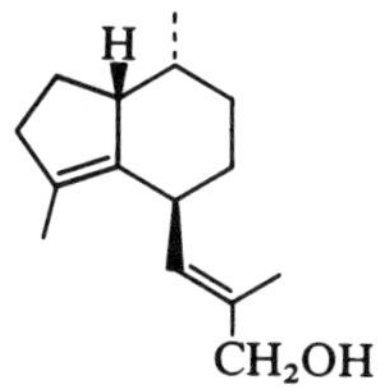

C₁₅H₂₄O M 220.354
$C_{15}H_{24}O$ M 220.354

See also Valerenol, V-30001 . Found as the free alcohol and as esters in *Valeriana officinalis*. Oil.

Bos, R. *et al, Phytochemistry*, 1986, **25**, 133.

Valilactone — V-70002

[113276-96-3]

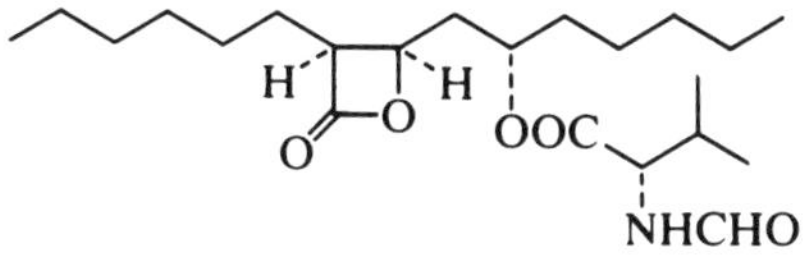

$C_{22}H_{39}NO_5$ M 397.554

Isol. from *Streptomyces* sp. Esterase inhibitor. Needles (MeOH aq.). Mp 57-58°. $[\alpha]_D^{28}$ −37° (c, 1 in CHCl₃).

Kitahara, M. *et al, J. Antibiot.*, 1987, **40**, 1647 (*isol, struct, props*)

Variabilin — V-70003

Updated Entry replacing V-00036

5-[13-(3-Furanyl)-2,6,10-trimethyl-6,10-tridecadienylidene]-4-hydroxy-3-methyl-2(5H)-furanone, 9CI

[51847-87-1]

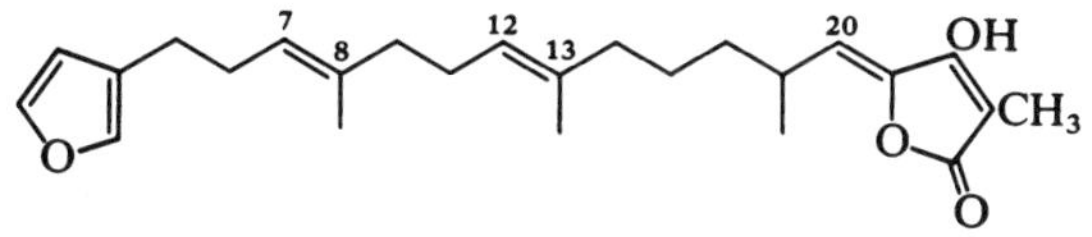

$C_{25}H_{34}O_4$ M 398.541

Isol. from the sponge *Ircinia strobilina* and a *Sarcotragus* sp. Plant phytoalexin with antibiotic props. Oil. $[\alpha]_D$ −4° (c, 1 in CHCl₃). λ_{max} 255 nm (MeOH).

20E-Isomer: [114761-90-9].
$C_{25}H_{34}O_4$ M 398.541
Constit. of a *S.* sp. Oil.

Δ¹³-Isomer (Z-): [114728-09-5].
$C_{25}H_{34}O_4$ M 398.541
Constit. of a *S.* sp. Oil. $[\alpha]_D$ −4.4° (c, 0.9 in CHCl₃).

7,8-Dihydro, 8-hydroxy: [82124-11-6]. *8-Hydroxyvariabilin*.
$C_{25}H_{36}O_5$ M 416.556
Constit. of a *S.* sp. Oil. $[\alpha]_D$ −24.5° (c, 0.22 in MeOH).

Faulkner, D.J., *Tetrahedron Lett.*, 1973, 3821 (*isol, uv, ir, nmr, struct*)

Rothberg, I. *et al, Tetrahedron Lett.*, 1975, 769 (*isol, ir, uv, nmr, struct*)
Barrow, C.J. *et al, J. Nat. Prod.*, 1988, **51**, 275 (*struct, deriv*)

Venustanol — V-70004

ent-3β-Bromo-13(16)-labdene-8β,14,15-triol

[115346-29-7]

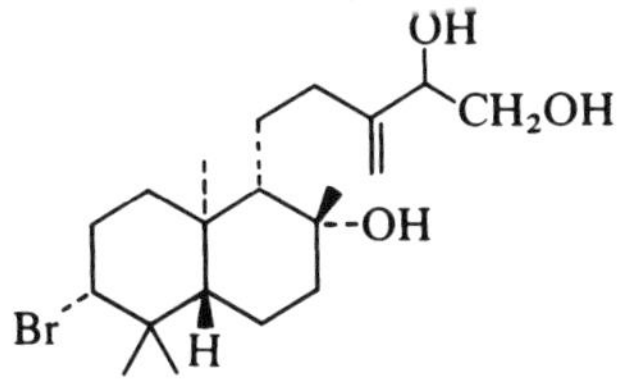

$C_{20}H_{35}BrO_3$ M 403.399

Constit. of *Laurencia venusta*. Cryst. Mp 97-99°. $[\alpha]_D^{24}$ −10.8° (c, 0.562 in MeOH).

Suzuki, M. *et al, Phytochemistry*, 1988, **27**, 1209.

Verboccidentafuran — V-70005

Updated Entry replacing V-50013

[67604-03-9]

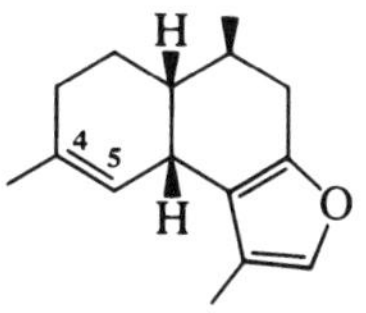

$C_{15}H_{20}O$ M 216.322

Constit. of *Verbesina occidentalis*. Oil. $[\alpha]_D^{24}$ +54.8° (c, 1 in CHCl₃).

4α,5α-Epoxide: **Verboccidentafuran 4α,5α-epoxide**.
$C_{15}H_{20}O_2$ M 232.322
Constit. of *Baccharis salificifolia*. $[\alpha]_D^{24}$ +24° (c, 0.25 in CHCl₃).

2-Oxo: **2-Oxoverboccidentafuran**.
$C_{15}H_{18}O_2$ M 230.306
Constit. of *Smallanthus uvedalia*. Oil.

Bohlmann, F. *et al, Phytochemistry*, 1978, **17**, 453 (*isol*)
Bohlmann, F. *et al, Justus Liebigs Ann. Chem.*, 1983, 1689 (*synth, struct*)
Zdero, C. *et al, Phytochemistry*, 1986, **25**, 2841 (*epoxide*)
Dominguez, X.A. *et al, Phytochemistry*, 1988, **27**, 1863 (*Oxoverboccidentafuran*)

Vescalagin V-70006

Updated Entry replacing V-10011

[36001-47-5]

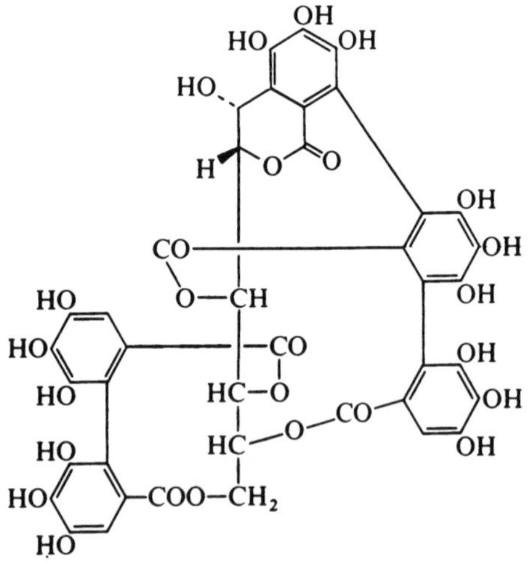

$C_{41}H_{26}O_{26}$ M 934.641

Compds. isol. from oak and chestnut having closely similar structs. include Castalagin (isomeric with Vescalagin) and Vescalin and Castalin (partial hydrol. prods.). Constit. of wood of chestnut (*Castanea sativa*) and oak (*Quercus sesseliflora*). Prisms (H_2O). Mp 200° dec. $[\alpha]_{578}^{25} -109.6°$ (c, 2 in H_2O).

3-Epimer, O^{25}-*(6-carboxy-2,3,4-trihydroxyphenyl)*: [60102-67-2]. **Castavaloninic acid**.
$C_{48}H_{30}O_{31}$ M 1102.746
Tannin from *Quercus valonea*, *Q. aegilops* and *Q. macrolepis*. Pale-beige glass + $2H_2O$. $[\alpha]_D^{20} -153°$ (c, 1 in H_2O).

Mayer, W. *et al*, *Justus Liebigs Ann. Chem.*, 1967, **707**, 177; 1969, **721**, 186; 1971, **751**, 60 (*isol, struct, pmr*)
Mayer, W. *et al*, *Justus Liebigs Ann. Chem.*, 1976, 876 (*Castavaloninic acid*)

Vesparione V-70007

[108834-58-8]

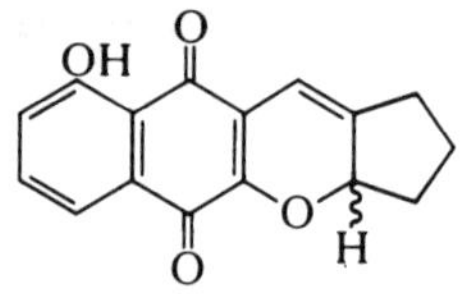

$C_{16}H_{12}O_4$ M 268.268

Pigment isol. from the slime mold *Metatrichia vesparium*. Red needles (MeOH/CH_2Cl_2). Mp 145-147° dec.

Kopanski, L. *et al*, *Justus Liebigs Ann. Chem.*, 1987, 793 (*isol, struct*)

Vestic acid V-70008

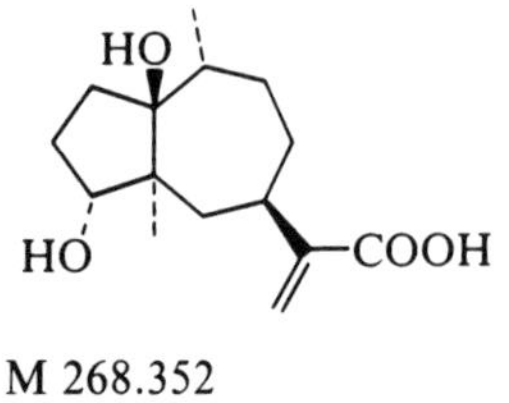

$C_{15}H_{24}O_4$ M 268.352

Constit. of *Vicoa vestita*. Needles (CHCl₃/Me₂CO). Mp 190-192°. $[\alpha]_D -18.6°$ (c, 1 in MeOH).

Sachdev, K. *et al*, *Indian J. Chem.*, *Sect. B*, 1987, **26**, 503.

Vinigrol V-70009

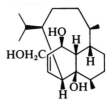

$C_{20}H_{34}O_3$ M 322.487

Metab. of *Virgaria nigra*. Antihypertensive and platelet aggregation-inhibiting substance. Cryst. Mp 108°. $[\alpha]_D -96.2°$ (c, 1.05 in CHCl₃).

Uchida, I. *et al*, *J. Org. Chem.*, 1987, **52**, 5292.

4-Vinyl-1,3-benzenediol V-70010

4-Ethenyl-1,3-benzenediol, *9CI*. *2,4-Dihydroxystyrene*. *4-Vinylresorcinol*

[6053-02-7]

$C_8H_8O_2$ M 136.150

Cryst. + 1MeOH (MeOH). Readily polymerises.

Di-Me ether: [40243-84-3]. *1-Ethenyl-2,4-dimethoxybenzene*, *9CI*. *2,4-Dimethoxystyrene*.
$C_{10}H_{12}O_2$ M 164.204
Constit. of *Dorstenia barnimiana*. Oil. Bp 235° dec.

Flood, S.A. *et al*, *J. Am. Chem. Soc.*, 1928, **50**, 2566 (*synth*)
Walda, Y. *et al*, *Phytochemistry*, 1988, **27**, 1227 (*isol, deriv*)

3-Vinyl-2-cyclopenten-1-one V-70011

3-Ethenyl-2-cyclopenten-1-one

[14252-62-1]

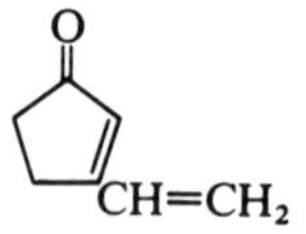

C_7H_8O M 108.140

Oil.

2,4-Dinitrophenylhydrazone: Cryst. (EtOH). Mp 165-166°.

Ishibashi, H. *et al*, *Synthesis*, 1986, 847 (*synth, ir, pmr*)

4(5)-Vinylimidazole V-70012

Updated Entry replacing V-10025

4-Ethenyl-1H-imidazole, *9CI*

[3718-04-5]

$C_5H_6N_2$ M 94.116

Used in polymerisations. Cryst. (H_2O or by subl.). Mp 84°.

1-Benzyl: [86803-30-7].
$C_{12}H_{12}N_2$ M 184.240
Oil.

1-Benzyl, picrate: Mp 183-185°.

1-Triphenylmethyl: [86803-29-4].
$C_{24}H_{20}N_2$ M 336.435
Cryst. (CHCl₃/pet. ether). Mp 205-207°.

3H-form

3-Benzyl: *1-Benzyl-5-vinylimidazole*.
$C_{12}H_{12}N_2$ M 184.240
Cryst. Mp 210° (glassifies >140°).

Overberger, C.G. *et al*, *Macromol. Synth.*, 1974, **5**, 43 (*synth*)
Kokosa, J.M. *et al*, *J. Org. Chem.*, 1983, **48**, 3605 (*synth, pmr, derivs*)
Griffith, R.K. *et al*, *Synth. Commun.*, 1986, **16**, 1761 (*synth, pmr*)
Altman, J. *et al*, *J. Heterocycl. Chem.*, 1988, **25**, 915 (*synth, pmr*)

2-Vinylindole   V-70013

2-Ethenyl-1H-indole, 9CI

[53654-35-6]

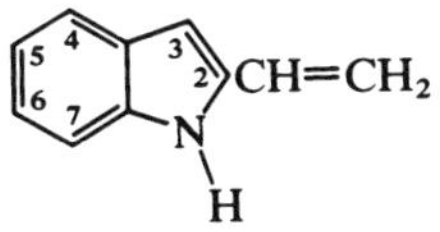

$C_{10}H_9N$ M 143.188
Cryst. (pet. ether). Mp 92°. Air- and light-sensitive.

Pindur, U. *et al*, *Helv. Chim. Acta*, 1988, **71**, 1060 (*synth, pmr, cmr*)

4-Vinylquinoline   V-70014

4-Ethenylquinoline, 9CI

[4945-29-3]

$C_{11}H_9N$ M 155.199
Sl.-yellow liq. d^{20} 1.093. Bp_3 120-123°. n_D^{20} 1.6450.
Picrate: [18121-46-5]. Mp 187°.

Kagan, E.Sh. *et al*, *Khim. Geterotsikl Soedin.*, 1967, 701; *CA*, **68**, 114408y (*synth*)
Rodriguez, J.G. *et al*, *J. Heterocycl. Chem.*, 1988, **25**, 819 (*synth, pmr, ms, bibl*)

5-Vinyluridine   V-70015

Updated Entry replacing V-00239
5-Ethenyluridine, 9CI

[55520-64-4]

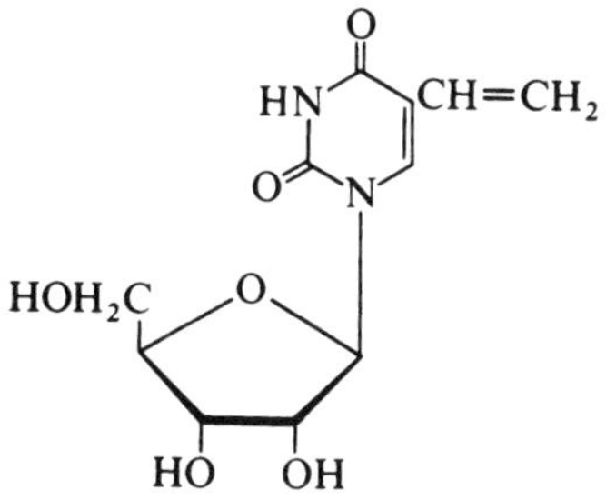

$C_{11}H_{14}N_2O_6$ M 270.241
Mp 255-260° dec. (sinters at 160-165°). λ_{max} 237, 291 nm (ϵ 11 285, 9 450) (MeOH).

2',3',5'-Tribenzoyl:
   $C_{32}H_{26}N_2O_9$ M 582.565
   Mp 141-142°.
2'-Deoxy: 2'-Deoxy-5-vinyluridine.
   $C_{11}H_{14}N_2O_5$ M 254.242
   Antiviral, antileukaemic in tissue culture, inactive *in vivo*. Needles ($CHCl_3$/MeOH).

Sharma, R.A. *et al*, *J. Org. Chem.*, 1975, **40**, 2377 (*synth, pmr*)
Lee, L. *et al*, *Biochemistry*, 1976, **15**, 3686.
Jones, A.S. *et al*, *J. Chem. Soc., Perkin Trans. 1*, 1987, 457 (*2'-Deoxy-5-vinyluridine*)

Vittadinal   V-70016

[117232-53-8]

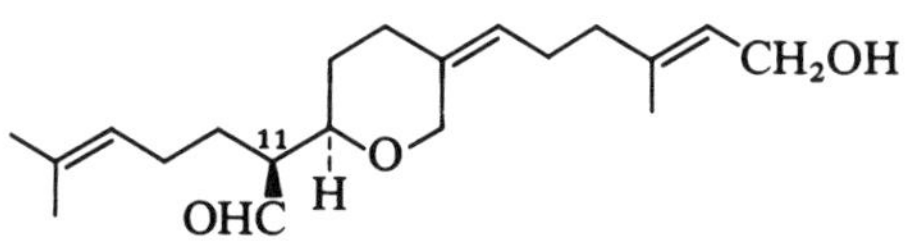

$C_{20}H_{32}O_3$ M 320.471
Constit. of *Vittadinia gracilis*. Oil.
11-Epimer: [117232-50-5]. **Epivittadinal**.
   $C_{20}H_{32}O_3$ M 320.471
   Constit. of *V. gracilis*. Oil.

Zdero, C. *et al*, *Phytochemistry*, 1988, **27**, 2251.

Vittagraciliolide   V-70017

[114703-26-3]

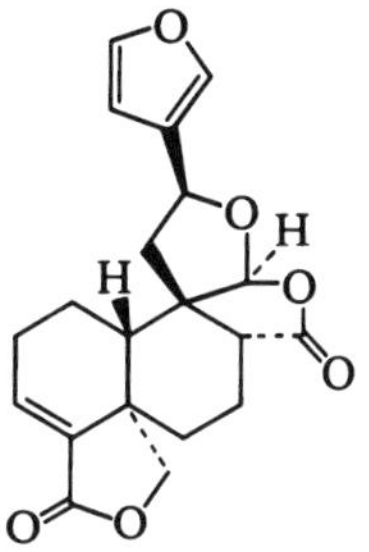

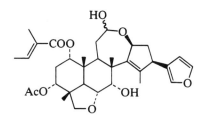

$C_{20}H_{20}O_6$ M 356.374
Constit. of *Vittadinia gracilis*. Cryst. Mp 221°.

Zdero, C. *et al*, *Phytochemistry*, 1988, **27**, 2251.

Volkensin   V-70018

$C_{33}H_{44}O_9$ M 584.705
Constit. of *Melia volkensii*. Insect antifeedant. Cryst. (Me_2CO/hexane). Mp 185-187°.

Rajab, M.S. *et al*, *J. Nat. Prod.*, 1988, **51**, 168.

W

Wasabidienone A W-70001

Updated Entry replacing W-60002

[90052-97-4]

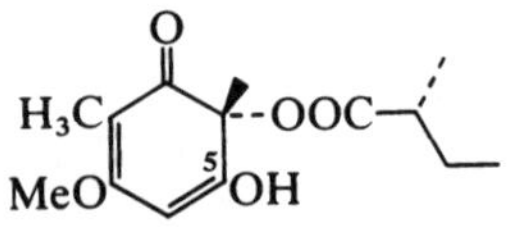

C₁₄H₂₀O₅   M 268.309

$C_{14}H_{20}O_5$ M 268.309

Tautomeric with the 5-one struct. Pigment from potato
culture solution of *Phoma wasabiae*. Oil.

p-*Nitrobenzoyl:* Pale-yellow prisms. Mp 114-114.5°.
[α]_D +159° (c, 0.55 in CHCl₃).

Soga, O. *et al, Chem. Lett.*, 1984, 339 (*isol*)
Soga, O. *et al, Agric. Biol. Chem.*, 1987, **51**, 283 (*cryst struct*)

Wasabidienone D W-70002

*2-Methylbutanoic acid 2-amino-4-methoxy-1,5-di-
methyl-6-oxo-2,4-cyclohexadien-1-yl ester, 9CI*

[100905-87-1]

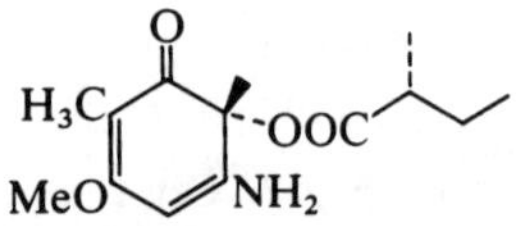

$C_{14}H_{21}NO_4$ M 267.324

Metab. of *Phoma wasabiae*.

N-*Hydroxyethyl:* [100905-88-2]. **Wasabidienone E**.
$C_{16}H_{25}NO_5$ M 311.377
From *P. wasabiae*. Orange-yellow prisms
(Et₂O/hexane). Mp 132-134°. [α]_D +50.5° (c, 1.1 in
CHCl₃).

Soga, O. *et al, CA*, 1986, **104**, 126212 (*isol, struct*)
Soga, O. *et al, Chem. Lett.*, 1987, 815 (*isol, struct*)

Wikstromol W-70003

Updated Entry replacing W-00019
Pinopalustrin. Nortrachelogenin

[34444-37-6]

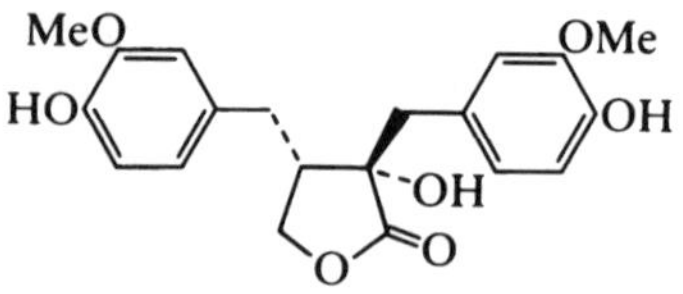

(+)-*form*

$C_{20}H_{22}O_7$ M 374.390

(+)-*form* [61521-74-2]
Isol. from *Wikstroemia* spp. Powder. [α]_D +72° (c, 0.37
in CHCl₃).

Di-Me ether: [61521-75-3]. [α]_D +35° (c, 1.47 in
CHCl₃).

(−)-*form* [63902-22-7]
Obt. from *Pinus palustris* and *Trachelospermum
asaticum*. Amorph. [α]_D^{17} −16.8° (c, 0.178 in EtOH).

Nishibe, S. *et al, Chem. Pharm. Bull.*, 1973, **21**, 1108 (*isol*)
Tandon, S. *et al, Phytochemistry*, 1976, **15**, 1789 (*isol, synth*)
Carnmalm, B. *et al, Acta Chem. Scand., Ser. B*, 1977, **31**, 433
(*struct*)
Torrance, S.J. *et al, J. Pharm. Sci.*, 1979, **68**, 664 (*isol*)
Kato, A. *et al, J. Nat. Prod.*, 1979, **42**, 159 (*isol*)
Belletire, J.L. *et al, J. Org. Chem.*, 1988, **53**, 4724 (*synth*)

X

Xanthorrhizol　　　　　　　　　　X-70001

Updated Entry replacing X-00041

5-(1,5-Dimethyl-4-hexenyl)-2-methylphenol, 9CI

[30199-26-9]

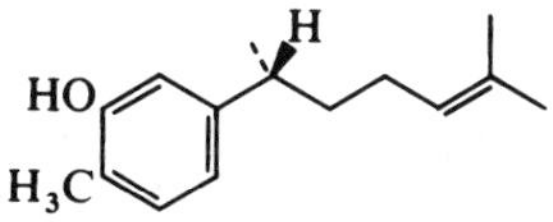

$C_{15}H_{22}O$　　M 218.338

Constit. of rhizomes of *Curcuma xanthorrhiza.* Oil. $[\alpha]_D^{20}$ −52.5°.

Rimpler, H. *et al, Z. Naturforsch., B,* 1970, **25**, 995 (*isol, struct*)
Mane, R.B. *et al, Indian J. Chem.,* 1974, **12**, 938 (*synth*)
John, T.K. *et al, Indian J. Chem., Sect. B,* 1985, **24**, 35 (*synth*)
Garcia, G.E. *et al, J. Nat. Prod.,* 1987, **50**, 1055 (*synth*)
Rane, R.K. *et al, Indian J. Chem., Sect. B,* 1987, **26**, 572 (*synth*)

Xanthostemone　　　　　　　　　X-70002

6,6-Dimethyl-2-(2-methyl-1-oxopropyl)-4-cyclohexene-1,3-dione, 9CI. *2-Isobutyryl-6,6-dimethyl-4-cyclohexene-1,3-dione,* 8CI

[22601-85-0]

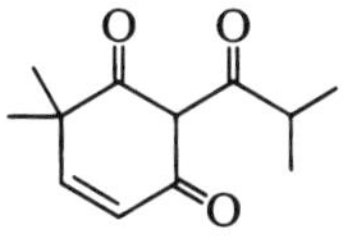

$C_{12}H_{16}O_3$　　M 208.257

Isol. from the oil of *Xanthostemon oppositifolius.* Oil. $Bp_{0.8}$ 95-102°. n_D^{24} 1.5112.

2,4-Dinitrophenylhydrazone: Orange-red needles (MeOH). Mp 239°.

Birch, A.J. *et al, Aust. J. Chem.,* 1956, **9**, 238 (*isol, uv, struct*)

Xanthylium(1+)　　　　　　　　　X-70003

[261-23-4]

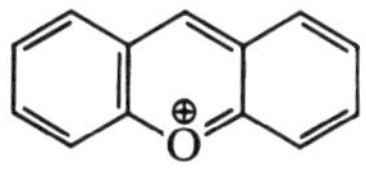

$C_{13}H_9O^{\oplus}$　　M 181.214 (ion)

Perchlorate: [2567-19-3].
　$C_{13}H_9ClO_5$　　M 280.664
　Bronze plates (AcOH). Mp 225-226°.

Bonthrone, W. *et al, J. Chem. Soc.,* 1959, 2774 (*synth*)

Xenicin　　　　　　　　　　　　　X-70004

Updated Entry replacing X-20011

[64504-52-5]

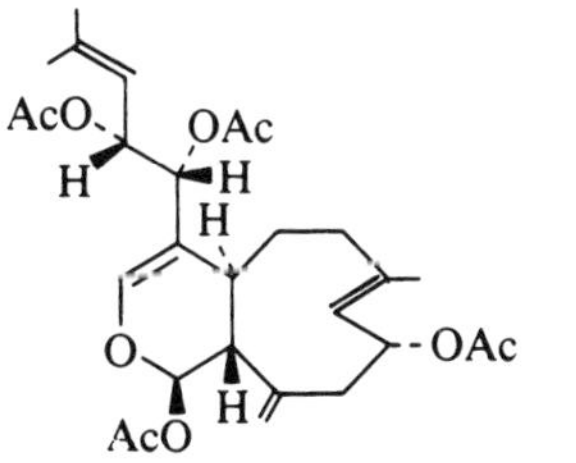

$C_{28}H_{38}O_9$　　M 518.603

Constit. of *Xenia elongata.* Cryst. Mp 141.5-142.3°. $[\alpha]_D^{23.5}$ −36.7° (c, 0.6 in $CHCl_3$).

9-Deacetoxy: **9-Deacetoxyxenicin.**
　$C_{26}H_{36}O_7$　　M 460.566
　Diterpene from *X. crassa* and *Anthelia glauca.* Oil.
　Mp 120-122°. $[\alpha]_D$ +26.2° (c, 0.15 in $CHCl_3$). Lacks
　the ring-OAc group.

9-Deacetyl, 9-acetoacetyl: [114933-24-3].
　$C_{30}H_{40}O_{10}$　　M 560.640
　Constit. of *A. glauca.* Oil.

Vanderah, D.J. *et al, J. Am. Chem. Soc.,* 1977, **99**, 5780 (*isol*)
Bowden, B.F. *et al, Aust. J. Chem.,* 1982, **35**, 997 (*isol*)
Green, D. *et al, Tetrahedron Lett.,* 1988, **29**, 1605 (*deriv*)

Xestenone　　　　　　　　　　　X-70005

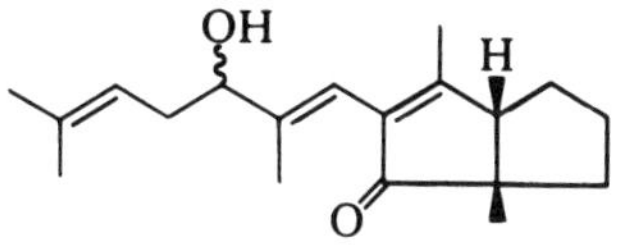

$C_{19}H_{28}O_2$　　M 288.429

Constit. of sponge *Xestospongia vanilla.* Clear oil. $[\alpha]_D$ 0° (c, 1.0 in MeOH). Positive Cotton effect.

Northcote, P.T. *et al, Tetrahedron Lett.,* 1988, **29**, 4357 (*struct*)

Xestodiol　　　　　　　　　　　X-70006

[112727-21-6]

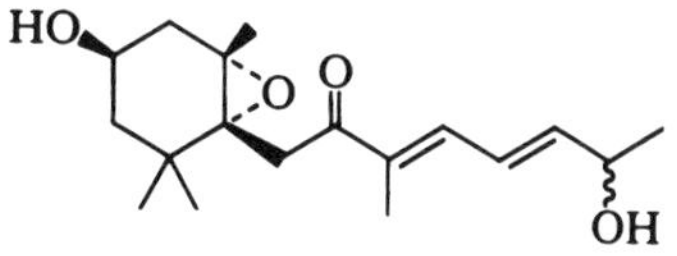

$C_{18}H_{28}O_4$　　M 308.417

Constit. of the sponge *Xestospongia vanilla.* Oil. $[\alpha]_D$ −27° (c, 0.5 in MeOH).

Northcote, P.T. *et al, J. Nat. Prod.,* 1987, **50**, 1174.

Y

Yomogin   Y-70001

Updated Entry replacing Y-00017

3-Oxo-1,4,11(13)-eudesmatrien-12,8β-olide

[10067-18-2]

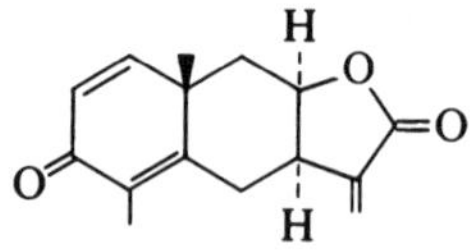

$C_{15}H_{16}O_3$ M 244.290

Constit. of *Artemesia princeps*. Cryst. (EtOAc). Mp 201-202°. $[\alpha]_D^{20}$ −88° (c, 0.11 in CHCl₃).

11α,13-Dihydro: 11α,13-Dihydroyomogin. 3-Oxo-1,4-eudesmadien-12,8β-olide.
$C_{15}H_{18}O_3$ M 246.305
Constit. of *Ferreyranthus fruticosus*. Cryst. Mp 167°.

1α,2α-Epoxide: 1α,2α-Epoxyyomogin. 1α,2α-Epoxy-4,11(13)-eudesmadien-12,8β-olide.
$C_{15}H_{16}O_4$ M 260.289
Constit. of *F. fruticosus*. Cryst. Mp 176°.

1α,2α-Epoxide, 4β,5α-dihydro: 1α,2α-Epoxy-4β,5-dihydroyomogin. 1α,2α-Epoxy-3-oxo-11(13)-eudesmen-12,8β-olide.
$C_{15}H_{18}O_4$ M 262.305
Constit. of *F. fruticosus*. Cryst. Mp 185°.

Geissman, T.A., *J. Org. Chem.*, 1966, **31**, 2523 (*isol, struct*)
Caine, D. *et al*, *Tetrahedron Lett.*, 1975, 743 (*synth*)
Marco, J.A. *et al*, *Can. J. Chem.*, 1987, **65**, 630 (*synth*)
Jakupovic, J. *et al*, *Phytochemistry*, 1988, **27**, 1113 (*derivs*)

Youlemycin   Y-70002

[110207-81-3]

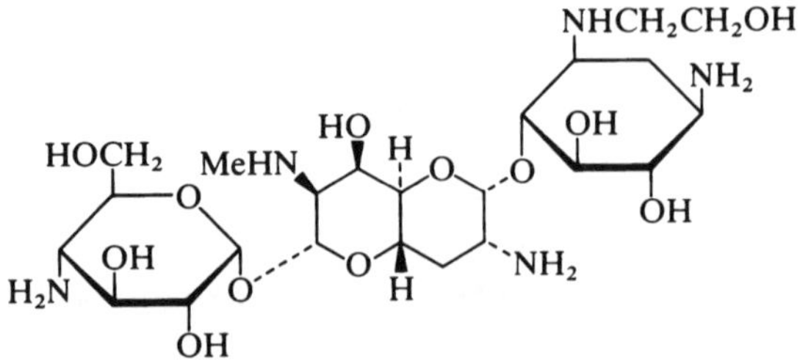

$C_{23}H_{45}N_5O_{12}$ M 583.635

Aminoglycoside antibiotic. Prod. by *Streptomyces* sp.

Ye, X. *et al*, *CA*, 1987, **107**, 112294 (*isol, struct*)

Z

Zafronic acid — Z-70001

3β,10α-Dihydroxy-4,11(13)-cadinadien-12-oic acid

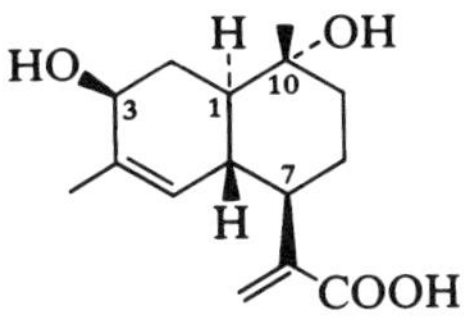

$C_{15}H_{22}O_4$ M 266.336

Compds. here have been renamed on the basis of the normal cadinane numbering system. Constit. of *Leucanthemopsis pulverulenta*. Not abstracted by *CA*.

Me ester: [91896-90-1]. Cryst. (Me$_2$CO). Mp 181-182°. $[\alpha]_D$ −77.5° (c, 1.6 in EtOH).

3-Ac: 3β-*Acetoxy-10α-hydroxy-4,11(13)-cadinadien-12-oic acid. Acetylzafronic acid.*
$C_{17}H_{24}O_5$ M 308.374
Constit. of *L. pulverulenta*.

3-Ac, Me ester: [91896-91-2]. Cryst. (hexane). Mp 104-105°. $[\alpha]_D$ −76° (c, 1.65 in CHCl$_3$).

O^{10}-Et: 10α-*Ethoxy-3β-hydroxy-4,11(13)-cadinadien-12-oic acid.*
$C_{17}H_{26}O_4$ M 294.390
Constit. of *L. pulverulenta*.

O^{10}-Et, Me ester: Oil. $[\alpha]_D$ −68.5° (c, 4.5 in CHCl$_3$).

O^{10}-Et, 3-Ac: 3β-*Acetoxy-10α-ethoxy-4,11(13)-cadinadien-12-oic acid.*
$C_{19}H_{28}O_5$ M 336.427
Constit. of *L. pulverulenta*.

O^{10}-Et, 3-Ac, Me ester: Oil. $[\alpha]_D$ −71° (c, 1.97 in CHCl$_3$).

De Pascual Teresa, J. *et al, Tetrahedron*, 1984, **40**, 2189.

Zebrinin — Z-70002

3-O-[6-O-(2,5-di-O-caffeoyl-α-L-arabinofuranosyl)-β-D-glucopyranosyl]-3′,7-di-O-(6-O-caffeoyl-β-D-glucopyranosyl)cyanidin

[110064-60-3]

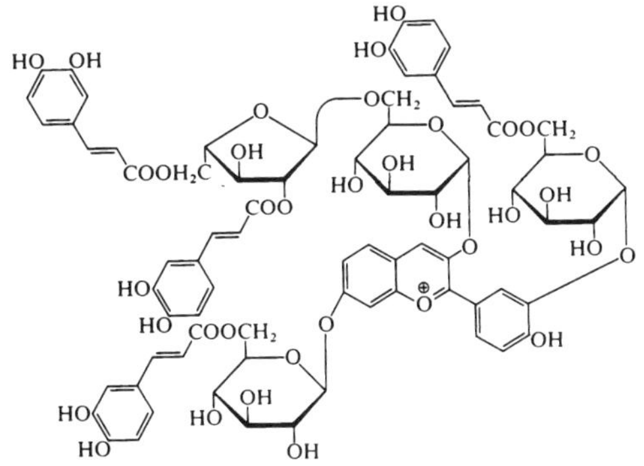

$C_{74}H_{73}O_{37}^{\oplus}$ M 1554.369 (ion)

Constit. of *Zebrina pendula*. Dark-red powder (MeOH/HCl/Et$_2$O) (as chloride). Mp 117-123° (chloride).

Idaka, E. *et al, Tetrahedron Lett.*, 1987, **28**, 1901.

Zederone — Z-70003

Updated Entry replacing Z-00010
[7727-79-9]

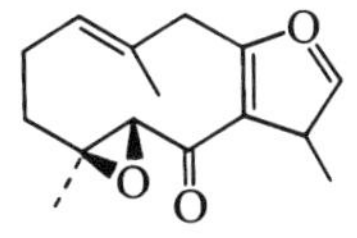

$C_{15}H_{18}O_3$ M 246.305

Constit. of the crude drug Zedoary, prepared from the rhizome of *Curcuma Zedoaria*. Cryst. (EtOAc). Mp 153.5-154°. $[\alpha]_D$ +265.8° (c, 0.5 in CHCl$_3$).

Hikino, H. *et al, J. Chem. Soc. (C)*, 1971, 688 (*isol*)
Shibuya, H. *et al, Chem. Pharm. Bull.*, 1987, **35**, 924 (*abs config, cryst struct*)

Zedoarondiol — Z-70004

Updated Entry replacing Z-40002
[98644-24-7]

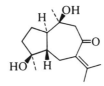

$C_{15}H_{24}O_3$ M 252.353

Constit. of *Curcuma zedoaria* and *C. aromatica*. Needles (CHCl$_3$). Mp 133-134°. $[\alpha]_D^{23}$ −44.0° (c, 1.0 in MeOH).

O^9-Me: **Methylzedoarondiol.**
$C_{16}H_{26}O_3$ M 266.380
Constit. of *C. aromatica*. Needles (hexane). Mp 83-85°. $[\alpha]_D^{23}$ −43.1° (c, 0.2 in MeOH).

Kouno, I. *et al, Phytochemistry*, 1985, **24**, 1845 (*isol*)
Kuroyanagi, M. *et al, Chem. Pharm. Bull.*, 1987, **35**, 53 (*cryst struct*)

Zexbrevin C — Z-70005

[50982-39-3]

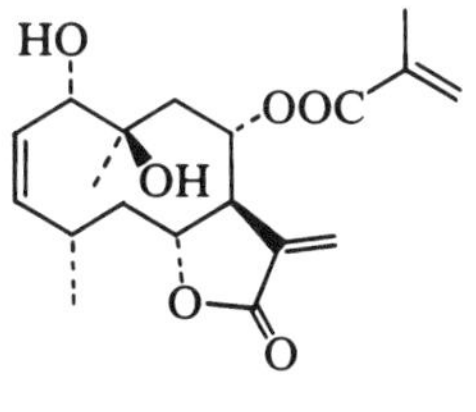

$C_{19}H_{26}O_6$ M 350.411

Isol. from *Zexmenia brevifolia*.

Ortega, A. *et al, Rev. Latinoamer. Quim.*, 1973, **4**, 1; *CA*, **79**, 123629a.

Zeylenol   Z-70006

Updated Entry replacing Z-20003

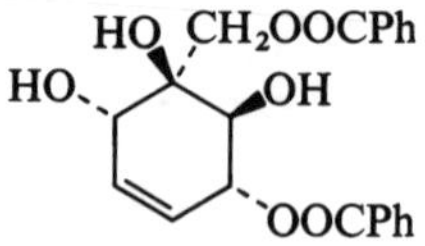

Absolute
configuration

$C_{21}H_{20}O_7$ M 384.385
(−)-form illus.

(+)-form

Constit. of a *Boesenbergia* sp. Needles (EtOAc). Mp
132-133°. $[\alpha]_D^{20}$ +113.5°.

(−)-form [78804-17-8]

Constit. of roots of *Uvaria zeylanica*. Needles (EtOAc).
Mp 144-145°. $[\alpha]_D^{20}$ −118° (c, 0.075 in $CHCl_3$).

Jolad, S.D. *et al*, *J. Org. Chem.*, 1981, **46**, 4267 (*struct*)
Schulte, G.R. *et al*, *Tetrahedron Lett.*, 1982, **23**, 4299 (*abs con-
    fig*)
Tuntiwachwuttikul, P. *et al*, *Aust. J. Chem.*, 1987, **40**, 2049
    (*isol*)

Zoapatanolide B   Z-70007

[84886-37-3]

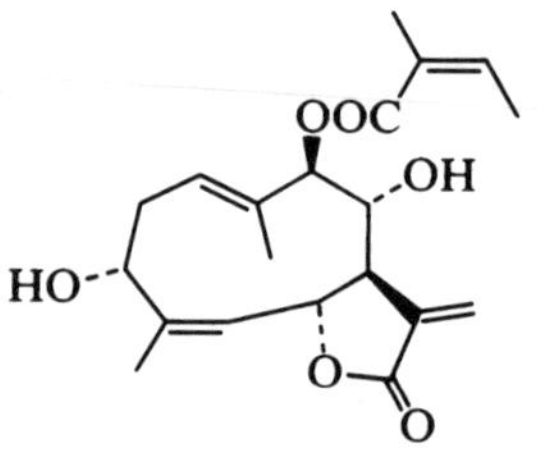

$C_{20}H_{26}O_6$ M 362.422

Constit. of *Montanoa tomentosa* and *Blainvillea latifolia*.

Quijano, L. *et al*, *Phytochemistry*, 1982, **21**, 2041 (*isol*)
Singh, P. *et al*, *Phytochemistry*, 1988, **27**, 609 (*isol*)

Name Index

This index becomes invalid after publication of the Eighth Supplement.

The Name Index lists in alphabetical order all names and synonyms contained in the Sixth and Seventh Supplements. Names contained in Supplements 1–5 are listed in the cumulative Index Volume published with the Fifth Supplement.

Each index term refers the user to a DOC Number consisting of a single letter of the alphabet followed by five digits. The letter is the first letter of the relevant DOC Name.

The first of the DOC Number (printed in bold type) indicates the number of the Supplement in which the entry is printed.

A DOC Number which follows immediately upon an index term means that the term is itself used as the Entry Name.

A DOC Number which is preceded by the word '*see*' means that the term is a synonym to an Entry Name.

A DOC Number which is preceded by the word '*in*' means that the term is embedded within an Entry, usually as a synonym to a particular stereoisomeric form or to a derivative.

The symbol ▷ preceding an index term indicates that the DOC Entry contains information on toxic or hazardous properties of the compound.

Name Index

A 40926, see A-70219
AAP, see A-60283
Abbeymycin, A-70001
Abbottin, A-70002
Abbreviatin PB, A-60001
(5S)-9(10→20)-Abeo-1(10),8,11,13-
abietatetraene-11,12-diol, see B-60007
Abeoanticopalic acid, A-60002
19(4→3)-Abeo-O-demethylcryptojaponol, see
A-60003
19(4→3)-Abeo-11,12-dihydroxy-4(18),-8,11,13-
abietatetraen-7-one, A-60003
19(4β→3β)-Abeo-6,11-epoxy-6,12-dihydroxy-
6,7-seco-4(18),8,11,13-abietatetraen-7-al,
A-60004
Abhexone, see E-60051
Abiesolidic acid, A-70003
Abiesonic acid, A-60005
6,8,11,13-Abietatetraene-11,12,14-triol,
A-60006
8,11,13-Abietatriene-7,18-diol, A-70004
8,11,13-Abietatriene-12,18-diol, A-70005
8,11,13-Abietatrien-3-ol, A-70006
8,11,13-Abietatrien-19-ol, A-60007
8,11,13-Abietatrien-12,16-oxide, A-60008
8(14)-Abieten-18-oic acid 9,13-endoperoxide,
A-60009
Abruslactone A, in H-70197
Absinthifolide, in H-60138
▷Acacetin, see D-70319
Acaciin, in D-70319
Acalycixeniolide A, A-60010
Acalycixeniolide B, A-60011
Acanthocerebroside A, A-70007
Acanthocerebroside B, A-70008
Acanthocerebroside C, A-70009
Acanthocerebroside D, A-70010
Acanthocerebroside E, A-70011
Acanthocerebroside F, A-70012
Acanthoglabrolide, see A-60012
Acanthospermal A, A-60012
1-Acenaphthenecarboxaldehyde, A-70013
1-Acenaphthenecarboxylic acid, A-70014
3,4-Acenaphthenedicarboxylic acid, A-60013
5,6-Acenaphthenedicarboxylic acid, A-70015
4-Acenaphthenol, see H-60099
Acenaphtho[1,2-b]benzo[d]thiophene, A-70016
Acenaphtho[5,4-b]furan, A-60014
1-Acenaphthoic acid, see A-70014
Acenaphtho[1,2-b]thiophene, A-70017
Acenaphtho[5,4-b]thiophene, A-60015
Acenaphthylene oxide, see D-70170
4-Acenaphthylenol, see H-70100
Acephenanthrylene, A-70018
5,6-Acepleiadylenedione, see C-60194
5,8-Acepleiadylenedione, see C-60195
▷Acetamide, A-70019
2-Acetamido-3-methylbiphenyl, in A-70164
5-Acetamido-2-methyl-2H-tetrazole, in A-70202
2-Acetamido-3H-phenoxazin-3-one, in A-60237
1-Acetamidophthalazine, in A-70184
Acetamidosalol, in H-70108
▷Acetamidoxime, A-60016
Acetaminosal, in H-70108
Acetaminosalol, in H-70108
S-Acetoacetylthiophenol, in O-70080
Acetoguanamine, see D-60040
Acetonediacetic acid, see O-60071
Acetonedicarboxylic acid, see O-60084
Acetoneglyceraldehyde, see D-70400
3-Acetonyl-5,8-dihydroxy-2-(hydroxymethyl)-6-
methoxy-1,4-naphthoquinone, see F-60108
Acetonylmercaptan, see M-60028
2-Acetonylpyridine, see P-60229

3-Acetonylpyridine, see P-60230
4-Acetonylpyridine, see P-60231
▷Acetophenone, A-60017
Acetophloroglucinol, see T-70248
▷5-Acetopyrogallol, see T-60315
3-Acetopyrrole, see A-70052
▷Acetosyringone, in T-60315
Acetothioacetic acid, see O-70080
1-Acetoxy-7-acetoxymethyl-3-acetoxymethyl-
ene-11,15-dimethyl-1,6,10,14-hexadecatetra-
ene, A-70020
4-Acetoxy-2-amino-3,5,14-trihydroxy-6-
eicosenoic acid, see T-60082
5-Acetoxy-9α-angeloyloxy-8β-hydroxy-10(14)-
oplopen-3-one, in T-70272
14-Acetoxy-7α-angeloyloxy-6β-hydroxy-8(10)-
oplopen-2-one (incorr.), in T-70272
2-Acetoxyaristotetralone, in A-70253
2-Acetoxy-1,3-benzenedicarboxylic acid, in
H-60101
ent-18-Acetoxy-15-beyeren-19-oic acid, in
H-60106
2-Acetoxy-1,11-bis(3-furanyl)-4,8-dimethyl-
3,6,8-undecatriene, in B-70143
4-Acetoxy-2-bromo-5,6-epoxy-2-cyclohexen-1-
one, in B-60239
14-Acetoxy-2-butanoyloxy-5,8(17)-briaradien-
18,7-olide, in E-70012
17α-Acetoxy-6α-butanoyloxy-15,17-oxido-16-
spongianone, in D-60365
3-Acetoxy-3-(2-chloro-2-isocyanovinyl)-2-
oxoindole, see I-60028
1-Acetoxy-2-chloromethoxyethane, see C-70088
7β-Acetoxy-13(17),15-cleistanthadien-18-oic
acid, in H-70115
9β-Acetoxy-4,5-dehydro-4(15)-dihydrocostic
acid, in H-60135
2-Acetoxy-2-desoxyperezone, in I-70075
4-Acetoxy-2,6-dibromo-1,5-dihydroxy-2-cyclo-
hexen-1-one, in D-60097
6-Acetoxy-3,10-dibromo-4,7-epoxy-12-penta-
decen-1-yne, in B-70217
4-Acetoxy-2,6-dibromo-5-hydroxy-2-cyclo-
hexen-1-one, in D-60097
21-Acetoxy-21,23:24,25-diepoxyapotirucall-14-
ene-3,7-diol, in T-60223
2α-Acetoxy-11α,13-dihydroconfertin, in
D-60211
7α-Acetoxy-12,20-dihydroxy-8,12-abietadiene-
11,14-dione, in R-60015
2'-Acetoxy-3,5-dihydroxy-7,8-dimethoxyflavone,
in P-60045
12-Acetoxy-5,8-dihydroxyfarnesol, in T-60363
3α-Acetoxy-15α,22S-dihydroxy-7,9(11),24E-
lanostatrien-26-oic acid, in T-70266
2'-Acetoxy-5,5'-dihydroxy-3,7,8-trimethoxy-
flavone, in H-70080
20-Acetoxy-2β,3α-dihydroxy-1(15),8(19)-
trinervitadiene, in T-70299
6-Acetoxy-3,7-dolabelladiene-12,16-diol, in
D-70541
3β-Acetoxydrimenin, in D-60519
2α-Acetoxy-8-epiivangustin, in I-70109
6α-Acetoxy-8-epiivangustin, in I-70109
6β-Acetoxy-1α,4α-epoxyeudesmane, in E-60019
6-Acetoxy-3,4-epoxy-12-hydroxy-7-dolabellen-
16-al, in D-70541
19-Acetoxy-15,16-epoxy-13,17-spatadien-5α-ol,
in E-60029
3β-Acetoxy-21,23-epoxy-7,24-tirucalladiene, in
E-60032
3β-Acetoxy-10α-ethoxy-4,11(13)-cadinadien-
12-oic acid, in Z-70001
9β-Acetoxy-4,11(13)-eudesmadien-12-oic acid,
in H-60135

3β-Acetoxy-4,11(13)-eudesmadien-12,8β-olide,
in H-60136
1β-Acetoxy-3,7(11),8-eudesmatrien-12,8-olide,
in H-60140
1β-Acetoxy-4(15),7(11),8-eudesmatrien-12,8-
olide, in H-60141
6β-Acetoxy-4(15)-eudesmene-1β,5α-diol, in
E-60068
4-Acetoxyflexilin, in F-70012
6α-Acetoxygedunin, in G-70006
11β-Acetoxygedunin, in G-70006
3β-Acetoxyhaageanolide acetate, in H-60001
2α-Acetoxyhardwickiic acid, in H-70002
6β-Acetoxyhebemacrophyllide, in H-60010
8-Acetoxy-2-(2,4-hexadiynylidene)-1,6-
dioxaspiro[4.5]dec-3-ene, in H-70042
1-Acetoxy-7-hydroperoxy-3,7-dimethyl-2E,5E-
octadien-4-one, in D-60318
3β-Acetoxy-10α-hydroxy-4,11(13)-cadinadien-
12-oic acid, in Z-70001
1-Acetoxy-7-hydroxy-3,7-dimethyl-2E,5E-
octadien-4-one, in D-60318
1β-Acetoxy-8β-hydroxy-3,7(11)-eudesmadien-
12,8-olide, in D-60322
3β-Acetoxy-2α-hydroxy-1(10),4,11(13)-
germacratrien-12,6α-olide, in H-60006
6α-Acetoxy-2α-hydroxy-1(10),4,11(13)-
germacratrien-12,8-olide, in H-70134
3α-Acetoxy-15α-hydroxy-7,9(11),24E-
lanostatrien-26-oic acid, in D-70314
15α-Acetoxy-3α-hydroxy-7,9(11),24E-
lanostatrien-26-oic acid, in D-70314
23-Acetoxy-30-hydroxy-20(29)-lupen-3-one, in
D-70318
3α-Acetoxy-15α-hydroxy-22-methoxy-8,24E-
lanostadien-26-oic acid, in T-60327
22-Acetoxy-3α-hydroxy-7α-methoxy-8,24E-
lanostadien-26-oic acid, in T-60328
4-Acetoxy-6-(4-hydroxy-4-methyl-2-cyclo-
hexenyl)-2-(4-methyl-3-pentenyl)-2-
heptenoic acid, A-60018
14-Acetoxy-9β-hydroxy-8β-(2-methyl-
propanoyloxy)-1(10),4,11(13)-germacratrien-
12,6α-olide, in T-60323
6α-Acetoxy-17β-hydroxy-15,17-oxido-16-
spongianone, in D-60365
3α-Acetoxy-15α-hydroxy-23-oxo-7,9(11),-24E-
lanostatrien-26-oic acid, in D-70330
15α-Acetoxy-3α-hydroxy-23-oxo-7,9(11),-24E-
lanostatrien-26-oic acid, in D-70330
3α-Acetoxy-15β-hydroxy-7,16-seco-7,11-
trinervitadiene, see S-60019
3β-Acetoxy-19-hydroxy-13(16),14-spongiadien-
2-one, in D-70345
2-Acetoxyisopropenyl-6-acetyl-5-
hydroxybenzofuran, see A-60065
ent-3β-Acetoxy-16β,17-kauranediol, in K-60005
ent-3β-Acetoxy-15-kauren-17-oic acid, in
H-60167
ent-3β-Acetoxy-15-kauren-17-ol, in K-60007
2β-Acetoxy-8(17),13E-labdadien-15-oic acid, in
H-70147
2β-Acetoxy-8(17),13Z-labdadien-15-oic acid, in
H-70147
ent-6β-Acetoxy-7,12E,14-labdatrien-17-oic acid,
in L-60004
ent-6β-Acetoxy-7,12E,14-labdatrien-17-ol, in
L-60004
15-Acetoxy-7-labden-17-oic acid, in H-60168
2α-Acetoxylaurenobiolide, in H-70134
3β-Acetoxy-14(26),17E,21-malabaricatriene, in
M-60008
19-Acetoxymanoyl oxide, in E-70023
1-Acetoxy-6-(2-methoxy-4,5-methyl-
enedioxyphenyl)-2-(3,4-methyl-
enedioxyphenoxy)-3,7-dioxabicyclo[3.3.0]-
octane, see P-60142

▷2-Acetoxy-1-methoxypropane, *in* P-70118
Acetoxyodontoschismenol, *in* D-70541
3β-Acetoxy-12-oleanene-2α,11α-diol, *in* O-60037
3β-Acetoxy-12-oleanene-1β,2α,11α-triol, *in* O-60034
3β-Acetoxy-4,10(14)-oplopadiene-8β,9α-diol, *in* O-70042
2β-Acetoxy-3(14),8(10)-oplopadiene-6β,7α-diol (incorr.), *in* O-70042
9β-Acetoxy-3-oxo-1,4(15),11(13)-eudesmatrien-12,6-olide, *in* H-60202
23-Acetoxy-3-oxo-20(30)-lupen-29-al, *in* D-70316
1-Acetoxypinoresinol, *in* H-60221
1-Acetoxypinoresinol 4′-*O*-β-D-glucopyranoside, *in* H-60221
▷3-Acetoxy-1,2-propanediol, *see* G-60029
6β-Acetoxypulchellin 4-*O*-angelate, *in* P-70138
4-Acetoxy-10-puteninone, *see* L-60009
7α-Acetoxyroyleanone, *in* D-60304
7β-Acetoxyroyleanone, *in* D-60304
3′-Acetoxy-4′-senecioyloxy-3′,4′-dihydroseselin, *in* K-70013
12-Acetoxysesquisabinene, *in* S-70037
13-Acetoxysesquisabinene, *in* S-70037
14-Acetoxytetraneurin *D*, *in* T-60158
3-Acetoxythiophene, *in* T-60212
17-Acetoxythymifodioic acid, *in* T-60218
9β-Acetoxytournefortiolide, *in* H-60137
6-Acetoxy-3,10,13-tribromo-4,7:9,12-diepoxy-1-pentadecyne, *in* B-70195
8-Acetoxy-2′,3,5-trihydroxy-7-methoxyflavone, *in* P-70029
15α-Acetoxy-5α,6β,14α-trihydroxy-1-oxo-20S,22R-witha-2,16,24-trienolide, *see* W-60006
16α-Acetoxy-2α,3β,12β-trihydroxy-4,7-pregnadien-20-one, *in* T-60124
15-Acetoxytubipofuran, *in* T-70332
3β-Acetoxy-28,20β-ursanolide, *in* H-60233
3β-Acetoxy-12-ursene-2α,11α-diol, *in* U-60008
3β-Acetoxy-12-ursene-1β,2α,11α,20β-tetrol, *in* U-60005
3β-Acetoxy-12-ursene-1β,2α,11α-triol, *in* U-60006
3β-Acetoxy-12-ursene-2α,11α,20β-triol, *in* U-60007
3β-Acetoxywedeliasecokaurenolide, *in* W-60004
▷Acetylacetone, *see* P-60059
Acetylacetone-thiourea, *see* D-60450
▷3-Acetylacrylic acid, *see* O-70098
2-Acetyl-3-aminobenzofuran, A-70021
2-Acetyl-3-aminopyridine, A-70022
2-Acetyl-5-aminopyridine, A-70023
3-Acetyl-2-aminopyridine, A-70024
3-Acetyl-4-aminopyridine, A-70025
4-Acetyl-2-aminopyridine, A-70026
4-Acetyl-3-aminopyridine, A-70027
5-Acetyl-2-aminopyridine, A-70028
6-Acetyl-2-amino-1,7,8,9-tetrahydro-4*H*-pyrimido[4,5-*b*][1,4]diazepin-4-one, *see* A-60019
6-Acetyl-2-amino-1,7,8,9-tetrahydro-4*H*-pyrimido[4,5-*b*][1,4]diazepin-4-one, *see* A-70043
2-Acetyl-3-aminothiophene, A-70029
1-Acetylanthracene, A-70030
2-Acetylanthracene, A-70031
[3][14-Acetyl-14-azacyclohexacosanone]-[25,26,53,54,55,56-hexa-acetoxytricyclo[49.3.1.1²⁴,²⁸]-hexapentaconta-1(55),24,26,28(56),-51,53-hexaene][14-acetyl-14-azacyclohexacosanone]-catenane, A-60020
1-Acetyl-2-azidobenzene, *see* A-60334
1-Acetyl-3-azidobenzene, *see* A-60335
1-Acetyl-4-azidobenzene, *see* A-60336
Acetylbarlerin, *in* S-60028
p-Acetylbenzenethiol, *see* M-60017
1-Acetyl-3-bromonaphthalene, A-60021
1-Acetyl-4-bromonaphthalene, A-60022
1-Acetyl-5-bromonaphthalene, A-60023
1-Acetyl-7-bromonaphthalene, A-60024
2-Acetyl-6-bromonaphthalene, A-60025
25-*O*-Acetylbryoamaride, *in* C-70204

2-Acetylbutadiene, *see* M-70075
Acetylbutadiyne, *see* H-60041
2-Acetylbutane, *see* M-70105
γ-Acetylbutyraldehyde, *see* O-70091
Acetylcedrene, A-70032
2-Acetyl-5-chloropyridine, A-70033
3-Acetyl-2-chloropyridine, A-70034
4-Acetyl-2-chloropyridine, A-70035
4-Acetyl-3-chloropyridine, A-70036
5-Acetyl-2-chloropyridine, A-70037
α-Acetylconstictic acid, *in* C-60163
1-Acetylcycloheptene, A-60026
1-Acetylcyclohexene, A-70038
4-Acetylcyclohexene, A-70039
1-Acetylcyclooctene, A-60027
1-Acetylcyclopentene, A-60028
4-Acetyl-2,3-dihydro-5-hydroxy-2-isopropenylbenzofuran, A-70040
N-Acetyl-*S*-[(5,8-dihydro-1-hydroxy-3-methoxy-7-methyl-5,8-dioxo-2-naphthalenyl)methyl]-L-cysteine, *in* F-70008
1-Acetyl-1,2-dihydro-3*H*-pyrrol-3-one, *in* D-70262
2-Acetyl-4,5-dihydrothiazole, A-70041
5-Acetyl-2,2-dimethyl-1,3-dioxane-4,6-dione, A-60029
4-Acetyl-2,2-dimethyl-1,3-dioxolane, A-70042
Acetyldinitromethane, *see* D-70454
Acetylenedicarboxaldehyde, *see* B-60351
Acetylenediurene, *see* T-70066
Acetyleumaitenol, *in* E-60072
6-Acetylferulidin, *in* F-70006
6-Acetylhomopterin, A-70043
3-Acetyl-4-hydroxy-2(5*H*)-furanone, A-60030
6-Acetyl-5-hydroxy-2-hydroxymethyl-2-methyl-chromene, A-60031
6-Acetyl-5-hydroxy-2-(1-hydroxymethylvinyl)-benzo[*b*]furan, *in* A-60065
6-Acetyl-5-hydroxy-2-isopropenylbenzo[*b*]-furan, *in* A-60065
5-Acetyl-2-(2-hydroxyisopropyl)benzofuran, *see* A-70044
5-Acetyl-2-(1-hydroxy-1-methylethyl)-benzofuran, A-70044
▷4-Acetyl-2-hydroxy-3-methyltetrahydrofuran, *see* B-60194
Acetylimbricatolol, *in* H-70150
1-Acetylindene, *see* A-70046
2-Acetyl-1*H*-indene, A-70045
3-Acetyl-1*H*-indene, A-70046
6-Acetyl-1*H*-indene, A-70047
2-Acetylindole, A-60032
4-Acetylindole, A-60033
▷5-Acetylindole, A-60034
6-Acetylindole, A-60035
7-Acetylindole, A-60036
1-Acetyl-7-iodonaphthalene, A-60037
1-Acetyl-8-iodonaphthalene, A-60038
Acetylisomontanolide, *in* M-70154
1-Acetyl-4-isopentenyl-6-methylphloroglucinol, *in* T-70248
4-Acetyl-2-isopropenyl-2,3-dihydro-5-benzofuranol, *see* A-70040
2-Acetyl-1-isopropylcyclopropaneacetic acid, *see* S-70027
1-Acetylisoquinoline, A-60039
3-Acetylisoquinoline, A-60040
4-Acetylisoquinoline, A-60041
5-Acetylisoquinoline, A-60042
Acetylmeldrum's acid, *see* A-60029
1-Acetyl-2-methylcyclopentene, A-60043
(Acetylmethylene)malonic acid, *see* O-70100
2-Acetyl-1-(1-methylethyl)-cyclopropaneacetic acid, *see* S-70027
2-Acetylnaphtho[1,8-*bc*]pyran, A-70048
3-Acetyl-4-nitrocatechol, *see* D-60357
4-Acetyl-5-nitrocatechol, *see* D-60359
2-Acetyl-3-nitroquinol, *see* D-60358
2-Acetyl-5-nitroquinol, *see* D-70322
12-(Acetyloxy)-10-[(acetyloxy)methylene]-6-methyl-2-(4-methyl-3-pentenyl)-2,6,11-dodecatrienal, A-70049
3-(Acetyloxy)-3-(2-chloro-3-isocyanoethenyl)-2-oxoindole, *see* I-60028

2-[2-(Acetyloxy)ethenyl]-6,10-dimethyl-2,5,9-undecatrienal, A-70050
19-(Acetyloxy)-13-methoxy-17-norkaur-15-en-18-oic acid, *in* H-60106
1-[2-[1-[(Acetyloxy)methyl]ethenyl]-5-hydroxy-6-benzofuranyl]ethanone, *see* A-60065
Acetylparaminosalol, *in* H-70108
2-Acetylpiperidine, A-70051
Acetylpringleine, *in* P-60172
2-Acetylpyrimidine, A-60044
4-Acetylpyrimidine, A-60045
5-Acetylpyrimidine, A-60046
5-Acetyl-2,4(1*H*,3*H*)-pyrimidinedione, A-60047
5-Acetyl-2(1*H*)-pyrimidinone, A-60048
3-Acetylpyrrole, A-70052
2-Acetylquinoline, A-60049
3-Acetylquinoline, A-60050
4-Acetylquinoline, A-60051
5-Acetylquinoline, A-60052
6-Acetylquinoline, A-60053
7-Acetylquinoline, A-60054
8-Acetylquinoline, A-60055
2-Acetylquinoxaline, A-60056
5-Acetylquinoxaline, A-60057
6-Acetylquinoxaline, A-60058
Acetylsanadaol, *in* S-70008
6-*O*-Acetylshanghiside methyl ester, *in* S-60028
2-Acetyl-1,2,3,4-tetrahydronaphthalene, A-70053
2-Acetyltetralin, *see* A-70053
1-Acetyl-2,3,4,5-tetramethyl-6-nitrobenzene, *see* T-60145
1-Acetyl-2,3,4,6-tetramethyl-5-nitrobenzene, *see* T-60146
1-Acetyl-2,3,5,6-tetramethyl-4-nitrobenzene, *see* T-60147
3-Acetyltetronic acid, *see* A-60030
3-Acetylteumicropin, *in* T-60181
2-Acetyl-2-thiazoline, *see* A-70041
3-(Acetylthio)cyclobutanol, *in* M-60019
21-*O*-Acetyltoosendantriol, *in* T-60223
6′-Acetyltrifolirhizin, *in* M-70001
1-Acetyltriphenylene, A-60059
2-Acetyltriphenylene, A-60060
5-Acetyluracil, *see* A-60047
Acetylzafronic acid, *in* Z-70001
Achalensolide, A-70054
▷Achilleaic acid, *see* A-70055
Achillolide *B*, A-60061
Achillolide *A*, *in* A-60061
Aciphen, *in* H-70108
▷Aconitic acid, A-70055
▷Acraldehyde, *see* P-70119
1-Acridinecarboxaldehyde, A-70056
2-Acridinecarboxaldehyde, A-70057
3-Acridinecarboxaldehyde, A-70058
4-Acridinecarboxaldehyde, A-70059
9-Acridinecarboxaldehyde, A-70060
9-Acridinecarboxylic acid, A-70061
▷2,9-Acridinediamine, *see* D-60027
▷3,6-Acridinediamine, *see* D-60028
▷3,9-Acridinediamine, *see* D-60029
Acriflavine, *in* D-60028
Acriflavinium chloride, *in* D-60028
▷Acrolein, *see* P-70119
▷Acrylamide, *in* P-70121
▷Acrylic acid, *see* P-70121
▷Acrylonitrile, *in* P-70121
β-Actinorhodin, A-70062
Actinostemmoside A, *in* D-70002
Actinostemmoside B, *in* D-70003
Actinostemmoside C, *in* D-70001
Actinostemmoside D, *in* D-70002
Actinoviridin A, *in* H-70193
Aculeatiside A, *in* N-60062
Aculeatiside B, *in* N-60062
Acuminatin, A-60062
Acuminatolide, A-70063
1,3,5,7-Adamantanetetraacetic acid, A-70064
Adamantylideneadamantane epoxide, *see* B-70091
1-Adamantylsulfinyl cyanide, *in* T-70177
Adamantyl thiocyanate, *see* T-70177
Adenanthin, A-70065
Adonirubin, A-70066

Adoxoside, A-70067
Adoxosidic acid, A-70068
Adriadysiolide, A-70069
▷Aesculetin 6-methyl ether, see H-70153
Afraglaucolide, A-70070
Afromosin, in T-60325
Afrormosin, in T-60325
Afzelechin, in T-60104
Agandencidial, see C-70177
▷Agaritine, A-70071
Agarospirene, in P-70114
Agasyllin, A-60064
Ageconyflavone C, in H-70081
Agerasanin, A-70072
Ageratone, A-60065
▷Agerite, see B-60074
Aglajne 2, A-70073
Aglajne 3, A-70074
Aglajne I, see H-70022
Aglycone A₃, in S-70051
Agrimophol, A-60066
Agropinic acid, A-70075
Agroskerin, A-70076
Agrostistachin, A-60067
Agrostophyllin, A-70077
AH₁₀, A-70078
AH₁₁, see D-70289
4a-Hydroxy-8-methoxychlorotetracycline, in M-70045
Aiapin, see M-70068
Ainsliaside A, in Z-60001
Ajafinin, A-60068
Ajoene, A-70079
Ajugasterone A, in H-60023
Akebonoic acid, in H-60199
3α-Akebonoic acid, in H-60199
[N-(β-Alanyl)-2-aminoethyl] disulfide, see A-60073
[N-(β-Alanyl)-3-aminopropyl]disulfide, see H-60084
N-(β-Alanyl)cystamine, see A-60073
N-Alanylcysteine, A-60069
N-β-Alanylhistidine, see C-60020
N-(β-Alanyl)homocystamine, see H-60084
Albidin, A-60070
Albomitomycin A, A-70080
Albrassitriol, in D-70548
Aldehydoacetic acid, see O-60090
6-Aldehydoisoophiopogone A, A-60071
6-Aldehydoisoophiopogone B, A-60072
9-Aldehydononanoic acid, see O-60063
Alectorialic acid, A-70081
Alesal, in H-70108
Alethine, A-60073
Algiospray, in A-60228
Alisol B, A-70082
Alkannin, in S-70038
Alkannin acetate, in S-70038
Alkhanol, in D-70299
2-Allenylbenzothiazole, see P-60174
Alliodorin, A-70083
Alliumoside A, in S-60045
Alloaromadendrane-4,10-diol, A-60074
(+)-Allocoronamic acid, in A-60151
(−)-Allocoronamic acid, in A-60151
(±)-Allocoronamic acid, in A-60151
L-Alloisothreonine, in A-60183
α-Allokainic acid, in K-60003
α-Allokaininic acid, in K-60003
Allophanamide, see B-60190
Allophanic methylamide, in B-60190
Allopteroxylin, A-60075
Allosamidin, A-60076
▷L-Allothreonine, in T-60217
Alloxanthine, see P-60202
4-Allyl-2-azetidinone, in A-60172
o-Allylbenzaldehyde, see P-70122
p-Allylbenzaldehyde, see P-60177
Allyl tert-butyl ketone, see D-60425
1-Allyl-2-chlorobenzene, see C-70152
1-Allyl-3-chlorobenzene, see C-70153
1-Allyl-4-chlorobenzene, see C-70154
1-Allylcyclohexene, see P-60179
3-Allylcyclohexene, see P-60180
4-Allylcyclohexene, see P-60182
3-Allylindole, see P-60183
Allylphenylcarbinol, see P-70061

α-Allylserine, see A-60190
1-Allyl-2,4,5-trimethoxybenzene, see T-70279
Alnusenol, in G-70020
Aloenin B, A-60077
Alogspray, in A-60228
Aloifol I, in T-60102
Aloifol II, in P-60044
Al R6-4, see K-60009
Al-R 2081, see P-70203
Altamisic acid, A-70084
Altamisin, in A-70084
Alteichin, see A-60078
Alterperylenol, in A-60078
Alterporriol B, A-60079
Alterporriol A, in A-60079
Altersolanol A, A-60080
Altersolanol B, in A-60080
Altersolanol C, in A-60080
Altertoxin I, in A-60078
▷Altertoxin II, A-60081
Altertoxin III, in A-60082
Altholactone, A-60083
▷Aluminon, in A-60318
Alyssin, in I-60140
▷Amanine, in A-60084
▷α-Amanitin, in A-60084
▷β-Amanitin, in A-60084
▷ε-Amanitin, in A-60084
▷γ-Amanitin, in A-60084
Amanitins, A-60084
Amantins, see A-60084
Amanullin, in A-60084
Amanullinic acid, in A-60084
Amarolide, A-60085
Ambocin, in T-60324
Ambonin, in D-60340
Ambrox, in D-70533
Ambroxide, in D-70533
Amentadione, A-60086
Amentaepoxide, A-60087
Amentol, A-60088
Amentol 1′-methyl ether, in A-60088
Aminoacetylene, see E-60056
2-(Aminoacetyl)pyridine, A-70085
3-(Aminoacetyl)pyridine, A-70086
4-(Aminoacetyl)pyridine, A-70087
2-Amino-6-acetyl-3,4,7,8-tetrahydro-4-oxo-3H,9H-pyrimido[4,5-b][1,4]diazepine, see A-60019
10-Amino-9-anthracenecarboxylic acid, A-70088
α-Amino-β-benzalpropionic acid, see A-60239
4-Amino-1,2-benzenedicarbonyl hydrazide, see A-70134
3-Amino-1,2-benzenediol, A-70089
▷4-Amino-1,2-benzenediol, A-70090
4-Amino-1H-1,5-benzodiazepine-3-carbonitrile, A-70091
1-(3-Amino-2-benzofuranyl)ethanone, see A-70021
3-Amino-2-benzofuryl methyl ketone, see A-70021
Amino-1,4-benzoquinone, A-60089
2-Amino-4H-benzo[d][1,3]thiazine, see A-70092
2-Amino-4H-3,1-benzothiazine, A-70092
α-Aminobenzo[b]thiophene-3-acetic acid, A-60090
▷3-Amino-2-benzo[b]thiophenecarboxylic acid, A-60091
4-Amino-1,2,3-benzotriazine, A-60092
▷2-Aminobenzoxazole, A-60093
3-Amino-3-benzoylpropanoic acid, see A-60236
2-Amino-2-benzylbutanedioic acid, A-60094
3-Amino-1-benzyl-4-cyanopyrazole, in A-70192
5-Amino-1-benzyl-4-cyanopyrazole, in A-70192
5-Amino-1-benzyl-1H-tetrazole, in A-70202
5-Amino-2-benzyl-2H-tetrazole, in A-70202
7-Amino-2,11-biaboladien-10R-ol, in A-70096
7-Aminobicyclo[4.1.0]heptane-7-carboxylic acid, A-60095
3-Aminobicyclo[2.2.1]heptane-2-methanol, A-70093
3-Aminobicyclo[2.2.1]hept-5-ene-2-methanol, A-70094
2-Aminobicyclo[2.1.1]hexane, A-70095
7-Amino-2,10-bisaboladiene, A-70096

7-Amino-3,10-bisaboladiene, A-70096
7-Amino-2,9-bisaboladien-11-ol, in A-70096
7-Amino-2,11-bisaboladien-10S-ol, in A-70096
7-Aminobisabolene, see A-70096
Aminobisabolenol, in A-70096
3-Amino-2-bromoanisole, in A-70099
3-Amino-2-bromobenzoic acid, A-70097
2-Amino-3-bromobutanedioic acid, A-70098
2-Amino-3-bromobutanoic acid, A-60097
2-Amino-4-bromobutanoic acid, A-60098
4-Amino-2-bromobutanoic acid, A-60099
6-Amino-2-bromohexanoic acid, A-60100
2-Amino-5-bromo-3-hydroxypyridine, A-60101
2-Amino-3-(2-bromo-1H-indol-3-yl)propanoic acid, see B-60332
2-Amino-8-bromoinosine, see B-70215
1-Amino-2-(bromomethyl)cyclohexane, see B-70239
1-Amino-2-(bromomethyl)cyclopentane, see B-70240
3-Amino-2-bromophenol, A-70099
2-Amino-5-bromo-3-pyridinol, see A-60101
2-Amino-4-bromopyrimidine, A-70100
2-Amino-5-bromopyrimidine, A-70101
4-Amino-5-bromopyrimidine, A-70102
5-Amino-5-bromopyrimidine, A-70103
2-Amino-8-bromo-9-β-D-ribofuranosyl-9H-purin-6(1H)-one, see B-70215
2-Amino-3-bromosuccinic acid, see A-70098
3-Amino-2-bromotoluene, see B-70234
▷4-Aminobutanoic acid lactam, see P-70188
2-Amino-4-butanolide, see A-60138
2-Amino-3-butylbutanedioic acid, A-60102
2-Amino-4-tert-butylcyclopentanol, A-70104
2-Amino-3-butylsuccinic acid, see A-60102
2-Amino-γ-butyrolactone, see A-60138
α-Aminobutyrolactone, see A-60138
1-Amino-β-carboline, A-70109
1-Amino-γ-carboline, A-70114
▷2-Amino-α-carboline, A-70105
3-Amino-α-carboline, A-70106
3-Amino-β-carboline, A-70110
3-Amino-γ-carboline, A-70115
4-Amino-α-carboline, A-70107
4-Amino-β-carboline, A-70111
5-Amino-γ-carboline, A-70116
6-Amino-α-carboline, A-70108
6-Amino-β-carboline, A-70112
6-Amino-γ-carboline, A-70117
8-Amino-β-carboline, A-70113
8-Amino-γ-carboline, A-70118
8-Amino-δ-carboline, A-70119
▷3-Amino-3-carboline (obsol.), see A-70105
4-Amino-3-carboline (obsol.), see A-70106
6-Amino-3-carboline (obsol.), see A-70107
10-Amino-3-carboline (obsol.), see A-70108
2-[(Aminocarbonyl)amino]-β-alanine, see A-60268
6-[[[3-(Aminocarbonyl)-2H-azirine-2-ylidene]-amino]oxy]-6-deoxy-1,2-O-(1-methyl-ethylidene)-α-D-glucofuranose 3,5-bis(4-methylbenzenesulfonate), see C-70014
N⁵-(Aminocarbonyl)ornithine, see C-60151
3-Aminocarbostyril, see A-60256
α-Amino-2-carboxy-5-oxo-1-pyrrolidinebutanoic acid, A-70120
S-(3-Amino-3-carboxypropyl)-S-methyl-sulfoximine, see M-70043
1-(3-Amino-3-carboxypropyl)-5-oxo-2-pyrrolidinecarboxyic acid, see A-70120
3-Aminocatechol, see A-70089
2-Amino-3-chlorobutanedioic acid, A-70121
2-Amino-3-chlorobutanoic acid, A-60103
2-Amino-4-chlorobutanoic acid, A-60104
4-Amino-2-chlorobutanoic acid, A-60105
4-Amino-3-chlorobutanoic acid, A-60106
4-Amino-1-(5-chloro-5-deoxy-β-D-arabinofuranosyl)-2(1H)-pyrimidinone, see C-70064
2-Amino-4-chloro-4,4-difluorobutanoic acid, A-60107
2-Amino-4-chloro-4-fluorobutanoic acid, A-60108
6-Amino-2-chlorohexanoic acid, A-60109
2-Amino-5-chloro-6-hydroxy-4-hexenoic acid, A-60110

2-Amino-5-chloro-3-hydroxypyridine, A-60111
2-Amino-3-(2-chloro-1*H*-indol-3-yl)propanoic
    acid, *see* C-60145
2-Amino-8-chloroinosine, *see* C-70068
2′-Amino-2-chloro-3′-methylacetophenone,
    A-60112
2′-Amino-2-chloro-4′-methylacetophenone,
    A-60113
2′-Amino-2-chloro-5′-methylacetophenone,
    A-60114
2′-Amino-2-chloro-6′-methylacetophenone,
    A-60115
▷2-Amino-3-chloro-1,4-naphthalenedione, *see*
    A-70122
▷2-Amino-3-chloro-1,4-naphthoquinone, A-70122
3-Amino-2-chloro-1-propene, *see* C-70151
▷2-Amino-6-chloro-1*H*-purine, A-60116
2-Amino-5-chloro-3-pyridinol, *see* A-60111
2-Amino-8-chloro-9-β-D-ribofuranosyl-9*H*-
    purin-6(1*H*)-one, *see* C-70068
2-Amino-3-chlorosuccinic acid, *see* A-70121
5-Aminocinchomeronic acid, *see* A-70195
4-Amino-3-cyano-1*H*-1,5-benzodiazepine, *see*
    A-70091
3-Amino-2-cyanobenzo[*b*]thiophene, *in* A-60091
3-Amino-4-cyano-1-methylpyrazole, *in* A-70187
5-Amino-4-cyano-1-methylpyrazole, *in* A-70192
2-Amino-5-cyanonicotinamide, *in* A-70193
3-Amino-4-cyanopyrazole, *in* A-70187
4-Amino-3-cyanopyrazole, *in* A-70189
2-Aminocycloheptanecarboxylic acid, A-70123
2-Aminocycloheptanemethanol, A-70124
2-Amino-2,5-cyclohexadiene-1,4-dione, *see*
    A-60089
1-Amino-1,4-cyclohexanedicarboxylic acid,
    A-60117
2-Amino-1,4-cyclohexanedicarboxylic acid,
    A-60118
3-Amino-1,2-cyclohexanedicarboxylic acid,
    A-60119
4-Amino-1,1-cyclohexanedicarboxylic acid,
    A-60120
4-Amino-1,3-cyclohexanedicarboxylic acid,
    A-60121
3-Amino-2-cyclohexen-1-one, A-70125
1-Aminocyclopropenecarboxylic acid, A-70126
1-Amino-2-cyclopropene-1-carboxylic acid,
    A-60122
3-(1-Aminocyclopropyl)-2-propenoic acid,
    A-60123
▷*N*⁴-Aminocytidine, A-70127
8-Amino-3-deazaguanine, *see* D-60034
2-Amino-3,4-dehydropimelic acid, *see* A-60164
2-Amino-4,5-dehydropimelic acid, *see* A-60165
3′-Amino-3′-deoxyadenosine, A-70128
9-Aminodeoxyclavulanic acid, *in* A-70139
9-(3-Amino-3-deoxy-β-D-ribofuranosyl)-
    adenine, *see* A-70128
6-Amino-2,2-dibromohexanoic acid, A-60124
1-Amino-6,7-dibromonaphthalene, *see* D-70097
2-Amino-4,4-dichloro-3-butenoic acid, A-60125
6-Amino-2,2-dichlorohexanoic acid, A-60126
4-Amino-2,5-dichlorophenol, A-70129
2-Amino-3-(2,3-dichlorophenyl)propanoic acid,
    A-60127
2-Amino-3-(2,4-dichlorophenyl)propanoic acid,
    A-60128
2-Amino-3-(2,5-dichlorophenyl)propanoic acid,
    A-60129
2-Amino-3-(2,6-dichlorophenyl)propanoic acid,
    A-60130
2-Amino-3-(3,4-dichlorophenyl)propanoic acid,
    A-60131
2-Amino-3-(3,5-dichlorophenyl)propanoic acid,
    A-60132
2-Amino-3,5-dicyanopyridine, *in* A-70193
2-Amino-3,3-difluorobutanoic acid, A-60133
3-Amino-4,4-difluorobutanoic acid, A-60134
2-Amino-3,3-difluoropropanoic acid, A-60135
9-Amino-9,10-dihydroanthracene, A-60136
7-Amino-7,8-dihydro-α-bisabolene, *see* A-60096
4-Amino-5,6-dihydro-4*H*-cyclopenta[*b*]-
    thiophene, A-60137

3-Aminodihydro-2(3*H*)-furanone, A-60138
2-Amino-7,8-dihydro-6-(2-hydroxy-1-
    oxopropyl)-4(1*H*)-pteridinone, *see* S-70035
4-Amino-2,5-dihydro-2-imino-5,5-dimethyl-1*H*-
    imidazol-1-yloxy, *see* P-70110
3-Amino-2,3-dihydro-1*H*-inden-1-one, *see*
    A-60211
4-Aminodihydro-3-methylene-2-(3*H*)furanone,
    A-70131
6-Amino-1,3-dihydro-1-methyl-2*H*-purine-2-
    one, *in* A-60144
1-Amino-1,4-dihydronaphthalene, A-60139
1-Amino-5,8-dihydronaphthalene, A-60140
2-Amino-4,7-dihydro-4-oxo-7β-D-ribofuranosyl-
    1*H*-pyrrolo[2,3-*d*]-pyrimidin-5-carbonitrile,
    A-70132
2-Amino-9,10-dihydrophenanthrene, A-60141
4-Amino-9,10-dihydrophenanthrene, A-60142
9-Amino-9,10-dihydrophenanthrene, A-60143
2-Amino-1,7-dihydro-7-phenyl-6*H*-purin-6-one,
    A-70133
6-Amino-2,3-dihydro-1,4-phthalazinedione,
    A-70134
6-Amino-1,3-dihydro-2*H*-purin-2-one, A-60144
5-Amino-1,4-dihydro-7*H*-pyrazolo[4,3-*d*]-
    pyrimidin-7-one, *see* A-60246
5-Amino-3,4-dihydro-2*H*-pyrrole-2-carboxylic
    acid, A-70135
2-Amino-1,7-dihydro-4*H*-pyrrolo[2,3-*d*]-
    pyrimidin-4-one, *see* A-60255
α-Amino-β,4-dihydroxybenzenebutanoic acid,
    see A-60185
2-Amino-5-[[(4,5-dihydroxy-2-cyclopenten-1-
    yl)amino]methyl]-1,7-dihydro-4*H*-
    pyrrolo[2,3-*d*]pyrimidin-4-one, *see* Q-70004
2-Amino-5-(4,5-dihydroxycyclopent-1-en-3-
    ylaminomethyl)pyrrolo[2,3-*d*]pyrimidine, *see*
    Q-70004
4-Amino-1-[2,3-dihydroxy-4-(hydroxymethyl)-
    cyclopentyl]-2(1*H*)-pyrimidinone, *see*
    C-70018
1-Amino-4,5-dihydroxy-7-methoxy-2-methyl-
    anthraquinone, A-60145
▷2-Amino-1,4-dihydroxy-18-methyl-4-aza-3,7,14-
    nonadecanetrione, *see* L-70033
▷2-Amino-*N*,3-dihydroxy-*N*-(14-methyl-3,10-
    dioxopentadecyl)propanamide, *see* L-70033
2-Amino-6-(1,2-dihydroxypropyl)-3-methyl-
    pterin-4-one, *in* B-70121
2-Amino-6-(1,2-dihydroxypropyl)-3-methyl-
    pterin-4-one, A-60146
2-Amino-6-(1,2-dihydroxypropyl)-4(1*H*)-
    pteridinone, *in* B-70121
3-Amino-1,2-dimethoxybenzene, *in* A-70089
▷4-Amino-1,2-dimethoxybenzene, *in* A-70090
2-Amino-3,3-dimethylbutanedioic acid, A-60147
2-Amino-2,3-dimethylbutanoic acid, A-60148
2-Amino-3,3-dimethylbutanoic acid, A-70136
2-Amino-4-(1,1-dimethylethyl)cyclopentanol,
    see A-70104
2-Amino-4,4-dimethylpentanoic acid, A-70137
2-Amino-1,3-diphenyl-1-propanone, A-60149
γ-Aminodipicolinic acid, *see* A-70194
Aminodithiocarbamic acid, *see* D-60509
2-Amino-5,6-epoxy-2-cyclohexene-1,4-dione, *in*
    A-60194
2-Amino-5,6-epoxy-3-hydroxy-2-cyclohexen-1-
    one, *see* A-60194
2-Amino-4,5-epoxy-3-hydroxy-6-oxo-1-cyclo-
    hexenecarboxamide, *see* A-70158
2-Amino-4-ethoxybutanoic acid, *in* A-70156
2-Amino-3-ethylbutanedioic acid, A-70138
2-Amino-2-ethyl-3-butenoic acid, A-60150
1-Amino-2-ethylcyclopropanecarboxylic acid,
    A-60151
S-(2-Aminoethyl)cysteine, A-60152
(2-Aminoethyl)hydrazine, *see* H-70096
3-(2-Aminoethylidene)-7-oxo-4-oxa-1-
    azabicyclo[3.2.0]heptane-2-carboxylic acid,
    A-70139
2-Amino-3-ethylsuccinic acid, *see* A-70138
5-Amino-1-ethyl-1*H*-tetrazole, *in* A-70202
5-Amino-2-ethyl-2*H*-tetrazole, *in* A-70202
3-[(2-Aminoethyl)thio]alanine, *see* A-60152

4-Amino-5-ethynyl-2(1*H*)-pyrimidinone,
    A-60153
2-Amino-3-fluorobenzoic acid, A-70140
2-Amino-4-fluorobenzoic acid, A-70141
2-Amino-5-fluorobenzoic acid, A-70142
2-Amino-6-fluorobenzoic acid, A-70143
3-Amino-4-fluorobenzoic acid, A-70144
4-Amino-2-fluorobenzoic acid, A-70145
▷4-Amino-3-fluorobenzoic acid, A-70146
5-Amino-2-fluorobenzoic acid, A-70147
2-Amino-3-fluorobutanoic acid, A-60154
2-Amino-4-fluorobutanoic acid, A-60155
3-Amino-4-fluorobutanoic acid, A-60156
4-Amino-2-fluorobutanoic acid, A-60157
4-Amino-3-fluorobutanoic acid, A-60158
2-Amino-3-fluoro-3-methylbutanoic acid,
    A-60159
2-Amino-3-fluoropentanoic acid, A-60160
4-Amino-5-fluoro-2-pentenoic acid, A-60161
3-Amino-2-fluoropropanoic acid, A-60162
8-Amino-6-fluoro-9*H*-purine, A-60163
3-Amino-6-fluoropyridazine, A-70148
α-Aminoglutaric acid lactam, *see* O-70102
6-Aminoguaiacol, *in* A-70089
2-Amino-5-guanidino-3-hydroxypentanoic acid,
    A-70149
8-Aminoguanine, *see* D-60035
8-Aminoguanosine, A-70150
3-Aminoguaiacol, *in* A-70089
4-Aminoguaiacol, *in* A-70090
5-Aminoguaiacol, *in* A-70090
2-Amino-3-heptenedioic acid, A-60164
2-Amino-5-heptenedioic acid, A-60165
4-Amino-2,5-hexadienoic acid, A-60166
6-Amino-2-hexene, *see* H-70088
2-Amino-2-hexenoic acid, A-60167
2-Amino-3-hexenoic acid, A-60168
2-Amino-5-hexenoic acid, A-60169
3-Amino-2-hexenoic acid, A-60170
3-Amino-4-hexenoic acid, A-60171
3-Amino-5-hexenoic acid, A-60172
6-Amino-2-hexenoic acid, A-60173
1-Amino-2-hydrazinoethane, *see* H-70096
2-Aminohydrosorbic acid, *see* A-60168
1-Amino-2-hydroxy-9,10-anthracenedione, *see*
    A-60174
▷1-Amino-4-hydroxy-9,10-anthracenedione, *see*
    A-60175
1-Amino-5-hydroxy-9,10-anthracenedione, *see*
    A-60176
1-Amino-8-hydroxy-9,10-anthracenedione, *see*
    A-60177
2-Amino-1-hydroxy-9,10-anthracenedione, *see*
    A-60178
2-Amino-3-hydroxy-9,10-anthracenedione, *see*
    A-60179
3-Amino-1-hydroxy-9,10-anthracenedione, *see*
    A-60180
9-Amino-10-hydroxy-1,4-anthracenedione, *see*
    A-60181
1-Amino-2-hydroxyanthraquinone, A-60174
▷1-Amino-4-hydroxyanthraquinone, A-60175
1-Amino-5-hydroxyanthraquinone, A-60176
1-Amino-8-hydroxyanthraquinone, A-60177
2-Amino-1-hydroxyanthraquinone, A-60178
2-Amino-3-hydroxyanthraquinone, A-60179
3-Amino-1-hydroxyanthraquinone, A-60180
9-Amino-10-hydroxy-1,4-anthraquinone,
    A-60181
3-Amino-4-hydroxy-1,2-benzenedicarboxylic
    acid, A-70151
4-Amino-5-hydroxy-1,2-benzenedicarboxylic
    acid, A-70152
4-Amino-6-hydroxy-1,3-benzenedicarboxylic
    acid, A-70153
5-Amino-2-hydroxy-1,3-benzenedicarboxylic
    acid, A-70154
5-Amino-4-hydroxy-1,3-benzenedicarboxylic
    acid, A-70155
3-Amino-5-hydroxybenzeneethanol, *see*
    A-60182
γ-Amino-β-hydroxybenzenepentanoic acid, *see*
    A-60199
3-Amino-5-hydroxybenzyl alcohol, A-60182
2-Amino-3-hydroxybutanoic acid, *see* T-60217
2-Amino-4-hydroxybutanoic acid, A-70156
3-Amino-2-hydroxybutanoic acid, A-60183
2-Amino-4-hydroxybutanoic acid lactone, *see*
    A-60138

2-Amino-1-hydroxy-1-cyclobutaneacetic acid, A-60184

2-Amino-4-hydroxy-6-(1,2-dihydroxypropyl)-pterin, see B-70121

2-Amino-3-hydroxy-4-(4-hydroxyphenyl)-butanoic acid, A-60185

2-Amino-4-hydroxy-α-imino-Δ^{7(3H)β}-pteridinepropionic acid, see L-60020

4-Amino-6-hydroxyisophthalic acid, see A-70153

5-Amino-2-hydroxyisophthalic acid, see A-70154

5-Amino-4-hydroxyisophthalic acid, see A-70155

2-Amino-4-hydroxy-2-methylbutanoic acid, A-60186

2-Amino-2-(hydroxymethyl)-3-butenoic acid, A-60187

1-Amino-3-(hydroxymethyl)-1-cyclobutanecarboxylic acid, A-70157

3-Amino-2-hydroxy-5-methylhexanoic acid, A-60188

2-Amino-3-hydroxymethylnorbornane, see A-70093

5-Amino-6-hydroxymethylnorborn-2-ene, see A-70094

2-Amino-3-hydroxy-4-methyl-6-octenoic acid, A-60189

2-Amino-2-(hydroxymethyl)-4-pentenoic acid, A-60190

2-Amino-2-(hydroxymethyl)-4-pentynoic acid, A-60191

2-Amino-3-hydroxy-6-methylpyridine, A-60192

5-Amino-3-hydroxy-2-methylpyridine, A-60193

3-Amino-5-hydroxy-7-oxabicyclo[4.1.0]hept-3-en-2-one, A-60194

5-Amino-3-hydroxy-2-(2-oxo-1-azetidinyl)-pentanoic acid, see P-70116

4-Amino-5-hydroxy-2-oxo-7-oxabicyclo[4.1.0]-hept-3-ene-3-carboxamide, A-70158

2-Amino-4-hydroxypentanedioic acid, A-60195

4-Amino-5-hydroxypentanoic acid, A-70159

5-Amino-4-hydroxypentanoic acid, A-60196

1-Amino-2-(4-hydroxyphenyl)-cyclopropanecarboxylic acid, A-60197

α-Amino-β-(p-hydroxyphenyl)isovaleric acid, see A-60198

2-Amino-3-(4-hydroxyphenyl)-3-methylbutanoic acid, A-60198

4-Amino-3-hydroxy-5-phenylpentanoic acid, A-60199

3-Amino-4-hydroxyphthalic acid, see A-70151

4-Amino-5-hydroxyphthalic acid, see A-70152

3-Amino-2-hydroxypropanoic acid, A-70160

α-(3-Amino-1-hydroxypropyl)-2-oxo-1-azetidineacetic acid, see P-70116

2-Amino-4-hydroxypteridine, see A-60242

3-[2-Amino-4-hydroxy-7(3H)-pteridinylidene]-2-iminopropanoic acid, see L-60020

2-Amino-3-hydroxypyridine, A-60200

2-Amino-4-hydroxypyridine, in A-60253

2-Amino-5-hydroxypyridine, A-60201

2-Amino-6-hydroxypyridine, in A-60252

3-Amino-2-hydroxypyridine, see A-60249

3-Amino-4-hydroxypyridine, see A-60254

4-Amino-2-hydroxypyridine, see A-60250

4-Amino-3-hydroxypyridine, A-60202

5-Amino-2-hydroxypyridine, see A-60251

5-Amino-3-hydroxypyridine, A-60203

5-Amino-4-hydroxypyrimidine, see A-70196

2-Amino-4-hydroxypyrrolo[2,3-d]pyrimidine, see A-60255

3-Amino-2-hydroxyquinoline, see A-60256

2-Amino-8-hydroxy-9-β-D-ribofuranosyl-9H-purin-6(1H)-one, see D-70238

α-Amino-1H-imidazole-1-propanoic acid, A-60204

α-Amino-1H-imidazole-2-propanoic acid, A-60205

4-Amino-1H-imidazo[4,5-c]pyridine, A-70161

8-Aminoimidazo[4,5-g]quinazoline, A-60207

9-Aminoimidazo[4,5-f]quinazoline, A-60206

7-Aminoimidazo[4,5-f]quinazolin-9(8H)-one, A-60208

6-Aminoimidazo[4,5-g]quinolin-8(7H)-one, A-60209

Aminoiminomethanesulfonic acid, A-60210

3-Amino-1-indanone, A-60211

8-Aminoinosine, see A-70198

2-Amino-8-iodo-9-β-D-ribofuranosyl-9H-purin-6(1H)-one, see I-70034

2-Amino-2-isopropylbutanedioic acid, A-60212

2-Amino-3-isopropylbutanedioic acid, A-70162

2-Amino-3-isopropylsuccinic acid, see A-70162

6-Aminoisoxanthopterine, see D-60047

3-Aminolactic acid, see A-70160

1-Amino-2-mercaptonaphthalene, see A-70172

1-Amino-8-mercaptonaphthalene, see A-70173

2-Amino-3-mercapto-1,4-naphthalenedione, see A-70163

2-Amino-3-mercapto-1,4-naphthoquinone, A-70163

2-Amino-3-mercapto-3-phenylpropanoic acid, A-60213

2-Amino-3-(4-mercaptophenyl)propanoic acid, A-60214

▷3-Amino-2-mercaptopyridine, see A-60247

2-Amino-8-mercapto-9-β-D-ribofuranosyl-9H-purin-6(1H)-one, see M-70029

3-Aminomesitoic acid, see A-60266

Aminomethanesulfonic acid, A-60215

▷1-Amino-2-methoxyanthraquinone, in A-60174

▷1-Amino-4-methoxyanthraquinone, in A-60175

1-Amino-5-methoxyanthraquinone, in A-60176

9-Amino-10-methoxy-1,4-anthraquinone, in A-60181

3-Amino-4-methoxy-1,2-benzenedicarboxylic acid, in A-70151

2-Amino-4-methoxy-6-nitrotoluene, in A-60222

2-Amino-6-methoxyphenol, in A-70089

3-Amino-2-methoxyphenol, in A-70089

4-Amino-2-methoxyphenol, in A-70090

5-Amino-2-methoxyphenol, in A-70090

2-Amino-3-(4-methoxyphenyl)-3-methylbutanoic acid, in A-60198

2-Amino-4-methoxypyridine, in A-60253

2-Amino-6-methoxypyridine, in A-60252

3-Amino-2-methoxypyridine, in A-60249

3-Amino-4-methoxypyridine, in A-60254

4-Amino-2-methoxypyridine, in A-60250

5-Amino-2-methoxypyridine, in A-60251

6-Amino-2-(methylamino)-4(3H)-pyrimidinone, in D-70043

2-Amino-3-methylbiphenyl, A-70164

5-Aminomethyl-3-bromoisoxazole, A-70165

3-Amino-3-methylbutanedioic acid, A-70166

1-Amino-3-methyl-2,3-butanediol, A-60216

2-Amino-2-methylbutanoic acid, A-70167

3-Amino-3-methyl-2-butanol, A-60217

2-Amino-3-methyl-3-butenoic acid, A-60218

2-(Aminomethyl)-4-(carboxymethyl)-1H-pyrrole-3-propanoic acid, see I-60114

2-(Aminomethyl)cycloheptanol, A-70168

2-(Aminomethyl)cyclohexanol, A-70169

2-(Aminomethyl)cyclooctanol, A-70170

5-(Aminomethyl)dihydro-2(3H)-furanone, in A-60196

β-Amino-α-methylene-γ-butyrolactone, see A-70131

2-Amino-3-methylenepentanedioic acid, A-60219

2-Amino-2-(1-methylethyl)butanedioic acid, see A-60212

2-Amino-2-methyl-3-(4-hydroxyphenyl)-propanoic acid, A-60220

4-Amino-5-methyl-3-isoxazolidinone, A-60221

2-Amino-3-methyl-1,4-naphthalenedione, see A-70171

2-Amino-3-methyl-1,4-naphthoquinone, A-70171

3-Amino-4-methyl-5-nitrophenol, A-60222

2-Amino-4-methylpentanedioic acid, A-60223

2-Aminomethyl-1,10-phenanthroline, A-60224

1-(2-Amino-3-methylphenyl)-2-chloroethanone, see A-60112

1-(2-Amino-4-methylphenyl)-2-chloroethanone, see A-60113

1-(2-Amino-5-methylphenyl)-2-chloroethanone, see A-60114

1-(2-Amino-6-methylphenyl)-2-chloroethanone, see A-60115

2-Amino-2-methyl-3-phenylpropanoic acid, A-60225

2-Amino-1-methyl-4(1H)-pteridinone, in A-60242

6-Amino-7-methylpurine, A-60226

8-Amino-9-methylpurine, in A-60244

3-Amino-1-methyl-4-pyrazolecarbonitrile, in A-70187

▷2-(Aminomethyl)pyridine, A-60227

3-(Aminomethyl)pyridine, A-60228

4-(Aminomethyl)pyridine, A-60229

2-Amino-6-methyl-3-pyridinol, see A-60192

5-Amino-2-methyl-3-pyridinol, see A-60193

3-Amino-1-methyl-2(1H)-pyridinone, in A-60249

3-Amino-1-methyl-4(1H)-pyridinone, in A-60254

5-Amino-1-methyl-2(1H)-pyridinone, in A-60251

6-Amino-1-methyl-2(1H)-pyridinone, in A-60252

Aminomethyl 2-pyridinyl ketone, see A-70085

Aminomethyl 3-pyridinyl ketone, see A-70086

Aminomethyl 4-pyridinyl ketone, see A-70087

2-Amino-6-methyl-4(1H)-pyrimidinone, A-60230

2-Amino-1-methyl-9-β-D-ribofuranosyl-9H-purin-6(1H)-one, see M-70084

2-Amino-8-methyl-9-β-D-ribofuranosyl-9H-purin-6(1H)-one, see M-70085

2-Amino-3-methylsuccinic acid, see A-70166

2-Amino-4-(S-methylsulfonimidoyl)butanoic acid, see M-70043

5-Amino-1-methyl-1H-tetrazole, in A-70202

5-Amino-1-methyl-2H-tetrazole, in A-70202

2-Amino-3-(methylthio)-1,4-naphthalenedione, in A-70163

2-Amino-5-(methylthio)pentanoic acid, A-60231

1-Amino-2-naphthalenethiol, A-70172

8-Amino-1-naphthalenethiol, A-70173

8-Amino-1-naphtholactam, see B-70027

2-Aminoneohexanoic acid, see A-70136

3-Amino-4-nitro-1,2-benzenedicarboxylic acid, A-70174

3-Amino-6-nitro-1,2-benzenedicarboxylic acid, A-70175

4-Amino-3-nitro-1,2-benzenedicarboxylic acid, A-70176

4-Amino-5-nitro-1,2-benzenedicarboxylic acid, A-70177

5-Amino-3-nitro-1,2-benzenedicarboxylic acid, A-70178

3-Amino-5-nitrobenzenemethanol, see A-60233

3-Amino-5-nitro-2,1-benzisothiazole, A-60232

3-Amino-5-nitrobenzyl alcohol, A-60233

α-Amino-5-nitro-1H-imidazole-4-propanoic acid, see N-60028

6-Amino-5-nitro-4-imino-β-D-ribofuranosylpyrimidine, see C-60155

3-Amino-4-nitrophthalic acid, see A-70174

3-Amino-6-nitrophthalic acid, see A-70175

4-Amino-3-nitrophthalic acid, see A-70176

4-Amino-5-nitrophthalic acid, see A-70177

5-Amino-3-nitrophthalic acid, see A-70178

N-(6-Amino-5-nitro-4-pyrimidinyl)-β-D-ribofuranosylamine, see C-60155

6-Amino-8-nitroquinoline, A-60234

8-Amino-6-nitroquinoline, A-60235

2-Aminooctane, see O-60031

3-Amino-7-oxabicyclo[4.1.0]hept-3-ene-2,5-dione, in A-60194

▷Aminooxamide, see S-60020

▷Aminooxoacetic acid hydrazide, see S-60020

3-Amino-4-oxo-4-phenylbutanoic acid, A-60236

2-(Aminooxy)ethanol, see H-60134

4-Amino-3-penten-2-one, in P-60059

2-Amino-1,10-phenanthroline, A-70179

4-Amino-1,10-phenanthroline, A-70181

5-Amino-1,10-phenanthroline, A-70182

8-Amino-1,7-phenanthroline, A-70180

▷2-Amino-3H-phenoxazin-3-one, A-60237

N-(4-Aminophenyl)-p-benzoquinone diimine, see I-60019

2-Amino-3-phenyl-3-butenoic acid, A-60238

2-Amino-4-phenyl-3-butenoic acid, A-60239

▷4-[(4-Aminophenyl)(4-imino-2,5-cyclohexadien-1-ylidene)methyl]-2-methylbenzenamine, see R-60010

5-Amino-3-phenylisoxazole, A-60240

5-Amino-1-(phenylmethyl)-1*H*-pyrazole-4-
    carboxamide, *in* A-70192
2-Amino-3-phenylpentanedioic acid, A-60241
1-Amino-1-phenyl-4-pentene, A-70183
3-Amino-1-phenylpropyne, *see* P-60129
3-Amino-4-phenyl-1,2,4-thiadiazole-5(4*H*)-
    imine, *see* H-60012
1-Aminophthalazine, A-70184
5-Aminophthalazine, A-70185
4-Aminophthalhydrazide, *see* A-70134
4-Aminophyscion, *see* A-60145
N-(3-Aminopropyl)-*N*′-[3-[(3-aminopropyl)-
    amino]propyl]-1,3-propanediamine, *see*
    P-60026
2-Amino-3-propylbutanedioic acid, A-70186
2-Amino-3-propylsuccinic acid, *see* A-70186
▷3-Aminopropyne, *see* P-60184
▷2-Amino-4,7(1*H*,8*H*)pteridinedione, *see* I-60141
2-Aminopteridinol, *see* A-60242
2-Amino-4(1*H*)-pteridinone, A-60242
▷2-Aminopurine, A-60243
8-Aminopurine, A-60244
9-Aminopurine, A-60245
5-Amino-4-pyrazolecarbonitrile, *in* A-70187
▷3-Amino-1*H*-pyrazole-4-carboxylic acid,
    A-70187
3-Amino-1*H*-pyrazole-5-carboxylic acid,
    A-70188
4-Amino-1*H*-pyrazole-3-carboxylic acid,
    A-70189
4-Amino-1*H*-pyrazole-5-carboxylic acid,
    A-70190
5-Amino-1*H*-pyrazole-3-carboxylic acid,
    A-70191
5-Amino-1*H*-pyrazole-4-carboxylic acid,
    A-70192
5-Amino-2*H*-pyrazole-3-carboxylic acid, *see*
    A-70188
5-Aminopyrazolo[4,3-*d*]pyrimidin-7(1*H*,6*H*)-
    one, A-60246
2-Amino-3,5-pyridinedicarboxylic acid, A-70193
4-Amino-2,6-pyridinedicarboxylic acid, A-70194
5-Amino-3,4-pyridinedicarboxylic acid, A-70195
▷3-Amino-2-pyridinethiol, *see* A-60247
▷3-Amino-2(1*H*)-pyridinethione, A-60247
2-Amino-3-pyridinol, *see* A-60200
2-Amino-4-pyridinol, *in* A-60253
3-Amino-2-pyridinol, *see* A-60249
3-Amino-4-pyridinol, *see* A-60254
4-Amino-2-pyridinol, *see* A-60250
4-Amino-3-pyridinol, *see* A-60202
5-Amino-2-pyridinol, *see* A-60251
5-Amino-3-pyridinol, *see* A-60203
6-Amino-2-pyridinol, *in* A-60252
6-Amino-3-pyridinol, *see* A-60201
1-Amino-2(1*H*)-pyridinone, A-60248
2-Amino-4(1*H*)-pyridinone, A-60253
3-Amino-2(1*H*)-pyridinone, A-60249
3-Amino-4(1*H*)-pyridinone, A-60254
4-Amino-2(1*H*)-pyridinone, A-60250
5-Amino-2(1*H*)-pyridinone, A-60251
6-Amino-2(1*H*)-pyridinone, A-60252
1-(2-Amino-3-pyridinyl)ethanone, *see* A-70024
1-(2-Amino-3-pyridinyl)ethanone, *see* A-70026
1-(3-Amino-2-pyridinyl)ethanone, *see* A-70022
1-(3-Amino-4-pyridinyl)ethanone, *see* A-70027
1-(4-Amino-3-pyridinyl)ethanone, *see* A-70025
1-(5-Amino-2-pyridinyl)ethanone, *see* A-70023
1-(6-Amino-3-pyridinyl)ethanone, *see* A-70028
2-Amino-1-(2-pyridinyl)ethanone, *see* A-70085
2-Amino-1-(3-pyridinyl)ethanone, *see* A-70086
2-Amino-1-(4-pyridinyl)ethanone, *see* A-70087
2-Amino-3-pyridinyl methyl ketone, *see*
    A-70024
2-Amino-4-pyridinyl methyl ketone, *see*
    A-70026
3-Amino-2-pyridinyl methyl ketone, *see*
    A-70022
3-Amino-4-pyridinyl methyl ketone, *see*
    A-70027
4-Amino-3-pyridinyl methyl ketone, *see*
    A-70025
5-Amino-2-pyridinyl methyl ketone, *see*
    A-70023

6-Amino-3-pyridinyl methyl ketone, *see*
    A-70028
1-Amino-2(1*H*)-pyridone, *see* A-60248
6-Amino-2(1*H*)-pyridone, *see* A-60252
5-Amino-4(1*H*)-pyrimidinone, A-70196
3-Aminopyrocatechol, *see* A-70089
▷4-Aminopyrocatechol, *see* A-70090
2-Amino-1-pyrroline-5-carboxylic acid, *see*
    A-70135
2-Aminopyrrolo[2,3-*d*]pyrimidin-4-one,
    A-60255
3-Amino-2(1*H*)quinolinethione, A-70197
3-Amino-2-quinolinol, *see* A-60256
3-Amino-2(1*H*)-quinolinone, A-60256
Aminoquinone, *see* A-60089
8-Amino-9-β-D-ribofuranosylguanine, *see*
    A-70150
2-Amino-9-β-D-ribofuranosylpurine, *in* A-60243
2-Amino-9-β-D-ribofuranosyl-6,8(1*H*,9*H*)-
    purinedione, *see* D-70238
8-Amino-9-β-D-ribofuranosyl-6*H*-purin-6-one,
    A-70198
4-Amino-2,3,5,6-tetrafluoropyridine, A-60257
2-Amino-4,5,6,7-tetrahydrobenzo[*b*]-thiophene-
    3-carboxylic acid, A-60258
3-Amino-1,2,3,4-tetrahydrocarbazole, A-60259
3-Aminotetrahydro-3-thiophenecarboxylic acid,
    A-70199
3-Aminotetrahydro-2*H*-thiopyran, A-70200
2-Amino-4,5-tetramethylene-3-
    thiophenecarboxylic acid, *see* A-60258
4-Amino-2,2,6,6-tetramethylpiperidine,
    A-70201
▷5-Amino-1*H*-tetrazole, A-70202
3-Aminothiacyclohexane, *see* A-70200
5-Amino-2-thiazolecarbothioamide, A-60260
5-Amino-4-thiazolecarbothioamide, A-60261
5-Aminothiazole-2-thiocarboxamide, *see*
    A-60260
5-Aminothiazole-4-thiocarboxamide, *see*
    A-60261
1-(3-Amino-2-thienyl)ethanone, *see* A-70029
5-Amino-2-thiocarbamoylthiazole, *see* A-60260
5-Amino-4-thiocarbamoylthiazole, *see* A-60261
5-Aminothio-2-thiazolecarboxamide, *see*
    A-60260
5-Aminothio-4-thiazolecarboxamide, *see*
    A-60261
▷3-Amino-1,2,4-triazine, A-70203
2-Amino-3-(2,3,4-trichlorophenyl)propanoic
    acid, A-60262
2-Amino-3-(2,3,6-trichlorophenyl)propanoic
    acid, A-60263
2-Amino-3-(2,4,5-trichlorophenyl)propanoic
    acid, A-60264
7-Amino-4-(trifluoromethyl)-2*H*-1-benzopyran-
    2-one, A-60265
7-Amino-4-(trifluoromethyl)coumarin, *see*
    A-60265
3-Amino-2,4,6-trimethylbenzoic acid, A-60266
5-Amino-2,3,4-trimethylbenzoic acid, A-60267
▷2-Amino-1,3,5-trinitrobenzene, *see* T-70301
3-Amino-2-ureidopropanoic acid, A-60268
2-Amino-5-ureidovaleric acid, *see* C-60151
3-Aminoveratrol, *in* A-70089
▷4-Aminoveratrol, *in* A-70090
2-Amino-5-vinylpyrimidine, A-60269
7-Aminoxanthopterine, *see* D-60046
▷Ammoidin, *see* X-60001
Ammothamnidin, A-70204
Amphidinolide *B*, A-70205
Amritoside, *in* E-70007
▷γ-Amylbutyrolactone, *see* D-70245
δ-Amyrin, *in* O-70036
δ-Amyrone, *in* O-70036
Anacardic acid, *in* H-60205
Anamarine, A-60270
▷Anametil, *see* C-70075
Anaphatol, *in* D-60339
Anastrephin, A-70206
Andalucin, A-70207
Andirolactone, A-60271
Andrimide, A-60272
▷α-Angelica lactone, *see* M-70081
▷β-Angelica lactone, *see* M-60083
▷Δ¹-Angelica lactone, *see* M-60083

▷Δ²-Angelica lactone, *see* M-70081
▷Angelic lactone, *see* M-70081
Angelicoidenol, *in* T-70280
Angelikoreanol, *see* B-60138
Angeloylprangeline, *in* P-60170
3-*O*-Angeloylsenecioodontol, *in* S-60022
Angoluvarin, A-70208
Angrendiol, *in* G-60014
Anguinomycin *A*, A-70209
Anguinomycin *B*, A-70210
Anhydrocumanin, *in* D-60211
3-Anhydro-6-epiophiobolin A, *in* O-70040
Anhydrofusarubin 9-methyl ether, *in* F-60108
Anhydronellinol, *see* T-60312
Anhydroscandenolide, A-70211
15-Anhydrothyrsiferol, A-60274
15-Anhydrothyrsiferyl diacetate, *in* A-60274
15(28)-Anhydrothyrsiferyl diacetate, *in*
    A-60274
4-Anilino-1,2,3-benzotriazine, *in* A-60092
1-Anilino-γ-carboline, *in* A-70114
4-Anilino-3-penten-2-one, *in* P-60059
4-Anilino-1,10-phenanthroline, *in* A-70181
Anisomelic acid, A-60275
Anisomelolide, *see* A-60275
Anisoylformic acid, *in* H-70212
Annonacin, A-70212
[4]Annulene, *see* C-70212
Annulin *A*, A-60276
Annulin *B*, A-60277
Anomalin, *in* K-70013
(−)-Anomalin, *in* K-70013
(+)-Anomalin, *in* K-70013
Ansadol, *in* H-70108
Antheliolide *A*, A-70213
Antheliolide B, *in* A-70213
Antheridic acid, A-60278
Antheridiogen Aₙ, *see* A-60278
Anthra[9,1,2-*cde*]benzo[*rst*]pentaphene,
    A-60279
Anthra[1,2-*b*]benzo[*d*]thiophene, A-70214
Anthra[2,1-*b*]benzo[*d*]thiophene, A-70215
Anthra[2,3-*b*]benzo[*d*]thiophene, A-70216
9-Anthraceneacetaldehyde, A-60280
1,8-Anthracenedicarboxylic acid, A-60281
1,1′-(1,5-Anthracenediyl)bisethanone, *see*
    D-70028
1,1′-(1,6-Anthracenediyl)bisethanone, *see*
    D-70029
1,1′-(1,8-Anthracenediyl)bisethanone, *see*
    D-70030
1,1′-(9,10-Anthracenediyl)bisethanone, *see*
    D-70031
9-Anthraceneethanol, *see* A-60282
2,2′,2″,2‴-(1,4,5,8-Anthracenetetraylidene)-
    tetrakispropanedinitrile, *see* O-70002
2-(9-Anthracenyl)ethanol, A-60282
1-(1-Anthracenyl)ethanone, *see* A-70030
1-(2-Anthracenyl)ethanone, *see* A-70031
9-Anthracenylpropanedinitrile, *see* D-60163
Anthra[1,2-*c*]furan, A-70217
9,10-Anthraquinodimethane, *see* D-70194
9-Anthrylacetaldehyde, *see* A-60280
2-(9-Anthryl)ethanol, *see* A-60282
1-Anthryl methyl ketone, *see* A-70030
2-Anthryl methyl ketone, *see* A-70031
Antiarrhythmic peptide (ox atrium), A-60283
Antibiotic 2061 *B*, *see* A-70158
Antibiotic 6270, *see* P-70111
Antibiotic 36531, *in* A-60194
Antibiotic X 14881C, *in* O-60003
▷Antibiotic 2061 A, *in* A-70158
Antibiotic 273a₁, *see* P-70002
Antibiotic 273a₁α, *in* P-70002
Antibiotic 273a₁β, *in* P-70002
Antibiotic 273a₂, A-70218
Antibiotic 273a₂α, *in* A-70218
Antibiotic 273a₂β, *in* A-70218
Antibiotic A 40926, A-70219
Antibiotic A 40926*A*, *in* A-70219
Antibiotic A 40926*B*, *in* A-70219
Antibiotic Al R6-4, *see* K-60009
Antibiotic Al-R 2081, *see* P-70203
Antibiotic B 1625*FA*₂β₋₁, A-70220
Antibiotic BMY 28100, A-70221
Antibiotic CA 146A, *in* C-70183

Antibiotic CA 146B, *in* C-70184
Antibiotic CP 54883, A-70222
Antibiotic FR 68504, *see* D-70465
Antibiotic FR 900359, A-70223
Antibiotic FR 900405, *in* E-60037
Antibiotic FR 900406, *in* E-60037
Antibiotic FR 900452, A-60284
Antibiotic FR 900482, A-60285
▷ Antibiotic G 7063-2, *in* A-70158
Antibiotic G 0069A, *in* C-70183
Antibiotic IM 8443T, *in* C-60178
Antibiotic K 818A, *see* C-60137
Antibiotic K 818B, *in* C-60137
Antibiotic L 660631, A-70224
Antibiotic LL C10037α, *in* A-60194
Antibiotic LL E33288γ₁, *see* C-70008
Antibiotic LL F42248α, *see* P-70203
Antibiotic M 167906, *in* P-60197
Antibiotic M 95464, *see* P-60197
▷ Antibiotic MA 321A₃, *see* P-60162
Antibiotic MM 14201, *in* A-60194
Antibiotic MT 35214, *in* A-60194
Antibiotic MT 36531, *in* A-60194
Antibiotic PB 5266A, A-70225
Antibiotic PB 5266B, A-70226
Antibiotic PB 5266C, A-70227
Antibiotic PC 766B, A-70228
Antibiotic PC 766B', *in* A-70228
Antibiotic PDE I, A-60286
Antibiotic PDE II, A-70229
Antibiotic PO1, *see* T-70262
Antibiotic Ro 09-0198, A-70231
Antibiotic R 20Y7, A-70230
Polyangium Antibiotics, A-60290
Antibiotic Sch 33256, *in* M-70045
Antibiotic Sch 34164, *in* M-70045
Antibiotic Sch 36969, *see* M-70045
Antibiotic Sch 38519, A-60287
Antibiotic SF 1993, *see* C-70015
Antibiotic SF 2312, A-60288
Antibiotic SF 2339, A-60289
Antibiotic SF 2415A₁, A-70232
Antibiotic SF 2415B₁, A-70235
Antibiotic SF 2415A₂, A-70233
Antibiotic SF 2415B₂, A-70236
Antibiotic SF 2415A₃, A-70234
Antibiotic SF 2415B₃, A-70237
Antibiotic SL 3440, *see* C-60038
Antibiotic SS 21020C, A-70238
Antibiotic SS 43405D, A-70239
Antibiotic SS 46506A, *see* P-70203
▷ Antibiotic T 42082, *see* C-60022
Antibiotic TAN 665A, *see* D-70063
Antibiotic TMF 518D, A-70240
Antibiotic W 341C, A-70241
Antibiotic WF 3405, *see* D-70465
Antibiotic WS 6049A, *in* E-60037
Antibiotic WS 43708A, *see* B-60114
Antibiotic WS 6049B, *in* E-60037
Antibiotic WS 43708B, *in* B-60114
Antibiotic YP 02908L-A, *see* O-70030
Anwuweizonic acid, *see* O-70096
Ap 2, *in* H-70108
Aparjitin, A-60291
Aphyllocladone, A-70242
Aphyllodenticulide, A-70243
▷ Apigenin 4'-methyl ether, *see* D-70319
Apionol, *see* B-60014
6'-O-Apiosylebuloside, *in* E-60001
Aplysiapyranoid A, A-70244
Aplysiapyranoid C, A-70245
Aplysiapyranoid B, *in* A-70244
Aplysiapyranoid D, *in* A-70245
α-APM, *see* A-60311
10'-Apo-β-carotene-3,10'-diol, *see* G-70002
10'-Apo-β-caroten-10'-ol, A-70246
Apotrisporin, *see* A-70247
Apotrisporin E, A-70247
Apotrisporin C, *in* A-70247
Apo-12'-violaxanthal, *in* P-60065
▷ Apresazide, *in* H-60092
▷ Apresoline, *in* H-60092
▷ Apyonin, *see* A-70268
▷ Aquedux, *in* C-70049
6-C-β-L-Arabinopyranosyl-8-C-β-D-
    galactopyranosylapigenin, *see* N-70021
Ara-doridosine, *in* A-60144

Araneol, *in* P-70031
Araneophthalide, A-60292
Aranochromanophthalide, A-60293
Arboreic acid, *in* T-60335
Arborone, A-60294
Arbutin, A-70248
Arbutoside, *see* A-70248
Arctigenin, A-70249
Arctiin, *in* A-70249
Arctodecurrolide, *in* D-60325
Argentilactone, *in* H-60025
3-L-Arginine-5-L-argininecyanoginosin LA, *see*
    C-70210
Argiopine, A-60295
Argiotoxin 636, *see* A-60295
Arguticinin, A-70250
Aristochilone, A-70251
Aristolactone, A-60296
1(10)-Aristolen-9-one, A-70252
Aristolignin, A-60297
Aristoligone, *in* A-70251
Aristosynone, *in* A-70251
Aristotetralol, A-70253
Aristotetralone, A-60298
Arizonin C₁, A-70254
Arizonin C₃, A-70255
Arizonin A₁, *in* A-70254
Arizonin A₂, *in* A-70255
Arizonin B₁, *in* A-70254
Arizonin B₂, *in* A-70255
Arjungenin, *in* T-70119
Arjunglucoside I, *in* T-70119
Arjunglucoside II, *in* T-70271
Arjunolic acid, *in* T-70271
Arnebin I, *in* S-70038
Arnebin III, *in* S-70038
Arnebin IV, *in* S-70038
1(5),3-Aromadendradiene, A-70256
1(10),4-Aromadendradiene, A-70257
1-Aromadendrene, A-70258
9-Aromadendrene, A-70259
10(14)-Aromadendrene-4,8-diol, A-60299
1(5)-Aromadendren-7-ol, A-70260
Artanomaloide, A-60300
Artapshin, A-60301
Artausin, A-70261
Arteannuic acid, *see* C-70004
Arteannuin B, A-70262
Arteanoflavone, *in* H-70081
Artecalin, A-60302
Artedouglasiaoxide, A-70263
Artelein, A-60303
Artelin, A-60304
Artemisiaglaucolide, *in* D-70303
Artemisic acid, *see* C-70004
Artemisininic acid, *see* C-70004
Artemone, A-60305
Artesovin, A-60306
Articulin, A-60307
Articulin acetate, *in* A-60307
Arvoside, A-70264
Asadanin, A-60308
Asadanol, *in* A-60308
Asalphen, *in* H-70108
γ-Asarone, *see* T-70279
Ascorbigen, A-60309
Ascorbigen A, *in* A-60309
Ascorbigen B, *in* A-60309
Asebogenin, *in* H-60219
Asebogenol, *see* H-60219
Asebotin, *in* H-60219
Asebotol, *in* H-60219
Asebotoside, *in* H-60219
ASK, *in* H-70108
N-Asparaginylalanine, A-60310
Aspartame, A-60311
▷ 1-L-Aspartic acid-α-amanitin, *in* A-60084
▷ 1-L-Aspartic acid-3-(S)-4-hydroxy-L-isoleucine-
    α-amanitin, *in* A-60084
1-L-Aspartic acid-3-isoleucine-α-amanitin, *in*
    A-60084
▷ 1-L-Aspartic acid-4-(2-mercapto-L-trytophan)-
    α-amanitin, *in* A-60084
N-α-Aspartylphenylalanine 1-methyl ester, *see*
    A-60311
N-α-Aspartyl-3-phenylalanine methyl ester, *see*
    A-60311

Aspercreme, *in* H-70108
Aspergel, *in* H-70108
Aspicilin, *see* A-60312
Aspicillin, A-60312
Assafoetidin, A-70265
Astilbin, *in* P-70027
Astrahygrol, *in* H-60169
Astrahygrone, *in* H-60169
ent-3β,16α,17-Atisanetriol, A-70266
Atlanton-6-ol, *see* H-60107
▷ Atletol, *see* C-70075
Atratogenin A, A-70267
Atratogenin B, *in* A-70267
Atratoside A, *in* A-70267
Atratoside B, *in* A-70267
Atratoside C, *in* A-70267
Atratoside D, *in* C-70257
Aurachin A, A-60313
Aurachin B, A-60314
Aurachin D, A-60315
Aurachin C, *in* A-60315
▷ Auramine, A-70268
Aurantioclavine, A-60316
Auraptenol, A-70269
Auricularic acid, *see* A-60317
Aurintricarboxylic acid, A-60318
Auropolin, A-60319
Austrobailignan-5, A-70270
Austrobailignan-6, A-60320
Austrobailignan-7, A-70271
Austrocortilutein, A-60321
Austrocortirubin, A-60322
Avellanin A, A-60323
Avellanin B, A-60324
Avenaciolide, A-60325
Avenacoside A, *in* N-60062
Avenacoside B, *in* N-60062
Avicennioside, A-60326
▷ AV toxin C, *see* A-60237
AV toxin D, *in* T-60353
AV toxin E, *see* T-60353
Axamide-1, A-70272
Axamide-2, A-60327
Axisonitrile-1, *in* A-60328
Axisonitrile-2, *in* A-60328
Axisonitrile-3, *in* A-60328
Axisonitrile-4, *in* A-60328
Axisonitriles, A-60328
Axisothiocyanate-1, *in* A-60329
Axisothiocyanate-2, *in* A-60329
Axisothiocyanate-3, *in* A-60329
Axisothiocyanate-4, *in* A-60329
Axisothiocyanates, A-60329
Ayapin, *see* M-70068
3a-Azaazulene, *see* P-70189
4-Azaazulene, *see* C-60225
1-Azaazulen-2-one, *see* C-60201
3a-Azaazulen-3-one, *see* P-70190
3a-Azaazulen-4-one, *see* P-70191
3a-Azaazulen-6-one, *see* P-70192
3a-Azaazulen-8-one, *see* P-70193
1-Azabenzo[h]cycl[3.2.2]azine, *see* B-70021
1-Azabicyclo[3.2.0]heptane-2,7-dione, *see* B-70330
1-Azabicyclo[3.2.0]heptane-2,7-dione, A-70273
8-Azabicyclo[5.2.0]nonane, A-70274
1-Azabicyclo[3.3.1]nonan-2-one, A-70275
2-Azabicyclo[3.3.1]nonan-7-one, A-70276
1-Azabicyclo[3.3.0]octa-5,7-dien-4-one, *see*
    D-60277
7-Azabicyclo[4.2.0]octane, A-70277
8-Azabicyclo[3.2.1]octane, A-70278
1-Azabicyclo[3.3.0]octan-2-one, *see* H-60058
1-Azabicyclo[3.3.0]octan-3-one, *see* T-60091
1-Azabicyclo[3.3.0]octan-4-one, *see* H-60057
2-Azabicyclo[2.2.2]oct-5-ene, A-70279
7-Azabicyclo[4.2.0]oct-3-en-8-one, A-60331
7-Azacoumaran, *see* D-70211
1-Azacycl[3.2.2]azine, *see* I-70004
▷ Azacyclooctane, *see* O-70010
1-Aza-4-cyclooctanone, *see* H-70047
1-Aza-5-cyclooctanone, *see* H-70048
5-Aza-2'-deoxycytidine, *see* D-70022
Azadirachtol, A-70280
Azadiradione, A-70281
4-Azahomoadamantane, *see* A-60333

▷4-Azaindole, *see* P-60249
7-Azaindoline, *see* D-70263
9-Azaindolizidine, *see* H-70067
5-Aza-6-oxo-1-thiacyclononane, *see* H-70075
1-Azaphenoxaselenine, *see* B-60054
14-Azaprostanoic acid, A-60332
8-Azapyrrolizidine, *see* T-70086
10-Azaquinolizidine, *see* O-70016
3-Azatetracyclo[5.3.1.1²,⁶.0⁴,⁹]dodecane, A-70282
9-Azatricyclo[4.3.1.0¹,⁶]deca-2,4,7-triene, *see* M-70041
4-Azatricyclo[4.3.1.1³,⁸]undecane, A-60333
1-Azatriphenylene, *see* D-70075
3-Azawurtzitane, *see* A-70282
2-Azaxanthone, *see* B-60042
Azelaoin, *see* H-60112
Azeloin, *see* H-60112
1*H*-Azepin-3(2*H*)-one, *see* D-70171
Azetidine, A-70283
2-Azetidinecarboxylic acid, A-70284
1-Azetine, *see* D-70172
2′-Azidoacetophenone, A-60334
3′-Azidoacetophenone, A-60335
4′-Azidoacetophenone, A-60336
2-Azidoadamantane, A-60337
2-Azidobenzophenone, A-60338
3-Azidobenzophenone, A-60339
4-Azidobenzophenone, A-60340
▷7-Azidobicyclo[2.2.1]hepta-2,5-diene, A-70285
4-Azidobutanal, A-70286
N-(Azidochloromethylene)-*N*-methyl-methanaminium, *see* D-70369
1-Azidodecahydro-3,5,1,7-[1,2,3,4]-butanetetra-ylnaphthalene, *see* A-60341
2-Azidodecahydro-3,5,1,7-[1,2,3,4]-butanetetra-ylnaphthalene, *see* A-60342
3-Azidodecahydro-3,5,1,7-[1,2,3,4]-butanetetra-ylnaphthalene, *see* A-60343
1-Azidodiamantane, A-60341
3-Azidodiamantane, A-60342
4-Azidodiamantane, A-60343
1-Azido-2,2-dimethylpropane, A-60344
▷2-Azidoethanol, A-70287
▷Azidoethyl alcohol, *see* A-70287
1-[3-Azido-4-(hydroxymethyl)cyclopentyl]-5-methyl-2,4(1*H*,3*H*)pyrimidinedione, A-70288
4-Azido-1*H*-imidazo[4,5-c]pyridine, A-60345
1-Azido-2-iodoethane, A-60346
1-Azido-3-iodopropane, A-60347
3-Azido-3-methylbutanoic acid, A-60348
1,1′,1″-(Azidomethylidyne)trisbenzene, *see* A-60352
2-Azido-2-nitropropane, A-60349
▷7-Azidonorbornadiene, *see* A-70285
3-Azidopentacyclo[7.3.1.1⁴,¹².0²,⁷.0⁶,¹¹]-tetra-decane, *see* A-60342
Azidopentafluorobenzene, A-60350
1-(2-Azidophenyl)ethanone, *see* A-60334
1-(3-Azidophenyl)ethanone, *see* A-60335
1-(4-Azidophenyl)ethanone, *see* A-60336
2-Azidophenyl methyl ketone, *see* A-60334
3-Azidophenyl methyl ketone, *see* A-60335
4-Azidophenyl methyl ketone, *see* A-60336
2-Azidophenyl phenyl ketone, *see* A-60338
3-Azidophenyl phenyl ketone, *see* A-60339
4-Azidophenyl phenyl ketone, *see* A-60340
(2-Azidophenyl)phenylmethanone, *see* A-60338
(3-Azidophenyl)phenylmethanone, *see* A-60339
(4-Azidophenyl)phenylmethanone, *see* A-60340
Azidophosgeniminium, *see* D-70369
2-Azidopropane, A-70289
3-Azido-2-propenoic acid, A-60290
3(5)-Azidopyrazole, A-70291
2-Azidopyrimidine, *see* T-60176
4-Azidopyrimidine, *see* T-60177
3-Azido-1,2,4-thiadiazole, A-60351
2-Azidotricyclo[3.3.1.1³,⁷]decane, *see* A-60337
Azidotriphenylmethane, A-60352
▷Azimidobenzene (obsol.), *see* B-70068
▷Aziminobenzene (obsol.), *see* B-70068
Azipyrazole, *see* P-60079
2*H*-Azirine, A-70292
Azlactone, *see* D-60485

2,2′-Azodiquinoxaline, A-60353
4-Azoniaspiro[3.3]heptane-2,6-diol, A-60354
1,4-Azulenedicarboxylic acid, A-70293
2,6-Azulenedicarboxylic acid, A-70294
5,7-Azulenedicarboxylic acid, A-70295
[2.2](2,6)Azulenophane, A-60355
Azuleno[1,2-d]thiazole, A-60356
Azuleno[2,1-d]thiazole, A-60357
Azulvalene, *see* T-70019
1,6-Azulylene, *see* D-60225
2,6-Azulylene, *see* D-60226
Azupyrene, *see* D-60167
B 1625FA₂ᵦ₋₁, *see* A-70220
Baccharinoid *B1*, B-70001
Baccharinoid *B7*, B-70002
Baccharinoid B2, *in* B-70001
Baccharinoid B3, *in* B-70002
Baccharinoid B4, *see* B-70003
Baccharinoid B6, *in* B-70003
Baccharinoid B9, B-60001
Baccharinoid B10, *in* B-60001
Baccharinoid B12, *in* R-70008
Baccharinoid B13, B-60002
Baccharinoid B14, *in* B-60002
Baccharinoid B16, B-60003
Baccharinoid B17, *in* R-70008
Baccharinoid B20, *in* B-60001
Baccharinoid B21, *in* R-70008
Baccharinoid B23, *in* B-60003
Baccharinoid B24, *in* B-60003
Baccharinoid B25, B-60004
Baccharinoid B27, *in* B-60002
Baccharinol, B-70003
Bacchariolide *B*, B-70004
Bacchariolide A, *in* B-70004
Bacteriochlorophyll f, B-70005
Baicalein, *see* T-60322
Balaenol, B-70006
Balaenonol, *in* B-70006
Balanitin 1, *in* S-60045
Balanitin 2, *in* S-60045
Balanitin 3, *in* S-60045
Balanitol, *in* E-70062
Balchanin, *see* S-60007
Balearone, B-60005
Ballotinone, *in* M-70013
ψ-Baptigenin, B-60006
ψ-Baptisin, *in* B-60006
▷Baratol, *in* T-70277
Barbaralane, *see* T-70230
Barbatic acid, B-70007
Barbatinic acid, *see* B-70007
Barbatusol, B-60007
Barbinervic acid, *in* T-60348
Barleriaquinone, *see* H-60170
Barlerin, *in* S-60028
▷Barrelin, *see* V-60017
Bartemidiolide, B-70008
▷Baycovin, *in* D-70111
Bayogenin, *in* T-60271
▷BCMO, *see* B-60149
Bedfordiolide, B-60008
Bellidifolin, *in* T-70123
Bellidin, *see* P-60052
▷Benoquin, *see* B-60074
Bentemazole, *in* I-70009
Benthamic acid, *in* D-70350
Benz[a]aceanthrylene, B-70009
Benz[d]aceanthrylene, B-60009
Benz[k]aceanthrylene, B-60010
▷Benz[e]acephenanthrylene, B-70010
Benz[j]acephenanthrylene, B-70011
Benzalchromanone, *see* D-60263
Benz[a]anthracene-1,2-dione, B-70012
Benz[a]anthracene-3,4-dione, B-70013
Benz[a]anthracene-5,6-dione, B-70014
Benz[a]anthracene-7,12-dione, B-70015
Benz[a]anthracene-8,9-dione, B-70016
Benz[a]anthracen-3-ol, *see* H-70104
1,2-Benzanthraquinone, B-70015
Benz[a]azulene-1,4-dione, B-60011
Benzenecarbo(diselenoperoxoic)acid phenyl ester, *see* B-70077
Benzenediazo-*p*-toluenesulfonate *N*-oxide, *in* H-70208
Benzene dimer, *see* T-60060
Benzeneethanethioic acid, B-70017

Benzeneethanethiol, *see* P-70069
▷Benzenehexamine, B-60012
Benzenenonanoic acid, *see* P-70084
Benzenesulfenyl thiocyanate, B-60013
▷Benzenesulfonyl peroxide, B-70018
Benzenetetramethylene diperoxide diamine, *see* T-60162
1,2,3,5-Benzenetetramine, *see* T-60012
1,2,4,5-Benzenetetramine, *see* T-60013
1,2,3,4-Benzenetetrol, B-60014
1,3,5-Benzenetrimethanethiol, B-70019
4,4′,4″-(1,3,5-Benzenetriyl)trismorpholine, *see* T-70294
1,1′,1″-(1,3,5-Benzenetriyl)trispiperidine, *see* T-70314
1,1′,1″-(1,3,5-Benzenetriyl)trispyrrolidine, *see* T-70319
Benzeneundecanoic acid, *see* P-70096
7,16[1′,2′]-Benzeno-7,16-dihydroheptacene, B-70020
Benz[b]homoheptalene, *see* B-60102
Benzilide, *see* T-60165
Benz[f]imidazo[5,1,2-cd]indolizine, B-70021
1*H*-Benzimidazole-2-sulfonic acid, B-70022
Benz[f]indane, B-60015
5,6-Benzindane, *see* B-60015
Benz[def]indeno[1,2,3-qr]chrysene, B-70023
Benz[def]indeno[1,2,3-hi]chrysene, B-70024
Benz[5,6]indeno[2,1-a]phenalene, B-70025
Benz[a]indeno[1,2,3-cd]pyrene, *see* F-70016
1*H*-Benz[e]indole-1,2(3*H*)-dione, B-70026
Benz[e]indoline-1,2-dione, *see* B-70026
Benz[cd]indol-2-(1*H*)-one, B-70027
Benz[e]isatin, *see* B-70026
lin-Benzoadenine, *see* A-60207
prox-Benzoadenine, *see* A-60206
lin-Benzoadenosine, *in* A-60207
lin-Benzoallopurinol, *see* P-60205
prox-Benzoallopurinol, *see* D-60271
▷Benzoaric acid, *see* E-70007
Benzo[c]-1-azabicyclo[4.3.0]nonane, *see* H-70072
Benzo[a]benzofuro[2,3-c]phenazine, *in* B-60033
Benzobenzvalene, *see* N-60012
2,3-Benzobicyclo[2.2.1]hepta-2,5-diene, *see* D-60256
2,3-Benzobicyclo[2.2.1]hept-2-ene, *see* M-60034
Benzo[a]biphenylene, B-60016
Benzo[b]biphenylene, B-60017
1,2-Benzobiphenylene, *see* B-60016
2,3-Benzobiphenylene, *see* B-60017
Benzo[1,2-b:4,5-b′]bis[1]benzothiophene, B-60018
Benzo[1,2-b:5,4-b′]bis[1]benzothiophene, B-60019
Benzo[1,2-d:4,5-d′]bis[1,3]dithiole-2,6-dione, B-70028
lin-Benzocaffeine, *in* I-60009
7,8-Benzochroman-4-one, *see* N-60010
Benzo[de]chromene, *see* N-70009
Benzo[h]chromone, *see* N-60010
Benzo[c]cinnoline, B-60020
Benzo[a]coronene, B-60021
1:2-Benzocoronene, *see* B-60021
Benzo[h]coumarin, *see* N-70011
▷Benzocyclobutene-4,5-quinone, *see* B-70110
Benzo[5,6]cycloocta[1,2-c]furan, B-70029
Benzocyclopropene, *see* C-70246
Benzo[1,2:3,4]dicyclopentene, *see* H-60094
Benzo[1,2:4,5]dicyclopentene, *see* H-60095
1,4-Benzodioxan-2-carbonitrile, *in* B-60024
1,4-Benzodioxan-2-carboxaldehyde, B-60022
1,4-Benzodioxan-6-carboxaldehyde, B-60023
1,4-Benzodioxan-2-carboxylic acid, B-60024
1,4-Benzodioxan-6-carboxylic acid, B-60025
1,4-Benzodioxan-2-methanol, *see* H-60172
4*H*-1,3-Benzodioxin-6-carboxaldehyde, B-60026
4*H*-1,3-Benzodioxin-6-carboxylic acid, B-60027
4*H*-1,3-Benzodioxin-2-one, B-60028
1,3-Benzodioxole-4-acetic acid, B-60029
1,3-Benzodioxole-4-acetonitrile, *in* B-60029
9-(1,3-Benzodioxol-5-yl)-1-(2,6-dihydroxyphenyl)-1-nonanone, *in* M-70007
3-(1,3-Benzodioxol-5-yl)-7-hydroxy-4*H*-1-benzopyran-4-one, *see* B-60006

▷9-(1,3-Benzodioxol-5-yl)-4-hydroxy-6,7-
  dimethoxynaphtho[2,3-c]furan-1(3H)-one,
  see D-60493
2-(1,3-Benzodioxol-5-yl)-1-(5-hydroxy-2,2-di-
  methyl-2H-1-benzopyran-6-yl)-2-propen-1-
  one, see G-70014
2-(1,3-Benzodioxol-5-yl)-7-methoxy-5-
  benzofuranpropanol, see E-70001
4-[5-(1,3-Benzodioxol-5-yl)tetrahydro-3,4-di-
  methyl-2-furanyl]-2-methoxyphenol, see
  A-70271
4-(1,3-Benzodioxol-5-yl)tetrahydro-1H,3H-
  furo[3,4-c]furan-1-one, see A-70063
Benzo[1,2-f:4,5-f']diquinoline, B-70030
1,3-Benzodiselenole, B-70031
1,3-Benzodiselenole-2-thione, B-70032
2-(1,3-Benzodiselenol-2-ylidene)-1,3-
  benzodiselenole, see D-70077
1,3,2-Benzodithiazol-1-ium(1+), B-70033
1,5-Benzodithiepan, see D-60195
3H-1,2-Benzodithiole-3-selone, B-70034
Benzo[1,2-b:4,5-b']dithiophene, B-60030
Benzo[1,2-b:4,5-b']dithiophene 4,8-dione,
  B-60031
Benzo[a]fluoranthene, see B-70009
▷Benzo[b]fluoranthene, see B-70010
▷3,4-Benzofluoranthene, see B-70010
3-Benzofurancarboxylic acid, B-60032
4-Benzofuranol, see H-70106
1,2-Benzo[5,6-c]furocyclooctatetraene, see
  B-70029
▷Benzoguanamine, see D-60041
lin-Benzoguanine, see A-60209
prox-Benzoguanine, see A-60208
5,6-Benzohydrindene, see B-60015
▷Benzohydroxamic acid, B-70035
dist-Benzohypoxanthine, see I-60013
lin-Benzohypoxanthine, see I-60012
prox-Benzohypoxanthine, see I-60011
prox-Benzoisoallopurinol, see P-60204
Benzo[f]isobenzofuran, see N-60008
2H-Benzo[cd]isoindol-2-one, see B-70027
lin-Benzolumazine, see P-60200
prox-Benzolumazine, see P-60201
1,2-Benzo[2.2]metaparacyclophan-9-ene,
  B-70036
Benzo[c]-1,7-methano[12]annulene, see
  B-60102
2,3-Benzo-6,7-naphthaanthraquinone, see
  H-60033
Benzo[b]naphtho[2,1-d]furan-5,6-dione,
  B-60033
Benzo[b]naphtho[2,3-d]furan-6,11-dione,
  B-60034
1H-Benzo[de][1,6]naphthyridine, B-70037
Benzo[f]ninhydrin, see D-60308
Benzonitrile N-oxide, B-60035
Benzonorbornadiene, see D-60256
Benzonorbornene, see M-60033
1,2-Benzo[2.2]paracyclophane, B-70038
Benzo[rst]phenaleno[1,2,3-de]pentaphene,
  B-60036
Benzo[rst]phenanthro[1,10,9-cde]pentaphene,
  B-60037
Benzo[rst]phenanthro[10,1,2-cde]pentaphene,
  B-60038
Benzo[b]phenanthro[1,2-d]thiophene, B-70039
Benzo[b]phenanthro[4,3-d]thiophene, B-70040
Benzo[3,4]phenanthro[1,2-b]thiophene,
  B-60039
Benzo[3,4]phenanthro[2,1-b]thiophene,
  B-60040
Benzo[a]phenathridizinium, see D-70076
Benzophenone-2,2'-dicarboxylic acid, B-70041
Benzophenone imine, see D-60482
Benzophenone nitrimine, in D-60482
β-Benzopinacolone, see T-60167
▷1H-2-Benzopyran-1-one, B-60041
1H-2-Benzopyran-4(3H)-one, B-70042
10H-[1]Benzopyrano[3,2-c]pyridin-10-one,
  B-60042
[1]-Benzopyrano[2,3-d]-1,2,3-triazol-9(1H)-
  one, B-60043
▷Benzoquin, see B-60074
Benzo[g]quinazoline-6,9-dione, B-60044
3H-Benzo[de]quinazolin-2-one, see P-70043
Benzo[g]quinoline-6,9-dione, B-60045

Benzo[a]quinolizinium(1+), B-70043
▷1,4-Benzoquinone, B-70044
▷p-Benzoquinone, see B-70044
1,2-Benzoquinone-3-carboxylic acid, B-70045
1,2-Benzoquinone-4-carboxylic acid, B-70046
1,4-Benzoquinone-2,3-dicarboxylic acid,
  B-70047
1,4-Benzoquinone-2,5-dicarboxylic acid,
  B-70048
1,4-Benzoquinone-2,6-dicarboxylic acid,
  B-70049
1,2-Benzoquinonedimethide, see D-70410
p-Benzoquinonemethide, see M-60071
Benzo[g]quinoxaline-6,9-dione, B-60046
Benzoselenazole, B-70050
1H-2-Benzoselenin-4(3H)-one, B-70051
[1]Benzoselenopheno[2,3-b][1]benzothiophene,
  B-70052
4,5-Benzo-3-selenoxo-1,2-dithiole, see B-70034
lin-Benzotheophylline, in I 60009
4H-3,1-Benzothiazin-2-amine, see A-70092
1,4-Benzothiazine, B-70053
2H-1,4-Benzothiazine-2,3(4H)-dione, B-70054
2H-1,4-Benzothiazine-2-thiol, see B-70055
2H-1,4-Benzothiazine-3(4H)-thione, B-70055
3-(2-Benzothiazolyl)-2H-1-benzopyran,
  B-70056
3-(2-Benzothiazolyl)coumarin, see B-70056
N-2-Benzothiazolyl-N,N',N''-triphenyl-
  guanidine, B-70057
6H-[1]Benzothieno[2,3-b]indole, B-70058
10H-[1]Benzothieno[3,2-b]indole, B-70059
[1]Benzothieno[2,3-c][1,5]naphthyridine,
  B-70060
[1]Benzothieno[2,3-c][1,7]naphthyridine,
  B-70061
[1]Benzothieno[2,3-c]quinoline, B-70062
[1]Benzothieno[3,2-b]quinoline, B-70063
[1]Benzothieno[3,2-c]quinoline, B-70064
3-Benzothienylglycine, see A-60090
1-Benzothiepin-5(4H)-one, B-60047
Benzothiet-2-one, B-70065
3,4-Benzothiocoumarin, see D-70080
Benzo[b]thiophen-3(2H)-one, B-60048
[1]Benzothiopyrano[6,5,4-def][1]-
  benzothiopyran, B-60049
1-Benzothiopyrylium(1+), B-70066
2-Benzothiopyrylium(1+), B-70067
1,2,3-Benzotriazin-4-amine, see A-60092
1,2,3-Benzotriazine-4-thiol, see B-60050
1,2,3-Benzotriazine-4(3H)-thione, B-60050
▷1H-Benzotriazole, B-70068
Benzo[1,2-b:3,4-b':5,6-b'']tripyrazine-
  2,3,6,7,10,11-tetracarboxylic acid, B-60051
Benzo[1,2-b:3,4-b':5,6-b'']tris[1]-
  benzothiophene, B-60052
Benzo[1,2-b:3,4-b':6,5-b'']tris[1]-
  benzothiophene, B-60053
Benzo[1,2-d:3,4-d':5,6-d'']tris[1,3]dithiole,
  B-70069
1,2,3-Benzoxadiazole, see D-70057
lin-Benzoxanthine, see I-60009
[1,4]Benzoxaselenino[3,2-b]pyridine, B-60054
1,3-Benzoxathiole-2-selone, B-70070
[1,4]Benzoxazino[3,2-b][1,4]benzoxazine,
  B-60055
[1,4]Benzoxazino[2,3-b]phenoxazine, see
  T-60378
▷2-Benzoxazolamine, see A-60093
1-[N-[3-(Benzoylamino)-2-hydroxy-4-phenyl-
  butyl]alanyl]proline, B-60056
Benzoylbenzylcarbinol, see H-70124
1-Benzoyl-2-butene, see P-60127
▷Benzoylcyclohexane, B-70071
1-Benzoylcyclopentene, B-70072
2-Benzoyl-1,2-dihydro-1-
  isoquinolinecarbonitrile, B-70073
1-Benzoyl-1,1-diphenylethane, see T-60389
1-Benzoyl-1,2-diphenylethane, see T-60390
2-Benzoyl-1,1-diphenylethane, see T-60391
6-Benzoyl-1-hexene, see P-70080
2-Benzoylimidazole, B-70074
4(5)-Benzoylimidazole, B-60057
6-Benzoylindole, B-60058

3-Benzoylisobutyric acid, see B-60064
1-Benzoylisoquinoline, B-60059
3-Benzoylisoquinoline, B-60060
4-Benzoylisoquinoline, B-60061
3-Benzoylisoxazole, B-70075
4-Benzoylisoxazole, B-60062
5-Benzoylisoxazole, B-60063
Benzoylmethylphenylmethanol, see H-70123
2-Benzoyl-2-methylpropane, see D-60443
3-Benzoyl-2-methylpropanoic acid, B-60064
2-Benzoylnaphtho[1,8-bc]pyran, B-70076
8α-Benzoyloxyacetylpringleine, in P-60172
1-[(Benzoyloxy)methyl]-7-oxabicyclo[4.1.0]-
  hept-4-ene-2,3-diol 3-benzoate, see P-60159
1-[(Benzoyloxy)methyl]-7-oxabicyclo[4.1.0]-
  hept-4-ene-2,3-diol diacetate, see S-60023
▷1-Benzoyl-1,3-pentadiyne, see P-60109
5-Benzoyl-1-pentene, see P-70081
Benzoyl phenyl diselenide, B-70077
3-Benzoylpropene, see P-60085
3-Benzoylpyrazole, B-70078
4-Benzoylpyrazole, B-60065
5-Benzoyl-2(1H)-pyrimidinone, B-60066
2-Benzoylthiophenol, see M-70020
3-Benzoylthiophenol, see M-70021
4-Benzoylthiophenol, see M-70022
4-Benzoyl-1,2,3-triazole, B-70079
Benzoyltrichloromethane, see T-70219
Benzoyltriphenylmethane, see T-60167
Benzvalenoquinoxaline, B-70080
Benzylamine-o-carboxylic lactam, see P-70100
5-Benzylamino-1H-tetrazole, in A-70202
α-Benzylaspartic acid, see A-60094
▷Benzyl chloroformate, see B-60071
1-Benzyl-3-chloropyrazole, in C-70155
1-Benzyl-5-chloropyrazole, in C-70155
Benzyldesoxybenzoin, see T-60390
1-Benzyl-1,4-dihydronicotinamide, B-60067
Benzyl hydroxycarbamate, see B-70086
Benzyl m-hydroxyphenyl ether, see B-60073
Benzyl o-hydroxyphenyl ether, see B-60072
▷Benzyl p-hydroxyphenyl ether, see B-60074
3-Benzylidene-4-chromanone, see D-60263
Benzylidenecyclobutane, B-70081
Benzylidenecycloheptane, B-60068
Benzylidenecyclohexane, B-70082
9-Benzylidene-1,3,5,7-cyclononatetraene,
  B-60069
Benzylidenecyclopentane, B-70083
1-Benzylidene-1H-cyclopropabenzene, B-70084
N-Benzylidenemethylamine, B-60070
4-Benzylidene-2-phenyloxazolin-5-one, see
  P-60128
1-Benzyl-5-imino-3-methyl-1H,3H-tetrazole, in
  A-70202
3-Benzyl-5-imino-1-methyl-1H,3H-tetrazole, in
  A-70202
1-Benzyl-5-methylamino-1H-tetrazole, in
  A-70202
2-Benzyloxyacetaldehyde, B-70085
m-Benzyloxyanisole, in B-60073
o-Benzyloxyanisole, in B-60072
p-Benzyloxyanisole, in B-60074
C-Benzyloxycarbohydroxamic acid, B-70086
▷Benzyloxycarbonyl chloride, B-60071
N-Benzyloxycarbonylhydroxylamine, see
  B-70086
2-(Benzyloxy)phenol, B-60072
3-(Benzyloxy)phenol, B-60073
▷4-(Benzyloxy)phenol, B-60074
1-Benzyloxy-2-phenyldiazene 2-oxide, in
  H-70208
1-Benzyloxy-2-phenyldiimide 2-oxide, in
  H-70208
2-Benzyloxy-1-propanol, in P-70118
1-Benzyl-5-vinylimidazole, in V-70012
Benzyne, in L-60087
Beyeriadiol, in L-60039
Bharangin, B-70088
Bharanginin, B-70089
Bhubaneswin, in B-70093
9,9'-Biacridylidene, B-70090
Biadamantylidene epoxide, B-70091
Biadamantylideneethane, B-60075

Bianthrone A2a, B-60076
Bianthrone A2b, B-60077
2,2′-Bibenzoxazoline, *in* T-60059
7,7′-Bi(bicyclo[2.2.1]heptylidene), B-60078
9,9′-Bi(bicyclo[3.3.1]nonylidene), B-60079
5,5′-Bibicyclo[2.1.0]pentane, B-70092
1,1′-Bibiphenylene, B-60080
2,2′-Bibiphenylene, B-60081
Bicoumol, B-70093
Bicyclo[1.1.0]butane-1-carboxylic acid, B-70094
Bicyclobutylidene, B-70095
Bicyclo[4.2.2]deca-6,8,9-triene, *see* P-60006
Bicyclo[4.3.1]deca-1(10),6,8-triene, *see* M-60029
Bicyclo[4.3.1]deca-1,3,5-trien-10-one, *see* T-70329
Bicyclo[4.4.0]dec-1(6)-ene-3,8-dione, *see* H-70063
Bicyclo[6.4.0]dodeca-1(8),2,4,6,10-pentaene, *see* D-70173
Bicyclo[6.2.2]dodeca-4,8,10,12-tetraene, *see* P-70010
Bicyclo[5.4.1]dodeca-7,9,11-trien-12-one, *see* T-70330
Bicyclo[8.8.2]eicosa-1(19),10(20),19-triene, B-70096
Bicyclofarnesyl epoxide, *see* D-70533
Bicyclo[3.2.0]hepta-2,6-diene, B-70097
Bicyclo[2.2.1]hepta-2,5-diene-2-carboxaldehyde, B-60083
Bicyclo[2.2.1]hepta-2,5-diene-2-carboxylic acid, B-60084
Bicyclo[2.2.1]heptane-1-carboxylic acid, B-60085
Bicyclo[3.1.1]heptane-1-carboxylic acid, B-60086
▷Bicyclo[2.2.1]heptan-2-one, B-70098
Bicyclo[4.1.0]hepta-1,3,5-triene, *see* C-70246
Bicyclo[3.1.1]hept-2-ene, B-60087
Bicyclo[3.2.0]hept-2-ene-6,7-dione, B-70099
7-Bicyclo[2.2.1]hept-7-ylidenebicyclo[2.2.1]-heptane, *see* B-60078
Bicyclo[2.1.1]hexan-2-amine, *see* A-70095
Bicyclo[2.1.1]hexan-2-ol, B-70100
Bicyclo[3.1.0]hexan-2-one, B-60088
[1,1′-Bicyclohexyl]-2,2′-dione, B-60089
(1,1′-Bicyclohexyl)-1,1′-diylbisbenzene, *see* D-70475
Bicyclo[12.4.1]nonadec-1,3,5,7,9,11,13,15,17-nonaene, B-70101
Bicyclo[3.3.1]nona-2,6-diene, B-70102
Bicyclo[3.3.1]nonane-2,4-dione, B-70103
Bi(cyclononatetraenylidene), B-70104
Bicyclo[3.3.1]non-3-en-2-one, B-70105
Bicyclo[4.2.1]non-2-en-9-one, B-70106
Bicyclo[4.2.1]non-3-en-9-one, B-70107
Bicyclo[4.2.1]non-7-en-9-one, B-70108
Bicyclo[4.3.0]non-8-en-7-one, *see* H-70058
Bicyclo[3.3.0]octa-1,3-diene, *see* T-70075
Bicyclo[3.3.0]octa-1,4-diene, *see* T-70077
Bicyclo[3.3.0]octa-1,5-diene, *see* T-70078
Bicyclo[3.3.0]octa-1,6-diene, *see* T-70079
Bicyclo[3.3.0]octa-1,7-diene, *see* T-70080
Bicyclo[3.3.0]octa-2,6-diene, *see* T-70081
Bicyclo[3.3.0]octa-2,7-diene, *see* T-70082
Bicyclo[4.1.1]octa-2,4-diene, B-60090
Bicyclo[3.3.0]octa-1(5),2-diene, *see* T-70076
Bicyclo[2.2.2]octa-5,7-diene-2,3-dione, B-70109
▷Bicyclo[4.2.0]octa-1,5-diene-3,4-dione, B-70110
Bicyclo[5.1.0]octa-3,5-dien-2-one, B-70111
Bicyclo[4.1.1]octane, B-60091
Bicyclo[2.2.2]octane-1-carboxylic acid, B-60092
Bicyclo[3.2.1]octane-1-carboxylic acid, B-70112
Bicyclo[4.2.0]octane-2,5-dione, B-60093
Bicyclo[3.2.1]octan-8-one, B-60094
Bicyclo[3.3.0]octa-1,3,5-triene, *see* D-70240
Bicyclo[3.3.0]octa-1,3,7-triene, *see* D-70244
Bicyclo[3.3.0]octa-1,4,6-triene, *see* D-70242
Bicyclo[3.3.0]octa-1(5),2,6-triene, *see* D-70241

Bicyclo[3.3.0]octa-1(5),2,7-triene, *see* D-70243
Bicyclo[3.3.0]oct-1-ene, *see* H-70064
Bicyclo[3.3.0]oct-2-ene, *see* H-70065
Bicyclo[4.1.1]oct-2-ene, B-60095
Bicyclo[4.1.1]oct-3-ene, B-60096
Bicyclo[3.3.0]oct-1(5)-ene, *see* H-70066
Bicyclo[2.2.2]oct-2-ene-1-carboxylic acid, B-70113
Bicyclo[4.2.0]oct-7-ene-2,5-dione, B-60097
Bicyclo[3.3.0]oct-1-en-3-one, *see* T-70084
Bicyclo[3.3.0]oct-6-en-2-one, *see* T-70083
Bicyclo[1.1.1]pentane-1-carboxylic acid, B-60101
[1,1′-Bicyclopropyl]-1,1′-dicarboxylic acid, B-70114
Bicyclo[4.4.1]undeca-1,6-diene, B-70115
Bicyclo[5.3.1]undeca-1,3,5,7,9-pentaene, B-60102
11-Bicyclo[4.4.1]undeca-1,3,5,7,9-pentaen-11-ylidenebicyclo[4.4.1]-undeca-1,3,5,7,9-pentaene, *see* B-70127
Bicyclo[5.3.1]undeca-1,4,6,8-tetraene-3,10-dione, *see* H-60086
Bicyclo[5.3.1]undeca-1,4,6,9-tetraene-3,8-dione, *see* H-60085
Bicyclo[5.3.1]undeca-3,6,8,10-tetraene-2,5-dione, *see* H-60087
Bi(1,3-diselenol-2-ylidene), B-70116
Bi(1,3-ditellurol-2-ylidene), B-70117
2,2′-Bi-1,3-dithiolane, B-60103
Bi(1,3-dithiolo[4,5-*b*][1,4]dithiin-2-ylidene), B-70118
Bi(1,3-dithiol-2-ylidene), B-70119
Δ⁹·⁹′-Bi-9*H*-fluorene, *see* B-60104
Δ⁹·⁹′-Bi-9-fluorenyl, *see* B-60104
9,9′-Bifluorenylidene, B-60104
Bifurcarenone, B-60105
8,8′-Bi[7-hydroxy-4-methoxy-5-methyl-coumarin], *see* O-70046
3,3′-Bi[5-hydroxy-2-methyl-1,4-naphthoquinone], B-60106
2,2′-Bi-1*H*-imidazole, B-60107
1,1′-Biindanylidene, B-70120
1,1′-Bi-1*H*-indene, B-60108
Δ¹·¹′-Biindene, B-60109
1,1′-Biindenyl, *see* B-60108
1,1′-Biindenylidene, *see* B-60109
2,2′-Biisonicotinic acid, *see* B-60126
3,3′-Biisonicotinic acid, *see* B-60134
2′,6″-Biluteolin, *see* D-70144
1,1′-Binaphthalene-4,4′,5,5′-tetrol, *see* T-70102
Bindschedler's green, B-60110
2,2′-Binicotinic acid, *see* B-60124
2,4′-Binicotinic acid, *see* B-60131
2,6-Binicotinic acid, *see* B-60125
4,4′-Binicotinic acid, *see* B-60137
6,6′-Binicotinic acid, *see* B-60127
7,7′-Binorbornylidene, *see* B-60078
Biopterin, B-70121
α-Biotol, B-60112
β-Biotol, B-60113
Biphenomycin *A*, B-60114
Biphenomycin B, *in* B-60114
[1,1′-Biphenyl]-2,2′-diacetic acid, B-60115
[1,1′-Biphenyl]-4,4′-diacetic acid, B-60116
[1,1′-Biphenyl]-2,2′-diamine, *see* D-60033
[1,1′-Biphenyl]-4,4′-diylbis[diphenylmethyl], B-60117
2,2′-[1,1′-Biphenyl]-2,2′-diylbis-1*H*-imidazole, *see* B-60148
1,8-Biphenylenedicarboxaldehyde, B-60118
[2.2](3,3′)Biphenylophane, *see* M-60031
2,2′,3,4-Biphenyltetrol, B-70122
2,3,4,4′-Biphenyltetrol, B-70123
2,3,4-Biphenyltriol, B-70124
3,3′-Bipicolinic acid, *see* B-60133
4,4′-Bipicolinic acid, *see* B-60136
6,6′-Bipicolinic acid, *see* B-60128
3,3′-Biplumbagin, *see* B-60106
1,3(5)-Bi-1*H*-pyrazole, B-60119
▷2,3′-Bipyridine, B-60120
2,4′-Bipyridine, B-60121
3,3′-Bipyridine, B-60122

3,4′-Bipyridine, B-60123
[2,2′-Bipyridine]-3,3′-dicarboxylic acid, B-60124
[2,2′-Bipyridine]-3,5-dicarboxylic acid, B-60125
[2,2′-Bipyridine]-4,4′-dicarboxylic acid, B-60126
[2,2′-Bipyridine]-5,5′-dicarboxylic acid, B-60127
[2,2′-Bipyridine]-6,6′-dicarboxylic acid, B-60128
[2,3′-Bipyridine]-2,3′-dicarboxylic acid, B-60129
[2,4′-Bipyridine]-2′,6′-dicarboxylic acid, B-60130
[2,4′-Bipyridine]-3,3′-dicarboxylic acid, B-60131
2,4′-Bipyridine-3,5-dicarboxylic acid, *see* B-60132
[2,4′-Bipyridine]-3′,5-dicarboxylic acid, B-60132
[3,3′-Bipyridine]-2,2′-dicarboxylic acid, B-60133
[3,3′-Bipyridine]-4,4′-dicarboxylic acid, B-60134
[3,4′-Bipyridine]-2′,6′-dicarboxylic acid, B-60135
[4,4′-Bipyridine]-2,2′-dicarboxylic acid, B-60136
[4,4′-Bipyridine]-3,3′-dicarboxylic acid, B-60137
2,4′-Bipyridyl, *see* B-60121
3,3′-Bipyridyl, *see* B-60122
3,4′-Bipyridyl, *see* B-60123
2,2′-Bipyridyl-3,3′-dicarboxylic acid, *see* B-60124
2,2′-Bipyridyl-3,5-dicarboxylic acid, *see* B-60125
2,2′-Bipyridyl-4,4′-dicarboxylic acid, *see* B-60126
2,2′-Bipyridyl-5,5′-dicarboxylic acid, *see* B-60127
2,2′-Bipyridyl-6,6′-dicarboxylic acid, *see* B-60128
2,3′-Bipyridyl-2′,3-dicarboxylic acid, *see* B-60129
2,4′-Bipyridyl-2′,6′-dicarboxylic acid, *see* B-60130
2,4′-Bipyridyl-3,3′-dicarboxylic acid, *see* B-60131
3,3′-Bipyridyl-2,2′-dicarboxylic acid, *see* B-60133
3,3′-Bipyridyl-4,4′-dicarboxylic acid, *see* B-60134
3,4′-Bipyridyl-2′,6′-dicarboxylic acid, *see* B-60135
4,4′-Bipyridyl-2,2′-dicarboxylic acid, *see* B-60136
4,4′-Bipyridyl-3,3′-dicarboxylic acid, *see* B-60137
[1,1′-Bipyrrolidine]-2,2′-dione, B-70125
4,4′-Biquinaldine, *see* D-60413
Bisabolangelone, B-60138
2-Bisabolene-1,12-diol, B-70126
5,6-Bis(acetyloxy)-10-chloro-12-hydroxy-9-oxoprosta-7,10,14,17-tetraen-1-oic acid methyl ester, *see* P-70143
1,3-Bis(allylthio)cyclobutane, *in* C-60184
1,2-Bis(2-aminoethoxy)ethane, B-60139
3,3-Bis(aminomethyl)oxetane, B-60140
1,1:2,2-Bis([10]annulene-1,6-diyl)ethylene, B-70127
2,3-Bis(9-anthryl)cyclopropenone, B-70128
Bis(1-azabicyclo[2.2.2]octane)bromine(1+), *see* B-60173
▷Bis(benzenesulfonyl)peroxide, *see* B-70018
6*H*,24*H*,26*H*,38*H*,56*H*,58*H*-16,65:48,66-Bis([1,3]benzenomethanothiomethano[1,3]-benzeno)-32,64-etheno-1,5;9,13:14,18:19,-23:27,31:33,37:41,45:46,50:51,55:59,63-decemetheno-8*H*,40*H*[1,7]dithiacyclotetra-triacontino[9,8-*h*][1,77]dithiocyclotetra-triacontin, B-70129
1,2-Bis(4-benzo-15-crown-5)diazene, B-60141
1,2-Bis(4-benzo-15-crown-5)ethene, B-60142
Bisbenzo[3,4]cyclobuta[1,2-*c*;1′,2′-*g*]-phenanthrene, B-60143

Name Index

Bis(benzothiazolo)[3,2-b:3′ . . . − 1,2-Bis(thiomethyl)cyclobut . . .

Bis(benzothiazolo)[3,2-b:3′,2″-e]-[1,2,4,5]tetra-zine, see T-60171
1,1′-Bisbiphenylenyl, see B-60080
2,2′-Bisbiphenylenyl, see B-60081
1,4-Bis[5-(4-biphenylyl)-2-oxazolyl]benzene, see P-60091
N,N-Bis([2,2′-bipyridin]-6-ylmethyl)-[2,2′-bipyridine]-6-methanamine, see T-70320
N,N′-Bis[3-(3-bromo-4-hydroxyphenyl)-2-oximidopropionyl]cystamine, B-70130
2,3-Bis(bromomethyl)-1,4-dibromobutane, B-60144
1,2-Bis(bromomethyl)-3-fluorobenzene, B-70131
1,2-Bis(bromomethyl)-4-fluorobenzene, B-70132
1,3-Bis(bromomethyl)-2-fluorobenzene, B-70133
1,3-Bis(bromomethyl)-5-nitrobenzene, B-60145
▷2,6-Bis(bromomethyl)pyridine, B-60146
3,5-Bis(bromomethyl)pyridine, B-60147
4,4-Bis(2-carboxyethyl)heptanedioic acid, see M-70036
2,2′-Bis(carboxymethyl)biphenyl, see B-60115
4,4′-Bis(carboxymethyl)biphenyl, see B-60116
▷3,12-Bis(carboxymethyl)-6,9-dioxa-3,12-diazatetradecanedioic acid, B-70134
2,3-Bis(chloromethyl)-1,4-dichlorobutane, B-60148
▷3,3-Bis(chloromethyl)oxetane, B-60149
2,2′-Bis(cyanomethyl)biphenyl, in B-60115
4,4′-Bis(cyanomethyl)biphenyl, in B-60116
1,1′-Bis(cyclopentadienylidene), see F-60080
2,3-Bis(dibromomethyl)pyrazine, B-60150
2,3-Bis(dibromomethyl)pyridine, B-60151
3,4-Bis(dibromomethyl)pyridine, B-60152
4,5-Bis(dibromomethyl)pyrimidine, B-60153
Bis(2,4-dichlorophenyl)ether, B-70135
Bis(2,6-dichlorophenyl) ether, B-70136
Bis(3,4-dichlorophenyl) ether, B-70137
Bis(1,1-dicyclopropylmethyl)diazene, B-60154
Bisdihydrotrifarin, in T-70236
1,5-Bis(3,4-dihydroxyphenyl)-1,2-dihydroxy-4-pentyne, see B-60155
1,2-Bis(3,4-dihydroxyphenyl)ethane, see T-60101
1,5-Bis(3,4-dihydroxyphenyl)-4-pentyne-1,2-diol, B-60155
3,6-Bis(3,4-dimethoxyphenyl)tetrahydro-1H,3H-furo[3,4-c]furan-1,4-diol, B-60156
▷4,4′-Bisdimethylaminobenzophenone imide, see A-70268
2,2′-Bis(dimethylamino)biphenyl, in D-60033
1,8-Bis(dimethylamino)naphthalene, in D-70042
▷Bis(4-dimethylaminophenyl)methyleneimine, see A-70268
1,1-Bis(1,1-dimethylethyl)cyclopropane, see D-60112
3,7-Bis(1,1-dimethylethyl)-9,10-dimethyl-2,6-anthracenedione, see D-60113
Bis[1-(1,1-dimethylethyl)-2,2-dimethylpropyl]-diazene, see T-60024
5-[3,5-Bis(1,1-dimethylethyl)-4-oxo-2,5-cyclo-hexadien-1-ylidene]-3,6-cycloheptadiene-1,2-dione, see D-70105
3,4-Bis(1,1-dimethylethyl)pyrazole, see D-70106
3,5-Bis(1,1-dimethylethyl)pyrazole, see D-70107
3,6-Bis(1,1-dimethylethyl)pyrrolo[3,2-b]-pyrrole, see D-70108
3,4-Bis(1,1-dimethylethyl)-2,2,5,5-tetramethyl-hexane, see D-70109
S,S-Bis(4,6-dimethyl-2-pyrimidinyl)-dithiocarbonate, B-70138
1,3-Bis(1,2-diphenyletnenyl)benzene, B-70139
1,4-Bis(1,2-diphenylethenyl)benzene, B-70140
Bis(diphenylmethyl)amine, see T-70150
Bis(diphenylmethyl) diselenide, B-70141
3,6-Bis(diphenylmethylene)-1,4-cyclohexadiene, B-60157
1,3-Bis(1,2-diphenylvinyl)benzene, see B-70139
1,4-Bis(1,2-diphenylvinyl)benzene, see B-70140

1,11-Bisepicaniojane, in C-70013
4,8-Bis-epi-inuviscolide, in I-70023
7,8,9,10-Bisepoxydispiro[2.0.2.4]decane, see D-60500
1,2;3,4-Bisepoxy-11(13)-eudesmen-12,8-olide, B-70142
Bis(ethylenedithio)benzene, see T-70039
Bis(ethylenedithio)tetrathiafulvalene, see D-60238
Bis(ethyleneoxythio)benzene, see T-70040
1,2-Bis(ethylidene)cyclohexane, see D-70154
4,6-Bis(ethylthio)-2,5-dihydropyrimidine, in D-70257
1,4-Bis(ethylthio)-3,6-diphenyl-2,5-diazapenta-lene, see B-60158
1,4-Bis(ethylthio)-3,6-diphenylpyrrolo[3,4-c]-pyrrole, B-60158
1,11-Bis(3-furanyl)-4,8-dimethyl-3,6,8-undeca-trien-2-ol, B-70143
N¹,N⁸-Bis(glutathionyl)spermidine, see T-70331
Bis(homoadamantane), see H-70031
Bishomoanthraquinone, see T-60264
1,4-Bishomohexaprismane, see H-70013
1,2-Bis(hydrazino)ethane, see D-60192
2,6-Bis(hydroxyacetyl)pyridine, B-70144
Bis(3-hydroxycyclobutyl)disulfide, in M-60019
S,S′-Bis(2-hydroxyethyl)-1,2-ethanedithiol, B-70145
1,2-Bis(2-hydroxyethylthio)ethane, see B-70145
2,3-Bis(hydroxymethyl)-1,4-butanediol, B-60159
2,5-Bis(hydroxymethyl)furan, B-60160
3,4-Bis(hydroxymethyl)furan, B-60161
1,7-Bis(hydroxymethyl)-7-methylbicyclo[2.2.1]-heptan-2-ol, B-70146
▷Bis(hydroxymethyl)nitromethane, see N-60038
3,3-Bis(hydroxymethyl)oxetane, see O-60056
3,6-Bis(hydroxymethyl)phenanthrene, see P-60069
4,5-Bis(hydroxymethyl)phenanthrene, see P-60070
2,5-Bis(4-hydroxyphenyl)-3,4-dimethyltetra-hydrofuran, see T-60061
1,5-Bis(4-hydroxyphenyl)-1,4-pentadiene, B-60162
1,2-Bis(4-hydroxyphenyl)-1,2-propanediol, B-70147
Bis(3-hydroxypropyl) ether, see O-60097
2,2′-Bis(2-imidazolyl)biphenyl, B-70148
Bis(2-imidazolyl)ketone, see D-70354
1,2-Bis(2-iodoethoxy)ethane, B-60163
Bis(2-iodoethyl) ether, see O-60096
▷1,4-Bis(1-isocyanato-1-methylethyl)benzene, B-70149
S,S′-Bis(2-mercaptoethyl)-1,2-ethanedithiol, B-70150
1,1-Bis(mercaptomethyl)cyclobutane, see C-70213
1,2-Bis(mercaptomethyl)cyclobutane, see C-70214
1,1-Bis(mercaptomethyl)cyclohexane, see C-70223
1,2-Bis(mercaptomethyl)cyclohexane, see C-70224
1,1-Bis(mercaptomethyl)cyclopentane, see C-70243
1,1-Bis(mercaptomethyl)cyclopropane, see C-70249
2,2-Bis(mercaptomethyl)-1,3-propanedithiol, B-60164
2,5-Bis(mercaptomethyl)thiophene, see T-70184
2,6-Bis(mercaptomethyl)toluene, see M-70050
1,6:7,12-Bismethano[14]annulene, B-60165
2,6-Bis(methoxyacetyl)pyridine, in B-70144
2,3-Bis(2-methoxy-4-nitro-5-sulfophenyl)-5-[(phenylamino)carbonyl]-2H-tetrazolium hydroxide inner salt, B-70151
2,2′-Bis(methylamino)biphenyl, in D-60033
1,8-Bis(methylamino)naphthalene, in D-70042
2,3-Bis(methylene)bicyclo[2.2.1]heptane, see D-70405
6,7-Bis(methylene)bicyclo[3.2.2]nonane, see D-70406

7,8-Bis(methylene)bicyclo[2.2.2]octa-2,5-diene, see D-70407
2,3-Bis(methylene)bicyclo[2.2.2]octane, see D-70408
Bis(methylene)bicyclo[1.1.1]pentanone, see D-60422
Bis(methylene)butanedinitrile, in D-70409
Bis(methylene)butanedioic acid, see D-70409
1,2-Bis(methylene)cyclohexane, B-60166
1,3-Bis(methylene)cyclohexane, B-60167
1,4-Bis(methylene)cyclohexane, B-60168
1,2-Bis(methylene)cyclopentane, B-60169
1,3-Bis(methylene)cyclopentane, B-60170
2,3-Bis(3,4-methylenedioxybenzyl)butane, see A-70270
3,4;3′,4′-Bis(methylenedioxy)bibenzyl, in T-60101
4,5;4′,5′-Bismethylenedioxypolemannone, in P-60166
9,10-Bis(methylene)tricyclo[5.3.0.0²,⁸]-deca-3,5-diene, B-70152
Bis(1-methylethyl)cyanamide, see D-60392
1,1′-[1,2-Bis(1-methylethylidene)-1,2-ethanediyl]bisbenzene, see D-60419
3,3-Bis(methylnitraminomethyl)oxetane, in B-60140
3,6-Bis(1-methylpropyl)-2,5-piperazinedione, B-70153
3,6-Bis(2-methylpropyl)-2,5-piperazinedione, B-60171
Bis(methylsulfonyl) peroxide, B-70154
2,3-Bis(methylsulfonyl)pyridine, in P-60218
2,5-Bis(methylsulfonyl)pyridine, in P-60220
3,5-Bis(methylsulfonyl)pyridine, in P-60222
2,3-Bis(methylthio)-1,4-benzenediamine, in D-70036
2,3-Bis(methylthio)-1,4-benzenediamine, in D-60030
[Bis(methylthio)methylene]propanedinitrile, in D-60395
2,3-Bis(methylthio)pyridine, in P-60218
2,4-Bis(methylthio)pyridine, in P-60219
2,5-Bis(methylthio)pyridine, in P-60220
3,5-Bis(methylthio)pyridine, in P-60222
2,5-Bis(methylthio)-1,3,4-thiadiazole, in M-70031
2,5-Bis(methylthio)-p-xylene, in D-60406
3,3-Bis(nitratomethyl)oxetane, in O-60056
Bis(2,3,5,6,8,9,11,12-octahydro-1,4,7,10,13-benzopentaoxacyclopentadecin-5-yl)diazene, see B-60141
1′,2′-Bis(2,5,8,11,14-pentaoxabicyclo[13.4.0]-nonadeca-1(15),-16,18-trien-17-yl)ethene, see B-60142
1,1′-Bis(1,3-phenylene)naphthalene, see D-60454
2,2′-Bis(1,3-phenylene)naphthalene, see D-60455
Bis(phenylmethyl)diselenide, see D-60082
1,2-Bis(phenylsulfinyl)cyclobutene, in B-70156
1,2-Bis(phenylsulfonyl)cyclobutene, in B-70156
▷Bis(phenylsulfonyl)peroxide, see B-70018
S,S′-Bis(1-phenyl-1H-tetrazol-5-yl) dithiocarbonate, B-70155
1,2-Bis(phenylthio)cyclobutene, B-70156
3,3-Bis(phenylthio)propanoic acid, B-70157
3,3-Bis(phenylthio)-1-propanol, B-70158
1,4-Bis(2-pyridylamino)phthalazine, B-60172
Bis(quinuclidine)bromine(1+), B-60173
Bissetone, B-60174
Bis(β-styryl)selenide, see D-60503
N,N′-Bis(3-sulfonatopropyl)-2,2′-bipyridinium, B-70159
N,N′-Bis(3-sulfonatopropyl)-4,4′-bipyridinium, B-70160
1,1′-Bis(3-sulfopropyl)-2,2′-bipyridinium dihydroxide bis(inner salt), see B-70159
1,1′-Bis(3-sulfopropyl)-4,4′-bipyridinium dihydroxide bis(inner salt), see B-70160
Bistheonellide A, see M-60138
Bistheonellide B, in M-60138
Bis[1,2,5-thiadiazolo][3,4-b:3′,4′-e]-pyrazine, B-70161
1,1-Bis(thiomethyl)cyclobutane, see C-70213
1,2-Bis(thiomethyl)cyclobutane, see C-70214

1,1-Bis(thiomethyl)cyclohexane, *see* C-70223
1,2-Bis(thiomethyl)cyclohexane, *see* C-70224
1,1-Bis(thiomethyl)cyclopropane, *see* C-70249
2,5-Bis(thiomethyl)thiophene, *see* T-70184
2,6-Bis(thiomethyl)toluene, *see* M-70050
Bis(2,4,6-tri-*tert*-butylphenyl)ditellane, *see* B-70162
Bis(2,4,6-tri-*tert*-butylphenyl)ditelluride, B-70162
Bis(trichloromethyl) carbonate, B-70163
1,2-Bis(trifluoromethyl)benzene, B-60175
1,3-Bis(trifluoromethyl)benzene, B-60176
1,4-Bis(trifluoromethyl)benzene, B-60177
3,3-Bis(trifluoromethyl)-3*H*-diazirine, B-60178
Bis-α-trifluoromethyldifluoroethane-β-sultone, *see* D-60181
4,5-Bis(trifluoromethyl)-1,3,2-dithiazol-2-yl, B-70164
3,4-Bis(trifluoromethyl)-1,2-dithiete, B-60179
N,*N*-Bis(trifluoromethyl)hydroxylamine, B-60180
2,3-Bis(trifluoromethyl)pyridine, B-60181
2,4-Bis(trifluoromethyl)pyridine, B-60182
2,5-Bis(trifluoromethyl)pyridine, B-60183
2,6-Bis(trifluoromethyl)pyridine, B-60184
3,4-Bis(trifluoromethyl)pyridine, B-60185
3,5-Bis(trifluoromethyl)pyridine, B-60186
Bis(trifluoromethyl)thioketene, B-60187
2,4-Bis[2,2,2-Trifluoro-1-(trifluoromethyl)-ethylidene]-1,3-dithietane, B-60188
1,2-Bis(3,4,5-trihydroxyphenyl)ethane, *see* H-60062
Bis[2,4,6-tris(1,1-dimethylethyl)phenyl]-ditelluride, *see* B-70162
Bis(vinylenedithio)tetrathiafulvalene, *see* B-70118
1,1′-Bitetralinylidene, B-70165
2,2′-Bithiazole, B-70166
2,4′-Bithiazole, B-70167
2,5′-Bithiazole, B-70168
4,4′-Bithiazole, B-70169
4,5′-Bithiazole, B-70170
5,5′-Bithiazole, B-70171
4,4′-Bi-4*H*-1,2,4-triazole, B-60189
3,3′-Bi(4′,5,7-trihydroxyflavanone), *see* I-70064
Biuret, B-60190
Blastmycetin *D*, B-60191
Blumealactone *C*, B-70172
Blumealactone A, *in* B-70172
Blumealactone B, *in* B-70172
BMY 28100, *see* A-70221
Boesenboxide, B-70173
Bomag, *in* O-70102
Bonducellin, B-60192
BOPOB, *see* P-60091
2,5-Bornanediol, *see* T-70280
2,3,10-Bornanetriol, *see* H-70173
2,8,10-Bornanetriol, *see* B-70146
Boronolide, B-60193
Botrylactone, B-70174
▷Botryodiplodin, B-60194
▷Bovolide, *see* D-60430
Bowdichione, B-70175
Boxazomycin *A*, B-70176
Boxazomycin B, *in* B-70176
Boxazomycin C, *in* B-70176
Brachycarpone, B-60195
Brachynereolide, B-60196
Brassicadiol, B-70177
Brassinolide, B-70178
Brasudol, B-60197
Brayleanin, *in* C-60026
α-Brazanquinone, *see* B-60033
β-Brazanquinone, *see* B-60034
Brevifloralactone, B-70179
Brevifloralactone acetate, *in* B-70179
Brevifolin, *in* T-70248
3-Bromoaspartic acid, *see* A-70098
2-Bromo-1,3,5-benzenetriol, B-60199
3-Bromo-1,2,4-benzenetriol, B-70180
4-Bromo-1,2,3-benzenetriol, B-60200
5-Bromo-1,2,3-benzenetriol, B-60201
5-Bromo-1,2,4-benzenetriol, B-70181

6-Bromo-1,2,4-benzenetriol, B-70182
p-Bromobenzhydrol, *see* B-60236
1-Bromo-2,3-benzobicyclo[2.2.1]hepta-2,5-diene, *see* B-60231
2-Bromobenzocyclopropene, B-60202
3-Bromobenzocyclopropene, B-60203
1-Bromobenzonorbornadiene, *see* B-60231
2-Bromobenzonorbornadiene, *see* B-60232
7-Bromobenzonorbornadiene, *see* B-60233
9-Bromobenzonorbornadiene, *see* B-60233
3-Bromo-4*H*-1-benzopyran-4-one, B-70183
2-Bromobenzo[*b*]thiophene, B-60204
3-Bromobenzo[*b*]thiophene, B-60205
4-Bromobenzo[*b*]thiophene, B-60206
5-Bromobenzo[*b*]thiophene, B-60207
6-Bromobenzo[*b*]thiophene, B-60208
7-Bromobenzo[*b*]thiophene, B-60209
2-Bromobicyclo[2.2.1]heptane-1-carboxylic acid, B-60210
2-Bromobicyclo[4.1.0]hepta-1,3,5-triene, *see* B-60202
3-Bromobicyclo[4.1.0]hepta-1,3,5-triene, *see* B-60203
1-Bromobicyclo[2.2.2]octane, B-70184
3-Bromo[1,1′-biphenyl]-4-carboxylic acid, B-60211
2-Bromo-3,3′-bipyridine, B-70185
4-Bromo-2,2′-bipyridine, B-70186
5-Bromo-2,3′-bipyridine, B-70187
5-Bromo-2,4′-bipyridine, B-70188
5-Bromo-3,3′-bipyridine, B-70189
5-Bromo-3,4′-bipyridine, B-70190
6-Bromo-2,2′-bipyridine, B-70191
6-Bromo-2,3′-bipyridine, B-70192
6-Bromo-2,4′-bipyridine, B-70193
3-Bromo-2-(3-bromo-4,5-dihydroxybenzyl)-4,5-dihydroxybenzyl alcohol, *see* B-70194
3-Bromo-4-[(3-bromo-4,5-dihydroxyphenyl)-methyl]-5-(hydroxymethyl)-1,2-benzenediol, B-70194
▷1-Bromo-2-(bromomethyl)naphthalene, B-60212
1-Bromo-4-(bromomethyl)naphthalene, B-60213
1-Bromo-5-(bromomethyl)naphthalene, B-60214
1-Bromo-7-(bromomethyl)naphthalene, B-60215
1-Bromo-8-(bromomethyl)naphthalene, B-60216
2-Bromo-3-(bromomethyl)naphthalene, B-60217
2-Bromo-6-(bromomethyl)naphthalene, B-60218
3-Bromo-1-(bromomethyl)naphthalene, B-60219
6-Bromo-1-(bromomethyl)naphthalene, B-60220
7-Bromo-1-(bromomethyl)naphthalene, B-60221
7-Bromo-4-(1-bromopropyl)-2-(2-penten-4-ynyl)-3,9-dioxabicyclo[4.2.1]nonane, *see* I-70067
2-[[3-Bromo-5-(1-bromopropyl)tetrahydro-2-furanyl]methyl]-5-(1-bromo-2-propynyl)-tetrahydro-3-furanol, B-70195
1-Bromo-1,2-butadiene, B-70196
4-Bromobutanal, B-70197
Bromo-*tert*-butylacetylene, *see* B-70207
2-(4-Bromobutyl)-1,3-dioxane, *in* B-70263
4-Bromo-1-butyne, B-70198
3-Bromo-1-butynylbenzene, *see* B-60314
3-Bromobutyrine, *see* A-60097
4-Bromobutyrine, *see* A-60098
2-Bromo-4-carboxybenzenethiol, *see* B-70228
3-Bromo-4-carboxybenzenethiol, *see* B-70226
4-Bromo-3-carboxybenzenethiol, *see* B-70227
5-Bromo-7-chlorocavernicolin, B-70199
1-Bromo-6-chlorohexane, B-70200
10*R*-Bromo-4-chloro-9*S*-hydroxy-3,7(14)-chamigradiene, *see* E-70002
3-(3-Bromo-4-chloro-4-methylcyclohexyl)-2-methylfuran, *see* F-60090
5-(3-Bromo-4-chloro-4-methylcyclohexyl)-5-methyl-2(5*H*)-furanone, B-60222
3-Bromo-5-chloropyridine, B-60223
4-Bromo-3-chloropyridine, B-60224
3-Bromochromone, *see* B-70183
3-Bromo-4-cyanobiphenyl, *in* B-60211
2-Bromo-1,3-cyclohexanedione, B-70201
5-Bromodecahydro-α,α,4a-trimethyl-8-methyl-ene-2-naphthalenemethanol, *see* B-60197
5-Bromo-5-decene, B-60225
4-Bromo-2,3-dichloroanisole, *in* B-60229
1-Bromo-2,3-dichloro-4-methoxybenzene, *in* B-60229

2-Bromo-3,5-dichlorophenol, B-60226
2-Bromo-4,5-dichlorophenol, B-60227
3-Bromo-2,4-dichlorophenol, B-60228
4-Bromo-2,3-dichlorophenol, B-60229
4-Bromo-3,5-dichlorophenol, B-60230
4-Bromo-4,4-difluoro-2-butynoic acid, B-70202
(Bromodifluoromethyl)phenylacetylene, *see* B-70203
3-Bromo-3,3-difluoro-1-phenylpropyne, B-70203
(3-Bromo-3,3-difluoro-1-propynyl)benzene, *see* B-70203
1-Bromo-1,4-dihydro-1,4-methanonaphthalene, B-60231
2-Bromo-1,4-dihydro-1,4-methanonaphthalene, B-60232
9-Bromo-1,4-dihydro-1,4-methanonaphthalene, B-60233
7-Bromo-3,4-dihydro-1(2*H*)-naphthalenone, B-60234
2-Bromodihydroresorcinol, *see* B-70201
3-Bromo-2,3-dihydrothiophene, B-70204
5-Bromo-3,4-dihydro-4-thioxo-2(1*H*)-pyrimidinone, B-70205
(4-Bromo-2,5-dihydroxyphenyl)(2,6-dimethyl-6-hydroxy-2-heptenyl)ketone, *see* B-70206
1-(4-Bromo-2,5-dihydroxyphenyl)-7-hydroxy-3,7-dimethyl-2-octen-1-one, B-70206
1-Bromo-(2-dimethoxymethyl)naphthalene, *in* B-60302
3-Bromo-2,6-dimethoxyphenol, *in* B-60200
4-Bromo-2,6-dimethoxyphenol, *in* B-60201
1-Bromo-3,3-dimethyl-1-butyne, B-70207
1-Bromo-2,3-dimethylcyclopropene, B-60235
3-Bromo-1,4-dimethylpyrazole, *in* B-70250
3-Bromo-1,5-dimethylpyrazole, *in* B-70252
4-Bromo-1,3-dimethylpyrazole, *in* B-70251
4-Bromo-1,5-dimethylpyrazole, *in* B-70251
4-Bromo-3,5-dimethylpyrazole, B-70208
5-Bromo-1,3-dimethylpyrazole, *in* B-70252
4-Bromodiphenylmethanol, B-60236
1-Bromo-1,2-diphenylpropene, B-60237
3-Bromo-1,2-diphenylpropene, B-70209
6′-Bromodisidein, *in* D-60499
6-Bromo-6-dodecene, B-60238
2-Bromo-5,6-epoxy-4-hydroxy-2-cyclohexen-1-one, B-60239
1,1′,1″-(1-Bromo-1,2,2-ethenediyl)-trisbenzene, *see* B-60276
2-Bromo-3-ethoxy-1,4-naphthoquinone, *in* B-60268
2-Bromo-1-ethoxy-3-nitrobenzene, *in* B-70257
2-Bromo-3-ethoxypyridine, *in* B-60275
2-(2-Bromoethyl)pyridine, B-70210
3-(2-Bromoethyl)pyridine, B-70211
1β-Bromo-4(15)-eudesmen-11-ol, *see* B-60197
2-Bromo-3-fluoroanisole, *in* B-60247
2-Bromo-4-fluoroanisole, *in* B-60248
4-Bromo-2-fluoroanisole, *in* B-60251
3-Bromo-4-fluorobenzenemethanol, *see* B-60244
5-Bromo-2-fluorobenzenemethanol, *see* B-60245
2-Bromo-4-fluorobenzoic acid, B-60240
2-Bromo-5-fluorobenzoic acid, B-60241
2-Bromo-6-fluorobenzoic acid, B-60242
3-Bromo-4-fluorobenzoic acid, B-60243
3-Bromo-4-fluorobenzyl alcohol, B-60244
5-Bromo-2-fluorobenzyl alcohol, B-60245
1-Bromo-2-fluoro-3,5-dinitrobenzene, B-60246
2-Bromo-1-fluoro-3-methoxybenzene, *in* B-60247
2-Bromo-4-fluoro-1-methoxybenzene, *in* B-60248
4-Bromo-2-fluoro-1-methoxybenzene, *in* B-60251
2-Bromo-3-fluorophenol, B-60247
2-Bromo-4-fluorophenol, B-60248
2-Bromo-6-fluorophenol, B-60249
3-Bromo-4-fluorophenol, B-60250
4-Bromo-2-fluorophenol, B-60251
1-Bromo-3-fluoropropane, B-70212
2-Bromo-3-fluoropyridine, B-70213
2-Bromo-5-fluoropyridine, B-60252
3-Bromo-2-fluoropyridine, B-60253
3-Bromo-4-fluoropyridine, B-70214

3-Bromo-5-fluoropyridine, B-60254
4-Bromo-3-fluoropyridine, B-60255
5-Bromo-2-fluoropyridine, B-60256
2-Bromo-5-formylfuran, see B-60259
3-Bromo-2-formylfuran, see B-60257
4-Bromo-2-formylfuran, see B-60258
3-Bromo-2-furancarboxaldehyde, B-60257
4-Bromo-2-furancarboxaldehyde, B-60258
5-Bromo-2-furancarboxaldehyde, B-60259
3-Bromofurfural, see B-60257
4-Bromofurfural, see B-60258
5-Bromofurfural, see B-60259
8-Bromoguanosine, B-70215
7-Bromo-1-heptene, B-60260
7-Bromo-5-heptynoic acid, B-60261
2-Bromo-1,5-hexadiene, B-70216
5-Bromo-2-hexanone, B-60262
5-Bromo-1-hexene, B-60263
6-Bromo-1-hexene, B-60264
α-(1-Bromo-3-hexenyl)-5-(1-bromo-2-
   propynyl)tetrahydro-3-hydroxy-2-
   furanethanol, B-70217
1-Bromo-1-hexyne, B-70218
1-Bromo-2-hexyne, B-70219
5′-Bromo-2′-hydroxyacetophenone, B-60265
3-Bromo-5-hydroxy-2-methyl-1,4-
   naphthalenedione, see B-60266
3-Bromo-5-hydroxy-2-methyl-1,4-
   naphthoquinone, B-60266
2-Bromo-3-hydroxy-6-methylpyridine, B-60267
2-Bromo-3-hydroxy-1,4-naphthalenedione, see
   B-60268
2-Bromo-5-hydroxy-1,4-naphthalenedione, see
   B-60269
2-Bromo-6-hydroxy-1,4-naphthalenedione, see
   B-60270
2-Bromo-8-hydroxy-1,4-naphthalenedione, see
   B-60272
7-Bromo-2-hydroxy-1,4-naphthalenedione, see
   B-60273
8-Bromo-2-hydroxy-1,4-naphthalenedione, see
   B-60274
2-Bromo-3-hydroxy-1,4-naphthoquinone,
   B-60268
2-Bromo-5-hydroxy-1,4-naphthoquinone,
   B-60269
2-Bromo-6-hydroxy-1,4-naphthoquinone,
   B-60270
2-Bromo-7-hydroxy-1,4-naphthoquinone,
   B-60271
2-Bromo-8-hydroxy-1,4-naphthoquinone,
   B-60272
7-Bromo-2-hydroxy-1,4-naphthoquinone,
   B-60273
8-Bromo-2-hydroxy-1,4-naphthoquinone,
   B-60274
3-Bromo-5-hydroxy-7-oxabicyclo[4.1.0]-hept-3-
   en-2-one, see B-60239
1-(5-Bromo-2-hydroxyphenyl)ethanone, see
   B-60265
6-Bromo-5-hydroxy-2-picoline, see B-60267
▷ 2-Bromo-3-hydroxypyridine, B-60275
2-Bromo-5-hydroxypyridine, B-60276
3-Bromo-5-hydroxypyridine, B-60277
4-Bromo-3-hydroxypyridine, B-60278
▷ 2-Bromo-1H-imidazole, B-70220
4(5)-Bromo-1H-imidazole, B-70221
2-Bromoinosine, see B-70269
8-Bromoinosine, see B-70270
5-Bromo-2-iodoaniline, B-60279
5-Bromo-2-iodobenzenamine, see B-60279
1-Bromo-2-iodo-3-nitrobenzene, B-70222
2-Bromo-1-iodo-3-nitrobenzene, B-70223
2-Bromo-4-iodopyridine, B-60280
2-Bromo-5-iodopyridine, B-60281
3-Bromo-4-iodopyridine, B-60282
3-Bromo-5-iodopyridine, B-60283
4-Bromo-2-iodopyridine, B-60284
4-Bromo-3-iodopyridine, B-60285
5-Bromo-2-iodopyridine, B-60286
2-Bromo-4-iodothiazole, B-60287
2-Bromo-5-iodothiazole, B-60288
4-Bromo-2-iodothiazole, B-60289
5-Bromo-2-iodothiazole, B-60290
4-Bromo-1(3H)-isobenzofuranone, B-70224
7-Bromo-1(3H)-isobenzofuranone, B-70225
3-Bromo-5-isoxazolemethanamine, see A-70165

2-Bromojuglone, see B-60269
3-Bromojuglone, see B-60272
ent-3β-Bromo-13(16)-labdene-8β,14,15-triol, see
   V-70004
2-Bromo-4-mercaptobenzoic acid, B-70226
2-Bromo-5-mercaptobenzoic acid, B-70227
3-Bromo-4-mercaptobenzoic acid, B-70228
3-Bromo-2-mercaptopyridine, in B-60315
3-Bromo-4-mercaptopyridine, see B-60316
5-Bromo-4-mercaptopyridine, see B-60317
3-Bromomesitoic acid, see B-60331
Bromomethanesulfonic acid, B-60291
5′-Bromo-2′-methoxyacetophenone, in B-60265
2-Bromo-3-methoxyaniline, in A-70099
2-Bromo-3-methoxy-2-cyclohexen-1-one, in
   B-70201
2-Bromo-3-methoxy-6-methylpyridine, in
   B-60267
2-Bromo-6-methoxy-1,4-naphthaquinone, in
   B-60270
2-Bromo-3-methoxy-1,4-naphthoquinone, in
   B-60268
2-Bromo-5-methoxy-1,4-naphthoquinone, in
   B-60269
2-Bromo-7-methoxy-1,4-naphthoquinone, in
   B-60271
2-Bromo-8-methoxy-1,4-naphthoquinone, in
   B-60272
2-Bromo-1-methoxy-3-nitrobenzene, in B-70257
1-(5-Bromo-2-methoxyphenyl)ethanone, in
   B-60265
6-Bromo-5-methoxy-2-picoline, in B-60267
2-Bromo-3-methoxypyridine, in B-60275
3-Bromo-5-methoxypyridine, in B-60277
1-(Bromomethyl)acridine, B-70229
2-(Bromomethyl)acridine, B-70230
3-(Bromomethyl)acridine, B-70231
4-(Bromomethyl)acridine, B-70232
9-(Bromomethyl)acridine, B-70233
α-Bromomethylacrylic acid, see B-70249
2-Bromo-3-methylaniline, B-70234
2-Bromo-3-methylbenzaldehyde, B-60292
2-Bromo-3-methylbenzenamine, see B-70234
2-Bromo-3-methylbenzoic acid, B-60293
3-Bromo-4-methylbenzoic acid, B-60294
2-(Bromomethyl)-6-chloropyridine, B-70235
3-(Bromomethyl)-2-chloropyridine, B-70236
4-(Bromomethyl)-2-chloropyridine, B-70237
5-Bromomethyl)-2-chloropyridine, B-70238
2-(Bromomethyl)cyclohexanamine, B-70239
2-(Bromomethyl)cyclopentanamine, B-70240
(Bromomethyl)cyclopentane, B-70241
1-Bromo-1-methylcyclopropane, B-70242
3-(Bromomethyl)-2,3-dihydrobenzofuran,
   B-60295
3-Bromo-2-methylenepropanoic acid, see
   B-70249
(Bromomethylene)tricyclo[3.3.1.1³,⁷]decane, see
   B-60296
1,1′-[1-(Bromomethyl)-1,2-ethenediyl]-
   bisbenzene, see B-70209
1,1′-(1-Bromo-2-methyl-1,2-ethenediyl)-
   bisbenzene, see B-60237
2-Bromomethyl)-6-fluoropyridine, B-70243
3-(Bromomethyl)-2-fluoropyridine, B-70244
4-(Bromomethyl)-2-fluoropyridine, B-70245
5-(Bromomethyl)-2-fluoropyridine, B-70246
(Bromomethylidene)adamantane, B-60296
4-Bromo-1-methyl-1H-imidazole, in B-70221
5-Bromo-1-methyl-1H-imidazole, in B-70221
Bromomethyl isobutyl ketone, see B-60298
5-Bromomethyl-2-(4-methyl-1,4-cyclo-
   hexadienyl)-2-methyl-1-methylenecyclo-
   pentane, see B-70247
3-(Bromomethyl)-1-methyl-2-methylene-1-(4-
   bromophenyl)cyclopentane, in B-70247
1-[3-(Bromomethyl)-1-methyl-2-methyl-
   enecyclopentyl]-4-methylbenzene, in B-70247
1-[3-(Bromomethyl)-1-methyl-2-methyl-
   enecyclopentyl]-4-methyl-1,4-cyclohexadiene,
   B-70247
4-(Bromomethyl)-2-nitrobenzoic acid, B-60297
2-Bromo-4-methylpentanoic acid, B-70248
1-Bromo-4-methyl-2-pentanone, B-60298
5-Bromo-3-methyl-1-pentene, B-60299
Bromomethyl phenyl selenide, B-60300

2-Bromomethyl-2-propenoic acid, B-70249
3(5)-Bromo-4-methylpyrazole, B-70250
4-Bromo-3(5)-methylpyrazole, B-70251
3(5)-Bromo-5(3)-methylpyrazole, B-70252
2-(Bromomethyl)pyridine, B-70253
3-(Bromomethyl)pyridine, B-70254
4-(Bromomethyl)pyridine, B-70255
2-Bromo-6-methyl-3-pyridinol, see B-60267
[(Bromomethyl)seleno]benzene, see B-60300
2-Bromo-4-(methylthio)benzoic acid, in
   B-70226
2-Bromo-5-(methylthio)benzoic acid, in
   B-70227
3-Bromo-4-(methylthio)benzoic acid, in
   B-70228
3-Bromo-2-(methylthio)pyridine, in B-60315
3-(Bromomethyl)-2,4,10-trioxadamantane, see
   B-60301
(3-Bromomethyl)-2,4,10-tri-
   oxatricyclo[3.3.1.1³,⁷]decane, B-60301
2-Bromo-4-methylvaleric acid, see B-70248
1-Bromo-2-naphthaldehyde, see B-60302
1-Bromo-2-naphthalenecarboxaldehyde,
   B-60302
1-(3-Bromo-1-naphthalenyl)ethanone, see
   A-60021
1-(4-Bromo-1-naphthalenyl)ethanone, see
   A-60022
1-(5-Bromo-1-naphthalenyl)ethanone, see
   A-60023
1-(6-Bromo-2-naphthalenyl)ethanone, see
   A-60025
1-(7-Bromo-1-naphthalenyl)ethanone, see
   A-60024
3-Bromo-1-naphthyl methyl ketone, see A-60021
4-Bromo-1-naphthyl methyl ketone, see A-60022
5-Bromo-1-naphthyl methyl ketone, see A-60023
6-Bromo-2-naphthyl methyl ketone, see A-60025
7-Bromo-1-naphthyl methyl ketone, see A-60024
2-Bromo-3-nitroanisole, in B-70257
2-Bromo-3-nitrobenzenemethanol, see B-70256
2-Bromo-5-nitrobenzotrifluoride, see B-60308
2-Bromo-5-nitrobenzotrifluoride, see B-60307
3-Bromo-2-nitrobenzotrifluoride, see B-60306
5-Bromo-2-nitrobenzotrifluoride, see B-60309
2-Bromo-3-nitrobenzyl alcohol, B-70256
1-Bromo-2-nitroethane, B-60303
2-Bromo-3-nitrophenetole, in B-70257
2-Bromo-3-nitrophenol, B-70257
1-Bromo-1-nitropropane, B-70258
2-Bromo-2-nitropropane, B-70259
4-Bromo-3(5)-nitro-1H-pyrazole, B-60304
3(5)-Bromo-5(3)-nitro-1H-pyrazole, B-60305
1-Bromo-2-nitro-3-(trifluoromethyl)benzene,
   B-60306
1-Bromo-4-nitro-2-(trifluoromethyl)benzene,
   B-60307
2-Bromo-1-nitro-3-(trifluoromethyl)benzene,
   B-60308
4-Bromo-1-nitro-2-(trifluoromethyl)benzene,
   B-60309
2-Bromo-5-nitro-α,α,α-trifluorotoluene, see
   B-60307
1-Bromo-3-nonene, B-60310
2-Bromo-1-norbornanecarboxylic acid, see
   B-60210
8-Bromo-1-octanol, B-70260
1-Bromo-1-octene, B-70261
(3-Bromo-2-oxopropylidene)propanedioic acid,
   B-70262
1-Bromo-1,3-pentadiene, B-60311
Bromopentamethylbenzene, B-60312
5-Bromopentanal, B-70263
5-Bromo-2-pentanol, B-60313
2-Bromo-1,10-phenanthroline, B-70264
8-Bromo-1,7-phenanthroline, B-70265
4-Bromo-α-phenylbenzenemethanol, see
   B-60236
2-Bromo-4-phenylbenzoic acid, see B-60211
3-Bromo-1-phenyl-1-butyne, B-60314
4-Bromophenylphenylmethanol, see B-60236
2-Bromophloroglucinol, see B-60199
1-Bromophthalazine, B-70266

5-Bromophthalazine, B-70267
6-Bromophthalazine, B-70268
4-Bromophthalide, *see* B-70224
7-Bromophthalide, *see* B-70225
3-Bromoplumbagin, *see* B-60266
3-Bromo-2-pyridinethiol, *in* B-60315
3-Bromo-4-pyridinethiol, *see* B-60316
5-Bromo-2-pyridinethiol, *see* B-60317
3-Bromo-2(1*H*)-pyridinethione, B-60315
3-Bromo-4(1*H*)-pyridinethione, B-60316
5-Bromo-2(1*H*)-pyridinethione, B-60317
▷2-Bromo-3-pyridinol, *see* B-60275
4-Bromo-3-pyridinol, *see* B-60278
5-Bromo-3-pyridinol, *see* B-60277
6-Bromo-3-pyridinol, *see* B-60276
2-Bromo-5-pyrimidinamine, *see* A-70103
4-Bromo-2-pyrimidinamine, *see* A-70100
5-Bromo-2-pyrimidinamine, *see* A-70101
5-Bromo-4-pyrimidinamine, *see* A-70102
▷5-Bromo-2,4-(1*H*,3*H*)-pyrimidinedione,
  B-60318
6-Bromo-2,4-(1*H*,3*H*)-pyrimidinedione,
  B-60319
4-Bromopyrogallol, *see* B-60200
5-Bromopyrogallol, *see* B-60201
8-Bromo-9-β-D-ribofuranosylguanine, *see*
  B-70215
2-Bromo-9-β-D-ribofuranosyl-6*H*-purin-6-one,
  B-70269
8-Bromo-9-β-D-ribofuranosyl-6*H*-purin-6-one,
  B-70270
4-Bromo-2-sulfolene, *in* B-70204
1-Bromo-2,3,4,5-tetrafluorobenzene, B-60320
1-Bromo-2,3,4,6-tetrafluorobenzene, *see*
  B-60321
1-Bromo-2,3,5,6-tetrafluorobenzene, *see*
  B-60322
2-Bromo-1,3,4,5-tetrafluorobenzene, B-60321
3-Bromo-1,2,4,5-tetrafluorobenzene, B-60322
1-Bromo-2,3,4,5-tetrafluoro-6-nitrobenzene,
  B-60323
1-Bromo-2,3,4,6-tetrafluoro-5-nitrobenzene,
  B-60324
1-Bromo-2,3,5,6-tetrafluoro-4-nitrobenzene,
  B-60325
2-Bromo-3,4,5,6-tetrafluoropyridine, B-60326
3-Bromo-2,4,5,6-tetrafluoropyridine, B-60327
4-Bromo-2,3,5,6-tetrafluoropyridine, B-60328
2-Bromotetrahydro-2*H*-pyran, B-70271
7-Bromo-1-tetralone, *see* B-60234
Bromotetrasphaerol, B-70272
5-Bromo-4-thiouracil, *see* B-70205
Bromothricin, *see* C-60137
2-Bromo-*m*-toluic acid, *see* B-60293
3-Bromo-*p*-toluic acid, *see* B-60294
2-Bromo-*m*-toluidine, *see* B-70234
5-Bromo-1,2,3-triazine, B-70273
3-Bromo[1,2,3]triazolo[1,5-*a*]pyridine, B-60329
7-Bromo[1,2,3]triazolo[1,5-*a*]pyridine, B-60330
1-Bromo-1,1,2,2,3,3,4,4,5,5,6,6,6-trideca-
  fluorohexane, B-70274
▷3-Bromo-3-(trifluoromethyl)-3*H*-diazirine,
  B-70275
(Bromotrifluoromethyl)propiolic acid, *see*
  B-70202
2-Bromo-α,α,α-trifluoro-3-nitrotoluene, *see*
  B-60308
3-Bromo-α,α,α-trifluoro-2-nitrotoluene, *see*
  B-60306
5-Bromo-α,α,α-trifluoro-2-nitrotoluene, *see*
  B-60309
1-Bromo-2,3,4-trimethoxybenzene, *in* B-60200
1-Bromo-2,3,5-trimethoxybenzene, *in* B-70182
1-Bromo-2,4,5-trimethoxybenzene, *in* B-70181
2-Bromo-1,3,4-trimethoxybenzene, *in* B-70180
2-Bromo-1,3,5-trimethoxybenzene, *in* B-60199
5-Bromo-1,2,3-trimethoxybenzene, *in* B-60201
3-Bromo-2,4,6-trimethylbenzoic acid, B-60331
2-Bromo-1,1,9-trimethylenespiro[5.5]-
  undeca-7,9-dien-3-ol, *see* I-70074
2-Bromo-1,1,9-trimethylspiro[5.5]undeca-7,9-
  dien-3-ol, *see* O-70001
Bromotriphenylethylene, B-70276
2-Bromotryptophan, B-60332
5-Bromotryptophan, B-60333
6-Bromotryptophan, B-60334

7-Bromotryptophan, B-60335
1-Bromo-2-undecene, B-70277
11-Bromo-1-undecene, B-60336
▷5-Bromouracil, *see* B-60318
6-Bromouracil, *see* B-60319
δ-Bromovaleraldehyde, *see* B-70263
Bromovulone I, B-70278
Brosimone *A*, B-70279
Broussonin E, *in* D-70336
Broussonin F, *in* D-70336
Bruceanol C, B-70280
Bruceantinoside C, B-70281
BR-Xanthone *A*, B-70282
BR-Xanthone B, *in* T-70122
Bryoamaride, *in* C-70204
Bryoflavone, B-70283
Bryostatin 1, B-70284
Bryostatin 2, *in* B-70284
Bryostatin 10, B-70285
Bryostatin 11, B-70286
Bryostatin 12, B-70287
Bryostatin 13, B-70288
Buchanaxanthone, *in* T-60349
▷Buddleoflavonol, *see* D-70319
Buddleoflavonoloside, *in* D-70319
▷Bufalin, B-70289
Bufalitoxin, *in* B-70289
Bug, *see* A-70136
Bullerone, B-60337
Bursatellin, B-60338
1,3-Butadiene-2,3-dicarboxylic acid, *see*
  D-70409
4a,8a-[1,3]Butadienonaphthalene, *see* T-70232
6-(1,3-Butadienyl)-1,4-cycloheptadiene,
  B-70290
1,1'-(1,3-Butadienylidene)bisbenzene, *see*
  D-70476
1,3-Butadienyl thiocyanate, *see* T-60210
1,1',1"-(1,3-Butadien-1-yl-4-ylidene)-
  trisbenzene, *see* T-70305
1,3-Butanedithiol, B-60339
2,2-Butanedithiol, B-60340
6α-Butanoyloxy-17β-hydroxy-15,17-oxido-16-
  spongianone, *in* D-60365
3-Butene-1,2-diol, B-70291
2-Butene-1,4-dithiol, B-70292
2-Butene-1,4-diyl dithiocyanate, *see* D-60510
▷1-Butene oxide, *see* E-70047
3-Butenophenone, *see* P-60085
α-3-Butenylbenzenemethanamine, *see* A-70183
4-(2-Butenyl)-4-methylthreonine, *see* A-60189
3-Buten-1-ynylcyclopropane, B-70293
3-Butylaspartic acid, *see* A-60102
3-*tert*-Butylbicyclo[3.1.0]hexane, B-70294
tert-Butyl chloromethyl ether, B-70295
1-*tert*-Butyl-1,3-cyclohexadiene, B-70296
1-*tert*-Butylcyclohexene, B-60341
3-*tert*-Butylcyclohexene, B-60342
4-*tert*-Butylcyclohexene, B-60343
3-*tert*-Butyl-2-cyclohexen-1-one, B-70297
1-*tert*-Butyl-1,2-cyclooctadiene, B-60344
1-*tert*-Butylcyclopentene, B-70298
3-*tert*-Butylcyclopentene, B-70299
4-*tert*-Butylcyclopentene, B-70300
4-*tert*-Butylcyclopentene-1,2-oxide *see* B-70294
1,1-*tert*-Butyl-3,3-diethoxy-2-azaallenium(1+),
  B-70301
2-*tert*-Butyl-2,4-dihydro-6-methyl-1,3-dioxol-4-
  one, B-60345
5-*tert*-Butyl-1,2,3,5-dithiadiazolyl, B-70302
5-*tert*-Butyl-1,3,2,4-dithiadiazolyl, B-70303
tert-Butylglycine, *see* A-70136
3-*sec*-Butylhexahydropyrrolo[1,2-*a*]-pyrazine-
  1,4-dione, *see* H-60052
3-*tert*-Butyl-4-hydroxybenzaldehyde, B-70304
3-Butylidene-4,5-dihydro-1(3*H*)-
  isobenzofuranone, *see* L-60026
3-Butylidene-4,5-dihydrophthalide, *see* L-60026
4-(Butylimino)-3,4-dihydro-1,2,3-benzotriazine
  (incorr.), *in* A-60092
Butyliodoacetylene, *see* I-70040
1-*tert*-Butyl-2-iodobenzene, B-60346
1-*tert*-Butyl-3-iodobenzene, B-60347
1-*tert*-Butyl-4-iodobenzene, B-60348
tert-Butyl iodomethyl ketone, *see* I-60044

tert-Butyl isobutenyl ketone, *see* T-60365
tert-Butyl isopropyl ketone, *see* T-60366
▷*tert*-Butyl isothiocyanate, *see* I-70105
sec-Butylmalonic acid, *see* M-70119
tert-Butylmethylacetylene, *see* D-60438
2-*tert*-Butyl-6-methyl-1,3-dioxan-4-one, *see*
  B-60345
▷*tert*-Butyl mustard oil, *see* I-70105
S-Butyl 3-oxobutanethioate, *in* O-70080
S-*tert*-Butyl 3-oxobutanethioate, *in* O-70080
2-*tert*-Butylphenanthrene, B-70305
3-*tert*-Butylphenanthrene, B-70306
9-*tert*-Butylphenanthrene, B-70307
tert-Butylphenylacetylene, *see* D-60441
tert-Butyl phenyl ketone, *see* D-60443
tert-Butyl-1-propenyl ketone, *see* D-70414
2-*tert*-Butyl-4(3*H*)-pyrimidinone, B-60349
3-*tert*-Butyl-2,2,4,5,5-tetramethyl-3-hexene,
  B-60350
2-*sec*-Butyl-2-thiazoline, *see* D-70224
tert-Butyl thiocyanate, *see* M-70135
2-Butynedial, B-60351
2-Butyne-1,4-dithiol, B-70308
3-Butynoic acid, B-70309
▷Butyrolactam, *see* P-70188
BW 55-5, *see* P-60202
CA 146A, *in* C-70183
CA 146B, *in* C-70184
Cachectin, *see* T-70335
Cacotheline, C-70001
Cadabine, *see* S-60050
Cadambagenic acid, C-70002
4(15),5-Cadinadiene, C-70003
4(14),10(15)-Cadinadiene, *see* C-60001
4,11(13)-Cadinadien-12-oic acid, C-70004
ε-Cadinene, C-60001
Caesalpin F, C-60002
Caffeidine, C-60003
▷3-*O*-Caffeoylquinic acid, C-70005
▷Caffetannic acid, *see* C-70005
▷Caffeylquinic acid, *see* C-70005
Cajaflavanone, C-60004
Cajeputene, *see* I-70087
Calaminthone, C-70006
Calaxin, C-70007
Calciopor, *in* O-70102
Calcium pidolate, *in* O-70102
Calcium red, *see* T-60059
Caleine E, C-60005
Caleine F, *in* C-60005
Calenduladiol, *in* L-60039
Calibragenin, *in* C-70172
Calichemicin γ₁¹, C-70008
Calipteryxin, *in* K-70013
Calopiptin, *in* A-70271
Calopogonium isoflavone *B*, C-60006
Caloverticillic acid *A*, C-60007
Caloverticillic acid C, C-60008
Caloverticillic acid B, *in* C-60007
Calpyrodil, *in* O-70102
Camellidin I, *in* D-60363
Camellidin II, *in* D-60363
▷Camphidorium, *in* T-70277
Camphonium, *in* T-70277
α-Camphorene, C-70009
p-Camphorene, *see* C-70009
Camporic acid, C-70010
Canaliculatol, C-70011
Candelabrone, *see* T-60313
▷Canderel, *in* A-60311
Candicanin, C-70012
Candidol, *in* P-60047
Candirone, *in* P-60048
Candollein, *in* E-70009
Caniojane, C-70013
▷Cannogenol, *see* C-60168
Cantabradienic acid, C-60009
Cantabrenolic acid, C-60010
Cantabrenonic acid, *in* C-60010
▷Capillin, *see* P-60109
Capsenone, C-60011
*O*⁶-[(3-Carbamoyl-2*H*-azirine-2-ylidene)-
  amino]-1,2-*O*-isopropylidene-3,5-di-*O*-tosyl-
  α-D-glucofuranoside, C-70014
N-Carbamoylglucosamine, C-70015
*N*⁶-(Carbamoylmethyl)-2'-deoxyadenosine,
  C-70016

N^5-Carbamoylornithine, see C-60151
1-Carbamoyl-2-pyrrolidone, see S-60049
▷Carbanilide, see D-70499
Carbapenam-3-carboxylic acid, see O-70079
1H-Carbazole-1,4(9H)-dione, C-60012
9H-Carbazol-2-ol, see H-60109
Carbazomycinal, see H-70154
Carbazomycin C, in T-70253
Carbazomycin D, in T-70253
Carbazomycin E, see H-70154
Carbazomycin F, in H-70154
1-Carbethoxybicyclo[1.1.0]butane, in B-70094
1-Carbethoxy-2-cyano-1,2-dihydropyridine,
    C-70017
▷Carbobenzoxy chloride, see B-60071
Carbocamphenilone, see D-60410
Carbocyclic 2′-deoxyuridine, see H-70140
Carbodine, see C-70018
2-Carbomethoxy-3-vinylcyclopentene, in
    V-60010
Carbonic acid 1-chloroethyl ethyl ester, see
    C-60067
Carbonimidic acid, C-70019
▷4,4′-Carbonimidoylbis[N,N-di-
    methylbenzenamine], see A-70268
▷Carbon monosulfide, C-70020
Carbonochloridothioic acid, C-60013
Carbonodithioic acid S,S-bis(1-phenyl-1H-tetra-
    zol-5-yl)ester, see B-70155
1,1′-Carbonothioylbis-2(1H)pyridinone,
    C-60014
2,2′-Carbonylbisbenzoic acid, see B-70041
O,O′-Carbonylbis(hydroxylamine), C-60015
N,N′-[Carbonylbis(oxy)]bis[2-methyl-2-
    propanamine], in C-60015
2,2′-Carbonyldibenzoic acid, see B-70041
3-Carboxyaspirin, in H-60101
2-Carboxy-α-[(3-carboxy-3-hydroxypropyl)-
    amino]-β-hydroxy-1-azetidinebutanoic acid,
    see M-60148
2-Carboxy-3-chlorobenzenethiol, see C-70082
2-Carboxy-4-chlorobenzenethiol, see C-70085
2-Carboxy-5-chlorobenzenethiol, see C-70084
3-Carboxy-2-chlorobenzenethiol, see C-70083
5-Carboxydecahydro-β-5,8a-trimethyl-2-
    methylene-1-naphthalenepentanoic acid, see
    O-60039
3-Carboxy-2,3-dihydro-8-hydroxy-5-methyl-
    thiazolo[3,2-a]pyridinium hydroxide inner
    salt, C-70021
2-Carboxy-1,1-dimethylpyrrolidinium hydroxide
    inner salt, see S-60050
o-Carboxydiphenylacetylene, see P-70074
p-Carboxydiphenylacetylene, see P-70075
N-(Carboxyethyl)glycine, see C-60016
2-Carboxy-3-fluorobenzenethiol, see F-70030
2-Carboxy-5-fluorobenzenethiol, see F-70031
4-Carboxy-3-fluorobenzenethiol, see F-70029
17-Carboxygrindelic acid, see E-70024
5-Carboxy-α-hydroxy-2-furanacetic acid, see
    O-70048
N-(N-1-Carboxy-6-hydroxy-3-oxoheptyl)-
    alanyltyrosine, C-70022
5-[(3-Carboxy-4-hydroxyphenyl)(3-carboxy-4-
    oxo-2,5-cyclohexadien-1-ylidene)-methyl]-2-
    hydroxybenzoic acid, see A-60318
3-Carboxy-2-hydroxy-N,N,N-trimethyl-1-
    propanaminium hydroxide inner salt, see
    C-70024
(1-Carboxy-2-imidazol-4-ylethyl)-trimethyl-
    ammonium hydroxide inner salt, see H-60083
4-Carboxy-2-iodobenzenethiol, see I-70046
N-(Carboxymethyl)-β-alanine, see C-60016
3-(Carboxymethylamino)propanoic acid,
    C-60016
▷3-(Carboxymethylamino)propionitrile, in
    C-60016
▷2-Carboxy-4-(1-methylethenyl)-3-
    pyrrolidineacetic acid, see K-60003
▷3-Carboxymethyl-4-isopropenylproline, see
    K-60003
4-Carboxy-3-nitrobenzyl bromide, see B-60297
1-(2-Carboxy-5-oxo-1-pyrrolidinyl)-1-deoxy-D-
    mannitol, see A-70075
2-Carboxyphenothiazine, see P-70056

5-Carboxyprotocatechuic acid, see D-60307
2-Carboxy-3-thiopheneacetic acid, C-60017
4-Carboxy-3-thiopheneacetic acid, C-60018
α-Carboxy-N,N,N-trimethyl-1H-imidazole-4-
    ethanaminium hydroxide inner salt, see
    H-60083
Cardiophyllidin, C-70023
3-Caren-5-one, C-60019
Carmenin, in F-70006
Carnitine, C-70024
Carnosiflogenin A, in C-60175
Carnosiflogenin B, in C-60175
Carnosiflogenin C, in C-60175
Carnosifloside I, in C-60175
Carnosifloside II, in C-60175
Carnosifloside III, in C-60175
Carnosifloside IV, in C-60175
Carnosifloside V, in C-60175
Carnosifloside VI, in C-60175
Carnosine, C-60020
ψ,ψ-Carotene, see I-60041
8-Carotene-4,6,10-triol, C-60021
▷Carriomycin, C-60022
Carvene, see I-70087
Caryoynencin A, in H-70195
Caryoynencin B, in H-70195
Caryoynencin C, in H-70195
Cassigarol A, C-70025
Castalagin, in V-70006
Castalin, in V-70006
Castasterone, in T-70116
Castavaloninic acid, in V-70006
Catalpol, C-70026
▷Catechin, in P-70026
[2′,2′]-Catechin-taxifolin, C-60023
▷Catechol, in P-70026
Catecholacetic acid, see H-70203
Catechol dichloromethylene ether, see D-70113
Cavoxinine, C-60024
Cavoxinone, C-60025
C-AZT, see A-70288
C₂-Bishomocuban-6-one, see P-60027
Cedrelopsin, C-60026
Cedronellone, C-60027
Cedronin, C-70027
Celastanhydride, C-70028
▷Celliton Fast Pink B, see A-60175
1,3,7,11-Cembratetraen-15-ol, C-70029
3,7,11,15(17)-Cembratetraen-16,2-olide,
    C-60028
2,7,11-Cembratriene-4,6-diol, C-70030
β-Cembrenediol, in C-70030
α-Cembrenediol, in C-70030
▷Centralite I, in D-70499
▷Centralite II, in D-70499
Centralite III, in D-70499
Centrohexaindane, C-70031
Cephalochromin, C-60029
Cerapicol, C-70032
Ceratenolone, C-60030
Ceratopicanol, C-70033
Cerevisterol, in M-70057
Cervicol, C-70034
Cerotone, see H-70009
Cestosal, in H-70108
Cetosalol, in H-70108
Cetraric acid, C-70035
CGA 134735, in D-70223
CGA 134736, see D-70223
Chaetochromin, in C-60029
Chaetochromin A, in C-60029
Chaetochromin B, in C-60029
Chaetochromin C, in C-60029
Chaetochromin D, in C-60029
Chalaurenol, C-70036
Chalcogran, see E-70043
Chalconaringenin, see H-60220
Chalepin, C-60031
Chalepin acetate, in C-60031
Chaloxone, see H-70155
Chamaecyparin, in H-60082
Chamaedroxide, C-60032
Chamaepitin, C-60033
Chapinolin, C-70037
Charamin, see A-60354
Chatferin, in F-60009

Chichibabin's hydrocarbon, see B-60117
Chiisanogenin, C-70038
Chiisanoside, in C-70038
Chilenone B, C-60034
Chimganidin, in G-60014
Chinensin I, C-60035
Chinensin II, C-60036
Chiratanin, C-60037
Chisocheton compound A, in T-60223
Chisocheton compound D, in T-60223
Chlamydocin, C-60038
Chloratranol, see C-60059
Chloretyl, see H-70053
▷Chlorisept, see C-70075
2-Chloroacetyl-3-methylaniline, see A-60115
2-Chloroacetyl-4-methylaniline, see A-60114
2-Chloroacetyl-5-methylaniline, see A-60113
2-Chloroacetyl-6-methylaniline, see A-60112
1-Chloroadamantane, C-60039
2-Chloroadamantane, C-60040
2-Chloroadenine, see C-70039
2-Chloroallylamine, see C-70151
2-Chloro-6-aminopurine, C-70039
2-Chloro-1,4,9,10-anthracenetetrone, C-60041
3-Chloroaspartic acid, see A-70121
Chloroatranol, see C-60059
2-Chloro-7H-benz[de]anthracen-7-one, C-70040
3-Chloro-7H-benz[de]anthracen-7-one, C-70041
4-Chloro-7H-benz[de]anthracen-7-one, C-70042
5-Chloro-7H-benz[de]anthracen-7-one, C-70043
6-Chloro-7H-benz[de]anthracen-7-one, C-70044
8-Chloro-7H-benz[de]anthracen-7-one, C-70045
9-Chloro-7H-benz[de]anthracen-7-one, C-70046
10-Chloro-7H-benz[de]anthracen-7-one,
    C-70047
11-Chloro-7H-benz[de]anthracen-7-one,
    C-70048
2-Chlorobenzanthrone, see C-70040
3-Chlorobenzanthrone, see C-70041
4-Chlorobenzanthrone, see C-70042
5-Chlorobenzanthrone, see C-70043
6-Chlorobenzanthrone, see C-70044
8-Chlorobenzanthrone, see C-70045
9-Chlorobenzanthrone, see C-70046
10-Chlorobenzanthrone, see C-70047
11-Chlorobenzanthrone, see C-70048
2-Chlorobenzanthrone (obsol.), see C-70042
2′-Chlorobenzanthrone (obsol.), see C-70040
6-Chlorobenzanthrone (obsol.), see C-70046
7-Chlorobenzanthrone (obsol.), see C-70047
13-Chlorobenzanthrone (obsol.), see C-70041
▷4-Chloro-1,3-benzenedisulfonamide, in C-70049
4-Chloro-1,3-benzenedisulfonic acid, C-70049
β-Chlorobenzeneethanol, see C-60128
p-Chlorobenzhydrol, see C-60065
4-Chloro-1,2-benzisothiazole, C-70050
5-Chloro-1,2-benzisothiazole, C-70051
6-Chloro-1,2-benzisothiazole, C-70052
7-Chloro-1,2-benzisothiazole, C-70053
1-Chlorobenzo[g]phthalazine, C-70054
4-Chlorobenzo[g]phthalazin-1(2H)-one,
    C-70055
2-Chlorobenzo[b]thiophene, C-60042
3-Chlorobenzo[b]thiophene, C-60043
4-Chlorobenzo[b]thiophene, C-60044
5-Chlorobenzo[b]thiophene, C-60045
6-Chlorobenzo[b]thiophene, C-60046
7-Chlorobenzo[b]thiophene, C-60047
▷2-Chlorobenzoxazole, C-60048
2-Chlorobicyclo[2.2.1]hepta-2,5-diene, C-70056
1-Chlorobicyclo[2.2.2]octane, C-70057
4-Chloro-1-butyne, C-70058
3-Chlorobutyrine, see A-60103
4-Chlorobutyrine, see A-60104
2[(Chlorocarbonyl)oxy]-3a,4,7,7a-tetrahydro-
    4,7-methano-1H-isoindole-1,3(2H)-dione,
    C-70059
1-Chloro-3-(chloromethyl)-7-methyl-2,6-
    octadiene, C-70060
3-Chloro-2-chloromethyl-1-propene, C-70061
α-Chloro-m-cresol, see C-70102
α-Chloro-o-cresol, see C-70101
α-Chloro-p-cresol, see C-70103
2-Chloro-6-cyanobenzenethiol, in C-70083
2-Chloro-4-cyanophenol, in C-60091

5-Chloro-1,3-cycloheptadiene, C-70062
2-Chloro-2-cyclohepten-1-one, C-60049
2-Chloro-2-cyclohexen-1-one, C-60050
2-Chloro-2-cyclopenten-1-one, C-60051
5-Chloro-2-cyclopenten-1-one, C-70063
5′-Chloro-5′-deoxyarabinosylcytosine, C-70064
2-Chloro-4,6-difluoro-1,3,5-triazine, C-60052
6-Chloro-3,5-difluoro-1,2,4-triazine, C-60053
2-Chloro-2,3-dihydro-1*H*-imidazole-4,5-dione, C-60054
4-Chloro-2,3-dihydro-1*H*-indole, C-60055
2-Chloro-3,6-dihydroxybenzaldehyde, C-60056
3-Chloro-2,5-dihydroxybenzaldehyde, C-60057
3-Chloro-2,6-dihydroxybenzaldehyde, C-60058
5-Chloro-2,3-dihydroxybenzaldehyde, C-70065
2-Chloro-1,8-dihydroxy-5-methoxy-6-methyl-xanthone, *in* T-70181
3-Chloro-2,6-dihydroxy-4-methylbenzaldehyde, C-60059
3-Chloro-4,6-dihydroxy-2-methylbenzaldehyde, C-60060
2-Chloro-4,6-dihydroxypyridine, *see* C-70074
4-Chloro-2,6-dihydroxypyridine, *see* C-70073
5-Chloro-2,3-dimethoxybenzaldehyde, *in* C-70065
6-Chloro-2,3-dimethoxy-1,4-naphthalenediol diacetate, *in* C-60118
2-Chloro-4,6-dimethoxypyridine, *in* C-70074
4-Chloro-2,6-dimethoxypyridine, *in* C-70073
2-Chloro-4,6-dimethylbenzaldehyde, C-60061
4-Chloro-2,6-dimethylbenzaldehyde, C-60062
4-Chloro-3,5-dimethylbenzaldehyde, C-60063
5-Chloro-2,4-dimethylbenzaldehyde, C-60064
4-Chlorodiphenylmethanol, C-60065
6′-Chlorodisidein, *in* D-60499
12-Chlorododecanoic acid, C-70066
2-Chloro-3,7-epoxy-9-chamigranone, C-60066
Chloroepoxytrifluoroethane, *see* C-60140
4-Chloro-2-ethoxy-1-nitrobenzene, *in* C-60119
(1-Chloroethyl) ethyl carbonate, C-60067
▷2-Chloroethyl isothiocyanate, *see* C-70081
▷β-Chloroethyl mustard oil, *see* C-70081
1-Chloro-1-ethynylcyclopropane, C-60068
2-Chloro-4-fluoroanisole, *in* C-60073
2-Chloro-5-fluoroanisole, *in* C-60074
4-Chloro-2-fluoroanisole, *in* C-60079
4-Chloro-3-fluoroanisole, *in* C-60080
3-Chloro-4-fluorobenzaldehyde, C-70067
3-Chloro-4-fluorobenzenethiol, C-60069
4-Chloro-2-fluorobenzenethiol, C-60070
4-Chloro-3-fluorobenzenethiol, C-60071
1-Chloro-2-fluoro-4-mercaptobenzene, *see* C-60071
2-Chloro-1-fluoro-4-mercaptobenzene, *see* C-60069
4-Chloro-2-fluoro-1-mercaptobenzene, *see* C-60070
1-Chloro-2-fluoro-4-methoxybenzene, *in* C-60080
1-Chloro-4-fluoro-2-methoxybenzene, *in* C-60074
2-Chloro-1-fluoro-4-methoxybenzene, *in* C-60073
4-Chloro-2-fluoro-1-methoxybenzene, *in* C-60079
2-Chloro-4-fluoro-5-nitrobenzoic acid, C-60072
2-Chloro-4-fluorophenol, C-60073
2-Chloro-5-fluorophenol, C-60074
2-Chloro-6-fluorophenol, C-60075
3-Chloro-2-fluorophenol, C-60076
▷3-Chloro-4-fluorophenol, C-60077
3-Chloro-5-fluorophenol, C-60078
4-Chloro-2-fluorophenol, C-60079
4-Chloro-3-fluorophenol, C-60080
2-Chloro-3-fluoropyridine, C-60081
2-Chloro-4-fluoropyridine, C-60082
2-Chloro-5-fluoropyridine, C-60083
2-Chloro-6-fluoropyridine, C-60084
3-Chloro-2-fluoropyridine, C-60085
4-Chloro-2-fluoropyridine, C-60086
4-Chloro-3-fluoropyridine, C-60087
5-Chloro-2-fluoropyridine, C-60088
3-Chloro-4-fluorothiophenol, *see* C-60069

4-Chloro-2-fluorothiophenol, *see* C-60070
4-Chloro-3-fluorothiophenol, *see* C-60071
▷Chloroflurazole, *see* D-60157
4-Chloro-6-formylorcinol, *see* C-60060
▷Chlorogenic acid, *see* C-70005
3-Chlorogentisaldehyde, *see* C-60057
6-Chlorogentisaldehyde, *see* C-60056
8-Chloroguanosine, C-70068
7-Chloro-2-heptenoic acid, C-70069
2-Chloro-1,5-hexadiene, C-70070
3-Chloro-4-hydroxybenzaldehyde, C-60089
4-Chloro-3-hydroxybenzaldehyde, C-60090
3-Chloro-4-hydroxybenzoic acid, C-60091
3-Chloro-5-hydroxybenzoic acid, C-60092
4-Chloro-3-hydroxybenzoic acid, C-60093
2-Chloro-3-hydroxy-7-chamigren-9-one, C-60094
3-Chloro-4-hydroxy-3-cyclobutene-1,2-dione, C-70071
2-Chloro-1-hydroxy-5,8-dimethoxy-6-methyl-xanthone, *in* T-70181
5-Chloro-8-hydroxy-1,4-dimethoxy-3-methyl-xanthone, *in* T-70181
3-Chloro-6-hydroxy-2-methoxybenzaldehyde, *in* C-60058
5-Chloro-2-hydroxy-3-methoxybenzaldehyde, *in* C-70065
2-Chloro-3-hydroxy-5-methylbenzoic acid, C-60095
3-Chloro-4-hydroxy-5-methylbenzoic acid, C-60096
5-Chloro-4-hydroxy-2-methylbenzoic acid, C-60097
3-Chloro-5-hydroxy-2-methyl-1,4-naphthalenedione, *see* C-60098
3-Chloro-5-hydroxy-2-methyl-1,4-naphthoquinone, C-60098
2-Chloro-5-hydroxy-6-methylpyridine, C-60099
6-Chloro-3-hydroxy-2-picoline, *see* C-60099
3-Chloro-2-hydroxypropanoic acid, C-70072
▷2-Chloro-3-hydroxypyridine, C-60100
2-Chloro-5-hydroxypyridine, C-60101
3-Chloro-5-hydroxypyridine, C-60102
4-Chloro-6-hydroxy-2(1*H*)-pyridinone, C-70073
6-Chloro-4-hydroxy-2(1*H*)pyridinone, C-70074
▷5-Chloro-8-hydroxyquinoline, *see* C-70075
8-Chloro-9-hydroxy-1,5,5,9-tetramethyl-spiro[5.5]undec-1-en-3-one, *see* C-60094
5-Chloro-4-hydroxy-*m*-toluic acid, *see* C-60096
5-Chloro-4-hydroxy-*o*-toluic acid, *see* C-60097
6-Chloro-5-hydroxy-*m*-toluic acid, *see* C-60095
2-Chloro-1*H*-imidazole, C-70076
4(5)-Chloro-1*H*-imidazole, C-70077
2-Chloro-4,5-imidazolidinedione, *see* C-60054
4-Chloro-1*H*-imidazo[4,5-*c*]pyridine, C-60104
6-Chloro-1*H*-imidazo[4,5-*c*]pyridine, C-60105
4-Chloroindoline, *see* C-60055
2-Chloroinosine, *see* C-70161
1-Chloro-1-iodoethane, C-60106
1-Chloro-2-iodoethane, C-70078
2-Chloro-3-iodopyridine, C-60107
2-Chloro-4-iodopyridine, C-60108
2-Chloro-5-iodopyridine, C-60109
3-Chloro-2-iodopyridine, C-60110
3-Chloro-5-iodopyridine, C-60111
4-Chloro-2-iodopyridine, C-60112
4-Chloro-3-iodopyridine, C-60113
4-Chloro-1(3*H*)-isobenzofuranone, C-70079
7-Chloro-1(3*H*)-isobenzofuranone, C-70080
▷1-Chloro-2-isothiocyanoethane, C-70081
β-Chlorolactic acid, *see* C-70072
2-Chloro-6-mercaptobenzoic acid, C-70082
3-Chloro-2-mercaptobenzoic acid, C-70083
4-Chloro-2-mercaptobenzoic acid, C-70084
5-Chloro-2-mercaptobenzoic acid, C-70085
3-Chloro-2-mercaptopyridine, *see* C-60129
Chloromethanesulfonic acid, C-70086
Chloromethanesulfonyl bromide, *in* C-70086
▷Chloromethanesulfonyl chloride, *in* C-70086
3-Chloro-4-methoxybenzaldehyde, *in* C-60089
4-Chloro-3-methoxybenzaldehyde, *in* C-60090
3-Chloro-4-methoxybenzoic acid, *in* C-60091
3-Chloro-5-methoxybenzoic acid, *in* C-60092

4-Chloro-3-methoxybenzoic acid, *in* C-60093
1-Chloro-4-methoxybenzo[*g*]phthalazine, *in* C-70055
3-Chloro-3-methoxy-3*H*-diazirine, C-70087
2-(Chloromethoxy)ethanol acetate, *see* C-70088
2-(Chloromethoxy)ethyl acetate, C-70088
3-Chloro-2-(methoxymethoxy)-1-propene, C-70089
2-Chloro-3-methoxy-5-methylbenzoic acid, *in* C-60095
3-Chloro-4-methoxy-5-methylbenzoic acid, *in* C-60096
5-Chloro-4-methoxy-2-methylbenzoic acid, *in* C-60097
1-Chloro-5-methoxy-2-methyl-3-nitrobenzene, *in* C-60116
2-Chloro-1-methoxy-3-methyl-4-nitrobenzene, *in* C-60115
2-(Chloromethoxy)-2-methylpropane, *see* B-70295
4-Chloro-6-methoxy-1-methyl-2(1*H*)-pyridinone, *in* C-70073
4-Chloro-2-methoxy-1-nitrobenzene, *in* C-60119
2-Chloro-3-methoxy-6-nitrotoluene, *in* C-60115
2-Chloro-4-methoxy-6-nitrotoluene, *in* C-60116
2-Chloro-3-methoxypyridine, *in* C-60100
3-Chloro-5-methoxypyridine, *in* C-60102
6-Chloro-4-methoxy-2(1*H*)-pyridinone, *in* C-70074
1-Chloromethylbenz[*cd*]indol-2(1*H*)-one, *in* B-70027
Chloromethyl *tert*-butyl ether, *see* B-70295
2-Chloro-3-methyl-2-cyclopenten-1-one, C-70090
5-Chloro-3-methyl-2-cyclopenten-1-one, C-70091
2-(Chloromethyl)-3,5-dioxa-1-hexene, *see* C-70089
4-Chloro-5-methylfurazan, C-60114
3-Chloro-4-methylfuroxan, *in* C-60114
4-Chloro-3-methylfuroxan, *in* C-60114
1-(Chloromethyl)-1*H*-indole-2,3-dione, C-70092
N-Chloromethylisatin, *see* C-70092
1-(Chloromethyl)-2-methoxybenzene, *in* C-70101
1-(Chloromethyl)-3-methoxybenzene, *in* C-70102
1-Chloromethyl-4-methoxybenzene, *in* C-70103
2-(Chloromethyl)-3-methylpyridine, C-70093
▷2-(Chloromethyl)-4-methylpyridine, *in* C-70094
2-(Chloromethyl)-5-methylpyridine, C-70095
▷2-(Chloromethyl)-6-methylpyridine, C-70096
3-(Chloromethyl)-2-methylpyridine, C-70097
3-(Chloromethyl)-4-methylpyridine, C-70098
▷4-(Chloromethyl)-2-methylpyridine, C-70099
5-(Chloromethyl)-2-methylpyridine, C-70100
2-Chloro-3-methyl-4-nitroanisole, *in* C-60115
2-Chloro-3-methyl-4-nitrophenol, C-60115
3-Chloro-4-methyl-5-nitrophenol, C-60116
4-Chloro-5-methyl-1,2,5-oxadiazole, *see* C-60114
5-Chloro-3-methyl-1-pentene, C-60117
2-(Chloromethyl)phenol, C-70101
3-(Chloromethyl)phenol, C-70102
4-(Chloromethyl)phenol, C-70103
1-Chloro-4-methylphthalazine, C-70104
2-Chloro-3-methylpyrazine, C-70105
2-Chloro-5-methylpyrazine, C-70106
2-Chloro-6-methylpyrazine, C-70107
3-Chloro-2-methylpyrazine 1-oxide, *in* C-70105
3-Chloro-1-methylpyrazole, *in* C-70155
4-Chloro-1-methylpyrazole, *in* C-70156
5-Chloro-1-methylpyrazole, *in* C-70155
▷4-Chloro-3(5)methylpyrazole, C-70108
3(5)-Chloro-5(3)-methylpyrazole, C-70109
3-Chloro-4-methylpyridazine, C-70110
3-Chloro-5-methylpyridazine, C-70111
3-Chloro-6-methylpyridazine, C-70112
6-Chloro-2-methyl-3-pyridinol, *see* C-60099
2-Chloro-4-methylpyrimidine, C-70113
2-Chloro-5-methylpyrimidine, C-70114

4-Chloro-2-methylpyrimidine, C-70115
4-Chloro-5-methylpyrimidine, C-70116
4-Chloro-6-methylpyrimidine, C-70117
5-Chloro-2-methylpyrimidine, C-70118
5-Chloro-4-methylpyrimidine, C-70119
2-(Chloromethyl)quinoxaline, C-70120
2-Chloro-3-methylquinoxaline, C-70121
2-Chloro-5-methylquinoxaline, C-70122
2-Chloro-6-methylquinoxaline, C-70123
2-Chloro-7-methylquinoxaline, C-70124
6-Chloro-7-methylquinoxaline, C-70125
3-Chloro-7-methylquinoxaline 1-oxide, *in* C-70123
3-Chloro-2-(methylsulfonyl)pyridine, *in* C-60129
5-Chloro-2-(methylthio)benzoic acid, *in* C-70085
2-Chloro-6-(methylthio)benzonitrile, *in* C-70082
3-Chloro-2-(methylthio)pyridine, *in* C-60129
6-Chloro-1,2,3,4-naphthalenetetrol, C-60118
2-Chloronebularine, *see* C-70159
6-Chloronebularine, *see* C-70160
1'-Chloronephroarctin, *in* N-70032
4'-Chloronephroarctin, *in* N-70032
5-Chloro-2-nitroanisole, *in* C-60119
2-Chloro-6-nitrobenzenemethanamine, *see* C-70126
4-Chloro-2-nitrobenzenemethanamine, *see* C-70127
4-Chloro-3-nitrobenzenemethanamine, *see* C-70128
5-Chloro-2-nitrobenzenemethanamine, *see* C-70129
2-Chloro-6-nitrobenzylamine, C-70126
4-Chloro-2-nitrobenzylamine, C-70127
4-Chloro-3-nitrobenzylamine, C-70128
5-Chloro-2-nitrobenzylamine, C-70129
2-Chloro-4-nitro-*m*-cresol, *see* C-60115
1-Chloro-1-nitrocyclobutane, C-70130
1-Chloro-1-nitrocyclohexane, C-70131
2-Chloro-3-nitro-9*H*-fluorene, C-70132
2-Chloro-7-nitro-9*H*-fluorene, C-70133
2-Chloro-9-nitro-9*H*-fluorene, C-70134
9-Chloro-3-nitro-9*H*-fluorene, C-70135
5-Chloro-2-nitrophenetole, *in* C-60119
▷5-Chloro-2-nitrophenol, C-60119
▷2-Chloro-2-nitropropane, C-60120
2-Chloronorbornadiene, *see* C-70056
5-Chloroorsellinaldehyde, *see* C-60060
5-Chloro-1,3-pentadiene, C-60121
6α-Chloro-4β,5β,14α,17β,20S-pentahydroxy-1-oxo-22R-witha-2,24-dienolide, *see* P-60148
Chloropentakis(dichloromethyl)benzene, C-70136
Chloropentamethylbenzene, C-60122
5-Chloropentanal, C-70137
1-Chloro-2,3-pentanedione, C-60123
1-Chloro-3-pentanol, C-70138
5-Chloro-1-pentyne, C-70139
1-Chloro-10*H*-phenothiazine, C-60124
▷2-Chloro-10*H*-phenothiazine, C-60125
3-Chloro-10*H*-phenothiazine, C-60126
4-Chloro-10*H*-phenothiazine, C-60127
4-Chloro-α-phenylbenzenemethanol, *see* C-60065
▷3-Chloro-3-phenyl-3*H*-diazirine, C-70140
2-Chloro-2-phenylethanol, C-70056
4-Chlorophenylphenylmethanol, *see* C-60065
1-Chloro-4-phenylphthalazine, C-70141
3-(2-Chlorophenyl)-1-propene, *see* C-70152
3-(3-Chlorophenyl)-1-propene, *see* C-70153
3-(4-Chlorophenyl)-1-propene, *see* C-70154
2-Chloro-3-phenylpyrazine, C-70142
2-Chloro-5-phenylpyrazine, C-70143
2-Chloro-6-phenylpyrazine, C-70144
2-Chloro-3-phenylquinoxaline, C-70145
6-Chloro-2-phenylquinoxaline, C-70146
7-Chloro-2-phenylquinoxaline, C-70147
1-Chlorophthalazine, C-70148
5-Chlorophthalazine, C-70149
6-Chlorophthalazine, C-70150
4-Chlorophthalide, *see* C-70079
7-Chlorophthalide, *see* C-70080
3-Chloroplumbagin, *see* C-60098
2-Chloro-2-propen-1-amine, C-70151

1-Chloro-2-(2-propenyl)benzene, C-70152
1-Chloro-3-(2-propenyl)benzene, C-70153
1-Chloro-4-(2-propenyl)benzene, C-70154
2-Chloropurin-6-amine, *see* C-70039
▷6-Chloro-1*H*-purin-2-amine, *see* A-60116
3-Chloro-1*H*-pyrazole, C-70155
▷4-Chloro-1*H*-pyrazole, C-70156
3-Chloro-2-pyridinethiol, *see* C-60129
3-Chloro-2(1*H*)-pyridinethione, C-60129
▷2-Chloro-3-pyridinol, *see* C-60100
5-Chloro-3-pyridinol, *see* C-60102
6-Chloro-3-pyridinol, *see* C-60101
1-(2-Chloro-3-pyridinyl)ethanone, *see* A-70034
1-(2-Chloro-4-pyridinyl)ethanone, *see* A-70035
1-(3-Chloro-4-pyridinyl)ethanone, *see* A-70036
1-(5-Chloro-2-pyridinyl)ethanone, *see* A-70033
1-(6-Chloro-3-pyridinyl)ethanone, *see* A-70037
2-Chloro-4-pyridinyl methyl ketone, *see* A-70035
3-Chloro-4-pyridinyl methyl ketone, *see* A-70036
5-Chloropyrimidine, C-70157
▷5-Chloro-2,4-(1*H*,3*H*)-pyrimidinedione, C-60130
6-Chloro-2,4-(1*H*,3*H*)-pyrimidinedione, C-70158
▷5-Chloro-8-quinolinol, *see* C-70075
8-Chloro-2(1*H*)-quinolinone, C-60131
8-Chloro-4(1*H*)-quinolinone, C-60132
Chlororepdiolide, C-60133
3-Chloro-γ-resorcylaldehyde, *see* C-60058
4-Chloro-9R-hydroxy-3,7(14)-chamigradiene, *in* E-70002
8-Chloro-9-β-D-ribofuranosylguanine, *see* C-70068
2-Chloro-9-β-D-ribofuranosyl-9*H*-purine, C-70159
6-Chloro-9-β-D-ribofuranosyl-9*H*-purine, C-70160
2-Chloro-9-β-D-ribofuranosyl-6*H*-purin-6-one, C-70161
▷Chlorosulfonyl isocyanate, *see* S-60062
6-Chloro-1,2,3,4-tetrahydroxynaphthalene, *see* C-60118
▷2-Chloro-3,4,5,6-tetranitroaniline, C-60134
▷2-Chloro-3,4,5,6-tetranitrobenzenamine, *see* C-60134
▷3-Chloro-1,2,4,5-tetranitrobenzene, C-60135
Chlorothioformic acid, *see* C-60013
3-Chlorothiophene, C-60136
5-Chloro-4-thiouracil, *see* C-70162
5-Chloro-4-thioxo-2(1*H*)-pyrimidinone, C-70162
Chlorothricin, C-60137
4-Chloro-1,3,5-triazaindene, *see* C-60104
6-Chloro-1,3,5-triazaindene, *see* C-60105
1-Chloro-1-(trichloroethenyl)cyclopropane, *see* C-60138
1-Chloro-1-(trichlorovinyl)cyclopropane, C-60138
1-Chlorotricyclo[3.3.1.1^{3,7}]decane, *see* C-60039
2-Chlorotricyclo[3.3.1.1^{3,7}]decane, *see* C-60040
1-Chloro-1,1,2,2,3,3,4,4,5,5,6,6,6-tridecafluorohexane, C-70163
2-Chloro-3-(trifluoromethyl)benzaldehyde, C-70164
2-Chloro-5-(trifluoromethyl)benzaldehyde, C-70165
2-Chloro-6-(trifluoromethyl)benzaldehyde, C-70166
4-Chloro-3-(trifluoromethyl)benzaldehyde, C-70167
5-Chloro-2-(trifluoromethyl)benzaldehyde, C-70168
▷3-Chloro-3-(trifluoromethyl)-3*H*-diazirine, C-70169
6-Chloro-2-(trifluoromethyl)-1*H*-imidazo[4,5-b]pyridine, C-60139
Chlorotrifluorooxirane, C-60140
2-Chloro-3,3,3-trifluoropropanoic acid, C-70170

2-Chloro-3,3,3-trifluoro-1-propene, C-60141
3-Chloro-4,5,6-trifluoropyridazine, C-60142
4-Chloro-3,5,6-trifluoropyridazine, C-60143
5-Chloro-2,4,6-trifluoropyrimidine, C-60144
2-Chlorotryptophan, C-60145
6-Chlorotryptophan, C-60146
▷5-Chlorouracil, *see* C-60130
6-Chlorouracil, *see* C-70158
Chokorin, C-70171
5-Cholestene-3,26-diol, C-60147
Cholest-5-ene-3,16,22-triol, C-70172
▷Choline salicylate, *in* H-70108
Chordarine, *see* T-70182
▷Chorismic acid, C-70173
2-Chromanol, *see* D-60198
3-Chromanol, *see* D-60199
4-Chromanol, *see* D-60200
5-Chromanol, *see* D-60201
6-Chromanol, *see* D-70175
7-Chromanol, *see* D-60202
8-Chromanol, *see* D-60203
3-Chromoneacetic acid, *see* O-60059
Chromophycadiol, C-60148
Chromophycadiol monoacetate, *in* C-60148
Chrysanthemine, *see* S-60050
Chrysarone, *in* T-70268
Chrysatropic acid, *see* H-70153
Chrysean, *see* A-60260
▷Chrysene-5,6-oxide, C-70174
Chrysophyllin *B*, C-70175
CI 911, *see* D-70260
▷C.I. Basic Violet 14, *see* R-60010
▷CI Basic Yellow 2, *in* A-70268
Ciclamycin 4, *in* C-70176
Ciclamycin O, C-70176
C.I. Disperse Blue 22, *in* A-60175
▷C.I. Disperse Red 15, *see* A-60175
Ciliapterin, *in* B-70121
Ciliarin, *in* C-70007
C.I. Mordant Violet 39, *in* A-60318
Cinene, *see* I-70087
Cinnamodial, C-70177
Cinnamoyl isothiocyanate, *see* P-70090
1-Cinnamoylmelianone, *in* M-60015
Cinnamoyl thiocarbimide, *see* P-70090
3-Cinnolinecarboxaldehyde, C-70178
4-Cinnolinecarboxaldehyde, C-70179
3,4-Cinnolinedicarboxaldehyde, C-70180
CI Pigment yellow 10, *see* D-70138
Circumanthracene, *see* P-60076
▷C.I. Solvent Red 53, *see* A-60175
▷CI Solvent yellow 34, *see* A-70268
Citrene, *see* I-70087
Citreomontanin, C-60149
Citreopyrone, *see* P-60213
Citreoviral, C-60150
Citreoviridinol A_1, C-70181
Citreoviridinol A_2, *in* C-70181
▷Citridinic acid, *see* A-70055
Citrulline, C-60151
Citrusinol, C-60152
Cladochrome, *in* P-70098
Cladochrome *A*, *in* P-70098
Clavaminic acid, *in* A-70139
Clavamycin C, C-70182
Clavamycin D, C-70183
Clavamycin E, C-70184
Clavularin *A*, C-60153
Clavularin *B*, *in* C-60153
13,15-Cleistanthadien-18-oic acid, A-60317
8,11,13-Cleistanthatrien-19-al, *in* C-60154
8,11,13-Cleistanthatrien-19-oic acid, *in* C-60154
8,11,13-Cleistanthatrien-19-ol, *in* C-60154
Cleistanthoside A, *in* D-60493
Cleistanthoside B, *in* D-60493
Cleomiscosin D, C-70185
ent-3,13-Clerodadien-15-oic acid, *see* K-60013
Clibadic acid, C-70186
Clibadiolide, C-70187
Clitocine, C-60155
▷Clofenamide, *in* C-70049
Clostomicin *A*, *see* T-70198
Clostomicin B_1, *in* L-60028
Clostomicin B_2, *see* T-70195

▷Cloxiquine, *see* C-70075
▷Cloxyquin, *see* C-70075
▷Cnestine, *in* M-70043
Coccellic acid, *see* B-70007
β-Coccinic acid, *see* H-70162
▷Cochliobolin, *see* O-70040
▷Cochliobolin *A*, *see* O-70040
Cochloxanthin, C-60156
Coestinol, *see* T-60110
Coflotriol, *in* O-70035
Coleon *C*, C-60157
Coleonol C, *in* E-70027
Colforsin, *in* E-70027
Colipase, C-70188
Colletodiol, C-60159
Colletoketol, *in* C-60159
Collettiside I, *in* S-60045
Colneleic acid, C-70189
▷Colocynthin, *in* C-70204
Coloradocin, *see* L-60038
Colubrinic acid, C-70190
Columbin, C-60160
Columbinyl glucoside, *in* C-60160
Comaparvin, C-70191
Combretastatin, C-70192
Combretastatin A1, *in* T-60344
Combretastatin B1, *in* T-60344
Commisterone, *in* H-60063
Confertin, *in* D-60211
Confertoside, *in* C-70193
Confusaridin, *in* H-60066
Confusarin, *in* P-60050
α-Conidendryl alcohol, *in* I-60139
▷Conocandin, C-70194
Conocarpan, C-60161
Conotoxin G1, C-70195
Conphysodalic acid, C-60162
Constictic acid, C-60163
Conyzic acid, *see* S-70077
Cookson's diketone, *see* P-70022
Cooperin, C-70196
α-Copaene, C-70197
α-Copaen-8-ol, *in* C-70197
α-Copaen-11-ol, *in* C-70197
15-Copaenol, *in* C-70197
Copazoline-2,4(1*H*,3*H*)-dione, *see* P-60237
Coptiside I, *in* D-70319
Coralloidin C, *in* E-60066
Coralloidin D, *in* E-60063
Coralloidin E, *in* E-60065
Coralloidolide *A*, C-60164
Coralloidolide *B*, C-60165
Corchorusin A, *in* O-60038
Corchorusin B, *in* S-60001
Corchorusin C, *in* O-60035
Cordatin, C-60166
Coriandrin, C-70198
(−)-Coriolic acid, *in* H-70196
(+)-Coriolic acid, *in* H-70196
▷Coriolin, C-70199
Coriolin B, *in* C-70199
Coriolin C, *in* C-70199
Cornubert's ketone, *see* S-70063
Cornudentanone, C-60167
▷Coroglaucigenin, C-60168
(−)-Coronamic acid, *in* A-60151
(+)-Coronamic acid, *in* A-60151
(±)-Coronamic acid, *in* A-60151
[4.5]Coronane, *see* P-70020
7,13-Corymbidienolide, C-60169
Corymbivillosol, C-60170
α-Corymbolol, *in* E-60067
β-Corymbolol, *in* E-60067
p-Coumaraldehyde, *see* H-60213
Coumarin-4-acetic acid, *see* O-60057
CP 54883, *see* A-70222
Cratystyolide, *in* T-70257
Crepiside A, *in* Z-60001
Crepiside B, *in* Z-60001
m-Cresol-4-aldehyde, *see* H-70161
m-Cresol-6-aldehyde, *see* H-70160
m-Cresol-4-carboxylic acid, *see* H-70164
m-Cresol-6-carboxylic acid, *see* H-70163
2,4-Cresotaldehyde, *see* H-70160
4,2-Cresotaldehyde, *see* H-70161
2,4-Cresotic acid, *see* H-70163
4,2-Cresotic acid, *see* H-70164

Croconate blue, *in* T-70321
Crombeone, C-70200
Crotalarin, C-60171
Crotarin, C-60172
2-Crotonyloxymethyl-4,5,6-trihydroxy-2-cyclo-
　hexenone, *see* T-60338
Croweacic acid, *in* H-60177
12-Crown-3, *see* T-70304
Crustecdysone, *in* H-60063
Cryptand 22, *see* T-70145
▷Cryptand 222, *see* H-60073
Cryptand 2.2.1, *see* P-60060
Cryptanol, *see* A-60006
Cryptolepine, *in* I-70016
Cryptomerin A, *in* H-60082
Cryptomerin B, *in* H-60082
Cryptoresinol, C-70201
Cryptotanshinone, C-60173
Crystal violet lactone, C-70202
CS 280, *see* S-60029
Cubanecarboxylic acid, C-60174
▷Cucurbitacin *D*, C-70203
▷Cucurbitacin *E*, C-70204
▷Cucurbitacin I, *in* C-70204
Cucurbitacin L, *in* C-70204
Cucurbitacin R, *in* C-70203
5,24-Cucurbitadiene-3,11,26-triol, C-60175
Cudraflavone *A*, C-70205
Cudraisoflavone *A*, C-70206
▷Cupferron, H-70208
Curcudiol, C-70207
Curculathyrane *A*, C-60176
Curculathyrane *B*, C-60177
Curcumene, *see* M-60112
Curcuphenol, C-70208
Curramycin B, *in* C-60178
Curromycin *A*, C-60178
1-Cyanoacenaphthene, *in* A-70014
9-Cyanoacridine, *in* A-70061
2-Cyano-1,4-benzodioxan, *in* B-60024
1-Cyano-2-benzoyl-1,2-dihydroisoquinoline, *see*
　B-70073
1-Cyanobicyclo[1.1.0]butane, *in* B-70094
2-Cyanobicyclo[2.2.1]hepta-2,5-diene, *in*
　B-60084
4-Cyanocyclohexanethiol, *in* M-60020
Cyanocyclooctatetraene, *in* C-70239
2-Cyanocyclopentanone, *in* O-70084
5-Cyano-7-deazaguanosine, *see* A-70132
2-Cyano-1,2-dihydrocyclobuta[*a*]naphthalene,
　in D-70177
1-Cyano-4,5-dimethoxy-2-nitrobenzene, *in*
　D-60360
4-Cyano-2,2-dimethylbutyraldehyde, *in*
　D-60434
5-Cyano-3,3-dimethyl-1-nitro-1-pentene, *in*
　D-60429
▷1-Cyano-3,5-dinitrobenzene, *in* D-60458
2-Cyano-1,3-dinitrobenzene, *in* D-60457
1-Cyano-1,1-diphenylethane, *in* D-70493
2-Cyanoethanethioamide, C-70209
▷Cyanoethylene, *in* P-70121
▷*N*-(2-Cyanoethyl)glycine, *in* C-60016
2-Cyano-3-fluoroaniline, *in* A-70143
3-Cyano-4-fluoroaniline, *in* A-70147
4-Cyano-2-fluoroaniline, *in* A-70146
4-Cyano-3-fluorobenzenethiol, *in* F-70029
2-Cyano-1-fluoro-3-iodobenzene, *in* F-70020
2-Cyanofuro[2,3-*b*]pyridine, *in* F-60098
2-Cyanofuro[2,3-*c*]pyridine, *in* F-60102
2-Cyanofuro[3,2-*b*]pyridine, *in* F-60100
2-Cyanofuro[3,2-*c*]pyridine, *in* F-60104
3-Cyanofuro[2,3-*b*]pyridine, *in* F-60099
3-Cyanofuro[2,3-*c*]pyridine, *in* F-60103
3-Cyanofuro[3,2-*b*]pyridine, *in* F-60101
3-Cyanofuro[3,2-*c*]pyridine, *in* F-60105
3-Cyano-2-hydroxy-5-nitrobenzoic acid, *in*
　H-70188
2-Cyano-3-hydroxyquinoline, *in* H-60228
3-Cyano-2-hydroxy-*N*,*N*,*N*-trimethyl-
　propanaminium, *in* C-70024
4(5)-Cyanoimidazole, *in* I-60004
2-Cyano-3-iodothiophene, *in* I-60075
2-Cyano-4-iodothiophene, *in* I-60076
2-Cyano-5-iodothiophene, *in* I-60078
3-Cyano-2-iodothiophene, *in* I-60074
3-Cyano-4-iodothiophene, *in* I-60077

4-Cyano-2-iodothiophene, *in* I-60079
5-Cyanoisoxazole, *in* I-60142
1-Cyano-3-methylbicyclo[1.1.0]butane, *in*
　M-70053
4-(Cyanomethyl)coumarin, *in* O-60057
Cyano methyl disulfide, *see* M-60032
▷3-Cyanomethyl-1*H*-indole, *in* I-70013
1-(Cyanomethyl)phenanthrene, *in* P-70046
9-Cyanomethylphenanthrene, *in* P-70047
3-Cyano-10-methylphenothiazine, *in* P-70057
3-Cyano-2-methyl-2-propenoic acid, *in* M-70103
N-(4-Cyano-1-methyl-1*H*-pyrazol-5-yl)-
　acetamide, *in* A-70192
3-Cyano-2-naphthoic acid, *in* N-60003
6-Cyano-2-pentene, *in* H-70024
2-Cyanophenol, *in* H-70108
2-Cyano-10*H*-phenothiazine, *in* P-70056
▷3-Cyano-10*H*-phenothiazine, *in* P-70057
Cyano phenyl disulfide, *see* B-60013
4-Cyano-1-phenyl-1,2,3-triazole, *in* P-60138
4-Cyano-2-phenyl-1,2,3-triazole, *in* P-60140
▷1-Cyano-2-propen-1-ol, *in* P-70119
1-Cyano-2,3,4,5-tetrafluorobenzene, *in* T-70032
3-Cyano-1,2,4,5-tetrafluorobenzene, *in* T-70034
2-Cyanotetrahydrothiophene, *in* T-70099
2-Cyanothioacetamide, *see* C-70209
4-Cyano-1,2,3-triazole, *in* T-60239
1-Cyano-2,4,5-trifluorobenzene, *in* T-70240
▷2-Cyano-3,3,3-trifluoropropene, *in* T-60295
1-Cyano-2,3,5-trimethoxybenzene, *in* T-70250
2-Cyanotriphenylene, *in* T-60384
Cyanoviridin *RR*, C-70210
▷Cyanuric acid, C-60179
▷Cyclic (L-asparaginyl-4-hydroxy-L-prolyl-(*R*)-
　4,5-dihydroxy-L-isoleucyl-6-hydroxy-2-
　mercapto-L-tryptophylglycyl-L-
　isoleucylglycyl-L-cysteinyl) cyclic (4→8)-
　sulfide (*R*)-*S*-oxide, *in* A-60084
Cyclic(*N*-methyl-L-alanyl-L-tyrosyl-D-
　tryptophyl-L-lysyl-L-valyl-L-phenylalanyl),
　C-60180
Cyclo(alanylaspartyl), *see* M-60064
Cycloanticopalic acid, C-60181
3,16,24,25,30-Cycloartanepentol, C-60182
Cycloaspeptide *A*, C-70211
Cycloaspeptide B, *in* C-70211
Cycloaspeptide C, *in* C-70211
▷4,5-Cyclobuta-1,2-benzoquinone, *see* B-70110
1,3-Cyclobutadiene, C-70212
1,1-Cyclobutanedimethanethiol, C-70213
1,2-Cyclobutanedimethanethiol, C-70214
1,2-Cyclobutanedione, C-60183
1,1-Cyclobutanedithiol, C-70215
1,3-Cyclobutanedithiol, C-60184
1,1′-(1,2-Cyclobutanediylidenedimethylidyne)-
　bisbenzene, *see* D-60083
Cyclobutylidenecyclobutane, *see* B-70095
(Cyclobutylidenemethyl)benzene, *see* B-70081
Cyclobutylidenephenylmethane, *see* B-70081
24,26-Cyclo-5,22-cholestadien-3β-ol, *see*
　P-60004
(24*S*,25*S*)-24,26-Cyclo-5α-cholest-22*E*-en-3β-
　ol, *in* P-60004
Cyclocratystyolide, C-70216
2,7-Cyclodecadien-1-one, C-60185
3,7-Cyclodecadien-1-one, C-60186
Cyclodeca[1,2,3-*de*:6,7,8-*d′e′*]dinaphthalene,
　C-60187
Cyclo[*d.e.e.e.d.e.d.e.e*]decakisbenzene, *see*
　D-60396
Cyclodehydromyoporone *A*, C-60188
Cyclodehydromyoporone B, C-60189
1,4,7,10-Cyclododecatetraene, C-70217
1,5,9-Cyclododecatrien-3-yne, C-60190
1-Cyclododecenecarboxaldehyde, C-60191
Cyclododecyne, C-60192
Cycloeuphordenol, C-70218
Cyclofoetigenin B, *in* C-60182
β-Cyclohallerin, *in* C-60193
α-Cyclohallerin, C-60193
Cyclohepta[1,2-*b*:1,7-*b′*]bis[1,4]benzoxazine,
　C-70219
Cyclohepta[*def*]carbazole, C-70220

Cyclohepta[*def*]carbazolium tetrafluoroborate, *in* C-70220
Cyclohept[*fg*]acenaphthylene-5,6-dione, C-60194
Cyclohept[*fg*]acenaphthylene-5,8-dione, C-60195
2,6-Cycloheptadien-4-yn-1-one, C-60196
Cyclohepta[*ef*]heptalene, C-60197
Cyclohepta[*de*]naphthalene, C-60198
peri-Cycloheptanaphthalene, *see* C-60198
Cyclohepta[*a*]phenalene, C-60199
5*H*-Cyclohepta[4,5]pyrrolo[2,3-*b*]indole, C-60200
Cyclohepta[*b*]pyrrol-2(1*H*)-one, C-60201
1,3,5-Cycloheptatriene-1,6-dicarboxylic acid, C-60202
2,4,6-Cycloheptatriene-1-thione, C-60203
2,4,6-Cycloheptatrien-1-one, C-60204
3-(1,3,6-Cycloheptatrien-1-yl-2,4,6-cyclo-heptatrien-1-ylidenemethyl)-1,3,5-cyclo-heptatriene, C-60205
2,4,6-Cycloheptatrienyl cyclopropyl ketone, *see* C-60206
2,4,6-Cycloheptatrien-1-ylcyclo-propylmethanone, C-60206
4-(2,4,6-Cycloheptatrien-1-ylidene)-bicyclo[5.4.1]dodeca-2,5,7,9,11-pentaene, C-60207
5-(2,4,6-Cycloheptatrien-1-ylidene)-3,6-cyclo-heptadiene-1,2-dione, *see* H-60022
2-(2,4,6-Cycloheptatrien-1-ylidene)-1,3-dithiane, C-70221
5-(2,4,6-Cycloheptatrien-1-ylideneethylidene)-3,6-cycloheptadiene-1,2-dione, C-70222
Cycloheptatrienylene(tetraphenylcyclo-pentadenylidene)ethylene, C-60208
2,3-Cycloheptenonaphthalene, *see* T-70047
1-(1-Cyclohepten-1-yl)ethanone, *see* A-60026
▷2,5-Cyclohexadiene-1,4-dione, B-70044
1,1′,1″,1‴-(2,5-Cyclohexadiene-1,4-diylidenedimethanetetrayl)-tetrakisbenzene, *see* B-60157
1,3-Cyclohexadien-5-yne, *see* B-70087
1*H*-Cyclohexa[4,5,6-*de*]methano[10]annulen-1-one, *see* M-70042
1,1-Cyclohexanedimethanethiol, C-70223
1,2-Cyclohexanedimethanethiol, C-70224
▷1,1-Cyclohexanedithiol, C-60209
1,2-Cyclohexanedithiol, C-70225
Cyclohexanehexone, C-60210
1,3,5-Cyclohexanetricarboxaldehyde, C-70226
Cyclohexanol-2-carboxylic acid, *see* H-70119
Cyclohexanone isoxime, *see* O-70052
Cyclohexano[*b*]-2-oxazolidone, *see* H-70052
3-Cyclohexene-1-carboxaldehyde, C-70227
1,2-Cyclohexenoacenaphthylene, *see* T-70061
3-Cyclohexen-1-one, C-60211
Cyclohexeno-1,2,3-thiadiazole, *see* T-70042
1-(1-Cyclohexen-1-yl)ethanone, *see* A-70038
3-Cyclohexenyl methyl ketone, *see* A-70039
α-Cyclohexylcyclohexanemethanol, *see* D-70150
1-Cyclohexylcyclohexene, *see* D-60009
3-Cyclohexylcyclohexene, *see* D-60008
4-Cyclohexylcyclohexene, *see* D-60007
Cyclohexylglycollic acid, *see* C-60212
2-Cyclohexyl-2-hydroxyacetic acid, C-60212
α-Cyclohexyl-α-hydroxybenzeneacetic acid (1,4,5,6-tetrahydro-1-methyl-2-pyrimidinyl)-methyl ester, *see* O-60102
▷*N*-Cyclohexylhydroxylamine, C-70228
(Cyclohexylidenemethyl)benzene, *see* B-70082
Cyclohexylidenephenylmethane, *see* B-70082
2-Cyclohexyl-2-methoxyacetic acid, *in* C-60212
▷Cyclohexyl phenyl ketone, *see* B-70071
▷Cyclohexylphenylmethanone, *see* B-70071
Cyclo(hydroxyprolylhydroxyprolyl), *see* O-60013
Cyclo(hydroxyprolylleucyl), C-60213
Cyclo(hydroxyprolylprolyl), *see* O-60017
3,12-Cycloiceane, *see* P-70019
Cyclo(isoleucylisoleucyl), *see* B-70153
Cyclo(leucylalanyl), *see* M-70118
Cyclo(leucylleucyl), *see* B-60171
Cyclolongipesin, C-60214
Cyclolycoserone, C-70229

3*H*-Cyclonona[*def*]biphenylene, C-60215
3*H*-Cyclonona[*def*]biphenylen-3-one, C-70230
Cyclonona[*c*]furan, *see* H-60047
1,5-Cyclooctadiene monoxide, *see* O-70058
Cycloocta[1,2-*b*:5,6-*b′*]dinaphthalene, C-60216
Cycloocta[2,1-*b*:3,4-*b′*]di[1,8]naphthyridine, C-70231
Cycloocta[2,1-*b*:3,4-*b′*]diquinoline, C-70232
Cycloocta[*def*]fluorene, C-70233
Cycloocta[*a*]naphthalene, C-70234
Cycloocta[*b*]naphthalene, C-70235
1*H*-Cyclooctapyrazole, C-70236
1,3,5,7-Cyclooctatetraene-1-acetic acid, C-70237
Cyclooctatetraenecarboxaldehyde, C-70238
Cyclooctatetraenecarboxylic acid, C-70239
1,3,5-Cyclooctatrien-7-yne, C-70240
1-Cyclooctene-1-acetic acid, C-70241
4-Cyclooctene-1-acetic acid, C-70242
1-Cyclooctenecarboxaldehyde, C-60217
1-(1-Cyclooctenyl)ethanone, *see* A-60027
(1-Cycloocten-1-yl) methyl ketone, *see* A-60027
Cycloorbicoside A, *in* C-60218
Cycloorbigenin, C-60218
▷Cyclopentadecanone, C-60219
Cyclopentadecyne, C-60220
1,1′,1″,1‴-(1,3-Cyclopentadiene-1,2,3,4-tetra-yl)tetrakisbenzene, *see* T-70149
5-(2,4-Cyclopentadien-1-ylidene)-1,3-cyclo-pentadiene, *see* F-60080
2-(2,4-Cyclopentadien-1-ylidene)-1,3-dithiolane, *in* C-60221
2-(2,4-Cyclopentadienylidene)-1,3-dithiole, C-60221
5*H*-Cyclopenta[2,1-*c*:3,4-*c′*]dipyridin-5-one, *see* D-70053
1,1-Cyclopentanedimethanethiol, C-70243
1,1-Cyclopentanedithiol, C-60222
Cyclopentanone-2-carboxylic acid, *see* O-70084
▷4*H*-Cyclopenta[*def*]phenanthrene, C-70244
▷Cyclopenta[*cd*]pyrene, C-60223
Cyclopenta[*b*]thiapyran, C-60224
Cyclopenta[*b*]thiophene, *see* D-70180
Cyclopenta[*c*]thiophene, *see* D-70181
4*H*-Cyclopenta[*c*]thiophen-5(6*H*)-one, C-70245
Cyclopent[*b*]azepine, C-60225
3-Cyclopentene-1-glyoxal, *see* O-60061
3-Cyclopentene-1-glyoxal, *see* O-70086
2,3-Cyclopentenonaphthalene, *see* B-60015
2-Cyclopenten-1-one, C-60226
2,3-Cyclopentenopyridine, *see* D-60275
2,3-Cyclopentenopyridine, D-70258
3,4-Cyclopentenopyridine, *see* D-60276
3,4-Cyclopentenopyridine, D-70259
Cyclopenteno-1,2,3-thiadiazole, *see* D-70179
1-(1-Cyclopenten-1-yl)ethanone, *see* A-60028
1-Cyclopenten-1-ylmethyl ketone, *see* A-60028
1-Cyclopentenyl phenyl ketone, *see* B-70072
1-Cyclopenten-1-ylphenylmethanone, *see* B-70072
2-Cyclopentylidenecyclopentanone, C-60227
(Cyclopentylidenemethyl)benzene, *see* B-70083
Cyclopentylidenephenylmethane, *see* B-70083
2-Cyclopentylpropane, I-70092
[2.2.2](1,2,3)Cyclophane, C-60228
[2.2.2](1,2,3)Cyclophane-1,9-diene, C-60229
13²,17³-Cyclopheophorbide enol, C-60230
Cyclo(prolylglycine), *see* H-70073
Cyclo(prolylisoleucyl), *see* H-60052
Cyclo(prolylprolyl), *see* O-60016
1*H*-Cyclopropabenzene, C-70246
5*H*-Cyclopropa[*f*][2]benzothiophene, C-70247
1,2-Cyclopropanedicarboxylic acid, C-70248
1,1-Cyclopropanedimethanethiol, C-70249
Cyclopropaneglyoxylic acid, *see* C-60231
Cyclopropanehexacarboxylic acid, C-70250
9,9′,9″-(1,2,3-Cyclopropanetriylidene)-tris-9*H*-fluorene, *see* T-70322
9,9′,9″-(1,2,3-Cyclopropanetriylidene)-tris-9*H*-thioxanthene, *see* T-70326
5*H*-Cycloprop[*f*]isobenzofuran, C-70251
Cyclopropylacetylene, *see* E-70045
4-Cyclopropyl-1-buten-3-yne, *see* B-70293
1,1′-Cyclopropylidenebisbenzene, *see* D-60478

1,1′-Cyclopropylidenebisethanone, *see* D-70032
2-Cyclopropyl-2-oxoacetic acid, C-60231
Cyclopropyltyrosine, *see* A-60197
1-Cyclopropyl-2-vinylacetylene, *see* B-70293
Cyclo (ProtProt), *see* O-70014
Cyclopterospermol, C-60232
▷Cyclotene, *see* H-60174
1,8-Cyclotetradecadiene, C-70252
Cyclotetradecane, C-70253
Cyclothreonine, *see* A-60221
Cyclotriveratrylene, C-70254
2-Cycloundecen-1-one, C-70255
1-Cycloundecen-3-yne, C-70256
Cymbodiacetal, C-60233
m-Cymene-2,5-diol, *see* I-60123
m-Cymene-5,6-diol, *see* I-60126
o-Cymene-4,6-diol, *see* I-60127
p-Cymene-2,3-diol, *see* I-60124
p-Cymene-2,6-diol, *see* I-60125
p-Cymene-3,5-diol, *see* I-60122
p-Cymene-2,5,7-triol, *see* D-70309
Cymopyrocatechol, *see* I-60124
Cymorcin, *see* I-60125
Cynajapogenin *A*, C-70257
Cynene, *see* I-70087
Cynomel, *in* T-60350
Cyparenoic acid, C-70258
Cypenamine, *see* P-60086
Cyperenal, *in* C-70259
Cyperenol, C-70259
Cyperenyl acetate, *in* C-70259
Cystofuranoquinol, C-70260
Cystoketal, C-60234
Cytobin, *in* T-60350
Cytochalasin *O*, C-60235
Cytochalasin *P*, C-60236
Cytochalasin N, *in* C-60235
Cytomel, *in* T-60350
Cytomine, *in* T-60350
▷Dabco, *see* D-70051
Dactylarin, *in* A-60080
Dactylariol, *in* A-60080
Daidzein, *see* D-60340
Daidzin, *in* D-60340
Daizeol, *see* D-60340
Dalzin, *see* D-60026
24-Dammarene-3,7,18,20,27-pentol, D-70001
25-Dammarene-3,12,17,20,24-pentol, D-60001
24-Dammarene-3,6,20,27-tetrol, D-70002
24-Dammarene-3,7,20,27-tetrol, D-70003
Damnacanthol, *in* D-70306
Daniellic acid, *in* L-60011
Danshenspiroketallactone, D-70004
Danshexinkun *A*, D-60002
Daphnodorin *D*, D-70005
Daphnoretin, D-70006
Daphnorin, *in* D-70006
▷Daricon, *in* O-60102
Datiscacin, *in* C-70204
Datiscoside, *in* C-70203
Datiscoside B, *in* C-70203
Datiscoside C, *in* C-70203
Datiscoside D, *in* C-70203
Datiscoside E, *in* C-70203
Datiscoside F, *in* C-70203
Datiscoside G, *in* C-70203
Datiscoside H, *in* C-70203
Daturilin, D-70007
Daturilinol, D-70007
4,8-Daucadiene, *see* D-70008
Daucene, D-70008
8-Daucene-2,4,6,10-tetrol, D-70009
8-Daucene-3,6,14-triol, D-60004
8-Daucene-4,6,10-triol, *see* C-60021
8(14)-Daucene-4,6,9-triol, D-60005
DDED, *in* E-60049
Deacetoxybrachycarpone, *in* B-60195
7-Deacetoxy-7-oxogedunin, *in* G-70006
9-Deacetoxyxenicin, *in* X-70004
7-Deacetyljadiradione, *in* A-70281
Deacetyleupaserrin, *in* D-70302
13-Deacetyl-11(9)-havannachlorohydrin, *in* H-70004
7-Deacetyl-17β-hydroxyazadiradione, *in* A-70281
Deacetylsessein, *in* S-60027

Deacetylteupyrenone, *in* T-60184
Deacetylviguiestin, *in* T-70003
7-Deaza-2'-deoxyguanosine, *in* A-60255
5-Deazaflavin, *see* P-60245
7-Deazaguanine, *see* A-60255
7-Deazaguanosine, *in* A-60255
3-Deazahypoxanthine, *see* H-60157
10-Deazariboflavin, *in* P-60245
3-Deazauracil, *see* H-60224
Debneyol, D-70010
1,4,7,10,13,16,19,22,25,28-Decaazacyclo-
    triacontane, D-70011
2,2,3,3,4,4,5,5,6,6-Decafluoropiperidine,
    D-60006
1,1',2,2',3,4,5,5',6,6'-Decahydrobiphenyl,
    D-60007
1,1',2,3,4,4',5,5',6,6'-Decahydrobiphenyl,
    D-60008
1,2,3,3',4,4',5,5',6,6'-Decahydrobiphenyl,
    D-60009
Decahydro-1,6-bis(methylene)-4-(1-methyl-
    ethyl)naphthalene, *see* C-60001
1,2,3,4,5,6,7,8,8a,9a-Decahydrocarbazole,
    D-70012
Decahydro-4,6-dihydroxy-3,5a-dimethyl-9-
    methylenenaphtho[1,2-*b*]furan-2(3*H*)-one,
    see A-60301
Decahydro-2,5:3,4-dimethanocyclopropa[*de*]-
    naphthalene, *see* P-70019
2,3,3a,4,5,8,9,12,13,15a-Decahydro-6,14-di-
    methyl-3-methylene-2-oxocyclotetradeca[*b*]-
    furan-10-carboxylic acid, *see* A-60275
Decahydro-4a,5-dimethyl-3-(1-methylethenyl)-
    1,2,7-naphthalenetriol, *see* E-70030
Decahydro-1,3,4,6-ethanediylidenecyclo-
    penta[*cd,gh*]-pentalene, *see* H-70012
Decahydro-5-hydroxy-4a,8-dimethyl-3-methyl-
    eneazuleno[6,5-*b*]furan-2(3*H*)-one, *see*
    D-60211
Decahydro-2,7-imino-3,6-methano-1,8-
    naphthyridine, *see* T-70206
Decahydro-5-(1*H*-indol-3-ylmethyl)-1,4a-di-
    methyl-6-methylene-1-(4-methyl-3-pentenyl)-
    2-naphthalenol, *see* E-60005
Decahydro-4-isopropyl-1,6-di-
    methylenenaphthalene, *see* C-60001
Decahydro-3,6-methanonaphthalen-2,7-imine,
    see A-70282
Decahydroquinazoline, D-70013
1,2,3,4,4a,5,6,6a,11b,13b-Decahydro-4,4,6a,9-
    tetramethyl-13*H*-benzo[*a*]-furo[2,3,4-*mn*]-
    xanthen-11-ol, *see* S-60029
Decahydro-2,5,5,8a-tetramethyl-1-methylene-2-
    naphthalenol, *see* D-60520
Decahydro-2,5,5,8a-tetramethyl-1-(4,8,12-
    trimethyl-3,7,11-tridecatrienyl)-2,6-
    naphthalenediol, *see* P-70107
5,5a,6,6a,7,12,12a,13,13a,14-Decahydro-
    5,14:6,13:7,12-trimethanopentacene,
    D-60010
Decahydro-5,5,8a-trimethyl-2-naphthalenol,
    D-70014
▷ γ-Decalactone, *see* H-60078
Decamethylcyclohexanone, D-60011
1,1,2,2,3,3,4,4,5,5-Decamethyl-6-methyl-
    enecyclohexane, D-60012
Decamethyl[5]radialene, *see* P-60053
5,6-Decanediol, D-60013
▷ 4-Decanolide, *see* H-60078
5-Decanone, D-60014
1,3,5,7,9-Decapentaene, D-70015
3-Decenal, D-70016
5-Decene, D-70017
5-Decene-2,8-diyne, D-60015
8-Decene-4,6-diynoic acid, D-70018
2-Decen-4-ol, D-60016
5-Decen-1-ol, D-70019
4-Decen-1-yne, D-70020
▷ Dechlorane, *see* D-70530
2-Dechloro-8-*O*-methylthiomelin, *in* T-70181
4-Dechloro-8-*O*-methylthiomelin, *in* T-70181
Dechloromicroline, *in* M-60135
Dechloromikrolin, *in* M-60135
Decumbeside C, D-60017

Decumbeside A, *in* G-60002
Decumbeside B, *in* G-60002
Decumbeside D, D-60017
Deferriviridomycin *A*, *see* H-60197
Dehydroabietinol, *see* A-60007
Dehydroabietinol acetate, *in* A-60007
2-Dehydroarcangelisinol, D-70021
Dehydroarsanin, *see* A-60302
Dehydro-*p*-asebotin, *in* H-60220
1,2-Dehydrobenzene, *see* B-70087
Dehydrochloroprepacifenol, D-60018
Dehydrocollinusin, *in* D-60493
Dehydrocurcuphenol, *in* C-70208
22-Dehydro-24,26-cyclocholesterol, *see* P-60004
10',11'-Dehydrocyclolycoserone, *in* C-70229
1,2-Dehydrocyclooctatetraene, *see* C-70240
2,3-Dehydro-11α,13-dihydroconfertin, *in*
    D-60211
Dehydrodihydrocostus lactone, *see* M-60142
2,3-Dehydro-1,2-dihydro-1,1-di-
    methylnaphthalene, *see* D-60168
11,13-Dehydroeriolin, *in* E-60036
Dehydroexobrevicomin, *see* E-60052
Dehydrohomocamphenilone, *see* D-60412
Dehydroisoerivanin, *in* I-70066
Dehydrojuncusol, *see* D-70294
Dehydromyoporone, D-60019
5-Dehydrooriciopsin, *in* O-70045
Dehydroosthol, D-60020
cis-Dehydroosthol, *in* D-60020
8,17*H*-7,8-Dehydropinifolic acid, *in* L-60007
Dehydrorabelomycin, *see* T-70270
6,7-Dehydroroyleanone, *in* R-60015
3-Dehydroshikimic acid, *in* D-70328
5-Dehydroshikimic acid (obsol.), *in* D-70328
Dehydrotectol, *see* T-60010
2,3-Dehydroteucrin E, *in* T-60180
Dehydrotrichostin, *in* T-70225
7'-Dehydrozearalenone, *in* Z-60002
3'-Deiodothyroxine, *see* T-60350
Deltic acid, *see* D-70290
30-Demethoxycurromycin *A*, *in* C-60178
Demethoxyegonol 2-methylbutanoate, *in*
    E-70001
Demethylbellidifolin, *in* T-70123
Demethyldaphnoretin, *in* D-70006
2-Demethyl-3''-*O*-demethyl-12-
    deoxyerythromycin, *in* N-70076
2-Demethyl-12-deoxyerythromycin, *in* N-70076
3''-*O*-Demethyl-12-deoxyerythromycin, *in*
    N-70076
2-Demethylerythromycin, *see* N-70076
2-Demethylerythromycin *B*, *in* N-70076
2-Demethylerythromycin *C*, *in* N-70076
2-Demethylerythromycin *D*, *in* N-70076
Demethylfruticulin *A*, *in* F-60076
Demethylpinusolide, *in* P-60156
Demethylpterocarpin, *see* M-70001
Demethylzeylasteral, *in* D-70027
Demethylzeylasterone, *see* D-70027
Demser, *in* A-60220
Dengibsin, *in* T-70261
Dengibsinin, *in* T-70109
De-*O*-methyldiaporthin, *in* D-70050
6-Deoxocastasterone, *in* T-70116
11-Deoxocucurbitacin I, *in* C-70204
7'-Deoxo-2'-deoxy-2',3'-epoxy-7'-(1-
    hydroxyethyl)verrucarin *A*, *see* R-70008
6-Deoxodihydrohomodolichosterone, *in* S-70075
6-Deoxodolichosterone, *in* D-70544
6-Deoxo-2α-hydroxytaxodione, *in* D-70282
Deoxyaustrocortilutein, *in* A-60321
Deoxyaustrocortirubin, *in* A-60327
2'-Deoxy-5-azacytidine, D-70022
Deoxybouvardin, *see* R-60002
2-Deoxycastasterone, *in* T-70116
6-Deoxycatalpol, *in* C-70026
2-Deoxychamaedroxide, *in* C-60032
3',4'-Deoxy-4'-chloropsorospermin-3'-ol, *in*
    D-60021
Deoxychondrine, *see* T-70182
1-Deoxy-1-[(1,3-dicarboxypropyl)amino]-
    mannitol, *see* M-70009
12-Deoxy-10,11-dihydro-4'-hydroxypicromycin,
    see K-60009
▷ 2'-Deoxy-5-ethyluridine, *in* E-60054

2'-Deoxy-5-ethynylcytidine, *in* A-60153
2'-Deoxy-5-ethynyluridine, *in* E-60060
Deoxyhavannahine, *in* H-60008
6-Deoxyilludin M, *in* I-60001
6-Deoxyilludin S, *in* I-60002
2'-Deoxyinosine, D-70023
3'-Deoxyinosine, D-70024
10-Deoxymelittoside, *in* M-70019
2'-Deoxymugeneic acid, *in* M-60148
5-Deoxymyricanone, *in* M-70162
8-Deoxyophiobolin J, *in* O-70041
9-(2-Deoxy-β-D-*erythro*-pentofuranosyl)-1,9-
    dihydro-6*H*-purin-6-one, *see* D-70023
9-(3-Deoxy-β-D-*erythro*-pentofuranosyl)-1,9-
    dihydro-6*H*-purin-6-one, *see* D-70024
3',4'-Deoxypsorospermin, D-60021
3',4'-Deoxypsorospermin-3',4'-diol, *in* D-60021
5-Deoxypulchelloside I, *in* P-70139
Deoxysepiapterin, *in* S-70035
Deoxytifruticin, *in* T-70199
Deoxyuvidin B, *in* D-70295
2'-Deoxy-5-vinyluridine, *in* V-70015
▷ Depigman, *see* B-60074
▷ Dermofongin A, *see* C-70075
Desacetoxylaurenobiolide, *see* G-70010
9-Desacetoxymelcanthin F, *in* M-60014
Desacetylacanthospermal A, *in* A-60012
13-Desacetyleudesmaafraglaucolide, *in* T-70258
13-Desacetyl-1α-hydroxyafraglaucolide, *in*
    A-70070
13-Desacetyl-1β-hydroxyafraglaucolide, *in*
    A-70070
13-Desacetyl-1β-hydroxyisoafraglaucolide, *in*
    I-70060
Desacetyllaurenobiolide, *in* H-70134
15-Desacetyltetraneurin C, *in* T-60158
Desacetyltetraneurin *D* 4-*O*-isobutyrate, *in*
    T-60158
15-Desacetyltetraneurin *C* isobutyrate, *in*
    T-60158
Desacetyltetraneurin *D* 15-*O*-isobutyrate, *in*
    T-60158
Desacylisoelephantopin senecioate, *in* I-60099
Desacylisoelephantopin tiglate, *in* I-60099
Desacylligulatin C, D-70025
Deschlorothricin, *in* C-60137
Desertorin *A*, D-60022
Desertorin B, *in* D-60022
Desertorin C, *in* D-60022
Desmarestene, *see* B-60290
10-Desmethyl-1-methyl-1,3,5(10),11(13)-
    eudesmatetraen-12,8-olide, D-70026
Desmethylzeylasterone, D-70027
3-Desoxoachalensolide, *in* A-70054
Desoxodehydrolaserpitin, *in* D-70009
Desoxyarticulin, *in* A-60307
3-Desoxyjuslimtetrol, *see* R-60012
Dess-Martin periodinane, *see* T-60228
Detrothyronine, *in* T-60350
Dewar furan, *see* O-70059
Dextrothyromine, *in* T-60350
1,3-Diacetin, *see* D-60023
12,13-Diacetoxy-1(10)-aromadendren-2-one, *in*
    D-70283
9α,14-Diacetoxy-1α-benzoyloxy-4β,6β,8β-
    trihydroxydihydro-β-agarofuran, *in* E-60018
ent-6β,17-Diacetoxy-14,15-bisnor-7,11*E*-
    labdadien-13-one, *in* D-60311
3,14-Diacetoxy-2-butanoyloxy-5,8(17)-
    briaradien-18,7-olide, *in* E-70012
4,14-Diacetoxy-2-butanoyloxy-5,8(17)-
    briaradien-18,7-olide, *in* E-70012
5-(3,4-Diacetoxy-1-butynyl)-5'-methyl-2,2'-
    bithiophene, *in* M-70054
9α,14-Diacetoxy-1α,8β-dibenzoyloxy-4β,8β-
    dihydroxydihydro-β-agarofuran, *in* E-60018
3,4-Diacetoxy-3,4-dihydro-2,5(2*H*,5*H*)-
    furandione, *in* T-70005
3α,22-Diacetoxy-7α,15α-dihydroxy-8,24*E*-
    lanostadien-26-oic acid, *in* T-60111
9β,20-Diacetoxy-2β,3α-dihydroxy-1(15),-8(19)-
    trinervitadiene, *in* T-70296

1,2-Diacetoxy-5,6-dimethoxyphenanthrene, *in* T-60115

6,16-Diacetoxy-3,7-dolabelladien-12-ol, *in* D-70541

6α,11β-Diacetoxygedunin, *in* G-70006

9β,15-Diacetoxy-1(10),4,11(13)-germacratrien-12,6α-olide, *in* D-60324

3,7a-Diacetoxyhexahydro-2-oxa-3a,7-dimethyl-1,4-indanedione, *in* H-70055

1β,6α-Diacetoxy-4α-hydroxy-11-eudesmen-12,8α-olide, *in* T-70259

3α,7α-Diacetoxy-15α-hydroxy-8,24E-lanostadien-26-oic acid, *in* T-60327

3α,22ξ-Diacetoxy-15α-hydroxy-7,9(11),24-lanostatrien-26-oic acid, *in* T-70266

3α,22S-Diacetoxy-15α-hydroxy-7,9(11),24E-lanostatrien-26-oic acid, *in* T-70266

15α,22S-Diacetoxy-3α-hydroxy-7,9(11),24E-lanostatrien-26-oic acid, *in* T-70266

15α,22S-Diacetoxy-3β-hydroxy-7,9(11),24E-lanostatrien-26-oic acid, *in* T-70266

3α,22-Diacetoxy-15α-hydroxy-7α-methoxy-8,24E-lanostadien-26-oic acid, *in* T-60111

2α,7α-Diacetoxy-6β-isovaleroyloxy-8,13E-labdadien-15-ol, *in* L-60002

ent-3β,17-Diacetoxy-16β-kauranol, *in* K-60005

ent-3β,19-Diacetoxy-15-kauren-17-oic acid, *in* D-60341

ent-6β,17-Diacetoxy-7,12E,14-labdatriene, *in* L-60004

ent-6β,17-Diacetoxy-7,11E,14-labdatrien-13ξ-ol, *in* L-60005

3α,15α-Diacetoxy-7,9(11),24E-lanostatrien-26-oic acid, *in* D-70314

3β,15α-Diacetoxy-7,9(11),24E-lanostatrien-26-oic acid, *in* D-70314

3β,23-Diacetoxy-20(30)-lupen-29-al, *in* D-70316

3α,22-Diacetoxy-7α-methoxy-8,24E-lanostadien-26-oic acid, *in* T-60328

6α,17α-Diacetoxy-spongian-15,17-oxido-16-spongianone, *in* D-60365

3α,15α-Diacetoxy-23-oxo-7,9(11),24E-lanostatrien-26-oic acid, *in* D-70330

4,14-Diacetoxy-2-propanoyloxy-5,8(17)-briaradien-18,7-olide, *in* E-70012

Diacetoxysuccinic acid, *in* T-70005

Diacetylacetylene, *see* H-70090

1,5-Diacetylanthracene, D-70028

1,6-Diacetylanthracene, D-70029

1,8-Diacetylanthracene, D-70030

9,10-Diacetylanthracene, D-70031

1,1-Diacetylcyclopropane, D-70032

1,2-Diacetyl-1,2-dibenzoylethane, *see* D-60081

1,3-Diacetylglycerol, D-60023

1,4-Diacetylindole, *in* A-60033

1,3-Diacetyl-5-iodobenzene, D-60024

▷Diacetylmethane, *see* P-60059

2,3-Di-*O*-acetyltartaric acid, *in* T-70005

7,7'-Diacetyl-1',8,8'-trihydroxy-6,6'-dimethyl[2,2'-binaphthalene]-1,4-dione, *see* I-70003

2,12-Di-9-acridinyl-7-phenyldibenz[*c,h*]acridine, D-70033

1,2-Di-1-adamantyl-1,2-di-*tert*-butylethane, *see* D-60025

Di-1-adamantyl selenoketone, D-70034

3,4-Di-1-adamantyl-2,2,5,5-tetramethylhexane, D-60025

Diallyl dicarbonate, *in* D-70111

1,6-Diallyl-2,5-dithiobiurea, D-60026

Diallyldithiocarbamidohydrazine, *see* D-60026

Diallyl pyrocarbonate, *in* D-70111

▷2,9-Diaminoacridine, D-60027

▷3,6-Diaminoacridine, D-60028

▷3,9-Diaminoacridine, D-60029

▷2,6-Diaminoacridine (obsol.), *see* D-60029

▷2,8-Diaminoacridine (obsol.), *see* D-60028

▷3,5-Diaminoacridine (obsol.), *see* D-60027

1,4-Diamino-2,3-benzenedithiol, D-60030

2,5-Diamino-1,4-benzenedithiol, D-70035

3,6-Diamino-1,2-benzenedithiol, D-60036

4,6-Diamino-1,3-benzenedithiol, D-70037

1,4-Diamino-2,3,5,6-benzenetetrathiol, D-60031

2,4-Diaminobenzoic acid, D-60032

2,2'-Diaminobiphenyl, D-60033

1,5-Diamino-1,5-dihydrobenzo[1,2-*d*:4,5-*d*']-bistriazole, D-70038

2,6-Diamino-1,5-dihydro-4*H*-imidazo[4,5-*c*]-pyridin-4-one, D-60034

2,7-Diamino-1,5-dihydro-4,6-pteridinedione, *see* D-60046

2,8-Diamino-1,7-dihydro-6*H*-purin-6-one, D-60035

1,4-Diamino-2,3-dimercaptobenzene, *see* D-60030

1,8-Diamino-3,6-dioxaoctane, *see* B-60139

1,2-Diamino-1,2-diphenylethane, *see* D-70480

2,6-Diamino-3-heptenedioic acid, D-60036

3,5-Diamino-1-heptyne, *see* H-60026

3,6-Diamino-1-hexyne, *see* H-60080

2,5-Diamino-3-hydroxypentanoic acid, D-70039

2,8-Diamino-6-hydroxypurine, *see* D-60035

2,4-Diamino-6-hydroxypyrimidine, *see* D-70043

2,8-Diaminohypoxanthine, *see* D-60035

2,8-Diaminoinosine, *see* A-70150

4,5-Diaminoisoquinoline, D-70040

5,8-Diaminoisoquinoline, D-70041

3,4-Diamino-1-methoxyisoquinoline, D-60037

1,2-Diamino-4,5-methylenedioxybenzene, D-60038

2,6-Diamino-4-methyleneheptanedioic acid, D-60039

2,4-Diamino-6-methyl-1,3,5-triazine, D-60040

1,8-Diaminononaphthalene, D-70014

2-[(2,5-Diamino-1-oxopentyl)amino]-ethanesulfonic acid, *see* O-60042

2,4-Diaminopentane, *see* P-70039

▷2,4-Diamino-6-phenyl-1,3,5-triazine, D-60041

2,4-Diaminopteridine, D-60042

4,6-Diaminopteridine, D-60043

4,7-Diaminopteridine, D-60044

6,7-Diaminopteridine, D-60045

2,6-Diamino-4,7(3*H*,8*H*)-pteridinedione, D-60047

2,7-Diamino-4,6(3*H*,5*H*)-pteridinedione, D-60046

2,6-Diamino-4(1*H*)-pyrimidinone, D-70043

2,4-Diaminopyrimido[4,5-*b*]pyrazine, *see* D-60042

2,4-Diaminoquinazoline, D-60048

4,5-Diaminoquinoline, D-70044

2,8-Diamino-9-β-D-ribofuranosyl-9*H*-purin-6(1*H*)-one, *see* A-70150

3,4-Diaminotetrahydrofuran, *see* T-70062

1,4-Diamino-2,3,5,6-tetramercaptobenzene, *see* D-60031

3,6-Diamino-1,2,4,5-tetrazine, D-70045

2,5-Diamino-1,3,4-thiadiazole, D-60049

4,6-Diaminothiorecorcinol, *see* D-70037

6,10-Diamino-2,3,5-trihydroxydecanoic acid, D-70046

3,5-Diamino-2,4,6-trinitrobenzoic acid, D-70047

1,2:8,9-Dianhydrononitol, *see* D-70465

3,5-Dianilino-1,2,4-thiadiazole, D-70048

Di-9-anthracenylmethanone, D-60050

Dianthramide *A*, D-70049

Dianthramide B, *in* D-70049

Di-9-anthrylacetylene, D-60051

Di-9-anthryl ketone, *see* D-60050

Diaporthin, D-70050

Diasterane, *see* T-60265

Diazaazupyrene, *see* D-70052

9,14-Diazabenz[*a,e*]fluoranthene, *see* D-60069

1,12-Diazabenzo[*a*]phenanthrene, *see* Q-70007

3,4-Diazabenzvalene, *see* D-70054

1,6-Diazabicyclo[4.4.0]decane, *see* O-70016

1,6-Diazabicyclo[4.4.0]decane-2,5-dione, *see* H-70070

2,5-Diazabicyclo[4.1.0]heptane, D-70053

1,4-Diazabicyclo[4.3.0]nonane, D-70054

1,5-Diazabicyclo[4.3.0]nonane, *see* H-70067

▷1,4-Diazabicyclo[2.2.2]octane, D-70051

1,5-Diazabicyclo[3.3.0]octane, *see* T-70086

3,7-Diazabicyclo[3.3.0]octane, *see* O-70017

1,5-Diazabicyclo[3.3.0]octane-4,8-dione, *see* T-70087

1,5-Diazabicyclo[5.2.2]undecane, D-60055

4,10-Diaza-15-crown-5, *see* T-60376

4,13-Diaza-18-crown-6, *see* T-60161

1,3-Diazacyclohexane, *see* H-60056

▷9,14-Diazadibenz[*a,e*]acephenanthrylene, *see* D-60074

5,12-Diazadibenz[*a,h*]anthracene, *see* B-70030

9,14-Diazadibenzo[*a,e*]aceanthrylene, *see* D-60069

10*b*,10*c*-Diazadicyclopenta[*ef,kl*]heptalene, D-70052

3,6-Diazafluorenone, D-70053

▷1,4-Diazaindene, *see* P-60249

2,4-Diazaindene, *see* P-70201

2,5-Diazaindene, *see* P-70200

3*a*,6-Diazaindene, *see* P-70197

17-Diaza-10-oxa-9-selenaanthracene, *see* O-70067

1,5-Diaza-9-oxa-10-selenaanthracene, *see* O-70065

1,6-Diaza-10-oxa-9-selenaanthracene, *see* O-70068

1,8-Diaza-10-oxa-9-selenaanthracene, *see* O-70066

9,10-Diazaphenanthrene, *see* B-60020

1,6-Diazaphenoxaselenine, *see* O-70065

1,7-Diazaphenoxaselenine, *see* O-70068

1,8-Diazaphenoxaselenine, *see* O-70067

1,9-Diazaphenoxaselenine, *see* O-70066

1,8-Diazaphenoxathiin, *see* O-60050

4,5-Diazapyrene, *see* N-70007

▷1,3-Diazaspiro[4.5]decane-2,4-dione, D-60056

1,7-Diazatricyclo[7.3.0.0^{3,7}]decane-2,8-dithione, *see* O-70014

3,4-Diazatricyclo[3.1.0.0^{2,6}]hex-3-ene, D-70054

6,13-Diazatricyclo[10.2.2.2^{5,8}]octadeca-5,7,12,14,15,17-hexaene, *see* P-60228

6,14-Diazatricyclo[10.2.2.2^{5,8}]octadeca-5,7,12,14,15,17-hexaene, *see* P-60227

1,3-Diazetidine-2,4-dione, D-60057

1,2-Diazetidin-3-one, D-70055

▷2,2-Diazido-1,3-indanedione, D-70056

▷2,2-Diazido-1*H*-indene-1,3(2*H*)-dione, *see* D-70056

▷2,2-Diazidopropane, D-60058

6-Diazo-2,4-cyclohexadien-1-one, D-70057

9-Diazo-9*H*-cyclopenta[1,2-*b*:4,3-*b*']-dipyridine, D-60059

Diazocyclopropane, D-60060

9-Diazo-1,8-diazafluorene, *see* D-60059

Diazodibenzoylmethane, *see* D-70058

2-Diazo-2,3-dihydro-1,1,3,3-tetramethyl-1*H*-indene, *see* D-60067

2-Diazo-1,3-diphenyl-1,3-propanedione, D-70058

1-(1-Diazoethyl)naphthalene, D-60061

9-Diazo-9*H*-fluorene, D-60062

3-Diazo-2,4(5*H*)-furandione, D-70059

▷2-Diazo-2*H*-imidazole, D-70060

4-Diazo-4*H*-imidazole, D-60063

▷4-Diazo-4*H*-imidazole, D-60061

▷1,2-Diazole, *see* P-70152

▷1-(Diazomethyl)-4-methoxybenzene, D-60064

1-(Diazomethyl)pyrene, D-70062

▷4-Diazo-2-nitro-2,5-cyclohexadien-1-one, *see* D-60065

▷4-Diazo-2-nitrophenol, D-60065

6-Diazo-5-oxo-1,3-cyclohexadiene-1-carboxylic acid, D-70063

o-Diazophenol, *see* D-70057

2-Diazo-1,1,3,3-tetramethylcyclopentane, D-60066

2-Diazo-2,2,6,6-tetramethyl-3,5-heptanedione, D-70064

2-Diazo-1,1,3,3-tetramethylindane, D-60067

3-Diazotetronic acid, *see* D-70059

9-Diazo-9*H*-thioxanthene, D-70065

Dibenz[*e,k*]acephenanthrylene, D-70066

Dibenz[*a,h*]anthracene-5,6:11,12-dioxide, *see* D-70151

Dibenz[*b,g*]azocine-5,7(6*H*,12*H*)-dione, D-60068

3,4:7,8-Dibenz-2,6,9-bisdioxan, *see* D-70505

1,2,3,4-Dibenzcyclohepta-1,3-diene-6-one, *see* D-70187

1,2,3,4-Dibenzcyclohepta-1,3-dien-5-one, *see* D-70188

o,o'-Dibenzene, *see* T-60060

sym-Dibenzfulvalene, *see* B-60109

Dibenzhydrylamine, see T-70150
13H-Dibenzimidazo[2,1-b:1',2'-e][1,3,5]-thiadiazine-13-thione, see T-60191
Dibenz[b,h]indeno[1,2,3-de][1,6]-naphthyridine, D-60069
2H-Dibenz[e,g]isoindole, D-70067
Dibenzobicyclo[2.2.2]octadiene, see D-70207
Dibenzo[a,cd]naphtho[1,2,3-lm]perylene, see B-60037
Dibenzo-54-crown-18, D-70068
Dibenzocyclohepta-1,3-diene, see D-70186
Dibenzo[a,c][1,3]cycloheptadien-5-one, see D-70188
Dibenzo[a,d]cycloheptenylium, D-60071
Dibenzo[jk,uv]dinaphtho[2,1,8,7-defg:2',1',8',7'-opqr]pentacene, D-60072
5H,7H-Dibenzo[b,g][1,5]dithiocin, D-70069
6H,12H-Dibenzo[b,f][1,5]dithiocin, D-70070
4,5:15,16-Dibenzo-3,6,14,17,23,24-hexa-azatricyclo[17.3.1.1^{8,12}]-tetracosa-1(23),4,8(24),-9,11,15,19,21-octaene, see D-60456
2,3,11,12-Dibenzo-1,4,7,10,13,16-hexathia-2,11-cyclooctadecadiene, D-60073
▷Dibenzo[c,f]indeno[1,2,3-ij][2,7]-naphthyridine, D-60074
1,2:9,10-Dibenzo[2.2]metaparacyclophane, D-70071
Dibenzo[a,rst]naphtho[8,1,2-cde]pentaphene, D-60075
1,2:7,8-Dibenzo[2.2]paracyclophane, D-70072
1,2:9,10-Dibenzo[2.2]paracyclophanediene, see D-70072
1,2:5,6-Dibenzopentafulvalene, see B-60109
Dibenzo[a,j]perylene-8,16-dione, D-70073
1,2:7,8-Dibenzoperylene-3,9-quinone, see D-70073
Dibenzo[2,3:10,11]perylo[1,12-bcd]thiophene, D-70074
Dibenzo[f,h]quinoline, D-70075
Dibenzo[a,h]quinolizinium(1+), D-70076
5H,7H-Dibenzo[c,e]selenepin, see D-70189
2,3:6,7-Dibenzosuberane, see D-60217
Dibenzotetraselenofulvalene, D-70077
Dibenzo[cd:c'd'][1,2,4,5]tetrazino[1,6-a:4,3-a']-diindole, D-60076
Dibenzo[b,i]thianthrene-5,7,12,14-tetrone, D-70078
Dibenzo[f,h]thieno[2,3-b]quinoline, D-70079
Dibenzo[b,f]thiepin-3-carboxylic acid, D-60077
Dibenzo[b,f]thiepin-3-methanol, see H-60176
2-Dibenzothiophenethiol, D-60078
4-Dibenzothiophenethiol, D-60079
6H-Dibenzo[b,d]thiopyran-6-one, D-70080
Dibenzo[b,d]thiopyrylium(1+), D-70081
2,3:7,8-Dibenzotricyclo[4.4.1.0^{1,6}]-undeca-4,9-diene, see M-70040
Dibenzo[b,m]triphenodithiazine-5,7,9,14,16,18(8H,17H)-tetrone, D-70082
7H-Dibenzo[c,h]xanthen-7-one, D-60080
1,3-Dibenzoylacetone, see D-60489
2,3-Dibenzoylbutanedioic acid, D-70083
2,3-Dibenzoyl-1,4-diphenyl-1,4-butanedione, see T-60020
2,3-Dibenzoyl-1,4-diphenyl-2-butene-1,4-dione, see T-60021
3,4-Dibenzoyl-2,5-hexanedione, see D-60081
2,3-Dibenzoylsuccinic acid, see D-70083
2,3-Di-O-benzoyltartaric acid, in T-70005
Dibenzyl dicarbonate, in D-70111
Dibenzyl diselenide, D-70077
1,2-Dibenzylidenecyclobutane, D-60083
Dibenzyl pyrocarbonate, in D-70111
1,3-Dibromoadamantane, D-70084
2,3-Dibromoanthracene, D-70085
9,10-Dibromoanthracene, D-60085
ms-Dibromoanthracene, see D-60085
2,3-Dibromobenzofuran, D-60086
5,7-Dibromobenzofuran, D-60087
1,2-Dibromobenzonorbornadiene, see D-60095
1,3-Dibromobenzonorbornadiene, see D-60096
2,6-Dibromo-1,4-benzoquinone, D-60088
α,α'-Dibromobibenzyl, see D-70091
1,4-Dibromobicyclo[2.2.2]octane, D-70086
2,2'-Dibromo-1,1'-binaphthalene, see D-60089

4,4'-Dibromo-1,1'-binaphthalene, see D-60090
2,2'-Dibromo-1,1'-binaphthyl, D-60089
4,4'-Dibromo-1,1'-binaphthyl, D-60090
1,5-Dibromo-3,3-bis(2-bromoethyl)pentane, see T-70124
▷1,4-Dibromo-2-butyne, D-70087
7,8-Dibromo-6-(chloromethylene)-2-methyl-2-octene, D-70088
2,3-Dibromocoumarone, see D-60086
5,7-Dibromocoumarone, see D-60087
1,1-Dibromocycloheptane, D-60091
2,6-Dibromo-2,5-cyclohexadiene-1,4-dione, see D-60088
2,2-Dibromocyclopropanecarboxaldehyde, D-60092
2,2-Dibromocyclopropanecarboxylic acid, D-60093
1,2-Dibromocyclopropene, D-60094
α,β-Dibromodibenzyl, see D-70091
9,13-Dibromo-6,12;7,10-diepoxypentadec-3-en-1-yne, see I-70067
1,2-Dibromo-1,4-dihydro-1,4-methanonaphthalene, D-60095
1,3-Dibromo-1,4-dihydro-1,4-methanonaphthalene, D-60096
2,6-Dibromo-4,5-dihydroxy-2-cyclohexen-1-one, D-60097
3,5-Dibromo-1,6-dihydroxy-4-oxo-2-cyclohexene-1-acetonitrile, D-60098
1,1-Dibromo-3,3-dimethoxy-2-methyl-1-butene, in D-70094
1,3-Dibromo-2,3-dimethylbutane, D-70089
1,5-Dibromo-3,3-dimethylpentane, D-70090
2,4-Dibromo-2,4-dimethyl-3-pentanone, D-60099
1,2-Dibromo-1,2-diphenylethane, D-70091
1,1'-(1,2-Dibromo-1,2-ethanediyl)bisbenzene, see D-70091
4,21-Dibromo-3-ethyl-2,19-dioxabicyclo[16.3.1]docosa-6,9,18(22),-21-tetraen-12-yn-20-one, in E-70042
4,5-Dibromo-2-formyl-1H-imidazole, see D-60101
1,20-Dibromoicosane, D-70092
4,5-Dibromo-1H-imidazole, D-60100
2,4(5)-Dibromo-1H-imidazole, D-60093
4,5-Dibromo-1H-imidazole-2-carboxaldehyde, D-60101
4,5-Dibromo-1H-imidazole-2-carboxylic acid, D-60102
1,3-Dibromo-2-methylbenzene, D-60103
1,1-Dibromo-3-methyl-1-butene, D-60104
4,4-Dibromo-3-methyl-3-buten-2-one, D-70094
6,7-Dibromo-1-naphthalenamine, see D-70097
6,7-Dibromo-1,4-naphthalenedione, see D-70096
6,7-Dibromo-1-naphthalenol, see D-70095
6,7-Dibromo-1-naphthol, D-70095
6,7-Dibromo-1,4-naphthoquinone, D-70096
6,7-Dibromo-1-naphthylamine, D-70097
6,7-Dibromo-1-nitronaphthalene, D-70098
2,3-Dibromo-5-nitronaphthalene (incorr.), see D-70098
2,6-Dibromo-4-nitroso-1-naphthol, in D-60088
ω,ω'-Dibromo-5-nitro-m-xylene, see B-60145
▷1,2-Dibromopropane, D-60105
1,8-Dibromopyrene, D-70099
1,2-Dibromo-3,4,5,6-tetrafluorobenzene, D-60106
1,3-Dibromo-2,4,5,6-tetrafluorobenzene, D-60107
1,4-Dibromo-2,3,5,6-tetrafluorobenzene, D-60108
2,4-Dibromothiazole, D-60109
2,5-Dibromothiazole, D-60110
4,5-Dibromothiazole, D-60111
2,6-Dibromotoluene, see D-60103
4,5-Dibromo-1,2,3-triazine, D-70100
1,3-Dibromotricyclo[3.3.1.1^{3,7}]decane, see D-70084
2,3-Di-tert-butoxycyclopropenone, in D-70290
1,3-Dibutylallene, see T-60272
1,1-Di-tert-butylcyclopropane, D-60112
3,6-Di-tert-butyl-1,4-diazapentalene, see D-70108
2,2-Di-tert-butyl-3-(di-tert-butylmethylene)-cyclopropanone, D-70101

Di-tert-butyl dicarbonate, in D-70111
2,7-Di-tert-butyldicyclopenta[a,e]-cyclooctene, D-70102
2,5-Di-tert-butyl-2,5-dihydrobenzo[e]-pyrrolo[3,4-g]isoindole, D-70103
2,7-Di-tert-butyl-2,7-dihydroisoindolo[5,4-e]-isoindole, D-70104
Di-sec-butyldiketopiperazine, see B-70153
3,7-Di-tert-butyl-9,10-dimethyl-2,6-anthraquinone, D-60113
5-(3,5-Di-tert-butyl-4-oxo-2,5-cyclo-hexadienylidene)-3,6-cycloheptadiene-1,2-dione, D-70105
Di-tert-butyl pentaketone, see T-60148
3,4-Di-tert-butylpyrazole, D-70106
3,5-Di-tert-butylpyrazole, D-70107
Di-tert-butyl pyrocarbonate, in D-70111
3,6-Di-tert-butylpyrrolo[3,2-b]pyrrole, D-70108
3,4-Di-tert-butyl-2,2,5,5-tetramethylhexane, D-70109
Di-tert-butylthioketene, D-60114
3-O-[6-O-(2,5-di-O-caffeoyl-α-L-arabinofuranosyl)-β-D-glucopyranosyl]-3',7-di-O-(6-O-caffeoyl-β-D-glucopyranosyl)-cyanidin, see Z-70002
3,4-Di-O-caffeoylquinic acid, D-70110
▷Dicarbamide, see T-70097
2,2'-Dicarbamoyldiphenylamine, in I-60014
1,4-Dicarbomethoxyazulene, in A-70293
2,3-Dicarbomethoxy-1,3-butadiene, in D-70409
Dicarbonic acid, D-70111
α-Diceroptene, D-70112
5,6-Dichloroacenaphthene, D-60115
4,5-Dichloroacenaphthene (obsol.), see D-60115
9,10-Dichloroanthracene, D-60116
ms-Dichloroanthracene, see D-60116
2,3-Dichloro-1,4,9,10-anthracenetetrone, D-60117
4,6-Dichloro-1,3-benzenedicarboxaldehyde, D-60118
2,3-Dichlorobenzenethiol, D-60119
2,4-Dichlorobenzenethiol, D-60120
2,5-Dichlorobenzenethiol, D-60121
2,6-Dichlorobenzenethiol, D-60122
3,4-Dichlorobenzenethiol, D-60123
3,5-Dichlorobenzenethiol, D-60124
2,2-Dichloro-1,3-benzodioxole, D-70113
2,3-Dichlorobenzofuran, D-60125
5,7-Dichlorobenzofuran, D-60126
1,4-Dichlorobenzo[g]phthalazine, D-70114
1,4-Dichlorobicyclo[2.2.2]octane, D-70115
1,8-Dichloro-6-chloromethyl-2-methyl-2,6-octadiene, D-70116
3,8-Dichloro-6-chloromethyl-2-methyl-1,6-octadiene, D-70117
1,3-Dichloro-2-(chloromethyl)propane, D-70118
2,3-Dichlorocoumarone, see D-60125
5,7-Dichlorocoumarone, see D-60126
2,2-Dichlorocyclopropanecarboxylic acid, D-60127
2,2-Dichloro-3,3-difluorooxirane, D-60128
5,6-Dichloro-1,2-dihydroacenaphthylene, see D-60115
2,4-Dichloro-1,8-dihydroxy-5-methoxy-6-methylxanthone, see T-70181
1-(3,5-Dichloro-2,6-dihydroxy-4-methoxy-phenyl)-1-hexanone, D-60129
3,5-Dichloro-2,4-dihydroxy-6-methylbenzoic acid 4-carbomethoxy-3-hydroxy-5-methyl-phenyl ester, see T-70334
4,6-Dichloro-5-[(dimethylamino)methylene]-3,6-cyclohexadiene-1,3-dicarboxaldehyde, D-70119
2,4-Dichloro-3,4-dimethyl-2-cyclobuten-1-one, D-60130
4,4-Dichloro-2,3-dimethyl-2-cyclobuten-1-one, D-60131
1,1-Dichloroepoxy-2,2-difluoroethane, see D-60128
2,3-Dichloro-N-(4-fluorophenyl)maleimide, see D-60132
3,4-Dichloro-1-(4-fluorophenyl)-1H-pyrrole-2,5-dione, D-60132

4,5-Dichloro-2-formylimidazole, see D-60134
2,5-Dichloro-4-hydroxybenzaldehyde, D-70120
2,6-Dichloro-4-hydroxybenzaldehyde, D-70121
2,3-Dichloro-4-hydroxybenzoic acid, D-60133
2,4-Dichloro-1-hydroxy-5,8-dimethoxyxanthone, in T-70181
4,5-Dichloro-1*H*-imidazole, D-70122
4,5-Dichloro-1*H*-imidazole-2-carboxaldehyde, D-60134
4,5-Dichloro-1*H*-imidazole-2-carboxylic acid, D-60135
5,6-Dichloro-1*H*-indole-3-acetic acid, D-70123
3,3'-Dichloroisobutylene, see C-70061
4,6-Dichloroisophthalaldehyde, see D-60118
6,7-Dichloroisoquinoline, D-60136
6,7-Dichloro-5,8-isoquinolinedione, D-60137
1,2-Dichloro-3-mercaptobenzene, see D-60119
1,2-Dichloro-4-mercaptobenzene, see D-60123
1,3-Dichloro-2-mercaptobenzene, see D-60122
1,3-Dichloro-5-mercaptobenzene, see D-60124
1,4-Dichloro-2-mercaptobenzene, see D-60121
2,4-Dichloro-1-mercaptobenzene, see D-60120
2,6-Dichloro-4-methoxybenzaldehyde, in D-70121
2,3-Dichloro-4-methoxybenzoic acid, in D-60133
2,3-Dichloro-1-methoxy-4-nitrobenzene, in D-70125
1,2-[(Dichloromethylene)dioxy]benzene, see D-70113
2,3-Dichloro-5-methylpyridine, D-60138
2,5-Dichloro-3-methylpyridine, D-60139
1,2-Dichloro-4-(methylthio)benzene, in D-60123
1,4-Dichloro-2-(methylthio)benzene, in D-60121
6,7-Dichloro-1,4-naphthalenedione, see D-70124
6,7-Dichloro-1,4-naphthoquinone, D-70124
2,3-Dichloro-4-nitroanisole, in D-70125
2,3-Dichloro-4-nitrophenol, D-70125
2,3-Dichloro-5-nitropyridine, D-60140
2,4-Dichloro-3-nitropyridine, D-60141
2,4-Dichloro-5-nitropyridine, D-60142
2,5-Dichloro-3-nitropyridine, D-60143
2,6-Dichloro-3-nitropyridine, D-60144
2,6-Dichloro-4-nitropyridine, D-60145
3,4-Dichloro-5-nitropyridine, D-60146
3,5-Dichloro-4-nitropyridine, D-60147
1,1-Dichlorooctane, D-70126
1,2-Dichlorooctane, D-70127
1,3-Dichlorooctane, D-70128
2,2-Dichlorooctane, D-70129
2,3-Dichlorooctane, D-70130
4,5-Dichlorooctane, D-70131
1,4-Dichlorophenanthrene, D-70132
2,4-Dichlorophenanthrene, D-70133
3,6-Dichlorophenanthrene, D-70134
3,9-Dichlorophenanthrene, D-70135
4,5-Dichlorophenanthrene, D-70136
9,10-Dichlorophenanthrene, D-70137
2,3-Dichlorophenylalanine, see A-60127
2,4-Dichlorophenylalanine, see A-60128
2,5-Dichlorophenylalanine, see A-60129
2,6-Dichlorophenylalanine, see A-60130
3,4-Dichlorophenylalanine, see A-60131
3,5-Dichlorophenylalanine, see A-60132
4-[(2,5-Dichlorophenyl)azo]-2,4-dihydro-5-methyl-2-phenyl-3*H*-pyrazol-3-one, see D-70138
4-(2,5-Dichlorophenylhydrazono)-5-methyl-2-phenyl-3*H*-pyrazol-3-one, D-70138
2,3-Dichlorophenyl mercaptan, see D-60119
2,4-Dichlorophenyl mercaptan, see D-60120
2,5-Dichlorophenyl mercaptan, see D-60121
2,6-Dichlorophenyl mercaptan, see D-60122
3,4-Dichlorophenyl mercaptan, see D-60123
3,5-Dichlorophenyl mercaptan, see D-60124
1,4-Dichlorophthalazine, D-60148
2,5-Dichloro-3-picoline, see D-60139
5,6-Dichloro-3-picoline, see D-60138
6,7-Dichloro-5,8-quinolinedione, D-60149
▷2,3-Dichloroquinoxaline, D-60150
1,2-Dichloro-3,4,5,6-tetrafluorobenzene, D-60151
1,3-Dichloro-2,4,5,6-tetrafluorobenzene, D-60152

1,4-Dichloro-2,3,5,6-tetrafluorobenzene, D-60153
2,4-Dichlorothiazole, D-60154
2,5-Dichlorothiazole, D-60155
4,5-Dichlorothiazole, D-60156
2,5-Dichlorothioanisole, in D-60121
3,4-Dichlorothioanisole, in D-60123
2,3-Dichlorothiophenol, see D-60119
2,4-Dichlorothiophenol, see D-60120
2,5-Dichlorothiophenol, see D-60121
2,6-Dichlorothiophenol, see D-60122
3,4-Dichlorothiophenol, see D-60123
3,5-Dichlorothiophenol, see D-60124
▷4,5-Dichloro-2-(trifluoromethyl)-1*H*-benzimidazole, D-60157
▷4,6-Dichloro-2-(trifluoromethyl)-1*H*-benzimidazole, D-60158
▷4,7-Dichloro-2-(trifluoromethyl)-1*H*-benzimidazole, D-60159
▷5,6-Dichloro-2-(trifluoromethyl)-1*H*-benzimidazole, D-60160
2,2-Dichloro-3,3,3-trifluoro-1-propanol, D-60161
1,5-Dichloro-9*H*-xanthen-9-one, see D-70139
1,7-Dichloro-9*H*-xanthen-9-one, see D-70140
2,6-Dichloro-9*H*-xanthen-9-one, see D-70141
2,7-Dichloro-9*H*-xanthen-9-one, see D-70142
3,6-Dichloro-9*H*-xanthen-9-one, see D-70143
1,5-Dichloroxanthone, D-70139
1,7-Dichloroxanthone, D-70140
2,6-Dichloroxanthone, D-70141
2,7-Dichloroxanthone, D-70142
3,6-Dichloroxanthone, D-70143
Dichotosinin, in T-60103
Dicranolomin, D-70144
Dictymal, D-60162
Dictyodial *A*, D-70145
Dictyone, D-70146
Dictyotriol *A*, D-70147
Dictytriene *A*, D-70148
Dictytriene B, in D-70148
Dictytriol, in D-70148
Dicyanic acid, see D-60057
5,6-Dicyanoacenaphthene, in A-70015
Dicyanoacetic acid, D-70149
1,8-Dicyanoanthracene, in A-60281
5,7-Dicyanoazulene, in A-70295
1,1'-Dicyanobicyclopropyl, in B-70114
5,5'-Dicyano-2,2'-bipyridine, in B-60127
2,3-Dicyano-1,3-butadiene, in D-70409
4,4'-Dicyanodiphenylamine, in I-60018
2,2-Dicyanoethylene-1,1-dithiol, in D-60395
2,3-Dicyano-1-methoxy-4-nitrobenzene, in H-70189
9-(Dicyanomethyl)anthracene, D-60163
10-(Dicyanomethyl)-9(10*H*)-anthracenone, see D-60164
[10-[(10-Dicyanomethylene)-9(10*H*)-anthracenylidene]-9(10*H*)-anthracenylidene]propanedinitrile, see T-60030
10-(Dicyanomethylene)anthrone, D-60164
2,3-Dicyanonaphthalene, in N-60003
1,5-Dicyano-3-pentanone, in O-60071
Dicyanotriselenide, D-60165
Dicyclobuta[1,4]dithiin-1,2,4,5-tetraone, see D-60506
Di(1,3,5-cycloheptatrien-3-yl)(2,4,6-cyclo-heptatrienylidene)methane, see C-60205
Di(2,4,6-cycloheptatrien-1-yl)ethanone, D-60166
Di(2,4,6-cycloheptatrien-1-yl) ketone, see D-60166
Dicyclohexylcarbinol, see D-70150
Dicyclohexylmethanol, D-70150
Dicyclopenta[*ef,kl*]heptalene, D-60167
2,3-Didehydro-1,2-dihydro-1,1-dimethylnaphthalene, D-60168
2,3-Didehydronorleucine, see A-60167
1,10-Didehydro-2-quinolizidone, see H-60059
N,6-Didehydro-1,2,3,6-tetrahydro-1-methyl-2-oxoadenosine, see D-60518
7,8-Didehydro-3,4,5,6-tetrahydro-2*H*-thiocin, see T-60189
5,6-Didehydro-1,4,7,10-tetramethyldibenzo[*ae*]cyclooctene, D-60169

4,5-Didehydrotropone, see C-60196
Didemethoxyaaptamine, see B-70037
▷6,27-Didemethoxyantibiotic *A* 204*A*, see C-60022
▷16,34-Didemethyl-21,27,29,31-tetradeoxy-12,13,32,33-tetrahydro-13-hydroxy-2-methyl-21,27-dioxodermostatin *A*, see R-70009
15,16-Dideoxy-15,17-dihydroxy-15,17-oxido-16-spongianoic acid, see T-60038
1,2-Dideoxy-2-(hydroxymethyl)-D-*chiro*-inositol, in H-70170
1,6-Dideoxy-1-(hydroxymethyl)-L-*chiro*-inositol, in H-70170
2,3-Dideoxy-2-(hydroxymethyl)-D-*epi*-inositol, in H-70170
2,3-Dideoxy-3-(hydroxymethyl)-D-*myo*-inositol, in H-70170
Dieckmann ester, in O-70084
8,13-diepi-Manoyl oxide, in E-70023
21,23*R*:24*S*,25-Diepoxyapotirucall-14-ene-3α,7α,21*S*-triol, see T-60223
2,16;7,8-Diepoxy-1(15),3,11-cembratriene, see S-70011
2,16:11,12-Diepoxy-1(15),8*E*,12*E*-cembratriene, see I-70100
4β,6β;15,16-Diepoxy-2β-hydroxy-13(16),14-clerodadiene-18,19;20,12*S*-diolide, see C-60032
1β,10α; 4α,5β-Diepoxy-8α-isobutoxyglechomanolide, in G-60022
1β,10α; 4α,5β-Diepoxy-8β-isobutoxyglechomanolide, in G-60022
5,6:12,13-Diepoxy-5,6,12,13-tetra-hydrodibenz[*a,h*]anthracene, D-70151
4,5:9,10-Diepoxy-4,5,9,10-tetrahydropyrene, D-70152
5,8:13,16-Diethenodibenzo[*a,g*]cyclododecene, see D-70072
1,3-Diethenyltricyclo[3.3.1.1³,⁷]decane, see D-70529
Diethoxyacetaldehyde, D-70153
4,4-Diethoxy-2-butynal, in B-60351
4,6-Diethoxy-2,5-dihydropyrimidine, in D-70256
1,5-Diethoxy-2,4-dinitrobenzene, in D-70450
3,4-Diethoxy-1,5-hexadiene, in H-70036
4,4-Diethoxy-2-methyl-2-butene, in M-70055
1,1-Diethoxy-3-nitropropane, in N-60037
2,4-Diethoxypyridine, in H-60224
2,5-Diethoxypyridine, in H-60225
2,6-Diethoxypyridine, in H-60226
3,4-Diethoxypyridine, in H-60227
2,2-Diethoxytetrahydro-5,6-dimethyl-2*H*-pyran, in T-70059
1-[4-[6-(Diethylamino)-2-benzofuranyl]-phenyl]-1*H*-pyrrole-2,5-dione, D-60170
2-(Diethylamino)purine, in A-60243
2,2-Diethylbutanoic acid, D-60171
▷Diethyl dicarbonate, in D-70111
8,14-Diethyl-16,17-dihydro-9,13,24-trimethyl-5,22:12,15-diimino-20,18-metheno-7,10-nitrilobenzo[*o*]-cyclopent[*b*]azacyclo-nonadecine, see E-60042
▷*N*,*N*'-Diethyl-*N*,*N*'-diphenylurea, in D-70499
1,2-Diethylidenecyclohexane, D-70154
N,*N*-Diethylnipecotamide, in P-70102
N,*N*-Diethyl-3-piperidinecarboxamide, in P-70102
1,1-Diethylpropanethiol, see E-60053
▷Diethyl pyrocarbonate, in D-70111
Diethyl tartrate, in T-70005
▷*N*,*N*'-Diethylthiocarbamide, see D-70155
▷*N*,*N*'-Diethylthiourea, D-70155
Diethyl(trichlorovinyl)amine, see T-70222
1,3-Diethynyladamantane, D-60172
1,3-Diethynylbenzene, D-70156
1,4-Diethynylbicyclo[2.2.2]octane, D-70157
2,7-Diethynylnaphthalene, D-70158
1,3-Diethynyl-2,4,5,6-tetrafluorobenzene, D-70159
1,4-Diethynyl-2,3,5,6-tetrafluorobenzene, D-70160
1,3-Diethynyltricyclo[3.3.1.1³,⁷]decane, see D-60172
▷Differenol *A*, see T-60324

Differentiation inducing factor, see T-70335
Diffutidin, in T-60103
Diffutin, in T-60103
3,3-Difluoroalanine, see A-60135
9,10-Difluoroanthracene, D-70161
2,5-Difluorobenzenethiol, D-60173
2,6-Difluorobenzenethiol, D-60174
3,4-Difluorobenzenethiol, D-60175
3,5-Difluorobenzenethiol, D-60176
7,7-Difluorobenzocyclopropene, see D-60177
7,7-Difluorobicyclo[4.1.0]hepta-1,3,5-triene, D-60177
4,4'-Difluoro-[1,1'-biphenyl]-3,3'-dicarboxylic acid, D-60178
4,5-Difluoro-[1,1'-biphenyl]-2,3-dicarboxylic acid, D-60179
6,6'-Difluoro-[1,1'-biphenyl]-2,2'-dicarboxylic acid, D-60180
4,4-Difluoro-3,3-bis(trifluoromethyl)-1,2-oxathietane 2,2-dioxide, D-60181
1,1-Difluoro-1,3-butadiene, D-70162
2,2-Difluorobutanedioic acid, D-70163
3,3-Difluorocyclobutanecarboxylic acid, D-70164
3,3-Difluorocyclobutene, D-70165
7,7-Difluorocyclopropabenzene, see D-60177
8,8-Difluoro-8,8-dihydro-4-methyl-2,2,6,6-tetrakis(trifluoromethyl)-2H,6H-[1,2]-iodoxolo[4,5,1-hi][1,2]-benziodoxole, D-60182
1,1-Difluoro-2,2-diiodoethylene, D-60183
6,6'-Difluorodiphenic acid, see D-60180
Difluorodiphenylmethane, D-60184
2,2-Difluoroethane β-sultone, see D-60186
1,2-Difluoroethylene oxide, see D-60187
1,2-Difluoro-4-mercaptobenzene, see D-60175
1,3-Difluoro-2-mercaptobenzene, see D-60174
1,3-Difluoro-5-mercaptobenzene, see D-60176
1,4-Difluoro-2-mercaptobenzene, see D-60173
1,1'-(Difluoromethylene)bisbenzene, see D-60184
▷ (Difluoromethylene)tetrafluorocyclopropane, D-60185
1,1-Difluoro-10-methyl-3,3,7,7-tetrakis-(trifluoromethyl)-4,5,6-benzo-1-ioda-2,8-dioxabicyclo[3.3.1]-octane, see D-60182
1,3-Difluoro-2-(methylthio)benzene, in D-60174
1,3-Difluoro-5-(methylthio)benzene, in D-60176
1,4-Difluoro-2-(methylthio)benzene, in D-60173
4,4-Difluoro-1,2-oxathietane 2,2-dioxide, D-60186
2,3-Difluorooxirane, D-60187
4,5-Difluorophenanthrene, D-70166
9,10-Difluorophenanthrene, D-70167
2,5-Difluorophenyl mercaptan, see D-60173
2,6-Difluorophenyl mercaptan, see D-60174
3,4-Difluorophenyl mercaptan, see D-60175
3,5-Difluorophenyl mercaptan, see D-60176
3,4-Difluoro-6-phenylphthalic acid, see D-60179
2,3-Difluoropyrazine, D-60188
2,6-Difluoropyrazine, D-60189
2,2-Difluorosuccinic acid, see D-70163
2,5-Difluorothioanisole, in D-60173
2,6-Difluorothioanisole, in D-60174
3,5-Difluorothioanisole, in D-60176
2,5-Difluorothiophenol, see D-60173
2,6-Difluorothiophenol, see D-60174
3,4-Difluorothiophenol, see D-60175
3,5-Difluorothiophenol, see D-60176
Difluoro(trifluoromethanethiolato)-(trifluoromethyl)sulfur, in T-60281
Diformylacetic acid, D-60190
3,4-Diformylcinnoline, see C-70180
3,5-Diformyl-2,4-dihydroxy-6-methylbenzoic acid 3-hydroxy-4-(methoxycarbonyl)-2,5,6-trimethylphenyl ester, see P-70054
3,5-Diformyl-2,4-dihydroxy-6-methylbenzoic acid 3-methoxy-2,5,6-trimethylphenyl ester, see N-70032
1,3-Diformylnaphthalene, see N-60002
3,4-Diformylthiadiazole, see T-70162
1,2-Di(2-furanyl)ethylene, D-60191

1,3-Di(2-furyl)benzene, D-70168
1,4-Di(2-furyl)benzene, D-70169
3,5-Digalloylepicatechin, in P-70026
▷ Digenic acid, see K-60003
▷ Digenin, see K-60003
1,2-Dihydrazinoethane, D-60192
6b,7a-Dihydroacenaphth[1,2-b]oxirene, D-70170
1,2-Dihydro-5-acenaphthylenecarboxylic acid, see A-70014
1,2-Dihydro-3,4-acenaphthylenedicarboxylic acid, see A-60013
1,2-Dihydro-5,6-acenaphthylenedicarboxylic acid, see A-70015
1,2-Dihydro-4-acenaphthylenol, see H-60099
Dihydroacepentalenediide(2−), see D-60216
9,20-Dihydro-9-(9(10H)-acridinylidene)-acridine, see B-70090
Dihydroalterperylenol, in A-60078
5,10-Dihydroanthra[9,1,2-cde]benzo[rst]-pentaphene, in A-60279
9,10-Dihydro-9-anthracenamine, see A-60136
9,10-Dihydro-9-anthrylamine, see A-60136
2,3-Dihydro-7-azaindole, see D-70263
1,2-Dihydro-3H-azepin-3-one, D-70171
2,3-Dihydroazete, D-70172
2,3-Dihydro-6(1H)-azulenone, D-60193
9,10-Dihydro-9,10[1',2']-benzenoanthracene-1,8,13-tricarboxylic acid, see T-60399
9,10-Dihydro-9,10[1',2']-benzenoanthracene-1,8,16-tricarboxylic acid, see T-60398
2,3-Dihydro-1H-benz[f]indene, see B-60015
1,4-Dihydrobenzocyclooctatetraene, D-70173
2,3-Dihydro-1,4-benzodioxin-2-carboxaldehyde, see B-60022
2,3-Dihydro-1,4-benzodioxin-6-carboxaldehyde, see B-60023
2,3-Dihydro-1,4-benzodioxin-2-carboxylic acid, see B-60024
2,3-Dihydro-1,4-benzodioxin-6-carboxylic acid, see B-60025
2,3-Dihydro-1,4-benzodioxin-2-methanol, see H-60172
1,8-Dihydrobenzo[2,1-b:3,4-b']dipyrrole, D-70174
1,5-Dihydro-2,4-benzodithiepin, D-60194
3,4-Dihydro-2H-1,5-benzodithiepin, D-60195
2,2'-(4,8-Dihydrobenzo[1,2-b:5,4-b']-dithiophene-4,8-diylidene)-bispropanedinitrile, D-60196
2,3-Dihydro-2-benzofurancarboxylic acid, D-60197
9,18-Dihydrobenzo[rst]phenanthro[10,1,2-cde]-pentaphene, in B-60038
3,4-Dihydro-2H-1-benzopyran-2-ol, D-60198
3,4-Dihydro-2H-1-benzopyran-3-ol, D-60199
3,4-Dihydro-2H-1-benzopyran-4-ol, D-60200
3,4-Dihydro-2H-1-benzopyran-5-ol, D-60201
3,4-Dihydro-2H-1-benzopyran-6-ol, D-70175
3,4-Dihydro-2H-1-benzopyran-7-ol, D-60202
3,4-Dihydro-2H-1-benzopyran-8-ol, D-60203
1,4-Dihydro-2(3H)-benzopyran-3-one, D-70176
1,4-Dihydro-3H-2-benzoselenin-3-one, D-60204
3,4-Dihydro-1H-2-benzoselenin-3-one, see D-60204
1,4-Dihydro-3H-2-benzotellurin-3-one, D-60205
3,4-Dihydro-1H-2-benzotellurin-3-one, see D-60205
3,4-Dihydro-1H-2-benzothiin-3-one, see D-60206
1,4-Dihydro-3H-2-benzothiopyran-3-one, D-60206
3,4-Dihydro-2H-1,5-benzoxathiepin-3-one, D-60207
2,3-Dihydro-2',6''-biluteolin, in D-70144
2,5-Dihydro-3,5-bis(4-hydroxyphenyl)-2-furanmethanol, see C-70201
9,10-Dihydro-9,10-bis(methylene)anthracene, see D-70194
1,6-Dihydro-1,6-bis(methylene)azulene, see D-60225
2,6-Dihydro-2,6-bis(methylene)azulene, see D-60226
2,3-Dihydro-2,3-bis(methylene)furan, D-60208

2,3-Dihydro-2,3-bis(methylene)thiophene, see D-70195
11β,13-Dihydrobrachynereolide, in B-60196
cis-9,10-Dihydrocapsenone, in C-60011
1,4-Dihydro-1,4-carbazoledione, see C-60012
2,3-Dihydrocarbazol-4(1H)-one, see O-70106
2,4-Dihydrocarbazol-3(1H)-one, see O-70105
3,4-Dihydrocarbazol-1(2H)-one, see O-70103
3,4-Dihydrocarbazol-2(1H)-one, see O-70104
3,4-Dihydro-β-carboline, D-60209
▷ 1a,11c-Dihydrochryseno[5,6-b]oxirene, see C-70174
8-Dihydrocinnamoyl-5,7-dihydroxy-4-phenyl-2H-benzopyran-2-one, D-60210
Dihydrocochloxanthin, in C-60156
Dihydroconfertin, D-60211
11β,13-Dihydroconfertin, in D-60211
Dihydrocoriandrin, in C-70198
Dihydrocuneane, see T-70022
4,5-Dihydrocyclobuta[b]furan, D-60212
1,2-Dihydrocyclobuta[a]naphthalene-2-carboxylic acid, D-70177
1,2-Dihydrocyclobuta[a]naphthalene-3,4-dione, D-60213
1,2-Dihydrocyclobuta[b]naphthalene-3,8-dione, D-60214
1,2-Dihydrocyclobuta[a]naphthalen-3-ol, D-60215
3,8-Dihydrocyclobuta[b]quinoxaline-1,2-dione, D-70178
Dihydrocyclopenta[c,d]pentalene(2−), D-60216
5,6-Dihydro-4H-cyclopenta-1,2,3-thiadiazole, D-70179
5,6-Dihydro-4H-cyclopentathiophen-4-amine, see A-60137
5,6-Dihydro-4H-cyclopenta[b]thiophene, D-70180
5,6-Dihydro-4H-cyclopenta[c]thiophene, D-70181
4,5-Dihydro-6H-cyclopenta[b]thiophen-6-one, D-70182
4,6-Dihydro-5H-cyclopenta[b]thiophen-5-one, D-70183
5,6-Dihydro-4H-cyclopenta[c]thiophen-4-one, D-70184
11β,13-Dihydrodesoxoachalensolide, in A-70054
2,3-Dihydro-2-diazo-3-oxobenzoic acid, see D-70063
▷ 10,11-Dihydro-5H-dibenz[b,f]azepine, D-70185
6,7-Dihydro-5H-dibenzo[a,c]cycloheptene, D-70186
10,11-Dihydro-5H-dibenzo[a,d]cycloheptene, D-60217
5,7-Dihydro-6H-dibenzo[a,c]cyclohepten-6-one, D-70187
6,7-Dihydro-5H-dibenzo[a,c]cyclohepten-5-one, D-70188
10,11-Dihydro-5H-dibenzo[a,d]cyclohepten-5-one, D-60218
5,7-Dihydrodibenzo[c,e]selenepin, D-70189
5,7-Dihydrodibenzo[c,e]thiepin, D-70190
2,3-Dihydrodicranolomin, in D-70144
9,10-Dihydrodicyclopenta[c,g]phenanthrene, D-60219
1,10-Dihydrodicyclopenta[a,h]naphthalene, D-60220
3,8-Dihydrodicyclopenta[a,h]naphthalene, D-60221
1,12-Dihydrodicyclopenta[c,g]phenanthrene, see D-60219
9,10-Dihydro-5,8:11,14-diethenobenzocyclo-dodecene, see B-70038
9,10-Dihydro-1,5-dihydroxy-2,7-dimethoxy-phenanthrene, in T-60116
9,10-Dihydro-2,7-dihydroxy-3,5-dimethoxy-phenanthrene, in T-60120
9,10-Dihydro-5,6-dihydroxy-2,4-dimethoxy-phenanthrene, in D-70247
9,10-Dihydro-9,10-dihydroxy-8,8-dimethyl-2H,8H-benzo[1,2-b:3,4-b']dipyran-2-one, see K-70013
1,3-Dihydro-1,4-dihydroxy-5-(hydroxymethyl)-10-methoxy-8-methyl-3,7-dioxo-7H-isobenzofuro[4,5-b][1,4]-benzodioxepin-11-carboxaldehyde, see C-60163

3,4-Dihydro-3,4-dihydroxy-2-(hydroxymethyl)-2*H*-pyrrole, D-60222

3,4-Dihydro-6,7-dihydroxy-3-(1-hydroxy-3-oxo-1-butenyl)-4-oxo-2*H*-1-benzopyran-5-carboxylic acid, *see* P-60167

2,3-Dihydro-5,7-dihydroxy-2-(4-hydroxyphenyl)-6,8-bis(3-methyl-2-butenyl)-4*H*-1-benzopyran-4-one, *see* L-60030

2,3-Dihydro-5,7-dihydroxy-2-(4-hydroxyphenyl)-6,8-bis(3-methyl-2-butenyl)-2*H*-1-benzopyran-4-one, *see* S-70034

2,3-Dihydro-5,7-dihydroxy-8-(3-methyl-2-butenyl)-2-phenyl-4*H*-1-benzopyran-4-one, *see* G-70015

2,3-Dihydro-3,5-dihydroxy-2-methyl-naphthoquinone, D-70191

3,4-Dihydro-2-(3,4-dihydroxyphenyl)-7-hydroxy-2*H*-1-benzopyran, *see* T-70260

2,3-Dihydro-2-(3,4-dihydroxyphenyl)-7-methoxy-3-methyl-5-propenylbenzofuran, *see* K-60001

3′,4′-Dihydro-3′,4′-dihydroxyseselin, *see* K-70013

▷ 2′,3′-Dihydro-3′,6′-dihydroxy-2′,2′,4′,6′-tetramethylspiro[cyclopropane-1,5′-[5*H*]inden]-7′(6′*H*)-one, *see* I-60001

9,10-Dihydro-2,7-dihydroxy-3,4,6-trimethoxyphenanthrene, *in* D-70246

9,10-Dihydro-5,6-dihydroxy-1,3,4-trimethoxyphenanthrene, *in* P-70034

9,10-Dihydro-6,7-dihydroxy-2,3,4-trimethoxyphenanthrene, *in* D-70246

6,11-Dihydro-5*H*-diindolo[2,3-*a*:2′,3′-*c*]-carbazole, *see* D-60384

3,8-Dihydro-6,6′;7,3′*a*-diliguetilide, D-60223

2,5-Dihydro-4,6-dimercaptopyrimidine, *see* D-70257

4-(3,4-Dihydro-5,7-dimethoxy-2*H*-1-benzopyran-3-yl)-1,3-benzenediol, *see* L-70038

9,10-Dihydro-2,7-dimethoxy-1,5-phenanthrenediol, *in* T-60116

9,10-Dihydro-5,7-dimethoxy-3,4-phenanthrenediol, *in* D-70247

2,5-Dihydro-4,6-dimethoxypyrimidine, *in* D-70256

1,4-Dihydro-3,6-dimethoxy-1,2,4,5-tetrazine, D-70192

9,10-Dihydro-9,9-dimethylanthracene, D-60224

4-[3,4-Dihydro-8,8-dimethyl-2*H*,8*H*-benzo[1,2-*b*:3,4-*b*′]dipyran-3-yl]-1,3-benzenediol, *see* G-70016

2,3-Dihydro-α²,α²-dimethyl-2,7-benzofurandimethanol, *see* B-70177

2,3-Dihydro-2,2-dimethyl-4*H*-1-benzothiopyran, D-70193

9,10-Dihydro-9,10-dimethyleneanthracene, D-70194

1,6-Dihydro-1,6-dimethyleneazulene, D-60225

2,6-Dihydro-2,6-dimethyleneazulene, D-60226

2,3-Dihydro-2,3-dimethylenethiophene, D-70195

Dihydro-4,4-dimethyl-2,3-furandione, D-70196

1,3-Dihydro-1,3-dimethyl-2*H*-imidazo[4,5-*b*]-pyrazine-2-thione, *in* D-60247

1,3-Dihydro-1,3-dimethyl-2*H*-imidazo[4,5-*b*]-pyrazin-2-one, *in* I-60006

1,4-Dihydro-1,4-dimethyl-2*H*-imidazo[4,5-*b*]-pyrazin-2-one, *in* I-60006

2,3-Dihydro-2,2-dimethyl-1*H*-inden-1-one, D-60227

Dihydro-4,4-dimethyl-5-methylene-2(3*H*)-furanone, D-70197

3,4-Dihydro-2,2-dimethyl-1(2*H*)-naphthalenone, D-60228

4,5-Dihydro-4,4-dimethyloxazole, D-70198

2,3-Dihydro-1,6-dimethyl-3-oxopyridazinium hydroxide, inner salt, *in* M-70124

3*a*,6*a*-Dihydro-3*a*,6*a*-dimethyl-1,6-pentalenedione, D-70199

3,7-Dihydro-3,7-dimethyl-6*H*-purin-6-amine, *see* D-60400

3,4-Dihydro-2,2-dimethyl-2*H*-pyrrole, D-60229

3,4-Dihydro-2,2-dimethyl-2*H*-pyrrole, D-70200

1,2-Dihydro-2,2-dimethyl-3*H*-pyrrol-3-one, D-60230

3,4-Dihydro-3,3-dimethyl-2(1*H*)quinolinone, D-60231

N-(1,4-Dihydro-1,4-dimethyl-5*H*-tetrazol-5-ylidene)methanamine, *in* A-70202

2,3-Dihydro-2,2-dimethylthiophene, D-60232

2,5-Dihydro-2,5-dimethylthiophene, D-60233

4,5-Dihydro-5,5-dimethyl-3*H*-1,2,4-triazol-3-one, D-70201

3,5-Dihydro-4*H*-dinaphth[2,1-*c*:1′,2′-*e*]-azepine, D-60234

(6*E*)-10,11-Dihydro-12,19-dioxogeranylnerol, *in* H-60186

6*a*,12*a*-Dihydro-6*H*-[1,3]dioxolo[5,6]-benzofuro[3,2-*c*][1]benzopyran-3-ol, *see* M-70001

4,9-Dihydro-4,9-dioxo-1*H*-naphtho[2,3-*d*]-*v*-triazole, *see* N-60011

5,6-Dihydro-11,12-diphenyldibenzo[*a,e*]-cyclooctene, D-70202

▷ 2,3-Dihydro-5,6-diphenyl-1,4-oxathiin, D-70203

1,2-Dihydro-4,6-diphenylpyrimidine, D-60235

2,5-Dihydro-3,6-diphenylpyrrolo[3,4-*c*]-pyrrole-1,4-dione, D-70204

2,5-Dihydro-3,4-diphenylselenophene, D-70205

2,5-Dihydro-2,2-diphenyl-1,3,4-thiadiazole, D-60236

2,5-Dihydro-1*H*-dipyrido[4,3-*b*:3′,4′-*d*]-pyrrol-1-one, D-60237

2-(5,6-Dihydro-1,3-dithiolo[4,5-*b*][1,4]-dithiin-2-ylidene)-5,6-dihydro-1,3-dithiolo[4,5-*b*][1,4]dithiin, D-60238

Dihydrodroserone, *see* D-70191

7,7-*O*-Dihydroebuloside, *in* E-60001

β-Dihydroentandrophragmin, *in* E-70009

11β,13-Dihydro-8-epi-confertin, *in* D-60211

9,10-Dihydro-9,10-epidithioanthracene, D-70206

11β,13-Dihydro-1-epi-inuviscolide, *in* I-70023

11β,13-Dihydroepiligustrin (incorr.), *in* D-60328

9,10-Dihydro-9,10-ethanoanthracene, D-70207

4,5-Dihydro-1,5-ethano-1*H*-1-benzazepin-2(3*H*)-one, D-60239

4,7-Dihydro-4,7-ethanoisobenzofuran, D-60240

9,10-Dihydro-9,10-ethenoanthracene-11,12-dicarboxylic acid, D-70208

5,6-Dihydro-2-(ethylthio)-4*H*-1,3-thiazine, *in* T-60094

3,4-Dihydro-3-fluoro-2*H*-1,5-benzodioxepin, D-70209

Dihydro-2(3*H*)-furanthione, D-70210

1-(4,5-Dihydro-3-furanyl)-2,2,2-trifluoroethanone, *see* D-70274

1,4-Dihydrofuro[3,4-*d*]pyridazine, D-60241

2,3-Dihydrofuro[2,3-*b*]pyridine, D-70211

Dihydrofusarubin A, *in* F-60108

Dihydrofusarubin B, *in* F-60108

▷ 1,2-Dihydrogedunin, *in* G-70006

5α,6-Dihydroglaucasterol, *in* P-60004

9,11-Dihydrogracilin *A*, *in* G-60033

Dihydroguaiaretic acid, *in* G-70033

10,15-Dihydro-2,3,7,8,12,13-hexamethoxy-5*H*-tribenzo[*a,d,g*]cyclononene, *see* C-70254

3,8-Dihydro-3,3,4,8,8,9-hexamethylbenzo[1,2-*c*:4,5-*c*′]dipyran-1,6-dione, *see* T-70190

Dihydrohomalicine, *in* H-70207

10,11-Dihydro-11-hydroxycurcuphenol, *in* C-70208

Dihydro-4-hydroxy-2(3*H*)-furanone, D-60242

4,5-Dihydro-3-hydroxy-2(3*H*)-furomone, *in* D-60313

3*a*,7*a*-Dihydro-4-hydroxy-4*H*-furo[2,3-*b*]-pyran-2(3*H*)-one, *see* O-70107

11β,13-Dihydro-8α-hydroxyglucozaluzanin C, *in* Z-60001

7,8-Dihydro-12β-hydroxyholothurinogenin, *in* T-70267

4,5-Dihydro-3-hydroxy-5-(hydroxymethyl)-2(3*H*)-furanone, D-70212

2,3-Dihydro-7-hydroxy-3-[(4-hydroxyphenyl)methylene]-4*H*-1-benzopyran-4-one, *see* H-60152

2,3-Dihydro-2-hydroxy-1*H*-imidazole-4,5-dione, D-60243

3,4-Dihydro-4-hydroxy-6-methoxy-2,7-dimethyl-1(2*H*)-naphthalenone, *see* F-60014

2,3-Dihydro-2-(4-hydroxy-3-methoxyphenyl)-5-(3-hydroxypropyl)-3-methylbenzofuran, *in* D-60244

2,3-Dihydro-7-hydroxy-3-[4-methoxyphenyl)methylene]-4*H*-1-benzopyran-4-one, *see* B-60192

2,3-Dihydro-2-(4-hydroxy-3-methoxyphenyl)-3-methyl-5-(1-propenyl)benzofuran, *in* C-60161

3,4-Dihydro-8-hydroxy-3-methylbenz[*a*]-anthracene-1,7,12(2*H*)-trione, *see* O-60003

2,3-Dihydro-3-hydroxy-2-(1-methylethenyl)-5-benzofurancarboxaldehyde, D-70213

1-[2,3-Dihydro-5-hydroxy-2-(1-methylethenyl)-4-benzofuranyl]ethanone, *see* A-70040

2,3-Dihydro-2-(1-hydroxy-1-methylethyl)-5-benzofurancarboxaldehyde, D-70214

2,3-Dihydro-7-hydroxymethyl-2-(2-hydroxy-2-propyl)-6-propylbenzofuran, *see* B-70177

7,12-Dihydro-11-hydroxy-2-(1-methylpropyl)-4,7,12-trioxo-4*H*-anthra[1,2-*b*]pyran-5-acetic acid, *see* A-70239

1,3-Dihydro-1-hydroxynaphtho[2,3-*c*]furan, *see* H-60188

4,5-Dihydro-6-hydroxy-3-oxo-8′-apo-ε-caroten-8′-oic acid, *in* C-60156

▷ 1,3-Dihydro-1-hydroxy-3-oxo-1,2-benziodoxole, *see* H-60102

3,4-Dihydro-3-(3-hydroxyphenyl)-1*H*-2-benzopyran-1-one, *in* H-70207

3,4-Dihydro-2-(4-hydroxyphenyl)-2*H*-1-benzopyran-3,5,7-triol, *see* T-60104

2,3-Dihydro-2-(4-hydroxyphenyl)-5-(3-hydroxypropyl)-3-methylbenzofuran, D-60244

2,3-Dihydro-2-(4-hydroxyphenyl)-7-methoxy-3-methyl-5-(1-propenyl)-benzofuran, *in* C-60161

2,3-Dihydro-2-(4-hydroxyphenyl)-3-methyl-5-benzofuranpropanol, *see* D-60244

2,3-Dihydro-2-(4-hydroxyphenyl)-3-methyl-5-(1-propenyl)benzofuran, *see* C-60161

9,10-Dihydro-7-hydroxy-2,3,4,6-tetramethoxyphenanthrene, *in* D-70246

11β,13-Dihydro-9α-hydroxyzaluzanin C, *in* Z-60001

4,5-Dihydro-1*H*-imidazole, D-60245

1-(4,5-Dihydro-1*H*-imidazol-2-yl)-2-imidazolidinethione, D-60246

1,3-Dihydro-2*H*-imidazo[4,5-*b*]pyrazine-2-thione, D-60247

1,5-Dihydro-6*H*-imidazo[4,5-*c*]pyridazine-6-thione, D-70215

1,5-Dihydro-6*H*-imidazo[4,5-*c*]pyridazin-6-one, D-60248

2,3-Dihydroimidazo[1,2-*a*]pyridine, D-70216

1,3-Dihydro-2*H*-imidazo[4,5-*b*]pyridin-2-one, D-60249

1,3-Dihydro-2*H*-imidazo[4,5-*c*]pyridin-2-one, D-60250

1,4-Dihydro-5-imino-3,4-dimethyltetrazole, *in* A-70202

4,5-Dihydro-5-imino-*N*,4-diphenyl-1,2,4-thiadiazol-3-amine, *see* H-60012

5,6-Dihydro-4*H*-indene, D-70217

1,3-Dihydro-1,1,3,3-2*H*-indene-2-selone, *see* T-60142

5,10-Dihydroindeno[2,1-*a*]indene, D-70218

4*b*,9*a*-Dihydroindeno[1,2-*a*]indene-9,10-dione, D-70219

4*b*,9*b*-Dihydroindeno[2,1-*a*]indene-5,10-dione, D-70220

1-(2,3-Dihydro-1*H*-inden-1-ylidene)-2,3-dihydro-1*H*-indene, *see* B-70120

2,3-Dihydro-1*H*-indole-2-sulfonic acid, D-70221

2,3-Dihydro-5(1*H*)-indolizinone, D-60251

11β,13-Dihydroinuviscolide, *in* I-70023

11α,13-Dihydroinuviscolide, *in* I-70023
Dihydro-3-iodo-2(3*H*)-furanone, D-60252
4,7-Dihydroisobenzofuran, D-60253
1,3-Dihydro-1-isobenzofuranol, *see* H-60171
2,3-Dihydro-1*H*-isoindol-1-one, *see* P-70100
3,4-Dihydro-8-isopropyl-5-methyl-2-
    naphthalenecarboxaldehyde, *in* I-60129
7,8-Dihydro-5(6*H*)-isoquinolinone, D-60254
2,3-Dihydrolinderazulene, *in* L-70031
11,12-Dihydrolinderazulene, *in* L-70031
Dihydromatricaria acid, *see* D-70018
7,12-Dihydro-5*H*-6,12-methanodibenz[*c,f*]-
    azocine, D-60255
1,4-Dihydro-1,4-methanonaphthalene, D-60256
2,3-Dihydro-1,2,3-metheno-1*H*-indene, *see*
    N-60012
5,6-Dihydro-4,5,6-metheno-4*H*-indene, *see*
    T-70019
3,4-Dihydro-6-methoxy-2*H*-1-benzopyran, *in*
    D-70175
3,4-Dihydro-8-methoxy-2*H*-1-benzopyran, *in*
    D-60203
3,4-Dihydro-5-methoxy-8,8-dimethyl-2-phenyl-
    2*H*,8*H*-benzo[1,2-*b*:3,4-*b'*]-dipyran-4-ol, *see*
    T-70007
4,5-Dihydro-3-(methoxymethylene)-2(3*H*)-
    furanone, D-70222
8,9-Dihydro-5-methoxy-8-(1-methylethenyl)-2-
    phenyl-2*H*-furo[2,3-*h*]-1-benzopyran, *see*
    A-70002
8,8*a*-Dihydro-3-methoxy-5-methyl-1*H*,6*H*-
    furo[3,4-*e*][1,3,2]dioxaphosphepin 3-oxide,
    D-70223
2,3-Dihydro-7-methoxy-3-methyl-5-(1-
    propenyl)-2-(3,4,5-trimethoxyphenyl)-
    benzofuran, *see* L-60024
15,16-Dihydro-15-methoxy-16-oxohardwickiic
    acid, *in* H-70002
15,16-Dihydro-15-methoxy-16-oxonidoresedic
    acid, *in* N-70035
3,4-Dihydro-5-methoxy-4-oxo-2-(1-pentenyl)-
    2*H*-benzopyran-7-acetic acid, *see* C-60025
15,16-Dihydro-15-methoxy-16-oxostrictic acid,
    in S-70077
9,10-Dihydro-7-methoxy-5*H*-phenanthro[4,5-
    bcd]pyran-2,8-diol, *see* F-70010
3,4-Dihydro-5-methoxy-2*H*-pyrrole, *in* P-70188
1,2-Dihydro-6-methylbenz[*j*]aceanthrylene, *see*
    M-70056
10,11-Dihydro-5-methyl-5*H*-dibenz[*b,f*]-
    azepine, *in* D-70185
2,3-Dihydro-1-methylene-1*H*-indene, *see*
    M-60078
3,4-Dihydro-1(2*H*)-methylenenaphthalene, *see*
    T-60073
Dihydro-3-methylene-4-octylfuro[3,4-*b*]-furan-
    2,6(3*H*,4*H*)-dione, *see* A-60325
1-[2,3-Dihydro-2-(1-methylethenyl)-5-
    benzofuranyl]-1,2-propanediol, *see* D-70340
1,3-Dihydro-3-(1-methylethyl)-2*H*-indol-2-one,
    see I-70097
5-(7,10-Dihydro-8-methylfuro[2,3-*g*][1]-
    benzoxepin-2-yl)-1,3-benzenediol, *see*
    M-60145
1,3-Dihydro-1-methyl-2*H*-imidazo[4,5-*b*]-
    pyrazine-2-thione, *in* D-60247
1,3-Dihydro-1-methyl-2*H*-imidazo[4,5-*b*]-
    pyrazin-2-one, *in* I-60006
Dihydro-4-methyl-5-(3-methyl-2-butenyl)-
    2(3*H*)-furanone, *see* E-70004
1,2-Dihydro-1-methyl-2-oxoadenosine, *see*
    D-60518
4-(1,5-Dihydro-3-methyl-5-oxo-1-phenyl-4*H*-
    pyrazol-4-ylidene)-2,4-dihydro-5-methyl-2-
    phenyl-3*H*-pyrazol-3-one, *see* P-70153
N-(2,9-Dihydro-1-methyl-2-oxo-1*H*-purin-6-yl)-
    acetamide, *in* A-60144
4-[2,3-Dihydro-3-methyl-5-(1-propenyl)-2-
    benzofuranyl]phenol, *see* C-60161
4,5-Dihydro-2-(1-methylpropyl)thiazole,
    D-70224
3,9-Dihydro-9-methyl-1*H*-purine-2,6-dione,
    D-60257
1,4-Dihydro-2-methylpyrimidine, D-70225

1,6-Dihydro-2-methylpyrimidine, *see* D-70225
3,4-Dihydro-5-methyl-2*H*-pyrrole, D-70226
2,3-Dihydro-2-methylthiophene, D-70227
2,3-Dihydro-4-methylthiophene, D-70228
2,3-Dihydro-5-methylthiophene, D-70229
2,5-Dihydro-2-methylthiophene, D-70230
2,5-Dihydro-3-methylthiophene, D-70231
5,6-Dihydro-2-(methylthio)-4*H*-1,3-thiazine, *in*
    T-60094
4,5-Dihydro-2-(methylthio)thiazole, *in* T-60197
▷ 2,3-Dihydro-6-methyl-2-thioxo-4(1*H*)-
    pyrimidinone, D-70232
3,4-Dihydro-5-methyl-4-thioxo-2(1*H*)-
    pyrimidinone, D-70233
3,4-Dihydro-6-methyl-4-thioxo-2(1*H*)-
    pyrimidinone, D-70234
10,11-Dihydromicroglossic acid, *in* M-60134
10,11-Dihydro-5-*O*-mycaminosylnarbonolide,
    see K-60009
Dihydromyoporone, *in* M-60156
1,4-Dihydro-1-naphthalenamine, *see* A-60139
5,8-Dihydro-1-naphthalenamine, *see* A-60140
1-(3,4-Dihydro-1(2*H*)-naphthalenylidene)-
    1,2,3,4-tetrahydronaphthalene, *see* B-70165
2,3-Dihydronaphtho[2,3-*b*]furan, D-70235
1*a*,7*a*-Dihydronaphth[2,3-*b*]oxirene-2,7-dione,
    see E-70011
1,4-Dihydro-1-naphthylamine, *see* A-60139
5,8-Dihydro-1-naphthylamine, *see* A-60140
Dihydronarigenin, *see* H-60219
Dihydroniloticin, *in* N-70036
2,3-Dihydro-4-nitro-2,3-dioxo-9,10-
    secostrychnidin-10-oic acid, *see* C-70001
2,5-Dihydro-3-nitrofuran, D-70236
11-Dihydro-12-norneoquassin, D-60258
9,11-Dihydro-22,25-oxido-11-
    oxoholothurinogenin, D-70237
7,8-Dihydro-8-oxoguanosine, D-70238
15,16-Dihydro-16-oxohardwickiic acid, *in*
    H-70002
2,3-Dihydro-2-oxo-1*H*-indole-3-acetic acid,
    D-60259
4,5-Dihydro-2-oxo-3*H*-oxazole-4-carboxylic
    acid, *see* O-60082
3,4-Dihydro-4-oxopyrido[2,3-*d*]pyrimidine-
    2(1*H*)-thione, *see* D-60291
5,6-Dihydro-4-oxopyrimidine, *see* D-60274
N-(1,4-Dihydro-4-oxo-5-pyrimidinyl)-
    formamide, *in* A-70196
2,3-Dihydro-3-oxo-1*H*-pyrrole-1-
    carboxaldehyde, *in* D-70262
Dihydropallescensin 2, *see* P-60015
Dihydroparthenolide, D-70239
9,10-Dihydropentafulvalene, *see* B-60100
9,10-Dihydro-2,3,4,6,7-penta-
    hydroxyphenanthrene, *see* D-70246
1,2-Dihydropentalene, D-70240
1,4-Dihydropentalene, D-70241
1,5-Dihydropentalene, D-70242
1,6-Dihydropentalene, D-70243
6*a*-Dihydropentalene, D-70244
▷ Dihydro-5-pentyl-2(3*H*)-furanone, D-70245
9,10-Dihydro-2-phenanthrenamine, *see* A-60141
9,10-Dihydro-4-phenanthrenamine, *see* A-60142
9,10-Dihydro-9-phenanthrenamine, *see* A-60143
9,10-Dihydrophenanthrene, D-60260
9,10-Dihydro-2,3,4,6,7-phenanthrenepentol,
    D-70246
9,10-Dihydro-2,4,5,6-phenanthrenetetrol,
    D-70247
9,10-Dihydro-2-phenanthrylamine, *see* A-60141
9,10-Dihydro-4-phenanthrylamine, *see* A-60142
9,10-Dihydro-9-phenanthrylamine, *see* A-60143
2,3-Dihydro-2-phenyl-1,2-benzisothiazole,
    D-60261
2,3-Dihydro-2-phenyl-2-benzofurancarboxylic
    acid, D-60262
2,3-Dihydro-2-phenyl-4*H*-benzo[*b*]thiin-4-one,
    see D-70248
2,3-Dihydro-2-phenyl-4*H*-1-benzothiopyran-4-
    one, D-70248
10,11-Dihydro-5-phenyl-5*H*-dibenz[*b,f*]-
    azepine, *in* D-70185
2,3-Dihydro-3-(phenylmethylene)-4*H*-1-
    benzopyran-4-one, D-60263

1,4-Dihydro-1-(phenylmethyl)-3-
    pyridinecarboxamide, *see* B-60067
1,2-Dihydro-1-phenyl-1-naphthalenecarboxylic
    acid, D-60264
3,4-Dihydro-2-phenyl-1(2*H*)-naphthalenone,
    D-60265
3,4-Dihydro-3-phenyl-1(2*H*)-naphthalenone,
    D-60266
3,4-Dihydro-4-phenyl-1(2*H*)-naphthalenone,
    D-60267
1,2-Dihydro-1-phenyl-1-naphthoic acid, *see*
    D-60264
1,4-Dihydro-2-phenylpyrimidine, D-70249
1,6-Dihydro-2-phenylpyrimidine, *see* D-70249
4,5-Dihydro-2-phenyl-6(1*H*)-pyrimidinone, *see*
    D-70250
5,6-Dihydro-2-phenyl-4(1*H*)-pyrimidinone,
    D-70250
3,4-Dihydro-5-phenyl-2*H*-pyrrole, D-70251
1,2-Dihydro-1-phenyl-3*H*-pyrrol-3-one, *in*
    D-70262
2,3-Dihydro-6-phenyl-2-thioxo-4(1*H*)-
    pyrimidinone, D-60268
Dihydropinosylvin, *see* P-60098
7,12-Dihydropleiadene, D-60269
11,13-Dihydropsilostachyin, *in* P-70131
1,5-Dihydro-2,4,6(3*H*)-pteridinetrione, *see*
    P-70134
▷ 1,7-Dihydro-6*H*-purine-6-thione, D-60270
3,4-Dihydro-2*H*-pyran-5-carboxaldehyde,
    D-70252
1-(3,4-Dihydro-2*H*-pyran-5-yl)-2,2,2-
    trifluoroethanone, *see* D-70275
4,5-Dihydro-1*H*-pyrazole, D-70253
1,8-Dihydro-5*H*-pyrazolo[4,3-*g*]quinazolin-
    5,7(6*H*)-dione, *see* P-60203
1,6-Dihydro-5*H*-pyrazolo[4,3-*g*]quinazolin-5-
    one, *see* P-60205
3,8-Dihydro-9*H*-pyrazolo[4,3-*f*]quinazolin-9-
    one, D-60271
4,5-Dihydro-3(2*H*)-pyridazinone, D-70272
7,12-Dihydropyrido[3,2-*b*:5,4-*b'*]diindole,
    D-70254
4,9-Dihydro-3*H*-pyrido[3,4-*b*]indole, *see*
    D-60209
Dihydropyrimidine, *see* D-70255
1,4-Dihydropyrimidine, D-70255
1,6-Dihydropyrimidine, *see* D-70255
2,5-Dihydro-4,6-pyrimidinediol, *see* D-70256
Dihydro-4,6-(1*H*,5*H*)-pyrimidinedione, D-70256
2,5-Dihydro-4,6-pyrimidinedithiol, D-70257
3,4-Dihydro-2(1*H*)-pyrimidinone, D-60273
5,6-Dihydro-4(1*H*)-pyrimidinone, D-60274
6,7-Dihydro-5*H*-1-pyrindine, D-60275
6,7-Dihydro-5*H*-1-pyrindine, D-70258
6,7-Dihydro-5*H*-2-pyrindine, D-60276
6,7-Dihydro-5*H*-2-pyrindine, D-70259
Dihydro-1*H*-pyrrolizine-3,5(2*H*,6*H*)-dione,
    D-70260
2,3-Dihydro-1*H*-pyrrolizin-1-ol, D-70261
2,3-Dihydro-1*H*-pyrrolizin-1-one, D-70277
2,5-Dihydro-1*H*-pyrrolo[3,2-*c*:4,5-*c'*]-dipyridin-
    1-one, *see* D-60237
1,2-Dihydro-3*H*-pyrrol-3-one, D-70262
2,3-Dihydro-1*H*-pyrrolo[2,3-*b*]pyridine,
    D-70263
▷ Dihydroquercetin, *see* P-70027
3,4-Dihydro-2(1*H*)-quinazolinone, D-60278
1,4-Dihydro-2,3-quinoxalinedione, D-60279
5,14-Dihydroquinoxalino[2,3-*b*]phenazine,
    D-70264
Dihydroridentin, *in* R-70007
Dihydrosamidin, *in* K-70013
2,3-Dihydro-2-selenoxo-4(1*H*)-quinazolinone,
    D-60281
3,4-Dihydro-4-selenoxo-2(1*H*)-quinazolinone,
    D-60280
Dihydroserruloside, D-70265
Dihydrosuberenol, *in* S-70080
6*a*,12*a*-Dihydro-2,3,8,10-tetrahydroxy[2]-
    benzopyrano[4,3-*b*][1]benzopyran-7(5*H*)-
    one, *see* C-70200
10,11-Dihydro-2,4,7,8-tetrahydroxy-10-(3,4-
    dihydroxyphenyl)-5-[(3,5-dihydroxyphenyl)-
    methyl]-5*H*-dibenzo[*a,d*]cycloheptene, *see*
    C-70025
5,6-Dihydro-6-(2,4,6,10-tetra-
    hydroxyheneicosyl)-2*H*-pyran-2-one, *in*
    D-70276

9,10-Dihydro-3,4,6,7-tetramethoxy-2-phenanthrenol, in D-70246

4,7-Dihydro-1,1,3,3-tetramethyl-1H,3H-3a,7a-episeleno-4,7-epoxyisobenzofuran, D-70266

Dihydro-3,3,4,4-tetramethyl-2(3H)-furanthione, D-60282

17,17a-Dihydro-4,4,8,17-tetramethyl-3′H-indeno[1′,2′:17,17a]-D-homoandrost-17-ene-4′,6′,7′-triol, see D-60499

1,3-Dihydro-1,1,3,3-tetramethyl-2H-inden-2-one, see T-60143

4,5-Dihydro-3,3,5,5-tetramethyl-4-methylene-3H-pyrazole, D-60283

2,5-Dihydro-2,2,5,5-tetramethyl-3-[[[(2-phenyl-3H-indol-3-ylidene)amino]oxy]-carbonyl]-1H-pyrrol-1-yloxy, D-60284

3,5-Dihydro-3,3,5,5-tetramethyl-4H-pyrazole-4-thione, D-60285

3,5-Dihydro-3,3,5,5-tetramethyl-4H-pyrazol-4-one, D-60286

3,6-Dihydro-3,3,6,6-tetramethyl-2(1H)-pyridinone, D-70267

4,5-Dihydro-3,3,4,4-tetramethyl-2(3H)-thiophenethione, D-60287

4,5-Dihydro-3,3,4,4-tetramethyl-2(3H)-thiophenone, D-60288

1,2-Dihydro-1,2,4,5-tetrazine-3,6-dione, D-70268

1,4-Dihydro-5H-tetrazole-5-thione, see T-60174

1,4-Dihydro-5H-tetrazol-5-one, see T-60175

1-(4,5-Dihydro-2-thiazolyl)ethanone, see A-70041

1,4-Dihydrothieno[3,4-d]pyridazine, D-60289

2,3-Dihydrothieno[2,3-b]pyridine, D-70269

2,3-Dihydrothiophene, D-60290

2,5-Dihydrothiophene, D-70270

7,8-Dihydro-8-thioxoguanosine, see M-70029

2,3-Dihydro-2-thioxopyrido[2,3-d]-pyrimidin-4(1H)-one, D-60291

2,3-Dihydro-2-thioxopyrido[3,2-d]-pyrimidin-4(1H)-one, D-60292

2,3-Dihydro-2-thioxopyrido[3,4-d]-pyrimidin-4(1H)-one, D-60293

2,3-Dihydro-2-thioxo-4H-pyrido[1,2-a]-1,3,5-triazin-4-one, D-60294

2,3-Dihydro-2-thioxo-4(1H)-quinazolinone, D-60296

2,3-Dihydro-4-thioxo-2(1H)-quinazolinone, D-60295

Dihydrotochuinyl acetate, in T-70202

Dihydro-1,3,5-triazine-2,4(1H,3H)-dione, D-60297

▷ 1,2-Dihydro-3H-1,2,4-triazole-3-thione, D-60298

2,3-Dihydro-4-(trichloroacetyl)furan, D-70271

3,4-Dihydro-5-(trichloroacetyl)-2H-pyran, D-70272

Dihydrotrichostin, D-70273

2,14-Dihydrotricyclo[7.3.2.0^{5.13}]-tetradeca-1,3,5,7,9,11-hexaen-2-one, see M-70042

2,3-Dihydro-4-(trifluoroacetyl)furan, D-70274

3,4-Dihydro-5-(trifluoroacetyl)-2H-pyran, D-70275

Dihydro-4-(trifluoromethyl)-2(3H)-furanone, D-60299

Dihydro-5-(trifluoromethyl)-2(3H)-furanone, D-60300

6a,12a-Dihydro-2,3,10-trihydroxy[2]-benzopyrano[4,3-b][1]benzopyran-7(5H)-one, C-70200

5,6-Dihydro-6-(2,4,6-trihydroxyheneicosyl)-2H-pyran-2-one, D-70276

3,4-Dihydro-3,5,7-trihydroxy-2-(4-hydroxyphenyl)-8-(3-methyl-2-butenyl)-2H-benzopyran, see T-60125

6a,12a-Dihydro-2,3,10-trihydroxy-8-methoxy[2]benzopyrano[4,3-b][1]-benzopyran-7(5H)-one, in C-70200

6a,12a-Dihydro-3,4,10-trihydroxy-8-methoxy[2]benzopyrano[4,3-b][1]-benzopyran-7(5H)-one, in C-70200

7,7a-Dihydro-3,6,7-trihydroxy-1a-(3-methyl-2-butenyl)naphth[2,3-b]oxiren-2(1aH)-one, D-70277

9,10-Dihydro-3,6,8-trihydroxy-1-methyl-9,10-dioxo-2-anthracenecarboxylic acid, see T-60332

4,10-Dihydro-3,7,8-trihydroxy-3-methyl-10-oxo-1H,3H-pyrano[4,3-b][1]-benzopyran-9-carboxylic acid, see F-60081

9,10-Dihydro-3,4,6-trimethoxy-2,7-phenanthrenediol, in D-70246

9,10-Dihydro-5,6,7-trimethoxy-2,3-phenanthrenediol, in D-70246

9,10-Dihydro-5,6,8-trimethoxy-3,4-phenanthrenediol, in P-70034

1,4-Dihydro-4,6,7-trimethyl-9H-imidazo[1,2-a]purin-9-one, D-70278

3,4-Dihydro-3,3,8a-trimethyl-1,6(2H,8aH)-naphthalenedione, D-60301

3,4-Dihydro-4,6,7-trimethyl-3-β-D-ribofuranosyl-9H-imidazo[1,2-a]purin-9-one, D-60302

3,5-Dihydro-3,5,5-trimethyl-4H-triazol-4-one, D-60303

5,6-Dihydro-6-[3,5,6-tris(acetyloxy)-1-heptenyl]-2H-pyran-2-one, see H-70236

5,6-Dihydro-6-[1,2,3-tris(acetyloxy)-heptyl]-2H-pyran-2-one, see B-60193

Dihydrotropidine, in A-70278

Dihydrotucumanoic acid, in T-60410

Dihydro-3-vinyl-2(3H)-furanone, D-70279

2β,12-Dihydroxy-8,12-abietadiene-11,14-dione, in R-60015

7,12-Dihydroxy-8,12-abietadiene-11,14-dione, D-60304

11,12-Dihydroxy-6,8,11,13-abietatetraen-20-oic acid, D-60305

2α,11-Dihydroxy-7,9(11),13-abietatriene-6,12-dione, in D-70282

7,15-Dihydroxy-8,11,13-abietatrien-18-oic acid, D-70280

2,12-Dihydroxy-8,11,13-abietatrien-7-one, D-70281

3,12-Dihydroxy-8,11,13-abietatrien-1-one, D-60306

2,11-Dihydroxy-7,9(11),13-abietatrien-12-one, D-70282

2,3-Dihydroxyaniline, see A-70089

▷ 3,4-Dihydroxyaniline, see A-70090

12,13-Dihydroxy-1(10)-aromadendren-2-one, D-70283

ent-16α,17-Dihydroxy-3-atisanone, D-70284

2,3-Dihydroxybenzaldehyde, D-70285

2,4-Dihydroxybenzeneacetaldehyde, see D-60369

2,5-Dihydroxybenzeneacetaldehyde, see D-60370

3,4-Dihydroxybenzeneacetaldehyde, see D-60371

4,5-Dihydroxy-1,3-benzenedicarboxylic acid, D-60307

2,5-Dihydroxybenzenemethanol, see D-70287

β,4-Dihydroxybenzenepentanoic acid, see H-70142

2,2-Dihydroxy-1H-benz[f]indene-1,3(2H)dione, D-60308

(6,7-Dihydroxy-1H-2-benzopyran-3-yl)(2,4-dihydroxyphenyl)methanone, see P-70017

(7,8-Dihydroxy-1H-2-benzopyran-3-yl)(2,4-dihydroxyphenyl)methanone, in P-70017

3,6-Dihydroxy-1,2-benzoquinone, D-70288

3-(2,4-Dihydroxybenzoyl)-6,7-dihydroxy-1H-2-benzopyran, see P-70017

3-(2,4-Dihydroxybenzoyl)-7,8-dihydroxy-1H-2-benzopyran, in P-70017

1-[3-(2,4-Dihydroxybenzoyl)-4,6-dihydroxy-2-(4-hydroxyphenyl)-7-benzofuranyl]-3-(4-hydroxyphenyl)-2-propen-1-one, see D-70286

2,5-Dihydroxybenzyl alcohol, D-70287

3-(3,4-Dihydroxybenzyl)-3,7-dihydroxy-4-chromanone, in D-70288

3-(3,4-Dihydroxybenzyl)-3,7-dihydroxy-4-methoxychroman, in D-70288

3-(3,4-Dihydroxybenzyl)-7-hydroxy-4-chromanone, D-70288

2-(3,4-Dihydroxybenzylidene)-6-hydroxy-3(2H)-benzofuranone, see S-70082

3-(3,4-Dihydroxybenzylidene)-7-hydroxy-4-chromanone, in D-70288

3-(3,4-Dihydroxybenzyl)-3,4,7-trihydroxychroman, in D-70288

7,7′-Dihydroxy-[6,8′-bi-2H-benzopyran]-2,2′-dione, see B-70093

2,3-Dihydroxybibenzyl, see P-60096

2,4-Dihydroxybibenzyl, see P-60097

2,5-Dihydroxybibenzyl, see P-60095

3,5-Dihydroxybibenzyl, see P-60098

6,7-Dihydroxy-3,7′-bicoumarin, see D-70006

7,7′-Dihydroxy-6,8′-bicoumarin, see B-70093

4,9-Dihydroxy-1,12-bis(2-hydroxypropyl)-2,6,7,11-tetramethoxy-3,10-perylenedione, see P-70098

1-[2,4-Dihydroxy-3,5-bis(3-methyl-2-butenyl)phenyl]-3-(4-hydroxyphenyl)-2-propen-1-one, D-60310

ent-6β,17-Dihydroxy-14,15-bisnor-7,11E-labdadien-13-one, D-60311

6,8′-Dihydroxy-2,2′-bis(2-phenylethyl)-[5,5′-bi-4H-1-benzopyran]-4,4′-dione, D-70289

6′,8-Dihydroxy-2,2′-bis(2-phenylethyl)-5,5′-bichromone, see D-70289

2,10-Dihydroxy-3-bornanone, in H-70173

3,10-Dihydroxyborneol, see H-70173

8,10-Dihydroxyborneol, see B-70146

▷ 3β,14β-Dihydroxy-5β-bufa-20,22-dienolide, see B-70289

2,3-Dihydroxybutanedioic acid, see T-70005

2,3-Dihydroxybutanoic acid, D-60312

3,4-Dihydroxybutanoic acid, D-60313

3,4-Dihydroxybutanoic acid γ-lactone, see D-60242

5-(3,4-Dihydroxy-1-butynyl)-5′-methyl-2,2′-bithiophene, see M-70054

3β,10α-Dihydroxy-4,11(13)-cadinadien-12-oic acid, see Z-70001

▷ 3-(3,4-Dihydroxycinnamoyl)quinic acid, see C-70005

ent-6β,15-Dihydroxy-3,13E-clerodadien-19-oic acid, see S-60003

3β,26-Dihydroxy-5,24E-cucurbitadien-11-one, in C-60175

3β,27-Dihydroxy-5,24Z-cucurbitadien-11-one, in C-60175

2,7-Dihydroxy-2,4,6-cycloheptatrien-1-one, D-60314

3,6-Dihydroxy-3,5-cyclohexadiene-1,2-dione, see D-60309

5,6-Dihydroxy-5-cyclohexene-1,2,3,4-tetrone, D-60315

7-(4,5-Dihydroxy-2-cyclopent-1-en-3-ylaminomethyl)-7-deazaguanine, see Q-70004

2,3-Dihydroxy-2-cyclopropen-1-one, D-70290

4β,6α-Dihydroxy-8-daucen-10-one, see L-60014

3,10-Dihydroxy-5,11-dielmenthadiene-4,9-dione, D-60316

4,6-Dihydroxy-2,5-dihydropyrimidine, see D-70256

4,6-Dihydroxy-3-(2,4-dihydroxybenzoyl)-7-(4-hydroxycinnamoyl)-2-(4-hydroxyphenyl)benzofuran, see D-70286

3′,4-Dihydroxy-3,5-dimethoxybibenzyl, in T-60102

2′,4-Dihydroxy-4′,6′-dimethoxychalcone, in H-60220

7,7′-Dihydroxy-4,4′-dimethoxy-5,5′-dimethyl[6,8′-bi-2H-1-benzopyran]-2,2′-dione, see D-60022

7,7′-Dihydroxy-4,4′-dimethoxy-5,5′-dimethyl-[8,8′-bi-2H-1-benzopyran]-2,2′-dione, see O-70046

5,7-Dihydroxy-3′,4′-dimethoxyflavan, in T-60103

5,7-Dihydroxy-2′,8-dimethoxyflavanone, in T-70104

4′,5-Dihydroxy-3′,7-dimethoxyflavone, D-70291

3,5-Dihydroxy-2,4-dimethoxy-9H-fluoren-9-one, in T-70109

2′,5-Dihydroxy-7,8-dimethoxyisoflavone, in T-70110

4′,5-Dihydroxy-3′,7-dimethoxyisoflavone, *in* T-60108

5,7-Dihydroxy-2′,6-dimethoxyisoflavone, *in* T-60107

5,7-Dihydroxy-3′,4′-dimethoxyisoflavone, *in* T-60108

5,8-Dihydroxy-2,3-dimethyl-6,7-methylenedioxy-1,4-naphthoquinone, *in* H-60065

2,3-Dihydroxy-5,7-dimethoxyphenanthrene, *in* T-60120

2,7-Dihydroxy-1,5-dimethoxyphenanthrene, *in* T-60116

5,8-Dihydroxy-6,10-dimethoxy-2-propyl-4H-naphtho[1,2-b]pyran-4-one, *in* C-70191

5,8-Dihydroxy-6,10-dimethoxy-2-propyl-4H-naphtho[2,3-b]pyran-4-one, D-70292

1,3-Dihydroxy-5,8-dimethoxyxanthone, *in* T-70123

1,8-Dihydroxy-3,5-dimethoxyxanthone, *in* T-70123

1,8-Dihydroxy-3,6-dimethoxyxanthone, *in* T-60127

2,6-Dihydroxy-3,5-dimethylbenzoic acid, D-70293

8,8′-Dihydroxy-3,3′-dimethyl[2,2′-binaphthalene]-1,1′,4,4′-tetrone, *see* B-60106

11,12-Dihydroxy-3,9-dimethyl-2,8-dioxacyclotetradeca-5,13-diene-1,7-dione, *see* C-60159

11,12-Dihydroxy-6,14-dimethyl-1,7-dioxacyclotetradeca-3,9-diene-2,8-dione, *see* C-60159

2,5-Dihydroxy-2,5-dimethyl-1,4-dithiane, *see* D-60421

3,3′-Dihydroxy-4,5;4′,5′-dimethylenedioxybibenzyl, *in* H-60062

5,8-Dihydroxy-2,3-dimethyl-1,4-naphthalenedione, *see* D-60317

5,8-Dihydroxy-2,3-dimethyl-1,4-naphthoquinone, D-60317

1,7-Dihydroxy-3,7-dimethyl-2,5-octadien-4-one, D-60318

2,7-Dihydroxy-1,6-dimethyl-5-vinylphenanthrene, D-70294

7β,8α-Dihydroxy-14,15-dinor-11-labden-13-one, *see* S-60052

3,5-Dihydroxy-1,7-diphenylheptane, *see* D-70481

1,5-Dihydroxy-1,7-diphenyl-3-heptanone, *in* D-60480

3,3-Dihydroxy-5,5-diphenyl-1,2,4,5-pentanetetrone, *in* D-60488

Dihydroxydiquinone, *see* D-60315

6,12-Dihydroxy-3,7-dolabelladiene, *see* D-70541

3,11-Dihydroxy-7-drimen-6-one, D-70295

5β,20R-Dihydroxyecdysone, *in* H-60023

5,20-Dihydroxy-6,8,11,14-eicosatetraenoic acid, D-70296

12,19-Dihydroxy-5,8,10,14-eicosatetraenoic acid, D-60319

15,20-Dihydroxy-5,8,11,13-eicosatetraenoic acid, D-70297

4,6-Dihydroxy-7,8-epoxy-20-nor-2-cembren-12-one, *in* D-70323

1,8-Dihydroxy-3,7(11)-eremophiladien-12,8-olide, D-60320

8α,9α-Dihydroxy-10βH-eremophil-11-en-2-one, *in* E-70030

1,5-Dihydroxyeriocephaloide, D-60321

▷ N,N′-Dihydroxyethanediamide, *see* O-60048

▷ 1,3-Dihydroxy-2-(ethoxymethyl)anthraquinone, *in* D-70306

1β,8β-Dihydroxy-4,11(13)eudesmadien-12-oic acid γ-lactone, *see* I-70109

1,8-Dihydroxy-3,7(11)-eudesmadien-12,8-olide, D-60322

1,5-Dihydroxy-2,4(15),11(13)-eudesmatrien-12,8-olide, D-70298

1α,3α-Dihydroxy-4-eudesmen-12,6α-olide, *see* I-70066

1β,3α-Dihydroxy-4-eudesmen-12,6α-olide, *in* I-70066

1,3-Dihydroxy-4-eudesmen-12,6-olide, D-70299

3β,19α-Dihydroxy-24-*trans*-feruryloxy-12-ursen-28-oic acid, *in* T-60348

2′,8-Dihydroxyflavone, D-70300

21,27-Dihydroxy-3-friedelanone, D-70301

3α,23R-Dihydroxy-17,13-friedo-9β-lanosta-7,12,24E-trien-26-oic acid, *see* M-60010

3α,23R-Dihydroxy-17,14-friedo-9β-lanosta-7,14,24E-trien-26-oic acid, *see* M-70012

21,27-Dihydroxy-D:A-friedooleanan-3-one, *see* D-70301

2,9-Dihydroxyfuranoeremophilane, *see* F-70047

2-(8,9-Dihydroxygeranylgeranyl)-5,6-dimethyl-1,4-benzoquinone, *in* S-60010

2-(8,9-Dihydroxygeranylgeranyl)-6-methyl-1,4-benzoquinone, *in* S-60010

1,4-Dihydroxy-5,10(14)-germacradiene, *see* G-70009

2,8-Dihydroxy-1(10),4,11(13)-germacratrien-12,6-olide, D-70302

3,13-Dihydroxy-1(10),4,7(11)-germacratrien-12,6-olide, D-70303

8,9-Dihydroxy-1(10),4,11(13)-germacratrien-12,6-olide, D-60323

9,15-Dihydroxy-1(10),4,11(13)-germacratrien-12,6-olide, D-60324

▷ Dihydroxyglyoxime, *see* O-60048

2,4-Dihydroxy-5,11(13)-guaiadien-12,8-olide, D-70304

4,10-Dihydroxy-2,11(13)-guaiadien-12,6-olide, D-60326

3,5-Dihydroxy-4(15),10(14)-guaiadien-12,8-olide, D-60325

3β,9β-Dihydroxy-4(15),10(14)-guaiadien-12,6α-olide, *in* D-70305

2,8-Dihydroxy-3,10(14),11(13)-guaiatrien-12,6-olide, D-60327

8,10-Dihydroxy-3-guaien-12,6-olide, D-60328

3,9-Dihydroxy-10(14)-guaien-12,6-olide, D-70305

2-(3,4-Dihydroxy-1,5-heptadienyl)-6-hydroxybenzaldehyde, *see* P-60216

5,7-Dihydroxy-2′,3,4′,5′,6,8-hexamethoxyflavone, *in* O-60022

5,7-Dihydroxy-3,3′,4′,5′,6,8-hexamethoxyflavone, *in* O-70018

5,6-Dihydroxyhexanoic acid, D-60329

3,4-Dihydroxyhydratropic acid, *see* D-60372

5,7-Dihydroxy-2-[4-hydroxy-3,5-bis(3-methyl-2-butenyl)phenyl]-4H-1-benzopyran-4-one, *see* H-70092

5,8-Dihydroxy-1-(1-hydroxyethyl)-7-methoxynaphtho[2,3-c]furan-4,9-dione, *see* N-60016

▷ 1,3-Dihydroxy-2-(hydroxymethyl)-9,10-anthracenedione, *see* D-70306

▷ 1,3-Dihydroxy-2-hydroxymethylanthraquinone, D-70306

3β,7β-Dihydroxy-4α-hydroxymethyl-4β,14α-dimethyl-11,15-dioxo-5α-chol-8-en-24-oic acid, *see* L-60036

3β,12β-Dihydroxy-4α-hydroxymethyl-4β,14α-dimethyl-7,11,15-trioxo-5α-chol-8-en-24-oic acid, *in* L-60036

5,8-Dihydroxy-2-(hydroxymethyl)-6-methoxy-3-(2-oxopropyl)-1,4-naphthalenedione, *see* F-60108

5,8-Dihydroxy-2-(1-hydroxy-4-methyl-3-pentenyl)-1,4-naphthalenedione, *see* S-70038

▷ 5,7-Dihydroxy-3-(4-hydroxyphenyl)-4H-1-benzopyran-4-one, *see* T-60324

7,8-Dihydroxy-3-(4-hydroxyphenyl)-4H-1-benzopyran-4-one, *see* T-70265

5,7-Dihydroxy-2-(4-hydroxyphenyl)-3-methoxy-6-(3-methyl-2-butenyl)-4H-1-benzopyran-4-one, *see* T-60224

1,4-Dihydroxy-10-imino-9(10H)-anthracenone, *see* A-60181

5,6-Dihydroxyindole, D-70307

5,6-Dihydroxy-1H-indole-2-carboxylic acid, D-70308

2,3-Dihydroxy-6-iodobenzoic acid, D-60330

2,4-Dihydroxy-3-iodobenzoic acid, D-60331

2,4-Dihydroxy-5-iodobenzoic acid, D-60332

2,5-Dihydroxy-4-iodobenzoic acid, D-60333

2,6-Dihydroxy-3-iodobenzoic acid, D-60334

3,4-Dihydroxy-5-iodobenzoic acid, D-60335

3,5-Dihydroxy-2-iodobenzoic acid, D-60336

3,5-Dihydroxy-4-iodobenzoic acid, D-60337

4,5-Dihydroxy-2-iodobenzoic acid, D-60338

5,7-Dihydroxy-1(3H)-isobenzofuranone, D-60339

5,7-Dihydroxy-8-isobutyryl-2,2,6-trimethylchromene, *see* D-70348

4′,7-Dihydroxyisoflavone, D-60340

4,5-Dihydroxyisophthalic acid, *see* D-60307

2,5-Dihydroxy-4-isopropylbenzyl alcohol, D-70309

3α,6β-Dihydroxyivangustin, *in* I-70109

ent-3β,19-Dihydroxy-15-kauren-17-oic acid, D-60341

3,15-Dihydroxy-8(17),13-labdadien-19-al, D-70310

8,15-Dihydroxy-13-labden-19-al, D-70311

6,15-Dihydroxy-7-labden-17-oic acid, D-60342

3β,15-Dihydroxy-8-labden-7-one, *in* L-70013

3,8-Dihydroxylactariusfuran, D-70312

4,8-Dihydroxylactariusfuran, D-70313

24,25-Dihydroxy-7,9(11)-lanostadien-3-one, D-60343

3,15-Dihydroxy-7,9(11),24-lanostatrien-26-oic acid, D-70314

7,13-Dihydroxy-2-longipinen-1-one, D-70315

3,23-Dihydroxy-20(29)-lupen-30-al, *see* D-70316

3,23-Dihydroxy-20(30)-lupen-29-al, D-70316

3,23-Dihydroxy-20(29)-lupen-28-oic acid, D-70317

1,11-Dihydroxy-20(29)-lupen-3-one, D-60345

23,30-Dihydroxy-20(29)-lupen-3-one, D-70318

9,10-Dihydroxy-7-marasmen-5,13-olide, D-60346

2,4-Dihydroxy-N-[3-[(3-mercaptopropyl)amino]-3-oxopropyl]-3,3-dimethylbutanamide, *see* H-60090

2′,6′-Dihydroxy-4′-methoxyacetophenone, *in* T-70248

3,4-Dihydroxy-5-methoxybenzeneacetic acid, *in* T-60343

3,5-Dihydroxy-4-methoxybenzeneacetic acid, *in* T-60343

3,9-Dihydroxy-2-methoxycoumestone, *in* T-70249

2′,7-Dihydroxy-4′-methoxy-4-(2′,7-dihydroxy-4′-methoxyisoflavan-5′-yl)-isoflavan, D-60347

7,10-Dihydroxy-11-methoxydracaenone, *in* D-70546

3β,5α-Dihydroxy-6β-methoxy-7,22-ergostadiene, *in* M-70057

5,8-Dihydroxy-7-methoxyflavanone, *in* T-60321

▷ 5,7-Dihydroxy-4′-methoxyflavone, D-70319

2,5-Dihydroxy-4-methoxy-9H-fluoren-9-one, *in* T-70261

3′,7-Dihydroxy-4′-methoxyisoflavanone, *in* T-70263

4′,6-Dihydroxy-7-methoxyisoflavone, *in* T-60325

7,8-Dihydroxy-4′-methoxyisoflavone, *in* T-70265

1,8-Dihydroxy-6-methoxy-2-methylanthraquinone, *in* T-70269

1,8-Dihydroxy-3-methoxy-7-methylanthraquinone (incorr.), *in* T-70269

2,5-Dihydroxy-8-methoxy-3-methyl-6,7-methylenedioxy-1,4-naphthoquinone, *in* T-60333

2,5-Dihydroxy-4-methoxyphenanthrene, *in* P-60071

3,4-Dihydroxy-5-methoxyphenylacetic acid, *in* T-60343

3,5-Dihydroxy-4-methoxyphenylacetic acid, *in* T-60343

▷ 5,7-Dihydroxy-2-(4-methoxyphenyl)-4H-1-benzopyran-4-one, *see* D-70319

1-(2,6-Dihydroxy-4-methoxyphenyl)-3-(4-hydroxyphenyl)-1-propanone, *in* H-60219

1-(2,3-Dihydroxy-4-methoxyphenyl)-2-(3,4,5-trimethoxyphenyl)ethane, *in* T-60344

1-(2,3-Dihydroxy-4-methoxyphenyl)-2-(3,4,5-trimethoxyphenyl)ethylene, *in* T-60344

5,8-Dihydroxy-10-methoxy-2-propyl-4H-naphtho[1,2-b]pyran-4-one, *see* C-70191

5,10-Dihydroxy-8-methoxy-2-propyl-4*H*-naphtho[1,2-*b*]pyran-4-one, *in* C-70191

9,10-Dihydroxy-5-methoxy-2*H*-pyrano[2,3,4-*kl*]xanthen-9-one, *in* T-70273

1,5-Dihydroxy-6-methoxyxanthone, *in* T-60349

1,6-Dihydroxy-5-methoxyxanthone, *in* T-60349

5,6-Dihydroxy-1-methoxyxanthone, *in* T-60349

3,4-Dihydroxy-α-methylbenzeneacetic acid, *see* D-60372

8,9-Dihydroxy-1-methyl-6*H*-benzofuro[3,2-*c*][1]benzopyran-6-one, *see* M-70161

8,11-Dihydroxy-12*b*-methyl-1*H*-benzo[6,7]-phenanthro[10,1-*bc*]furan-3,6(2*H*,12*bH*)-dione, *see* H-60002

4,7-Dihydroxy-5-methyl-2*H*-1-benzopyran-2-one, D-60348

2,3-Dihydroxy-5-methyl-1,4-benzoquinone, D-70320

3,6-Dihydroxy-1*a*-(3-methyl-2-butenyl)-naphth[2,3-*b*]-2,7(1*aH*,7*aH*)-dione, *in* D-70277

2-[3,4-Dihydroxy-5-(3-methyl-2-butenyl)-phenyl]-2,3-dihydro-5,7-dihydroxy-4*H*-1-benzopyran-4-one, *see* S-70041

N-(2,3-Dihydroxy-3-methylbutyl)acetamide, *in* A-60216

4,7-Dihydroxy-5-methylcoumarin, *see* D-60348

2,3-Dihydroxy-5-methyl-2,5-cyclohexadiene-1,4-dione, *see* D-70320

1,3-Dihydroxy-4-methyl-6,8-decadien-5-one, D-70321

3,5-Dihydroxy-6,7-methylenedioxyflavanone, *in* T-60105

2′,5-Dihydroxy-6,7-methylenedioxyisoflavone, *in* T-60107

2-(1,2-Dihydroxy-1-methylethyl)-2,3-dihydro-5-benzofurancarboxaldehyde, *in* D-70214

2,4-Dihydroxy-3-methyl-2-hexenoic acid γ-lactone, *see* E-60051

4,4′-Dihydroxy-α-methylhydrobenzoin, *see* B-70147

5,7-Dihydroxy-6-methyl-1(3*H*)-isobenzofuranone, D-60349

1,2-Dihydroxy-3-methylnaphthalene, *see* M-60093

1,2-Dihydroxy-6-methylnaphthalene, *see* M-60095

1,3-Dihydroxy-2-methylnaphthalene, *see* M-60091

1,5-Dihydroxy-2-methylnaphthalene, *see* M-60092

2,4-Dihydroxy-1-methylnaphthalene, *see* M-60094

3,5-Dihydroxy-2-methyl-1,4-naphthalenedione, *see* D-60350

5,6-Dihydroxy-2-methyl-1,4-naphthalenedione, *see* D-60351

5,8-Dihydroxy-2-methyl-1,4-naphthalenedione, *see* D-60352

3,5-Dihydroxy-2-methyl-1,4-naphthoquinone, D-60350

5,6-Dihydroxy-2-methyl-1,4-naphthoquinone, D-60351

5,8-Dihydroxy-2-methyl-1,4-naphthoquinone, D-60352

4,5-Dihydroxy-3-methyl-7-oxabicyclo[4.1.0]-hept-3-en-2-one, *see* T-70009

1-[4,5-Dihydroxy-*N*-(12-methyl-1-oxotetra-decyl)ornithine]-5-serineechinocandin B, *see* M-70157

2,5-Dihydroxy-3-methylpentanoic acid, D-60353

16-(2,5-Dihydroxy-3-methylphenyl)-2,6,10,19-tetramethyl-2,6,10,14-hexadecatetraene-4,12-dione, *in* H-70225

5,7-Dihydroxy-6-methylphthalide, *see* D-60349

8′,9′-Dihydroxy-5-methylsargaquinone, *in* S-60010

2,5-Dihydroxy-1,4-naphthalenedione, *see* D-60354

2,8-Dihydroxy-1,4-naphthalenedione, *see* D-60355

▷5,8-Dihydroxy-1,4-naphthalenedione, *see* D-60356

2,5-Dihydroxy-1,4-naphthoquinone, D-60354

2,8-Dihydroxy-1,4-naphthoquinone, D-60355

3,5-Dihydroxy-1,4-naphthoquinone, *see* D-60355

▷5,8-Dihydroxy-1,4-naphthoquinone, D-60356

18,19-Dihydroxynerylgeraniol, *see* H-70179

2′,3′-Dihydroxy-6′-nitroacetophenone, D-60357

2′,5′-Dihydroxy-4′-nitroacetophenone, D-70322

3′,6′-Dihydroxy-2′-nitroacetophenone, D-60358

4′,5′-Dihydroxy-2′-nitroacetophenone, D-60359

4,5-Dihydroxy-2-nitrobenzoic acid, D-60360

1-(2,3-Dihydroxy-6-nitrophenyl)ethanone, *see* D-60357

1-(2,5-Dihydroxy-4-nitrophenyl)ethanone, *see* D-70322

1-(3,6-Dihydroxy-2-nitrophenyl)ethanone, *see* D-60358

1-(4,5-Dihydroxy-2-nitrophenyl)ethanone, *see* D-60359

12,16-Dihydroxy-20-nor-5(10),6,8,12-abietatetraene-11,14-dione, *see* N-60017

4,6-Dihydroxy-20-nor-2,7-cembradien-12-one, D-70323

ent-10β,13-Dihydroxy-20-nor-16-gibberellene-7,19-dioic acid 19,10-lactone, *see* G-70011

5,7-Dihydroxy-13-nor-8-marasmanone, D-60361

5,8-Dihydroxy-13-nor-7-marasmanone, D-60362

3,23-Dihydroxy-30-nor-12,20(29)-oleanadien-28-oic acid, D-70324

3,18-Dihydroxy-28-nor-12-oleanen-16-one, D-60363

ent-15,16-Dihydroxy-19-nor-4-rosen-3-one, *see* N-60061

1,3-Dihydroxy-12-oleanen-29-oic acid, D-70325

3,6-Dihydroxy-12-oleanen-29-oic acid, D-70326

3,19-Dihydroxy-12-oleanen-28-oic acid, D-70327

1,11-Dihydroxy-18-oleanen-3-one, D-60364

8β,9α-Dihydroxy-4,10(14)-oplopadien-3-one, *in* O-70042

3,5-Dihydroxy-1,2,4-oxadiazole, *see* O-70060

▷*N*,*N*′-Dihydroxyoxamide, *see* O-60048

6,17-Dihydroxy-15,17-oxido-16-spongianone, D-60365

11,12-Dihydroxy-7-oxo-8,11,13-abietatrien-20-oic acid, *in* T-60337

6,7-Dihydroxy-3-[(2-oxo-2*H*-1-benzopyran-7-yl)oxy]-2*H*-1-benzopyran-2-one, *in* D-70006

6-[4-(5,7-Dihydroxy-4-oxo-4*H*-1-benzopyran-2-yl)phenoxy]-5,7-dihydroxy-2-(4-hydroxyphenyl)-4*H*-1-benzopyran-4-one, *see* H-60082

4,5-Dihydroxy-3-oxo-1-cyclohexenecarboxylic acid, D-70328

1,5-Dihydroxy-2-oxo-3,11(13)-eudesmadien-12,8-olide, D-70329

2,3-Dihydroxy-6-oxo-D:A-friedo-24-noroleana-1,3,5(10),7-tetraen-23,29-dioic acid, *see* D-70027

2,9-Dihydroxy-8-oxo-1(10),4,11(13)-germacratrien-12,6-olide, D-60366

3,15-Dihydroxy-23-oxo-7,9(11),24-lanostatrien-26-oic acid, D-70330

19,29-Dihydroxy-3-oxo-12-oleanen-28-oic acid, D-70331

11,15-Dihydroxy-9-oxo-5,13-prostadienoic acid, D-60367

3,5-Dihydroxy-4-oxo-4*H*-pyran-2-carboxylic acid, D-60368

(1,5-Dihydroxy-2-oxo-3-pyrrolidinyl)-phosphonic acid, *see* A-60288

3,6-Dihydroxy-2-(1-oxo-10-tetradecenyl)-2-cyclohexen-1-one, D-70332

22,23-Dihydroxy-3-oxo-12-ursen-30-oic acid, D-70333

3,5-Dihydroxy-3′,4′,5′,7,8-pentamethoxyflavone, *in* H-70021

5,7-Dihydroxy-3,3′,4′,5′,8-pentamethoxyflavone, *in* H-70021

4,5-Dihydroxyphenanthrene, *see* P-70048

2,4-Dihydroxyphenylacetaldehyde, D-60369

2,5-Dihydroxyphenylacetaldehyde, D-60370

3,4-Dihydroxyphenylacetaldehyde, D-60371

2-(3,4-Dihydroxyphenyl)-3,4-dihydro-2*H*-1-benzopyran-5,7-diol, *see* T-60103

2-(3,4-Dihydroxyphenyl)-2,3-dihydro-5,7-dihydroxy-4*H*-1-benzopyran, *see* T-70105

2-(2,3-Dihydroxyphenyl)-5,7-dihydroxy-4*H*-1-benzopyran-4-one, *see* T-60106

2-(2,6-Dihydroxyphenyl)-5,6-dihydroxy-4*H*-1-benzopyran-4-one, *see* T-70107

3-(3,4-Dihydroxyphenyl)-6,7-dihydroxy-4*H*-1-benzopyran-4-one, *see* T-70111

4-(3,4-Dihydroxyphenyl)-5,7-dihydroxy-2*H*-1-benzopyran-2-one, D-70334

4-(3,4-Dihydroxyphenyl)-5,7-dihydroxycoumarin, *see* D-70334

1-(2,4-Dihydroxyphenyl)-2-(3,5-dihydroxyphenyl)ethylene, D-70335

1-(2,6-Dihydroxyphenyl)-9-(3,4-dihydroxyphenyl)-1-nonanone, *in* M-70007

1-(2,4-Dihydroxyphenyl)-3-(3,4-dihydroxyphenyl)propane, D-70336

2-(3,5-Dihydroxyphenyl)-5,6-dimethoxy-benzofuran, *in* M-60146

4-[2-(3,5-Dihydroxyphenyl)ethenyl]-1,3-benzenediol, *see* D-70335

5-[2-(3,4-Dihydroxyphenyl)ethyl]-1,2,3-benzenetriol, *see* P-60044

6-(2,3-Dihydroxyphenylethyl)-2-[(4-hydroxymethyl)-3-pentenyl]-2-methyl-chromene, *see* T-70008

2,4-Dihydroxyphenyl 1-hydro-7,8-dihydroxy-2-oxa-3-naphthyl ketone, *in* P-70017

2-(3,5-Dihydroxyphenyl)-6-hydroxybenzofuran, *see* M-60146

3-(2,4-Dihydroxyphenyl)-7-hydroxy-4*H*-1-benzopyran-4-one, *see* T-70264

2-(3,5-Dihydroxyphenyl)-5-hydroxy-6-methoxy-benzofuran, *in* M-60146

1-(3,4-Dihydroxyphenyl)-2-(3-hydroxyphenyl)-ethane, *see* T-60316

1-(2,6-Dihydroxyphenyl)-9-(4-hydroxyphenyl)-1-nonanone, *in* M-70007

3,5-Dihydroxy-4-phenylisoxazole, D-70337

2-(2,4-Dihydroxyphenyl)-7-methoxy-5-(1-propenyl)benzofuran, *in* H-60214

3-[(3,4-Dihydroxyphenyl)methyl]-2,3-dihydro-7-hydroxy-4*H*-1-benzopyran-4-one, *see* D-70288

9-(3,4-Dihydroxyphenyl)-1-(3,4-methyl-enedioxyphenyl)-1-nonanone, *in* M-70007

2-[(3,4-Dihydroxyphenyl)methylene]-6-hydroxy-3(2*H*)-benzofuranone, *see* S-70082

1-(3,4-Dihydroxyphenyl)-2-nitroethylene, *see* N-70049

2-(3,4-Dihydroxyphenyl)-3,5,6,7,8-penta-hydroxy-4*H*-1-benzopyran-4-one, *see* H-60024

1-(2,6-Dihydroxyphenyl)-9-phenyl-1-nonanone, *see* M-70007

2-(3,4-Dihydroxyphenyl)propanoic acid, D-60372

2-(2,4-Dihydroxyphenyl)-5-(1-propenyl)-benzofuran, *in* H-60214

2-(3,4-Dihydroxyphenyl)propionic acid, *see* D-60372

4-[3-[(2,4-Dihydroxyphenyl)propyl]]-1,2-benzenediol, *see* D-70336

1-(2,6-Dihydroxyphenyl)-1-tetradecanone, D-70338

1-(3,4-Dihydroxyphenyl)-1,2,3,4-tetrahydro-7-hydroxy-6-methoxy-2,3-naphthalenedimethanol, *see* I-60139

1-(3,4-Dihydroxyphenyl)-2-(3,4,5-trihydroxyphenyl)ethane, *see* P-60044

3-(3,4-Dihydroxyphenyl)-1-(2,4,5-trihydroxyphenyl)-2-propen-1-one, D-70339

3,4-Dihydroxyphthalaldehydic acid, *see* F-60070

5,6-Dihydroxyphthalaldehydic acid, *see* F-60073

5,7-Dihydroxyphthalide, *see* D-60339

1α,5α-Dihydroxypinnatifidin, *in* D-70329

5,7-Dihydroxy-8-prenylflavanone, *see* G-70015
2-(3,5-Dihydroxy-4-prenylphenyl)-6-hydroxybenzofuran, *in* M-60146
3,4-Dihydroxyproline, *see* D-70342
5-(1,3-Dihydroxypropyl)-2-isopropenyl-2,3-dihydrobenzofuran, D-70340
3,9-Dihydroxypterocarpan, D-60373
2,3-Dihydroxypyridine, *in* H-60223
2,4-Dihydroxypyridine, *in* H-60224
2,5-Dihydroxypyridine, *in* H-60225
2,6-Dihydroxypyridine, *in* H-60226
3,4-Dihydroxypyridine, *in* H-60227
4,6-Dihydroxy-2(1*H*)-pyridinone, D-70341
2,4-Dihydroxypyrido[2,3-*d*]pyrimidine, *see* P-60236
2,4-Dihydroxypyrimido[4,5-*b*]quinoline, *see* P-60245
3,4-Dihydroxy-2-pyrrolidinecarboxylic acid, D-70342
5,8-Dihydroxyquinazoline, *see* Q-60002
2,3-Dihydroxyquinoxaline, *see* D-60279
5,8-Dihydroxyquinoxaline, *see* Q-60008
6,7-Dihydroxyquinoxaline, *see* Q-60009
8′,9′-Dihydroxysargaquinone, *in* S-60010
20,24-Dihydroxy-3,4-seco-4(28),23-dammaradien-3-oic acid, D-70343
20,25-Dihydroxy-3,4-seco-4(28),23-dammaradien-3-oic acid, D-70344
2α,19-Dihydroxy-13(16),14-spongiadien-3-one, *see* I-60137
2α,19-Dihydroxy-13(16),14-spongiadien-3-one, *in* H-70220
3,19-Dihydroxy-13(16),14-spongiadien-2-one, D-70345
7,12-Dihydroxysterpurene, D-70346
9,12-Dihydroxysterpurene, D-70347
2,4-Dihydroxystyrene, *see* V-70010
Di-(*p*-hydroxystyryl)methane, *see* B-60162
2α,7-Dihydroxytaxodone, *in* D-70282
2,3-Dihydroxy-3,4,4′,5-tetramethoxybibenzyl, *in* T-60344
3′,5-Dihydroxy-4′,6,7,8-tetramethoxyflavone, D-60375
4′,5-Dihydroxy-3′,5′,6,7-tetramethoxyflavone, *in* H-70081
5,7-Dihydroxy-3′,4′,5′,6-tetramethoxyflavone, *in* H-70081
5,7-Dihydroxy-3′,4′,6,8-tetramethoxyflavone, D-60376
5′,5-Dihydroxy-2′,3,7,8-tetramethoxyflavone, *in* H-70080
3,7-Dihydroxy-1,2,5,6-tetramethoxyphenanthrene, *in* H-60066
2-(5,16-Dihydroxy-3,7,11,15-tetramethyl-2,6,10,14-hexadecatetraenyl)-6-methyl-1,4-benzenediol, *in* H-70225
5,8-Dihydroxy-2,3,6,7-tetramethyl-1,4-naphthalenedione, *see* D-60377
5,8-Dihydroxy-2,3,6,7-tetramethyl-1,4-naphthoquinone, D-60377
7,12-Dihydroxy-3,11,15,23-tetraoxo-8,20(22)-lanostadien-26-oic acid, D-60378
7,20-Dihydroxy-3,11,15,23-tetraoxo-8-lanosten-26-oic acid, D-60379
3,6-Dihydroxy-1,2,4,5-tetrazine, *see* D-70268
5,6-Dihydroxy-*p*-toluquinone, *see* D-70320
3,4-Dihydroxy-2,5-toluquinone, *see* D-70320
5,8-Dihydroxy-2,3,6-trimethoxy-1,4-benzoquinone, *in* P-70033
4,4′-Dihydroxy-3,3′,5-trimethoxybibenzyl, *in* P-60044
3,3′-Dihydroxy-4′,5,7-trimethoxyflavan, *in* P-70026
3,4′-Dihydroxy-3′,5,7-trimethoxyflavan, *in* P-70026
3′,5-Dihydroxy-4′,6,7-trimethoxyflavanone, *in* P-70028
3,4′-Dihydroxy-5,6,7-trimethoxyflavone, *in* P-60047
3,5-Dihydroxy-4′,6,7-trimethoxyflavone, *in* P-60047
4′,5-Dihydroxy-3,6,7-trimethoxyflavone, *in* P-60047
4′,5-Dihydroxy-3,6,8-trimethoxyflavone, *in* P-60048
5,6-Dihydroxy-3,4′,8-trimethoxyflavone, *in* P-60048

5,7-Dihydroxy-3,6,8-trimethoxyflavone, *in* P-70031
5′,6-Dihydroxy-2′,3,5,7-trimethoxyflavone, *in* H-60064
2,7-Dihydroxy-1,5,6-trimethoxyphenanthrene, *in* P-60050
7,9-Dihydroxy-2,3,4-trimethoxyphenanthrene, *in* P-70035
1,8-Dihydroxy-2,3,6-trimethoxyxanthone, *in* P-70036
3,6-Dihydroxy-1,2,3-trimethoxyxanthone, *in* P-70036
3,8-Dihydroxy-1,2,4-trimethoxyxanthone, *in* P-60051
6,8-Dihydroxy-1,2,5-trimethoxyxanthone, *in* P-60052
1-(5,7-Dihydroxy-2,2,6-trimethyl-2*H*-1-benzopyran-8-yl)-2-methyl-1-propanone, D-70348
3,23-Dihydroxy-4,4,17-trimethylcholesta-7,14,24-trien-26-oic acid, *see* M-70012
4-(2,4-Dihydroxy-2,6,6-trimethylcyclohexyl-idene)-3-buten-2-one, *see* G-70028
1-(5,7-Dihydroxy-2,2,6-trimethyl-2H-1-benzopyran-8-yl)-2-methyl-1-butanone, *in* D-70348
5,8-Dihydroxy-2,3,6-trimethyl-1,4-naphthalenedione, *see* D-60380
5,8-Dihydroxy-2,3,6-trimethyl-1,4-naphthoquinone, D-60380
3β,12β-Dihydroxy-4,4,14α-trimethyl-7,11,15-trioxo-5α-chol-8-en-24-oic acid, *in* L-60037
3β,7β-Dihydroxy-11,15,23-trioxo-8,20(22)-*E*-lanostadien-26-oic acid, *in* T-70254
7β,15α-Dihydroxy-3,11,23-trioxo-8,20(22)-*E*-lanostadien-26-oic acid, *in* T-70254
3β,12β-Dihydroxy-7,11,15-trioxo-25,26,27-trisnor-8-lanosten-24-oic acid, *in* L-60037
3β,28-Dihydroxy-7,11,15-trioxo-25,26,27-trisnor-8-lanosten-24-oic acid, *in* L-60036
2,7-Dihydroxytropone, *see* D-60314
3,5-Dihydroxytyrosyl-α,β-didehydro-3,5-dihydroxy-*N*-[2-(3,4,5-trihydroxyphenyl)-ethenyl]tyrosinamide, *see* T-70336
2,3-Dihydroxy-12,20(30)-ursadien-28-oic acid, D-60381
3,19-Dihydroxy-12-ursene-24,28-dioic acid, D-60382
1,3-Dihydroxy-12-ursen-28-oic acid, D-60383
2,3-Dihydroxy-12-ursen-28-oic acid, D-70349
3,19-Dihydroxy-12-ursen-28-oic acid, D-70350
3,23-Dihydroxy-12-ursen-28-oic acid, D-70351
4,6-Dihydroxy-5-vinylbenzofuran, D-70352
1,2-Dihydroxy-9*H*-xanthen-9-one, *see* D-70353
1,2-Dihydroxyxanthone, D-70353
Dihydroyashabushiketol, *in* D-70481
11α,13-Dihydroyomogin, *in* Y-70001
11β,13-Dihydrozaluzanin C, *in* Z-60001
11α,13-Dihydrozaluzanin C, *in* Z-60001
Di-2-imidazolylmethane, *see* M-70064
Di-1*H*-imidazol-2-ylmethanone, D-70354
2,2′-Diiminazole, *see* B-60107
2,4-Diimino-5,5-dimethyl-1-imidazolinyloxy, *see* P-70110
Diindolo[3,2-a:3′,2′-*c*]carbazole, D-60384
9,10-Diiodoanthracene, D-70355
2,6-Diiodobenzoic acid, D-70356
1,3-Diiodobicyclo[1.1.1]pentane, D-70357
2,2′-Diiodo-1,1′-binaphthalene, *see* D-60385
4,4′-Diiodo-1,1′-binaphthalene, *see* D-60386
2,2′-Diiodo-1,1′-binaphthyl, D-60385
4,4′-Diiodo-1,1′-binaphthyl, D-60386
1,4-Diiodocubane, *see* D-70359
1,3-Diiodo-2,2-diazidopropane, D-60387
β,β′-Diiododiethyl ether, *see* O-60096
1,3-Diiodo-2-(iodomethyl)propane, D-70358
1,4-Diiodopentacyclo[4.2.0.0²⋅⁵.0³⋅⁸.0⁴⋅⁷]octane, D-70359
2,7-Diiodophenanthrene, D-70360
3,6-Diiodophenanthrene, D-70361
2,4-Diiodothiazole, D-60388
2,5-Diiodothiazole, D-60389

3,6-Diisobutylpiperazinedione, *see* B-60171
1,3-Diisocyanatocyclopentane, D-60390
2,2′-Diisocyano-1,1′-binaphthalene, *see* D-60391
2,2′-Diisocyano-1,1′-binaphthyl, D-60391
Diisocyanogen, D-70362
3′,5′-Diisopentenyl-2′,4,4′-trihydroxychalcone, *see* D-60310
Diisoprene, *see* I-70087
2,3-Diisopropoxycyclopropenone, *in* D-70290
Diisopropylcyanamide, D-60392
3,4-Diisopropyl-2,5-dimethylhexane, D-60393
6,8′;7,3′-Diligustide, *see* R-60009
6,6′;7,3a′-Diligustilide, D-70363
Dilophic acid, D-60394
▷Dimazine, *see* D-70416
1,3-Dimercaptobutane, *see* B-60339
2,2-Dimercaptobutane, *see* B-60340
1,4-Dimercapto-2-butene, *see* B-70292
1,4-Dimercapto-2-butyne, *see* B-70308
1,1-Dimercaptocyclobutane, *see* C-70215
1,3-Dimercaptocyclobutane, *see* C-60184
3,4-Dimercapto-3-cyclobutene-1,2-dithione, D-70364
▷1,1-Dimercaptocyclohexane, *see* C-60209
1,2-Dimercaptocyclohexane, *see* C-70225
1,1-Dimercaptocyclopentane, *see* C-60222
1,5-Dimercapto-3,3-dimethylpentane, *see* D-70427
1,3-Dimercapto-2,2-dimethylpropane, *see* D-70433
2-(4,5-Dimercapto-1,3-dithiol-2-ylidene)-4,5-dimercapto-1,3-dithiole, *see* T-70130
1,7-Dimercaptoheptane, *see* H-70023
(Dimercaptomethylene)malonic acid, *see* D-60395
(Dimercaptomethylene)malononitrile, *in* D-60395
(Dimercaptomethylene)propanedioic acid, D-60395
2,4-Dimercapto-6-methylpyrimidine, *see* M-70128
1,2-Dimercaptonaphthalene, *see* N-70002
1,8-Dimercaptonaphthalene, *see* N-70003
2,3-Dimercaptonaphthalene, *see* N-70004
2,3-Dimercapto-*p*-phenylenediamine, *see* D-60030
2,5-Dimercapto-*p*-phenylenediamine, *see* D-70035
4,6-Dimercapto-*m*-phenylenediamine, *see* D-70037
2,2-Dimercaptopropane, *see* P-60175
2,3-Dimercaptopyridine, *see* P-60218
2,4-Dimercaptopyridine, *see* P-60219
2,5-Dimercaptopyridine, *see* P-60220
3,4-Dimercaptopyridine, *see* P-60221
3,5-Dimercaptopyridine, *see* P-60222
2,5-Dimercapto-*p*-xylene, *see* D-60406
1,6:7,12-Dimethano[14]annulene, *see* B-60165
13,19:14,18-Dimethenoanthra[1,2-*a*]benzo[*o*]pentaphene, D-60396
2,3-Dimethoxyaniline, *in* A-70089
▷3,4-Dimethoxyaniline, *in* A-70090
2,3-Dimethoxybenzaldehyde, *in* D-70285
2,3-Dimethoxybenzenamine, *in* A-70089
▷3,4-Dimethoxybenzenamine, *in* A-70090
2,5-Dimethoxybenzeneacetaldehyde, *in* D-60370
3,4-Dimethoxybenzeneacetaldehyde, *in* D-60371
4,5-Dimethoxy-1,3-benzenedicarboxylic acid, *in* D-60307
3,6-Dimethoxy-1,2-benzenediol, *in* B-60014
5-(5,6-Dimethoxy-2-benzofuranyl)-1,3-benzenediol, *in* M-60146
3,6-Dimethoxy-1,2-benzoquinone, *in* D-60309
7,7′-Dimethoxy-[6,8′-bi-2*H*-benzopyran]-2,2′-dione, *in* B-70093
7,7′-Dimethoxy-6,8′-bicoumarin, *in* B-70093
1,1-Dimethoxy-2-butanol, *in* H-70112
5,6-Dimethoxy-1,3-cyclohexadiene, D-60397
4,4-Dimethoxycyclohexene, *in* C-60211
Dimethoxycyclopropenone, *in* D-70290
1,1-Dimethoxy-3-decene, *in* D-70016
5,7-Dimethoxy-2′,4′-dihydroxyisoflavan, *see* L-70038
2,6-Dimethoxy-3,5-dimethylbenzoic acid, *in* D-70293
6,7-Dimethoxy-2,2-dimethyl-2*H*-1-benzopyran, D-70365

6,7-Dimethoxy-2,2-dimethyl-3-chromene, *see* D-70365

5,7-Dimethoxy-α,α-dimethyl-2-oxo-2*H*-1-benzopyran-8-propanoic acid, *see* M-70158

1,2-Dimethoxy-4,5-dinitrobenzene, *in* D-70449

1,5-Dimethoxy-2,4-dinitrobenzene, *in* D-70450

5,6-Dimethoxyindole, *in* D-70307

4′,7-Dimethoxyisoflavone, *in* D-60340

5,5′-Dimethoxylariciresinol, *in* L-70019

5,6-Dimethoxy-3-(4-methoxybenzyl)phthalide, D-60398

5,6-Dimethoxy-3-[(4-methoxyphenyl)methyl]-1(3*H*)-isobenzofuranone, *see* D-60398

2-(Dimethoxymethyl)benzenemethanol, *in* H-60171

α-(Dimethoxymethyl)benzenemethanol, *in* H-60210

4,7-Dimethoxy-5-methyl-2*H*-1-benzopyran-2-one, *in* D-60348

2,3-Dimethoxy-5-methyl-1,4-benzoquinone, *in* D-70320

2-(Dimethoxymethyl)benzyl alcohol, *in* H-60171

4,7-Dimethoxy-5-methylcoumarin, *in* D-60348

3′,5-Dimethoxy-3,4-methylenedioxybibenzyl, *in* T-60102

6,7-Dimethoxy-3′,4′-methylenedioxyflavanone, *in* T-70106

3′,4′-Dimethoxy-7,8-methylenedioxyisoflavone, *in* T-70112

6,7-Dimethoxy-3′,4′-methylenedioxyisoflavone, *in* T-70111

2,3-Dimethoxy-6,7-methylenedioxynaphthazarin, *in* H-60065

4,5-Dimethoxy-4′,5′-methylenedioxypolemannone, *in* P-60166

▷4,9-Dimethoxy-7-methyl-5*H*-furo[3,2-*g*][1]-benzopyran-5-one, *see* K-60010

2-(Dimethoxymethyl)-1-naphthaldehyde, *in* N-60001

2-(Dimethoxymethyl)-1-naphthalenecarboxaldehyde, *in* N-60001

5,6-Dimethoxy-2-methyl-1,4-naphthoquinone, *in* D-60351

5,7-Dimethoxy-6-methylphthalide, *in* D-60349

1-(Dimethoxymethyl)-2,3,5,6-tetrakis(methylene)-7-oxabicyclo[2.2.1]heptane, D-60399

2,4-Dimethoxy-6-methyl-1,3,5-triazine, *in* M-60132

4,5-Dimethoxy-1-naphthol, *in* N-60005

4,8-Dimethoxy-1-naphthol, *in* N-60005

2′,3′-Dimethoxy-6′-nitroacetophenone, *in* D-60357

2′,5′-Dimethoxy-4′-nitroacetophenone, *in* D-70322

3′,6′-Dimethoxy-2′-nitroacetophenone, *in* D-60358

4′,5′-Dimethoxy-2′-nitroacetophenone, *in* D-60359

4,5-Dimethoxy-2-nitrobenzoic acid, *in* D-60360

1,1-Dimethoxy-3-nitropropane, *in* N-60037

6,7-Dimethoxy-8-[(2-oxo-2*H*-1-benzopyran-7-yl)oxy]-2*H*-1-benzopyran-2-one, *see* O-70044

5,5-Dimethoxy-2-pentanone, *in* O-60083

4,5-Dimethoxyphenanthrene, *in* P-70048

1,5-Dimethoxy-2,7-phenanthrenediol, *in* T-60116

5,7-Dimethoxy-2,3-phenanthrenediol, *in* T-60120

3,7-Dimethoxy-2-phenanthrenol, *in* T-63339

2,6-Dimethoxy-5*H*-phenanthro-[4,5-*bcd*]-pyran-7-ol, *see* A-70077

1,2-Dimethoxy-3*H*-phenoxazin-3-one, *in* T-60353

2,5-Dimethoxyphenylacetaldehyde, *in* D-60370

3,4-Dimethoxyphenylacetaldehyde, *in* D-60371

2-(3,4-Dimethoxyphenyl)-5,7-dihydroxy-6,8-dimethoxy-4*H*-1-benzopyran-4-one, *see* D-60376

2,2-Dimethoxy-1-phenylethanol, *in* H-60210

6-[2-(3,4-Dimethoxyphenyl)ethenyl]-4-methoxy-2*H*-pyran-2-one, *in* H-70122

4-[4-(3,4-Dimethoxyphenyl)-1*H*,3*H*-furo[3,4-*c*]-furan-1-yl]-2-methoxyphenol, *see* P-60141

4-[(3,4-Dimethoxyphenyl)methyl]dihydro-3-[(4-hydroxy-3-methoxyphenyl)methyl]-2(3*H*)-furanone, *see* A-70249

3,4-Dimethoxyphthalaldehydic acid, *in* F-60070

5,6-Dimethoxyphthalaldehydic acid, *in* F-60073

5,7-Dimethoxyphthalide, *in* D-60339

▷3,3-Dimethoxypropene, *in* P-70119

2,3-Dimethoxypyridine, *in* H-60223

2,4-Dimethoxypyridine, *in* H-60224

2,6-Dimethoxypyridine, *in* H-60226

3,4-Dimethoxypyridine, *in* H-60227

4,6-Dimethoxy-2(1*H*)-pyridinone, *in* D-70341

2,3-Dimethoxyquinoxaline, *in* D-60279

5,8-Dimethoxyquinoxaline, *in* Q-60002

6,7-Dimethoxyquinoxaline, *in* Q-60009

2,4-Dimethoxystyrene, *in* V-70010

6-(3,4-Dimethoxystyryl)-4-methoxy-2-pyrone, *in* H-70122

3,6-Dimethoxy-1,2,4,5-tetrazine, *in* D-70268

16,17-Dimethoxytricyclo[12.3.1.1²,⁶]-nonadeca-1(18),2,4,6(19),14,16-hexaene-3,9,15-triol, *see* M-70162

24,24-Dimethoxy-25,26,27-trisnor-3-cycloartanol, *in* H-60232

4,6-Dimethoxy-5-vinylbenzofuran, *in* D-70352

3,3-Dimethylacrolein, *see* M-70055

β,β-Dimethylacrylalkannin, *in* S-70038

β,β-Dimethylacrylshikonin, *in* S-70038

1,3-Dimethyladamantane, D-70366

3,7-Dimethyladenine, D-60400

8-Dimethylally-7-hydroxy-6-methylcoumarin, *see* C-60026

3-(1,1-Dimethylallyl)-7-hydroxy-6-(3-methyl-2-butenyl)coumarin, *see* G-60037

8-(3,3-Dimethylallyl)-5-methoxy-3,4′,7-trihydroxyflavan, *in* T-60125

2-(Dimethylamino)benzaldehyde, D-60401

N-[4-[2-(6-Dimethylaminobenzofuranyl)-phenyl]]maleimide, *see* D-60170

4-Dimethylamino-1,2,3-benzotriazine, *in* A-60092

6-(Dimethylamino)-3,3-bis[4-(dimethylamino)-phenyl]-1(3*H*)-isobenzofuranone, *see* C-70202

1-(Dimethylamino)-γ-carboline, *in* A-70114

2-Dimethylamino-3,3-dimethylazirine, D-60402

8-(Dimethylamino)guanosine, *in* A-70150

8-(Dimethylamino)inosine, *in* A-70198

(Dimethylamino)methanesulfonic acid, *in* A-60215

Dimethyl 4-amino-5-methoxyphthalate, *in* A-70152

1-Dimethylamino-4-nitro-1,3-butadiene, D-70367

N-[4-[[4-(Dimethylamino)phenyl]imino]-2,5-cyclohexadien-1-ylidene]-*N*-methyl-methanaminium, *see* B-60110

2-(Dimethylamino)purine, *in* A-60243

8-(Dimethylamino)purine, *in* A-60244

5-Dimethylamino-1*H*-tetrazole, *in* A-70202

2-Dimethylamino-6-(trihydroxypropyl)-4(3*H*)-pteridinone, *see* E-60070

10,10-Dimethyl-9(10*H*)-anthracenone, D-60403

10,10-Dimethylanthrone, *see* D-60403

3,3-Dimethylaspartic acid, *see* A-60147

N-(7,7-Dimethyl-3-azabicyclo[4.1.0]hepta-1,3,5-trien-4-yl)benzamide, D-70368

1,6-Dimethyl-5-azauracil, *in* M-60132

3,6-Dimethyl-5-azauracil, *in* M-60132

N,*N*-Dimethylazidochloromethyl-eniminium(1+), D-70369

1,6-Dimethylazulene, D-60404

4,7-Di-*O*-methylbellidin, *in* P-60052

▷7,12-Dimethylbenz[*a*]anthracene, D-60405

▷9,10-Dimethyl-1,2-benzanthracene (obsol.), *see* D-60405

α,α-Dimethylbenzeneacetic acid, *see* M-60113

α,4-Dimethylbenzeneacetic acid, *see* M-60114

2,5-Dimethyl-1,4-benzenedithiol, D-60406

3,6-Dimethyl-1,2,4,5-benzenetetracarboxylic acid, D-60407

2,3-Dimethyl-1,4-benzodioxin, D-70370

2,2-Dimethyl-2*H*-1-benzopyran-6-carboxylic acid, D-60408

2,5-Dimethyl-4*H*-1-benzopyran-4-one, D-70371

3-(2,2-Dimethyl-2*H*-1-benzopyran-6-yl)-2-propenoic acid, D-70372

2,2-Dimethyl-2*H*-1-benzothiopyran, D-70373

7,7-Dimethylbicyclo[4.1.0]hept-3-ene, D-60409

1,5-Dimethylbicyclo[3.3.0]octa-3,6-diene-2,8-dione, *see* D-70199

4,4-Dimethylbicyclo[3.2.1]octane-2,3-dione, D-60410

3,3-Dimethylbicyclo[2.2.2]octan-2-one, D-60411

3,3-Dimethylbicyclo[2.2.2]oct-5-en-2-one, D-60412

Dimethylbiochanin B, *see* D-60340

2,2′-Dimethyl-4,4′-biquinoline, D-60413

2,5-Dimethyl-3,4-bis(1-methylethyl)hexane, *see* D-60393

1,4-Dimethyl-2,5-bis(methylthio)benzene, *in* D-60406

5,5′-(2,3-Dimethyl-1,4-butanediyl)bis-1,3-benzodioxole, *see* A-70270

1,1′-(2,2-Dimethyl-3-butenylidene)-bisbenzene, *see* D-60418

3,3-Dimethyl-1-butynylbenzene, *see* D-60441

N,*N*-Dimethylcarbamohydroxamic acid, D-60414

1,2-Dimethylcarbazole, D-70374

1,3-Dimethylcarbazole, D-70375

1,4-Dimethylcarbazole, D-70376

1,5-Dimethylcarbazole, D-70377

1,6-Dimethylcarbazole, D-70378

1,7-Dimethylcarbazole, D-70379

1,8-Dimethylcarbazole, D-70380

2,3-Dimethylcarbazole, D-70381

2,4-Dimethylcarbazole, D-70382

2,5-Dimethylcarbazole, D-70383

2,7-Dimethylcarbazole, D-70384

3,4-Dimethylcarbazole, D-70385

3,5-Dimethylcarbazole, D-70386

3,6-Dimethylcarbazole, D-70387

4,5-Dimethylcarbazole, D-70388

1,2-Dimethyl-9*H*-carbazole-3,4,6-triol, *see* T-70253

24,24-Dimethyl-3-cholestanol, D-70389

2,5-Dimethylchromone, *see* D-70371

▷4,6-Dimethylcoumalin, *see* D-70435

Dimethyl cyanocarbonimidate, *see* D-70390

Dimethyl *N*-cyanoimidocarbonate, D-70390

2,3-Dimethyl-2-cyclobuten-1-one, D-60415

4,14-Dimethyl-9,19-cycloergost-20-en-3β-ol, *see* C-70218

2,7-Dimethyl-2,4,6-cycloheptatrien-1-one, D-70391

2,2-Dimethyl-5-cycloheptene-1,3-dione, D-60416

5,5-Dimethyl-2-cyclohexene-1,4-dione, D-70392

2,2-Dimethyl-1,3-cyclopentanedione, D-70393

4,4-Dimethyl-1,3-cyclopentanedione, D-70394

2,2-Dimethyl-4-cyclopentene-1,3-dione, D-70395

2,2-Dimethylcyclopropanemethanol, D-70396

2,3-Dimethylcyclopropanol, D-70397

3,3-Dimethyl-1-cyclopropene-1-carboxylic acid, D-60417

4,8-Dimethyl-3*E*,8*E*-decadien-10-olide, *see* S-70084

Dimethyl dicarbonate, *in* D-70111

2,3-Dimethyl-3,4-dihydro-2*H*-pyran, D-70398

▷8,16-Dimethyl-1,9-dioxacyclohexadeca-3,11-diene-2,5,10,13-tetrone, *see* P-70167

Dimethyldioxirane, D-70399

2,2-Dimethyl-1,3-dioxolane-4-carboxaldehyde, D-70400

1-(2,2-Dimethyl-1,3-dioxolan-4-yl)ethanone, *see* A-70042

3,3′-Dimethyl-1,1′-diphenyl[Δ⁴,⁴′-bi-2-pyrazoline]-5,5′-dione, *see* P-70153

2,2-Dimethyl-4,4-diphenyl-3-butenal, D-70401

3,3-Dimethyl-4,4-diphenyl-1-butene, D-60418

2,2-Dimethyl-4,4-diphenyl-3-butenoic acid, D-70402

2,5-Dimethyl-3,4-diphenyl-2,4-cyclopentadien-1-one, D-70403

2,5-Dimethyl-3,4-diphenyl-2,4-hexadiene, D-60419

3,3-Dimethyl-5,5-diphenyl-4-pentenoic acid, D-60420

▷*N,N'*-Dimethyl-*N,N'*-diphenylurea, *in* D-70499

2,3-Dimethyl-1,4-dipiperonylbutane, *see* A-70270

9,18-Dimethyl-2,11-diselena[3.3]-metacyclophane, D-70404

2,5-Dimethyl-1,4-dithiane-2,5-diol, D-60421

2,3-Dimethylenebicyclo[2.2.1]heptane, D-70405

2,3-Dimethylenebicyclo[2.2.3]nonane, D-70406

7,8-Dimethylenebicyclo[2.2.2]octa-2,5-diene, D-70407

2,3-Dimethylenebicyclo[2.2.2]octane, D-70408

Dimethylenebicyclo[1.1.1]pentanone, D-60422

Dimethylenebutanedioic acid, D-70409

5,6-Dimethylene-1,3-cyclohexadiene, D-70410

1,2-Dimethylenecyclohexane, *see* B-60166

1,3-Dimethylenecyclohexane, *see* B-60167

1,4-Dimethylenecyclohexane, *see* B-60168

1,2-Dimethylenecyclopentane, *see* B-60169

1,2-Dimethylenecyclopentane, *see* B-60170

3,4;3',4'-Dimethylenedioxybibenzyl, *in* T-60101

Dimethylenediurea, *see* T-70098

2,3-Dimethylenenorbornane, *see* D-70405

3,4-Dimethyleneoxolane, *see* T-70046

Dimethylenesuccinic acid, *see* D-70409

3,4-Dimethylenetetrahydrofuran, *see* T-70046

6,6'-(1,2-Dimethyl-1,2-ethanediyl)bis[1-phenazinecarboxylic acid], D-70411

2-(1,1-Dimethylethyl)-1-butene-1-thione, *see* D-60114

1-(1,1-Dimethylethyl)-1,3-cyclohexadiene, *see* B-70296

1-(1,1-Dimethylethyl)cyclohexene, *see* B-60341

3-(1,1-Dimethylethyl)cyclohexene, *see* B-60342

4-(1,1-Dimethylethyl)cyclohexene, *see* B-60343

3-(1,1-Dimethylethyl)-2-cyclohexen-1-one, B-70297

1-(1,1-Dimethylethyl)-1,2-cyclooctadiene, *see* B-60344

1-(1,1-Dimethylethyl)cyclopentene, *see* B-70298

3-(1,1-Dimethylethyl)cyclopentene, *see* B-70299

4-(1,1-Dimethylethyl)cyclopentene, *see* B-70300

5-(1,1-Dimethylethyl)-1,3,2,4-dithiadiazol-2-yl, *see* B-70303

4-(1,1-Dimethylethyl)-3*H*-1,2,3,5-dithiadiazol-3-yl, *see* B-70302

3-(1,1-Dimethylethyl)-4-hydroxybenzaldehyde, *see* B-70304

4-(1,1-Dimethylethyl)-2-hydroxycyclopentanecarboxylic acid, *see* H-70113

▷Dimethylethylideneurea, *in* I-70005

1-(1,1-Dimethylethyl)-2-iodobenzene, *see* B-60346

1-(1,1-Dimethylethyl)-3-iodobenzene, *see* B-60347

1-(1,1-Dimethylethyl)-4-iodobenzene, *see* B-60348

3-(1,1-Dimethylethyl)-6-oxabicyclo[3.1.0]-hexane, *see* B-70294

3-(1,1-Dimethylethyl)-2,2,4,5,5-pentamethyl-3-hexene, *see* B-60350

2-(1,1-Dimethylethyl)phenanthrene, *see* B-70305

3-(1,1-Dimethylethyl)phenanthrene, *see* B-70306

9-(1,1-Dimethylethyl)phenanthrene, *see* B-70307

2,5-Dimethyl-3-furancarboxaldehyde, D-70412

4,5-Dimethyl-2-furancarboxylic acid, D-70413

3,4-Dimethyl-2(5*H*)-furanone, D-60423

4,4-Dimethylglutaraldehydic acid, *see* D-60434

4,4-Dimethylglutaraldehydonitrile, *in* D-60434

4'-(3,6-Dimethyl-2-heptenyloxy-5-hydroxy-3',7-dimethoxyflavanone, *in* T-70105

4-(1,5-Dimethyl-1,3-hexadienyl)-1-isocyano-1-methylcyclohexane, *see* I-60091

4-(1,5-Dimethyl-1,4-hexadienyl)-1-methyl-7-oxabicyclo[4.1.0]heptane, *see* E-60009

2,5-Dimethyl-1,3,5-hexatriene, D-60424

2,2-Dimethyl-4-hexen-3-one, D-70414

2,2-Dimethyl-5-hexen-3-one, D-60425

2-(1,5-Dimethyl-4-hexenyl)-*p*-cresol, *see* E-60004

9-(1,5-Dimethyl-4-hexenyl)-2,6-dimethyl-2,5-cyclononadiene-1-carboxylic acid, *see* D-60394

5-(1,5-Dimethyl-4-hexenyl)-3-hydroxy-2-methyl-2,5-cyclohexadiene-1,4-dione, *see* I-70075

1-(1,5-Dimethyl-4-hexenyl)-4-methylenebicyclo[3.1.0]hexane, *see* S-70037

2-(1,5-Dimethyl-4-hexenyl)-4-methylphenol, *see* E-60004

2-(1,5-Dimethyl-4-hexenyl)-5-methylphenol, *see* C-70208

5-(1,5-Dimethyl-4-hexenyl)-2-methylphenol, *see* X-70001

5,5-Dimethyl-2-hexynedioic acid, D-70415

▷1,1-Dimethylhydrazine, D-70416

5,5-Dimethyl-1-hydroxy-1-cyclopenten-3-one, *see* D-70394

1,4-Dimethyl-4-[4-hydroxy-2-(hydroxymethyl)-1-methyl-2-cyclopentenyl]-2-cyclohexen-1-ol, D-70417

O,N-Dimethylhydroxylamine, D-70418

1,2-Dimethylidenecyclohexane, *see* B-60166

1,3-Dimethylidenecyclohexane, *see* B-60167

1,4-Dimethylidenecyclohexane, *see* B-60168

1,2-Dimethylidenecyclopentane, *see* B-60169

1,2-Dimethylidenecyclopentane, *see* B-60170

▷1,3-Dimethylimidazolidinone, *in* I-70005

2,2-Dimethyl-1-indanone, *see* D-60227

1,6-Dimethyl-1*H*-indole, *in* M-60088

1,3-Dimethyl-1*H*-indole-2-carboxaldehyde, *in* M-70090

5,13-Dimethyl[2.2]metacyclophane, D-60426

▷*N,N*-Dimethylmethanamine *N*-oxide, *see* T-60354

N,N-Dimethylmethaneselenoamide, *in* M-70035

N,1-Dimethyl-4-(methylamino)-1*H*-imidazole-5-carboxamide, *see* C-60003

2,2-Dimethyl-3-methylenebicyclo[2.2.2]-octane, D-60427

4,4-Dimethyl-5-methylene-γ-butyrolactone, *see* D-70197

4-(2,2-Dimethyl-6-methylenecyclohexyl)-3-buten-2-one, *see* I-70058

3-[2-(2,2-Dimethyl-6-methylenecyclohexyl)-ethyl]furan, *see* P-60015

2,2-Dimethyl-4-methylene-1,3-dioxolane, D-70419

5,5-Dimethyl-4-methylene-1,2-dioxolan-3-one, D-60428

1,3-Dimethyl-2-methyleneimidazolidine, D-70420

6,10-Dimethyl-2-(1-methylethenyl)-spiro[4.5]-dec-6-ene, *see* P-70114

1,7-Dimethyl-4-(1-methylethyl)-bicyclo[3.2.1]-oct-6-ene-6,8-dicarboxaldehyde, *see* H-60016

1,3-Dimethyl-8-(1-methylethyl)-tricyclo[4.4.0.0^{2,7}]dec-3-ene, *see* C-70197

6,6-Dimethyl-2-(2-methyl-1-oxopropyl)-4-cyclohexene-1,3-dione, *see* X-70002

1,2-Dimethyl-2-(4-methylphenyl)-cyclopentanemethanol, *see* T-70202

8,8-Dimethyl-1,4,5(8*H*)-naphthalenetrione, D-70421

2,3-Dimethylnaphthazarin, *see* D-60317

2,2-Dimethyl-1-(1-naphthyl)propane, *see* D-60446

2,2-Dimethyl-1-(2-naphthyl)propane, *see* D-60447

N,N-Dimethyl-4-nitro-1,3-butadien-1-amine, *see* D-70367

4,4-Dimethyl-6-nitro-5-hexenenitrile, *in* D-60429

4,4-Dimethyl-6-nitro-5-hexenoic acid, D-60429

α-(4,8-Dimethyl-3,7-nonadienyl)-1,2-dihydro-α,4-dimethylfuro[2,3-c]-quinoline-2-methanol 5-oxide, *see* A-60313

α-(4,8-Dimethyl-3,7-nonadienyl)-2-formyl-3-(hydroxymethyl)-2-cyclopentene-1-acetaldehyde, *see* F-70038

2,6-Dimethyl-3,7-octadiene-2,6-diol, D-70422

3-(3,7-Dimethyl-2,6-octadienyl)-4-hydroxybenzoic acid, D-70423

3,7-Dimethyl-6-octenal, D-60430

3,7-Dimethyl-6-octen-4-olide, *see* E-70004

4,6-Dimethyl-4-octen-3-one, D-70424

6,6-Dimethyl-3-oxabicyclo[3.1.0]hexan-2-one, D-70425

3,5-Dimethyl-2-oxazolidinone, *in* M-60107

4,4-Dimethyl-2-oxazoline, *see* D-70198

β,β-Dimethyloxiraneethanol, D-60431

3,3-Dimethyloxiranemethanol, D-70426

2-(3,3-Dimethyloxiranyl)-4-ethenyl-3,4-dihydro-2-hydroxy-4,10-dimethyl-2*H*,5*H*-pyrano[3,2-c][1]benzopyran-5-one, *see* E-60045

3-[1-(1,1-Dimethyl-3-oxobutyl)imidazol-4-yl]-2-propenoic acid, *in* U-70005

4,4-Dimethyl-3-oxo-1-cyclopentene-1-carboxaldehyde, D-60432

4,4-Dimethyl-17-oxo-18-nor-16-oxaandrostane-8-carboxaldehyde, *see* O-60094

3,3-Dimethyl-4-oxopentanoic acid, D-60433

4,4-Dimethyl-5-oxopentanoic acid, D-60434

2,2-Dimethyl-4-oxothiochroman, *see* D-70193

3,3-Dimethyl-1,5-pentanedithiol, D-70427

3,4-Dimethylpentanoic acid, D-60435

4,4-Dimethylpentanoic acid, D-70428

3,4-Dimethyl-1-pentanol, D-60436

3,3-Dimethyl-4-pentenoic acid, D-60437

3,3-Dimethyl-4-penten-2-one, D-70429

7-(1,3-Dimethyl-1-pentenyl)-7,8-dihydro-4,8-dihydroxy-3,8-dimethyl-2*H*,5*H*-pyrano[4,3-b]pyran-2-one, *see* N-70029

▷3,4-Dimethyl-5-pentylidene-2(5*H*)-furanone, D-70430

4,4-Dimethyl-2-pentyne, D-70438

4,5-Dimethylphenanthrene, D-60439

2,2-Dimethyl-3-phenyl-2*H*-azirine, D-60440

3,3-Dimethyl-1-phenyl-1-butyne, D-60441

4,5-Dimethyl-2-phenyl-1,3-dioxol-1-ium, D-70431

2,5-Dimethyl-3-(2-phenylethenyl)pyrazine, D-60442

2-(2,4-Dimethylphenyl)-5-hydroxy-6-methoxy-benzofuran, *in* S-60002

4,4-Dimethyl-2-phenyl-4*H*-imidazole, D-70432

2,2-Dimethyl-1-phenylpropane, *see* D-60445

2,2-Dimethyl-1-phenyl-1-propanone, D-60443

4,6-Dimethyl-2-phenylpyrimidine, D-60444

2,2-Dimethyl-1,3-propanedithiol, D-70433

6-(1,1-Dimethyl-2-propenyl)-2,3-dihydro-2-(1-hydroxy-1-methylethyl)-7*H*-furo[3,2-g][1]-benzopyran-7-one, *see* C-60031

6-(1,1-Dimethyl-2-propenyl)-8*H*-1,3-dioxolo[4,5-h][1]benzopyran-8-one, *see* R-70015

3-(1,1-Dimethyl-2-propenyl)-7-hydroxy-6-(3-methyl-2-butenyl)-2*H*-1-benzopyran-2-one, *see* G-60037

3-(1,1-Dimethyl-2-propenyl)-4-hydroxy-6-phenyl-2*H*-pyran-2-one, D-70434

[[2-(1,2-Dimethyl-2-propenyl)-1*H*-indol-3-yl]-methylene]piperazinetrione, *see* N-70023

4-(1,1-Dimethylpropenyl)-7,8-methyl-enedioxycoumarin, *see* R-70015

(2,2-Dimethylpropyl)benzene, D-60445

▷Dimethylpropyleneurea, *in* T-60088

1-(2,2-Dimethylpropyl)naphthalene, D-60446

2-(2,2-Dimethylpropyl)naphthalene, D-60447

7,8-(2,2-Dimethylpyrano)-3,4'-dihydroxy-5-methoxyflavan, *in* D-60448

▷4,6-Dimethyl-2*H*-pyran-2-one, D-70435

7,8-(2,2-Dimethylpyrano)-3,4',5-trihydroxyflavan, D-60448

7,6-(2,2-Dimethylpyrano)-3,4',5-trihydroxyflavone, D-60449

4,5-Dimethylpyrimidine, D-70436

4,6-Dimethyl-2(1*H*)-pyrimidinethione, D-60450

Dimethyl pyrocarbonate, *in* D-70111

▷4,6-Dimethyl-α-pyrone, *see* D-70435

1,5-Dimethyl-1*H*-pyrrole-2-carboxaldehyde, *in* M-70129
2,5-Dimethylpyrrolidine, D-70437
5,5-Dimethyl-Δ¹-pyrroline, *see* D-70200
5,5-Dimethyl-1-pyrroline, *see* D-60229
3,3-Dimethyl-2,4(1*H*,3*H*)-quinolinedione, D-60451
3,5-Di-*O*-methylsenecioodontol, *in* S-60022
2,5-Dimethyl-3-styrylpyrazine, *see* D-60442
Dimethyl tartrate, *in* T-70005
2,2-Dimethyltetralone, *see* D-60228
2,2-Dimethyl-2*H*-thiochromene, *see* D-70373
4,6-Dimethyl-2-thio-2,3-dihydropyrimidine, *see* D-60450
2,5-Dimethyl-3-thiophenecarboxaldehyde, D-60452
2,5-Dimethyl-3-thiophenecarboxaldehyde, D-70438
3,5-Dimethyl-2-thiophenecarboxaldehyde, D-70439
1,2-Dimethyl-9*H*-thioxanthen-9-one, *see* D-70440
1,3-Dimethyl-9*H*-thioxanthen-9-one, *see* D-70441
1,4-Dimethyl-9*H*-thioxanthen-9-one, *see* D-70442
2,3-Dimethyl-9*H*-thioxanthen-9-one, *see* D-70443
2,4-Dimethyl-9*H*-thioxanthen-9-one, *see* D-70444
3,4-Dimethyl-9*H*-thioxanthen-9-one, *see* D-70445
1,2-Dimethylthioxanthone, D-70440
1,3-Dimethylthioxanthone, D-70441
1,4-Dimethylthioxanthone, D-70442
2,3-Dimethylthioxanthone, D-70443
2,4-Dimethylthioxanthone, D-70444
3,4-Dimethylthioxanthone, D-70445
5,5-Dimethyl-Δ¹-1,2,4-triazolin-3-one, D-70201
1,3-Dimethyltricyclo[3.3.1.1³,⁷]decane, *see* D-70366
6,13-Dimethyltricyclo[9.3.1.1⁴,⁸]-hexadeca-1(15),4,6,8(16),11,13-hexaene, *see* D-60426
2,3-Dimethyl-5-(2,6,10-trimethylundecyl)-thiophene, D-70446
2,7-Dimethyltropone, *see* D-70391
β,β-Dimethyltyrosine, *see* A-60198
4,4-Dimethylvaleric acid, *see* D-70428
γ,δ-Dimethyl-δ-valerolactone, *see* T-70059
2,7-Dimethyl-9*H*-xanthen-9-one, *see* D-60453
2,7-Dimethylxanthone, D-60453
1,1-Di-4-morpholinylethene, D-70447
1,2;3;6,7,8-Di(1′,8′-naphth)[10]annulene, *see* C-60187
Dinaphtho[1,8-*ab*,1′,8′-*fg*]cyclodecene, *see* C-60187
Dinaphtho[2,3-*a*:2′,3′-*e*]cyclooctene, *see* C-60216
7*H*,9*H*,16*H*,18*H*-Dinaphtho[1,8-*cd*:1′,8′-*ij*]-[1,7]dithiacyclododecin, D-70448
Dinaphtho[7′:1′-1:13][1″:7″-6:8]peropyrene, *see* D-60072
Dinaphtho[1,2,3-*cd*:1′,2′,3′-*lm*]perylene, *see* B-60038
Dinaphtho[1,2,3-*cd*:3′,2′,1′-*lm*]perylene, *see* A-60279
1,3-Di-(1-naphthyl)benzene, D-60454
1,3-Di(2-naphthyl)benzene, D-60455
7,11:20,24-Dinitrilodibenzo[*b*,*m*]-[1,4,12,15]-tetraazacyclodocosine, D-60456
1,1-Dinitroacetone, *see* D-70454
3,5-Dinitrobenzamide, *in* D-60458
4,5-Dinitro-1,2-benzenediol, D-70449
4,6-Dinitro-1,3-benzenediol, D-70450
2,6-Dinitrobenzenemethanol, *see* D-60459
2,6-Dinitrobenzoic acid, D-60457
3,5-Dinitrobenzoic acid, D-60458
2,4-Dinitrobenzotrifluoride, *see* D-70456
3,5-Dinitrobenzotrifluoride, *see* D-70455
2,6-Dinitrobenzyl alcohol, D-60459
2,2′-Dinitrobiphenyl, D-60460
4,5-Dinitrocatechol, *see* D-70449
1,2-Dinitrocyclohexene, D-70451
4,5-Dinitroguaiacol, *in* D-70449

3,5-Dinitroisoxazole, D-70452
▷2,4-Dinitropentane, D-70453
1,1-Dinitro-2-propanone, D-70454
2,3-Dinitro-1*H*-pyrrole, D-60461
2,4-Dinitro-1*H*-pyrrole, D-60462
2,5-Dinitro-1*H*-pyrrole, D-60463
3,4-Dinitro-1*H*-pyrrole, D-60464
4,6-Dinitroresorcinol, *see* D-70450
▷2,6-Dinitrotoluene, *see* M-60061
1,3-Dinitro-5-(trifluoromethyl)benzene, D-70455
2,4-Dinitro-1-(trifluoromethyl)benzene, D-70456
4,5-Dinitroveratrole, *in* D-70449
▷Dinopron *EM*, *in* D-60367
▷Dinoprostone, *in* D-60367
14,15-Dinor-13-oxo-7-labden-17-oic acid, D-70457
Dioctadecylacetic acid, *see* O-70005
▷1,3-Diodo-2-methylenepropane, *see* I-60047
4,6-Di-*O*-galloylarbutin, *in* A-70248
Diomelquinone A, *in* D-60351
Dioscin, *in* S-60045
▷Diosgenin, *in* S-60045
1,4-Dioxa-8-azaspiro[4.7]dodecane, *in* H-70047
1,4-Dioxa-9-azaspiro[4.7]dodecane, *in* H-70048
Dioxa-1,3-bishomopentaprismane, *see* D-70461
1,6-Dioxacyclodeca-3,8-diene, D-70458
1,7-Dioxa-4,10-diazacyclododecane, D-60465
5,12-Dioxa-7,14-diazapentacene, *see* T-60378
2*H*-1,5,2,4-Dioxadiazine-3,6(4*H*)dione, D-60466
1,3-Dioxahexaaza-24-crown-8, *see* D-70460
1,13-Dioxa-4,7,10,16,20,24-hexaazacyclo-hexacosane, D-70459
1,13-Dioxa-4,7,10,16,19,22-hexaazacyclotetra-cosane, D-70460
11,12-Dioxahexacyclo[6.2.1.1³,⁶.0²,⁷.0⁴,¹⁰.0⁵,⁹]-dodecane, D-70461
2,2′-(1,3-Dioxan-2-ylidene)bispyridine, *in* D-70502
8,13-Dioxapentacyclo[6.5.0.0²,⁶.0⁵,¹⁰.0³,¹¹]-tridecane-9,12-dione, D-60467
Dioxapyrrolomycin, *see* P-70203
7,10-Dioxaspiro[5.4]deca-2,5-dien-4-one, *in* B-70044
1,4-Dioxaspiro[4.5]decane-7-thiol, *in* M-70026
1,4-Dioxaspiro[4.5]dec-7-ene, *in* C-60211
2,6-Dioxaspiro[3.3]heptane, D-60468
3,9-Dioxaspiro[5.5]undecane, D-70462
5,12-Dioxatetracyclo[7.2.1.0⁴,¹¹.0⁶,¹⁰]-dodeca-2,7-diene, D-70463
2*H*-1,5,2-Dioxazine-3,6(4*H*)-dione, D-70464
▷1,2-Dioxetane, D-60469
1,5-Dioxiranyl-1,2,3,4,5-pentanepentol, D-70465
3,23-Dioxo-8(14→13*R*)-abeo-17,13-friedo-9β-lanosta-7,14,24*E*-trien-26-oic acid, *see* I-60108
3,23-Dioxo-8(14→13*R*)-abeo-17,13-friedo-9β-lanosta-7,14(30),24*E*-trien-26-oic acid, *see* M-60011
2,2′-Dioxo-*N*,*N*′-bipyrrolidinyl, *see* B-70125
3,6-Dioxo-4,7,11,15-cembratetraen-10,20-olide, D-60470
3,4-Dioxo-1,5-cyclohexadiene-1-carboxylic acid, *see* B-70046
5,6-Dioxo-1,3-cyclohexadiene-1-carboxylic acid, *see* B-70045
2,5-Dioxo-3,6-cyclohexadiene-1,3-dicarboxylic acid, *see* B-70049
3,6-Dioxo-1,4-cyclohexadiene-1,2-dicarboxylic acid, *see* B-70047
3,6-Dioxo-1,4-cyclohexadiene-1,4-dicarboxylic acid, *see* B-70048
9,12-Dioxododecanoic acid, D-70471
1,1′-(1,2-Dioxo-1,2-ethanediyl)bis-1*H*-indole-2,3-dione, D-70466
3,6-Dioxohexahydropyrrolo[1,2-*a*]pyrazine (incorr.), *see* H-70073
1,2-Dioxolane, D-70467
1,3-Dioxolane-2,2-diacetic acid, *in* O-60084
1,3-Dioxolane-2-methanol, D-70468
3,23-Dioxo-7,24-lanostadien-26-oic acid, D-60472

3,23-Dioxo-7,25(27)-lanostadien-26-oic acid, *in* D-60472
1,3-Dioxole, D-60473
6*H*-1,3-Dioxolo[4,5-*g*][1]benzopyran-6-one, *see* M-70068
3,7-Dioxo-12-oleanen-28-oic acid, D-70469
3,5-Dioxo-1,2,4-oxadiazolidine, *see* O-70060
3,4-Dioxopyrrolizidine, *see* D-70260
4,5-Dioxo-4,5-seco-11(13)-eudesmen-12,8β-olide, *see* U-60001
2,4-Dioxo-1,2,3,4-tetrahydropyrimido[4,5-*b*]-quinoline, *see* P-60245
1,3-Dioxototaryl methyl ether, *in* T-70204
4,5-Dioxo-1(10)-xanthen-12,8-olide, D-60474
▷Dipentene, *in* I-70087
Diphenanthro[2,1-*b*:1′,2′-*d*]furan, D-70470
Diphenanthro[9,10-*b*:9′,10′-*d*]furan, D-70471
Diphenanthro[5,4,3-*abcd*:5′,4′,3′-*jklm*]-perylene, *see* D-60072
Di-9-phenanthrylamine, D-70472
Diphenospiropyran, *see* S-70062
2,3-Diphenylallyl bromide, *see* B-70209
▷Diphenylamine-2,2′-dicarboxylic acid, *see* I-60014
Diphenylamine-2,3′-dicarboxylic acid, *see* I-60015
Diphenylamine-2,4′-dicarboxylic acid, *see* I-60016
Diphenylamine-3,3′-dicarboxylic acid, *see* I-60017
Diphenylamine-4,4′-dicarboxylic acid, *see* I-60018
7,7-Diphenylbenzo[*c*]fluorene, D-70473
1,3-Diphenylbenzo[*c*]thiophene, D-70474
1,1′-Diphenylbicyclohexyl, D-70475
1,1-Diphenyl-1,3-butadiene, D-70476
2,3-Diphenyl-2-butenal, D-60475
2,3-Diphenylcrotonaldehyde, *see* D-60475
3,3-Diphenylcyclobutanone, D-60476
1,2-Diphenyl-1,4-cyclohexadiene, D-60477
1,1-Diphenylcyclopropane, D-60478
Diphenylcyclopropenone anil, *see* D-70491
N-(2,3-Diphenyl-2-cyclopropen-1-ylidene)-benzenamine, *see* D-70491
3,6-Diphenyl-2,5-diaza-1,4(2*H*,5*H*)-penta-lenedione, *see* D-70204
1,2-Diphenyl-1,2-diazetidin-3-one, *in* D-70055
4,6-Diphenyl-1,2-dihydropyrimidine, *see* D-60235
3,6-Diphenyl-2,5-dihydropyrrolo[3,4-*c*]-pyrrole-1,4-dione, D-70477
2,5-Diphenyl-1,3-dioxol-1-ium, D-70478
2,11-Diphenyldipyrrolo[1,2-*a*:2′,1′-*c*]-quinoxaline, D-70479
1,2-Diphenyl-1,2-ethanediamine, D-70480
α-(2,2-Diphenylethenyl)-α-phenyl-benzeneacetaldehyde, *see* T-70148
1,2-Diphenylethylenediamine, *see* D-70480
N,*N*′-Diphenylformamidine, D-60479
1,7-Diphenyl-3,5-heptanediol, D-70481
1,7-Diphenyl-1,3,5-heptanetriol, D-70480
7,7-Diphenyl[2.2.1]hericene, *see* D-70488
Diphenyl imidocarbonate, *in* C-70019
1,3-Diphenyl-1*H*-indene-2-carboxylic acid, D-70481
2,3-Diphenyl-1*H*-inden-1-one, D-70482
2,3-Diphenylindole, D-70483
2,3-Diphenylindone, *see* D-70482
Diphenylmethaneimine, D-60482
N,*N*′-Diphenylmethanimidamide, *see* D-60479
Diphenylmethanone-2,2′-dicarboxylic acid, *see* B-70041
Diphenylmethylamine, *see* M-70060
Diphenylmethyl diselenide, *see* B-70141
N-(Diphenylmethylene)acetamide, *in* D-60482
7-(Diphenylmethylene)bicyclo[4.1.0]hepta-1,3,5-triene, *see* D-70484
1-(Diphenylmethylene)-1*H*-cyclopropabenzene, D-70484
1-(Diphenylmethylene)-1*H*-cyclopropa[*b*]-naphthalene, D-70485
2,2-Diphenyl-4-methylene-1,3-dioxolane, D-70486
4-(Diphenylmethylene)-1(4*H*)-naphthalenone, D-70487
7-(Diphenylmethylene)-2,3,5,6-tetramethyl-enebicyclo[2.2.1]heptane, D-70488

1,1-Diphenylmethylenimine, *see* D-60482
N-(Diphenylmethyl)-α-phenyl-
benzenemethanamine, *see* T-70150
1,8-Diphenylnaphthalene, D-60483
3,5-Diphenyl-1,2,4-oxadiazole, D-60484
3,4-Diphenyl-2-oxazolidinone, *in* P-60126
3,4-Diphenyl-1,3-oxazolidin-2-one, *in* P-60126
2,4-Diphenyl-2-oxazolin-5-one, *see* D-60485
2,4-Diphenyl-5(4*H*)-oxazolone, D-60485
2,7-Diphenyloxepin, D-60486
1,1-Diphenyl-1,3-pentadiene, D-70489
Diphenyl pentaketone, *see* D-60488
1,5-Diphenyl-1,3-pentanediol, D-70490
1,5-Diphenylpentanepentone, D-60488
1,5-Diphenyl-1,3,5-pentanetrione, D-60489
9,18-Diphenylphenanthro[9,10-*b*]triphenylene,
D-60490
1,2-Diphenyl-3-(phenylimino)cyclopropene,
D-70491
(Diphenylpropadienylidene)propanedioic acid,
D-70492
2,2-Diphenylpropanoic acid, D-70493
1,1-Diphenylpropene, D-60491
2,2-Diphenylpropiophenone, *see* T-60389
2,3-Diphenylpropiophenone, *see* T-60390
3,3-Diphenylpropiophenone, *see* T-60391
2,4-Diphenylpyrimidine, D-70494
4,5-Diphenylpyrimidine, D-70495
4,6-Diphenylpyrimidine, D-70496
3,4-Diphenyl-1,2,5-selenadiazole, D-70497
▷Diphenylstyrylcarbinol, *see* T-70313
Diphenyl sulfide-2,2′-dicarboxylic acid, *see*
T-70174
Diphenyl sulfide 4,4′-dicarboxylic acid, *see*
T-70175
Diphenyl tartrate, *in* T-70005
2,5-Diphenyl-1,2,4,5-tetraazabicyclo[2.2.1]-
heptane, D-70498
9,18-Diphenyltetrabenz[*a,c,h,j*]anthracene, *see*
D-60490
N,N′-Diphenyl-1,2,4-thiadiazole-3,5-diamine,
see D-70048
2,2-Diphenyl-1,3,4-thiadiazoline, *see* D-60236
2,6-Diphenyl-4*H*-thiopyran-4-one, D-60492
▷*N,N*′-Diphenylurea, D-70499
▷Diphosgene, *see* T-70223
▷Diphyllin, D-60493
▷Di-1-piperidinyl disulfide, *see* D-70523
1,1-Di-1-piperidinylethene, D-70500
N,N′-Di-2-propenyl-1,2-
hydrazinedicarbothioamide, *see* D-60026
▷*N,N*-Di-2-propenyl-2-propen-1-amine, *see*
T-70315
Dipropyl 4,4′-bipyridinium disulfonate, *see*
B-70160
Dipteryxin, *in* P-70032
Dipyridiniomethane, *see* M-70065
Di-2-pyridinylmethanone, *see* D-70502
Di-3-pyridinylmethanone, *see* D-70503
Di-4-pyridinylmethanone, *see* D-70504
N,N′-Di-2-pyridinyl-1,4-phthalazinediamine, *see*
B-60172
Dipyrido[1,2-*a*:1′,2′-*c*]imidazol-10-ium(1+),
D-70501
Dipyrido[1,2-*a*:1′,2′-*e*]-1,3,4,6-tetraazapenta-
lene, *see* P-60234
Dipyrido[1,2-*b*:1′,2′-*e*][1,2,4]tetrazine,
D-60494
6*H*-Dipyrido[1,2-*a*:2′,1′-*d*][1,3,5]triazin-5-ium,
D-60495
▷2,3′-Dipyridyl, *see* B-60120
1,4-Di(2-pyridylamino)phthalazine, *see* B-60172
2,2′-Dipyridyl ether, *see* O-60098
2,3′-Dipyridyl ether, *see* O-60099
3,3′-Dipyridyl ether, *see* O-60100
4,4′-Dipyridyl ether, *see* O-60101
Di-2-pyridyl ketone, D-70502
Di-3-pyridyl ketone, D-70503
Di-4-pyridyl ketone, D-70504
Di-2-pyridylmethane, D-60496
2,2′-Dipyridyl oxide, *see* O-60098
2,3′-Dipyridyl oxide, *see* O-60099
3,3′-Dipyridyl oxide, *see* O-60100
4,4′-Dipyridyl oxide, *see* O-60101

Di-2-pyridyl sulfite, D-60497
▷Disacryl, *in* P-70119
Disalicylaldehyde, D-70505
1,2-Diselenete, D-60498
2-(1,3-Diselenol-2-ylidene)-1,3-diselenole, *see*
B-70116
3′,4′-Disenecioyloxy-3′,4′-dihydroseselin, *in*
K-70013
Disidein, D-60499
▷Disnogamycin, D-70506
Disogluside, *in* S-60045
Dispiro[cyclopropane-1,5′-[3,8]-
dioxatricyclo[5.1.0.0^{2,4}]octane-6′,1″-cyclo-
propane], D-60500
Dispiro[2.0.2.4]dec-8-ene-7,10-dione, D-60501
Dispiro[tricyclo[3.3.1.1^{3,7}]decane-2,2′-oxirane-
3,2″-tricyclo[3.3.1.1^{3,7}]decane], *see* B-70091
Distichol, D-60502
▷Distylin, *see* P-70027
Distyryl selenide, D-60503
2-(1,3-Ditellurol-2-ylidene)-1,3-tellurole, *see*
B-70117
1,2-Dithiacyclodecane, *see* D-60507
1,2-Dithiacycloheptane, *see* D-70517
1,5-Dithiacyclononane, *see* D-70526
9,10-Dithia-1,6-diazaanthracene, *see* D-70518
9,10-Dithia-1,7-diazaanthracene, *see* D-70520
9,10-Dithia-1,8-diazaanthracene, *see* D-70519
1,4-Dithiafulvalene, *see* C-60221
8,8-Dithiaheptafulvene, *see* C-70221
1,3-Dithiane-2-selone, D-70507
2-(1,3-Dithian-2-yl)pyridine, D-60504
3,6-Dithia-1,8-octanediol, *see* B-70145
3,6-Dithia-1,8-octanedithiol, *see* B-70150
2,11-Dithia[3.3]paracyclophane, D-60505
3,10-Dithiaperylene, *see* P-60072
1,6-Dithiapyrene, *see* B-60049
Dithiatopazine, D-70508
2,7-Dithiatricyclo[6.2.0.0^{3,6}]deca-1(8),-3(6)-
diene-4,5,9,10-tetrone, D-60506
3,10-Dithiatricyclo[10.2.2.2^{5,8}]octadeca-
5,7,12,14,15,17-hexaene, *see* D-60505
2,2′-Dithiazolyl, *see* B-70166
2,4′-Dithiazolyl, *see* B-70167
2,5′-Dithiazolyl, *see* B-70168
4,4′-Dithiazolyl, *see* B-70169
4,5′-Dithiazolyl, *see* B-70170
5,5′-Dithiazolyl, *see* B-70171
1,2-Dithiecane, D-60507
Dithieno[2,3-*b*:2′,3′-*d*]pyridine, D-70509
Dithieno[2,3-*b*:3′,2′-*d*]pyridine, D-70510
Dithieno[2,3-*b*:3′,4′-*d*]pyridine, D-70511
Dithieno[3,2-*b*:2′,3′-*d*]pyridine, D-70512
Dithieno[3,2-*b*:3′,4′-*d*]pyridine, D-70513
Dithieno[3,4-*b*:2′,3′-*d*]pyridine, D-70514
Dithieno[3,4-*b*:3′2′-*d*]pyridine, D-70515
Dithieno[3,4-*b*:3′,4′-*d*]pyridine, D-70516
1,2-Dithiepane, D-70517
1,4-Dithiino[2,3-*b*:5,6-*c*′]dipyridine, D-70518
[1,4]Dithiino[2,3-*b*:6,5-*b*′]dipyridine, D-70519
1,4-Dithiino[2,3-*b*:6,5-*c*′]dipyridine, D-70520
[1,4]Dithiino[2,3-*b*]-1,4-dithiin, D-70521
3,3′-Dithiobiscyclobutanol, *in* M-60019
2,2′-Dithiobis-1*H*-isoindole-1,3(2*H*)-dione,
D-70522
Dithiobisphthalimide, D-60508
Dithiobisphthalimide, *see* D-70522
▷1,1′-Dithiobispiperidine, D-70523
1,1′-Dithiobis-2,5-pyrrolidinedione, D-70524
N,N′-Dithiobissuccinimide, *see* D-70524
2,2′-Dithiobisthiazole, *in* T-60200
N,N′-[Dithiobis(trimethylene)]bis[3-
aminopropionamide], *see* H-60084
Dithiocarbazic acid, D-60509
Dithiocarbazinic acid, *see* D-60509
1,4-Dithiocyano-2-butene, D-60510
1,1-Dithiocyclopentane, *see* C-60222
N,N′-(Dithiodi-2,1-ethanediyl)bis[3-
aminopropanamide], *see* A-60073
N,N′-(Dithiodi-2,1-ethanediyl)bis[3-bromo-4-
hydroxy-α-(hydroxyimino)-
benzenepropanamide], B-70130

N,N′-(Dithiodiethylene)bis[3-
aminopropionamide], *see* A-60073
N,N′-(Dithiodi-3,1-propanediyl)bis[3-
aminopropanamide], *see* H-60084
1,3-Dithiolane-2-selone, D-70525
2-(1,3-Dithiolo[4,8-*b*][1,4]dithiin-2-ylidene)-
1,3-dithiolo[4,5-*b*][1,4]dithiin, *see* B-70118
2-(1,3-Dithiol-2-ylidene)-1,3-dithiole, *see*
B-70119
1,5-Dithionane, D-70526
1,4-Dithioniabicyclo[2.2.0]hexane, D-70527
▷Dithiooxamide, *see* E-60041
Dithiosilvatin, D-60511
▷Dithiothymine, *see* M-70127
2,6-Dithioxobenzo[1,2-*d*:4,5-*d*′]bis[1,3]-dithiole-
4,8-dione, D-60512
Di-*p*-toluoyl tartrate, *in* T-70005
▷Diurea, *see* T-70097
Divaronic acid, D-70528
Divaroside, *in* C-70038
1,3-Divinyladamantane, D-70529
▷Divinylethylene glycol, *see* H-70036
▷1,2-Divinylglycol, *see* H-70036
Divinylglyoxal, *see* H-60039
▷DMI, *in* I-70005
▷DMPU, *in* T-60088
▷1,1*a*,2,2,3,3*a*,4,5,5,5*a*,5*b*,6-Dodecachlorocyclo-
octahydro-1,3,4-metheno-1*H*-cyclobuta[*cd*]-
pentalene, D-70530
▷Dodecachlorodihomocubane, *see* D-70530
▷Dodecachloropentacyclo[5.3.0.0^{2,6}.0^{3,9}.0^{5,8}]-
decane, D-70530
2,6-Dodecadien-5-olide, *see* H-60025
Dodecafluorobicyclobutylidene, D-60513
Dodecahedrane, D-60514
1,3,5,7,9,11-Dodecahexaene, D-70531
Dodecahydrocarbazole, D-70532
7,8,10,11,13,14,23,24,26,27,29,30-Dodecahydro-
5,3,2[1′,2′]:16,21[1″,2″]-
dibenzenodibenzo[*l,z*]-[1,4,7,10,15,18,21,24]-
octaoxacyclooctacosin, *see* H-70085
Dodecahydro-6*H*,13*H*-5*a*,12*a*-
epidithiopyrano[3,2-*b*]-pyrano[2′,3′:6,7]-
oxepino[2,3-*f*]oxepin, *see* D-70508
Dodecahydro-2,6,3,5-ethanediylidenecyclo-
but[*jkl*]-*as*-indacene, *see* H-70013
Dodecahydropentaleno[1,6-*cd*]pentalene, *see*
T-60035
1,2,2*a*,3,4,4*a*,5,7,8,9,10,10*a*-Dodecahydro-
3,3,4*a*,7,10*a*-pentamethylpentaleno[1,6-*cd*]-
azulene, *see* L-60018
Dodecahydro-1,7*a*:2,3*b*:3*a*,5:6,7*b*-tetra-
methanodicyclobuta[*b,h*]biphenylene, *see*
N-70070
Dodecahydro-3*a*,6,6,9*a*-tetramethylnaphtho[2,1-
b]furan, D-70533
1,2,3,4,5,6,7,8,9,10,11,12-Dodeca-
hydrotriphenylene, D-60516
2,6-Dodecanedione, D-70534
4,7,10,15,18,21,24,27,30,33,36,39-Dodeca-
oxatricyclo[11.9.9.9^{2,12}]-tetraconta-1,12-
diene, D-70535
▷1-*O*-(1,3,5,7,9-dodecapentaenyl)-*sn*-glycerol, *in*
D-70536
3-(1,3,5,7,9-Dodecapentaenyloxy)-1,2-
propanediol, D-70536
9-Dodecen-1-ol, D-70537
10-Dodecen-12-olide, *in* H-60126
5-Dodecen-7-yne, D-70538
5-Dodecyn-1-ol, D-70539
11-Dodecyn-1-ol, D-70540
3,7-Dolabelladiene-6,12-diol, D-70541
Dolastatin 10, D-70542
1(15)-Dolastene-4,8,9,14-tetrol, D-70543
Dolichol, D-60517
Dolichosterone, D-70544
Dologel, *in* H-70108
Dolunguent, *in* H-70108
Domesticoside, *in* T-70248
Dopal, *see* D-60371
Dopaldehyde, *see* D-60371
α-Doradexanthin, D-70545

Doridosine, D-60518
Dost's base, see D-70048
dota, see T-60015
DR 4003, in A-60228
Dracaenone, D-70546
Drechslerol A, see N-70068
6-Drimene-8,9,11-triol, D-70547
7-Drimene-6,9,11-triol, D-70548
Drimenin, D-60519
9(11)-Drimen-8-ol, D-60520
Droserone, see D-60350
Droserone 5-methyl ether, in D-60350
Drosopterin, D-60521
Dukunolide A, D-60522
Dukunolide D, D-60523
Dukunolide B, in D-60522
Dukunolide C, in D-60522
Dukunolide E, in D-60523
Dukunolide F, in D-60523
Duramycin, D-70549
Duryne, D-70550
4,8,13-Duvatriene-1,3 diol, see C-70030
Dypnopinacol, D-60524
Dypnopinacone, see D-60524
Dysoxylin, D-60525
E-Bifurcarenone, in B-60105
Ebuloside, E-60001
β-Ecdysone, in H-60063
β-Ecdysone 2-cinnamate, in H-60063
β-Ecdysone 3-p-coumarate, in H-60063
Ecdysterone, in H-60063
Ecdysterone 22-O-benzoate, in H-60063
Echinofuran, E-60002
Echinofuran B, in E-60002
Echinosides, in T-70267
Edgeworthin, in D-70006
1(10)E,8E-Millerdienolide, in M-60136
Efical, in O-70102
Egonol, E-70001
Egonol glucoside, in E-70001
Egonol 2-methylbutanoate, in E-70001
▷egta, see B-70134
6-Eicosyl-4-methyl-2H-pyran-2-one, see
  A-60291
(3E)-Isolaureatin, in I-70067
▷Elagostasine, see E-70007
▷Elatericin A, see C-70203
▷Elatericin B, in C-70204
▷α-Elaterin, see C-70204
▷Elaterinide, in C-70204
Elatol, E-70002
Eldanolide, E-70004
Eleganonal, in T-60150
Elemasteiractinolide, E-70005
1,3,7(11)-Elematriene, E-70006
γ-Elemene, see E-70006
▷Ellagic acid, E-70007
Elvirol, E-60004
(E)-Methylsuberenol, in S-70080
Emindole DA, E-60005
Emindole SA, in E-60005
Enarax, in O-60102
Encecanescin, E-70008
2,5-Endomethylenebenzoic acid, see B-60084
Entandrophragmin, E-70009
ent-1(10)13E-Halimadiene-15,18-dioic acid, in
  O-60069
ent-Oplopanone, in O-70043
ent-1(10),13Z-Halimadiene-15,18-dioic acid, in
  O-60069
▷Enzaprost E, in D-60367
Ephemeric acid, E-60006
Ephemeroside, in E-60006
5-epi-6β-acetoxyhebemacrophyllide, in H-60010
(−)-Epiafzelechin, in T-60104
(+)-Epiafzelechin, in T-60104
1-Epialkhanol, in D-70299
Epianastrephin, in A-70206
3-epi-Astrahygrol, in H-60169
17-epi-Azadiradione, in A-70281
Epicatechin, in P-70026
Epicatechol, in P-70026
7-Epicedronin, in C-70027
Epidanshenspiroketallactone, in D-70004
7-Epidebneyol, in D-70010
Epidermin, E-70010

6-Epidesacetyllaurenobiolide, in H-70134
11-Epidihydroridentin, in R-70007
3-epi-Diosgenin, in S-60045
3,6-Epidioxy-6-methoxy-4,16,18-eicosatrienoic
  acid, E-60007
3,6-Epidioxy-6-methoxy-4-octadecenoic acid,
  E-60008
9,10-Epidithio-9,10-dihydroanthracene, see
  D-70206
9′-Epiencecanescin, in E-70008
Epientphenmenthol, in M-60090
1-Epierivanin, in E-70033
23-epi-Firmanolide, in F-60013
▷Epifluorohydrin, see F-70032
1-Epi-α-gurjunene, in G-70034
3-epi-20-Hydroxyecdysone, in H-60063
8′-epi-Hydroxyzearalenone, in Z-60002
5-Epiilimaquinone, in I-70002
1-epi-Inuviscolide, in I-70023
8-epi-Inuviscolide, in I-70023
1-Episoerivanin, in I-70066
8-Epijateorin, in C-60160
2-Epijatrogrossidione, in J-70002
Epilophodione, in D-60470
1′-Epilycoserone, in L-70045
8-epi-Manoyl oxide, in E-70023
13-epi-Manoyl oxide, in E-70023
6-Epimonomelittoside, in M-70019
4-Epineopulchellin, in N-70028
13-Epi-5β-neoverrucosanol, in N-70030
10-Epinidoresedic acid, in N-70035
6-Epiophiobolin A, in O-70040
4-Epipulchellin, in P-70138
Epirosmanol, in R-70010
13-Episclareol, in S-70018
3-Episiaresinolic acid, in D-70327
Episongiadiol, in D-70345
Epitaxifolin, in P-70027
12-Epiteupolin II, in T-70159
22-Epitokorogenin, in S-60043
Epiverrucarinolactone, in D-60353
Epivittadinal, in V-70016
Epivolkenin, in T-60007
4-Epivulgarin, in V-60017
3-Epizaluzanin C, in Z-60001
12,16-Epoxy-8,11,13-abietatriene, see A-60008
1,2-Epoxyacenaphthene, see D-70170
1,2-Epoxyacenaphthylene, see D-70170
2,3-Epoxy-7,10-bisaboladiene, E-60009
1,5-Epoxy-5-bromopentane, see B-70271
▷1,2-Epoxybutane, see E-70047
Epoxycantabronic acid, C-60010
7,8-Epoxy-2,11-cembradiene-4,6-diol, E-60010
3,4-Epoxy-7,11,15(17)-cembratrien-16,2-olide,
  see I-60106
4,5-Epoxycyclohexene, see O-60047
3,6-Epoxy-1-cyclohexene, see O-70054
20,24-Epoxy-9,19-cyclolanostane-3,16,25-triol,
  see Q-70011
1,2-Epoxy-5-cyclooctene, see O-70058
8α,9α-Epoxy-2α,4β,6α,10α-daucanetetrol, see
  L-60015
6,12-Epoxy-6H,12H-dibenzo[b,f][1,5]dioxocin,
  see D-70505
5,6-Epoxy-5,6-dihydro-12′-apo-β-carotene-3,12′-
  diol, see P-60065
▷5,6-Epoxy-5,6-dihydrochrysene, see C-70174
1,2-Epoxy-1,2-dihydrolycopene, in L-60041
5,6-Epoxy-5,6-dihydrolycopene, in L-60041
2,3-Epoxy-2,3-dihydro-1,4-naphthoquinone,
  E-70011
1,2-Epoxy-1,2-dihydrotriphenylene, see T-60385
12,16-Epoxy-11,14-dihydroxy-5,8,11,13-
  abietatetraen-7-one, E-60011
11,12-Epoxy-2,14-dihydroxy-5,8(17)-briaradien-
  18,7-olide, E-70012
24R,25-Epoxy-11β,23S-dihydroxydammar-
  13(17)-en-3-one, see A-70082
2,3-Epoxy-1,4-dihydroxy-7(11),8-eudesmadien-
  12,8-olide, E-60012
4,5-Epoxy-2,8-dihydroxy-1(10),11(13)-
  germacradien-12,6-olide, E-60013
24,25-Epoxy-7,26-dihydroxy-8-lanosten-3-one,
  E-70013
2α,3α-Epoxy-1β,4α-dihydroxy-8β-methoxy-
  7(11)-eudesmen-12,8-olide, in E-60033

5,6-Epoxy-3,4-dihydroxy-2-methyl-2-cyclo-
  hexen-1-one, see T-70009
22,25-Epoxy-3β,17α-dihydroxy-11-oxo-7-
  lanosten-18,20S-olide, see D-70237
6α,7α-Epoxy-5α,20R-dihydroxy-1-oxo-
  22R,24S,25R-with-2-enolide, see I-60143
24R,25-Epoxy-11β,23S-dihydroxy-13(17)-
  protosten-3-one, see A-70082
6,11-Epoxy-6,12-dihydroxy-6,7-seco-8,11,13-
  abietatrien-7-al, E-60014
15,20-Epoxy-2α,3β-dihydroxy-14,15-seco-
  5,15,17(20)-pregnatrien-14-one, see A-70267
1α,2α-Epoxy-4β,5-dihydroyomogin, in Y-70001
3,4-Epoxy-2,2-dimethyl-1-butanol, see D-60431
8,16-Epoxy-8H,16H-dinaphtho[2,1-b:2′,1′-
  f][1,5]dioxocin, E-70014
4β,15-Epoxy-1(10)E,8E-millerdienolide, in
  M-60136
5,6-Epoxy-7,9,11,14-eicosatetraenoic acid,
  F-60015
8,9-Epoxy-5,11,14-eicosatrienoic acid, E-60016
11,12-Epoxy-5,8,14-eicosatrienoic acid, E-60017
4β,15-Epoxy-9E-millerenolide, in M-60137
4β,15-Epoxy-1β-ethoxy-4Z-millerenolide, in
  M-60137
1α,2α-Epoxy-4,11(13)-eudesmadien-12,8β-
  olide, in Y-70001
ent-5α,11-Epoxy-1β,4α,6α,8α,9β,14-
  eudesmanehexol, E-60018
1,4-Epoxy-6-eudesmanol, E-60019
8,12-Epoxy-3,7,11-eudesmatrien-1-ol, E-60020
3,4-Epoxy-11(13)-eudesmen-12-oic acid,
  E-60021
▷1,2-Epoxy-3-fluoropropane, see F-70032
22,25-Epoxy-2,3,6,26-furostanetetrol, E-70015
22S,25S-Epoxy-5-furostene-3β,26-diol, see
  N-60062
22,25-Epoxy-5-furostene-2α,3β,26-triol, in
  E-70015
Epoxyganoderiol A, in E-70013
Epoxyganoderiol B, in E-70025
Epoxyganoderiol C, in E-70025
1,10-Epoxy-4,11(13)-germacradien-12,8-olide,
  E-60022
4,5-Epoxy-1(10),7(11)-germacradien-8-one,
  E-60023
4α,5α-Epoxy-1(10)E-germacrene-6β,8β-diol,
  see S-70039
1β,10α-Epoxyhaageanolide, in H-60001
1β,10β-Epoxy-8α-hydroxyachillin, in H-60100
14,15-Epoxy-5-hydroxy-6,8,10,12-eicosatetra-
  enoic acid, E-70016
5,6-Epoxy-20-hydroxy-8,11,14-eicosatrienoic
  acid, E-70017
14,15-Epoxy-20-hydroxy-5,8,11-eicosatrienoic
  acid, E-70018
1,2-Epoxy-3-hydroxy-4,11(13)-eudesmadien-
  12,8-olide, E-70019
4α,5α-Epoxy-3α-hydroxy-11(13)-eudesmen-
  12,8β-olide, in H-60136
4α,5α-Epoxy-3β-hydroxy-11(13)-eudesmen-
  12,8β-olide, in H-60136
24S,25S-Epoxy-26-hydroxy-7,9(11)-
  lanostadien-3-one, in E-70025
5,6-Epoxy-4-hydroxy-2-methoxy-2-cyclohexen-
  1-one, see H-70155
5,6-Epoxy-2-hydroxymethyl-2-cyclohexene-1,4-
  dione, see H-70181
12,16-Epoxy-18-hydroxy-20-nor-5(10),-
  6,8,12,15-abietapentaene-11,14-dione, see
  I-60138
15,20-Epoxy-3β-hydroxy-14,15-seco-
  5,15,17(20)-pregnatriene-2,14-dione, in
  A-70267
24,25-Epoxy-23-hydroxy-7-tirucallen-3-one, see
  N-70036
Epoxyisodihydrorhodophytin, E-60024
Epoxyjaeschkeanadiol p-hydroxybenzoate, in
  J-70001
Epoxyjaeschkeanadiol vanillate, in J-70001
1α,10β-Epoxykurubaschic acid benzoate, in
  K-60014

1β,10α-Epoxykurubashic acid angelate, *in* K-60014

1β,10α-Epoxykurubashic acid benzoate, *in* K-60014

14,15-Epoxy-8(17),12-labdadien-16-oic acid, E-60025

ent-8,13-Epoxy-14,15,19-labdanetriol, E-70020

8,13-Epoxy-3-labdanol, E-70021

ent-15,16-Epoxy-7,13(16),14-labdatriene, E-70022

15,16-Epoxy-8(17),13(16),14-labdatrien-19-oic acid, *see* L-60011

8,13-Epoxy-14-labdene, E-70023

ent-9,13-Epoxy-7-labdene-15,17-dioic acid, E-70024

21,23-Epoxylanosta-7,24-diene-3,21-diol, *see* F-70013

24,25-Epoxy-7,9(11)-lanostadiene-3,26-diol, E-70025

Epoxylaphodione, *in* D-60470

20(29)-Epoxy-3-lupanol, E-60026

4β,15-Epoxy-1β-methoxy-9Z-millerenolide, *in* M-60137

2,3-Epoxy-3-methyl-1-butanol, *see* D-70426

3,4-Epoxy-2-methyl-1-butene, *see* M-70080

2-(3,4-Epoxy-4-methylcyclohexyl)-6-methyl-2,5-heptadiene, *see* E-60009

24,25-Epoxy-29-nor-3-cycloartanol, E-70026

13β,28-Epoxy-11-oleanene-3β,16β,23-triol, *see* S-60001

1α,2α-Epoxy-3-oxo-11(13)-eudesmen-12,8β-olide, *in* Y-70001

15,16-Epoxy-12-oxo-8(17),13(16),14-labdatrien-19-oic acid, *in* L-60011

(20S,22S)-21,24S-Epoxy-1-oxo-2,5,25-withatrienolide, *see* D-60003

▷1,2-Epoxypropane, *see* M-70100

5-(1,2-Epoxypropyl)benzofuran, *see* M-70101

Epoxypteryxin, *in* K-70013

5,6-Epoxyquinoline, E-60027

7,8-Epoxyquinoline, E-60028

15,16-Epoxy-13,17-spatadiene-5,19-diol, E-60029

3,4-Epoxy-13(15),16-sphenolobadiene-5,18-diol, E-60030

3,4-Epoxy-13(15),16,18-sphenolobatrien-5-ol, E-60031

▷Epoxytetrafluoroethane, *see* T-60049

9,10-Epoxy-7,8,9,10-tetrahydro-7,8-dihydroxybenzo[a]pyrene, *see* T-70041

8,13-Epoxy-1,6,7,9-tetrahydroxy-14-labden-11-one, E-70027

21,23-Epoxy-7,24-tirucalladien-3-ol, E-60032

21,23-Epoxy-7,24-tirucalladien-3-one, *in* E-60032

12,13-Epoxytrichothec-9-en-3-ol, E-70028

Epoxytrifluoroethane, *see* T-60305

2,3-Epoxy-1,1,1-trifluoropropane, *see* T-60294

2,3-Epoxy-1,4,8-trihydroxy-7(11)-eudesmen-12,8-olide, E-60033

5β,6β-Epoxy-4β,17α,20R-trihydroxy-3β-methoxy-1-oxo-8(14),24-withadienolide, *see* P-60144

6,7-Epoxy-5,20,22-trihydroxy-1-oxoergost-2-en-26-oic acid δ-lactone, *see* I-60143

5α,6α-Epoxy-7α,17α,20R-trihydroxy-1-oxo-22R-6α,7α-witha-2,24-dienolide, *see* W-60007

5β,6β-Epoxy-4β,16β,20R-trihydroxy-1-oxo-22R,24R,25R-with-2-enolide, *see* W-60008

6α,7α-Epoxy-5α,14α,20R-trihydroxy-1-oxo-22R,24S,25R-with-2-enolide, *in* I-60143

2,3-Epoxy-4,5,8-trihydroxy-2-prenyl-1-tetralone, *see* D-70277

17a,21-Epoxy-4,4,8-trimethyl-D(17a)-homopregna-17,20-diene-12α,16β-diol 12-acetate, *see* F-70057

1α,2α-Epoxyyomogin, *in* Y-70001

4β,15-Epoxy-9Z-millerenolide, *in* M-60137

▷Equisetic acid, *see* A-70055

Eranthemoside, E-60034

9,11(13)-Eremophiladien-12-oic acid, E-70029

11-Eremophilene-2,8,9-triol, E-70030

1(10)-Eremophilen-7-ol, E-70031

Ergolide, E-60035

5,24(28)-Ergostadiene-3,4,7,20-tetrol, *see* M-60070

7,22-Ergostadiene-3,5,6-triol, *see* M-70057

Ericin, *in* H-70108

Eriocephaloside, *in* H-70133

Eriodermic acid, E-70032

Eriodictyol, *see* T-70105

Eriolin, E-60036

Erivanin, E-70033

Erycristin, E-70034

Erythrisenegalone, *see* C-60004

Erythrol, *see* B-70291

Erzinine, *see* H-70083

Esperamicin, E-60037

Esperamicin X, E-70035

Esperamicin A₁, *in* E-60037

Esperamicin A₂, *in* E-60037

Esperamicin A₁b, *in* E-60037

Espirosal, *in* H-70108

Esulone A, E-60038

Esulone C, E-60039

Esulone B, *in* E-60038

▷Etenzamide, *in* H-70108

Ethanebis(dithioic)acid, E-60040

▷Ethanedihydroxamic acid, *see* O-60048

▷Ethanedithioamide, E-60041

4,4'-(1,2-Ethanediyl)bis[1,2-benzenediol], *see* T-60101

5,5'-(1,2-Ethanediyl)bis[1,2,3-benzenetriol], *see* H-60062

5,5'-(1,2-Ethanediyl)bis-1,3-benzodioxole, *in* T-60101

6,6'-(1,2-Ethanediyl)bis-1,3-benzodioxol-4-ol, *in* H-60062

1,1'-(1,2-Ethanediyl)bishydrazine, *see* D-60192

2,2'-[1,2-Ethanediylbis(oxy)]bisethanamine, *see* B-60139

2,2'-[1,2-Ethanediylbis(thio)]-bisethanethiol, *see* B-70150

2,2'-[1,2-Ethanediylbis(thio)]bisethanol, *see* B-70145

2,2'-(1,2-Ethanediylidene)-bistricyclo[3.3.1.1³,⁷]decane, *see* B-60075

1,1',1'',1'''-1,2-Ethanediylidenetetrakiscyclopropane, *see* T-60034

S₆-Ethano-18, *see* H-60075

S₆-Ethano-18, *see* H-70087

1,10-Ethanodibenzo[a,e]cyclooctene, *see* C-60229

13,15-Ethano-3,17-diethyl-2,12,18-trimethylmonobenzo[g]porphyrin, *see* E-60042

13,15-Ethano-17-ethyl-2,3,12,18-tetramethylmonobenzo[g]porphyrin, E-60043

2,2'-(1,2-Ethenediyl)bisfuran, *see* D-60191

15,15'-(1,2-Ethenediyl)-bis[2,3,5,6,8,9,11,12-octahydro-1,4,7,10,13-benzopentaoxacyclopentadecin], *see* B-60142

2,2',2'',2'''-(1,2-Ethenediylidene)-tetrakis[9,10-dihydro-1',2'-benzenoanthracene], *see* T-60169

▷Ethenetetracarbonitrile, *see* T-60031

Ethenetetracarboxylic acid, *see* E-60049

2,6-(Etheno[1,4]benzenoetheno)naphthalene, *see* N-70006

N²,3-Ethenoguanine, *see* I-70006

N²,3-Ethenoguanosine, *see* I-70006

12,15-Etheno-5,9-metheno-9H-benzocyclotridecene, *see* B-70036

5,8-Etheno-17,13-metheno-13H-dibenzo[a,g]cyclotridecene, *see* D-70071

4-Ethenyl-1,3-benzenediol, *see* V-70010

5-Ethenyl-4,6-benzofurandiol, *see* D-70352

5-Ethenyl-6-benzofuranol, *see* H-70231

▷5-Ethenylbicyclo[2.2.1]hept-2-ene, E-70036

3-Ethenylcyclohexene, *see* V-60009

5-Ethenyl-1-cyclopentenecarboxylic acid, *see* V-60010

3-Ethenyl-2-cyclopenten-1-one, *see* V-70011

2-Ethenyl-1,1-cyclopropanedicarboxylic acid, *see* V-60011

3-Ethenylcyclopropene, *see* V-60012

3-Ethenyldihydro-2(3H)-furanone, *see* D-70279

1-Ethenyl-2,4-dimethoxybenzene, *in* V-70010

5-Ethenyl-4,6-dimethoxybenzofuran, *in* D-70352

5-Ethenyl-1,6-dimethyl-2,7-phenanthrenediol, *see* D-70294

7-Ethenyldodecahydro-1,1,4a,7-tetramethyl-8a(2H)-phenanthrenol, *see* H-60163

6-Ethenyl-11-ethyl-15,19,20,21-tetrahydro-18-hydroxy-5,10,22,23-tetramethyl-4,7-imino-2,21-methano-14,16-metheno-9,12-nitrilo-17H-azuleno[1,8-bc][1,5]-diazacyclooctadecin-17-one, *see* C-60230

4-Ethenylhexahydro-4,7a-dimethyl-2(3H)-benzofuranone, *see* A-70206

2-Ethenyl-4-hydroxybenzenepropanoic acid, *see* H-70232

4,4'-Ethenylidenebismorpholine, *see* D-70447

1,1'-Ethenylidenebispiperidine, *see* D-70500

4-Ethenyl-1H-imidazole, *see* V-70012

2-Ethenyl-1H-indole, *see* V-70013

1-Ethenyl-4-methoxybenzene, *in* H-70209

5-Ethenyl-5-methyl-1,3-cyclopentadiene, *see* M-70146

▷4-Ethenylphenol, *see* H-70209

5-Ethenyl-2-pyrimidinamine, *see* A-60269

2-Ethenyl-1H-pyrrole, *see* V-60013

4-Ethenylquinoline, *see* V-70014

2-(5-Ethenyltetrahydro-5-methyl-2-furanyl)-4,4-dimethyl-5-hexen-3-one, *see* A-60305

4-Ethenyl-1,2,3-thiadiazole, *see* V-60014

5-Ethenyl-1,2,3-thiadiazole, *see* V-60015

5-Ethenyluridine, *see* V-70015

▷Ethenzamide, *in* H-70108

Ethosalamide, *in* H-70108

α-Ethoxyacrylic acid, *see* E-70037

▷2-Ethoxybenzamide, *in* H-70108

2-Ethoxybenzoic acid, *in* H-70108

2-Ethoxycarbazole, *in* H-60109

2-Ethoxycarbonylcyclopentanone, *in* O-70084

Ethoxycarbonylmalonaldehyde, *in* D-60190

5-Ethoxy-3,4-dihydro-2H-pyrrole, *in* P-70188

7-Ethoxy-3,4-dimethylcoumarin, *in* H-60119

5-Ethoxy-2,4-dinitrophenol, *in* D-70450

2-(2-Ethoxyethoxy)benzamide, *in* H-70108

▷(1-Ethoxyethyl) hydroperoxide, *see* E-60044

N-Ethoxy-1,1,1,1',1',1'-hexafluorodimethylamine, *in* B-60180

▷1-Ethoxy-1-hydroperoxyethane, E-60044

10α-Ethoxy-3β-hydroxy-4,11(13)-cadinadien-12-oic acid, *in* Z-70001

2-Ethoxy-3-hydroxyquinoxaline, *in* D-60279

4-Ethoxy-2-hydroxy-N,N,N-trimethyl-4-oxo-1-butanaminium, *in* C-70024

3-Ethoxy-1H-isoindole, *in* P-70100

6-Ethoxy-7-methoxy-2,2-dimethyl-2H-1-benzopyran, *in* D-70365

9-(Ethoxymethyl)-4-formyl-3,8-dihydroxy-1,6-dimethyl-11-oxo-11H-dibenzo[b,e]-[1,4]-dioxepin-7-carboxylic acid, *see* C-70035

▷2-Ethoxymethyl-3-hydroxy-1-methoxy-anthraquinone, *in* D-70306

4-Ethoxy-1,2-naphthoquinone, *in* H-60194

4-Ethoxy-2-nitropyridine, *in* N-70063

1-Ethoxy-1-phenylethylene, *in* P-60092

2-Ethoxy-2-propenoic acid, E-70037

3-Ethoxy-4(1H)-pyridinone, *in* H-60227

2-Ethoxy-1-pyrroline, *in* P-70188

3-Ethoxy-2-quinoxalinol, *in* D-60279

3-Ethoxy-2(1H)quinoxalinone, *in* D-60279

α-Ethoxystyrene, *in* P-60092

3-Ethoxy-4,5,6,7-tetrahydroisoxazolo[4,5-c]pyridine, *in* T-60071

1-Ethoxy-9Z-millerenolide, *in* M-60137

Ethuliacoumarin, E-60045

Ethyl 3-acetoxyacrylate, *in* O-60090

▷Ethyl acrylate, *in* P-70121

5-Ethylamino-1H-tetrazole, *in* A-70202

3-Ethylaspartic acid, *see* A-70138

3-(2-Ethyl-2-butenyl)-9,10-dihydro-1,6,8-trihydroxy-9,10-dioxo-2-anthracenecarboxylic acid, E-70038

1-Ethylcyclohexene, E-60046

3-Ethylcyclohexene, E-70039

4-Ethylcyclohexene, E-60048
Ethyl dicyanoacetate, *in* D-70149
Ethyl 3,3-diethoxypropionate, *in* O-60090
Ethyl diformylacetate, *in* D-60190
3-*O*-Ethyldihydrofusarubin B, *in* F-60108
14-Ethyl-16,17-dihydro-8,9,13,24-tetramethyl-5,22:12,15-diimino-20,18-metheno-7,10-nitrilobenzo[*o*]-cyclopent[*b*]azacyclo-nonadecine, *see* E-60043
3-Ethyl-5,7-dihydroxy-4*H*-1-benzopyran-4-one, E-70040
3-Ethyl-5,7-dihydroxychromone, *see* E-70040
6-Ethyl-2,4-dihydroxy-3-methylbenzoic acid, E-70041
3-Ethyl-2,19-dioxabicyclo[16.3.1]docosa-3,6,9,18(22),21-pentaen-12-yn-20-one, E-70042
2-Ethyl-1,6-dioxaspiro[4.4]nonane, E-70043
▷ Ethylenebis(oxyethylenenitrilo)tetraacetic acid, *see* B-70134
Ethylenedihydrazine, *see* D-60192
3,4-Ethylenedioxybenzaldehyde, *see* B-60023
3,4-Ethylenedioxybenzoic acid, *see* B-60025
3,3-(Ethylenedioxy)pentanedioic acid, *in* O-60084
2,2′-(Ethylenedithio)diethanol, *see* B-70145
▷ Ethylene glycol bis(2-aminoethyl ether)-*N*-tetra-acetic acid, *see* B-70134
▷ Ethylene glycol bis[2-[bis(carboxymethyl)-amino]ethyl] ether, *see* B-70134
Ethylene 1,3-propylene orthocarbonate, *see* T-60164
Ethylenetetracarboxylic acid, E-60049
▷ Ethyleneurea, *see* I-70005
21-Ethyl-2,6-epoxy-17-hydroxy-1-oxacyclo-henicosa-2,5,14,18,20-pentaen-11-yn-4-one, E-60050
Ethyl 2-ethoxycarbonyl-5,5-diphenyl-2,3,4-pentatrienoate, *in* D-70492
5-Ethyl-2-hydroxy-2,4,6-cycloheptatrien-1-one, E-70044
9-Ethyl-4-hydroxy-1,7-dioxaspiro[5.5]-undecane-3-methanol, *see* T-70004
5-Ethyl-3-hydroxy-4-methyl-2(5*H*)-furanone, E-60051
24-Ethylidene-2,3,22,23-cholestanetetrol, *see* S-70075
Ethylidenedimercaptofulvene, *in* C-60221
α-Ethylidenediphenylmethane, *see* D-60491
1-Ethylideneoctahydro-4-methylene-7-(1-methylethyl)-1*H*-indene-2,5,6-triol, *see* O-70042
3-Ethyl-3-mercaptopentane, *see* E-60053
▷ 3-Ethyl-2-methylacraldehyde, *see* M-70106
7-Ethyl-5-methyl-6,8-dioxabicyclo[3.2.1]-oct-3-ene, E-60052
N-Ethyl-*N*′-methyl-*N*,*N*′-diphenylurea, *in* D-70499
5-Ethyl-4-methylisotetronic acid, *see* E-60051
23-Ethyl-24-methyl-27-nor-5,25-cholestadien-3β-ol, *see* F-60010
Ethylnylcyclopropane, E-70045
1-Ethylnylnaphthalene, E-70046
▷ 2-Ethyloxirane, E-70047
O-Ethyl 3-oxobutanethioate, *in* O-70080
S-Ethyl 3-oxobutanethiol, *in* O-70080
3-Ethyl-3-pentanethiol, E-60053
Ethyl *N*-phenacylcarbamate, *in* O-70099
(1-Ethyl-2-propynyl)benzene, *see* P-70089
5-Ethyl-2,4(1*H*,3*H*)-pyrimidinedione, E-60054
▷ Ethyl salicylate, *in* H-70108
5-Ethylsotolone, *see* E-60051
3-Ethylthioacrylic acid, *see* E-60055
3-(Ethylthio)-2-propenoic acid, E-60055
2-(Ethylthio)thiazole, *in* T-60200
Ethyl trifluoromethanesulfenate, *in* T-60279
7-Ethyl-2,8,9-trihydroxy-2,4,4,6-tetramethyl-1,3(2*H*,4*H*)anthracenedione, *see* H-60148
5-Ethyltropolone, *see* E-70044
5-Ethyluracil, *see* E-60054
5-Ethyluridine, *in* E-60054
2-Ethyl-2-vinylglycine, *see* A-60150
▷ Ethyl vinyl ketone, *see* P-60062
Ethynamine, E-60056

9,9′-(1,2-Ethynediyl)bisanthracene, *see* D-60051
4-Ethynylbenzaldehyde, E-70048
1-Ethynylbicyclo[2.2.2]octane, E-70049
Ethynylbutadiene, *see* H-70041
1-Ethynylcycloheptene, E-70050
1-Ethynylcyclohexene, E-70051
1-Ethynylcyclopentene, E-70052
5-Ethynylcytidine, *in* A-60153
5-Ethynylcytosine, *see* A-60153
2-Ethynylindole, E-60057
3-Ethynylindole, E-60058
3-Ethynyl-3-methyl-4-pentenoic acid, E-60059
Ethynylpentafluorobenzene, E-70053
2-Ethynyl-1,3-propanediol, E-70054
Ethynyl propargyl ketone, *see* H-60040
5-Ethynyl-2,4(1*H*,3*H*)-pyrimidinedione, E-60060
2-Ethynylthiazole, E-60061
4-Ethynylthiazole, E-60062
2-Ethynylthiophene, E-70055
5-Ethynyluracil, *see* E-60060
5-Ethynyluridine, *in* E-60060
Etosalamide, *in* H-70108
Euchrenone A_1, E-70056
Euchrenone A_2, E-70057
Euchrenone b_1, E-70058
Euchrenone b_3, E-70058
Eucrenone b_2, *in* E-70058
Eudesmaafraglaucolide, *in* T-70258
4,7(11)-Eudesmadiene-12,13-diol, E-60063
▷ 3,11(13)-Eudesmadien-12-oic acid, E-70060
4,11(13)-Eudesmadien-12-oic acid, E-70061
3,5-Eudesmadien-1-ol, E-60064
5,7-Eudesmadien-11-ol, E-60065
5,7(11)-Eudesmadien-15-ol, E-60066
1,11-Eudesmanediol, E-70062
5,11-Eudesmanediol, E-70063
11-Eudesmene-1,5-diol, E-60067
4(15)-Eudesmene-1,11-diol, E-70064
4(15)-Eudesmene-1,5,6-triol, E-60068
3-Eudesmen-6-ol, E-70065
11-Eudesmen-5-ol, E-60069
4(15)-Eudesmen-11-ol, E-70066
β-Eudesmol, *see* E-70066
5-Eudesmol, *see* E-60069
Euglenapterin, E-60070
Eulophiol, *in* T-60116
Eumaitenin, E-60071
Eumaitenol, E-60072
Eumorphistonol, E-60073
Eupalin, *in* P-60047
Eupalitin, *in* P-60047
Eupaserrin, *in* D-70302
Euphorianin, E-60074
Eupomatenoid 2, *in* E-60075
Eupomatenoid 4, *in* E-60075
Eupomatenoid 5, *in* E-60075
Eupomatenoid 6, E-60075
Eupomatenoid 7, *in* E-60075
Eupomatenoid 12, *in* E-60075
Eupomatenoid 13, *in* K-60001
▷ Euprex, *in* T-70277
Euthroid, *in* T-60350
▷ Exaltone, *see* C-60219
Eximin, *in* A-70248
E2Z4, *see* D-70321
F-5-3, *in* Z-60002
F-5-4, *in* Z-60002
Fabiatrin, *in* H-70153
Faranal, *see* T-60156
Farnesenic acid, *see* T-60364
Farnesic acid, *see* T-60364
3-(1-Farnesyl)-4-hydroxycoumarin, *see* F-60005
Farnesylic acid, *see* T-60364
Fasciculatin, F-70001
Fauronol, F-70002
Fauronyl acetate, *in* F-70002
Fecapentaene-12, *see* D-70536
Fecapentaene-14, *see* T-70024
▷ Fenazaflor, *in* D-60160
[4.4.5.5]Fenestrane, *see* T-60037
[5.5.5.5]Fenestrane, *see* T-60035
[5.5.5.5]Fenestratetraene, *see* T-60036
Fenestrindane, *see* T-60019
▷ Fenoflurazole, *in* D-60160

Feraginidin, *in* D-60004
Fercoperol, F-60001
9(11)-Fernene-3,7,19-triol, F-60002
Ferolin, *in* G-60014
Ferprenin, F-70003
Ferreyrantholide, F-70004
Ferrioxamine B, F-60003
Ferrioxamine D_1, *in* F-60003
Ferugin, *in* D-60005
Feruginin, F-60004
Ferulenol, F-60005
Ferulide, F-70005
Ferulidene, *in* X-60002
Ferulidin, F-70006
Ferulinolone, F-60006
Feruone, F-70007
Fervanol, F-60007
Fervanol benzoate, *in* F-60007
Fervanol *p*-hydroxybenzoate, *in* F-60007
Fervanol vanillate, *in* F-60007
Fevicordin *A*, F-60008
Fevicordin *A* glucoside, *in* F-60008
Fexerol, F-60009
Fibrostatin *A*, *in* F-70008
Fibrostatin *B*, *in* F-70008
Fibrostatin *C*, *in* F-70008
Fibrostatin *D*, *in* F-70008
Fibrostatin *E*, *in* F-70008
Fibrostatin *F*, *in* F-70008
Fibrostatins, F-70008
Ficisterol, F-60010
Ficulinic acid A, F-60011
Ficulinic acid B, F-60012
Firmanoic acid, *in* D-60472
Firmanolide, F-60013
Flabellata secoclerodane, F-70009
Flaccidin, F-70010
Flavanthrin, F-70011
Flavokawin A, *in* H-60220
Flavokermesic acid, *see* T-60332
3-Flavonecarboxylic acid, *see* O-60085
Flavophene, *see* D-70074
Flavoyadorigenin B, *see* D-70291
Flavoyadorinin B, *in* D-70291
Flexilin, F-70012
Flindissol, F-70013
Flindissol lactone, *in* F-70013
Flindissone, *in* F-70013
Flindissone lactone, *in* F-70013
Floionolic acid, *in* T-60334
Floroselin, *in* K-70013
Flossonol, F-60014
Fluorantheno[3,4-*cd*]-1,2-diselenole, F-60015
Fluorantheno[3,4-*cd*]-1,2-ditellurole, F-60016
Fluorantheno[3,4-*cd*]-1,2-dithiole, F-60017
Fluorazone, *see* P-60247
trans-Fluorenacenedione, *see* I-70011
9*H*-Fluorene-1-carboxaldehyde, F-70014
▷ 9*H*-Fluorene-9-carboxaldehyde, F-70015
▷ 9*H*-Fluorene-1-methanol, *see* H-70176
9*H*-Fluorene-9-selenone, *see* F-60018
Fluoreno[3,2,1,9-*defg*]chrysene, F-70016
Fluoreno[9,1,2,3-*cdef*]chrysene, F-70017
9-(9*H*-Fluoren-9-ylidene)-9*H*-fluorene, *see* B-60104
▷ 9*H*-Fluoren-9-ylidenemethanol, *see* F-70015
9-Fluorenylmethyl pentafluorophenyl carbonate, F-60019
Fluorindine, *see* D-70264
2-Fluoro-β-alanine, *see* A-60162
3-Fluoroanthranilic acid, *see* A-70140
4-Fluoroanthranilic acid, *see* A-70141
5-Fluoroanthranilic acid, *see* A-70142
6-Fluoroanthranilic acid, *see* A-70143
α-Fluorobenzeneacetaldehyde, *see* F-60062
4-Fluoro-1,2-benzenediol, F-60020
4-Fluoro-1,3-benzenediol, F-70018
α-Fluorobenzenepropanal, *see* F-60064
(4-Fluoro-1-butenyl)benzene, *see* F-70035
3-Fluorobutyrine, *see* A-60154
4-Fluorobutyrine, *see* A-60155
4-Fluorocatechol, *see* F-60020
2-Fluoro-*m*-cresol, *see* F-60052
2-Fluoro-*p*-cresol, *see* F-60053
3-Fluoro-*o*-cresol, *see* F-60056
3-Fluoro-*p*-cresol, *see* F-60057

4-Fluoro-*m*-cresol, *see* F-60059
4-Fluoro-*o*-cresol, *see* F-60058
5-Fluoro-*o*-cresol, *see* F-60060
6-Fluoro-*m*-cresol, *see* F-60054
6-Fluoro-*o*-cresol, *see* F-60055
2-Fluorocyclohexanone, F-60021
2-Fluorocyclopentanone, F-60022
5-Fluoro-3,4-dihydro-4-thioxo-2(1*H*)-
pyrimidinone, F-70019
3-Fluoro-2,6-dihydroxybenzoic acid, F-60023
4-Fluoro-3,5-dihydroxybenzoic acid, F-60024
5-Fluoro-2,3-dihydroxybenzoic acid, F-60025
1-Fluoro-2,4-dimethoxybenzene, *in* F-70018
4-Fluoro-1,2-dimethoxybenzene, *in* F-60020
3-Fluoro-2,6-dimethoxybenzoic acid, *in* F-60023
5-Fluoro-2,3-dimethoxybenzoic acid, *in* F-60025
2-Fluoro-1,3-dimethylbenzene, F-60026
2-Fluoro-3,5-dinitroaniline, F-60027
2-Fluoro-4,6-dinitroaniline, F-60028
4-Fluoro-2,6-dinitroaniline, F-60029
4-Fluoro-3,5-dinitroaniline, F-60030
5-Fluoro-2,4-dinitroaniline, F-60031
6-Fluoro-3,4-dinitroaniline, F-60032
2-Fluoro-3,5-dinitrobenzenamine, *see* F-60027
2-Fluoro-4,6-dinitrobenzenamine, *see* F-60028
4-Fluoro-2,6-dinitrobenzenamine, *see* F-60029
4-Fluoro-3,5-dinitrobenzenamine, *see* F-60030
5-Fluoro-2,4-dinitrobenzenamine, *see* F-60031
6-Fluoro-3,4-dinitrobenzenamine, *see* F-60032
4-Fluoro-2,6-dinitrotoluene, *see* F-60051
2-Fluoro-2,2-diphenylacetaldehyde, F-60033
2-Fluoroheptanal, F-60034
2-Fluoro-3-hydroxybenzaldehyde, F-60035
2-Fluoro-5-hydroxybenzaldehyde, F-60036
4-Fluoro-3-hydroxybenzaldehyde, F-60037
2-Fluoro-5-hydroxybenzoic acid, F-60038
2-Fluoro-6-hydroxybenzoic acid, F-60039
3-Fluoro-2-hydroxybenzoic acid, F-60040
3-Fluoro-4-hydroxybenzoic acid, F-60041
4-Fluoro-2-hydroxybenzoic acid, F-60042
4-Fluoro-3-hydroxybenzoic acid, F-60043
5-Fluoro-2-hydroxybenzoic acid, F-60044
2-Fluoro-3-hydroxytoluene, *see* F-60052
2-Fluoro-4-hydroxytoluene, *see* F-60057
2-Fluoro-5-hydroxytoluene, *see* F-60059
2-Fluoro-6-hydroxytoluene, *see* F-60056
3-Fluoro-2-hydroxytoluene, *see* F-60055
3-Fluoro-4-hydroxytoluene, *see* F-60053
4-Fluoro-2-hydroxytoluene, *see* F-60060
4-Fluoro-3-hydroxytoluene, *see* F-60054
5-Fluoro-2-hydroxytoluene, *see* F-60058
2-Fluoro-6-iodobenzoic acid, F-70020
3-Fluoro-2-iodobenzoic acid, F-70021
3-Fluoro-4-iodobenzoic acid, F-70022
4-Fluoro-2-iodobenzoic acid, F-70023
4-Fluoro-3-iodobenzoic acid, F-70024
5-Fluoro-2-iodobenzoic acid, F-70025
1-Fluoro-3-iodopropane, F-70026
2-Fluoro-4-iodopyridine, F-60045
3-Fluoro-4-iodopyridine, F-60046
4-Fluoro-1(3*H*)-isobenzofuranone, F-70027
7-Fluoro-1(3*H*)-isobenzofuranone, F-70028
▷1-Fluoro-2-isocyanatobenzene, F-60047
▷1-Fluoro-3-isocyanatobenzene, F-60048
▷1-Fluoro-4-isocyanatobenzene, F-60049
2-Fluoro-4-mercaptobenzoic acid, F-70029
2-Fluoro-6-mercaptobenzoic acid, F-70030
4-Fluoro-2-mercaptobenzoic acid, F-70031
2-Fluoro-5-methoxybenzoic acid, *in* F-60038
3-Fluoro-2-methoxybenzoic acid, *in* F-60041
5-Fluoro-2-methoxybenzoic acid, *in* F-60044
▷3-Fluoro-3-methoxy-3*H*-diazirine, F-60050
1-Fluoro-2-methoxy-4-methylbenzene, *in*
F-60054
1-Fluoro-3-methoxy-6-methylbenzene, *in*
F-60057
1-Fluoro-4-methoxy-3-methylbenzene, *in*
F-60059
2-Fluoro-1-methoxy-4-methylbenzene, *in*
F-60053
4-Fluoro-2-methoxy-1-nitrobenzene, *in* F-60061
2-Fluoro-4-methoxytoluene, *in* F-60057
2-Fluoro-5-methoxytoluene, *in* F-60059
3-Fluoro-4-methoxytoluene, *in* F-60053

4-Fluoro-3-methoxytoluene, *in* F-60054
2-Fluoro-4-methylanisole, *in* F-60053
2-Fluoro-5-methylanisole, *in* F-60054
3-Fluoro-4-methylanisole, *in* F-60057
4-Fluoro-3-methylanisole, *in* F-60059
α-Fluoro-α-methylbenzeneacetaldehyde, *see*
F-60063
5-Fluoro-2-methyl-1,3-dinitrobenzene, F-60051
▷(Fluoromethyl)oxirane, F-70032
2-Fluoro-3-methylphenol, F-60052
2-Fluoro-4-methylphenol, F-60053
2-Fluoro-5-methylphenol, F-60054
2-Fluoro-6-methylphenol, F-60055
3-Fluoro-2-methylphenol, F-60056
3-Fluoro-4-methylphenol, F-60057
4-Fluoro-2-methylphenol, F-60058
4-Fluoro-3-methylphenol, F-60059
5-Fluoro-2-methylphenol, F-60060
2-Fluoro-6-(methylthio)benzonitrile, *in* F-70030
Fluoromide, *see* D-60132
Fluoromidine, *see* C-60139
5-Fluoro-2-nitroanisole, *in* F-60061
3-Fluoro-4-nitrobenzoic acid, F-70033
▷5-Fluoro-2-nitrophenol, F-60061
3-Fluoronorvaline, *see* A-60160
1-Fluoro-1-octene, F-70034
2-Fluoro-2-phenylacetaldehyde, F-60062
α-Fluoro-α-phenylbenzeneacetaldehyde, *see*
F-60033
4-Fluoro-1-phenyl-1-butene, F-70035
▷3-Fluoro-3-phenyl-3*H*-diazirine, F-70036
▷2-Fluorophenyl isocyanate, *see* F-60047
▷3-Fluorophenyl isocyanate, *see* F-60048
▷4-Fluorophenyl isocyanate, *see* F-60049
2-Fluoro-2-phenylpropanal, F-60063
2-Fluoro-3-phenylpropanal, F-60064
2-Fluoro-1-phenyl-1-propanone, F-60065
4-Fluorophthalide, *see* F-70027
7-Fluorophthalide, *see* F-70028
2-Fluoropropiophenone, *see* F-60065
6-Fluoro-3-pyridazinamine, *see* A-70148
1-Fluoropyridinium, F-60066
4-Fluoropyrocatechol, *see* F-60020
4-Fluororesorcinol, *see* F-70018
3-Fluorosalicylic acid, *see* F-60040
4-Fluorosalicylic acid, *see* F-60042
5-Fluorosalicylic acid, *see* F-60044
6-Fluorosalicylic acid, *see* F-60039
5-Fluoro-4-thiouracil, *see* F-70019
▷3-Fluoro-3-(trifluoromethyl)-3*H*-diazirine,
F-70037
3-Fluorovaline, *see* A-60159
4-Fluoroveratrole, *in* F-60020
2-Fluoro-*m*-xylene, *see* F-60026
Fluromidine, *see* C-60139
Fontonamide, F-60067
10α-Formamidoalloaromadendrane, *in* A-60327
11-Formamido-5-eudesmene, F-60068
6-Formamido-4(15)-eudesmene, F-60069
1-Formylacenaphthene, *see* A-70013
Formylacetic acid, *see* O-60090
1-Formylacridine, *see* A-70056
2-Formylacridine, *see* A-70057
3-Formylacridine, *see* A-70058
4-Formylacridine, *see* A-70059
5-Formyl-6-azauracil, *see* F-60075
2-Formyl-1,4-benzodioxan, *see* B-60022
6-Formyl-1,4-benzodioxan, *see* B-60023
6-Formyl-4*H*-1,3-benzodioxin, *see* B-60026
o-Formylbenzyl alcohol, *see* H-60171
3-Formyl-2-butanone, *see* M-70102
2-Formyl-2-butenenitrile, *in* M-70103
3-Formyl-2-butenoic acid, *see* M-70103
2-Formylcinnamaldehyde, *see* O-60091
3-Formylcinnoline, *see* C-70178
4-Formylcinnoline, *see* C-70179
3-Formylcrotonic acid, *see* M-70103
1-Formylcyclododecene, *see* C-60191
4-Formylcyclohexene, *see* C-70227
3-Formyl-2-cyclohexen-1-one, *see* O-60060
Formylcyclooctatetraene, *see* C-70238
1-Formylcyclooctene, *see* C-60217
3-Formyl-2-cyclopentenone, *see* O-60062
2-Formyl-3,4-dihydroxybenzoic acid, F-60070
2-Formyl-3,5-dihydroxybenzoic acid, F-60071

4-Formyl-2,5-dihydroxybenzoic acid, F-60072
6-Formyl-2,3-dihydroxybenzoic acid, F-60073
5-Formyl-2-(1,2-dihydroxyisopropyl)-2,3-
dihydrobenzofuran, *in* D-70214
10-Formyl-2,16-dihydroxy-2,6,14-trimethyl-
3,10,14-hexacatrien-5-one, *in* H-60186
2-Formyl-3,4-dimethoxybenzoic acid, *in*
F-60070
4-Formyl-2,5-dimethoxybenzoic acid, *in*
F-60072
6-Formyl-2,3-dimethoxybenzoic acid, *in*
F-60073
4-Formyl-4,4-dimethylbutanoic acid, *see*
D-60434
2-Formyl-α,3-dimethylcyclo-
pentaneacetaldehyde, *see* I-60084
3-Formyl-2,5-dimethylfuran, *see* D-70412
2-Formyl-3,5-dimethylthiophene, *see* D-70439
3-Formyl-2,5-dimethylthiophene, *see* D-60452
3-Formyl-2,5-dimethylthiophene, *see* D-70438
Formylethanoic acid, *see* O-60090
1-Formylfluorene, *see* F-70014
▷9-Formylfluorene, *see* F-70015
2-Formylfuro[2,3-*b*]pyridine, *see* F-60093
2-Formylfuro[2,3-*c*]pyridine, *see* F-60096
2-Formylfuro[3,2-*b*]pyridine, *see* F-60095
2-Formylfuro[3,2-*c*]pyridine, *see* F-60097
3-Formylfuro[2,3-*b*]pyridine, *see* F-60094
▷3-Formylguaiacol, *in* D-70285
5-Formyl-3-hydroxy-2-isopropenyl-2,3-
dihydrobenzofuran, *in* D-70213
5-Formyl-2-(1-hydroxyisopropyl)benzofuran, *in*
D-70214
5-Formyl-2-(1-hydroxyisopropyl)-2,3-
dihydrobenzofuran, *see* D-70214
2-(2-Formyl-3-hydroxymethyl-2-cyclopentenyl)-
6,10-dimethyl-5,9-undecadienal, F-70038
▷2-Formyl-5-hydroxymethylfuran, *see* H-70177
3-Formyl-2-(4-hydroxyphenyl)-7-methoxy-5-(1-
propenyl)benzofuran, *see* H-60212
2-Formyl-3-hydroxy-2-propenoic acid, *see*
D-60190
10-Formyl-16-hydroxy-2,6,14-trimethyl-
2,10,14-hexacatrien-5-one, *see* H-60186
15-Formylimbricatolal, *in* H-70150
5-Formylindole, *see* I-60029
6-Formylindole, *see* I-60030
7-Formylindole, *see* I-60031
2-Formyl-3-iodothiophene, *see* I-60069
2-Formyl-4-iodothiophene, *see* I-60070
2-Formyl-5-iodothiophene, *see* I-60072
3-Formyl-4-iodothiophene, *see* I-60071
2-Formylmalonaldehydic acid, *see* D-60190
3-Formyl-3-methylacrylic acid, *see* M-70103
2-Formyl-3-methylindole, *see* M-70090
2-Formyl-5-methylphenol, *see* H-70160
4-Formyl-3-methylphenol, *see* H-70161
2-Formyl-5-methylpyrrole, *see* M-70129
2-Formyl-3-methylthiophene, *see* M-70140
2-Formyl-4-methylthiophene, *see* M-70141
2-Formyl-5-methylthiophene, *see* M-70143
3-Formyl-2-methylthiophene, *see* M-70139
3-Formyl-4-methylthiophene, *see* M-70142
4-Formyl-2-methythiophene, *see* M-70144
9-Formylnonanoic acid, *see* O-60063
2-Formylnorbornadiene, *see* B-60083
15-Formyloxyimbricatalal (incorr.), *in* H-70150
2-Formyl-1,4-pentadiene, *see* M-70074
5-Formyl-4-phenanthrenecarboxylic acid,
F-60074
2-Formylpropanoic acid, F-70039
3-Formylpyrrole, *see* P-70182
2-Formylpyrrolidine, *see* P-70184
▷2-Formylthiophene, *see* T-70183
6-Formyl-1,2,4-triazine-3,5(2*H*,4*H*)-dione,
F-60075
2-Formyltriphenylene, *see* T-60382
10-Formylundecanoic acid, *see* O-60066
5-Formyluracil, *see* T-60064
5-Formylzearalenone, *see* Z-60002
Fortunellin, *in* D-70319
FR 68504, *see* D-70465
FR 900359, *see* A-70223
FR 900405, *in* E-60037

FR 900406, *in* E-60037
FR 900452, *see* A-60284
FR 900482, *see* A-60285
Fragransin D_1, F-70040
Fragransin D_2, *in* F-70040
Fragransin D_3, *in* F-70040
Fragransin E_1, *in* A-70271
Fragransol *A*, F-70041
Fragransol *B*, F-70042
Fridamycin *A*, F-70043
Fridamycin *D*, F-70044
Fridamycin B, *in* F-70043
Fridamycin E, *in* F-70043
D:B-Friedoolean-5-en-3-ol, *see* G-70020
Fritschiellaxanthin, *in* D-70545
▷Frugoside, *in* C-60168
Fruticolide, F-70045
Fruticulin *A*, F-60076
Fruticulin *B*, F-60077
FS-2, F-60078
FS 2, *see* D-70417
▷Fuchsine, *in* R-60010
Fucoserratene, *in* O-70022
Fujikinetin, *in* T-70111
Fujikinin, *in* T-70111
Fulgenic acid, *see* D-70409
Fuligorubin *A*, F-60079
Fulvalene, F-60080
Fulvic acid, F-60081
Fumifungin, F-60082
Fumigatin, *in* D-70320
Fumigatin epoxide, *in* D-70320
Fumigatin methyl ether, *in* D-70320
Fumigatin oxide, *in* D-70320
Funadonin, F-70046
Funkioside *A*, *in* S-60045
▷Funkioside C, *in* S-60045
▷Funkioside D, *in* S-60045
▷Funkioside E, *in* S-60045
Funkioside F, *in* S-60045
▷Funkioside G, *in* S-60045
3-Furanacetaldehyde, F-60083
2,5-Furandimethanol, *see* B-60160
3,4-Furandimethanol, *see* B-60161
Furaneol, *see* H-60120
Furanoeremophilane-2,9-diol, F-70047
Furanopetasin, *in* F-70047
Furanopetasol, *in* F-70047
5(7*H*)-Furano[2,3-*d*]pyrimidin-6-one, *see*
   F-60106
2-(2-Furanyl)-4*H*-1-benzopyran-4-one, F-60084
2-(3-Furanyl)-4*H*-1-benzopyran-4-one, F-60085
4-(2-Furanyl)-3-buten-2-one, F-60086
6-(3-Furanyl)-5,6-dihydro-3-[2-(2,6,6-
   trimethyl-4-oxo-2-cyclohexen-1-yl)-ethyl]-
   2*H*-pyran-2-one, *see* H-60009
1-(3-Furanyl)-4,8-dimethyl-1,6-nonanedione,
   see M-60156
α-[9-(3-Furanyl)-2,6-dimethyl-1,4,6-
   nonatrienyl]-3-furanethanol, *see* B-70143
1-(3-Furanyl)-4,8-dimethyl-7-nonene-1,6-dione,
   see D-60019
2-[2-(2-Furanyl)ethenyl]-1*H*-pyrrole, *see*
   F-60088
5-[2-(3-Furanyl)ethyl]decahydro-1,4a-dimethyl-
   6-methylene-1-naphthalenecarboxylic acid,
   see L-60011
3-Furanyl[4-methyl-2-(2-methylpropyl)-1-cyclo-
   penten-1-yl]methanone, *see* M-60155
▷1-(3-Furanyl)-4-methyl-1-pentanone, F-60087
5-(2-Furanyl)oxazole, F-70048
6-[3-(3-Furanyl)propyl]-2-(4-methyl-3-
   pentenyl)-2-heptenedioic acid 1-methyl ester,
   see M-60134
1-(2-Furanyl)-2-(2-pyrrolyl)ethylene, F-60088
2-(2-Furanyl)quinoxaline, F-60089
14-(3-Furanyl)-3,7,11-trimethyl-7,11-tetra-
   decadienoic acid, F-70049
5-[13-(3-Furanyl)-2,6,10-trimethyl-6,8-trideca-
   dienyl]-4-hydroxy-3-methyl-2(5*H*)-furanone,
   F-70050
5-[13-(3-Furanyl)-2,6,10-trimethyl-6,10-trideca-
   dienylidene]-4-hydroxy-3-methyl-2(5*H*)-
   furanone, *see* V-70003
Furazano[3,4-*b*]quinoxaline, F-70051
Furcellataepoxylactone, F-70052

2-Furfuralacetone, *see* F-60086
3-Furfuraldehyde, *see* F-60083
Furfurylethylene oxide, *see* O-70078
Furlone yellow, F-70053
Furocaespitane, F-60090
Furodysinin hydroperoxide, F-70054
3-(Furo[3,4-*b*]furan-4-yl)-2-propenenitrile,
   F-70055
Furoixiolal, F-70056
Furonic acid, *see* O-60072
Furo[3,4-*c*]octalene, *see* O-70019
Furo[2,3-*d*]pyridazine, F-60091
Furo[3,4-*d*]pyridazine, F-60092
Furo[2,3-*d*]pyridazin-4(5*H*)-one, *see* H-60146
Furo[2,3-*d*]pyridazin-7(6*H*)-one, *see* H-60147
Furo[2,3-*b*]pyridine-2-carboxaldehyde, F-60093
Furo[2,3-*c*]pyridine-2-carboxaldehyde, F-60096
Furo[2,3-*b*]pyridine-3-carboxaldehyde, F-60094
Furo[3,2-*b*]pyridine-2-carboxaldehyde, F-60095
Furo[3,2-*c*]pyridine-2-carboxaldehyde, F-60097
Furo[2,3-*b*]pyridine-2-carboxylic acid, F-60098
Furo[2,3-*c*]pyridine-2-carboxylic acid, F-60102
Furo[2,3-*b*]pyridine-3-carboxylic acid, F-60099
Furo[2,3-*c*]pyridine-3-carboxylic acid, F-60103
Furo[3,2-*b*]pyridine-2-carboxylic acid, F-60100
Furo[3,2-*c*]pyridine-2-carboxylic acid, F-60104
Furo[3,2-*b*]pyridine-3-carboxylic acid, F-60101
Furo[3,2-*c*]pyridine-3-carboxylic acid, F-60105
Furo[2,3-*d*]pyrimidin-2(1*H*)-one, F-60106
Furoscalarol, F-70057
Furoxano[3,4-*b*]quinoxaline, *in* F-70051
2-(3-Furyl)acetaldehyde, *see* F-60083
2-(2-Furyl)chromone, *see* F-60084
2-(3-Furyl)chromone, *see* F-60085
2-Furyloxirane, *see* O-70078
2-(2-Furyl)quinoxaline, *see* F-60089
Fusarin *A*, F-60107
Fusarin D, *in* F-60107
Fusarubin, F-60108
Fusarubin ethyl acetal, *in* F-60108
Fusarubin methyl acetal, *in* F-60108
Fusarubinoic acid, F-70058
▷G 7063-2, *in* A-70158
G 0069A, *in* C-70183
Gaboxadol, *see* T-60072
Gaillardoside, *in* T-70284
Galantinamic acid, *see* D-70046
Galipein, G-60001
Galiridoside, G-60002
Gallicadiol, G-70001
Gallisal, *in* H-70108
Galloxanthin, G-70002
3-Galloylcatechin, *in* P-70026
Galtamycin, G-70003
Galtamycinone, *in* G-70003
Ganderic acid Me, *in* D-70314
Gandodermic acid *P2*, *in* T-70266
Ganoderal B, *in* H-70200
Ganoderenic acid A, *in* T-70254
Ganoderenic acid B, *in* T-70254
Ganoderenic acid C, *in* T-70254
Ganoderenic acid D, *in* T-70254
Ganoderenic acid E, *in* D-60378
Ganoderic acid *Mb*, *in* T-60111
Ganoderic acid *Mc*, *in* T-60111
Ganoderic acid *Mg*, *in* T-60111
Ganoderic acid *Mh*, *in* T-60111
Ganoderic acid *Mk*, *in* T-70266
Ganoderic acid *N*, *in* D-60379
Ganoderic acid *O*, *in* D-60379
Ganoderic acid *O*, *in* T-60111
Ganoderic acid *P*, *in* T-70266
Ganoderic acid *Q*, *in* T-70266
Ganoderic acid *T*, *in* T-70266

Ganoderic acid X, *in* D-70314
Ganoderic acid Ma, *in* T-60327
Ganoderic acid Md, *in* T-60328
Ganoderic acid Mf, *in* D-70314
Ganoderic acid Mi, *in* T-60327
Ganoderic acid Mj, *in* T-60328
Ganoderic acid R, *in* D-70314
Ganoderic acid S, *in* D-70314
Ganoderiol A, *in* L-60012
Ganoderiol B, *in* T-60330
Ganodermanondiol, *in* D-60343
Ganodermanontriol, *in* L-60012
Ganodermic acid *Pl*, *in* T-70266
Ganodermic acid Ja, *in* D-70314
Ganodermic acid Jb, *in* D-70314
Gansongone, *see* A-70252
Garcinone D, G-70004
Garcinone E, G-60003
Gardenin D, *see* D-60375
Garudane, *see* H-70013
Garvalone *A*, G-60004
Garvalone *B*, G-60005
Garveatin D, G-60007
Garveatin *A* quinone, G-60006
Garvin *A*, G-60008
Garvin B, G-60009
GB 1, G-70005
GB-2, *in* G-70005
GB1a, *in* G-70005
GB2a, *in* G-70005
▷Gedunin, G-70006
Gelsemide, G-60010
Gelsemide 7-glucoside, *in* G-60010
▷Gelseminic acid, *see* H-70153
Gelsemiol, G-60011
Gelsemiol 1-glucoside, *in* G-60011
Gelsemiol 3-glucoside, *in* G-60011
▷Genistein, *see* T-60324
Genistin, *in* T-60324
Gentisin alcohol, *see* D-70287
Gentisyl alcohol, *see* D-70287
Geodiamolide *A*, G-60012
Geodiamolide B, *in* G-60012
Geodoxin, G-70007
2-Geranylgeranyl-6-methylbenzoquinone, *see*
   S-60010
3-Geranyl-4-hydroxybenzoic acid, *see* D-70423
8-Geranyloxypsoralen, *in* X-60002
4-Geranyl-3,4′,5-trihydroxystilbene, G-70008
Gerberinol, *see* G-60013
Gerberinol 1, G-60013
1(10),4-Germacradiene-6,8-diol, G-60014
5,10(14)-Germacradiene-1,4-diol, G-70009
1(10),4,11(13)-Germacratrien-8,12-olide,
   G-70010
Germacrone 4,5-epoxide, *see* E-60023
Gersemolide, G-60015
Gersolide, G-60016
Gibberellin A_2, G-60017
Gibberellin A_9, G-60018
Gibberellin A_{20}, G-70011
Gibberellin A_{24}, G-60019
Gibberellin A_{29}, *in* G-70011
Gibberellin A_{36}, *in* G-60019
Gibberellin A_{40}, *in* G-60018
Gibberellin A_{45}, *in* G-60018
Gibberellin A_{51}, *in* G-60018
Gibberellin A_{55}, *in* G-60018
Gibberellin A_{63}, *in* G-60018
Gibberellin A_{64}, G-70012
Gibberellin A_{65}, G-70013
Gibberellin A_{66}, *in* G-70013
Gibberellin A_{67}, *in* G-60018
Gibberellin A_{72}, *in* G-70011
Gibberellin A_{73}, *in* G-60018
Gibboside, G-60020
Giffordene, *in* U-70003
Gigantanolide A, *in* T-60323
Gigantanolide B, *in* T-60323
Gigantanolide C, *in* T-60323
Ginkgoic acid, *in* H-60205
Ginkgolic acid, *in* H-60205
GL 7, *in* H-70108
Glabone, G-60021
Glabrachromene II, G-70014
Glabranin, G-70015
Glabranin 7-methyl ether, *in* G-70015

Glabridin, G-70016
▷Glabrin, *in* M-70043
Glaucasterol, *see* P-60004
Glaucin *B*, G-70017
Glechomanolide, G-60022
Gleinadiene, *in* G-60023
Gleinene, G-60023
Gliotoxin *E*, G-60024
Globularicisin, *in* C-70026
Globularidin, *in* C-70026
Globularin, *in* C-70026
Gloeosporone, G-60025
Glomelliferonic acid, G-70018
Glomellonic acid, *in* G-70018
Glucazidone, *see* F-60089
Glucodistylin, *in* P-70027
1-Glucopyranosyl-2-(2-hydroxyhexa-
    decanoylamino)-1,3,4-docesanetriol, *see*
    A-70008
1-Glucopyranosyl-2-(2-hydroxyhexa-
    decanoylamino)-13-docosene-1,3,4-triol, *see*
    A-70009
1-Glucopyranosyl-2-(2-hydroxytetra-
    cosanoylamino)-1,3,4-hexadecanetriol, *see*
    A-70007
(1-β-D-Glucopyranosyloxy)-1,4*a*,5,6,7,7*a*-hexa-
    hydro-5,7-dihydroxy-7-methylcyclopenta[*c*]-
    pyran-4-carboxylic acid, *see* S-60028
1-(β-D-Glucopyranosyloxy)-4-hydroxy-2-cyclo-
    pentene-1-carbonitrile, *see* T-60007
2-[[15-(β-D-Glucopyranosyloxy)-8-hydroxy-1-
    oxohexadecyl]amino]ethanesulfonic acid, *see*
    G-70019
N-[15-(β-D-Glucopyranosyloxy)-8-
    hydroxypalmitoyl]taurine, G-70019
Glucozaluzanin *C*, *in* Z-60001
Glutaconimide, *see* H-60226
34-Glutamic acid-43-histidinethymopoietin II, *in*
    T-70192
▷Glutamic acid 5-2-(α-hydroxy-*p*-tolyl)-
    hydrazide, *see* A-70071
Glutamic acid lactam, *see* O-70102
L-γ-Glutamyl-L-cysteinyl-*N*-[3-[4-[[[*N*-(*N*-L-γ-
    glutamyl-L-cysteinyl)glycyl]-amino]butyl]-
    amino]propyl]glycinamide cyclic(2→3)-
    disulfide, *see* T-70331
▷β-*N*-(γ-Glutamyl)-4-hydroxymethyl-
    phenylhydrazine, *see* A-70071
γ-Glutamylmarasmine, G-60026
N-γ-Glutamyl-3-[[(methylthio)methyl]-
    sulfinyl]alanine, *see* G-60026
γ-Glutamyltaurine, G-60027
Glutaroin, *see* H-60114
Glutaurine, *see* G-60027
5-Gluten-3-ol, G-70020
5(10)-Gluten-3-one, G-60028
Glutimic acid, *see* O-70102
Glutiminic acid, *see* O-70102
Glutinol, *in* G-70020
Glutinopallal, G-70021
Glyceraldehyde acetonide, *see* D-70400
▷Glycerol 1-acetate, G-60029
Glycerol 1,3-diacetate, *see* D-60023
Glycerol 2-(3-methyl-2-butenoate) 1-(2,4,11-
    tridecatrienoate), *see* U-70001
Glycerol 2-(3-methylthio-2-propenoate) 1-
    (2,4,11-tridecatrienoate), *see* U-70002
1-Glycine-2-glutamine-43-histidinethymopoietin
    II, *in* T-70192
Glycinoeclepin *A*, G-70022
▷Glycol acetal, *see* M-60063
Glycol salicylate, *in* H-70108
Glycoluril, *see* T-70066
Glycosine, *see* B-60107
N-[*N*-[*N*-[1-(1-Glycylproline)-4-
    hydroxyprolyl]glycyl]alanyl]glycine, *see*
    A-60283
Glyoxal bis 2-hydroxyanil, *in* T-60059
Glyoxal diethylacetal, *see* D-70153
Glyoxaldiurene, *see* T-70066
Glysal, *in* H-70108
Gnetin A, G-70023
Gnididione, G-60030
Gochnatolide, G-70024
Gomisin *D*, G-70025

Gomphidic acid, G-70026
Goniothalamicin, G-70027
▷Goniothalamin, G-60031
Goniothalamin oxide, *in* G-60031
Goniothalenol, *see* A-60083
Gosferol, *see* P-60170
▷Gossypol, G-60032
Gracilin *A*, G-60033
Gracilin *B*, G-60034
Gracilin *F*, G-60035
Gracilin C, *in* G-60034
Gracilin D, *in* G-60034
Gracilin E, *in* G-60035
Gracillin, *in* S-60045
Grahamimycin A₁, *in* C-60159
Grasshopper ketone, G-70028
Graucin *A*, G-60036
Gravelliferone, G-60037
▷Green Cross, *see* T-70223
Green Pigment, *see* M-60123
Grenoblone, G-70029
Grevilline *A*, G-60038
Grevilline B, *in* G-60038
Grevilline C, *in* G-60038
Grevilline D, *in* G-60038
Griffonin, *in* L-70034
Grimaldone, G-70030
GSN (as disodium salt), *in* G-60032
1(5),6-Guaiadiene, G-70031
1(10),3-Guaiadien-12,8-olide, G-70032
4,11(13)-Guaiadien-12,8β-olide, *in* A-70054
Guaianin D, *in* H-60199
Guaianin E, *in* H-60199
1,3,5,7(11),9-Guaiapentaen-14-al, G-60039
Guaiaretic acid, G-70033
1(10),2,4-Guaiatrien-12,8β-olide, *see* S-70073
4-Guaien-12,8β-olide, *in* A-70054
Guanylsulfonic acid, *see* A-60210
Guayulin C, *in* A-60299
Guayulin D, *in* A-60299
α-Gurjunene, G-70034
▷α-Guttiferin, G-60040
Gymnomitrol, G-70035
Gymnopusin, *in* P-70035
Gymnorhizol, *in* O-70036
Gyplure, *in* O-60006
Gypopinifolone, G-70036
Gypothamniol, G-70037
▷HA 106, *in* T-70277
Haageanolide, H-60001
Halenaquinol, H-60002
Halenaquinone, *in* H-60002
Halichondramide, H-60003
Halleridone, H-60004
Hallerone, H-60005
Hamachilobene *A*, H-70001
Hamachilobene B, *in* H-70001
Hamachilobene C, *in* H-70001
Hamachilobene D, *in* H-70001
Hamachilobene E, *in* H-70001
Hanphyllin, H-60006
Hardwickiic acid, H-70002
Hautriwaic acid, *in* H-70002
Hautriwaic acid acetate, *in* H-70002
7(18)-Havannachlorohydrin, H-70003
11(19)-Havannachlorohydrin, H-70004
7(18),11(19)-Havannadichlorohydrin, H-70005
Havannahine, H-60008
Havardic acid A, *in* L-60006
Havardic acid B, *in* H-60168
Havardic acid C, *in* D-60342
Havardic acid D, *in* O-60077
Havardic acid E, *see* D-70457
Havardic acid F, *in* H-60198
Havardiol, *in* H-60198
Heavenly blue anthocyanin, H-70006
Hebeclinolide, H-60009
Hebemacrophyllide, H-60010
Hebesterol, H-60011
Hector's base, H-60012
Heleniumlactone 1, H-60013
Heleniumlactone 2, *in* H-60013
Heleniumlactone 3, H-60014
Heliantriol A₁, *in* O-70035
Helichrysin, *in* H-60220

Helicquinone, H-60015
Heliettin, *see* C-60031
Heliopsolide, H-70007
Helminthosporal, H-60016
Helogynic acid, H-70008
Hemicyclone, *see* D-70403
sym-Hemimellitenol, *see* T-60371
m-Hemipic acid, *in* D-60307
26-Henpentacontanone, H-70009
Henricine, H-60017
1,14,34,35,36,37,38-Heptaazaheptacyclo-
    [12.12.7.1³,⁷.1⁸,¹².1¹⁶,²⁰.1²¹,²⁵.1²⁸,³²]octatri-
    aconta-3,5,7(38),8,10,12(37),16,18,20(36),-
    21,23,25(35),28,30,32(34)pentadecane,
    H-70010
7,16-Heptacenedione, H-60018
7,16-Heptacenequinone, *see* H-60018
13-Heptacosene, H-60019
Heptacyclo[7.7.0.0²,⁶.0³,¹⁵.0⁴,¹².0⁵,¹⁰.0¹¹,¹⁶]-
    hexadeca-7,13-diene, H-70011
Heptacyclo[7.7.0.0²,⁶.0³,¹⁵.0⁴,¹².0⁵,¹⁰.0¹¹,¹⁶]-
    hexadecane, *in* H-70011
Heptacyclo[5.5.1.1⁴,¹⁰.0²,⁶.0³,¹¹.0⁵,⁹.0⁸,¹²]tetra-
    decane, *see* H-70012
Heptacyclo[6.6.0.0²,⁶.0³,¹³.0⁴,¹¹.0⁵,⁹.0¹⁰,¹⁴]tetra-
    decane, H-70012
Heptacyclo[9.3.0.0²,⁵.0³,¹³.0⁴,⁸.0⁶,¹⁰.0⁹,¹²]tetra-
    decane, H-70013
2,4-Heptadienal, H-60020
2,4-Heptadienoic acid, H-60021
2,4-Heptadien-1-ol, H-70014
1,5-Heptadien-3-one, H-70015
1,5-Heptadien-4-one, H-70016
1,6-Heptadien-3-one, H-70017
4,5-Heptadien-3-one, H-70018
1*H*-Heptafluoronaphthalene, *see* H-70019
2*H*-Heptafluoronaphthalene, *see* H-70020
1,2,3,4,5,6,7-Heptafluoronaphthalene, H-70019
1,2,3,4,5,6,8-Heptafluoronaphthalene, H-70020
3,4-Heptafulvalenedione, H-60022
8-(8-Heptafulvenyl)-*p*-tropoquinone methide,
    see C-70222
3″,4′,4‴,5,5″,7,7″-Heptahydroxy-3,8‴-
    biflavanone, *see* G-70005
2,3,5,14,20,22,25-Heptahydroxycholest-7-en-6-
    one, H-60023
3,3′,4′,5,5′,7,8-Heptahydroxyflavone, H-70021
3,3′,4′,5,6,7,8-Heptahydroxyflavone, H-60024
3,3′,4′,5,5′,7,8-Heptamethoxyflavone, *in*
    H-70021
3,4-Heptamethylenefuran, *see* H-60047
▷Heptamethyleneimine, *see* O-70010
4,6,8,10,12,14,16-Heptamethyl-6,8,11-octadeca-
    triene-3,5,13-trione, H-70022
1,7-Heptanedithiol, H-70023
5-Heptenenitrile, *in* H-70024
5-Heptenoic acid, H-70024
6-(1-Heptenyl)-5,6-dihydro-2*H*-pyran-2-one,
    H-60025
2-Heptyl-12-oxo-13-henicosenoic acid, *see*
    F-60012
2-Heptyl-10-oxo-11-nonadecenoic acid, *see*
    F-60011
6-Heptyne-2,5-diamine, H-60026
6-Heptyn-2-one, H-70025
Herbacin, H-70134
Herbicidal substance 1328-2, *see* M-70134
Herbicidal substance 1328-3, *in* M-70134
Hercycin, *see* H-60083
Heritol, H-60028
Hermosillol, H-60029
Herpetetrone, H-70026
Herzynine, *see* H-60083
Hesperidene, *see* I-70087
heta, *see* T-60016
5-HETE, *see* H-60127
8-HETE, *see* H-60128
9-HETE, *see* H-60129
11-HETE, *see* H-60130
12-HETE, *see* H-60131
▷Heteroauxin, *see* I-70013
Heterobryoflavone, H-70027
Heterocoerdianthrone, *see* D-70073
Heteronemin, H-60030

▷Hexaaminobenzene, see B-60012

1,4,7,12,15,18-Hexaazacyclodocosane, H-70028

1,4,7,13,16,19-Hexaazacyclotetracosane, H-70029

1,5,6,10,10a,10c-Hexaazadibenzo[a,f]-pentalene, see P-70169

1,14,29,30,31,32-Hexaazahexacyclo[12.7.7.1³,⁷.1⁸,¹².1¹⁶,²⁰.1²³,²⁷]dotriaconta-3,5,7(32),8,10,12(31),16,18,20(30),23,25,-27(29)-dodecaene, H-70030

1,4,5,8,9,12-Hexaazatriphenylenehexacarboxylic acid, see B-60051

Hexabenzo[bc,ef,hi,kl,no,qr]coronene, H-60031

Hexabenzohexacyclo[5.5.2.2⁴,¹⁰.1¹,⁷.0⁴,¹⁷.0¹⁰,¹⁷]-heptadecane, see C-70031

Hexabutylbenzene, H-60032

6,15-Hexacenedione, H-60033

▷2,2′,4,4′,5,5′-Hexachlorobiphenyl, H-60034

Hexacyclo[11.3.1.1³,⁷.1⁵,⁹.1¹¹,¹⁵.0²,¹⁰]-eicos-2-ene, H-70031

Hexacyclo[4.4.0.0⁷,⁵.0²,⁹.O⁴,⁸.0⁷,¹⁰]decane, H-70032

Hexacyclo[5.4.0.0²,⁶.0³,¹⁰.0⁵,⁹.0⁸,¹¹]-undecan-4-one, H-60035

11,13-Hexadecadienal, H-70033

8,9-Hexadecanedione, H-70034

N-Hexadecanoylhomoserine lactone, in A-60138

2-Hexadecenal, H-60036

7-Hexadecenal, H-70035

5-Hexadecyne, H-60037

1,2,9,10,17,18-Hexadehydro[2.2.2]-paracyclophane, H-60038

▷1,5-Hexadiene-3,4-diol, H-70036

1,5-Hexadiene-3,4-dione, H-60039

3,4-Hexadienoic acid, H-70037

3,5-Hexadienoic acid, H-70038

3,5-Hexadien-1-ol, H-70039

1,2-Hexadien-5-yne, H-70040

1,3-Hexadien-5-yne, H-70041

1,5-Hexadiyn-3-one, H-60040

3,5-Hexadiyn-2-one, H-60041

▷2,4-Hexadiynophenone, see P-60109

2-(2,4-Hexadiynylidene)-1,6-dioxaspiro[4.5]dec-3-en-8-ol, H-70042

2-(2,4-Hexadiynylidene)-1,6-dioxaspiro[4.5]dec-3-ene, in H-70042

7-(2,4-Hexadiynylidene)-1,6-dioxaspiro[4.4]-nona-2,8-dien-4-ol, see M-60154

2-(2,4-Hexadiynylidene)-3,4-epoxy-1,6-dioxaspiro[4.5]decane, in H-70042

▷Hexaethylbenzene, H-70043

Hexafluorobiacetyl, see H-60043

1,1,1,4,4,4-Hexafluoro-2,3-butanedione, H-60043

1,2,3,4,5,5-Hexafluoro-1,3-cyclopentadiene, H-60044

Hexafluoro(hexafluorocyclobutylidene)-cyclobutane, see D-60513

Hexafluoroisobutylidene sulfate, see T-60311

1,1,1,1′,1′,1′-Hexafluoro-N-methoxydimethyl-amine, in B-60180

▷Hexafluoro(methylenecyclopropane), see D-60185

1,2,3,4,5,6-Hexafluoronaphthalene, H-70044

1,2,4,5,6,8-Hexafluoronaphthalene, H-70045

α,α,α′,α′,α′-Hexafluoro-m-xylene, see B-60176

α,α,α,α′,α′,α′-Hexafluoro-o-xylene, see B-60175

α,α,α,α′,α′,α′-Hexafluoro-p-xylene, see B-60177

Hexahydroazirino[2,3,1-hi]indol-2(1H)-one, H-60046

Hexahydro-4(1H)-azocinone, H-70047

Hexahydro-5(2H)-azocinone, H-70048

1,2,3,5,10,10a-Hexahydrobenz[f]indolizine, see H-70072

Hexahydro-2(3H)-benzofuranone, H-70049

▷Hexahydrobenzophenone, see B-70071

2,3,4,4a,5,6-Hexahydro-7H-1-benzopyran-7-one, H-60045

2,3,6,7,10,11-Hexahydrobenzo[1,2-b:3,4-b′:5,6-b″]tris[1,4]dithiin, H-70050

2,3,6,7,10,11-Hexahydrobenzo[1,2-b:3,4-b′:5,6-b″]tris[1,4]oxathiin, H-70051

Hexahydro-2(3H)-benzoxazolone, H-70052

Hexahydro-2,5-bis(trichloromethyl)-furo[3,2-b]furan-3a,6a-diol, H-70053

Hexahydrochroman-7-one, see O-60011

5,6,7,8,9,10-Hexahydro-4H-cyclonona[c]furan, H-60047

Hexahydro-1H-cyclopenta[c]furan-1-one, H-70054

Hexahydrocyclopenta[cd]pentalene-1,3,5(2H)-trione, H-60048

Hexahydro-2H-cyclopenta[b]thiophene, H-60049

1,5,6,8,12,13-Hexahydro-4,14:7,11-diethenodicyclopenta[a,g]cyclododecene, see I-60021

Hexahydro-3,7a-dihydroxy-3a,7-dimethyl-1,4-isobenzofurandione, H-70055

▷3,4,5,6,9,10-Hexahydro-14,16-dihydroxy-3-methyl-1H-2-benzoxacyclotetradecin-1,7(8H)-dione, see Z-60002

2-[(2,3,3a,4,5,7a-Hexahydro-3,6-dimethyl-2-benzofuranyl)ethylidene]-6-methyl-5-heptenoic acid, H-60050

2-[(3,4,4a,5,6,8a-Hexahydro-4,7-dimethyl-2H-1-benzopyran-2-yl)ethylidene]-6-methyl-5-heptenoic acid, H-60046

1,2,3,4,5,6-Hexahydro-1,4-dimethyl-7-(1-methylethyl)azulene, see G-70031

1,2,5,7,8,8a-Hexahydro-4,8a-dimethyl-6-(1-methylethyl)-1-naphthalenol, see E-60064

4,4a,5,6,7,8-Hexahydro-2,5-dimethyl-5-(4-methyl-3-pentenyl)-1,8a(3H)-naphthalenedicarboxaldehyde, see P-60064

2a,3a,5a,6a,6b,6c-Hexahydro-3,6-dioxacyclopenta[cd,gh]pentalene, see D-70463

Hexahydro-2,6-dioxoimidazolo[4,5-d]-imidazole, see T-70066

Hexahydro-1,4-dithiino[2,3-b]-1,4-dithiin, H-60051

Hexahydro-1,2,3,5-ethanediylidene-1H-cyclobuta[cd]pentalen-4(1aH)one, see H-60035

Hexahydro-2,6,3,5-ethanediylidene-2H,3H-1,4-dioxacyclopenta[cd]pentalene, see D-70461

Hexahydro-1,2,4-ethanylylidene-1H-cyclobuta[cd]pentalene-5,7(1aH)dione, see P-70022

Hexahydro-1,2,4-ethanylylidene-1H-cyclobuta[cd]pentalene-3,5,7-trione, see P-70023

1,2,3,6,7,8-Hexahydro-3a,5a-ethenopyrene, H-70056

Hexahydrofarnesol, see T-70282

3a,4,5,8,9,11a-Hexahydro-9-hydroxy-6,10-di-methyl-3-methylenecyclodeca[b]-furan-2(3H)-one, see H-60006

5-[Hexahydro-5-hydroxy-4-(3-hydroxy-4-methyl-1-octen-6-ynyl)-2(1H)-penta-lenylidene]pentanoic acid, see I-60003

1,2,3,10,11,11a-Hexahydro-2-hydroxy-11-methoxy-5H-pyrrolo[2,1-c][1,4]-benzodiazepin-5-one, see A-70001

2,3,4,7,8,8a-Hexahydro-8-(5-hydroxy-4-methyl-pentyl)-6,8-dimethyl-1H-3a,7-methanoazulene-3-carboxylic acid, H-70057

Hexahydro-7-hydroxy-3-(2-methylpropyl)-pyrrolo[1,2-a]pyrazine-1,4-dione, see C-60213

3,3a,4,5,8a,8b-Hexahydro-4-hydroxy-3,6,9-trimethylazuleno[4,5-b]furan-2,7-dione, see H-60100

1,2,3,5,6,7-Hexahydro-s-indacene, see H-60095

1,2,3,6,7,8-Hexahydro-as-indacene, see H-60094

3a,4,5,6,7,7a-Hexahydro-1H-inden-1-one, H-70058

2,3,5,6,7,8-Hexahydro-1H-indolizinium(1+), H-70059

1,3,3a,4,5,7a-Hexahydro-2H-indol-2-one, H-70060

Hexahydro-1(3H)-isobenzofuranone, H-70061

1,2,3,4,5,6-Hexahydro-7-isopropyl-1,4-di-methylazulene, see G-70031

1,2,6,7,8,8a-Hexahydro-6-isopropyl-4,8a-di-methyl-1-naphthol, see E-60064

Hexahydromandelic acid, see C-60212

2,3,4,5,6,7-Hexahydro-4,7-methano-1H-indene, see T-60260

3a,4,6a,7,9a,9b-Hexahydro-1,4,7-methano-1H-phenalene, see P-70021

4b,9b,10,14b,14c,15-Hexahydro-5,10,15-metheno-5H-benzo[a]naphth[1,2,3-de]-anthracene, see T-70210

Hexahydro-1,3,5-methenocyclopenta[cd]-penta-lene-2,4,6(1H)-trione, see P-60031

Hexahydro-7-methylcyclopenta[c]pyran-1(3H)-one, see M-70150

3a,4,5,6,7,7a-Hexahydro-3-methylene-2-(3H)-benzofuranone, H-70062

Hexahydro-3-(1-methylpropyl)pyrrolo[1,2-a]-pyrazine-1,4-dione, H-60052

1,2,3,5,8,8a-Hexahydronaphthalene, H-60053

1,2,3,7,8,8a-Hexahydronaphthalene, H-60054

1,3,4,5,7,8-Hexahydro-2,6-naphthalenedione, H-70063

Hexahydronezukone, see I-70091

Hexahydronicotinic acid, see P-70102

1,2,3,3a,4,5-Hexahydropentalene, H-70064

1,2,3,3a,4,6a-Hexahydropentalene, H-70065

1,2,3,4,5,6-Hexahydropentalene, H-70066

Hexahydrophthalide, see H-70061

Hexahydro-6,1,3[1]-propan[3]ylidene-1H-cyclo-penta[c]triazirino[a]pyrazole, see T-70205

3,4,4a,5,6,7-Hexahydro-2H-pyrano[2,3-b]-pyrylium, H-60055

Hexahydro-2,5-pyrazinedicarboxylic acid, see P-60157

Hexahydro-1H-pyrazolo[1,2-a]pyridazine, H-70067

1,2,3,3a,4,5-Hexahydropyrene, H-70068

1,2,3,6,7,8-Hexahydropyrene, H-70069

as-Hexahydropyrene, see H-70068

s-Hexahydropyrene, see H-70069

Hexahydropyridazino[1,2-a]pyridazine-1,4-dione, H-70070

4,5,6,8,9,10-Hexahydropyrido[3,4,5-ij]-quinolizine, see T-60086

Hexahydropyrimidine, H-60056

1,2,3,5,6,7-Hexahydropyrrolizinium(1+), H-70071

Hexahydro-1H-pyrrolizin-1-one, H-60057

Hexahydro-3H-pyrrolizin-3-one, H-60058

1,2,3,5,10,10a-Hexahydropyrrolo[1,2-b]-isoquinoline, H-70072

Hexahydropyrrolo[1,2-a]pyrazine-1,4-dione, H-70073

1,2,3,6,7,9a-Hexahydro-4(1H)-quinolizinone, H-70074

3,4,6,7,8,9-Hexahydro-2H-quinolizin-2-one, H-60059

Hexahydrosalicylic acid, see H-70119

Hexahydro-1,3,4,5-tetrahydroxybenzoic acid, see Q-70005

4a,5,6,7,7a,8a-Hexahydro-4,4a,7,7a-tetramethyl-5,8-epoxy-4H-indeno[5,6-b]furan, see P-60155

2,3,4,7,8,8a-Hexahydro-3,6,8,8-tetramethyl-1H-3a,7-methanoazulen-1-ol, see B-60112

▷Hexahydro-s-tetrazine-3,6-dione, see T-70097

Hexahydro-1,5-thiazonin-6(7H)-one, H-70075

2,3,4,5,8,9-Hexahydrothionin, H-70076

6bH-2a,4,4a,6a-Hexahydrotriazacyclopenta[cd]-pentalene, H-70077

1H,4H,7H,9bH-2,3,5,6,8,9-Hexahydro-3a,6a,9a-triazaphenalene, H-70078

Hexahydro-1,3,5-triazine, H-60060

Hexahydro-1,2,4-triazine-3,6-dione, see T-60099

1,3,4,8,9,11b-Hexahydro-5,7,11-trihydroxy-4,4,8,11b-tetramethylphenanthro[3,2-b]-furan-6(2H)-one, see L-70046

4,5,5a,6,7,9a-Hexahydro-5a,6,9a-trimethyl-naphtho[1,2-b]furan, see H-60027

1,2,6,7,8,9-Hexahydro-1,6,6-trimethyl-phenanthro[1,2-b]furan-10,11-dione, see C-60173

▷Hexahydro-1,3,5-trimethyl-1,3,5-triazine, *in* H-60060
2,3,6,7,10,11-Hexahydrotrisimidazo[1,2-*a*;1′,2′-*c*;1″,2″-*e*][1,3,5]-triazine, H-60061
3,3′,4,4′,5,5′-Hexahydroxybibenzyl, H-60062
▷2,3,14,20,22,25-Hexahydroxy-7-cholesten-6-one, H-60063
2β,3β,14α,20ξ,22ξ,24ξ-Hexahydroxy-5β-cholest-7-en-6-one, *see* P-60192
1α,4β,6β,8β,9α,14-Hexahydroxydihydro-β-agarofuran, *see* E-60018
▷1,1′,6,6′,7,7′-Hexahydroxy-3,3′-dimethyl-5,5′-bis(1-methylethyl)-[2,2′-binaphthalene]-8,8′-dicarboxaldehyde, *see* G-60032
2′,3,3′,5,7,8-Hexahydroxyflavone, H-70079
2′,3,5,5′,6,7-Hexahydroxyflavone, H-60064
2′,3,5,5′,7,8-Hexahydroxyflavone, H-70080
3′,4′,5,5′,6,7-Hexahydroxyflavone, H-70081
2,3,5,6,7,8-Hexahydroxy-1,4-naphthalenedione, *see* H-60065
Hexahydroxy-1,4-naphthoquinone, H-60065
1,2,3,5,6,7-Hexahydroxyphenanthrene, H-60066
2,3,3′,4,4′,5′-Hexahydroxystilbene, *see* T-60344
Hexaketocyclohexane, *see* C-60210
Hexakis(dichloromethyl)benzene, H-60067
3′,4′,5,5′,6,7-Hexamethoxyflavone, *in* H-70081
1,1,2,2,4,4-Hexamethyl-3,5-bis(methylene)-cyclopentane, H-70082
2,2,3,3,4,4-Hexamethylcyclobutanol, H-60068
Hexamethylcyclobutanone, H-60069
1,1,2,2,3,3-Hexamethylcyclohexane, H-60070
Hexamethylene diperoxide diamine, *see* T-70144
2,2,3,4,5,5-Hexamethyl-3-hexene, H-70083
2,2,3,3,4,4-Hexamethylpentane, H-60071
2-[3,7,11,15,19,23-Hexamethyl-25-(2,6,6-trimethyl-2-cyclohexenyl)pentacosa-2,14,18,22-tetraenyl]-3-methyl-1,4-naphthoquinone, H-70084
1,1,3,3,5,5-Hexamethyl-2,4,6-tris(methylene)-cyclohexane, H-60072
1,1′,1″-(1,3,6-Hexanetriyl)trisbenzene, *see* T-60386
2,2′,4,4′,6,6′-Hexanitro-[1,1′-biphenyl]-3,3′,5,5′-tetramine, *see* T-60014
1,4,7,10,25,28,31,34-Hexaoxa[10.10](9,10)-anthracenophane, H-70085
1,4,7,22,25,28-Hexaoxa[7.7](9,10)-anthracenophane, H-70086
▷4,7,13,16,21,24-Hexaoxa-1,10-diazabicyclo[8.8.8]hexacosane, H-60073
Hexaperibenzocoronene, *see* H-60031
Hexaphenylbenzene, H-60074
Hexathia-18-crown-6, *see* H-70087
1,4,7,10,13,16-Hexathiacyclooctadecane, H-60075
1,4,7,10,13,16-Hexathiacyclooctadecane, H-70087
Hexathiobenzo[18]crown-6, *see* D-60073
▷2-Hexenal, H-60076
4-Hexen-1-amine, H-70088
4-Hexen-3-ol, H-60077
1-Hexen-5-yn-3-ol, H-70089
▷Hexone, *see* M-60109
▷γ-Hexylbutyrolactone, *see* H-60078
▷5-Hexyldihydro-2(3H)-furanone, H-60078
δ-Hexylene oxide, *see* T-60075
2-Hexyl-5-methyl-3(2H)furanone, H-60079
2-Hexyl-5-methyl-3-oxo-2H-furan, *see* H-60079
Hexylvinylcarbinol, *see* N-60058
5-Hexyne-1,4-diamine, H-60080
3-Hexyne-2,5-dione, H-70090
Hibiscetin, *see* H-70021
Hibiscitin, *in* H-70021
Hildecarpidin, H-60081
▷1H-Indole-3-acetonitrile, *in* I-70013
Hinokiflavone, H-60082
Hipposterol, H-70091
Hirsudiol, *in* O-60033
Hispidol A, *in* T-70201
Hispidol B, *in* T-70201
Histidine trimethylbetaine, H-60083
Homalicine, *in* H-70207

Homoadamantane-2,7-dione, *see* T-70233
Homoadamantane-4,5-dione, *see* T-70234
Homoalethine, H-60084
Homoasaronic acid, *in* T-60342
Homoazulene-1,5-quinone, H-60085
Homoazulene-1,7-quinone, H-60086
Homoazulene-4,7-quinone, H-60087
Homocamphene, *see* D-60427
Homocamphenilone, *see* D-60411
Homocyclolongipesin, H-60088
Homocytidine, *in* A-70130
Homofarnesene, *see* T-70135
Homofulvene, *see* M-60069
Homofuran, *see* O-70056
Homoheveadride, H-60089
Homoisochroman, *see* T-70044
Homoisovanillin, *in* D-60371
Homomethionine, *see* A-60231
Homopantetheine, H-60090
Homopantethine, *in* H-60090
Homopentaprismanone, *see* H-60035
Homophthalimide, *see* I-70099
Homoprotocatechualdehyde, *see* D-60371
Homoserine, *see* A-70156
Homoserine lactone, *see* A-60138
Homotetrahydroquinoline, *see* T-70036
Homothiophene, *see* T-70161
Homotropone, *see* B-70111
Homovanillin, *in* D-60371
Honyucitrin, H-70092
Honyudisin, H-70093
Horminone, *in* D-60304
Hugershoff's base, *see* B-70057
Hyanilid, *in* H-70108
Hydnowightin, H-70094
2-Hydoxypiperonal, *see* H-70175
▷Hydral, *in* H-60092
▷Hydralazine, *see* H-60092
▷Hydralazine hydrochloride, *in* H-60092
Hydrazinecarbodithioic acid, *see* D-60509
2-Hydrazinobenzothiazole, H-70095
Hydrazinodithioformic acid, *see* D-60509
2-Hydrazinoethanamine, *see* H-70096
2-Hydrazinoethylamine, H-70096
2-[(2-Hydrazinoethyl)amino]ethanol, H-60091
▷1-Hydrazinophthalazine, H-60092
4-Hydrazino-2(1H)-pyridinone, H-60093
Hydrazinotriphenylmethane, *see* T-60388
as-Hydrindacene, H-60094
s-Hydrindacene, H-60095
as-Hydrindacene-1,8-dione, *see* T-70069
s-Hydrindacene-1,5-dione, *see* T-70068
s-Hydrindacene-1,7-dione, *see* T-70067
Hydrochelidonic acid, *see* O-60071
7-Hydro-8-methylpteroylglutamylglutamic acid, H-60096
4α-Hydroperoxydesoxyvulgarin, *in* V-60017
3α-Hydroperoxy-4,11(13)-eudesmadien-12,8β-olide, *in* H-60136
5-Hydroperoxy-4(15)-eudesmen-11-ol, H-70097
2-Hydroperoxy-2-methyl-6-methylene-3,7-octadiene, H-60097
3-Hydroperoxy-2-methyl-6-methylene-1,7-octadiene, H-60098
Hydroquinone-glucose, *see* A-70248
12-Hydroxy-8,12-abietadiene-11,14-dione, *see* R-60015
12-Hydroxy-8,11,13-abietatrien-18-al, *in* A-70005
12-Hydroxy-6,8,12-abietatriene-11,14-dione, *in* R-60015
7′-Hydroxyabscisic acid, H-70098
8′-Hydroxyabscisic acid, H-70099
4-Hydroxyacenaphthene, H-60099
4-Hydroxyacenaphthylene, H-70100
2-(Hydroxyacetyl)pyridine, H-70101
3-(Hydroxyacetyl)pyridine, H-70102
2-(Hydroxyacetyl)thiophene, H-70103
8-Hydroxyachillin, H-60100
3-Hydroxyacrylic acid, *in* O-60090
6β-Hydroxyadoxoside, *in* A-70067
1α-Hydroxyafraglaucolide, *in* A-70070
1β-Hydroxyafraglaucolide, *in* A-70070
2-Hydroxy-β-alanine, *see* A-70160
16β-Hydroxyalisol B monoacetate, *in* A-70082
15α-Hydroxy-β-amyrin, *in* O-70033
6-Hydroxyarcangelisin, *in* D-70021

3-Hydroxyarginine, *see* A-70149
2-Hydroxyaristotetralone, *in* A-70253
2-Hydroxy-4-azabenzimidazole, *see* D-60249
3-Hydroxybaikiain, *see* T-60068
8α-Hydroxybalchanin, *in* S-60007
▷2-Hydroxybenzamide, *in* H-70108
▷N-Hydroxybenzamide, *see* B-70035
3-Hydroxybenz[a]anthracene, H-70104
2-Hydroxybenzeneacetaldehyde, *see* H-60207
3-Hydroxybenzeneacetaldehyde, *see* H-60208
4-Hydroxybenzeneacetaldehyde, *see* H-60209
α-Hydroxybenzeneacetaldehyde, *see* H-70206
α-Hydroxybenzeneacetaldehyde, *see* H-60210
N-Hydroxybenzeneacetamide, *see* P-70059
2-Hydroxybenzenediazonium hydroxide inner salt, *see* D-70057
2-Hydroxy-1,3-benzenedicarboxylic acid, H-60101
2-Hydroxy-1,4-benzenedicarboxylic acid, H-70105
▷1-Hydroxy-1,2-benziodoxol-3(1H)-one, H-60102
6-Hydroxy-1,3-benzodioxole-5-carboxaldehyde, *see* H-70175
4-Hydroxy-1,3-benzodioxole-5-carboxylic acid, *see* H-60177
5-Hydroxy-1,3-benzodioxole-4-carboxylic acid, *see* H-60180
6-Hydroxy-1,3-benzodioxole-5-carboxylic acid, *see* H-60181
7-Hydroxy-1,3-benzodioxole-4-carboxylic acid, *see* H-60179
7-Hydroxy-1,3-benzodioxole-5-carboxylic acid, *see* H-60178
4-Hydroxybenzofuran, H-70106
4-Hydroxy-5-benzofurancarboxylic acid, H-60103
5-(6-Hydroxy-2-benzofuranyl)-1,3-benzenediol, *see* M-60146
5-(6-Hydroxy-2-benzofuranyl)-2,2-dimethyl-2H-1-benzopyran-7-ol, *see* M-60144
7-(6-Hydroxy-2-benzofuranyl)-2,2-dimethyl-2H-1-benzopyran-5-ol, *see* M-60143
5-(6-Hydroxy-2-benzofuranyl)-2-(3-methyl-2-butenyl)-1,3-benzenediol, *in* M-60146
1-(4-Hydroxy-5-benzofuranyl)-3-phenyl-2-propen-1-one, H-70107
▷2-Hydroxybenzoic acid, H-70108
6-Hydroxy-2H-1-benzopyran-2-one, H-70109
2-Hydroxy-4H-1-benzothiopyran-4-one, *see* H-70110
4-Hydroxy-2H-1-benzothiopyran-2-one, H-70110
2-[(2-Hydroxybenzoyl)amino]-4-methoxy-benzoic acid, *see* D-70049
2-Hydroxybenzoylformic acid, *see* H-70210
3-Hydroxybenzoylformic acid, *see* H-70211
4-Hydroxybenzoylformic acid, *see* H-70212
β-Hydroxy-β-benzylacetophenone, *see* H-70124
2-Hydroxybenzyl chloride, *see* C-70101
3-Hydroxybenzyl chloride, *see* C-70102
4-Hydroxybenzyl chloride, *see* C-70103
ent-7α-Hydroxy-15-beyeren-19-oic acid, H-60104
ent-12β-Hydroxy-15-beyeren-19-oic acid, H-60105
ent-18-Hydroxy-15-beyeren-19-oic acid, H-60106
2-Hydroxybibenzyl, *see* P-70071
3-Hydroxybibenzyl, *see* P-70072
4-Hydroxybibenzyl, *see* P-70073
4′-Hydroxy-2-biphenylcarboxylic acid, H-70111
6-Hydroxy-2,7,10-bisabolatrien-9-one, H-60107
5-Hydroxyborneol, *see* T-70280
5-Hydroxybowdichione, *in* B-70175
12-Hydroxybromosphaerol, H-60108
2-Hydroxybutanal, H-70112
3-Hydroxy-4-butanolide, *see* D-60242
▷2-Hydroxy-3-butenenitrile, *in* P-70119
2-Hydroxy-4-*tert*-butylcyclopentanecarboxylic acid, H-70113
α-Hydroxybutyraldehyde, *see* H-70112
β-Hydroxy-γ-butyrotrimethylbetaine, *see* C-70024

8-Hydroxy-1(6),2,4,7(11)-cadinatetraen-12,8-olide, H-70114

2-Hydroxycarbazole, H-60109

α-Hydroxy-β-carboxyisocaproic acid, *see* H-60166

β-Hydroxy-β-carboxyisocaproic acid, *see* H-60165

3-Hydroxy-β,β-carotene-4,4′-dione, *see* A-70066

5-Hydroxycarvacrol, *see* I-60125

2‴-Hydroxychlorothricin, *in* C-60137

(25*R*)-26-Hydroxycholesterol, *in* C-60147

(25*S*)-26-Hydroxycholesterol, *in* C-60147

2-Hydroxychroman, *see* D-60198

3-Hydroxychroman, *see* D-60199

4-Hydroxychroman, *see* D-60200

5-Hydroxychroman, *see* D-60201

6-Hydroxychroman, *see* D-70175

7-Hydroxychroman, *see* D-60202

8-Hydroxychroman, *see* D-60203

3′-Hydroxycinerubin *X*, *see* C-70176

p-Hydroxycinnamaldehyde, *see* H-60213

l-Hydroxy-13(17),15-cleistanthadien-18-oic acid, H-70115

7-Hydroxy-3,13-clerodadiene-16,15;18,19-diolide, H-70116

ent-7β-Hydroxy-3,13-clerodadiene-15,16;18,19-diolide, H-70117

16-Hydroxy-3,13-clerodadien-15,16-olide, H-70118

ent-7β-Hydroxy-3-clerodene-15,16:18,19-diolide, *in* H-70117

8β-Hydroxycolumbin, *in* C-60160

10α-Hydroxycolumbin, *in* C-60160

7-Hydroxy-8(17),13-corymbidienolide, H-60110

6-Hydroxycoumarin, *see* H-70109

5β-Hydroxycrustecdysone, *in* H-60023

5′-Hydroxycudraflavone A, *in* C-70205

▷ *N*-Hydroxycyclohexanamine, *see* C-70228

α-Hydroxycyclohexaneacetic acid, *see* C-60212

2-Hydroxycyclohexaneacetic acid lactone, *see* H-70049

2-Hydroxycyclohexanecarboxylic acid, H-70119

2-Hydroxycyclohexanemethanol, H-70120

2-(2-Hydroxycyclohexyl)propenoic acid lactone, *see* H-70062

10′-Hydroxycyclolycoserone, *in* C-70229

6-Hydroxycyclonerolidol, H-60111

2-Hydroxycyclononanone, H-60112

3-Hydroxycyclononanone, H-60113

2-Hydroxycyclooctanecarboxylic acid, H-70121

2-Hydroxycyclopentanone, H-60114

3-Hydroxycyclopentanone, H-60115

4-Hydroxy-2-cyclopentenone, H-60116

1-Hydroxycyclopropanecarboxylic acid, H-60117

5-Hydroxycystofuranoquinol, *in* C-70260

8-Hydroxydatriscetin, *see* H-70029

6-Hydroxy-3-deazapurine, *see* H-60157

1-Hydroxydebneyol, *in* D-70010

8-Hydroxydebneyol, *in* D-70010

▷ 4-Hydroxydecanoic acid lactone, *see* H-60078

4′-Hydroxydehydrokawain, H-70122

8β-Hydroxydehydrozaluzanin C, *in* Z-60001

3β-Hydroxydemethylcryptojaponol, *in* T-60314

7α-Hydroxy-3-deoxyzaluzanin C, *in* Z-60001

3α-Hydroxydesoxoachalensolide, *in* H-70135

3β-Hydroxydesoxoachalensolide, *in* H-70135

1-Hydroxy-1,3-dihydrobenzofuran, *see* H-60171

2α-Hydroxy-11α,13-dihydroconfertin, *in* D-60211

3β-Hydroxy-11β,13-dihydrodesoxoachalensolide, *in* H-70135

11β-Hydroxy-11,13-dihydro-8-epi-confertin, *in* D-60211

14-Hydroxy-11,13-dihydrohypocretenolide, *in* H-70235

3α-Hydroxy-14,15-dihydromanoyl oxide, *in* E-70021

2α-Hydroxydihydroparthenolide, *in* D-70239

9α-Hydroxydihydroparthenolide, *in* D-70239

7-Hydroxy-6,7-dihydro-5*H*-pyrrolo[1,2-*a*]-imidazole, H-60118

2-Hydroxy-6-(3,4-dihydroxy-1-pentenyl)-benzyl alcohol, *see* S-70056

7-Hydroxy-4-(3,4-dihydroxyphenyl)-2′,5-oxidocoumarin, *see* T-70273

9α-Hydroxy-11β,13-dihydrozaluzanin *C*, *in* Z-60001

9β-Hydroxy-11β,13-dihydrozaluzanin *C*, *in* D-70305

8α-Hydroxy-11α,13-dihydrozaluzanin C, *in* Z-60001

9α-Hydroxy-11β,13-dihyrosantamarine, *in* S-60007

2′-Hydroxy-4′,6′-dimethoxyacetophenone, *in* T-70248

▷ 4′-Hydroxy-3′,5′-dimethoxyacetophenone, *in* T-60315

6-Hydroxy-4,7-dimethoxy-5-acetylcoumarone, *see* K-60011

1-(6-Hydroxy-4,7-dimethoxy-5-benzofuranyl)-ethanone, *see* K-60011

5-Hydroxy-2,3-dimethoxybenzoic acid, *in* T-70250

3-Hydroxy-3′,4-dimethoxybibenzyl, *in* T-60316

4-Hydroxy-3,6-dimethoxy-1,2-dimethylcarbazole, *in* T-70253

4′-Hydroxy-3′,7-dimethoxyflavan, *in* T-70260

5-Hydroxy-6,7-dimethoxyflavone, *in* T-60322

5-Hydroxy-7,8-dimethoxyflavone, *in* T-60321

6-Hydroxy-4′,7-dimethoxyisoflavone, *in* T-60325

4-Hydroxy-3,6-dimethoxy-2-methyl-9*H*-carbazole-1-carboxaldehyde, *in* H-70154

4′-Hydroxy-3′,5-dimethoxy-3,4-methylenedioxybibenzyl, *in* P-60044

7-Hydroxy-5,6-dimethoxy-3′,4′-methylenedioxyisoflavone, *in* P-70032

▷ 4-Hydroxy-6,7-dimethoxy-9-(3,4-methylenedioxyphenyl)naphtho[2,3-*c*]-furan-1(3*H*)-one, *see* D-60493

2-Hydroxy-3,7-dimethoxyphenanthrene, *in* T-60339

▷ 1-(4-Hydroxy-3,5-dimethoxyphenyl)ethanone, *in* T-60315

3-(4-Hydroxy-3,5-dimethoxyphenyl)-2-propen-1-ol, *see* S-70046

1-Hydroxy-5,6-dimethoxyxanthone, *in* T-60349

7-Hydroxy-3,4-dimethyl-2*H*-1-benzopyran-2-one, H-60119

6-(4-Hydroxy-1,3-dimethyl-2-butenylamino)-purine, *see* M-60133

7-Hydroxy-3,4-dimethylcoumarin, *see* H-60119

12-Hydroxy-6,14-dimethyl-1,7-dioxacyclotetradeca-3,9-diene-2,8,11-trione, *in* C-60159

4-Hydroxy-2,5-dimethyl-3(2*H*)-furanone, H-60120

6-Hydroxy-2,4-dimethylheptanoic acid, H-60121

5-Hydroxy-8,8-dimethyl-6-(3-methyl-2-butenyl)-2*H*,8*H*-benzo[1,2-*b*:3,4-*b*′]-dipyran-2-one, *see* H-70093

5-Hydroxy-2,2-dimethyl-6-(3,4-methylenedioxystyryl)chromene, *see* G-70014

3-(2-Hydroxy-4,8-dimethyl-3,7-nonadienyl)-benzaldehyde, H-60122

2-(7-Hydroxy-3,7-dimethyl-2-octenyl)-6-methoxy-1,4-benzoquinone, H-60123

4-Hydroxy-3,3-dimethyl-2-oxobutanoic acid, H-60124

α-Hydroxy-2,6-dinitrotoluene, *see* D-60459

15-Hydroxy-12,19-dioxo-13,14-dehydro-10,11,14,15-tetrahydrogeranylnerol, *in* H-60186

5-Hydroxy-1,7-diphenyl-3-heptanone, *in* D-70481

5-Hydroxy-1,7-diphenyl-1-hepten-3-one, *in* D-70481

2-Hydroxy-1,2-diphenyl-1-propanone, H-70123

2-Hydroxy-1,3-diphenyl-1-propanone, H-70124

10-Hydroxy-7,11,13,16,19-docosapentaenoic acid, H-70125

2-Hydroxy-2-dodecenal, *see* O-70087

10-Hydroxy-11-dodecenoic acid, H-60125

12-Hydroxy-10-dodecenoic acid, H-60126

20-Hydroxyecdysone, *in* H-60063

5β-Hydroxyecdysterone, *in* H-60023

8-Hydroxy-5,9,11,14,17-eicosapentaenoic acid, H-70126

5-Hydroxy-6,8,11,14-eicosatetraenoic acid, H-60127

8-Hydroxy-5,9,11,14-eicosatetraenoic acid, H-60128

12-Hydroxy-5,8,10,14-eicosatetraenoic acid, H-60131

20-Hydroxyelemajurinelloide, H-60132

6α-Hydroxy-8-epiivangustin, *in* I-70109

10′-Hydroxy-1′-epilycoserone, *in* L-70045

12α-Hydroxy-13-epi-(+)-manoyl oxide, *in* E-70023

11-Hydroxy-12,13-epoxy-9,15-octadecadienoic acid, H-70127

▷ *N*-Hydroxyethanimidamide, *see* A-60016

1-(2-Hydroxyethyl)-1,4-cyclohexanediol, H-60133

O-(2-Hydroxyethyl)hydroxylamine, H-60134

5-(1-Hydroxyethyl)-2-(1-hydroxy-1-methylethyl)benzofuran, *in* A-70044

α-Hydroxy-1-ethylnaphthalene, *see* N-70015

3-(2-Hydroxyethyl)piperidine, *see* P-70103

3-(2-Hydroxyethyl)thiophene, *see* I-60211

9-Hydroxy-4,11(13)-eudesmadien-12-oic acid, H-60135

1β-Hydroxy-3,11(13)-eudesmadien-12,6α-olide, *see* S-60007

1β-Hydroxy-4,11(13)-eudesmadien-12,8β-olide, *see* I-70109

3-Hydroxy-4,11(13)-eudesmadien-12,8-olide, H-60136

7-Hydroxy-4,11(13)-eudesmadien-12,6-olide, H-70128

9-Hydroxy-4,11(13)-eudesmadien-12,6-olide, H-60137

15-Hydroxy-4,11(13)-eudesmadien-12,8-olide, H-60138

6-Hydroxy-1,4-eudesmadien-3-one, H-60139

1-Hydroxy-3,7(11),8-eudesmatrien-12,8-olide, H-60140

1-Hydroxy-2,4(15),11(13)-eudesmatrien-12,8-olide, H-70129

1-Hydroxy-4(15),7(11),8-eudesmatrien-12,8-olide, H-60141

4-Hydroxy-11(13)-eudesmen-12,8-olide, H-60142

18-Hydroxyferruginol, *see* A-70005

ω-Hydroxyferulenol, *in* F-60005

6-Hydroxyflavone, H-70130

7-Hydroxyflavone, H-70131

▷ 8-Hydroxyflavone, H-70132

3β-Hydroxy-27-friedelanal, *see* T-70218

27-Hydroxy-3,21-friedelanedione, *in* D-70301

3β-Hydroxy-27-friedelanoic acid, *in* T-70218

3α-Hydroxy-27-friedelanoic acid, *in* T-70218

27-Hydroxy-3-friedelanone, *in* D-70301

▷ 9-Hydroxy-7*H*-furo[3,2-*g*][1]benzopyran-7-one, *see* X-60002

10-Hydroxy-2*H*,13*H*-furo[3,2-*c*:5,4-*h*′]-bis[1]-benzopyran-2,13-dione, H-70133

4-Hydroxyfuro[2,3-*d*]pyridazine, H-60146

7-Hydroxyfuro[2,3-*d*]pyridazine, H-60147

2-Hydroxygarveatin *B*, H-60148

2-Hydroxygarveatin A, *in* H-60148

2-Hydroxygarvin *B*, H-60009

6α-Hydroxygedunin, *in* G-70006

3β-Hydroxy-1(10),4,11(13)-germacratrien-12,6α-olide, *see* H-60006

6-Hydroxy-1(10),4,11(13)-germacratrien-12,8-olide, H-70134

4-Hydroxyglutamic acid, *see* A-60195

4-Hydroxygrenoblone, *in* G-70029

3-Hydroxy-4,11(13)-guaiadien-12,8-olide, H-70135

15-Hydroxy-3,10(14)-guaiadien-12,8-olide, H-70136

5-Hydroxy-3,11(13)-guaiatrien-12,8-olide, H-70137

8-Hydroxy-4(15),11(13),10(14)-guaiatrien-12,6-olide, H-70138

3β-Hydroxy-4-guaien-12,8β-olide, *in* H-70135

11-Hydroxy-4-guaien-3-one, H-60149

8-Hydroxyguanosine, *see* D-70238

9α-Hydroxygymnomitryl acetate, *in* G-70035

9α-Hydroxygymnomitryl cinnamate, *in* G-70035

2α-Hydroxyhanphyllin-3-*O*-acetate, *in* H-60006

2β-Hydroxyhardwickiic acid, *in* H-70002

2-Hydroxyhautriwaic acid, *in* H-70002

3β-Hydroxyhebeclinolide, *in* H-60009

17-Hydroxyhebemacrophyllide, *in* H-60010

5-Hydroxyhemimellitene, *see* T-60371

5-Hydroxy-3,3′,4′,5′,6,7,8-heptamethoxyflavone, *in* O-70018

5-(1-Hydroxy-2,4,6-heptatriynyl)-2-oxo-1,3-dioxolane-4-heptanoic acid, *see* A-70224

7-Hydroxy-5-heptynoic acid, H-60150

16-Hydroxy-9-hexadecenoic acid *o*-lactone, *see* I-60086

3-Hydroxy-3′,4′,5,6,7,8-hexamethoxyflavone, *in* H-60024

5-Hydroxy-3,3′,4′,6,7,8-hexamethoxyflavone, *in* H-60024

3-Hydroxy-5-hexanolide, *see* T-60067

3-Hydroxy-22,23,24,25,26,27-hexanor-20-dammaranone, H-60151

3-Hydroxyhomotyrosine, *see* A-60185

5-Hydroxyhydantoin, *see* H-60156

7-Hydroxy-3-(4-hydroxybenzylidene)-4-chromanone, H-60152

3-Hydroxy-4-(1-hydroxy-1,5-dimethylhexyl)-benzoic acid, *see* S-60065

7-Hydroxy-1-hydroxy-3,7-dimethyl-2*E*,5*E*-octadien-4-one, *in* D-60318

4-Hydroxy-4-(2-hydroxyethyl)cyclohexanone, H-60153

4-Hydroxy-5-(1-hydroxyethyl)-3-(3-methyl-2-butenyl)-2*H*-1-benzopyran-2-one, *see* M-70091

5-Hydroxy-2-[5-hydroxy-2-(hydroxymethyl)-4-pyrimidinyl]-7-methyl-4-benzoxazolecarboxamide, *see* B-70176

2-Hydroxy-4-[(2-hydroxy-4-methoxy-3,6-dimethylbenzoyl)oxy]-3,6-dimethylbenzoic acid, *see* B-70007

2-Hydroxy-4-[[2-hydroxy-4-methoxy-6-(2-oxoheptyl)benzoyl]oxy]-6-(2-oxoheptyl)-benzoic acid, *see* M-70147

5-Hydroxy-2-(4-hydroxy-3-methoxyphenyl)-7-methoxy-4*H*-1-benzopyran-4-one, *see* D-70291

5-Hydroxy-2-(3-hydroxy-4-methoxyphenyl)-6,7,8-trimethoxy-4*H*-1-benzopyran-4-one, *see* D-60375

6-Hydroxy-2-(5-hydroxy-3-methoxy-2-prenylphenyl)benzofuran, *in* M-60146

7-Hydroxy-6-(hydroxymethyl)-2*H*-1-benzopyran-2-one, H-70139

2-Hydroxy-4-hydroxymethyl-4-butanolide, *see* D-70212

4-Hydroxy-3-(2-hydroxy-3-methyl-3-butenyl)benzoic acid, H-60154

7-Hydroxy-6-hydroxymethylcoumarin, *see* H-70139

1-[3-Hydroxy-4-(hydroxymethyl)-cyclopentyl]-2,4(1*H*,3*H*)-pyrimidinedione, H-70140

3-Hydroxy-4-(hydroxymethyl)-7,7-dimethylbicyclo[2.2.1]heptan-2-one, *in* H-70173

3β-Hydroxy-4α-hydroxymethyl-4β,14α-dimethyl-7,11,15-trioxo-5α-chol-8-en-24-oic acid, *in* L-60036

10-Hydroxy-2-(6-hydroxy-4-methyl-4-hexylidene)-6,10-dimethyl-7-oxo-8-undecenal, *in* H-60186

3-Hydroxy-2-hydroxymethyl-1-methoxy-anthraquinone, *in* D-70306

5-Hydroxy-6-(hydroxymethyl)-7-methoxy-2-methyl-4*H*-1-benzopyran-4-one, H-60155

5-Hydroxy-6-hydroxymethyl-7-methoxy-2-methylchromone, *see* H-60155

1-[5-Hydroxy-2-(hydroxymethyl)-2-methyl-2*H*-1-benzopyran-6-yl]ethanone, *see* A-60031

4-Hydroxy-5-(hydroxymethyl)-3-(3-methyl-2-butenyl)-2*H*-1-benzopyran-2-one, *see* L-60034

12-Hydroxy-6-(hydroxymethyl)-10-methyl-2-(4-methyl-3-pentenyl)-2,6,10-dodecatrienal, *in* H-70179

5-[3-Hydroxy-2-(hydroxymethyl)phenyl]-4-pentene-2,3-diol, *see* S-70056

4-Hydroxy-5-hydroxymethyl-3-prenylcoumarin, *see* L-60034

2-Hydroxy-2-(hydroxymethyl)-2*H*-pyran-3(6*H*)-one, H-70141

4-Hydroxy-3-(9-hydroxy-8-oxofarnesyl)-5-methylcoumarin, *in* H-70183

4-Hydroxy-α-[3-hydroxy-5-oxo-4-(3,4,5-trihydroxyphenyl)-2(5*H*)-furanylidene]-benzeneacetic acid, *see* G-70026

3-Hydroxy-4-(4-hydroxyphenyl)-6-[(4-hydroxyphenylmethylene)]-2*H*-pyran-2,5(6*H*)-dione, *see* G-60038

3-Hydroxy-5-(4-hydroxyphenyl)pentanoic acid, H-70142

8-Hydroxy-3-(2-hydroxypropyl)-6-methoxy-1*H*-2-benzopyran-1-one, *see* D-70050

8-Hydroxy-3-(2-hydroxypropyl)-6-methoxy-isocoumarin, *see* D-70050

14-Hydroxyhypocretenolide, *in* H-70235

8-Hydroxy-5,9,11,14-icosatetraenoic acid, *in* H-70126

9-Hydroxy-5,7,11,14-icosatetraenoic acid, H-60129

11-Hydroxy-5,8,12,14-icosatetraenoic acid, H-60130

15-Hydroxyimbricatolal (incorr.), *in* H-70150

2-Hydroxy-4,5-imidazolidinedione, *see* D-60243

5-Hydroxy-2,4-imidazolidinedione, H-60156

2-Hydroxyimidazo[*b*]pyrazine, *see* I-60006

6-Hydroxyimidazo[4,5-*c*]pyridazine, *see* D-60248

2-Hydroxyimidazo[*b*]pyridine, *see* D-60249

4-Hydroxyimidazo[4,5-*b*]pyridine, H-60157

3-Hydroxyiminoindole, *in* I-60035

▷1-Hydroxyindeno[1,2,3-*cd*]pyrene, *see* I-60022

▷2-Hydroxyindeno[1,2,3-*cd*]pyrene, *see* I-60023

▷6-Hydroxyindeno[1,2,3-*cd*]pyrene, *see* I-60024

▷7-Hydroxyindeno[1,2,3-*cd*]pyrene, *see* I-60025

▷8-Hydroxyindeno[1,2,3-*cd*]pyrene, *see* I-60026

6-Hydroxyindole, H-60158

2-Hydroxy-5-iodobenzaldehyde, H-70143

3-Hydroxy-2-iodo-6-methylpyridine, H-60159

3-[4-(4-Hydroxy-3-iodophenoxy)-3,5-diiodophenyl]alanine, *see* T-60350

O-(4-Hydroxy-3-iodophenyl)-3,5-diiodotyrosine, *see* T-60350

5-Hydroxy-6-iodo-2-picoline, *see* H-60159

3-Hydroxy-2-iodopyridine, H-60160

2-Hydroxyislandicin, *see* T-60112

1α-Hydroxyisoafraglaucolide, *in* I-70060

3α-Hydroxyisoagatholal, *in* D-70310

3β-Hydroxyisoagatholal, *in* D-70310

5′-Hydroxyisoavrainvilleol, *see* B-70194

4-Hydroxyisobacchasmacranone, H-60161

10α-Hydroxy-6-isodaucen-14-al, *in* O-70093

▷3-(4-Hydroxy-L-isoleucine)-α-amanitin, *in* A-60084

10′-Hydroxyisolycoserone, *in* I-70068

2-Hydroxyisophthalic acid, *see* H-60101

18-Hydroxy-8,15-isopimaradien-7-one, H-60162

8-Hydroxyisopimar-15-ene, H-60163

4-Hydroxy-2-isopropyl-5-benzofurancarboxylic acid, H-60164

2-Hydroxy-2-isopropylbutanedioic acid, H-60165

2-Hydroxy-3-isopropylbutanedioic acid, H-60166

7-Hydroxy-4-isopropyl-3-methoxy-6-methylcoumarin, H-70144

2-Hydroxy-2-isopropylsuccinamic acid, *in* H-60165

2-Hydroxy-2-isopropylsuccinamide, *in* H-60165

2-Hydroxy-2-isopropylsuccinic acid, *see* H-60165

2-Hydroxy-3-isopropylsuccinic acid, *see* H-60166

7-Hydroxy-3-isopropyl-5,5,9-trimethyl-1,2,6(5*H*)-anthracenetrione, *see* P-70144

β-Hydroxyisovalerylshikonin, *in* S-70038

4-Hydroxyisovaline, *see* A-60186

2α-Hydroxyivangustin, *in* I-70109

14α-Hydroxyixocarpanolide, *in* I-60143

Hydroxyjavanicin, *see* F-60108

2-Hydroxyjuglone, *see* D-60354

3-Hydroxyjuglone, *see* D-60355

20-Hydroxyjurinelloide, *in* J-60003

ent-19-Hydroxy-16*S*-kauran-17-al, *in* K-70008

ent-3β-Hydroxy-15-kauren-17-oic acid, H-60167

15-Hydroxy-8(17),12-labdadien-16-al, H-70145

19-Hydroxy-8(17),13-labdadien-15-al, H-70146

2-Hydroxy-8(17),13-labdadien-15-oic acid, H-70147

ent-6β-Hydroxy-8(17),13-labdadien-15-oic acid, H-70148

ent-15-Hydroxy-8,13-labdadien-7-one, *in* L-70004

ent-6β-Hydroxy-7,12*E*,14-labdatrien-17-oic acid, *in* L-60004

8-Hydroxy-13-labden-15-al, H-70149

15-Hydroxy-8(17)-labden-19-al, H-70150

15-Hydroxy-7-labden-17-oic acid, H-60168

15-Hydroxy-7-labden-3-one, H-70151

4-Hydroxylamino-1,2,3-benzotriazine, *in* A-60092

▷Hydroxylaminocyclohexane, *see* C-70228

3-Hydroxy-8,24-lanostadien-21-oic acid, H-70152

3-Hydroxy-8-lanosten-26,22-olide, H-60169

21-Hydroxylanosterol, *in* L-70017

2α-Hydroxylaurenobiolide, *in* H-70134

Hydroxyleptomycin *A*, *see* K-70007

23-Hydroxylongispinogenin, *in* O-60035

5-Hydroxy-14,15-LTA₄, *see* E-70016

15-Hydroxylycopersene, *in* L-70044

3α-Hydroxymanool, *in* L-70002

19-Hydroxymanoyl oxide, *in* E-70023

9α-Hydroxymedioresinol, *in* M-70016

3-Hydroxy-2-mercaptopyridine, *see* H-60222

▷2-Hydroxy-3-methoxybenzaldehyde, *in* D-70285

3-Hydroxy-2-methoxybenzaldehyde, *in* D-70285

5-(5-Hydroxy-6-methoxy-2-benzofuranyl)-1,3-benzenediol, *in* M-60146

▷7-Hydroxy-6-methoxy-2*H*-1-benzopyran-2-one, H-70153

▷7-Hydroxy-6-methoxycoumarin, *see* H-70153

11-Hydroxy-2-methoxy-19,20-dinor-1,3,5(10),7,9(11),13-abietahexaene-6,12-dione, *see* F-60077

10-Hydroxy-11-methoxydracaenone, *in* D-70546

5-Hydroxy-6-methoxyindole, *in* D-70307

6-Hydroxy-5-methoxyindole, *in* D-70307

3-Hydroxy-1-methoxy-2-(methoxymethyl)-anthraquinone, *in* D-70306

7-Hydroxy-4-methoxy-5-methyl-2*H*-1-benzopyran-2-one, *in* D-60348

2-Hydroxy-3-methoxy-5-methyl-1,4-benzoquinone, *in* D-70320

3-Hydroxy-2-methoxy-5-methyl-1,4-benzoquinone, *in* D-70320

7-Hydroxy-6-methoxy-8-(3-methyl-2-butenyl)-2*H*-1-benzopyran-1-one, *see* C-60026

4-Hydroxy-3-methoxy-2-methyl-9*H*-carbazole-1-carboxaldehyde, H-70154

7-Hydroxy-4-methoxy-5-methylcoumarin, *in* D-60348

5-Hydroxy-2′-methoxy-6,7-methylenedioxyisoflavone, *in* T-60107

7-Hydroxy-6-methoxy-3′,4′-methylenedioxyisoflavone, *in* T-70111

7-Hydroxy-8-methoxy-3′,4′-methylenedioxyisoflavone, *in* T-70112

3-Hydroxy-4-methoxy-8,9-methylenedioxypterocarpan, *see* S-60038

7-Hydroxy-3-methoxy-6-methyl-4-(1-methylethyl)-2*H*-1-benzopyran-2-one, *see* H-70144

3-Hydroxy-5-methoxy-2-methyl-1,4-naphthoquinone, *in* D-60350

6-Hydroxy-5-methoxy-2-methyl-1,4-naphthoquinone, *in* D-60351

3-Hydroxy-4-methoxy-1-methyl-7-oxabicyclo[4.1.0]hept-3-ene-2,5-dione, *in* D-70320

5-Hydroxy-7-methoxy-6-methylphthalide, *in* D-60349

2-Hydroxy-5-methoxy-1,4-naphthoquinone, *in* D-60354

2-Hydroxy-8-methoxy-1,4-naphthoquinone, *in* D-60355

4-Hydroxy-5-methoxy-2-nitrobenzoic acid, *in* D-60360

5-Hydroxy-4-methoxy-2-nitrobenzoic acid, *in* D-60360

5-Hydroxy-3-methoxy-7-oxabicyclo[4.1.0]-hept-3-en-2-one, H-70155

7-Hydroxy-6-methoxy-3-[(2-oxo-2*H*-1-benzopyran-7-yl)oxy]-2*H*-1-benzopyran-2-one, *see* D-70006

3-Hydroxy-5-methoxy-4-(1-oxo-2,4-octadienyl)-benzeneacetic acid, *see* C-60024

2-Hydroxy-5-methoxy-3-(10-pentadecenyl)-1,4-benzoquinone, *see* M-60004

3-Hydroxy-7-methoxy-9-pentyl-1-propyldibenzofuran-2-carboxylic acid, *see* I-60098

3-Hydroxy-4-methoxyphenylacetaldehyde, *in* D-60371

4-Hydroxy-3-methoxyphenylacetaldehyde, *in* D-60371

7-Hydroxy-6-methoxy-4-phenyl-2*H*-1-benzopyran-2-one, I-60097

7-Hydroxy-6-methoxy-4-phenylcoumarin, *see* I-60097

6-[2-(4-Hydroxy-3-methoxyphenyl)ethenyl]-4-methoxy-2*H*-pyran-2-one, *in* H-70122

2-(2-Hydroxy-4-methoxyphenyl)-5-hydroxy-6-methoxybenzofuran, *see* S-60002

1-(2-Hydroxy-4-methoxyphenyl)-3-(3-hydroxy-4-methoxyphenyl)propane, *in* D-70336

1-(4-Hydroxy-2-methoxyphenyl)-3-(4-hydroxy-3-methoxyphenyl)propane, *in* D-70336

1-(2-Hydroxy-4-methoxyphenyl)-3-(4-hydroxy-3-methoxyphenyl)propene, *in* D-70336

1-(6-Hydroxy-2-methoxyphenyl)-9-(4-hydroxyphenyl)-1-nonanone, *in* M-70007

2-(2-Hydroxy-4-methoxyphenyl)-6-methoxy-3-methylbenzofuran, H-70156

2-(2-Hydroxy-4-methoxyphenyl)-7-methoxy-5-(1-propenyl)benzofuran, *in* H-60214

1-(6-Hydroxy-2-methoxyphenyl)-9-(3,4-methyl-enedioxyphenyl)-1-nonanone, *in* M-70007

2-(2-Hydroxy-4-methoxyphenyl)-3-methyl-5,6-methylenedioxybenzofuran, H-70157

2-(2-Hydroxy-4-methoxyphenyl)-5-(1-propenyl)benzofuran, *in* H-60214

1-(2-Hydroxy-6-methoxyphenyl)-1-tetra-decanone, *in* D-70338

2-(3-Hydroxy-4-methoxyphenyl)-1-(3,4,5-trimethoxyphenyl)ethanol, *see* C-70192

5-Hydroxy-7-methoxyphthalide, *in* D-60339

7-Hydroxy-5-methoxyphthalide, *in* D-60339

9-Hydroxy-3-methoxypterocarpan, *in* D-60373

6-Hydroxy-7-methoxyquinoxaline, *in* Q-60009

6-(4-Hydroxy-3-methoxystyryl)-4-methoxy-2-pyrone, *in* H-70122

3-Hydroxy-4-methoxy-α-(3,4,5-trimethoxy-phenyl)benzenemethanol, *see* C-70192

8-Hydroxy-3-methoxy-1,4,6-trimethyl-11-oxo-11*H*-dibenzo[*b,e*][1,4]dioxepin-7-carboxylic acid, *see* N-70081

2-Hydroxy-1-methoxyxanthone, *in* D-70353

2′-(Hydroxymethyl)acetophenone, H-70158

4′-(Hydroxymethyl)acetophenone, H-70159

1-Hydroxy-4-(methylamino)anthraquinone, *in* A-60175

1-Hydroxy-7-methyl-9,10-anthracenedione, *see* H-60170

1-Hydroxy-7-methylanthraquinone, H-60170

2-(Hydroxymethyl)benzaldehyde, H-60171

2-Hydroxy-4-methylbenzaldehyde, H-70160

4-Hydroxy-2-methylbenzaldehyde, H-70161

5-Hydroxy-3-methyl-1,2-benzenedicarboxylic acid, H-70162

2-Hydroxymethyl-1,4-benzenediol, *see* D-70287

2-(Hydroxymethyl)-1,4-benzodioxan, H-60172

2-Hydroxy-4-methylbenzoic acid, H-70163

4-Hydroxy-2-methylbenzoic acid, H-70164

2-(Hydroxymethyl)benzonitrile, *see* I-70062

5-Hydroxy-4-methyl-2*H*-1-benzopyran-2-one, H-60173

8-Hydroxy-4-methyl-2*H*-1-benzopyran-2-one, H-70165

3-Hydroxymethylbicyclo[2.2.1]heptan-2-amine, *see* A-70093

3-Hydroxymethylbicyclo[2.2.1]hept-5-en-2-amine, *see* A-70094

Hydroxymethylbilane, H-70166

1-[4-Hydroxy-3-(3-methyl-1,3-butadienyl)-phenyl]-2-(3,5-dihydroxyphenyl)ethylene, H-70167

5-[2-[4-Hydroxy-3-(3-methyl-1,3-butadienyl)-phenyl]ethenyl]-1,3-benzenediol, *see* H-70167

4-(3-Hydroxy-3-methylbutanoyloxy)-3-(1,1-di-methyl-2-propenyl)-6-phenyl-2*H*-pyran-2-one, *in* D-70434

5-Hydroxymethyl-2-butenolide, *see* H-60182

4-Hydroxy-3-methylbut-2-enolide, *in* M-70103

6-(3-Hydroxy-3-methyl-1-butenyl)-7-methoxy-2*H*-1-benzopyran-2-one, *see* S-70080

8-(2-Hydroxy-3-methyl-3-butenyl)-7-methoxy-2*H*-1-benzopyran-2-one, *see* A-70269

8-(2-Hydroxy-3-methyl-3-butenyl)-7-methoxy-coumarin, *see* A-70269

2-(Hydroxymethyl)-3-butyn-1-ol, *see* E-70054

2-Hydroxy-9-methylcarbazole, *in* H-60109

2-[(Hydroxymethyl)carbonyl]thiophene, *see* H-70103

5-Hydroxy-4-methylcoumarin, *see* H-60173

8-Hydroxy-4-methylcoumarin, *see* H-70165

2-Hydroxymethylcyclobutanecarboxylic acid lactone, *see* O-70053

2-Hydroxy-2-methylcyclobutanone, H-70168

4-Hydroxy-4-methyl-2,5-cyclohexadien-1-one, H-70169

5-(Hydroxymethyl)-1,2,3,4-cyclohexanetetrol, H-70170

1L-(1,2,4/3,5)-1-*C*-Hydroxymethyl-2,3,4,5-cyclohexanetetrol, *in* H-70170

1D-(1,2,4/3,5)-5-*C*-Hydroxymethyl-1,2,3,4-cyclohexanetetrol, *in* H-70170

1D-(1,2,4,5/3)-5-*C*-Hydroxymethyl-1,2,3,4-cyclohexanetetrol, *in* H-70170

1L-(1,3,5/2,4)-5-*C*-Hydroxymethyl-1,2,3,4-cyclohexanetetrol, *in* H-70170

2-Hydroxymethylcyclohexanol, *see* H-70120

5-Hydroxy-2-methylcyclohexanone, H-70171

5-(Hydroxymethyl)-5-cyclohexene-1,2,3,4-tetrol, H-70172

2-Hydroxymethylcyclopentanecarboxylic acid lactone, *see* H-70054

▷2-Hydroxy-3-methyl-2-cyclopenten-1-one, H-60174

4-Hydroxy-2-methyl-2-cyclopenten-1-one, H-60175

2-Hydroxymethylcyclopropanecarboxylic acid lactone, *see* O-70055

3-Hydroxymethyldibenzo[*b,f*]thiepin, H-60176

8-Hydroxy-5-methyldihydrothiazolo[3,2-*a*]-pyridinium-3-carboxylate, *see* C-70021

1-(Hydroxymethyl)-7,7-dimethylbicyclo[2.2.1]-heptane-2,3-diol, H-70173

2-Hydroxymethyl-1,1-dimethylcyclopropane, *see* D-70396

3-Hydroxymethyl-7,11-dimethyl-2,6,11-dodeca-triene-1,5,10-triol, H-70174

2-Hydroxymethyl-1,1-dimethyloxirane, *see* D-70426

2-Hydroxymethyl-1,3-dioxolane, *see* D-70468

12α-Hydroxy-24-methyl-24,25-dioxo-14-scalaren-22-oic acid, *in* H-60191

2-Hydroxy-4,5-methylenedioxybenzaldehyde, H-70175

2-Hydroxy-3,4-methylenedioxybenzoic acid, H-60177

3-Hydroxy-4,5-methylenedioxybenzoic acid, H-60178

4-Hydroxy-2,3-methylenedioxybenzoic acid, H-60179

6-Hydroxy-2,3-methylenedioxybenzoic acid, H-60180

6-Hydroxy-3,4-methylenedioxybenzoic acid, H-60181

5-Hydroxy-6,7-methylenedioxyflavone, *in* T-60322

7-Hydroxy-3-(3,4-methylenedioxyphenyl)-chromone, *see* B-60006

3-Hydroxy-8,9-methylenedioxypterocarpan, *see* M-70001

7-Hydroxy-4′,5′-methylenedioxypterocarpan (obsol.), *see* M-70001

2-(1-Hydroxy-1-methylethyl)-5-benzofurancarboxaldehyde, *in* D-70214

4-Hydroxy-2-(1-methylethyl)-5-benzofurancarboxylic acid, *see* H-60164

1-[2-(1-Hydroxy-1-methylethyl)-5-benzofuranyl]ethanone, *see* A-70044

2-Hydroxy-2-(1-methylethyl)butanedioic acid, *see* H-60165

2-Hydroxy-3-(1-methylethyl)butanedioic acid, *see* H-60166

2-(1-Hydroxy-1-methylethyl)-5-methyltetra-hydrofuran, *see* P-60160

6-Hydroxymethyleugenin, *see* H-60155

▷1-(Hydroxymethyl)fluorene, H-70176

▷5-Hydroxymethyl-2-furancarboxaldehyde, H-70177

5-Hydroxymethyl-2(5*H*)-furanone, H-60182

5-Hydroxy-4-methyl-2(5*H*)-furanone, *in* M-70103

▷5-Hydroxy-5-methyl-2(5*H*)-furanone, *see* O-70098

▷5-Hydroxymethylfurfural, *see* H-70177

5-(Hydroxymethyl)furfuryl alcohol, *see* B-60160

12-(Hydroxymethyl)-10-hentetraconten-4-ol, *see* N-70068

13-Hydroxymethyl-2,6,10,18,22,26,30-heptamethyl-14-methylene-10-hentriacontene, *see* T-70138

5-Hydroxy-4-methyl-3-heptanone, H-60183

6-Hydroxymethylherniarin, *in* H-70139

2-Hydroxy-2-methylhexanoic acid, H-60184

2-Hydroxy-3-methylhexanoic acid, H-60185

2-(6-Hydroxy-4-methyl-4-hexenylidene)-6,10-dimethyl-7-oxo-9-undecenal, H-60186

6-Hydroxymethylindole, *see* I-60032

6-(Hydroxymethyl)-7-methoxy-2*H*-1-benzopyran-2-one, *in* H-70139

1-[2-(Hydroxymethyl)-3-methoxyphenyl]-1,5-heptadiene-3,4-diol, H-70178

4-Hydroxymethyl-3-methyl-2-buten-1-olide, *see* H-60187

4a-Hydroxy-1-methyl-8-methyleneglobane-1,10-dicarboxylic acid 1,4a-lactone, *see* G-60018

2-(Hydroxymethyl)-5-(1-methylethyl)-1,4-benzenediol, *see* D-70309

5-Hydroxymethyl-4-methyl-2(5*H*)-furanone, H-60187

6-(Hydroxymethyl)-10-methyl-2-(4-methyl-3-pentenyl)-2,6,10-dodecatriene-1,12-diol, H-70179

11-Hydroxy-5-methyl-2-(1-methyl-1,3-penta-dienyl)-8-[2,3,6-trideoxy-3-(dimethylamino)-β-D-*arabino*-hexopyranosyl]-10-[2,3,6-trideoxy-3-(dimethylamino)-3-*C*-methyl-α-L-*lyxo*-hexopyranosyl]-4*H*-anthra[1,2-*b*]-pyran-4,7,12-trione, *see* R-70011

3-(Hydroxymethyl)-2-naphthaldehyde, H-60188

2-(Hydroxymethyl)naphthalene, *see* N-60004

3-(Hydroxymethyl)-2-naphthalenecarboxaldehyde, H-60188

5-Hydroxy-7-methyl-1,2-naphthalenedione, *see* H-60189

8-Hydroxy-3-methyl-1,2-naphthalenedione, *see* H-60190

5-Hydroxy-7-methyl-1,2-naphthoquinone, H-60189

8-Hydroxy-3-methyl-1,2-naphthoquinone, H-60190

2-Hydroxy-5-methyl-3-nitrobenzaldehyde, H-70180

3-(Hydroxymethyl)-7-oxabicyclo[4.1.0]-hept-3-ene-2,5-dione, H-70181

4-Hydroxy-5-methyl-3-(8-oxofarnesyl)-coumarin, *see* H-70183

12-Hydroxy-24-methyl-24-oxo-16-scalarene-22,25-dial, H-60191

3-Hydroxy-2-methylpentanoic acid, H-60192

5-Hydroxy-3-methylpentanoic acid, H-70182

5-Hydroxy-3-methyl-2-penten-4-olide, *see* H-60187

3-[2-[2-(5-Hydroxy-4-methyl-3-pentenyl)-2-methyl-2*H*-1-benzopyran-6-yl]ethyl]-1,2-benzenediol, *see* T-70008

Hydroxymethylphenylacetylene, see P-60130
4-Hydroxy-α-methylphenylalanine, see A-60220
1-[2-(Hydroxymethyl)phenyl]ethanone, see H-70158
1-[4-(Hydroxymethyl)phenyl]ethanone, see H-70159
5-Hydroxy-3-methylphthalic acid, see H-70162
3-Hydroxy-6-methylpyridazine, in M-70124
6-(Hydroxymethyl)tetrahydro-2H-pyran-2-one, in D-60329
2-Hydroxymethyl-1-tetralol, see T-70064
4-Hydroxy-5-methyl-3-(3,8,11-trimethyl-8-oxo-2,6,10-dodecatrienyl)-2H-1-benzopyran-2-one, H-70183
3-Hydroxy-5-methyl-2,4,6-trinitrobenzoic acid, H-60193
(2-Hydroxy-6-methyl-2,4,6-triphenyl-3-cyclohexen-1-yl)phenylmethanone, see D-60524
3-Hydroxy-2-methylvaleric acid, see H-60192
5-Hydroxy-3-methylvaleric acid, see H-70182
5-(Hydroxymethyl)-δ-valerolactone, in D-60329
3-Hydroxymugeneic acid, in M-60148
6-Hydroxy-1-naphthaldehyde, H-70184
6-Hydroxy-2-naphthaldehyde, H-70185
7-Hydroxy-2-naphthaldehyde, H-70186
1-Hydroxy-2-naphthaleneacrylic acid δ-lactone, see N-70011
6-Hydroxy-1-naphthalenecarboxaldehyde, see H-70184
6-Hydroxy-2-naphthalenecarboxaldehyde, see H-70185
7-Hydroxy-2-naphthalenecarboxaldehyde, see H-70186
4-Hydroxy-1,2-naphthalenedione, see H-60194
3-Hydroxynaphtho[a]cyclobutene, see D-60215
4-Hydroxy-1,2-naphthoquinone, H-60194
12-Hydroxynerolidol, in T-70283
13-Hydroxynerolidol, in T-70283
Hydroxyniranthin, H-70187
2-Hydroxy-5-nitro-1,3-benzenedicarboxylic acid, H-70188
3-Hydroxy-6-nitro-1,2-benzenedicarboxylic acid, H-70189
4-Hydroxy-3-nitro-1,2-benzenedicarboxylic acid, H-70190
4-Hydroxy-5-nitro-1,2-benzenedicarboxylic acid, H-70191
4-Hydroxy-5-nitro-1,3-benzenedicarboxylic acid, H-70192
2-Hydroxy-3-nitrobenzotrifluoride, see N-60050
2-Hydroxy-5-nitrobenzotrifluoride, see N-60053
3-Hydroxy-2-nitrobenzotrifluoride, see N-60047
3-Hydroxy-4-nitrobenzotrifluoride, see N-60049
4-Hydroxy-2-nitrobenzotrifluoride, see N-60051
▷4-Hydroxy-3-nitrobenzotrifluoride, see N-60048
2-Hydroxy-5-nitroisophthalic acid, see H-70188
4-Hydroxy-5-nitroisophthalic acid, see H-70192
2-Hydroxy-3-nitro-1,4-naphthalenedione, see H-60195
2-Hydroxy-3-nitro-1,4-naphthoquinone, H-60195
3-Hydroxy-6-nitrophthalic acid, see H-70189
4-Hydroxy-3-nitrophthalic acid, see H-70190
4-Hydroxy-5-nitrophthalic acid, see H-70191
4-Hydroxy-3-nitropyridine, see N-70063
▷N-Hydroxy-N-nitrosoaniline, see H-70208
4-Hydroxy-3-nitrosobenzaldehyde, H-60197
▷N-Hydroxy-N-nitrosobenzenamine, see H-70208
4-Hydroxy-3-nitrosobenzoic acid, H-70193
6-Hydroxy-5-nitro-m-tolualdehyde, see H-70180
▷4-Hydroxynonanoic acid γ-lactone, see D-70245
9-Hydroxy-5-nonenoic acid, H-70194
ent-10β-Hydroxy-20-nor-16-gibberellene-7,19-dioic acid 19,10-lactone, see G-60018

6β-Hydroxynorjuslimdiolone, in N-60061
15-Hydroxy-17-nor-8-labden-7-one, H-60198
3-Hydroxy-30-nor-12,20(29)-oleanadien-28-oic acid, H-60199
3β-Hydroxy-28-noroleana-12,17-dien-16-one, in M-70011
18β-Hydroxy-28-nor-3,16-oleanenedione, in D-60363
3β-Hydroxy-28-norolean-12-en-16-one, see M-70011
6-Hydroxy-7,9-octadecadiene-11,13,15,17-tetraynoic acid, H-70195
13-Hydroxy-9,11-octadecadienoic acid, H-70196
3-Hydroxy-28,13-oleananolide, H-60200
3β-Hydroxy-18α-olean-12-ene-27,28-dioic acid, see C-70002
3-Hydroxy-12-oleanen-29,22-olide, H-70197
Hydroxyoleanonic lactone, in H-70201
3-Hydroxyornithine, see D-70039
3-Hydroxy-1,2,4-oxadiazolidin-5-one, see O-70060
17-Hydroxy-15,17-oxido-16-spongianone, H-60201
6-Hydroxy-3-oxo-8′-apo-ε-caroten-8′-oic acid, see C-60156
2-Hydroxy-α-oxobenzeneacetic acid, see H-70210
3-Hydroxy-α-oxobenzeneacetic acid, see H-70211
4-Hydroxy-α-oxobenzeneacetic acid, see H-70212
2-(7-Hydroxy-4-oxo-4H-1-benzopyran-3-yl)-5-methoxy-2,5-cyclohexadiene-1,4-dione, see B-70175
10-Hydroxy-3-oxoborneol, in H-70173
6-(2-Hydroxy-3-oxobutyl)-7-methoxy-2H-1-benzopyran-2-one, see F-70046
7-Hydroxy-6-oxocarnosic acid, see T-60337
2-[[[(1-Hydroxy-6-oxo-2-cyclohexen-1-yl)carbonyl]oxy]methyl]phenyl β-D-glucopyranoside, see S-70002
4-Hydroxy-1-oxo-2,11(13)-eudesmadien-12,8-olide, H-70198
8-Hydroxy-3-oxo-1,4,11(13)-eudesmatrien-12-oic acid, H-70199
9-Hydroxy-3-oxo-1,4(15),11(13)-eudesmatrien-12,6-olide, H-60202
6-Hydroxy-5-oxo-7,15-fusicoccadien-15-al, H-60203
5β-Hydroxy-2-oxo-1(10),3,11(13)-guaiatrien-12-oic acid, see H-70235
9β-Hydroxy-3-oxo-10(14)-guaien-12,6α-olide, in D-70305
7-Hydroxy-3-oxo-8,24-lanostadien-26-al, H-70200
11-Hydroxy-3-oxo-13-nor-7(11)-eudesmen-12,6-olide, H-60204
12-Hydroxy-3-oxo-28,13-oleananolide, H-70201
19α-Hydroxy-3-oxo-12-oleanen-28-oic acid, in D-70327
26-Hydroxy-28-oxoooligomycin A, see O-70039
N-(5-Hydroxy-2-oxo-7-oxobicyclo[4.1.0]-hept-3-en-3-yl)acetamide, in A-60194
3-Hydroxy-1-oxo-1H-pyrazolo[1,2-a]-pyrazol-4-ium hydroxide inner salt, see O-60092
19α-Hydroxy-3-oxo-12-ursen-28-oic acid, in D-70350
22-Hydroxy-3-oxo-12-ursen-30-oic acid, H-70202
2-Hydroxy-6-(8,11-pentadecadienyl)benzoic acid, H-60205
2-Hydroxy-6-(8,11,14-pentadecatrienyl)-benzoic acid, in H-60205
2-Hydroxy-6-(8-pentadecenyl)benzoic acid, in H-60205
2-Hydroxy-6-(10-pentadecenyl)benzoic acid, in H-60205
4′-Hydroxy-3′,5,5′,6,7-pentamethoxyflavone, H-70081
6-Hydroxy-2′,3,5,5′,7-pentamethoxyflavone, in H-60064
Hydroxypentamethylbenzene, see P-60058
10-Hydroxy-1,2,5,7,9-pentamethyl-3,11,12-trioxatricyclo[5.3.1.1²,⁶]dodecan-4-one, B-70174

20ξ-Hydroxy-3,7,11,15,23-pentaoxo-8-lanosten-26-oic acid, in D-60379
▷4-Hydroxy-2-pentenoic acid γ-lactone, see M-60083
▷4-Hydroxy-3-pentenoic acid γ-lactone, see M-70081
4-Hydroxy-3-penten-2-one, in P-60059
11-Hydroxy-1-[3-(2-pentenyl)oxiranyl]-9-undecenoic acid, see H-70127
1-Hydroxy-6-pentyl-5,15-dioxabicyclo[10.2.1]-pentadecane-4,13-dione, see G-60025
5-Hydroxy-13(3-pentyloxiranyl)-6,8,10,12-tridecatetraenoic acid, see E-70016
13-[3-(5-Hydroxypentyl)oxiranyl]-5,8,11-tridecatrienoic acid, see E-70018
4-Hydroxy-2-pentyne, see P-70041
2-Hydroxy-3H-phenoxazin-3-one, H-60206
(2-Hydroxyphenoxy)acetic acid, H-70203
(3-Hydroxyphenoxy)acetic acid, H-70204
(4-Hydroxyphenoxy)acetic acid, H-70205
2-Hydroxyphenylacetaldehyde, H-60207
3-Hydroxyphenylacetaldehyde, H-60208
4-Hydroxyphenylacetaldehyde, H-60209
2-Hydroxy-2-phenylacetaldehyde, H-70206
2-Hydroxy-2-phenylacetaldehyde, H-60210
4-Hydroxyphenylacetaldoxime, in H-60209
N-Hydroxyphenylacetamide, see P-70059
2-Hydroxy-N-phenylbenzamide, see H-70108
2-(4-Hydroxyphenyl)benzoic acid, see H-70111
3-(3-Hydroxyphenyl)-1H-2-benzopyran-1-one, H-70207
6-Hydroxy-2-phenyl-4H-1-benzopyran-4-one, see H-70130
7-Hydroxy-2-phenyl-4H-1-benzopyran-4-one, see H-70131
6-Hydroxy-2-phenylchromone, see H-70130
7-Hydroxy-2-phenylchromone, see H-70131
▷1-Hydroxy-2-phenyldiazene 2-oxide, H-70208
3-(4-Hydroxyphenyl)-1-(2,4-dihydroxy-6-methoxyphenyl)-1-propanone, in H-60220
5-[2-(3-Hydroxyphenyl)ethenyl]-1,2,4-benzenetriol, see H-70215
6-[2-(4-Hydroxyphenyl)ethenyl]-4-methoxy-2H-pyran-2-one, see H-70122
4-[2-(3-Hydroxyphenyl)ethyl]-1,2-benzenediol, see T-60316
5-[2-[(3-Hydroxyphenyl)ethyl]]-1,2,3-benzenetriol, see T-60102
4-[2-(3-Hydroxyphenyl)ethyl]-2,6-dimethoxyphenol, in T-60102
▷(4-Hydroxyphenyl)ethylene, H-70209
4-Hydroxyphenyl β-D-glucopyranoside, see A-70248
(m-Hydroxyphenyl)glyoxylic acid, see H-70211
(p-Hydroxyphenyl)glyoxylic acid, see H-70212
o-Hydroxyphenylglyoxylic acid, see H-70210
5-(4-Hydroxyphenyl)-2-(4-hydroxy-3-methoxyphenyl)-3,4-dimethyltetrahydrofuran, in T-60061
2-(4-Hydroxyphenyl)-5-(3-hydroxypropyl)-3-methylbenzofuran, see O-60040
3-(3-Hydroxyphenyl)isocoumarin, see H-70207
3-Hydroxy-4-phenyl-5(2H)-isoxazolone, see D-70337
5-Hydroxy-4-phenyl-3(2H)-isoxazolone, see D-70337
5-(2-Hydroxyphenyl)-7-methoxy-2-(3,4-methylenedioxyphenyl)benzofuran, see E-70001
2-(4-Hydroxyphenyl)-7-methoxy-3-methyl-5-propenylbenzofuran, in K-60001
2-(4-Hydroxyphenyl)-7-methoxy-5-(1-propenyl)benzofuran, in H-60214
2-(4-Hydroxyphenyl)-7-methoxy-5-(1-propenyl)-3-benzofurancarboxaldehyde, H-60212
1-(4-Hydroxyphenyl)-2-(2-methoxy-4-propenylphenoxy)-1-propanol, in H-60215
2-(4-Hydroxyphenyl)-3-methyl-5-benzofuranpropanol, see O-60040
2-(4-Hydroxyphenyl)-3-methyl-5-(1-propenyl)benzofuran, see E-60075
2-(2-Hydroxyphenyl)-2-oxoacetic acid, H-70210
2-(3-Hydroxyphenyl)-2-oxoacetic acid, H-70211

2-(4-Hydroxyphenyl)-2-oxoacetic acid, H-70212
1-(2-Hydroxyphenyl)-2-phenylethane, *see* P-70071
1-(3-Hydroxyphenyl)-2-phenylethane, *see* P-70072
1-(4-Hydroxyphenyl)-2-phenylethane, *see* P-70073
3-(4-Hydroxyphenyl)-2-propenal, H-60213
2-(4-Hydroxyphenyl)-5-(1-propenyl)-benzofuran, H-60214
1-(4-Hydroxyphenyl)-2-(4-propenylphenoxy)-1-propanol, H-60215
2-Hydroxy-2-phenylpropiophenone, *see* H-70123
2-Hydroxy-3-phenylpropiophenone, *see* H-70124
1-(4-Hydroxyphenyl)pyrrole, H-60216
2-(2-Hydroxyphenyl)pyrrole, H-60217
2-Hydroxy-4-phenylquinoline, *see* P-60131
4-Hydroxy-2-phenylquinoline, *in* P-60132
8-Hydroxy-7-phenylquinoline, H-60218
5-Hydroxy-2-(phenylthio)-1,4-naphthalenedione, *see* H-70213
5-Hydroxy-3-(phenylthio)-1,4-naphthalenedione, *see* H-70214
5-Hydroxy-2-(phenylthio)-1,4-naphthoquinone, H-70213
5-Hydroxy-3-(phenylthio)-1,4-naphthoquinone, H-70214
1-(3-Hydroxyphenyl)-2-(2,4,5-trihydroxyphenyl)ethylene, H-70215
3-(4-Hydroxyphenyl)-1-(2,4,6-trihydroxyphenyl)-1-propanone, H-60219
3-(4-Hydroxyphenyl)-1-(2,4,6-trihydroxyphenyl)-2-propen-1-one, H-60220
1-Hydroxypinoresinol, H-60221
9-Hydroxypinoresinol, H-70216
2-Hydroxypiperonylic acid, *see* H-60177
6-Hydroxypiperonylic acid, *see* H-60181
Hydroxyproline leucine anhydride, *see* C-60213
Hydroxyprolylproline anhydride, *see* O-60017
3-Hydroxy-2-propenoic acid, *in* O-60090
4-(3-Hydroxy-1-propenyl)-2,6-dimethoxyphenol, *see* S-70046
6-(1-Hydroxypropyl)-1-methyllumazine, *see* L-70029
5′-Hydroxypsorospermin, *in* P-70132
11-Hydroxyptilosarcenone, *in* P-60193
6α-Hydroxypulchellin, *in* P-70138
6β-Hydroxypulchellin 2-O-acetate, *in* P-70138
6α-Hydroxypulchellin 4-O-angelate, *in* P-70138
6β-Hydroxypulchellin 2-O-isovalerate, *in* P-70138
3-Hydroxy-2-pyridinamine, *see* A-60200
3-Hydroxy-4-pyridinamine, *see* A-60202
5-Hydroxy-2-pyridinamine, *see* A-60201
5-Hydroxy-3-pyridinamine, *see* A-60203
▷1-Hydroxy-2(1*H*)-pyridinethione, H-70217
3-Hydroxy-2(1*H*)-pyridinethione, H-60222
2-Hydroxy-4(1*H*)-pyridinone, *in* H-60224
3-Hydroxy-2(1*H*)-pyridinone, H-60223
3-Hydroxy-4(1*H*)-pyridinone, H-60227
4-Hydroxy-2(1*H*)-pyridinone, H-60224
5-Hydroxy-2(1*H*)-pyridinone, H-60225
6-Hydroxy-2(1*H*)-pyridinone, H-60226
2-Hydroxy-1-(2-pyridinyl)ethanone, *see* H-70101
2-Hydroxy-1-(3-pyridinyl)ethanone, *see* H-70102
3-Hydroxypyrrole, *see* D-70262
3-Hydroxypyrrolidine, *see* P-70187
3-Hydroxyquinaldic acid, *see* H-60228
2-Hydroxy-4-quinazolinethione, *see* D-60295
3-Hydroxy-2-quinolinecarboxylic acid, H-60228
8-Hydroxy-9-β-D-ribofuranosylguanine, *see* D-70238
2β-Hydroxyroyleanone, *in* R-60015
7-Hydroxyroyleanone, *see* D-60304
8α-Hydroxysambucoin, *in* S-70007
8β-Hydroxysambucoin, *in* S-70007
8-Hydroxy-15-sandaracopimarene, *see* H-60163
9α-Hydroxysantamarine, *in* S-60007
ent-15-Hydroxy-8,9-seco-13-labdene-8,9-dione, H-70218

1*R*-Hydroxy-3,4-seco-4(23),20(29)-lupadiene-3,28-dioic acid 3,11α-lactone, *see* C-70038
9-Hydroxysemperoside, *in* S-60021
15-Hydroxysolavetivone, *in* S-70051
6-Hydroxy-3-spirostanone, H-70219
19-Hydroxy-13(16),14-spongiadien-3-one, H-70220
13-Hydroxysterpuric acid, *in* S-60053
▷4-Hydroxystyrene, *see* H-70209
α-Hydroxystyrene, *see* P-60092
6-(4-Hydroxystyryl)-4-methoxy-2-pyrone, *see* H-70122
4-Hydroxystyryl 2,4,6-trihydroxyphenyl ketone, *see* H-60220
2α-Hydroxysugiol, *in* D-70281
1-Hydroxysyringaresinol, *in* H-60221
3-Hydroxytaraxastan-28,20-olide, H-70221
2α-Hydroxytaxodione, *in* D-70282
2α-Hydroxytaxodone, *in* D-70282
Hydroxyterephthalic acid, *see* H-70105
6-Hydroxy-2,4,8-tetradecatrienoic acid, H-70222
3-(14-Hydroxy-2,5,8-tetradecatrienyl)-oxiranebutanoic acid, *see* E-70017
12-Hydroxy-13-tetradecenoic acid, H-60229
14-Hydroxy-12-tetradecenoic acid, H-60230
9β-Hydroxy-4β,11β,13,15-tetra-hydrodehydrozaluzanin C, *in* D-70305
3-Hydroxytetrahydrofuranone, *see* D-60242
4-Hydroxy-1,2,3,4-tetrahydroisoquinoline, *see* T-60066
9β-Hydroxy-4β,11β,13,15-tetrahydrozaluzanin C, *in* D-70305
9β-Hydroxy-4β,11α,13,15-tetrahydrozaluzanin C, *in* D-70305
5-Hydroxy-2′,3,7,8-tetramethoxyflavone, *in* P-70029
5-Hydroxy-3,4′,6,7-tetramethoxyflavone, *in* P-60047
5-Hydroxy-3,4′,6,7-tetramethoxyflavone, *in* P-60047
6-Hydroxy-3,4′,5,7-tetramethoxyflavone, *in* P-60047
12-Hydroxy-3,7,11,15-tetramethyl-2,6,10,14-hexadecatetraenoic acid, H-70223
16-Hydroxy-2,6,10,14-tetramethyl-2,6,10,14-hexadecatetraenoic acid, H-70224
2-(5-Hydroxy-3,7,11,15-tetramethyl-2,6,10,14-hexadecatetraenyl)-6-methyl-1,2-benzenediol, H-70225
ent-7β-Hydroxy-13,14,15,16-tetranor-3-cleroden-12-oic acid 18,19-lactone, H-70226
ent-12-Hydroxy-13,14,15,16-tetranor-1(10)-halimen-18-oic acid, H-60231
7β-Hydroxy-3,11,15,23-tetraoxo-8,20(22)-lanostadien-26-oic acid, *in* T-70254
12α-Hydroxy-3,7,11,15-tetraoxo-25,26,27-trisnor-8-lanosten-24-oic acid, *in* L-60037
6-Hydroxy-1,2,4,5-tetrazin-3(2*H*)-one, *see* D-70268
2-Hydroxythiochromone, *see* H-70110
4-Hydroxythiocoumarin, *see* H-70110
3-Hydroxythiophene, *see* T-60212
22-Hydroxy-7,24-tirucalladiene-3,23-dione, H-70227
2-Hydroxy-*p*-tolualdehyde, *see* H-70160
4-Hydroxy-*o*-tolualdehyde, *see* H-70161
α-Hydroxy-*o*-tolualdehyde, *see* H-60171
2-Hydroxy-*p*-toluic acid, *see* H-70163
4-Hydroxy-*o*-toluic acid, *see* H-70164
9β-Hydroxytournefortiolide, *in* H-60137
2-Hydroxy-1,3,5-triazaindene, *see* D-60250
3-Hydroxytrichothecene, *see* E-70028
4-Hydroxy-2-(trifluoromethyl)benzaldehyde, H-70228
4-Hydroxy-3-(trifluoromethyl)benzaldehyde, H-70229
2′-Hydroxy-4,4′,6′-trimethoxychalcone, *in* H-60220
4-Hydroxy-2′,4′,6′-trimethoxychalcone, *in* H-60220
7-Hydroxy-2′,5,8-trimethoxyflavanone, *in* T-70104

5-Hydroxy-2′,6,6′-trimethoxyflavone, *in* T-70107
2-Hydroxy-3,5,7-trimethoxyphenanthrene, *in* T-60120
8-Hydroxy-5,6,10-trimethoxy-2-propyl-4*H*-naphtho[1,2-*b*]pyran-4-one, *in* C-70191
3-Hydroxy 6,8,9-trimethoxypterocarpan, *see* S-60037
2-Hydroxy-5,6,7-trimethoxyxanthone, *in* T-60128
3-Hydroxy-1,2,4-trimethoxyxanthone, *in* T-60126
7-Hydroxy-2,3,4-trimethoxyxanthone, *in* T-60128
8-Hydroxy-1,3,5-trimethoxyxanthone, *in* T-70123
5-Hydroxy-2,8,8-trimethyl-4*H*,8*H*-benzo[1,2-*b*:3,4-*b*′]dipyran-4-one, *see* A-60075
4-Hydroxy-3-[(3,7,11-trimethyl-2,6,10-dodecatrienyl)]-2*H*-1-benzopyran-2-one, *see* F-60005
7-Hydroxy-5,5,9-trimethyl-(1-methylethyl)-1,2,6(5*H*)anthracenetrione, *see* P-70144
12α-Hydroxy-4,4,14α-trimethyl-3,7,11,15-tetra-oxo-5α-chol-8-en-24-oic acid, *in* L-60037
3-Hydroxy-25,26,27-trisnor-24-cycloartanal, H-60232
▷1-Hydroxytropane, *see* P-60151
3-Hydroxytropolone, *see* D-60314
3-Hydroxy-28,20-ursanolide, H-60233
2α-Hydroxyursolic acid, *in* D-70349
11-Hydroxy-1(10)-valencen-2-one, H-60234
8-Hydroxyvariabilin, *in* V-70003
Hydroxyversicolorone, H-70230
6-Hydroxy-5-vinylbenzofuran, H-70231
3-(4-Hydroxy-2-vinylphenyl)propanoic acid, H-70232
9α-Hydroxyzaluzanin B, *in* Z-60001
8′-Hydroxyzearalenone, *in* Z-60002
Hygric acid methylbetaine, *see* S-60050
Hymatoxin *A*, H-60235
Hymenoxin, *see* D-60376
Hyperevoline, H-70233
Hypericorin, H-70234
Hyperlatolic acid, H-60236
▷Hypnone, *see* A-60017
Hypocretenoic acid, H-70235
Hypocretenolide, *in* H-70235
Hypoxanthine 2′-deoxyriboside, *see* D-70023
Hypoxanthine 3′-deoxyriboside, *see* D-70024
Hyptolide, H-70236
▷Hyserp, *in* H-60092
Hythizine, *see* D-60509
Ibaacid, *see* D-60434
Ibanitrile, *in* D-60429
▷Ibericin, *in* D-70306
Icariside B₁, *in* G-70028
Icariside C₁, *in* T-60362
Icariside C₂, *in* T-60362
Icariside C₃, *in* T-60362
Icariside C₄, *in* T-60362
Ichangensin, I-70001
6-Icosyl-4-methyl-2*H*-pyran-2-one, *see* A-60291
Idomain, *in* D-60324
Ilexgenin A, *in* D-60382
Ilexoside A, *in* D-70327
Ilexoside B, *in* D-70350
Ilexsaponin A1, *in* D-60382
Ilexside I, *in* D-70350
Ilexside II, *in* D-70350
Ilimaquinone, I-70002
▷Illudin *M*, I-60001
▷Illudin *S*, I-60002
Illurinic acid, *in* L-60011
Iloprost, I-60003
IM 8443*T*, *in* C-60178
Imberbic acid, *in* D-70325
Imbricatolal, *in* H-70150
Imbricatonol, I-70003
Imidazate, *in* H-70108
Imidazo[*a*,*c*]dipyridinium, *see* D-70501
Imidazo[5,1,2-*cd*]indolizine, I-70004
▷4-Imidazoleacrylic acid, *see* U-70005
1-Imidazolealanine, *see* A-60204

2-Imidazolealanine, *see* A-60205
1*H*-Imidazole-4-carboxylic acid, I-60004
Imidazole salicylate, *in* H-70108
▷2-Imidazolidinone, I-70005
2-Imidazolidone-4-carboxylic acid, *see* O-60074
2-Imidazoline, *see* D-60245
β-(Imidazol-1-yl)-α-alanine, *see* A-60204
β-(Imidazol-2-yl)-α-alanine, *see* A-60205
1*H*-Imidazol-2-ylphenylmethanone, *see* B-70074
1*H*-Imidazol-4-ylphenylmethanone, *see* B-60057
▷3-(1*H*-Imidazol-4-yl)-2-propenoic acid, *see*
  U-70005
1*H*-Imidazo[2,1-*b*]purin-4(5*H*)-one, I-70006
Imidazo[1,2-*a*]pyrazine, I-60005
2*H*-Imidazo[4,5-*b*]pyrazin-2-one, I-60006
Imidazo[1,2-*b*]pyridazine, I-60007
Imidazo[4,5-*c*]pyridazine-6-thiol, *see* D-70215
1*H*-Imidazo[4,5-*c*]pyridin-4-amine, *see* A-70161
▷Imidazo[4,5-*c*]pyridine, I-60008
Imidazo[4,5-*b*]pyridin-2-one, *see* D-60249
1*H*-Imidazo[4,5-*c*]pyridin-4(5*H*)-one, *see* .
  H-60157
1*H*-Imidazo[4,5-*g*]quinazolin-8-amine, *see*
  A-60207
1*H*-Imidazo[4,5-*f*]quinazolin-9-amine, *see*
  A-60206
Imidazo[4,5-*g*]quinazoline-6,8(5*H*,7*H*)-dione,
  I-60009
Imidazo[4,5-*g*]quinazoline-4,8,9(3*H*,7*H*)-trione,
  I-60010
Imidazo[4,5-*h*]quinazolin-6-one, I-60013
Imidazo[4,5-*g*]quinazolin-8-one, I-60012
Imidazo[4,5-*f*]quinazolin-9(8*H*)-one, *in* I-60011
1*H*-Imidazo[4,5-*f*]quinoline, I-70007
1*H*-Imidazo[4,5-*h*]quinoline, I-70008
1*H*-Imidazo[4,5-*c*]tetrazolo[1,5-*a*]pyridine, *see*
  A-60345
5-(1*H*-Imidaz-2-yl)-1*H*-tetrazole, I-70009
Imidocarbonic acid, *see* C-70019
▷*o*-Imidodibenzyl, *see* D-70185
Imidodicarbonic diamide, *see* B-60190
2,2′-Iminobisbenzamide, *in* I-60014
▷2,2′-Iminobisbenzoic acid, *see* I-60014
2,3′-Iminobisbenzoic acid, *see* I-60015
2,4′-Iminobisbenzoic acid, *see* I-60016
3,3′-Iminobisbenzoic acid, *see* I-60017
4,4′-Iminobisbenzoic acid, *see* I-60018
4,4′-Iminobisbenzonitrile, *in* I-60018
N-(4-Imino-2,5-cyclohexadien-1-ylidene)-1,4-
  benzenediamine, *see* I-60019
N-(4-Imino-2,5-cyclohexadien-1-ylidene)-*p*-
  phenylenediamine, *see* I-60019
▷2,2′-Iminodibenzoic acid, I-60014
2,3′-Iminodibenzoic acid, I-60015
2,4′-Iminodibenzoic acid, I-60016
3,3′-Iminodibenzoic acid, I-60017
4,4′-Iminodibenzoic acid, I-60018
▷Iminodibenzyl, *see* D-70185
5-Imino-1,3-dimethyl-1*H*,3*H*-tetrazole, *in*
  A-70202
4-Imino-2-pentanone, *in* P-60059
Iminopropionicacetic acid, *see* C-60016
Indamine, I-60019
Indazolo[3,2-*a*]isoquinoline, I-60020
Indeno[1,2,3-*hi*]chrysene, I-70010
Indeno[1,2-*b*]fluorene-6,12-dione, I-70011
[2.2](4,7)(7,4)Indenophane, I-60021
▷Indeno[1,2,3-*cd*]pyrene, I-70012
▷Indeno[1,2,3-*cd*]pyren-1-ol, I-60022
▷Indeno[1,2,3-*cd*]pyren-2-ol, I-60023
▷Indeno[1,2,3-*cd*]pyren-6-ol, I-60024
▷Indeno[1,2,3-*cd*]pyren-7-ol, I-60025
▷Indeno[1,2,3-*cd*]pyren-8-ol, I-60026
1-(1*H*-Inden-2-yl)ethanone, *see* A-70045
1-(1*H*-Inden-3-yl)ethanone, *see* A-70046
1-(1*H*-Inden-6-yl)ethanone, *see* A-70047
Inden-2-yl methyl ketone, *see* A-70045
Inden-3-yl methyl ketone, *see* A-70046
Inden-6-yl methyl ketone, *see* A-70047
Indicoside *A*, I-60027
Indisocin, I-60028
▷1*H*-Indole-3-acetic acid, I-70013
1*H*-Indole-5-carboxaldehyde, I-60029
1*H*-Indole-6-carboxaldehyde, I-60030

1*H*-Indole-7-carboxaldehyde, I-60031
1*H*-Indole-5,6-diol, *see* D-70307
1*H*-Indole-6-methanol, I-60032
Indoline-2-sulfonic acid, *see* D-70221
Indolizine, I-60033
8*H*-Indolo[3,2,1-*de*]acridine, I-70014
Indolo[2,3-*b*][1]azaazulene, *see* C-60200
Indolo[1,7-*ab*][1]benzazepine, I-60034
Indolo[2,3-*b*][1,4]benzodiazepin-12(6*H*)-one,
  I-70015
1*H*-Indol-6-ol, *see* H-60158
3*H*-Indol-3-one, I-60035
10*H*-Indolo[3,2-*b*]quinoline, I-70016
5*H*-Indolo[2,3-*b*]quinoxaline, I-60036
▷3-Indolylacetic acid, *see* I-70013
2-Indolylacetylene, *see* E-60057
3-Indolylacetylene, *see* E-60058
1-(1*H*-Indol-2-yl)ethanone, *see* A-60032
1-(1*H*-Indol-4-yl)ethanone, *see* A-60033
▷1-(1*H*-Indol-5-yl)ethanone, *see* A-60034
1-(1*H*-Indol-6-yl)ethanone, *see* A-60035
1-(1*H*-Indol-7-yl)ethanone, *see* A-60036
2-*C*-(1*H*-Indol-3-ylmethyl)-β-L-*lyxo*-3-
  hexulofuranosonic acid γ-lactone, *in* A-60309
2-*C*-(1*H*-Indol-3-ylmethyl)-β-L-*xylo*-3-
  hexulofuranosonic acid γ-lactone, *in* A-60309
2-Indolyl methyl ketone, *see* A-60032
4-Indolyl methyl ketone, *see* A-60033
▷5-Indolyl methyl ketone, *see* A-60034
6-Indolyl methyl ketone, *see* A-60035
7-Indolyl methyl ketone, *see* A-60036
▷1*H*-Indol-3-ylphenyliodonium hydroxide inner
  salt, I-70017
6-Indolyl phenyl ketone, *see* B-60058
1*H*-Indol-6-ylphenylmethanone, *see* B-60058
Indophenazine, *see* I-60036
Inermin, *see* M-70001
Ingol, I-70018
myo-Inositol-1,3,4,5-tetrakis(dihydrogen
  phosphate), *see* I-70019
myo-Inositol-1,3,4,5-tetraphosphate, I-70019
myo-Inositol-1,3,4-triphosphate, I-70020
myo-Inositol-2,4,5-triphosphate, I-70021
myo-Inositol-2,4,5-tris(dihydrogen phosphate),
  see I-70021
Inunal, I-70022
Inuviscolide, I-70023
Invictolide, I-60038
2-Iodoacetophenone, I-60039
ω-Iodoacetophenone, *see* I-60039
2-Iodobutanal, I-60040
3-Iodo-2-butanone, I-60041
4-Iodo-1-butyne, I-70024
4-Iodo-*m*-cresol, *see* I-60050
2-Iodocycloheptanone, I-70025
2-Iodocyclohexanone, I-60042
1-Iodocyclohexene, I-70026
2-Iodocyclooctanone, I-70027
2-Iodocyclopentanone, I-60043
2-Iodo-4,5-dimethoxybenzoic acid, *in* D-60338
3-Iodo-2,4-dimethoxybenzoic acid, *in* D-60331
3-Iodo-2,6-dimethoxybenzoic acid, *in* D-60334
3-Iodo-4,5-dimethoxybenzoic acid, *in* D-60335
4-Iodo-2,5-dimethoxybenzoic acid, *in* D-60333
5-Iodo-2,4-dimethoxybenzoic acid, *in* D-60332
6-Iodo-2,3-dimethoxybenzoic acid, *in* D-60330
1-Iodo-5,5-dimethoxypentene, *in* I-70052
2-Iodo-4,5-dimethylaniline, I-70028
4-Iodo-2,3-dimethylaniline, I-70029
4-Iodo-2,5-dimethylaniline, I-70030
4-Iodo-3,5-dimethylaniline, I-70031
2-Iodo-4,5-dimethylbenzenamine, *see* I-70028
4-Iodo-2,3-dimethylbenzenamine, *see* I-70029
4-Iodo-2,5-dimethylbenzenamine, *see* I-70030
4-Iodo-3,5-dimethylbenzenamine, *see* I-70031
1-Iodo-3,3-dimethyl-2-butanone, I-60044
3-Iodo-4,5-dimethylpyrazole, I-70032
4-Iodo-1,3-dimethylpyrazole, *in* I-70049
4-Iodo-1,5-dimethylpyrazole, *in* I-70049
4-Iodo-3,5-dimethylpyrazole, I-70033
5-Iodo-1,3-dimethylpyrazole, *in* I-70050
6-Iodo-1,3-dimethyluracil, *in* I-70057
2-Iododiphenylmethane, I-60045
4-Iododiphenylmethane, I-60046
2-Iodoethyl azide, *see* A-60346

▷Iodoethynylbenzene, *see* I-70055
3-Iodoflavone, *see* I-60059
2-Iodo-3-formylthiophene, *see* I-60068
2-Iodo-4-formylthiophene, *see* I-60073
8-Iodoguanosine, I-70034
6-Iodo-1-heptene, I-70035
7-Iodo-1-heptene, I-70036
1-Iodo-1-hexene, I-70037
6-Iodo-1-hexene, I-70038
6-Iodo-5-hexen-1-ol, I-70039
1-Iodo-1-hexyne, I-70040
6-Iodo-5-hexyn-1-ol, I-70041
4(5)-Iodo-1*H*-imidazole, I-70042
7-Iodo-1*H*-indole, I-70043
▷3-Iodo-2-(iodomethyl)-1-propene, I-60047
4-Iodo-1(3*H*)-isobenzofuranone, I-70044
7-Iodo-1(3*H*)-isobenzofuranone, I-70045
3-Iodo-4-mercaptobenzoic acid, I-70046
Iodomethanesulfonic acid, I-60048
5-Iodo-2-methoxybenzaldehyde, *in* H-70143
1-Iodo-4-methoxy-2-methylbenzene, *in* I-60050
2-Iodo-5-methoxytoluene, *in* I-60050
4-Iodo-3-methylanisole, *in* I-60050
2-Iodo-3-methylbutanal, I-60049
(Iodomethyl)cyclopentane, I-70047
4-Iodo-3-methylphenol, I-60050
Iodomethyl phenyl ketone, *see* I-60039
2-Iodo-2-methylpropanal, I-60051
3(5)-Iodo-4-methylpyrazole, I-70048
4-Iodo-3(5)-methylpyrazole, I-70049
3(5)-Iodo-5(3)-methylpyrazole, I-70050
2-Iodo-6-methyl-3-pyridinol, *see* H-60159
3-Iodo-4-(methylthio)benzoic acid, *in* I-70046
1-(7-Iodo-1-naphthalenyl)ethanone, *see*
  A-60037
1-(8-Iodo-1-naphthalenyl)ethanone, *see*
  A-60038
7-Iodo-1-naphthyl methyl ketone, *see* A-60037
8-Iodo-1-naphthyl methyl ketone, *see* A-60038
5-Iodo-2-nitrobenzaldehyde, I-60052
3-Iodo-2-octanone, I-60053
1-Iodo-1-octene, I-70051
2-Iodopentanal, I-60054
1-Iodo-3-pentanol, I-60055
5-Iodo-2-pentanol, I-60056
5-Iodo-2-pentanone, I-60057
5-Iodo-4-pentenal, I-70052
5-Iodo-4-penten-1-ol, I-60058
5-Iodo-1-penten-4-yne, I-70053
5-Iodo-1-pentyne, I-70054
▷1-Iodo-2-phenylacetylene, I-70055
3-Iodo-2-phenyl-4*H*-1-benzopyran-4-one,
  I-60059
2-Iodo-1-phenyl-1-butanone, I-60060
1,1′-(5-Iodo-1,3-phenylene)bisethanone, *see*
  D-60024
2-Iodo-1-phenylethanone, *see* I-60039
▷1-Iodo-2-phenylethyne, *see* I-70055
1-Iodo-2-(phenylmethyl)benzene, *see* I-60045
1-Iodo-4-(phenylmethyl)benzene, *see* I-60046
(2-Iodophenyl)phenylmethane, *see* I-60045
(4-Iodophenyl)phenylmethane, *see* I-60046
2-Iodo-1-phenyl-2-(phenylsulfonyl)ethanone,
  I-60061
4-Iodophthalide, *see* I-70044
7-Iodophthalide, *see* I-70045
3-Iodopropyl azide, *see* A-60347
2-(3-Iodopropyl)-2-methyl-1,3-dioxolane, *in*
  I-60057
2-Iodo-3-pyridinol, *see* H-60160
▷5-Iodo-2,4(1*H*,3*H*)-pyrimidinedione, I-70056
▷5-Iodo-2,4(1*H*,3*H*)-pyrimidinedione, I-60062
6-Iodo-2,4-(1*H*,3*H*)-pyrimidinedione, I-70057
8-Iodo-9-β-D-ribofuranosylguanine, *see* I-70034
5-Iodosalicylaldehyde, *see* H-70143
2-Iodosobenzeneacetic acid, *see* I-60063
2-Iodosophenylacetic acid, *see* I-60063
2-Iodothiazole, I-60064
4-Iodothiazole, I-60065
5-Iodothiazole, I-60066
3-Iodothiophene, I-60067
2-Iodo-3-thiophenecarboxaldehyde, I-60068
3-Iodo-2-thiophenecarboxaldehyde, I-60069
4-Iodo-2-thiophenecarboxaldehyde, I-60070
4-Iodo-3-thiophenecarboxaldehyde, I-60071

5-Iodo-2-thiophenecarboxaldehyde, I-60072
5-Iodo-3-thiophenecarboxaldehyde, I-60073
2-Iodo-3-thiophenecarboxylic acid, I-60074
3-Iodo-2-thiophenecarboxylic acid, I-60075
4-Iodo-2,3-thiophenecarboxylic acid, I-60076
4-Iodo-3-thiophenecarboxylic acid, I-60077
5-Iodo-2-thiophenecarboxylic acid, I-60078
5-Iodo-3-thiophenecarboxylic acid, I-60079
▷5-Iodouracil, see I-70056
▷5-Iodouracil, see I-60062
6-Iodouracil, see I-70057
5-Iodoveratric acid, in D-60335
6-Iodo-o-veratric acid, in D-60330
Iodovulone I, in B-70278
4-Iodo-2,3-xylidine, see I-70029
4-Iodo-2,5-xylidine, see I-70030
4-Iodo-3,5-xylidine, see I-70031
6-Iodo-3,4-xylidine, see I-70028
γ-Ionone, I-70058
Ipsdienol, see M-70093
Ipsenol, in M-70093
Ircinianin, I-60081
Ircinic acid, I-60082
Ircinin 1, I-60083
Ircinin 2, in I-60083
Iridodial, I-60084
Irisone A, in T-60107
Irisone B, in T-60107
Isatogen, in I-60035
Isatronic acid, see D-60264
ε-Isoactinorhodin, I-70059
▷Isoadenine, see A-60243
Isoafraglaucolide, I-70060
12-Isoagathen-15-oic acid, I-60085
Isoalbizziine, see A-60268
Isoalbrassitriol, in D-70547
Isoambrettolide, I-60086
Δ⁹-Isoambrettolide, see I-60086
Isoambrox, in D-70533
Isoamericanin A, I-60087
Isoaminobisabolenol a, in A-70096
Isoaminobisabolenol b, in A-70096
Isoapressin, I-70061
Isobaccharinol, in B-70003
Isobalearone, in B-60005
Isobellidifolin, in T-70123
1(3H)-Isobenzofuranimine, I-70062
1(3H)-Isobenzofuranthione, I-70063
Isobicyclogermacrenal, I-60088
Isobrasudol, in B-60197
▷Isobutyl methyl ketone, see M-60109
Isobutyric acid 4-isopropyl-2,5-dimethoxybenzyl
   ester, in D-70309
2-Isobutyryl-6,6-dimethyl-4-cyclohexene-1,3-
   dione, see X-70002
1-Isobutyryloxymethyl-4-isopropyl-2,5-
   dimethoxybenzene, in D-70309
Isobutyrylshikonin, in S-70038
Isoceroptene, see D-70112
Isochamaejasmin, I-70064
3-Isochromanone, see D-70176
4-Isochromanone, see B-70042
Isochrysean, see A-60261
Isocostic acid, see E-70061
▷Isocoumarin, see B-60041
Isocoumarin-3-acetic acid, see O-60058
Isocryptomerin, in H-60082
7-Isocyanato-2,10-bisaboladiene, I-60089
7-Isocyanato-7,8-dihydro-α-bisabolene, see
   I-60089
3-Isocyanatopyridine, I-70065
▷Isocyanic acid m-fluorophenyl ester, see F-60048
▷Isocyanic acid o-fluorophenyl ester, see F-60047
▷Isocyanic acid p-fluorophenyl ester, see F-60049
Isocyanic acid pentafluorophenyl ester, see
   P-60035
10α-Isocyanoalloaromadendrane, in A-60328
8-Isocyano-10,14-amphilectadiene, I-60090
3-Isocyano-7,9-bisaboladiene, I-60091
7-Isocyano-3,10-bisaboladiene, I-60092
7-Isocyano-1-cycloamphilectene, I-60093
7-Isocyano-11-cycloamphilectene, I-60094
8-Isocyano-1(12)-cycloamphilectrene, I-60095
7-Isocyano-7,8-dihydro-α-bisabolene, see
   I-60092

11-Isocyano-5-eudesmene, in F-60068
6-Isocyano-4(15)-eudesmene, in F-60069
3-Isocyanotheonellin, see I-60091
▷Isocyanuric acid, see C-60179
Isocyclocalamin, I-60096
Isocymorcin, see I-60122
Isocystoketal, in C-60234
Isodalbergin, I-60097
Isodehydrovaline, see A-60218
Isodictytriol, in D-70148
Isodidymic acid, I-60098
Isodrosopterin, in D-60521
α-Isodurylic acid, see T-60355
Isoeleganonal, in T-60150
Isoelephantopin, I-60099
Isoerivanin, I-70066
Isofasciculatin, in F-70001
Isofirmanoic acid, in D-60472
▷Isoflav, see D-60028
Isoflavone-2-carboxylic acid, see O-60086
Isofloroseselin, in K-70013
Isoglucodistylin, in P-70027
Isogosferol, in X-60002
Isogospherol, in X-60002
Isoguanine, see A-60144
Isohallerin, I-60100
Isohemipic acid, in D-60307
1-Isohistidine, see A-60204
2-Isohistidine, see A-60205
Isohydroxylycopersene, see O-70020
Isoiloprost, in I-60003
1-Isoindolinone, see P-70100
1H-Isoindolin-1-one-3-carboxylic acid, I-60102
Isoinunal, in I-70022
7-Isojasmonic acid, in J-60002
Isojateorin, in C-60160
Isojateorinyl glucoside, in C-60160
Isokidamycin, see R-70012
Isolariciresinol, in I-60139
ent-Isolariciresinol, in I-60139
Isolariciresinol 4′-methyl ether, in I-60139
Isolaserpitin, in D-70009
Isolaureatin, I-70067
Isolecanoric acid, I-60103
Isoligustroside, I-60104
Isolimbolide, I-60105
Isolinderatone, in L-70030
Isolobophytolide, I-60106
Isolophodione, in D-60470
Isoluminol, see A-70134
Isolycoserone, I-70068
Isomammeigin, I-70069
Isomarchantin C, in I-60107
Isomaresiic acid C, I-60108
Isomedicarpin, in D-60373
Isomitomycin A, I-70070
Isomontanolide, in M-70154
Isomugeneic acid, in M-60148
Isomurralonginol, I-70071
Isomyomontanone, in M-60155
Isoneonepetalactone, in N-70027
Isoneriucoumaric acid, in D-70349
▷Isonicoteine, see B-60120
Isonicotinylaminomethane, see A-70087
Isonimbinolide, I-70072
Isonimolicinolide, I-60109
Isonimolide, I-60110
Isonitrosoisopropanoic acid, in O-60090
Isonuatigenin, in S-60044
Isonuezhenide, I-70073
Isoobtusadiene, I-70074
10-Isopentenylemodinanthran-10-ol, I-60111
Isoperezone, I-70075
Isophleichrome, in P-70098
Isophloionolic acid, in T-60334
Isophthalanil, see P-60112
Isopicropolin, I-60112
7,15-Isopimaradiene-3,18-diol, I-70076
7,15-Isopimaradiene-3,19-diol, I-70077
7,15-Isopimaradiene-18,19-diol, I-70078
8,15-Isopimaradiene-7,18-diol, I-70079
8(14),15-Isopimaradiene-2,18-diol, I-70113
8(14),15-Isopimaradiene-3,18-diol, I-70080
7,15-Isopimaradiene-2,3,19-triol, I-70081
8,15-Isopimaradiene-3,7,19-triol, I-70082

Isopiperitenone, see I-70088
Isoplatanin, see T-70117
Isoporphobilinogen, I-60114
Isoprene cyclic sulfone, in D-70231
α-Isoprene sulfone, in D-70228
Isopristimerin III, I-70083
24-Isopropenyl-7-cholestene-3,6-diol, I-70084
24-Isopropenyl-7-cholesten-3-ol, I-70085
1-Isopropenylcyclohexene, I-60115
3-Isopropenylcyclohexene, I-60116
4-Isopropenylcyclohexene, I-60117
2-Isopropenyl-6,10-dimethylspiro[4.5]dec-6-ene,
   see P-70114
4-Isopropenyl-1-methyl-1,2-cyclohexanediol,
   I-70086
4-Isopropenyl-1-methylcyclohexene, I-70087
6-Isopropenyl-3-methyl-2-cyclohexen-1-one,
   I-70088
6-Isopropenyl-3-methyl-3,9-decadien-1-ol
   acetate, in I-70089
6-Isopropenyl-3-methyl-9-decen-1-ol, I-70089
Isopropenyloxirane, see M-70080
▷Isopropylacetone, see M-60109
α-Isopropylaspartic acid, see A-60212
β-Isopropylaspartic acid, see A-70162
Isopropyl azide, see A-70289
2-Isopropylcycloheptanone, I-70090
4-Isopropylcycloheptanone, I-70091
1-Isopropylcyclohexene, I-60118
3-Isopropylcyclohexene, I-60119
4-Isopropylcyclohexene, I-60120
Isopropylcyclopentane, I-70092
4-Isopropyl-1,7-dimethylbicyclo[3.2.1]-oct-6-
   ene-6,8-dicarboxaldehyde, see H-60016
8-Isopropyl-1,3-dimethyltricyclo[4.4.0.0²·⁷]dec-
   3-ene, see C-70197
2-Isopropylideneadamantane, I-60121
4-Isopropylidenecrotonic acid, see M-70086
2,3-O-Isopropylideneglyceraldehyde, see
   D-70400
2-Isopropylmalic acid, see H-60165
3-Isopropylmalic acid, see H-60166
2-Isopropyl-5-methyl-1,3-benzenediol, I-60122
2-Isopropyl-6-methyl-1,4-benzenediol, I-60123
3-Isopropyl-6-methyl-1,2-benzenediol, I-60124
5-Isopropyl-2-methyl-1,3-benzenediol, I-60125
5-Isopropyl-3-methyl-1,2-benzenediol, I-60126
5-Isopropyl-4-methyl-1,3-benzenediol, I-60127
3-Isopropyl-6-methylcatechol, see I-60124
4-Isopropyl-1-methyl-4-cyclohexene-1,2-diol,
   I-70094
4-Isopropyl-1-methyl-3-cyclohexenene-1,2-diol,
   I-70093
2-Isopropyl-5-methyl-4-cyclohexen-1-one,
   I-60128
7-Isopropyl-4-methyl-10-methylene-5-cyclo-
   decene-1,4-diol, see G-70009
8-Isopropyl-5-methyl-2-
   naphthalenecarboxaldehyde, I-60129
3-Isopropyl-2-naphthol, I-70095
4-Isopropyl-1-naphthol, I-70096
3-Isopropyloxindole, I-70097
14-Isopropyl-8,11,13-podocarpatrien-13-ol, see
   T-70204
2-Isopropyl-1,3,5-trimethylbenzene, I-60130
7-Isopropyl-3,3,4-trimethylnaphtho[2,3-b]-
   furan-2,5,6(3H)-trione, see B-70089
Isopropyl vinyl carbinol, see M-70107
Isopseudocyphellarin A, I-70098
Isopteryxin, in K-70013
4,5-Isoquinolinediamine, see D-70040
5,8-Isoquinolinediamine, see D-70041
▷5,8-Isoquinolinedione, see I-60131
1,3(2H,4H)-Isoquinolinedione, I-70099
▷5,8-Isoquinolinequinone, I-60131
Isoquinoline Reissert compound, see B-70073
3-Isoquinolinethiol, see I-60133
1(2H)-Isoquinolinethione, I-60132

3(2*H*)-Isoquinolinethione, I-60133
1-(1-Isoquinolinyl)ethanone, *see* A-60039
1-(3-Isoquinolinyl)ethanone, *see* A-60040
1-(4-Isoquinolinyl)ethanone, *see* A-60041
1-(5-Isoquinolinyl)ethanone, *see* A-60042
1-Isoquinolinyl methyl ketone, *see* A-60039
3-Isoquinolinyl methyl ketone, *see* A-60040
4-Isoquinolinyl methyl ketone, *see* A-60041
5-Isoquinolinyl methyl ketone, *see* A-60042
1-(1-Isoquinolinyl)-1-(2-pyridinyl)ethanol,
　I-60134
1-Isoquinolyl phenyl ketone, *see* B-60059
1-Isoquinolylphenylmethanone, *see* B-60059
3-Isoquinolylphenylmethanone, *see* B-60060
4-Isoquinolylphenylmethanone, *see* B-60061
Isorengyol, *in* H-60133
Isoriccardin C, I-60135
Isosamidin, *in* K-70013
Isosarcophytoxide, I-70100
Isoschizandrin, I-70101
3-Isoselenochromanone, *see* D-60204
Isosepiapterin, *in* S-70035
Isoserine, *see* A-70160
Isosilybin, I-70102
Isosilychristin, I-70103
Isosphaeric acid, I-60136
Isospongiadiol, I-60137
Isosulochrin, *in* S-70083
Isotanshinone IIB, I-60138
Isotaxiresinol, I-60139
3-Isotellurochromanone, *see* D-60205
Isoterebentine, *see* I-70087
Isothamnosin A, I-70104
Isothamnosin B, *in* I-70104
3-Isothiochromanone, *see* D-60206
10α-Isothiocyanatoalloaromadendrane, *in*
　A-60329
▷2-Isothiocyanato-2-methylpropane, I-70105
1-Isothiocyanato-5-(methylsulfinyl)pentene, *in*
　I-60140
1-Isothiocyanato-5-(methylthio)pentane,
　I-60140
2-Isothiocyanatopyridine, I-70106
3-Isothiocyanatopyridine, I-70107
▷Isothiocyanic acid, *tert*-butyl ester, *see* I-70105
11-Isothiocyano-5-eudesmene, *in* F-60068
6-Isothiocyano-4(15)-eudesmene, *in* F-60069
(±)-Isothreonine, *in* A-60183
D-Isothreonine, *in* A-60183
Isotrichodermin, *in* E-70028
Isotubaic acid, *see* H-60164
6β-Isovaleroyloxy-8,13*E*-labdadiene-7α,15-diol,
　in L-60003
8-Isovaleryloxy-2-(2,4-hexadiynylidene)-1,6-
　dioxaspiro[4.5]dec-3-ene, *in* H-70042
Isovaline, *see* A-70167
Isoviolanthrene, *in* B-60038
Isoviolanthrene *A*, *see* B-60038
Isoviolanthrene *B*, *see* B-60037
Isoxantholumazine, *see* P-70135
▷Isoxanthopterin, I-60141
5-Isoxazolecarboxylic acid, I-60142
3-Isoxazolylphenylmethanone, *see* B-70075
4-Isoxazolylphenylmethanone, *see* B-60062
5-Isoxazolylphenylmethanone, *see* B-60063
Isozaluzanin C, *in* Z-60001
Istanbulin F, *in* D-60320
ITF 182, *in* H-70108
Ivangulic acid, I-70108
Ivangustin, I-70109
Ixocarpanolide, I-60143
Jaborol, J-60001
Jacoumaric acid, *in* D-70349
Jaeschkeanadiol, J-70001
Jaeschkeanadiol, *in* D-60004
Jaffe's base, *see* D-60246
Jasmonic acid, J-60002
Jateorin, *in* C-60160
Jateorinyl glucoside, *in* C-60160
Jatrogrossidione, J-70002
Jewenol A, J-70003
Jewenol B, J-70004
▷Judaicin, *see* V-60017
Juglorin, J-70005
Jurinelloide, J-60003
Juslimtetrol, *see* R-60011

4-*epi*-Juslimtetrol, *in* R-60011
Justicidin A, *in* D-60493
Justicidin B, *in* D-60493
Juvenile hormone III, *in* T-60364
K 13, K-70001
K 818*A*, *see* C-60137
K 818*B*, *in* C-60137
K 259-2, *see* E-70038
Kachirachirol B, K-60001
Kachirachirol A, *in* K-60001
Kahweol, K-60002
▷Kainic acid, K-60003
▷α-Kaininic acid, *see* K-60003
Kakkatin, *in* T-60325
Kamebacetal A, K-70002
Kamebacetal B, *in* K-70002
Kamebakaurin, *see* T-70114
Kamebakaurinin, *see* T-70115
Kaneric acid, *in* D-60383
Kanjone, K-70003
Kanshone *A*, K-70004
Kanshone *B*, K-70005
Kanshone *C*, K-70006
Karanjic acid, *see* H-60103
Karanjol, *see* H-70106
Kasuzamycin B, K-70007
ent-16*S*-Kauran-17-al, K-70008
ent-16*S*-Kaurane-17,19-dial, *in* K-70008
ent-2α,16β,17-Kauranetriol, K-60004
ent-3β,16β,17-Kauranetriol, K-60005
ent-16*R*-Kauran-17-oic acid, *in* K-70008
ent-16*S*-Kauran-17-oic acid, *in* K-70008
ent-11-Kaurene-16β,18-diol, K-60006
ent-15-Kaurene-3β,17-diol, K-60007
ent-16-Kaurene-7α,15β-diol, K-70009
ent-15-Kauren-17-oic acid, K-60008
ent-16-Kauren-3β-ol, K-70010
ent-Kauren-15-one-18-oic acid, *see* O-60075
Kautschine, *see* I-70087
Kayamycin, K-60009
Kerlinic acid, K-70011
3-Keto-2,3-dihydrothianaphthene, *see* B-60048
α-Keto-β,β-dimethyl-γ-butyrolactone, *see*
　D-70196
Ketopantoic acid, *see* H-60124
Ketopantolactone, *see* D-70196
4-Ketopipecolic acid, *see* O-60089
2-Ketopiperazine, *see* P-70101
1-Ketopyrrolizidine, *see* H-60057
2-Ketopyrrolizidine, *see* T-60091
3-Ketopyrrolizidine, *see* H-60058
Ketosantalic acid, K-70012
Khellactone, K-70013
▷Khellin, K-60010
Khellinone, K-60011
Kickxioside, K-70014
Kielcorin B, K-60012
Koanoadmantic acid, *in* H-70224
Kokoondiol, *in* D-70301
Kokoonol, *in* D-70301
Kokoononol, *in* D-70301
Kolaflavanone, *in* G-70005
Kolavenic acid, K-60013
Kolaviron, *in* G-70005
▷Kriptofix 222, *see* H-60073
KS 619-1, *see* T-70095
Kurospongin, K-70015
Kurubasch aldehyde, K-60014
Kurubasch aldehyde benzoate, *in* K-60014
Kurubasch aldehyde vanillate, *in* K-60014
Kurubashic acid angelate, *in* K-60014
Kurubashic acid benzoate, *in* K-60014
Kuwanon *J*, K-60015
Kuwanone *L*, K-70016
Kuwanone K, *in* K-70016
Kuwanon Q, *in* K-60015
Kuwanon R, *in* K-60015
Kuwanon V, *in* K-60015
Kwakhurin, K-60016
Kytta-Gel, *in* H-70108
L 363586, *see* C-60180
L 660631, *see* A-70224
8(17),13-Labdadiene-3,15-diol, L-70001
8(17),14-Labdadiene-3,13-diol, L-70002
ent-7,13*E*-Labdadiene-3β,15-diol, L-60001
ent-7,14-Labdadiene-2α,13-diol, L-70003
ent-8,13-Labdadiene-7α,15-diol, L-70004

ent-8,13-Labdadiene-7β,15-diol, *in* L-70004
ent-8,13(16)-Labdadiene-14,15-diol, L-70005
ent-8(17),13-Labdadiene-7α,15-diol, L-70006
ent-8(17),13-Labdadiene-9α,15-diol, L-70007
11,13(16)-Labdadiene-6,7,8,14,15-pentol,
　L-70008
8,13-Labdadiene-2,6,7,15-tetrol, L-60002
11,13-Labdadiene-6,7,8,15-tetrol, L-70009
8,13-Labdadiene-6,7,15-triol, L-60003
ent-7,14-Labdadiene-2α,3α,13-triol, L-70010
ent-7,14-Labdadiene-2α,13,20-triol, L-70011
8(17),12-Labdadien-3-ol, L-70012
ent-7,12*E*,14-Labdatriene-6β,17-diol, L-60004
ent-7,11*E*,14-Labdatriene-6β,13ξ,17-triol,
　L-60005
7-Labdene-15,17-dioic acid, L-60006
7-Labdene-15,18-dioic acid, L-60007
13*S*-Labd-8(17)-ene-15,19-dioic acid, *see*
　O-60039
14-Labdene-8α,13*R*-diol, *in* S-70018
14-Labdene-8α,13*S*-diol, *see* S-70018
8-Labdene-3,7,15-triol, L-70013
ent-14-Labden-8β-ol, L-70014
Laccaic acid *D*, *see* T-60332
Lachnellulone, L-70015
Lactarorufin *D*, L-70016
Lactarorufin E, *in* L-70016
6-Lactoyl-7,8-dihydropterin, *see* S-70035
Ladibranolide, L-60008
Laferin, L-60009
Lagerstronolide, L-60010
▷Lagiotase, *see* E-70007
Lambertianic acid, L-60011
▷Lampterol, *see* I-60002
Lanceolin, *in* P-60052
Lanceoside, *in* P-60052
8,24-Lanostadiene-3,21-diol, L-60017
7,9(11)-Lanostadiene-3,24,25,26-tetrol, L-60012
Lansilactone, L-60013
Lansisterol A, *in* M-60070
Lapidin, *in* L-60014
Lapidol, L-60014
Lapidolin, *in* L-60015
Lapidolinin, *in* L-60015
Lapidolinol, L-60015
Lappaphen a, L-70018
Lappaphen b, *in* L-70018
Lardolure, *in* T-60372
Lariciresinol, L-70019
Larreantin, L-70020
Lascrol, *in* L-60016
Laserpitine, L-60016
Laserpitinol, *in* L-60016
Lasianthin, L-60017
Lateriflorol, *in* E-70030
Lathodoratin, *see* E-70040
Latrunculin *A*, L-70021
Latrunculin *B*, L-70022
Latrunculin *D*, L-70023
Latrunculin C, *in* L-70023
Laurencenone *A*, L-70024
Laurencenone *B*, L-70025
Laurencenone *D*, L-70026
Laurencenone C, *in* L-70025
Laurenene, L-60018
Laurenobiolide, *in* H-70134
▷Leaf aldehyde, *in* H-60076
Lehmanin, *see* L-70027
Lehmannin, L-70027
Leiopathic acid, *in* H-70125
Lepidopterans, L-70028
Lepidopterin, L-60020
Lespedazaflavone *B*, L-60021
Lespedezaflavanone *A*, L-60022
Leucettidine, L-70029
tert-Leucine, *see* A-70136
N-Leucyl-3-nitroglutamic acid, *see* N-70055
▷Leukerin, *see* D-60270
Leukotriene A₄, *in* E-60015
▷Levocarmitine, *in* C-70024
Levulinaldehyde, *see* O-60083
Levulinic aldehyde, *see* O-60083
Licarin C, L-60024
Licoricidin, L-60025
Ligulatin A, *in* D-70025
Ligulatin C, *in* D-70025

Ligustilide, L-60026
Lilacin, *in* S-70046
Limonene, *see* I-70087
▷ Linarigenin, *see* D-70319
Linarin, *in* D-70319
Linderatone, L-70030
Linderazulene, L-70031
Lineatin, L-60027
▷ Liothyronine, *in* T-60350
Liothyronine sodium, *in* T-60350
Liotrix, *in* T-60350
Lipase cofactor, *see* C-70188
Lipiarmycin, L-60028
Lipiarmycin A_3, *in* L-60028
Lipiarmycin A_4, *in* L-60028
Lipid A, L-70032
▷ Lipoxamycin, L-70033
Lipoxin B, *in* T-60320
8-*trans*-Lipoxin B, *in* T-60320
(14S)-8-*trans*-Lipoxin B, *in* T-60320
Lipoxin A_5, *in* T-70255
Lipoxin B_5, *in* T-70256
Liquidambin, L-60029
Lircal, *in* O-70102
Lithospermoside, L-70034
Litoralon, *see* G-60027
Lixetone, *see* A-70032
LL C10037α, *in* A-60194
LL E33288γ_1, *see* C-70008
LL F42248α, *see* P-70203
Lonapalene, *in* C-60118
Lonchocarpol A, L-60030
Lonchocarpol C, L-60031
Lonchocarpol D, L-60032
Lonchocarpol E, L-60033
Lonchocarpol B, *in* L-60030
Longikaurin A, L-70035
Longikaurin B, *in* L-70035
Longikaurin C, *in* L-70035
Longikaurin D, *in* L-70035
Longikaurin E, *in* L-70035
Longikaurin F, *in* L-70035
Longipesin, L-60034
Longirabdosin, L-70036
Longispinogenin, *in* O-60038
Lophirone A, L-60035
Lophodione, *in* D-60470
▷ Lopress, *in* H-60092
Loroglossin, L-70037
Lotisoflavan, L-70038
Loxodellonic acid, L-70039
Lp-2, *see* A-70120
LTA₄, *in* E-60015
Lucidenic acid H, L-60036
Lucidenic acid M, L-60037
Lucidenic acid I, *in* L-60036
Lucidenic acid J, *in* L-60036
Lucidenic acid K, *in* L-60037
Lucidenic acid L, *in* L-60037
▷ Lucidin, *see* D-70306
Luffariellolide, L-70040
Lukes-Šorm dilactam, *see* D-70260
Lumazine, *see* P-60191
Luminamicin, L-60038
Lumiyomogin, L-70041
12,20(29)-Lupadiene-3,27,28-triol, L-70042
20(29)-Lupene-3,16-diol, L-60039
13(18)-Lupen-3-ol, L-70043
LXA₅, *in* T-70255
LX-B, *in* T-60320
LXB₅, *in* T-70256
8-*trans*-LXB, *in* T-60320
(14S)-8-*trans*-LXB, *in* T-60320
Lycopadiene, L-60040
Lycopene, L-60041
Lycopersene, L-70044
Lycopersiconolide, L-60042
Lycopine, L-60041
Lycoserone, L-70045
Lycoxanthol, L-70046
M 167906, *in* P-60197
M 95464, *see* P-60197
▷ MA 321A₃, *see* P-60162
Maackiain, M-70001
Maackiasin, M-60001
Macelignan, *in* A-60320
Machilin B, M-60002

Machilin C, M-60003
Machilin H, M-70002
Machilin I, M-70003
Machilin A, *in* A-70270
Machilin D, *in* M-60003
Machilin E, *in* M-60003
Machilin F, *in* A-70271
Machilin G, *in* A-70271
Macrocliniside B, *in* Z-60001
Macrocliniside I, *in* Z-60001
Macrophylloside C, M-70004
Macrophylloside A, *in* M-70004
Macrophylloside B, *in* M-70004
Maesanin, M-60004
MAG 2, *in* O-70102
▷ Magenta I, *in* R-60010
Magireol A, M-60005
Magireol B, M-60006
Magireol C, *in* M-60006
Magnesone, *in* O-70102
Magnoshinin, M-60007
Majucin, M-70005
Majusculone, M-70006
14(26),17,21-Malabaricatrien-3-ol, M-60008
14(26),17E,21-Malabaricatrien-3-one, *in*
     M-60008
Malabaricone A, M-70007
Malabaricone B, *in* M-70007
Malabaricone C, *in* M-70007
Malabaricone D, *in* M-70007
Malabarolide, M-70008
Malonaldehydic acid, *see* O-60090
Malyngamide C, M-60009
Mandelaldehyde, *see* H-70206
Mandelaldehyde, *see* H-60210
Manicone, *in* D-70424
Mannopine, *in* M-70009
Mannopinic acid, M-70009
Manwuweizic acid, M-70010
Maquiroside A, *in* C-60168
Maragenin I, M-70011
Maragenin II, *in* M-70011
Maragenin III, *in* M-70011
Mariesiic acid A, M-70012
Mariesiic acid B, M-60010
Mariesiic acid C, M-60011
Marrubiin, M-70013
Marsupol, *see* B-70147
Matsukaze lactone, *in* B-70093
Maximaisoflavone B, *in* B-60006
Maximaisoflavone D, *in* T-70112
Maximaisoflavone E, *in* T-70112
Mbamichalcone, M-70014
Meciadanol, *in* P-70026
Medigenin, M-70015
Medinin, *in* M-70015
Medioresinol, M-70016
▷ Medullin, *in* D-60367
7-Megastigmene-5,6,9-triol, M-60012
Meglumine salicylate, *in* H-70108
Meijicoccene, M-70017
Melampodin D, *in* M-60013
Melcanthin F, M-60014
Melianolone, M-60015
Melisodoric acid, M-70018
Melissodoric acid, *see* M-70018
Melittoside, M-70019
Melongoside B, *in* S-60045
▷ Meloxine, *see* X-60001
Membranolide, M-60016
Menisdaurin, *in* L-70034
1,8-p-Menthadiene, *see* I-70087
p-Mentha-1,8-dien-3-one, *see* I-70088
p-Menth-3-ene-1,2-diol, *see* I-70093
p-Menth-4-ene-1,2-diol, *see* I-70094
p-Menth-8-ene-1,2-diol, *see* I-70086
p-Menth-1(6)-en-3-one, *see* I-60128
Mercaptoacetone, *see* M-60028
4′-Mercaptoacetophenone, *see* M-60017
α-Mercaptobenzeneacetic acid, *see* M-60024
α-Mercaptobenzenepropanoic acid, *see* M-60026
2-Mercaptobenzophenone, M-70020
3-Mercaptobenzophenone, M-70021
4-Mercaptobenzophenone, M-70022
2-Mercapto-1,4-benzothiazine, *see* B-70055
4-Mercapto-1,2,3-benzotriazine, *see* B-60050

3-Mercaptobutanoic acid, M-70023
4-Mercapto-2-butenoic acid, M-60018
3-Mercaptocyclobutanol, M-60019
4-Mercaptocyclohexanecarbonitrile, *in* M-60020
4-Mercaptocyclohexanecarboxylic acid,
     M-60020
2-Mercaptocyclohexanol, M-70024
2-Mercaptocyclohexanone, M-70025
3-Mercaptocyclohexanone, M-70026
4-Mercaptocyclohexanone, M-70027
3-Mercaptocyclopentanecarboxylic acid,
     M-60021
2-Mercaptodibenzothiophene, *see* D-60078
4-Mercaptodibenzothiophene, *see* D-60079
2-Mercapto-4,6-dimethylpyrimidine, *see*
     D-60450
2-Mercapto-1,1-diphenylethanol, M-70028
2-Mercaptoethanesulfinic acid, M-60022
2-(2-Mercaptoethyl)pyridine, *see* P-60239
4-(2-Mercaptoethyl)pyridine, *see* P-60240
3-Mercaptoflavone, *see* M-60025
8-Mercaptoguanosine, M-70029
α-Mercaptohydrocinnamic acid, *see* M-60026
6-Mercaptoimidazo[4,5-c]pyridazine, *see*
     D-70215
1-Mercaptoisoquinoline, *see* I-60132
3-Mercaptoisoquinoline, *see* I-60133
3-Mercapto-2-(mercaptomethyl)-1-propene, *see*
     M-70076
4-Mercapto-2-methyl-1,2,3-benzotriazinium
     hydroxide inner salt, *in* B-60050
2-Mercapto-3-methylbutanoic acid, M-60023
▷ 3-Mercapto-2-methylfuran, *see* M-70082
2-Mercaptomethyl-2-methyl-1,3-propanedithiol,
     M-70030
α-(Mercaptomethyl)-α-phenylbenzenemethanol,
     see M-70028
2-(Mercaptomethyl)-1-propene-3-thiol, *see*
     M-70076
2-(Mercaptomethyl)pyrrolidine, *see* P-70185
2-Mercapto-5-(methylthio)-1,3,4-thiadiazole, *in*
     M-70031
2-Mercapto-1-naphthylamine, *see* A-70172
8-Mercapto-1-naphthylamine, *see* A-70173
9-Mercaptophenanthrene, *see* P-70049
2-Mercapto-2-phenylacetic acid, M-60024
α-Mercaptophenylacetic acid, *see* M-60024
4-Mercaptophenylalanine, *see* A-60214
β-Mercaptophenylalanine, *see* A-60213
3-Mercapto-2-phenyl-4H-1-benzopyran-4-one,
     M-60025
1-Mercapto-2-phenylethane, *see* P-70069
1-(4-Mercaptophenyl)ethanone, *see* M-60017
(2-Mercaptophenyl)phenylmethanone, *see*
     M-70020
(3-Mercaptophenyl)phenylmethanone, *see*
     M-70021
(4-Mercaptophenyl)phenylmethanone, *see*
     M-70022
2-Mercapto-3-phenylpropanoic acid, M-60026
▷ 2-Mercaptopropanoic acid, M-60027
1-Mercapto-2-propanone, M-60028
▷ Mercaptopurine, *see* D-60270
6-Mercaptopurine, *in* D-60270
2-Mercaptopyridine, *see* P-60223
3-Mercapto-2(1H)-pyridinethione, *see* P-60218
3-Mercapto-4(1H)-pyridinethione, *see* P-60221
4-Mercapto-2(1H)-pyridinethione, *see* P-60219
5-Mercapto-2(1H)-pyridinethione, *see* P-60220
3-Mercaptoquinoline, *see* Q-60003
4-Mercaptoquinoline, *see* Q-60006
5-Mercaptoquinoline, *see* Q-60004
6-Mercaptoquinoline, *see* Q-60005
8-Mercapto-9-β-D-ribofuranosylguanine, *see*
     M-70029
▷ 6-Mercapto-9-β-D-ribofuranosyl-9H-purine, *see*
     T-70180
5-Mercapto-1H-tetrazole, *in* T-60174
▷ 2,5-Mercapto-1,3,4-thiadiazole, *see* M-70031
▷ 5-Mercapto-1,3,4-thiadiazoline-2-thione,
     M-70031
2-Mercaptothiazole, *see* T-60200
▷ 3-Mercapto-1,2,4-thiazole, *see* D-60298
1-Mercapto-2-(trifluoromethyl)benzene, *see*
     T-60284
1-Mercapto-3-(trifluoromethyl)benzene, *see*
     T-60285

1-Mercapto-4-(trifluoromethyl)benzene, *see*
　T-60286
▷Mesitene lactone, *see* D-70435
　Mesotan, *in* H-70108
▷Mesotol, *in* H-70108
　Mesuaferrol, M-70032
　Mesuxanthone B, *see* T-60349
▷Mesyl azide, *see* M-60033
　Mesyltriflone, *see* M-70133
　Metachromin A, M-70033
　Metachromin B, M-70034
　Metacrolein, *in* P-70119
　[4]Metacyclophane, M-60029
　[3⁴,¹⁰][7]Metacyclophane, M-60030
　[2.0.2.0]Metacyclophane, M-60031
　Metaphin (as disodium salt), *in* G-60032
▷Methaneselenal, *see* S-70029
　Methaneselenoamide, *in* M-70035
　Methaneselenoic acid, M-70035
　Methanesulfenyl thiocyanate, M-60032
▷Methanesulfonyl azide, M-60033
　Methanesulfonyl peroxide, *see* B-70154
　Methanetetrapropanoic acid, M-70036
　Methanethial, M-70037
▷Methanimine, M-70038
　1,6-Methano[18]annulene, *see* B-70101
　4,10b-Methano-8H-benzo[ab]cyclodecen-8-one,
　　see M-70042
　10,11-Methano-1H-benzo[5,6]-cyclo-
　　octa[1,2,3,4-def]fluorene-1,14-dione,
　　M-70039
　4b,10b-Methanochrysene, M-70040
　2H,5H-(Methanodioxymethano)-3,4,1,6-
　　benzodioxadiazocine, *see* T-60162
　15,20-Methano-1,5-(ethano[1,6]-cyclo-
　　decethano)naphthalene, *see* N-70005
　3a,7a-Methano-1H-indole, M-70041
　6a,9b-Methano-1H-phenalen-1-one, M-70042
　1,4-Methano-1,2,3,4-tetrahydronaphthalene,
　　M-60034
▷Methiodal sodium, *in* I-60048
　Methionine sulfoximine, M-70043
　N-Methionylalanine, M-60035
▷Methoxa-Dome, *see* X-60001
▷Methoxsalen, *see* X-60001
　12-Methoxy-8,11,13-abietatrien-20-oic acid,
　　M-60036
　4-Methoxyacenaphthene, *in* H-60099
　4-Methoxyacenaphthylene, *in* H-70100
　2-(Methoxyacetyl)pyridine, *in* H-70101
　2-(Methoxyacetyl)thiophene, *in* H-70103
　16β-Methoxyalisol B monoacetate, *in* A-70082
　3-Methoxybenz[a]anthracene, *in* H-70104
　2-Methoxybenzeneacetaldehyde, *in* H-60207
　α-Methoxybenzeneacetaldehyde, *in* H-60210
　2-Methoxy-1,3-benzenedicarboxylic acid, *in*
　　H-60101
　2-Methoxy-1,4-benzenedicarboxylic acid, *in*
　　H-70105
　1-Methoxy-1,2-benziodoxol-3(1H)one, *in*
　　H-60102
　6-Methoxy-1,3-benzodioxole-5-carboxaldehyde,
　　in H-70175
　5-Methoxy-1,3-benzodioxole-4-carboxylic acid,
　　in H-60180
　6-Methoxy-1,3-benzodioxole-5-carboxylic acid,
　　in H-60181
　7-Methoxy-1,3-benzodioxole-4-carboxylic acid,
　　in H-60179
　3-(6-Methoxy-1,3-benzodioxol-5-yl)-2-propenal,
　　see M-70048
　4-Methoxybenzofuran, *in* H-70106
　4-Methoxy-5-benzofurancarboxylic acid, *in*
　　H-60103
　6-Methoxy-2H-1-benzopyran-2-one, *in* H-70109
　4-Methoxybenzoyl cyanide, *in* H-70212
　2-Methoxybenzyl chloride, *in* C-70101
　m-Methoxybenzyl chloride, *in* C-70102
　p-Methoxybenzyl chloride, *in* C-70103
　4′-Methoxy-2-biphenylcarboxylic acid, *in*
　　H-70111
　8-Methoxybonducellin, *in* B-60192
　2-Methoxybutanal, *in* H-70112
　1-Methoxy-3-buten-2-ol, *in* B-70291
　8-Methoxy-1(6),2,4,7(11)-cadinatetraen-12,8-
　　olide, *in* H-70114
　6-Methoxycarbazomycinal, *in* H-70154

C-Methoxycarbohydroxamic acid, M-70044
Methoxycarbonylmalonaldehyde, *in* D-60190
N-Methoxycarbonyl-(2,3,4,5-tetrachloro-1-
　thiophenio)amide, *see* T-60028
8-Methoxychlorotetracycline, M-70045
6-Methoxychroman, *in* D-70175
8-Methoxychroman, *in* D-60203
p-Methoxycinnamaldehyde, *in* H-60213
6-Methoxycoelonin, *in* T-60120
6-Methoxycomaparvin, *in* C-70191
6-Methoxycomaparvin 5-methyl ether, *in*
　C-70191
Methoxyconiferin, *in* S-70046
6-Methoxycoumarin, *in* H-70109
1-Methoxycyclohexene, M-60037
2-Methoxycyclopentanone, *in* H-60114
1-Methoxycyclopropanecarboxylic acid, *in*
　H-60117
2-Methoxy-6-(3,4-dihydroxy-1,5-heptadienyl)-
　benzyl alcohol, *see* H-70178
1-Methoxy-2,3-dimethylcyclopropane, *in*
　D-70397
4-Methoxy-2,5-dimethyl-3(2H)-furanone, *in*
　H-60120
6-Methoxy-2-(3,7-dimethyl-2,6-octadienyl)-1,4-
　benzoquinone, M-60038
4-Methoxy-1,6-dimethyl-1,3,5-triazin-2(1H)-
　one, *in* M-60132
2-Methoxy-4,5-dinitrophenol, *in* D-70449
5-Methoxy-2,4-dinitrophenol, *in* D-70450
6β-Methoxy-7,22-ergostadiene-3β,5α-diol, *in*
　M-70057
2-(2-Methoxyethoxy)-N,N-bis[2-(2-methoxy-
　ethoxy)ethyl]ethanamine, *in* T-70323
6-Methoxyflavone, *in* H-70130
7-Methoxyflavone, *in* H-70131
8-Methoxyflavone, *in* H-70132
6-Methoxyfurano[4″,5″:8,7]flavone, *see*
　K-70003
▷9-Methoxy-7H-furo[3,2-g][1]benzopyran-7-one,
　see X-60001
▷9-Methoxyfuro[3,2-g]chromen-7-one, *see*
　X-60001
▷8-Methoxy-4′,5′:6,7-furocoumarin, *see* X-60001
2-(9-Methoxygeranylgeranyl)-6-methyl-1,4-
　benzoquinone, *in* S-60010
2-(11-Methoxygeranylgeranyl)-6-methyl-1,4-
　benzoquinone, *in* S-60010
3′-Methoxyglabridin, *in* G-70016
8-Methoxyguanosine, *in* D-70238
1-Methoxy-1-hexene, M-70046
4-Methoxy-2-hexene, *in* H-60077
7-Methoxyindeno[1,2,3-cd]pyrene, *in* I-60025
8-Methoxyindeno[1,2,3-cd]pyrene, *in* I-60026
6-Methoxyindole, *in* H-60158
5-Methoxy-1H-indol-6-ol, *in* D-70307
6-Methoxy-1H-indol-5-ol, *in* D-70307
3-Methoxy-1H-isoindole, *in* P-70100
1-Methoxy-3,4-isoquinolinediamine, *see*
　D-60037
7α-Methoxy-8-labdene-3β,15-diol, *in* L-70013
α-Methoxylidene-γ-butyrolactone, *see* D-70222
6β-Methoxymaackiain, *see* S-60038
N-Methoxymethanamine, *see* D-70418
5-Methoxy-2-(6-methoxy-3-methyl-2-
　benzofuranyl)phenol, *see* H-70156
6-Methoxy-2-[2-(4-methoxyphenyl)ethyl]-4H-1-
　benzopyran-4-one, M-60039
6-Methoxy-2-[2-(4-methoxyphenyl)ethyl]-
　chromone, *see* M-60039
2-Methoxy-4-methylbenzaldehyde, *in* H-70160
4-Methoxy-2-methylbenzaldehyde, *in* H-70161
2-Methoxy-4-methylbenzoic acid, *in* H-70163
4-Methoxy-2-methylbenzoic acid, *in* H-70164
8-Methoxy-4-methyl-2H-1-benzopyran-2-one, *in*
　H-70165
4-Methoxy-3-(3-methyl-2-butenyl)-5-phenyl-
　2(5H)-furanone, M-70047
5-Methoxy-4-methylcoumarin, *in* H-60173
8-Methoxy-4-methylcoumarin, *in* H-70165
2-Methoxy-3-methyl-2-cyclopenten-1-one, *in*
　H-60174
2-Methoxy-4,5-methylenedioxybenzaldehyde, *in*
　H-70175
2-Methoxy-3,4-methylenedioxybenzoic acid, *in*
　H-60177

3-Methoxy-4,5-methylenedioxybenzoic acid, *in*
　H-60178
4-Methoxy-2,3-methylenedioxybenzoic acid, *in*
　H-60179
6-Methoxy-2,3-methylenedioxybenzoic acid, *in*
　H-60180
6-Methoxy-3,4-methylenedioxybenzoic acid, *in*
　H-60181
3′-Methoxy-3,4-methylenedioxybibenzyl, *in*
　T-60316
3-Methoxy-4,5-methylenedioxycinnamaldehyde,
　see M-70048
3-(2-Methoxy-4,5-methylenedioxyphenyl)-
　propenal, M-70048
5-Methoxy-2-(7-methylfuro[2,3-f]-1,3-
　benzodioxol-6-yl)phenol, *see* H-70157
4-Methoxy-7-methyl-5H-furo[2,3-g][2]-
　benzopyran-5-one, *see* C-70198
2-Methoxy-4-methyl-4H-imidazo[4,5-b]-
　pyrazine, *in* I-60006
6-Methoxy-3-methyl-4,5-isobenzofurandione,
　see A-60070
5-Methoxy-7-methyl-1,2-naphthoquinone, *in*
　H-60189
8-Methoxy-3-methyl-1,2-naphthoquinone, *in*
　H-60190
3-Methoxy-4,5-methylnedioxybenzonitrile, *in*
　H-60178
5-Methoxy-2-methyl-3-nitroaniline, *in* A-60222
3-Methoxy-2-methyl-6-nitrophenol, *in* M-70097
3-Methoxy-4-methyl-1,2,4-oxadiazolidin-5(4H)-
　one, *in* O-70060
4-Methoxy-6-methyl-5-(1-oxo-2-butenyl)-2H-
　pyran-2-one, *see* P-60213
3-Methoxy-6-methylpyridazine, *in* M-70124
2-Methoxy-1-methyl-4(1H)-pyridinone, *in*
　H-60224
3-Methoxy-1-methyl-4(1H)-pyridinone, *in*
　H-60227
4-Methoxy-1-methyl-2(1H)-pyridinone, *in*
　H-60224
Methoxymethyl salicylate, *in* H-70108
4-Methoxy-5-methyl-6-(7,9,11-trimethyl-
　1,3,5,7,9,11-tridecahexaenyl)-2H-pyran-2-
　one, *see* C-60149
6-Methoxy-1-naphthaldehyde, *in* H-70184
6-Methoxy-2-naphthaldehyde, *in* H-70185
7-Methoxy-2-naphthaldehyde, *in* H-70186
4-Methoxy-1,2-naphthoquinone, *in* H-60194
3-Methoxy-6-nitro-1,2-benzenedicarboxylic
　acid, *in* H-70189
4-Methoxy-3-nitro-1,2-benzenedicarboxylic
　acid, *in* H-70190
4-Methoxy-5-nitro-1,2-benzenedicarboxylic
　acid, *in* H-70191
4-Methoxy-5-nitro-1,3-benzenedicarboxylic
　acid, *in* H-70192
2-Methoxy-5-nitrobenzotrifluoride, *in* N-60053
4-Methoxy-3-nitropyridine, *in* N-70063
1-Methoxy-2-nitro-4-(trifluoromethyl)-benzene,
　in N-60048
1-Methoxy-3-nitro-5-(trifluoromethyl)-benzene,
　in N-60052
1-Methoxy-4-nitro-2-(trifluoromethyl)-benzene,
　in N-60053
4-Methoxy-1-nitro-2-(trifluoromethyl)-benzene,
　in N-60054
8-Methoxy-N-methylchlorotetracycline, *in*
　M-70045
11-Methoxynoryangonin, *in* H-70122
6-Methoxyorobol 7-methyl ether, *in* P-70032
4-Methoxy-α-oxobenzeneacetic acid, *in* H-70212
4-Methoxy-α-oxobenzeneacetonitrile, *in*
　H-70212
5-Methoxy-3-(8,11,14)pentadecatrienyl)-1,2,4-
　benzenetriol, M-70049
Methoxypentamethylbenzene, *in* P-60058
4-Methoxy-3-penten-2-one, *in* P-60059
4-Methoxy-2,5-phenanthrenediol, *in* P-60071
2-Methoxy-3H-phenoxazin-3-one, *in* H-60206
(2-Methoxyphenoxy)acetic acid, *in* H-70203
(3-Methoxyphenoxy)acetic acid, *in* H-70204
(4-Methoxyphenoxy)acetic acid, *in* H-70205
(2-Methoxyphenyl)acetaldehyde, *in* H-60207

4-Methoxyphenylacetaldehyde, *in* H-60209
2-Methoxy-2-phenylacetaldehyde, *in* H-60210
5-Methoxy-2-phenyl-7*H*-1-benzopyran-7-one, *see* N-70074
4-Methoxy-4-phenyl-1-butene, *in* P-70061
1-Methoxy-2-phenyldiazene 2-oxide, *in* H-70208
▷4-Methoxyphenyldiazomethane, *see* D-60064
1-Methoxy-2-phenyldiimine 2-oxide, *in* H-70208
5-[2-(3-Methoxyphenyl)ethyl]-1,3-benzodioxole, *in* T-60316
(4-Methoxyphenyl)ethylene, *in* H-70209
1-Methoxy-1-phenylethylene, *in* P-60092
6-Methoxy-2-phenyl-4*H*-furo[2,3-*h*]-1-benzopyran-4-one, *see* K-70003
1-Methoxy-2-(phenylmethoxy)benzene, *in* B-60072
1-Methoxy-3-(phenylmethoxy)benzene, *in* B-60073
1-Methoxy-4-(phenylmethoxy)benzene, *in* B-60074
2-Methoxy-1-phenylnaphthalene, *in* P-60115
2-Methoxy-5-phenylnaphthalene, *in* P-60118
7-Methoxy-1-phenylnaphthalene, *in* P-60119
3-Methoxy-4-phenyl-2-prenyl-2-butenolide, *see* M-70047
3-(4-Methoxyphenyl)-2-propenal, *in* H-60213
1-(4-Methoxyphenyl)pyrrole, *in* H-60216
2-(2-Methoxyphenyl)pyrrole, *in* H-60217
2-Methoxy-4-phenylquinoline, *in* P-60131
4-Methoxy-2-phenylquinoline, *in* P-60132
5-Methoxy-2-(phenylthio)-1,4-naphthoquinone, *in* H-70213
5-Methoxy-3-(phenylthio)-1,4-naphthoquinone, *in* H-70214
6-Methoxy-4-phenylumbelliferone, *see* I-60097
2-Methoxypiperonylic acid, *in* H-60177
5-Methoxypiperonylic acid, *in* H-60178
▷1-Methoxy-2-propanol, *in* P-70118
4-[2-[2-Methoxy-5-(1-propenyl)phenyl]-2-propenyl]phenol, *see* H-60029
▷9-Methoxypsoralen, *see* X-60001
3-Methoxy-4(1*H*)-pyridinone, *in* H-60227
6-Methoxy-2(1*H*)-pyridinone, *in* H-60226
2-Methoxy-1-(2-pyridinyl)ethanone, *in* H-70101
1-[1-Methoxy-1-(2-pyridinyl)ethyl]-isoquinoline, *in* I-60134
2-Methoxy-1-pyrroline, *in* P-70188
7-Methoxyrosmanol, *in* R-70010
9′-Methoxysargaquinone, *in* S-60010
11′-Methoxysargaquinone, *in* S-60010
α-Methoxystyrene, *in* P-60092
p-Methoxystyrene, *in* H-70209
2-Methoxy-1-(2-thienyl)ethanone, *in* H-70103
3-Methoxythiophene, *in* T-60212
4-Methoxy-2-(trifluoromethyl)benzaldehyde, *in* H-70228
4-Methoxy-3-(trifluoromethyl)benzaldehyde, *in* H-70229
3-Methoxy-6-[2-(3,4-5-trimethoxyphenyl)-ethenyl]-1,2-benzenediol, *in* T-60344
3-Methoxy-6-[2-(3,4,5-trimethoxyphenyl)-ethyl]1,2-benzenediol, *in* T-60344
5-Methoxy-1,2,3-trimethylbenzene, *in* T-60371
3-Methoxy-2,5,6-trimethylphenyl 3,5-diformyl-2,4-dihydroxy-6-methylbenzoate, *see* N-70032
6-Methoxy-5-vinylbenzofuran, *in* H-70231
11-Methoxyyangonin, *in* H-70122
▷Methyl acrylate, *in* P-70121
1-Methyladamantanone, M-60040
5-Methyladamantanone, M-60041
7-Methyladenine, *see* A-60226
▷3β-Methylaesculetin, *in* H-70153
O-Methylalloptaeroxylin, *in* A-60075
Methylallosamidin, *in* A-60076
Methyl altamisate, *in* A-70084
▷4-(Methylamino)cycloheptanone, *see* P-60151
8-(Methylamino)guanosine, *in* A-70150
(Methylamino)methanesulfonic acid, *in* A-60215
2-(Methylamino)-4(1*H*)-pteridinone, *in* A-60242
2-(Methylamino)purine, *in* A-60243
8-(Methylamino)purine, *in* A-60244

2-(Methylamino)-1*H*-pyrido[2,3-*b*]indole, *in* A-70105
5-Methylamino-1*H*-tetrazole, *in* A-70202
2-Methyl-*p*-anisaldehyde, *in* H-70161
3-Methylaspartic acid, *see* A-70166
Methyl aspartylphenylalanine, *see* A-60311
4-Methylayapin, *see* M-70092
▷8-Methyl-8-azabicyclo[3.2.1]octan-1-ol, *see* P-60151
6-Methyl-5-azauracil, *see* M-60132
4-Methyl-1-azulenecarboxylic acid, M-60042
1-Methylazupyrene, M-60043
9*b*-Methyl-9*bH*-benz[*cd*]azulene, M-60044
N-Methylbenzenecarboximidic acid, M-60045
2-Methyl-1,3-benzenedicarboxaldehyde, M-60046
2-Methyl-1,4-benzenedicarboxaldehyde, M-60047
4-Methyl-1,2-benzenedicarboxaldehyde, M-60048
4-Methyl-1,3-benzenedicarboxaldehyde, M-60049
5-Methyl-1,3-benzenedicarboxaldehyde, M-60050
2-Methyl-1,3-benzenedimethanethiol, M-70050
α-Methylbenzeneethanesulfonic acid, *see* P-70068
N-Methylbenzimidic acid, *see* M-60045
8-Methyl-6*H*-1,3-benzodioxolo[4,5-*g*][1]-benzopyran-6-one, *see* M-70092
3-Methyl-4,5-benzofurandione, M-70051
α-Methylbenzoin, *see* H-70123
2-Methylbenzo[*gh*]perimidine, M-60051
3-Methyl-1*H*-2-benzopyran-1-one, M-70052
1-Methyl-8*H*-benzo[*cd*]triazirino[*a*]indazole, M-60052
Methyl 3-benzoyloxyacrylate, *in* O-60090
Methyl 1-benzyl-1*H*-1,2,3-triazole-5-carboxylate, *in* T-60239
3-Methylbicyclo[1.1.0]butane-1-carboxylic acid, M-70053
3-Methylbiphenyl, M-60053
3-Methyl-2-biphenylamine, *see* A-70164
3-Methyl-[1,1′-biphenyl]-2-amine, *see* A-70164
1′-Methyl-1,5′-bi-1*H*-pyrazole, *in* B-60119
2-Methyl-*N*,*N*-bis(2-methylphenyl)-benzenamine, *see* T-70288
3-Methyl-*N*,*N*-bis(3-methylphenyl)-benzenamine, *see* T-70289
4-Methyl-*N*,*N*-bis(4-methylphenyl)-benzenamine, *see* T-70290
4-[5′-Methyl[2,2′-bithiophen]-5-yl]-3-butyne-1,2-diol, M-70054
Methyl 3-(3-bromo-4-chloro-4-methyl-cyclohexyl)-4-oxo-2-pentenoate, M-60054
2-Methylbutane-1,1-dicarboxylic acid, *see* M-70119
2-Methylbutanoic acid 2-amino-4-methoxy-1,5-dimethyl-6-oxo-2,4-cyclohexadien-1-yl ester, *see* W-70002
3-Methyl-2-butenal, M-70055
6-(3-Methyl-2-butenyl)allopteroxylin, *in* A-60075
6-(3-Methyl-2-butenyl)allopteroxylin methyl ether, *in* A-60075
3-(3-Methyl-2-butenyl)furo[2,3,4-*de*]-1-benzopyran-2(5*H*)-one, *see* C-60214
Methyl *sec*-butyl ketone, *see* M-70105
▷3-[(3-Methylbutyl)nitrosoamino]-2-butanone, M-60055
9-Methyl-9*H*-carbazole, M-60056
N-Methylcarbazole, *see* M-60056
9-Methyl-9*H*-carbazol-2-ol, *in* H-60109
Methylcatalpol, *in* C-70026
Methyl 5-chloro-4-*O*-demethylbarbatate, *in* B-70007
Methyl 2-(3-chloro-2,6-dihydroxy-4-methyl-benzoyl)-5-hydroxy-3-methoxybenzoate, *in* S-70083
▷6-Methylcholanthrene, M-70056
24-Methyl-7,22-cholestadiene-3,5,6-triol, M-70057
24-Methyl-5,22,25-cholestatrien-3-ol, M-60057
3-Methylcrotonaldehyde, *see* M-70055

2-Methyl-4-cyanopyrrole, *in* M-60125
Methyl-α-cyclohallerin, *in* C-60193
Methyl-β-cyclohallerin, *in* C-60193
6-Methyl-2-cyclohexen-1-ol, M-70058
6-Methyl-2-cyclohexen-1-one, M-70059
9-Methylcyclolongipesin, *in* C-60214
Methylcyclononane, M-60058
2-Methylcyclooctanone, M-60059
11*b*-Methyl-11*bH*-Cyclooct[*cd*]azulene, M-60060
1-(2-Methyl-1-cyclopenten-1-yl)ethanone, *see* A-60043
23-(2-Methylcyclopropyl)-24-norchola-5,22-dien-3-ol, *see* P-60004
3-*O*-Methyl-2,5-dehydrosenecioodentol, *in* S-60022
2-*O*-Methyl-1,4-dehydrosenecioodontol, *in* S-60022
*O*⁵-Methyl-3′,4′-deoxypsorospermin-3′-ol, *in* D-60021
Methyl 15α,17β-diacetoxy-6α-butanoyloxy-15,16-dideoxy-15,17-oxido-16-spongianate, *in* T-60038
Methyl 15α,17β-diacetoxy-15,16-dideoxy-15,17-oxido-16-spongianoate, *in* T-60038
2-Methyl-1,3-diazapyrene, *see* M-60051
Methyl 3,5-dichlorolecanorate, *see* T-70334
Methyl dicyanoacetate, *in* D-70149
1-Methyldicyclopenta[*ef,kl*]heptalene, *see* M-60043
Methyl 3,3-diethoxypropionate, *in* O-60090
Methyl diformylacetate, *in* D-60190
Methyl 3,4-dihydro-8-isopropyl-5-methyl-2-naphalenecarboxylate, *in* I-60129
2-Methyl-1,4-dihydropyrimidine, *see* D-70225
2-Methyl-1,6-dihydropyrimidine, *see* D-70225
1-Methyl-1,2-dihydro-3*H*-pyrrol-3-one, *in* D-70262
Methyl 2-(2,6-dihydroxy-4-methylbenzoyl)-5-hydroxy-3-methoxybenzoate, *see* S-70083
Methyl 11,12-dimethoxy-6,8,11,13-abietatrien-20-oate, *in* D-60305
Methyl 2,2′-di-*O*-methyleriodermate, *in* E-70032
▷2-Methyl-1,3-dinitrobenzene, M-60061
1-Methyl-2,3-dinitro-1*H*-pyrrole, *in* D-60461
1-Methyl-2,4-dinitro-1*H*-pyrrole, *in* D-60462
1-Methyl-2,5-dinitro-1*H*-pyrrole, *in* D-60463
1-Methyl-3,4-dinitro-1*H*-pyrrole, *in* D-60464
2-Methyl-1,6-dioxaspiro[4.5]decane, M-60062
▷2-Methyl-1,3-dioxolane, M-60063
5-Methyl-3,6-dioxo-2-piperazineacetic acid, M-60064
N-Methyldiphenylamine, M-70060
1-Methyl-4,4-diphenylcyclohexene, M-60065
3-Methyl-4,4-diphenylcyclohexene, M-60066
2-Methyl-1,5-diphenyl-1*H*-pyrrole, *in* M-60115
5-Methyl-1,3-dithiane, M-60067
5-Methyl-1,3,2,4-dithiazolium(1+), M-70061
▷5-Methyl-2,4-dithiouracil, *see* M-70127
6-Methyl-2,4-dithiouracil, *see* M-70128
Methyleneadamantane, M-60068
N,*N*′-Methylene-2,2′-azapyridocyanine, *see* D-60495
α-Methylenebenzenemethanol, *see* P-60092
4-Methylenebicyclo[3.1.0]hex-2-ene, M-60069
7-Methylenebicyclo[3.3.1]nonan-3-one, M-70062
3,3′-Methylenebis[4-hydroxy-5-methyl-2*H*-1-benzopyran-2-one], *see* G-60013
2,2′-Methylenebis[8-hydroxy-3-methyl-1,4-naphthalenedione], M-70063
2,2′-Methylenebis-1*H*-imidazole, M-70064
2,2′-Methylenebispyridine, *see* D-60496
1,1′-Methylenebispyridinium(1+), M-70065
24-Methylene-5-cholestene-3,4,7,20-tetrol, M-60070
22-Methylene-3β-cycloartanol, *see* C-60232
Methylenecyclobutane, M-70066
4-Methylene-2,5-cyclohexadien-1-one, M-70071
6-Methylene-2,4-cyclohexadien-1-one, M-60072
9-Methylene-1,3,5,7-cyclononatetraene, M-60073
7-Methylene-1,3,5-cyclooctatriene, M-60074
2-Methylenecyclopropaneacetic acid, M-70067
2-Methylene-1,3-dioxane, M-60075
4-Methylene-1,3-dioxolane, M-60076
6,7-Methylenedioxy-2*H*-1-benzopyran-2-one, M-70068

6,7-Methylenedioxycoumarin, *see* M-70068
2,3-Methylenedioxyphenylacetic acid, *see* B-60029
2,3-Methylenedioxyphenylacetonitrile, *in* B-60029
3-(3,4-Methylenedioxyphenyl)-8,8-dimethyl-4*H*,8*H*-benzo[1,2-*b*:3,4-*b*′]-dipyran-4-one, *see* C-60006
1-(3,4-Methylenedioxyphenyl)-1-tetradecene, M-60077
4,5-Methylenedioxysalicylaldehyde, *see* H-70175
3,3′-Methylenediplumbagin, *see* M-70063
1,1′-Methylenedipyridinium, *see* M-70065
2-Methylene-1,3-diselenole, M-70069
3-Methyleneglutamic acid, *see* A-60219
4-Methylene-1,2,5,6-heptatetraene, M-70070
4-Methylene-1,2,5-hexatriene, M-70071
1-Methyleneindane, M-60078
24-Methylene-7,9(11)-lanostadien-3-ol, M-70072
22-Methylene-30-nor-3β-cycloartanol, *see* N-60059
22-Methylene-30-nor-3-cycloartanone, *in* N-60059
2-Methylene-7-oxabicyclo[2.2.1]heptane, M-70073
4-Methylene-2-oxabicyclo[3.3.0]octan-3-one, *see* H-70062
2-Methylene-4-pentenal, M-70074
3-Methylene-4-penten-2-one, M-70075
3-Methyleneperhydrobenzofuran-2-one, *see* H-70062
β-Methylenephenylalanine, *see* A-60238
2-Methylene-1,3-propanedithiol, M-70076
1-Methylenepyrrolidinium(1+), M-70077
24-Methylene-3,4-seco-4(28)-cycloarten-3-oic acid, M-70078
4-Methylene-1,2,3,5-tetraphenylbicyclo[3.1.0]-hex-2-ene, M-60079
11-Methylene-1,5,9-triazabicyclo[7.3.3]-pentadecane, M-60080
Methylenetricyclo[3.3.1.1³·⁷]decane, *see* M-70068
Methylene-1,3,5-trioxane, M-60081
Methylenolactocin, M-70079
Methyl 3,6-epidioxy-6-methoxy-4,16,18-eicosatrienoate, *in* E-60007
Methyl 3,6-epidioxy-6-methoxy-4,16-octadecadienoate, *in* E-60008
Methyl 3,6-epidioxy-6-methoxy-4,14,16-octadecatrienoate, *in* E-60008
Methyl 3,6-epidioxy-6-methoxy-4-octadecenoate, *in* E-60008
Methyl 14ξ,15-epoxy-8(17),12*E*-labdadien-16-oate, *in* E-60025
Methyl eriodermate, *in* E-70032
13¹-Methyl-13,15-ethano-13²,17-prop-13²(15²)-enoporphyrin, M-60082
1-(1-Methylethenyl)cyclohexene, *see* I-60115
3-(1-Methylethenyl)cyclohexene, *see* I-60116
4-(1-Methylethenyl)cyclohexene, *see* I-60117
(1-Methylethenyl)oxirane, *see* M-70080
5-Methylethuliacoumarin, *in* E-60045
3-(1-Methylethyl)aspartic acid, *see* A-70162
α-(1-Methylethyl)benzeneacetaldehyde, *see* M-70109
2-(1-Methylethyl)cycloheptanone, *see* I-70090
4-(1-Methylethyl)cycloheptanone, *see* I-70091
1-(1-Methylethyl)cyclohexene, *see* I-60118
3-(1-Methylethyl)cyclohexene, *see* I-60119
4-(1-Methylethyl)cyclohexene, *see* I-60120
(1-Methylethyl)cyclopentane, *see* I-70092
Methyl 2-ethyl-1,2,3,4,6,11-hexahydro-2,4,5,7,12-pentahydroxy-6,11-dioxo-1-naphthacenecarboxylate, *see* R-70002
2-(1-Methylethylidene)-tricyclo[3.3.1.1³·⁷]-decane, *see* I-60121
4-(Methylethyl)-1-naphthalenol, *see* I-70096
3-(1-Methylethyl)-2-naphthalenol, *see* I-70095
2-(1-Methylethyl)-1,3,5-trimethylbenzene, *see* I-60130
Methyl 4-fluoro-3,5-dimethoxybenzoate, *in* F-60024
Methyl 3-formylcrotonate, *in* M-70103

Methyl 3-formyl-3-methylacrylate, *in* M-70103
▷5-Methyl-2(3*H*)-furanone, M-70081
▷5-Methyl-2(5*H*)-furanone, M-60083
▷2-Methyl-3-furanthiol, M-70082
2-Methylfuro[2,3-*b*]pyridine, M-60084
3-Methylfuro[2,3-*b*]pyridine, M-60085
2-Methylfuro[3,2-*c*]quinoline, M-70083
O-Methylganoderic acid *O*, *in* T-60111
4-Methylglutamic acid, *see* A-60223
γ-Methylglutamic acid, *see* A-60223
2-Methylglyceric acid, *see* D-60312
1-Methylguanosine, M-70084
8-Methylguanosine, M-70085
3-Methyl-2-heptenoic acid, M-60086
4-Methyl-3-hepten-2-one, M-60087
5-Methyl-2,4-hexadienoic acid, M-70086
5-Methyl-3,4-hexadienoic acid, M-70087
Methylhomoserine, *see* A-60186
p-Methylhydratropic acid, *see* M-60114
Methyl hydrogen 2-(*tert*-butoxymethyl)-2-methylmalonate, M-70088
Methyl hydrogen 3-hydroxyglutarate, M-70089
Methyl 2-[(2-hydroxybenzoyl)amino]-4-hydroxybenzoate, *in* D-70049
Methyl hydroxycarbamate, *see* M-70044
7-Methylidenebicyclo[3.3.1]nonan-3-one, *see* M-70062
6-Methylidenetetrahydro-2-pyrone, *see* T-70074
1,1′,1″-Methylidynetris[2,3,4,5,6-penta-chlorobenzene], *see* T-60405
6-Methyl-1*H*-indole, M-60088
3-Methyl-1*H*-indole-2-carboxaldehyde, M-70090
Methyl 2-iodo-3,5-dimethoxybenzoate, *in* D-60336
Methyl 4-iodo-3,5-dimethoxybenzoate, *in* D-60337
3-Methylisocoumarin, *see* M-70052
6-Methylisocytosine, *see* A-60230
1-Methylisoguanosine, *see* D-60518
Methylisohallerin, *in* I-60100
2-Methylisophthalaldehyde, *see* M-60046
4-Methylisophthalaldehyde, *see* M-60049
5-Methylisophthalaldehyde, *see* M-60050
Methyl 8-isopropyl-5-methyl-2-naphthalenecarboxylate, *in* I-60129
Methylisoselenourea hydrogen sulfate, *in* S-70031
Methylisoselenourea iodide, *in* S-70031
3-Methylisoserine, *see* A-60183
Methyl jasmonate, *in* J-60002
Methyllanceolin, *in* P-60052
4-Methylleucine, *see* A-70137
5-*O*-Methyllicoricidin, *in* L-60025
Methyllinderatone, *in* L-70030
9-Methyllongipesin, M-70091
1-Methyl-4-mercaptohistidine, *see* O-70050
Methyl 3-(4-methoxy-2-vinylphenyl)-propanoate, *in* H-70232
1-Methyl-4-methylamino-5-(methyl-aminocarbonyl)imidazole, *see* C-60003
1-Methyl-5-methylamino-1*H*-tetrazole, *in* A-70202
2-Methyl-5-methylamino-2*H*-tetrazole, *in* A-70202
2-Methyl-6-(4-methylenebicyclo[3.1.0]hex-1-yl)-2-heptenal, *in* S-70037
4-Methyl-6,7-methylenedioxy-2*H*-1-benzopyran-2-one, M-70092
4-Methyl-6,7-methylenedioxycoumarin, *see* M-70092
4-(5-Methyl-1-methylene-4-hexenyl)-1-(4-methyl-3-pentenyl)cyclohexene, *see* C-70009
4-Methyl-10-methylene-7-(1-methylethyl)-5-cyclodecene-1,4-diol, *see* G-70009
2-Methyl-6-methylene-2,7-octadien-4-ol, M-70093
2-Methyl-6-methylene-1,3,7-octatriene, M-60089
2-Methyl-6-methylene-7-octen-4-ol, *in* M-70093
Methyl 2-*O*-methyleriodermate, *in* E-70032
Methyl 2′-*O*-methyleriodermate, *in* E-70032
Methyl 4-*O*-methyleriodermate, *in* E-70032
1-Methyl-4-(1-methylethenyl)-1,2-cyclo-hexanediol, *see* I-70086

3-Methyl-6-(1-methylethenyl)-2-cyclohexen-1-one, *see* I-70088
3-Methyl-6-(1-methylethenyl)-9-decen-1-ol, *see* I-70089
5-Methyl-8-(1-methylethenyl)-10-oxabicyclo[7.2.1]dodeca-1(12),4-dien-11-one, *see* A-60296
2-Methyl-5-(1-methylethyl)-1,3-benzenediol, *see* I-60125
2-Methyl-6-(1-methylethyl)-1,4-benzenediol, *see* I-60123
3-Methyl-5-(1-methylethyl)-1,2-benzenediol, *see* I-60126
3-Methyl-6-(1-methylethyl)-1,2-benzenediol, *see* I-60124
4-Methyl-5-(1-methylethyl)-1,3-benzenediol, *see* I-60127
5-Methyl-2-(1-methylethyl)-1,3-benzenediol, *see* I-60122
1-Methyl-4-(1-methylethyl)-3-cyclohexane-1,2-diol, *see* I-70093
1-Methyl-4-(1-methylethyl)-4-cyclohexene-1,2-diol, *see* I-70094
5-Methyl-8-(1-methylethyl)-2-naphthalenecarboxaldehyde, *see* I-60129
5-Methyl-2-(1-methyl-1-phenylethyl)-cyclohexanol, M-60090
Methyl 4-methylphenyl sulfoxide, M-70094
1-Methyl-4-(methylsulfinyl)benzene, *see* M-70094
1-Methyl-2-methylthio-1*H*-imidazo[4,5-*b*]pyrazine, *in* D-60247
4-Methyl-2-methylthio-4*H*-imidazo[4,5-*b*]pyrazine, *in* D-60247
1-Methyl-4-(methylthio)-2(1*H*)-pyridinethione, *in* P-60219
3-Methyl-5-(methylthio)-1,3,4-thiadiazole-2(3*H*)-thione, *in* M-70031
Methyl 1-methyl-1*H*-1,2,3-triazole-5-carboxylate, *in* T-60239
2-Methyl-1,3-naphthalenediol, M-60091
2-Methyl-1,5-naphthalenediol, M-60092
3-Methyl-1,2-naphthalenediol, M-60093
4-Methyl-1,3-naphthalenediol, M-60094
6-Methyl-1,2-naphthalenediol, M-60095
α-Methyl-1-naphthaleneethanol, *see* N-60015
α-Methyl-1-naphthalenemethanol, *see* N-70015
Methylnaphthazarin, *see* D-60352
1-Methylnaphtho[2,1-*b*]thiophene, M-60096
2-Methylnaphtho[2,1-*b*]thiophene, M-60097
4-Methylnaphtho[2,1-*b*]thiophene, M-60098
5-Methylnaphtho[2,1-*b*]thiophene, M-60099
6-Methylnaphtho[2,1-*b*]thiophene, M-60100
7-Methylnaphtho[2,1-*b*]thiophene, M-60101
8-Methylnaphtho[2,1-*b*]thiophene, M-60102
9-Methylnaphtho[2,1-*b*]thiophene, M-60103
2-Methyl-2*H*-naphtho[1,8-*de*]triazine, M-70095
2-Methyl-1*H*-naphtho[1,8-*de*]-1,2,3-triazinium hydroxide inner salt, *see* M-70095
Methyl-1-naphthylcarbinol, *see* N-70015
7-Methylnapyradiomycin *A*, *see* A-70237
2-Methylnebularine, *see* M-70130
6-Methylnebularine, *see* M-70131
Methyl nidoresedate, *in* N-70035
2-Methyl-4-nitrobenzaldehyde, M-70096
2-Methyl-4-nitro-1,3-benzenediol, M-70097
3-Methyl-2-nitrobiphenyl, M-70098
3-Methyl-4-nitropyridine, M-70099
4-Methyl-5-nitropyrimidine, M-60104
2-Methyl-4-nitroresorcinol, *see* M-70097
24-Methyl-25-nor-12,24-dioxo-16-scalaren-22-oic acid, M-60105
8-*O*-Methylochramycinone, *in* O-60003
2-Methyl-2-octenal, M-60106
5-Methyl-2-oxazolidinone, M-60107
▷Methyloxirane, M-70100
5-(2-Methyloxiranyl)benzofuran, M-70101
α-Methyl-γ-oxobenzenebutanoic acid, *see* B-60064
2-Methyl-3-oxobutanal, M-70102
3-Methyl-oxo-2-butenoic acid, M-70103
6-Methyl-7-(3-oxobutyl)-2-cyclohepten-1-one, *see* C-60153
Methyl 7-oxo-8(14),15-isopimaradien-18-oate, *in* O-70094
5-Methyl-2-oxo-4-oxazolidinecarboxylic acid, M-60108

Name Index

3-Methyl-4-oxo-2-pentenal − 1-Methyl-2,3,4-trinitro-1*H*- . . .

3-Methyl-4-oxo-2-pentenal, M-70104
3-Methyl-4-[(5-oxo-3-phenyl-2-pyrazolin-4-ylidene)methyl]-1-phenyl-2-pyrazolin-5-one, *see* P-70153
4-Methyl-1,3-pentadiene-1-carboxylic acid, *see* M-70086
3-Methyl-5-pentanolide, *in* H-70182
3-Methyl-2-pentanone, M-70105
▷4-Methyl-2-pentanone, M-60109
▷2-Methyl-2-pentenal, M-70106
4-Methyl-1-penten-3-ol, M-70107
Methyl *N*-phenacylcarbamate, *in* O-70099
3-Methylphenanthrene, M-70108
N-Methylphenylacetohydroxamic acid, *in* P-70059
2-Methyl-3-phenylalanine, *see* A-60225
α-Methylphenylalanine, *see* A-60225
N-Methyl-*N*-phenylbenzenamine, *see* M-70060
α-Methyl-α-phenylbenzeneacetic acid, *see* D-70493
5-Methyl-2-phenyl-1,5-benzothiazepin-4(5*H*)-one, *in* P-60081
3-Methyl 2-phenylbutanal, M-70109
2-Methyl-4-phenyl-3-butyn-2-ol, M-70110
1-Methyl-2-phenyl-1*H*-imidazole-4-carboxaldehyde, *in* P-60111
1-Methyl-2-phenyl-1*H*-imidazole-5-carboxaldehyde, *in* P-60111
▷Methyl phenyl ketone, *see* A-60017
5-Methyl-3-phenyl-1,2,4-oxadiazole, M-60110
4-Methyl-5-phenyl-2-oxazolidinethione, M-60111
5-Methyl-3-phenyl-2-oxazolidinone, *in* M-60107
4-(4-Methylphenyl)-2-pentanone, M-60112
2-(2-Methylphenyl)propanoic acid, M-60113
2-(4-Methylphenyl)propanoic acid, M-60114
3-Methyl-1-phenylpyrazole-4,5-dione 4-(2,5-dichlorophenyl)hydrazone, *see* D-70138
2-Methyl-4-phenylpyrimidine, M-70111
2-Methyl-5-phenylpyrimidine, M-70112
4-Methyl-2-phenylpyrimidine, M-70113
4-Methyl-5-phenylpyrimidine, M-70114
4-Methyl-6-phenylpyrimidine, M-70115
5-Methyl-2-phenylpyrimidine, M-70116
5-Methyl-4-phenylpyrimidine, M-70117
▷2-Methyl-5-phenyl-1*H*-pyrrole, M-60115
1-Methyl-2-phenyl-4(1*H*)-quinolinone, *in* P-60132
▷1-Methyl-4-phenyl-1,2,3,6-tetrahydropyridine, *in* T-60081
4-Methyl-5-phenyl-1,2,3-thiadiazole, M-60116
5-Methyl-4-phenyl-1,2,3-thiadiazole, M-60117
6-Methyl-3-phenyl-1,2,4-triazine, M-60118
4-*O*-Methylphloracetophenone, *in* T-70248
4-Methylphthalaldehyde, *see* M-60048
α-Methylpipecolic acid, *see* M-60119
2-Methyl-2-piperidinecarboxylic acid, M-60119
O-Methylpisiferic acid, *see* M-60036
3′-*O*-Methylpratensein, *in* T-60108
2-Methylproline, *see* M-60126
N-Methylproline methylbetaine, *see* S-60050
2-Methyl-1,3-propanediol, M-60120
2-Methyl-1,3-propanedithiol, M-60121
4-[3-Methyl-5-(1-propenyl)-2-benzofuranyl]phenol, *see* E-60075
3-(2-Methylpropyl)-6-methyl-2,5-piperazinedione, M-70118
2-Methyl-4-propyl-1,3-oxathiane, M-60122
(1-Methylpropyl)propanedioic acid, M-70119
N-Methylprotoporphyrin IX, M-60123
2-Methylpteridine, M-70120
4-Methylpteridine, M-70121
6-Methylpteridine, M-70122
7-Methylpteridine, M-70123
7-Methyl-7*H*-purin-6-amine, *see* A-60226
9-Methyl-9*H*-purin-8-amine, *in* A-60244
2-Methyl-4-(1*H*-purin-6-ylamino)-2-penten-1-ol, *see* M-60133
6-Methyl-3-pyridazinol, *in* M-70124
6-Methyl-3-pyridazinone, M-70124
4-Methyl-3(2*H*)-pyridazinone, M-70125
5-Methyl-3(2*H*)-pyridazinone, M-70126
α-Methyl-α-2-pyridinylisoquinolinemethanol, *see* I-60134

▷5-Methyl-2,4-(1*H*,3*H*)-pyrimidinedithione, M-70127
6-Methyl-2,4(1*H*,3*H*)-pyrimidinedithione, M-70128
Methyl 2-pyrimidinyl ketone, *see* A-60044
Methyl 4-pyrimidinyl ketone, *see* A-60045
Methyl 5-pyrimidinyl ketone, *see* A-60046
5-Methyl-1*H*-pyrrole-2-carboxaldehyde, M-70129
5-Methyl-1*H*-pyrrole-2-carboxylic acid, M-60124
5-Methyl-1*H*-pyrrole-3-carboxylic acid, M-60125
2-Methyl-2-pyrrolidinecarboxylic acid, M-60126
2-Methyl-1-pyrroline, *see* D-70226
Methyl 3-pyrryl ketone, *see* A-70052
5-Methyl-5*H*-quindoline, *in* I-70016
Methyl 2-quinolinyl ketone, *see* A-60049
Methyl 3-quinolinyl ketone, *see* A-60050
Methyl 4-quinolinyl ketone, *see* A-60051
Methyl 5-quinolinyl ketone, *see* A-60052
Methyl 6-quinolinyl ketone, *see* A-60053
Methyl 7-quinolinyl ketone, *see* A-60054
Methyl 8-quinolinyl ketone, *see* A-60055
Methyl 2-quinoxalinyl ketone, *see* A-60056
Methyl 5-quinoxalinyl ketone, *see* A-60057
Methyl 6-quinoxalinyl ketone, *see* A-60058
1-Methyl-9-β-D-ribofuranosylguanine, *see* M-70084
8-Methyl-9-β-D-ribofuranosylguanine, *see* M-70085
2-Methyl-9-β-D-ribofuranosyl-9*H*-purine, M-70130
6-Methyl-9-β-D-ribofuranosyl-9*H*-purine, M-70131
Methylsainfuran, *in* S-60002
4-Methylsalicylaldehyde, *see* H-70160
Methyl *N*-salicyl-4-hydroxyanthranilate, *in* D-70049
4-Methylsalicylic acid, *see* H-70163
[2-(Methylseleninyl)ethenyl]benzene, *in* M-70132
Methyl selenocyanate, M-60127
[2-(Methylseleno)ethenyl]benzene, M-70132
[2-(Methylselenonyl)ethenyl]benzene, *in* M-70132
2-(Methylseleno)pyridine, *in* P-70170
3-*O*-Methylsenecioodentol, *in* S-60022
5-Methylsorbic acid, *see* M-70086
Methyl styryl selenide, *see* M-70132
Methyl styryl selenoxide, *in* M-70132
Methyl styryl sulfide, *see* M-70137
2-Methyl-3-sulfolene, *in* D-70230
2-(Methylsulfonyl)cyclohexanone, *in* M-70025
8-(Methylsulfonyl)guanosine, *in* M-70029
(Methylsulfonyl)[(trifluoromethyl)-sulfonyl]methane, M-70133
Methylsulochrin, *in* S-70083
Methylterephthalaldehyde, *see* M-60047
3-Methyl-*N*-(5*a*,6*a*,7,8-tetrahydro-4-methyl-1,5-dioxo-1*H*,5*H*-pyrrolo[1,2-*c*]-[1,3]oxazepin-3-yl)butanamide, M-70134
10-Methyl-3,3,7,7-tetrakis(trifluoromethyl)-4,5,6-benzo-1-bromo-2,8-dioxabicyclo[3.3.1]-octane, *see* M-60128
4-Methyl-2,2,6,6-tetrakis(trifluoromethyl)-2*H*,6*H*-[1,2]-bromoxolo[4,5,1-*hi*][1,2]-benzobromoxole, M-60128
2-Methyl-6-(3,7,11,15-tetramethyl-2,6,10,14-hexadecatetraenyl)-1,4-benzoquinone, *see* S-60010
▷2-Methyl-3,4,5,6-tetranitroaniline, M-60129
▷3-Methyl-2,4,5,6-tetranitroaniline, M-60130
▷4-Methyl-2,3,5,6-tetranitroaniline, M-60131
▷2-Methyl-3,4,5,6-tetranitrobenzenamine, *see* M-60129
▷3-Methyl-2,4,5,6-tetranitrobenzenamine, *see* M-60130
▷4-Methyl-2,3,5,6-tetranitrobenzeneamine, *see* M-60131
▷1-Methyl-2,3,4,5-tetranitro-1*H*-pyrrole, *in* T-60160
3-Methyl-1,3,4-thiadiazolidine-2,5-dithione, M-70031

2-Methyl-Δ²-1,3-thiazoline, *in* T-60197
Methyl 2-thiazolin-2-yl ketone, *see* A-70041
2-(Methylthio)benzophenone, *in* M-70020
3-(Methylthio)benzophenone, *in* M-70021
4-(Methylthio)benzophenone, *in* M-70022
2-Methyl-2-thiocyanatopropane, M-70135
2-(Methylthio)cyclohexanone, *in* M-70025
3-(Methylthio)cyclohexanone, *in* M-70026
4-Methylthiocyclohexanone, *in* M-70027
2-(Methylthio)cyclopentanone, M-70136
[2-(Methylthio)ethenyl]benzene, M-70137
8-(Methylthio)guanosine, *in* M-70029
2-(Methylthio)-1*H*-imidazo[4,5-*b*]pyrazine, *in* D-60247
6-(Methylthio)-5*H*-imidazo[4,5-*c*]pyridazine, *in* D-70215
▷6-(Methylthio)inosine, *in* T-70180
1-Methylthioisoquinoline, *in* I-60132
8-*O*-Methylthiomelin, *in* T-70181
(Methylthio)methanol, M-70138
2-[(Methylthio)methyl]pyrrolidine, *in* P-70185
5-Methylthionorvaline, *see* A-60231
5-(Methylthio)pentyl isothiocyanate, *see* I-60140
2-Methyl-3-thiophenecarboxaldehyde, M-70139
3-Methyl-2-thiophenecarboxaldehyde, M-70140
4-Methyl-2-thiophenecarboxaldehyde, M-70141
4-Methyl-3-thiophenecarboxaldehyde, M-70142
5-Methyl-2-thiophenecarboxaldehyde, M-70143
5-Methyl-3-thiophenecarboxaldehyde, M-70144
α-Methyl-2-thiophenemethanol, *see* T-60207
3-Methylthio-2-phenyl-4*H*-1-benzopyran-4-one, *in* M-60025
1-[4-(Methylthio)phenyl]ethanone, *in* M-60017
[2-(Methylthio)phenyl]phenylmethanone, *in* M-70020
[3-(Methylthio)phenyl]phenylmethanone, *in* M-70021
[4-(Methylthio)phenyl]phenylmethanone, *in* M-70022
2-(Methylthio)propanoic acid, *in* M-60027
4-(Methylthio)-2(1*H*)-pyridinethione, *in* P-60219
4-Methylthioquinoline, *in* Q-60006
5-Methylthioquinoline, *in* Q-60004
6-Methylthioquinoline, *in* Q-60005
▷6-(Methylthio)-9-β-D-ribofuranosyl-9*H*-purine, *in* T-70180
β-(Methylthio)styrene, *see* M-70137
5-Methylthio-1,3,4-thiadiazole-2(3*H*)-thione, M-70031
2-(Methylthio)thiazole, *in* T-60200
1-(Methylthio)-2-(trifluoromethyl)benzene, *in* T-60284
1-(Methylthio)-3-(trifluoromethyl)benzene, *in* T-60285
1-(Methylthio)-4-(trifluoromethyl)benzene, *in* T-60286
▷Methylthiouracil, *see* D-70232
5-Methyl-4-thiouracil, *see* D-70233
▷6-Methyl-2-thiouracil, *see* D-70232
6-Methyl-4-thiouracil, *see* D-70234
Methyl-*p*-tolylacetic acid, *see* M-60114
Methyl *p*-tolyl sulfoxide, *see* M-70094
Methyl 6α,15α,17β-triacetoxy-15,16-dideoxy-15,17-oxido-16-spongianoate, *in* T-60038
6-Methyl-1,3,5-triazine-2,4-diamine, *see* D-60040
6-Methyl-1,3,5-triazine-2,4-diol, *see* M-60132
6-Methyl-1,3,5-triazine-2,4(1*H*,3*H*)-dione, M-60132
1-Methyltricyclo[3.3.1.1³,⁷]decan-2-one, *see* M-60040
5-Methyltricyclo[3.3.1.1³,⁷]decan-2-one, *see* M-60041
15-Methyltricyclo[6.5.2¹³,¹⁴]pentadeca-1,3,5,7,9,11,13-heptene, *see* M-60060
Methyl trifluoromethanesulfenate, *in* T-60279
Methyl trifluoromethyl sulfide, M-70145
Methyl trifluoromethyl sulfone, *in* M-70145
Methyl trifluoromethyl sulfoxide, *in* M-70145
3-Methyl-2-(3,7,11-trimethyl-2,6,10-dodeca-trienyl)furan, *see* M-70148
▷1-Methyl-2,3,4-trinitro-1*H*-pyrrole, *in* T-60373

▷1-Methyl-2,3,5-trinitro-1*H*-pyrrole, *in* T-60374
Methyl 1-triphenylenyl ketone, *see* A-60059
Methyl 2-triphenylenyl ketone, *see* A-60060
Methyltriphenylhydroxycyclohexenophenone, *see* D-60524
α-Methyltyrosine, *see* A-60220
5-(*N*-Methyl-L-tyrosine)bouvardin, *see* R-60002
3-Methyl-δ-valerolactone, *in* H-70182
2-Methylvaline, *see* A-60148
3-Methylvaline, *see* A-70136
5-Methyl-5-vinyl-1,3-cyclopentadiene, M-70146
Methylvinylglyoxal, *see* P-60061
3-Methyl-3-vinyl-4-pentynoic acid, *see* E-60059
7-Methylwye, *see* D-70278
9-Methylxanthine, *see* D-60257
Methyl yomoginate, *in* H-70199
1'-Methylzeatin, M-60133
Methylzedoarondiol, *in* Z-70004
Metoxal, *in* H-70108
Metyrosine, *in* A-60220
Mexolide, *see* T-70203
Microglossic acid, M-60134
Microline, *see* M-60135
Microphyllic acid, *see* M-70147
Microphyllinic acid, M-70147
Microthecin, *see* H-70141
Mikanifuraн, M-70148
Mikanin, *in* P-60047
Mikrolin, M-60135
9-Millerenolide, M-60137
1(10)-Millerenolide, M-60136
Milletenin C, *in* T-70111
Miltiodiol, M-70149
mimG, *see* D-60302
▷Minprostin E₂, *in* D-60367
▷Mirex, *see* D-70530
Misakinolide *A*, M-60138
▷Miserotoxin, *in* N-60039
Mitsugashiwalactone, M-70150
Miyabenol *A*, M-60139
Miyabenol *B*, M-60140
Miyabenol *C*, M-60141
MK 781, *in* A-60220
MM 14201, *in* A-60194
Mobisyl, *in* H-70108
Mokkolactone, M-60142
Mollisorin A, *in* D-70302
Mollisorin B, *in* D-70302
Moluccanin, M-70151
▷α-Monoacetin, *see* G-60029
Monoacetylarbutin, *in* A-70248
Monoalide, M-70152
Monoalide 25-acetate, *in* M-70152
▷Monobenzone, *see* B-60074
Monochlorosulochrin, *in* S-70083
Monocillin II, *in* N-60060
Monocillin III, *in* N-60060
Monocillin IV, *in* N-60060
Monocillin V, *in* N-60060
Monocillin I, *in* N-60060
Monomelittoside, *in* M-70019
Monomethyl 3-hydroxypentanedioate, *see* M-70089
Monomethylsulochrin, *in* S-70083
▷Monopicolylamine, *see* A-60227
Montanin *F*, *see* T-60178
Montanin *G*, M-70153
Montanin C, *in* T-70159
Montanolide, M-70154
Moracin *D*, M-60143
Moracin *E*, M-60144
Moracin *G*, M-60145
Moracin *M*, M-60146
Moracin C, *in* M-60146
Moracin F, *in* M-60146
Moracin H, *in* M-60145
Moracin I, *in* M-60146
Moracin J, *in* M-60146
3-Morpholinecarboxylic acid, M-70155
3-Morpholinone, M-60147
2-Morpholinotetrahydrofuran, *see* T-60065
Morpholone, *see* M-60147
Moscatilin, *in* P-60044

Moscatin, *in* P-60071
▷Mozuku toxin *A*, *see* E-60044
▷MPTP, *in* T-60081
MT 35214, *in* A-60194
MT 36531, *in* A-60194
Muamvatin, M-70156
Mugineic acid, M-60148
Mulberrofuran *R*, M-60149
Mulberroside A, *in* D-70335
Mulundocandin, M-70157
▷Muracin, *see* D-70232
Murraculatin, M-70158
Murralongin, M-70159
Murraol, M-60150
Murraxocin, M-60151
Murraxonin, M-60152
Murrayanone, M-70160
Muscalure, *in* T-60259
▷Muscarine, M-60153
Mutisifurocoumarin, M-70161
Mycosinol, M-60154
▷Mycotoxin F2, *see* Z-60002
Myoflex, *in* H-70108
Myomontanone, M-60155
Myoporone, M-60156
α-Myrcene hydroperoxide, *see* H-60098
β-Myrcene hydroperoxide, *see* H-60097
Myrianthinic acid, *in* D-70326
Myricanol, M-70162
Myricanone, *in* M-70162
Myristargenol *A*, M-70163
Myristargenol *B*, M-70164
Myristicic acid, *in* H-60178
Myristicin acid, *in* H-60178
Myrocin *C*, M-70165
Mysal, *in* H-70108
Mzikonone, M-70166
▷Naematolin, M-70001
Naematolin B, *in* N-70001
Naematolone, *in* N-70001
Namakochrome, *in* H-60065
Naphthalene-1,3-dialdehyde, *see* N-60002
1,8-Naphthalenediamine, *see* D-70042
1,2-Naphthalenedicarboxaldehyde, N-60001
1,3-Naphthalenedicarboxaldehyde, N-60002
2,3-Naphthalenedicarboxylic acid, N-60003
1,2-Naphthalenedithiol, N-70002
1,8-Naphthalenedithiol, N-70003
2,3-Naphthalenedithiol, N-70004
2-Naphthalenemethanol, N-60004
1,4,5-Naphthalenetriol, N-60005
[2](1,5)Naphthaleno[2](2,7)(1,6-methano[10]-annuleno)phane, N-70005
[2](2,6)-Naphthaleno[2]paracyclophane-1,11-diene, N-70006
1-(1-Naphthalenyl)piperazine, *see* N-60013
1-(2-Naphthalenyl)piperazine, *see* N-60014
Naphthanthraquinone, *see* B-70015
▷Naphthazarin, *see* D-60356
Naphth[2,3-*a*]azulene-5,12-dione, N-60006
β-Naphthisatin, *see* B-70026
Naphtho[2,1-*b*:6,5-*b*′]bis[1]benzothiophene, N-60007
Naphtho[1,8-*bc*:5,4-*b*′,*c*′]bisthiopyran, *see* B-60049
Naphtho[8,1,2-*cde*]cinnoline, N-70007
Naphtho[2,3-*b*]fluoranthene, *see* D-70066
p-Naphthofuchsone, *see* D-70487
Naphtho[1,2-*c*]furan, N-70008
Naphtho[2,3-*c*]furan, N-60008
Naphtho[1,2,3,4-*ghi*]perylene, N-60009
Naphtho[1,8-*bc*]pyran, N-70009
Naphtho[1,8-*bc*]pyran-2-carboxylic acid, N-70010
2*H*-Naphtho[1,2-*b*]pyran-2-one, N-70011
4*H*-Naphtho[1,2-*b*]pyran-4-one, N-60010
Naphtho[1,2-*a*]quinolizinium(1+), N-70012
Naphthostyril, *see* B-70027
Naphtho[2,1-*b*]thiet-1-one, N-70013
Naphtho[2,1-*b*]thiet-2-one, *see* N-70013
Naphtho[2,3-*b*]thiophene-2,3-dione, N-70014
1*H*-Naphtho[2,3-*d*]triazole-4,9-dione, N-60011
α-Naphthoylformic acid, *see* O-60078
β-Naphthoylformic acid, *see* O-60079
Naphthvalene, N-60012
(1-Naphthyl)acetylene, *see* E-70046
1-(1-Naphthyl)ethanol, N-70015

1-Naphthylglyoxylic acid, *see* O-60078
2-Naphthylglyoxylic acid, *see* O-60079
1-(1-Naphthyl)piperazine, N-60013
1-(2-Naphthyl)piperazine, N-60014
1-(1-Naphthyl)-2-propanol, N-60015
Napyradiomycin *A₂*, N-70016
Napyradiomycin *B₄*, N-70017
Nardonoxide, N-70018
Nardostachin, N-70019
▷Naridan, *in* O-60102
Nasutin B, *in* E-70007
Nasutin C, *in* E-70007
Natsudaidan, *in* H-60024
Nectandrin B, *in* M-70003
Nectriafurone, N-60016
Nemorosonol, N-70020
Neobacin, *in* T-60324
Neobanin, *in* D-60340
Neocorymboside, N-70021
Neocryptomerin, *in* H-60082
Neocryptotanshinone, N-60017
Neodolabellenol, N-70022
Neodrosopterin, *in* D-60521
Neoechinulin *E*, N-70023
▷Neoenactin *M₁*, *see* L-70033
Neokadsuranin, N-70024
Neokyotorphin, N-70025
Neoleuropein, N-60018
Neoliacinic acid, N-70026
Neomajucin, *in* M-70005
Neonepetalactone, N-70027
Neopentylbenzene, *see* D-60445
Neopentylglycine, *see* A-70137
1-Neopentylnaphthalene, *see* D-60446
2-Neopentylnaphthalene, *see* D-60447
Neoplathymenin, *see* D-70339
Neopulchellin, N-70028
Neotokorigenin, *in* S-60043
Neotokoronin, *in* S-60043
Neotyrosine, *see* A-60198
Neovasinin, N-70029
Neovasinone, N-60019
Neovasinone, *in* N-70029
Neoverrucosanol, N-70030
Nepenthone A, *in* T-60333
Nepetalactone, N-60020
Nepetaside, N-70031
Nephroarctin, N-70032
Nephthene, *see* C-70003
Neriucoumaric acid, *in* D-70349
Neurotoxin NSTX 3, N-70033
Nezukol, *in* H-60163
Nicotinamide salicylate, *in* H-70108
N-Nicotinoylglycine, *see* N-70034
Nicotinuric acid, N-70034
Nicotinylaminomethane, *see* A-70086
Nidoresedaic acid, *see* N-70035
Nidoresedic acid, N-70035
Niduloic acid, *see* H-70142
Nidulol, *in* D-60349
Nigellic acid, *see* H-70098
Niksalin, *in* H-70108
Niloticin, N-70036
Niloticin acetate, *in* N-70036
Nimbidiol, N-60021
Nimbinone, N-70037
Nimbione, N-70038
Nimbocinol, *in* A-70281
Nimolicinoic acid, N-60022
Nipecotamide, *in* P-70102
Nipecotic acid, *see* P-70102
Nitraldin, *see* N-70056
2,2′,2″-[Nitrilotris(2,1-ethanediyloxy)]-trisethanol, *see* T-70323
2,2′,2″-Nitrilotrisphenol, N-70039
4,4′,4″-Nitrilotrisphenol, N-70040
2-Nitrobenzil, *see* N-70058
4-Nitrobenzil, *see* N-70059
5-Nitro-2,1-benzisothiazol-3-amine, *see* A-60232
5-Nitro-1,2-benzisothiazole, N-70041
7-Nitro-1,2-benzisothiazole, N-70042
▷*m*-Nitrobenzotrifluoride, *see* N-70065
▷*o*-Nitrobenzotrifluoride, *see* N-70064
▷*p*-Nitrobenzotrifluoride, *see* N-70066
1-Nitro-1,3-butadiene, N-60023
1-Nitro-2-butanone, N-70043

3-Nitro-2-butanone, N-70044
4-Nitro-2-butanone, N-70045
Nitrococussic acid, see H-60193
3-Nitro-2,5-cresotaldehyde, see H-70180
Nitrocyclobutane, N-60024
2-Nitrocyclodecanone, N-70046
Nitrocycloheptane, N-60025
1-Nitro-1,3-cyclohexadiene, N-70047
2-Nitro-3,4-dihydro-1(2*H*)naphthalenone, N-60026
2-Nitro-1,3-diphenylpropane, N-70048
4-(2-Nitroethenyl)-1,2-benzenediol, N-70049
[(2-Nitroethenyl)sulfinyl]benzene, in N-70061
[(2-Nitroethenyl)sulfonyl]benzene, in N-70061
[(2-Nitroethenyl)thio]benzene, see N-70061
4-(2-Nitroethyl)phenol, N-70050
2-Nitroethyl phenyl sulfide, see N-70060
2-Nitroethyl phenyl sulfoxide, in N-70060
[(2-Nitroethyl)sulfinyl]benzene, in N-70060
[(2-Nitroethyl)thio]benzene, see N-70060
▷Nitrogenin, in S-60045
4-Nitroheptane, N-60027
1-Nitro-3-hexene, N-70051
5-Nitrohistidine, N-60028
▷2-Nitro-5-hydroxybenzotrifluoride, see N-60054
3-Nitro-5-hydroxybenzotrifluoride, see N-60052
5-Nitro-1*H*-indole, N-70052
1-Nitroisoquinoline, N-60029
4-Nitroisoquinoline, N-60030
5-Nitroisoquinoline, N-60031
6-Nitroisoquinoline, N-60032
7-Nitroisoquinoline, N-60033
8-Nitroisoquinoline, N-60034
6-Nitroisovanillic acid, in D-60360
3-Nitromesidine, see T-70285
3-Nitromesitoic acid, see T-60370
(Nitromethyl)benzene, see N-70053
Nitromethyl phenyl sulfide, see P-60137
Nitromethyl phenyl sulfone, in P-60137
(Nitromethyl)sulfonylbenzene, in P-60137
(Nitromethyl)thiobenzene, see P-60137
Nitromide, in D-60458
3-Nitro-2-naphthaldehyde, N-70054
3-Nitro-2-naphthalenecarboxaldehyde, see N-70054
6-Nitro-1,4-naphthalenedione, see N-60035
7-Nitro-1,2-naphthalenedione, see N-60036
6-Nitro-1,4-naphthoquinone, N-60035
7-Nitro-1,2-naphthoquinone, N-60036
7-Nitro-4-nitroso-1-naphthol, in N-60035
2-Nitro-1-oxo-1,2,3,4-tetrahydronaphthalene, see N-60026
Nitropeptin, N-70055
m-Nitrophenylethylene oxide, see N-70057
o-Nitrophenylethylene oxide, see N-70056
Nitrophenylmethane, see N-70053
(2-Nitrophenyl)oxirane, see N-70056
(3-Nitrophenyl)oxirane, see N-70057
(2-Nitrophenyl)phenylethanedione, N-70058
(4-Nitrophenyl)phenylethanedione, N-70059
α-2-Nitrophenyl-β-phenylglyoxal, see N-70058
α-4-Nitrophenyl-β-phenylglyoxal, see N-70059
1-Nitro-2-(phenylsulfinyl)ethane, in N-70060
1-Nitro-2-(phenylsulfinyl)ethylene, in N-70061
1-Nitro-2-(phenylsulfonyl)ethylene, in N-70061
1-Nitro-2-(phenylthio)ethane, N-70060
1-Nitro-2-(phenylthio)ethylene, N-70061
5-Nitrophthalazine, N-70062
4-Nitro-β-picoline, see M-70099
4-Nitro-3-picoline, see M-70099
▷3-Nitropropanal, N-60037
▷2-Nitro-1,3-propanediol, N-60038
1,1′-(2-Nitro-1,3-propanediyl)bisbenzene, see N-70048
▷3-Nitro-1-propanol, N-60039
▷1-Nitropyrene, N-60040
2-Nitropyrene, N-60041
4-Nitropyrene, N-60042
3-Nitro-4-pyridinol, see N-70063
3-Nitro-4(1*H*)-pyridone, N-70063
2-Nitropyrimidine, N-60043

6-Nitropyrocatechuic acid, see D-60360
3-Nitropyrrole, N-60044
6-Nitro-8-quinolinamine, see A-60235
8-Nitro-6-quinolinamine, see A-60234
Nitrosin, N-60045
4-Nitrosobenzaldehyde, N-60046
O-Nitroso-*N*,*N*-bis(trifluoromethyl)-hydroxylamine, in B-60180
3-Nitrosoindole, in I-60035
▷*N*-Nitroso-*N*-(1-methylacetonyl)-3-methyl-butylamine, see M-60055
▷*N*-Nitroso-*N*-(1-methyl-2-oxopropyl)-3-methyl-1-butanamine, see M-60055
▷*N*-Nitroso-*N*-phenylhydroxylamine, see H-70208
3-Nitroso-2-phenylindole, in P-60114
m-Nitrostyrene oxide, see N-70057
o-Nitrostyrene oxide, see N-70056
2-Nitro-1-tetralone, see N-60026
4-Nitro-*o*-tolualdehyde, see M-70096
α-Nitrotoluene, see N-70053
ω-Nitrotoluene, see N-70053
2-Nitro-4-(trifluoromethyl)anisole, in N-60048
3-Nitro-5-(trifluoromethyl)anisole, in N-60052
4-Nitro-2-(trifluoromethyl)anisole, in N-60053
4-Nitro-3-(trifluoromethyl)anisole, in N-60054
▷1-Nitro-2-(trifluoromethyl)benzene, N-70064
▷1-Nitro-3-(trifluoromethyl)benzene, N-70065
▷1-Nitro-4-(trifluoromethyl)benzene, N-70066
2-Nitro-3-(trifluoromethyl)phenol, N-60047
▷2-Nitro-4-(trifluoromethyl)phenol, N-60048
2-Nitro-5-(trifluoromethyl)phenol, N-60049
2-Nitro-6-(trifluoromethyl)phenol, N-60050
3-Nitro-4-(trifluoromethyl)phenol, N-60051
3-Nitro-5-(trifluoromethyl)phenol, N-60052
4-Nitro-2-(trifluoromethyl)phenol, N-60053
▷4-Nitro-3-(trifluoromethyl)phenol, N-60054
2-Nitrotryptophan, N-70067
6-Nitrovanillic acid, in D-60360
▷Nogamycin, see D-70506
Nonachlorophenalenyl, N-60055
2-Nonacosyl-3-tridecene-1,10-diol, N-70068
Nonactic acid, see N-70069
Nonactinic acid, N-70069
Nonacyclo[10.8.0.0²,¹¹.0⁴,⁹.0⁴,¹⁹.0⁶,¹⁷.0⁷,¹⁶.0⁹,¹⁴-.0¹⁴,¹⁹]icosane, N-70070
9-(2,4-Nonadienyloxy)-8-nonenoic acid, see C-70189
Nonafluoromorpholine, N-60056
Nonafluorotrimethylamine, see T-60406
Nonafulvalene, see B-70104
Nonafulvene, see M-60073
▷γ-Nonalactone, see D-70245
Nonamethylcyclopentanol, N-60057
▷1,4-Nonanolide, see D-70245
1,3,5,7-Nonatetraene, N-70071
1-Nonen-3-ol, N-60058
4-Nonen-2-ynoic acid, N-70072
1′-Noraltamisin, in A-70084
2,5-Norbornadiene-2-carboxylic acid, see B-60084
Norbornane-1-carboxylic acid, see B-60085
▷2-Norbornanone, see B-70098
5-Norbornene-2,3-dicarboximido carbonochloridate, see C-70059
▷Norcamphor, see B-70098
▷Norcepanone, see H-60079
29-Nor-23-cycloartene-3,25-diol, N-70073
30-Norcyclopterospermol, N-60059
30-Norcyclopterospermone, in N-60059
Nordehydro-α-matrinidine, see T-60086
Nordinone, N-60060
Nordinonediol, in N-60060
Nordracorhodin, N-70074
Nordracorubin, N-70075
2-Norerythromycin, N-70076
2-Norerythromycin B, in N-70076
2-Norerythromycin C, in N-70076
2-Norerythromycin D, in N-70076
Norhardwickiic acid, N-70077
Norharmalan, see D-60209
Norjuslimdiolone, N-60061
▷Normuscone, see C-60219

Nornotatic acid, in N-70081
17-Nor-7-oxo-8-labden-15-oic acid, in H-60198
28-Nor-16-oxo-17-oleanen-3β-ol, in M-70011
Norpatchoulenol, N-70078
Norpinene, see B-60087
Norsalvioxide, N-70079
Norsantal, see T-60108
Norstrictic acid, N-70080
Nortrachelogenin, see W-70003
3-Nortricyclanone, see T-70228
Nortropane, see A-70278
Noryangonin, see H-70122
Notatic acid, N-70081
Novain, see C-70024
▷NSC 755, see D-60270
NSC 339555, see B-70284
NSC 60719, in D-60458
NSC 76239, see P-60202
NSTX 3, see N-70033
Nuatigenin, N-70062
Nuciferol, N-70082
Nucleoside Q, in Q-70004
Nyasicaside, in B-60155
6-*O*-Acetylarbutin, in A-70248
7-*O*-Acetyldaphnoretin, in D-70006
19-*O*-Acetyl-1,2-dehydrohautriwaic acid, in H-70002
O-Angeloylalkannin, in S-70038
Oblongolide, O-60001
Obscuronatin, O-60002
Obtusinin, in H-70153
Obtusadiene, O-70001
3-*O*-Caffeoyl-4-*O*-sinapoylquinic acid, in C-70005
Ochromycinone, O-60003
Ochromycinone methyl ether, in O-60003
11,11,12,12,13,13,14,14-Octacyano-1,4:5,8-anthradiquinotetramethane, O-70002
2,4-Octadecadienoic acid, O-70003
2,4-Octadecadien-1-ol, O-60004
Octadecahydrotriphenylene, O-70004
3-Octadecene-1,2-diol, O-60005
9-Octadecene-1,12-diol, O-60006
2-Octadecen-1-ol, O-60007
2-Octadecylicosanoic acid, O-70005
5,6,10,11,16,17,21,22-Octadehydro-7,9,18,20-tetrahydrodibenzo[*e*,*n*]-[1,10]dioxacyclo-octadecin, O-70006
2,4-Octadien-1-ol, O-70007
N,*N*,*N*′,*N*′,*N*″,*N*″,*N*‴,*N*‴-Octaethylbenzo[1,2-*c*:4,5-*c*′]dipyrrole-1,3,5,7-tetramine, see T-70125
Octaethyltetramethyl[26]porphyrin[3.3.3.3], O-60008
2,2,3,3,5,5,6,6-Octafluoromorpholine, O-60009
Octafluoronaphthalene, O-70008
Octafluoro[2.2]paracyclophane, O-70009
4,5,7,8,12,13,15,16-Octafluoro[2.2]-paracyclo-phane, O-60010
5,6,11,12,13,14,15,16-Octafluorotricyclo[8.2.2.2⁴,⁷]-hexadeca-4,6,10,12,13,15-hexaene, see O-70009
▷Octahydroazocine, O-70010
5,7,9,14,16,18,28,33-Octahydro-28,33[1,2′]-benzeno-7,16[2′,3′]-anthraceno-5,18[1,2′]:9,14[1″,2″]-dibenzenoheptacene, see T-60407
Octahydro-7*H*-1-benzopyran-7-one, O-60011
Octahydrobenzo[*b*]thiophene, O-60012
Octahydro-2*H*-1,3-benzoxazine, O-70011
Octahydro-2*H*-3,1-benzoxazine, O-70012
▷6*H*,13*H*-Octahydrobis[pyridazino[1,2-*a*:1′,2′-*d*]-*s*-tetrazine, see O-70013
Octahydro-1*H*,4*H*,7*H*,10*H*-cyclo-buta[1,2:1,4:2,3:3,4]-tetracyclopentene, see P-70020
2,3,4,5,6,7,8,9-Octahydro-1*H*-cyclopent[*e*]-*as*-indacene, see O-60021
7,8,10,11,20,21,23,24-Octahydro-5,26[1′,2′]:13,18[1″,2″]-dibenzenodibenzo[*i*,*t*][1,4,7,12,15,18]-hexa-oxacyclodocosin, see H-70086
6,7,9,10,17,18,20,21-Octahydrodibenzo[*b*,*k*][1,4,7,10,13,16]-hexa-thiacyclooctadecin, see D-60073

2a,2b,4a,4b,4c,4d,4e,4f-Octahydrodicyclo-
buta[def,jkl]-biphenylene, see P-60028

Octahydrodicyclopropa[cd,gh]pentalene, see
T-70022

Octahydro-2,7-dihydroxy-5H,10H-
dipyrrolo[1,2-a:1′,2′-d]pyrazine-5,10-dione,
O-60013

Octahydro-5a,10a-dihydroxy-5H,10H-
dipyrrolo[1,2-a:1′,2′-d]pyrazine-5,10-dione,
O-60014

Octahydro-2,7-dihydroxypyrocoll, see O-60013

Octahydro-5a,10a-dihydroxypyrocoll, see
O-60014

Octahydro-2,7:3,6-dimethano-2H-1-benzopyran,
see O-70069

1,2,3,4,5,6,7,8-Octahydro-1,4:5,8-
dimethanonaphthalene, see T-70021

1,2,3,4,4a,5,6,8a-Octahydro-4,7-dimethyl-α-
methylene-1-naphthaleneacetic acid, see
C-70004

1,2,3,4,4a,5,6,7-Octahydro-4a,5-dimethyl-2-(1-
methylethenyl)naphthalene, O-60015

▷6H,13H-Octahydrodipyridazino[1,2-a:1′,2′-
d][1,2,4,5]tetrazine, O-70013

Octahydro-5H,10H-dipyrrolo[1,2-a:1′,2′-d]-
pyrazine-5,10-dione, O-60016

Octahydro-5H,10H-dipyrrolo[1,2-a:1′,2′-d]-
pyrazine-5,10-dithione, O-70014

Octahydro-1,2,4-(epoxyethanylylidene)-5H-6-
oxacyclobut[cd]indene-5,8-dione, see
D-60467

3,5,6,6a,7,8,9,10-Octahydro-9-hydroxy-7,8-di-
methyl-3-oxo-1H-naphtho[1,8a-c]-furan-7-
acetic acid, see H-70226

Octahydro-2-hydroxy-5H,10H-dipyrrolo[1,2-
a:1′,2′-d]pyrazine-5,10-dione, O-60017

1,2,3,4,4a,5,6,6a-Octahydro-11-hydroxy-
4,4,6a,9-tetramethyl-12bH-benzo[a]-
xanthene-12b-methanol, see S-60030

1,4,4a,5,6,7,8,8a-Octahydro-1-hydroxy-5,5,8a-
trimethyl-1,2-naphthalenedicarboxaldehyde,
see W-60001

Octahydroimidazo[4,5-d]imidazole, O-70015

Octahydro-1-methyl-2H-3,1-benzoxazine, in
O-70012

1,2,3,3a,6,7,8,8a-Octahydro-8a-methyl-3-(1-
methylethyl)-8-oxo-5-
azulenecarboxaldehyde, see O-70093

Octahydropyridazino[1,2-a]pyridazine, O-70016

1,2,3,4,5,6,7,8-Octahydropyridazino[4,5-d]-
pyridazine, O-60018

Octahydropyrocoll, see O-60016

Octahydropyrrolo[3,4-c]pyrrole, O-70017

Octahydro-4H-quinolizin-4-one, O-60019

1,2,3,3a,5a,6,7,8-Octahydro-1,4,7,7-tetra-
methylcyclopenta[c]pentalene, see P-70037

1,2,3,4,5,6,7,8-Octahydro-1,4,9,9-tetramethyl-
4,7-methanoazulene, see P-60010

Octahydro-1,2,5,6-tetrazocine, O-60020

Octahydrothianaphthene, see O-60012

2,3,4,5,6,7,8,9-Octahydro-1H-triindene,
O-60021

Octahydro-3,8,8-trimethyl-6-methylene-1H-
3a,7-methanoazulen-1-ol, see B-60113

Octahydro-5,5,8a-trimethyl-2(1H)-
naphthalenone, in D-70014

1,3a,4,5,5a,6,7,8-Octahydro-1,3a,6-trimethyl-8-
oxocyclopenta[c]-pentalene-2-carboxylic
acid, see S-70081

3″,3‴,4′,4‴,5,5″,7,7″-Octahydroxy-3,8″-
biflavanone, in G-70005

2′,3,4′,5,5′,6,7,8-Octahydroxyflavone, O-60022

3,3′,4′,5,5′,6,7,8-Octahydroxyflavone, O-70018

Octaleno[3,4-c]furan, O-70019

Δ^{4a(8a)}-Octalin-2,6-dione, see H-70063

2,2,3,3,4,4,5,5-Octamethylcyclopentanol,
O-60024

Octamethylcyclopentanone, O-60025

Octamethylcyclopentene, O-60026

2,6,10,14,19,23,27,31-Octamethyl-
2,6,10,14,18,22,26,30-dotriacontaoctaene, see
L-70044

2,6,10,14,19,23,27,31-Octamethyl-
2,6,10,14,17,22,26,30-dotriacontaoctaen-19-
ol, O-70020

1,1,2,2,3,3,4,4-Octamethyl-5-methylenecyclo-
pentane, O-60027

1,2,2,3,3,4,4,5-Octamethyl-6-oxabicyclo[3.1.0]-
hexane, O-70021

N,N,N′,N′,N″,N″,N‴,N‴-Octamethyl-
pyrrolo[3,4-c]pyrroleteramine, see T-70127

2-Octanamine, see O-60031

Octaphenyl[4]radialene, see T-70128

1,3,5-Octatriene, O-70022

2,4,6-Octatriene, O-70023

3-Octenal, O-70024

4-Octenedial, O-70025

4-Octene-1,8-diol, O-70026

4-Octene-1,7-diyne, O-60028

2-Octen-1-ol, O-60029

3-Octen-2-one, O-70027

4-Octen-2-one, O-60030

5-Octen-2-one, O-70028

1-Octenyl bromide, see B-70261

1-Octenyl iodide, see I-70051

10-[3-(2-Octenyl)oxiranyl]-5,8-decadienoic
acid, see E-60017

3-Octen-1-yne, O-70029

2-Octylamine, O-60031

4-O-Demethylmicrophyllinic acid, in M-70147

1-O-Demethylpsorospermindiol, in P-70133

3-O-Demethylsulochrin, in S-70083

Odoratin, in P-70032

3-O-Ethyldihydrofusarubin A, in F-60108

3-O-Feruloylquinic acid, in C-70005

7-O-Formylhorminone, in D-60304

6-O-Galloylarbutin, in A-70248

3-O-Glucosylgibberellin A₂₉, in G-70011

Okilactomycin, O-70030

9(11),12-Oleanadien-3-ol, O-70031

11,13(18)-Oleanadien-3-ol, O-70032

9(11),12-Oleanadien-3-one, O-60032

Oleanderol, in L-70042

12-Oleanene-3,15-diol, O-70033

13(18)-Oleanene-2,3-diol, O-60033

12-Oleanene-1,2,3,11-tetrol, O-60034

12-Oleanene-3,16,23,28-tetrol, O-60035

11(12)-Oleanene-3,13,23,28-tetrol, O-70034

12-Oleanene-1,3,11-triol, O-60036

12-Oleanene-2,3,11-triol, O-60037

12-Oleanene-3,16,28-triol, O-60038

13(18)-Oleanene-3,16,28-triol, O-70035

13(18)-Oleanen-3-ol, O-70036

13(18)-Oleanen-3-one, in O-70036

Olearyl oxide, in E-70023

Oleonuezhenide, O-70037

Oleuroside, O-70038

Oligomycin E, O-70039

Oliveric acid, O-60039

Olmecol, O-60040

O-Methylcedrelopsin, in C-60026

3-O-Methyldihydrofusarubin A, in F-60108

4′-O-Methylglabridin, in G-70016

2′-O-Methylisopseudocyphellarin A, in I-70098

2′-O-Methylmicrophyllinic acid, in M-70147

2′-O-Methylphenarctin, in P-70054

2′-O-Methylpseudocyphellarin A, in P-70127

4′-O-Methylpsorospermindiol, in P-70133

5-O-Methylpsorospermindiol, in P-70133

5-O-Methylsulochrin, in S-70083

Onikulactone, in M-70150

▷Ophiobalin, see O-70040

▷Ophiobolin, see O-70040

▷Ophiobolin A, see O-70040

Ophiobolin J, O-70041

Opianic acid, in F-60073

3(14),8(10)-Oplapadiene-2,6,7-triol (incorr.),
see O-70042

4,10(14)-Oplopadiene-3,8,9-triol, O-70042

Oplopanone, O-70043

Oreojasmin, O-70044

Oriciopsin, O-70045

Orlandin, O-70046

N-Ornithyl-β-alanine, O-60041

N-Ornithylglycine, O-70047

N-Ornithyltaurine, O-60042

Oroboside, in T-60108

Orotinichalcone, O-60043

Orotinin, O-60044

Orthopappolide, O-60045

Orthopappolide methacrylate, in O-60045

Orthopappolide senecioate, in O-60045

Orthopappolide tiglate, in O-60045

[2₆](Orthopara)₃cyclophanehexaene, see
T-60273

Osbeckic acid, O-70048

▷Ostensin, in T-70277

▷Ostensol, in T-70277

Osthenone, O-70049

Ovalene, O-60046

Ovothiol A, O-70050

Ovothiol B, in O-70050

Ovothiol C, in O-70050

2-Oxaadamantane, see O-60051

2-Oxa-3-azabicyclo[2.2.2]octane, O-70051

1-Oxa-2-azaspiro[2,5]octane, O-70052

3-Oxabenzocycloheptene, see T-70045

5-Oxabenzocycloheptene, see T-70043

3-Oxabicyclo[3.2.0]heptan-2-one, O-70053

7-Oxabicyclo[2.2.1]hept-2-ene, O-70054

7-Oxabicyclo[4.1.0]hept-3-ene, O-60047

3-Oxabicyclo[3.1.0]hexane-2,4-dione, in
C-70248

3-Oxabicyclo[3.1.0]hexan-2-one, O-70055

2-Oxabicyclo[3.1.0]hex-3-ene, O-70056

8-Oxabicyclo[4.3.0]nonan-7-one, see H-70061

9-Oxabicyclo[4.2.1]nona-2,4,7-triene, O-70057

9-Oxabicyclo[6.1.0]non-4-ene, O-70058

3-Oxabicyclo[4.3.0]non-7-en-2-one, see T-70072

3-Oxabicyclo[3.3.0]octan-2-one, see H-70054

5-Oxabicyclo[2.1.0]pent-2-ene, O-70059

Oxacycloheptadec-10-en-2-one, see I-60086

1,2,4-Oxadiazolidine-3,5-dione, O-70060

1,3,4-Oxadiazolidine-2,5-dione, O-70061

▷2H-[1,2,4]Oxadiazolo[2,3-a]pyridine-2-thione,
O-70062

1,3,4-Oxadiazolo[3,2-a]pyridin-2(3H)-one,
O-70063

2H-[1,2,4]Oxadiazolo[2,3-a]pyridin-2-one,
O-70064

[1,2,5]Oxadiazolo[3,4-b]quinoxaline, see
F-70051

3-Oxa-1,7-heptanediol, see O-60097

7-Oxa[2.2.1]hericene, see T-70136

Oxaiceane, see O-70069

▷Oxalodihydroxamic acid, see O-60048

▷Oxalohydroxamic acid, O-60048

1,1′-Oxalylbisisatin, see D-70466

O-Oxalylhomoserine, in A-70156

▷Oxalylhydroxamic acid, see O-60048

▷Oxamic hydrazide, see S-60020

1-Oxaphenalene, see N-70009

4-Oxapipecolic acid, see M-70155

10-Oxa-9-selena-1-azaanthracene, see B-60054

[1,4]Oxaselenino[2,3-b:5,6-b′]dipyridine,
O-70065

[1,4]Oxaselenino[3,2-b:5,6-b′]dipyridine,
O-70066

[1,4]Oxaselenino[3,2-b:5,6-c′]dipyridine,
O-70067

[1,4]Oxaselenino[3,2-b:6,5-c′]dipyridine,
O-70068

3-Oxatetracyclo[5.3.1.1^{2,6}.0^{4,9}]dodecane,
O-70069

2-(1,3-Oxathian-2-yl)pyridine, O-60049

[1,4]Oxathiino[3,2-b:5,6-c′]dipyridine, O-60050

[1,4]Oxathiino[3,2-c:5,6-b′]dipyridine (incorr.),
see O-60050

2-Oxatricyclo[3.3.1.1^{3,7}]decane, O-60051

2-Oxatricyclo[4.1.0^{1,6}.0^{3,5}]heptane, O-60052

7-Oxatricyclo[4.1.1.0^{2,5}]octane, O-60053

8-Oxatricyclo[3.3.0.0^{2,7}]octane, O-60054

4-Oxatricyclo[4.3.1.1^{3,8}]undecan-5-one,
O-60055

3-Oxawurtzitane, see O-70069

1,2,4-Oxazolidine-3,5-dione, O-70070

2-(5-Oxazolyl)furan, see F-70048

2-(2-Oxazolyl)pyridine, O-70071

2-(5-Oxazolyl)pyridine, O-70072

3-(2-Oxazolyl)pyridine, O-70073

3-(5-Oxazolyl)pyridine, O-70074
4-(2-Oxazolyl)pyridine, O-70075
4-(5-Oxazolyl)pyridine, O-70076
2-(5-Oxazolyl)thiophene, see T-70171
2-Oxepanethione, O-70077
3,3-Oxetanebis(methylamine), see B-60140
3,3-Oxetanedimethanamine, see B-60140
3,3-Oxetanedimethanol, O-60056
8,14-Oxido-9-oxo-8,9-secodolast-1(15)-ene-7,8-
diol, see C-70034
Oxindole-3-acetic acid, see D-60259
Oxipurinol, see P-60202
2-Oxiranylfuran, O-70078
2-Oxiranyl-2-methyl-1-propanol, see D-60431
Oxoallopurinol, see P-60202
10-Oxo-Δ⁹⁽¹⁰⁻ᴴ⁾,ᵃ-anthracenemalononitrile, see
D-60164
ent-3-Oxo-16α,17-atisanediol, see D-70284
7-Oxo-1-azabicyclo[3.2.0]heptane-2-carboxylic
acid, O-70079
2-Oxo-2H-benzopyran-4-acetic acid, O-60057
1-Oxo-1H-2-benzopyran-3-acetic acid, O-60058
4-Oxo-4H-1-benzopyran-3-acetic acid, O-60059
9-Oxo-1H,9H-benzopyrano[2,3-d]-v-triazole,
see B-60043
μ-Oxobis[trifluoroacetato(phenyl)iodine], see
O-60065
3-Oxobutanethioic acid, O-70080
2-(3-Oxo-1-butenyl)furan, see F-60086
7-Oxocarnosic acid, in T-60337
6-Oxo-3,13-clerodadien-15-oic acid, O-70081
ent-7-Oxo-3-clerodene-15,16:18,19-diolide, in
H-70117
4-Oxo-1-cycloheptene-1-carboxylic acid,
O-70082
3-Oxo-1-cyclohexene-1-carboxaldehyde,
O-60060
α-(2-Oxocyclohexyl)acetophenone, see P-70045
6-Oxocyclonerolidol, in H-60111
▷Oxocyclopentadecane, see C-60219
3-Oxocyclopentaneacetic acid, O-70083
2-Oxocyclopentanecarboxylic acid, O-70084
2-Oxo-1,3-cyclopentanediglyoxylic acid,
O-70085
2-Oxocyclopentanediylbisglyoxylic acid, see
O-70085
4-Oxocyclopenta[c]thiophene, see D-70184
5-Oxocyclopenta[c]thiophene, see D-70183
6-Oxocyclopenta[b]thiophene, see D-70182
3-Oxocyclopentene, see C-60226
α-Oxo-3-cyclopenteneacetaldehyde, O-70086
α-Oxo-3-cyclopentene-1-acetaldehyde, O-60061
3-Oxo-1-cyclopentenecarboxaldehyde, O-60062
α-Oxocyclopropaneacetic acid, see C-60231
5-Oxocystofuranoquinol, in C-70260
2-Oxodecanal, O-70087
4-Oxodecanal, O-70088
10-Oxodecanoic acid, O-60063
9-Oxo-2-decenoic acid, O-70089
9-Oxo-4,5-dehydro-4(15)-dihydrocostic acid, in
H-60135
2-Oxodesoxyligustrin, in O-70090
3-Oxo-1,2-diazetidine, see D-70055
4-Oxo-4,5-dihydrofuro[2,3-d]pyridazine, see
H-60146
7-Oxo-6,7-dihydrofuro[2,3-d]pyridazine, see
H-60147
7-Oxodihydrogmelinol, O-60064
1-Oxo-3H-1,2-dihydropyrrolo[1,2-a]pyrrole, see
D-60277
μ-Oxodiphenylbis(trifluoroacetato-O)-diiodine,
O-60065
12-Oxododecanoic acid, O-60066
3-Oxo-4,11(13)-eudesmadien-12-oic acid, in
E-70061
9-Oxo-4,11(13)-eudesmadien-12-oic acid, in
H-60135
3-Oxo-1,4-eudesmadien-12,8β-olide, in Y-70001
3-Oxo-4,11(13)-eudesmadien-12,8β-olide, in
H-60136
9-Oxo-4,11(13)-eudesmadien-12,16β-olide, in
H-60137
3-Oxo-1,4,11(13)-eudesmatrien-12,8β-olide, see
Y-70001
1-Oxo-7(11)-eudesmen-12,8-olide, O-60067
18-Oxoferruginol, in A-70005

3-Oxo-27-friedelanoic acid, in T-70218
1-Oxo-4,10(14)-germacradien-12,6α-olide, in
O-60068
1-Oxo-4-germacren-12,6-olide, O-60068
ent-20-Oxo-16-gibberellene-7,19-dioic acid, see
G-60019
3-Oxoglutaric acid, see O-60084
3-Oxo-4,11(13)-guaiadien-12,8β-olide, see
A-70054
2-Oxo-3,10(14),11(13)-guaiatrien-12,6-olide,
O-70090
9-Oxogymnomitryl acetate, in G-70035
ent-15-Oxo-1(10),13-halimadien-18-oic acid,
O-60069
17-Oxohebemacrophyllide, in H-60010
4-Oxoheptanal, O-60070
4-Oxoheptanedinitrile, in O-60071
4-Oxoheptanedioic acid, O-60071
4-Oxo-2-heptenedioic acid, O-60072
4-Oxohexanal, O-60073
5-Oxohexanal, O-70091
5-Oxo-3-hexenoic acid, O-70092
1-Oxohinokiol, in D-60306
18-Oxo-19-hydroxynerylgeraniol, in H-70179
2-Oxo-4-imidazolidinecarboxylic acid, O-60074
▷4-Oxo-2-imidazolidinethione, see T-70188
2-Oxo-1H,3H-imidazo[4,5-b]pyridine, see
D-60249
3-Oxoindole, see I-60035
2-(2-Oxo-3-indolinyl)acetic acid, see D-60259
3-Oxoisocostic acid, in E-70061
5-Oxoisocystofuranoquinol, in C-70260
10-Oxo-6-isodaucen-14-al, O-70093
7-Oxo-8(14),15-isopimaradien-18-oic acid,
O-70094
23-Oxoisopristimerin III, in I-70083
ent-15-Oxo-16-kauren-18-oic acid, O-60075
2-Oxokolavenic acid, in K-60013
7-Oxo-8,13-labdadien-15-oic acid, O-70095
ent-7-Oxo-7,12E,14-labdatrien-17,11α-olide, in
O-60076
6-Oxo-7-labdene-15,17-dioic acid, O-60077
12-Oxolambertianic acid, in L-60011
3-Oxo-8,24-lanostadien-26-oic acid, O-70096
3-Oxo-8-lanosten-26,22-olide, in H-60169
6-Oxolumazine, see P-70134
7-Oxolumazine, see P-70135
23-Oxomariesiic acid A, in M-70012
23-Oxomariesiic acid B, in M-60010
3-Oxomorpholine, see M-60147
α-Oxo-1-naphthaleneacetic acid, O-60078
α-Oxo-2-naphthaleneacetic acid, O-60079
4-Oxo-2-nonenal, O-70097
▷2-Oxonorbornane, see B-70098
7-Oxo-11-nordrim-8-en-12-oic acid, O-60080
4-Oxooctanal, O-60081
2-Oxo-4-oxazolidinecarboxylic acid, O-60082
4-Oxopentanal, O-60083
3-Oxo-1-pentene-1,5-dicarboxylic acid, see
O-60072
▷4-Oxo-2-pentenoic acid, O-70098
3-Oxo-2-(2-pentenyl)-cyclopentaneacetic acid,
see J-60002
N-(3-Oxo-3H-phenoxazin-2-yl)acetamide, in
A-60237
4-Oxo-2-phenyl-4H-1-benzopyran-3-carboxylic
acid, O-60085
4-Oxo-3-phenyl-4H-1-benzopyran-2-carboxylic
acid, O-60086
4-Oxo-2-phenyl-4H-1-benzothiopyran-3-
carboxylic acid, O-60087
2-Oxo-4-phenyl-2,5-dihydrofuran, see P-60103
(2-Oxo-2-phenylethyl)carbamic acid, O-70099
2-(2-Oxo-2-phenylethyl)cyclohexanone, see
P-70045
3-Oxo-2-phenylindolenine, see P-60114
1-Oxo-2-phenyl-1,2,3,4-tetrahydronaphthalene,
see D-60265
1-Oxo-3-phenyl-1,2,3,4-tetrahydronaphthalene,
see D-60266
1-Oxo-4-phenyl-1,2,3,4-tetrahydronaphthalene,
see D-60267
6-Oxo-2-phenyl-1,4,5,6-tetrahydropyrimidine,
see D-70250

2-Oxo-4-phenylthio-3-butenoic acid, O-60088
γ-Oxopimelic acid, see O-60071
4-Oxopipecolic acid, see O-60089
4-Oxo-2-piperidinecarboxylic acid, O-60089
5-Oxoproline, see O-70102
2-Oxopropanethiol, see M-60028
3-Oxopropanoic acid, O-60090
2-(3-Oxo-1-propenyl)benzaldehyde, O-60091
(2-Oxopropylidene)propanedioic acid, O-70100
2-(2-Oxopropyl)pyridine, see P-60229
3-(2-Oxopropyl)pyridine, see P-60230
4-(2-Oxopropyl)pyridine, see P-60231
3-Oxopyrazolo[1,2-a]pyrazol-8-ylium-1-olate,
O-60092
α-Oxo-1H-pyrrole-3-acetic acid, O-70101
▷2-Oxopyrrolidine, see P-70188
2-Oxo-1-pyrrolidinecarboxamide, see S-60049
5-Oxo-2-pyrrolidinecarboxylic acid, O-70102
1-Oxopyrrolizidine, see H-60057
2-Oxopyrrolizidine, see T-60091
3-Oxopyrrolizidine, see H-60058
4-Oxoquinolizidine, see O-60019
7-Oxosandaracopimaric acid, see O-70094
4-Oxo-3,4-secoambrosan-12,6-olid-3-oic acid,
O-60093
3-Oxosilphinene, see S-60033
16-Oxo-17-spongianal, O-60094
6-Oxo-1,2,3,6-tetrahydroazulene, see D-60193
1-Oxo-1,2,3,4-tetrahydrocarbazole, O-70103
2-Oxo-1,2,3,4-tetrahydrocarbazole, O-70104
3-Oxo-1,2,3,4-tetrahydrocarbazole, O-70105
4-Oxo-1,2,3,4-tetrahydrocarbazole, O-70106
5-Oxo-5,6,7,8-tetrahydroisoquinoline, see
D-60254
2-Oxo-1,2,3,4-tetrahydropyrimidine, see
D-60273
3-Oxo-2-thiabicyclo[2.2.2]oct-5-ene, see
T-60188
4-Oxo-2-thionodihydro-2H-pyrido[1,2-a]-1,3,5-
triazine, see D-60294
9-Oxotournefortiolide, in H-60137
2-Oxoverboccidentafuran, in V-70005
18-Oxo-3-virgene, O-60095
1,1'-Oxybis[2,4-dichlorobenzene], see B-70135
1,1'-Oxybis[3,4-dichlorobenzene], see B-70137
2,2'-Oxybis[1,3-dichlorobenzene], see B-70136
1,1'-Oxybis[2-iodoethane], O-60096
3,3'-Oxybis-1-propanol, O-60097
2,2'-Oxybispyridine, O-60098
2,3'-Oxybispyridine, O-60099
3,3'-Oxybispyridine, O-60100
4,4'-Oxybispyridine, O-60101
6,6'-(Oxydiethylidene)bis[7-methoxy-2,2-di-
methyl-2H-1-benzopyran], see E-70008
Oxydiformic acid, see D-70111
Oxyjavanicin, see F-60108
3[(1-Oxyl-2,2,5,5-tetramethyl-2,5-
dihydropyrrole-3-carbonyloxy)imino]-2-
phenyl-3H-indole, see D-60284
Oxyphencyclimine, O-60102
▷Oxyphencyclimine hydrochloride, in O-60102
Oxypurinol, see P-60202
Oxyresveratrol, see D-70335
Oxysolavetivone, in S-70051
Oxysporone, O-70107
P 23924A, in F-70008
P 23924B, in F-70008
P 23924C, in F-70008
P 23924D, in F-70008
P 23924E, in F-70008
P 23924F, in F-70008
Pabulenol, see P-60170
Pabulenone, in P-60170
Pachytriol, P-70001
Palasitrin, in S-70082
Palauolide, P-60001
Paldimycin, P-70002
Paldimycin A, in P-70002
Paldimycin B, in P-70002
Pallescensone, P-70003
Palliferin, in L-60014
Palliferinin, in L-60014
Pallinin, in C-60021
Pallinol, in C-60021

Palmitylglutinopallal, in G-70021
Pamamycin-607, P-70004
Panasinsanol A, P-70005
Panasinsanol B, in P-70005
>Pancridine, see D-60028
Pandoxide, P-60002
Panduratin B, P-60003
Pangesic, in A-60228
Panial, P-70006
pap, see B-60172
Papakusterol, P-60004
papf, in B-60172
papfs, in B-60172
paps, in B-60172
Paracamphorene, see C-70009
[2.2]Paracyclo(4,8)[2.2]metaparacyclophane, P-60005
[2.2]Paracyclophadiene, P-70007
[2.2]Paracyclophan-1,9-diene, see P-70007
[4]Paracyclophane, P-60006
[14.0]Paracyclophane, P-70008
[2.2][2.2]Paracyclophane-5,8-quinone, P-60007
[2.2][2.2]Paracyclophane-12,15-quinone, P-60008
[2₃]Paracyclophanetriene, see H-60038
[2.2.2]Paracyclophane-1,9,17-triene, see H-60038
[2₆]Paracyclophene, P-70009
[6]-Paracycloph-3-ene, P-70010
[2.2.2]Paracyclophene, see H-60038
Paralycolin A, P-60009
Paramicholide, P-70011
Parasorboside, in T-60067
Paristerone, in H-60063
Parvodicin, P-70012
Parvodicin A, in P-70012
Parvodicin B_1, in P-70012
Parvodicin C_1, in P-70012
Parvodicin B_2, in P-70012
Parvodicin C_2, in P-70012
Parvodicin C_3, in P-70012
Parvodicin C_4, in P-70012
Patagonic acid, P-70013
β-Patchoulene, P-60010
Patellamide A, P-60011
Patellamide B, P-60012
Patellamide C, P-60013
Patrinalloside, P-70014
PB 5266A, see A-70225
PB 5266B, see A-70226
PB 5266C, see A-70227
PC 766B, see A-70228
PC 766B', in A-70228
PD 105587, see D-70260
PDE I, see A-60286
PDE II, see A-70229
Pedicellosine, P-70015
Pedonin, P-60014
Peltigerin, P-70016
Peltochalcone, P-70017
Penduletin, in P-60047
Penlanpallescensin, P-60015
Penstebioside, P-60016
7H-2,3,4,6,7-Pentaazabenz[de]anthracene, P-60017
1,5,9,13,17-Pentaazacycloeicosane, P-60018
1,4,7,10,14-Pentaazacycloheptadecane, P-60019
1,4,7,11,14-Pentaazacycloheptadecane, P-60020
1,4,7,10,13-Pentaazacyclohexadecane, P-60021
1,4,8,12,16-Pentaazacyclononadecane, P-60022
1,4,7,11,15-Pentaazacyclooctadecane, P-60023
1,4,8,11,15-Pentaazacyclooctadecane, P-60024
1,4,7,10,13-Pentaazacyclopentadecane, P-60025
1,5,9,13,17-Pentaazaheptadecane, P-60026
Pentacarbon dioxide, see P-70040
Pentachlorocyclopropane, P-70018
Pentacyclo[5.3.0.0^{2,5}.0^{3,9}.0^{4,8}]decan-6-one, P-60027
Pentacyclo[6.4.0.0^{2,7}.0^{3,12}.0^{6,9}]dodeca-4,10-diene, P-60028
Pentacyclo[6.3.1.0^{2,4}.0^{5,10}.0^{7,8}]dodecane, P-70019
Pentacyclo[18.2.2.2^{8,11}.0^{4,14}.0^{5,17}]-hexacosa-4,8,10,14,16,20,22,23,25-nonaene, see P-60005

Pentacyclo[18.2.2.2^{9,12}.0^{4,15}.0^{6,17}]-hexacosa-4(15),6(17),-9,11,20,22,23,25-octaene-5,16-dione, see P-60007
Pentacyclo[18.2.2.2^{9,12}.0^{4,15}.0^{6,17}]-hexacosa-4,6(17),9(26),-11,15,20,22,23-octaene-10,25-dione, see P-60008
Pentacyclo[11.3.0.0^{1,5}.0^{5,9}.0^{9,13}]-hexadecane, P-70020
Pentacyclo[19.3.1.1^{2,6}.1^{9,13}.1^{14,18}]-octacosa-1(25),2,4,6(28),9,11,13(27),-14,16,18(26),21,23-dodecaene, see M-60031
Pentacyclo[4.2.0.0^{2,5}.0^{3,8}.0^{4,7}]octane-1-carboxylic acid, see C-60174
Pentacyclo[10.4.4.4^{4,9}.0^{6,23}.0^{14,19}]-tetracosa-4,6,8,12,14,16,17,19,21,23-decaene, see A-60355
Pentacyclo[12.2.2.2^{2,5}.2^{6,9}.2^{10,13}]-tetracosa-1,5,9,13-tetraene, P-60029
Pentacyclo[8.4.0.0^{2,7}.0^{3,12}.0^{6,11}]-tetradeca-4,8,13-triene, P-70021
Pentacyclo[5,4.0.0^{2,6}.0^{3,10}.0^{5,9}]-undecane-1,11-dione, P-60030
Pentacyclo[6.2.1.0^{2,7}.0^{4,10}.0^{5,9}]-undecane-3,6-dione, P-70022
Pentacyclo[5.4.0.0^{2,6}.0^{3,10}.0^{5,9}]-undecane-4,8,11-trione, P-70023
Pentacyclo[6.3.0.0^{2,6}.0^{5,9}]undecane-4,7,11-trione, P-60031
2,4-Pentadecadienal, P-70024
6-Pentadecadienylsalicylic acid, see H-60205
6-(8-Pentadecenyl)salicylic acid, in H-60205
6-(10-Pentadecenyl)salicylic acid, in H-60205
1,2-Pentadienylbenzene, see P-70085
1,1'-(1,3-Pentadienylidene)bisbenzene, see D-70489
Pentaerythritol tetramercaptan, see B-60164
3,3,4,4,4-Pentafluoro-1-butyne, P-60032
1,3,3,4,4-Pentafluorocyclobutene, P-60033
(Pentafluoroethyl)acetylene, see P-60032
Pentafluoroiodosobenzene, P-60034
Pentafluoroiodosylbenzene, see P-60034
Pentafluoroisocyanatobenzene, P-60035
Pentafluoroisocyanobenzene, P-60036
Pentafluoroisothiocyanatobenzene, P-60037
Pentafluoromethylbenzene, P-60038
2,3,4,5,6-Pentafluoro-α-methyl-benzenemethanol, see P-60040
2,3,4,5,6-Pentafluoro-α-methylbenzyl alcohol, see P-60040
Pentafluoronitrosobenzene, P-60039
Pentafluorophenol formate, see P-70025
(Pentafluorophenyl)acetylene, see E-70053
1-(Pentafluorophenyl)ethanol, P-60040
Pentafluorophenyl formate, see P-70025
Pentafluorophenyl isocyanide, see P-60036
Pentafluorophenylisothiocyanate, see P-60037
2,3,4,5,6-Pentafluorotoluene, see P-60038
Pentafluoro(trifluoromethyl)sulfur, P-60041
Pentafulvalene, see F-60080
Pentahydroxybenzaldehyde, P-60042
Pentahydroxybenzoic acid, P-60043
3,3',4,4',5-Pentahydroxybibenzyl, P-60044
2',3,4,4',5'-Pentahydroxychalcone, see D-70339
2,3,4,6,7-Pentahydroxy-9,10-dihydrophenanthrene, see D-70246
3,3',4',5,7-Pentahydroxyflavan, P-70026
2',3,5,7,8-Pentahydroxyflavanone, P-60045
>3,3',4',5,7-Pentahydroxyflavanone, P-70027
3,3',4',5,6,7-Pentahydroxyflavanone, P-70028
2',3,5,7,8-Pentahydroxyflavone, P-70029
2',3,4',5,6-Pentahydroxyflavone, P-70030
3,4',5,6,7-Pentahydroxyflavone, P-60047
3,4',5,6,8-Pentahydroxyflavone, P-60048
3,5,6,7,8-Pentahydroxyflavone, P-70031
3',4',5,6,7-Pentahydroxyisoflavone, P-70032
6,7,8,14,15-Pentahydroxy-11,13(16)-labdadiene, see L-70008
2,3,5,6,8-Pentahydroxy-7-methoxy-1,4-naphthoquinone, in H-60065
2,5,6,7,8-Pentahydroxy-3-methoxy-1,4-naphthoquinone, in H-60065
2,3,5,6,8-Pentahydroxy-1,4-naphthalenedione, see P-70033

2,3,5,6,8-Pentahydroxy-1,4-naphthoquinone, P-70033
1,2,5,6,7-Pentahydroxyphenanthrene, P-60050
1,3,4,5,6-Pentahydroxyphenanthrene, P-70034
2,3,4,7,9-Pentahydroxyphenanthrene, P-70035
3,8,9,12,17-Pentahydroxytricyclo[12.3.1.1^{2,6}]-nonadeca-1(18),2,4,6(19),14,16-hexaen-10-one, see A-60308
3',4',5,5',7-Pentahydroxy-3,6,8-trimethoxy-flavone, in O-70018
1,2,5,6,8-Pentahydroxy-9H-xanthen-9-one, see P-60052
1,2,3,4,8-Pentahydroxyxanthone, P-60051
1,2,3,6,8-Pentahydroxyxanthone, P-70036
1,2,5,6,8-Pentahydroxyxanthone, P-60052
1,3,4,7,8-Pentahydroxyxanthone (incorrect), see P-60052
Pentaisopropylidenecyclopentane, P-60053
Pentakis(1-methylethylidene)cyclopentane, see P-60053
Pentalenene, P-70037
Pentalenic acid, P-60054
Pentalenolactone E, P-60055
Pentamethoxybenzaldehyde, in P-60042
Pentamethoxybenzoic acid, in P-60043
3,3',4',5,7-Pentamethoxyflavan, in P-70026
2',3,5,7,8-Pentamethoxyflavone, in P-70029
2',3',4',5,6-Pentamethoxyflavone, in P-70030
3,4',5,6,7-Pentamethoxyflavone, in P-60047
3,5,6,7,8-Pentamethoxyflavone, in P-70031
2,3,4,7,9-Pentamethoxyphenanthrene, in P-70035
2-(Pentamethoxyphenyl)-5-(1-propenyl)-benzofuran, see R-60001
1,2,3,4,8-Pentamethoxyxanthone, in P-60051
Pentamethylanisole, in P-60058
2,3,5,6,7-Pentamethylenebicyclo[2.2.2]-octane, P-60056
3,3-Pentamethyleneoxaziridine, see O-70052
2,3,5,6,7-Pentamethylidenebicyclo[2.2.2]-octane, see P-60056
1,2,3,4,5-Pentamethyl-6-nitrobenzene, P-60057
Pentamethylphenol, P-60058
(Pentamethylphenyl)acetic acid, P-70038
2,4-Pentanediamine, P-70039
▷2,4-Pentanedione, P-60059
4,7,13,16,21-Pentaoxa-1,10-diazabicyclo[8.8.5]-tricosane, P-60060
Pentaprismane, see H-70032
1,2,3,4-Pentatetraene-1,5-dione, P-70040
4-Pentene-2,3-dione, P-60061
▷1-Penten-3-one, P-60062
▷Penthiazolidine, see T-60093
2-Penthiazolidone, see T-60096
▷5-Pentyldihydro-2(3H)-furanone, see D-70245
3-Pentyn-2-ol, P-70041
▷Perchlordecone, see D-70530
▷Perchloromethyl chloroformate, see T-70223
▷Perchloropentacyclo[5.3.0.0^{2,6}.0^{3,9}.0^{5,8}]-decane, see D-70530
Perchlorophenalenyl, see N-60055
Perchlorotriphenylmethane, see T-60405
Peregrinin, in M-70013
Peregrinone, in M-70013
Perfluoroazapropene, see T-60288
Perfluorobiacetyl, see H-60043
Perfluorobicyclobutylidene, see D-60513
Perfluoro-2,3-butanedione, see H-60043
Perfluorocyclopentadiene, see H-60044
Perfluorohexyl bromide, see B-70274
Perfluorohexyl chloride, see C-70163
Perfluoromethanesulfonimide, P-70042
▷Perfluoro(methylenecyclopropane), see D-60185
Perfluoro(methylenemethylamine), see T-60288
Perfluoro(2-methyl-1,2-oxazetidine), see T-60051
Perfluoromorpholine, see N-60056
Perfluoropiperidine, see U-60002
Perfluorotrimethylamine, see T-60406
Perforatin A, in A-60075
Perforenone, P-60063
▷Perhydroazocine, see O-70010
Perhydro-3a,6a-diazapentalene, see T-70086
Perhydroquinazoline, see D-70013

Perhydro-1,2,5,6-tetrazocine, see O-60020
Perhydrotriphenylene, see O-70004
▷Perilla ketone, see F-60087
1H-Perimidin-2(3H)-one, P-70043
Perimidone, see P-70043
Peroxyauraptenol, in A-70269
3α-Peroxy-4,11(13)-eudesmadien-12,8β-olide, in H-60136
Perrottetianal A, P-60064
Persicaxanthin, P-60065
Petrostanol, P-60066
Petrosynol, P-70044
Petrosynone, in P-70044
Peuformosin, in K-70013
PFA 186, in H-70108
PFIB, see P-60034
▷PGE₂, in D-60367
▷(5E)-PGE₂, in D-60367
(±)-8-iso-PGE₂, in D-60367
11-epi-PGE₂, in D-60367
(±)-15-epi-PGE₂, in D-60367
Pharbitis gibberellin, see G-70011
Phaseolinic acid, P-60067
Phenacylcarbamic acid, see O-70099
2-Phenacylcyclohexanone, P-70045
Phenacyl iodide, see I-60039
2-Phenacylpropionic acid, see B-60064
ψ-Phenalenone, see M-70042
Phenanthrane, see D-60260
1-Phenanthreneacetic acid, P-70046
9-Phenanthreneacetic acid, P-70047
3,4-Phenanthrenedicarboxylic acid, P-60068
3,6-Phenanthrenedimethanol, P-60069
4,5-Phenanthrenedimethanol, P-60070
4,5-Phenanthrenediol, P-70048
1,2,3,5,6,7-Phenanthrenehexol, see H-60066
1,2,5,6,7-Phenanthrenepentol, see P-60050
1,3,4,5,6-Phenanthrenepentol, see P-70034
2,3,4,7,9-Phenanthrenepentol, see P-70035
1,2,5,6-Phenanthrenetetrol, see T-60115
1,2,5,7-Phenanthrenetetrol, see T-60116
1,2,6,7-Phenanthrenetetrol, see T-60117
1,3,5,6-Phenanthrenetetrol, see T-60118
1,3,6,7-Phenanthrenetetrol, see T-60119
2,3,5,7-Phenanthrenetetrol, see T-60120
2,3,6,7-Phenanthrenetetrol, see T-60121
2,4,5,6-Phenanthrenetetrol, see T-60122
3,4,5,6-Phenanthrenetetrol, see T-60123
9-Phenanthrenethiol, P-70049
2,3,7-Phenanthrenetriol, T-60339
2,4,5-Phenanthrenetriol, P-60071
N-9-Phenanthrenyl-9-phenanthrenamine, see D-70472
▷Phenanthridene, see C-70244
Phenanthro[1,10-bc:8,9-b',c']bisthiopyran, P-60072
Phenanthro[1,2-c]furan, P-70050
Phenanthro[3,4-c]furan, P-70051
Phenanthro[9,10-b]furan, P-60073
Phenanthro[9,10-c]furan, P-60074
Phenanthro[9,10-g]isoquinoline, P-60075
1,7-Phenanthrolin-8-amine, see A-70180
1,10-Phenanthrolin-2-amine, see A-70179
1,10-Phenanthrolin-4-amine, see A-70181
1,10-Phenanthrolin-5-amine, see A-70182
1,7-Phenanthroline-5,6-dione, P-70052
1,10-Phenanthroline-2-methanamine, see A-60224
Phenanthro[3,4,5,6-bcdef]ovalene, P-60076
Phenanthro[9,10-c][1,2,5]selenadiazole, P-70053
Phenanthro[9,10-e]thieno[2,3-b]pyridine, see D-70079
Phenanthro[4,5-bcd]thiophene, P-60077
Phenarctin, P-70054
Phenazone, see B-60020
Phenethylhydroquinone, see P-60095
m-Phenethylphenol, see P-70072
o-Phenethylphenol, see P-70071
p-Phenethylphenol, see P-70073
4-Phenethylpyrocatechol, see P-60096
4-Phenethylresorcinol, see P-60097
5-Phenethylresorcinol, see P-60098
Phenetrol, see B-60014
Phenetsal, in H-70108
▷Phenidone, see P-70091
Phenmenthol, in M-60090

Phenosol, in H-70108
10H-Phenothiazine-1-carboxylic acid, P-70055
10H-Phenothiazine-2-carboxylic acid, P-70056
10H-Phenothiazine-3-carboxylic acid, P-70057
10H-Phenothiazine-4-carboxylic acid, P-70058
1H-Phenothiazin-1-one, P-60078
1-Phenoxy-1-phenylethylene, in P-60092
α-Phenoxystyrene, in P-60092
Phenylacetohydroxamic acid, P-70059
2-Phenyl-2-adamantanol, P-70060
3,3'-[1-[(Phenylamino)carbonyl-3,4-tetra-zolium]bis(4-methoxy-6-nitrobenzenesulfonic acid)], see B-70151
3-Phenyl-2-aza-1,4-naphthoquinone, see P-70083
3-(Phenylazo)-2-butenenitrile, P-60079
3-(Phenylazo)crotononitrile, see P-60079
α-Phenylbenzenemethanimine, see D-60482
2-Phenyl-4H-3,1,2-benzooxathiazine, P-60080
2-Phenyl-1-benzosuberone, see T-60080
▷2-Phenyl-1,5-benzothiazepin-4(5H)-one, P-60081
2-Phenyl-4H-benzo[b]thiin-4-one, see P-60082
2-Phenyl-4H-1-benzothiopyran-4-one, see P-60082
2-Phenyl-2H-benzotriazole, P-60083
2-Phenyl-1,3-benzoxathiol-1-ium, P-60084
1-Phenyl-3-buten-1-ol, P-70061
5-Phenyl-2(3H)-butenolide, see P-70079
β-Phenyl-Δ^{α,β}-butenolide, see P-60103
1-Phenyl-3-buten-1-one, P-60085
1-Phenyl-2-butyn-1-ol, P-70062
4-Phenyl-3-butyn-2-ol, P-70063
4-Phenylcarbostyril, see P-60131
▷3-Phenyl-2-cyclobuten-1-one, P-70064
2-Phenylcyclohexanecarboxylic acid, P-70065
α-Phenylcyclohexaneglycolic acid (1,4,5,6-tetra-hydro-1-methyl-2-pyrimidinyl)methyl ester, see O-60102
2-Phenylcyclopentanamine, P-60086
2-Phenylcyclopentylamine, P-60086
2-Phenylcyclopropanecarboxylic acid, P-60087
1-Phenylcyclopropanol, P-70066
β-Phenylcysteine, see A-60213
3-Phenyl-3H-diazirine, P-70067
▷Phenyl 5,6-dichloro-2-(trifluoromethyl)-1H-benzimidazole-1-carboxylate, in D-60160
2-Phenyl-1,4-dihydropyrimidine, see D-70249
2-Phenyl-1,6-dihydropyrimidine, see D-70249
4-Phenyldisic acid, see D-70337
3-Phenyl-1,4,2-dithiazole-5-thione, P-60088
[4]Phenylene, P-60089
[5]Phenylene, P-60090
2,2'-p-Phenylenebis[5-(4-biphenylyl)]-oxazole, see P-60091
2,2'-(1,4-Phenylene)bis[5,1-[1,1'-biphenyl]-4-yl]oxazole, P-60091
2,2'-(1,3-Phenylene)bisfuran, see D-70168
2,2'-(1,4-Phenylene)bisfuran, see D-70169
1,2-Phenylenebismethyl, see D-70410
1,1'-(1,3-Phenylene)bisnaphthalene, see D-60454
2,2'-(1,3-Phenylene)bisnaphthalene, see D-60455
Phenylene blue, see I-60019
1:12-o-Phenyleneperylene, see N-60009
2,2'-o-Phenylene-1,1'-spirobiindan, see T-70058
1-Phenylethanesulfonic acid, P-70068
2-Phenylethanethiol, P-70069
▷1-Phenylethanone, see A-60017
1-Phenylethenol, P-60092
2-(2-Phenylethenyl)benzothiazole, P-60093
2-(2-Phenylethenyl)-1H-pyrrole, P-60094
[(2-Phenylethenyl)seleno]benzene, P-70070
[(2-Phenylethenyl)selenonyl]benzene, in P-70070
2-(2-Phenylethyl)-1,4-benzenediol, P-60095
4-(2-Phenylethyl)-1,2-benzenediol, P-60096
4-(2-Phenylethyl)-1,3-benzenediol, P-60097
5-(2-Phenylethyl)-1,3-benzenediol, P-60098
α-(1-Phenylethylidene)benzeneacetaldehyde, see D-60475
β-Phenylethylmercaptan, see P-70069
2-(2-Phenylethyl)phenol, P-70071
3-(2-Phenylethyl)phenol, P-70072
4-(2-Phenylethyl)phenol, P-70073

2-(Phenylethynyl)benzoic acid, P-70074
4-(Phenylethynyl)benzoic acid, P-70075
2-(Phenylethynyl)thiazole, P-60099
4-(Phenylethynyl)thiazole, P-60100
N-Phenylformamidine, P-60101
2-Phenyl-3-furancarboxylic acid, P-70076
4-Phenyl-2-furancarboxylic acid, P-70077
5-Phenyl-2-furancarboxylic acid, P-70078
3-Phenyl-2(5H)furanone, P-60102
4-Phenyl-2(5H)-furanone, P-60103
5-Phenyl-2(3H)-furanone, P-70079
2-Phenyl-3-furoic acid, see P-70076
4-Phenyl-2-furoic acid, see P-70077
5-Phenyl-2-furoic acid, see P-70078
2-Phenylfuro[2,3-b]pyridine, P-60104
2-Phenylfuro[3,2-c]pyridine, P-60105
3-Phenylglutamic acid, see A-60241
Phenylglycolaldehyde, see H-70206
Phenylglycolaldehyde, see H-60210
7-Phenylguanine, see A-70133
1-Phenyl-6-hepten-1-one, P-70080
7-Phenyl-5-heptynoic acid, P-70106
7-Phenyl-3-heptyn-2-ol, P-60107
1-Phenyl-1,4-hexadiyn-3-one, P-60108
▷1-Phenyl-2,4-hexadiyn-1-one, P-60109
6-Phenyl-3,5-hexadiyn-2-one, P-60110
1-Phenyl-5-hexen-1-one, P-60081
α-Phenylhydratropic acid, see D-70493
2-Phenyl-1H-imidazole-4(5)-carboxaldehyde, P-60111
Phenyl 2-imidazolyl ketone, see B-70074
3-(Phenylimino)-1(3H)-isobenzofuranone, P-60112
4-(Phenylimino)-2-pentanone, in P-60059
3-(Phenylimino)phthalide, see P-60112
1-Phenyl-1H-indole, P-70082
6-Phenyl-1H-indole, P-60113
N-Phenylindole, see P-70082
2-Phenyl-3H-indol-3-one, P-60114
▷Phenyl(β-indolyl)iodone, see I-70017
2-Phenylisatogen, in P-60114
2-Phenylisatogen oxime, in P-60114
8-Phenylisomenthol, in M-60090
Phenylisonitromethane, in N-70053
N-Phenylisophthalimide, see P-60112
3-Phenyl-1,4-isoquinolinedione, P-70083
3-Phenyl-5-isoxazolamine, see A-60240
4-Phenyl-3,5-isoxazolediol, see D-70337
2-Phenyl-3,5-isoxazolidinedione, see D-70337
(−)-8-Phenylmenthol, in M-60090
(+)-8-Phenylmenthol, in M-60090
N-Phenylmethanimidamide, see P-60101
(Phenylmethoxy)acetaldehyde, see B-70085
1-(Phenylmethoxy)-3-buten-2-ol, in B-70291
2-(Phenylmethoxy)phenol, see B-60072
3-(Phenylmethoxy)phenol, see B-60073
▷4-(Phenylmethoxy)phenol, see B-60074
2-(Phenylmethyl)aspartic acid, see A-60094
▷Phenylmethyl carbonochloridate, see B-60071
7-(Phenylmethylene)bicyclo[4.1.0]hepta-1,3,5-triene, see B-70084
(Phenylmethylene)cycloheptane, see B-60068
9-(Phenylmethylene)-1,3,5,7-cyclononatetraene, see B-60069
1-(Phenylmethylene)-1H-cyclopropabenzene, see B-70084
N-(Phenylmethylene)methanamine, see B-60070
5-(1-Phenylmethyl-1H-imidaz-2-yl)-1H-tetra-zole, in I-70009
1-(Phenylmethyl)-1H-tetrazol-5-amine, in A-70202
N-(Phenylmethyl)-1H-tetrazol-5-amine, in A-70202
4-Phenyl-3-morpholinone, in M-60147
3-Phenyl-1,2-naphthalenedione, see P-60120
4-Phenyl-1,2-naphthalenedione, see P-60121
1-Phenyl-2-naphthalenol, see P-60115
3-Phenyl-2-naphthalenol, see P-60116
4-Phenyl-2-naphthalenol, see P-60117
5-Phenyl-2-naphthalenol, see P-60118
8-Phenyl-2-naphthalenol, see P-60119
1-Phenyl-2-naphthol, P-60115
3-Phenyl-2-naphthol, P-60116
4-Phenyl-2-naphthol, P-60117
5-Phenyl-2-naphthol, P-60118
8-Phenyl-2-naphthol, P-60119
3-Phenyl-1,2-naphthoquinone, P-60120

4-Phenyl-1,2-naphthoquinone, P-60121
(+)-8-Phenylneomenthol, *in* M-60090
Phenylnitromethane, *see* N-70053
Phenylnitrosohydroxylamine tosylate, *in*
    H-70208
10-Phenylnonafulvene, *see* B-60069
9-Phenylnonanoic acid, P-70084
5-Phenyl-1,2,4-oxadiazole-3-carboxaldehyde,
    P-60122
2-Phenyl-1,3-oxazepine, P-60123
5-Phenyl-1,4-oxazepine, P-60124
5-Phenyloxazole, P-60125
4-Phenyl-1,3-oxazolidine-2-one, *see* P-60126
4-Phenyl-2-oxazolidinone, P-60126
S-Phenyl 3-oxobutanethioate, *in* O-70080
1-Phenyl-4-oxo-2-pyrroline, *in* D-70262
1-Phenyl-1,2-pentadiene, P-70085
5-Phenyl-2,4-pentadienoic acid, P-70086
1-Phenyl-1,4-pentadiyn-3-one, P-70087
1-Phenyl-3-penten-1-one, P-60127
5-Phenyl-4-penten-2-one, P-70088
3-Phenyl-1-pentyne, P-70089
2N-Phenyl-N-[[(phenylamino)(phenylimino)-
    methyl]amino]benzothiazole, *see* B-70057
▷α-Phenyl-α-(2-phenylethenyl)benzenemethanol,
    see T-70313
2-Phenyl-4-(phenylmethylene)-5(4H)-
    oxazolone, P-60128
▷N-Phenylphthalamidine, *in* P-70100
N-Phenylphthalisoimide, *see* P-60112
1-Phenyl-2-piperazinone, *in* P-70101
4-Phenyl-Δ³-piperideine, *see* T-60081
Phenylpropargyl alcohol, *see* P-60130
Phenylpropargylamine, *see* P-60129
3-Phenyl-2-propenoyl isothiocyanate, P-70090
3-Phenyl-2-propyn-1-amine, P-60129
3-Phenyl-2-propyn-1-ol, *see* P-60130
▷1-Phenyl-3-pyrazolidinone, P-70091
1-Phenyl-1H-pyrazolo[3,4-e]indolizine, P-70092
Phenyl 3-pyrazolyl ketone, *see* B-70078
Phenyl-1H-pyrazol-3-ylmethanone, *see* B-70078
Phenyl-1H-pyrazol-4-ylmethanone, *see* B-60065
5-Phenylpyrimidine, P-70093
4-Phenylpyrogallol, *see* B-70124
2-Phenyl-1-pyrroline, *see* D-70251
1-Phenyl-2-(2-pyrrolyl)ethylene, *see* P-60094
2-Phenyl-4-quinolinol, *in* P-60132
4-Phenyl-2-quinolinol, *see* P-60131
7-Phenyl-8-quinolinol, *see* H-60218
2-Phenyl-4(1H)-quinolinone, P-60132
4-Phenyl-2(1H)-quinolinone, P-60131
Phenylselenium tribromide, *see* T-70213
Phenylselenium trichloride, *see* T-70224
9-(Phenylseleno)phenanthrene, P-60133
1-Phenylseleno-2-phenylethene, *see* P-70070
Phenyl styryl selenide, *see* P-70070
Phenyl styryl selenone, *in* P-70070
Phenyl styryl selenoxide, *in* P-70070
3-(Phenylsulfinyl)butanoic acid, *in* M-70023
1-(Phenylsulfinyl)-2-(phenylsulfonyl)-cyclo-
    butene, *in* B-70156
1-(Phenylsulfinyl)-2-(phenylthio)-cyclobutene,
    in B-70156
2-Phenylsulfonyl-1,3-butadiene, P-70094
2-(Phenylsulfonyl)-1,3-cyclohexadiene, P-70095
4-Phenyltetralone, *see* D-60267
2-Phenyl-1-tetralone, *see* D-60265
3-Phenyl-1-tetralone, *see* D-60266
3-Phenyltetrazolo[1,5-a]pyridinium, P-60134
4-Phenylthieno[3,4-b]furan, P-60135
2-Phenylthieno[2,3-b]pyridine, P-60136
Phenylthioacetic acid, *see* B-70017
3-(Phenylthio)butanoic acid, *in* M-70023
2-(Phenylthio)juglone, *see* H-70213
3-(Phenylthio)juglone, *see* H-70214
Phenylthiolacetamide, *in* B-70017
2-[(Phenylthio)methyl]pyrrolidine, *in* P-70185
(Phenylthio)nitromethane, P-60137
9-(Phenylthio)phenanthrene, *in* P-70049
N-Phenyl-2-thiophenesulfonamide, *in* T-60213
N-Phenyl-3-thiophenesulfonamide, *in* T-60214
5-Phenyl-2-thiouracil, *see* D-60268
m-Phenyltoluene, *see* M-60053

▷6-Phenyl-1,3,5-triazine-2,4-diamine, *see*
    D-60041
1-Phenyl-1H-1,2,3-triazole-4-carboxylic acid,
    P-60138
1-Phenyl-1H-1,2,3-triazole-5-carboxylic acid,
    P-60139
2-Phenyl-2H-1,2,3-triazole-4-carboxylic acid,
    P-60140
3-Phenyl-3H-1,2,3-triazole-4-carboxylic acid,
    see P-60139
Phenyl tribromomethyl sulfide, *see* T-70212
Phenyl tribromomethyl sulfone, *in* T-70212
Phenyl tribromomethyl sulfoxide, *in* T-70212
Phenyl trichloromethyl ketone, *see* T-70219
2-Phenyltricyclo[3.3.1.1³,⁷]decan-2-ol, *see*
    P-70060
Phenyl(triphenylmethyl)diazene, *see* T-70312
Phenyl triphenylmethyl ketone, *see* T-60167
Phenyl trityl ketone, *see* T-60167
11-Phenylundecanoic acid, P-70096
1-Phenylvinyl alcohol, *see* P-60092
Phillygenin, P-60141
Phillyrin, P-60141
Phlebiakauranol alcohol, *in* P-70097
Phlebiakauranol aldehyde, P-70097
Phleichrome, P-70098
Phlogantholide A, P-70099
Phloganthoside, *in* P-70099
Phloinolic acid, *in* T-60334
Phloionolic acid, *in* T-60334
Phloracetophenone, *see* T-70248
Phloretin, *see* H-60219
▷Phlorhizin, *in* H-60219
▷Phloridzin, *in* H-60219
▷Phlorrhizin, *in* H-60219
Phoenicoxanthin, *see* A-70066
O-Phosphohomoserine, *in* A-70156
Phosphorylhomoserine, *in* A-70156
Phrymarolin II, P-60142
Phrymarolin I, *in* P-60142
1-Phthalazinamine, *see* A-70184
5-Phthalazinamine, *see* A-70185
▷Phthalidanil, *in* P-70100
Phthalidochromene, P-60143
Phthalimidine, P-70100
ψ-Phthalimidine, *see* I-70062
8-p-Hydroxybenzoylshiromodiol, *in* S-70039
Phyllostine, *see* H-70181
Physalactone, P-60144
Physalin D, P-60145
Physalin G, P-60146
Physalin L, P-60147
Physalin I, *in* P-60145
Physalolactone, P-60148
Physanolide, P-60149
Physarochrome A, P-60150
▷Physoperuvine, P-60151
Phytal, *in* P-60152
Phytol, P-60152
Picolamine, *see* A-60228
Picolamine salicylate, *in* A-60228
Picolinylaminomethane, *see* A-70085
▷α-Picolylamine, *see* A-60227
β-Picolylamine, *see* A-60228
γ-Picolylamine, *see* A-60229
2-Picolyl bromide, *see* B-70253
3-Picolyl bromide, *see* B-70254
4-Picolyl bromide, *see* B-70255
▷Picramide, *see* T-70301
Picroside I, *in* C-70026
Picroside II, *in* C-70026
Picroside III, *in* C-70026
Pidolic acid, *see* O-70102
▷Pigmex, *see* B-70074
Pilosanone A, P-60153
Pilosanone B, *in* P-60153
ent-8(14)-Pimarene-2α,3α,15R,16-tetrol,
    P-60154
Pinguisanin, P-60155
Pinopalustrin, *see* W-70003
Pinusolide, P-60156
2,5-Piperazinedicarboxylic acid, P-60157
Piperazinone, P-70101
Δ¹-Piperideine, *see* T-70090
▷Δ³-Piperideine, *see* T-70089
3-Piperidinecarboxamide, *in* P-70102
3-Piperidinecarboxylic acid, P-70102

3-Piperidineethanol, P-70103
▷Piperidinic lactam, *see* P-70188
2-Piperidinotetrahydrofuran, *see* T-60082
1-(2-Piperidinyl)ethanone, *see* A-70051
4-Piperidone-2-carboxylic acid, *see* O-60089
Piperimidine, *see* H-60056
α-Pipitzol, P-60158
Pipoxide, P-60159
Pitumbin, P-70104
Pityol, P-60160
Pivalophenone, *see* D-60443
Platanetin, P-70105
Platanin, *see* T-70118
Platypterophthalide, P-60161
Pleiadene, *see* C-60198
Pleiadiene, *see* C-60198
Pleiaheptalene, *see* C-60197
Pleichrome, *see* P-70098
Plumbazeylanone, P-70106
▷Pluramycin A, P-60162
Pluviatide, *in* A-70063
PO 1, *see* T-70262
Podocarpusflavanone, P-60163
Podoverine A, P-60164
Podoverine C, P-60165
Podoverine B, *in* P-60165
Polemannone, P-60166
Polivione, P-60167
Polydine, *in* P-70026
Polygonatoside A, *in* S-60045
Laurencia Polyketal, P-60168
Polyphyllin A, *in* S-60045
Polyphyllin D, *in* S-60045
13,17,21-Polypodatriene-3,8-diol, P-70107
Polypodine A, *in* H-60063
Polypodine B, *in* H-60023
Polypodoaurein, *in* H-60063
Pomolic acid, *in* D-70350
Pomonic acid, *in* D-70350
Pongone, P-70108
Ponostop, *in* H-70108
21H,23H-Porphyrazine, P-70109
Porphyrexide, P-70110
Portmicin, P-70111
Praderin, P-70112
Praecansone B, P-70113
Praecansone A, *in* P-70113
Praeruptorin A, P-60169
Prangeline, P-60170
Prasinic acid, *see* S-60063
Pratensein, *in* T-60108
Precapnelladiene, P-60171
Precocene II, *see* D-70365
Precocene III, *in* D-70365
Premnaspiral, *in* P-70114
Premnaspirodiene, P-70114
Prephytoene alcohol, P-70115
Preuroporphyrinogen, *see* H-70166
Primulagenin A, *in* O-60038
Pringleine, P-60172
Proclavaminic acid, P-70116
▷Proflavine, *see* D-60028
Prolinal, *see* P-70184
Prolylproline anhydride, *see* O-60016
N-Prolylserine, P-60173
1,2-Propadiene-1-thione, P-70117
2-(1,2-Propadienyl)benzothiazole, P-60174
▷1,2-Propanediol, P-70118
2,2-Propanedithiol, P-60175
▷1,2,3-Propanetriol 1-acetate, *see* G-60029
1,2,3-Propanetriol 1,3-diacetate, *see* D-60023
Propargylallene, *see* H-70040
▷Propargylamine, *see* P-60184
4-Propargyl-2-azetidinone, *see* P-60185
α-Propargylserine, *see* A-60191
Propargylvinylcarbinol, *see* H-70089
[4.2.2]Propella-7,9-diene, *see* T-70226
[4.4.4]Propellahexaene, *see* T-70232
▷Propenal, P-70119
1-Propene-1-sulfenic acid, *see* P-70123
2-Propenethioamide, P-70120
▷1-Propene-1,2,3-tricarboxylic acid, *see* A-70055
▷2-Propenoic acid, P-70121
4-(2-Propenyl)-2-azetidinone, *in* A-60172
2-(2-Propenyl)benzaldehyde, P-70122

4-(2-Propenyl)benzaldehyde, P-60177
α-2-Propenylbenzenemethanol, *see* P-70061
4-[5-(1-Propenyl)-2-benzofuranyl]-2,3-
   benzenediol, *in* H-60214
4-[5-(1-Propenyl)-2-benzofuranyl]phenol, *see*
   H-60214
1-(1-Propenyl)cyclohexene, P-60178
1-Propen-2-ylcyclohexene, *see* I-60115
1-(2-Propenyl)cyclohexene, P-60179
3-Propen-2-ylcyclohexene, *see* I-60116
3-(2-Propenyl)cyclohexene, P-60180
4-(1-Propenyl)cyclohexene, P-60181
4-Propen-2-ylcyclohexene, *see* I-60117
4-(2-Propenyl)cyclohexene, P-60182
1,1′-(1-Propenylidene)bisbenzene, *see* D-60491
3-(2-Propenyl)indole, P-60183
2-Propenyl 3-(2-propenylsulfinyl)-1-propenyl
   disulfide, *see* A-70079
Propenylsulfenic acid, P-70123
8,1-[1]Propen[1]yl[3]ylidene-1*H*-benzocyclo-
   hepten-4(9*H*)-one, *see* M-70042
▷Propionylethylene, *see* P-60062
▷3-Propylacrolein, *see* H-60076
3-Propylaspartic acid, *see* A-70186
▷Propylene dibromide, *see* D-60105
▷α-Propylene glycol, *see* P-70118
▷Propylene oxide, *see* M-70100
Propyleneurea, *see* T-60088
▷2-Propylidenepropionaldehyde, *see* M-70106
Propyl viologensulfonate, *see* B-70160
▷2-Propyn-1-amine, P-60184
4-(2-Propynyl)-2-azetidinone, P-60185
α-1-Propynylbenzenemethanol, *see* P-70062
Prosapogenin *D*′₃, *in* S-60045
▷Prostaglandin E₂, *in* D-60367
▷Prostarmon *E*, *in* D-60367
▷Prostenon, *in* D-60367
▷Prostin *E*₂, *in* D-60367
▷Prostrumyl, *see* D-70232
Protetrone, P-70124
Proton sponge, *in* D-70042
Protosappanin *A*, P-60186
▷Prunetol, *see* T-60324
Przewanoic acid *A*, P-70125
Przewanoic acid *B*, P-70126
Pseudilin, *see* H-60217
Pseudobaptigenin, *see* B-60006
Pseudobaptisin, *in* B-60006
Pseudocyphellarin *B*, P-70127
Pseudocyphellarin A, *in* P-70127
Pseudo-α-D-glucopyranose, *in* H-70170
Pseudo-β-D-glucopyranose, *in* H-70170
Pseudo-α-DL-glucopyranose, *in* H-70170
Pseudo-α-L-glucopyranose, *in* H-70170
Pseudo-α-DL-idopyranose, *in* H-70170
Pseudo-α-L-idopyranose, *in* H-70170
Pseudo-β-L-idopyranose, *in* H-70170
3-Pseudoindolone, *see* I-60035
Pseudoleucine, *see* A-70136
Pseudoopianic acid, *in* F-60070
Pseudophenalenone, *see* M-70042
Pseudophthalimidine, *see* I-70062
Pseudopterogorgia diterpenoid *B*, P-70128
Pseudopterogorgia diterpenoid *C*, P-70129
Pseudopterosin *A*, P-60187
Pseudopterosin B, *in* P-60187
Pseudopterosin C, *in* P-60187
Pseudopterosin D, *in* P-60187
Pseudrelone *B*, P-70130
Psilostachyin, P-70131
Psorolactone, P-60188
Psorospermin, P-70132
Psorospermindiol, P-70133
▷Ptaquiloside, P-60190
Ptelatoside A, *in* H-70209
Ptelatoside B, *in* H-70209
2,4-Pteridinediamine, *see* D-60042
4,6-Pteridinediamine, *see* D-60043
4,7-Pteridinediamine, *see* D-60044
6,7-Pteridinediamine, *see* D-60045
2,4(1*H*,3*H*)-Pteridinedione, P-60191
2,4,7-Pteridinetriamine, *see* T-60229
4,6,7-Pteridinetriamine, *see* T-60230
2,4,6(1*H*,3*H*,5*H*)-Pteridinetrione, P-70134
2,4,7(1*H*,3*H*,8*H*)-Pteridinetrione, P-70135
Pterin, *see* A-60242

Pterocarpin, *in* M-70001
Pterosterone, P-60192
Pteryxin, *in* K-70013
12-Ptilosarcenol, P-60193
Ptilosarcenone, *in* P-60193
Ptilosarcenone, P-70136
Ptilosarcol, P-60195
Ptilosarcone, *in* P-60195
Puerol *A*, P-70137
Puerol B, *in* P-70137
Pueroside A, *in* P-70137
Pueroside B, *in* P-70137
Pulchellin, *see* P-70138
Pulchellin *A*, P-70138
Pulchellin diacetate, *in* P-70138
Pulchellin 2-O-isovalerate, *in* P-70138
Pulchellin 2-O-tiglate, *in* P-70138
Pulchelloside I, P-70139
Pumilaisoflavone *A*, P-70140
Pumilaisoflavone *B*, P-70141
Punaglandin 1, P-70142
Punaglandin 2, *in* P-70142
Punaglandin 3, P-70143
Punaglandin 4, P-70143
(*E*)-Punaglandin 4, *in* P-70143
Punctaporonin *A*, P-60197
Punctaporonin D, *in* P-60197
Punctatin *A*, *see* P-60197
Punctatin D, *in* P-60197
▷1*H*-Purin-2-amine, *see* A-60243
9*H*-Purin-8-amine, *see* A-60244
9*H*-Purin-9-amine, *see* A-60245
▷6-Purinethiol, *see* D-60270
6-Purinethiol, *in* D-60270
▷Purinethiol, *see* D-60270
Pygmaeocine *E*, P-70144
Pygmaeoherin, P-70145
Pyoverdin *C*, P-70146
Pyoverdin *D*, P-70147
Pyoverdin *Pa*, *in* P-70147
Pyoverdin I, P-70148
Pyoverdin II, P-70149
Pyoverdin III, *in* P-70149
Pyoverdin E, *in* P-70147
Pyracanthoside, *in* T-70105
Pyraceheptylene, *see* D-60167
4,4-Pyrandiacetic acid, P-70150
Pyranthrene, P-70151
Pyrazino[2′,3′:3,4]cyclobuta[1,2-*g*]-quinoxaline,
   P-60198
Pyrazino[2,3-*b*]pyrido[3′,2′-*e*][1,4]thiazine,
   P-60199
Pyrazino[2,3-*g*]quinazoline-2,4-(1*H*,3*H*)-dione,
   P-60200
Pyrazino[2,3-*f*]quinazoline-8,10-(7*H*,9*H*)-dione,
   P-60201
▷1*H*-Pyrazole, P-70152
Pyrazole blue, P-70153
2-Pyrazoline, *see* D-70253
10*H*-Pyrazolo[5,1-*c*][1,4]benzodiazepine,
   P-70154
1*H*-Pyrazolo[3,4-*b*]pyridine, P-70155
1*H*-Pyrazolo[3,4-*d*]pyrimidine, P-70156
1*H*-Pyrazolo[3,4-*d*]pyrimidine-4,6(5*H*,7*H*)-
   dione, P-60202
1*H*-Pyrazolo[4,3-*g*]quinazoline-5,7(6*H*,8*H*)-
   dione, P-60203
Pyrazolo[3,4-*f*]quinazolin-9(8*H*)-one, P-60204
Pyrazolo[4,3-*g*]quinazolin-5(6*H*)-one, P-60205
Pyrazolo[4,3-*f*]quinazolin-9(8*H*)-one, *see*
   D-60271
Pyrazolo[3,4-*c*]quinoline, P-60206
Pyrazolo[1,5-*b*][1,2,4]triazine, P-60207
1*H*-Pyrazolo[5,1-*c*]-1,2,4-triazole, P-70157
2-[(1*H*-Pyrazol-1-yl)methyl]pyridine, *see*
   P-70178
2-(1*H*-Pyrazol-1-yl)pyridine, P-70158
2-(1*H*-Pyrazol-3-yl)pyridine, P-70159
3-(1*H*-Pyrazol-1-yl)pyridine, P-70160
3-(1*H*-Pyrazol-3-yl)pyridine, P-70161
4-(1*H*-Pyrazol-1-yl)pyridine, P-70162
4-(1*H*-Pyrazol-3-yl)pyridine, P-70163
4-(1*H*-Pyrazol-4-yl)pyridine, P-70164
1-Pyreneacetic acid, P-60208
4-Pyreneacetic acid, P-60209
Pyrene-4,5:9,10-dioxide, *see* D-70152
1-Pyrenesulfonic acid, P-60210

2-Pyrenesulfonic acid, P-60211
4-Pyrenesulfonic acid, P-60212
Pyrenocin *A*, P-60213
Pyrenocin B, *in* P-60213
Pyreno[1,2-*c*]furan, P-70165
Pyreno[4,5-*b*]furan, P-60214
Pyreno[4,5-*c*]furan, P-70166
▷Pyrenophorin, P-70167
1-Pyrenyldiazomethane, *see* D-70062
Pyrichalasin *H*, P-60215
Pyriculol, P-60216
2,3-Pyridane, *see* D-60275
3,4-Pyridane, *see* D-60276
3,6-Pyridazinedicarboxylic acid, P-70168
Pyridazino[1″,6″:1′,2′]imidazo[4′,5′:4,5]-
   imidazo[1,2-*b*]pyridazine, P-70169
3,5-Pyridinebis(propanoic acid), *see* P-60217
2,3-Pyridinediol, *in* H-60223
2,4-Pyridinediol, *in* H-60224
2,5-Pyridinediol, *in* H-60225
2,6-Pyridinediol, *in* H-60226
3,4-Pyridinediol, *in* H-60227
3,5-Pyridinedipropanoic acid, P-60217
2,3-Pyridinedithiol, P-60218
2,4-Pyridinedithiol, P-60219
2,5-Pyridinedithiol, P-60220
3,4-Pyridinedithiol, P-60221
3,5-Pyridinedithiol, P-60222
1,1′-(2,6-Pyridinediyl)bis[2-hydroxyethanone],
   see B-70144
2-Pyridineethanethiol, *see* P-60239
4-Pyridineethanethiol, *see* P-60240
▷2-Pyridinemethanamine, *see* A-60227
3-Pyridinemethanamine, *see* A-60228
4-Pyridinemethanamine, *see* A-60229
2(1*H*)-Pyridineselone, P-70170
2-Pyridinesulfonic acid, P-70171
2-Pyridinethiol, P-60223
▷2(1*H*)-Pyridinethione, P-60224
▷4(1*H*)-Pyridinethione, P-60225
2,4,6-Pyridinetriol, *see* D-70341
N-Pyridinium-2-benzimidazole, P-70172
Pyridinium 2-benzimidazolylide, *see* P-70172
2-Pyridinol sulfite (2:1) (ester), *see* D-60497
[3.3][2.6]Pyridinophane, P-60226
[3](2.2)[3](5.5)Pyridinophane, P-60227
[3](2.5)[3](5.2)Pyridinophane, P-60228
N-(3-Pyridinylcarbonyl)glycine, *see* N-70034
2-(3(4*H*)-Pyridinylidene)-1*H*-indene-1,3(2*H*)-
   dione, *see* P-60242
2-(4(1*H*)-Pyridinylidene)-1*H*-indene-1,3(2*H*)-
   dione, *see* P-60243
2-(2-Pyridinyl)-1*H*-indene-1,3(2*H*)-dione, *see*
   P-60241
2-(3-Pyridinyl)-1*H*-indene-1,3(2*H*)-dione, *see*
   P-60242
2-(4-Pyridinyl)-1*H*-indene-1,3(2*H*)-dione, *see*
   P-60243
2-(3-Pyridinyloxy)pyridine, *see* O-60099
1-(2-Pyridinyl)-2-propanone, P-60229
1-(3-Pyridinyl)-2-propanone, P-60230
1-(4-Pyridinyl)-2-propanone, P-60231
2-Pyridinyl-3-pyridinylmethanone, *see* P-70179
2-Pyridinyl-4-pyridinylmethanone, *see* P-70180
3-Pyridinyl-4-pyridinylmethanone, *see* P-70181
2-Pyridinyl trifluoromethanesulfonate, P-70173
Pyrido[2′,3′:3,4]cyclobuta[1,2-*g*]quinoline,
   P-60232
Pyrido[3′,2′:3,4]cyclobuta[1,2-*g*]quinoline,
   P-60233
Pyrido[3′,4′:4,5]furo[3,2-*b*]indole, P-70174
Pyrido[1″,2″:1′,2′]imidazo[4′,5′:4,5]-
   imidazo[1,2-*a*]pyridine, P-70175
Pyrido[2″,1″:2′,3′]imidazo[4′,5′:4,5]-
   imidazo[1,2-*a*]pyridine, P-60234
Pyrido[2′,1′:2,3]imidazo[4,5-*c*]isoquinoline,
   P-60235
Pyrido[1′,2′:3,4]imidazo[1,2-*a*]pyrimidin-5-
   ium(1+), P-70176
1*H*-Pyrido[2,3-*b*]indol-4-amine, *see* A-70107
5*H*-Pyrido[3,2-*b*]indol-8-amine, *see* A-70119
5*H*-Pyrido[4,3-*b*]indol-1-amine, *see* A-70114

5*H*-Pyrido[4,3-*b*]indol-3-amine, *see* A-70115
5*H*-Pyrido[4,3-*b*]indol-5-amine, *see* A-70116
5*H*-Pyrido[4,3-*b*]indol-6-amine, *see* A-70117
5*H*-Pyrido[4,3-*b*]indol-8-amine, *see* A-70118
▷ 9*H*-Pyrido[2,3-*b*]indol-2-amine, *see* A-70105
9*H*-Pyrido[2,3-*b*]indol-3-amine, *see* A-70106
9*H*-Pyrido[2,3-*b*]indol-6-amine, *see* A-70108
9*H*-Pyrido[3,4-*b*]indol-1-amine, *see* A-70109
9*H*-Pyrido[3,4-*b*]indol-3-amine, *see* A-70110
9*H*-Pyrido[3,4-*b*]indol-4-amine, *see* A-70111
9*H*-Pyrido[3,4-*b*]indol-6-amine, *see* A-70112
9*H*-Pyrido[3,4-*b*]indol-8-amine, *see* A-70113
Pyridooxadiazolone, *see* O-70064
Pyrido[2,3-*d*]pyrimidine-2,4(1*H*,3*H*)-dione, P-60236
Pyrido[3,4-*d*]pyrimidine-2,4(1*H*,3*H*)-dione, P-60237
Pyrido[3,2-*e*]thiazole, *see* T-70168
6*H*-Pyrido[3′,4′:4,5]thieno[2,3-*b*]indole, P-70177
Pyrido[3,2-*e*]-2-thiouracil, *see* D-60291
Pyrido[3,4-*d*]-1,2,3-triazin-4(3*H*)-one, P-60238
2,3,4-Pyridotriazole, *see* T-60241
2-Pyridylacetone, *see* P-60229
3-Pyridylacetone, *see* P-60230
4-Pyridylacetone, *see* P-60231
2-(2-Pyridyl)-1,3-dithiane, *see* D-60504
2-(2-Pyridyl)ethanethiol, P-60239
2-(4-Pyridyl)ethanethiol, P-60240
3-(Pyridylformamido)acetic acid, *see* N-70034
2-(2-Pyridyl)-1,3-indanedione, P-60241
2-(3-Pyridyl)-1,3-indanedione, P-60242
2-(4-Pyridyl)-1,3-indanedione, P-60243
3-Pyridyl isocyanate, *see* I-70065
2-Pyridyl isothiocyanate, *see* I-70106
3-Pyridyl isothiocyanate, *see* I-70107
2-Pyridyl ketone, *see* D-70502
3-Pyridyl ketone, *see* D-70503
4-Pyridyl ketone, *see* D-70504
2-Pyridyl mercaptan, *see* P-60223
▷ 2-Pyridylmethylamine, *see* A-60227
3-Pyridylmethylamine, *see* A-60228
4-Pyridylmethylamine, *see* A-60229
2-Pyridylmethyl bromide, *see* B-70253
3-Pyridylmethyl bromide, *see* B-70254
4-Pyridylmethyl bromide, *see* B-70255
2-(2-Pyridyl)oxazole, *see* O-70071
2-(3-Pyridyl)oxazole, *see* O-70073
2-(4-Pyridyl)oxazole, *see* O-70075
5-(2-Pyridyl)oxazole, *see* O-70072
5-(3-Pyridyl)oxazole, *see* O-70074
5-(4-Pyridyl)oxazole, *see* O-70076
1-(2-Pyridyl)pyrazole, *see* P-70158
1-(3-Pyridyl)pyrazole, *see* P-70160
1-(4-Pyridyl)pyrazole, *see* P-70162
3-(2-Pyridyl)pyrazole, *see* P-70159
3-(3-Pyridyl)pyrazole, *see* P-70161
3-(4-Pyridyl)pyrazole, *see* P-70163
4-(4-Pyridyl)pyrazole, *see* P-70164
(2-Pyridyl)(1-pyrazolyl)methane, P-70178
2-Pyridyl 3-pyridyl ketone, P-70179
2-Pyridyl 4-pyridyl ketone, P-70180
3-Pyridyl 4-pyridyl ketone, P-70181
2-(2-Pyridyl)tetrahydrothiophene, *see* T-60097
2-(2-Pyridyl)-1,3,4-thiadiazole, *see* T-60192
2-(4-Pyridyl)-1,3,4-thiadiazole, *see* T-60193
1-(2-Pyrimidinyl)ethanone, *see* A-60044
1-(4-Pyrimidinyl)ethanone, *see* A-60045
1-(5-Pyrimidinyl)ethanone, *see* A-60046
Pyrimido[4,5-*i*]imidazo[4,5-*g*]cinnoline, P-60244
Pyrimido[4,5-*b*]quinoline-2,4(3*H*,10*H*)-dione, P-60245
Pyrindane, *see* D-70258
1-Pyrindane, *see* D-70258
2-Pyrindane, *see* D-70259
2,3-Pyrindane, *see* D-70258
3,4-Pyrindane, *see* D-70259
▷ Pyrithione, *see* H-70217
Pyrocarbonic acid, *see* D-70111
▷ Pyrocitric acid, *see* A-70055
Pyroglutamic acid, *see* O-70102
Pyrophthalone, *see* P-60241
β-Pyrophthalone, *see* P-60242
γ-Pyrophthalone, *see* P-60243

Pyrrocoline, *see* I-60033
Pyrrole-3-aldehyde, *see* P-70182
1*H*-Pyrrole-3-carboxaldehyde, P-70182
1*H*-Pyrrole-3,4-dicarboxylic acid, P-70183
Pyrrole-3-glyoxylic acid, *see* O-70101
2-Pyrrolidinecarboxaldehyde, P-70184
2-Pyrrolidinemethanethiol, P-70185
2-Pyrrolidinethione, P-70186
3-Pyrrolidinol, P-70187
▷ 2-Pyrrolidinone, P-70188
2-Pyrrolidinotetrahydrofuran, *see* T-60090
▷ 2-Pyrrolidone, P-70188
2-Pyrrolidone-5-carboxylic acid, *see* O-70102
2,5-Pyrrolizidinedione, *see* D-70260
1-Pyrrolizidinone, *see* H-60057
2-Pyrrolizidinone, *see* T-60091
3-Pyrrolizidinone, *see* H-60058
5*H*-Pyrrolo[1,2-*a*]azepine, P-70189
3*H*-Pyrrolo[1,2-*a*]azepin-3-one, P-70190
5*H*-Pyrrolo[1,2-*a*]azepin-5-one, P-70191
7*H*-Pyrrolo[1,2-*a*]azepin-7-one, P-70192
9*H*-Pyrrolo[1,2-*a*]azepin-9-one, P-70193
4*H*,6*H*-Pyrrolo[1,2-*a*][4,1]benzoxazepine, P-60246
5*H*-Pyrrolo[1,2-*a*]imidazol-5-one, P-70194
5*H*-Pyrrolo[1,2-*c*]imidazol-5-one, P-70195
9*H*-Pyrrolo[1,2-*a*]indol-9-one, P-60247
1*H*-Pyrrol-3-ol, *see* D-70262
3(2*H*)-Pyrrolone, *see* D-70262
Pyrrolo[3,2,1-*kl*]phenothiazine, P-70196
Pyrrolo[1,2-*a*]pyrazine, P-70197
5*H*-Pyrrolo[2,3-*b*]pyrazine, P-60248
Pyrrolo[1,2-*a*]pyrazin-1(2*H*)-one, P-70198
6*H*-Pyrrolo[1,2-*b*]pyrazol-6-one, P-70199
Pyrrolo[1,2-*a*]pyridine, *see* I-60033
▷ 1*H*-Pyrrolo[3,2-*b*]pyridine, P-60249
2*H*-Pyrrolo[3,4-*c*]pyridine, P-70200
6*H*-Pyrrolo[3,4-*b*]pyridine, P-70201
1-(1*H*-Pyrrol-3-yl)ethanone, *see* A-70052
2-(1*H*-Pyrrol-2-yl)phenol, *see* H-60217
4-(1*H*-Pyrrol-1-yl)phenol, *see* H-60216
3-(1*H*-Pyrrol-2-yl)-2-propenal, P-70202
1-(2-Pyrrolyl)-2-(2-thienyl)ethylene, P-60250
▷ 2-Pyrrolyl thiocyanate, *see* T-60209
Pyrroxamycin, P-70203
Pyruvic acid ethyl enol ether, *see* E-70037
PZ 5, *see* D-70191
Q base, *see* Q-70004
Qing Hau acid, *see* C-70004
Quadrangolide, Q-60001
Queen substance, *in* O-70089
Quench spot, *see* A-60019
Quercetol *A*, Q-70001
Quercetol *B*, Q-70002
Quercetol *C*, Q-70003
▷ Questiomycin *A*, *see* A-60237
Queuine, Q-70004
Queuosine, *in* Q-70004
2,4-Quinazolinediamine, *see* D-60048
5,8-Quinazolinediol, Q-60002
Quindoline, *see* I-70016
▷ Quinhydrone, *in* B-70044
Quinic acid, Q-70005
Quinide, *in* Q-70005
5*H*,9*H*-Quino[3,2,1-*de*]acridine, Q-70006
o-Quinodimethane, *see* D-70410
4,5-Quinolinediamine, *see* D-70044
Quinoline 5,6-oxide, *see* E-60027
Quinoline 7,8-oxide, *see* E-60028
3-Quinolinethiol, Q-60003
4-Quinolinethiol, *see* Q-60006
5-Quinolinethiol, Q-60004
6-Quinolinethiol, Q-60005
4(1*H*)-Quinolinethione, Q-60006
1-(2-Quinolinyl)ethanone, *see* A-60049
1-(3-Quinolinyl)ethanone, *see* A-60050
1-(4-Quinolinyl)ethanone, *see* A-60051
1-(5-Quinolinyl)ethanone, *see* A-60052
1-(6-Quinolinyl)ethanone, *see* A-60053
1-(7-Quinolinyl)ethanone, *see* A-60054
1-(8-Quinolinyl)ethanone, *see* A-60055
4-Quinolizidinone, *see* O-60019
Quinomethane, *see* M-60071
o-Quinomethane, *see* M-60072
▷ Quinone, *see* B-70044
o-Quinonemethide, *see* M-60072

p-Quinonemethide, *see* M-60071
Quino[7,8-*h*]quinoline, Q-70007
Quino[7,8-*h*]quinoline-4,9(1*H*,12*H*)dione, Q-60007
2,3-Quinoxalinediol, *see* D-60279
5,8-Quinoxalinediol, Q-60008
6,7-Quinoxalinediol, Q-60009
5,8-Quinoxalinedione, Q-60010
Quinoxaline 5,8-quinone, *see* Q-60010
12*H*-Quinoxalino[2,3-*b*][1,4]benzothiazine, Q-70008
12*H*-Quinoxalino[2,3-*b*][1,4]benzoxazine, Q-70009
1-(2-Quinoxalinyl)ethanone, *see* A-60056
1-(5-Quinoxalinyl)ethanone, *see* A-60057
1-(6-Quinoxalinyl)ethanone, *see* A-60058
2-(2-Quinoxalinyl)furan, *see* F-60089
2,2′:5′,2″:5″,2‴:5‴,2‴′-Quinquethiophene, Q-70010
α-Quinquethiophene, *see* Q-70010
Quinrhodine, *see* T-60201
Quisquagenin, Q-70011
Racemic acid, *in* T-70005
RA-I, *in* R-60002
RA-II, *in* R-60002
RA-III, *in* R-60002
RA-IV, *in* R-60002
Ramentone, *see* D-60352
Ramosissin, R-60001
Ranunculin, *in* H-60182
Ratanhiaphenol II, *see* E-60075
Ratanhiaphenol I, *in* H-60214
Rathyronine, *in* T-60350
RA-V, R-60002
RA-VII, *in* R-60002
Regelide, *in* H-70197
Regelin, *in* H-70202
Regelinol, *in* D-70333
Rehmaglutin *C*, R-60003
Reiswigin *A*, R-60004
Reiswigin B, *in* R-60004
Rengyol, *see* H-60133
Repensolide, R-60005
Resinoside, R-60006
Resorcinolacetic acid, *see* H-70204
Resorcinol benzyl ether, *see* B-60073
Resorcylide, R-70001
Rheumacyl, *in* H-70108
Rheumatidermol, *in* H-70108
Rhinocerotinoic acid, *see* O-70095
Rhizoic acid, *see* B-70007
Rhodizonic acid, *see* D-60315
ζ-Rhodomycinone, *in* R-70002
ε-Rhodomycinone, R-70002
τ-Rhodomycinone, *in* R-70002
Rhodonocardin *A*, R-70003
Rhodonocardin B, *in* R-70003
Rhodotorucin *A*, R-70004
13R-Hydroxy-8(17),14-labdadien-3-one, *in* L-70002
Rhynchosperin *A*, R-60007
Rhynchosperin B, *in* R-60007
Rhynchosperin C, *in* R-60007
Rhynchospermoside *A*, R-60008
Rhynchospermoside B, *in* R-60008
9-β-D-Ribofuranosyl-9*H*-purin-2-amine, *in* A-60243
▷ 9-β-D-Ribofuranosyl-9*H*-purine-6-thiol, *see* T-70180
Riccardin D, R-70005
Riccardin E, *in* R-70005
Richardianidin 1, R-70006
Richardianidin 2, *in* R-70006
Ricinelaidyl alcohol, *in* O-60006
Ricinoleyl alcohol, *in* O-60006
Ridentin, R-70007
Riligustilide, R-60009
Ro 2-5959, *in* T-60350
Ro 09-0198, *see* A-70231
Rolziracetam, *see* D-70260
Rolziracetanal, *see* D-70260
Roridin D, R-70008
▷ Rosaniline, R-60010
ent-5-Rosene-3α,15,16,18-tetrol, *in* R-60011
ent-5-Rosene-3α,15,16,19-tetrol, R-60011
ent-5-Rosene-15,16,18-triol, *in* R-60012
ent-5-Rosene-15,16,19-triol, R-60012

▷Roseofungin, R-70009
Rosmanol, R-70010
Rosmanoyl carnosate, R-60013
[3]Rotane, *see* T-70325
Rotenic acid, *see* H-60164
Dalbergia Rotenolone, R-60014
Roxburghin, *see* H-70215
Royflex, *in* H-70108
Royleanone, R-60015
RS 43179, *in* C-60118
▷Rubeanic acid, *see* E-60041
Rubiatriol, *in* F-60002
Rubiflavin F, R-70012
Rubiflavin C_1, R-70011
Rubiflavin A, *in* P-60162
Rubiflavin C_2, *in* R-70011
Rubiflavin D, *in* R-70011
Rubiflavin E, *in* R-70011
Rubifolide, R-60016
Rubiginic acid, *see* D-60368
Rubonic acid, *see* D-70469
Rudbeckin A, R-70013
Ruptalgor, *in* A-60228
Rutaevinexic acid, R-70014
Rutalpinin, R-70015
Rutamarin, *in* C-60031
Rutamarin alcohol, *see* C-60031
R 20Y7, *see* A-70230
Rzedowskin A, R-70016
Rzedowskin B, *in* R-70016
Rzedowskin C, *in* R-70016
S 604, *in* C-70075
Safflor Yellow B, S-70001
Saikogenin F, S-60001
Saikosaponin A, *in* S-60001
Sainfuran, S-60002
▷Sal ethyl, *in* H-70108
Salicortin, S-70002
▷Salicylamide, *in* H-70108
Salicylanilide, *in* H-70108
▷Salicylic acid, *see* H-70108
N-Salicyl-4-methoxyanthranilic acid, *see* D-70049
Salifebrin, *in* H-70108
Saliment, *in* H-70108
Salimethyl, *in* H-70108
Salinidol, *in* H-70108
Salirepin, *in* D-70287
Salirepol, *see* D-70287
Salireposide, *in* D-70287
Salizolo, *in* H-70108
Salmester, *in* H-70108
Salocolum, *in* H-70108
Salophen, *in* H-70108
Sal-Rub, *in* H-70108
▷Salstan, *in* H-70108
▷Saltron, *in* C-70049
Salvianolic acid B, S-70003
Salvianolic acid C, *in* S-70003
Salvicanaric acid, S-70004
Salvicin, S-60003
Salvileucolidone, S-70005
Salviolone, S-70006
Salvisyriacolide, S-60004
Sambucoin, S-70007
▷Samidin, *in* K-70013
Sanadaol, S-70008
8(14),15-Sandaracopimaradiene-2,18-diol, *see* I-60113
8(14),15-Sandaracopimaradiene-3,18-diol, *see* I-70080
Sanguinone A, S-60006
▷Sanoflavin, *see* S-60028
Santal, *in* T-60108
Santamarine, S-60007
▷Saponaceolide A, S-60008
Saponin A, *in* C-70002
Saptarangiquinone A, S-70009
Sarcodictyin A, S-70010
Sarcodictyin B, *in* S-70010
Sarcophytoxide, S-70011
Sargahydroquinoic acid, S-60009
Sargaquinone, S-60010
Saroaspidin A, S-70012
Saroaspidin B, S-70013
Saroaspidin C, S-70014
Sarocol, *in* H-70108

Sarothamnoside, *in* T-60324
Sarothralen B, S-60011
Sauvagine, S-70015
Sawaranin, S-70016
Saxosterol, *in* C-70172
Scapaniapyrone A, S-70017
Sch 33256, *in* M-70045
Sch 34164, *in* M-70045
Sch 36969, *see* M-70045
Sch 38519, *see* A-60287
Schkuhridin B, S-60012
Schkuhridin A, *in* S-60012
Sclareol, S-70018
Scleroderolide, S-60013
Scleroderris green, S-60014
Scopadulcic acid A, S-60015
Scopadulcic acid B, *in* S-60015
▷Scopoletin, *see* H-70153
Scutellarioside I, *in* C-70026
Scutellarioside II, *in* C-70026
Scutellone A, S-60017
Scutellone C, S-70019
Scutellone F, S-70020
Scuterivulactone C_1, *see* S-60017
Scuterivulactone C_2, *in* S-60017
Scuterivulactone D, *in* T-70010
Sebacic semialdehyde, *see* O-60063
Secocubane, *see* T-70023
3,4-Seco-4(28),20,24-dammaratrien-3,26-dioic acid, S-70021
Secofloribundione, S-70022
Seco-4-hydroxylintetralin, S-70023
Secoisoerivanin pseudoacid, S-70024
4,5-Seconeopulchell-5-ene, S-70025
Seconidoresedic acid, *see* S-70077
Secopseudopterosin A, S-70026
Secopseudopterosin B, *in* S-70026
Secopseudopterosin C, *in* S-70026
Secopseudopterosin D, *in* S-70026
3,4-Seco-4(23),14-taraxeradien-3-oic acid, S-60018
Secothujene, S-70027
Secotrinervitane, S-60019
Seglitide, *see* C-60180
Seglitide acetate, *in* C-60180
Seiricuprolide, S-70028
Selenium diselenocyanate, *see* D-60165
2-Selenochroman-4-one, *see* B-70051
Selenocyanic acid methyl ester, *see* M-60127
1,1′-(Selenodi-2,1-ethenediyl)bisbenzene, *see* D-60503
Selenofluorenone, *see* F-60018
▷Selenoformaldehyde, S-70029
Selenoformic acid, *see* M-70035
Selenolo[3,4-*b*]thiophene, S-70030
2-Selenopyridine, *see* P-70170
▷Selenourea, S-70031
Selezen, *in* H-70108
1,11-Selinanediol, *see* E-70062
4(15)-Selinene-1,11-diol, *see* E-70064
4(15)-Selinen-11-ol, *see* E-70066
Semecarpetin, S-70032
▷Semioxamazide, S-60020
Semperoside, S-60021
Senburiside II, S-70033
Senecialdehyde, *see* M-70055
Senecioaldehyde, *see* M-70055
Senecioodontol, S-60022
Senegalensein, S-70034
Senepoxide, S-60023
β-Senepoxide, *in* S-60023
Senkyunolide F, *in* L-60026
Senkyunolide H, *in* L-60026
Sepiapterin, S-70035
Septuplinolide, *in* H-60142
▷Seragen, *in* H-60092
N-Serylmethionine, S-60024
N-Serylproline, S-60025
Sesamoside, S-70036
Sesebrinic acid, S-60026
Sesquinorbornatriene, *see* T-60032
Sesquinorbornene, *see* T-70021
12-Sesquisabinenal, *in* S-70037
Sesquisabinene, S-70037
Sesquisabinene hydrate, *in* S-70037
12-Sesquisabinenol acetate, *in* S-70037
13-Sesquisabinenol acetate, *in* S-70037

Sesquithujene, *in* S-70037
Sessein, S-60027
SF 1993, *see* C-70015
SF 2312, *see* A-60288
SF 2339, *see* A-60289
SF 2415A_1, *see* A-70232
SF 2415B_1, *see* A-70235
SF 2415A_2, *see* A-70233
SF 2415B_2, *see* A-70236
SF 2415A_3, *see* A-70234
SF 2415B_3, *see* A-70237
Shanzhiside, S-60028
Shanzhisin methyl ester gentiobioside, *in* S-60028
Shikalkin, *in* S-70038
Shikonin, S-70038
Shikonin acetate, *in* S-70038
Shikonin isobutyrate, *in* S-70038
Shiromodiol, S-70039
Shonachalin D, S-70040
SI 23548, *see* S-60029
Siaresinolic acid, *in* D-70327
Siccanin, S-60029
Siccanochromene E, S-60030
Sidendrodiol, *see* K-60006
Siderin, *in* D-60348
Sigmoidin B, S-70041
Sigmoidin D, S-60031
Sigmoidin A, *in* S-70041
Sikkimotoxin, S-60032
Silandrin, S-70042
Sileneoside, *in* H-60063
Silidianin, *see* S-70044
6-Siliphiperfolene, S-70043
Silital, *in* C-70075
3-Silphinenone, S-60033
Silydianin, S-70044
Silyhermin, S-70045
Silymonin, *in* S-70044
Sinapyl alcohol, S-70046
Sintenin, S-70047
Sinulareone, S-60034
Sirutekkone, S-70048
Sitophilure, *in* H-60183
SKFD 2623, *in* T-60350
Skimmial, *in* D-70316
Skimmianone, *in* D-70316
Skimminin, S-60035
Skimmiol, *in* D-70318
▷Skiodan sodium, *in* I-60048
Skutchiolide A, S-70049
Skutchiolide B, S-70050
SL 3440, *see* C-60038
SL 1 pigment, *in* N-70049
Smenospongine, S-60036
Solavetivone, S-70051
Solenolide A, S-70052
Solenolide B, *in* S-70052
Solenolide C, S-70053
Solenolide D, *in* S-70053
Solenolide E, S-70054
Solenolide F, S-70055
Solumag, *in* O-70102
▷Soluran, *in* C-70049
Sophicoroside, *in* T-60324
Sophojaponicin, *in* M-70001
Sophoraside A, *in* P-70137
▷Sophoricol, *see* T-60324
Sophorocarpan A, S-60037
Sophorocarpan B, S-60038
Sordariol, S-70056
Sorguyl, *see* T-70096
Soulattrone A, S-70057
Soyanal, *see* O-70091
13,17-Spatadien-10-ol, S-70058
Specionin, S-70059
Specionin, S-60039
Sphaerobioside, *in* T-60324
Sphaeropyrane, S-70060
Sphaeroxetane, S-70061
Spinochrome D, *see* P-70033
Spinochrome E, *see* H-60065
Spinosic acid, *in* D-70327
2,2′-Spirobi[2*H*-1-benzopyran], S-70062
2,2′-Spirobichroman, *see* T-70094
2,2′-Spirobichromene, *see* S-70062
Spiro[bicyclo[3.2.1]octane-8,2′-[1,3]dioxolane], *in* B-60094

Spirobi[9H-fluorene], S-60040
Spirobrassinin, S-60041
Spiro[3,4-cyclohexano-4-hydroxybicyclo[3.3.1]-
 nonan-9-one-2,1'-cyclohexane], S-70063
▷Spirohydantoin, see D-60056
Spiroketal enol ether polyyne, see H-70042
Spiroketal enol ether polyyne, in H-70042
Spiro[4.4]non-2-ene-1,4-dione, S-60042
Spirosal, in H-70108
1,2,3-Spirostanetriol, S-60043
5-Spirostene-3,25-diol, S-60044
25(27)-Spirostene-2,3,6-triol, S-70064
5-Spirosten-3-ol, S-60045
25(27)-Spirosten-3-ol, S-70065
Spiro[5.5]undeca-1,3-dien-7-one, S-60046
Spiro[5.5]undecane-1,9-dione, S-70066
Spiro[5.5]undecane-3,8-dione, S-70067
Spiro[5.5]undecane-3,9-dione, S-70068
1(10),11-Spirovetivadien-2-one, see S-70051
Spirovetivene, in P-70114
Spongiadiol, in D-70345
17-Spongianal-16-one, see O-60094
Spongiolactone, S-60047
Spongionellin, S-60048
Sporol, S-70069
Squamolone, S-60049
Squaric acid monochloride, see C-70071
Squarrofuric acid, S-70070
SS 21020C, see A-70238
SS 43405D, see A-70239
SS 46506A, see P-70203
SS 7313A, see D-70321
Stachydrine, S-60050
Stachydrine ethyl ester, in S-60050
Stacopin P1, S-60051
Stacopin P2, in S-60051
Staurane, see T-60035
2,5,8,11-Stauranetetraene, see T-60036
Stearylglutinopallal, in G-70021
Stemodin, S-70071
Stemodinone, in S-70071
Stemphylin, see A-60080
Stenosporonic acid, S-70072
Sterebin D, S-60052
Sterebin A, in S-60052
Sterebin B, in S-60052
Sterebin C, in S-60052
Sterebin E, in L-70009
Sterebin F, in L-70009
Sterebin G, in L-70008
Sterebin H, in L-70008
Sterophyllolide, in D-60324
1-Sterpurene-9,12-diol, see D-70347
Sterpuric acid, S-60053
Stevisamolide, S-70073
3,6-Stigmastanediol, S-70074
4-Stigmastene-3,6-diol, S-60054
24(28)-Stigmastene-2,3,22,23-tetrol, S-70075
Stilbenediamine, see D-70480
Stilbene dibromide, see D-70091
2,3,3',4,4',5'-Stilbenehexol, see T-60344
2,3',4,5-Stilbenetetrol, see H-70215
2,3',4,5'-Stilbenetetrol, see D-70335
Stilbenobis-15-crown-5, see B-60142
Stillopsidin, see D-70339
Stillopsin, in D-70339
Stolonidiol, S-60055
Stolonidiol acetate, in S-60055
Streptol, in H-70172
Strictaepoxide, S-60056
Strictaketal, S-70076
Strictic acid, S-70077
Strigol, S-60057
▷Strumacil, see D-70232
2-Styrylbenzothiazole, see P-60093
2-Styrylpyrrole, see P-60094
Subcordatolide D, S-70078
Subcordatolide E, S-70079
Suberenol, S-70080
Subergorgic acid, S-70081
Subexpinnatin, in H-70138
Subexpinnatin C, S-60058
Subexpinnatin B, in S-60058
Substance P, S-60059
Substictic acid, S-60060
1,2-Succinoylpiperidazine, see H-70070

Suksdorfin, in K-70013
Sulcatine, S-60061
N-(2-Sulfoethyl)glutamine, see G-60027
4,4'-Sulfonylbisbenzoic acid, in T-70175
4,4'-Sulfonyldibenzoic acid, in T-70175
Sulfurein, in S-70082
Sulfuretin, S-70082
▷Sulfuryl chloride isocyanate, S-60062
Sulochrin, S-70083
Sulphurein, in S-70082
Sulphuretin, in S-70082
Superlatolic acid, S-60063
▷Superpalite, see T-70223
Suspensolide, S-70084
Suvanine, S-70085
Swalpamycin, S-70086
Swerchirin, in T-70123
Swertialactone D, S-60064
Swertialactone C, in S-60064
Sydonic acid, S-60065
Sylvestroside I, S-70087
Sylvestroside II, in S-70087
Sylvestroside III, S-70088
Sylvestroside IV, S-70089
Synrotolide, S-60066
Syoyualdehyde, see M-70104
Syringenin, in S-70046
Syringin, in S-70046
Syringopicrogenin A, S-60067
Syringopicrogenin B, in S-60067
Syringopicrogenin C, in S-60067
Syringopicroside B, in S-60067
Syringopicroside C, in S-60067
▷T₃, in T-60350
▷T 42082, see C-60022
Taccalonolide B, T-70001
Taccalonolide C, T-70002
Taccalonolide A, in T-70001
Taccanolide D, in T-70001
Tackle, see S-60029
Tagitinin A, T-60002
Tagitinin E, T-70003
Talaromycin A, T-70004
Talaromycin B, in T-70004
Talaromycin C, in T-70004
Talaromycin D, in T-70004
Talaromycin E, in T-70004
Talaromycin F, in T-70004
Talaromycin G, in T-70004
TAN 665A, see D-70063
Tanavulgarol, T-60004
Tannunolide A, T-60005
Tannunolide B, in T-70005
Taonianone, T-60006
Taraktophyllin, T-60007
14-Taraxerene-3,24-diol, T-60008
Tarchonanthus lactone, T-60009
Tartaric acid, T-70005
▷Tauremisin A, see V-60017
▷Taxifolin, see P-70027
Taxifolin 4'-glucoside, in P-70027
Taxillusin, in P-70027
Taxoquinone, in D-60304
Tecomaquinone I, T-60010
Tecomaquinone III, T-70006
Tenuiorin, see P-70016
Tephrinone, in G-70015
Tephrobbottin, T-70007
Teracrylshikonin, in S-70038
2,3':2',3''-Ter-1H-indole, T-60011
Ternatin, T-70008
[2](4,4'')-1,1':2',1''-Terphenylophane, see
 B-70038
Terpilene, see I-70087
Terremutin, T-70009
Tertroxin, in T-60350
teta, see T-60017
Tethracene, see T-60053
Tethracenequinone, see T-60054
6β,8β,9α,14-Tetraacetoxy-1α-benzoyloxy-4β-
 hydroxydihydro-β-agarofuran, in E-60018
1,2,3,5-Tetraaminobenzene, T-60012
1,2,4,5-Tetraaminobenzene, T-60013
3,3',5,5'-Tetraamino-2,2',4,4',6,6'-hexa-
 nitrobiphenyl, T-60014
▷Tetraazabenzo[a]naphth[1,2,3-de]anthracene,
 see T-60271

3,4,8,9-Tetraazabicyclo[4.4.0]dec-1(6)-ene, see
 O-60018
2,4,6,8-Tetraazabicyclo[3.3.0]octane, see
 O-70175
2,4,6,8-Tetraazabicyclooctane-3,7-dione, see
 T-70066
1,4,7,10-Tetraazacyclododecane-1,4,7,10-tetra-
 acetic acid, T-60015
1,5,9,13-Tetraazacyclohexadecane-1,5,9,13-
 tetraacetic acid, T-60016
1,2,5,6-Tetraazacyclooctane, see O-60020
1,4,8,11-Tetraazacyclotetradecane-1,4,8,11-
 tetraacetic acid, T-60017
5,5b,10,10b-Tetraazadibenzo[a,e]pentalene, see
 P-60234
5,6,10a,10c-Tetraazadibenzo[a,f]pentalene, see
 P-70175
5,6,8,9-Tetraaza[3.3]paracyclophane, T-60018
25,26,27,28-Tetraazapentacyclo[19.3.1.1³,⁷-
 .1⁹,¹³.1¹⁵,¹⁹]octacosa-1(25),3,5,7(28),9,11,-
 13(27),15,17,19(26),21,23-dodecaene-
 2,8,14,20-tetrone, T-70010
Tetraazaporphin, see P-70109
6,7,17,18-Tetraazatricyclo[10.2.2.2⁵,⁸]-
 octadeca-5,7,12,14,15,17-hexaene, see
 T-60018
9H-Tetrabenzo[a,c,g,i]carbazole, T-70011
Tetrabenzotetracyclo[5.5.1.0⁴,¹³.0¹⁰,¹³]-trideca-
 ne, T-60019
Tetrabenzo[b,h,n,t]tetraphenylene, T-70012
1,1,2,2-Tetrabenzoylethane, T-60020
Tetrabenzoylethylene, T-60021
1,2,4,5-Tetrabromo-3-chloro-6-methylbenzene,
 T-60022
2,3,5,6-Tetrabromo-4-chlorotoluene, see
 T-60022
Tetrabromofuran, T-60023
Tetra-tert-butoxycyclopentadienone, T-70013
1,1,1',1'-Tetra-tert-butylazomethane, T-60024
Tetra-tert-butylcyclobutadiene, T-70014
1,1,2,2-Tetra-tert-butylethane, see D-70109
1,3,5,7-Tetra-tert-butyl-s-indacene, T-60025
Tetra-tert-butyltetrahedrane, T-70015
1,4,5,8-Tetrachloro-9,10-anthraquinodimethane,
 T-60026
4,5,6,7-Tetrachlorobenzotriazole, T-60027
▷1,2,3,4-Tetrachlorodibenzo-p-dioxin, T-70016
1,4,5,8-Tetrachloro-9,10-dihydro-9,10-bis-
 (methylene)anthracene, see T-60026
2,3,4,5-Tetrachloro-1,1-dihydro-1-[(methoxycarbonyl)imino]thiophene, T-60028
2,2',4,4'-Tetrachlorodiphenyl ether, see B-70135
2,2',6,6'-Tetrachlorodiphenyl ether, see B-70136
3,3',4,4'-Tetrachlorodiphenyl ether, see B-70137
▷1,1,2,3-Tetrachloro-1-propene, T-60029
1-Tetracosyne, T-70017
12-Tetracosyne, T-70018
29,29,30,30-Tetra-
 cyanobianthraquinodimethane, T-60030
1,2,4,5-Tetracyano-3,6-dimethylbenzene, in
 D-60407
▷Tetracyanoethene, see T-60031
▷Tetracyanoethylene, T-60031
Tetracyano-p-xylene, in D-60407
Tetracyclo[5.3.0.0²,⁴.0³,⁵]deca-6,8,10-triene,
 T-70019
Tetracyclo[6.2.1.1³,⁶.0²,⁷]dodeca-2(7),-4,9-
 triene, T-60032
Tetracyclo[6.4.0.0⁴,¹².0⁵,⁹]dodec-10-ene,
 T-70020
Tetracyclo[6.2.1.1³,⁶.0²,⁷]dodec-2(7)-ene,
 T-70021
Tetracyclo[3.3.0.0²,⁸.0⁴,⁶]octane, T-70022
Tetracyclo[4.2.0.0²,⁵.0³,⁸]octane, T-70023
1,1,2,2-Tetracyclopropylethane, T-60034
Tetracyclo[14.2.2.2⁴,⁷.2¹⁰,¹³]tetracosa-
 2,4,6,8,10,12,14,16,18,19,21,23-dodecaene,
 see H-60038
Tetracyclo[5.5.1.0⁴,¹³.0¹⁰,¹³]tridecane, T-60035
Tetracyclo[5.5.1.0⁴,¹³.0¹⁰,¹³]trideca-2,5,8,11-
 tetraene, T-60036

Tetracyclo[4.4.1.0³,¹¹.0⁹,¹¹]undecane, T-60037

Tetradecahydro-4,6-dihydroxy-9,9,12a-trimethyl-6H-phenanthro[1,10a-c]-furan-3-carboxylic acid, T-60038

1,2,3,4,5,6,7,8,9,10,11,12,13,14-Tetradecahydro-1,5:3,7:8,12:10,14-tetramethanononalene, see H-70031

4,4'-Tetradecamethylenebiphenyl, see P-70008

5-Tetradecanone, T-60039

2-Tetradecanoyl-1,3-benzenediol, see D-70338

1-O-(1,3,5,7,9-Tetradecapentaenyl)glycerol, see T-70024

3-(1,3,5,7,9-Tetradecapentaenyloxy)-1,2-propanediol, T-70024

3-(1,3,5,8-Tetradecatetraenyl)-oxiranebutanoic acid, see E-60015

12-Tetradecen-14-olide, in H-60230

5-(1-Tetradecenyl)-1,3-benzodioxole, see M-60077

1-Tetradecylcyclopropanecarboxylic acid, T-60040

2-Tetradecyloxiranecarboxylic acid, T-60041

7-Tetradecyn-1-ol, T-60042

9,10,23,24-Tetradecahydro-5,8:11,14:19,22:25,28-tetraethenodibenzo[a,m]cyclotetracosene, see T-70012

1,3,5,7-Tetraethenyladamantane, see T-70154

1,3,5,7-Tetraethenyltricyclo[3.3.1.1³,⁷]-decane, see T-70154

1,1,4,4-Tetraethoxy-2-butyne, in B-60351

Tetraethylammonium tris(pentachlorophenyl)-methylide, in T-60405

1,3,5,7-Tetraethynyladamantane, T-70025

1,3,5,7-Tetraethynyltricyclo[3.3.1.1³,⁷]-decane, see T-70025

2,3,4,5-Tetrafluorobenzaldehyde, T-70026

2,3,4,6-Tetrafluorobenzaldehyde, T-70027

2,3,5,6-Tetrafluorobenzaldehyde, T-70028

2,3,5,6-Tetrafluoro-1,4-benzenedicarboxylic acid, T-70029

2,4,5,6-Tetrafluoro-1,3-benzenedicarboxylic acid, T-70030

3,4,5,6-Tetrafluoro-1,2-benzenedicarboxylic acid, T-70031

2,3,4,5-Tetrafluorobenzoic acid, T-70032

2,3,4,6-Tetrafluorobenzoic acid, T-70033

2,3,5,6-Tetrafluorobenzoic acid, T-70034

2,3,4,5-Tetrafluorobenzonitrile, in T-70032

2,3,5,6-Tetrafluorobenzonitrile, in T-70034

4,5,6,7-Tetrafluoro-1H-benzotriazole, T-70035

3,3,4,4-Tetrafluorocyclobutene, T-60043

1,2,3,4-Tetrafluoro-5,6-diiodobenzene, T-60044

1,2,3,5-Tetrafluoro-4,6-diiodobenzene, T-60045

▷1,2,4,5-Tetrafluoro-3,6-diiodobenzene, T-60046

1,1,2,2-Tetrafluoro-3,4-dimethylcyclobutane, T-60047

2,2,4,4-Tetrafluoro-1,3-dithietane, T-60048

▷Tetrafluoroethylene oxide, see T-60049

4,5,6,7-Tetrafluoro-1,3-isobenzofurandione, in T-70031

Tetrafluoroisophthalic acid, see T-70030

▷Tetrafluorooxirane, T-60049

4,5,7,8-Tetrafluoro[2.2]paracyclophane, T-60050

Tetrafluorophthalic acid, see T-70031

2,3,5,6-Tetrafluoro-4-pyridinamine, see A-60257

Tetrafluoroterephthalic acid, see T-70029

5,6,15,16-Tetrafluorotricyclo[8.2.2.2⁴,⁷]-hexadeca-4,6,10,12,13,15-hexaene, see T-60050

3,3,4,4-Tetrafluoro-2-(trifluoromethyl)-1,2-oxazetidine, T-60051

N,1,1,1-Tetrafluoro-N-[(trifluoromethyl)-sulfonyl]methanesulfonamide, see P-70042

1',2',3',6'-Tetrahydroacetophenone, see A-70039

3,4,5,6-Tetrahydroacetophenone, see A-70038

1,3,4,10-Tetrahydro-9(2H)-acridinone, T-60052

1,2,3,4-Tetrahydroacridone, see T-60052

1,2,3,4-Tetrahydroanthracene, T-60053

1,2,3,4-Tetrahydro-9,10-anthracenedione, see T-60054

Tetrahydroanthranil-3-one, see T-70038

1,2,3,4-Tetrahydroanthraquinone, T-60054

1,2,3,6-Tetrahydrobenzaldehyde, see C-70227

2,3,4,5-Tetrahydro-1H-1-benzazepine, T-70036

4,5,6,7-Tetrahydro-1,2-benzisoxazol-3-ol, see T-70037

4,5,6,7-Tetrahydro-1,2-benzisoxazol-3(2H)-one, T-70037

4,5,6,7-Tetrahydro-2,1-benzisoxazol-3(1H)-one, T-70038

4,5,6,7-Tetrahydro-2,1-benzisoxazol-3(3aH)-one, see T-70038

2,3,8,9-Tetrahydrobenzo[1,2-b:3,4-b']-bis[1,4]-dithiin, T-70039

2,3,8,9-Tetrahydrobenzo[2,1-b:3,4-b']-bis[1,4]-oxathiin, T-70040

7,8,8a,9a-Tetrahydrobenzo[10,11]-chryseno[3,4-b]oxirene-7,8-diol, T-70041

6,7,8,9-Tetrahydro-5H-benzocycloheptene-6-carboxaldehyde, T-60055

6,7,8,9-Tetrahydro-5H-benzocycloheptene-7-carboxaldehyde, T-60056

1,2,3,4-Tetrahydro-7H-benzocyclohepten-7-one, T-60057

1,3,4,5-Tetrahydro-2H-1,3-benzodiazepin-2-one, T-60058

4,5,6,7-Tetrahydrobenzothiadiazole, T-70042

5a,6,11a,12-Tetrahydro[1,4]-benzoxazino[3,2-b][1,4]benzoxazine, T-60059

1,2,4,5-Tetrahydro-3-benzoxepin, T-70043

1,3,4,5-Tetrahydro-2-benzoxepin, T-70044

2,3,4,5-Tetrahydro-1-benzoxepin, T-70045

4a,4b,8a,8b-Tetrahydrobiphenylene, T-60060

Tetrahydro-2,5-bis(4-hydroxy-3-methoxy-phenyl)-3,4-dimethylfuran, see M-70003

Tetrahydrobis(4-hydroxyphenyl)-3,4-dimethylfuran, T-60061

Tetrahydro-3,4-bis(methylene)furan, T-70046

3,4,5,6-Tetrahydro-4,5-bis(methylene)-pyridazine, T-60062

Tetrahydro-2-bromo-2H-pyran, see B-70271

1,2,3,4-Tetrahydro-9H-carbazol-3-amine, see A-60259

1,2,3,9-Tetrahydro-4H-carbazol-4-one, see O-70106

1,2,4,9-Tetrahydro-3H-carbazol-3-one, see O-70105

1,3,4,9-Tetrahydro-2H-carbazol-2-one, see O-70104

2,3,4,9-Tetrahydro-1H-carbazol-1-one, see O-70103

4a,5,6,7-Tetrahydrochroman-7-one, see H-60045

5a,5b,11a,11b-Tetrahydrocyclobuta[1,2-b:4,3-b']benzothiopyran-11,12-dione, T-60063

1,2,3,8-Tetrahydrocyclobuta[b]-quinoxaline-1,2-dione, see D-70178

7,8,9,10-Tetrahydro-6H-cyclohepta[b]-naphthalene, T-70047

3,3a,4,6a-Tetrahydrocyclopenta[b]pyrrol-2(1H)-one, T-70048

1,2,3,4-Tetrahydrocyclopent[b]indole, T-70049

3,4,5,6-Tetrahydro-1H-cyclopropa[b]-naphthalene, T-70050

4,5,6,7-Tetrahydro-1H-cyclopropa[a]-naphthalene, T-70051

1,3,4,5-Tetrahydrocycloprop[f]indene, T-70052

1,4,5,6-Tetrahydrocycloprop[e]indene, T-70053

1a,6b,7a,12b-Tetrahydrodibenz[3,4:7,8]-anthra[1,2-b:5,6-b']bisoxirene, see D-70151

5,6,7,8-Tetrahydrodibenz[c,e]azocine, T-70054

5,6,7,12-Tetrahydrodibenz[b,e]azocine, T-70055

5,6,7,12-Tetrahydrodibenz[b,g]azocine, T-70056

5,6,11,12-Tetrahydrodibenz[b,f]azocine, T-70057

4b,8b,13,14-Tetrahydrodiindeno[1,2-a:2',1'-b]-indene, T-70058

1,4,5,8-Tetrahydro-1,4:5,8-dimethanonaphthalene, see T-70032

3,3a,4,5-Tetrahydro-5,7-dimethoxy-2-methyl-3a-(2-propenyl)-3-(3,4,5-trimethoxyphenyl)-6(2H)-benzofuranone, see C-70175

5,6,7,7a-Tetrahydro-4,7-dimethylcyclopenta[c]-pyran-1(4aH)-one, see N-60020

3,4,5,6-Tetrahydro-4,5-dimethylenepyridazine, see T-60062

4,4'-(Tetrahydro-3,4-dimethyl-2,5-furandiyl)-bis[2-methoxyphenol], see M-70003

4,4'-(Tetrahydro-3,4-dimethyl-2,5-furandiyl)-bisphenol, see T-60061

Tetrahydro-3,4-dimethyl-6-(1-methylbutyl)-2H-pyran-2-one, see I-60038

Tetrahydro-5,6-dimethyl-2H-pyran-2-one, T-70059

1,2,3,4-Tetrahydro-2,4-dioxopyrido[2,3-d]-pyrimidine, see P-60236

1,2,3,4-Tetrahydro-2,4-dioxopyrido[3,4-d]-pyrimidine, see P-60237

1,2,3,6-Tetrahydro-2,6-dioxo-4-pyrimidineacetic acid, T-70060

1,2,3,4-Tetrahydro-2,4-dioxo-5-pyrimidinecarboxaldehyde, T-60064

Tetrahydrodiphenospiropyran, see T-70094

5,6,11,12-Tetrahydro-1,10-ethanodibenzo[a,e]-cyclooctene, see C-60228

Tetrahydro-4a,8a-ethanonaphthalene-1,5(2H,6H)-dione, see T-60261

7,8,9,10-Tetrahydrofluoranthene, T-70061

Tetrahydro-3,4-furandiamine, T-70062

4-(Tetrahydro-2-furanyl)morpholine, T-60065

5,6,7,8-Tetrahydro-4,4',5,6,6',7,8-heptahydroxy-2,2'-dimethoxy-7,7'-dimethyl-[1,1'-bianthracene]-9,9',10,10'-tetrone, see A-60079

2,2',3,3'-Tetrahydro-5,5',6,6',8,8'-hexahydroxy-2,2'-dimethyl[9,9'-bi-4H-naphtho[2,3-b]-pyran]-4,4'-dione, see C-60029

3,3a,7,7a-Tetrahydro-3-hydroxy-3,6-dimethyl-2-(3-methyl-2-butenylidene)-4(2H)-benzofuranone, see B-60138

5,6,7,8-Tetrahydro-3-hydroxy-2-(2-hydroxy-1-methylethyl)-8,8-dimethyl-1,4-phenanthrenedione, see N-60017

1,2,3,4-Tetrahydro-4-hydroxyisoquinoline, T-60066

1,3,4,9-Tetrahydro-8-hydroxy-5-methoxy-6,7-methylenedioxy-1-methylnaphtho[2,3-c]-furan-4,9-dione, see V-60004

Tetrahydro-2-(4-hydroxy-3-methoxyphenyl)-4-[(4-hydroxy-3-methoxyphenyl)methyl]-3-furanmethanol, see L-70019

▷1-(Tetrahydro-5-hydroxy-4-methyl-3-furanyl)-ethanone, see B-60194

Tetrahydro-3-hydroxy-4-methyl-2H-pyran-2-one, in D-60353

3,4,5,6-Tetrahydro-4-hydroxy-6-methyl-2H-pyran-2-one, T-60067

1,2,3,4-Tetrahydro-1-hydroxy-2-naphthalenecarboxylic acid, T-70063

1,2,3,4-Tetrahydro-1-hydroxy-2-naphthalenemethanol, T-70064

1,2,3,4-Tetrahydro-1-hydroxy-2-naphthoic acid, see T-70063

Tetrahydro-2-(4-hydroxyphenyl)-5-(4-hydroxy-3-methoxyphenyl)-3,4-dimethylfuran, in T-60061

1,2,3,6-Tetrahydro-3-hydroxypicolinic acid, see T-60068

Tetrahydro-5-(2-hydroxypropyl)-α-methyl-2-furanacetic acid, see N-70069

1,2,3,6-Tetrahydro-3-hydroxy-2-pyridinecarboxylic acid, T-60068

1,2,5,6-Tetrahydro-5-hydroxy-3-pyridinecarboxylic acid, T-70065

1,2,5,6-Tetrahydro-4-hydroxypyrimidine, *see* T-60089

▷Tetrahydro-4-hydroxy-*N,N,N*,5-tetramethyl-2-furanmethanaminium, *see* M-60153

Tetrahydroimidazo[4,5-*d*]imidazole-2,5(1*H*,3*H*)-dione, T-70066

2,3,5,6-Tetrahydro-*s*-indacene-1,7-dione, T-70067

2,3,6,7-Tetrahydro-*s*-indacene-1,5-dione, T-70068

2,3,6,7-Tetrahydro-*as*-indacene-1,8-dione, T-70069

3,5,6,7-Tetrahydro-4*H*-inden-4-one, T-70070

1,5,6,8*a*-Tetrahydro-3(2*H*)-indolizinone, T-70071

Tetrahydroindoxazen-3-one, *see* T-70037

Tetrahydro-3-iodo-2*H*-pyran-2-one, T-60069

3*a*,4,7,7*a*-Tetrahydro-1(3*H*)-isobenzofuranone, T-70072

1,2,3,4-Tetrahydro-1-isoquinolinecarboxylic acid, T-60070

1,2,3,4-Tetrahydro-4-isoquinolinol, *see* T-60066

4,5,6,7-Tetrahydroisoxazolo[4,5-*c*]-pyridin-3-ol, T-60071

4,5,6,7-Tetrahydroisoxazolo[5,4-*c*]-pyridin-3-ol, T-60072

4,5,6,7-Tetrahydroisoxazolo[4,5-*c*]-pyridin-3(2*H*)-one, *see* T-60071

4,5,6,7-Tetrahydroisoxazolo[5,4-*c*]-pyridin-3(2*H*)-one, *see* T-60072

4,5,6,7-Tetrahydro-3-methoxyisoxazolo[4,5-*c*]-pyridine, *in* T-60071

N-(5*a*,6,7,8-Tetrahydro-4-methyl-1,5-dioxo-1*H*,5*H*-pyrrolo[1,2-*c*][1,3]-oxazepin-3-yl)-hexanamide, *in* M-70134

4,5,6,7-Tetrahydro-1-methylene-1*H*-indene, T-70073

1,2,3,4-Tetrahydro-1-methylenenaphthalene, T-60073

Tetrahydro-4-methylene-5-oxo-2-pentyl-3-furancarboxylic acid, *see* M-70079

Tetrahydro-6-methylene-2*H*-pyran-2-one, T-70074

1,2,3,4-Tetrahydro-3-methyl-3-isoquinolinecarboxylic acid, T-60074

Tetrahydro-6-methyl-3-methylene-6-(4-oxopentyl)-2,5-(3*H*,4*H*)-benzofurandione, *see* U-60001

7,8,9,9*a*-Tetrahydro-3-[(3-methyl-1-oxobutyl)-amino]-4-methyl-1*H*,5*H*-pyrrolo[1,2-*c*][1,3]-oxazepine-1,5-dione, *see* M-70134

Tetrahydro-4-methyl-5-oxo-2-pentyl-3-furancarboxylic acid, *see* P-60067

▷1,2,3,6-Tetrahydro-1-methyl-4-phenylpyridine, *in* T-60081

3,4,5,6-Tetrahydro-6-(2-methyl-1-propenyl)-1*H*-azepino[5,4,3-*cd*]indole, *see* A-60316

Tetrahydro-2-methylpyran, T-60075

Tetrahydro-4-methyl-2*H*-pyran-2-one, *in* H-70182

Tetrahydro-2-(4-morpholinyl)furan, *see* T-60065

3,4,7,8-Tetrahydro-2,6(1*H*,5*H*)-naphthalenedione, *see* H-70063

1-(1,2,3,4-Tetrahydro-2-naphthalenyl)-ethanone, *see* A-70053

1,1′-(1,2,3,4-Tetrahydronaphthylidene), *see* B-70165

3,6,7,8-Tetrahydro-2*H*-oxocin, T-60076

2,3,4,5-Tetrahydro-2-oxo-1,5-ethanobenzazepine, *see* D-60239

N-(Tetrahydro-2-oxo-3-furanyl)-hexadecanamide, *in* A-60138

▷Tetrahydro-2-oxoglyoxaline, *see* I-70005

7,8,9,9*a*-Tetrahydro-3-[(1-oxohexyl)amino]-4-methyl-1*H*,5*H*-pyrrolo[1,2-*c*][1,3]-oxazepine-1,5-dione, *in* M-70134

1,2,5,6-Tetrahydro-4-oxopyrimidine, *see* T-60089

1,2,3,4-Tetrahydro-2-oxo-4-quinolinecarboxylic acid, T-60077

1,2,3,4-Tetrahydro-4-oxo-6-quinolinecarboxylic acid, T-60078

1,2,3,4-Tetrahydro-4-oxo-7-quinolinecarboxylic acid, T-60079

2,2′,3,3′-Tetrahydro-3′,5,5′,7,7′-pentahydroxy-2,2′-bis(4-hydroxyphenyl)[3,8′-bi-4*H*-1-benzopyran]-4,4′-dione, *see* G-70005

1,2,3,3*a*-Tetrahydropentalene, T-70075

1,2,3,4-Tetrahydropentalene, T-70076

1,2,3,5-Tetrahydropentalene, T-70077

1,2,4,5-Tetrahydropentalene, T-70078

1,2,4,6*a*-Tetrahydropentalene, T-70079

1,2,6,6*a*-Tetrahydropentalene, T-70080

1,3*a*,4,6*a*-Tetrahydropentalene, T-70081

1,3*a*,6,6*a*-Tetrahydropentalene, T-70082

3,3*a*,6,6*a*-Tetrahydro-1(2*H*)-pentalenone, T-70083

4,5,6,6*a*-Tetrahydro-2(1*H*)-pentalenone, T-70084

6,7,8,9-Tetrahydro-6-phenyl-5*H*-benzocyclo-hepten-5-one, T-60080

1,2,3,6-Tetrahydro-4-phenylpyridine, T-60081

Tetrahydro-2-phenyl-2*H*-1,2-thiazin-3-one, *in* T-60095

Δ⁴-Tetrahydrophthalide, *see* T-70072

Tetrahydro-2-(2-piperidinyl)furan, T-60082

Tetrahydro-2*H*-pyran-2-thione, T-70085

Tetrahydro-1*H*,5*H*-pyrazolo[1,2-*a*]pyrazole, T-70086

Tetrahydro-1*H*,5*H*-pyrazolo[1,2-*a*]pyrazole-1,5-dione, T-70087

Tetrahydro-1*H*-pyrazolo[1,2-*a*]pyridazine-1,3(2*H*)-dione, T-70088

4,5,9,10-Tetrahydropyrene, T-60083

3,4,8,9-Tetrahydropyrene (obsol.), *see* T-60083

3*b*,4*a*,7*b*,8*a*-Tetrahydropyreno[4,5-*b*:9,10-*b*′]-bisoxirene, *see* D-70152

▷1,2,3,6-Tetrahydropyridine, T-70089

2,3,4,5-Tetrahydropyridine, T-70090

1,2,3,6-Tetrahydro-3-pyridinecarboxylic acid, T-60084

6,7,8,9-Tetrahydro-5*H*-pyrido[2,3-*b*]azepine, T-70091

6,7,8,9-Tetrahydro-5*H*-pyrido[3,2-*b*]azepine, T-70092

7,8,9,10-Tetrahydropyrido[1,2-*a*]azepin-4(6*H*)-one, T-70085

5,6,9,10-Tetrahydro-4*H*,8*H*-pyrido[3,2,1-*ij*][1,6]naphthyridine, T-60086

5-(3,4,5,6-Tetrahydro-3-pyridylidenemethyl)-2-furanmethanol, T-60087

Tetrahydro-2(1*H*)-pyrimidinone, T-60088

Tetrahydro-4(1*H*)-pyrimidinone, T-60089

▷3,4,5,6-Tetrahydro-2(1*H*)-pyrimidinone, *in* T-60088

Tetrahydro-2-(2-pyrrolidinyl)furan, T-60090

Tetrahydro-1*H*-pyrrolizin-2(3*H*)-one, T-60091

1,2,3,4-Tetrahydroquinoline, T-70093

6,7,8,9-Tetrahydro-4*H*-quinolizin-4-one, T-60092

3,3′,4,4′-Tetrahydro-2,2′-spirobi[2*H*-1-benzopyran], T-70094

2,2′,3,3′-Tetrahydro-3′,5,7,8′-tetrahydroxy-2′,2′-dimethyl[2,6′-bi-4*H*-1-benzopyran]-4-one, *see* S-60031

5,6,8,13-Tetrahydro-1,7,9,11-tetrahydroxy-8,13-dioxo-3-(2-oxopropyl)benzo[*a*]naphthacene-2-carboxylic acid, T-70095

2,3,4,4*a*-Tetrahydro-5,6,8,10-tetrahydroxy-7-(2-hydroxy-1-methylethyl)-1,1,4*a*-trimethyl-9(1*H*)-phenanthrenone, *see* C-60157

5,6,7,8-Tetrahydro-5,7,9,10-tetrahydroxy-2-methoxy-7-methyl-1,4-anthracenedione, *see* A-60322

1,2,12*a*,12*b*-Tetrahydro-1,4,9,12*a*-tetrahydroxy-3,10-perylenedione, *see* A-60078

Tetrahydro-1,3,4,6-tetranitroimidazo[4,5-*d*]-imidazole-2,5-(1*H*,3*H*)-dione, T-70096

▷Tetrahydro-1,2,4,5-tetrazine-3,6-dione, T-70097

Tetrahydro-1,3,5,7-tetrazocine-2,6(1*H*,3*H*)-dione, T-70098

▷Tetrahydro-2*H*-1,3-thiazine, T-60093

Tetrahydro-2*H*-1,4-thiazine-3-carboxylic acid, *see* T-70182

Tetrahydro-2*H*-1,3-thiazine-2-thione, T-60094

Tetrahydro-2*H*-1,2-thiazin-3-one, T-60095

Tetrahydro-2*H*-1,3-thiazin-2-one, T-60096

2-(Tetrahydro-2-thienyl)pyridine, T-60097

2,3,4,5-Tetrahydrothiepino[2,3-*b*]pyridine, T-60098

Tetrahydro-2-thiophenecarboxylic acid, T-70099

Tetrahydro-2*H*-thiopyran-3-amine, *see* A-70200

1,2,3,4-Tetrahydro-2-thioxo-5*H*-1,4-benzodiazepin-5-one, T-70100

Tetrahydro-1,2,4-triazine-3,6-dione, T-60099

1,2,3,4-Tetrahydro-1,3,8-trihydroxy-6-methoxy-3-methyl-9,10-anthracenedione, *see* A-60321

5,6,7,8-Tetrahydro-6,7,8-trihydroxy-5-[[4-oxo-2-(2-phenylethyl)-4*H*-1-benzopyran-6-yl]-oxy]-2-(2-phenylethyl)-4*H*-benzopyran-4-one, *see* A-70078

7,8,11,11*a*-Tetrahydro-3,6,10-trimethyl-cyclodeca[*b*]furan-2(4*H*)-one, *see* G-60022

Tetrahydro-*α,α*,5-trimethyl-2-furanmethanol, *see* P-60160

Tetrahydro-1,6,7-tris(methylene)-1*H*,4*H*-3*a*,6*a*-propanopentalene, T-70101

2*α*,6*α*,7,11-Tetrahydro-7,9(11),13-abietatrien-12-one, *in* D-70282

1,2,3,4-Tetrahydroxybenzene, *see* B-60014

2,3,5,6-Tetrahydroxybenzoic acid, T-60100

▷2,3,7,8-Tetrahydroxy[1]benzopyrano[5,4,3-*cde*][1]benzopyran-5,10-dione, *see* E-70007

3,3′,4,4′-Tetrahydroxybibenzyl, T-60101

3,3′,4,5′-Tetrahydroxybibenzyl, T-60102

4,4′,5,5′-Tetrahydroxy-1,1′-binaphthyl, T-70102

2,2′,3,4-Tetrahydroxybiphenyl, *see* B-70122

2,3,4,4′-Tetrahydroxybiphenyl, *see* B-70123

2′,4,4′,6′-Tetrahydroxychalcone, *see* H-60220

3,4,6,8-Tetrahydroxy-11,13-clerodadien-15,16-olide, T-70103

(1*α*,3*α*,4*α*,5*β*)-1,3,4,5-Tetrahydroxycyclo-hexanecarboxylic acid, *see* Q-70005

3,4,5,6-Tetrahydroxy-1-cyclohexene-1-methanol, *see* H-70172

2,4,6,10-Tetrahydroxy-8-daucene, *see* D-70009

2,4,4′,6-Tetrahydroxydihydrochalcone, *see* H-60219

2,4,5,6-Tetrahydroxy-9,10-dihydrophenanthrene, *see* D-70247

3′,4′,5,5′-Tetrahydroxy-6,7-dimethoxyflavone, *in* H-70081

3′,4′,5,7-Tetrahydroxy-5′,6-dimethoxyflavone, *in* H-70081

2,5,6,8-Tetrahydroxy-3,7-dimethoxy-1,4-naphthoquinone, *in* H-60065

2,5,7,8-Tetrahydroxy-3,6-dimethoxy-1,4-naphthoquinone, *in* H-60065

5,8,10,11-Tetrahydroxy-2,6-dodecadienoic acid *δ*-lactone triacetate, *see* H-70236

2,3,22,23-Tetrahydroxy-6-ergostanone, *see* T-70116

2*α*,3*α*,22*R*,23*R*-Tetrahydroxy-5*α*-ergost-24(28)-en-6-one, *see* D-70544

3′,4′,5,7-Tetrahydroxyflavan, T-60103

2′,5,7,8-Tetrahydroxyflavanone, T-70104

3,4′,5,7-Tetrahydroxyflavanone, T-60104

3,5,6,7-Tetrahydroxyflavanone, T-60105

3′,4′,5,7-Tetrahydroxyflavanone, T-70105

3′,4′,6,7-Tetrahydroxyflavanone, T-70106

2′,3′,5,7-Tetrahydroxyflavone, T-60106

2′,5,6,6′-Tetrahydroxyflavone, T-70107

3′,4′,5,6-Tetrahydroxyflavone, T-70108

2,3,4,5-Tetrahydroxy-9*H*-fluoren-9-one, T-70109

7,9,11,15-Tetrahydroxy-2-hexacosen-1,5-olide, *in* D-70276

2,3,3*a*,7*a*-Tetrahydroxy-3-hydroxy-2-phenyl-5*H*-furo[3,2-*b*]pyran-5-one, *see* A-60083

2′,5,6,7-Tetrahydroxyisoflavone, T-60107

2′,5,7,8-Tetrahydroxyisoflavone, T-70110

3′,4′,5,7-Tetrahydroxyisoflavone, T-60108
3′,4′,6,7-Tetrahydroxyisoflavone, T-70111
3′,4′,7,8-Tetrahydroxyisoflavone, T-70112
4′,6,7,8-Tetrahydroxyisoflavone, T-70113
ent-1β,3α,7α,11α-Tetrahydroxy-16-kaurene-6,15-dione, T-60109
ent-1β,7β,14S,20-Tetrahydroxy-16-kauren-15-one, T-70114
ent-3α,7β,14α,20-Tetrahydroxy-16-kauren-15-one, T-60110
ent-7β,11α,14S,20-Tetrahydroxy-16-kauren-15-one, T-70115
6,7,18,15-Tetrahydroxy-11,13-labdadiene, see L-70009
3,7,15,22-Tetrahydroxy-8,24-lanostadien-26-oic acid, T-60111
3,3′,5,7-Tetrahydroxy-4′-methoxyflavan, in P-70026
3′,4′,5,6-Tetrahydroxy-7-methoxyflavanone, in P-70028
3′,4′,5,7-Tetrahydroxy-6-methoxyflavanone, in P-70028
2′,3,5,8-Tetrahydroxy-7-methoxyflavone, in P-70029
3′,4′,5,7-Tetrahydroxy-3-methoxy-2′-prenylflavone, see P-60164
1,3,5,8-Tetrahydroxy-2-methyl-9,10-anthracenedione, see T-60113
1,3,4,5-Tetrahydroxy-2-methylanthraquinone, T-60112
1,3,5,8-Tetrahydroxy-2-methylanthraquinone, T-60113
3,5,7,8-Tetrahydroxy-6-(3-methyl-2-butenyl)-2-phenyl-4H-1-benzopyran-4-one, see P-70105
1,3,5,8-Tetrahydroxy-2-(3-methyl-2-butenyl)-xanthone, T-60114
2,3,22,23-Tetrahydroxy-24-methyl-6-cholestanone, T-70116
3,5,6,7-Tetrahydroxy-8-methylflavone, T-70117
3,5,7,8-Tetrahydroxy-6-methylflavone, T-70118
2α,3α,22R,23R-Tetrahydroxy-24S-methyl-B-homo-7-oxa-5α-cholestan-6-one, see B-70178
▷2,16,20,25-Tetrahydroxy-9-methyl-19-norlanosta-5,23-diene-3,11,22-trione, see C-70203
3,5,6,7-Tetrahydroxy-8-methyl-2-phenyl-4H-1-benzopyran-4-one, see T-70117
3,5,7,8-Tetrahydroxy-6-methyl-2-phenyl-4H-1-benzopyran-4-one, see T-70118
2,3,19,23-Tetrahydroxy-12-oleanen-28-oic acid, T-70119
9,12,13,16-Tetrahydroxy-11-oxokauran-17-al, see P-70097
4β,14α,17β,20R-Tetrahydroxy-1-oxo-20R-witha-2,5,24-trienolide, see W-60010
3′,4′,5,8-Tetrahydroxypeltogynan-4-one, see C-70200
1,2,5,6-Tetrahydroxyphenanthrene, T-60115
1,2,5,7-Tetrahydroxyphenanthrene, T-60116
1,2,6,7-Tetrahydroxyphenanthrene, T-60117
1,3,5,6-Tetrahydroxyphenanthrene, T-60118
1,3,6,7-Tetrahydroxyphenanthrene, T-60119
2,3,5,7-Tetrahydroxyphenanthrene, T-60120
2,3,6,7-Tetrahydroxyphenanthrene, T-60121
2,4,5,6-Tetrahydroxyphenanthrene, T-60122
3,4,5,6-Tetrahydroxyphenanthrene, T-60123
1-(2,3,4,5-Tetrahydroxyphenyl)-3-(2,4,5-trihydroxyphenyl)-1,3-propanedione, T-70120
2,3,12,16-Tetrahydroxy-4,7-pregnadien-20-one, T-60124
3,4′,5,7-Tetrahydroxy-8-prenylflavan, T-60125
3′,4′,5,7-Tetrahydroxy-5′-C-prenylflavanone, see S-70041
3,5,7,8-Tetrahydroxy-6-prenylflavone, see P-70105
2,3′,4,5-Tetrahydroxystilbene, see H-70215
2,3′,4,5′-Tetrahydroxystilbene, see D-70335
2′,4′,5,7-Tetrahydroxy-3,5′,6,8-tetramethoxyflavone, in O-60022
2′,5,5′,7-Tetrahydroxy-3,4′,6,8-tetramethoxyflavone, in O-60022

3′,5,5′,6-Tetrahydroxy-3,4′,7,8-tetramethoxyflavone, in O-70018
4′,5,5′,7-Tetrahydroxy-2′,3,6,8-tetramethoxyflavone, in O-60022
3,5,7,8-Tetrahydroxy-2-(3,4,5-trihydroxyphenyl)-4H-1-benzopyran-4-one, see H-70021
3′,5,5′,7-Tetrahydroxy-3,4′,8-trimethoxyflavone, in H-70021
2′,4′,5,7-Tetrahydroxy-3′,6,8-triprenylisoflavone, in E-70058
1,3,6,7-Tetrahydroxy-2,5,8-triprenylxanthone, see G-60003
1,2,3,19-Tetrahydroxy-12-ursen-28-oic acid, T-70121
1,2,3,4-Tetrahydroxy-9H-xanthen-9-one, see T-60126
1,2,4,5-Tetrahydroxy-9H-xanthen-9-one, see T-70122
1,3,5,8-Tetrahydroxy-9H-xanthen-9-one, see T-70123
1,3,6,8-Tetrahydroxy-9H-xanthen-9-one, see T-60127
1,2,3,4-Tetrahydroxyxanthone, T-60126
1,2,4,5-Tetrahydroxyxanthone, T-70122
1,3,5,8-Tetrahydroxyxanthone, T-70123
1,3,6,8-Tetrahydroxyxanthone, T-60127
2,3,4,7-Tetrahydroxyxanthone, T-60128
1,1,2,2-Tetraisopropylethane, see D-60393
Tetrakis(2-bromoethyl)methane, T-70124
1,1,2,2-Tetrakis(bromomethyl)ethane, see B-60144
1,3,4,6-Tetrakis(tert-butylthio)-thieno[3,4-c]-thiophene, in T-70170
Tetrakis(2-carboxyethyl)methane, see M-70036
1,3,5,7-Tetrakis(carboxymethyl)adamantane, see A-70064
1,1,2,2-Tetrakis(chloromethyl)ethane, see B-60148
Tetrakis(2-cyanoethyl)methane, in M-70036
1,3,5,7-Tetrakis(cyanomethyl)adamantane, in A-70064
1,3,5,7-Tetrakis(diethylamino)-2,6-diaza-s-indacene, see T-70125
1,3,5,7-Tetrakis(diethylamino)-pyrrolo[3,4-f]-isoindole, T-70125
1,2,4,5-Tetrakis(dimethylamino)benzene, in T-60012
1,2,4,5-Tetrakis(dimethylamino)benzene, in T-60013
Tetrakis(dimethylamino)-2,5-diazapentalene, see T-70127
5,8,14,17-Tetrakis(dimethylamino)[3.3]-paracyclophene, T-70126
1,3,4,6-Tetrakis(dimethylamino)-pyrrolo[3,4-c]-pyrrole, T-70127
2,3,4,5-Tetrakis(1,1-dimethylethoxy)-2,4-cyclopentadien-1-one, see T-70013
1,2,3,4-Tetrakis(1,1-dimethylethyl)-1,3-cyclobutadiene, see T-70014
1,3,5,7-Tetrakis(1,1-dimethylethyl)-s-indacene, see T-60025
Tetrakis(1,1-dimethylethyl)-tricyclo[1.1.0²,⁴]-butane, see T-70015
Tetrakis(diphenylmethylene)cyclobutane, T-70128
1,3,4,6-Tetrakis(ethylthio)thieno[3,4-b]-thiophene, in T-70170
1,1,2,2-Tetrakis(hydroxymethyl)ethane, see B-60159
1,3,4,6-Tetrakis(isospropylthio)-thieno[3,4-c]-thiophene, in T-70170
Tetrakis(mercaptomethyl)methane, see B-60164
Tetrakis(methylsulfinylmethyl)methane, in B-60164
2,3,5,6-Tetrakis(methylthio)-1,4-benzenediamine, see D-60031
Tetrakis(methylthiomethyl)methane, in B-60164
▷2,2,4,4-Tetrakis(trifluoromethyl)-1,3-dithietane, T-60129
2,3,9,10-Tetrakis(trimethylsilyl)[5]-phenylene, T-60130
1,1,2,2-Tetrakis(2-triptycyl)ethene, see T-60169
1,3,5,7-Tetrakis[2-D₃-trishomocubanyl-1,3-butadiynyl]adamantane, T-70129

1-(1-Tetralinylidene)tetralin, see B-70165
2,3,5,6-Tetramercapto-p-phenylenediamine, see D-60031
Tetramercaptotetrathiafulvalene, T-70130
1,3,4,6-Tetramercaptothieno[3,4-c]thiophene, see T-70170
1,2,3,4-Tetramethoxybenzene, in B-60014
2,3,5,6-Tetramethoxybenzoic acid, in T-60100
2,2′,3,4-Tetramethoxybiphenyl, in B-60122
2,3,4,4′-Tetramethoxybiphenyl, in B-60123
2′,5,6,6′-Tetramethoxyflavone, in T-70107
3′,4′,5,6-Tetramethoxyflavone, in T-70108
2′,5,6,7-Tetramethoxyisoflavone, in T-60107
3′,4′,6,7-Tetramethoxyisoflavone, in T-70111
5,5′,6,7-Tetramethoxy-3′,4′-methylenedioxyflavone, in H-70081
1,1,8,8-Tetramethoxy-4-octene, in O-70025
1,2,5,6-Tetramethoxyphenanthrene, in T-60115
1,2,5,7-Tetramethoxyphenanthrene, in T-60116
1,2,6,7-Tetramethoxyphenanthrene, in T-60117
1,3,5,6-Tetramethoxyphenanthrene, in T-60118
1,3,6,7-Tetramethoxyphenanthrene, in T-60119
2,3,5,7-Tetramethoxyphenanthrene, in T-60120
2,3,6,7-Tetramethoxyphenanthrene, in T-60121
2,4,5,6-Tetramethoxyphenanthrene, in T-60122
3,4,5,6-Tetramethoxyphenanthrene, in T-60123
3,4,7,8-Tetramethoxy-2,6-phenanthrenediol, in H-60066
1-(2,3,4,5-Tetramethoxyphenyl)-3-(2,4,5-trimethoxyphenyl)-1,3-propanedione, in T-70120
4,4′,5,5′-Tetramethoxypolemannone, in P-60166
1,2,3,4-Tetramethoxyxanthone, in T-60126
1,3,5,8-Tetramethoxyxanthone, in T-60123
1,3,5,7-Tetramethyl-2,4-adamantanedione, T-60131
1,3,5,7-Tetramethyladamantanone, T-60132
N,N,2,2-Tetramethyl-2H-azirin-3-amine, see D-60402
2,2,5,5-Tetramethylbicyclo[4.1.0]hept-1(6)-en-7-one, T-70131
2,2,6,6-Tetramethylcyclohexanone, T-60133
3,3,6,6-Tetramethyl-4-cyclohexene-1,2-dione, T-60134
5,5,6,6-Tetramethyl-2-cyclohexene-1,4-dione, T-60135
2,2,5,5-Tetramethyl-3-cyclohexen-1-one, T-60136
3,3,6,6-Tetramethylcyclohexyne, T-70132
2,2,5,5-Tetramethylcyclopentaneselone, T-60137
2,2,5,5-Tetramethylcyclopentanone, T-60138
2,2,5,5-Tetramethyl-3-cyclopentene-1-thione, T-60139
2,2,5,5-Tetramethyl-3-cyclopenten-1-one, T-60140
2,2,3,3-Tetramethyl-cyclopropanecarboxaldehyde, T-70133
2,2,3,3-Tetramethylcyclopropanemethanol, T-70134
2,2,8,8-Tetramethyl-5,5-dihydroxy-3,4,6,7-nonanetetrone, in T-60148
Tetramethyldimethylenediurea, in T-70098
2-(3,7,11,15-Tetramethyl-5,13-dioxo-2,6,10,14-hexadecatetraenyl)-6-methylhydroquinone, in H-70225
3,4,7,11-Tetramethyl-1,3,6,10-dodectetraene, T-70135
2,3,5,6-Tetramethylenebicyclo[2.2.1]-heptan-7-one, T-70136
3,4-Tetramethyleneisoxazolin-5-one, see T-70038
4,5-Tetramethylenetetrahydro-1,3-oxazine, see O-70012
5,6-Tetramethylenetetrahydro-1,3-oxazine, see O-70011
1,1,4a,6-Tetramethyl-5-ethyl-6,5-oxidodeca-hydronaphthalene, see D-70533
3,7,11,15-Tetramethyl-2,6,10,13,15-hexadecapentaen-1-ol, T-70137
3,7,11,15-Tetramethyl-2-hexadecenal, in P-60152
3,7,11,15-Tetramethyl-2-hexadecen-1-ol, see P-60152
2,2,5,5-Tetramethyl-3-hexenedioic acid, T-60141

1,1,3,3-Tetramethyl-2-indaneselone, T-60142
1,1,3,3-Tetramethyl-2-indanone, T-60143
1,1,3,3-Tetramethyl-2-methylenecyclohexane,
    T-60144
3,3,5,5-Tetramethyl-4-methylene-1-pyrazoline,
    see D-60283
7,11,15,19-Tetramethyl-3-methylene-2-(3,7,11-
    trimethyl-2-dodecenyl)-1-eicosanol, T-70138
Tetramethylnaphthazarin, see D-60377
2′,3′,4′,5′-Tetramethyl-6′-nitroacetophenone,
    T-60145
2′,3′,4′,6′-Tetramethyl-5′-nitroacetophenone,
    T-60146
2′,3′,5′,6′-Tetramethyl-4′-nitroacetophenone,
    T-60147
1-(2,3,4,5-Tetramethyl-6-nitrophenyl)-ethanone,
    see T-60145
1-(2,3,4,6-Tetramethyl-5-nitrophenyl)-ethanone,
    see T-60146
1-(2,3,5,6-Tetramethyl-4-nitrophenyl)-ethanone,
    see T-60147
2,2,8,8-Tetramethyl-3,4,5,6,7-nonanepentone,
    T-60148
2,5,7,7-Tetramethyloctanal, T-70139
3,3,4,4-Tetramethyl-1,2-oxathietane, T-60149
3,7,11,15-Tetramethyl-13-oxo-2,6,10,14-hexa-
    decatetraenal, T-60150
2,5,8,11-Tetramethylperylene, T-70140
3,4,9,10-Tetramethylperylene, T-70141
1,3,6,8-Tetramethylphenanthrene, T-60151
2,4,5,7-Tetramethylphenanthrene, T-60152
2,2,6,6-Tetramethyl-4-piperidinamine, see
    A-70201
2,4,6,7-Tetramethylpteridine, T-70142
3,3,5,5-Tetramethyl-4-pyrazolidinone, T-60153
3,3,5,5-Tetramethyl-4-pyrazolinethione, see
    D-60285
3,3,5,5-Tetramethyl-1-pyrazolin-4-one, see
    D-60286
5,5,9,9-Tetramethyl-5H,9H-quino[3,2,1-de]-
    acridine, T-70143
1,5,5,9-Tetramethylspiro[5.5]undeca-1,8-dien-3-
    one, in L-70025
2,4,6,8-Tetramethyl-2,4,6,8-tetra-
    azabicyclo[3.3.0]octane, in O-70015
1,3,5,7-Tetramethyltetrahydro-1,3,5,7-tetra-
    zocine-2,6(1H,3H)dione, in T-70098
2,2,4,4-Tetramethyl-3-thietanone, T-60154
3,3,4,4-Tetramethyl-2-thietanone, T-60155
1,3,5,7-Tetramethyltricyclo[3.3.1.1³,⁷]-decane-
    2,4-dione, see T-60131
1,3,5,7-Tetramethyltricyclo[3.3.1.1³,⁷]-decan-2-
    one, see T-60132
1,4,4,8-Tetramethyltricyclo[6.2.1.0²,⁶]-undecan-
    11-ol, see C-70032
3,4,7,11-Tetramethyl-6,10-tridecadienal,
    T-60156
1,3,8,8-Tetramethyl-3-[3-(trimethylammonio)-
    propyl]-3-azoniabicyclo[3.2.1]octane, see
    T-70277
2,4,6,8-Tetramethylundecanoic acid, T-60157
▷p-Tetramethylxylylene diisocyanate, see
    B-70149
Tetraneurin D, T-60158
Tetraneurin C, in T-60158
2,3,4,6-Tetranitroacetanilide, in T-60159
▷2,3,4,6-Tetranitroaniline, T-60159
▷2,3,4,6-Tetranitrobenzenamine, see T-60159
Tetranitroglycoluril, see T-70096
2,3,4,5-Tetranitro-1H-pyrrole, T-60160
3,4,8,9-Tetraoxa-1,6-diazabicyclo[4.4.2]-do-
    decane, T-70144
1,4,10,13-Tetraoxa-7,16-diazacyclooctadecane,
    T-70145
1,7,10,16-Tetraoxa-4,13-diazacyclooctadecane,
    T-60161
10,11,14,15-Tetraoxa-1,8-
    diazatricyclo[6.4.4.0²,⁷]hexadeca-2(7),3,5-
    triene, T-60162
21,22,23,24-Tetraoxapenta-
    cyclo[16.2.1.1²,⁵,1⁸,¹¹,1¹²,¹⁵]tetracosa-
    1,3,5,7,9,1,13,15,17,19-decaene, see T-70146
21,22,23,24-Tetraoxapenta-
    cyclo[16.2.1.1³,⁶,1⁸,¹¹,1¹³,¹⁶]tetracosa-
    3,5,8,10,13,15,18,20-octaene, see T-70147

Tetraoxaporphycene, T-70146
Tetraoxaporphyrinogen, T-70147
1,3,7,9-Tetraoxaspiro[4,5]decane, T-70163
1,4,6,10-Tetraoxaspiro[4,5]decane, T-60164
Tetraphene-7,12-quinone, see B-70015
sym-Tetraphenylacetone, see T-60168
2,2,4,4-Tetraphenyl-3-butenal, T-70148
1,2,3,4-Tetraphenylcyclopentadiene, T-70149
7-[(2,3,4,5-Tetraphenyl-2,4-cyclopentadien-1-
    ylidene)-ethenylidene]-1,3,5-cyclo-
    heptatriene, see C-60208
1,1,1′,1′-Tetraphenyldimethylamine, T-70150
3,3,6,6-Tetraphenyl-1,4-dioxane-2,5-dione,
    T-60165
4,4,5,5-Tetraphenyl-1,3-dithiolane, T-60166
α,α,α′,α′-Tetraphenyldi-p-xylylene, see B-60117
Tetraphenylenefurfuran, see D-70471
Tetraphenylethanone, T-60167
Tetraphenylglycolide, see T-60165
Tetraphenylhomofulvene, see M-60079
1,1,3,3-Tetraphenyl-2-propanone, T-60168
2′,4′,5′,6′-Tetraphenyl-m-terphenyl, see
    H-60074
3′,4′,5′,6′-Tetraphenyl-1,1′:2′,1″-terphenyl, see
    H-60074
9,11,20,22-Tetraphenyltetrabenzo[a,c,l,n]-
    pentacene-10,21-dione, T-70151
4,4,5,5-Tetraphenyltetramethylene disulfide, see
    T-60166
1,3,4,6-Tetraphenylthieno[3,4-c]-thiophene-5-
    S^IV, T-70152
3,3,5,5-Tetraphenyl-1,2,4-trithiolane, T-70153
α,α,α′,α′-Tetraphenyl-p-xylylene, see B-60157
Tetraselenafulvalene, see B-70116
Tetratellurafulvalene, see B-70117
2,5,7,10-Tetrathiabicyclo[4.4.0]decane, see
    H-60051
1,4,5,8-Tetrathiadecalin, see H-60051
Tetrathiafulvalene, see B-70119
1,3,5,7-Tetrathia-s-indacene-2,6-dione, see
    B-70028
1,4,5,8-Tetrathianaphthalene, see D-70521
Tetrathiooxalic acid, see E-60040
Tetrathioquadratic acid, see D-70364
Tetrathiosquaric acid, see D-70364
Tetra(2-triptycyl)ethylene, T-60169
1,3,5,7-Tetravinyladamantane, T-70154
1,2,4,5-Tetrazine, T-60170
1,2,4,5-Tetrazine-3,6-diamine, see D-70045
1,2,4,5-Tetrazine-3,6-diol, see D-70268
[1,2,4,5]Tetrazino[1,6-a:4,3-a′]bis[1]-
    azaacenaphthylene, see D-60076
[1,2,4,5]Tetrazino[3,4-b:6,1-b′]-
    bisbenzothiazole, T-60171
[1,2,4,5]Tetrazino[1,6-a:4,3-a′]-diisoquinoline,
    T-60172
[1,2,4,5]Tetrazino[1,6-a:4,3-a′]diquinoline,
    T-60173
1,2,4,5-Tetrazino[1,6-a:4,3-a′]-diquinoxaline, in
    A-60353
▷1H-Tetrazol-5-amine, see A-70202
1H-Tetrazole-5-thiol, in T-60174
Tetrazole-5-thione, T-60174
2-Tetrazolin-5-one, T-60175
Tetrazolo[5,1-a]phthalazine, T-70155
Tetrazolo[1,5-a]pyrimidine, T-60176
Tetrazolo[1,5-c]pyrimidine, T-60177
Tetrazolo[1,5-b]quinoline-5,10-dione, T-70156
Teuchamaedryn A, T-70157
Teuchamaedryn B, in T-60180
Teucjaponin A, T-60178
Teucjaponin B, in T-60178
Teucretol, T-60179
Teucrin E, T-60180
Teucrin H2, in T-60180
Teulamifin B, T-70158
Teumicropin, T-60181
Teumicropodin, T-60182
Teupolin I, T-70159
Teupolin II, in T-70159
Teupyrenone, T-60184
Thalicticoside, in N-70050

Theaspirone, T-60185
▷α-Thenaldehyde, see T-70183
▷Thespesin, see G-60032
2-Thiaadamantane, see T-60195
2-Thiabicyclo[2.2.1]heptan-3-one, T-60186
2-Thiabicyclo[2.2.1]hept-5-ene, T-70160
2-Thiabicyclo[3.1.0]hex-3-ene, T-70161
7-Thiabicyclo[4.3.0]nonane, see O-60012
2-Thiabicyclo[3.3.0]octane, see H-60049
2-Thiabicyclo[2.2.2]octan-3-one, T-60187
7-Thiabicyclo[4.2.0]octa-1,3,5-trien-8-one, see
    B-70065
2-Thiabicyclo[2.2.2]oct-5-en-3-one, T-60188
Thiacyclonon-4-ene, see H-70076
1-Thia-2-cyclooctyne, T-60189
1-Thia-3-cyclooctyne, T-60190
13H-[1,3,5]Thiadiazino[3,2-a:5,6-a′]-
    bisbenzimidazole-13-thione, T-60191
1,3,4-Thiadiazole-2,5-diamine, see D-60049
1,2,5-Thiadiazole-3,4-dicarboxaldehyde,
    T-70162
▷1,3,4-Thiadiazolidine-2,5-dithione, see M-70031
[1,2,5]Thiadiazolo[3,4-d]pyridazine, T-70163
[1,2,5]Thiadiazolo[3,4-d]pyridazin-4(5H)-one,
    T-70164
2-(1,3,4-Thiadiazol-2-yl)pyridine, T-60192
4-(1,3,4-Thiadiazol-2-yl)pyridine, T-60193
1-Thiahydrindane, see O-60012
Thialen, see C-60224
Thialysine, see A-60152
Thia[2.2]metacyclophane, T-60194
1-Thianaphthalenium, see B-70066
2-Thianaphthalenium, see B-70067
1-Thianaphthen-3-one, see B-60048
3(2H)-Thianaphthenone, see B-60048
9H-10-Thia-1,4,5,9-tetraazaanthracene, see
    P-60199
2-Thiatricyclo[3.3.1.1³,⁷]decane, T-60195
3-Thiatricyclo[2.2.1.0²,⁶]heptane, T-70165
2-Thiatricyclo[9.3.1.1⁴,⁸]hexadeca-1(15),-
    4,6,8(16),11,13-hexaene, see T-60194
▷1,3-Thiazane, see T-60093
1,4-Thiazane-3-carboxylic acid, see T-70182
1,2-Thiazan-3-one, see T-60095
1,4-Thiazepine, T-60196
2-Thiazoledinethione, T-60197
4-Thiazolesulfonic acid, T-60198
5-Thiazolesulfonic acid, T-60199
2-Thiazolethiol, see T-60200
2(3H)-Thiazolethione, T-60200
4-Thiazoline-2-thione, see T-60200
Thiazolo[4,5-b]pyridine, T-70166
Thiazolo[4,5-c]pyridine, T-70167
Thiazolo[5,4-b]pyridine, T-70168
Thiazolo[5,4-c]pyridine, T-70169
Thiazolo[4,5-b]quinoline-2(3H)-thione, T-60201
Thiele's hydrocarbon, see B-60157
▷2-Thienal, see T-70183
Thieno[2,3-c]cinnoline, T-60202
Thieno[3,2-c]cinnoline, T-60203
Thieno[3,4-b]furan, T-60204
3-Thienol, see T-60212
4H-Thieno[3,4-c]pyran-4,6(7H)-dione, in
    C-60018
5H-Thieno[2,3-c]pyran-5,7(4H)dione, in
    C-60017
Thieno[3,2-d]pyrimidine, T-60205
5H-Thieno[2,3-c]pyrrole, T-60206
Thieno[3,4-c]thiophene-1,3,4,6-tetrathiol,
    T-70170
3-(2-Thienyl)acrylic acid, see T-60208
▷2-Thienylaldehyde, see T-70183
1-(2-Thienyl)ethanol, T-60207
2-(3-Thienyl)ethanol, see T-60211
2-[2-(2-Thienyl)ethenyl]-1H-pyrrole, see
    P-60250
5-(2-Thienyl)oxazole, T-70171
3-(2-Thienyl)-2-propenoic acid, T-60208
3-(3-Thienyl)-2-propenoic acid, T-70172
2-Thienylpropiolic acid, see T-70173
3-(2-Thienyl)-2-propynoic acid, see T-70173
Thiepano[2,3-b]pyridine, see T-60098
▷Thimecil, see D-70232
Thioacetoacetic acid, see O-70080
Thioacrylamide, see P-70120

2-Thiobenzpropiolactone, *see* B-70065
2,2′-Thiobisbenzoic acid, T-70174
4,4′-Thiobisbenzoic acid, T-70175
2,2′-Thiobis-1*H*-isoindole-1,3(2*H*)-dione, T-70176
N,N′-Thiobisphthalimide, *see* T-70176
Thiobis(trifluoromethyl)ketene, *see* B-60187
1,1′-Thiocarbonyl-2,2′-pyridone, *see* C-60014
Thiochloroformic acid, *see* C-60013
1-Thiocyanatoadamantane, T-70177
▷2-Thiocyanato-1*H*-pyrrole, T-60209
Thiocyanic acid 1,1-dimethylethyl ester, *see* M-70135
▷Thiocyanic acid 1*H*-pyrrol-2-yl ester, *see* T-60209
Thiocyanic acid tricyclo [3.3.1.1³,⁷]decyl ester, *see* T-70177
1-Thiocyano-1,3-butadiene, T-60210
2,2′-Thiodibenzoic acid, *see* T-70174
4,4′-Thiodibenzoic acid, *see* T-70175
Thioflavanone, *see* D-70248
Thioflavone, *see* P-60082
1-Thioflavone, *see* P-60082
Thioformaldehyde, *see* M-70037
Thiofurodysin, T-70178
Thiofurodysin acetate, *in* T-70178
Thiofurodysinin, T-70179
Thiofurodysinin acetate, *in* T-70179
▷2-Thiohydantoin, *see* T-70188
2-Thioimidazolidinetrione, *see* T-70187
Thioindoxyl, *see* B-60048
▷Thioinosine, *see* T-70180
▷6-Thioinosine, T-70180
▷Thiolactic acid, *see* M-60027
2-Thiolanecarboxylic acid, *see* T-70099
Thiomandelic acid, *see* M-60024
Thiomelin, T-70181
3-Thiomorpholinecarboxylic acid, T-70182
1-Thionianaphthalene, *see* B-70066
2-Thionianaphthalene, *see* B-70067
Thionobutyrolactone, *see* D-70210
ε-Thionocaprolactone, *see* O-70077
2-Thiono-4-oxotetrahydroquinazoline, *see* D-60296
4-Thiono-2-oxotetrahydroquinazoline, *see* D-60295
δ-Thionovalerolactone, *see* T-70085
Thioparabanic acid, *see* T-70187
3-Thiopheneacrylic acid, *see* T-60208
3-Thiopheneacrylic acid, *see* T-70172
▷Thiophene-2-aldehyde, *see* T-70183
▷2-Thiophenecarboxaldehyde, T-70183
2,5-Thiophenedimethanethiol, T-70184
2,4(3*H,5H*)-Thiophenedione, T-70185
3-Thiopheneethanol, T-60211
Thiophene-3-ol, T-60212
2-Thiophenesulfonic acid, T-60213
3-Thiophenesulfonic acid, T-60214
Thiophenylacetic acid, *see* B-70017
1-Thiophthalide, *see* I-70063
▷2-Thiopyridone, *see* P-60224
▷4-Thiopyridone, *see* P-60225
Thiopyrrolidone, *see* P-70186
2-Thio-2,4(1*H,3H*)-quinazolinedione, *see* D-60296
4-Thio-2,4(1*H,3H*)-quinazolinedione, *see* D-60295
Thiosine, *see* A-60152
2-Thiotetrahydro-1,3-thiazine, *see* T-60094
Thiotetronic acid, *see* T-70185
2-Thiothiazolone, *see* T-60200
▷Thiothymin, *see* D-70232
4-Thiothymine, *see* D-70233
Thio-α-toluic acid, *see* B-70017
Thiourea *S*-trioxide, *see* A-60210
Thioxanthylium(1+), T-70186
4-Thioxo-2-hexanone, T-60215
2-Thioxo-4,5-imidazolidinedione, T-70187
▷2-Thioxo-4-imidazolidinone, *see* T-70188
2-Thioxo-6-phenyl-4-pyrimidinone, *see* D-60268
3-Thioxo-1,2-propadien-1-one, T-60216
THIP, *see* T-60072
Thorectolide, T-70189
Thorectolide 25-acetate, *in* T-70189
D-Threaric acid, *in* T-70005
▷L-Threaric acid, *in* T-70005

Threonine, T-60217
Thujin, T-70190
Thurberin, *in* L-60039
Thurfyl salicylate, *in* H-70108
Thuriferic acid, T-70191
Thybon, *in* T-60350
Thymifodioic acid, T-60218
Thymin (hormone), *see* T-70192
Thymopoietin, T-70192
Thymopoietin I, *in* T-70192
Thymopoietin III, *in* T-70192
Thymopoietin II, 9CI, *in* T-70192
▷Thyreostat, *see* D-70232
Thyrolar, *in* T-60350
Thyrsiferol, T-70193
Tiacumicin A, T-70194
Tiacumicin B, *in* L-60028
Tiacumicin C, T-70195
Tiacumicin D, T-70196
Tiacumicin E, T-70197
Tiacumicin F, T-70198
Tifruticin, T-70199
Tilianin, *in* D-70319
Tingitanol, *in* D-70009
Tinospora clerodane, T-70200
▷Tioinosine, *see* T-70180
7-Tirucallene-3,23,24,25-tetrol, T-70201
TMF 518D, *see* A-70240
TNF, *see* T-70335
Tochuinol, T-70202
Tochuinyl acetate, *in* T-70202
Toddasin, T-70203
Tokorigenin, *in* S-60043
Tokoronin, *in* S-60043
Toltecol, *in* H-60214
4-*p*-Tolyl-2-pentanone, *see* M-60112
α-*o*-Tolylpropionic acid, *see* M-60113
α-*p*-Tolylpropionic acid, *see* M-60114
Tomenphantopin B, T-60222
Tomenphantopin A, *in* T-60222
Toosendantriol, T-60223
Topazolin, T-60224
Topazolin hydrate, *in* T-60224
α-Torosol, T-60225
β-Torosol, T-60226
Totarol, T-70204
Totarolone, *in* T-70204
Tovarol, *in* G-60014
Trametenolic acid, *in* H-70152
Trechonolide A, T-60227
Trechonolide B, *in* T-60227
Tremulacin, *in* S-70002
6β,9α,14-Triacetoxy-1α-benzoyloxy-4β,8β-dihydroxydihydro-β-agarofuran, *in* E-60018
6β,9α,14-Triacetoxy-1α-benzoyloxy-4β-hydroxy-8-oxodihydro-β-agarofuran, *in* E-60018
1,1,1-Triacetoxy-1,1-dihydro-1,2-benziodoxol-3(1*H*)-one, T-60228
ent-1β,7α,11α-Triacetoxy-3α-hydroxy-16-kaurene-6,15-dione, *in* H-60109
3α,7α,22-Triacetoxy-15α-hydroxy-8,24*E*-lanostadien-26-oic acid, *in* T-61011
3α,15α,22-Triacetoxy-7α-hydroxy-8,24*E*-lanostadien-26-oic acid, *in* T-61011
3α,15α,22*S*-Triacetoxy-7α-hydroxy-8,24*E*-lanostadien-26-oic acid, *in* T-61011
3α,15α,22*S*-Triacetoxy-7,9(11),24-lanostatrien-26-oic acid, *in* T-70266
3β,15α,22*S*-Triacetoxy-7,9(11),24*E*-lanostatrien-26-oic acid, *in* T-70266
3α,15α,22*S*-Triacetoxy-7α-methoxy-8,24*E*-lanostadien-26-oic acid, *in* T-61011
4,15,26-Triacontatriene-1,29-diyne-3,28-diol, *see* D-70550
4,15,26-Triacontatriene-1,12,18,29-tetrayne-3,14,17,28-tetrol, *see* P-70044
▷Triallylamine, *see* T-70315
2,4,7-Triaminopteridine, T-60229
4,6,7-Triaminopteridine, T-60230
3*H*-2,3,5-Triazabenz[*e*]indene, *see* P-60206
1,2,6-Triaza-4,5-benzindene (obsol.), *see* P-60206
1,4,7-Triazacyclododecane, T-60231
1,5,9-Triazacyclododecane, T-60232
1,3,5-Triazacyclohexane, *see* H-60060

1,5,9-Triazacyclotridecane, T-60233
1,4,7-Triazacycloundecane, T-60234
1,4,8-Triazacycloundecane, T-60235
1,2,7-Triazaindene, *see* P-70155
1,3,4-Triazaindene, *see* D-60249
▷1,3,5-Triazaindene, *see* I-60008
1,2,12-Triazapentacyclo[6.4.0.0²,¹⁷.0³,⁷.0⁴,¹¹]-dodecane, T-70205
1,4,6-Triazaphenothiazine, *see* P-60199
3,5,12-Triazatetracyclo[5.3.1.1²,⁶.0⁴,⁹]-dodecane, T-70206
19,20,21-Triazatetracyclo[13.3.1.1³,⁷.1⁹,¹³]-heneicosa-1(19),3,5,7(21),9,11,13(20),15,17-nonaene-2,8,14-trione, T-70236
▷1,2,4-Triazin-3-amine, *see* A-70203
1,3,5-Triazine-2,4,6-tricarboxaldehyde, T-60237
▷1,3,5-Triazine-2,4,6(1*H,3H,5H*)-trione, *see* C-60179
[1,3,5]Triazino[1,2-*a*.3,4-*a*′:5,6-*a*″]-trisbenzimidazole, T-60238
1*H*-1,2,3-Triazole-4-carboxylic acid, T-60239
1*H*-1,2,3-Triazole-5-carboxylic acid, *in* T-60239
v-Triazole-4-carboxylic acid, *see* T-60239
▷*s*-Triazole-3-thiol, *see* D-60298
[1,2,4]Triazolo[1,5-*b*]pyridazine, T-60240
[1,2,4]Triazolo[2,3-*b*]pyridazine, *see* T-60240
[1,2,3]Triazolo[1,5-*a*]pyridine, T-60241
[1,2,4]Triazolo[1,5-*a*]pyrimidine, T-60242
1,2,4-Triazolo[4,3-*c*]pyrimidine, T-60243
1*H*-1,2,3-Triazolo[4,5-*f*]quinoline, T-70207
1*H*-1,2,3-Triazolo[4,5-*h*]quinoline, T-70208
[1,2,4]Triazolo[1,2-*a*][1,2,4]triazole-1,3,5,7(2*H,6H*)-tetrone, T-70209
Tribenzimidazo[1,2-*a*:1′,2′-*c*:1″,2″-*e*]-[1,3,5]-triazine, *see* T-60238
Tribenzo[*b,n,pqr*]perylene, T-60244
Tribenzotricyclo[6.3.0.0¹,⁵]undecane, *see* T-70058
Tribenzotritwistatriene, T-70210
1,2,3-Tribromocyclopropane, T-70211
▷Tribromoglyoxaline, *see* T-60245
▷2,4,5-Tribromo-1*H*-imidazole, T-60245
[(Tribromomethyl)sulfinyl]benzene, *in* T-70212
[(Tribromomethyl)sulfonyl]benzene, *in* T-70212
[(Tribromomethyl)thio]benzene, *in* T-70212
Tribromophenylselenium, T-70213
3,4,5-Tribromo-1*H*-pyrazole, T-60246
2*a*,4*a*,6*b*-Tribromo-2*a*,4*a*,6*b*-tetrahydrocyclopenta[*cd*]pentalene, *see* T-70248
2,3,5-Tribromothiophene, T-60247
1,3,5-Tribromo-2,4,6-trifluorobenzene, T-70214
1,4,7-Tribromotriquinacene, T-60248
Tri-*tert*-butylazete, T-60249
1,3,5-Tributylbenzene, T-70215
Tri-*tert*-butylmethylethylene, *see* B-60350
1,3,5-Tri-*tert*-butyl-2-nitrobenzene, T-70216
2,4,6-Tri-*tert*-butylpyridine, T-70217
Tri-*tert*-butyl-1,2,3-triazine, T-60250
Tricarbon oxide sulfide, *see* T-60216
Trichadenal, T-70218
Trichadenic acid A, *in* T-70218
Trichadenic acid B, *in* T-70218
Trichadonic acid, *in* T-70218
Trichilinin, T-60251
2,2,3-Trichloroacetophenone, T-70219
1,2,3-Trichlorocyclopropane, T-70220
Trichlorocyclopropenylium, T-70221
1,1,2-Trichloro-2-(diethylamino)ethylene, T-70222
1,2,2-Trichloro-*N,N*-diethylethanamine, *see* T-70222
2,2,2-Trichloro-1-(4,5-dihydro-3-furanyl)-ethanone, *see* D-70271
2,2,2-Trichloro-1-(3,4-dihydro-2*H*-pyan-5-yl)-ethanone, *see* D-70272
▷1,1,1-Trichloro-3,4-epoxybutane, *see* T-60252
▷(2,2,2-Trichloroethyl)oxirane, T-60252
▷Trichloromethyl carbonochloridate, *see* T-70223
▷Trichloromethyl chloroformate, T-70223

5-(Trichloromethyl)isoxazole, T-60253
2,3,4-Trichlorophenylalanine, *see* A-60262
2,3,6-Trichlorophenylalanine, *see* A-60263
2,4,5-Trichlorophenylalanine, *see* A-60264
2,2,2-Trichloro-1-phenylethanone, *see* T-70219
Trichlorophenylselenium, T-70224
1,1,2-Trichloro-1-propene, T-60254
▷2,4,6-Trichloropyridine, T-60255
1,2,3-Trichloro-4,5,6-trifluorobenzene, T-60256
▷1,3,5-Trichloro-2,4,6-trifluorobenzene, T-60257
Trichostin, T-70225
Trichotriol, T-60258
4,15,26-Tricontatrien-1,12,18,29-tetrayne-3,14,17,28-tetrone, *in* P-70044
9-Tricosene, T-60259
Tricrozarin A, *in* H-60065
Tricrozarin B, *in* P-70033
Tricyclo[4.2.2.0^{1,6}]deca-7,9-diene, T-70226
Tricyclo[3.3.1.1^{3,7}]decane-1,3,5,7-tetraacetic acid, *see* A-70064
Tricyclo[3.3.1.1^{3,7}]decane-1,3,5,7-tetra-acetonitrile, *in* A-70064
Tricyclo[5.2.1.0^{4,10}]decane-2,5,8-trione, *see* H-60048
Tricyclo[5.3.0.0^{2,8}]deca-3,5,9-triene, T-70227
Tricyclo[5.2.1.0^{2,6}]dec-2(6)-ene, T-60260
Tricyclo[3.3.1.1^{3,7}]decyl-1-sulfinylcyanide, *in* T-70177
Tricyclo[7.3.2.0^{5,13}]dodeca-3,5,7,9,11(13)-hexaen-2-one, *see* M-70042
Tricyclo[4.4.2.0^{1,6}]dodecane-2,7-dione, T-60261
Tricyclo[6.4.0.0^{2,7}]dodeca-3,5,9,11-tetraene, *see* T-60060
Tricyclo[4.1.0.0^{1,3}]heptane, T-60262
Tricyclo[2.2.1.0^{2,6}]heptan-3-one, *see* T-70228
Tricyclo[18.4.1.1^{8,13}]hexacosa-2,4,6,8,10,12,14,16,18,20,22,24-dodecaene, T-60263
Tricyclo[18.2.2.2^{2,5}]hexacosa-2,4,20,22,23,25-hexaene, *see* P-70008
Tricyclo[8.4.1.1^{2,7}]hexadeca-2,4,6,8,10,12,14-heptaene, *see* B-60165
5,6,11,12,13,14,15,16-Tricyclo[8.2.2.2^{4,7}]-hexadeca-4,6,10,12,13,15-hexaene, *see* O-60010
Tricyclo[8.4.1.1^{3,8}]hexadeca-3,5,7,10,12,14-hexaene-2,9-dione, T-60264
Tricyclo[8.2.2.2^{4,7}]hexadeca-2,4,6,8,10,12,13,15-octaene, *see* P-70007
Tricyclo[3.1.0.0^{2,6}]hexanedione, T-70229
Tricyclo[3.3.1.0^{2,8}]nona-3,6-diene, T-70230
Tricyclo[3.1.1.1^{2,4}]octane, T-60265
Tricyclo[5.1.0.0^{2,8}]octane, T-60266
Tricyclo[2.2.2.0^{2,6}]oct-7-ene, *see* T-60267
Tricyclo[3.2.1.0^{2,7}]oct-3-ene, T-60267
Tricyclo[5.1.0.0^{2,8}]oct-3-ene, T-60268
Tricyclo[5.1.0.0^{2,8}]oct-4-ene, T-60269
Tricyclo[2.1.0.0^{1,3}]pentane, T-70231
Tricyclopropylcyclopropenium(1+), T-60270
Tricyclopropylidene, *see* T-70325
▷Tricycloquinazoline, T-60271
Tricyclo[4.4.4.0^{1,6}]tetradeca-2,4,7,9,10,12-hexaene, T-70232
Tricyclotrimethylenebenzene, *see* O-60021
Tricyclo[4.3.1.1^{3,8}]undecane-2,7-dione, T-70233
Tricyclo[4.3.1.1^{3,8}]undecane-4,5-dione, T-70234
6,7-Tridecadiene, T-60272
7,10:19,22:31,34-Triethenotribenzo[*a,k,u*]-cyclotriacontane, T-60273
Triethylacetic acid, *see* D-60171
▷Triethylenediamine, *see* D-70051
Triethyl isocyanurate, *in* C-60179
Triethylthiocarbinol, *see* E-60053
2,4,6-Triethyl-1,3,5-triazine, T-70235
Trifarin, T-70236
Triflic acid, *see* T-60241
Triflorestevione, T-70237
▷Trifluoroacetic acid, anhydride with nitrous acid, *see* T-60274

▷Trifluoroacetyl nitrite, T-60274
Trifluoroacetyl trifluoromethanesulfonate, *in* T-70241
2,3,5-Trifluorobenzoic acid, T-70238
2,3,6-Trifluorobenzoic acid, T-70239
2,4,5-Trifluorobenzoic acid, T-70240
2,4,5-Trifluorobenzonitrile, *in* T-70240
1,1,1-Trifluoro-*N,N*-bis(trifluoromethyl)-methanamine, *see* T-60406
1,1,1-Trifluoro-2-butyne, T-60275
1,4,4-Trifluorocyclobutene, T-60276
α,α,α-Trifluoro-2,4-dinitrotoluene, *see* D-70456
α,α,α-Trifluoro-3,5-dinitrotoluene, *see* D-70455
1,1,1-Trifluoro-2,3-epoxypropane, *see* T-60294
Trifluoroethenesulfonic acid, T-60277
Trifluoroethylene oxide, *see* T-60305
1,1,1-Trifluoro-*N*-hydroxy-*N*-(trifluoromethyl)-methanamine, *see* B-60180
Trifluoroisocyanomethane, T-60278
Trifluoromethanesulfenic acid, T-60279
Trifluoromethanesulfenyl bromide, T-60280
▷Trifluoromethanesulfenyl fluoride, T-60281
Trifluoromethanesulfenyl iodide, T-60282
Trifluoromethanesulfonic acid, T-70241
2-Trifluoromethylacrylic acid, *see* T-60295
▷2-Trifluoromethylacrylonitrile, *in* T-60295
2-(Trifluoromethyl)anisaldehyde, *in* H-70228
3-(Trifluoromethyl)anisaldehyde, *in* H-70229
2-(Trifluoromethyl)benzenesulfonic acid, T-70242
3-(Trifluoromethyl)benzenesulfonic acid, T-60283
4-(Trifluoromethyl)benzenesulfonic acid, T-70243
2-(Trifluoromethyl)benzenethiol, T-60284
3-(Trifluoromethyl)benzenethiol, T-60285
4-(Trifluoromethyl)benzenethiol, T-60286
2-(Trifluoromethyl)-1*H*-benzimidazole, T-60287
3-Trifluoromethyl-γ-butyrolactone, *see* D-60299
(Trifluoromethyl)carbonimidic difluoride, T-60288
2-Trifluoromethyl-1*H*-imidazole, T-60289
4(5)-Trifluoromethyl-1*H*-imidazole, T-60290
(Trifluoromethyl)imidocarbonyl fluoride, *see* T-60288
2-Trifluoromethyl-1*H*-indole, T-60291
3-Trifluoromethyl-1*H*-indole, T-60292
6-Trifluoromethyl-1*H*-indole, T-60293
Trifluoromethyl isocyanide, *see* T-60278
(Trifluoromethyl)oxirane, T-60294
2-(Trifluoromethyl)phenyl methyl sulfide, *in* T-60284
3-(Trifluoromethyl)phenyl methyl sulfide, *in* T-60285
4-(Trifluoromethyl)phenyl methyl sulfide, *in* T-60286
2-(Trifluoromethyl)propenoic acid, T-60295
2-(Trifluoromethyl)pyridine, T-60296
3-(Trifluoromethyl)pyridine, T-60297
4-(Trifluoromethyl)pyridine, T-60298
Trifluoro(methylsulfinyl)methane, *in* M-70145
Trifluoro(methylsulfonyl)methane, *in* M-70145
2-(Trifluoromethylsulfonyloxy)pyridine, P-70173
(Trifluoromethyl)sulfur pentafluoride, *see* P-60041
Trifluoro(methylthio)methane, *see* M-70145
2-(Trifluoromethyl)thiophenol, *see* T-60284
3-(Trifluoromethyl)thiophenol, *see* T-60285
4-(Trifluoromethyl)thiophenol, *see* T-60286
Trifluoromethyl trifluoromethanesulfonate, *in* T-70241
1,2,3-Trifluoro-4-nitrobenzene, T-60299
1,2,3-Trifluoro-5-nitrobenzene, T-60300
▷1,2,4-Trifluoro-5-nitrobenzene, T-60301
1,2,5-Trifluoro-3-nitrobenzene, T-60302
1,2,5-Trifluoro-4-nitrobenzene, T-60303
1,3,5-Trifluoro-2-nitrobenzene, T-60304
α,α,α-Trifluoro-2-nitro-*m*-cresol, *see* N-60047
▷α,α,α-Trifluoro-2-nitro-*p*-cresol, *see* N-60048

α,α,α-Trifluoro-3-nitro-*p*-cresol, *see* N-60051
▷α,α,α-Trifluoro-4-nitro-*m*-cresol, *see* N-60054
α,α,α-Trifluoro-4-nitro-*o*-cresol, *see* N-60053
α,α,α-Trifluoro-5-nitro-*m*-cresol, *see* N-60052
α,α,α-Trifluoro-6-nitro-*m*-cresol, *see* N-60049
α,α,α-Trifluoro-6-nitro-*o*-cresol, *see* N-60050
α,α,α-Trifluoro-3-nitro-4-methoxytoluene, *in* N-60048
1,1,1-Trifluoro-*N*-(nitrosooxy)-*N*-(tri-fluoromethyl)methanamine, *see* T-60180
▷α,α,α-Trifluoro-*m*-nitrotoluene, *see* N-70065
▷α,α,α-Trifluoro-*o*-nitrotoluene, *see* N-70064
▷α,α,α-Trifluoro-*p*-nitrotoluene, *see* N-70066
Trifluorooxirane, T-60305
1,1,1-Trifluoro-4-phenyl-3-butyn-2-one, T-70244
Trifluoropyrazine, T-60306
2,4,6-Trifluoropyrimidine, T-60307
4,5,6-Trifluoropyrimidine, T-60308
α,α,α-Trifluoro-*m*-toluenesulfonic acid, *see* T-60283
α,α,α-Trifluoro-*o*-toluenesulfonic acid, *see* T-70242
α,α,α-Trifluoro-*p*-toluenesulfonic acid, *see* T-70243
α,α,α-Trifluoro-*m*-toluenethiol, *see* T-60285
α,α,α-Trifluoro-*o*-toluenethiol, *see* T-60284
α,α,α-Trifluoro-*p*-toluenethiol, *see* T-60286
Trifluoro-1,2,3-triazine, T-70245
4,4,4-Trifluoro-3-(trifluoromethyl)-2-butenal, T-60309
1,1,1-Trifluoro-2-(trifluoromethyl)-3-butyn-2-ol, T-60310
4-[2,2,2-Trifluoro-1-(trifluoromethyl)-ethyl-idene]-1,3,2-dioxathietane 2,2-dioxide, T-60311
3,3,3-Trifluoro-2-(trifluoromethyl)-1-propene-1-thione, *see* B-60187
5,5,5-Trifluoro-γ-valerolactone, *see* D-60300
Trifluorovinylsulfonic acid, *see* T-60277
Trifolirhizin, *in* M-70001
1,3,5-Triformylcyclohexane, *see* C-70226
Triformyl-*s*-triazine, *see* T-60237
1,3,5-Triheptylbenzene, T-70246
1,3,5-Trihexylbenzene, T-70247
11,12,16-Trihydroxy-5,8,11,13-abietatetraen-7-one, T-60312
11,12,14-Trihydroxy-8,11,13-abietatriene-3,7-dione, T-60313
7β,11,12-Trihydroxy-8,11,13-abietatrien-20,6β-olide, *see* R-70010
3,11,12-Trihydroxy-8,11,13-abietatrien-7-one, T-60314
2α,6α,11-Trihydroxy-7,9(11),13-abietatrien-12-one, *in* D-70282
2′,4′,6′-Trihydroxyacetophenone, T-70248
▷3′,4′,5′-Trihydroxyacetophenone, T-60315
3′,4′,6-Trihydroxyaurone, *see* S-70082
3′,4′,6-Trihydroxybenzalcoumaranone, *see* S-70082
2,3,4-Trihydroxybenzeneacetic acid, *see* T-60340
2,3,5-Trihydroxybenzeneacetic acid, *see* T-60341
2,4,5-Trihydroxybenzeneacetic acid, *see* T-60342
3,4,5-Trihydroxybenzeneacetic acid, *see* T-60343
2,3,9-Trihydroxy-6*H*-benzofuro[3,2-*c*][1]-benzopyran-6-one, T-70249
2,3,5-Trihydroxybenzoic acid, T-70250
3,3′,4-Trihydroxybibenzyl, T-60316
2,3,4-Trihydroxybiphenyl, B-70124
▷3β,14β,19-Trihydroxy-5α-card-20(22)-enolide, *see* C-60168
3,11,15-Trihydroxycholanic acid, T-60317
3,15,18-Trihydroxycholanic acid, T-60318
ent-2β,3β,4α-Trihydroxy-15-clerodanoic acid, *in* T-60410
ent-2β,3β,4α-Trihydroxy-13-cleroden-15-oic acid, *see* T-60410

2,3,9-Trihydroxycoumestone, *see* T-70249

2,3,24-Trihydroxy-13,27-cyclo-11-oleanen-28-oic acid, T-70251

2,3,24-Trihydroxy-12,27-cyclo-14-taraxeren-28-oic acid, T-70252

12,20,25-Trihydroxy-23-dammaren-3-one, T-60319

3,10,11-Trihydroxy-7,8-dihydro-6*H*-dibenz[*b*,*d*]oxcin-7-one, *see* P-60186

3,3′,5-Trihydroxy-4′,7-dihydroxyflavan, *in* P-70026

2′,3,5-Trihydroxy-7,8-dimethoxyflavanone, *in* P-60045

4′,5,7-Trihydroxy-3′,6-dimethoxyflavanone, *in* P-70028

3,4′,5-Trihydroxy-6,7-dimethoxyflavone, *in* P-60047

3,4′,7-Trihydroxy-5,6-dimethoxyflavone, *in* P-60047

3′,4′,5-Trihydroxy-6,7-dimethoxyisoflavone, *in* P-70032

1,3,8-Trihydroxy-2,6-dimethoxyxanthone, *in* P-70036

1,3,8-Trihydroxy-4,7-dimethoxyxanthone, *in* P-60052

1,5,8-Trihydroxy-2,6-dimethoxyxanthone, *in* P-60052

3,4,6-Trihydroxy-1,2-dimethylcarbazole, T-70253

2,4*a*,8-Trihydroxy-1,8-dimethylgibbane-1,10-dicarboxylic acid 1,4*a*-lactone, *see* G-60017

6α,7β,8α-Trihydroxy-14,15-dinor-11-labden-13-one, *in* S-60052

2,3,10-Trihydroxy-6,12-dioxabenz[*a*]-anthracen-7(5*H*,6*H*,12*aH*)-one, *in* C-70200

3,7,15-Trihydroxy-11,23-dioxo-8,20(22)-lanostadien-26-oic acid, T-70254

3β,7β,28-Trihydroxy-11,15-dioxo-25,26,27-trisnor-8-lanosten-24-oic acid, *see* L-60036

14α,17β,20*R*-Trihydroxy-1,4-dioxo-22*R*-witha-5,24-dienolide, *see* P-60149

4′,5,7-Trihydroxy-3′,8-diprenylflavanone, *see* L-60021

4′,5,7-Trihydroxy-6,8-diprenylflavanone, *see* L-60030

4′,5,7-Trihydroxy-6,8-diprenylflavanone, *see* S-70034

4′,5,7-Trihydroxy-3′,5′-diprenylflavone, *see* H-70092

6,9,11-Trihydroxy-7-drimene, *see* D-70548

8,9,11-Trihydroxy-6-drimene, *see* D-70547

5,6,15-Trihydroxy-7,9,11,13,17-eicosapenta-enoic acid, T-70255

5,14,15-Trihydroxy-6,8,10,12,17-eicosapenta-enoic acid, T-70256

5,14,15-Trihydroxy-6,8,10,12-eicosatetraenoic acid, T-60320

2,8,9-Trihydroxy-11-eremophilene, *see* E-70030

1β,3α,6β-Trihydroxy-4,11(13)-eudesmadien-12,8β-olide, *in* I-70109

1,3,9-Trihydroxy-4(15),11(13)-eudesmadien-12,6-olide, T-70257

1,3,13-Trihydroxy-4(15),7(11)-eudesmadien-12,6-olide, T-70258

1,4,6-Trihydroxy-11(13)-eudesmen-12,8-olide, T-70259

5,8,12-Trihydroxyfarnesol, *in* T-60363

3′,4′,7-Trihydroxyflavan, T-70260

5,7,8-Trihydroxyflavanone, T-60321

5,6,7-Trihydroxyflavone, T-60322

2,4,5-Trihydroxy-9*H*-fluoren-9-one, T-70261

8,9,14-Trihydroxy-1(10),4,11(13)-germacratrien-12,6-olide, T-60323

7,9,11-Trihydroxy-2-hexacosen-1,5-olide, *see* D-70276

2,5,7-Trihydroxy-3-(5-hydroxyhexyl)-1,4-naphthoquinone, T-70262

3′,4′,7-Trihydroxyisoflavanone, T-70263

2′,4′,7-Trihydroxyisoflavone, T-70264

▷4′,5,7-Trihydroxyisoflavone, T-60324

4′,6,7-Trihydroxyisoflavone, T-70265

4′,7,8-Trihydroxyisoflavone, T-70265

3′,4,5′-Trihydroxy-3-isopentadienylstilbene, *see* H-70167

ent-3α,7β,14α-Trihydroxy-16-kauren-15-one, T-60326

3,7,15-Trihydroxy-8,24-lanostadien-26-oic acid, T-60327

3,7,22-Trihydroxy-8,24-lanostadien-26-oic acid, T-60328

24,25,26-Trihydroxy-7,9(11)-lanostadien-3-one, *in* L-60012

3,15,22-Trihydroxy-7,9(11),24-lanostatrien-26-oic acid, T-70266

15,26,27-Trihydroxy-7,9(11),24-lanostatrien-3-one, T-60330

3,12,17-Trihydroxy-9(11)-lanosten-18,20-olide, T-70267

1,11,20-Trihydroxy-3-lupanone, T-60331

ent-14,15,19-Trihydroxymanoyl oxide, *see* E-70020

2′,4,4′-Trihydroxy-6′-methoxydihydrochalcone, *in* H-60220

2,3,10-Trihydroxy-8-methoxy-6,12-dioxabenz[*a*]anthracen-7(5*H*,6*aH*,12*aH*)-one, *in* C-70200

3,4,10-Trihydroxy-8-methoxy-6,12-dioxabenz[*a*]anthracen-7(5*H*,6*H*,12*aH*)-one, *in* C-70200

2′,5,7-Trihydroxy-4′-methoxy-6,8-diprenylflavanone, *see* L-60022

2′,5,8-Trihydroxy-7-methoxyflavanone, *in* T-70104

3′,4′,5-Trihydroxy-7-methoxyisoflavone, *in* T-60108

3′,5,7-Trihydroxy-4′-methoxyisoflavone, *in* T-60108

3′,7,8-Trihydroxy-4′-methoxyisoflavone, *in* T-70112

4′,5,7-Trihydroxy-3′-methoxyisoflavone, *in* T-60108

4′,7,8-Trihydroxy-6-methoxyisoflavone, *in* T-70113

1,3,5-Trihydroxy-4-methoxy-2-methyl-anthraquinone, *in* T-60112

1,3,8-Trihydroxy-5-methoxy-2-methyl-anthraquinone, *in* T-60113

2′,4,6′-Trihydroxy-6-methoxy-4′-methyl-2-benzophenonecarboxylic acid methyl ester, *see* S-70083

1,2,4-Trihydroxy-5-methoxy-3-(8,11,14-penta-decatrienyl)benzene, *see* M-70049

4′,5,7-Trihydroxy-3′-methoxy-5′-prenylflavanone, *in* S-70041

4′,5,7-Trihydroxy-3-methoxy-6-prenylflavone, *see* T-60224

4′,6′,7-Trihydroxy-3′-methoxy-2′-prenylisoflavone, *see* K-60016

1,3,8-Trihydroxy-5-methoxyxanthone, *in* T-70123

1,5,8-Trihydroxy-3-methoxyxanthone, *in* T-70123

2,4,5-Trihydroxy-1-methoxyxanthone, *in* T-70122

1,2,7-Trihydroxy-6-methyl-9,10-anthracenedione, *see* T-70268

1,2,7-Trihydroxy-6-methylanthraquinone, T-70268

1,6,8-Trihydroxy-2-methylanthraquinone, T-70269

3,5,6-Trihydroxy-2-methylanthraquinone, *see* T-70268

3,6,8-Trihydroxy-1-methylanthraquinone-2-carboxylic acid, T-60332

1,3,8-Trihydroxy-7-methylanthraquinone (incorr.), *see* T-70269

1,6,8-Trihydroxy-3-methylbenz[*a*]-anthracene-7,12-dione, T-70270

3α,22*R*,23*R*-Trihydroxy-24*S*-methyl-5α-cholestan-6-one, *in* T-70116

1,5,8-Trihydroxy-3-methyl-2-(3-methyl-2-butenyl)xanthone, *in* T-60114

2,5,8-Trihydroxy-3-methyl-6,7-methylenedioxy-1,4-naphthoquinone, T-60333

4,6,9-Trihydroxy-7-methylnaphtho[2,3-*d*]-1,3-dioxole-5,8-dione, T-60333

5,6,7-Trihydroxy-18-methyloxacyclooctadec-3-en-2-one, *see* A-60312

1,4,5-Trihydroxynaphthalene, *see* N-60005

ent-3α,10β,15β-Trihydroxy-20-norgibberellane-7,19-dioic acid 19,10-lactone, *see* G-60017

ent-6α,15,16-Trihydroxy-19-nor-4-rosen-3-one, *in* N-60061

9,10,18-Trihydroxyoctadecanoic acid, T-60334

4,5,6-Trihydroxyoctadec-2-en-1,17-olide, *see* A-60312

3,16,28-Trihydroxy-13(18)-oleanene, *see* O-70035

2,3,23-Trihydroxy-12-oleanen-28-oic acid, T-70271

2,3,24-Trihydroxy-12-oleanen-28-oic acid, T-60335

2,3,24-Trihydroxy-11,13(18)-oleonadien-28-oic acid, T-60336

5,8,9-Trihydroxy-10(14)-oplopen-3-one, T-70272

6,7,14-Trihydroxy-8(10)-oplopen-2-one (incorr.), *see* T-70272

1,3,4-Trihydroxy-6-oxabicyclo[3.2.1]-octan-7-one, *in* Q-70005

7,11,12-Trihydroxy-6-oxo-8,11,13-abietatrien-20-oic acid, T-60337

(3,4,5-Trihydroxy-6-oxo-1-cyclohexen-1-yl)-methyl 2-butenoate, T-60338

3β,7α,15α-Trihydroxy-11-oxo-25,26,27-trisnor-8-lanosten-24-oic acid, *see* L-60037

2′,5,7-Trihydroxy-3,4′,5′,6,8-penta-methoxyflavone, *in* O-60022

3′,5,7-Trihydroxy-3,4′,5′,6,8-penta-methoxyflavone, *in* O-70018

4′,5,5′-Trihydroxy-2′,3,6,7,8-penta-methoxyflavone, *in* O-60022

4′,5,7-Trihydroxy-2′,3,5′,6,8-penta-methoxyflavone, *in* O-60022

4′,5,7-Trihydroxy-3,3′,5′,6,8-penta-methoxyflavone, *in* O-70018

2,3,7-Trihydroxyphenanthrene, *see* T-60339

2,4,5-Trihydroxyphenanthrene, *see* P-60071

2,3,4-Trihydroxyphenylacetic acid, T-60340

2,3,5-Trihydroxyphenylacetic acid, T-60341

2,4,5-Trihydroxyphenylacetic acid, T-60342

3,4,5-Trihydroxyphenylacetic acid, T-60343

5,6,7-Trihydroxy-2-phenyl-4*H*-1-benzopyran-4-one, *see* T-60322

1-(2,4,6-Trihydroxyphenyl)ethanone, *see* T-70248

▷1-(3,4,5-Trihydroxyphenyl)ethanone, *see* T-60315

1-(2,3,4-Trihydroxyphenyl)-2-(3,4,5-trihydroxyphenyl)ethylene, T-60344

ent-2α,3α,16-Trihydroxy-8(14)-pimaren-15-one, *in* P-60154

2,3,12-Trihydroxy-4,7-pregnadien-20-one, T-60345

3β,16β,20*S*-Trihydroxy-5α-pregnane-20-carboxylic acid 22,16-lactone, *see* L-60042

2,3,12-Trihydroxy-4,7,16-pregnatrien-20-one, T-60346

5,9,10-Trihydroxy-2*H*-pyrano[2,3,4-*kl*]-xanthen-2-one, T-70273

2,4,6-Trihydroxypyridine, *see* D-70341

12,20,25-Trihydroxy-3,4-seco-4(28),23-dammaradien-3-oic acid, T-70274

20,25,26-Trihydroxy-3,4-seco-4(28),23-dammaradien-3-oic acid, T-70275

3β,23,27-Trihydroxy-28,20-taraxastanolide, *in* H-70221

3,4′,5-Trihydroxy-3′,6,7,8-tetramethoxyflavone, *in* H-60024

3′,5,5′-Trihydroxy-3,4′,7,8-tetramethoxyflavone, *in* H-70021

3′,5,7-Trihydroxy-3,4′,5′,8-tetramethoxyflavone, *in* H-70021

3,23,25-Trihydroxy-7-tiracallen-24-one, T-60347

α,2,5-Trihydroxytoluene, *see* D-70287

2′,3′,5-Trihydroxy-3,7,8-trimethoxyflavone, *in* H-70079

2′,5,5′-Trihydroxy-3,7,8-trimethoxyflavone, *in* H-70080

2′,5′,6-Trihydroxy-3,5,7-trimethoxyflavone, *in* H-60064

3′,4′,5-Trihydroxy-5′,6,7-trimethoxyflavone, *in* H-70081

3',5,7-Trihydroxy-4',5',6-trimethoxyflavone, *in* H-70081

3β,7α,15α-Trihydroxy-4,4,14α-trimethyl-11-oxo-5α-chol-8-en-24-oic acid, *see* L-60037

3,9,13-Trihydroxy-11,15(17)-trinervitadiene, *see* T-70297

2,3,20-Trihydroxy-1(15),8(19)-trinervitadiene, *see* T-70299

3β,12β,28-Trihydroxy-7,11,15-trioxo-25,26,27-trisnor-8-lanosten-24-oic acid, *in* L-60036

2,2',2''-Trihydroxytriphenylamine, *see* N-70039

4,4',4''-Trihydroxytriphenylamine, *see* N-70040

4',5,7-Trihydroxy-3',6,8-triprenylisoflavone, *see* E-70058

2α,3α,24-Trihydroxy-12,20(30)-ursadien-28-oic acid, *in* D-60381

3,19,24-Trihydroxy-12-ursen-28-oic acid, T-60348

1,5,6-Trihydroxy-9*H*-xanthen-9-one, *see* T-60349

1,5,6-Trihydroxyxanthone, T-60349

2,3'-2',3''-Triindolyl, *see* T-60011

▷2,4,5-Triiodo-1*H*-imidazole, T-70276

3,3',5-Triiodothyronine, T-60350

Trijugin *A*, T-60351

Trijugin *B*, T-60352

Trillin, *in* S-60045

Trimethidinium(2+), T-70277

▷Trimethidinium methosulphate, *in* T-70277

2',4',6'-Trimethoxyacetophenone, *in* T-70248

3',4',5'-Trimethoxyacetophenone, *in* T-60315

2,3,4-Trimethoxybenzeneacetic acid, *in* T-60340

2,4,5-Trimethoxybenzeneacetic acid, *in* T-60342

3,4,5-Trimethoxybenzeneacetic acid, *in* T-60343

▷3,4,5-Trimethoxybenzeneacetonitrile, *in* T-60343

2,3,5-Trimethoxybenzoic acid, *in* T-70250

2,3,5-Trimethoxybenzonitrile, *in* T-70250

2,3,4-Trimethoxybiphenyl, *in* B-70124

2,4,5-Trimethoxycinnamaldehyde, *see* T-70278

3,4,6-Trimethoxy-1,2-dimethylcarbazole, *in* T-70253

5,6,7-Trimethoxyflavone, *in* T-60322

5,7,8-Trimethoxyflavone, *in* T-60321

2',4',7-Trimethoxyisoflavone, *in* T-70264

4',6,7-Trimethoxyisoflavone, *in* T-60325

4',7,8-Trimethoxyisoflavone, *in* T-70265

5,6,7-Trimethoxy-3',4'-methylenedioxyisoflavone, *in* P-70032

5,6,7-Trimethoxy-8-(3-methyl-2-oxobutyl)-2*H*-1-benzopyran-2-one, *see* M-70160

1,4,5-Trimethoxynaphthalene, *in* N-60005

2,3,7-Trimethoxyphenanthrene, *in* T-60339

1,5,6-Trimethoxy-2,7-phenanthrenediol, *in* P-60050

5,6,7-Trimethoxy-2,10-phenanthrenediol, *in* P-70035

3,5,7-Trimethoxy-2-phenanthrenol, *in* T-60120

2,3,6-Trimethoxyphenol, *in* B-60014

1,2,4-Trimethoxy-3*H*-phenoxazin-3-one, T-60353

2,3,4-Trimethoxyphenylacetic acid, *in* T-60340

2,4,5-Trimethoxyphenylacetic acid, *in* T-60342

3,4,5-Trimethoxyphenylacetic acid, *in* T-60343

▷3,4,5-Trimethoxyphenylacetonitrile, *in* T-60343

3-(2,4,5-Trimethoxyphenyl)-2-propenal, T-70278

1,2,4-Trimethoxy-5-(2-propenyl)benzene, T-70279

2,4,6-Trimethoxypyridine, *in* D-70341

ω-Trimethylacetophenone, *see* D-60443

▷Trimethylamine oxide, T-60354

N-(γ-Trimethylammoniopropyl)-*N*-methyl-camphidinium, *see* T-70277

1,3,6-Trimethyl-5-azauracil, *in* M-60132

3,5,8-Trimethylazuleno[5,6-*b*]furan, *see* L-70031

3,4,5-Trimethylbenzoic acid, T-60355

2,5,8-Trimethylbenzotriimidazole, T-60356

1,7,7-Trimethylbicyclo[2.2.1]heptane-2,5-diol, T-70280

4,7,7-Trimethylbicyclo[4.1.0]heptan-2-one, *see* C-60019

7,11,11-Trimethylbicyclo[8.1.0]undeca-2,6-diene-3-carboxaldehyde, *see* I-60088

1,3,9-Trimethylcarbazole, *in* D-70375

2,4,9-Trimethylcarbazole, *in* D-70382

1,2,3-Trimethyl-9*H*-carbazole, T-60357

1,2,4-Trimethyl-9*H*-carbazole, T-60358

4α,14α,24*R*-Trimethyl-9β,19-cyclo-20-cholesten-3β-ol, *see* C-70218

2,3,4-Trimethyl-2-cyclohexen-1-one, T-60359

1,2,5-Trimethylcyclopentanecarboxylic acid, T-60361

2,2,5-Trimethylcyclopentanecarboxylic acid, T-60360

2,5,5-Trimethyl-2-cyclopenten-1-one, T-70281

3,3,7-Trimethyl-2,9-dioxatricyclo[3.3.1.0⁴·⁷]nonane, *see* L-60027

2,6,10-Trimethyl-6,11-dodecadiene-2,3,10-triol, T-60362

3,7,11-Trimethyl-1-dodecanol, T-70282

2,6,10-Trimethyl-2,6,11-dodecatriene-1,10-diol, T-70283

3,7,11-Trimethyl-1,6,10-dodecatriene-3,9-diol, T-70284

3,7,11-Trimethyl-2,6,10-dodecatrienoic acid, T-60364

2,6,10-Trimethyl-2,6,10-dodecatrien-1,5,8,12-tetrol, T-60363

2,3,5-Trimethylenebicyclo[2.2.1]heptane, *see* T-60403

Trimethyleneimine, *see* A-70283

2,8,9-Trimethylene[3.3.3]propellane, *see* T-70101

1,2-Trimethylenepyrazolidine, *see* T-70086

2,3-Trimethylenepyridine, *see* D-60275

2,3-Trimethylenepyridine, *see* D-70258

3,4-Trimethylenepyridine, *see* D-60276

3,4-Trimethylenepyridine, *see* D-70259

Trimethylenetriamine, *see* H-60060

Trimethyleneurea, *see* T-60088

2,2,6-Trimethyl-5-hepten-3-one, *see* T-60365

2,2,5-Trimethyl-4-hexen-3-one, T-60366

2,3,5-Trimethylidenebicyclo[2.2.1]heptane, *see* T-60403

Trimethyl isocyanurate, *in* C-60179

3,4,4-Trimethyl-5(4*H*)-isoxazolone, T-60367

7,11,15-Trimethyl-3-methylene-1,2-hexadecanediol, T-60368

Trimethylnaphthazarin, *see* D-60380

2,4,6-Trimethyl-3-nitroaniline, T-70285

2,4,6-Trimethyl-3-nitrobenzenamine, *see* T-70285

2,3,4-Trimethyl-5-nitrobenzoic acid, T-60369

2,4,6-Trimethyl-3-nitrobenzoic acid, T-60370

▷Trimethyloxamine, *see* T-60354

2,6,6-Trimethyl-1-(3-oxo-1-butenylidene)-2,4-cyclohexanediol, *see* G-70028

3,4,5-Trimethylphenol, T-60371

2',4',6'-Trimethyl-5'-phenyl-1,1':3',1''-terphenyl, *see* T-70293

3',4',6'-Trimethyl-5'-phenyl-1,1':3',1''-terphenyl, *see* T-70292

4',5',6'-Trimethyl-2'-phenyl-1,1':3',1''-terphenyl, *see* T-70291

2,6,7-Trimethylpteridine, T-70286

4,6,7-Trimethylpteridine, T-70287

1,2,5-Trimethylpyrrolidine, *in* D-70437

1,5,5-Trimethylspiro[5.5]undeca-1,7-diene-3,9-dione, *see* M-70006

2,2',2''-Trimethyltriphenylamine, T-70288

3,3',3''-Trimethyltriphenylamine, T-70289

4,4',4''-Trimethyltriphenylamine, T-70290

1,2,3-Trimethyl-4,5,6-triphenylbenzene, T-70291

1,2,4-Trimethyl-3,5,6-triphenylbenzene, T-70292

1,3,5-Trimethyl-2,4,6-triphenylbenzene, T-70293

N,*N*',*N*''-Trimethyl-*N*,*N*',*N*''-triphenylmethanetriamine, *see* T-70324

4,6,8-Trimethyl-2-undecanol, T-60372

1,3,5-Trimorpholinobenzene, T-70294

Trindan, *see* O-60021

2,3-Trinervidiol, *see* T-70295

Trinerviol, *see* T-70300

1(15),8(19)-Trinervitadiene-2,3-diol, T-70295

1(15),8(19)-Trinervitadiene-2,3,9,20-tetrol, T-70296

1(15),8(19)-Trinervitadiene-2β,3α,9β,20-tetrol 9,20-diacetate, *in* T-70296

11,15(17)-Trinervitadiene-3,9,13-triol, T-70297

1(15),8(19)-Trinervitadiene-2,3,9-triol, T-70298

1(15),8(19)-Trinervitadiene-2,3,20-triol, T-70299

1(15),8(19)-Trinervitadien-9-ol, T-70300

2,3,9-Trinervitriol, *see* T-70298

2,3,9-Trinerviriol, *see* T-70298

2,4,6-Trinitroacetanilide, *in* T-70301

▷2,4,6-Trinitroaniline, T-70301

▷2,4,6-Trinitrobenzenamine, *see* T-70301

▷2,4,6-Trinitrobenzenesulfonic acid, T-70302

▷2,4,6-Trinitrobenzoic acid, T-70303

2,4,6-Trinitro-3,5-cresotic acid, *see* H-60193

2,3,4-Trinitro-1*H*-pyrrole, T-60373

2,3,5-Trinitro-1*H*-pyrrole, T-60374

20,29,30-Trinorlupane-3,19-diol, *see* T-60404

Triomet-125, *in* T-60350

Trionine, *in* T-60350

Triothyrone, *see* T-60350

2,4,10-Trioxaadamantane, *see* T-60377

1,5,9-Trioxacyclododecane, T-70304

4,7,13-Trioxa-1,10-diazabicyclo[8.5.5]-icosane, T-60375

1,4,10-Trioxa-7,13-diazacyclopentadecane, T-60376

1,4,8-Trioxaspiro[4.5]decan-7,9-dione, *in* O-60084

2,4,10-Trioxatricyclo[3.3.1.1³·⁷]decane, T-60377

Triphenodioxazine, T-60378

2,2,2-Triphenylacetophenone, *see* T-60167

1,2,3-Triphenylbenzopentalene, *see* T-70306

1,1,4-Triphenyl-1,3-butadiene, T-70305

3,4,4-Triphenyl-2-cyclohexen-1-one, T-60379

3,5,5-Triphenyl-2-cyclohexen-1-one, T-60380

4,5,5-Triphenyl-2-cyclohexen-1-one, T-60381

1,2,3-Triphenylcyclopent[*a*]indene, T-70306

2,3,*N*-Triphenylcyclopropenoneimine, *see* D-70491

2,4,5-Triphenyl-1,3-dioxol-1-ium, T-70307

2-Triphenylenecarboxaldehyde, T-60382

1-Triphenylenecarboxylic acid, T-60383

2-Triphenylenecarboxylic acid, T-60384

1,2-Triphenylenedione, T-70308

1,4-Triphenylenedione, T-70309

Triphenylene-1,2-oxide, T-60385

1,2-Triphenylenequinone, *see* T-70308

1,4-Triphenylenequinone, *see* T-70309

Triphenyleno[1,12-*bcd*]thiophene, T-70310

Triphenyleno[4,5-*bcd*]thiophene, *see* T-70310

1-(1-Triphenylenyl)ethanone, *see* A-60059

1-(2-Triphenylenyl)ethanone, *see* A-60060

1,1,2-Triphenyl-1,2-ethanediol, T-70311

Triphenylethyleneglycol, *see* T-70311

1,3,6-Triphenylhexane, T-60386

Triphenylidene (obsol.), *see* D-70075

1,2,3-Triphenyl-1*H*-indene, T-60387

Triphenylmethyl azide, *see* A-60352

(Triphenylmethyl)hydrazine, T-60388

4-[(Triphenylmethyl)thio]-cyclohexanecarboxylic acid, *in* M-60020

3-[(Triphenylmethyl)thio]-cyclopentanecarboxylic acid, *in* M-60021

Triphenylphenylazomethane, T-70312

1,2,2-Triphenyl-1-propanone, T-60389

1,2,3-Triphenyl-1-propanone, T-60390

1,3,3-Triphenyl-1-propanone, T-60391

▷1,1,3-Triphenyl-2-propen-1-ol, T-70313

Triphosgene, *see* B-70163

1,3,5-Tripiperidinobenzene, T-70314

▷Tri(2-propenyl)amine, *see* T-70315

Triptofordin *A*, T-60392

Triptofordin *B*, T-60393

Triptofordin *E*, T-60397

Triptofordin *D*-1, T-60395

Triptofordin *C*-2, T-60394
Triptofordin *D*-2, T-60396
Triptofordin C-1, *in* T-60394
1,8,11-Triptycenetricarboxylic acid, T-60398
1,8,14-Triptycenetricarboxylic acid, T-60399
1,3,5-Tri[2,6]pyridacyclohexaphane-2,4,6-
trione, *see* T-60236
2,4,6-Tri-2-pyridinyl-1,3,5-triazine, T-70316
2,4,6-Tri-3-pyridinyl-1,3,5-triazine, T-70317
2,4,6-Tri-4-pyridinyl-1,3,5-triazine, T-70318
1,3,5-Tripyrrolidinobenzene, T-70319
Triquinoyl, *see* C-60210
Trisabbreviatin BBB, T-60400
5,6,7-Tris(acetyloxy)-10-chloro-12-hydroxy-9-
oxoprosta-10,14,17-trien-1-oic acid methyl
ester, *see* P-70142
Tris[(2,2'-bipyridyl-6-yl)methyl]amine, T-70320
1,3,5-Tris(bromomethyl)benzene, T-60401
1,3,5-Tris(chloromethyl)benzene, T-60402
Tris(chloromethyl)methane, *see* D-70118
2,4,5-Tris(dicyanomethylene)-1,3-cyclo-
pentanedione, T-70321
1,3,5-Tris(dimethoxymethyl)cyclohexane, *in*
C-70226
Tris(1,1-dimethylethyl)azete, *see* T-60249
1,3,5-Tris(1,1-dimethylethyl)-2-nitrobenzene,
see T-70216
2,4,6-Tris(1,1-dimethylethyl)pyridine, *see*
T-70217
Tris(1,1-dimethylethyl)-1,2,3-triazine, *see*
T-60250
Tris(3,6-dioxaheptyl)amine, *in* T-70323
Tris(3,6-dioxahexyl)amine, *see* T-70323
Tris(ethylenedithio)benzene, *see* H-70050
Tris(9-fluorenylidene)cyclopropane, T-70322
*D*₃-Trishomocubanetrione, P-60031
Tris[(2-hydroxyethoxy)ethyl]amine, T-70323
Tris(iodomethyl)methane, *see* D-70358
1,3,5-Tris(mercaptomethyl)benzene, *see*
B-70019
1,1,1-Tris(mercaptomethyl)ethane, *see* M-70030
Tris(2-methoxyethoxyethyl)amine, *in* T-70323
2,3,5-Tris(methylene)bicyclo[2.2.1]heptane,
T-60403
Tris(methylenedithio)benzene, *see* B-70069
Tris(methylphenylamino)methane, T-70324
20,29,30-Trisnor-3,19-lupanediol, T-60404
Tris(pentachlorophenyl)methane, T-60405
Tris(pentachlorophenyl)methyl, *in* T-60405
Tris(pentachlorophenyl)methyl hexa-
chloroantimonate, *in* T-60405
Trispiro[2.0.2.0.2.0]nonane, T-70325
s-Tris-2,3-thiocoumaronobenzene (obsol.), *see*
B-60052
1,3,5-Tris(thiomethyl)benzene, *see* B-70019
Tris(thioxanthen-9-ylidene)cyclopropane,
T-70326
Tris(*m*-tolyl)amine, *see* T-70289
Tris(*o*-tolyl)amine, *see* T-70288
Tris(*p*-tolyl)amine, *see* T-70290
Tris(trifluoromethyl)amine, T-60406
Tristrimethylenebenzene, *see* O-60021
Tritetralin, *see* D-60516
1,3λ⁴,δ²,5,2,4-Trithiadiazine, T-70327
4,5,9-Trithia-1,6,11-dodecatriene 9-oxide, *see*
A-70079
2,8,17-Trithia[4⁵,¹²][9]metacyclophane,
T-70328
5,11,16-Trithia[6.6.4.1³,¹³]nonadeca-1,3(19),13-
triene, *see* T-70328
7,14,21-Trithiatrispiro[5.1.5.1.5.1]-heneicosane,
in C-60209
Tri-Thyrotope, *in* T-60350
β-Triticene, *in* P-70024
Tri-*m*-tolylamine, *see* T-70289
Tri-*o*-tolylamine, *see* T-70288
Tri-*p*-tolylamine, *see* T-70290
b,*b*',*b*"-Tritriptycene, T-60407
*D*₃-Tritwistatriene, *see* P-70021
Tritylhydrazine, *see* T-60388
Tropane, *in* A-70278
▷1-Tropanol, *see* P-60151

Tropone, *see* C-60204
Troponethione, *see* C-60203
[3](2,7)Troponophane, T-70329
[5](2,7)Troponophane, T-70330
Trunculin *A*, T-60408
Trunculin *B*, T-60409
Trypanothione, T-70331
Tubipofuran, T-70332
Tubocapside A, *in* T-70333
Tubocapside B, *in* T-70333
Tubocapsigenin *A*, T-70333
Tucumanoic acid, T-60410
Tumidulin, T-70334
Tumour necrosis factor (human), T-70335
Tunichrome B-1, T-70336
Typhasterol, *in* T-70116
N-Tyrosylalanine, T-60411
N-Tyrosylphenylalanine, T-60412
▷U 12062, *in* D-60367
▷UDMH, *see* D-70416
Ugandensidial, *see* C-70117
Umbelactone, *see* H-60187
Umbellifolide, U-60001
Umbraculumin *A*, U-70001
Umbraculumin *B*, U-70002
Undecacyclo[9.9.0.0².⁹.0³,⁷.0⁴,²⁰.0⁵,¹⁸.0⁶,¹⁵.0⁸,¹⁵
.0¹⁰,¹⁴.0¹²,¹⁹.0¹³,¹⁷]eicosane, *see* D-60514
7-[3-(2,5-Undecadienyl)oxiranyl]-5-heptenoic
acid, *see* E-60016
Undecafluoropiperidine, U-60002
Undecamethylcyclohexanol, U-60003
2,4,6,8-Undecatetraene, U-70003
1,3,5-Undecatriene, U-70004
6-Undecyn-1-ol, U-60004
▷Unipres, *in* H-60092
Uracil-4-acetic acid, *see* T-70060
▷*p*-Urazine, *see* T-70097
Urazourazole, *see* T-70209
Ureidoformamide, B-60190
5-Ureidoornithine, *see* C-60151
2,4-Uretidinedione, *see* D-60057
▷Uridine 4-hydrazone, *see* A-70127
▷Urocanic acid, U-70005
▷Urocaninic acid, *see* U-70005
Urodiolenone, U-70006
9(11),12-Ursadien-3-ol, U-70007
12-Ursene-1,2,3,11,20-pentol, U-60005
12-Ursene-1,2,3,11-tetrol, U-60006
12-Ursene-2,3,11,20-tetrol, U-60007
12-Ursene-2,3,11-triol, U-60008
Uvitaldehyde, *see* M-60050
1(10)-Valencen-7-ol, *see* E-70031
Valerenol, V-70001
Valilactone, V-70002
Valyl-*N*-(1,2-dicarboxy-2-hydroxyethoxy)-
valinamide, *see* A-60289
N-Valylleucine, V-60001
Valylvalylvalyl-*N*-(1-formyl-2-phenylethyl)-
valinamide, *see* S-60051
▷*o*-Vanillin, *in* D-70285
8-Vanilloylshiromodiol, *in* S-70039
Variabilin, V-70003
Vaticaffinol, V-60002
Vebraside, V-60003
Velcorin, *in* D-70111
Velutin, *see* D-70291
Ventilone *C*, V-60004
Ventilone D, *in* V-60004
Ventilone E, *in* V-60004
Venustanol, V-70004
Venustatriol, *in* T-70193
Verapliquinone A, *in* M-60038
Verapliquinone B, *in* M-60038
Verapliquinone C, *in* H-60123
Verapliquinone D, *in* H-60123
o-Veratraldehyde, *in* D-70285
Verbascogenin, *in* O-70034
Verbascosaponin, *in* O-70034
Verboccidentafuran, V-70005
Verboccidentafuran 4α,5α-epoxide, *in* V-70005
Verecynarmin *A*, V-60005
Vernoflexin, *in* Z-60001
Vernoflexuoside, *in* Z-60001
Verproside, *in* C-70026
Verpyran, *in* H-70108
Verrucarinic acid, *in* D-60353

Verrucarinolactone, *in* D-60353
Verrucosidin, V-60006
Vertofix, *see* A-70032
Vescalagin, V-70006
Vescalin, *in* V-70006
Vesparione, V-70007
Vestic acid, V-70008
Vetidiol, V-60007
Viguiestin, *in* T-70003
Vinigrol, V-70009
4-Vinyl-2-azetidinone, V-60008
4-Vinyl-1,3-benzenediol, V-70010
2-Vinyl-γ-butyrolactone, *see* D-70279
▷Vinyl cyanide, *in* P-70121
3-Vinylcyclohexene, V-60009
5-Vinyl-1-cyclopentenecarboxylic acid, V-60010
3-Vinyl-2-cyclopenten-1-one, V-70011
2-Vinyl-1,1-cyclopropanedicarboxylic acid,
V-60011
3-Vinylcyclopropene, V-60012
2-Vinyl-1,3-dioxolane, *in* P-70119
2-Vinyl-1,3-dithiolane, *in* P-70119
2,2'-Vinylenedifuran, *see* D-60191
2-Vinylfuran oxide, *see* O-70078
▷Vinylglycollic nitrile, *in* P-70119
4(5)-Vinylimidazole, V-70012
2-Vinylindole, V-70013
▷5-Vinyl-2-norbornene, *see* E-70036
▷*p*-Vinylphenol, *see* H-70209
2-Vinyl-1*H*-pyrrole, V-60013
4-Vinylquinoline, V-70014
4-Vinylresorcinol, *see* V-70010
α-Vinylserine, A-60187
4-Vinyl-1,2,3-thiadiazole, V-60014
5-Vinyl-1,2,3-thiadiazole, V-60015
5-Vinyluridine, V-70015
Violanthrene, *in* A-60279
Violanthrene A, *see* A-60279
Violanthrene *B*, *see* D-60075
Violanthrene *C*, *see* B-60036
Violapterin, *see* P-70135
Virescenol A, *in* I-70081
Virescenol *B*, *in* I-70077
Virescenoside A, *in* I-70081
Virescenoside B, *in* I-70077
Virescenoside E, *in* I-70081
Virescenoside F, *in* I-70081
Virescenoside G, *in* I-70077
Virescenoside L, *in* I-70081
Viridomycin A, *in* H-60197
Viscosic acid, *in* E-60021
Viscutin 1, *in* T-60103
Viscutin 2, *in* T-60103
Viscutin 3, *in* T-60103
▷Visnadin, *in* K-70013
Vistrax, *in* O-60102
Vitamin B_T, *see* C-70024
Viticosterone, *in* H-60063
Viticosterone E, *in* H-60063
Vittadinal, V-70016
Vittagraciliolide, V-70017
Volkensiachromone, V-60016
Volkensin, V-70018
▷Vulgarin, V-60017
W 341*C*, *see* A-70241
WA 185, *in* S-60045
Warburganal, W-60001
Wasabidienone *A*, W-70001
Wasabidienone *D*, W-70002
Wasabidienone *E*, W-60003
Wasabidienone E, *in* W-70002
Wedeliasecokaurenolide, W-60004
WF 3405, *see* D-70465
WF 10129, *see* C-70022
Wikstromol, W-70003
Wilforlide *A*, *in* H-70197
Wistin, *in* T-60325
Withametelin, W-60005
Withaminimin, W-60006
Withanolide *Y*, W-60007
Withaphysacarpin, W-60008
Withaphysalin *E*, W-60009
Withaphysanolide, W-60010
WS 6049A, *in* E-60037
WS 43708A, *see* B-60114
WS 6049B, *in* E-60037

WS 43708B, *in* B-60114
▷Wy 1395, *in* T-70277
Xanthokermesic acid, *see* T-60332
Xanthorrhizol, X-70001
Xanthostemone, X-70002
▷Xanthotoxin, X-60001
▷Xanthotoxol, X-60002
Xanthoxylin, *in* T-70248
Xanthylium(1+), X-70003
X 14881C, *in* O-60003
Xenicin, X-70004
Xestenone, X-70005
Xestodiol, X-70006
XTT, *see* B-70151
o-Xylylene, *see* D-70410
1,8-*o*-Xylylenenaphthalene, *see* D-60269
Yadanzioside *K*, Y-60001

Yamogenin, *in* S-60045
Yangonin, *in* H-70122
Yashabushidiol A, *in* D-70481
Yashabushidiol B, *in* D-70481
Yashabushiketodiol A, *in* D-60480
Yashabushiketodiol B, *in* D-60480
Yashabushiketol, *in* D-70481
Yashabushitriol, *in* D-60480
▷Yellow pyoctenin, *see* A-70268
▷Yessotoxin, Y-60002
Yezoquinolide, Y-60003
Yomogin, Y-70001
Yomoginic acid, *in* H-70199
Youlemycin, Y-70002
YP 02908L-*A*, *see* O-70030
Zafronic acid, Z-70001
Zaluzanin *C*, Z-60001
Zaluzanin D, *in* Z-60001

▷Zanthotoxin, *see* X-60001
Zapotin, *in* T-70107
Zapotinin, *in* T-70107
Z-Asn-Ala-OH, *in* A-60310
Z-Bifurcarenone, *in* B-60105
▷Zearalenone, Z-60002
Zebrinin, Z-70002
Zederone, Z-70003
Zedoarondiol, Z-70004
(Z)-6,7-Epoxyligustilide, *in* L-60026
Zexbrevin *C*, Z-70005
Zeylasteral, *in* D-70027
Zeylenol, Z-70006
ZK 36374, *see* I-60003
ZK 36375, *in* I-60003
(Z)-Methylsuberenol, *in* S-70080
Zoapatanolide *B*, Z-70007
Zuonin *A*, Z-60003

Molecular Formula Index

This index becomes invalid after publication of the Eighth Supplement.

The Molecular Formula Index lists the molecular formulae of compounds in the Sixth and Seventh Supplements which occur as Entry Names or as important derivatives. Molecular formulae of compounds contained in Supplements 1 to 5 are listed in the cumulative Index Volume published with the Fifth Supplement.

The first digit of the DOC Number (printed in bold type) refers to the number of the Supplement in which the Entry appears.

Where a molecular formula applies to a derivative the DOC Number is prefixed by the word '*in*'.

The symbol ▷ preceding an Index Entry indicates that the DOC Entry contains information on toxic or hazardous properties of the compound.

Molecular Formula Index

CBrF₃S
Trifluoromethanesulfenyl bromide, T-60280

CClF₃O₂S
Trifluoromethanesulfonic acid; Chloride, *in* T-70241

CClNO₃S
▷Sulfuryl chloride isocyanate, S-60062

CF₃IS
Trifluoromethanesulfenyl iodide, T-60282

CF₄O₂S
Trifluoromethanesulfonic acid; Fluoride, *in* T-70241

CF₄S
▷Trifluoromethanesulfenyl fluoride, T-60281

CF₈S
Pentafluoro(trifluoromethyl)sulfur, P-60041

CHClOS
Carbonochloridothioic acid, C-60013

CHF₃OS
Trifluoromethanesulfenic acid, T-60279

CHF₃O₃S
Trifluoromethanesulfonic acid, T-70241

CH₂BrClO₂S
Bromomethanesulfonic acid; Chloride, *in* B-60291
Chloromethanesulfonyl bromide, *in* C-70086

CH₂BrIO₂S
Iodomethanesulfonic acid; Bromide, *in* I-60048

CH₂Br₂O₂S
Bromomethanesulfonic acid; Bromide, *in* B-60291

CH₂Cl₂O₂S
▷Chloromethanesulfonyl chloride, *in* C-70086

CH₂F₃NO₂S
▷Trifluoromethanesulfonic acid; Me ester, *in* T-70241
Trifluoromethanesulfonic acid; Amide, *in* T-70241

CH₂N₂OS₃
1,3λ⁴,δ²,5,2,4-Trithiadiazine; 1-Oxide, *in* T-70327

CH₂N₂S₃
1,3λ⁴,δ²,5,2,4-Trithiadiazine, T-70327

CH₂N₄O
2-Tetrazolin-5-one, T-60175

CH₂N₄S
Tetrazole-5-thione, T-60174

CH₂OS
Methanethial; *S*-Oxide, *in* M-70037

CH₂OSe
Methaneselenoic acid, M-70035

CH₂S
Methanethial, M-70037

CH₂Se
▷Selenoformaldehyde, S-70029

CH₃BrO₃S
Bromomethanesulfonic acid, B-60291

CH₃ClO₃S
Chloromethanesulfonic acid, C-70086

CH₃IO₃S
Iodomethanesulfonic acid, I-60048

CH₃N
▷Methanimine, M-70038

CH₃NO₂
Carbonimidic acid, C-70019

CH₃NSe
Methaneselenoamide, *in* M-70035

CH₃N₃O₂S
▷Methanesulfonyl azide, M-60033

CH₃N₅
▷5-Amino-1*H*-tetrazole, A-70202

CH₄N₂O₃
O,*O*′-Carbonylbis(hydroxylamine), C-60015

CH₄N₂O₃S
Aminoiminomethanesulfonic acid, A-60210

CH₄N₂S₂
Dithiocarbazic acid, D-60509

CH₄N₂Se
▷Selenourea, S-70031

CH₅NO₃S
Aminomethanesulfonic acid, A-60215

CS
▷Carbon monosulfide, C-70020

C₂BrF₃N₂
▷3-Bromo-3-(trifluoromethyl)-3*H*-diazirine, B-70275

C₂ClF₃N₂
▷3-Chloro-3-(trifluoromethyl)-3*H*-diazirine, C-70169

C₂ClF₃O
Chlorotrifluorooxirane, C-60140

C₂Cl₂F₂O
2,2-Dichloro-3,3-difluorooxirane, D-60128

C₂Cl₄O₂
▷Trichloromethyl chloroformate, T-70223

C₂F₂I₂
1,1-Difluoro-2,2-diiodoethylene, D-60183

C₂F₃N
Trifluoroisocyanomethane, T-60278

C₂F₃NO₃
▷Trifluoroacetyl nitrite, T-60274

C₂F₄N₂
▷3-Fluoro-3-(trifluoromethyl)-3*H*-diazirine, F-70037

C₂F₄O
▷Tetrafluorooxirane, T-60049

C₂F₄OS₂
2,2,4,4-Tetrafluoro-1,3-dithietane; 1-Oxide, *in* T-60048

C₂F₄O₂S
Trifluoroethenesulfonic acid; Fluoride, *in* T-60277

C₂F₄O₂S₂
2,2,4,4-Tetrafluoro-1,3-dithietane; 1,1-Dioxide, *in* T-60048

C₂F₄O₃S₂
2,2,4,4-Tetrafluoro-1,3-dithietane; 1,1,3-Trioxide, *in* T-60048

C₂F₄O₄S₂
2,2,4,4-Tetrafluoro-1,3-dithietane; 1,1,3,3-Tetroxide, *in* T-60048

C₂F₄S₂
2,2,4,4-Tetrafluoro-1,3-dithietane, T-60048

C₂F₅N
(Trifluoromethyl)carbonimidic difluoride, T-60288

C₂F₆N₂O₂
1,1,1-Trifluoro-*N*-(nitrosooxy)-*N*-(trifluoromethyl)methanamine, *in* B-60180

C₂F₆O₃S
Trifluoromethyl trifluoromethanesulfonate, *in* T-70241

C₂F₆O₅S₂
Trifluoromethanesulfonic acid; Anhydride, *in* T-70241

C₂F₇NO₄S₂
Perfluoromethanesulfonimide, P-70042

C₂F₈S₂
Difluoro(trifluoromethanethiolato)-(trifluoromethyl)sulfur, *in* T-60281

C₂HF₃O
Trifluorooxirane, T-60305

C₂HF₃O₃S
Trifluoroethenesulfonic acid, T-60277

C₂HF₆NO
N,*N*-Biss(trifluoromethyl)hydroxylamine, B-60180

C₂HN₅S
3-Azido-1,2,4-thiadiazole, A-60351

C₂H₂F₂O
2,3-Difluorooxirane, D-60187

C₂H₂F₂O₃S
4,4-Difluoro-1,2-oxathietane 2,2-dioxide, D-60186

C₂H₂N₂O₂
1,3-Diazetidine-2,4-dione, D-60057

C₂H₂N₂O₃
1,2,4-Oxadiazolidine-3,5-dione, O-70060
1,3,4-Oxadiazolidine-2,5-dione, O-70061
1,2,4-Oxazolidine-3,5-dione, O-70070

C₂H₂N₂O₄
2*H*-1,5,2,4-Dioxadiazine-3,6(4*H*)dione, D-60466

C₂H₂N₂S₃
▷5-Mercapto-1,3,4-thiadiazoline-2-thione, M-70031

C₂H₂N₄
1,2,4,5-Tetrazine, T-60170

C₂H₂N₄O₂
1,2-Dihydro-1,2,4,5-tetrazine-3,6-dione, D-70268

C₂H₂O₅
Dicarbonic acid, D-70111

C₂H₂S₄
Ethanebis(dithioic)acid, E-60040

C₂H₂Se₂
1,2-Diselenete, D-60498

C₂H₃AsF₆N₂S₂
5-Methyl-1,3,2,4-dithiazolium(1+); Hexafluoroarsenate, *in* M-70061

C₂H₃ClN₂O
3-Chloro-3-methoxy-3*H*-diazirine, C-70087

C₂H₃ClN₂S₂
5-Methyl-1,3,2,4-dithiazolium(1+); Chloride, *in* M-70061

C₂H₃ClOS
Carbonochloridothioic acid; *O*-Me, *in* C-60013

C₂H₃FN₂O
▷3-Fluoro-3-methoxy-3*H*-diazirine, F-60050

C₂H₃F₃OS
Methyl trifluoromethanesulfenate, *in* T-60279
Trifluoro(ethylsulfinyl)methane, *in* M-70145

C₂H₃F₃O₂S
Trifluoro(ethylsulfonyl)methane, *in* M-70145

C₂H₃F₃S
Methyl trifluoromethyl sulfide, M-70145

C₂H₃N
2*H*-Azirine, A-70292
Ethynamine, E-60056

C₂H₃NS₂
Methanesulfenyl thiocyanate, M-60032

C₂H₃NSe
Methyl selenocyanate, M-60127

C₂H₃N₂S₂⁺
5-Methyl-1,3,2,4-dithiazolium(1+), M-70061

C₂H₃N₃S
▷1,2-Dihydro-3*H*-1,2,4-triazole-3-thione, D-60298

C₂H₄BrNO₂
1-Bromo-2-nitroethane, B-60303

C₂H₄ClI
1-Chloro-1-iodoethane, C-60106
1-Chloro-2-iodoethane, C-70078

C₂H₄IN₃
1-Azido-2-iodoethane, A-60346

C₂H₄N₂O
1,2-Diazetidin-3-one, D-70055

C₂H₄N₂O₄
▷Oxalohydroxamic acid, O-60048

C₂H₄N₂S₂
▷Ethanedithioamide, E-60041

C₂H₄N₄O₂
▷Tetrahydro-1,2,4,5-tetrazine-3,6-dione, T-70097

C₂H₄N₄S
2,5-Diamino-1,3,4-thiadiazole, D-60049
Tetrazole-5-thione; 1-Me, *in* T-60174
Tetrazole-5-thione; *S*-Me, *in* T-60174

C₂H₄N₆
3,6-Diamino-1,2,4,5-tetrazine, D-70045

C₂H₄O₂
▷1,2-Dioxetane, D-60469

C₂H₅NO
▷Acetamide, A-70019

C₂H₅NO₃
C-Methoxycarbohydroxamic acid, M-70044

C₂H₅N₃O
▷2-Azidoethanol, A-70287

C₂H₅N₃O₂
Biuret, B-60190
▷Semioxamazide, S-60020

C₂H₅N₅
5-Amino-1-methyl-1*H*-tetrazole, *in* A-70202
5-Amino-2-methyl-2*H*-tetrazole, *in* A-70202
5-Methylamino-1*H*-tetrazole, *in* A-70202

C₂H₆N₂O
▷Acetamidoxime, A-60016

C₂H₆N₂O₃S
Aminoiminomethanesulfonic acid; *N*-Me, *in* A-60210

C₂H₆N₂S₂
Dithiocarbazic acid; Me ester, *in* D-60509

C₂H₆N₂Se
Selenourea; *N*-Me, *in* S-70031

C₂H₆OS
(Methylthio)methanol, M-70138

C₂H₆O₂S₂
2-Mercaptoethanesulfinic acid, M-60022

C₂H₆O₆S₂
Biss(ethylsulfonyl) peroxide, B-70154

C₂H₇NO
O,*N*-Dimethylhydroxylamine, D-70418

C₂H₇NO₂
O-(2-Hydroxyethyl)hydroxylamine, H-60134

C₂H₇NO₃S
(Methylamino)methanesulfonic acid, *in* A-60215

C₂H₈N₂
▷1,1-Dimethylhydrazine, D-70416

C₂H₉N₃
2-Hydrazinoethylamine, H-70096 '

C₂H₁₀N₄
1,2-Dihydrazinoethane, D-60192

C₂N₂
Diisocyanogen, D-70362

C₂N₂Se₃
Dicyanotriselenide, D-60165

C₃AlCl₇
Trichlorocyclopropenylium; Tetrachloroaluminate, *in* T-70221

C₃Br₃N₃O₂
3,4,5-Tribromo-1*H*-pyrazole; 1-Nitro, *in* T-60246

C₃ClF₂N₃
2-Chloro-4,6-difluoro-1,3,5-triazine, C-60052
6-Chloro-3,5-difluoro-1,2,4-triazine, C-60053

C₃Cl₃⁺
Trichlorocyclopropenylium, T-70221

C₃Cl₆O₃
Biss(richloromethyl) carbonate, B-70163

C₃Cl₉Sb
Trichlorocyclopropenylium;
Hexachloroantimonate, *in* T-70221

C₃F₃N₃
Trifluoro-1,2,3-triazine, T-70245

C₃F₆N₂
3,3-Biss(rifluoromethyl)-3*H*-diazirine, B-60178

C₃F₆O₄S
Trifluoroacetyl trifluoromethanesulfonate, *in* T-70241

C₃F₇NO
3,3,4,4-Tetrafluoro-2-(trifluoromethyl)-1,2-oxazetidine, T-60051

C₃F₉N
Triss(rifluoromethyl)amine, T-60406

C₃HBrINS
2-Bromo-4-iodothiazole, B-60287
2-Bromo-5-iodothiazole, B-60288
4-Bromo-2-iodothiazole, B-60289
5-Bromo-2-iodothiazole, B-60290

C₃HBr₂NS
2,4-Dibromothiazole, D-60109
2,5-Dibromothiazole, D-60110
4,5-Dibromothiazole, D-60111

C₃HBr₂N₃
4,5-Dibromo-1,2,3-triazine, D-70100

C₃HBr₃N₂
▷2,4,5-Tribromo-1*H*-imidazole, T-60245
3,4,5-Tribromo-1*H*-pyrazole, T-60246

C₃HCl₂NS
2,4-Dichlorothiazole, D-60154
2,5-Dichlorothiazole, D-60155
4,5-Dichlorothiazole, D-60156

C₃HCl₅
Pentachlorocyclopropane, P-70018

C₃HI₂NS
2,4-Diiodothiazole, D-60388
2,5-Diiodothiazole, D-60389

C₃HI₃N₂
▷2,4,5-Triiodo-1*H*-imidazole, T-70276

C₃HN₃O₅
3,5-Dinitroisoxazole, D-70452

C₃H₂BrN₃
5-Bromo-1,2,3-triazine, B-70273

C₃H₂BrN₃O₂
4-Bromo-3(5)-nitro-1*H*-pyrazole, B-60304
3(5)-Bromo-5(3)-nitro-1*H*-pyrazole, B-60305

C₃H₂Br₂
1,2-Dibromocyclopropene, D-60094

C₃H₂Br₂N₂
4,5-Dibromo-1*H*-imidazole, D-60100
2,4(5)-Dibromo-1*H*-imidazole, D-70093

C₃H₂ClF₃
2-Chloro-3,3,3-trifluoro-1-propene, C-60141

C₃H₂ClF₃O₂
2-Chloro-3,3,3-trifluoropropanoic acid, C-70170

C₃H₂Cl₂N₂
4,5-Dichloro-1*H*-imidazole, D-70122

C₃H₂Cl₄
▷1,1,2,3-Tetrachloro-1-propene, T-60029

C₃H₂INS
2-Iodothiazole, I-60064
4-Iodothiazole, I-60065
5-Iodothiazole, I-60066

C₃H₂N₂O₂S
2-Thioxo-4,5-imidazolidinedione, T-70187

C₃H₂N₃
▷2-Diazo-2*H*-imidazole, D-70060

C₃H₂N₄
4-Cyano-1,2,3-triazole, *in* T-60239
4-Diazo-4*H*-imidazole, D-60063
▷4-Diazo-4*H*-imidazole, D-70061

C₃H₂O₃
2,3-Dihydroxy-2-cyclopropen-1-one, D-70290

C₃H₂S
1,2-Propadiene-1-thione, P-70117

C₃H₃BrN₂
▷2-Bromo-1*H*-imidazole, B-70220
4(5)-Bromo-1*H*-imidazole, B-70221

C₃H₃Br₃
1,2,3-Tribromocyclopropane, T-70211

C₃H₃ClN₂
2-Chloro-1*H*-imidazole, C-70076
4(5)-Chloro-1*H*-imidazole, C-70077
3-Chloro-1*H*-pyrazole, C-70155
▷4-Chloro-1*H*-pyrazole, C-70156

C₃H₃ClN₂O
4-Chloro-5-methylfurazan, C-60114

C₃H₃ClN₂O₂
2-Chloro-2,3-dihydro-1*H*-imidazole-4,5-dione, C-60054
4-Chloro-3-methylfuroxan, *in* C-60114

C₃H₃ClO
▷2-Propenoic acid; Chloride, *in* P-70121

C₃H₃Cl₂F₃O
2,2-Dichloro-3,3,3-trifluoro-1-propanol, D-60161

C₃H₃Cl₂N₂O₂
3-Chloro-4-methylfuroxan, *in* C-60114

C₃H₃Cl₃
1,2,3-Trichlorocyclopropane, T-70220
1,1,2-Trichloro-1-propene, T-60254

C₃H₃F₃O
(Trifluoromethyl)oxirane, T-60294

C₃H₃F₆NO
1,1,1,1′,1′,1′-Hexafluoro-*N*-methoxydimethyl-amine, *in* B-60180

C₃H₃IN₂
4(5)-Iodo-1*H*-imidazole, I-70042

C₃H₃N
▷Vinyl cyanide, *in* P-70121

C₃H₃NO₃S₂
4-Thiazolesulfonic acid, T-60198
5-Thiazolesulfonic acid, T-60199

C₃H₃NO₄
2*H*-1,5,2-Dioxazine-3,6(4*H*)-dione, D-70464

C₃H₃NS₂
2(3*H*)-Thiazolethione, T-60200

$C_3H_3N_3O_2$
3-Azido-2-propenoic acid, A-70290
1*H*-1,2,3-Triazole-4-carboxylic acid, T-60239

$C_3H_3N_3O_3$
▷Cyanuric acid, C-60179

$C_3H_3N_5$
3(5)-Azidopyrazole, A-70291

C_3H_4ClNS
▷1-Chloro-2-isothiocyanoethane, C-70081

$C_3H_4I_2N_6$
1,3-Diiodo-2,2-diazidopropane, D-60387

$C_3H_4N_2$
Diazocyclopropane, D-60060
▷1*H*-Pyrazole, P-70152

$C_3H_4N_2OS$
▷2-Thioxo-4-imidazolidinone, T-70188

$C_3H_4N_2O_2$
2-Propyn-1-amine; *N*-Nitro, *in* P-60184

$C_3H_4N_2O_3$
2,3-Dihydro-2-hydroxy-1*H*-imidazole-4,5-dione, D-60243
5-Hydroxy-2,4-imidazolidinedione, H-60156
1,2,4-Oxadiazolidine-3,5-dione; 4-Me, *in* O-70060

$C_3H_4N_2O_5$
1,1-Dinitro-2-propanone, D-70454

$C_3H_4N_2S$
2-Cyanoethanethioamide, C-70209

$C_3H_4N_2S_3$
3-Methyl-1,3,4-thiadiazolidine-2,5-dithione, *in* M-70031
5-Methylthio-1,3,4-thiadiazole-2(3*H*)-thione, *in* M-70031

$C_3H_4N_4$
▷3-Amino-1,2,4-triazine, A-70203

$C_3H_4N_4O$
1*H*-1,2,3-Triazole-4-carboxylic acid; Amide, *in* T-60239

C_3H_4O
▷Propenal, P-70119

$C_3H_4O_2$
1,3-Dioxole, D-60473
▷2-Propenoic acid, P-70121

$C_3H_4O_3$
3-Oxopropanoic acid, O-60090

$C_3H_4S_2Se$
1,3-Dithiolane-2-selone, D-70525

C_3H_5ClOS
▷Carbonochloridothioic acid; *O*-Et, *in* C-60013
Carbonochloridothioic acid; *S*-Et, *in* C-60013

$C_3H_5ClO_3$
3-Chloro-2-hydroxypropanoic acid, C-70072

C_3H_5FO
▷(Fluoromethyl)oxirane, F-70032

$C_3H_5F_2NO_2$
2-Amino-3,3-difluoropropanoic acid, A-60135

$C_3H_5F_3OS$
Ethyl trifluoromethanesulfenate, *in* T-60279

$C_3H_5F_3O_3S$
Trifluoromethanesulfonic acid; Et ester, *in* T-70241

$C_3H_5F_3O_4S_2$
(Methylsulfonyl)[(trifluoromethyl)sulfonyl]-methane, M-70133

C_3H_5N
2,3-Dihydroazete, D-70172
▷2-Propyn-1-amine, P-60184

C_3H_5NO
▷Acrylamide, *in* P-70121
Propenal; Oxime, *in* P-70119

$C_3H_5NO_3$
Isonitrosopropanoic acid, *in* O-60090
▷3-Nitropropanal, N-60037

C_3H_5NS
2-Propenethioamide, P-70120

$C_3H_5NS_2$
2-Thiazolidinethione, T-60197

$C_3H_5N_3O_2$
Dihydro-1,3,5-triazine-2,4(1*H*,3*H*)-dione, D-60297
▷2-Imidazolidinone; 1-Nitroso, *in* I-70005
Tetrahydro-1,2,4-triazine-3,6-dione, T-60099

C_3H_6BrF
1-Bromo-3-fluoropropane, B-70212

$C_3H_6BrNO_2$
1-Bromo-1-nitropropane, B-70258
2-Bromo-2-nitropropane, B-70259

$C_3H_6Br_2$
▷1,2-Dibromopropane, D-60105

C_3H_6ClN
2-Chloro-2-propen-1-amine, C-70151

$C_3H_6ClNO_2$
▷2-Chloro-2-nitropropane, C-60120

$C_3H_6ClN_4^\oplus$
N,*N*-Dimethylazidochloromethyl-eniminium(1+), D-70369

$C_3H_6Cl_2N_4$
N,*N*-Dimethylazidochloromethyl-eniminium(1+); Chloride, *in* D-70369

C_3H_6FI
1-Fluoro-3-iodopropane, F-70026

$C_3H_6FNO_2$
3-Amino-2-fluoropropanoic acid, A-60162

$C_3H_6F_3NO_2S$
Trifluoromethanesulfonic acid; Diethylamide, *in* T-70241

$C_3H_6IN_3$
1-Azido-3-iodopropane, A-60347

$C_3H_6N_2$
4,5-Dihydro-1*H*-imidazole, D-60245
4,5-Dihydro-1*H*-pyrazole, D-70253

$C_3H_6N_2O$
▷Azetidine; *N*-Nitroso, *in* A-70283
▷2-Imidazolidinone, I-70005

$C_3H_6N_2O_2$
Acetamidoxime; *O*-Formyl, *in* A-60016

$C_3H_6N_4O$
2-Tetrazolin-5-one; 1-Et, *in* T-60175

$C_3H_6N_4O_2$
2-Azido-2-nitropropane, A-60349

$C_3H_6N_4S$
Tetrazole-5-thione; 1-Et, *in* T-60174
Tetrazole-5-thione; 1,4-Di-Me, *in* T-60174
Tetrazole-5-thione; *S*-Et, *in* T-60174

$C_3H_6N_6$
▷2,2-Diazidopropane, D-60058

$C_3H_6N_6O_3$
▷Hexahydro-1,3,5-triazine; 1,3,5-Trinitroso, *in* H-60060

C_3H_6O
▷Methyloxirane, M-70100

C_3H_6OS
1-Mercapto-2-propanone, M-60028
Propenylsulfenic acid, P-70123

$C_3H_6O_2$
Dimethyldioxirane, D-70399
1,2-Dioxolane, D-70467

$C_3H_6O_2S$
▷2-Mercaptopropanoic acid, M-60027

C_3H_7N
Azetidine, A-70283

C_3H_7NO
▷Acetamide; *N*-Me, *in* A-70019

C_3H_7NOS
2-Mercaptopropanoic acid; Amide, *in* M-60027

$C_3H_7NO_3$
3-Amino-2-hydroxypropanoic acid, A-70160
▷3-Nitro-1-propanol, N-60039

$C_3H_7NO_4$
▷2-Nitro-1,3-propanediol, N-60038

C_3H_7NSe
N,*N*-Dimethylmethaneselenoamide, *in* M-70035

$C_3H_7N_3$
2-Azidopropane, A-70289

$C_3H_7N_3O_2$
Allophanic methylamide, *in* B-60190

$C_3H_7N_5$
5-Amino-1-ethyl-1*H*-tetrazole, *in* A-70202
5-Amino-2-ethyl-2*H*-tetrazole, *in* A-70202
1,4-Dihydro-5-imino-3,4-dimethyltetrazole, *in* A-70202
5-Dimethylamino-1*H*-tetrazole, *in* A-70202
5-Ethylamino-1*H*-tetrazole, *in* A-70202
5-Imino-1,3-dimethyl-1*H*,3*H*-tetrazole, *in* A-70202
1-Methyl-5-methylamino-1*H*-tetrazole, *in* A-70202
2-Methyl-5-methylamino-2*H*-tetrazole, *in* A-70202

$C_3H_8N_2O_2$
N,*N*-Dimethylcarbamohydroxamic acid, D-60414

$C_3H_8N_2O_3S$
Aminoiminomethanesulfonic acid; *N*,*N*-Di-Me, *in* A-60210

$C_3H_8N_2Se$
Selenourea; N^1,N^1-Di-Me, *in* S-70031
Selenourea; N^1,N^3-Di-Me, *in* S-70031
Selenourea; *N*-Et, *in* S-70031

$C_3H_8O_2$
▷1,2-Propanediol, P-70118

$C_3H_8S_2$
2,2-Propanedithiol, P-60175

C_3H_9NO
▷Trimethylamine oxide, T-60354

$C_3H_9NO_3S$
(Dimethylamino)methanesulfonic acid, *in* A-60215

$C_3H_9N_3$
Hexahydro-1,3,5-triazine, H-60060

C_3OS
3-Thioxo-1,2-propadien-1-one, T-60216

C_4Br_4O
Tetrabromofuran, T-60023

$C_4ClF_3N_2$
3-Chloro-4,5,6-trifluoropyridazine, C-60142
4-Chloro-3,5,6-trifluoropyridazine, C-60143
5-Chloro-2,4,6-trifluoropyrimidine, C-60144

C_4F_6
▷(Difluoromethylene)tetrafluorocyclopropane, D-60185

$C_4F_6NS_2$
4,5-Biss(rifluoromethyl)-1,3,2-dithiazol-2-yl, B-70164

$C_4F_6O_2$
1,1,1,4,4,4-Hexafluoro-2,3-butanedione, H-60043

$C_4F_6O_4S$
4-[2,2,2-Trifluoro-1-(trifluoromethyl)-ethylidene]-1,3,2-dioxathietane 2,2-dioxide, T-60311

C_4F_6S
Biss(rifluoromethyl)thioketene, B-60187

C₄F₆S₂
  3,4-Biss(rifluoromethyl)-1,2-dithiete, B-60179

C₄F₈O₃S
  4,4-Difluoro-3,3-biss(rifluoromethyl)-1,2-
    oxathietane 2,2-dioxide, D-60181

C₄F₉NO
  Nonafluoromorpholine, N-60056

C₄HBrF₂O₂
  4-Bromo-4,4-difluoro-2-butynoic acid, B-70202

C₄HBr₃S
  2,3,5-Tribromothiophene, T-60247

C₄HClO₃
  3-Chloro-4-hydroxy-3-cyclobutene-1,2-dione,
    C-70071

C₄HF₃N₂
  Trifluoropyrazine, T-60306
  2,4,6-Trifluoropyrimidine, T-60307
  4,5,6-Trifluoropyrimidine, T-60308

C₄HF₅
  3,3,4,4,4-Pentafluoro-1-butyne, P-60032
  1,3,3,4,4-Pentafluorocyclobutene, P-60033

C₄HF₈NO
  2,2,3,3,5,5,6,6-Octafluoromorpholine, O-60009

C₄HN₅O₈
  2,3,4,5-Tetranitro-1H-pyrrole, T-60160

C₄H₂Br₂N₂O
  4,5-Dibromo-1H-imidazole-2-carboxaldehyde,
    D-60101

C₄H₂Br₂N₂O₂
  4,5-Dibromo-1H-imidazole-2-carboxylic acid,
    D-60102

C₄H₂ClNO₂
  5-Isoxazolecarboxylic acid; Chloride, in I-60142

C₄H₂Cl₂N₂O
  4,5-Dichloro-1H-imidazole-2-carboxaldehyde,
    D-60134

C₄H₂Cl₂N₂O₂
  4,5-Dichloro-1H-imidazole-2-carboxylic acid,
    D-60135

C₄H₂Cl₃NO
  5-(Trichloromethyl)isoxazole, T-60253

C₄H₂F₂N₂
  2,3-Difluoropyrazine, D-60188
  2,6-Difluoropyrazine, D-60189

C₄H₂F₃N
  ▷2-Trifluoromethylacrylonitrile, in T-60295

C₄H₂F₄
  3,3,4,4-Tetrafluorocyclobutene, T-60043

C₄H₂N₂O
  5-Cyanoisoxazole, in I-60142

C₄H₂N₂O₂
  Dicyanoacetic acid, D-70149

C₄H₂N₂O₂S
  1,2,5-Thiadiazole-3,4-dicarboxaldehyde,
    T-70162

C₄H₂N₂O₃
  3-Diazo-2,4(5H)-furandione, D-70059

C₄H₂N₂S₂
  (Dimercaptomethylene)malononitrile, in
    D-60395

C₄H₂N₄OS
  [1,2,5]Thiadiazolo[3,4-d]pyridazin-4(5H)-one,
    T-70164

C₄H₂N₄O₄
  [1,2,4]Triazolo[1,2-a][1,2,4]triazole-
    1,3,5,7(2H,6H)-tetrone, T-70209

C₄H₂N₄O₆
  2,3,4-Trinitro-1H-pyrrole, T-60373
  2,3,5-Trinitro-1H-pyrrole, T-60374

C₄H₂N₄S
  [1,2,5]Thiadiazolo[3,4-d]pyridazine, T-70163

C₄H₂N₈O₁₀
  Tetrahydro-1,3,4,6-tetranitroimidazo[4,5-d]-
    imidazole-2,5-(1H,3H)-dione, T-70096

C₄H₂O₂
  2-Butynedial, B-60351

C₄H₂S₄
  3,4-Dimercapto-3-cyclobutene-1,2-dithione,
    D-70364

C₄H₃BrN₂OS
  5-Bromo-3,4-dihydro-4-thioxo-2(1H)-
    pyrimidinone, B-70205

C₄H₃BrN₂O₂
  ▷5-Bromo-2,4-(1H,3H)-pyrimidinedione,
    B-60318
  6-Bromo-2,4-(1H,3H)-pyrimidinedione,
    B-60319

C₄H₃ClN₂
  5-Chloropyrimidine, C-70157

C₄H₃ClN₂O
  5-Chloropyrimidine; N-Oxide, in C-70157

C₄H₃ClN₂OS
  5-Chloro-4-thioxo-2(1H)-pyrimidinone,
    C-70162

C₄H₃ClN₂O₂
  ▷5-Chloro-2,4-(1H,3H)-pyrimidinedione,
    C-60130
  6-Chloro-2,4-(1H,3H)-pyrimidinedione,
    C-70158

C₄H₃ClO₂S₂
  2-Thiophenesulfonic acid; Chloride, in T-60213
  3-Thiophenesulfonic acid; Chloride, in T-60214

C₄H₃ClS
  3-Chlorothiophene, C-60136

C₄H₃FN₂OS
  5-Fluoro-3,4-dihydro-4-thioxo-2(1H)-
    pyrimidinone, F-70019

C₄H₃F₃
  1,1,1-Trifluoro-2-butyne, T-60275
  1,4,4-Trifluorocyclobutene, T-60276

C₄H₃F₃N₂
  2-Trifluoromethyl-1H-imidazole, T-60289
  4(5)-Trifluoromethyl-1H-imidazole, T-60290

C₄H₃F₃O₂
  2-(Trifluoromethyl)propenoic acid, T-60295

C₄H₃IN₂O₂
  ▷5-Iodo-2,4(1H,3H)-pyrimidinedione, I-70056
  ▷5-Iodo-2,4(1H,3H)-pyrimidinedione, I-60062
  6-Iodo-2,4-(1H,3H)-pyrimidinedione, I-70057

C₄H₃IS
  3-Iodothiophene, I-60067

C₄H₃I₃N₂
  2,4,5-Triiodo-1H-imidazole; 1-Me, in T-70276

C₄H₃NO₃
  5-Isoxazolecarboxylic acid, I-60142

C₄H₃N₃
  4(5)-Cyanoimidazole, in I-60004

C₄H₃N₃O₂
  2-Nitropyrimidine, N-60043

C₄H₃N₃O₃
  6-Formyl-1,2,4-triazine-3,5(2H,4H)-dione,
    F-60075

C₄H₃N₃O₄
  2,3-Dinitro-1H-pyrrole, D-60461
  2,4-Dinitro-1H-pyrrole, D-60462

  2,5-Dinitro-1H-pyrrole, D-60463
  3,4-Dinitro-1H-pyrrole, D-60464

C₄H₃N₅
  Tetrazolo[1,5-a]pyrimidine, T-60176
  Tetrazolo[1,5-c]pyrimidine, T-60177

C₄H₄
  1,3-Cyclobutadiene, C-70212

C₄H₄BrN₃
  2-Amino-4-bromopyrimidine, A-70100
  2-Amino-5-bromopyrimidine, A-70101
  4-Amino-5-bromopyrimidine, A-70102
  5-Amino-2-bromopyrimidine, A-70103

C₄H₄BrN₃O₂
  4-Bromo-3(5)-nitro-1H-pyrazole; 1-Me, in
    B-60304

C₄H₄Br₂
  ▷1,4-Dibromo-2-butyne, D-70087

C₄H₄Br₂O
  2,2-Dibromocyclopropanecarboxaldehyde,
    D-60092

C₄H₄Br₂O₂
  2,2-Dibromocyclopropanecarboxylic acid,
    D-60093

C₄H₄Cl₂O₂
  2,2-Dichlorocyclopropanecarboxylic acid,
    D-60127

C₄H₄FN₃
  3-Amino-6-fluoropyridazine, A-70148

C₄H₄F₂
  1,1-Difluoro-1,3-butadiene, D-70162
  3,3-Difluorocyclobutene, D-70165

C₄H₄F₂O₄
  2,2-Difluorobutanedioic acid, D-70163

C₄H₄F₃NO
  2-(Trifluoromethyl)propenoic acid; Amide, in
    T-60295

C₄H₄N₂
  3,4-Diazatricyclo[3.1.0.0²,⁶]hex-3-ene, D-70054

C₄H₄N₂O₂
  1H-Imidazole-4-carboxylic acid, I-60004
  5-Isoxazolecarboxylic acid; Amide, in I-60142
  3-Nitropyrrole, N-60044
  Tartaric acid; Dinitrile, in T-70005

C₄H₄N₂S
  4-Vinyl-1,2,3-thiadiazole, V-60014
  5-Vinyl-1,2,3-thiadiazole, V-60015

C₄H₄N₄
  4-Amino-3-cyanopyrazole, in A-70189
  5-Amino-4-pyrazolecarbonitrile, in A-70187
  1H-Pyrazolo[5,1-c]-1,2,4-triazole, P-70157

C₄H₄N₆
  4,4′-Bi-4H-1,2,4-triazole, B-60189
  5-(1H-Imidaz-2-yl)-1H-tetrazole, I-70009

C₄H₄O
  5-Oxabicyclo[2.1.0]pent-2-ene, O-70059

C₄H₄OS
  Thiophene-3-ol, T-60212

C₄H₄O₂
  3-Butynoic acid, B-70309
  1,2-Cyclobutanedione, C-60183

C₄H₄O₂S
  2,4(3H,5H)-Thiophenedione, T-70185

C₄H₄O₃S₂
  2-Thiophenesulfonic acid, T-60213
  3-Thiophenesulfonic acid, T-60214

C₄H₄O₄
  Diformylacetic acid, D-60190

C₄H₄O₄S₂
  (Dimercaptomethylene)propanedioic acid,
    D-60395

$C_4H_4Se_2$
2-Methylene-1,3-diselenole, M-70069

C_4H_5Br
1-Bromo-1,2-butadiene, B-70196
4-Bromo-1-butyne, B-70198

$C_4H_5BrN_2$
3-Bromo-1,4-dimethylpyrazole, *in* B-70250
4-Bromo-1-methyl-1H-imidazole, *in* B-70221
5-Bromo-1-methyl-1H-imidazole, *in* B-70221
3(5)-Bromo-4-methylpyrazole, B-70250
4-Bromo-3(5)-methylpyrazole, B-70251
3(5)-Bromo-5(3)-methylpyrazole, B-70252

$C_4H_5BrN_2O$
5-Aminomethyl-3-bromoisoxazole, A-70165

$C_4H_5BrO_2$
2-Bromomethyl-2-propenoic acid, B-70249

$C_4H_5BrO_2S$
4-Bromo-2-sulfolene, *in* B-70204

C_4H_5BrS
3-Bromo-2,3-dihydrothiophene, B-70204

C_4H_5Cl
4-Chloro-1-butyne, C-70058

$C_4H_5ClN_2$
3-Chloro-1-methylpyrazole, *in* C-70155
4-Chloro-1-methylpyrazole, *in* C-70156
5-Chloro-1-methylpyrazole, *in* C-70155
▷4-Chloro-3(5)methylpyrazole, C-70108
3(5)-Chloro-5(3)-methylpyrazole, C-70109

$C_4H_5Cl_2NO_2$
2-Amino-4,4-dichloro-3-butenoic acid, A-60125

$C_4H_5Cl_3O$
▷(2,2,2-Trichloroethyl)oxirane, T-60252

$C_4H_5F_6NO$
N-Ethoxy-1,1,1,1′,1′,1′-hexafluorodi-
methylamine, *in* B-60180

C_4H_5I
4-Iodo-1-butyne, I-70024

$C_4H_5IN_2$
3(5)-Iodo-4-methylpyrazole, I-70048
4-Iodo-3(5)-methylpyrazole, I-70049
3(5)-Iodo-5(3)-methylpyrazole, I-70050

$C_4H_5IO_2$
Dihydro-3-iodo-2(3H)-furanone, D-60252

C_4H_5NO
1,2-Dihydro-3H-pyrrol-3-one, D-70262
▷2-Hydroxy-3-butenenitrile, *in* P-70119

$C_4H_5NO_2$
1-Aminocyclopropenecarboxylic acid, A-70126
1-Amino-2-cyclopropene-1-carboxylic acid,
A-60122
1-Nitro-1,3-butadiene, N-60023

$C_4H_5NO_2S_2$
2-Thiophenesulfonic acid; Amide, *in* T-60213
3-Thiophenesulfonic acid; Amide, *in* T-60214

$C_4H_5NO_3$
2,5-Dihydro-3-nitrofuran, D-70236

$C_4H_5NO_4$
2-Oxo-4-oxazolidinecarboxylic acid, O-60082

C_4H_5NS
4-Mercapto-2-butenoic acid; Nitrile, *in*
M-60018

$C_4H_5NS_2$
2-(Methylthio)thiazole, *in* T-60200

$C_4H_5N_3O$
5-Amino-4(1H)-pyrimidinone, A-70196
1H-Imidazole-4-carboxylic acid; Amide, *in*
I-60004

$C_4H_5N_3O_2$
▷3-Amino-1H-pyrazole-4-carboxylic acid,
A-70187
3-Amino-1H-pyrazole-5-carboxylic acid,
A-70188

4-Amino-1H-pyrazole-3-carboxylic acid,
A-70189
4-Amino-1H-pyrazole-5-carboxylic acid,
A-70190
5-Amino-1H-pyrazole-3-carboxylic acid,
A-70191
5-Amino-1H-pyrazole-4-carboxylic acid,
A-70192
3-Azido-2-propenoic acid; Me ester, *in* A-70290
6-Methyl-1,3,5-triazine-2,4(1H,3H)-dione,
M-60132
1H-1,2,3-Triazole-4-carboxylic acid; 1-Me, *in*
T-60239
1H-1,2,3-Triazole-4-carboxylic acid; Me ester,
in T-60239
1H-1,2,3-Triazole-4-carboxylic acid; 2-Me, *in*
T-60239

$C_4H_5N_3S_2$
5-Amino-2-thiazolecarbothioamide, A-60260
5-Amino-4-thiazolecarbothioamide, A-60261

$C_4H_6BrNO_4$
2-Amino-3-bromobutanedioic acid, A-70098

$C_4H_6ClF_2NO_2$
2-Amino-4-chloro-4,4-difluorobutanoic acid,
A-60107

$C_4H_6ClNO_2$
1-Chloro-1-nitrocyclobutane, C-70130

$C_4H_6ClNO_4$
2-Amino-3-chlorobutanedioic acid, A-70121

$C_4H_6Cl_2$
3-Chloro-2-chloromethyl-1-propene, C-70061

$C_4H_6I_2$
▷3-Iodo-2-(iodomethyl)-1-propene, I-60047

$C_4H_6N_2$
1,4-Dihydropyrimidine, D-70255

$C_4H_6N_2O$
4,5-Dihydro-3(2H)-pyridazinone, D-60272
3,4-Dihydro-2(1H)-pyrimidinone, D-60273
5,6-Dihydro-4(1H)-pyrimidinone, D-60274

$C_4H_6N_2OS$
2-Thioxo-4-imidazolidinone; 1-N-Me, *in*
T-70188
2-Thioxo-4-imidazolidinone; 3-N-Me, *in*
T-70188

$C_4H_6N_2O_2$
1,3-Diazetidine-2,4-dione; 1,3-Di-Me, *in*
D-60057
Dihydro-4,6-(1H,5H)-pyrimidinedione,
D-70256
Dimethyl N-cyanoimidocarbonate, D-70390

$C_4H_6N_2O_3$
3-Methoxy-4-methyl-1,2,4-oxadiazolidin-
5(4H)-one, *in* O-70060
1,2,4-Oxadiazolidine-3,5-dione; 4-Et, *in*
O-70060
1,2,4-Oxadiazolidine-3,5-dione; 2,4-Di-Me, *in*
O-70060
2-Oxo-4-imidazolidinecarboxylic acid, O-60074

$C_4H_6N_2S_2$
2,5-Dihydro-4,6-pyrimidinedithiol, D-70257

$C_4H_6N_2S_3$
2,5-Biss(ethylthio)-1,3,4-thiadiazole, *in*
M-70031
3-Methyl-5-(methylthio)-1,3,4-thiadiazole-
2(3H)-thione, *in* M-70031

$C_4H_6N_4$
3-Amino-1,2,4-triazine; N-Me, *in* A-70203

$C_4H_6N_4O$
3-Amino-1H-pyrazole-4-carboxylic acid;
Amide, *in* A-70187
4-Amino-1H-pyrazole-3-carboxylic acid;
Amide, *in* A-70189
4-Amino-1H-pyrazole-5-carboxylic acid; N(1)-
Me, Amide, *in* A-70190

2,6-Diamino-4(1H)-pyrimidinone, D-70043
1H-1,2,3-Triazole-4-carboxylic acid; 1-Me,
amide, *in* T-60239

$C_4H_6N_4O_2$
3,6-Dimethoxy-1,2,4,5-tetrazine, *in* D-70268
Tetrahydroimidazo[4,5-d]imidazole-
2,5(1H,3H)-dione, T-70066

C_4H_6OS
Dihydro-2(3H)-furanthione, D-70210

$C_4H_6O_2$
▷Methyl acrylate, *in* P-70121
2-Methylene-1,3-dioxolane, M-60075
4-Methylene-1,3-dioxolane, M-60076

$C_4H_6O_2S$
4-Mercapto-2-butenoic acid, M-60018
3-Oxobutanethioic acid, O-70080

$C_4H_6O_3$
Dihydro-4-hydroxy-2(3H)-furanone, D-60242
4,5-Dihydro-3-hydroxy-2(3H)-furomone, *in*
D-60313
2-Formylpropanoic acid, F-70039
1-Hydroxycyclopropanecarboxylic acid,
H-60117
Methylene-1,3,5-trioxane, M-60081
3-Oxopropanoic acid; Me ester, *in* O-60090

$C_4H_6O_5$
Dimethyl dicarbonate, *in* D-70111

$C_4H_6O_6$
Tartaric acid, T-70005

C_4H_6S
2,3-Dihydrothiophene, D-60290
2,5-Dihydrothiophene, D-70270

$C_4H_6S_2$
2-Butyne-1,4-dithiol, B-70308

$C_4H_6S_2Se$
1,3-Dithiane-2-selone, D-70507

$C_4H_6S_4$
Ethanebis(dithioic)acid; Di-Me ester, *in*
E-60040

C_4H_7Br
1-Bromo-1-methylcyclopropane, B-70242

C_4H_7BrO
4-Bromobutanal, B-70197

$C_4H_7ClFNO_2$
2-Amino-4-chloro-4-fluorobutanoic acid,
A-60108

C_4H_7ClOS
Carbonochloridothioic acid; O-Propyl, *in*
C-60013
Carbonochloridothioic acid; O-Isopropyl, *in*
C-60013
▷Carbonochloridothioic acid; S-Propyl, *in*
C-60013

$C_4H_7ClO_3$
3-Chloro-2-hydroxypropanoic acid; Me ester, *in*
C-70072

$C_4H_7Cl_3$
1,3-Dichloro-2-(chloromethyl)propane,
D-70118

$C_4H_7F_2NO_2$
2-Amino-3,3-difluorobutanoic acid, A-60133
3-Amino-4,4-difluorobutanoic acid, A-60134

$C_4H_7F_3OS$
Trifluoromethanesulfenic acid; Isopropyl ester,
in T-60279

C_4H_7IO
2-Iodobutanal, I-60040
3-Iodo-2-butanone, I-60041
2-Iodo-2-methylpropanal, I-60051

$C_4H_7I_3$
1,3-Diiodo-2-(iodomethyl)propane, D-70358

C_4H_7NO
▷2-Pyrrolidinone, P-70188

C₄H₇NOS

Tetrahydro-2H-1,2-thiazin-3-one, T-60095
Tetrahydro-2H-1,3-thiazin-2-one, T-60096

C₄H₇NO₂

3-Aminodihydro-2(3H)-furanone, A-60138
2-Azetidinecarboxylic acid, A-70284
5-Methyl-2-oxazolidinone, M-60107
3-Morpholinone, M-60147
Nitrocyclobutane, N-60024

C₄H₇NO₃

1-Nitro-2-butanone, N-70043
3-Nitro-2-butanone, N-70044
4-Nitro-2-butanone, N-70045

C₄H₇NS

2-Propenethioamide; N-Me, in P-70120
2-Pyrrolidinethione, P-70186

C₄H₇NS₂

4,5-Dihydro-2-(methylthio)thiazole, in T-60197
Tetrahydro-2H-1,3-thiazine-2-thione, T-60094
2-Thiazoledinethione; N-Me, in T-60197

C₄H₇N₃O

4-Azidobutanal, A-70286
4,5-Dihydro-5,5-dimethyl-3H-1,2,4-triazol-3-one, D-70201

C₄H₇N₃O₂

Biuret; N-Ac, in B-60190

C₄H₇N₅

2,4-Diamino-6-methyl-1,3,5-triazine, D-60040

C₄H₇N₅O

5-Acetamido-2-methyl-2H-tetrazole, in A-70202
2,4-Diamino-6-methyl-1,3,5-triazine; N³-Oxide, in D-60040
2,4-Diamino-6-methyl-1,3,5-triazine; N⁵-Oxide, in D-60040

C₄H₈BrNO₂

2-Amino-3-bromobutanoic acid, A-60097
2-Amino-4-bromobutanoic acid, A-60098
4-Amino-2-bromobutanoic acid, A-60099

C₄H₈ClNO₂

2-Amino-3-chlorobutanoic acid, A-60103
2-Amino-4-chlorobutanoic acid, A-60104
4-Amino-2-chlorobutanoic acid, A-60105
4-Amino-3-chlorobutanoic acid, A-60106

C₄H₈FNO₂

2-Amino-3-fluorobutanoic acid, A-60154
2-Amino-4-fluorobutanoic acid, A-60155
3-Amino-4-fluorobutanoic acid, A-60156
4-Amino-2-fluorobutanoic acid, A-60157
4-Amino-3-fluorobutanoic acid, A-60158

C₄H₈IN

2-Propyn-1-amine; N-Me; B,HI, in P-60184

C₄H₈I₂O

1,1'-Oxybis[2-iodoethane], O-60096

C₄H₈NO₆P

Antibiotic SF 2312, A-60288

C₄H₈N₂O

2-Imidazolidinone; 1-Me, in I-70005
Piperazinone, P-70101
Tetrahydro-2(1H)-pyrimidinone, T-60088
Tetrahydro-4(1H)-pyrimidinone, T-60089

C₄H₈N₂O₂

Acetamidoxime; O-Ac, in A-60016
4-Amino-5-methyl-3-isoxazolidinone, A-60221

C₄H₈N₂O₄

Tartaric acid; Diamide, in T-70005

C₄H₈N₄O₂

1,4-Dihydro-3,6-dimethoxy-1,2,4,5-tetrazine, D-70192
▷Hexahydropyrimidine; N,N'-Dinitroso, in H-60056
Tetrahydro-1,3,5,7-tetrazocine-2,6(1H,3H)-dione, T-70098

C₄H₈N₄O₄

▷Hexahydropyrimidine; N,N'-Dinitro, in H-60056

C₄H₈O

▷2-Ethyloxirane, E-70047

C₄H₈OS

3-Mercaptocyclobutanol, M-60019

C₄H₈O₂

3-Butene-1,2-diol, B-70291
2-Hydroxybutanal, H-70112
▷2-Methyl-1,3-dioxolane, M-60063

C₄H₈O₂S

3-Mercaptobutanoic acid, M-70023
2-(Methylthio)propanoic acid, in M-60027

C₄H₈O₃

1,3-Dioxolane-2-methanol, D-70468

C₄H₈O₄

2,3-Dihydroxybutanoic acid, D-60312
3,4-Dihydroxybutanoic acid, D-60313

C₄H₈S₂

2-Butene-1,4-dithiol, B-70292
1,1-Cyclobutanedithiol, C-70215
1,3-Cyclobutanedithiol, C-60184
2-Methylene-1,3-propanedithiol, M-70076

C₄H₈S₂⊕⊕

1,4-Dithioniabicyclo[2.2.0]hexane, D-70527

C₄H₉F₃O₃SSi

Trifluoromethanesulfonic acid; Trimethylsilyl ester, in T-70241

C₄H₉NO

▷Acetamide; N-Di-Me, in A-70019
3-Pyrrolidinol, P-70187

C₄H₉NO₃

2-Amino-4-hydroxybutanoic acid, A-70156
3-Amino-2-hydroxybutanoic acid, A-60183
Threonine, T-60217

C₄H₉NS

▷Tetrahydro-2H-1,3-thiazine, T-60093

C₄H₉N₃O₃

3-Amino-2-ureidopropanoic acid, A-60268

C₄H₉N₅

N-(1,4-Dihydro-1,4-dimethyl-5H-tetrazol-5-ylidene)methanamine, in A-70202

C₄H₁₀NO₆P

O-Phosphohomoserine, in A-70156

C₄H₁₀N₂

Hexahydropyrimidine, H-60056

C₄H₁₀N₂O

Tetrahydro-3,4-furandiamine, T-70062

C₄H₁₀N₄

Octahydroimidazo[4,5-d]imidazole, O-70015

C₄H₁₀O₂

▷1-Methoxy-2-propanol, in P-70118
2-Methyl-1,3-propanediol, M-60120

C₄H₁₀O₃

▷1-Ethoxy-1-hydroperoxyethane, E-60044

C₄H₁₀S₂

1,3-Butanedithiol, B-60339
2,2-Butanedithiol, B-60340
2-Methyl-1,3-propanedithiol, M-60121

C₄H₁₂N₄

Octahydro-1,2,5,6-tetrazocine, O-60020

C₄H₁₃N₃O

2-[(2-Hydrazinoethyl)amino]ethanol, H-60091

C₄N₆S₂

Bis[1,2,5-thiadiazolo][3,4-b:3',4'-e]pyrazine, B-70161

C₅BrF₄N

4-Bromo-2,3,5,6-tetrafluoropyridine, B-60328

C₅BrF₅N

2-Bromo-3,4,5,6-tetrafluoropyridine, B-60326

3-Bromo-2,4,5,6-tetrafluoropyridine, B-60327

C₅F₆

1,2,3,4,5,5-Hexafluoro-1,3-cyclopentadiene, H-60044

C₅F₁₀N₂O₂

2,2,3,3,4,4,5,5,6,6-Decafluoropiperidine; N-Nitro, in D-60006

C₅F₁₁N

Undecafluoropiperidine, U-60002

C₅HF₁₀N

2,2,3,3,4,4,5,5,6,6-Decafluoropiperidine, D-60006

C₅H₂Cl₂N₂O₂

2,3-Dichloro-5-nitropyridine, D-60140
2,4-Dichloro-5-nitropyridine, D-60141
2,4-Dichloro-5-nitropyridine, D-60142
2,5-Dichloro-3-nitropyridine, D-60143
2,6-Dichloro-3-nitropyridine, D-60144
2,6-Dichloro-4-nitropyridine, D-60145
3,4-Dichloro-5-nitropyridine, D-60146
3,5-Dichloro-4-nitropyridine, D-60147

C₅H₂Cl₂N₂O₃

2,6-Dichloro-4-nitropyridine; 1-Oxide, in D-60145
3,5-Dichloro-4-nitropyridine; 1-Oxide, in D-60147

C₅H₂Cl₃N

▷2,4,6-Trichloropyridine, T-60255

C₅H₂F₄N₂

4-Amino-2,3,5,6-tetrafluoropyridine, A-60257

C₅H₂F₆O

4,4,4-Trifluoro-3-(trifluoromethyl)-2-butenal, T-60309
1,1,1-Trifluoro-2-(trifluoromethyl)-3-butyn-2-ol, T-60310

C₅H₂INS

2-Cyano-3-iodothiophene, in I-60075
2-Cyano-4-iodothiophene, in I-60076
2-Cyano-5-iodothiophene, in I-60078
3-Cyano-2-iodothiophene, in I-60074
3-Cyano-4-iodothiophene, in I-60077
4-Cyano-2-iodothiophene, in I-60079

C₅H₃BrClN

3-Bromo-5-chloropyridine, B-60223
4-Bromo-3-chloropyridine, B-60224

C₅H₃BrFN

2-Bromo-3-fluoropyridine, B-70213
2-Bromo-5-fluoropyridine, B-60252
3-Bromo-2-fluoropyridine, B-60253
3-Bromo-4-fluoropyridine, B-70214
3-Bromo-5-fluoropyridine, B-60254
4-Bromo-3-fluoropyridine, B-60255
5-Bromo-2-fluoropyridine, B-60256

C₅H₃BrIN

2-Bromo-4-iodopyridine, B-60280
2-Bromo-5-iodopyridine, B-60281
3-Bromo-4-iodopyridine, B-60282
3-Bromo-5-iodopyridine, B-60283
4-Bromo-2-iodopyridine, B-60284
4-Bromo-3-iodopyridine, B-60285
5-Bromo-2-iodopyridine, B-60286

C₅H₃BrO₂

3-Bromo-2-furancarboxaldehyde, B-60257
4-Bromo-2-furancarboxaldehyde, B-60258
5-Bromo-2-furancarboxaldehyde, B-60259

C₅H₃ClFN

2-Chloro-3-fluoropyridine, C-60081
2-Chloro-4-fluoropyridine, C-60082
2-Chloro-5-fluoropyridine, C-60083
2-Chloro-6-fluoropyridine, C-60084
3-Chloro-2-fluoropyridine, C-60085
4-Chloro-2-fluoropyridine, C-60086
4-Chloro-3-fluoropyridine, C-60087
5-Chloro-2-fluoropyridine, C-60088

C₅H₃ClFNO

2-Chloro-3-fluoropyridine; N-Oxide, in C-60081

C₅H₃ClIN

2-Chloro-3-iodopyridine, C-60107
2-Chloro-4-iodopyridine, C-60108
2-Chloro-5-iodopyridine, C-60109
3-Chloro-2-iodopyridine, C-60110
3-Chloro-5-iodopyridine, C-60111
4-Chloro-2-iodopyridine, C-60112
4-Chloro-3-iodopyridine, C-60113

C₅H₃FIN

2-Fluoro-4-iodopyridine, F-60045
3-Fluoro-4-iodopyridine, F-60046

C₅H₃IOS

2-Iodo-3-thiophenecarboxaldehyde, I-60068
3-Iodo-2-thiophenecarboxaldehyde, I-60069
4-Iodo-2-thiophenecarboxaldehyde, I-60070
4-Iodo-3-thiophenecarboxaldehyde, I-60071
5-Iodo-2-thiophenecarboxaldehyde, I-60072
5-Iodo-3-thiophenecarboxaldehyde, I-60073

C₅H₃IO₂S

2-Iodo-3-thiophenecarboxylic acid, I-60074
3-Iodo-2-thiophenecarboxylic acid, I-60075
4-Iodo-2-thiophenecarboxylic acid, I-60076
4-Iodo-3-thiophenecarboxylic acid, I-60077
5-Iodo-2-thiophenecarboxylic acid, I-60078
5-Iodo-3-thiophenecarboxylic acid, I-60079

C₅H₃NS

2-Ethynylthiazole, E-60061
4-Ethynylthiazole, E-60062

C₅H₃N₅O₈

▷1-Methyl-2,3,4,5-tetranitro-1H-pyrrole, in
T-60160

C₅H₄BrNO

▷2-Bromo-3-hydroxypyridine, B-60275
2-Bromo-5-hydroxypyridine, B-60276
3-Bromo-5-hydroxypyridine, B-60277
4-Bromo-3-hydroxypyridine, B-60278

C₅H₄BrNO₂

2-Bromo-3-hydroxypyridine; N-Oxide, in
B-60275

C₅H₄BrNS

3-Bromo-2(1H)-pyridinethione, B-60315
3-Bromo-4(1H)-pyridinethione, B-60316
5-Bromo-2(1H)-pyridinethione, B-60317

C₅H₄ClNO

▷2-Chloro-3-hydroxypyridine, C-60100
2-Chloro-5-hydroxypyridine, C-60101
3-Chloro-5-hydroxypyridine, C-60102

C₅H₄ClNO₂

4-Chloro-6-hydroxy-2(1H)-pyridinone,
C-70073
6-Chloro-4-hydroxy-2(1H)pyridinone, C-70074

C₅H₄ClNO₂S

2-Pyridinesulfonic acid; Chloride, in P-70171

C₅H₄ClNS

3-Chloro-2(1H)-pyridinethione, C-60129

C₅H₄ClN₅

▷2-Amino-6-chloro-1H-purine, A-60116
2-Chloro-6-aminopurine, C-70039

C₅H₄Cl₄

1-Chloro-1-(trichlorovinyl)cyclopropane,
C-60138

C₅H₄FN₅

8-Amino-6-fluoro-9H-purine, A-60163

C₅H₄INO

3-Hydroxy-2-iodopyridine, H-60160

C₅H₄INOS

2-Iodo-3-thiophenecarboxaldehyde; Oxime, in
I-60068
3-Iodo-2-thiophenecarboxaldehyde; Oxime, in
I-60069
4-Iodo-2-thiophenecarboxaldehyde; Oxime, in
I-60070

4-Iodo-3-thiophenecarboxaldehyde; Oxime, in
I-60071
5-Iodo-2-thiophenecarboxaldehyde; Oxime, in
I-60072
5-Iodo-3-thiophenecarboxaldehyde; Oxime, in
I-60073
2-Iodo-3-thiophenecarboxylic acid; Amide, in
I-60074
3-Iodo-2-thiophenecarboxylic acid; Amide, in
I-60075
4-Iodo-2-thiophenecarboxylic acid; Amide, in
I-60076
4-Iodo-3-thiophenecarboxylic acid; Amide, in
I-60077
5-Iodo-2-thiophenecarboxylic acid; Amide, in
I-60078
5-Iodo-3-thiophenecarboxylic acid; Amide, in
I-60079

C₅H₄N₂O₂

Methyl dicyanoacetate, in D-70149

C₅H₄N₂O₃

3-Nitro-4(1H)-pyridone, N-70063
1,2,3,4-Tetrahydro-2,4-dioxo-5-
pyrimidinecarboxaldehyde, T-60064

C₅H₄N₂S

▷2-Thiocyanato-1H-pyrrole, T-60209

C₅H₄N₄

1H-Pyrazolo[3,4-d]pyrimidine, P-70156
Pyrazolo[1,5-b][1,2,4]triazine, P-60207
[1,2,4]Triazolo[1,5-b]pyridazine, T-60240
[1,2,4]Triazolo[1,5-a]pyrimidine, T-60242
1,2,4-Triazolo[4,3-c]pyrimidine, T-60243

C₅H₄N₄O

1,5-Dihydro-6H-imidazo[4,5-c]pyridazin-6-one,
D-60248
2H-Imidazo[4,5-b]pyrazin-2-one, I-60006

C₅H₄N₄O₂

1H-Pyrazolo[3,4-d]pyrimidine-4,6(5H,7H)-
dione, P-60202

C₅H₄N₄O₆

▷1-Methyl-2,3,4-trinitro-1H-pyrrole, in T-60373
▷1-Methyl-2,3,5-trinitro-1H-pyrrole, in T-60374

C₅H₄N₄S

1,3-Dihydro-2H-imidazo[4,5-b]pyrazine-2-
thione, D-60247
1,5-Dihydro-6H-imidazo[4,5-c]pyridazine-6-
thione, D-70215
▷1,7-Dihydro-6H-purine-6-thione, D-60270

C₅H₄OS

▷2-Thiophenecarboxaldehyde, T-70183

C₅H₄O₃

3-Oxabicyclo[3.1.0]hexane-2,4-dione, in
C-70248

C₅H₅BF₅N

1-Fluoropyridinium; Tetrafluoroborate, in
F-60066

C₅H₅BrN₂O

2-Amino-5-bromo-3-hydroxypyridine, A-60101

C₅H₅BrO₅

(3-Bromo-2-oxopropylidene)propanedioic acid,
B-70262

C₅H₅Cl

1-Chloro-1-ethynylcyclopropane, C-60068

C₅H₅ClFNO₄

1-Fluoropyridinium; Perchlorate, in F-60066

C₅H₅ClN₂

2-Chloro-3-methylpyrazine, C-70105
2-Chloro-5-methylpyrazine, C-70106
2-Chloro-6-methylpyrazine, C-70107
3-Chloro-4-methylpyridazine, C-70110
3-Chloro-5-methylpyridazine, C-70111
3-Chloro-6-methylpyridazine, C-70112
2-Chloro-4-methylpyrimidine, C-70113
2-Chloro-5-methylpyrimidine, C-70114
4-Chloro-2-methylpyrimidine, C-70115

4-Chloro-5-methylpyrimidine, C-70116
4-Chloro-6-methylpyrimidine, C-70117
5-Chloro-2-methylpyrimidine, C-70118
5-Chloro-4-methylpyrimidine, C-70119

C₅H₅ClN₂O

2-Amino-5-chloro-3-hydroxypyridine, A-60111
▷2-Chloro-3-methylpyrazine; 1-Oxide, in
C-70105
2-Chloro-6-methylpyrazine; 4-Oxide, in
C-70107
3-Chloro-2-methylpyrazine 1-oxide, in C-70105
3-Chloro-4-methylpyridazine; 1-Oxide, in
C-70110
3-Chloro-5-methylpyridazine; 1-Oxide, in
C-70111
3-Chloro-6-methylpyridazine; 1-Oxide, in
C-70112
4-Chloro-6-methylpyrimidine; 1-Oxide, in
C-70117

C₅H₅ClO

2-Chloro-2-cyclopenten-1-one, C-60051
5-Chloro-2-cyclopenten-1-one, C-70063

C₅H₅Cl₂F₃O₂

2,2-Dichloro-3,3,3-trifluoro-1-propanol; Ac, in
D-60161

C₅H₅FN⊕

1-Fluoropyridinium, F-60066

C₅H₅F₃O₂

Dihydro-4-(trifluoromethyl)-2(3H)-furanone,
D-60299
Dihydro-5-(trifluoromethyl)-2(3H)-furanone,
D-60300
▷2-(Trifluoromethyl)propenoic acid; Me ester, in
T-60295

C₅H₅F₇NSb

1-Fluoropyridinium; Hexafluoroantimonate, in
F-60066

C₅H₅I

5-Iodo-1-penten-4-yne, I-70053

C₅H₅I₃N₂

2,4,5-Triiodo-1H-imidazole; 1-Et, in T-70276

C₅H₅N

1-Cyanobicyclo[1.1.0]butane, in B-70094

C₅H₅NO

3-Formyl-2-butenenitrile, in M-70103
1H-Pyrrole-3-carboxaldehyde, P-70182

C₅H₅NOS

▷1-Hydroxy-2(1H)-pyridinethione, H-70217
3-Hydroxy-2(1H)-pyridinethione, H-60222
▷2(1H)-Pyridinethione; 1-Hydroxy, in P-60224
4(1H)-Pyridinethione; 1-Hydroxy, in P-60225
2-Thiophenecarboxaldehyde; Oxime, in
T-70183

C₅H₅NOSe

2(1H)-Pyridineselone; N-Oxide, in P-70170

C₅H₅NO₂

2,3-Dihydro-3-oxo-1H-pyrrole-1-
carboxaldehyde, in D-70262
3-Hydroxy-2(1H)-pyridinone, H-60223
3-Hydroxy-4(1H)-pyridinone, H-60227
4-Hydroxy-2(1H)-pyridinone, H-60224
5-Hydroxy-2(1H)-pyridinone, H-60225
6-Hydroxy-2(1H)-pyridinone, H-60226

C₅H₅NO₃

4,6-Dihydroxy-2(1H)-pyridinone, D-70341
5-Isoxazolecarboxylic acid; Me ester, in I-60142

C₅H₅NO₃S

2-Pyridinesulfonic acid, P-70171

C₅H₅NS

2-Pyridinethiol, P-60223
▷2(1H)-Pyridinethione, P-60224
▷4(1H)-Pyridinethione, P-60225

1,4-Thiazepine, T-60196
1-Thiocyano-1,3-butadiene, T-60210

C₅H₅NS₂

2,3-Pyridinedithiol, P-60218
2,4-Pyridinedithiol, P-60219
2,5-Pyridinedithiol, P-60220
3,4-Pyridinedithiol, P-60221
3,5-Pyridinedithiol, P-60222

C₅H₅NSe

2(1*H*)-Pyridineselone, P-70170

C₅H₅N₃O₂

N-(1,4-Dihydro-4-oxo-5-pyrimidinyl)-formamide, *in* A-70196
4-Methyl-5-nitropyrimidine, M-60104

C₅H₅N₃O₄

1-Methyl-2,3-dinitro-1*H*-pyrrole, *in* D-60461
1-Methyl-2,4-dinitro-1*H*-pyrrole, *in* D-60462
1-Methyl-2,5-dinitro-1*H*-pyrrole, *in* D-60463
1-Methyl-3,4-dinitro-1*H*-pyrrole, *in* D-60464

C₅H₅N₅

▷2-Aminopurine, A-60243
8-Aminopurine, A-60244
9-Aminopurine, A-60245

C₅H₅N₅O

6-Amino-1,3-dihydro-2*H*-purin-2-one, A-60144
5-Aminopyrazolo[4,3-*d*]pyrimidin-7(1*H*,6*H*)-one, A-60246

C₅H₆

Ethynylcyclopropane, E-70045
Tricyclo[2.1.0.0¹,³]pentane, T-70231
3-Vinylcyclopropene, V-60012

C₅H₆Br₂O

4,4-Dibromo-3-methyl-3-buten-2-one, D-70094

C₅H₆Br₂O₂

2,2-Dibromocyclopropanecarboxylic acid; Me ester, *in* D-60093

C₅H₆Cl₂O₂

2,2-Dichlorocyclopropanecarboxylic acid; Me ester, *in* D-60127

C₅H₆F₂O₂

3,3-Difluorocyclobutanecarboxylic acid, D-70164

C₅H₆I₂

1,3-Diiodobicyclo[1.1.1]pentane, D-70357

C₅H₆N₂

4(5)-Vinylimidazole, V-70012

C₅H₆N₂O

2-Amino-3-hydroxypyridine, A-60200
2-Amino-5-hydroxypyridine, A-60201
4-Amino-3-hydroxypyridine, A-60202
5-Amino-3-hydroxypyridine, A-60203
1-Amino-2(1*H*)-pyridinone, A-60248
2-Amino-4(1*H*)-pyridinone, A-60253
3-Amino-2(1*H*)-pyridinone, A-60249
3-Amino-4(1*H*)-pyridinone, A-60254
4-Amino-2(1*H*)-pyridinone, A-60250
5-Amino-2(1*H*)-pyridinone, A-60251
6-Amino-2(1*H*)-pyridinone, A-60252
6-Methyl-3(2*H*)-pyridazinone, M-70124
4-Methyl-3(2*H*)-pyridazinone, M-70125
5-Methyl-3(2*H*)-pyridazinone, M-70126
1*H*-Pyrazole; *N*-Ac, *in* P-70152

C₅H₆N₂OS

5,6-Dihydro-4*H*-cyclopenta-1,2,3-thiadiazole; 2-Oxide, *in* D-70179
▷2,3-Dihydro-6-methyl-2-thioxo-4(1*H*)-pyrimidinone, D-70232
3,4-Dihydro-5-methyl-4-thioxo-2(1*H*)-pyrimidinone, D-70233

C₅H₆N₂O₂

4-Amino-2(1*H*)-pyridinone; *N*-Oxide, *in* A-60250
1*H*-Imidazole-4-carboxylic acid; Me ester, *in* I-60004

C₅H₆N₂O₂S

▷2-Thioxo-4-imidazolidinone; 1-*N*-Ac, *in* T-70188

C₅H₆N₂S

▷3-Amino-2(1*H*)-pyridinethione, A-60247
5,6-Dihydro-4*H*-cyclopenta-1,2,3-thiadiazole, D-70179

C₅H₆N₂S₂

▷5-Methyl-2,4-(1*H*,3*H*)-pyrimidinedithione, M-70127
6-Methyl-2,4(1*H*,3*H*)-pyrimidinedithione, M-70128

C₅H₆N₄

5-Amino-4-cyano-1-methylpyrazole, *in* A-70192
3-Amino-1-methyl-4-pyrazolecarbonitrile, *in* A-70187

C₅H₆N₆O

2,8-Diamino-1,7-dihydro-6*H*-purin-6-one, D-60035

C₅H₆O

2-Cyclopenten-1-one, C-60226
2-Oxabicyclo[3.1.0]hex-3-ene, O-70056

C₅H₆OS

3-Methoxythiophene, *in* T-60212
▷2-Methyl-3-furanthiol, M-70082

C₅H₆O₂

Bicyclo[1.1.0]butane-1-carboxylic acid, B-70094
3-Butynoic acid; Me ester, *in* B-70309
4-Hydroxy-2-cyclopentenone, H-60116
▷5-Methyl-2(3*H*)-furanone, M-70081
▷5-Methyl-2(5*H*)-furanone, M-60083
3-Oxabicyclo[3.1.0]hexan-2-one, O-70055
4-Pentene-2,3-dione, P-60061

C₅H₆O₃

2-Cyclopropyl-2-oxoacetic acid, C-60231
Dimethoxycyclopropenone, *in* D-70290
5-Hydroxymethyl-2(5*H*)-furanone, H-60182
3-Methyl-4-oxo-2-butenoic acid, M-70103
▷4-Oxo-2-pentenoic acid, O-70098

C₅H₆O₄

1,2-Cyclopropanedicarboxylic acid, C-70248
Methyl diformylacetate, *in* D-60190

C₅H₆O₅

3-Oxopentanedioic acid, O-60084

C₅H₆S

2-Thiabicyclo[3.1.0]hex-3-ene, T-70161

C₅H₇Br

1-Bromo-2,3-dimethylcyclopropene, B-60235
1-Bromo-1,3-pentadiene, B-60311

C₅H₇BrN₂

3-Bromo-1,5-dimethylpyrazole, *in* B-70252
4-Bromo-1,3-dimethylpyrazole, *in* B-70251
4-Bromo-1,5-dimethylpyrazole, *in* B-70251
4-Bromo-3,5-dimethylpyrazole, B-70208
5-Bromo-1,3-dimethylpyrazole, *in* B-70252

C₅H₇BrO₂

2-Bromomethyl-2-propenoic acid; Me ester, *in* B-70249

C₅H₇Cl

5-Chloro-1,3-pentadiene, C-60121
5-Chloro-1-pentyne, C-70139

C₅H₇ClOS

Tetrahydro-2-thiophenecarboxylic acid; Chloride, *in* T-70099

C₅H₇ClO₂

1-Chloro-2,3-pentanedione, C-60123

C₅H₇FO

2-Fluorocyclopentanone, F-60022

C₅H₇I

5-Iodo-1-pentyne, I-70054

C₅H₇IN₂

3-Iodo-4,5-dimethylpyrazole, I-70032
4-Iodo-1,3-dimethylpyrazole, *in* I-70049
4-Iodo-1,5-dimethylpyrazole, *in* I-70049

4-Iodo-3,5-dimethylpyrazole, I-70033
5-Iodo-1,3-dimethylpyrazole, *in* I-70050

C₅H₇IO

2-Iodocyclopentanone, I-60043
5-Iodo-4-pentenal, I-70052

C₅H₇IO₂

Tetrahydro-3-iodo-2*H*-pyran-2-one, T-60069

C₅H₇NO

2-Cyclopenten-1-one; Oxime, *in* C-60226
1-Methyl-1,2-dihydro-3*H*-pyrrol-3-one, *in* D-70262
4-Vinyl-2-azetidinone, V-60008

C₅H₇NOS

2-Acetyl-4,5-dihydrothiazole, A-70041

C₅H₇NO₂

4-Aminodihydro-3-methylene-2-(3*H*)furanone, A-70131

C₅H₇NO₂S

3-(Ethylthio)-2-propenoic acid; Nitrile, *S*-dioxide, *in* E-60055
Tetrahydro-2-thiophenecarboxylic acid; Nitrile, 1,1-dioxide, *in* T-70099

C₅H₇NO₃

5-Oxo-2-pyrrolidinecarboxylic acid, O-70102

C₅H₇NO₄

2-Amino-4-hydroxypentanedioic acid; Lactone, *in* A-60195
5-Methyl-2-oxo-4-oxazolidinecarboxylic acid, M-60108
3-Oxopentanedioic acid; Amide, *in* O-60084

C₅H₇NO₅

3-Oxopentanedioic acid; Oxime, *in* O-60084

C₅H₇NS

2-Cyanotetrahydrothiophene, *in* T-70099
3-(Ethylthio)-2-propenoic acid; Nitrile, *in* E-60055

C₅H₇NS₂

2-(Ethylthio)thiazole, *in* T-60200

C₅H₇N₃O

4-Amino-1,7-dihydro-2*H*-1,3-diazepin-2-one, A-70130
2-Amino-6-methyl-4(1*H*)-pyrimidinone, A-60230
4-Hydrazino-2(1*H*)-pyridinone, H-60093
1*H*-Imidazole-4-carboxylic acid; Methylamide, *in* I-60004

C₅H₇N₃O₂

3-Amino-1*H*-pyrazole-4-carboxylic acid; Me ester, *in* A-70187
3-Amino-1*H*-pyrazole-5-carboxylic acid; *N*(1)-Me, *in* A-70188
5-Amino-1*H*-pyrazole-3-carboxylic acid; *N*(1)-Me, *in* A-70191
3-Azido-2-propenoic acid; Et ester, *in* A-70290
1,6-Dimethyl-5-azauracil, *in* M-60132
3,6-Dimethyl-5-azauracil, *in* M-60132
Methyl 1-methyl-1*H*-1,2,3-triazole-5-carboxylate, *in* T-60239
1*H*-1,2,3-Triazole-4-carboxylic acid; 1-Me, Me ester, *in* T-60239
1*H*-1,2,3-Triazole-4-carboxylic acid; Et ester, *in* T-60239
1*H*-1,2,3-Triazole-4-carboxylic acid; 2-Me, Me ester, *in* T-60239

C₅H₇N₃O₂S

2-Pyridinesulfonic acid; Hydrazide, *in* P-70171

C₅H₇N₃O₃

Cyanuric acid; 1,3-Di-Me, *in* C-60179

C₅H₇N₃O₆

1,1-Dinitro-2-propanone; *O*-Acetyloxime, *in* D-70454

C₅H₈

Methylenecyclobutane, M-70066

C₅H₈Br₂

1,1-Dibromo-3-methyl-1-butene, D-60104

C$_5$H$_8$ClNO$_2$
1-Chloro-2,3-pentanedione; 2-Oxime, *in* C-60123

C$_5$H$_8$Cl$_2$O
▷3,3-Biss(hloromethyl)oxetane, B-60149

C$_5$H$_8$FNO$_2$
4-Amino-5-fluoro-2-pentenoic acid, A-60161

C$_5$H$_8$N$_2$
1,4-Dihydro-2-methylpyrimidine, D-70225

C$_5$H$_8$N$_2$O
3,4-Dihydro-2(1*H*)-pyrimidinone; 1-Me, *in* D-60273
5,6-Dihydro-4(1*H*)-pyrimidinone; 3-Me, *in* D-60274

C$_5$H$_8$N$_2$OS
2-Acetyl-4,5-dihydrothiazole; Oxime, *in* A-70041
2-Thioxo-4-imidazolidinone; 1,3-*N*-Di-Me, *in* T-70188

C$_5$H$_8$N$_2$O$_2$
5-Amino-3,4-dihydro-2*H*-pyrrole-2-carboxylic acid, A-70135
▷*N*-(2-Cyanoethyl)glycine, *in* C-60016
5-Oxo-2-pyrrolidinecarboxylic acid; Amide, *in* O-70102
Squamolone, S-60049

C$_5$H$_8$N$_2$O$_7$
3,3-Biss(itratomethyl)oxetane, *in* O-60056

C$_5$H$_8$N$_4$
3-Amino-1,2,4-triazine; *N*,*N*-Di-Me, *in* A-70203

C$_5$H$_8$N$_4$O
6-Amino-2-(methylamino)-4(3*H*)-pyrimidinone, *in* D-70043
5-Amino-1*H*-pyrazole-4-carboxylic acid; *N*(1)-Me, amide, *in* A-70192
2,6-Diamino-4(1*H*)-pyrimidinone; 1-Me, *in* D-70043

C$_5$H$_8$O
3-Methyl-2-butenal, M-70055
(1-Methylethenyl)oxirane, M-70080
▷1-Penten-3-one, P-60062
3-Pentyn-2-ol, P-70041

C$_5$H$_8$OS
Tetrahydro-2*H*-pyran-2-thione, T-70085

C$_5$H$_8$O$_2$
2,6-Dioxaspiro[3.3]heptane, D-60468
2-Ethynyl-1,3-propanediol, E-70054
2-Hydroxycyclopentanone, H-60114
3-Hydroxycyclopentanone, H-60115
2-Hydroxy-2-methylcyclobutanone, H-70168
2-Methyl-3-oxobutanal, M-70102
4-Oxopentanal, O-60083
▷2,4-Pentanedione, P-60059
2-Vinyl-1,3-dioxolane, *in* P-70119

C$_5$H$_8$O$_2$S
2,3-Dihydro-2-methylthiophene; 1,1-Dioxide, *in* D-70227
2,3-Dihydro-5-methylthiophene; 1,1-Dioxide, *in* D-70229
3-(Ethylthio)-2-propenoic acid, E-60055
Isoprene cyclic sulfone, *in* D-70231
α-Isoprene sulfone, *in* D-70228
4-Mercapto-2-butenoic acid; Me ester, *in* M-60018
2-Methyl-3-sulfolene, *in* D-70230
Tetrahydro-2-thiophenecarboxylic acid, T-70099

C$_5$H$_8$O$_3$
2-Ethoxy-2-propenoic acid, E-70037

C$_5$H$_8$O$_3$S
2-Mercaptopropanoic acid; *S*-Ac, *in* M-60027

C$_5$H$_8$O$_4$
4,5-Dihydro-3-hydroxy-5-(hydroxymethyl)-2(3*H*)-furanone, D-70212

C$_5$H$_8$S
2,3-Dihydro-2-methylthiophene, D-70227

2,3-Dihydro-4-methylthiophene, D-70228
2,3-Dihydro-5-methylthiophene, D-70229
2,5-Dihydro-2-methylthiophene, D-70230
2,5-Dihydro-3-methylthiophene, D-70231

C$_5$H$_8$S$_2$
2-Vinyl-1,3-dithiolane, *in* P-70119

C$_5$H$_9$BrO
5-Bromopentanal, B-70263
2-Bromotetrahydro-2*H*-pyran, B-70271

C$_5$H$_9$ClO
5-Chloropentanal, C-70137

C$_5$H$_9$ClOS
Carbonochloridothioic acid; *O*-Butyl, *in* C-60013
Carbonochloridothioic acid; *O*-*tert*-Butyl, *in* C-60013
Carbonochloridothioic acid; *S*-Butyl, *in* C-60013

C$_5$H$_9$ClO$_2$
3-Chloro-2-(methoxymethyl)-1-propene, C-70089

C$_5$H$_9$ClO$_3$
(1-Chloroethyl) ethyl carbonate, C-60067
2-(Chloromethoxy)ethyl acetate, C-70088

C$_5$H$_9$FN$_2$
2-Amino-3-fluoro-3-methylbutanoic acid; Nitrile, *in* A-60159

C$_5$H$_9$IO
2-Iodo-3-methylbutanal, I-60049
2-Iodopentanal, I-60054
5-Iodo-2-pentanone, I-60057
5-Iodo-4-penten-1-ol, I-60058

C$_5$H$_9$N
3,4-Dihydro-5-methyl-2*H*-pyrrole, D-70226
2-Propyn-1-amine; *N*-Di-Me, *in* P-60184
▷1,2,3,6-Tetrahydropyridine, T-70089
2,3,4,5-Tetrahydropyridine, T-70090

C$_5$H$_9$NO
4,5-Dihydro-4,4-dimethyloxazole, D-70198
3,4-Dihydro-5-methoxy-2*H*-pyrrole, *in* P-70188
4-Imino-2-pentanone, *in* P-60059
2-Pyrrolidinecarboxaldehyde, P-70184

C$_5$H$_9$NO$_2$
2-Amino-3-methyl-3-butenoic acid, A-60218
5-(Aminomethyl)dihydro-2(3*H*)-furanone, *in* A-60196
3,5-Dimethyl-2-oxazolidinone, *in* M-60107
2-Ethoxy-2-propenoic acid; Amide, *in* E-70037
2-Hydroxycyclopentanone; Oxime, *in* H-60114

C$_5$H$_9$NO$_2$S
3-Aminotetrahydro-3-thiophenecarboxylic acid, A-70199
3-Thiomorpholinecarboxylic acid, T-70182

C$_5$H$_9$NO$_3$
2-Amino-2-(hydroxymethyl)-3-butenoic acid, A-60187
3,4-Dihydro-3,4-dihydroxy-2-(hydroxymethyl)-2*H*-pyrrole, D-60222
3-Morpholinecarboxylic acid, M-70155

C$_5$H$_9$NO$_4$
2-Amino-3-methylbutanedioic acid, A-70166
3-(Carboxymethylamino)propanoic acid, C-60016
3,4-Dihydroxy-2-pyrrolidinecarboxylic acid, D-70342
3-Nitro-1-propanol; Ac, *in* N-60039
Threonine; *N*-Formyl, *in* T-60217

C$_5$H$_9$NO$_5$
2-Amino-4-hydroxypentanedioic acid, A-60195

C$_5$H$_9$NS
▷2-Isothiocyanato-2-methylpropane, I-70105
2-Methyl-2-thiocyanatopropane, M-70135
2-Propenethioamide; *N*,*N*-Di-Me, *in* P-70120

2-Pyrrolidinethione; *N*-Me, *in* P-70186

C$_5$H$_9$NS$_2$
5,6-Dihydro-2-(methylthio)-4*H*-1,3-thiazine, *in* T-60094
Tetrahydro-2*H*-1,3-thiazine-2-thione; *N*-Me, *in* T-60094

C$_5$H$_9$N$_2$S$_2$
5-*tert*-Butyl-1,2,3,5-dithiadiazolyl, B-70302
5-*tert*-Butyl-1,3,2,4-dithiadiazolyl, B-70303

C$_5$H$_9$N$_3$O
3,5-Dihydro-3,5,5-trimethyl-4*H*-triazol-4-one, D-60303

C$_5$H$_9$N$_3$O$_2$
3-Azido-3-methylbutanoic acid, A-60348

C$_5$H$_9$N$_4$O
Porphyrexide, P-70110

C$_5$H$_{10}$ClN
1-Methylenepyrrolidinium(1+); Chloride, *in* M-70077

C$_5$H$_{10}$ClNO$_2$
2-Amino-3-chlorobutanoic acid; Me ester, *in* A-60103
2-Amino-4-chlorobutanoic acid; Me ester, *in* A-60104

C$_5$H$_{10}$FNO$_2$
2-Amino-3-fluoro-3-methylbutanoic acid, A-60159
2-Amino-3-fluoropentanoic acid, A-60160

C$_5$H$_{10}$N$^⊕$
1-Methylenepyrrolidinium(1+), M-70077

C$_5$H$_{10}$N$_2$
2-Amino-2-methylbutanoic acid; Nitrile, *in* A-70167
2,5-Diazabicyclo[4.1.0]heptane, D-60053

C$_5$H$_{10}$N$_2$O
▷1,3-Dimethylimidazolidinone, *in* I-70005
Tetrahydro-2(1*H*)-pyrimidinone; 1-Me, *in* T-60088
Tetrahydro-4(1*H*)-pyrimidinone; 3-Me, *in* T-60089

C$_5$H$_{10}$N$_2$O$_2$
4-Oxopentanal; Dioxime, *in* O-60083
2,4-Pentanedione; Dioxime, *in* P-60059

C$_5$H$_{10}$N$_2$O$_4$
▷2,4-Dinitropentane, D-70453

C$_5$H$_{10}$O
2,3-Dimethylcyclopropanol, D-70397

C$_5$H$_{10}$OSe
Methaneselenoic acid; *tert*-Butyl ester, *in* M-70035

C$_5$H$_{10}$O$_2$
▷3,3-Dimethoxypropene, *in* P-70119
3,3-Dimethyloxiranemethanol, D-70426
2-Methoxybutanal, *in* H-70112
1-Methoxy-3-buten-2-ol, *in* B-70291

C$_5$H$_{10}$O$_2$S
2-Mercapto-3-methylbutanoic acid, M-60023
2-Mercaptopropanoic acid; Et ester, *in* M-60027
2-Mercaptopropanoic acid; Me ester, *S*-Me ether, *in* M-60027

C$_5$H$_{10}$O$_2$S$_2$
1,2-Dithiepane; 1,1-Dioxide, *in* D-70517

C$_5$H$_{10}$O$_3$
3,3-Oxetanedimethanol, O-60056

C$_5$H$_{10}$O$_4$
2,3-Dihydroxybutanoic acid; Me ester, *in* D-60312
▷Glycerol 1-acetate, G-60029

C$_5$H$_{10}$O$_4$S$_2$
1,2-Dithiepane; 1,1,2,2-Tetraoxide, *in* D-70517

C$_5$H$_{10}$S$_2$
1,1-Cyclopentanedithiol, C-60222
1,1-Cyclopropanedimethanethiol, C-70249

1,2-Dithiepane, D-70517
5-Methyl-1,3-dithiane, M-60067

C₅H₁₁BrO
5-Bromo-2-pentanol, B-60313

C₅H₁₁ClO
tert-Butyl chloromethyl ether, B-70295
1-Chloro-3-pentanol, C-70138

C₅H₁₁IO
1-Iodo-3-pentanol, I-60055
5-Iodo-2-pentanol, I-60056

C₅H₁₁NOS
3-Aminotetrahydro-2*H*-thiopyran; *S*-Oxide, *in*
A-70200

C₅H₁₁NO₂
2-Amino-2-methylbutanoic acid, A-70167

C₅H₁₁NO₂S
3-Aminotetrahydro-2*H*-thiopyran; *S,S*-Dioxide,
in A-70200

C₅H₁₁NO₃
2-Amino-4-hydroxy-2-methylbutanoic acid,
A-60186
4-Amino-5-hydroxypentanoic acid, A-70159
5-Amino-4-hydroxypentanoic acid, A-60196
3-Amino-2-hydroxypropanoic acid; Et ester, *in*
A-70160
Threonine; Me ether, *in* T-60217
Threonine; *N*-Me, *in* T-60217
Threonine; Me ester, *in* T-60217

C₅H₁₁NO₄
1,1-Dimethoxy-3-nitropropane, *in* N-60037

C₅H₁₁NS
3-Aminotetrahydro-2*H*-thiopyran, A-70200
2-Pyrrolidinemethanethiol, P-70185

C₅H₁₁N₃
1-Azido-2,2-dimethylpropane, A-60344

C₅H₁₁N₃O₂
Biuret; 1,3,5-Tri-Me, *in* B-60190

C₅H₁₂N₂O
3,3-Biss(minomethyl)oxetane, B-60140

C₅H₁₂N₂O₂S
S-(2-Aminoethyl)cysteine, A-60152

C₅H₁₂N₂O₃
2,5-Diamino-3-hydroxypentanoic acid, D-70039

C₅H₁₂N₂O₃S
S-(2-Aminoethyl)cysteine; *S*-Oxide, *in* A-60152
Aminoiminomethanesulfonic acid; *N-tert*-Butyl,
in A-60210
Methionine sulfoximine, M-70043

C₅H₁₂N₂O₄S
S-(2-Aminoethyl)cysteine; *S,S*-Dioxide, *in*
A-60152

C₅H₁₂N₂S
▷*N,N′*-Diethylthiourea, D-70155

C₅H₁₂N₂Se
Selenourea; *N*-Tetra-Me, *in* S-70031

C₅H₁₂S₂
2,2-Dimethyl-1,3-propanedithiol, D-70433

C₅H₁₂S₃
2-Mercaptomethyl-2-methyl-1,3-
propanedithiol, M-70030

C₅H₁₂S₄
2,2-Biss(ercaptomethyl)-1,3-propanedithiol,
B-60164

C₅H₁₃NO
3-Amino-3-methyl-2-butanol, A-60217

C₅H₁₃NO₂
1-Amino-3-methyl-2,3-butanediol, A-60216

C₅H₁₄N₂
2,4-Pentanediamine, P-70039

C₅O₂
1,2,3,4-Pentatetraene-1,5-dione, P-70040

C₆BrF₄NO₂
1-Bromo-2,3,4,5-tetrafluoro-6-nitrobenzene,
B-60323

1-Bromo-2,3,4,6-tetrafluoro-5-nitrobenzene,
B-60324
1-Bromo-2,3,5,6-tetrafluoro-4-nitrobenzene,
B-60325

C₆BrF₁₃
1-Bromo-1,1,2,2,3,3,4,4,5,5,6,6,6-trideca-
fluorohexane, B-70274

C₆Br₂F₄
1,2-Dibromo-3,4,5,6-tetrafluorobenzene,
D-60106
1,3-Dibromo-2,4,5,6-tetrafluorobenzene,
D-60107
1,4-Dibromo-2,3,5,6-tetrafluorobenzene,
D-60108

C₆Br₃F₃
1,3,5-Tribromo-2,4,6-trifluorobenzene, T-70214

C₆ClF₁₃
1-Chloro-1,1,2,2,3,3,4,4,5,5,6,6,6-trideca-
fluorohexane, C-70163

C₆Cl₂F₄
1,2-Dichloro-3,4,5,6-tetrafluorobenzene,
D-60151
1,3-Dichloro-2,4,5,6-tetrafluorobenzene,
D-60152
1,4-Dichloro-2,3,5,6-tetrafluorobenzene,
D-60153

C₆Cl₃F₃
1,2,3-Trichloro-4,5,6-trifluorobenzene, T-60256
▷1,3,5-Trichloro-2,4,6-trifluorobenzene, T-60257

C₆F₄I₂
1,2,3,4-Tetrafluoro-5,6-diiodobenzene, T-60044
1,2,3,5-Tetrafluoro-4,6-diiodobenzene, T-60045
▷1,2,4,5-Tetrafluoro-3,6-diiodobenzene, T-60046

C₆F₅IO
Pentafluoroiodosobenzene, P-60034

C₆F₅NO
Pentafluoronitrosobenzene, P-60039

C₆F₅N₃
Azidopentafluorobenzene, A-60350

C₆F₁₂OS₂
2,2,4,4-Tetrakiss(rifluoromethyl)-1,3-
dithietane; 1-Oxide, *in* T-60129

C₆F₁₂O₂S₂
2,2,4,4-Tetrakiss(rifluoromethyl)-1,3-
dithietane; 1,1-Dioxide, *in* T-60129
2,2,4,4-Tetrakiss(rifluoromethyl)-1,3-
dithietane; 1,3-Dioxide, *in* T-60129

C₆F₁₂O₃S₂
2,2,4,4-Tetrakiss(rifluoromethyl)-1,3-
dithietane; 1,1,3-Trioxide, *in* T-60129

C₆F₁₂O₄S₂
2,2,4,4-Tetrakiss(rifluoromethyl)-1,3-
dithietane; 1,1,3,3-Tetraoxide, *in* T-60129

C₆F₁₂S₂
▷2,2,4,4-Tetrakiss(rifluoromethyl)-1,3-
dithietane, T-60129

C₆HBrF₄
1-Bromo-2,3,4,5-tetrafluorobenzene, B-60320
2-Bromo-1,3,4,5-tetrafluorobenzene, B-60321
3-Bromo-1,2,4,5-tetrafluorobenzene, B-60322

C₆HClN₄O₈
▷3-Chloro-1,2,4,5-tetranitrobenzene, C-60135

C₆HCl₄N₃
4,5,6,7-Tetrachlorobenzotriazole, T-60027

C₆HF₄N₃
4,5,6,7-Tetrafluoro-1*H*-benzotriazole, T-70035

C₆H₂BrFN₂O₄
1-Bromo-2-fluoro-3,5-dinitrobenzene, B-60246

C₆H₂Br₂O₂
2,6-Dibromo-1,4-benzoquinone, D-60088

C₆H₂ClN₅O₈
▷2-Chloro-3,4,5,6-tetranitroaniline, C-60134

C₆H₂F₃NO₂
1,2,3-Trifluoro-4-nitrobenzene, T-60299
1,2,3-Trifluoro-5-nitrobenzene, T-60300
▷1,2,4-Trifluoro-5-nitrobenzene, T-60301
1,2,5-Trifluoro-3-nitrobenzene, T-60302
1,2,5-Trifluoro-4-nitrobenzene, T-60303
1,3,5-Trifluoro-2-nitrobenzene, T-60304

C₆H₂O₆
5,6-Dihydroxy-5-cyclohexene-1,2,3,4-tetrone,
D-60315

C₆H₃BrCl₂O
2-Bromo-3,5-dichlorophenol, B-60226
2-Bromo-4,5-dichlorophenol, B-60227
3-Bromo-2,4-dichlorophenol, B-60228
4-Bromo-2,3-dichlorophenol, B-60229
4-Bromo-3,5-dichlorophenol, B-60230

C₆H₃BrINO₂
1-Bromo-2-iodo-3-nitrobenzene, B-70222
2-Bromo-1-iodo-3-nitrobenzene, B-70223

C₆H₃Br₂NO₂
2,6-Dibromo-4-nitroso-1-naphthol, *in* D-60088

C₆H₃Cl₂NO₃
2,3-Dichloro-4-nitrophenol, D-70125

C₆H₃Cl₄NO₂S
2,3,4,5-Tetrachloro-1,1-dihydro-1-[(methoxycarbonyl)imino]thiophene, T-60028

C₆H₃N₃O₃
▷4-Diazo-2-nitrophenol, D-60065
1,3,5-Triazine-2,4,6-tricarboxaldehyde,
T-60237

C₆H₃N₃O₉S
▷2,4,6-Trinitrobenzenesulfonic acid, T-70302

C₆H₃N₅O₈
▷2,3,4,6-Tetranitroaniline, T-60159

C₆H₄
Benzyne, B-70087

C₆H₄BrFO
2-Bromo-3-fluorophenol, B-60247
2-Bromo-4-fluorophenol, B-60248
2-Bromo-6-fluorophenol, B-60249
3-Bromo-4-fluorophenol, B-60250
4-Bromo-2-fluorophenol, B-60251

C₆H₄BrNO₃
2-Bromo-3-nitrophenol, B-70257

C₆H₄BrNS₂
1,3,2-Benzodithiazol-1-ium(1+); Bromide, *in*
B-70033

C₆H₄BrN₃
3-Bromo[1,2,3]triazolo[1,5-*a*]pyridine, B-60329
7-Bromo[1,2,3]triazolo[1,5-*a*]pyridine, B-60330

C₆H₄Br₄N₂
2,3-Biss(ibromomethyl)pyrazine, B-60150
4,5-Biss(ibromomethyl)pyrimidine, B-60153

C₆H₄ClFO
2-Chloro-4-fluorophenol, C-60073
2-Chloro-5-fluorophenol, C-60074
2-Chloro-6-fluorophenol, C-60075
3-Chloro-2-fluorophenol, C-60076
▷3-Chloro-4-fluorophenol, C-60077
3-Chloro-5-fluorophenol, C-60078
4-Chloro-2-fluorophenol, C-60079
4-Chloro-3-fluorophenol, C-60080

C₆H₄ClFS
3-Chloro-4-fluorobenzenethiol, C-60069
4-Chloro-2-fluorobenzenethiol, C-60070

4-Chloro-3-fluorobenzenethiol, C-60071

$C_6H_4ClNO_3$
▷5-Chloro-2-nitrophenol, C-60119

$C_6H_4ClNS_2$
1,3,2-Benzodithiazol-1-ium(1+); Chloride, *in* B-70033

$C_6H_4ClN_3$
4-Chloro-1H-imidazo[4,5-c]pyridine, C-60104
6-Chloro-1H-imidazo[4,5-c]pyridine, C-60105

$C_6H_4Cl_2S$
2,3-Dichlorobenzenethiol, D-60119
2,4-Dichlorobenzenethiol, D-60120
2,5-Dichlorobenzenethiol, D-60121
2,6-Dichlorobenzenethiol, D-60122
3,4-Dichlorobenzenethiol, D-60123
3,5-Dichlorobenzenethiol, D-60124

$C_6H_4FNO_3$
▷5-Fluoro-2-nitrophenol, F-60061

$C_6H_4FN_3O_4$
2-Fluoro-3,5-dinitroaniline, F-60027
2-Fluoro-4,6-dinitroaniline, F-60028
4-Fluoro-2,6-dinitroaniline, F-60029
4-Fluoro-3,5-dinitroaniline, F-60030
5-Fluoro-2,4-dinitroaniline, F-60031
6-Fluoro-3,4-dinitroaniline, F-60032

$C_6H_4F_2S$
2,5-Difluorobenzenethiol, D-60173
2,6-Difluorobenzenethiol, D-60174
3,4-Difluorobenzenethiol, D-60175
3,5-Difluorobenzenethiol, D-60176

$C_6H_4F_3N$
2-(Trifluoromethyl)pyridine, T-60296
3-(Trifluoromethyl)pyridine, T-60297
4-(Trifluoromethyl)pyridine, T-60298

$C_6H_4F_3NO$
2-(Trifluoromethyl)pyridine; 1-Oxide, *in* T-60296
3-(Trifluoromethyl)pyridine; 1-Oxide, *in* T-60297
4-(Trifluoromethyl)pyridine; 1-Oxide, *in* T-60298

$C_6H_4F_3NO_3S$
2-Pyridinyl trifluoromethanesulfonate, P-70173

$C_6H_4NS_2^{\oplus}$
1,3,2-Benzodithiazol-1-ium(1+), B-70033

$C_6H_4N_2$
Biss(ethylene)butanedinitrile, *in* D-70409

$C_6H_4N_2O$
6-Diazo-2,4-cyclohexadien-1-one, D-70057
Furo[2,3-d]pyridazine, F-60091
Furo[3,4-d]pyridazine, F-60092
3-Isocyanatopyridine, I-70065
5H-Pyrrolo[1,2-a]imidazol-5-one, P-70194
5H-Pyrrolo[1,2-c]imidazol-5-one, P-70195
6H-Pyrrolo[1,2-b]pyrazol-6-one, P-70199

$C_6H_4N_2OS$
▷2H-[1,2,4]Oxadiazolo[2,3-a]pyridine-2-thione, O-70062

$C_6H_4N_2O_2$
5-Ethynyl-2,4(1H,3H)-pyrimidinedione, E-60060
Furo[2,3-d]pyrimidin-2(1H)-one, F-60106
4-Hydroxyfuro[2,3-d]pyridazine, H-60146
7-Hydroxyfuro[2,3-d]pyridazine, H-60147
1,3,4-Oxadiazolo[3,2-a]pyridin-2(3H)-one, O-70063
2H-[1,2,4]Oxadiazolo[2,3-a]pyridin-2-one, O-70064
3-Oxopyrazolo[1,2-a]pyrazol-8-ylium-1-olate, O-60092

$C_6H_4N_2O_4$
3,6-Pyridazinedicarboxylic acid, P-70168

$C_6H_4N_2O_6$
4,5-Dinitro-1,2-benzenediol, D-70449
4,6-Dinitro-1,3-benzenediol, D-70450

$C_6H_4N_2S$
2-Isothiocyanatopyridine, I-70106
3-Isothiocyanatopyridine, I-70107
Thiazolo[4,5-b]pyridine, T-70166
Thiazolo[4,5-c]pyridine, T-70167
Thiazolo[5,4-b]pyridine, T-70168
Thiazolo[5,4-c]pyridine, T-70169
Thieno[3,2-d]pyrimidine, T-60205

$C_6H_4N_2S_2$
2,2'-Bithiazole, B-70166
2,4'-Bithiazole, B-70167
2,5'-Bithiazole, B-70168
4,4'-Bithiazole, B-70169
4,5'-Bithiazole, B-70170
5,5'-Bithiazole, B-70171

$C_6H_4N_2S_4$
2,2'-Dithiobisthiazole, *in* T-60200

$C_6H_4N_4O$
Pyrido[3,4-d]-1,2,3-triazin-4(3H)-one, P-60238

$C_6H_4N_4O_2$
2,4(1H,3H)-Pteridinedione, P-60191

$C_6H_4N_4O_3$
2,4,6(1H,3H,5H)-Pteridinetrione, P-70134
2,4,7(1H,3H,8H)-Pteridinetrione, P-70135

$C_6H_4N_4O_6$
▷2,4,6-Trinitroaniline, T-70301

$C_6H_4N_6$
4-Azido-1H-imidazo[4,5-c]pyridine, A-60345

C_6H_4O
1,5-Hexadiyn-3-one, H-60040
3,5-Hexadiyn-2-one, H-60041

C_6H_4OS
Thieno[3,4-b]furan, T-60204

$C_6H_4O_2$
▷1,4-Benzoquinone, B-70044
Tricyclo[3.1.0.0^{2,6}]hexanedione, T-70229

$C_6H_4O_4$
3,6-Dihydroxy-1,2-benzoquinone, D-60309

$C_6H_4O_5$
Aconitic acid; Anhydride, *in* A-70055

$C_6H_4O_6$
3,5-Dihydroxy-4-oxo-4H-pyran-2-carboxylic acid, D-60368

$C_6H_4O_8$
Ethylenetetracarboxylic acid, E-60049

C_6H_4S
2-Ethynylthiophene, E-70055

C_6H_4SSe
Selenolo[3,4-b]thiophene, S-70030

$C_6H_4S_4$
Bi(1,3-dithiol-2-ylidene), B-70119
[1,4]Dithiino[2,3-b]-1,4-dithiin, D-70521

$C_6H_4S_6$
Thieno[3,4-c]thiophene-1,3,4,6-tetrathiol, T-70170

$C_6H_4S_8$
Tetramercaptotetrathiafulvalene, T-70130

$C_6H_4Se_4$
Bi(1,3-diselenol-2-ylidene), B-70116

$C_6H_4Te_4$
Bi(1,3-ditellurol-2-ylidene), B-70117

C_6H_5BrClN
2-(Bromomethyl)-6-chloropyridine, B-70235
3-(Bromomethyl)-2-chloropyridine, B-70236
4-(Bromomethyl)-2-chloropyridine, B-70237
5-Bromomethyl)-2-chloropyridine, B-70238

$C_6H_5BrClNO$
4-(Bromomethyl)-2-chloropyridine; 1-Oxide, *in* B-70237

C_6H_5BrFN
2-Bromomethyl-6-fluoropyridine, B-70243
3-(Bromomethyl)-2-fluoropyridine, B-70244
4-(Bromomethyl)-2-fluoropyridine, B-70245
5-(Bromomethyl)-2-fluoropyridine, B-70246

C_6H_5BrIN
5-Bromo-2-iodoaniline, B-60279

$C_6H_5BrO_3$
2-Bromo-1,3,5-benzenetriol, B-60199
3-Bromo-1,2,4-benzenetriol, B-70180
4-Bromo-1,2,3-benzenetriol, B-60200
5-Bromo-1,2,3-benzenetriol, B-60201
5-Bromo-1,2,4-benzenetriol, B-70181
6-Bromo-1,2,4-benzenetriol, B-70182
2-Bromo-5,6-epoxy-4-hydroxy-2-cyclohexen-1-one, B-60239

$C_6H_5Br_3Se$
Tribromophenylselenium, T-70213

$C_6H_5ClO_6S_2$
4-Chloro-1,3-benzenedisulfonic acid, C-70049

$C_6H_5Cl_2N$
2,3-Dichloro-5-methylpyridine, D-60138
2,5-Dichloro-3-methylpyridine, D-60139

$C_6H_5Cl_2NO$
4-Amino-2,5-dichlorophenol, A-70129

$C_6H_5Cl_3O_2$
2,3-Dihydro-4-(trichloroacetyl)furan, D-70271

$C_6H_5Cl_3Se$
Trichlorophenylselenium, T-70224

$C_6H_5FO_2$
4-Fluoro-1,2-benzenediol, F-60020
4-Fluoro-1,3-benzenediol, F-70018

$C_6H_5F_3O_2$
2,3-Dihydro-4-(trifluoroacetyl)furan, D-70274

$C_6H_5F_4NO_3S$
1-Fluoropyridinium;
Trifluoromethanesulfonate, *in* F-60066

$C_6H_5NO_2$
Amino-1,4-benzoquinone, A-60089

$C_6H_5NO_3$
α-Oxo-1H-pyrrole-3-acetic acid, O-70101

$C_6H_5NO_4$
1H-Pyrrole-3,4-dicarboxylic acid, P-70183

C_6H_5NS
5H-Thieno[2,3-c]pyrrole, T-60206

$C_6H_5N_2OS$
3,4-Dihydro-6-methyl-4-thioxo-2(1H)-pyrimidinone, D-70234

$C_6H_5N_3$
▷1H-Benzotriazole, B-70068
Imidazo[1,2-a]pyrazine, I-60005
Imidazo[1,2-b]pyridazine, I-60007
▷Imidazo[4,5-c]pyridine, I-60008
1H-Pyrazolo[3,4-b]pyridine, P-70155
5H-Pyrrolo[2,3-b]pyrazine, P-60248
[1,2,3]Triazolo[1,5-a]pyridine, T-60241

$C_6H_5N_3O$
4-Amino-5-ethynyl-2(1H)-pyrimidinone, A-60153
1,3-Dihydro-2H-imidazo[4,5-b]pyridin-2-one, D-60249
1,3-Dihydro-2H-imidazo[4,5-c]pyridin-2-one, D-60250
4-Hydroxyimidazo[4,5-b]pyridine, H-60157

$C_6H_5N_5O$
2-Amino-4(1H)-pteridinone, A-60242

$C_6H_5N_5O_2$
2-Amino-4(1H)-pteridinone; 8-Oxide, *in* A-60242
▷Isoxanthopterin, I-60141

C_6H_6
1,2-Hexadien-5-yne, H-70040
1,3-Hexadien-5-yne, H-70041

C_6H_6BrN
2-(Bromomethyl)pyridine, B-70253
3-(Bromomethyl)pyridine, B-70254
4-(Bromomethyl)pyridine, B-70255

C_6H_6BrNO
3-Amino-2-bromophenol, A-70099

2-Bromo-3-hydroxy-6-methylpyridine, B-60267
2-Bromo-3-methoxypyridine, *in* B-60275
3-Bromo-5-methoxypyridine, *in* B-60277

$C_6H_6BrNO_2$

2-Bromo-3-hydroxy-6-methylpyridine; *N*-Oxide, *in* B-60267
2-Bromo-3-hydroxypyridine; Me ether, *N*-Oxide, *in* B-60275

C_6H_6BrNS

3-Bromo-2-(methylthio)pyridine, *in* B-60315

$C_6H_6Br_2N_2O_2$

4,5-Dibromo-1*H*-imidazole-2-carboxylic acid; Et ester, *in* D-60102

$C_6H_6Br_2O_3$

2,6-Dibromo-4,5-dihydroxy-2-cyclohexen-1-one, D-60097

C_6H_6ClNO

2-Chloro-5-hydroxy-6-methylpyridine, C-60099
2-Chloro-3-methoxypyridine, *in* C-60100
3-Chloro-5-methoxypyridine, *in* C-60102

$C_6H_6ClNO_2$

6-Chloro-4-methoxy-2(1*H*)-pyridinone, *in* C-70074

$C_6H_6ClNO_2S$

3-Chloro-2-(methylsulfonyl)pyridine, *in* C-60129

C_6H_6ClNS

3-Chloro-2-(methylthio)pyridine, *in* C-60129

$C_6H_6Cl_2O$

2,4-Dichloro-3,4-dimethyl-2-cyclobuten-1-one, D-60130
4,4-Dichloro-2,3-dimethyl-2-cyclobuten-1-one, D-60131

C_6H_6INO

3-Hydroxy-2-iodo-6-methylpyridine, H-60159

$C_6H_6N_2$

2-Methyl-4-cyanopyrrole, *in* M-60125

$C_6H_6N_2O$

2-Acetylpyrimidine, A-60044
4-Acetylpyrimidine, A-60045
5-Acetylpyrimidine, A-60046
1,4-Dihydrofuro[3,4-*d*]pyridazine, D-60241

$C_6H_6N_2O_2$

5-Acetyl-2(1*H*)-pyrimidinone, A-60048
▷1,4-Benzoquinone; Dioxime, *in* B-70044
Ethyl dicyanoacetate, *in* D-70149
▷1-Hydroxy-2-phenyldiazene 2-oxide, H-70208
3-Methyl-4-nitropyridine, M-70099
▷Urocanic acid, U-70005

$C_6H_6N_2O_3$

5-Acetyl-2,4(1*H*,3*H*)-pyrimidinedione, A-60047
4-Methoxy-3-nitropyridine, *in* N-70063
▷3-Methyl-4-nitropyridine; *N*-Oxide, *in* M-70099
3-Nitro-4(1*H*)-pyridone; *N*-Me, *in* N-70063

$C_6H_6N_2O_4$

1,2,3,6-Tetrahydro-2,6-dioxo-4-pyrimidineacetic acid, T-70060

$C_6H_6N_2S$

1,4-Dihydrothieno[3,4-*d*]pyridazine, D-60289

$C_6H_6N_2S_2$

[Biss(ethylthio)methylene]propanedinitrile, *in* D-60395
1,4-Dithiocyano-2-butene, D-60510

$C_6H_6N_4$

4-Amino-1*H*-imidazo[4,5-*c*]pyridine, A-70161
2,2′-Bi-1*H*-imidazole, B-60107
1,3(5)-Bi-1*H*-pyrazole, B-60119
1*H*-Pyrazolo[3,4-*d*]pyrimidine; 1-Me, *in* P-70156

$C_6H_6N_4O$

3-Amino-1*H*-pyrazole-4-carboxylic acid; Nitrile, *N*(3)-Ac, *in* A-70187
2-Aminopyrrolo[2,3-*d*]pyrimidin-4-one, A-60255
1,3-Dihydro-1-methyl-2*H*-imidazo[4,5-*b*]pyrazin-2-one, *in* I-60006
1*H*-Pyrazolo[3,4-*d*]pyrimidine; 1-Me, 5-oxide, *in* P-70156

$C_6H_6N_4O_2$

3,9-Dihydro-9-methyl-1*H*-purine-2,6-dione, D-60257

$C_6H_6N_4S$

1,3-Dihydro-1-methyl-2*H*-imidazo[4,5-*b*]pyrazine-2-thione, *in* D-60247
1,7-Dihydro-6*H*-purine-6-thione; 1-Me, *in* D-60270
1,7-Dihydro-6*H*-purine-6-thione; 7-Me, *in* D-60270
1,7-Dihydro-6*H*-purine-6-thione; 9-Me, *in* D-60270
1,7-Dihydro-6*H*-purine-6-thione; 3-Me, *in* D-60270
▷1,7-Dihydro-6*H*-purine-6-thione; *S*-Me, *in* D-60270
2-(Methylthio)-1*H*-imidazo[4,5-*b*]pyrazine, *in* D-60247
6-(Methylthio)-5*H*-imidazo[4,5-*c*]pyridazine, *in* D-70215

$C_6H_6N_6$

2,4-Diaminopteridine, D-60042
4,6-Diaminopteridine, D-60043
4,7-Diaminopteridine, D-60044
6,7-Diaminopteridine, D-60045

$C_6H_6N_6O_2$

2,4-Diaminopteridine; 5,8-Dioxide, *in* D-60042
2,6-Diamino-4,7(3*H*,8*H*)-pteridinedione, D-60047
2,7-Diamino-4,6(3*H*,5*H*)-pteridinedione, D-60046

$C_6H_6N_6O_3$

1,3,5-Triazine-2,4,6-tricarboxaldehyde; Trioxime, *in* T-60237

$C_6H_6N_8$

1,5-Diamino-1,5-dihydrobenzo[1,2-*d*:4,5-*d*′]-bistriazole, D-70038

C_6H_6O

2,3-Dihydro-2,3-biss(ethylene)furan, D-60208
4,5-Dihydrocyclobuta[*b*]furan, D-60212

C_6H_6OS

2-Methyl-3-thiophenecarboxaldehyde, M-70139
3-Methyl-2-thiophenecarboxaldehyde, M-70140
4-Methyl-2-thiophenecarboxaldehyde, M-70141
4-Methyl-3-thiophenecarboxaldehyde, M-70142
5-Methyl-2-thiophenecarboxaldehyde, M-70143
5-Methyl-3-thiophenecarboxaldehyde, M-70144

$C_6H_6O_2$

3-Furanacetaldehyde, F-60083
1,5-Hexadiene-3,4-dione, H-60039
3-Hexyne-2,5-dione, H-70090
2-Oxiranylfuran, O-70078
3-Oxo-1-cyclopentenecarboxaldehyde, O-60062

$C_6H_6O_2S$

3-Acetoxythiophene, *in* T-60212
2-(Hydroxyacetyl)thiophene, H-70103

$C_6H_6O_3$

▷5-Hydroxymethyl-2-furancarboxaldehyde, H-70177
2-Propenoic acid; Anhydride, *in* P-70121

$C_6H_6O_4$

3-Acetyl-4-hydroxy-2(5*H*)-furanone, A-60030
1,2,3,4-Benzenetetrol, B-60014
Dimethylenebutanedioic acid, D-70409

$C_6H_6O_5$

(2-Oxopropylidene)propanedioic acid, O-70100

$C_6H_6O_6$

▷Aconitic acid, A-70055

C_6H_6S

2,3-Dihydro-2,3-dimethylenethiophene, D-70195

$C_6H_7BrO_2$

2-Bromo-1,3-cyclohexanedione, B-70201

$C_6H_7ClN_2O_4S_2$

▷Clofenamide, *in* C-70049

C_6H_7ClO

2-Chloro-2-cyclohexen-1-one, C-60050
2-Chloro-3-methyl-2-cyclopenten-1-one, C-70090
5-Chloro-3-methyl-2-cyclopenten-1-one, C-70091
3,5-Hexadienoic acid; Chloride, *in* H-70038

$C_6H_7IN_2O_2$

6-Iodo-1,3-dimethyluracil, *in* I-70057

C_6H_7N

1-Cyano-3-methylbicyclo[1.1.0]butane, *in* M-70053
2-Vinyl-1*H*-pyrrole, V-60013

C_6H_7NO

3-Acetylpyrrole, A-70052
2-Cyanocyclopentanone, *in* O-70084
1,2-Dihydro-3*H*-azepin-3-one, D-70171
5-Methyl-1*H*-pyrrole-2-carboxaldehyde, M-70129
4-(2-Propynyl)-2-azetidinone, P-60185

C_6H_7NOS

2-Acetyl-3-aminothiophene, A-70029
2-Pyridinethiol; *S*-Me, *N*-Oxide, *in* P-60223

$C_6H_7NO_2$

1-Acetyl-1,2-dihydro-3*H*-pyrrol-3-one, *in* D-70262
3-Amino-1,2-benzenediol, A-70089
▷4-Amino-1,2-benzenediol, A-70090
1-Azabicyclo[3.2.0]heptane-2,7-dione, A-60330
1-Azabicyclo[3.2.0]heptane-2,7-dione, A-70273
3-Hydroxy-2(1*H*)-pyridinone; *N*-Me, *in* H-60223
3-Hydroxy-4(1*H*)-pyridinone; *N*-Me, *in* H-60227
4-Hydroxy-2(1*H*)-pyridinone; *N*-Me, *in* H-60224
5-Hydroxy-2(1*H*)-pyridinone; *N*-Me, *in* H-60225
6-Hydroxy-2(1*H*)-pyridinone; *N*-Me, *in* H-60226
3-Methoxy-4(1*H*)-pyridinone, *in* H-60227
6-Methoxy-2(1*H*)-pyridinone, *in* H-60226
5-Methyl-1*H*-pyrrole-2-carboxylic acid, M-60124
5-Methyl-1*H*-pyrrole-3-carboxylic acid, M-60125
1-Nitro-1,3-cyclohexadiene, N-70047

$C_6H_7NO_3$

3-Amino-5-hydroxy-7-oxabicyclo[4.1.0]hept-3-en-2-one, A-60194
5-Hydroxymethyl-2-furancarboxaldehyde; Oxime, *in* H-70177
5-Isoxazolecarboxylic acid; Et ester, *in* I-60142

C_6H_7NS

▷2-Pyridinethiol; *S*-Me, *in* P-60223
▷2(1*H*)-Pyridinethione; *N*-Me, *in* P-60224
4(1*H*)-Pyridinethione; *N*-Me, *in* P-60225

$C_6H_7NS_2$

4-(Methylthio)-2(1*H*)-pyridinethione, *in* P-60219

C_6H_7NSe

2-(Methylseleno)pyridine, *in* P-70170
2(1*H*)-Pyridineselone; *N*-Me, *in* P-70170

$C_6H_7N_3$

2-Amino-5-vinylpyrimidine, A-60269

$C_6H_7N_3OS_2$

5-Amino-2-thiazolecarbothioamide; N^5-Ac, *in* A-60260
5-Amino-4-thiazolecarbothioamide; N^5-Ac, *in* A-60261

$C_6H_7N_5$

6-Amino-7-methylpurine, A-60226
2-(Methylamino)purine, *in* A-60243
8-(Methylamino)purine, *in* A-60244
9-Methyl-9*H*-purin-8-amine, *in* A-60244

$C_6H_7N_5O$

6-Amino-1,3-dihydro-1-methyl-2*H*-purine-2-one, *in* A-60144
2,6-Diamino-1,5-dihydro-4*H*-imidazo[4,5-*c*]-pyridin-4-one, D-60034

$C_6H_7N_7$

2,4,7-Triaminopteridine, T-60229
4,6,7-Triaminopteridine, T-60230

$C_6H_8Cl_2N_2$

3-Chloro-6-methylpyridazine; 2-Methochloride, *in* C-70112

$C_6H_8F_4$

1,1,2,2-Tetrafluoro-3,4-dimethylcyclobutane, T-60047

$C_6H_8F_6O_6S_4$

1,4-Dithioniabicyclo[2.2.0]hexane; Bis-s(rifluoromethanesulfonate), *in* D-70527

$C_6H_8N_2$

▷2-(Aminomethyl)pyridine, A-60227
3-(Aminomethyl)pyridine, A-60228
4-(Aminomethyl)pyridine, A-60229
4,5-Dimethylpyrimidine, D-70436
3,4,5,6-Tetrahydro-4,5-biss(ethylene)-pyridazine, T-60062

$C_6H_8N_2O$

2-Amino-3-hydroxy-6-methylpyridine, A-60192
5-Amino-3-hydroxy-2-methylpyridine, A-60193
2-Amino-4-methoxypyridine, *in* A-60253
2-Amino-6-methoxypyridine, *in* A-60252
3-Amino-2-methoxypyridine, *in* A-60249
3-Amino-4-methoxypyridine, *in* A-60254
4-Amino-2-methoxypyridine, *in* A-60250
5-Amino-2-methoxypyridine, *in* A-60251
3-Amino-1-methyl-2(1*H*)-pyridinone, *in* A-60249
3-Amino-1-methyl-4(1*H*)-pyridinone, *in* A-60254
5-Amino-1-methyl-2(1*H*)-pyridinone, *in* A-60251
6-Amino-1-methyl-2(1*H*)-pyridinone, *in* A-60252
2,3-Dihydro-1,6-dimethyl-3-oxopyridazinium hydroxide, inner salt, *in* M-70124
4,5-Dimethylpyrimidine; 1-Oxide, *in* D-70436
4,5-Dimethylpyrimidine; 3-Oxide, *in* D-70436
7-Hydroxy-6,7-dihydro-5*H*-pyrrolo[1,2-*a*]-imidazole, H-60118
3-Methoxy-6-methylpyridazine, *in* M-70124
6-Methyl-3-pyridazinone; 2-Me, *in* M-70124
5-Methyl-1*H*-pyrrole-2-carboxaldehyde; Oxime, *in* M-70129

$C_6H_8N_2OS$

4,5,6,7-Tetrahydrobenzothiadiazole; 2-Oxide, *in* T-70042

$C_6H_8N_2O_2$

5-Ethyl-2,4(1*H*,3*H*)-pyrimidinedione, E-60054
1*H*-Imidazole-4-carboxylic acid; Et ester, *in* I-60004
6-Methyl-3-pyridazinone; Me ether, 1-Oxide, *in* M-70124
4,5,6,7-Tetrahydroisoxazolo[4,5-*c*]pyridin-3-ol, T-60071

4,5,6,7-Tetrahydroisoxazolo[5,4-*c*]pyridin-3-ol, T-60072
Tetrahydro-1*H*,5*H*-pyrazolo[1,2-*a*]pyrazole-1,5-dione, T-70087

$C_6H_8N_2O_3S$

4,5,6,7-Tetrahydrobenzothiadiazole; 1,1,2-Trioxide, *in* T-70042

$C_6H_8N_2O_4$

1,2-Dinitrocyclohexene, D-70451

$C_6H_8N_2S$

4,6-Dimethyl-2(1*H*)-pyrimidinethione, D-60450
4,5,6,7-Tetrahydrobenzothiadiazole, T-70042

$C_6H_8N_2S_2$

1,4-Diamino-2,3-benzenedithiol, D-60030
2,5-Diamino-1,4-benzenedithiol, D-70035
3,6-Diamino-1,2-benzenedithiol, D-70036

$C_6H_8N_2S_4$

1,4-Diamino-2,3,5,6-benzenetetrathiol, D-60031

$C_6H_8N_4O_4$

5-Nitrohistidine, N-60028
Tetrahydro-1,2,4,5-tetrazine-3,6-dione; 1,5-Di-Ac, *in* T-70097

$C_6H_8N_6O_2$

3,6-Diamino-1,2,4,5-tetrazine; Di-*N*-Ac, *in* D-70045
3,6-Pyridazinedicarboxylic acid; Dihydrazide, *in* P-70168

C_6H_8O

Bicyclo[3.1.0]hexan-2-one, B-60088
3-Cyclohexen-1-one, C-60211
2,3-Dimethyl-2-cyclobuten-1-one, D-60415
1-Hexen-5-yn-3-ol, H-70089
2-Methylene-4-pentenal, M-70074
3-Methylene-4-penten-2-one, M-70075
7-Oxabicyclo[2.2.1]hept-2-ene, O-70054
7-Oxabicyclo[4.1.0]hept-3-ene, O-60047
2-Oxatricyclo[4.1.0^{1,6}.0^{3,5}]heptane, O-60052
Tetrahydro-3,4-biss(ethylene)furan, T-70046

C_6H_8OS

2-Methyl-3-furanthiol; *S*-Me, *in* M-70082
2-Thiabicyclo[2.2.1]heptan-3-one, T-60186
2-Thiabicyclo[2.2.1]hept-5-ene; *exo*-2-Oxide, *in* T-70160
2-Thiabicyclo[2.2.1]hept-5-ene; *endo*-2-Oxide, *in* T-70160
1-(2-Thienyl)ethanol, T-60207
3-Thiopheneethanol, T-60211

$C_6H_8O_2$

Bicyclo[1.1.1]pentane-1-carboxylic acid, B-60101
3,4-Dihydro-2*H*-pyran-5-carboxaldehyde, D-70252
Dihydro-3-vinyl-2(3*H*)-furanone, D-70279
3,3-Dimethyl-1-cyclopropene-1-carboxylic acid, D-60417
3,4-Dimethyl-2(5*H*)-furanone, D-60423
3,4-Hexadienoic acid, H-70037
3,5-Hexadienoic acid, H-70038
▷2-Hydroxy-3-methyl-2-cyclopenten-1-one, H-60174
4-Hydroxy-2-methyl-2-cyclopenten-1-one, H-60175
3-Methylbicyclo[1.1.0]butane-1-carboxylic acid, M-70053
2-Methylenecyclopropaneacetic acid, M-70067
3-Methyl-4-oxo-2-pentenal, M-70104
3-Oxabicyclo[3.2.0]heptan-2-one, O-70053
Tetrahydro-6-methylene-2*H*-pyran-2-one, T-70074

$C_6H_8O_2S$

2-Thiabicyclo[2.2.1]hept-5-ene; 2,2-Dioxide, *in* T-70160
3-Thiatricyclo[2.2.1.0^{2,6}]heptane; *S,S*-Dioxide, *in* T-70165

$C_6H_8O_3$

2,5-Biss(ydroxymethyl)furan, B-60160
3,4-Biss(ydroxymethyl)furan, B-60161
2-Cyclopropyl-2-oxoacetic acid; Me ester, *in* C-60231
Dihydro-4,4-dimethyl-2,3-furandione, D-70196
4,5-Dihydro-3-(methoxymethylene)-2(3*H*)-furanone, D-70222
5,5-Dimethyl-4-methylene-1,2-dioxolan-3-one, D-60428
4-Hydroxy-2,5-dimethyl-3(2*H*)-furanone, H-60120
5-Hydroxymethyl-4-methyl-2(5*H*)-furanone, H-60187
2-Oxocyclopentanecarboxylic acid, O-70084
5-Oxo-3-hexenoic acid, O-70092
4-Oxo-2-pentenoic acid; Me ester, *in* O-70098

$C_6H_8O_3S_2$

2-Thiophenesulfonic acid; Et ester, *in* T-60213

$C_6H_8O_4$

Ethyl diformylacetate, *in* D-60190
2-Hydroxy-2-(hydroxymethyl)-2*H*-pyran-3(6*H*)-one, H-70141

C_6H_8S

2-Thiabicyclo[2.2.1]hept-5-ene, T-70160
3-Thiatricyclo[2.2.1.0^{2,6}]heptane, T-70165

$C_6H_8S_3$

2,5-Thiophenedimethanethiol, T-70184

C_6H_9Br

1-Bromo-3,3-dimethyl-1-butyne, B-70207
2-Bromo-1,5-hexadiene, B-70216
1-Bromo-1-hexyne, B-70218
1-Bromo-2-hexyne, B-70219

$C_6H_9BrO_2$

2-Bromomethyl-2-propenoic acid; Et ester, *in* B-70249

C_6H_9Cl

2-Chloro-1,5-hexadiene, C-70070

C_6H_9FO

2-Fluorocyclohexanone, F-60021

C_6H_9I

1-Iodocyclohexene, I-70026
1-Iodo-1-hexyne, I-70040

C_6H_9IO

2-Iodocyclohexanone, I-60042
6-Iodo-5-hexyn-1-ol, I-70041

C_6H_9NO

3-Amino-2-cyclohexen-1-one, A-70125
3-Cyclohexen-1-one; Oxime, *in* C-60211
1,2-Dihydro-2,2-dimethyl-3*H*-pyrrol-3-one, D-60230
3-Methylbicyclo[1.1.0]butane-1-carboxylic acid; Amide, *in* M-70053
4-(2-Propenyl)-2-azetidinone, *in* A-60172

$C_6H_9NO_2$

3-(1-Aminocyclopropyl)-2-propenoic acid, A-60123
4-Amino-2,5-hexadienoic acid, A-60166
▷2-Pyrrolidinone; *N*-Ac, *in* P-70188
1,2,3,6-Tetrahydro-3-pyridinecarboxylic acid, T-60084
3,4,4-Trimethyl-5(4*H*)-isoxazolone, T-60367

$C_6H_9NO_3$

3-Aminodihydro-2(3*H*)-furanone; *N*-Ac, *in* A-60138
2-Amino-2-(hydroxymethyl)-4-pentynoic acid, A-60191
4-Oxo-2-piperidinecarboxylic acid, O-60089
5-Oxo-2-pyrrolidinecarboxylic acid; Me ester, *in* O-70102
1,2,3,6-Tetrahydro-3-hydroxy-2-pyridinecarboxylic acid, T-60068
1,2,5,6-Tetrahydro-5-hydroxy-3-pyridinecarboxylic acid, T-70065

C$_6$H$_9$NO$_4$

2-Amino-3-methylenepentanedioic acid,
A-60219

C$_6$H$_9$NO$_6$

O-Oxalylhomoserine, *in* A-70156

C$_6$H$_9$NS$_2$

2(3H)-Thiazolethione; S-Isopropyl, *in* T-60200

C$_6$H$_9$N$_3$O

2-Amino-6-methyl-4(1H)-pyrimidinone; 2-N-
Me, *in* A-60230

C$_6$H$_9$N$_3$O$_2$

α-Amino-1H-imidazole-1-propanoic acid,
A-60204
α-Amino-1H-imidazole-2-propanoic acid,
A-60205
3-Amino-1H-pyrazole-4-carboxylic acid; Et
ester, *in* A-70187
3-Amino-1H-pyrazole-5-carboxylic acid; N(1)-
Me, Me ester, *in* A-70188
4-Amino-1H-pyrazole-3-carboxylic acid; Et
ester, *in* A-70189
5-Amino-1H-pyrazole-3-carboxylic acid; N(1)-
Me, Me ester, *in* A-70191
▷Cupferron, *in* H-70208
2,4-Dimethoxy-6-methyl-1,3,5-triazine, *in*
M-60132
4-Methoxy-1,6-dimethyl-1,3,5-triazin-2(1H)-
one, *in* M-60132
1H-1,2,3-Triazole-4-carboxylic acid; 1-Me, Et
ester, *in* T-60239
1H-1,2,3-Triazole-4-carboxylic acid; 2-Me, Et
ˈester, *in* T-60239
1,3,6-Trimethyl-5-azauracil, *in* M-60132

C$_6$H$_9$N$_3$O$_3$

Aconitic acid; Triamide, *in* A-70055
Hexahydro-1,3,5-triazine; 1,3,5-Triformyl, *in*
H-60060
Trimethyl isocyanurate, *in* C-60179

C$_6$H$_{10}$Br$_4$

2,3-Biss(romomethyl)-1,4-dibromobutane,
B-60144

C$_6$H$_{10}$ClF$_2$NO$_2$

2-Amino-4-chloro-4,4-difluorobutanoic acid; Et
ester, *in* A-60107

C$_6$H$_{10}$ClNO$_2$

1-Chloro-1-nitrocyclohexane, C-70131

C$_6$H$_{10}$ClNO$_3$

2-Amino-5-chloro-6-hydroxy-4-hexenoic acid,
A-60110

C$_6$H$_{10}$Cl$_3$N

1,1,2-Trichloro-2-(diethylamino)ethylene,
T-70222

C$_6$H$_{10}$Cl$_4$

2,3-Biss(hloromethyl)-1,4-dichlorobutane,
B-60148

C$_6$H$_{10}$N$_2$

2-Amino-5-hexenoic acid; Nitrile, *in* A-60169
3-Amino-2-hexenoic acid; Nitrile, *in* A-60170

C$_6$H$_{10}$N$_2$O$_2$

5-Amino-3,4-dihydro-2H-pyrrole-2-carboxylic
acid; Me ester, *in* A-70135
2,5-Dihydro-4,6-dimethoxypyrimidine, *in*
D-70256
Dihydro-4,4-dimethyl-2,3-furandione;
Hydrazone, *in* D-70196
1-Dimethylamino-4-nitro-1,3-butadiene,
D-70367

C$_6$H$_{10}$N$_2$O$_3$

2-Amino-3-methylenepentanedioic acid; Amide,
in A-60219
3-Piperidinecarboxylic acid; N-Nitroso, *in*
P-70102

C$_6$H$_{10}$N$_2$O$_4$

2,5-Piperazinedicarboxylic acid, P-60157

C$_6$H$_{10}$N$_4$

1,2,3,5-Tetraaminobenzene, T-60012

1,2,4,5-Tetraaminobenzene, T-60013

C$_6$H$_{10}$N$_4$O$_2$

Tetrahydroimidazo[4,5-d]imidazole-
2,5(1H,3H)-dione; 1,4-Di-Me, *in* T-70066 ʼ
Tetrahydroimidazo[4,5-d]imidazole-
2,5(1H,3H)-dione; 1,6-Di-Me, *in* T-70066

C$_6$H$_{10}$N$_4$S

1-(4,5-Dihydro-1H-imidazol-2-yl)-2-
imidazolidinethione, D-60246

C$_6$H$_{10}$O

Bicyclo[2.1.1]hexan-2-ol, B-70100
3,5-Hexadien-1-ol, H-70039
▷2-Hexenal, H-60076
▷2-Methyl-2-pentenal, M-70106

C$_6$H$_{10}$OS

2-Mercaptocyclohexanone, M-70025
3-Mercaptocyclohexanone, M-70026
4-Mercaptocyclohexanone, M-70027
2-(Methylthio)cyclopentanone, M-70136
2-Oxepanethione, O-70077
4-Thioxo-2-hexanone, T-60215

C$_6$H$_{10}$O$_2$

2,2-Dimethyl-4-methylene-1,3-dioxolane,
D-70419
▷1,5-Hexadiene-3,4-diol, H-70036
2-Methoxycyclopentanone, *in* H-60114
4-Methoxy-3-penten-2-one, *in* P-60059
4-Oxohexanal, O-60073
5-Oxohexanal, O-70091
Tetrahydro-4-methyl-2H-pyran-2-one, *in*
H-70182

C$_6$H$_{10}$O$_2$S

3-(Acetylthio)cyclobutanol, *in* M-60019
2,5-Dihydro-2,5-dimethylthiophene; 1,1-
Dioxide, *in* D-60233
O-Ethyl 3-oxobutanethioate, *in* O-70080
S-Ethyl 3-oxobutanethioate, *in* O-70080
3-Mercaptocyclopentanecarboxylic acid,
M-60021

C$_6$H$_{10}$O$_3$

3-Butene-1,2-diol; Ac, *in* B-70291
2,2-Dimethyl-1,3-dioxolane-4-carboxaldehyde,
D-70400
Epiverrucarinolactone, *in* D-60353
2-Ethoxy-2-propenoic acid; Me ester, *in*
E-70037
6-(Hydroxymethyl)tetrahydro-2H-pyran-2-one,
in D-60329
3-Oxopropanoic acid; Isopropyl ester, *in*
O-60090
Tetrahydro-3-hydroxy-4-methyl-2H-pyran-2-
one, *in* D-60353
3,4,5,6-Tetrahydro-4-hydroxy-6-methyl-2H-
pyran-2-one, T-60067

C$_6$H$_{10}$O$_4$

4-Hydroxy-3,3-dimethyl-2-oxobutanoic acid,
H-60124
1,3,7,9-Tetraoxaspiro[4,5]decane, T-60163
1,4,6,10-Tetraoxaspiro[4,5]decane, T-60164

C$_6$H$_{10}$O$_5$

▷Diethyl dicarbonate, *in* D-70111
Methyl hydrogen 3-hydroxyglutarate, M-70089

C$_6$H$_{10}$O$_6$

Dimethyl tartrate, *in* T-70005

C$_6$H$_{10}$S

2,3-Dihydro-2,2-dimethylthiophene, D-60232
2,5-Dihydro-2,5-dimethylthiophene, D-60233

C$_6$H$_{10}$S$_4$

2,2′-Bi-1,3-dithiolane, B-60103
Hexahydro-1,4-dithiino[2,3-b]-1,4-dithiin,
H-60051

C$_6$H$_{11}$Br

5-Bromo-1-hexene, B-60263
6-Bromo-1-hexene, B-60264
(Bromomethyl)cyclopentane, B-70241
5-Bromo-3-methyl-1-pentene, B-60299

C$_6$H$_{11}$BrO

5-Bromo-2-hexanone, B-60262
1-Bromo-4-methyl-2-pentanone, B-60298

C$_6$H$_{11}$BrO$_2$

2-Bromo-4-methylpentanoic acid, B-70248

C$_6$H$_{11}$Br$_2$NO$_2$

6-Amino-2,2-dibromohexanoic acid, A-60124

C$_6$H$_{11}$Cl

5-Chloro-3-methyl-1-pentene, C-60117

C$_6$H$_{11}$ClFNO$_2$

2-Amino-4-chloro-4-fluorobutanoic acid; Et
ester, *in* A-60108

C$_6$H$_{11}$Cl$_2$NO$_2$

6-Amino-2,2-dichlorohexanoic acid, A-60126

C$_6$H$_{11}$F$_2$NO$_2$

2-Amino-3,3-difluorobutanoic acid; Et ester, *in*
A-60133

C$_6$H$_{11}$I

1-Iodo-1-hexene, I-70037
6-Iodo-1-hexene, I-70038
(Iodomethyl)cyclopentane, I-70047

C$_6$H$_{11}$IO

1-Iodo-3,3-dimethyl-2-butanone, I-60044
6-Iodo-5-hexen-1-ol, I-70039

C$_6$H$_{11}$N

2-Aminobicyclo[2.1.1]hexane, A-70095
3,4-Dihydro-2,2-dimethyl-2H-pyrrole, D-60229
3,4-Dihydro-2,2-dimethyl-2H-pyrrole, D-70200

C$_6$H$_{11}$NO

3,4-Dihydro-2,2-dimethyl-2H-pyrrole; N-Oxide,
in D-60229
3,4-Dihydro-2,2-dimethyl-2H-pyrrole; 1-Oxide,
in D-70200
5-Ethoxy-3,4-dihydro-2H-pyrrole, *in* P-70188
2-Oxa-3-azabicyclo[2.2.2]octane, O-70051
1-Oxa-2-azaspiro[2,5]octane, O-70052

C$_6$H$_{11}$NOS

3-Mercaptocyclopentanecarboxylic acid;
Amide, *in* M-60021

C$_6$H$_{11}$NO$_2$

2-Amino-2-ethyl-3-butenoic acid, A-60150
1-Amino-2-ethylcyclopropanecarboxylic acid,
A-60151
2-Amino-2-hexenoic acid, A-60167
2-Amino-3-hexenoic acid, A-60168
2-Amino-5-hexenoic acid, A-60169
3-Amino-2-hexenoic acid, A-60170
3-Amino-4-hexenoic acid, A-60171
3-Amino-5-hexenoic acid, A-60172
6-Amino-2-hexenoic acid, A-60173
2-Amino-3-methyl-3-butenoic acid; Me ester, *in*
A-60218
2-Methyl-2-pyrrolidinecarboxylic acid,
M-60126
1-Nitro-3-hexene, N-70051
3-Piperidinecarboxylic acid, P-70102

C$_6$H$_{11}$NO$_3$

2-Amino-1-hydroxy-1-cyclobutaneacetic acid,
A-60184
1-Amino-3-(hydroxymethyl)-1-cyclo-
butanecarboxylic acid, A-70157
2-Amino-2-(hydroxymethyl)-4-pentenoic acid,
A-60190

C$_6$H$_{11}$NO$_4$

2-Amino-3,3-dimethylbutanedioic acid,
A-60147
2-Amino-3-ethylbutanedioic acid, A-70138
2-Amino-4-methylpentanedioic acid, A-60223
Threonine; N-Ac, *in* T-60217

C$_6$H$_{11}$NS$_2$

5,6-Dihydro-2-(ethylthio)-4H-1,3-thiazine, *in*
T-60094
Tetrahydro-2H-1,3-thiazine-2-thione; N-Et, *in*
T-60094

$C_6H_{11}N_3O_2$

Dihydro-1,3,5-triazine-2,4(1H,3H)-dione; 1,3,5-Tri-Me, *in* D-60297

$C_6H_{12}BrCl$

1-Bromo-6-chlorohexane, B-70200

$C_6H_{12}BrN$

2-(Bromomethyl)cyclopentanamine, B-70240

$C_6H_{12}BrNO_2$

6-Amino-2-bromohexanoic acid, A-60100

$C_6H_{12}Br_2$

1,3-Dibromo-2,3-dimethylbutane, D-70089

$C_6H_{12}ClNO_2$

6-Amino-2-chlorohexanoic acid, A-60109

$C_6H_{12}I_2O_2$

1,2-Bis(2-iodoethoxy)ethane, B-60163

$C_6H_{12}NO_2^{\oplus}$

4-Azoniaspiro[3.3]heptane-2,6-diol, A-60354

$C_6H_{12}N_2$

▷1,4-Diazabicyclo[2.2.2]octane, D-70051
2-Dimethylamino-3,3-dimethylazirine, D-60402
1,3-Dimethyl-2-methyleneimidazolidine, D-70420
5-Hexyne-1,4-diamine, H-60080
Octahydropyrrolo[3,4-c]pyrrole, O-70017
Tetrahydro-1H,5H-pyrazolo[1,2-a]pyrazole, T-70086

$C_6H_{12}N_2O$

3-Amino-2-hexenoic acid; Amide, *in* A-60170
3-Piperidinecarboxamide, *in* P-70102
▷3,4,5,6-Tetrahydro-2(1H)-pyrimidinone, *in* T-60088

$C_6H_{12}N_2O_3S$

N-Alanylcysteine, A-60069

$C_6H_{12}N_2O_4$

3,4,8,9-Tetraoxa-1,6-diazabicyclo[4.4.2]-dodecane, T-70144

$C_6H_{12}N_4$

1,2,3,4,5,6,7,8-Octahydropyridazino[4,5-d]-pyridazine, O-60018

$C_6H_{12}N_4O_2$

2,5-Piperazinedicarboxylic acid; Diamide, *in* P-60157

$C_6H_{12}N_6$

▷Benzenehexamine, B-60012

$C_6H_{12}O$

2,2-Dimethylcyclopropanemethanol, D-70396
4-Hexen-3-ol, H-60077
1-Methoxy-2,3-dimethylcyclopropane, *in* D-70397
3-Methyl-2-pentanone, M-70105
▷4-Methyl-2-pentanone, M-60109
4-Methyl-1-penten-3-ol, M-70107
Tetrahydro-2-methylpyran, T-60075

$C_6H_{12}OS$

2-Mercaptocyclohexanol, M-70024
3,3,4,4-Tetramethyl-1,2-oxathietane, T-60149

$C_6H_{12}O_2$

β,β-Dimethyloxiraneethanol, D-60431

$C_6H_{12}O_2S$

3-Mercaptobutanoic acid; Et ester, *in* M-70023

$C_6H_{12}O_2S_2$

2,5-Dimethyl-1,4-dithiane-2,5-diol, D-60421

$C_6H_{12}O_3$

▷2-Acetoxy-1-methoxypropane, *in* P-70118
Diethoxyacetaldehyde, D-70153
3-Hydroxy-2-methylpentanoic acid, H-60192
5-Hydroxy-3-methylpentanoic acid, H-70182

$C_6H_{12}O_4$

3,4-Dihydroxybutanoic acid; Et ester, *in* D-60313
5,6-Dihydroxyhexanoic acid, D-60329

2,5-Dihydroxy-3-methylpentanoic acid, D-60353

$C_6H_{12}S_2$

1,1-Cyclobutanedimethanethiol, C-70213
1,2-Cyclobutanedimethanethiol, C-70214
▷1,1-Cyclohexanedithiol, C-60209
1,2-Cyclohexanedithiol, C-70225

$C_6H_{13}N$

2,5-Dimethylpyrrolidine, D-70437
4-Hexen-1-amine, H-70088

$C_6H_{13}NO$

▷N-Cyclohexylhydroxylamine, C-70228

$C_6H_{13}NO_2$

2-Amino-2,3-dimethylbutanoic acid, A-60148
2-Amino-3,3-dimethylbutanoic acid, A-70136

$C_6H_{13}NO_2S$

2-Amino-5-(methylthio)pentanoic acid, A-60231

$C_6H_{13}NO_3$

2-Amino-4-ethoxybutanoic acid, *in* A-70156

$C_6H_{13}NS$

2-[(Methylthio)methyl]pyrrolidine, *in* P-70185

$C_6H_{13}N_3O_2$

Biuret; 1,1,3,5-Tetra-Me, *in* B-60190

$C_6H_{13}N_3O_3$

Citrulline, C-60151

$C_6H_{14}N_2$

Hexahydropyrimidine; 1,3-Di-Me, *in* H-60056

$C_6H_{14}N_2O$

2-Amino-3,3-dimethylbutanoic acid; Amide, *in* A-70136

$C_6H_{14}N_4O_3$

2-Amino-5-guanidino-3-hydroxypentanoic acid, A-70149

$C_6H_{14}O_2S_2$

S,S'-Bis(2-hydroxyethyl)-1,2-ethanedithiol, B-70145

$C_6H_{14}O_3$

1,1-Dimethoxy-2-butanol, *in* H-70112
3,3'-Oxybis-1-propanol, O-60097

$C_6H_{14}O_4$

2,3-Biss(ydroxymethyl)-1,4-butanediol, B-60159

$C_6H_{14}O_6S_2$

S,S'-Bis(2-hydroxyethyl)-1,2-ethanedithiol; S-Tetroxide, *in* B-70145

$C_6H_{14}S_4$

S,S'-Bis(2-mercaptoethyl)-1,2-ethanedithiol, B-70150

$C_6H_{15}N_3$

▷Hexahydro-1,3,5-trimethyl-1,3,5-triazine, *in* H-60060

$C_6H_{15}O_{15}P_3$

myo-Inositol-1,3,4-triphosphate, I-70020
myo-Inositol-2,4,5-triphosphate, I-70021

$C_6H_{16}N_2O_2$

1,2-Bis(2-aminoethoxy)ethane, B-60139

$C_6H_{16}O_{18}P_4$

myo-Inositol-1,3,4,5-tetraphosphate, I-70019

$C_6H_{18}N_4$

1,2-Dihydrazinoethane; $N^{\beta},N^{\beta},N^{\beta'},N^{\beta'}$-Tetra-Me, *in* D-60192

C_6N_4

▷Tetracyanoethylene, T-60031

C_6O_6

Cyclohexanehexone, C-60210

C_7F_5N

Pentafluoroisocyanobenzene, P-60036

C_7F_5NS

Pentafluoroisothiocyanatobenzene, P-60037

C_7HF_4N

2,3,4,5-Tetrafluorobenzonitrile, *in* T-70032
2,3,5,6-Tetrafluorobenzonitrile, *in* T-70034

C_7HF_5

Ethynylpentafluorobenzene, E-70053

$C_7HF_5O_2$

Pentafluorophenyl formate, P-70025

$C_7H_2ClN_3O_7$

▷2,4,6-Trinitrobenzoic acid; Chloride, *in* T-70303

$C_7H_2F_3N$

2,4,5-Trifluorobenzonitrile, *in* T-70240

$C_7H_2F_4O$

2,3,4,5-Tetrafluorobenzaldehyde, T-70026
2,3,4,6-Tetrafluorobenzaldehyde, T-70027
2,3,5,6-Tetrafluorobenzaldehyde, T-70028

$C_7H_2F_4O_2$

2,3,4,5-Tetrafluorobenzoic acid, T-70032
2,3,4,6-Tetrafluorobenzoic acid, T-70033
2,3,5,6-Tetrafluorobenzoic acid, T-70034

$C_7H_3BrF_3NO_2$

1-Bromo-2-nitro-3-(trifluoromethyl)benzene, B-60306
1-Bromo-4-nitro-2-(trifluoromethyl)benzene, B-60307
2-Bromo-1-nitro-3-(trifluoromethyl)benzene, B-60308

$C_7H_3Br_4Cl$

1,2,4,5-Tetrabromo-3-chloro-6-methylbenzene, T-60022

$C_7H_3ClFNO_4$

2-Chloro-4-fluoro-5-nitrobenzoic acid, C-60072

$C_7H_3ClF_3N_3$

6-Chloro-2-(trifluoromethyl)-1H-imidazo[4,5-b]pyridine, C-60139

$C_7H_3ClF_3N_3O$

6-Chloro-2-(trifluoromethyl)-1H-imidazo[4,5-b]pyridine; 4-Oxide, *in* C-60139

$C_7H_3ClN_2O_5$

2,6-Dinitrobenzoic acid; Chloride, *in* D-60457
▷3,5-Dinitrobenzoic acid; Chloride, *in* D-60458

$C_7H_3Cl_4N_3$

4,5,6,7-Tetrachlorobenzotriazole; 1-Me, *in* T-60027
4,5,6,7-Tetrachlorobenzotriazole; 2-Me, *in* T-60027

C_7H_3FIN

2-Cyano-1-fluoro-3-iodobenzene, *in* F-70020

$C_7H_3F_3N_2O_4$

1,3-Dinitro-5-(trifluoromethyl)benzene, D-70455
2,4-Dinitro-1-(trifluoromethyl)benzene, D-70456

$C_7H_3F_3O_2$

2,3,5-Trifluorobenzoic acid, T-70238
2,3,6-Trifluorobenzoic acid, T-70239
2,4,5-Trifluorobenzoic acid, T-70240

$C_7H_3F_5$

Pentafluoromethylbenzene, P-60038

$C_7H_3F_6N$

2,3-Biss(rifluoromethyl)pyridine, B-60181
2,4-Biss(rifluoromethyl)pyridine, B-60182
2,5-Biss(rifluoromethyl)pyridine, B-60183
2,6-Biss(rifluoromethyl)pyridine, B-60184
3,4-Biss(rifluoromethyl)pyridine, B-60185
3,5-Biss(rifluoromethyl)pyridine, B-60186

$C_7H_3F_6NO$

2,4-Biss(rifluoromethyl)pyridine; 1-Oxide, *in* B-60182

2,5-Biss(rifluoromethyl)pyridine; 1-Oxide, *in* B-60183

2,6-Biss(rifluoromethyl)pyridine; 1-Oxide, *in* B-60184

C₇H₃N₃O₄

▷1-Cyano-3,5-dinitrobenzene, *in* D-60458

2-Cyano-1,3-dinitrobenzene, *in* D-60457

C₇H₃N₃O₈

▷2,4,6-Trinitrobenzoic acid, T-70303

C₇H₄BrFO₂

2-Bromo-4-fluorobenzoic acid, B-60240

2-Bromo-5-fluorobenzoic acid, B-60241

2-Bromo-6-fluorobenzoic acid, B-60242

3-Bromo-4-fluorobenzoic acid, B-60243

C₇H₄ClFO

3-Chloro-4-fluorobenzaldehyde, C-70067

C₇H₄ClF₃O₂S

2-(Trifluoromethyl)benzenesulfonic acid; Chloride, *in* T-70242

▷3-(Trifluoromethyl)benzenesulfonic acid; Chloride, *in* T-60283

4-(Trifluoromethyl)benzenesulfonic acid; Chloride, *in* T-70243

C₇H₄ClNO

▷2-Chlorobenzoxazole, C-60048

2-Chloro-4-cyanophenol, *in* C-60091

C₇H₄ClNS

4-Chloro-1,2-benzisothiazole, C-70050

5-Chloro-1,2-benzisothiazole, C-70051

6-Chloro-1,2-benzisothiazole, C-70052

7-Chloro-1,2-benzisothiazole, C-70053

2-Chloro-6-cyanobenzenethiol, *in* C-70083

C₇H₄Cl₂O₂

2,2-Dichloro-1,3-benzodioxole, D-70113

2,5-Dichloro-4-hydroxybenzaldehyde, D-70120

2,6-Dichloro-4-hydroxybenzaldehyde, D-70121

C₇H₄Cl₂O₃

2,3-Dichloro-4-hydroxybenzoic acid, D-60133

C₇H₄FIO₂

2-Fluoro-6-iodobenzoic acid, F-70020

3-Fluoro-2-iodobenzoic acid, F-70021

3-Fluoro-4-iodobenzoic acid, F-70022

4-Fluoro-2-iodobenzoic acid, F-70023

4-Fluoro-3-iodobenzoic acid, F-70024

5-Fluoro-2-iodobenzoic acid, F-70025

C₇H₄FNO

▷1-Fluoro-2-isocyanatobenzene, F-60047

▷1-Fluoro-3-isocyanatobenzene, F-60048

▷1-Fluoro-4-isocyanatobenzene, F-60049

C₇H₄FNO₄

3-Fluoro-4-nitrobenzoic acid, F-70033

C₇H₄FNS

4-Cyano-3-fluorobenzenethiol, *in* F-70029

C₇H₄F₂

7,7-Difluorobicyclo[4.1.0]hepta-1,3,5-triene, D-60177

C₇H₄F₃NO₂

▷1-Nitro-2-(trifluoromethyl)benzene, N-70064

▷1-Nitro-3-(trifluoromethyl)benzene, N-70065

▷1-Nitro-4-(trifluoromethyl)benzene, N-70066

C₇H₄F₃NO₃

2-Nitro-3-(trifluoromethyl)phenol, N-60047

▷2-Nitro-4-(trifluoromethyl)phenol, N-60048

2-Nitro-5-(trifluoromethyl)phenol, N-60049

2-Nitro-6-(trifluoromethyl)phenol, N-60050

3-Nitro-4-(trifluoromethyl)phenol, N-60051

3-Nitro-5-(trifluoromethyl)phenol, N-60052

4-Nitro-2-(trifluoromethyl)phenol, N-60053

▷4-Nitro-3-(trifluoromethyl)phenol, N-60054

C₇H₄INO₃

5-Iodo-2-nitrobenzaldehyde, I-60052

C₇H₄N₂O₂S

5-Nitro-1,2-benzisothiazole, N-70041

7-Nitro-1,2-benzisothiazole, N-70042

C₇H₄N₂O₃

6-Diazo-5-oxo-1,3-cyclohexadiene-1-carboxylic acid, D-70063

C₇H₄N₂O₆

2,6-Dinitrobenzoic acid, D-60457

3,5-Dinitrobenzoic acid, D-60458

C₇H₄N₄

2-Amino-3,5-dicyanopyridine, *in* A-70193

C₇H₄N₄O₇

2,4,6-Trinitrobenzoic acid; Amide, *in* T-70303

C₇H₄N₆O₆

3,5-Diamino-2,4,6-trinitrobenzoic acid; Nitrile, *in* D-70047

C₇H₄O

2,6-Cycloheptadien-4-yn-1-one, C-60196

C₇H₄OS

Benzothiet-2-one, B-70065

C₇H₄OSSe

1,3-Benzoxathiole-2-selone, B-70070

C₇H₄O₂S

3-(2-Thienyl)-2-propynoic acid, T-70173

C₇H₄O₃S

4H-Thieno[3,4-c]pyran-4,6(7H)-dione, *in* C-60018

5H-Thieno[2,3-c]pyran-5,7(4H)dione, *in* C-60017

C₇H₄O₄

1,2-Benzoquinone-3-carboxylic acid, B-70045

1,2-Benzoquinone-4-carboxylic acid, B-70046

C₇H₄SSe₂

1,3-Benzodiselenole-2-thione, B-70032

C₇H₄S₂Se

3H-1,2-Benzodithiole-3-selone, B-70034

C₇H₅Br

2-Bromobenzocyclopropene, B-60202

3-Bromobenzocyclopropene, B-60203

C₇H₅BrCl₂O

1-Bromo-2,3-dichloro-4-methoxybenzene, *in* B-60229

C₇H₅BrO₂S

2-Bromo-4-mercaptobenzoic acid, B-70226

2-Bromo-5-mercaptobenzoic acid, B-70227

3-Bromo-4-mercaptobenzoic acid, B-70228

C₇H₅Br₃OS

[(Tribromomethyl)sulfinyl]benzene, *in* T-70212

C₇H₅Br₃O₂S

[(Tribromomethyl)sulfonyl]benzene, *in* T-70212

C₇H₅Br₃S

[(Tribromomethyl)thio]benzene, T-70212

C₇H₅Br₄N

2,3-Biss(ibromomethyl)pyridine, B-60151

3,4-Biss(ibromomethyl)pyridine, B-60152

C₇H₅ClN₂

▷3-Chloro-3-phenyl-3H-diazirine, C-70140

C₇H₅ClOS

Carbonochloridothioic acid; *S*-Ph, *in* C-60013

C₇H₅ClO₂

3-Chloro-4-hydroxybenzaldehyde, C-60089

4-Chloro-3-hydroxybenzaldehyde, C-60090

2-Hydroxybenzoic acid; Chloride, *in* H-70108

C₇H₅ClO₂S

2-Chloro-6-mercaptobenzoic acid, C-70082

3-Chloro-2-mercaptobenzoic acid, C-70083

4-Chloro-2-mercaptobenzoic acid, C-70084

5-Chloro-2-mercaptobenzoic acid, C-70085

C₇H₅ClO₃

2-Chloro-3,6-dihydroxybenzaldehyde, C-60056

3-Chloro-2,5-dihydroxybenzaldehyde, C-60057

3-Chloro-2,6-dihydroxybenzaldehyde, C-60058

5-Chloro-2,3-dihydroxybenzaldehyde, C-70065

3-Chloro-4-hydroxybenzoic acid, C-60091

3-Chloro-5-hydroxybenzoic acid, C-60092

4-Chloro-3-hydroxybenzoic acid, C-60093

C₇H₅Cl₂NO₃

2,3-Dichloro-1-methoxy-4-nitrobenzene, *in* D-70125

C₇H₅FN₂

2-Cyano-3-fluoroaniline, *in* A-70143

3-Cyano-4-fluoroaniline, *in* A-70147

4-Cyano-2-fluoroaniline, *in* A-70146

▷3-Fluoro-3-phenyl-3H-diazirine, F-70036

C₇H₅FN₂O₄

5-Fluoro-2-methyl-1,3-dinitrobenzene, F-60051

C₇H₅FO₂

2-Fluoro-3-hydroxybenzaldehyde, F-60035

2-Fluoro-5-hydroxybenzaldehyde, F-60036

4-Fluoro-3-hydroxybenzaldehyde, F-60037

C₇H₅FO₂S

2-Fluoro-4-mercaptobenzoic acid, F-70029

2-Fluoro-6-mercaptobenzoic acid, F-70030

4-Fluoro-2-mercaptobenzoic acid, F-70031

C₇H₅FO₃

2-Fluoro-5-hydroxybenzoic acid, F-60038

2-Fluoro-6-hydroxybenzoic acid, F-60039

3-Fluoro-2-hydroxybenzoic acid, F-60040

3-Fluoro-4-hydroxybenzoic acid, F-60041

4-Fluoro-2-hydroxybenzoic acid, F-60042

4-Fluoro-3-hydroxybenzoic acid, F-60043

5-Fluoro-2-hydroxybenzoic acid, F-60044

C₇H₅FO₄

3-Fluoro-2,6-dihydroxybenzoic acid, F-60023

4-Fluoro-3,5-dihydroxybenzoic acid, F-60024

5-Fluoro-2,3-dihydroxybenzoic acid, F-60025

C₇H₅F₃O₃S

2-(Trifluoromethyl)benzenesulfonic acid, T-70242

3-(Trifluoromethyl)benzenesulfonic acid, T-60283

4-(Trifluoromethyl)benzenesulfonic acid, T-70243

C₇H₅F₃S

2-(Trifluoromethyl)benzenethiol, T-60284

3-(Trifluoromethyl)benzenethiol, T-60285

4-(Trifluoromethyl)benzenethiol, T-60286

C₇H₅IO₂

2-Hydroxy-5-iodobenzaldehyde, H-70143

C₇H₅IO₂S

3-Iodo-4-mercaptobenzoic acid, I-70046

C₇H₅IO₃

▷1-Hydroxy-1,2-benziodoxol-3(1H)-one, H-60102

C₇H₅IO₄

2,3-Dihydroxy-6-iodobenzoic acid, D-60330

2,4-Dihydroxy-3-iodobenzoic acid, D-60331

2,4-Dihydroxy-5-iodobenzoic acid, D-60332

2,5-Dihydroxy-4-iodobenzoic acid, D-60333

2,6-Dihydroxy-3-iodobenzoic acid, D-60334

3,4-Dihydroxy-5-iodobenzoic acid, D-60335

3,5-Dihydroxy-2-iodobenzoic acid, D-60336

3,5-Dihydroxy-4-iodobenzoic acid, D-60337

4,5-Dihydroxy-2-iodobenzoic acid, D-60338

$C_7H_5I_2O_2$

2,6-Diiodobenzoic acid, D-70356

C_7H_5NO

Benzonitrile *N*-oxide, B-60035
2-Cyanophenol, *in* H-70108
Pentafluoroisocyanatobenzene, P-60035

C_7H_5NOS

5-(2-Thienyl)oxazole, T-70171

$C_7H_5NO_2$

5-(2-Furanyl)oxazole, F-70048
4-Nitrosobenzaldehyde, N-60046

$C_7H_5NO_3$

4-Hydroxy-3-nitrosobenzaldehyde, H-60197

$C_7H_5NO_4$

4-Hydroxy-3-nitrosobenzoic acid, H-70193

$C_7H_5NO_6$

4,5-Dihydroxy-2-nitrobenzoic acid, D-60360

$C_7H_5NS_2$

Benzenesulfenyl thiocyanate, B-60013

C_7H_5NSe

Benzoselenazole, B-70050

$C_7H_5N_3OS$

2,3-Dihydro-2-thioxopyrido[2,3-*d*]pyrimidin-4(1*H*)-one, D-60291
2,3-Dihydro-2-thioxopyrido[3,2-*d*]pyrimidin-4(1*H*)-one, D-60292
2,3-Dihydro-2-thioxopyrido[3,4-*d*]pyrimidin-4(1*H*)-one, D-60293
2,3-Dihydro-2-thioxo-4*H*-pyrido[1,2-*a*]-1,3,5-triazin-4-one, D-60294

$C_7H_5N_3O_2$

Pyrido[2,3-*d*]pyrimidine-2,4(1*H*,3*H*)-dione, P-60236
Pyrido[3,4-*d*]pyrimidine-2,4(1*H*,3*H*)-dione, P-60237

$C_7H_5N_3O_2S$

3-Amino-5-nitro-2,1-benzisothiazole, A-60232

$C_7H_5N_3O_5$

3,5-Dinitrobenzamide, *in* D-60458

$C_7H_5N_3S$

1,2,3-Benzotriazine-4(3*H*)-thione, B-60050
2-(1,3,4-Thiadiazol-2-yl)pyridine, T-60192
4-(1,3,4-Thiadiazol-2-yl)pyridine, T-60193

$C_7H_5N_5O$

1*H*-Imidazo[2,1-*b*]purin-4(5*H*)-one, I-70006

$C_7H_5N_5O_8$

3,5-Diamino-2,4,6-trinitrobenzoic acid, D-70047
▷2-Methyl-3,4,5,6-tetranitroaniline, M-60129
▷3-Methyl-2,4,5,6-tetranitroaniline, M-60130
▷4-Methyl-2,3,5,6-tetranitroaniline, M-60131
2,3,4,6-Tetranitroaniline; *N*-Me, *in* T-60159

C_7H_6

1*H*-Cyclopropabenzene, C-70246

C_7H_6BrFO

3-Bromo-4-fluorobenzyl alcohol, B-60244
5-Bromo-2-fluorobenzyl alcohol, B-60245
2-Bromo-1-fluoro-3-methoxybenzene, *in* B-60247
2-Bromo-4-fluoro-1-methoxybenzene, *in* B-60248
4-Bromo-2-fluoro-1-methoxybenzene, *in* B-60251

$C_7H_6BrNO_2$

3-Amino-2-bromobenzoic acid, A-70097

$C_7H_6BrNO_3$

2-Bromo-1-methoxy-3-nitrobenzene, *in* B-70257
2-Bromo-3-nitrobenzyl alcohol, B-70256

$C_7H_6Br_2$

1,3-Dibromo-2-methylbenzene, D-60103

C_7H_6ClFO

1-Chloro-2-fluoro-4-methoxybenzene, *in* C-60080
1-Chloro-4-fluoro-2-methoxybenzene, *in* C-60074
2-Chloro-4-fluoro-1-methoxybenzene, *in* C-60073
4-Chloro-2-fluoro-1-methoxybenzene, *in* C-60079

C_7H_6ClNO

2-Acetyl-5-chloropyridine, A-70033
3-Acetyl-2-chloropyridine, A-70034
4-Acetyl-2-chloropyridine, A-70035
4-Acetyl-3-chloropyridine, A-70036
5-Acetyl-2-chloropyridine, A-70037

$C_7H_6ClNO_2$

3-Chloro-4-hydroxybenzoic acid; Amide, *in* C-60091

$C_7H_6ClNO_3$

4-Chloro-2-methoxy-1-nitrobenzene, *in* C-60119
2-Chloro-3-methyl-4-nitrophenol, C-60115
3-Chloro-4-methyl-5-nitrophenol, C-60116

$C_7H_6Cl_2S$

1,2-Dichloro-4-(methylthio)benzene, *in* D-60123
1,4-Dichloro-2-(methylthio)benzene, *in* D-60121

$C_7H_6FNO_2$

2-Amino-3-fluorobenzoic acid, A-70140
2-Amino-4-fluorobenzoic acid, A-70141
2-Amino-5-fluorobenzoic acid, A-70142
2-Amino-6-fluorobenzoic acid, A-70143
3-Amino-4-fluorobenzoic acid, A-70144
4-Amino-2-fluorobenzoic acid, A-70145
▷4-Amino-3-fluorobenzoic acid, A-70146
5-Amino-2-fluorobenzoic acid, A-70147
4-Fluoro-2-hydroxybenzoic acid; Amide, *in* F-60042

$C_7H_6FNO_3$

4-Fluoro-2-methoxy-1-nitrobenzene, *in* F-60061

$C_7H_6F_2S$

1,3-Difluoro-2-(methylthio)benzene, *in* D-60174
1,3-Difluoro-5-(methylthio)benzene, *in* D-60176
1,4-Difluoro-2-(methylthio)benzene, *in* D-60173

$C_7H_6F_3NO_2S$

2-(Trifluoromethyl)benzenesulfonic acid; Amide, *in* T-70242
3-(Trifluoromethyl)benzenesulfonic acid; Amide, *in* T-60283
4-(Trifluoromethyl)benzenesulfonic acid; Amide, *in* T-70243

$C_7H_6INO_2$

2-Hydroxy-5-iodobenzaldehyde; Oxime, *in* H-70143

$C_7H_6N_2$

3-Phenyl-3*H*-diazirine, P-70067
Pyrrolo[1,2-*a*]pyrazine, P-70197
▷1*H*-Pyrrolo[3,2-*b*]pyridine, P-60249
2*H*-Pyrrolo[3,4-*c*]pyridine, P-70200
6*H*-Pyrrolo[3,4-*b*]pyridine, P-70201

$C_7H_6N_2O$

▷2-Aminobenzoxazole, A-60093
Pyrrolo[1,2-*a*]pyrazin-1(2*H*)-one, P-70198

$C_7H_6N_2O_3S$

1*H*-Benzimidazole-2-sulfonic acid, B-70022

$C_7H_6N_2O_4$

2-Amino-3,5-pyridinedicarboxylic acid, A-70193
4-Amino-2,6-pyridinedicarboxylic acid, A-70194
5-Amino-3,4-pyridinedicarboxylic acid, A-70195
▷Antibiotic 2061 A, *in* A-70158

▷2-Methyl-1,3-dinitrobenzene, M-60061

$C_7H_6N_2O_5$

2,6-Dinitrobenzyl alcohol, D-60459

$C_7H_6N_2O_6$

2-Methoxy-4,5-dinitrophenol, *in* D-70449
5-Methoxy-2,4-dinitrophenol, *in* D-70450

$C_7H_6N_4$

4-Amino-1,2,3-benzotriazine, A-60092
2-Methylpteridine, M-70120
4-Methylpteridine, M-70121
6-Methylpteridine, M-70122
7-Methylpteridine, M-70123

$C_7H_6N_4O$

4-Amino-1,2,3-benzotriazine; 2-Oxide, *in* A-60092
4-Amino-1,2,3-benzotriazine; 3-Oxide, *in* A-60092
2-Amino-5-cyanonicotinamide, *in* A-70193
Di-1*H*-imidazol-2-ylmethanone, D-70354
4-Hydroxylamino-1,2,3-benzotriazine, *in* A-60092

$C_7H_6N_4O_2$

2,4(1*H*,3*H*)-Pteridinedione; 1-Me, *in* P-60191
2,4(1*H*,3*H*)-Pteridinedione; 3-Me, *in* P-60191

$C_7H_6N_4O_3$

2,4,6(1*H*,3*H*,5*H*)-Pteridinetrione; 1-Me, *in* P-70134
2,4,6(1*H*,3*H*,5*H*)-Pteridinetrione; 3-Me, *in* P-70134
2,4,6(1*H*,3*H*,5*H*)-Pteridinetrione; 7-Me, *in* P-70134
2,4,7(1*H*,3*H*,8*H*)-Pteridinetrione; 1-Me, *in* P-70135
2,4,7(1*H*,3*H*,8*H*)-Pteridinetrione; 3-Me, *in* P-70135
2,4,7(1*H*,3*H*,8*H*)-Pteridinetrione; 6-Me, *in* P-70135

$C_7H_6N_4O_6$

2,4,6-Trinitroaniline; *N*-Me, *in* T-70301

$C_7H_6N_4S$

▷Tetrazole-5-thione; 1-Ph, *in* T-60174
Tetrazole-5-thione; *S*-Ph, *in* T-60174

C_7H_6O

2,4,6-Cycloheptatrien-1-one, C-60204
Dimethylenebicyclo[1.1.1]pentanone, D-60422
4-Methylene-2,5-cyclohexadien-1-one, M-60071
6-Methylene-2,4-cyclohexadien-1-one, M-60072

C_7H_6OS

4*H*-Cyclopenta[*c*]thiophen-5(6*H*)-one, C-70245
4,5-Dihydro-6*H*-cyclopenta[*b*]thiophen-6-one, D-70182
4,6-Dihydro-5*H*-cyclopenta[*b*]thiophen-5-one, D-70183
5,6-Dihydro-4*H*-cyclopenta[*c*]thiophen-4-one, D-70184

$C_7H_6O_2$

Bicyclo[3.2.0]hept-2-ene-6,7-dione, B-70099

$C_7H_6O_2S$

3-(2-Thienyl)-2-propenoic acid, T-60208
3-(3-Thienyl)-2-propenoic acid, T-70172

$C_7H_6O_3$

2,3-Dihydroxybenzaldehyde, D-70285
2,7-Dihydroxy-2,4,6-cycloheptatrien-1-one, D-60314
▷2-Hydroxybenzoic acid, H-70108

$C_7H_6O_4$

2,3-Dihydroxy-5-methyl-1,4-benzoquinone, D-70320
3-(Hydroxymethyl)-7-oxabicyclo[4.1.0]hept-3-ene-2,5-dione, H-70181

$C_7H_6O_4S$

2-Carboxy-3-thiopheneacetic acid, C-60017

4-Carboxy-3-thiopheneacetic acid, C-**60018**

C$_7$H$_6$O$_5$

2,3,5-Trihydroxybenzoic acid, T-70250

C$_7$H$_6$O$_6$

Osbeckic acid, O-**70048**
Pentahydroxybenzaldehyde, P-**60042**
2,3,5,6-Tetrahydroxybenzoic acid, T-**60100**

C$_7$H$_6$O$_7$

Pentahydroxybenzoic acid, P-**60043**

C$_7$H$_6$S

2,4,6-Cycloheptatriene-1-thione, C-**60203**

C$_7$H$_6$Se$_2$

1,3-Benzodiselenole, B-**70031**

C$_7$H$_7$BrO$_3$

4-Bromo-2,6-dimethoxyphenol, *in* B-**60201**

C$_7$H$_7$BrSe

Bromomethyl phenyl selenide, B-**60300**

C$_7$H$_7$Br$_2$N

▷2,6-Biss(romomethyl)pyridine, B-**60146**
3,5-Biss(romomethyl)pyridine, B-**60147**

C$_7$H$_7$Cl

2-Chlorobicyclo[2.2.1]hepta-2,5-diene, C-**70056**

C$_7$H$_7$ClN$_2$O$_2$

2-Chloro-6-nitrobenzylamine, C-**70126**
4-Chloro-2-nitrobenzylamine, C-**70127**
4-Chloro-3-nitrobenzylamine, C-**70128**
5-Chloro-2-nitrobenzylamine, C-**70129**

C$_7$H$_7$ClO

2-(Chloromethyl)phenol, C-**70101**
3-(Chloromethyl)phenol, C-**70102**
4-(Chloromethyl)phenol, C-**70103**

C$_7$H$_7$Cl$_3$O$_2$

3,4-Dihydro-5-(trichloroacetyl)-2*H*-pyran,
D-**70272**

C$_7$H$_7$FO

2-Fluoro-3-methylphenol, F-**60052**
2-Fluoro-4-methylphenol, F-**60053**
2-Fluoro-5-methylphenol, F-**60054**
2-Fluoro-6-methylphenol, F-**60055**
3-Fluoro-2-methylphenol, F-**60056**
3-Fluoro-4-methylphenol, F-**60057**
4-Fluoro-2-methylphenol, F-**60058**
4-Fluoro-3-methylphenol, F-**60059**
5-Fluoro-2-methylphenol, F-**60060**

C$_7$H$_7$F$_3$O$_2$

3,4-Dihydro-5-(trifluoroacetyl)-2*H*-pyran,
D-**70275**

C$_7$H$_7$IO

4-Iodo-3-methylphenol, I-**60050**

C$_7$H$_7$NO

2,3-Dihydrofuro[2,3-*b*]pyridine, D-**70211**
2,3-Dihydro-1*H*-pyrrolizin-1-one, D-**60277**
3-(1*H*-Pyrrol-2-yl)-2-propenal, P-**70202**

C$_7$H$_7$NOS

2,3-Dihydrothieno[2,3-*b*]pyridine; 1-Oxide, *in*
D-**70269**

C$_7$H$_7$NO$_2$

▷Benzohydroxamic acid, B-**70035**
2-(Hydroxyacetyl)pyridine, H-**70101**
3-(Hydroxyacetyl)pyridine, H-**70102**
▷2-Hydroxybenzamide, *in* H-**70108**
(Nitromethyl)benzene, N-**70053**

C$_7$H$_7$NO$_2$S

(Phenylthio)nitromethane, P-**60137**
2(1*H*)-Pyridinethione; 1-Acetoxy, *in* P-**60224**

C$_7$H$_7$NO$_3$

3-Hydroxy-2(1*H*)-pyridinone; 3-Ac, *in* H-**60223**
5-Hydroxy-2(1*H*)-pyridinone; 5-Ac, *in* H-**60225**

C$_7$H$_7$NO$_3$S

4-Carboxy-3-thiopheneacetic acid; Acetamide,
in C-**60018**

C$_7$H$_7$NO$_4$

2-Methyl-4-nitro-1,3-benzenediol, M-**70097**

1*H*-Pyrrole-3,4-dicarboxylic acid; Me ester, *in*
P-**70183**

C$_7$H$_7$NO$_4$S

(Nitromethyl)sulfonylbenzene, *in* P-**60137**

C$_7$H$_7$NS

2,3-Dihydrothieno[2,3-*b*]pyridine, D-**70269**

C$_7$H$_7$N$_3$

▷7-Azidobicyclo[2.2.1]hepta-2,5-diene, A-**70285**
5*H*-Pyrrolo[2,3-*b*]pyrazine; 5-Me, *in* P-**60248**

C$_7$H$_7$N$_3$O

1,3-Dihydro-2*H*-imidazo[4,5-*c*]pyridin-2-one;
1-Me, *in* D-**60250**

C$_7$H$_7$N$_3$S

2-Hydrazinobenzothiazole, H-**70095**

C$_7$H$_7$N$_5$O

2-Amino-1-methyl-4(1*H*)-pteridinone, *in*
A-**60242**
2-(Methylamino)-4(1*H*)-pteridinone, *in*
A-**60242**

C$_7$H$_8$

Bicyclo[3.2.0]hepta-2,6-diene, B-**70097**
3-Buten-1-ynylcyclopropane, B-**70293**
1-Ethynylcyclopentene, E-**70052**
4-Methylenebicyclo[3.1.0]hex-2-ene, M-**60069**
4-Methylene-1,2,5-hexatriene, M-**70071**

C$_7$H$_8$BrF$_3$NO$_2$

4-Bromo-1-nitro-2-(trifluoromethyl)benzene,
B-**60309**

C$_7$H$_8$BrN

2-(2-Bromoethyl)pyridine, B-**70210**
3-(2-Bromoethyl)pyridine, B-**70211**
2-Bromo-3-methylaniline, B-**70234**

C$_7$H$_8$BrNO

2-Bromo-3-ethoxypyridine, *in* B-**60275**
2-Bromo-3-methoxyaniline, *in* A-**70099**
2-Bromo-3-methoxy-6-methylpyridine, *in*
B-**60267**

C$_7$H$_8$ClN

2-(Chloromethyl)-3-methylpyridine, C-**70093**
▷2-(Chloromethyl)-4-methylpyridine, C-**70094**
2-(Chloromethyl)-5-methylpyridine, C-**70095**
▷2-(Chloromethyl)-6-methylpyridine, C-**70096**
3-(Chloromethyl)-2-methylpyridine, C-**70097**
3-(Chloromethyl)-4-methylpyridine, C-**70098**
▷4-(Chloromethyl)-2-methylpyridine, C-**70099**
5-(Chloromethyl)-2-methylpyridine, C-**70100**

C$_7$H$_8$ClNO$_2$

2-Chloro-4,6-dimethoxypyridine, *in* C-**70074**
4-Chloro-2,6-dimethoxypyridine, *in* C-**70073**
4-Chloro-6-methoxy-1-methyl-2(1*H*)-
pyridinone, *in* C-**70073**

C$_7$H$_8$N$_2$

2,3-Dihydroimidazo[1,2-*a*]pyridine, D-**70216**
2,3-Dihydro-1*H*-pyrrolo[2,3-*b*]pyridine,
D-**70263**
N-Phenylformamidine, P-**60101**

C$_7$H$_8$N$_2$O

2-Acetyl-3-aminopyridine, A-**70022**
2-Acetyl-5-aminopyridine, A-**70023**
3-Acetyl-2-aminopyridine, A-**70024**
3-Acetyl-4-aminopyridine, A-**70025**
4-Acetyl-2-aminopyridine, A-**70026**
4-Acetyl-3-aminopyridine, A-**70027**
5-Acetyl-2-aminopyridine, A-**70028**
2-(Aminoacetyl)pyridine, A-**70085**
3-(Aminoacetyl)pyridine, A-**70086**

4-(Aminoacetyl)pyridine, A-**70087**
2,3-Dihydro-1*H*-pyrrolo[2,3-*b*]pyridine; 7-
Oxide, *in* D-**70263**
4-Oxoheptanedinitrile, *in* O-**60071**

C$_7$H$_8$N$_2$O$_2$

5-Acetyl-2-aminopyridine; *N*-Oxide, *in* A-**70028**
1-Amino-2(1*H*)-pyridinone; *N*-Ac, *in* A-**60248**
2,4-Diaminobenzoic acid, D-**60032**
1,2-Diamino-4,5-methylenedioxybenzene,
D-**60038**
1,3-Diisocyanatocyclopentane, D-**60390**
1-Methoxy-2-phenyldiazene 2-oxide, *in*
H-**70208**
Urocanic acid; Me ester, *in* U-**70005**

C$_7$H$_8$N$_2$O$_3$

3-Amino-4-methyl-5-nitrophenol, A-**60222**
3-Amino-5-nitrobenzyl alcohol, A-**60233**
4-Ethoxy-2-nitropyridine, *in* N-**70063**

C$_7$H$_8$N$_2$O$_3$S

Aminoiminomethanesulfonic acid; *N*-Ph, *in*
A-**60210**

C$_7$H$_8$N$_2$O$_4$

4-Amino-5-hydroxy-2-oxo-7-oxabicyclo[4.1.0]-
hept-3-ene-3-carboxamide, A-**70158**
1,2,3,6-Tetrahydro-2,6-dioxo-4-
pyrimidineacetic acid; Me ester, *in* T-**70060**

C$_7$H$_8$N$_2$O$_5$

3-Amino-2(1*H*)-pyridinethione; N^3-Ac, *in*
A-**60247**

C$_7$H$_8$N$_2$Se

Selenourea; *N*-Ph, *in* S-**70031**

C$_7$H$_8$N$_4$

1′-Methyl-1,5′-bi-1*H*-pyrazole, *in* B-**60119**
2,2′-Methylenebis-1*H*-imidazole, M-**70064**

C$_7$H$_8$N$_4$O

N-(4-Cyano-1-methyl-1*H*-pyrazol-5-yl)-
acetamide, *in* A-**70192**
1,3-Dihydro-1,3-dimethyl-2*H*-imidazo[4,5-*b*]-
pyrazin-2-one, *in* I-**60006**
1,4-Dihydro-1,4-dimethyl-2*H*-imidazo[4,5-*b*]-
pyrazin-2-one, *in* I-**60006**
2-Methoxy-4-methyl-4*H*-imidazo[4,5-*b*]-
pyrazine, *in* I-**60006**

C$_7$H$_8$N$_4$O$_2$

2-Amino-3,5-pyridinedicarboxylic acid;
Diamide, *in* A-**70193**
1*H*-Pyrazolo[3,4-*d*]pyrimidine-4,6(5*H*,7*H*)-
dione; 1,5-Di-Me, *in* P-**60202**
1*H*-Pyrazolo[3,4-*d*]pyrimidine-4,6(5*H*,7*H*)-
dione; 5,7-Di-Me, *in* P-**60202**

C$_7$H$_8$N$_4$S

1,3-Dihydro-1,3-dimethyl-2*H*-imidazo[4,5-*b*]-
pyrazine-2-thione, *in* D-**60247**
1,5-Dihydro-6*H*-imidazo[4,5-*c*]pyridazine-6-
thione; N^1,*S*-Di-Me, *in* D-**70215**
1,5-Dihydro-6*H*-imidazo[4,5-*c*]pyridazine-6-
thione; N^2,*S*-Di-Me, *in* D-**70215**
1,5-Dihydro-6*H*-imidazo[4,5-*c*]pyridazine-6-
thione; N^5,*S*-Di-Me, *in* D-**70215**
1,5-Dihydro-6*H*-imidazo[4,5-*c*]pyridazine-6-
thione; N^7,*S*-Di-Me, *in* D-**70215**
1,7-Dihydro-6*H*-purine-6-thione; 1,9-Di-Me, *in*
D-**60270**
1,7-Dihydro-6*H*-purine-6-thione; 3,7-Di-Me, *in*
D-**60270**
1,7-Dihydro-6*H*-purine-6-thione; 3,9-Di-Me, *in*
D-**60270**
1,7-Dihydro-6*H*-purine-6-thione; *S*,3*N*-Di-Me,
in D-**60270**
1,7-Dihydro-6*H*-purine-6-thione; *S*,7*N*-Di-Me,
in D-**60270**
1,7-Dihydro-6*H*-purine-6-thione; *S*,9*N*-Di-Me,
in D-**60270**
1-Methyl-2-methylthio-1*H*-imidazo[4,5-*b*]-
pyrazine, *in* D-**60247**
4-Methyl-2-methylthio-4*H*-imidazo[4,5-*b*]-
pyrazine, *in* D-**60247**

C_7H_8O

Tricyclo[2.2.1.0^{2,6}]heptan-3-one, T-70228

3-Vinyl-2-cyclopenten-1-one, V-70011

C_7H_8OS

2,5-Dimethyl-3-thiophenecarboxaldehyde,
D-60452

2,5-Dimethyl-3-thiophenecarboxaldehyde,
D-70438

3,5-Dimethyl-2-thiophenecarboxaldehyde,
D-70439

2-Thiabicyclo[2.2.2]oct-5-en-3-one, T-60188

$C_7H_8O_2$

2,2-Dimethyl-4-cyclopentene-1,3-dione,
D-70395

2,5-Dimethyl-3-furancarboxaldehyde, D-70412

▷4,6-Dimethyl-2H-pyran-2-one, D-70435

4-Hydroxy-4-methyl-2,5-cyclohexadien-1-one,
H-70169

3-Oxo-1-cyclohexene-1-carboxaldehyde,
O-60060

α-Oxo-3-cyclopenteneacetaldehyde, O-70086

α-Oxo-3-cyclopentene-1-acetaldehyde, O-60061

$C_7H_8O_2S$

2-(Methoxyacetyl)thiophene, in H-70103

$C_7H_8O_3$

2,5-Dihydroxybenzyl alcohol, D-70287

4,5-Dimethyl-2-furancarboxylic acid, D-70413

4-Hydroxy-2-cyclopentenone; Ac, in H-60116

Methyl 3-formylcrotonate, in M-70103

$C_7H_8O_4$

5-Hydroxy-3-methoxy-7-oxabicyclo[4.1.0]hept-
3-en-2-one, H-70155

5-Hydroxymethyl-2(5H)-furanone; Ac, in
H-60182

Oxysporone, O-70107

Terremutin, T-70009

2-Vinyl-1,1-cyclopropanedicarboxylic acid,
V-60011

$C_7H_8O_5$

4,5-Dihydroxy-3-oxo-1-cyclohexenecarboxylic
acid, D-70328

4-Oxo-2-heptenedioic acid, O-60072

1,4,8-Trioxaspiro[4.5]decan-7,9-dione, in
O-60084

$C_7H_8O_6$

Aconitic acid; α-Mono-Me ester, in A-70055

Aconitic acid; β-Mono-Me ester, in A-70055

Aconitic acid; γ-Mono-Me ester, in A-70055

C_7H_8S

5,6-Dihydro-4H-cyclopenta[b]thiophene,
D-70180

5,6-Dihydro-4H-cyclopenta[c]thiophene,
D-70181

$C_7H_9BrO_2$

7-Bromo-5-heptynoic acid, B-60261

2-Bromo-3-methoxy-2-cyclohexen-1-one, in
B-70201

C_7H_9Cl

5-Chloro-1,3-cycloheptadiene, C-70062

C_7H_9ClO

2-Chloro-2-cyclohepten-1-one, C-60049

C_7H_9N

2-Vinyl-1H-pyrrole; N-Me, in V-60013

C_7H_9NO

7-Azabicyclo[4.2.0]oct-3-en-8-one, A-60331

1,2-Dihydro-3H-azepin-3-one; 1-Me, in
D-70171

2,3-Dihydro-1H-pyrrolizin-1-ol, D-70261

1,5-Dimethyl-1H-pyrrole-2-carboxaldehyde, in
M-70129

3,3a,4,6a-Tetrahydrocyclopenta[b]pyrrol-
2(1H)-one, T-70048

Tricyclo[2.2.1.0^{2,6}]heptan-3-one; Oxime, in
T-70228

$C_7H_9NO_2$

3-Amino-5-hydroxybenzyl alcohol, A-60182

2-Amino-6-methoxyphenol, in A-70089

5-Amino-2-methoxyphenol, in A-70090

Dihydro-1H-pyrrolizine-3,5(2H,6H)-dione,
D-70260

2,3-Dimethoxypyridine, in H-60223

2,4-Dimethoxypyridine, in H-60224

2,6-Dimethoxypyridine, in H-60226

3,4-Dimethoxypyridine, in H-60227

3-Ethoxy-4(1H)-pyridinone, in H-60227

6-Hydroxy-2(1H)-pyridinone; Me ether, N-Me,
in H-60226

2-Methoxy-1-methyl-4(1H)-pyridinone, in
H-60224

3-Methoxy-1-methyl-4(1H)-pyridinone, in
H-60227

4-Methoxy-1-methyl-2(1H)-pyridinone, in
H-60224

5-Methyl-1H-pyrrole-2-carboxylic acid; Me
ester, in M-60124

5-Methyl-1H-pyrrole-3-carboxylic acid; Me
ester, in M-60125

4,5,6,7-Tetrahydro-1,2-benzisoxazol-3(2H)-one,
T-70037

4,5,6,7-Tetrahydro-2,1-benzisoxazol-3(1H)-one,
T-70038

$C_7H_9NO_3$

4-Aminodihydro-3-methylene-2-(3H)furanone;
N-Ac, in A-70131

4,6-Dimethoxy-2(1H)-pyridinone, in D-70341

3-Hydroxy-4(1H)-pyridinone; Di-Me ether, 1-
oxide, in H-60227

4-Hydroxy-2(1H)-pyridinone; Di-Me ether, 1-
oxide, in H-60224

7-Oxo-1-azabicyclo[3.2.0]heptane-2-carboxylic
acid, O-70079

$C_7H_9NO_4$

5-Oxo-2-pyrrolidinecarboxylic acid; Me ester,
N-formyl, in O-70102

$C_7H_9NO_4S_2$

2,3-Biss(ethylsulfonyl)pyridine, in P-60218

2,5-Biss(ethylsulfonyl)pyridine, in P-60220

3,5-Biss(ethylsulfonyl)pyridine, in P-60222

C_7H_9NS

4-Amino-5,6-dihydro-4H-cyclopenta[b]-
thiophene, A-60137

2-(2-Pyridyl)ethanethiol, P-60239

2-(4-Pyridyl)ethanethiol, P-60240

$C_7H_9NS_2$

2,3-Biss(ethylthio)pyridine, in P-60218

2,4-Biss(ethylthio)pyridine, in P-60219

2,5-Biss(ethylthio)pyridine, in P-60220

3,5-Biss(ethylthio)pyridine, in P-60222

1-Methyl-4-(methylthio)-2(1H)-pyridinethione,
in P-60219

$C_7H_9N_2S$

3,5-Dimethyl-2-thiophenecarboxaldehyde;
Hydrazone, in D-70439

$C_7H_9N_3O$

2-(Hydroxyacetyl)pyridine; Hydrazone, in
H-70101

$C_7H_9N_3O_3$

5-Amino-1H-pyrazole-4-carboxylic acid; N(1)-
Me, N(5)-Ac, in A-70192

$C_7H_9N_5$

3,7-Dimethyladenine, D-60400

2-(Dimethylamino)purine, in A-60243

8-(Dimethylamino)purine, in A-60244

C_7H_{10}

Bicyclo[3.1.1]hept-2-ene, B-60087

1,2-Biss(ethylene)cyclopentane, B-60169

1,3-Biss(ethylene)cyclopentane, B-60170

Tricyclo[4.1.0.0^{1,3}]heptane, T-60262

$C_7H_{10}N_2O$

2-Oxa-3-azabicyclo[2.2.2]octane; N-Cyano, in
O-70051

$C_7H_{10}N_2O_2$

Hexahydropyrrolo[1,2-a]pyrazine-1,4-dione,
H-70073

4,5,6,7-Tetrahydro-3-methoxyisoxazolo[4,5-c]-
pyridine, in T-60071

Tetrahydro-1H-pyrazolo[1,2-a]pyridazine-
1,3(2H)-dione, T-70088

$C_7H_{10}N_2O_3$

2-Imidazolidinone; 1,3-Di-Ac, in I-70005

$C_7H_{10}N_2O_4$

5-Methyl-3,6-dioxo-2-piperazineacetic acid,
M-60064

$C_7H_{10}N_2S$

4,6-Dimethyl-2(1H) pyrimidinethione; 1-Me, in
D-60450

$C_7H_{10}O$

1-Acetylcyclopentene, A-60028

▷Bicyclo[2.2.1]heptan-2-one, B-70098

3-Cyclohexene-1-carboxaldehyde, C-70227

2,4-Heptadienal, H-60020

1,5-Heptadien-3-one, H-70015

1,5-Heptadien-4-one, H-70016

1,6-Heptadien-3-one, H-70017

4,5-Heptadien-3-one, H-70018

6-Heptyn-2-one, H-70025

6-Methyl-2-cyclohexen-1-one, M-70059

2-Methylene-7-oxabicyclo[2.2.1]heptane,
M-70073

7-Oxatricyclo[4.1.1.0^{2,5}]octane, O-60053

8-Oxatricyclo[3.3.0.0^{2,7}]octane, O-60054

$C_7H_{10}OS$

2-Thiabicyclo[2.2.2]octan-3-one, T-60187

$C_7H_{10}O_2$

1-Carbethoxybicyclo[1.1.0]butane, in B-70094

1,1-Diacetylcyclopropane, D-70032

Dihydro-4,4-dimethyl-5-methylene-2(3H)-
furanone, D-70197

2,2-Dimethyl-1,3-cyclopentanedione, D-70393

4,4-Dimethyl-1,3-cyclopentanedione, D-70394

3,3-Dimethyl-1-cyclopropene-1-carboxylic acid;
Me ester, in D-60417

6,6-Dimethyl-3-oxabicyclo[3.1.0]hexan-2-one,
D-70425

2,4-Heptadienoic acid, H-60021

3,4-Hexadienoic acid; Me ester, in H-70037

3,5-Hexadienoic acid; Me ester, in H-70038

Hexahydro-1H-cyclopenta[c]furan-1-one,
H-70054

2-Methoxy-3-methyl-2-cyclopenten-1-one, in
H-60174

5-Methyl-2,4-hexadienoic acid, M-70086

5-Methyl-3,4-hexadienoic acid, M-70087

3-Penten-2-ol; Ac, in P-70041

$C_7H_{10}O_3$

5-Ethyl-3-hydroxy-4-methyl-2(5H)-furanone,
E-60051

2-Hydroxycyclopentanone; Ac, in H-60114

7-Hydroxy-5-heptynoic acid, H-60150

4-Methoxy-2,5-dimethyl-3(2H)-furanone, in
H-60120

3-Methyl-4-oxo-2-butenoic acid; Et ester, in
M-70103

3-Oxocyclopentaneacetic acid, O-70083

2-Oxocyclopentanecarboxylic acid; Me ester, in
O-70084

4-Oxo-2-pentenoic acid; Et ester, in O-70098

2,4-Pentanedione; Ac, in P-60059

2,4,10-Trioxatricyclo[3.3.1.1^{3,7}]decane,
T-60377

$C_7H_{10}O_4$

1,2-Cyclopropanedicarboxylic acid; Di-Me
ester, in C-70248

Ethyl 3-acetoxyacrylate, in O-60090

$C_7H_{10}O_5$

4-Oxoheptanedioic acid, O-60071
3-Oxopentanedioic acid; Di-Me ester, *in* O-60084
1,3,4-Trihydroxy-6-oxabicyclo[3.2.1]octan-7-one, *in* Q-70005

$C_7H_{10}O_6$

1,3-Dioxolane-2,2-diacetic acid, *in* O-60084

$C_7H_{10}S$

1-Thia-2-cyclooctyne, T-60189
1-Thia-3-cyclooctyne, T-60190

$C_7H_{11}ClO$

3,3-Dimethyl-4-pentenoic acid; Chloride, *in* D-60437

$C_7H_{11}ClO_2$

7-Chloro-2-heptenoic acid, C-70069

$C_7H_{11}IO$

2-Iodocycloheptanone, I-70025

$C_7H_{11}N$

2-Azabicyclo[2.2.2]oct-5-ene, A-70279
5-Heptenenitrile, *in* H-70024

$C_7H_{11}NO$

Bicyclo[2.2.1]heptan-2-one; Oxime, *in* B-70098
3-Cyclohexene-1-carboxaldehyde; Oxime, *in* C-70227
4,4-Dimethylglutaraldehydonitrile, *in* D-60434
Hexahydro-1H-pyrrolizin-1-one, H-60057
Hexahydro-3H-pyrrolizin-3-one, H-60058
Tetrahydro-1H-pyrrolizin-2(3H)-one, T-60091

$C_7H_{11}NO_2$

Hexahydro-2(3H)-benzoxazolone, H-70052
2-Oxa-3-azabicyclo[2.2.2]octane; N-Formyl, *in* O-70051
1,2,3,6-Tetrahydro-3-pyridinecarboxylic acid; N-Me, *in* T-60084

$C_7H_{11}NO_3$

5-Oxo-2-pyrrolidinecarboxylic acid; Et ester, *in* O-70102

$C_7H_{11}NO_4$

2-Amino-3-heptenedioic acid, A-60164
2-Amino-5-heptenedioic acid, A-60165

$C_7H_{11}NO_5$

4-Oxoheptanedioic acid; Oxime, *in* O-60071

$C_7H_{11}NS$

4-Mercaptocyclohexanecarbonitrile, *in* M-60020

$C_7H_{11}N_3O_2$

3-Amino-1H-pyrazole-4-carboxylic acid; N(1)-Me, Et ester, *in* A-70187
3-Amino-1H-pyrazole-5-carboxylic acid; N(1)-Me, Et ester, *in* A-70188
5-Amino-1H-pyrazole-4-carboxylic acid; N(1)-Me, Et ester, *in* A-70192

$C_7H_{11}N_3O_2S$

Ovothiol A, O-70050

C_7H_{12}

4,4-Dimethyl-2-pentyne, D-60438

$C_7H_{12}BrO_2$

2-(4-Bromobutyl)-1,3-dioxole, *in* B-70263

$C_7H_{12}Br_2$

1,1-Dibromocycloheptane, D-60091

$C_7H_{12}Br_2O$

2,4-Dibromo-2,4-dimethyl-3-pentanone, D-60099

$C_7H_{12}Br_2O_2$

1,1-Dibromo-3,3-dimethoxy-2-methyl-1-butene, *in* D-70094

$C_7H_{12}ClN$

8-Azabicyclo[3.2.1]octane; N-Chloro, *in* A-70278

$C_7H_{12}ClNO_4$

1,2,3,5,6,7-Hexahydropyrrolizinium(1+); Perchlorate, *in* H-70071

$C_7H_{12}N^{\oplus}$

1,2,3,5,6,7-Hexahydropyrrolizinium(1+), H-70071

$C_7H_{12}N_2O$

3,5-Dihydro-3,3,5,5-tetramethyl-4H-pyrazol-4-one, D-60286
Hexahydro-1H-pyrrolizin-1-one; Oxime, *in* H-60057

$C_7H_{12}N_2OS$

3,5-Dihydro-3,3,5,5-tetramethyl-4H-pyrazole-4-thione; S-Oxide, *in* D-60285

$C_7H_{12}N_2O_2$

5-Amino-3,4-dihydro-2H-pyrrole-2-carboxylic acid; Et ester, *in* A-70135
1-Oxa-2-azaspiro[2,5]octane; N-Carbamoyl, *in* O-70052

$C_7H_{12}N_2O_4$

2,6-Diamino-3-heptenedioic acid, D-60036

$C_7H_{12}N_2S$

3,5-Dihydro-3,3,5,5-tetramethyl-4H-pyrazole-4-thione, D-60285

$C_7H_{12}N_4O$

Caffeidine, C-60003

$C_7H_{12}O$

2,3-Dimethyl-3,4-dihydro-2H-pyran, D-70398
3,3-Dimethyl-4-penten-2-one, D-70429
2,4-Heptadien-1-ol, H-70014
1-Methoxycyclohexene, M-60037
6-Methyl-2-cyclohexen-1-ol, M-70058
3,6,7,8-Tetrahydro-2H-oxocin, T-60076

$C_7H_{12}OS$

2-Hexyl-5-methyl-3(2H)furanone; S-Oxide (exo-), *in* H-60079
2-(Methylthio)cyclohexanone, *in* M-70025
3-(Methylthio)cyclohexanone, *in* M-70026
4-Methylthiocyclohexanone, *in* M-70027
2,2,4,4-Tetramethyl-3-thietanone, T-60154
3,3,4,4-Tetramethyl-2-thietanone, T-60155

$C_7H_{12}O_2$

2,3-Dimethylcyclopropanol; Ac, *in* D-70397
3,3-Dimethyl-4-pentenoic acid, D-60437
5-Heptenoic acid, H-70024
5-Hydroxy-2-methylcyclohexanone, H-70171
4-Oxoheptanal, O-60070
Tetrahydro-5,6-dimethyl-2H-pyran-2-one, T-70059

$C_7H_{12}O_2S$

3-(Ethylthio)-2-propenoic acid; Et ester, *in* E-60055
4-Mercaptocyclohexanecarboxylic acid, M-60020
3-Oxobutanethioic acid; Isopropyl ester, *in* O-70080

$C_7H_{12}O_3$

4-Acetyl-2,2-dimethyl-1,3-dioxolane, A-70042
▷Botryodiplodin, B-60194
3,3-Dimethyl-4-oxopentanoic acid, D-60433
4,4-Dimethyl-5-oxopentanoic acid, D-60434
2-Hydroxycyclohexanecarboxylic acid, H-70119
3-Oxopropanoic acid; tert-Butyl ester, *in* O-60090

$C_7H_{12}O_3S$

2-(Methylsulfonyl)cyclohexanone, *in* M-70025
2,2,4,4-Tetramethyl-3-thietanone; 1,1-Dioxide, *in* T-60154

$C_7H_{12}O_4$

(1-Methylpropyl)propanedioic acid, M-70119
▷1,2-Propanediol; Di-Ac, *in* P-70118

$C_7H_{12}O_5$

1,3-Diacetylglycerol, D-60023
2-Hydroxy-2-isopropylbutanedioic acid, H-60165
2-Hydroxy-3-isopropylbutanedioic acid, H-60166
5-(Hydroxymethyl)-5-cyclohexene-1,2,3,4-tetrol, H-70172

$C_7H_{12}O_6$

Quinic acid, Q-70005

$C_7H_{12}S$

Hexahydro-2H-cyclopenta[b]thiophene, H-60049

$C_7H_{13}Br$

7-Bromo-1-heptene, B-60260

$C_7H_{13}ClO$

4,4-Dimethylpentanoic acid; Chloride, *in* D-70428

$C_7H_{13}ClO_2$

1-Chloro-3-pentanol; Ac, *in* C-70138

$C_7H_{13}FO$

2-Fluoroheptanal, F-60034

$C_7H_{13}I$

6-Iodo-1-heptene, I-70035
7-Iodo-1-heptene, I-70036

$C_7H_{13}IO_2$

1-Iodo-5,5-dimethoxypentene, *in* I-70052
5-Iodo-2-pentanol; Ac, *in* I-60056
2-(3-Iodopropyl)-2-methyl-1,3-dioxolane, *in* I-60057

$C_7H_{13}N$

7-Azabicyclo[4.2.0]octane, A-70277
8-Azabicyclo[3.2.1]octane, A-70278
4,4-Dimethylpentanoic acid; Nitrile, *in* D-70428
▷2-Propyn-1-amine; N-Di-Et, *in* P-60184

$C_7H_{13}NO$

2-Acetylpiperidine, A-70051
4,4-Dimethyl-5-oxopentanoic acid; Oxime, *in* D-60434
3,3-Dimethyl-4-penten-2-one; Oxime, *in* D-70429
Hexahydro-4(1H)-azocinone, H-70047
Hexahydro-5(2H)-azocinone, H-70048

$C_7H_{13}NOS$

Hexahydro-1,5-thiazonin-6(7H)-one, H-70075
4-Mercaptocyclohexanecarboxylic acid; Amide, *in* M-60020

$C_7H_{13}NOS_2$

1-Isothiocyanato-5-(methylsulfinyl)pentene, *in* I-60140

$C_7H_{13}NO_2$

2-Amino-2-hexenoic acid; Me ester, *in* A-60167
3-Amino-4-hexenoic acid; Me ester, *in* A-60171
2-Hydroxycyclohexanecarboxylic acid; Amide, *in* H-70119
2-Methyl-2-piperidinecarboxylic acid, M-60119
Nitrocycloheptane, N-60025
3-Piperidinecarboxylic acid; Me ester, *in* P-70102
Stachydrine, S-60050

$C_7H_{13}NO_4$

2-Amino-2-isopropylbutanedioic acid, A-60212
2-Amino-3-isopropylbutanedioic acid, A-70162
2-Amino-3-propylbutanedioic acid, A-70186
2-Hydroxy-2-isopropylsuccinamic acid, *in* H-60165
Threonine; Ac, Me ether, *in* T-60217

$C_7H_{13}NS$

4,5-Dihydro-2-(1-methylpropyl)thiazole, D-70224

$C_7H_{13}NS_2$

1-Isothiocyanato-5-(methylthio)pentane, I-60140

$C_7H_{13}N_3$

6bH-2a,4a,6a-Hexahydrotriazacyclopenta[cd]pentalene, H-70077

$C_7H_{13}N_3O_4$

N-Asparaginylalanine, A-60310

C₇H₁₄BrN

2-(Bromomethyl)cyclohexanamine, B-70239

C₇H₁₄Br₂

1,5-Dibromo-3,3-dimethylpentane, D-70090

C₇H₁₄N₂

1,4-Diazabicyclo[4.3.0]nonane, D-60054
Diisopropylcyanamide, D-60392
6-Heptyne-2,5-diamine, H-60026
Hexahydro-1*H*-pyrazolo[1,2-*a*]pyridazine, H-70067
Octahydropyrrolo[3,4-*c*]pyrrole; *N*-Me, *in* O-70017

C₇H₁₄N₂O

2-Acetylpiperidine; Oxime (*E*-), *in* A-70051
2-Imidazolidinone; 1-*tert*-Butyl, *in* I-70005
3,3,5,5-Tetramethyl-4-pyrazolidinone, T-60153

C₇H₁₄N₂O₂

2-Hydroxycyclohexanecarboxylic acid; Hydrazide, *in* H-70119

C₇H₁₄N₂O₃

2-Hydroxy-2-isopropylsuccinamide, *in* H-60165

C₇H₁₄N₂O₃S

S-(2-Aminoethyl)cysteine; *N*^α^-Ac, *in* A-60152
S-(2-Aminoethyl)cysteine; *N*^ε^-Ac, *in* A-60152

C₇H₁₄N₂O₆

N-Carbamoylglucosamine, C-70015

C₇H₁₄N₂O₆S

γ-Glutamyltaurine, G-60027

C₇H₁₄N₄O₅

3,3-Biss(ethylnitraminomethyl)oxetane, *in* B-60140

C₇H₁₄O

1-Methoxy-1-hexene, M-70046
4-Methoxy-2-hexene, *in* H-60077

C₇H₁₄OS₂

1,5-Dithionane; 1-Oxide, *in* D-70526

C₇H₁₄O₂

3,4-Dimethylpentanoic acid, D-60435
4,4-Dimethylpentanoic acid, D-70428
2-Hydroxycyclohexanemethanol, H-70120

C₇H₁₄O₃

5,5-Dimethoxy-2-pentanone, *in* O-60083
2-Hydroxy-2-methylhexanoic acid, H-60184
2-Hydroxy-3-methylhexanoic acid, H-60185
3-Hydroxy-2-methylpentanoic acid; Me ester, *in* H-60192

C₇H₁₄O₃S

2-Mercaptopropanoic acid; Et ester, *S*-Ac, *in* M-60027

C₇H₁₄O₄

5,6-Dihydroxyhexanoic acid; Me ester, *in* D-60329
Glycerol 1-acetate; Di-Me ether, *in* G-60029

C₇H₁₄O₅

5-(Hydroxymethyl)-1,2,3,4-cyclohexanetetrol, H-70170

C₇H₁₄S₂

1,1-Cyclopentanedimethanethiol, C-70243
1,5-Dithionane, D-70526

C₇H₁₅N

▷Octahydroazocine, O-70010
1,2,5-Trimethylpyrrolidine, *in* D-70437

C₇H₁₅NO

2-(Aminomethyl)cyclohexanol, A-70169
4,4-Dimethylpentanoic acid; Amide, *in* D-70428
3-Piperidineethanol, P-70103

C₇H₁₅NO₂

2-Amino-2,3-dimethylbutanoic acid; Me ester, *in* A-60148

2-Amino-3,3-dimethylbutanoic acid; Me ester, *in* A-70136
2-Amino-4,4-dimethylpentanoic acid, A-70137
2-Amino-2-methylbutanoic acid; Et ester, *in* A-70167
4-Nitroheptane, N-60027

C₇H₁₅NO₃

3-Amino-2-hydroxy-5-methylhexanoic acid, A-60188
Carnitine, C-70024
N-(2,3-Dihydroxy-3-methylbutyl)acetamide, *in* A-60216
3-Nitro-1-propanol; *tert*-Butyl ether, *in* N-60039

C₇H₁₅NO₄

1,1-Diethoxy-3-nitropropane, *in* N-60037

C₇H₁₅NS

3-Aminotetrahydro-2*H*-thiopyran; *N*,*N*-Di-Me, *in* A-70200
2-Pyrrolidinemethanethiol; 1-Et, *in* P-70185

C₇H₁₅N₂O⊕

3-Cyano-2-hydroxy-*N*,*N*,*N*-trimethyl-propanaminium, *in* C-70024

C₇H₁₅N₃O₃

N-Ornithylglycine, O-70047

C₇H₁₆N₂O₃S

Aminoiminomethanesulfonic acid; *N*,*N*-Diisopropyl, *in* A-60210

C₇H₁₆N₂O₄

2-Amino-4-hydroxybutanoic acid; *O*-(2-Amino-3-hydroxypropyl), *in* A-70156

C₇H₁₆O

3,4-Dimethyl-1-pentanol, D-60436

C₇H₁₆S

3-Ethyl-3-pentanethiol, E-60053

C₇H₁₆S₂

3,3-Dimethyl-1,5-pentanedithiol, D-70427
1,7-Heptanedithiol, H-70023

C₇H₁₇N₃O₄S

N-Ornithyltaurine, O-60042

C₈F₄O₃

4,5,6,7-Tetrafluoro-1,3-isobenzofurandione, *in* T-70031

C₈F₁₂

Dodecafluorobicyclobutylidene, D-60513

C₈F₁₂S₂

2,4-Bis[2,2,2-Trifluoro-1-(trifluoromethyl)-ethylidene]-1,3-dithietane, B-60188

C₈H₂F₄O₄

2,3,5,6-Tetrafluoro-1,4-benzenedicarboxylic acid, T-70029
2,4,5,6-Tetrafluoro-1,3-benzenedicarboxylic acid, T-70030
3,4,5,6-Tetrafluoro-1,2-benzenedicarboxylic acid, T-70031

C₈H₂O₂S₄

Benzo[1,2-*d*:4,5-*d*′]bis[1,3]dithiole-2,6-dione, B-70028

C₈H₃Cl₂F₃N₂

▷4,5-Dichloro-2-(trifluoromethyl)-1*H*-benzimidazole, D-60157
▷4,6-Dichloro-2-(trifluoromethyl)-1*H*-benzimidazole, D-60158
▷4,7-Dichloro-2-(trifluoromethyl)-1*H*-benzimidazole, D-60159
▷5,6-Dichloro-2-(trifluoromethyl)-1*H*-benzimidazole, D-60160

C₈H₃F₄N₃O

4,5,6,7-Tetrafluoro-1*H*-benzotriazole; 1-Ac, *in* T-70035

C₈H₃NO₆

3-Hydroxy-6-nitro-1,2-benzenedicarboxylic acid; Anhydride, *in* H-70189
4-Hydroxy-5-nitro-1,2-benzenedicarboxylic acid; Anhydride, *in* H-70191

C₈H₄Br₂O

2,3-Dibromobenzofuran, D-60086
5,7-Dibromobenzofuran, D-60087

C₈H₄ClF₃O

2-Chloro-3-(trifluoromethyl)benzaldehyde, C-70164
2-Chloro-5-(trifluoromethyl)benzaldehyde, C-70165
2-Chloro-6-(trifluoromethyl)benzaldehyde, C-70166
4-Chloro-3-(trifluoromethyl)benzaldehyde, C-70167
5-Chloro-2-(trifluoromethyl)benzaldehyde, C-70168

C₈H₄Cl₂N₂

1,4-Dichlorophthalazine, D-60148
▷2,3-Dichloroquinoxaline, D-60150

C₈H₄Cl₂N₂O

2,3-Dichloroquinoxaline; 1-Oxide, *in* D-60150

C₈H₄Cl₂O

2,3-Dichlorobenzofuran, D-60125
5,7-Dichlorobenzofuran, D-60126

C₈H₄Cl₂O₂

4,6-Dichloro-1,3-benzenedicarboxaldehyde, D-60118

C₈H₄FN₃O₅

5-Fluoro-2,4-dinitroaniline; *N*-Ac, *in* F-60031

C₈H₄F₄O₂

2,3,4,5-Tetrafluorobenzoic acid; Me ester, *in* T-70032

C₈H₄F₆

1,2-Biss(rifluoromethyl)benzene, B-60175
1,3-Biss(rifluoromethyl)benzene, B-60176
1,4-Biss(rifluoromethyl)benzene, B-60177

C₈H₄N₂O

2-Cyanofuro[2,3-*b*]pyridine, *in* F-60098
2-Cyanofuro[2,3-*c*]pyridine, *in* F-60102
2-Cyanofuro[3,2-*b*]pyridine, *in* F-60100
2-Cyanofuro[3,2-*c*]pyridine, *in* F-60104
3-Cyanofuro[2,3-*b*]pyridine, *in* F-60099
3-Cyanofuro[2,3-*c*]pyridine, *in* F-60103
3-Cyanofuro[3,2-*b*]pyridine, *in* F-60101
3-Cyanofuro[3,2-*c*]pyridine, *in* F-60105

C₈H₄N₂O₂

5,8-Quinoxalinedione, Q-60010

C₈H₄N₂O₅

4-Amino-5-nitro-1,2-benzenedicarboxylic acid; Anhydride, *in* A-70177
3-Cyano-2-hydroxy-5-nitrobenzoic acid, *in* H-70188

C₈H₄N₄O

Furazano[3,4-*b*]quinoxaline, F-70051

C₈H₄N₄O₂

Furoxano[3,4-*b*]quinoxaline, *in* F-70051

C₈H₄O₆

1,4-Benzoquinone-2,3-dicarboxylic acid, B-70047
1,4-Benzoquinone-2,5-dicarboxylic acid, B-70048
1,4-Benzoquinone-2,6-dicarboxylic acid, B-70049

C₈H₅BrN₂

1-Bromophthalazine, B-70266
5-Bromophthalazine, B-70267
6-Bromophthalazine, B-70268

C₈H₅BrOS

2-Bromobenzo[*b*]thiophene; 1-Oxide, *in* B-60204
3-Bromobenzo[*b*]thiophene; 1-Oxide, *in* B-60205

C₈H₅BrO₂

4-Bromo-1(3*H*)-isobenzofuranone, B-70224
7-Bromo-1(3*H*)-isobenzofuranone, B-70225

C₈H₅BrO₂S

2-Bromobenzo[*b*]thiophene; 1,1-Dioxide, *in* B-60204

3-Bromobenzo[*b*]thiophene; 1,1-Dioxide, *in* B-60205

4-Bromobenzo[*b*]thiophene; 1,1-Dioxide, *in* B-60206

5-Bromobenzo[*b*]thiophene; 1,1-Dioxide, *in* B-60207

6-Bromobenzo[*b*]thiophene; 1,1-Dioxide, *in* B-60208

C$_8$H$_5$BrS

2-Bromobenzo[*b*]thiophene, B-60204
3-Bromobenzo[*b*]thiophene, B-60205
4-Bromobenzo[*b*]thiophene, B-60206
5-Bromobenzo[*b*]thiophene, B-60207
6-Bromobenzo[*b*]thiophene, B-60208
7-Bromobenzo[*b*]thiophene, B-60209

C$_8$H$_5$ClN$_2$

1-Chlorophthalazine, C-70148
5-Chlorophthalazine, C-70149
6-Chlorophthalazine, C-70150

C$_8$H$_5$ClOS

2-Chlorobenzo[*b*]thiophene; 1-Oxide, *in* C-60042

3-Chlorobenzo[*b*]thiophene; 1-Oxide, *in* C-60043

C$_8$H$_5$ClO$_2$

4-Chloro-1(3*H*)-isobenzofuranone, C-70079
7-Chloro-1(3*H*)-isobenzofuranone, C-70080

C$_8$H$_5$ClO$_2$S

2-Chlorobenzo[*b*]thiophene; 1,1-Dioxide, *in* C-60042

3-Chlorobenzo[*b*]thiophene; 1,1-Dioxide, *in* C-60043

7-Chlorobenzo[*b*]thiophene; 1,1-Dioxide, *in* C-60047

C$_8$H$_5$ClS

2-Chlorobenzo[*b*]thiophene, C-60042
3-Chlorobenzo[*b*]thiophene, C-60043
4-Chlorobenzo[*b*]thiophene, C-60044
5-Chlorobenzo[*b*]thiophene, C-60045
6-Chlorobenzo[*b*]thiophene, C-60046
7-Chlorobenzo[*b*]thiophene, C-60047

C$_8$H$_5$Cl$_3$O

2,2,2-Trichloroacetophenone, T-70219

C$_8$H$_5$FO$_2$

4-Fluoro-1(3*H*)-isobenzofuranone, F-70027
7-Fluoro-1(3*H*)-isobenzofuranone, F-70028

C$_8$H$_5$F$_3$N$_2$

2-(Trifluoromethyl)-1*H*-benzimidazole, T-60287

C$_8$H$_5$F$_3$O$_2$

4-Hydroxy-2-(trifluoromethyl)benzaldehyde, H-70228

4-Hydroxy-3-(trifluoromethyl)benzaldehyde, H-70229

2,4,5-Trifluorobenzoic acid; Me ester, *in* T-70240

C$_8$H$_5$F$_5$O

1-(Pentafluorophenyl)ethanol, P-60040

C$_8$H$_5$I

▷1-Iodo-2-phenylacetylene, I-70055

C$_8$H$_5$IO$_2$

4-Iodo-1(3*H*)-isobenzofuranone, I-70044
7-Iodo-1(3*H*)-isobenzofuranone, I-70045

C$_8$H$_5$NO

3*H*-Indol-3-one, I-60035

C$_8$H$_5$NO$_2$

Furo[2,3-*b*]pyridine-2-carboxaldehyde, F-60093
Furo[2,3-*c*]pyridine-2-carboxaldehyde, F-60096
Furo[2,3-*b*]pyridine-3-carboxaldehyde, F-60094
Furo[3,2-*b*]pyridine-2-carboxaldehyde, F-60095
Furo[3,2-*c*]pyridine-2-carboxaldehyde, F-60097
2-(2-Hydroxyphenyl)-2-oxoacetic acid; Nitrile, *in* H-70210
Isatogen, *in* I-60035

C$_8$H$_5$NO$_2$S

2*H*-1,4-Benzothiazine-2,3(4*H*)-dione, B-70054

C$_8$H$_5$NO$_3$

Furo[2,3-*b*]pyridine-2-carboxylic acid, F-60098
Furo[2,3-*c*]pyridine-2-carboxylic acid, F-60102
Furo[2,3-*b*]pyridine-3-carboxylic acid, F-60099
Furo[2,3-*c*]pyridine-3-carboxylic acid, F-60103
Furo[3,2-*b*]pyridine-2-carboxylic acid, F-60100
Furo[3,2-*c*]pyridine-2-carboxylic acid, F-60104
Furo[3,2-*b*]pyridine-3-carboxylic acid, F-60101
Furo[3,2-*c*]pyridine-3-carboxylic acid, F-60105

C$_8$H$_5$NO$_7$

2-Hydroxy-5-nitro-1,3-benzenedicarboxylic acid, H-70188

3-Hydroxy-6-nitro-1,2-benzenedicarboxylic acid, H-70189

4-Hydroxy-3-nitro-1,2-benzenedicarboxylic acid, H-70190

4-Hydroxy-5-nitro-1,2-benzenedicarboxylic acid, H-70191

4-Hydroxy-5-nitro-1,3-benzenedicarboxylic acid, H-70192

C$_8$H$_5$NS$_3$

3-Phenyl-1,4,2-dithiazole-5-thione, P-60088

C$_8$H$_5$N$_2$O

3-Hydroxyiminoindole, *in* I-60035

C$_8$H$_5$N$_3$O$_2$

5-Nitrophthalazine, N-70062

C$_8$H$_5$N$_3$O$_8$

2,4,6-Trinitrobenzoic acid; Me ester, *in* T-70303

C$_8$H$_5$N$_3$O$_9$

3-Hydroxy-5-methyl-2,4,6-trinitrobenzoic acid, H-60193

C$_8$H$_5$N$_5$

Tetrazolo[5,1-*a*]phthalazine, T-70155

C$_8$H$_5$N$_5$O$_9$

2,3,4,6-Tetranitroacetanilide, *in* T-60159

C$_8$H$_6$

1,3,5-Cyclooctatrien-7-yne, C-70240

C$_8$H$_6$BrFO$_2$

2-Bromo-4-fluorobenzoic acid; Me ester, *in* B-60240

C$_8$H$_6$BrNO$_4$

4-(Bromomethyl)-2-nitrobenzoic acid, B-60297

C$_8$H$_6$ClNS

2-Chloro-6-(methylthio)benzonitrile, *in* C-70082

C$_8$H$_6$Cl$_2$O$_2$

2,6-Dichloro-4-methoxybenzaldehyde, *in* D-70121

C$_8$H$_6$Cl$_2$O$_3$

2,3-Dichloro-4-methoxybenzoic acid, *in* D-60133

C$_8$H$_6$FIO$_2$

3-Fluoro-2-iodobenzoic acid; Me ester, *in* F-70021

C$_8$H$_6$FNS

2-Fluoro-6-(methylthio)benzonitrile, *in* F-70030

C$_8$H$_6$FN$_3$O$_5$

6-Fluoro-3,4-dinitroaniline; *N*-Ac, *in* F-60032

C$_8$H$_6$F$_3$NO$_3$

1-Methoxy-2-nitro-4-(trifluoromethyl)benzene, *in* N-60048

1-Methoxy-3-nitro-5-(trifluoromethyl)benzene, *in* N-60052

1-Methoxy-4-nitro-2-(trifluoromethyl)benzene, *in* N-60053

4-Methoxy-1-nitro-2-(trifluoromethyl)benzene, *in* N-60054

C$_8$H$_6$IN

7-Iodo-1*H*-indole, I-70043

C$_8$H$_6$I$_2$

1,4-Diiodopentacyclo[4.2.0.0^{2,5}.0^{3,8}.0^{4,7}]octane, D-70359

C$_8$H$_6$N$_2$O

2-(2-Oxazolyl)pyridine, O-70071
2-(5-Oxazolyl)pyridine, O-70072
3-(2-Oxazolyl)pyridine, O-70073
3-(5-Oxazolyl)pyridine, O-70074
4-(2-Oxazolyl)pyridine, O-70075
4-(5-Oxazolyl)pyridine, O-70076

C$_8$H$_6$N$_2$OS

2,3-Dihydro-2-thioxo-4(1*H*)-quinazolinone, D-60296

2,3-Dihydro-4-thioxo-2(1*H*)-quinazolinone, D-60295

C$_8$H$_6$N$_2$OSe

2,3-Dihydro-2-selenoxo-4(1*H*)-quinazolinone, D-60281

3,4-Dihydro-4-selenoxo-2(1*H*)-quinazolinone, D-60280

C$_8$H$_6$N$_2$O$_2$

1,4-Dihydro-2,3-quinoxalinedione, D-60279
Furo[2,3-*b*]pyridine-2-carboxaldehyde; Oxime, *in* F-60093
Furo[2,3-*c*]pyridine-2-carboxaldehyde; Oxime, *in* F-60096
Furo[3,2-*b*]pyridine-2-carboxaldehyde; Oxime, *in* F-60095
Furo[3,2-*c*]pyridine-2-carboxaldehyde; Oxime, *in* F-60097
Furo[2,3-*b*]pyridine-3-carboxylic acid; Amide, *in* F-60099
Furo[2,3-*c*]pyridine-3-carboxylic acid; Amide, *in* F-60103
Furo[3,2-*b*]pyridine-3-carboxylic acid; Amide, *in* F-60101
Furo[3,2-*c*]pyridine-3-carboxylic acid; Amide, *in* F-60105
3*H*-Indol-3-one; 1-Oxide, oxime, *in* I-60035
5-Nitro-1*H*-indole, N-70052
5,8-Quinazolinediol, Q-60002
5,8-Quinoxalinediol, Q-60008
6,7-Quinoxalinediol, Q-60009

C$_8$H$_6$N$_2$O$_3$

1,2,4-Oxadiazolidine-3,5-dione; 2-Ph, *in* O-70060

1,2,4-Oxadiazolidine-3,5-dione; 4-Ph, *in* O-70060

C$_8$H$_6$N$_2$O$_4$

DDED, *in* E-60049

C$_8$H$_6$N$_2$O$_6$

3-Amino-4-nitro-1,2-benzenedicarboxylic acid, A-70174

3-Amino-6-nitro-1,2-benzenedicarboxylic acid, A-70175

4-Amino-3-nitro-1,2-benzenedicarboxylic acid, A-70176

4-Amino-5-nitro-1,2-benzenedicarboxylic acid, A-70177

5-Amino-3-nitro-1,2-benzenedicarboxylic acid, A-70178

2,6-Dinitrobenzoic acid; Me ester, *in* D-60457
3,5-Dinitrobenzoic acid; Me ester, *in* D-60458

C$_8$H$_6$N$_4$O$_7$

2,4,6-Trinitroacetanilide, *in* T-70301

C$_8$H$_6$OS

Benzo[*b*]thiophen-3(2*H*)-one, B-60048
1(3*H*)-Isobenzofuranthione, I-70063

C$_8$H$_6$O$_2$

Bicyclo[2.2.2]octa-5,7-diene-2,3-dione, B-70109
▷Bicyclo[4.2.0]octa-1,5-diene-3,4-dione, B-70110

$C_8H_6O_3$

4H-1,3-Benzodioxin-2-one, B-60028

$C_8H_6O_3S$

Benzo[b]thiophen-3(2H)-one; 1,1-Dioxide, in B-60048

$C_8H_6O_4$

1,2-Benzoquinone-3-carboxylic acid; Me ester, in B-70045

1,2-Benzoquinone-4-carboxylic acid; Me ester, in B-70046

5,7-Dihydroxy-1(3H)-isobenzofuranone, D-60339

2-Hydroxy-4,5-methylenedioxybenzaldehyde, H-70175

2-(2-Hydroxyphenyl)-2-oxoacetic acid, H-70210

2-(3-Hydroxyphenyl)-2-oxoacetic acid, H-70211

2-(4-Hydroxyphenyl)-2-oxoacetic acid, H-70212

$C_8H_6O_5$

2-Formyl-3,4-dihydroxybenzoic acid, F-60070

2-Formyl-3,5-dihydroxybenzoic acid, F-60071

4-Formyl-2,5-dihydroxybenzoic acid, F-60072

6-Formyl-2,3-dihydroxybenzoic acid, F-60073

2-Hydroxy-1,3-benzenedicarboxylic acid, H-60101

2-Hydroxy-1,4-benzenedicarboxylic acid, H-70105

2-Hydroxy-3,4-methylenedioxybenzoic acid, H-60177

3-Hydroxy-4,5-methylenedioxybenzoic acid, H-60178

4-Hydroxy-2,3-methylenedioxybenzoic acid, H-60179

6-Hydroxy-2,3-methylenedioxybenzoic acid, H-60180

6-Hydroxy-3,4-methylenedioxybenzoic acid, H-60181

$C_8H_6O_6$

4,5-Dihydroxy-1,3-benzenedicarboxylic acid, D-60307

C_8H_6S

Cyclopenta[b]thiapyran, C-60224

$C_8H_6S_2$

2-(2,4-Cyclopentadienylidene)-1,3-dithiole, C-60221

$C_8H_7BrClNO_3$

5-Bromo-7-chlorocavernicolin, B-70199

C_8H_7BrO

2-Bromo-3-methylbenzaldehyde, B-60292

$C_8H_7BrO_2$

5'-Bromo-2'-hydroxyacetophenone, B-60265

2-Bromo-3-methylbenzoic acid, B-60293

3-Bromo-4-methylbenzoic acid, B-60294

$C_8H_7BrO_2S$

2-Bromo-4-(methylthio)benzoic acid, in B-70226

2-Bromo-5-(methylthio)benzoic acid, in B-70227

3-Bromo-4-(methylthio)benzoic acid, in B-70228

$C_8H_7BrO_4$

4-Acetoxy-2-bromo-5,6-epoxy-2-cyclohexen-1-one, in B-60239

$C_8H_7Br_2F$

1,2-Bis(bromomethyl)-3-fluorobenzene, B-70131

1,2-Bis(bromomethyl)-4-fluorobenzene, B-70132

1,3-Bis(bromomethyl)-2-fluorobenzene, B-70133

$C_8H_7Br_2NO_2$

1,3-Bis(bromomethyl)-5-nitrobenzene, B-60145

$C_8H_7Br_2NO_3$

3,5-Dibromo-1,6-dihydroxy-4-oxo-2-cyclohexene-1-acetonitrile, D-60098

$C_8H_7ClO_2$

▷Benzyloxycarbonyl chloride, B-60071

3-Chloro-4-methoxybenzaldehyde, in C-60089

4-Chloro-3-methoxybenzaldehyde, in C-60090

$C_8H_7ClO_2S$

5-Chloro-2-(methylthio)benzoic acid, in C-70085

$C_8H_7ClO_3$

3-Chloro-2,6-dihydroxy-4-methylbenzaldehyde, C-60059

3-Chloro-4,6-dihydroxy-2-methylbenzaldehyde, C-60060

3-Chloro-4-hydroxybenzoic acid; Me ester, in C-60091

3-Chloro-6-hydroxy-2-methoxybenzaldehyde, in C-60058

5-Chloro-2-hydroxy-3-methoxybenzaldehyde, in C-70065

2-Chloro-3-hydroxy-5-methylbenzoic acid, C-60095

3-Chloro-4-hydroxy-5-methylbenzoic acid, C-60096

5-Chloro-4-hydroxy-2-methylbenzoic acid, C-60097

3-Chloro-4-methoxybenzoic acid, in C-60091

3-Chloro-5-methoxybenzoic acid, in C-60092

4-Chloro-3-methoxybenzoic acid, in C-60093

C_8H_7FO

2-Fluoro-2-phenylacetaldehyde, F-60062

$C_8H_7FO_2S$

4-Fluoro-2-mercaptobenzoic acid; Me ester, in F-70031

$C_8H_7FO_3$

3-Fluoro-2-hydroxybenzoic acid; Me ester, in F-60040

3-Fluoro-4-hydroxybenzoic acid; Me ester, in F-60041

4-Fluoro-2-hydroxybenzoic acid; Me ester, in F-60042

5-Fluoro-2-hydroxybenzoic acid; Me ester, in F-60044

2-Fluoro-5-methoxybenzoic acid, in F-60038

3-Fluoro-4-methoxybenzoic acid, in F-60041

5-Fluoro-2-methoxybenzoic acid, in F-60044

$C_8H_7F_3S$

1-(Methylthio)-4-(trifluoromethyl)benzene, in T-60286

C_8H_7IO

2-Iodoacetophenone, I-60039

$C_8H_7IO_2$

5-Iodo-2-methoxybenzaldehyde, in H-70143

$C_8H_7IO_2S$

3-Iodo-4-(methylthio)benzoic acid, in I-70046

$C_8H_7IO_3$

2-Iodosophenylacetic acid, I-60063

1-Methoxy-1,2-benziodoxol-3(1H)one, in H-60102

C_8H_7N

Cyanocyclooctatetraene, in C-70239

Indolizine, I-60033

C_8H_7NO

6-Hydroxyindole, H-60158

1(3H)-Isobenzofuranimine, I-70062

2-Methylfuro[2,3-b]pyridine, M-60084

3-Methylfuro[2,3-b]pyridine, M-60085

Phthalimidine, P-70100

$C_8H_7NO_2$

5,6-Dihydroxyindole, D-70307

$C_8H_7NO_2S$

1-Nitro-2-(phenylthio)ethylene, N-70061

$C_8H_7NO_2S_2$

2H-1,4-Benzothiazine-3(4H)-thione; 1,1-Dioxide, in B-70055

$C_8H_7NO_3$

Amino-1,4-benzoquinone; N-Ac, in A-60089

2-(2-Hydroxyphenyl)-2-oxoacetic acid; Amide, in H-70210

2-Methyl-4-nitrobenzaldehyde, M-70096

(2-Nitrophenyl)oxirane, N-70056

(3-Nitrophenyl)oxirane, N-70057

$C_8H_7NO_3S$

1-Nitro-2-(phenylsulfinyl)ethylene, in N-70061

$C_8H_7NO_4$

3-Amino-7-oxabicyclo[4.1.0]hept-3-ene-2,5-dione, in A-60194

2-Hydroxy-5-methyl-3-nitrobenzaldehyde, H-70180

4-(2-Nitroethenyl)-1,2-benzenediol, N-70049

$C_8H_7NO_4S$

1-Nitro-2-(phenylsulfonyl)ethylene, in N-70061

$C_8H_7NO_5$

3-Amino-4-hydroxy-1,2-benzenedicarboxylic acid, A-70151

4-Amino-5-hydroxy-1,2-benzenedicarboxylic acid, A-70152

4-Amino-6-hydroxy-1,3-benzenedicarboxylic acid, A-70153

5-Amino-2-hydroxy-1,3-benzenedicarboxylic acid, A-70154

5-Amino-4-hydroxy-1,3-benzenedicarboxylic acid, A-70155

2',3'-Dihydroxy-6'-nitroacetophenone, D-60357

2',5'-Dihydroxy-4'-nitroacetophenone, D-70322

3',6'-Dihydroxy-2'-nitroacetophenone, D-60358

4',6'-Dihydroxy-2'-nitroacetophenone, D-60359

$C_8H_7NO_6$

4-Hydroxy-5-methoxy-2-nitrobenzoic acid, in D-60360

5-Hydroxy-4-methoxy-2-nitrobenzoic acid, in D-60360

C_8H_7NS

1,4-Benzothiazine, B-70053

$C_8H_7NS_2$

2H-1,4-Benzothiazine-3(4H)-thione, B-70055

$C_8H_7N_3$

1-Aminophthalazine, A-70184

5-Aminophthalazine, A-70185

2-(1H-Pyrazol-1-yl)pyridine, P-70158

2-(1H-Pyrazol-3-yl)pyridine, P-70159

3-(1H-Pyrazol-1-yl)pyridine, P-70160

3-(1H-Pyrazol-3-yl)pyridine, P-70161

4-(1H-Pyrazol-1-yl)pyridine, P-70162

4-(1H-Pyrazol-3-yl)pyridine, P-70163

4-(1H-Pyrazol-4-yl)pyridine, P-70164

$C_8H_7N_3O$

2'-Azidoacetophenone, A-60334

3'-Azidoacetophenone, A-60335

4'-Azidoacetophenone, A-60336

1H-Benzotriazole; N-Ac, in B-70068

Furo[2,3-b]pyridine-3-carboxaldehyde; Hydrazone, in F-60094

$C_8H_7N_3O_2$

6-Amino-2,3-dihydro-1,4-phthalazinedione, A-70134

$C_8H_7N_3S$

1,2,3-Benzotriazine-4(3H)-thione; 3-N-Me, in B-60050

1,2,3-Benzotriazine-4(3H)-thione; S-Me, in B-60050

4-Mercapto-2-methyl-1,2,3-benzotriazinium hydroxide inner salt, in B-60050

C₈H₇N₅O₂

2-Amino-4(1H)-pteridinone; N²-Ac, in A-60242

C₈H₇N₅O₈

2,3,4,6-Tetranitroaniline; N-Di-Me, in T-60159

C₈H₈

1,2-Dihydropentalene, D-70240
1,4-Dihydropentalene, D-70241
1,5-Dihydropentalene, D-70242
1,6-Dihydropentalene, D-70243
1,6a-Dihydropentalene, D-70244
5,6-Dimethylene-1,3-cyclohexadiene, D-70410
4-Methylene-1,2,5,6-heptatetraene, M-70070
4-Octene-1,7-diyne, O-60028

C₈H₈BrNO₂

3-Amino-2-bromobenzoic acid; Me ester, in A-70097

C₈H₈BrNO₃

2-Bromo-1-ethoxy-3-nitrobenzene, in B-70257

C₈H₈Br₂O₄

4-Acetoxy-2,6-dibromo-5-hydroxy-2-cyclo-hexen-1-one, in D-60097
2,6-Dibromo-4,5-dihydroxy-2-cyclohexen-1-one; 4-Ac, in D-60097

C₈H₈ClN

4-Chloro-2,3-dihydro-1H-indole, C-60055

C₈H₈ClNO₃

4-Chloro-2-ethoxy-1-nitrobenzene, in C-60119
1-Chloro-5-methoxy-2-methyl-3-nitrobenzene, in C-60116
2-Chloro-1-methoxy-3-methyl-4-nitrobenzene, in C-60115

C₈H₈Cl₆O₄

Hexahydro-2,5-biss(richloromethyl)furo[3,2-b]-furan-3a,6a-diol, H-70053

C₈H₈FNO₂

2-Amino-5-fluorobenzoic acid; Me ester, in A-70142
3-Amino-4-fluorobenzoic acid; Me ester, in A-70144
5-Amino-2-fluorobenzoic acid; Me ester, in A-70147

C₈H₈INO

2-Iodoacetophenone; Oxime (Z-), in I-60039

C₈H₈INO₂

3-Hydroxy-2-iodo-6-methylpyridine; O-Ac, in H-60159

C₈H₈N₂

1,1'-Dicyanobicyclopropyl, in B-70114
2H-Pyrrolo[3,4-c]pyridine; 2-Me, in P-70200
6H-Pyrrolo[3,4-b]pyridine; 6-Me, in P-70201

C₈H₈N₂O

▷1-(Diazomethyl)-4-methoxybenzene, D-60064
3,4-Dihydro-2(1H)-quinazolinone, D-60278
Pyrrolo[1,2-a]pyrazin-1(2H)-one; N-Me, in P-70198

C₈H₈N₂O₂

Bicyclo[2.2.2]octa-5,7-diene-2,3-dione; Dioxime, in B-70109

C₈H₈N₂O₂S

Benzo[b]thiophen-3(2H)-one; Hydrazone, 1,1-dioxide, in B-60048

C₈H₈N₂O₃

Nicotinuric acid, N-70034

C₈H₈N₂O₄

2-Hydroxy-5-methyl-3-nitrobenzaldehyde; Oxime, in H-70180
3,6-Pyridazinedicarboxylic acid; Di-Me ester, in P-70168

C₈H₈N₂O₄S₂

1,1'-Dithiobis-2,5-pyrrolidinedione, D-70524

C₈H₈N₂O₆

1,2-Dimethoxy-4,5-dinitrobenzene, in D-70449
1,5-Dimethoxy-2,4-dinitrobenzene, in D-70450
5-Ethoxy-2,4-dinitrophenol, in D-70450
Tartaric acid; Dinitrile, Di-Ac, in T-70005

C₈H₈N₂S

2-Amino-4H-3,1-benzothiazine, A-70092

C₈H₈N₄

2,4-Diaminoquinazoline, D-60048
▷1-Hydrazinophthalazine, H-60092

C₈H₈N₄O

2-Tetrazolin-5-one; 1-Benzyl, in T-60175

C₈H₈N₄O₂

2,4(1H,3H)-Pteridinedione; 1,3-Di-Me, in P-60191

C₈H₈N₄O₃

2,4,6(1H,3H,5H)-Pteridinetrione; 1,3-Di-Me, in P-70134
2,4,6(1H,3H,5H)-Pteridinetrione; 1,7-Di-Me, in P-70134
2,4,6(1H,3H,5H)-Pteridinetrione; 3,7-Di-Me, in P-70134
2,4,7(1H,3H,8H)-Pteridinetrione; 1,3-Di-Me, in P-70135
2,4,7(1H,3H,8H)-Pteridinetrione; 1,6-Di-Me, in P-70135
2,4,7(1H,3H,8H)-Pteridinetrione; 1,8-Di-Me, in P-70135
2,4,7(1H,3H,8H)-Pteridinetrione; 3,6-Di-Me, in P-70135
2,4,7(1H,3H,8H)-Pteridinetrione; 3,8-Di-Me, in P-70135
2,4,7(1H,3H,8H)-Pteridinetrione; 6,8-Di-Me, in P-70135

C₈H₈N₄O₆

2,4,6-Trinitroaniline; N-Di-Me, in T-70301

C₈H₈N₄S

Tetrazole-5-thione; 1-Me, 4-Ph, in T-60174
Tetrazole-5-thione; 1-N-Ph, S-Me, in T-60174

C₈H₈O

▷Acetophenone, A-60017
Bicyclo[2.2.1]hepta-2,5-diene-2-carboxaldehyde, B-60083
Bicyclo[5.1.0]octa-3,5-dien-2-one, B-70111
4,7-Dihydroisobenzofuran, D-60253
▷(4-Hydroxyphenyl)ethylene, H-70209
9-Oxabicyclo[4.2.1]nona-2,4,7-triene, O-70057
1-Phenylethenol, P-60092

C₈H₈OS

Benzeneethanethioic acid, B-70017
4'-Mercaptoacetophenone, M-60017

C₈H₈O₂

Bicyclo[2.2.1]hepta-2,5-diene-2-carboxylic acid, B-60084
Bicyclo[4.2.0]oct-7-ene-2,5-dione, B-60097
4-(2-Furanyl)-3-buten-2-one, F-60086
2-(Hydroxymethyl)benzaldehyde, H-60171
2-Hydroxy-4-methylbenzaldehyde, H-70160
4-Hydroxy-2-methylbenzaldehyde, H-70161
2-Hydroxyphenylacetaldehyde, H-60207
3-Hydroxyphenylacetaldehyde, H-60208
4-Hydroxyphenylacetaldehyde, H-60209
2-Hydroxy-2-phenylacetaldehyde, H-70206
2-Hydroxy-2-phenylacetaldehyde, H-60210
4-Vinyl-1,3-benzenediol, V-70010

C₈H₈O₂S

2-Mercapto-2-phenylacetic acid, M-60024

C₈H₈O₃

2,4-Dihydroxyphenylacetaldehyde, D-60369

2,5-Dihydroxyphenylacetaldehyde, D-60370
3,4-Dihydroxyphenylacetaldehyde, D-60371
7,10-Dioxaspiro[5.4]deca-2,5-dien-4-one, in B-70044
▷2-Hydroxy-3-methoxybenzaldehyde, in D-70285
3-Hydroxy-2-methoxybenzaldehyde, in D-70285
2-Hydroxy-4-methylbenzoic acid, H-70163
4-Hydroxy-2-methylbenzoic acid, H-70164

C₈H₈O₄

3,6-Dimethoxy-1,2-benzoquinone, in D-60309
2-Hydroxy-3-methoxy-5-methyl-1,4-benzoquinone, in D-70320
3-Hydroxy-2-methoxy-5-methyl-1,4-benzoquinone, in D-70320
(2-Hydroxyphenoxy)acetic acid, H-70203
(3-Hydroxyphenoxy)acetic acid, H-70204
(4-Hydroxyphenoxy)acetic acid, H-70205
2',4',6'-Trihydroxyacetophenone, T-70248
▷3',4',5'-Trihydroxyacetophenone, T-60315

C₈H₈O₅

3-Hydroxy-4-methoxy-1-methyl-7-oxabicyclo[4.1.0]hept-3-ene-2,5-dione, in D-70320
2,3,5-Trihydroxybenzoic acid; Me ester, in T-70250
2,3,4-Trihydroxyphenylacetic acid, T-60340
2,3,5-Trihydroxyphenylacetic acid, T-60341
2,4,5-Trihydroxyphenylacetic acid, T-60342
3,4,5-Trihydroxyphenylacetic acid, T-60343

C₈H₈O₇

3,4-Diacetoxy-3,4-dihydro-2,5(2H,5H)-furandione, in T-70005

C₈H₈S₂

2-(2,4-Cyclopentadien-1-ylidene)-1,3-dithiolane, in C-60221

C₈H₉BrO₃

3-Bromo-2,6-dimethoxyphenol, in B-60200

C₈H₉BrO₅

(3-Bromo-2-oxopropylidene)propanedioic acid; Di-Me ester, in B-70262

C₈H₉ClO

1-(Chloromethyl)-2-methoxybenzene, in C-70101
1-(Chloromethyl)-3-methoxybenzene, in C-70102
1-Chloromethyl-4-methoxybenzene, in C-70103
2-Chloro-2-phenylethanol, C-60128

C₈H₉F

2-Fluoro-1,3-dimethylbenzene, F-60026

C₈H₉FO

1-Fluoro-2-methoxy-4-methylbenzene, in F-60054
1-Fluoro-3-methoxy-6-methylbenzene, in F-60057
1-Fluoro-4-methoxy-3-methylbenzene, in F-60059
2-Fluoro-1-methoxy-4-methylbenzene, in F-60053

C₈H₉FO₂

1-Fluoro-2,4-dimethoxybenzene, in F-70018
4-Fluoro-1,2-dimethoxybenzene, in F-60020

C₈H₉N

N-Benzylidenemethylamine, B-60070
2-Cyanobicyclo[2.2.1]hepta-2,5-diene, in B-60084
6,7-Dihydro-5H-1-pyrindine, D-60275
6,7-Dihydro-5H-1-pyrindine, D-70258
6,7-Dihydro-5H-2-pyrindine, D-60276
6,7-Dihydro-5H-2-pyrindine, D-70259

C₈H₉NO

Acetophenone; (E)-Oxime, in A-60017
2,3-Dihydro-5(1H)-indolizinone, D-60251
6,7-Dihydro-5H-1-pyrindine; N-Oxide, in D-60275

N-Methylbenzenecarboximidic acid, M-60045
1-(2-Pyridinyl)-2-propanone, P-60229
1-(3-Pyridinyl)-2-propanone, P-60230
1-(4-Pyridinyl)-2-propanone, P-60231

$C_8H_9NO_2$
2-Hydroxy-4-methylbenzaldehyde; Oxime, *in* H-70160
4-Hydroxy-2-methylbenzaldehyde; Oxime, *in* H-70161
4-Hydroxyphenylacetaldoxime, *in* H-60209
2-(Methoxyacetyl)pyridine, *in* H-70101
Phenylacetohydroxamic acid, P-70059

$C_8H_9NO_2S$
1-Nitro-2-(phenylthio)ethane, N-70060

$C_8H_9NO_3$
C-Benzyloxycarbohydroxamic acid, B-70086
(2-Hydroxyphenoxy)acetic acid; Amide, *in* H-70203
4-(2-Nitroethyl)phenol, N-70050
1*H*-Pyrrole-3-carboxaldehyde; *N*-Ethoxycarbonyl, *in* P-70182

$C_8H_9NO_3S$
4-Carboxy-3-thiopheneacetic acid; *N*-Methylacetamide, *in* C-60018
2,3-Dihydro-1*H*-indole-2-sulfonic acid, D-70221
1-Nitro-2-(phenylsulfinyl)ethane, *in* N-70060

$C_8H_9NO_4$
Antibiotic MT 35214, *in* A-60194
N-(5-Hydroxy-2-oxo-7-oxobicyclo[4.1.0]hept-3-en-3-yl)acetamide, *in* A-60194
3-Methoxy-2-methyl-6-nitrophenol, *in* M-70097
1*H*-Pyrrole-3,4-dicarboxylic acid; Di-Me ester, *in* P-70183

$C_8H_9NO_5$
1*H*-Pyrrole-3,4-dicarboxylic acid; 1-Hydroxy, Di-Me ether, *in* P-70183

C_8H_9NS
Phenylthiolacetamide, *in* B-70017

$C_8H_9N_3$
1*H*-Benzotriazole; 1-Et, *in* B-70068

$C_8H_9N_5$
5-Amino-2-benzyl-2*H*-tetrazole, *in* A-70202
1-(Phenylmethyl)-1*H*-tetrazol-5-amine, *in* A-70202
N-(Phenylmethyl)-1*H*-tetrazol-5-amine, *in* A-70202

$C_8H_9N_5O_2$
N-(2,9-Dihydro-1-methyl-2-oxo-1*H*-purin-6-yl)acetamide, *in* A-60144

C_8H_{10}
Bicyclo[4.1.1]octa-2,4-diene, B-60090
1-Ethynylcyclohexene, E-70051
5-Methyl-5-vinyl-1,3-cyclopentadiene, M-70146
Tetracyclo[3.3.0.0^{2,8}.0^{4,6}]octane, T-70022
Tetracyclo[4.2.0.0^{2,5}.0^{3,8}]octane, T-70023
1,2,3,3*a*-Tetrahydropentalene, T-70075
1,2,3,4-Tetrahydropentalene, T-70076
1,2,3,5-Tetrahydropentalene, T-70077
1,2,4,5-Tetrahydropentalene, T-70078
1,2,4,6*a*-Tetrahydropentalene, T-70079
1,2,6,6*a*-Tetrahydropentalene, T-70080
1,3*a*,4,6*a*-Tetrahydropentalene, T-70081
1,3*a*,6,6*a*-Tetrahydropentalene, T-70082
Tricyclo[3.2.1.0^{2,7}]oct-3-ene, T-60267
Tricyclo[5.1.0.0^{2,8}]oct-3-ene, T-60268
Tricyclo[5.1.0.0^{2,8}]oct-4-ene, T-60269

$C_8H_{10}Br_2O_4$
4-Acetoxy-2,6-dibromo-1,5-dihydroxy-2-cyclohexen-1-one, *in* D-60097

$C_8H_{10}IN$
2-Iodo-4,5-dimethylaniline, I-70028
4-Iodo-2,3-dimethylaniline, I-70029
4-Iodo-2,5-dimethylaniline, I-70030
4-Iodo-3,5-dimethylaniline, I-70031

$C_8H_{10}N_2O$
Acetamidoxime; *N*-Ph, *in* A-60016
1-(3-Pyridinyl)-2-propanone; Oxime, *in* P-60230

$C_8H_{10}N_2O_2$
3-Amino-4(1*H*)-pyridinone; 1-Me, *N*^{3}-Ac, *in* A-60254

$C_8H_{10}N_2O_3$
5-Ethyl-2,4(1*H*,3*H*)-pyrimidinedione; 1-Ac, *in* E-60054
5-Methoxy-2-methyl-3-nitroaniline, *in* A-60222

$C_8H_{10}N_2O_3S$
Aminoiminomethanesulfonic acid; *N*-Benzyl, *in* A-60210

$C_8H_{10}N_2O_4$
3-(2-Aminoethylidene)-7-oxo-4-oxa-1-azabicyclo[3.2.0]heptane-2-carboxylic acid, A-70139
1,2,3,6-Tetrahydro-2,6-dioxo-4-pyrimidineacetic acid; Et ester, *in* T-70060

$C_8H_{10}N_4$
4-Amino-1*H*-imidazo[4,5-*c*]pyridine; 4-*N*-Di-Me, *in* A-70161

$C_8H_{10}N_4O_2$
1*H*-Pyrazolo[3,4-*d*]pyrimidine-4,6(5*H*,7*H*)-dione; 1,5,7-Tri-Me, *in* P-60202
1*H*-Pyrazolo[3,4-*d*]pyrimidine-4,6(5*H*,7*H*)-dione; 2,5,7-Tri-Me, *in* P-60202

$C_8H_{10}N_4O_5$
5-Nitrohistidine; *N*$^\alpha$-Ac, *in* N-60028

$C_8H_{10}O$
3,3*a*,6,6*a*-Tetrahydro-1(2*H*)-pentalenone, T-70083
4,5,6,6*a*-Tetrahydro-2(1*H*)-pentalenone, T-70084

$C_8H_{10}OS$
Methyl 4-methylphenyl sulfoxide, M-70094

$C_8H_{10}O_2$
Bicyclo[4.2.0]octane-2,5-dione, B-60093
5,5-Dimethyl-2-cyclohexene-1,4-dione, D-70392
4,4-Dimethyl-3-oxo-1-cyclopentene-1-carboxaldehyde, D-60432
3-Ethynyl-3-methyl-4-pentenoic acid, E-60059
3*a*,4,7,7*a*-Tetrahydro-1(3*H*)-isobenzofuranone, T-70072
5-Vinyl-1-cyclopentenecarboxylic acid, V-60010

$C_8H_{10}O_2S$
4-Methyl-2-thiophenecarboxaldehyde; Ethylene acetal, *in* M-70141

$C_8H_{10}O_3$
Halleridone, H-60004
4-Oxo-1-cycloheptene-1-carboxylic acid, O-70082

$C_8H_{10}O_3S$
1-Phenylethanesulfonic acid, P-70068

$C_8H_{10}O_4$
[1,1'-Bicyclopropyl]-1,1'-dicarboxylic acid, B-70114
2,3-Dicarbomethoxy-1,3-butadiene, *in* D-70409
3,6-Dimethoxy-1,2-benzenediol, *in* B-60014
5,5-Dimethyl-2-hexynedioic acid, D-70415
4-Hydroxy-2,5-dimethyl-3(2*H*)-furanone; Ac, *in* H-60120

$C_8H_{10}O_5$
5-Acetyl-2,2-dimethyl-1,3-dioxane-4,6-dione, A-60029
Diallyl dicarbonate, *in* D-70111
(2-Oxopropylidene)propanedioic acid; Di-Me ester, *in* O-70100

$C_8H_{10}O_8$
2,3-Di-*O*-acetyltartaric acid, *in* T-70005

$C_8H_{10}S$
2-Phenylethanethiol, P-70069

$C_8H_{10}S_2$
2,5-Dimethyl-1,4-benzenedithiol, D-60406

$C_8H_{11}BrO_2$
2-Bromobicyclo[2.2.1]heptane-1-carboxylic acid, B-60210
7-Bromo-5-heptynoic acid; Me ester, *in* B-60261

$C_8H_{11}BrO_3$
(3-Bromomethyl)-2,4,10-trioxatricyclo[3.3.1.1^{3,7}]decane, B-60301

$C_8H_{11}ClO_5$
2,3-Dihydroxybutanoic acid; Di-Ac, chloride, *in* D-60312

$C_8H_{11}NO$
Hexahydroazirino[2,3,1-*hi*]indol-2(1*H*)-one, H-70046
1,3,3*a*,4,5,7*a*-Hexahydro-2*H*-indol-2-one, H-70060
1,5,6,8*a*-Tetrahydro-3(2*H*)-indolizinone, T-70071

$C_8H_{11}NO_2$
3-Amino-2-cyclohexen-1-one; *N*-Ac, *in* A-70125
2,3-Dimethoxyaniline, *in* A-70089
▷ 3,4-Dimethoxyaniline, *in* A-70090
5-Methyl-1*H*-pyrrole-3-carboxylic acid; Et ester, *in* M-60125

$C_8H_{11}NO_3$
4-Oxoheptanedioic acid; Me ester, mononitrile, *in* O-60071
2,4,6-Trimethoxypyridine, *in* D-70341

$C_8H_{11}NO_4$
5-Oxo-2-pyrrolidinecarboxylic acid; Me ester, *N*-Ac, *in* O-70102
5-Oxo-2-pyrrolidinecarboxylic acid; Me ester, *N*-benzoyl, *in* O-70102

$C_8H_{11}N_2$
Acetophenone; Hydrazone, *in* A-60017

$C_8H_{11}N_3$
2-Amino-5-vinylpyrimidine; *N*^{2},*N*^{2}-Di-Me, *in* A-60269

$C_8H_{11}N_3O_3$
α-Amino-1*H*-imidazole-2-propanoic acid; *N*$^\alpha$-Ac, *in* A-60205
3-Amino-1*H*-pyrazole-4-carboxylic acid; Et ester, *N*(3)-Ac, *in* A-70187
5-Amino-1*H*-pyrazole-4-carboxylic acid; *N*(1)-Ac, Et ester, *in* A-70192
1,2,3,6-Tetrahydro-2,6-dioxo-4-pyrimidineacetic acid; *N*-Ethylamide, *in* T-70060

C_8H_{12}
Bicyclobutylidene, B-70095
Bicyclo[4.1.1]oct-2-ene, B-60095
Bicyclo[4.1.1]oct-3-ene, B-60096
1,2-Biss(ethylene)cyclohexane, B-60166
1,3-Biss(ethylene)cyclohexane, B-60167
1,4-Biss(ethylene)cyclohexane, B-60168
2,5-Dimethyl-1,3,5-hexatriene, D-60424
1,2,3,3*a*,4,5-Hexahydropentalene, H-70064
1,2,3,3*a*,4,6*a*-Hexahydropentalene, H-70065
1,2,3,4,5,6-Hexahydropentalene, H-70066
1,3,5-Octatriene, O-70022
2,4,6-Octatriene, O-70023
3-Octen-1-yne, O-70029
Tricyclo[3.1.1.1^{2,4}]octane, T-60265
Tricyclo[5.1.0.0^{2,8}]octane, T-60266
3-Vinylcyclohexene, V-60009

$C_8H_{12}Br_2$
1,4-Dibromobicyclo[2.2.2]octane, D-70086

$C_8H_{12}Cl_2$
1,4-Dichlorobicyclo[2.2.2]octane, D-70115

$C_8H_{12}Cl_2N_2O_2$
4,5-Dichloro-1*H*-imidazole-2-carboxaldehyde; Di-Et acetal, *in* D-60134

$C_8H_{12}N_2O$
2-*tert*-Butyl-4(3*H*)-pyrimidinone, B-60349

$C_8H_{12}N_2O_2$
[1,1'-Bipyrrolidine]-2,2'-dione, B-70125
▷ 1,3-Diazaspiro[4.5]decane-2,4-dione, D-60056

4,4-Dimethyl-6-nitro-5-hexenenitrile, *in*
D-60429
3-Ethoxy-4,5,6,7-tetrahydroisoxazolo[4,5-*c*]-
pyridine, *in* T-60071
Hexahydropyridazino[1,2-*a*]pyridazine-1,4-
dione, H-70070

C₈H₁₂N₂S₂
2,3-Biss(ethylthio)-1,4-benzenediamine, *in*
D-70036
2,3-Biss(ethylthio)-1,4-benzenediamine, *in*
D-60030
4,6-Diamino-1,3-benzenedithiol; Di-*S*-Me, *in*
D-70037

C₈H₁₂N₄O₄
2′-Deoxy-5-azacytidine, D-70022

C₈H₁₂O
1-Acetylcyclohexene, A-70038
4-Acetylcyclohexene, A-70039
1-Acetyl-2-methylcyclopentene, A-60043
Bicyclo[3.2.1]octan-8-one, B-60094
9-Oxabicyclo[6.1.0]non-4-ene, O-70058
2,5,5-Trimethyl-2-cyclopenten-1-one, T-70281

C₈H₁₂O₂
Bicyclo[2.2.1]heptane-1-carboxylic acid,
B-60085
Bicyclo[3.1.1]heptane-1-carboxylic acid,
B-60086
5,6-Dimethoxy-1,3-cyclohexadiene, D-60397
1,6-Dioxacyclodeca-3,8-diene, D-70458
1,4-Dioxaspiro[4,5]dec-7-ene, *in* C-60211
3,5-Hexadienoic acid; Et ester, *in* H-70038
Hexahydro-2(3*H*)-benzofuranone, H-70049
Hexahydro-1(3*H*)-isobenzofuranone, H-70061
5-Methyl-3,4-hexadienoic acid; Me ester, *in*
M-70087
4-Octenedial, O-70025

C₈H₁₂O₃
4,4-Diethoxy-2-butynal, *in* B-60351
2-Ethoxycarbonylcyclopentanone, *in* O-70084
7-Hydroxy-5-heptynoic acid; Me ester, *in*
H-60150
3-Oxocyclopentaneacetic acid; Me ester, *in*
O-70083

C₈H₁₂O₄
3-Butene-1,2-diol; Di-Ac, *in* B-70291

C₈H₁₂O₄S₂
(Dimercaptomethylene)propanedioic acid; Di-
S-Me, di-Me ester, *in* D-60395

C₈H₁₂O₆
2,3-Dihydroxybutanoic acid; Di-Ac, *in* D-60312

C₈H₁₃BF₄O₂
3,4,4*a*,5,6,7-Hexahydro-2*H*-pyrano[2,3-*b*]-
pyrilium; Tetrafluoroborate, *in* H-60055

C₈H₁₃Br
1-Bromobicyclo[2.2.2]octane, B-70184

C₈H₁₃Br₂NO₃
6-Amino-2,2-dibromohexanoic acid; *N*-Ac, *in*
A-60124

C₈H₁₃Cl
1-Chlorobicyclo[2.2.2]octane, C-70057

C₈H₁₃Cl₂NO₃
6-Amino-2,2-dichlorohexanoic acid; *N*-Ac, *in*
A-60126

C₈H₁₃IO
2-Iodocyclooctanone, I-70027

C₈H₁₃NO
1-Acetylcyclohexene; Oxime, *in* A-70038
3-Aminobicyclo[2.2.1]hept-5-ene-2-methanol,
A-70094
1-Azabicyclo[3.3.1]nonan-2-one, A-70275
2-Azabicyclo[3.3.1]nonan-7-one, A-70276
Bicyclo[2.2.1]heptane-1-carboxylic acid;
Amide, *in* B-60085

C₈H₁₃NO₂
7-Aminobicyclo[4.1.0]heptane-7-carboxylic
acid, A-60095

Hexahydro-2(3*H*)-benzoxazolone; *N*-Me, *in*
H-70052
1-Oxa-2-azaspiro[2,5]octane; *N*-Ac, *in* O-70052

C₈H₁₃NO₃
4-Oxo-2-piperidinecarboxylic acid; *N*-Me, Me
ester, *in* O-60089
3-Piperidinecarboxylic acid; *N*-Ac, *in* P-70102

C₈H₁₃NO₄
1-Amino-1,4-cyclohexanedicarboxylic acid,
A-60117
2-Amino-1,4-cyclohexanedicarboxylic acid,
A-60118
3-Amino-1,2-cyclohexanedicarboxylic acid,
A-60119
4-Amino-1,1-cyclohexanedicarboxylic acid,
A-60120
4-Amino-1,3-cyclohexanedicarboxylic acid,
A-60121
4,4-Dimethyl-6-nitro-5-hexenoic acid, D-60429

C₈H₁₃N₃O₂S
Ovothiol B, *in* O-70050

C₈H₁₃O₂⊕
3,4,4*a*,5,6,7-Hexahydro-2*H*-pyrano[2,3-*b*]-
pyrilium, H-60055

C₈H₁₃O₅P
8,8*a*-Dihydro-3-methoxy-5-methyl-1*H*,6*H*-
furo[3,4-*e*][1,3,2]dioxaphosphepin 3-oxide,
D-70223

C₈H₁₄
Bicyclo[4.1.1]octane, B-60091
1-Ethylcyclohexene, E-60046
3-Ethylcyclohexene, E-70039
4-Ethylcyclohexene, E-60048

C₈H₁₄BrNO₃
6-Amino-2-bromohexanoic acid; *N*-Ac, *in*
A-60100

C₈H₁₄ClNO₃
6-Amino-2-chlorohexanoic acid; *N*-Ac, *in*
A-60109

C₈H₁₄ClNO₄
2,3,5,6,7,8-Hexahydro-1*H*-indolizinium(1+);
Perchlorate, *in* H-70059

C₈H₁₄N⊕
2,3,5,6,7,8-Hexahydro-1*H*-indolizinium(1+),
H-70059

C₈H₁₄N₂
4,5-Dihydro-3,3,5,5-tetramethyl-4-methylene-
3*H*-pyrazole, D-60283

C₈H₁₄N₂O₂
1,3-Diazetidine-2,4-dione; 1,3-Diisopropyl, *in*
D-60057
4,6-Diethoxy-2,5-dihydropyrimidine, *in*
D-70256

C₈H₁₄N₂O₄
2,6-Diamino-4-methyleneheptanedioic acid,
D-60039
2,5-Piperazinedicarboxylic acid; Di-Me ester, *in*
P-60157
Proclavaminic acid, P-70116
N-Prolylserine, P-60173
N-Serylproline, S-60025

C₈H₁₄N₂S₂
4,6-Biss(thylthio)-2,5-dihydropyrimidine, *in*
D-70257

C₈H₁₄N₄O₂
Tetrahydroimidazo[4,5-*d*]imidazole-
2,5(1*H*,3*H*)-dione; 1,3,4,6-Tetra-Me, *in*
T-70066

C₈H₁₄N₄S₂
1,6-Diallyl-2,5-dithiobiurea, D-60026

C₈H₁₄O
2,2-Dimethyl-4-hexen-3-one, D-70414
2,2-Dimethyl-5-hexen-3-one, D-60425
4-Methyl-3-hepten-2-one, M-60087
2,4-Octadien-1-ol, O-70007
3-Octenal, O-70024
3-Octen-2-one, O-70027

4-Octen-2-one, O-60030
5-Octen-2-one, O-70028
2,2,3,3-Tetramethyl-
cyclopropanecarboxaldehyde, T-70133

C₈H₁₄OS
Dihydro-3,3,4,4-tetramethyl-2(3*H*)-
furanthione, D-60282
4,5-Dihydro-3,3,4,4-tetramethyl-2(3*H*)-
thiophenone, D-60288
2,3,4,5,8,9-Hexahydrothionin; 1-Oxide, *in*
H-70076

C₈H₁₄O₂
4,4-Dimethoxycyclohexene, *in* C-60211
4-Hexen-3-ol; Ac, *in* H-60077
3-Methyl-2-heptenoic acid, M-60086
4-Oxooctanal, O-60081

C₈H₁₄O₂S
S-Butyl 3-oxobutanethioate, *in* O-70080
S-*tert*-Butyl 3-oxobutanethioate, *in* O-70080
1,4-Dioxaspiro[4.5]decane-7-thiol, *in* M-70026
4-Mercapto-2-butenoic acid; *tert*-Butyl ester, *in*
M-60018
Octahydrobenzo[*b*]thiophene; 1,1-Dioxide, *in*
O-60012

C₈H₁₄O₂S₂
Bis(3-hydroxycyclobutyl)disulfide, *in* M-60019

C₈H₁₄O₃
2-Cyclohexyl-2-hydroxyacetic acid, C-60212
4,4-Dimethyl-5-oxopentanoic acid; Me ester, *in*
D-60434
2-Formylpropanoic acid; *tert*-Butyl ester, *in*
F-70039
2-Hydroxycyclohexanecarboxylic acid; Me
ester, *in* H-70119
4-Hydroxy-4-(2-hydroxyethyl)cyclohexanone,
H-60153

C₈H₁₄O₄
Tartaric acid; Mono-*tert*-butyl ester, *in* T-70005
Tartaric acid; Mono-*tert-butyl ether*, *in*
T-70005

C₈H₁₄O₆
Diethyl tartrate, *in* T-70005

C₈H₁₄S
2,3,4,5,8,9-Hexahydrothionin, H-70076
Octahydrobenzo[*b*]thiophene, O-60012

C₈H₁₄S₂
4,5-Dihydro-3,3,4,4-tetramethyl-2(3*H*)-
thiophenethione, D-60287

C₈H₁₅Br
1-Bromo-1-octene, B-70261

C₈H₁₅BrO₂
5-Bromo-2-hexanone; Ethylene acetal, *in*
B-60262

C₈H₁₅ClO
2,2-Diethylbutanoic acid; Chloride, *in* D-60171

C₈H₁₅F
1-Fluoro-1-octene, F-70034

C₈H₁₅I
1-Iodo-1-octene, I-70051

C₈H₁₅IO
3-Iodo-2-octanone, I-60053

C₈H₁₅N
8-Azabicyclo[5.2.0]nonane, A-70274
1,2,3,6-Tetrahydropyridine; 1-Propyl, *in*
T-70089
Tropane, *in* A-70278

C₈H₁₅NO
3-Aminobicyclo[2.2.1]heptane-2-methanol,
A-70093
Hexahydro-5(2*H*)-azocinone; 1-Me, *in* H-70048
Octahydro-2*H*-1,3-benzoxazine, O-70011
Octahydro-2*H*-3,1-benzoxazine, O-70012

▷Physoperuvine, P-60151
Tetrahydro-2-(2-pyrrolidinyl)furan, T-60090

$C_8H_{15}NO_2$

2-Aminocycloheptanecarboxylic acid, A-70123
3-Amino-2-hexenoic acid; Et ester, in A-60170
2-Methyl-2-piperidinecarboxylic acid; Me ester, in M-60119
3-Piperidinecarboxylic acid; Et ester, in P-70102
4-(Tetrahydro-2-furanyl)morpholine, T-60065

$C_8H_{15}NO_3$

2-Amino-4,4-dimethylpentanoic acid; N-Formyl, in A-70137

$C_8H_{15}NO_4$

2-Amino-3-butylbutanedioic acid, A-60102

C_8H_{16}

Isopropylcyclopentane, I-70092

$C_8H_{16}BrNO_2$

2-Amino-4-bromobutanoic acid; tert-Butyl ester, in A-60098

$C_8H_{16}Cl_2$

1,1-Dichlorooctane, D-70126
1,2-Dichlorooctane, D-70127
1,3-Dichlorooctane, D-70128
2,2-Dichlorooctane, D-70129
2,3-Dichlorooctane, D-70130
4,5-Dichlorooctane, D-70131

$C_8H_{16}N_2$

Decahydroquinazoline, D-70013
Octahydropyridazino[1,2-a]pyridazine, O-70016

$C_8H_{16}N_2O$

▷Tetrahydro-2(1H)-pyrimidinone; 1,3-Di-Et, in T-60088

$C_8H_{16}N_2O_3S$

N-Methionylalanine, M-60035

$C_8H_{16}N_2O_4S$

N-Serylmethionine, S-60024

$C_8H_{16}N_4O_2$

1,3,5,7-Tetramethyltetrahydro-1,3,5,7-tetrazocine-2,6(1H,3H)dione, in T-70098

$C_8H_{16}O$

2-Octen-1-ol, O-60029
2,2,3,3-Tetramethylcyclopropanemethanol, T-70134

$C_8H_{16}OS$

2-Methyl-4-propyl-1,3-oxathiane, M-60122

$C_8H_{16}O_2$

2,2-Diethylbutanoic acid, D-60171
3,4-Dimethylpentanoic acid; Me ester, in D-60435
4,4-Dimethylpentanoic acid; Me ester, in D-70428
5-Hydroxy-4-methyl-3-heptanone, H-60183
4-Octene-1,8-diol, O-70026
Pityol, P-60160

$C_8H_{16}O_3$

1-(2-Hydroxyethyl)-1,4-cyclohexanediol, H-60133
Isorengyol, in H-60133

$C_8H_{16}O_4$

Methyl 3,3-diethoxypropionate, in O-60090

$C_8H_{16}O_5$

5,7,8-Trimethoxyflavone, in T-60321

$C_8H_{16}S_2$

1,1-Cyclohexanedimethanethiol, C-70223
1,2-Cyclohexanedimethanethiol, C-70224
1,2-Dithiecane, D-60507

$C_8H_{17}BrO$

8-Bromo-1-octanol, B-70260

$C_8H_{17}N$

Octahydroazocine; 1-Me, in O-70010

$C_8H_{17}NO$

2-Aminocycloheptanemethanol, A-70124
2-(Aminomethyl)cycloheptanol, A-70168
2,2-Diethylbutanoic acid; Amide, in D-60171

$C_8H_{17}NO_3$

Carnitine; Me ether, in C-70024

$C_8H_{17}N_3O_3$

N-Ornithyl-β-alanine, O-60041

$C_8H_{18}N_2$

Hexahydropyrimidine; 1,3-Di-Et, in H-60056

$C_8H_{18}N_2O_2$

1,7-Dioxa-4,10-diazacyclododecane, D-60465

$C_8H_{18}N_4$

2,4,6,8-Tetramethyl-2,4,6,8-tetraazabicyclo[3.3.0]octane, in O-70015

$C_8H_{19}N$

2 Octylamine, O-60031

$C_8H_{19}N_3$

1,4,7-Triazacycloundecane, T-60234
1,4,8-Triazacycloundecane, T-60235

$C_8H_{22}N_4$

1,2-Dihydrazinoethane; $N^\alpha,N^{\alpha'},N^\beta,N^\beta,N^{\beta'},N^{\beta'}$-Hexa-Me, in D-60192

$C_8O_2S_6$

2,6-Dithioxobenzo[1,2-d:4,5-d']bis[1,3]dithiole-4,8-dione, D-60512

$C_8O_4S_2$

2,7-Dithiatricyclo[6.2.0.0^{3,6}]deca-1(8),3(6)-diene-4,5,9,10-tetrone, D-60506

$C_9H_3Cl_2NO_2$

6,7-Dichloro-5,8-isoquinolinedione, D-60137
6,7-Dichloro-5,8-quinolinedione, D-60149

$C_9H_3Cl_2NO_3$

6,7-Dichloro-5,8-quinolinedione; N-Oxide, in D-60149

$C_9H_4N_4O_2$

Tetrazolo[1,5-b]quinoline-5,10-dione, T-70156

$C_9H_4N_4O_3$

Imidazo[4,5-g]quinazoline-4,8,9(3H,7H)-trione, I-60010

$C_9H_4N_6O_2$

▷2,2-Diazido-1,3-indanedione, D-70056

$C_9H_5BrF_2$

3-Bromo-3,3-difluoro-1-phenylpropyne, B-70203

$C_9H_5BrO_2$

3-Bromo-4H-1-benzopyran-4-one, B-70183

$C_9H_5Cl_2F_3N_2$

5,6-Dichloro-2-(trifluoromethyl)-1H-benzimidazole; N-Me, in D-60160

$C_9H_5Cl_2N$

6,7-Dichloroisoquinoline, D-60136

$C_9H_5NO_2$

3-(Furo[3,4-b]furan-4-yl)-2-propenenitrile, F-70055
▷5,8-Isoquinolinequinone, I-60131

$C_9H_5NO_6$

4-Hydroxy-5-nitro-1,2-benzenedicarboxylic acid; Me ether, anhydride, in H-70191

$C_9H_5NS_2$

Dithieno[2,3-b:2',3'-d]pyridine, D-70509
Dithieno[2,3-b:3',2'-d]pyridine, D-70510
Dithieno[2,3-b:3',4'-d]pyridine, D-70511
Dithieno[3,2-b:2',3'-d]pyridine, D-70512
Dithieno[3,2-b:3',4'-d]pyridine, D-70513
Dithieno[3,4-b:2',3'-d]pyridine, D-70514
Dithieno[3,4-b:3'2'-d]pyridine, D-70515
Dithieno[3,4-b:3',4'-d]pyridine, D-70516

$C_9H_5N_3O_2$

[1]-Benzopyrano[2,3-d]-1,2,3-triazol-9(1H)-one, B-60043

$C_9H_5N_3O_3$

2,3-Dicyano-1-methoxy-4-nitrobenzene, in H-70189

C_9H_6ClNO

▷5-Chloro-8-hydroxyquinoline, C-70075
8-Chloro-2(1H)-quinolinone, C-60131
8-Chloro-4(1H)-quinolinone, C-60132
1H-Indole-3-acetic acid; Chloride, in I-70013

$C_9H_6ClNO_2$

5-Chloro-8-hydroxyquinoline; 1-Oxide, in C-70075
1-(Chloromethyl)-1H-indole-2,3-dione, C-70092

$C_9H_6Cl_2O_2$

1,3,5-Cycloheptatriene-1,6-dicarboxylic acid; Dichloride, in C-60202

$C_9H_6F_3N$

2-Trifluoromethyl-1H-indole, T-60291
3-Trifluoromethyl-1H-indole, T-60292
6-Trifluoromethyl-1H-indole, T-60293

$C_9H_6N_2$

Imidazo[5,1,2-cd]indolizine, I-70004

$C_9H_6N_2O$

3-Cinnolinecarboxaldehyde, C-70178
4-Cinnolinecarboxaldehyde, C-70179

$C_9H_6N_2O_2$

5,8-Isoquinolinequinone; 8-Oxime, in I-60131
1-Nitroisoquinoline, N-60029
4-Nitroisoquinoline, N-60030
5-Nitroisoquinoline, N-60031
6-Nitroisoquinoline, N-60032
7-Nitroisoquinoline, N-60033
8-Nitroisoquinoline, N-60034
5-Phenyl-1,2,4-oxadiazole-3-carboxaldehyde, P-60122

$C_9H_6N_2O_3$

5-Nitroisoquinoline; 2-Oxide, in N-60031
8-Nitroisoquinoline; 2-Oxide, in N-60034

$C_9H_6N_2S$

3-Amino-2-cyanobenzo[b]thiophene, in A-60091

$C_9H_6N_4$

4-Cyano-1-phenyl-1,2,3-triazole, in P-60138
4-Cyano-2-phenyl-1,2,3-triazole, in P-60140
1H-1,2,3-Triazolo[4,5-f]quinoline, T-70207
1H-1,2,3-Triazolo[4,5-h]quinoline, T-70208

$C_9H_6N_4O$

3,8-Dihydro-9H-pyrazolo[4,3-f]quinazolin-9-one, P-60271
Imidazo[4,5-h]quinazolin-6-one, I-60013
Imidazo[4,5-g]quinazolin-8-one, I-60012
Imidazo[4,5-f]quinazolin-9(8H)-one, I-60011
Pyrazolo[3,4-f]quinazolin-9(8H)-one, P-60204
Pyrazolo[4,3-g]quinazolin-5(6H)-one, P-60205

$C_9H_6N_4O_2$

Imidazo[4,5-g]quinazoline-6,8(5H,7H)-dione, I-60009
1H-Pyrazolo[4,3-g]quinazoline-5,7(6H,8H)-dione, P-60203

$C_9H_6N_4S$

Pyrazino[2,3-b]pyrido[3',2'-e][1,4]thiazine, P-60199

C_9H_6O

5H-Cycloprop[f]isobenzofuran, C-70251
4-Ethynylbenzaldehyde, E-70048

$C_9H_6O_2$

▷1H-2-Benzopyran-1-one, B-60041

$C_9H_6O_2S$

4-Hydroxy-2H-1-benzothiopyran-2-one, H-70110

C₉H₆O₃

3-Benzofurancarboxylic acid, B-**60032**

6-Hydroxy-2*H*-1-benzopyran-2-one, H-**70109**

3-Methyl-4,5-benzofurandione, M-**70051**

C₉H₆O₄

4-Hydroxy-5-benzofurancarboxylic acid, H-**60103**

5-Hydroxy-3-methyl-1,2-benzenedicarboxylic acid; Anhydride, *in* H-**70162**

C₉H₆O₁₂

Cyclopropanehexacarboxylic acid, C-**70250**

C₉H₆S

5*H*-Cyclopropa[*f*][2]benzothiophene, C-**70247**

C₉H₆S₆

Benzo[1,2-*d*:3,4-*d*':5,6-*d*″]tris[1,3]dithiole, B-**70069**

C₉H₇ClN₂

1-Chloro-4-methylphthalazine, C-**70104**

2-(Chloromethyl)quinoxaline, C-**70120**

2-Chloro-3-methylquinoxaline, C-**70121**

2-Chloro-5-methylquinoxaline, C-**70122**

2-Chloro-6-methylquinoxaline, C-**70123**

2-Chloro-7-methylquinoxaline, C-**70124**

6-Chloro-7-methylquinoxaline, C-**70125**

C₉H₇ClN₂O

2-Chloro-3-methylquinoxaline; 1-Oxide, *in* C-**70121**

2-Chloro-3-methylquinoxaline; 4-Oxide, *in* C-**70121**

6-Chloro-7-methylquinoxaline; 1-Oxide, *in* C-**70125**

3-Chloro-7-methylquinoxaline 1-oxide, *in* C-**70123**

C₉H₇ClN₂O₂

2-Chloro-3-methylquinoxaline; 1,4-Dioxide, *in* C-**70121**

6-Chloro-7-methylquinoxaline; 1,4-Dioxide, *in* C-**70125**

C₉H₇ClO₃

1,4-Benzodioxan-2-carboxylic acid; Chloride, *in* B-**60024**

1,4-Benzodioxan-6-carboxylic acid; Chloride, *in* B-**60025**

C₉H₇ClO₄

3-Hydroxy-4,5-methylenedioxybenzoic acid; Me ether, chloride, *in* H-**60178**

6-Hydroxy-3,4-methylenedioxybenzoic acid; Me ether, chloride, *in* H-**60181**

C₉H₇ClO₄S

1-Benzothiopyrylium(1+); Perchlorate, *in* B-**70066**

2-Benzothiopyrylium(1+); Perchlorate, *in* B-**70067**

C₉H₇F₃O₂

4-Methoxy-2-(trifluoromethyl)benzaldehyde, *in* H-**70228**

4-Methoxy-3-(trifluoromethyl)benzaldehyde, *in* H-**70229**

C₉H₇F₃S

1-(Methylthio)-3-(trifluoromethyl)benzene, *in* T-**60285**

C₉H₇IO₄

1-Hydroxy-1,2-benziodoxol-3(1*H*)-one; Ac, *in* H-**60102**

C₉H₇N

Cyclopent[*b*]azepine, C-**60225**

C₉H₇NO

Cyclohepta[*b*]pyrrol-2(1*H*)-one, C-**60201**

Cyclopent[*b*]azepine; *N*-Oxide, *in* C-**60225**

5,6-Epoxyquinoline, E-**60027**

7,8-Epoxyquinoline, E-**60028**

1*H*-Indole-5-carboxaldehyde, I-**60029**

1*H*-Indole-6-carboxaldehyde, I-**60030**

1*H*-Indole-7-carboxaldehyde, I-**60031**

5-Phenyloxazole, P-**60125**

3*H*-Pyrrolo[1,2-*a*]azepin-3-one, P-**70190**

5*H*-Pyrrolo[1,2-*a*]azepin-5-one, P-**70191**

7*H*-Pyrrolo[1,2-*a*]azepin-7-one, P-**70192**

9*H*-Pyrrolo[1,2-*a*]azepin-9-one, P-**70193**

C₉H₇NO₂

1,4-Benzodioxan-2-carbonitrile, *in* B-**60024**

1,3-Benzodioxole-4-acetonitrile, *in* B-**60029**

2-(3-Hydroxyphenyl)-2-oxoacetic acid; Me ether, nitrile, *in* H-**70211**

1,3(2*H*,4*H*)-Isoquinolinedione, I-**70099**

4-Methoxy-α-oxobenzeneacetonitrile, *in* H-**70212**

C₉H₇NO₂S

▷3-Amino-2-benzo[*b*]thiophenecarboxylic acid, A-**60091**

C₉H₇NO₃

3,5-Dihydroxy-4-phenylisoxazole, D-**70337**

1*H*-Isoindol-1-one-3-carboxylic acid, I-**60102**

3-Methoxy-4,5-methylnedioxybenzonitrile, *in* H-**60178**

C₉H₇NO₄

3-Amino-4-hydroxy-1,2-benzenedicarboxylic acid; Me ether, anhydride, *in* A-**70151**

5,6-Dihydroxy-1*H*-indole-2-carboxylic acid, D-**70308**

C₉H₇NO₇

3-Methoxy-6-nitro-1,2-benzenedicarboxylic acid, *in* H-**70189**

4-Methoxy-3-nitro-1,2-benzenedicarboxylic acid, *in* H-**70190**

4-Methoxy-5-nitro-1,2-benzenedicarboxylic acid, *in* H-**70191**

4-Methoxy-5-nitro-1,3-benzenedicarboxylic acid, *in* H-**70192**

C₉H₇NS

1(2*H*)-Isoquinolinethione, I-**60132**

3(2*H*)-Isoquinolinethione, I-**60133**

3-Quinolinethiol, Q-**60003**

5-Quinolinethiol, Q-**60004**

6-Quinolinethiol, Q-**60005**

4(1*H*)-Quinolinethione, Q-**60006**

C₉H₇N₃O

4-Benzoyl-1,2,3-triazole, B-**70079**

4-Cinnolinecarboxaldehyde; Oxime, *in* C-**70179**

C₉H₇N₃OS

1,2,3-Benzotriazine-4(3*H*)-thione; *N*-Ac, *in* B-**60050**

C₉H₇N₃O₂

6-Amino-8-nitroquinoline, A-**60234**

8-Amino-6-nitroquinoline, A-**60235**

5-Phenyl-1,2,4-oxadiazole-3-carboxaldehyde; Oxime, *in* P-**60122**

1-Phenyl-1*H*-1,2,3-triazole-4-carboxylic acid, P-**60138**

1-Phenyl-1*H*-1,2,3-triazole-5-carboxylic acid, P-**60139**

2-Phenyl-2*H*-1,2,3-triazole-4-carboxylic acid, P-**60140**

C₉H₇N₅

8-Aminoimidazo[4,5-*g*]quinazoline, A-**60207**

9-Aminoimidazo[4,5-*f*]quinazoline, A-**60206**

C₉H₇N₅O

7-Aminoimidazo[4,5-*f*]quinazolin-9(8*H*)-one, A-**60208**

6-Aminoimidazo[4,5-*g*]quinolin-8(7*H*)-one, A-**60209**

C₉H₇S⊕

1-Benzothiopyrylium(1+), B-**70066**

2-Benzothiopyrylium(1+), B-**70067**

C₉H₈ClNO₅

4,5-Dihydroxy-2-nitrobenzoic acid; Di-Me ether, chloride, *in* D-**60360**

C₉H₈Cl₃NO₂

2-Amino-3-(2,3,4-trichlorophenyl)propanoic acid, A-**60262**

2-Amino-3-(2,3,6-trichlorophenyl)propanoic acid, A-**60263**

2-Amino-3-(2,4,5-trichlorophenyl)propanoic acid, A-**60264**

C₉H₈FNO₃

2-Amino-4-fluorobenzoic acid; *N*-Ac, *in* A-**70141**

2-Amino-5-fluorobenzoic acid; *N*-Ac, *in* A-**70142**

3-Amino-4-fluorobenzoic acid; *N*-Ac, *in* A-**70144**

4-Amino-2-fluorobenzoic acid; *N*-Ac, *in* A-**70145**

4-Amino-3-fluorobenzoic acid; *N*-Ac, *in* A-**70146**

C₉H₈N₂

1*H*-Cyclooctapyrazole, C-**70236**

C₉H₈N₂O

5-Amino-3-phenylisoxazole, A-**60240**

3-Amino-2(1*H*)-quinolinone, A-**60256**

5-Methyl-3-phenyl-1,2,4-oxadiazole, M-**60110**

1*H*-Pyrrolo[3,2-*b*]pyridine; *N*-Ac, *in* P-**60249**

C₉H₈N₂OS

3-Amino-2-benzo[*b*]thiophenecarboxylic acid; Amide, *in* A-**60091**

1,2,3,4-Tetrahydro-2-thioxo-5*H*-1,4-benzodiazepin-5-one, T-**70100**

▷2-Thioxo-4-imidazolidinone; 3-*N*-Ph, *in* T-**70188**

C₉H₈N₂O₂

1,4-Dihydro-2,3-quinoxalinedione; 1-Me, *in* D-**60279**

6-Hydroxy-7-methoxyquinoxaline, *in* Q-**60009**

C₉H₈N₂O₃

5-Phenyl-1,2,4-oxadiazole-3-carboxaldehyde; Covalent hydrate, *in* P-**60122**

C₉H₈N₂O₄

1-Cyano-4,5-dimethoxy-2-nitrobenzene, *in* D-**60360**

C₉H₈N₂O₆

3,5-Dinitrobenzoic acid; Et ester, *in* D-**60458**

C₉H₈N₂O₇

4,5-Dinitro-1,2-benzenediol; Mono-Me ether, Ac, *in* D-**70449**

C₉H₈N₂S

3-Amino-2(1*H*)quinolinethione, A-**70197**

4-Methyl-5-phenyl-1,2,3-thiadiazole, M-**60116**

5-Methyl-4-phenyl-1,2,3-thiadiazole, M-**60117**

C₉H₈N₄

4-Cinnolinecarboxaldehyde; Hydrazone, *in* C-**70179**

C₉H₈N₄O

1-Phenyl-1*H*-1,2,3-triazole-4-carboxylic acid; Amide, *in* P-**60138**

1-Phenyl-1*H*-1,2,3-triazole-5-carboxylic acid; Amide, *in* P-**60139**

2-Phenyl-2*H*-1,2,3-triazole-4-carboxylic acid; Amide, *in* P-**60140**

C₉H₈N₆O₃

Lepidopterin, L-**60020**

C₉H₈O

Cyclooctatetraenecarboxaldehyde, C-**70238**

3-Phenyl-2-propyn-1-ol, P-**60130**

C₉H₈OS

1,4-Dihydro-3*H*-2-benzothiopyran-3-one, D-**60206**

C₉H₈OSe

1*H*-2-Benzoselenin-4(3*H*)-one, B-**60051**

1,4-Dihydro-3*H*-2-benzoselenin-3-one, D-**60204**

C₉H₈OTe

1,4-Dihydro-3*H*-2-benzotellurin-3-one, D-**60205**

C₉H₈O₂

1*H*-2-Benzopyran-4(3*H*)-one, B-**70042**

Cubanecarboxylic acid, C-**60174**

Cyclooctatetraenecarboxylic acid, C-**70239**

1,4-Dihydro-2(3H)-benzopyran-3-one, D-70176
4-Hydroxybenzofuran, H-70106
3-(4-Hydroxyphenyl)-2-propenal, H-60213
4-Methoxybenzofuran, in H-70106
2-Methyl-1,3-benzenedicarboxaldehyde, M-60046
2-Methyl-1,4-benzenedicarboxaldehyde, M-60047
4-Methyl-1,2-benzenedicarboxaldehyde, M-60048
4-Methyl-1,3-benzenedicarboxaldehyde, M-60049
5-Methyl-1,3-benzenedicarboxaldehyde, M-60050

$C_9H_8O_2S$

3,4-Dihydro-2H-1,5-benzoxathiepin-3-one, D-60207
3-(2-Thienyl)-2-propenoic acid; Me ester, in T-60208

$C_9H_8O_3$

1,4-Benzodioxan-2-carboxaldehyde, B-60022
1,4-Benzodioxan-6-carboxaldehyde, B-60023
4H-1,3-Benzodioxin-6-carboxaldehyde, B-60026
2,3-Dihydro-2-benzofurancarboxylic acid, D-60197

$C_9H_8O_4$

1,4-Benzodioxan-2-carboxylic acid, B-60024
1,4-Benzodioxan-6-carboxylic acid, B-60025
4H-1,3-Benzodioxin-6-carboxylic acid, B-60027
1,3-Benzodioxole-4-acetic acid, B-60029
1,3,5-Cycloheptatriene-1,6-dicarboxylic acid, C-60202
5,7-Dihydroxy-6-methyl-1(3H)-isobenzofuranone, D-60349
5-Hydroxy-7-methoxyphthalide, in D-60339
7-Hydroxy-5-methoxyphthalide, in D-60339
2-Methoxy-4,5-methylenedioxybenzaldehyde, in H-70175
4-Methoxy-α-oxobenzeneacetic acid, in H-70212

$C_9H_8O_5$

2-Formyl-3,5-dihydroxybenzoic acid; Me ester, in F-60071
2-Hydroxy-1,3-benzenedicarboxylic acid; Mono-Me ester, in H-60101
5-Hydroxy-3-methyl-1,2-benzenedicarboxylic acid, H-70162
2-Methoxy-1,3-benzenedicarboxylic acid, in H-60101
2-Methoxy-1,4-benzenedicarboxylic acid, in H-70105
5-Methoxy-1,3-benzodioxole-4-carboxylic acid, in H-60180
6-Methoxy-1,3-benzodioxole-5-carboxylic acid, in H-60181
7-Methoxy-1,3-benzodioxole-4-carboxylic acid, in H-60179
2-Methoxy-3,4-methylenedioxybenzoic acid, in H-60177
3-Methoxy-4,5-methylenedioxybenzoic acid, in H-60178

$C_9H_8O_7$

2-Oxo-1,3-cyclopentanediglyoxylic acid, O-70085

C_9H_9BrO

3-(Bromomethyl)-2,3-dihydrobenzofuran, B-60295

$C_9H_9BrO_2$

1-(5-Bromo-2-methoxyphenyl)ethanone, in B-60265

$C_9H_9Br_3$

1,3,5-Triss(romomethyl)benzene, T-60401

C_9H_9Cl

1-Chloro-2-(2-propenyl)benzene, C-70152
1-Chloro-3-(2-propenyl)benzene, C-70153
1-Chloro-4-(2-propenyl)benzene, C-70154

C_9H_9ClO

2-Chloro-4,6-dimethylbenzaldehyde, C-60061
4-Chloro-2,6-dimethylbenzaldehyde, C-60062
4-Chloro-3,5-dimethylbenzaldehyde, C-60063
5-Chloro-2,4-dimethylbenzaldehyde, C-60064

$C_9H_9ClO_2$

3-(Chloromethyl)phenol; Ac, in C-70102

$C_9H_9ClO_3$

5-Chloro-2,3-dimethoxybenzaldehyde, in C-70065
3-Chloro-4-hydroxybenzoic acid; Me ester, Me ether, in C-60091
2-Chloro-3-methoxy-5-methylbenzoic acid, in C-60095
3-Chloro-4-methoxy-5-methylbenzoic acid, in C-60096
5-Chloro-4-methoxy-2-methylbenzoic acid, in C-60097
4-(Chloromethyl)phenol; Ac, in C-70103

$C_9H_9Cl_2NO_2$

2-Amino-3-(2,3-dichlorophenyl)propanoic acid, A-60127
2-Amino-3-(2,4-dichlorophenyl)propanoic acid, A-60128
2-Amino-3-(2,5-dichlorophenyl)propanoic acid, A-60129
2-Amino-3-(2,6-dichlorophenyl)propanoic acid, A-60130
2-Amino-3-(3,4-dichlorophenyl)propanoic acid, A-60131
2-Amino-3-(3,5-dichlorophenyl)propanoic acid, A-60132

$C_9H_9Cl_3$

1,3,5-Triss(hloromethyl)benzene, T-60402

C_9H_9FO

2-Fluoro-2-phenylpropanal, F-60063
2-Fluoro-3-phenylpropanal, F-60064
2-Fluoro-1-phenyl-1-propanone, F-60065

$C_9H_9FO_2$

3,4-Dihydro-3-fluoro-2H-1,5-benzodioxepin, D-70209
3-Fluoro-2-methylphenol; Ac, in F-60056

$C_9H_9FO_4$

3-Fluoro-2,6-dimethoxybenzoic acid, in F-60023
5-Fluoro-2,3-dimethoxybenzoic acid, in F-60025

$C_9H_9F_3O_3S$

3-(Trifluoromethyl)benzenesulfonic acid; Et ester, in T-60283

$C_9H_9IO_4$

2-Iodo-4,5-dimethoxybenzoic acid, in D-60338
3-Iodo-2,4-dimethoxybenzoic acid, in D-60331
3-Iodo-2,6-dimethoxybenzoic acid, in D-60334
3-Iodo-4,5-dimethoxybenzoic acid, in D-60335
4-Iodo-2,5-dimethoxybenzoic acid, in D-60333
5-Iodo-2,4-dimethoxybenzoic acid, in D-60332
6-Iodo-2,3-dimethoxybenzoic acid, in D-60330

C_9H_9N

3a,7a-Methano-1H-indole, M-70041
6-Methyl-1H-indole, M-60088
3-Phenyl-2-propyn-1-amine, P-60129
5H-Pyrrolo[1,2-a]azepine, P-70189

C_9H_9NO

3-Amino-1-indanone, A-60211
7,8-Dihydro-5(6H)-isoquinolinone, D-60254
1H-Indole-6-methanol, I-60032
6-Methoxyindole, in H-60158
3-Methoxy-1H-isoindole, in P-70100
Phthalimidine; N-Me, in P-70100

$C_9H_9NO_2$

5,6-Dihydroxyindole; N-Me, in D-70307
5-Methoxy-1H-indol-6-ol, in D-70307
6-Methoxy-1H-indol-5-ol, in D-70307
4-Phenyl-2-oxazolidinone, P-60126

$C_9H_9NO_3$

1,4-Benzodioxan-6-carboxaldehyde; Oxime, in B-60023
1,4-Benzodioxan-2-carboxylic acid; Amide, in B-60024
2-(4-Hydroxyphenyl)-2-oxoacetic acid; Me ether, amide, in H-70212
(2-Oxo-2-phenylethyl)carbamic acid, O-70099

$C_9H_9NO_3S$

3-Carboxy-2,3-dihydro-8-hydroxy-5-methyl-thiazolo[3,2-a]pyridinium hydroxide inner salt, C-70021

$C_9H_9NO_4$

2,6-Biss(ydroxyacetyl)pyridine, B-70144
3-Hydroxy-4,5-methylenedioxybenzoic acid; Me ether, amide, in H-60178
2-(4-Hydroxyphenyl)-2-oxoacetic acid; Me ether, oxime, in H-70212

$C_9H_9NO_5$

3-Amino-4-methoxy-1,2-benzenedicarboxylic acid, in A-70151

$C_9H_9NO_6$

4,5-Dihydroxy-2-nitrobenzoic acid; 4-Me ether, Me ester, in D-60360
4,5-Dimethoxy-2-nitrobenzoic acid, in D-60360

$C_9H_9N_3$

1-Aminophthalazine; N-Me, in A-70184
4,5-Diaminoisoquinoline, D-70040
5,8-Diaminoisoquinoline, D-70041
4,5-Diaminoquinoline, D-70044
(2-Pyridyl)(1-pyrazolyl)methane, P-70178

$C_9H_9N_3O_2$

1,3(2H,4H)-Isoquinolinedione; Dioxime, in I-70099

$C_9H_9N_3O_3$

Biuret; N-Benzoyl, in B-60190

$C_9H_9N_3S$

1,2,3-Benzotriazine-4(3H)-thione; 2-N-Et, in B-60050

$C_9H_9N_5$

4-Benzoyl-1,2,3-triazole; Hydrazone, in B-70079
▷2,4-Diamino-6-phenyl-1,3,5-triazine, D-60041

C_9H_{10}

5,6-Dihydro-4H-indene, D-70217
7-Methylene-1,3,5-cyclooctatriene, M-60074
Tricyclo[3.3.1.0^{2,8}]nona-3,6-diene, T-70230

$C_9H_{10}BrNO$

2-Bromo-3-methylaniline; N-Ac, in B-70234

$C_9H_{10}ClN$

4-Chloro-2,3-dihydro-1H-indole; 1-Me, in C-60055

$C_9H_{10}ClNO$

2'-Amino-2-chloro-3'-methylacetophenone, A-60112
2'-Amino-2-chloro-4'-methylacetophenone, A-60113
2'-Amino-2-chloro-5'-methylacetophenone, A-60114
2'-Amino-2-chloro-6'-methylacetophenone, A-60115
2-Chloro-4,6-dimethylbenzaldehyde; Oxime, in C-60061

$C_9H_{10}N_2$

4,5-Dihydro-1H-pyrazole; 1-Ph, in D-70253

$C_9H_{10}N_2O$

3,4-Dihydro-2(1H)-quinazolinone; 3-Me, in D-60278

▷1-Phenyl-3-pyrazolidinone, P-70091
1,3,4,5-Tetrahydro-2*H*-1,3-benzodiazepin-2-
one, T-60058

C₉H₁₀N₂O₂

2-Acetyl-5-aminopyridine; *N*-Ac, *in* A-70023
1-Carbethoxy-2-cyano-1,2-dihydropyridine,
C-70017

C₉H₁₀N₂O₃

1,4-Benzodioxan-6-carboxylic acid; Hydrazide,
in B-60025
2,4-Diaminobenzoic acid; 2-*N*-Ac, *in* D-60032
2,4-Diaminobenzoic acid; 4-*N*-Ac, *in* D-60032

C₉H₁₀N₂O₄

2-Amino-3,5-pyridinedicarboxylic acid; 3-
Mono-Et ester, *in* A-70193
2-Amino-3,5-pyridinedicarboxylic acid; 5-
Mono-Et ester, *in* A-70193
5-Amino-3,4-pyridinedicarboxylic acid; Di-Me
ester, *in* A-70195

C₉H₁₀N₂S

2-Amino-4,5,6,7-tetrahydrobenzo[*b*]thiophene-
3-carboxylic acid; Nitrile, *in* A-60258

C₉H₁₀N₄

4-Dimethylamino-1,2,3-benzotriazine, *in*
A-60092
2,6,7-Trimethylpteridine, T-70286
4,6,7-Trimethylpteridine, T-70287

C₉H₁₀N₄O

4-Amino-1,2,3-benzotriazine; *N*,*N*(4)-Di-Me, 2-
Oxide, *in* A-60092
Di-1*H*-imidazol-2-ylmethanone; 1,1′-Di-Me, *in*
D-70354

C₉H₁₀N₄O₃

2,4,6(1*H*,3*H*,5*H*)-Pteridinetrione; 1,3,5-Tri-Me,
in P-70134
2,4,7(1*H*,3*H*,8*H*)-Pteridinetrione; 1,3,6-Tri-Me,
in P-70135
2,4,7(1*H*,3*H*,8*H*)-Pteridinetrione; 1,3,8-Tri-Me,
in P-70135
2,4,7(1*H*,3*H*,8*H*)-Pteridinetrione; 1,6,8-Tri-Me,
in P-70135
2,4,7(1*H*,3*H*,8*H*)-Pteridinetrione; 3,6,8-Tri-Me,
in P-70135

C₉H₁₀N₄S

Tetrazole-5-thione; 3-Et, 1-Ph, *in* T-60174

C₉H₁₀O

2,7-Dimethyl-2,4,6-cycloheptatrien-1-one,
D-70391
1-Ethenyl-4-methoxybenzene, *in* H-70209
1-Methoxy-1-phenylethylene, *in* P-60092
1-Phenylcyclopropanol, P-70066
3,5,6,7-Tetrahydro-4*H*-inden-4-one, T-70070

C₉H₁₀OS

1-[4-(Methylthio)phenyl]ethanone, *in* M-60017

C₉H₁₀OSe

[2-(Methylseleninyl)ethenyl]benzene, *in*
M-70132

C₉H₁₀O₂

2-Benzyloxyacetaldehyde, B-70085
Bicyclo[2.2.1]hepta-2,5-diene-2-carboxylic acid;
Me ester, *in* B-60084
3,4-Dihydro-2*H*-1-benzopyran-2-ol, D-60198
3,4-Dihydro-2*H*-1-benzopyran-3-ol, D-60199
3,4-Dihydro-2*H*-1-benzopyran-4-ol, D-60200
3,4-Dihydro-2*H*-1-benzopyran-5-ol, D-60201
3,4-Dihydro-2*H*-1-benzopyran-6-ol, D-70175
3,4-Dihydro-2*H*-1-benzopyran-7-ol, D-60202
3,4-Dihydro-2*H*-1-benzopyran-8-ol, D-60203
5-Ethyl-2-hydroxy-2,4,6-cycloheptatrien-1-one,
E-70044
2′-(Hydroxymethyl)acetophenone, H-70158
4′-(Hydroxymethyl)acetophenone, H-70159
2-Methoxybenzeneacetaldehyde, *in* H-60207
α-Methoxybenzeneacetaldehyde, *in* H-60210
2-Methoxy-4-methylbenzaldehyde, *in* H-70160

4-Methoxy-2-methylbenzaldehyde, *in* H-70161
4-Methoxyphenylacetaldehyde, *in* H-60209
Spiro[4.4]non-2-ene-1,4-dione, S-60042

C₉H₁₀O₂S

2-Mercapto-2-phenylacetic acid; Me ester, *in*
M-60024
2-Mercapto-3-phenylpropanoic acid, M-60026
3-(3-Thienyl)-2-propenoic acid; Et ester, *in*
T-70172

C₉H₁₀O₂Se

[2-(Methylselenonyl)ethenyl]benzene, *in*
M-70132

C₉H₁₀O₃

2-Ethoxybenzoic acid, *in* H-70108
▷Ethyl salicylate, *in* H-70108
3-Hydroxy-4-methoxyphenylacetaldehyde, *in*
D-60371
4-Hydroxy-3-methoxyphenylacetaldehyde, *in*
D-60371
2-(Hydroxymethyl)-1,4-benzodioxan, H-60172
2-Hydroxy-4-methylbenzoic acid; Me ester, *in*
H-70163
2-Methoxy-4-methylbenzoic acid, *in* H-70163
4-Methoxy-2-methylbenzoic acid, *in* H-70164

C₉H₁₀O₄

2,6-Dihydroxy-3,5-dimethylbenzoic acid,
D-70293
2′,6′-Dihydroxy-4′-methoxyacetophenone, *in*
T-70248
2-(3,4-Dihydroxyphenyl)propanoic acid,
D-60372
2,3-Dimethoxy-5-methyl-1,4-benzoquinone, *in*
D-70320
Glycol salicylate, *in* H-70108
(2-Hydroxyphenoxy)acetic acid; Me ester, *in*
H-70203
Methoxymethyl salicylate, *in* H-70108
(2-Methoxyphenoxy)acetic acid, *in* H-70203
(3-Methoxyphenoxy)acetic acid, *in* H-70204
(4-Methoxyphenoxy)acetic acid, *in* H-70205

C₉H₁₀O₅

3,4-Dihydroxy-5-methoxybenzeneacetic acid, *in*
T-60343
3,5-Dihydroxy-4-methoxybenzeneacetic acid, *in*
T-60343
5-Hydroxy-2,3-dimethoxybenzoic acid, *in*
T-70250
3,4,5-Trihydroxyphenylacetic acid; Me ester, *in*
T-60343

C₉H₁₀O₆

3,5-Dihydroxy-4-oxo-4*H*-pyran-2-carboxylic
acid; Di-Me ether, Me ester, *in* D-60368

C₉H₁₀S

[2-(Methylthio)ethenyl]benzene, M-70137

C₉H₁₀S₂

1,5-Dihydro-2,4-benzodithiepin, D-60194
3,4-Dihydro-2*H*-1,5-benzodithiepin, D-60195

C₉H₁₀Se

[2-(Methylseleno)ethenyl]benzene, M-70132

C₉H₁₁BrO₃

1-Bromo-2,3,4-trimethoxybenzene, *in* B-60200
1-Bromo-2,3,5-trimethoxybenzene, *in* B-70182
1-Bromo-2,4,5-trimethoxybenzene, *in* B-70181
2-Bromo-1,3,4-trimethoxybenzene, *in* B-70180
2-Bromo-1,3,5-trimethoxybenzene, *in* B-60199
5-Bromo-1,2,3-trimethoxybenzene, *in* B-60201

C₉H₁₁N

Azetidine; *N*-Ph, *in* A-70283
Bicyclo[2.2.2]oct-2-ene-1-carboxylic acid;
Nitrile, *in* B-70113

1,2,3,4-Tetrahydroquinoline, T-70093

C₉H₁₁NO

2-(Dimethylamino)benzaldehyde, D-60401
N-Methylbenzenecarboximidic acid; Me ester,
in M-60045
1,2,3,4-Tetrahydro-4-hydroxyisoquinoline,
T-60066
6,7,8,9-Tetrahydro-4*H*-quinolizin-4-one,
T-60092

C₉H₁₁NOS

2-(1,3-Oxathian-2-yl)pyridine, O-60049

C₉H₁₁NO₂

▷2-Ethoxybenzamide, *in* H-70108
2-Hydroxyphenylacetaldehyde; Me ether,
oxime, *in* H-60207
3-Hydroxyphenylacetaldehyde; Me ether,
oxime, *in* H-60208
4-Hydroxyphenylacetaldehyde; Me ether,
oxime, *in* H-60209
N-Methylphenylacetohydroxamic acid, *in*
P-70059

C₉H₁₁NO₂S

2-Amino-3-mercapto-3-phenylpropanoic acid,
A-60213
2-Amino-3-(4-mercaptophenyl)propanoic acid,
A-60214
2-Amino-4,5,6,7-tetrahydrobenzo[*b*]thiophene-
3-carboxylic acid, A-60258

C₉H₁₁NO₃

4-Amino-1,2-benzenediol; *O*¹-Me, 2-Ac, *in*
A-70090
4-Amino-1,2-benzenediol; *O*²-Me, 1-Ac, *in*
A-70090
2,3-Dihydroxybenzaldehyde; Di-Me ether,
oxime, *in* D-70285
2,4-Dihydroxyphenylacetaldehyde; Di-Me
ether, oxime, *in* D-60369
3,4-Dihydroxyphenylacetaldehyde; 3-Me ether,
oxime, *in* D-60371
(2-Hydroxyphenoxy)acetic acid; Me ether,
amide, *in* H-70203

C₉H₁₁NO₅

1*H*-Pyrrole-3,4-dicarboxylic acid; 1-Methoxy,
Di-Me ester, *in* P-70183

C₉H₁₁NS

2-(Tetrahydro-2-thienyl)pyridine, T-60097
2,3,4,5-Tetrahydrothiepino[2,3-*b*]pyridine,
T-60098

C₉H₁₁NS₂

2-(1,3-Dithian-2-yl)pyridine, D-60504

C₉H₁₁N₅

5-Amino-1*H*-tetrazole; 2-Me, 5-*N*-Benzyl, *in*
A-70202
5-Amino-1*H*-tetrazole; 1-Me, 4-Benzyl, *in*
A-70202
1-Benzyl-5-imino-3-methyl-1*H*,3*H*-tetrazole, *in*
A-70202
3-Benzyl-5-imino-1-methyl-1*H*,3*H*-tetrazole, *in*
A-70202
1-Benzyl-5-methylamino-1*H*-tetrazole, *in*
A-70202

C₉H₁₁N₅O₂

6-Acetyl-2-amino-1,7,8,9-tetrahydro-4*H*-
pyrimido[4,5-*b*][1,4]diazepin-4-one, A-60019
6-Acetylhomopterin, A-70043
Deoxysepiapterin, *in* S-70035

C₉H₁₁N₅O₃

Biopterin, B-70121
Sepiapterin, S-70035

C₉H₁₂

Bicyclo[3.3.1]nona-2,6-diene, B-70102
2,3-Dimethylenebicyclo[2.2.1]heptane,
D-70405
▷5-Ethenylbicyclo[2.2.1]hept-2-ene, E-70036
1-Ethynylcycloheptene, E-70050
1,3,5,7-Nonatetraene, N-70071
Trispiro[2.0.2.0.2.0]nonane, T-70325

C₉H₁₂ClN₃O₄

5′-Chloro-5′-deoxyarabinosylcytosine, C-70064

$C_9H_{12}N_2$

6,7,8,9-Tetrahydro-5H-pyrido[2,3-b]azepine, T-70091

6,7,8,9-Tetrahydro-5H-pyrido[3,2-b]azepine, T-70092

$C_9H_{12}N_2O$

Acetamidoxime; O-Benzyl, in A-60016

4-Acetyl-2-aminopyridine; N^2,N^2-Di-Me, in A-70026

2-(Dimethylamino)benzaldehyde; Oxime, in D-60401

$C_9H_{12}N_2OS$

2-Amino-4,5,6,7-tetrahydrobenzo[b]thiophene-3-carboxylic acid; Amide, in A-60258

$C_9H_{12}N_2O_2$

2,4,6-Trimethyl-3-nitroaniline, T-70285

$C_9H_{12}N_4O_5$

5-Nitrohistidine; Me ester, N^α-Ac, in N-60028

$C_9H_{12}N_6$

2,3,6,7,10,11-Hexahydrotrisimidazo[1,2-a;1',2'-c;1'',2''-e][1,3,5]triazine, H-60061

$C_9H_{12}O$

Bicyclo[3.3.1]non-3-en-2-one, B-70105

Bicyclo[4.2.1]non-2-en-9-one, B-70106

Bicyclo[4.2.1]non-3-en-9-one, B-70107

Bicyclo[4.2.1]non-7-en-9-one, B-70108

3a,4,5,6,7,7a-Hexahydro-1H-inden-1-one, H-70058

3,4,5-Trimethylphenol, T-60371

$C_9H_{12}O_2$

Bicyclo[3.3.1]nonane-2,4-dione, B-70103

Bicyclo[2.2.2]oct-2-ene-1-carboxylic acid, B-70113

2-Carbomethoxy-3-vinylcyclopentene, in V-60010

2,2-Dimethyl-5-cycloheptene-1,3-dione, D-60416

3-Ethynyl-3-methyl-4-pentenoic acid; Me ester, in E-60059

2,3,4,4a,5,6-Hexahydro-7H-1-benzopyran-7-one, H-60045

3a,4,5,6,7,7a-Hexahydro-3-methylene-2-(3H)-benzofuranone, H-70062

4-Nonen-2-ynoic acid, N-70072

$C_9H_{12}O_3$

1,3,5-Cyclohexanetricarboxaldehyde, C-70226

Metacrolein, in P-70119

4-Oxo-1-cycloheptene-1-carboxylic acid; Me ester, in O-70082

$C_9H_{12}O_4$

5,5-Dimethyl-2-hexynedioic acid; Di-Me ester, in D-70415

5-Ethyl-3-hydroxy-4-methyl-2(5H)-furanone; Ac, in E-60051

2,3,6-Trimethoxyphenol, in B-60014

2-Vinyl-1,1-cyclopropanedicarboxylic acid; Di-Me ester, in V-60011

$C_9H_{12}O_5$

4-Oxo-2-heptenedioic acid; Di-Me ester, in O-60072

Rehmaglutin C, R-60003

$C_9H_{12}O_6$

Aconitic acid; Tri-Me ester, in A-70055

$C_9H_{12}S_2$

2-Methyl-1,3-benzenedimethanethiol, M-70050

$C_9H_{12}S_3$

1,3,5-Benzenetrimethanethiol, B-70019

$C_9H_{13}NO$

1,2,3,6,7,9a-Hexahydro-4(1H)-quinolizinone, H-70074

3,4,6,7,8,9-Hexahydro-2H-quinolizin-2-one, H-60059

$C_9H_{13}NO_2$

2,4-Diethoxypyridine, in H-60224

2,5-Diethoxypyridine, in H-60225

2,6-Diethoxypyridine, in H-60226

3,4-Diethoxypyridine, in H-60227

$C_9H_{13}N_3$

1,2,12-Triazapentacyclo[6.4.0.0^{2,17}.0^{3,7}.0^{4,11}]-dodecane, T-70205

$C_9H_{13}N_3O_3$

5-Amino-1H-pyrazole-4-carboxylic acid; N(1)-Me, N(5)-Ac, Et ester, in A-70192

$C_9H_{13}N_5$

2-(Diethylamino)purine, in A-60243

$C_9H_{13}N_5O_6$

Clitocine, C-60155

C_9H_{14}

7,7-Dimethylbicyclo[4.1.0]hept-3-ene, D-60409

1-Isopropenylcyclohexene, I-60115

3-Isopropenylcyclohexene, I-60116

4-Isopropenylcyclohexene, I-60117

1-(1-Propenyl)cyclohexene, P-60178

1-(2-Propenyl)cyclohexene, P-60179

3-(2-Propenyl)cyclohexene, P-60180

4-(1-Propenyl)cyclohexene, P-60181

4-(2-Propenyl)cyclohexene, P-60182

$C_9H_{14}N_2O_5$

α-Amino-2-carboxy-5-oxo-1-pyrrolidinebutanoic acid, A-70120

3,4-Dihydro-2(1H)-pyrimidinone; 1-β-D-Ribofuranosyl, in D-60273

3,4-Dihydro-2(1H)-pyrimidinone; 3-β-D-Ribofuranosyl, in D-60273

$C_9H_{14}N_4O_3$

Caffeidine; Ac, in C-60003

Carnosine, C-60020

$C_9H_{14}N_4O_5$

▷N^4-Aminocytidine, A-70127

$C_9H_{14}O$

1-Acetylcycloheptene, A-60026

1-Cyclooctenecarboxaldehyde, C-60217

2-Oxatricyclo[3.3.1.1^{3,7}]decane, O-60051

2,2,5,5-Tetramethyl-3-cyclopenten-1-one, T-60140

2,3,4-Trimethyl-2-cyclohexen-1-one, T-60359

$C_9H_{14}OS_3$

Ajoene, A-70079

$C_9H_{14}O_2$

Bicyclo[2.2.1]heptane-1-carboxylic acid; Me ester, in B-60085

Bicyclo[2.2.2]octane-1-carboxylic acid, B-60092

Bicyclo[3.2.1]octane-1-carboxylic acid, B-70112

7-Ethyl-5-methyl-6,8-dioxabicyclo[3.2.1]oct-3-ene, E-60052

5-Methyl-2,4-hexadienoic acid; Et ester, in M-70086

Mitsugashiwalactone, M-70150

Octahydro-7H-1-benzopyran-7-one, O-60011

4-Oxo-2-nonenal, O-70097

$C_9H_{14}O_2S$

2-Thiatricyclo[3.3.1.1^{3,7}]decane; 2,2-Dioxide, in T-60195

$C_9H_{14}O_3$

2-tert-Butyl-2,4-dihydro-6-methyl-1,3-dioxol-4-one, B-60345

2,3-Diisopropoxycyclopropenone, in D-70290

3-Oxocyclopentaneacetic acid; Et ester, in O-70083

$C_9H_{14}O_4$

2-Hydroxycyclohexanecarboxylic acid; Ac, in H-70119

$C_9H_{14}O_5$

Bissetone, B-60174

4-Oxoheptanedioic acid; Di-Me ester, in O-60071

4,4-Pyrandiacetic acid, P-70150

$C_9H_{14}S$

2,2,5,5-Tetramethyl-3-cyclopentene-1-thione, T-60139

2-Thiatricyclo[3.3.1.1^{3,7}]decane, T-60195

$C_9H_{15}ClO_2$

7-Chloro-2-heptenoic acid; Et ester, in C-70069

$C_9H_{15}N$

▷Tri(2-propenyl)amine, T-70315

$C_9H_{15}NO$

3-Aminobicyclo[2.2.1]hept-5-ene-2-methanol; N-Me, in A-70094

3,6-Dihydro-3,3,6,6-tetramethyl-2(1H)-pyridinone, D-70267

Octahydro-4H-quinolizin-4-one, O-60019

$C_9H_{15}NO_3$

2-Amino-2-hexenoic acid; N-Ac, Me ester, in A-60167

5-Oxo-2-pyrrolidinecarboxylic acid; tert-Butyl ester, in O-70102

$C_9H_{15}NO_5$

4-Oxoheptanedioic acid; Di-Me ester, oxime, in O-60071

$C_9H_{15}N_3$

3,5,12-Triazatetracyclo[5.3.1.1^{2,6}.0^{4,9}]dodecane, T-70206

2,4,6-Triethyl-1,3,5-triazine, T-70235

$C_9H_{15}N_3O_2$

Histidine trimethylbetaine, H-60083

$C_9H_{15}N_3O_2S$

Ovothiol C, in O-70050

$C_9H_{15}N_3O_3$

Triethyl isocyanurate, in C-60179

$C_9H_{15}N_3O_8$

Hexahydro-1,3,5-triazine; 1,3,5-Tri-Ac, in H-60060

C_9H_{16}

1-tert-Butylcyclopentene, B-70298

3-tert-Butylcyclopentene, B-70299

4-tert-Butylcyclopentene, B-70300

1-Isopropylcyclohexene, I-60118

3-Isopropylcyclohexene, I-60119

4-Isopropylcyclohexene, I-60120

$C_9H_{16}Br_4$

Tetrakis(2-bromoethyl)methane, T-70124

$C_9H_{16}N_2$

2-Diazo-1,1,3,3-tetramethylcyclopentane, D-60066

2,2,5,5-Tetramethyl-3-cyclopenten-1-one; Hydrazone, in T-60140

$C_9H_{16}N_2O_2$

3-(2-Methylpropyl)-6-methyl-2,5-piperazinedione, M-70118

$C_9H_{16}N_2O_4S$

S-(2-Aminoethyl)cysteine; N^α,N'-Di-Ac, in A-60152

$C_9H_{16}N_2O_5$

Tetrahydro-2(1H)-pyrimidinone; 1-β-D-Ribofuranosyl, in T-60088

Tetrahydro-2(1H)-pyrimidinone; 1-β-D-Ribopyranosyl, in T-60088

$C_9H_{16}O$

3-tert-Butylbicyclo[3.1.0]hexane, B-70294

2-Methylclooctanone, M-60059

2-Methyl-2-octenal, M-60106

2,2,5,5-Tetramethylcyclopentanone, T-60138

2,2,5-Trimethyl-4-hexen-3-one, T-60366

$C_9H_{16}O_2$

▷Dihydro-5-pentyl-2(3H)-furanone, D-70245

3,3-Dimethyl-4-pentenoic acid; Et ester, in D-60437

3,9-Dioxaspiro[5.5]undecane, D-70462

2-Ethyl-1,6-dioxaspiro[4.4]nonane, E-70043

2-Hydroxycyclononanone, H-60112

3-Hydroxycyclononanone, H-60113

2-Methyl-1,6-dioxaspiro[4.5]decane, M-60062

1,2,5-Trimethylcyclopentanecarboxylic acid, T-60361

2,2,5-Trimethylcyclopentanecarboxylic acid, T-60360

$C_9H_{16}O_3$

2-Cyclohexyl-2-hydroxyacetic acid; Me ester, *in* C-60212

2-Cyclohexyl-2-methoxyacetic acid, *in* C-60212

3,3-Dimethyl-4-oxopentanoic acid; Et ester, *in* D-60433

2-Hydroxycyclooctanecarboxylic acid, H-70121

9-Hydroxy-5-nonenoic acid, H-70194

$C_9H_{16}O_4$

(1-Methylpropyl)propanedioic acid; Di-Me ester, *in* M-70119

$C_9H_{16}O_7$

1,5-Dioxiranyl-1,2,3,4,5-pentanepentol, D-70465

$C_9H_{16}Se$

2,2,5,5-Tetramethylcyclopentaneselone, T-60137

$C_9H_{17}Br$

1-Bromo-3-nonene, B-60310

$C_9H_{17}NO$

3-Aminobicyclo[2.2.1]heptane-2-methanol; *N*-Me, *in* A-70093

Hexahydro-5(2*H*)-azocinone; 1-Et, *in* H-70048

Octahydro-2*H*-1,3-benzoxazine; *N*-Me, *in* O-70011

Octahydro-1-methyl-2*H*-3,1-benzoxazine, *in* O-70012

Tetrahydro-2-(2-piperidinyl)furan, T-60082

2,2,5-Trimethylcyclopentanecarboxylic acid; Amide, *in* T-60360

$C_9H_{17}NO_2$

2-Acetylpiperidine; Ethylene ketal, *in* A-70051

1,4-Dioxa-8-azaspiro[4.7]dodecane, *in* H-70047

1,4-Dioxa-9-azaspiro[4.7]dodecane, *in* H-70048

2-Methyl-2-piperidinecarboxylic acid; Et ester, *in* M-60119

Stachydrine ethyl ester, *in* S-60050

$C_9H_{17}NO_3$

2-Amino-4,4-dimethylpentanoic acid; *N*-Ac, *in* A-70137

2-Amino-3-hydroxy-4-methyl-6-octenoic acid, A-60189

$C_9H_{17}NO_4$

2-Amino-3-methylbutanedioic acid; Di-Et ester, *in* A-70166

3-(Carboxymethylamino)propanoic acid; Di-Et ester, *in* C-60016

▷Carnitine; *O*-Ac, *in* C-70024

$C_9H_{17}NO_8$

▷Miserotoxin, *in* N-60039

$C_9H_{18}N_2$

1,5-Diazabicyclo[5.2.2]undecane, D-60055

$C_9H_{18}N_2O_2$

▷3-[(3-Methylbutyl)nitrosoamino]-2-butanone, M-60055

2,4-Pentanediamine; N^2,N^4-Di-Ac, *in* P-70039

$C_9H_{18}N_2O_4S$

N-Serylmethionine; Me ester, *in* S-60024

$C_9H_{18}N_6$

1,3,5-Cyclohexanetricarboxaldehyde; Trishydrazone, *in* C-70226

$C_9H_{18}O$

1-Nonen-3-ol, N-60058

$C_9H_{18}O_2$

4,4-Diethoxy-2-methyl-2-butene, *in* M-70055

4,4-Dimethylpentanoic acid; Et ester, *in* D-70428

$C_9H_{18}O_3$

6-Hydroxy-2,4-dimethylheptanoic acid, H-60121

1,5,9-Trioxacyclododecane, T-70304

$C_9H_{18}O_4$

Ethyl 3,3-diethoxypropionate, *in* O-60090

$C_9H_{19}N$

Octahydroazocine; 1-Et, *in* O-70010

$C_9H_{19}NO$

2-Amino-4-*tert*-butylcyclopentanol, A-70104

2-(Aminomethyl)cyclooctanol, A-70170

$C_9H_{19}NS_2$

2-Pyrrolidinecarboxaldehyde; Diethyl dithioacetal, *in* P-70184

$C_9H_{20}ClNO_2$

▷Muscarine; Chloride, *in* M-60153

$C_9H_{20}INO_2$

Muscarine; Iodide, *in* M-60153

$C_9H_{20}NO_2^\oplus$

▷Muscarine, M-60153

$C_9H_{20}NO_3^\oplus$

4-Ethoxy-2-hydroxy-*N,N,N*-trimethyl-4-oxo-1-butanaminium, *in* C-70024

$C_9H_{20}N_2$

4-Amino-2,2,6,6-tetramethylpiperidine, A-70201

$C_9H_{20}N_2O_3$

N,N′-[Carbonylbis(oxy)]bis[2-methyl-2-propanamine], *in* C-60015

$C_9H_{20}O_4S_4$

Tetrakiss(ethylsulfinylmethyl)methane, *in* B-60164

$C_9H_{20}S_4$

Tetrakiss(ethylthiomethyl)methane, *in* B-60164

$C_9H_{21}N_3$

▷Hexahydro-1,3,5-triazine; 1,3,5-Tri-Et, *in* H-60060

1,4,7-Triazacyclododecane, T-60231

1,5,9-Triazacyclododecane, T-60232

$C_{10}Cl_{12}$

▷Dodecachloropentacyclo[5.3.0.0^{2,6}.0^{3,9}.0^{5,8}]-decane, D-70530

$C_{10}F_8$

Octafluoronaphthalene, O-70008

$C_{10}HF_7$

1,2,3,4,5,6,7-Heptafluoronaphthalene, H-70019

1,2,3,4,5,6,8-Heptafluoronaphthalene, H-70020

$C_{10}H_2F_4$

1,3-Diethynyl-2,4,5,6-tetrafluorobenzene, D-70159

1,4-Diethynyl-2,3,5,6-tetrafluorobenzene, D-70160

$C_{10}H_2F_6$

1,2,3,4,5,6-Hexafluoronaphthalene, H-70044

1,2,4,5,6,8-Hexafluoronaphthalene, H-70045

$C_{10}H_4^{\ominus\ominus}$

Dihydrocyclopenta[c,d]pentalene(2−), D-60216

$C_{10}H_4Br_2O_2$

6,7-Dibromo-1,4-naphthoquinone, D-70096

$C_{10}H_4Cl_2FNO_2$

3,4-Dichloro-1-(4-fluorophenyl)-1*H*-pyrrole-2,5-dione, D-60132

$C_{10}H_4Cl_2O_2$

6,7-Dichloro-1,4-naphthoquinone, D-70124

$C_{10}H_4K_2$

Dihydrocyclopenta[c,d]pentalene(2−); Di-K salt, *in* D-60216

$C_{10}H_4O_2S_2$

Benzo[1,2-*b*:4,5-*b′*]dithiophene-4,8-dione, B-60031

$C_{10}H_4S_8$

Bi(1,3-dithiolo[4,5-*b*][1,4]dithiin-2-ylidene), B-70118

$C_{10}H_5BrO_3$

2-Bromo-3-hydroxy-1,4-naphthoquinone, B-60268

2-Bromo-5-hydroxy-1,4-naphthoquinone, B-60269

2-Bromo-6-hydroxy-1,4-naphthoquinone, B-60270

2-Bromo-7-hydroxy-1,4-naphthoquinone, B-60271

2-Bromo-8-hydroxy-1,4-naphthoquinone, B-60272

7-Bromo-2-hydroxy-1,4-naphthoquinone, B-60273

8-Bromo-2-hydroxy-1,4-naphthoquinone, B-60274

$C_{10}H_5Br_2NO_2$

6,7-Dibromo-1-nitronaphthalene, D-70098

$C_{10}H_5F_3O$

1,1,1-Trifluoro-4-phenyl-3-butyn-2-one, T-70244

$C_{10}H_5NO_4$

6-Nitro-1,4-naphthoquinone, N-60035

7-Nitro-1,2-naphthoquinone, N-60036

$C_{10}H_5NO_5$

2-Hydroxy-3-nitro-1,4-naphthoquinone, H-60195

$C_{10}H_5N_3O_2$

1*H*-Naphtho[2,3-*d*]triazole-4,9-dione, N-60011

$C_{10}H_6$

1,3-Diethynylbenzene, D-70156

$C_{10}H_6Br_2O$

6,7-Dibromo-1-naphthol, D-70095

$C_{10}H_6ClNO_2$

▷2-Amino-3-chloro-1,4-naphthoquinone, A-70122

$C_{10}H_6F_3NO_2$

7-Amino-4-(trifluoromethyl)-2*H*-1-benzopyran-2-one, A-60265

$C_{10}H_6F_4O_4$

2,3,5,6-Tetrafluoro-1,4-benzenedicarboxylic acid; Di-Me ester, *in* T-70029

$C_{10}H_6N_2O$

2-Cyano-3-hydroxyquinoline, *in* H-60228

$C_{10}H_6N_2OS$

[1,4]Oxathiino[3,2-*b*:5,6-*c′*]dipyridine, O-60050

$C_{10}H_6N_2OSe$

[1,4]Oxaselenino[2,3-*b*:5,6-*b′*]dipyridine, O-70065

[1,4]Oxaselenino[3,2-*b*:5,6-*b′*]dipyridine, O-70066

[1,4]Oxaselenino[3,2-*b*:5,6-*c′*]dipyridine, O-70067

[1,4]Oxaselenino[3,2-*b*:6,5-*c′*]dipyridine, O-70068

$C_{10}H_6N_2O_2$

3,4-Cinnolinedicarboxaldehyde, C-70180

3,8-Dihydrocyclobuta[*b*]quinoxaline-1,2-dione, D-70178

$C_{10}H_6N_2O_2S$

[1,4]Oxathiino[3,2-*b*:5,6-*c′*]dipyridine; 8-Oxide, *in* O-60050

$C_{10}H_6N_2O_2Se$

[1,4]Oxaselenino[3,2-*b*:5,6-*c′*]dipyridine; 8-Oxide, *in* O-70067

[1,4]Oxaselenino[3,2-*b*:6,5-*c′*]dipyridine; 7-Oxide, *in* O-70068

$C_{10}H_6N_2O_4$

7-Nitro-4-nitroso-1-naphthol, *in* N-60035

$C_{10}H_6N_2S$

Thieno[2,3-*c*]cinnoline, T-60202

Thieno[3,2-*c*]cinnoline, T-60203

$C_{10}H_6N_2S_2$

1,4-Dithiino[2,3-*b*:5,6-*c′*]dipyridine, D-70518

[1,4]Dithiino[2,3-*b*:6,5-*b'*]dipyridine, D-70519
1,4-Dithiino[2,3-*b*:6,5-*c'*]dipyridine, D-70520
Thiazolo[4,5-*b*]quinoline-2(3*H*)-thione, T-60201

$C_{10}H_6N_4O_2$

Pyrazino[2,3-*g*]quinazoline-2,4-(1*H*,3*H*)-dione, P-60200
Pyrazino[2,3-*f*]quinazoline-8,10-(7*H*,9*H*)-dione, P-60201

$C_{10}H_6N_6$

Pyridazino[1″,6″:1′,2′]imidazo[4′,5′:4,5]-imidazo[1,2-*b*]pyridazine, P-70169

$C_{10}H_6O_3$

2,3-Epoxy-2,3-dihydro-1,4-naphthoquinone, E-70011
4-Hydroxy-1,2-naphthoquinone, H-60194

$C_{10}H_6O_4$

2,5-Dihydroxy-1,4-naphthoquinone, D-60354
2,8-Dihydroxy-1,4-naphthoquinone, D-60355
▷5,8-Dihydroxy-1,4-naphthoquinone, D-60356
6,7-Methylenedioxy-2*H*-1-benzopyran-2-one, M-70068

$C_{10}H_6O_7$

2,3,5,6,8-Pentahydroxy-1,4-naphthoquinone, P-70033

$C_{10}H_6O_8$

Hexahydroxy-1,4-naphthoquinone, H-60065

$C_{10}H_6S_2$

Benzo[1,2-*b*:4,5-*b'*]dithiophene, B-60030

$C_{10}H_7BrN_2$

2-Bromo-3,3′-bipyridine, B-70185
4-Bromo-2,2′-bipyridine, B-70186
5-Bromo-2,3′-bipyridine, B-70187
5-Bromo-2,4′-bipyridine, B-70188
5-Bromo-3,3′-bipyridine, B-70189
5-Bromo-3,4′-bipyridine, B-70190
6-Bromo-2,2′-bipyridine, B-70191
6-Bromo-2,3′-bipyridine, B-70192
6-Bromo-2,4′-bipyridine, B-70193

$C_{10}H_7Br_2N$

6,7-Dibromo-1-naphthylamine, D-70097

$C_{10}H_7Br_3$

1,4,7-Tribromotriquinacene, T-60248

$C_{10}H_7ClN_2$

2-Chloro-3-phenylpyrazine, C-70142
2-Chloro-5-phenylpyrazine, C-70143
2-Chloro-6-phenylpyrazine, C-70144

$C_{10}H_7ClN_2O$

2-Chloro-3-phenylpyrazine; 1-Oxide, *in* C-70142
2-Chloro-3-phenylpyrazine; 4-Oxide, *in* C-70142
2-Chloro-5-phenylpyrazine; 1-Oxide, *in* C-70143
2-Chloro-5-phenylpyrazine; 4-Oxide, *in* C-70143
2-Chloro-6-phenylpyrazine; 1-Oxide, *in* C-70144
2-Chloro-6-phenylpyrazine; 4-Oxide, *in* C-70144

$C_{10}H_7ClN_2O_2$

2-Chloro-3-phenylpyrazine; 1,4-Dioxide, *in* C-70142
2-Chloro-5-phenylpyrazine; 1,4-Dioxide, *in* C-70143

$C_{10}H_7ClO_3$

4-Hydroxy-5-benzofurancarboxylic acid; Me ether, chloride, *in* H-60103

$C_{10}H_7ClO_4$

6-Chloro-1,2,3,4-naphthalenetetrol, C-60118

$C_{10}H_7Cl_2NO_2$

5,6-Dichloro-1*H*-indole-3-acetic acid, D-70123

$C_{10}H_7N$

2-Ethynylindole, E-60057

3-Ethynylindole, E-60058

$C_{10}H_7NOS$

3-Phenyl-2-propenoyl isothiocyanate, P-70090

$C_{10}H_7NO_2$

3-Benzoylisoxazole, B-70075
4-Benzoylisoxazole, B-60062
5-Benzoylisoxazole, B-60063

$C_{10}H_7NO_2S$

2-Amino-3-mercapto-1,4-naphthoquinone, A-70163

$C_{10}H_7NO_3$

3-Hydroxy-2-quinolinecarboxylic acid, H-60228

$C_{10}H_7NS$

2-(1,2-Propadienyl)benzothiazole, P-60174

$C_{10}H_7N_3$

1*H*-Imidazo[4,5 *f*]quinoline, I-70007
1*H*-Imidazo[4,5-*h*]quinoline, I-70008
Pyrazolo[3,4-*c*]quinoline, P-60206

$C_{10}H_7N_3O$

2,5-Dihydro-1*H*-dipyrido[4,3-*b*:3′,4′-*d*]pyrrol-1-one, D-60237

$C_{10}H_8$

Fulvalene, F-60080
Naphthvalene, N-60012
Tetracyclo[5.3.0.0^{2,4}.0^{3,5}]deca-6,8,10-triene, T-70019

$C_{10}H_8ClNO_4$

2[(Chlorocarbonyl)oxy]-3*a*,4,7,7*a*-tetrahydro-4,7-methano-1*H*-isoindole-1,3(2*H*)-dione, C-70059

$C_{10}H_8ClN_3O_4$

Pyrido[1′,2′:3,4]imidazo[1,2-*a*]pyrimidin-5-ium(1+); Perchlorate, *in* P-70176

$C_{10}H_8N_2$

▷2,3′-Bipyridine, B-60120
2,4′-Bipyridine, B-60121
3,3′-Bipyridine, B-60122
3,4′-Bipyridine, B-60123
1,8-Dihydrobenzo[2,1-*b*:3,4-*b'*]dipyrrole, D-70174
▷1H-Indole-3-acetonitrile, *in* I-70013
5-Phenylpyrimidine, P-70093

$C_{10}H_8N_2O$

2-Acetylquinoxaline, A-60056
5-Acetylquinoxaline, A-60057
6-Acetylquinoxaline, A-60058
2-Benzoylimidazole, B-70074
4(5)-Benzoylimidazole, B-60057
3-Benzoylpyrazole, B-70078
4-Benzoylpyrazole, B-60065
2,3′-Bipyridine; 1′-Oxide, *in* B-60120
2,4′-Bipyridine; 1-Oxide, *in* B-60121
3,3′-Bipyridine; Mono-*N*-oxide, *in* B-60122
2,2′-Oxybispyridine, O-60098
2,3′-Oxybispyridine, O-60099
3,3′-Oxybispyridine, O-60100
4,4′-Oxybispyridine, O-60101
2-Phenyl-1*H*-imidazole-4(5)-carboxaldehyde, P-60111
5-Phenylpyrimidine; *N*-Oxide, *in* P-70093
1*H*-Pyrazole; *N*-Benzoyl, *in* P-70152

$C_{10}H_8N_2OS$

2,3-Dihydro-6-phenyl-2-thioxo-4(1*H*)-pyrimidinone, D-60268

$C_{10}H_8N_2O_2$

5-Amino-3-phenylisoxazole; *N*-Formyl, *in* A-60240
4-Benzoylisoxazole; Oxime, *in* B-60062
2,3′-Bipyridine; 1,1′-Dioxide, *in* B-60120
2,4′-Bipyridine; 1,1′-Dioxide, *in* B-60121
3-Hydroxy-2-quinolinecarboxylic acid; Amide, *in* H-60228
3,3′-Oxybispyridine; *N*-Oxide, *in* O-60100

$C_{10}H_8N_2O_2S$

▷2-Thioxo-4-imidazolidinone; 1-*N*-Benzoyl, *in* T-70188

$C_{10}H_8N_2O_3$

2-Acetylquinoxaline; 1,4-Dioxide, *in* A-60056
3,3′-Oxybispyridine; *N*,*N*′-Dioxide, *in* O-60100

$C_{10}H_8N_2O_3S$

Di-2-pyridyl sulfite, D-60497

$C_{10}H_8N_2O_8$

4,6-Dinitro-1,3-benzenediol; Di-Ac, *in* D-70450

$C_{10}H_8N_3^{\oplus}$

Pyrido[1′,2′:3,4]imidazo[1,2-*a*]pyrimidin-5-ium(1+), P-70176

$C_{10}H_8N_4$

4-Amino-1*H*-1,5-benzodiazepine-3-carbonitrile, A-70091
5-Amino-1*H*-pyrazole-4-carboxylic acid; *N*(1)-Ph, nitrile, *in* A-70192
Dipyrido[1,2-*b*:1′,2′-*e*][1,2,4,5]tetrazine, D-60494

$C_{10}H_8O$

▷3-Phenyl-2-cyclobuten-1-one, P-70064

$C_{10}H_8OS$

1-Benzothiepin-5(4*H*)-one, B-60047

$C_{10}H_8O_2$

1,2-Di(2-furanyl)ethylene, D-60191
6-Hydroxy-5-vinylbenzofuran, H-70231
3-Methyl-1*H*-2-benzopyran-1-one, M-70052
2-(3-Oxo-1-propenyl)benzaldehyde, O-60091
3-Phenyl-2(5*H*)furanone, P-60102
4-Phenyl-2(5*H*)-furanone, P-60103
5-Phenyl-2(3*H*)-furanone, P-70079

$C_{10}H_8O_3$

3-Benzofurancarboxylic acid; Me ester, *in* B-60032
4,6-Dihydroxy-5-vinylbenzofuran, D-70352
5-Hydroxy-4-methyl-2*H*-1-benzopyran-2-one, H-60173
8-Hydroxy-4-methyl-2*H*-1-benzopyran-2-one, H-70165
6-Methoxy-2*H*-1-benzopyran-2-one, *in* H-70109
1,4,5-Naphthalenetriol, N-60005

$C_{10}H_8O_3S$

2-Oxo-4-phenylthio-3-butenoic acid, O-60088

$C_{10}H_8O_4$

Albidin, A-60070
4,7-Dihydroxy-5-methyl-2*H*-1-benzopyran-2-one, D-60348
4-Hydroxy-5-benzofurancarboxylic acid; Me ester, *in* H-60103
7-Hydroxy-6-(hydroxymethyl)-2*H*-1-benzopyran-2-one, H-70139
▷7-Hydroxy-6-methoxy-2*H*-1-benzopyran-2-one, H-70153
4-Methoxy-5-benzofurancarboxylic acid, *in* H-60103

$C_{10}H_8O_5$

2-Hydroxy-4,5-methylenedioxybenzaldehyde; Ac, *in* H-70175
2-(2-Hydroxyphenyl)-2-oxoacetic acid; Ac, *in* H-70210

$C_{10}H_8O_6$

2-Acetoxy-1,3-benzenedicarboxylic acid, *in* H-60101
1,4-Benzoquinone-2,3-dicarboxylic acid; Di-Me ester, *in* B-70047
1,4-Benzoquinone-2,5-dicarboxylic acid; Di-Me ester, *in* B-70048

$C_{10}H_8S_2$

1,2-Naphthalenedithiol, N-70002
1,8-Naphthalenedithiol, N-70003
2,3-Naphthalenedithiol, N-70004

$C_{10}H_8S_8$

2-(5,6-Dihydro-1,3-dithiolo[4,5-*b*][1,4]dithiin-2-ylidene)-5,6-dihydro-1,3-dithiolo[4,5-*b*][1,4]dithiin, D-60238

C₁₀H₉Br

3-Bromo-1-phenyl-1-butyne, B-60314

C₁₀H₉BrN₂

4(5)-Bromo-1H-imidazole; 1-Benzyl, *in* B-70221

C₁₀H₉BrO

7-Bromo-3,4-dihydro-1(2H)-naphthalenone, B-60234

C₁₀H₉ClN₂

1-Benzyl-3-chloropyrazole, *in* C-70155
1-Benzyl-5-chloropyrazole, *in* C-70155

C₁₀H₉IO₂

1,3-Diacetyl-5-iodobenzene, D-60024

C₁₀H₉N

2-Vinylindole, V-70013

C₁₀H₉NO

2-Acetylindole, A-60032
4-Acetylindole, A-60033
▷5-Acetylindole, A-60034
6-Acetylindole, A-60035
7-Acetylindole, A-60036
1,2-Dihydro-1-phenyl-3H-pyrrol-3-one, *in* D-70262
1-(2-Furanyl)-2-(2-pyrrolyl)ethylene, F-60088
1-(4-Hydroxyphenyl)pyrrole, H-60216
2-(2-Hydroxyphenyl)pyrrole, H-60217
3-Methyl-1H-indole-2-carboxaldehyde, M-70090

C₁₀H₉NO₂

2-Acetyl-3-aminobenzofuran, A-70021
4-Hydroxy-5-nitro-1,2-benzenedicarboxylic acid; Di-Me ester, *in* H-70191
▷1H-Indole-3-acetic acid, I-70013
Phthalimidine; N-Ac, *in* P-70100

C₁₀H₉NO₂S

α-Aminobenzo[b]thiophene-3-acetic acid, A-60090
3-Amino-2-benzo[b]thiophenecarboxylic acid; Me ester, *in* A-60091

C₁₀H₉NO₂S₂

N-Phenyl-2-thiophenesulfonamide, *in* T-60213
N-Phenyl-3-thiophenesulfonamide, *in* T-60214

C₁₀H₉NO₃

2,3-Dihydro-2-oxo-1H-indole-3-acetic acid, D-60259
2-Nitro-3,4-dihydro-1(2H)naphthalenone, N-60026
1,2,3,4-Tetrahydro-2-oxo-4-quinolinecarboxylic acid, T-60077
1,2,3,4-Tetrahydro-4-oxo-6-quinolinecarboxylic acid, T-60078
1,2,3,4-Tetrahydro-4-oxo-7-quinolinecarboxylic acid, T-60079

C₁₀H₉NO₇

4-Hydroxy-3-nitro-1,2-benzenedicarboxylic acid; Me ether, 2-Me ester, *in* H-70190
4-Hydroxy-5-nitro-1,3-benzenedicarboxylic acid; Di-Me ester, *in* H-70192

C₁₀H₉NS

1-Amino-2-naphthalenethiol, A-70172
8-Amino-1-naphthalenethiol, A-70173
3(2H)-Isoquinolinethione; S-Me; B,HCl, *in* I-60133
1-Methylthioisoquinoline, *in* I-60132
4-Methylthioquinoline, *in* Q-60006
5-Methylthioquinoline, *in* Q-60004
6-Methylthioquinoline, *in* Q-60005
1-(2-Pyrrolyl)-2-(2-thienyl)ethylene, P-60250
3-Quinolinethiol; S-Me, *in* Q-60003
4(1H)-Quinolinethione; N-Me, *in* Q-60006

C₁₀H₉NS₂

2-Thiazoledinethione; N-Ph, *in* T-60197

C₁₀H₉N₃

6-Methyl-3-phenyl-1,2,4-triazine, M-60118
3-(Phenylazo)-2-butenenitrile, P-60079

C₁₀H₉N₃O

1-Acetamidophthalazine, *in* A-70184
2-Acetylquinoxaline; Oxime, *in* A-60056
4-Benzoyl-1,2,3-triazole; 1-Me, *in* B-70079
4-Benzoyl-1,2,3-triazole; 2-Me, *in* B-70079
4-Benzoyl-1,2,3-triazole; 3-Me, *in* B-70079

C₁₀H₉N₃O₂

4-Amino-1H-pyrazole-3-carboxylic acid; 1-Ph, *in* A-70189
1-Phenyl-1H-1,2,3-triazole-4-carboxylic acid; Me ester, *in* P-60138
1-Phenyl-1H-1,2,3-triazole-5-carboxylic acid; Me ester, *in* P-60139
2-Phenyl-2H-1,2,3-triazole-4-carboxylic acid; Me ester, *in* P-60140

C₁₀H₉N₃O₉

3-Hydroxy-5-methyl-2,4,6-trinitrobenzoic acid; Me ether, Me ester, *in* H-60193

C₁₀H₁₀

Bi-2,4-cyclopentadien-1-yl, B-60100
7,8-Dimethylenebicyclo[2.2.2]octa-2,5-diene, D-70407
Hexacyclo[4.4.0.0²·⁵.0³·⁹.0⁴·⁸.0⁷·¹⁰]decane, H-70032
9-Methylene-1,3,5,7-cyclononatetraene, M-60073
1-Methyleneindane, M-60078
1,3,4,5-Tetrahydrocycloprop[f]indene, T-70052
1,4,5,6-Tetrahydrocycloprop[e]indene, T-70053
Tricyclo[5.3.0.0²·⁸]deca-3,5,9-triene, T-70227

C₁₀H₁₀N₂

1,8-Diaminonaphthalene, D-70042
1,4-Dihydro-2-phenylpyrimidine, D-70249

C₁₀H₁₀N₂O

5-Acetylindole; Oxime, *in* A-60034
5-Amino-3-phenylisoxazole; N-Me, *in* A-60240
5,6-Dihydro-2-phenyl-4(1H)-pyrimidinone, D-70250
1H-Indole-3-acetic acid; Amide, *in* I-70013

C₁₀H₁₀N₂O₂

1,2-Diazetidin-3-one; 1-Ac, 2-Ph, *in* D-70055
1,4-Dihydro-2,3-quinoxalinedione; 1,4-Di-Me, *in* D-60279
2,3-Dimethoxyquinoxaline, *in* D-60279
5,8-Dimethoxyquinoxaline, *in* Q-60002
6,7-Dimethoxyquinoxaline, *in* Q-60009
3-Ethoxy-2-quinoxalinol, *in* D-60279
3-Ethoxy-2(1H)quinoxalinone, *in* D-60279
5,8-Quinoxalinediol; Di-Me ether, *in* Q-60008

C₁₀H₁₀N₂O₆

3-Amino-4-nitro-1,2-benzenedicarboxylic acid; Di-Me ester, *in* A-70174
3-Amino-6-nitro-1,2-benzenedicarboxylic acid; Di-Me ester, *in* A-70175
4-Amino-3-nitro-1,2-benzenedicarboxylic acid; Di-Me ester, *in* A-70176

C₁₀H₁₀N₄O

3-Amino-1H-pyrazole-4-carboxylic acid; Anilide, *in* A-70187
4-Amino-1H-pyrazole-3-carboxylic acid; 1-Ph, amide, *in* A-70189
5-Amino-1H-pyrazole-4-carboxylic acid; N(1)-Ph, amide, *in* A-70192

C₁₀H₁₀O

2,3-Dihydro-6(1H)-azulenone, D-60193
4,7-Dihydro-4,7-ethanoisobenzofuran, D-60240
Pentacyclo[5.3.0.0²·⁵.0³·⁹.0⁴·⁸]decan-6-one, P-60027
1-Phenyl-3-buten-1-one, P-60085
1-Phenyl-2-butyn-1-ol, P-70062
4-Phenyl-3-butyn-2-ol, P-70063
2-(2-Propenyl)benzaldehyde, P-70122

4-(2-Propenyl)benzaldehyde, P-60177
[3](2,7)Troponophane, T-70329

C₁₀H₁₀O₂

1,3,5,7-Cyclooctatetraene-1-acetic acid, C-70237
Cyclooctatetraenecarboxylic acid; Me ester, *in* C-70239
8-Decene-4,6-diynoic acid, D-70018
3a,6a-Dihydro-3a,6a-dimethyl-1,6-pentalenedione, D-70199
2,3-Dimethyl-1,4-benzodioxin, D-70370
11,12-Dioxahexacyclo[6.2.1.1³·⁶.0²·⁷.0⁴·¹⁰.0⁵·⁹]-dodecane, D-70461
5,12-Dioxatetracyclo[7.2.1.0⁴·¹¹.0⁶·¹⁰]dodeca-2,7-diene, D-70463
Dispiro[2.0.2.4]dec-8-ene-7,10-dione, D-60501
(4-Hydroxyphenyl)ethylene; Ac, *in* H-70209
3-(4-Methoxyphenyl)-2-propenal, *in* H-60213
2-Phenylcyclopropanecarboxylic acid, P-60087

C₁₀H₁₀O₂S

4′-Mercaptoacetophenone; Ac, *in* M-60017
S-Phenyl 3-oxobutanethioate, *in* O-70080
2-Phenylsulfonyl-1,3-butadiene, P-70094

C₁₀H₁₀O₂S₂

2,3,8,9-Tetrahydrobenzo[2,1-b:3,4-b′]bis[1,4]-oxathiin, T-70040

C₁₀H₁₀O₃

Hexahydrocyclopenta[cd]pentalene-1,3,5(2H)-trione, H-60048
2-Hydroxy-2-phenylacetaldehyde; Ac, *in* H-60210
3-Oxopropanoic acid; Benzyl ester, *in* O-60090

C₁₀H₁₀O₃S

2-Mercaptopropanoic acid; S-Benzoyl, *in* M-60027

C₁₀H₁₀O₄

1,4-Benzodioxan-2-carboxylic acid; Me ester, *in* B-60024
1,4-Benzodioxan-6-carboxylic acid; Me ester, *in* B-60025
4H-1,3-Benzodioxin-6-carboxylic acid; Me ester, *in* B-60027
1,3,5-Cycloheptatriene-1,6-dicarboxylic acid; Mono-Me ester, *in* C-60202
3,4-Dihydroxyphenylacetaldehyde; 3-Me ether, Ac, *in* D-60371
5,7-Dimethoxyphthalide, *in* D-60339
5-Hydroxy-7-methoxy-6-methylphthalide, *in* D-60349
2-Hydroxy-4-methylbenzoic acid; Ac, *in* H-70163
2-(2-Hydroxyphenyl)-2-oxoacetic acid; Et ester, *in* H-70210

C₁₀H₁₀O₅

2,3-Dihydroxy-5-methyl-1,4-benzoquinone; 2-Me ether, 3-Ac, *in* D-70320
2,3-Dihydroxy-5-methyl-1,4-benzoquinone; 3-Me ether, 2-Ac, *in* D-70320
2-Formyl-3,4-dimethoxybenzoic acid, *in* F-60070
4-Formyl-2,5-dimethoxybenzoic acid, *in* F-60072
6-Formyl-2,3-dimethoxybenzoic acid, *in* F-60073
2-Hydroxy-1,3-benzenedicarboxylic acid; Di-Me ester, *in* H-60101
2-Hydroxy-1,4-benzenedicarboxylic acid; Di-Me ester, *in* H-70105
5-Hydroxy-3-methyl-1,2-benzenedicarboxylic acid; Me ether, *in* H-70162
2-Hydroxy-3,4-methylenedioxybenzoic acid; Me ether, Me ester, *in* H-60177
3-Hydroxy-4,5-methylenedioxybenzoic acid; Me ether, Me ester, *in* H-60178
4-Hydroxy-2,3-methylenedioxybenzoic acid; Me ether, Me ester, *in* H-60179
(2-Hydroxyphenoxy)acetic acid; O-Ac, *in* H-70203

$C_{10}H_{10}O_6$

Chorismic acid, C-70173
4,5-Dihydroxy-1,3-benzenedicarboxylic acid;
Di-Me ester, *in* D-60307
4,5-Dimethoxy-1,3-benzenedicarboxylic acid, *in* D-60307

$C_{10}H_{10}S_4$

2,3,8,9-Tetrahydrobenzo[1,2-*b*:3,4-*b'*]bis[1,4]-dithiin, T-70039

$C_{10}H_{11}BrN_4O_5$

2-Bromo-9-β-D-ribofuranosyl-6*H*-purin-6-one, B-70269
8-Bromo-9-β-D-ribofuranosyl-6*H*-purin-6-one, B-70270

$C_{10}H_{11}BrO_2$

3-Bromo-2,4,6-trimethylbenzoic acid, B-60331

$C_{10}H_{11}ClN_4O_4$

2-Chloro-9-β-D-ribofuranosyl-9*H*-purine, C-70159
6-Chloro-9-β-D-ribofuranosyl-9*H*-purine, C-70160

$C_{10}H_{11}ClN_4O_5$

2-Chloro-9-β-D-ribofuranosyl-6*H*-purin-6-one, C-70161

$C_{10}H_{11}ClO_2$

2-Chloro-2-phenylethanol; Ac, *in* C-60128

$C_{10}H_{11}ClO_3$

5-Chloro-4-hydroxy-2-methylbenzoic acid; Me ether, Me ester, *in* C-60097

$C_{10}H_{11}F$

4-Fluoro-1-phenyl-1-butene, F-70035

$C_{10}H_{11}FO_4$

Methyl 4-fluoro-3,5-dimethoxybenzoate, *in* F-60024

$C_{10}H_{11}IO$

2-Iodo-1-phenyl-1-butanone, I-60060

$C_{10}H_{11}IO_4$

2,3-Dihydroxy-6-iodobenzoic acid; Di-Me ether, Me ester, *in* D-60330
2,4-Dihydroxy-5-iodobenzoic acid; Di-Me ether, Me ester, *in* D-60332
2,6-Dihydroxy-3-iodobenzoic acid; Di-Me ether, Me ester, *in* D-60334
3,4-Dihydroxy-5-iodobenzoic acid; Di-Me ether, Me ester, *in* D-60335
Methyl 2-iodo-3,5-dimethoxybenzoate, *in* D-60336
Methyl 4-iodo-3,5-dimethoxybenzoate, *in* D-60337

$C_{10}H_{11}N$

1-Amino-1,4-dihydronaphthalene, A-60139
1-Amino-5,8-dihydronaphthalene, A-60140
3,4-Dihydro-5-phenyl-2*H*-pyrrole, D-70251
1,6-Dimethyl-1*H*-indole, *in* M-60088
2,2-Dimethyl-3-phenyl-2*H*-azirine, D-60440
2-(4-Methylphenyl)propanoic acid; Nitrile, *in* M-60114

$C_{10}H_{11}NO$

3-Ethoxy-1*H*-isoindole, *in* P-70100
Phthalimidine; *N*-Et, *in* P-70100

$C_{10}H_{11}NOS$

4-Methyl-5-phenyl-2-oxazolidinethione, M-60111
Tetrahydro-2-phenyl-2*H*-1,2-thiazin-3-one, *in* T-60095

$C_{10}H_{11}NO_2$

2-Amino-3-phenyl-3-butenoic acid, A-60238
2-Amino-4-phenyl-3-butenoic acid, A-60239
5,6-Dimethoxyindole, *in* D-70307
5-Methyl-3-phenyl-2-oxazolidinone, *in* M-60107
4-Phenyl-3-morpholinone, *in* M-60147
1,2,3,4-Tetrahydro-1-isoquinolinecarboxylic acid, T-60070

$C_{10}H_{11}NO_3$

1-Amino-2-(4-hydroxyphenyl)-cyclopropanecarboxylic acid, A-60197
3-Amino-4-oxo-4-phenylbutanoic acid, A-60236

4-Hydroxyphenylacetaldehyde; Ac, oxime, *in* H-60209
Methyl *N*-phenacylcarbamate, *in* O-70099
2,3,5-Trimethoxybenzonitrile, *in* T-70250

$C_{10}H_{11}NO_4$

3-Amino-2-hydroxypropanoic acid; Benzoyl, *in* A-70160
2,3,4-Trimethyl-5-nitrobenzoic acid, T-60369
2,4,6-Trimethyl-3-nitrobenzoic acid, T-60370

$C_{10}H_{11}NO_4S$

2,3-Dihydro-1*H*-indole-2-sulfonic acid; *N*-Ac, *in* D-70221

$C_{10}H_{11}NO_5$

4-Amino-6-hydroxy-1,3-benzenedicarboxylic acid; Di-Me ester, *in* A-70153
5-Amino-2-hydroxy-1,3-benzenedicarboxylic acid; Di-Me ester, *in* A-70154
5-Amino-4-hydroxy-1,3-benzenedicarboxylic acid; Di-Me ester, *in* A-70155
2',3'-Dimethoxy-6'-nitroacetophenone, *in* D-60357
2',5'-Dimethoxy-4'-nitroacetophenone, *in* D-70322
3',6'-Dimethoxy-2'-nitroacetophenone, *in* D-60358
4',5'-Dimethoxy-2'-nitroacetophenone, *in* D-60359

$C_{10}H_{11}N_3$

1-Aminophthalazine; *N,N*-Di-Me, *in* A-70184

$C_{10}H_{11}N_3O$

3,4-Diamino-1-methoxyisoquinoline, D-60037
▷1*H*-Indole-3-acetic acid; Hydrazide, *in* I-70013

$C_{10}H_{11}N_5$

2,4-Diamino-6-methyl-1,3,5-triazine; *N*-Ph, *in* D-60040

$C_{10}H_{11}N_5O$

1,4-Dihydro-4,6,7-trimethyl-9*H*-imidazo[1,2-*a*]purin-9-one, D-70278

$C_{10}H_{11}N_5O_3$

6-Amino-1,3-dihydro-2*H*-purin-2-one; 6,9-Di-Ac, 1-Me, *in* A-60144

$C_{10}H_{12}$

1,3,5,7,9-Decapentaene, D-70015
5-Decene-2,8-diyne, D-60015
[4]Metacyclophane, M-60029
[4]Paracyclophane, P-60006
4,5,6,7-Tetrahydro-1-methylene-1*H*-indene, T-70073
Tricyclo[4.2.2.0^{1,6}]deca-7,9-diene, T-70226
2,3,5-Triss(ethylene)bicyclo[2.2.1]heptane, T-60403

$C_{10}H_{12}BrN_5O_5$

8-Bromoguanosine, B-70215

$C_{10}H_{12}ClN_5O_5$

8-Chloroguanosine, C-70068

$C_{10}H_{12}FN_5O_4$

8-Amino-6-fluoro-9*H*-purine; 9-β-D-Ribofuranosyl, *in* A-60163

$C_{10}H_{12}IN_5O_5$

8-Iodoguanosine, I-70034

$C_{10}H_{12}N_2O$

1-Phenyl-2-piperazinone, *in* P-70101

$C_{10}H_{12}N_2O_3$

3-Amino-4(1*H*)-pyridinone; 1-*N*-Me, N^3,N^3-di-Ac, *in* A-60254

$C_{10}H_{12}N_2O_4$

3,6-Pyridazinedicarboxylic acid; Di-Et ester, *in* P-70168
10,11,14,15-Tetraoxa-1,8-diazatricyclo[6.4.4.0^{2,7}]hexadeca-2(7),3,5-triene, T-60162

$C_{10}H_{12}N_2O_6$

1,5-Diethoxy-2,4-dinitrobenzene, D-70450

$C_{10}H_{12}N_2S$

2-Amino-4*H*-3,1-benzothiazine; *N*-Et, *in* A-70092

$C_{10}H_{12}N_4$

2,4,6,7-Tetramethylpteridine, T-70142

$C_{10}H_{12}N_4O_3$

Leucettidine, L-70029
2,4,7(1*H*,3*H*,8*H*)-Pteridinetrione; 1,3,6,8-Tetra-Me, *in* P-70135

$C_{10}H_{12}N_4O_4$

2'-Deoxyinosine, D-70023
3'-Deoxyinosine, D-70024

$C_{10}H_{12}N_4O_4S$

▷1,7-Dihydro-6*H*-purine-6-thione; 9-β-D-Ribofuranosyl, *in* D-60270
▷6-Thioinosine, T-70180

$C_{10}H_{12}N_4O_6$

Tetrahydro-1,2,4,5-tetrazine-3,6-dione; 1,2,4,5-Tetra-Ac, *in* T-70097

$C_{10}H_{12}O$

1-Ethoxy-1-phenylethylene, *in* P-60092
1-Phenyl-3-buten-1-ol, P-70061
1,2,4,5-Tetrahydro-3-benzoxepin, T-70043
1,3,4,5-Tetrahydro-2-benzoxepin, T-70044
2,3,4,5-Tetrahydro-1-benzoxepin, T-70045

$C_{10}H_{12}OS$

Benzeneethanethioic acid; *O*-Et ester, *in* B-70017
Benzeneethanethioic acid; *S*-Et ester, *in* B-70017

$C_{10}H_{12}O_2$

3,4-Dihydro-2*H*-1-benzopyran-7-ol; Me ether, *in* D-60202
3,4-Dihydro-6-methoxy-2*H*-1-benzopyran, *in* D-70175
3,4-Dihydro-8-methoxy-2*H*-1-benzopyran, *in* D-60203
Dispiro[cyclopropane-1,5'-[3,8]-dioxatricyclo[5.1.0.0^{2,4}]octane-6',1''-cyclopropane], D-60500
1-Ethenyl-2,4-dimethoxybenzene, *in* V-70010
1,3,4,5,7,8-Hexahydro-2,6-naphthalenedione, H-70063
2-(2-Methylphenyl)propanoic acid, M-60113
2-(4-Methylphenyl)propanoic acid, M-60114
3,4,5-Trimethylbenzoic acid, T-60355

$C_{10}H_{12}O_2S$

2-Mercapto-3-phenylpropanoic acid; Me ester, *in* M-60026
3-(Phenylthio)butanoic acid, *in* M-70023

$C_{10}H_{12}O_3$

Adriadysiolide, A-70069
2,5-Dimethoxybenzeneacetaldehyde, *in* D-60370
3,4-Dimethoxybenzeneacetaldehyde, *in* D-60371
2-Hydroxy-4-methylbenzoic acid; Et ester, *in* H-70163
4-Hydroxy-2-methylbenzoic acid; Et ester, *in* H-70164

$C_{10}H_{12}O_3S$

3-(Phenylsulfinyl)butanoic acid, *in* M-70023

$C_{10}H_{12}O_4$

1,4-Benzoquinone; Biss(thylene ketal), *in* B-70044
6-Ethyl-2,4-dihydroxy-3-methylbenzoic acid, E-70041
Hallerone, H-60005
2'-Hydroxy-4',6'-dimethoxyacetophenone, *in* T-70248
▷1-(4-Hydroxy-3,5-dimethoxyphenyl)ethanone, *in* T-60315
(3-Hydroxyphenoxy)acetic acid; Et ester, *in* H-70204

$C_{10}H_{12}O_5$

2,5-Biss(ydroxymethyl)furan; Di-Ac, *in* B-60160
3,4-Biss(ydroxymethyl)furan; Di-Ac, *in* B-60161

Gelsemide, G-60010

2,3,5-Trimethoxybenzoic acid, *in* T-70250

$C_{10}H_{12}S_8$

Tetramercaptotetrathiafulvalene; Tetrakis(*S*-Me), *in* T-70130

$C_{10}H_{13}I$

1-*tert*-Butyl-2-iodobenzene, B-60346

1-*tert*-Butyl-3-iodobenzene, B-60347

1-*tert*-Butyl-4-iodobenzene, B-60348

$C_{10}H_{13}N$

Azetidine; *N*-Benzyl, *in* A-70283

2,3,4,5-Tetrahydro-1*H*-1-benzazepine, T-70036

$C_{10}H_{13}NO$

2-(4-Methylphenyl)propanoic acid; Amide, *in* M-60114

7,8,9,10-Tetrahydropyrido[1,2-*a*]azepin-4(6*H*)-one, T-60085

$C_{10}H_{13}NO_2$

2-Amino-2-methyl-3-phenylpropanoic acid, A-60225

3-Amino-2,4,6-trimethylbenzoic acid, A-60266

5-Amino-2,3,4-trimethylbenzoic acid, A-60267

$C_{10}H_{13}NO_3$

4-Amino-1,2-benzenediol; Di-Me ether, *N*-Ac, *in* A-70090

2-Amino-2-methyl-3-(4-hydroxyphenyl)-propanoic acid, A-60220

$C_{10}H_{13}NO_4$

2-Amino-3-hydroxy-4-(4-hydroxyphenyl)-butanoic acid, A-60185

1*H*-Pyrrole-3,4-dicarboxylic acid; Di-Et ester, *in* P-70183

2′,4′,6′-Trihydroxyacetophenone; 2′,4′-Di-Me ether, oxime, *in* T-70248

$C_{10}H_{13}N_5O_3$

2-Amino-6-(1,2-dihydroxypropyl)-3-methyl-pterin-4-one, *in* B-70121

2-Amino-6-(1,2-dihydroxypropyl)-3-methyl-pterin-4-one, A-60146

$C_{10}H_{13}N_5O_4$

9-β-D-Ribofuranosyl-9*H*-purin-2-amine, *in* A-60243

$C_{10}H_{13}N_5O_5$

6-Amino-1,3-dihydro-2*H*-purin-2-one; 9-(β-D-Arabinofuranosyl), *in* A-60144

8-Amino-9-β-D-ribofuranosyl-6*H*-purin-6-one, A-70198

$C_{10}H_{13}N_5O_5S$

8-Mercaptoguanosine, M-70029

$C_{10}H_{13}N_5O_6$

7,8-Dihydro-8-oxoguanosine, D-70238

$C_{10}H_{14}$

5,5′-Bibicyclo[2.1.0]pentane, B-70092

2,3-Dimethylenebicyclo[2.2.2]octane, D-70408

1-Ethynylbicyclo[2.2.2]octane, E-70049

1,2,3,5,8,8*a*-Hexahydronaphthalene, H-60053

1,2,3,7,8,8*a*-Hexahydronaphthalene, H-60054

2-Methyl-6-methylene-1,3,7-octatriene, M-60089

Tricyclo[5.2.1.0²·⁶]dec-2(6)-ene, T-60260

$C_{10}H_{14}Br_2$

1,3-Dibromoadamantane, D-70084

$C_{10}H_{14}IN$

4-Iodo-3,5-dimethylaniline; *N*,*N*-Di-Me, *in* I-70031

$C_{10}H_{14}N_2O_2$

2,4-Dihydroxyphenylacetaldehyde; Dimethyl-hydrazone, *in* D-60369

Octahydro-5*H*,10*H*-dipyrrolo[1,2-*a*:1′,2′-*d*]-pyrazine-5,10-dione, O-60016

$C_{10}H_{14}N_2O_3$

Octahydro-2-hydroxy-5*H*,10*H*-dipyrrolo[1,2-*a*:1′,2′-*d*]pyrazine-5,10-dione, O-60017

$C_{10}H_{14}N_2O_4$

1-[3-Hydroxy-4-(hydroxymethyl)cyclopentyl]-2,4(1*H*,3*H*)-pyrimidinedione, H-70140

Isoporphobilinogen, I-60114

Octahydro-2,7-dihydroxy-5*H*,10*H*-dipyrrolo[1,2-*a*:1′,2′-*d*]pyrazine-5,10-dione, O-60013

Octahydro-5*a*,10*a*-dihydroxy-5*H*,10*H*-dipyrrolo[1,2-*a*:1′,2′-*d*]pyrazine-5,10-dione, O-60014

$C_{10}H_{14}N_2S_2$

Octahydro-5*H*,10*H*-dipyrrolo[1,2-*a*:1′,2′-*d*]-pyrazine-5,10-dithione, O-70014

$C_{10}H_{14}N_6O_3$

3′-Amino-3′-deoxyadenosine, A-70128

$C_{10}H_{14}N_6O_5$

8-Aminoguanosine, A-70150

$C_{10}H_{14}O$

3-Caren-5-one, C-60019

2,7-Cyclodecadien-1-one, C-60185

3,7-Cyclodecadien-1-one, C-60186

2-Cyclopentylidenecyclopentanone, C-60227

3,3-Dimethylbicyclo[2.2.2]oct-5-en-2-one, D-60412

6-Isopropenyl-3-methyl-2-cyclohexen-1-one, I-70088

5-Methoxy-1,2,3-trimethylbenzene, *in* T-60371

7-Methylenebicyclo[3.3.1]nonan-3-one, M-70062

$C_{10}H_{14}O_2$

2-Benzyloxy-1-propanol, *in* P-70118

4,4-Dimethylbicyclo[3.2.1]octane-2,3-dione, D-60410

3-Ethynyl-3-methyl-4-pentenoic acid; Et ester, *in* E-60059

▷1-(3-Furanyl)-4-methyl-1-pentanone, F-60087

Isoneonepetalactone, *in* N-70027

2-Isopropyl-5-methyl-1,3-benzenediol, I-60122

2-Isopropyl-6-methyl-1,4-benzenediol, I-60123

3-Isopropyl-6-methyl-1,2-benzenediol, I-60124

5-Isopropyl-2-methyl-1,3-benzenediol, I-60125

5-Isopropyl-3-methyl-1,2-benzenediol, I-60126

5-Isopropyl-4-methyl-1,3-benzenediol, I-60127

Neonepetalactone, N-70027

Nepetalactone, N-60020

4-Oxatricyclo[4.3.1.1³·⁸]undecan-5-one, O-60055

3,3,6,6-Tetramethyl-4-cyclohexene-1,2-dione, T-60134

5,5,6,6-Tetramethyl-2-cyclohexene-1,4-dione, T-60135

$C_{10}H_{14}O_3$

2,5-Dihydroxy-4-isopropylbenzyl alcohol, D-70309

2-(Dimethoxymethyl)benzenemethanol, *in* H-60171

α-(Dimethoxymethyl)benzenemethanol, *in* H-60210

$C_{10}H_{14}O_4$

1,2,3,4-Tetramethoxybenzene, *in* B-60014

$C_{10}H_{14}O_5$

Hexahydro-3,7*a*-dihydroxy-3*a*,7-dimethyl-1,4-isobenzofurandione, H-70055

$C_{10}H_{14}S_2$

1,4-Dimethyl-2,5-biss(ethylthio)benzene, *in* D-60406

$C_{10}H_{15}BrCl_2O$

Aplysiapyranoid *C*, A-70245

Aplysiapyranoid D, *in* A-70245

$C_{10}H_{15}Br_2Cl$

7,8-Dibromo-6-(chloromethylene)-2-methyl-2-octene, D-70088

$C_{10}H_{15}Br_2ClO$

Aplysiapyranoid *A*, A-70244

Aplysiapyranoid B, *in* A-70244

$C_{10}H_{15}Cl$

1-Chloroadamantane, C-60039

2-Chloroadamantane, C-60040

$C_{10}H_{15}Cl_3$

1,8-Dichloro-6-chloromethyl-2-methyl-2,6-octadiene, D-70116

3,8-Dichloro-6-chloromethyl-2-methyl-1,6-octadiene, D-70117

$C_{10}H_{15}NO$

2-Cyclopentylidenecyclopentanone; Oxime, *in* C-60227

3-Ethynyl-3-methyl-4-pentenoic acid; *N*,*N*-Dimethylamide, *in* E-60059

$C_{10}H_{15}NO_4$

α-Allokainic acid, *in* K-60003

▷Kainic acid, K-60003

$C_{10}H_{15}N_3$

2-Azidoadamantane, A-60337

$C_{10}H_{15}N_3O_2$

1*H*-Pyrrole-3,4-dicarboxylic acid; Dimethyl-amide, *in* P-70183

$C_{10}H_{15}N_3O_4$

Carbodine, C-70018

$C_{10}H_{15}N_3O_5$

Homocytidine, *in* A-70130

$C_{10}H_{16}$

1-*tert*-Butyl-1,3-cyclohexadiene, B-70296

4-Decen-1-yne, D-70020

1,2-Diethylidenecyclohexane, D-70154

4-Isopropenyl-1-methylcyclohexene, I-70087

3,3,6,6-Tetramethylcyclohexyne, T-70132

$C_{10}H_{16}Cl_2$

1-Chloro-3-(chloromethyl)-7-methyl-2,6-octadiene, C-70060

$C_{10}H_{16}N_2S_2$

4,6-Diamino-1,3-benzenedithiol; Di-*S*-Et, *in* D-70037

$C_{10}H_{16}N_2S_4$

2,3,5,6-Tetrakiss(ethylthio)-1,4-benzenediamine, *in* D-60031

$C_{10}H_{16}O$

1-Acetylcyclooctene, A-60027

3-*tert*-Butyl-2-cyclohexen-1-one, B-70297

3,3-Dimethylbicyclo[2.2.2]octan-2-one, D-60411

2-Isopropyl-5-methyl-4-cyclohexen-1-one, I-60128

2-Methyl-6-methylene-2,7-octadien-4-ol, M-70093

2,2,5,5-Tetramethyl-3-cyclohexen-1-one, T-60136

$C_{10}H_{16}O_2$

Bicyclo[3.2.1]octane-1-carboxylic acid; Me ester, *in* B-70112

1-Cyclooctene-1-acetic acid, C-70241

4-Cyclooctene-1-acetic acid, C-70242

Eldanolide, E-70004

2-Hydroperoxy-2-methyl-6-methylene-3,7-octadiene, H-60097

3-Hydroperoxy-2-methyl-6-methylene-1,7-octadiene, H-60098

Iridodial, I-60084

Lineatin, L-60027

Spiro[bicyclo[3.2.1]octane-8,2′-[1,3]dioxolane], *in* B-60094

$C_{10}H_{16}O_3$

1,7-Dihydroxy-3,7-dimethyl-2,5-octadien-4-one, D-60318

3-Hydroxy-4-(hydroxymethyl)-7,7-di-methylbicyclo[2.2.1]heptan-2-one, *in* H-70173

9-Oxo-2-decenoic acid, O-70089

Secothujene, S-70027

$C_{10}H_{16}O_4$

Gelsemiol, G-60011

7-Hydroxy-1-hydroxy-3,7-dimethyl-2*E*,5*E*-octadien-4-one, *in* D-60318

2,2,5,5-Tetramethyl-3-hexenedioic acid,
T-60141

$C_{10}H_{16}S_2$
1,3-Biss(llylthio)cyclobutane, *in* C-60184

$C_{10}H_{17}N$
4-Azatricyclo[4.3.1.1^{3,8}]undecane, A-60333

$C_{10}H_{17}NO_3$
2-Nitrocyclodecanone, N-70046

$C_{10}H_{18}$
1-*tert*-Butylcyclohexene, B-60341
3-*tert*-Butylcyclohexene, B-60342
4-*tert*-Butylcyclohexene, B-60343

$C_{10}H_{18}N_2O_2$
1,1-Di-4-morpholinylethene, D-70447

$C_{10}H_{18}N_2O_4$
2*H*-1,5,2,4-Dioxadiazine-3,6(4*H*)dione; 2,4-Di-
tert-butyl, *in* D-60466

$C_{10}H_{18}N_2O_6S_2$
γ-Glutamylmarasmine, G-60026

$C_{10}H_{18}N_4O_4$
1,2-Dihydrazinoethane; $N^α,N^{α'},N^β,N^{β'}$-Tetra-
Ac, *in* D-60192

$C_{10}H_{18}O$
3-Decenal, D-70016
3,7-Dimethyl-6-octenal, D-60430
4,6-Dimethyl-4-octen-3-one, D-70424
Hexamethylcyclobutanone, H-60069
2-Isopropylcyclohoptanone, I-70090
4-Isopropylcycloheptanone, I-70091
2-Methyl-6-methylene-7-octen-4-ol, *in*
M-70093
2,2,6,6-Tetramethylcyclohexanone, T-60133
2,2,6-Trimethyl-5-hepten-3-one, T-60365

$C_{10}H_{18}OS$
Di-*tert*-butylthioketene; *S*-Oxide, *in* D-60114

$C_{10}H_{18}O_2$
3,4-Diethoxy-1,5-hexadiene, *in* H-70036
2,6-Dimethyl-3,7-octadiene-2,6-diol, D-70422
▷5-Hexyldihydro-2(3*H*)-furanone, H-60078
4-Isopropenyl-1-methyl-1,2-cyclohexanediol,
I-70086
4-Isopropyl-1-methyl-4-cyclohexene-1,2-diol,
I-70094
4-Isopropyl-1-methyl-3-cyclohexenene-1,2-diol,
I-70093
2-Octen-1-ol; Ac, *in* O-60029
4-Oxodecanal, O-70088
1,7,7-Trimethylbicyclo[2.2.1]heptane-2,5-diol,
T-70280

$C_{10}H_{18}O_3$
1,7-Biss(ydroxymethyl)-7-methylbicyclo[2.2.1]-
heptan-2-ol, B-70146
2-Hydroxy-4-*tert*-butylcyclopentanecarboxylic
acid, H-70113
1-(Hydroxymethyl)-7,7-dimethylbicyclo[2.2.1]-
heptane-2,3-diol, H-70173
9-Hydroxy-5-nonenoic acid; Me ester, *in*
H-70194
10-Oxodecanoic acid, O-60063

$C_{10}H_{18}O_4$
3-Methyl-4-oxo-2-butenoic acid; Me ester, di-Et
acetal, *in* M-70103
Nonactinic acid, N-70069

$C_{10}H_{18}O_5$
Methyl hydrogen 2-(*tert*-butoxymethyl)-2-
methylmalonate, M-70088

$C_{10}H_{18}S$
Di-*tert*-butylthioketene, D-60114

$C_{10}H_{19}Br$
5-Bromo-5-decene, B-60225

$C_{10}H_{19}NO$
1-Oxa-2-azaspiro[2,5]octane; *N-tert*-Butyl, *in*
O-70052
2,2,6,6-Tetramethylcyclohexanone; Oxime, *in*
T-60133

$C_{10}H_{19}NO_2$
3-Amino-2-hexenoic acid; *tert*-Butyl ester, *in*
A-60170

$C_{10}H_{19}NO_3$
2-Amino-1-hydroxy-1-cyclobutaneacetic acid;
tert-Butyl ester, *in* A-60184
2-Amino-3-hydroxy-4-methyl-6-octenoic acid;
N-Me, *in* A-60189
10-Oxodecanoic acid; Oxime, *in* O-60063

$C_{10}H_{19}N_3$
1*H*,4*H*,7*H*,9b*H*-2,3,5,6,8,9-Hexahydro-
3*a*,6*a*,9*a*-triazaphenalene, H-70078

$C_{10}H_{20}$
5-Decene, D-70017
Methylcyclononane, M-60058

$C_{10}H_{20}N_2$
1,5-Diazabicyclo[5.2.2]undecane; *N*-Me, *in*
D-60055

$C_{10}H_{20}N_2O$
N,*N*-Diethyl-3-piperidinecarboxamide, *in*
P-70102

$C_{10}H_{20}N_2S_2$
▷1,1'-Dithiobispiperidine, D-70523

$C_{10}H_{20}N_4$
▷6*H*,13*H*-Octahydrodipyridazino[1,2-*a*:1',2'-
d][1,2,4,5]tetrazine, O-70013

$C_{10}H_{20}O$
5-Decanone, D-60014
2-Decen-4-ol, D-60016
5-Decen-1-ol, D-70019
2,2,3,3,4,4-Hexamethylcyclobutanol, H-60068

$C_{10}H_{20}O_4$
3,4-Dihydroxybutanoic acid; O^4-*tert*-Butyl, Et
ester, *in* D-60313

$C_{10}H_{20}O_5$
Di-*tert*-butyl dicarbonate, *in* D-70111

$C_{10}H_{21}N$
Octahydroazocine; 1-Isopropyl, *in* O-70010

$C_{10}H_{22}N_2$
Hexahydropyrimidine; 1,3-Diisopropyl, *in*
H-60056

$C_{10}H_{22}N_2O_3$
1,4,10-Trioxa-7,13-diazacyclopentadecane,
T-60376

$C_{10}H_{22}N_2O_5$
6,10-Diamino-2,3,5-trihydroxydecanoic acid,
D-70046

$C_{10}H_{22}N_4O_2S_2$
Alethine, A-60073

$C_{10}H_{22}O_2$
5,6-Decanediol, D-60013

$C_{10}H_{23}N_3$
1,5,9-Triazacyclotridecane, T-60233

$C_{10}H_{25}N_5$
1,4,7,10,13-Pentaazacyclopentadecane, P-60025

$C_{11}H_5Cl_{11}$
Chloropentakiss(ichloromethyl)benzene,
C-70136

$C_{11}H_6N_2O$
3,6-Diazafluorenone, D-70053

$C_{11}H_6N_4$
9-Diazo-9*H*-cyclopenta[1,2-*b*:4,3-*b'*]dipyridine,
D-60059

$C_{11}H_6N_6$
Pyrimido[4,5-*i*]imidazo[4,5-*g*]cinnoline,
P-60244

$C_{11}H_6O$
1-Phenyl-1,4-pentadiyn-3-one, P-70087

$C_{11}H_6OS$
Naphtho[2,1-*b*]thiet-1-one, N-70013

$C_{11}H_7BrO$
1-Bromo-2-naphthalenecarboxaldehyde,
B-60302

$C_{11}H_7BrO_3$
3-Bromo-5-hydroxy-2-methyl-1,4-
naphthoquinone, B-60266
2-Bromo-6-methoxy-1,4-naphthaquinone, *in*
B-60270
2-Bromo-3-methoxy-1,4-naphthoquinone, *in*
B-60268
2-Bromo-5-methoxy-1,4-naphthoquinone, *in*
B-60269
2-Bromo-7-methoxy-1,4-naphthoquinone, *in*
B-60271
2-Bromo-8-methoxy-1,4-naphthoquinone, *in*
B-60272

$C_{11}H_7ClO_3$
3-Chloro-5-hydroxy-2-methyl-1,4-
naphthoquinone, C-60098
2-Oxo-2*H*-benzopyran-4-acetic acid; Chloride,
in O-60057

$C_{11}H_7NO$
Benz[*cd*]indol-2-(1*H*)-one, B-70027
9*H*-Pyrrolo[1,2-*a*]indol-9-one, P-60247

$C_{11}H_7NOSe$
[1,4]Benzoxaselenino[3,2-*b*]pyridine, B-60054

$C_{11}H_7NO_2$
4-(Cyanomethyl)coumarin, *in* O-60057
4-Oxo-4*H*-1-benzopyran-3-acetic acid; Nitrile,
in O-60059

$C_{11}H_7NO_3$
3-Nitro-2-naphthaldehyde, N-70054

$C_{11}H_7NS$
Azuleno[1,2-*d*]thiazole, A-60356
Azuleno[2,1-*d*]thiazole, A-60357
2-(Phenylethynyl)thiazole, P-60099
4-(Phenylethynyl)thiazole, P-60100

$C_{11}H_7N_3O_2$
Pyrimido[4,5-*b*]quinoline-2,4(3*H*,10*H*)-dione,
P-60245

$C_{11}H_8Br_2$
▷1-Bromo-2-(bromomethyl)naphthalene,
B-60212
1-Bromo-4-(bromomethyl)naphthalene,
B-60213
1-Bromo-5-(bromomethyl)naphthalene,
B-60214
1-Bromo-7-(bromomethyl)naphthalene,
B-60215
1-Bromo-8-(bromomethyl)naphthalene,
B-60216
2-Bromo-1-(bromomethyl)naphthalene,
B-60217
2-Bromo-6-(bromomethyl)naphthalene,
B-60218
3-Bromo-1-(bromomethyl)naphthalene,
B-60219
6-Bromo-1-(bromomethyl)naphthalene,
B-60220
7-Bromo-1-(bromomethyl)naphthalene,
B-60221
1,2-Dibromo-1,4-dihydro-1,4-
methanonaphthalene, D-60095
1,3-Dibromo-1,4-dihydro-1,4-
methanonaphthalene, D-60096

$C_{11}H_8ClNO_2$
Silital, *in* C-70075

$C_{11}H_8N_2$
1*H*-Benzo[*de*][1,6]naphthyridine, B-70037

$C_{11}H_8N_2O$
Di-2-pyridyl ketone, D-70502
Di-3-pyridyl ketone, D-70503
Di-4-pyridyl ketone, D-70504
1*H*-Perimidin-2(3*H*)-one, P-70043
2-Pyridyl 3-pyridyl ketone, P-70179
2-Pyridyl 4-pyridyl ketone, P-70180
3-Pyridyl 4-pyridyl ketone, P-70181

$C_{11}H_8N_2O_2$
5-Benzoyl-2(1*H*)-pyrimidinone, B-60066

$C_{11}H_8N_2O_2S$
1,1'-Carbonothioylbis-2(1*H*)pyridinone,
C-60014

C₁₁H₈N₄

1*H*-Pyrazolo[3,4-*d*]pyrimidine; 1-Ph, *in* P-70156

C₁₁H₈N₄O

1*H*-Pyrazolo[3,4-*d*]pyrimidine; 1-Ph, 5-oxide, *in* P-70156

C₁₁H₈O₂

Homoazulene-1,5-quinone, H-60085
Homoazulene-1,7-quinone, H-60086
Homoazulene-4,7-quinone, H-60087
6-Hydroxy-1-naphthaldehyde, H-70184
6-Hydroxy-2-naphthaldehyde, H-70185
7-Hydroxy-2-naphthaldehyde, H-70186

C₁₁H₈O₃

5-Hydroxy-7-methyl-1,2-naphthoquinone, H-60189
8-Hydroxy-3-methyl-1,2-naphthoquinone, H-60190
4-Methoxy-1,2-naphthoquinone, *in* H-60194
Pentacyclo[5.4.0.0²,⁶.0³,¹⁰.0⁵,⁹]undecane-4,8,11-trione, P-70023
Pentacyclo[6.3.0.0²,⁶.0⁵,⁹]undecane-4,7,11-trione, P-60031
2-Phenyl-3-furancarboxylic acid, P-70076
4-Phenyl-2-furancarboxylic acid, P-70077
5-Phenyl-2-furancarboxylic acid, P-70078

C₁₁H₈O₄

3,5-Dihydroxy-2-methyl-1,4-naphthoquinone, D-60350
5,6-Dihydroxy-2-methyl-1,4-naphthoquinone, D-60351
5,8-Dihydroxy-2-methyl-1,4-naphthoquinone, D-60352
2-Hydroxy-8-methoxy-1,4-naphthoquinone, *in* D-60355
4-Methyl-6,7-methylenedioxy-2*H*-1-benzopyran-2-one, M-70092
2-Oxo-2*H*-benzopyran-4-acetic acid, O-60057
1-Oxo-1*H*-2-benzopyran-3-acetic acid, O-60058
4-Oxo-4*H*-1-benzopyran-3-acetic acid, O-60059

C₁₁H₈O₅

4-Hydroxy-5-benzofurancarboxylic acid; Ac, *in* H-60103
2-Hydroxy-5-methoxy-1,4-naphthoquinone, *in* D-60354

C₁₁H₈O₈

2,3,5,6,8-Pentahydroxy-7-methoxy-1,4-naphthoquinone, *in* H-60065

C₁₁H₉BF₄N₄

3-Phenyltetrazolo[1,5-*a*]pyridinium; Tetrafluoroborate, *in* P-60134

C₁₁H₉Br

1-Bromo-1,4-dihydro-1,4-methanonaphthalene, B-60231
2-Bromo-1,4-dihydro-1,4-methanonaphthalene, B-60232
9-Bromo-1,4-dihydro-1,4-methanonaphthalene, B-60233

C₁₁H₉BrN₂

Dipyrido[1,2-*a*:1′,2′-*c*]imidazol-10-ium(1+); Bromide, *in* D-70501

C₁₁H₉BrN₄

3-Phenyltetrazolo[1,5-*a*]pyridinium; Bromide, *in* P-60134

C₁₁H₉ClN₂O₄

Dipyrido[1,2-*a*:1′,2′-*c*]imidazol-10-ium(1+); Perchlorate, *in* D-70501

C₁₁H₉IN₂

Dipyrido[1,2-*a*:1′,2′-*c*]imidazol-10-ium(1+); Iodide, *in* D-70501

C₁₁H₉N

2-Ethynylindole; 1-Me, *in* E-60057
4-Vinylquinoline, V-70014

C₁₁H₉NO

1-Acetylisoquinoline, A-60039
3-Acetylisoquinoline, A-60040
4-Acetylisoquinoline, A-60041

5-Acetylisoquinoline, A-60042
2-Acetylquinoline, A-60049
3-Acetylquinoline, A-60050
4-Acetylquinoline, A-60051
5-Acetylquinoline, A-60052
6-Acetylquinoline, A-60053
7-Acetylquinoline, A-60054
8-Acetylquinoline, A-60055
2-Phenyl-1,3-oxazepine, P-60123
5-Phenyl-1,4-oxazepine, P-60124

C₁₁H₉NO₂

4-Acetylisoquinoline; 1-Oxide, *in* A-60051
2-Amino-3-methyl-1,4-naphthoquinone, A-70171
Cyclohepta[*b*]pyrrol-2(1*H*)-one; *N*-Ac, *in* C-60201

C₁₁H₉NO₂S

2-Amino-3-(methylthio)-1,4-naphthalenedione, *in* A-70163

C₁₁H₉NO₃

2-Oxo-2*H*-benzopyran-4-acetic acid; Amide, *in* O-60057
4-Oxo-4*H*-1-benzopyran-3-acetic acid; Amide, *in* O-60059

C₁₁H₉N₂⊕

Dipyrido[1,2-*a*:1′,2′-*c*]imidazol-10-ium(1+), D-70501

C₁₁H₉N₃

1-Amino-β-carboline, A-70109
1-Amino-γ-carboline, A-70114
▷2-Amino-α-carboline, A-70105
3-Amino-α-carboline, A-70106
3-Amino-β-carboline, A-70110
3-Amino-γ-carboline, A-70115
4-Amino-α-carboline, A-70107
4-Amino-β-carboline, A-70111
5-Amino-γ-carboline, A-70116
6-Amino-α-carboline, A-70108
6-Amino-β-carboline, A-70112
6-Amino-γ-carboline, A-70117
8-Amino-β-carboline, A-70113
8-Amino-γ-carboline, A-70118
8-Amino-δ-carboline, A-70119
1-Methyl-8*H*-benzo[*cd*]triazirino[*a*]indazole, M-60052
2-Methyl-2*H*-naphtho[1,8-*de*]triazine, M-70095
10*H*-Pyrazolo[5,1-*c*][1,4]benzodiazepine, P-70154
9*H*-Pyrrolo[1,2-*a*]indol-9-one; Hydrazone, *in* P-60247

C₁₁H₉N₃O

Di-2-pyridyl ketone; Oxime, *in* D-70502
Di-3-pyridyl ketone; Oxime, *in* D-70503
Di-4-pyridyl ketone; Oxime, *in* D-70504
2-Pyridyl 3-pyridyl ketone; Oxime, *in* P-70179
2-Pyridyl 4-pyridyl ketone; Oxime, *in* P-70180
3-Pyridyl 4-pyridyl ketone; Oxime, *in* P-70181

C₁₁H₉N₃O₃

8-Amino-6-nitroquinoline; 8-*N*-Ac, *in* A-60235

C₁₁H₉N₄⊕

3-Phenyltetrazolo[1,5-*a*]pyridinium, P-60134

C₁₁H₉N₅O

2-Amino-1,7-dihydro-7-phenyl-6*H*-purin-6-one, A-70133

C₁₁H₁₀

1,4-Dihydro-1,4-methanonaphthalene, D-60256

C₁₁H₁₀IN₃

6*H*-Dipyrido[1,2-*a*:2′,1′-*d*][1,3,5]triazin-5-ium; Iodide, *in* D-60495

C₁₁H₁₀N₂

3,4-Dihydro-β-carboline, D-60209
Di-2-pyridylmethane, D-60496
2-Methyl-4-phenylpyrimidine, M-70111
2-Methyl-5-phenylpyrimidine, M-70112

4-Methyl-2-phenylpyrimidine, M-70113
4-Methyl-5-phenylpyrimidine, M-70114
4-Methyl-6-phenylpyrimidine, M-70115
5-Methyl-2-phenylpyrimidine, M-70116
5-Methyl-4-phenylpyrimidine, M-70117

C₁₁H₁₀N₂O

1-Acetylisoquinoline; Oxime, *in* A-60039
4-Acetylisoquinoline; Oxime, *in* A-60041
2-Acetylquinoline; Oxime, *in* A-60049
5-Acetylquinoline; Oxime, *in* A-60052
6-Acetylquinoline; Oxime, *in* A-60053
7-Acetylquinoline; Oxime, *in* A-60054
8-Acetylquinoline; Oxime, *in* A-60055
3,4-Dihydro-β-carboline; 2-Oxide, *in* D-60209
1-Methyl-2-phenyl-1*H*-imidazole-4-carboxaldehyde, *in* P-60111
1-Methyl-2-phenyl-1*H*-imidazole-5-carboxaldehyde, *in* P-60111
2-Methyl-4-phenylpyrimidine; 1-Oxide, *in* M-70111
4-Methyl-5-phenylpyrimidine; 1-Oxide, *in* M-70114
4-Methyl-5-phenylpyrimidine; 3-Oxide, *in* M-70114
4-Methyl-6-phenylpyrimidine; 1-Oxide, *in* M-70115
4-Methyl-6-phenylpyrimidine; 3-Oxide, *in* M-70115
5-Methyl-4-phenylpyrimidine; 1-Oxide, *in* M-70117
5-Methyl-4-phenylpyrimidine; 3-Oxide, *in* M-70117
6-Methyl-3-pyridazinone; 2-Ph, *in* M-70124
5-Methyl-3(2*H*)-pyridazinone; 2-Ph, *in* M-70126

C₁₁H₁₀N₂OS₂

Spirobrassinin, S-60041

C₁₁H₁₀N₂O₂

5-Amino-3-phenylisoxazole; *N*-Ac, *in* A-60240
3-Amino-2(1*H*)-quinolinone; 3-*N*-Ac, *in* A-60256
5-Methyl-4-phenylpyrimidine; 1,3-Dioxide, *in* M-70117

C₁₁H₁₀N₃⊕

6*H*-Dipyrido[1,2-*a*:2′,1′-*d*][1,3,5]triazin-5-ium, D-60495

C₁₁H₁₀N₄

5-Amino-1-benzyl-4-cyanopyrazole, *in* A-70192

C₁₁H₁₀N₄O₂

lin-Benzotheophylline, *in* I-60009

C₁₁H₁₀N₆

Bentemazole, *in* I-70009

C₁₁H₁₀O

2-Acetyl-1*H*-indene, A-70045
3-Acetyl-1*H*-indene, A-70046
6-Acetyl-1*H*-indene, A-70047
Hexacyclo[5.4.0.0²,⁶.0³,¹⁰.0⁵,⁹.0⁸,¹¹]undecan-4-one, H-60035
2-Naphthalenemethanol, N-60004
2,3,5,6-Tetramethylenebicyclo[2.2.1]heptan-7-one, T-70136

C₁₁H₁₀O₂

2,5-Dimethyl-4*H*-1-benzopyran-4-one, D-70371
6-Methoxy-5-vinylbenzofuran, *in* H-70231
2-Methyl-1,3-naphthalenediol, M-60091
2-Methyl-1,5-naphthalenediol, M-60092
3-Methyl-1,2-naphthalenediol, M-60093
4-Methyl-1,3-naphthalenediol, M-60094
6-Methyl-1,2-naphthalenediol, M-60095
5-(2-Methyloxiranyl)benzofuran, M-70101
Pentacyclo[5.4.0.0²,⁶.0³,¹⁰.0⁵,⁹]undecane-1,11-dione, P-60030
Pentacyclo[6.2.1.0²,⁷.0⁴,¹⁰.0⁵,⁹]undecane-3,6-dione, P-70022
5-Phenyl-2,4-pentadienoic acid, P-70086

C₁₁H₁₀O₃

7-Hydroxy-3,4-dimethyl-2*H*-1-benzopyran-2-one, H-60119

8-Methoxy-4-methyl-2*H*-1-benzopyran-2-one, *in* H-70165
5-Methoxy-4-methylcoumarin, *in* H-60173

$C_{11}H_{10}O_4$

2,3-Dihydro-3,5-dihydroxy-2-methyl-naphthoquinone, D-70191
8,13-Dioxapentacyclo[6.5.0.0^{2,6}.0^{5,10}.0^{3,11}]-tridecane-9,12-dione, D-60467
3-Ethyl-5,7-dihydroxy-4*H*-1-benzopyran-4-one, E-70040
7-Hydroxy-4-methoxy-5-methyl-2*H*-1-benzopyran-2-one, *in* D-60348
6-(Hydroxymethyl)-7-methoxy-2*H*-1-benzopyran-2-one, *in* H-70139
3-(2-Methoxy-4,5-methylenedioxyphenyl)-propenal, M-70048
Methyl 3-benzoyloxyacrylate, *in* O-60090

$C_{11}H_{10}O_5$

2,7-Dihydroxy-2,4,6-cycloheptatrien-1-one; Di-Ac, *in* D-60314

$C_{11}H_{11}BrN_2O_2$

2-Bromotryptophan, B-60332
5-Bromotryptophan, B-60333
6-Bromotryptophan, B-60334
7-Bromotryptophan, B-60335

$C_{11}H_{11}ClN_2O_2$

2-Chlorotryptophan, C-60145
6-Chlorotryptophan, C-60146

$C_{11}H_{11}Cl_2NO_2$

4,6-Dichloro-5-[(dimethylamino)methylene]-3,6-cyclohexadiene-1,3-dicarboxaldehyde, D-70119

$C_{11}H_{11}F_4O_2$

2,3,5,6-Tetrafluorobenzaldehyde; Di-Et acetal, *in* T-70028

$C_{11}H_{11}N$

▷2-Methyl-5-phenyl-1*H*-pyrrole, M-60115
3-(2-Propenyl)indole, P-60183
1,2,3,4-Tetrahydrocyclopent[*b*]indole, T-70049

$C_{11}H_{11}NO$

1,3-Dimethyl-1*H*-indole-2-carboxaldehyde, *in* M-70090
1-(2-Furanyl)-2-(2-pyrrolyl)ethylene; *N*-Me, *in* F-60088
3a,7a-Methano-1*H*-indole; *N*-Ac, *in* M-70041
1-(4-Methoxyphenyl)pyrrole, *in* H-60216
2-(2-Methoxyphenyl)pyrrole, *in* H-60217

$C_{11}H_{11}NO_2$

3,3-Dimethyl-2,4(1*H*,3*H*)-quinolinedione, D-60451
1*H*-Indole-3-acetic acid; *N*-Me, *in* I-70013

$C_{11}H_{11}NO_2S$

α-Aminobenzo[*b*]thiophene-3-acetic acid; Me ester, *in* A-60090
3-Amino-2-benzo[*b*]thiophenecarboxylic acid; Et ester, *in* A-60091

$C_{11}H_{11}NO_3$

3-Aminodihydro-2(3*H*)-furanone; *N*-Benzoyl, *in* A-60138
2,3-Dihydro-2-oxo-1*H*-indole-3-acetic acid; Me ester, *in* D-60259
3-Morpholinone; 4-Benzoyl, *in* M-60147
1,2,3,4-Tetrahydro-2-oxo-4-quinolinecarboxylic acid; Me ester, *in* T-60077
1,2,3,4-Tetrahydro-4-oxo-6-quinolinecarboxylic acid; Me ester, *in* T-60078

$C_{11}H_{11}NO_4$

5,6-Dihydroxy-1*H*-indole-2-carboxylic acid; Et ester, *in* D-70308
5,6-Dihydroxy-1*H*-indole-2-carboxylic acid; Di-Me ether, *in* D-70308

$C_{11}H_{11}NO_7$

4-Hydroxy-5-nitro-1,2-benzenedicarboxylic acid; Me ether, di-Me ester, *in* H-70191
4-Hydroxy-5-nitro-1,3-benzenedicarboxylic acid; Me ether, di-Me ester, *in* H-70192

$C_{11}H_{11}NS$

1-(2-Pyrrolyl)-2-(2-thienyl)ethylene; *N*-Me, *in* P-60250

$C_{11}H_{11}N_3O_2$

3-Amino-1*H*-pyrazole-4-carboxylic acid; *N*(1)-Benzyl, *in* A-70187
Methyl 1-benzyl-1*H*-1,2,3-triazole-5-carboxylate, *in* T-60239
1-Phenyl-1*H*-1,2,3-triazole-4-carboxylic acid; Et ester, *in* P-60138
1-Phenyl-1*H*-1,2,3-triazole-5-carboxylic acid; Et ester, *in* P-60139
2-Phenyl-2*H*-1,2,3-triazole-4-carboxylic acid; Et ester, *in* P-60140
1*H*-1,2,3-Triazole-4-carboxylic acid; 1-Benzyl, Me ester, *in* T-60239

$C_{11}H_{11}N_3O_4$

2-Nitrotryptophan, N-70067

$C_{11}H_{11}O_2^{⊕}$

4,5-Dimethyl-2-phenyl-1,3-dioxol-1-ium, D-70431

$C_{11}H_{17}$

Benzylidenecyclobutane, B-70081
1,4-Methano-1,2,3,4-tetrahydronaphthalene, M-60034
1-Phenyl-1,2-pentadiene, P-70085
3-Phenyl-1-pentyne, P-70089
3,4,5,6-Tetrahydro-1*H*-cyclopropa[*b*]-naphthalene, T-70050
4,5,6,7-Tetrahydro-1*H*-cyclopropa[*a*]-naphthalene, T-70051
1,2,3,4-Tetrahydro-1-methylenenaphthalene, T-60073

$C_{11}H_{12}Br_2N_2$

1,1′-Methylenebispyridinium(1+); Dibromide, *in* M-70065

$C_{11}H_{12}I_2N_2$

1,1′-Methylenebispyridinium(1+); Diiodide, *in* M-70065

$C_{11}H_{12}N_2$

4,4-Dimethyl-2-phenyl-4*H*-imidazole, D-70432

$C_{11}H_{12}N_2^{⊕}$

1,1′-Methylenebispyridinium(1+), M-70065

$C_{11}H_{12}N_2O$

5-Amino-3-phenylisoxazole; *N*-Et, *in* A-60240

$C_{11}H_{12}N_2O_2$

5-Oxo-2-pyrrolidinecarboxylic acid; Anilide, *in* O-70102

$C_{11}H_{12}N_2O_4$

2,4-Diaminobenzoic acid; 2,4-*N*-Di-Ac, *in* D-60032

$C_{11}H_{12}N_2O_5$

2′-Deoxy-5-ethynyluridine, *in* E-60060

$C_{11}H_{12}N_2O_6$

5-Ethynyluridine, *in* E-60060

$C_{11}H_{12}N_4O$

5-Amino-1-(phenylmethyl)-1*H*-pyrazole-4-carboxamide, *in* A-70192

$C_{11}H_{12}O$

2,4,6-Cycloheptatrien-1-ylcyclo-propylmethanone, C-60206
2,3-Dihydro-2,2-dimethyl-1*H*-inden-1-one, D-60227
2-Methyl-4-phenyl-3-butyn-2-ol, M-70110
1-Phenyl-3-penten-1-one, P-60127
5-Phenyl-4-penten-2-one, P-70088
1,2,3,4-Tetrahydro-7*H*-benzocyclohepten-7-one, T-60057

$C_{11}H_{12}OS$

2,3-Dihydro-2,2-dimethyl-4*H*-1-benzothiopyran, D-70193

$C_{11}H_{12}O_2$

8-Decene-4,6-diynoic acid; Me ester, *in* D-70018
2-Phenylcyclopropanecarboxylic acid; Me ester, *in* P-60087

$C_{11}H_{12}O_2S$

4-Mercapto-2-butenoic acid; Benzyl ester, *in* M-60018

$C_{11}H_{12}O_3$

3-Benzoyl-2-methylpropanoic acid, B-60064
3,4-Dihydro-2*H*-1-benzopyran-2-ol; Ac, *in* D-60198
3,4-Dihydro-2*H*-1-benzopyran-4-ol; Ac, *in* D-60200
3,4-Dihydro-2*H*-1-benzopyran-6-ol; Ac, *in* D-70175
3-(4-Hydroxy-2-vinylphenyl)propanoic acid, H-70232
1,2,3,4-Tetrahydro-1-hydroxy-2-naphthalenecarboxylic acid, T-70063

$C_{11}H_{12}O_4$

1,3,5-Cycloheptatriene-1,6-dicarboxylic acid; Di-Me ester, *in* C-60202
5,7-Dimethoxy-6-methylphthalide, *in* D-60349
Pyrenocin *A*, P-60213

$C_{11}H_{12}O_5$

2-Formyl-3,5-dihydroxybenzoic acid; Di-Me ether, Me ester, *in* F-60071
5-Hydroxy-3-methyl-1,2-benzenedicarboxylic acid; Me ether, 2-Me ester, *in* H-70162

$C_{11}H_{12}S$

2,2-Dimethyl-2*H*-1-benzothiopyran, D-70373

$C_{11}H_{12}S_2$

2-(2,4,6-Cycloheptatrien-1-ylidene)-1,3-dithiane, C-70221

$C_{11}H_{13}BrO_2$

3-Bromo-2,4,6-trimethylbenzoic acid; Me ester, *in* B-60331

$C_{11}H_{13}IO_4$

2,5-Dihydroxy-4-iodobenzoic acid; Di-Me ether, Et ester, *in* D-60333

$C_{11}H_{13}I_2NO$

2,6-Diiodobenzoic acid; Diethylamide, *in* D-70356

$C_{11}H_{13}N$

1,2,3,6-Tetrahydro-4-phenylpyridine, T-60081
1,2,3,6-Tetrahydropyridine; 1-Ph, *in* T-70089

$C_{11}H_{13}NO$

3,4-Dihydro-3,3-dimethyl-2(1*H*)quinolinone, D-60231
3-Isopropyloxindole, I-70097
4-(Phenylimino)-2-pentanone, *in* P-60059
1,2,3,4-Tetrahydroquinoline; *N*-Ac, *in* T-70093

$C_{11}H_{13}NO_2$

2-Amino-3-phenyl-3-butenoic acid; Me ester, *in* A-60238
5,6-Dihydroxyindole; Di-Me ether, *N*-Me, *in* D-70307
1,2,3,4-Tetrahydro-4-hydroxyisoquinoline; *N*-Ac, *in* T-60066
1,2,3,4-Tetrahydro-3-methyl-3-isoquinolinecarboxylic acid, T-60074
5-(3,4,5,6-Tetrahydro-3-pyridylidenemethyl)-2-furanmethanol, T-60087

$C_{11}H_{13}NO_3$

1-Amino-2-(4-hydroxyphenyl)-cyclopropanecarboxylic acid; Me ester, *in* A-60197
1-Amino-2-(4-hydroxyphenyl)-cyclopropanecarboxylic acid; Me ether, *in* A-60197
Ethyl *N*-phenacylcarbamate, *in* O-70099
2,3,4-Trihydroxyphenylacetic acid; Tri-Me ether, nitrile, *in* T-60340
2,4,5-Trihydroxyphenylacetic acid; Tri-Me ether, nitrile, *in* T-60342
▷3,4,5-Trimethoxybenzeneacetonitrile, *in* T-60343

$C_{11}H_{13}NO_4$

2-Amino-2-benzylbutanedioic acid, A-60094
2-Amino-4-hydroxybutanoic acid; *N*-Benzoyl, *in* A-70156
3-Amino-2-hydroxybutanoic acid; *N*-Benzoyl, *in* A-60183
2-Amino-3-phenylpentanedioic acid, A-60241
2,6-Biss(ethoxyacetyl)pyridine, *in* B-70144

3,5-Pyridinedipropanoic acid, P-60217
Threonine; N-Benzoyl, in T-60217

C$_{11}$H$_{13}$NO$_5$
5-Amino-4-hydroxy-1,3-benzenedicarboxylic
acid; Me ether, di-Me ester, in A-70155
Dimethyl 4-amino-5-methoxyphthalate, in
A-70152

C$_{11}$H$_{13}$N$_3$O$_4$
2′-Deoxy-5-ethynylcytidine, in A-60153
Imidazo[4,5-c]pyridine; 1-β-D-Ribofuranosyl, in
I-60008

C$_{11}$H$_{13}$N$_3$O$_5$
5-Ethynylcytidine, in A-60153

C$_{11}$H$_{14}$
6-(1,3-Butadienyl)-1,4-cycloheptadiene,
B-70290

C$_{11}$H$_{14}$N$_2$
5,6,9,10-Tetrahydro-4H,8H-pyrido[3,2,1-
ij][1,6]naphthyridine, T-60086

C$_{11}$H$_{14}$N$_2$O
Piperazinone; N^4-Benzyl, in P-70101

C$_{11}$H$_{14}$N$_2$O$_3$
2,4,6-Trimethyl-3-nitroaniline; N-Ac, in
T-70285

C$_{11}$H$_{14}$N$_2$O$_4$
4-Amino-2,6-pyridinedicarboxylic acid; Di-Et
ester, in A-70194

C$_{11}$H$_{14}$N$_2$O$_5$
2′-Deoxy-5-vinyluridine, in V-70015

C$_{11}$H$_{14}$N$_2$O$_6$
2-Amino-3,5-pyridinedicarboxylic acid; Di-Et
ester, in A-70193
5-Vinyluridine, V-70015

C$_{11}$H$_{14}$N$_4$
4-(Butylimino)-3,4-dihydro-1,2,3-benzotriazine
(incorr.), in A-60092

C$_{11}$H$_{14}$N$_4$O$_3$
4-Amino-1H-imidazo[4,5-c]pyridine; 1-(2-
Deoxy-β-D-ribofuranosyl), in A-70161
4-Amino-1H-imidazo[4,5-c]pyridine; 1-(5-
Deoxy-β-D-ribofuranosyl), in A-70161

C$_{11}$H$_{14}$N$_4$O$_4$
4-Amino-1H-imidazo[4,5-c]pyridine; 1-β-D-
Ribofuranosyl, in A-70161
7-Deaza-2′-deoxyguanosine, in A-60255
2-Methyl-9-β-D-ribofuranosyl-9H-purine,
M-70130
6-Methyl-9-β-D-ribofuranosyl-9H-purine,
M-70131

C$_{11}$H$_{14}$N$_4$O$_4$S
▷1,7-Dihydro-6H-purine-6-thione; 5-Me, 9-β-D-
ribofuranosyl, in D-60270
▷6-(Methylthio)-9-β-D-ribofuranosyl-9H-purine,
in T-70180

C$_{11}$H$_{14}$N$_4$O$_5$
7-Deazaguanosine, in A-60255

C$_{11}$H$_{14}$O
2,2-Dimethyl-1-phenyl-1-propanone, D-60443
(3-Hydroxyphenoxy)acetic acid; Me ether, Et
ester, in H-70204
4-Methoxy-4-phenyl-1-butene, in P-70061
3-Methyl-2-phenylbutanal, M-70109
Spiro[5.5]undeca-1,3-dien-7-one, S-60046

C$_{11}$H$_{14}$O$_2$
Andirolactone, A-60271
3-tert-Butyl-4-hydroxybenzaldehyde, B-70304
1-(Phenylmethoxy)-3-buten-2-ol, in B-70291
1,2,3,4-Tetrahydro-1-hydroxy-2-
naphthalenemethanol, T-70064
Tricyclo[4.3.1.1^{3,8}]undecane-2,7-dione,
T-70233
Tricyclo[4.3.1.1^{3,8}]undecane-4,5-dione,
T-70234
3,4,5-Trimethylbenzoic acid; Me ester, in
T-60355

3,4,5-Trimethylphenol; Ac, in T-60371

C$_{11}$H$_{14}$O$_4$
2-(3,4-Dihydroxyphenyl)propanoic acid; Di-Me
ether, in D-60372
2,6-Dimethoxy-3,5-dimethylbenzoic acid, in
D-70293
6-Ethyl-2,4-dihydroxy-3-methylbenzoic acid;
Me ester, in E-70041
3-Hydroxy-5-(4-hydroxyphenyl)pentanoic acid,
H-70142
(2-Hydroxyphenoxy)acetic acid; Me ether, Et
ester, in H-70203
(4-Hydroxyphenoxy)acetic acid; Me ether, Et
ester, in H-70205
Sinapyl alcohol, S-70046
2′,4′,6′-Trimethoxyacetophenone, in T-70248
3′,4′,5′-Trimethoxyacetophenone, in T-60315

C$_{11}$H$_{14}$O$_5$
Pyrenocin B, in P-60213
2,3,4-Trimethoxybenzeneacetic acid, in
T-60340
2,4,5-Trimethoxybenzeneacetic acid, in
T-60342
3,4,5-Trimethoxybenzeneacetic acid, in
T-60343

C$_{11}$H$_{14}$O$_6$
2,3,5,6-Tetramethoxybenzoic acid, in T-60100
(3,4,5-Trihydroxy-6-oxo-1-cyclohexen-1-yl)-
methyl 2-butenoate, T-60338

C$_{11}$H$_{15}$Br
(Bromomethylidene)adamantane, B-60296
Bromopentamethylbenzene, B-60312

C$_{11}$H$_{15}$Cl
Chloropentamethylbenzene, C-60122

C$_{11}$H$_{15}$N
1-Amino-1-phenyl-4-pentene, A-70183
2-Phenylcyclopentylamine, P-60086
2,3,4,5-Tetrahydro-1H-1-benzazepine; 1-Me, in
T-70036

C$_{11}$H$_{15}$NOS
Tetrahydro-2-thiophenecarboxylic acid; Anilide,
in T-70099
Tricyclo[3.3.1.1^{3,7}]decyl-1-sulfinylcyanide, in
T-70177

C$_{11}$H$_{15}$NO$_2$
2-Amino-2-methyl-3-phenylpropanoic acid; Me
ester, in A-60225
1,2,3,4,5-Pentamethyl-6-nitrobenzene, P-60057

C$_{11}$H$_{15}$NO$_2$S
2-Amino-4,5,6,7-tetrahydrobenzo[b]thiophene-
3-carboxylic acid; Et ester, in A-60258

C$_{11}$H$_{15}$NO$_3$
2-Amino-3-(4-hydroxyphenyl)-3-methyl-
butanoic acid, A-60198
4-Amino-3-hydroxy-5-phenylpentanoic acid,
A-60199
2-(2-Ethoxyethoxy)benzamide, in H-70108

C$_{11}$H$_{15}$NO$_4$
3′,4′,5′-Trihydroxyacetophenone; Tri-Me ether,
oxime, in T-60315

C$_{11}$H$_{15}$NO$_5$
1H-Pyrrole-3,4-dicarboxylic acid; 1-Methoxy,
di-Et ester, in P-70183

C$_{11}$H$_{15}$NS
3-Aminotetrahydro-2H-thiopyran; N-Ph, in
A-70200
1-Thiocyanatoadamantane, T-70177

C$_{11}$H$_{15}$N$_5$O
1′-Methylzeatin, M-60133

C$_{11}$H$_{15}$N$_5$O$_3$
1-[3-Azido-4-(hydroxymethyl)cyclopentyl]-5-
methyl-2,4(1H,3H)pyrimidinedione, A-70288

C$_{11}$H$_{15}$N$_5$O$_4$
Euglenapterin, E-60070

C$_{11}$H$_{15}$N$_5$O$_5$
Ara-doridosine, in A-60144
Doridosine, D-60518
1-Methylguanosine, M-70084
8-Methylguanosine, M-70085

C$_{11}$H$_{15}$N$_5$O$_5$S
8-(Methylthio)guanosine, in M-70029

C$_{11}$H$_{15}$N$_5$O$_6$
8-Methoxyguanosine, in D-70238

C$_{11}$H$_{15}$N$_5$O$_7$S
8-(Methylsulfonyl)guanosine, in M-70029

C$_{11}$H$_{16}$
Bicyclo[4.4.1]undeca-1,6-diene, B-70115
1-Cycloundecen-3-yne, C-70256
2,3-Dimethylenebicyclo[2.2.3]nonane, D-70406
(2,2-Dimethylpropyl)benzene, D-60445
Methyleneadamantane, M-60068
Tetracyclo[4.4.1.0^{3,11}.0^{9,11}]undecane, T-60037
2,4,6,8-Undecatetraene, U-70003

C$_{11}$H$_{16}$N$_2$O$_5$
▷2′-Deoxy-5-ethyluridine, in E-60054

C$_{11}$H$_{16}$N$_2$O$_6$
5-Ethyluridine, in E-60054

C$_{11}$H$_{16}$N$_6$O$_3$
3′-Amino-3′-deoxyadenosine; 3′N-Me, in
A-70128

C$_{11}$H$_{16}$N$_6$O$_5$
8-(Methylamino)guanosine, in A-70150

C$_{11}$H$_{16}$O
5,6,7,8,9,10-Hexahydro-4H-cyclonona[c]furan,
H-60047
1-Methyladamantanone, M-60040
5-Methyladamantanone, M-60041
3-Oxatetracyclo[5.3.1.1^{2,6}.0^{4,9}]dodecane,
O-70069
Pentamethylphenol, P-60058
2,2,5,5-Tetramethylbicyclo[4.1.0]hept-1(6)-en-
7-one, T-70131

C$_{11}$H$_{16}$O$_2$
Bicyclo[2.2.2]oct-2-ene-1-carboxylic acid; Et
ester, in B-70113
▷3,4-Dimethyl-5-pentylidene-2(5H)-furanone,
D-70430
Spiro[5.5]undecane-1,9-dione, S-70066
Spiro[5.5]undecane-3,8-dione, S-70067
Spiro[5.5]undecane-3,9-dione, S-70068

C$_{11}$H$_{16}$O$_4$
Methylenolactocin, M-70079
▷Xanthotoxol, X-60002

C$_{11}$H$_{16}$O$_8$
Ranunculin, in H-60182

C$_{11}$H$_{17}$N
3-Azatetracyclo[5.3.1.1^{2,6}.0^{4,9}]dodecane,
A-70282

C$_{11}$H$_{17}$NO$_7$
2-(Dimethylamino)benzaldehyde; Di-Me acetal,
in D-60401

C$_{11}$H$_{17}$N$_3$O$_6$
Antibiotic CA 146B, in C-70184
Clavamycin E, C-70184

C$_{11}$H$_{18}$
2,2-Dimethyl-3-methylenebicyclo[2.2.2]octane,
D-60427
1,3,5-Undecatriene, U-70004

C$_{11}$H$_{18}$N$_2$O$_2$
2-Diazo-2,2,6,6-tetramethyl-3,5-heptanedione,
D-70064
Hexahydro-3-(1-methylpropyl)pyrrolo[1,2-a]-
pyrazine-1,4-dione, H-60052
Spiro[5.5]undecane-3,9-dione; Dioxime, in
S-70068

C$_{11}$H$_{18}$N$_2$O$_3$
Cycloo(ydroxyprolylleucyl), C-60213

C$_{11}$H$_{18}$O
2-Cycloundecen-1-one, C-70255

$C_{11}H_{18}O_2$

Bicyclo[2.2.2]octane-1-carboxylic acid; Et ester, *in* B-60092
2-Hexyl-5-methyl-3(2*H*)furanone, H-60079

$C_{11}H_{18}O_3$

2,3-Di-*tert*-butoxycyclopropenone, *in* D-70290
1,3-Dihydroxy-4-methyl-6,8-decadien-5-one, D-70321
9-Oxo-2-decenoic acid; Me ester, *in* O-70089

$C_{11}H_{18}O_4$

Citreoviral, C-60150
Phaseolinic acid, P-60067

$C_{11}H_{18}O_5$

4-Oxoheptanedioic acid; Di-Et ester, *in* O-60071
4,4-Pyrandiacetic acid; Di-Me ester, *in* P-70150

$C_{11}H_{19}NO_8$

Agropinic acid, A-70075

$C_{11}H_{19}N_3O_7$

Nitropeptin, N-70055

$C_{11}H_{20}$

1,1,3,3-Tetramethyl-2-methylenecyclohexane, T-60144

$C_{11}H_{20}N_2$

3,4-Di-*tert*-butylpyrazole, D-70106
3,5-Di-*tert*-butylpyrazole, D-70107

$C_{11}H_{20}O$

6-Undecyn-1-ol, U-60004

$C_{11}H_{20}O_2$

2,4-Heptadienal; Di-Et acetal, *in* H-60020

$C_{11}H_{20}O_3$

2-Hydroxycyclooctanecarboxylic acid; Me ether, Me ester, *in* H-70121
2-Hydroxycyclooctanecarboxylic acid; Et ester, *in* H-70121
10-Oxodecanoic acid; Me ester, *in* O-60063

$C_{11}H_{21}Br$

1-Bromo-2-undecene, B-70277
11-Bromo-1-undecene, B-60336

$C_{11}H_{21}NO$

Hexahydro-5(2*H*)-azocinone; 1-*tert*-Butyl, *in* H-70048

$C_{11}H_{21}NO_9$

Mannopinic acid, M-70009

$C_{11}H_{22}$

1,1-Di-*tert*-butylcyclopropane, D-60112

$C_{11}H_{22}N_2O$

4-Amino-2,2,6,6-tetramethylpiperidine; *N*-Ac, *in* A-70201

$C_{11}H_{22}N_2O_3$

N-Valylleucine, V-60001

$C_{11}H_{22}N_2O_8$

Mannopine, *in* M-70009

$C_{11}H_{22}O_3$

2,2-Diethoxytetrahydro-5,6-dimethyl-2*H*-pyran, *in* T-70059

$C_{11}H_{23}N$

Octahydroazocine; 1-*tert*-Butyl, *in* O-70010

$C_{11}H_{24}$

2,2,3,3,4,4-Hexamethylpentane, H-60071

$C_{11}H_{27}N_5$

1,4,7,10,13-Pentaazacyclohexadecane, P-60021

$C_{12}H_4Cl_4O_2$

▷1,2,3,4-Tetrachlorodibenzo-*p*-dioxin, T-70016

$C_{12}H_4Cl_6$

▷2,2′,4,4′,5,5′-Hexachlorobiphenyl, H-60034

$C_{12}H_6Cl_2N_2$

1,4-Dichlorobenzo[g]phthalazine, D-70114

$C_{12}H_6Cl_2N_2O_2$

[2,2′-Bipyridine]-4,4′-dicarboxylic acid; Dichloride, *in* B-60126
[2,2′-Bipyridine]-6,6′-dicarboxylic acid; Dichloride, *in* B-60128

$C_{12}H_6Cl_4N_2O_4$

Pyrroxamycin, P-70203

$C_{12}H_6Cl_4O$

Bis(2,4-dichlorophenyl)ether, B-70135
Bis(2,6-dichlorophenyl) ether, B-70136
Bis(3,4-dichlorophenyl) ether, B-70137

$C_{12}H_6Cl_{12}$

Hexakiss(ichloromethyl)benzene, H-60067

$C_{12}H_6N_2$

5,7-Dicyanoazulene, *in* A-70295
2,3-Dicyanonaphthalene, *in* N-60003

$C_{12}H_6N_2O_2$

Benzo[g]quinazoline-6,9-dione, B-60044
Benzo[g]quinoxaline-6,9-dione, B-60046
1,7-Phenanthroline-5,6-dione, P-70052

$C_{12}H_6N_4$

5,5′-Dicyano-2,2′-bipyridine, *in* B-60127
Pyrazino[2′,3′:3,4]cyclobuta[1,2-g]quinoxaline, P-60198
1,2,4,5-Tetracyano-3,6-dimethylbenzene, *in* D-60407

$C_{12}H_6O_2S$

Naphtho[2,3-b]thiophene-2,3-dione, N-70014

$C_{12}H_6O_3$

2,3-Naphthalenedicarboxylic acid; Anhydride, *in* N-60003

$C_{12}H_7BrN_2$

2-Bromo-1,10-phenanthroline, B-70264
8-Bromo-1,7-phenanthroline, B-70265

$C_{12}H_7BrO_4$

2-Bromo-5-hydroxy-1,4-naphthoquinone; Ac, *in* B-60269
2-Bromo-6-hydroxy-1,4-naphthoquinone; Ac, *in* B-60270
2-Bromo-8-hydroxy-1,4-naphthoquinone; Ac, *in* B-60272

$C_{12}H_7ClN_2$

1-Chlorobenzo[g]phthalazine, C-70054

$C_{12}H_7ClN_2O$

4-Chlorobenzo[g]phthalazin-1(2*H*)-one, C-70055

$C_{12}H_7NOS$

1*H*-Phenothiazin-1-one, P-60078

$C_{12}H_7NO_2$

1*H*-Benz[e]indole-1,2(3*H*)-dione, B-70026
10*H*-[1]Benzopyrano[3,2-c]pyridin-10-one, B-60042
1*H*-Carbazole-1,4(9*H*)-dione, C-60012
3-Cyano-2-naphthoic acid, *in* N-60003
2,3-Naphthalenedicarboxylic acid; Imide, *in* N-60003

$C_{12}H_7NO_3$

2-Hydroxy-3*H*-phenoxazin-3-one, H-60206

$C_{12}H_7N_5$

7*H*-2,3,4,6,7-Pentaazabenz[de]anthracene, P-60017

$C_{12}H_8$

1-Ethylnylnaphthalene, E-70046

$C_{12}H_8ClNO$

1-Chloromethylbenz[cd]indol-2(1*H*)-one, *in* B-70027

$C_{12}H_8ClNOS$

3-Chloro-10*H*-phenothiazine; 5-Oxide, *in* C-60126
4-Chloro-10*H*-phenothiazine; 5-Oxide, *in* C-60127

$C_{12}H_8ClNO_2S$

3-Chloro-10*H*-phenothiazine; 5,5-Dioxide, *in* C-60126
4-Chloro-10*H*-phenothiazine; 5,5-Dioxide, *in* C-60127

$C_{12}H_8ClNO_3$

▷2-Amino-3-chloro-1,4-naphthoquinone; *N*-Ac, *in* A-70122

$C_{12}H_8ClNS$

1-Chloro-10*H*-phenothiazine, C-60124
▷2-Chloro-10*H*-phenothiazine, C-60125
3-Chloro-10*H*-phenothiazine, C-60126

$C_{12}H_8Cl_2$

5,6-Dichloroacenaphthene, D-60115

$C_{12}H_8F_3NO_3$

7-Amino-4-(trifluoromethyl)-2*H*-1-benzopyran-2-one; *N*-Ac, *in* A-60265

$C_{12}H_8N_2$

Benzo[c]cinnoline, B-60020
Benzvalenoquinoxaline, B-70080

$C_{12}H_8N_2O$

Benzo[c]cinnoline; *N*-Oxide, *in* B-60020
2-(2-Furanyl)quinoxaline, F-60089

$C_{12}H_8N_2O_2$

▷2-Amino-3*H*-phenoxazin-3-one, A-60237
1*H*-Benz[e]indole-1,2(3*H*)-dione; 1-Oxime, *in* B-70026
Benzo[c]cinnoline; 5,6-Di-*N*-oxide, *in* B-60020

$C_{12}H_8N_2O_4$

[2,2′-Bipyridine]-3,3′-dicarboxylic acid, B-60124
[2,2′-Bipyridine]-3,5′-dicarboxylic acid, B-60125
[2,2′-Bipyridine]-4,4′-dicarboxylic acid, B-60126
[2,2′-Bipyridine]-5,5′-dicarboxylic acid, B-60127
[2,2′-Bipyridine]-6,6′-dicarboxylic acid, B-60128
[2,3′-Bipyridine]-2,3′-dicarboxylic acid, B-60129
[2,4′-Bipyridine]-2′,6′-dicarboxylic acid, B-60130
[2,4′-Bipyridine]-3,3′-dicarboxylic acid, B-60131
[2,4′-Bipyridine]-3′,5-dicarboxylic acid, B-60132
[3,3′-Bipyridine]-2,2′-dicarboxylic acid, B-60133
[3,3′-Bipyridine]-4,4′-dicarboxylic acid, B-60134
[3,4′-Bipyridine]-2′,6′-dicarboxylic acid, B-60135
[4,4′-Bipyridine]-2,2′-dicarboxylic acid, B-60136
[4,4′-Bipyridine]-3,3′-dicarboxylic acid, B-60137
2,2′-Dinitrobiphenyl, D-60460

$C_{12}H_8N_4$

· Pyrido[1″,2″:1′,2′]imidazo[4′,5′:4,5]-imidazo[1,2-a]pyridine, P-70175
Pyrido[2″,1″:2′,3′]imidazo[4′,5′:4,5]-imidazo[1,2-a]pyridine, P-60234

$C_{12}H_8N_{10}O_{12}$

3,3′,5,5′-Tetraamino-2,2′,4,4′,6,6′-hexa-nitrobiphenyl, T-60014

$C_{12}H_8O$

6b,7a-Dihydroacenaphth[1,2-b]oxirene, D-70170
4-Hydroxyacenaphthylene, H-70100
Naphtho[1,2-c]furan, N-70008
Naphtho[2,3-c]furan, N-60008
Naphtho[1,8-bc]pyran, N-70009
1-Phenyl-1,4-hexadiyn-3-one, P-60108
▷1-Phenyl-2,4-hexadiyn-1-one, P-60109
6-Phenyl-3,5-hexadiyn-2-one, P-60110

$C_{12}H_8OS$

4-Phenylthieno[3,4-b]furan, P-60135

$C_{12}H_8O_2$

1,2-Dihydrocyclobuta[a]naphthalene-3,4-dione, D-60213

1,2-Dihydrocyclobuta[*b*]naphthalene-3,8-dione, D-60214
1,2-Naphthalenedicarboxaldehyde, N-60001
1,3-Naphthalenedicarboxaldehyde, N-60002

C₁₂H₈O₃

α-Oxo-1-naphthaleneacetic acid, O-60078
α-Oxo-2-naphthaleneacetic acid, O-60079

C₁₂H₈O₄

1,4-Azulenedicarboxylic acid, A-70293
2,6-Azulenedicarboxylic acid, A-70294
5,7-Azulenedicarboxylic acid, A-70295
2,3-Naphthalenedicarboxylic acid, N-60003
▷Xanthotoxin, X-60001

C₁₂H₈O₇

2,5,8-Trihydroxy-3-methyl-6,7-methylenedioxy-1,4-naphthoquinone, T-60333

C₁₂H₈S₂

2-Dibenzothiophenethiol, D-60078
4-Dibenzothiophenethiol, D-60079

C₁₂H₉BrO

1-Acetyl-3-bromonaphthalene, A-60021
1-Acetyl-4-bromonaphthalene, A-60022
1-Acetyl-5-bromonaphthalene, A-60023
1-Acetyl-7-bromonaphthalene, A-60024
2-Acetyl-6-bromonaphthalene, A-60025

C₁₂H₉BrO₃

2-Bromo-3-ethoxy-1,4-naphthoquinone, *in* B-60268

C₁₂H₉IO

1-Acetyl-7-iodonaphthalene, A-60037
1-Acetyl-8-iodonaphthalene, A-60038

C₁₂H₉NO

Benz[*cd*]indol-2-(1*H*)-one; *N*-Me, *in* B-70027
2-Hydroxycarbazole, H-60109
2-Methylfuro[3,2-*c*]quinoline, M-70083

C₁₂H₉N₃

2-Amino-1,10-phenanthroline, A-70179
4-Amino-1,10-phenanthroline, A-70181
5-Amino-1,10-phenanthroline, A-70182
8-Amino-1,7-phenanthroline, A-70180
2-Phenyl-2*H*-benzotriazole, P-60083
N-Pyridinium-2-benzimidazole, P-70172

C₁₂H₉N₃O

3-Amino-β-carboline; *N³*-Formyl, *in* A-70110
2-Phenyl-2*H*-benzotriazole; 1-*N*-Oxide, *in* P-60083

C₁₂H₁₀

1,6-Dihydro-1,6-dimethyleneazulene, D-60225
2,6-Dihydro-2,6-dimethyleneazulene, D-60226

C₁₂H₁₀F₃NO₂

7-Amino-4-(trifluoromethyl)-2*H*-1-benzopyran-2-one; *N*-Et, *in* A-60265

C₁₂H₁₀N₂

1-(1-Diazoethyl)naphthalene, D-60061

C₁₂H₁₀N₂O₃

5-Acetyl-2,4(1*H*,3*H*)-pyrimidinedione; 1-Ph, *in* A-60047
3-Nitro-4(1*H*)-pyridone; *N*-Benzyl, *in* N-70063

C₁₂H₁₀N₄O₂

[2,2′-Bipyridine]-4,4′-dicarboxylic acid; Diamide, *in* B-60126
[2,2′-Bipyridine]-5,5′-dicarboxylic acid; Diamide, *in* B-60127

C₁₂H₁₀O

1,2-Dihydrocyclobuta[*a*]naphthalen-3-ol, D-60215
2,3-Dihydronaphtho[2,3-*b*]furan, D-70235
4-Hydroxyacenaphthene, H-60099

C₁₂H₁₀O₂

3-(Hydroxymethyl)-2-naphthalenecarboxaldehyde, H-60188
6-Methoxy-1-naphthaldehyde, *in* H-70184
6-Methoxy-2-naphthaldehyde, *in* H-70185
7-Methoxy-2-naphthaldehyde, *in* H-70186

4-Methyl-1-azulenecarboxylic acid, M-60042
2,3,5,6-Tetrahydro-*s*-indacene-1,7-dione, T-70067
2,3,6,7-Tetrahydro-*s*-indacene-1,5-dione, T-70068
2,3,6,7-Tetrahydro-*as*-indacene-1,8-dione, T-70069

C₁₂H₁₀O₃

2,3,4-Biphenyltriol, B-70124
8,8-Dimethyl-1,4,5(8*H*)-naphthalenetrione, D-70421
4-Ethoxy-1,2-naphthoquinone, *in* H-60194
4-Hydroxy-2-cyclopentenone; Benzoyl, *in* H-60116
5-Methoxy-7-methyl-1,2-naphthoquinone, *in* H-60189
8-Methoxy-3-methyl-1,2-naphthoquinone, *in* H-60190
5-Phenyl-2-furancarboxylic acid; Me ester, *in* P-70078

C₁₂H₁₀O₄

2,2′,3,4-Biphenyltetrol, B-70122
2,3,4,4′-Biphenyltetrol, B-70123
5,8-Dihydroxy-2,3-dimethyl-1,4-naphthoquinone, D-60317
3-Hydroxy-5-methoxy-2-methyl-1,4-naphthoquinone, *in* D-60350
6-Hydroxy-5-methoxy-2-methyl-1,4-naphthoquinone, *in* D-60351
1-Oxo-1*H*-2-benzopyran-3-acetic acid; Me ester, *in* O-60058

C₁₂H₁₀O₅

Murraxonin, M-60152

C₁₂H₁₀O₆S₂

▷Benzenesulfonyl peroxide, B-70018

C₁₂H₁₀O₈

3,6-Dimethyl-1,2,4,5-benzenetetracarboxylic acid, D-60407
2,5,6,8-Tetrahydroxy-3,7-dimethoxy-1,4-naphthoquinone, *in* H-60065
2,5,7,8-Tetrahydroxy-3,6-dimethoxy-1,4-naphthoquinone, *in* H-60065

C₁₂H₁₁BrO₃

5-Bromo-1,2,4-benzenetriol; Tri-Ac, *in* B-70181
6-Bromo-1,2,4-benzenetriol; Tri-Ac, *in* B-70182

C₁₂H₁₁BrO₆

2-Bromo-1,3,5-benzenetriol; Tri-Ac, *in* B-60199

C₁₂H₁₁F₃O₅S

4,5-Dimethyl-2-phenyl-1,3-dioxol-1-ium; Trifluoromethanesulfonate, *in* D-70431

C₁₂H₁₁N

2-(2-Phenylethenyl)-1*H*-pyrrole, P-60094

C₁₂H₁₁NO

1,2-Dihydro-3*H*-azepin-3-one; 1-Ph, *in* D-70171
1-Oxo-1,2,3,4-tetrahydrocarbazole, O-70103
2-Oxo-1,2,3,4-tetrahydrocarbazole, O-70104
3-Oxo-1,2,3,4-tetrahydrocarbazole, O-70105
4-Oxo-1,2,3,4-tetrahydrocarbazole, O-70106
4*H*,6*H*-Pyrrolo[1,2-*a*][4,1]benzoxazepine, P-60246

C₁₂H₁₁NO₂

1,4-Diacetylindole, *in* A-60033
6-Hydroxy-1-naphthaldehyde; Me ether, oxime, *in* H-70184
6-Hydroxy-2-naphthaldehyde; Me ether, oxime, *in* H-70185

C₁₂H₁₁NO₃

3-Hydroxy-2-quinolinecarboxylic acid; Me ether, Me ester, *in* H-60228

C₁₂H₁₁NO₄

5,6-Dihydroxyindole; *O,O*-Di-Ac, *in* D-70307

C₁₂H₁₁N₃

Indamine, I-60019
2-(Methylamino)-1*H*-pyrido[2,3-*b*]indole, *in* A-70105

C₁₂H₁₂

9,10-Biss(ethylene)tricyclo[5.3.0.0²,⁸]deca-3,5-diene, B-70152
2,3-Didehydro-1,2-dihydro-1,1-dimethylnaphthalene, D-60168
1,4-Dihydrobenzocyclooctatetraene, D-70173
1,6-Dimethylazulene, D-60404
Pentacyclo[6.4.0.0²,⁷.0³,¹².0⁶,⁹]dodeca-4,10-diene, P-60028
Tetracyclo[6.2.1.1³,⁶.0²,⁷]dodeca-2(7),4,9-triene, T-60032
4*a*,4*b*,8*a*,8*b*-Tetrahydrobiphenylene, T-60060

C₁₂H₁₂N₂

1-Benzyl-5-vinylimidazole, *in* V-70012
2,2′-Diaminobiphenyl, D-60033
4,6-Dimethyl-2-phenylpyrimidine, D-60444
4(5)-Vinylimidazole; 1-Benzyl, *in* V-70012

C₁₂H₁₂N₂O

1-Oxo-1,2,3,4-tetrahydrocarbazole; (*E*)-Oxime, *in* O-70103
1-Oxo-1,2,3,4-tetrahydrocarbazole; (*Z*)-oxime, *in* O-70103
2-Oxo-1,2,3,4-tetrahydrocarbazole; Oxime, *in* O-70104
3-Oxo-1,2,3,4-tetrahydrocarbazole; Oxime, *in* O-70105
4-Oxo-1,2,3,4-tetrahydrocarbazole; Oxime, *in* O-70106

C₁₂H₁₂N₂O₇

3-Amino-4-nitro-1,2-benzenedicarboxylic acid; Di-Me ester, Ac, *in* A-70174
4-Amino-3-nitro-1,2-benzenedicarboxylic acid; Di-Me ester, Ac, *in* A-70176
4-Amino-5-nitro-1,2-benzenedicarboxylic acid; Di Me ester, Ac, *in* A-70177

C₁₂H₁₂N₄O₂

lin-Benzocaffeine, *in* I-60009
Imidazo[4,5-*g*]quinazoline-6,8(5*H*,7*H*)-dione; 3,5,7-Tri-Me, *in* I-60009

C₁₂H₁₂N₆

2,5,8-Trimethylbenzotriimidazole, T-60356

C₁₂H₁₂N₆O₂

[2,2′-Bipyridine]-5,5′-dicarboxylic acid; Dihydrazide, *in* B-60127
[3,3′-Bipyridine]-2,2′-dicarboxylic acid; Dihydrazide, *in* B-60133

C₁₂H₁₂O

1-Benzoylcyclopentene, B-70072
1-(1-Naphthyl)ethanol, N-70015

C₁₂H₁₂O₂

5-Phenyl-2,4-pentadienoic acid; Me ester, *in* P-70086

C₁₂H₁₂O₂S

2-(Phenylsulfonyl)-1,3-cyclohexadiene, P-70095

C₁₂H₁₂O₃

2,3-Dihydro-3-hydroxy-2-(1-methylethenyl)-5-benzofurancarboxaldehyde, D-70213
4,5-Dimethoxy-1-naphthol, *in* N-60005
4,8-Dimethoxy-1-naphthol, *in* N-60005
4,6-Dimethoxy-5-vinylbenzofuran, *in* D-70352
2,2-Dimethyl-2*H*-1-benzopyran-6-carboxylic acid, D-60408
2-Hydroxycyclopentanone; Benzoyl, *in* H-60114
2-(1-Hydroxy-1-methylethyl)-5-benzofurancarboxaldehyde, *in* D-70214

C₁₂H₁₂O₃S

2-Oxo-4-phenylthio-3-butenoic acid; Et ester, *in* O-60088

C₁₂H₁₂O₃S₃

2,3,6,7,10,11-Hexahydrobenzo[1,2-*b*:3,4-*b*′:5,6-*b*″]tris[1,4]oxathiin, H-70051

$C_{12}H_{12}O_4$

4,7-Dimethoxy-5-methyl-2H-1-benzopyran-2-
one, in D-60348
4-Hydroxy-2-isopropyl-5-benzofurancarboxylic
acid, H-60164

$C_{12}H_{12}O_5$

De-O-methyldiaporthin, in D-70050
5-Hydroxy-6-(hydroxymethyl)-7-methoxy-2-
methyl-4H-1-benzopyran-4-one, H-60155
Khellinone, K-60011

$C_{12}H_{12}O_6$

1,4-Benzoquinone-2,6-dicarboxylic acid; Di-Et
ester, in B-70049
2-Hydroxy-1,4-benzenedicarboxylic acid; Ac,
di-Me ester, in H-70105

$C_{12}H_{12}S_6$

2,3,6,7,10,11-Hexahydrobenzo[1,2-b:3,4-b':5,6-
b'']tris[1,4]dithiin, H-70050

$C_{12}H_{13}BrN_2O_2$

5-Bromotryptophan; Me ester, in B-60333

$C_{12}H_{13}N$

1,2,3,4-Tetrahydrocyclopent[b]indole; N-Me, in
T-70049

$C_{12}H_{13}NO$

1-Amino-1,4-dihydronaphthalene; N-Ac, in
A-60139
1-Amino-5,8-dihydronaphthalene; N-Ac, in
A-60140
4,5-Dihydro-1,5-ethano-1H-1-benzazepin-
2(3H)-one, D-60239

$C_{12}H_{13}NO_3$

2,3-Dihydro-2-oxo-1H-indole-3-acetic acid; Et
ester, in D-60259
5,6-Dihydroxyindole; Di-Me ether, N-Ac, in
D-70307
5-Oxo-2-pyrrolidinecarboxylic acid; Benzyl
ester, in O-70102
1,2,3,4-Tetrahydro-2-oxo-4-quinolinecarboxylic
acid; Et ester, in T-60077
1,2,3,4-Tetrahydro-4-oxo-7-quinolinecarboxylic
acid; Et ester, in T-60079

$C_{12}H_{13}NO_7$

2-Hydroxy-5-nitro-1,3-benzenedicarboxylic
acid; Di-Et ester, in H-70188
4-Hydroxy-3-nitro-1,2-benzenedicarboxylic
acid; Di-Et ester, in H-70190

$C_{12}H_{13}N_5O_5$

2-Amino-4,7-dihydro-4-oxo-7β-D-ribofuranosyl-
1H-pyrrolo[2,3-d]pyrimidin-5-carbonitrile,
A-70132
N^2,3-Ethenoguanosine, in I-70006

$C_{12}H_{14}$

Benzylidenecyclopentane, B-70083
1,5,9-Cyclododecatrien-3-yne, C-60190
1,4-Diethynylbicyclo[2.2.2]octane, D-70157
3,3-Dimethyl-1-phenyl-1-butyne, D-60441
1,3,5,7,9,11-Dodecahexaene, D-70531
as-Hydrindacene, H-60094
s-Hydrindacene, H-60095
[6]Paracycloph-3-ene, P-70010

$C_{12}H_{14}BrNO_4$

4-(Bromomethyl)-2-nitrobenzoic acid; tert-
Butyl ester, in B-60297

$C_{12}H_{14}N_2$

3-Amino-1,2,3,4-tetrahydrocarbazole, A-60259
1,8-Biss(ethylamino)naphthalene, in D-70042

$C_{12}H_{14}N_2O_3$

1,2,3,4-Tetrahydro-4-oxo-7-quinolinecarboxylic
acid; Et ester, oxime, in T-60079

$C_{12}H_{14}N_4O_6$

Tetrahydroimidazo[4,5-d]imidazole-
2,5(1H,3H)-dione; 1,3,4,6-Tetra-Ac, in
T-70066

$C_{12}H_{14}O$

2-Acetyl-1,2,3,4-tetrahydronaphthalene,
A-70053
3,4-Dihydro-2,2-dimethyl-1(2H)-
naphthalenone, D-60228

1-Phenyl-5-hexen-1-one, P-70081
6,7,8,9-Tetrahydro-5H-benzocycloheptene-6-
carboxaldehyde, T-60055
6,7,8,9-Tetrahydro-5H-benzocycloheptene-7-
carboxaldehyde, T-60056
[5](2,7)Troponophane, T-70330

$C_{12}H_{14}O_2$

Ligustilide, L-60026
1-Phenyl-3-buten-1-ol; Ac, in P-70061

$C_{12}H_{14}O_3$

2,3-Dihydro-2-(1-hydroxy-1-methylethyl)-5-
benzofurancarboxaldehyde, D-70214
Senkyunolide F, in L-60026
1,2,3,4-Tetrahydro-1-hydroxy-2-
naphthalenecarboxylic acid; Me ester, in
T-70063
(Z)-6,7-Epoxyligustilide, in L-60026

$C_{12}H_{14}O_3S$

2-Mercapto-3-methylbutanoic acid; S-Benzoyl,
in M-60023

$C_{12}H_{14}O_4$

Calaminthone, C-70006
2-(1,2-Dihydroxy-1-methylethyl)-2,3-dihydro-
5-benzofurancarboxaldehyde, in D-70214
4-Hydroxy-3-(hydroxy-3-methyl-3-butenyl)-
benzoic acid, H-60154
Thurfyl salicylate, in H-70108
3-(2,4,5-Trimethoxyphenyl)-2-propenal,
T-70278

$C_{12}H_{14}O_5$

2-Hydroxy-1,3-benzenedicarboxylic acid; Di-Et
ester, in H-60101
5-Hydroxy-3-methyl-1,2-benzenedicarboxylic
acid; Me ether, di-Me ester, in H-70162
2',4',6'-Trihydroxyacetophenone; 2',4'-Di-Me
ether, Ac, in T-70248

$C_{12}H_{14}O_6$

4,5-Dihydroxy-1,3-benzenedicarboxylic acid;
Di-Me ether, di-Me ester, in D-60307

$C_{12}H_{15}^{\oplus}$

Tricyclopropylcyclopropenium(1+), T-60270

$C_{12}H_{15}BF_4$

Tricyclopropylcyclopropenium(1+); Tetrafluoroborate, in T-60270

$C_{12}H_{15}Cl$

Tricyclopropylcyclopropenium(1+); Chloride,
in T-60270

$C_{12}H_{15}F_6Sb$

Tricyclopropylcyclopropenium(1+);
Hexafluoroantimonate, in T-60270

$C_{12}H_{15}N$

1,2,3,5,10,10a-Hexahydropyrrolo[1,2-b]-
isoquinoline, H-70072
▷1,2,3,6-Tetrahydro-1-methyl-4-phenylpyridine,
in T-60081

$C_{12}H_{15}NO_2$

2-Amino-3-methyl-3-butenoic acid; Benzyl
ester, in A-60218
1,2,3,4-Tetrahydro-1-isoquinolinecarboxylic
acid; Et ester, in T-60070

$C_{12}H_{15}NO_3$

2-Amino-2-methyl-3-phenylpropanoic acid; N-
Ac, in A-60225
2',3',4',5'-Tetramethyl-6'-nitroacetophenone,
T-60145
2',3',4',6'-Tetramethyl-5'-nitroacetophenone,
T-60146
2',3',5',6'-Tetramethyl-4'-nitroacetophenone,
T-60147

$C_{12}H_{15}N_5O_3$

Queuine, Q-70004

$C_{12}H_{16}$

1,4,7,10-Cyclododecatetraene, C-70217
Pentacyclo[6.3.1.0^{2,4}.0^{5,10}.0^{7,8}]dodecane,
P-70019
Tetracyclo[6.4.0.0^{4,12}.0^{5,9}]dodec-10-ene,
T-70020
Tetracyclo[6.2.1.1^{3,6}.0^{2,7}]dodec-2(7)-ene,
T-70021

$C_{12}H_{16}BrClO$

Furocaespitane, F-60090

$C_{12}H_{16}BrClO_2$

5-(3-Bromo-4-chloro-4-methylcyclohexyl)-5-
methyl-2(5H)-furanone, B-60222

$C_{12}H_{16}N_2$

Octahydropyrrolo[3,4-c]pyrrole; N-Ph, in
O-70017

$C_{12}H_{16}N_2O_3$

3-[1-(1,1-Dimethyl-3-oxobutyl)imidazol-4-yl]-
2-propenoic acid, in U-70005

$C_{12}H_{16}N_2O_4$

N-Tyrosylalanine, T-60411

$C_{12}H_{16}N_6O_4$

N^6-(Carbamoylmethyl)-2'-deoxyadenosine,
C-70016

$C_{12}H_{16}O$

4-(4-Methylphenyl)-2-pentanone, M-60112

$C_{12}H_{16}O_2$

2-(4-Methylphenyl)propanoic acid; Et ester, in
M-60114
Tricyclo[4.4.2.0^{1,6}]dodecane-2,7-dione, T-60261

$C_{12}H_{16}O_2Se$

4,7-Dihydro-1,1,3,3-tetramethyl-1H,3H-3a,7a-
episeleno-4,7-epoxyisobenzofuran, D-70266

$C_{12}H_{16}O_3$

1,2,4-Trimethoxy-5-(2-propenyl)benzene,
T-70279
Xanthostemone, X-70002

$C_{12}H_{16}O_4$

6-Ethyl-2,4-dihydroxy-3-methylbenzoic acid; Et
ester, in E-70041
Senkyunolide H, in L-60026
Sordariol, S-70056

$C_{12}H_{16}O_5$

2,3,4-Trihydroxyphenylacetic acid; Tri-Me
ether, Me ester, in T-60340
2,4,5-Trihydroxyphenylacetic acid; Tri-Me
ether, Me ester, in T-60342
3,4,5-Trihydroxyphenylacetic acid; tert-Butyl
ester, in T-60343

$C_{12}H_{16}O_6$

Pentamethoxybenzaldehyde, in P-60042

$C_{12}H_{16}O_7$

Arbutin, A-70248
Pentamethoxybenzoic acid, in P-60043

$C_{12}H_{17}NO_2$

2-Amino-2-methyl-3-phenylpropanoic acid; Et
ester, in A-60225

$C_{12}H_{17}NO_3$

2-Amino-3-(4-methoxyphenyl)-3-methyl-
butanoic acid, in A-60198

$C_{12}H_{17}NO_7$

Epivolkenin, in T-60007
Taraktophyllin, T-60007

$C_{12}H_{17}N_3O_4$

▷Agaritine, A-70071

$C_{12}H_{17}N_5O_5$

8-(Dimethylamino)inosine, in A-70198

$C_{12}H_{17}N_5O_{10}S$

Antibiotic PB 5266C, A-70227

$C_{12}H_{18}$

2-Isopropyl-1,3,5-trimethylbenzene, I-60130

$C_{12}H_{18}N_4O_4$

Octahydroimidazo[4,5-d]imidazole; 1,3,4,6-
Tetra-Ac, in O-70015

$C_{12}H_{18}N_6O_5$

8-(Dimethylamino)guanosine, in A-70150

$C_{12}H_{18}O$

Methoxypentamethylbenzene, in P-60058

$C_{12}H_{18}O_2$

Anastrephin, A-70206
[1,1'-Bicyclohexyl]-2,2'-dione, B-60089
Clavularin A, C-60153
Clavularin B, in C-60153

Epianastrephin, *in* A-70206
6-(1-Heptenyl)-5,6-dihydro-2*H*-pyran-2-one, H-60025
5-Isopropyl-2-methyl-1,3-benzenediol; Di-Me ether, *in* I-60125
Suspensolide, S-70084

$C_{12}H_{18}O_3$
7-Isojasmonic acid, *in* J-60002
Jasmonic acid, J-60002

$C_{12}H_{18}O_4$
1-Acetoxy-7-hydroxy-3,7-dimethyl-2*E*,5*E*-octadien-4-one, *in* D-60318
3,4-Dihydroxyphenylacetaldehyde; Di-Me ether, di-Me acetal, *in* D-60371

$C_{12}H_{18}O_5$
1-Acetoxy-7-hydroperoxy-3,7-dimethyl-2*E*,5*E*-octadien-4-one, *in* D-60318

$C_{12}H_{19}N$
3-Azatetracyclo[5.3.1.1^{2,6}.0^{4,9}]dodecane; *N*-Me, *in* A-70282
1,2,3,4,5,6,7,8,8*a*,9*a*-Decahydrocarbazole, D-70012

$C_{12}H_{19}NO_4$
▷Choline salicylate, *in* H-70108

$C_{12}H_{20}$
1-*tert*-Butyl-1,2-cyclooctadiene, B-60344
Cyclododecyne, C-60192
1,1′,2,2′,3,4,5,5′,6,6′-Decahydrobiphenyl, D-60007
1,1′,2,3,4,4′,5,5′,6,6′-Decahydrobiphenyl, D-60008
1,2,3,3′,4,4′,5,5′,6,6′-Decahydrobiphenyl, D-60009
1,3-Dimethyladamantane, D-70366
5-Dodecen-7-yne, D-70538

$C_{12}H_{20}N_2O_7$
2′-Deoxymugeneic acid, *in* M-60148

$C_{12}H_{20}N_2O_8$
Isomugeneic acid, *in* M-60148
Mugineic acid, M-60148

$C_{12}H_{20}N_2O_9$
3-Hydroxymugeneic acid, *in* M-60148

$C_{12}H_{20}N_4O_4$
Octahydro-1,2,5,6-tetrazocine; Tetra-Ac, *in* O-60020

$C_{12}H_{20}O_2$
10-Dodecen-12-olide, *in* H-60126

$C_{12}H_{20}O_4$
9,12-Dioxododecanoic acid, D-60471
4-Octene-1,8-diol; Di-Ac, *in* O-70026

$C_{12}H_{20}O_8$
Parasorboside, *in* T-60067

$C_{12}H_{21}N$
Dodecahydrocarbazole, D-70532

$C_{12}H_{21}NO_4$
2-Amino-1,4-cyclohexanedicarboxylic acid; Di-Et ester, *in* A-60118
3-Amino-1,2-cyclohexanedicarboxylic acid; Di-Et ester, *in* A-60119
4-Amino-1,1-cyclohexanedicarboxylic acid; Di-Et ester, *in* A-60120
4-Amino-1,3-cyclohexanedicarboxylic acid; Di-Et ester, *in* A-60121

$C_{12}H_{21}N_3$
3,5,12-Triazatetracyclo[5.3.1.1^{2,6}.0^{4,9}]dodecane; 3,5,12-Tri-Me, *in* T-70206

$C_{12}H_{22}Cl_2O$
12-Chlorododecanoic acid; Chloride, *in* C-70066

$C_{12}H_{22}N_2$
1,1-Di-1-piperidinylethene, D-70500

$C_{12}H_{22}N_2O_2$
3,6-Bis(1-methylpropyl)-2,5-piperazinedione, B-70153
3,6-Bis(2-methylpropyl)-2,5-piperazinedione, B-60171
Dihydro-4,6-(1*H*,5*H*)-pyrimidinedione; Di-*tert*-butyl ether, *in* D-70256

$C_{12}H_{22}O$
5-Dodecyn-1-ol, D-70539
11-Dodecyn-1-ol, D-70540

$C_{12}H_{22}O_2$
5-Decen-1-ol; Ac, *in* D-70019
2,6-Dodecanedione, D-70534
Invictolide, I-60038
2-Oxodecanal, O-70087

$C_{12}H_{22}O_3$
10-Hydroxy-11-dodecenoic acid, H-60125
12-Hydroxy-10-dodecenoic acid, H-60126
12-Oxododecanoic acid, O-60066

$C_{12}H_{22}O_4$
Talaromycin *A*, T-70004
Talaromycin B, *in* T-70004
Talaromycin C, *in* T-70004
Talaromycin D, *in* T-70004
Talaromycin E, *in* T-70004
Talaromycin F, *in* T-70004
Talaromycin G, *in* T-70004
Tartaric acid; Di-*tert*-butyl ester, *in* T-70005
Tartaric acid; Di-*tert*-butyl ether, *in* T-70005
1,1,4,4-Tetraethoxy-2-butyne, *in* B-60351

$C_{12}H_{23}Br$
6-Bromo-6-dodecene, B-60238

$C_{12}H_{23}ClO_2$
12-Chlorododecanoic acid, C-70066

$C_{12}H_{24}$
1,1,2,2,3,3-Hexamethylcyclohexane, H-60070
2,2,3,4,5,5-Hexamethyl-3-hexene, H-70083

$C_{12}H_{24}N_2O_4S$
Homopantetheine, H-60090

$C_{12}H_{24}O$
9-Dodecen-1-ol, D-70537
2,5,7,7-Tetramethyloctanal, T-70139

$C_{12}H_{24}O_2$
1,1-Dimethoxy-3-decene, *in* D-70016

$C_{12}H_{24}O_4$
1,1,8,8-Tetramethoxy-4-octene, *in* O-70025

$C_{12}H_{24}S_6$
1,4,7,10,13,16-Hexathiacyclooctadecane, H-60075
1,4,7,10,13,16-Hexathiacyclooctadecane, H-70087

$C_{12}H_{26}N_2$
Hexahydropyrimidine; 1,3-Di-*tert*-butyl, *in* H-60056

$C_{12}H_{26}N_2O_4$
1,4,10,13-Tetraoxa-7,16-diazacyclooctadecane, T-70145
1,7,10,16-Tetraoxa-4,13-diazacyclooctadecane, T-60161

$C_{12}H_{26}N_4O_2S_2$
Homoalethine, H-60084

$C_{12}H_{27}NO_6$
Tris[(2-hydroxyethoxy)ethyl]amine, T-70323

$C_{12}H_{27}N_3$
Hexahydro-1,3,5-triazine; 1,3,5-Triisopropyl, *in* H-60060

$C_{12}H_{29}N_5$
1,4,7,10,14-Pentaazacycloheptadecane, P-60019
1,4,7,11,14-Pentaazacycloheptadecane, P-60020

$C_{12}H_{31}N_5$
1,5,9,13,17-Pentaazaheptadecane, P-60026

$C_{13}Cl_9$
Nonachlorophenalenyl, N-60055

$C_{13}H_5BrF_{12}O_2$
4-Methyl-2,2,6,6-tetrakiss(trifluoromethyl)-2*H*,6*H*-[1,2]bromoxolo[4,5,1-*hi*][1,2]-benzobromoxole, M-60128

$C_{13}H_5F_{14}IO_2$
8,8-Difluoro-8,8-dihydro-4-methyl-2,2,6,6-tetrakiss(trifluoromethyl)-2*H*,6*H*-[1,2]-iodoxolo[4,5,1-*hi*][1,2]benziodoxole, D-60182

$C_{13}H_6Cl_2O_2$
1,5-Dichloroxanthone, D-70139
1,7-Dichloroxanthone, D-70140
2,6-Dichloroxanthone, D-70141
2,7-Dichloroxanthone, D-70142
3,6-Dichloroxanthone, D-70143

$C_{13}H_7NO_2$
Benzo[*g*]quinoline-6,9-dione, B-60045

$C_{13}H_8BrN$
3-Bromo-4-cyanobiphenyl, *in* B-60211

$C_{13}H_8ClNO_2$
2-Chloro-3-nitro-9*H*-fluorene, C-70132

$C_{13}H_8FNO_4$
5-Fluoro-2-nitrophenol; *O*-Benzoyl, *in* F-60061

$C_{13}H_8N_2$
Benz[*f*]imidazo[5,1,2-*cd*]indolizine, B-70021
9-Diazo-9*H*-fluorene, D-60062

$C_{13}H_8N_2O$
Pyrido[3′,4′:4,5]furo[3,2-*b*]indole, P-70174

$C_{13}H_8N_2O_2S$
9-Diazo-9*H*-thioxanthene; 10,10-Dioxide, *in* D-70065

$C_{13}H_8N_2S$
2-Cyano-10*H*-phenothiazine, *in* P-70056
▷3-Cyano-10*H*-phenothiazine, *in* P-70057
9-Diazo-9*H*-thioxanthene, D-70065
6*H*-Pyrido[3′,4′:4,5]thieno[2,3-*b*]indole, P-70177

$C_{13}H_8OS$
6*H*-Dibenzo[*b*,*d*]thiopyran-6-one, D-70080

$C_{13}H_8O_2$
2*H*-Naphtho[1,2-*b*]pyran-2-one, N-70011
4*H*-Naphtho[1,2-*b*]pyran-4-one, N-60010

$C_{13}H_8O_3$
2-(2-Furanyl)-4*H*-1-benzopyran-4-one, F-60084
2-(3-Furanyl)-4*H*-1-benzopyran-4-one, F-60085
Naphtho[1,8-*bc*]pyran-2-carboxylic acid, N-70010

$C_{13}H_8O_4$
2,2-Dihydroxy-1*H*-benz[*f*]indene-1,3(2*H*)dione, D-60308
1,2-Dihydroxyxanthone, D-70353
2,4,5-Trihydroxy-9*H*-fluoren-9-one, T-70261

$C_{13}H_8O_5$
2,3,4,5-Tetrahydroxy-9*H*-fluoren-9-one, T-70109
1,5,6-Trihydroxyxanthone, T-60349

$C_{13}H_8O_6$
1,2,3,4-Tetrahydroxyxanthone, T-60126
1,2,4,5-Tetrahydroxyxanthone, T-70122
1,3,5,8-Tetrahydroxyxanthone, T-70123
1,3,6,8-Tetrahydroxyxanthone, T-60127
2,3,4,7-Tetrahydroxyxanthone, T-60128

$C_{13}H_8O_7$
1,2,3,4,8-Pentahydroxyxanthone, P-60051
1,2,3,6,8-Pentahydroxyxanthone, P-70036

$C_{13}H_8Se$
9*H*-Fluorene-9-selenone, F-60018

$C_{13}H_9BF_4OS$
2-Phenyl-1,3-benzoxathiol-1-ium; Tetra-fluoroborate, *in* P-60084

$C_{13}H_9BrO_2$
3-Bromo[1,1′-biphenyl]-4-carboxylic acid, B-60211

$C_{13}H_9ClN_2O_3$
Indisocin, I-60028

$C_{13}H_9ClO_4S$

Dibenzo[*b*,*d*]thiopyrylium(1+); Perchlorate, *in* D-70081

Thioxanthylium(1+); Perchlorate, *in* T-70186

$C_{13}H_9ClO_5$

Xanthylium(1+); Perchlorate, *in* X-70003

$C_{13}H_9ClO_5S$

2-Phenyl-1,3-benzoxathiol-1-ium; Perchlorate, *in* P-60084

$C_{13}H_9F_3O_3S$

3-(Trifluoromethyl)benzenesulfonic acid; Ph ester, *in* T-60283

$C_{13}H_9N$

1-Cyanoacenaphthene, *in* A-70014

2-Cyano-1,2-dihydrocyclobuta[*a*]naphthalene, *in* D-70177

$C_{13}H_9NO$

2-Phenylfuro[2,3-*b*]pyridine, P-60104

2-Phenylfuro[3,2-*c*]pyridine, P-60105

$C_{13}H_9NO_2$

1*H*-Benz[*e*]indole-1,2(3*H*)-dione; *N*-Me, *in* B-70026

Benz[*cd*]indol-2-(1*H*)-one; *N*-Ac, *in* B-70027

$C_{13}H_9NO_2S$

10*H*-Phenothiazine-1-carboxylic acid, P-70055

10*H*-Phenothiazine-2-carboxylic acid, P-70056

10*H*-Phenothiazine-3-carboxylic acid, P-70057

10*H*-Phenothiazine-4-carboxylic acid, P-70058

$C_{13}H_9NO_3$

2-Methoxy-3*H*-phenoxazin-3-one, *in* H-60206

$C_{13}H_9NO_4S$

10*H*-Phenothiazine-2-carboxylic acid; 5,5-Dioxide, *in* P-70056

$C_{13}H_9NS$

2-Phenylthieno[2,3-*b*]pyridine, P-60136

$C_{13}H_9N_3O$

2-Azidobenzophenone, A-60338

3-Azidobenzophenone, A-60339

4-Azidobenzophenone, A-60340

1*H*-Benzotriazole; *N*-Benzoyl, *in* B-70068

$C_{13}H_9O^{\oplus}$

Xanthylium(1+), X-70003

$C_{13}H_9OS^{\oplus}$

2-Phenyl-1,3-benzoxathiol-1-ium, P-60084

$C_{13}H_9S^{\oplus}$

Dibenzo[*b*,*d*]thiopyrylium(1+), D-70081

Thioxanthylium(1+), T-70186

$C_{13}H_{10}BrN$

Benzo[*a*]quinolizinium(1+); Bromide, *in* B-70043

$C_{13}H_{10}ClNO_4$

Benzo[*a*]quinolizinium(1+); Perchlorate, *in* B-70043

$C_{13}H_{10}F_2$

Difluorodiphenylmethane, D-60184

$C_{13}H_{10}N^{\oplus}$

Benzo[*a*]quinolizinium(1+), B-70043

$C_{13}H_{10}N_2OS$

10*H*-Phenothiazine-1-carboxylic acid; Amide, *in* P-70055

10*H*-Phenothiazine-2-carboxylic acid; Amide, *in* P-70056

$C_{13}H_{10}N_2O_2$

Benzophenone nitrimine, *in* D-60482

$C_{13}H_{10}N_2O_4$

[2,2'-Bipyridine]-3,3'-dicarboxylic acid; Mono-Me ester, *in* B-60124

$C_{13}H_{10}N_4$

4-Amino-1,2,3-benzotriazine; 3-Ph, *in* A-60092

4-Anilino-1,2,3-benzotriazine, *in* A-60092

$C_{13}H_{10}O$

1-Acenaphthenecarboxaldehyde, A-70013

4-Methoxyacenaphthylene, *in* H-70100

$C_{13}H_{10}OS$

2-Mercaptobenzophenone, M-70020

3-Mercaptobenzophenone, M-70021

4-Mercaptobenzophenone, M-70022

$C_{13}H_{10}OSe_2$

Benzoyl phenyl diselenide, B-70077

$C_{13}H_{10}O_2$

1-Acenaphthenecarboxylic acid, A-70014

1,2-Dihydrocyclobuta[*a*]naphthalene-2-carboxylic acid, D-70177

$C_{13}H_{10}O_3$

4'-Hydroxy-2-biphenylcarboxylic acid, H-70111

Mycosinol, M-60154

$C_{13}H_{10}O_4$

Coriandrin, C-70198

$C_{13}H_{10}O_7$

Nepenthone A, *in* T-60333

$C_{13}H_{10}O_8$

5,8-Dihydroxy-2,3-dimethoxy-6,7-methyl-enedioxy-1,4-naphthoquinone, *in* H-60065

$C_{13}H_{10}S$

1-Methylnaphtho[2,1-*b*]thiophene, M-60096

2-Methylnaphtho[2,1-*b*]thiophene, M-60097

4-Methylnaphtho[2,1-*b*]thiophene, M-60098

5-Methylnaphtho[2,1-*b*]thiophene, M-60099

6-Methylnaphtho[2,1-*b*]thiophene, M-60100

7-Methylnaphtho[2,1-*b*]thiophene, M-60101

8-Methylnaphtho[2,1-*b*]thiophene, M-60102

9-Methylnaphtho[2,1-*b*]thiophene, M-60103

$C_{13}H_{10}S_2$

2-Dibenzothiophenethiol; *S*-Me, *in* D-60078

4-Dibenzothiophenethiol; *S*-Me, *in* D-60079

$C_{13}H_{11}BrO$

4-Bromodiphenylmethanol, B-60236

$C_{13}H_{11}ClO$

4-Chlorodiphenylmethanol, C-60065

$C_{13}H_{11}I$

2-Iododiphenylmethane, I-60045

4-Iododiphenylmethane, I-60046

$C_{13}H_{11}N$

Diphenylmethaneimine, D-60482

9-Methyl-9*H*-carbazole, M-60056

$C_{13}H_{11}NO$

1-Acenaphthenecarboxaldehyde; Oxime, *in* A-70013

1,2-Dihydrocyclobuta[*a*]naphthalene-2-carboxylic acid; Amide, *in* D-70177

9-Methyl-9*H*-carbazol-2-ol, *in* H-60109

$C_{13}H_{11}NOS$

2-Phenyl-4*H*-3,1,2-benzooxathiazine, P-60080

$C_{13}H_{11}NO_2$

Diphenyl imidocarbonate, *in* C-70019

2-Hydroxy-*N*-phenylbenzamide, *in* H-70108

3-Methyl-2-nitrobiphenyl, M-70098

$C_{13}H_{11}NO_3$

2-Amino-3-methyl-1,4-naphthoquinone; Ac, *in* A-70171

$C_{13}H_{11}NS$

2,3-Dihydro-2-phenyl-1,2-benzisothiazole, D-60261

$C_{13}H_{11}N_3$

2-Aminomethyl-1,10-phenanthroline, A-60224

1*H*-Benzotriazole; 1-Benzyl, *in* B-70068

▷2,9-Diaminoacridine, D-60027

▷3,6-Diaminoacridine, D-60028

▷3,9-Diaminoacridine, D-60029

1-(Dimethylamino)-γ-carboline, *in* A-70114

$C_{13}H_{11}N_3O$

3-Amino-β-carboline; N^3-Ac, *in* A-70110

1,3-Dihydro-2*H*-imidazo[4,5-*c*]pyridin-2-one; 1-Benzyl, *in* D-60250

$C_{13}H_{12}$

Benz[*f*]indane, B-60015

3-Methylbiphenyl, M-60053

Tetracyclo[5.5.1.0^{4,13}.0^{10,13}]trideca-2,5,8,11-tetraene, T-60036

$C_{13}H_{12}N_2$

N,*N*'-Diphenylformamidine, D-60479

$C_{13}H_{12}N_2O$

2-(Aminomethyl)pyridine; *N*-Benzoyl, *in* A-60227

4-(Aminomethyl)pyridine; *N*-Benzoyl, *in* A-60229

▷*N*,*N*'-Diphenylurea, D 70499

$C_{13}H_{12}N_2O_2$

1-Benzyloxy-2-phenyldiazene 2-oxide, *in* H-70208

2,2'-(1,3-Dioxan-2-ylidene)bispyridine, *in* D-70502

$C_{13}H_{12}N_2O_3S$

Aminoiminomethanesulfonic acid; *N*,*N*-Di-Ph, *in* A-60210

$C_{13}H_{12}N_2O_4S$

Benzenediazo-*p*-toluenesulfonate *N*-oxide, *in* H-70208

$C_{13}H_{12}N_2Se$

Selenourea; N^1,N^1-Di-Ph, *in* S-70031

Selenourea; N^1,N^3-Di-Ph, *in* S-70031

$C_{13}H_{12}N_4$

4-Amino-1*H*-imidazo[4,5-*c*]pyridine; 4-*N*-Benzyl, *in* A-70161

$C_{13}H_{12}O$

4-Methoxyacenaphthene, *in* H-60099

$C_{13}H_{12}O_2$

2-(Benzyloxy)phenol, B-60072

3-(Benzyloxy)phenol, B-60073

▷4-(Benzyloxy)phenol, B-60074

▷Goniothalamin, G-60031

4-Methyl-1-azulenecarboxylic acid; Me ester, *in* M-60042

$C_{13}H_{12}O_2S_2$

4-[5'-Methyl[2,2'-bithiophen]-5-yl]-3-butyne-1,2-diol, M-70054

$C_{13}H_{12}O_3$

6-Acetyl-5-hydroxy-2-isopropenylbenzo[*b*]-furan, A-60065

Goniothalamin oxide, *in* G-60031

$C_{13}H_{12}O_4$

6-Acetyl-5-hydroxy-2-(1-hydroxymethylvinyl)-benzo[*b*]furan, *in* A-60065

Altholactone, A-60083

Dihydrocoriandrin, *in* C-70198

5,8-Dihydroxy-2,3,6-trimethyl-1,4-naphthoquinone, D-60380

5,6-Dimethoxy-2-methyl-1,4-naphthoquinone, *in* D-60351

1,4,5-Naphthalenetriol; 5-Me ether, 1-Ac, *in* N-60005

2-Oxo-2*H*-benzopyran-4-acetic acid; Et ester, *in* O-60057

Platypterophthalide, P-60161

$C_{13}H_{12}O_5$

Acuminatolide, A-70063

$C_{13}H_{12}O_7$

5,8-Dihydroxy-2,3,6-trimethoxy-1,4-benzoquinone, *in* P-70033

$C_{13}H_{13}BrN_2O_3$

5-Bromotryptophan; N^α-Ac, *in* B-60333

6-Bromotryptophan; N^α-Ac, *in* B-60334

7-Bromotryptophan; N^α-Ac, *in* B-60335

$C_{13}H_{13}BrO_2$

1-Bromo-(2-dimethoxymethyl)naphthalene, *in*
B-60302

$C_{13}H_{13}IO_8$

1,1,1-Triacetoxy-1,1-dihydro-1,2-benziodoxol-
3(1*H*)-one, T-60228

$C_{13}H_{13}N$

2-Amino-3-methylbiphenyl, A-70164
N-Methyldiphenylamine, M-70060
2-(2-Phenylethenyl)-1*H*-pyrrole; *N*-Me, *in*
P-60094
2-(2-Phenylethenyl)-1*H*-pyrrole; *N*-Me, *in*
P-60094

$C_{13}H_{13}NO$

1-Oxo-1,2,3,4-tetrahydrocarbazole; *N*-Me, *in*
O-70103
1,3,4,10-Tetrahydro-9(2*H*)-acridinone, T-60052

$C_{13}H_{13}NO_2$

5-Methyl-1*H*-pyrrole-2-carboxylic acid; Benzyl
ester, *in* M-60124

$C_{13}H_{13}NO_4$

5,6-Dihydroxyindole; *N*-Me, di-*O*-Ac, *in*
D-70307

$C_{13}H_{13}N_3$

3-Amino-β-carboline; *N*³-Et, *in* A-70110

$C_{13}H_{13}N_3O_5$

Antibiotic PDE I, A-60286

$C_{13}H_{14}$

2,3,5,6,7-Pentamethylenebicyclo[2.2.2]octane,
P-60056

$C_{13}H_{14}N_2O$

1-Benzyl-1,4-dihydronicotinamide, B-60067
1-Oxo-1,2,3,4-tetrahydrocarbazole; *N*-Me,
oxime, *in* O-70103

$C_{13}H_{14}N_2O_4S_3$

Gliotoxin *E*, G-60024

$C_{13}H_{14}N_4OS_2$

S,*S*-Bis(4,6-dimethyl-2-pyrimidinyl)-
dithiocarbonate, B-70138

$C_{13}H_{14}O$

3-Isopropyl-2-naphthol, I-70095
4-Isopropyl-1-naphthol, I-70096
1-(1-Naphthyl)-2-propanol, N-60015

$C_{13}H_{14}O_2$

2-Methyl-4-phenyl-3-butyn-2-ol; Ac, *in*
M-70110
7-Phenyl-5-heptynoic acid, P-60106

$C_{13}H_{14}O_3$

4-Acetyl-2,3-dihydro-5-hydroxy-2-
isopropenylbenzofuran, A-70040
5-Acetyl-2-(1-hydroxy-1-methylethyl)-
benzofuran, A-70044
2,2-Dimethyl-2*H*-1-benzopyran-6-carboxylic
acid; Me ester, *in* D-60408
7-Ethoxy-3,4-dimethylcoumarin, *in* H-60119
1,4,5-Trimethoxynaphthalene, *in* N-60005

$C_{13}H_{14}O_4$

6-Acetyl-5-hydroxy-2-hydroxymethyl-2-methyl-
chromene, A-60031
Anaphatol, *in* D-60339
4-Hydroxy-2-isopropyl-5-benzofurancarboxylic
acid; Me ester, *in* H-60164

$C_{13}H_{14}O_5$

Diaporthin, D-70050

$C_{13}H_{15}ClO$

2-Phenylcyclohexanecarboxylic acid; Chloride,
in P-70065

$C_{13}H_{15}Cl_2NO_3$

6-Amino-2,2-dichlorohexanoic acid; *N*-Benzoyl,
in A-60126

$C_{13}H_{15}NO_2$

1-Oxa-2-azaspiro[2,5]octane; *N*-Benzoyl, *in*
O-70052

$C_{13}H_{15}NO_3$

2-Pyrrolidinecarboxaldehyde; *N*-
Benzyloxycarbonyl, *in* P-70184

$C_{13}H_{15}N_3O_2$

3-Amino-1*H*-pyrazole-4-carboxylic acid; *N*(1)-
Benzyl, Et ester, *in* A-70187

$C_{13}H_{16}$

Benzylidenecyclohexane, B-70082

$C_{13}H_{16}Cl_2O_4$

1-(3,5-Dichloro-2,6-dihydroxy-4-methoxy-
phenyl)-1-hexanone, D-60129

$C_{13}H_{16}NO_3$

3-Benzoyl-2-methylpropanoic acid; Et ester, *in*
B-60064

$C_{13}H_{16}N_2$

1,8-Diaminonaphthalene; *N*,*N*,*N*′-Tri-Me, *in*
D-70042
2-Diazo-1,1,3,3-tetramethylindane, D-60067

$C_{13}H_{16}N_2O_3$

Abbeymycin, A-70001

$C_{13}H_{16}N_2O_4$

Bursatellin, B-60338

$C_{13}H_{16}N_4$

Tetrakis(2-cyanoethyl)methane, *in* M-70036

$C_{13}H_{16}O$

▷Benzoylcyclohexane, B-70071
1-Phenyl-6-hepten-1-one, P-70080
7-Phenyl-3-heptyn-2-ol, P-60107
1,1,3,3-Tetramethyl-2-indanone, T-60143

$C_{13}H_{16}O_2$

3,4-Dihydro-3,3,8a-trimethyl-1,6(2*H*,8a*H*)-
naphthalenedione, D-60301
2-Phenylcyclohexanecarboxylic acid, P-70065

$C_{13}H_{16}O_3$

6,7-Dimethoxy-2,2-dimethyl-2*H*-1-benzopyran,
D-70365
1-(Dimethoxymethyl)-2,3,5,6-tetrakiss(ethylene)-7-oxabicyclo[2.2.1]heptane, D-60399
Flossonol, F-60014
5-(1-Hydroxyethyl)-2-(1-hydroxy-1-methyl-
ethyl)benzofuran, *in* A-70044
Methyl 3-(4-methoxy-2-vinylphenyl)-
propanoate, *in* H-70232

$C_{13}H_{16}O_4$

4-Hydroxy-3-(2-hydroxy-3-methyl-3-butenyl)-
benzoic acid; Me ester, *in* H-60154

$C_{13}H_{16}O_7$

2-Oxo-1,3-cyclopentanediglyoxylic acid; Di-Et
ester, *in* O-70085

$C_{13}H_{16}Se$

1,1,3,3-Tetramethyl-2-indaneselone, T-60142

$C_{13}H_{17}NO$

Benzoylcyclohexane; (*E*)-Oxime, *in* B-70071

$C_{13}H_{17}NO_3S$

2-Amino-4,5,6,7-tetrahydrobenzo[*b*]thiophene-
3-carboxylic acid; Et ester, *N*-Ac, *in* A-60258

$C_{13}H_{17}NO_4$

2-Amino-3-(4-hydroxyphenyl)-3-methyl-
butanoic acid; *N*-Ac, *in* A-60198

$C_{13}H_{18}BrClO_3$

Methyl 3-(3-bromo-4-chloro-4-methyl-
cyclohexyl)-4-oxo-2-pentenoate, M-60054

$C_{13}H_{18}N_2$

Octahydropyrrolo[3,4-*c*]pyrrole; *N*-Benzyl, *in*
O-70017
1,1,3,3-Tetramethyl-2-indanone; Hydrazone, *in*
T-60143

$C_{13}H_{18}N_4O_3$

Citrulline; α-*N*-Benzoyl, amide, *in* C-60151

$C_{13}H_{18}O_2$

Pentamethylphenol; Ac, *in* P-60058
(Pentamethylphenyl)acetic acid, P-70038

$C_{13}H_{18}O_4$

3,4-Dihydroxybutanoic acid; *O*⁴-Benzyl, Et
ester, *in* D-60313

$C_{13}H_{18}O_5$

2,2,8,8-Tetramethyl-3,4,5,6,7-nonanepentone,
T-60148

$C_{13}H_{18}O_7$

Pentahydroxybenzoic acid; Penta-Me ether, Me
ester, *in* P-60043

$C_{13}H_{18}O_8$

Salirepin, *in* D-70287

$C_{13}H_{19}N_5O_{10}S$

Antibiotic PB 5266*A*, A-70225

$C_{13}H_{19}N_5O_{11}S$

Antibiotic PB 5266*B*, A-70226

$C_{13}H_{20}$

2-Isopropylideneadamantane, I-60121
Tetracyclo[5.5.1.0⁴·¹³.0¹⁰·¹³]tridecane, T-60035

$C_{13}H_{20}N_6O_3$

3′-Amino-3′-deoxyadenosine; 3′,6,6-Tri-*N*-Me,
in A-70128

$C_{13}H_{20}O$

γ-Ionone, I-70058

$C_{13}H_{20}O_3$

Grasshopper ketone, G-70028
Methyl jasmonate, *in* J-60002

$C_{13}H_{20}O_6$

2,2,8,8-Tetramethyl-5,5-dihydroxy-3,4,6,7-
nonanetetrone, *in* T-60148

$C_{13}H_{20}O_8$

Methanetetrapropanoic acid, M-70036

$C_{13}H_{21}N$

1,2,3,4,5,6,7,8,8a,9a-Decahydrocarbazole; *N*-
Me, *in* D-70012

$C_{13}H_{21}NO_6$

2,6-Biss(ydroxyacetyl)pyridine; Bis di-Me ketal,
in B-70144

$C_{13}H_{21}N_3O_6$

Antibiotic CA 146A, *in* C-70183
Clavamycin *D*, C-70183

$C_{13}H_{22}$

1,1,2,2,4,4-Hexamethyl-3,5-biss(ethylene)-
cyclopentane, H-70082

$C_{13}H_{22}N_4O_8$

Clavamycin *C*, C-70182

$C_{13}H_{22}O$

1-Cyclododecenecarboxaldehyde, C-60191
Octahydro-5,5,8a-trimethyl-2(1*H*)-
naphthalenone, *in* D-70014
Theaspirone, T-60185

$C_{13}H_{22}O_4$

9,12-Dioxododecanoic acid; Me ester, *in*
D-60471

$C_{13}H_{22}O_5$

4,4-Pyrandiacetic acid; Di-Et ester, *in* P-70150

$C_{13}H_{24}$

Octamethylcyclopentene, O-60026
6,7-Tridecadiene, T-60272

$C_{13}H_{24}O$

Decahydro-5,5,8a-trimethyl-2-naphthalenol,
D-70014
Dicyclohexylmethanol, D-70150
Octamethylcyclopentanone, O-60025
1,2,2,3,3,4,4,5-Octamethyl-6-oxabicyclo[3.1.0]-
hexane, O-70021

$C_{13}H_{24}O_2$

9-Dodecen-1-ol; Formyl, *in* D-70537

$C_{13}H_{24}O_3$

10-Hydroxy-11-dodecenoic acid; Me ester, *in*
H-60125
12-Hydroxy-10-dodecenoic acid; Me ester, *in*
H-60126
7-Megastigmene-5,6,9-triol, M-60012
12-Oxododecanoic acid; Me ester, *in* O-60066

$C_{13}H_{25}N_3$

11-Methylene-1,5,9-triazabicyclo[7.3.3]-
pentadecane, M-60080

$C_{13}H_{26}O$

2,2,3,3,4,4,5,5-Octamethylcyclopentanol,
O-60024

$C_{13}H_{26}O_4$

10-Oxodecanoic acid; Me ester, Di-Me acetal,
in O-60063

$C_{13}H_{31}N_5$

1,4,7,11,15-Pentaazacyclooctadecane, P-60023
1,4,8,11,15-Pentaazacyclooctadecane, P-60024

$C_{14}H_4Cl_2O_4$

2,3-Dichloro-1,4,9,10-anthracenetetrone,
D-60117

$C_{14}H_4N_6O_{15}$

2,4,6-Trinitrobenzoic acid; Anhydride, *in*
T-70303

$C_{14}H_5ClO_4$

2-Chloro-1,4,9,10-anthracenetetrone, C-60041

$C_{14}H_6N_4O_{11}$

3,5-Dinitrobenzoic acid; Anhydride, *in* D-60458

$C_{14}H_6O_8$

▷Ellagic acid, E-70007

$C_{14}H_8$

2,7-Diethynylnaphthalene, D-70158

$C_{14}H_8Br_2$

2,3-Dibromoanthracene, D-70085
9,10-Dibromoanthracene, D-60085

$C_{14}H_8ClNO$

9-Acridinecarboxylic acid; Chloride, *in* A-70061

$C_{14}H_8Cl_2$

9,10-Dichloroanthracene, D-60116
1,4-Dichlorophenanthrene, D-70132
2,4-Dichlorophenanthrene, D-70133
3,6-Dichlorophenanthrene, D-70134
3,9-Dichlorophenanthrene, D-70135
4,5-Dichlorophenanthrene, D-70136
9,10-Dichlorophenanthrene, D-70137

$C_{14}H_8Cl_2O_2S$

2,2'-Thiobisbenzoic acid; Dichloride, *in* T-70174
4,4'-Thiobisbenzoic acid; Dichloride, *in* T-70175

$C_{14}H_8F_2$

9,10-Difluoroanthracene, D-70161
4,5-Difluorophenanthrene, D-70166
9,10-Difluorophenanthrene, D-70167

$C_{14}H_8F_2O_4$

4,4'-Difluoro-[1,1'-biphenyl]-3,3'-dicarboxylic
acid, D-60178
4,5-Difluoro-[1,1'-biphenyl]-2,3-dicarboxylic
acid, D-60179
6,6'-Difluoro-[1,1'-biphenyl]-2,2'-dicarboxylic
acid, D-60180

$C_{14}H_8F_6O_5S$

3-(Trifluoromethyl)benzenesulfonic acid;
Anhydride, *in* T-60283

$C_{14}H_8FeN_2O_8$

Actinoviridin A, *in* H-70193

$C_{14}H_8I_2$

9,10-Diiodoanthracene, D-70355
2,7-Diiodophenanthrene, D-70360
3,6-Diiodophenanthrene, D-70361

$C_{14}H_8N_2$

9-Cyanoacridine, *in* A-70061
5,6-Dicyanoacenaphthene, *in* A-70015
Naphtho[8,1,2-cde]cinnoline, N-70007
Pyrido[2',3':3,4]cyclobuta[1,2-g]quinoline,
P-62232
Pyrido[3',2':3,4]cyclobuta[1,2-g]quinoline,
P-62233

$C_{14}H_8N_2O$

Naphtho[8,1,2-cde]cinnoline; N-Oxide, *in*
N-70007

$C_{14}H_8N_2O_2$

[1,4]Benzoxazino[3,2-b][1,4]benzoxazine,
B-60055

$C_{14}H_8N_2S$

[1]Benzothieno[2,3-c][1,5]naphthyridine,
B-70060
[1]Benzothieno[2,3-c][1,7]naphthyridine,
B-70061

$C_{14}H_8N_2Se$

Phenanthro[9,10-c][1,2,5]selenadiazole,
P-70053

$C_{14}H_8N_4S_2$

[1,2,4,5]Tetrazino[3,4-b:6,1-b']-
bisbenzothiazole, T-60171

$C_{14}H_8O$

Acenaphtho[5,4-b]furan, A-60014

$C_{14}H_8OS$

Phenanthro[4,5-bcd]thiophene; 4-Oxide, *in*
P-60077

$C_{14}H_8O_2$

Benz[a]azulene-1,4-dione, B-60011
1,8-Biphenylenedicarboxaldehyde, B-60118

$C_{14}H_8O_2S$

Phenanthro[4,5-bcd]thiophene; 4,4-Dioxide, *in*
P-60077

$C_{14}H_8O_3$

3,4-Acenaphthenedicarboxylic acid; Anhydride,
in A-60013

$C_{14}H_8S$

Acenaphtho[1,2-b]thiophene, A-70017
Acenaphtho[5,4-b]thiophene, A-60015
Phenanthro[4,5-bcd]thiophene, P-60077

$C_{14}H_8SSe$

[1]Benzoselenopheno[2,3-b][1]benzothiophene,
B-70052

$C_{14}H_8S_2$

[1]Benzothiopyrano[6,5,4-def][1]-
benzothiopyran, B-60049

$C_{14}H_8Se_4$

Dibenzotetraselenofulvalene, D-70077

$C_{14}H_9ClN_2$

1-Chloro-4-phenylphthalazine, C-70141
2-Chloro-3-phenylquinoxaline, C-70145
6-Chloro-2-phenylquinoxaline, C-70146
7-Chloro-2-phenylquinoxaline, C-70147

$C_{14}H_9ClN_2O$

2-Chloro-3-phenylquinoxaline; 4-Oxide, *in*
C-70145
6-Chloro-2-phenylquinoxaline; 4-Oxide, *in*
C-70146
7-Chloro-2-phenylquinoxaline; 4-Oxide, *in*
C-70147

$C_{14}H_9ClN_2O_2$

6-Chloro-2-phenylquinoxaline; 1,4-Dioxide (?),
in C-70146

$C_{14}H_9Cl_2NO_2$

2,2'-Iminodibenzoic acid; Dichloride, *in* I-60014

$C_{14}H_9NO$

1-Acridinecarboxaldehyde, A-70056
2-Acridinecarboxaldehyde, A-70057
3-Acridinecarboxaldehyde, A-70058
4-Acridinecarboxaldehyde, A-70059
9-Acridinecarboxaldehyde, A-70060
2-Phenyl-3H-indol-3-one, P-60114

$C_{14}H_9NO_2$

9-Acridinecarboxaldehyde; 10-Oxide, *in*
A-70060
9-Acridinecarboxylic acid, A-70061
3-(Phenylimino)-1(3H)-isobenzofuranone,
P-60112
2-Phenylisatogen, *in* P-60114
2-(2-Pyridyl)-1,3-indanedione, P-60241
2-(3-Pyridyl)-1,3-indanedione, P-60242
2-(4-Pyridyl)-1,3-indanedione, P-60243

$C_{14}H_9NO_3$

1-Amino-2-hydroxyanthraquinone, A-60174
▷1-Amino-4-hydroxyanthraquinone, A-60175
1-Amino-5-hydroxyanthraquinone, A-60176
1-Amino-6-hydroxyanthraquinone, A-60177
2-Amino-1-hydroxyanthraquinone, A-60178
2-Amino-3-hydroxyanthraquinone, A-60179
3-Amino-1-hydroxyanthraquinone, A-60180
9-Amino-10-hydroxy-1,4-anthraquinone,
A-60181
1H-Benz[e]indole-1,2(3H)-dione; N-Ac, *in*
B-70026

$C_{14}H_9NO_4$

(2-Nitrophenyl)phenylethanedione, N-70058
(4-Nitrophenyl)phenylethanedione, N-70059

$C_{14}H_9NS$

6H-[1]Benzothieno[2,3-b]indole, B-70058
10H-[1]Benzothieno[3,2-b]indole, B-70059
Pyrrolo[3,2,1-kl]phenothiazine, P-70196

$C_{14}H_9N_3$

4,4'-Iminobisbenzonitrile, *in* I-60018
5H-Indolo[2,3-b]quinoxaline, I-60036
Pyrido[2',1':2,3]imidazo[4,5-c]isoquinoline,
P-60235

$C_{14}H_9N_3O$

12H-Quinoxalino[2,3-b][1,4]benzoxazine,
Q-70009

$C_{14}H_9N_3OS$

1,2,3-Benzotriazine-4(3H)-thione; N-Benzoyl,
in B-60050

$C_{14}H_9N_3O_2$

5H-Indolo[2,3-b]quinoxaline; 5,11-Dioxide, *in*
I-60036

$C_{14}H_9N_3S$

12H-Quinoxalino[2,3-b][1,4]benzothiazine,
Q-70008

$C_{14}H_{10}$

1-Benzylidene-1H-cyclopropabenzene, B-70084
Cyclohepta[de]naphthalene, C-60198

$C_{14}H_{10}BrN$

1-(Bromomethyl)acridine, B-70229
2-(Bromomethyl)acridine, B-70230
3-(Bromomethyl)acridine, B-70231
4-(Bromomethyl)acridine, B-70232
9-(Bromomethyl)acridine, B-70233

$C_{14}H_{10}IN$

▷1H-Indol-3-ylphenyliodonium hydroxide inner
salt, I-70017

$C_{14}H_{10}N_2$

10b,10c-Diazadicyclopenta[ef,kl]heptalene,
D-70052

$C_{14}H_{10}N_2O$

9-Acridinecarboxylic acid; Amide, *in* A-70061
3,5-Diphenyl-1,2,4-oxadiazole, D-60484
3-Nitroso-2-phenylindole, *in* P-60114
1H-Pyrrolo[3,2-b]pyridine; N-Benzoyl, *in*
P-60249

$C_{14}H_{10}N_2O_2$

1,3-Diazetidine-2,4-dione; 1,3-Di-Ph, *in*
D-60057
3,5-Diphenyl-1,2,4-oxadiazole; 4-Oxide, *in*
D-60484
2-Phenylisatogen oxime, *in* P-60114

$C_{14}H_{10}N_2O_3$

N-(3-Oxo-3H-phenoxazin-2-yl)acetamide, *in*
A-60237

$C_{14}H_{10}N_2O_4$

(2-Nitrophenyl)phenylethanedione; 1-Oxime, *in*
N-70058
(2-Nitrophenyl)phenylethanedione; 2-Oxime, *in*
N-70058
(4-Nitrophenyl)phenylethanedione; Monoxime,
in N-70059

$C_{14}H_{10}N_2O_5$

4-Amino-2,6-pyridinedicarboxylic acid; N-
Benzoyl, *in* A-70194

C₁₄H₁₀N₂S

3-Cyano-10-methylphenothiazine, *in* P-70057

C₁₄H₁₀N₂Se

3,4-Diphenyl-1,2,5-selenadiazole, D-70497

C₁₄H₁₀O

Benzo[5,6]cycloocta[1,2-*c*]furan, B-70029

9*H*-Fluorene-1-carboxaldehyde, F-70014

▷9*H*-Fluorene-9-carboxaldehyde, F-70015

6*a*,9*b*-Methano-1*H*-phenalen-1-one, M-70042

C₁₄H₁₀O₂

2-Acetylnaphtho[1,8-*bc*]pyran, A-70048

1,3-Di(2-furyl)benzene, D-70168

1,4-Di(2-furyl)benzene, D-70169

3,4-Heptafulvalenedione, H-60022

4-Hydroxyacenaphthylene; Ac, *in* H-70100

4,5-Phenanthrenediol, P-70048

C₁₄H₁₀O₃

Disalicylaldehyde, D-70505

Naphtho[1,8-*bc*]pyran-2-carboxylic acid; Me ester, *in* N-70010

2,3,7-Phenanthrenetriol, T-60339

2,4,5-Phenanthrenetriol, P-60071

C₁₄H₁₀O₄

3,4-Acenaphthenedicarboxylic acid, A-60013

5,6-Acenaphthenedicarboxylic acid, A-70015

2,5-Dihydroxy-4-methoxy-9*H*-fluoren-9-one, *in* T-70261

2-Hydroxybenzoic acid; Benzoyl, *in* H-70108

2-Hydroxy-1-methoxyxanthone, *in* D-70353

Moracin *M*, M-60146

1,2,5,6-Tetrahydroxyphenanthrene, T-60115

1,2,5,7-Tetrahydroxyphenanthrene, T-60116

1,2,6,7-Tetrahydroxyphenanthrene, T-60117

1,3,5,6-Tetrahydroxyphenanthrene, T-60118

1,3,6,7-Tetrahydroxyphenanthrene, T-60119

2,3,5,7-Tetrahydroxyphenanthrene, T-60120

2,3,6,7-Tetrahydroxyphenanthrene, T-60121

2,4,5,6-Tetrahydroxyphenanthrene, T-60122

3,4,5,6-Tetrahydroxyphenanthrene, T-60123

C₁₄H₁₀O₄S

2,2′-Thiobisbenzoic acid, T-70174

4,4′-Thiobisbenzoic acid, T-70175

C₁₄H₁₀O₅

1,5-Dihydroxy-6-methoxyxanthone, *in* T-60349

5,6-Dihydroxy-1-methoxyxanthone, *in* T-60349

1,2,5,6,7-Pentahydroxyphenanthrene, P-60050

1,3,4,5,6-Pentahydroxyphenanthrene, P-70034

2,3,4,7,9-Pentahydroxyphenanthrene, P-70035

C₁₄H₁₀O₆

2,5-Dihydroxy-1,4-naphthoquinone; Di-Ac, *in* D-60354

2,8-Dihydroxy-1,4-naphthoquinone; Di-Ac, *in* D-60355

5,8-Dihydroxy-1,4-naphthoquinone; Di-Ac, *in* D-60356

1,2,3,5,6,7-Hexahydroxyphenanthrene, H-60066

2,4,5-Trihydroxy-1-methoxyxanthone, *in* T-70122

C₁₄H₁₀O₆S

4,4′-Sulfonylbisbenzoic acid, *in* T-70175

C₁₄H₁₀S

9-Phenanthrenethiol, P-70049

C₁₄H₁₀S₂

9,10-Dihydro-9,10-epidithioanthracene, D-70206

C₁₄H₁₁FO

2-Fluoro-2,2-diphenylacetaldehyde, F-60033

3-Fluoro-4-methylphenol; Benzoyl, *in* F-60057

C₁₄H₁₁FO₂

2-Fluoro-3-hydroxybenzaldehyde; Benzyl ether, *in* F-60035

2-Fluoro-5-hydroxybenzaldehyde; Benzyl ether, *in* F-60036

4-Fluoro-3-hydroxybenzaldehyde; Benzyl ether, *in* F-60037

C₁₄H₁₁IO₃S

2-Iodo-1-phenyl-2-(phenylsulfonyl)ethanone, I-60061

C₁₄H₁₁N

1-Phenyl-1*H*-indole, P-70082

6-Phenyl-1*H*-indole, P-60113

C₁₄H₁₁NO

9*H*-Fluorene-1-carboxaldehyde; Oxime, *in* F-70014

▷*N*-Phenylphthalamidine, *in* P-70100

C₁₄H₁₁NO₂

1*H*-Benz[*e*]indole-1,2(3*H*)-dione; *N*-Et, *in* B-70026

2-Hydroxycarbazole; Ac, *in* H-60109

C₁₄H₁₁NO₂S

10*H*-Phenothiazine-1-carboxylic acid; Me ester, *in* P-70055

10*H*-Phenothiazine-1-carboxylic acid; *N*-Me, *in* P-70055

10*H*-Phenothiazine-2-carboxylic acid; Me ester, *in* P-70056

10*H*-Phenothiazine-2-carboxylic acid; *N*-Me, *in* P-70056

10*H*-Phenothiazine-3-carboxylic acid; *N*-Me, *in* P-70057

10*H*-Phenothiazine-4-carboxylic acid; *N*-Me, *in* P-70058

C₁₄H₁₁NO₃S

10*H*-Phenothiazine-1-carboxylic acid; Me ester, 5-oxide, *in* P-70055

C₁₄H₁₁NO₄

1,2-Dimethoxy-3*H*-phenoxazin-3-one, *in* T-60353

▷2,2′-Iminodibenzoic acid, I-60014

2,3′-Iminodibenzoic acid, I-60015

2,4′-Iminodibenzoic acid, I-60016

3,3′-Iminodibenzoic acid, I-60017

4,4′-Iminodibenzoic acid, I-60018

C₁₄H₁₁NO₄S

10*H*-Phenothiazine-1-carboxylic acid; *N*-Me, 5,5-dioxide, *in* P-70055

10*H*-Phenothiazine-2-carboxylic acid; Me ester, 5,5-dioxide, *in* P-70056

10*H*-Phenothiazine-2-carboxylic acid; *N*-Me, 5,5-dioxide, *in* P-70056

10*H*-Phenothiazine-3-carboxylic acid; *N*-Me, 5,5-dioxide, *in* P-70057

10*H*-Phenothiazine-4-carboxylic acid; *N*-Me, 5,5-dioxide, *in* P-70058

C₁₄H₁₁N₃O₄

(2-Nitrophenyl)phenylethanedione; Dioxime, *in* N-70058

(4-Nitrophenyl)phenylethanedione; Dioxime, *in* N-70059

C₁₄H₁₂

9,10-Dihydrophenanthrene, D-60260

9*b*-Methyl-9*bH*-benz[*cd*]azulene, M-60044

Tricyclo[4.4.4.0¹,⁶]tetradeca-2,4,7,9,10,12-hexaene, T-70232

C₁₄H₁₂Br₂

1,2-Dibromo-1,2-diphenylethane, D-70091

C₁₄H₁₂Br₂O₅

3-Bromo-4-[(3-bromo-4,5-dihydroxyphenyl)-methyl]-5-(hydroxymethyl)-1,2-benzenediol, B-70194

C₁₄H₁₂N₂O

2,4-Dimethylcarbazole; *N*-Nitroso, *in* D-70382

3,6-Dimethylcarbazole; *N*-Nitroso, *in* D-70387

1,2-Diphenyl-1,2-diazetidin-3-one, *in* D-70055

C₁₄H₁₂N₂O₂

5*a*,6,11*a*,12-Tetrahydro[1,4]benzoxazino[3,2-*b*][1,4]benzoxazine, T-60059

C₁₄H₁₂N₂O₄

[2,2′-Bipyridine]-3,3′-dicarboxylic acid; Di-Me ester, *in* B-60124

[2,2′-Bipyridine]-3,5′-dicarboxylic acid; Di-Me ester, *in* B-60125

[2,2′-Bipyridine]-4,4′-dicarboxylic acid; Di-Me ester, *in* B-60126

[2,2′-Bipyridine]-5,5′-dicarboxylic acid; Di-Me ester, *in* B-60127

[3,3′-Bipyridine]-2,2′-dicarboxylic acid; Di-Me ester, *in* B-60133

[3,3′-Bipyridine]-4,4′-dicarboxylic acid; Di-Me ester, *in* B-60134

C₁₄H₁₂N₂S

2-Amino-4*H*-3,1-benzothiazine; *N*-Ph, *in* A-70092

2,5-Dihydro-2,2-diphenyl-1,3,4-thiadiazole, D-60236

C₁₄H₁₂N₄

4-Amino-1,2,3-benzotriazine; *N*(4)-Benzyl, *in* A-60092

4-Amino-1,2,3-benzotriazine; 2-Me, *N*(4)-Ph, *in* A-60092

4-Amino-1,2,3-benzotriazine; 3-Me, *N*(4)-Ph, *in* A-60092

4-Amino-1,2,3-benzotriazine; 3-Benzyl, *in* A-60092

C₁₄H₁₂N₄O₄

Boxazomycin B, *in* B-70176

Boxazomycin C, *in* B-70176

C₁₄H₁₂N₄O₅

Boxazomycin *A*, B-70176

C₁₄H₁₂N₄S

3,5-Dianilino-1,2,4-thiadiazole, D-70048

Hector's base, H-60012

C₁₄H₁₂O

▷1-(Hydroxymethyl)fluorene, H-70176

1-Phenoxy-1-phenylethylene, *in* P-60092

C₁₄H₁₂OS

5,7-Dihydrodibenzo[*c,e*]thiepin; *S*-Oxide, *in* D-70190

2-(Methylthio)benzophenone, *in* M-70020

3-(Methylthio)benzophenone, *in* M-70021

4-(Methylthio)benzophenone, *in* M-70022

C₁₄H₁₂OS₂

5*H*,7*H*-Dibenzo[*b,g*][1,5]dithiocin; 6-Oxide, *in* D-70069

5*H*,7*H*-Dibenzo[*b,g*][1,5]dithiocin; 12-Oxide, *in* D-70069

6*H*,12*H*-Dibenzo[*b,f*][1,5]dithiocin; *S*-Oxide, *in* D-70070

C₁₄H₁₂OSe

Phenyl styryl selenoxide, *in* P-70070

C₁₄H₁₂O₂

1,2-Dihydrocyclobuta[*a*]naphthalene-2-carboxylic acid; Me ester, *in* D-70177

4-Hydroxyacenaphthene; Ac, *in* H-60099

2,3-Naphthalenedicarboxylic acid; Di-Me ester, *in* N-60003

1,2,3,4-Tetrahydroanthraquinone, T-60054

C₁₄H₁₂O₂S

5,7-Dihydrodibenzo[*c,e*]thiepin; *S*-Dioxide, *in* D-70190

C₁₄H₁₂O₂S₂

5*H*,7*H*-Dibenzo[*b,g*][1,5]dithiocin; 12,12-Dioxide, *in* D-70069

C₁₄H₁₂O₂Se

Phenyl styryl selenone, *in* P-70070

C₁₄H₁₂O₃

4′-Methoxy-2-biphenylcarboxylic acid, *in* H-70111

α-Oxo-1-naphthaleneacetic acid; Et ester, *in* O-60078

α-Oxo-2-naphthaleneacetic acid; Et ester, *in* O-60079

$C_{14}H_{12}O_4$

2,6-Azulenedicarboxylic acid; Di-Me ester, *in* A-70294

1,4-Dicarbomethoxyazulene, *in* A-70293

9,10-Dihydro-2,4,5,6-phenanthrenetetrol, D-70247

1-(2,4-Dihydroxyphenyl)-2-(3,5-dihydroxyphenyl)ethylene, D-70335

4′-Hydroxydehydrokawain, H-70122

1-(3-Hydroxyphenyl)-2-(2,4,5-trihydroxyphenyl)ethylene, H-70215

Osthenone, O-70049

$C_{14}H_{12}O_4S_2$

$5H,7H$-Dibenzo[b,g][1,5]dithiocin; 6,6,12,12-Tetroxide, *in* D-70069

$C_{14}H_{12}O_5$

9,10-Dihydro-2,3,4,6,7-phenanthrenepentol, D-70246

▷ Khellin, K-60010

$C_{14}H_{12}O_6$

1-(2,3,4-Trihydroxyphenyl)-2-(3,4,5-trihydroxyphenyl)ethylene, T-60344

$C_{14}H_{12}O_8$

Fulvic acid, F-60081

Polivione, P-60167

$C_{14}H_{12}S$

5,7-Dihydrodibenzo[c,e]thiepin, D-70190

$C_{14}H_{12}S_2$

$5H,7H$-Dibenzo[b,g][1,5]dithiocin, D-70069

$6H,12H$-Dibenzo[b,f][1,5]dithiocin, D-70070

$C_{14}H_{12}Se$

5,7-Dihydrodibenzo[c,e]selenepin, D-70189

[(2-Phenylethenyl)seleno]benzene, P-70070

$C_{14}H_{13}N$

9-Amino-9,10-dihydroanthracene, A-60136

2-Amino-9,10-dihydrophenanthrene, A-60141

4-Amino-9,10-dihydrophenanthrene, A-60142

9-Amino-9,10-dihydrophenanthrene, A-60143

▷ 10,11-Dihydro-5H-dibenz[b,f]azepine, D-70185

1,2-Dimethylcarbazole, D-70374

1,3-Dimethylcarbazole, D-70375

1,4-Dimethylcarbazole, D-70376

1,5-Dimethylcarbazole, D-70377

1,6-Dimethylcarbazole, D-70378

1,7-Dimethylcarbazole, D-70379

1,8-Dimethylcarbazole, D-70380

2,3-Dimethylcarbazole, D-70381

2,4-Dimethylcarbazole, D-70382

2,5-Dimethylcarbazole, D-70383

2,7-Dimethylcarbazole, D-70384

3,4-Dimethylcarbazole, D-70385

3,5-Dimethylcarbazole, D-70386

3,6-Dimethylcarbazole, D-70387

4,5-Dimethylcarbazole, D-70388

$C_{14}H_{13}NO$

2-Ethoxycarbazole, *in* H-60109

$C_{14}H_{13}NO_2$

4-Oxo-1,2,3,4-tetrahydrocarbazole; N-Ac, *in* O-70106

$C_{14}H_{13}NO_3$

(4-Hydroxyphenoxy)acetic acid; Anilide, *in* H-70205

3,4,6-Trihydroxy-1,2-dimethylcarbazole, T-70253

$C_{14}H_{13}NO_4$

6-Amino-4-hexenoic acid; N-Phthalimido, *in* A-60173

$C_{14}H_{13}N_3O_2$

2,2′-Iminobisbenzamide, *in* I-60014

$C_{14}H_{14}$

Pentacyclo[8.4.0.0^{2,7}.0^{3,12}.0^{6,11}]tetradeca-4,8,13-triene, P-70021

1,2,3,4-Tetrahydroanthracene, T-60053

$C_{14}H_{14}N_2$

2,5-Dimethyl-3-(2-phenylethenyl)pyrazine, D-60442

$N,N′$-Diphenylformamidine; N-Me, *in* D-60479

$C_{14}H_{14}N_2O$

2,2′-Diaminobiphenyl; 2-N-Ac, *in* D-60033

$N,N′$-Diphenylurea; N-Me, *in* D-70499

$C_{14}H_{14}N_2O_5$

Antibiotic PDE II, A-70229

$C_{14}H_{14}O$

2-(2-Phenylethyl)phenol, P-70071

3-(2-Phenylethyl)phenol, P-70072

4-(2-Phenylethyl)phenol, P-70073

$C_{14}H_{14}OS$

2-Mercapto-1,1-diphenylethanol, M-70028

$C_{14}H_{14}O_2$

2-(2,4-Hexadiynylidene)-1,6-dioxaspiro[4.5]-dec-3-ene, *in* H-70042

1-Methoxy-2-(phenylmethoxy)benzene, *in* B-60072

1-Methoxy-3-(phenylmethoxy)benzene, *in* B-60073

1-Methoxy-4-(phenylmethoxy)benzene, *in* B-60074

2-(2-Phenylethyl)-1,4-benzenediol, P-60095

4-(2-Phenylethyl)-1,2-benzenediol, P-60096

4-(2-Phenylethyl)-1,3-benzenediol, P-60097

5-(2-Phenylethyl)-1,3-benzenediol, P-60098

$C_{14}H_{14}O_3$

2-(Dimethoxymethyl)-1-naphthalenecarboxaldehyde, *in* N-60001

3-(2,2-Dimethyl-2H-1-benzopyran-6-yl)-2-propenoic acid, D-70372

2-(2,4-Hexadiynylidene)-1,6-dioxaspira[4.5]-dec-3-en-8-ol, H-70042

2-(2,4-Hexadiynylidene)-3,4-epoxy-1,6-dioxaspiro[4.5]decane, *in* H-70042

3,3′,4-Trihydroxybibenzyl, T-60316

$C_{14}H_{14}O_4$

5,8-Dihydroxy-2,3,6,7-tetramethyl-1,4-naphthoquinone, D-60377

1,4,5-Naphthalenetriol; 1,5-Di-Me ether, Ac, *in* N-60005

1,4,5-Naphthalenetriol; 4,5-Di-Me ether, Ac, *in* N-60005

Phthalidochromene, P-60143

3,3′,4,4′-Tetrahydroxybibenzyl, T-70101

3,3′,4,5-Tetrahydroxybibenzyl, T-60102

$C_{14}H_{14}O_5$

Funadonin, F-70046

Khellactone, K-70013

3,3′,4,4′,5-Pentahydroxybibenzyl, P-60044

$C_{14}H_{14}O_6$

3,3′,4,4′,5,5′-Hexahydroxybibenzyl, H-60062

$C_{14}H_{14}O_7$

3′,4′,5′-Trihydroxyacetophenone; Tri-Ac, *in* T-60315

$C_{14}H_{14}O_8$

1,2,3,4-Benzenetetrol; Tetra-Ac, *in* B-60014

Hexahydroxy-1,4-naphthoquinone; Tetra-Me ether, *in* H-60065

3,4,5-Trihydroxyphenylacetic acid; Tri-Ac, *in* T-60343

$C_{14}H_{14}Se_2$

Dibenzyl diselenide, D-60082

$C_{14}H_{15}ClO_5$

Mikrolin, M-60135

$C_{14}H_{15}NO$

1,3,4,10-Tetrahydro-9(2H)-acridinone; N-Me, *in* T-60052

$C_{14}H_{15}N_3O_6$

Antibiotic FR 900482, A-60285

$C_{14}H_{15}N_5O_4$

8-Aminoimidazo[4,5-g]quinazoline; 1-($β$-D-Ribofuranosyl), *in* A-60207

lin-Benzoadenosine, *in* A-60207

$C_{14}H_{16}$

1,3-Diethynyladamantane, D-60172

Heptacyclo[6.6.0.0^{2,6}.0^{3,13}.0^{4,11}.0^{5,9}.0^{10,14}]tetradecane, H-70012

Heptacyclo[9.3.0.0^{2,5}.0^{3,13}.0^{4,8}.0^{6,10}.0^{9,12}]tetradecane, H-70013

$C_{14}H_{16}N_2$

2,2′-Biss(ethylamino)biphenyl, *in* D-60033

1,2-Diphenyl-1,2-ethanediamine, D-70480

1-(1-Naphthyl)piperazine, N-60013

1-(2-Naphthyl)piperazine, N-60014

$C_{14}H_{16}N_2O_2$

▷ 1,4-Bis(1-isocyanato-1-methylethyl)benzene, B-70149

$C_{14}H_{16}N_4$

5,6,8,9-Tetraaza[3.3]paracyclophane, T-60018

$C_{14}H_{16}O_2$

2-Phenacylcyclohexanone, P-70045

7-Phenyl-5-heptynoic acid; Me ester, *in* P-60106

$C_{14}H_{16}O_4$

7-Hydroxy-4-isopropyl-3-methoxy-6-methyl-coumarin, H-70144

Pyriculol, P-60216

$C_{14}H_{16}O_5$

Dechloromikrolin, *in* M-60135

$C_{14}H_{16}O_6$

1,5,8-Trihydroxy-3-methoxyxanthone, *in* T-70123

$C_{14}H_{16}O_7$

Sawaranin, S-70016

$C_{14}H_{17}NO$

8-Azabicyclo[3.2.1]octane; N-Benzoyl, *in* A-70278

$C_{14}H_{17}NO_3$

3-Amino-4-hexenoic acid; N-Benzoyl, Me ester, *in* A-60171

$C_{14}H_{18}$

Benzylidenecycloheptane, B-60068

Tetrahydro-1,6,7-triss(ethylene)-1H,4H-3a,6a-propanopentalene, T-70101

$C_{14}H_{18}N_2$

1,8-Biss(imethylamino)naphthalene, *in* D-70042

$C_{14}H_{18}N_2O_5$

Aspartame, A-60311

$C_{14}H_{18}N_4O_4$

1,2,3,5-Tetraaminobenzene; N-Tetra-Ac, *in* T-60012

1,2,4,5-Tetraaminobenzene; 1,2,4,5-N-Tetra-Ac, *in* T-60013

$C_{14}H_{18}O_2$

Majusculone, M-70006

$C_{14}H_{18}O_3$

5-(1,3-Dihydroxypropyl)-2-isopropenyl-2,3-dihydrobenzofuran, D-70340

6-Ethoxy-7-methoxy-2,2-dimethyl-2H-1-benzopyran, *in* D-70365

$C_{14}H_{18}O_4$

1-Acetyl-4-isopentenyl-6-methylphloroglucinol, *in* T-70248

11-Hydroxy-3-oxo-13-nor-7(11)-eudesmen-12,6-olide, H-60204

3-Isopropyl-6-methyl-1,2-benzenediol; Di-Ac, *in* I-60124

$C_{14}H_{18}O_6$

Colletoketol, *in* C-60159

Grahamimycin A$_1$, *in* C-60159

$C_{14}H_{18}O_7$

3,7a-Diacetoxyhexahydro-2-oxa-3a,7-dimethyl-1,4-indanedione, *in* H-70055

$C_{14}H_{18}O_8$

6-O-Acetylarbutin, *in* A-70248

$C_{14}H_{19}NO_2$

2-(Aminomethyl)cyclohexanol; N-Benzoyl, *in* A-70169

C14H19NO7
Menisdaurin, *in* L-70034

C14H19NO8
Griffonin, *in* L-70034
Lithospermoside, L-70034
Lithospermoside; 5-Epimer, *in* L-70034
Thalictoside, *in* N-70050

C14H19N3
1-Azidodiamantane, A-60341
3-Azidodiamantane, A-60342
4-Azidodiamantane, A-60343

C14H19N3O4
Citrulline; α-*N*-Benzoyl, Me ester, *in* C-60151

C14H20
7,7′-Bii(icyclo[2.2.1]heptylidene), B-60078
1,3-Divinyladamantane, D-70529

C14H20N2
3,6-Di-*tert*-butylpyrrolo[3,2-*b*]pyrrole, D-70108

C14H20N2O4
3-Methyl-*N*-(5a,6a,7,8-tetrahydro-4-methyl-1,5-dioxo-1*H*,5*H*-pyrrolo[1,2-*c*][1,3]-oxazepin-3-yl)butanamide, M-70134

C14H20O2
Oblongolide, O-60001
1,3,5,7-Tetramethyl-2,4-adamantanedione, T-60131

C14H20O3
Ceratenolone, C-60030
7-Oxo-11-nordrim-8-en-12-oic acid, O-60080

C14H20O5
Seiricuprolide, S-70028
Wasabidienone *A*, W-70001

C14H20O6
Colletodiol, C-60159

C14H20O8
Ethylenetetracarboxylic acid; Tetra-Et ester, *in* E-60049

C14H20S6
1,3,4,6-Tetrakiss(thylthio)thieno[3,4-*b*]-thiophene, *in* T-70170

C14H21NO
1,2,3,4,5,6,7,8,8a,9a-Decahydrocarbazole; *N*-Ac, *in* D-70012

C14H21NO4
Wasabidienone *D*, W-70002

C14H22
1,1,2,2-Tetracyclopropylethane, T-60034

C14H22N2
Bis(1,1-dicyclopropylmethyl)diazene, B-60154

C14H22N6O3
3′-Amino-3′-deoxyadenosine; 3′,3′,6,6-Tetra-*N*-Me, *in* A-70128

C14H22O
Norpatchoulenol, N-70078
1,3,5,7-Tetramethyladamantanone, T-60132

C14H22O2
Sulcatine, S-60061

C14H22O3
5,7-Dihydroxy-13-nor-8-marasmanone, D-60361
5,8-Dihydroxy-13-nor-7-marasmanone, D-60362
6-Hydroxy-2,4,8-tetradecatrienoic acid, H-70222

C14H22O5
Botrylactone, B-70174

C14H23N3O2
1*H*-Pyrrole-3,4-dicarboxylic acid; Diethyl-amide, *in* P-70183

C14H24
1,8-Cyclotetradecadiene, C-70252

C14H24N2O10
▷3,12-Biss(arboxymethyl)-6,9-dioxa-3,12-diazatetradecanedioic acid, B-70134

C14H24N6
1,3,4,6-Tetrakiss(imethylamino)pyrrolo[3,4-*c*]-pyrrole, T-70127

C14H24O2
11-Dodecyn-1-ol; Ac, *in* D-70540
12-Tetradecen-14-olide, *in* H-60230

C14H25N3O8
Antibiotic SF 2339, A-60289

C14H26
1,1,2,2,3,3,4,4-Octamethyl-5-methylenecyclo-pentane, O-60027

C14H26BBrF4N2
Biss(uinuclidine)bromine(1+); Tetra-fluoroborate, *in* B-60173

C14H26BrN2⊕
Biss(uinuclidine)bromine(1+), B-60173

C14H26Br2N2
Biss(uinuclidine)bromine(1+); Bromide, *in* B-60173

C14H26N4
1,2,4,5-Tetrakiss(imethylamino)benzene, *in* T-60012
1,2,4,5-Tetrakiss(imethylamino)benzene, *in* T-60013

C14H26O
6-Isopropenyl-3-methyl-9-decen-1-ol, I-70089
7-Tetradecyn-1-ol, T-60042

C14H26O2
9-Dodecen-1-ol; Ac, *in* D-70537

C14H26O3
12-Hydroxy-13-tetradecenoic acid, H-60229
14-Hydroxy-12-tetradecenoic acid, H-60230

C14H28
Cyclotetradecane, C-70253

C14H28NO2⊕
1,1-*tert*-Butyl-3,3-diethoxy-2-azaallenium(1+), B-70301

C14H28N2O5
Carnitine; Dimeric intermolecular ester, *in* C-70024

C14H28O
Nonamethylcyclopentanol, N-60057
5-Tetradecanone, T-60039

C14H30
3,4-Diisopropyl-2,5-dimethylhexane, D-60393

C14H30O
4,6,8-Trimethyl-2-undecanol, T-60372

C14H33N5
1,4,8,12,16-Pentaazacyclononadecane, P-60022

C14K2N6O2
Croconate blue, *in* T-70321

C14N6O2
2,4,5-Triss(icyanomethyl)-1,3-cyclo-pentanedione, T-70321

C15H7Cl2F3N2O2
▷Phenyl 5,6-dichloro-2-(trifluoromethyl)-1*H*-benzimidazole-1-carboxylate, *in* D-60160

C15H8ClNS
4-Chloro-10*H*-phenothiazine, C-60127

C15H8N4S2
13*H*-[1,3,5]Thiadiazino[3,2-*a*:5,6-*a*′]-bisbenzimidazole-13-thione, T-60191

C15H8O4
Benzophenone-2,2′-dicarboxylic acid; Anhydride, *in* B-70041

C15H8O6
2,3,9-Trihydroxy-6*H*-benzofuro[3,2-*c*][1]-benzopyran-6-one, T-70249

C15H9IO2
5,9,10-Trihydroxy-2*H*-pyrano[2,3,4-*kl*]-xanthen-2-one, T-70273

C15H9IO2
3-Iodo-2-phenyl-4*H*-1-benzopyran-4-one, I-60059

C15H9N
Cyclohepta[*def*]carbazole, C-70220
2-(Phenylethynyl)benzoic acid; Nitrile, *in* P-70074

C15H9NO2
3-Phenyl-1,4-isoquinolinedione, P-70083

C15H9NO3
Benzophenone-2,2′-dicarboxylic acid; Imide, *in* B-70041

C15H9NS
[1]Benzothieno[2,3-*c*]quinoline, B-70062
[1]Benzothieno[3,2-*b*]quinoline, B-70063
[1]Benzothieno[3,2-*c*]quinoline, B-70064

C15H9N3O
Indolo[2,3-*b*][1,4]benzodiazepin-12(6*H*)-one, I-70015

C15H10
▷4*H*-Cyclopenta[*def*]phenanthrene, C-70244

C15H10BF4N
Cyclohepta[*def*]carbazolium tetrafluoroborate, *in* C-70220

C15H10Cl2O5
Thiomelin, T-70181

C15H10N2
5*H*-Cyclohepta[4,5]pyrrolo[2,3-*b*]indole, C-60200
Indazolo[3,2-*a*]isoquinoline, I-60020
10*H*-Indolo[3,2-*b*]quinoline, I-70016
2-Methylbenzo[*gh*]perimidine, M-60051

C15H10N2O
10*H*-Indolo[3,2-*b*]quinoline; 5-Oxide, *in* I-70016

C15H10N2O2
2-Diazo-1,3-diphenyl-1,3-propanedione, D-70058

C15H10N8OS2
S,*S*′-Bis(1-phenyl-1*H*-tetrazol-5-yl) dithiocarbonate, B-70155

C15H10O2
2-(Phenylethynyl)benzoic acid, P-70074
4-(Phenylethynyl)benzoic acid, P-70075

C15H10O2S
Dibenzo[*b*,*f*]thiepin-3-carboxylic acid, D-60077
3-Mercapto-2-phenyl-4*H*-1-benzopyran-4-one, M-60025
2-Phenyl-4*H*-1-benzothiopyran-4-one; *S*-Oxide, *in* P-60082

C15H10O3
6-Hydroxyflavone, H-70130
7-Hydroxyflavone, H-70131
▷8-Hydroxyflavone, H-70132
1-Hydroxy-7-methylanthraquinone, H-60170
3-(3-Hydroxyphenyl)-1*H*-2-benzopyran-1-one, H-70207

C15H10O3S
2-Phenyl-4*H*-1-benzothiopyran-4-one; *S*,*S*-Dioxide, *in* P-60082

C15H10O4
2′,8-Dihydroxyflavone, D-70300
4′,7-Dihydroxyisoflavone, D-60340

C15H10O4S
Dibenzo[*b*,*f*]thiepin-3-carboxylic acid; 5,5-Dioxide, *in* D-60077

C15H10O5
Benzophenone-2,2′-dicarboxylic acid, B-70041
Chalaurenol, C-70036
▷1,3-Dihydroxy-2-hydroxymethylanthraquinone, D-70306
Sulfuretin, S-70082
5,6,7-Trihydroxyflavone, T-60322
2′,4′,7-Trihydroxyisoflavone, T-70264

▷4′,5,7-Trihydroxyisoflavone, T-60324
4′,6,7-Trihydroxyisoflavone, T-60325
4′,7,8-Trihydroxyisoflavone, T-70265
1,2,7-Trihydroxy-6-methylanthraquinone, T-70268
1,6,8-Trihydroxy-2-methylanthraquinone, T-70269

$C_{15}H_{10}O_6$

4-(3,4-Dihydroxyphenyl)-5,7-dihydroxy-2H-1-benzopyran-2-one, D-70334
2′,3′,5,7-Tetrahydroxyflavone, T-60106
2′,5,6,6′-Tetrahydroxyflavone, T-70107
3′,4′,5,6-Tetrahydroxyflavone, T-70108
2′,5,6,7-Tetrahydroxyisoflavone, T-60107
2′,5,7,8-Tetrahydroxyisoflavone, T-70110
3′,4′,5,7-Tetrahydroxyisoflavone, T-60108
3′,4′,7,8-Tetrahydroxyisoflavone, T-70112
4′,6,7,8-Tetrahydroxyisoflavone, T-70113
1,3,4,5-Tetrahydroxy-2-methylanthraquinone, T-60112
1,3,5,8-Tetrahydroxy-2-methylanthraquinone, T-60113

$C_{15}H_{10}O_7$

2′,3,5,7,8-Pentahydroxyflavone, P-70029
2′,3′,4′,5,6-Pentahydroxyflavone, P-70030
3,4′,5,6,7-Pentahydroxyflavone, P-60047
3,4′,5,6,8-Pentahydroxyflavone, P-60048
3,5,6,7,8-Pentahydroxyflavone, P-70031
3′,4′,5,6,7-Pentahydroxyisoflavone, P-70032

$C_{15}H_{10}O_8$

2′,3,3′,5,7,8-Hexahydroxyflavone, H-70079
2′,3,5,5′,6,7-Hexahydroxyflavone, H-60064
2′,3,5,5′,7,8-Hexahydroxyflavone, H-70080
3′,4′,5,5′,6,7-Hexahydroxyflavone, H-70081

$C_{15}H_{10}O_9$

3,3′,4′,5,5′,7,8-Heptahydroxyflavone, H-70021
3,3′,4′,5,6,7,8-Heptahydroxyflavone, H-60024

$C_{15}H_{10}O_{10}$

2′,3,4′,5,5′,6,7,8-Octahydroxyflavone, O-60022
3,3′,4′,5,5′,6,7,8-Octahydroxyflavone, O-70018

$C_{15}H_{11}^{\oplus}$

Dibenzo[a,d]cycloheptenylium, D-60071

$C_{15}H_{11}ClN_2O_2$

2-Chloro-2,3-dihydro-1H-imidazole-4,5-dione; 1,3-Di-Ph, in C-60054

$C_{15}H_{11}ClO_4$

Dibenzo[a,d]cycloheptenylium; Perchlorate, in D-60071

$C_{15}H_{11}ClO_5$

2-Chloro-1,8-dihydroxy-5-methoxy-6-methylxanthone, in T-70181

$C_{15}H_{11}NO$

6-Benzoylindole, B-60058
8-Hydroxy-7-phenylquinoline, H-60218
2-(Phenylethynyl)benzoic acid; Amide, in P-70074
2-Phenyl-4(1H)-quinolinone, P-60132
4-Phenyl-2(1H)-quinolinone, P-60131

$C_{15}H_{11}NOS$

▷2-Phenyl-1,5-benzothiazepin-4(5H)-one, P-60081

$C_{15}H_{11}NO_2$

▷9-Acridinecarboxylic acid; Me ester, in A-70061
10-Amino-9-anthracenecarboxylic acid, A-70088
Dibenz[b,g]azocine-5,7(6H,12H)-dione, D-60068
2,4-Diphenyl-5(4H)-oxazolone, D-60485

$C_{15}H_{11}NO_3$

▷1-Amino-2-methoxyanthraquinone, in A-60174
▷1-Amino-4-methoxyanthraquinone, in A-60175
1-Amino-5-methoxyanthraquinone, in A-60176

9-Amino-10-methoxy-1,4-anthraquinone, in A-60181
1-Hydroxy-4-(methylamino)anthraquinone, in A-60175

$C_{15}H_{11}NO_3S$

10H-Phenothiazine-2-carboxylic acid; N-Ac, in P-70056

$C_{15}H_{11}NS$

10H-[1]Benzothieno[3,2-b]indole; N-Me, in B-70059
2-(2-Phenylethenyl)benzothiazole, P-60093

$C_{15}H_{11}N_3$

1-Phenyl-1H-pyrazolo[3,4-e]indolizine, P-70092

$C_{15}H_{11}O_2^{\oplus}$

2,5-Diphenyl-1,3-dioxol-1-ium, D-70478

$C_{15}H_{12}$

3-Methylphenanthrene, M-70108

$C_{15}H_{12}I_3NO_4$

3,3′,5-Triiodothyronine, T-60350

$C_{15}H_{12}N_2O$

5-Amino-3-phenylisoxazole; N-Ph, in A-60240

$C_{15}H_{12}N_2O_3$

2,3-Dihydro-2-hydroxy-1H-imidazole-4,5-dione; 1,3-Di-Ph, in D-60243

$C_{15}H_{12}O$

5,7-Dihydro-6H-dibenzo[a,c]cyclohepten-6-one, D-70187
6,7-Dihydro-5H-dibenzo[a,c]cyclohepten-5-one, D-70188
10,11-Dihydro-5H-dibenzo[a,d]cyclohepten-5-one, D-60218

$C_{15}H_{12}OS$

2,3-Dihydro-2-phenyl-4H-1-benzothiopyran-4-one, D-70248
1,2-Dimethylthioxanthone, D-70440
1,3-Dimethylthioxanthone, D-70441
1,4-Dimethylthioxanthone, D-70442
2,3-Dimethylthioxanthone, D-70443
2,4-Dimethylthioxanthone, D-70444
3,4-Dimethylthioxanthone, D-70445
3-Hydroxymethyldibenzo[b,f]thiepin, H-60176

$C_{15}H_{12}O_2$

2,7-Dimethylxanthone, D-60453
(4-Hydroxyphenyl)ethylene; Benzoyl, in H-70209

$C_{15}H_{12}O_2S$

2,3-Dihydro-2-phenyl-4H-1-benzothiopyran-4-one; 1-Oxide, in D-70248

$C_{15}H_{12}O_3$

3,4-Dihydro-3-(3-hydroxyphenyl)-1H-2-benzopyran-1-one, in H-70207
2,3-Dihydro-2-phenyl-2-benzofurancarboxylic acid, D-60262
4-Methoxy-2,5-phenanthrenediol, in P-60071
Naphtho[1,8-bc]pyran-2-carboxylic acid; Et ester, in N-70010

$C_{15}H_{12}O_3S$

2,3-Dihydro-2-phenyl-4H-1-benzothiopyran-4-one; 1,1-Dioxide, in D-70248
1,4-Dimethylthioxanthone; Dioxide, in D-70442
2,4-Dimethylthioxanthone; Dioxide, in D-70444
3-Hydroxymethyldibenzo[b,f]thiepin; 5,5-Dioxide, in H-60176

$C_{15}H_{12}O_4$

3,9-Dihydroxypterocarpan, D-60373

$C_{15}H_{12}O_5$

3,5-Dihydroxy-2,4-dimethoxy-9H-fluoren-9-one, in T-70109
1,6-Dihydroxy-5-methoxyxanthone, in T-60349
1-Hydroxy-5,6-dimethoxyxanthone, in T-60349
3-(4-Hydroxyphenyl)-1-(2,4,6-trihydroxyphenyl)-2-propen-1-one, H-60220

Protosappanin A, P-60186
5,7,8-Trihydroxyflavanone, T-60321
3′,4′,7-Trihydroxyisoflavanone, T-70263

$C_{15}H_{12}O_6$

1,3-Dihydroxy-5,8-dimethoxyxanthone, in T-70123
1,8-Dihydroxy-3,5-dimethoxyxanthone, in T-70123
1,8-Dihydroxy-3,6-dimethoxyxanthone, in T-60127
3,5-Dihydroxy-2-methyl-1,4-naphthoquinone; Di-Ac, in D-60350
3-(3,4-Dihydroxyphenyl)-1-(2,4,5-trihydroxyphenyl)-2-propen-1-one, D-70339
2′,5,7,8-Tetrahydroxyflavanone, T-70104
3,5,6,7-Tetrahydroxyflavanone, T-60105
3′,4′,5,7-Tetrahydroxyflavanone, T-70105
3′,4′,5,7-Tetrahydroxyflavanone, T-70106

$C_{15}H_{12}O_7$

Nectriafurone, N-60016
2′,3,5,7,8-Pentahydroxyflavanone, P-60045
▷3,3′,4′,5,7-Pentahydroxyflavanone, P-70027
3′,4′,5,6,7-Pentahydroxyflavanone, P-70028
1,2,5,6,8-Pentahydroxyxanthone, P-60052
1,3,8-Trihydroxy-2,6-dimethoxyxanthone, in P-70036
1,5,8-Trihydroxy-2,6-dimethoxyxanthone, in P-60052
Ventilone C, V-60004
Ventilone D, in V-60004

$C_{15}H_{12}O_8$

Fusarubinoic acid, F-70058

$C_{15}H_{12}O_9$

1-(2,3,4,5-Tetrahydroxyphenyl)-3-(2,4,5-trihydroxyphenyl)-1,3-propanedione, T-70120

$C_{15}H_{13}Br$

1-Bromo-1,2-diphenylpropene, B-60237
3-Bromo-1,2-diphenylpropene, B-70209

$C_{15}H_{13}ClO$

2,2-Diphenylpropanoic acid; Chloride, in D-70493

$C_{15}H_{13}N$

1-Cyano-1,1-diphenylethane, in D-70493

$C_{15}H_{13}NO$

N-(Diphenylmethylene)acetamide, in D-60482
6-Hydroxyindole; Benzyl ether, in H-60158
Phthalimidine; N-Benzyl, in P-70100

$C_{15}H_{13}NOS$

2-Phenyl-1,5-benzothiazepin-4(5H)-one; 2,3-Dihydro, in P-60081

$C_{15}H_{13}NO_2$

3,4-Diphenyl-2-oxazolidinone, in P-60126

$C_{15}H_{13}NO_2S$

10H-Phenothiazine-1-carboxylic acid; N-Me, Me ester, in P-70055
10H-Phenothiazine-1-carboxylic acid; N-Et, in P-70055
10H-Phenothiazine-2-carboxylic acid; Et ester, in P-70056
10H-Phenothiazine-2-carboxylic acid; N-Me, Me ester, in P-70056
10H-Phenothiazine-3-carboxylic acid; N-Me, Me ester, in P-70057
10H-Phenothiazine-3-carboxylic acid; N-Et, in P-70057
10H-Phenothiazine-4-carboxylic acid; N-Me, Me ester, in P-70058
10H-Phenothiazine-4-carboxylic acid; N-Et, in P-70058

$C_{15}H_{13}NO_3$

4-Hydroxy-3-methoxy-2-methyl-9H-carbazole-1-carboxaldehyde, H-70154

$C_{15}H_{13}NO_4$

Acetaminosalol, in H-70108

$C_{15}H_{13}NO_4S$

10H-Phenothiazine-3-carboxylic acid; N-Me, Me ester, 5,5-dioxide, in P-70057

C₁₅H₁₃NO₄S

10*H*-Phenothiazine-3-carboxylic acid; *N*-Et,
5,5-dioxide, *in* P-70057

10*H*-Phenothiazine-4-carboxylic acid; *N*-Me,
5,5-dioxide, Me ester, *in* P-70058

C₁₅H₁₃NO₅

Dianthramide *A*, D-70049

Dianthramide B, *in* D-70049

1,2,4-Trimethoxy-3*H*-phenoxazin-3-one,
T-60353

C₁₅H₁₃NS₂

2*H*-1,4-Benzothiazine-3(4*H*)-thione; 4-Benzyl,
in B-70055

C₁₅H₁₄

6,7-Dihydro-5*H*-dibenzo[*a,c*]cycloheptene,
D-70186

10,11-Dihydro-5*H*-dibenzo[*a,d*]cycloheptene,
D-60217

1,1-Diphenylcyclopropane, D-60478

1,1-Diphenylpropene, D-60491

C₁₅H₁₄ClN

4-Chloro-2,3-dihydro-1*H*-indole; 1-Benzyl, *in*
C-60055

C₁₅H₁₄N₂

10,11-Dihydro-5*H*-dibenzo[*a,d*]cyclohepten-5-
one; Hydrazone, *in* D-60218

C₁₅H₁₄N₂O

N-(7,7-Dimethyl-3-azabicyclo[4.1.0]hepta-
1,3,5-trien-4-yl)benzamide, D-70368

C₁₅H₁₄N₂OS

10*H*-Phenothiazine-2-carboxylic acid;
Dimethylamide, *in* P-70056

C₁₅H₁₄N₂O₂

N,N'-Diphenylurea; Mono-*N*-Ac, *in* D-70499

C₁₅H₁₄N₂S

2-Amino-4*H*-3,1-benzothiazine; *N*-Benzyl, *in*
A-70092

C₁₅H₁₄N₄

4-Amino-1,2,3-benzotriazine; 2-Et, *N*(4)-Ph, *in*
A-60092

4-Amino-1,2,3-benzotriazine; 3-Et, *N*(4)-Ph, *in*
A-60092

C₁₅H₁₄O

Di(2,4,6-cycloheptatrien-1-yl)ethanone,
D-60166

Linderazulene, L-70031

C₁₅H₁₄OS

Thia[2.2]metacyclophane; *S*-Oxide, *in* T-60194

C₁₅H₁₄O₂

2,2-Diphenylpropanoic acid, D-70493

2-Hydroxy-1,2-diphenyl-1-propanone, H-70123

2-Hydroxy-1,3-diphenyl-1-propanone, H-70124

C₁₅H₁₄O₂S

2-Mercapto-2-phenylacetic acid; *S*-Benzyl, *in*
M-60024

Thia[2.2]metacyclophane; *S*-Dioxide, *in*
T-60194

C₁₅H₁₄O₂S₂

3,3-Biss(henylthio)propanoic acid, B-70157

C₁₅H₁₄O₃

4-(Benzyloxy)phenol; Ac, *in* B-60074

Cyclolongipesin, C-60214

Dehydroosthol, D-60020

cis-Dehydroosthol, *in* D-60020

4'-Hydroxy-2-biphenylcarboxylic acid; Me
ether, Me ester, *in* H-70111

C₁₅H₁₄O₄

Allopteroxylin, A-60075

Helicquinone, H-60015

2-Methyl-1,3-naphthalenediol; Di-Ac, *in*
M-60091

4-Methyl-1,3-naphthalenediol; Di-Ac, *in*
M-60094

Murralongin, M-70159

Rutalpinin, R-70015

3',4',7-Trihydroxyflavan, T-70260

Yangonin, *in* H-70122

C₁₅H₁₄O₅

Ageratone, A-60065

Altholactone; Ac, *in* A-60083

Anhydroscandenolide, A-70211

3,6-Dihydroxy-1*a*-(3-methyl-2-butenyl)-
naphth[2,3-*b*],2,7(1*aH*,7*aH*)-dione, *in*
D-70277

6-[2-(4-Hydroxy-3-methoxyphenyl)ethenyl]-4-
methoxy-2*H*-pyran-2-one, *in* H-70122

3-(4-Hydroxyphenyl)-1-(2,4,6-
trihydroxyphenyl)-1-propanone, H-60219

Panial, P-70006

3',4',5,7-Tetrahydroxyflavan, T-60103

3,4',5,7-Tetrahydroxyflavanone, T-60104

C₁₅H₁₄O₆

3,3',4',5,7-Pentahydroxyflavan, P-70026

C₁₅H₁₄O₇

Fusarubin, F-60108

2,5,8-Trihydroxy-3-methyl-6,7-methylenedioxy-
1,4-naphthoquinone; Tri-Me ether, *in*
T-60333

C₁₅H₁₄S

Thia[2.2]metacyclophane, T-60194

C₁₅H₁₅N

10,11-Dihydro-5-methyl-5*H*-dibenz[*b,f*]-
azepine, *in* D-70185

5,6,7,8-Tetrahydrodibenz[*c,e*]azocine, T-70054

5,6,7,12-Tetrahydrodibenz[*b,e*]azocine,
T-70055

5,6,7,12-Tetrahydrodibenz[*b,g*]azocine,
T-70056

5,6,11,12-Tetrahydrodibenz[*b,f*]azocine,
T-70057

1,3,9-Trimethylcarbazole, *in* D-70375

2,4,9-Trimethylcarbazole, *in* D-70382

1,2,3-Trimethyl-9*H*-carbazole, T-60357

1,2,4-Trimethyl-9*H*-carbazole, T-60358

C₁₅H₁₅NO

2-Acetamido-3-methylbiphenyl, *in* A-70164

2-Amino-1,3-diphenyl-1-propanone, A-60149

2,2-Diphenylpropanoic acid; Amide, *in* D-70493

C₁₅H₁₅NO₂

2-Hydroxy-1,3-diphenyl-1-propanone; Oxime,
in H-70124

2-Nitro-1,3-diphenylpropane, N-70048

C₁₅H₁₅NO₄

6-Amino-2-hexenoic acid; *N*-Phthalimido, Me
ester, *in* A-60173

C₁₅H₁₅NO₆

Ascorbigen, A-60309

C₁₅H₁₅N₅

5-Amino-1*H*-tetrazole; 1,5(*N*)-Dibenzyl, *in*
A-70202

5-Amino-1*H*-tetrazole; 2,5(*N*)-Dibenzyl, *in*
A-70202

C₁₅H₁₆

7,8,9,10-Tetrahydro-6*H*-cyclohepta[*b*]-
naphthalene, T-70047

C₁₅H₁₆N₂O

▷*N,N'*-Dimethyl-*N,N'*-diphenylurea, *in* D-70499

C₁₅H₁₆N₄

2,5-Diphenyl-1,2,4,5-tetraazabicyclo[2.2.1]-
heptane, D-70498

C₁₅H₁₆N₁₀O₂

Drosopterin, D-60521

Isodrosopterin, *in* D-60521

Neodrosopterin, *in* D-60521

C₁₅H₁₆O

2,3-Dihydrolinderazulene, *in* L-70031

1,3,5,7(11),9-Guaiapentaen-14-al, G-60039

8-Isopropyl-5-methyl-2-
naphthalenecarboxaldehyde, I-60129

4-(2-Phenylethyl)phenol; Me ether, *in* P-70073

C₁₅H₁₆OS₂

3,3-Biss(henylthio)-1-propanol, B-70158

C₁₅H₁₆O₂

10-Desmethyl-1-methyl-1,3,5(10),11(13)-
eudesmatetraen-12,8-olide, D-70026

Furoixiolal, F-70056

C₁₅H₁₆O₃

3-(2,2-Dimethyl-2*H*-1-benzopyran-6-yl)-2-
propenoic acid; Me ester, *in* D-70372

Gnididione, G-60030

Heritol, H-60028

8-Hydroxy-1(6),2,4,7(11)-cadinatetraen-12,8-
olide, H-70114

Hypocretenolide, *in* H-70235

Lumiyomogin, L-70041

2-Oxo-3,10(14),11(13)-guaiatrien-12,6-olide,
O-70090

2,3,4-Trimethoxybiphenyl, *in* B-70124

Yomogin, Y-70001

C₁₅H₁₆O₄

Auraptenol, A-70269

1,2-Bis(4-hydroxyphenyl)-1,2-propanediol,
B-70147

Cedrelopsin, C-60026

1-(2,4-Dihydroxyphenyl)-3-(3,4-
dihydroxyphenyl)propane, D-70336

1α,2α-Epoxyyomogin, *in* Y-70001

Ferreyrantholide, F-70004

8β-Hydroxydehydrozaluzanin C, *in* Z-60001

9-Hydroxy-3-oxo-1,4(15),11(13)-eudesmatrien-
12,6-olide, H-60202

14-Hydroxyhypocretenolide, *in* H-70235

Isomurralonginol, I-70071

Longipesin, L-60034

Murraol, M-60150

Suberenol, S-70080

C₁₅H₁₆O₅

7,7a-Dihydro-3,6,7-trihydroxy-1*a*-(3-methyl-2-
butenyl)naphth[2,3-*b*]oxiren-2(1*aH*)-one,
D-70277

Peroxyauraptenol, *in* A-70269

C₁₅H₁₆O₇

Dihydrofusarubin A, *in* F-60108

Dihydrofusarubin B, *in* F-60108

C₁₅H₁₆O₈

3,4,5-Trihydroxyphenylacetic acid; Tri-Ac, Me
ester, *in* T-60343

C₁₅H₁₇BrN₂O₃

6-Bromotryptophan; *N*ᵅ-Ac, Et ester, *in*
B-60334

C₁₅H₁₈

1-(2,2-Dimethylpropyl)naphthalene, D-60446

2-(2,2-Dimethylpropyl)naphthalene, D-60447

2,3,4,5,6,7,8,9-Octahydro-1*H*-triindene,
O-60021

C₁₅H₁₈N₂

Aurantioclavine, A-60316

C₁₅H₁₈O

3,4-Dihydro-8-isopropyl-5-methyl-2-
naphthalenecarboxaldehyde, *in* I-60129

Tubipofuran, T-70332

C₁₅H₁₈O₂

2-Oxoverboccidentafuran, *in* V-70005

Stevisamolide, S-70073

Tannunolide *A*, T-60005

Tannunolide B, *in* T-60005

C₁₅H₁₈O₃

6α-Acetoxy-2α-hydroxy-1(10),4,11(13)-
germacratrien-12,8-olide, *in* H-70134

Achalensolide, A-70054

Bullerone, B-60337

11α,13-Dihydroyomogin, *in* Y-70001

3-Epizaluzanin C, *in* Z-60001

7α-Hydroxy-3-deoxyzaluzanin C, *in* Z-60001

1-Hydroxy-3,7(11),8-eudesmatrien-12,8-olide,
H-60140

1-Hydroxy-2,4(15),11(13)-eudesmatrien-12,8-
olide, H-70129

1-Hydroxy-4(15),7(11),8-eudesmatrien-12,8-
olide, H-60141

8-Hydroxy-4(15),11(13),10(14)-guaiatrien-
12,6-olide, H-70138

Inunal, I-70022
Isoinunal, *in* I-70022
3-Oxo-4,11(13)-eudesmadien-12,8β-olide, *in*
    H-60136
9-Oxo-4,11(13)-eudesmadien-12,16β-olide, *in*
    H-60137
Zaluzanin C, Z-60001
Zederone, Z-70003

$C_{15}H_{18}O_4$

1,2;3,4-Bisepoxy-11(13)-eudesmen-12,8-olide,
    B-70142
Carmenin, *in* F-70006
Cooperin, C-70196
Dihydrosuberenol, *in* S-70080
1,5-Dihydroxy-2,4(15),11(13)-eudesmatrien-
    12,8-olide, D-70298
2,8-Dihydroxy-3,10(14),11(13)-guaiatrien-
    12,6-olide, D-60327
1α,2α-Epoxy-4β,5-dihydroyomogin, *in* Y-70001
1β,10β-Epoxy-8α-hydroxyachillin, *in* H-60100
1,2-Epoxy-3-hydroxy-4,11(13)-eudesmadien-
    12,8-olide, E-70019
Ferulidin, F-70006
8-Hydroxyachillin, H-60100
14-Hydroxy-11,13-dihydrohypocretenolide, *in*
    H-70235
4-Hydroxy-1-oxo-2,11(13)-eudesmadien-12,8-
    olide, H-70198
8-Hydroxy-3-oxo-1,4,11(13)-eudesmatrien-12-
    oic acid, H-70199
Hypocretenoic acid, H-70235
Pentalenolactone E, P-60055

$C_{15}H_{18}O_5$

Artelin, A-60304
1,5-Dihydroxy-2-oxo-3,11(13)-eudesmadien-
    12,8-olide, D-70329
2,9-Dihydroxy-8-oxo-1(10),4,11(13)-
    germacratrien-12,6-olide, D-60366
2,3-Epoxy-1,4-dihydroxy-7(11),8-eudesmadien-
    12,8-olide, E-60012

$C_{15}H_{18}O_6$

Araneophthalide, A-60292
Chilenone B, C-60034
Obtsusinin, *in* H-70153

$C_{15}H_{18}O_8$

Neoliacinic acid, N-70026

$C_{15}H_{18}O_9$

5,7-Dihydroxy-1(3H)-isobenzofuranone; 5-Me
    ether, glucoside, *in* D-60339

$C_{15}H_{19}Br$

1-[3-(Bromomethyl)-1-methyl-2-methyl-
    enecyclopentyl]-4-methylbenzene, *in* B-70247

$C_{15}H_{19}ClO_5$

Andalucin, A-70207

$C_{15}H_{19}N_5O_5$

3,4-Dihydro-4,6,7-trimethyl-3-β-D-
    ribofuranosyl-9H-imidazo[1,2-a]purin-9-one,
    D-60302

$C_{15}H_{20}BrClO_2$

Epoxyisodihydrorhodophytin, E-60024

$C_{15}H_{20}Br_2O_2$

Dehydrochloroprepacifenol, D-60018
(3E)-Isolaureatin, *in* I-70067
Isolaureatin, I-70067

$C_{15}H_{20}O$

Dehydrocurcuphenol, *in* C-70208
Herbacin, H-60027
Verboccidentafuran, V-70005

$C_{15}H_{20}OS$

Thiofurodysin, T-70178
Thiofurodysinin, T-70179

$C_{15}H_{20}O_2$

Aristolactone, A-60296
Cantabradienic acid, C-60009
6-Deoxyilludin M, *in* I-60001
Elemasteiractinolide, E-70005
8,12-Epoxy-3,7,11-eudesmatrien-1-ol, E-60020
1(10),4,11(13)-Germacratrien-8,12-olide,
    G-70010

Glechomanolide, G-60022
1(10),3-Guaiadien-12,8-olide, G-70032
4,11(13)-Guaiadien-12,8β-olide, *in* A-70054
Isomyomontanone, *in* M-60155
Mokkolactone, M-60142
Myomontanone, M-60155
Nardonoxide, N-70018
Pallescensone, P-70003
Pinguisanin, P-60155
Verboccidentafuran 4α,5α-epoxide, *in* V-70005

$C_{15}H_{20}O_3$

Arteannuin B, A-70262
Artesovin, A-60306
Bisabolangelone, B-60138
4,8-Bis-epi-inuviscolide, *in* I-70023
Cantabrenonic acid, *in* C-60010
Cyclodehydromyoporone A, C-60188
Cyclodehydromyoporone B, C-60189
Dehydromyoporone, D-60019
6-Deoxyilludin S, *in* I-60002
11β,13-Dihydrozaluzanin C, *in* Z-60001
11α,13-Dihydrozaluzanin C, *in* Z-60001
Elemasteiractinolide; 15-Hydroxy, *in* E-70005
1-epi-Inuviscolide, *in* I-70023
8-epi-Inuviscolide, *in* I-70023
1,10-Epoxy-4,11(13)-germacradien-12,8-olide,
    E-60022
9,11(13)-Eremophiladien-12-oic acid; 3-Oxo, *in*
    E-70029
3,11(13)-Eudesmadien-12-oic acid; 2-Oxo, *in*
    E-70060
Eumorphistonol, E-60073
Haageanolide, H-60001
Hanphyllin, H-60006
3-Hydroxy-4,11(13)-eudesmadien-12,8-olide,
    H-60136
7-Hydroxy-4,11(13)-eudesmadien-12,6-olide,
    H-70128
9-Hydroxy-4,11(13)-eudesmadien-12,6-olide,
    H-60137
15-Hydroxy-4,11(13)-eudesmadien-12,8-olide,
    H-60138
6-Hydroxy-1(10),4,11(13)-germacratrien-12,8-
    olide, H-70134
3-Hydroxy-4,11(13)-guaiadien-12,8-olide,
    H-70135
15-Hydroxy-3,10(14)-guaiadien-12,8-olide,
    H-70136
5-Hydroxy-3,11(13)-guaiatrien-12,8-olide,
    H-70137
▷Illudin M, I-60001
Inunal; Δ⁴-Isomer, 15-alcohol, *in* I-70022
Inuviscolide, I-70023
Isoperezone, I-70075
Ivangustin, I-70109
Kanshone C, K-70006
3-Oxo-4,11(13)-eudesmadien-12-oic acid, *in*
    E-70061
9-Oxo-4,11(13)-eudesmadien-12-oic acid, *in*
    H-60135
1-Oxo-7(11)-eudesmen-12,8-olide, O-60067
1-Oxo-4,10(14)-germacradien-12,6α-olide, *in*
    O-60068
α-Pipitzol, P-60158
Quadrangolide, Q-60001
Santamarine, S-60007
Subergorgic acid, S-70081

$C_{15}H_{20}O_4$

Artecalin, A-60302
Confertin, *in* D-60211
2,3-Dehydro-11α,13-dihydroconfertin, *in*
    D-60211
11,13-Dehydroeriolin, *in* E-60036
11β,13-Dihydro-9α-hydroxyzaluzanin C, *in*
    Z-60001
1,8-Dihydroxy-3,7(11)-eremophiladien-12,8-
    olide, D-60320
1,5-Dihydroxyeriocephaloide, D-60321
1,8-Dihydroxy-3,7(11)-eudesmadien-12,8-olide,
    D-60322
2,8-Dihydroxy-1(10),4,11(13)-germacratrien-
    12,6-olide, D-70302

3,13-Dihydroxy-1(10),4,7(11)-germacratrien-
    12,6-olide, D-70303
8,9-Dihydroxy-1(10),4,11(13)-germacratrien-
    12,6-olide, D-60323
9,15-Dihydroxy-1(10),4,11(13)-germacratrien-
    12,6-olide, D-60324
2,4-Dihydroxy-5,11(13)-guaiadien-12,8-olide,
    D-70304
4,10-Dihydroxy-2,11(13)-guaiadien-12,6-olide,
    D-60326
3,5-Dihydroxy-4(15),10(14)-guaiadien-12,8-
    olide, D-60325
3β,9β-Dihydroxy-4(15),10(14)-guaiadien-
    12,6α-olide, *in* D-70305
9,10-Dihydroxy-7-marasmen-5,13-olide,
    D-60346
4,5-Dioxo-1(10)-xanthen-12,8-olide, D-60474
4-Epivulgarin, *in* V-60017
Epoxycantabronic acid, *in* C-60010
1β,10α-Epoxyhaageanolide, *in* H-60001
4α,5α-Epoxy-3α-hydroxy-11(13)-eudesmen-
    12,8β-olide, *in* H-60136
4α,5α-Epoxy-3β-hydroxy-11(13)-eudesmen-
    12,8β-olide, *in* H-60136
3α-Hydroperoxy-4,11(13)-eudesmadien-12,8β-
    olide, *in* H-60136
8α-Hydroxybalchanin, *in* S-60007
8α-Hydroxy-11α,13-dihydrozaluzanin C, *in*
    Z-60001
6α-Hydroxy-8-epiivangustin, *in* I-70109
2α-Hydroxyivangustin, *in* I-70109
1-[2-(Hydroxymethyl)-3-methoxyphenyl]-1,5-
    heptadiene-3,4-diol, H-70178
9β-Hydroxy-3-oxo-10(14)-guaien-12,6α-olide,
    in D-70305
9α-Hydroxysantamarine, *in* S-60007
▷Illudin S, I-60002
Ivangulic acid, I-70108
Ridentin, R-70007
Schkuhridin B, S-60012
8,9,14-Trihydroxy-1(10),4,11(13)-
    germacratrien-12,6-olide, T-60323
Umbellifolide, U-60001
▷Vulgarin, V-60017

$C_{15}H_{20}O_5$

Altamisic acid, A-70084
▷Coriolin, C-70199
Dehydroisoerivanin, *in* I-70066
3,5-Dihydroxy-4(15),10(14)-guaiadien-12,8-
    olide; 10α,14-Epoxide, *in* D-60325
4,5-Epoxy-2,8-dihydroxy-1(10),11(13)-
    germacradien-12,6-olide, E-60013
4α-Hydroperoxydesoxyvulgarin, *in* V-60017
7'-Hydroxyabscisic acid, H-70098
8'-Hydroxyabscisic acid, H-70099
Psilostachyin, P-70131
Secoisoerivanin pseudoacid, S-70024
1β,3α,6β-Trihydroxy-4,11(13)-eudesmadien-
    12,8β-olide, *in* I-70109
1,3,9-Trihydroxy-4(15),11(13)-eudesmadien-
    12,6-olide, T-70257
1,3,13-Trihydroxy-4(15),7(11)-eudesmadien-
    12,6-olide, T-70258

$C_{15}H_{20}O_6$

Ajafinin, A-60068
2,3-Epoxy-1,4,8-trihydroxy-7(11)-eudesmen-
    12,8-olide, E-60033

$C_{15}H_{20}O_7$

Domesticoside, *in* T-70248
Neomajucin, *in* M-70005

$C_{15}H_{20}O_8$

Majucin, M-70005
Laurencia Polyketal, P-60168

$C_{15}H_{20}O_{10}$

Quinic acid; Tetra-Ac, *in* Q-70005

$C_{15}H_{21}Br$

1-[3-(Bromomethyl)-1-methyl-2-methyl-
    enecyclopentyl]-4-methyl-1,4-cyclohexadiene,
    B-70247

$C_{15}H_{21}BrO$

Isoobtusadiene, I-70074
Obtusadiene, O-70001

C₁₅H₂₁Br₃O₂

2-[[3-Bromo-5-(1-bromopropyl)tetrahydro-2-
furanyl]methyl]-5-(1-bromo-2-propynyl)-
tetrahydro-3-furanol, B-70195

C₁₅H₂₁ClO

Laurencenone B, L-70025

C₁₅H₂₁NO

Octahydro-2H-1,3-benzoxazine; N-Benzyl, in
O-70011
Octahydro-2H-3,1-benzoxazine; N-Benzyl, in
O-70012

C₁₅H₂₁NO₄

3,5-Pyridinedipropanoic acid; Di-Et ester, in
P-60217

C₁₅H₂₂

1(5),3-Aromadendradiene, A-70256
1(10),4-Aromadendradiene, A-70257

C₁₅H₂₂BrClO

Elatol, E-70002
Laurencenone A, L-70024
Laurencenone D, L-70026

C₁₅H₂₂Br₂O₃

α-(1-Bromo-3-hexenyl)-5-(1-bromo-2-
propynyl)tetrahydro-3-hydroxy-2-
furanethanol, B-70217

C₁₅H₂₂N₂O₄

N-(5a,6,7,8-Tetrahydro-4-methyl-1,5-dioxo-
1H,5H-pyrrolo[1,2-c][1,3]oxazepin-3-yl)-
hexanamide, in M-70134

C₁₅H₂₂O

1(10)-Aristolen-9-one, A-70252
Curcuphenol, C-70208
Cyperenal, in C-70259
Elvirol, E-60004
Grimaldone, G-70030
Isobicyclogermacrenal, I-60088
Laurencenone C, in L-70025
Nuciferol, N-70082
Penlanpallescensin, P-60015
Perforenone, P-60063
Premnaspiral, in P-70114
12-Sesquisabinenal, in S-70037
3-Silphinenone, S-60033
Solavetivone, S-70051
Tochuinol, T-70202
Xanthorrhizol, X-70001

C₁₅H₂₂O₂

Aglycone A₃, in S-70051
Apotrisporin C, in A-70247
4,11(13)-Cadinadien-12-oic acid, C-70004
Capsenone, C-60011
Cyparenoic acid, C-70258
Drimenin, D-60519
4,5-Epoxy-1(10),7(11)-germacradien-8-one,
E-60023
9,11(13)-Eremophiladien-12-oic acid, E-70029
▷3,11(13)-Eudesmadien-12-oic acid, E-70060
4,11(13)-Eudesmadien-12-oic acid, E-70061
4-Guaien-12,8β-olide, in A-70054
Helminthosporal, H-60016
6-Hydroxy-2,7,10-bisabolatrien-9-one, H-60107
6-Hydroxy-1,4-eudesmadien-3-one, H-60139
Kanshone A, K-70004
10-Oxo-6-isodaucen-14-al, O-70093
Oxysolavetivone, in S-70051
9-Phenylnonanoic acid, P-70084

C₁₅H₂₂O₃

Apotrisporin E, A-70247
Artausin, A-70261
Artedouglasiaoxide, A-70263
Bedfordiolide, B-60008
Brassicadiol, B-70177
Cantabrenolic acid, C-60010
Dihydroconfertin, D-60211
11β,13-Dihydroconfertin, in D-60211
11β,13-Dihydro-8-epi-confertin, in D-60211
11β,13-Dihydro-1-epi-inuviscolide, in I-70023

11β,13-Dihydroinuviscolide, in I-70023
11α,13-Dihydroinuviscolide, in I-70023
Dihydroparthenolide, D-70239
12,13-Dihydroxy-1(10)-aromadendren-2-one,
D-70283
3,8-Dihydroxylactariusfuran, D-70312
4,8-Dihydroxylactariusfuran, D-70313
7,13-Dihydroxy-2-longipinen-1-one, D-70315
8β,9α-Dihydroxy-4,10(14)-oplopadien-3-one, in
O-70042
3-(1,3,5,7,9-Dodecapentaenyloxy)-1,2-
propanediol, D-70536
Elemasteiractinolide; 11β,13-Dihydro, 15-
hydroxy, in E-70005
3,4-Epoxy-11(13)-eudesmen-12-oic acid,
E-60021
12,13-Epoxytrichothec-9-en-3-ol, E-70028
9,11(13)-Eremophiladien-12-oic acid; 2β-
Hydroxy, in E-70029
9,11(13)-Eremophiladien-12-oic acid; 3β-
Hydroxy, in E-70029
3,11(13)-Eudesmadien-12-oic acid; 2α-
Hydroxy, in E-70060
4,11(13)-Eudesmadien-12-oic acid; 3β-
Hydroxy, in E-70061
Furanoeremophilane-2,9-diol, F-70047
3β-Hydroxy-11β,13-
dihydrodesoxoachalensolide, in H-70135
9-Hydroxy-4,11(13)-eudesmadien-12-oic acid,
H-60135
4-Hydroxy-11(13)-eudesmen-12,8-olide,
H-60142
Ketosantalic acid, K-70012
Myoporone, M-60156
1-Oxo-4-germacren-12,6-olide, O-60068
7-Oxo-11-nordrim-8-en-12-oic acid; Me ester,
in O-60080
Pentalenic acid, P-60054
Sambucoin, S-70007
Santamarine; 11β,13-Dihydro, in S-60007
Shonachalin D, S-70040
Sterpuric acid, S-60053
Warburganal, W-60001

C₁₅H₂₂O₄

Artapshin, A-60301
Avenaciolide, A-60325
Desacylligulatin C, D-70025
Dihydroridentin, in R-70007
1β,3α-Dihydroxy-4-eudesmen-12,6α-olide, in
I-70066
1,3-Dihydroxy-4-eudesmen-12,6-olide, D-70299
8,10-Dihydroxy-3-guaien-12,6-olide, D-60328
3,9-Dihydroxy-10(14)-guaien-12,6-olide,
D-70305
11-Epidihydroridentin, in R-70007
1-Epierivanin, in E-70033
4-Epineopulchellin, in N-70028
Eriolin, E-60036
Erivanin, E-70033
Gallicadiol, G-70001
11β-Hydroxy-11,13-dihydro-8-epi-confertin, in
D-60211
2α-Hydroxydihydroparthenolide, in D-70239
9α-Hydroxydihydroparthenolide, in D-70239
9α-Hydroxy-11β,13-dihyrosantamarine, in
S-60007
8α-Hydroxysambucoin, in S-70007
8β-Hydroxysambucoin, in S-70007
13-Hydroxysterpuric acid, in S-60053
Isoerivanin, I-70066
Kanshone B, K-70005
Lactarorufin D, L-70016
Lactarorufin E, in L-70016
Neopulchellin, N-70028
Pulchellin A, P-70138
4,5-Seconeopulchell-5-ene, S-70025
Sporol, S-70069
Sydonic acid, S-60065
Zafronic acid, Z-70001

C₁₅H₂₂O₅

11,13-Dihydropsilostachyin, in P-70131
2α-Hydroxy-11α,13-dihydroconfertin, in
D-60211

6α-Hydroxypulchellin, in P-70138
4-Oxo-3,4-secoambrosan-12,6-olid-3-oic acid,
O-60093
1,4,6-Trihydroxy-11(13)-eudesmen-12,8-olide,
T-70259

C₁₅H₂₂O₉

6-Deoxycatalpol, in C-70026
Eranthemoside, E-60034
Galiridoside, G-60002

C₁₅H₂₂O₁₀

Catalpol, C-70026
6-Epimonomelittoside, in M-70019
Monomelittoside, in M-70019

C₁₅H₂₃ClO

4-Chloro-9R-hydroxy-3,7(14)-chamigradiene,
in E-70002

C₁₅H₂₃ClO₂

2-Chloro-3,7-epoxy-9-chamigranone, C-60066
2-Chloro-3-hydroxy-7-chamigren-9-one,
C-60094

C₁₅H₂₃NO

2-Octylamine; N-Benzoyl, in O-60031

C₁₅H₂₄

Agarospirene, in P-70114
1-Aromadendrene, A-70258
9-Aromadendrene, A-70259
4(15),5-Cadinadiene, C-70003
ε-Cadinene, C-60001
α-Copaene, C-70197
Daucene, D-70008
1,3,7(11)-Elematriene, E-70006
1-Epi-α-gurjunene, in G-70034
1(5),6-Guaiadiene, G-70031
α-Gurjunene, G-70034
1,1,3,3,5,5-Hexamethyl-2,4,6-triss(ethylene)-
cyclohexane, H-60072
1,2,3,4,4a,5,6,7-Octahydro-4a,5-dimethyl-2-(1-
methylethenyl)naphthalene, O-60015
β-Patchoulene, P-60010
Pentalenene, P-70037
Precapnelladiene, P-60171
Premnaspirodiene, P-70114
Sesquisabinene, S-70037
Sesquithujene, in S-70037
6-Siliphiperfolene, S-70043
Sinularene, S-60034

C₁₅H₂₄O

1(5)-Aromadendren-7-ol, A-70260
α-Biotol, B-60112
β-Biotol, B-60113
α-Copaene, C-70197
α-Copaen-8-ol, in C-70197
α-Copaen-11-ol, in C-70197
15-Copaenol, in C-70197
Cyperenol, C-70259
2,3-Epoxy-7,10-bisaboladiene, E-60009
3,5-Eudesmadien-1-ol, E-60064
5,7-Eudesmadien-11-ol, E-60065
5,7(11)-Eudesmadien-15-ol, E-60066
Fervanol, F-60007
Gymnomitrol, G-70035
Sesquisabinene hydrate, in S-70037
α-Torosol, T-60225
β-Torosol, T-60226
Valerenol, V-70001

C₁₅H₂₄O₂

10(14)-Aromadendrene-4,8-diol, A-60299
Artemone, A-60305
Curcudiol, C-70207
cis-9,10-Dihydrocapsenone, in C-60011
10,11-Dihydro-11-hydroxycurcuphenol, in
C-70208
7,12-Dihydroxysterpurene, D-70346
9,12-Dihydroxysterpurene, D-70347
4,7(11)-Eudesmadiene-12,13-diol, E-60063
11-Hydroxy-4-guaien-3-one, H-60149
10α-Hydroxy-6-isodaucen-14-al, in O-70093
11-Hydroxy-1(10)-valencen-2-one, H-60234
Kurubasch aldehyde, K-60014
6-Oxocyclonerolidol, in H-60111
Tanavulgarol, T-60004
3,7,11-Trimethyl-2,6,10-dodecatrienoic acid,
T-60364

Vetidiol, V-**60007**

$C_{15}H_{24}O_3$
Dihydromyoporone, in M-**60156**
3,11-Dihydroxy-7-drimen-6-one, D-**70295**
8α,9α-Dihydroxy-10βH-eremophil-11-en-2-one,
  in E-**70030**
1,4-Dimethyl-4-[4-hydroxy-2-(hydroxymethyl)-
  1-methyl-2-cyclopentenyl]-2-cyclohexen-1-ol,
  D-**70417**
Feruone, F-**70007**
FS-2, F-**60078**
Lapidol, L-**60014**
4,10(14)-Oplopadiene-3,8,9-triol, O-**70042**
Punctaporonin A, P-**60197**
Punctaporonin D, in P-**60197**
Urodiolenone, U-**70006**
Zedoarondiol, Z-**70004**

$C_{15}H_{24}O_4$
4-Epipulchellin, in P-**70138**
Secofloribundione, S-**70022**
Trichotriol, T-**60258**
5,8,9-Trihydroxy-10(14)-oplopen-3-one,
  T-**70272**
Vestic acid, V-**70008**

$C_{15}H_{24}O_5$
Rudbeckin A, R-**70013**

$C_{15}H_{24}O_{11}$
Avicennioside, A-**60326**

$C_{15}H_{25}BrO$
Brasudol, B-**60197**
Isobrasudol, in B-**60197**

$C_{15}H_{25}NS$
6-Isothiocyano-4(15)-eudesmene, in F-**60069**

$C_{15}H_{26}$
Cyclopentadecyne, C-**60220**

$C_{15}H_{26}O$
Cerapicol, C-**70032**
Ceratopicanol, C-**70033**
9(11)-Drimen-8-ol, D-**60520**
1(10)-Eremophilen-7-ol, E-**70031**
3-Eudesmen-6-ol, E-**70065**
11-Eudesmen-5-ol, E-**60069**
4(15)-Eudesmen-11-ol, E-**70066**
Panasinsanol A, P-**70005**
Panasinsanol B, in P-**70005**
2,4-Pentadecadienal, P-**70024**

$C_{15}H_{26}O_2$
Alloaromadendrane-4,10-diol, A-**60074**
Debneyol, D-**70010**
Decahydro-5,5,8a-trimethyl-2-naphthalenol;
  Ac, in D-**70014**
7-Epidebneyol, in D-**70010**
1,4-Epoxy-6-eudesmanol, E-**60019**
11-Eudesmene-1,5-diol, E-**60067**
4(15)-Eudesmene-1,11-diol, E-**70064**
Fauronol, F-**70002**
1(10),4-Germacradiene-6,8-diol, G-**60014**
5,10(14)-Germacradiene-1,4-diol, G-**70009**
6-Hydroxycyclonerolidol, H-**60111**
Jaeschkeanadiol, J-**70001**
Oplopanone, O-**70043**
2,6,10-Trimethyl-2,6,11-dodecatriene-1,10-diol,
  T-**70283**
3,7,11-Trimethyl-1,6,10-dodecatriene-3,9-diol,
  T-**70284**

$C_{15}H_{26}O_3$
8-Carotene-4,6,10-triol, C-**60021**
8-Daucene-3,6,14-triol, D-**70011**
8(14)-Daucene-4,6,9-triol, D-**60005**
6-Drimene-8,9,11-triol, D-**70547**
7-Drimene-6,9,11-triol, D-**70548**
11-Eremophilene-2,8,9-triol, E-**70030**
4(15)-Eudesmene-1,5,6-triol, E-**60068**
Fexerol, F-**60009**
5-Hydroperoxy-4(15)-eudesmen-11-ol,
  H-**70097**
1-Hydroxydebneyol, in D-**70010**
8-Hydroxydebneyol, in D-**70010**
Shiromodiol, S-**70039**

$C_{15}H_{26}O_4$
8-Daucene-2,4,6,10-tetrol, D-**70009**

Fercoperol, F-**60001**
3-Hydroxymethyl-7,11-dimethyl-2,6,11-dodeca-
  triene-1,5,10-triol, H-**70174**
2,6,10-Trimethyl-2,6,10-dodecatrien-1,5,8,12-
  tetrol, T-**60363**

$C_{15}H_{26}O_5$
9,12-Dioxododecanoic acid; Me ester, ethylene
  acetal, in D-**60471**
Lapidolinol, L-**60015**

$C_{15}H_{26}O_7$
ent-5α,11-Epoxy-1β,4α,6α,8α,9β,14-
  eudesmanehexol, E-**60018**

$C_{15}H_{27}N$
7-Amino-2,10-bisaboladiene, A-**70096**
7-Amino-3,10-bisaboladiene, A-**60096**
Tri-tert-butylazete, T-**60249**

$C_{15}H_{27}NO$
7-Amino-2,11-biaboladien-10R-ol, in A-**70096**
7-Amino-2,9-bisaboladien-11-ol, in A-**70096**
7-Amino-2,11-bisaboladien-10S-ol, in A-**70096**

$C_{15}H_{27}N_3$
3,5,12-Triazatetracyclo[5.3.1.1²,⁶.0⁴,⁹]dodecane;
  3,5,12-Tri-Et, in T-**70206**
Tri-tert-butyl-1,2,3-triazine, T-**60250**

$C_{15}H_{28}O$
▷Cyclopentadecanone, C-**60219**

$C_{15}H_{28}O_2$
2-Bisabolene-1,12-diol, B-**70126**
5-Decen-1-ol; 3-Methylbutanoyl, in D-**70019**
9-Dodecen-1-ol; Propanoyl, in D-**70537**
1,11-Eudesmanediol, E-**70062**
5,11-Eudesmanediol, E-**70063**

$C_{15}H_{28}O_3$
12-Hydroxy-13-tetradecenoic acid; Me ester, in
  H-**60229**
2,6,10-Trimethyl-6,11-dodecadiene-2,3,10-triol,
  T-**60362**

$C_{15}H_{29}NO$
Cyclopentadecanone; Oxime, in C-**60219**

$C_{15}H_{30}$
3-tert-Butyl-2,2,4,5,5-tetramethyl-3-hexene,
  B-**60350**

$C_{15}H_{30}N_2O_3$
4,7,13-Trioxa-1,10-diazabicyclo[8.5.5]icosane,
  T-**60375**

$C_{15}H_{30}O_2$
Lardolure, in T-**60372**
2,4,6,8-Tetramethylundecanoic acid, T-**60157**

$C_{15}H_{30}O_4$
12-Oxododecanoic acid; Me ester, Di-Me
  acetal, in O-**60066**

$C_{15}H_{30}O_6$
1,3,5-Triss(imethoxymethyl)cyclohexane, in
  C-**70226**

$C_{15}H_{32}O$
3,7,11-Trimethyl-1-dodecanol, T-**70282**

$C_{15}H_{33}NO_6$
2-(2-Methoxyethoxy)-N,N-bis[2-(2-methoxy-
  ethoxy)ethyl]ethanamine, in T-**70323**

$C_{15}H_{33}N_3$
Hexahydro-1,3,5-triazine; 1,3,5-Tri-tert-butyl
  , in H-**60060**

$C_{15}H_{35}N_5$
1,5,9,13,17-Pentaazacycloeicosane, P-**60018**

$C_{16}H_4N_4S_2$
2,2'-(4,8-Dihydrobenzo[1,2-b:5,4-b']-
  dithiophene-4,8-diylidene)bispropanedinitrile,
  D-**60196**

$C_{16}H_8Br_2$
1,8-Dibromopyrene, D-**70099**

$C_{16}H_8Cl_2O_2$
1,8-Anthracenedicarboxylic acid; Dichloride, in
  A-**60281**

$C_{16}H_8Cl_4$
1,4,5,8-Tetrachloro-9,10-
  anthraquinodimethane, T-**60026**

$C_{16}H_8F_8$
Octafluoro[2.2]paracyclophane, O-**70009**
4,5,7,8,12,13,15,16-Octafluoro[2.2]paracyclo-
  phane, O-**60010**

$C_{16}H_8N_2$
1,8-Dicyanoanthracene, in A-**60281**

$C_{16}H_8N_2O_4S$
2,2'-Thiobis-1H-isoindole-1,3(2H)-dione,
  T-**70176**

$C_{16}H_8N_2O_4S_2$
2,2'-Dithiobis-1H-isoindole-1,3(2H)-dione,
  D-**70522**
Dithiobisphthalimide, D-**60508**

$C_{16}H_8O_2$
Cyclohept[fg]acenaphthylene-5,6-dione,
  C-**60194**
Cyclohept[fg]acenaphthylene-5,8-dione,
  C-**60195**

$C_{16}H_8O_3$
Benzo[b]naphtho[2,1-d]furan-5,6-dione,
  B-**60033**
Benzo[b]naphtho[2,3-d]furan-6,11-dione,
  B-**60034**
3,4-Phenanthrenedicarboxylic acid; Anhydride,
  in P-**60068**

$C_{16}H_8S_2$
Fluorantheno[3,4-cd]-1,2-dithiole, F-**60017**

$C_{16}H_8Se_2$
Fluorantheno[3,4-cd]-1,2-diselenole, F-**60015**

$C_{16}H_8Te_2$
Fluorantheno[3,4-cd]-1,2-ditellurole, F-**60016**

$C_{16}H_9ClO_2S$
1-Pyrenesulfonic acid; Chloride, in P-**60210**

$C_{16}H_9NO_2$
▷1-Nitropyrene, N-**60040**
2-Nitropyrene, N-**60041**
4-Nitropyrene, N-**60042**

$C_{16}H_9NO_2S$
3-(2-Benzothiazolyl)-2H-1-benzopyran,
  B-**70056**

$C_{16}H_{10}$
Acephenanthrylene, A-**70018**
Benzo[a]biphenylene, B-**60016**
Benzo[b]biphenylene, B-**60017**
Dicyclopenta[ef,kl]heptalene, D-**60167**

$C_{16}H_{10}F_3NO_2$
7-Amino-4-(trifluoromethyl)-2H-1-benzopyran-
  2-one; N-Ph, in A-**60265**

$C_{16}H_{10}F_6I_2O_5$
μ-Oxodiphenylbiss(rifluoroacetato-O)diiodine,
  O-**60065**

$C_{16}H_{10}N_2$
Quino[7,8-h]quinoline, Q-**70007**

$C_{16}H_{10}N_2O_2$
Quino[7,8-h]quinoline-4,9(1H,12H)dione,
  Q-**60007**

$C_{16}H_{10}N_2S_2$
Thiazolo[4,5-b]quinoline-2(3H)-thione; 3-Ph, in
  T-**60201**

$C_{16}H_{10}N_6$
2,2'-Azodiquinoxaline, A-**60353**

$C_{16}H_{10}N_8$
21H,23H-Porphyrazine, P-**70109**

$C_{16}H_{10}O$
Anthra[1,2-c]furan, A-**70217**
Phenanthro[1,2-c]furan, P-**70050**

Phenanthro[3,4-*c*]furan, P-70051
Phenanthro[9,10-*b*]furan, P-60073
Phenanthro[9,10-*c*]furan, P-60074

$C_{16}H_{10}O_2$
4,5:9,10-Diepoxy-4,5,9,10-tetrahydropyrene,
   D-70152
4*b*,9*a*-Dihydroindeno[1,2-*a*]indene-9,10-dione,
   D-70219
4*b*,9*b*-Dihydroindeno[2,1-*a*]indene-5,10-dione,
   D-70220
3-Phenyl-1,2-naphthoquinone, P-60120
4-Phenyl-1,2-naphthoquinone, P-60121

$C_{16}H_{10}O_3$
5-Formyl-4-phenanthrenecarboxylic acid,
   F-60074

$C_{16}H_{10}O_3S$
5-Hydroxy-2-(phenylthio)-1,4-naphthoquinone,
   H-70213
5-Hydroxy-3-(phenylthio)-1,4-naphthoquinone,
   H-70214
4-Oxo-2-phenyl-4*H*-1-benzothiopyran-3-
   carboxylic acid, O-60087
1-Pyrenesulfonic acid, P-60210
2-Pyrenesulfonic acid, P-60211
4-Pyrenesulfonic acid, P-60212

$C_{16}H_{10}O_4$
1,8-Anthracenedicarboxylic acid, A-60281
4-Oxo-2-phenyl-4*H*-1-benzopyran-3-carboxylic
   acid, O-60085
4-Oxo-3-phenyl-4*H*-1-benzopyran-2-carboxylic
   ·acid, O-60086
3,4-Phenanthrenedicarboxylic acid, P-60068

$C_{16}H_{10}O_5$
ψ-Baptigenin, B-60006
5-Hydroxy-6,7-methylenedioxyflavone, *in*
   T-63322
Mutisifurocoumarin, M-70161

$C_{16}H_{10}O_6$
Bowdichione, B-70175
3,9-Dihydroxy-2-methoxycoumestone, *in*
   T-70249
9,10-Dihydroxy-5-methoxy-2*H*-pyrano[2,3,4-
   kl]xanthen-9-one, *in* T-70273
2′,5-Dihydroxy-6,7-methylenedioxyisoflavone,
   in T-60107

$C_{16}H_{10}O_7$
5-Hydroxybowdichione, *in* B-70175
3,6,8-Trihydroxy-1-methylanthraquinone-2-
   carboxylic acid, T-60332

$C_{16}H_{10}O_8$
Nasutin C, *in* E-70007

$C_{16}H_{11}ClO$
1-Phenanthreneacetic acid; Chloride, *in*
   P-70046

$C_{16}H_{11}F_3O_5S$
2,5-Diphenyl-1,3-dioxol-1-ium;
   Trifluoromethanesulfonate, *in* D-70478

$C_{16}H_{11}N$
1-(Cyanomethyl)phenanthrene, *in* P-70046
9-Cyanomethylphenanthrene, *in* P-70047
2*H*-Dibenz[*e,g*]isoindole, D-70067
Indolo[1,7-*ab*][1]benzazepine, I-60034

$C_{16}H_{11}NO$
1-Benzoylisoquinoline, B-60059
3-Benzoylisoquinoline, B-60060
4-Benzoylisoquinoline, B-60061

$C_{16}H_{11}NO_2$
2-Phenyl-4-(phenylmethylene)-5(4*H*)-
   oxazolone, P-60128

$C_{16}H_{11}NO_3$
1-Amino-8-hydroxyanthraquinone; *N*-Ac, *in*
   A-60177

$C_{16}H_{11}NO_4$
1-Amino-2-hydroxyanthraquinone; *N*-Ac, *in*
   A-60174
1-Amino-5-hydroxyanthraquinone; *N*-Ac, *in*
   A-60176

$C_{16}H_{11}N_3OS$
12*H*-Quinoxalino[2,3-*b*][1,4]benzothiazine; *N*-
   Ac, *in* Q-70008

$C_{16}H_{11}N_3O_2$
12*H*-Quinoxalino[2,3-*b*][1,4]benzoxazine; *N*-
   Ac, *in* Q-70009

$C_{16}H_{12}$
Cyclohepta[*ef*]heptalene, C-60197
Cycloocta[*a*]naphthalene, C-70234
Cycloocta[*b*]naphthalene, C-70235
1,10-Dihydrodicyclopenta[*a,h*]naphthalene,
   D-60220
3,8-Dihydrodicyclopenta[*a,h*]naphthalene,
   D-60221
9,10-Dihydro-9,10-dimethyleneanthracene,
   D-70194
5,10-Dihydroindeno[2,1-*a*]indene, D-70218
[2.2]Paracyclophadiene, P-70007

$C_{16}H_{12}Cl_2N_4O$
4-(2,5-Dichlorophenylhydrazono)-5-methyl-2-
   phenyl-3*H*-pyrazol-3-one, D-70138

$C_{16}H_{12}Cl_2O_5$
2,4-Dichloro-1-hydroxy-5,8-dimethoxy-
   xanthone, *in* T-70181

$C_{16}H_{12}F_2O_4$
4,4′-Difluoro-[1,1′-biphenyl]-3,3′-dicarboxylic
   acid; Di-Me ester, *in* D-60178
4,5-Difluoro-[1,1′-biphenyl]-2,3-dicarboxylic
   acid; Di-Me ester, *in* D-60179
6,6′-Difluoro-[1,1′-biphenyl]-2,2′-dicarboxylic
   acid; Di-Me ester, *in* D-60180

$C_{16}H_{12}F_4$
4,5,7,8-Tetrafluoro[2.2]paracyclophane,
   T-60050

$C_{16}H_{12}N_2$
2,2′-Biss(yanomethyl)biphenyl, *in* B-60115
4,4′-Biss(yanomethyl)biphenyl, *in* B-60116
2,4-Diphenylpyrimidine, D-70494
4,5-Diphenylpyrimidine, D-70495
4,6-Diphenylpyrimidine, D-70496

$C_{16}H_{12}N_2O$
1-Benzoylisoquinoline; Oxime, *in* B-60059
4,5-Diphenylpyrimidine; 1-Oxide, *in* D-70495
4,6-Diphenylpyrimidine; *N*-Oxide, *in* D-70496

$C_{16}H_{12}N_2O_2$
4*b*,9*b*-Dihydroindeno[2,1-*a*]indene-5,10-dione;
   Dioxime, *in* D-70220

$C_{16}H_{12}O$
1-Acetylanthracene, A-70030
2-Acetylanthracene, A-70031
9-Anthraceneacetaldehyde, A-60280
Octaleno[3,4-*c*]furan, O-70019
1-Phenyl-2-naphthol, P-60115
3-Phenyl-2-naphthol, P-60116
4-Phenyl-2-naphthol, P-60117
5-Phenyl-2-naphthol, P-60118
8-Phenyl-2-naphthol, P-60119

$C_{16}H_{12}O_2$
5-(2,4,6-Cycloheptatrien-1-ylideneethylidene)-
   3,6-cycloheptadiene-1,2-dione, C-70222
2,3-Dihydro-3-(phenylmethylene)-4*H*-1-
   benzopyran-4-one, D-60263
1-Phenanthreneacetic acid, P-70046
9-Phenanthreneacetic acid, P-70047
Tricyclo[8.4.1.1³,⁸]hexadeca-3,5,7,10,12,14-
   hexaene-2,9-dione, T-60264

$C_{16}H_{12}O_2S$
3-Methylthio-2-phenyl-4*H*-1-benzopyran-4-one,
   in M-60025

$C_{16}H_{12}O_3$
6-Methoxyflavone, *in* H-70130
7-Methoxyflavone, *in* H-70131
8-Methoxyflavone, *in* H-70132
Nordracorhodin, N-70074

$C_{16}H_{12}O_4$
7-Hydroxy-3-(4-hydroxybenzylidene)-4-
   chromanone, H-60152
Isodalbergin, I-60097
Vesparione, V-70007

$C_{16}H_{12}O_5$
3-(3,4-Dihydroxybenzylidene)-7-hydroxy-4-
   chromanone, *in* D-70288
▷5,7-Dihydroxy-4′-methoxyflavone, D-70319
4′,6-Dihydroxy-7-methoxyisoflavone, *in*
   T-60325
7,8-Dihydroxy-4′-methoxyisoflavone, *in*
   T-70265
1,8-Dihydroxy-6-methoxy-2-methyl-
   anthraquinone, *in* T-70269
3-Hydroxy-2-hydroxymethyl-1-methoxy-
   anthraquinone, *in* D-70306
Maackiain, M-70001
Pabulenone, *in* P-60170

$C_{16}H_{12}O_6$
6*a*,12*a*-Dihydro-2,3,10-trihydroxy[2]-
   benzopyrano[4,3-*b*][1]benzopyran-7(5*H*)-
   one, *in* C-70200
3-(2,4-Dihydroxybenzoyl)-7,8-dihydroxy-1*H*-2-
   benzopyran, *in* P-70017
3,5-Dihydroxy-6,7-methylenedioxyflavanone, *in*
   T-60105
Peltochalcone, P-70017
3,5,6,7-Tetrahydroxy-8-methylflavone, T-70117
3,5,7,8-Tetrahydroxy-6-methylflavone, T-70118
3′,4′,5-Trihydroxy-7-methoxyisoflavone, *in*
   T-60108
3′,5,7-Trihydroxy-4′-methoxyisoflavone, *in*
   T-60108
3′,7,8-Trihydroxy-4′-methoxyisoflavone, *in*
   T-70112
4′,5,7-Trihydroxy-3′-methoxyisoflavone, *in*
   T-60108
4′,7,8-Trihydroxy-6-methoxyisoflavone, *in*
   T-70113
1,3,5-Trihydroxy-4-methoxy-2-methyl-
   anthraquinone, *in* T-60112
1,3,8-Trihydroxy-5-methoxy-2-methyl-
   anthraquinone, *in* T-60113

$C_{16}H_{12}O_7$
Crombeone, C-70200
2′,3,5,8-Tetrahydroxy-7-methoxyflavone, *in*
   P-70029

$C_{16}H_{13}ClO_5$
2-Chloro-1-hydroxy-5,8-dimethoxy-6-methyl-
   xanthone, *in* T-70181
5-Chloro-8-hydroxy-1,4-dimethoxy-3-methyl-
   xanthone, *in* T-70181

$C_{16}H_{13}NO$
2-Methoxy-4-phenylquinoline, *in* P-60131
4-Methoxy-2-phenylquinoline, *in* P-60132
1-Methyl-2-phenyl-4(1*H*)-quinolinone, *in*
   P-60132
9-Phenanthreneacetic acid; Amide, *in* P-70047

$C_{16}H_{13}NOS$
5-Methyl-2-phenyl-1,5-benzothiazepin-4(5*H*)-
   one, *in* P-60081

$C_{16}H_{13}NO_2$
10-Amino-9-anthracenecarboxylic acid; Me
   ester, *in* A-70088
Dibenz[*b,g*]azocine-5,7(6*H*,12*H*)-dione; *N*-Me,
   in D-60068

$C_{16}H_{13}NO_5$
1-Amino-4,5-dihydroxy-7-methoxy-2-methyl-
   anthraquinone, A-60145

$C_{16}H_{13}NS$
10*H*-[1]Benzothieno[3,2-*b*]indole; *N*-Et, *in*
   B-70059

$C_{16}H_{14}$
9-Benzylidene-1,3,5,7-cyclononatetraene,
   B-60069
1,6:7,12-Bismethano[14]annulene, B-60165
9,10-Dihydro-9,10-ethanoanthracene, D-70207
4,5-Dimethylphenanthrene, D-60439
1,1-Diphenyl-1,3-butadiene, D-70476
11*b*-Methyl-11*bH*-Cyclooct[*cd*]azulene,
   M-60060
7,8,9,10-Tetrahydrofluoranthene, T-70061
4,5,9,10-Tetrahydropyrene, T-60083

$C_{16}H_{14}N_2$

1,2-Dihydro-4,6-diphenylpyrimidine, D-60235

$C_{16}H_{14}N_2O$

1-(1-Isoquinolinyl)-1-(2-pyridinyl)ethanol, I-60134

$C_{16}H_{14}N_2O_2$

2-Phenyl-3*H*-indol-3-one; *N*-Oxide, oxime, Et ether, *in* P-60114

$C_{16}H_{14}N_2O_3$

2,3-Dihydro-2-hydroxy-1*H*-imidazole-4,5-dione; 1,3-Di-Ph, Me ether, *in* D-60243

$C_{16}H_{14}N_2S_2$

2,5-Dihydro-4,6-pyrimidinedithiol; Di-Ph thioether, *in* D-70257

$C_{16}H_{14}O$

2-(9-Anthracenyl)ethanol, A-60282

3,4-Dihydro-2-phenyl-1(2*H*)-naphthalenone, D-60265

3,4-Dihydro-3-phenyl-1(2*H*)-naphthalenone, D-60266

3,4-Dihydro-4-phenyl-1(2*H*)-naphthalenone, D-60267

10,10-Dimethyl-9(10*H*)-anthracenone, D-60403

2,3-Diphenyl-2-butenal, D-60475

3,3-Diphenylcyclobutanone, D-60476

$C_{16}H_{14}OS$

▷2,3-Dihydro-5,6-diphenyl-1,4-oxathiin, D-70203

$C_{16}H_{14}OS_2$

1-(Phenylsulfinyl)-2-(phenylthio)cyclobutene, *in* B-70156

$C_{16}H_{14}O_2$

4,5-Dimethoxyphenanthrene, *in* P-70048

2,2-Diphenyl-4-methylene-1,3-dioxolane, D-70486

Dracaenone, D-70546

3,6-Phenanthrenedimethanol, P-60069

4,5-Phenanthrenedimethanol, P-60070

$C_{16}H_{14}O_2S_2$

1,2-Biss(henylsulfinyl)cyclobutene, *in* B-70156

$C_{16}H_{14}O_3$

3,7-Dimethoxy-2-phenanthrenol, *in* T-60339

$C_{16}H_{14}O_3S_2$

1-(Phenylsulfinyl)-2-(phenylsulfonyl)-cyclobutene, *in* B-70156

$C_{16}H_{14}O_4$

[1,1′-Biphenyl]-2,2′-diacetic acid, B-60115

[1,1′-Biphenyl]-4,4′-diacetic acid, B-60116

1,5-Dimethoxy-2,7-phenanthrenediol, *in* T-60116

5,7-Dimethoxy-2,3-phenanthrenediol, *in* T-60120

5,5′-(1,2-Ethanediyl)bis-1,3-benzodioxole, *in* T-60101

Ferulidene, *in* X-60002

Flaccidin, F-70010

9-Hydroxy-3-methoxypterocarpan, *in* D-60373

$C_{16}H_{14}O_4S$

2,2′-Thiobisbenzoic acid; Di-Me ester, *in* T-70174

4,4′-Thiobisbenzoic acid; Di-Me ester, *in* T-70175

$C_{16}H_{14}O_4S_2$

1,2-Biss(henylsulfonyl)cyclobutene, *in* B-70156

$C_{16}H_{14}O_5$

Dibenzyl dicarbonate, D-70111

3-(3,4-Dihydroxybenzyl)-7-hydroxy-4-chromanone, D-70288

5,8-Dihydroxy-7-methoxyflavanone, *in* T-60321

3′,7-Dihydroxy-4′-methoxyisoflavanone, *in* T-70263

Isogosferol, *in* X-60002

Moracin F, *in* M-60146

Prangeline, P-60170

Sainfuran, S-60002

$C_{16}H_{14}O_6$

Anhydrofusarubin 9-methyl ether, *in* F-60108

3-(3,4-Dihydroxybenzyl)-3,7-dihydroxy-4-chromanone, *in* D-70288

5,8-Dihydroxy-2,3-dimethyl-1,4-naphthoquinone; Di-Ac, *in* D-60317

Diphenyl tartrate, *in* T-70005

6,6′-(1,2-Ethanediyl)bis-1,3-benzodioxol-4-ol, *in* H-60062

3-Hydroxy-1,2,4-trimethoxyxanthone, *in* T-60126

7-Hydroxy-2,3,4-trimethoxyxanthone, *in* T-60128

8-Hydroxy-1,3,5-trimethoxyxanthone, *in* T-70123

1,4,5-Naphthalenetriol; Tri-Ac, *in* N-60005

2′,5,8-Trihydroxy-7-methoxyflavanone, *in* T-70104

$C_{16}H_{14}O_6S$

4,4′-Thiobisbenzoic acid; Di-Me ester, *S*-dioxide, *in* T-70175

$C_{16}H_{14}O_7$

3′,5-Dihydroxy-4′,6,7-trimethoxyflavanone, *in* P-70028

1,8-Dihydroxy-2,3,6-trimethoxyxanthone, *in* P-70036

3,6-Dihydroxy-1,2,3-trimethoxyxanthone, *in* P-70036

3,8-Dihydroxy-1,2,4-trimethoxyxanthone, *in* P-60051

6,8-Dihydroxy-1,2,5-trimethoxyxanthone, *in* P-60052

Isolecanoric acid, I-60103

Nectriafurone; 8-Me ether, *in* N-60016

3-*O*-Demethylsulochrin, *in* S-70083

3′,4′,5,7-Tetrahydroxy-6-methoxyflavanone, *in* P-70028

Ventilone E, *in* V-60004

$C_{16}H_{14}S_2$

1,2-Biss(henylthio)cyclobutene, B-70156

$C_{16}H_{14}Se$

2,5-Dihydro-3,4-diphenylselenophene, D-70205

Distyryl selenide, D-60503

$C_{16}H_{15}ClO_6$

Lonapalene, *in* C-60118

$C_{16}H_{15}N$

7,12-Dihydro-5*H*-6,12-methanodibenz[*c,f*]-azocine, D-60255

$C_{16}H_{15}NO$

9-Amino-9,10-dihydroanthracene; *N*-Ac, *in* A-60136

2-Amino-9,10-dihydrophenanthrene; *N*-Ac, *in* A-60141

9-Amino-9,10-dihydrophenanthrene; *N*-Ac, *in* A-60143

10,11-Dihydro-5*H*-dibenz[*b,f*]azepine; 5-Ac, *in* D-70185

3,4-Dihydro-4-phenyl-1(2*H*)-naphthalenone; Oxime, *in* D-60267

3,6-Dimethylcarbazole; *N*-Ac, *in* D-70387

$C_{16}H_{15}NOS$

2-Phenyl-1,5-benzothiazepin-4(5*H*)-one; 2,3-Dihydro, *N*-Me, *in* P-60081

$C_{16}H_{15}NO_2$

2-Amino-1,3-diphenyl-1-propanone; *N*-Formyl, *in* A-60149

$C_{16}H_{15}NO_2S$

10*H*-Phenothiazine-1-carboxylic acid; *N*-Et, Me ester, *in* P-70055

10*H*-Phenothiazine-3-carboxylic acid; *N*-Et, Me ester, *in* P-70057

10*H*-Phenothiazine-4-carboxylic acid; *N*-Et, Me ester, *in* P-70058

$C_{16}H_{15}NO_4$

4-Hydroxy-3,6-dimethoxy-2-methyl-9*H*-carbazole-1-carboxaldehyde, *in* H-70154

2,2′-Iminodibenzoic acid; Di-Me ester, *in* I-60014

2,4′-Iminodibenzoic acid; Di-Me ester, *in* I-60016

4,4′-Iminodibenzoic acid; Di-Me ester, *in* I-60018

$C_{16}H_{15}N_5$

2,4-Diamino-6-methyl-1,3,5-triazine; *N,N*-Di-Ph, *in* D-60040

2,4-Diamino-6-methyl-1,3,5-triazine; *N,N′*-Di-Ph, *in* D-60040

$C_{16}H_{16}$

9,10-Dihydro-9,9-dimethylanthracene, D-60224

4,5-Dimethylphenanthrene; 9,10-Dihydro, *in* D-60439

Heptacyclo[7.7.0.0²,⁶.0³,¹⁵.0⁴,¹².0⁵,¹⁰.0¹¹,¹⁶]-hexadeca-7,13-diene, H-70011

1,2,3,3*a*,4,5-Hexahydropyrene, H-70068

1,2,3,6,7,8-Hexahydropyrene, H-70069

$C_{16}H_{16}N_2$

10,10-Dimethyl-9(10*H*)-anthracenone; Hydrazone, *in* D-60403

$C_{16}H_{16}N_2O_2$

2,2′-Diaminobiphenyl; 2,2′-*N*-Di-Ac, *in* D-60033

$C_{16}H_{16}N_2O_3$

2,4,6-Trimethyl-3-nitroaniline; *N*-Benzoyl, *in* T-70285

$C_{16}H_{16}N_2O_4$

[2,2′-Bipyridine]-6,6′-dicarboxylic acid; Di-Et ester, *in* B-60128

$C_{16}H_{16}O$

6,7,8,9-Tetrahydro-6-phenyl-5*H*-benzocyclohepten-5-one, T-60080

$C_{16}H_{16}O_2$

2,2-Diphenylpropanoic acid; Me ester, *in* D-70493

2,3-Naphthalenedicarboxylic acid; Di-Et ester, *in* N-60003

$C_{16}H_{16}O_3$

3-(1,1-Dimethyl-2-propenyl)-4-hydroxy-6-phenyl-2*H*-pyran-2-one, D-70434

Echinofuran B, *in* E-60002

5-[2-(3-Methoxyphenyl)ethyl]-1,3-benzodioxole, *in* T-60316

9-Methylcyclolongipesin, *in* C-60214

$C_{16}H_{16}O_4$

8-Acetoxy-2-(2,4-hexadiynylidene)-1,6-dioxaspiro[4.5]dec-3-ene, *in* H-70042

5,7-Azulenedicarboxylic acid; Di-Et ester, *in* A-70295

9,10-Dihydro-5,6-dihydroxy-2,4-dimethoxy-phenanthrene, *in* D-70247

9,10-Dihydro-2,7-dimethoxy-1,5-phenanthrenediol, *in* T-60116

Gleinadiene, *in* G-60023

Homocyclolongipesin, H-60088

Perforatin A, *in* A-60075

$C_{16}H_{16}O_5$

Deoxyaustrocortilutein, *in* A-60321

1-(2,6-Dihydroxy-4-methoxyphenyl)-3-(4-hydroxyphenyl)-1-propanone, *in* H-60219

6-[2-(3,4-Dimethoxyphenyl)ethenyl]-4-methoxy-2*H*-pyran-2-one, *in* H-70122

3-(4-Hydroxyphenyl)-1-(2,4-dihydroxy-6-methoxyphenyl)-1-propanone, *in* H-60220

Shikonin, S-70038

$C_{16}H_{16}O_6$

Altersolanol B, *in* A-60080

Austrocortilutein, A-60321

Deoxyaustrocortirubin, *in* A-60322

3-(3,4-Dihydroxybenzyl-3,4,7-trihydroxychroman, *in* D-70288

Meciadanol, *in* P-70026

3,3′,4′,5,7-Pentahydroxyflavan; *O³*-Me, *in* P-70026

3,3′,5,7-Tetrahydroxy-4′-methoxyflavan, *in* P-70026

$C_{16}H_{16}O_7$
Altersolanol C, *in* A-60080
Austrocortirubin, A-60322
Fusarubin; O^9-Me, *in* F-60108
Fusarubin methyl acetal, *in* F-60108

$C_{16}H_{16}O_8$
Altersolanol *A*, A-60080

$C_{16}H_{16}S_2$
2,11-Dithia[3.3]paracyclophane, D-60505

$C_{16}H_{17}N$
Azetidine; *N*-Benzhydryl, *in* A-70283
1,2,3,4-Tetrahydroquinoline; *N*-Benzyl, *in* T-70093

$C_{16}H_{17}NO_3$
4-Hydroxy-3,6-dimethoxy-1,2-dimethylcarbazole, *in* T-70253

$C_{16}H_{17}O_4$
9,10-Dihydro-2,7-dihydroxy-3,5-dimethoxyphenanthrene, *in* T-60120

$C_{16}H_{18}N_2$
[3.3][2.6]Pyridinophane, P-60226
[3](2.2)[3](5.5)Pyridinophane, P-60227
[3](2.5)[3](5.2)Pyridinophane, P-60228

$C_{16}H_{18}N_2O$
N-Ethyl-*N'*-methyl-*N*,*N'*-diphenylurea, *in* D-70499

$C_{16}H_{18}O_2$
Methyl 8-isopropyl-5-methyl-2-naphthalenecarboxylate, *in* I-60129

$C_{16}H_{18}O_3$
3-Hydroxy-3',4-dimethoxybibenzyl, T-60316
8-Methoxy-1(6),2,4,7(11)-cadinatetraen-12,8-olide, *in* H-70114
4-Methoxy-3-(3-methyl-2-butenyl)-5-phenyl-2(5*H*)-furanone, M-70047

$C_{16}H_{18}O_4$
(E)-Methylsuberenol, *in* S-70080
Gleinene, G-60023
4-[2-(3-Hydroxyphenyl)ethyl]-2,6-dimethoxyphenol, *in* T-60102
9-Methyllongipesin, M-70091
O-Methylcedrelopsin, *in* C-60026
2,2',3,4-Tetramethoxybiphenyl, *in* B-70122
2,3,4,4'-Tetramethoxybiphenyl, *in* B-70123
(Z)-Methylsuberenol, *in* S-70080

$C_{16}H_{18}O_5$
Resorcylide, R-70001
Skimminin, S-60035

$C_{16}H_{18}O_6$
Murraculatin, M-70158
2,5,7-Trihydroxy-3-(5-hydroxyhexyl)-1,4-naphthoquinone, T-70262

$C_{16}H_{18}O_7$
3-O-Methyldihydrofusarubin A, *in* F-60108

$C_{16}H_{18}O_8$
3,6-Dimethyl-1,2,4,5-benzenetetracarboxylic acid; Tetra-Me ester, *in* D-60407

$C_{16}H_{18}O_9$
▷3-O-Caffeoylquinic acid, C-70005
Fabiatrin, *in* H-70153

$C_{16}H_{19}N_3O_6$
Albomitomycin *A*, A-70080
Isomitomycin *A*, I-70070

$C_{16}H_{20}$
Heptacyclo[7.7.0.0^{2,6}.0^{3,15}.0^{4,12}.0^{5,10}.0^{11,16}]-hexadecane, *in* H-70011

$C_{16}H_{20}ClN_3$
Bindschedler's green; Chloride, *in* B-60110

$C_{16}H_{20}N_2$
2,2'-Biss(imethylamino)biphenyl, *in* D-60033
1,2-Diphenyl-1,2-ethanediamine; *N*,*N'*-Di-Me, *in* D-70480

$C_{16}H_{20}N_2O_6S_2$
N,*N'*-Bis(3-sulfonatopropyl)-2,2'-bipyridinium, B-70159

N,*N'*-Bis(3-sulfonatopropyl)-4,4'-bipyridinium, B-70160

$C_{16}H_{20}N_3^{\oplus}$
Bindschedler's green, B-60110

$C_{16}H_{20}O$
2-Phenyl-2-adamantanol, P-70060

$C_{16}H_{20}O_2$
Methyl 3,4-dihydro-8-isopropyl-5-methyl-2-naphalenecarboxylate, *in* I-60129

$C_{16}H_{20}O_3$
Alliodorin, A-70083

$C_{16}H_{20}O_4$
1-(5,7-Dihydroxy-2,2,6-trimethyl-2*H*-1-benzopyran-8-yl)-2-methyl-1-propanone, D-70348
Hypocretenoic acid; Me ester, *in* H-70235
Methyl yomoginate, *in* H-70199

$C_{16}H_{20}O_5$
Glutinopallal, G-70021

$C_{16}H_{20}O_6$
Orthopappolide, O-60045
▷Pyrenophorin, P-70167

$C_{16}H_{21}BrO_4$
1-(4-Bromo-2,5-dihydroxyphenyl)-7-hydroxy-3,7-dimethyl-2-octen-1-one, B-70206

$C_{16}H_{22}$
[3^{4,10}][7]Metacyclophane, M-60030

$C_{16}H_{22}O_3$
3-Hydroxy-4,11(13)-guaiadien-12,8-olide; 3-Me ether, *in* H-70135
3-Hydroxy-4,11(13)-guaiadien-12,8-olide; Me ether, *in* H-70135

$C_{16}H_{22}O_5$
ent-7β-Hydroxy-13,14,15,16-tetranor-3-cleroden-12-oic acid 18,19-lactone, H-70226
Methyl altamisate, *in* A-70084

$C_{16}H_{22}O_6$
2α,3α-Epoxy-1β,4α-dihydroxy-8β-methoxy-7(11)-eudesmen-12,8-olide, *in* E-60033

$C_{16}H_{22}O_6S_3$
2,8,17-Trithia[4^{5,12}][9]metacyclophane; Trisulfone, *in* T-70328

$C_{16}H_{22}O_8$
Synrotolide, S-60066

$C_{16}H_{22}O_{10}$
Gelsemide 7-glucoside, *in* G-60010

$C_{16}H_{22}S_3$
2,8,17-Trithia[4^{5,12}][9]metacyclophane, T-70328

$C_{16}H_{23}N$
Axisonitrile-4, *in* A-60328

$C_{16}H_{23}NO$
2-(Aminomethyl)cyclooctanol; *N*-Benzoyl, *in* A-70170

$C_{16}H_{23}NO_2$
2-Amino-4-*tert*-butylcyclopentanol; Benzoyl, *in* A-70104

$C_{16}H_{23}NS$
Axisothiocyanate-4, *in* A-60329

$C_{16}H_{23}N_5O_5$
1'-Methylzeatin; 9-β-D-Ribofuranosyl, *in* M-60133

$C_{16}H_{24}$
Pentacyclo[11.3.0.0^{1,5}.0^{5,9}.0^{9,13}]hexadecane, P-70020

$C_{16}H_{24}O$
5-Methyl-2-(1-methyl-1-phenylethyl)cyclohexanol, M-60090

$C_{16}H_{24}O_4$
Furodysinin hydroperoxide, F-70054
1-Isobutyryloxymethyl-4-isopropyl-2,5-dimethoxybenzene, *in* D-70309

$C_{16}H_{24}O_4S_2$
Dithiatopazine, D-70508

$C_{16}H_{24}O_9$
Semperoside, S-60021

$C_{16}H_{24}O_{10}$
Adoxosidic acid, A-70068
9-Hydroxysemperoside, *in* S-60021
Methylcatalpol, *in* C-70026
Vebraside, V-60003

$C_{16}H_{24}O_{11}$
Shanzhiside, S-60028

$C_{16}H_{25}N$
Axisonitrile-1, *in* A-60328
Axisonitrile-2, *in* A-60328
Axisonitrile-3, *in* A-60328
10α-Isocyanoalloaromadendrane, *in* A-60328
3-Isocyano-7,9-bisaboladiene, I-60091
7-Isocyano-3,10-bisaboladiene, I-60092
11-Isocyano-5-eudesmene, *in* F-60068
6-Isocyano-4(15)-eudesmene, *in* F-60069

$C_{16}H_{25}NO$
7-Isocyanato-2,10-bisaboladiene, I-60089

$C_{16}H_{25}NO_5$
Wasabidienone *E*, W-60003
Wasabidienone E, *in* W-70002

$C_{16}H_{25}NS$
Axisothiocyanate-1, *in* A-60329
Axisothiocyanate-2, *in* A-60329
Axisothiocyanate-3, *in* A-60329
10α-Isothiocyanatoalloaromadendrane, *in* A-60329
11-Isothiocyano-5-eudesmene, *in* F-60068

$C_{16}H_{26}$
3,4,7,11-Tetramethyl-1,3,6,10-dodectetraene, T-70135

$C_{16}H_{26}O_2$
6-Isopropenyl-3-methyl-3,9-decadien-1-ol acetate, *in* I-70089
3,7,11-Trimethyl-2,6,10-dodecatrienoic acid; Me ester, *in* T-60364

$C_{16}H_{26}O_3$
6-Hydroxy-2,4,8-tetradecatrienoic acid; Et ester, *in* H-70222
ent-12-Hydroxy-13,14,15,16-tetranor-1(10)-halimen-18-oic acid, H-60231
Juvenile hormone III, *in* T-60364
Methylzedoarondiol, *in* Z-70004

$C_{16}H_{26}O_8$
Nepetaside, N-70031

$C_{16}H_{26}O_9$
Gelsemiol 1-glucoside, *in* G-60011
Gelsemiol 3-glucoside, *in* G-60011
Gibboside, G-60020

$C_{16}H_{27}NO$
Axamide-1, A-70272
Axamide-2, A-60327
10α-Formamidoalloaromadendrane, *in* A-60327
11-Formamido-5-eudesmene, F-60068
6-Formamido-4(15)-eudesmene, F-60069

$C_{16}H_{28}N_4O_8$
1,4,7,10-Tetraazacyclododecane-1,4,7,10-tetraacetic acid, T-60015

$C_{16}H_{28}O$
Dodecahydro-3*a*,6,6,9*a*-tetramethylnaphtho[2,1-*b*]furan, D-70533
11,13-Hexadecadienal, H-70033

$C_{16}H_{28}O_2$
Isoambrettolide, I-60086
6-Isopropenyl-3-methyl-9-decen-1-ol; Ac, *in* I-70089

$C_{16}H_{30}$
5-Hexadecyne, H-60037

$C_{16}H_{30}N_4O_2$
Tetrahydroimidazo[4,5-*d*]imidazole-2,5(1*H*,3*H*)-dione; 1,3,4,6-Tetraisopropyl, *in* T-70066

$C_{16}H_{30}O$
Decamethylcyclohexanone, D-60011
2-Hexadecenal, H-60036

7-Hexadecenal, H-70035

$C_{16}H_{30}O_2$

8,9-Hexadecanedione, H-70034

$C_{16}H_{32}N_2O_5$

4,7,13,16,21-Pentaoxa-1,10-diazabicyclo[8.8.5]tricosane, P-60060

$C_{16}H_{34}N_4$

Octahydroimidazo[4,5-*d*]imidazole; 1,3,4,6-Tetraisopropyl, *in* O-70015

$C_{16}H_{38}N_6$

1,4,7,12,15,18-Hexaazacyclodocosane, H-70028

$C_{16}H_{38}N_6O_2$

1,13-Dioxa-4,7,10,16,19,22-hexaazacyclotetracosane, D-70460

$C_{17}H_8N_2O$

10-(Dicyanomethylene)anthrone, D-60164

$C_{17}H_9ClO$

2-Chloro-7*H*-benz[*de*]anthracen-7-one, C-70040

3-Chloro-7*H*-benz[*de*]anthracen-7-one, C-70041

4-Chloro-7*H*-benz[*de*]anthracen-7-one, C-70042

5-Chloro-7*H*-benz[*de*]anthracen-7-one, C-70043

6-Chloro-7*H*-benz[*de*]anthracen-7-one, C-70044

8-Chloro-7*H*-benz[*de*]anthracen-7-one, C-70045

9-Chloro-7*H*-benz[*de*]anthracen-7-one, C-70046

10-Chloro-7*H*-benz[*de*]anthracen-7-one, C-70047

11-Chloro-7*H*-benz[*de*]anthracen-7-one, C-70048

$C_{17}H_{10}F_3NO_3$

7-Amino-4-(trifluoromethyl)-2*H*-1-benzopyran-2-one; *N*-Benzoyl, *in* A-60265

$C_{17}H_{10}N_2$

1-(Diazomethyl)pyrene, D-70062

9-(Dicyanomethyl)anthracene, D-60163

$C_{17}H_{10}O$

3*H*-Cyclonona[*def*]biphenylen-3-one, C-70230

$C_{17}H_{10}O_5$

1,5-Diphenylpentanepentone, D-60488

$C_{17}H_{10}O_8$

Scapaniapyrone *A*, S-70017

$C_{17}H_{11}N$

Dibenzo[*f,h*]quinoline, D-70075

$C_{17}H_{11}NO$

Benz[*cd*]indol-2-(1*H*)-one; *N*-Ph, *in* B-70027

$C_{17}H_{11}N_3$

7,12-Dihydropyrido[3,2-*b*:5,4-*b'*]diindole, D-70254

$C_{17}H_{12}$

3*H*-Cyclonona[*def*]biphenylene, C-60215

Cycloocta[*def*]fluorene, C-70233

1-Methylazupyrene, M-60043

$C_{17}H_{12}BrN$

Dibenzo[*a,h*]quinolizinium(1+); Bromide, *in* D-70076

Naphtho[1,2-*a*]quinolizinium(1+); Bromide, *in* N-70012

$C_{17}H_{12}ClNO_4$

Dibenzo[*a,h*]quinolizinium(1+); Perchlorate, *in* D-70076

Naphtho[1,2-*a*]quinolizinium(1+); Perchlorate, *in* N-70012

$C_{17}H_{12}Cl_2O_8$

Geodoxin, G-70007

$C_{17}H_{12}N^{\oplus}$

Dibenzo[*a,h*]quinolizinium(1+), D-70076

Naphtho[1,2-*a*]quinolizinium(1+), N-70012

$C_{17}H_{12}N_2O$

2-Benzoyl-1,2-dihydro-1-isoquinolinecarbonitrile, B-70073

$C_{17}H_{12}OS$

2,6-Diphenyl-4*H*-thiopyran-4-one, D-60492

$C_{17}H_{12}O_2$

2,2'-Spirobi[2*H*-1-benzopyran], S-70062

$C_{17}H_{12}O_3$

1-(4-Hydroxy-5-benzofuranyl)-3-phenyl-2-propen-1-one, H-70107

$C_{17}H_{12}O_3S$

2,6-Diphenyl-4*H*-thiopyran-4-one; 1,1-Dioxide, *in* D-60492

5-Methoxy-2-(phenylthio)-1,4-naphthoquinone, *in* H-70213

5-Methoxy-3-(phenylthio)-1,4-naphthoquinone, *in* H-70214

$C_{17}H_{12}O_6$

3,3-Dihydroxy-5,5-diphenyl-1,2,4,5-pentanetetrone, *in* D-60488

5-Hydroxy-2'-methoxy-6,7-methylenedioxyisoflavone, *in* T-60107

7-Hydroxy-6-methoxy-3',4'-methylenedioxyisoflavone, *in* T-70111

7-Hydroxy-8-methoxy-3',4'-methylenedioxyisoflavone, *in* T-70112

$C_{17}H_{12}O_7$

3,6,8-Trihydroxy-1-methylanthraquinone-2-carboxylic acid; Me ester, *in* T-60332

$C_{17}H_{12}O_8$

Nasutin B, *in* E-70007

$C_{17}H_{13}NO_2S$

2-Phenyl-1,5-benzothiazepin-4(5*H*)-one; *N*-Ac, *in* P-60081

$C_{17}H_{13}N_3$

1-Anilino-γ-carboline, *in* A-70114

$C_{17}H_{14}$

Bicyclo[5.3.1]undeca-1,3,5,7,9-pentaene, B-60102

$C_{17}H_{14}Cl_2O_7$

Tumidulin, T-70334

$C_{17}H_{14}N_2O_4$

2,3-Dihydro-2-hydroxy-1*H*-imidazole-4,5-dione; 1,3-Di-Ph, *O*-Ac, *in* D-60243

$C_{17}H_{14}O$

2-Methoxy-1-phenylnaphthalene, *in* P-60115

2-Methoxy-5-phenylnaphthalene, *in* P-60118

7-Methoxy-1-phenylnaphthalene, *in* P-60119

$C_{17}H_{14}O_2$

1,2-Dihydro-1-phenyl-1-naphthalenecarboxylic acid, D-60264

2-(4-Hydroxyphenyl)-5-(1-propenyl)benzofuran, H-60214

9-Phenanthreneacetic acid; Me ester, *in* P-70047

2-(Phenylethynyl)benzoic acid; Et ester, *in* P-70074

$C_{17}H_{14}O_3$

1,5-Diphenyl-1,3,5-pentanetrione, D-60489

4-[5-(1-Propenyl)-2-benzofuranyl]-2,3-benzenediol, *in* H-60214

$C_{17}H_{14}O_4$

Agrostophyllin, A-70077

Bonducellin, B-60192

$C_{17}H_{14}O_5$

Benzophenone-2,2'-dicarboxylic acid; Di-Me ester, *in* B-70041

▷ 1,3-Dihydroxy-2-(ethoxymethyl)anthraquinone, *in* D-70306

5-Hydroxy-6,7-dimethoxyflavone, *in* T-60322

6-Hydroxy-4',7-dimethoxyisoflavone, *in* T-60325

3-Hydroxy-1-methoxy-2-(methoxymethyl)anthraquinone, *in* D-70306

2-(2-Hydroxy-4-methoxyphenyl)-3-methyl-5,6-methylenedioxybenzofuran, H-70157

Pterocarpin, *in* M-70001

Puerol *A*, P-70137

$C_{17}H_{14}O_6$

4',5-Dihydroxy-3',7-dimethoxyflavone, D-70291

2',5-Dihydroxy-7,8-dimethoxyisoflavone, *in* T-70110

4',5-Dihydroxy-3',7-dimethoxyisoflavone, *in* T-60108

5,7-Dihydroxy-2',6-dimethoxyisoflavone, *in* T-60107

5,7-Dihydroxy-3',4'-dimethoxyisoflavone, *in* T-60108

Sophorocarpan *B*, S-60038

$C_{17}H_{14}O_7$

Arizonin A_1, *in* A-70254

Arizonin B_1, *in* A-70254

6*a*,12*a*-Dihydro-2,3,10-trihydroxy-8-methoxy[2]benzopyrano[4,3-*b*][1]benzopyran-7(5*H*)-one, *in* C-70200

6*a*,12*a*-Dihydro-3,4,10-trihydroxy-8-methoxy[2]benzopyrano[4,3-*b*][1]benzopyran-7(5*H*)-one, *in* C-70200

Nornotatic acid, *in* N-70081

3,4',5-Trihydroxy-6,7-dimethoxyflavone, *in* P-60047

3,4',7-Trihydroxy-5,6-dimethoxyflavone, *in* P-60047

3',4',5-Trihydroxy-6,7-dimethoxyisoflavone, *in* P-70032

$C_{17}H_{14}O_8$

3',4',5,5'-Tetrahydroxy-6,7-dimethoxyflavone, *in* H-70081

3',4',5,7-Tetrahydroxy-5',6-dimethoxyflavone, *in* H-70081

$C_{17}H_{15}ClO_7$

Methyl 2-(3-chloro-2,6-dihydroxy-4-methylbenzoyl)-5-hydroxy-3-methoxybenzoate, *in* S-70083

$C_{17}H_{15}N$

2-Methyl-1,5-diphenyl-1*H*-pyrrole, *in* M-60115

$C_{17}H_{15}NO_2S$

2-Phenyl-1,5-benzothiazepin-4(5*H*)-one; 2,3-Dihydro, *N*-Ac, *in* P-60081

$C_{17}H_{16}$

1,1-Diphenyl-1,3-pentadiene, D-70489

$C_{17}H_{16}N_2O$

1-[1-Methoxy-1-(2-pyridinyl)ethyl]isoquinoline, *in* I-60134

$C_{17}H_{16}N_2O_3$

3-Oxopentanedioic acid; Dianilide, *in* O-60084

$C_{17}H_{16}O_2$

1,5-Bis(4-hydroxyphenyl)-1,4-pentadiene, B-60162

3,3',4,4'-Tetrahydro-2,2'-spirobi[2*H*-1-benzopyran], T-70094

$C_{17}H_{16}O_3$

Danshenspiroketallactone, D-70004

Epidanshenspiroketallactone, *in* D-70004

2,3,7-Trimethoxyphenanthrene, *in* T-60339

$C_{17}H_{16}O_4$

Cryptoresinol, C-70201

10-Hydroxy-11-methoxydracaenone, *in* D-70546

2-(2-Hydroxy-4-methoxyphenyl)-6-methoxy-3-methylbenzofuran, H-70156

3,5,7-Trimethoxy-2-phenanthrenol, *in* T-60120

$C_{17}H_{16}O_4S_2$

5-(3,4-Diacetoxy-1-butynyl)-5'-methyl-2,2'-bithiophene, *in* M-70054

$C_{17}H_{16}O_5$

Comaparvin, C-70191

2',4-Dihydroxy-4',6'-dimethoxychalcone, *in* H-60220

7,10-Dihydroxy-11-methoxydracaenone, *in* D-70546

5,10-Dihydroxy-8-methoxy-2-propyl-4*H*-naphtho[1,2-*b*]pyran-4-one, *in* C-70191

7,9-Dihydroxy-2,3,4-trimethoxyphenanthrene, *in* P-70035

5-Hydroxy-7,8-dimethoxyflavone, *in* T-60321
Methylsainfuran, *in* S-60002
Sophorocarpan *A*, S-60037
1,5,6-Trimethoxy-2,7-phenanthrenediol, *in* P-60050

C$_{17}$H$_{16}$O$_6$

1,5-Bis(3,4-dihydroxyphenyl)-4-pentyne-1,2-diol, B-60155
5,7-Dihydroxy-2′,8-dimethoxyflavanone, *in* T-70104
5,8-Dihydroxy-2,3,6-trimethyl-1,4-naphthoquinone; Di-Ac, *in* D-60380
1,2,3,4-Tetramethoxyxanthone, *in* T-60126
1,3,5,8-Tetramethoxyxanthone, *in* T-70123

C$_{17}$H$_{16}$O$_7$

Isosulochrin, *in* S-70083
Sulochrin, S-70083
2′,3,5-Trihydroxy-7,8-dimethoxyflavanone, *in* P-60045
4′,5,7-Trihydroxy-3′,6-dimethoxyflavanone, *in* P-70028
1,3,8-Trihydroxy-4,7-dimethoxyxanthone, *in* P-60052

C$_{17}$H$_{17}$NO

5,6,11,12-Tetrahydrodibenz[*b,f*]azocine; *N*-Ac, *in* T-70057

C$_{17}$H$_{18}$O$_2$S$_2$

3,3-Biss(henylthio)propanoic acid; Et ester, *in* B-70157

C$_{17}$H$_{18}$O$_4$

3′,5-Dimethoxy-3,4-methylenedioxybibenzyl, *in* T-60102
4′-Hydroxy-3′,7-dimethoxyflavan, *in* T-70260
Pygmaeoherin, P-70145

C$_{17}$H$_{18}$O$_5$

9β-Acetoxy-3-oxo-1,4(15),11(13)-eudesmatrien-12,6-olide, *in* H-60202
9,10-Dihydro-2,7-dihydroxy-3,4,6-trimethoxyphenanthrene, *in* D-70246
9,10-Dihydro-5,6-dihydroxy-1,3,4-trimethoxyphenanthrene, *in* P-70034
9,10-Dihydro-5,6,7-trimethoxy-2,3-phenanthrenediol, *in* D-70246
5,7-Dihydroxy-3′,4′-dimethoxyflavan, *in* T-60103
4′-Hydroxy-3′,5-dimethoxy-3,4-methylenedioxybibenzyl, *in* P-60044
Isomurralonginol; Ac, *in* I-70071
Longipesin; 9-Ac, *in* L-60034
Lotisoflavan, L-70038

C$_{17}$H$_{18}$O$_6$

Antibiotic L 660631, A-70224
3-(3,4-Dihydroxybenzyl)-3,7-dihydroxy-4-methoxychroman, *in* D-70288
Pandoxide, P-60002
3,3′,4′,5,7-Pentahydroxyflavan; *O*³,*O*⁴-Di-Me, *in* P-70026
3,3′,4′,5,7-Pentahydroxyflavan; *O*³,*O*⁵-Di-Me, *in* P-70026
3,3′,5-Trihydroxy-4′,7-dihydroxyflavan, P-70026

C$_{17}$H$_{18}$O$_7$

Fusarubin; *O*³-Et, *in* F-60108
Fusarubin ethyl acetal, *in* F-60108

C$_{17}$H$_{19}$N

2,3,4,5-Tetrahydro-1*H*-1-benzazepine; *N*-Benzyl, *in* T-70036

C$_{17}$H$_{19}$NO$_3$

3,4,6-Trimethoxy-1,2-dimethylcarbazole, *in* T-70253

C$_{17}$H$_{20}$N$_2$O

▷*N,N′*-Diethyl-*N,N′*-diphenylurea, *in* D-70499

C$_{17}$H$_{20}$O$_2$

1,5-Diphenyl-1,3-pentanediol, D-70490

C$_{17}$H$_{20}$O$_3$

1β-Acetoxy-4(15),7(11),8-eudesmatrien-12,8-olide, *in* H-60141

15-Acetoxytubipofuran, *in* T-70332

C$_{17}$H$_{20}$O$_4$

1β-Acetoxy-3,7(11),8-eudesmatrien-12,8-olide, *in* H-60140
1-(2-Hydroxy-4-methoxyphenyl)-3-(3-hydroxy-4-methoxyphenyl)propane, *in* D-70336
1-(4-Hydroxy-2-methoxyphenyl)-3-(4-hydroxy-3-methoxyphenyl)propane, *in* D-70336
1-(2-Hydroxy-4-methoxyphenyl)-3-(4-hydroxy-3-methoxyphenyl)propene, *in* D-70336
Zaluzanin D, *in* Z-60001

C$_{17}$H$_{20}$O$_5$

6-Acetylferulidin, *in* F-70006
Cavoxinine, C-60024
Cavoxinone, C-60025
4,4′-Dihydroxy-3,3′,5-trimethoxybibenzyl, *in* P-60044
8-Hydroxyachillin; Ac, *in* H-60100
Murraxocin, M-60151

C$_{17}$H$_{20}$O$_6$

Achillolide A, *in* A-60061
Murrayanone, M-70160
Tarchonanthus lactone, T-60009

C$_{17}$H$_{20}$O$_7$

Isoapressin, I-70061
3-*O*-Ethyldihydrofusarubin A, *in* F-60108

C$_{17}$H$_{20}$O$_9$

4,7-Dihydroxy-5-methyl-2*H*-1-benzopyran-2-one; 4-Me ether, 7-*O*-β-D-glucopyranoside, *in* D-60348
3-*O*-Feruloylquinic acid, *in* C-70005

C$_{17}$H$_{21}$ClO$_6$

Arctodecurrolide, *in* D-60325

C$_{17}$H$_{21}$N$_3$

▷Auramine, A-70268

C$_{17}$H$_{22}$N$_6$O$_6$

3′-Amino-3′-deoxyadenosine; 3′*N*-Ac, 3′*N*-Me, 2′,5′-di-Ac, *in* A-70128

C$_{17}$H$_{22}$O$_2$S

Thiofurodysin acetate, *in* T-70178
Thiofurodysinin acetate, *in* T-70179

C$_{17}$H$_{22}$O$_3$

3-(3,7-Dimethyl-2,6-octadienyl)-4-hydroxybenzoic acid, D-70423
8,12-Epoxy-3,7,11-eudesmatrien-1-ol; Ac, *in* E-60020
6-Methoxy-2-(3,7-dimethyl-2,6-octadienyl)-1,4-benzoquinone, M-60038
Nimbidiol, N-60021

C$_{17}$H$_{22}$O$_4$

2-Acetoxy-2-desoxyperezone, *in* I-70075
3β-Acetoxy-4,11(13)-eudesmadien-12,8β-olide, *in* H-60136
9β-Acetoxytournefortiolide, *in* H-60137
1-(5,7-Dihydroxy-2,2,6-trimethyl-2*H*-1-benzopyran-8-yl)-2-methyl-1-butanone, *in* D-70348
3-Hydroxy-4,11(13)-guaiadien-12,8-olide; Ac, *in* H-70135
15-Hydroxy-3,10(14)-guaiadien-12,8-olide; Ac, *in* H-70136
Laurenobiolide, *in* H-70134
Subexpinnatin *C*, S-60058

C$_{17}$H$_{22}$O$_5$

2α-Acetoxy-8-epiivangustin, *in* I-70109
6α-Acetoxy-8-epiivangustin, *in* I-70109
1β-Acetoxy-8β-hydroxy-3,7(11)-eudesmadien-12,8-olide, *in* D-60322
3β-Acetoxy-2α-hydroxy-1(10),4,11(13)-germacratrien-12,6α-olide, *in* H-60006
Ergolide, E-60035
Naematolone, *in* N-70001

C$_{17}$H$_{22}$O$_6$

13-Desacetyleudesmaafraglaucolide, *in* T-70258
13-Desacetyl-1α-hydroxyafraglaucolide, A-70070
13-Desacetyl-1β-hydroxyafraglaucolide, A-70070
13-Desacetyl-1β-hydroxyisoafraglaucolide, I-70060

Neovasinone, N-60019
Neovasinone, *in* N-70029
1,3,9-Trihydroxy-4(15),11(13)-eudesmadien-12,6-olide; 3-Ac, *in* T-70257

C$_{17}$H$_{23}$Br$_3$O$_4$

6-Acetoxy-3,10,13-tribromo-4,7:9,12-diepoxy-1-pentadecyne, *in* B-70195

C$_{17}$H$_{23}$N$_5$O$_7$

Queuosine, *in* Q-70004

C$_{17}$H$_{24}$Br$_2$O$_4$

6-Acetoxy-3,10-dibromo-4,7-epoxy-12-pentadecen-1-yne, *in* B-70217

C$_{17}$H$_{24}$O$_2$

Tochuinyl acetate, *in* T-70202

C$_{17}$H$_{24}$O$_3$

2-[2-(Acetyloxy)ethenyl]-6,10-dimethyl-2,5,9-undecatrienal, A-70050
9-Oxogymnomitryl acetate, *in* G-70035

C$_{17}$H$_{24}$O$_4$

3β-Acetoxydrimenin, *in* D-60519
9β-Acetoxy-4,11(13)-eudesmadien-12-oic acid, *in* H-60135
2-(7-Hydroxy-3,7-dimethyl-2-octenyl)-6-methoxy-1,4-benzoquinone, H-60123
Isotrichodermin, *in* E-70028
Norstrictic acid, N-70080
Sterpuric acid; 13-Hydroxy, 3,13-ethylidene acetal, *in* S-60053

C$_{17}$H$_{24}$O$_5$

3β-Acetoxy-10α-hydroxy-4,11(13)-cadinadien-12-oic acid, *in* Z-70001
Altamisin, *in* A-70084
Cinnamodial, C-70177
▷Naematolin, N-70001
Neovasinin, N-70029
Nitrosin, N-60045
4,5-Seco-neopulchell-5-ene; 2-Ac, *in* S-70025
4,5-Seco-neopulchell-5-ene; 4-Ac, *in* S-70025

C$_{17}$H$_{24}$O$_6$

2α-Acetoxy-11α,13-dihydroconfertin, *in* D-60211
Blumealactone *C*, B-70172
15-Desacetyltetraneurin *C*, *in* T-60158
6β-Hydroxypulchellin 2-*O*-acetate, *in* P-70138
Naematolin B, *in* N-70001
Paramicholide, P-70011
Tetraneurin *D*, T-60158

C$_{17}$H$_{24}$O$_9$

Syringin, *in* S-70046

C$_{17}$H$_{24}$O$_{10}$

5-(Hydroxymethyl)-1,2,3,4-cyclohexanetetrol; Penta-Ac, *in* H-70170

C$_{17}$H$_{24}$O$_{12}$

Sesamoside, S-70036

C$_{17}$H$_{26}$O

Acetylcedrene, A-70032

C$_{17}$H$_{26}$O$_2$

Coralloidin C, *in* E-60066
Coralloidin E, *in* E-60065
Cyperenyl acetate, *in* C-70259
Dihydrotochuinyl acetate, *in* T-70202
11-Phenylundecanoic acid, P-70096
12-Sesquisabinenol acetate, *in* S-70037
13-Sesquisabinenol acetate, *in* S-70037

C$_{17}$H$_{26}$O$_3$

9α-Hydroxygymnomitryl acetate, *in* G-70035
3-(1,3,5,7,9-Tetradecapentaenyloxy)-1,2-propanediol, T-70024

C$_{17}$H$_{26}$O$_4$

3β-Acetoxy-4,10(14)-oplopadiene-8β,9α-diol, *in* O-70042
10α-Ethoxy-3β-hydroxy-4,11(13)-cadinadien-12-oic acid, *in* Z-70001
Norhardwickiic acid, N-70077

C$_{17}$H$_{26}$O$_{10}$

Adoxoside, A-70067

$C_{17}H_{26}O_{11}$
5-Deoxypulchelloside I, *in* P-70139
6β-Hydroxyadoxoside, *in* A-70067
Shanzhiside; Me ester, *in* S-60028

$C_{17}H_{26}O_{12}$
Pulchelloside I, P-70139

$C_{17}H_{28}O_2$
5,6-Decanediol; Benzyl ether, *in* D-60013

$C_{17}H_{28}O_3$
6β-Acetoxy-1α,4α-epoxyeudesmane, *in* E-60019
Fauronyl acetate, *in* F-70002
5,10(14)-Germacradiene-1,4-diol; 1-Ac, *in* G-70009

$C_{17}H_{28}O_4$
6β-Acetoxy-4(15)-eudesmene-1β,5α-diol, *in* E-60068
Shiromodiol; 8-Ac, *in* S-70039

$C_{17}H_{28}O_5$
12-Acetoxy-5,8-dihydroxyfarnesol, *in* T-60363

$C_{17}H_{29}N$
2,4,6-Tri-*tert*-butylpyridine, T-70217

$C_{17}H_{30}O$
3,4,7,11-Tetramethyl-6,10-tridecadienal, T-60156

$C_{17}H_{32}$
1,1,2,2,3,3,4,4,5,5-Decamethyl-6-methyl-enecyclohexane, D-60012

$C_{17}H_{32}O_2$
Cyclopentadecanone; Ethylene acetal, *in* C-60219

$C_{17}H_{32}O_3$
2-Tetradecyloxiranecarboxylic acid, T-60041

$C_{17}H_{34}O$
Undecamethylcyclohexanol, U-60003

$C_{17}H_{36}I_2N_2$
Camphonium, *in* T-70277

$C_{17}H_{36}N_2^{\oplus\oplus}$
Trimethidinium(2+), T-70277

$C_{18}H_6N_6O_{12}$
Benzo[1,2-*b*:3,4-*b'*:5,6-*b''*]tripyrazine-2,3,6,7,10,11-tetracarboxylic acid, B-60051

$C_{18}H_8N_2O_2$
1,1'-(1,2-Dioxo-1,2-ethanediyl)bis-1*H*-indole-2,3-dione, D-70466

$C_{18}H_8O_6$
10-Hydroxy-2*H*,13*H*-furo[3,2-*c*:5,4-*h'*]bis[1]-benzopyran-2,13-dione, H-70133

$C_{18}H_9N_3O_3$
19,20,21-Triazatetracyclo[13.3.1.1^{3,7}.1^{9,13}]-heneicosa-1(19),3,5,7(21),9,11,13(20),15,17-nonaene-2,8,14-trione, T-60236

$C_{18}H_{10}$
▷Cyclopenta[*cd*]pyrene, C-60223

$C_{18}H_{10}N_2$
9,10-Dihydro-9,10-ethenoanthracene-11,12-dicarboxylic acid; Dinitrile, *in* D-70208

$C_{18}H_{10}N_2O_2$
Triphenodioxazine, T-60378

$C_{18}H_{10}O$
Pyreno[1,2-*c*]furan, P-70165
Pyreno[4,5-*b*]furan, P-60214
Pyreno[4,5-*c*]furan, P-70166

$C_{18}H_{10}OS$
Triphenyleno[1,12-*bcd*]thiophene; *S*-Oxide, *in* T-70310

$C_{18}H_{10}O_2$
Benz[*a*]anthracene-1,2-dione, B-70012
Benz[*a*]anthracene-3,4-dione, B-70013
Benz[*a*]anthracene-5,6-dione, B-70014
Benz[*a*]anthracene-7,12-dione, B-70015
Benz[*a*]anthracene-8,9-dione, B-70016
Naphth[2,3-*a*]azulene-5,12-dione, N-60006
1,2-Triphenylenedione, T-70308

1,4-Triphenylenedione, T-70309

$C_{18}H_{10}O_2S$
Triphenyleno[1,12-*bcd*]thiophene; *S*-Dioxide, *in* T-70310

$C_{18}H_{10}O_6$
Bicoumol, B-70093

$C_{18}H_{10}O_7$
Edgeworthin, *in* D-70006

$C_{18}H_{10}S$
Acenaphtho[1,2-*b*]benzo[*d*]thiophene, A-70016
Triphenyleno[1,12-*bcd*]thiophene, T-70310

$C_{18}H_{10}S_2$
Benzo[1,2-*b*:4,5-*b'*]bis[1]benzothiophene, B-60018
Benzo[1,2-*b*:5,4-*b'*]bis[1]benzothiophene, B-60019
Phenanthro[1,10-*bc*:8,9-*b',c'*]bisthiopyran, P-60072

$C_{18}H_{11}NO_2$
1*H*-Benz[*e*]indole-1,2(3*H*)-dione; *N*-Ph, *in* B-70026
Benz[*cd*]indol-2-(1*H*)-one; *N*-Benzoyl, *in* B-70027

$C_{18}H_{12}$
$\Delta^{1,1'}$-Biindene, B-60109
Cyclohepta[*a*]phenalene, C-60199

$C_{18}H_{12}N_2O_2$
2,5-Dihydro-3,6-diphenylpyrrolo[3,4-*c*]pyrrole-1,4-dione, D-70204
3,6-Diphenyl-2,5-dihydropyrrolo[3,4-*c*]pyrrole-1,4-dione, D-70477

$C_{18}H_{12}N_4$
5,14-Dihydroquinoxalino[2,3-*b*]phenazine, D-70264
[1,2,4,5]Tetrazino[1,6-*a*:4,3-*a'*]diisoquinoline, T-60172
[1,2,4,5]Tetrazino[1,6-*a*:4,3-*a'*]diquinoline, T-60173

$C_{18}H_{12}N_6$
2,4,6-Tri-2-pyridinyl-1,3,5-triazine, T-70316
2,4,6-Tri-3-pyridinyl-1,3,5-triazine, T-70317
2,4,6-Tri-4-pyridinyl-1,3,5-triazine, T-70318

$C_{18}H_{12}N_{12}O_6$
Benzo[1,2-*b*:3,4-*b'*:5,6-*b''*]tripyrazine-2,3,6,7,10,11-tetracarboxylic acid; Hexaamide, *in* B-60051

$C_{18}H_{12}O$
▷Chrysene-5,6-oxide, C-70174
3-Hydroxybenz[*a*]anthracene, H-70104
Triphenylene-1,2-oxide, T-60385

$C_{18}H_{12}O_2$
1-Pyreneacetic acid, P-60208
4-Pyreneacetic acid, P-60209

$C_{18}H_{12}O_2S_2$
5a,5b,11a,11b-Tetrahydrocyclobuta[1,2-*b*:4,3-*b'*]benzothiopyran-11,12-dione, T-70063

$C_{18}H_{12}O_4$
9,10-Dihydro-9,10-ethenoanthracene-11,12-dicarboxylic acid, D-70208
(Diphenylpropadienylidene)propanedioic acid, D-70492
Glabone, G-60021
Kanjone, K-70003
Pongone, P-70108

$C_{18}H_{12}O_4S$
5-Hydroxy-2-(phenylthio)-1,4-naphthoquinone; Ac, *in* H-70213
5-Hydroxy-3-(phenylthio)-1,4-naphthoquinone; Ac, *in* H-70214

$C_{18}H_{12}O_6$
Grevilline *A*, G-60038

$C_{18}H_{12}O_7$
Grevilline B, *in* G-60038
Tartaric acid; Anhydride, dibenzoyl, *in* T-70005

$C_{18}H_{12}O_8$
Grevilline C, *in* G-60038
Grevilline D, *in* G-60038

$C_{18}H_{12}O_9$
Gomphidic acid, G-70026
Substictic acid, S-60060

$C_{18}H_{13}N_3$
4-Anilino-1,10-phenanthroline, *in* A-70181

$C_{18}H_{14}$
1,1'-Bi-1*H*-indene, B-60108
[2.2.2](1,2,3)Cyclophane-1,9-diene, C-60229
7,12-Dihydropleiadene, D-60269

$C_{18}H_{14}N_4$
2,2'-Bis(2-imidazolyl)biphenyl, B-70148

$C_{18}H_{14}N_6$
1,4-Bis(2-pyridylamino)phthalazine, B-60172

$C_{18}H_{14}O$
2,7-Diphenyloxepin, D-60486

$C_{18}H_{14}O_2$
1,5-Diacetylanthracene, D-70028
1,6-Diacetylanthracene, D-70029
1,8-Diacetylanthracene, D-70030
9,10-Diacetylanthracene, D-70031

$C_{18}H_{14}O_4$
3,4-Acenaphthenedicarboxylic acid; Di-Et ester, *in* A-60013
1,8-Anthracenedicarboxylic acid; Di-Me ester, *in* A-60281
4-Oxo-2-phenyl-4*H*-1-benzopyran-3-carboxylic acid; Et ester, *in* O-60085
3,4-Phenanthrenedicarboxylic acid; Di-Me ester, *in* P-60068

$C_{18}H_{14}O_6$
2,3-Dibenzoylbutanedioic acid, D-70083
3',4'-Dimethoxy-7,8-methylenedioxyisoflavone, *in* T-70112
6,7-Dimethoxy-3',4'-methylenedioxyisoflavone, *in* T-70111

$C_{18}H_{14}O_7$
7-Hydroxy-5,6-dimethoxy-3',4'-methyl-enedioxyisoflavone, *in* P-70032

$C_{18}H_{14}O_8$
8-Acetoxy-2',3,5-trihydroxy-7-methoxyflavone, *in* P-70029
2,3-Di-*O*-benzoyltartaric acid, *in* T-70005

$C_{18}H_{15}NO_3$
10-Amino-9-anthracenecarboxylic acid; *N*-Ac, Me ester, *in* A-70088
2,2',2''-Nitrilotrisphenol, N-70039
4,4',4''-Nitrilotrisphenol, N-70040

$C_{18}H_{15}N_3$
Amino-1,4-benzoquinone; Dianil, *in* A-60089

$C_{18}H_{16}$
Bii(yclononatetraenylidene), B-70104
1,1'-Biindanylidene, B-70120
1,2-Dibenzylidenecyclobutane, D-60083
1,2-Diphenyl-1,4-cyclohexadiene, D-60477
1,3,5,7-Tetraethynyladamantane, T-70025

$C_{18}H_{16}N_2S_2$
4,6-Diamino-1,3-benzenedithiol; Di-*S*-Ph, *in* D-70037

$C_{18}H_{16}O_2$
2,7-Dihydroxy-1,6-dimethyl-5-vinylphenanthrene, D-70294
Eupomatenoid 6, E-60075

$C_{18}H_{16}O_3$
2-(2-Hydroxy-4-methoxyphenyl)-5-(1-propenyl)benzofuran, *in* H-60214
6-Hydroxy-7,9-octadecadiene-11,13,15,17-tetraynoic acid, H-70195
2-(4-Hydroxyphenyl)-7-methoxy-5-(1-propenyl)benzofuran, *in* H-60214

$C_{18}H_{16}O_4$
3-Butene-1,2-diol; Dibenzoyl, *in* B-70291
Danshexinkun *A*, D-60002

2-(2,4-Dihydroxyphenyl)-7-methoxy-5-(1-propenyl)benzofuran, *in* H-60214

$C_{18}H_{16}O_5$

1,3-Dihydroxy-2-hydroxymethylanthraquinone; Tri-Me ether, *in* D-70306

▷ 2-Ethoxymethyl-3-hydroxy-1-methoxyanthraquinone, *in* D-70306

8-Methoxybonducellin, *in* B-60192

Puerol *B*, *in* P-70137

5,6,7-Trimethoxyflavone, *in* T-60322

2',4',7-Trimethoxyisoflavone, *in* T-70264

4',6,7-Trimethoxyisoflavone, *in* T-60325

4',7,8-Trimethoxyisoflavone, *in* T-70265

$C_{18}H_{16}O_6$

2,3,4-Biphenyltriol; Tri-Ac, *in* B-70124

6,7-Dimethoxy-3',4'-methylenedioxyflavanone, *in* T-70106

5-Hydroxy-2',6,6'-trimethoxyflavone, *in* T-70107

Isoamericanin *A*, I-60087

Pedicellosine, P-70015

Scleroderolide, S-60013

1,3,5,8-Tetrahydroxy-2-(3-methyl-2-butenyl)-xanthone, T-60114

2',5',6-Trihydroxy-3,5,7-trimethoxyflavone, *in* H-60064

$C_{18}H_{16}O_7$

Arizonin C_1, A-70254

3,4'-Dihydroxy-5,6,7-trimethoxyflavone, *in* P-60047

3,5-Dihydroxy-4',6,7-trimethoxyflavone, *in* P-60047

4',5-Dihydroxy-3,6,7-trimethoxyflavone, *in* P-60047

4',5-Dihydroxy-3,6,8-trimethoxyflavone, *in* P-60048

5,6-Dihydroxy-3,4',8-trimethoxyflavone, *in* P-60048

5,7-Dihydroxy-3,6,8-trimethoxyflavone, *in* P-70031

Notatic acid, N-70081

1-O-Demethylpsorospermindiol, *in* P-70133

Psorospermindiol; O-De-Me, epimer, *in* P-70133

$C_{18}H_{16}O_8$

2',3',5-Trihydroxy-3,7,8-trimethoxyflavone, *in* H-70079

2',5,5'-Trihydroxy-3,7,8-trimethoxyflavone, *in* H-70080

3',4',5-Trihydroxy-5',6,7-trimethoxyflavone, *in* H-70081

3',5,7-Trihydroxy-4',5',6-trimethoxyflavone, *in* H-70081

$C_{18}H_{16}O_9$

Alectorialic acid, A-70081

3',5,5',7-Tetrahydroxy-3,4',8-trimethoxyflavone, *in* H-70021

$C_{18}H_{16}O_{10}$

3',4',5,5',7-Pentahydroxy-3,6,8-trimethoxyflavone, *in* O-70018

$C_{18}H_{17}N$

2,2-Dimethyl-4,4-diphenyl-3-butenoic acid; Nitrile, *in* D-70402

$C_{18}H_{17}NO_2$

10-Amino-9-anthracenecarboxylic acid; *N*-Di-Me, Me ester, *in* A-70088

$C_{18}H_{17}NO_5$

2-Amino-4-hydroxybutanoic acid; *O,N*-Dibenzoyl, *in* A-70156

2-Amino-3-phenylpentanedioic acid; *N*-Benzoyl, *in* A-60241

$C_{18}H_{17}N_3O_3$

Neoechinulin *E*, N-70023

$C_{18}H_{18}$

2-*tert*-Butylphenanthrene, B-70305

3-*tert*-Butylphenanthrene, B-70306

9-*tert*-Butylphenanthrene, B-70307

[2.2.2](1,2,3)Cyclophane, C-60228

1,2,3,6,7,8-Hexahydro-3a,5a-ethenopyrene, H-70056

1,3,6,8-Tetramethylphenanthrene, T-60151

2,4,5,7-Tetramethylphenanthrene, T-60152

$C_{18}H_{18}O$

2,2-Dimethyl-4,4-diphenyl-3-butenal, D-70401

$C_{18}H_{18}O_2$

Conocarpan, C-60161

2,2-Dimethyl-4,4-diphenyl-3-butenoic acid, D-70402

$C_{18}H_{18}O_3$

Olmecol, O-60040

$C_{18}H_{18}O_4$

Bharanginin, B-70089

1,2,5,6-Tetramethoxyphenanthrene, *in* T-60115

1,2,5,7-Tetramethoxyphenanthrene, *in* T-60116

1,2,6,7-Tetramethoxyphenanthrene, *in* T-60117

1,3,5,6-Tetramethoxyphenanthrene, *in* T-60118

1,3,6,7-Tetramethoxyphenanthrene, *in* T-60119

2,3,5,7-Tetramethoxyphenanthrene, *in* T-60120

2,3,6,7-Tetramethoxyphenanthrene, *in* T-60121

2,4,5,6-Tetramethoxyphenanthrene, *in* T-60122

3,4,5,6-Tetramethoxyphenanthrene, *in* T-60123

$C_{18}H_{18}O_5$

5,6-Dimethoxy-3-(4-methoxybenzyl)phthalide, D-60398

Echinofuran, E-60002

Homocyclolongipesin; Ac, *in* H-60088

2'-Hydroxy-4,4',6'-trimethoxychalcone, *in* H-60220

4-Hydroxy-2',4',6'-trimethoxychalcone, *in* H-60220

$C_{18}H_{18}O_6$

Alkannin acetate, *in* S-70038

5,8-Dihydroxy-6,10-dimethoxy-2-propyl-4*H*-naphtho[2,3-*b*]pyran-4-one, D-70292

5,8-Dihydroxy-2,3,6,7-tetramethyl-1,4-naphthoquinone; Di-Ac, *in* D-60377

7-Hydroxy-2',5,8-trimethoxyflavanone, *in* T-70104

Monocillin I, *in* N-60060

Shikonin acetate, *in* S-70038

3,4,7,8-Tetramethoxy-2,6-phenanthrenediol, *in* H-60066

$C_{18}H_{18}O_7$

Monomethylsulochrin, *in* S-70083

5-O-Methylsulochrin, *in* S-70083

1,2,3,4,8-Pentamethoxyxanthone, *in* P-60051

Senepoxide, S-60023

β-Senepoxide, *in* S-60023

3',4',5,6-Tetrahydroxy-7-methoxyflavanone, *in* P-70028

$C_{18}H_{18}O_8$

Arizonin A_2, *in* A-70255

Arizonin B_2, *in* A-70255

$C_{18}H_{19}NO$

2,2-Dimethyl-4,4-diphenyl-3-butenal; Oxime, *in* D-70401

2-Phenylcyclopentylamine; *N*-Benzoyl, *in* P-60086

2-Phenylcyclopentylamine; Benzoyl, *in* P-60086

$C_{18}H_{19}NO_3S$

2-Amino-4,5,6,7-tetrahydrobenzo[*b*]thiophene-3-carboxylic acid; Et ester, *N*-benzoyl, *in* A-60258

$C_{18}H_{19}NO_7S$

Fibrostatin *A*, *in* F-70008

$C_{18}H_{19}NO_8S$

Fibrostatin *C*, *in* F-70008

Fibrostatin *D*, *in* F-70008

Fibrostatin *E*, *in* F-70008

$C_{18}H_{19}N_3O_5S$

Antibiotic BMY 28100, A-70221

$C_{18}H_{20}$

3,3-Dimethyl-4,4-diphenyl-1-butene, D-60418

5,13-Dimethyl[2.2]metacyclophane, D-60426

$C_{18}H_{20}N_2O_2$

5,6-Acenaphthenedicarboxylic acid; Bis(imethylamide), *in* A-70015

$C_{18}H_{20}N_2O_4$

N-Tyrosylphenylalanine, T-60412

$C_{18}H_{20}N_4$

Tricyclo[3.3.1.1^{3,7}]decane-1,3,5,7-tetra-acetonitrile, *in* A-70064

$C_{18}H_{20}O_2$

Pentamethylphenol; Benzoyl, *in* P-60058

Salviolone, S-70006

$C_{18}H_{20}O_3$

2,3-Dihydro-2-(4-hydroxyphenyl)-5-(3-hydroxypropyl)-3-methylbenzofuran, D-60244

1-(4-Hydroxyphenyl)-2-(4-propenylphenoxy)-1-propanol, H-60215

Tetrahydrobis(4-hydroxyphenyl)-3,4-dimethylfuran, T-60061

$C_{18}H_{20}O_4$

Thujin, T-70190

$C_{18}H_{20}O_5$

9,10-Dihydro-7-hydroxy-2,3,4,6-tetramethoxyphenanthrene, *in* D-70246

Longipesin; Propanoyl, *in* L-60034

9-Methyllongipesin; O^9-Ac, *in* M-70091

Monocillin II, *in* N-60060

$C_{18}H_{20}O_6$

3,3'-Dihydroxy-4',5,7-trimethoxyflavan, *in* P-70026

3,4'-Dihydroxy-3',5,7-trimethoxyflavan, *in* P-70026

3-Methoxy-6-[2-(3,4-5-trimethoxyphenyl)-ethenyl]-1,2-benzenediol, *in* T-60344

Monocillin III, *in* N-60060

3,3',4',5,7-Pentahydroxyflavan; $O^{3'},O^5,O^7$-Tri-Me, *in* P-70026

Syringopicrogenin *A*, S-60067

$C_{18}H_{20}O_7$

Syringopicrogenin B, *in* S-60067

$C_{18}H_{20}Se_2$

9,18-Dimethyl-2,11-diselena[3.3]metacyclophane, D-70404

$C_{18}H_{21}N_3O_6$

10-Deazariboflavin, *in* P-60245

$C_{18}H_{22}N_2O_3S_2$

Dithiosilvatin, D-60511

$C_{18}H_{22}O_3$

Nimbinone, N-70037

Nimbione, N-70038

$C_{18}H_{22}O_5$

Monocillin IV, *in* N-60060

▷ Zearalenone, Z-60002

$C_{18}H_{22}O_6$

Combretastatin, C-70192

8'-Hydroxyzearalenone, *in* Z-60002

3-Methoxy-6-[2-(3,4,5-trimethoxyphenyl)-ethyl]1,2-benzenediol, *in* T-60344

Monocillin V, *in* N-60060

$C_{18}H_{24}$

1,2,3,4,5,6,7,8,9,10,11,12-Dodecahydrotriphenylene, D-60516

1,3,5,7-Tetravinyladamantane, T-70154

$C_{18}H_{24}N_2$

1,2-Diphenyl-1,2-ethanediamine; *N*-Tetra-Me, *in* D-70480

$C_{18}H_{24}O_2$

3-(2-Hydroxy-4,8-dimethyl-3,7-nonadienyl)-benzaldehyde, H-60122

Norsalvioxide, N-70079

$C_{18}H_{24}O_5$

Nordinone, N-60060

$C_{18}H_{24}O_7$
Malabarolide, M-70008
Nordinonediol, *in* N-60060

$C_{18}H_{24}O_8$
1,3,5,7-Adamantanetetraacetic acid, A-70064
Hyptolide, H-70236

$C_{18}H_{26}O_8$
Boronolide, B-60193

$C_{18}H_{27}N_3$
1,3,5-Tripyrrolidinobenzene, T-70319

$C_{18}H_{27}N_3O_3$
1,3,5-Trimorpholinobenzene, T-70294 ·

$C_{18}H_{28}$
9,9'-Bii(icyclo[3.3.1]nonylidene), B-60079

$C_{18}H_{28}BF_4NO_2$
1,1-*tert*-Butyl-3,3-diethoxy-2-azaallenium(1+);
Tetrafluoroborate, *in* B-70301

$C_{18}H_{28}O_2$
Spiro[3,4-cyclohexano-4-hydroxybicyclo[3.3.1]-
nonan-9-one-2,1'-cyclohexane], S-70063

$C_{18}H_{28}O_3$
ent-6β,17-Dihydroxy-14,15-bisnor-7,11E-
labdadien-13-one, D-60311
14,15-Dinor-13-oxo-7-labden-17-oic acid,
D-70457

$C_{18}H_{28}O_4$
ent-12-Hydroxy-13,14,15,16-tetranor-1(10)-
halimen-18-oic acid; Ac, *in* H-60231
Xestodiol, X-70006

$C_{18}H_{28}O_5$
Lachnellulone, L-70015

$C_{18}H_{28}S_6$
1,3,4,6-Tetrakiss(sospropylthio)thieno[3,4-c]-
thiophene, *in* T-70170

$C_{18}H_{29}NO_2$
1,3,5-Tri-*tert*-butyl-2-nitrobenzene, T-70216

$C_{18}H_{30}$
▷Hexaethylbenzene, H-70043
Octadecahydrotriphenylene, O-70004
1,3,5-Tributylbenzene, T-70215

$C_{18}H_{30}O_3$
Colneleic acid, C-70189
▷Conocandin, C-70194
Sterebin D, S-60052

$C_{18}H_{30}O_4$
11-Hydroxy-12,13-epoxy-9,15-octadecadienoic
acid, H-70127
Sterebin A, *in* S-60052
3,7,11-Trimethyl-2,6,10-dodecatrienoic acid;
2,3-Dihydroxypropyl ester, *in* T-60364

$C_{18}H_{30}O_5$
Gloeosporone, G-60025

$C_{18}H_{30}S_3$
7,14,21-Trithiatrispiro[5.1.5.1.5.1]heneicosane,
in C-60209

$C_{18}H_{32}N_4O_8$
1,4,8,11-Tetraazacyclotetradecane-1,4,8,11-
tetraacetic acid, T-60017

$C_{18}H_{32}O_2$
2,4-Octadecadienoic acid, O-70003

$C_{18}H_{32}O_3$
13-Hydroxy-9,11-octadecadienoic acid,
H-70196

$C_{18}H_{32}O_5$
Aspicillin, A-60312

$C_{18}H_{34}O$
2,4-Octadecadien-1-ol, O-60004

$C_{18}H_{34}O_2$
1-Tetradecylcyclopropanecarboxylic acid,
T-60040

$C_{18}H_{34}O_3$
2-Tetradecyloxiranecarboxylic acid; Me ester,
in T-60041

$C_{18}H_{36}N_2O_6$
▷4,7,13,16,21,24-Hexaoxa-1,10-
diazabicyclo[8.8.8]hexacosane, H-60073

$C_{18}H_{36}O$
2-Octadecen-1-ol, O-60007

$C_{18}H_{36}O_2$
3-Octadecene-1,2-diol, O-60005
9-Octadecene-1,12-diol, O-60006

$C_{18}H_{36}O_5$
9,10,18-Trihydroxyoctadecanoic acid, T-60334

$C_{18}H_{38}$
3,4-Di-*tert*-butyl-2,2,5,5-tetramethylhexane,
D-70109

$C_{18}H_{38}N_2$
1,1,1,1',1'-Tetra-*tert*-butylazomethane, T-60024

$C_{18}H_{42}N_6$
1,4,7,13,16,19-Hexaazacyclotetracosane,
H-70029

$C_{18}H_{42}N_6O_2$
1,13-Dioxa-4,7,10,16,20,24-hexaazacyclo-
hexacosane, D-70459

$C_{18}N_6O_9$
Benzo[1,2-b:3,4-b':5,6-b'']tripyrazine-
2,3,6,7,10,11-tetracarboxylic acid;
Trianhydride, *in* B-60051

$C_{18}N_{12}$
Benzo[1,2-b:3,4-b':5,6-b'']tripyrazine-
2,3,6,7,10,11-tetracarboxylic acid;
Hexanitrile, *in* B-60051

$C_{19}Cl_{15}$
Triss(entachlorophenyl)methyl, *in* T-60405

$C_{19}Cl_{21}Sb$
Triss(entachlorophenyl)methyl hexachloroantimonate, *in* T-60405

$C_{19}HCl_{15}$
Triss(entachlorophenyl)methane, T-60405

$C_{19}H_{11}ClO$
2-Triphenylenecarboxylic acid; Chloride, *in*
T-60384

$C_{19}H_{11}N$
2-Cyanotriphenylene, *in* T-60384

$C_{19}H_{11}NOS$
Dibenzo[f,h]thieno[2,3-b]quinoline; 1-Oxide, *in*
D-70079

$C_{19}H_{11}NS$
Dibenzo[f,h]thieno[2,3-b]quinoline, D-70079

$C_{19}H_{12}N_2O_2$
Cyclohepta[1,2-b:1,7-b']bis[1,4]benzoxazine,
C-70219

$C_{19}H_{12}N_4O_7$
Benzo[a]quinolizinium(1+); Picrate, *in*
B-70043

$C_{19}H_{12}O$
2-Triphenylenecarboxaldehyde, T-60382

$C_{19}H_{12}O_2$
2-Benzoylnaphtho[1,8-bc]pyran, B-70076
1-Triphenylenecarboxylic acid, T-60383
2-Triphenylenecarboxylic acid, T-60384

$C_{19}H_{12}O_5$
1,6,8-Trihydroxy-3-methylbenz[a]anthracene-
7,12-dione, T-70270

$C_{19}H_{12}O_6$
Bhubaneswin, *in* B-70093

$C_{19}H_{12}O_7$
Daphnoretin, D-70006

$C_{19}H_{13}N$
8H-Indolo[3,2,1-de]acridine, I-70014

$C_{19}H_{13}NO_2S$
10H-Phenothiazine-3-carboxylic acid; N-Ph, *in*
P-70057

$C_{19}H_{13}NO_4S$
10H-Phenothiazine-3-carboxylic acid; N-Ph,
5,5-dioxide, *in* P-70057

$C_{19}H_{13}NO_8$
Protetrone, P-70124

$C_{19}H_{14}$
4b,10b-Methanochrysene, M-70040

$C_{19}H_{14}N_4$
4-Amino-1,2,3-benzotriazine; 3,N(4)-Di-Ph, *in*
A-60092

$C_{19}H_{14}O$
3-Methoxybenz[a]anthracene, *in* H-70104

$C_{19}H_{14}O_2$
1-Pyreneacetic acid; Me ester, *in* P-60208
4-Pyreneacetic acid; Me ester, *in* P-60209

$C_{19}H_{14}O_4$
4',7-Dimethoxyisoflavone, *in* D-60340
Ochromycinone, O-60003

$C_{19}H_{14}O_6$
4',7-Dihydroxyisoflavone; Di-Ac, *in* D-60340

$C_{19}H_{14}O_7$
6-Aldehydoisoophiopogone A, A-60071

$C_{19}H_{14}O_{10}$
Constictic acid, C-60163

$C_{19}H_{14}S_2$
2-Dibenzothiophenethiol; S-Benzyl, *in* D-60078
4-Dibenzothiophenethiol; S-Benzyl, *in* D-60079

$C_{19}H_{15}N_3$
Azidotriphenylmethane, A-60352

$C_{19}H_{16}$
4-(2,4,6-Cycloheptatrien-1-ylidene)-
bicyclo[5.4.1]dodeca-2,5,7,9,11-pentaene,
C-60207

$C_{19}H_{16}O$
2,5-Dimethyl-3,4-diphenyl-2,4-cyclopentadien-
1-one, D-70403

$C_{19}H_{16}O_4$
2-(4-Hydroxyphenyl)-7-methoxy-5-(1-
propenyl)-3-benzofurancarboxaldehyde,
H-60212
Moracin D, M-60143
Moracin E, M-60144
Moracin G, M-60145

$C_{19}H_{16}O_5$
3',4'-Deoxypsorospermin, D-60021

$C_{19}H_{16}O_6$
6-Aldehydoisoophiopogone B, A-60072
Psorospermin, P-70132

$C_{19}H_{16}O_7$
Fridamycin E, *in* F-70043
5'-Hydroxypsorospermin, *in* P-70132
5,6,7-Trimethoxy-3',4'-methyl-
enedioxyisoflavone, *in* P-70032

$C_{19}H_{17}ClO_6$
3',4'-Deoxy-4'-chlorosorospermin-3'-ol, *in*
D-60021

$C_{19}H_{17}NO_2$
2-Hydroxycyclononanone; Oxime, *in* H-60112

$C_{19}H_{18}$
Bicyclo[12.4.1]nonadec-1,3,5,7,9,11,13,15,17-
nonaene, B-70101

$C_{19}H_{18}N_2$
(Triphenylmethyl)hydrazine, T-60388

$C_{19}H_{18}O_3$
Eupomatenoid 5, *in* E-60075
1-[4-Hydroxy-3-(3-methyl-1,3-butadienyl)-
phenyl]-2-(3,5-dihydroxyphenyl)ethylene,
H-70167
Kachirachirol A, *in* K-60001

$C_{19}H_{18}O_4$
Demethylfruticulin A, *in* F-60076
2-(2-Hydroxy-4-methoxyphenyl)-7-methoxy-5-
(1-propenyl)benzofuran, *in* H-60214
Isotanshinone IIB, I-60138
6-Methoxy-2-[2-(4-methoxyphenyl)ethyl]-4H-
1-benzopyran-4-one, M-60039
Moracin C, *in* M-60146

C$_{19}$H$_{18}$O$_5$
Benzophenone-2,2'-dicarboxylic acid; Di-Et ester, *in* B-70041
Egonol, E-70001

C$_{19}$H$_{18}$O$_6$
Ciliarin, *in* C-70007
Fruticulin *B*, F-60077
2',5,6,6'-Tetramethoxyflavone, *in* T-70107
3',4',5,6-Tetramethoxyflavone, *in* T-70108
3',4',6,7-Tetramethoxyisoflavone, *in* T-70111
1,5,8-Trihydroxy-3-methyl-2-(3-methyl-2-butenyl)xanthone, *in* T-60114

C$_{19}$H$_{18}$O$_7$
3',4'-Deoxypsorospermin-3',4'-diol, *in* D-60021
5-Hydroxy-2',3,7,8-tetramethoxyflavone, *in* P-70029
5-Hydroxy-3,4',6,7-tetramethoxyflavone, *in* P-60047
5-Hydroxy-3,4',6,7-tetramethoxyflavone, *in* P-60047
6-Hydroxy-3,4',5,7-tetramethoxyflavone, *in* P-60047
Psorospermindiol, P-70133

C$_{19}$H$_{18}$O$_8$
2'-Acetoxy-3,5-dihydroxy-7,8-dimethoxyflavone, *in* P-60045
3',5-Dihydroxy-4',6,7,8-tetramethoxyflavone, D-60375
4',5-Dihydroxy-3',5',6,7-tetramethoxyflavone, *in* H-70081
5,7-Dihydroxy-3',4',5',6-tetramethoxyflavone, *in* H-70081
5,7-Dihydroxy-3',4',6,8-tetramethoxyflavone, D-60376
5',5'-Dihydroxy-2',3,7,8-tetramethoxyflavone, *in* H-70080
5',6-Dihydroxy-2',3,5,7-trimethoxyflavone, *in* H-60064

C$_{19}$H$_{18}$O$_9$
3,4',5-Trihydroxy-3',6,7,8-tetramethoxyflavone, *in* H-60024
3',5,5'-Trihydroxy-3,4',7,8-tetramethoxyflavone, *in* H-70021
3',5,7-Trihydroxy-3,4',5',8-tetramethoxyflavone, *in* H-70021

C$_{19}$H$_{18}$O$_{10}$
2',4',5,7-Tetrahydroxy-3,5',6,8-tetramethoxyflavone, *in* O-60022
2',5,5',7-Tetrahydroxy-3,4',6,8-tetramethoxyflavone, *in* O-60022
3',5,5',6-Tetrahydroxy-3,4',7,8-tetramethoxyflavone, *in* O-70018
2',5,7-Trihydroxy-3,4',5',6,8-pentamethoxyflavone, *in* O-60022

C$_{19}$H$_{18}$O$_{11}$
1,3,5,8-Tetrahydroxyxanthone; 8-Glucoside, *in* T-70123

C$_{19}$H$_{19}$ClO$_7$
Eriodermic acid, E-70032
Methyl 5-chloro-4-*O*-demethylbarbatate, *in* B-70007

C$_{19}$H$_{20}$
1-Methyl-4,4-diphenylcyclohexene, M-60065
3-Methyl-4,4-diphenylcyclohexene, M-60066

C$_{19}$H$_{20}$O$_2$
3,3-Dimethyl-5,5-diphenyl-4-pentenoic acid, D-60420
Hermosillol, H-60029

C$_{19}$H$_{20}$O$_3$
Cryptotanshinone, C-60173
2,3-Dihydro-2-(4-hydroxy-3-methoxyphenyl)-3-methyl-5-(1-propenyl)benzofuran, *in* C-60161
2,3-Dihydro-2-(4-hydroxyphenyl)-7-methoxy-3-methyl-5-(1-propenyl)benzofuran, *in* C-60161

C$_{19}$H$_{20}$O$_4$
Galipein, G-60001
Honyudisin, H-70093
Kachirachirol *B*, K-60001

C$_{19}$H$_{20}$O$_5$
Agasyllin, A-60064
Homocyclolongipesin; Propanoyl, *in* H-60088
3-(4-Hydroxyphenyl)-1-(2,4,6-trihydroxyphenyl)-2-propen-1-one; Tetra-Me ether, *in* H-60220
2,3,4,7,9-Pentamethoxyphenanthrene, *in* P-70035
Teuchamaedryn *A*, T-70157

C$_{19}$H$_{20}$O$_6$
Asadanin, A-60308
Calaxin, C-70007
1(10)E,8E-Millerdienolide, *in* M-60136
Khellactone; O^{10}-Angeloyl, *in* K-70013
Khellactone; O^9-Angeloyl, *in* K-70013

C$_{19}$H$_{20}$O$_7$
Annulin *A*, A-60276
Barbatic acid, B-70007
4β,15-Epoxy-1(10)E,8E-millerdienolide, *in* M-60136
Isoelephantopin, I-60099

C$_{19}$H$_{20}$O$_8$
Eximin, *in* A-70248

C$_{19}$H$_{20}$O$_{11}$
6-O-Galloylarbutin, *in* A-70248

C$_{19}$H$_{21}$NO$_8$S
Fibrostatin *B*, *in* F-70008

C$_{19}$H$_{21}$NO$_9$S
Fibrostatin *F*, *in* F-70008

C$_{19}$H$_{22}$N$_2$O$_2$
2,4-Pentanediamine; N^2,N^4-Dibenzoyl, *in* P-70039

C$_{19}$H$_{22}$O$_2$
5-Hydroxy-1,7-diphenyl-3-heptanone, *in* D-70481

C$_{19}$H$_{22}$O$_3$
Gravelliferone, G-60037
Miltiodiol, M-70149

C$_{19}$H$_{22}$O$_4$
Chalepin, C-60031
2,3-Dihydro-2-(4-hydroxy-3-methoxyphenyl)-5-(3-hydroxypropyl)-3-methylbenzofuran, *in* D-60244
1,5-Dihydroxy-1,7-diphenyl-3-heptanone, *in* D-60480
Gibberellin A$_{73}$, *in* G-60018
1-(4-Hydroxyphenyl)-2-(2-methoxy-4-propenylphenoxy)-1-propanol, *in* H-60215
8-Isovaleryloxy-2-(2,4-hexadiynylidene)-1,6-dioxaspiro[4.5]dec-3-ene, *in* H-70042
Neocryptotansohinone, N-60017
Tetrahydrobis(4-hydroxyphenyl)-3,4-dimethylfuran; 3'-Methoxy, *in* T-60061
Tetrahydro-2-(4-hydroxyphenyl)-5-(4-hydroxy-3-methoxyphenyl)-3,4-dimethylfuran, *in* T-60061
Yashabushiketodiol *B*, *in* D-60480

C$_{19}$H$_{22}$O$_5$
Fragransol *B*, F-70042
9-Methyllongipesin; 9-Propanoyl, *in* M-70091
Subexpinnatin, *in* H-70138

C$_{19}$H$_{22}$O$_6$
Antheridic acid, A-60278
Asadanol, *in* A-60308
5-Formylzearalenone, *in* Z-60002
Isotaxiresinol, I-60139
9-Millerenolide, M-60137
1(10)-Millerenolide, M-60136
Strigol, S-60057

C$_{19}$H$_{22}$O$_7$
4β,15-Epoxy-9E-millerenolide, *in* M-60137
4β,15-Epoxy-9Z-millerenolide, *in* M-60137
Tomenphantopin *B*, T-60222

C$_{19}$H$_{22}$O$_8$
Syringopicrogenin *C*, *in* S-60067

C$_{19}$H$_{22}$O$_{10}$
Aranochromanophthalide, A-60293

C$_{19}$H$_{23}$ClO$_7$
Chlororepdiolide, C-60133
Repensolide, R-60005

C$_{19}$H$_{23}$NO
1,2,3,4,5,6,7,8,8a,9a-Decahydrocarbazole; *N*-Benzoyl, *in* D-70012

C$_{19}$H$_{24}$O$_2$
1,7-Diphenyl-3,5-heptanediol, D-70481

C$_{19}$H$_{24}$O$_3$
1,7-Diphenyl-1,3,5-heptanetriol, D-60480

C$_{19}$H$_{24}$O$_4$
Gibberellin A$_9$, G-60018

C$_{19}$H$_{24}$O$_5$
Gibberellin A$_{20}$, G-70011
Gibberellin A$_{40}$, *in* G-60018
Gibberellin A$_{45}$, *in* G-60018
Gibberellin A$_{51}$, *in* G-60018

C$_{19}$H$_{24}$O$_6$
3β-Acetoxyhaageanolide acetate, *in* H-60001
2α-Acetoxylaurenobiolide, *in* H-70134
Afraglaucolide, A-70070
Artemisiaglaucolide, *in* D-70303
Caleine *E*, C-60005
Desacetylacanthospermal A, *in* A-60012
9β,15-Diacetoxy-1(10),4,11(13)-germacratrien-12,6α-olide, *in* D-60324
Gibberellin A$_{29}$, *in* G-70011
Gibberellin A$_{55}$, *in* G-60018
Gibberellin A$_{63}$, *in* G-60018
Gibberellin A$_{67}$, *in* G-70011
Gochnatolide, G-70024
Isoafraglaucolide, I-70060
Ivangustin; 8-Epimer, 6α-acetoxy, Ac, *in* I-70109

C$_{19}$H$_{24}$O$_7$
Achillolide *B*, A-60061
Cedronin, C-70027
7-Epicedronin, *in* C-70027
Eudesmaafraglaucolide, *in* T-70258
Gibberellin A$_{72}$, *in* G-70011
1α-Hydroxyafraglaucolide, *in* A-70070
1β-Hydroxyafraglaucolide, *in* A-70070
1α-Hydroxyisoafraglaucolide, *in* I-70060
Ligulatin A, *in* D-70025
1,3,9-Trihydroxy-4(15),11(13)-eudesmadien-12,6-olide; 1,9-Di-Ac, *in* T-70257

C$_{19}$H$_{26}$O$_2$
Acalycixeniolide *B*, A-60011

C$_{19}$H$_{26}$O$_4$
Subcordatolide *E*, S-70079

C$_{19}$H$_{26}$O$_5$
12,13-Diacetoxy-1(10)-aromadendren-2-one, *in* D-70283
Salvicanaric acid, S-70004
Subcordatolide *D*, S-70078

C$_{19}$H$_{26}$O$_6$
Deacetylviguiestin, *in* T-70003
Gibberellin A$_2$, G-60017
Ligulatin C, *in* D-70025
Pulchellin diacetate, *in* P-70138
Sintenin, S-70047
Tagitinin *E*, T-70003
Zexbrevin C, Z-70005

C$_{19}$H$_{26}$O$_7$
14-Acetoxytetraneurin D, *in* T-60158
1β,6α-Diacetoxy-4α-hydroxy-11-eudesmen-12,8α-olide, *in* T-70259
Resinoside, R-60006
Tetraneurin C, *in* T-60158

C$_{19}$H$_{26}$O$_{10}$
Ptelatoside A, *in* H-70209

C$_{19}$H$_{28}$O$_2$
Acalycixeniolide *A*, A-60010
Xestenone, X-70005

C$_{19}$H$_{28}$O$_4$
Coralloidin D, *in* E-60063

Flexilin, F-70012

$C_{19}H_{28}O_5$

3β-Acetoxy-10α-ethoxy-4,11(13)-cadinadien-12-oic acid, *in* Z-70001

1β,10α; 4α,5β-Diepoxy-8α-isobutoxyglechomanolide, *in* G-60022

1β,10α; 4α,5β-Diepoxy-8β-isobutoxyglechomanolide, *in* G-60022

$C_{19}H_{28}O_6$

Desacetyltetraneurin *D* 4-*O*-isobutyrate, *in* T-60158

Desacetyltetraneurin *D* 15-*O*-isobutyrate, *in* T-60158

$C_{19}H_{28}O_7$

Tagitinin *A*, T-70002

$C_{19}H_{28}O_{12}$

6-*O*-Acetylshanghiside methyl ester, *in* S-60028

Barlerin, *in* S-60028

$C_{19}I I_{30}N_6O_8$

Antiarrhythmic peptide (ox atrium), A-60283

$C_{19}H_{30}O_2$

Gracilin *F*, G-60035

$C_{19}H_{30}O_3$

Norjuslimdiolone, N-60061

17-Nor-7-oxo-8-labden-15-oic acid, *in* H-60198

$C_{19}H_{30}O_4$

ent-6α,15,16-Trihydroxy-19-nor-4-rosen-3-one, *in* N-60061

$C_{19}H_{30}O_5$

Shiromodiol; Di-Ac, *in* S-70039

$C_{19}H_{30}O_6$

3-Hydroxymethyl-7,11-dimethyl-2,6,11-dodecatriene-1,5,10-triol; 1,1'-Di-Ac, *in* H-70174

$C_{19}H_{30}O_8$

Icariside B$_1$, *in* G-70028

$C_{19}H_{32}O_2$

15-Hydroxy-17-nor-8-labden-7-one, H-60198

$C_{19}H_{32}O_3$

4,6-Dihydroxy-20-nor-2,7-cembradien-12-one, D-70323

$C_{19}H_{32}O_4$

4,6-Dihydroxy-7,8-epoxy-20-nor-2-cembren-12-one, *in* D-70323

$C_{19}H_{34}O_2$

2,4-Octadecadienoic acid; Me ester, *in* O-70003

$C_{19}H_{34}O_3$

13-Hydroxy-9,11-octadecadienoic acid; Me ester, *in* H-70196

$C_{19}H_{34}O_5$

3,6-Epidioxy-6-methoxy-4-octadecenoic acid, E-60008

$C_{19}H_{36}N_2O_5$

▷Lipoxamycin, L-70033

$C_{19}H_{37}NO_2$

14-Azaprostanoic acid, A-60332

$C_{19}H_{38}O_5$

9,10,18-Trihydroxyoctadecanoic acid; Me ester, *in* T-60334

$C_{19}H_{42}N_2O_8S_2$

▷Trimethidinium methosulphate, *in* T-70277

$C_{20}H_8O_4S_2$

Dibenzo[*b,i*]thianthrene-5,7,12,14-tetrone, D-70078

$C_{20}H_{10}O_2$

Indeno[1,2-*b*]fluorene-6,12-dione, I-70011

$C_{20}H_{12}$

Benz[*a*]aceanthrylene, B-70009

Benz[*d*]aceanthrylene, B-60009

Benz[*k*]aceanthrylene, B-60010

▷Benz[*e*]acephenanthrylene, B-70010

Benz[*j*]acephenanthrylene, B-70011

$C_{20}H_{12}Br_2$

2,2'-Dibromo-1,1'-binaphthyl, D-60089

4,4'-Dibromo-1,1'-binaphthyl, D-60090

$C_{20}H_{12}I_2$

2,2'-Diiodo-1,1'-binaphthyl, D-60385

4,4'-Diiodo-1,1'-binaphthyl, D-60386

$C_{20}H_{12}N_2$

Benzo[1,2-*f*:4,5-*f'*]diquinoline, B-70030

$C_{20}H_{12}N_4$

Cycloocta[2,1-*b*:3,4-*b'*]di[1,8]naphthyridine, C-70231

$C_{20}H_{12}O_4$

Tetraoxaporphycene, T-70146

$C_{20}H_{12}O_5$

Halenaquinone, *in* H-60002

$C_{20}H_{12}O_6$

Altertoxin III, A-60082

$C_{20}H_{12}S$

Anthra[1,2-*b*]benzo[*d*]thiophene, A-70214

Anthra[2,1-*b*]benzo[*d*]thiophene, A-70215

Anthra[2,3-*b*]benzo[*d*]thiophene, A-70216

Benzo[*b*]phenanthro[1,2-*d*]thiophene, B-70039

Benzo[*b*]phenanthro[4,3-*d*]thiophene, B-70040

Benzo[3,4]phenanthro[1,2-*b*]thiophene, B-60039

Benzo[3,4]phenanthro[2,1-*b*]thiophene, B-60040

$C_{20}H_{12}S_5$

2,2':5',2'':5'',2''':5''',2''''-Quinquethiophene, Q-70010

$C_{20}H_{13}NO_2$

9-Acridinecarboxylic acid; Ph ester, *in* A-70061

$C_{20}H_{14}$

1,2-Benzo[2.2]metaparacyclophan-9-ene, B-70036

9,10-Dihydrodicyclopenta[*c,g*]phenanthrene, D-60219

1-(Diphenylmethylene)-1*H*-cyclopropabenzene, D-70484

[2](2,6)-Naphthaleno[2]paracyclophane-1,11-diene, N-70006

$C_{20}H_{14}O$

1-Acetyltriphenylene, A-60059

2-Acetyltriphenylene, A-60060

$C_{20}H_{14}O_2$

3-Hydroxybenz[*a*]anthracene; Ac, *in* H-70104

1-Triphenylenecarboxylic acid; Me ester, *in* T-60383

2-Triphenylenecarboxylic acid; Me ester, *in* T-60384

$C_{20}H_{14}O_3$

7,8,8*a*,9*a*-Tetrahydrobenzo[10,11]chryseno[3,4-*b*]oxirene-7,8-diol, T-70041

$C_{20}H_{14}O_4$

4,4',5,5'-Tetrahydroxy-1,1'-binaphthyl, T-70102

$C_{20}H_{14}O_5$

Halenaquinol, H-60002

$C_{20}H_{14}O_6$

Alterperylenol, A-60078

▷Altertoxin II, A-60081

7,7'-Dimethoxy-[6,8'-bi-2*H*-benzopyran]-2,2'-dione, *in* B-70093

$C_{20}H_{14}O_7$

Oreojasmin, O-70044

$C_{20}H_{14}O_8S$

Halenaquinol; O^{16}-Sulfate, *in* H-60002

$C_{20}H_{14}S$

1,3-Diphenylbenzo[*c*]thiophene, D-70474

9-(Phenylthio)phenanthrene, *in* P-70049

$C_{20}H_{14}Se$

9-(Phenylseleno)phenanthrene, P-60133

$C_{20}H_{15}Br$

Bromotriphenylethylene, B-70276

$C_{20}H_{15}N$

10,11-Dihydro-5-phenyl-5*H*-dibenz[*b,f*]azepine, *in* D-70185

2,3-Diphenylindole, D-70483

5*H*,9*H*-Quino[3,2,1-*de*]acridine, Q-70006

$C_{20}H_{15}NO$

2-Acetyltriphenylene; Oxime, *in* A-60060

$C_{20}H_{15}NO_3S$

10*H*-Phenothiazine-3-carboxylic acid; *N*-Ph, Me ester, *in* P-70057

$C_{20}H_{15}NO_4S$

10*H*-Phenothiazine-3-carboxylic acid; *N*-Ph, 5,5-dioxide, Me ester, *in* P-70057

$C_{20}H_{16}$

1,2-Benzo[2.2]paracyclophane, B-70038

▷7,12-Dimethylbenz[*a*]anthracene, D-60405

$C_{20}H_{16}N_2$

9,10-Dihydro-9,10-ethenoanthracene-11,12-dicarboxylic acid; Di-Me ester, *in* D-70208

2,2'-Dimethyl-4,4'-biquinoline, D-60413

$C_{20}H_{16}N_2O_3$

1-[4-[6-(Diethylamino)-2-benzofuranyl]-phenyl]-1*H*-pyrrole-2,5-dione, D-60170

$C_{20}H_{16}N_4O_2$

Pyrazole blue, P-70153

$C_{20}H_{16}O_3$

3-(Benzyloxy)phenol; Benzoyl, *in* B-60073

$C_{20}H_{16}O_4$

Ochromycinone methyl ether, *in* O-60003

Tetraoxaporphyrinogen, T-70147

$C_{20}H_{16}O_5$

2,5-Biss(ydroxymethyl)furan; Dibenzoyl, *in* B-60160

Citrusinol, C-60152

$C_{20}H_{16}O_6$

Crotarin, C-60172

Dihydroalterperylenol, *in* A-60078

7,6-(2,2-Dimethylpyrano)-3,4',5-trihydroxyflavone, D-60449

$C_{20}H_{16}O_8$

Hydroxyversicolorone, H-70230

$C_{20}H_{16}O_{12}$

2,3,5,6,8-Pentahydroxy-1,4-naphthoquinone; Penta-Ac, *in* P-70033

$C_{20}H_{18}$

5,6-Didehydro-1,4,7,10-tetramethyl-dibenzo[*ae*]cyclooctene, D-60169

$C_{20}H_{18}N_2O$

3-Amino-1,2,3,4-tetrahydrocarbazole; N^3-Benzoyl, *in* A-60259

$C_{20}H_{18}O_2$

1,1,2-Triphenyl-1,2-ethanediol, T-70311

$C_{20}H_{18}O_4$

3,4-Dibenzoyl-2,5-hexanedione, D-60081

4,5-Phenanthrenedimethanol; Di-Ac, *in* P-60070

$C_{20}H_{18}O_5$

Crotalarin, C-60171

Moracin H, *in* M-60145

$C_{20}H_{18}O_6$

1,2-Diacetoxy-5,6-dimethoxyphenanthrene, *in* T-60115

Garveatin *A* quinone, G-60006

Platanetin, P-70105

$C_{20}H_{18}O_8$

Arborone, A-60294

Di-*p*-toluoyl tartrate, *in* T-70005

Moluccanin, M-70151

5,5',6,7-Tetramethoxy-3',4'-methylenedioxyflavone, *in* H-70081

$C_{20}H_{18}O_9$

2'-Acetoxy-5,5'-dihydroxy-3,7,8-trimethoxy-flavone, *in* H-70080

Cetraric acid, C-70035

C$_{20}$H$_{18}$O$_{10}$
Conphysodalic acid, C-60162

C$_{20}$H$_{19}$ClO$_7$
4'-Chloronephroarctin, in N-70032

C$_{20}$H$_{19}$N
2*H*-Dibenz[*e,g*]isoindole; *N-tert*-Butyl, *in*
D-70067

C$_{20}$H$_{19}$N$_3$
▷Rosaniline, R-60010

C$_{20}$H$_{20}$
1,1'-Bitetralinylidene, B-70165
▷7,12-Dimethylbenz[*a*]anthracene; 1,2,3,4-
Tetrahydro, *in* D-60405
Dodecahedrane, D-60514

C$_{20}$H$_{20}$O$_3$
Eupomatenoid 4, *in* E-60075

C$_{20}$H$_{20}$O$_4$
Eupomatenoid 7, *in* E-60075
Fruticulin *A*, F-60076
Glabranin, G-70015
Glabridin, G-70016
Moracin I, *in* M-60146
Psorolactone, P-60188
Pygmaeocine *E*, P-70144

C$_{20}$H$_{20}$O$_5$
Aphyllodenticulide, A-70243
7,8-(2,2-Dimethylpyrano)-3,4',5-
trihydroxyflavan, D-60448
3-(4-Hydroxyphenyl)-1-(2,4,6-
trihydroxyphenyl)-2-propen-1-one; 4'-(3-
Methyl-2-butenyl), *in* H-60220
10-Isopentenylemodinanthran-10-ol, I-60111
Machilin *B*, M-60002
Moracin J, *in* M-60146
Zuonin *A*, Z-60003

C$_{20}$H$_{20}$O$_6$
Cardiophyllidin, C-70023
2-Hydroxygarveatin A, *in* H-60148
3-(4-Hydroxyphenyl)-1-(2,4,6-
trihydroxyphenyl)-2-propen-1-one; 2',4',6'-
Tri-Me, 4-Ac, *in* H-60220
3-(4-Hydroxyphenyl)-1-(2,4,6-
trihydroxyphenyl)-2-propen-1-one; 4,4',6'-
Tri-Me, 2-Ac, *in* H-60220
O^5-Methyl-3',4'-deoxypsorospermin-3'-ol, *in*
D-60021
Rhynchosperin *A*, R-60007
Sigmoidin *B*, S-70041
Vittagraciliolide, V-70017

C$_{20}$H$_{20}$O$_7$
Juglorin, J-70005
Nephroarctin, N-70032
4'-*O*-Methylpsorospermindiol, *in* P-70133
5-*O*-Methylpsorospermindiol, *in* P-70133
2',3,5,7,8-Pentamethoxyflavone, *in* P-70029
2',3',4',5,6-Pentamethoxyflavone, *in* P-70030
3,4',5,6,7-Pentamethoxyflavone, *in* P-60047
3,5,6,7,8-Pentamethoxyflavone, *in* P-70031
Rhynchosperin B, *in* R-60007
Sigmoidin *D*, S-60031

C$_{20}$H$_{20}$O$_8$
6-Hydroxy-2',3,5,5',7-pentamethoxyflavone, *in*
H-60064

C$_{20}$H$_{20}$O$_9$
3,5-Dihydroxy-3',4',5',7,8-penta-
methoxyflavone, *in* H-70021
5,7-Dihydroxy-3,3',4',5',8-penta-
methoxyflavone, *in* H-70021

C$_{20}$H$_{20}$O$_{10}$
4',5,5',7-Tetrahydroxy-2',3,6,8-tetra-
methoxyflavone, *in* O-60022
3',5,7-Trihydroxy-3,4',5',6,8-penta-
methoxyflavone, *in* O-70018
4',5,5'-Trihydroxy-2',3,6,7,8-penta-
methoxyflavone, *in* O-60022
4',5,7-Trihydroxy-2',3,5',6,8-penta-
methoxyflavone, *in* O-60022

4',5,7-Trihydroxy-3,3',5',6,8-penta-
methoxyflavone, *in* O-70018

C$_{20}$H$_{20}$O$_{11}$
3,3',4',5,7-Pentahydroxyflavanone; 3-*O*-β-D-
Xylopyranoside, *in* P-70027
1,3,5,8-Tetrahydroxyxanthone; 3-Me ether, 8-
glucoside, *in* T-70123

C$_{20}$H$_{21}$ClO$_7$
Methyl eriodermate, *in* E-70032

C$_{20}$H$_{22}$
2,5-Dimethyl-3,4-diphenyl-2,4-hexadiene,
D-60419

C$_{20}$H$_{22}$ClNO$_2$
Fontonamide, F-60067

C$_{20}$H$_{22}$O$_2$
3,3-Dimethyl-5,5-diphenyl-4-pentenoic acid;
Me ester, *in* D-60420

C$_{20}$H$_{22}$O$_4$
Austrobailignan-5, A-70270
6-(3-Methyl-2-butenyl)allopteroxylin, *in*
A-60075
Polemannone, P-60166

C$_{20}$H$_{22}$O$_5$
Austrobailignan-7, A-70271
Ethuliacoumarin, E-60045
Fragransin E$_1$, *in* A-70271
2-Hydroxygarveatin *B*, H-60148
3,4',5,7-Tetrahydroxy-8-prenylflavan, T-60125
Volkensiachromone, V-60016

C$_{20}$H$_{22}$O$_6$
Acuminatin, A-60062
Bartemidiolide, B-70008
Columbin, C-60160
2,3-Dehydroteucrin E, *in* T-60180
2-Deoxychamaedroxide, *in* C-60032
Isobutyrylshikonin, *in* S-70038

C$_{20}$H$_{22}$O$_7$
Chamaedroxide, C-60032
2-Dehydroarcangelisinol, D-70021
Desacylisoelephantopin senecioate, *in* I-60099
Desacylisoelephantopin tiglate, *in* I-60099
8β-Hydroxycolumbin, C-60160
10α-Hydroxycolumbin, C-60160
1-Hydroxypinoresinol, H-60221
9-Hydroxypinoresinol, H-70216
Isojateorin, *in* C-60160
Jateorin, *in* C-60160
Wikstromol, W-70003

C$_{20}$H$_{22}$O$_8$
Arizonin C$_3$, A-70255
2-Dehydroarcangelisinol; 2β,3β-Epoxide, 6,12-
diepimer, *in* D-70021
6-Hydroxyarcangelisin, *in* D-70021
4'-Hydroxy-3',5,5',6,7-pentamethoxyflavone, *in*
H-70081
Tinospora clerodane, T-70200

C$_{20}$H$_{22}$O$_9$
Salireposide, *in* D-70287
3,4',5,7-Tetrahydroxyflavanone; 7-*O*-β-D-
Apioside, *in* T-60104
Viscutin 3, *in* T-60103

C$_{20}$H$_{22}$O$_{10}$
Polydine, *in* P-70026

C$_{20}$H$_{23}$NO$_5$
Fuligorubin *A*, F-60079

C$_{20}$H$_{24}$
Nonacyclo[10.8.0.0^{2,11}.0^{4,9}.0^{4,19}.0^{6,17}.0^{7,16}.0^{9,14}.-
0^{14,19}]icosane, N-70070

C$_{20}$H$_{24}$O$_3$
2-Epijatrogrossidione, *in* J-70002
Gravelliferone; Me ether, *in* G-60037
Jatrogrossidione, J-70002
Rubifolide, R-60016

C$_{20}$H$_{24}$O$_4$
Austrobailignan-6, A-60320
Bharangin, B-70088

Brayleanin, *in* C-60026
Coralloidolide *A*, C-60164
3,6-Dioxo-4,7,11,15-cembratetraen-10,20-olide,
D-60470
12,16-Epoxy-11,14-dihydroxy-5,8,11,13-
abietatetraen-7-one, E-60011
Gersemolide, G-60015
Gersolide, G-60016
Gravelliferone; 8-Methoxy, *in* G-60037
Guaiaretic acid, G-70033
Hebeclinolide, H-60009
Sirutekkone, S-70048
Vernoflexin, *in* Z-60001
Zaluzanin *C*; Angeloyl, *in* Z-60001

C$_{20}$H$_{24}$O$_5$
1,11-Bisepicaniojane, *in* C-70013
Caniojane, C-70013
Epoxylaphodione, *in* D-60470
Ferulide, F-70005
3β-Hydroxyhebeclinolide, *in* H-60009
4-Hydroxyisobacchasmacranone, H-60161
Lycoxanthol, L-70046
Machilin *C*, M-60003
Machilin *I*, M-70003
Machilin D, *in* M-60003
Myristargenol *A*, M-70163
Myrocin *C*, M-70165
Nectandrin B, *in* M-70003

C$_{20}$H$_{24}$O$_6$
Bacchariolide *B*, B-70004
Deacetylsessein, *in* S-60027
Deacetylteupyrenone, *in* T-60184
Heliopsolide, H-70007
Isolariciresinol, I-60139
ent-Isolariciresinol, *in* I-60139
Isotaxiresinol; 7-Me ether, *in* I-60139
Ladibranolide, L-60008
Lariciresinol, L-70019
3,3',4',5,7-Pentamethoxyflavan, *in* P-70026
Sanguinone *A*, S-60006
Teucrin *E*, T-60180
Teucrin H2, *in* T-60180

C$_{20}$H$_{24}$O$_7$
Bacchariolide A, *in* B-70004
Melampodin *D*, M-60013
Orthopappolide methacrylate, *in* O-60045
Tomenphantopin A, *in* T-62222

C$_{20}$H$_{24}$O$_8$
4β,15-Epoxy-1β-methoxy-9Z-millerenolide, *in*
M-60137

C$_{20}$H$_{24}$O$_{10}$
Salicortin, S-70002

C$_{20}$H$_{24}$S$_6$
2,3,11,12-Dibenzo-1,4,7,10,13,16-hexathia-
2,11-cyclooctadecadiene, D-60073

C$_{20}$H$_{26}$O$_3$
19(4→3)-Abeo-11,12-dihydroxy-4(18),8,11,13-
abietatetraen-7-one, A-60003
6,7-Dehydroroyleanone, *in* R-60015
10-Epinidoresedic acid, *in* N-70035
3-(2-Hydroxy-4,8-dimethyl-3,7-nonadienyl)-
benzaldehyde; Ac, *in* H-60122
Kahweol, K-60002
Nidoresedic acid, N-70035
ent-6-Oxo-7,12E,14-labdatrien-17,11α-olide,
O-60076
Strictic acid, S-70077

C$_{20}$H$_{26}$O$_4$
19(4β→3β)-Abeo-6,11-epoxy-6,12-dihydroxy-
6,7-seco-4(18),8,11,13-abietatetraen-7-al,
A-60004
Anisomelic acid, A-60275
Cynajapogenin *A*, C-70257
Desoxyarticulin, *in* A-60307
Dihydroguaiaretic acid, *in* G-70033
11,12-Dihydroxy-6,8,11,13-abietatetraen-20-oic
acid, D-60305
2α,11-Dihydroxy-7,9(11),13-abietatriene-6,12-
dione, *in* D-70282
15,16-Epoxy-12-oxo-8(17),13(16),14-
labdatrien-19-oic acid, *in* L-60011
Hardwickiic acid; 19-Oxo, *in* H-70002

17-Oxohebemacrophyllide, *in* H-60010
11,12,16-Trihydroxy-5,8,11,13-abietatetraen-7-
one, T-60312

$C_{20}H_{26}O_5$
Articulin, A-60307
Ballotinone, *in* M-70013
Chapinolin, C-70037
8,9-Dihydroxy-1(10),4,11(13)-germacratrien-
12,6-olide; 9-(2*R*,3*R*-Epoxy-2-methyl-
butanoyl), *in* D-60323
11,12-Dihydroxy-7-oxo-8,11,13-abietatrien-20-
oic acid, *in* T-60337
Epirosmanol, *in* R-70010
Feruginin, F-60004
Gibberellin A_{24}, G-60019
Gibberellin A_{64}, G-70012
7-Hydroxy-3,13-clerodadiene-16,15;18,19-
diolide, H-70116
ent-7β-Hydroxy-3,13-clerodadiene-15,16;18,19-
diolide, H-70117
Mollisorin A, *in* D-70302
Myristargenol *B*, M-70164
ent-7-Oxo-3-clerodene-15,16:18,19-diolide, *in*
H-70117
Peregrinone, *in* M-70013
Rosmanol, R-70010
Senecioodontol, S-60022
Thymifodioic acid, T-60218
11,12,14-Trihydroxy-8,11,13-abietatriene-3,7-
dione, T-60313

$C_{20}H_{26}O_6$
Caleine F, *in* C-60005
Coleon *C*, C-60157
Coralloidolide *B*, C-60165
Deacetyleupaserrin, *in* D-70302
Deoxytifruticin, *in* T-70199
2,8-Dihydroxy-1(10),4,11(13)-germacratrien-
12,6-olide; 8-(3*S*-Hydroxy-2-methyl-
enebutanoyl), *in* D-70302
4,5-Epoxy-2,8-dihydroxy-1(10),11(13)-
germacradien-12,6-olide; 8-(3-Methyl-2-
butenoyl), *in* E-60013
Gibberellin A_{36}, *in* G-60019
Gibberellin A_{65}, G-70013
Gigantanolide A, *in* T-60323
Gigantanolide B, *in* T-60323
Jurinelloide, J-60003
Mollisorin B, *in* D-70302
Sesebrinic acid, S-60026
Teulamifin *B*, T-70158
Teumicropin, T-60181
7,11,12-Trihydroxy-6-oxo-8,11,13-abietatrien-
20-oic acid, T-60337
Zoapatanolide *B*, Z-70007

$C_{20}H_{26}O_7$
Cyclocratystyolide, C-70216
2,8-Dihydroxy-1(10),4,11(13)-germacratrien-
12,6-olide; 8-(2*R*,3*R*-Epoxy-2-
hydroxymethylbutanoyl), *in* D-70302
Gibberellin A_{66}, *in* G-70013
20-Hydroxyelemajurinelloide, H-60132
20-Hydroxyjurinelloide, *in* J-60003
Specionin, S-70059
Tifruticin, T-70199

$C_{20}H_{26}O_8$
Specionin, S-60039

$C_{20}H_{26}O_{10}$
Anamarine, A-60270

$C_{20}H_{28}$
Hexacyclo[11.3.1.1³,⁷.1⁵,⁹.1¹¹,¹⁵.0²,¹⁰]eicos-2-
ene, H-70031

$C_{20}H_{28}N_2O_3$
Oxyphencyclimine, O-60102

$C_{20}H_{28}N_2O_8$
N-(*N*-1-Carboxy-6-hydroxy-3-oxoheptyl)-
alanyltyrosine, C-70022

$C_{20}H_{28}O$
8,11,13-Abietatrien-12,16-oxide, A-60008
Biadamantylidene epoxide, B-70091
8,11,13-Cleistanthrien-19-al, *in* C-60154

$C_{20}H_{28}O_2$
Barbatusol, B-60007
Cedronellone, C-60027
3,7,11,15(17)-Cembratetraen-16,2-olide,
C-60028
8,11,13-Cleistanthatrien-19-oic acid, *in*
C-60154
12-Hydroxy-8,11,13-abietatrien-18-al, *in*
A-70005
Taonianone, T-60006
Totarolone, *in* T-70204

$C_{20}H_{28}O_3$
6,8,11,13-Abietatetraene-11,12,14-triol,
A-60006
2,12-Dihydroxy-8,11,13-abietatrien-7-one,
D-70281
3,12-Dihydroxy-8,11,13-abietatrien-1-one,
D-60306
2,11-Dihydroxy-7,9(11),13-abietatrien-12-one,
D-70282
Hardwickiic acid, H-70002
Hebemacrophyllide, H-60010
6-Hydroxy-5-oxo-7,15-fusicoccadien-15-al,
H-60203
19-Hydroxy-13(16),14-spongiadien-3-one,
H-70220
Isolobophytolide, I-60106
Lambertianic acid, L-60011
7-Oxo-8(14),15-isopimaradien-18-oic acid,
O-70094
ent-15-Oxo-16-kauren-18-oic acid, O-60075
Royleanone, R-60015
Wedeliasecokaurenolide, W-60004

$C_{20}H_{28}O_4$
Brevifloralactone, B-70179
Curculathyrane *A*, C-60176
Curculathyrane *B*, C-60177
Demethylpinusolide, *in* P-60156
15,16-Dihydro-16-oxohardwickiic acid, *in*
H-70002
2β,12-Dihydroxy-8,12-abietadiene-11,14-dione,
in R-60015
7,12-Dihydroxy-8,12-abietadiene-11,14-dione,
D-60304
7,15-Dihydroxy-8,11,13-abietatrien-18-oic acid,
D-70280
3,10-Dihydroxy-5,11-dielmenthadiene-4,9-
dione, D-60316
2α,19-Dihydroxy-13(16),14-spongiadien-3-one,
in H-70220
3,19-Dihydroxy-13(16),14-spongiadien-2-one,
D-70345
Ephemeric acid, E-60006
6,11-Epoxy-6,12-dihydroxy-6,7-seco-8,11,13-
abietatrien-7-al, E-60014
Furanopetasin, *in* F-70047
Hautriwaic acid, *in* H-70002
2β-Hydroxyhardwickiic acid, *in* H-70002
17-Hydroxyhebemacrophyllide, *in* H-60010
Isospongiadiol, I-60137
Ivangustin; *O*-(2-Methylbutanoyl), *in* I-70109
Kerlinic acid, K-70011
Marrubiin, M-70013
Patagonic acid, P-70013
Pseudopterogorgia diterpenoid *C*, P-70129
3,11,12-Trihydroxy-8,11,13-abietatrien-7-one,
T-60314

$C_{20}H_{28}O_5$
8,9-Dihydroxy-1(10),4,11(13)-germacratrien-
12,6-olide; 8-(2-Methylbutanoyl), *in* D-60323
11,12-Epoxy-2,14-dihydroxy-5,8(17)-
briaradien-18,7-olide, E-70012
ent-7β-Hydroxy-3-clerodene-15,16:18,19-
diolide, *in* H-70117
2-Hydroxyhautriwaic acid, *in* H-70002
Longikaurin *A*, L-70035
Melisodoric acid, M-70018
Pulchellin 2-O-tiglate, *in* P-70138
2α,6α,7,11-Tetrahydroxy-7,9(11),13-
abietatrien-12-one, *in* D-70282

$C_{20}H_{28}O_6$
Amarolide, A-60085
Blumealactone A, *in* B-70172

Blumealactone B, *in* B-70172
Gigantanolide C, *in* T-60323
6α-Hydroxypulchellin 4-O-angelate, *in* P-70138
Schkuhridin A, *in* S-60012
Sigmoidin A, *in* S-70041
ent-1β,3α,7α,11α-Tetrahydroxy-16-kaurene-
6,15-dione, T-60109

$C_{20}H_{28}O_7S$
2,8-Dihydroxy-1(10),4,11(13)-germacratrien-
12,6-olide; 8-(2*S*-Hydroxy-2-hydroxymethyl-
3*S*-mercaptobutanoyl), *in* D-70302

$C_{20}H_{28}O_{10}$
Ptelatoside B, *in* H-70209

$C_{20}H_{29}NO_5S$
Latrunculin B, L-70022

$C_{20}H_{30}$
Pentaisopropylidenecyclopentane, P-60053

$C_{20}H_{30}O$
8,11,13-Abietatrien-3-ol, A-70006
8,11,13-Abietatrien-19-ol, A-60007
8,11,13-Cleistanthatrien-19-ol, C-60154
ent-15,16-Epoxy-7,13(16),14-labdatriene,
E-70022
Mikanifuran, M-70148
Totarol, T-70204

$C_{20}H_{30}O_2$
Abeoanticopalic acid, A-60002
8,11,13-Abietatriene-12,18-diol, A-70005
13,15-Cleistanthadien-18-oic acid, A-60317
Cycloanticopalic acid, C-60181
Dictyodial *A*, D-70145
3,4-Epoxy-13(15),16,18-sphenolobatrien-5-ol,
E-60031
18-Hydroxy-8,15-isopimaradien-7-one,
H-60162
Isosarcophytoxide, I-70100
ent-16*S*-Kaurane-17,19-dial, *in* K-70008
ent-15-Kauren-17-oic acid, K-60008
Perrottetianal *A*, P-60064
Reiswigin B, *in* R-60004
Sanadaol, S-70008
Sarcophytoxide, S-70011
3,7,11,15-Tetramethyl-13-oxo-2,6,10,14-hexa-
decatetraenal, T-60150

$C_{20}H_{30}O_3$
8,11,13-Abietatriene-7,18-diol, A-70004
Agroskerin, A-70076
Agrostistachin, A-60067
Dictyodial *A*; 4β-Hydroxy, *in* D-70145
5,6-Epoxy-7,9,11,14-eicosatetraenoic acid,
E-60015
14,15-Epoxy-8(17),12-labdadien-16-oic acid,
E-60025
15,16-Epoxy-13,17-spatadiene-5,19-diol,
E-60029
2-(2-Formyl-3-hydroxymethyl-2-cyclo-
pentenyl)-6,10-dimethyl-5,9-undecadienal,
F-70038
2-[(2,3,3*a*,4,5,7*a*-Hexahydro-3,6-dimethyl-2-
benzofuranyl)ethylidene]-6-methyl-5-
heptenoic acid, H-60050
2-[(3,4,4*a*,5,6,8*a*-Hexahydro-4,7-dimethyl-2*H*-
1-benzopyran-2-yl)ethylidene]-6-methyl-5-
heptenoic acid, H-60046
ent-7α-Hydroxy-15-beyeren-19-oic acid,
H-60104
ent-12β-Hydroxy-15-beyeren-19-oic acid,
H-60105
ent-18-Hydroxy-15-beyeren-19-oic acid,
H-60106
7-Hydroxy-13(17),15-cleistanthadien-18-oic
acid, H-70115
16-Hydroxy-3,13-clerodadien-15,16-olide,
H-70118
8-Hydroxy-5,9,11,14,17-eicosapentaenoic acid,
H-70126
ent-3β-Hydroxy-15-kauren-17-oic acid,
H-60167
Isosarcophytoxide; 3,4-Epoxide, *in* I-70100
Medigenin, M-70015
6-Oxo-3,13-clerodadien-15-oic acid, O-70081

ent-15-Oxo-1(10),13-halimadien-18-oic acid, O-60069
2-Oxokolavenic acid, *in* K-60013
7-Oxo-8,13-labdadien-15-oic acid, O-70095
16-Oxo-17-spongianal, O-60094

C$_{20}$H$_{30}$O$_4$
8(14)-Abieten-18-oic acid 9,13-endoperoxide, A-60009
β-Cyclohallerin, *in* C-60193
α-Cyclohallerin, C-60193
Cymbodiacetal, C-60233
ent-3β,19-Dihydroxy-15-kauren-17-oic acid, D-60341
ent-1(10)13E-Halimadiene-15,18-dioic acid, *in* O-60069
ent-1(10),13Z-Halimadiene-15,18-dioic acid, *in* O-60069
14,15-Epoxy-5-hydroxy-6,8,10,12-eicosatetraenoic acid, E-70016
17-Hydroxy-15,17-oxido-16-spongianone, H-60201
Isohallerin, I-60100
Kurubashic acid angelate, *in* K-60014
Lapidin, *in* L-60014
Phlogantholide A, P-70099
ent-3α,7β,14α-Trihydroxy-16-kauren-15-one, T-60326

C$_{20}$H$_{30}$O$_5$
6,17-Dihydroxy-15,17-oxido-16-spongianone, D-60365
1β,10α-Epoxykurubashic acid angelate, *in* K-60014
ent-9,13-Epoxy-7-labdene-15,17-dioic acid, E-70024
Ivangustin; 4,5α-Dihydro,4α-hydroxy, O^1-(2-methylbutanoyl), *in* I-70109
6-Oxo-7-labdene-15,17-dioic acid, O-60077
ent-1β,7β,14S,20-Tetrahydroxy-16-kauren-15-one, T-70114
ent-3α,7β,14α,20-Tetrahydroxy-16-kauren-15-one, T-60110
ent-7β,11α,14S,20-Tetrahydroxy-16-kauren-15-one, T-70115
5,6,15-Trihydroxy-7,9,11,13,17-eicosapentaenoic acid, T-70255
5,14,15-Trihydroxy-6,8,10,12,17-eicosapentaenoic acid, T-70256

C$_{20}$H$_{30}$O$_5$S
Umbraculumin B, U-70002

C$_{20}$H$_{30}$O$_6$
6β-Hydroxypulchellin 2-O-isovalerate, *in* P-70138
Ingol, I-70018
Phlebiakauranol aldehyde, P-70097
3,4,6,8-Tetrahydroxy-11,13-clerodadien-15,16-olide, T-70103

C$_{20}$H$_{30}$O$_7$S
Hymatoxin A, H-60235

C$_{20}$H$_{30}$O$_8$
▷Ptaquiloside, P-60190

C$_{20}$H$_{31}$BrO
Sphaeroxetane, S-70061

C$_{20}$H$_{31}$NO$_5$S
Latrunculin D, L-70023

C$_{20}$H$_{32}$
Bicyclo[8.8.2]eicosa-1(19),10(20),19-triene, B-70096
α-Camphorene, C-70009
Dictytriene A, D-70148
Dictytriene B, *in* D-70148
Laurenene, L-60018

C$_{20}$H$_{32}$Br$_2$O$_2$
12-Hydroxybromosphaerol, H-60108

C$_{20}$H$_{32}$O
1,3,7,11-Cembratetraen-15-ol, C-70029
Dictymal, D-60162
ent-16S-Kauran-17-al, K-70008
ent-11-Kaurene-16β,18-diol, K-60006
ent-16-Kauren-3β-ol, K-70010
Neodolabellenol, N-70022
Neodolabellenol; 5-Epimer, *in* N-70022

18-Oxo-3-virgene, O-60095
13,17-Spatadien-10-ol, S-70058
Sphaeropyrane, S-70060
3,7,11,15-Tetramethyl-2,6,10,13,15-hexadecapentaen-1-ol, T-70137
1(15),8(19)-Trinervitadien-9-ol, T-70300

C$_{20}$H$_{32}$O$_2$
Dictyone, D-70146
Dilophic acid, D-60394
ent-19-Hydroxy-16S-kauran-17-al, *in* K-70008
15-Hydroxy-8(17),12-labdadien-16-al, H-70145
19-Hydroxy-8(17),13-labdadien-15-al, H-70146
ent-15-Hydroxy-8,13-labdadien-7-one, *in* L-70004
12-Isoagathen-15-oic acid, I-60085
7,15-Isopimaradiene-3,18-diol, I-70076
7,15-Isopimaradiene-3,19-diol, I-70077
7,15-Isopimaradiene-18,19-diol, I-70078
8,15-Isopimaradiene-7,18-diol, I-70079
8(14),15-Isopimaradiene-2,18-diol, I-60113
8(14),15-Isopimaradiene-3,18-diol, I-70080
ent-16R-Kauran-17-oic acid, *in* K-70008
ent-16S-Kauran-17-oic acid, *in* K-70008
ent-15-Kaurene-3β,17-diol, K-70007
ent-16-Kaurene-7α,15β-diol, K-70009
Kolavenic acid, K-60013
ent-7,12E,14-Labdatriene-6β,17-diol, L-60004
Reiswigin A, R-60004
13R-Hydroxy-8(17),14-labdadien-3-one, *in* L-70002
Stemodinone, *in* S-70071
1(15),8(19)-Trinervitadiene-2,3-diol, T-70295

C$_{20}$H$_{32}$O$_3$
Dictyotriol A, D-70147
ent-16α,17-Dihydroxy-3-atisanone, D-70284
3,15-Dihydroxy-8(17),13-labdadien-19-al, D-70310
1-(2,6-Dihydroxyphenyl)-1-tetradecanone, D-70338
Epivittadinal, *in* V-70016
8,9-Epoxy-5,11,14-eicosatrienoic acid, E-60016
11,12-Epoxy-5,8,14-eicosatrienoic acid, E-60017
3,4-Epoxy-13(15),16-sphenolobadiene-5,18-diol, E-60030
2,3,4,7,8,8a-Hexahydro-8-(5-hydroxy-4-methylpentyl)-6,8-dimethyl-1H-3a,7-methanoazulene-3-carboxylic acid, H-70057
5-Hydroxy-6,8,11,14-eicosatetraenoic acid, H-60127
8-Hydroxy-5,9,11,14-eicosatetraenoic acid, H-60128
12-Hydroxy-5,8,10,14-eicosatetraenoic acid, H-60131
12-Hydroxy-6-(hydroxymethyl)-10-methyl-2-(4-methyl-3-pentenyl)-2,6,10-dodecatrienal, *in* H-70179
8-Hydroxy-5,9,11,14-icosatetraenoic acid, *in* H-70126
9-Hydroxy-5,7,11,14-icosatetraenoic acid, H-60129
11-Hydroxy-5,8,12,14-icosatetraenoic acid, H-60130
2-Hydroxy-8(17),13-labdadien-15-oic acid, H-70147
ent-6β-Hydroxy-8(17),13-labdadien-15-oic acid, H-70148
2-(6-Hydroxy-4-methyl-4-hexenylidene)-6,10-dimethyl-7-oxo-9-undecenal, H-60186
12-Hydroxy-3,7,11,15-tetramethyl-2,6,10,14-hexadecatetraenoic acid, H-70223
16-Hydroxy-2,6,10,14-tetramethyl-2,6,10,14-hexadecatetraenoic acid, H-70224
7,15-Isopimaradiene-2,3,19-triol, I-70081
8,15-Isopimaradiene-3,7,19-triol, I-70082
ent-7,11E,14-Labdatriene-6β,13ξ,17-triol, L-60005
Pilosanone A, P-60153

ent-5-Rosene-3α,15,16,18-tetrol, *in* R-60011
11,15(17)-Trinervitadiene-3,9,13-triol, T-70297
1(15),8(19)-Trinervitadiene-2,3,9-triol, T-70298
1(15),8(19)-Trinervitadiene-2,3,20-triol, T-70299
Vittadinal, V-70016

C$_{20}$H$_{32}$O$_4$
Cervicol, C-70034
Chatferin, *in* F-60009
5,20-Dihydroxy-6,8,11,14-eicosatetraenoic acid, D-70296
12,19-Dihydroxy-5,8,10,14-eicosatetraenoic acid, D-60319
15,20-Dihydroxy-5,8,11,13-eicosatetraenoic acid, D-70297
3,6-Dihydroxy-2-(1-oxo-10-tetradecenyl)-2-cyclohexen-1-one, D-70332
5,6-Epoxy-20-hydroxy-8,11,14-eicosatrienoic acid, E-70017
14,15-Epoxy-20-hydroxy-5,8,11-eicosatrienoic acid, E-70018
10-Hydroxy-2-(6-hydroxy-4-methyl-4-hexylidene)-6,10-dimethyl-7-oxo-8-undecenal, *in* H-60186
7-Labdene-15,17-dioic acid, L-60006
7-Labdene-15,18-dioic acid, L-60007
Oliveric acid, O-60039
Pilosanone B, *in* P-60153
Salvicin, S-60003
Stolonidiol, S-60055
ent-2α,3α,16-Trihydroxy-8(14)-pimaren-15-one, *in* P-60154
1(15),8(19)-Trinervitadiene-2,3,9,20-tetrol, T-70296

C$_{20}$H$_{32}$O$_5$
Camporic acid, C-70010
11,15-Dihydroxy-9-oxo-5,13-prostadienoic acid, D-60367
Methyl 3,6-epidioxy-6-methoxy-4,14,16-octadecatrienoate, *in* E-60008
Pulchellin 2-O-isovalerate, *in* P-70138
Sterebin B, *in* S-60052
Sterebin C, *in* S-60052
Tetradecahydro-4,6-dihydroxy-9,9,12a-trimethyl-6H-phenanthro[1,10a-c]furan-3-carboxylic acid, T-60038
5,14,15-Trihydroxy-6,8,10,12-eicosatetraenoic acid, T-60320
3,7,11-Trimethyl-2,6,10-dodecatrienoic acid; 2-Acetoxy-3-hydroxypropyl ester, *in* T-60364

C$_{20}$H$_{32}$O$_6$
8,13-Epoxy-1,6,7,9-tetrahydroxy-14-labden-11-one, E-70027
Lascrol, *in* L-60016
Nardostachin, N-70019
Phlebiakauranol alcohol, *in* P-70097

C$_{20}$H$_{33}$BrO$_2$
Bromotetrasphaerol, B-70272

C$_{20}$H$_{34}$O
13-Epi-5β-neoverrucosanol, *in* N-70030
8,13-Epoxy-14-labdene, E-70023
8-Hydroxyisopimar-15-ene, H-60163
8(17),12-Labdadien-3-ol, L-70012
Neoverrucosanol, N-70030
Obscuronatin, O-60002

C$_{20}$H$_{34}$O$_2$
2,7,11-Cembratriene-4,6-diol, C-70030
Chromophycadiol, C-60148
3,7-Dolabelladiene-6,12-diol, D-70541
12α-Hydroxy-13-epi-(+)-manoyl oxide, *in* E-70023
8-Hydroxy-13-labden-15-al, H-70149
15-Hydroxy-7-labden-3-one, H-70151
19-Hydroxymanoyl oxide, *in* E-70023
8(17),13-Labdadiene-3,15-diol, L-70001
8(17),14-Labdadiene-3,13-diol, L-70001
ent-7,13E-Labdadiene-3β,15-diol, L-60001
ent-7,14-Labdadiene-2α,13-diol, L-70003
ent-8,13-Labdadiene-7α,15-diol, L-70004
ent-8,13-Labdadiene-7β,15-diol, *in* L-70004

$C_{20}H_{34}O_2$ (continued)

ent-8,13(16)-Labdadiene-14,15-diol, L-70005
ent-8(17),13-Labdadiene-7α,15-diol, L-70006
ent-8(17),13-Labdadiene-9α,15-diol, L-70007
Stemodin, S-70071

$C_{20}H_{34}O_3$

ent-3β,16α,17-Atisanetriol, A-70266
Dictytriol, *in* D-70148
8,15-Dihydroxy-13-labden-19-al, D-70311
3β,15-Dihydroxy-8-labden-7-one, *in* L-70013
7,8-Epoxy-2,11-cembradiene-4,6-diol, E-60010
Gypopinifolone, G-70036
15-Hydroxy-8(17)-labden-19-al, H-70150
15-Hydroxy-7-labden-17-oic acid, H-60168
6-(Hydroxymethyl)-10-methyl-2-(4-methyl-3-
   pentenyl)-2,6,10-dodecatriene-1,12-diol,
   H-70179
ent-15-Hydroxy-8,9-seco-13-labdene-8,9-dione,
   H-70218
Isodictytriol, *in* D-70148
ent-2α,16β,17-Kauranetriol, K-60004
ent-3β,16β,17-Kauranetriol, K-60005
8,13-Labdadiene-6,7,15-triol, L-60003
ent-7,14-Labdadiene-2α,3α,13-triol, L-70010
ent-7,14-Labdadiene-2α,13,20-triol, L-70011
Pachytriol, P-70001
ent-5-Rosene-15,16,18-triol, *in* R-60012
ent-5-Rosene-15,16,19-triol, R-60012
Vinigrol, V-70009

$C_{20}H_{34}O_4$

6,15-Dihydroxy-7-labden-17-oic acid, D-60342
1(15)-Dolastene-4,8,9,14-tetrol, D-70543
Jewenol A, J-70003
Jewenol B, J-70004
8,13-Labdadiene-2,6,7,15-tetrol, L-60002
11,13-Labdadiene-6,7,8,15-tetrol, L-70009
ent-8(14)-Pimarene-2α,3α,15R,16-tetrol,
   P-60154
ent-5-Rosene-3α,15,16,19-tetrol, R-60011

$C_{20}H_{34}O_5$

4(15)-Eudesmen-11-ol; α-L-Arabopyranoside,
   in E-70066
11,13(16)-Labdadiene-6,7,8,14,15-pentol,
   L-70008
Methyl 3,6-epidioxy-6-methoxy-4,16-octadeca-
   dienoate, *in* E-60008
Sterebin H, *in* L-70008
Tucumanoic acid, T-60410

$C_{20}H_{35}BrO_3$

Venustanol, V-70004

$C_{20}H_{36}$

Tetra-*tert*-butylcyclobutadiene, T-70014
Tetra-*tert*-butyltetrahedrane, T-70015

$C_{20}H_{36}Cl_2O_{10}$

7(18),11(19)-Havannadichlorohydrin, H-70005

$C_{20}H_{36}N_4O_8$

1,5,9,13-Tetraazacyclohexadecane-1,5,9,13-
   tetraacetic acid, T-60016

$C_{20}H_{36}O$

2,2-Di-*tert*-butyl-3-(di-*tert*-butylmethylene)-
   cyclopropanone, D-70101
ent-14-Labden-8β-ol, L-70014

$C_{20}H_{36}O_2$

13-Episclareol, *in* S-70018
8,13-Epoxy-3-labdanol, E-70021
Sclareol, S-70018

$C_{20}H_{36}O_3$

8-Labdene-3,7,15-triol, L-70013

$C_{20}H_{36}O_4$

ent-8,13-Epoxy-14,15,19-labdanetriol, E-70020

$C_{20}H_{36}O_5$

Methyl 3,6-epidioxy-6-methoxy-4-
   octadecenoate, *in* E-60008

---

ent-2β,3β,4α-Trihydroxy-15-clerodanoic acid,
   in T-60410

$C_{20}H_{36}S$

2,3-Dimethyl-5-(2,6,10-trimethylundecyl)-
   thiophene, D-70446

$C_{20}H_{37}NO_3$

N-(Tetrahydro-2-oxo-3-furanyl)hexadeca-
   namide, *in* A-60138

$C_{20}H_{38}O$

3,7,11,15-Tetramethyl-2-hexadecenal, *in*
   P-60152

$C_{20}H_{38}O_3$

Gyplure, *in* O-60006

$C_{20}H_{40}Br_2$

1,20-Dibromoicosane, D-70092

$C_{20}H_{40}O$

Phytol, P-60152

$C_{20}H_{40}O_2$

7,11,15-Trimethyl-3-methylene-1,2-hexa-
   decanediol, T-60368

$C_{20}H_{50}N_{10}$

1,4,7,10,13,16,19,22,25,28-Decaazacyclo-
   triacontane, D-70011

$C_{21}H_{11}F_5O_3$

9-Fluorenylmethyl pentafluorophenyl
   carbonate, F-60019

$C_{21}H_{12}N_4$

▷Tricycloquinazoline, T-60271

$C_{21}H_{12}N_6$

[1,3,5]Triazino[1,2-*a*:3,4-*a'*:5,6-*a''*]-
   trisbenzimidazole, T-60238

$C_{21}H_{12}O_2$

7*H*-Dibenzo[*c,h*]xanthen-7-one, D-60080

$C_{21}H_{13}N$

Phenanthro[9,10-*g*]isoquinoline, P-60075

$C_{21}H_{13}N_3OS$

12*H*-Quinoxalino[2,3-*b*][1,4]benzothiazine; *N*-
   Benzoyl, *in* Q-70008

$C_{21}H_{14}O$

2,3-Diphenyl-1*H*-inden-1-one, D-70482

$C_{21}H_{14}O_8$

7-*O*-Acetyldaphnoretin, *in* D-70006

$C_{21}H_{15}N$

1,2-Diphenyl-3-(phenylimino)cyclopropene,
   D-70491

$C_{21}H_{15}NO_2$

Dibenz[*b,g*]azocine-5,7(6*H*,12*H*)-dione; *N*-Ph,
   in D-60068

$C_{21}H_{15}O_2^\oplus$

2,4,5-Triphenyl-1,3-dioxol-1-ium, T-70307

$C_{21}H_{16}$

▷6-Methylcholanthrene, M-70056

$C_{21}H_{16}O_5$

Calopogonium isoflavone B, C-60006

$C_{21}H_{16}O_6$

Gerberinol 1, G-60013
Justicidin B, *in* D-60493

$C_{21}H_{16}O_7$

▷Diphyllin, D-60493

$C_{21}H_{16}O_8$

1,3-Dihydroxy-2-hydroxymethylanthraquinone;
   Tri-Ac, *in* D-70306

$C_{21}H_{16}O_{11}$

α-Acetylconstictic acid, *in* C-60163

$C_{21}H_{18}O$

Maximaisoflavone B, *in* B-60006
1,2,2-Triphenyl-1-propanone, T-60389
1,2,3-Triphenyl-1-propanone, T-60390
1,3,3-Triphenyl-1-propanone, T-60391
▷1,1,3-Triphenyl-2-propen-1-ol, T-70313

$C_{21}H_{18}O_3$

6-Methylene-2,4-cyclohexadien-1-one; Trimer,
   in M-60072

$C_{21}H_{18}O_5$

Glabrachromene II, G-70014

$C_{21}H_{18}O_6$

Pipoxide, P-60159

$C_{21}H_{18}O_7$

3-(2-Ethyl-2-butenyl)-9,10-dihydro-1,6,8-
   trihydroxy-9,10-dioxo-2-
   anthracenecarboxylic acid, E-70038
Hildecarpidin, H-60081

$C_{21}H_{19}NO$

1,3,3-Triphenyl-1-propanone; Oxime, *in*
   T-60391

$C_{21}H_{19}NO_5$

Isomurralonginol; 3-Pyridinecarboxylate, *in*
   I-70071

$C_{21}H_{20}O_3$

Abbottin, A-70002

$C_{21}H_{20}O_6$

Angeloylprangeline, *in* P-60170
Egonol; Ac, *in* E-70001
Garvin B, G-60009
Kwakhurin, K-60016
Topazolin, T-60224

$C_{21}H_{20}O_7$

Dehydrotrichostin, *in* T-70225
2-Hydroxygarvin B, *in* G-60009
3-(4-Hydroxyphenyl)-1-(2,4,6-
   trihydroxyphenyl)-2-propen-1-one; 4',6'-Di-
   Me, 2',4-di-Ac, *in* H-60220
Podoverine A, P-60164
Zeylenol, Z-70006

$C_{21}H_{20}O_8$

Homalicine, *in* H-70207

$C_{21}H_{20}O_9$

Cleomiscosin D, C-70185
Daidzin, *in* D-60340
Phenarctin, P-70054

$C_{21}H_{20}O_{10}$

Genistin, *in* T-60324
Sophicoroside, *in* T-60324

$C_{21}H_{20}O_{11}$

Sulfurein, *in* S-70082

$C_{21}H_{20}O_{14}$

Hibiscitin, *in* H-70021

$C_{21}H_{21}N$

2,2',2''-Trimethyltriphenylamine, T-70288
3,3',3''-Trimethyltriphenylamine, T-70289
4,4',4''-Trimethyltriphenylamine, T-70290

$C_{21}H_{21}N_3O_7$

Cacotheline, C-70001

$C_{21}H_{22}O_4$

Eupomatenoid 12, *in* E-60075
Kurospongin, K-70015
4'-O-Methylglabridin, *in* G-70016
Tephrinone, *in* G-70015
Tephrobbottin, T-70007

$C_{21}H_{22}O_5$

Aristotetralone, A-60298
7,8-(2,2-Dimethylpyrano)-3,4'-dihydroxy-5-
   methoxyflavan, *in* D-60448
Garveatin D, G-60007
3'-Methoxyglabridin, *in* G-70016
Quercetol A, Q-70001

$C_{21}H_{22}O_6$

β,β-Dimethylacrylalkannin, *in* S-70038
β,β-Dimethylacrylshikonin, *in* S-70038
2-Hydroxyaristotetralone, *in* A-70253
O-Angeloylalkannin, *in* S-70038
4',5,7-Trihydroxy-3'-methoxy-5'-
   prenylflavanone, *in* S-70041

$C_{21}H_{22}O_7$

Annulin B, A-60277
Divaronic acid, D-70528
Isopteryxin, *in* K-70013
Isosamidin, *in* K-70013
Praeruptorin A, P-60169
Pteryxin, *in* K-70013
▷Samidin, *in* K-70013

Topazolin hydrate, *in* T-60224
Trichostin, T-70225

$C_{21}H_{22}O_8$

Dihydrohomalicine, *in* H-70207
Epoxypteryxin, *in* K-70013
3′,4′,5,5′,6,7-Hexamethoxyflavone, *in* H-70081
Isopseudocyphellarin *A*, I-70098
Pseudocyphellarin A, *in* P-70127

$C_{21}H_{22}O_9$

3-Hydroxy-3′,4′,5,6,7,8-hexamethoxyflavone, *in* H-60024
5-Hydroxy-3,3′,4′,6,7,8-hexamethoxyflavone, *in* H-60024

$C_{21}H_{22}O_{10}$

5,7-Dihydroxy-2′,3,4′,5′,6,8-hexamethoxyflavone, *in* O-60022
5,7-Dihydroxy-3,3′,4′,5′,6,8-hexamethoxyflavone, *in* O-70018

$C_{21}H_{22}O_{11}$

Pyracanthoside, *in* T-70105
Stillopsin, *in* D-70339

$C_{21}H_{22}O_{12}$

Astilbin, *in* P-70027
Glucodistylin, *in* P-70027
Isoglucodistylin, *in* P-70027
Lanceoside, *in* P-60052
Taxifolin 4′-glucoside, *in* P-70027
1,3,5,8-Tetrahydroxyxanthone; 3,5-Di-Me ether, 8-O-β-D-glucopyranoside, *in* T-70123

$C_{21}H_{23}ClO_7$

Methyl 2-O-methyleriodermate, *in* E-70032
Methyl 2′-O-methyleriodermate, *in* E-70032
Methyl 4-O-methyleriodermate, *in* E-70032

$C_{21}H_{24}O_3$

5-(3,5-Di-*tert*-butyl-4-oxo-2,5-cyclohexadienylidene)-3,6-cycloheptadiene-1,2-dione, D-70105

$C_{21}H_{24}O_4$

5-Deoxymyricanone, *in* M-70162
8-Geranyloxypsoralen, *in* X-60002
6-(3-Methyl-2-butenyl)allopteroxylin methyl ether, *in* A-60075

$C_{21}H_{24}O_5$

Aristochilone, A-70251
Aristotetralol, A-70253
Calopiptin, *in* A-70271
8-(3,3-Dimethylallyl)-5-methoxy-3,4′,7-trihydroxyflavan, *in* T-60125
4-(3-Hydroxy-3-methylbutanoyloxy)-3-(1,1-dimethyl-2-propenyl)-6-phenyl-2H-pyran-2-one, *in* D-70434
Machilin G, *in* A-70271
5-Methylethuliacoumarin, *in* E-60045
Myricanone, *in* M-70162
Rutamarin, *in* C-60031

$C_{21}H_{24}O_6$

Arctigenin, A-70249
Phillygenin, P-60141
▷Phloridzin, *in* H-60219
Shikonin isobutyrate, *in* S-70038

$C_{21}H_{24}O_7$

Dihydrosamidin, *in* K-70013
Dihydrotrichostin, D-70273
β-Hydroxyisovalerylshikonin, *in* S-70038
1-Hydroxypinoresinol; 4″-Me ether, *in* H-60221
Medioresinol, M-70016
Suksdorfin, *in* K-70013
▷Visnadin, *in* K-70013

$C_{21}H_{24}O_8$

9α-Hydroxymedioresinol, *in* M-70016
Pseudocyphellarin B, P-70127

$C_{21}H_{24}O_{10}$

7,7a-Dihydro-3,6,7-trihydroxy-1a-(3-methyl-2-butenyl)naphth[2,3-b]oxiren-2(1aH)-one; 7-Ketone, 3-O-β-D-glucopyranoside, *in* D-70277

$C_{21}H_{26}O_3$

1,11-Bis(3-furanyl)-4,8-dimethyl-3,6,8-undecatrien-2-ol, B-70143

Malabaricone *A*, M-70007

$C_{21}H_{26}O_4$

1-(2,6-Dihydroxyphenyl)-9-(4-hydroxyphenyl)-1-nonanone, *in* M-70007
15,20-Epoxy-3β-hydroxy-14,15-seco-5,15,17(20)-pregnatriene-2,14-dione, *in* A-70267

$C_{21}H_{26}O_5$

Aristolignin, A-60297
1-(2,6-Dihydroxyphenyl)-9-(3,4-dihydroxyphenyl)-1-nonanone, *in* M-70007
3-O-Methyl-2,5-dehydrosenecioodentol, *in* S-60022
Myricanol, M-70162

$C_{21}H_{26}O_6$

Cordatin, C-60166
Fragransol *A*, F-70041
Isolariciresinol 4′-methyl ether, *in* I-60139
Machilin *H*, M-70002
Subexpinnatin B, *in* S-60058

$C_{21}H_{26}O_7$

1-Ethoxy-9Z-millerenolide, *in* M-60137
Orthopappolide senecioate, *in* O-60045
Orthopappolide tiglate, *in* O-60045

$C_{21}H_{26}O_8$

4β,15-Epoxy-1β-ethoxy-4Z-millerenolide, *in* M-60137
1,3,9-Trihydroxy-4(15),11(13)-eudesmadien-12,6-olide; Tri-Ac, *in* T-70257

$C_{21}H_{26}O_9$

Hypocretenoic acid; Lactone, 14-β-D-Glucopyranosyloxy, *in* H-70235

$C_{21}H_{26}O_{10}$

7,7a-Dihydro-3,6,7-trihydroxy-1a-(3-methyl-2-butenyl)naphth[2,3-b]oxiren-2(1aH)-one; 3-O-β-D-Glucopyranoside, *in* D-70277

$C_{21}H_{28}O_3$

1,3-Dioxototaryl methyl ether, *in* T-70204
Methyl nidoresedate, *in* N-70035

$C_{21}H_{28}O_4$

Atratogenin *A*, A-70267
Membranolide, M-60016
2,3,12-Trihydroxy-4,7,16-pregnatrien-20-one, T-60346

$C_{21}H_{28}O_5$

15,16-Dihydro-15-methoxy-16-oxonidoresedic acid, *in* N-70035
15,16-Dihydro-15-methoxy-16-oxostrictic acid, *in* S-70077
7-Methoxyrosmanol, *in* R-70010
3-O-Methylsenecioodentol, *in* S-60022
7-O-Formylhorminone, *in* D-60304

$C_{21}H_{28}O_7$

14-Acetoxy-9β-hydroxy-8β-(2-methylpropanoyloxy)-1(10),4,11(13)-germacratrien-12,6α-olide, *in* T-60323
9-Desacetoxymelcanthin F, *in* M-60014
Viguiestin, *in* T-70003

$C_{21}H_{28}O_8$

Vernoflexuoside, *in* Z-60001

$C_{21}H_{28}O_9$

Hypocretenoic acid; Lactone, 11α,13-Dihydro, 14-β-D-glucopyranosyloxy, *in* H-70235

$C_{21}H_{29}BrO_4$

Bromovulone I, B-70278

$C_{21}H_{29}IO_4$

Iodovulone I, *in* B-70278

$C_{21}H_{29}NO_3$

Smenospongine, S-60036

$C_{21}H_{30}O_3$

12-Methoxy-8,11,13-abietatrien-20-oic acid, M-60036
Methyl 7-oxo-8(14),15-isopimaradien-18-oate, *in* O-70094

$C_{21}H_{30}O_4$

Pinusolide, P-60156
2,3,12-Trihydroxy-4,7-pregnadien-20-one, T-60345

$C_{21}H_{30}O_5$

15,16-Dihydro-15-methoxy-16-oxohardwickiic acid, *in* H-70002
Kamebacetal *A*, K-70002
Kamebacetal B, *in* K-70002
Microglossic acid, M-60134
Spongionellin, S-60048
2,3,12,16-Tetrahydroxy-4,7-pregnadien-20-one, T-60124

$C_{21}H_{30}O_6$

4-Acetoxyflexilin, *in* F-70012
11-Dihydro-12-norneoquassin, D-60258

$C_{21}H_{30}O_7$

15-Desacetyltetraneurin *C* isobutyrate, *in* T-60158

$C_{21}H_{30}O_8$

Brachynereolide, B-60196
Zaluzanin *C*; 11β,13-Dihydro, β-D-Glucoside, *in* Z-60001

$C_{21}H_{30}O_9$

11β,13-Dihydro-8α-hydroxyglucozaluzanin C, *in* Z-60001

$C_{21}H_{30}O_{12}$

Cyclopropanehexacarboxylic acid; Hexa-Et ester, *in* C-70250

$C_{21}H_{30}O_{13}$

Acetylbarlerin, *in* S-60028

$C_{21}H_{30}Se$

Di-1-adamantyl selenoketone, D-70034

$C_{21}H_{31}N$

8-Isocyano-10,14-amphilectadiene, I-60090
7-Isocyano-1-cycloamphilectene, I-60093
7-Isocyano-11-cycloamphilectene, I-60094
8-Isocyano-1(12)-cycloamphilectrene, I-60095

$C_{21}H_{31}NO_5S$

Latrunculin C, *in* L-70023

$C_{21}H_{32}O_2$

1-(3,4-Methylenedioxyphenyl)-1-tetradecene, M-60077

$C_{21}H_{32}O_3$

14-(3-Furanyl)-3,7,11-trimethyl-7,11-tetradecadienoic acid, F-70049
Gracilin E, *in* G-60035
Methyl 14ξ,15-epoxy-8(17),12E-labdadien-16-oate, *in* E-60025

$C_{21}H_{32}O_4$

Lycopersiconolide, L-60042

$C_{21}H_{32}O_5$

10,11-Dihydromicroglossic acid, *in* M-60134
Umbraculumin *A*, U-70001

$C_{21}H_{32}O_8$

11β,13-Dihydrobrachynereolide, *in* B-60196

$C_{21}H_{32}O_{10}$

Dihydroserruloside, D-70265
Ebuloside, E-60001

$C_{21}H_{32}O_{14}$

10-Deoxymelittoside, *in* M-70019

$C_{21}H_{32}O_{15}$

Melittoside, M-70019

$C_{21}H_{33}N_3$

1,3,5-Tripiperidinobenzene, T-70314

$C_{21}H_{34}O_2$

Kolavenic acid; Me ester, *in* K-60013

$C_{21}H_{34}O_3$

15-Formylimbricatolal, *in* H-70150
5-Hydroxy-6,8,11,14-eicosatetraenoic acid; Me ester, *in* H-60127
8-Hydroxy-5,9,11,14-eicosatetraenoic acid; Me ester, *in* H-60128
12-Hydroxy-5,8,10,14-eicosatetraenoic acid; Me ester, *in* H-60131
9-Hydroxy-5,7,11,14-icosatetraenoic acid; Me ester, *in* H-60129
11-Hydroxy-5,8,12,14-icosatetraenoic acid; Me ester, *in* H-60130

1-(2-Hydroxy-6-methoxyphenyl)-1-tetra-
decanone, *in* D-70338

C₂₁H₃₄O₅
3,6-Epidioxy-6-methoxy-4,16,18-eicosatrienoic
acid, E-60007

C₂₁H₃₄O₁₀
7,7-*O*-Dihydroebuloside, *in* E-60001

C₂₁H₃₄O₁₁
Patrinalloside, P-70014

C₂₁H₃₆O₅
Arvoside, A-70264
Tetra-*tert*-butoxycyclopentadienone, T-70013

C₂₁H₃₆O₈
Methanetetrapropanoic acid; Tetra-Et ester, *in*
M-70036

C₂₁H₃₈O₃
7α-Methoxy-8-labdene-3β,15-diol, *in* L-70013

C₂₁H₃₈O₈
Icariside C₁, *in* T-60362
Icariside C₂, *in* T-60362
Icariside C₃, *in* T-60362
Icariside C₄, *in* T-60362

C₂₂H₁₀O₂
10,11-Methano-1*H*-benzo[5,6]-
cycloocta[1,2,3,4-*def*]fluorene-1,14-dione,
M-70039

C₂₂H₁₂
▷Indeno[1,2,3-*cd*]pyrene, I-70012

C₂₂H₁₂N₂
Dibenz[*b,h*]indeno[1,2,3-*de*][1,6]-
naphthyridine, D-60069
▷Dibenzo[*c,f*]indeno[1,2,3-*ij*][2,7]naphthyridine,
D-60074
2,2′-Diisocyano-1,1′-binaphthyl, D-60391

C₂₂H₁₂N₂O
Benzo[*a*]benzofuro[2,3-*c*]phenazine, *in*
B-60033

C₂₂H₁₂N₄
Dibenzo[*cd:c′d′*][1,2,4,5]tetrazino[1,6-*a*:4,3-
a′]diindole, D-60076

C₂₂H₁₂O
▷Indeno[1,2,3-*cd*]pyren-1-ol, I-60022
▷Indeno[1,2,3-*cd*]pyren-2-ol, I-60023
▷Indeno[1,2,3-*cd*]pyren-6-ol, I-60024
▷Indeno[1,2,3-*cd*]pyren-7-ol, I-60025
▷Indeno[1,2,3-*cd*]pyren-8-ol, I-60026

C₂₂H₁₂S₂
Naphtho[2,1-*b*:6,5-*b′*]bis[1]benzothiophene,
N-60007

C₂₂H₁₄N₂
Cycloocta[2,1-*b*:3,4-*b′*]diquinoline, C-70232

C₂₂H₁₄O₂
5,6:12,13-Diepoxy-5,6,12,13-tetra-
hydrodibenz[*a,h*]anthracene, D-70151

C₂₂H₁₄O₃
8,16-Epoxy-8*H*,16*H*-dinaphtho[2,1-*b*:2′,1′-
f][1,5]dioxocin, E-70014

C₂₂H₁₄O₆
3,3′-Bi[5-hydroxy-2-methyl-1,4-
naphthoquinone], B-60106

C₂₂H₁₄O₉
Aurintricarboxylic acid, A-60318

C₂₂H₁₅F₃O₅S
2,4,5-Triphenyl-1,3-dioxol-1-ium;
Trifluoromethanesulfonate, *in* T-70307

C₂₂H₁₆
1,1:2,2-Bis([10]annulene-1,6-diyl)ethylene,
B-70127
1,8-Diphenylnaphthalene, D-60483

C₂₂H₁₆O₂
1,3-Diphenyl-1*H*-indene-2-carboxylic acid,
D-60481

C₂₂H₁₇N
3,5-Dihydro-4*H*-dinaphth[2,1-*c*:1′,2′-*e*]azepine,
D-60234

C₂₂H₁₇NO₂
Dibenz[*b,g*]azocine-5,7(6*H*,12*H*)-dione; *N*-
Benzyl, *in* D-60068

C₂₂H₁₇N₇O₁₃S₂
2,3-Bis(2-methoxy-4-nitro-5-sulfophenyl)-5-
[(phenylamino)carbonyl]-2*H*-tetrazolium
hydroxide inner salt, B-70151

C₂₂H₁₈
1,1,4-Triphenyl-1,3-butadiene, T-70305

C₂₂H₁₈O₃
2-Hydroxy-1,3-diphenyl-1-propanone; Benzoyl,
in H-70124

C₂₂H₁₈O₇
Justicidin A, *in* D-60493

C₂₂H₁₈O₈
Desertorin A, D-60022
Orlandin, O-70046
2,4,5,6-Tetrahydroxyphenanthrene; Tetra-Ac,
in T-60122

C₂₂H₁₈O₁₀
3-Galloylcatechin, *in* P-70026

C₂₂H₁₈O₁₄
Hexahydroxy-1,4-naphthoquinone; Hexa-Ac, *in*
H-60065

C₂₂H₂₀
3-(1,3,6-Cycloheptatrien-1-yl-2,4,6-cyclo-
heptatrien-1-ylidenemethyl)-1,3,5-cyclo-
heptatriene, C-60205
[2.2](4,7)(7,4)Indenophane, I-60021

C₂₂H₂₀N₂S₂
1,4-Biss(thylthio)-3,6-diphenylpyrrolo[3,4-*c*]-
pyrrole, B-60158

C₂₂H₂₀O₃
1,1,2-Triphenyl-1,2-ethanediol; *O²*-Ac, *in*
T-70311

C₂₂H₂₀O₄
Ethyl 2-ethoxycarbonyl-5,5-diphenyl-2,3,4-
pentatrienoate, *in* D-70492

C₂₂H₂₀O₈
ζ-Rhodomycinone, *in* R-70002
Thuriferic acid, T-70191

C₂₂H₂₀O₉
ε-Rhodomycinone, R-70002
τ-Rhodomycinone, *in* R-70002

C₂₂H₂₀O₁₀
ψ-Baptisin, *in* B-60006

C₂₂H₂₀O₁₃
Ellagic acid; 3,3′-Di-Me ether, 4-glucoside, *in*
E-70007

C₂₂H₂₂O₄
Paralycolin A, P-60009

C₂₂H₂₂O₅
Praecansone B, P-70113

C₂₂H₂₂O₆
2,3-Dibenzoylbutanedioic acid; Di-Et ester, *in*
D-70083

C₂₂H₂₂O₈
4,5;4′,5′-Bismethylenedioxypolemannone, *in*
P-60166
Dalbergia Rotenolone, R-60014

C₂₂H₂₂O₉
2′-O-Methylphenarctin, *in* P-70054

C₂₂H₂₂O₁₀
5,7-Dihydroxy-4′-methoxyflavone; 7-β-D-
Galactoside, *in* D-70319
Sophojaponicin, *in* M-70001
1,3,5,8-Tetrahydroxy-2-methylanthraquinone;
5-Me ether, 8-*O*-α-L-rhamnopyranoside, *in*
T-60113
Tilianin, *in* D-70319
Trifolirhizin, *in* M-70001

C₂₂H₂₂O₁₁
4-(3,4-Dihydroxyphenyl)-5,7-dihydroxy-2*H*-1-
benzopyran-2-one; 7-Me ether, 5-*O*-β-D-
galactopyranosyl, *in* D-70334

C₂₂H₂₄Br₂N₄O₆S₂
N,N′-Bis[3-(3-bromo-4-hydroxyphenyl)-2-
oximidopropionyl]cystamine, B-70130

C₂₂H₂₄O₅
Quercetol C, Q-70003

C₂₂H₂₄O₆
Austrobailignan-7; Ac, *in* A-70271
Ramosissin, R-60001

C₂₂H₂₄O₇
Machilin E, *in* M-60003
Richardianidin 1, R-70006
Richardianidin 2, *in* R-70006

C₂₂H₂₄O₈
1-Acetoxypinoresinol, *in* H-60221
2′-*O*-Methylisopseudocyphellarin A, *in* I-70098
2′-*O*-Methylpseudocyphellarin A, *in* P-70127
Skutchiolide B, S-70050

C₂₂H₂₄O₉
3,3′,4′,5,5′,7,8-Heptamethoxyflavone, *in*
H-70021

C₂₂H₂₄O₁₀
Helichrysin, *in* H-60220
5-Hydroxy-3,3′,4′,5′,6,7,8-heptamethoxy-
flavone, *in* O-70018

C₂₂H₂₅ClO₇
Methyl 2,2′-di-*O*-methyleriodermate, *in*
E-70032

C₂₂H₂₅N₃
Triss(ethylphenylamino)methane, T-70324

C₂₂H₂₅N₃O₃S
Antibiotic FR 900452, A-60284

C₂₂H₂₅N₃O₁₀
▷Aluminon, *in* A-60318

C₂₂H₂₆
2,7-Di-*tert*-butyldicyclopenta[*a,e*]cyclooctene,
D-70102

C₂₂H₂₆Br₂O₃
4,21-Dibromo-3-ethyl-2,19-
dioxabicyclo[16.3.1]docosa-6,9,18(22),21-
tetraen-12-yn-20-one, *in* E-70042

C₂₂H₂₆N₂
2,5-Di-*tert*-butyl-2,5-dihydrobenzo[*e*]-
pyrrolo[3,4-*g*]isoindole, D-70103
2,7-Di-*tert*-butyl-2,7-dihydroisoindolo[5,4-*e*]-
isoindole, D-70104

C₂₂H₂₆O₃
Acetylimbricatolol, *in* H-70150
3-Ethyl-2,19-dioxabicyclo[16.3.1]docosa-
3,6,9,18(22),21-pentaen-12-yn-20-one,
E-70042

C₂₂H₂₆O₄
21-Ethyl-2,6-epoxy-17-hydroxy-1-oxacyclo-
henicosa-2,5,14,18,20-pentaen-11-yn-4-one,
E-60050

C₂₂H₂₆O₅
Aristoligone, *in* A-70251
Aristosynone, *in* A-70251
9-(1,3-Benzodioxol-5-yl)-1-(2,6-
dihydroxyphenyl)-1-nonanone, *in* M-70007
Isodidymic acid, I-60098
Licarin C, L-60024

C₂₂H₂₆O₆
Asebotin, *in* H-60219
Henricine, H-60017

C₂₂H₂₆O₇
Heliopsolide; 8-Ac, *in* H-70007
Heliopsolide; 4-Epimer, 8-Ac, *in* H-70007
Laferin, L-60009
Praderin, P-70112
Sessein, S-60027
Teupyrenone, T-60184

C₂₂H₂₆O₈
Abbreviatin PB, A-60001
3,6-Bis(3,4-dimethoxyphenyl)tetrahydro-
1*H*,3*H*-furo[3,4-*c*]furan-1,4-diol, B-60156
Heliopsolide; 2α,3α-Epoxide, 8-Ac, *in* H-70007

Isopicropolin, I-60112
7-Oxodihydrogmelinol, O-60064

C$_{22}$H$_{26}$O$_9$

1-Hydroxysyringaresinol, in H-60221
Skutchiolide A, S-70049
1-(2,3,4,5-Tetramethoxyphenyl)-3-(2,4,5-
trimethoxyphenyl)-1,3-propanedione, in
T-70120

C$_{22}$H$_{26}$O$_{13}$

Verproside, in C-70026

C$_{22}$H$_{28}$O$_2$

Fervanol benzoate, in F-60007

C$_{22}$H$_{28}$O$_3$

Fervanol p-hydroxybenzoate, in F-60007
Kurubasch aldehyde benzoate, in K-60014

C$_{22}$H$_{28}$O$_4$

1-(6-Hydroxy-2-methoxyphenyl)-9-(4-
hydroxyphenyl)-1-nonanone, in M-70007
Kurubashic acid benzoate, in K-60014
Verecynarmin A, V-60005

C$_{22}$H$_{28}$O$_5$

1α,10β-Epoxykurubaschic acid benzoate, in
K-60014
1β,10α-Epoxykurubashic acid benzoate, in
K-60014
Flabellata secoclerodane, F-70009
1-(6-Hydroxy-2-methoxyphenyl)-9-(3,4-
methylenedioxyphenyl)-1-nonanone, in
M-70007
19-O-Acetyl-1,2-dehydrohautriwaic acid, in
H-70002

C$_{22}$H$_{28}$O$_6$

Articulin acetate, in A-60307
Chrysophyllin B, C-70175
8,9-Dihydroxy-1(10),4,11(13)-germacratrien-
12,6-olide; 9-Ac, 8-(2R,3R-epoxy-2-methyl-
butanoyl), in D-60323
Fragransin D$_1$, F-70040
Fragransin D$_2$, in F-70040
Fragransin D$_3$, in F-70040
Homoheveadride, H-60089

C$_{22}$H$_{28}$O$_7$

17-Acetoxythymifodioic acid, in T-60218
3-Acetylteumicropin, in T-60181
12-Epiteupolin II, in T-70159
Eupaserrin, in D-70302
Teucjaponin A, T-60178
Teucjaponin B, in T-60178
Teupolin I, T-70159
Teupolin II, in T-70159

C$_{22}$H$_{28}$O$_8$

Citreoviridinol A$_1$, C-70181
Citreoviridinol A$_2$, in C-70181
5,5'-Dimethoxylariciresinol, in L-70019
Gracilin B, G-60034
Gracilin C, in G-60034

C$_{22}$H$_{29}$ClO$_7$

Solenolide E, S-70054

C$_{22}$H$_{30}$

Biadamantylideneethane, B-60075

C$_{22}$H$_{30}$O$_3$

2-Hydroxy-6-(8,11,14-pentadecatrienyl)benzoic
acid, in H-60205
Pseudopterosin A; Aglycone, O-Ac, in P-60187
Siccanin, S-60029
Siccanochromene E, S-60030

C$_{22}$H$_{30}$O$_4$

5-Epiilimaquinone, in I-70002
Ferolin, in G-60014
Ilimaquinone, I-70002
Metachromin A, M-70033

C$_{22}$H$_{30}$O$_5$

2α-Acetoxyhardwickiic acid, in H-70002
6β-Acetoxyhebemacrophyllide, in H-60010
3β-Acetoxy-19-hydroxy-13(16),14-spongiadien-
2-one, in D-70345
7α-Acetoxyroyleanone, in D-60304
7β-Acetoxyroyleanone, in D-60304
3β-Acetoxywedeliasecokaurenolide, in W-60004

Brevifloralactone acetate, in B-70179
3,5-Di-O-methylsenecioodontol, in S-60022
5-epi-6β-acetoxyhebemacrophyllide, in
H-60010
Epoxyjaeschkeanadiol p-hydroxybenzoate, in
J-70001
Feraginidin, in D-60004
Ferugin, in D-60005
Hautriwaic acid acetate, in H-70002
Lasianthin, L-60017
8-p-Hydroxybenzoylshiromodiol, in S-70039
Pseudopterogorgia diterpenoid B, P-70128

C$_{22}$H$_{30}$O$_6$

7α-Acetoxy-12,20-dihydroxy-8,12-abietadiene-
11,14-dione, in R-60015
Longikaurin C, in L-70035
Longikaurin E, in L-70035

C$_{22}$H$_{30}$O$_7$

6β-Acetoxypulchellin 4-O-angelate, in P-70138
Isomontanolide, in M-70154
Longikaurin B, in L-70035
Longikaurin D, in L-70035
Montanolide, M-70154

C$_{22}$H$_{31}$NO$_5$S

Latrunculin A, L-70021

C$_{22}$H$_{32}$N$_2$

1,2-Diphenyl-1,2-ethanediamine; N-Tetra-Et, in
D-70480

C$_{22}$H$_{32}$O$_2$

Dehydroabietinol acetate, in A-60007

C$_{22}$H$_{32}$O$_3$

Acetylsanadaol, in S-70008
3,4-Epoxy-13(15),16,18-sphenolobatrien-5-ol;
Ac, in E-60031
2-Hydroxy-6-(8,11-pentadecadienyl)benzoic
acid, in H-60205

C$_{22}$H$_{32}$O$_4$

ent-18-Acetoxy-15-beyeren-19-oic acid, in
H-60106
7β-Acetoxy-13(17),15-cleistanthadien-18-oic
acid, in H-70115
19-Acetoxy-15,16-epoxy-13,17-spatadien-5α-ol,
in E-60029
ent-3β-Acetoxy-15-kauren-17-oic acid, in
H-60167
ent-6β-Acetoxy-7,12E,14-labdatrien-17-oic
acid, in L-60004
Iloprost, I-60003
Isoiloprost, in I-60003
Lagerstronolide, L-60010
5-Methoxy-3-(8,11,14)pentadecatrienyl)-1,2,4-
benzenetriol, M-70049

C$_{22}$H$_{32}$O$_5$

ent-6β,17-Diacetoxy-14,15-bisnor-7,11E-
labdadien-13-one, in D-60311
7-Hydroxy-8(17),13-corymbidienolide, in
H-60110
4,10(14)-Oplopadiene-3,8,9-triol; 3-Ac, 8-
angeloyl, in O-70042
4,10(14)-Oplopadiene-3,8,9-triol; 3-Ac, 9-
angeloyl, in O-70042

C$_{22}$H$_{32}$O$_6$

5-Acetoxy-9α-angeloyloxy-8β-hydroxy-10(14)-
oplopen-3-one, in T-70272
6α-Acetoxy-17β-hydroxy-15,17-oxido-16-
spongianone, in D-60365

C$_{22}$H$_{32}$O$_7$

Arguticinin, A-70250
Arguticinin; 4-Epimer, in A-70250

C$_{22}$H$_{34}$O$_3$

ent-3β-Acetoxy-15-kauren-17-ol, in K-60007
ent-6β-Acetoxy-7,12E,14-labdatrien-17-ol, in
L-60004
10-Hydroxy-7,11,13,16,19-docosapentaenoic
acid, H-70125
8-Hydroxy-5,9,11,14,17-eicosapentaenoic acid;
Et ester, in H-70126
2-Hydroxy-6-(8-pentadecenyl)benzoic acid, in
H-60205
2-Hydroxy-6-(10-pentadecenyl)benzoic acid, in
H-60205

C$_{22}$H$_{34}$O$_4$

20-Acetoxy-2β,3α-dihydroxy-1(15),8(19)-
trinervitadiene, in T-70299
2β-Acetoxy-8(17),13E-labdadien-15-oic acid, in
H-70147
2β-Acetoxy-8(17),13Z-labdadien-15-oic acid, in
H-70147
Dictyodial A; 4β-Hydroxy, 18,O-Dihydro, 18-
Ac, in D-70145
Dictyotriol A; 12-Ac, in D-70147
3,15-Dihydroxy-8(17),13-labdadien-19-al; 3-
Ac, in D-70310
Koanoadmantic acid, in H-70224
Maesanin, M-60004
1(15),8(19)-Trinervitadiene-2,3,9-triol; 9-Ac, in
T-70298

C$_{22}$H$_{34}$O$_5$

6-Acetoxy-3,4-epoxy-12-hydroxy-7-dolabellen-
16-al, in D-70541
4-Acetoxy-6-(4-hydroxy-4-methyl-2-cyclo-
hexenyl)-2-(4-methyl-3-pentenyl)-2-
heptenoic acid, A-60018
Cornudentanone, C-60167
3,4-Epoxy-13(15),16-sphenolobadiene-5,18-
diol; 5-Ac, in E-60030
Stolonidiol acetate, in S-60055

C$_{22}$H$_{34}$O$_7$

8,13-Epoxy-1,6,7,9-tetrahydroxy-14-labden-11-
one; 6-Ac, in E-70027
8,13-Epoxy-1,6,7,9-tetrahydroxy-14-labden-11-
one; 7-Ac, in E-70027

C$_{22}$H$_{36}$O$_3$

19-Acetoxymanoyl oxide, in E-70023
Acetoxyodontoschismenol, in D-70541
Chromophycadiol monoacetate, in C-60148
8-Hydroxy-5,9,11,14,17-eicosapentaenoic acid;
17,18-Dihydro, Et ester, in H-70126
ent-8(17),13-Labdadiene-7α,15-diol; 7-Ac, in
L-70006
Secotrinervitane, S-60019

C$_{22}$H$_{36}$O$_4$

6-Acetoxy-3,7-dolabelladiene-12,16-diol, in
D-70541
ent-3β-Acetoxy-16β,17-kauranediol, in K-60005
15-Acetoxy-7-labden-17-oic acid, in H-60168
Hamachilobene E, in H-70001

C$_{22}$H$_{36}$O$_5$

Methyl 3,6-epidioxy-6-methoxy-4,16,18-
eicosatrienoate, in E-60007

C$_{22}$H$_{36}$O$_{16}$

Shanzhisin methyl ester gentiobioside, in
S-60028

C$_{22}$H$_{36}$S$_6$

1,3,4,6-Tetrakis(tert-butylthio)thieno[3,4-c]-
thiophene, in T-70170

C$_{22}$H$_{39}$NO$_5$

Valilactone, V-70002

C$_{22}$H$_{40}$O$_4$

9-Octadecene-1,12-diol; Di-Ac, in O-60006

C$_{22}$H$_{41}$NO$_7$

Fumifungin, F-60082

C$_{23}$H$_{14}$O

7-Methoxyindeno[1,2,3-cd]pyrene, in I-60025
8-Methoxyindeno[1,2,3-cd]pyrene, in I-60026

C$_{23}$H$_{14}$O$_6$

1,8,11-Triptycenetricarboxylic acid, T-60398
1,8,14-Triptycenetricarboxylic acid, T-60399

C$_{23}$H$_{16}$O

4-(Diphenylmethylene)-1(4H)-naphthalenone,
D-70487

C$_{23}$H$_{16}$O$_6$

2,2'-Methylenebis[8-hydroxy-3-methyl-1,4-
naphthalenedione], M-70063

$C_{23}H_{17}N$
2H-Dibenz[e,g]isoindole; N-Benzyl, in D-70067

$C_{23}H_{18}$
4b,8b,13,14-Tetrahydrodiindeno[1,2-a:2',1'-b]-
indene, T-70058

$C_{23}H_{18}O_2$
1,3-Diphenyl-1H-indene-2-carboxylic acid; Me
ester, in D-60481

$C_{23}H_{18}O_7$
Antibiotic SS 43405D, A-70239

$C_{23}H_{20}O_8$
Boesenboxide, B-70173
Desertorin B, in D-60022

$C_{23}H_{22}N_3O_3$
2,5-Dihydro-2,2,5,5-tetramethyl-3-[[[(2-
phenyl-3H-indol-3-ylidene)amino]oxy]-
carbonyl]-1H-pyrrol-1-yloxy, D-60284

$C_{23}H_{22}O_{10}$
Phrymarolin II, P-60142

$C_{23}H_{22}O_{11}$
Fujikinin, in T-70111

$C_{23}H_{24}O_5$
Demethoxyegonol 2-methylbutanoate, in
E-70001
Praecansone A, in P-70113

$C_{23}H_{24}O_6$
BR-Xanthone A, B-70282

$C_{23}H_{24}O_7$
2-Acetoxyaristotetralone, in A-70253

$C_{23}H_{24}O_7S$
Floroselin, in K-70013
Isofloroseselin, in K-70013

$C_{23}H_{24}O_8$
Loxodellonic acid, L-70039

$C_{23}H_{24}O_9$
Wistin, in T-60325

$C_{23}H_{24}O_{11}$
Eupalin, in P-60047
Flavoyadorinin B, in D-70291
3,4',5,6,7-Pentahydroxyflavone; 6,7-Di-Me
ether, 3-rhamnoside, in P-60047

$C_{23}H_{25}ClN_2O_9$
8-Methoxychlorotetracycline, M-70045

$C_{23}H_{25}ClN_2O_{10}$
4a-Hydroxy-8-methoxychlorotetracycline, in
M-70045

$C_{23}H_{26}O_6$
Garvin A, G-60008
Teracrylshikonin, in S-70038

$C_{23}H_{26}O_7$
Neokadsuranin, N-70024
Stenosporonic acid, S-70072

$C_{23}H_{26}O_8$
4,5-Dimethoxy-4',5'-methyl-
enedioxypolemannone, in P-60166
1-Hydroxypinoresinol; 4''-Me ether, 1-Ac, in
H-60221
Sikkimotoxin, S-60032

$C_{23}H_{26}O_{11}$
Macrophylloside C, M-70004
Macrophylloside B, in M-70004
Nyasicaside, in B-60155

$C_{23}H_{28}N_4O_7$
Biphenomycin B, in B-60114

$C_{23}H_{28}N_4O_8$
Biphenomycin A, B-60114

$C_{23}H_{28}O_3$
Citreomontanin, C-60149

$C_{23}H_{28}O_4$
2-Acetoxy-1,11-bis(3-furanyl)-4,8-dimethyl-
3,6,8-undecatriene, in B-70143
Quercetol B, Q-70002

$C_{23}H_{28}O_7$
Isosphaeric acid, I-60136

Palliferinin, in L-60014

$C_{23}H_{28}O_{10}$
Diffutin, in T-60103

$C_{23}H_{28}O_{13}$
Picroside II, in C-70026

$C_{23}H_{29}NO_6$
Fusarin A, F-60107

$C_{23}H_{29}NO_7$
Fusarin D, in F-60107

$C_{23}H_{30}O_4$
Fervanol vanillate, in F-60007
Guayulin D, in A-60299

$C_{23}H_{30}O_5$
Kurubasch aldehyde vanillate, in K-60014

$C_{23}H_{30}O_7$
Seco-4-hydroxylintetralin, S-70023

$C_{23}H_{30}O_8$
Acanthospermal A, A-60012
Gracilin D, in G-60034

$C_{23}H_{32}O_4$
8,11,13-Abietatrien-19-ol; 19-(Carboxyacetyl),
in A-60007
Metachromin B, M-70034
Methyl 11,12-dimethoxy-6,8,11,13-abietatrien-
20-oate, in D-60305

$C_{23}H_{32}O_5$
Chimganidin, in G-60014

$C_{23}H_{32}O_6$
16α-Acetoxy-2α,3β,12β-trihydroxy-4,7-
pregnadien-20-one, in T-60124
Epoxyjaeschkeanadiol vanillate, in J-70001
8-Vanilloylshiromodiol, in S-70039

$C_{23}H_{32}O_9$
Absinthifolide, in H-60138

$C_{23}H_{34}O_4$
7,13-Corymbidienolide, C-60169

$C_{23}H_{34}O_5$
▷Coroglaucigenin, C-60168
Gracilin A, G-60033
4,10(14)-Oplopadiene-3,8,9-triol; 3-Ac, 9-(3-
Methyl-2-pentenoyl), in O-70042

$C_{23}H_{34}O_6$
Coriolin B, in C-70199

$C_{23}H_{34}O_7$
Coriolin C, in C-70199

$C_{23}H_{36}O_5$
9,11-Dihydrogracilin A, in G-60033

$C_{23}H_{36}O_6$
Helogynic acid, H-70008

$C_{23}H_{38}O_4$
15-Hydroxy-7-labden-17-oic acid; 15-Ac, Me
ester, in H-60168
12-Isoagathen-15-oic acid; 2,3-Dihydroxypropyl
ester, in I-60085

$C_{23}H_{38}O_5$
Muamvatin, M-70156

$C_{23}H_{38}O_8$
Gaillardoside, in T-70284

$C_{23}H_{45}N_5O_{12}$
Youlemycin, Y-70002

$C_{23}H_{46}$
9-Tricosene, T-60259

$C_{24}H_{12}$
[4]Phenylene, P-60089

$C_{24}H_{12}N_4O_4$
25,26,27,28-Tetraazapentacyclo[19.3.1.1³,⁷.-
1⁹,¹³.1¹⁵,¹⁹]octacosa-1(25),3,5,7(28),-
9,11,13(27),15,17,19(26),21,23-dodecaene-
2,8,14,20-tetrone, T-70010

$C_{24}H_{12}S_3$
Benzo[1,2-b:3,4-b':5,6-b'']tris[1]-
benzothiophene, B-60052
Benzo[1,2-b:3,4-b':6,5-b'']tris[1]-
benzothiophene, B-60053

$C_{24}H_{14}$
Benz[5,6]indeno[2,1-a]phenalene, B-70025
1,1'-Bibiphenylene, B-60080
2,2'-Bibiphenylene, B-60081
Dibenz[e,k]acephenanthrylene, D-70066
Indeno[1,2,3-hi]chrysene, I-70010

$C_{24}H_{14}O_2$
Indeno[1,2,3-cd]pyren-1-ol; Ac, in I-60022
Indeno[1,2,3-cd]pyren-2-ol; Ac, in I-60023
Indeno[1,2,3-cd]pyren-6-ol; Ac, in I-60024

$C_{24}H_{15}N_3$
Diindolo[3,2-a:3',2'-c]carbazole, D-60384

$C_{24}H_{16}$
Cyclodeca[1,2,3-de:6,7,8-d'e']dinaphthalene,
C-60187
Cycloocta[1,2-b:5,6-b']dinaphthalene, C-60216
1,2:9,10-Dibenzo[2.2]metaparacyclophane,
D-70071
1,2:7,8-Dibenzo[2.2]paracyclophane, D-70072
1-(Diphenylmethylene)-1H-cyclopropa[b]-
naphthalene, D-70485

$C_{24}H_{16}O_2$
5,6,10,11,16,17,21,22-Octadehydro-7,9,18,20-
tetrahydrodibenzo[e,n][1,10]dioxacyclo-
octadecin, O-70006

$C_{24}H_{17}N_3$
2,3':2',3''-Ter-1H-indole, T-60011

$C_{24}H_{18}$
1,2,9,10,17,18-Hexahydro[2.2.2]paracyclo-
phane, H-60038

$C_{24}H_{18}N_6O_{12}$
Benzo[1,2-b:3,4-b':5,6-b'']tripyrazine-
2,3,6,7,10,11-tetracarboxylic acid; Hexa-Me
ester, in B-60051

$C_{24}H_{18}O_{10}$
Eriocephaloside, in H-70133

$C_{24}H_{20}$
[2.2](2,6)Azulenophane, A-60355
7-(Diphenylmethylene)-2,3,5,6-tetramethyl-
enebicyclo[2.2.1]heptane, D-70488
2,5,8,11-Tetramethylperylene, T-70140
3,4,9,10-Tetramethylperylene, T-70141

$C_{24}H_{20}N_2$
4(5)-Vinylimidazole; 1-Triphenylmethyl, in
V-70012

$C_{24}H_{20}O$
3,4,4-Triphenyl-2-cyclohexen-1-one, T-60379
3,5,5-Triphenyl-2-cyclohexen-1-one, T-60380
4,5,5-Triphenyl-2-cyclohexen-1-one, T-60381

$C_{24}H_{20}O_5$
8-Dihydrocinnamoyl-5,7-dihydroxy-4-phenyl-
2H-benzopyran-2-one, D-60210

$C_{24}H_{20}O_8$
Kielcorin B, K-60012

$C_{24}H_{20}S_2$
7H,9H,16H,18H-Dinaphtho[1,8-cd:1',8'-
ij][1,7]dithiacyclododecin, D-70448
2,3-Naphthalenedithiol; Di-S-benzyl, in
N-70004

$C_{24}H_{22}O_8$
Desertorin C, in D-60022

$C_{24}H_{23}N$
5,5,9,9-Tetramethyl-5H,9H-quino[3,2,1-de]-
acridine, T-70143

$C_{24}H_{24}O_{11}$
Phrymarolin I, in P-60142

$C_{24}H_{25}NO_8$
Antibiotic Sch 38519, A-60287

$C_{24}H_{26}$
1,3,6-Triphenylhexane, T-60386

C$_{24}$H$_{26}$O$_6$
Egonol 2-methylbutanoate, *in* E-70001

C$_{24}$H$_{26}$O$_7$
Anomalin, *in* K-70013
(−)-Anomalin, *in* K-70013
(+)-Anomalin, *in* K-70013
Calipteryxin, *in* K-70013
Khellactone; O^9-(3-Methyl-2-butenoyl), O^{10}-angeloyl, *in* K-70013
Khellactone; Bis(3-methyl-3-butenoyl), *in* K-70013
Khellactone; Di-O-(3-Methyl-2-butenoyl), *in* K-70013
Peuformosin, *in* K-70013

C$_{24}$H$_{26}$O$_8$
Khellactone; O^9-Angeloyl, O^{10}-(2,3-Epoxy-2-methylbutanoyl), *in* K-70013

C$_{24}$H$_{26}$O$_9$
Khellactone; Di-O-(2,3-Epoxy-2-methylbutanoyl), *in* K-70013

C$_{24}$H$_{26}$O$_{10}$
Sophoraside A, *in* P-70137

C$_{24}$H$_{27}$ClN$_2$O$_9$
8-Methoxy-N-methylchlorotetracycline, *in* M-70045

C$_{24}$H$_{27}$N$_3$
Hexahydro-1,3,5-triazine; 1,3,5-Tribenzyl, *in* H-60060

C$_{24}$H$_{27}$N$_3$O$_6$
Physarochrome A, P-60150

C$_{24}$H$_{28}$O$_2$
3,7-Di-*tert*-butyl-9,10-dimethyl-2,6-anthraquinone, D-60113

C$_{24}$H$_{28}$O$_3$
Ferprenin, F-70003
4-Geranyl-3,4',5-trihydroxystilbene, G-70008

C$_{24}$H$_{28}$O$_4$
6,6';7,3a'-Diligustilide, D-70363
Riligustilide, R-60009
Ternatin, T-70008

C$_{24}$H$_{28}$O$_5$
Aphyllocladone, A-70242

C$_{24}$H$_{28}$O$_7$
Garcinone D, G-70004
Khellactone; O-Angeloyl, O-2-methylpropanoyl, *in* K-70013
Khellactone; O^9-(3-Methylbutanoyl), O^{10}-angeloyl, *in* K-70013
Khellactone; O-Angeloyl, O-3-methylbutanoyl, *in* K-70013

C$_{24}$H$_{28}$O$_{11}$
Decumbeside A, *in* G-60002
Decumbeside B, *in* G-60002
Globularicisin, *in* C-70026
Globularin, *in* C-70026
Macrophylloside A, *in* M-70004

C$_{24}$H$_{28}$O$_{12}$
Scutellarioside II, *in* C-70026

C$_{24}$H$_{29}$ClO$_8$
Ptilosarcenone, *in* P-60193
Ptilosarcenone, P-70136

C$_{24}$H$_{29}$ClO$_9$
11-Hydroxyptilosarcenone, *in* P-60193

C$_{24}$H$_{30}$
1,1'-Diphenylbicyclohexyl, D-70475

C$_{24}$H$_{30}$O$_3$
Ferulenol, F-60005
Guayulin C, *in* A-60299
9α-Hydroxygymnomitryl cinnamate, *in* G-70035

C$_{24}$H$_{30}$O$_4$
Assafoetidin, A-70265
3,8-Dihydro-6,6';7,3'a-diliguetilide, D-60223
ω-Hydroxyferulenol, *in* F-60005

C$_{24}$H$_{30}$O$_6$
Magnoshinin, M-60007

C$_{24}$H$_{30}$O$_7$
Khellactone; Bis(3-methylbutanoyl), *in* K-70013

C$_{24}$H$_{30}$O$_8$
Montanin C, *in* T-70159
Saroaspidin A, S-70012
4,4',5,5'-Tetramethoxypolemannone, *in* P-60166

C$_{24}$H$_{30}$O$_9$
Auropolin, A-60319
Montanin G, M-70153
Teumicropodin, T-60182

C$_{24}$H$_{30}$O$_{10}$
Dichotosinin, *in* T-60103

C$_{24}$H$_{30}$O$_{11}$
Globularidin, *in* C-70026

C$_{24}$H$_{30}$O$_{12}$
Syringopicroside B, *in* S-60067

C$_{24}$H$_{31}$ClO$_8$
12-Ptilosarcenol, P-60193

C$_{24}$H$_{31}$ClO$_{10}$
Solenolide C, S-70053

C$_{24}$H$_{32}$
Pentacyclo[12.2.2.2^{2,5}.2^{6,9}.2^{10,13}]tetracosa-1,5,9,13-tetraene, P-60029

C$_{24}$H$_{32}$O$_3$
Antheliolide A, A-70213

C$_{24}$H$_{32}$O$_4$
Antheliolide B, *in* A-70213

C$_{24}$H$_{32}$O$_6$
3,19-Dihydroxy-13(16),14-spongiadien-2-one; Di-Ac, *in* D-70345
Okilactomycin, O-70030
Verrucosidin, V-60006

C$_{24}$H$_{32}$O$_7$
Furcellataepoxylactone, F-70052
Isoschizandrin, I-70101
Palliferin, *in* L-60014

C$_{24}$H$_{32}$O$_8$
Acetylisomontanolide, *in* M-70154
Hydroxyniranthin, H-70187
Longikaurin F, *in* L-70035

C$_{24}$H$_{33}$ClO$_9$
13-Deacetyl-11(9)-havannachlorohydrin, *in* H-70004
Solenolide B, *in* S-70052

C$_{24}$H$_{34}$N$_2$O$_3$
1,4,10-Trioxa-7,13-diazacyclopentadecane; 4,10-Dibenzyl, *in* T-60376

C$_{24}$H$_{34}$O$_4$
▷Bufalin, B-70289

C$_{24}$H$_{34}$O$_5$
12-(Acetyloxy)-10-[(acetyloxy)methylene]-6-methyl-2-(4-methyl-3-pentenyl)-2,6,11-dodecatrienal, A-70049

C$_{24}$H$_{34}$O$_6$
ent-3β,19-Diacetoxy-15-kauren-17-oic acid, *in* D-60341

C$_{24}$H$_{34}$O$_7$
6α,17α-Diacetoxy-15,17-oxido-16-spongianone, *in* D-60365

C$_{24}$H$_{34}$O$_8$
Solenolide F, S-70055
Teucretol, T-60179

C$_{24}$H$_{36}$O$_4$
ent-6β,17-Diacetoxy-7,12E,14-labdatriene, *in* L-60004
Soulattrone A, S-70057

C$_{24}$H$_{36}$O$_5$
ent-6β,17-Diacetoxy-7,11E,14-labdatrien-13ξ-ol, *in* L-60005
1(15),8(19)-Trinervitadiene-2,3,9-triol; 2,3-Di-Ac, *in* T-70298

C$_{24}$H$_{36}$O$_6$
6α-Butanoyloxy-17β-hydroxy-15,17-oxido-16-spongianone, *in* D-60365
Hamachilobene B, *in* H-70001
1(15),8(19)-Trinervitadiene-2β,3α,9β,20-tetrol 9,20-diacetate, *in* T-70296

C$_{24}$H$_{36}$O$_8$
Lapidolin, *in* L-60015

C$_{24}$H$_{38}$ClNO$_5$
Malyngamide C, M-60009

C$_{24}$H$_{38}$O$_3$
10-Hydroxy-7,11,13,16,19-docosapentaenoic acid; Et ester, *in* H-70125

C$_{24}$H$_{38}$O$_4$
Trifarin, T-70236
Trunculin A, T-60408

C$_{24}$H$_{38}$O$_5$
6,16-Diacetoxy-3,7-dolabelladien-12-ol, *in* D-70541
ent-3β,17-Diacetoxy-16β-kauranol, *in* K-60005
Trunculin B, T-60409

C$_{24}$H$_{38}$O$_6$
Hamachilobene A, H-70001
Hamachilobene C, *in* H-70001
Hamachilobene D, *in* H-70001

C$_{24}$H$_{40}$O$_2$
3-Hydroxy-22,23,24,25,26,27-hexanor-20-dammaranone, H-60151

C$_{24}$H$_{40}$O$_5$
3,11,15-Trihydroxycholanic acid, T-60317
3,15,18-Trihydroxycholanic acid, T-60318

C$_{24}$H$_{42}$
1,3,5-Trihexylbenzene, T-70247

C$_{24}$H$_{42}$O$_4$
Bisdihydrotrifarin, *in* T-70236

C$_{24}$H$_{44}$O$_4$
7,11,15-Trimethyl-3-methylene-1,2-hexadecanediol; Di-Ac, *in* T-60368

C$_{24}$H$_{46}$
1-Tetracosyne, T-70017
12-Tetracosyne, T-70018

C$_{24}$H$_{46}$N$_4$O$_8$S$_2$
Homopantethine, *in* H-60090

C$_{24}$H$_{47}$NO$_{11}$S
N-[15-(β-D-Glucopyranosyloxy)-8-hydroxypalmitoyl]taurine, G-70019

C$_{25}$H$_{16}$
Spirobi[9H-fluorene], S-60040

C$_{25}$H$_{20}$N$_2$
Triphenylphenylazomethane, T-70312

C$_{25}$H$_{22}$
[2](1,5)Naphthaleno[2](2,7)(1,6-methano[10]-annuleno)phane, N-70005

C$_{25}$H$_{22}$O$_6$
Cudraflavone A, C-70205

C$_{25}$H$_{22}$O$_7$
5'-Hydroxycudraflavone A, *in* C-70205

C$_{25}$H$_{22}$O$_8$
Galtamycinone, *in* G-70003

C$_{25}$H$_{22}$O$_9$
Hypericorin, H-70234
Silandrin, S-70042
Silyhermin, S-70045
Silymonin, *in* S-70044

C$_{25}$H$_{22}$O$_{10}$
Isosilybin, I-70102
Isosilychristin, I-70103
Silydianin, S-70044

C$_{25}$H$_{22}$O$_{12}$
Daphnorin, *in* D-70006

C$_{25}$H$_{24}$
5,5a,6,6a,7,12,12a,13,13a,14-Decahydro-5,14:6,13:7,12-trimethanopentacene, D-60010

$C_{25}H_{24}O_2S$
3-[(Triphenylmethyl)thio]cyclopenta-
necarboxylic acid, *in* M-60021

$C_{25}H_{24}O_5$
Euchrenone A_1, E-70056
Isomammeigin, I-70069

$C_{25}H_{24}O_6$
Cudraisoflavone A, C-70206

$C_{25}H_{24}O_7$
Laserpitinol, *in* L-60016

$C_{25}H_{24}O_{12}$
3,4-Di-*O*-caffeoylquinic acid, D-70110

$C_{25}H_{26}O_5$
Cajaflavanone, C-60004
Euchrenone A_2, E-70057
Honyucitrin, H-70092

$C_{25}H_{26}O_6$
Orotinin, O-60044

$C_{25}H_{26}O_7$
Garvalone B, G-60005

$C_{25}H_{26}O_8$
Rhynchosperin C, *in* R-60007

$C_{25}H_{26}O_9$
Glomellonic acid, *in* G-70018

$C_{25}H_{26}O_{10}$
Fridamycin A, F-70043
Fridamycin B, *in* F-70043

$C_{25}H_{28}O_4$
1-[2,4-Dihydroxy-3,5-bis(3-methyl-2-butenyl)-
phenyl]-3-(4-hydroxyphenyl)-2-propen-1-one,
D-60310
Isolinderatone, *in* L-70030
Linderatone, L-70030

$C_{25}H_{28}O_5$
Ammothamnidin, A-70204
10′,11′-Dehydrocyclolycoserone, *in* C-70229
Lehmannin, L-70027
Lespedazaflavone B, L-60021
Lonchocarpol A, L-60030
Senegalensein, S-70034

$C_{25}H_{28}O_6$
Lonchocarpol C, L-60031
Lonchocarpol D, L-60032

$C_{25}H_{28}O_7$
Lonchocarpol E, L-60033

$C_{25}H_{28}O_8$
Glomelliferonic acid. G-70018

$C_{25}H_{28}O_{10}$
Egonol glucoside, *in* E-70001

$C_{25}H_{30}Cl_2O_6$
Napyradiomycin A_2, N-70016

$C_{25}H_{30}N_8O_9$
7-Hydro-8-methylpteroylglutamylglutamic
acid, H-60096

$C_{25}H_{30}O_4$
4-Hydroxy-5-methyl-3-(3,8,11-trimethyl-8-oxo-
2,6,10-dodecatrienyl)-2*H*-1-benzopyran-2-
one, H-70183

$C_{25}H_{30}O_5$
Cyclolycoserone, C-70229
Cyclolycoserone; 3′-Epimer, *in* C-70229
1′-Epilycoserone, *in* L-70045
Gypothamniol, G-70037
4-Hydroxy-3-(9-hydroxy-8-oxofarnesyl)-5-
methylcoumarin, *in* H-70183
Ircinin 1, I-60083
Ircinin 2, *in* I-60083
Isolycoserone, I-70068
Lycoserone, L-70045

$C_{25}H_{30}O_6$
10′-Hydroxycyclolycoserone, *in* C-70229
10′-Hydroxy-1′-epilycoserone, *in* L-70045
10′-Hydroxyisolycoserone, *in* I-70068

$C_{25}H_{30}O_7$
Lonchocarpol B, *in* L-60030

$C_{25}H_{30}O_{13}$
Picroside III, *in* C-70026

$C_{25}H_{31}Cl_3O_6$
Napyradiomycin B_4, N-70017

$C_{25}H_{31}N_3O_5$
1-[*N*-[3-(Benzoylamino)-2-hydroxy-4-phenyl-
butyl]alanyl]proline, B-60056

$C_{25}H_{32}O_4$
Ircinianin, I-60081
Ircinic acid, I-60082

$C_{25}H_{32}O_6$
3-*O*-Angeloylsenecioodontol, *in* S-60022

$C_{25}H_{32}O_7$
Ichangensin, I-70001

$C_{25}H_{32}O_8$
Saroaspidin B, S-70013

$C_{25}H_{32}O_{11}$
Melcanthin Γ, M-60014

$C_{25}H_{32}O_{12}$
Isoligustroside, I-60104

$C_{25}H_{32}O_{13}$
Oleuroside, O-70038
Syringopicroside C, *in* S-60067

$C_{25}H_{33}ClO_8$
Punaglandin 3, P-70143

$C_{25}H_{33}NO$
Aurachin D, A-60315

$C_{25}H_{33}NO_2$
Aurachin B, A-60314
Aurachin C, *in* A-60315

$C_{25}H_{33}NO_3$
Aurachin A, A-60313

$C_{25}H_{34}O_3$
3-Anhydro-6-epiophiobolin A, *in* O-70040
Apo-12′-violaxanthal, *in* P-60065

$C_{25}H_{34}O_4$
Fasciculatin, F-70001
Isofasciculatin, *in* F-70001
Variabilin, V-70003
Variabilin; 20*E*-Isomer, *in* V-70003
Variabilin; Δ^{13}-Isomer (*Z*-), *in* V-70003

$C_{25}H_{34}O_7$
Glycinoeclepin A, G-70022

$C_{25}H_{34}O_{11}$
3-O-Glucosylgibberellin A_{29}, *in* G-70011

$C_{25}H_{35}ClO_8$
Punaglandin 4, *in* P-70143

$C_{25}H_{36}O_3$
8-Deoxyophiobolin J, *in* O-70041
Persicaxanthin, P-60065

$C_{25}H_{36}O_4$
6-Epiophiobolin A, *in* O-70040
5-[13-(3-Furanyl)-2,6,10-trimethyl-6,8-trideca-
dienyl]-4-hydroxy-3-methyl-2(5*H*)-furanone,
F-70050
24-Methyl-25-nor-12,24-dioxo-16-scalaren-22-
oic acid, M-60105
▷Ophiobolin *A*, O-70040
Ophiobolin J, O-70041

$C_{25}H_{36}O_5$
8-Hydroxyvariabilin, *in* V-70003
Monoalide, M-70152
Salvileucolidone, S-70005
Thorectolide, T-70189

$C_{25}H_{36}O_6$
Pseudopterosin A, P-60187
Triflorestevione, T-70237

$C_{25}H_{36}O_{12}$
Kickxioside, K-70014

$C_{25}H_{38}O_3$
Luffariellolide, L-70040
Palauolide, P-60001

$C_{25}H_{38}O_4$
Aglajne 2, A-70073

Spongiolactone, S-60047

$C_{25}H_{38}O_4S$
Suvanine, S-70085

$C_{25}H_{38}O_5$
Pallinin, *in* C-60021

$C_{25}H_{38}O_6$
Clibadic acid, C-70186
Desoxodehydrolaserpitin, *in* D-70009
Secopseudopterosin A, S-70026
Tingitanol, *in* D-70009

$C_{25}H_{38}O_7$
Isolaserpitin, *in* D-70009
Laserpitine, L-70016
Methyl 15α,17β-diacetoxy-15,16-dideoxy-
15,17-oxido-16-spongianoate, *in* T-60038

$C_{25}H_{40}O_3$
4,6,8,10,12,14,16-Heptamethyl-6,8,11-
octadecatriene-3,5,13-trione, H-70022
3-Octadecene-1,2-diol; 1-Benzoyl, *in* O-60005

$C_{25}H_{40}O_6$
Salvisyriacolide, S-60004

$C_{25}H_{42}N_4O_{14}$
Allosamidin, A-60076

$C_{25}H_{42}O_3$
Sclareol; 6α-Angeloyloxy, *in* S-70018

$C_{25}H_{42}O_4$
6β-Isovaleroyloxy-8,13*E*-labdadiene-7α,15-diol,
in L-60003

$C_{25}H_{45}FeN_6O_8$
Ferrioxamine B, F-60003

$C_{26}H_6N_8$
11,11,12,12,13,13,14,14-Octacyano-1,4:5,8-
anthradiquinotetramethane, O-70002

$C_{26}H_{10}N_2O_6S_2$
Dibenzo[*b,m*]triphenodithiazine-
5,7,9,14,16,18(8*H*,17*H*)-tetrone, D-70082

$C_{26}H_{14}$
Benz[*def*]indeno[1,2,3-*qr*]chrysene, B-70023
Benz[*def*]indeno[1,2,3-*hi*]chrysene, B-70024
Bisbenzo[3,4]cyclobuta[1,2-*c*;1′,2′-*g*]-
phenanthrene, B-60143
Fluoreno[3,2,1,9-*defg*]chrysene, F-70016
Fluoreno[9,1,2,3-*cdef*]chrysene, F-70017
Naphtho[1,2,3,4-*ghi*]perylene, N-60009

$C_{26}H_{14}O_2$
6,15-Hexacenedione, H-60033

$C_{26}H_{16}$
9,9′-Bifluorenylidene, B-60104

$C_{26}H_{18}$
1,3-Di-(1-naphthyl)benzene, D-60454
1,3-Di(2-naphthyl)benzene, D-60455

$C_{26}H_{18}N_2$
9,9′-Biacridylidene, B-70090
2,11-Diphenyldipyrrolo[1,2-*a*:2′,1′-*c*]-
quinoxaline, D-70479

$C_{26}H_{18}N_6$
7,11:20,24-Dinitrilodibenzo[*b,m*][1,4,12,15]-
tetraazacyclodocosine, D-60456

$C_{26}H_{18}O_9$
5,6,8,13-Tetrahydro-1,7,9,11-tetrahydroxy-
8,13-dioxo-3-(2-oxopropyl)benzo[*a*]-
naphthacene-2-carboxylic acid, T-70095

$C_{26}H_{20}$
Tribenzotritwistatriene, T-70210

$C_{26}H_{20}N_4S$
N-2-Benzothiazolyl-*N*,*N*′,*N*″-triphenyl-
guanidine, B-70057

$C_{26}H_{20}O$
Tetraphenylethanone, T-60167

$C_{26}H_{20}O_6$
1,8,11-Triptycenetricarboxylic acid; Tri-Me
ester, *in* T-60398
1,8,14-Triptycenetricarboxylic acid; Tri-Me
ester, *in* T-60399

C$_{26}$H$_{20}$O$_7$
Imbricatonol, I-70003

C$_{26}$H$_{20}$O$_{10}$
Salvianolic acid C, *in* S-70003

C$_{26}$H$_{20}$S$_3$
3,3,5,5-Tetraphenyl-1,2,4-trithiolane, T-70153

C$_{26}$H$_{22}$N$_2$O
1,1,1′,1′-Tetraphenyldimethylamine; *N*-Nitroso, *in* T-70150

C$_{26}$H$_{22}$Se$_2$
Biss(iphenylmethyl) diselenide, B-70141

C$_{26}$H$_{23}$N
1,1,1′,1′-Tetraphenyldimethylamine, T-70150

C$_{26}$H$_{24}$
Tricyclo[18.4.1.1^{8,13}]hexacosa-
2,4,6,8,10,12,14,16,18,20,22,24-dodecaene,
T-60263

C$_{26}$H$_{24}$N$_6$
1,14,29,30,31,32-Hexaazahexacyclo-
[12.7.7.1^{3,7}.1^{8,12}.1^{16,20}.1^{23,27}]dotriaconta-
3,5,7(32),8,10,12(31),16,18,20(30),-
23,25,27(29)-dodecaene, H-70030

C$_{26}$H$_{24}$O$_2$
[2.2][2.2]Paracyclophane-5,8-quinone, P-60007
[2.2][2.2]Paracyclophane-12,15-quinone,
P-60008

C$_{26}$H$_{24}$O$_{10}$
Peltigerin, P-70016

C$_{26}$H$_{24}$O$_{15}$
4,6-Di-O-galloylarbutin, *in* A-70248

C$_{26}$H$_{25}$N$_3$O$_{11}$
Tunichrome B-1, T-70336

C$_{26}$H$_{26}$
[2.2]Paracyclo(4,8)[2.2]metaparacyclophane,
P-60005

C$_{26}$H$_{26}$O$_2$S
4-[(Triphenylmethyl)thio]cyclohexa-
necarboxylic acid, *in* M-60020

C$_{26}$H$_{26}$O$_9$
Dukunolide *A*, D-60522

C$_{26}$H$_{26}$O$_{10}$
Dukunolide B, *in* D-60522

C$_{26}$H$_{26}$O$_{18}$
Amritoside, *in* E-70007

C$_{26}$H$_{28}$O$_6$
Orotinichalcone, O-60043
Orotinin; 5-Me ether, *in* O-60044

C$_{26}$H$_{28}$O$_8$
Dukunolide D, D-60523

C$_{26}$H$_{28}$O$_9$
Dukunolide E, *in* D-60523
Dukunolide F, *in* D-60523
Ephemeroside, *in* E-60006

C$_{26}$H$_{28}$O$_{13}$
Ambonin, *in* D-60340
Neobanin, *in* D-60340

C$_{26}$H$_{28}$O$_{14}$
Ambocin, *in* T-60324
Neobacin, *in* T-60324
Neocorymboside, N-70021
1,2,7-Trihydroxy-6-methylanthraquinone; O^2-β-
Primeveroside, *in* T-70268

C$_{26}$H$_{29}$NO$_9$
Antibiotic R 20Y7, A-70230

C$_{26}$H$_{29}$N$_3$O$_2$
Crystal violet lactone, C-70202

C$_{26}$H$_{29}$N$_3$O$_{11}$S$_2$
O^6-[(3-Carbamoyl-2*H*-azirine-2-ylidene)-
amino]-1,2-*O*-isopropylidene-3,5-di-*O*-tosyl-
α-D-glucofuranoside, C-70014

C$_{26}$H$_{30}$Cl$_2$N$_2$O$_5$
Antibiotic SF 2415*A*$_3$, A-70234

C$_{26}$H$_{30}$N$_2$O$_5$
Antibiotic SF 2415*A*$_2$, A-70233

C$_{26}$H$_{30}$O$_4$
Erycristin, E-70034
Grenoblone, G-70029
Methyllinderatone, *in* L-70030

C$_{26}$H$_{30}$O$_5$
4-Hydroxygrenoblone, *in* G-70029

C$_{26}$H$_{30}$O$_6$
7-Deacetoxy-7-oxogedunin, *in* G-70006
2,3-Dibenzoylbutanedioic acid; Dibutyl ester, *in*
D-70083
Lespedezaflavanone *A*, L-60022

C$_{26}$H$_{30}$O$_8$
Dysoxylin, D-60525

C$_{26}$H$_{30}$O$_{10}$
Graucin *A*, G-60036

C$_{26}$H$_{30}$O$_{11}$
Rutaevinexic acid, R-70014

C$_{26}$H$_{30}$O$_{13}$
3-(4-Hydroxyphenyl)-1-(2,4,6-
trihydroxyphenyl)-2-propen-1-one; 2′-(*O*-
Rhamnosyl(1→4)xyloside), *in* H-60220

C$_{26}$H$_{31}$ClN$_2$O$_5$
Antibiotic SF 2415*A*$_1$, A-70232

C$_{26}$H$_{32}$Cl$_2$O$_5$
Antibiotic SF 2415*B*$_3$, A-70237

C$_{26}$H$_{32}$O$_4$
Nimbocinol, *in* A-70281

C$_{26}$H$_{32}$O$_5$
Antibiotic SF 2415*B*$_2$, A-70236
7-Deacetyl-17β-hydroxyazadiradione, *in*
A-70281
Licoricidin, L-60025

C$_{26}$H$_{32}$O$_6$
4′-(3,6-Dimethyl-2-heptenyloxy-5-hydroxy-3′,7-
dimethoxyflavanone, *in* T-70105

C$_{26}$H$_{32}$O$_8$
Longirabdosin, L-70036

C$_{26}$H$_{32}$O$_{11}$
Columbinyl glucoside, *in* C-60160

C$_{26}$H$_{32}$O$_{12}$
1-Hydroxypinoresinol; 1-*O*-β-D-
Glucopyranoside, *in* H-60221
1-Hydroxypinoresinol; 4-*O*-β-D-
Glucopyranoside, *in* H-60221
Isojateorinyl glucoside, *in* C-60160
Jateorinyl glucoside, *in* C-60160

C$_{26}$H$_{32}$O$_{13}$
Decumbeside *C*, D-60017
Decumbeside D, *in* D-60017

C$_{26}$H$_{32}$O$_{14}$
Mulberroside A, *in* D-70335

C$_{26}$H$_{33}$ClO$_5$
Antibiotic SF 2415*B*$_1$, A-70235

C$_{26}$H$_{33}$ClO$_8$
Ptilosarcenone; 2-De-Ac, 2-butanoyl, *in*
P-70136

C$_{26}$H$_{33}$ClO$_9$
12-Ptilosarcenol; 12-Ac, *in* P-60193

C$_{26}$H$_{33}$ClO$_{11}$
Solenolide D, *in* S-70053

C$_{26}$H$_{34}$O$_6$
Nimolicinoic acid, N-60022

C$_{26}$H$_{34}$O$_8$
Agrimophol, A-60066
Saroaspidin *C*, S-70014

C$_{26}$H$_{34}$O$_9$
Adenanthin, A-70065
Deoxyhavannahine, *in* H-60008
ent-1β,7α,11α-Triacetoxy-3α-hydroxy-16-
kaurene-6,15-dione, *in* T-60109

C$_{26}$H$_{34}$O$_{10}$
9α,14-Diacetoxy-1α-benzoyloxy-4β,6β,8β-
trihydroxydihydro-β-agarofuran, *in* E-60018

Eumaitenin, E-60071
Havannahine, H-60008
Rhynchospermoside *A*, R-60008
Rhynchospermoside B, *in* R-60008

C$_{26}$H$_{35}$ClO$_{10}$
7(18)-Havannachlorohydrin, H-70003
11(19)-Havannachlorohydrin, H-70004

C$_{26}$H$_{36}$
[14.0]Paracyclophane, P-70008

C$_{26}$H$_{36}$O$_7$
14-Acetoxy-2-butanoyloxy-5,8(17)-briaradien-
18,7-olide, *in* E-70012
9-Deacetoxyxenicin, *in* X-70004

C$_{26}$H$_{36}$O$_9$
Caesalpin F, C-60002

C$_{26}$H$_{36}$O$_{10}$
Melianolone, M-60015

C$_{26}$H$_{38}$N$_2$O$_4$
1,4,10,13-Tetraoxa-7,16-diazacyclooctadecane;
N,*N*′-Dibenzyl, *in* T-70145

C$_{26}$H$_{38}$O$_4$
Aglajne 3, A-70074
12-Hydroxy-24-methyl-24-oxo-16-scalarene-
22,25-dial, H-60191

C$_{26}$H$_{38}$O$_5$
Chinensin II, C-60036
12α-Hydroxy-24-methyl-24,25-dioxo-14-
scalaren-22-oic acid, *in* H-60191

C$_{26}$H$_{38}$O$_6$
1-Acetoxy-7-acetoxymethyl-3-acetoxymethyl-
ene-11,15-dimethyl-1,6,10,14-hexadecatetra-
ene, A-70020

C$_{26}$H$_{38}$O$_7$
17α-Acetoxy-6α-butanoyloxy-15,17-oxido-16-
spongianone, *in* D-60365
Virescenoside E, *in* I-70081

C$_{26}$H$_{40}$N$_4$
5,8,14,17-Tetrakiss(imethylamino)[3.3]-
paracyclophene, T-70126

C$_{26}$H$_{40}$O$_7$
Virescenoside L, *in* I-70081

C$_{26}$H$_{40}$O$_8$
Virescenoside G, *in* I-70077

C$_{26}$H$_{40}$O$_9$
Phloganthoside, *in* P-70099

C$_{26}$H$_{40}$O$_{14}$
6′-*O*-Apiosylebuloside, *in* E-60001

C$_{26}$H$_{42}$N$_6$
1,3,5,7-Tetrakiss(iethylamino)pyrrolo[3,4-*f*]-
isoindole, T-70125

C$_{26}$H$_{42}$O$_8$
Virescenoside A, *in* I-70081

C$_{26}$H$_{44}$N$_4$O$_{14}$
Methylallosamidin, *in* A-60076

C$_{26}$H$_{44}$O$_4$N$_4$
1,3,5,7-Adamantanetetraacetic acid; Tetrakis-
s(imethylamide), *in* A-70064

C$_{26}$H$_{46}$O$_6$
Sclareol; 6′-Deoxy-α-L-idopyranoside, *in*
S-70018

C$_{26}$H$_{48}$O$_3$
Ficulinic acid A, F-60011

C$_{26}$H$_{48}$O$_5$
5,6-Dihydro-6-(2,4,6-trihydroxyheneicosyl)-2*H*-
pyran-2-one, D-70276

C$_{26}$H$_{48}$O$_6$
5,6-Dihydro-6-(2,4,6,10-tetra-
hydroxyheneicosyl)-2*H*-pyran-2-one, *in*
D-70276

C$_{26}$H$_{50}$O$_2$
Aparjitin, A-60291

C$_{27}$H$_{20}$
1,2,3-Triphenyl-1*H*-indene, T-60387

C$_{27}$H$_{20}$Cl$_{15}$N
Tetraethylammonium triss(entachlorophenyl)-
methylide, *in* T-60405

$C_{27}H_{20}O_7$
Mulberrofuran R, M-60149

$C_{27}H_{22}O$
1,1,3,3-Tetraphenyl-2-propanone, T-60168

$C_{27}H_{22}S_2$
4,4,5,5-Tetraphenyl-1,3-dithiolane, T-60166

$C_{27}H_{24}$
1,2,3-Trimethyl-4,5,6-triphenylbenzene, T-70291
1,2,4-Trimethyl-3,5,6-triphenylbenzene, T-70292
1,3,5-Trimethyl-2,4,6-triphenylbenzene, T-70293

$C_{27}H_{24}O_7$
Larreantin, L-70020

$C_{27}H_{26}O_4S_2$
Lappaphen a, L-70018
Lappaphen b, in L-70018

$C_{27}H_{26}O_7$
Euchrenone b₃, E-70059

$C_{27}H_{26}O_{11}$
Cleistanthoside B, in D-60493
Viscutin 1, in T-60103

$C_{27}H_{28}O_7$
Pumilaisoflavone A, P-70140
Pumilaisoflavone B, P-70141

$C_{27}H_{28}O_{11}$
Tremulacin, in S-70002

$C_{27}H_{28}O_{13}$
3-O-Caffeoyl-4-O-sinapoylquinic acid, in C-70005

$C_{27}H_{30}O_6$
Cyclotriveratrylene, C-70254

$C_{27}H_{30}O_8$
5-Dehydrooriciopsin, in O-70045

$C_{27}H_{30}O_{14}$
4′,7-Dihydroxyisoflavone; 4′,7-Di-O-β-D-glucopyranosyl, in D-60340
Sphaerobioside, in T-60324

$C_{27}H_{30}O_{15}$
Palasitrin, in S-70082

$C_{27}H_{32}O_6$
Scutellone F, S-70020

$C_{27}H_{32}O_7$
Garvalone A, G-60004

$C_{27}H_{32}O_8$
Oriciopsin, O-70045

$C_{27}H_{32}O_9$
Pedonin, P-60014
Trijugin B, T-60352

$C_{27}H_{32}O_{10}$
Baccharinoid B25, B-60004
Eumaitenol, E-60072

$C_{27}H_{32}O_{16}$
3′,4′,5,7-Tetrahydroxyflavanone; 3′,5-Di-O-β-D-Glucoside, in T-70105

$C_{27}H_{33}N_3O_5$
Andrimide, A-60272

$C_{27}H_{34}O_4$
5-Oxocystofuranoquinol, in C-70260
5-Oxoisocystofuranoquinol, in C-70260
Yezoquinolide, Y-60003

$C_{27}H_{34}O_5$
5-O-Methyllicoricidin, in L-60025
Scopadulcic acid B, in S-60015

$C_{27}H_{34}O_6$
Scopadulcic acid A, S-60015

$C_{27}H_{34}O_7$
Scuterivulactone D, in T-70103

$C_{27}H_{34}O_9$
Isocyclocalamin, I-60096

$C_{27}H_{34}O_{11}$
Arctiin, in A-70249

Phillyrin, in P-60141

$C_{27}H_{35}ClO_9$
12-Ptilosarcenol; 12-Propanoyl, in P-60193

$C_{27}H_{36}O_3$
Cystofuranoquinol, C-70260

$C_{27}H_{36}O_4$
16-(2,5-Dihydroxy-3-methylphenyl)-2,6,10,19-tetramethyl-2,6,10,14-hexadecatetraene-4,12-dione, in H-70225
5-Hydroxycystofuranoquinol, in C-70260

$C_{27}H_{36}O_7$
12α-Hydroxy-4,4,14α-trimethyl-3,7,11,15-tetraoxo-5α-chol-8-en-24-oic acid, in L-60037
Hyperlatolic acid, H-60236
Isohyperlatolic acid, I-60101

$C_{27}H_{36}O_9$
4,14-Diacetoxy-2-propanoyloxy-5,8(17)-briaradien-18,7-olide, in E-70012

$C_{27}H_{36}O_{14}$
Sylvestroside III, S-70088
Sylvestroside IV, S-70089

$C_{27}H_{37}ClO_{10}$
Punaglandin 1, P-70142

$C_{27}H_{38}O$
10′-Apo-β-caroten-10′-ol, A-70246

$C_{27}H_{38}O_2$
Galloxanthin, G-70002
Sargaquinone, S-60010

$C_{27}H_{38}O_4$
8′,9′-Dihydroxysargaquinone, in S-60010
Sargahydroquinoic acid, S-60009

$C_{27}H_{38}O_5$
Amentadione, A-60086
Amentaepoxide, A-60087
Amentol, A-60088
Bifurcarenone, B-60105

$C_{27}H_{38}O_6$
Monoalide 25-acetate, in M-70152
Thorectolide 25-acetate, in T-70189

$C_{27}H_{38}O_7$
3β,12β-Dihydroxy-4,4,14α-trimethyl-7,11,15-trioxo-5α-chol-8-en-24-oic acid, in L-60037
3β-Hydroxy-4α-hydroxymethyl-4β,14α-dimethyl-7,11,15-trioxo-5α-chol-8-en-24-oic acid, in L-60036
Pseudopterosin B, in P-60187
Pseudopterosin C, in P-60187
Pseudopterosin D, in P-60187

$C_{27}H_{38}O_8$
3β,12β-Dihydroxy-4α-hydroxymethyl-4β,14α-dimethyl-7,11,15-trioxo-5α-chol-8-en-24-oic acid, in L-60036

$C_{27}H_{38}O_{13}$
Macrocliniside B, in Z-60001

$C_{27}H_{39}ClO_{10}$
Punaglandin 2, in P-70142

$C_{27}H_{40}O_3$
2-(5-Hydroxy-3,7,11,15-tetramethyl-2,6,10,14-hexadecatetraenyl)-6-methyl-1,2-benzenediol, H-70225

$C_{27}H_{40}O_4$
2-(5,16-Dihydroxy-3,7,11,15-tetramethyl-2,6,10,14-hexadecatetraenyl)-6-methyl-1,4-benzenediol, in H-70225
Furoscalarol, F-70057

$C_{27}H_{40}O_5$
Chinensin I, C-60035

$C_{27}H_{40}O_7$
Lucidenic acid H, L-60036
Secopseudopterosin B, in S-70026
Secopseudopterosin C, in S-70026
Secopseudopterosin D, in S-70026

$C_{27}H_{40}O_9$
Methyl 6α,15α,17β-triacetoxy-15,16-dideoxy-15,17-oxido-16-spongianoate, in T-60038

$C_{27}H_{42}O$
Papakusterol, P-60004

$C_{27}H_{42}O_3$
5-Spirosten-3-ol, S-60045
25(27)-Spirosten-3-ol, S-70065

$C_{27}H_{42}O_4$
Deacetoxybrachycarpone, in B-60195
6-Hydroxy-3-spirostanone, H-70219
Nuatigenin, N-60062
5-Spirostene-3,25-diol, S-60044

$C_{27}H_{42}O_5$
22,25-Epoxy-5-furostene-2α,3β,26-triol, in E-70015
25(27)-Spirostene-2,3,6-triol, S-70064

$C_{27}H_{42}O_6$
Lucidenic acid M, L-60037

$C_{27}H_{42}O_{15}$
Penstebioside, P-60016

$C_{27}H_{44}O$
(24S,25S)-24,26-Cyclo-5α-cholest-22E-en-3β-ol, in P-60004

$C_{27}H_{44}O_2$
3-Hydroxy-25,26,27-trisnor-24-cycloartanal, H-60232

$C_{27}H_{44}O_3$
Lansilactone, L-60013

$C_{27}H_{44}O_5$
1,2,3-Spirostanetriol, S-60043

$C_{27}H_{44}O_6$
22,25-Epoxy-2,3,6,26-furostanetetrol, E-70015

$C_{27}H_{44}O_7$
▷ 2,3,14,20,22,25-Hexahydroxy-7-cholesten-6-one, H-60063
Pterosterone, P-60192

$C_{27}H_{44}O_8$
2,3,5,14,20,22,25-Heptahydroxycholest-7-en-6-one, H-60023

$C_{27}H_{44}O_{15}$
Confertoside, C-70193

$C_{27}H_{46}O_2$
5-Cholestene-3,26-diol, C-60147
20,29,30-Trisnor-3,19-lupanediol, T-60404

$C_{27}H_{46}O_3$
Cholest-5-ene-3,16,22-triol, C-70172

$C_{27}H_{47}FeN_6O_9$
Ferrioxamine D₁, in F-60003

$C_{27}H_{47}N_9O_{10}S_2$
Trypanothione, T-70331

$C_{27}H_{48}$
1,3,5-Triheptylbenzene, T-70246

$C_{27}H_{48}O_3$
Hipposterol, H-70091

$C_{27}H_{54}$
13-Heptacosene, H-60019

$C_{28}H_{14}$
Benzo[a]coronene, B-60021

$C_{28}H_{14}O_2$
Dibenzo[a,j]perylene-8,16-dione, D-70073

$C_{28}H_{14}S$
Dibenzo[2,3:10,11]perylo[1,12-bcd]thiophene, D-70074

$C_{28}H_{16}O$
Diphenanthro[2,1-b:1′,2′-d]furan, D-70470
Diphenanthro[9,10-b:9′,10′-d]furan, D-70471

$C_{28}H_{17}N$
9H-Tetrabenzo[a,c,g,i]carbazole, T-70011

$C_{28}H_{19}N$
Di-9-phenanthrylamine, D-70472

$C_{28}H_{20}O_4$
3,3,6,6-Tetraphenyl-1,4-dioxane-2,5-dione, T-60165

$C_{28}H_{22}$
5,6-Dihydro-11,12-diphenyldibenzo[*a,e*]-cyclooctene, D-70202

$C_{28}H_{22}N_2$
9,9'-Biacridylidene; *N,N'*-Di-Me, *in* B-70090

$C_{28}H_{22}O$
2,2,4,4-Tetraphenyl-3-butenal, T-70148

$C_{28}H_{22}O_6$
Gnetin A, G-70023

$C_{28}H_{22}O_8$
4,4',5,5'-Tetrahydroxy-1,1'-binaphthyl; Tetra-Ac, *in* T-70102

$C_{28}H_{22}O_{10}$
Cephalochromin, C-60029

$C_{28}H_{24}$
[2.0.2.0]Metacyclophane, M-60031

$C_{28}H_{24}O_4$
Isomarchantin C, I-60107
Isoriccardin C, I-60135
Riccardin D, R-70005

$C_{28}H_{24}O_8$
Cassigarol A, C-70025

$C_{28}H_{26}O_{16}$
Taxillusin, *in* P-70027

$C_{28}H_{28}O_4$
2α,6α,11-Trihydroxy-7,9(11),13-abietatrien-12-one, *in* D-70282

$C_{28}H_{28}O_{11}$
Dukunolide C, *in* D-60522

$C_{28}H_{28}O_{15}$
ε-Rhodomycinone; 7-Glucoside, *in* R-70002

$C_{28}H_{30}O_{10}$
Physalin G, P-60146

$C_{28}H_{32}O_4$
Agerasanin, A-70072

$C_{28}H_{32}O_6$
Garcinone E, G-60003

$C_{28}H_{32}O_{10}$
Glaucin B, G-70017
Physalin L, P-60147

$C_{28}H_{32}O_{11}$
Physalin D, P-60145

$C_{28}H_{32}O_{14}$
Fortunellin, *in* D-70319
Linarin, *in* D-70319

$C_{28}H_{34}O_4$
Balaenonol, *in* B-70006

$C_{28}H_{34}O_5$
Azadiradione, A-70281
Encecanescin, E-70008
17-epi-Azadiradione, *in* A-70281
9'-Epiencecanescin, *in* E-70008

$C_{28}H_{34}O_7$
▷Gedunin, G-70006
Withaphysalin E, W-60009

$C_{28}H_{34}O_8$
6α-Hydroxygedunin, *in* G-70006

$C_{28}H_{34}O_9$
4-O-Demethylmicrophyllinic acid, *in* M-70147

$C_{28}H_{34}O_{10}$
Gomisin D, G-70025

$C_{28}H_{34}O_{11}$
6β,9α,14-Triacetoxy-1α-benzoyloxy-4β-hydroxy-8-oxodihydro-β-agarofuran, *in* E-60018

$C_{28}H_{34}O_{13}$
1-Acetoxypinoresinol 4'-O-β-D-glucopyranoside, *in* H-60221

$C_{28}H_{36}N_2O_6$
Sarcodictyin A, S-70010

$C_{28}H_{36}O_3$
Balaenol, B-70006

$C_{28}H_{36}O_4$
Daturilin, D-60003
Withametelin, W-60005

$C_{28}H_{36}O_5$
Celastanhydride, C-70028

$C_{28}H_{36}O_6$
Jaborol, J-60001

$C_{28}H_{36}O_7$
▷1,2-Dihydrogedunin, *in* G-70006
Trechonolide A, T-60227

$C_{28}H_{36}O_{10}$
Rzedowskin A, R-70016

$C_{28}H_{36}O_{11}$
6β,9α,14-Triacetoxy-1α-benzoyloxy-4β,8β-dihydroxydihydro-β-agarofuran, *in* E-60018

$C_{28}H_{36}O_{16}$
Shanzhiside; 6-(4-Hydroxy-3,5-dimethoxy-benzoyl), 8-Ac, Me ester, *in* S-60028

$C_{28}H_{37}ClO_{10}$
Ptilosarcone, *in* P-60195

$C_{28}H_{37}NO_4$
Cytochalasin O, C-60235

$C_{28}H_{38}N_2O_{10}$
1,2-Bis(4'-benzo-15-crown-5)diazene, B-60141

$C_{28}H_{38}N_4O_6$
Chlamydocin, C-60038

$C_{28}H_{38}O_4$
Cystoketal, C-60234
Isocystoketal, *in* C-60234

$C_{28}H_{38}O_5$
Daturilinol, D-70007
Mzikonone, M-70166

$C_{28}H_{38}O_7$
Physanolide, P-60149
Withanolide Y, W-60007
Withaphysanolide, W-60010

$C_{28}H_{38}O_9$
3,14-Diacetoxy-2-butanoyloxy-5,8(17)-briaradien-18,7-olide, *in* E-70012
4,14-Diacetoxy-2-butanoyloxy-5,8(17)-briaradien-18,7-olide, *in* E-70012
Xenicin, X-70004

$C_{28}H_{38}O_{10}$
Ingol; Tetra-Ac, *in* I-70018
Lapidolinin, *in* L-60015

$C_{28}H_{39}ClO_8$
Physalolactone, P-60148

$C_{28}H_{39}ClO_{10}$
Ptilosarcol, P-60195

$C_{28}H_{39}NO$
Emindole DA, E-60005
Emindole SA, *in* E-60005

$C_{28}H_{40}$
1,3,5,7-Tetra-*tert*-butyl-*s*-indacene, T-60025

$C_{28}H_{40}BrN_3O_6$
Geodiamolide B, *in* G-60012

$C_{28}H_{40}IN_3O_6$
Geodiamolide A, G-60012

$C_{28}H_{40}O_3$
9'-Methoxysargaquinone, *in* S-60010
11'-Methoxysargaquinone, *in* S-60010

$C_{28}H_{40}O_4$
8',9'-Dihydroxy-5-methylsargaquinone, *in* S-60010

$C_{28}H_{40}O_5$
Amentol 1'-methyl ether, *in* A-60088
Balearone, B-60005
Bifurcarenone; 2'-Me ether, *in* B-60105
12-Hydroxy-24-methyl-24-oxo-16-scalarene-22,25-dial; O¹²-Ac, *in* H-60191
Isobalearone, *in* B-60005
Strictaepoxide, S-60056
Strictaketal, S-70076

$C_{28}H_{40}O_6$
Ixocarpanolide, I-60143

$C_{28}H_{40}O_7$
6α,7α-Epoxy-5α,14α,20R-trihydroxy-1-oxo-22R,24S,25R-with-2-enolide, *in* I-60143
Withaphysacarpin, W-60008

$C_{28}H_{41}ClO_9$
Solenolide A, S-70052

$C_{28}H_{42}O_5$
Tubocapsigenin *A*, T-70333

$C_{28}H_{42}O_6$
11,15(17)-Trinervitadiene-3,9,13-triol; 9-Ac, 3,13-dipropanoyl, *in* T-70297

$C_{28}H_{44}O$
24-Methyl-5,22,25-cholestatrien-3-ol, M-60057

$C_{28}H_{46}O_3$
24-Methyl-7,22-cholestadiene-3,5,6-triol, M-70057

$C_{28}H_{46}O_4$
24-Methylene-5-cholestene-3,4,7,20-tetrol, M-60070

$C_{28}H_{46}O_5$
Dolichosterone, D-70544

$C_{28}H_{46}O_7$
Polypodoaurein, *in* H-60063

$C_{28}H_{47}N_9O_9$
Neokyotorphin, N-70025

$C_{28}H_{48}O_4$
6-Deoxodolichosterone, *in* D-70544
3α,22R,23R-Trihydroxy-24S-methyl-5α-cholestan-6-one, *in* T-70116

$C_{28}H_{48}O_5$
2,3,22,23-Tetrahydroxy-24-methyl-6-cholestanone, T-70116

$C_{28}H_{48}O_6$
Brassinolide, B-70178

$C_{28}H_{48}O_{12}$
4,7,10,15,18,21,24,27,30,33,36,39-Dodeca-oxatricyclo[11.9.9.9²,¹²]tetraconta-1,12-diene, D-70535

$C_{28}H_{49}NO_8$
Kayamycin, K-60009

$C_{28}H_{50}O_4$
6-Deoxocastasterone, *in* T-70116

$C_{28}H_{52}O_3$
Ficulinic acid B, F-60012

$C_{29}H_{18}O$
Di-9-anthracenylmethanone, D-60050

$C_{29}H_{20}$
7,7-Diphenylbenzo[*c*]fluorene, D-70473
Tetrabenzotetracyclo[5.5.1.0⁴,¹³.0¹⁰,¹³]tridecane, T-60019

$C_{29}H_{22}$
1,2,3,4-Tetraphenylcyclopentadiene, T-70149

$C_{29}H_{22}O_{14}$
3,5-Digalloylepicatechin, *in* P-70026

$C_{29}H_{24}O_{10}$
Chaetochromin C, *in* C-60029

$C_{29}H_{26}O_4$
Riccardin E, *in* R-70005

$C_{29}H_{28}O_{12}$
Viscutin 2, *in* T-60103

$C_{29}H_{29}N_3O_8$
K 13, K-70001

$C_{29}H_{34}O_7$
Triptofordin B, T-60393

$C_{29}H_{34}O_{10}$
Crepiside A, *in* Z-60001
Crepiside B, *in* Z-60001

$C_{29}H_{34}O_{11}$
Physalin I, *in* P-60145
Trijugin *A*, T-60351

$C_{29}H_{34}O_{12}$
Acetyleumaitenol, *in* E-60072

$C_{29}H_{34}O_{14}$
Pueroside A, *in* P-70137

$C_{29}H_{36}O_6$
Demethylzeylasteral, *in* D-70027

$C_{29}H_{36}O_7$
Desmethylzeylasterone, D-70027

$C_{29}H_{36}O_9$
Microphyllinic acid, M-70147

$C_{29}H_{36}O_{10}$
Baccharinoid B27, *in* B-60002

$C_{29}H_{36}O_{13}$
1-Hydroxypinoresinol; 4″-Me ether, 1-Ac, 4′-O-
β-D-glucopyranoside, *in* H-60221

$C_{29}H_{38}N_2O_6$
Sarcodictyin B, *in* S-70010

$C_{29}H_{38}O_7$
Isonimolide, I-60110
Trechonolide *B*, *in* T-60227

$C_{29}H_{38}O_9$
Roridin *D*, R-70008
Scutellone *A*, S-60017
Scutellone *C*, S-70019
Scuterivulactone C_2, *in* S-60017

$C_{29}H_{38}O_{10}$
Baccharinoid B9, B-60001
Baccharinoid B10, *in* B-60001
Baccharinoid B12, *in* R-70008
Baccharinoid B13, B-60002
Baccharinoid B14, *in* B-60002
Baccharinoid B16, B-60003
Baccharinoid B17, *in* R-70008
Baccharinoid B21, *in* R-70008

$C_{29}H_{38}O_{11}$
Baccharinol, B-70003
Isobaccharinol, *in* B-70003

$C_{29}H_{40}O_7$
Superlatolic acid, S-60063

$C_{29}H_{40}O_8$
Physalactone, P-60144

$C_{29}H_{40}O_9$
Ingol; 3,7-Di-Ac, 12-tigloyl, *in* I-70018

$C_{29}H_{40}O_{10}$
Baccharinoid *B1*, B-70001
Baccharinoid *B7*, B-70002
Baccharinoid B2, *in* B-70001
Baccharinoid B3, *in* B-70002
Baccharinoid B20, *in* B-60001
Baccharinoid B23, *in* B-60003
Baccharinoid B24, *in* B-60003

$C_{29}H_{42}O_4$
Przewanoic acid *B*, P-70126

$C_{29}H_{44}O_2$
3β-Hydroxy-28-noroleana-12,17-dien-16-one, *in*
M-70011

$C_{29}H_{44}O_3$
3-Hydroxy-30-nor-12,20(29)-oleanadien-28-oic
acid, H-60199
18β-Hydroxy-28-nor-3,16-oleanenedione, *in*
D-60363

$C_{29}H_{44}O_4$
3,23-Dihydroxy-30-nor-12,20(29)-oleanadien-
28-oic acid, D-70324

$C_{29}H_{44}O_6$
Brachycarpone, B-60195
Heteronemin, H-60030
11,15(17)-Trinervitadiene-3,9,13-triol; 3,9,13-
Tripropanoyl, *in* T-70297

$C_{29}H_{44}O_9$
Coroglaucigenin; 3-O-Rhamnoside, *in* C-60168
▷Frugoside, *in* C-60168
Methyl 15α,17β-diacetoxy-6α-butanoyloxy-
15,16-dideoxy-15,17-oxido-16-spongianate, *in*
T-60038

$C_{29}H_{45}N_3O_3$
Blastmycetin *D*, B-60191

$C_{29}H_{46}O_2$
Maragenin I, M-70011

$C_{29}H_{46}O_3$
3,18-Dihydroxy-28-nor-12-oleanen-16-one,
D-60363

$C_{29}H_{46}O_7$
2α,7α-Diacetoxy-6β-isovaleroyloxy-8,13E-
labdadien-15-ol, *in* L-60002

$C_{29}H_{46}O_8$
Viticosterone E, *in* H-60063

$C_{29}H_{47}N_5O_5$
Stacopin P1, S-60051

$C_{29}H_{47}N_5O_6$
Stacopin P2, *in* S-60051

$C_{29}H_{48}O$
Ficisterol, F-60010
Hebesterol, H-60011

$C_{29}H_{48}O_2$
24,25-Epoxy-29-nor-3-cycloartanol, E-70026
29-Nor-23-cycloartene-3,25-diol, N-70073
28-Nor-16-oxo-17-oleanen-3β-ol, *in* M-70011

$C_{29}H_{48}O_3$
6β-Methoxy-7,22-ergostadiene-3β,5α-diol, *in*
M-70057

$C_{29}H_{50}O$
Petrostanol, P-60066

$C_{29}H_{50}O_2$
4-Stigmastene-3,6-diol, S-60054

$C_{29}H_{50}O_3$
24,24-Dimethoxy-25,26,27-trisnor-3-cyclo-
artanol, *in* H-60232

$C_{29}H_{50}O_4$
24(28)-Stigmastene-2,3,22,23-tetrol, S-70075

$C_{29}H_{52}N_{10}O_6$
Argiopine, A-60295

$C_{29}H_{52}O$
24,24-Dimethyl-3-cholestanol, D-70389

$C_{29}H_{52}O_2$
3,6-Stigmastanediol, S-70074

$C_{30}H_{14}$
[5]Phenylene, P-60090

$C_{30}H_{16}$
Pyranthrene, P-70151
Tribenzo[*b,n,pqr*]perylene, T-60244

$C_{30}H_{16}O_2$
7,16-Heptacenedione, H-60018

$C_{30}H_{18}$
Di-9-anthrylacetylene, D-60051

$C_{30}H_{18}O_{10}$
Hinokiflavone, H-60082

$C_{30}H_{18}O_{12}$
Bryoflavone, B-70283
Dicranolomin, D-70144
Heterobryoflavone, H-70027

$C_{30}H_{20}$
1,2,3-Triphenylcyclopent[*a*]indene, T-70306

$C_{30}H_{20}O_4$
Tetrabenzoylethylene, T-60021

$C_{30}H_{20}O_9$
1-[3-(2,4-Dihydroxybenzoyl)-4,6-dihydroxy-2-
(4-hydroxyphenyl)-7-benzofuranyl]-3-(4-
hydroxyphenyl)-2-propen-1-one, D-70286

$C_{30}H_{20}O_{12}$
2,3-Dihydrodicranolomin, *in* D-70144

$C_{30}H_{20}S_2$
1,3,4,6-Tetraphenylthieno[3,4-*c*]thiophene-5-
S^{IV}, T-70152

$C_{30}H_{20}S_8$
Tetramercaptotetrathiafulvalene; Tetrakis (*S*-
Ph), *in* T-70130

$C_{30}H_{22}N_4O_4$
6,6′-(1,2-Dimethyl-1,2-ethanediyl)bis[1-
phenazinecarboxylic acid], D-70411

$C_{30}H_{22}O_4$
1,1,2,2-Tetrabenzoylethane, T-60020

$C_{30}H_{22}O_8$
Lophirone *A*, L-60035

$C_{30}H_{22}O_9$
Daphnodorin *D*, D-70005
1-[3-(2,4-Dihydroxybenzoyl)-4,6-dihydroxy-2-
(4-hydroxyphenyl)-7-benzofuranyl]-3-(4-
hydroxyphenyl)-2-propen-1-one; 2,3-Dihydro
(trans-), *in* D-70286
Maackiasin, M-60001

$C_{30}H_{22}O_{10}$
GB1a, *in* G-70005
Isochamaejasmin, I-70064

$C_{30}H_{22}O_{11}$
GB 1, G-70005
GB2a, *in* G-70005

$C_{30}H_{22}O_{12}$
3″,3‴,4′,4‴,5,5″,7,7″-Octahydroxy-3,8″-
biflavanone, *in* G-70005

$C_{30}H_{22}O_{13}$
Chiratanin, C-60037

$C_{30}H_{24}N_6O_3$
Furlone yellow, F-70053

$C_{30}H_{24}O_4$
Tecomaquinone I, T-60010

$C_{30}H_{24}O_{10}$
Chaetochromin D, *in* C-60029

$C_{30}H_{24}O_{13}$
[2′,2′]-Catechin-taxifolin, C-60023

$C_{30}H_{26}O_5$
Tecomaquinone III, T-70006

$C_{30}H_{26}O_6$
Flavanthrin, F-70011

$C_{30}H_{26}O_8$
Mbamichalcone, M-70014

$C_{30}H_{26}O_{10}$
Chaetochromin, *in* C-60029
Chaetochromin B, *in* C-60029

$C_{30}H_{28}N_4O_2$
Tetrahydro-1,2,4,5-tetrazine-3,6-dione; 1,2,4,5-
Tetrabenzyl, *in* T-70097

$C_{30}H_{28}O_6$
Angoluvarin, A-70208
Isothamnosin A, I-70104
Isothamnosin B, *in* I-70104

$C_{30}H_{30}O_8$
▷Gossypol, G-60032

$C_{30}H_{30}O_{10}$
Phleichrome, P-70098

$C_{30}H_{32}O_4$
4,15,26-Tricontatrien-1,12,18,29-tetrayne-
3,14,17,28-tetrone, *in* P-70044

$C_{30}H_{32}O_6$
Fruticolide, F-70045

$C_{30}H_{32}O_{14}$
Senburiside II, S-70033

$C_{30}H_{34}O_5$
Euchrenone b_1, E-70058

$C_{30}H_{34}O_6$
Eucrenone b_2, *in* E-70058

$C_{30}H_{34}O_{11}$
Ainsliaside A, *in* Z-60001

$C_{30}H_{36}O_8$
Artelein, A-60303

$C_{30}H_{36}O_9$
6α-Acetoxygedunin, *in* G-70006
11β-Acetoxygedunin, *in* G-70006

$C_{30}H_{36}O_9$

Isonimolicinolide, I-60109

$C_{30}H_{36}O_{11}$

Isonimbinolide, I-70072

$C_{30}H_{36}O_{15}$

Pueroside B, *in* P-70137

$C_{30}H_{37}N_5O_5$

Avellanin *B*, A-60324

$C_{30}H_{38}O_4$

Cochloxanthin, C-60156

$C_{30}H_{38}O_5$

23-Oxoisopristimerin III, *in* I-70083

$C_{30}H_{38}O_6$

Zeylasteral, *in* D-70027

$C_{30}H_{38}O_9$

Isolimbolide, I-60105
2'-O-Methylmicrophyllinic acid, *in* M-70147

$C_{30}H_{38}O_{12}$

$6\beta,8\beta,9\alpha,14$-Tetraacetoxy-1α-benzoyloxy-4β-
    hydroxydihydro-β-agarofuran, *in* E-60018

$C_{30}H_{38}O_{13}$

Bruceanol *C*, B-70280

$C_{30}H_{39}NO_5$

Cytochalasin N, *in* C-60235

$C_{30}H_{40}O_4$

4,5-Dihydro-6-hydroxy-3-oxo-8'-apo-ϵ-caroten-
    8'-oic acid, *in* C-60156
Isopristimerin III, I-70083
Petrosynol, P-70044

$C_{30}H_{40}O_7$

7β-Hydroxy-3,11,15,23-tetraoxo-8,20(22)-
    lanostadien-26-oic acid, *in* T-70254

$C_{30}H_{40}O_8$

7,12-Dihydroxy-3,11,15,23-tetraoxo-8,20(22)-
    lanostadien-26-oic acid, D-60378
Ganoderic acid *O*, *in* D-60379
Trichilinin, T-60251
Withaminimin, W-60006

$C_{30}H_{40}O_{10}$

1,2-Bis(4'-benzo-15-crown-5)ethene, B-60142
Xenicin; 9-Deacetyl, 9-acetoacetyl, *in* X-70004

$C_{30}H_{41}NO_6$

Cytochalasin P, C-60236

$C_{30}H_{42}O_3$

Heleniumlactone 1, H-60013
Heleniumlactone 2, *in* H-60013
Heleniumlactone 3, H-60014

$C_{30}H_{42}O_4$

23-epi-Firmanolide, *in* F-60013
Firmanolide, F-60013
Isomaresiic acid *C*, I-60108
Mariesiic acid *C*, M-60011

$C_{30}H_{42}O_5$

Abiesonic acid, A-60005

$C_{30}H_{42}O_7$

▷Cucurbitacin I, *in* C-70204
$3\beta,7\beta$-Dihydroxy-11,15,23-trioxo-8,20(22)*E*-
    lanostadien-26-oic acid, *in* T-70254
$7\beta,15\alpha$-Dihydroxy-3,11,23-trioxo-8,20(22)*E*-
    lanostadien-26-oic acid, *in* T-70254

$C_{30}H_{42}O_8$

7,20-Dihydroxy-3,11,15,23-tetraoxo-8-lanosten-
    26-oic acid, D-60379

$C_{30}H_{42}O_9$

Ingol; 7-Tigloyl, O^8-Me, 3,12-di-Ac, *in* I-70018
Ingol; 7-Angeloyl, O^8-Me, 3,12-Di-Ac, *in*
    I-70018

$C_{30}H_{44}O_4$

3,23-Dioxo-7,24-lanostadien-26-oic acid,
    D-60472
3,23-Dioxo-7,25(27)-lanostadien-26-oic acid, *in*
    D-60472
3,7-Dioxo-12-oleanen-28-oic acid, D-70469
23-Oxomariesiic acid *A*, *in* M-70012

23-Oxomariesiic acid B, *in* M-60010

$C_{30}H_{44}O_5$

Chiisanogenin, C-70038
3,15-Dihydroxy-23-oxo-7,9(11),24-lanostatrien-
    26-oic acid, D-70330

$C_{30}H_{44}O_6$

11-Deoxocucurbitacin I, *in* C-70204
9,11-Dihydro-22,25-oxido-11-
    oxoholothurinogenin, D-70237

$C_{30}H_{44}O_7$

▷Cucurbitacin D, C-70203
Cucurbitacin L, *in* C-70204
3,7,15-Trihydroxy-11,23-dioxo-8,20(22)-
    lanostadien-26-oic acid, T-70254

$C_{30}H_{46}O$

9(11),12-Oleanadien-3-one, O-60032

$C_{30}H_{46}O_2$

21,23-Epoxy-7,24-tirucalladien-3-one, *in*
    E-60032

$C_{30}H_{46}O_3$

24*S*,25*S*-Epoxy-26-hydroxy-7,9(11)-
    lanostadien-3-one, *in* E-70025
Flindissol lactone, *in* F-70013
Flindissone, *in* F-70013
3-Hydroxy-12-oleanen-29,22-olide, H-70197
7-Hydroxy-3-oxo-8,24-lanostadien-26-al,
    H-70200
22-Hydroxy-7,24-tirucalladiene-3,23-dione,
    H-70227
3-Oxo-8,24-lanostadien-26-oic acid, O-70096
3-Oxo-8-lanosten-26,22-olide, *in* H-60169
Swertialactone *D*, S-60064
Swertialactone *C*, *in* S-60064

$C_{30}H_{46}O_4$

Abiesolidic acid, A-70003
Colubrinic acid, C-70190
3,15-Dihydroxy-7,9(11),24-lanostatrien-26-oic
    acid, D-70314
2,3-Dihydroxy-12,20(30)-ursadien-28-oic acid,
    D-60381
12-Hydroxy-3-oxo-28,13-oleananolide,
    H-70201
19α-Hydroxy-3-oxo-12-oleanen-28-oic acid, *in*
    D-70327
19α-Hydroxy-3-oxo-12-ursen-28-oic acid, *in*
    D-70350
22-Hydroxy-3-oxo-12-ursen-30-oic acid,
    H-70202
Manwuweizic acid, M-70010
Mariesiic acid *A*, M-70012
Mariesiic acid *B*, M-60010
Przewanoic acid *A*, P-70125
3,4-Seco-4(28),20,24-dammaratrien-3,26-dioic
    acid, S-70021
15,26,27-Trihydroxy-7,9(11),24-lanostatrien-3-
    one, T-60330

$C_{30}H_{46}O_5$

Cadambagenic acid, C-70002
19,29-Dihydroxy-3-oxo-12-oleanen-28-oic acid,
    D-70331
22,23-Dihydroxy-3-oxo-12-ursen-30-oic acid,
    D-70333
2,3,24-Trihydroxy-13,27-cyclo-11-oleanen-28-
    oic acid, T-70251
2,3,24-Trihydroxy-12,27-cyclo-14-taraxeren-28-
    oic acid, T-70252
3,15,22-Trihydroxy-7,9(11),24-lanostatrien-26-
    oic acid, T-70266
2,3,24-Trihydroxy-11,13(18)-oleonadien-28-oic
    acid, T-60336
2α,3α,24-Trihydroxy-12,20(30)-ursadien-28-oic
    acid, *in* D-60381

$C_{30}H_{46}O_6$

3,19-Dihydroxy-12-ursene-24,28-dioic acid,
    D-60382

$C_{30}H_{46}O_7$

Cucurbitacin R, *in* C-70203
▷Saponaceolide *A*, S-60008

$C_{30}H_{46}O_8$

Maquiroside A, *in* C-60168

$C_{30}H_{48}O$

5(10)-Gluten-3-one, G-60028
14(26),17*E*,21-Malabaricatrien-3-one, *in*
    M-60008
9(11),12-Oleanadien-3-ol, O-70031
11,13(18)-Oleanadien-3-ol, O-70032
9(11),12-Ursadien-3-ol, U-70007

$C_{30}H_{48}O_2$

Duryne, D-70550
21,23-Epoxy-7,24-tirucalladien-3-ol, E-60032
30-Norcyclopterospermone, *in* N-60059
3,4-Seco-4(23),14-taraxeradien-3-oic acid,
    S-60018

$C_{30}H_{48}O_3$

3β,26-Dihydroxy-5,24*E*-cucurbitadien-11-one,
    in C-60175
3β,27-Dihydroxy-5,24*Z*-cucurbitadien-11-one,
    in C-60175
24,25-Dihydroxy-7,9(11)-lanostadien-3-one,
    D-60343
3,23-Dihydroxy-20(30)-lupen-29-al, D-70316
1,11-Dihydroxy-20(29)-lupen-3-one, D-60345
23,30-Dihydroxy-20(29)-lupen-3-one, D-70318
1,11-Dihydroxy-18-oleanen-3-one, D-60364
24,25-Epoxy-7,9(11)-lanostadiene-3,26-diol,
    E-70025
Flindissol, F-70013
27-Hydroxy-3,21-friedelanedione, *in* D-70301
3-Hydroxy-8,24-lanostadien-21-oic acid,
    H-70152
3-Hydroxy-8-lanosten-26,22-olide, H-60169
3-Hydroxy-28,13-oleananolide, H-60200
3-Hydroxytaraxastan-28,20-olide, H-70221
3-Hydroxy-28,20-ursanolide, H-60233
12,20(29)-Lupadiene-3,27,28-triol, L-70042
Niloticin, N-70036
3-Oxo-27-friedelanoic acid, *in* T-70218

$C_{30}H_{48}O_4$

Alisol *B*, A-70082
3,23-Dihydroxy-20(29)-lupen-28-oic acid,
    D-70317
1,3-Dihydroxy-12-oleanen-29-oic acid, D-70325
3,6-Dihydroxy-12-oleanen-29-oic acid, D-70326
3,19-Dihydroxy-12-oleanen-28-oic acid,
    D-70327
1,3-Dihydroxy-12-ursen-28-oic acid, D-60383
2,3-Dihydroxy-12-ursen-28-oic acid, D-70349
3,19-Dihydroxy-12-ursen-28-oic acid, D-70350
3,23-Dihydroxy-12-ursen-28-oic acid, D-70351
24,25-Epoxy-7,26-dihydroxy-8-lanosten-3-one,
    E-70013
Saikogenin *F*, S-60001
24,25,26-Trihydroxy-7,9(11)-lanostadien-3-one,
    in L-60012

$C_{30}H_{48}O_5$

Cycloorbigenin, C-60218
Squarrofuric acid, S-70070
Toosendantriol, T-60223
3,7,15-Trihydroxy-8,24-lanostadien-26-oic acid,
    T-60327
3,7,22-Trihydroxy-8,24-lanostadien-26-oic acid,
    T-60328
3,12,17-Trihydroxy-9(11)-lanosten-18,20-olide,
    T-70267
2,3,23-Trihydroxy-12-oleanen-28-oic acid,
    T-70271
2,3,24-Trihydroxy-12-oleanen-28-oic acid,
    T-60335
3β,23,27-Trihydroxy-28,20-taraxastanolide, *in*
    H-70221
3,19,24-Trihydroxy-12-ursen-28-oic acid,
    T-60348

$C_{30}H_{48}O_6$

3,7,15,22-Tetrahydroxy-8,24-lanostadien-26-oic acid, T-60111

2,3,19,23-Tetrahydroxy-12-oleanen-28-oic acid, T-70119

1,2,3,19-Tetrahydroxy-12-ursen-28-oic acid, T-70121

$C_{30}H_{50}$

3,4-Di-1-adamantyl-2,2,5,5-tetramethylhexane, D-60025

$C_{30}H_{50}O$

Cycloeuphordenol, C-70218

5-Gluten-3-ol, G-70020

24-Isopropenyl-7-cholesten-3-ol, I-70085

13(18)-Lupen-3-ol, L-70043

14(26),17,21-Malabaricatrien-3-ol, M-60008

13(18)-Oleanen-3-ol, O-70036

$C_{30}H_{50}O_2$

20(29)-Epoxy-3-lupanol, E-60026

27-Hydroxy-3-triedelanone, in D-70301

24-Isopropenyl-7-cholestene-3,6-diol, I-70084

8,24-Lanostadiene-3,21-diol, L-70017

20(29)-Lupene-3,16-diol, L-60039

30-Norcyclopterospermol, N-60059

12-Oleanene-3,15-diol, O-70033

13(18)-Oleanene-2,3-diol, O-60033

14-Taraxerene-3,24-diol, T-60008

Trichadenal, T-70218

$C_{30}H_{50}O_3$

5,24-Cucurbitadiene-3,11,26-triol, C-60175

Dihydroniloticin, in N-70036

21,27-Dihydroxy-3-friedelanone, D-70301

9(11)-Fernene-3,7,19-triol, F-60002

3β-Hydroxy-27-friedelanoic acid, in T-70218

3α-Hydroxy-27-friedelanoic acid, in T-70218

12-Oleanene-1,3,11-triol, O-60036

12-Oleanene-2,3,11-triol, O-60037

12-Oleanene-3,16,28-triol, O-60038

13(18)-Oleanene-3,16,28-triol, O-70035

12-Ursene-2,3,11-triol, U-60008

$C_{30}H_{50}O_4$

20,24-Dihydroxy-3,4-seco-4(28),23-dammaradien-3-oic acid, D-70343

20,25-Dihydroxy-3,4-seco-4(28),23-dammaradien-3-oic acid, D-70344

7,9(11)-Lanostadiene-3,24,25,26-tetrol, L-60012

12-Oleanene-1,2,3,11-tetrol, O-60034

12-Oleanene-3,16,23,28-tetrol, O-60035

11(12)-Oleanene-3,13,23,28-tetrol, O-70034

Quisquagenin, Q-70011

12,20,25-Trihydroxy-23-dammaren-3-one, T-60319

1,11,20-Trihydroxy-3-lupanone, T-60331

3,23,25-Trihydroxy-7-tiracallen-24-one, T-60347

12-Ursene-1,2,3,11-tetrol, U-60006

12-Ursene-2,3,11,20-tetrol, U-60007

$C_{30}H_{50}O_5$

12,20,25-Trihydroxy-3,4-seco-4(28),23-dammaradien-3-oic acid, T-70274

20,25,26-Trihydroxy-3,4-seco-4(28),23-dammaradien-3-oic acid, T-70275

12-Ursene-1,2,3,11,20-pentol, U-60005

$C_{30}H_{51}BrO_5$

Magireol B, M-60006

Magireol C, in M-60006

$C_{30}H_{51}BrO_6$

15-Anhydrothyrsiferol, A-60274

$C_{30}H_{52}N_{10}O_7$

Neurotoxin NSTX 3, N-70033

$C_{30}H_{52}O_2$

13,17,21-Polypodatriene-3,8-diol, P-70107

$C_{30}H_{52}O_4$

24-Dammarene-3,6,20,27-tetrol, D-70002

24-Dammarene-3,7,20,27-tetrol, D-70003

7-Tirucallene-3,23,24,25-tetrol, T-70201

$C_{30}H_{52}O_5$

3,16,24,25,30-Cycloartanepentol, C-60182

24-Dammarene-3,7,18,20,27-pentol, D-70001

25-Dammarene-3,12,17,20,24-pentol, D-60001

$C_{30}H_{53}BrO_6$

Magireol A, M-60005

$C_{30}H_{53}BrO_7$

Thyrsiferol, T-70193

Venustatriol, in T-70193

$C_{30}H_{54}$

Hexabutylbenzene, H-60032

$C_{31}H_{18}O$

2,3-Bis(9-anthryl)cyclopropenone, B-70128

$C_{31}H_{20}O_{10}$

Cryptomerin A, in H-60082

Isocryptomerin, in H-60082

Neocryptomerin, in H-60082

$C_{31}H_{22}O_5$

Nordracorubin, N-70075

$C_{31}H_{24}$

4-Methylene-1,2,3,5-tetraphenylbicyclo[3.1.0]hex-2-ene, M-60079

$C_{31}H_{24}O_{12}$

Kolaflavanone, in G-70005

$C_{31}H_{27}N_7$

1,14,34,35,36,37,38-Heptaazaheptacyclo-[12.12.7.1^{3,7}.1^{8,12}.1^{16,20}.1^{21,25}.1^{28,32}]octatriaconta-3,5,7(38),8,10,12(37),16,18,20(36),-21,23,25(35),28,30,32(34)pentadecane, H-70010

$C_{31}H_{32}O_8$

Gossypol; 6-Me ether, in G-60032

$C_{31}H_{32}O_{12}$

Fridamycin D, F-70044

$C_{31}H_{32}O_{16}$

3,4-Di-O-caffeoylquinic acid; O^5-(3-Hydroxy-3-methylglutaroyl), in D-70110

$C_{31}H_{36}O_6$

Triptofordin A, T-60392

$C_{31}H_{36}O_{18}S$

Rhodonocardin B, in R-70003

$C_{31}H_{39}N_5O_5$

Avellanin A, A-60323

$C_{31}H_{41}NO_6$

Pyrichalasin H, P-60215

$C_{31}H_{42}O_8$

Fevicordin A, F-60008

$C_{31}H_{42}O_{10}$

Ingol; 3,7,8-Tri-Ac, 12-tigloyl, in I-70018

Ingol; 7-Tigloyl, 3,8,12-tri-Ac, in I-70018

Ingol; 7-Angeloyl, 3,8,12-tri-Ac, in I-70018

$C_{31}H_{42}O_{11}$

Euphorianin, E-60074

$C_{31}H_{44}O_6$

Anguinomycin A, A-70209

$C_{31}H_{46}O_3$

Disidein, D-60499

$C_{31}H_{46}O_7$

Heteronemin; 12-Ac, in H-60030

Heteronemin; 12-Epimer, 12-Ac, in H-60030

$C_{31}H_{46}O_{13}$

Chamaepitin, C-60033

$C_{31}H_{48}O_4$

Regelin, in H-70202

$C_{31}H_{48}O_5$

Regelinol, in D-70333

$C_{31}H_{50}O$

24-Methylene-7,9(11)-lanostadien-3-ol, M-70072

$C_{31}H_{50}O_2$

24-Methylene-3,4-seco-4(28)-cycloarten-3-oic acid, M-70078

$C_{31}H_{50}O_4$

2,3-Dihydroxy-12-ursen-28-oic acid; Me ester, in D-70349

$C_{31}H_{50}O_5$

3,12,17-Trihydroxy-9(11)-lanosten-18,20-olide; 12-Me ether, in T-70267

$C_{31}H_{52}O$

Cyclopterospermol, C-60232

$C_{32}H_{14}$

Ovalene, O-60046

$C_{32}H_{22}O_{10}$

Chamaecyparin, in H-60082

Cryptomerin B, in H-60082

$C_{32}H_{24}$

3,6-Biss(iphenylmethylene)-1,4-cyclohexadiene, B-60157

$C_{32}H_{24}O_{15}$

ε-Isoactinorhodin, I-70059

$C_{32}H_{26}N_2O_9$

5-Vinyluridine; 2',3',5'-Tribenzoyl, in V-70015

$C_{32}H_{26}N_4O_4$

6,6'-(1,2-Dimethyl-1,2-ethanediyl)bis[1-phenazinecarboxylic acid]; Di-Me ester, in D-70411

$C_{32}H_{26}O_{13}$

Alterporriol B, A-60079

Alterporriol A, in A-60079

$C_{32}H_{28}O_2$

Dypnopinacol, D-60524

$C_{32}H_{28}O_{10}$

Candicanin, C-70012

$C_{32}H_{30}N_4$

13,15-Ethano-17-ethyl-2,3,12,18-tetramethyl-monobenzo[g]porphyrin, E-60043

$C_{32}H_{30}O_8$

2',7-Dihydroxy-4'-methoxy-4-(2',7-dihydroxy-4'-methoxyisoflavan-5'-yl)isoflavan, D-60347

$C_{32}H_{30}O_9$

2',7-Dihydroxy-4'-methoxy-4-(2',7-dihydroxy-4'-methoxyisoflavan-5'-yl)isoflavan; 3'-Hydroxy, in D-60347

$C_{32}H_{32}O_8$

Toddasin, T-70203

$C_{32}H_{34}O_8$

Gossypol; 6,6'-Di-Me ether, in G-60032

$C_{32}H_{36}O_8$

Artanomaloide, A-60300

$C_{32}H_{38}O_{11}$

6α,11β-Diacetoxygedunin, in G-70006

$C_{32}H_{38}O_{12}$

Luminamicin, L-60038

$C_{32}H_{38}O_{15}$

Neoleuropein, N-60018

$C_{32}H_{40}O_8$

Hyperevoline, H-70233

$C_{32}H_{40}O_9$

Ingol; O^7-Benzoyl, O^8-Me, 3,12-di-Ac, in I-70018

$C_{32}H_{44}O_4$

3-Octadecene-1,2-diol; Dibenzoyl, in O-60005

$C_{32}H_{44}O_8$

▷Cucurbitacin E, C-70204

Datiscacin, in C-70204

$C_{32}H_{44}O_{18}$

Isonuezhenide, I-70073

$C_{32}H_{46}O_6$

3α-Acetoxy-15α-hydroxy-23-oxo-7,9(11),24E-lanostatrien-26-oic acid, in D-70330

15α-Acetoxy-3α-hydroxy-23-oxo-7,9(11),24E-lanostatrien-26-oic acid, *in* D-70330
Anguinomycin *B*, A-70210
Azadirachtol, A-70280

C₃₂H₄₆O₇
Bufalin; 3-(Hydrogen suberoyl), *in* B-70289
Kasuzamycin *B*, K-70007

C₃₂H₄₈O₄
23-Acetoxy-3-oxo-20(30)-lupen-29-al, *in* D-70316
3,4-Seco-4(28),20,24-dammaratrien-3,26-dioic acid; 3-Me ester, *in* S-70021

C₃₂H₄₈O₅
3α-Acetoxy-15α-hydroxy-7,9(11),24E-lanostatrien-26-oic acid, *in* D-70314
15α-Acetoxy-3α-hydroxy-7,9(11),24E-lanostatrien-26-oic acid, *in* D-70314

C₃₂H₄₈O₆
3α-Acetoxy-15α,22S-dihydroxy-7,9(11),24E-lanostatrien-26-oic acid, *in* T-70266
Chisocheton compound A, *in* T-60223

C₃₂H₅₀O₃
3β-Acetoxy-21,23-epoxy-7,24-tirucalladiene, *in* E-60032

C₃₂H₅₀O₄
23-Acetoxy-30-hydroxy-20(29)-lupen-3-one, *in* D-70318
3β-Acetoxy-28,20β-ursanolide, *in* H-60233
Niloticin acetate, *in* N-70036

C₃₂H₅₀O₅
Alisol *B*; 23-Ac, *in* A-70082

C₃₂H₅₀O₆
21-Acetoxy-21,23:24,25-diepoxyapotirucall-14-ene-3,7-diol, *in* T-60223
16β-Hydroxyalisol B monoacetate, *in* A-70082
Palmitylglutinopallal, *in* G-70021

C₃₂H₅₀O₈
Amphidinolide *B*, A-70205

C₃₂H₅₀O₁₃
Medinin, *in* M-70015

C₃₂H₅₂O₂
3β-Acetoxy-14(26),17E,21-malabaricatriene, *in* M-60008
5-Gluten-3-ol; Ac, *in* G-70020

C₃₂H₅₂O₄
3β-Acetoxy-12-oleanene-2α,11α-diol, *in* O-60037
3β-Acetoxy-12-ursene-2α,11α-diol, *in* U-60008
Trichadenal; 27-Carboxylic acid, Ac, *in* T-70218

C₃₂H₅₂O₅
3β-Acetoxy-12-oleanene-1β,2α,11α-triol, *in* O-60034
3β-Acetoxy-12-ursene-1β,2α,11α-triol, *in* U-60006
3β-Acetoxy-12-ursene-2α,11α,20β-triol, *in* U-60007

C₃₂H₅₂O₆
3β-Acetoxy-12-ursene-1β,2α,11α,20β-tetrol, *in* U-60005

C₃₂H₅₂O₉
Neotokoronin, *in* S-60043
1,2,3-Spirostanetriol; 1-O-α-L-Arabinopyranoside, *in* S-60043
Tokoronin, *in* S-60043

C₃₂H₅₄
Meijicoccene, M-70017

C₃₂H₅₄O₁₃
ent-2α,16β,17-Kauranetriol; O²,O¹⁷-Bis-β-D-glucopyranoside, *in* K-60004

C₃₂H₅₅BrO₈
Thyrsiferol; 23-Ac, *in* T-70193

C₃₃H₂₇N₇
Tris[(2,2'-bipyridyl-6-yl)methyl]amine, T-70320

C₃₃H₂₈O₁₀
Podocarpusflavanone, P-60163

C₃₃H₃₂N₄
13,15-Ethano-3,17-diethyl-2,12,18-trimethyl-monobenzo[g]porphyrin, E-60042

C₃₃H₃₂N₄O₂
13²,17³-Cyclopheophorbide enol, C-60230

C₃₃H₃₂O₉
2',7-Dihydroxy-4'-methoxy-4-(2',7-dihydroxy-4'-methoxyisoflavan-5'-yl)isoflavan; 4'-Methoxy, *in* D-60347
2',7-Dihydroxy-4'-methoxy-4-(2',7-dihydroxy-4'-methoxyisoflavan-5'-yl)isoflavan; 3'-Hydroxy, 2'-Me ether, *in* D-60347

C₃₃H₃₆O₁₁
Triptofordin C-1, *in* T-60394

C₃₃H₃₈O₂
▷α-Guttiferin, G-60040

C₃₃H₃₈O₁₁
9α,14-Diacetoxy-1α,8β-dibenzoyloxy-4β,8β-dihydroxydihydro-β-agarofuran, *in* E-60018
Pringleine, P-60172
Triptofordin C-2, T-60394

C₃₃H₄₀BNO₂
Muscarine; Tetraphenylborate, *in* M-60153

C₃₃H₄₂O₄
Nemorosonol, N-70020

C₃₃H₄₂O₈
Sarothralen *B*, S-60011

C₃₃H₄₄O₉
Volkensin, V-70018

C₃₃H₄₄O₁₆
Arctigenin; 4-β-Gentiobioside, *in* A-70249

C₃₃H₄₈O₁₈
Macrocliniside I, *in* Z-60001

C₃₃H₄₈O₁₉
Sylvestroside I, S-70087

C₃₃H₅₂O₅
Saptarangiquinone *A*, S-70009

C₃₃H₅₂O₆
3α-Acetoxy-15α-hydroxy-22-methoxy-8,24E-lanostadien-26-oic acid, *in* T-60327
22-Acetoxy-3α-hydroxy-7α-methoxy-8,24E-lanostadien-26-oic acid, *in* T-60328
16β-Methoxyalisol B monoacetate, *in* A-70082

C₃₃H₅₂O₈
5-Spirosten-3-ol; 3-O-β-D-Glucopyranoside, *in* S-60045
Trillin, *in* S-60045

C₃₃H₅₄O₁₂
Sileneoside, *in* H-60063

C₃₄H₁₆N₄
29,29,30,30-Tetra-cyanobianthraquinodimethane, T-60030

C₃₄H₁₈
Anthra[9,1,2-cde]benzo[rst]pentaphene, A-60279
Benzo[rst]phenaleno[1,2,3-de]pentaphene, B-60036
Benzo[rst]phenanthro[1,10,9-cde]pentaphene, B-60037
Benzo[rst]phenanthro[10,1,2-cde]pentaphene, B-60038
Dibenzo[a,rst]naphtho[8,1,2-cde]pentaphene, D-60075

C₃₄H₂₀
5,10-Dihydroanthra[9,1,2-cde]benzo[rst]-pentaphene, *in* A-60279
9,18-Dihydrobenzo[rst]phenanthro[10,1,2-cde]-pentaphene, *in* B-60038

C₃₄H₂₄O₉
Plumbazeylanone, P-70106

C₃₄H₂₆
1,3-Bis(1,2-diphenylethenyl)benzene, B-70139

C₃₄H₂₆O₆
6,8'-Dihydroxy-2,2'-bis(2-phenylethyl)[5,5'-bi-4H-1-benzopyran]-4,4'-dione, D-70289

C₃₄H₃₀O₈
AH₁₀, A-70078

C₃₄H₃₀O₉
Semecarpetin, S-70032

C₃₄H₃₆N₄
13¹-Methyl-13,15-ethano-13²,17-prop-13²(15²)-enoporphyrin, M-60082

C₃₄H₃₈O₁₆
Cleistanthoside A, *in* D-60493

C₃₄H₃₈O₁₇
Aloenin *B*, A-60077

C₃₄H₄₀O₁₂
Trisabbreviatin BBB, T-60400

C₃₄H₄₄O₁₃
Taccalonolide *B*, T-70001

C₃₄H₄₆O₁₈
Loroglossin, L-70037

C₃₄H₄₈O₇
3α,15α-Diacetoxy-23-oxo-7,9(11),24E-lanostatrien-26-oic acid, *in* D-70330

C₃₄H₄₈O₈
Ecdysterone 22-O-benzoate, *in* H-60063
Pitumbin, P-70104

C₃₄H₅₀O₆
Caloverticillic acid *A*, C-60007
Caloverticillic acid *C*, C-60008
Caloverticillic acid B, *in* C-60007
3α,15α-Diacetoxy-7,9(11),24E-lanostatrien-26-oic acid, *in* D-70314
3β,15α-Diacetoxy-7,9(11),24E-lanostatrien-26-oic acid, *in* D-70314

C₃₄H₅₀O₇
3α,22ξ-Diacetoxy-15α-hydroxy-7,9(11),24-lanostatrien-26-oic acid, *in* T-70266
3α,22S-Diacetoxy-15α-hydroxy-7,9(11),24E-lanostatrien-26-oic acid, *in* T-70266
15α,22S-Diacetoxy-3α-hydroxy-7,9(11),24E-lanostatrien-26-oic acid, *in* T-70266
15α,22S-Diacetoxy-3β-hydroxy-7,9(11),24E-lanostatrien-26-oic acid, *in* T-70266

C₃₄H₅₂O₅
3β,23-Diacetoxy-20(30)-lupen-29-al, *in* D-70316

C₃₄H₅₂O₇
3α,7α-Diacetoxy-15α-hydroxy-8,24E-lanostadien-26-oic acid, *in* T-60327

C₃₄H₅₂O₈
3α,22-Diacetoxy-7α,15α-dihydroxy-8,24E-lanostadien-26-oic acid, *in* T-60111

C₃₄H₅₂O₉
Tiacumicin *A*, T-70194

C₃₄H₅₄O₆
Stearylglutinopallal, *in* G-70021

C₃₄H₅₅BrO₈
15-Anhydrothyrsiferyl diacetate, *in* A-60274
15(28)-Anhydrothyrsiferyl diacetate, *in* A-60274

C₃₅H₂₈O₁₄
β-Actinorhodin, A-70062

C₃₅H₃₀O₁₁
Kuwanone *L*, K-70016

C₃₅H₃₀O₁₂
Hydnowightin, H-70094

C₃₅H₃₆N₄O₄
N-Methylprotoporphyrin IX, M-60123

C₃₅H₃₈O₁₁
Triptofordin D-1, T-60395

C₃₅H₃₈O₁₃
Triptofordin E, T-60397

$C_{35}H_{40}O_{12}$
Acetylpringleine, in P-60172

$C_{35}H_{41}N_5O_6$
Cycloaspeptide B, in C-70211
Cycloaspeptide C, in C-70211

$C_{35}H_{42}O_{11}$
1-Cinnamoylmelianone, in M-60015

$C_{35}H_{46}O_6$
Mesuaferrol, M-70032

$C_{35}H_{50}N_8O_6S_2$
Patellamide A, P-60011

$C_{35}H_{50}O_{20}$
Sylvestroside II, in S-70087

$C_{35}H_{54}O_7$
3α,22-Diacetoxy-7α-methoxy-8,24E-lanostadien-26-oic acid, in T-60328

$C_{35}H_{54}O_8$
3α,22-Diacetoxy-15α-hydroxy-7α-methoxy-8,24E-lanostadien-26-oic acid, in T-60111

$C_{35}H_{56}O_8$
Ilexoside A, in D-70327
Ilexoside B, in D-70350

$C_{35}H_{56}O_9$
Cycloorbicoside A, in C-60218

$C_{35}H_{61}NO_7$
Pamamycin-607, P-70004

$C_{35}H_{62}O_{12}$
Indicoside A, I-60027

$C_{35}H_{63}NO_{12}$
2-Norerythromycin D, in N-70076

$C_{35}H_{63}NO_{13}$
2-Norerythromycin C, in N-70076

$C_{35}H_{64}O_7$
Annonacin, A-70212
Goniothalamicin, G-70027

$C_{36}H_{22}$
7,16[1′,2′]-Benzeno-7,16-dihydroheptacene, B-70020

$C_{36}H_{24}N_2O_2$
2,2′-(1,4-Phenylene)bis[5,1-[1,1′-biphenyl]-4-yl]oxazole, P-60091

$C_{36}H_{30}O_{14}$
Podoverine C, P-60165

$C_{36}H_{30}O_{15}$
Podoverine B, in P-60165

$C_{36}H_{30}O_{16}$
Salvianolic acid B, S-70003

$C_{36}H_{32}O_6$
1,4,7,22,25,28-Hexaoxa[7.7](9,10)-anthracenophane, H-70086

$C_{36}H_{36}O_8$
α-Diceroptene, D-70112

$C_{36}H_{40}N_2O_{14}S$
Esperamicin X, E-70035

$C_{36}H_{43}N_5O_6$
Cycloaspeptide A, C-70211

$C_{36}H_{44}O_4$
Panduratin B, P-60003

$C_{36}H_{46}O_{14}$
Taccalonolide C, T-70002
Taccalonolide A, in T-70001
Taccanolide D, in T-70001

$C_{36}H_{48}O_{18}$
Bruceantinoside C, B-70281
Yadanzioside K, Y-60001

$C_{36}H_{50}O_8$
β-Ecdyson 2-cinnamate, in H-60063

$C_{36}H_{50}O_9$
β-Ecdyson 3-p-coumarate, in H-60063
2,3,5,14,20,22,25-Heptahydroxycholest-7-en-6-one; 2-Cinnamoyl, in H-60023

$C_{36}H_{52}O_{11}$
3α,15α,22S-Triacetoxy-7,9(11),24-lanostatrien-26-oic acid, in T-70266

3β,15α,22S-Triacetoxy-7,9(11),24E-lanostatrien-26-oic acid, in T-70266

$C_{36}H_{52}O_{12}$
▷Cucurbitacin E; Deacetyl, 2-O-β-D-glucopyranosyl, in C-70204

$C_{36}H_{54}O_9$
3α,7α,22-Triacetoxy-15α-hydroxy-8,24E-lanostadien-26-oic acid, in T-60111
3α,15α,22-Triacetoxy-7α-hydroxy-8,24E-lanostadien-26-oic acid, in T-60111
3α,15α,22S-Triacetoxy-7α-hydroxy-8,24E-lanostadien-26-oic acid, in T-60111

$C_{36}H_{54}O_{12}$
Bryoamaride, in C-70204

$C_{36}H_{56}O_{11}$
Ilexsaponin A1, in D-60382

$C_{36}H_{58}O_9$
Corchorusin B, in S-60001

$C_{36}H_{58}O_{10}$
Arjunglucoside II, in T-70271

$C_{36}H_{58}Te_2$
Bis(2,4,6-tri-tert-butylphenyl)ditelluride, B-70162

$C_{36}H_{60}O_8$
Corchorusin A, in O-60038

$C_{36}H_{60}O_9$
Corchorusin C, in O-60035

$C_{36}H_{62}O_9$
Actinostemmoside A, in D-70002
Actinostemmoside B, in D-70003

$C_{36}H_{62}O_{10}$
Actinostemmoside C, in D-70001

$C_{36}H_{65}NO_{12}$
2-Norerythromycin B, in N-70076

$C_{36}H_{65}NO_{13}$
2-Norerythromycin, N-70076

$C_{37}H_{42}O_{12}$
Triptofordin D-2, T-60396

$C_{37}H_{46}N_8O_6S_2$
Patellamide C, P-60013

$C_{37}H_{46}O_{20}S$
Rhodonocardin A, R-70003

$C_{37}H_{46}O_{23}$
Sarothamnoside, in T-60324

$C_{37}H_{47}NO_{14}$
▷Disnogamycin, D-70506

$C_{37}H_{49}N_7O_{12}$
Antibiotic B 1625FA_{2β-1}, A-70220

$C_{37}H_{51}BrO_6$
6′-Bromodisidein, in D-60499

$C_{37}H_{51}ClO_6$
6′-Chlorodisidein, in D-60499

$C_{37}H_{52}O_{13}$
Fevicordin A glucoside, in F-60008

$C_{37}H_{53}N_3O_9$
30-Demethoxycurromycin A, in C-60178

$C_{37}H_{56}O_9$
3α,15α,22S-Triacetoxy-7α-methoxy-8,24E-lanostadien-26-oic acid, in T-60111

$C_{37}H_{56}O_{14}$
Swalpamycin, S-70086

$C_{38}H_{18}$
Dibenzo[jk,uv]dinaphtho[2,1,8,7-defg:2′,1′,8′,7′-opqr]pentacene, D-60072

$C_{38}H_{26}$
Cycloheptatrienylidenee(etraphenylcyclopentadenylidene)ethylene, C-60208

$C_{38}H_{28}$
[1,1′-Biphenyl]-4,4′-diylbis[diphenylmethyl], B-60117

$C_{38}H_{32}O_2$
2,5-Dimethyl-3,4-diphenyl-2,4-cyclopentadien-1-one; Dimer, in D-70403

$C_{38}H_{35}NO_{10}$
Scleroderris green, S-60014

$C_{38}H_{42}O_{12}$
Esulone A, E-60038
Esulone C, E-60039

$C_{38}H_{42}O_{14}$
Cladochrome, in P-70098

$C_{38}H_{46}O_{16}$
Pseudrelone B, P-70130

$C_{38}H_{48}N_8O_6S_2$
Patellamide B, P-60012

$C_{38}H_{53}N_3O_{20}S_2$
Antibiotic 273a_{2β}, in A-70218

$C_{38}H_{54}O_{12}$
Datiscoside, in C-70203
Datiscoside B, in C-70203

$C_{38}H_{54}O_{13}$
▷Elaterinide, in C-70204

$C_{38}H_{55}N_3O_{10}$
Curromycin A, C-60178

$C_{38}H_{56}O_{11}$
Datiscoside H, in C-70203

$C_{38}H_{56}O_{13}$
25-O-Acetylbryoamaride, in C-70204
Datiscoside G, in C-70203

$C_{38}H_{58}N_4O_8$
Bufalitoxin, in B-70289

$C_{38}H_{76}O_2$
2-Octadecylicosanoic acid, O-70005

$C_{39}H_{48}N_2O_9$
Rubiflavin F, R-70012

$C_{39}H_{53}N_9O_{13}S$
Amanullinic acid, in A-60084

$C_{39}H_{53}N_9O_{14}S$
▷Amanine, in A-60084
▷ε-Amanitin, in A-60084

$C_{39}H_{53}N_9O_{15}S$
▷β-Amanitin, in A-60084

$C_{39}H_{54}N_{10}O_{12}S$
Amanullin, in A-60084

$C_{39}H_{54}N_{10}O_{13}S$
▷γ-Amanitin, in A-60084

$C_{39}H_{54}N_{10}O_{14}S$
▷α-Amanitin, in A-60084

$C_{39}H_{54}O_6$
Isoneriucoumaric acid, in D-70349
Neriucoumaric acid, in D-70349

$C_{39}H_{55}N_3O_{20}S_2$
Antibiotic 273a_{2α}, in A-70218

$C_{39}H_{58}O_{15}$
Bryostatin 11, B-70286

$C_{39}H_{62}O_{10}$
▷Roseofungin, R-70009

$C_{39}H_{62}O_{12}$
5-Spirosten-3-ol; 3-O-α-L-Rhamnopyranosyl(1→2)-β-D-glucopyranoside, in S-60045

$C_{39}H_{62}O_{13}$
▷Funkioside C, in S-60045
Nuatigenin; 3-O-α-L-Rhamnopyranosyl(1→2)-β-D-glucopyranoside, in N-60062
5-Spirostene-3,25-diol; 3-O-α-L-Rhamnopyranosyl(1→2)-β-D-glucopyranoside, in S-60044

$C_{40}H_{16}$
Phenanthro[3,4,5,6-bcdef]ovalene, P-60076

$C_{40}H_{20}$
13,19:14,18-Dimethenoanthra[1,2-a]benzo[o]pentaphene, D-60396

$C_{40}H_{24}$
Tetrabenzo[b,h,n,t]tetraphenylene, T-70012

$C_{40}H_{36}O_{10}$
Brosimone *A*, B-70279

$C_{40}H_{36}O_{11}$
Kuwanone K, *in* K-70016

$C_{40}H_{38}O_8$
Bianthrone A2b, B-60077
Kuwanon V, *in* K-60015

$C_{40}H_{38}O_9$
Kuwanon Q, *in* K-60015
Kuwanon R, *in* K-60015

$C_{40}H_{38}O_{10}$
Kuwanon J, K-60015

$C_{40}H_{40}O_8$
1,4,7,10,25,28,31,34-Hexaoxa[10.10](9,10)-
anthracenophane, H-70085

$C_{40}H_{42}N_4O_{17}$
Hydroxymethylbilane, H-70166

$C_{40}H_{42}O_{13}$
Herpetetrone, H-70026

$C_{40}H_{44}O_{13}$
Esulone B, *in* E-60038

$C_{40}H_{48}N_6O_9$
RA-II, *in* R-60002
RA-V, R-60002

$C_{40}H_{48}N_6O_{10}$
RA-I, *in* R-60002

$C_{40}H_{48}O_{17}$
Ciclamycin O, C-70176

$C_{40}H_{49}NO_{13}$
Antibiotic TMF 518*D*, A-70240

$C_{40}H_{50}O_{16}$
Ciclamycin 4, *in* C-70176

$C_{40}H_{52}O_3$
Adonirubin, A-70066

$C_{40}H_{52}O_8$
Rosmanoyl carnosate, R-60013

$C_{40}H_{54}O_3$
α-Doradexanthin, D-70545
Fritschiellaxanthin, *in* D-70545

$C_{40}H_{56}$
Lycopene, L-60041

$C_{40}H_{56}O$
1,2-Epoxy-1,2-dihydrolycopene, *in* L-60041
5,6-Epoxy-5,6-dihydrolycopene, *in* L-60041

$C_{40}H_{56}O_8$
3β,19α-Dihydroxy-24-*trans*-ferulyloxy-12-
ursen-28-oic acid, *in* T-60348

$C_{40}H_{58}O_{12}$
Datiscoside F, *in* C-70203

$C_{40}H_{58}O_{13}$
Datiscoside C, *in* C-70203
Datiscoside E, *in* C-70203

$C_{40}H_{60}O_{13}$
Atratoside D, *in* C-70257
Datiscoside D, *in* C-70203

$C_{40}H_{66}$
Lycopersene, L-70044

$C_{40}H_{66}O$
15-Hydroxylycopersene, *in* L-70044
2,6,10,14,19,23,27,31-Octamethyl-
2,6,10,14,17,22,26,30-dotriacontaoctaen-19-
ol, O-70020
Prephytoene alcohol, P-70115

$C_{40}H_{78}$
Lycopadiene, L-60040

$C_{40}H_{78}O$
7,11,15,19-Tetramethyl-3-methylene-2-(3,7,11-
trimethyl-2-dodecenyl)-1-eicosanol, T-70138

$C_{41}H_{24}$
Centrohexaindane, C-70031

$C_{41}H_{26}O_{26}$
Vescalagin, V-70006

$C_{41}H_{28}O_{26}$
Liquidambin, L-60029

$C_{41}H_{50}N_2O_9$
Rubiflavin C_1, R-70011
Rubiflavin C_2, *in* R-70011

$C_{41}H_{50}N_2O_{10}$
Rubiflavin A, *in* P-60162

$C_{41}H_{50}N_6O_9$
RA-VII, *in* R-60002

$C_{41}H_{50}N_6O_{10}$
RA-III, *in* R-60002
RA-IV, *in* R-60002

$C_{41}H_{52}N_2O_9$
Rubiflavin D, *in* R-70011

$C_{41}H_{52}N_2O_{10}$
Antibiotic SS 21020*C*, A-70238
Rubiflavin E, *in* R-70011

$C_{41}H_{52}O_5$
Ferulinolone, F-60006

$C_{41}H_{62}Cl_2O_{12}$
Antibiotic CP 54883, A-70222

$C_{41}H_{62}O_{15}$
Bryostatin 13, B-70288

$C_{41}H_{66}O_{13}$
Ilexside I, *in* D-70350

$C_{42}H_{18}$
Hexabenzo[*bc,ef,hi,kl,no,qr*]coronene, H-60031

$C_{42}H_{24}$
Tris(9-fluorenylidene)cyclopropane, T-70322

$C_{42}H_{24}S_3$
Triss(hioxanthen-9-ylidene)cyclopropane,
T-70326

$C_{42}H_{26}$
9,18-Diphenylphenanthro[9,10-*b*]triphenylene,
D-60490

$C_{42}H_{30}$
Hexaphenylbenzene, H-60074

$C_{42}H_{32}O_9$
Canaliculatol, C-70011
Distichol, D-60502
Miyabenol C, M-60141

$C_{42}H_{40}O_{14}$
Chokorin, C-70171

$C_{42}H_{44}O_{14}$
8α-Benzoyloxyacetylpringleine, *in* P-60172

$C_{42}H_{46}Si_4$
2,3,9,10-Tetrakiss(rimethylsilyl)[5]phenylene,
T-60130

$C_{42}H_{64}O_{13}$
Atratoside A, *in* A-70267

$C_{42}H_{64}O_{15}$
Bryostatin 10, B-70285
Divaroside, *in* C-70038

$C_{42}H_{64}O_{16}$
Tubocapside A, *in* T-70333

$C_{42}H_{68}N_6O_6S$
Dolastatin 10, D-70542

$C_{42}H_{68}O_{13}$
Carnosifloside I, *in* C-60175
Saikosaponin A, *in* S-60001

$C_{42}H_{68}O_{16}$
2,3,19,23-Tetrahydroxy-12-oleanen-28-oic acid;
3,28-*O*-Bis-β-D-glucopyranosyl, *in* T-70119

$C_{42}H_{72}O_{13}$
Actinostemmoside D, *in* D-70002

$C_{42}H_{84}O_2$
2-Nonacosyl-3-tridecene-1,10-diol, N-70068

$C_{43}H_{52}N_2O_{11}$
▷Pluramycin *A*, P-60162

$C_{43}H_{56}O_{17}$
Entandrophragmin, E-70009

$C_{43}H_{58}O_{16}$
Candollein, *in* E-70009

$C_{43}H_{58}O_{17}$
β-Dihydroentandrophragmin, *in* E-70009

$C_{43}H_{62}N_4O_{23}S_3$
Paldimycin *B*, *in* P-70002

$C_{43}H_{68}O_6$
Corymbivillosol, C-60170

$C_{43}H_{68}O_{12}$
Antibiotic PC 766*B*, A-70228

$C_{44}H_{56}N_8O_7$
Cyclic(*N*-methyl-L-alanyl-L-tyrosyl-D-
tryptophyl-L-lysyl-L-valyl-L-phenylalanyl),
C-60180

$C_{44}H_{58}O_{12}$
Clibadiolide, C-70187

$C_{44}H_{60}N_4O_{12}$
Halichondramide, H-60003

$C_{44}H_{64}N_4O_{23}S_3$
Paldimycin *A*, *in* P-70002

$C_{44}H_{70}O_{12}$
Antibiotic PC 766B', *in* A-70228

$C_{44}H_{70}O_{16}$
Polyphyllin D, *in* S-60045

$C_{44}H_{72}O_{18}$
Dibenzo-54-crown-18, D-70068

$C_{44}H_{76}O_3$
20(29)-Lupene-3,16-diol; 3-Tetradecanoyl, *in*
L-60039

$C_{44}H_{76}O_{14}$
Portmicin, P-70111

$C_{44}H_{85}NO_{10}$
Acanthocerebroside C, A-70009

$C_{44}H_{87}NO_{10}$
Acanthocerebroside B, A-70008

$C_{45}H_{46}O_8$
Bianthrone A2a, B-60076

$C_{45}H_{66}O_{16}$
Bryostatin 2, *in* B-70284

$C_{45}H_{70}O_7$
Corymbivillosol; 3-Ac, *in* C-60170

$C_{45}H_{72}O_{13}$
Oligomycin E, O-70039

$C_{45}H_{72}O_{16}$
Dioscin, *in* S-60045
5-Spirosten-3-ol; 3-*O*-[α-L-
Rhamnopyranosyl(1→2)][β-D-
glucopyranosyl(1→4)-α-L-
rhamnopyranosyl(1→4)]-β-D-
glucopyranoside, *in* S-60045

$C_{45}H_{72}O_{17}$
Balanitin 3, *in* S-60045
Gracillin, *in* S-60045
5-Spirosten-3-ol; 3-*O*-[α-L-
Rhamnopyranosyl(1→2)][β-D-
glucopyranosyl(1→4)]-β-D-glucopyranoside,
in S-60045
5-Spirosten-3-ol; 3-*O*-β-D-
Glucopyranosyl(1→4)-α-L-
rhamnopyranosyl(1→4)-β-D-
glucopyranoside, *in* S-60045

$C_{45}H_{72}O_{18}$
▷Funkioside D, *in* S-60045

$C_{46}H_{56}MgN_4O_5$
Bacteriochlorophyll f, B-70005

$C_{46}H_{58}O_{27}$
Coptiside I, *in* D-70319

$C_{46}H_{80}O_3$
20(29)-Lupene-3,16-diol; 3-Hexadecanoyl, *in*
L-60039

$C_{46}H_{87}NO_9$
Acanthocerebroside D, A-70010

$C_{46}H_{91}NO_{10}$
Acanthocerebroside *A*, A-70007

$C_{47}H_{68}O_{17}$
Bryostatin 1, B-70284

$C_{47}H_{76}O_{18}$
Ilexside II, in D-70350

$C_{47}H_{80}O_{15}$
▷Carriomycin, C-60022

$C_{47}H_{80}O_{16}$
Antibiotic W 341C, A-70241

$C_{47}H_{89}NO_9$
Acanthocerebroside E, A-70011

$C_{48}H_{30}O_{31}$
Castavaloninic acid, in V-70006

$C_{48}H_{36}$
[2_6]Paracyclophene, P-70009
7,10:19,22:31,34-Triethenotribenzo[a,k,u]-cyclotriacontane, T-60273

$C_{48}H_{54}O_{27}$
Safflor Yellow B, S-70001

$C_{48}H_{64}Br_2N_4$
Octaethyltetramethyl[26]porphyrin[3.3.3.3];
Dibromide, in O-60008

$C_{48}H_{64}N_4^{\oplus\oplus}$
Octaethyltetramethyl[26]porphyrin[3.3.3.3],
O-60008

$C_{48}H_{72}O_{18}$
Atratoside C, in A-70267

$C_{48}H_{74}O_{18}$
Atratoside B, in A-70267

$C_{48}H_{74}O_{19}$
Chiisanoside, in C-70038

$C_{48}H_{74}O_{21}$
Tubocapside B, in T-70333

$C_{48}H_{76}O_{19}$
Saponin A, in C-70002

$C_{48}H_{77}N_7O_{16}$
Mulundocandin, M-70157

$C_{48}H_{78}O_{18}$
Carnosifloside II, in C-60175
Carnosifloside III, in C-60175
Carnosifloside IV, in C-60175

$C_{48}H_{80}O_{18}$
Carnosifloside V, in C-60175
Carnosifloside VI, in C-60175

$C_{48}H_{91}NO_9$
Acanthocerebroside F, A-70012

$C_{49}H_{66}O_{28}$
Oleonuezhenide, O-70037

$C_{49}H_{72}O_{17}$
Bryostatin 12, B-70287

$C_{49}H_{75}N_7O_{15}$
Antibiotic FR 900359, A-70223

$C_{49}H_{75}N_{13}O_{12}$
Cyanoviridin RR, C-70210

$C_{50}H_{63}BrO_{16}$
Bromothricin, in C-60137

$C_{50}H_{63}ClO_{16}$
Chlorothricin, C-60137

$C_{50}H_{63}ClO_{17}$
2‴-Hydroxychlorothricin, in C-60137

$C_{50}H_{80}O_{21}$
Balanitin 2, S-60045

$C_{50}H_{80}O_{22}$
▷Funkioside F, in S-60045

$C_{51}H_{72}Cl_2O_{18}$
Lipiarmycin A_4, in L-60028
Tiacumicin E, T-70197

$C_{51}H_{76}O_2$
2-[3,7,11,15,19,23-Hexamethyl-25-(2,6,6-trimethyl-2-cyclohexenyl)pentacosa-2,14,18,22-tetraenyl]-3-methyl-1,4-naphthoquinone, H-70084

$C_{51}H_{82}O_{22}$
Aculeatiside A, in N-60062
Balanitin 1, in S-60045
Funkioside E, in S-60045

$C_{51}H_{82}O_{23}$
Aculeatiside B, in N-60062
Avenacoside A, in N-60062

$C_{51}H_{102}O$
26-Henpentacontanone, H-70009

$C_{52}H_{74}Cl_2O_{18}$
Lipiarmycin A_3, in L-60028
Tiacumicin C, T-70195
Tiacumicin D, T-70196
Tiacumicin F, T-70198

$C_{52}H_{82}O_{20}$
Guaianin D, in H-60199

$C_{53}H_{31}N_3$
2,12-Di-9-acridinyl-7-phenyldibenz[c,h]-acridine, D-70033

$C_{53}H_{84}O_{24}$
Camellidin II, in D-60363

$C_{54}H_{90}O_{22}$
Verbascosaponin, in O-70034

$C_{55}H_{74}IN_3O_{21}S_4$
Calichemicin γ_1^I, C-70008

$C_{55}H_{82}O_{21}S_2$
▷Yessotoxin, Y-60002

$C_{55}H_{84}N_{16}O_{20}$
Pyoverdin II, P-70149

$C_{55}H_{85}N_{17}O_{19}$
Pyoverdin III, in P-70149

$C_{55}H_{85}N_{17}O_{22}$
Pyoverdin D, P-70147

$C_{55}H_{86}N_{18}O_{21}$
Pyoverdin E, in P-70147

$C_{55}H_{86}O_{25}$
Camellidin I, in D-60363

$C_{56}H_{40}$
Tetrakiss(iphenylmethylene)cyclobutane, T-70128

$C_{56}H_{40}O_{12}$
Miyabenol B, M-60140

$C_{56}H_{42}O_{12}$
Miyabenol A, M-60139
Vaticaffinol, V-60002

$C_{56}H_{84}N_{16}O_{21}$
Pyoverdin I, P-70148

$C_{56}H_{85}N_{17}O_{23}$
Pyoverdin C, P-70146

$C_{57}H_{92}O_{28}$
Avenacoside B, in N-60062

$C_{58}H_{78}N_4O_{22}S_4$
Esperamicin A_{1b}, in E-60037

$C_{58}H_{92}O_{52}$
Guaianin E, in H-60199

$C_{59}H_{80}N_4O_{22}S_4$
Esperamicin A_1, in E-60037
Esperamicin A_2, in E-60037

$C_{59}H_{90}O_{26}$
▷Funkioside G, in S-60045

$C_{62}H_{36}O_2$
9,11,20,22-Tetraphenyltetrabenzo[a,c,l,n]-pentacene-10,21-dione, T-70151

$C_{62}H_{38}$
b,b′,b″-Tritriptycene, T-60407

$C_{62}H_{40}N_6O_2P_2$
2,4,5-Triss(icyanomethylene)-1,3-cyclo-pentanedione; Biss(etraphenylphosphonium salt), in T-70321

$C_{63}H_{98}N_{18}O_{13}S$
Substance P, S-60059

$C_{68}H_{130}N_2O_{23}P_2$
Lipid A, L-70032

$C_{70}H_{68}$
1,3,5,7-Tetrakis[2-D_3-trishomocubanyl-1,3-butadiynyl]adamantane, T-70129

$C_{73}H_{108}N_{16}O_{18}S$
Rhodotorucin A, R-70004

$C_{73}H_{126}O_{20}$
Bistheonellide B, in M-60138

$C_{74}H_{73}O_{37}^{\oplus}$
Zebrinin, Z-70002

$C_{74}H_{128}O_{20}$
Misakinolide A, M-60138

$C_{79}H_{91}O_{45}^{\oplus}$
Heavenly blue anthocyanin, H-70006

$C_{81}H_{84}Cl_2N_8O_{29}$
Parvodicin A, in P-70012

$C_{82}H_{52}$
Tetra(2-triptycyl)ethylene, T-60169

$C_{82}H_{86}Cl_2N_8O_{29}$
Antibiotic A 40926A, in A-70219
Parvodicin B_1, in P-70012
Parvodicin B_2, in P-70012

$C_{83}H_{88}Cl_2N_8O_{29}$
Antibiotic A 40926B, in A-70219
Parvodicin C_1, in P-70012
Parvodicin C_2, in P-70012

$C_{85}H_{90}Cl_2N_8O_{30}$
Parvodicin C_3, in P-70012
Parvodicin C_4, in P-70012

$C_{98}H_{141}N_{25}O_{23}S_4$
Epidermin, E-70010

$C_{100}H_{164}O$
Dolichol, D-60517

$C_{102}H_{78}S_6$
6H,24H,26H,38H,56H,58H-16,65:48,66-Bis-([1,3]benzenomethanothiomethano[1,3]-benzeno)-32,64-etheno-1,5:9,13:14,18:19,23:-27,31:33,37:41,45:46,50:51,55:59,63-decemetheno-8H,40H[1,7]dithiacyclotetra-triacontino[9,8-h][1,77]dithiocyclotetra-triacontin, B-70129

$C_{122}H_{210}N_2O_{16}$
[3][14-Acetyl-14-azacyclo-hexacosanone][25,26,53,54,55,56-hexa-acetoxytricyclo[49.3.1.1^{24,28}]hexapentaconta-1(55),24,26,28(56),51,53-hexaene][14-acetyl-14-azacyclohexacosanone]catenane, A-60020

$C_{250}H_{410}N_{68}O_{75}$
Thymopoietin I, in T-70192

$C_{251}H_{412}N_{64}O_{78}$
Thymopoietin II, 9CI, in T-70192

$C_{255}H_{416}N_{66}O_{77}$
Thymopoietin III, in T-70192

Chemical Abstracts Service Registry Number Index

This index becomes invalid after publication of the Eighth Supplement.

The CAS Registry Number Index lists in ascending numerical order all CAS Registry Numbers recorded in the Sixth and Seventh Supplements. CAS Registry numbers of compounds contained in Supplements 1 to 5 are listed in the cumulative Index Volume published with the Fifth Supplement.

Each CAS Registry Number listed refers the user to a chemical name (with stereochemical or derivative descriptors where relevant) and a DOC number.

The first digit of the DOC Number (printed in bold type) refers to the number of the Supplement in which the Entry appears.

A DOC Number which follows immediately upon a chemical name means that the name is the DOC Name and it is to this name that the CAS Registry Number refers.

A DOC Number which is preceded by the word '*in*' means that the CAS Registry Number refers to the specified stereoisomer or derivative which is to be found embedded within the particular Entry.

The symbol ▷ preceding an index term indicates that the DOC Entry contains information on toxic or hazardous properties of the compound.

CAS Registry Number Index

50-44-2 ▷1,7-Dihydro-6*H*-purine-6-thione, D-60270

50-46-2 ▷5,8-Isoquinolinequinone, I-60131

50-66-8 ▷1,7-Dihydro-6*H*-purine-6-thione; 6-*Thiol-form, S*-Me, *in* D-60270

50-85-1 2-Hydroxy-4-methylbenzoic acid, H-70163

51-20-7 ▷5-Bromo-2,4-(1*H*,3*H*)pyrimidinedione, B-60318

54-21-7 ▷2-Hydroxybenzoic acid; Na salt, *in* H-70108

55-06-1 Liothyronine sodium, *in* T-60350

56-04-2 ▷2,3-Dihydro-6-methyl-2-thioxo4(1*H*)-pyrimidinone, D-70232

56-06-4 2,6-Diamino-4(1*H*)-pyrimidinone, D-70043

57-14-7 ▷1,1-Dimethylhydrazine, D-70416

57-55-6 ▷1,2-Propanediol, P-70118

57-97-6 ▷7,12-Dimethylbenz[*a*]anthracene, D-60405

60-35-5 ▷Acetamide, A-70019

60-81-1 ▷Phloridzin, *in* H-60219

60-82-2 3-(4-Hydroxyphenyl)-1-(2,4,6trihydroxyphenyl)-1propanone, H-60219

65-45-2 ▷2-Hydroxybenzamide, *in* H-70108

67-42-5 ▷3,12-Biss(arboxymethyl)-6,9dioxa-3,12-diazatetradecanedioic acid, B-70134

67-47-0 ▷5-Hydroxymethyl-2furancarboxaldehyde, H-70177

69-72-7 ▷2-Hydroxybenzoic acid, H-70108

72-19-5 ▷Threonine; (2*S*,3*R*)-*form, in* T-60217

74-89-5 ▷Methanamine, M-70038

75-56-9 ▷Methyloxirane, M-70100

77-95-2 ▷Quinic acid; (−)-*form, in* Q-70005

78-71-7 ▷3,3-Biss(hloromethyl)oxetane, B-60149

78-75-1 ▷1,2-Dibromopropane, D-60105

79-06-1 ▷Acrylamide, *in* P-70121

79-10-7 ▷2-Propenoic acid, P-70121

79-16-3 ▷Acetamide; *N*-Me, *in* A-70019

79-40-3 ▷Ethanedithioamide, E-60041

79-42-5 ▷2-Mercaptopropanoic acid, M-60027

79-76-5 γ-Ionone, I-70058

79-84-5 2-Thiophenesulfonic acid, T-60213

80-68-2 Threonine; (2*RS*,3*SR*)-*form, in* T-60217

80-71-7 ▷2-Hydroxy-3-methyl-2cyclopenten-1-one, H-60174

80-73-9 ▷1,3-Dimethylimidazolidinone, *in* I-70005

81-31-2 5,10-Dihydroanthra[9,1,2-*cde*]benzo[*rst*]pentaphene, *in* A-60279

81-97-0 4-Chloro-7*H*-benz[*de*]anthracen-7-one, C-70042

82-02-0 ▷Khellin, K-60010

82-04-2 2-Chloro-7*H*-benz[*de*]anthracen-7-one, C-70040

85-98-3 ▷*N,N*′-Diethyl-*N,N*′-diphenylurea, *in* D-70499

86-54-4 ▷1-Hydrazinophthalazine, H-60092

86-79-3 2-Hydroxycarbazole, H-60109

86-93-1 ▷Tetrazole-5-thione; 1,4-*Dihydro-form,* 1-Ph, *in* T-60174

87-17-2 2-Hydroxy-*N*-phenylbenzamide, *in* H-70108

87-28-5 Glycol salicylate, *in* H-70108

87-51-4 ▷1*H*-Indole-3-acetic acid, I-70013

87-69-4 ▷Tartaric acid; (2*R*,3*R*)-*form, in* T-70005

87-91-2 Diethyl tartrate, *in* T-70005

88-30-2 ▷4-Nitro-3-(trifluoromethyl)phenol, N-60054

90-24-4 2′-Hydroxy-4′,6′dimethoxy-acetophenone, *in* T-70248

90-29-9 ψ-Baptigenin, B-60006

91-12-3 Fulvalene, F-60080

91-76-9 ▷2,4-Diamino-6-phenyl-1,3,5triazine, D-60041

92-39-7 ▷2-Chloro-10*H*-phenothiazine, C-60125

92-43-3 ▷1-Phenyl-3-pyrazolidinone, P-70091

92-61-5 ▷7-Hydroxy-6-methoxy-2*H*-1benzopyran-2-one, H-70153

92-62-6 ▷3,6-Diaminoacridine, D-60028

94-77-9 4-(2-Phenylethyl)-1,3-benzenediol, P-60097

95-58-9 2-Chloro-3-methylpyrazine, C-70105

96-33-3 ▷Methyl acrylate, *in* P-70121

98-03-3 ▷2-Thiophenecarboxylic acid, T-70183

98-46-4 ▷1-Nitro-3-(trifluoromethyl)benzene, N-70065

98-79-5 5-Oxo-2-pyrrolidinecarboxylic acid; (*S*)-*form, in* O-70102

98-86-2 ▷Acetophenone, A-60017

99-33-2 ▷3,5-Dinitrobenzoic acid; Chloride, *in* D-60458

99-34-3 3,5-Dinitrobenzoic acid, D-60458

100-50-5 ▷3-Cyclohexene-1-carboxaldehyde; (±)-*form, in* C-70227

101-78-0 Indamine, I-60019

102-07-8 ▷*N,N*′-Diphenylurea, D-70499

102-70-5 ▷Tri(2-propenyl)amine, T-70315

103-16-2 ▷4-(Benzyloxy)phenol, B-60074

104-61-0 ▷Dihydro-5-pentyl-2(3*H*)-furanone, D-70245

104-98-3 ▷Urocanic acid, U-70005

105-11-3 ▷1,4-Benzoquinone; Dioxime, *in* B-70044

105-55-5 ▷*N,N*′-Diethylthiourea, D-70155

105-70-4 1,3-Diacetylglycerol, D-60023

106-23-0 3,7-Dimethyl-6-octenal, D-60430

106-34-3 ▷Quinhydrone, *in* B-70044

106-51-4 ▷1,4-Benzoquinone, B-70044

106-61-6 ▷Glycerol 1-acetate, G-60029

107-02-8 ▷Propenal, P-70119

107-13-1 ▷Vinyl cyanide, *in* P-70121

107-86-8 3-Methyl-2-butenal, M-70055

107-98-2 ▷1-Methoxy-2-propanol, *in* P-70118

108-10-1 ▷4-Methyl-2-pentanone, M-60109

108-19-0 Biuret, B-60190

108-65-6 ▷2-Acetoxy-1-methoxypropane, *in* P-70118

108-74-7 ▷Hexahydro-1,3,5-trimethyl1,3,5-triazine, *in* H-60060

108-80-5 ▷Cyanuric acid, C-60179

109-11-5 3-Morpholinone, M-60147

109-98-8 4,5-Dihydro-1*H*-pyrazole, D-70253

110-90-7 Hexahydro-1,3,5-triazine, H-60060

116-83-6 ▷1-Amino-4-methoxyanthraquinone, *in* A-60175

116-85-8 ▷1-Amino-4-hydroxyanthraquinone, A-60175

117-77-1 2-Amino-3-hydroxyanthraquinone, A-60179

118-34-3 Syringin, *in* S-70046

118-57-0 Acetaminosalol, H-70108

118-61-6 ▷Ethyl salicylate, *in* H-70108

118-76-3 5,6-Dihydroxy-5-cyclohexene1,2,3,4-tetrone, D-60315

119-56-2 4-Chlorodiphenylmethanol, C-60065

120-05-8 Sulfuretin, S-70082

120-93-4 ▷2-Imidazolidinone, I-70005

121-81-3 3,5-Dinitrobenzamide, *in* D-60458

123-54-6 ▷2,4-Pentanedione, P-60059

125-52-0 ▷Oxyphencyclimine hydrochloride, *in* O-60102

125-53-1 Oxyphencyclimine, O-60102

126-31-8 ▷Methiodal sodium, I-60048

127-19-5 ▷Acetamide; *N*-Di-Me, *in* A-70019

128-70-1 Pyranthrene, P-70151

129-66-8 ▷2,4,6-Trinitrobenzoic acid, T-70303

130-00-7 Benz[*cd*]indol-2-(1*H*)-one, B-70027

130-03-0 Benzo[*b*]thiophen-3(2*H*)-one, B-60048

130-16-5 ▷5-Chloro-8-hydroxyquinoline, C-70075

131-62-4 ▷1-Hydroxy-1,2-benziodoxol-3(1*H*)one, H-60102

133-37-9 Tartaric acid; (2*RS*,3*RS*)-*form, in* T-70005

134-11-2 2-Ethoxybenzoic acid, *in* H-70108

135-20-6 ▷Cupferron, *in* H-70208

140-88-5 ▷Ethyl acrylate, *in* P-70121

143-47-5 Iodomethanesulfonic acid, I-60048

144-98-9 Threonine; (2*RS*,3*SR*)-*form, in* T-60217

147-71-7 Tartaric acid; (2*S*,3*S*)-*form, in* T-70005

147-73-9 Tartaric acid; (2*RS*,3*SR*)-*form, in* T-70005

148-53-8 ▷2-Hydroxy-3-methoxybenzaldehyde, *in* D-70285

148-97-0 ▷1-Hydroxy-2-phenyldiazene 2-oxide, H-70208

149-87-1 5-Oxo-2-pyrrolidinecarboxylic acid; (±)-*form, in* O-70102

150-86-7 ▷Phytol; (2*E*,7*R*,11*R*)-*form, in* P-60152

151-11-1 5-Hydroxy-3-methylpentanoic acid, H-70182

152-95-4 Sophicoroside, *in* T-60324

154-23-4 ▷3,3′,4′,5,7-Pentahydroxyflavan; (2*R*,3*S*)-*form, in* T-70026

157-16-4 2*H*-Azirine, A-70292

159-66-0 Spirobi[9*H*-fluorene], S-60040

174-73-2 Tricyclo[4.1.0.0^{1,3}]heptane, T-60262

174-79-8 2,6-Dioxaspiro[3.3]heptane, D-60468

177-58-2 7,14,21Trithiatrispiro[5.1.5.1.5.1]-heneicosane, *in* C-60209

178-30-3 2,2′-Spirobi[2*H*-1-benzopyran], S-70062

180-47-2 3,9-Dioxaspiro[5.5]undecane, D-70462

185-80-8 1-Oxa-2-azaspiro[2,5]octane, O-70052

188-84-1 Benzo[*rst*]phenanthro[10,1,2-*cde*]pentaphene, B-60038

188-87-4 Anthra[9,1,2-*cde*]benzo[*rst*]penta-phene, A-60279

190-24-9 Hexabenzo[*bc,ef,hi,kl,no,qr*]coronene, H-60031

190-26-1 Ovalene, O-60046

190-28-3 Phenanthro[3,4,5,6-*bcdef*]ovalene, P-60076

190-70-5 Benzo[*a*]coronene, B-60021

190-81-8 Tribenzo[*b,n,pqr*]perylene, T-60244

190-84-1 Naphtho[1,2,3,4-*ghi*]perylene, N-60009

190-89-6 Dibenzo[*jk,uv*]-dinaphtho[2,1,8,7*defg*:2′,1′,8′,7′-*opqr*]pentacene, D-60072

190-93-2 Benzo[*rst*]phenanthro[1,10,9-*cde*]pentaphene, B-60037

191-46-8 Dibenzo[*a,rst*]naphtho[8,1,2-*cde*]pentaphene, D-60075

192-35-8 Fluoreno[3,2,1,9-*defg*]chrysene, F-70016

192-88-1 Benzo[*b*]phenanthro[4,3-*d*]thiophene, B-70040

193-37-3 Fluoreno[9,1,2,3-*cdef*]chrysene, F-70017

193-39-5 ▷Indeno[1,2,3-*cd*]pyrene, I-70012

193-40-8 ▷Dibenzo[*c,f*]indeno[1,2,3-*ij*][2,7]-naphthyridine, D-60074

193-85-1 Dicyclopenta[*ef,kl*]heptalene, D-60167

194-07-0 [1]Benzothiopyrano[6,5,4-*def*][1]benzothiopyran, B-60049

194-54-7 Diphenanthro[2,1-*b*:1′,2′-*d*]furan, D-70470

195-40-4 Naphtho[1,2-*a*]quinolizinium(1+), N-70012

195-41-5 Quino[7,8-*h*]quinoline, Q-70007

195-84-6 ▷Tricycloquinazoline, T-60271

196-23-6 Dibenzo[2,3:10,11]perylo[1,12*bcd*]-thiophene, D-70074

198-11-8 Benzo[*b*]phenanthro[1,2-*d*]thiophene, B-70039

199-14-4 Benzo[1,2-*b*:3,4-*b*′:5,6-*b*″]tris[1]-benzothiophene, B-60052

201-06-9 Acephenanthrylene, A-70018

202-01-7 Indolo[1,7-*ab*][1]benzazepine, I-60034

202-71-1 Diphenanthro[9,10-*b*:9′,10′-*d*]furan, D-70471

203-33-8 Benz[*a*]aceanthrylene, B-70009

203-35-0 8*H*-Indolo[3,2,1-*de*]acridine, I-70014

203-64-5 ▷4*H*-Cyclopenta[*def*]phenanthrene, C-70244

203-91-8 Naphtho[1,8-*bc*]pyran, N-70009

205-99-2 ▷Benz[*e*]acephenanthrylene, B-70010

206-06-4 Dibenz[*e,k*]acephenanthrylene, D-70066

206-76-8 [1]Benzoselenopheno[2,3-*b*][1]benzothiophene, B-70052

208-20-8 Cyclohepta[*de*]naphthalene, C-60198

209-83-6 Imidazo[5,1,2-*cd*]indolizine, I-70004

216-48-8 Benz[*j*]acephenanthrylene, B-70011

217-65-2 Dibenzo[*f,h*]quinoline, D-70075

219-20-5 Phenanthro[1,2-*c*]furan, P-70050

223-47-2 Anthra[1,2-*b*]benzo[*d*]thiophene, A-70214

226-65-3 [1,2,4,5]Tetrazino[1,6-*a*:4,3-*a*′]di-isoquinoline, T-60172

227-53-2 Anthra[1,2-*c*]furan, A-70217

230-17-1 Benzo[*c*]cinnoline, B-60020

231-02-7 Benzo[*a*]quinolizinium(1+), B-70043

231-95-8 Cycloocta[*a*]naphthalene, C-70234

232-74-6 Naphtho[1,2-*c*]furan, N-70008

233-55-6 1*H*-Imidazo[4,5-*f*]quinoline, I-70007

233-62-5 1*H*-1,2,3-Triazolo[4,5-*f*]quinoline, T-70207

233-92-1 1*H*-1,2,3-Triazolo[4,5-*h*]quinoline, T-70208

234-82-2 Tetrazolo[5,1-*a*]phthalazine, T-70155

235-94-9 Phenanthro[9,10-*c*]furan, P-60074

235-98-3 Phenanthro[9,10-*b*]furan, P-60073

236-08-8 Phenanthro[9,10-*c*][1,2,5]selenadiazole, P-70053

241-34-9 Benzo[1,2-*b*:4,5-*b*′]-bis[1]benzothiophene, B-60018

241-37-2 Benzo[1,2-*b*:5,4-*b*′]-bis[1]benzothiophene, B-60019

243-47-0 [1]Benzothieno[3,2-*b*]quinoline, B-70063

243-58-3 10*H*-Indolo[3,2-*b*]quinoline, I-70016

243-59-4 5*H*-Indolo[2,3-*b*]quinoxaline, I-60036

245-75-0 Dipyrido[1,2-*a*:1′,2′-*c*]imidazol10-ium(1+), D-70501

248-67-9 10*H*-[1]Benzothieno[3,2-*b*]indole, B-70059

249-05-8 Anthra[2,3-*b*]benzo[*d*]thiophene, A-70216

250-63-5 5*H*-Thieno[2,3-*c*]pyrrole, T-60206

252-47-1 Benzo[*a*]biphenylene, B-70016

252-72-2 Disalicylaldehyde, D-70505

254-38-6 1-Benzothiopyrylium(1+), B-70066

255-55-0 [1,4]Dithiino[2,3-*b*]-1,4-dithiin, D-70521

258-16-2 12*H*-Quinoxalino[2,3-*b*][1,4]benzoxazine, Q-70009

258-17-3 12*H*-Quinoxalino[2,3-*b*][1,4]benzothiazine, Q-70008

258-19-5 [1,4]Benzoxazino[3,2-*b*][1,4]benzoxazine, B-60055

258-72-0 Triphenodioxazine, T-60017

259-56-3 Benzo[*b*]biphenylene, B-60017

261-23-4 Xanthylium(1+), X-70003

261-32-5 Thioxanthylium(1+), T-70186

262-83-9 Cycloocta[*b*]naphthalene, C-70235

263-06-9 6*H*,12*H*-Dibenzo[*b,f*][1,5]dithiocin, D-70070

267-65-2 Benzo[1,2-*b*:4,5-*b*′]dithiophene, B-60030

270-70-2 2*H*-Pyrrolo[3,4-*c*]pyridine, P-70200

270-80-4 Furo[3,4-*d*]pyridazine, F-60092

271-01-2 6*H*-Pyrrolo[3,4-*b*]pyridine, P-70201

271-17-0 Cyclopenta[*b*]thiapyran, C-60224

271-73-8 1*H*-Pyrazolo[3,4-*b*]pyridine, P-70155

271-80-7 1*H*-Pyrazolo[3,4-*d*]pyrimidine, P-70156

271-93-2 Furo[2,3-*d*]pyridazine, F-60091

272-49-1 ▷1*H*-Pyrrolo[3,2-*b*]pyridine, P-60249

272-68-4 Thieno[3,2-*d*]pyrimidine, T-60205

272-97-9 ▷Imidazo[4,5-*c*]pyridine, I-60008

273-14-3 [1,2,5]Thiadiazolo[3,4-*d*]pyridazine, T-70163

273-70-1 Thiazolo[5,4-*c*]pyridine, T-70169

273-75-6 Thiazolo[4,5-*c*]pyridine, T-70167

273-84-7 Thiazolo[5,4-*b*]pyridine, T-70168

273-91-6 Benzoselenazole, B-70050

273-98-3 Thiazolo[4,5-*b*]pyridine, T-70166

274-40-8 Indolizine, I-60033

274-45-3 Pyrrolo[1,2-*a*]pyrazine, P-70197

274-59-9 [1,2,3]Triazolo[1,5-*a*]pyridine, T-60241

274-79-3 Imidazo[1,2-*a*]pyrazine, I-60005

274-81-7 1,2,4-Triazolo[4,3-*c*]pyrimidine, T-60243

274-88-4 Tetrazolo[1,5-*c*]pyrimidine, T-60177

275-02-5 [1,2,4]Triazolo[1,5-*a*]pyrimidine, T-60242

275-03-6 Tetrazolo[1,5-*a*]pyrimidine, T-60176

275-61-6 Cyclopent[*b*]azepine, C-60225

276-16-4 5*H*-Pyrrolo[1,2-*a*]azepine, P-70189

278-36-4 7-Azabicyclo[4.2.0]octane, A-70277

280-05-7 8-Azabicyclo[3.2.1]octane, A-70278

280-50-2 2-Oxa-3-azabicyclo[2.2.2]octane, O-70051

280-57-9 ▷1,4-Diazabicyclo[2.2.2]octane, D-70051

281-24-3 2-Oxatricyclo[3.3.1.1³,⁷]decane, O-60051

281-25-4 2-Thiatricyclo[3.3.1.1³,⁷]decane, T-60195

281-32-3 2,4,10-Trioxatricyclo[3.3.1.1³,⁷]-decane, T-60377

281-97-0 Tetraoxaporphyrinogen, T-70147

288-13-1 ▷1*H*-Pyrazole, P-70152

288-53-9 1,3-Dioxole, D-60473

290-96-0 1,2,4,5-Tetrazine, T-60170

294-80-4 1,5,9-Triazacyclododecane, T-60232

294-92-8 1,7-Dioxa-4,10-diazacyclododecane, D-60465

295-17-0 Cyclotetradecane, C-70253

295-64-7 1,4,7,10,13Pentaazacyclopentadecane, P-60025

296-41-3 1,4,7,10,13,16Hexathiacyclooctadecane, H-60075

296-41-3 1,4,7,10,13,16Hexathiacyclooctadecane, H-70087

298-81-7 ▷Xanthotoxin, X-60001

300-54-9 ▷Muscarine; (2*S*,4*R*,5*S*)-*form, in* M-60153

303-45-7 ▷Gossypol, G-60032

304-20-1 ▷Hydralazine hydrochloride, *in* H-60092

305-84-0 ▷Carnosine; (*S*)-*form, in* C-60020

306-79-6 Trifluoromethanesulfenic acid, T-60279

312-73-2 2-(Trifluoromethyl)-1*H*benzimidazole, T-60287

313-72-4 Octafluoronaphthalene, O-70008

315-14-0 1,3,5-Trifluoro-2-nitrobenzene, T-60304

319-24-4 2-Amino-5-fluorobenzoic acid; Me ester, *in* A-70142

319-88-0 ▷1,3,5-Trichloro-2,4,6trifluorobenzene, T-60257

322-58-7 1-(Methylthio)-2-(trifluoromethyl)-benzene, *in* T-60284

327-86-6 3,3′,5-Triiodothyronine, T-60350

327-97-9 ▷3-*O*-Caffeoylquinic acid, C-70005

328-98-3 1-(Methylthio)-3(trifluoromethyl)-benzene, *in* T-60285

333-27-7 ▷Trifluoromethanesulfonic acid; Me ester, *in* T-70241

334-20-3 9-Oxo-2-decenoic acid; (*E*)-*form, in* O-70089

335-05-7 Trifluoromethanesulfonic acid; Fluoride, *in* T-70241

335-56-8 1-Bromo1,1,2,2,3,3,4,4,5,5,6,6,6-tridecafluorohexane, B-70274

337-65-5 Spiro[bicyclo[3.2.1]octane-8,2′[1,3]dioxolane], *in* B-60094

340-45-4 Dibenzo[*a,h*]quinolizinium(1+), D-70076

341-27-5 3-Fluoro-2-hydroxybenzoic acid, F-60040

342-69-8 ▷1,7-Dihydro-6*H*-purine-6-thione; 6-*Thiol-form*, 5-Me, 9-β-D-ribofuranosyl, *in* D-60270

342-69-8 ▷6-(Methylthio)-9-β-Dribofuranosyl-9*H*-purine, *in* T-70180

344-03-6 1,4-Dibromo-2,3,5,6tetra-fluorobenzene, D-60108

344-38-7 4-Bromo-1-nitro-2(trifluoromethyl)-benzene, B-60309

344-39-8 4-Methoxy-1-nitro-2(tri-fluoromethyl)benzene, *in* N-60054

345-16-4 5-Fluoro-2-hydroxybenzoic acid, F-60044

345-29-9 4-Fluoro-2-hydroxybenzoic acid, F-60042

348-62-9 4-Chloro-2-fluorophenol, C-60079

350-29-8 3-Fluoro-4-hydroxybenzoic acid, F-60041

352-91-0 1-Bromo-3-fluoropropane, B-70212

355-41-9 1-Chloro1,1,2,2,3,3,4,4,5,5,6,6,6-tridecafluorohexane, C-70163

358-23-6 Trifluoromethanesulfonic acid; Anhydride, *in* T-70241

359-41-1 (Trifluoromethyl)oxirane, T-60294

359-63-7 *N,N*-Bis-s(rifluoromethyl)hydroxylamine, B-60180

359-75-1 1,1,1-Trifluoro-*N*-(nitrosooxy)*N*-(tri-fluoromethyl)methanamine, *in* B-60180

360-11-2 Difluorodiphenylmethane, D-60184

360-91-8 3,4-Biss(rifluoromethyl)-1,2dithiete, B-60179

363-24-6 ▷11,15-Dihydroxy-9-oxo-5,13prostadienoic acid; (5*Z*,8*R*,11*R*,12*R*,13*E*,15*S*)-*form, in* D-60367

367-32-8 4-Fluoro-1,2-benzenediol, F-60020

367-67-9 1-Bromo-4-nitro-2(trifluoromethyl)-benzene, B-60307

367-78-2 2-Fluoro-4,6-dinitroaniline, F-60028

367-81-7 5-Fluoro-2,4-dinitroaniline, F-60031

367-83-9 2-Fluoro-5-methoxybenzoic acid, *in* F-60038

368-48-9 2-(Trifluoromethyl)pyridine, T-60296

369-26-6 3-Amino-4-fluorobenzoic acid; Me ester, *in* A-70144

371-71-1 (Trifluoromethyl)carbonimidic difluoride, T-60288

372-75-8 Citrulline; (*S*)-*form, in* C-60151

373-80-8 Pentafluoroo(rifluoromethyl)sulfur, P-60041

374-31-2 1,3,3,4,4-Pentafluorocyclobutene, P-60033

378-94-9 Nonafluoromorpholine, N-60056

381-84-0 ▷2-Trifluoromethylacrylonitrile, *in* T-60295

381-98-6 2-(Trifluoromethyl)propenoic acid, T-60295

382-90-1 ▷2-(Trifluoromethyl)propenoic acid; Me ester, *in* T-60295

384-22-5 ▷1-Nitro-2-(trifluoromethyl)benzene, N-70064

386-72-1 2-Nitro-3-(trifluoromethyl)phenol, N-60047

391-92-4 5-Fluoro-2-hydroxybenzoic acid; Me ester, *in* F-60044

392-57-4 ▷1,2,4,5-Tetrafluoro-3,6diiodobenzene, T-60046

394-04-7 5-Fluoro-2-methoxybenzoic acid, *in* F-60044

394-25-2 1-Methoxy-2-nitro-4-(trifluoromethyl)benzene, *in* N-60048

394-28-5 2-Bromo-5-fluorobenzoic acid, B-60241

398-62-9 4-Fluoro-1,2-dimethoxybenzene, *in* F-60020

399-55-3 2-Fluoro-1-methoxy-4-methylbenzene, *in* F-60053

400-99-7 ▷2-Nitro-4-(trifluoromethyl)phenol, N-60048

401-53-6 2-Amino-4-fluorobutanoic acid, A-60155

401-99-0 1,3-Dinitro-5-(trifluoromethyl)benzene, D-70455

402-31-3 1,3-Biss(rifluoromethyl)benzene, B-60176

402-54-0 ▷1-Nitro-4-(trifluoromethyl)benzene, N-70066

403-01-0 3-Fluoro-4-hydroxybenzoic acid; Me ester, *in* F-60041

403-20-3 3-Fluoro-4-methoxybenzoic acid, *in* F-60041

403-21-4 3-Fluoro-4-nitrobenzoic acid, F-70033

404-71-7 ▷1-Fluoro-3-isocyanatobenzene, F-60048

406-41-7 1,1,1-Trifluoro-2-butyne, T-60275

406-76-8 ▷Carnitine; (±)-*form*, *in* C-70024

421-16-9 Methyl trifluoromethyl sulfide, M-70145

421-82-9 Trifluoroo(ethylsulfonyl)methane, *in* M-70145

421-83-0 Trifluoromethanesulfonic acid; Chloride, *in* T-70241

421-85-2 Trifluoromethanesulfonic acid; Amide, *in* T-70241

425-75-2 Trifluoromethanesulfonic acid; Et ester, *in* T-70241

427-77-0 Gibberellin A_9, G-60018

428-21-7 Geodoxin, G-70007

432-03-1 Triss(rifluoromethyl)amine, T-60406

433-19-2 1,4-Biss(rifluoromethyl)benzene, B-60177

433-69-2 4-Amino-5-methyl-3-isoxazolidinone, A-60221

433-95-4 1,2-Biss(rifluoromethyl)benzene, B-60175

434-76-4 2-Amino-6-fluorobenzoic acid, A-70143

443-88-9 2-Fluoro-1,3-dimethylbenzene, F-60026

443-90-3 2-Fluoro-6-methylphenol, F-60055

446-08-2 2-Amino-5-fluorobenzoic acid, A-70142

446-17-3 2,4,5-Trifluorobenzoic acid, T-70240

446-31-1 4-Amino-2-fluorobenzoic acid, A-70145

446-32-2 2-Amino-4-fluorobenzoic acid, A-70141

446-36-6 ▷5-Fluoro-2-nitrophenol, F-60061

446-72-0 ▷4′,5,7-Trihydroxyisoflavone, T-60324

450-89-5 1-Chloro-4-fluoro-2-methoxybenzene, *in* C-60074

451-25-2 4-(4-Methylphenyl)-2-pentanone, M-60112

452-06-2 ▷2-Aminopurine, A-60243

452-08-4 2-Bromo-4-fluoro-1-methoxybenzene, *in* B-60248

452-09-5 4-Chloro-2-fluoro-1-methoxybenzene, *in* C-60079

452-70-0 4-Fluoro-3-methylphenol, F-60059

452-72-2 4-Fluoro-2-methylphenol, F-60058

452-78-8 3-Fluoro-4-methylphenol, F-60057

452-81-3 2-Fluoro-4-methylphenol, F-60053

452-85-7 5-Fluoro-2-methylphenol, F-60060

454-99-9 2,4-Biss(rifluoromethyl)pyridine, B-60182

455-00-5 2,6-Biss(rifluoromethyl)pyridine, B-60184

455-87-8 ▷4-Amino-3-fluorobenzoic acid, A-70146

461-06-3 Carnitine, C-70024

462-40-8 1-Fluoro-3-iodopropane, F-70026

462-80-6 Benzyne, B-70087

465-00-9 2,3,23-Trihydroxy-12-oleanen-28oic acid; (2α,3β)-*form*, *in* T-70271

465-21-4 ▷Bufalin, B-70289

465-58-7 2-Amino-2-methylbutanoic acid, A-70167

465-92-9 Marrubiin, M-70013

465-94-1 12-Oleanene-3,16,28-triol; (3β,16β)-*form*, *in* O-60038

465-95-2 12-Oleanene-3,16,28-triol; (3β,16α)-*form*, *in* O-60038

466-37-5 Tetraphenylethanone, T-60167

467-32-3 3,3,6,6-Tetraphenyl-1,4-dioxane2,5-dione, T-60165

468-19-9 ▷Coroglaucigenin, C-60168

470-30-4 4-Hydroxy-3,3-dimethyl-2oxobutanoic acid, H-60124

471-32-9 Dithiocarbazic acid, D-60509

471-87-4 Stachydrine, S-60050

473-15-4 4(15)-Eudesmen-11-ol, E-70066

473-84-7 2-Hydroxycyclopentanone, H-60114

475-38-7 ▷5,8-Dihydroxy-1,4-naphthoquinone, D-60356

476-37-9 Hexahydroxy-1,4-naphthoquinone, H-60065

476-66-4 ▷Ellagic acid, E-70007

477-32-7 ▷Visnadin, *in* K-70013

477-33-8 ▷Samidin, *in* K-70013

477-83-8 3-Hydroxy-2-hydroxymethyl-1methoxyanthraquinone, *in* D-70306

478-08-0 ▷1,3-Dihydroxy-2hydroxymethylanthraquinone, D-70306

478-40-0 3,5-Dihydroxy-2-methyl-1,4naphthoquinone, D-60350

479-11-8 Benzo[*b*]naphtho[2,3-*d*]furan6,11-dione, B-60034

479-27-6 1,8-Diaminonaphthalene, D-70042

479-66-3 Fulvic acid, F-60081

480-18-2 ▷3,3′,4′,5,7-Pentahydroxyflavanone, P-70027

480-23-9 3′,4′,5,7-Tetrahydroxyisoflavone, T-60108

480-26-2 5-Methyl-5*H*-quindoline, *in* I-70016

480-44-4 ▷5,7-Dihydroxy-4′-methoxyflavone, D-70319

480-66-0 2′,4′,6′-Trihydroxyacetophenone, T-70248

480-91-1 Phthalimidine, P-70100

480-97-7 4-Hydroxybenzofuran, H-70106

481-40-3 1,4,5-Naphthalenetriol, N-60005

483-85-2 2-Formyl-3,4-dimethoxybenzoic acid, *in* F-60070

484-32-2 2-Methoxy-3,4methylenedioxybenzoic acid, *in* H-60177

484-51-5 Khellinone, K-60011

484-89-9 3-Hydroxy-2-methoxy-5-methyl1,4-benzoquinone, *in* D-7032C

484-92-4 1-(4,5-Dihydro-1*H*-imidazol-2-yl)2-imidazolidinethione, D-60246

485-38-1 4,5-Dimethoxy-1,3benzenedicarboxylic acid, *in* D-60307

486-66-8 4′,7-Dihydroxyisoflavone, D-60340

487-21-8 2,4(1*H*,3*H*)-Pteridinedione, P-60191

487-39-8 Phillygenin; (+)-*form*, *in* P-60141

487-41-2 Phillyrin, *in* P-60141

487-42-3 3-(Phenylimino)-1(3*H*)isobenzofuranone, P-60112

487-56-9 4-Hydroxy-5-benzofurancarboxylic acid, H-60103

487-79-6 ▷Kainic acid, K-60003

488-20-0 Dimethylenebutanedioic acid, D-70409

489-35-0 3,3′,4′,5,5′,7,8Heptahydroxyflavone, H-70021

489-40-7 α-Gurjunene, G-70034

489-49-6 Cetraric acid, C-70035

489-79-2 Linderazulene, L-70031

489-98-5 ▷2,4,6-Trinitroaniline, T-70301

490-06-2 3-Isopropyl-6-methyl-1,2benzenediol, I-60124

491-31-6 ▷1*H*-2-Benzopyran-1-one, B-60041

491-40-7 Octahydro-4*H*-quinolizin-4-one, O-60019

491-46-3 Microphyllinic acid, M-70147

491-67-8 5,6,7-Trihydroxyflavone, T-60322

492-80-8 ▷Auramine, A-70268

492-98-8 2,2′-Bi-1*H*-imidazole, B-60107

494-19-9 ▷10,11-Dihydro-5*H*-dibenz[*b,f*]azepine, D-70185

494-21-3 2-(2-Furanyl)quinoxaline, F-60089

494-49-5 Palasitrin, *in* S-70082

494-56-4 6,7-Methylenedioxy-2*H*-1benzopyran-2-one, M-70068

494-67-7 2,2′-(1,4-Phenylene)bis[5,1[1,1′-biphenyl]-4-yl]oxazole, P-60091

495-08-9 2,5-Dihydroxybenzyl alcohol, D-70287

495-18-1 ▷Benzohydroxamic acid, B-70035

495-35-2 Thiazolo[4,5-*b*]quinoline-2(3*H*)thione, T-60201

495-52-3 *s*-Hydrindacene, H-60095

495-74-9 ▷1-Phenyl-2,4-hexadiyn-1-one, P-60109

496-46-8 Tetrahydroimidazo[4,5-*d*]imidazole-2,5(1*H*,3*H*)dione, T-70066

496-69-5 2-Bromo-4-fluorophenol, B-60248

496-83-3 2-Hydroxycyclononanone, H-60112

496-86-6 9,10,18-Trihydroxyoctadecanoic acid, T-60334

496-89-9 2-Thioxo-4,5-imidazolidinedione, T-70187

497-06-3 3-Butene-1,2-diol, B-70291

497-26-7 ▷2-Methyl-1,3-dioxolane, M-60063

497-38-1 ▷Bicyclo[2.2.1]heptan-2-one, B-70098

497-76-7 Arbutin, A-70248

498-19-1 2-Amino-4-hydroxybutanoic acid, A-70156

498-95-3 3-Piperidinecarboxylic acid, P-70102

499-12-7 ▷Aconitic acid, A-70055

500-40-3 Guaiaretic acid, G-70033

500-77-6 21*H*,23*H*-Porphyrazine, P-70109

501-32-6 α-Amino-1*H*-imidazole-1-propanoic acid, A-60204

501-53-1 ▷Benzyloxycarbonyl chloride, B-60071

502-50-1 4-Oxoheptanedioic acid, O-60071

502-51-2 4-Oxo-2-heptenedioic acid, O-60072

502-62-5 Lycopersene, L-70044

502-65-8 Lycopene, L-60041

502-72-7 ▷Cyclopentadecanone, C-60219

502-87-4 4-Methylene-2,5-cyclohexadien-1one, M-60071

503-09-3 ▷(Fluoromethyl)oxirane, F-70032

503-29-7 Azetidine, A-70283

503-38-8 ▷Trichloromethyl chloroformate, T-70223

503-53-7 4,4-Dimethyl-5-oxopentanoic acid, D-60434

503-81-1 Dicarbonic acid, D-70111

503-87-7 ▷2-Thioxo-4-imidazolidinone, T-70188

504-75-6 4,5-Dihydro-1*H*-imidazole, D-60245

505-03-3 5-Oxohexanal, O-70091

505-18-0 2,3,4,5-Tetrahydropyridine, T-70090

505-19-1 Hexahydropyrimidine, H-60056

505-57-7 ▷2-Hexenal, H-60076

505-72-6 3-(Carboxymethylamino)propanoic acid, C-60016

507-29-9 Histidine trimethylbetaine, H-60083

508-04-3 13(18)-Oleanen-3-ol; 3β-*form*, *in* O-70036

511-15-9   Totarol, T-70204

511-77-3   3,19-Dihydroxy-12-oleanen-28-oic acid; (3β,19α)-*form, in* D-70327

512-04-9 ▷5-Spirosten-3-ol; (3β,25R)-*form, in* S-60045

514-51-2   β-Patchoulene, P-60010

515-03-7   Sclareol, S-70018

515-85-5   3,3,4,4-Tetrafluoro-2(trifluoromethyl)-1,2oxazetidine, T-60051

515-96-8 ▷Semioxamazide, S-60020

516-37-0   24-Methyl-7,22-cholestadiene3,5,6-triol; (3β,5α,6β,22E)-*form, in* M-70057

517-88-4   Shikonin; (S)-*form, in* S-70038

517-89-5 ▷Shikonin; (R)-*form, in* S-70038

519-05-1   6-Formyl-2,3-dimethoxybenzoic acid, *in* F-60073

519-57-3   Sulochrin, S-70083

520-42-3   1-(2,6-Dihydroxy-4methoxyphenyl)-3-(4hydroxyphenyl)-1propanone, *in* H-60219

521-65-3   1,8-Dihydroxy-3,5dimethoxy-xanthone, *in* T-70123

523-21-7   5,6-Dihydroxy-5-cyclohexene1,2,3,4-tetrone; Di-Na salt, *in* D-60315

523-27-3   9,10-Dibromoanthracene, D-60085

524-97-0   Pterocarpin, *in* M-70001

525-24-6   9H-Pyrrolo[1,2-a]indol-9-one, P-60247

526-34-1   3-Methoxy-4,5methylenedioxybenzoic acid, *in* H-60178

526-49-8   4-Hydroxy-2-isopropyl-5benzofurancarboxylic acid, H-60164

526-51-2   Isoporphobilinogen, I-60114

526-94-3   Tartaric acid; (2R,3R)-*form*, Mono-Na salt, *in* T-70005

527-31-1   Cyclohexanehexone, C-60210

527-54-8   3,4,5-Trimethylphenol, T-60371

528-22-3   3,5-Dihydroxy-4-oxo-4H-pyran-2carboxylic acid, D-60368

529-01-1   6-Isopropenyl-3-methyl-2cyclohexen-1-one, I-70088

529-17-9   Tropane, *in* A-70278

529-59-9   Genistin, *in* T-60324

529-60-2   3',4',5-Trihydroxy-7methoxy-isoflavone, *in* T-60108

529-69-1 ▷Isoxanthopterin, I-60141

530-22-3   Egonol, E-70001

531-47-5   5,14-Dihydroquinoxalino[2,3-b]phenazine, D-70264

531-63-5   Sulfurein, *in* S-70082

532-87-6   α-Camphorene, C-70009

533-35-7   6,7-Dihydro-5H-2-pyridine, D-60276

533-35-7   6,7-Dihydro-5H-2-pyridine, D-70259

533-37-9   6,7-Dihydro-5H-1-pyridine, D-60275

533-37-9   6,7-Dihydro-5H-1-pyridine, D-70258

534-30-5   Histidine trimethylbetaine; (S)-*form*, *in* H-60083

535-67-1   5-Amino-2-thiazolecarbothioamide, A-60260

537-33-7   Sinapyl alcohol, S-70046

539-80-0   2,4,6-Cycloheptatrien-1-one, C-60204

539-97-9   1,6-Diallyl-2,5-dithiobiurea, D-60026

540-11-4   9-Octadecene-1,12-diol; (R,Z)-*form*, *in* O-60006

541-14-0   Carnitine; (S)-*form, in* C-70024

541-15-1 ▷Carnitine; (R)-*form, in* C-70024

542-02-9   2,4-Diamino-6-methyl-1,3,5triazine, D-60040

542-05-2   3-Oxopentanedioic acid, O-60084

542-48-3   26-Henpentacontanone, H-70009

543-71-5 ▷Tetrahydro-2H-1,3-thiazine, T-60093

545-24-4   5-Gluten-3-ol; 3β-*form, in* G-70020

546-02-1 ▷Frugoside, *in* C-60168

546-97-4   Columbin, C-60160

547-01-3   1,2,3-Spirostanetriol; (1β,2β,3α,5β,20S,22R,25R)-*form, in* S-60043

548-29-8   Isolariciresinol, *in* I-60139

550-45-8   Iridodial, I-60084

550-79-8   6-Hydroxy-4',7-dimethoxyisoflavone, *in* T-60325

552-00-1   1,3,8-Trihydroxy-5-methoxyxanthone, *in* T-70123

552-58-9   3',4',5,7-Tetrahydroxyflavanone; (S)-*form, in* T-70105

552-66-9   Daidzin, *in* D-60340

552-82-9   N-Methyldiphenylamine, M-70060

553-30-0 ▷3,6-Diaminoacridine; B,H₂SO₄, *in* D-60028

553-84-4 ▷1-(3-Furanyl)-4-methyl-1pentanone, F-60087

559-31-9   2,2,3,3,4,4,5,5,6,6Decafluoropiperidine, D-60006

561-20-6   Cacotheline, C-70001

561-68-2   Gibberellin A₂, G-60017

565-12-8   2-Amino-3-(4-hydroxyphenyl)-3methylbutanoic acid, A-60198

565-61-7   3-Methyl-2-pentanone, M-70105

565-71-9   3-Amino-2-hydroxypropanoic acid, A-70160

565-81-1   3-Amino-2-hydroxybutanoic acid, A-60183

567-85-1   6,6'-Difluoro-[1,1'-biphenyl]2,2'-dicarboxylic acid, D-60180

568-98-9   1-Amino-2-hydroxyanthraquinone, A-60174

568-99-0   2-Amino-1-hydroxyanthraquinone, A-60178

569-58-4 ▷Aluminon, *in* A-60318

573-32-0   Benzophenone-2,2'-dicarboxylic acid, B-70041

574-25-4 ▷1,7-Dihydro-6H-purine-6-thione; 6-Thiol-*form*, 9-β-D-Ribofuranosyl, *in* D-60270

574-25-4 ▷6-Thioinosine, T-70180

575-82-6   Methoxymethyl salicylate, *in* H-70108

577-47-9 ▷2-Thioxo-4-imidazolidinone; 1-N-Benzoyl, *in* T-70188

577-67-3   6,7-Dichloro-1,4-naphthoquinone, D-70124

578-39-2   4-Hydroxy-2-methylbenzoic acid, H-70164

579-92-0 ▷2,2'-Iminodibenzoic acid, I-60014

581-46-4   3,3'-Bipyridine, B-60122

581-47-5   2,4'-Bipyridine, B-60121

581-50-0 ▷2,3'-Bipyridine, B-60120

583-08-4   Nicotinuric acid, N-70034

583-11-9   Acetophenone; Phenylhydrazone, *in* A-60017

583-86-8   9,10,18-Trihydroxyoctadecanoic acid; (9RS,10RS)-*form, in* T-60334

584-07-6   C-Methoxycarbohydroxamic acid, M-70044

584-26-9 ▷2-Thioxo-4-imidazolidinone; 1-N-Ac, *in* T-70188

585-84-2   Aconitic acid; (Z)-*form, in* A-70055

590-42-1 ▷2-Isothiocyanato-2-methylpropane, I-70105

590-91-0   1,1-Difluoro-1,3-butadiene, D-70162

591-05-9   2,4-Pentanediamine, P-70039

591-11-7 ▷5-Methyl-2(5H)-furanone, M-60083

591-12-8   5-Methyl-2(3H)-furanone, M-70081

593-82-8 ▷1,1-Dimethylhydrazine; B,HCl, *in* D-70416

594-00-3   1-Chloro-1-iodoethane, C-60106

594-71-8 ▷2-Chloro-2-nitropropane, C-60120

595-39-1   2-Amino-2-methylbutanoic acid; (±)-*form, in* A-70167

595-40-4   2-Amino-2-methylbutanoic acid; (S)-*form, in* A-70167

596-84-9   8,13-Epoxy-14-labdene; (8α,13R)-*form, in* E-70023

602-14-2   3-Hydroxy-5-methyl-2,4,6trinitrobenzoic acid, H-60193

603-12-3   2,6-Dinitrobenzoic acid, D-60457

604-88-6 ▷Hexaethylbenzene, H-70043

605-09-4   Isodalbergin, I-60097

605-48-1   9,10-Dichloroanthracene, D-60116

605-94-7   2,3-Dimethoxy-5-methyl-1,4benzoquinone, *in* D-70320

606-19-9   2-Hydroxy-1,3-benzenedicarboxylic acid, H-60101

606-20-2 ▷2-Methyl-1,3-dinitrobenzene, M-60061

606-86-0   1,3,3-Triphenyl-1-propanone, T-60391

607-32-9   5-Nitroisoquinoline, N-60031

608-68-4   Dimethyl tartrate, *in* T-70005

609-69-8   2-Hydroxycyclohexanecarboxylic acid, H-70119

611-03-0   2,4-Diaminobenzoic acid, D-60032

611-07-4 ▷5-Chloro-2-nitrophenol, C-60119

611-20-1   2-Cyanophenol, *in* H-70108

611-92-7 ▷N,N'-Dimethyl-N,N'-diphenylurea, *in* D-70499

612-01-1   N,N'-Diphenylurea; N-Me, *in* D-70499

615-18-9 ▷2-Chlorobenzoxazole, C-60048

616-45-5 ▷2-Pyrrolidinone, P-70188

616-74-0   4,6-Dinitro-1,3-benzenediol, D-70450

617-12-9   Chorismic acid, C-70173

617-71-0   S-(2-Aminoethyl)cysteine, A-60152

618-71-3   3,5-Dinitrobenzoic acid; Et ester, *in* D-60458

620-30-4   2-Amino-2-methyl-3-(4hydroxyphenyl)propanoic acid; (±)-*form, in* A-60220

622-15-1   N,N'-Diphenylformamidine, D-60479

622-29-7   N-Benzylidenemethylamine, B-60070

622-42-4   (Nitromethyl)benzene, N-70053

622-43-5   (Nitromethyl)benzene; aci-*form, in* N-70053

623-15-4   4-(2-Furanyl)-3-buten-2-one, F-60086

623-36-9 ▷2-Methyl-2-pentenal, M-70106

623-84-7 ▷1,2-Propanediol; (±)-*form*, Di-Ac, *in* P-70118

624-40-8 ▷Tetrahydro-1,2,4,5-tetrazine3,6-dione, T-70097

624-70-4   1-Chloro-2-iodoethane, C-70078

626-06-2   6-Hydroxy-2(1H)-pyridinone, H-60226

626-47-1   4,6-Dihydroxy-2(1H)-pyridinone, D-70341

626-96-0   4-Oxopentanal, O-60083

630-10-4 ▷Selenourea, S-70031

632-11-1   3-Amino-2-hydroxypropanoic acid; (R)-*form, in* A-70160

632-12-2   3-Amino-2-hydroxypropanoic acid; (±)-*form, in* A-70160

632-13-3   3-Amino-2-hydroxypropanoic acid; (S)-*form, in* A-70160

632-20-2 ▷Threonine; (2R,3S)-*form, in* T-60217

632-99-5 ▷Fuchsine, *in* R-60010

634-63-9   Tartaric acid; (2R,3R)-*form*, Diamide, *in* T-70005

635-46-1   1,2,3,4-Tetrahydroquinoline, T-70093

636-94-2   2-Hydroxy-1,4-benzenedicarboxylic acid, H-70105

637-69-4   1-Ethenyl-4-methoxybenzene, *in* H-70209

637-90-1   9-Oxabicyclo[6.1.0]non-4-ene, O-70058

638-13-1   3,4-Dihydro-6-methyl-4-thioxo2(1H)-pyrimidinone, D-70234

641-63-4   2-(2-Pyridyl)-1,3-indanedione, P-60241

642-96-6   1,2,3,4-Benzenetetrol, B-60014

643-93-6   3-Methylbiphenyl, M-60053

644-06-4   6,7-Dimethoxy-2,2-dimethyl-2H-1benzopyran, D-70365

644-69-9   Ranunculin, in H-60182

645-54-5   Phenylthiolacetamide, in B-70017

646-08-2   Alethine, A-60073

646-23-1   1-Isothiocyanato-5(methylsulfinyl)-pentene, in I-60140

646-72-0   1,1,1-Trifluoro-2(trifluoromethyl)-3butyn-2-ol, T-60310

652-03-9   3,4,5,6-Tetrafluoro-1,2benzenedicarboxylic acid, T-70031

652-12-0   4,5,6,7-Tetrafluoro-1,3isobenzofurandione, in T-70031

652-18-6   2,3,5,6-Tetrafluorobenzoic acid, T-70034

652-36-8   2,3,5,6-Tetrafluoro-1,4benzenedicarboxylic acid, T-70029

653-92-9   2-Bromo-4-fluorobenzoic acid; Me ester, in B-60240

654-76-2   1-Methoxy-4-nitro-2(trifluoromethyl)benzene, in N-60053

658-48-0   2-Amino-2-methyl-3-(4hydroxyphenyl)propanoic acid, A-60220

665-31-6   2,2-Difluorobutanedioic acid, D-70163

667-29-8 ▷ Trifluoroacetyl nitrite, T-60274

670-54-2 ▷ Tetracyanoethylene, T-60031

671-41-0   5-Fluoro-3,4-dihydro-4-thioxo2(1H)-pyrimidinone, F-70019

671-95-4 ▷ Clofenamide, in C-70049

672-15-1   2-Amino-4-hydroxybutanoic acid; (S)-form, in A-70156

672-87-7   2-Amino-2-methyl-3-(4hydroxyphenyl)propanoic acid; (S)-form, in A-60220

675-09-2 ▷ 4,6-Dimethyl-2H-pyran-2-one, D-70435

683-79-4   1,1-Difluoro-2,2-diiodoethylene, D-60183

684-10-6   Trifluoroethenesulfonic acid; Fluoride, in T-60277

685-24-5   1,1,1,4,4,4-Hexafluoro-2,3butanedione, H-60043

686-10-2   1,1,2-Trichloro-2-(diethylamino)ethylene, T-70222

691-04-3   Ethyl trifluoromethanesulfenate, in T-60279

691-57-6   2-Azidopropane, A-70289

693-16-3   2-Octylamine, O-60031

694-05-3 ▷ 1,2,3,6-Tetrahydropyridine, T-70089

694-17-7 ▷ Tetrafluorooxirane, T-60049

694-25-7   1-Cyano-3-methylbicyclo[1.1.0]butane, in M-70053

694-32-6   2-Imidazolidinone; 1-Me, in I-70005

694-68-8   2-Thioxo-4-imidazolidinone; 3-N-Me, in T-70188

694-73-5   1,2,3,3a,4,5-Hexahydropentalene, H-70064

694-81-5   4,5-Dimethylpyrimidine, D-70436

694-82-6   2-Fluorocyclohexanone, F-60021

695-05-6   Tricyclo[2.2.1.0²·⁶]heptan-3-one, T-70228

695-70-5   1,1-Diacetylcyclopropane, D-70032

696-07-1 ▷ 5-Iodo-2,4(1H,3H)-pyrimidinedione, I-70056

696-07-1 ▷ 5-Iodo-2,4(1H,3H)-pyrimidinedione, I-60062

696-74-2   1,2-Cyclopropanedicarboxylic acid; (1RS,2SR)-form, in C-70248

696-82-2   2,4,6-Trifluoropyrimidine, T-60307

696-85-5   2-Chloro-4,6-difluoro-1,3,5triazine, C-60052

697-83-6   5-Chloro-2,4,6-trifluoropyrimidine, C-60144

698-10-2   5-Ethyl-3-hydroxy-4-methyl-2(5H)furanone, E-60051

698-27-1   2-Hydroxy-4-methylbenzaldehyde, H-70160

698-40-8   Bicyclo[2.2.1]hepta-2,5-diene-2carboxylic acid, B-60084

699-39-8   1,2,3,4,5,5-Hexafluoro-1,3cyclopentadiene, H-60044

699-55-8   Bicyclo[2.2.2]octane-1-carboxylic acid, B-60092

700-91-4   3,4-Dihydro-5-phenyl-2H-pyrrole, D-70251

702-62-5 ▷ 1,3-Diazaspiro[4.5]decane-2,4dione, D-60056

702-79-4   1,3-Dimethyladamantane, D-70366

704-45-0   2-Methoxy-4-methylbenzoic acid, in H-70163

706-14-9 ▷ 5 Hexyldihydro-2(3H)-furanone, H-60078

708-57-6   5-Methyl-3-phenyl-2-oxazolidinone, in M-60107

712-50-5 ▷ Benzoylcyclohexane, B-70071

715-63-9   Triethyl isocyanurate, in C-60179

716-39-2   2,3-Naphthalenedicarboxylic acid; Anhydride, in N-60003

723-61-5   Helminthosporal, H-60016

727-55-9   2,3,5,6-Tetrafluoro-1,4benzenedicarboxylic acid; Di-Me ester, in T-70029

730-08-5   N-Tyrosylalanine; L-L-form, in T-60411

740-33-0   5-Hydroxy-6,7-dimethoxyflavone, in T-60322

746-47-4   9,9'-Bifluorenylidene, B-60104

753-92-4   Trifluoromethanesulfenyl bromide, T-60280

764-75-0   2,4,6-Octatriene, O-70023

765-72-0   Tetracyclo[3.3.0.0²·⁸.0⁴·⁶]octane, T-70022

766-03-0   3-Vinylcyclohexene, V-60009

766-11-0   5-Bromo-2-fluoropyridine, B-60256

766-55-2   Imidazo[1,2-b]pyridazine, I-60007

767-01-1   2-(Chloromethyl)-5-methylpyridine, C-70095

767-45-3   2,2,4,4-Tetrafluoro-1,3-dithietane; 1,1-Dioxide, in T-60048

769-71-1   3-Hydroxy-4(1H)-pyridinone; Diol-form, Di-Me ether, 1-oxide, in H-60227

771-41-5   4,7-Diaminopteridine, D-60044

771-51-7 ▷ 1H-Indole-3-acetonitrile, in I-70013

771-56-2   Pentafluoromethylbenzene, P-60038

771-69-7   1,2,3-Trifluoro-4-nitrobenzene, T-60299

774-53-8   5-Bromo-2,3'-bipyridine, B-70187

774-64-1 ▷ 3,4-Dimethyl-5-pentylidene2(5H)-furanone, D-70430

776-35-2   9,10-Dihydrophenanthrene, D-60260

777-44-6 ▷ 3-(Trifluoromethyl)benzenesulfonic acid; Chloride, in T-60283

778-66-5   1,1-Diphenylpropene, D-60491

781-17-9   4,5,9,10-Tetrahydropyrene, T-60083

784-00-9   1,2,3,4,5,6,8Heptafluoronaphthalene, H-70020

784-62-3   2-Phenyl-4H-1-benzothiopyran-4-one, P-60082

789-07-1   2-Nitropyrene, N-60041

789-99-1   2-Cyclopenten-1-one; 2,4-Dinitrophenylhydrazone, in C-60226

791-50-4 ▷ 2,2,4,4Tetrakiss(rifluoromethyl)-1,3-dithietane, T-60129

795-31-3   2,2,4,4Tetrakiss(rifluoromethyl)-1,3-dithietane; 1,1-Dioxide, in T-60129

798-61-8   4',6,7-Trimethoxyisoflavone, in T-60325

813-58-1   2,2-Diethylbutanoic acid, D-60171

814-68-6 ▷ 2-Propenoic acid; Chloride, in P-70121

816-00-2   Biuret; 1,3,5-Tri-Me, in B-60190

820-29-1   5-Decanone, D-60014

824-94-2   1-Chloromethyl-4-methoxybenzene, in C-70103

824-98-6   1-(Chloromethyl)-3-methoxybenzene, in C-70102

825-22-9   2-Amino-3-fluorobenzoic acid, A-70140

825-25-2   2-Cyclopentylidenecyclopentanone, C-60227

825-33-2   2,3-Dihydroxy-5-methyl-1,4benzoquinone, D-70320

825-83-2   4-(Trifluoromethyl)benzenethiol, T-60286

826-34-6   1,2-Cyclopropanedicarboxylic acid; (1RS,2SR)-form, Di-Me ester, in C-70248

827-08-7   1,2-Dibromo-3,4,5,6tetrafluorobenzene, D-60106

827-12-3   1,2,3-Trichloro-4,5,6trifluorobenzene, T-60256

827-16-7   Trimethyl isocyanurate, in C-60179

830-50-2   1-(Pentafluorophenyl)ethanol, P-60040

832-58-6   2',4',6'-Trimethoxyacetophenone, in T-70248

832-71-3   3-Methylphenanthrene, M-70108

832-80-4   9-Diazo-9H-fluorene, D-60062

833-48-7   10,11-Dihydro-5H-dibenzo[a,d]cycloheptene, D-60217

835-05-2   3,6-Diiodophenanthrene, D-70361

835-79-0   1-Methoxy-2-(phenylmethoxy)-benzene, in B-60072

836-77-1   Undecafluoropiperidine, U-60002

839-97-4   Benzoylcyclohexane; Semicarbazone, in B-70071

841-07-6   1,3,5-Tributylbenzene, T-70215

842-74-0   2-Phenyl-4-(phenylmethylene)5(4H)-oxazolone, P-60128

844-25-7   2-Benzoyl-1,2-dihydro-1isoquinolinecarbonitrile, B-70073

862-28-2   1,4,7,10,13,16,19,22,25,28Decaazacyclotriacontane, D-70011

865-36-1   Methanethiol, M-70037

870-74-6   4-Imino-2-pentanone, in P-60059

870-77-9   Carnitine; (±)-form, O-Ac, in C-70024

872-32-2   3,4-Dihydro-5-methyl-2H-pyrrole, D-70226

872-94-6   α-Isoprene sulfone, in D-70228

873-67-6   Benzonitrile N-oxide, B-60035

873-92-7   1-Chloro-1-nitrocyclohexane, C-70131

874-99-7   Methylcyclononane, M-60058

875-72-9   Methyleneadamantane, M-60068

876-53-9   1,3-Dibromoadamantane, D-70084

879-37-8   1H-Indole-3-acetic acid; Amide, in I-70013

880-12-6   4-(Phenylimino)-2-pentanone, in P-60059

881-64-1   Dipyrido[1,2-b:1',2'-e][1,2,4,5]tetrazine, D-60494

881-70-9   4-Amino-1,2-benzenediol; Di-Me ether, N-Ac, in A-70090

885-23-4   9-Acridinecarboxaldehyde, A-70060

888-35-7   4-Anilino-1,2,3-benzotriazine, in A-60092

888-71-1   3,5-Diphenyl-1,2,4-oxadiazole, D-60484

890-38-0   2'-Deoxyinosine, D-70023

908-21-4   Hyptolide, H-70236

926-61-4   3-Oxopropanoic acid, O-60090

927-66-2   1-Nitro-1,3-butadiene, N-60023

928-39-2   6-Heptyn-2-one, H-70025

929-59-9   1,2-Bis(2-aminoethoxy)ethane, B-60139

930-30-3   2-Cyclopenten-1-one, C-60226

930-50-7   2,2-Dimethylcyclopropanemethanol, D-70396

930-97-2   3-Iodo-2-thiophenecarboxaldehyde, I-60069

931-46-4   5-Ethoxy-3,4-dihydro-2H-pyrrole, in P-70188

931-49-7   1-Ethynylcyclohexene, E-70051

931-57-7   1-Methoxycyclohexene, M-60037

932-17-2 ▷2-Pyrrolidinone; *N*-Ac, *in* P-70188
932-66-1 1-Acetylcyclohexene, A-70038
932-88-7 ▷1-Iodo-2-phenylacetylene, I-70055
933-19-7 6-Methyl-1,3,5-triazine2,4(1*H*,3*H*)-dione, M-60132
934-72-5 Methyl 4-methylphenyl sulfoxide, M-70094
935-56-8 1-Chloroadamantane, C-60039
935-69-3 6-Amino-7-methylpurine, A-60226
935-72-8 1*H*-Pyrrole-3,4-dicarboxylic acid, P-70183
936-04-9 2-Hydroxycyclohexanecarboxylic acid; (1*RS*,2*RS*)-*form*, Me ester, *in* H-70119
936-40-3 7-Methylpteridine, M-70123
936-53-8 4,5-Dihydro-1*H*-pyrazole; 1-Ph, *in* D-70253
936-58-3 1-Phenyl-3-buten-1-ol, P-70061
937-00-8 3-(Trifluoromethyl)benzenethiol, T-60285
938-16-9 2,2-Dimethyl-1-phenyl-1-propanone, D-60443
938-73-8 ▷2-Ethoxybenzamide, *in* H-70108
938-94-3 2-(4-Methylphenyl)propanoic acid, M-60114
951-80-4 ▷3,9-Diaminoacridine, D-60029
951-82-6 3,4,5-Trimethoxybenzeneacetic acid, *in* T-60343
951-87-1 1,2-Diphenyl-1,2-ethanediamine; (1*RS*,2*SR*)-*form*, *in* D-70480
952-45-4 3,6-Bis(2-methylpropyl)-2,5piperazinedione; (3*S*,6*S*)-*form*, *in* B-60171
952-92-1 1-Benzyl-1,4-dihydronicotinamide, B-60067
954-63-2 4-Amino-1,2,3-benzotriazine; 3*H*-form, 3-Ph, *in* A-60092
958-17-8 4-Amino-1,2,3-benzotriazine; 3*H*-form, 3-Benzyl, *in* A-60092
973-67-1 5,6,7-Trimethoxyflavone, *in* T-60322
973-82-0 2-Hydroxy-4-methylbenzaldehyde; 2,4-Dinitrophenylhydrazone, *in* H-70160
981-18-0 Triphenylphenylazomethane, T-70312
992-04-1 Hexaphenylbenzene, H-60074
999-78-0 4,4-Dimethyl-2-pentyne, D-60438
1001-62-3 Biss(ethylsulfonyl) peroxide, B-70154
1004-99-5 2-Chloro-2-phenylethanol, C-60128
1006-12-8 1,7-Dihydro-6*H*-purine-6-thione; 3,7-*Dihydro-form*, 3-Me, *in* D-60270
1006-20-8 1,7-Dihydro-6*H*-purine-6-thione; 1,9-*Dihydro-form*, 9-Me, *in* D-60270
1006-22-0 1,7-Dihydro-6*H*-purine-6-thione; 1,7-*Dihydro-form*, 1-Me, *in* D-60270
1006-41-3 2-Bromo-4-fluorobenzoic acid, B-60240
1006-68-4 5-Phenyloxazole, P-60125
1007-16-5 3-Bromo-4-fluorobenzoic acid, B-60243
1007-26-7 (2,2-Dimethylpropyl)benzene, D-60445
1008-01-1 1,7-Dihydro-6*H*-purine-6-thione; 6-*Thiol-form*, *S*,7*N*-Di-Me, *in* D-60270
1008-08-8 1,7-Dihydro-6*H*-purine-6-thione; 6-*Thiol-form*, *S*,3*N*-Di-Me, *in* D-60270
1008-92-0 1,4-Benzodioxan-2-carbonitrile, *in* B-60024
1009-74-1 2,4,6-Triethyl-1,3,5-triazine, T-70235
1011-47-8 2-Acetylquinoline, A-60049
1013-88-3 Diphenylmethaneimine, D-60482
1015-80-1 6,7-Dihydro-5*H*-dibenzo[*a*,*c*]cycloheptene, D-70186
1022-07-7 2,4,6-Trinitroaniline; *N*-Me, *in* T-70301
1025-36-1 1,3-Diazetidine-2,4-dione; 1,3-Di-Ph, *in* D-60057

1029-96-5 2,6-Diphenyl-4*H*-thiopyran-4-one, D-60492
1032-98-0 3-(2-Benzothiazolyl)-2*H*-1benzopyran, B-70056
1038-67-1 1,8-Diphenylnaphthalene, D-60483
1061-93-4 3',4',5,7-Tetrahydroxyisoflavone; Tetra-Ac, *in* T-60108
1069-23-4 ▷1,5-Hexadiene-3,4-diol, H-70036
1072-70-4 5-Methyl-2-oxazolidinone, M-60107
1072-71-5 ▷5-Mercapto-1,3,4thiadiazoline-2-thione, M-70031
1072-82-8 3-Acetylpyrrole, A-70052
1072-84-0 1*H*-Imidazole-4-carboxylic acid, I-60004
1074-48-2 6-Methyl-3-pyridazinone; *OH-form*, Me ether, 1-Oxide, *in* M-70124
1074-88-0 1*H*-Indole-7-carboxaldehyde, I-60031
1074-91-5 1-Bromo-2,3,4,5tetrafluorobenzene, B-60320
1074-98-2 ▷3-Methyl-4-nitropyridine; *N*-Oxide, *in* M-70099
1075-26-9 1*H*-Indole-6-methanol, I-60032
1076-17-1 *as*-Hydrindacene, H-60094
1076-88-6 3,4,5-Trimethylbenzoic acid, T-60355
1084-95-3 2-(Phenylethynyl)benzoic acid, P-70074
1094-12-8 Kanjone, K-70003
1110-02-7 Cucurbitacin L, *in* C-70204
1110-58-3 Gossypol; (±)-*form*, 6,6'-Di-Me ether, *in* G-60032
1117-97-1 *O*,*N*-Dimethylhydroxylamine, D-70418
1118-47-4 4,4-Dimethylpentanoic acid, D-70428
1119-64-8 1-Bromo-1-hexyne, B-70218
1119-67-1 1-Iodo-1-hexyne, I-70040
1120-53-2 1,3-Cyclobutadiene, C-70212
1120-56-5 Methylenecyclobutane, M-70066
1120-59-8 2,3-Dihydrothiophene, D-60290
1120-99-6 ▷3-Amino-1,2,4-triazine, A-70203
1121-30-8 ▷2(1*H*)-Pyridinethione; 1-Hydroxy, *in* P-60224
1121-30-8 ▷1-Hydroxy-2(1*H*)-pyridinethione, H-70217
1121-79-5 3-Chloro-6-methylpyridazine, C-70112
1121-92-2 ▷Octahydroazocine, O-70010
1122-21-0 Tetrahydro-3-hydroxy-4-methyl2*H*-pyran-2-one, *in* D-60353
1122-41-4 2,4-Dichlorobenzenethiol, D-60120
1123-39-3 1,4-Dichlorobicyclo[2.2.2]octane, D-70115
1124-65-8 3-(2-Thienyl)-2-propenoic acid, T-60208
1127-35-1 Benzo[*b*]thiophen-3(2*H*)-one; 1,1-Dioxide, *in* B-60048
1127-49-7 5,8-Diaminoisoquinoline, D-70041
1127-75-9 1,7-Dihydro-6*H*-purine-6-thione; 6-*Thiol-form*, *S*,9*N*-Di-Me, *in* D-60270
1127-93-1 2,4-Diaminopteridine, D-60042
1129-90-4 Cyclododecyne, C-60192
1131-40-4 2-Bromo-1,3,5-trimethoxybenzene, *in* B-60199
1132-14-5 3,5-Di-*tert*-butylpyrazole, D-70107
1132-26-9 2-Amino-2-methyl-3phenylpropanoic acid; (±)-*form*, *in* A-60225
1132-37-2 Di-2-pyridylmethane, D-60496
1136-86-3 3',4',5'-Trimethoxyacetophenone, *in* T-60315
1139-82-8 5,7-Dihydro-6*H*-dibenzo[*a*,*c*]cyclohepten-6-one, D-70187
1143-11-9 2,3,5,6,8-Pentahydroxy-1,4naphthoquinone, P-70033
1144-20-3 2-Phenyl-4(1*H*)-quinolinone; *OH-form*, *in* P-60132
1146-04-9 ▷Illudin M, I-60001
1149-99-1 ▷Illudin S, I-60002
1157-39-7 4',7-Dimethoxyisoflavone, *in* D-60340
1159-53-1 4,4',4''-Trimethyltriphenylamine, T-70290

1176-88-1 5-Hydroxy-3,3',4',6,7,8hexamethoxyflavone, *in* H-60024
1180-60-5 Cyclotriveratrylene, C-70254
1184-78-7 ▷Trimethylamine oxide, T-60354
1184-90-3 Aminoiminomethanesulfonic acid, A-60210
1189-64-6 9-Oxo-2-decenoic acid; (*E*)-*form*, Me ester, *in* O-70089
1189-71-5 ▷Sulfuryl chloride isocyanate, S-60062
1191-30-6 5-Bromopentanal, B-70263
1192-20-7 3-Aminodihydro-2(3*H*)-furanone, A-60138
1192-79-6 5-Methyl-1*H*-pyrrole-2carboxaldehyde, M-70129
1193-10-8 Isoprene cyclic sulfone, *in* D-70231
1193-59-5 1,5-Dimethyl-1*H*-pyrrole-2carboxaldehyde, *in* M-70129
1194-56-5 1-Methyleneindane, M-60078
1195-08-0 1,2,3,4-Tetrahydro-2,4-dioxo-5pyrimidinecarboxaldehyde, T-60064
1195-19-3 5-Methyl-2-oxo-4oxazolidinecarboxylic acid, M-60108
1195-20-6 5-Methyl-2-oxo-4oxazolidinecarboxylic acid; (4*S*,5*R*)-*form*, *in* M-60108
1195-45-5 ▷1-Fluoro-4-isocyanatobenzene, F-60049
1195-93-3 2,2,6,6-Tetramethylcyclohexanone, T-60133
1196-69-6 1*H*-Indole-5-carboxaldehyde, I-60029
1196-70-9 1*H*-Indole-6-carboxaldehyde, I-60030
1198-33-0 3,9-Dihydro-9-methyl-1*H*-purine2,6-dione, D-60257
1198-59-0 1,2-Dichloro-3,4,5,6tetrafluorobenzene, D-60151
1198-61-4 1,3-Dichloro-2,4,5,6tetrafluorobenzene, D-60152
1198-62-5 1,4-Dichloro-2,3,5,6tetrafluorobenzene, D-60153
1198-98-7 5-Methyl-3-phenyl-1,2,4oxadiazole, M-60110
1201-31-6 2,3,4,5-Tetrafluorobenzoic acid, T-70032
1203-39-0 2-Bromo-3-hydroxy-1,4naphthoquinone, B-60268
1206-79-7 2,3,4,5,6,7,8,9-Octahydro-1*H*triindene, O-60021
1207-99-4 3-Chloro-10*H*-phenothiazine, C-60126
1210-35-1 10,11-Dihydro-5*H*-dibenzo[*a*,*d*]cyclohepten-5-one, D-60218
1210-96-4 1,5-Dimethoxy-2,4-dinitrobenzene, *in* D-70450
1221-43-8 Auraptenol; (*S*)-*form*, *in* A-70269
1227-93-6 8,13-Epoxy-14-labdene; (8*α*,13*S*)-*form*, *in* E-70023
1235-77-4 Lambertianic acid; (−)-*form*, *in* L-60011
1248-51-7 Benzoylcyclohexane; 2,4-Dinitrophenylhydrazone, *in* B-70071
1391-36-2 Duramycin, D-70549
1398-78-3 ▷Elaterinide, *in* C-70204
1399-49-1 Globularin, *in* C-70026
1402-69-3 Albidin, A-60070
1423-13-8 Pentafluoronitrosobenzene, P-60039
1423-15-0 Azidopentafluorobenzene, A-60350
1423-61-6 7-Bromobenzo[*b*]thiophene, B-60209
1436-27-7 3,6-Bis(2-methylpropyl)-2,5piperazinedione, B-60171
1437-88-3 2-Thiazoledinethione, T-60197
1439-10-7 4-Amino-5-bromopyrimidine, A-70102
1439-19-6 1,2-Di(2-furanyl)ethylene; (*E*)-*form*, *in* D-60191
1441-87-8 2-Hydroxybenzoic acid; Chloride, *in* H-70108

1453-24-3	1-Ethylcyclohexene, E-60046	
1454-80-4	2,2′-Diaminobiphenyl, D-60033	
1455-91-0	Tetrazole-5-thione; 1,4-*Dihydro-form*, 1-Me, 4-Ph, *in* T-60174	
1455-92-1	Tetrazole-5-thione; (*SH*)-*form*, 1-*N*-Ph, *S*-Me, *in* T-60174	
1467-40-9	1,5-Diphenyl-1,3,5-pentanetrione, D-60489	
1480-64-4	3-Chloro-2-fluoropyridine, C-60085	
1480-65-5	5-Chloro-2-fluoropyridine, C-60088	
1481-93-2	3,4-Dihydro-2*H*-benzopyran-4-ol, D-60200	
1482-82-2	Dibenzyl diselenide, D-60082	
1483-30-3	2-(2-Phenylethenyl)benzothiazole, P-60093	
1484-12-4	9-Methyl-9*H*-carbazole, M-60056	
1485-19-4	1-Oxo-1,2,3,4-tetrahydrocarbazole; *N*-Me, *in* O-70103	
1493-13-6	Trifluoromethanesulfonic acid, T-70241	
1504-58-1	3-Phenyl-2-propyn-1-ol, P-60130	
1516-70-7	▷Methanesulfonyl azide, M-60033	
1517-05-1	▷2-Azidoethanol, A-70287	
1517-72-2	1-(1-Naphthyl)ethanol, N-70015	
1518-61-2	3,4-Dihydroxybutanoic acid, D-60313	
1519-39-7	Methyl 4-methylphenyl sulfoxide; (*R*)-*form*, *in* M-70094	
1521-60-4	2,4,6-Trimethyl-3-nitroaniline, T-70285	
1522-20-9	2,4-Pentanedione; *Enol-form, in* P-60059	
1527-07-7	5,6,7,12-Tetra-hydrodibenz[*b,e*]azocine, T-70055	
1527-08-8	5,6,7,12-Tetra-hydrodibenz[*b,e*]azocine; B,HCl, *in* T-70055	
1544-50-9	Methyl trifluoromethanesulfenate, *in* T-60279	
1545-69-3	9,10-Difluoroanthracene, D-70161	
1548-61-4	4-Nitro-2-(trifluoromethyl)phenol, N-60053	
1548-62-5	2-Nitro-6-(trifluoromethyl)phenol, N-60050	
1551-39-9	2,4,5,6-Tetrafluoro-1,3benzenedicarboxylic acid, T-70030	
1552-42-7	Crystal violet lactone, C-70202	
1552-94-9	5-Phenyl-2,4-pentadienoic acid, P-70086	
1556-34-9	9-(Bromomethyl)acridine, B-70233	
1556-62-3	2,6-Dichloroxanthone, D-70141	
1556-78-1	Dibenzo[*a,h*]quinolizinium(1+); Bromide, *in* D-70076	
1556-79-2	Dibenzo[*a,h*]quinolizinium(1+); Perchlorate, *in* D-70076	
1559-86-0	2-Bromo-1,3,4,5tetrafluorobenzene, B-60321	
1559-87-1	1,3-Dibromo-2,4,5,6tetra-fluorobenzene, D-60107	
1559-88-2	3-Bromo-1,2,4,5tetrafluorobenzene, B-60322	
1562-95-4	Di-2-pyridyl ketone; Oxime, *in* D-70502	
1572-64-1	2-Methyl-2-pentenal; (*E*)-*form*, 2,4-Dinitrophenylhydrazone, *in* M-70106	
1573-66-6	1-Hexen-5-yn-3-ol, H-70089	
1575-46-8	3,4-Dimethyl-2(5*H*)-furanone, D-60423	
1575-47-9	4-Phenyl-2(5*H*)-furanone, P-60103	
1587-07-1	1-Chloro-2-(2-propenyl)benzene, C-70152	
1590-23-4	1-Acetyl-7-iodonaphthalene, A-60037	
1590-24-5	1-Acetyl-7-bromonaphthalene, A-60024	
1590-25-6	2-Acetyl-6-bromonaphthalene, A-60025	
1591-95-3	Pentafluoroisocyanatobenzene, P-60035	
1592-38-7	2-Naphthalenemethanol, N-60004	
1595-73-9	6-Phenyl-3,5-hexadiyn-2-one, P-60110	
1607-57-4	Bromotriphenylethylene, B-70276	
1608-31-7	Benzylidenecyclohexane, B-70082	

1609-47-8	▷Diethyl dicarbonate, *in* D-70111	
1610-13-5	1-Ethynylcyclopentene, E-70052	
1610-39-5	1,2,3,4,5,6,7,8,9,10,11,12Dodeca-hydrotriphenylene, D-60516	
1620-94-6	4-Mercaptobenzophenone, M-70022	
1620-95-7	2-(Methylthio)benzophenone, *in* M-70020	
1624-26-6	Benz[*f*]indane, B-60015	
1625-81-6	9,10-Dihydro-9,10ethenoanthracene-11,12dicarboxylic acid, D-70208	
1625-82-7	9,10-Dihydro-9,10ethenoanthracene-11,12dicarboxylic acid; Di-Me ester, *in* D-70208	
1629-58-9	▷1-Penten-3-one, P-60062	
1639-73-2	5,6,7,12-Tetra-hydrodibenz[*b,g*]azocine, T-70056	
1643-69-2	3-(Trifluoromethyl)benzenesulfonic acid, T-60283	
1644-68-4	2,3-Biss(trifluoromethyl)pyridine, B-60181	
1654-67-7	2-Hydroxycyclohexanecarboxylic acid; (1*R*,2*R*)-*form, in* H-70119	
1655-00-1	2-Hydroxycyclohexanecarboxylic acid; (1*R*,2*S*)-*form, in* H-70119	
1655-01-2	2-Hydroxycyclohexanecarboxylic acid; (1*S*,2*R*)-*form, in* H-70119	
1669-44-9	3-Octen-2-one, O-70027	
1669-73-4	1,3,3-Triphenyl-1-propanone; Oxime, *in* T-60391	
1678-53-1	3-Methyl-4-nitropyridine, M-70099	
1682-20-8	4-Amino-2,3,5,6tetrafluoropyridine, A-60257	
1687-46-3	1,1-Cyclopentanedithiol, C-60222	
1687-47-4	2,2-Propanedithiol, P-60175	
1687-53-2	5-Amino-2-methoxyphenol, *in* A-70090	
1687-60-1	▷Oxalohydroxamic acid, O-60048	
1701-57-1	2,3,4,5-Tetrahydro-1*H*-1benzazepine, T-70036	
1702-77-8	Fusarubin, F-60108	
1703-85-1	Hexamethylcyclobutanone, H-60069	
1708-32-3	2,5-Dihydrothiophene, D-70270	
1710-20-9	Benz[*cd*]indol-2-(1*H*)-one; *N*-Me, *in* B-70027	
1713-85-5	3-Chloro-2-hydroxypropanoic acid, C-70072	
1716-19-4	3-Hydroxy-4-methoxy-1-methyl-7oxabicyclo[4.1.0]hept3-ene-2,5-dione, *in* D-70320	
1717-50-6	2,2,4,4-Tetrafluoro-1,3dithietane, T-60048	
1719-19-3	2-Methyl-4-phenyl-3-butyn-2-ol, M-70110	
1732-13-4	1,2,3,6,7,8-Hexahydropyrene, H-70069	
1740-74-5	4,4-Diethoxy-2-methyl-2-butene, *in* M-70055	
1745-18-2	1-Chloro-4-(2-propenyl)benzene, C-70154	
1746-25-4	▷2,4,5-Triiodo-1*H*-imidazole, T-70276	
1755-12-0	2-Fluorocyclopentanone, F-60022	
1761-62-2	2-Hydroxy-5-iodobenzaldehyde, H-70143	
1762-45-4	[2,2′-Bipyridine]-5,5′dicarboxylic acid; Di-Me ester, *in* B-60127	
1771-65-9	3-Benzoyl-2-methylpropanoic acid, B-60064	
1778-09-2	1-[4-(Methylthio)phenyl]ethanone, *in* M-60017	
1782-65-6	Hardwickiic acid; (−)-*form, in* H-70002	
1785-61-1	1,3-Diethynylbenzene, D-70156	
1785-64-4	Octafluoro[2.2]paracyclophane, O-70009	
1785-64-4	4,5,7,8,12,13,15,16-Octafluoro[2.2]paracyclophane, O-60010	
1794-90-7	▷2-Nitro-1,3-propanediol, N-60038	

1796-84-5	4-Ethoxy-2-nitropyridine, *in* N-70063	
1797-87-1	Deoxysepiapterin, *in* S-70035	
1799-37-7	4-Bromo-3-fluoropyridine; Picrate, *in* B-60255	
1799-38-8	4-Chloro-3-fluoropyridine; Picrate, *in* C-60087	
1801-42-9	2,3-Diphenyl-1*H*-inden-1-one, D-70482	
1801-62-3	2-Thioxo-4-imidazolidinone; 1,3-*N*-Di-Me, *in* T-70188	
1802-29-5	5,5′-Dicyano-2,2′-bipyridine, *in* B-60127	
1802-30-8	[2,2′-Bipyridine]-5,5′dicarboxylic acid, B-60127	
1803-60-7	3-(2-Methylpropyl)-6-methyl2,5-piperazinedione, M-70118	
1805-67-0	5-Methyl-1,3benzenedicarboxaldehyde, M-60050	
1807-47-2	9-Diazo-9*H*-cyclopenta[1,2*b*:4,3-*b′*]-dipyridine, D-60059	
1808-09-9	1,1′,2,3,4,4′,5,5′,6,6′Decahydrobi-phenyl, D-60008	
1820-81-1	▷5-Chloro-2,4-(1*H*,3*H*)pyrimidinedione, C-60130	
1823-76-3	2-(4-Hydroxyphenyl)-2-oxoacetic acid; Me ether, amide, *in* H-70212	
1830-54-2	3-Oxopentanedioic acid; Di-Me ester, *in* O-60084	
1830-57-5	Benz[*cd*]indol-2-(1*H*)-one; *N*-Ph, *in* B-70027	
1839-18-5	2-Chloro-6-aminopurine, C-70039	
1840-07-9	2,2,3,3,4,4,5,5,6,6-Decafluoropiperidine; *N*-Nitro, *in* D-60006	
1852-17-1	Tetrahydro-2(1*H*)-pyrimidinone, T-60088	
1871-57-4	3-Chloro-2-chloromethyl-1-propene, C-70061	
1877-75-4	(4-Methoxyphenoxy)acetic acid, *in* H-70205	
1878-83-7	(3-Hydroxyphenoxy)acetic acid, H-70204	
1878-84-8	(4-Hydroxyphenoxy)acetic acid, H-70205	
1878-85-9	(2-Methoxyphenoxy)acetic acid, *in* H-70203	
1883-75-6	2,5-Biss(ydroxymethyl)furan, B-60160	
1899-24-7	5-Bromo-2-furancarboxaldehyde, B-60259	
1899-48-5	2,4-Diaminoquinazoline, D-60048	
1910-85-6	1-Chloro-10*H*-phenothiazine, C-60124	
1911-78-0	Oplopanone; (−)-*form, in* O-70043	
1912-33-0	1*H*-Indole-3-acetic acid; Me ester, picrate, *in* I-70013	
1912-48-7	1*H*-Indole-3-acetic acid; *N*-Me, *in* I-70013	
1914-60-9	2,3-Dihydro-2benzofurancarboxylic acid, D-60197	
1915-20-4	3,4-Dihydro-2*H*-1-benzopyran-8-ol, D-60203	
1915-49-7	2-Hydroxy-3*H*-phenoxazin-3-one, H-60206	
1916-55-8	*N*-(3-Oxo-3*H*-phenoxazin-2-yl)acetamide, *in* A-60237	
1916-59-2	▷2-Amino-3*H*-phenoxazin-3-one, A-60237	
1916-72-9	2-Phenyl-2*H*-benzotriazole, P-60083	
1927-25-9	2-Amino-4-hydroxybutanoic acid; (±)-*form, in* A-70156	
1928-98-9	3-Oxopropanoic acid; *Oxo-form*, Me ester, 2,4-dinitrophenylhydrazone, *in* O-60090	
1931-01-7	2-(Methylamino)purine, *in* A-60243	
1931-67-5	12-Oxododecanoic acid; Me ester, Di-Me acetal, *in* O-60066	
1936-10-3	5-Methoxy-7-methyl-1,2naphthoquinone, *in* H-60189	
1940-28-9	4-Bromo-3,5-dichlorophenol, B-60230	

1940-44-9   4-Bromo-2,3-dichlorophenol, B-50229

1955-39-1   5-Phenyl-2(3*H*)-furanone, P-70079

1963-36-6   3-(4-Methoxyphenyl)-2-propenal, *in* H-60213

1969-60-4   Acenaphtho[1,2-*b*]thiophene, A-70017

1969-74-0   2-Phenylisatogen, *in* P-60114

1971-44-4   1,3-Dimethyl-1*H*-indole-2carboxaldehyde, *in* M-70090

1996-41-4   2-Chloro-4-fluorophenol, C-60073

2009-24-7   ▷Xanthotoxol, X-60002

2009-59-8   12-Oxododecanoic acid; Me ester, *in* O-60066

2010-15-3   ▷2-Thioxo-4-imidazolidinone; 3-*N*-Ph, *in* T-70188

2013-43-6   [14.0]Paracyclophane, P-70008

2016-36-6   ▷Choline salicylate, *in* H-70108

2032-75-9   2,2-Dichloro-1,3-benzodioxole, D-70113

2033-45-6   4-Iodo-3,5-dimethylpyrazole, I-70033

2034-22-2   ▷2,4,5-Tribromo-1*H*-imidazole, T-60245

2034-69-7   Daphnoretin, D-70006

2035-15-6   Maackiain; (−)-*form*, *in* M-70001

2040-70-2   Methyl dicyanoacetate, *in* D-70149

2040-89-3   2-Bromo-6-fluorophenol, B-60249

2040-90-6   2-Chloro-6-fluorophenol, C-60075

2044-26-0   2(1*H*)-Pyridineselone, P-70170

2044-27-1   ▷2(1*H*)-Pyridinethione; *N*-Me, *in* P-60224

2044-28-2   2-(2-Pyridyl)ethanethiol, P-60239

2045-76-3   3,4-Dihydro-2,2-dimethyl-2*H*pyrrole, D-60229

2045-76-3   3,4-Dihydro-2,2-dimethyl-2*H*pyrrole, D-70200

2047-91-8   1,2,3,4-Tetrahydrocyclopent[*b*]indole, T-70049

2055-52-9   1-Phenyl-1*H*-1,2,3-triazole-4carboxylic acid; Me ester, *in* P-60138

2055-53-0   1-Phenyl-1*H*-1,2,3-triazole-4carboxylic acid; Amide, *in* P-60138

2064-03-1   1-Chlorobicyclo[2.2.2]octane, C-70057

2067-66-5   Dolichol, D-60517

2085-31-6   2-Diazo-1,3-diphenyl-1,3propanedione, D-70058

2088-24-6   (3-Methoxyphenoxy)acetic acid, *in* H-70204

2089-76-1   4-Methyl-1,3-naphthalenediol, M-60094

2104-66-7   8-(Methylthio)guanosine, *in* M-70029

2104-68-9   8-Chloroguanosine, C-70068

2105-61-5   ▷1,2,4-Trifluoro-5-nitrobenzene, T-60301

2105-94-4   4-Bromo-2-fluorophenol, B-60251

2107-78-0   7-Hydroxy-3,4-dimethyl-2*H*-1benzopyran-2-one, H-60119

2122-02-3   2-Hexenal; (*E*)-*form*, 2,4-Dinitrophenylhydrazone, *in* H-60076

2127-05-1   2-(4-Pyridyl)ethanethiol, P-60240

2132-34-5   Cyclohepta[*b*]pyrrol-2(1*H*)-one, C-60201

2133-34-8   ▷2-Azetidinecarboxylic acid; (*S*)-*form*, *in* A-70284

2140-65-0   1-Methylguanosine, M-70084

2141-42-6   1,2,3,4-Tetrahydroanthracene, T-60053

2147-34-4   2*H*-Naphtho[1,2-*b*]pyran-2-one, N-70011

2157-56-4   2,4-Pentanedione; *Oxo-form*, Dioxime, *in* P-60059

2163-42-0   2-Methyl-1,3-propanediol, M-60120

2166-24-7   3,6-Pyridazinedicarboxylic acid; Di-Me ester, *in* P-70168

2169-87-1   2,3-Naphthalenedicarboxylic acid, N-60003

2174-16-5   Aspercreme, *in* H-70108

2177-49-3   1,1′-Bi-1*H*-indene, B-60108

2179-80-8   Methyl selenocyanate, M-60127

2179-96-6   Dicyanotriselenide, D-60165

2185-02-6   3-Aminodihydro-2(3*H*)-furanone; (*S*)-*form*, *in* A-60138

2196-20-5   3,5-Hexadienoic acid; (*Z*)-*form*, *in* H-70038

2211-20-3   α-Pipitzol, P-60158

2211-64-5   ▷*N*-Cyclohexylhydroxylamine, C-70228

2211-81-6   4-Fluoro-1(3*H*)-isobenzofuranone, F-70027

2211-82-7   7-Fluoro-1(3*H*)-isobenzofuranone, F-70028

2213-63-0   ▷2,3-Dichloroquinoxaline, D-60150

2217-35-8   Thurfyl salicylate, *in* H-70108

2219-62-7   3-Thioxo-1,2-propadien-1-one, T-60216

2219-66-1   ▷1,4-Dibromo-2-butyne, D-70087

2222-07-3   ▷Cucurbitacin I, *in* C-70204

2228-98-0   4-*tert*-Butylcyclohexene, B-60343

2236-60-4   2-Amino-4(1*H*)-pteridinone, A-60242

2242-71-9   1,2-Cyclohexanedithiol; (1*RS*,2*SR*)-*form*, *in* C-70225

2252-37-1   2-Bromo-6-fluorobenzoic acid, B-60242

2267-25-6   2-Chloro-4-fluoro-1methoxybenzene, *in* C-60073

2271-12-7   3,4-Di-*O*-caffeoylquinic acid, D-70110

2284-31-3   3′,5,7-Trihydroxy-4′methoxyisoflavone, *in* T-60108

2287-57-2   Bicyclo[2.2.1]heptane-1carboxylic acid; Me ester, *in* B-60085

2289-83-0   1-Oxa-2-azaspiro[2,5]octane; *N*-Benzoyl, *in* O-70052

2289-95-4   1-Oxa-2-azaspiro[2,5]octane; *N*-Carbamoyl, *in* O-70052

2294-41-9   2-Chlorobicyclo[2.2.1]hepta2,5-diene, C-70056

2295-35-4   2-Pyrrolidinethione, P-70186

2302-25-2   4(5)-Bromo-1*H*-imidazole, B-70221

2302-30-9   4,5-Dibromo-1*H*-imidazole, D-60100

2303-35-7   ▷Muscarine; (2*S*,4*R*,5*S*)-*form*, Chloride, *in* M-60153

2309-47-9   1-(2-Thienyl)ethanol, T-60207

2316-56-5   2-Bromo-4,5-dichlorophenol, B-60227

2326-89-8   Drimenin, D-60519

2327-97-1   1-Phenyl-1,2-pentadiene, P-70085

2338-10-5   4,5,6,7-Tetrachlorobenzotriazole, T-60027

2338-25-2   ▷5,6-Dichloro-2(trifluoromethyl)-1*H*benzimidazole, D-60160

2338-54-7   1-Fluoro-4-methoxy-3methylbenzene, *in* F-60059

2342-60-1   3-(Trifluoromethyl)benzenesulfonic acid; *S*-Benzythiuronium salt, *in* T-60283

2345-51-9   3-Butynoic acid, B-70309

2353-33-5   ▷2′-Deoxy-5-azacytidine; β-*form*, *in* D-70022

2357-52-0   4-Bromo-2-fluoro-1-methoxybenzene, *in* B-60251

2358-29-4   2,3,6-Trifluorobenzoic acid, T-70239

2365-85-7   3-Amino-4-fluorobenzoic acid, A-70144

2367-75-1   1-Bromo-2-fluoro-3,5dinitrobenzene, B-60246

2368-49-2   1,3,5-Tribromo-2,4,6trifluorobenzene, T-70214

2373-34-4   5-Hydroxy-4-methyl-2*H*-1benzopyran-2-one, H-60173

2380-82-7   5-Methoxy-1*H*-indol-6-ol, *in* D-70307

2380-83-8   6-Methoxy-1*H*-indol-5-ol, *in* D-70307

2380-86-1   6-Hydroxyindole, H-60158

2385-85-5   ▷Dodecachloropentacyclo[5.3.0.0²,⁶.0³,⁹.0⁵,⁸]decane, D-70530

2396-61-4   3,3′-Oxybis-1-propanol, O-60097

2411-95-2   Cyclooctatetraenecarboxylic acid, C-70239

2415-96-5   2,2,3,3Tetramethyl-cyclopropanemethanol, T-70134

2420-16-8   3-Chloro-4-hydroxybenzaldehyde, C-60089

2422-86-8   Bicyclo[3.2.0]hepta-2,6-diene, B-70097

2423-91-8   1,3,5,7,9-Decapentaene, D-70015

2423-92-9   1,3,5,7,9,11-Dodecahexaene, D-70531

2432-20-4   2-Methylpteridine, M-70120

2432-21-5   4-Methylpteridine, M-70121

2433-58-1   2-(2-Phenylethenyl)-1*H*-pyrrole; (*Z*)-*form*, *in* P-60094

2433-66-1   2-Vinyl-1*H*-pyrrole, V-60013

2433-77-4   2-(2-Phenylethenyl)-1*H*-pyrrole; (*Z*)-*form*, *N*-Me, *in* P-60094

2436-96-6   2,2′-Dinitrobiphenyl, D-60460

2439-44-3   3-Methyl-2-phenylbutanal, M-70109

2446-67-5   2-Hydroxy-3-methoxy-5-methyl1,4-benzoquinone, *in* D-70320

2449-35-6   4,4′-Sulfonylbisbenzoic acid, *in* T-70175

2450-71-7   ▷2-Propyn-1-amine, P-60184

2465-27-2   ▷CI Basic Yellow 2, *in* A-70268

2465-59-0   1*H*-Pyrazolo[3,4-*d*]pyrimidine4,6(5*H*,7*H*)-dione, P-60202

2468-55-5   5-Iodo-1-pentyne, I-70054

2478-38-8   ▷1-(4-Hydroxy-3,5dimethoxyphenyl)-ethanone, *in* T-60315

2481-26-7   *N*-Prolylserine; L-L-*form*, Z-Pro-SerOMe, *in* P-60173

2485-33-8   2-Amino-4-hydroxypentanedioic acid; (2*S*,4*R*)-*form*, *in* A-60195

2490-55-3   2,4,6,8-Tetramethylundecanoic acid; (2*R*,4*R*,6*R*,8*R*)-*form*, Me ester, *in* T-60157

2490-72-4   *N*-Alanylcysteine; L-L-*form*, *in* A-60069

2493-31-4   2,4,6-Trinitroaniline; *N*-Di-Me, *in* T-70301

2498-66-0   Benz[*a*]anthracene-7,12-dione, B-70015

2504-55-4   3′-Amino-3′-deoxyadenosine, A-70128

2508-19-2   ▷2,4,6-Trinitrobenzenesulfonic acid, T-70302

2517-04-6   2-Azetidinecarboxylic acid, A-70284

2534-80-7   Bicyclo[2.2.2]oct-2-ene1carboxylic acid, B-70113

2534-83-0   Bicyclo[3.2.1]octane-1-carboxylic acid, B-70112

2536-88-1   1,2,3-Benzotriazine-4(3*H*)-thione; Thiol-*form*, *in* B-60050

2538-87-6   3-(4-Hydroxyphenyl)-2-propenal, H-60213

2540-06-9   2-Vinyl-1*H*-pyrrole; *N*-Me, *in* V-60013

2545-00-8   3,4′,5,7-Tetrahydroxyflavanone; (2*R*,3*S*)-*form*, *in* T-60104

2546-52-3   4-Bromo-3-fluoropyridine, B-60255

2546-56-7   4-Chloro-3-fluoropyridine, C-60087

2562-40-5   Nitrocycloheptane, N-60025

2567-19-3   Xanthylium(1+); Perchlorate, *in* X-70003

2567-20-6   Thioxanthylium(1+); Perchlorate, *in* T-70186

2570-13-0   7-Methylene-1,3,5-cyclooctatriene, M-60074

2575-01-1   9-Oxo-2-decenoic acid, O-70089

2577-35-7   2,4,6(1*H*,3*H*,5*H*)-Pteridinetrione, P-70134

2577-38-0   2,4,7(1*H*,3*H*,8*H*)-Pteridinetrione, P-70135

2581-27-3   Gyplure, *in* O-60006

2586-62-1   Naphtho[2,3-*c*]furan, N-60008

2587-01-1	2,6-Dichloro-4-nitropyridine; 1-Oxide, *in* D-60145	
2594-64-1	1,2-Dihydrazinoethane; $N^\beta,N^\beta,N^{\beta'},N^{\beta'}$-Tetra-Me, *in* D-60192	
2596-04-5	2-Amino-4-methylpentanedioic acid, A-60223	
2611-01-0	Hexahydro-1(3H)-isobenzofuranone, H-70061	
2613-23-2	▷3-Chloro-4-fluorophenol, C-60077	
2614-43-9	2,4,7(1H,3H,8H)-Pteridinetrione; 1,3-Di-Me, *in* P-70135	
2614-44-0	2,4,7(1H,3H,8H)-Pteridinetrione; 1-Me, *in* P-70135	
2622-65-3	2,4,7(1H,3H,8H)-Pteridinetrione; 3-Me, *in* P-70135	
2624-50-2	Trimethidinium(2+), T-70277	
2625-21-0	2,4,7(1H,3H,8H)-Pteridinetrione; 1,3,6-Tri-Me, *in* P-70135	
2625-37-8	4-Nitroheptane, N-60027	
2625-41-4	Nitrocyclobutane, N-60024	
2628-16-2	(4-Hydroxyphenyl)ethylene; Ac, *in* H-70209	
2628-17-3	▷(4-Hydroxyphenyl)ethylene, H-70209	
2629-11-0	▷1,2-Dihydrogedunin, *in* G-70006	
2630-37-7	3-Oxocyclopentaneacetic acid; (S)-form, *in* O-70083	
2630-38-8	3-Oxocyclopentaneacetic acid; (S)-form, Me ester, *in* O-70083	
2630-41-3	Bicyclo[2.2.1]heptan-2-one; (1S,5R)-form, *in* B-70098	
2637-34-5	▷2(1H)-Pyridinethione, P-60224	
2637-76-5	3-Piperidinecarboxylic acid; (±)-form, N-Ac, *in* P-70102	
2638-15-5	2,4,5-Trihydroxyphenylacetic acid; Tri-Me ether, Me ester, *in* T-60342	
2652-16-6	5-Hydroxy-2'-methoxy-6,7methylenedioxyisoflavone, *in* T-60107	
2658-82-4	Fauronyl acetate, *in* F-70002	
2675-79-8	5-Bromo-1,2,3-trimethoxybenzene, *in* B-60201	
2683-49-0	4-Amino-2,6-pyridinedicarboxylic acid, A-70194	
2695-47-8	6-Bromo-1-hexene, B-60264	
2695-53-6	3-Methyl-2-pentanone; (S)-form, *in* M-70105	
2697-49-6	Trifluoroo(ethylsulfinyl)methane, *in* M-70145	
2698-71-7	2,3-Dihydroxy-5-methyl-1,4benzoquinone; 3-Me ether, 2-Ac, *in* D-70320	
2702-58-1	3,5-Dinitrobenzoic acid; Me ester, *in* D-60458	
2708-97-6	1,2,3,4-Tetrafluoro-5,6diiodobenzene, T-60044	
2714-38-7	3,3,4,4-Tetrafluorocyclobutene, T-60043	
2730-62-3	2-Chloro-3,3,3-trifluoro-1propene, C-60141	
2730-96-3	3,4-Dihydro-5-methyl-2H-pyrrole; B,MeClO₄, *in* D-70226	
2739-17-5	1,2,3,4-Tetrahydroquinoline; B,HCl, *in* T-70093	
2744-05-0	3a,4,7,7a-Tetrahydro-1(3H)isobenzofuranone; (3aRS,7aSR)-form, *in* T-70072	
2745-17-7	2-Oxiranylfuran, O-70078	
2753-30-2	▷Gedunin, G-70006	
2754-18-9	3,3-Oxetanedimethanol, O-60056	
2757-90-6	▷Agaritine, A-70071	
2761-76-4	2-(2-Phenylethenyl)-1H-pyrrole; (E)-form, *in* P-60094	
2761-79-7	2-(2-Phenylethenyl)-1H-pyrrole; (E)-form, N-Me, *in* P-60094	
2763-93-1	5-Aminomethyl-3-bromoisoxazole, A-70165	
2770-01-6	4-Chloro-1H-imidazo[4,5-c]pyridine, C-60104	
2778-41-8	▷1,4-Bis(1-isocyanato-1methylethyl)benzene, B-70149	
2783-57-5	Amino-1,4-benzoquinone, A-60089	
2797-51-5	▷2-Amino-3-chloro-1,4naphthoquinone, A-70122	
2798-25-6	1,5,8-Trihydroxy-3methoxyxanthone, *in* T-70123	
2808-71-1	3-Ethylcyclohexene, E-70039	
2809-83-8	1-Ethynylcycloheptene, E-70050	
2812-28-4	Threonine; (2S,3R)-form, N-Me, *in* T-60217	
2812-66-0	1,3,7,9-Tetraoxaspiro[4,5]decane, T-60163	
2812-72-8	Carbonochloridothioic acid; O-Me, *in* C-60013	
2812-73-9	▷Carbonochloridothioic acid; O-Et, *in* C-60013	
2812-74-0	Carbonochloridothioic acid; O-Propyl, *in* C-60013	
2812-75-1	Carbonochloridothioic acid; O-Butyl, *in* C-60013	
2818-06-6	1H-Pyrrole-3,4-dicarboxylic acid; Di-Me ester, *in* P-70183	
2819-48-9	1,2-Biss(ethylene)cyclohexane, B-60166	
2819-86-5	Pentamethylphenol, P-60058	
2831-86-9	2-Octadecen-1-ol; (Z)-form, *in* O-60007	
2833-24-1	4-Oxo-2-pentenoic acid; (E)-form, Me ester, *in* O-70098	
2833-26-3	4-Oxo-2-pentenoic acid; (Z)-form, Et ester, *in* O-70098	
2833-28-5	4-Oxo-2-pentenoic acid; (E)-form, *in* O-70098	
2833-34-3	1-Phenyl-3-buten-1-ol; (±)-form, Ac, *in* P-70061	
2845-83-2	4-Methoxy-3-penten-2-one, *in* P-60059	
2902-69-4	2,2,2-Trichloroacetophenone, T-70219	
2922-42-1	4,5-Dihydroxy-3-oxo-1cyclohexenecarboxylic acid; (4S,5R)-form, *in* D-70328	
2925-24-8	Trifluorooxirane, T-60305	
2936-69-8	S-(2-Aminoethyl)cysteine; (R)-form, *in* A-60152	
2937-81-7	2,5-Diamino-1,3,4-thiadiazole, D-60049	
2941-29-9	2-Cyanocyclopentanone, *in* O-70084	
2941-64-2	Carbonochloridothioic acid; S-Et, *in* C-60013	
2944-05-0	▷Carbon monosulfide, C-70020	
2948-69-8	1-Methyl-2,4-dinitro-1H-pyrrole, *in* D-60462	
2951-15-7	N-Prolylserine; L-L-form, Z-Pro-Ser-NHNH₂, *in* P-60173	
2955-50-2	2-Amino-3-methylbutanedioic acid, A-70166	
2958-72-7	Pentacyclo[5,4.0.0²,⁶.0³,¹⁰.0⁵,⁹]undecane-1,11-dione, P-60030	
2958-72-7	Pentacyclo[6.2.1.0²,⁷.0⁴,¹⁰.0⁵,⁹]undecane-3,6-dione, P-70022	
2971-31-5	2,3-Dihydro-2-oxo-1H-indole-3acetic acid, D-60259	
2977-45-9	3,4-Dihydro-2,2-dimethyl-1(2H)naphthalenone, D-60228	
2980-32-7	1,3,5,8-Tetrahydroxyxanthone, T-70123	
2983-95-1	3-Oxabicyclo[3.2.0]heptan-2-one, O-70053	
2989-10-8	3,4-Dihydroxy-5methoxybenzeneacetic acid, *in* T-60343	
2989-63-1	2-Phenyl-3H-indol-3-one, P-60114	
2991-42-6	4-(Trifluoromethyl)benzenesulfonic acid; Chloride, *in* T-70243	
3006-64-2	Phthalimidine; N-Ac, *in* P-70100	
3024-50-8	3,3-Biss(rifluoromethyl)-3Hdiazirine, B-60178	
3034-14-8	2-(2-Furanyl)-4H-1-benzopyran4-one, F-60084	
3034-54-6	2-Iodothiazole, I-60064	
3037-84-1	Bicyclo[2.2.2]oct-2-ene-1carboxylic acid; Et ester, *in* B-70113	
3040-38-8	▷Carnitine; (R)-form, O-Ac, *in* C-70024	
3042-21-5	▷2-Methyl-5-phenyl-1H-pyrrole, M-60115	
3048-64-4	▷5-Ethenylbicyclo[2.2.1]hept-2ene, E-70036	
3056-03-9	5-Quinolinethiol, Q-60004	
3059-97-0	2-Amino-2-methylbutanoic acid; (R)-form, *in* A-70167	
3061-96-9	N-Methionylalanine; L-L-form, *in* M-60035	
3075-23-8	S-Ethyl 3-oxobutanethioate, *in* O-70080	
3080-67-9	6-Amino-2,2-dichlorohexanoic acid; B,HCl, *in* A-60126	
3083-25-8	▷(2,2,2-Trichloroethyl)oxirane, T-60252	
3085-76-5	Diisopropylcyanamide, D-60392	
3088-42-4	▷N-(2-Cyanoethyl)glycine, *in* C-60016	
3099-29-4	▷2-(Chloromethyl)-6methylpyridine, C-70096	
3099-30-7	2-(Chloromethyl)-6-methylpyridine; B,HCl, *in* C-70096	
3104-90-3	Tetracyclo[4.2.0.0²,⁵.0³,⁸]octane, T-70023	
3128-05-0	3-Oxocyclopentaneacetic acid, O-70083	
3130-54-9	2,4-Dinitro-1H-pyrrole, D-60462	
3130-96-9	3,3',5-Triiodothyronine; (±)-form, *in* T-60350	
3131-52-0	5,6-Dihydroxyindole, D-70307	
3141-24-0	2,3,5-Tribromothiophene, T-60247	
3153-52-4	Benzenesulfenyl thiocyanate, B-60013	
3153-73-9	7,7'-Dimethoxy-[6,8'-bi-2Hbenzopyran]-2,2'-dione, *in* B-70093	
3153-77-3	5-Iodo-2,4-dimethoxybenzoic acid, *in* D-60332	
3153-79-5	2,4-Dihydroxy-5-iodobenzoic acid; Di-Me ether, Me ester, *in* D-60332	
3157-50-4	2-Amino-4-chlorobutanoic acid, A-60104	
3162-56-9	▷Vulgarin, V-60017	
3164-29-2	▷Tartaric acid; (2R,3R)-form, Di-NH₄ salt, *in* T-70005	
3166-17-4	9-Diazo-9H-thioxanthene; 10,10-Dioxide, *in* D-70065	
3168-90-9	1-Acetyl-2-methylcyclopentene, A-60043	
3179-31-5	▷1,2-Dihydro-3H-1,2,4-triazole3-thione, D-60298	
3189-13-7	6-Methoxyindole, *in* H-60158	
3204-61-3	1,2,4,5-Tetraaminobenzene, T-60013	
3207-47-4	2-(2-Hydroxy-4-methoxyphenyl)6-methoxy-3-methylbenzofuran, H-70156	
3207-48-5	2-(2-Hydroxy-4-methoxyphenyl)3-methyl-5,6-methylenedioxybenzofuran, H-70157	
3209-78-7	5,5-Dimethoxy-2-pentanone, *in* O-60083	
3220-72-2	1-Benzothiopyrylium(1+); Perchlorate, *in* B-70066	
3227-09-6	N-Serylmethionine; L-L-form, *in* S-60024	
3237-44-3	2-Hydroxy-2-isopropylbutanedioic acid, H-60165	
3243-24-1	4-Hydroxyimidazo[4,5-b]pyridine, H-60157	
3248-93-9	▷Rosaniline, R-60010	
3264-24-2	7H-Dibenzo[c,h]xanthen-7-one, D-60080	
3279-95-6	O-(2-Hydroxyethyl)hydroxylamine, H-60134	
3281-03-6	Bicyclo[2.2.1]heptan-2-one; (1RS,4SR)-form, 2,4-Dinitrophenylhydrazone, *in* B-70098	

3282-18-6 1,1-Diphenylcyclopropane, D-60478
3282-54-0 1,2,3,3',4,4',5,5',6,6'Decahydrobi-phenyl, D-60009
3302-06-5 3,4-Dimethylpentanoic acid, D-60435
3306-40-9 2-(2,4-Hexadiynylidene)-1,6dioxaspira[4.5]dec-3en-8-ol; (E)-form, in H-70042
3306-40-9 2-(2,4-Hexadiynylidene)-1,6dioxaspiro[4.5]dec-3-ene, in H-70042
3317-61-1 3,4-Dihydro-2,2-dimethyl-2Hpyrrole; N-Oxide, in D-60229
3317-61-1 3,4-Dihydro-2,2-dimethyl-2Hpyrrole; 1-Oxide, in D-70200
3324-79-6 1,7-Dihydro-6H-purine-6-thione; 1,7-Dihydro-form, 7-Me, in D-60270
3334-89-2 Azetidine; N-Ph, in A-70283
3343-24-6 11-Phenylundecanoic acid, P-70096
3357-53-7 2-(2,4-Cyclopentadien-1ylidene)-1,3-dithiolane, in C-60221
3367-95-1 N,N-Diethyl-3piperidinecarboxamide, in P-70102
3373-53-3 6-Amino-1,3-dihydro-2H-purin-2one, A-60144
3378-71-0 2,5-Dimethylpyrrolidine, D-70437
3378-82-3 1-Bromo-2-naphthalenecarboxaldehyde, B-60302
3395-03-7 1,2-Dimethoxy-4,5-dinitrobenzene, in D-70449
3398-16-1 4-Bromo-3,5-dimethylpyrazole, B-70208
3400-88-2 2-Chloro-2-cyclohexen-1-one, C-60050
3400-89-3 2-Chloro-2-cyclopenten-1-one, C-60051
3413-97-6 2,3-Dihydroxybutanoic acid, D-60312
3419-66-7 1-tert-Butylcyclohexene, B-60341
3419-67-8 1-tert-Butylcyclopentene, B-70298
3420-02-8 6-Methyl-1H-indole, M-60088
3420-72-2 2'-Hydroxy-4,4',6trimethoxy-chalcone, in H-60220
3426-71-9 C-Benzyloxycarbohydroxamic acid, B-70086
3435-25-4 4-Chloro-6-methylpyrimidine, C-70117
3449-55-6 1-Oxo-1,2,3,4-tetrahydrocarbazole; N-Me, oxime, in O-70103
3453-33-6 6-Methoxy-2-naphthaldehyde, in H-70185
3456-99-3 1-Oxo-1,2,3,4-tetrahydrocarbazole, O-70103
3465-55-2 5-Hydroxy-7-methoxyphthalide, in D-60339
3465-69-8 5,7-Dimethoxyphthalide, in D-60339
3465-72-3 Urocanic acid; (E)-form, in U-70005
3469-20-3 2,3-Diphenylindole, D-70483
3471-10-1 2-Phenylcyclopropanecarboxylic acid; (1R,2R)-form, in P-60087
3511-90-8 4-Bromo-2,3,5,6tetrafluoropyridine, B-60328
3518-65-8 ▷Chloromethanesulfonyl chloride, in C-70086
3526-27-0 2,2'-Biss(yanomethyl)biphenyl, in B-60115
3528-23-2 4H-Naphtho[1,2-b]pyran-4-one, N-60010
3533-47-9 Psilostachyin, P-70131
3543-01-9 5-Amino-3-hydroxypyridine, A-60203
3560-71-2 5,8-Dihydroxy-2,3,6-trimethoxy1,4-benzoquinone, in P-70033
3582-05-6 Trifluoromethyl trifluoromethanesulfonate, in T-70241
3591-47-7 2,2-Dichlorocyclopropanecarboxylic acid; (±)-form, Me ester, in D-60127

3593-00-8 5-Heptenoic acid, H-70024
3597-42-0 6-Methoxy-1-naphthaldehyde, in H-70184
3604-36-2 Bicyclo[2.2.1]hepta-2,5-diene2-carboxylic acid; (1R)-form, Me ester, in B-60084
3615-21-2 ▷4,5-Dichloro-2(trifluoromethyl)-1Hbenzimidazole, D-60157
3658-77-3 4-Hydroxy-2,5-dimethyl-3(2H)furanone, H-60120
3661-10-7 Hexahydropyridazino[1,2-a]pyridazine-1,4-dione, H-70070
3661-15-2 Octahydropyridazino[1,2-a]pyridazine, O-70016
3663-79-4 1,4-Benzodioxan-2-carboxylic acid; (±)-form, Me ester, in B-60024
3663-80-7 1,4-Benzodioxan-2-carboxylic acid, B-60024
3663-81-8 1,4-Benzodioxan-2-carboxylic acid; (±)-form, Chloride, in B-60024
3663-82-9 2-(Hydroxymethyl)-1,4-benzodioxan, H-60172
3674-69-9 4,5-Dimethylphenanthrene, D-60439
3682-01-7 4',7-Dihydroxyisoflavone; Di-Ac, in D-60340
3682-14-2 6-Amino-2,3-dihydro-1,4phthalazinedione, A-70134
3682-35-7 2,4,6-Tri-2-pyridinyl-1,3,5triazine, T-70316
3693-58-1 2-Azabicyclo[2.2.2]oct-5-ene, A-70279
3695-28-1 2-(3-Iodopropyl)-2-methyl-1,3dioxolane, in I-60057
3695-29-2 5-Iodo-2-pentanone, I-60057
3698-54-2 ▷2,3,4,6-Tetranitroaniline, T-60159
3705-27-9 Hexahydropyrrolo[1,2-a]pyrazine-1,4-dione; (S)-form, in H-70073
3718-04-5 4(5)-Vinylimidazole, V-70012
3718-56-7 6-Methyl-2-cyclohexen-1-ol, M-70058
3720-98-7 Tetrahydroimidazo[4,5-d]imidazole-2,5(1H,3H)dione; 1,6-Di-Me, in T-70066
3723-32-8 [3,3'-Bipyridine]-2,2'dicarboxylic acid, B-60133
3725-23-3 Tricyclo[3.2.1.0²,⁷]oct-3-ene, T-60267
3727-53-5 3a,4,5,6,7,7a-Hexahydro-3methyl-ene-2-(3H)benzofuranone; (3aRS,7aSR)-form, in H-70062
3731-51-9 ▷2-(Aminomethyl)pyridine, A-60227
3731-52-0 3-(Aminomethyl)pyridine, A-60228
3731-53-1 4-(Aminomethyl)pyridine, A-60229
3736-99-0 Propenylsulfenic acid, P-70123
3742-42-5 4-Ethylcyclohexene, E-60048
3747-69-1 6,7-Diaminopteridine, D-60045
3749-17-5 2-Hydroxycyclohexanecarboxylic acid; (1RS,2SR)-form, in H-70119
3757-53-7 5-Methyl-1H-pyrrole-2-carboxylic acid, M-60124
3760-95-0 ▷2-Ethyloxirane, E-70047
3769-41-3 3-(Benzyloxy)phenol, B-60073
3771-57-1 2-Methyl-1,5-diphenyl-1H-pyrrole, in M-60115
3783-61-7 3-(Carboxymethylamino)propanoic acid; Di-Et ester, in C-60016
3796-23-4 3-(Trifluoromethyl)pyridine, T-60297
3796-24-5 4-(Trifluoromethyl)pyridine, T-60298
3814-20-8 4'-Mercaptoacetophenone, M-60017
3814-30-0 (Bromomethyl)cyclopentane, B-70241
3818-97-1 Bicyclo[5.1.0]octa-3,5-dien-2-one, B-70111
3820-26-6 4-(Aminomethyl)pyridine; N-Benzoyl, in A-60229

3821-81-6 3-Amino-2-fluoropropanoic acid, A-60162
3840-17-3 1-Chloro-3-(2-propenyl)benzene, C-70153
3843-55-8 4,8-Dimethoxy-1-naphthol, in N-60005
3844-63-1 ▷2-Imidazolidinone; 1-Nitroso, in I-70005
3855-24-1 ▷1,1-Cyclohexanedithiol, C-60209
3856-25-5 α-Copaene, C-70197
3868-31-3 7,8-Dihydro-8-oxoguanosine, D-70238
3868-32-4 8-Aminoguanosine, A-70150
3868-36-8 7,8-Dihydro-8-oxoguanosine; 8-Benzyl ether, in D-70238
3875-51-2 Isopropylcyclopentane, I-70092
3877-86-9 ▷Cucurbitacin D, C-70203
3883-58-7 2,2-Dimethyl-1,3cyclopentanedione, D-70393
3913-68-6 2-Amino-4-hydroxypentanedioic acid; (2S,4S)-form, in A-60195
3913-80-2 2-Octen-1-ol; (E)-form, Ac, in O-60029
3915-98-8 1H-Naphtho[2,3-d]triazole-4,9dione, N-60011
3929-62-2 N-Valylleucine; DL-DL-form, in V-60001
3932-66-9 1,4,4-Trifluorocyclobutene, T-60276
3935-49-7 Chlorotrifluorooxirane, C-60140
3956-80-7 12-Oxododecanoic acid, O-60066
3964-57-6 3-Chloro-4-hydroxybenzoic acid; Me ester, in C-60091
3964-58-7 3-Chloro-4-hydroxybenzoic acid, C-60091
3965-53-5 4,4'-Thiobisbenzoic acid; Di-Me ester, S-dioxide, in T-70175
3975-08-4 1,1,4,4-Tetraethoxy-2-butyne, in B-60351
3977-29-5 2-Amino-6-methyl-4(1H)pyrimidinone, A-60230
3977-48-8 4,6-Diphenylpyrimidine, D-70496
3983-08-2 3-Isopropylcyclohexene, I-60119
3984-22-3 2-Vinyl-1,3-dioxolane, in P-70119
3989-97-7 N-Valylleucine; L-L-form, in V-60001
4000-78-6 2,4-Dimethoxy-6-methyl-1,3,5triazine, in M-60132
4007-81-2 4,4-Dimethyl-5-oxopentanoic acid; Me ester, in D-60434
4016-63-1 8-Bromoguanosine, B-70215
4023-65-8 Aconitic acid; (E)-form, in A-70055
4024-72-0 6-Diazo-2,4-cyclohexadien-1-one, D-70057
4042-36-8 5-Oxo-2-pyrrolidinecarboxylic acid; (R)-form, in O-70102
4049-38-1 3',4',5,7-Tetrahydroxyflavanone, T-70105
4063-33-6 10H-Phenothiazine-1-carboxylic acid; Me ester, in P-70055
4070-01-3 Triss(entachlorophenyl)methyl, in T-60405
4071-37-8 Kainic acid; Di-Me ester, in K-60003
4071-39-0 α-Allokainic acid, in K-60003
4074-25-3 1,3,5-Tri-tert-butyl-2nitrobenzene, T-70216
4077-47-8 4-Methoxy-2,5-dimethyl-3(2H)furanone, in H-60120
4079-68-9 ▷2-Propyn-1-amine; N-Di-Et, in P-60184
4082-19-3 ▷Bicyclo[4.2.0]octa-1,5-diene3,4-dione, B-70110
4096-34-8 3-Cyclohexen-1-one, C-60211
4096-48-4 O-Oxalylhomoserine, in A-70156
4099-35-8 ▷S-(2-Aminoethyl)cysteine; (R)-form, B,HCl, in A-60152
4101-32-0 4',5'-Dimethoxy-2'nitro-acetophenone, in D-60359
4110-35-4 ▷1-Cyano-3,5-dinitrobenzene, in D-60458

4115-26-8 Hector's base, H-60012
4117-09-3 7-Bromo-1-heptene, B-60260
4138-26-5 3-Piperidinecarboxamide, in P-70102
4144-02-9 Threonine; (2S,3R)-form, Me ether, in T-60217
4146-32-1 Benzo[a]quinolizinium(1+); Perchlorate, in B-70043
4147-27-7 5,6-Dimethoxy-2-methyl-1,4naphthoquinone, in D-60351
4160-49-0 Bicyclo[3.1.0]hexan-2-one, B-60088
4161-50-6 8-Decene-4,6-diynoic acid; (Z)-form, Me ester, in D-70018
4165-81-5 1,1-Diphenyl-1,3-butadiene, D-70476
4169-19-1 1,2,3,4-Tetrahydroquinoline; N-Ac, in T-70093
4175-76-2 2,4-Dichlorothiazole, D-60154
4175-77-3 2,4-Dibromothiazole, D-60109
4175-78-4 2,5-Dibromothiazole, D-60110
4181-07-1 3,3-Dimethyl-4-penten-2-one, D-70429
4182-55-2 10H-Phenothiazine-1-carboxylic acid, P-70055
4187-59-1 2-Hydroxycyclohexanemethanol; (1RS,2SR)-form, in H-70120
4187-60-4 2-Hydroxycyclohexanemethanol; (1RS,2RS)-form, in H-70120
4188-88-9 4-Oxo-2-pentenoic acid; (Z)-form, in O-70098
4198-12-3 3,3,5,5-Tetraphenyl-1,2,4trithiolane, T-70153
4208-97-3 5,6-Dichloroacenaphthene, D-60115
4210-66-6 O-Phosphohomoserine, in A-70156
4212-49-1 5-Ethyl-2,4(1H,3H)pyrimidinedione, E-60054
4219-23-2 4-Amino-2-chlorobutanoic acid, A-60105
4222-27-9 3-Chloro-3-methoxy-3H-diazirine, C-70087
4224-69-5 2-Bromomethyl-2-propenoic acid; Me ester, in B-70249
4228-88-0 ▷4,6-Dichloro-2(trifluoromethyl)-1Hbenzimidazole, D-60158
4228-89-1 ▷4,7-Dichloro-2(trifluoromethyl)-1Hbenzimidazole, D-60159
4231-62-3 1H-Benzotriazole; N-Benzoyl, in B-70068
4250-82-2 3,3-Dimethyl-1-phenyl-1-butyne, D-60441
4254-14-2 1,2-Propanediol; (R)-form, in P-70118
4254-15-3 1,2-Propanediol; (S)-form, in P-70118
4254-16-4 ▷1,2-Propanediol; (±)-form, in P-70118
4264-35-1 2-Acetylindole, A-60032
4269-93-6 6-Bromo-2,4-(1H,3H)pyrimidinedione, B-60319
4269-94-7 6-Iodo-2,4-(1H,3H)pyrimidinedione, I-70057
4270-27-3 6-Chloro-2,4-(1H,3H)pyrimidinedione, C-70158
4271-99-2 Aconitic acid; (E)-form, Tri-Me ester, in A-70055
4290-13-5 Santamarine, S-60007
4292-04-0 1-Isopropylcyclohexene, I-60118
4298-52-6 2-Ethynylthiophene, E-70055
4308-14-9 1,8-Cyclotetradecadiene, C-70252
4313-03-5 2,4-Heptadienal, H-60020
4318-52-9 1H-Pyrazolo[3,4-d]-pyrimidine4,6(5H,7H)-dione; 1H-form, 1,5,7-Tri-Me, in P-60202
4324-53-2 3,5-Dihydroxy-4',6,7trimethoxy-flavone, in P-60047
4324-55-4 3,4',5,6,7-Pentahydroxyflavone, P-60047
4362-13-4 1,2-Dioxolane, D-70467
4362-23-6 2-Methylene-1,3-dioxolane, M-60075

4362-24-7 4-Methylene-1,3-dioxolane, M-60076
4363-44-4 Ethylenetetracarboxylic acid, E-60049
4369-55-5 5-Amino-3-phenylisoxazole, A-60240
4370-22-3 2-(Chloromethyl)-3-methylpyridine; B,HCl, in C-70093
4371-55-5 2-Amino-3-mercapto-3phenyl-propanoic acid, A-60213
4371-81-7 Benzophenone nitrimine, in D-60482
4377-43-9 2-(Chloromethyl)-3-methylpyridine, C-70093
4378-19-2 2-Amino-2,3-dimethylbutanoic acid, A-60148
4383-15-7 1-Phenylethenol, P-60092
4385-35-7 1,4-Dihydro-2(3H)-benzopyran-3one, D-70176
4389-62-2 5-Isopropyl-2-methyl-1,3benzenediol, I-60125
4389-63-3 2-Isopropyl-5-methyl-1,3benzenediol, I-60122
4394-11-0 3,4'-Bipyridine, B-60123
4410-77-9 Benzylidenecyclopentane, B-70083
4410-99-5 2-Phenylethanethiol, P-70069
4418-61-5 ▷5-Amino-1H-tetrazole, A-70202
4418-72-8 Adonirubin, A-70066
4419-92-5 Aciphen, in H-70108
4430-29-9 9,18-Dihydrobenzo[rst]phenanthro-[10,1,2-cde]pentaphene, in B-60038
4430-42-6 1-Isothiocyanato-5-(methyl-thio)pentane, I-60140
4431-00-9 Aurintricarboxylic acid, A-60318
4431-01-0 Ligustilide, L-60026
4433-01-6 [2,2'-Bipyridine]-3,3'dicarboxylic acid, B-60124
4440-93-1 1,1,2,2-Tetrabenzoylethane, T-60020
4442-54-0 1,4-Benzodioxan-6-carboxylic acid, B-60025
4442-94-8 2-Cyclohexyl-2-hydroxyacetic acid, C-60212
4444-26-2 ▷Benzenehexamine, B-60012
4444-36-4 [2,2'-Bipyridine]-5,5'dicarboxylic acid; Diamide, in B-60127
4445-36-7 6H-Dibenzo[b,d]thiopyran-6-one, D-70080
4453-82-1 Dicyclohexylmethanol, D-70150
4453-90-1 1,4-Dihydro-1,4methanonaphthalene, D-60256
4455-27-0 1,3-Diazetidine-2,4-dione, D-60057
4456-77-3 1,3(2H,4H)-Isoquinolinedione, I-70099
4460-46-2 ▷3-Chloro-3-phenyl-3H-diazirine, C-70140
4463-16-5 2,4,5-Trimethoxybenzeneacetic acid, in T-60342
4474-03-7 N-Ethyl-N'-methyl-N,N'diphenyl-urea, in D-70499
4475-95-0 2-Amino-2-methylbutanoic acid; (±)-form, Nitrile, in A-70167
4479-74-7 [2,2'-Bipyridine]-6,6'dicarboxylic acid, B-60128
4486-05-9 Bindschedler's green; Chloride, in B-60110
4486-29-7 1,4-Methano-1,2,3,4tetra-hydronaphthalene, M-60034
4487-56-3 2,4-Dichloro-5-nitropyridine, D-60142
4488-18-0 Homopantetheine, H-60090
4493-23-6 Dodecahedrane, D-60514
4506-66-5 1,2,4,5-Tetraaminobenzene; B,4HCl, in T-60013
4506-71-2 2-Amino-4,5,6,7tetra-hydrobenzo[b]thiophene-3carboxylic acid; Et ester, in A-60258
4513-01-3 10,11-Dihydro-5-methyl-5Hdibenz[b,f]azepine, in D-70185
4513-71-7 Diphenyl imidocarbonate, in C-70019
4514-69-6 1,3,5-Cyclooctatrien-7-yne, C-70240

4525-33-1 Dimethyl dicarbonate, in D-70111
4530-44-3 3-(Chloromethyl)phenol; Ac, in C-70102
4541-35-9 2,2,5,5-Tetramethylcyclopentanone, T-60138
4546-54-7 9-β-D-Ribofuranosyl-9H-purin-2amine, in A-60243
4547-24-4 2,3-Dihydroxy-12-ursen-28-oic acid; (2α,3β)-form, in D-70349
4558-27-4 5-Bromo-1-hexene, B-60263
4567-34-4 9-Methylnaphtho[2,1-b]thiophene, M-60103
4567-36-6 8-Methylnaphtho[2,1-b]thiophene, M-60102
4567-37-7 6-Methylnaphtho[2,1-b]thiophene, M-60100
4567-39-9 7-Methylnaphtho[2,1-b]thiophene, M-60101
4570-41-6 ▷2-Aminobenzoxazole, A-60093
4572-17-2 Hexacyclo[4.4.0.0^{2,5}.0^{3,9}.0^{4,8}.0^{7,10}]-decane, H-70032
4576-48-1 Bicyclo[2.2.1]heptan-2-one; (1RS,4SR)-form, Oxime, in B-70098
4578-66-9 2-Hydroxybenzoic acid; Benzoyl, in H-70108
4580-70-5 7,12-Dihydropleiadene, D-60269
4600-04-8 1-Phenyl-1H-1,2,3-triazole-4carboxylic acid, P-60138
4611-05-6 ▷Ophiobolin A, O-70040
4628-39-1 1,2,3,6-Tetrahydro-2,6-dioxo-4pyrimidineacetic acid, T-70060
4628-94-8 5-Mercapto-1,3,4-thiadiazoline2-thione; Di-K salt, in M-70031
4636-16-2 2-Iodoacetophenone, I-60039
4646-69-9 1H-Cyclopropabenzene, C-70246
4651-91-6 2-Amino-4,5,6,7tetra-hydrobenzo[b]thiophene-3carboxylic acid; Nitrile, in A-60258
4663-36-9 ▷1,1,3-Triphenyl-2-propen-1-ol, T-70313
4670-29-5 Alloperoxylin, A-60075
4670-56-8 2-Hydroxy-4-methylbenzoic acid; Me ester, in H-70163
4676-99-7 3-Nitroso-2-phenylindole, in P-60114
4680-51-7 1H-Pyrazolo[3,4-d]-pyrimidine4,6(5H,7H)-dione; 1H-form, 5,7-Di-Me, in P-60202
4683-51-6 4,4-Dimethyl-1,3cyclopentanedione, D-70394
4687-24-5 3-Benzofurancarboxylic acid; Me ester, in B-60032
4695-09-4 2-Mercapto-2-phenylacetic acid, M-60024
4700-82-7 Secothujene; (1S,2S)-form, in S-70027
4702-25-4 1(2H)-Isoquinolinethione, I-60132
4707-77-1 4,5-Dihydroxy-1,3benzenedicarboxylic acid, D-60307
4720-60-9 2,2-Biss(ercaptomethyl)-1,3propanedithiol, B-60164
4720-68-7 2-Hydroxy-4,5methyl-enedioxybenzaldehyde, H-70175
4721-07-7 1-(2,4-Dihydroxyphenyl)-2-(3,5dihydroxyphenyl)ethylene, D-70335
4736-13-4 3-Cyclohexene-1-carboxaldehyde; (±)-form, Oxime, in C-70227
4741-68-8 1(3H)-Isobenzofuranthione, I-70063
4743-82-2 ▷4-Oxo-2-pentenoic acid, O-70098
4745-93-1 5H-Pyrrolo[2,3-b]pyrazine, P-60248
4752-10-7 1,4-Dichlorophthalazine, D-60148
4759-45-9 Homoalethine, H-60084
4783-35-1 Cryptotanshinone, C-60173
4790-08-3 5,6-Dihydroxy-1H-indole-2carboxylic acid, D-70308
4798-45-2 4-Methyl-1-penten-3-ol, M-70107
4798-58-7 4-Hexen-3-ol, H-60077

4815-28-5 2-Amino-4,5,6,7tetra-
hydrobenzo[*b*]thiophene-3-
carboxylic acid; Amide, *in*
A-60258

4817-93-0 *N*-Valylleucine; L-L-*form*, Z-Val-
Leu-OMe, *in* V-60001

4821-00-5 5,6-Dihydroxyindole; *N*-Me, *in*
H-70307

4823-41-0 ▷3-Fluoro-3-methoxy-3*H*-diazirine,
F-60050

4834-35-9 3-Hydroxybenz[*a*]anthracene,
H-70104

4842-45-9 1,2,3-Triphenyl-1-propanone,
T-60390

4843-44-1 3-(2-Thienyl)-2-propynoic acid,
T-70173

4881-00-9 2-Hydroxy-2-phenylacetaldehyde;
(±)-*form*, Me ether, 2,4-dinitro-
phenylhydrazone, *in* H-60210

4885-93-2 (Dimercaptomethylene)malononitrile, *in* D-60395

4890-01-1 6-Hydroxy-3,4methyl-
enedioxybenzoic acid, H-60181

4894-26-2 3,4-Dihydro-β-carboline, D-60209

4903-09-7 3-Chloro-4-methoxybenzaldehyde, *in*
C-60089

4915-97-3 1-Phenyl-1*H*-1,2,3-triazole-
4carboxylic acid; Et ester, *in*
P-60218

4916-55-6 3-(Bromomethyl)pyridine; B,HBr, *in*
B-70254

4919-48-6 4,4′-Thiobisbenzoic acid, T-70175

4919-96-4 1-Phenyl-2-naphthol, P-60115

4923-55-1 2,5-Dihydroxy-1,4-naphthoquinone,
D-60354

4923-58-4 2,8-Dihydroxy-1,4-naphthoquinone,
D-60355

4923-66-4 1,2,3,4-Tetrahydroanthraquinone,
T-60054

4923-87-9 5-Bromobenzo[*b*]thiophene, B-60207

4945-29-3 4-Vinylquinoline, V-70014

4955-03-7 Mokkolactone, M-60142

4956-02-9 4-Amino-1,2-benzenediol; *O*²-Me;
B,HCl, *in* A-70090

4956-56-3 4-Amino-1,2-benzenediol; B,HCl, *in*
A-70090

4960-31-0 Triss(ethylphenylamino)methane,
T-70324

4966-13-6 Lambertianic acid; (+)-*form*, *in*
L-60011

4970-26-7 Anomalin, *in* K-70013

4970-26-7 (−)-Anomalin, *in* K-70013

4994-86-9 4-Chloro-2-methylpyrimidine,
C-70115

4997-39-1 10*H*-Phenothiazine-3-carboxylic
acid; *N*-Me, 5,5-dioxide, *in*
P-70057

4998-07-6 4,5-Dimethoxy-2-nitrobenzoic acid,
in D-60360

5014-22-2 Barbatic acid; Me ester, *in* B-70007

5020-69-9 2-Acetyl-6-bromonaphthalene; 2,4-
Dinitrophenylhydrazone, *in*
A-60025

5041-99-6 1,3,8-Trihydroxy-4,7dimethoxy-
xanthone, *in* P-60052

5042-03-5 1,5,6-Trihydroxyxanthone, T-60349

5042-07-9 1-Hydroxy-5,6-dimethoxyxanthone,
in T-60349

5053-24-7 2-(Methylthio)thiazole, *in* T-60200

5056-07-5 Methyl 4-methylphenyl sulfoxide;
(*S*)-*form*, *in* M-70094

5057-96-5 Tartaric acid; (2*RS*,3*SR*)-*form*, Di-
Me ester, *in* T-70005

5058-15-1 Pabulenone, *in* P-60170

5062-80-6 Bicyclo[2.1.1]hexan-2-ol, B-70100

5076-10-8 2,5-Biss(ydroxymethyl)furan; Di-Ac,
in B-60160

5117-16-8 Selenourea; *N*¹,*N*¹-Di-Me, *in*
S-70031

5118-13-8 4-Bromobenzo[*b*]thiophene, B-60206

5122-07-6 Ethynylpentafluorobenzene, E-70053

5130-17-6 4-Amino-2-fluorobutanoic acid,
A-60157

5132-80-9 9-Acridinecarboxylic acid; Chloride,
in A-70061

5132-81-0 ▷9-Acridinecarboxylic acid; Me ester,
in A-70061

5145-71-1 1-(4-Methoxyphenyl)pyrrole, *in*
H-60216

5147-80-8 [Biss(ethylthio)methyl-
ene]propanedinitrile, *in* D-60395

5153-39-9 Chloropentamethylbenzene, C-60122

5153-40-2 Bromopentamethylbenzene, B-60312

5154-00-7 6-Amino-2(1*H*)-pyridinone; *NH*-
form, *in* A-60252

5154-01-8 5-Hydroxy-2(1*H*)-pyridinone,
H-60225

5156-00-3 2-Methoxy-1,4-benzenedicarboxylic
acid, *in* H-70105

5157-11-9 1*H*-Perimidin-2(3*H*)-one, P-70043

5162-01-6 β,β-Dimethylacrylshikonin, *in*
S-70038

5162-01-6 β,β-Dimethylacrylalkannin, *in*
S-70038

5167-17-9 2-(Diethylamino)purine, *in* A-60243

5204-88-6 13-Hydroxy-9,11-octadecadienoic
acid, H-70196

5212-50-0 Bicyclo[2.2.1]hepta-2,5-diene2-
carboxaldehyde, B-60083

5216-17-1 2,3,5,6-Tetrafluorobenzonitrile, *in*
T-70034

5220-49-5 3-Amino-2-cyclohexen-1-one,
A-70125

5244-34-8 *S*,*S*′-Bis(2-hydroxyethyl)-
1,2ethanedithiol, B-70145

5244-75-7 Benzylidenecyclobutane, B-70081

5257-24-9 3-Methyl-1*H*-indole-
2carboxaldehyde, M-70090

5264-35-7 3,4-Dihydro-5-methoxy-2*H*-pyrrole,
in P-70188

5266-64-8 1,4,5,6-Tetrahydrocyclo-
prop[*e*]indene, T-70053

5289-74-7 ▷2,3,14,20,22,25-Hexahydroxy-
7cholesten-6-one, H-60063

5291-91-8 2,3-Dihydro-6(1*H*)-azulenone,
D-60193

5292-42-2 2,3,4,5-Tetrafluorobenzoic acid; Me
ester, *in* T-70032

5302-11-4 1,2,4-Oxadiazolidine-3,5-dione;
Dioxo-*form*, 2,4-Di-Me, *in*
O-70060

5302-20-5 1,2,4-Oxadiazolidine-3,5-dione;
Dioxo-*form*, 4-Ph, *in* O-70060

5314-33-0 Propenal; Oxime, *in* P-70119

5319-67-5 Diphenylmethaneimine; B,HCl, *in*
D-60482

5330-97-2 Phenylacetohydroxamic acid,
P-70059

5333-13-1 3-Bromo-2,4,6-trimethylbenzoic
acid, B-60331

5334-41-8 5-Amino-4-cyano-1-methylpyrazole,
in A-70192

5334-42-9 *N*-(4-Cyano-1-methyl-1*H*-pyrazol5-
yl)acetamide, *in* A-70192

5334-43-0 5-Amino-1*H*-pyrazole-4-carboxylic
acid; *N*(1)-Ph, nitrile, *in* A-70192

5336-90-3 9-Acridinecarboxylic acid, A-70061

5337-95-1 2-Methyl-1,3-propanedithiol,
M-60121

5342-91-6 Phthalimidine; *N*-Me, *in* P-70100

5344-23-0 Diethoxyacetaldehyde, D-70153

5350-05-0 2-Bromobenzo[*b*]thiophene; 1,1-
Dioxide, *in* B-60204

5350-71-0 1,1,1′,1′Tetraphenyldimethylamine,
T-70150

5353-15-1 1,2,4-Trimethoxy-5-(2-
propenyl)benzene, T-70279

5365-14-0 2,2-Dichlorocyclopropanecarboxylic
acid, D-60127

5365-17-3 2,2-Dibromocyclopropanecarboxylic
acid, D-60093

5370-19-4 5-Iodo-2-thiophenecarboxaldehyde,
I-60072

5373-21-7 2,3-Biss(ydroxymethyl)-
1,4butanediol, B-60159

5381-83-9 2-Hydroxy-1,3-diphenyl-1propanone,
H-70124

5385-37-5 1,2,3,3*a*,4,5-Hexahydropyrene,
H-70068

5388-42-1 ▷*N*-Phenylphthalamidine, *in* P-70100

5391-40-2 2-Imidazolidinone; 1,3-Di-Ac, *in*
I-70005

5394-13-8 2-Bromobenzo[*b*]thiophene, B-60204

5397-03-5 Dithiocarbazic acid; Me ester, *in*
D-60509

5397-67-1 Tetrahydro-1*H*,5*H*-pyrazolo[1,2*a*]-
pyrazole, T-70086

5397-78-4 ▷2-Amino-3-chloro-
1,4naphthoquinone; *N*-Ac, *in*
A-70122

5399-87-1 6-Chloro-9-β-D-ribofuranosyl9*H*-
purine, C-70160

5416-79-5 1,2-Di(2-furanyl)ethylene, D-60191

5422-44-6 5-Amino-1-methyl-1*H*-tetrazole, *in*
A-70202

5422-45-7 5-Dimethylamino-1*H*-tetrazole, *in*
A-70202

5426-04-0 Acetamidoxime; B,HCl, *in* A-60016

5447-86-9 10,10-Dimethyl-
9(10*H*)anthracenone, D-60403

5447-96-1 1-Bromo-1-nitropropane, B-70258

5447-97-2 2-Bromo-2-nitropropane, B-70259

5448-47-5 ▷1*H*-Indole-3-acetic acid; Hydrazide,
in I-70013

5464-89-1 1,3,4,10-Tetrahydro-
9(2*H*)acridinone; *N*-Me, *in*
T-60052

5466-11-5 2-Chloro-9-β-D-ribofuranosyl9*H*-
purine, C-70159

5469-16-9 Dihydro-4-hydroxy-2(3*H*)-furanone,
D-60242

5470-95-1 2,3-Dihydroxybenzaldehyde; Di-Me
ether, oxime, *in* D-70285

5472-67-3 2-Amino-3-(3,4-
dichlorophenyl)propanoic acid;
(±)-*form*, *in* A-60131

5472-68-4 2-Amino-3-(2,4-
dichlorophenyl)propanoic acid;
(±)-*form*, *in* A-60128

5492-30-8 Phytol; (2*Z*,7*R*,11*R*)-*form*, *in*
P-60152

5522-43-0 ▷1-Nitropyrene, N-60040

5533-46-0 Selenourea; *N*¹,*N*³-Di-Me, *in*
S-70031

5533-49-3 Selenourea; *N*-Me, *in* S-70031

5535-87-5 2-(2,4-Hexadiynylidene)-
1,6dioxaspira[4.5]dec-3-en-8-ol;
(*Z*)-*form*, *in* H-70042

5535-87-5 2-(2,4-Hexadiynylidene)-
1,6dioxaspira[4.5]dec-3-en-8-ol;
(*Z*)-*form*, Deoxy, *in* H-70042

5549-09-7 1,2,3,3*a*,4,6*a*-Hexahydropentalene,
H-70065

5554-48-3 Tetrahydro-2*H*-1,3-thiazine-2thione,
T-60094

5554-52-9 Tetrahydro-2*H*-1,3-thiazine-2thione;
Thione-form, *N*-Me, *in* T-60094

5557-27-7 8-Hydroxy-1,3,5trimethoxyxanthone,
in T-70123

5558-66-7 2,2-Diphenylpropanoic acid,
D-70493

5578-80-3 10-Oxodecanoic acid, O-60063

5580-83-6 1-Bromo-2,3,4,5-tetrafluoro-6nitro-
benzene, B-60323

5581-97-5 4-*tert*-Butylcyclopentene, B-70300

5581-98-6 3-*tert*-Butylbicyclo[3.1.0]hexane;
(1α,3α,5β)-*form*, *in* B-70294

5588-87-4 1*H*-Benz[*e*]indole-1,2(3*H*)-dione,
B-70026

5590-96-5 3-*tert*-Butylbicyclo[3.1.0]hexane;
(1α,3α,5α)-*form*, *in* B-70294

5599-50-8 3,6-Dimethylcarbazole, D-70387

5607-35-2 Spiro[5.5]undecane-3,9-dione,
S-70068

5614-78-8 3,4-Dihydro-2*H*-1-benzopyran-6-ol,
D-70175

5617-74-3   3-Oxabicyclo[3.1.0]hexane-2,4dione, *in* C-70248

5621-15-8   1,6-Dimethyl-1*H*-indole, *in* M-60088

5623-26-7   2-Hydroxy-1,2-diphenyl-1propanone, H-70123

5625-67-2   Piperazinone, P-70101

5626-52-8   5-Oxo-2-pyrrolidinecarboxylic acid; (±)-*form*, Amide, *in* O-70102

5632-95-1   2-Amino-5-(methylthio)pentanoic acid, A-60231

5636-66-8   Tetrahydro-6-methylene-2*H*pyran-2-one, T-70074

5636-70-4   4,4-Dimethyl-6-nitro-5hexenenitrile, *in* D-60429

5649-86-5   1,3,4-Oxadiazolidine-2,5-dione, O-70061

5650-50-0   5,6-Dihydro-4*H*-cyclopenta[*b*]thiophene, D-70180

5650-52-2   4,5-Dihydro-6*H*-cyclopenta[*b*]thiophen-6-one, D-70182

5660-45-7   2,2′:5′,2″:5″,2‴:5‴,2⁗-Quinquethiophene, Q-70010

5661-30-3   Acetamidoxime; *N*-Ph, *in* A-60016

5664-20-0   1,4-Biss(ethylene)cyclohexane, B-60168

5667-03-8   ▷2-Phenyl-1,5-benzothiazepin4(5*H*)-one, P-60081

5675-64-9   9,10-Dihydro-9,10ethanoanthracene, D-70207

5678-33-1   2*H*-[1,2,4]Oxadiazolo[2,3-*a*]pyridin-2-one, O-70064

5680-41-1   1-(1-Propenyl)cyclohexene; (*Z*)-*form*, *in* P-60178

5684-15-1   5-Formyl-4-phenanthrenecarboxylic acid, F-60074

5685-05-2   2(3*H*)-Thiazolethione, T-60200

5685-38-1   2-Phenylcyclopropanecarboxylic acid, P-60087

5689-12-3   1,1,3,3-Tetramethyl-2-indanone, T-60143

5690-41-5   4-(Diphenylmethylene)-1(4*H*)naphthalenone, D-70487

5691-09-8   2-(Aminomethyl)cyclohexanol; (1*RS*,2*SR*)-*form*, *in* A-70169

5691-13-4   7-Azabicyclo[4.2.0]octane; (1*RS*,6*RS*)-*form*, Picrate, *in* A-70277

5691-38-3   7-Azabicyclo[4.2.0]octane; (1*RS*,6*SR*)-*form*, Picrate, *in* A-70277

5694-68-8   1,3-Dioxolane-2-methanol, D-70468

5694-91-7   1,3-Dioxolane-2,2-diacetic acid, *in* O-60084

5695-13-6   Indeno[1,2-*b*]fluorene-6,12-dione, I-70011

5697-88-1   5,6,11,12-Tetrahydrodibenz[*b*,*f*]azocine, T-70057

5698-85-1   1,3,4,5-Tetrahydro-2-benzoxepin, T-70044

5698-99-7   5,6-Acenaphthenedicarboxylic acid, A-70015

5699-78-5   4,4-Dimethylpentanoic acid; Chloride, *in* D-70428

5700-60-7   1,2-Diphenyl-1,2-ethanediamine, D-70480

5703-21-9   3,4-Dimethoxybenzeneacetaldehyde, *in* D-60371

5703-24-2   4-Hydroxy-3methoxyphenylacetaldehyde, *in* D-60371

5703-26-4   4-Methoxyphenylacetaldehyde, *in* H-60209

5707-55-1   3,4-Dihydroxyphenylacetaldehyde, D-60371

5714-08-9   3,3′,5-Triiodothyronine; (*R*)-*form*, *in* T-60350

5721-43-7   Hexahydro-1*H*-pyrazolo[1,2-*a*]pyridazine, H-70067

5723-54-6   3,4-Phenanthrenedicarboxylic acid; Anhydride, *in* P-60068

5732-37-6   3,3′,4,4′-Tetrahydro-2,2′spirobi[2*H*-1-benzopyran], T-70094

5736-03-8   2,2-Dimethyl-1,3-dioxolane-4carboxaldehyde, D-70400

5737-94-0   Dibenzo[*a*,*j*]perylene-8,16-dione, D-70073

5739-85-5   ▷Pyrenophorin, P-70167

5744-70-7   5-Bromo-1,3-dimethylpyrazole, *in* B-70252

5744-80-9   3-Bromo-1,5-dimethylpyrazole, *in* B-70252

5745-52-8   Octahydrobenzo[*b*]thiophene, O-60012

5747-07-9   3,5-Hexadien-1-ol, H-70039

5754-89-2   ▷Hexahydropyrimidine; *N*,*N*′-Dinitro, *in* H-60056

5759-60-4   1,7-Dihydro-6*H*-purine-6-thione; 3,7-*Dihydro-form*, 3,7-Di-Me, *in* D-60270

5759-62-6   1,7-Dihydro-6*H*-purine-6-thione; 1,9-*Dihydro-form*, 1,9-Di-Me, *in* D-60270

5767-20-4   ▷6*H*,13*H*Octahydrodipyridazino[1,2-*a*:1′,2′-*d*][1,2,4,5]tetrazine, O-70013

5770-70-7   1,4-Diazabicyclo[2.2.2]octane; Br₂ complex, *in* D-70051

5775-82-6   4-Bromo-1,3-dimethylpyrazole, *in* B-70251

5775-86-0   4-Bromo-1,5-dimethylpyrazole, *in* B-70251

5780-00-7   2-Methoxy-4,5methylenedioxybenzaldehyde, *in* H-70175

5784-45-2   1-Chlorophthalazine, C-70148

5789-30-0   1,2-Dibromo-1,2-diphenylethane, D-70091

5794-98-9   Tetrakis(2-bromoethyl)methane, T-70124

5799-71-3   1,3-Dichlorooctane, D-70128

5809-59-6   ▷2-Hydroxy-3-butenenitrile, *in* P-70119

5814-98-2   Isatogen, *in* I-60035

5820-02-0   2,2-Dimethyl-4,4-diphenyl-3butenal, D-70401

5827-80-5   3-Ethyl-3-pentanethiol, E-60053

5834-10-6   2-Cyclopentylidenecyclopentanone; Oxime, *in* C-60227

5834-16-2   3-Methyl-2thiophenecarboxaldehyde, M-70140

5852-98-2   Carnosine; (*S*)-*form*, B,HNO₃, *in* C-60020

5852-99-3   Carnosine; (*S*)-*form*, B,HCl, *in* C-60020

5853-00-9   Carnosine; (*R*)-*form*, *in* C-60020

5855-57-2   4-Phenyl-2(1*H*)-quinolinone, P-60131

5856-47-3   2-Amino-3-chlorobutanoic acid; (2*RS*,3*SR*)-*form*, *in* A-60103

5858-17-3   3,4-Dichlorobenzenethiol, D-60123

5858-18-4   2,5-Dichlorobenzenethiol, D-60121

5860-38-8   Tetrabenzoylethylene, T-60021

5860-48-0   1,2-Dihydro-3*H*-pyrrol-3-one, D-70262

5873-00-7   3-Amino-2(1*H*)-quinolinone, A-60256

5876-76-6   4-Phenyl-3-butyn-2-ol, P-70063

5892-39-7   5,7-Dihydroxy-4′-methoxyflavone; Di-Ac, *in* D-70319

5897-94-9   3-Chloro-2(1*H*)-pyridinethione, C-60129

5910-52-1   Aspartame; L-L-*form*, B,HCl, *in* A-60311

5919-29-9   2-Amino-4,5,6,7tetrahydrobenzo[*b*]thiophene-3-carboxylic acid; Et ester, *N*-Ac, *in* A-60258

5930-94-9   3-Nitropyrrole, N-60044

5932-20-7   3(5)-Bromo-4-methylpyrazole, B-70250

5936-58-3   2-Amino-4,5,6,7tetrahydrobenzo[*b*]thiophene-3-carboxylic acid, A-60258

5943-53-3   Selenourea; *N*-Tetra-Me, *in* S-70031

5962-00-5   2,3-Dihydro-2-phenyl-4*H*-1benzothiopyran-4-one, D-70248

5975-12-2   2,4-Dichloro-3-nitropyridine, D-60141

5980-96-1   2-Isopropyl-1,3,5trimethylbenzene, I-60130

5989-27-5   ▷4-Isopropenyl-1methylcyclohexene; (*R*)-*form*, *in* I-70087

5989-54-8   ▷4-Isopropenyl-1methylcyclohexene; (*S*)-*form*, *in* I-70087

5998-92-5   4-(2-Oxazolyl)pyridine, O-70075

6007-50-7   Furlone yellow, F-70053

6007-71-2   2-Methyl-3-sulfolene, *in* D-70230

6008-51-1   1,2-Dithiepane, D-70517

6027-21-0   2-Amino-4-hydroxybutanoic acid; (*R*)-*form*, *in* A-70156

6030-36-0   4-Methyl-2thiophenecarboxaldehyde, M-70141

6032-87-7   Muscarine; (2*RS*,4*SR*,5*RS*)-*form*, Tetrachloroaurate, *in* M-60153

6038-12-6   1-Cyclooctenecarboxaldehyde, C-60217

6042-73-5   11β-Acetoxygedunin, *in* G-70006

6044-68-4   ▷3,3-Dimethoxypropene, *in* P-70119

6051-03-2   Hexahydro-2(3*H*)-benzofuranone, H-70049

6052-72-8   5,6,9,10-Tetrahydro-4*H*,8*H*pyrido[3,2,1-*ij*][1,6]naphthyridine, T-60086

6053-02-7   4-Vinyl-1,3-benzenediol, V-70010

6068-98-0   1,2-Dihydrazinoethane, D-60192

6093-68-1   6-Hydroxy-2*H*-1-benzopyran-2-one, H-70109

6094-76-4   2-Amino-5-(methylthio)pentanoic acid; (±)-*form*, *in* A-60231

6099-88-3   ▷1-Chloro-2-isothiocyanoethane, C-70081

6124-02-3   Selenourea; *N*-Ph, *in* S-70031

6138-47-2   Thybon, *in* T-60350

6140-61-0   4,4-Dimethylglutaraldehydonitrile, *in* D-60434

6141-98-6   Benzo[*c*]cinnoline; *N*-Oxide, *in* B-60020

6154-04-7   5-Amino-2-methyl-2*H*-tetrazole, *in* A-70202

6169-78-4   2,3,4,5-Tetrahydro-1-benzoxepin, T-70045

6172-65-2   2,3,5,6-Tetramethoxybenzoic acid, *in* T-60100

6174-95-4   Ethylenetetracarboxylic acid; Tetra-Et ester, *in* E-60049

6189-88-4   3-*tert*-Butylcyclopentene, B-70299

6196-36-7   5,6,7,8-Tetrahydrodibenz[*c*,*e*]azocine; B,HCl, *in* T-70054

6196-54-9   5,6,7,8-Tetrahydrodibenz[*c*,*e*]azocine, T-70054

6214-65-9   5-Acetyl-2,4(1*H*,3*H*)pyrimidinedione, A-60047

6217-61-4   ▷5-Methyl-2,4-(1*H*,3*H*)pyrimidinedithione, M-70127

6231-16-9   6-Hydroxy-2(1*H*)-pyridinone; Me ether, *N*-Me, *in* H-60226

6231-17-0   6-Hydroxy-2(1*H*)-pyridinone; *N*-Me, *in* H-60226

6231-18-1   2,6-Dimethoxypyridine, *in* H-60226

6231-26-1   1,3,5-Trimethyl-2,4,6triphenyl-benzene, T-70293

6245-57-4   4-Methoxy-2-methylbenzoic acid, *in* H-70164

6249-29-2   3-(1*H*-Pyrrol-2-yl)-2-propenal, P-70202

6249-80-5   1-Phenyl-3-buten-1-one, P-60085

6252-18-2   1-Isopropenylcyclohexene, I-60115

6253-27-6   7-Oxabicyclo[4.1.0]hept-3-ene, O-60047

6262-51-7   Pentachlorocyclopropane, P-70018

6264-40-0   5-Methylthio-1,3,4-thiadiazole2(3*H*)-thione, *in* M-70031

6272-38-4   2-(Benzyloxy)phenol, B-60072

6288-86-4   1*H*-Pyrazolo[3,4-*d*]pyrimidine; 1-Me, *in* P-70156

6294-17-3 1-Bromo-6-chlorohexane, B-70200
6295-29-0 6,7-Dimethoxyquinoxaline, *in* Q-60009
6296-95-3 1,1,2-Triphenyl-1,2-ethanediol, T-70311
6296-98-6 4-Hydroxyacenaphthene, H-60099
6298-11-9 2-(4-Pyridyl)ethanethiol; B,HCl, *in* P-60240
6299-67-8 2,3-Dimethoxyaniline, *in* A-70089
6302-02-9 1-(2-Pyridinyl)-2-propanone, P-60229
6302-03-0 1-(3-Pyridinyl)-2-propanone, P-60230
6304-16-1 1-(4-Pyridinyl)-2-propanone, P-60231
6305-38-0 3-Aminodihydro-2(3H)-furanone; (S)-form, B,HBr, *in* A-60138
6308-38-9 6-Methyl-2,4(1H,3H)pyrimidinedithione, M-70128
6313-13-9 9-Aminopurine, A-60245
6315-89-5 ▷3,4-Dimethoxyaniline, *in* A-70090
6320-39-4 3-Amino-4(1H)-pyridinone; OH-form, *in* A-60254
6324-11-4 (2-Hydroxyphenoxy)acetic acid, H-70203
6333-43-3 2,3-Dimethoxyquinoxaline, *in* D-60279
6334-24-3 Pyrazole blue, P-70153
6335-83-7 4-(2-Phenylethyl)phenol, P-70073
6339-87-3 2-Thiophenesulfonic acid; Amide, *in* T-60213
6342-72-9 2-Hydroxy-1,4-benzenedicarboxylic acid; Di-Me ester, *in* H-70105
6373-16-6 1-Hydroxy-4-(methylamino)anthraquinone, *in* A-60175
6395-79-5 2-Cyclopropyl-2-oxoacetic acid; Me ester, *in* C-60231
6407-75-6 4-(2,5-Dichlorophenylhydrazono)5-methyl-2-phenyl-3Hpyrazol-3-one, D-70138
6409-44-5 3-Chloro-7H-benz[de]anthracen7-one, C-70041
6418-52-6 [1,1'-Biphenyl]-4,4'diylbis[diphenyl-methyl], B-60117
6426-84-2 1,2,3,6-Tetrahydro-2,6-dioxo-4pyrimidineacetic acid; Et ester, *in* T-70060
6431-92-1 2-Mercaptopropanoic acid; (±)-form, S-Ac, *in* M-60027
6443-68-1 3-Methoxy-4,5methylnedioxybenzonitrile, *in* H-60178
6453-27-6 6-Amino-β-carboline, A-70112
6468-80-0 Peuformosin, *in* K-70013
6491-93-6 1,2,3,4,5,6-Hexahydropentalene, H-70066
6535-93-9 1,2,3,6-Tetrahydro-2,6-dioxo-4pyrimidineacetic acid; Me ester, *in* T-70060
6541-19-1 6,7-Dichloro-5,8-quinolinedione, D-60149
6543-29-9 5,10-Dihydroindeno[2,1-a]indene, D-70218
6548-09-0 5-Bromotryptophan; (±)-form, *in* B-60333
6558-83-4 1,8-Dimethylcarbazole, D-70380
6570-87-2 3,4-Dimethyl-1-pentanol, D-60436
6572-60-7 [2.2]Paracyclophadiene, P-70007
6573-48-4 1,5-Dithionane, D-70526
6573-66-6 1,2-Dithiecane, D-60507
6573-73-5 Cyclopentadecyne, C-60220
6573-95-1 2-Thiabicyclo[3.1.0]hex-3-ene, T-70161
6577-17-9 Octahydro-1,2,5,6-tetrazocine, O-60020
6596-50-5 ▷Selenoformaldehyde, S-70029
6602-29-5 2-Bromo-3-hydroxypyridine; N-Oxide, *in* B-60275
6602-32-0 ▷2-Bromo-3-hydroxypyridine, B-60275
6604-06-4 2-Phenylcyclopentylamine; (1RS,2SR)-form, *in* P-60086

6610-21-5 6-Methyl-2-cyclohexen-1-one, M-70059
6627-53-8 4-Chloro-2-methoxy-1-nitrobenzene, *in* C-60119
6628-77-9 5-Amino-2-methoxypyridine, *in* A-60251
6630-18-8 1-Methoxy-4-(phenylmethoxy)benzene, *in* B-60074
6636-78-8 ▷2-Chloro-3-hydroxypyridine, C-60100
6638-79-5 O,N-Dimethylhydroxylamine; B,HCl, *in* D-70418
6647-96-7 4-Iodo-1,5-dimethylpyrazole, *in* I-70049
6647-97-8 4-Iodo-1,3-dimethylpyrazole, *in* I-70049
6664-26-2 2-Oxabicyclo[3.1.0]hex-3-ene, O-70056
6665-83-4 6-Hydroxyflavone, H-70130
6665-86-7 7-Hydroxyflavone, H-70131
6667-26-1 2-Bromotetrahydro-2H-pyran, B-70271
6672-64-6 5,7-Dihydrodibenzo[c,e]thiepin, D-70190
6681-01-2 4,5-Dihydrocyclobuta[b]furan, D-60212
6685-67-2 3,3',4',5,7-Pentahydroxyflavanone; (2R,3R)-form, Penta-Ac, *in* P-70027
6703-46-4 Imidazo[4,5-c]pyridine; 1-β-D-Ribofuranosyl, *in* I-60008
6703-56-6 Carbonimidic acid, C-70019
6705-50-6 7-Oxabicyclo[2.2.1]hept-2-ene, O-70054
6708-06-1 Octahydro-5H,10Hdipyrrolo[1,2a:1',2'-d]-pyrazine5,10-dione, O-60016
6708-14-1 Bicyclobutylidene, B-70095
6719-74-0 3,4,5-Trimethylphenol; Ac, *in* T-60371
6725-69-5 2-Butene-1,4-dithiol, B-70292
6726-48-3 Cyanuric acid; 1,3-Di-Me, *in* C-60179
6728-26-3 ▷2-Hexenal; (E)-form, *in* H-60076
6736-58-9 4-Amino-1H-imidazo[4,5-c]pyridine; 1-β-D-Ribofuranosyl, *in* A-70161
6741-90-8 6-Thioinosine; 2',3',5'-Tribenzoyl, *in* T-70180
6743-25-5 2,4,7(1H,3H,8H)-Pteridinetrione; 6,8-Di-Me, *in* P-70135
6743-26-6 2,4,7(1H,3H,8H)-Pteridinetrione; 3,6,8-Tri-Me, *in* P-70135
6746-94-7 Ethynylcyclopropane, E-70045
6750-34-1 3,7,11-Trimethyl-1-dodecanol, T-70282
6754-35-4 Pulchellin A, P-70138
6755-93-7 Totarolone, *in* T-70204
6760-99-2 8-Azabicyclo[3.2.1]octane; B,HCl, *in* A-70278
6761-70-2 1,4-Benzodioxan-6-carboxylic acid; Chloride, *in* B-60025
6784-47-0 2,2'-Bi-1,3-dithiolane, B-60103
6788-84-7 ▷1,2-Dioxetane, D-60469
6788-85-8 2,3-Dihydroazete, D-70172
6790-58-5 Dodecahydro-3a,6,6,9atetramethyl-naphtho[2,1b]furan, D-70533
6790-58-5 Dodecahydro-3a,6,6,9atetramethyl-naphtho[2,1b]furan; (3aα,5aβ,6α,9aα,9bβ)-form, *in* D-70533
6790-85-8 Aristolactone, A-60296
6805-37-4 Flindissol, F-70013
6807-83-6 Trifolirhizin, *in* M-70001
6811-35-4 Nuatigenin, N-60062
6811-77-4 4-Amino-1H-imidazo[4,5-c]pyridine, A-70161
6812-87-9 Royleanone, R-60015
6812-88-0 7β-Acetoxyroyleanone, *in* D-60304
6813-38-3 [2,2'-Bipyridine]-4,4'dicarboxylic acid, B-60126
6833-51-8 1-Acenaphthenecarboxylic acid, A-70014

6841-59-4 2-Thiabicyclo[2.2.1]hept-5-ene, T-70160
6855-99-8 6,7-Dehydroroyleanone, *in* R-60015
6887-59-8 4(1H)-Pyridinethione; N-Me, *in* P-60225
6893-02-3 ▷3,3',5-Triiodothyronine; (S)-form, *in* T-60350
6894-38-8 Jasmonic acid, J-60002
6894-43-5 Kahweol, K-60002
6902-62-1 Furanopetasin, F-70047
6918-15-6 Di-4-pyridyl ketone, D-70504
6937-91-3 Allophanic methylamide, *in* B-60190
6938-44-9 3-Benzoyl-2-methylpropanoic acid; (±)-form, Et ester, *in* B-60064
6939-05-5 2-Chloro-7-nitro-9H-fluorene, C-70133
6939-71-5 Hexahydro-1(3H)-isobenzofuranone; (3aRS,7aSR)-form, *in* H-70061
6989-24-8 2,3,23-Trihydroxy-12-oleanen28-oic acid; (2β,3β)-form, *in* T-70271
6994-25-8 3-Amino-1H-pyrazole-4-carboxylic acid; Et ester, *in* A-70187
7007-40-1 2,2,6,6-Tetramethylcyclohexanone; Oxime, *in* T-60133
7035-02-1 1-(Chloromethyl)-2-methoxybenzene, *in* C-70101
7057-50-3 8-(Methylsulfonyl)guanosine, *in* M-70029
7057-52-5 8-(Dimethylamino)guanosine, *in* A-70150
7057-53-6 8-Methoxyguanosine, *in* D-70238
7061-35-0 Homopantethine, *in* H-60090
7065-92-1 2-Chloro-3-phenylquinoxaline, C-70145
7067-12-1 Laserpitine, L-60016
7078-34-4 Bicyclo[4.1.1]octane, B-60091
7092-24-2 1,4-Dioxaspiro[4,5]dec-7-ene, *in* C-60211
7092-91-3 7,8,9,10-Tetrahydro-6Hcyclo-hepta[b]naphthalene, T-70047
7095-82-1 Bicyclo[3.1.1]hept-2-ene, B-60087
7096-51-7 3,3,4,4,4-Pentafluoro-1-butyne, P-60032
7099-91-4 4-Methoxy-α-oxobenzeneacetic acid, *in* H-70212
7124-96-1 [4]Paracyclophane, P-60006
7126-39-8 1H-Pyrrole-3-carboxaldehyde, P-70182
7140-63-8 9-Oxabicyclo[4.2.1]nona-2,4,7triene, O-70057
7149-65-7 5-Oxo-2-pyrrolidinecarboxylic acid; (S)-form, Et ester, *in* O-70102
7159-85-5 5-Ethyl-2-hydroxy-2,4,6cyclo-heptatrien-1-one, E-70044
7168-93-6 6-Methoxy-1,3-benzodioxole-5carboxylic acid, *in* H-60181
7193-15-9 2,3-Dihydronaphtho[2,3-b]furan, D-70235
7216-19-5 1,5-Dihydro-2,4-benzodithiepin, D-60194
7223-38-3 2-Propyn-1-amine; N-Di-Me, *in* P-60184
7226-23-5 ▷3,4,5,6-Tetrahydro-2(1H)pyrimidinone, *in* T-60088
7254-33-3 1H-Pyrazolo[3,4-d]-pyrimidine4,6(5H,7H)-dione; 1H-form, 1,5-Di-Me, *in* P-60202
7255-83-6 4,4'-Biss(yanomethyl)biphenyl, *in* B-60116
7287-73-2 2,2',2''-Trimethyltriphenylamine, T-70288
7288-07-5 2,2',2''-Nitrilotrisphenol, N-70039
7294-84-0 2-(2-Phenylethyl)phenol, P-70071
7299-38-9 2,4-Dichlorophenanthrene, D-70133
7331-52-4 Dihydro-4-hydroxy-2(3H)-furanone; (S)-form, *in* D-60242
7339-87-9 4-Hydroxyphenylacetaldehyde, H-60209
7342-82-7 3-Bromobenzo[b]thiophene, B-60205
7342-85-0 2-Chlorobenzo[b]thiophene, C-60042

7342-86-1   3-Chlorobenzo[b]thiophene, C-60043

7346-41-0   2-Chloroadamantane, C-60040

7355-55-7   2-Aminopyrrolo[2,3-d]pyrimidin4-one, A-60255

7357-70-2   2-Cyanoethanethioamide, C-70209

7369-69-9   4-Chloro-10H-phenothiazine, C-60127

7396-21-6   1-Acetylanthracene, A-70030

7396-38-5   2,4,5,7-Tetramethylphenanthrene, T-60152

7397-68-4   1,3-Dihydro-2H-imidazo[4,5-c]pyridin-2-one, D-60250

7427-09-0   2-Amino-3-methyl-1,4naphthoquinone, A-70171

7432-88-4   2-Benzothiopyrylium(1+); Perchlorate, in B-70067

7432-94-2   Dibenzo[b,d]thiopyrylium(1+); Perchlorate, in D-70081

7433-56-9   ▷5-Decene; (E)-form, in D-70017

7433-78-5   5-Decene; (Z)-form, in D-70017

7437-61-8   (1-Methylethenyl)oxirane, M-70080

7445-60-5   Biss(rifluoromethyl)thioketene, B-60187

7445-61-6   2,4-Bis[2,2,2-Trifluoro-1(trifluoromethyl)ethylidene]-1,3dithietane, B-60188

7451-95-8   2-Hydroxyphenylacetaldehyde, H-60207

7473-19-0   4-Ethoxy-1,2-naphthoquinone, in H-60194

7476-11-1   1,1,3,3-Tetraphenyl-2-propanone, T-60168

7480-30-0   Hexahydro-2(3H)-benzoxazolone; (3aRS,7aSR)-form, in H-70052

7480-31-1   Hexahydro-2(3H)-benzoxazolone; (3aRS,7aRS)-form, in H-70052

7480-32-2   4-Phenyl-2-oxazolidinone, P-60126

7486-94-4   4-Vinyl-2-azetidinone, V-60008

7498-87-5   3,4-Dihydro-2-phenyl-1(2H)naphthalenone, D-60265

7500-86-9   ▷2,4,6-Trinitrobenzoic acid; Chloride, in T-70303

7507-89-3   2',6'-Dihydroxy-4'methoxy-acetophenone, in T-70248

7539-68-6   1,2,3,4,5,6,7Heptafluoronaphthalene, H-70019

7539-90-4   6-Hydroxy-5-methoxy-2-methyl1,4-naphthoquinone, in D-60351

7548-13-2   3,7,11-Trimethyl-2,6,10dodeca-trienoic acid, T-60364

7573-15-1   2,7-Dimethylxanthone, D-60453

7605-10-9   9-Bromo-1,4-dihydro-1,4methanonaphthalene; anti-form, in B-60233

7619-12-7   ▷Muscarine, M-60153

7622-78-8   4-Oxo-3-phenyl-4H-1-benzopyran2-carboxylic acid, O-60086

7639-27-2   4'-Hydroxydehydrokawain, H-70122

7651-88-9   Trimethylamine oxide; B,HCl, in T-60354

7653-69-2   2,5-Biss(ethylthio)-1,3,4thiadiazole, in M-70031

7678-85-5   2',4',7-Trihydroxyisoflavone, T-70264

7686-78-4   2-Vinyl-1,1cyclopropanedicarboxylic acid, V-60011

7687-82-3   5,6-Dihydro-4H-cyclopenta[c]thiophen-4-one, D-70184

7690-98-4   5,6-Dihydro-4H-cyclopenta[c]thiophene, D-70181

7697-09-8   1-Bromobicyclo[2.2.2]octane, B-70184

7697-26-9   3-Bromo-4-methylbenzoic acid, B-60294

7699-35-6   Urocanic acid; (Z)-form, in U-70005

7703-74-4   ▷2,6-Biss(romomethyl)pyridine, B-60146

7705-14-8   ▷4-Isopropenyl-1methylcyclohexene; (±)-form, in I-70087

7715-02-8   [2-(Methylthio)ethenyl]benzene, M-70137

7727-79-9   Zederone, Z-70003

7729-30-8   2-Azetidinecarboxylic acid; (R)-form, in A-70284

7730-39-4   Azetidine; N-Benzyl, in A-70283

7730-45-2   Azetidine; N-(4-Methyl-benzenesulfonyl), in A-70283

7740-05-8   5-Chloro-2-hydroxy-3methoxy-benzaldehyde, in C-70065

7740-66-1   2-Fluoroheptanal, F-60034

7741-54-0   2,4-Dimethylthioxanthone; Dioxide, in D-70444

7752-82-1   2-Amino-5-bromopyrimidine, A-70101

7764-29-6   2,2'-Thiobis-1H-isoindole1,3(2H)-dione, T-70176

7764-30-9   2,2'-Dithiobis-1H-isoindole1,3(2H)-dione, D-70522

7766-50-9   11-Bromo-1-undecene, B-60336

7770-78-7   Arctigenin, A-70249

7779-27-3   ▷Hexahydro-1,3,5-triazine; 1,3,5-Tri-Et, in H-60060

7796-72-7   3,3-Dimethyl-4-pentenoic acid; Et ester, in D-60437

7796-73-8   3,3-Dimethyl-4-pentenoic acid, D-60437

8063-24-9   Acriflavinium chloride, in D-60028

8075-98-7   Ascorbigen, A-60309

10036-12-1   Bicyclo[3.3.1]non-3-en-2-one, B-70105

10067-18-2   Yomogin, Y-70001

10070-48-1   20(29)-Lupene-3,16-diol; (3β,16β)-form, in L-60039

10095-06-4   Tetrahydroimidazo[4,5-d]imidazole-2,5(1H,3H)dione; 1,3,4,6-Tetra-Me, in T-70066

10099-08-8   Bromomethanesulfonic acid; Chloride, in B-60291

10127-54-3   4,5-Dimethoxyphenanthrene, in P-70048

10127-55-6   4,5-Phenanthrenediol, P-70048

10132-01-1   1-Chloro-4-phenylphthalazine, C-70141

10133-41-2   2-Chlorobenzo[b]thiophene; 1,1-Dioxide, in C-60042

10141-72-7   Tetrahydro-2-methylpyran, T-60075

10165-33-0   ▷1-Amino-2-methoxyanthraquinone, in A-60174

10167-06-3   5,6-Dihydro-4(1H)-pyrimidinone, D-60274

10167-09-6   Tetrahydro-4(1H)-pyrimidinone, T-60089

10167-11-0   3,4-Dihydro-2(1H)-pyrimidinone, D-60273

10168-50-0   2,3'-Oxybispyridine, O-60099

10173-02-1   Silital, in C-70075

10182-48-6   3-Hydroxy-4(1H)-pyridinone, H-60227

10201-73-7   2-Amino-4-methoxypyridine, in A-60253

10208-16-9   1,8-Diacetylanthracene, D-70030

10210-32-9   2-Acetylanthracene, A-70031

10210-34-1   1,5-Diacetylanthracene, D-70028

10220-20-9   ▷1,1'-Dithiobispiperidine, D-70523

10228-97-4   9-Acridinecarboxaldehyde; 10-Oxide, in A-70060

10228-99-6   4,4-Dimethylpentanoic acid; Et ester, in D-70428

10233-78-0   N-Serylmethionine, L-L-form, Me ester, in S-60024

10266-89-4   19-Hydroxy-8(17),13-labdadien15-al; (E)-form, in H-70146

10310-21-1   ▷2-Amino-6-chloro-1H-purine, A-60116

10325-70-9   5-Acetylpyrimidine, A-60046

10338-69-9   1,2,3,6-Tetrahydro-4phenyl-pyridine, T-60081

10338-88-2   6-O-Acetylarbutin, in A-70248

10341-75-0   Acetophenone; (E)-Oxime, in A-60017

10352-22-4   1H-2-Benzoselenin-4(3H)-one, B-70051

10352-63-3   Chloromethanesulfonic acid; Na salt, in C-70086

10363-27-6   2-Methylcyclooctanone, M-60059

10364-04-2   1,4-Dibromobicyclo[2.2.2]octane, D-70086

10385-36-1   1-Bromo-2,3,4-trimethoxybenzene, in B-60200

10395-02-5   10-(Dicyanomethylene)anthrone, D-60164

10420-90-3   1,3-Hexadien-5-yne, H-70041

10421-73-5   Trichlorocyclopropenylium; Hexachloroantimonate, in T-70221

10428 68-9   Tetrakls(2-cyanoethyl)methane, in M-70036

10428-69-0   Methanetetrapropanoic acid, M-70036

10428-70-3   Methanetetrapropanoic acid; Tetra-Et ester, in M-70036

10428-73-6   Spiro[5.5]undecane-3,9-dione; Dioxime, in S-70068

10436-39-2   ▷1,1,2,3-Tetrachloro-1-propene, T-60029

10438-65-0   Trichlorocyclopropenylium; Tetra-chloroaluminate, in T-70221

10441-57-3   2-Pyrrolidinethione; N-Me, in P-70186

10467-11-5   1,2,4-Trimethyl-3,5,6triphenyl-benzene, T-70292

10472-24-9   2-Oxocyclopentanecarboxylic acid; (±)-form, Me ester, in O-70084

10478-89-4   2-(3-Pyridyl)-1,3-indanedione, P-60242

10478-90-7   2-(3-Pyridyl)-1,3-indanedione; B,½ HCl, in P-60242

10478-94-1   2-(3-Pyridyl)-1,3-indanedione; Hemipicrate, in P-60242

10478-99-6   2-(4-Pyridyl)-1,3-indanedione, P-60243

10485-70-8   3,7,11-Trimethyl-2,6,10dodeca-trienoic acid; Me ester, in T-60364

10486-61-0   3-Iodothiophene, I-60067

10489-28-8   2,3-Dihydro-2,2-dimethyl-1Hinden-1-one, D-60227

10495-73-5   6-Bromo-2,2'-bipyridine, B-70191

10498-47-2   2-Cyclohexyl-2-hydroxyacetic acid; (±)-form, in C-60212

10505-26-7   1H-Pyrazolo[3,4-d]pyrimidine4,6(5H,7H)-dione; 2H-form, 2,5,7-Tri-Me, in P-60202

10524-56-8   1-Bromo-2-nitroethane, B-60303

10527-10-3   4-Methyl-1-azulenecarboxylic acid, M-60042

10532-39-5   Diaporthin; (S)-form, in D-70050

10543-60-9   Tetrahydroimidazo[4,5-d]imidazole-2,5(1H,3H)dione; 1,3,4,6-Tetra-Ac, in T-70066

10553-03-4   2,4,6,8-Tetramethylundecanoic acid; (2R,4R,6R,8R)-form, in T-60157

10556-96-4   Hexahydropyrimidine; 1,3-Di-Me, in H-60056

10556-98-6   Hexahydro-1,3,5-triazine; 1,3,5-Triisopropyl, in H-60060

10560-39-1   Hexahydro-1,3,5-triazine; 1,3,5-Tri-tert-butyl, in H-60060

10568-83-9   3-(Ethylthio)-2-propenoic acid; Nitrile, in E-60055

10592-27-5   2,3-Dihydro-1H-pyrrolo[2,3-b]pyridine, D-70263

10601-80-6   Ethyl 3,3-diethoxypropionate, in O-60090

10602-65-0   Benzeneethanethioic acid; O-Et ester, in B-70017

11003-34-2   Cholest-5-ene-3,16,22-triol; (3β,16ξ,22ξ)-form, in C-70172

11013-05-1   Entandrophragmin, E-70009

11016-27-6 ▷ Pluramycin A, P-60162

11017-56-4   Strigol, S-60057

11023-64-6   Cladochrome, in P-70098

11031-58-6   Shikonin; (±)-form, in S-70038

11048-67-2   Candollein, in E-70009

11048-92-3 ▷ α-Guttiferin, G-60040

11054-16-3 ▷ Naematolin, N-70001

11075-15-3   Asebotin, in H-60219

12687-98-8 ▷ Roseofungin, R-70009

13005-84-0   2-(Methylamino)-4(1H)pteridinone, in A-60242

13005-86-2   2-Amino-1-methyl-4(1H)pteridinone, in A-60242

13019-52-8   Biopterin; (1′R,2′R)-form, in B-70121

13021-40-4   5,6-Dihydroxy-5-cyclohexene1,2,3,4-tetrone; Di-K salt, in D-60315

13027-48-0   1,2-Dibromo-1,2-diphenylethane; (1RS,2RS)-form, in D-70091

13031-04-4   Dihydro-4,4-dimethyl-2,3furandione, D-70196

13036-57-2   2-Chloro-4-methylpyrimidine, C-70113

13039-62-8   Biopterin; (1′S,2′R)-form, in B-70121

13039-82-2   Biopterin; (1′S,2′S)-form, in B-70121

13047-04-6 ▷ 4-Amino-1,2-benzenediol, A-70090

13055-36-2   3,4-Acenaphthenedicarboxylic acid, A-60013

13058-70-3   3-Nitro-2-butanone, N-70044

13058-73-6   7-Nitroisoquinoline, N-60033

13072-74-7   7-Deacetoxy-7-oxogedunin, in G-70006

13080-75-6   10,11-Dihydro-5H-dibenz[b,f]azepine; 5-Ac, in D-70185

13090-26-1   [1,2,4,5]Tetrazino[1,6-a:4,3a′]-diquinoline, T-60173

13120-66-6   Isoperezone, I-70075

13136-51-1   2-Mercapto-2-phenylacetic acid; (S)-form, S-Benzyl, in M-60024

13136-52-2   2-Mercapto-2-phenylacetic acid; (R)-form, S-Benzyl, in M-60024

13146-72-0   3′-Deoxyinosine, D-70024

13161-75-6   Pteryxin, in K-70013

13161-85-8   1,3,4,10-Tetrahydro-9(2H)acridinone, T-60052

13164-04-0   Chalepin, C-60031

13171-59-0   1,2,3,4,5-Pentamethyl-6nitro-benzene, P-60057

13177-73-6   7-Hydroxyfuro[2,3-d]pyridazine, H-60147

13186-45-3   C.I. Mordant Violet 39, in A-60318

13200-02-7   7,16-Heptacenedione, H-60018

13210-25-8   2-Acetyl-3-aminopyridine, A-70022

13210-52-1   4-Acetyl-3-aminopyridine, A-70027

13210-54-3   4-Acetyl-3-aminopyridine; 2,4-Dinitrophenylhydrazone; B,HCl, in A-70027

13214-71-6   6,15-Hexacenedione, H-60033

13220-52-5   10H-Indolo[3,2-b]quinoline; 5-Oxide, in I-70016

13261-50-2   2-Hydroxy-8-methoxy-1,4naphthoquinone, in D-60355

13274-42-5   1,5,9,13,17-Pentaazaheptadecane, P-60026

13276-43-2   2-Chloro-9-β-D-ribofuranosyl6H-purin-6-one, C-70161

13306-99-5   2-Phenyl-2H-1,2,3-triazole-4carboxylic acid, P-60140

13315-72-5   3-Amino-α-carboline, A-70106

13325-14-9   3-Amino-3-methyl-2-butanol, A-60217

13327-27-0   6-Methyl-3-pyridazinone; 2H-form, in M-70124

13333-97-6   2-(Trifluoromethyl)benzenethiol, T-60284

13338-63-1 ▷ 3,4,5Trimethoxy-benzeneacetonitrile, in T-60343

13344-76-8   2-Fluoro-2-phenylacetaldehyde, F-60062

13375-11-6   2-Hydroxycyclohexanecarboxylic acid; (1S,2R)-form, Me ester, in H-70119

13375-12-7   2-Hydroxycyclohexanecarboxylic acid; (1R,2S)-form, Me ester, in H-70119

13380-32-0   Phthalimidine; N-Benzyl, in P-70100

13385-49-4   2-Acetyl-3-aminopyridine; 2,4-Dinitrophenylhydrazone; B,HCl, in A-70022

13389-05-4   8-(Methylamino)guanosine, in A-70150

13389-16-7   8-Amino-9-β-D-ribofuranosyl6H-purin-6-one, A-70198

13389-17-8   8-(Dimethylamino)inosine, in A-70198

13389-36-1   6-Iodo-1-heptene, I-70035

13401-18-8   2,4(1H,3H)-Pteridinedione; 1,3-Di-Me, in P-60191

13431-29-3   3-Carboxy-2,3-dihydro-8hydroxy-5-methylthiazolo[3,2-a]pyridinium hydroxide inner salt, C-70021

13440-24-9   1,2-Dibromo-1,2-diphenylethane; (1RS,2SR)-form, in D-70091

13464-19-2   Carbonochloridothioic acid; S-Ph, in C-60013

13466-30-3   Acetophenone; Hydrazone, in A-60017

13471-31-3   1H-Benz[e]indole-1,2(3H)-dione; N-Ph, in B-70026

13472-57-6   2,6-Diethoxypyridine, in H-60226

13484-76-9   N-Phenylformamidine, P-60101

13511-13-2   1-(2-Propenyl)cyclohexene, P-60179

13534-07-1   Bicyclo[3.3.1]nona-2,6-diene, B-70102

13538-66-4   9-Methylene-1,3,5,7cyclononatetra-ene, M-60073

13544-43-9   6-Trifluoromethyl-1H-indole, T-60293

13544-66-6   3,4,5-Trimethylbenzoic acid; Me ester, in T-60355

13577-71-4   6-Chloro-2-(trifluoromethyl)1H-imidazo[4,5-b]pyridine, C-60139

13580-54-6   2,2,3,3,5,5,6,6Octafluoromorphol-ine, O-60009

13601-86-0   1-Bromo-3,3-dimethyl-1-butyne, B-70207

13602-68-1   4-Amino-2(1H)-pyridinone; NH-form, N-Oxide; B,HCl, in A-60250

13602-69-2   4-Amino-2(1H)-pyridinone; NH-form, N-Oxide, in A-60250

13606-71-8   3,4-Diphenyl-2-oxazolidinone, in P-60126

13656-84-3   4-Isopropylcycloheptanone, I-70091

13659-21-7   3-Bromo-2,4-dichlorophenol, B-60228

13659-22-8   2-Bromo-3,5-dichlorophenol, B-60226

13667-30-6   2,3,4-Trimethyl-5-nitrobenzoic acid, T-60369

13679-70-4   5-Methyl-2thiophenecarboxaldehyde, M-70143

13726-16-4   4-Chloro-3-methoxybenzaldehyde, in C-60090

13728-34-2   2,3-Naphthalenedicarboxylic acid; Di-Me ester, in N-60003

13741-21-4   2,6-Dimethyl-3,7-octadiene2,6-diol, D-70422

13745-59-0   3-Bromo-1,4-dimethylpyrazole, in B-70250

13754-41-1   Piperazinone; N4-Benzyl, in P-70101

13754-69-3   3,7,11,15-Tetramethyl-2hexa-decenal, in P-60152

13781-67-4   3-Thiopheneethanol, T-60211

13808-64-5   4-Bromo-3(5)-methylpyrazole, B-70251

13816-21-2   2,2′-Bithiazole, B-70166

13830-35-8   2,3-Dimethylcyclopropanol, D-70397

13830-35-8   2,3-Dimethylcyclopropanol; (2RS,3RS)-form, in D-70397

13849-32-6   3,4-Dihydro-2H-1-benzopyran-5-ol, D-60201

13849-91-7   3,19-Dihydroxy-12-ursen-28-oic acid; (3β,19α)-form, in D-70350

13855-90-8   2,2,5,5-Tetramethyl-3cyclohexen-1-one, T-60136

13881-91-9   Aminomethanesulfonic acid, A-60215

13885-13-7   2-Cyclopropyl-2-oxoacetic acid, C-60231

13889-92-4 ▷ Carbonochloridothioic acid; S-Propyl, in C-60013

13889-94-6   Carbonochloridothioic acid; S-Butyl, in C-60013

13906-09-7   2,3-Dihydro-2-thioxo-4(1H)quinazolinone, D-60296

13980-04-6 ▷ Hexahydro-1,3,5-triazine; 1,3,5-Trinitroso, in H-60060

13988-19-7   5,6-Dihydroxyindole; N-Me, di-O-Ac, in D-70307

14017-71-1   Isopteryxin, in K-70013

14017-72-2   Calipteryxin, in K-70013

14033-65-9   4-(1-Propenyl)cyclohexene, P-60181

14072-82-3   4-Isopropylcyclohexene, I-60120

14072-87-8   3-tert-Butylcyclohexene, B-60342

14090-99-4   2,5,7,8-Tetrahydroxy-3,6dimethoxy-1,4-naphthoquinone, in H-60065

14097-60-0   4-Bromo-2,2′-bipyridine; Picrate, in B-70186

14144-06-0   Trillin, in S-60045

14149-43-0 ▷ Trimethidinium methosulphate, in T-70277

14161-43-4   4-Methyl-6-phenylpyrimidine; 3-Oxide, in M-70115

14162-95-9   4-Bromo-2,2′-bipyridine, B-70186

14164-34-2   4,6-Dimethyl-2-phenylpyrimidine, D-60444

14164-59-1   Ivangustin, I-70109

14174-83-5   Hexahydro-1H-pyrrolizin-1-one, H-60057

14174-86-8   Tetrahydro-1H-pyrrolizin-2(3H)one, T-60091

14189-85-6   2-Methoxy-3-methyl-2cyclopenten-1-one, in H-60174

14212-87-4   1,1-Di-4-morpholinylethene, D-70447

14250-96-5   2-Methyl-2-pentenal; (E)-form, in M-70106

14252-62-1   3-Vinyl-2-cyclopenten-1-one, V-70011

14252-67-6   2-Chloro-2-phenylethanol; (S)-form, in C-60128

14255-88-0 ▷ Phenyl 5,6-dichloro-2(trifluoromethyl)-1Hbenzimidazole-1carboxylate, in D-60160

14267-92-6   5-Chloro-1-pentyne, C-70139

14271-83-1   4-Methoxy-αoxobenzeneacetonitrile, in H-70212

14282-62-3   2,5-Thiophenedimethanethiol, T-70184

14289-45-3   α-Oxo-2-naphthaleneacetic acid, O-60079

14294-33-8   2-Amino-3-methylbiphenyl, A-70164

14296-16-3   1,20-Dibromoicosane, D-70092

14307-88-1   Alethine; B,2HCl, in A-60073

14309-25-2   Azidotriphenylmethane, A-60352

14333-81-4   5-Amino-1H-pyrazole-4-carboxylic acid; N(1)-Ac, Et ester, in A-70192

14339-33-4 3-Chloro-1*H*-pyrazole, C-70155

14346-19-1 3-Amino-5-nitro-2,1benzisothiazole, A-60232

14352-55-7 2-Hydroxycyclopentanone; (±)-*form*, Oxime, *in* H-60114

14353-90-3 Pentafluoroiodosobenzene, P-60034

14356-59-3 Saikogenin *F*, S-60001

14362-50-6 4-Oxo-1,2,3,4tetrahydrocarbazole; Oxime, *in* O-70106

14377-11-8 1-Acetylcycloheptene, A-60026

14387-31-6 4,4′-Thiobisbenzoic acid; Di-Me ester, *in* T-70175

14408-81-2 Tetrahydro-1*H*-pyrrolizin-2(3*H*)one; (±)-*form*, Picrate, *in* T-60091

14430-23-0 5,6-Dimethoxyindole, *in* D-70307

14439-13-5 2,4,7-Triaminopteridine, T-60229

14476-63-2 Benzeneethanethioic acid; *S*-Et ester, *in* B-70017

14478-61-6 2-Hydrazinoethylamine, H-70096

14491-02-2 2,2-Dimethyl-3-phenyl-2*H*-azirine, D-60440

14496-24-3 3,4-Biss(ydroxymethyl)furan, B-60161

14504-07-5 2-Hydroxy-4-methylbenzoic acid; Ac, *in* H-70163

14527-48-1 2-(Ethylthio)thiazole, *in* T-60200

14531-52-3 5-(2-Phenylethyl)-1,3benzenediol, P-60098

14554-09-7 5,8-Dihydroxy-2-methyl-1,4naphthoquinone, D-60352

14554-13-3 5-Gluten-3-ol; 3α-*form*, *in* G-70020

14561-37-6 2-Amino-3-chlorobutanoic acid; (2*RS*,3*RS*)-form, *in* A-60103

14561-56-9 2-Amino-3-chlorobutanoic acid, A-60103

14578-68-8 3,4-Dihydro-4-phenyl-1(2*H*)naphthalenone, D-60267

14590-54-6 1,2-Cyclopropanedicarboxylic acid; (1*S*,2*S*)-*form*, *in* C-70248

14594-57-1 2,3,5,6,7,8-Hexahydro-1*H*indolizinium(1+); Perchlorate, *in* H-70059

14671-09-1 Hexahydro-1*H*-pyrrolizin-1-one; (±)-*form*, Picrate, *in* H-60057

14675-48-0 6-Methyl-9-β-D-ribofuranosyl9*H*-purine, M-70131

14677-10-2 4-Benzoyl-1,2,3-triazole, B-70079

14679-41-5 3*a*,4,7,7*a*-Tetrahydro-1(3*H*)isobenzofuranone; (3*aRS*,7*aRS*)-*form*, *in* T-70072

14682-29-2 Roridin *D*, R-70008

14690-58-5 5,6-Dimethylene-1,3cyclohexadiene, D-70410

14693-11-9 Tricyclo[3.3.1.0²,⁸]nona-3,6diene, T-70230

14696-36-7 *ent*-15-Kauren-17-oic acid, K-60008

14698-56-7 Arbutin; Penta-Ac, *in* A-70248

14699-32-2 8-Hydroxyisopimar-15-ene; 8β-*form*, *in* H-60163

14705-30-7 2,2,5-Trimethyl-4-hexen-3-one, T-60366

14730-25-7 3-Phenyl-1,4,2-dithiazole-5thione, P-60088

14736-58-4 GB 1, G-70005

14757-77-8 4-Hydroxyfuro[2,3-*d*]pyridazine, H-60146

14757-78-9 3-Bromo-2-furancarboxaldehyde, B-60257

14764-52-4 3-Oxabicyclo[3.2.0]heptan-2-one; (1*RS*,5*SR*)-*form*, *in* O-70053

14776-33-1 1-Phenyl-3-pyrazolidinone; B,HCl, *in* P-70091

14787-34-9 5-Hydroxy-3′,4′,6,7tetramethoxyflavone, *in* P-60047

14790-51-3 1,2-Diphenyl-1,2-diazetidin-3-one, *in* D-70055

14802-18-7 2-Phenyl-4(1*H*)-quinolinone, P-60132

14804-37-6 Methoxypentamethylbenzene, *in* P-60058

14811-73-5 10-Oxodecanoic acid; Me ester, *in* O-60063

14813-19-5 2′,5,6,6′-Tetrahydroxyflavone, T-70107

14813-19-5 2′,5,6,6′-Tetramethoxyflavone, *in* T-70107

14813-20-8 5-Hydroxy-2′,6,6′trimethoxyflavone, *in* T-70107

14832-58-7 *N*-(Phenylmethyl)-1*H*-tetrazol5-amine, *in* A-70202

14836-73-8 Ferrioxamine *B*, F-60003

14882-94-1 Rutamarin, *in* C-60031

14889-64-6 Tetrahydro-2*H*-1,3-thiazin-2-one, T-60096

14936-99-3 1,4-Dimethylthioxanthone; Dioxide, *in* D-70442

14944-26-4 3,4-Dihydro-3-phenyl-1(2*H*)naphthalenone, D-60266

14958-06-6 2-Methyl-1,3-naphthalenediol; Di-Ac, *in* M-60091

14985-78-5 6-Methyl-9-β-D-ribofuranosyl9*H*-purine; 2′,3′,5′-Tribenzoyl, *in* M-70131

14988-20-6 Sphaerobioside, *in* T-60324

14993-03-4 1*H*-Imidazo[4,5-*h*]quinoline, I-70008

15012-38-1 2,4,6-Trinitrobenzoic acid; Me ester, *in* T-70303

15055-49-9 5-Amino-3-phenylisoxazole; *N*-Ph, *in* A-60240

15055-81-9 5-Isoxazolecarboxylic acid; Me ester, *in* I-60142

15068-08-3 2-(Chloromethyl)phenol; Ac, *in* C-70101

15069-79-1 5,6-Dihydroxyindole; *O*,*O*-Di-Ac, *in* D-70307

15071-04-2 4-Methyl-6,7-methylenedioxy2*H*-1-benzopyran-2-one, M-70092

15074-91-6 Octadecahydrotriphenylene, O-70004

15089-43-7 2,2-Butanedithiol, B-60340

15103-48-7 2-Pyridinesulfonic acid, P-70171

15123-40-7 Pyreno[4,5-*c*]furan, P-70166

15123-45-2 13,19:14,18Dimethenoanthra[1,2-*a*]benzo[*o*]pentaphene, D-60396

15128-52-6 4-Oxo-1,2,3,4tetrahydrocarbazole, O-70106

15131-84-7 ▷Chrysene-5,6-oxide, C-70174

15144-23-7 1,1-Cyclobutanedithiol, C-70215

15149-00-5 4-Bromo-2-fluorophenol; Methanesulfonate, *in* B-60251

15158-36-8 4-Methyl-1,2benzenedicarboxaldehyde, M-60048

15159-65-6 2-Amino-4-bromobutanoic acid; (*S*)-*form*, B,HBr, *in* A-60098

15176-29-1 ▷2′-Deoxy-5-ethyluridine, *in* E-60054

15177-05-6 1-Phenyl-6-hepten-1-one, P-70080

15186-48-8 2,2-Dimethyl-1,3-dioxolane-4carboxaldehyde; (*R*)-*form*, *in* D-70400

15216-10-1 ▷Azetidine; *N*-Nitroso, *in* A-70283

15232-95-8 3-(2-Propenyl)cyclohexene, P-60180

15250-36-9 3-Amino-1*H*-pyrazole-4-carboxylic acid; Et ester, *N*(3)-Ac, *in* A-70187

15250-38-1 5,8-Quinoxalinedione, Q-60010

15268-31-2 3-Isocyanatopyridine, I-70065

15295-32-6 1,1-Diphenyl-1,3-pentadiene; (*E*)-*form*, *in* D-70489

15300-62-6 Benzo[*a*]biphenylene; 2,4,7-Trinitrofluorenone complex, *in* B-60016

15301-54-9 2-Phenylcyclopentylamine, P-60086

15302-15-5 2-(2-Ethoxyethoxy)benzamide, *in* H-70108

15308-22-2 2,5,6,8-Tetrahydroxy-3,7dimethoxy-1,4-naphthoquinone, *in* H-60065

15308-24-4 2,3,5,6,8-Pentahydroxy-7methoxy-1,4-naphthoquinone, *in* H-60065

15383-88-7 2-Amino-3-ethylbutanedioic acid; (2*RS*,3*RS*)-(?)-*form*, *in* A-70138

15383-89-8 2-Amino-3-propylbutanedioic acid; (2*RS*,3*RS*)-(?)-*form*, *in* A-70186

15383-90-1 2-Amino-3-isopropylbutanedioic acid; (2*RS*,3*RS*)-(?)-*form*, *in* A-70162

15383-91-2 2-Amino-3-butylbutanedioic acid; (±)-*form*, *in* A-60102

15414-36-5 4-(2-Propenyl)cyclohexene, P-60182

15430-52-1 2-Propyn-1-amine; B,HCl, *in* P-60184

15434-00-1 Egonol; Ac, *in* E-70001

15448-47-2 Methyloxirane; (*R*)-*form*, *in* M-70100

15448-58-5 2,3-Epoxy-2,3-dihydro-1,4naphthoquinone, E-70011

15454-54-3 5-Amino-1*H*-tetrazole; Monohydrate, *in* A-70202

15462-43-8 2-Cyano-3-hydroxyquinoline, *in* H-60228

15462-44-9 3-Hydroxy-2-quinolinecarboxylic acid; Amide, *in* H-60228

15462-45-0 3-Hydroxy-2-quinolinecarboxylic acid, H-60228

15537-53-8 Benzylidenecycloheptane, B-60068

15540-18-8 2-Cyclohexyl-2-methoxyacetic acid, *in* C-60212

15540-79-1 4-Amino-6-hydroxy-1,3benzenedicarboxylic acid, A-70153

15545-30-9 2(3*H*)-Thiazolethione; *SH-form*, *S*-Isopropyl, *in* T-60200

15567-77-8 2-(Hydroxymethyl)-1,4benzodioxan; Carbamate, *in* H-60172

15570-45-3 1,2,3,4Tetraphenylcyclopentadiene, T-70149

15573-67-8 2-(4-Hydroxyphenyl)-2-oxoacetic acid, H-70212

15575-52-7 Bicoumol, B-70093

15575-69-6 Khellactone; (9*RS*,10*SR*)-*form*, Di-Ac, *in* K-70013

15590-89-3 3-Amino-4(1*H*)-pyridinone; *NH-form*, *in* A-60254

15590-90-6 3-Nitro-4(1*H*)-pyridone, N-70063

15622-62-5 Porphyrexide, P-70110

15657-79-1 2,5-Diamino-1,4-benzenedithiol, D-70035

15672-96-5 4,4-Dimethylpentanoic acid; Amide, *in* D-70428

15673-05-9 4,4-Dimethylpentanoic acid; Nitrile, *in* D-70428

15673-17-3 4,4-Dimethylpentanoic acid; Me ester, *in* D-70428

15690-25-2 3-(2-Thienyl)-2-propenoic acid; (*E*)-*form*, *in* T-60208

15727-65-8 1-Ethylnylnaphthalene, E-70046

15731-98-3 3,4,4-Trimethyl-5(4*H*)isoxazolone, T-60367

15732-43-1 2-Phenyl-4-(phenylmethylene)5(4*H*)-oxazolone; (*E*)-*form*, *in* P-60128

15780-62-8 3-Oxobutanethioic acid; (*SH*)-*form*, Isopropyl ester, *in* O-70080

15780-63-9 *S*-Butyl 3-oxobutanethioate, *in* O-70080

15787-60-7 4-(Bromomethyl)acridine, B-70232

15802-75-2 4-Iodo-3(5)-methylpyrazole, I-70049

15804-19-0 ▷1,4-Dihydro-2,3quinoxalinedione; Dione-*form*, *in* D-60279

15833-10-0	3,5-Dimethyl-2-oxazolidinone, *in* M-60107
15851-57-7	2,3-Dihydroxybutanoic acid; (2*R*,3*R*)-*form*, *in* D-60312
15851-58-8	2,3-Dihydroxybutanoic acid; (2*S*,3*R*)-*form*, *in* D-60312
15862-22-3	5-Bromo-3,3′-bipyridine, B-70189
15878-00-9	▷4-Chloro-1*H*-pyrazole, C-70156
15878-08-7	▷4-Chloro-3(5)methylpyrazole, C-70108
15903-94-3	6-Hydroxyindole; Benzyl ether, *in* H-60158
15912-21-7	Hexahydro-1*H*-pyrrolizin-1-one; (±)-*form*, Oxime; B,HCl, *in* H-60057
15912-22-8	Hexahydro-1*H*-pyrrolizin-1-one; (±)-*form*, Oxime, *in* H-60057
15914-84-8	1-(1-Naphthyl)ethanol; (*S*)-*form*, *in* N-70015
15918-19-1	10*H*-Phenothiazine-2-carboxylic acid; Me ester, *in* P-70056
15925-47-0	*S-tert*-Butyl 3-oxobutanethioate, *in* O-70080
15953-45-4	3(5)-Chloro-5(3)-methylpyrazole, C-70109
15965-30-7	4,5-Dichloro-1*H*-imidazole, D-70122
15965-31-8	4(5)-Chloro-1*H*-imidazole, C-70077
15966-72-0	1-Phenyl-1*H*-1,2,3-triazole-5carboxylic acid, P-60139
15973-99-6	▷Hexahydropyrimidine; *N,N′*-Dinitroso, *in* H-60056
16013-85-7	2,6-Dichloro-3-nitropyridine, D-60144
16024-30-9	5-Methoxy-2-methyl-3nitroaniline, *in* A-60222
16033-71-9	Methyloxirane; (±)-*form*, *in* M-70100
16046-28-9	Benzenediazo-*p*-toluenesulfonate *N*-oxide, *in* H-70208
16048-89-8	2-Hydroxy-3-isopropylbutanedioic acid, H-60166
16063-69-7	▷2,4,6-Trichloropyridine, T-60255
16069-13-9	Octadecahydrotriphenylene; (*R*, all-*trans*)-*form*, *in* O-70004
16077-41-1	Hexahydropyrimidine; 1,3-Di-*tert*-butyl, *in* H-60056
16088-62-3	Methyloxirane; (*S*)-*form*, *in* M-70100
16096-33-6	1-Phenyl-1*H*-indole, P-70082
16096-71-2	2-(Methylsulfonyl)cyclohexanone, *in* M-70025
16112-10-0	1-Acetylcyclopentene, A-60028
16192-49-7	1-Oxa-2-azaspiro[2,5]octane; *N*-Ac, *in* O-70052
16201-50-6	2-Mercapto-2-phenylacetic acid; (±)-*form*, *in* M-60024
16201-51-7	2-Mercapto-2-phenylacetic acid; (*R*)-*form*, *in* M-60024
16201-52-8	2-Mercapto-2-phenylacetic acid; (*R*)-*form*, Me ester, *in* M-60024
16205-72-4	2-Phenylcyclopropanecarboxylic acid; (1*S*,2*S*)-*form*, Me ester, *in* P-60087
16227-15-9	4,4′-Bi-4*H*-1,2,4-triazole, B-60189
16258-05-2	2-Amino-5-hexenoic acid, A-60169
16265-04-6	2-Chloro-1*H*-imidazole, C-70076
16269-06-0	9-Phenylnonanoic acid, P-70084
16269-40-2	Di-9-phenanthrylamine, D-70472
16272-83-6	1-Phenyl-1,4-hexadiyn-3-one, P-60108
16291-08-0	▷2,2-Diazido-1,3-indanedione, D-70056
16293-80-4	4*b*,9*b*-Dihydroindeno[2,1-*a*]indene-5,10-dione; (4*bRS*,9*bRS*)-*form*, *in* D-70220
16328-62-4	1,3-Dihydro-2*H*-imidazo[4,5-*b*]pyridin-2-one, D-60249

16328-63-5	2*H*-Imidazo[4,5-*b*]pyrazin-2-one, I-60006
16336-55-3	9-(Phenylthio)phenanthrene, *in* P-70049
16336-82-6	5-Dodecen-7-yne; (*E*)-*form*, *in* D-70538
16337-16-9	1,2,9,10,17,18Hexadehydro[2.2.2]-paracyclophane, H-60038
16340-68-4	2,2,3,3Tetramethyl-cyclopropanecarboxaldehyde, T-70133
16395-57-6	5-Oxo-2-pyrrolidinecarboxylic acid; (*S*)-*form*, Amide, *in* O-70102
16400-86-5	2,4,6-Trinitroacetanilide, *in* T-70301
16401-28-8	Octaleno[3,4-*c*]furan, O-70019
16421-52-6	2-Tetrazolin-5-one, T-60175
16492-75-4	5-Bromo-1,2,3-benzenetriol, B-60201
16519-43-0	Selenourea; *N*[1],*N*[3]-Di-Ph, *in* S-70031
16538-47-9	1-Iodo-1-hexene; (*Z*)-*form*, *in* I-70037
16544-26-6	Bicyclo[4.1.1]oct-2-ene, B-60095
16576-23-1	1-Benzoylisoquinoline, B-60059
16583-06-5	2,3,4,5-Tetrafluorobenzaldehyde, T-70026
16587-35-2	2-Methylnaphtho[2,1-*b*]thiophene, M-60097
16587-39-6	1,3-Diphenylbenzo[*c*]thiophene, D-70474
16611-84-0	2-Hydroxy-6-(8,11penta-decadienyl)benzoic acid; (all-*Z*)-*form*, *in* H-60205
16617-46-2	5-Amino-4-pyrazolecarbonitrile, *in* A-70187
16629-14-4	2,5-Dichlorothiazole, D-60155
16629-16-6	4,5-Dichlorothiazole, D-60156
16629-19-9	2-Thiophenesulfonic acid; Chloride, *in* T-60213
16635-95-3	1,2-Diphenyl-1,2-ethanediamine; (1*RS*,2*RS*)-*form*, *in* D-70480
16644-98-7	1-Iodo-1-hexene; (*E*)-*form*, *in* I-70037
16652-35-0	2-Amino-4-fluorobutanoic acid; (±)-*form*, *in* A-60155
16661-00-0	Daucene, D-70008
16679-66-6	3,6-Bis(2-methylpropyl)-2,5piperazinedione; (3*RS*,6*SR*)-*form*, *in* B-60171
16681-56-4	▷2-Bromo-1*H*-imidazole, B-70220
16681-70-2	1*H*-1,2,3-Triazole-4-carboxylic acid, T-60239
16681-71-3	1*H*-1,2,3-Triazole-4-carboxylic acid; (1*H*)-*form*, 1-Me, *in* T-60239
16683-64-0	Benz[*k*]aceanthrylene, B-60010
16699-76-6	2-Amino-4-fluorobutanoic acid; (±)-*form*, B,HCl, *in* A-60155
16714-26-4	2′-Azidoacetophenone, A-60334
16714-27-5	2-Azidobenzophenone, A-60338
16740-73-1	1-(5-Bromo-2-methoxy-phenyl)ethanone, *in* B-60265
16744-98-2	▷1-Fluoro-2-isocyanatobenzene, F-60047
16750-82-6	6-Isopropenyl-3-methyl-2cyclohexen-1-one; (*S*)-*form*, *in* I-70088
16767-46-7	1*H*-Cyclooctapyrazole, C-70236
16781-33-2	2-Amino-4*H*-3,1-benzothiazine; *N*-Ph, *in* A-70092
16797-75-4	Di-*tert*-butylthioketene, D-60114
16797-76-5	Di-*tert*-butylthioketene; *S*-Oxide, *in* D-60114
16803-13-7	1,4-Dioxa-9-azaspiro[4.7]dodecane, *in* H-70048
16803-16-0	Hexahydro-4(1*H*)-azocinone; B,HClO₄, *in* H-70047
16822-06-3	3a,4,5,6,7,7a-Hexahydro-3methyl-ene-2-(3*H*)benzofuranone; (3*aRS*,7*aRS*)-*form*, *in* H-70062

16831-48-4	4,4-Dimethoxycyclohexene, *in* C-60211
16836-36-5	Isolaserpitin, *in* D-70009
16836-38-7	Laserpitinol, *in* L-60016
16838-85-0	Zaluzanin D, *in* Z-60001
16838-87-2	Zaluzanin C, Z-60001
16849-78-8	2-Formyl-3,5-dihydroxybenzoic acid; Me ester, *in* F-60071
16853-10-4	1,2,3,5,6,7Hexahydropyrrolizinium-(1+); Perchlorate, *in* H-70071
16853-11-5	1,4-Dioxa-8-azaspiro[4.7]dodecane, *in* H-70047
16857-93-5	1,3,5-Tripyrrolidinobenzene, T-70319
16857-95-7	1,3,5-Tripiperidinobenzene, T-70314
16857-97-9	1,3,5-Trimorpholinobenzene, T-70294
16867-03-1	2-Amino-3-hydroxypyridine, A-60200
16867-04-2	3-Hydroxy-2(1*H*)-pyridinone, H-60223
16867-95-1	3,4-Di-*tert*-butylpyrazole, D-70106
16886-09-2	3-(2-Propenyl)indole, P-60183
16890-85-0	Carbonochloridothioic acid, C-60013
16914-66-2	1,2-Diselenete, D-60498
16936-63-3	[(Tribromomethyl)thio]benzene, T-70212
16955-35-4	1-Cyanobicyclo[1.1.0]butane, *in* B-70094
16955-55-8	Salireposide, *in* D-70287
16957-97-4	3-Bromobenzo[*b*]thiophene; 1,1-Dioxide, *in* B-60205
16958-22-8	2-Methyl-2-pentenal; (*Z*)-*form*, *in* M-70106
16974-11-1	9-Dodecen-1-ol; (*Z*)-*form*, Ac, *in* D-70537
16981-80-9	Cyperenol, C-70259
16993-42-3	Avenaciolide, A-60325
17024-05-4	9-*tert*-Butylphenanthrene, B-70307
17025-47-7	[(Tribromomethyl)-sulfonyl]benzene, *in* T-70212
17078-60-3	4,5-Dihydroxy-1,3benzenedicarboxylic acid; Di-Me ether, di-Me ester, *in* D-60307
17093-75-3	2-Amino-4-hydroxypentanedioic acid; (2*RS*,4*RS*)-*form*, *in* A-60195
17094-01-8	Sepiapterin; (*S*)-*form*, *in* S-70035
17104-81-3	4,4′-Iminodibenzoic acid; Di-Me ester, *in* I-60018
17137-03-0	*N*-Valylleucine; L-L-*form*, Z-Val-Leu-NHNH₂, *in* V-60001
17175-86-9	2,4-Heptadienoic acid, H-60021
17180-94-8	5-Chloropyrimidine, C-70157
17194-58-0	Curcuphenol, C-70208
17219-94-2	9,10-Dichlorophenanthrene, D-70137
17231-94-6	3,5-Dichlorobenzenethiol, D-60124
17231-95-7	2,3-Dichlorobenzenethiol, D-60119
17236-59-8	Thiophene-3-ol, T-60212
17249-80-8	3-Chlorothiophene, C-60136
17257-15-7	Nepetalactone; (4*aS*,7*S*,7*aS*)-*form*, *in* N-60020
17266-64-7	2,3-Dihydro-1*H*-pyrrolizin-1-one, D-60277
17267-51-5	1-Methyl-5-methylamino-1*H*tetrazole, *in* A-70202
17268-93-8	2-Amino-4-ethoxybutanoic acid, *in* A-70156
17273-30-2	Gossypol; (+)-*form*, Hexa-Me ether, *in* G-60032
17282-04-1	2-Chloro-3-fluoropyridine, C-60081
17299-35-3	3-*tert*-Butyl-2-cyclohexen-1-one, B-70297
17299-97-7	1,3,5-Triss(hloromethyl)benzene, T-60402
17303-67-2	▷Goniothalamin, G-60031
17325-26-7	1*H*-Imidazole-4-carboxylic acid; Me ester, *in* I-60004

17332-57-9 2,4'-Iminodibenzoic acid, I-60016
17339-74-1 1-Acetylcyclooctene, A-60027
17345-72-1 4-Bromo-1,2,3-benzenetriol, B-60200
17346-16-6 2,4-Dibromo-2,4-dimethyl-3pentanone, D-60099
17347-32-9 6-Bromobenzo[b]thiophene, B-60208
17351-29-0 1,2-Diphenyl-1,4-cyclohexadiene, D-60477
17355-11-2 N-Tyrosylphenylalanine; (S,S)-form, in T-60412
17366-20-0 4,6-Diamino-1,3-benzenedithiol, D-70037
17372-53-1 6-Methoxy-2H-1-benzopyran-2-one, in H-70109
17392-16-4 2-(2-Hydroxyphenyl)-2-oxoacetic acid, H-70210
17422-92-3 2,3,4,4a,5,6-Hexahydro-7H-1benzopyran-7-one, H-60045
17425-64-8 2,3,6,7-Tetrahydroxyphenanthrene, T-60121
17435-72-2 2-Bromomethyl-2-propenoic acid; Et ester, in B-70249
17452-27-6 3-Isothiocyanatopyridine, I-70107
17497-53-9 1-Iodocyclohexene, I-70026
17502-32-8 2-Hydroxycyclohexanecarboxylic acid; (1RS,2RS)-form, in H-70119
17526-17-9 ▷1,3-Dihydroxy-2-(ethoxymethyl)-anthraquinone, in D-70306
17539-96-7 Hexahydro-2(3H)-benzoxazolone, H-70052
17550-38-8 Senepoxide; (−)-form, in S-60023
17573-78-3 4,5,6-Trifluoropyrimidine, T-60308
17573-92-1 3-Methoxythiophene, in T-60212
17606-70-1 2-Phenyl-4-(phenylmethylene)5(4H)-oxazolone; (Z)-form, in P-60128
17616-43-2 ▷3-Iodo-2-(iodomethyl)-1-propene, I-60047
17623-69-7 1,2-Dihydroxyxanthone, D-70353
17635-44-8 3,4,5-Tribromo-1H-pyrazole, T-60246
17636-16-7 Barbatic acid, B-70007
17644-83-6 3-Methoxy-6-methylpyridazine, in M-70124
17645-02-2 1,5-Dihydro-6H-imidazo[4,5-c]pyridazine-6-thione, D-70215
17645-03-3 6-(Methylthio)-5H-imidazo[4,5c]pyridazine, in D-70215
17654-26-1 3,3',4',5,7Pentahydroxyflavanone; (2R,3R)-form, in P-70027
17666-93-3 1,1'-Biindanylidene, B-70120
17673-81-3 9,10,18-Trihydroxyoctadecanoic acid; (9RS,10SR)-form, in T-60334
17687-76-2 1-Methoxy-3-buten-2-ol, in B-70291
17705-68-9 9,10,18-Trihydroxyoctadecanoic acid; (9S,10S)-form, in T-60334
17708-79-1 N-Valylleucine; L-L-form, Z-Val-Leu-OH, in V-60001
17715-70-7 1-Fluoro-2,4-dimethoxybenzene, in F-70018
17733-23-2 1,2-Dichloro-4-(methylthio)benzene, in D-60123
17733-24-3 1,4-Dichloro-2-(methylthio)benzene, in D-60121
17734-85-9 1-Amino-5-methoxyanthraquinone, in A-60176
17742-04-0 ▷Trifluoromethanesulfenyl fluoride, T-60281
17754-76-6 Tetrahydroimidazo[4,5-d]imidazole-2,5(1H,3H)dione; 1,4-Di-Me, in T-70066
17759-10-3 4-Methyl-6-phenylpyrimidine; 1-Oxide, in M-70115
17759-27-2 4-Methyl-6-phenylpyrimidine, M-70115

17762-03-7 3-Chloro-6-methylpyridazine; 1-Oxide, in C-70112
17784-60-0 4-(1H-Pyrazol-3-yl)pyridine, P-70163
17796-47-3 2,3-Dihydro-4-thioxo-2(1H)quinazolinone, D-60295
17808-21-8 6-Chlorotryptophan; (±)-form, in C-60146
17811-38-0 5-Hydroxy-7-methoxy-6methylphthalide, in D-60349
17811-39-1 5,7-Dimethoxy-6-methylphthalide, in D-60349
17817-31-1 4',6,7-Trihydroxyisoflavone, T-60325
17823-37-9 1-Bromo-2,3,5,6-tetrafluoro-4nitrobenzene, B-60325
17826-68-5 1-Bromo-2,3,4,6-tetrafluoro-5nitrobenzene, B-60324
17833-66-8 2,3,6,7-Tetrahydro-asindacene-1,8-dione, T-70069
17839-51-9 3-Chloro-2-mercaptobenzoic acid, C-70083
17854-46-5 2-Methylene-4-pentenal, M-70074
17872-39-8 Heptacyclo[6.6.0.0^{2,6}.0^{3,13}.0^{4,11}.0^{5,9}.0^{10,14}]tetradecane, H-70012
17920-35-3 2-Amino-6-methoxypyridine, in A-60252
17920-37-5 6-Amino-1-methyl-2(1H)pyridinone, in A-60252
17924-92-4 ▷Zearalenone; (S)-form, in Z-60002
17933-29-8 7-Methylenebicyclo[3.3.1]nonan-3-one, M-70062
17944-05-7 5,8-Dimethoxyquinoxaline, in Q-60002
17951-19-8 Justicidin B, in D-60493
18014-96-5 4-Methoxybenzofuran, in H-70106
18028-55-2 1,4-Dimethylcarbazole, D-70376
18069-14-2 2,3,5,14,20,22,25Heptahydroxy-cholest-7-en-6-one; (2β,3β,5β,14α,20R,22R)-form, in H-60023
18085-01-3 3-Butene-1,2-diol; (±)-form, Ac, in B-70291
18085-02-4 3-Butene-1,2-diol; (±)-form, Di-Ac, in B-70291
18087-70-2 Imidazo[1,2-b]pyridazine; B,HCl, in I-60007
18089-44-6 Pterosterone, P-60192
18110-71-9 4'-Methoxy-2-biphenylcarboxylic acid, in H-70111
18111-34-7 3-Bromo-2,6-dimethoxyphenol, in B-60200
18121-46-5 4-Vinylquinoline; Picrate, in V-70014
18202-10-3 11-Dodecyn-1-ol, D-70540
18212-30-1 4-Methyl-5-phenyl-1,2,3thiadiazole, M-60116
18213-75-7 5-Amino-1H-pyrazole-4-carboxylic acid; N(1)-Me, amide, in A-70192
18226-42-1 1,3,5-Triss(romomethyl)benzene, T-60401
18229-29-3 8,9-Hexadecanedione, H-70034
18238-55-6 7,7-Difluorobicyclo[4.1.0]hepta-1,3,5-triene, D-60177
18266-93-8 1,2-Di(2-furanyl)ethylene; (Z)-form, in D-60191
18304-03-5 Bicoumol; Di-Ac, in B-70093
18321-17-0 1,2-Dithiepane; 1,1-Dioxide, in D-70517
18321-20-5 1,2-Dithiepane; 1,1,2,2-Tetraoxide, in D-70517
18344-58-6 3,5-Dichloro-4-nitropyridine; 1-Oxide, in D-60147
18356-28-0 Dihydro-1H-pyrrolizine3,5(2H,6H)-dione, D-70260
18377-79-2 2,3-Dihydro-1H-pyrrolizin-1-ol, D-70261
18396-83-3 ▷4-Diazo-2-nitrophenol, D-60065
18398-35-1 7-Nitro-1,2-naphthoquinone, N-60036
18402-82-9 3-Octen-2-one; (E)-form, in O-70027
18409-17-1 2-Octen-1-ol; (E)-form, in O-60029

18409-20-6 2,4-Octadien-1-ol; (2E,4E)-form, in O-70007
18411-75-1 Hautriwaic acid, in H-70002
18412-96-9 GB2a, in G-70005
18438-38-5 ▷2-Pyridinethiol; S-Me, in P-60223
18438-99-8 8-Iodoguanosine, I-70034
18444-66-1 ▷Cucurbitacin E, C-70204
18468-57-0 Hexahydropyrimidine; 1,3-Di-Et, in H-60056
18495-25-5 1-Bromo-2-hexyne, B-70219
18499-84-8 3,6,8-Trihydroxy-1methylanthraquinone-2carboxylic acid, T-60332
18499-99-5 1,3,6,8-Tetramethylphenanthrene, T-60151
18500-00-0 1,3,6,8-Tetramethylphenanthrene; Monopicrate, in T-60151
18508-00-4 Benz[a]anthracene-5,6-dione, D-70014
18513-79-6 1,2,3,6-Tetrahydropyridine; B,HCl, in T-70089
18529-47-0 Suberenol; (E)-form, in S-70080
18546-38-8 Thieno[3,4-b]furan, T-60204
18646-02-1 2-Fluoro-3,5-dinitroaniline, F-60027
18649-93-9 Alisol B, A-70082
18651-67-7 Sikkimotoxin, S-60032
18654-17-6 2-Hydroxy-6-(8,11pentadecadienyl)benzoic acid, H-60205
18670-99-0 N-Methionylalanine; L-L-form, Boc-Met-Ala-OMe, in M-60035
18677-43-5 2,4-Dimethoxypyridine, in H-60224
18686-81-2 Tetrazole-5-thione; (SH)-form, in T-60174
18720-11-1 1,1,4-Triphenyl-1,3-butadiene, T-70305
18720-30-4 Bicyclo[2.2.1]heptane-1carboxylic acid, B-60085
18746-82-2 ▷Terremutin; (−)-form, in T-70009
18755-49-2 4-Cyano-1,2,3-triazole, in T-60239
18773-93-8 1H-Benzotriazole; N-Ac, in B-70068
18776-90-4 5-Heptenoic acid; (E)-form, in H-70024
18799-84-3 4-Iodo-3-thiophenecarboxaldehyde, I-60071
18799-85-4 5-Iodo-2-thiophenecarboxaldehyde, I-60073
18799-94-5 3-Iodo-2-thiophenecarboxaldehyde; Oxime, in I-60069
18799-95-6 4-Iodo-2-thiophenecarboxaldehyde; Oxime, in I-60070
18799-96-7 5-Iodo-2-thiophenecarboxaldehyde; Oxime, in I-60072
18799-97-8 2-Iodo-3-thiophenecarboxaldehyde; Oxime, in I-60068
18799-98-9 4-Iodo-3-thiophenecarboxaldehyde; Oxime, in I-60071
18799-99-0 5-Iodo-3-thiophenecarboxaldehyde; Oxime, in I-60073
18800-00-5 2-Cyano-3-iodothiophene, in I-60075
18800-01-6 3-Cyano-2-iodothiophene, in I-60074
18800-02-7 4-Cyano-2-iodothiophene, in I-60079
18800-03-8 3-Iodo-2-thiophenecarboxylic acid; Amide, in I-60075
18800-04-9 4-Iodo-2-thiophenecarboxylic acid; Amide, in I-60076
18800-05-0 5-Iodo-2-thiophenecarboxylic acid; Amide, in I-60078
18800-06-1 2-Iodo-3-thiophenecarboxylic acid; Amide, in I-60074
18800-07-2 4-Iodo-3-thiophenecarboxylic acid; Amide, in I-60077
18800-08-3 5-Iodo-3-thiophenecarboxylic acid; Amide, in I-60079
18812-38-9 4-Iodo-2-thiophenecarboxaldehyde, I-60070
18812-40-3 2-Iodo-3-thiophenecarboxaldehyde, I-60068

18881-07-7 Hexahydro-1*H*-pyrrolizin-1-one; (*S*)-*form*, *in* H-60057

18887-22-4 1,2,5-Trimethylpyrrolidine, *in* D-70437

18887-24-6 2,5-Dimethylpyrrolidine; (2*RS*,5*SR*)-*form*, *N*-Me, *in* D-70437

18887-25-7 2,5-Dimethylpyrrolidine; (2*RS*,5*SR*)-*form*, *N*-Me, picrate, *in* D-70437

18894-98-9 2-Cyano-4-iodothiophene, *in* I-60076

18894-99-0 3-Cyano-4-iodothiophene, *in* I-60077

18895-00-6 2-Iodo-3-thiophenecarboxylic acid, I-60074

18895-01-7 5-Iodo-3-thiophenecarboxylic acid, I-60079

18913-18-3 3″,3‴,4′,4‴,5,5″,7,7″Octahydroxy-3,8″biflavanone, *in* G-70005

18916-57-9 4-Methoxy-1,2-naphthoquinone, *in* H-60194

18922-04-8 6-Iodo-1-hexene, I-70038

18933-33-0 3-Cyano-2-hydroxy-*N*,*N*,*N*trimethylpropanaminium, *in* C-70024

18945-81-8 2-Cyano-5-iodothiophene, *in* I-60078

18992-65-9 2,7-Dimethylcarbazole, D-70384

18992-66-0 4,5-Dimethylcarbazole, D-70388

18992-67-1 1,2-Dimethylcarbazole, D-70374

18992-68-2 1,3-Dimethylcarbazole, D-70375

18992-70-6 2,3-Dimethylcarbazole, D-70381

18992-71-7 2,4-Dimethylcarbazole, D-70382

18992-72-8 3,4-Dimethylcarbazole, D-70385

19027-02-2 Tricyclo[4.3.1.1^{3,8}]undecane2,7-dione, T-70233

19039-48-6 3,3′-Iminodibenzoic acid, I-60017

19046-26-5 Wistin, *in* T-60325

19057-60-4 Dioscin, *in* S-60045

19064-68-7 1-Chloro-4-methylphthalazine, C-70104

19064-69-8 1-Aminophthalazine, A-70184

19064-74-5 6-Bromophthalazine, B-70268

19067-49-3 9,10-Dihydro-9,10ethenoanthracene-11,12dicarboxylic acid; Dinitrile, *in* D-70208

19083-00-2 Gracillin, *in* S-60045

19087-65-1 Di-2-pyridylmethane; B,2HCl, *in* D-60496

19093-20-0 5-Octen-2-one; (*E*)-*form*, *in* O-70028

19114-85-3 2,2-Dimethyl-2*H*-1-benzothiopyran, D-70373

19143-87-4 Gibberellin A_{20}, G-70011

19149-48-5 Tetrahydro-2(1*H*)-pyrimidinone; 1-β-D-Ribofuranosyl, *in* T-60088

19167-60-3 4,6-Diaminopteridine, D-60043

19167-62-5 4,6,7-Triaminopteridine, T-60230

19179-12-5 Hexahydropyrrolo[1,2-*a*]pyrazine-1,4-dione, H-70073

19190-53-5 α-Methoxybenzeneacetaldehyde, *in* H-60210

19202-36-9 Hinokiflavone, H-60082

19279-79-9 Tetrahydro-1,2,4-triazine-3,6dione, T-60099

19297-13-3 2-Hydroxycyclooctanecarboxylic acid; (1*RS*,2*SR*)-*form*, Et ester, *in* H-70121

19297-14-4 2-Hydroxycyclooctanecarboxylic acid; (1*RS*,2*SR*)-*form*, *in* H-70121

19312-07-3 2-(2-Phenylethyl)-1,4benzenediol, P-60095

19342-73-5 10-Deazariboflavin, *in* P-60245

19358-05-5 2,2-Dimethyl-4-methylene-1,3dioxolane, D-70419

19360-72-6 GB1a, *in* G-70005

19365-01-6 3-Hydroxy-2(1*H*)-pyridinone; *N*-Me, *in* H-60223

19384-65-7 1,2,3,4-Tetrahydro-4-oxo-7quinolinecarboxylic acid, T-60079

19393-86-3 4-Hexen-3-ol; (±)-(*E*)-*form*, Ac, *in* H-60077

19397-28-5 Cedrelopsin, C-60026

19405-19-7 4-Methylenebicyclo[3.1.0]hex2-ene, M-60069

19418-11-2 Tetrahydro-2-thiophenecarboxylic acid, T-70099

19427-32-8 Gibberellin A_{24}, G-60019

19437-26-4 Di-2-pyridyl ketone, D-70502

19437-32-2 2,5-Diphenyl-1,2,4,5tetra-azabicyclo[2.2.1]heptane, D-70498

19467-03-9 Melittoside, M-70019

19479-15-3 Myoporone; (*S*)-*form*, *in* M-60156

19480-01-4 Trifluoroisocyanomethane, T-60278

19495-08-0 3-Phenyl-2-propenoyl isothiocyanate, P-70090

19506-18-4 5,8-Quinoxalinediol, Q-60008

19506-19-5 5,8-Quinoxalinediol; Di-Me ether, *in* Q-60008

19506-20-8 6,7-Quinoxalinediol, Q-60009

19516-14-4 Octahydrobenzo[*b*]thiophene; (3*aRS*,7*aRS*)-*form*, *in* O-60012

19517-16-9 5,6-Dichloro-2(trifluoromethyl)-1*H*benzimidazole; *N*-Me, *in* D-60160

19525-95-2 2-Oxo-4-oxazolidinecarboxylic acid; (*S*)-*form*, *in* O-60082

19536-50-6 Acenaphtho[5,4-*b*]furan, A-60014

19556-33-3 Protetrone, P-70124

19607-08-0 (Dimercaptomethylene)propanedioic acid; Di-*S*-Me, di-Me ester, *in* D-60395

19609-07-5 Pyrrolo[3,2,1-*kl*]phenothiazine, P-70196

19617-90-4 3,6-Diamino-1,2,4,5-tetrazine, D-70045

19643-45-9 2,6-Dibromo-1,4-benzoquinone, D-60088

19652-57-4 Biss(ethylene)butanedinitrile, *in* D-70409

19658-76-5 1-Nitroisoquinoline, N-60029

19689-19-1 5-Decene, D-70017

19700-96-0 1,5-Hexadiene-3,4-diol; (3*RS*,4*SR*)-*form*, *in* H-70036

19700-97-1 1,5-Hexadiene-3,4-diol; (3*RS*,4*RS*)-*form*, *in* H-70036

19719-81-4 Octahydroazocine; 1-Me, *in* O-70010

19720-72-0 Tetrahydro-1*H*,5*H*-pyrazolo[1,2*a*]pyrazole-1,5-dione, T-70087

19740-90-0 9-Oxabicyclo[6.1.0]non-4-ene; (1*RS*,2*SR*,4*Z*)-*form*, *in* O-70058

19756-06-0 1,2-Cyclohexanedithiol; (1*RS*,2*RS*)-*form*, *in* C-70225

19760-94-2 2,3-Biss(romomethyl)-1,4dibromobutane, B-60144

19768-00-4 3,4-Diphenyl-1,2,5-selenadiazole, D-70497

19770-52-6 Benz[*d*]aceanthrylene, B-60009

19806-14-5 [1,1′-Biphenyl]-4,4′-diacetic acid, B-60116

19828-20-7 5-Acetyl-2-aminopyridine, A-70028

19828-21-8 5-Acetyl-2-aminopyridine; B,HCl, *in* A-70028

19829-29-9 ▷4(1*H*)-Pyridinethione, P-60225

19842-75-2 2,3,4,5-Tetrafluorobenzaldehyde; 2,4-Dinitrophenylhydrazone, *in* T-70026

19842-76-3 2,3,5,6-Tetrafluorobenzaldehyde, T-70028

19842-78-5 2,3,4,6-Tetrafluorobenzaldehyde, T-70027

19842-78-5 2,3,4,6-Tetrafluorobenzaldehyde; 2,4-Dinitrophenylhydrazone, *in* T-70027

19842-79-6 2,3,5,6-Tetrafluorobenzaldehyde; 2,4-Dinitrophenylhydrazone, *in* T-70028

19854-92-3 3-Benzoylpyrazole, B-70078

19865-76-0 Alisol *B*; 23-Ac, *in* A-70082

19869-19-3 2,3-Biss(hloromethyl)-1,4dichlorobutane, B-60148

19883-26-2 1,3,5-Undecatriene; (3*Z*,5*Z*)-*form*, *in* U-70004

19883-29-5 1,3,5-Undecatriene; (3*E*,5*E*)-*form*, *in* U-70004

19885-32-6 Octahydropyrrolo[3,4-*c*]pyrrole; (3*aRS*,6*aSR*)-*form*, *N*-(3,5-Dinitrobenzoyl), *in* O-70017

19885-60-0 Octahydropyrrolo[3,4-*c*]pyrrole; (3*aRS*,6*aSR*)-*form*, *in* O-70017

19885-61-1 Octahydropyrrolo[3,4-*c*]pyrrole; (3*aRS*,6*aSR*)-*form*, Dipicrate, *in* O-70017

19891-40-8 *ent*-16-Kauren-3β-ol, K-70010

19897-64-4 Isolaureatin, I-70067

19898-90-9 Peregrinone, *in* M-70013

19899-61-7 4,6,7-Trimethylpteridine, T-70287

19899-62-8 2,4,6,7-Tetramethylpteridine, T-70142

19901-58-7 Homopantetheine; (*R*)-*form*, *in* H-60090

19902-26-2 β-Biotol, B-60113

19902-30-8 α-Biotol, B-60112

19908-48-6 Maackiain; (±)-*form*, *in* M-70001

19908-69-1 Confertin, *in* D-60211

19918-38-8 6-Chloro-2-(trifluoromethyl)1*H*-imidazo[4,5-*b*]pyridine; 4-Oxide, *in* C-60139

19931-06-7 5-Chloro-2-cyclopenten-1-one, C-70063

19932-33-3 5-Methyl-2,4-hexadienoic acid, M-70086

19943-27-2 Octahydro-5*H*,10*H*dipyrrolo[1,2-*a*:1′,2′*d*]pyrazine-5,10-dione; (5*aS*,10*aS*)-*form*, *in* O-60016

19949-46-3 8-Decene-4,6-diynoic acid; (*Z*)-*form*, *in* D-70018

19950-67-5 Calopiptin, *in* A-70271

19959-71-8 4-(1*H*-Pyrazol-4-yl)pyridine, P-70164

19960-71-5 2,5′-Bithiazole, B-70168

19960-72-6 5,5′-Bithiazole, B-70171

19975-56-5 4,5-Dihydro-2-(methylthio)thiazole, *in* T-60197

19991-86-7 6-Amino-2-hexenoic acid; (*E*)-*form*, *N*-Phthalimido, *in* A-60173

19991-87-8 6-Amino-2-hexenoic acid; (*E*)-*form*, *N*-Phthalimido, Me ester, *in* A-60173

19991-88-9 6-Amino-2-hexenoic acid; (*E*)-*form*, B,HCl, *in* A-60173

20024-89-9 4-(Tetrahydro-2-furanyl)morpholine, T-60065

20025-03-0 1,2,3-Trimethyl-4,5,6triphenyl-benzene, T-70291

20034-31-5 2-Methyl-1,3-naphthalenediol, M-60091

20036-72-0 2-Isopropylcycloheptanone, I-70090

20041-90-1 Caffeidine, C-60003

20051-55-2 1,4-Dioxaspiro[4.5]decane-7-thiol, *in* M-70026

20060-07-5 4-Octene-1,7-diyne; (*E*)-*form*, *in* O-60028

20062-24-2 4′-Azidoacetophenone, A-60336

20063-89-2 2-Azetidinecarboxylic acid; (±)-*form*, *in* A-70284

20071-58-3 Shiromodiol; Di-Ac, *in* S-70039

20071-59-4 Shiromodiol; 8-Ac, *in* S-70039

20071-60-7 Shiromodiol, S-70039

20073-13-6 Methyl jasmonate, *in* J-60002

20074-80-0 5-Chloropentanal, C-70137

20081-65-6 1,6-Dihydroxy-5-methoxyxanthone, *in* T-60349

20081-69-0 1,5-Dihydroxy-6-methoxyxanthone, *in* T-60349

20129-11-7 1-Bromo-2,4,5-trimethoxybenzene, *in* B-70181

20145-38-4 8-Oxatricyclo[3.3.0.0²,⁷]octane, O-60054

20175-17-1 2-Iodobutanal; (±)-*form, in* I-60040

20175-18-2 2-Iodo-2-methylpropanal, I-60051

20180-65-8 3,4-Dimethylpentanoic acid; (*S*)-*form, in* D-60435

20180-66-9 3,4-Dimethyl-1-pentanol; (*S*)-*form, in* D-60436

20197-75-5 1,4-Benzodioxan-6-carboxylic acid; Me ester, *in* B-60025

20199-19-3 Di-9-anthrylacetylene, D-60051

20223-76-1 Avenaciolide; (−)-*form, in* A-60325

20235-61-4 1,1,4-Triphenyl-1,3-butadiene; (*E*)-*form, in* T-70305

20247-93-2 1,1′,2,2′,3,4,5,5′,6,6′Decahydrobiphenyl, D-60007

20248-08-2 13(18)-Oleanen-3-one, *in* O-70036

20265-38-7 3-Amino-2-methoxypyridine, *in* A-60249

20265-39-8 4-Amino-2-methoxypyridine, *in* A-60250

20296-09-7 8-Aminopurine, A-60244

20300-26-9 ▷Gossypol; (+)-*form, in* G-60032

20312-78-1 Neotokoronin, *in* S-60043

20315-06-4 Cyclodeca[1,2,3-*de*:6,7,8-*d′e′*]dinaphthalene, C-60187

20324-49-6 4-Chloro-2-mercaptobenzoic acid, C-70084

20324-50-9 5-Chloro-2-mercaptobenzoic acid, C-70085

20324-51-0 2-Chloro-6-mercaptobenzoic acid, C-70082

20335-71-1 4-Benzoylisoquinoline, B-60061

20336-15-6 2,4,6-Tri-*tert*-butylpyridine, T-70217

20348-16-7 2-Amino-3-hydroxy-6methylpyridine, A-60192

20355-59-3 1,8-Dihydroxy-3,6dimethoxyxanthone, *in* T-60127

20362-31-6 Arctiin, *in* A-70249

20362-54-3 2,2′-Dithiobisthiazole, *in* T-60200

20372-66-1 2,4,5-Trifluorobenzoic acid; Me ester, *in* T-70240

20380-11-4 5-Cholestene-3,26-diol; (3*β*,25*R*)-*form, in* C-60147

20395-24-8 1,1-Dichlorooctane, D-70126

20411-45-4 1-(2,2-Dimethylpropyl)naphthalene, D-60446

20411-84-1 2,2-Dichloro-3,3,3-trifluoro1-propanol, D-60161

20420-52-4 3,8-Dihydrocyclobuta[*b*]quinoxaline-1,2-dione, D-70178

20441-18-3 2-Isopropylideneadamantane, I-60121

20456-88-6 5,6-Dihydro-2-phenyl-4(1*H*)pyrimidinone, D-70250

20457-80-1 1-(1-Propenyl)cyclohexene, P-60178

20482-28-4 11-Hydroxy-4-guaien-3-one, H-60149

20511-09-5 Dypnopinacol, D-60524

20532-33-6 5-Chlorobenzo[*b*]thiophene, C-60045

20542-67-0 Cyclohepta[*a*]phenalene, C-60199

20594-92-7 3,5-Diphenyl-1,2,4-oxadiazole; 4-Oxide, *in* D-60484

20615-64-9 ▷9*H*-Fluorene-9-carboxaldehyde, F-70015

20617-33-8 2-(Aminomethyl)cyclohexanol; (1*RS*,2*SR*)-*form, N*-Benzoyl, *in* A-70169

20627-78-5 2,2′-Biss(imethylamino)biphenyl, *in* D-60033

20633-72-1 Monomelittoside, *in* M-70019

20633-93-6 Fortunellin, *in* D-70319

20676-79-3 3,3′,3″-Trimethyltriphenylamine, T-70289

20697-05-6 (3-Nitrophenyl)oxirane, N-70057

20731-46-8 Acetamide; B,½ HBr, *in* A-70019

20734-56-9 1,8-Biss(ethylamino)naphthalene, *in* D-70042

20734-57-0 1,8-Diaminonaphthalene; *N*,*N*,*N*′-Tri-Me, *in* D-70042

20734-58-1 1,8-Biss(imethylamino)naphthalene, *in* D-70042

20734-66-1 3-Amino-1,2-benzenediol, A-70089

20736-09-8 Saikosaponin A, *in* S-60001

20800-00-4 4,4′-Iminodibenzoic acid, I-60018

20804-96-0 5-Amino-2,3,4-trimethylbenzoic acid, A-60267

20851-90-5 3,6-Dichlorophenanthrene, D-70134

20857-44-7 2,5-Biss(rifluoromethyl)pyridine, B 60183

20857-46-9 3,4-Biss(rifluoromethyl)pyridine, B-60185

20857-47-0 3,5-Biss(rifluoromethyl)pyridine, B-60186

20859-02-3 2-Amino-3,3-dimethylbutanoic acid, A-70136

20859-02-3 2-Amino-3,3-dimethylbutanoic acid; (*S*)-*form, in* A-70136

20885-12-5 2-Chloro-6-fluoropyridine, C-60084

20917-57-1 *N*-Asparaginylalanine; L-L-*form, in* A-60310

20924-56-5 1*H*-2-Benzopyran-4(3*H*)-one, B-70042

20931-35-5 Chamaecyparin, *in* H-60082

20931-36-6 Neocryptomerin, *in* H-60082

20931-58-2 Isocryptomerin, *in* H-60082

20934-51-4 1,4-Dihydro-2,3-quinoxalinedione; Dione-*form*, 1-Me, *in* D-60279

20958-62-7 Gravelliferone; Me ether, *in* G-60037

20960-92-3 3-Thiomorpholinecarboxylic acid, T-70182

20968-70-1 1,2-Biss(ethylene)cyclopentane, B-60169

20973-46-0 2-Bromo-3,4,5,6tetrafluoropyridine, B-60326

20987-33-1 Bufalin; 3-(Methylsuberoyl), *in* B-70289

21038-66-4 Pyrido[2,3-*d*]pyrimidine2,4(1*H*,3*H*)-dione, P-60236

21038-67-5 Pyrido[3,4-*d*]pyrimidine2,4(1*H*,3*H*)-dione, P-60237

21083-36-3 Benzothiet-2-one, B-70065

21120-36-5 2-Fluoro-1-phenyl-1-propanone, F-60065

21142-67-6 Colletodiol, C-60159

21144-16-1 1-Methoxy-3-(phenylmethoxy)benzene, *in* B-60073

21146-26-9 1-Amino-1,4cyclohexanedicarboxylic acid; (1*RS*,4*SR*)-*form*, B,HCl, *in* A-60117

21146-28-1 1-Amino-1,4cyclohexanedicarboxylic acid; (1*RS*,4*SR*)-*form, in* A-60117

21146-34-9 1-Amino-1,4cyclohexanedicarboxylic acid; (1*RS*,4*SR*)-*form*, Di-Me ester; B,HCl, *in* A-60117

21150-21-0 ▷Amanine, *in* A-60084

21150-22-1 ▷Amanitins; *β*-Amanitin, *in* A-60084

21150-23-2 ▷Amanitins; *γ*-Amanitin, *in* A-60084

21168-34-3 5-Chloro-8-hydroxyquinoline; 1-Oxide, *in* C-70075

21169-71-1 5-Isoxazolecarboxylic acid, I-60142

21179-17-9 *ζ*-Rhodomycinone, *in* R-70002

21203-79-2 2-Methyl-4-phenylpyrimidine, M-70111

21203-80-5 2-Methyl-4-phenylpyrimidine; Monopicrate, *in* M-70111

21211-29-0 3-Chlorobenzo[*b*]thiophene; 1,1-Dioxide, *in* C-60043

21215-00-9 Dihydro-4,6-(1*H*,5*H*)pyrimidinedione, D-70256

21230-43-3 3-Amino-1*H*-pyrazole-4-carboxylic acid; *N*(1)-Me, Et ester, *in* A-70187

21230-50-2 3-Amino-1-methyl-4pyrazolecarbonitrile, *in* A-70187

21251-20-7 2-Butynedial, B-60351

21251-69-4 1(3*H*)-Isobenzofuranimine, I-70062

21272-85-5 (Nitromethyl)sulfonylbenzene, *in* P-60137

21277-16-7 2-Oxo-4-imidazolidinecarboxylic acid, O-60074

21288-60-8 *ε*-Rhodomycinone, R-70002

21316-80-3 Gravelliferone, G-60037

21347-28-4 Selenourea; *N*¹,*N*¹-Di-Ph, *in* S-70031

21365-84-4 5-Methyladamantanone, M-60041

21377-11-7 3-Amino-1*H*-pyrazole-4-carboxylic acid; *N*(1)-Benzyl, Et ester, *in* A-70187

21400-25-9 1,1,2-Trichloro-1-propene, T-60254

21418-10-0 5,8-Dihydroxy-2,3-dimethyl1,4-naphthoquinone, D-60317

21423-86-9 Bi-2,4-cyclopentadien-1-yl, B-60100

21427-62-3 2,5-Dichloro-3-nitropyridine, D-60143

21428-19-3 6-Iodo-1,3-dimethyluracil, *in* I-70057

21454-46-6 1-Penten-3-one; 2,4-Dinitrophenylhydrazone, *in* P-60062

21461-49-4 3-Nitro-1-propanol; Ac, *in* N-60039

21504-23-4 *α*-(Dimethoxymethyl)benzenemethanol, *in* H-60210

21573-41-1 5-Methoxy-1,2,3-trimethylbenzene, *in* T-70371

21573-70-6 1-Benzoylcyclopentene, B-70072

21615-76-9 4(15)-Eudesmen-11-ol; *α*-L-Arabopyranoside, *in* E-70066

21634-42-4 Dibenzo[*b*,*i*]thianthrene5,7,12,14-tetrone, D-70078

21634-47-9 3′,5,7-Trihydroxy-3,4′,5′,8tetramethoxyflavone, *in* H-70021

21634-52-6 3,3′,4′,5,5′,7,8Heptamethoxyflavone, *in* H-70021

21651-53-6 Nepetalactone; (4*aR*,7*S*,7*aS*)-*form, in* N-60020

21651-62-7 Nepetalactone; (4*aS*,7*S*,7*aR*)-*form, in* N-60020

21657-71-6 4*a*,4*b*,8*a*,8*b*Tetrahydrobiphenylene; (4*aα*,4*bβ*,8*aβ*,8*bα*)-*form, in* T-60060

21705-02-2 ▷Amanitins; *ε*-Amanitin, *in* A-60084

21752-31-8 Methionine sulfoximine; (*S*)ᴄ(*R*)ₛ-*form, in* M-70043

21752-32-9 ▷Methionine sulfoximine; (*S*)ᴄ(*S*)ₛ-*form, in* M-70043

21764-41-0 7,12-Dihydroxy-8,12abietadiene-11,14-dione; 7*β*-*form, in* D-60304

21802-18-6 9-Phenanthreneacetic acid; Me ester, *in* P-70047

21834-60-6 3,4-Dihydro-2*H*-1-benzopyran-3-ol, D-60199

21863-32-1 1,2,3,4-Tetrahydroquinoline; *N*-Benzyl, *in* T-70093

21887-01-4 7,12-Dihydroxy-8,12abietadiene-11,14-dione; 7*α*-*form, in* D-60304

21890-57-3   3-Chloro-5-hydroxy-2-methyl1,4-naphthoquinone, C-60098

21896-87-7   4-Amino-1,1cyclohexanedicarboxylic acid; Di-Et ester; B,HCl, *in* A-60120

21898-84-0   4-Oxatricyclo[4.3.1.1³,⁸]undecan-5-one, O-60055

21917-86-2   7,8-Dihydro-5(6H)-isoquinolinone, D-60254

21921-76-6   4-Bromo-2-furancarboxaldehyde, B-60258

21948-46-9   1,2-Dichlorooctane, D-70127

21964-44-3   1-Nonen-3-ol, N-60058

21977-71-9   1H-1,2,3-Triazole-4-carboxylic acid; (2H)-form, Me ester, *in* T-60239

21985-91-1   3,3',5,5'-Tetraamino2,2',4,4',6,6'-hexanitrobiphenyl, T-60014

22012-97-1   Cryptomerin A, *in* H-60082

22012-98-2   Cryptomerin B, *in* H-60082

22013-58-7   6H-Dipyrido[1,2-a:2',1'-d][1,3,5]triazin-5-ium; Iodide, *in* D-60495

22033-96-1   Viticosterone E, *in* H-60063

22048-84-6   1,2,3,4-Tetrahydro-4-oxo-7quinolinecarboxylic acid; Et ester, *in* T-60079

22048-85-7   1,2,3,4-Tetrahydro-4-oxo-7quinolinecarboxylic acid; Et ester, oxime, *in* T-60079

22055-22-7   ▷Diphyllin, D-60493

22058-69-1   6b,7a-Dihydroacenaphth[1,2-b]oxirene, D-70170

22059-22-9   ▷Acetamidoxime, A-60016

22070-24-2   2,3,4-Trimethyl-2-cyclohexen1-one, T-60359

22075-86-1   12-Chlorododecanoic acid, C-70066

22094-21-9   2-Hydroxy-2-phenylacetaldehyde; (±)-form, Ac, *in* H-60210

22104-78-5   2-Octen-1-ol, O-60029

22104-84-3   2-Octadecen-1-ol, O-60007

22150-76-1   Biopterin; (1'R,2'S)-form, *in* B-70121

22191-26-0   2H-1,4-Benzothiazine-3(4H)thione; 1,1-Dioxide, *in* B-70055

22191-30-6   2H-1,4-Benzothiazine-3(4H)thione, B-70055

22191-31-7   2H-1,4-Benzothiazine-3(4H)thione; 4-Benzyl, *in* B-70055

22219-02-9   2,2'-Thiobisbenzoic acid, T-70174

22235-27-4   4-Hydroxy-5-nitro-1,3benzenedicarboxylic acid, H-70192

22235-28-5   4-Hydroxy-5-nitro-1,3benzenedicarboxylic acid; Di-Me ester, *in* H-70192

22245-55-2   4-Pyreneacetic acid, P-60209

22245-56-3   4-Pyreneacetic acid; Me ester, *in* P-60209

22245-81-4   2,6-Biss(rifluoromethyl)pyridine; 1-Oxide, *in* B-60184

22253-59-4   2-(Trifluoromethyl)pyridine; 1-Oxide, *in* T-60296

22253-59-4   4-(Trifluoromethyl)pyridine; 1-Oxide, *in* T-60298

22253-60-7   2,5-Biss(rifluoromethyl)pyridine; 1-Oxide, *in* B-60183

22253-61-8   2,4-Biss(rifluoromethyl)pyridine; 1-Oxide, *in* B-60182

22253-72-1   3-(Trifluoromethyl)pyridine; 1-Oxide, *in* T-60297

22264-96-6   1,1,1,1',1',1'-Hexafluoro-Nmethoxydimethylamine, *in* B-60180

22267-00-1   3-Hydroxy-5-methoxy-2-methyl1,4-naphthoquinone, *in* D-60350

22267-03-4   8-Methoxy-3-methyl-1,2naphthoquinone, *in* H-60190

22268-16-2   Altersolanol A, A-60080

22270-13-9   Bicyclo[2.2.1]heptan-2-one; (1RS,4SR)-form, *in* B-70098

22273-47-8   5,6-Dihydroxy-2-methyl-1,4naphthoquinone, D-60351

22282-70-8   2-Fluoro-4-iodopyridine, F-60045

22282-75-3   3-Fluoro-4-iodopyridine, F-60046

22282-76-4   3-Fluoro-4-iodopyridine; Picrate, *in* F-60046

22287-28-1   Bicyclo[1.1.1]pentane-1carboxylic acid, B-60101

22298-37-9   1,2,3-Benzotriazine-4(3H)-thione; (2H)-form, 2-N-Et, *in* B-60050

22305-48-2   1,2,3-Benzotriazine-4(3H)-thione; 3H-thione-form, 3-N-Me, *in* B-60050

22305-49-3   4-Mercapto-2-methyl-1,2,3benzotriazinium hydroxide inner salt, *in* B-60050

22305-56-2   1,2,3-Benzotriazine-4(3H)-thione; Thiol-form, S-Me, *in* B-60050

22318-84-9   2,3,5,7-Tetramethoxyphenanthrene, *in* T-60120

22323-80-4   2,2-Dimethyl-1,3-dioxolane-4carboxaldehyde; (S)-form, *in* D-70400

22325-27-5   4,6-Dimethyl-2(1H)pyrimidinethione, D-60450

22329-74-4   3,4-Dihydroxybutanoic acid; (±)-form, *in* D-60313

22343-46-0   7,15-Isopimaradiene-2,3,19-triol; (2α,3β)-form, *in* I-70081

22343-47-1   7,15-Isopimaradiene-3,19-diol; 3β-form, *in* I-70077

22350-90-9   Altersolanol B, *in* A-60080

22353-40-8   2,3-Dichloro-5-nitropyridine, D-60140

22358-21-0   2-Oxo-1,3cyclopentanediglyoxylic acid; Di-Et ester, *in* O-70085

22362-40-9   2-Bromo-5-(methylthio)benzoic acid, *in* B-70227

22395-22-8   7-Methoxyflavone, *in* H-70131

22426-14-8   2-Bromo-1,10-phenanthroline, B-70264

22426-18-2   2-Amino-1,10-phenanthroline, A-70179

22426-19-3   5-Amino-1,10-phenanthroline; B,2HBr, *in* A-70182

22426-20-6   5-Amino-1,10-phenanthroline; B,HBr, *in* A-70182

22428-91-7   2-Methyl-3-oxobutanal, M-70102

22432-95-7   2'-Deoxy-5-azacytidine; α-form, *in* D-70022

22436-26-6   9-Bromo-1,4-dihydro-1,4methanonaphthalene; syn-form, *in* B-60233

22480-88-2   2,3,4-Trihydroxyphenylacetic acid; Tri-Me ether, Me ester, *in* T-60340

22480-91-7   2,3,4-Trimethoxybenzeneacetic acid, *in* T-60340

22503-68-0   4-Oxo-2-heptenedioic acid; Di-Me ester, 2,4-Dinitrophenylhydrazone, *in* O-60072

22503-69-1   4-Oxo-2-heptenedioic acid; Di-Me ester, *in* O-60072

22524-25-0   1-Phenyl-5-hexen-1-one, P-70081

22536-61-4   2-Chloro-5-methylpyrimidine, C-70114

22557-12-6   4,5,7,8-Tetrafluoro[2.2]paracyclophane, T-60050

22601-85-0   Xanthostemone, X-70002

22610-86-2   5-Octen-2-one; (Z)-form, *in* O-70028

22621-26-7   3-Acetyl-4-hydroxy-2(5H)furanone, A-60030

22627-00-5   Diphenylmethaneimine; N-Me, *in* D-60482

22634-92-0   4-Oxoheptanedioic acid; Di-Me ester, *in* O-60071

22644-96-8   2-Hexadecenal; (E)-form, *in* H-60036

22680-62-2   4-Methoxy-2-phenylquinoline, *in* P-60132

22711-24-6   (4-Nitrophenyl)phenylethanedione, N-70059

22713-97-9   3,5-Dianilino-1,2,4-thiadiazole, D-70048

22721-28-4   2-Amino-4,5,6,7tetrahydrobenzo[b]thiophene-3-carboxylic acid; Hydrazide, *in* A-60258

22733-60-4   Siccanin, S-60029

22741-45-3   Methyl N-phenacylcarbamate, *in* O-70099

22756-44-1   Asadanin, A-60308

22776-74-5   4-Azatricyclo[4.3.1.1³,⁸]undecane, A-60333

22800-71-1   N-(Diphenylmethylene)acetamide, *in* D-60482

22804-63-3   1,2,3,4,8-Pentamethoxyxanthone, *in* P-60051

22810-28-2   2-Phenyl-4H-1-benzothiopyran4-one; S,S-Dioxide, *in* P-60082

22839-47-0   ▷Aspartame; L-L-form, *in* A-60311

22839-65-2   Aspartame; L-D-form, *in* A-60311

22842-45-1   3-(Methylthio)cyclohexanone, *in* M-70026

22856-30-0   2,3-Dicyanonaphthalene, *in* N-60003

22856-34-4   4-Oxo-2-heptenedioic acid; 2,4-Dinitrophenylhydrazone, *in* O-60072

22906-55-4   N-Valylleucine; L-D-form, *in* V-60001

22910-60-7   2-Hydroxy-6-(8-pentadecenyl)benzoic acid, *in* H-60205

22918-03-2   4-Chloro-2-iodopyridine, C-60112

22920-77-0   1-Nitro-2-butanone, N-70043

22934-58-3   3-Hydroxy-4,5methylenedioxybenzoic acid; Me ether, Me ester, *in* H-60178

22940-91-6   2,2-Dichloro-3,3-difluorooxirane, D-60128

22963-93-5   Juvenile hormone III, *in* T-60364

22993-77-7   6-Amino-2,2-dichlorohexanoic acid, A-60126

23003-22-7   3-Hydroxy-2(1H)-pyridinethione, H-60222

23003-30-7   3-Hydroxy-2-iodo-6methylpyridine, H-60159

23003-35-2   2-Bromo-3-hydroxy-6methylpyridine, B-60267

23004-60-6   Nephroarctin, N-70032

23007-85-4   1,2,3,6-Tetrahydro-4phenylpyridine; N-Me; B,HCl, *in* T-60081

23020-15-7   2-Phenylcyclopropanecarboxylic acid; (1S,2S)-form, *in* P-60087

23020-18-0   2-Phenylcyclopropanecarboxylic acid; (1S,2R)-form, *in* P-60087

23030-39-9   1-Bromo-2,3,5-trimethoxybenzene, *in* B-70182

23038-58-6   4-Methyl-1,3benzenedicarboxaldehyde, M-60049

23042-98-0   1-Acetylcyclohexene; Oxime, *in* A-70038

23043-62-1   ▷2,9-Diaminoacridine, D-60027

23046-47-1   5-Bromo-1,2,4-benzenetriol; Tri-Ac, *in* B-70181

23109-05-9   ▷Amanitins; α-Amanitin, *in* A-60084

23130-58-7   3-Oxopropanoic acid; Oxo-form, 2,4-Dinitrophenylhydrazone, *in* O-60090

23141-17-5   Isotaxiresinol; 7-Me ether, *in* I-60139

23145-06-4   5,7-Dichlorobenzofuran, D-60126

23145-08-6   5,7-Dibromobenzofuran, D-60087

23156-68-5   O-(2-Hydroxyethyl)hydroxylamine; B,HCl, *in* H-60134

23161-63-9   3,4-Dihydroxy-2pyrrolidinecarboxylic acid; (2S,3S,4S)-form, *in* D-70342

23194-66-3   2-(3-Hydroxyphenyl)-2-oxoacetic acid; Me ether, nitrile, *in* H-70211

23239-35-2   2-Amino-2-methyl-3phenyl-propanoic acid; (*S*)-*form*, *in* A-60225

23240-49-5   1-Oxo-1,2,3,4tetrahydrocarbazole; (*Z*)-oxime, *in* O-70103

23240-52-0   1-Oxo-1,2,3,4tetrahydrocarbazole; (*E*)-Oxime, *in* O-70103

23245-77-4   [4,4′-Bipyridine]-3,3′dicarboxylic acid, B-60137

23245-99-0   Acetophenone; (*E*)-2,4-Dinitrophenylhydrazone, *in* A-60017

23277-50-1   Meglumine salicylate, *in* H-70108

23304-25-8   ▷1-(Diazomethyl)-4methoxybenzene, D-60064

23311-86-6   2-Amino-3-methyl-3-butenoic acid, A-60218

23334-72-7   2,3-Dihydroxybutanoic acid; (2*S*,3*S*)-*form*, *in* D-60312

23351-09-9   1-(4-Hydroxyphenyl)pyrrole, H-60216

23365-04-0   *N*-Valylleucine; L-L-*form*, H-Val-Leu-OMe; B,HCl, *in* V-60001

23369-74-6   Jateorin, *in* C-60160

23386-93-8   1,3,4,6-Tetraphenylthieno[3,4*c*]-thiophene-5-*S*^{IV}, T-70152

23402-19-9   Agasyllin, A-60064

23405-48-3   4-(Methylthio)benzophenone, *in* M-70022

23438-23-5   4-Hydroxy-4-methyl-2,5cyclohexadien-1-one, H-70169

23445-00-3   1,3,5,8-Tetrahydroxyxanthone; 3-Me ether, 8-glucoside, *in* T-70123

23458-04-0   Khellactone; (9*R*,10*S*)-*form*, *in* K-70013

23461-67-8   3,3-Dimethyl-4-oxopentanoic acid, D-60433

23479-73-4   Elvirol, E-60004

23500-57-4   3,3-Biss(minomethyl)oxetane, B-60140

23504-97-4   2-Thiatricyclo[3.3.1.1^{3,7}]decane; 2,2-Dioxide, *in* T-60195

23510-98-7   4-Methylthiocyclohexanone, *in* M-70027

23517-35-3   Benzoylcyclohexane; (*Z*)-Oxime, *in* B-70071

23527-24-4   Kolavenic acid; Me ester, *in* K-60013

23535-17-3   4-Hydroxy-2-methyl-2cyclopenten-1-one, H-60175

23537-79-3   2-Bromo-1,4-dihydro-1,4methanonaphthalene, B-60232

23537-80-6   1-Bromo-1,4-dihydro-1,4methanonaphthalene, B-60231

23554-71-4   Benzoylcyclohexane; (*E*)-Oxime, *in* B-70071

23592-45-2   (Methylamino)methanesulfonic acid, *in* A-60215

23599-45-3   Cinnamodial, C-70177

23619-77-4   9-Diazo-9*H*-thioxanthene, D-70065

23635-83-8   1,4,7-Triazacyclododecane, T-60231

23658-61-9   2-(Dimethylamino)purine, *in* A-60243

23658-67-5   8-(Methylamino)purine, *in* A-60244

23663-77-6   9,9′-Biacridylidene; *N*,*N*′-Di-Me, *in* B-70090

23664-28-0   4,7-Dihydroxy-5-methyl-2*H*-1benzopyran-2-one, D-60348

23674-56-8   3-Amino-2-cyclohexen-1-one; *N*-Ac, *in* A-70125

23687-23-2   8-(Dimethylamino)purine, *in* A-60244

23709-78-6   1,4-Dihydrobenzocyclooctatetraene, D-70173

23720-86-7   Phrymarolin II, P-60142

23724-57-4   7-Methoxy-1,3-benzodioxole-4carboxylic acid, *in* H-60179

23731-78-4   2-Hydroxy-3,4methylenedioxybenzoic acid; Me ether, Me ester, *in* H-60177

23732-22-1   4-Methyl-3-hepten-2-one; (*E*)-form, *in* M-60087

23732-23-2   4-Methyl-3-hepten-2-one; (*Z*)-form, *in* M-60087

23745-82-6   4-Hydroxyphenylacetaldoxime, *in* H-60209

23754-36-1   Pulchellin diacetate, *in* P-70138

23767-00-2   2,6,7-Trimethylpteridine, T-70286

23785-21-9   1*H*-Imidazole-4-carboxylic acid; Et ester, *in* I-60004

23794-15-2   4-Acetyl-2-chloropyridine, A-70035

23812-55-7   4-Hydroxy-2,3methylenedioxybenzoic acid; Me ether, Me ester, *in* H-60179

23827-93-2   *N*-Serylproline, S-60025

23833-96-7   8-Chloro-4(1*H*)-quinolinone, C-60132

23890-32-6   5-Ethenylbicyclo[2.2.1]hept-2ene; (1*RS*,5*SR*)-*form*, *in* E-70036

23967-95-5   Phthalimidine; *N*-Et, *in* P-70100

23978-09-8   ▷4,7,13,16,21,24-Hexaoxa-1,10diazabicyclo[8.8.8]hexacosane, H-60073

23978-55-4   1,7,10,16-Tetraoxa-4,13diazacyclooctadecane, T-60161

23978-55-4   1,4,10,13-Tetraoxa-7,16diazacyclooctadecane, T-70145

23981-25-1   8-Chloro-2(1*H*)-quinolinone, C-60131

23994-23-2   3-Phenyl-1,4-isoquinolinedione, P-70083

24007-54-3   2-Hydroxyphenylacetaldehyde; 2,4-Dinitrophenylhydrazone, *in* H-60207

24034-22-8   2-Bromo-1-nitro-3(trifluoromethyl)benzene, B-60308

24034-72-8   6-Epiophiobolin A, *in* O-70040

24035-43-6   8,11,13-Abietatrien-19-ol, A-60007

24086-18-8   3(5)-Iodo-4-methylpyrazole, I-70048

24092-47-5   9-Chloro-7*H*-benz[*de*]anthracen7-one, C-70046

24100-18-3   2-Bromo-3-methoxypyridine, *in* B-60275

24126-93-0   3′,4′,6,7-Tetramethoxyisoflavone, *in* T-70111

24144-59-0   Khellactone; (9*R*,10*S*)-*form*, Di-Ac, *in* K-70013

24144-61-4   Khellactone; (9*R*,10*R*)-*form*, *in* K-70013

24160-36-9   3-Hydroxy-8,24-lanostadien-21oic acid; 3β-*form*, *in* H-70152

24190-25-8   Neonepetalactone, N-70027

24190-32-7   γ-Ionone; (*R*)-*form*, *in* I-70058

24192-00-5   5-Hydroxy-1,7-diphenyl-1hepten-3-one, *in* D-70481

24192-01-6   5-Hydroxy-1,7-diphenyl-3heptanone, *in* D-70481

24192-73-2   6-Methylpteridine, M-70122

24196-39-2   5-Phenyl-2,4-pentadienoic acid; (*E*,*E*)-*form*, Me ester, *in* P-70086

24207-04-3   2-Bromo-3-hydroxy-6methylpyridine; *N*-Oxide, *in* B-60267

24207-22-5   2-Bromo-3-methoxy-6methylpyridine, *in* B-60267

24211-27-6   Siccanochromene E, S-60030

24259-39-0   3-(Phenylimino)-1(3*H*)isobenzofuranone; B,HClO$_4$, *in* P-60112

24271-83-8   5,8-Quinazolinediol, Q-60002

24280-74-8   4-Azatricyclo[4.3.1.1^{3,8}]undecane; B,HCl, *in* A-60333

24308-04-1   1,4,5-Naphthalenetriol; Tri-Ac, *in* N-60005

24330-52-7   1,3-Butanedithiol, B-60339

24330-53-8   2-Methyl-1,3-propanediol; Bis(4-methylbenzenesulfonyl), *in* M-60120

24383-66-2   2-Thiophenecarboxaldehyde; 2,4-Dinitrophenylhydrazone, *in* T-70183

24405-15-0   Tetrahydro-5,6-dimethyl-2*H*pyran-2-one; (5*RS*,6*RS*)-*form*, *in* T-70059

24462-15-5   Dehydroabietinol acetate, *in* A-60007

24470-47-1   Hardwickiic acid; (+)-*form*, *in* H-70002

24472-05-7   1,4,6,10-Tetraoxaspiro[4,5]decane, T-60164

24502-76-9   ▷Miserotoxin, *in* N-60039

24509-73-7   1*H*-Pyrrolo[3,2-*b*]pyridine; *N*-Ac, *in* P-60249

24513-65-3   2,3-Dihydro-3(phenylmethylene)-4*H*1benzopyran-4-one; (*Z*)-*form*, *in* D-60263

24513-66-4   2,3-Dihydro-3(phenylmethylene)-4*H*1benzopyran-4-one; (*E*)-*form*, *in* D-60263

24514-98-5   3-(Phenylazo)-2-butenenitrile; (*E*,*E*)-*form*, *in* P-60079

24515-82-0   3-(Phenylazo)-2-butenenitrile; (*Z*,*E*)-*form*, P-60079

24526-70-3   2-Oxo-2*H*-benzopyran-4-acetic acid; Chloride, *in* O-60057

24526-71-4   2-Oxo-2*H*-benzopyran-4-acetic acid; Et ester, *in* O-60057

24526-73-6   2-Oxo-2*H*-benzopyran-4-acetic acid, O-60057

24526-74-7   4-(Cyanomethyl)coumarin, *in* O-60057

24536-68-3   1,1′-Biindanylidene; *E*-*form*, *in* B-70210

24570-49-8   Muscarine; (2*S*,4*R*,5*S*)-*form*, Iodide, *in* M-60153

24572-14-3   Biss(iphenylmethyl) diselenide, B-70141

24587-97-1   Domesticoside, *in* T-70248

24593-67-7   2-Cycloundecen-1-one; (*Z*)-*form*, *in* C-70255

24593-68-8   2-Cycloundecen-1-one; (*E*)-*form*, *in* C-70255

24603-68-7   1,2,4-Oxadiazolidine-3,5-dione, O-70060

24653-75-6   1-Mercapto-2-propanone, M-60028

24656-54-0   [2.0.2.0]Metacyclophane, M-60031

24676-83-3   3-(2-Methylpropyl)-6-methyl2,5-piperazinedione; (3*S*,6*S*)-*form*, *in* M-70118

24677-78-9   2,3-Dihydroxybenzaldehyde, D-70285

24716-98-1   Decahydroquinazoline; (4*aR*,8*aS*)-*form*, *in* D-70013

24717-06-4   2-(Bromomethyl)cyclohexanamine; (1*RS*,2*SR*)-*form*, B,HBr, *in* B-70239

24769-97-9   6-Amino-2,2-dichlorohexanoic acid; *N*-Benzoyl, *in* A-60126

24771-25-3   Dimethyl *N*-cyanoimidocarbonate, D-70390

24778-20-9   Artecalin, A-60302

24808-04-6   3,4′,5,7-Tetrahydroxyflavanone; (2*R*,3*R*)-*form*, *in* T-60104

24868-59-5   Pyriculol, P-60216

24871-12-3   Hexahydro-2(3*H*)-benzofuranone; (3*aRS*,7*aRS*)-*form*, *in* H-70049

24915-65-9   Avenacoside A, *in* N-60062

24947-68-0   2-(Aminomethyl)cyclohexanol; (1*RS*,2*RS*)-*form*, B,HCl, *in* A-70169

24948-05-8   2-(Aminomethyl)cyclohexanol; (1*RS*,2*SR*)-*form*, B,HCl, *in* A-70169

24953-77-3   7-Hydroxy-5-methoxyphthalide, *in* D-60339

24966-39-0   2,6-Dichlorobenzenethiol, D-60122

25001-57-4   Justicidin A, *in* D-60493

25006-86-4   1,3-Biss(romomethyl)-2fluorobenzene, B-70133

25079-77-0   1,8-Naphthalenedithiol, N-70003

25090-33-9   3,4-Dihydro-2*H*-pyran-5carboxaldehyde, D-70252

25093-48-5   5-Ethenylbicyclo[2.2.1]hept-2ene; (1*RS*,5*RS*)-*form*, *in* E-70036

25095-48-1   2,4-Diphenylpyrimidine, D-70494

25108-63-8   1,2,3,4-Tetrahydro-1methyl-enenaphthalene, T-60073

25110-76-3   5-Ethyluridine, *in* E-60054

25132-36-9   5,8-Isoquinolinequinone; 8-Oxime, *in* I-60131

25137-00-2   3-Piperidinecarboxylic acid; (*R*)-*form*, *in* P-70102

25146-17-2   Fauronol, F-70002

25148-30-5   2-Amino-5-(methylthio)pentanoic acid; (*S*)-*form*, *in* A-60231

25163-67-1   3-(4-Hydroxyphenyl)-1-(2,4,6trihydroxyphenyl)-2propen-1-one; Tetra-Me ether, *in* H-60220

25177-46-2   9-Phenanthreneacetic acid, P-70047

25181-82-2   5-Ethyl-2,4(1*H*,3*H*)pyrimidinedione; Mono-Hg salt, *in* E-60054

25182-84-7 ▷3-Nitro-1-propanol, N-60039

25187-19-3   6-Chloro-2-phenylquinoxaline, C-70146

25194-01-8   2,6-Dichloro-4-nitropyridine, D-60145

25197-99-3   5-Bromotryptophan; (*S*)-*form*, *in* B-60333

25208-34-8   4-Amino-α-carboline, A-70107

25234-50-8   10*H*-Phenothiazine-2-carboxylic acid, P-70056

25244-60-4   10*H*-Phenothiazine-2-carboxylic acid; *N*-Me, *in* P-70056

25245-41-4   2-Bromo-1,3,4-trimethoxybenzene, *in* B-70180

25252-67-9   4,4-Dimethyl-5-oxopentanoic acid; Nitrile, 2,4-dinitro-phenylhydrazone, *in* D-60434

25328-77-2   Gomphidic acid, G-70026

25343-47-9   Aminoiminomethanesulfonic acid; *N*-Me, *in* A-60210

25343-52-6   Aminoiminomethanesulfonic acid; *N*-Ph, *in* A-60210

25343-55-9   Aminoiminomethanesulfonic acid; *N*,*N*-Di-Me, *in* A-60210

25343-56-0   Aminoiminomethanesulfonic acid; *N*-*tert*-Butyl, *in* A-60210

25346-59-2   4-Oxohexanal, O-60073

25348-84-9   Aminoiminomethanesulfonic acid; *N*,*N*-Diisopropyl, *in* A-60210

25370-92-7   1-Benzyloxy-2-phenyldiazene 2-oxide, *in* H-70208

25370-94-9   1-Methoxy-2-phenyldiazene 2-oxide, *in* H-70208

25380-61-4   4-Chloro-1,2-benzisothiazole, C-70050

25395-13-5   5-Chloro-8-hydroxyquinoline; B,HCl, *in* C-70075

25423-55-6   S,S'-Bis(2-mercaptoethyl)-1,2ethanedithiol, B-70150

25436-90-2   Kolavenic acid, K-60013

25465-34-3   1,2,3-Benzotriazine-4(3*H*)-thione, B-60050

25465-37-6   4-(Butylimino)-3,4-dihydro1,2,3-benzotriazine (incorr.), *in* A-60092

25535-53-9   1,4-Bis(2-pyridylamino)phthalazine, B-60172

25548-04-3   ε-Cadinene, C-60001

25594-62-1   2-Acetylquinoxaline, A-60056

25635-44-3   5-Amino-4-hydroxypentanoic acid, A-60196

25645-08-3   Erivanin, E-70033

25700-11-2   2-(1*H*-Pyrazol-1-yl)pyridine, P-70158

25700-12-3   3-(1*H*-Pyrazol-1-yl)pyridine, P-70160

25700-13-4   4-(1*H*-Pyrazol-1-yl)pyridine, P-70162

25733-27-1   2-Hydroxy-2-methylcyclobutanone, H-70168

25739-23-5   4-(Phenylethynyl)benzoic acid, P-70075

25739-41-7   4',5-Dihydroxy-3',7dimethoxy-flavone, D-70291

25787-42-2   3'-Amino-3'-deoxyadenosine; 3',3',6,6-Tetra-*N*-Me, *in* A-70128

25787-43-3   3'-Amino-3'-deoxyadenosine; 3'*N*-Me, *in* A-70128

25834-71-3   3'-Amino-3'-deoxyadenosine; 3'*N*-Ac, 3'*N*-Me, 2',5'-di-Ac, *in* A-70128

25841-53-6   8'-Hydroxyabscisic acid, H-70099

25844-73-9   2-Chloro-5-phenylpyrazine, C-70143

25889-36-5   3-Nitro-4-(trifluoromethyl)phenol, N-60051

25908-26-3   Cephalochromin, C-60029

25922-93-4   4,5-Diphenylpyrimidine, D-70495

25926-14-1   4,4',4''-Nitrilotrisphenol, N-70040

25942-61-4   1,4-Dimethylthioxanthone, D-70442

26001-38-7   8-Mercaptoguanosine, M-70029

26001-58-1   2-Octen-1-ol; (*Z*)-*form*, *in* O-60029

26028-46-6   Hexahydro-1,3,5-triazine; 1,3,5-Tri-Ac, *in* H-60060

26037-61-6   2-Bromo-3-methoxy-1,4naphthoquinone, *in* B-60268

26057-70-5   Avenaciolide; (±)-*form*, *in* A-60325

26114-57-8   5-Methylthioquinoline, *in* Q-60004

26148-68-5 ▷2-Amino-α-carboline, A-70105

26148-70-9   2-(Methylamino)-1*H*-pyrido[2,3*b*]-indole, *in* A-70105

26153-26-4   α-Oxo-1-naphthaleneacetic acid, O-60078

26154-22-3   2,2-Dimethyl-4-cyclopentene1,3-dione, D-70395

26194-57-0   Isotaxiresinol, I-60139

26241-51-0   Azadiradione, A-70281

26261-80-3   Benzo[5,6]cycloocta[1,2-*c*]furan, B-70029

26279-50-5   4-Chloro-3,5,6trifluoropyridazine, C-60143

26307-17-5   2,5-Dimethyl-3,4-diphenyl-2,4cyclopentadien-1-one, D-70403

26325-89-3   4-Isopropenylcyclohexene, I-60117

26391-89-9   Difluoro-o(rifluoromethanethiolato)-(trifluoromethyl)sulfur, *in* T-60281

26392-12-1   3,6-Biss(iphenylmethylene)1,4-cyclohexadiene, B-60157

26421-44-3   2,5-Dimethyl-3thiophenecarboxaldehyde, D-60452

26421-44-3   2,5-Dimethyl-3thiophenecarboxaldehyde, D-70438

26473-49-4   3-Mercaptobutanoic acid, M-70023

26511-45-5   1,2-Dihydrocyclo-buta[*b*]naphthalene-3,8-dione, D-60214

26537-68-8   3-Benzofurancarboxylic acid, B-60021

26540-64-7   13(18)-Oleanene-3,16,28-triol; (3β,16β)-*form*, *in* O-70035

26548-49-2   Ascorbigen; (2*R*)-*form*, *in* A-60309

26574-86-7   3-Carboxy-2,3-dihydro-8hydroxy-5-methylthiazolo[3,2-*a*]pyridinium hydroxide inner salt; (*S*)-*form*, *in* C-70021

26584-20-3   3-Bromo-2,4,6-trimethylbenzoic acid; Me ester, *in* B-60331

26586-00-5   3-Methyl-4-oxo-2-butenoic acid; (*E*)-*form*, Me ester, di-Et acetal, *in* M-70103

26651-23-0   1-Pyrenesulfonic acid, P-60210

26652-12-0   Salirepin, *in* D-70287

26676-89-1   Ascorbigen; (2*S*)-*form*, *in* A-60309

26685-89-2   Decahydroquinazoline; (4*sS*,8a*R*)-*form*, *in* D-70013

26693-40-3   Decahydroquinazoline; (4*aR*,8a*R*)-*form*, *in* D-70013

26693-41-4   Decahydroquinazoline; (4*aS*,8a*S*)-*form*, *in* D-70013

26775-76-8   Tricyclo[4.3.1.1³,⁸]undecane4,5-dione, T-70234

26807-68-1   2,3-Dihydro-1*H*-indole-2-sulfonic acid; Na salt, *in* D-70221

26807-69-2   2,3-Dihydro-1*H*-indole-2-sulfonic acid; *N*-Ac, *in* D-70221

26831-63-0   3-Hydroxycyclopentanone, H-60115

26832-08-6   1*H*-Imidazole-4-carboxylic acid; Amide, *in* I-60004

26832-19-9   1-Methyladamantanone, M-60040

26888-72-2   4,5,6,7-Tetrafluoro-1*H*benzotriazole, T-70035

26888-73-3   4,5,6,7-Tetrafluoro-1*H*benzotriazole; 1-Ac, *in* T-70035

26908-38-3   Pyrimido[4,5-*b*]quinoline2,4(3*H*,10*H*)-dione, P-60245

26944-65-0   4-Amino-1,2,3-benzotriazine; Amine-*form*, *N*(4)-Benzyl, *in* A-60092

26964-24-9   6-Methoxyflavone, *in* H-70130

26964-26-1   8-Methoxyflavone, *in* H-70132

27003-73-2   Lariciresinol, L-70019

27032-78-6   Dihydro-1,3,5-triazine2,4(1*H*,3*H*)-dione, D-60297

27038-07-9   2-Amino-4-phenyl-3-butenoic acid; (−)-(*E*)-*form*, *in* A-60239

27038-50-2   2,3-Dihydrofuro[2,3-*b*]pyridine, D-70211

27098-03-9 ▷Botryodiplodin, B-60194

27164-10-9   2,5-Dichloro-4hydroxybenzaldehyde, D-70120

27165-32-8   3-Amino-1-hydroxyanthraquinone, A-60180

27208-37-3 ▷Cyclopenta[*cd*]pyrene, C-60223

27268-68-4   1,2,4-Oxadiazolidine-3,5-dione; Dioxo-*form*, 4-Me, *in* O-70060

27268-89-9   3-Methoxy-4-methyl-1,2,4oxadiazolidin-5(4*H*)-one, *in* O-70060

27270-89-9   3-(Hydroxymethyl)-7oxabicyclo[4.1.0]hept3-ene-2,5-dione; (−)-*form*, *in* H-70181

27280-21-3   1,2,4-Oxadiazolidine-3,5-dione; Dioxo-*form*, 4-Et, *in* O-70060

27301-54-8   3-Pentyn-2-ol, P-70041

27305-38-0   Allopteroxylin; Ac, *in* A-60075

27331-38-0   7-Methoxy-1-phenylnaphthalene, *in* P-60119

27331-47-1   2-Methoxy-5-phenylnaphthalene, *in* P-60118

27345-71-7   Hexahydro-2(3*H*)-benzofuranone; (3*aRS*,7*aSR*)-*form*, *in* H-70049

27407-11-0   3-Bromo-4-fluorophenol, B-60250

27455-93-2   4,4-Dimethylbicyclo[3.2.1]octane-2,3-dione, D-60410

27509-30-4   3-(1*H*-Pyrazol-3-yl)pyridine; Monopicrate, *in* P-70161

27519-02-4   9-Tricosene; (*Z*)-*form*, *in* T-60259

27542-21-8   Eriolin, E-60036

27578-93-4   2,3-Dihydroimidazo[1,2-*a*]pyridine, D-70216

27579-06-2   4,4-Dimethyl-5-oxopentanoic acid; Oxime, *in* D-60434

27583-43-3 2-Hydroxycyclohexanemethanol, H-70120

27587-17-3 2-Methyl-1,4benzenedicarboxaldehyde, M-60047

27607-77-8 Trifluoromethanesulfonic acid; Trimethylsilyl ester, *in* T-70241

27642-41-7 8,13-Epoxy-14-labdene; (*ent*-8α,13S)-*form, in* E-70023

27653-13-0 1-(1-Naphthyl)-2-propanol, N-60015

27693-67-0 2,3′-Iminodibenzoic acid, I-60015

27783-00-2 1,3,4-Trihydroxy-6oxabicyclo[3.2.1]octan7-one, *in* Q-70005

27790-74-5 1,4-Dihydropyrimidine, D-70255

27830-00-8 4,4′-Difluoro-[1,1′-biphenyl]3,3′-dicarboxylic acid, D-60178

27830-01-9 4,4′-Difluoro-[1,1′-biphenyl]3,3′-dicarboxylic acid; Di Me ester, *in* D-60178

27839-39-0 Phenarctin, P-70054

27886-55-1 4-Chloro-1,3-benzenedisulfonic acid, C-70049

27890-67-1 6-Methylene-2,4-cyclohexadien1-one, M-60072

27935-87-1 (Iodomethyl)cyclopentane, I-70047

27949-41-3 Δ¹,¹′-Biindene, B-60109

27968-82-7 Isopicropolin, I-60112

27975-80-0 3-Bromo-1-phenyl-1-butyne, B-60314

27979-58-4 5,7-Dihydroxy-1(3H)isobenzofuranone, D-60339

28010-12-0 5-Phenyl-2,4-pentadienoic acid; (E,E)-*form, in* P-70086

28035-02-1 2,3-Dihydro-2,2-dimethyl-4H-1benzothiopyran, D-70193

28048-17-1 Trifluoromethanesulfonic acid; Diethylamide, *in* T-70241

28053-21-6 1,2,3,4-Tetrahydro-1-hydroxy2-naphthalenemethanol; (1RS,2RS)-*form, in* T-70064

28060-21-1 1,2,3,4-Tetrahydro-1-hydroxy2-naphthalenemethanol; (1RS,2SR)-*form, in* T-70064

28070-18-0 5-Chloro-1,3-pentadiene; (E)-*form, in* C-60121

28076-73-5 Bis(2,4-dichlorophenyl)ether, B-70135

28128-41-8 2,8-Diamino-1,7-dihydro-6Hpurin-6-one, D-60035

28140-93-4 ▷3-Cyano-10H-phenothiazine, *in* P-70057

28148-84-7 Ridentin, R-70007

28168-96-9 Pluviatide, *in* A-70063

28230-81-1 Neopulchellin, N-70028

28235-38-3 8,13-Epoxy-14-labdene; (8β,13S)-*form, in* E-70023

28250-37-5 2-(Aminomethyl)cyclohexanol; (1RS,2RS)-*form, in* A-70169

28251-73-2 Virescenoside A, *in* I-70081

28251-74-3 Virescenoside B, *in* I-70077

28286-13-7 Hexahydropyrimidine; 1,3-Diisopropyl, *in* H-60056

28289-54-5 ▷1,2,3,6-Tetrahydro-1-methyl4-phenylpyridine, *in* T-60081

28385-45-7 2-Amino-2-methyl-3phenyl-propanoic acid; (S)-*form*, Me ester, *in* A-60225

28399-17-9 ▷3,11(13)-Eudesmadien-12-oic acid, E-70060

28508-52-3 2,4-Pyridinedithiol, P-60219

28529-48-8 3-Acetyl-1H-indene, A-70046

28584-89-6 2,3,6,7,10,11Hexahydrotrisimidazo-[1,2-*a*;1′,2′-*c*;1″,2″-*e*][1,3,5]-triazine, H-60061

28587-46-4 Tetraneurin C, T-60158

28587-47-5 Tetraneurin *D*, T-60158

28588-74-1 ▷2-Methyl-3-furanthiol, M-70082

28645-51-4 Isoambrettolide, I-60086

28659-87-2 2-Bromo-4-methylpentanoic acid; (S)-*form, in* B-70248

28667-63-2 2,11-Dithia[3.3]paracyclophane, D-60505

28687-81-2 2,4-Diphenyl-5(4H)-oxazolone, D-60485

28845-69-4 5-Oxo-3-hexenoic acid; (E)-*form*, 2,4-Dinitrophenylhydrazone, *in* O-70092

28882-53-3 Ochromycinone, O-60003

28887-47-0 Anthra[2,1-*b*]benzo[*d*]thiophene, A-70215

28892-73-1 3-Hydroxy-2-methylpentanoic acid, H-60192

28915-02-8 Ageratone, A-60065

28954-12-3 ▷Threonine; (2S,3S)-*form, in* T-60217

28966-95-2 Terremutin; (±)-*form, in* T-70009

28986-48-3 Tetrazole-5-thione; (SH)-*form*, S-Ph, *in* T-60174

29043-07-0 3′,4′,5,5′,6,7Hexamethoxyflavone, *in* H-70081

29067-92-3 Lepidopterin, L-60020

29094-75-5 5-Hydroxy-2(1H)-pyridinone; N-Me, *in* H-60225

29097-00-5 3-Amino-1H-pyrazole-4-carboxylic acid; Me ester, *in* A-70187

29125-77-7 α-Doradexanthin, D-70545

29176-18-9 1H-Pyrazolo[5,1-*c*]-1,2,4triazole, P-70157

29176-21-4 3(5)-Azidopyrazole, A-70291

29186-07-0 Biadamantylidene epoxide, B-70091

29202-00-4 3′,5-Dihydroxy-4′,6,7,8tetra-methoxyflavone, D-60375

29282-32-4 9-Oxo-2-decenoic acid; (Z)-*form, in* O-70089

29289-02-9 Polydine, *in* P-70026

29334-16-5 4-Bromodiphenylmethanol, B-60236

29342-61-8 ▷Benzenesulfonyl peroxide, B-70018

29372-44-9 4-Amino-1,2,3-benzotriazine; 3H-*form*, 3-Me, N(4)-Ph, *in* A-60092

29372-45-0 4-Amino-1,2,3-benzotriazine; 3H-*form*, 3-Et, N(4)-Ph, *in* A-60092

29399-87-9 Pipoxide; (S)-*form, in* P-60159

29410-13-7 5-Hydroxy-2,4-imidazolidinedione, H-60156

29421-72-5 5-Methyl-3thiophenecarboxaldehyde, M-70144

29430-70-4 8-Phenyl-2-naphthol, P-60119

29455-26-3 Oliveric acid; (−)-*form, in* O-60039

29476-22-0 2-Phenyl-1,5-benzothiazepin(5H)-one; 2,3-Dihydro, *in* P-60081

29478-27-1 4-Hexen-3-ol; (±)-(E)-*form, in* H-60077

29478-30-6 4-Hexen-3-ol; (±)-(Z)-*form, in* H-60077

29480-18-0 2-Phenyl-2-adamantanol, P-70060

29507-09-3 3,3-Difluorocyclobutene, D-70165

29511-09-9 3,4,5-Trihydroxyphenylacetic acid, T-60343

29518-11-4 4-Phenyl-3-morpholinone, *in* M-60147

29526-96-3 1-Phenylcyclopropanol, P-70066

29536-28-5 1,3,5-Trihexylbenzene, T-70247

29536-29-6 1,3,5-Triheptylbenzene, T-70246

29538-78-1 Apotrisporin *E*, A-70247

29539-21-7 3-Methyl-1H-2-benzopyran-1-one, M-70052

29546-26-7 3-Methyl-1,3,4thiadiazolidine-2,5dithione, *in* M-70031

29555-14-4 4-Hydroxy-2-cyclopentenone; (±)-*form*, Benzoyl, *in* H-60116

29568-78-3 4,5-Dihydroxy-2-nitrobenzoic acid; Di-Me ether, chloride, *in* D-60360

29583-35-5 Bicyclo[2.2.1]heptan-2-one; (1R,4S)-*form, in* B-70098

29585-02-2 1-Bromo-4-methyl-2-pentanone, B-60298

29586-62-7 2-(Bromomethyl)cyclohexanamine; (1R*,2R*)-*form*, B,HBr, *in* B-70239

29598-09-2 4,5,6,7-Tetrahydro-1,2benzisoxazol-3(2H)-one, T-70037

29604-25-9 4-Chloro-2-ethoxy-1-nitrobenzene, *in* C-60119

29623-28-7 13-Hydroxy-9,11-octadecadienoic acid; (9Z,11E,13S)-*form, in* H-70196

29668-44-8 1,4-Benzodioxan-6-carboxaldehyde, B-60023

29700-22-9 1-(2,4-Dihydroxyphenyl)-2(3,5-dihydroxyphenyl)ethylene; (E)-*form, in* D-70335

29745-96-8 2,4-Pentanediamine; (2RS,4SR)-*form, in* P-70039

29745-97-9 2,4-Pentanediamine; (2RS,4RS)-*form, in* P-70039

29746-51-8 4b,9a-Dihydroindeno[1,2-*a*]indene-9,10-dione; *cis*-form, *in* D-70219

29774-53-6 Gibberellin A₂₉, *in* G-70011

29779-78-0 1,1,3,3-Tetramethyl-2methyl-enecyclohexane, T-60144

29782-68-1 Silydianin, S-70044

29783-72-0 1,4,7,10,13Pentaazacyclohexadeca-ne, P-60021

29803-94-9 ▷Cucurbitacin E; Deacetyl, 2-O-β-D-glucopyranosyl, *in* C-70204

29820-55-1 1-Carbethoxybicyclo[1.1.0]butane, *in* B-70094

29836-27-9 Shanzhiside, S-60028

29836-40-6 Tremulacin, *in* S-70002

29836-41-7 Salicortin, S-70002

29838-67-3 Astilbin, *in* P-70027

29841-69-8 1,2-Diphenyl-1,2-ethanediamine; (1R,2R)-*form, in* D-70480

29873-25-4 ▷6-Methylcholanthrene, M-70056

29873-99-2 1,3,7(11)-Elematriene, E-70006

29879-50-3 4,5,6,7-Tetrahydro-2,1benzisoxazol-3(1H)-one, T-70038

29913-86-8 Amarolide, A-60085

29914-17-8 Tetrazole-5-thione; (SH)-*form*, S-Me, *in* T-60174

29926-41-8 2-Acetyl-4,5-dihydrothiazole, A-70041

29949-29-9 3,5-Hexadienoic acid, H-70038

29980-84-5 4-Amino-1,2,3-benzotriazine; 3H-*form*, 3,N(4)-Di-Ph, *in* A-60092

29984-72-3 4-Amino-1,2,3-benzotriazine; 2H-*form*, 2-Me, N(4)-Ph, *in* A-60092

29984-74-5 4-Amino-1,2,3-benzotriazine; 2H-*form*, 2-Et, N(4)-Ph, *in* A-60092

30011-36-0 3-Phenyl-2-propyn-1-amine; B,HCl, *in* P-60129

30013-84-4 10-Chloro-7H-benz[*de*]anthracen-7-one, C-70047

30046-29-8 3-O-Glucosylgibberellin A₂₉, *in* G-70011

30074-90-9 5-Acetylquinoline; Oxime, *in* A-60052

30074-91-0 6-Acetylquinoline; Oxime, *in* A-60053

30093-99-3 4,5-Dihydro-4,4-dimethyloxazole, D-70198

30197-64-9 8-Acetylquinoline; Oxime, *in* A-**60055**

30199-26-9 Xanthorrhizol, X-**70001**

30219-13-7 Bufalin; 3-(Hydrogen suberoyl), *in* B-**70289**

30269-34-2 2,3,6,7-Tetramethoxyphenanthrene, *in* T-**60121**

30271-21-7 Flavoyadorinin B, *in* D-**70291**

30287-05-9 Constictic acid, C-**60163**

30287-26-4 2,4-Dinitro-1(trifluoromethyl)-benzene, D-**70456**

30294-54-3 1,2-Dihydropentalene, D-**70240**

30412-86-3 Calaxin, C-**70007**

30412-87-4 Ciliarin, *in* C-**70007**

30417-04-0 Hexacyclo[5.4.0.0^{2,6}.0^{3,10}.0^{5,9}.0^{8,11}]-undecan-4-one, H-**60035**

30430-91-2 Gravelliferone; 8-Methoxy, *in* G-**60037**

30456-90-7 2,4,6-Cycloheptatriene-1-thione, C-**60203**

30467-62-0 3,5-Dihydro-3,3,5,5tetramethyl-4*H*-pyrazol4-one, D-**60286**

30468-02-1 11-Chloro-7*H*-benz[*de*]anthracen-7-one, C-**70048**

30468-03-2 8-Chloro-7*H*-benz[*de*]anthracen7-one, C-**70045**

30482-37-2 Triss(entachlorophenyl)methyl hexachloroantimonate, *in* T-**60405**

30493-99-3 Bicyclo[1.1.0]butane-1carboxylic acid, B-**70094**

30508-27-1 Licoricidin, L-**60025**

30557-67-6 Montanolide, M-**70154**

30557-81-4 Bisabolangelone, B-**60138**

30614-34-7 Hexacyclo[11.3.1.1^{3,7}.1^{5,9}.1^{11,15}.-0^{2,10}]eicos-2-ene, H-**70031**

30614-73-4 3,4-Biss(ydroxymethyl)furan; Di-Ac, *in* B-**60161**

30684-41-4 1-Amino-β-carboline, A-**70109**

30684-51-6 8-Amino-β-carboline, A-**70113**

30688-55-2 Galiridoside, G-**60002**

30719-18-7 3-Iodo-2-butanone; (±)-*form*, *in* I-**60041**

30719-67-6 ▷Gossypol; (±)-*form*, Hexa-Ac, *in* G-**60032**

30746-58-8 ▷1,2,3,4-Tetrachlorodibenzo-*p*dioxin, T-**70016**

30752-15-9 5-Methyl-2-phenyl-1,5benzothiazepin-4(5*H*)-one, *in* P-**60081**

30759-13-8 Crombeone, C-**70200**

30779-90-9 2,3-Dihydro-3(phenylmethylene)-4*H*-1benzopyran-4-one, D-**60263**

30796-92-0 Phenanthro[4,5-*bcd*]thiophene, P-**60077**

30796-93-1 Phenanthro[4,5-*bcd*]thiophene; 4-Oxide, *in* P-**60077**

30796-94-2 Phenanthro[4,5-*bcd*]thiophene; 4,4-Dioxide, *in* P-**60077**

30800-67-0 1,4-Dichlorobenzo[*g*]phthalazine, D-**70114**

30800-68-1 1-Chlorobenzo[*g*]phthalazine, C-**70054**

30800-69-2 1-Chloro-4-methoxy-benzo[*g*]phthalazine, *in* C-**70055**

30801-90-2 3-Methyl-2-heptenoic acid; (*E*)-*form*, *in* M-**60086**

30826-85-8 ▷Tetrahydro-2(1*H*)-pyrimidinone; 1,3-Di-Et, *in* T-**60088**

30844-12-3 Cyclooctatetraenecarboxaldehyde, C-**70238**

30860-27-6 Ferulidin, F-**70006**

30889-48-6 3-Phenyl-2-naphthol, P-**60116**

30925-48-5 Artemone, A-**60305**

30954-70-2 1,6,8-Trihydroxy-3methylbenz[*a*]anthracene-7,12-dione, T-**70270**

30990-90-0 2-Amino-2-methyl-3phenyl-propanoic acid; (*S*)-*form*, B,HCl, *in* A-**60225**

31020-17-4 1,4-Bis(1,2-diphenyl-ethenyl)benzene; (*E*,*E*)-*form*, *in* B-**70140**

31020-18-5 1,4-Bis(1,2-diphenyl-ethenyl)benzene; (*E*,*Z*)-*form*, *in* B-**70140**

31024-80-3 1,4-Bis(1,2-diphenyl-ethenyl)benzene; (*Z*,*Z*)-*form*, *in* B-**70140**

31037-02-2 5-Amino-1*H*-pyrazole-4-carboxylic acid; *N*(1)-Me, Et ester, *in* A-**70192**

31053-46-0 2,4,7(1*H*,3*H*,8*H*)-Pteridinetrione; 6-Me, *in* P-**70135**

31097-80-0 3,5-Hexadiyn-2-one, H-**60041**

31106-82-8 2-(Bromomethyl)pyridine; B,HBr, *in* B-**70253**

31127-39-6 1,4-Benzodioxan-6-carboxaldehyde; Oxime, *in* B-**60023**

31139-36-3 Dibenzyl dicarbonate, *in* D-**70111**

31140-59-7 3-(Bromomethyl)-2-fluoropyridine; B,HBr, *in* B-**70244**

31140-60-0 4-(Bromomethyl)-2-fluoropyridine; B,HBr, *in* B-**70245**

31140-61-1 5-(Bromomethyl)-2-fluoropyridine; B,HBr, *in* B-**70246**

31140-62-2 2-Bromomethyl)-6-fluoropyridine; B,HBr, *in* B-**70243**

31167-92-7 Dihydro-3-iodo-2(3*H*)-furanone, D-**60252**

31189-39-6 7-Acetylquinoline; Oxime, *in* A-**60054**

31249-95-3 1,4,10-Trioxa-7,13diazacyclo-pentadecane, T-**60376**

31301-40-3 5-Amino-3-phenylisoxazole; *N*-Ac, *in* A-**60240**

31301-51-6 2-Chloro-5-fluoropyridine, C-**60083**

31341-59-0 7,11:20,24Dinitrilodibenzo[*b*,*m*]-[1,4,12,15]tetraazacyclodocosine, D-**60456**

31364-42-8 4,7,13,16,21-Pentaoxa-1,10diazabicyclo[8.8.5]tricosane, P-**60000**

31366-25-3 Bi(1,3-dithiol-2-ylidene), B-**70119**

31377-05-6 Calciopor, *in* O-**70102**

31390-84-8 (4-Nitrophenyl)phenylethanedione; Monoxime, *in* N-**70059**

31415-00-6 6-Methoxy-2-(3,7-dimethyl-2,6octadienyl)-1,4benzoquinone; (*E*)-*form*, *in* M-**60038**

31458-06-7 5,6-Acenaphthenedicarboxylic acid; Biss(imethylamide), *in* A-**70015**

31486-46-1 10*H*-[1]Benzothieno[3,2-*b*]indole; *N*-Me, *in* B-**70059**

31486-47-2 10*H*-[1]Benzothieno[3,2-*b*]indole; *N*-Et, *in* B-**70059**

31511-11-2 Aconitic acid; (*Z*)-*form*, Anhydride, *in* A-**70055**

31557-77-4 2-Acetyl-5-aminopyridine; *N*-Ac, *in* A-**70023**

31561-59-8 Trispiro[2.0.2.0.2.0]nonane, T-**70325**

31571-69-4 2-Amino-3-methylbutanedioic acid; (2*S*,3*S*)-*form*, *in* A-**70166**

31646-15-8 *N*,*N*-Dimethylmethaneselenoamide, *in* M-**70035**

31660-13-6 11,15-Dihydroxy-9-oxo-5,13prostadienoic acid; (5*Z*,8*RS*,-11*RS*,12*RS*,13*E*,15*RS*)-*form*, *in* D-**60367**

31660-17-0 11,15-Dihydroxy-9-oxo-5,13prostadienoic acid; (5*Z*,8*SR*,-11*RS*,12*RS*,13*E*,15*SR*)-*form*, *in* D-**60367**

31685-80-0 Pinusolide, P-**60156**

31694-90-3 1-(Phenylmethyl)-1*H*-tetrazol5-amine, *in* A-**70202**

31694-91-4 5-Amino-2-benzyl-2*H*-tetrazole, *in* A-**70202**

31699-37-3 2,4,6,8-Undecatetraene, U-**70003**

31704-42-4 1-(Chloromethyl)-1*H*-indole2,3-dione, C-**70092**

31806-66-3 Benzo[*a*]quinolizinium(1+); Bromide, *in* B-**70043**

31818-12-9 Bicyclo[2.2.2]octane-1carboxylic acid; Et ester, *in* B-**60092**

31823-03-7 5-(2-Methyloxiranyl)benzofuran, M-**70101**

31839-20-0 5-Hydroxy-4-methoxy-2nitro-benzoic acid, *in* D-**60360**

31857-89-3 5-Tetradecanone, T-**60039**

31872-62-5 4-Methoxy-3-nitropyridine, *in* N-**70063**

31917-92-7 2-Phenylisatogen oxime, *in* P-**60114**

31968-33-9 2-Acetyl-3-aminothiophene, A-**70029**

32093-23-5 1,2,3,4-Tetrahydro-1-hydroxy2-naphthalenecarboxylic acid; (1*RS*,2*RS*)-*form*, Me ester, *in* T-**70063**

32093-24-6 1,2,3,4-Tetrahydro-1-hydroxy2-naphthalenecarboxylic acid; (1*RS*,2*SR*)-*form*, Me ester, *in* T-**70063**

32141-12-1 1-Acetyl-8-iodonaphthalene, A-**60038**

32180-43-1 3-(Hydroxymethyl)-7oxabicyclo[4.1.0]hept-3-ene-2,5-dione; (±)-*form*, *in* H-**70181**

32189-59-6 *N*,*N*'-Diphenylformamidine; *N*-Me, *in* D-**60479**

32203-46-6 1-Epierivanin, *in* E-**70033**

32205-23-5 3,19-Dihydroxy-12-oleanen-28-oic acid; (3β,19β)-*form*, *in* D-**70327**

32257-52-6 Spiro[5.5]undecane-1,9-dione, S-**70066**

32273-62-4 Bicyclo[2.2.2]octa-5,7-diene2,3-dione, B-**70109**

32276-00-9 *S*,*S*'-Bis(1-phenyl-1*H*-tetrazol5-yl) dithiocarbonate, B-**70155**

32281-36-0 Benzo[1,2-*b*:4,5-*b*']dithiophene-4,8-dione, B-**70169**

32281-97-3 7-Bromo-3,4-dihydro-1(2*H*)naphthalenone, B-**60234**

32296-88-1 1,4,8-Trioxaspiro[4.5]decan7,9-dione, O-**60084**

32315-10-9 Biss(richloromethyl) carbonate, B-**70163**

32337-95-4 2-Bromo-1-iodo-3-nitrobenzene, B-**70223**

32337-96-5 1-Bromo-2-iodo-3-nitrobenzene, B-**70222**

32363-92-1 1,1-Dibromo-3-methyl-1-butene, D-**60104**

32388-55-9 Acetylcedrene, A-**70032**

32398-66-6 1-Phenyl-2-butyn-1-ol, P-**70062**

32405-50-8 1,1'-Methylenebispyridinium(1+); Diiodide, *in* M-**70065**

32460-09-6 Tetrabromofuran, T-**60023**

32479-45-1 Asadanol, *in* A-**60308**

32492-74-3 Myricanone, *in* M-**70162**

32521-21-4 2'-(Hydroxymethyl)acetophenone, H-**70158**

32541-33-6 1-Chloro-3-pentanol, C-**70138**

32548-24-6 Hexahydro-3*H*-pyrrolizin-3-one, H-**60058**

32560-26-2 3,4-Dihydro-2*H*-1-benzopyran-2-ol, D-**60198**

32568-59-5 (4-Hydroxyphenyl)ethylene; Benzoyl, *in* H-**70209**

32601-86-8 2-Chloro-3-methylquinoxaline, C-**70121**

32602-96-3 2,5-Dinitro-1*H*-pyrrole, D-**60463**

32615-84-2 2,2'-Iminobisbenzamide, *in* I-**60014**

32621-46-8 2,2'-Iminodibenzoic acid; Dichloride, *in* I-**60014**

32659-31-7 5'-Chloro-5'deoxyarabinosylcytosine, C-**70064**

32666-56-1 1-Amino-5,8-dihydronaphthalene, A-**60140**

32673-76-0 [1,1'-Bicyclohexyl]-2,2'-dione, B-60089

32686-09-2 Floroselin, in K-70013

32744-83-5 3-Chloro-2,5dihydroxybenzaldehyde, C-60057

32744-84-6 2-Chloro-3,6dihydroxybenzaldehyde, C-60056

32749-97-6 4-Hydroxy-3-nitrosobenzoic acid, H-70193

32775-94-3 3,5-Hexadienoic acid; (E)-form, Me ester, in H-70038

32777-04-1 3-Chloro-2-hydroxypropanoic acid; (±)-form, Me ester, in C-70072

32804-66-3 3-Butynoic acid; Me ester, in B-70309

32819-31-1 13-Hydroxy-9,11-octadecadienoic acid; (±)-(9E,11E)-form, in H-70196

32821-76-4 2-Ethoxy-2-propenoic acid, E-70037

32833-13-9 [1,3,5]Triazino[1,2-a:3,4a':5,6-a'']-trisbenzimidazole, T-60238

32861-73-7 5H-Indolo[2,3-b]quinoxaline; 5,11-Dioxide, in I-60036

32872-29-0 N-Ethoxy-1,1,1,1',1',1'hexa-fluorodimethylamine, in B-60180

32886-15-0 ▷Lipoxamycin, L-70033

32890-92-9 2,3,4,6-Tetrafluorobenzoic acid, T-70033

32918-38-0 3-(Bromomethyl)-2-chloropyridine; B,HBr, in B-70236

32918-40-4 5-Bromomethyl)-2-chloropyridine; B,HBr, in B-70238

32936-71-3 6a,9b-Methano-1H-phenalen-1-one, M-70042

32938-48-0 4-(Bromomethyl)-2-chloropyridine; B,HBr, in B-70237

32938-50-4 2-(Bromomethyl)-6-chloropyridine; B,HBr, in B-70235

32954-39-5 4-Amino-3-chlorobutanoic acid, A-60106

32988-67-3 Saptarangiquinone A, S-70009

33011-97-1 Dibenzo[a,d]cycloheptenylium, D-60071

33021-53-3 3-Acetylquinoline, A-60050

33023-90-4 Tetrazolo[1,5-b]quinoline5,10-dione, T-70156

33059-38-0 5H-Cycloprop[f]isobenzofuran, C-70251

33070-04-1 1,4-Benzodioxan-2-carboxylic acid; (±)-form, Amide, in B-60024

33080-41-0 5H,9H-Quino[3,2,1-de]acridine, Q-70006

33097-28-8 [1,2,5]Thiadiazolo[3,4-d]pyridazin-4(5H)-one, T-70164

33101-88-1 1,2,4-Oxadiazolidine-3,5-dione; Dioxo-form, 2-Ph, in O-70060

33105-81-6 2-Amino-3,3-dimethylbutanoic acid; (±)-form, in A-70136

33142-15-3 1,2-Hexadien-5-yne, H-70040

33178-96-0 2-Mercaptopropanoic acid; (R)-form, in M-60027

33179-02-1 2-Mercaptopropanoic acid; (R)-form, S-Benzoyl, in M-60027

33193-41-8 [1]Benzothieno[3,2-c]quinoline, B-70064

33240-33-4 3-tert-Butylphenanthrene, B-70306

33240-61-8 Triss(entachlorophenyl)methane, T-60405

33251-42-2 Selenourea; N-Et, in S-70031

33275-54-6 3,4,5,6-Tetrahydro-4-hydroxy6-methyl-2H-pyran-2-one; (4S,6S)-form, in T-60067

33276-04-9 Parasorboside, in T-60067

33284-11-6 1,5-Dihydropentalene, D-70242

33368-92-2 6-Chloro-7-methylquinoxaline; 1,4-Dioxide, in C-70125

33400-89-4 Coriolin B, in C-70199

33400-90-7 Coriolin C, in C-70199

33404-85-2 ▷Coriolin; (−)-form, in C-70199

33442-56-7 2-Norerythromycin C, in N-70076

33449-51-3 4,6-Dihydro-5H-cyclopenta[b]thiophen-5-one, D-70183

33449-52-4 3-Mercaptocyclohexanone, M-70026

33466-46-5 Drosopterin, D-60521

33466-47-6 Isodrosopterin, D-60521

33467-79-7 2,4-Heptadien-1-ol; (2E,4E)-form, in H-70014

33468-69-8 4(5)-Trifluoromethyl-1Himidazole, T-60290

33471-40-8 4-Methyl-3(2H)-pyridazinone, M-70125

33507-63-0 Substance P, S-60059

33511-69-2 2-Amino-4-methylpentanedioic acid; (2S,4S)-form, in A-60223

33511-70-5 2-Amino-4-methylpentanedioic acid; (2S,4R)-form, in A-60223

33512-01-5 N-Alanylcysteine; D-L-form, in A-60069

33553-23-0 2-Phenacylcyclohexanone, P-70045

33567-59-8 2-Methoxybenzeneacetaldehyde, in H-60207

33567-62-3 2,5-Dimethoxy-benzeneacetaldehyde, in D-60370

33580-60-8 2,3,5-Trihydroxybenzoic acid, T-70250

33599-61-0 6-Bromotryptophan; (±)-form, in B-60334

33606-81-4 Myricanol, M-70162

33623-18-6 2-Amino-4(1H)-pyridinone, A-60253

33624-92-9 Hexakiss(ichloromethyl)benzene, H-60067

33630-94-3 5-Amino-2(1H)-pyridinone; NH-form, in A-60251

33630-96-5 5-Amino-1-methyl-2(1H)pyridinone, in A-60251

33630-97-6 4-Amino-2(1H)-pyridinone; OH-form, in A-60250

33630-99-8 3-Amino-2(1H)-pyridinone; NH-form, in A-60249

33631-01-5 3-Amino-1-methyl-2(1H)-pyridinone, in A-60249

33631-05-9 2-Amino-4(1H)-pyridinone; OH-form, in A-60253

33631-09-3 3-Amino-4-methoxypyridine, in A-60254

33631-10-6 3-Amino-1-methyl-4(1H)pyridinone, in A-60254

33631-21-9 3-Amino-2(1H)-pyridinone; NH-form, B,HCl, in A-60249

33631-22-0 3-Amino-2(1H)-pyridinone; NH-form, Picrate, in A-60249

33646-78-5 5,8-Dihydroxy-6,10-dimethoxy2-propyl-4H-naphtho[1,2-b]pyran-4-one, in C-70191

33646-79-6 8-Hydroxy-5,6,10-trimethoxy-2propyl-4H-naphtho[1,2b]pyran-4-one, in C-70191

33656-65-4 α-Oxo-1-naphthaleneacetic acid; Et ester, in O-60078

33665-27-9 4-Octen-2-one, O-60030

33675-75-1 3-(2-Phenylethyl)phenol, P-70072

33682-80-3 3-Methyl-5-(methylthio)-1,3,4thiadiazole-2(3H)-thione, in M-70031

33689-28-0 1,2-Cyclobutanedione, C-60183

33698-60-1 1,6-Heptadien-3-one, H-70017

33698-63-4 1,5-Heptadien-4-one; (E)-form, in H-70016

33698-68-9 1,5-Heptadien-3-one; (Z)-form, in H-70015

33709-29-4 ▷3',4',5'-Trihydroxyacetophenone, T-60315

33739-50-3 2,7-Dihydroxy-2,4,6cyclo-heptatrien-1-one, D-60314

33835-87-9 4H-1,3-Benzodioxin-6-carboxylic acid, B-60027

33835-89-1 4H-1,3-Benzodioxin-6-carboxylic acid; Me ester, in B-60027

33842-22-7 4,5-Dihydroxy-1,3benzenedicarboxylic acid; Di-Me ester, in D-60307

33853-96-2 Lycoxanthol, L-70046

33873-09-5 2,6-Difluoropyrazine, D-60189

33884-53-6 Angeloylprangeline, in P-60170

33889-70-2 Prangeline; (R)-form, in P-60170

33889-70-2 Prangeline; (S)-form, in P-60170

33903-83-2 Boronolide, B-60193

33922-66-6 2-Hexyl-5-methyl-3(2H)furanone, H-60079

33942-00-6 2-Chloro-2-phenylethanol; (S)-form, Ac, in C-60128

33948-83-3 5,7-Dihydrodibenzo[c,e]selenepin, D-70189

34021-73-3 3,5-Dihydroxy-4methoxy-benzeneacetic acid, in T-60343

34025-29-1 2-Hydroxy-2-phenylacetaldehyde, H-70206

34025-29-1 2-Hydroxy-2-phenylacetaldehyde, H-60210

34068-59-2 2-Ethoxy-2-propenoic acid; Amide, in E-70037

34069-89-1 2,2'-Iminodibenzoic acid; Di-Me ester, in I-60014

34121-64-7 1,4-Di(2-furyl)benzene, D-70169

34175-33-2 α-Amino-1H-imidazole-2-propanoic acid, A-60205

34177-31-6 2-Cyanotriphenylene, in T-60384

34197-88-1 2-Azidoadamantane, A-60337

34202-45-4 1,2-Cyclopropanedicarboxylic acid; (1R,2R)-form, in C-70248

34212-88-9 Virescenoside F, in I-70081

34212-90-3 Virescenoside G, in I-70077

34212-94-7 Virescenoside E, in I-70081

34239-78-6 Bromomethanesulfonic acid; Na salt, in B-60291

34251-41-7 Dithiobisphthalimide, D-60508

34251-41-7 1,1'-Dithiobis-2,5pyrrolidinedione, D-70524

34270-90-1 1,1'-Oxybis[2-iodoethane], O-60096

34272-56-5 1-Isobutyryloxymethyl-4isopropyl-2,5-dimethoxybenzene, in D-70309

34305-47-0 Naphthvalene, N-60012

34318-16-6 1,8-Dihydroxy-2,3,6trimethoxy-xanthone, in P-70036

34328-46-6 4-Chloro-3-(tri-fluoromethyl)benzaldehyde, C-70167

34328-61-5 3-Chloro-4-fluorobenzaldehyde, C-70067

34341-05-4 Cyclopentadecanone; Oxime, in C-60219

34341-27-0 3,3'-Bi[5-hydroxy-2-methyl1,4-naphthoquinone], B-60106

34385-93-8 1,4-Benzodioxan-2-carboxylic acid; (±)-form, in B-60024

34444-37-6 Wikstromol, W-70003

34522-43-5 3'-Amino-3'-deoxyadenosine; 3',6,6-Tri-N-Me, in A-70128

34566-04-6 2-Octylamine; (S)-form, in O-60031

34566-05-7 2-Octylamine; (R)-form, in O-60031

34668-26-3 Furo[2,3-b]pyridine-2-carboxylic acid, F-60098

34673-69-3 [1]Benzoselenopheno[2,3-b][1]benzothiophene; Picrate, in B-60052

34673-70-6 [1]Benzoselenopheno[2,3-b][1]benzothiophene; Trinitro-fluorenone complex, in B-70052

34678-70-1 1,4-Dimethyl-2,5biss(ethylthio)-benzene, in D-60406

34689-84-4 2-Acetyl-5-aminopyridine; B,HCl, in A-70023

34707-92-1 Chlorothricin, C-60137

34713-97-8 2-Phenylcyclohexanecarboxylic acid; (1*S*,2*S*)-*form*, Chloride, in P-70065

34729-49-2 4,4′,4″-Trimethyltriphenylamine; B,HClO₄, in T-70290

34761-14-3 3-Amino-2-cyanobenzo[*b*]-thiophene, in A-60091

34771-45-4 5-Phenylpyrimidine, P-70093

34771-47-6 2-Methyl-5-phenylpyrimidine, M-70112

34771-48-7 4-Methyl-2-phenylpyrimidine, M-70113

34804-98-3 *N*-Alanylcysteine; L-L-*form*, Z-Ala-Cys-OMe, in A-60069

34818-57-0 2,2-Dimethyl-2*H*-1-benzopyran6-carboxylic acid; Me ester, in D-60408

34858-80-5 13*H*-[1,3,5]Thiadiazino[3,2*a*:5,6-*a*′]bisbenzimidazole-13-thione, T-60191

34897-33-1 3,4-Dimethylpentanoic acid; (*S*)-*form*, Me ester, in D-60435

34941-91-8 2-Chloro-4-fluoropyridine, C-60082

34941-92-9 4-Chloro-2-fluoropyridine, C-60086

34998-98-6 2,4-Pentanediamine; (2*R*,4*R*)-*form*, in P-70039

34998-99-7 2,4-Pentanediamine; (2*S*,4*S*)-*form*, in P-70039

35001-24-2 6-Hydroxy-1(10),4,11(13)germacratrien-12,8-olide; (6α,8α)-*form*, in H-70134

35001-25-3 6-Hydroxy-1(10),4,11(13)germacratrien-12,8-olide, H-70134

35001-25-3 Laurenobiolide, in H-70134

35010-10-7 (2-Nitrophenyl)phenylethanedione, N-70058

35013-09-3 5,7-Dihydroxy-4′-methoxyflavone; 7-β-D-Galactoside, in D-70319

35060-50-5 2-Acetyl-1,2,3,4tetra-hydronaphthalene, A-70053

35065-27-1 ▷2,2′,4,4′,5,5′Hexachlorobiphenyl, H-60034

35070-01-0 2-(Phenylethynyl)thiazole, P-60099

35132-20-8 1,2-Diphenyl-1,2-ethanediamine; (1*S*,2*S*)-*form*, in D-70480

35148-18-6 9-Dodecen-1-ol; (*Z*)-*form*, in D-70537

35148-19-7 9-Dodecen-1-ol; (*E*)-*form*, Ac, in D-70537

35151-69-0 *N*-(1,4-Dihydro-1,4-dimethyl5*H*-tetrazol-5-ylidene)methanamine, in A-70202

35154-55-3 3-Hydroxy-3′,4′,5,6,7,8hexamethoxyflavone, in H-60024

35154-56-4 3,3′,4′,5,6,7,8Heptahydroxyflavone; 3′,4,5,6,7,8-Hexa-Me ether, Ac, in H-60024

35193-37-4 2-Hydroxycyclooctanecarboxylic acid; (1*RS*,2*RS*)-*form*, Me ester, in H-70121

35200-11-4 3,3*a*,6,6*a*-Tetrahydro-1(2*H*)penta-lenone, T-70083

35212-85-2 3-Amino-2-benzo[*b*]thiophenecarboxylic acid; Me ester, in A-60091

35213-00-4 2-Cyano-1,3-dinitrobenzene, in D-60457

35237-62-8 9-Dodecen-1-ol; (*E*)-*form*, in D-70537

35290-21-2 Nordracorhodin, N-70074

35290-22-3 Nordracorubin, N-70075

35290-22-3 Nordracorubin; (−)-*form*, in N-70075

35298-85-2 Coleon *C*, C-60157

35354-15-5 2,2-Diethylbutanoic acid; Chloride, in D-60171

35357-33-6 1,4-Benzoquinone; Biss(thylene ketal), in B-70044

35357-34-7 7,10-Dioxaspiro[5.4]deca-2,5dien-4-one, in B-70044

35365-19-6 2-Iodocyclohexanone, I-60042

35394-09-3 2-Methoxycyclopentanone, in H-60114

35417-96-0 9-Acridinecarboxylic acid; Amide, in A-70061

35421-08-0 4-(Chloromethyl)phenol, C-70103

35444-93-0 2-Iododiphenylmethane, I-60045

35444-94-1 4-Iododiphenylmethane, I-60046

35455-33-5 Bufalitoxin, in B-70289

35455-79-9 3,4-Dihydro-5-methyl-4-thioxo2(1*H*)-pyrimidinone, D-70233

35552-67-1 1-Aminophthalazine; *N*,*N*-Di-Me, in A-70184

35558-00-0 Bicyclo[2.2.2]octa-5,7-diene2,3-dione; Dioxime, in B-70109

35589-31-2 4-Amino-1,2-benzenediol; *O*¹-Me; B,HCl, in A-70090

35628-00-3 2-Methyl-6-methylene-2,7octadien-4-ol; (*S*)-*form*, in M-70093

35628-05-8 2-Methyl-6-methylene-7-octen4-ol, in M-70093

35721-53-0 1-Cyclododecenecarboxaldehyde; (*E*)-*form*, in C-60191

35731-89-6 Ircinin 1, I-60083

35732-93-5 9-Octadecene-1,12-diol; (*R*,*E*)-*form*, in O-60006

35761-52-5 Ircinin 2, in I-60083

35779-04-5 1-*tert*-Butyl-4-iodobenzene, B-60348

35779-35-2 Di-3-pyridyl ketone, D-70503

35822-50-5 [2-(Methylthio)ethenyl]benzene; (*Z*)-*form*, in M-70137

35825-57-1 Cryptotanshinone; (*R*)-*form*, in C-60173

35852-81-4 4-Chloro-1-methylpyrazole, in C-70156

35854-47-8 2-(Aminomethyl)pyridine; *N*-Benzoyl, in A-60227

35857-62-6 9-Tricosene; (*E*)-*form*, in T-60259

35866-60-5 8-Hydroxyachillin; 8α-*form*, Ac, in H-60100

35879-92-6 8-Hydroxyachillin; 8α-*form*, in H-60100

35920-91-3 Avenacoside B, in N-60062

35923-79-6 Pentafluoroisothiocyanatobenzene, P-60037

35930-25-7 Khellactone; (9*RS*,10*RS*)-*form*, Di-Ac, in K-70013

35930-31-5 Perforatin A, in A-60075

35980-61-1 1,4,8-Triazacycloundecane, T-60235

35980-62-2 1,5,9-Triazacyclododecane; B,3HBr, in T-60232

35991-75-4 3,5-Biss(romomethyl)pyridine, B-60147

36001-47-5 Vescalagin, V-70006

36059-14-0 6-Chloro-2-phenylquinoxaline; 1,4-Dioxide (?), in C-70146

36067-56-8 Datiscoside, in C-70203

36073-93-5 4-Nitroisoquinoline, N-60030

36099-80-6 Octahydro-2-hydroxy-5*H*,10*H*dipyrrolo[1,2-*a*:1′,2′*d*]-pyrazine-5,10-dione; (2*R*,5*aS*,10*aS*)-*form*, in O-60017

36112-60-4 6-Hydroxy-1-naphthaldehyde; Me ether, oxime, in H-70184

36149-85-6 Candicanin, C-70012

36150-00-2 ▷11,15-Dihydroxy-9-oxo-5,13prostadienoic acid; (5*E*,8*R*,11*R*,12*R*,13*E*,15*S*)-*form*, in D-60367

36159-74-7 4-Phenyl-2-naphthol, P-60117

36168-41-9 7,7-Dimethylbicyclo[4.1.0]hept-3-ene; (1*RS*,6*SR*)-*form*, in D-60409

36178-05-9 3-Bromo-2-fluoropyridine, B-60253

36190-95-1 4′,5,7-Trihydroxy-3′methoxy-isoflavone, in T-60108

36210-71-6 4-Azidobenzophenone, A-60340

36294-90-3 2,3-Dihydroxybutanoic acid; (2*RS*,3*RS*)-*form*, in D-60312

36328-29-7 Tricyclo[5.1.0.0²,⁸]octane, T-60266

36359-70-3 5-Octen-2-one, O-70028

36364-49-5 Imidazole salicylate, in H-70108

36383-38-7 4-Amino-1,2-benzenediol; 2-Benzenesulfonyl, in A-70090

36383-41-2 4-Amino-1,2-benzenediol; 1-Ac, 2-benzenesulfonyl, in A-70090

36383-42-3 3-Amino-1,2-benzenediol; 1-Benzenesulfonyl, in A-70089

36383-45-6 3-Amino-1,2-benzenediol; 2-Ac, 1-benzenesulfonyl, in A-70089

36386-83-1 4-Cyano-2-phenyl-1,2,3-triazole, in P-60140

36413-60-2 Quinic acid, Q-70005

36431-72-8 Theaspirone, T-60185

36439-78-8 2,3-Dimethyl-enebicyclo[2.2.1]heptane, D-70405

36439-79-9 2,3-Dimethyl-enebicyclo[2.2.2]octane, D-70408

36439-81-3 2,3-Dimethyl-enebicyclo[2.2.3]nonane, D-70406

36504-01-5 1,2,2-Triphenyl-1-propanone, T-60389

36602-05-8 4,4′-Iminobisbenzonitrile, in I-60018

36638-46-7 4-[2,2,2-Trifluoro-1(tri-fluoromethyl)ethylidene]-1,3,2dioxathietane 2,2-dioxide, T-60311

36669-06-4 2-Hydroxy-1,3benzenedicarboxylic acid; Di-Me ester, in H-60101

36676-50-3 Threonine, T-60217

36693-05-7 9,10,18-Trihydroxyoctadecanoic acid; (9*S*,10*S*)-*form*, Me ester, in T-60334

36727-13-6 2-Methoxy-1,3benzenedicarboxylic acid, in H-60101

36730-68-4 5,7-Dihydroxy-4′-methoxyflavone; 7-β-D-Glucurono-β(1→2)-D-glucuronide, in D-70319

36768-62-4 4-Amino-2,2,6,6tetramethyl-piperidine, A-70201

36771-92-3 Aspartame; L-L-*form*, B,HBr, in A-60311

36799-17-4 8-Methylguanosine, M-70085

36822-11-4 2,3-Dihydro-6-phenyl-2-thioxo4(1*H*)-pyrimidinone, D-60268

36839-55-1 1,2-Bis(2-iodoethoxy)ethane, B-60163

36873-96-8 2,3,5-Trimethoxybenzoic acid, in T-70250

36909-44-1 1,3-Diazetidine-2,4-dione; 1,3-Di-Me, in D-60057

36913-17-4 3′-Amino-3′-deoxyadenosine; 3′*N*-Ac, 3′*N*-Me, 2′,5′-di-Ac, 6*N*-benzoyl, in A-70128

36917-96-1 1*H*-Benzo[*de*][1,6]naphthyridine, B-70037

36980-84-4 5-Acetyl-2,4(1*H*,3*H*)pyrimidinedione; 1-Ph, in A-60047

37011-94-2 7-Tetradecyn-1-ol, T-60042

37042-21-0 5-Hydroxy-6-(hydroxymethyl)-7methoxy-2-methyl-4*H*-1benzopyran-4-one, H-60155

37067-97-3 2,4,5-Triiodo-1*H*-imidazole; 1-Me, in T-70276

37110-03-5 5-Hydroxymethyl-2furancarboxaldehyde; Oxime, *in* H-70177

37112-89-3 2-Acetyl-4,5-dihydrothiazole; Oxime, *in* A-70041

37164-17-3 Cyanocyclooctatetraene, *in* C-70239

37428-67-4 2-Hydroxybutanal, H-70112

37464-71-4 2,3-Dichlorooctane; (2*RS*,3*RS*)-form, *in* D-70130

37464-73-6 Cyclooctatetraenecarboxylic acid; Me ester, *in* C-70239

37494-11-4 3,3,6,6-Tetramethylcyclohexyne, T-70132

37551-62-5 Prangeline, P-60170

37567-58-1 4-(2-Nitroethyl)phenol, N-70050

37616-04-9 5-Decen-1-ol; (*Z*)-form, 3-Methylbutanoyl, *in* D-70019

37617-02-0 1-Methoxy-1-hexene, M-70046

37662-02-5 2-Amino-4-hydroxybutanoic acid; (*S*)-form, *O*-(2-Amino-3-hydroxypropyl), *in* A-70156

37675-18-6 3-Piperidinecarboxylic acid; (*S*)-form, Et ester, *in* P-70102

37687-16-4 4-Benzoylpyrazole, B-60065

37697-50-0 2-(3,4-Dihydroxyphenyl)propanoic acid, D-60372

37715-16-5 2-Hydroxy-4-*tert*butylcyclopentanecarboxylic acid; (1*RS*,2*SR*,4*SR*)-form, *in* H-70113

37715-19-8 2-Amino-4-*tert*butylcyclopentanol; (1*RS*,2*SR*,4*RS*)-form, *in* A-70104

37715-20-1 2-Hydroxy-4-*tert*butylcyclopentanecarboxylic acid; (1*RS*,2*RS*,4*RS*)-form, *in* H-70113

37720-70-0 4-Octen-2-one; (*E*)-form, *in* O-60030

37727-90-5 Hexahydro-5(2*H*)-azocinone; 1-Et, *in* H-70048

37745-11-2 2-Amino-4-*tert*butylcyclopentanol; (1*RS*,2*RS*,4*SR*)-form, *in* A-70104

37745-18-9 2-Amino-4-*tert*butylcyclopentanol; (1*RS*,2*RS*,4*RS*)-form, *in* A-70104

37745-21-4 2-Amino-4-*tert*butylcyclopentanol; (1*RS*,2*SR*,4*RS*)-form, *in* A-70104

37746-17-1 1,5-Dibromo-3,3-dimethylpentane, D-70090

37763-43-2 ▷1-Bromo-2-(bromomethyl)naphthalene, B-60212

37791-33-6 5-Methyl-2-oxo-4oxazolidinecarboxylic acid; (4*RS*,5*SR*)-form, *N*-Benzoyl, *in* M-60108

37791-34-7 5-Methyl-2-oxo-4oxazolidinecarboxylic acid; (4*RS*,5*SR*)-form, *N*-Benzoyl, Me ester, *in* M-60108

37791-36-9 5-Methyl-2-oxo-4oxazolidinecarboxylic acid; (4*RS*,5*SR*)-form, *in* M-60108

37816-19-6 7,8-Dihydroxy-4′methoxyisoflavone, *in* T-70265

37839-59-1 3-Amino-2-benzo[*b*]thiophenecarboxylic acid; Amide, *in* A-60091

37842-29-8 8,11,13-Abietatrien-12,16-oxide, A-60008

37890-96-3 2,3-Dihydro-2-thioxo-4*H*pyrido[1,2-*a*]-1,3,5triazin-4-one, D-60294

37891-04-6 2,3-Dihydro-2thioxopyrido[2,3-*d*]pyrimidin-4(1*H*)-one, D-60291

37891-05-7 2,3-Dihydro-2thioxopyrido[3,2-*d*]pyrimidin-4(1*H*)-one, D-60292

37905-12-7 Fasciculatin, F-70001

37908-96-6 3-Chloro-4-methoxybenzoic acid, *in* C-60091

37908-98-8 3-Chloro-4-hydroxybenzoic acid; Me ester, Me ether, *in* C-60091

37967-82-1 6-Methyl-3-phenyl-1,2,4-triazine, M-60118

37977-48-3 1-Acenaphthenecarboxaldehyde, A-70013

37982-24-4 2-Phenylcyclohexanecarboxylic acid; (1*S*,2*S*)-form, *in* P-70065

37985-18-5 2-Methyl-2-thiocyanatopropane, M-70135

38002-32-3 1,4-Diethynyl-2,3,5,6tetrafluorobenzene, D-70160

38005-60-6 Prephytoene alcohol, P-70115

38070-65-4 6*H*-Pyrrolo[3,4-*b*]pyridine; 6-Me, *in* P-70201

38070-66-5 6*H*-Pyrrolo[3,4-*b*]pyridine; 6-Me, picrate, *in* P-70201

38070-67-6 2*H*-Pyrrolo[3,4-*c*]pyridine; 2-Me, *in* P-70200

38070-68-7 2*H*-Pyrrolo[3,4-*c*]pyridine; 2-Me, picrate, *in* P-70200

38070-91-6 Dehydroosthol, D-60020

38076-35-6 Kolavenic acid; 2-Oxo, Me ester, *in* K-60013

38076-57-2 Gibberellin A₃₆, *in* G-60019

38114-47-5 Isomontanolide, *in* M-70154

38146-68-8 Glechomanolide, G-60022

38147-15-8 β-Ecdysone 2-cinnamate, *in* H-60063

38147-16-9 β-Ecdysone 3-p-coumarate, *in* H-60063

38147-16-9 2,3,5,14,20,22,25Heptahydroxycholest-7en-6-one; 2-Cinnamoyl, *in* H-60023

38154-50-6 6-Methyl-3-pyridazinone; 2*H*-form, 2-Ph, *in* M-70124

38157-12-9 1,2,3,4-Tetrahydro-1-hydroxy2-naphthalenemethanol; (1*R*,2*S*)-form, *in* T-70064

38157-17-4 1,2,3,4-Tetrahydro-1-hydroxy2-naphthalenecarboxylic acid; (1*R*,2*S*)-form, *in* T-70063

38157-19-6 1,2,3,4-Tetrahydro-1-hydroxy2-naphthalenecarboxylic acid; (1*R*,2*R*)-form, *in* T-70063

38157-20-9 1,2,3,4-Tetrahydro-1-hydroxy2-naphthalenecarboxylic acid; (1*R*,2*R*)-form, Me ester, *in* T-70063

38198-16-2 ▷2-(Chloromethyl)-4methylpyridine, C-70094

38222-13-8 Alkannin acetate, *in* S-70038

38226-08-3 3-Fluoro-2-methylphenol; Ac, *in* F-60056

38240-21-0 ▷3-Amino-2(1*H*)-pyridinethione, A-60247

38261-87-9 1,2-Cyclohexanedimethanethiol; (1*S*,2*S*)-form, *in* C-70224

38274-35-0 1,2,3-Triphenyl-1*H*-indene, T-70061

38279-47-9 6*a*,12*a*-Dihydro-2,3,10trihydroxy-8methoxy[2]benzopyrano[4,3-*b*][1]benzopyran-7(5*H*)-one, *in* C-70200

38279-50-4 6*a*,12*a*-Dihydro-3,4,10trihydroxy-8methoxy[2]benzopyrano[4,3-*b*][1]benzopyran-7(5*H*)-one, *in* C-70200

38279-52-6 6*a*,12*a*-Dihydro-2,3,10trihydroxy[2]benzopyrano[4,3-*b*][1]benzopyran-7(5*H*)-one, *in* C-70200

38279-57-1 3-(2,4-Dihydroxybenzoyl)-7,8dihydroxy-1*H*-2-benzopyran, *in* P-70017

38303-35-4 1,8-Dibromopyrene, D-70099

38303-95-6 Phrymarolin I, *in* P-60142

38308-89-3 Datiscacin, *in* C-70204

38310-90-6 11,15-Dihydroxy-9-oxo-5,13prostadienoic acid; (5*Z*,8*R*,11*S*,12*R*,13*E*,15*S*)-form, *in* D-60367

38353-02-5 2-Benzoylimidazole, B-70074

38363-18-7 Carbonochloridothioic acid; *O*-Isopropyl, *in* C-60013

38378-77-7 1,8-Anthracenedicarboxylic acid, A-60281

38387-33-6 1-Chloro-1-ethynylcyclopropane, C-60068

38399-20-1 2-Bromo-3-(bromomethyl)naphthalene, B-60217

38401-84-2 2-Ethyl-1,6-dioxaspiro[4.4]nonane, E-70043

38409-30-2 Suberenol, S-70080

38425-47-7 ▷3-Phenyl-2-cyclobuten-1-one, P-70064

38428-17-0 *N*-Asparaginylalanine; L-L-form, *N*²-(*p*-Methoxybenzyloxycarbonyl), Me ester, *in* A-60310

38444-50-7 2,4,5-Trihydroxyphenylacetic acid; Tri-Me ether, nitrile, *in* T-60342

38447-47-1 4*H*-Cyclopenta[*c*]thiophen-5(6*H*)one, C-70245

38456-36-9 Eupaserrin, *in* D-70302

38456-39-2 Deacetyleupaserrin, *in* D-70302

38460-57-0 1,3,5-Benzenetrimethanethiol, B-70019

38462-36-1 Dehydromyoporone, D-60019

38475-38-6 6-Bromo-1,2,4-benzenetriol; Tri-Ac, *in* B-70182

38523-30-7 2-Amino-4-hydroxypentanedioic acid; (2*RS*,4*SR*)-form, *in* A-60195

38557-71-0 2-Chloro-6-methylpyrazine, C-70107

38627-57-5 2,2-Dihydroxy-1*H*-benz[*f*]indene-1,3(2*H*)dione, D-60308

38629-30-0 Notatic acid; Me ester, *in* N-70081

38631-03-7 1-(4,5-Dihydro-1*H*-imidazol-2yl)-2-imidazolidinethione; B,HI, *in* D-60246

38636-86-1 Notatic acid, N-70081

38642-74-9 2-Cyano-10*H*-phenothiazine, *in* P-70056

38694-47-2 4-Bromobutanal, B-70197

38729-96-3 2-Chloro-2-propen-1-amine, C-70151

38746-33-7 1,2,3,4-Tetrahydro-1-hydroxy2-naphthalenemethanol; (1*R*,2*S*)-form, 1′-(4-Methylbenzenesulfonyl), *in* T-70064

38761-25-0 5-Methyl-1,3-dithiane, M-60067

38767-72-5 4-Amino-2(1*H*)-pyridinone; *NH*-form, *in* A-60250

38771-21-0 4-Bromo-1-butyne, B-70198

38786-97-9 3,6,7,8-Tetrahydro-2*H*-oxocin; (*Z*)-form, *in* T-60076

38791-02-5 3-Benzoylisoxazole, B-70075

38818-30-3 2,3-Dicarbomethoxy-1,3-butadiene, *in* D-70409

38819-11-3 6-Amino-1,3-dihydro-2*H*-purin2-one; 9-(β-D-Arabinofuranosyl), *in* A-60144

38822-00-3 9-Hydroxy-3-methoxypterocarpan, *in* D-60373

38883-84-0 2,5-Dimethyl-3,4-diphenyl-2,4cyclopentadien-1-one; Dimer, *in* D-70403

38910-53-1 Octahydro-4*H*-quinolizin-4-one; (±)-form, B,HCl, *in* O-60019

38927-01-4 Tribromophenylselenium, T-70213
38965-66-1 7-Hydroxy-6-methoxy-3',4'methyl-enedioxyisoflavone, in T-70111
38965-67-2 Fujikinin, in T-70111
39012-20-9 Picroside II, in C-70026
39039-48-0 1H-1,2,3-Triazole-4-carboxylic acid; (1H)-form, 1-Me, Et ester, in T-60239
39066-80-3 Methyl 4-methylphenyl sulfoxide; (±)-form, in M-70094
39129-27-6 1,1-Diphenyl-1,3-pentadiene, D-70489
39153-95-2 1-Oxo-1H-2-benzopyran-3-acetic acid, O-60058
39153-96-3 1-Oxo-1H-2-benzopyran-3-acetic acid; Me ester, in O-60058
39232-04-7 2-(2-Bromoethyl)pyridine, B-70210
39266-96-1 6-Chloro-7-methylquinoxaline; 1-Oxide, in C-70125
39267-00-0 3-Chloro-7-methylquinoxaline 1-oxide, in C-70123
39380-12-6 Ferolin, in G-60014
39499-84-8 3,5,7-Trimethoxy-2-phenanthrenol, in T-60120
39499-93-9 9,10-Dihydro-2,7-dihydroxy3,4,6-trimethoxyphenanthrene, in D-70246
39500-00-0 9,10-Dihydro-5,6,7-trimethoxy2,3-phenanthrenediol, in D-70246
39520-23-5 (1-Methylpropyl)propanedioic acid; (±)-form, Di-Me ester, in M-70119
39576-78-8 2-Chloro-3-methylquinoxaline; 1,4-Dioxide, in C-70121
39597-87-0 2-Mercaptomethyl-2-methyl-1,3propanedithiol, M-70030
39599-18-3 Nuciferol, N-70082
39623-20-6 6a,9b-Methano-1H-phenalen-1-one; (±)-form, in M-70042
39677-96-8 2-Thiophenecarboxaldehyde; Phenylhydrazone, in T-70183
39700-44-2 Dihydro-2(3H)-furanthione, D-70210
39713-71-8 2,5-Dimethylpyrrolidine; (2RS,5SR)-form, in D-70437
39713-72-9 2,5-Dimethylpyrrolidine; (2RS,5RS)-form, in D-70437
39720-27-9 4-(Chloromethyl)phenol; Ac, in C-70103
39724-61-3 6α-Acetoxygedunin, in G-70006
39731-51-6 1,3,6,8-Tetrahydroxyxanthone, T-60127
39770-63-3 [3,3'-Bipyridine]-2,2'dicarboxylic acid; Di-Me ester, in B-60133
39770-64-4 [3,3'-Bipyridine]-2,2'dicarboxylic acid; Dihydrazide, in B-60133
39775-31-0 [2,2'-Bipyridine]-3,3'dicarboxylic acid; Di-Me ester, in B-60124
39801-53-1 2-Mercapto-3-methylbutanoic acid; (R)-form, in M-60023
39810-46-3 N-Phenyl-2-thiophenesulfonamide, in T-60213
39825-84-8 1-Thiocyanatoadamantane, T-70177
39830-70-1 (2-Nitrophenyl)oxirane, N-70056
39838-58-9 6α-Hydroxygedunin, in G-70006
39863-20-2 2-Cyanobicyclo[2.2.1]hepta2,5-diene, in B-60084
39870-05-8 4-Acetylpyrimidine, A-60045
39872-95-2 8-Azabicyclo[5.2.0]nonane, A-70274
39897-59-1 3,6-Bis(1-methylpropyl)-2,5piperazinedione, B-70153
39903-01-0 2-Amino-5-bromo-3hydroxypyridine, A-60101
39919-02-3 4-Amino-2-chlorobutanoic acid; (±)-form, in A-60105
39947-70-1 Methyl diformylacetate, in D-60190

39986-86-2 4'-Hydroxydehydrokawain; (E)-form, in H-70122
39998-73-7 1-Aminophthalazine; N-Me, in A-70184
40030-53-3 Octahydro-5a,10a-dihydroxy5H,10H-dipyrrolo[1,2a:1',2'-d]-pyrazine5,10-dione; (5aRS,10aSR)-form, in O-60014
40030-54-4 Octahydro-5a,10a-dihydroxy5H,10H-dipyrrolo[1,2a:1',2'-d]-pyrazine5,10-dione; (5aRS,10aRS)-form, in O-60014
40032-49-3 1,1'-Methylenebispyridinium(1+); Dibromide, in M-70065
40053-29-0 S-Phenyl 3-oxobutanethioate, in O-70080
40053-98-3 2-(Chloromethyl)phenol, C-70101
40087-61-4 1,3,5-Octatriene; (3E,5Z)-form, in O-70022
40100-16-1 Methanethial; S-Oxide, in M-70037
40104-07-2 Chloromethanesulfonic acid, C-70086
40107-00-4 4,5-Diaminoquinoline; B,AcOH, in D-70044
40112-23-0 Gossypol; (±)-form, in G-60032
40139-58-0 3-Amino-2-benzo[b]thiophenecarboxylic acid; K salt, in A-60091
40142-71-0 ▷3-Amino-2-benzo[b]thiophenecarboxylic acid, A-60091
40168-97-6 2-Thiabicyclo[2.2.2]oct-5-en3-one, T-60188
40217-11-6 4-Amino-5-hydroxypentanoic acid, A-70159
40237-01-2 2-Methyl-2H-naphtho[1,8-de]triazine, M-70095
40243-84-3 1-Ethenyl-2,4-dimethoxybenzene, in V-70010
40263-57-8 3-Hydroxy-2-iodopyridine, H-60160
40264-04-8 2-Phenylcyclopentylamine; (1RS,2RS)-form, in P-60086
40265-80-3 1,6-Dimethyl-5-azauracil, in M-60132
40265-81-4 1,3,6-Trimethyl-5-azauracil, in M-60132
40273-45-8 2-Bromo-3-fluoropyridine, B-70213
40315-17-1 3-Oxopentanedioic acid; Dianilide, in O-60084
40325-85-7 4,4-Difluoro-3,3bis-s(rifluoromethyl)1,2-oxathietane 2,2-dioxide, D-60181
40357-87-7 4-Hydroxy-2(1H)-pyridinone; N-Me, in H-60224
40369-38-8 [1,2,4]Triazolo[1,5-b]pyridazine, T-60240
40392-86-7 3-Bromo-2,4,5,6tetrafluoropyridine, B-60327
40420-22-2 4-Oxoheptanedioic acid; Di-Et ester, in O-60071
40423-52-7 1,3-Dihydro-2H-imidazo[4,5-c]pyridin-2-one; 1-Me, in D-60250
40429-00-3 2-Oxo-1,2,3,4tetrahydrocarbazole, O-70104
40451-67-0 Squamolone, S-60049
40499-83-0 3-Pyrrolidinol, P-70187
40510-88-1 2-(Chloromethoxy)ethyl acetate, C-70088
40523-67-9 6-Methyl-2-cyclohexen-1-ol; (1RS,6SR)-form, in M-70058
40556-01-2 tert-Butyl chloromethyl ether, B-70295
40570-74-9 4-Acetylisoquinoline, A-60041
40594-98-7 1H-1,2,3-Triazole-4-carboxylic acid; (2H)-form, Et ester, in T-60239
40596-30-3 5-Chloro-1,3-pentadiene, C-60121

40611-76-5 5-Methyl-1H-pyrrole-3-carboxylic acid; Me ester, in M-60125
40620-23-3 2-Propenethioamide, P-70120
40785-55-5 2-(2-Hydroxyphenyl)-2-oxoacetic acid; Et ester, in H-70210
40785-64-6 8'-Hydroxyzearalenone, in Z-60002
40785-65-7 8'-epi-Hydroxyzearalenone, in Z-60002
40835-18-5 Methyl 3-formylcrotonate, in M-70103
40836-94-0 2-Thiazoledinethione; Thiol-form, S-Me; B,HI, in T-60197
40863-57-8 Bicyclo[4.2.1]non-3-en-9-one, B-70107
40908-37-0 4-Amino-2,2,6,6tetramethyl-piperidine; N-Ac, in A-70201
40915-13-7 4-Ethoxy-2-hydroxy-N,N,Ntrimethyl-4-oxo-1butanaminium, in C-70024
40917-92-8 N,N'-Diphenylformamidine; N-Me; B,HCl, in D-60479
40925-71-1 2-Amino-6-methoxyphenol, in A-70089
40940-16-7 2,3-Dihydro-2-oxo-1H-indole-3acetic acid; (±)-form, Et ester, in D-60259
40957-99-1 Medioresinol, M-70016
40966-87-8 2-Amino-5-chloro-3hydroxypyridine, A-60111
40993-10-0 3,5-Dinitrobenzoic acid; Anhydride, in D-60458
40997-78-2 2,2-Diphenylpropanoic acid; Chloride, in D-70493
41002-50-0 2,4-Pentanedione; Enol-form, Ac, in P-60059
41034-52-0 1,2,3,4-Tetrahydro-1isoquinolinecarboxylic acid, T-60070
41039-91-2 3-(2-Bromoethyl)pyridine; B,HBr, in B-70211
41068-60-4 2,6-Diphenyl-4H-thiopyran-4-one; 1,1-Dioxide, in D-60492
41164-14-1 1,3a,6,6a-Tetrahydropentalene, T-70082
41205-21-4 3,4-Dichloro-1-(4fluorophenyl)-1Hpyrrole-2,5-dione, D-60132
41207-34-5 2-Octadecen-1-ol; (E)-form, in O-60007
41221-01-6 Dihydro-1,3,5-triazine2,4(1H,3H)-dione; 1,3,5-Tri-Me, in D-60297
41233-74-3 2,5-Dimethyl-1,3,5-hexatriene; (E)-form, in D-60424
41270-62-6 2-Chloro-6-phenylpyrazine, C-70144
41270-65-9 2-Chloro-3-phenylpyrazine, C-70142
41288-96-4 2-Chloro-5-hydroxypyridine, C-60101
41300-59-8 7-Hydroxy-5-heptynoic acid, H-60150
41300-60-1 7-Bromo-5-heptynoic acid, B-60261
41326-74-3 2,2,4,4-Tetraphenyl-3-butenal, T-70148
41349-38-6 7-Bromo-5-heptynoic acid; Me ester, in B-60261
41365-37-1 Eupomatenoid 12, in E-60075
41367-00-4 5-Nitrohistidine; (S)-form, Nα-Ac, in N-60028
41370-56-3 α-Copaen-11-ol, in C-70197
41371-53-3 2-Oxo-4-imidazolidinecarboxylic acid; (S)-form, in O-60074
41404-58-4 2-Bromo-5-fluoropyridine, B-60252
41410-53-1 Gymnomitrol, G-70035
41429-52-1 Norpatchoulenol, N-70078
41429-88-3 5-Nitrohistidine; (S)-form, Me ester, Nα-Ac, in N-60028
41438-18-0 4-Hydroxy-2-methylbenzaldehyde, H-70161
41443-48-5 2-Methoxy-4-phenylquinoline, in P-60131

41500-02-1	O-Ethyl 3-oxobutanethioate, in O-70080	42216-62-6	3,5-Dinitroisoxazole, D-70452	43084-62-4	(4-Nitrophenyl)phenylethanedione; Dioxime, in N-70059
41505-89-9	4,4-Difluoro-1,2-oxathietane 2,2-dioxide, D-60186	42245-90-9	Tricyclo[4.4.2.0^{1,6}]dodecane2,7-dione, T-60261	43090-52-4	1,13-Dioxa-4,7,10,16,19,22hexa-azacyclotetracosane, D-70460
41527-66-6	1,3a,4,6a-Tetrahydropentalene, T-70081	42259-31-4	1,1-Di-1-piperidinylethene, D-70500	43119-27-3	Hexahydro-2(3H)-benzofuranone; (3aR,7aR)-form, in H-70049
41535-76-6	5-Aminopyrazolo[4,3-d]pyrimidin-7(1H,6H)-one, A-60246	42270-91-7	3-Phenyl-3H-diazirine, P-70067	43119-99-9	2-Hydroxy-2-isopropylbutanedioic acid; (−)-form, in H-60165
41536-78-1	1,3(2H,4H)-Isoquinolinedione; Dioxime, in I-70099	42298-41-9	5-Iodo-2-methoxybenzaldehyde, in H-70143	43134-76-5	2,4-Diaminobenzoic acid; 4-N-Ac, in D-60032
41563-67-1	2-Methyl-1,3benzenedimethanethiol, M-70050	42332-94-5	9,10-Dihydro-9,9di-methylanthracene, D-60224	43142-60-5	Dihydro-3-vinyl-2(3H)-furanone, D-70279
41610-70-2	15-Copaenol, in C-70197	42333-76-6	2,4,6-Tri-3-pyridinyl-1,3,5triazine, T-70317	43160-78-7	4-Oxodecanal, O-70088
41623-91-0	2-Bromo-9-β-D-ribofuranosyl6H-purin-6-one; 2′,3′,5′-Tri-Ac, in B-70269	42333-78-8	2,4,6-Tri-4-pyridinyl-1,3,5triazine, T-70318	43163-94-6	2-Amino-3-fluoro-3methylbutanoic acid, A-60159
41680-12-0	7-Hydroxy-4-methoxy-5-methyl2H-1-benzopyran-2-one, in D-60348	42346-89-4	2-Amino-4(1H)-pteridinone; 8-Oxide, in A-60242	43163-95-7	2-Amino-3-fluoropentanoic acid, A-60160
41680-34-6	▷3-Amino-1H-pyrazole-4carboxylic acid, A-70187	42346-97-4	2,4-Diaminopteridine; 5,8-Dioxide, in D-60042	44855-57-4	2-Octylamine, (±)-form, in O-60031
41682-24-0	Arctigenin; 4-β-Gentiobioside, in A-70249	42417-39-0	3-Aminodihydro-2(3H)-furanone; (±)-form, B,HCl, in A-60138	45731-99-5	5,5-Dimethyl-2-cyclohexene1,4-dione, D-70392
41690-67-9	Jaeschkeanadiol, J-70001	42418-76-8	2-(Aminomethyl)cycloheptanol; (1RS,2RS)-form, B,HCl, in A-70168	45887-08-9	3-(1H-Pyrazol-3-yl)pyridine, P-70161
41703-38-2	Grasshopper ketone, G-70028			45990-28-1	4,5-Diaminoquinoline, D-70044
41720-93-8	Capsenone, C-60011	42418-77-9	2-(Aminomethyl)cycloheptanol; (1RS,2SR)-form, B,HCl, in A-70168	46000-25-3	4-Methylthioquinoline, in Q-60006
41731-34-4	2-Bromo-4-iodothiazole, B-60287			46258-62-2	1-Acetyl-4-bromonaphthalene, A-60022
41744-25-6	Eupomatenoid 2, in E-60075	42418-78-0	2-Aminocycloheptanemethanol; (1RS,2SR)-form, B,HCl, in A-70124	46898-00-4	1,2-Dihydro-4,6diphenylpyrimidine, D-60235
41744-26-7	Eupomatenoid 6, in E-60075				
41744-28-9	Eupomatenoid 5, in E-60075	42418-79-1	2-Aminocycloheptanemethanol; (1RS,2RS)-form, B,HCl, in A-70124	48126-51-8	2-Phenylcyclopropanecarboxylic acid; (1R,2S)-form, in P-60087
41744-29-0	Eupomatenoid 4, in E-60075				
41744-30-3	Eupomatenoid 7, in E-60075	42418-83-7	2-Aminocycloheptanecarboxylic acid; (1RS,2SR)-form, in A-70123	48194-29-2	1,2-Diphenyl-3-(phenylimino)cyclo-propene, D-70491
41744-32-5	Grevilline A, G-60038				
41744-33-6	Grevilline B, in G-60038	42418-84-8	2-Aminocycloheptanecarboxylic acid; (1RS,2RS)-form, in A-70123	49576-57-0	2-Methyl-2-octenal; (E)-form, in M-60106
41744-34-7	Grevilline C, in G-60038				
41753-44-0	9(11),12-Ursadien-3-ol; 3β-form, in U-70007	42429-92-5	7,8,9,10-Tetrahydrofluoranthene, T-70061	49583-10-0	4-Methyl-1,3-naphthalenediol; Di-Ac, in M-60094
41759-19-7	4-Methoxy-1-methyl-2(1H)pyridinone, in H-60224	42510-46-3	4-Amino-1H-1,5-benzodiazepine3-carbonitrile; B,HCl, in A-70091	49590-24-1	2,2′-Thiobisbenzoic acid; Di-Me ester, in T-70174
41787-69-3	Oliveric acid; (+)-form, in O-60039	42572-42-9	Trichlorophenylselenium, T-70224	49610-35-7	4,4′-Dibromo-1,1′-binaphthyl, D-60090
41840-27-1	4,6-Dimethyl-2(1H)pyrimidinethione; Na salt, in D-60450	42599-17-7	1-Iodo-1-octene; (E)-form, in I-70051	49615-96-5	2-Acetyl-3-aminobenzofuran, A-70021
41910-64-9	4-Chloro-2,3-dihydro-1H-indole, C-60055	42715-03-7	Norjuslimdiolone, N-60061	49616-64-0	3-(1H-Pyrrol-2-yl)-2-propenal; (E)-form, in P-70202
41918-07-4	5-Chloro-1,2-benzisothiazole, C-70051	42762-56-1	5-Phenyl-4-penten-2-one; (E)-form, in P-70088	49619-48-9	3-Methoxy-1H-isoindole, in P-70100
41921-63-5	1,3,4,5-Tetrahydro-2H-1,3benzodiazepin-2-one, T-60058	42772-86-1	2,2′-(1,3-Dioxan-2-ylidene)bispyridine, in D-70502	49619-49-0	3-Ethoxy-1H-isoindole, in P-70100
41929-21-9	Withaphysacarpin, W-60008	42793-20-4	5H-Pyrrolo[1,2-a]azepin-5-one, P-70191	49619-82-1	3-Bromo-4H-1-benzopyran-4-one, B-70183
41934-74-1	5-Nitrohistidine; (S)-form, in N-60028	42830-26-2	Adoxoside, A-70067	49623-54-3	1,3-Dibromo-2,3-dimethylbutane, D-70089
41943-79-7	Stemodin, S-70071	42836-42-0	Di(2,4,6-cycloheptatrien-1-yl)ethanone, D-60166	49628-52-6	2-Bromo-4-methylpentanoic acid, B-70248
41943-80-0	Stemodinone, in S-70071	42842-41-1	3,3-Dimethyl-4,4-diphenyl-1butene, D-60418	49634-78-8	7-Chloro-2-phenylquinoxaline, C-70147
41969-71-5	1H-Pyrrole-3,4-dicarboxylic acid; Di-Et ester, in P-70183	42843-49-2	1-Bromo-1-octene; (Z)-form, in B-70261	49647-13-4	5-Acetyl-2-aminopyridine; N-Oxide, in A-70028
41969-74-8	1H-Pyrrole-3,4-dicarboxylic acid; Me ester, in P-70183	42851-22-9	10b,10cDiazadicyclopenta[ef,kl]-heptalene, D-70052	49664-20-2	6-Methylene-2,4-cyclohexadien1-one; Trimer, in M-60072
41983-91-9	Glabranin, G-70015	42855-50-5	2,5-Dihydro-3-methylthiophene, D-70231	49679-57-4	2-Phenyl-1,3-oxazepine, P-60123
42041-50-9	2-(2-Hydroxyphenyl)pyrrole, H-60217	42856-71-3	2-Methyl-2-pyrrolidinecarboxylic acid, M-60126	49747-09-3	2,7-Dimethyl-2,4,6cycloheptatrien-1-one, D-70391
42046-60-6	2-Methyl-4-cyanopyrrole, in M-60125	42890-34-6	2-Methyl-9-β-D-ribofuranosyl9H-purine, M-70130	49840-62-2	1,2-Dihydrazinoethane; N^α,N^{α′},N^β,N^β,N^{β′},N^{β′}-Hexa-Me, in D-60192
42050-06-6	1-Phenanthreneacetic acid, P-70046	42904-05-2	2-Mercaptocyclohexanone, M-70025		
42050-16-8	5,7-Dimethoxy-2,3phenanthrenediol, in T-60120	42906-89-3	3,3,4,4-Tetramethyl-2-thietanone, T-60155	50256-18-3	2,4(1H,3H)-Pteridinedione; 1-Me, in P-60191
42053-19-0	5,13-Dimethyl[2.2]metacyclo-phane; anti-form, in D-60426	42916-14-3	2-Phenylcyclopropanecarboxylic acid; (1RS,2RS)-form, in P-60087	50256-19-4	2,4(1H,3H)-Pteridinedione; 3-Me, in P-60191
42059-30-3	1-Azido-2-iodoethane, A-60346				
42087-82-1	2,6-Dinitrobenzoic acid; Me ester, in D-60457	42948-91-4	Bicyclo[4.2.1]non-7-en-9-one, B-70108	50266-44-9	9,12-Dioxododecanoic acid; Me ester, in D-60471
42088-41-5	1-Methylthioisoquinoline, in I-60132	42978-64-3	9-Cyanoacridine, in A-70061	50298-93-6	Stachydrine; (±)-form, in S-60050
42104-42-7	3-Octen-1-yne; (E)-form, in O-70029	42990-28-3	2-Bromo-4-methylpentanoic acid; (R)-form, in B-70248	50314-86-8	Acetophenone; (Z)-Oxime, in A-60017
42110-76-9	5-Chloro-1-methylpyrazole, in C-70155	43001-25-8	4-Iodo-1-butyne, I-70024	50354-48-8	2,2-Diphenylpropanoic acid; Me ester, in D-70493
42155-49-7	1-(Cyanomethyl)phenanthrene, in P-70046	43064-12-6	1,2,3,6-Tetrahydro-4phenyl-pyridine; B,HCl, in T-60081		
42177-25-3	1-(1-Naphthyl)ethanol; (R)-form, in N-70015	43073-11-6	2-Acetyl-1H-indene, A-70045		
42182-25-2	4-Acetyl-2-aminopyridine, A-70026	43084-29-3	6-Amino-2-chlorohexanoic acid; (±)-form, B,HCl, in A-60109		

50376-39-1 Colletoketol, *in* C-60159

50392-39-7 4-Amino-2,5-dichlorophenol,
            A-70129

50427-77-5 5-Amino-1*H*-pyrazole-4-carboxylic
            acid; *N*(1)-Ph, amide, *in* A-70192

50463-74-6 2-(3,4-Dihydroxyphenyl)propanoic
            acid; (±)-*form*, Di-Me ether, *in*
            D-60372

50464-85-2 Trichadenal; 27-Carboxylic acid,
            Ac, *in* T-70218

50465-22-0 Trichadenal, T-70218

50489-46-8 Nornotatic acid, *in* N-70081

50500-59-9 6-Methyl-3-pyridazinone; 2*H*-*form*,
            2-Me, *in* M-70124

50521-40-9 Tetrahydro-3,4-biss(ethylene)furan,
            T-70046

50585-13-2 4-Cyclooctene-1-acetic acid,
            C-70242

50585-89-2 3-Piperidinecarboxylic acid; (±)-
            form, Me ester, *in* P-70102

50592-48-8 1-Methyl-4,4-diphenylcyclohexene,
            M-60065

50614-86-3 4-Acetylindole, A-60033

50656-69-4 3α-Hydroxy-27-friedelanoic acid, *in*
            T-70218

50656-70-7 3β-Hydroxy-27-friedelanoic acid, *in*
            T-70218

50682-96-7 3,3-Dimethylbicyclo[2.2.2]octan-2-
            one, D-60411

50700-60-2 3-Methoxy-4(1*H*)-pyridinone, *in*
            H-60227

50703-54-3 4*b*,9*b*-Dihydroindeno[2,1-*a*]indene-
            5,10-dione, D-70220

50720-05-3 1*H*-Indole-3-acetic acid; Chloride,
            in I-70013

50720-12-2 3-Bromo-5-methoxypyridine, *in*
            B-60277

50720-19-9 6,7,8,9-Tetrahydro-4*H*quinolizin-4-
            one, T-60092

50722-63-9 [2,4′-Bipyridine]-3′,5dicarboxylic
            acid, B-60132

50772-81-1 3,4-Dihydroxy-5-iodobenzoic acid;
            Di-Me ether, Me ester, *in*
            D-60335

50781-52-7 9-Cyanomethylphenanthrene, *in*
            P-70047

50781-91-4 7-Hydroxy-5-heptynoic acid; Me
            ester, *in* H-60150

50816-19-8 8-Bromo-1-octanol, B-70260

50823-87-5 4-Methoxy-3-(tri-
            fluoromethyl)benzaldehyde, *in*
            H-70229

50845-35-7 3,4-Dihydro-4-phenyl-
            1(2*H*)naphthalenone; (±)-*form*,
            Oxime, *in* D-60267

50846-98-5 ▷2-Diazo-2*H*-imidazole, D-70060

50874-54-9 1,2,3,3*a*-Tetrahydropentalene,
            T-70075

50874-54-9 1,2,3,4-Tetrahydropentalene,
            T-70076

50877-39-9 1-Benzyl-5-chloropyrazole, *in*
            C-70155

50877-43-5 1-Benzyl-3-chloropyrazole, *in*
            C-70155

50878-08-5 4-Oxo-4*H*-1-benzopyran-3-acetic
            acid; Nitrile, *in* O-60059

50878-09-6 4-Oxo-4*H*-1-benzopyran-3-acetic
            acid, O-60059

50878-10-9 4-Oxo-4*H*-1-benzopyran-3-acetic
            acid; Amide, *in* O-60059

50882-16-1 2-Oxocyclopentanecarboxylic acid,
            O-70084

50885-01-3 2-Amino-3-fluorobutanoic acid,
            A-60154

50893-36-2 (1-Chloroethyl) ethyl carbonate,
            C-60067

50896-26-9 5-Methyl-2-oxo-
            4oxazolidinecarboxylic acid;
            (4*RS*,5*RS*)-*form*, *in* M-60108

50906-02-0 3-Chloro-2-hydroxypropanoic acid;
            (±)-*form*, *in* C-70072

50906-56-4 Arteannuin *B*, A-70262

50906-59-7 Alliodorin, A-70083

50915-27-0 1-Bromo-1-methylcyclopropane,
            B-70242

50915-66-7 2,4-Heptadienoic acid; (2*E*,4*Z*)-
            form, *in* H-60021

50919-54-5 2,3-Naphthalenedicarboxylic acid;
            Di-Et ester, *in* N-60003

50932-20-2 Verproside, *in* C-70026

50982-39-3 Zexbrevin *C*, Z-70005

50996-37-7 2,4,6(1*H*,3*H*,5*H*)-Pteridinetrione;
            I-Me, *in* P-70134

51019-83-1 3-(2-Thienyl)-2-propenoic acid;
            (*Z*)-*form*, *in* T-60208

51024-97-6 3-Oxo-27-friedelanoic acid, *in*
            T-70218

51073-57-5 3-Cinnolinecarboxaldehyde,
            C-70178

51082-19-0 3,6-Dimethyl-1,2,4,5benzenetetra-
            carboxylic acid; Tetra-Me ester,
            in D-60407

51093-34-6 2,6-Diamino-4(1*H*)-pyrimidinone;
            1-Me, *in* D-70043

51097-60-0 4-Octenedial, O-70025

51109-27-4 2,4,5-Trihydroxyphenylacetic acid,
            T-60342

51137-13-4 Tetrahydro-1,3,5,7tetrazocine-
            2,6(1*H*,3*H*)dione, T-70098

51145-61-0 3-Oxo-1,2,3,4tetrahydrocarbazole,
            O-70105

51149-23-6 4,5-Dichlorooctane; (4*RS*,5*SR*)-
            form, *in* D-70131

51174-45-9 5-Chloro-3-methyl-1-pentene,
            C-60117

51175-66-7 *N*-Phenyl-3-thiophenesulfonamide,
            in T-60214

51175-71-4 3-Thiophenesulfonic acid; Chloride,
            in T-60214

51175-72-5 3-Thiophenesulfonic acid; Na salt,
            in T-60214

51193-77-2 3-Octen-2-one; (*Z*)-*form*, *in*
            O-70027

51197-89-8 6,7,8,9-Tetrahydro-6-phenyl5*H*-
            benzocyclohepten-5-one, T-60080

51220-97-4 3-Amino-1,2-benzenediol; B,HCl,
            in A-70089

51226-42-7 2,4,6-Trinitrobenzoic acid; Amide,
            in T-70303

51271-34-2 2-Chloro-6-(methyl-
            thio)benzonitrile, *in* C-70082

51273-49-5 Tricyclo[3.1.1.1^{2,4}]octane, T-60265

51310-54-4 2-Trifluoromethyl-1*H*-indole,
            T-60291

51310-55-5 3-Trifluoromethyl-1*H*-indole,
            T-60292

51338-33-1 2,4(3*H*,5*H*)-Thiophenedione,
            T-70185

51364-89-7 1-Carbethoxy-2-cyano-
            1,2dihydropyridine, C-70017

51385-49-0 2′-Deoxyinosine; Oxime, *in*
            D-70023

51439-47-5 3,4-Dibenzoyl-2,5-hexanedione,
            D-60081

51445-93-3 Tetrakiss(iphenylmethylene)cyclo-
            butane, T-70128

51446-30-1 2-Fluoro-5-hydroxybenzoic acid,
            F-60038

51446-31-2 4-Fluoro-3-hydroxybenzoic acid,
            F-60043

51460-32-3 2-Acetyl-5-aminopyridine, A-70023

51460-34-5 2-Acetyl-3-aminopyridine; Picrate,
            in A-70022

51460-35-6 4-Acetyl-3-aminopyridine; Picrate,
            in A-70027

51460-36-7 2-Acetyl-5-aminopyridine; Picrate,
            in A-70023

51460-81-2 5,6-Dihydroxyhexanoic acid,
            D-60329

51497-33-7 1-*tert*-Butyl-1,3-cyclohexadiene,
            B-70294

51501-77-0 Tetramercaptotetrathiafulvalene;
            Tetrakis(*S*-Me), *in* T-70130

51502-05-7 ▷(Difluoromethylene)tetra-
            fluorocyclopropane, D-60185

51551-01-0 9,12-Dioxododecanoic acid,
            D-60471

51552-75-1 1,2,3,4-Tetrahydro-4-oxo-
            6quinolinecarboxylic acid,
            T-60078

51552-76-2 1,2,3,4-Tetrahydro-4-oxo-
            6quinolinecarboxylic acid; Me
            ester, *in* T-60078

51591-38-9 2,3-Di-*O*-acetyltartaric acid, *in*
            T-70005

51599-09-8 1,2-Epoxy-1,2-dihydrolycopene, *in*
            L-60041

51599-10-1 5,6-Epoxy-5,6-dihydrolycopene, *in*
            L-60041

51606-50-9 2-Hydroxycyclohexanemethanol;
            (1*R*,2*R*)-*form*, *in* H-70120

51640-60-9 1,5-Dimethylcarbazole, D-70377

51641-23-7 1,2,3,4-Tetrahydro-
            4hydroxyisoquinoline, T-60066

51652-35-8 5-Methoxy-2,4-dinitrophenol, *in*
            D-70450

51652-47-2 5-Decen-1-ol; (*Z*)-*form*, *in* D-70019

51670-51-0 3-Phenyl-1,2-naphthoquinone,
            P-60120

51689-29-3 7,7′-Bii(icyclo[2.2.1]heptylidene),
            B-60078

51689-50-0 2-Methyl-
            1,3benzenedicarboxaldehyde,
            M-60046

51698-73-8 7,8-Dimethylenebicyclo[2.2.2]octa-
            2,5-diene, D-70407

51719-64-3 4-Chloro-3,5dimethylbenzaldehyde,
            C-60063

51746-83-9 4-(Aminoacetyl)pyridine; B,2HCl,
            in A-70087

51747-02-5 3-(1*H*-Pyrazol-3-yl)pyridine;
            B,HCl, *in* P-70161

51750-18-6 2-Phenyl-2*H*-benzotriazole; 1-*N*-
            Oxide, *in* P-60083

51751-87-2 1-Bromo-1-octene; (*E*)-*form*, *in*
            B-70261

51760-20-4 1,3-Biss(romomethyl)-5nitro-
            benzene, B-60145

51789-99-2 5*H*-Pyrrolo[1,2-*a*]imidazol-5-one,
            P-70194

51799-36-1 1,2,3,4-Pentatetraene-1,5-dione,
            P-70040

51846-33-4 3*H*-1,2-Benzodithiole-3-selone,
            B-70034

51846-67-4 9-Methyl-9*H*-carbazol-2-ol, *in*
            H-60109

51847-78-0 Furocaespitane, F-60090

51847-86-0 Ingol, I-70018

51847-87-1 Variabilin, V-70003

51848-09-0 Glabrachromene II, G-70014

51906-02-6 Ingol; Tetra-Ac, *in* I-70018

51908-64-6 4-Chloro-1-butyne, C-70058

51933-77-8 3-Bromo-2-(methylthio)pyridine, *in*
            B-60315

51934-46-4 3-Quinolinethiol; *S*-Me, *in* Q-60003

51941-15-2 3-(Aminoacetyl)pyridine, A-70086

51957-32-5 4-Chloro-5-methylpyrimidine,
            C-70116

51958-72-6 6-Chloro-7*H*-benz[*de*]anthracen7-
            one, C-70044

51958-76-0 Benzo[*rst*]phenaleno[1,2,3-
            de]pentaphene, B-60036

52010-22-7 4-Chloro-1(3*H*)-isobenzofuranone,
            C-70079

52066-62-3 5,5,9,9-Tetramethyl-
            5*H*,9*H*quino[3,2,1-*de*]acridine,
            T-70143

52078-48-5 9-Tricosene, T-60259

52086-82-5 1,3-Biss(ethylene)cyclohexane,
            B-60167

52089-56-2 Bicyclo[4.2.1]non-2-en-9-one,
            B-70106

52096-54-5 4-Octenedial; (*Z*)-*form*, Bis-2,4-
            dinitrophenylhydrazone, *in*
            O-70025

52148-91-1 5-Methyl-2,4-hexadienoic acid;
            (*E*)-*form*, Me ester, *in* M-70086

52153-48-7 4-Octenedial; (Z)-form, in O-70025

52189-67-0 3-Fluoro-2,6-dimethoxybenzoic acid, in F-60023

52190-34-8 2-(Methylthio)cyclopentanone, M-70136

52190-35-9 2-(Methylthio)cyclohexanone, in M-70025

52191-71-6 2,2'-Dimethyl-4,4'-biquinoline, D-60413

52200-90-5 4-Amino-2-methoxyphenol, in A-70090

52213-27-1 2,3-Dihydroxy-12-ursen-28-oic acid; (2α,3α)-form, in D-70349

52243-98-8 8,16-Epoxy-8H,16Hdinaphtho[2,1-b:2',1'f][1,5]dioxocin, E-70014

52289-54-0 4-Methoxy-2-methylbenzaldehyde, in H-70161

52311-30-5 2,4-Diethoxypyridine, in H-60224

52324-04-6 Ethynamine, E-60056

52334-53-9 4-Amino-3-hydroxypyridine, A-60202

52342-51-5 2-Methylene-1,3-propanedithiol, M-70076

52344-93-1 2-Formyl-3,5-dihydroxybenzoic acid; Di-Me ether, Me ester, in F-60071

52345-24-1 1,6-Dimethylazulene, D-60404

52355-85-8 5-Bromo-2-hexanone, B-60262

52356-93-1 1-Iodo-1-octene; (Z)-form, in I-70051

52363-44-7 Octahydro-2,7-dihydroxy5H,10H-dipyrrolo[1,2a:1',2'-d]-pyrazine5,10-dione; (2R,5aS,7R,10aS)-form, in O-60013

52393-53-0 1,7-Diphenyl-3,5-heptanediol; (3RS,5SR)-form, in D-70481

52412-55-2 4-Amino-5-nitro-1,2benzenedicarboxylic acid; Di Me ester, Ac, in A-70177

52412-56-3 4-Amino-3-nitro-1,2benzenedicarboxylic acid; Di-Me ester, Ac, in A-70176

52412-59-6 4-Amino-3-nitro-1,2benzenedicarboxylic acid; Di-Me ester, in A-70176

52412-66-5 3-Amino-4-nitro-1,2benzenedicarboxylic acid; Di-Me ester, in A-70174

52431-65-9 2-Bromo-8-hydroxy-1,4naphthoquinone, B-60272

52448-17-6 6-Bromotryptophan; (S)-form, in B-60334

52456-88-9 6-Methyl-2-cyclohexen-1-one; (±)-form, 2,4-Dinitro-phenylhydrazone, in M-70059

52461-05-9 Aspicillin, A-60312

52461-07-1 Aspicillin; Tri-Ac, in A-60312

52461-08-2 Aspicillin; Dihydro, in A-60312

52535-73-6 2-Amino-4,5,6,7tetra-hydrobenzo[b]thiophene-3-carboxylic acid; Et ester, N-benzoyl, in A-60258

52538-26-8 2-Bromo-9-β-D-ribofuranosyl6H-purin-6-one; Mono-NH₄ salt, in B-70269

52548-63-7 5-Fluoro-2-iodobenzoic acid, F-70025

52601-74-8 1,4,7,10,13Pentaazacyclo-pentadecane; Pentakis(4-methyl-benzenesulfonyl), in P-60025

52605-96-6 2-Chloro-3-methoxypyridine, in C-60100

52605-97-7 2,3-Dimethoxypyridine, in H-60223

52631-78-4 Khellinone; Oxime, in K-60011

52647-56-0 13(18)-Oleanen-3-ol; 3α-form, in O-70036

52648-45-0 2-Isothiocyanatopyridine, I-70106

52677-91-5 Polypodoaurein, in H-60063

52703-34-1 [(Tribromomethyl)-sulfinyl]benzene, in T-70212

52719-86-5 1,3,5,7-Tetramethyladamantanone, T-60132

52719-88-7 1,3,5,7-Tetramethyl-2,4adamantanedione, T-60131

52745-08-1 4-Amino-1,2,3-benzotriazine; Amine-form, 3-Oxide, in A-60092

52751-15-2 2,3-Difluoropyrazine, D-60188

52751-31-2 1,2,3,4-Tetrahydrocyclo-pent[b]indole; N-Me, in T-70049

52761-34-9 Colneleic acid; (2'E,4'Z,8E)-form, in C-70189

52789-75-0 2-Hydroxycyclopentanone; (±)-form, Ac, in H-60114

52793-99-4 8-Hydroxy-7-phenylquinoline, H-60218

52805-44-4 3-Cyano-10-methylphenothiazine, in P-70057

52809-70-8 1-Nitro-2-(phenylthio)ethane, N-70060

52816-76-9 4-Chloro-6-methylpyrimidine; 1-Oxide, in C-70117

52840-38-7 7-Amino-4-(trifluoromethyl)2H-1-benzopyran-2-one; N-Et, in A-60265

52853-03-9 1-Methylenepyrrolidinium(1+); Chloride, in M-70077

52879-18-2 Cyclohepta[def]carbazole, C-70220

52903-50-1 [2.2]Paracyclo(4,8)-[2.2]metaparacyclophane, P-60005

52906-16-8 3-Amino-1H-pyrazole-4-carboxylic acid; Amide, in A-70187

52938-97-3 5-Phenyl-2-furancarboxylic acid, P-70078

52939-03-4 5-Phenyl-2-furancarboxylic acid; Me ester, in P-70078

53000-40-1 ▷1H-Indol-3-ylphenyliodonium hydroxide inner salt, I-70017

53010-00-7 1-Benzyl-5-methylamino-1Htetra-zole, in A-70202

53010-03-0 5-Methylamino-1H-tetrazole, in A-70202

53011-72-6 Murralongin, M-70159

53023-17-9 Suksdorfin, in K-70013

53023-18-0 Isosamidin, in K-70013

53081-42-8 8-Hydroxy-4-methyl-2H-1benzopyran-2-one, H-70165

53137-51-2 6,7-Dihydro-5H-dibenzo[a,c]cyclo-hepten-5-one, D-70188

53216-06-1 Threonine; (2RS,3SR)-form, Me ester, in T-60217

53229-92-8 2-Ethoxycarbonylcyclopentanone, in O-70084

53242-29-8 2-Hydroxycyclooctanecarboxylic acid; (1RS,2RS)-form, in H-70121

53254-85-6 3,6,8-Trihydroxy-1methyl-anthraquinone-2carboxylic acid; Me ester, in T-60332

53258-94-9 2,2'-Oxybispyridine, O-60098

53258-95-0 3,3'-Oxybispyridine, O-60100

53258-96-1 4,4'-Oxybispyridine, O-60101

53277-43-3 3-Acetyl-4-aminopyridine, A-70025

53300-17-7 3-Amino-1,2,4-triazine; N,N-Di-Me, in A-70203

53312-81-5 3-Cyano-4-fluoroaniline, in A-70147

53318-10-8 Cadambagenic acid, C-70002

53319-52-1 Isogosferol, in X-60002

53330-94-2 ▷5-Acetylindole, A-60034

53336-42-8 4-Bromo-2-sulfolene, in B-70204

53342-16-8 Chlamydocin, C-60038

53342-27-1 2-Acetylpyrimidine, A-60044

53377-54-1 4,7-Dimethoxy-5-methyl-2H-1benzopyran-2-one, in D-60348

53385-78-7 1,2,3,6-Tetrahydropyridine; 1-Propyl, in T-70089

53387-38-5 3a,4,5,6,7,7a-Hexahydro-3methyl-ene-2-(3H)benzofuranone, H-70062

53449-12-0 8-Aminoimidazo[4,5-g]quinazoline, A-60207

53449-13-1 8-Aminoimidazo[4,5-g]quinazoline; B,HCl, in A-60207

53449-18-6 Imidazo[4,5-g]quinazolin-8-one, I-60012

53449-43-7 9-Aminoimidazo[4,5-f]quinazoline, A-60206

53449-49-3 Imidazo[4,5-h]quinazolin-6-one, I-60013

53449-52-8 Imidazo[4,5-f]quinazolin-9(8H)one, I-60011

53515-94-9 4-(2-Phenylethyl)-1,2benzenediol, P-60096

53518-15-3 7-Amino-4-(trifluoromethyl)2H-1-benzopyran-2-one, A-60265

53537-91-0 4,4-Dimethylbicyclo[3.2.1]octane-2,3-dione; (±)-form, in D-60410

53549 12 5 4,4-Dimethylbicyclo[3.2.1]octane-2,3-dione; (1S)-form, in D-60410

53555-42-3 2,2-Dimethyl-1,3-propanedithiol, D-70433

53578-15-7 Cubanecarboxylic acid, C-60174

53579-96-7 Hexahydro-1,5-thiazonin-6(7H)one, H-70075

53585-93-6 2-Cyclohexyl-2-hydroxyacetic acid; (R)-form, in C-60212

53601-99-3 4,5,6,7Tetrahydroisoxazolo[4,5-c]-pyridin-3-ol; B,HBr, in T-60071

53602-00-9 4,5,6,7Tetrahydroisoxazolo[4,5-c]-pyridin-3-ol, T-60071

53606-43-2 4-Amino-1,2-benzenediol; O¹-Me, 2-Ac, in A-70090

53636-08-1 Dithiocarbazic acid; Hydrazine salt, in D-60509

53645-79-7 1H-Pyrazolo[3,4-d]pyrimidine; 1-Ph, in P-70156

53654-35-6 2-Vinylindole, V-70013

53658-78-9 3,3-Dimethylbicyclo[2.2.2]oct5-en-2-one, D-60412

53681-67-7 4',7-Dihydroxyisoflavone; 4',7-Di-O-β-D-glucopyranosyl, in D-60340

53703-55-2 3-Thiatricyclo[2.2.1.0²,⁶]heptane, T-70165

53755-85-4 2-Phenyl-1,3-benzoxathiol-1-ium; Perchlorate, in P-60084

53766-30-6 Peltochalcone, P-70017

53774-75-7 Bowdichione, B-70175

53798-51-9 2,5-Dihydroxy-3-methylpentanoic acid; (2S,3R)-form, in D-60353

53798-95-1 2-Tetrazolin-5-one; 1-Benzyl, in T-60175

53802-73-6 Peltochalcone; 4',6,7-Tri-Me ether, in P-70017

53820-53-4 11-Eremophilene-2,8,9-triol; (2α,8α,9α,10αH)-form, in E-70030

53822-96-1 Axisonitriles; Axisonitrile-1, in A-60328

53822-97-2 Axisothiocyanates; Axisothiocyanate-1, in A-60329

53870-24-9 2,3-Dichloroquinoxaline; 1-Oxide, in D-60150

53873-12-4 1-(3-Pyridinyl)-2-propanone; Picrate, in P-60230

53897-99-7 1H-1,2,3-Triazole-4-carboxylic acid; (2H)-form, Amide, in T-60239

53922-55-7 Tricyclo[4.2.2.0¹,⁶]deca-7,9diene, T-70226

53944-71-1 1,2-Naphthalenediol, N-70002

53947-90-3 Edgeworthin, in D-70006

53978-11-3 2-Amino-1,4cyclo-hexanedicarboxylic acid; (1RS,2RS,4RS)-form, Di-Et ester, in A-60118

53978-12-4 2-Amino-1,4cyclo-hexanedicarboxylic acid; (1RS,2RS,4SR)-form, Di-Et ester, in A-60118

53978-13-5 2-Amino-1,4cyclo-
hexanedicarboxylic acid;
(1*RS*,2*SR*,4*SR*)-*form*, Di-Et
ester, *in* A-60118

53978-17-9 2-Amino-1,4cyclo-
hexanedicarboxylic acid;
(1*RS*,2*SR*,4*RS*)-*form*, Di-Et
ester, *N*-benzoyl, *in* A-60118

53978-31-7 2-Amino-1,4cyclo-
hexanedicarboxylic acid;
(1*RS*,2*RS*,4*RS*)-*form*, Di-Et
ester, B,HCl, *in* A-60118

53978-32-8 2-Amino-1,4cyclo-
hexanedicarboxylic acid;
(1*RS*,2*RS*,4*SR*)-*form*, Di-Et
ester, B,HCl, *in* A-60118

53978-33-9 2-Amino-1,4cyclo-
hexanedicarboxylic acid;
(1*RS*,2*SR*,4*SR*)-*form*, Di-Et
ester, B,HCl, *in* A-60118

53984-36-4 3-Chloro-5-hydroxybenzoic acid,
C-60092

53990-67-3 Octahydro-2,7-dihydroxy5*H*,10*H*-
dipyrrolo[1,2a:1′,2′-*d*]-
pyrazine5,10-dione;
(2*R*,5a*S*,7*R*,10a*S*)-*form*, *in*
O-60013

53990-68-4 Octahydro-2,7-dihydroxy5*H*,10*H*-
dipyrrolo[1,2a:1′,2′-*d*]-
pyrazine5,10-dione;
(2*R*,5a*S*,7*R*,10a*R*)-*form*, *in*
O-60013

53990-71-9 Octahydro-5*H*,10*H*dipyrrolo[1,2-
a:1′,2′*d*]pyrazine-5,10-dione;
(5a*R*,10a*R*)-*form*, *in* O-60016

53990-72-0 Octahydro-5*H*,10*H*dipyrrolo[1,2-
a:1′,2′*d*]pyrazine-5,10-dione;
(5a*R*,10a*S*)-*form*, *in* O-60016

54002-75-4 Octahydro-2,7-dihydroxy5*H*,10*H*-
dipyrrolo[1,2a:1′,2′-*d*]-
pyrazine5,10-dione;
(2*R*,5a*S*,7*S*,10a*R*)-*form*, *in*
O-60013

54022-19-4 Diazocyclopropane, D-60060

54053-93-9 2,3,14,20,22,25-Hexahydroxy-
7cholesten-6-one;
(2β,3α,5β,20*R*,22*R*)-*form*, *in*
H-60063

54060-73-0 2-(9-Anthracenyl)ethanol, A-60282

54070-65-4 Biuret; 1,1,3,5-Tetra-Me, *in*
B-60190

54092-45-4 5,7-Dicyanoazulene, *in* A-70295

54092-46-5 5,7-Azulenedicarboxylic acid; Di-Et
ester, *in* A-70295

54147-74-9 2,2′-Diaminobiphenyl; 2-*N*-Ac, *in*
D-60033

54153-70-7 Melisodoric acid, M-70018

54154-26-6 3,7,11-Trimethyl-1-dodecanol;
(3*R*,7*R*)-*form*, *in* T-70282

54166-20-0 3,3-Diphenylcyclobutanone,
D-60476

54168-84-2 3-Methyl-4-oxo-2-butenoic acid;
(*E*)-*form*, *in* M-70103

54198-82-2 5-Chloro-4-methylpyrimidine,
C-70119

54198-89-9 5-Chloro-2-methylpyrimidine,
C-70118

54255-13-9 1-(4,5-Dihydro-1*H*-imidazol-2yl)-2-
imidazolidinethione; B,2HCl, *in*
D-60246

54258-41-2 5-Amino-1,10-phenanthroline,
A-70182

54283-14-6 2,2,5,5Tetramethylbicyclo[4.1.0]-
hept-1(6)-en-7-one, T-70131

54290-13-0 2,5-Dimethyl-3-(2phenylethenyl)-
pyrazine; (*E*)-*form*, *in* D-60442

54290-40-3 2,2,3,4,5,5-Hexamethyl-3-hexene;
(*E*)-*form*, *in* H-70083

54302-42-0 Gossypol; (±)-*form*, 6-Me ether, *in*
G-60032

54303-97-8 10*H*-Phenothiazine-2-carboxylic
acid; 5,5-Dioxide, *in* P-70056

54303-98-9 10*H*-Phenothiazine-2-carboxylic
acid; *N*-Me, 5,5-dioxide, *in*
P-70056

54304-03-9 10*H*-Phenothiazine-2-carboxylic
acid; Me ester, 5,5-dioxide, *in*
P-70056

54304-04-0 10*H*-Phenothiazine-2-carboxylic
acid; *N*-Me, Me ester, *in* P-70056

54338-82-8 Bicyclo[4.2.0]octane-2,5-dione;
(1*RS*,2*SR*)-*form*, *in* B-60093

54338-83-9 Bicyclo[4.2.0]oct-7-ene-2,5dione;
(1*RS*,5*SR*)-*form*, *in* B-60097

54354-35-7 1-(1-Propenyl)cyclohexene; (*E*)-
form, *in* P-60178

54365-83-2 1,5,9-Triazacyclotridecane,
T-60233

54378-77-7 1,3-Difluoro-5-(methylthio)benzene, *in* D-60176

54378-78-8 1,4-Difluoro-2-(methylthio)benzene, *in* D-60173

54384-48-4 Tris[(2-hydroxyethoxy)-
ethyl]amine, T-70323

54396-36-0 2-Thiabicyclo[2.2.1]heptan-3-one,
T-60186

54397-83-0 12-Hydroxy-5,8,10,14eicosatetra-
enoic acid;
(5*Z*,8*Z*,10*E*,12*S*,14*Z*)-*form*, *in*
H-60131

54415-31-5 3-Hexyne-2,5-dione, H-70090

54415-44-0 5-Acetylisoquinoline, A-60042

54429-93-5 2,2,3,4,5,5-Hexamethyl-3-hexene;
(*Z*)-*form*, *in* H-70083

54436-60-1 1,2,3,5,10,10aHexahydropyrrolo-
[1,2-*b*]isoquinoline, H-70072

54489-01-9 Bi(1,3-diselenol-2-ylidene),
B-70116

54532-45-5 Amanitins; Amanullinic acid, *in*
A-60084

54561-74-9 2,2-Diphenylpropanoic acid; Amide,
in D-70493

54571-66-3 5-Oxo-2-pyrrolidinecarboxylic acid;
(±)-*form*, Me ester, *in* O-70102

54583-69-6 2,5-Dimethyl-
3furancarboxaldehyde, D-70412

54629-30-0 10*H*-[1]Benzopyrano[3,2-*c*]pyridin-
10-one, B-60042

54630-44-3 1-Methyl-1,2-dihydro-3*H*pyrrol-3-
one, *in* D-70262

54660-00-3 2,5-Dihydro-3,6diphenyl-
pyrrolo[3,4-*c*]pyrrole-1,4-dione,
D-70204

54660-00-3 3,6-Diphenyl-
2,5dihydropyrrolo[3,4-*c*]pyrrole-
1,4-dione, D-70477

54668-28-9 1-Phenyl-1,4-pentadiyn-3-one,
P-70087

54676-87-8 Khellactone; (9*S*,10*S*)-*form*, Bis(3-
methylbutanoyl), *in* K-70013

54676-88-9 Khellactone; (9*S*,10*S*)-*form*, Bis(3-
methyl-3-butenoyl), *in* K-70013

54707-49-2 Grevilline D, *in* G-60038

54709-94-3 5-Methyl-3(2*H*)-pyridazinone,
M-70126

54730-18-6 Bromomethanesulfonic acid;
Bromide, *in* B-60291

54751-01-8 4-(Bromomethyl)pyridine, B-70255

54808-16-1 8-Bromo-2-hydroxy-
1,4naphthoquinone, B-60274

54808-30-9 2-Hydroxy-3-nitro-
1,4naphthoquinone, H-60195

54808-90-1 Decahydro-5,5,8a-trimethyl-
2naphthalenol; (2*S*,4a*S*,8a*R*)-
form, 2-Ketone, *in* D-70014

54819-17-9 1*H*-Phenothiazin-1-one, P-60078

54826-91-4 2,3-Dihydroxy-2-cyclopropen-1one,
D-70290

54856-83-6 2-Dimethylamino-3,3di-
methylazirine, D-60402

54863-75-1 13-Heptacosene; (*Z*)-*form*, *in*
H-60019

54878-25-0 Solavetivone, S-70051

54879-19-5 2-Bromo-3-methylaniline; *N*-Ac, *in*
B-70234

54879-20-8 2-Bromo-3-methylaniline, B-70234

54927-87-6 Teucrin *E*, T-60180

54931-11-2 1-Amino-2(1*H*)-pyridinone,
A-60248

54954-12-0 1,3,5,8-Tetrahydroxyxanthone; 8-
Glucoside, *in* T-70123

54954-13-1 1,3,5,8-Tetramethoxyxanthone, *in*
T-70123

54963-31-4 Isofloroseselin, *in* K-70013

54980-22-2 Physalin *D*, P-60145

54986-14-0 Tetrazole-5-thione; 1,4-*Dihydro*-
form, 1,4-Di-Me, *in* T-60174

55035-85-3 Gibberellin A₅₅, *in* G-60018

55092-18-7 3-Oxatetracyclo[5.3.1.1²,⁶.0⁴,⁹]-
dodecane, O-70069

55097-80-8 7,8,8a,9aTetra-
hydrobenzo[10,11]chryseno[3,4-
b]oxirene7,8-diol, T-70041

55103-01-0 2,7-Dichloroxanthone, D-70142

55108-01-5 4-Hexen-1-amine; (*E*)-*form*, *in*
H-70088

55126-92-6 Colipase, C-70188

55156-16-6 3-Methyl-2-pentanone; (±)-*form*,
in M-70105

55215-60-6 Trifluoropyrazine, T-60306

55220-86-5 Nonactinic acid; (±)-*form*, *in*
N-70069

55226-63-6 2,3,23-Trihydroxy-12-oleanen28-oic
acid; (2α,3α)-*form*, Me ester, *in*
T-70271

55273-87-5 2,5-Dihydro-2-methylthiophene,
D-70230

55303-89-4 6,7-Dimethoxy-3′,4′methyl-
enedioxyisoflavone, *in* T-70111

55303-97-4 Elatol, E-70002

55304-68-2 Senepoxide; (±)-*form*, *in* S-60023

55305-24-3 Tricyclo[3.3.1.1³,⁷]decyl-
1sulfinylcyanide, *in* T-70177

55400-60-7 3-(Trifluoromethyl)benzenesulfonic
acid; Ph ester, *in* T-60283

55400-68-5 3-(Trifluoromethyl)benzenesulfonic
acid; Et ester, *in* T-60283

55400-87-8 5,6:12,13-Diepoxy-5,6,12,13tetra-
hydrodibenz[*a*,*h*]anthracene,
D-70151

55400-88-9 4,5:9,10-Diepoxy-4,5,9,10tetra-
hydropyrene, D-70152

55401-97-3 2-(Bromomethyl)pyridine, B-70253

55443-61-3 2-Butene-1,4-dithiol; (*Z*)-*form*, *in*
B-70292

55479-94-2 2-(Hydroxymethyl)benzaldehyde,
H-60171

55483-01-7 5,7-Dihydroxy-6-methyl-
1(3*H*)isobenzofuranone, D-60349

55504-08-0 *N*-Methylbenzenecarboximidic
acid; (*E*)-*form*, Me ester, *in*
M-60045

55510-03-7 Tetrahydro-1,3,4,6tetra-
nitroimidazo[4,5*d*]imidazole-2,5-
(1*H*,3*H*)dione, T-70096

55520-64-4 5-Vinyluridine, V-70015

55532-07-5 2,2-Dimethyl-5-hexen-3-one,
D-60425

55627-73-1 8-Bromo-9-β-D-ribofuranosyl6*H*-
purin-6-one, B-70270

55627-78-6 8-Amino-9-β-D-ribofuranosyl6*H*-
purin-6-one; 6-Hydrazone, *in*
A-70198

55676-21-6 3-Acetyl-2-chloropyridine, A-70034
55676-22-7 5-Acetyl-2-chloropyridine, A-70037
55679-31-7 Bicyclo[3.2.1]octan-8-one, B-60094
55687-00-8 2-Chloro-6-methylquinoxaline, C-70123
55704-78-4 2,5-Dimethyl-1,4-dithiane-2,5diol, D-60421
55717-45-8 2-Bromo-5-hydroxypyridine, B-60276
55717-46-9 2-Amino-5-hydroxypyridine, A-60201
55730-99-9 Austrobailignan-7; Ac, in A-70271
55747-78-9 Ferulidene, in X-60002
55790-78-8 4,5-Dihydro-3,3,5,5tetramethyl-4-methylene-3H-pyrazole, D-60283
55790-79-9 3,3,5,5-Tetramethyl-4pyrazolidinone, T-60153
55806-40-1 Daphnorin, in D-70006
55812-47-0 Gibberellin A$_{45}$, in G-60018
55831-04-4 1,2-Biss(romomethyl)-4fluorobenzene, B-70132
55890-23-8 Austrobailignan-5, A-70270
55890-24-9 Austrobailignan-6; (2R,3R)-form, in A-60320
55890-25-0 Austrobailignan-7, A-70271
55901-80-9 2,3-Dichloro-4-methoxybenzoic acid, in D-60133
55903-92-9 Cucurbitacin R, in C-70203
55903-97-4 Octahydro-2-hydroxy-5H,10Hdipyrrolo[1,2-a:1',2'd]-pyrazine-5,10-dione; (2S,5aR,10aS)-form, in O-60017
55904-61-5 4-Amino-1H-pyrazole-3-carboxylic acid; Et ester, in A-70189
55907-33-0 Axisonitriles; Axisonitrile-2, in A-60328
55993-21-0 9,9'-Bii(icyclo[3.3.1]nonylidene), B-60079
56003-01-1 5,7-Dihydroxy-3',4',6,8tetra-methoxyflavone, D-60376
56003-87-3 Bi(1,3-diselenol-2-ylidene); Tetra-cyanoquinodimethane salt, in B-70116
56012-79-4 Disidein, D-60499
56012-88-5 Axamide-1, A-70272
56012-89-6 Axamide-2, A-60327
56012-90-9 Axisothiocyanates; Axisothiocyanate-2, in A-60329
56096-89-0 4-Fluoro-2-iodobenzoic acid, F-70023
56156-22-0 5-Amino-1-(phenylmethyl)-1Hpyrazole-4-carboxamide, in A-70192
56187-03-2 1,4,7-Triazacyclododecane; B,3HCl, in T-60231
56187-15-6 1,4,7,13,16,19Hexaazacyclotetra-cosane, H-70029
56218-77-0 9-Dodecen-1-ol; (Z)-form, Formyl, in D-70537
56218-84-9 9-Dodecen-1-ol; (Z)-form, Propanoyl, in D-70537
56221-26-2 2,3-Dihydroxy-6-iodobenzoic acid; Di-Me ether, Me ester, in D-60330
56221-41-1 6-Iodo-2,3-dimethoxybenzoic acid, in D-60330
56234-20-9 8-Acetylquinoline, A-60055
56257-59-1 ▷Altertoxin II, A-60081
56283-44-4 1(10),4-Germacradiene-6,8-diol; (1E,4E,6R,8R)-form, in G-60014
56284-07-2 3,4-Dimercapto-3-cyclobutene1,2-dithione; Di-K salt, in D-70364
56317-16-9 Strictic acid, S-70077
56317-21-6 Moracin M, M-60146
56319-02-9 Conocarpan, C-60161
56348-72-2 Bis(3,4-dichlorophenyl) ether, B-70137
56350-95-9 1-Methyl-2,5-dinitro-1H-pyrrole, in D-60463

56367-27-2 4,5-Dihydro-2-(1-methyl-propyl)thiazole, D-70224
56377-63-0 Deoxytifruticin, in T-70199
56377-69-6 Tifruticin, T-70199
56382-72-0 4,5,6,7Tetrahydrobenzothiadiazole, T-70042
56382-73-1 5,6-Dihydro-4H-cyclopenta1,2,3-thiadiazole, D-70179
56382-76-4 4,5,6,7Tetrahydrobenzothiadiazole; 2-Oxide, in T-70042
56382-80-0 4,5,6,7Tetrahydrobenzothiadiazole; 1,1,2-Trioxide, in T-70042
56383-84-7 3-(3-Hydroxyphenyl)-1H-2benzopyran-1-one, H-70207
56383-84-7 Homalicine, in H-70207
56405-97-1 3,3-Dimethyl-5,5-diphenyl-4pentenoic acid; Me ester, in D-60420
56472-18-5 1,1-Cyclopentanedimethanethiol, C-70243
56472-19-6 1,1-Cyclohexanedimethanethiol, C-70223
56488-60-9 γ-Glutamyltaurine; (S)-form, in G-60027
56506-08-2 4-Octene-1,8-diol, O-70026
56512-18-6 2-Cyclopropyl-2-oxoacetic acid; K salt, in C-60231
56555-55-6 Pyreno[1,2-c]furan, P-70165
56578-18-8 5-Decen-1-ol; (E)-form, in D-70019
56582-03-7 4,7-Dihydroisobenzofuran, D-60253
56586-60-8 3,19-Dihydroxy-12-oleanen-28-oic acid; (3α,19α)-form, in D-70327
56621-91-1 5-Amino-2-bromopyrimidine, A-70103
56632-86-1 6-Chlorotryptophan; (R)-form, in C-60146
56642-52-5 5-Methyl-4-phenylpyrimidine; 1,3-Dioxide, in M-70117
56645-60-4 Lipiarmycin, L-60028
56673-34-8 5-Bromo-2(1H)-pyridinethione, B-60317
56689-33-9 Acanthospermal A, A-60012
56711-41-2 Neodrosopterin, in D-60521
56741-33-4 5-Amino-2-fluorobenzoic acid, A-70147
56741-34-5 5-Amino-2-fluorobenzoic acid; Me ester, in A-70147
56797-40-1 7-Hexadecenal; (Z)-form, in H-70035
56798-34-6 2',4-Dihydroxy-4',6'dimethoxy-chalcone, in H-60220
56809-84-8 3,4-Dichloro-5-nitropyridine, D-60460
56814-40-5 3-Isopropenylcyclohexene, I-60116
56830-34-3 9,10-Difluorophenanthrene, D-70167
56845-83-1 5-Cholestene-3,26-diol; (3β,25S)-form, in C-60147
56900-24-4 1-Cyclooctene-1-acetic acid, C-70241
56900-31-3 1,3,5,7-Cyclooctatetraene-1acetic acid, C-70237
56904-85-9 2,4-Octadien-1-ol; (2E,4Z)-form, in O-70007
56962-12-0 4-Chloro-3-hydroxybenzaldehyde, C-60090
56970-91-3 2-Pyridyl 3-pyridyl ketone, P-70179
56970-92-4 2-Pyridyl 4-pyridyl ketone, P-70180
56970-93-5 3-Pyridyl 4-pyridyl ketone, P-70181
56974-44-8 Phleichrome, P-70098
56978-14-4 Gibberellin A$_{51}$, in G-60018
56986-80-2 3-Methylthio-2-phenyl-4H-1benzopyran-4-one, in M-60025
56996-26-0 Thymopoietin; Thymopoietin II, 9Cl, in T-70192
57020-32-3 N,N-Dimethylazidochloromethyl-eniminium(1+); Chloride, in D-70369

57052-72-9 3,4-Dihydro-2H-1-benzopyran-7-ol, D-60202
57072-36-3 Queuosine, in Q-70004
57074-21-2 3-Chloro-2,6-dihydroxy-4methyl-benzaldehyde, C-60059
57081-00-2 4-Hydroxy-2-methylbenzoic acid; Et ester, in H-70164
57090-88-7 4(5)-Cyanoimidazole, in I-60004
57097-81-1 3(5)-Bromo-5(3)-methylpyrazole, B-70252
57113-93-6 3-Amino-6-nitro-1,2benzenedicarboxylic acid; Di-Me ester, in A-70175
57147-26-9 3-Bromobenzo[b]thiophene; 1-Oxide, in B-60205
57147-27-0 2-Bromobenzo[b]thiophene; 1-Oxide, in B-60204
57147-28-1 2-Chlorobenzo[b]thiophene; 1-Oxide, in C-60042
57154-98-0 2,3,5-Trihydroxyphenylacetic acid, T-60341
57183-27-4 5-Amino-3-hydroxy-2methyl-pyridine, A-60193
57200-23-4 3-Phenyl-2(5H)furanone, P-60102
57224-50-7 2-Amino-4,4-dimethylpentanoic acid; (S)-form, in A-70137
57233-85-9 6-Chlorotryptophan; (±)-form, N^α-Formyl, in C-60146
57233-89-3 6-Chlorotryptophan; (R)-form, N^α-Formyl, in C-60146
57266-70-3 3,6-Pyridazinedicarboxylic acid, P-70168
57270-07-2 5-Methyl-4-phenylpyrimidine, M-70117
57270-08-3 5-Methyl-4-phenylpyrimidine; 1-Oxide, in M-70117
57270-09-4 5-Methyl-4-phenylpyrimidine; 3-Oxide, in M-70117
57279-38-6 4H-Thieno[3,4-c]pyran-4,6(7H)dione, in C-60018
57279-42-2 2-Carboxy-3-thiopheneacetic acid, C-60017
57279-54-6 4-Carboxy-3-thiopheneacetic acid, C-60018
57282-48-1 Verpyran, in H-70108
57289-92-6 [1]Benzothieno[2,3-c]quinoline, B-70062
57294-74-3 ▷N^4-Aminocytidine, A-70127
57350-38-6 4-Hydroxy-3-nitrosobenzaldehyde, H-60197
57362-82-0 1H-1,2,3-Triazole-4-carboxylic acid; (1H)-form, 1-Me, Me ester, in T-60239
57362-84-2 1H-1,2,3-Triazole-4-carboxylic acid; (1H)-form, 1-Me, amide, in T-60239
57415-35-7 2-Methoxy-4-methylbenzaldehyde, in H-70160
57416-04-3 Glycerol 1-acetate; (R)-form, in G-60029
57420-45-8 Acetylbarlerin, in S-60028
57420-46-9 Barlerin, in S-60028
57430-10-1 1-(4-Hydroxy-2-methoxyphenyl)3-(4-hydroxy-3-methoxyphenyl)-propane, in D-70336
57433-56-4 6-Chloro-2-phenylquinoxaline; 4-Oxide, in C-70146
57433-57-5 7-Chloro-2-phenylquinoxaline; 4-Oxide, in C-70147
57449-30-6 (3,4,5-Trihydroxy-6-oxo-1cyclo-hexen-1-yl)methyl 2-butenoate, T-60338
57459-06-0 3-O-Demethylsulochrin, in S-70083
57468-58-3 [2.2][2.2]Paracyclophane12,15-quinone, P-60008
57468-59-4 [2.2][2.2]Paracyclophane-5,8quinone, P-60007
57488-06-9 4-Octene-1,7-diyne, O-60028
57536-86-4 1-(1-Naphthyl)piperazine, N-60013
57536-91-1 1-(2-Naphthyl)piperazine, N-60014
57554-78-6 5-Nitroisoquinoline; 2-Oxide, in N-60031
57560-02-8 1,3-Dithiolane-2-selone, D-70525
57562-58-0 4-Methyl-5-phenylpyrimidine, M-70114

57576-33-7 Vernoflexuoside, *in* Z-60001
57576-43-9 Vernoflexin, *in* Z-60001
57576-82-6 Sargaquinone, S-60010
57605-80-8 2,7,11-Cembratriene-4,6-diol, C-70030
57605-95-5 1-(1-Naphthyl)ethanol; (±)-*form*, *in* N-70015
57606-91-4 3,3-Dimethyl-4-penten-2-one; Oxime, *in* D-70429
57672-81-8 Gibberellin A$_{40}$, *in* G-60018
57678-40-7 2-(Dimethylamino)benzaldehyde; Oxime, *in* D-60401
57697-76-4 2-Phenyl-3-furancarboxylic acid, P-70076
57707-91-2 5-Methyl-2-(1-methyl-1phenyl-ethyl)cyclohexanol; (1*S*,2*R*,5*S*)-*form*, *in* M-60090
57743-92-7 Phlebiakauranol aldehyde, P-70097
57766-27-5 2,3-Dihydro-1,6-dimethyl-3oxopyridazinium hydroxide, inner salt, *in* M-70124
57835-92-4 4-Nitropyrene, N-60042
57872-14-7 12-Hydroxy-5,8,10,14eicosatetra-enoic acid; (5*Z*,8*Z*,10*E*,12*S*,14*Z*)-*form*, Me ester, *in* H-60131
57899-68-0 4-Amino-1,1cyclo-hexanedicarboxylic acid; Di-Et ester, *in* A-60120
57899-74-8 4-Amino-1,1cyclo-hexanedicarboxylic acid, A-60120
57906-83-9 2-(Aminomethyl)cyclooctanol; (1*RS*,2*RS*)-*form*, *in* A-70170
57906-84-0 2-(Aminomethyl)cyclooctanol; (1*RS*,2*RS*)-*form*, B,HCl, *in* A-70170
57944-03-3 Neoechinulin *E*, N-70023
57965-30-7 2-Mercaptopropanoic acid; (*S*)-*form*, *in* M-60027
57984-70-0 3-Pentyn-2-ol; (*R*)-*form*, *in* P-70041
58001-78-8 Bicyclo[3.1.0]hexan-2-one; (1*R*,5*R*)-*form*, *in* B-60088
58022-21-2 1-Acetylisoquinoline, A-60039
58022-24-5 1-Benzoylisoquinoline; Hydrazone, *in* B-60059
58041-19-3 ▷2,3-Dihydro-5,6-diphenyl-1,4oxathiin, D-70203
58066-59-4 1,4,7,10,13Pentaazacyclo-pentadecane; B,5HCl, *in* P-60025
58081-05-3 Dihydro-4-hydroxy-2(3*H*)-furanone; (*R*)-*form*, *in* D-60242
58088-16-7 2-Pyridyl 3-pyridyl ketone; Oxime, *in* P-70179
58088-17-8 2-Pyridyl 4-pyridyl ketone; Oxime, *in* P-70180
58088-20-3 3-Pyridyl 4-pyridyl ketone; Oxime, *in* P-70181
58088-22-5 Di-4-pyridyl ketone; Oxime, *in* D-70504
58096-02-9 3-Amino-2-hexenoic acid; Et ester, *in* A-60170
58109-92-5 2,3-Dihydro-2-phenyl-4*H*-1benzothiopyran-4-one; (±)-*form*, 1-Oxide, *in* D-70248
58109-94-7 2,3-Dihydro-2-phenyl-4*H*-1benzothiopyran-4-one; (±)-*form*, 1,1-Dioxide, *in* D-70248
58115-32-5 9,10-Dihydro-5,6-dihydroxy1,3,4-trimethoxyphenanthrene, *in* P-70034
58115-33-6 9,10-Dihydro-5,6-dihydroxy2,4-dimethoxyphenanthrene, *in* D-70247
58123-70-9 3-Bromo-4-mercaptobenzoic acid, B-70228
58123-71-0 3-Bromo-4-(methylthio)benzoic acid, *in* B-70228

58123-73-2 3-Iodo-4-mercaptobenzoic acid, I-70046
58139-22-3 Loroglossin, L-70037
58149-65-8 1-Acetyl-3-bromonaphthalene, A-60021
58149-87-4 1-Acetyl-5-bromonaphthalene, A-60023
58164-02-6 1-*tert*-Butyl-3-iodobenzene, B-60347
58175-07-8 1,4-Dihydro-2,3-quinoxalinedione; Dione-*form*, 1,4-Di-Me, *in* D-60279
58190-98-0 2,7,11-Cembratriene-4,6-diol; (1*S*,2*E*,4*R*,6*R*,7*E*,11*E*)-*form*, *in* C-70030
58200-82-1 6-Nitro-1,4-naphthoquinone, N-60035
58239-51-3 Decahydro-5,5,8a-trimethyl-2naphthalenol; (2*S*,4a*S*,8a*R*)-*form*, *in* D-70014
58242-80-1 3,4-Dimethyl-1-pentanol; (*R*)-*form*, *in* D-60436
58275-00-6 2,2'-Methylenebis[8-hydroxy-3methyl-1,4-naphthalenedione], M-70063
58286-53-6 Scutellarioside II, *in* C-70026
58319-04-3 Sesquisabinene, S-70037
58319-06-5 Sesquithujene, *in* S-70037
58374-94-0 Imidazo[5,1,2-*cd*]indolizine; B,MeI, *in* I-70004
58427-00-2 Acenaphtho[5,4-*b*]thiophene, A-60015
58461-87-3 Tetrahydro-1,6,7triss(ethylene)-1*H*,4*H*3a,6a-propanopentalene, T-70101
58472-34-7 7-Bromo-2-hydroxy-1,4naphthoquinone, B-60273
58474-80-9 3-Decenal, D-70016
58503-62-1 1-Azido-3-iodopropane, A-60347
58512-71-3 1,4,7,12,15,18Hexaazacyclo-docosane, H-70028
58518-08-4 1,1,1-Trifluoro-4-phenyl-3butyn-2-one, T-70244
58539-77-8 3-(Chloromethyl)-2methylpyridine; B,HCl, *in* C-70097
58561-94-7 Laferin, L-60009
58569-25-8 α-Copaen-8-ol, *in* C-70197
58594-48-2 Camphonium, *in* T-70277
58616-95-8 1,2-Cyclopropanedicarboxylic acid; (1*RS*,2*RS*)-*form*, *in* C-70248
58657-26-4 ▷3-Nitropropanal, N-60037
58660-74-5 4-Amino-1,3cyclo-hexanedicarboxylic acid; (1*RS*,3*RS*,4*SR*)-*form*, Di-Et ester, *in* A-60121
58660-75-6 4-Amino-1,3cyclo-hexanedicarboxylic acid; (1*RS*,3*RS*,4*RS*)-*form*, Di-Et ester, *in* A-60121
58721-01-0 2,2,4,4-Tetramethyl-3-thietanone, T-60154
58721-02-1 2,2,4,4-Tetramethyl-3-thietanone; 1,1-Dioxide, *in* T-60154
58746-94-4 1,8-Biphenylenedicarboxaldehyde, B-60118
58751-09-0 Pentafluoroisocyanobenzene, P-60036
58779-65-0 Benz[*cd*]indol-2-(1*H*)-one; N-Ac, *in* B-70027
58785-54-9 3,6-Dihydroxy-1,2,3trimethoxy-xanthone, P-70036
58793-59-2 Hexahydro-1,3,5-triazine; 1,3,5-Triformyl, *in* H-60060
58809-18-0 Rubiflavin *F*, R-70012
58809-73-7 2-(Methylthio)propanoic acid, *in* M-60027
58842-19-6 5,6-Dihydro-2-(methylthio)-4*H*1,3-thiazine, *in* T-60094

58861-87-3 2-Chloro-3-phenylpyrazine; 4-Oxide, *in* C-70142
58861-88-4 2-Chloro-3-phenylpyrazine; 1,4-Dioxide, *in* C-70142
58880-25-4 2,3,19,23-Tetrahydroxy-12oleanen-28-oic acid; (2α,3β,19α)-*form*, *in* T-70119
58898-04-7 2-Mercapto-1,1-diphenylethanol, M-70028
58911-30-1 ▷3-Chloro-3-(trifluoromethyl)3*H*-diazirine, C-70169
58914-34-4 4-Hydroxy-2-(tri-fluoromethyl)benzaldehyde, H-70228
58917-56-9 Hexahydro-3-(1-methyl-propyl)pyrrolo[1,2-*a*]pyrazine1,4-dione, H-60052
58921-44-1 3-Methyl-4-oxo-2-pentenal; (*Z*)-*form*, *in* M-70104
58930-57-7 5-(3,4-Diacetoxy-1-butynyl)5'-methyl-2,2'-bithiophene, *in* M-70054
58935-95-8 4-Nitro-2-butanone, N-70045
58937-88-5 2-Azabicyclo[3.3.1]nonan-7-one, A-70276
58947-87-8 2,4,6(1*H*,3*H*,5*H*)-Pteridinetrione; 3-Me, *in* P-70134
58948-35-9 2-Phenyl-1,3-benzoxathiol-1-ium; Tetrafluoroborate, *in* P-60084
58960-35-3 2-Amino-3-fluorobutanoic acid; (2*R*,3*R*)-*form*, *in* A-60154
59045-82-8 3-Piperidinecarboxylic acid; (*S*)-*form*, *in* P-70102
59058-16-1 2-Hydroxycyclopentanone; (±)-*form*, Benzoyl, *in* H-60114
59066-61-4 2-(2-Phenylethenyl)benzothiazole; (*E*)-*form*, *in* P-60093
59096-04-7 Oreojasmin, O-70044
59156-43-3 2,4-Diaminobenzoic acid; 2-*N*-Ac, *in* D-60032
59219-48-6 1,3-Biss(ethylene)cyclopentane, B-60170
59219-64-6 8(14),15-Isopimaradiene-3,18diol; 3β-*form*, *in* I-70080
59263-72-8 2',4',6'-Trihydroxyacetophenone; 2',4'-Di-Me ether, Ac, *in* T-70248
59303-10-5 2-Chloro-5-methylpyrazine, C-70106
59315-44-5 3-Amino-2(1*H*)-pyridinone; *OH*-*form*, *in* A-60249
59331-62-3 2-Amino-3-(3,4-dichlorophenyl)propanoic acid, A-60131
59358-71-3 1,1'-Diphenylbicyclohexyl, D-70475
59384-57-5 2,3-Dichloro-4-nitrophenol, D-70125
59404-49-8 2,4-Octadecadienoic acid; (*E*,*E*)-*form*, *in* O-70003
59416-97-6 Octahydro-1,2,5,6-tetrazocine; Tetra-Ac, *in* O-60020
59416-98-7 Octahydro-1,2,5,6-tetrazocine; B,2HCl, *in* O-60020
59416-99-8 Octahydro-1,2,5,6-tetrazocine; B,H$_2$SO$_4$, *in* O-60020
59417-01-5 Octahydro-1,2,5,6-tetrazocine; Dipicrate, *in* O-60020
59632-76-7 Anisomelic acid, A-60275
59633-81-7 Axisothiocyanates; Axisothiocyanate-3, *in* A-60329
59633-83-9 Axisonitriles; Axisonitrile-3, *in* A-60328
59684-34-3 3,8-Dihydroxylactariusfuran, D-70312
59752-74-8 2-Amino-3-fluoro-3methylbutanoic acid; (*R*)-*form*, *in* A-60159
59779-79-2 2-Amino-3-mercapto-3phenyl-propanoic acid; (±)-*form*, B,HCl, *in* A-60213

59782-88-6 2,5-Dichloro-3-methylpyridine, D-60139

59782-90-0 2,3-Dichloro-5-methylpyridine, D-60138

59822-10-5 5-Bromo-3-methyl-1-pentene; (±)-form, in B-60299

59830-27-2 2,2-Dimethyl-1-phenyl-1propanone; 2,4-Dinitrophenylhydrazone, in D-60443

59840-70-9 Ethyl N-phenacylcarbamate, in O-70099

59845-92-0 Kainic acid; N-Ac, in K-60003

59858-28-5 2,6-Dinitrobenzoic acid; Chloride, in D-60457

59865-23-5 2-Amino-3-hydroxy-4-methyl-6octenoic acid; (2S,3R,4R,6E)-form, N-Me, in A-60189

59870-68-7 Glabridin, G-70016

59870-70-1 Glabridin; Di-Me ether, in G-70016

59979-58-7 Tagitinin E, T-70003

59979-61-2 Tagitinin A, T-60002

59980-98-2 2-Amino-3-chlorobutanoic acid; (2R,3S)-form, in A-60103

59985-28-3 12-Hydroxy-5,8,10,14eicosatetra-enoic acid, H-60131

59985-51-2 Coptiside I, in D-70319

59989-18-3 5-Ethynyl-2,4(1H,3H)pyrimidinedione, E-60060

59995-47-0 4-Hydroxy-2-cyclopentenone; (R)-form, in H-60116

59995-48-1 4-Hydroxy-2-cyclopentenone; (R)-form, Ac, in H-60116

59995-49-2 4-Hydroxy-2-cyclopentenone; (S)-form, in H-60116

60047-82-7 Cycloocta[def]fluorene, C-70233

60049-36-7 2-Amino-3-methyl-3-butenoic acid; (±)-form, in A-60218

60060-44-8 2-Bromo-1,3-cyclohexanedione, B-70201

60064-29-1 6-Aminoimidazo[4,5-g]-quinolin8(7H)-one, A-60209

60077-57-8 4-Hydroxy-5-benzofurancarboxylic acid; Me ester, in H-60103

60102-67-2 Castavaloninic acid, in V-70006

60103-01-7 2-Amino-3-methyl-3-butenoic acid; (R)-form, in A-60218

60122-72-7 2-Amino-4,4-dimethylpentanoic acid; (±)-form, in A-70137

60132-36-7 4,6-Dimethyl-4-octen-3-one; (S,E)-form, in D-70424

60168-05-0 4-Hexen-1-amine, H-70088

60189-62-0 lin-Benzoadenosine, in A-60207

60189-64-2 Imidazo[4,5-g]-quinazoline6,8(5H,7H)-dione, I-60009

60189-88-0 8-Aminoimidazo[4,5-g]quinazoline; 1-(β-D-Ribofuranosyl), in A-60207

60239-18-1 1,4,7,10Tetraazacyclododeca-ne1,4,7,10-tetraacetic acid, T-60015

60239-22-7 1,4,8,11Tetraazacyclotetradecane-1,4,8,11-tetraacetic acid, T-60017

60252-41-7 3-Piperidinecarboxylic acid; (±)-form, in P-70102

60268-40-8 Hanphyllin, H-60006

60297-83-8 Licarin C, L-60024

60302-27-4 2,2,3,3,4,4-Hexamethylpentane, H-60071

60338-65-0 Thieno[3,2-c]cinnoline, T-60203

60363-04-4 Onikulactone, in M-70150

60363-05-5 Mitsugashiwalactone, M-70150

60418-10-2 5H,7H-Dibenzo[b,g][1,5]dithiocin, D-70069

60418-11-3 5H,7H-Dibenzo[b,g][1,5]dithiocin; 12,12-Dioxide, in D-70069

60418-12-4 5H,7H-Dibenzo[b,g][1,5]dithiocin; 6,6,12,12-Tetroxide, in D-70069

60454-77-5 ▷Funkioside C, in S-60045

60454-78-6 ▷Funkioside D, in S-60045

60454-79-7 Funkioside E, in S-60045

60454-80-0 ▷Funkioside F, in S-60045

60454-81-1 ▷Funkioside G, in S-60045

60466-40-2 [(2-Phenylethenyl)seleno]benzene; (E)-form, in P-70070

60466-50-4 Bromomethyl phenyl selenide, B-60300

60498-89-7 Gnididione, G-60030

60508-97-6 1,2-Diphenyl-1,2-ethanediamine; (1RS,2RS)-form, N,N'-Di-Me, in D-70480

60509-62-8 1,2-Diphenyl-1,2-ethanediamine; (1RS,2SR)-form, N,N'-Di-Me, in D-70480

60529-76-2 Thymopoietin, T-70192

60545-98-4 4,6-Diphenylpyrimidine; N-Oxide, in D-70496

60546-10-3 Gomisin D, G-70025

60549-48-6 [2.2](2,6)Azulenophane; anti-form, in A-60355

60576-03-6 2-(2,4,6-Cycloheptatrien-1ylidene)-1,3-dithiane, C-70221

60583-16-6 5,8-Dihydroxy-2,3,6,7tetramethyl-1,4naphthoquinone, D-60377

60595-16-6 (Phenylthio)nitromethane, P-60137

60611-22-5 2-Chloro-6-(tri-fluoromethyl)benzaldehyde, C-70166

60622-99-3 2-(Aminomethyl)cyclooctanol; (1RS,2SR)-form, B,HCl, in A-70170

60631-76-7 3-Cyclohexene-1-carboxaldehyde; (R)-form, in C-70227

60656-87-3 2-Benzyloxyacetaldehyde, B-70085

60658-78-8 Comaparvin; 6-Methoxy, di-Me ether, in C-70191

60659-10-1 2-(Aminomethyl)cyclohexanol, A-70169

60671-71-8 3-Octenal, O-70024

60679-70-1 6'-Acetyltrifolirhizin, in M-70001

60681-09-6 2-Diazo-2,2,6,6-tetramethyl3,5-heptanedione, D-70064

60683-46-7 3-Isopropyl-2-naphthol, I-70095

60705-54-6 1-Bromo-3-nonene; (Z)-form, in B-60310

60723-27-5 ▷Physoperuvine, P-60151

60727-70-0 1-Acetylcyclooctene; (E)-form, in A-60027

60740-67-2 2,3,4,5-Tetrahydro-1H-1benzazepine; N-Benzyl, in T-70036

60753-26-6 1(15),8(19)-Trinervitadiene2,3-diol; (2β,3α)-form, in T-70295

60753-28-8 1(15),8(19)-Trinervitadien-9-ol; 9β-form, in T-70300

60760-06-7 3-(Chloromethyl)phenol, C-70102

60761-12-8 Nonactinic acid, N-70069

60768-66-3 5-Nitro-1,2-benzisothiazole, N-70041

60769-64-4 [1,1'-Bipyrrolidine]-2,2'-dione, B-70125

60770-00-5 2-Hydroxy-4-methylbenzoic acid; Et ester, in H-70163

60791-29-9 1(15),8(19)-Trinervitadiene2,3,9-triol; (2β,3α,9α)-form, 2,3-Di-Ac, in T-70298

60791-30-2 1(15),8(19)-Trinervitadiene2,3,9-triol; (2β,3α,9α)-form, 9-Ac, in T-70298

60811-22-5 4-Chloro-3-fluorobenzenethiol, C-60071

60811-23-6 3-Chloro-4-fluorobenzenethiol, C-60069

60811-24-7 3,4-Difluorobenzenethiol, D-60175

60814-30-4 4-Acetylquinoline, A-60051

60828-09-3 13(18)-Oleanene-2,3-diol; (2β,3β)-form, in O-60033

60861-06-5 2-Mercaptocyclohexanol; (1RS,2RS)-form, in M-70024

60894-97-5 2-Methyl-6-methylene-2,7octadien-4-ol; (R)-form, in M-70093

60958-71-6 Mikrolin, M-60135

60958-72-7 Dechloromikrolin, in M-60135

60964-09-2 2,6-Dichloro-4hydroxybenzaldehyde, D-70121

61014-18-4 25-O-Acetylbryoamaride, in C-70204

61047-38-9 10,11-Dihydro-5H-dibenzo[a,d]cyclohepten-5-one; Hydrazone, in D-60218

61062-50-8 4-tert-Butylcyclohexene; (R)-form, in B-60343

61082-23-3 3-Vinylcyclopropene, V-60012

61105-51-9 Bryoamaride, in C-70204

61117-54-2 Hexahydro-3-(1-methyl-propyl)pyrrolo[1,2-a]pyrazine1,4-dione; (3R,8aS,1'S)-form, in H-60052

61117-55-3 Hexahydro-3-(1-methyl-propyl)pyrrolo[1,2-a]pyrazine1,4-dione; (3R,8aS,1'R)-form, in H-60052

61135-33-9 2'-Deoxy-5-ethynyluridine, in E-60060

61148-17-2 2-Chloro-5-methylquinoxaline, C-70122

61149-88-0 Bicyclo[3.2.0]hept-2-ene-6,7dione, B-70099

61203-48-3 2-Iodo-4,5-dimethoxybenzoic acid, in D-60338

61210-21-7 2-Amino-4,7-dihydro-4-oxo-7βD-ribofuranosyl-1Hpyrrolo[2,3-d]-pyrimidin-5-carbonitrile, A-70132

61235-25-4 Auraptenol; (±)-form, in A-70269

61276-36-6 2,8-Dihydroxy-1,4-naphthoquinone; Di-Ac, in D-60355

61278-86-2 9,10-Biss(ethyl-ene)tricyclo[5.3.0.0²,⁸]deca-3,5-diene, B-70152

61289-05-2 Ballotinone, in M-70013

61305-27-9 4-Hydroxy-2-cyclopentenone, H-60116

61348-75-2 2-Amino-3-methyl-3-butenoic acid; (±)-form, B,HCl, in A-60218

61348-76-3 2-Amino-3-methyl-3-butenoic acid; (R)-form, B,HCl, in A-60218

61348-78-5 2-Amino-3-methyl-3-butenoic acid; (R)-form, Me ester, in A-60218

61366-76-5 2-Mercaptopropanoic acid; (±)-form, Me ester, S-Me ether, in M-60027

61371-61-7 ▷Conocandin, C-70194

61376-23-6 2-Amino-3-methyl-3-butenoic acid; (S)-form, in A-60218

61376-24-7 2-Amino-3-methyl-3-butenoic acid; (S)-form, B,HCl, in A-60218

61456-87-9 4-Cyano-1-phenyl-1,2,3-triazole, in P-60138

61468-81-3 4,5-Dihydro-3(2H)-pyridazinone, D-60272

61474-12-2 1,2,3,4-Tetrahydro-1-hydroxy2-naphthalenecarboxylic acid; (1R,2S)-form, Me ester, in T-70063

61474-13-3 1,2,3,4-Tetrahydro-1-hydroxy2-naphthalenemethanol; (1*R*,2*R*)-*form, in* T-70064

61480-98-6 2-Pyrrolidinecarboxaldehyde, P-70184

61481-40-1 γ-Glutamylmarasmine, G-60026

61485-47-0 Ethanebis(dithioic)acid; Di-Me ester, *in* E-60040

61494-52-8 1-Pyrenesulfonic acid; Chloride, *in* P-60210

61505-41-7 3-Chloro-2-hydroxypropanoic acid; (*R*)-*form, in* C-70072

61521-39-9 2-Amino-3,3-dimethylbutanedioic acid; (±)-*form, in* A-60147

61521-74-2 Wikstromol; (+)-*form, in* W-70003

61521-75-3 Wikstromol; (+)-*form*, Di-Me ether, *in* W-70003

61557-94-6 Decahydroquinazoline, D-70013

61561-22-6 2-(3-Hydroxyphenyl)-2-oxoacetic acid, H-70211

61566-58-3 2,4-Diaminobenzoic acid; B,HCl, *in* D-60032

61578-11-8 2-Chloro-5-phenylpyrazine; 1-Oxide, *in* C-70143

61578-12-9 2-Chloro-5-phenylpyrazine; 4-Oxide, *in* C-70143

61582-73-8 2,8,17-Trithia[4^{5,12}][9]metacyclophane, T-70328

61650-52-0 2-(3-Oxo-1-propenyl)benzaldehyde; (*E*)-*form, in* O-60091

61661-37-8 4-Chloro-9R-hydroxy-3,7(14)chamigradiene, *in* E-70002

61689-41-6 3-Chloro-2-methylpyrazine 1-oxide, *in* C-70105

61689-42-7 ▷2-Chloro-3-methylpyrazine; 1-Oxide, *in* C-70105

61689-44-9 2-Chloro-3-methylquinoxaline; 1-Oxide, *in* C-70121

61689-45-0 2-Chloro-3-methylquinoxaline; 4-Oxide, *in* C-70121

61708-79-0 9-Phenanthrenethiol, P-70049

61719-58-2 1,3-Dihydro-2*H*-imidazo[4,5-*c*]pyridin-2-one; 1-Benzyl, *in* D-60250

61740-29-2 4-Hydroxy-2-cyclopentenone; (±)-*form, in* H-60116

61752-45-2 1-Phenyl-3-penten-1-one; (*Z*)-*form, in* P-60127

61760-11-0 2-(2,2-Dimethylpropyl)naphthalene, D-60447

61771-84-4 1,4-Dihydropentalene, D-70241

61775-04-0 Saponin A, *in* C-70002

61782-63-6 2,2-Dibromocyclopropanecarboxaldehyde, D-60092

61826-89-9 Helichrysin, *in* H-60220

61836-40-6 4,5-Dimethoxy-1-naphthol, *in* N-60005

61836-58-6 3,5-Dihydroxy-2-methyl-1,4naphthoquinone; Di-Ac, *in* D-60350

61846-18-2 2,4,6(1*H*,3*H*,5*H*)-Pteridinetrione; 1,3-Di-Me, *in* P-70134

61847-84-5 1-Tetracosyne, T-70017

61871-80-5 2′-Amino-2-chloro-5′methylacetophenone, A-60114

61883-14-5 3-Amino-1,2cyclohexanedicarboxylic acid; (1*RS*,2*RS*,3*RS*)-*form*, Di-Et ester, *in* A-60119

61883-62-3 3-Morpholinone; 4-Benzoyl, *in* M-60147

61885-53-8 Bicyclo[4.1.1]octa-2,4-diene, B-60090

61885-54-9 Bicyclo[4.1.1]oct-3-ene, B-60096

61898-55-3 Tetrahydro-4-methyl-2*H*-pyran2-one, *in* H-70182

61898-56-4 5-Hydroxy-3-methylpentanoic acid; (*S*)-*form*, Lactone, *in* H-70182

61906-95-4 13-Heptacosene, H-60019

61915-20-6 3-Amino-1,2cyclohexanedicarboxylic acid; (1*RS*,2*SR*,3*RS*)-*form*, Di-Et ester, *in* A-60119

61915-22-8 3-Amino-1,2cyclohexanedicarboxylic acid; (1*RS*,2*SR*,3*SR*)-*form*, Di-Et ester, *in* A-60119

61936-90-1 Cyclopropanehexacarboxylic acid; Hexa-Et ester, *in* C-70250

61947-27-1 1,2-Cyclohexanedithiol, C-70225

61978-16-3 3-Methyl-1,2-naphthalenediol, M-60093

61978-33-4 6-Methyl-1,2-naphthalenediol, M-60095

61985-32-8 4(5)-Benzoylimidazole, B-60057

62003-27-4 Bomag, *in* O-70102

62008-04-2 Heteronemin, H-60030

62012-57-1 4,4′-Diiodo-1,1′-binaphthyl, D-60386

62013-43-8 5-Nitrohistidine; (*S*)-*form*, Me ester; B,HCl, *in* N-60028

62023-30-7 3-Amino-2-hydroxy-5methylhexanoic acid, A-60188

62054-49-3 3-Methyl-4-oxo-2-butenoic acid; (*E*)-*form*, Et ester, *in* M-70103

62076-67-9 2-Amino-3-chlorobutanoic acid; (2*RS*,3*SR*)-*form*, Me ester, *in* A-60103

62076-72-6 2-Amino-3-chlorobutanoic acid; (2*RS*,3*SR*)-*form*, N-Benzoyl, *in* A-60103

62076-81-7 2-Amino-3-chlorobutanoic acid; (2*RS*,3*RS*)-*form*, Me ester, *in* A-60103

62078-10-8 Axisonitriles; Axisonitrile-4, *in* A-60328

62078-11-9 Axisothiocyanates; Axisothiocyanate-4, *in* A-60329

62102-28-7 1-[3-Hydroxy-4-(hydroxymethyl)cyclopentyl]-2,4(1*H*,3*H*)pyrimidinedione, H-70140

62114-58-3 Dibenzo[*a,d*]cycloheptenylium; Perchlorate, *in* D-60071

62160-23-0 7-Deazaguanosine, *in* A-60255

62167-00-4 Benzeneethanethioic acid, B-70017

62171-59-9 1-*tert*-Butyl-2-iodobenzene, B-60346

62176-39-0 5-Chloro-2-(methylthio)benzoic acid, *in* C-70085

62222-77-9 6,6-Dimethyl-3oxabicyclo[3.1.0]-hexan2-one; (1*RS*,5*SR*)-*form*, *in* D-70425

62224-02-6 1,7-Heptanedithiol, H-70023

62289-79-6 2-Phenyl-2*H*-1,2,3-triazole-4carboxylic acid; Me ester, *in* P-60140

62316-29-4 3-Chloro-4-methoxy-5methylbenzoic acid, *in* C-60096

62319-70-4 Arjunglucoside I, *in* T-70119

62322-48-9 5-Methyl-2(5*H*)-furanone; (*R*)-*form, in* M-60083

62325-31-9 2,7-Diiodophenanthrene, D-70360

62348-13-4 5-Isoxazolecarboxylic acid; Chloride, *in* I-60142

62365-78-0 3-Ethynylindole, E-60058

62366-40-9 Di-1*H*-imidazol-2-ylmethanone; 1,1′-Di-Me, *in* D-70354

62369-72-6 Arjunglucoside II, *in* T-70271

62392-84-1 6-Methyl-2-cyclohexen-1-one; (*R*)-*form, in* M-70059

62393-64-0 4-Isopropenylcyclohexene; (*R*)-*form, in* I-60117

62422-45-1 4-Octene-1,8-diol; (*Z*)-*form, in* O-70026

62424-07-1 Aconitic acid; (*E*)-*form*, γ-Mono-Me ester, *in* A-70055

62438-05-5 1-Amino-2(1*H*)-pyridinone; B,HCl, *in* A-60248

62452-74-8 2-Bromo-3-ethoxy-1,4naphthoquinone, *in* B-60268

62458-42-8 3-Oxo-4,11(13)-eudesmadien-12oic acid, *in* E-70061

62488-55-5 2,4-Heptadien-1-ol, H-70014

62497-62-5 Antibiotic PDE I, A-60286

62502-14-1 Calopogonium isoflavone *B*, C-60006

62564-63-0 1*H*-Pyrazolo[3,4-*d*]pyrimidine; 1-Me, 5-oxide, *in* P-70156

62564-64-1 1*H*-Pyrazolo[3,4-*d*]pyrimidine; 1-Ph, 5-oxide, *in* P-70156

62590-16-3 1,2-Biss(romomethyl)-3fluorobenzene, B-70131

62616-12-0 4-Chloro-6-hydroxy-2(1*H*)pyridinone, C-70073

62616-13-1 4-Chloro-6-methoxy-1-methyl2(1*H*)-pyridinone, *in* C-70073

62616-14-2 4-Chloro-2,6-dimethoxypyridine, *in* C-70073

62665-02-5 1,2,3,6,7,8-Hexahydro-3*a*,5aethenopyrene, H-70056

62681-14-5 Tetrazole-5-thione; 1,3-*Dihydro-form*, 3-Et, 1-Ph, *in* T-60174

62690-65-7 1,2,3,5,8,8*a*Hexahydronaphthalene, H-60053

62690-66-8 1,2,3,7,8,8*a*Hexahydronaphthalene, H-60054

62708-42-3 2-(2-Propenyl)benzaldehyde, P-70122

62748-09-8 3,3-Dimethyloxiranemethanol; (*R*)-*form, in* D-70426

62785-91-5 Furo[2,3-*d*]pyrimidin-2(1*H*)-one, F-60106

62805-43-0 Carbodine, C-70018

62835-95-4 2-(2-Methylphenyl)propanoic acid, M-60113

62850-21-9 3,4-Di-*tert*-butyl-2,2,5,5tetramethylhexane, D-70109

62874-94-6 Antibiotic PDE II, A-70229

62928-74-9 Pentacyclo[5.3.0.0^{2,5}.0^{3,9}.0^{4,8}]-decan-6-one; (+)-*form, in* P-60027

62952-40-3 3-Oxo-1-cyclohexene-1carboxaldehyde, O-60060

62953-00-8 5,7-Dihydroxy-3,3′,4′,5′,8penta-methoxyflavone, *in* H-70021

62957-46-4 5-Bromo-2-pentanol, B-60313

62962-42-9 1*H*-Imidazo[2,1-*b*]purin-4(5*H*)-one, I-70006

62965-57-5 2-Amino-3,3-dimethylbutanoic acid; (*S*)-*form*, Amide, *in* A-70136

62989-38-2 5-Hydroxy-3-methylpentanoic acid; (±)-*form*, Lactone, *in* H-70182

62994-47-2 Warburganal, W-60001

63012-97-5 2-Methyl-3-furanthiol; *S*-Me, *in* M-70082

63027-86-1 9-Chloro-2-nitro-9*H*-fluorene, C-70134

63038-23-3 2-Amino-4-bromobutanoic acid; (±)-*form, in* A-60098

63038-26-6 2-Amino-3,3-dimethylbutanoic acid; (*S*)-*form*, Me ester, *in* A-70136

63062-88-4 1*H*-1,2,3-Triazole-4-carboxylic acid; (2*H*)-*form*, 2-Me, *in* T-60239

63069-50-1 4-Cyano-2-fluoroaniline, *in* A-70146

63109-30-8 Inuviscolide, I-70023

63121-50-6 3,4-Dihydroxy-2pyrrolidinecarboxylic acid, D-70342

63147-18-2 Hebeclinolide, H-60009

63181-39-5 5,10(14)-Germacradiene-1,4-diol; (1*S*,3*S*)-*form, in* G-70009

63181-40-8 5,10(14)-Germacradiene-1,4-diol; (1*S*,3*S*)-*form*, 1-Ac, *in* G-70009

63284-78-6 1,1,8,8-Tetramethoxy-4-octene, *in* O-70025

63285-40-5 Naphtho[8,1,2-*cde*]cinnoline, N-70007

63285-41-6 Naphtho[8,1,2-*cde*]cinnoline; *N*-Oxide, *in* N-70007

63286-42-0 Isoambrettolide; (*E*)-*form, in* I-60086

63303-42-4 Jacoumaric acid, *in* D-70349

63323-31-9 7,8,8a,9aTetra-hydrobenzo[10,11]chryseno[3,4-*b*]oxirene7,8-diol; (7*R*,8*S*,9*R*,10*S*)-*form, in* T-70041

63331-62-4 5-Chloro-4-thioxo-2(1*H*)pyrimidinone, C-70162

63335-23-9 Malabaricone *A*, M-70007

63335-24-0 1-(2,6-Dihydroxyphenyl)-9-(4hydroxyphenyl)-1-nonanone, *in* M-70007

63335-25-1 1-(2,6-Dihydroxyphenyl)-9(3,4-dihydroxyphenyl)1-nonanone, *in* M-70007

63335-26-2 9-(1,3-Benzodioxol-5-yl)-1(2,6-dihydroxyphenyl)1-nonanone, *in* M-70007

63351-80-4 Gibberellin A$_{63}$, *in* G-60018

63357-09-5 7,8,8a,9aTetra-hydrobenzo[10,11]chryseno[3,4-*b*]oxirene7,8-diol; (7*R*,8*S*,9*S*,10*R*)-*form, in* T-70041

63357-96-0 Dihydro-5-pentyl-2(3*H*)-furanone; (*R*)-*form, in* D-70245

63357-97-1 Dihydro-5-pentyl-2(3*H*)-furanone; (*S*)-*form, in* D-70245

63361-67-1 [2,2′-Bipyridine]-5,5′dicarboxylic acid; Dihydrazide, *in* B-60127

63364-52-3 2-Vinyl-1,1cyclo-propanedicarboxylic acid; (±)-*form*, Me ester, *in* V-60011

63364-56-7 1-Amino-2ethyl-cyclopropanecarboxylic acid; (1*RS*,2*RS*)-*form, in* A-60151

63367-96-4 3-Hydroxy-3′,4-dimethoxybibenzyl, *in* T-60316

63370-07-0 3-Bromobenzocyclopropene, B-60203

63393-56-6 1-Amino-2ethyl-cyclopropanecarboxylic acid; (1*S*,2*S*)-*form, in* A-60151

63393-57-7 1-Amino-2ethyl-cyclopropanecarboxylic acid; (1*R*,2*R*)-*form, in* A-60151

63399-77-9 2-Methyl-2-pyrrolidinecarboxylic acid; (*R*)-*form, in* M-60126

63425-54-7 3-Chloro-1-methylpyrazole, *in* C-70155

63452-69-7 1-Iodo-4-methoxy-2-methyl-benzene, *in* I-60050

63492-69-3 Lithospermoside, L-70034

63512-54-9 1-Amino-2-naphthalenethiol, A-70172

63555-48-6 Ircinianin, I-60081

63572-76-9 1-Amino-8-hydroxyanthraquinone, A-60177

63582-80-9 ▷Antibiotic 2061 A, *in* A-70158

63641-49-6 2-Iodocyclooctanone, I-70027

63641-53-2 Tetrahydro-3-iodo-2*H*-pyran-2-one, T-60069

63646-27-5 3,4-Dihydro-4-phenyl-1(2*H*)naphthalenone; (±)-*form, in* D-60267

63697-96-1 4-Ethynylbenzaldehyde, E-70048

63724-95-8 3-Chlorobenzo[*b*]thiophene; 1-Oxide, *in* C-60043

63744-96-7 4-O-Demethylmicrophyllinic acid, *in* M-70147

63758-60-1 Virescenoside L, *in* I-70081

63762-78-7 1-Fluoro-2-methoxy-4methyl-benzene, *in* F-60054

63762-79-8 2-Fluoro-5-methylphenol, F-60054

63763-79-1 2-(Bromomethyl)-6-chloropyridine, B-70235

63783-94-8 Baccharinol, B-70003

63814-58-4 Isobaccharinol, *in* B-70003

63857-17-0 3-Oxopropanoic acid; *Oxo-form*, Me ester, *in* O-60090

63893-02-7 Ethuliacoumarin, E-60045

63902-22-7 Wikstromol; (−)-*form, in* W-70003

63904-66-5 2,3-Diphenyl-2-butenal; (*Z*)-*form, in* D-60475

63904-67-6 2,3-Diphenyl-2-butenal; (*E*)-*form, in* D-60475

63904-72-3 1-Bromo-1,2-diphenylpropene; (*E*)-*form, in* B-60237

63904-73-4 1-Bromo-1,2-diphenylpropene; (*Z*)-*form, in* B-60237

63927-95-7 Bentemazole, *in* I-70009

63927-96-8 5-(1*H*-Imidaz-2-yl)-1*H*-tetrazole, I-70009

64067-72-7 Tetrahydro-2*H*-1,3-thiazine-2thione; *Thione-form*, *N*-Et, *in* T-60094

64129-96-0 γ-Ionone; (*S*)-*form, in* I-70058

64150-61-4 2,3-Dibromobenzofuran, D-60086

64165-99-7 5-Acetyl-2-(1-hydroxy-1methyl-ethyl)benzofuran, A-70044

64179-67-5 1,4-Benzodioxan-2-carboxaldehyde, B-60022

64181-95-9 1,5,8-Trihydroxy-2,6dimethoxy-xanthone, *in* P-60052

64187-50-4 2-Amino-3-bromobutanoic acid; (2*R*,3*S*)-*form, in* A-60097

64188-96-1 Pyrido[3,4-*d*]-1,2,3-triazin4(3*H*)-one, P-60238

64199-78-6 3,19,24-Trihydroxy-12-ursen28-oic acid; (3α,19α)-*form, in* T-60348

64255-63-6 3-Thiophenesulfonic acid; Amide, *in* T-60214

64267-46-5 ▷Disnogamycin, D-70506

64269-79-0 Di-1*H*-imidazol-2-ylmethanone, D-70354

64269-81-4 2,2′-Methylenebis-1*H*-imidazole, M-70064

64273-28-5 5-Methyl-4-phenyl-1,2,3thiadiazole, M-60117

64285-85-4 Furoscalarol, F-70057

64299-25-8 4-Amino-1*H*-pyrazole-3-carboxylic acid; 1-Ph, amide, *in* A-70189

64299-26-9 4-Amino-1*H*-pyrazole-3-carboxylic acid; 1-Ph, *in* A-70189

64329-58-4 12*H*-Quinoxalino[2,3-*b*][1,4]benzothiazine; *N*-Ac, *in* Q-70008

64350-83-0 2-Pyrenesulfonic acid, P-60211

64396-70-9 2-Phenylcyclohexanecarboxylic acid; (1*RS*,2*RS*)-*form, in* P-70065

64421-28-9 Shanzhiside; Me ester, *in* S-60028

64461-95-6 Picroside III, *in* C-70026

64481-14-7 2′,5′-Dihydroxy-4′nitro-acetophenone, D-70322

64504-48-9 β-Dihydroentandrophragmin, *in* E-70027

64504-52-5 Xenicin, X-70004

64543-31-3 6-(1-Heptenyl)-5,6-dihydro-2*H*pyran-2-one; (*R*,*Z*)-*form, in* H-60025

64586-13-6 1,2-Benzo[2.2]paracyclophane, B-70038

64591-03-3 2,4(5)-Dibromo-1*H*-imidazole, D-70093

64597-86-0 Ptilosarcone, *in* P-60195

64597-87-1 Ptilosarcenone, P-70136

64603-91-4 4,5,6,7Tetrahydroisoxazolo[5,4-*c*]pyridin-3-ol, T-60072

64636-39-1 1,4,5-Trimethoxynaphthalene, *in* N-60005

64657-20-1 8,13-Epoxy-1,6,7,9tetrahydroxy-14-labden11-one; (1α,6β,7β,8α,9α,13*R*)-*form, in* E-70027

64657-21-2 8,13-Epoxy-1,6,7,9tetrahydroxy-14-labden11-one; (1α,6β,7β,8α,9α,13*R*)-*form*, 6-Ac, *in* E-70027

64677-33-4 (2-Oxopropylidene)propanedioic acid; Di-Me ester, *in* O-70100

64701-00-4 2,3,7-Trimethoxyphenanthrene, *in* T-60339

64709-55-3 1-Pyreneacetic acid, P-60208

64725-24-2 RA-V, R-60002

64725-43-5 3,3-Dimethyl-4-oxopentanoic acid; Et ester, *in* D-60433

64736-53-4 4,5-Dichloro-1*H*-imidazole-2carboxylic acid, D-60135

64760-88-9 1,1′-Dicyanobicyclopropyl, *in* B-70114

64761-30-4 3-Nitro-4(1*H*)-pyridone; *N*-Me, *in* N-70063

64766-16-1 Physalactone, P-60144

64817-94-3 2,5-Diamino-3-hydroxypentanoic acid; (2*RS*,3*RS*)-*form, in* D-70039

64818-17-3 2,5-Diamino-3-hydroxypentanoic acid; (2*RS*,3*SR*)-*form, in* D-70039

64844-66-2 1,13-Dioxa-4,7,10,16,19,22hexa-azacyclotetracosane; 4,10,16,22-Tetrakis(4-methyl-benzenesulfonyl), *in* D-70460

64845-75-6 Sinularene, S-60034

64853-55-0 2′,3′,5′,6′-Tetramethyl-4′nitro-acetophenone, T-60147

64911-68-8 1,7,7-Trimethyl-bicyclo[2.2.1]heptane-2,5-diol; (1*S*,2*R*,5*S*)-*form, in* T-70280

64992-03-6 4-(Bromomethyl)-2-fluoropyridine, B-70245

65007-00-3 2-Pyridinyl trifluoromethanesulfonate, P-70173

65035-34-9 Lineatin; (+)-*form, in* L-60027

65043-59-6 20(29)-Lupene-3,16-diol; (3β,16β)-*form*, Di-Ac, *in* L-60039

65060-18-6 4-Oxo-2-piperidinecarboxylic acid; (*S*)-*form, in* O-60089

65075-32-3 4-(4-Methylphenyl)-2-pentanone; (±)-*form, in* M-60112

65092-35-5 5-Nitrohistidine; (*S*)-*form*, B,HCl, *in* N-60028

65129-90-0 Ligulatin A, *in* D-70025

65146-85-2 Aconitic acid; (*Z*)-*form*, α-Mono-Me ester, *in* A-70055

65146-86-3 Aconitic acid; (*Z*)-*form*, β-Mono-Me ester, *in* A-70055

65146-87-4 Aconitic acid; (*Z*)-*form*, γ-Mono-Me ester, *in* A-70055

65146-88-5 Aconitic acid; (*E*)-*form*, α-Mono-Me ester, *in* A-70055

65146-89-6 Aconitic acid; (*E*)-*form*, β-Mono-Me ester, *in* A-70055

65157-90-6 10-Oxodecanoic acid; Me ester, Di-Me acetal, *in* O-60063

65160-04-5 2,6-Dithioxobenzo[1,2-*d*:4,5*d*′]-bis[1,3]dithiole4,8-dione, D-60512

65172-10-3 1,2-Dihydro-1-phenyl-3*H*pyrrol-3-one, *in* D-70262

65179-91-1 Ligulatin C, *in* D-70025

65202-63-3 4,5,6,7Tetrahydroisoxazolo[5,4-*c*]pyridin-3-ol; B,HBr, *in* T-60072

65223-78-1 5-Ethynylcytidine, *in* A-60153

65223-79-2 4-Amino-5-ethynyl-2(1*H*)pyrimidinone, A-60153

65253-04-5 5-Methyl-2-(1-methyl-1phenylethyl)cyclohexanol; (1*R*,2*S*,5*R*)-*form, in* M-60090

65258-53-9 5-Amino-1-ethyl-1*H*-tetrazole, *in* A-70202

65326-33-2 3-Acetyl-2-aminopyridine, A-70024

65350-86-9 Meciadanol, *in* P-70026

65371-43-9 Dihydro-4,4-dimethyl-5methylene-2(3*H*)-furanone, D-70197

65373-82-2 2-Phenyl-4*H*-1-benzothiopyran4-one; *S*-Oxide, *in* P-60082

65383-74-6 6-Ethoxy-7-methoxy-2,2dimethyl-2*H*-1-benzopyran, *in* D-70365

65395-77-9 3,4,7,11-Tetramethyl-6,10tridecadienal, T-60156

65408-91-5 Altholactone, A-60083

65409-11-2 Altholactone; Ac, *in* A-60083

65464-36-0 8-Nitroisoquinoline; 2-Oxide, *in* N-60034

65478-59-3 ▷2*H*-[1,2,4]Oxadiazolo[2,3-*a*]pyridine-2-thione, O-70062

65490-23-5 9-(Phenylseleno)phenanthrene, P-60133

65514-04-7 3-Chloro-4-methylfuroxan, *in* C-60114

65518-46-9 2,4-Heptadienoic acid; (*E,E*)-*form, in* H-60021

65527-54-0 3-Thiomorpholinecarboxylic acid; (*R*)-*form, in* T-70182

65539-69-7 3-(Hydroxymethyl)-2naphthalenecarboxaldehyde, H-60188

65566-14-5 Methyl 4-fluoro-3,5dimethoxybenzoate, *in* F-60024

65566-16-7 Methyl 4-iodo-3,5dimethoxybenzoate, *in* D-60337

65594-65-2 Benzophenone-2,2'-dicarboxylic acid; Di-Me ester, *in* B-70041

65654-51-5 Tetrakiss(ethylthiomethyl)methane, *in* B-60164

65678-11-7 3-*tert*-Butyl-4hydroxybenzaldehyde, B-70304

65682-45-3 [1]Benzothieno[2,3-*c*][1,5]naphthyridine, B-70060

65682-47-5 [1]Benzothieno[2,3-*c*][1,7]naphthyridine, B-70061

65739-39-1 [2,2'-Bipyridine]-6,6'dicarboxylic acid; Dichloride, *in* B-60128

65739-40-4 [2,2'-Bipyridine]-6,6'dicarboxylic acid; Di-Et ester, *in* B-60128

65754-71-4 Bicyclo[5.3.1]undeca1,3,5,7,9-pentaene, B-60102

65788-55-8 3-(2-Aminoethylidene)-7-oxo-4oxa-1-azabicyclo[3.2.0]heptane-2-carboxylic acid; (2*R*,3*Z*,5*R*)-*form, in* A-70139

65792-05-4 Agrimophol, A-60066

65878-52-6 1-Amino-2ethylcyclopropanecarboxylic acid; (1*S*,2*R*)-*form, in* A-60151

65878-53-7 1-Amino-2ethylcyclopropanecarboxylic acid; (1*R*,2*S*)-*form, in* A-60151

65878-54-8 1-Amino-2ethylcyclopropanecarboxylic acid; (1*RS*,2*SR*)-*form, in* A-60151

65907-75-7 Danshexinkun *A*, D-60002

65915-07-3 3-Amino-1,2,4-triazine; *N*-Me, *in* A-70203

65927-08-4 3,5-Dihydro-3,3,5tetramethyl-4*H*-pyrazole-4-thione, D-60285

65927-19-7 3,5-Dihydro-3,3,5,5tetramethyl-4*H*-pyrazole-4-thione; *S*-Oxide, *in* D-60285

65938-86-5 3-Bromo-2(1*H*)-pyridinethione; *NH*-form, *in* B-60315

65967-73-9 Sydonic acid, S-60065

65978-43-0 ▷Carriomycin, C-60022

66039-62-1 Tricyclo[5.3.0.0²,⁸]deca3,5,9-triene, T-70227

66085-83-4 1-Dimethylamino-4-nitro-1,3butadiene, D-70367

66106-15-8 Rhodotorucin *A*, R-70004

66113-28-8 Perforenone, P-60063

66173-52-2 Artelin, A-60304

66193-78-0 1-Amino-2(1*H*)-pyridinone; *N*-Ac, *in* A-60248

66242-97-5 3,4-Pyridinedithiol, P-60221

66275-29-4 Isolobophytolide, I-60106

66289-51-8 5-Spirosten-3-ol; (3α,25*R*)-*form*, 3-*O*-β-D-Glucopyranoside, *in* S-60045

66289-52-9 5-Spirosten-3-ol; (3α,25*R*)-*form, in* S-60045

66289-87-0 4-Epivulgarin, *in* V-60017

66397-64-6 Dihydroparthenolide, D-70239

66428-84-0 8,13-Epoxy-1,6,7,9tetrahydroxy-14-labden11-one, E-70027

66478-35-1 [2](1,5)Naphthaleno[2](2,7)(1,6-methano[10]annuleno)phane, N-70005

66478-67-9 2,3-Di-*tert*-butoxycyclopropenone, *in* D-70290

66490-20-8 6-Chlorobenzo[*b*]thiophene, C-60046

66490-33-3 4-Chlorobenzo[*b*]thiophene, C-60044

66553-04-6 2-*tert*-Butylphenanthrene, B-70305

66575-29-9 Colforsin, *in* E-70027

66587-78-8 1,3-Diiodo-2-(iodomethyl)propane, D-70358

66619-32-7 4-Oxoheptanedinitrile, *in* O-60071

66620-31-3 2-Hydroxy-5-methyl-3nitrobenzaldehyde, H-70180

66634-12-6 Nicotinamide salicylate, *in* H-70108

66640-91-3 5-Hexyne-1,4-diamine; (*R*)-*form, in* H-60080

66655-67-2 3,4-Dihydro-2(1*H*)-quinazolinone, D-60278

66656-93-7 Taxillusin, *in* P-70027

66662-22-4 4-Oxooctanal, O-60081

66675-22-7 2-Trifluoromethyl-1*H*-imidazole, T-60289

66684-57-9 1,2,5-Trifluoro-3-nitrobenzene, T-60302

66684-58-0 1,2,3-Trifluoro-5-nitrobenzene, T-60300

66703-69-3 1,3-Dichloro-2-(chloromethyl)propane, D-70118

66707-26-4 2-Mercaptopropanoic acid; (±)-*form*, Et ester, *in* M-60027

66726-67-8 7,10:19,22:31,34Triethenotribenzo[*a*,*k*,*u*]cyclotriacontane; (*all-Z*)-*form, in* T-60273

66737-76-6 1,5-Hexadiyn-3-one, H-60040

66769-49-1 2,3-Dimethylcyclopropanol; (2*R**,3*R**)-*form*, Ac, *in* D-70397

66769-58-2 2-Chloro-3-phenylpyrazine; 1-Oxide, *in* C-70142

66769-59-3 2-Chloro-6-phenylpyrazine; 1-Oxide, *in* C-70144

66769-60-6 2-Chloro-6-phenylpyrazine; 4-Oxide, *in* C-70144

66786-05-8 ▷2-Thiocyanato-1*H*-pyrrole, T-60209

66791-94-4 1-Methoxy-2,3dimethylcyclopropane, *in* D-70397

66802-95-7 1,4,7,11,14Pentaazacycloheptadecane, P-60020

66809-05-0 Tetra-*tert*-butylcyclobutadiene, T-70014

66809-06-1 Tetra-*tert*-butyltetrahedrane, T-70015

66873-39-0 Thyrsiferol, T-70193

66888-85-5 3,6-Phenanthrenedimethanol, P-60069

66901-51-7 5-Hydroxy-2-methylcyclohexanone; (2*R*,5*S*)-*form, in* H-70171

66907-78-6 5-Amino-1*H*-tetrazole; 1*H*-Amino-(A)-*form*, 1,5(*N*)-Dibenzyl, *in* A-70202

66943-28-0 Thymopoietin; Thymopoietin I, *in* T-70192

66946-48-3 2-(5,6-Dihydro-1,3dithiolo[4,5-*b*][1,4]dithiin-2-ylidene)-5,6dihydro-1,3-dithiolo[4,5-*b*][1,4]dithiin, D-60238

66954-82-3 2,5-Diamino-3-hydroxypentanoic acid; (2*RS*,3*RS*)-*form*, B,HCl, *in* D-70039

66954-83-4 2,5-Diamino-3-hydroxypentanoic acid; (2*RS*,3*SR*)-*form*, B,HCl, *in* D-70039

66965-46-6 4-Isopropyl-1-methyl-4cyclohexene-1,2-diol; (1*R**,2*R**)-*form, in* I-70094

67022-41-7 Altersolanol C, *in* A-60080

67155-45-7 2-Iodoacetophenone; Oxime (*Z*-), *in* I-60039

67161-27-7 4-Octene-1,8-diol; (*Z*)-*form*, Di-Ac, *in* O-70026

67203-48-9 6,7,8,9-Tetrahydro-5*H*pyrido[3,2-*b*]azepine, T-70092

67221-50-5 4-Amino-1*H*-pyrazole-3-carboxylic acid; Amide, *in* A-70189

67242-54-0 ▷7,12-Dimethylbenz[*a*]anthracene; 1,2,3,4-Tetrahydro, *in* D-60405

67244-49-9 Pulchelloside I, P-70139

67263-73-4 9,10-Diacetylanthracene, D-70031

67280-33-5 7,12-Dihydro-5*H*-6,12methanodibenz[*c*,*f*]azocine, D-60255

67285-12-5 3*H*-Indol-3-one, I-60035

67294-53-5 2-Hydroxy-5-nitro-1,3benzenedicarboxylic acid, H-70188

67294-57-9 5-Amino-2-hydroxy-1,3benzenedicarboxylic acid; Di-Me ester, *in* A-70154

67323-02-8 3-(2-Methoxy-4,5methylenedioxyphenyl)propenal, M-70048

67382-69-8 2-Chloro-2-cyclohepten-1-one, C-60049

67388-22-1 4-Methylnaphtho[2,1-*b*]thiophene, M-60098

67446-07-5 5-Decen-1-ol; (*Z*)-*form*, Ac, *in* D-70019

67463-80-3 1,3-Diazetidine-2,4-dione; 1,3-Diisopropyl, *in* D-60057

67463-84-7 1,5-Dithionane; 1-Oxide, *in* D-70526

67490-05-5 Tetracyclo[5.5.1.0⁴,¹³.0¹⁰,¹³]tridecane, T-60035

67492-31-3 3',7-Dihydroxy-4'methoxyisoflavanone, *in* T-70263

67498-21-9 3-Hydroxy-2-methylpentanoic acid; (2*RS*,3*SR*)-*form*, Me ester, *in* H-60192

67502-64-1 Distyryl selenide; (*E,E*)-*form, in* D-60503

67502-78-7 Distyryl selenide; (*Z,Z*)-*form, in* D-60503

67506-48-3 Furazano[3,4-*b*]quinoxaline, F-70051

67528-62-5 2-Phenylcyclopropanecarboxylic acid; (1*R*,2*S*)-*form*, Me ester, *in* P-60087

67528-63-6 2-Phenylcyclopropanecarboxylic acid; (1*S*,2*R*)-*form*, Me ester, *in* P-60087

67528-68-1　2-Phenylcyclopropanecarboxylic acid; (1*RS*,2*SR*)-*form, in* P-60087

67531-86-6　2-Fluoro-6-hydroxybenzoic acid, F-60039

67533-03-3　Dihydrofusarubin A, *in* F-60108

67542-95-4　9*H*-Pyrrolo[1,2-*a*]azepin-9-one, P-70193

67560-61-6　2,3-Dibenzoylbutanedioic acid; (2*RS*,3*SR*)-*form,* Di-Et ester, *in* D-70083

67567-43-5　4-Chloro-2-nitrobenzylamine, C-70127

67567-44-6　5-Chloro-2-nitrobenzylamine, C-70129

67576-71-0　Dihydrofusarubin B, *in* F-60108

67594-67-6　4,5-Dibromothiazole, D-60111

67594-73-4　Colubrinic acid, C-70190

67601-06-3　6-Isopropenyl-3-methyl-9decen-1-ol; (3*S*,6*R*)-*form,* Ac, *in* I-70089

67604-03-9　Verboccidentafuran, V-70005

67604-04-0　Altamisin, *in* A-70084

67667-64-5　3-Epizaluzanin C, *in* Z-60001

67705-38-8　6*bH*-2*a*,4*a*,6*a*Hexahydrotriazacyclopenta[*cd*]pentalene, H-70077

67705-41-3　1*H*,4*H*,7*H*,9*bH*-2,3,5,6,8,9Hexahydro-3*a*,6*a*,9atriazaphenalene, H-70078

67715-80-4　2-Methyl-4-propyl-1,3-oxathiane, M-60122

67765-58-6　Menisdaurin, *in* L-70034

67815-57-0　1,2,3,5-Tetrafluoro-4,6diiodobenzene, T-60045

67845-24-3　1,3-Diphenyl-1*H*-indene-2carboxylic acid, D-60481

67853-37-6　2-Bromo-1-methoxy-3-nitrobenzene, *in* B-70257

67921-06-6　Coleonol C, *in* E-70027

67952-61-8　1-Bromo-2-undecene, B-70277

67965-45-1　Isothamnosin B, *in* I-70104

67979-24-2　Isothamnosin A, I-70104

67987-83-1　Montanin C, *in* T-70159

68013-79-6　2,3,4,5,8,9-Hexahydrothionin; (*E*)-*form, in* H-70076

68057-72-7　3-Chloro-4-hydroxy-3cyclobutene-1,2-dione, C-70071

68068-70-2　α-Amino-1*H*-imidazole-1-propanoic acid; (±)-*form, in* A-60204

68108-92-9　2-Methyl-1,6-dioxaspiro[4.5]decane, M-60062

68143-02-2　5-Methyl-3(2*H*)-pyridazinone; 2-Ph, *in* M-70126

68170-97-8　2-Tetradecyloxiranecarboxylic acid, T-60041

68206-04-2　3-Chloro-4-methylpyridazine, C-70110

68230-59-1　10*H*-Phenothiazine-3-carboxylic acid, P-70057

68236-11-3　Lonchocarpol A; (*S*)-*form, in* L-60030

68236-12-4　Cajaflavanone, C-60004

68251-94-5　(Bromomethylidene)adamantane, B-60296

68252-05-1　Dodecafluorobicyclobutylidene, D-60513

68282-47-3　2-Phenyl-1*H*-imidazole-4(5)carboxaldehyde, P-60111

68365-88-8　Dodecahydro-3*a*,6,6,9atetramethylnaphtho[2,1*b*]furan; (3*aα*,5*aα*,9*aβ*,9*bα*)-*form, in* D-70533

68373-50-2　3,4-Dihydro-2,2-dimethyl-2*H*pyrrole; B,HCl, *in* D-60229

68395-57-3　1,2,3,4,5,6Hexafluoronaphthalene, H-70044

68474-14-6　Fritschiellaxanthin, *in* D-70545

68507-34-6　(Dimethylamino)methanesulfonic acid, *in* A-60215

68541-99-1　Isoporphobilinogen; Di-Et ester; B,HCl, *in* I-60114

68558-73-6　Triphenyleno[1,12-*bcd*]thiophene, T-70310

68570-34-3　2,4,5,6-Tetrahydroxyphenanthrene, T-60122

68570-35-4　2,4,5,6-Tetrahydroxyphenanthrene; Tetra-Ac, *in* T-60122

68602-57-3　Trifluoroacetyl trifluoromethanesulfonate, *in* T-70241

68612-45-3　Melampodin D, M-60013

68631-48-1　3-(3,7-Dimethyl-2,6octadienyl)-4-hydroxybenzoic acid, D-70423

68676-85-7　3-Decenal; (*E*)-*form, in* D-70016

68681-08-3　17-epi-Azadiradione, *in* A-70281

68682-03-1　Cajaflavanone; (±)-*form, in* C-60004

68703-67-3　4-Amino-3-cyanopyrazole, *in* A-70189

68705-66-8　Kolaflavanone, *in* G-70005

68710-17-8　5,7-Dihydroxy-3′,4′,5′,6tetramethoxyflavone, *in* H-70081

68712-54-9　1-Methyl-3,4-dinitro-1*H*-pyrrole, *in* D-60464

68715-67-3　Haageanolide, H-60001

68738-47-6　1,3-Dimethyl-2methyleneimidazolidine, D-70420

68766-96-1　5-Oxo-2-pyrrolidinecarboxylic acid; (*R*)-*form,* Et ester, *in* O-70102

68776-18-1　Eumorphistonol, E-60073

68776-59-0　5-Cyanoisoxazole, *in* I-60142

68781-12-4　4-Amino-3-fluorobutanoic acid; (±)-*form,* B,HCl, *in* A-60158

68781-14-6　2-Amino-3-fluorobutanoic acid; (2*R*,3*S*)-*form, in* A-60154

68781-15-7　2-Amino-3-fluorobutanoic acid; (2*S*,3*R*)-*form, in* A-60154

68781-16-8　2-Amino-3-fluorobutanoic acid; (2*S*,3*S*)-*form, in* A-60154

68799-58-6　Arguticinin; 4-Epimer, *in* A-70250

68803-47-4　5-Methoxy-1,3-benzodioxole-4carboxylic acid, *in* H-60180

68832-40-6　Dihydroconfertin, D-60211

68871-27-2　19,20,21Triazatetracyclo-[13.3.1.1^{3,7}.1^{9,13}]heneicosa-1(19),3,5,7(21),9,11,13(20),-15,17-nonaene-2,8,14-trione, T-60236

68882-69-9　Bicyclo[4.2.0]octane-2,5-dione, B-60093

68913-85-9　2,3,4-Trihydroxyphenylacetic acid; Tri-Me ether, nitrile, *in* T-60340

68942-37-0　2-Amino-3-bromobutanoic acid, A-60097

68963-23-5　1-Hydroxy-7-methylanthraquinone, H-60170

68964-81-8　Austrobailignan-5; (2*RS*,3*RS*)-*form, in* A-70270

68964-82-9　Austrobailignan-6; (2*RS*,3*RS*)-*form, in* A-60320

68972-89-4　2-Mercaptocyclohexanol; (1*RS*,2*SR*)-*form, in* M-70024

68978-05-2　Neodolabellenol; 5-Epimer, *in* N-70022

68978-09-6　4′-O-Methylglabridin, *in* G-70016

69008-03-3　2-Bromo-5-hydroxy-1,4naphthoquinone, B-60269

69008-15-7　2-Bromo-6-methoxy-1,4naphthaquinone, *in* B-60270

69008-16-8　2-Bromo-7-methoxy-1,4naphthoquinone, *in* B-60271

69010-14-6　Neodolabellenol, N-70022

69032-72-0　3,6-Dimethyl-5-azauracil, *in* M-60132

69045-79-0　2-Chloro-5-iodopyridine, C-60109

69048-98-2　2-Tetrazolin-5-one; 1-Et, *in* T-60175

69075-42-9　5-Ethynyluridine, *in* E-60060

69075-47-4　2′-Deoxy-5-ethynylcytidine, *in* A-60153

69095-03-0　Bicyclo[2.2.1]heptane-1carboxylic acid; Amide, *in* B-60085

69098-69-7　2-Methyl-5-phenylpyrimidine; 1-Oxide, *in* M-70111

69111-91-7　3*H*-Indol-3-one; 1-Oxide, oxime, *in* I-60035

69120-06-5　Moracin C, *in* M-60146

69120-07-6　Moracin D, M-60143

69199-37-7　Mugineic acid, M-60148

69207-52-9　2-Tetradecyloxiranecarboxylic acid; (±)-*form,* Me ester, *in* T-60041

69212-29-9　2,3-Pyridinedithiol, P-60218

69212-36-8　2,3-Biss(ethylthio)pyridine, *in* P-60218

69222-10-2　6-Undecyn-1-ol, U-60004

69301-27-5　Curcuphenol; (*R*)-*form, in* C-70208

69314-18-7　1,1′-(1,2-Dioxo-1,2ethanediyl)bis-1Hindole-2,3-dione, D-70466

69314-26-7　2,3-Dichlorobenzofuran, D-60125

69321-60-4　1,3-Dibromo-2-methylbenzene, D-60103

69355-11-9　5-Hexyne-1,4-diamine, H-60080

69380-42-3　4-Anilino-1,10-phenanthroline, *in* A-70181

69380-48-9　4-Amino-1,10-phenanthroline, A-70181

69381-32-4　2-Iodocyclopentanone, I-60043

69386-75-0　Tetrahydro-1*H*-pyrazolo[1,2-*a*]pyridazine-1,3(2*H*)-dione, T-70088

69394-19-0　Pentalenic acid, P-60054

69433-66-5　Resorcylide; (*S*,*Z*)-*form, in* R-70001

69440-10-4　Viguiestin, *in* T-70003

69459-93-4　5,6-Didehydro-1,4,7,10tetramethyl-dibenzo[*ae*]cyclooctene, D-60169

69483-32-5　Resorcylide; (*S*,*E*)-*form, in* R-70001

69534-86-7　5-Hydroxymethyl-4-methyl-2(5*H*)furanone; (*R*)-*form, in* H-60187

69616-80-4　5,6-Dihydro-6-(2,4,6,10tetrahydroxyheneicosyl)2*H*-pyran-2-one, *in* D-70276

69625-33-8　Flexilin, F-70012

69625-34-9　Trifarin, T-70236

69640-32-0　2-(2-Methoxyphenyl)pyrrole, *in* H-60217

69657-27-8　4-(4-Methylphenyl)-2-pentanone; (*S*)-*form, in* M-60112

69703-25-9　1,4,10,13-Tetraoxa-7,16diazacyclo-octadecane; *N*,*N*′-Dibenzyl, *in* T-70145

69726-47-2　▷1-Methyl-2,3,5-trinitro-1*H*pyrrole, *in* T-60374

69726-53-0　▷1-Methyl-2,3,4-trinitro-1*H*pyrrole, *in* T-60373

69726-67-6　▷1-Methyl-2,3,4,5-tetranitro1*H*-pyrrole, *in* T-60160

69736-21-6　1-Methylnaphtho[2,1-*b*]thiophene, M-60096

69779-95-9　2-Amino-3-propylbutanedioic acid, A-70186

69785-94-0　5-Amino-4(1*H*)-pyrimidinone, A-70196

69833-09-6　2-Bromo-5-methoxy-1,4naphthoquinone, *in* B-60269

69833-10-9　2-Bromo-8-methoxy-1,4naphthoquinone, *in* B-60272

69847-25-2　3-Methoxybenz[*a*]anthracene, *in* H-70104

69868-02-6　Senecioodontol, S-60022

69891-94-7　3-Decenal; (*Z*)-*form, in* D-70016

69897-44-5　2-Hydroxy-1,3-diphenyl-1propanone; (*R*)-*form, in* H-70124

69897-45-6　2-Hydroxy-1,3-diphenyl-1propanone; (*S*)-*form*, *in* H-70124

69901-70-8　3-Amino-γ-carboline, A-70115

69905-01-7　9,11(13)-Eremophiladien-12-oic acid, E-70029

69905-05-1　3-*O*-Methyl-2,5dehydrosenecioodentol, *in* S-60022

69905-06-2　3-*O*-Methylsenecioodentol, *in* S-60022

69905-08-4　3,5-Di-*O*-methylsenecioodontol, *in* S-60022

69905-11-9　3-*O*-Angeloylsenecioodontol, *in* S-60022

69957-35-3　2-Nitro-1,3-diphenylpropane, N-70048

69966-55-8　3-(Bromomethyl)pyridine, B-70254

69975-77-5　Orlandin, O-70046

69978-82-1　4,11(13)-Eudesmadien-12-oic acid, E-70061

70092-10-3　3-Hydroxybenz[*a*]anthracene; Ac, *in* H-70104

70097-45-9　7-Chloro-1(3*H*)-isobenzofuranone, C-70080

70143-04-3　3-Methyl-4-oxo-2-butenoic acid; (*Z*)-*form*, *in* M-70103

70144-18-2　2,5-Dimethylpyrrolidine; (2*R*,5*R*)-*form*, B,HCl, *in* D-70437

70163-98-3　3-Fluoro-2-hydroxybenzoic acid; Me ester, *in* F-60040

70205-58-2　2,4,5-Phenanthrenetriol, P-60071

70256-08-5　Globularidin, *in* C-70026

70258-20-7　4-(Bromomethyl)-2-chloropyridine; 1-Oxide, *in* B-70237

70313-21-2　2′,5′-Dimethoxy-4′nitro-acetophenone, *in* D-70322

70334-60-0　3′-Azidoacetophenone, A-60335

70346-51-9　Urocanic acid; (*E*)-*form*, Me ester, *in* U-70005

70377-05-8　1*H*-Carbazole-1,4(9*H*)-dione, C-60012

70377-99-0　Nidoresedic acid, N-70035

70380-67-5　5-(2-Furanyl)oxazole, F-70048

70380-70-0　5-(2-Thienyl)oxazole, T-70171

70380-73-3　2-(5-Oxazolyl)pyridine, O-70072

70380-74-4　3-(5-Oxazolyl)pyridine, O-70074

70380-75-5　4-(5-Oxazolyl)pyridine, O-70076

70384-51-9　2-(2-Methoxyethoxy)-*N*,*N*-bis[2(2-methoxyethoxy)-ethyl]ethanamine, *in* T-70323

70398-11-7　2-Amino-2-benzylbutanedioic acid; (±)-*form*, *in* A-60094

70428-45-4　5-Methyl-2(5*H*)-furanone; (±)-*form*, *in* M-60083

70538-57-7　6-Nitroisoquinoline, N-60032

70547-85-2　1,3,6-Triphenylhexane, T-60386

70552-61-3　Dictyodial *A*, D-70145

70583-30-1　Triphenyleno[1,12-*bcd*]thiophene; *S*-Oxide, *in* T-70310

70583-31-2　Triphenyleno[1,12-*bcd*]thiophene; *S*-Dioxide, *in* T-70310

70597-76-1　2,7-Dithiatricyclo[6.2.0.0^{3,6}]deca-1(8),3(6)-diene4,5,9,10-tetrone, D-60506

70608-72-9　5-Hydroxy-6,8,11,14eicosatetra-enoic acid; (5*S*,6*E*,8*Z*,11*Z*,14*Z*)-*form*, *in* H-60127

70622-52-5　Malyngamide *C*, M-60009

70639-65-5　Doridosine, D-60518

70645-59-9　Isosarcophytoxide; (2*R*,11*R*,12*R*)-*form*, 3,4-Epoxide, *in* I-70100

70645-61-3　Isosarcophytoxide, I-70100

70654-71-6　4-Bromo-2,6-dimethoxyphenol, *in* B-60201

70656-98-3　1,5,9-Triazacyclotridecane; B,3HCl, *in* T-60233

70671-47-5　3-Amino-2-hydroxybutanoic acid; (2*R*,3*S*)-*form*, *in* A-60183

70671-91-9　3-Bromo-1,2-diphenylpropene, B-70209

70674-02-1　2,4,7(1*H*,3*H*,8*H*)-Pteridinetrione; 1,3,8-Tri-Me, *in* P-70135

70748-49-1　Sarcophytoxide, S-70011

70815-31-5　Silymonin, *in* S-70044

70815-32-6　Silandrin, S-70042

70840-66-3　RA-III, *in* R-60002

70853-11-1　3-Amino-2-hydroxy-5methyl-hexanoic acid; (2*S*,3*R*)-*form*, *in* A-60188

70853-17-7　3-Amino-2-hydroxy-5methyl-hexanoic acid; (2*R*,3*R*)-*form*, *in* A-60188

70858-10-5　1-Tetradecylcyclo-propanecarboxylic acid, T-60040

70871-28-2　1,2-Biss(henylthio)cyclobutene, B-70156

70908-63-3　2-Hydroxy-2-methylhexanoic acid; (±)-*form*, *in* H-60184

70913-80-3　Thiofurodysinin, T-70179

70913-81-4　Thiofurodysin, T-70178

70916-39-1　2,4,7(1*H*,3*H*,8*H*)-Pteridinetrione; 3,8-Di-Me, *in* P-70135

70916-41-5　2,4,7(1*H*,3*H*,8*H*)-Pteridinetrione; 1,6,8-Tri-Me, *in* P-70135

70916-42-6　2,4,7(1*H*,3*H*,8*H*)-Pteridinetrione; 1,8-Di-Me, *in* P-70135

70918-53-5　1,4-Benzodioxan-2-carboxylic acid; (*R*)-*form*, *in* B-60024

70954-68-6　2-Hydroxy-2-methylhexanoic acid; (*R*)-*form*, *in* H-60184

70979-88-3　2,4-Heptadien-1-ol; (2*E*,4*Z*)-*form*, *in* H-70014

70999-07-4　3,5-Pyridinedithiol, P-60222

70999-08-5　3,5-Biss(ethylthio)pyridine, *in* P-60222

71014-14-7　2,7-Diamino-4,6(3*H*,5*H*)pteridinedione, D-60046

71016-15-4　▷3-[(3-Methylbutyl)nitrosoamino]-2-butanone, M-60055

71028-59-6　2-Amino-3-phenyl-3-butenoic acid, A-60238

71030-39-2　5-Hydroxy-6,8,11,14eicosatetra-enoic acid, H-60127

71042-53-0　4,5-Biss(rifluoromethyl)1,3,2-dithiazol-2-yl, B-70164

71050-05-0　Queuine; 7-β-D-Ribofuranosyl; B,HCl, *in* Q-70004

71071-46-0　[2,2′-Bipyridine]-4,4′dicarboxylic acid; Di-Me ester, *in* B-60126

71149-60-5　3′,5,5′-Trihydroxy-3,4′,7,8tetra-methoxyflavone, *in* H-70021

71186-96-4　2-Hydroxy-5-methoxy-1,4naphthoquinone, *in* D-60354

71202-99-8　Eximin, *in* A-70248

71221-73-3　2,5,5-Trimethyl-2-cyclopenten1-one, T-70281

71239-86-6　α-Amino-1*H*-imidazole-2-propanoic acid; (±)-*form*, *N*^{α}-Ac, *N*^{1}-benzyl, *in* A-60205

71239-87-7　α-Amino-1*H*-imidazole-2-propanoic acid; (±)-*form*, *N*^{α}-Ac, *in* A-60205

71239-88-8　α-Amino-1*H*-imidazole-2-propanoic acid; (±)-*form*, *N*^{α}-*tert*-Butoxycarbonyl, *N*^{1}-benzyl, *in* A-60205

71241-93-5　Isoerivanin, I-70066

71241-93-5　1,3-Dihydroxy-4-eudesmen-12,6olide; (1α,3α,6α)-*form*, *in* D-70299

71243-11-3　4-Hydroxy-2-isopropyl-5benzofurancarboxylic acid; Me ester, *in* H-60164

71302-24-4　3,19-Dihydroxy-13(16),14spongiadien-2-one; 3α-*form*, Di-Ac, *in* D-70345

71302-26-6　3,19-Dihydroxy-13(16),14spongiadien-2-one; 3β-*form*, *in* D-70345

71302-28-8　3,19-Dihydroxy-13(16),14spongiadien-2-one; 3α-*form*, *in* D-70345

71302-30-2　3,19-Dihydroxy-13(16),14spongiadien-2-one; 3β-*form*, Di-Ac, *in* D-70345

71316-38-6　3-Amino-2(1*H*)quinolinethione, A-70197

71317-73-2　11,13-Hexadecadienal; (*Z*,*Z*)-*form*, *in* H-70033

71327-31-6　Dehydroisoerivanin, *in* I-70066

71339-25-8　Physalolactone, P-60148

71388-73-3　4-Amino-3-chlorobutanoic acid; (±)-*form*, *in* A-60106

71392-06-8　Hydnowightin, H-70094

71407-99-3　4-Amino-6-hydroxy-1,3benzenedicarboxylic acid; Di-Me ester, *in* A-70153

71426-39-6　1,3(5)-Bi-1*H*-pyrazole, B-60119

71431-22-6　Sylvestroside I, S-70087

71431-23-7　Sylvestroside II, *in* S-70087

71431-24-8　Sylvestroside III, S-70088

71431-25-9　Sylvestroside IV, S-70089

71452-17-0　3-Buten-1-ynylcyclopropane, B-70293

71461-30-8　2-Pyrrolidinecarboxaldehyde; (*S*)-*form*, *N*-Benzyloxycarbonyl, *in* P-70184

71474-60-7　Dibenzo[*b*,*f*]thiepin-3-carboxylic acid, D-60077

71474-64-1　Dibenzo[*b*,*f*]thiepin-3-carboxylic acid; 5,5-Dioxide, *in* D-60077

71479-92-0　5,7-Dihydroxy-3,6,8trimethoxy-flavone, *in* P-70031

71481-01-1　Pipoxide; (*R*)-*form*, *in* P-60159

71502-46-0　1-Amino-5-hydroxyanthraquinone, A-60176

71506-84-8　4-(Methylthio)-2(1*H*)pyridinethione, *in* P-60219

71506-85-9　2,4-Biss(ethylthio)pyridine, *in* P-60219

71506-86-0　1-Methyl-4-(methylthio)-2(1*H*)pyridinethione, *in* P-60219

71512-38-4　Hexahydro-5(2*H*)-azocinone; 1-Me, *in* H-70048

71545-19-2　Maragenin I, M-70011

71545-20-5　3β-Hydroxy-28-noroleana-12,17dien-16-one, *in* M-70011

71545-21-6　28-Nor-16-oxo-17-oleanen-3β-ol, *in* M-70011

71548-17-9　5,6-Epoxy-7,9,11,14eicosatetra-enoic acid, E-60015

71554-62-6　2,5-Pyridinedithiol, P-60220

71563-86-5　2-Mercaptopropanoic acid; (±)-*form*, *in* M-60027

71628-20-1　1,10-Dihydrodicyclo-penta[*a*,*h*]naphthalene, D-60220

71628-26-7　3,8-Dihydrodicyclo-penta[*a*,*h*]naphthalene, D-60221

71628-27-8　4,5-Phenanthrenedimethanol, P-60070

71634-96-3　9,10-Dihydrodicyclo-penta[*c*,*g*]phenanthrene, D-60219

71666-01-8　2,2-Dibromocyclo-propanecarboxylic acid; (±)-*form*, Me ester, *in* D-60093

71678-03-0　Ilimaquinone, I-70002

71690-19-2　2,3-Dihydro-2,2dimethylthiophene, D-60232

71690-98-7　2-Diazo-1,1,3,3tetramethyl-cyclopentane, D-60066

71700-92-0　5-Hydroxy-2-(phenylthio)-1,4naphthoquinone; Ac, *in* H-70213

71700-93-1　5-Hydroxy-2-(phenylthio)-1,4naphthoquinone, H-70213

71724-91-9　Fusarubin; *O*^{3}-Et, *in* F-60108

71748-38-4   1,11-Eudesmanediol; 1β-*form*, *in* E-70062

71779-54-9   1,5,6,8a-Tetrahydro-3(2H)indolizinone; (±)-*form*, *in* T-70071

71785-46-1   Pyrazolo[4,3-g]quinazolin5(6H)-one, P-60205

71785-68-7   2-Amino-2,3-dimethylbutanoic acid; (R)-*form*, Me ester, *in* A-60148

71835-80-8   N-Prolylserine; L-L-*form*, *in* P-60173

71868-25-2   N-Carbamoylglucosamine, C-70015

71899-16-6   Lineatin; (±)-*form*, *in* L-60027

71899-37-1   5-Hexadecyne, H-60037

71939-67-8   Tarchonanthus lactone, T-60009

71949-03-6   4b,10b-Methanochrysene, M-70040

72005-84-6   2-Methyl-4-nitrobenzaldehyde, M-70096

72011-63-3   1-Fluoro-1-octene, F-70034

72037-37-7   Tetrahydro-2H-pyran-2-thione, T-70085

72037-38-8   2-Oxepanethione, O-70077

72052-64-3   8-Azabicyclo[5.2.0]nonane; (1RS,7RS)-*form*, Picrate, *in* A-70274

72052-66-5   8-Azabicyclo[5.2.0]nonane; (1RS,7SR)-*form*, Picrate, *in* A-70274

72053-11-3   2-(Bromomethyl)cyclopenta-namine; (1RS,2SR)-*form*, B,HBr, *in* B-70240

72053-12-4   2-(Bromomethyl)cyclopenta-namine; (1RS,2RS)-*form*, B,HBr, *in* B-70240

72053-13-5   2-(Bromomethyl)cyclohexanamine; (1RS,2RS)-*form*, B,HBr, *in* B-70239

72059-45-1   5,6-Epoxy-7,9,11,14eicosatetra-enoic acid; (5S,6S,7E,9E,11Z,14Z)-*form*, *in* E-60015

72078-90-1   2-Hydroxy-5-nitro-1,3benzenedicarboxylic acid; Di-Et ester, *in* H-70188

72145-16-5   Dihydromyoporone, *in* M-60156

72154-34-8   Isobrasudol, *in* B-60197

72154-35-9   Brasudol, B-60197

72155-50-1   4-Amino-3-hydroxy-5phenyl-pentanoic acid; (3S,4S)-*form*, *in* A-60199

72155-51-2   4-Amino-3-hydroxy-5phenyl-pentanoic acid; (3R,4S)-*form*, *in* A-60199

72183-90-5   27-Hydroxy-3,21-friedelanedione, *in* D-70301

72183-91-6   21,27-Dihydroxy-3-friedelanone; 21α-*form*, *in* D-70301

72183-92-7   27-Hydroxy-3-friedelanone, *in* D-70301

72200-41-0   4,5,6,6a-Tetrahydro-2(1H)penta-lenone, T-70084

72229-33-5   2,8-Dihydroxy-1(10),4,11(13)germacratrien-12,6-olide; (1(10)E,2α,4E,6α,8β)-*form*, *in* D-70302

72229-34-6   Mollisorin B, *in* D-70302

72229-35-7   2,8-Dihydroxy-1(10),4,11(13)germacratrien-12,6-olide; (1(10)E,2α,4E,6α,8β)-*form*, 8-(2R,3R-Epoxy-2-hydroxymethyl-butanoyl), *in* D-70302

72229-36-8   2,8-Dihydroxy-1(10),4,11(13)germacratrien-12,6-olide; (1(10)E,2α,4E,6α,8β)-*form*, 8-(2S-Hydroxy-2-hydroxymethyl-3S-mercaptobutanoyl), *in* D-70302

72229-37-9   2,8-Dihydroxy-1(10),4,11(13)germacratrien-12,6-olide; (1(10)E,2α,4E,6α,8β)-*form*, 8-(3S-Hydroxy-2-methyl-enebutanoyl), *in* D-70302

72229-75-5   N-Methylphenylacetohydroxamic acid, *in* P-70059

72239-42-0   9'-Methoxysargaquinone, *in* S-60010

72239-43-1   11'-Methoxysargaquinone, *in* S-60010

72239-44-2   8',9'-Dihydroxysargaquinone, *in* S-60010

72239-45-3   8',9'-Dihydroxy-5methyl-sargaquinone, *in* S-60010

72241-26-0   Stillopsin, *in* D-70339

72284-72-1   4,4-Dichloro-2,3-dimethyl-2cyclo-buten-1-one, D-60131

72319-64-3   8-Amino-1-naphthalenethiol, A-70173

72324-39-1   5-Acetyl-2,2-dimethyl-1,3dioxane-4,6-dione, A-60029

72330-50-8   1-[2-(Hydroxymethyl)-3methoxy-phenyl]-1,5heptadiene-3,4-diol, H-70178

72404-94-5   2-Acetylquinoline; Phenylhydrazone, *in* A-60049

72447-81-5   1,1-Dimethoxy-3-nitropropane, *in* N-60037

72450-73-8   2-(Aminomethyl)cycloheptanol; (1RS,2RS)-*form*, *in* A-70168

72450-74-9   2-(Aminomethyl)cycloheptanol; (1RS,2SR)-*form*, *in* A-70168

72450-75-0   2-Aminocycloheptanemethanol; (1RS,2SR)-*form*, *in* A-70124

72450-76-1   2-Aminocycloheptanemethanol; (1RS,2RS)-*form*, *in* A-70124

72460-28-7   [2,2'-Bipyridine]-4,4'dicarboxylic acid; Dichloride, *in* B-60126

72496-59-4   Queuine, Q-70004

72505-77-2   4α-Hydroperoxydesoxyvulgarin, *in* V-60017

72518-41-3   2-Methyl-2-piperidinecarboxylic acid, M-60119

72579-04-5   Botrylactone; Ac, *in* B-70174

72579-05-6   Botrylactone, B-70174

72581-71-6   Isosilybin, I-70102

72612-74-9   5-Bromo-5-decene; (E)-*form*, *in* B-60225

72612-75-0   5-Bromo-5-decene; (Z)-*form*, *in* B-60225

72620-08-7   3,5-Dihydroxy-3',4',5',7,8penta-methoxyflavone, *in* H-70021

72638-72-3   8,9-Dihydroxy-1(10),4,11(13)germacratrien-12,6-olide; (1(10)E,4E,6β,8α,9β)-*form*, 9-(2R,3R-Epoxy-2-methyl-butanoyl), *in* D-60323

72640-82-5   4,7,13-Trioxa-1,10diazabicyclo[8.5.5]icosane, T-60375

72653-76-0   2-Amino-3-(methylthio)-1,4naphthalenedione, *in* A-70163

72656-14-5   Deschlorothricin, *in* C-60137

72674-29-4   Pyrenocin B, *in* P-60213

72690-77-8   8,9-Dihydroxy-1(10),4,11(13)germacratrien-12,6-olide; (1(10)E,4E,6β,8α,9β)-*form*, 9-Ac, 8-(2R,3R-epoxy-2-methyl-butanoyl), *in* D-60323

72704-04-2   Mollisorin A, *in* D-70302

72707-66-5   2-Bromomethyl-2-propenoic acid, B-70249

72711-84-3   Bharangin, B-70088

72715-03-8   Pentalenolactone E, P-60055

72715-05-0   Precapnelladiene, P-60171

72758-17-9   1-Bromo-8-(bromomethyl)naphthalene, B-60216

72764-82-0   2-Amino-4-phenyl-3-butenoic acid, A-60239

72772-24-8   2-Mercapto-3-phenylpropanoic acid; (±)-*form*, *in* M-60026

72779-23-8   Laurenene, L-60018

72795-78-9   1-Methyl-2,3-dinitro-1H-pyrrole, *in* D-60461

72812-45-4   5-Bromo-3,4-dihydro-4-thioxo2(1H)-pyrimidinone, B-70205

72893-87-9   Croconate blue, *in* T-70321

72948-20-0   Teucjaponin B, *in* T-60178

72948-20-0   Teupolin I, T-70159

72948-21-1   Teupolin II, *in* T-70159

72962-43-7   Brassinolide, B-70178

72996-65-7   2-(2-Bromoethyl)pyridine; B,HBr, *in* B-70210

73013-68-0   6-Acetylquinoline, A-60053

73020-90-3   Pinguisanin, P-60155

73021-14-4   8,9-Dihydroxy-1(10),4,11(13)germacratrien-12,6-olide; (1(10)E,4E,6α,8β,9ξ)-*form*, 8-(2-Methylbutanoyl), *in* D-60323

73023-76-4   Hydroxymethylbilane, H-70166

73027-11-9   Kachirachirol A, *in* K-60001

73033-01-9   2-Hydroxy-2-(hydroxymethyl)2H-pyran-3(6H)-one, H-70141

73042-64-5   2H-1,4-Benzothiazine-2,3(4H)dione, B-70054

73045-98-4   2-Hydroxycyclohexanecarboxylic acid; (1RS,2SR)-*form*, Amide, *in* H-70119

73050-40-5   2,2-Dimethyl-4,4-diphenyl-3butenoic acid; Nitrile, *in* D-70402

73069-27-9   Praeruptorin A, P-60169

73075-60-2   6,7-Dichloroisoquinoline; B,HCl, *in* D-60136

73090-23-0   2,2,4,4-Tetrafluoro-1,3dithietane; 1,1,3,3-Tetroxide, *in* T-60048

73114-31-5   Chisocheton compound A, *in* T-60223

73114-34-8   21-Acetoxy-21,23:24,25diepoxyapotirucall-14ene-3,7-diol, *in* T-60223

73129-12-1   4-Chloro-2-fluorobenzenethiol, C-60070

73217-63-7   5-Phenyl-1,2,4-oxadiazole-3carboxaldehyde; Covalent hydrate, *in* P-60122

73217-79-5   5-Phenyl-1,2,4-oxadiazole-3carboxaldehyde, P-60122

73256-82-3   Isobicyclogermacrenal; (−)-*form*, *in* I-60088

73264-91-2   11,13-Hexadecadienal; (E,E)-*form*, *in* H-70033

73275-96-4   5-Chloro-2,3dihydroxybenzaldehyde, C-70065

73290-22-9   2-Bromo-5-iodopyridine, B-60281

73306-73-7   Pentalenene, P-70037

73318-93-1   3-Butene-1,2-diol; (±)-*form*, Dibenzoyl, *in* B-70291

73338-84-8   Moracin E, M-60144

73338-85-9   Moracin F, *in* M-60146

73338-86-0   Moracin G, M-60145

73338-87-1   Moracin H, M-60145

73338-88-2   Moracin I, *in* M-60146

73338-89-3   Moracin J, *in* M-60146

73343-08-5   Oxysporone, O-70107

73347-43-0   11-Hydroxy-5,8,12,14icosatetra-enoic acid; (5Z,8Z,11R,12E,14Z)-*form*, *in* H-60130

73396-43-7   Pentamethylphenol; Ac, *in* P-60058

73397-20-3   3-Amino-2-hydroxy-5methyl-hexanoic acid; (2S,3S)-*form*, *in* A-60188

73397-21-4   3-Amino-2-hydroxy-5methyl-hexanoic acid; (2R,3S)-*form*, *in* A-60188

73413-69-1   Anamarine, A-60270

73420-44-7   6-Methylthioquinoline, *in* Q-60005

73428-05-4   3-Nitro-2-naphthaldehyde, N-70054

73453-34-6   4,4'-Dibromo-1,1'-binaphthyl; (+)-*form*, *in* D-60090

73473-42-4 2-Amino-2,3-dimethylbutanoic acid; (S)-form, B,HCl, in A-60148

73483-87-1 Perrottetianal A, P-60064

73491-37-9 1,2-Bis(4′-benzo-15-crown-5)diazene; (E)-form, in B-60141

73513-77-6 Benzo[1,2-f:4,5-f′]diquinoline, B-70030

73567-98-3 2,3-Dihydro-2,3-biss(ethylene)furan, D-60208

73583-39-8 3-Bromo-5-chloropyridine, B-60223

73583-41-2 4-Bromo-3-chloropyridine, B-60224

73618-70-9 Fusarubin; O⁹-Me, in F-60108

73625-00-0 Cycloheptatrienylidenee(etraphenylcyclopentadenylidene)-ethylene, C-60208

73642-95-2 2,2-Dichlorooctane, D-70129

73649-05-5 6,6′-(1,2-Dimethyl-1,2ethanediyl)-bis[1phenazinecarboxylic acid]; Di-Me ester, in D-70411

73654-19-0 1-Pyreneacetic acid; Me ester, in P-60208

73670-84-5 3,5-Hexadienoic acid; (E)-form, Chloride, in H-70038

73670-87-8 3,5-Hexadienoic acid; (E)-form, in H-70038

73671-07-5 4-Phenyl-1,2-naphthoquinone, P-60121

73676-23-0 Octahydroazocine; 1-Et, in O-70010

73679-39-7 Tetracyclo[6.2.1.1³,⁶.0²,⁷]dodec-2(7)-ene; (1α,3β,6β,8α)-form, in T-70021

73691-67-5 6-Amino-1,3-dihydro-1-methyl2H-purine-2-one, in A-60144

73692-50-9 3-(4-Hydroxyphenyl)-1-(2,4,6trihydroxyphenyl)-2propen-1-one, H-60220

73702-69-9 Pseudrelone B, P-70130

▷73728-55-9 1-(Hydroxymethyl)fluorene, H-70176

73746-49-3 3-Iodo-2-octanone, I-60053

73748-77-3 9-Chloro-3-nitro-9H-fluorene, C-70135

73748-78-4 2,4-Diaminobenzoic acid; 2,4-N-Di-Ac, in D-60032

73757-45-6 2-Amino-3,3-difluorobutanoic acid; (±)-form, Et ester, in A-60133

73773-07-6 4-Bromodiphenylmethanol; (S)-form, in B-60236

73789-39-6 Euglenapterin, E-60070

73790-09-7 α-Oxo-2-naphthaleneacetic acid; Et ester, in O-60079

73815-13-1 O-Methylcedrelopsin, in C-60026

73834-77-2 3-Amino-β-carboline, A-70110

73870-24-3 4-(Bromomethyl)pyridine; B,HBr, in B-70255

73873-87-7 Iloprost; (E)-form, in I-60003

73875-02-2 1,3,5-Cycloheptatriene-1,6dicarboxylic acid; Dichloride, in C-60202

73907-90-1 Pyrazolo[3,4-f]quinazolin9(8H)-one, P-60204

73908-01-7 1H-Pyrazolo[4,3-g]-quinazoline5,7(6H,8H)-dione, P-60203

73922-81-3 4-Phenyl-3-butyn-2-ol; (R)-form, in P-70063

73923-18-9 2-Chloro-3-methyl-2cyclopenten-1-one, C-70090

73981-34-7 ent-1β,7β,14S,20-Tetrahydroxy16-kauren-15-one, T-70114

73981-35-8 Kamebacetal B, in K-70002

73981-36-9 Kamebacetal A, K-70002

74045-97-9 Psorospermin, P-70132

74046-00-7 3′,4′-Deoxy-4′chloropsorospermin-3′-ol, in D-60021

74046-05-2 3′-Methoxyglabridin, in G-70016

74054-58-3 3,5-Hexadienoic acid; (E)-form, Et ester, in H-70038

74057-36-6 1,2-Naphthalenedicarboxaldehyde, N-60001

74067-77-9 2-Iodo-3-methylbutanal, I-60049

74070-00-1 Selenolo[3,4-b]thiophene, S-70030

74087-85-7 Dimethyldioxirane, D-70399

74115-12-1 3-Chloro-5-hydroxypyridine, C-60102

74115-13-2 3-Bromo-5-hydroxypyridine, B-60277

74144-54-0 ent-7β,11α,14S,20Tetrahydroxy-16-kauren15-one, T-70115

74144-55-1 3-Eudesmen-6-ol; 6β-form, in E-70065

74149-25-0 4,4-Diethoxy-2-butynal, in B-60351

74157-93-0 1-Phenyl-3-penten-1-one; (E)-form, in P-60127

74163-66-9 Verbascosaponin, in O-70034

74176-06-0 Obscuronatin, O-60002

74183-96-3 Coriolin; (±)-form, in C-70199

74213-96-0 Thalictoside, in N-70050

74214-39-4 1-Amino-2-(4-hydroxyphenyl)cyclopropanecarboxylic acid; (1RS,2SR)-form, in A-60197

74216-21-0 Z-Asn-Ala-OH, in A-60310

74235-23-7 3-Hydroxymugeneic acid, in M-60148

74281-81-5 Isomugeneic acid, in M-60148

74284-56-3 6-Siliphiperfolene, S-70043

74284-66-5 Norhardwickiic acid; Di-Me ester, in N-70077

74321-33-8 2,4-Dihydroxyphenylacetaldehyde; Dimethylhydrazone, in D-60369

74327-28-9 4-Oxoheptanal, O-60070

74327-95-0 8-Azabicyclo[3.2.1]octane; N-Chloro, in A-70278

74339-76-7 1,1′-Bi-1H-indene; (1RS,1′SR)-form, in B-60108

74357-37-2 [2,2′-Bipyridine]-3,5′dicarboxylic acid, B-60125

74373-21-0 2,6-Cycloheptadien-4-yn-1-one, C-60196

74404-32-3 1,2,3-Trimethyl-9H-carbazole, T-60357

74431-35-9 4b,9b-Dihydroindeno[2,1-a]indene-5,10-dione; (4bS,9bS)-form, in D-70220

74434-59-6 Sauvagine, S-70015

74451-57-3 Pyrazole blue; (E)-form, in P-70153

74474-66-1 Citreomontanin, C-60149

74474-75-2 Agropinic acid, A-70075

74496-19-8 5-Oxabicyclo[2.1.0]pent-2-ene, O-70059

74510-43-3 2-Chloro-3-methoxy-5methylbenzoic acid, in C-60095

74517-75-2 Praecansone B, P-70113

74517-76-3 Praecansone A, in P-70113

74524-18-8 Mannopinic acid, M-70009

74560-70-6 5-(Hydroxymethyl)-1,2,3,4cyclohexanetetrol; (1RS,2SR,3SR,4RS,5SR)-form, Penta-Ac, in H-70170

74560-71-7 5-(Hydroxymethyl)-1,2,3,4cyclohexanetetrol; (1RS,2SR,3SR,4RS,5SR)-form, in H-70170

74635-75-9 ent-8(17),13-Labdadiene-7α,15diol; (13E)-form, 7-Ac, in L-70006

74644-88-5 3,4-Dihydroxy-2pyrrolidinecarboxylic acid; (2S,3R,4R)-form, in D-70342

74663-99-3 4-Nitrosobenzaldehyde, N-60046

74683-16-2 Nectandrin B, in M-70003

74708-14-8 Hexahydro-2(3H)-benzofuranone; (3aS,7aR)-form, in H-70049

74708-16-0 Hexahydro-2(3H)-benzofuranone; (3aS,7aS)-form, in H-70049

74725-44-3 1,2-Dibenzylidenecyclobutane, D-60083

74732-99-3 1-Acetyltriphenylene, A-60059

74733-00-9 2-Acetyltriphenylene, A-60060

74741-45-0 Hexahydro-2(3H)-benzofuranone; (3aR,7aS)-form, in H-70049

74768-64-2 1,1,3,3-Tetramethyl-2indaneselone, T-60142

74768-84-6 1,1,3,3-Tetramethyl-2-indanone; Hydrazone, in T-60143

74783-36-1 4-(Aminoacetyl)pyridine; Semicarbazone; B,HCl, in A-70087

74799-65-8 Withaphysanolide, W-60010

74805-70-2 2′,3′,5,7-Tetrahydroxyflavone, T-60106

74838-13-4 Grahamimycin A₁, in C-60159

74840-91-8 4,5-Dibromo-1H-imidazole-2carboxylic acid, D-60102

74840-99-6 4,5-Dibromo-1H-imidazole-2carboxylic acid; Et ester, in D-60102

74866-28-7 2,2′-Dibromo-1,1′-binaphthyl, D-60089

74877-25-1 Benz[a]anthracene-3,4-dione, B-70013

74892-41-4 Nonactinic acid; (±)-form, Me ester, in N-70069

74908-84-2 Ethyl dicyanoacetate, in D-70149

74912-54-2 2-Butyne-1,4-dithiol, B-70308

74958-11-5 Ficisterol, F-60010

75020-20-1 [1]-Benzopyrano[2,3-d]-1,2,3triazol-9(1H)-one, B-60043

75078-18-1 7-Chloro-2-heptenoic acid; (E)-form, in C-70069

75088-80-1 Monoalide, M-70152

75094-00-7 Tricyclopropylcyclopropenium(1+); Chloride, in T-60270

75140-33-9 2-(Aminoacetyl)pyridine, A-70085

75140-34-0 4-(Aminoacetyl)pyridine, A-70087

75207-11-3 Monocillin III, in N-60060

75207-12-4 Monocillin V, in N-60060

75207-13-5 Monocillin I, in N-60060

75207-14-6 Monocillin IV, in N-60060

75207-15-7 Monocillin II, in N-60060

75206-68-8 Longikaurin B, in L-70035

75207-67-9 Longikaurin A, L-70035

75240-08-3 3-Amino-γ-carboline; B,HCl, in A-70115

75240-38-9 1,2,3,4-Tetrahydro-4hydroxyisoquinoline; (±)-form, in T-60066

75258-11-6 2-Ethynylindole, E-60057

75282-01-8 2,7,11-Cembratriene-4,6-diol; (1S,2E,4S,6R,7E,11E)-form, in C-70030

75332-26-2 2-Methylfuro[2,3-b]pyridine, M-60084

75347-13-6 Bursatellin, B-60338

75350-44-6 Tephrinone, in G-70015

75359-38-5 Tricyclopropylcyclopropenium(1+); Tetrafluoroborate, in T-60270

75415-03-1 2-(1H-Pyrazol-3-yl)pyridine, P-70159

75435-17-5 1,2-Benzoquinone-4-carboxylic acid, B-70046

75443-74-2 Teucrin H2, in T-60180

75443-75-3 Teuchamaedryn A, T-70157

75464-52-7 2,5-Diamino-1,4-benzenedithiol; B,2HCl, in D-70035

75478-98-7 Articulin, A-60307

75514-37-3 Chaetochromin, in C-60029

75520-53-5 2-Amino-3-methyl-3-butenoic acid; (R)-form, Benzyl ester, in A-60218

▷75523-42-1 4-(Chloromethyl)-2methylpyridine, C-70099

75538-94-2 1,3,5-Cycloheptatriene-1,6dicarboxylic acid; Mono-Me ester, in C-60202

75556-30-8 3-Amino-2-hydroxy-5methyl-hexanoic acid; (2*RS*,3*SR*)-*form*, *in* A-**60**188

75599-97-2 4,5-Difluoro-[1,1'-biphenyl]2,3-dicarboxylic acid; Di-Me ester, *in* D-**60**179

75612-51-0 3,4-Dihydro-3,3,8*a*-trimethyl1,6(2*H*,8*aH*)-naphthalenedione, D-**60**301

75619-06-6 2-Methylene-1,3-diselenole, M-**70**069

75628-00-1 2,6,10-Trimethyl-2,6,11dodeca-triene-1,10-diol; (2*Z*,6*E*)-*form*, 1-Angelyl, *in* T-**70**283

75628-01-2 2,6,10-Trimethyl-2,6,11dodeca-triene-1,10-diol; (2*Z*,6*E*)-*form*, *in* T-**70**283

75628-02-3 2,6,10-Trimethyl-2,6,11dodeca-triene-1,10-diol; (2*E*,6*E*)-*form*, *in* T-**70**283

75628-34-1 Nepenthone A, *in* T-**60**333

75628-93-2 2,5,8-Trihydroxy-3-methyl-6,7methylenedioxy-1,4naphthoquinone; Tri-Me ether, *in* T-**60**333

75658-84-3 Hexahydro-1*H*-cyclopenta[*c*]furan-1-one; (3*aR*,6*aS*)-*form*, *in* H-**70**054

75658-85-4 3-Oxabicyclo[3.2.0]heptan-2-one; (1*S*,5*R*)-*form*, *in* O-**70**053

75658-86-5 3-Oxabicyclo[3.1.0]hexan-2-one; (1*S*,5*R*)-*form*, *in* O-**70**055

75775-35-8 Toddasin, T-**70**203

75812-86-1 3-Amino-4-hexenoic acid; (3*S*,4*E*)-*form*, *N*-Benzoyl, Me ester, *in* A-**60**171

75816-15-8 5-Bromotryptophan; (±)-*form*, *N*α-Ac, *in* B-**60**333

75816-16-9 5-Bromotryptophan; (*R*)-*form*, *N*α-Ac, *in* B-**60**333

75816-18-1 7-Bromotryptophan; (±)-*form*, *N*α-Ac, *in* B-**60**335

75816-19-2 7-Bromotryptophan; (*S*)-*form*, *in* B-**60**335

75822-15-0 3-Piperidinecarboxylic acid; (±)-*form*, *N*-Nitroso, *in* P-**70**102

75853-08-6 1-(Pentafluorophenyl)ethanol; (±)-*form*, *in* P-**60**040

75872-68-3 Bifurcarenone; (2*Z*)-*form*, *in* B-**60**105

75907-52-7 2-Methoxy-1-phenylnaphthalene, *in* P-**60**115

75964-05-5 9-Hydroxy-5-nonenoic acid; (*Z*)-*form*, Me ester, *in* H-**70**194

75967-78-1 3,4,4*a*,5,6,7-Hexahydro-2*H*pyrano[2,3-*b*]pyrilium; Tetra-fluoroborate, *in* H-**60**055

76003-76-4 1*H*-1,2,3-Triazole-4-carboxylic acid; (1*H*)-*form*, 1-Benzyl, Me ester, *in* T-**60**239

76015-47-9 3-Hydroxy-4,5methyl-enedioxybenzoic acid; Me ether, chloride, *in* H-**60**178

76045-38-0 Physalin *G*, P-**60**146

76076-27-2 4(1*H*)-Quinolinethione, Q-**60**006

76076-35-2 3-Quinolinethiol, Q-**60**003

76093-26-0 1,5-Dichloroxanthone, D-**70**139

76093-30-6 1,7-Dichloroxanthone, D-**70**140

76148-76-0 2-Amino-3-mercapto-1,4naphthoquinone, A-**70**163

76152-63-1 3-Vinylcyclohexene; (*S*)-*form*, *in* V-**60**009

76152-65-3 3-Ethylcyclohexene; (*R*)-*form*, *in* E-**70**039

76167-53-8 3,3'-Oxybispyridine; *N*-Oxide, *in* O-**60**100

76167-54-9 3,3'-Oxybispyridine; *N*,*N*'-Dioxide, *in* O-**60**100

76220-92-3 4-Methyl-2,2,6,6tetrakis-s(rifluoromethyl)-2*H*,6*H*-[1,2]-bromoxolo[4,5,1-*hi*][1,2]-benzobromoxole, M-**60**128

76235-09-1 Neoverrucosanol; 5β-*form*, *in* N-**70**030

76235-94-4 12-Methoxy-8,11,13abietatrien-20-oic acid, M-**60**036

76248-14-1 Globularicisin, *in* C-**70**026

76282-17-2 2,4-Octadecadienoic acid, O-**70**003

76284-65-6 2,2'-Dibromo-1,1'-binaphthyl; (±)-*form*, *in* D-**60**089

76293-13-5 2,4-Dimethylthioxanthone, D-**70**444

76296-72-5 Polyphyllin D, *in* S-**60**045

76302-58-4 2-Amino-9,10-dihydrophenanthrene, A-**60**141

76319-75-0 Naphth[2,3-*a*]azulene-5,12-dione, N-**60**006

76343-93-6 Latrunculin A, L-**70**021

76343-94-7 Latrunculin B, L-**70**022

76343-94-7 Latrunculin D, L-**70**023

76343-96-9 3-O-Ethyldihydrofusarubin A, *in* F-**60**108

76356-92-8 1-Nitro-1,3-cyclohexadiene, N-**70**047

76376-32-4 Latrunculin C, *in* L-**70**023

76376-33-5 3-*O*-Ethyldihydrofusarubin B, *in* F-**60**108

76382-73-5 Anaphatol, *in* D-**60**339

76402-03-4 *N*-2-Benzothiazolyl-*N*,*N*',*N*″triphenylguanidine, B-**70**057

76419-75-5 Threonine; (2*RS*,3*RS*)-*form*, Me ester; B,HCl, *in* T-**60**217

76447-13-7 3,5-Dihydroxy-4-iodobenzoic acid, D-**60**337

76488-06-7 2-Amino-4,5,6,7tetra-hydrobenzo[*b*]thiophene-3-carboxylic acid; Et ester; B,HCl, *in* A-**60**258

76529-35-6 1,3-Dimethylthioxanthone, D-**70**441

76549-18-3 Isoneonepetalactone, *in* N-**70**027

76595-71-6 3-Octenal; (*E*)-*form*, *in* O-**70**024

76693-50-0 3-Ethyl-5,7-dihydroxy-4*H*-1benzopyran-4-one, E-**70**040

76695-73-3 2,3-Dibenzoylbutanedioic acid; (2*RS*,3*SR*)-*form*, Dibutyl ester, *in* D-**70**083

76695-74-4 2,3-Dibenzoylbutanedioic acid; (2*RS*,3*RS*)-*form*, Dibutyl ester, *in* D-**70**083

76713-39-8 2-Hydroxy-3-methylhexanoic acid; (2*S*,3*R*)-*form*, *in* H-**60**185

76713-40-1 2-Hydroxy-3-methylhexanoic acid; (2*R*,3*S*)-*form*, *in* H-**60**185

76727-31-6 7*H*,9*H*,16*H*,18*H*-Dinaphtho[1,8*cd*:1',8'-*ij*][1,7]dithiacyclododecin, D-**70**448

76727-41-8 Benz[5,6]indeno[2,1-*a*]phenalene, B-**70**025

76732-76-8 2-(Tetrahydro-2-thienyl)pyridine, T-**60**097

76822-71-4 *lin*-Benzotheophylline, *in* I-**60**009

76822-74-7 Imidazo[4,5-*g*]-quinazoline6,8(5*H*,7*H*)-dione; 3,5,7-Tri-Me, *in* I-**60**009

76832-42-3 *lin*-Benzocaffeine, *in* I-**60**009

76850-13-0 2-Chloro-6-methylpyrazine; 4-Oxide, *in* C-**70**107

76862-65-2 Conotoxin G1, C-**70**195

76868-97-8 Pyrenocin *A*, P-**60**213

76902-44-8 1,4-Triphenylenedione, T-**70**309

76905-80-1 2,2'-Diiodo-1,1'-binaphthyl, D-**60**385

76969-87-4 2-Amino-4-bromobutanoic acid; (*S*)-*form*, *N*-*tert*-Butyloxycarbonyl, Me ester, *in* A-**60**098

76983-82-9 3-Hydroxyiminoindole, *in* I-**60**035

77037-45-7 4-Oxo-2-phenyl-4*H*-1benzopyran-3-carboxylic acid, O-**60**085

77037-46-8 4-Oxo-2-phenyl-4*H*-1benzopyran-3-carboxylic acid; Et ester, *in* O-**60**085

77118-87-7 1-Phenyl-3-buten-1-ol; (*S*)-*form*, *in* P-**70**061

77162-46-0 3-Amino-4,4-difluorobutanoic acid, A-**60**134

77162-47-1 3-Amino-4-fluorobutanoic acid, A-**60**156

77167-91-0 3-Hydroxymethyl-dibenzo[*b*,*f*]thiepin, H-**60**176

77167-93-2 3-Hydroxymethyl-dibenzo[*b*,*f*]thiepin; 5,5-Dioxide, *in* H-**60**176

77182-66-2 Isosilychristin, I-**70**103

77189-69-6 2-Bromo-5-hydroxy-1,4naphthoquinone; Ac, *in* B-**60**269

77197-58-1 2-Bromo-8-hydroxy-1,4naphthoquinone; Ac, *in* B-**60**272

77209-45-1 5-(Hydroxymethyl)-1,2,3,4cyclo-hexanetetrol; (1*RS*,2*RS*,3*RS*,4*SR*,5*SR*)-*form*, Penta-Ac, *in* H-**70**170

77210-71-0 3*a*,4,7,7*a*-Tetrahydro-1(3*H*)isobenzofuranone, T-**70**072

77220-15-6 1,4-Benzoquinone-2,3dicarboxylic acid; Di-Me ester, *in* B-**70**047

77232-48-5 5-Methyl-2-phenylpyrimidine, M-**70**116

77242-40-1 4-*tert*-Butylcyclohexene; (*S*)-*form*, *in* B-**60**343

77242-41-2 3-*tert*-Butylcyclohexene; (*S*)-*form*, *in* B-**60**342

77256-26-9 2-Iodocycloheptanone, I-**70**025

77282-68-9 Isosulochrin, *in* S-**70**083

77282-69-0 5-O-Methylsulochrin, *in* S-**70**083

77284-05-0 Longikaurin C, *in* L-**70**035

77298-64-7 ▷8-Hydroxyflavone, H-**70**132

77300-37-9 6-Amino-2-bromohexanoic acid; (±)-*form*, B,HBr, *in* A-**60**100

77320-30-0 1,4,7,10,13Pentaazacyclohexadeca-ne; Pentakis(4-methyl-benzenesulfonyl), *in* P-**60**021

77326-36-4 2-Cyano-3-fluoroaniline, *in* A-**70**143

77332-89-9 3-Chloro-2-iodopyridine, C-**60**110

77332-90-2 3-Chloro-5-iodopyridine, C-**60**111

77370-02-6 Lotisoflavan, L-**70**038

77380-28-0 2,5-Difluorobenzenethiol, D-**60**173

77400-30-7 4,5-Dinitro-1,2-benzenediol, D-**70**449

77405-43-7 3-Hydroxy-2-methylpentanoic acid; (2*R*,3*S*)-*form*, *in* H-**60**192

77451-20-8 Sclareol; 6'-Deoxy-α-L-idopyranoside, *in* S-**70**018

77483-35-3 3-Azatetracyclo[5.3.1.1²,⁶.0⁴,⁹]-dodecane; B,HCl, *in* A-**70**282

77483-36-4 3-Azatetracyclo[5.3.1.1²,⁶.0⁴,⁹]-dodecane; *N*-Me, *in* A-**70**282

77553-75-4 *N*-Methylprotoporphyrin IX; Di-Me ester, *in* M-**60**123

77670-93-0 Epianastrephin, *in* A-**70**206

77670-94-1 Anastrephin, A-**70**206

77690-83-6 Egonol glucoside, *in* E-**70**001

77766-56-4 4-Chlorobenzo[*g*]phthalazin1(2*H*)-one, C-**70**055

77771-03-0 3-Bromo-4-fluorobenzyl alcohol, B-**60**244

77772-72-6 2-Fluoro-3-methylphenol, F-**60**052

77785-94-5 4-(2-Propenyl)benzaldehyde, P-**60**177

77849-09-3 Rubiflavin A, *in* P-**60**162

77856-33-8 Ara-doridosine, *in* A-**60**144

77873-76-8 3-Morpholinecarboxylic acid, M-**70**155

77935-54-7 5-Methyl-3,6-dioxo-2piperazineacetic acid; (2*S*,5*S*)-form, Benzyl ester, *in* M-60064

77949-42-9 Longikaurin E, *in* L-70035

77951-49-6 *N*,*N*′-Bis(3-sulfonatopropyl)4,4′-bipyridinium, B-70160

77967-61-4 Longikaurin D, *in* L-70035

77967-62-5 Longikaurin F, *in* L-70035

77977-04-9 2-Bromo-9-β-D-ribofuranosyl6*H*-purin-6-one, B-70269

77984-53-3 Auropolin, A-60319

78012-02-9 9β-Hydroxy-3-oxo-10(14)guaien-12,6α-olide, *in* D-70305

78032-07-2 6-Chlorophthalazine, C-70150

78032-08-3 5-Chlorophthalazine, C-70149

78037-99-7 5-Hydroxy-6,8,11,14eicosatetra-enoic acid; (5*S*,6*E*,8*Z*,11*Z*,14*Z*)-form, Me ester, *in* H-60127

78078-41-8 8,11,13-Abietatrien-3-ol; 3β-form, *in* A-70006

78168-48-6 Hexahydrocyclopenta[*cd*]penta-lene-1,3,5(2*H*)trione, H-60048

78168-74-8 3-Phenyl-2-propyn-1-amine, P-60129

78269-38-2 3,4-Dihydro-2*H*-1,5benzodithiepin, D-60195

78277-34-6 7-Amino-4-(trifluoromethyl)2*H*-1-benzopyran-2-one; *N*-Ph, *in* A-60265

78277-38-0 7-Amino-4-(trifluoromethyl)2*H*-1-benzopyran-2-one; *N*-Ac, *in* A-60265

78277-39-1 7-Amino-4-(trifluoromethyl)2*H*-1-benzopyran-2-one; *N*-Benzoyl, *in* A-60265

78347-59-8 3-Amino-4-fluorobutanoic acid; (±)-form, B,HCl, *in* A-60156

78365-98-7 3-Aminobicyclo[2.2.1]heptane2-methanol; (1*RS*,2*SR*,3*RS*)-form, *in* A-70093

78377-23-8 1-(Diazomethyl)pyrene, D-70062

78416-44-1 3,8-Dihydro-9*H*-pyrazolo[4,3-*f*]quinazolin-9-one, D-60271

78472-10-3 5-Hydroxy-3-methoxy-7oxabicyclo[4.1.0]hept3-en-2-one, H-70155

78472-11-4 3-Hydroxy-5-(4-hydroxyphenyl)pentanoic acid; (+)-form, *in* H-70142

78473-10-6 2-Amino-3,5-dicyanopyridine, *in* A-70193

78476-69-4 12-Chlorododecanoic acid; Chloride, *in* C-70066

78508-96-0 5-Hydroxymethyl-2(5*H*)-furanone; (*S*)-form, *in* H-60182

78516-78-6 3′,5,5′,6-Tetrahydroxy3,4′,7,8-tetramethoxyflavone, *in* O-70018

78549-00-5 Spiro[3,4-cyclohexano-4hydroxybicyclo[3.3.1]nonan-9-one-2,1′cyclohexane], S-70063

78560-77-7 2-Mercaptopropanoic acid; (*R*)-form, Et ester, *S*-Ac, *in* M-60027

78582-15-7 4-Amino-1*H*-imidazo[4,5-*c*]pyridine; 1-(2-Deoxy-β-D-ribofuranosyl), *in* A-70161

78583-89-8 2-Acetylquinoxaline; Oxime, *in* A-60056

78594-10-2 2,3-Bis(9-anthryl)cyclopropenone, B-70128

78607-36-0 2-Chloro-3-iodopyridine, C-60107

78680-95-2 5*a*,5*b*,11*a*,11*b*Tetrahydrocyclo-buta[1,2-*b*:4,3-*b*′]benzothiopyran-11,12-dione; (5*aα*,5*bβ*,11*aβ*,11*bα*)-form, *in* T-60063

78693-34-2 3-Octenal; (*Z*)-form, *in* O-70024

78693-95-5 Sarothamnoside, *in* T-60324

78738-44-0 5*a*,5*b*,11*a*,11*b*Tetrahydrocyclo-buta[1,2-*b*:4,3-*b*′]benzothiopyran-11,12-dione; (5*aα*,5*bα*,11*aα*,11*bα*)-form, *in* T-60063

78739-37-4 7-Tirucallene-3,23,24,25-tetrol; (3α,23*R*,24*R*)-form, *in* T-70201

78739-39-6 7-Tirucallene-3,23,24,25-tetrol; (3β,23*R*,24*R*)-form, *in* T-70201

78739-85-2 3-(Bromomethyl)-2,3dihydrobenzofuran, B-60295

78740-45-1 2′,3′,4′,6′-Tetramethyl-5′nitro-acetophenone, T-60146

78754-84-4 Pyrazino[2,3-*f*]quinazoline8,10-(7*H*,9*H*)-dione, P-60201

78787-77-6 1,6-Dimethylcarbazole, D-70378

78787-78-7 1,7-Dimethylcarbazole, D-70379

78787-79-8 2,5-Dimethylcarbazole, D-70383

78787-81-2 3,5-Dimethylcarbazole, D-70386

78790-82-6 4-Acetyl-3-chloropyridine, A-70036

78795-11-6 Pyrazino[2,3-*g*]quinazoline2,4-(1*H*,3*H*)-dione, P-60200

78804-17-8 Zeylenol; (−)-form, *in* Z-70006

78812-65-4 2,6-Biss(ydroxyacetyl)pyridine, B-70144

78843-78-4 3-Formyl-2-butenenitrile, *in* M-70103

78919-13-8 Iloprost, I-60003

78943-60-9 10,11-Dihydro-5-phenyl-5*H*dibenz[*b*,*f*]azepine, *in* D-70185

78959-46-3 2-Amino-4*H*-3,1-benzothiazine, A-70092

78981-67-4 *N*-Valylleucine; D-L-form, Boc-Val-Leu-OMe, *in* V-60001

79083-18-4 11-Hydroxy-5,8,12,14icosatetra-enoic acid; (5*Z*,8*Z*,11*R*,12*E*,14*Z*)-form, Me ester, *in* H-60130

79100-18-8 4*H*,6*H*-Pyrrolo[1,2-*a*][4,1]benzoxazepine, P-60246

79100-21-3 1,3-Dihydro-2*H*-imidazo[4,5-*b*]pyrazine-2-thione, D-60247

79103-34-7 Thymopoietin; Thymopoietin III, *in* T-70192

79120-58-4 Zuonin *A*, Z-60003

79121-29-2 Leucettidine, L-70029

79128-34-0 Tetrahydro-2*H*-1,3-thiazine; B,HCl, *in* T-60093

79144-85-7 1-(1-Naphthyl)piperazine; B,HBr, *in* N-60013

79199-28-3 Physanolide, P-60149

79203-37-5 Taonianone, T-60006

79205-57-5 2-Amino-3-fluoro-3methylbutanoic acid; (±)-form, Nitrile, *in* A-60159

79214-54-3 2-(2-Hydroxy-4-methoxyphenyl)5-(1-propenyl)benzofuran, *in* H-60214

79289-49-9 1,2-Diazetidin-3-one; 4-Methyl-benzenesulfonate salt, *in* D-70055

79314-57-1 5-Hydroxy-4-methyl-3-heptanone, H-60183

79357-44-1 2-Amino-2-hexenoic acid; (*Z*)-form, *N*-Ac, Me ester, *in* A-60167

79357-45-2 2-Amino-2-hexenoic acid; (*E*)-form, *N*-Ac, Me ester, *in* A-60167

79367-59-2 Sterpuric acid, S-60053

79367-60-5 13-Hydroxysterpuric acid, *in* S-60053

79367-61-6 Sterpuric acid; 13-Hydroxy, 3,13-ethylidene acetal, *in* S-60053

79383-33-8 3,4,7,11-Tetramethyl-1,3,6,10dodectetraene; (3*Z*,6*Z*)-form, *in* T-70135

79383-34-9 3,4,7,11-Tetramethyl-1,3,6,10dodectetraene; (3*Z*,6*E*)-form, *in* T-70135

79405-85-9 Melcanthin F, M-60014

79421-50-4 Furoxano[3,4-*b*]quinoxaline, *in* F-70051

79435-73-7 2-Amino-3-phenyl-3-butenoic acid; (*R*)-form, Me ester, *in* A-60238

79441-12-6 2-Acetylquinoxaline; 1,4-Dioxide, *in* A-60056

79483-43-5 8-Bromo-9-β-D-ribofuranosyl6*H*-purin-6-one; 2′-(4-Methyl-benzenesulfonyl), *in* B-70270

79544-29-9 2-Cyano-1-fluoro-3-iodobenzene, *in* F-70020

79579-62-7 2′-O-Methylmicrophyllinic acid, *in* M-70147

79605-61-1 1-Nonen-3-ol; (±)-form, *in* N-60058

79620-69-2 Heavenly blue anthocyanin, H-70006

79642-24-3 8-Amino-γ-carboline, A-70118

79642-25-4 5-Amino-γ-carboline, A-70116

79642-26-5 6-Amino-γ-carboline, A-70117

79642-27-6 1-Amino-γ-carboline, A-70114

79690-90-7 1,5-Dihydro-6*H*-imidazo[4,5-*c*]pyridazine-6-thione; *N*², *S*-Di-Me, *in* D-70215

79690-91-8 1,5-Dihydro-6*H*-imidazo[4,5-*c*]pyridazine-6-thione; *N*⁵, *S*-Di-Me, *in* D-70215

79690-92-9 1,5-Dihydro-6*H*-imidazo[4,5-*c*]pyridazine-6-thione; *N*⁷, *S*-Di-Me, *in* D-70215

79690-93-0 1,5-Dihydro-6*H*-imidazo[4,5-*c*]pyridazine-6-thione; *N*¹, *S*-Di-Me, *in* D-70215

79690-94-1 1,5-Dihydro-6*H*-imidazo[4,5-*c*]pyridazin-6-one, D-60248

79693-15-5 10-Amino-9-anthracenecarboxylic acid; Me ester, *in* A-70088

79693-35-9 10-Amino-9-anthracenecarboxylic acid; *N*-Di-Me, Me ester, *in* A-70088

79695-53-7 5-Methyl-2,4-hexadienoic acid; (*E*)-form, *in* M-70086

79786-99-5 2-Methyl-1,5-naphthalenediol, M-60092

79792-92-0 2,3-Dimethyl-1,4-benzodioxin, D-70370

79816-98-1 Octamethylcyclopentene, O-60026

79863-23-3 Lapidol, L-60014

79863-24-4 Lapidin, *in* L-60014

79917-54-7 2-Nitropyrimidine, N-60043

79958-32-0 4-Amino-3-chlorobutanoic acid; (±)-form, B,HCl, *in* A-60106

79975-20-5 Balanitin 2, *in* S-60045

79975-21-6 Balanitin 1, *in* S-60045

79996-99-9 1-Bromo-4-(bromomethyl)naphthalene, B-60213

79999-60-3 2,5-Dihydro-2,2-diphenyl1,3,4-thiadiazole, D-60236

80003-63-0 6-Acetyl-2-amino-1,7,8,9tetra-hydro-4*H*-pyrimido[4,5-*b*][1,4]-diazepin-4-one, A-60019

80003-63-0 6-Acetylhomopterin, A-70043

80063-50-9 3-Bromo-1-(bromomethyl)naphthalene, B-60219

80085-67-2 2-(1,3-Dithian-2-yl)pyridine, D-60504

80113-53-7 9(11),12-Oleanadien-3-one, O-60032

80127-31-7 Pyrazino[2,3-*b*]pyrido[3′,2′-*e*][1,4]-thiazine, P-60199

80152-02-9 Aurantioclavine, A-60316

80221-08-5 2-(Phenylethynyl)benzoic acid; Amide, *in* P-70074

80225-53-2 Rosmanol, R-70010

80262-04-0 2-Amino-4-chloro-4,4difluorobutanoic acid; (±)-form, Et ester, *in* A-60107

80262-07-3   2-Amino-4-chloro-4,4difluorobutanoic acid; (±)-*form, in* A-**60**107

80286-58-4   4,11(13)-Cadinadien-12-oic acid, C-**70**004

80300-23-8   2-Amino-4,4-dichloro-3-butenoic acid; (±)-*form, in* A-**60**125

80344-66-7   α-Oxo-3-cyclopentene-1acetaldehyde, O-**60**061

80344-66-7   α-Oxo-3-cyclopenteneacetaldehyde, O-**70**086

80348-64-7   Echinofuran, E-**60**002

80357-95-5   5-[2-(3-Methoxyphenyl)ethyl]1,3-benzodioxole, *in* T-**60**316

80357-96-6   3′,5-Dimethoxy-3,4methyl-enedioxybibenzyl, *in* T-**60**102

80357-98-8   6,6′-(1,2-Ethanediyl)bis-1,3benzodioxol-4-ol, *in* H **60**062

80360-39-0   8,8-Difluoro-8,8-dihydro-4methyl-2,2,6,6-tetrakiss(rifluoromethyl)-2*H*,6*H*-[1,2]iodoxolo[4,5,1-*hi*][1,2]benziodoxole, D-**60**182

80368-54-3   Koanoadmantic acid, *in* H-**70**224

80374-27-2   Panasinsanol *A*, P-**70**005

80377-69-1   Artapshin, A-**60**301

80386-32-9   2-Amino-4-chloro-4,4difluorobutanoic acid; (±)-*form*, B,HCl, *in* A-**60**107

80442-35-9   Mikanifuran, M-**70**148

80443-18-1   4-Amino-1*H*-imidazo[4,5-*c*]pyridine; 4-*N*-Benzyl, *in* A-**70**161

80535-88-2   Palliferinin, *in* L-**60**014

80535-89-3   Palliferin, *in* L-**60**014

80639-40-3   5,6-Dihydroxyindole; Di-Me ether, *N*-Me, *in* D-**70**307

80639-85-6   4-Amino-1*H*-imidazo[4,5-*c*]pyridine; B,2HCl, *in* A-**70**161

80680-23-5   Eriocephaloside, *in* H-**70**133

80707-70-6   8,13-Dioxapenta-cyclo[6.5.0.0²,⁶.0⁵,¹⁰.0³,¹¹]trideca-ne-9,12-dione, D-**60**467

80717-49-3   1,2,4,5-Tetracyano-3,6di-methylbenzene, *in* D-**60**407

80733-35-3   Ethanebis(dithioic)acid; Biss(etra-ethylammonium) salt, *in* E-**60**040

80735-94-0   1-Phenyl-3-buten-1-ol; (±)-*form, in* P-**70**061

80736-41-0   2,3,22,23-Tetrahydroxy-24methyl-6-cholestanone; (2α,3α,5α,22*R*,23*R*,24*S*)-*form, in* T-**70**116

80744-88-3   2-Amino-3-hexenoic acid; (±)-(*E*)-*form*, B,HCl, *in* A-**60**168

80745-00-2   2-Amino-3-hexenoic acid; (±)-(*E*)-*form, in* A-**60**168

80745-01-3   2-Amino-4,4-dichloro-3-butenoic acid, A-**60**125

80784-19-6   5,5′-(1,2-Ethanediyl)bis-1,3benzodioxole, *in* T-**60**101

80784-20-9   4′-Hydroxy-3′,5-dimethoxy-3,4methylenedioxybibenzyl, *in* P-**60**044

80816-10-0   1,1-Di-*tert*-butylcyclopropane, D-**60**112

80875-71-4   2-Amino-3-phenyl-3-butenoic acid; (±)-*form, in* A-**60**238

80875-72-5   2-Amino-3-phenyl-3-butenoic acid; (±)-*form*, B,HCl, *in* A-**60**238

80882-45-7   4-Acetyl-2-aminopyridine; Oxime, *in* A-**70**026

80882-53-7   4-Acetyl-2-aminopyridine; *N*²,*N*²-Di-Me, *in* A-**70**026

80882-54-8   4-Acetyl-2-aminopyridine; *N*²,*N*²-Di-Me, oxime, *in* A-**70**026

80931-30-2   2α-Acetoxy-8-epiivangustin, *in* I-**70**109

80952-82-5   Persicaxanthin, P-**60**065

80980-56-9   3,3-Dimethyl-1,5-pentanedithiol, D-**70**427

81098-23-9   Patellamide *B*, P-**60**012

81104-39-4   3-Hydroxyphenylacetaldehyde, H-**60**208

81120-73-2   Patellamide *A*, P-**60**011

81120-74-3   Patellamide *C*, P-**60**013

81129-96-6   1,3,5,7-Nonatetraene; (*E*,*E*,*E*)-*form, in* N-**70**071

81223-43-0   4-Fluoro-2-mercaptobenzoic acid, F-**70**031

81223-44-1   4-Fluoro-2-mercaptobenzoic acid; Me ester, *in* F-**70**031

81223-73-6   6-Acetylindole, A-**60**035

81255-91-6   3*a*,4,5,6,7,7*a*-Hexahydro-1*H*inden-1-one; (3*aRS*,7*aSR*)-*form, in* H-**70**058

81276-02-0   11,12-Epoxy-5,8,14eicosatrienoic acid, E-**60**017

81293-97-2   4,5-Dichloro-1*H*-imidazole-2carboxaldehyde, D-**60**134

81315-59-5   4,5-Dichloro-1*H*-imidazole-2carboxaldehyde; Di-Et acetal, *in* D-**60**134

81331-45-5   2-Diazo-1,1,3,3tetramethylindane, D-**60**067

81344-96-9   Vaticaffinol, V-**60**002

81355-47-7   4-(2-Propynyl)-2-azetidinone, P-**60**185

81377-02-8   Cyclic(*N*-methyl-L-alanyl-Ltyrosyl-D-tryptophyl-Llysyl-L-valyl-Lphenyl-alanyl), C-**60**180

81396-36-3   2,2,5,5-Tetramethyl-3cyclopenten-1-one, T-**60**140

81396-39-6   2,2,5,5-Tetramethyl-3-cyclopentene-1-thione, T-**60**139

81421-79-6   Subexpinnatin, *in* H-**70**138

81422-44-8   Teucjaponin *A*, T-**60**178

81423-50-9   2,3,6,7-Tetrahydro-*s*-indacene1,5-dione, T-**70**068

81426-14-4   1-Acetoxypinoresinol, *in* H-**60**221

81426-17-7   1-Hydroxypinoresinol, H-**60**221

81426-18-8   1-Hydroxypinoresinol; 4″-Me ether, *in* H-**60**221

81426-90-6   18β-Hydroxy-28-nor-3,16oleanenedione, *in* D-**60**363

81427-04-5   Ingol; 3,7,8-Tri-Ac, 12-tigloyl, *in* I-**70**018

81509-31-1   12-Hydroxybromosphaerol; 12β-*form, in* H-**60**108

81523-14-0   1,1′-Bi-1*H*-indene; (1*RS*,1′*RS*)-*form, in* B-**60**108

81523-46-8   Bicyclo[3.2.1]octane-1carboxylic acid; Me ester, *in* B-**70**112

81566-86-1   1,4-Benzoquinone-2,5dicarboxylic acid, B-**70**048

81566-86-1   1,4-Benzoquinone-2,5dicarboxylic acid; Di-Me ester, *in* B-**70**048

81581-30-8   2,3-Dibenzoylbutanedioic acid; (2*RS*,3*RS*)-*form*, Et ester, *in* D-**70**083

81623-72-5   12-Hydroxy-5,8,10,14eicosatetra-enoic acid; (5*Z*,8*Z*,10*Z*,12*S*,14*Z*)-*form*, Me ester, *in* H-**60**131

81633-96-7   5-Isopropyl-4-methyl-1,3butanediol, I-**60**127

81645-70-7   6-Heptyne-2,5-diamine, H-**60**026

81674-19-3   4-Thioxo-2-hexanone, T-**60**215

81771-37-1   Antiarrhythmic peptide (ox atrium); L-L-L-*form, in* A-**60**283

81830-68-4   7-Bromo-1-(bromomethyl)naphthalene, B-**60**221

81877-47-6   3,4-Dimethylthioxanthone, D-**70**445

81877-48-7   2,3-Dimethylthioxanthone, D-**70**443

81892-79-7   Dihydrosuberenol, *in* S-**70**080

81892-89-9   *N*-(2,3-Dihydroxy-3methylbutyl)-acetamide, *in* A-**60**216

81925-83-9   2,2-Dimethyl-4-hexen-3-one, D-**70**414

81930-23-6   Tetrahydro-1,2,4,5-tetrazine3,6-dione; 1,2,4,5-Tetrabenzyl, *in* T-**70**097

81930-26-9   Tetrahydro-1,2,4,5-tetrazine3,6-dione; 1,5-Di-Ac, *in* T-**70**097

81930-28-1   1,2-Dihydro-1,2,4,5-tetrazine3,6-dione, D-**70**268

81930-31-6   3,6-Dimethoxy-1,2,4,5-tetrazine, *in* D-**70**268

81930-32-7   1,4-Dihydro-3,6-dimethoxy1,2,4,5-tetrazine, D-**70**192

81980-37-2   Ingol; 3,7-Di-Ac, 12-tigloyl, *in* I-**70**018

81981-05-7   3,4-Hexadienoic acid; Me ester, *in* H-**70**037

81992-00-9   Lanceoside, *in* P-**60**052

82045-40-7   Tetrahydro-5,6-dimethyl-2*H*pyran-2-one; (5*RS*,6*SR*)-*form, in* T-**70**059

82079-44-5   3-Chloro-2-hydroxypropanoic acid; (*S*)-*form, in* C-**70**072

82081-51-4   2-(1,3-Oxathian-2-yl)pyridine, O-**60**049

82084-87-5   Gnetin A, G-**70**023

82120-26-1   Benz[*a*]anthracene-8,9-dione, B-**70**016

82120-27-2   Benz[*a*]anthracene-1,2-dione, B-**70**012

82120-28-3   1,2-Triphenylenedione, T-**70**308

82124-06-9   14-(3-Furanyl)-3,7,11trimethyl-7,11-tetradecadienoic acid, F-**70**049

82124-11-6   8-Hydroxyvariabilin, *in* V-**70**003

82131-06-4   5-Iodo-2-pentanol; (±)-*form*, Ac, *in* I-**60**056

82177-66-0   2-Formyl-3,4-dihydroxybenzoic acid, F-**60**070

82178-33-4   Premnaspiral, *in* P-**70**114

82189-85-3   Premnaspirodiene, P-**70**114

82204-36-2   2-Hydroxyphenylacetaldehyde; Me ether, oxime, *in* H-**60**207

82205-22-9   Palauolide, P-**60**001

82208-38-6   8,8-Dimethyl-1,4,5(8*H*)naphthalenetrione, D-**70**421

82218-01-7   7*H*-Pyrrolo[1,2-*a*]azepin-7-one, P-**70**192

82257-10-1   3-Bromo-4(1*H*)-pyridinethione, B-**60**316

82326-33-8   2,4′-Bithiazole, B-**70**167

82344-84-1   3-(4-Hydroxyphenyl)-1-(2,4,6trihydroxyphenyl)-2propen-1-one; 2′-(*O*-Rhamnosyl(1→4)-xyloside), *in* H-**60**220

82358-27-8   Balanitin 3, *in* S-**60**045

82366-44-7   4-Fluoro-2,6-dinitroaniline, F-**60**029

82386-89-8   2-Chloro-5-(tri-fluoromethyl)benzaldehyde, C-**70**165

82390-19-0   Spiro[5.5]undeca-1,3-dien-7-one, S-**60**046

82442-61-3   Hexahydro-1(3*H*)-isobenzofuranone; (3*aR*,7*aS*)-*form, in* H-**70**061

82442-62-4   3*a*,4,7,7*a*-Tetrahydro-1(3*H*)isobenzofuranone; (3*aR*,7*aS*)-*form, in* T-**70**072

82442-63-5   Hexahydro-1*H*-cyclopenta[*c*]furan-1-one; (3*aRS*,6*aSR*)-*form, in* H-**70**054

82442-65-7   3-Oxabicyclo[3.1.0]hexan-2-one; (1*RS*,5*SR*)-*form, in* O-**70**055

82442-72-6   6,6-Dimethyl-3oxabicyclo[3.1.0]-hexan-2one; (1*R*,5*S*)-*form, in* D-**70**425

82443-67-2 3,6-Dioxo-4,7,11,15cembratetraen-10,20-olide; (4*Z*,7*E*,10β)-*form, in* D-60470

82448-92-8 3-Amino-5-hexenoic acid; (*R*)-*form, in* A-60172

82452-81-1 Dibenzotetraselenofulvalene, D-70077

82467-25-2 Tetrahydro-5,6-dimethyl-2*H*pyran-2-one; (5*S*,6*R*)-*form, in* T-70059

82468-77-7 4-(2-Propenyl)-2-azetidinone, *in* A-60172

82477-42-7 4-Bromo-4,4-difluoro-2-butynoic acid, B-70202

82477-43-8 3-Bromo-3,3-difluoro-1phenyl-propyne, B-70203

82477-67-6 3-Chloro-5-methoxybenzoic acid, *in* C-60092

82493-54-7 6-Hydroxy-2,4,8tetradecatrienoic acid; (2*E*,4*E*,6*R*,8*Z*)-*form,* Et ester, *in* H-70222

82498-75-7 6-Hydroxy-2,4,8tetradecatrienoic acid; (2*E*,4*E*,6*R*,8*Z*)-*form,* Et ester, *tert*-butyldiphenylsilyl ether, *in* H-70222

82617-45-6 3-Methyl-2-nitrobiphenyl, M-70098

82679-43-4 Chamaedroxide, C-60032

82701-91-5 1,7-Phenanthroline-5,6-dione, P-70052

82733-10-6 1,2-Benzo[2.2]metaparacyclophan-9-ene, B-70036

82736-94-5 6-Chloro-3,5-difluoro-1,2,4triazine, C-60053

82749-44-8 Ethanebis(dithioic)acid; Biss(etra-phenylphosphonium) salt, *in* E-60040

82766-65-2 Ethanebis(dithioic)acid, E-60040

82772-93-8 2,6-Dichloro-4methoxy-benzaldehyde, *in* D-70121

82828-86-2 [3,3′-Bipyridine]-4,4′dicarboxylic acid; Di-Me ester, *in* B-60134

82843-99-0 2,3,19,23-Tetrahydroxy-12oleanen-28-oic acid; (2α,3β,19α)-*form,* 3,28-*O*-Bis-β-D-glucopyranosyl, *in* T-70119

82849-49-8 DDED, *in* E-60049

82855-09-2 Combretastatin, C-70192

82864-43-5 8,9-Epoxy-5,11,14-eicosatrienoic acid; (5*Z*,8*R*,*9*R**,11*Z*,14*Z*)-*form, in* E-60016

82971-12-8 4-Thiazolesulfonic acid, T-60198

82979-27-9 1-Chloro-1-(trichlorovinyl)cyclo-propane, C-60138

82988-65-6 4,5,6,7-Tetrahydro-3methoxy-isoxazolo[4,5-*c*]pyridine, *in* T-60071

83004-15-3 4-(Bromomethyl)-2-chloropyridine, B-70237

83013-90-5 6-(1,3-Butadienyl)-1,4cyclo-heptadiene, B-70290

83124-80-5 2,3-Dihydro-4(trichloroacetyl)-furan, D-70271

83124-87-2 3,4-Dihydro-5(trichloroacetyl)-2*H*pyran, D-70272

83133-17-9 4′,5-Dihydroxy-3′,5′,6,7tetra-methoxyflavone, *in* H-70081

83161-51-7 12-Sesquisabinenal, *in* S-70037

83162-84-9 Bonducellin, B-60192

83188-13-0 1,4-Diacetylindole, *in* A-60033

83216-10-8 Glycinoeclepin *A*, G-70022

83217-89-4 4(15)-Eudesmene-1,11-diol; 1β-*form, in* E-70064

83220-72-8 3-Pyrrolidinol; (±)-*form, in* P-70187

83248-46-8 ▷3-(1,3,5,7,9Dodeca-pentaenyloxy)1,2-propanediol; (*S*,all-*E*)-*form, in* D-70536

83298-49-1 2,7-Cyclodecadien-1-one; (*Z*,*Z*)-*form, in* C-60185

83298-50-4 3,7-Cyclodecadien-1-one; (*Z*,*Z*)-*form, in* C-60186

83314-01-6 Bryostatin 1, B-70284

83324-59-8 3-Chloro-4,6-dihydroxy-2methyl-benzaldehyde, C-60060

83474-31-1 4,6,8-Trimethyl-2-undecanol, T-60372

83527-88-2 4-Amino-9,10-dihydrophenanthrene, A-60142

83540-84-5 Lardolure, *in* T-60372

83570-42-7 6-Acetylquinoxaline, A-60058

83629-95-2 3-Benzoylisoquinoline, B-60060

83631-17-8 Penlanpallescensin, P-60015

83631-17-8 Penlanpallescensin; (*S*)-*form, in* P-60015

83643-92-9 Sanadaol, S-70008

83643-93-0 Acetylsanadaol, *in* S-70008

83648-98-0 Isoglucodistylin, *in* P-70027

83680-48-2 Glucodistylin, *in* P-70027

83709-26-6 Kuwanon *J*, K-60015

83711-16-4 3,3-Biss(henylthio)-1-propanol, B-70158

83720-10-9 Talaromycin *A*, T-70004

83725-19-3 Ilexside I, *in* D-70350

83725-52-4 3β-Acetoxyhaageanolide acetate, *in* H-60001

83725-59-1 2-Oxokolavenic acid, *in* K-60013

83759-54-0 Neokyotorphin, N-70025

83780-27-2 Talaromycin B, *in* T-70004

83797-23-3 1,2-Propadiene-1-thione, P-70117

83897-48-7 2,3-Dimethyl-2-cyclobuten-1-one, D-60415

84000-58-8 3-(1,3,5,7,9Tetradecapenta-enyloxy)1,2-propanediol, T-70024

84000-59-9 3-(1,3,5,7,9Dodeca-pentaenyloxy)1,2-propanediol, D-70536

84030-65-9 1,5,9,13,17Pentaazacycloeicosane, P-60018

84053-10-1 2-Amino-3-(4-mercaptophenyl)propanoic acid; (*S*)-*form, in* A-60214

84093-50-5 11β,13-Dihydroinuviscolide, *in* I-70023

84104-71-2 3-Hydroxy-12-oleanen-29,22-olide; (3β,22α)-*form, in* H-70197

84107-93-7 Eldanolide, E-70004

84153-78-6 Gerberinol 1, G-60013

84164-85-2 Dictyodial *A*; 4β-Hydroxy, 18,*O*-Dihydro, 18-Ac, *in* D-70145

84164-86-3 Dictytriene *A*, D-70148

84164-87-4 Dictytriene B, *in* D-70148

84164-88-5 Dictyone, D-70146

84180-01-8 1,2,9,10,17,18Hexadehydro[2.2.2]-paracyclophane; (*Z*,*Z*,*Z*)-*form, in* H-60038

84192-00-7 *N*-Methylprotoporphyrin IX, M-60123

84213-17-2 1,3,4,5,7,8-Hexahydro-2,6naphthalenedione, H-70063

84249-42-3 Valerenol, V-70001

84289-01-0 6,7-Dichloro-5,8-quinolinedione; *N*-Oxide, *in* D-60149

84289-03-2 6,7-Dichloro-5,8isoquinolinedione, D-60137

84290-29-9 3-Chloro-6-hydroxy-2methoxy-benzaldehyde, *in* C-60058

84294-86-0 ent-3α,7β,14α-Trihydroxy-16kauren-15-one, T-60326

84305-05-5 8α-Hydroxybalchanin, *in* S-60007

84323-33-1 1,2-Bis(4-hydroxyphenyl)-1,2propanediol, B-70147

84332-58-1 9-Anthraceneacetaldehyde, A-60280

84352-58-9 2,2,6-Trimethyl-5-hepten-3-one, T-60365

84412-93-1 5-Hydroxymethyl-4-methyl-2(5*H*)furanone; (±)-*form, in* H-60187

84432-53-1 ▷4-Methyl-2,3,5,6tetranitroaniline, M-60131

84432-56-4 ▷3-Methyl-2,4,5,6tetranitroaniline, M-60130

84432-57-5 ▷2-Methyl-3,4,5,6tetranitroaniline, M-60129

84434-78-6 1,3-Benzodioxole-4-acetonitrile, *in* B-60029

84580-28-9 2,3,14,20,22,25-Hexahydroxy-7cholesten-6-one; (2α,3β,5β,20*R*,22*R*)-*form, in* H-60063

84602-80-2 9-Methyl-9*H*-purin-8-amine, *in* A-60244

84633-05-6 Trisabbreviatin BBB, T-60400

84633-06-7 Abbreviatin PB, A-60001

84673-59-6 2-Amino-5-hexenoic acid; Nitrile, *in* A-60169

84743-77-1 2-Bromo-1,3,5-benzenetriol, B-60199

84782-75-2 Tricyclo[3.3.1.1^{3,7}]decane1,3,5,7-tetraacetonitrile, *in* A-70064

84782-76-3 1,3,5,7-Adamantanetetraacetic acid, A-70064

84782-80-9 1,3,5,7-Tetravinyladamantane, T-70154

84782-82-1 1,3,5,7-Tetraethynyladamantane, T-70025

84799-31-5 Lithospermoside; 5-Epimer, *in* L-70034

84800-12-4 2-Mercapto-3-phenylpropanoic acid; (*R*)-*form, in* M-60026

84808-91-3 Octahydro-2*H*-1,3-benzoxazine; (4a*RS*,8a*RS*)-*form,* *N*-Me, *in* O-70011

84813-70-7 Hypocretenolide, *in* H-70235

84813-80-9 Hypocretenoic acid; Me ester, *in* H-70235

84815-20-3 2-Methyl-3thiophenecarboxaldehyde, M-70139

84871-06-7 Papakusterol, P-60004

84886-37-3 Zoapatanolide B, Z-70007

84889-60-1 2-Bromo-5-mercaptobenzoic acid, B-70227

84890-11-9 Tetra-*tert*butoxycyclopentadienone, T-70013

84909-66-0 Octahydro-2*H*-1,3-benzoxazine; (4a*RS*,8a*SR*)-*form,* *N*-Me, *in* O-70011

84909-67-1 Octahydro-2*H*-1,3-benzoxazine; (4a*RS*,8a*RS*)-*form,* *N*-Benzyl, *in* O-70011

84996-45-2 1,3-Dihydro-1-methyl-2*H*imidazo[4,5-*b*]pyrazin2-one, *in* I-60006

84996-46-3 1,3-Dihydro-1-methyl-2*H*imidazo[4,5-*b*]pyrazine2-thione, *in* D-60247

84996-47-4 1-Methyl-2-methylthio-1*H*imidazo[4,5-*b*]pyrazine, *in* D-60247

84996-48-5 2-(Methylthio)-1*H*-imidazo[4,5*b*]-pyrazine, *in* D-60247

84996-49-6 4-Methyl-2-methylthio-4*H*imidazo[4,5-*b*]pyrazine, *in* D-60247

84996-50-9 1,3-Dihydro-1-methyl-2*H*imidazo[4,5-*b*]pyrazine2-thione, *in* D-60247

84996-53-2 1,3-Dihydro-1,3-dimethyl-2*H*imidazo[4,5-*b*]pyrazin-2-one, *in* I-60006

84996-54-3 2-Methoxy-4-methyl-4*H*imidazo[4,5-*b*]pyrazine, *in* I-60006

84996-55-4 1,4-Dihydro-1,4-dimethyl-2*H*imidazo[4,5-*b*]pyrazin2-one, *in* I-60006

85093-93-2 2-Chloro-5-phenylpyrazine; 1,4-Dioxide, *in* C-70143

85122-38-9 1,2,5-Thiadiazole-3,4dicarboxaldehyde, T-70162

85179-07-3   Lapidolinin, *in* L-60015
85179-08-4   Lapidolin, *in* L-60015
85180-89-8   1,2-Dihydrocyclo-
buta[*a*]naphthalene-2-carboxylic
acid; (±)-*form*, Amide, *in*
D-70177
85202-11-5   Ochromycinone methyl ether, *in*
O-60003
85202-12-6   Lapidolinol, L-60015
85216-79-1   2-Tetradecyloxiranecarboxylic acid;
(±)-*form*, Na salt, *in* T-60041
85282-84-4   Biss(uinuclidine)bromine(1+);
Bromide, *in* B-60173
85282-86-6   Biss(uinuclidine)bromine(1+);
Tetrafluoroborate, *in* B-60173
85288-97-7   10-Hydroxy-11-dodecenoic acid,
H-60125
85288-98-8   12-Hydroxy-13-tetradecenoic acid,
H-60229
85288-99-9   12-Hydroxy-10-dodecenoic acid,
H-60126
85289-00-5   14-Hydroxy-12-tetradecenoic acid,
H-60230
85330-62-7   2,5-Biss(ethylthio)pyridine, *in*
P-60220
85330-63-8   2,5-Biss(ethylsulfonyl)pyridine, *in*
P-60220
85330-79-6   2,3-Biss(ethylsulfonyl)pyridine, *in*
P-60218
85359-61-1   Chalaurenol, C-70036
85385-68-8   1,6:7,12-Bismethano[14]annulene;
syn-form, *in* B-60165
85386-94-3   2-Chloro-3-fluoropyridine; *N*-
Oxide, *in* C-60081
85440-62-6   1,6:7,12-Bismethano[14]annulene;
anti-form, *in* B-60165
85452-72-8   3,4-Dihydroxyphenylacetaldehyde;
Di-Me ether, di-Me acetal, *in*
D-60371
85493-96-5   2-Bromo-3-methoxy-2-cyclohexen-1-
one, *in* B-70201
85515-06-6   6-Bromotryptophan; (±)-*form*, *N*^α-
Ac, *in* B-60334
85526-81-4   10(14)-Aromadendrene-4,8-diol;
(4β,8α)-*form*, *in* A-60299
85531-49-3   [4,4′-Bipyridine]-2,2′dicarboxylic
acid, B-60136
85541-04-4   Chrysophyllin *B*, C-70175
85545-06-8   2,2,5Trimethylcyclopenta-
necarboxylic acid; (1*RS*,5*SR*)-
form, *in* T-60360
85545-07-9   2,2,5Trimethylcyclopenta-
necarboxylic acid; (1*RS*,5*SR*)-
form, Amide, *in* T-60360
85551-57-1   1-Phenyl-3-buten-1-ol; (*R*)-*form*, *in*
P-70061
85552-23-4   Wedeliasecokaurenolide, W-60004
85564-03-0   Teupyrenone, T-60184
85570-50-9   Dimethylenebicyclo[1.1.1]penta-
none, D-60422
85601-43-0   Naphtho[2,1-*b*]thiet-1-one,
N-70013
85620-94-6   2-Azido-2-nitropropane, A-60349
85620-95-7   ▷2,2-Diazidopropane, D-60058
85657-09-6   2-Pyrrolidinemethanethiol; (*S*)-
form, *in* P-70185
85769-29-5   2,4-Pentadecadienal; (2*E*,4*Z*)-
form, *in* P-70024
85779-98-2   5-Iodo-1,3-dimethylpyrazole, *in*
I-70050
85814-86-4   3-Hydroxycyclononanone, H-60113
85846-83-9   5-Hydroxymethyl-2(5*H*)-furanone;
(*S*)-*form*, Ac, *in* H-60182
85847-69-4   1β,3α-Dihydroxy-4-
eudesmen12,6α-olide, *in* I-70066
85847-69-4   1,3-Dihydroxy-4-eudesmen-
12,6olide; (1β,3α,6α)-*form*, *in*
D-70299

85895-83-6   3,5-Dimethyl-
2thiophenecarboxaldehyde,
D-70439
85960-33-4   Cyclohept[*fg*]acenaphthylene5,6-
dione, C-60194
85960-34-5   Cyclohept[*fg*]acenaphthylene5,8-
dione, C-60195
85963-75-3   2,2,4,4-Tetrafluoro-1,3dithietane;
1-Oxide, *in* T-60048
85963-76-4   2,2,4,4-Tetrafluoro-1,3dithietane;
1,1,3-Trioxide, *in* T-60048
86117-82-0   5-(Hydroxymethyl)-1,2,3,4cyclo-
hexanetetrol; (1*R*,2*S*,3*S*,4*R*,5*R*)-
form, *in* H-70170
86120-41-4   2,4-Octadecadienoic acid; (*E*,*E*)-
form, Me ester, *in* O-70003
86195-44-0   5-(Hydroxymethyl)-5cyclohexene-
1,2,3,4tetrol;
(1*RS*,2*RS*,3*RS*,4*SR*)-*form*, *in*
H-70172
86195-45-1   5-(Hydroxymethyl)-5cyclohexene-
1,2,3,4tetrol;
(1*RS*,2*SR*,3*SR*,4*SR*)-*form*, *in*
H-70172
86229-97-2   RA-VII, *in* R-60002
86232-28-2   5-Chloro-2,3dimethoxy-
benzaldehyde, *in* C-70065
86238-50-8   3,7,11-Trimethyl-1-dodecanol;
(3*R*,7*S*)-*form*, *in* T-70282
86377-53-9   15α-Acetoxy-3α-hydroxy-
7,9(11),24*E*-lanostatrien-26-oic
acid, *in* D-70314
86379-88-6   Octahydro-7*H*-1-benzopyran-7-one;
(4*aRS*,8*aRS*)-*form*, *in* O-60011
86402-39-3   Cleistanthoside A, *in* D-60493
86447-23-6   1,2,3,6-Tetrahydro-
3pyridinecarboxylic acid,
T-60084
86447-28-1   1,2,5,6-Tetrahydro-5-hydroxy3-
pyridinecarboxylic acid; (±)-
form, B,HBr, *in* T-70065
86447-29-2   1,2,3,6-Tetrahydro-
3pyridinecarboxylic acid; (±)-
form, *N*-Me, *in* T-60084
86450-76-2   Shanzhiside; 6-(4-Hydroxy-3,5-
dimethoxybenzoyl), 8-Ac, Me
ester, *in* S-60028
86456-69-1   6-Bromo-1-
(bromomethyl)naphthalene,
B-60220
86471-28-5   4-Phenyl-2-furancarboxylic acid,
P-70077
86474-77-3   5-Methylnaphtho[2,1-*b*]thiophene,
M-60099
86474-78-4   5-Methylnaphtho[2,1-*b*]thiophene;
Picrate, *in* M-60099
86474-80-8   4-Methylnaphtho[2,1-*b*]thiophene;
Picrate, *in* M-60098
86474-81-9   2-Methylnaphtho[2,1-*b*]thiophene;
Picrate, *in* M-60097
86475-22-1   Pentamethoxybenzaldehyde, *in*
P-60042
86492-07-1   1*H*-Pyrrole-3,4-dicarboxylic acid;
Diethylamide, *in* P-70183
86492-08-2   1*H*-Pyrrole-3,4-dicarboxylic acid;
Dimethylamide, *in* P-70183
86496-18-6   Queuine; B,2HCl, *in* Q-70004
86528-79-2   5,6-Dicyanoacenaphthene, *in*
A-70050
86530-73-6   1-Nitro-2-(phenylsulfinyl)ethane, *in*
N-70060
86538-52-5   8-Bromo-1,7-phenanthroline,
B-70265
86538-53-6   8-Amino-1,7-phenanthroline,
A-70180
86538-55-8   8-Amino-1,7-phenanthroline;
Picrate, *in* A-70180

86544-37-8   2-Hydroxy-2-phenylacetaldehyde;
(*S*)-*form*, Me ether, *in* H-60210
86546-83-0   9(11)-Drimen-8-ol; (5*R*,8*S*,10*R*)-
form, *in* D-60520
86546-84-1   9(11)-Drimen-8-ol; (5*R*,8*R*,10*R*)-
form, *in* D-60520
86575-88-4   Artedouglasiaoxide, A-70263
86582-92-5   Clavularin *A*, C-60153
86630-47-9   1,5-Dimethoxy-
2,7phenanthrenediol, *in* T-60116
86632-29-3   2,2′-Diiodo-1,1′-binaphthyl; (±)-
form, *in* D-60385
86635-85-0   2,4-Dihydroxy-3-iodobenzoic acid,
D-60331
86685-13-4   5-Amino-3-phenylisoxazole; *N*-
Formyl, *in* A-60240
86685-14-5   5-Amino-3-phenylisoxazole, *N*-Me,
in A-60240
86685-93-0   5-Amino-3-phenylisoxazole; *N*-Et,
in A-60240
86688-06-4   2,2′-Diiodo-1,1′-binaphthyl; (*R*)-
form, *in* D-60385
86688-08-6   2,2′-Dibromo-1,1′-binaphthyl; (*R*)-
form, *in* D-60089
86690-04-2   *N*,*N*′-Bis(3-sulfonatopropyl)2,2′-
bipyridinium, B-70159
86690-14-4   Halenaquinone, *in* H-60002
86702-02-5   Magnoshinin, M-60007
86732-28-7   Octahydropyrrolo[3,4-*c*]pyrrole;
(3*aRS*,6*aSR*)-*form*, *N*-Me, *in*
O-70017
86746-90-9   Clavularin B, *in* C-60153
86747-02-6   β-Senepoxide, *in* S-60023
86766-46-3   4-Methoxy-4-phenyl-1-butene, *in*
P-70061
86803-29-4   4(5)-Vinylimidazole; 1-Triphenyl-
methyl, *in* V-70012
86803-30-7   4(5)-Vinylimidazole; 1-Benzyl, *in*
V-70012
86810-04-0   *N*-Alanylcysteine; L-L-*form*, Boc-
Ala-Cys-OMe, *in* A-60069
86831-55-2   2,5-Diamino-3-hydroxypentanoic
acid; (2*R*,3*R*)-*form*, *in* D-70039
86849-13-0   RA-IV, *in* R-60002
86963-40-8   1-Phenylethanesulfonic acid; (+)-
form, *in* P-70068
86963-85-1   2,3,5-Triss(ethylene)bicyclo[2.2.1]-
heptane, T-60403
86988-90-1   4-Chloro-3-methylfuroxan, *in*
C-60114
86989-08-4   Isomyomontanone, *in* M-60155
86989-09-5   Myomontanone, M-60155
87013-24-9   2-Oxa-3-azabicyclo[2.2.2]octane;
N-Formyl, *in* O-70051
87037-69-2   2-Benzyloxy-1-propanol, *in* P-70118
87064-16-2   Zeylasteral, *in* D-70027
87064-40-2   Desmethylzeylasterone, D-70027
87084-52-4   Mannopine, *in* M-70009
87108-78-9   2,2,4,4Tetrakis(rifluoromethyl)-
1,3-dithietane; 1,3-Dioxide, *in*
T-60129
87118-53-4   Methyl hydrogen 3-
hydroxyglutarate; (*R*)-*form*, *in*
M-70089
87121-83-3   Benz[*a*]azulene-1,4-dione, B-60011
87143-01-9   1,3-Benzodiselenole-2-thione,
B-70032
87161-46-4   Isopristimerin III, I-70083
87174-13-8   23-Oxoisopristimerin III, *in*
I-70083
87206-01-7   *N*-(Tetrahydro-2-oxo-3-
furanyl)hexadecanamide, *in*
A-60138
87216-46-4   1,4-Dihydro-3*H*-2benzothiopyran-
3-one, D-60206
87216-47-5   1,4-Dihydro-3*H*-2-benzoselenin3-
one, D-60204

87216-48-6   1,4-Dihydro-3H-2benzotellurin-3-one, D-60205

87219-30-5   3-Aminodihydro-2(3H)-furanone; (S)-*form*, N-Benzoyl, *in* A-60138

87227-41-6   2-(4-Bromobutyl)-1,3-dioxole, *in* B-70263

87238-75-3   Tricyclo[5.2.1.0^{2,6}]dec-2(6)-ene, T-60260

87255-31-0   3-Amino-5-hexenoic acid, A-60172

87282-19-7   ▷3-Fluoro-3-phenyl-3H-diazirine, F-70036

87302-42-9   Achalensolide, A-70054

87402-72-0   9,10-Dihydro-2,7-dimethoxy1,5-phenanthrenediol, *in* T-60116

87413-09-0   1,1,1-Triacetoxy-1,1-dihydro1,2-benziodoxol-3(1H)-one, T-60228

87441-73-4   11,13-Dehydroeriolin, *in* E-60036

87462-15-5   5-Methyl-1H-pyrrole-2-carboxylic acid; Benzyl ester, *in* M-60124

87480-58-8   1-Amino-2ethylcyclopropanecarboxylic acid, A-60151

87512-32-1   3-Amino-2-hexenoic acid; *tert*-Butyl ester, *in* A-60170

87533-53-7   1,3-Diacetyl-5-iodobenzene, D-60024

87573-88-4   5-Acetyl-2(1H)-pyrimidinone, A-60048

87592-08-3   10H-Pyrazolo[5,1-c][1,4]benzodiazepine, P-70154

87592-85-6   Encecanescin, E-70008

87592-86-7   9′-Epiencecanescin, *in* E-70008

87596-55-2   Nectriafurone, N-60016

87606-98-2   GSN (as disodium salt), *in* G-60032

87619-34-9   1-Bromo-2,3-dimethylcyclopropene, B-60235

87625-62-5   ▷Ptaquiloside, P-60190

87656-32-4   2-(Dimethoxymethyl)benzenemethanol, *in* H-60171

87667-46-7   Fusarubin methyl acetal, *in* F-60108

87682-63-1   1,2,3,6-Tetrahydropyridine; 1-Ph, *in* T-70089

87682-75-5   1,2,3,6,7,9a-Hexahydro-4(1H)quinolizinone; (±)-*form*, *in* H-70074

87701-33-5   Ptaquiloside; Tetra-Ac, *in* P-60190

87710-32-5   Datiscoside F, *in* C-70203

87710-33-6   Datiscoside H, *in* C-70203

87710-34-7   Datiscoside B, *in* C-70203

87718-77-2   Datiscoside E, *in* C-70203

87734-68-7   3α,22R,23R-Trihydroxy-24Smethyl-5α-cholestan-6-one, *in* T-70116

87745-28-6   Bryostatin 2, *in* B-70284

87746-47-2   Sigmoidin B, S-70041

87746-48-3   Sigmoidin A, *in* S-70041

87768-94-3   3-Aminobicyclo[2.2.1]hept-5ene-2-methanol; (1RS,2SR,3SR)-*form*, *in* A-70094

87785-65-7   3,5-Hexadienoic acid; (E)-*form*, *p*-Nitrobenzyl ester, *in* H-70038

87827-49-4   Datiscoside C, *in* C-70203

87827-50-7   Datiscoside D, *in* C-70203

87833-54-3   6-Deoxocastasterone, *in* T-70116

87842-36-2   2,2,5,5Tetramethylcyclopentaneselone, T-60137

87865-19-8   5,9,10-Trihydroxy-2Hpyrano[2,3,4-kl]xanthen-2-one, T-70273

87999-44-8   1-Azidodiamantane, A-60341

87999-45-9   4-Azidodiamantane, A-60343

88010-68-8   3,9-Dihydroxy-10(14)-guaien12,6-olide; (3β,4α,6α,9β,11α)-*form*, *in* D-70305

88015-28-5   9-(Dicyanomethyl)anthracene, D-60163

88047-46-5   5-Methyl-1,3,2,4-dithiazolium(1+); Hexafluoroarsenate, *in* M-70061

88070-48-8 N-Methylbenzenecarboximidic acid, M-60045

88090-59-9   9H-Tetrabenzo[a,c,g,i]carbazole, T-70011

88099-66-5   3-Amino-2-fluoropropanoic acid; (R)-*form*, *in* A-60162

88109-70-0   5-Dodecyn-1-ol, D-70539

88153-64-4   3,11(13)-Eudesmadien-12-oic acid; 2α-Hydroxy, *in* E-70060

88153-65-5   3,11(13)-Eudesmadien-12-oic acid; 2-Oxo, *in* E-70060

88174-22-5   2-Chloro-4,6dimethylbenzaldehyde, C-60061

88181-49-1   Urocanic acid; (Z)-*form*, Me ester, *in* U-70005

88182-33-6   6,6′;7,3a'-Diligustilide, D-70363

88192-22-7   6-Heptyne-2,5-diamine; (2R,5R)-*form*, *in* H-60026

88218-12-6   2-Pyrrolidinecarboxaldehyde; (S)-*form*, *in* P-70184

88255-07-6   7-Phenyl-5-heptynoic acid, P-60106

88255-19-0   7-Phenyl-5-heptynoic acid; Me ester, *in* P-60106

88255-95-2   Ilexside II, *in* D-70350

88283-10-7   5-(Trichloromethyl)isoxazole, T-60253

88303-65-5   2-[(2-Hydrazinoethyl)-amino]ethanol, H-60091

88320-47-2   5-Amino-1H-pyrazole-4-carboxylic acid; N(1)-Me, N(5)-Ac, Et ester, *in* A-70192

88330-30-7   3-Aminobicyclo[2.2.1]hept-5ene-2-methanol; (1RS,2SR,3SR)-*form*, Picrate, *in* A-70094

88330-31-8   3-Aminobicyclo[2.2.1]heptane2-methanol; (1RS,2SR,3RS)-*form*, Picrate, *in* A-70093

88335-95-9   3-Oxabicyclo[3.2.0]heptan-2-one; (1R,5S)-*form*, *in* O-70053

88378-97-6   19-O-Acetyl-1,2dehydrohautriwaic acid, *in* H-70002

88389-71-3   Verrucosidin, V-60006

88488-54-4   1,2-Dimethylthioxanthone, D-70440

88489-55-8   1,2,3-Trichlorocyclopropane; (1α,2α,3α)-*form*, *in* T-70220

88524-65-6   Kuwanone L, K-70016

88524-66-7   Kuwanone K, *in* K-70016

88525-42-2   7-Ethyl-5-methyl-6,8dioxabicyclo[3.2.1]oct3-ene, E-60052

88544-98-3   2-Hydroxy-1,3benzenedicarboxylic acid; Di-Et ester, *in* H-60101

88609-21-6   8-Amino-6-nitroquinoline, A-60235

88609-21-6   8-Amino-6-nitroquinoline; 8-N-Ac, *in* A-60235

88640-94-2   Ammothamnidin, A-70204

88682-15-9   Bi(1,3-dithiolo[4,5-b][1,4]dithiin-2-ylidene), B-70118

88691-94-5   4,4,5,5-Tetraphenyl-1,3dithiolane, T-60166

88692-18-6   3-Chloro-4,5,6trifluoropyridazine, C-60142

88744-04-1   9-Fluorenylmethyl pentafluorophenyl carbonate, F-60019

88819-78-7   3,3-Dimethyl-4-pentenoic acid; Chloride, *in* D-60437

88841-60-5   2,3-Dihydro-2-phenyl-1,2benzisothiazole, D-60261

88895-06-1   Esperamicin; Esperamicin A$_{1b}$, *in* E-60037

88899-62-1   Alterperylenol, A-60078

88899-63-2   Dihydroalterperylenol, *in* A-60078

88901-69-3   Alterporriol B, A-60079

88901-97-7   1-[3-(2,4-Dihydroxybenzoyl)4,6-dihydroxy-2-(4hydroxyphenyl)-7benzofuranyl]-3-(4hydroxyphenyl)-2-propen-1-one, D-70286

88901-98-8   1-[3-(2,4-Dihydroxybenzoyl)4,6-dihydroxy-2-(4hydroxyphenyl)-7benzofuranyl]-3-(4hydroxyphenyl)-2-propen-1-one; 2,3-Dihydro (trans-), *in* D-70286

88907-88-4   1,2,5Trimethylcyclopentanecarboxylic acid, T-60361

88919-86-2   2,3′:2′,3″-Ter-1H-indole, T-60011

88936-02-1   Gloeosporone, G-60025

88966-86-3   Pyoverdin E, *in* P-60147

89002-63-1   4,4-Dimethyl-2-phenyl-4Himidazole, D-70432

89002-85-7   2-(3-Furanyl)-4H-1-benzopyran4-one, F-60085

89020-30-4   3-Amino-5-hydroxy-7oxabicyclo[4.1.0]hept3-en-2-one, A-60194

89020-30-4   3-Amino-5-hydroxy-7oxabicyclo[4.1.0]hept3-en-2-one; (1S,5S,6S)-*form*, *in* A-60194

89020-31-5   Antibiotic 36531, *in* A-60194

89026-40-4   Umbellifolide, U-60001

89029-10-7   5,5′,6,7-Tetramethoxy-3′,4′methylenedioxyflavone, *in* H-70081

89033-27-2   2-Oxo-4-oxazolidinecarboxylic acid, O-60082

89046-74-2   3-*tert*-Butyl-2,2,4,5,5tetramethyl-3-hexene, B-60350

89088-56-2   3-Amino-1H-pyrazole-5-carboxylic acid; N(1)-Me, Me ester, *in* A-70188

89088-57-3   3-Amino-1H-pyrazole-5-carboxylic acid; N(1)-Me, Et ester, *in* A-70188

89108-47-4   4-Diazo-4H-imidazole, D-60063

89108-47-4   ▷4-Diazo-4H-imidazole, D-70061

89181-81-7   6-Amino-2-(methylamino)-4(3H)pyrimidinone, *in* D-70043

89186-13-0   4-Oxo-2-nonenal, O-70097

89199-95-1   1-Hydroxysyringaresinol, *in* H-60221

89212-61-3   Pyrazolo[1,5-b][1,2,4]triazine, P-60207

89289-91-8   Diffutin, *in* T-60103

89289-92-9   5,7-Dihydroxy-3′,4′dimethoxyflavan, *in* T-60103

89311-52-4   2-Bromotryptophan; (S)-*form*, *in* B-60332

89311-53-5   2-Chlorotryptophan; (S)-*form*, *in* C-60145

89317-93-1   1,4,7,22,25,28-Hexaoxa[7.7](9,10)anthracenophane, H-70086

89317-94-2   1,4,7,10,25,28,31,34Hexaoxa-[10.10](9,10)anthracenophane, H-70085

89320-82-1   Bhubaneswin, *in* B-70093

89324-31-2   4,4′-Bithiazole, B-70169

89334-34-9   5-Acetylquinoxaline, A-60057

89353-99-1   Chiisanogenin, C-70038

89354-01-8   Chiisanoside, *in* C-70038

89354-45-0   Riligustilide, R-60009

89355-02-2   Tecomaquinone I, T-60010

89362-84-5   13,17,21-Polypodatriene-3,8-diol, P-70107

89395-28-8   3-Oxabicyclo[3.1.0]hexan-2-one; (1R,5S)-*form*, *in* O-70055

89408-86-6   6-(Hydroxymethyl)tetrahydro2H-pyran-2-one, *in* D-60329

89482-11-1   Dictyodial A; 4β-Hydroxy, *in* D-70145

89576-50-1   8-Hydroxy-13-labden-15-al; (8α,E)-*form*, *in* H-70149

89576-51-2 8-Hydroxy-13-labden-15-al; (8α,Z)-form, in H-70149
89583-90-4 7-Chloro-1,2-benzisothiazole, C-70053
89607-12-5 4-Bromo-3(5)-nitro-1H-pyrazole; 1-Me, in B-60304
89639-83-8 4,5-Dimethyl-2-furancarboxylic acid, D-70413
89641-97-4 7-Nitro-1,2-benzisothiazole, N-70042
89648-25-9 2-Amino-3-heptenedioic acid; (R,E)-form, in A-60164
89654-25-1 1,6-Dihydropentalene, D-70243
89701-80-4 1,3,4,5-Tetrahydroxy-2methyl-anthraquinone, T-60112
89717-64-6 4-Bromo-3(5)-nitro-1H-pyrazole, B-60304
89783-74-4 9,10-Dihydroxy-5-methoxy-2Hpyrano[2,3,4-kl]xanthen-9-one, in T-70273
89803-84-9 Kuwanon V, in K-60015
89803-85-0 Kuwanon R, in K-60015
89803-86-1 Kuwanon Q, in K-60015
89803-99-6 Chatferin, in F-60009
89885-39-2 3,7-Dimethyladenine; B,HI, in D-60400
89885-86-9 Talaromycin C, in T-70004
89889-13-4 3-Amino-4-hexenoic acid; (±)-(E)-form, Me ester, in A-60171
89898-86-2 5-Nitrophthalazine, N-70062
89921-31-3 3-Amino-4-hexenoic acid; (±)-(E)-form, N-Benzoyl, Me ester, in A-60171
89936-87-8 4-Mercapto-2-butenoic acid; (E)-form, tert-Butyl ester, in M-60018
89942-33-6 2-Iodosophenylacetic acid, I-60063
89976-15-8 7-Iodo-1H-indole, I-70043
89984-05-4 Ingol; O7-Benzoyl, O8-Me, 3,12-di-Ac, in I-70018
89984-06-5 Ingol; 7-Tigloyl, O8-Me, 3,12-di-Ac, in I-70018
90014-13-4 Bicyclo[2.2.2]oct-2-ene-1carboxylic acid; Nitrile, in B-70113
90027-10-4 Ingol; 7-Angeloyl, O8-Me, 3,12-Di-Ac, in I-70018
90044-34-1 Sambucoin, S-70007
90052-97-4 Wasabidienone A, W-70001
90134-81-9 Cycloocta[2,1-b:3,4-b']-di[1,8]naphthyridine, C-70231
90171-34-9 5,5-Dimethyl-2-hexynedioic acid; Di-Me ester, in D-70415
90243-46-2 2-Methyl-4-propyl-1,3-oxathiane; (2R,4S)-form, in M-60122
90243-47-3 2-Methyl-4-propyl-1,3-oxathiane; (2S,4R)-form, in M-60122
90332-19-7 Isohyperlatolic acid, I-60101
90332-20-0 Hyperlatolic acid, H-60236
90332-21-1 Superlatolic acid, S-60063
90332-22-2 Isophaeric acid, I-60136
90347-70-9 3-Iodo-2,6-dimethoxybenzoic acid, in D-60334
90375-51-2 Acetoxyodontoschismenol, in D-70541
90381-07-0 5-Chloro-2-(trifluoromethyl)benzaldehyde, C-70168
90390-46-8 3-Amino-5-nitrobenzyl alcohol, A-60233
90397-87-8 5-Iodo-2-pentanol, I-60056
90399-89-6 3-Amino-1H-pyrazole-4-carboxylic acid; Anilide, in A-70187
90407-14-0 7-Chlorobenzo[b]thiophene, C-60047
90407-20-8 2-Bromo-3-nitrobenzyl alcohol, B-70256
90411-20-4 3-Anhydro-6-epiophiobolin A, in O-70040
90522-71-7 4-Amino-1,2,3-benzotriazine; 2H-form, 2-Propyl, N(4)-butyl; B,HI, in A-60092

90528-92-0 2-Amino-4-phenyl-3-butenoic acid; (±)-(E)-form, in A-60239
90536-15-5 2-Mercapto-3-phenylpropanoic acid, M-60026
90539-42-7 2,3,6-Trimethoxyphenol, in B-60014
90624-29-6 10,10-Dimethyl-9(10H)anthracenone; Hydrazone, in D-60403
90632-49-8 Indisocin, I-60028
90685-55-5 Echinofuran B, in E-60002
90685-58-8 6,7-Dihydro-5H-1-pyridine; N-Oxide, in D-60275
90685-95-3 Pseudocyphellarin A, in P-70127
90685-96-4 Pseudocyphellarin B, P-70127
90695-04-8 Pallinin, in C-60021
90695-22-0 5-(Hydroxymethyl)-1,2,3,4cyclo-hexanetetrol; (1R,2S,3S,4R,5R)-form, Penta-Ac, in H-70170
90760-94-4 Agarospirene, in P-70114
90772-05-7 2-Acetoxy-1,3benzenedicarboxylic acid, in H-60101
90852-99-6 Ptelatoside B, in H-70209
90885-92-0 1,8-Anthracenedicarboxylic acid; Dichloride, in A-60281
90899-20-0 Ptelatoside A, in H-70209
90902-21-9 1-(2-Hydroxy-4-methoxyphenyl)3-(3-hydroxy-4-methoxyphenyl)-propane, in D-70336
90970-33-5 N-Ornithylglycine; (S)-form, B,HCl, in O-70047
90970-63-1 N-Ornithyl-β-alanine; (S)-form, B,HCl, in O-60041
90970-64-2 N-Ornithyltaurine; (S)-form, B,HCl, in O-60042
90990-60-6 N-Ornithyltaurine; (S)-form, Nα,Nδ-Biss(enzyloxycarbonyl), in O-60042
91091-13-3 5-Amino-1-benzyl-4-cyanopyrazole, in A-70192
91158-94-0 4b,8b,13,14Tetra-hydrodiindeno[1,2a:2',1'-b]-indene, T-70058
91161-74-9 Punctaporonin A, P-60197
91164-64-6 3,5-Biss(ethylsulfonyl)pyridine, in P-60222
91196-41-7 Lepidopterans, L-70028
91238-45-8 4-Pentene-2,3-dione, P-60061
91239-72-4 Bicyclo[3.1.1]heptane-1carboxylic acid, B-60086
91265-96-2 1-Chloro-2,3-pentanedione, C-60123
91279-91-3 4a,4b,8a,8bTetrahydrobiphenylene; (4aα,4bα,8aα,8bα)-form, in T-70060
91407-43-1 5-Fluoro-2,3-dimethoxybenzoic acid, in F-60025
91423-90-4 Isotrichodermin, in E-70028
91431-42-4 Lonapalene, in C-60118
91477-70-2 1,5-Diamino-1,5dihydrobenzo[1,2-d:4,5d']bistriazole, D-70038
91511-06-7 4-(2,4,6-Cycloheptatrien-1ylidene)-bicyclo[5.4.1]dodeca-2,5,7,9,11pentaene, C-60207
91524-69-5 1,3-Difluoro-2-(methyl-thio)benzene, in D-60174
91538-70-4 Naphtho[2,1-b:6,5-b']-bis[1]benzothiophene, N-60007
91544-03-5 3-Acetylisoquinoline, A-60040
91573-04-5 1,1'-Bitetralinylidene; (Z)-form, in B-70165
91574-92-4 Safflor Yellow B, S-70001
91586-90-2 Iodomethanesulfonic acid; Bromide, in I-60048
91590-50-0 1,1'-Bitetralinylidene; (E)-form, in B-70165
91591-88-7 2,4,6-Trimethoxypyridine, in D-70341

91682-00-7 1,4,7-Tribromotriquinacene, T-60248
91717-64-5 Isomitomycin A, I-70070
91776-42-0 Fibrostatins; Fibrostatin A, in F-70008
91776-44-2 Fibrostatins; Fibrostatin E, in F-70008
91776-45-3 Fibrostatins; Fibrostatin F, in F-70008
91776-46-4 Fibrostatins; Fibrostatin D, in F-70008
91776-47-5 Fibrostatins; Fibrostatin C, in F-70008
91776-48-6 Fibrostatins; Fibrostatin B, in F-70008
91790-20-4 5,5a,6,6a,7,12,12a,13,13a,14Deca-hydro-5,14:6,13:7,12-trimeth-anopentacene, D-60010
91794-28-4 4-Methyl-5-phenyl-2oxazolidinethione, M-60111
91840-07-2 myo-Inositol-2,4,5-triphosphate, I-70021
91879-79-7 μ-Oxodiphenylbiss(rifluoroacetato-O)diiodine, O-60065
91896-90-1 Zafronic acid; Me ester, in Z-70001
91896-91-2 Zafronic acid; 3-Ac, Me ester, in Z-70001
91897-25-5 7'-Hydroxyabscisic acid, H-70098
91914-33-9 7,11,15,19-Tetramethyl-3-methyl-ene-2-(3,7,11trimethyl-2-dodecenyl)1-eicosanol, T-70138
91934-41-7 1,3,2-Benzodithiazol-1-ium(1+); Chloride, in B-70033
91971-25-4 2,6-Dimethoxy-3,5dimethylbenzoic acid, in D-70293
91985-75-0 3-Amino-β-carboline; N3-Ac, in A-70110
92012-54-9 1-(Diphenylmethylene)-1Hcyclo-propabenzene, D-70484
92012-57-2 1-(Diphenylmethylene)-1Hcyclo-propa[b]naphthalene, D-70485
92013-62-2 1,2-Diethylidenecyclohexane; (E,E)-form, in D-70154
92075-16-6 2-Imidazolidinone; 1-tert-Butyl, in I-70005
92121-62-5 Naematolone, in N-70001
92136-58-8 2-Amino-4-bromobutanoic acid; (S)-form, in A-60098
92285-01-3 Ajoene, A-70079
92367-07-2 4-Fluoro-3,5-dinitroaniline, F-60030
92367-59-4 3,4-Dihydro-3,3-dimethyl-2(1H)quinolinone, D-60231
92406-53-6 5-Amino-1H-pyrazole-3-carboxylic acid; N(1)-Me, Me ester, in A-70191
92406-77-4 Decamethylcyclohexanone, D-60011
92406-82-1 Octamethylcyclopentanone, O-60025
92534-70-8 4-Amino-1H-pyrazole-5-carboxylic acid; N(1)-Me; B,HCl, in A-70190
92534-73-1 4-Amino-1H-pyrazole-5-carboxylic acid; N(1)-Me, Amide, in A-70190
92587-21-8 2-Cyclohexyl-2-hydroxyacetic acid; (R)-form, Me ester, in C-60212
92597-03-0 Octahydro-2H-1,3-benzoxazine; (4aRS,8aSR)-form, N-Benzyl, in O-70011
92597-05-2 Octahydro-2H-3,1-benzoxazine; (4aRS,8aSR)-form, N-Me, in O-70012
92597-07-4 Octahydro-1-methyl-2H-3,1benzoxazine, in O-70012
92597-09-6 Octahydro-2H-3,1-benzoxazine; (4aRS,8aSR)-form, N-Benzyl, in O-70012
92597-11-0 Octahydro-2H-3,1-benzoxazine; (4aRS,8aRS)-form, N-Benzyl, in O-70012

92607-80-2 4*H*-1,3-Benzodioxin-6carboxaldehyde, B-60026

92609-97-7 Methaneselenoamide, *in* M-70035

92619-42-6 4-*tert*-Butylcyclohexene; (±)-*form*, *in* B-60343

92622-71-4 Tetracyclo[5.3.0.0^{2,4}.0^{3,5}]deca-6,8,10-triene, T-70019

92632-00-3 3-(2,2-Dimethyl-2*H*-1benzopyran-6-yl)-2propenoic acid; Me ester, *in* D-70372

92665-29-7 Antibiotic BMY 28100, A-70221

92675-09-7 Balearone, B-60005

92694-51-4 5-Methyl-2(5*H*)-furanone; (*S*)-*form*, *in* M-60083

92810-15-6 2-Formyl-3,5-dihydroxybenzoic acid, F-60071

92910-93-5 Ingol; 7-Tigloyl, 3,8,12-tri-Ac, *in* I-70018

92914-76-6 5-Amino-1*H*-pyrazole-4-carboxylic acid; *N*(1)-Me, *N*(5)-Ac, *in* A-70192

92950-25-9 5,14,15-Trihydroxy-6,8,10,12eicosatetraenoic acid, T-60320

92967-66-3 1,8-Dicyanoanthracene, *in* A-60281

92998-77-1 Ingol; 7-Angeloyl, 3,8,12-tri-Ac, *in* I-70018

93081-07-3 1,2-Benzoquinone-3-carboxylic acid; Me ester, *in* B-70045

93103-00-5 3-(Aminoacetyl)pyridine; B,HCl, *in* A-70086

93118-03-7 2-Chloro-3-(trifluoromethyl)benzaldehyde, C-70164

93124-90-4 4′-Hydroxy-3′,5,5′,6,7pentamethoxyflavone, *in* H-70081

93127-02-7 1,2-Benzoquinone-4-carboxylic acid; Me ester, *in* B-70046

93198-79-9 3-Methylene-4-penten-2-one, M-70075

93233-21-7 3(5)-Iodo-5(3)-methylpyrazole, I-70050

93236-48-7 Ainsliaside A, *in* Z-60001

93255-83-5 2,4-Octadecadien-1-ol; (*E,E*)-*form*, *in* O-60004

93273-84-8 2-(3-Oxo-1-propenyl)benzaldehyde; (*Z*)-*form*, *in* O-60091

93278-93-4 Cyclohepta[1,2-*b*:1,7-*b*′]bis[1,4]benzoxazine; (±)-*form*, *in* C-70219

93289-90-8 Dianthramide *A*, D-70049

93289-91-9 Dianthramide B, *in* D-70049

93299-38-8 5-Bromotryptophan; (*S*)-*form*, Me ester, *in* B-60333

93299-39-9 5-Bromotryptophan; (*R*)-*form*, Me ester, *in* B-60333

93299-40-2 5-Bromotryptophan; (*R*)-*form*, *in* B-60333

93373-41-2 4,5-Seconeopulchell-5-ene; 2-Ac, *in* S-70025

93380-12-2 Epirosmanol, *in* R-70010

93380-74-6 Cyclohepta[1,2-*b*:1,7-*b*′]bis[1,4]benzoxazine; (*S*)-*form*, *in* C-70219

93395-09-6 1-[3-(Bromomethyl)-1-methyl2methylenecyclopentyl]4-methyl-1,4-cyclohexadiene, B-70247

93395-10-9 1-[3-(Bromomethyl)-1-methyl2methylenecyclopentyl]4-methylbenzene, B-70247

93446-15-2 1,3,5-Trihydroxy-4-methoxy-2methylanthraquinone, *in* T-60112

93569-21-2 Heptacyclo[9.3.0.0^{2,5}.0^{3,13}.0^{4,8}.-0^{6,10}.0^{9,12}]tetradecane, H-70013

93588-88-6 6,7-Tridecadiene; (*R*)-*form*, *in* T-60272

93635-21-3 Isoambrettolide; (*Z*)-*form*, *in* I-60086

93643-24-4 1,4-Diazabicyclo[4.3.0]nonane; (*S*)-*form*, *in* D-60054

93655-34-6 1,8-Anthracenedicarboxylic acid; Di-Me ester, *in* A-60281

93675-85-5 1-(2-Hydroxyethyl)-1,4cyclohexanediol, H-60133

93710-24-8 Dictyotriol *A*, D-70147

93713-40-7 Glycerol 1-acetate; (±)-*form*, *in* G-60029

93752-54-6 *N*-(5-Hydroxy-2-oxo-7oxobicyclo[4.1.0]hept3-en-3-yl)-acetamide, *in* A-60194

93764-49-9 2,2′-Azodiquinoxaline, A-60353

93782-93-5 5-Iodo-4-penten-1-ol; (*E*)-*form*, *in* I-60058

93859-63-3 Isochamaejasmin, I-70064

93888-62-1 2-(2-Formyl-3-hydroxymethyl-2cyclopentenyl)-6,10dimethyl-5,9-undecadienal, F-70038

93888-65-4 4-Acetoxyflexilin, *in* F-70012

93888-67-6 2-[2-(Acetyloxy)ethenyl]-6,10dimethyl-2,5,9-undecatrienal, A-70050

93916-15-5 (Methylsulfonyl)[(trifluoromethyl)sulfonyl]methane, M-70133

93930-21-3 5,6-Dihydroxy-1-methoxyxanthone, *in* T-60349

93930-25-7 2-Amino-3-(3,5-dichlorophenyl)propanoic acid; (±)-*form*, *in* A-60132

94052-77-4 Biadamantylideneethane, B-60075

94122-10-8 Citreoviridinol A$_2$, *in* C-70181

94137-75-4 1-Bromo-1,2-butadiene; (*R*)-*form*, *in* B-70196

94161-12-3 Citreoviridinol A$_1$, C-70181

94195-16-1 1,4,10-Trioxa-7,13diazacyclopentadecane; 4,10-Dibenzyl, *in* T-60376

94203-53-9 Suvanine, S-70085

94285-21-9 5,7-Dihydroxy-2′,6dimethoxyisoflavone, *in* T-60107

94286-87-0 Lactarorufin E, *in* L-70016

94303-31-8 3′,4′,5,6-Tetramethoxyflavone, *in* T-70108

94346-70-0 Lactarorufin D, L-70016

94410-15-8 Plumbazeylanone, P-70106

94414-19-4 3,23-Dihydroxy-12-ursen-28-oic acid; 3β-*form*, *in* D-70351

94444-24-3 6-Acetoxy-3,10-dibromo-4,7epoxy-12-pentadecen-1-yne, B-70217

94444-25-4 6-Acetoxy-3,10,13-tribromo4,7:9,12-diepoxy-1pentadecyne, B-70195

94465-43-7 2,5-Biss(ydroxymethyl)furan; Dibenzoyl, *in* B-60160

94474-62-1 Macrocliniside B, *in* Z-60001

94513-61-8 Kachirachirol *B*, K-60001

94530-84-4 Senkyunolide F, *in* L-60026

94533-51-4 1,2-Diphenyl-1,2-ethanediamine; (1*RS*,2*SR*)-*form*, *N*-Tetra-Me, *in* D-70480

94533-52-5 1,2-Diphenyl-1,2-ethanediamine; (1*RS*,2*RS*)-*form*, *N*-Tetra-Me, *in* D-70480

94533-53-6 1,2-Diphenyl-1,2-ethanediamine; (1*RS*,2*SR*)-*form*, *N*-Tetra-Et, *in* D-70480

94533-54-7 1,2-Diphenyl-1,2-ethanediamine; (1*RS*,2*RS*)-*form*, *N*-Tetra-Et, *in* D-70480

94596-27-7 Senkyunolide H, *in* L-60026

94618-65-2 1,2-Dibenzylidenecyclobutane; (*E,E*)-*form*, *in* D-60083

94785-96-3 9*H*-Pyrrolo[1,2-*a*]indol-9-one; Hydrazone, *in* P-60247

94930-47-9 4-Formyl-2,5-dimethoxybenzoic acid, *in* F-60072

94936-00-2 Isofasciculatin, *in* F-70001

94938-02-0 1-Methyl-2-phenyl-1*H*imidazole-4-carboxaldehyde, *in* P-60111

94938-03-1 1-Methyl-2-phenyl-1*H*imidazole-5-carboxaldehyde, *in* P-60111

94943-06-3 5-(Hydroxymethyl)-1,2,3,4cyclohexanetetrol; (1*S*,2*S*,3*S*,4*R*,5*R*)-*form*, Penta-Ac, *in* H-70170

94943-08-5 5-(Hydroxymethyl)-1,2,3,4cyclohexanetetrol; (1*S*,2*S*,3*S*,4*R*,5*R*)-*form*, *in* H-70170

94948-59-1 Tumour necrosis factor (human), T-70335

94952-46-2 2-Acetyl-5-chloropyridine, A-70033

94971-81-0 3-Phenyltetrazolo[1,5-*a*]pyridinium; Bromide, *in* P-60134

95043-48-4 5-(Hydroxymethyl)-1,2,3,4cyclohexanetetrol; (1*RS*,2*RS*,3*RS*,4*SR*,5*SR*)-*form*, *in* H-70170

95061-46-4 1,1,2-Triphenyl-1,2-ethanediol; (*R*)-*form*, *in* T-70311

95061-47-5 1,1,2-Triphenyl-1,2-ethanediol; (*R*)-*form*, O^2-Ac, *in* T-70311

95127-11-0 2,3,4-Trimethoxybiphenyl, *in* B-70124

95165-17-6 5-Hydroxy-3,3′,4′,5′,6,7,8heptamethoxyflavone, *in* O-70018

95172-28-4 Cycloocta[2,1-*b*:3,4-*b*′]diquinoline, C-70232

95264-32-7 9*H*-Fluorene-1-carboxaldehyde, F-70014

95303-77-8 *N*-Valylleucine; L-L-*form*, Z-Val-Leu-OBzl, *in* V-60001

95342-43-1 Tingitanol, *in* D-70009

95391-78-9 [2-(Methylseleno)ethenyl]benzene; (*E*)-*form*, *in* M-70132

95391-81-4 [2-(Methylseleno)ethenyl]benzene; (*Z*)-*form*, *in* M-70132

95395-77-0 3-Amino-2-hexenoic acid; Amide, *in* A-60170

95421-88-8 3-Vinylcyclohexene; (*R*)-*form*, *in* V-60009

95457-17-3 3,5-Eudesmadien-1-ol; 1β-*form*, *in* E-60064

95462-97-8 α-Amino-1*H*-imidazole-2-propanoic acid; (±)-*form*, *in* A-60205

95474-41-2 1-Amino-2-(4-hydroxyphenyl)cyclopropanecarboxylic acid; (1*RS*,2*SR*)-*form*, Me ether, *in* A-60197

95485-50-0 Biphenomycin *A*, B-60114

95509-70-9 5-Isopropyl-3-methyl-1,2benzenediol, I-60126

95549-39-6 1-[*N*-[3-(Benzoylamino)-2hydroxy-4-phenylbutyl]alanyl]proline, B-60056

95589-34-7 1-[*N*-[3-(Benzoylamino)-2hydroxy-4-phenylbutyl]alanyl]proline; (2*S*,8*S*,11*R*,12*S*)-*form*, *in* B-60056

95589-35-8 1-[*N*-[3-(Benzoylamino)-2hydroxy-4-phenylbutyl]alanyl]proline; (2*S*,8*S*,11*S*,12*S*)-*form*, *in* B-60056

95597-32-3 3-Ethoxy-4,5,6,7tetrahydroisoxazolo[4,5-*c*]pyridine, *in* T-60071

95597-35-6 4,5,6,7Tetrahydroisoxazolo[4,5-*c*]-pyridin-3-ol; Me ether; B,MeCl, *in* T-60071

95630-78-7 3-Aminobicyclo[2.2.1]hept-5-ene-2methanol; (1*RS*,2*RS*,3*SR*)-*form*, *in* A-70094

95630-79-8 3-Aminobicyclo[2.2.1]heptane2-methanol; (1*RS*,2*RS*,3*SR*)-*form*, *in* A-70093

95630-80-1 3-Aminobicyclo[2.2.1]heptane2-methanol; (1*RS*,2*RS*,3*SR*)-*form*, *N*-Me, *in* A-70093

95630-81-2 3-Aminobicyclo[2.2.1]heptane2-methanol; (1*RS*,2*SR*,3*RS*)-*form*, *N*-Me, *in* A-70093

95824-80-9 2-Amino-2-hexenoic acid; (*Z*)-*form*, Me ester, *in* A-60167

95834-55-2 α-Aminobenzo[*b*]thiophene-3acetic acid; (±)-*form*, *in* A-60090

95834-94-9 α-Aminobenzo[*b*]thiophene-3acetic acid, A-60090

95835-03-3 α-Aminobenzo[*b*]thiophene-3acetic acid; (±)-*form*, Me ester, *in* A-60090

95836-52-5 2-(Hydroxyacetyl)pyridine, H-70101

95836-53-6 2-(Hydroxyacetyl)pyridine; Hydrazone, *in* H-70101

95881-83-7 3-Chloro-5-methoxypyridine, *in* C-60102

95882-39-6 3-Amino-2-hexenoic acid; Nitrile, *in* A-60170

95909-97-0 α-Aminobenzo[*b*]thiophene-3acetic acid; (*R*)-*form*, Me ester, *in* A-60090

95909-98-1 α-Aminobenzo[*b*]thiophene-3acetic acid; (*R*)-*form*, *in* A-60090

95935-50-5 3-Amino-β-carboline; *N*³-Formyl, *in* A-70110

95935-52-7 3-Amino-β-carboline; *N*³-Et, *in* A-70110

95968-93-7 4-Mercapto-2-butenoic acid; (*E*)-*form*, Me ester, *in* M-60018

96012-97-4 2-Methylbenzo[*gh*]perimidine, M-60051

96025-84-2 2,2,4,4Tetrakiss(rifluoromethyl)-1,3-dithietane; 1-Oxide, *in* T-60129

96025-87-5 2,2,4,4Tetrakiss(rifluoromethyl)-1,3-dithietane; 1,1,3-Trioxide, *in* T-60129

96055-63-9 Punaglandin 1, P-70142

96055-64-0 Punaglandin 2, *in* P-70142

96055-65-1 Punaglandin 3, P-70143

96055-66-2 Punaglandin 4, *in* P-70143

96055-66-2 (*E*)-Punaglandin 4, *in* P-70143

96120-85-3 10*H*-Phenothiazine-1-carboxylic acid; Me ester, 5-oxide, *in* P-70055

96202-77-6 9,12-Dioxododecanoic acid; Me ester, ethylene acetal, *in* D-60471

96225-88-6 2,2,4,4Tetrakiss(rifluoromethyl)-1,3-dithietane; 1,1,3,3-Tetraoxide, *in* T-60129

96238-88-9 Silyhermin, S-70045

96245-29-3 Pyrimido[4,5-*i*]imidazo[4,5-*g*]cinnoline, P-60244

96253-60-0 Cystoketal, C-60234

96304-42-6 Trypanothione, T-70331

96304-95-9 Thyrsiferol; 23-Ac, *in* T-70193

96308-48-4 1-Cyclooctenecarboxaldehyde; (*E*)-*form*, *in* C-60217

96313-95-0 Gracilin *B*, G-60034

96385-81-8 Ventilone *C*, V-60004

96385-82-9 Ventilone D, *in* V-60004

96385-83-0 Ventilone E, *in* V-60004

96404-79-4 2-Triphenylenecarboxaldehyde, T-60382

96423-70-0 9-Amino-10-methoxy-1,4anthraquinone, *in* A-60181

96423-72-2 9-Amino-10-hydroxy-1,4anthraquinone, A-60181

96454-73-8 1-Ethynylbicyclo[2.2.2]octane, E-70049

96523-73-8 *N*-Tyrosylalanine; L-L-*form*, Boc-Tyr-Ala-OBzl, *in* T-60411

96552-87-3 6-Hydroxyarcangelisin, *in* D-70021

96552-88-4 2-Dehydroarcangelisinol, D-70021

96574-27-5 2,5-Dimethyl-4*H*-1-benzopyran4-one, D-70371

96603-02-0 Halenaquinol, H-60002

96694-90-5 1,1'-Bibiphenylene, B-60080

96694-92-7 2,2'-Bibiphenylene, B-60081

96695-04-4 Bisbenzo[3,4]cyclobuta[1,2*c*;1',2'-*g*]phenanthrene, B-60143

96705-91-8 2,5-Piperazinedicarboxylic acid; (2*RS*,5*RS*)-*form*, *in* P-60157

96705-92-9 2,5-Piperazinedicarboxylic acid; (2*RS*,5*SR*)-*form*, *in* P-60157

96705-93-0 2,5-Piperazinedicarboxylic acid; (2*RS*,5*RS*)-*form*, Di-Me ester, *in* P-60157

96705-94-1 2,5-Piperazinedicarboxylic acid; (2*RS*,5*RS*)-*form*, Diamide, *in* P-60157

96705-95-2 2,5-Piperazinedicarboxylic acid; (2*RS*,5*SR*)-*form*, Diamide, *in* P-60157

96728-88-0 2,5-Piperazinedicarboxylic acid; (2*RS*,5*SR*)-*form*, Di-Me ester, *in* P-60157

96738-89-5 9-Hydroxypinoresinol; 9β-*form*, *in* H-70216

96754-01-7 1,2,5,7-Tetramethoxyphenanthrene, *in* T-60116

96806-31-4 5-Epiilimaquinone, *in* I-70002

96816-28-3 *N*-2-Benzothiazolyl-*N*,*N*',*N*″triphenylguanidine; (*Z*)-*form*, *in* B-70057

96820-10-9 2-Bromo-1,3,5-benzenetriol; Tri-Ac, *in* B-60199

96827-22-4 Camellidin I, *in* D-60363

96827-23-5 Camellidin II, *in* D-60363

96839-34-8 2,6-Dinitrobenzyl alcohol, D-60459

96843-73-1 Cudraflavone *A*, C-70205

96853-63-3 8-p-Hydroxybenzoylshiromodiol, *in* S-70039

96853-63-3 Epoxyjaeschkeanadiol p-hydroxybenzoate, *in* J-70001

96861-15-3 3,18-Dihydroxy-28-nor-12oleanen-16-one, D-60363

96881-39-9 1,4-Dihydro-4,6,7-trimethyl9*H*-imidazo[1,2-*a*]purin9-one, D-70278

96887-18-2 5,7-Dihydroxy-3,3',4',5',6,8hexamethoxyflavone, *in* O-70018

96887-19-3 4',5,7-Trihydroxy-3,3',5',6,8pentamethoxyflavone, *in* O-70018

96887-20-6 3',5,7-Trihydroxy-3,4',5',6,8pentamethoxyflavone, *in* O-70018

96888-55-0 Anhydrofusarubin 9-methyl ether, *in* F-60108

96919-50-5 Thieno[2,3-*c*]cinnoline, T-60202

96944-53-5 Specionin, S-60039

97038-01-2 2-Phenyl-1,5-benzothiazepin4(5*H*)-one; 2,3-Dihydro, *N*-Me, *in* P-60081

97165-43-0 Tucumanoic acid, T-60410

97217-74-8 3,5-Diamino-2,4,6trinitrobenzoic acid, D-70047

97218-74-1 1,2,4-Trimethyl-9*H*-carbazole, T-60358

97229-08-8 1,4-Diiodopentacyclo[4.2.0.0²,⁵.0³,⁸.0⁴,⁷]octane, D-70359

97253-51-5 Tetracyclo[6.2.1.1³,⁶.0²,⁷]dodeca-2(7),4,9-triene; *syn-form*, *in* T-60032

97274-08-3 2,2'-Azodiquinoxaline; (*E*)-*form*, *in* A-60353

97315-26-9 1,2:7,8-Dibenzo[2.2]paracyclophane, D-70072

97344-05-3 Oblongolide, O-60001

97350-56-6 Dodecahedrane; C_6H_6 complex, *in* D-60514

97359-75-6 2',5-Dihydroxy-6,7methylenedioxyisoflavone, *in* T-60107

97400-69-6 Austrocortirubin, A-60322

97400-70-9 Austrocortilutein, A-60321

97402-98-7 2-Amino-3-methylenepentanedioic acid; (±)-*form*, *in* A-60219

97402-99-8 2-Amino-3-methylenepentanedioic acid; (±)-*form*, Amide, *in* A-60219

97444-11-6 6-Bromotryptophan; (±)-*form*, *N*α-Ac, Et ester, *in* B-60334

97444-12-7 6-Bromotryptophan; (*S*)-*form*, *N*α-*tert*-Butyloxycarbonyl, *in* B-60334

97530-63-7 2-Acetoxy-1,11-bis(3-furanyl)4,8-dimethyl-3,6,8undecatriene, *in* B-70143

97551-09-2 3,5-Dihydro-4*H*-dinaphth[2,1*c*:1',2'-*e*]azepine; (*S*)-*form*, *in* D-60234

97563-40-1 3,3*a*,6,6*a*-Tetrahydro-1(2*H*)pentalenone; (3*aRS*,6*aSR*)-*form*, *in* T-70083

97567-50-5 2,6-Diiodobenzoic acid; Diethylamide, *in* D-70356

97605-30-6 Pulchellin 2-O-tiglate, *in* P-70138

97605-31-7 6β-Hydroxypulchellin 2-Oisovalerate, *in* P-70138

97643-92-0 6β-Hydroxypulchellin 2-O-acetate, *in* P-70138

97689-87-7 Tunichrome B-1, T-70336

97718-45-1 Subergorgic acid, S-70081

97776-60-8 6-Formyl-1,2,4-triazine3,5(2*H*,4*H*)-dione, F-60075

97804-04-1 Dukunolide *A*, D-60522

97806-44-5 Tetracyclo[6.2.1.1³,⁶.0²,⁷]dodeca-2(7),4,9-triene; *anti-form*, *in* T-60032

97825-83-7 9-Amino-9,10-dihydrophenanthrene; (±)-*form*, *in* A-60143

97825-91-7 9-Amino-9,10-dihydroanthracene, A-60136

97825-92-8 1-Amino-1,4-dihydronaphthalene; (±)-*form*, *in* A-60139

97847-61-5 2,4,6-Cycloheptatrien-1ylcyclopropylmethanone, C-60206

97915-33-8 2,5-Dihydroxy-4-methoxy-9*H*fluoren-9-one, *in* T-70261

97938-47-1 3-Oxopyrazolo[1,2-*a*]pyrazol-8ylium-1-olate, O-60092

97984-28-6 2,5-Dihydro-4,6dimethoxypyrimidine, *in* D-70256

97984-29-7 4,6-Diethoxy-2,5dihydropyrimidine, *in* D-70256

97984-32-2 4,6-Biss(thylthio)-2,5dihydropyrimidine, *in* D-70257

97995-82-9 2-Amino-3,3-difluoropropanoic acid, A-60135

98026-98-3 3-Diazo-2,4(5*H*)-furandione, D-70059

98036-41-0 Lachnellulone, L-70015

98094-08-7 Cyclolycoserone, C-70229

98094-09-8 Lycoserone, L-70045

98094-16-7 α-Diceroptene, D-70112

98149-05-4 *N*-Methionylalanine; L-L-*form*, Boc-Met-Ala-OPh, *in* M-60035

98153-12-9 3-Iodo-2-phenyl-4*H*-1benzopyran-4-one, I-60059

98153-13-0 3-Mercapto-2-phenyl-4*H*-1benzopyran-4-one, M-60025

98153-25-4 4-Oxo-2-phenyl-4*H*-1benzothiopyran-3carboxylic acid, O-60087

98155-84-1 Linderatone, L-70030

98202-88-1 1-(Dimethoxymethyl)-2,3,5,6tetra-kiss(ethylene)-7oxabicyclo[2.2.1]heptane, D-60399

98211-30-4 1'-Methylzeatin; 9-β-D-Ribofuranosyl, in M-60133

98212-93-2 3-(1-Aminocyclopropyl)-2propenoic acid; (E)-form, in A-60123

98243-92-6 Chrysene-5,6-oxide; (5S,6R)-form, in C-70174

98243-93-7 Chrysene-5,6-oxide; (5R,6S)-form, in C-70174

98263-44-6 4-Amino-β-carboline, A-70111

98263-45-7 7,12-Dihydropyrido[3,2-b:5,4b']-diindole, D-70254

98273-52-0 2-(1,3,4-Thiadiazol-2-yl)pyridine, T-60192

98300-40-4 2',3'-Dimethoxy-6'nitro-acetophenone, in D-60357

98329-63-6 3,5-Pyridinedipropanoic acid; B,HCl, in P-60217

98331-27-2 1-Bromo-7-(bromomethyl)naphthalene, B-60215

98349-22-5 2,4,5-Trifluorobenzonitrile, in T-70240

98442-42-3 Bicyclo[3.3.1]nonane-2,4-dione, B-70103

98462-03-4 8-Hydroxy-5,9,11,14eicosatetra-enoic acid; (8S,5Z,9E,11Z,14Z)-form, in H-60128

98571-24-5 4,6,8,10,12,14,16-Heptamethyl6,8,11-octadeca-triene3,5,13-trione, H-70022

98599-90-7 Polivione, P-60167

98626-97-2 3-Chloro-2-(methylthio)pyridine, in C-60129

98626-98-3 3-Chloro-2-(methyl-sulfonyl)pyridine, in C-60129

98640-00-7 5,6-Dichloro-1H-indole-3-acetic acid, D-70123

98644-24-7 Zedoarondiol, Z-70004

98665-64-6 3,5-Dihydroxy-2,4-dimethoxy9H-fluoren-9-one, in T-70109

98676-96-1 Decahydro-5,5,8a-trimethyl-2naphthalenol; (2S,4aS,8aR)-form, Ac, in D-70014

98760-30-6 α-Aminobenzo[b]thiophene-3acetic acid; (±)-form, Me ester; B,HCl, in A-60090

98816-35-4 4-Chloro-3(5)methylpyrazole; B,HCl, in C-70108

98819-33-1 α-Aminobenzo[b]thiophene-3acetic acid; (S)-form, Me ester, in A-60090

98819-35-3 α-Aminobenzo[b]thiophene-3acetic acid; (S)-form, Isopropyl ester, in A-60090

98819-36-4 α-Aminobenzo[b]thiophene-3acetic acid; (R)-form, Isopropyl ester, in A-60090

98856-65-6 7-Azabicyclo[4.2.0]oct-3-en-8one; (1S,6R)-form, in A-60331

98857-38-6 4-Isopropyl-1-methyl-3cyclo-hexenene-1,2-diol; (1R*,2S*)-form, in I-70093

98858-06-1 4-Azido-1H-imidazo[4,5-c]pyridine, A-60345

98858-07-2 4-Amino-1H-imidazo[4,5-c]pyridine; 4-N-Di-Me, in A-70161

98858-09-4 4-Amino-1H-imidazo[4,5-c]pyridine; 1-(5-Deoxy-β-D-ribofuranosyl), in A-70161

98933-69-8 6-Amino-1,3-dihydro-2H-purin2-one; 6,9-Di-Ac, 1-Me, in A-60144

98933-71-2 N-(2,9-Dihydro-1-methyl-2-oxo1H-purin-6-yl)acetamide, in A-60144

98941-66-3 1(10),4-Germacradiene-6,8-diol; (1E,4E,6S,8S)-form, in G-60014

98959-58-1 3,4,6,7,8,9-Hexahydro-2Hquinolizin-2-one, H-60059

99007-16-6 3-Amino-1H-pyrazole-4-carboxylic acid; N(1)-Benzyl, in A-70187

99103-26-1 Inunal, I-70022

99165-17-0 Epidermin, E-70010

99165-99-8 [2,2'-Bipyridine]-3,3'dicarboxylic acid; Mono-Me ester, in B-60124

99211-67-3 Aurantioclavine; (±)-form, in A-60316

99233-89-3 Dibenz[b,g]azocine-5,7(6H,12H)dione; N-Ph, in D-60068

99248-33-6 Seglitide acetate, in C-60180

99291-28-8 4-Amino-5-hydroxy-2-oxo-7oxabicyclo[4.1.0]hept3-ene-3-carboxamide, A-70158

99305-07-4 Subepinnatin C, S-60058

99310-21-1 Tricyclopropylcyclopropenium(1+); Hexafluoroantimonate, in T-60270

99343-73-4 Dukunolide B, in D-60522

99343-74-5 Dukunolide C, in D-60522

99354-18-4 Bis(2,4,6-tri-tertbutylphenyl)-ditelluride, B-70162

99389-26-1 3,5-Difluorobenzenethiol, D-60176

99493-88-6 2-Hydroxycyclopentanone; (±)-form, in H-60114

99499-99-7 6-Deoxycatalpol, in C-70026

99502-89-3 2[(Chlorocarbonyl)oxy]3a,4,7,7a-tetrahydro4,7-methano-1H-isoindole-1,3(2H)-dione, C-70059

99520-58-8 ▷Indeno[1,2,3-cd]pyren-8-ol, I-60026

99520-65-7 ▷Indeno[1,2,3-cd]pyren-1-ol, I-60022

99520-66-8 ▷Indeno[1,2,3-cd]pyren-2-ol, I-60023

99520-67-9 ▷Indeno[1,2,3-cd]pyren-6-ol, I-60024

99528-87-7 Halenaquinol; O16-Sulfate, in H-60002

99545-47-8 4-Azidobutanal, A-70286

99566-52-6 4-Acetyl-2,2-dimethyl-1,3dioxolane; (S)-form, in A-70042

99593-51-8 4-Methyl-5-nitropyrimidine, M-60104

99605-65-9 Bicyclo[8.2.2]eicosa-1(19),10(20),19-triene, B-70096

99624-26-7 Mariesiic acid A, M-70012

99624-47-2 Methyl 3-(4-methoxy-2vinylphenyl)propanoate, in H-70232

99643-51-3 2,3-Naphthalenedithiol; Di-S-benzyl, in N-70004

99643-52-4 2,3-Naphthalenedithiol, N-70004

99646-78-3 2,3-Dihydro-2,3di-methylenethiophene, D-70195

99651-86-2 2-Amino-5-bromopyrimidine; B,HCl, in A-70101

99674-26-7 Esperamicin; Esperamicin A₁, in E-60037

99684-26-1 Dimethyl 4-amino-5methoxy-phthalate, in A-70152

99694-82-3 Debneyol, D-70010

99702-05-3 1-(2-Furanyl)-2-(2-pyrrolyl)ethyl-ene; (E)-form, N-Me, in F-60088

99702-07-5 1-(Pyrrolyl)-2-(2-thienyl)ethyl-ene; (E)-form, N-Me, in P-60250

99705-97-2 Tetrahydro-2-(2-piperidinyl)furan, T-60082

99706-00-0 Tetrahydro-2-(2-pyrrolidinyl)furan, T-60090

99714-13-3 1-Iodo-3,3-dimethyl-2-butanone, I-60044

99725-13-0 5-Bromo-2-fluorobenzyl alcohol, B-60245

99809-68-4 5,6-Decanediol; (5S,6S)-form, Benzyl ether, in D-60013

99816-52-1 3',4',5,5',7-Pentahydroxy3,6,8-trimethoxyflavone, in O-70018

99816-54-3 3',5,5',7-Tetrahydroxy-3,4',8trimethoxyflavone, in H-70021

99820-21-0 Luminamicin, L-60038

99835-10-6 2,3-Dimethyl-5-(2,6,10trimethyl-undecyl)thiophene, D-70446

99881-77-3 5,6-Decanediol; (5S,6S)-form, in D-60013

99895-14-4 Cyclopent[b]azepine; N-Oxide, in C-60225

99910-88-0 3-Bromo-1,2,4-benzenetriol, B-70180

99922-00-6 Pentacyclo[12.2.2.2²,⁵.2⁶,⁹.2¹⁰,¹³]-tetracosa-1,5,9,13-tetraene, P-60029

99946-16-4 Bicyclo[2.2.1]hepta-2,5-diene2-carboxylic acid; (±)-form, Me ester, in B-60084

99947-39-4 3-Ethynyl-3-methyl-4-pentenoic acid; (±)-form, Et ester, in E-60059

99947-40-7 3-Ethynyl-3-methyl-4-pentenoic acid; (±)-form, in E-60059

99947-42-9 3-Ethynyl-3-methyl-4-pentenoic acid; (S)-form, N,N-Dimethyl-amide, in E-60059

99947-54-3 3-Ethynyl-3-methyl-4-pentenoic acid; (R)-form, Me ester, in E-60059

99948-48-8 3-Furanacetaldehyde, F-60083

100018-84-6 3-Ethynyl-3-methyl-4-pentenoic acid; (S)-form, in E-60059

100018-85-7 3-Ethynyl-3-methyl-4-pentenoic acid; (R)-form, in E-60059

100020-51-7 1-[3-Azido-4-(hydroxymethyl)cyclopentyl]-5methyl2,4(1H,3H)-pyrimidinedione; (±)-form, in A-70288

100045-39-4 7,13-Dihydroxy-2-longipinen1-one; 7β-form, in D-70315

100045-45-2 Elemasteiractinolide; 15-Hydroxy, in E-70005

100045-46-3 Elemasteiractinolide; 11β,13-Dihydro, 15-hydroxy, in E-70005

100045-52-1 4,11(13)-Eudesmadien-12-oic acid; 3β-Hydroxy, in E-70061

100045-53-2 9,11(13)-Eremophiladien-12-oic acid; 2β-Hydroxy, in E-70029

100045-54-3 9,11(13)-Eremophiladien-12-oic acid; 3β-Hydroxy, in E-70029

100045-55-4 9,11(13)-Eremophiladien-12-oic acid; 3-Oxo, in E-70029

100077-49-4 1,3-Benzodioxole-4-acetic acid, B-60029

100079-33-2 5,7-Dihydroxy-2',8dimethoxy-flavanone, in T-70104

100079-34-3 7-Hydroxy-2',5,8trimethoxy-flavanone, in T-70104

100079-50-3 Fusarin A, F-60107

100101-42-6 5-Methyl-2-(1-methyl-1phenyl-ethyl)cyclohexanol; (1S,2R,5R)-form, in M-60090

100187-63-1 Macrocliniside I, in Z-60001

100198-26-3 Esulone B, in E-60038

100202-26-4 Crepiside B, in Z-60001

100202-28-6   Crepiside A, *in* Z-60001

100202-78-6   2-Bromomethyl)-6-fluoropyridine, B-70243

100215-74-5   Esulone *A*, E-60038

100217-74-1   Biphenomycin B, *in* B-60114

100243-39-8   3-Pyrrolidinol; (*S*)-*form, in* P-70187

100324-68-3   8-Hydroxy-1(6),2,4,7(11)cadinatetraen-12,8-olide, H-70114

100324-69-4   8-Methoxy-1(6),2,4,7(11)cadinatetraen-12,8-olide, *in* H-70114

100324-71-8   10-Oxo-6-isodaucen-14-al, O-70093

100343-67-7   ent-7β-Hydroxy-13,14,15,16tetranor-3-cleroden-12oic acid 18,19-lactone, H-70226

100343-68-8   ent-7β-Hydroxy-3,13clerodadiene-15,16;18,19-diolide, H-70117

100343-69-9   ent-7-Oxo-3-clerodene15,16:18,19-diolide, *in* H-70117

100343-70-2   ent-7β-Hydroxy-3-clerodene15,16,18,19-diolide, *in* H-70117

100367-36-0   3-Amino-2-bromophenol, A-70099

100414-80-0   Danshenspiroketallactone, D-70004

100428-90-8   10α-Hydroxy-6-isodaucen-14-al, *in* O-70093

100460-12-6   Pyrido[1″,2″:1′,2′]imidazo[4′,5′:-4,5]imidazo[1,2-*a*]pyridine, P-70175

100461-65-2   1,1,1′,1′-Tetraphenyldimethylamine; *N*-Nitroso, *in* T-70150

100466-27-1   4-(Bromomethyl)-2-nitrobenzoic acid, B-60297

100483-73-6   5-Hydroxy-4-methyl-3-heptanone; (4*S*,5*S*)-*form, in* H-60183

100486-34-8   1,2,3,4-Tetrahydro-3-methyl3-isoquinolinecarboxylic acid; (±)-*form*, B,HCl, *in* T-60074

100515-65-9   Bis(1,1-dicyclopropylmethyl)diazene; (*Z*)-*form, in* B-60154

100515-66-0   Bis(1,1-dicyclopropylmethyl)diazene; (*E*)-*form, in* B-60154

100515-67-1   1,1,1′,1′-Tetra-*tert*butyl-azomethane; (*Z*)-*form, in* T-60024

100515-68-2   1,1,1′,1′-Tetra-*tert*butyl-azomethane; (*E*)-*form, in* T-60024

100605-94-5   7,11,15-Trimethyl-3methylene-1,2-hexadecanediol, T-60368

100633-68-9   2-Methylfuro[3,2-*c*]quinoline, M-70083

100656-07-3   2-(Dimethylamino)benzaldehyde; Di-Me acetal, *in* D-60401

100665-40-5   7β,15α-Dihydroxy-3,11,23trioxo-8,20(22)*E*lanostadien-26-oic acid, *in* T-70254

100665-41-6   3β,7β-Dihydroxy-11,15,23trioxo-8,20(22)*E*lanostadien-26-oic acid, *in* T-70254

100665-42-7   3,7,15-Trihydroxy-11,23dioxo-8,20(22)-lanostadien-26-oic acid; (3β,7β,15α,20(22)*E*)-*form, in* T-70254

100665-43-8   7β-Hydroxy-3,11,15,23tetraoxo-8,20(22)lanostadien-26-oic acid, *in* T-70254

100683-08-7   1-Methoxycyclopropanecarboxylic acid, *in* H-60117

100692-52-2   Pueroside A, *in* P-70137

100692-54-4   Pueroside B, *in* P-70137

100758-27-8   4-Octene-1,7-diyne; (*Z*)-*form, in* O-60028

100758-30-3   5-Decene-2,8-diyne; (*Z*)-*form, in* D-60015

100813-32-9   13,15-Ethano-17-ethyl2,3,12,18-tetramethylmonobenzo[*g*]-porphyrin, E-60043

100813-35-2   13,15-Ethano-3,17-diethyl2,12,18-trimethylmonobenzo[*g*]-porphyrin, E-60042

100905-87-1   Wasabidienone *D*, W-70002

100905-88-2   Wasabidienone E, *in* W-70002

100905-89-3   Pamamycin-607, P-70004

100910-16-5   2,6-Diamino-3-heptenedioic acid; (*E*)-*form, in* D-60036

100910-32-5   2,6-Diamino-4methyl-eneheptanedioic acid; (2*S**,6*S**)-*form*, B,2HCl, *in* D-60039

100910-33-6   2,6-Diamino-4methyl-eneheptanedioic acid; (2*R**,6*R**)-*form*, B,2HCl, *in* D-60039

100910-34-7   2,6-Diamino-4methyl-eneheptanedioic acid; (2*RS*,6*SR*)-*form, in* D-60039

100910-60-9   2,6-Diamino-4methyl-eneheptanedioic acid, D-60039

100910-61-0   2,6-Diamino-4methyl-eneheptanedioic acid; (2*R**,6*R**)-*form, in* D-60039

100928-36-7   2,6-Diamino-4methyl-eneheptanedioic acid; (2*S**,6*S**)-*form, in* D-60039

100939-69-3   2,3,5,6-Tetrahydro-sindacene-1,7-dione, T-70067

100994-20-5   [3](2.2)[3](5.5)Pyridinophane; High-melting-*form, in* P-60227

100994-22-7   [3](2.5)[3](5.2)Pyridinophane; Low-melting-*form, in* P-60228

100994-29-4   [3.3][2.6]Pyridinophane, P-60226

101002-03-3   2-Bromo-3,3′-bipyridine, B-70185

101023-30-7   10,10-Dimethyl-9(10*H*)anthracenone; Hydrazone, picrate, *in* D-60403

101030-88-0   Aminoiminomethanesulfonic acid; *N*-Benzyl, *in* A-60210

101046-20-2   1-[4-[6-(Diethylamino)-2benzofuranyl]phenyl]1*H*-pyrrole-2,5-dione, D-60170

101053-45-6   [3](2.2)[3](5.5)Pyridinophane; Low-melting-*form, in* P-60227

101053-47-8   [3](2.5)[3](5.2)Pyridinophane; High-melting-*form, in* P-60228

101055-48-5   [6]-Paracycloph-3-ene, P-70010

101055-88-3   5-Vinyl-1,2,3-thiadiazole, V-60015

101187-91-1   4-Cyano-3-fluorobenzenethiol, *in* F-70029

101210-95-1   3-Amino-3-methyl-2-butanol; (±)-*form*, Picrate, *in* A-60217

101392-95-4   4,6-Dichloro-5[(dimethylamino)methylene]-3,6-cyclohexadiene-1,3-dicarboxaldehyde, D-70119

101401-88-1   Sporol, S-70069

101402-04-4   4-Chlorodiphenylmethanol; (*S*)-*form, in* C-60065

101411-70-5   Paldimycin; Paldimycin *A, in* P-70002

101411-71-6   Paldimycin; Paldimycin *B, in* P-70002

101467-55-4   2,3,4,7,8,8*a*-Hexahydro-8-(5hydroxy-4-methylpentyl)6,8-dimethyl-1*H*-3*a*,7methanoazulene-3carboxylic acid, H-70057

101489-38-7   Isorengyol, *in* H-60133

101508-37-6   Azadirachtol, A-70280

101509-58-4   Nimbocinol, *in* A-70281

101512-26-9   1′-Methylzeatin, M-60133

101541-87-1   4-Vinyl-1,2,3-thiadiazole, V-60014

101541-97-3   3-(Ethylthio)-2-propenoic acid; (*E*)-*form, in* E-60055

101559-98-2   Yadanzioside *K*, Y-60001

101623-21-6   Gliotoxin E, G-60024

101670-85-3   2-Hydroxy-1,3benzenedicarboxylic acid; Mono-Me ester, *in* H-60101

101706-33-6   Antibiotic FR 900452, A-60284

101773-62-0   2,3-Dihydro-5(1*H*)-indolizinone, D-60251

101773-63-1   7,8,9,10Tetrahydropyrido[1,2-*a*]azepin-4(6*H*)-one, T-60085

101916-65-8   1,2-Dibromopropane; (*S*)-*form, in* D-60105

101933-27-1   1-Nitro-2-(phenylthio)ethylene; (*E*)-*form, in* N-70061

101933-28-2   1-Nitro-2-(phenylsulfinyl)ethylene, *in* N-70061

101933-29-3   1-Nitro-2-(phenylsulfonyl)ethylene, *in* N-70061

101933-87-3   2-Bromo-1,5-hexadiene, B-70216

101933-88-4   2-Chloro-1,5-hexadiene, C-70070

101934-24-1   7,7-Dimethylbicyclo[4.1.0]hept-3-ene; (1*RS*,6*RS*)-*form, in* D-60409

101934-33-3   5-Methyl-3,4-hexadienoic acid; Me ester, *in* M-70087

101955-04-8   6,7,8,9-Tetrahydro-5*H*benzocyclo-heptene-7carboxaldehyde, T-60056

101955-05-9   6,7,8,9-Tetrahydro-5*H*benzocyclo-heptene-6carboxaldehyde, T-60055

101977-88-2   2,2-Diethoxytetrahydro-5,6dimethyl-2*H*-pyran, *in* T-70059

101978-00-1   2,3-Dimethyl-3,4-dihydro-2*H*pyran; (2*R*,3*S*)-*form, in* D-70398

101998-46-3   9*b*-Methyl-9*bH*-benz[*cd*]azulene, M-60044

101998-70-3   1,3,5,7-Tetra-*tert*-butyl-*s*-indacene, T-60025

102036-28-2   Protosappanin *A*, P-60186

102067-87-8   3-(3,4-Dihydroxybenzyl-3,4,7trihydroxychroman, *in* D-70288

102067-88-9   3-(3,4-Dihydroxybenzyl)-7hydroxy-4-chromanone, D-70288

102072-84-4   5-Aminophthalazine, A-70185

102127-62-8   Trifluoromethanesulfenyl iodide, T-60282

102167-68-0   Naematolin B, *in* N-70001

102208-92-4   Tetrahydro-2-methylpyran; (*S*)-*form, in* T-60075

102283-96-5   Phenanthro[1,10-*bc*:8,9-*b*′,*c*′]bisthiopyran, P-60072

102301-13-3   Antibiotic 273a₂; Antibiotic 273a$_{2β}$, *in* A-70218

102308-43-0   4-Bromo-1(3*H*)-isobenzofuranone, B-70224

102326-69-2   Antibiotic 273a₂; Antibiotic 273a$_{2α}$, *in* A-70218

102342-18-7   2,7-Diphenyloxepin, D-60486

102363-08-6   Antibiotic FR 900482, A-60285

102367-93-1   ▷2-Chloro-3,4,5,6tetranitroaniline, C-60134

102367-94-2   ▷3-Chloro-1,2,4,5tetranitrobenzene, C-60135

102368-13-8   1,1′-Carbonothioylbis-2(1*H*)pyridinone, C-60014

102420-40-6   4-Amino-2,5-hexadienoic acid; (±)-(*E*)-*form, in* A-60166

102420-57-5   Indeno[1,2,3-*cd*]pyren-1-ol; Ac, *in* I-60022

102420-58-6   Indeno[1,2,3-*cd*]pyren-2-ol; Ac, *in* I-60023

102420-62-2 Indeno[1,2,3-*cd*]pyren-6-ol; Ac, *in* I-60024

102420-63-3 ▷Indeno[1,2,3-*cd*]pyren-7-ol, I-60025

102420-65-5 8-Methoxyindeno[1,2,3-*cd*]pyrene, *in* I-60026

102420-67-7 7-Methoxyindeno[1,2,3-*cd*]pyrene, *in* I-60025

102421-45-4 1-(1-Diazoethyl)naphthalene, D-60061

102427-32-7 2,5-Dimethyl-3,4-diphenyl2,4-hexadiene, D-60419

102435-72-3 2,2-Dimethyl-3methylenebicyclo[2.2.2]octane, D-60427

102450-37-3 1,1-Dibromocycloheptane, D-60091

102491-83-8 4-Amino-5-fluoro-2-pentenoic acid; (±)-(*E*)-*form*, *in* A-60161

102493-70-9 4-Chloro-2,3-dihydro-1*H*-indole; 1-Benzyl, *in* C-60055

102505-81-7 3,3,4,4-Tetramethyl-1,2oxathietane, T-60149

102518-95-6 3,5-Dihydro-4*H*-dinaphth[2,1*c*:1′,2′-*e*]azepine; (±)-*form*, *in* D-60234

102521-04-0 11*b*-Methyl-11*bH*-Cyclooct[*cd*]azulene, M-60060

102521-12-0 2,2,5,5-Tetramethyl-3cyclopenten-1-one; 2,4-Dinitrophenylhydrazone, *in* T-60140

102573-92-2 2-Ethynyl-1,3-propanediol, E-70054

102574-14-1 3-Oxo-1-cyclopentenecarboxaldehyde, O-60062

102575-25-7 Tricyclo[5.1.0.0^{2,8}]oct-4-ene, T-60269

102575-26-8 Tricyclo[5.1.0.0^{2,8}]oct-3-ene, T-60268

102580-63-2 Bryostatin 11, B-70286

102580-65-4 Bryostatin 10, B-70285

102586-55-0 1,3-Diiodo-2,2-diazidopropane, D-60387

102586-88-9 4,5-Dihydro-1,5-ethano-1*H*-1benzazepin-2(3*H*)-one, D-60239

102652-07-3 5,6,7,8,9,10-Hexahydro-4*H*cyclonona[*c*]furan, H-60047

102652-65-3 1,1,2,2-Tetracyclopropylethane, T-60034

102652-67-5 3,4-Diisopropyl-2,5dimethylhexane, D-60393

102696-71-9 3-(3-Thienyl)-2-propenoic acid; (*E*)-*form*, *in* T-70172

102698-33-9 5,6-Dimethoxy-1,3cyclohexadiene, D-60397

102711-08-0 1-Azido-2,2-dimethylpropane, A-60344

102714-71-6 1-Cyano-4,5-dimethoxy-2nitrobenzene, *in* D-60360

102719-89-1 3-Methyl-*N*-(5*a*,6*a*,7,8tetrahydro-4-methyl1,5-dioxo-1*H*,5*H*pyrrolo[1,2-*c*][1,3]oxazepin-3-yl)-butanamide, M-70134

102719-90-4 *N*-(5*a*,6,7,8-Tetrahydro-4methyl-1,5-dioxo-1*H*,5*H*pyrrolo[1,2-*c*][1,3]oxazepin-3-yl)-hexanamide, *in* M-70134

102725-09-7 Di-9-anthracenylmethanone, D-60050

102735-88-6 5-Fluoro-2-methyl-1,3dinitrobenzene, F-60051

102735-89-7 1-Chloro-5-methoxy-2-methyl3-nitrobenzene, *in* C-60116

102737-62-2 Methaneselenoic acid; *tert*-Butyl ester, *in* M-70035

102774-90-3 5-Amino-4-hydroxypentanoic acid; (*S*)-*form*, *in* A-60196

102774-96-9 5-(Aminomethyl)dihydro-2(3*H*)furanone, *in* A-60196

102778-41-6 2-Phenylcyclopentylamine; (1*R*,2*R*)-*form*, B,HCl, *in* P-60086

102778-42-7 2-Phenylcyclopentylamine; (1*S*,2*S*)-*form*, B,HCl, *in* P-60086

102781-57-7 2-Aminobicyclo[2.1.1]hexane; (±)-*form*, B,HCl, *in* A-70095

102786-89-0 1,3,5,8-Tetrahydroxy-2methylanthraquinone; 5-Me ether, 8-*O*-α-L-rhamnopyranoside, *in* T-60113

102786-92-5 1,3,8-Trihydroxy-5-methoxy-2methylanthraquinone, *in* T-60113

102791-30-0 Ficulinic acid A; (*E*)-*form*, *in* F-60011

102791-31-1 Ficulinic acid B; (*E*)-*form*, *in* F-60012

102795-79-9 Homoazulene-1,5-quinone, H-60085

102795-80-2 Homoazulene-1,7-quinone, H-60086

102795-81-3 Homoazulene-4,7-quinone, H-60087

102819-46-5 3-(2-Ethyl-2-butenyl)-9,10dihydro-1,6,8-trihydroxy-9,10-dioxo-2-anthracenecarboxylic acid, E-70038

102830-05-7 1-Methylazupyrene, M-60043

102831-11-8 2-Amino-1,3-diphenyl1propanone; (±)-*form*, B,HBr, *in* A-60149

102831-18-5 2-Amino-1,3-diphenyl1propanone; (±)-*form*, *N*-Formyl, *in* A-60149

102831-40-3 2-Amino-3-methylenepentanedioic acid; (±)-*form*, B,HCl, *in* A-60219

102831-41-4 2-Amino-3-methylenepentanedioic acid; (±)-*form*, Amide; B,HCl, *in* A-60219

102841-42-9 Mulberroside A, *in* D-70335

102860-22-0 2-(Phenylsulfonyl)-1,3cyclohexadiene, P-70095

102907-96-0 Kayamycin, K-60009

102920-05-8 3-Azidodiamantane, A-60342

102922-02-1 3,4-Dihydro-2(1*H*)-pyrimidinone; 3-β-D-Ribofuranosyl, *in* D-60273

102922-03-2 3,4-Dihydro-2(1*H*)-pyrimidinone; 1-β-D-Ribofuranosyl, *in* D-60273

102961-72-8 Antibiotic A 40926, A-70219

102961-73-9 Antibiotic A 40926; Antibiotic A 40926*A*, *in* A-70219

102961-74-0 Antibiotic A 40926; Antibiotic A 40926*B*, *in* A-70219

102979-48-6 2-Carbomethoxy-3vinylcyclopentene, *in* V-60010

103020-20-8 2,3-Dihydrothieno[2,3-*b*]pyridine, D-70269

103059-96-7 Antibiotic CA 146B, *in* C-70184

103068-16-2 1,3-Di-(1-naphthyl)benzene, D-60454

103068-17-3 1,3-Di(2-naphthyl)benzene, D-60455

103094-97-9 1,5-Diazabicyclo[5.2.2]undecane; *N*-Me, *in* D-60055

103119-78-4 5-Bromophthalazine, B-70267

103129-23-3 ▷(Fluoromethyl)oxirane; (±)-*form*, *in* F-70032

103147-87-1 Maackiasin, M-60001

103202-15-9 Furodysinin hydroperoxide, F-70054

103223-12-7 Brachycarpone, B-60195

103224-59-5 2′,3′,4′,5′-Tetramethyl-6′nitroacetophenone, T-60145

103258-83-9 *N,N′*-[Carbonylbis(oxy)]-bis[2methyl-2-propanamine], *in* C-60015

103258-84-0 2*H*-1,5,2,4-Dioxadiazine-3,6(4*H*)-dione; 2,4-Di-*tert*-butyl, *in* D-60466

103259-13-8 1,8,14-Triptycenetricarboxylic acid, T-60399

103259-15-0 1,8,11-Triptycenetricarboxylic acid, T-60398

103259-19-4 1,8,11-Triptycenetricarboxylic acid; Tri-Me ester, *in* T-60398

103259-20-7 1,8,14-Triptycenetricarboxylic acid; Tri-Me ester, *in* T-60399

103322-21-0 1-Benzylidene-1*H*cyclopropabenzene, B-70084

103322-24-3 2,6-Dihydro-2,6dimethyleneazulene, D-60226

103348-91-0 7-Aminobicyclo[4.1.0]heptane7-carboxylic acid; *trans-form*, *in* A-60095

103367-32-4 5,6-Dihydroxyhexanoic acid; (*S*)-*form*, Me ester, *in* D-60329

103367-47-1 3,4,4-Triphenyl-2-cyclohexen1-one, T-60379

103367-48-2 3,4,4-Triphenyl-2-cyclohexen1-one; Tosylhydrazone, *in* T-60379

103367-50-6 4,5,5-Triphenyl-2-cyclohexen1-one; (±)-*form*, *in* T-60381

103367-54-0 3,5,5-Triphenyl-2-cyclohexen1-one, T-60380

103367-60-8 3-Methyl-4,4diphenylcyclohexene; (±)-*form*, *in* M-60066

103367-81-3 4,4-Dibromo-3-methyl-3-buten2-one, D-70094

103367-82-4 1,1-Dibromo-3,3-dimethoxy2methyl-1-butene, *in* D-70094

103370-21-4 5,6,8,13-Tetrahydro-1,7,9,11tetrahydroxy-8,13dioxo-3-(2-oxopropyl)benzo[*a*]naphthacene-2carboxylic acid, T-70095

103438-45-5 2,6-Diamino-1,5-dihydro-4*H*imidazo[4,5-*c*]pyridin4-one, D-60034

103438-47-7 2,6-Diamino-1,5-dihydro-4*H*imidazo[4,5-*c*]pyridin4-one; Methanesulfonyl, *in* D-60034

103438-84-2 2-Fluoro-5-hydroxybenzaldehyde, F-60036

103438-85-3 4-Fluoro-3-hydroxybenzaldehyde, F-60037

103438-86-4 2-Fluoro-3-hydroxybenzaldehyde, F-60035

103438-90-0 2-Fluoro-3-hydroxybenzaldehyde; Benzyl ether, *in* F-60035

103438-91-1 4-Fluoro-3-hydroxybenzaldehyde; Benzyl ether, *in* F-60037

103438-92-2 2-Fluoro-5-hydroxybenzaldehyde; Benzyl ether, *in* F-60036

103457-83-6 Nonamethylcyclopentanol, N-60057

103457-84-7 Undecamethylcyclohexanol, U-60003

103457-85-8 1,1,2,2,3,3,4,4-Octamethyl-5methylenecyclopentane, O-60027

103457-86-9 1,1,2,2,3,3,4,4,5,5Decamethyl-6-methylenecyclohexane, D-60012

103495-77-8 1,1,3,3,5,5-Hexamethyl-2,4,6triss(ethylene)cyclohexane, H-60072

103495-83-6 1,1,2,2,3,3Hexamethylcyclohexane, H-60070

103499-08-7 5-Phenyl-1,2,4-oxadiazole-3carboxaldehyde; Oxime, *in* P-60122

103499-59-8 2-Mercapto-3-phenylpropanoic acid; (*R*)-*form*, Me ester, *in* M-60026

103499-61-2 2-Mercapto-3-methylbutanoic acid; (*R*)-*form*, S-Benzoyl, *in* M-60023

103500-26-1 3-(1-Aminocyclopropyl)-2propenoic acid; (*E*)-*form*, *N*-Benzyloxycarbonyl, *in* A-60123

103500-27-2 3-(1-Aminocyclopropyl)-2propenoic acid; (*E*)-*form*, B,HCl, *in* A-60123

103521-25-1 Portmicin, P-70111
103528-06-9 Antibiotic SF 2339, A-60289
103528-07-0 Antibiotic TMF 518*D*, A-70240
103538-04-1 Stenosporonic acid, S-70072
103538-05-2 Divaronic acid, D-70528
103538-56-3 13²,17³-Cyclopheophorbide enol, C-60230
103547-83-7 2,2,3,3,4,4Hexamethyl-cyclobutanol, H-60068
103547-87-1 2,2,3,3,4,4,5,5Octamethyl-cyclopentanol, O-60024
103553-93-1 Ebuloside, E-60001
103560-62-9 4-Oxo-2-nonenal; (*E*)-*form, in* O-70097
103562-47-6 3-Mercaptocyclobutanol; *trans-form, in* M-60019
103562-48-7 3-Mercaptocyclobutanol; *cis-form, in* M-60019
103562-49-8 3-Mercaptocyclobutanol; *trans-form, O*-(4-Methyl-benzenesulfonyl), *in* M-60019
103562-50-1 3-Mercaptocyclobutanol; *cis-form, O*-(4-Methylbenzenesulfonyl), *in* M-60019
103562-67-0 3-(Acetylthio)cyclobutanol, *in* M-60019
103562-70-5 Bis(3-hydroxycyclobutyl)disulfide, *in* M-60019
103562-75-0 1,3-Biss(llylthio)cyclobutane, *in* C-60184
103562-76-1 1,3-Cyclobutanedithiol; *cis-form, in* C-60184
103573-06-4 Oxysolavetivone, *in* S-70051
103616-07-5 2-Mercaptopropanoic acid; (*R*)-form, Et ester, *in* M-60027
103616-08-6 2-Mercapto-2-phenylacetic acid; (*S*)-*form, in* M-60024
103620-96-8 4,5-Dihydro-3,3,4,4tetramethyl-2(3*H*)thiophenone, D-60288
103620-99-1 Dihydro-3,3,4,4-tetra-methyl2(3*H*)-furanthione, D-60282
103621-02-9 4,5-Dihydro-3,3,4,4tetramethyl-2(3*H*)thiophenethione, D-60287
103654-23-5 1,5-Dihydroxy-1,7-diphenyl-3heptanone, *in* D-60480
103654-24-6 Yashabushiketodiol B, *in* D-60480
103654-25-7 1,7-Diphenyl-1,3,5-heptanetriol; (1*R*,3*R*,5*S*)-form, *in* D-60480
103654-30-4 Artelein, A-60303
103654-31-5 8-Carotene-4,6,10-triol, C-60021
103665-30-1 [2.2](4,7)(7,4)Indenophane, I-60021
103668-59-3 1-Bromo-(2-dimethoxy-methyl)naphthalene, *in* B-60302
103668-60-6 2-(Dimethoxymethyl)-1naphthalenecarboxaldehyde, *in* N-60001
103692-62-2 9,18-Diphenylphenanthro[9,10*b*]-triphenylene, D-60490
103722-81-2 5-Decene-2,8-diyne; (*E*)-*form, in* D-60015
103724-51-2 1,3-Divinyladamantane, D-70529
103729-38-0 1,7-Diphenyl-3,5-heptanediol; (3*R*,5*R*)-form, *in* D-70481
103729-39-1 1,7-Diphenyl-3,5-heptanediol; (3*S*,5*S*)-form, *in* D-70481
103733-33-1 1,2,3,4-Tetrahydro-1isoquinolinecarboxylic acid; (±)-*form*, Et ester; B, HCl, *in* T-60070
103735-89-3 Galtamycin, G-70003
103744-20-3 4-Hydroxy-3-methoxy-2-methyl9*H*-carbazole-1-carboxaldehyde, H-70154

103744-21-4 4-Hydroxy-3,6-dimethoxy-2methyl-9*H*-carbazole-1carboxaldehyde, *in* H-70154
103744-76-9 3,4-Di-*O*-caffeoylquinic acid; *O*⁵-(3-Hydroxy-3-methylglutaroyl), *in* D-70110
103744-81-6 Rehmaglutin *C*, R-60003
103772-42-5 Cervicol, C-70034
103782-08-7 Allosamidin, A-60076
103794-87-2 Tri-*tert*-butylazete, T-60249
103794-88-3 Tri-*tert*-butyl-1,2,3-triazine, T-60250
103794-91-8 1-Thia-2-cyclooctyne, T-60189
103884-21-5 7-Aminoimidazo[4,5-*f*]quinazolin-9(8*H*)-one, A-60208
103884-30-6 7-Aminoimidazo[4,5-*f*]quinazolin-9(8*H*)-one; B,2HCl, *in* A-60208
103896-92-0 5*H*,7*H*-Dibenzo[*b,g*][1,5]dithiocin; 6-Oxide, *in* D-70069
103896-93-1 5*H*,7*H*-Dibenzo[*b,g*][1,5]dithiocin; 12-Oxide, *in* D-70069
103960-12-9 *b,b′,b″*-Tritriptycene, T-60407
103980-84-3 1-Bromo-1,3-pentadiene; (*E,E*)-*form, in* B-60311
103992-91-2 3α,15α,22S-Triacetoxy-7,9(11),24-lanostatrien-26-oic acid, *in* T-70266
104007-33-2 1,6-Dihydro-1,6dimethyleneazulene, D-60225
104013-99-2 Muamvatin, M-70156
104014-34-8 Dibenz[*b,g*]azocine5,7(6*H*,12*H*)-dione; *N*-Benzyl, *in* D-60068
104014-40-6 Dibenz[*b,g*]azocine5,7(6*H*,12*H*)-dione; *N*-Me, *in* D-60068
104014-78-0 3,5-Dihydro-3,5,5-trimethyl4*H*-triazol-4-one, D-60303
104014-86-0 2,5-Dimethyl-1,4-benzenedithiol, D-60406
104019-20-7 7-Acetylindole, A-60036
104021-41-2 Grenoblone, G-70029
104022-78-8 Pyoverdin *C*, P-70146
104022-79-9 Pyoverdin *D*, P-70147
104051-86-7 Tetrahydro-2(1*H*)-pyrimidinone; 1-β-D-Ribopyranosyl, *in* T-60088
104070-31-7 5-Chloro-3-methyl-1-pentene; (±)-*form, in* C-60117
104090-87-1 1-(2-Naphthyl)piperazine; B,HCl, *in* N-60014
104091-52-3 2,2-Dimethyl-5-cycloheptene1,3-dione, D-60416
104113-37-3 2-Bromobicyclo[2.2.1]heptane1-carboxylic acid; (1*R*,2*R*)-*form, in* B-60210
104113-38-4 2-Bromobicyclo[2.2.1]heptane1-carboxylic acid; (1*S*,2*S*)-*form, in* B-60210
104113-47-5 5a,5b,11a,11bTetrahydrocyclo-buta[1,2-*b*:4,3-*b′*]-benzothiopyran-11,12-dione; (5aα,5bα,11aα,11bβ)-*form, in* T-60063
104113-71-5 1-(1-Naphthyl)piperazine; B,HCl, *in* N-60013
104144-79-8 2,2′-Diisocyano-1,1′-binaphthyl; (*R*)-*form, in* D-60391
104144-80-1 2,2′-Diisocyano-1,1′-binaphthyl; (*S*)-*form, in* D-60391
104155-10-4 12,13-Epoxytrichothec-9-en-3-ol, E-70028
104157-40-6 Methanesulfenyl thiocyanate, M-60032
104169-29-1 Benz[*f*]imidazo[5,1,2-*cd*]indolizine, B-70021
104170-63-0 9-Benzylidene-1,3,5,7cyclo-nonatetraene, B-60069

104172-76-1 ▷7-Azidobicyclo[2.2.1]hepta2,5-diene, A-70285
104184-80-7 7,10:19,22:31,34Triethenotri-benzo[*a,k,u*]cyclotriacontane; (*all-E*)-*form, in* T-60273
104241-29-4 2-Amino-3-hydroxy-4-(4hydroxyphenyl)butanoic acid; (2*S*,3*R*)-*form, in* A-60185
104286-26-2 Pentacyclo[6.4.0.0²,⁷.0³,¹².0⁶,⁹]-dodeca-4,10-diene, P-60028
104291-39-6 4,4,4-Trifluoro-3(trifluoromethyl)-2butenal, T-60309
104292-92-4 5*H*-Thieno[2,3-*c*]pyran-5,7(4*H*)dione, *in* C-60017
104370-29-8 (Triphenylmethyl)hydrazine; B,HCl, *in* T-60388
104371-20-2 1 (Pentafluorophenyl)ethanol; (*S*)-*form, in* P-60040
104371-21-3 1-(Pentafluorophenyl)ethanol; (*R*)-*form, in* P-60040
104384-54-5 1′-Methyl-1,5′-bi-1*H*-pyrazole, *in* B-60119
104387-05-5 3-(2,2-Dimethyl-2*H*-1benzopyran-6-yl)-2propenoic acid, D-70372
104394-22-1 3,3-Dimethyl-5,5-diphenyl-4pentenoic acid, D-60420
104422-32-4 3-(1,3,6-Cycloheptatrien-1yl-2,4,6-cycloheptatrien-1-ylidenemethyl)-1,3,5-cyclo-heptatriene, C-60205
104462-81-9 5a,6,11a,12-Tetra-hydro[1,4]benzoxazino[3,2-*b*][1,4]benzoxazine, T-60059
104487-60-7 Muscarine; (2*S*,4*R*,5*S*)-*form*, Tetraphenylborate, *in* M-60153
104501-59-9 3-(Hydroxyacetyl)pyridine, H-70102
104505-67-1 2-Phenyl-1,5-benzothiazepin4(5*H*)-one; 2,3-Dihydro, *N*-Ac, *in* P-60081
104505-68-2 2-Phenyl-1,5-benzothiazepin4(5*H*)-one; *N*-Ac, *in* P-60081
104505-71-7 2-(2-Phenylethenyl)benzothiazole; (*Z*)-*form, in* P-60093
104519-25-7 3-Bromo-1-phenyl-1-butyne; (±)-*form, in* B-60314
104597-02-6 6,7-Tridecadiene; (±)-*form, in* T-60272
104599-38-4 3(5)-Bromo-5(3)-nitro-1*H*pyrazole, B-60305
104599-40-8 3,4,5-Tribromo-1*H*-pyrazole; 1-Nitro, *in* T-60246
104642-21-9 3*H*-Cyclonona[*def*]biphenylene; (*Z,Z*)-*form, in* C-60215
104690-12-2 Pyreno[4,5-*b*]furan, P-60214
104700-94-9 Salvicin, S-60003
104713-91-9 Tricyclo[8.4.1.1³,⁸]-hexadeca3,5,7,10,12,14-hexa-ene2,9-dione; *syn-form, in* T-60264
104713-92-0 Tricyclo[8.4.1.1³,⁸]-hexadeca3,5,7,10,12,14-hexa-ene2,9-dione; *anti-form, in* T-60264
104716-12-3 Pyrido[2″,1″:2′,3′]imidazo[4′,5′:-4,5]imidazo[1,2-*a*]pyridine; Tetracyano-1,4-quinodimethane complex, *in* P-60234
104716-50-9 Pyrido[2″,1″:2′,3′]imidazo[4′,5′:-4,5]imidazo[1,2-*a*]pyridine, P-60234
104716-55-4 Pyrido[2″,1″:2′,3′]imidazo[4′,5′:-4,5]imidazo[1,2-*a*]pyridine; Picrate, *in* P-60234

104716-56-5 Pyrido[2″,1″:2′,3′]imidazo[4′,5′:-4,5]imidazo[1,2-a]pyridine; B,EtBr, in P-60234

104716-60-1 Pyrido[2″,1″:2′,3′]imidazo[4′,5′:-4,5]imidazo[1,2-a]pyridine; 2,4,7-Trinitrofluorenone complex, in P-60234

104716-67-8 2-Chloro-2,3-dihydro-1Himidazole-4,5-dione; (±)-form, 1,3-Di-Ph, in C-60054

104716-68-9 2,3-Dihydro-2-hydroxy-1Himidazole-4,5-dione; (±)-form, 1,3-Di-Ph, in D-60243

104716-69-0 2,3-Dihydro-2-hydroxy-.1Himidazole-4,5-dione; (±)-form, 1,3-Di-Ph, Me ether, in D-60243

104716-70-3 2,3-Dihydro-2-hydroxy-1Himidazole-4,5-dione; (±)-form, 1,3-Di-Ph, O-Ac, in D-60243

104770-08-3 3-Aminobicyclo[2.2.1]hept-5ene-2-methanol; (1RS,2SR,3SR)-form, N-Me, in A-70094

104770-09-4 3-Aminobicyclo[2.2.1]hept-5ene-2-methanol; (1RS,2RS,3SR)-form, N-Me, in A-70094

104777-60-8 3-Hydroxy-30-nor-12,20(29)oleanadien-28-oic acid; 3β-form, in H-60199

104777-98-2 Tephrobbottin, T-70007

104778-14-5 3-(3,4-Dihydroxybenzylidene)7-hydroxy-4-chromanone, in D-70288

104778-15-6 3-(3,4-Dihydroxybenzyl)-3,7dihydroxy-4-chromanone, in D-70288

104779-70-6 1,2,4,5Tetrakiss(imethylamino)-benzene, in T-60012

104779-77-3 2,2,8,8-Tetramethyl-5,5dihydroxy-3,4,6,7nonanetetrone, in T-60148

104779-79-5 2,2,8,8-Tetramethyl3,4,5,6,7-nonanepentone, T-60148

104779-80-8 1,5-Diphenylpentanepentone, D-60488

104779-82-0 3,3-Dihydroxy-5,5-diphenyl1,2,4,5-pentanetetrone, in D-60488

104793-17-1 1(15),8(19)-Trinervitadiene2β,3α,9β,20-tetrol 9,20-diacetate, in T-70296

104808-79-9 2,3,4,5,8,9-Hexahydrothionin; (E)-form, 1-Oxide, in H-70076

104808-81-3 2-Hexyl-5-methyl-3(2H)furanone; (±)-form, S-Oxide (exo-), in H-60079

104808-82-4 Hexahydro-2H-cyclopenta[b]thiophene; (3aRS,6aRS)-form, in H-60049

104835-53-2 1,2,3,4,5,6,7,8Octahydropyridazino[4,5-d]pyridazine, O-60018

104835-54-3 3,4,5,6-Tetrahydro-4,5biss(ethylene)pyridazine, T-60062

104848-77-3 5,5-Dimethyl-4-methylene-1,2dioxolan-3-one, D-60428

104855-20-1 Pseudopterosin A, P-60187

104855-21-2 Pseudopterosin B, in P-60187

104855-82-5 Rubiflavin E, in R-70011

104870-75-9 5-Methyl-2-(1-methyl-1phenylethyl)cyclohexanol; (1S,2S,5R)-form, in M-60090

104872-75-5 2-Hexyl-5-methyl-3(2H)furanone; (±)-form, S-Oxide (endo-), in H-60079

104875-17-4 11-Methylene-1,5,9triazabicyclo[7.3.3]pentadecane; B,HI, in M-60080

104875-18-5 11-Methylene-1,5,9triazabicyclo[7.3.3]pentadecane, M-60080

104881-78-9 Pseudopterosin C, in P-60187

104900-64-3 Bromotetrasphaerol, B-70272

104910-78-3 1,5-Hexadiene-3,4-dione, H-60039

104933-75-7 (Triphenylmethyl)hydrazine, T-60388

104975-37-3 2-Oxo-4-oxazolidinecarboxylic acid; (±)-form, in O-60082

104984-21-6 Dihydrocyclopenta[c,d]pentalene(2−), D-60216

104993-16-0 20-Acetoxy-2β,3α-dihydroxy1(15),8(19)-trinervitadiene, in T-70299

104995-43-9 Dihydro-4,6-(1H,5H)pyrimidinedione; Di-tert-butyl ether, in D-70256

104995-46-2 2,5-Dihydro-4,6pyrimidinedithiol; Di-Ph thioether, in D-70257

105017-73-0 5-(Hydroxymethyl)-1,2,3,4cyclohexanetetrol; (1R,2S,3S,4R,5S)-form, Penta-Ac, in H-70170

105017-74-1 5-(Hydroxymethyl)-1,2,3,4cyclohexanetetrol; (1R,2S,3S,4R,5S)-form, in H-70170

105017-80-9 5-(Hydroxymethyl)-1,2,3,4cyclohexanetetrol; (1S,2S,3R,4R,5S)-form, Penta-Ac, in H-70170

105018-77-7 5-Methoxy-3-(8,11,14)pentadecatrienyl)1,2,4-benzenetriol; (8Z,11Z)-form, in M-70049

105020-38-0 Methyl 1-methyl-1H-1,2,3triazole-5-carboxylate, in T-60239

105020-39-1 1H-1,2,3-Triazole-4-carboxylic acid; (2H)-form, 2-Me, Me ester, in T-60239

105020-57-3 9,10-Dihydro-9,10dimethyleneanthracene, D-70194

105020-57-3 1,4,5,8-Tetrachloro-9,10anthraquinodimethane, T-60026

105029-41-2 Argiopine, A-60295

105064-77-5 1,4-Dihydrofuro[3,4-d]pyridazine, D-60241

105089-88-1 2-Pyrrolidinecarboxaldehyde; (S)-form, Diethyl dithioacetal, in P-70184

105090-72-0 3-Azido-3-methylbutanoic acid, A-60348

105104-40-3 3-Chloro-2-(methoxymethoxy)1-propene, C-70089

105114-21-4 4-Amino-3-hydroxy-5phenylpentanoic acid, A-60199

105118-17-0 3,4-Dihydroxy-2pyrrolidinecarboxylic acid; (2R,3S,4R)-form, in D-70342

105121-97-9 1,2,3,4-Tetrahydro-4hydroxyisoquinoline; (±)-form, B,HCl, in T-60066

105121-98-0 1,2,3,4-Tetrahydro-4hydroxyisoquinoline; (±)-form, N-Ac, in T-60066

105125-43-7 Di-2-pyridyl sulfite, D-60497

105141-61-5 2-Methyl-2-piperidinecarboxylic acid; (R)-form, in M-60119

105162-50-3 Octaethyltetramethyl[26]porphyrin[3.3.3.3]; Dibromide, in O-60008

105181-84-8 1,2,3,4-Tetrahydro-4hydroxyisoquinoline; (S)-form, in T-60066

105181-85-9 1,2,3,4-Tetrahydro-4hydroxyisoquinoline; (R)-form, in T-60066

105181-86-0 1,2,3,4-Tetrahydro-4hydroxyisoquinoline; (S)-form, B,HCl, in T-60066

105205-50-3 6-Benzoylindole, B-60058

105226-66-2 1,2,3,4-Tetrahydro-3-methyl3-isoquinolinecarboxylic acid; (R)-form, in T-60074

105239-72-3 Sarothralen B, S-60011

105245-47-4 5-Methoxy-3-(phenylthio)-1,4naphthoquinone, in H-70214

105259-49-2 5-Methoxy-2-(phenylthio)-1,4naphthoquinone, in H-70213

105281-33-2 7-Amino-3,10-bisaboladiene; (6R,7R)-form, B,HCl, in A-60096

105281-34-3 7-Amino-3,10-bisaboladiene; (6R,7R)-form, in A-60096

105281-35-4 7-Isocyanato-2,10-bisaboladiene; (6R,7R)-form, in I-60089

105281-36-5 7-Isocyano-3,10-bisaboladiene; (6R,7R)-form, in I-60092

105281-40-1 3-Isocyano-7,9-bisaboladiene; (7E,9E)-form, in I-60091

105281-43-4 7-Amino-3,10-bisaboladiene; (6R,7S)-form, in A-60096

105282-83-5 2,3-Didehydro-1,2-dihydro1,1-dimethylnaphthalene, D-60168

105333-29-7 4-Nitrosobenzaldehyde; Dimer, in N-60046

105335-73-7 Annulin A, A-60276

105335-74-8 Annulin B, A-60277

105343-03-1 Iodovulone I, in B-70278

105343-04-2 Bromovulone I, B-70278

105363-23-3 Tricyclo[18.4.1.1^{8,13}]hexacosa-2,4,6,8,10,12,14,16,18,20,22,24-dodecaene, T-60263

105363-51-7 3,5-Diamino-2,4,6trinitrobenzoic acid; Nitrile, in D-70047

105365-16-0 Spiro[4.4]non-2-ene-1,4-dione, S-60042

105376-48-5 3-Azido-1,2,4-thiadiazole, A-60351

105382-66-9 10,11,14,15-Tetraoxa-1,8diazatricyclo[6.4.4.0^{2,7}]hexadeca-2(7),3,5-triene, T-60162

105400-44-0 2,3,11,12-Dibenzo1,4,7,10,13,16-hexathia-2,11-cyclooctadecadiene, D-60073

105400-80-4 1H-Benzo[de][1,6]naphthyridine; B,HCl, in B-70037

105440-04-8 6H,12H-Dibenzo[b,f][1,5]dithiocin; S-Oxide, in D-70070

105456-38-0 2-Mercaptoethanesulfinic acid; Li salt, in M-60022

105456-49-3 2-Mercaptoethanesulfinic acid, M-60022

105496-34-2 Ovothiol C, in O-70050

105500-09-2 8-Hydroxy-5,9,11,14icosatetraenoic acid, in H-70126

105514-98-5 Antibiotic SS 43405D, A-70239

105542-98-1 1,3-Diiodobicyclo[1.1.1]pentane, D-70357

105553-45-5 2-Methyl-6-methylene-1,3,7octatriene, M-60089

105555-58-6 4,5,5-Triphenyl-2-cyclohexen1-one, T-60381

105562-32-1 5,8,14,17Tetrakiss(imethylamino)[3.3]paracyclophene, T-70126

105575-75-5 (2-Pyridyl)(1-pyrazolyl)methane, P-70178
105579-74-6 Altertoxin III, A-60082
105590-36-1 Distyryl selenide; (*E*,*Z*)-*form*, *in* D-60503
105593-16-6 Gibberellin A$_{67}$, *in* G-70011
105598-27-4 Benzo[1,2-*b*:3,4-*b*′:5,6-*b*″]tripyrazine-2,3,6,7,10,11-tetracarboxylic acid; Hexanitrile, *in* B-60051
105598-29-6 Benzo[1,2-*b*:3,4-*b*′:5,6-*b*″]tripyrazine-2,3,6,7,10,11-tetracarboxylic acid, B-60051
105601-68-1 1,2,3,3*a*,4,5-Hexahydropentalene; (±)-*form*, *in* H-70064
105618-29-9 Benzo[1,2-*b*:3,4-*b*′:5,6-*b*″]tripyrazine-2,3,6,7,10,11-tetracarboxylic acid; Trianhydride, *in* B-60051
105630-54-4 2,3-Dichloro-1-methoxy-4-nitrobenzene, *in* D 70125
105633-41-8 2-Oxo-4-phenylthio-3-butenoic acid; (*Z*)-*form*, *in* O-60088
105633-54-3 2-Oxo-4-phenylthio-3-butenoic acid; (*Z*)-*form*, Et ester, *in* O-60088
105660-81-9 Nepetalactone; (4*aR*,7*R*,7*aS*)-*form*, *in* N-60020
105664-72-0 Imidazo[4,5-*g*]quinazoline4,8,9(3*H*,7*H*)-trione, I-60010
105675-99-8 4-Mercaptocyclohexanecarbonitrile, *in* M-60020
105676-05-9 3-Mercaptocyclopentanecarboxylic acid, M-60021
105676-06-0 4-Mercaptocyclohexanecarboxylic acid; *cis*-*form*, *in* M-60020
105676-07-1 3-Mercaptocyclopentanecarboxylic acid; (1*RS*,3*SR*)-*form*, Amide, *in* M-60021
105676-08-2 4-Mercaptocyclohexanecarboxylic acid; *cis*-*form*, Amide, *in* M-60020
105676-09-3 3-[(Triphenylmethyl)thio]cyclopentanecarboxylic acid, *in* M-60021
105676-10-6 4-[(Triphenylmethyl)thio]cyclohexanecarboxylic acid, *in* M-60020
105676-16-2 2-Thiabicyclo[2.2.2]octan-3-one, T-60187
105694-44-8 7-Bromo-1(3*H*)-isobenzofuranone, B-70225
105694-45-9 4-Iodo-1(3*H*)-isobenzofuranone, I-70044
105694-46-0 7-Iodo-1(3*H*)-isobenzofuranone, I-70045
105728-31-2 5-Iodo-2-nitrobenzaldehyde, I-60052
105729-06-4 2-(Methoxyacetyl)pyridine, *in* H-70101
105729-07-5 2,6-Biss(ethoxyacetyl)pyridine, *in* B-70144
105729-09-7 2-(Methoxyacetyl)thiophene, *in* H-70103
105754-75-4 29,29,30,30Tetracyanobianthraquinodimethane, T-60030
105763-01-7 1,13-Dioxa-4,7,10,16,20,24hexaazacyclohexacosane, D-70459
105798-67-2 12,20,25-Trihydroxy-23dammaren-3-one; (12β,20*S*,23*E*)-*form*, *in* T-60319
105798-74-1 Clitocine, C-60155
105814-93-5 Pityol; (2*S*,5*R*)-*form*, *in* P-60160
105814-94-6 Pityol; (2*R*,5*S*)-*form*, *in* P-60160
105822-01-3 25-Dammarene-3,12,17,20,24pentol; (3α,12β,17*R*,20*S*,24ξ)-*form*, *in* D-60001

105858-44-4 1,4-Diethynylbicyclo[2.2.2.]octane, D-70157
105858-62-6 3,7-Di-*tert*-butyl-9,10dimethyl-2,6-anthraquinone, D-60113
105858-77-3 1,2-Biss(henylsulfonyl)cyclobutene, *in* B-70156
105858-78-4 1-(Phenylsulfinyl)-2(phenylsulfonyl)cyclobutene, *in* B-70156
105858-81-9 1-(Phenylsulfinyl)-2(phenylthio)cyclobutene, *in* B-70156
105880-10-2 Venustatriol, *in* T-70193
105888-54-8 Pyrroxamycin, P-70203
105918-54-5 6-Diazo-5-oxo-1,3cyclohexadiene-1carboxylic acid, D-70063
105958-17-6 *N*-Pyridinium-2-benzimidazole, P-70172
105958 33-6 *N*-Pyridinium-2-benzimidazole; B,HCl, *in* P-70172
105969-64-0 Trunculin *A*, T-70408
105969-65-1 Trunculin *B*, T-70409
105977-35-3 11-Hydroxy-12,13-epoxy-9,15octadecadienoic acid; (9*Z*,11*S*,12*S*,13*S*,12*Z*)-*form*, *in* H-70127
105997-04-4 Galtamycinone, *in* G-70003
106001-28-9 3-Ethyl-2,19dioxabicyclo[16.3.1]docosa-3,6,9,18(22),21pentaen-12-yn-20-one, E-70042
106008-69-9 Tiacumicin *F*, T-70198
106008-70-2 Tiacumicin *C*, T-70195
106009-81-8 17-Hydroxy-15,17-oxido-16spongianone; 17β-*form*, *in* H-60201
106009-90-9 4,21-Dibromo-3-ethyl-2,19dioxabicyclo[16.3.1]docosa-6,9,18(22),21tetraen-12-yn-20-one, *in* E-70042
106009-91-0 Cycloorbigenin, C-60218
106016-50-6 2,4,5-Triphenyl-1,3-dioxol-1ium; Trifluoromethanesulfonate, *in* T-70307
106016-52-8 4,5-Dimethyl-2-phenyl-1,3dioxol-1-ium; Trifluoromethanesulfonate, *in* D-70431
106016-54-0 2,5-Diphenyl-1,3-dioxol-1-ium; Trifluoromethanesulfonate, *in* D-70478
106020-39-7 Benzo[1,2-*b*:3,4-*b*′:6,5-*b*″]tris[1]benzothiophene, B-60053
106034-49-5 11-Hydroxy-12,13-epoxy-9,15octadecadienoic acid; (9*Z*,11*R*,12*S*,13*S*,15*Z*)-*form*, *in* H-70127
106047-28-3 6-Bromo-2,3′-bipyridine, B-70192
106047-29-4 6-Bromo-2,4′-bipyridine, B-70193
106047-33-0 5-Bromo-2,4′-bipyridine, B-70188
106047-38-5 5-Bromo-3,4′-bipyridine, B-70190
106058-89-3 3,4-Dihydroxybutanoic acid; (*R*)-*form*, *O*^{4}-*tert*-Butyl, Et ester, *in* D-60313
106058-93-9 3,4-Dihydroxybutanoic acid; (*R*)-*form*, *O*^{4}-*tert*-Butyl, Et ester, 3,5-dinitrobenzoyl, *in* D-60313
106062-83-3 5,14,15-Trihydroxy-6,8,10,12eicosatetraenoic acid; (5*S*,6*E*,8*Z*,10*E*,12*E*,14*R*,15*S*)-*form*, *in* T-60320
106073-70-5 2-Amino-3-chlorobutanedioic acid; (2*R*,3*S*)-*form*, *in* A-70121
106128-88-5 3-(2,4,5-Trimethoxyphenyl)-2propenal, T-70278
106139-18-8 11,12Dioxahexacyclo[6.2.1.1^{3,6}.0^{2,7}.0^{4,10}.0^{5,9}]dodecane, D-70461

106139-20-2 5,12Dioxatetracyclo[7.2.1.0^{4,11}.0^{6,10}]dodeca-2,7-diene, D-70463
106140-37-8 2-Azabicyclo[3.3.1]nonan-7-one; B,HCl, *in* A-70276
106212-04-8 3′,4′-Deoxypsorospermin; (*R*)-*form*, *in* D-60021
106212-05-9 3′,4′-Deoxypsorospermin3′,4′-diol, *in* D-60021
106230-68-6 *O*^{5}-Methyl-3′,4′deoxypsorospermin-3′-ol, *in* D-60021
106231-25-8 9,11-Dihydrogracilin *A*, *in* G-60033
106231-26-9 Membranolide, M-60016
106231-29-2 Rubifolide, R-60016
106289-05-8 *N*-(1,4-Dihydro-4-oxo-5pyrimidinyl)formamide, *in* A-70196
106293-82-7 Gersemolide, G-60015
106310-35-4 Repensolide, R-60005
106310-45-6 Arctodecurrolide, *in* D-60325
106310-46-7 3,5-Dihydroxy-4(15),10(14)guaiadien-12,8olide; (1α,3β,5α,8α,11*R*)-*form*, *in* D-60325
106310-47-8 3,5-Dihydroxy-4(15),10(14)guaiadien-12,8olide; (1α,3β,5α,8α,11*R*)-*form*, 10α,14-Epoxide, *in* D-60325
106312-36-1 4-Methoxy-2-(trifluoromethyl)benzaldehyde, *in* H-70228
106318-66-5 2-Acetylpiperidine; (±)-*form*, B,HCl, *in* A-70051
106318-67-6 2-Acetylpiperidine; (±)-*form*, Oxime (*E*-), *in* A-70051
106318-86-9 2-Acetylpiperidine; (±)-*form*, A-70051
106318-87-0 2-Acetylpiperidine; (±)-*form*, Ethylene ketal, *in* A-70051
106325-28-4 1,3-Diethynyladamantane, D-60172
106327-62-2 Abbottin, A-70002
106335-91-5 6-Iodo-5-hexen-1-ol; (*Z*)-*form*, *in* I-70039
106335-92-6 6-Iodo-5-hexyn-1-ol, I-70041
106367-37-7 2,5-Dihydro-2,2,5tetramethyl-3-[[[(2phenyl-3*H*-indol-3ylidene)amino]oxy]carbonyl]-1*H*-pyrrol-1yloxy, D-60284
106387-02-4 Cassigarol *A*, C-70025
106403-98-9 3-Bromo-1,2-diphenylpropene; (*Z*)-*form*, *in* B-70209
106435-53-4 2-(Chloromethyl)quinoxaline, C-70120
106484-46-2 Thia[2.2]metacyclophane, T-60194
106518-62-1 15,26,27-Trihydroxy-7,9(11),24lanostatrien-3-one; 15α-*form*, *in* T-60330
106533-38-4 3,8-Dihydro-6,6′;7,3′adiliguetilide, D-60223
106533-41-9 Aloenin *B*, A-60077
106533-43-1 Fercoperol, F-60001
106533-44-2 Sigmoidin *D*, S-60031
106533-72-6 Isodidymic acid, I-60098
106534-60-5 Ircinic acid, I-60082
106539-42-8 6-Heptyne-2,5-diamine; (2*R*,5*S*)-*form*, *in* H-60026
106539-45-1 6-Heptyne-2,5-diamine; (2*R*,5*R*)-*form*, B,2HCl, *in* H-60026
106539-46-2 6-Heptyne-2,5-diamine; (2*R*,5*S*)-*form*, B,2HCl, *in* H-60026
106541-96-2 14-Hydroxy-12-tetradecenoic acid; (*Z*)-*form*, *in* H-60230
106550-54-3 2,3,4,5-Tetrachloro-1,1dihydro-1-[(methoxycarbonyl)imino]thiophene, T-60028

106551-87-5 1-(1-Isoquinolinyl)-1-(2pyridinyl)-ethanol; (±)-*form*, Me ether, *in* I-60134

106556-70-1 1,2-Dihydro-2,2-dimethyl-3*H*pyrrol-3-one, D-60230

106565-95-1 γ-Glutamylmarasmine; (2*R*,2'*S*,4*S*)-*form*, *in* G-60026

106565-99-5 γ-Glutamylmarasmine; (2*R*,2'*S*,4*R*)-*form*, *in* G-60026

106566-98-7 Chlororepdiolide, C-60133

106567-40-2 3-Amino-2,4,6-trimethylbenzoic acid, A-60266

106567-41-3 2,4,6-Trimethyl-3-nitrobenzoic acid, T-60370

106575-40-0 7-Phenyl-3-heptyn-2-ol, P-60107

106611-74-9 2-[3,7,11,15,19,23Hexamethyl-25-(2,6,6trimethyl-2-cyclohexenyl)-pentacosa-2,14,18,22-tetraenyl]-3-methyl-1,4-naphthoquinone, H-70084

106621-85-6 Gracilin C, *in* G-60034

106623-23-8 3,10-Dihydroxy-5,11dielmenthadiene-4,9-dione, D-60316

106623-38-5 1-(1-Isoquinolinyl)-1-(2pyridinyl)-ethanol; (*R*)-*form*, *in* I-60134

106623-39-6 1-(1-Isoquinolinyl)-1-(2pyridinyl)-ethanol; (*S*)-*form*, Me ether, *in* I-60134

106623-40-9 1-[1-Methoxy-1-(2-pyridinyl)ethyl]isoquinoline, *in* I-60134

106624-74-2 1-(1-Isoquinolinyl)-1-(2pyridinyl)-ethanol; (*S*)-*form*, *in* I-60134

106680-46-0 Nonachlorophenalenyl, N-60055

106689-41-2 4-Hydrazino-2(1*H*)-pyridinone, H-60093

106689-46-7 2,5-Dihydro-1*H*-dipyrido[4,3*b*:3',4'-*d*]pyrrol-1-one, D-60237

106709-91-5 Deacetoxybrachycarpone, *in* B-60195

106726-77-6 Tetra(2-triptycyl)ethylene, T-60169

106753-87-1 10-Hydroxy-11-dodecenoic acid; (±)-*form*, Me ester, *in* H-60125

106753-88-2 12-Hydroxy-13-tetradecenoic acid; (±)-*form*, Me ester, *in* H-60229

106753-90-6 12-Hydroxy-10-dodecenoic acid; (*E*)-*form*, Me ester, *in* H-60126

106753-93-9 14-Hydroxy-12-tetradecenoic acid; (*E*)-*form*, *in* H-60230

106780-28-3 2,4,6,8-Tetramethyl-2,4,6,8tetra-azabicyclo[3.3.0]octane, *in* O-70015

106780-30-7 Octahydroimidazo[4,5-*d*]imidazole; (3a*RS*,6a*SR*)-*form*, 1,3,4,6-Tetraisopropyl, *in* O-70015

106780-32-9 Octahydroimidazo[4,5-*d*]imidazole; (3a*RS*,6a*SR*)-*form*, 1,3,4,6-Tetra-Ac, *in* O-70015

106780-46-5 1,2,3,6-Tetrahydro-2,6-dioxo4-pyrimidineacetic acid; *N*-Ethylamide, *in* T-70060

106821-88-9 Hexabutylbenzene, H-60032

106848-38-8 4(5)-Bromo-1*H*-imidazole; 1-Benzyl, *in* B-70221

106848-46-8 4,5-Dibromo-1*H*-imidazole-2carboxaldehyde, D-60101

106849-36-9 Gracilin D, *in* G-60034

106851-31-4 6-Phenyl-1*H*-indole, P-60113

106865-52-5 2-[(Phenylthio)-methyl]pyrrolidine, *in* P-70185

106865-55-8 2-[(Methylthio)-methyl]pyrrolidine, *in* P-70185

106868-03-5 4-Methylene-1,2,3,5tetra-phenylbicyclo[3.1.0]hex-2-ene, M-60079

106875-07-4 3,3-Dimethyl-2,4(1*H*,3*H*)quinolinedione, D-60451

106880-56-2 10,11-Methano-1*H*-benzo[5,6]cycloocta[1,2,3,4-*def*]fluorene-1,14-dione, M-70039

106881-87-2 3*a*,7*a*-Methano-1*H*-indole; *N*-Ac, *in* M-70041

106896-48-4 3-Amino-2-bromobenzoic acid; Me ester, *in* A-70097

106911-04-0 3-Bromo[1,2,3]triazolo[1,5-*a*]pyridine, B-60329

106913-64-8 5-Amino-4(1*H*)-pyrimidinone; B,HCl, *in* A-70196

106924-75-8 6-Bromo-6-dodecene; (*E*)-*form*, *in* B-60238

106924-82-7 6-Bromo-6-dodecene; (*Z*)-*form*, *in* B-60238

106941-33-7 Cantabradienic acid; Me ester, *in* C-60009

106941-34-8 Cantabrenolic acid; 5-Ketone, 6α,7α-epoxide, Me ester, *in* C-60010

106941-35-9 Cantabrenolic acid; 5-Ketone, Me ester, *in* C-60010

106941-36-0 Cantabrenolic acid; Me ester, *in* C-60010

106941-40-6 Cantabrenonic acid, *in* C-60010

106941-41-7 Cantabradienic acid, C-60009

106941-42-8 Epoxycantabronic acid, *in* C-60010

106941-43-9 Cantabrenolic acid, C-60010

106982-90-5 Ternatin, T-70008

107010-23-1 1,2,2,3,3,4,4,5-Octamethyl-6oxabicyclo[3.1.0]hexane, O-70021

107010-25-3 1,1,2,2,4,4-Hexamethyl-3,5bis-s(ethylene)cyclopentane, H-70082

107011-08-5 2,2'-Bis(2-imidazolyl)biphenyl, B-70148

107021-10-3 Bryostatin 12, B-70287

107021-11-4 Bryostatin 13, B-70288

107021-64-7 Boxazomycin *A*, B-70176

107021-65-8 Boxazomycin B, *in* B-70176

107021-66-9 Boxazomycin C, *in* B-70176

107040-99-3 15-Anhydrothyrsiferyl diacetate, *in* A-60274

107065-86-1 15(28)-Anhydrothyrsiferyl diacetate, *in* A-60274

107098-23-7 1,13-Dioxa-4,7,10,16,19,22hexa-azacyclotetracosane; 7,19-Bis-s(minoethyl); B,HCl, *in* D-70460

107128-58-5 3-Azidobenzophenone, A-60339

107135-39-7 3,4-Dihydro-4-selenoxo-2(1*H*)quinazolinone, D-60280

107140-30-7 Kasuzamycin *B*, K-70007

107159-83-1 13(18)-Lupen-3-ol; 3β-*form*, *in* L-70043

107170-30-9 1,5,9-Cyclododecatrien-3-yne; (1*Z*,5*E*,9*Z*)-*form*, *in* C-60190

107175-47-3 Esperamicin *X*, E-70035

107175-49-5 7-Iodo-1-heptene, I-70036

107182-80-9 2,3-Dihydro-2-selenoxo-4(1*H*)quinazolinone, D-60281

107191-86-6 Stacopin P2, *in* S-60051

107191-87-7 Stacopin P1, S-60051

107195-81-3 Abiesonic acid; Di-Me ester, *in* A-60005

107195-86-8 Abiesonic acid, A-60005

107221-65-8 6α,7α-Epoxy-5α,14α,20*R*trihydroxy-1-oxo22*R*,24*S*,25*R*-with-2enolide, *in* I-60143

107263-95-6 1-Fluoropyridinium; Trifluoromethanesulfonate, *in* F-60066

107264-09-5 1-Fluoropyridinium; Tetra-fluoroborate, *in* F-60066

107264-11-9 1-Fluoropyridinium; Perchlorate, *in* F-60066

107264-12-0 1-Fluoropyridinium; Hexafluoroantimonate, *in* F-60066

107264-13-1 1-Fluoropyridinium; Perfluorobutanesulfonate, *in* F-60066

107288-22-2 Neurotoxin NSTX 3, N-70033

107299-36-5 5,13-Dimethyl[2.2]metacyclo-phane; *syn-form*, *in* D-60426

107316-88-1 Demethylzeylasteral, *in* D-70027

107321-16-4 1,5,9,13Tetraazacyclohexadecane-1,5,9,13-tetraacetic acid, T-60016

107354-37-0 1,3,5Cyclohexa-netricarboxaldehyde; (1α,3α,5α)-*form*, *in* C-70226

107365-21-9 2-Fluoro-2-phenylpropanal, F-60063

107365-22-0 2-Fluoro-2,2diphenylacetaldehyde, F-60033

107365-23-1 2-Fluoro-3-phenylpropanal, F-60064

107373-23-9 5,20-Dihydroxy-6,8,11,14eicosatetraenoic acid; (5*S*,6*E*,8*Z*,11*Z*,14*Z*)-*form*, *in* D-70296

107383-92-6 3,5,12Triazatetra-cyclo[5.3.1.1[2,6].0[4,9]]dodecane; 3,5,12-Tri-Me, *in* T-70206

107383-93-7 3,5,12Triazatetra-cyclo[5.3.1.1[2,6].0[4,9]]dodecane; 3,5,12-Tri-Et, *in* T-70206

107383-98-2 1,3,5Cyclohexa-netricarboxaldehyde; (1α,3α,5α)-*form*, Tris-s(henylhydrazone), *in* C-70226

107383-99-3 1,3,5Cyclohexa-netricarboxaldehyde; (1α,3α,5α)-*form*, Trishydrazone, *in* C-70226

107384-03-2 1,3,5-Triss(imethoxymethyl)cyclo-hexane, *in* C-70226

107389-91-3 4-Hydroxy-4-(2-hydroxyethyl)cyclohexanone, H-60151

107390-08-9 Garcinone *D*, G-70004

107395-29-9 Methylallosamidin, *in* A-60076

107400-83-9 1,2-Dinitrocyclohexene, D-70451

107408-20-8 1-Iodo-5,5-dimethoxypentene, *in* I-70052

107408-21-9 5-Iodo-4-pentenal; (*Z*)-*form*, *in* I-70052

107408-35-5 7-Chloro-2-heptenoic acid; (*E*)-*form*, Et ester, *in* C-70069

107408-92-4 3,4-Dihydro-3-fluoro-2*H*-1,5benzodioxepin, D-70209

107427-51-0 2,3-Biss(ibromomethyl)pyridine, B-60151

107427-52-1 2,3-Biss(ibromomethyl)pyrazine, B-60150

107427-53-2 4,5-Bis-s(ibromomethyl)pyrimidine, B-60153

107427-57-6 Pyrido[2',3':3,4]cyclobuta[1,2-*g*]-quinoline, P-60232

107427-58-7 Pyrido[3',2':3,4]cyclobuta[1,2-*g*]-quinoline, P-60233

107427-60-1 Pyrazino[2',3':3,4]cyclobuta[1,2-*g*]quinoxaline, P-60198

107427-65-6 Benzo[g]quinoline-6,9-dione, B-60045

107427-66-7 Benzo[g]quinoxaline-6,9-dione, B-60046

107427-67-8 Benzo[g]quinazoline-6,9-dione, B-60044

107445-08-9 3,4-Biss(ibromomethyl)pyridine, B-60152

107465-26-9 7-Bromo[1,2,3]triazolo[1,5-*a*]pyridine, B-60330

107474-49-7 3,6-Diamino-1,2-benzenedithiol; B,2HI, *in* D-70036

107474-49-7 1,4-Diamino-2,3-benzenedithiol; B,2HI, *in* D-60030

107474-50-0 2,3-Biss(ethylthio)-1,4benzenediamine, *in* D-70036

107474-50-0 2,3-Biss(ethylthio)-1,4benzenediamine, *in* D-60030

107474-53-3 2,3,5,6-Tetrakiss(ethylthio)1,4-benzenediamine, *in* D-60031

107484-53-7 Cynajapogenin *A*, C-70257

107484-59-3 Tubocapside A, *in* T-70333

107484-60-6 Tubocapside B, *in* T-70333

107496-54-8 3,3-Difluorocyclobutanecarboxylic acid, D-70164

107514-60-3 4-Hydroxy-2*H*-1benzothiopyran-2-one, H-70110

107530-18-7 Antibiotic FR 900359, A-70223

107550-91-4 [1,2,4,5]Tetrazino[3,4-*b*:6,1*b'*]-bisbenzothiazole, T-60171

107550-92-5 Dibenzo[*cd*:*c'd'*][1,2,4,5]tetra-zino[1,6-*a*:4,3-*a'*]diindole, D-60076

107575-48-4 [4]Metacyclophane, M-60029

107575-50-8 4,5,6,7-Tetrahydro-1methylene-1*H*-indene, T-70073

107584-83-8 3,23-Dioxo-7,24-lanostadien26-oic acid; (9β,24*E*)-*form*, *in* D-60472

107584-84-9 3,23-Dioxo-7,24-lanostadien26-oic acid; (9β,24*Z*)-*form*, *in* D-60472

107602-74-4 6β,8β,9α,14-Tetraacetoxy-1αbenzoyloxy-4β-hydroxydihydro-β-agarofuran, *in* E-60018

107602-75-5 9α,14-Diacetoxy-1αbenzoyloxy-4β,6β,8β,trihydroxydihydro-βagarofuran, *in* E-60018

107602-76-6 6β,9α,14-Triacetoxy-1αbenzoyloxy-4β-hydroxy8-oxodihydro-β-agarofuran, *in* E-60018

107602-77-7 6β,9α,14-Triacetoxy-1αbenzoyloxy-4β,8β,dihydroxydihydro-βagarofuran, *in* E-60018

107602-78-8 9α,14-Diacetoxy-1α,8β,dibenzoyloxy-4β,8β,dihydroxydihydro-βagarofuran, *in* E-60018

107616-06-8 5-Chloro-1,3-cycloheptadiene, C-70062

107617-00-5 2-Methylenecyclopropaneacetic acid; (±)-*form*, *in* M-70067

107678-82-0 6-Amino-8-nitroquinoline, A-60234

107678-85-3 Aminoiminomethanesulfonic acid; *N*,*N*-Di-Ph, *in* A-60210

107693-87-8 1,3-Dihydroxy-12-ursen-28-oic acid; (1β,3β)-*form*, *in* D-60383

107708-98-5 4-Hydroxy-2-methyl-2cyclopenten-1-one; (±)-*form*, *in* H-60175

107711-30-8 Homocytidine, *in* A-70130

107748-88-9 Coralloidolide *A*, C-60164

107748-93-6 Heteronemin; 12-Ac, *in* H-60030

107748-94-7 Heteronemin; 12-Epimer, 12-Ac, *in* H-60030

107783-46-0 Henricine, H-60017

107784-07-6 7-Oxatricyclo[4.1.1.0²,⁵]octane; (1α,2β,5β,6α)-*form*, *in* O-60053

107798-57-2 Quino[7,8-*h*]-quinoline4,9(1*H*,12*H*)dione, Q-60007

107798-86-7 1-Benzothiepin-5(4*H*)-one, B-60047

107812-57-7 Guayulin C, *in* A-60299

107812-58-8 Guayulin D, *in* A-60299

107825-31-0 1,2-Bis(4'-benzo-15-crown-5)ethene; (*E*)-*form*, *in* B-60142

107825-33-2 1,2-Bis(4'-benzo-15-crown-5)ethene; (*Z*)-*form*, *in* B-60142

107826-05-1 4,7-Dihydro-4,7ethanoisobenzofuran, D-60240

107833-73-8 1,1-Diethoxy-3-nitropropane, *in* N-60037

107841-93-0 2-Methyl-6-methylene-1,3,7octatriene; (*E*)-*form*, *in* M-60089

107847-11-0 2-Methoxybutanal, *in* H-70112

107847-15-4 3-Methyl-2-phenylbutanal; (±)-*form*, *in* M-70109

107869-22-7 3,16,24,25,30-Cycloartanepentol; (3β,16β,24*S*)-*form*, *in* C-60182

107882-40-6 Podoverine C, P-60165

107882-41-7 Podoverine B, *in* P-60165

107882-43-9 Podoverine *A*, P-60164

107882-65-5 6'-*O*-Apiosylebuloside, *in* E-60001

107900-70-9 Ferulinolone, F-60006

107902-33-0 1-(2-Furanyl)-2-(2-pyrrolyl)ethyl-ene; (*E*)-*form*, *in* F-60088

107902-35-2 1-(2-Pyrrolyl)-2-(2-thienyl)ethyl-ene; (*E*)-*form*, *in* P-60250

107902-36-3 1-(2-Pyrrolyl)-2-(2-thienyl)ethyl-ene; (*Z*)-*form*, *in* P-60250

107942-05-2 2-Amino-5-guanidino-3hydroxypentanoic acid; (2*S*,3*R*)-*form*, B,HCl, *in* A-70149

107942-06-3 2-Amino-5-guanidino-3hydroxypentanoic acid; (2*S*,3*S*)-*form*, B,HCl, *in* A-70149

107959-84-2 *N*-[15-(β-D-Glucopyranosyloxy)8-hydroxypalmitoyl]taurine, G-70019

107971-15-3 Benzo[3,4]phenanthro[1,2-*b*]thiophene, B-60039

107971-19-7 Benzo[3,4]phenanthro[2,1-*b*]thiophene, B-60040

107976-45-4 2,7-Dihydroxy-2,4,6cyclo-heptatrien-1-one; Di-Ac, *in* D-60314

108005-21-6 2,3,4,5Tetrahydrothiepino[2,3*b*]-pyridine, T-60098

108026-93-3 3α,15α-Diacetoxy-7,9(11),24*E*lanostatrien-26-oic acid, *in* D-70314

108026-94-4 3α-Acetoxy-15α-hydroxy7,9(11),24*E*-lanostatrien-26-oic acid, *in* D-70314

108027-02-7 1,3,5,8-Tetrahydroxy-2methyl-anthraquinone, T-60113

108044-04-8 Ambonin, *in* D-60340

108044-05-9 Ambocin, *in* T-60324

108044-10-6 7-Oxodihydromelinol, O-60064

108044-16-2 Isonimbinolide, I-70072

108044-18-4 8,12-Epoxy-3,7,11eudesmatrien-1-ol; 1β-*form*, Ac, *in* E-60020

108044-19-5 1β-Acetoxy-8β-hydroxy-3,7(11)eudesmadien-12,8-olide, *in* D-60322

108044-20-8 1-Oxo-7(11)-eudesmen-12,8-olide; 8α-*form*, *in* O-60067

108046-24-8 4-Amino-5,6-dihydro-4*H*cyclopenta[*b*]thiophene, A-60137

108046-27-1 4-Amino-5,6-dihydro-4*H*cyclopenta[*b*]thiophene; (±)-*form*, B,HCl, *in* A-60137

108058-71-5 3*H*-Cyclonona[*def*]biphenylen3-one; (*Z*,*Z*)-*form*, *in* C-70230

108060-85-1 *N²*,3-Ethenoguanosine, *in* I-70006

108065-95-8 Antibiotic PB 5266*B*, A-70226

108065-96-9 Antibiotic PB 5266*C*, A-70227

108069-00-7 Neobacin, *in* T-60324

108069-01-8 Neobanin, *in* D-60340

108069-03-0 Arborone, A-60294

108073-62-7 4-Hydroxy-3,6-dimethoxy-1,2dimethylcarbazole, *in* T-70253

108073-63-8 3,4,6-Trimethoxy-1,2dimethylcarbazole, *in* T-70253

108073-64-9 Abbeymycin, A-70001

108074-44-8 4-(2-Nitroethenyl)-1,2benzenediol; (*E*)-*form*, *in* N-70049

108079-34-1 Fluorantheno[3,4-*cd*]-1,2dithiole, F-60017

108079-35-2 Fluorantheno[3,4-*cd*]-1,2diselenole, F-60015

108079-36-3 Fluorantheno[3,4-*cd*]-1,2ditellurole, F-60016

108101-63-9 2-Amino-5-chloro-6-hydroxy-4hexenoic acid; (*S*,*Z*)-*form*, *in* A-60110

108112-82-9 Antibiotic SS 21020*C*, A-70238

108120-13-4 2,6-Diamino-4,7(3*H*,8*H*)pteridinedione, D-60047

108157-59-1 Ixocarpanolide, I-60143

108161-72-4 1-Iodo-3-pentanol, I-60055

108176-13-2 3,3-Dimethyl-1-cyclopropene1-carboxylic acid, D-60417

108185-85-9 1,2,3,4-Tetrahydro-7*H*benzocyclo-hepten-7-one, T-60057

108186-05-6 1,2-Dibromocyclopropene, D-60094

108186-13-6 1-*tert*-Butyl-1,2-cyclooctadiene, B-60344

108189-46-4 Tricyclo[4.4.4.0¹,⁶]tetradeca-2,4,7,9,10,12-hexaene, T-70232

108195-55-7 Abiesolidic acid; Me ester, *in* A-70003

108211-37-6 Dihydro-5-(tri-fluoromethyl)2(3*H*)-furanone; (*R*)-*form*, *in* D-60300

108212-12-0 Aplysiapyranoid B, *in* A-70244

108212-75-5 Calichemicin γ₁¹, C-70008

108213-73-6 3,6-Dimethoxy-1,2-benzoquinone, *in* D-60309

108263-91-8 2,2-Di-*tert*-butyl-3-(di-*tert*butyl-methylene)cyclopropanone, D-70101

108274-04-0 3,4-Dihydro-4,6,7-trimethyl3-β-D-ribofuranosyl-9*H*imidazo[1,2-*a*]purin-9-one, D-60302

108279-65-8 6-Chloro-4-methoxy-2(1*H*)pyridinone, *in* C-70074

108279-66-9 4,6-Dimethoxy-2(1*H*)-pyridinone, *in* D-70341

108279-89-6 2-Chloro-4,6-dimethoxypyridine, *in* C-70074

108288-44-4 2-Amino-4*H*-3,1-benzothiazine; *N*-Et, *in* A-70092

108288-46-6 2-Amino-4*H*-3,1-benzothiazine; *N*-Benzyl, *in* A-70092

108288-67-1 2-Phenyl-4*H*-3,1,2benzoxathiazine, P-60080

108293-08-9 2-Octadecylicosanoic acid, O-70005

108295-47-2 Heritol, H-60028

108298-35-7 5-Hydroxy-3-methyl-1,2benzenedicarboxylic acid; Me ether, di-Me ester, *in* H-70162

108306-55-4 2,4-Diiodothiazole, D-60388

108306-56-5 4-Bromo-2-iodothiazole, B-60289

108306-60-1 4-Iodothiazole, I-60065

108306-61-2 5-Iodothiazole, I-60066

108306-62-3 2,5-Diiodothiazole, D-60389

108306-63-4 2-Bromo-5-iodothiazole, B-60288

108306-64-5 5-Bromo-2-iodothiazole, B-60290

108332-47-4 Dibenzo-54-crown-18, D-70068

108335-06-4 4-Methoxy-2,5-phenanthrenediol, *in* P-60071

108343-55-1 Fuligorubin *A*, F-60079

108350-37-4 2-Iodopentanal, I-60054

108350-39-6 2-Iodo-1-phenyl-1-butanone, I-60060

108351-20-8 Mulundocandin, M-70157

108351-33-3 Antibiotic PC 766*B'*, *in* A-70228

108354-12-7 Aurachin B, A-60314

108354-13-8 Aurachin *D*, A-**60315**
108354-14-9 Aurachin C, *in* A-**60315**
108354-15-0 Aurachin *A*, A-**60313**
108354-43-4 1,5-Dioxiranyl-1,2,3,4,5penta-
 nepentol, D-**70465**
108375-77-5 Antibiotic PC 766*B*, A-**70228**
108388-06-3 Perfluoromethanesulfonimide,
 P-**70042**
108402-49-9 7-Hydro-8methyl-
 pteroylglutamylglutamic acid,
 H-**60096**
108403-43-6 Viscutin 1, *in* T-**60103**
108403-44-7 Viscutin 2, *in* T-**60103**
108403-45-8 Viscutin 3, *in* T-**60103**
108404-40-6 Pentaisopropylidenecyclopentane,
 P-**60053**
108415-79-8 Phenyl styryl selenoxide, *in*
 P-**70070**
108415-80-1 [2-(Methylseleninyl)-
 ethenyl]benzene, *in* M-**70132**
108415-81-2 [(2-Phenylethenyl)seleno]benzene;
 (*Z*)-*form*, *S*-Oxide, *in* P-**70070**
108415-82-3 [(2-(Methylseleno)-
 ethenyl]benzene; (*Z*)-*form*, *S*-
 Oxide, *in* M-**70132**
108415-83-4 Phenyl styryl selenone, *in* P-**70070**
108415-84-5 [2-(Methylselenonyl)-
 ethenyl]benzene, *in* M-**70132**
108415-85-6 [(2-Phenylethenyl)seleno]benzene;
 (*Z*)-*form*, *S*-Dioxide, *in* P-**70070**
108415-86-7 [2-(Methylseleno)-
 ethenyl]benzene; (*Z*)-*form*, *S*-
 Dioxide, *in* M-**70132**
108418-13-9 Ovothiol *A*, O-**70050**
108418-14-0 Ovothiol B, *in* O-**70050**
108418-48-0 Bi(1,3-ditellurol-2-ylidene),
 B-**70117**
108425-53-2 4,5-Heptadien-3-one, H-**70018**
108428-42-8 2-Amino-1-hydroxy-1cyclo-
 butaneacetic acid; (1*S*,2*S*)-
 form, *in* A-**60184**
108444-56-0 2-Amino-5-vinylpyrimidine,
 A-**60269**
108451-23-6 Pentacyclo[11.3.0.0¹,⁵.0⁵,⁹.0⁹,¹³]-
 hexadecane, P-**70020**
108460-21-5 5*H*-Cyclohepta[4,5]pyrrolo[2,3-
 b]indole, C-**60200**
108461-92-3 2-Amino-5-vinylpyrimidine;
 N^2,N^2-Di-Me, *in* A-**60269**
108472-40-8 1,4-Biss(thylthio)-3,6diphenyl-
 pyrrolo[3,4-*c*]pyrrole, B-**60158**
108507-92-2 2-Amino-1-hydroxy-1cyclo-
 butaneacetic acid; (1*RS*,2*RS*)-
 form, *in* A-**60184**
108507-93-3 2-Amino-1-hydroxy-1cyclo-
 butaneacetic acid; (1*RS*,2*SR*)-
 form, *in* A-**60184**
108511-44-0 1,2-Diazetidin-3-one; 1-Ac, 2-Ph,
 in D-**70055**
108524-93-2 Ilexsaponin A1, *in* D-**60382**
108524-94-3 3,19-Dihydroxy-12-ursene24,28-
 dioic acid; (3β,19α)-*form*, *in*
 D-**60382**
108526-92-7 3,5-Dibromo-1,6-dihydroxy-4oxo-
 2-cyclohexene-1acetonitrile;
 (1*R*,5*S*,6*S*)-*form*, *in* D-**60098**
108529-23-3 Ilexoside A, *in* D-**70327**
108544-40-7 Ilexoside B, *in* D-**70350**
108560-05-0 Inunal; Δ⁴-Isomer, 15-alcohol, *in*
 I-**70022**
108560-93-6 1,1-Dinitro-2-propanone; *O*-
 Acetyloxime, *in* D-**70454**
108572-36-7 12,16-Epoxy-11,14-
 dihydroxy5,8,11,13-abietatetra-
 en-7-one, E-**60011**
108577-36-2 1,4,8,12,16Pentaazacyclo-
 nonadecane, P-**60022**

108577-37-3 1,4,8,11,15Pentaazacyclooctadeca-
 ne, P-**60024**
108577-38-4 1,4,7,10,14Pentaazacyclo-
 heptadecane, P-**60019**
108577-39-5 1,4,7,11,15Pentaazacyclooctadeca-
 ne, P-**60023**
108577-40-8 1,4,7,10,13Pentaazacyclo-
 hexadecane; B,5HBr, *in* P-**60021**
108577-41-9 1,4,7,10,14Pentaazacyclo-
 heptadecane; B,5HBr, *in*
 P-**60019**
108577-42-0 1,4,7,11,14Pentaazacyclo-
 heptadecane; B,5HBr, *in*
 P-**60020**
108577-44-2 1,4,7,11,15Pentaazacyclooctadeca-
 ne; B,5HBr, *in* P-**60023**
108577-47-5 1,4,8,12,16Pentaazacyclo-
 nonadecane; B,5HBr, *in*
 P-**60022**
108577-49-7 1,5,9,13,17Pentaazacycloeicosane;
 B,5HBr, *in* P-**60018**
108586-93-2 2,2,5,5-Tetramethyl-3hexenedioic
 acid, T-**60141**
108586-95-4 3,3,6,6-Tetramethyl-4cyclohexene-
 1,2-dione, T-**60134**
108590-99-4 3,5-Dibromo-1,6-dihydroxy-4oxo-
 2-cyclohexene-1acetonitrile;
 (1*R*,5*R*,6*S*)-*form*, *in* D-**60098**
108593-59-5 1,1-*tert*-Butyl-3,3-diethoxy2-
 azaallenium(1+); Tetra-
 fluoroborate, *in* B-**70301**
108603-00-5 1,4,7,10,13Pentaazacyclo-
 pentadecane; B,5HBr, *in*
 P-**60025**
108603-02-7 1,4,8,11,15Pentaazacyclooctadeca-
 ne; B,5HBr, *in* P-**60024**
108605-55-6 1,3-Dihydroxy-4-methyl-6,8deca-
 dien-5-one, D-**70321**
108605-66-9 6-Hydroxy-5-oxo-
 7,15fusicoccadien-15-al,
 H-**60203**
108635-94-5 Pentacyclo[6.3.0.0²,⁶.0⁵,⁹]undeca-
 ne-4,7,11-trione; (±)-*form*, *in*
 P-**60031**
108641-87-8 Duryne, D-**70550**
108643-62-5 Oxyphencyclimine; (*S*)-*form*,
 B,HCl, *in* O-**60102**
108643-89-6 Pentacyclo[6.3.1.0²,⁴.0⁵,¹⁰.0⁷,⁸]-
 dodecane, P-**70019**
108645-28-9 Acuminatolide, A-**70063**
108646-47-5 11,12,16-Trihydroxy5,8,11,13-
 abietatetraen-7-one, T-**60312**
108665-25-4 3,4-Acenaphthenedicarboxylic
 acid; Anhydride, *in* A-**60013**
108665-29-8 3,4-Acenaphthenedicarboxylic
 acid; Di-Et ester, *in* A-**60013**
108665-38-9 4,5-Difluorophenanthrene,
 D-**70166**
108665-39-0 4,5-Dichlorophenanthrene,
 D-**70136**
108665-93-6 3-Isopropyloxindole; (±)-*form*, *in*
 I-**70097**
108666-51-9 Tetramercaptotetrathiafulvalene;
 Tetrakis (*S*-Ph), *in* T-**70130**
108692-07-5 Oxyphencyclimine; (*R*)-*form*,
 B,HCl, *in* O-**60102**
108692-47-3 3,4-Dihydro-3,4-dihydroxy-
 2(hydroxymethyl)-2*H*pyrrole,
 D-**60222**
108695-79-0 7-Isocyano-1-cycloamphilectene,
 I-**60093**
108695-80-3 7-Isocyano-11-cycloamphilectene,
 I-**60094**
108695-81-4 8-Isocyano-10,14amphilectadiene,
 I-**60090**

108695-82-5 8-Isocyano-1(12)cyclo-
 amphilectrene, I-**60095**
108696-10-2 Hexahydro-5(2*H*)-azocinone; 1-
 tert-Butyl, *in* H-**70048**
108696-11-3 Octahydroazocine; 1-Isopropyl, *in*
 O-**70010**
108696-12-4 Octahydroazocine; 1-*tert*-Butyl, *in*
 O-**70010**
108708-02-7 Phlebiakauranol alcohol, *in*
 P-**70097**
108739-39-5 10α-Isocyanoalloaromadendrane,
 in A-**60328**
108739-40-8 10α-Isothiocyanatoalloaromad-
 endrane, *in* A-**60329**
108739-41-9 10α-
 Formamidoalloaromadendrane,
 in A-**60327**
108789-15-7 Isoligustroside; Penta-Ac, *in*
 I-**60104**
108789-16-8 Neoleuropein, N-**60018**
108789-18-0 Isoligustroside, I-**60104**
108793-61-9 2-Nonacosyl-3-tridecene-1,10diol,
 N-**70068**
108816-39-3 2,4,5-Triss(icyanomethylene)1,3-
 cyclopentanedione; Biss(etra-
 phenylphosphonium salt), *in*
 T-**70321**
108818-44-6 2-Ethoxy-2-propenoic acid; 4-
 Nitrobenzyl ester, *in* E-**70037**
108834-58-8 Vesparione, V-**70007**
108835-41-2 3,5,6,7-Tetrahydro-4*H*-inden4-
 one, T-**70070**
108835-42-3 5,6-Dihydro-4*H*-indene, D-**70217**
108853-12-9 4-[2-(3-Hydroxyphenyl)ethyl]2,6-
 dimethoxyphenol, *in* T-**60102**
108864-15-9 Dilophic acid, D-**60394**
108864-22-8 Tomenphantopin A, *in* T-**60222**
108864-23-9 Tomenphantopin *B*, T-**60222**
108864-26-2 3,4-Epoxy-
 13(15),16sphenolobadiene-5,18-
 diol; (3α,4α,5α,13(15)*E*,16*E*)-
 form, *in* E-**60030**
108869-44-9 *N*-(7,7-Dimethyl-
 3azabicyclo[4.1.0]hepta1,3,5-
 trien-4-yl)benzamide, D-**70368**
108885-55-8 Bedfordiolide, B-**60008**
108885-57-0 8-epi-Inuviscolide, *in* I-**70023**
108885-58-1 4,8-Bis-epi-inuviscolide, *in* I-**70023**
108885-63-8 Taccalonolide A, *in* T-**70001**
108885-69-4 Taccalonolide *B*, T-**70001**
108885-72-9 Neovasinone, *in* N-**70029**
108890-86-4 Arizonin C_1, A-**70254**
108890-87-5 Arizonin A_1, *in* A-**70254**
108890-88-6 Arizonin A_2, *in* A-**70255**
108890-89-7 Arizonin B_1, *in* A-**70254**
108890-90-0 K 13, K-**70001**
108905-76-6 Arizonin B_2, *in* A-**70255**
108906-50-9 Icariside C_1, *in* T-**60362**
108906-61-2 Actinostemmoside D, *in* D-**70002**
108906-62-3 Actinostemmoside C, *in* D-**70001**
108906-63-4 Actinostemmoside B, *in* D-**70003**
108906-64-5 Actinostemmoside A, *in* D-**70002**
108906-79-2 2-Chloro-3,7-epoxy-
 9chamigranone, C-**60066**
108906-99-6 20(29)-Lupene-3,16-diol;
 (3β,16β)-*form*, 3-Tetra-
 decanoyl, *in* L-**60039**
108907-00-2 3,4-Epoxy-11(13)-eudesmen-12oic
 acid; (3α,4α)-*form*, *in* E-**60021**
108907-02-4 5,6-Dimethoxy-3-(4methoxy-
 benzyl)phthalide, D-**60398**
108907-23-9 12-Hydroxybromosphaerol; 12α-
 form, *in* H-**60108**
108909-02-0 1,5,6-Trimethoxy-
 2,7phenanthrenediol, *in* P-**60050**
108925-06-0 Arizonin C_3, A-**70255**
108925-14-0 2-Chloro-3-hydroxy-7chamigren-
 9-one, C-**60094**
108925-15-1 Laurencenone *B*, L-**70025**
108925-82-2 3,4,7,8-Tetramethoxy-
 2,6phenanthrenediol, *in*
 H-**60066**

108970-88-3   Benzo[1,2-*d*:4,5-*d'*]-bis[1,3]dithiole-2,6-dione, B-70028
109007-67-2   1,5-Diphenyl-1,3-pentanediol; (1*S*,3*S*)-form, in D-70490
109007-68-3   1,5-Diphenyl-1,3-pentanediol; (1*S*,3*R*)-form, in D-70490
109028-21-9   Magireol C, in M-60006
109062-00-2   Icariside B₁, in G-70028
109063-83-4   Magireol *A*, M-60005
109063-84-5   Magireol *B*, M-60006
109063-85-6   Isoamericanin *A*, I-60087
109070-40-8   3,6-Dihydro-3,3,6,6tetramethyl-2(1*H*)pyridinone, D-70267
109075-64-1   *N*-(*N*-1-Carboxy-6-hydroxy-3oxoheptyl)alanyltyrosine, C-70022
109138-61-6   2,4,6,8-Undecatetraene; (2*Z*,4*Z*,6*E*,8*Z*)-form, in U-70003
109141-97-1   2-Chloro-1,4,9,10anthracenetetrone, C-60041
109141-98-2   2,3-Dichloro-1,4,9,10anthracenetetrone, D-60117
109145-64-4   2-(2-Hydroxy-4-methoxyphenyl)7-methoxy-5-(1-propenyl)-benzofuran, in H-60214
109145-65-5   Olmecol, O-60040
109151-44-2   Naphtho[2,3-*b*]thiophene-2,3dione, N-70014
109152-40-1   6α,17α-Diacetoxy-15,17oxido16-spongianone, in D-60365
109152-41-2   17α-Acetoxy-6α-butanoyloxy15,17-oxido-16-spongianone, in D-60365
109152-42-3   16-Oxo-17-spongianal, O-60094
109152-43-4   Methyl 15α,17β-diacetoxy-6αbutanoyloxy-15,16dideoxy-15,17-oxido-16spongianate, in T-60038
109152-44-5   Methyl 15α,17β-diacetoxy15,16-dideoxy-15,17oxido-16-spongianoate, in T-60038
109163-57-7   *ent*-12-Hydroxy-13,14,15,16tetranor-1(10)-halimen18-oic acid; Ac, Me ester, in H-60231
109163-58-8   *ent*-15-Oxo-1(10),13halimadien-18-oic acid; (13*Z*)-form, Me ester, in O-60069
109163-59-9   *ent*-15-Oxo-1(10),13halimadien-18-oic acid; (13*E*)-form, Me ester, in O-60069
109163-61-3   ent-1(10),13Z-Halimadiene15,18-dioic acid, in O-60069
109163-62-4   ent-1(10)13E-Halimadiene15,18-dioic acid, in O-60069
109163-64-6   *ent*-12-Hydroxy-13,14,15,16tetranor-1(10)-halimen18-oic acid; Ac, in H-60231
109163-65-7   *ent*-15-Oxo-1(10),13halimadien-18-oic acid; (13*Z*)-form, in O-60069
109171-13-3   Cycloaspeptide *A*, C-70211
109171-14-4   Cycloaspeptide B, in C-70211
109171-15-5   Cycloaspeptide C, in C-70211
109178-30-5   1-Chloro-1-nitrocyclobutane, C-70130
109178-63-4   7*H*-2,3,4,6,7-Pentaazabenz[*de*]anthracene, P-60017
109178-83-8   1-Nitro-3-hexene; (*Z*)-form, in N-70051
109179-31-9   2-Bromo-3-methylbenzaldehyde, B-60292
109181-77-3   *ent*-15-Oxo-1(10),13halimadien-18-oic acid; (13*E*)-form, in O-60069
109181-97-7   6α-Butanoyloxy-17β-hydroxy15,17-oxido-16-spongianone, in D-60365
109181-98-8   6α-Acetoxy-17β-hydroxy-15,17oxido-16-spongianone, in D-60365

109181-99-9   Methyl 6α,15α,17β-triacetoxy15,16-dideoxy-15,17oxido-16-spongianoate, in T-60038
109182-83-4   3-Phenyl-1-pentyne; (±)-form, in P-70089
109194-67-4   2,3-Dihydro-2-(4hydroxyphenyl)-5-(3hydroxypropyl)-3methyl-benzofuran; (2*R*,3*R*)-form, in D-60244
109194-68-5   2-(4-Hydroxyphenyl)-7methoxy-5-(1propenyl)3-benzofurancarboxaldehyde, H-60212
109194-69-6   2-(4-Hydroxyphenyl)-5-(1propenyl)benzofuran, H-60214
109194-70-9   2-(4-Hydroxyphenyl)-7methoxy-5-(1propenyl)benzofuran, in H-60214
109194-71-0   4-[5-(1-Propenyl)-2benzofuranyl]-2,3benzenediol, in H-60214
109194-72-1   2-(2,4-Dihydroxyphenyl)-7methoxy-5-(1-propenyl)benzofuran, in H-60214
109194-73-2   1-(4-Hydroxyphenyl)-2-(4propenylphenoxy)-1propanol, H-60215
109194-74-3   Tetrahydrobis(4hydroxyphenyl)-3,4dimethylfuran; (2*R**,3*R**,4*S**,5*R**)-form, in T-60061
109194-75-4   Tetrahydro-2-(4hydroxyphenyl)-5-(4hydroxy-3-methoxyphenyl)-3,4-dimethylfuran, in T-60061
109202-01-9   4-Mercapto-2-butenoic acid; (*E*)-form, Benzyl ester, in M-60018
109202-04-2   4-Mercapto-2-butenoic acid; (*E*)-form, Nitrile, in M-60018
109216-74-2   2,3-Dihydrothieno[2,3-*b*]pyridine; 1-Oxide, in D-70269
109216-79-7   Dibenzo[*f*,*h*]thieno[2,3-*b*]quinoline, D-70079
109216-80-0   Dibenzo[*f*,*h*]thieno[2,3-*b*]quinoline; 1-Oxide, in D-70079
109217-15-4   Fontonamide, F-60067
109225-42-5   2,3-Dihydro-2-(4-hydroxy-3methoxyphenyl)-5-(3hydroxypropyl)-3methyl-benzofuran, in D-60244
109225-43-6   1-(4-Hydroxyphenyl)-2-(2methoxy-4-propenylphenoxy)-1-propanol, in H-60215
109237-00-5   2,3,12-Trihydroxy-4,7,16pregnatrien-20-one; (2α,3β,12β)-form, in T-60346
109237-01-6   2,3,12-Trihydroxy-4,7pregnadien-20-one; (2α,3β,12β)-form, in T-60345
109237-38-9   Chiratanin, C-60037
109274-90-0   3-Methylfuro[2,3-*b*]pyridine, M-60085
109274-92-2   Furo[2,3-*b*]pyridine-2carboxaldehyde, F-60093
109274-95-5   2-Cyanofuro[2,3-*b*]pyridine, in F-60098
109274-96-6   3-Cyanofuro[2,3-*b*]pyridine, in F-60099
109274-98-8   Furo[2,3-*b*]pyridine-3carboxylic acid; Amide, in F-60099
109274-99-9   Furo[2,3-*b*]pyridine-3carboxaldehyde, F-60094
109275-00-5   Furo[2,3-*b*]pyridine-3carboxaldehyde; Hydrazone, in F-60094
109279-75-6   1-Bromo-1,2-butadiene; (±)-form, in B-70196

109280-45-7   Tetrahydrobis(4hydroxyphenyl)-3,4dimethylfuran; (2*R**,3*R**,4*S**,5*S**)-form, 3'-Methoxy, in T-60061
109291-64-7   15-Desacetyltetraneurin C, in T-60158
109291-66-9   15-Desacetyltetraneurin *C* isobutyrate, in T-60158
109296-83-5   3,4-Heptafulvalenedione, H-60022
109305-25-1   5,6,8,9-Tetraaza[3.3]paracyclophane, T-60018
109308-71-6   3*a*,6*a*-Dihydro-3*a*,6*a*-dimethyl1,6-pentalenedione; (3*a*R*S*,6*a*S*R*)-form, in D-70199
109317-74-0   3,4-Dihydro-5(trifluoroacetyl)-2*H*pyran, D-70275
109317-75-1   2,3-Dihydro-4(trifluoroacetyl)-furan, D-70274
109360-94-3   Alloaromadendrane-4,10-diol; (4α,10β)-form, in A-60074
109428-38-8   6-Methyl-2-cyclohexen-1-ol; (1*R*,6*R*)-form, in M-70058
109517-68-2   Crotalarin, C-60171
109517-69-3   Crotarin, C-60172
109517-70-6   Cavoxinone, C-60025
109517-71-7   Cavoxinine, C-60024
109517-72-8   Ferugin, in D-60005
109521-81-5   Juglorin, J-70005
109528-43-0   3,6-Diazafluorenone, D-70053
109532-22-1   2'-Amino-2-chloro-3'methyl-acetophenone, A-60112
109532-23-2   2'-Amino-2-chloro-4'methyl-acetophenone, A-60113
109532-24-3   2'-Amino-2-chloro-6'methyl-acetophenone, A-60115
109534-58-9   2-Triphenylenecarboxylic acid, T-60384
109575-71-5   Indicoside *A*, I-60027
109576-20-7   24-Methylene-3,4-seco-4(28)cycloarten-3-oic acid, M-70078
109576-21-8   24-Methylene-3,4-seco-4(28)cycloarten-3-oic acid; Me ester, in M-70078
109605-79-0   Topazolin, T-60224
109605-84-7   Topazolin hydrate, in T-60224
109621-33-2   Hymatoxin *A*, H-60235
109664-01-9   Isotanshinone IIB, I-60138
109664-02-0   Neocryptotanshinone, N-60017
109667-12-1   Phaseolinic acid, P-60067
109679-44-9   Tiacumicin E, T-70197
109679-45-0   Tiacumicin D, T-70196
109681-73-4   [1,4]Benzoxaselenino[3,2-*b*]pyridine, B-60054
109703-33-5   4,4-Dimethyl-3-oxo-1cyclopentene-1-carboxaldehyde, D-60432
109713-94-2   Tiacumicin *A*, T-70194
109719-06-4   Dihydro-4-(trifluoromethyl)2(3*H*)-furanone; (*S*)-form, in D-60299
109719-07-5   Dihydro-4-(trifluoromethyl)2(3*H*)-furanone; (*R*)-form, in D-60299
109720-41-4   1,2,3,6-Tetrahydropyridine; 1-Propyl; B, (COOH)₂, in T-70089
109720-43-6   1,2,3,6-Tetrahydropyridine; 1-Ph; B, (COOH)₂, in T-70089
109741-39-1   Peroxyauraptenol, in A-70269
109764-44-5   1,3-Bis(1,2-diphenyl-ethenyl)benzene; (*Z*,*Z*)-form, in B-70139
109765-60-8   3,6-Dimethoxy-1,2-benzenediol, in B-60014
109785-99-1   Panasinsanol B, in P-70005
109792-56-5   Nitropeptin, N-70055
109794-96-9   1-Amino-3-(hydroxymethyl)-1cyclobutanecarboxylic acid; *cis*-form, in A-70157

109795-03-1 20,29,30-Trisnor-3,19lupanediol; (3β,19α)-form, in T-60404

109802-71-3 2-Phenylsulfonyl-1,3-butadiene, P-70094

109803-47-6 5-Chloro-4-methoxy-2methylbenzoic acid, in C-60097

109803-48-7 5-Chloro-4-hydroxy-2methylbenzoic acid; Me ether, Me ester, in C-60097

109803-52-3 1-Bromo-2,3-dichloro-4methoxybenzene, in B-60229

109872-63-1 Nordinone, N-60060

109872-64-2 Nordinonediol, in N-60060

109890-37-1 Trichotriol, T-60258

109894-10-2 2-Hydroxygarveatin A, in H-60148

109894-11-3 2-Hydroxygarveatin B, H-60148

109905-56-8 Epivolkenin, in T-60007

109905-96-6 2-Amino-3-phenylpentanedioic acid; (2S,3S)-form, in A-60241

109918-65-2 2-Amino-4-hydroxy-2methylbutanoic acid; (R)-form, in A-60186

109918-71-0 2-Amino-2-ethyl-3-butenoic acid; (R)-form, in A-60150

109925-99-7 1-Amino-1-phenyl-4-pentene, A-70183

109927-30-2 Aplysiapyranoid A, A-70244

109927-31-3 Aplysiapyranoid C, A-70245

109927-32-4 Aplysiapyranoid D, in A-70245

109948-61-0 2-(1,2-Propadienyl)benzothiazole, P-60174

109954-46-3 2-(7-Hydroxy-3,7-dimethyl2octenyl)-6-methoxy-1,4benzoquinone; (Z)-form, in H-60123

109954-47-4 2-(7-Hydroxy-3,7-dimethyl2octenyl)-6-methoxy-1,4benzoquinone; (E)-form, in H-60123

109954-48-5 6-Methoxy-2-(3,7-dimethyl2,6-octadienyl)-1,4benzoquinone; (Z)-form, in M-60038

109971-64-4 3-Methoxy-6-[2-(3,4,5trimethoxyphenyl)ethyl]1,2-benzenediol, in T-60344

109972-13-6 Shonachalin D, S-70040

109974-21-2 Regelin, in H-70202

109974-22-3 Regelinol, in D-70333

109974-30-3 Adenanthin, A-70065

109978-12-3 6H,24H,26H,38H,56H,58H16,-65:48,66-Bis([1,3]benzomethanothiomethano[1,3]benzeno)-32,64-etheno-1,5:9,13:14,18:-19,23:27,31:33,37:41,45:46,50:-51,55:59,63-decemetheno-8H,40H[1,7]dithiacyclotetratriacontino[9,8-h][1,77]dithiacyclotetratriacontin, B-70129

109978-15-6 β-Actinorhodin, A-70062

109978-16-7 ε-Isoactinorhodin, I-70059

109986-01-8 cis-9,10-Dihydrocapsenone, in C-60011

109988-40-1 1,3,2-Benzodithiazol-1-ium(1+); Bromide, in B-70033

109997-30-0 9-Hydroxy-5-nonenoic acid; (Z)-form, in H-70194

110007-10-8 5-Deoxymyricanone, in M-70162

110024-14-1 3α,22S-Diacetoxy-15αhydroxy7,9(11),24Elanostatrien-26-oic acid, in T-70266

110024-18-5 Volkensiachromone, V-60016

110024-19-6 Cyclolongipesin, C-60214

110024-20-9 Homocyclolongipesin, H-60088

110024-35-6 Longipesin; 9-Ac, in L-60034

110024-38-9 9-Methyllongipesin, M-70091

110024-40-3 9-Methyllongipesin; O⁹-Ac, in M-70091

110024-43-6 9-Methyllongipesin; 9-Propanoyl, in M-70091

110024-46-9 9-Methylcyclolongipesin, in C-60214

110024-47-0 Homocyclolongipesin; Ac, in H-60088

110024-48-1 Homocyclolongipesin; Propanoyl, in H-60088

110028-19-8 2H-Dibenz[e,g]isoindole; N-tert-Butyl, in D-70067

110028-20-1 2H-Dibenz[e,g]isoindole; N-Benzyl, in D-70067

110042-12-1 Longipesin; 9-Ac, 4-Me ether, in L-60034

110044-90-1 6-Isocyano-4(15)-eudesmene, in F-60069

110044-92-3 6-Formamido-4(15)-eudesmene; (6α,10α)-form, in F-60069

110045-38-0 2-Oxodecanal, O-70087

110064-50-1 7-Hydroxy-3-(4hydroxybenzylidene)-4chromanone, H-60152

110064-60-3 Zebrinin, Z-70002

110064-64-7 Blastmycetin D, B-60191

110064-65-8 Isolecanoric acid, I-60103

110066-02-9 12,19-Dihydroxy-5,8,10,14eicosatetraenoic acid; (5Z,8Z,10E,12S,14Z,19R)-form, in D-60319

110066-03-0 12,19-Dihydroxy-5,8,10,14eicosatetraenoic acid; (5Z,8Z,10E,12S,14Z,19S)-form, in D-60319

110115-55-4 Taraktophyllin, T-60007

110128-59-1 3,4-Diamino-1methoxyisoquinoline, D-60037

110129-67-4 4-Oxo-1-cycloheptene-1carboxylic acid; Me ester, in O-70082

110167-42-5 Pyrido[3',4':4,5]furo[3,2-b]indole, P-70174

110186-13-5 6-O-Acetylshanghiside methyl ester, in S-60028

110187-11-6 1,5,8-Trihydroxy-3-methyl-2(3-methyl-2-butenyl)xanthone, in T-60114

110187-31-0 ent-3β-Acetoxy-15-kauren-17-oic acid, in H-60167

110187-32-1 ent-3β,19-Diacetoxy-15kauren-17-oic acid, in D-60341

110187-33-2 ent-18-Acetoxy-15-beyeren-19oic acid, in H-60106

110187-34-3 ent-15-Kauren-17-oic acid; Me ester, in K-60008

110187-35-4 ent-3β-Hydroxy-15-kauren-17-oic acid; 3-Ac, Me ester, in H-60167

110187-38-7 3,8-Dihydroxy-1,2,4trimethoxyxanthone, in P-60051

110191-89-4 4,5-Diaminoisoquinoline, D-70040

110200-29-8 Antibiotic SF 2415B₃, A-70237

110200-30-1 Antibiotic SF 2415B₂, A-70236

110200-31-2 Antibiotic SF 2415A₂, A-70233

110200-32-3 Antibiotic SF 2415A₁, A-70232

110200-33-4 Antibiotic SF 2415A₃, A-70234

110200-34-5 Antibiotic SF 2415B₁, A-70235

110201-59-7 Havannahine, H-60008

110201-60-0 Deoxyhavannahine, in H-60008

110201-62-2 18-Hydroxy-8,15isopimaradien-7-one, H-60162

110201-66-6 7,11,12-Trihydroxy-6-oxo8,11,13-abietatrien-20oic acid; 7β-form, in T-60337

110201-67-7 7,11,12-Trihydroxy-6-oxo8,11,13-abietatrien-20oic acid; 7α-form, in T-60337

110201-83-7 ent-6-Oxo-7,12E,14labdatrien-17,11α-olide, O-60076

110202-83-0 3,7-Dimethoxy-2-phenanthrenol, in T-60339

110204-44-9 3,5-Dihydroxy-6,7methylenedioxyflavanone, in T-60105

110206-37-6 1,1-Cyclobutanedimethanethiol, C-70213

110206-40-1 1,2-Cyclobutanedimethanethiol; (1RS,2RS)-form, in C-70214

110207-64-2 2,3-Dihydrolinderazulene, in L-70031

110207-81-3 Youlemycin, Y-70002

110209-90-0 Machilin B, M-60002

110209-97-7 6,8,11,13-Abietatetraene11,12,14-triol, A-60006

110209-99-9 Tannunolide A, T-60005

110210-00-9 Tannunolide B, in T-60005

110219-85-7 Sulcatine, S-60061

110230-36-9 2-Chloro-3,3,3trifluoropropanoic acid, C-70170

110231-33-9 Fumifungin, F-60082

110231-34-0 Oligomycin E, O-70039

110241-11-7 Hermosillol, H-60029

110241-17-3 Sesebrinic acid, S-60026

110241-35-5 3-O-Caffeoyl-4-O-sinapoylquinic acid, in C-70005

110243-21-5 Pentacyclo[5.4.0.0²,⁶.0³,¹⁰.0⁵,⁹]undecane-4,8,11-trione, P-70023

110268-34-3 Aristolignin, A-60297

110268-36-5 Isobicyclogermacrenal; (+)-form, in I-60088

110268-39-8 20(29)-Lupene-3,16-diol; (3β,16β)-form, 3-Hexadecanoyl, in L-60039

110269-50-6 Austrobailignan-5; (2RS,3SR)-form, in A-70270

110269-51-7 Machilin C, M-60003

110269-53-9 Machilin D, in M-60003

110297-46-6 Avellanin B, A-60324

110297-47-7 Avellanin A, A-60323

110298-63-0 8-Methoxychlorotetracycline, M-70045

110298-64-1 4a-Hydroxy-8methoxychlorotetracycline, in M-70045

110298-65-2 8-Methoxy-Nmethylchlorotetracycline, in M-70045

110299-91-7 Coralloidin C, in E-60066

110299-97-3 5,7(11)-Eudesmadien-15-ol; (4R,10S)-form, in E-60066

110300-00-0 2-Amino-3-(2,4,5trichlorophenyl)propanoic acid; (±)-form, in A-60264

110300-01-1 2-Amino-3-(2,3,6trichlorophenyl)propanoic acid; (±)-form, in A-60263

110300-02-2 2-Amino-3-(2,3,4trichlorophenyl)propanoic acid; (±)-form, in A-60262

110300-03-3 2-Amino-3-(2,6dichlorophenyl)propanoic acid; (±)-form, in A-60130

110300-04-4 2-Amino-3-(2,3dichlorophenyl)propanoic acid; (±)-form, in A-60127

110300-20-4 Synrotolide, S-60066

110309-27-8 Gelsemide, G-60010

110309-28-9 9-Hydroxysemperoside, in S-60021

110309-32-5 Gelsemiol 1-glucoside, in G-60011

110309-33-6 Gelsemiol 3-glucoside, in G-60011

110318-31-5 2,3-Dihydro-2-phenyl-4H1benzothiopyran-4-one; (±)-form, in D-70248

110318-34-8 2,3-Dihydro-2-phenyl-4H1benzothiopyran-4-one; (+)-form, in D-70248

110318-35-9 2,3-Dihydro-2-phenyl-4H1benzothiopyran-4-one; (−)-form, in D-70248

110318-58-6 Nonacyclo[10.8.0.0^{2,11}.0^{4,9}.0^{4,19}.-
0^{6,17}.0^{7,16}.0^{9,14}.0^{14,19}]icosane,
N-70070

110321-62-5 9,11,20,22Tetraphenyltetra-
benzo[a,c,l,n]pentacene-10,21-
dione, T-70151

110322-49-1 Gelsemide 7-glucoside, *in* G-60010

110325-63-8 Desertorin A, D-60022

110325-64-9 Desertorin B, *in* D-60022

110325-65-0 Desertorin C, *in* D-60022

110325-70-7 Lappaphen a, L-70018

110344-57-5 Semperoside, S-60021

110345-06-7 24(28)-Stigmastene-
2,3,22,23tetrol;
(2α,3α,5α,22R,23R,24(28)E)-
form, *in* S-70075

110351-36-5 2-Amino-3-(2,5dichlorophenyl)-
propanoic acid; (±)-*form, in*
A-60129

110351-75-2 Isobalearone, *in* B-60005

110361-76-7 2,3,5-Trihydroxybenzoic acid; Me
ester, *in* T-70250

110368-36-0 Antibiotic W 341C, A-70241

110372-75-3 7,16[1′,2′]-Benzeno-
7,16dihydroheptacene, B-70020

110374-51-1 1-Amino-2-cyclopropene-
1carboxylic acid; N-tert-
Butyloxycarbonyl, Me ester, *in*
A-60122

110374-52-2 1-Amino-2-cyclopropene-
1carboxylic acid; N-tert-
Butyloxycarbonyl, *in* A-60122

110374-53-3 1-Amino-2-cyclopropene-
1carboxylic acid; B,HCl, *in*
A-60122

110374-53-3 1-Aminocyclopropenecarboxylic
acid; B,HCl, *in* A-70126

110374-54-4 1-Amino-2-cyclopropene-
1carboxylic acid, A-60122

110374-54-4 1-Aminocyclopropenecarboxylic
acid, A-70126

110397-95-0 6-Hydroxy-2,4-dimethylheptanoic
acid; (2R,4S,6S)-*form, in*
H-60121

110414-77-2 Gelsemiol, G-60011

110415-32-2 Lappaphen b, *in* L-70018

110417-88-4 Dolastatin 10, D-70542

110429-26-0 3-Hydroxy-2-quinolinecarboxylic
acid; Me ether, Me ester, *in*
H-60228

110429-27-1 3-Hydroxy-2-quinolinecarboxylic
acid; Ca salt, *in* H-60228

110429-58-8 Rhodonocardin A, R-70003

110429-59-9 Rhodonocardin B, *in* R-70003

110431-54-4 2,3,6,7,10,11Hexahydrobenzo-
[1,2b:3,4-$b′$:5,6-$b″$]tris[1,4]-
dithiin, H-70050

110431-55-5 Benzo[1,2-d:3,4-$d′$:5,6-
$d″$]tris[1,3]dithiole, B-70069

110431-56-6 2,3,8,9-Tetrahydrobenzo[1,2b:3,4-
$b′$]bis[1,4]dithiin, T-70039

110431-57-7 2,3,8,9-Tetrahydrobenzo[2,1b:3,4-
$b′$]bis[1,4]oxathiin, T-70040

110431-67-9 2,3,6,7,10,11Hexahydrobenzo-
[1,2b:3,4-$b′$:5,6-$b″$]tris[1,4]-
dithiin; Tetrafluoroborate, *in*
H-70050

110434-71-4 2H-1,5,2-Dioxazine-3,6(4H)dione,
D-70464

110448-11-8 2,3,6,7,10,11Hexahydrobenzo-
[1,2b:3,4-$b′$:5,6-$b″$]tris[1,4]-
oxathiin, H-70051

110454-83-6 6-Hydroxy-2,4-dimethylheptanoic
acid; (2S,4R,6R)-*form, in*
H-60121

110505-66-3 6-Acetoxy-3,7-
dolabelladiene12,16-diol, *in*
D-70541

110512-27-1 4,7,10,15,18,21,24,27,30,33,36,39-
Dodecaoxatricyclo[11.9.9.9^{2,12}]-
tetraconta-1,12-diene, D-70535

110520-17-7 Phenanthro[9,10-g]isoquinoline,
P-60075

110538-24-4 Brassicadiol, B-70177

110559-86-9 6β-Hydroxyadoxoside, *in* A-70067

110561-67-6 1,2-Dihydro-3H-azepin-3-one; 1-
Me, *in* D-70171

110561-68-7 1,2-Dihydro-3H-azepin-3-one; 1-
Ph, *in* D-70171

110568-96-2 1,3,5,7Tetrakiss(iethylamino)pyrrolo[3,4-f]isoindole,
T-70125

110590-01-7 2-Oxa-3-azabicyclo[2.2.2]octane;
N-Cyano, *in* O-70051

110590-22-2 2-Oxa-3-azabicyclo[2.2.2]octane;
B,HCl, *in* O-70051

110601-24-6 1(10)-Millerenolide; (1(10)E)-
form, *in* M-60136

110601-29-1 9-Millerenolide; (9E)-form, *in*
M-60137

110601-38-2 ent-8(14)-Pimarene2α,3α,15R,16-
tetrol, P-60154

110601-39-3 ent-2α,3α,16-Trihydroxy-
8(14)pimaren-15-one, *in*
P-60154

110601-42-8 2-(6-Hydroxy-4-methyl-
4hexenylidene)-6,10dimethyl-7-
oxo-9undecenal; (10E,14Z)-
form, *in* H-60186

110601-43-9 10-Hydroxy-2-(6-hydroxy-
4methyl-4-hexylidene)6,10-di-
methyl-7-oxo-8undecenal, *in*
H-60186

110653-03-7 4-Fluoro-1-phenyl-1-butene; (E)-
form, *in* F-70035

110655-89-5 2,4-Dichloro-3,4-dimethyl-2cyclo-
buten-1-one, D-60130

110657-98-2 5,6,15-Trihydroxy7,9,11,13,17-
eicosapentaenoic acid; (5S,6R,-
7E,9E,11Z,13E,15S,17Z)-form,
in T-70255

110657-99-3 5,14,15-Trihydroxy6,8,10,12,17-
eicosapentaenoic acid; (5S,6E,-
8Z,10E,12E,14R,15S,17Z)-
form, *in* T-70256

110659-91-1 $N,N′$-Bis[3-(3-bromo-
4hydroxyphenyl)-
2oximidopropionyl]cystamine,
B-70130

110661-69-3 Cyclooctatetraenecarboxaldehyde;
2,4-Dinitrophenylhydrazone, *in*
C-70238

110661-70-6 Cyclooctatetraenecarboxaldehyde;
4-Methyl-
benzenesulfonylhydrazone, *in*
C-70238

110661-83-1 1,13-Dioxa-4,7,10,16,20,24hexa-
azacyclohexacosane; B,6HCl, *in*
D-70459

110668-26-3 Brevifloralactone, B-70179

110694-79-6 4-Acetyl-2,3-dihydro-5hydroxy-2-
isopropenylbenzofuran, A-70040

110718-94-0 2-Amino-1,7-dihydro-7-phenyl6H-
purin-6-one, A-70133

110744-41-7 12H-Quinoxalino[2,3-
b][1,4]benzoxazine; N-Ac, *in*
Q-70009

110744-44-0 12H-Quinoxalino[2,3-
b][1,4]benzothiazine; N-
Benzoyl, *in* Q-70008

110744-71-3 Acanthocerebroside A, A-70007

110744-72-4 Acanthocerebroside B, A-70008

110744-73-5 Acanthocerebroside C, A-70009

110744-74-6 Acanthocerebroside D, A-70010

110744-75-7 Acanthocerebroside E, A-70011

110744-76-8 Acanthocerebroside F, A-70012

110784-15-1 Kielcorin B, K-60012

110784-16-2 Ramosissin, R-60001

110786-60-2 4-Acetoxy-2,6-dibromo-5hydroxy-
2-cyclohexen-1one, *in* D-60097

110786-62-4 4-Acetoxy-2-bromo-5,6-epoxy2-
cyclohexen-1-one, *in* B-60239

110786-63-5 2-Bromo-5,6-epoxy-4-hydroxy2-
cyclohexen-1-one; (4S,5R,6R)-
form, in B-60239

110786-78-2 Amphidinolide B, A-70205

110796-49-1 4,4-Pyrandiacetic acid, P-70150

110796-53-7 4,4-Pyrandiacetic acid; Di-Et
ester, *in* P-70150

110796-54-8 4,4-Pyrandiacetic acid; Di-Me
ester, *in* P-70150

110835-94-4 2,2-Dimethyl-4,4-diphenyl-
3butenoic acid, D-70402

110847-11-5 Dihydro-4-(trifluoromethyl)2(3H)-furanone,
D-60299

110850-10-7 2,6-Dibromo-4,5-dihydroxy-
2cyclohexen-1-one; (4S,5R,6R)-
form, 4-Ac, *in* D-60097

110874-85-6 5-Chloro-3-methyl-2cyclopenten-
1-one, C-70091

110882-81-0 Parvodicin; Parvodicin A, *in*
P-70012

110882-82-1 Parvodicin; Parvodicin B_1, *in*
P-70012

110882-83-2 Parvodicin; Parvodicin B_2, *in*
P-70012

110882-84-3 Parvodicin; Parvodicin C_1, *in*
P-70012

110882-85-4 Parvodicin; Parvodicin C_2, *in*
P-70012

110882-86-5 Parvodicin; Parvodicin C_3, *in*
P-70012

110883-36-8 Emindole DA, E-60005

110883-37-9 Emindole SA, *in* E-60005

110901-51-4 Hypericorin, H-70234

110906-83-7 Kickxioside, K-70014

110925-96-7 6-Ethyl-2,4-dihydroxy-3methyl-
benzoic acid; Et ester, *in*
E-70041

110925-97-8 6-Ethyl-2,4-dihydroxy-3methyl-
benzoic acid; Me ester, *in*
E-70041

110934-24-2 Albomitomycin A, A-70080

110935-41-6 1,4-Benzoquinone-2,6dicarboxylic
acid; Di-Et ester, *in* B-70049

110941-52-1 Pedicellosine, P-70015

110954-19-3 5-Methyl-3,6-dioxo-
2piperazineacetic acid; (2S,5S)-
form, *in* M-60064

110954-32-0 Rubiflavin C_2, *in* R-70011

110954-33-1 Rubiflavin D, *in* R-70011

110971-80-7 5-Amino-3,4-dihydro-2Hpyrrole-
2-carboxylic acid; (S)-*form*, Me
ester, *in* A-70135

110971-81-8 5-Amino-3,4-dihydro-2Hpyrrole-
2-carboxylic acid; (S)-*form*,
B,HCl, *in* A-70135

110971-82-9 5-Amino-3,4-dihydro-2Hpyrrole-
2-carboxylic acid; (S)-*form*, Et
ester, *in* A-70135

110977-28-1 Bicyclo[12.4.1]nonadec1,3,5,7,-
9,11,13,15,17nonaene, B-70101

110978-89-7 2-Norerythromycin D, *in* N-70076

110978-90-0 2-Norerythromycin B, *in* N-70076

111004-32-1 Isocyclocalamin, I-60096

111010-24-3 2-Norerythromycin, N-70076

111013-09-3 4-Hydroxyacenaphthylene,
H-70100

111013-13-9 4-Methoxyacenaphthene, *in*
H-60099

111013-14-0 4-Methoxyacenaphthylene, *in*
H-70100

111013-17-3 4-Hydroxyacenaphthene; Ac, *in*
H-60099

111013-18-4 4-Hydroxyacenaphthylene; Ac, *in*
H-70100

111035-65-5 Annonacin, A-70212

111051-87-7 6-Hydroxy-7,9-octadeca-
diene11,13,15,17-tetraynoic
acid; (7E,9E)-*form, in* H-70195

111058-14-1 Rubiflavin C_1, R-70011

111079-50-6   1,7-Dihydro-6*H*-purine-6-thione; 1,9-*Dihydro-form, in* D-60270

111080-51-4   2,3-Dihydro-2-phenyl-2benzofurancarboxylic acid, D-60262

111136-25-5   5-(Hydroxymethyl)-5cyclohexene-1,2,3,4tetrol; (1*S*,2*S*,3*S*,4*R*)-*form, in* H-70172

111138-74-0   6-Hydroxy-7,9-octadecadiene11,13,15,17-tetraynoic acid; (7*Z*,9*E*)-*form, in* H-70195

111138-75-1   6-Hydroxy-7,9-octadecadiene11,13,15,17-tetraynoic acid; (7*E*,9*Z*)-*form, in* H-70195

111149-87-2   Luffariellolide, L-70040

111150-38-0   10,11-Dihydromicroglossic acid, *in* M-60134

111150-39-1   Microglossic acid, M-60134

111150-42-6   8,11,13-Cleistanthatrien-19-oic acid, *in* C-60154

111150-44-8   8,11,13-Cleistanthatrien-19-al, *in* C-60154

111170-37-7   1,2-Dihydrocyclobuta[*a*]naphthalen-3-ol, D-60215

111170-38-8   1,2-Dihydrocyclobuta[*a*]naphthalene-3,4-dione, D-60213

111179-61-4   8,11,13-Cleistanthatrien-19-ol, C-60154

111185-06-9   4,5′-Bithiazole, B-70170

111188-75-1   Aristotetralone, A-60298

111189-32-3   Indeno[1,2,3-*hi*]chrysene, I-70010

111189-33-4   Benz[*def*]indeno[1,2,3-*hi*]chrysene, B-70024

111189-34-5   Benz[*def*]indeno[1,2,3-*qr*]chrysene, B-70023

111192-70-2   Dispiro[cyclopropane-1,5′[3,8]-dioxatricyclo[5.1.0.0²,⁴]octane-6′,1″-cyclopropane], D-60500

111192-77-9   Dispiro[2.0.2.4]dec-8-ene7,10-dione, D-60501

111192-78-0   5,5,6,6-Tetramethyl-2cyclohexene-1,4-dione, T-60135

111200-22-7   Bryoflavone, B-70283

111200-23-8   Heterobryoflavone, H-70027

111216-62-7   Napyradiomycin A_2, N-70016

111220-88-3   2,2-Dimethyl-4,4-diphenyl-3butenal; Oxime, *in* D-70401

111228-19-4   1,8-Dihydroxy-6-methoxy-2methylanthraquinone, *in* T-70269

111233-39-7   Ciclamycin 4, *in* C-70176

111233-40-0   Ciclamycin 0, C-70176

111258-41-4   Methyl eriodermate, *in* E-70032

111258-42-5   Methyl 2′-*O*-methyleriodermate, *in* E-70032

111258-43-6   Methyl 2-*O*-methyleriodermate, *in* E-70032

111258-44-7   Methyl 4-*O*-methyleriodermate, *in* E-70032

111258-45-8   Methyl 2,2′-di-*O*methyleriodermate, *in* E-70032

111261-12-2   3,6-Dihydroxy-1*a*-(3-methyl-2butenyl)naphth[2,3-*b*]2,7(1*aH*,7*aH*)-dione, *in* D-70277

111261-13-3   7,7*a*-Dihydro-3,6,7trihydroxy-1*a*-(3methyl-2-butenyl)naphth[2,3-*b*]oxiren2(1*aH*)-one, D-70277

111261-14-4   7,7*a*-Dihydro-3,6,7trihydroxy-1*a*-(3methyl-2-butenyl)naphth[2,3-*b*]oxiren2(1*aH*)-one; 7-Ketone, 3-*O*-β-D-glucopyranoside, *in* D-70277

111261-15-5   7,7*a*-Dihydro-3,6,7trihydroxy-1*a*-(3methyl-2-butenyl)naphth[2,3-*b*]oxiren2(1*aH*)-one; 3-*O*-β-D-Glucopyranoside, *in* D-70277

111278-00-3   Anguinomycin *B*, A-70210

111278-01-4   Anguinomycin *A*, A-70209

111286-05-6   6-Methoxy-2-[2-(4methoxyphenyl)ethyl]4*H*-1-benzopyran-4-one, M-60039

111286-34-1   2,5-Dihydro-3-nitrofuran, D-70236

111299-76-4   5-*tert*-Butyl-1,3,2,4dithiadiazolyl, B-70303

111314-74-0   1-Amino-2-(4-hydroxyphenyl)cyclopropanecarboxylic acid; (1*RS*,2*SR*)-*form*, B,HCl, *in* A-60197

111314-77-3   1-Amino-2-(4-hydroxyphenyl)cyclopropanecarboxylic acid; (1*RS*,2*SR*)-*form*, Me ester, *in* A-60197

111317-37-4   2-Amino-6-(1,2dihydroxypropyl)-3methylpterin-4-one, *in* B-70121

111333-97-2   Pentafluorophenyl formate, P-70025

111351-51-0   Chloropentakiss(ichloromethyl)-benzene, C-70136

111367-04-5   Okilactomycin, O-70030

111394-30-0   Decumbeside D, *in* D-60017

111394-32-2   7-Hydroxy-4-isopropyl-3methoxy-6-methylcoumarin, H-70144

111394-33-3   9-Hydroxy-4,11(13)eudesmadien-12,6-olide; (6β,9β)-*form, in* H-60137

111394-34-4   9β-Acetoxytournefortiolide, *in* H-60137

111394-35-5   9-Oxo-4,11(13)-eudesmadien12-oic acid, *in* H-60135

111394-37-7   9β-Acetoxy-4,11(13)eudesmadien-12-oic acid, *in* H-60135

111394-39-9   1-Acetoxy-7-hydroxy-3,7dimethyl-2*E*,5*E*-octadien-4-one, *in* D-60318

111394-41-3   1-Acetoxy-7-hydroperoxy-3,7dimethyl-2*E*,5*E*-octadien-4-one, *in* D-60318

111397-47-8   5,8-Dihydroxy-6,10-dimethoxy2-propyl-4*H*-naphtho[2,3-*b*]pyran-4-one, D-70292

111397-51-4   Secopseudoterosin B, *in* S-70026

111397-55-8   4,15,26-Tricontatrien1,12,18,29-tetrayne3,14,17,28-tetrone, *in* P-70044

111397-58-1   5,10-Dihydroxy-8-methoxy-2propyl-4*H*-naphtho[1,2*b*]pyran-4-one, *in* C-70191

111398-15-3   6,6′-Difluoro-[1,1′-biphenyl]2,2′-dicarboxylic acid; Di-Me ester, *in* D-60180

111408-27-6   2-Hydroxy-3isopropylbutanedioic acid; (2*RS*,3*RS*)-*form, in* H-60166

111409-81-5   2,3,9,10Tetrakiss(rimethylsilyl)-[5]phenylene, T-60130

111420-30-5   Antibiotic B 1625*FA*₂β-1, A-70220

111420-61-2   9-Oxo-4,11(13)-eudesmadien12,16β-olide, *in* H-60137

111424-67-0   1,2,3,4-Tetrahydro-2-thioxo5*H*-1,4-benzodiazepin-5one, T-70100

111465-42-0   Talaromycin E, *in* T-70004

111465-43-1   Talaromycin D, *in* T-70004

111465-44-2   Talaromycin F, *in* T-70004

111466-65-0   Secopseudopterosin *A*, S-70026

111466-66-1   Secopseudopterosin C, *in* S-70026

111466-67-2   Secopseudopterosin D, *in* S-70026

111467-42-6   Decumbeside C, D-60017

111480-78-5   7-Hydroxy-1-hydroxy-3,7dimethyl-2*E*,5*E*-octadien-4-one, *in* D-60318

111508-79-3   19(4β→3β)-Abeo-6,11-epoxy6,12-dihydroxy-6,7seco-4(18),8,11,13abietatetraen-7-al; 6α-*form, in* A-60004

111508-81-7   19(4β→3β)-Abeo-6,11-epoxy6,12-dihydroxy-6,7seco-4(18),8,11,13abietatetraen-7-al; 6α-*form*, Di-Ac, *in* A-60004

111508-89-5   6β-Acetoxyhebemacrophyllide, *in* H-60010

111508-90-8   5-epi-6βacetoxyhebemacrophyllide, *in* H-60010

111509-25-2   6,8-Dihydroxy-1,2,5trimethoxyxanthone, *in* P-60052

111515-49-2   2-Methylene-7oxabicyclo[2.2.1]-heptane, M-70073

111524-31-3   6,11-Epoxy-6,12-dihydroxy6,7-seco-8,11,13abietatrien-7-al; 6α-*form, in* E-60014

111534-64-6   Cymbodiacetal, C-60233

111537-53-2   3-Bromo-4-[(3-bromo-4,5dihydroxyphenyl)methyl]5-(hydroxymethyl)-1,2benzenediol, B-70194

111540-00-2   Ochromycinone; (±)-*form, in* O-60003

111540-01-3   Ochromycinone; (±)-*form*, Ac, *in* O-60003

111545-12-1   Lonchocarpol B, *in* L-60030

111545-13-2   Lonchocarpol C, L-60031

111545-14-3   Lonchocarpol D, L-60032

111545-46-1   Blumealactone A, *in* B-70172

111545-47-2   Blumealactone B, *in* B-70172

111545-48-3   Blumealactone C, B-70172

111550-86-8   6*H*-[1]Benzothieno[2,3-*b*]indole, B-70058

111550-91-5   6*H*-Pyrido[3′,4′:4,5]thieno[2,3-*b*]-indole, P-70177

111554-19-9   Petrosynol, P-70044

111557-80-3   2,11-Diphenyldipyrrolo[1,2a:2′,1′-*c*]quinoxaline, D-70479

111558-01-1   2,7-Di-*tert*-butyl-2,7dihydroisoindolo[5,4-*e*]isoindole, D-70104

111558-02-2   2,5-Di-*tert*-butyl-2,5dihydrobenzo[*e*]pyrrolo[3,4-*g*]isoindole, D-70103

111567-20-5   Lonchocarpol E, L-60033

111573-52-5   5*H*-Pyrrolo[1,2-*c*]imidazol-5-one, P-70195

111573-53-6   6*H*-Pyrrolo[1,2-*b*]pyrazol-6-one, P-70199

111576-31-9   19-Acetoxy-15,16-epoxy-13,17spatadien-5α-ol, *in* E-60029

111576-37-5   2,8-Dihydroxy-3,10(14),11(13)guaiatrien-12,6-olide; (1α,2β,5α,6β,8β)-*form, in* D-60327

111576-39-7   *ent*-6α,15,16-Trihydroxy-19nor-4-rosen-3-one, *in* N-60061

111600-85-2   2-Ethynylthiazole, E-60061

111600-88-5   4-(Phenylethynyl)thiazole, P-60100

111600-89-6   4-Ethynylthiazole, E-60062

111614-78-9   3,6-Bis(3,4-dimethoxyphenyl)tetrahydro-1*H*,3*H*furo[3,4-*c*]furan-1,4-diol, B-60156

111615-79-3   Tetrabenzo[*b,h,n,t*]tetraphenylene, T-70012

111621-35-3   Tochuinyl acetate, *in* T-70202

111621-36-4   Dihydrotochuinyl acetate, *in* T-70202

111622-25-4   2-Aminomethyl-1,10phenanthroline, A-60224

111631-97-1   Pyrichalasin H, P-60215

111660-59-4   Nectriafurone; 5-Me ether, *in* N-60016

111660-60-7   Nectriafurone; 8-Me ether, *in* N-60016

111689-53-3 1-Cycloundecen-3-yne; (*E*)-*form*, *in* C-70256

111727-91-4 Hexahydro-1,4-dithiino[2,3-*b*]1,4-dithiin; *trans-form*, *in* H-60051

111755-37-4 Cyanoviridin *RR*, C-70210

111773-09-2 2,12-Di-9-acridinyl-7phenyl-dibenz[*c,h*]acridine, D-70033

111778-06-4 1(5),3-Aromadendradiene, A-70256

111786-26-6 Methyllinderatone, *in* L-70030

111821-79-5 1-Aromadendrene, A-70258

111822-11-8 Isolinderatone, *in* L-70030

111853-12-4 1-Phenylethanesulfonic acid; (±)-*form*, Hemi-Ba salt, *in* P-70068

111870-18-9 2,3-Diisopropoxycyclopropenone, *in* D-70290

111897-37-1 Neovasinin, N-70029

111897-38-2 2,5,7 Trihydroxy-3-(5hydroxyhexyl)-1,4naphthoquinone, T-70262

111900-50-6 1(5),6-Guaiadiene; (4β,10α)-*form*, *in* G-70031

111900-51-7 1(5),6-Guaiadiene; (4α,10α)-*form*, *in* G-70031

111929-99-8 *myo*-Inositol-1,3,4,5tetra-phosphate, I-70019

111931-75-0 Benzoyl phenyl diselenide, B-70077

111983-96-1 1-(5,7-Dihydroxy-2,2,6trimethyl-2*H*-1-benzopyran-8-yl)-2-methyl-1-propanone, D-70348

111983-97-2 1-(5,7-Dihydroxy-2,2,6trimethyl-2H-1-benzopyran-8-yl)-2-methyl-1-butanone, *in* D-70348

112008-27-2 Swalpamycin, S-70086

112038-99-0 [3](2,7)Troponophane, T-70329

112039-01-7 [5](2,7)Troponophane, T-70330

112078-67-8 4-(3,4-Dihydroxyphenyl)-5,7dihydroxy-2*H*-1-benzopyran-2-one; 7-Me ether, 5-*O*-β-D-galactopyranosyl, *in* D-70334

112107-61-6 Naphtho[1,2-*a*]quinolizinium(1+); Perchlorate, *in* N-70012

112107-62-7 Naphtho[1,2-*a*]quinolizinium(1+); Bromide, *in* N-70012

112149-08-3 1,8-Dihydrobenzo[2,1-*b*:3,4*b'*]dipyrrole, D-70174

112168-43-1 5,6-Epoxy-20-hydroxy-8,11,14eicosatrienoic acid; (5*S*,6*R*,8*Z*,11*Z*,14*Z*)-*form*, *in* E-70017

112168-44-2 14,15-Epoxy-20-hydroxy5,8,11-eicosatrienoic acid; (5*Z*,8*Z*,11*Z*,14*R*,15*S*)-*form*, *in* E-70018

112168-45-3 15,20-Dihydroxy-5,8,11,13eicosatetraenoic acid; (5*Z*,8*Z*,11*Z*,13*E*,15*S*)-*form*, *in* D-70297

112204-35-0 3,4,8,9-Tetraoxa-1,6diazabicyclo[4.4.2]dodecane, T-70144

112240-59-2 Proclavaminic acid, P-70116

112283-40-6 7-Oxo-1-azabicyclo[3.2.0]heptane-2-carboxylic acid; (2*S*,5*R*)-*form*, *in* O-70079

112291-51-7 Trifluoro-1,2,3-triazine, T-70245

112292-35-0 5-(3,5-Di-*tert*-butyl-4-oxo2,5-cyclohexadienylidene)-3,6-cyclo-heptadiene-1,2-dione, D-70105

112296-12-5 3-(2-Aminoethylidene)-7-oxo4-oxa-1-azabicyclo[3.2.0]heptane-2-carboxylic acid; (2*S*,3*Z*,5*S*)-*form*, *in* A-70139

112343-16-5 Sophoraside A, *in* P-70137

112343-17-6 Puerol *B*, *in* P-70137

112362-74-0 1(10),4-Aromadendradiene, A-70257

112396-63-1 Antibiotic CP 54883, A-70222

112419-10-0 7-Oxo-1-azabicyclo[3.2.0]heptane-2-carboxylic acid; (2*R*,5*R*)-*form*, *in* O-70079

112421-19-9 1-Epi-α-gurjunene, *in* G-70034

112421-20-2 9-Aromadendrene, A-70259

112430-63-4 3β,15α-Diacetoxy-7,9(11),24*E*lanostatrien-26-oic acid, *in* D-70314

112430-67-8 3,15-Dihydroxy-7,9(11),24lanostatrien-26-oic acid; (3α,5α,15α,24*E*)-*form*, *in* D-70314

112430-68-9 3,15-Dihydroxy-7,9(11),24lanostatrien-26-oic acid; (3β,15α,24*E*)-*form*, *in* D-70314

112430-69-0 15α,22S-Diacetoxy-3β-hydroxy7,9(11),24*E*-lanostatrien-26-oic acid, *in* T-70266

112430-70-3 3β,15α,22S-Triacetoxy-7,9(11),24*E*-lanostatrien-26-oic acid, *in* T-70266

112448-38-1 5-Hydroxybowdichione, *in* B-70175

112468-60-7 8α-Hydroxysambucoin, *in* S-70007

112470-96-9 Salvicanaric acid, S-70004

112497-53-7 2,6-Dodecanedione, D-70534

112501-40-3 8-Vanilloylshiromodiol, *in* S-70039

112501-40-3 Epoxyjaeschkeanadiol vanillate, *in* J-70001

112503-91-0 Machilin *H*, M-70002

112504-80-0 4,5,6,7-Tetrahydro-1*H*cyclo-propa[*a*]naphthalene, T-70051

112504-81-1 1,3,4,5Tetrahydrocycloprop[*f*]-indene, T-70052

112504-82-2 3,4,5,6-Tetrahydro-1*H*cyclo-propa[*b*]naphthalene, T-70050

112520-11-3 4,7-Dihydro-1,1,3,3tetramethyl-1*H*,3*H*3*a*,7*a*-episeleno-4,7epoxyisobenzofuran, D-70266

112531-13-2 8β-Hydroxysambucoin, *in* S-70007

112543-27-8 Arvoside, A-70264

112570-87-3 9,11-Dihydro-22,25-oxido-11oxoholothurinogenin, D-70237

112606-14-1 Kerlinic acid, K-70011

112613-42-0 Octahydro-5*H*,10*H*dipyrrolo[1,2-*a*:1',2'*d*]pyrazine-5,10-dithione; (*S*,*S*)-*form*, *in* O-70014

112613-52-2 Lehmannin, L-70027

112633-64-4 1-Phenyl-1*H*-pyrazolo[3,4-*e*]indolizine, P-70092

112634-75-0 25,26,27,28Tetraazapenta-cyclo[19.3.1.1^{3,7}.1^{9,13}.1^{15,19}]octa-cosa-1(25),3,5,7(28)-9,11,13(27),15,17,19(26),21,23-dodecaene-2,8,14,20-tetrone, T-70010

112642-49-6 Majusculone, M-70006

112642-59-8 1,8-Dichloro-6-chloromethyl2-methyl-2,6-octadiene, D-70116

112642-60-1 7,8-Dibromo-6-(chloromethylene)-2methyl-2-octene, D-70088

112642-61-2 1-Chloro-3-(chloromethyl)-7methyl-2,6-octadiene; (*Z*)-*form*, *in* C-70060

112642-62-3 3,8-Dichloro-6-chloromethyl2-methyl-1,6-octadiene, D-70117

112649-07-7 6,8'-Dihydroxy-2,2'-bis(2phenyl-ethyl)[5,5'-bi4*H*-1-benzopyran]-4,4'dione, D-70289

112649-48-6 BR-Xanthone *A*, B-70282

112663-68-0 Saroaspidin *B*, S-70013

112663-69-1 Saroaspidin *A*, S-70012

112663-70-4 Saroaspidin *C*, S-70014

112664-27-4 AH_{10}, A-70078

112667-14-8 15α,22S-Diacetoxy-3α-hydroxy7,9(11),24*E*-lanostatrien-26-oic acid, *in* T-70266

112681-65-9 1,4,7,10-Cyclododecatetraene; (*Z*,*Z*,*Z*,*Z*)-*form*, *in* C-70217

112693-17-1 Neriucoumaric acid, *in* D-70349

112693-21-7 Oleonuezhenide, O-70037

112693-22-8 Isonuezhenide, I-70073

112700-85-3 4,5-Dihydro-5,5-dimethyl3*H*1,2,4-triazol-3-one, D-70201

112709-68-9 Quisquagenin, Q-70011

112710-66-4 (3*E*)-Isolaureatin, *in* I-70067

112727-21-6 Xestodiol, X-70006

112727-26-1 3,6-Dihydroxy-2-(1-oxo-10tetra-decenyl)-2cyclohexen-1-one, D-70332

112740-62-2 5-Hydroxy-3-(phenylthio)-1,4naphthoquinone, H-70214

112740-63-3 5-Hydroxy-3-(phenylthio)-1,4naphthoquinone; Ac, *in* H-70214

112757-06-9 Daphnodorin *D*; (*S*)-*form*, *in* D-70005

112781-23-4 3,14-Diacetoxy-2-butanoyloxy5,8(17)-briaradien18,7-olide, *in* E-70012

112781-24-5 14-Acetoxy-2-butanoyloxy5,8(17)-briaradien18,7-olide, *in* E-70012

112781-25-6 4,14-Diacetoxy-2-butanoyloxy5,8(17)-briaradien18,7-olide, *in* E-70012

112781-26-7 4,14-Diacetoxy-2propanoyloxy-5,8(17)briaradien-18,7-olide, *in* E-70012

112790-14-4 1,4,7,10-Cyclododecatetraene; (*E*,*Z*,*Z*,*Z*)-*form*, *in* C-70217

112863-43-1 3,4-Dihydro-β-carboline; 2-Oxide, *in* D-60209

112897-61-7 3-Thiatricyclo[2.2.1.0^{2,6}]heptane; *S*,*S*-Dioxide, *in* T-70165

112899-36-2 Herpetetrone, H-70026

112923-64-5 Osbeckic acid, O-70048

112970-44-2 2-Bromo-3-methoxyaniline, *in* A-70099

113055-86-0 Bis[1,2,5-thiadiazolo][3,4*b*:3',4'-*e*]pyrazine, B-70161

113085-62-4 7-Methoxyrosmanol, *in* R-70010

113088-07-6 Tetraoxaporphycene, T-70146

113105-30-9 23-Oxomariesiic acid *A*, *in* M-70012

113122-50-2 Myrocin *C*, M-70165

113139-04-1 1-(Bromomethyl)acridine, B-70229

113139-05-2 2-(Bromomethyl)acridine, B-70230

113139-06-3 3-(Bromomethyl)acridine, B-70231

113139-13-2 1-Acridinecarboxaldehyde, A-70056

113139-14-3 2-Acridinecarboxaldehyde, A-70057

113139-15-4 3-Acridinecarboxaldehyde, A-70058

113139-16-5 4-Acridinecarboxaldehyde, A-70059

113139-56-3 Tris[(2,2'-bipyridyl-6-yl)methyl]-amine, T-70320

113201-68-6 1-(2,6-Dihydroxyphenyl)-1tetra-decanone, D-70338

113215-84-2 2-Phenylcyclohexanecarboxylic acid; (1*R*,2*S*)-*form*, *in* P-70065

113215-85-3 2-Phenylcyclohexanecarboxylic acid; (1*S*,2*R*)-*form*, *in* P-70065

113219-94-6 5-Oxo-3-hexenoic acid; (*Z*)-*form*, *in* O-70092

113236-00-3 4'-Hydroxy-3',7-dimethoxyflavan, *in* T-70260

113262-62-7 Obtusadiene, O-70001

113262-62-7 Isoobtusadiene, I-70074

113266-71-0 8,8*a*-Dihydro-3-methoxy-5methyl-1*H*,6*H*-furo[3,4*e*][1,3,2]dioxaphosphepin 3-oxide, D-70223

113270-95-4 Macrophylloside A, *in* M-70004
113270-96-5 Macrophylloside *C*, M-70004
113276-96-3 Valilactone, V-70002
113290-67-8 3,4-Cinnolinedicarboxaldehyde, C-70180
113296-38-1 Macrophylloside B, *in* M-70004
113297-21-5 3-Methyl-4,5-benzofurandione, M-70051
113308-92-2 12-Tetracosyne, T-70018
113332-14-2 Dihydroserruloside, D-70265
113332-15-3 Confertoside, C-70193
113333-46-3 5-Iodo-1-penten-4-yne, I-70053
113340-30-0 2-Cyano-1,2dihydrocyclobuta[*a*]-naphthalene, *in* D-70177
113340-31-1 1,2-Dihydrocyclobuta[*a*]naphthalene-2-carboxylic acid; (±)-*form, in* D-70177
113340-32-2 1,2-Dihydrocyclobuta[*a*]naphthalene-2-carboxylic acid; (±)-*form*, Me ester, *in* D-70177
113372-44-4 [1,4]Dithiino[2,3-*b*:6,5-*b*']dipyridine, D-70519
113372-47-7 1,4-Dithiino[2,3-*b*:5,6-*c*']dipyridine, D-70518
113372-48-8 1,4-Dithiino[2,3-*b*:6,5-*c*']dipyridine, D-70520
113426-94-1 1-(Phenylmethoxy)-3-buten-2-ol, *in* B-70291
113446-75-6 11,11,12,12,13,13,14,14Octacyano-1,4:5,8anthradiquinotetramethane, O-70002
113446-81-4 11,11,12,12,13,13,14,14Octacyano-1,4:5,8anthradiquinotetramethane; Tetraethylammonium salt, *in* O-70002
113459-56-6 10-Hydroxy-11-methoxydracaenone, *in* D-70546
113464-53-2 Cardiophyllidin, C-70023
113466-87-8 1,3,3*a*,4,5,7*a*-Hexahydro-2*H*indol-2-one; (3*aRS*,7*aSR*)-*form, in* H-70060
113466-88-9 3,3*a*,4,6*a*Tetrahydrocyclopenta[*b*]pyrrol-2(1*H*)-one; (3*aRS*,6*aSR*)-*form, in* T-70048
113466-89-0 Hexahydroazirino[2,3,1-*hi*]indol-2(1*H*)-one, H-70046
113476-61-2 7,9-Dihydroxy-2,3,4trimethoxyphenanthrene, *in* P-70035
113476-63-4 2,3,4,7,9Pentamethoxyphenanthrene, *in* P-70035
113477-35-3 7,10-Dihydroxy-11methoxydracaenone, *in* D-70546
113495-60-6 2,5-Dihydro-3,4diphenylselenophene, D-70205
113504-10-2 1,4-Dithioniabicyclo[2.2.0]hexane; Biss(rifluoromethanesulfonate), *in* D-70527
113532-14-2 Boesenboxide, B-70173
113540-73-1 Adriadysiolide, A-70069
113540-81-1 Sarcodictyin *A*, S-70010
113540-84-4 Imbricatonol, I-70003
113555-26-3 Sarcodictyin B, *in* S-70010
113605-05-3 5*H*-Cyclopropa[*f*][2]benzothiophene, C-70247
113631-50-8 3,6-Pyridazinedicarboxylic acid; Di-Et ester, *in* P-70168
113631-52-0 3,6-Pyridazinedicarboxylic acid; Dihydrazide, *in* P-70168
113667-19-9 Decahydro-5,5,8*a*-trimethyl-2naphthalenol; (2*R*,4*aR*,8*aS*)-*form*, Ac, *in* D-70014
113667-24-6 Decahydro-5,5,8*a*-trimethyl-2naphthalenol; (2*R*,4*aR*,8*aS*)-*form, in* D-70014

113667-25-7 Octahydro-5,5,8*a*-trimethyl2(1*H*)-naphthalenone, *in* D-70014
113689-49-9 Loxodellonic acid, L-70039
113689-51-3 2'-*O*-Methylpseudocyphellarin A, *in* P-70127
113689-52-4 2'-*O*-Methylphenarctin, *in* P-70054
113689-53-5 4'-Chloronephroarctin, *in* N-70032
113705-11-6 9,10-Diiodoanthracene, D-70355
113705-24-1 1,3-Diethynyl-2,4,5,6tetrafluorobenzene, D-70159
113705-27-4 2,7-Diethynylnaphthalene, D-70158
113706-22-2 Glomellonic acid, *in* G-70018
113706-23-3 Glomelliferonic acid, G-70018
113706-24-4 Isopseudocyphellarin *A*, I-70098
113724-67-7 Galloxanthin; (*R*)-*form, in* G-70002
113728-07-7 Cyclohepta[*def*]carbazolium tetrafluoroborate, *in* C-70220
113734-83-1 Thiomelin, T-70181
113734-84-2 2-Chloro-1,8-dihydroxy-5methoxy-6-methylxanthone, *in* T-70181
113734-85-3 2,4-Dichloro-1-hydroxy-5,8dimethoxyxanthone, *in* T-70181
113734-86-4 2-Chloro-1-hydroxy-5,8dimethoxy-6-methylxanthone, *in* T-70181
113738-79-7 3,7-Dioxo-12-oleanen-28-oic acid, D-70469
113762-83-7 3,6-Dihydroxy-12-oleanen-29-oic acid; (3*β*,6*β*)-*form, in* D-70326
113789-85-8 5-(2,4,6-Cycloheptatrien-1ylideneethylidene)-3,6cycloheptadiene-1,2-dione, C-70222
113817-64-4 Goniothalamicin, G-70027
113866-80-1 1-(3-Hydroxyphenyl)-2-(2,4,5trihydroxyphenyl)ethylene; (*E*)-*form, in* H-70215
113880-22-1 1,2-Diphenyl-3-(phenylimino)cyclopropene; B,HBF₄, *in* D-70491
113893-26-8 [1,2,4]Triazolo[1,2-*a*][1,2,4]triazole-1,3,5,7(2*H*,6*H*)tetrone, T-70209
113893-50-8 3-(Furo[3,4-*b*]furan-4-yl)-2propenenitrile; (*E*)-*form, in* F-70055
113893-52-0 3-(Furo[3,4-*b*]furan-4-yl)-2propenenitrile; (*Z*)-*form, in* F-70055
113996-94-4 5-Chloro-8-hydroxy-1,4dimethoxy-3-methylxanthone, *in* T-70181
114020-56-3 24,25-Epoxy-7,26-dihydroxy-8lanosten-3-one; (7*α*,24*S*,25*S*)-*form, in* E-70013
114020-57-4 24*S*,25*S*-Epoxy-26hydroxy7,9(11)-lanostadien-3-one, *in* E-70025
114020-58-5 24,25-Epoxy-7,9(11)lanostadiene-3,26-diol; (3*β*,24*S*,25*S*)-*form, in* E-70025
114020-67-6 Patagonic acid, P-70013
114032-09-6 1,2:9,10-Dibenzo[2.2]metaparacyclophane, D-70071
114044-23-4 1-Chloromethylbenz[*cd*]indol2(1*H*)-one, *in* B-70027
114049-72-8 2,3-Dihydro-3-oxo-1*H*-pyrrole1-carboxaldehyde, *in* D-70262
114049-73-9 1-Acetyl-1,2-dihydro-3*H*pyrrol-3-one, *in* D-70262
114058-42-3 Solenolide B, *in* S-70052
114058-43-4 Solenolide D, *in* S-70053
114058-44-5 Solenolide E, S-70054
114066-81-8 Meijicoccene, M-70017
114068-57-4 3,4-Diazatricyclo[3.1.0.0²,⁶]hex-3-ene, D-70054

114076-53-8 1,3-Dihydroxy-12-oleanen-29-oic acid; (1*α*,3*β*)-*form, in* D-70325
114076-72-1 2*α*-Hydroxydihydroparthenolide, *in* D-70239
114076-73-2 9*α*-Hydroxydihydroparthenolide, *in* D-70239
114078-88-5 5-Bromo-1,2,3-triazine, B-70273
114078-89-6 4,5-Dibromo-1,2,3-triazine, D-70100
114078-94-3 1,3λ⁴,δ²,5,2,4-Trithiadiazine, T-70327
114094-31-4 Solenolide A, S-70052
114094-32-5 Solenolide C, S-70053
114094-33-6 Solenolide F, S-70055
114094-46-1 Larreantin, L-70020
114102-66-8 1,3λ⁴,δ²,5,2,4-Trithiadiazine; 1-Oxide, *in* T-70327
114144-28-4 14,15-Epoxy-5-hydroxy6,8,10,12-eicosatetraenoic acid; (5*S*,6*E*,8*Z*,10*E*,12*E*)-*form, in* E-70016
114191-55-8 Thujin, T-70190
114191-58-1 8,15-Isopimaradiene-7,18-diol; 7*α*-*form, in* I-70079
114191-60-5 8,15-Isopimaradiene-7,18-diol; 7*β*-*form, in* I-70079
114216-80-7 Tecomaquinone III, T-70006
114216-82-9 Rutalpinin, R-70015
114216-99-8 3,9-Dihydroxy-2methoxycoumestone, *in* T-70249
114226-24-3 1-(2-Hydroxy-6-methoxyphenyl)1-tetradecanone, *in* D-70338
114246-84-3 11,13-Dihydropsilostachyin, *in* P-70131
114263-84-2 9*H*-Fluorene-9-selenone, F-60018
114298-64-5 Neonepetalactone; 4*a*-Epimer, *in* N-70027
114339-93-4 1(10)-Aristolen-9-one, A-70252
114339-95-6 5'-Hydroxycudraflavone A, *in* C-70205
114340-00-4 4',5,7-Trihydroxy-3'-methoxy5'-prenylflavanone, *in* S-70041
114359-99-8 7*α*-Methoxy-8-labdene-3*β*,15-diol, *in* L-70013
114360-00-8 3*β*,15-Dihydroxy-8-labden-7-one, *in* L-70013
114394-19-3 Fragransol *A*, F-70041
114394-20-6 Fragransol B, F-70042
114394-21-7 Fragransin *D₁*, F-70040
114420-29-0 *ent*-9,13-Epoxy-7-labdene15,17-dioic acid; Di-Me ester, *in* E-70024
114422-15-0 Cleomiscosin D, C-70185
114422-18-3 Isoschizandrin, I-70101
114422-19-4 Machilin G, *in* A-70271
114422-21-8 Machilin *I*, M-70003
114422-23-0 Fragransin D₂, *in* F-70040
114422-24-1 Fragransin D₃, *in* F-70040
114437-17-1 1-(Hydroxymethyl)-7,7dimethylbicyclo[2.2.1]heptane-2,3-diol; (1*S*,2*S*,3*R*)-*form, in* H-70173
114437-19-3 3-Hydroxy-4-(hydroxymethyl)7,7dimethylbicyclo[2.2.1]heptan-2-one, *in* H-70173
114466-74-9 Metachromin A, M-70033
114466-75-0 Metachromin B, M-70034
114489-64-1 1-[3-Azido-4-(hydroxymethyl)cyclopentyl]-5-methyl2,4(1*H*,3*H*)-pyrimidinedione; (+)-*form, in* A-70288
114489-73-5 Scutellone C, S-70019
114489-85-9 1-(Hydroxymethyl)-7,7dimethylbicyclo[2.2.1]heptane-2,3-diol; (1*S*,2*S*,3*S*)-*form, in* H-70173
114491-65-5 5,6-Dihydro-11,12diphenyldibenzo[*a,e*]cyclooctene; (*E*)-*form, in* D-70202
114491-81-5 1,3,4,6Tetrakiss(imethylamino)pyrrolo[3,4-*c*]pyrrole, T-70127

114504-92-6   5,6-Dihydro-11,12diphenyl-
dibenzo[a,e]cyclooctene; (Z)-
form, *in* D-70202
114542-44-8   Honyucitrin, H-70092
114542-45-9   Honyudisin, H-70093
114567-38-3   3,5,6,7-Tetrahydroxy-8methyl-
flavone, T-70117
114567-42-9   1,3-Dihydroxy-5,8dimethoxy-
xanthone, *in* T-70123
114587-58-5   Pyoverdin II, P-70149
114587-59-6   Pyoverdin III, *in* P-70149
114616-35-2   Pyoverdin I, P-70148
114687-97-7   Majucin, M-70005
114687-98-8   Neomajucin, *in* M-70005
114703-26-3   Vittagraciliolide, V-70017
114709-99-8   3H-Pyrrolo[1,2-a]azepin-3-one,
P-70190
114715-53-6   3-Aminotetrahydro-
3thiophenecarboxylic acid; (S)-
form, *in* A-70199
114727-96-7   Ferprenin, F-70003
114727-97-8   Cudraisoflavone A, C-70206
114728-09-5   Variabilin; Δ^{13}-Isomer (Z-), *in*
V-70003
114728-44-8   4,8-Dihydroxylactariusfuran,
D-70313
114742-71-1   Anhydroscandenolide, A-70211
114742-92-6   7-Hydroxy-
13(17),15cleistanthadien-18-oic
acid; 7β-form, *in* H-70115
114742-93-7   7β-Acetoxy-
13(17),15cleistanthadien-18-oic
acid, *in* H-70115
114761-90-9   Variabilin; 20E-Isomer, *in*
V-70003
114769-49-2   4,5-Diphenylpyrimidine; 1-Oxide,
in D-70495
114861-40-4   Diisocyanogen, D-70362
114906-01-3   12,20(29)-Lupadiene-3,27,28triol;
3β-form, *in* L-70042
114915-36-5   Antheliolide A, A-70213
114933-19-6   Antheliolide B, *in* A-70213
114933-24-3   Xenicin; 9-Deacetyl, 9-acetoacetyl,
in X-70004
114969-48-1   5-Chloropyrimidine; N-Oxide, *in*
C-70157
114969-53-8   4,5-Dimethylpyrimidine; 1-Oxide,
in D-70436
114969-95-8   4-Methyl-5-phenylpyrimidine; 3-
Oxide, *in* M-70114
115022-87-2   1-Azabicyclo[3.2.0]heptane2,7-
dione, A-70273
115028-52-9   Balaenonol, *in* B-70006
115031-67-9   Balaenol, B-70006
115055-41-9   [1,4]Oxaselenino[3,2-b:6,5c']-
dipyridine, O-70068
115125-38-7   5-Methyl-
1,3,2,4dithiazolium(1+);
Chloride, *in* M-70061
115216-83-6   Antibiotic L 660631, A-70224
115271-20-0   3,6-Di-*tert*-butylpyrrolo[3,2b]-
pyrrole, D-70108
115319-22-7 O^6-[(3-Carbamoyl-2H-azirine2-
ylidene)amino]-1,2-
Oisopropylidene-3,5-di$O$-tosyl-
α-D-glucofuranoside, C-70014
115333-90-9   16β-Methoxyalisol B monoacetate,
in A-70082
115333-92-1   Pygmaeocine E, P-70144
115333-98-7   6-Methoxy-5-vinylbenzofuran, *in*
H-70231
115334-05-9   Dihydroniloticin, *in* N-70036
115334-06-0   Egonol 2-methylbutanoate, *in*
E-70001
115334-07-1   Demethoxyegonol 2-methyl-
butanoate, *in* E-70001
115334-13-9   Cyperenyl acetate, *in* C-70259
115334-16-2   Cyperenal, *in* C-70259
115334-17-3   Tinospora clerodane, T-70200
115334-52-6   6,16-Diacetoxy-3,7dolabelladien-
12-ol, *in* D-70541

115334-66-2   13-Desacetyl-
1αhydroxyafraglaucolide, *in*
A-70070
115338-10-8   13-Desacetyleudesmaafraglau-
colide, *in* T-70258
115338-11-9   13-Desacetyl-
1βhydroxyisoafraglaucolide, *in*
I-70060
115346-25-3   16β-Hydroxyalisol B monoacetate,
in A-70082
115346-29-7   Venustanol, V-70004
115346-31-1   Artemisiaglaucolide, *in* D-70303
115346-35-5   Afraglaucolide, A-70070
115346-36-6   Eudesmaafraglaucolide, *in*
T-70258
115346-38-8   1α-Hydroxyisoafraglaucolide, *in*
I-70060
115346-39-9   1β-Hydroxyafraglaucolide, *in*
A-70070
115356-18-8   Kanshone A, K-70004
115361-83-6   Assafoetidin, A-70265
115367-47-0   1α-Hydroxyafraglaucolide, *in*
A-70070
115370-61-1   Kanshone B, K-70005
115374-33-9   7-Oxo-8,13-labdadien-15-oic acid,
O-70095
115404-57-4   Niloticin, N-70036
115404-62-1   3,7-Dolabelladiene-6,12-diol;
(1R,3E,6R,7E,11R,12R)-form,
in D-70541
115458-72-5   Glaucin B, G-70017
115491-93-5   Diallyl dicarbonate, *in* D-70111
115492-54-1   1,2,12Triazapenta-
cyclo[6.4.0.0^{2,17}.0^{3,7}.0^{4,11}]-
dodecane, T-70205
115532-08-6   Mutisifurocoumarin; Di-Ac, *in*
M-70161
115533-27-2   Tris(9-fluorenylidene)cyclo-
propane, T-70322
115547-12-1   Chapinolin, C-70037
115547-14-3   Skutchiolide B, S-70050
115553-99-6 α-Amino-2-carboxy-5-oxo-
1pyrrolidinebutanoic acid,
A-70120
115591-79-2   Triss(hioxanthen-9-ylidene)cyclo-
propane, T-70326
115610-51-0   Sphaeropyrane, S-70060
115699-93-9   2,3-Dihydro-3-hydroxy-2-
(1methylethenyl)-
5benzofurancarboxaldehyde,
D-70213
115699-94-0   2,3-Dihydro-2-(1-hydroxy-
1methylethyl)-5-
benzofurancarboxaldehyde,
D-70214
115699-95-1   2-(1,2-Dihydroxy-1methylethyl)-
2,3dihydro-5-
benzofurancarboxaldehyde, *in*
D-70214
115699-96-2   5-(1,3-Dihydroxypropyl)-
2isopropenyl-
2,3dihydrobenzofuran, D-70340
115712-89-5   Pumilaisoflavone A, P-70140
115712-90-8   Pumilaisoflavone B, P-70141
115713-07-0   2,4,5-Trihydroxy-1methoxy-
xanthone, *in* T-70122
115713-11-6   Cryptoresinol, C-70201
115713-23-0   2′,5-Dihydroxy-7,8dimethoxy-
isoflavone, *in* T-70110
115713-34-3   Trichostin, T-70225
115713-37-6   Dihydrotrichostin, D-70273
115722-54-8   Kurospongin, K-70019
115783-40-9   Norstrictic acid, N-70080
115783-44-3   ent-3β,16α,17-Atisanetriol,
A-70266
115783-47-6   3,7a-Diacetoxyhexahydro-2oxa-
3a,7-dimethyl-1,4indanedione,
in H-70055
115783-48-7   Hexahydro-3,7a-dihydroxy3a,7-
dimethyl-1,4isobenzofurandione;
(3S,3aR,7R,7aR)-form, Di-Ac,
in H-70055
115783-49-8   Hexahydro-3,7a-dihydroxy3a,7-
dimethyl-1,4isobenzofurandione;
(3R,3aR,7S,7aR)-form, Di-Ac,
in H-70055

115783-50-1   Hexahydro-3,7a-dihydroxy3a,7-
dimethyl-1,4isobenzofurandione;
(3S,3aR,7S,7aR)-form, Di-Ac,
in H-70055
115787-98-9   5-Oxoisocystofuranoquinol, *in*
C-70260
115787-99-0   5-Oxocystofuranoquinol, *in*
C-70260
115788-00-6   5-Hydroxycystofuranoquinol, *in*
C-70260
115788-01-7   2-(5-Hydroxy-3,7,11,15tetra-
methyl-2,6,10,14hexadecatetra-
enyl)-6methyl-1,2-benzenediol,
H-70225
115788-02-8   2-(5,16-Dihydroxy-3,7,11,15tetra-
methyl-2,6,10,14hexadecatetra-
enyl)-6methyl-1,4-benzenediol,
in H-70225
115788-03-9   16-(2,5-Dihydroxy-3methyl-
phenyl)-2,6,10,19-tetramethyl-
2,6,10,14-hexadecatetraene-
4,12-dione, *in* H-70225
115794-38-2   Dehydrotrichostin, *in* T-70225
115812-55-0   Chokorin, C-70171
115841-09-3   Salvianolic acid C, *in* S-70003
115855-37-3   [1,4]Oxaselenino[3,2-b:5,6b']-
dipyridine, O-70066
115855-39-5   [1,4]Oxaselenino[3,2-b:5,6c']-
dipyridine; 8-Oxide, *in* O-70067
115855-40-8   [1,4]Oxaselenino[3,2-b:5,6c']-
dipyridine; 7-Oxide, *in* O-70068
115857-55-1   [1,4]Oxaselenino[3,2-b:5,6c']-
dipyridine, O-70067
115873-03-5   Sordariol, S-70056
115879-75-9   Pyrido[1′,2′:3,4]imidazo[1,2a]-
pyrimidin-5-ium(1+);
Perchlorate, *in* P-70176
115939-25-8   Salvianolic acid B, S-70003
116070-47-4   Indolo[2,3-b][1,4]benzodiazepin-
12(6H)-one, I-70015
116120-54-8   Fridamycin E, *in* F-70043
116174-99-3   Agrostophyllin, A-70077
116175-04-3   23-Acetoxy-30-hydroxy-
20(29)lupen-3-one, *in* D-70318
116175-10-1   23-Acetoxy-3-oxo-20(30)lupen-
29-al, *in* D-70316
116195-01-8   9,10-Dihydro-
9,10epidithioanthracene,
D-70206
116199-61-2   3β,23-Diacetoxy-20(30)-lupen29-
al, *in* D-70316
116204-21-8   1,1:2,2-Bis([10]annulene-1,6diyl)-
ethylene, B-70127
116212-47-6   Pyrrolo[1,2-a]pyrazin-1(2H)-one;
N-Me, *in* P-70198
116280-37-6   Pentacyclo[8.4.0.0^{2,7}.0^{3,12}.0^{6,11}]-
tetradeca-4,8,13-triene; ($\pm$)-
form, *in* P-70021
116310-62-4   Euchrenone A$_2$, E-70057
116310-63-5   Clibadic acid, C-70186
116310-64-6   Clibadiolide, C-70187
116310-65-7   Sphaeroxetane, S-70061
116360-06-6   3,9-Dihydroxy-10(14)-guaien12,6-
olide; (3β,4α,6α,9β,11β)-form,
in D-70305
116360-07-7   3β,9β-Dihydroxy-
4(15),10(14)guaiadien-12,6α-
olide, *in* D-70305
116360-08-3   3,9-Dihydroxy-10(14)-guaien12,6-
olide; (3β,4α,6α,9β,11β)-form,
4,15-Didehydro, *in* D-70305
116368-97-9   1,3,8-Trihydroxy-2,6dimethoxy-
xanthone, *in* P-70036
116369-02-9   Flabellata secoclerodane, F-70009
116374-12-0   Secofloribundione, S-70022
116383-31-4   Oleuroside, O-70038
116383-99-4   Dihydrocoriandrin, *in* C-70198

116384-09-9 Isolycoserone, I-70068
116384-12-6 10′-Hydroxyisolycoserone, *in* I-70068
116384-28-2 Laurencenone *A*, L-70024
116384-28-2 Laurencenone *D*, L-70026
116397-92-3 Agerasanin, A-70072
116404-82-1 3,9-Dihydroxy-10(14)-guaien12,6-olide; (3β,4α,6α,9β,11β)-*form*, 3-Ketone, *in* D-70305
116408-25-4 Salvileucolidone, S-70005
116408-80-1 Coriandrin, C-70198
116425-02-6 Longirabdosin, L-70036
116425-75-3 2,11-Dihydroxy-7,9(11),13abietatrien-12-one; 2α-*form*, *in* D-70282
116425-76-4 2α,11-Dihydroxy-7,9(11),13abietatriene-6,12-dione, *in* D-70282
116425-77-5 2α,6α,11-Trihydroxy-7,9(11),13-abietatrien-12-one, *in* D-70282
116425-78-6 2α,6α,7,11-Tetrahydroxy7,9(11),13-abietatrien12-one, *in* D-70282
116425-92-4 Nimbinone, N-70037
116425-93-5 Nimbione, N-70038
116425-97-9 Niloticin acetate, *in* N-70036
116428-57-0 Neoliacinic acid; Me ester, *in* N-70026
116460-77-6 Bicyclo[5.1.0]octa-3,5-dien2-one; (1*RS*,7*RS*)-*form*, *in* B-70111
116481-69-7 Praderin, P-70112
116489-73-7 Dibenzo[*b,m*]triphenodithiazine-5,7,9,14,16,18(8*H*,17*H*)tetrone, D-70082
116497-04-2 Arguticinin, A-70250
116498-58-9 5,5′-Dimethoxylariciresinol, *in* L-70019
116504-33-7 [2](2,6)-Naphthaleno[2]paracyclophane-1,11-diene; (±)-*form*, *in* N-70006
116521-73-4 Moluccanin, M-70151
116528-02-0 Pitumbin, P-70104
116561-97-8 5,6-Dihydro-6-(2,4,6trihydroxyheneicosyl)2*H*-pyran-2-one, D-70276
116857-23-9 2-Nitrotryptophan; (*S*)-*form*, *in* N-70067
116913-58-7 5-Hydroxy-3-methyl-1,2benzenedicarboxylic acid; Me ether, 2-Me ester, *in* H-70162
116919-57-4 2-(1-Hydroxy-1-methylethyl)5-benzofurancarboxaldehyde, *in* D-70214
116922-60-2 3-Bromo-4-fluoropyridine, B-70214
116943-38-5 Methylenolactocin, M-70079
116944-99-1 Umbraculumin *A*, U-70001
116963-50-9 Umbraculumin *B*, U-70002
116963-91-8 9,10-Dihydro-7-hydroxy2,3,4,6-tetramethoxyphenanthrene, *in* D-70246
116965-35-6 Isoinunal, *in* I-70022
116988-15-9 Sintenin, S-70047
116988-16-0 3,7,11,15-Tetramethyl2,6,10,13,15-hexadecapentaen-1-ol, T-70137
117005-44-4 1(10)-Eremophilen-7-ol, E-70031
117007-26-8 Murraculatin, M-70158
117007-28-0 12-Hydroxy-3,7,11,15tetramethyl-2,6,10,14hexadecatetraenoic acid, H-70223
117038-70-7 2,3-Bis(2-methoxy-4-nitro-5sulfophenyl)-5-[(phenylamino)carbonyl]-2*H*-tetrazolium hydroxide inner salt, B-70151
117047-27-5 Rudbeckin *A*, R-70013
117052-59-2 Pyridazino[1″,6″:1′,2′]imidazo-[4′,5′:4,5]imidazo[1,2-*b*]pyridazine, P-70169
117054-70-3 [2](2,6)-Naphthaleno[2]paracyclophane-1,11-diene, N-70006

117065-26-6 Neocorymboside, N-70021
117073-97-9 22-Hydroxy-7,24tirucalladiene-3,23-dione, H-70227
117073-98-0 11,13(18)-Oleanadien-3-ol; 3β-*form*, *in* O-70032
117082-14-1 3-[1-(1,1-Dimethyl-3oxobutyl)-imidazol-4-yl]2-propenoic acid, *in* U-70005
117101-04-9 *ent*-14-Labden-8β-ol, L-70014
117113-32-3 ▷3-Fluoro-3-(trifluoromethyl)3*H*-diazirine, F-70037
117113-33-4 ▷3-Bromo-3-(trifluoromethyl)3*H*-diazirine, B-70275
117157-35-4 6,7-Dibromo-1-nitronaphthalene, D-70098
117157-36-5 6,7-Dibromo-1-naphthylamine, D-70097
117157-37-6 6,7-Dibromo-1-naphthol, D-70095
117176-24-6 4,10(14)-Oplopadiene-3,8,9triol; (3β,4*Z*,8β,9α)-*form*, 3-Ac, 8-angeloyl, *in* O-70042
117176-25-7 4,10(14)-Oplopadiene-3,8,9triol; (3β,4*Z*,8β,9α)-*form*, 3-Ac, 9-(3-Methyl-2-pentenoyl), *in* O-70042
117176-26-8 3β-Acetoxy-4,10(14)oplopadiene-8β,9α-diol, *in* O-70042
117180-87-7 Tetrahydro-3,4-furandiamine; (3*RS*,4*RS*)-*form*, *in* T-70062
117180-88-8 Tetrahydro-3,4-furandiamine; (3*RS*,4*RS*)-*form*, B,2HBr, *in* T-70062
117183-11-6 4,10(14)-Oplopadiene-3,8,9triol; (3β,4*Z*,8β,9α)-*form*, 3-Ac, 9-angeloyl, *in* O-70042
117193-18-7 Cycloeuphordenol, C-70218
117193-22-3 Nemorosonol, N-70020
117210-47-6 Richardianidin 2, *in* R-70006
117232-44-7 3,15-Dihydroxy-8(17),13labdadien-19-al; (3β,13*E*)-*form*, 3-Ac, *in* D-70310
117232-45-8 8,15-Dihydroxy-13-labden-19-al; (8α,*E*)-*form*, *in* D-70311
117232-46-9 7,15-Isopimaradiene-3,18-diol; 3β-*form*, *in* I-70076
117232-47-0 7,15-Isopimaradiene-18,19-diol, I-70078
117232-50-5 Epivittadinal, *in* V-70016
117232-52-7 2-Bisabolene-1,12-diol, B-70126
117232-53-8 Vittadinal, V-70016
117232-54-9 6-(Hydroxymethyl)-10-methyl2-(4-methyl-3-pentenyl)2,6,10-dodecatriene1,12-diol, H-70179
117232-58-3 11-Epidihydroridentin, *in* R-70007
117232-60-7 6-Acetylferulidin, *in* F-70006
117232-63-0 Richardianidin 1, R-70006
117249-10-2 *ent*-15,16-Epoxy-7,13(16),14labdatriene, E-70022
117254-99-6 3,15-Dihydroxy-8(17),13labdadien-19-al; (3α,13*E*)-*form*, *in* D-70310
117255-00-2 8,15-Isopimaradiene-3,7,19triol; (3β,7α)-*form*, *in* I-70082
117255-04-6 12-Hydroxy-6-(hydroxymethyl)10-methyl-2-(4-methyl3-pentenyl)-2,6,10dodecatrienal, *in* H-70179
117255-05-7 3,15-Dihydroxy-8(17),13labdadien-19-al; (3β,13*E*)-*form*, *in* D-70310
117255-06-8 Dihydroridentin, *in* R-70007
117255-07-9 Carmenin, *in* F-70006
117255-08-0 Andalucin, A-70207
117274-04-1 Heliopsolide; 8-Ac, *in* H-70007
117274-10-9 Skutchiolide A, S-70049

117274-11-0 Heliopsolide; 4-Epimer, 8-Ac, *in* H-70007
117274-13-2 Heliopsolide, H-70007
117278-48-5 Mzikonone, M-70166
117332-99-7 10-Hydroxy-7,11,13,16,19docosapentaenoic acid; (7*Z*,10*R*,11*E*,13*Z*,16*Z*,19*Z*)-*form*, *in* H-70125
117333-00-3 10-Hydroxy-7,11,13,16,19docosapentaenoic acid; (7*Z*,10*R*,11*E*,13*Z*,16*Z*,19*Z*)-*form*, Et ester, *in* H-70125
117333-01-4 8-Hydroxy-5,9,11,14,17eicosapentaenoic acid; (5*Z*,8*R*,9*E*,11*Z*,14*Z*,17*Z*)-*form*, Et ester, *in* H-70126
117333-02-5 8-Hydroxy-5,9,11,14,17eicosapentaenoic acid; (5*Z*,8*R*,9*E*,11*Z*,14*Z*,17*Z*)-*form*, 17,18-Dihydro, Et ester, *in* H-70126
117383-35-4 15α-Acetoxy-3α-hydroxy-23oxo-7,9(11),24*E*lanostatrien-26-oic acid, *in* D-70330
117383-36-5 3α,15α-Diacetoxy-23oxo7,9(11),24*E*-lanostatrien-26-oic acid, *in* D-70330
117383-37-6 3α-Acetoxy-15α-hydroxy-23oxo-7,9(11),24*E*lanostatrien-26-oic acid, *in* D-70330
117404-55-4 19-Hydroxy-8(17),13labdadien-15-al; (*Z*)-*form*, *in* H-70146
117407-06-4 8-Hydroxy-5,9,11,14,17eicosapentaenoic acid; (5*Z*,8*R*,9*E*,11*Z*,14*Z*,17*Z*)-*form*, *in* H-70126
117472-88-5 1,14,29,30,31,32Hexaazahexacyclo[12.7.7.1³,⁷.1⁸,¹².1¹⁶,²⁰.-1²³,²⁷]dotriaconta-3,5,7(32),-8,10,12(31),16,18,20(30),-23,25,27(29)-dodecaene; LiBr complex, *in* H-70030
117479-91-1 Gaillardoside, *in* T-70284
117500-97-7 1,14,34,35,36,37,38Heptaazaheptacyclo[12.12.7.1³,⁷.1⁸,¹².-1¹⁶,²⁰.1²¹,²⁵.1²⁸,³²]octatriaconta-3,5,7(38),8,10,12(37),-16,18,20(36),21,23,25(35),-28,30,32(34)pentadecane; NaBr complex, *in* H-70010
117585-50-9 6β-Methoxy-7,22ergostadiene3,5α-diol, *in* M-70057
117590-96-2 Jewenol A, J-70003
117590-97-3 Jewenol B, J-70004
117590-99-5 Sirutekkone, S-70048
117593-38-1 3-Aminotetrahydro-2*H*-thiopyran, A-70200
117593-43-8 3-Aminotetrahydro-2*H*-thiopyran; (±)-*form*, *S,S*-Dioxide, *in* A-70200
117593-44-9 3-Aminotetrahydro-2*H*-thiopyran; (±)-*form*, *N,N*-Di-Me, *in* A-70200
117593-45-0 3-Aminotetrahydro-2*H*-thiopyran; (±)-*form*, *N*-Ph, *in* A-70200
117593-51-8 3-Aminotetrahydro-2*H*-thiopyran; (±)-*form*, Oxalate, *in* A-70200
117597-78-1 Funadonin, F-70046
117597-79-2 6-(Hydroxymethyl)-7methoxy2*H*-1-benzopyran-2-one, *in* H-70139
117597-80-5 (E)-Methylsuberenol, *in* S-70080
117597-81-6 (Z)-Methylsuberenol, *in* S-70080
117610-40-9 Rzedowskin *A*, R-70016
117615-11-9 19,29-Dihydroxy-3-oxo-12oleanen-28-oic acid; 19α-*form*, Me ester, *in* D-70331
117634-61-4 2,4-Dihydroxy-5,11(13)guaiadien-12,8-olide; (2α,4α,8α)-*form*, *in* D-70304

117634-64-7 Kanshone *C*, K-70006

117634-70-5 4,5-Seconeopulchell-5-ene; 4-Ac, *in* S-70025

117654-06-5 3,23-Dihydroxy-30-nor12,20(29)-oleanadien28-oic acid; 3β-*form*, *in* D-70324

117659-70-8 7-Deacetyl-17βhydroxyazadiradione, *in* A-70281

117732-40-8 3,4-Seco-4(28),20,24dammaratrien-3,26-dioic acid, S-70021

117752-92-8 4-Aminodihydro-3-methylene-2(3*H*)furanone; (*R*)-*form*, *in* A-70131

117752-93-9 4-Aminodihydro-3-methylene-2(3*H*)furanone; (*R*)-*form*, *N*-Ac, *in* A-70131

117860-53-4 5-Amino-1*H*-pyrazole-3carboxylic acid; *N*(1)-Me, *in* A-70191

117860-54-5 3-Amino-1*H*-pyrazole-5carboxylic acid; *N*(1)-Me, *in* A-70188

118002-94-1 Atratoside D, *in* C-70257

555999-04-7 5-Heptenenitrile, *in* H-70024

117732-43-1 20,25,26-Trihydroxy-3,4-seco4(28),23-dammaradien-3oic acid; (20*S*,23*E*)-*form*, *in* T-70275

117732-44-2 12,20,25-Trihydroxy-3,4-seco4(28),23-dammaradien-3oic acid; (12β,20*S*,23*E*)-*form*, *in* T-70274

117752-87-1 4-Aminodihydro-3-methylene-2(3*H*)furanone; (*S*)-*form*, *N*-Ac, *in* A-70131